8th International School and Conference "Saint Petersburg OPEN 2021": Optoelectronics, Photonics, Engineering and Nanostructures (SPbOPEN 2021)

Journal of Physics: Conference Series Volume 2086

St. Petersburg, Russia
25 – 28 May 2021

Part 1 of 2

ISBN: 978-1-7138-5057-1
ISSN: 1742-6588

Printed from e-media with permission by:

Curran Associates, Inc.
57 Morehouse Lane
Red Hook, NY 12571

Some format issues inherent in the e-media version may also appear in this print version.

Institute of Physics
Dirac House, Temple Back
Bristol BS1 6BE UK

Phone: 44 1 17 929 7481
Fax: 44 1 17 920 0979

techtracking@iop.org

Additional copies of this publication are available from:

Curran Associates, Inc.
57 Morehouse Lane
Red Hook, NY 12571 USA
Phone: 845-758-0400
Fax: 845-758-2633
Email: curran@proceedings.com
Web: www.proceedings.com

TABLE OF CONTENTS

PART 1

PART 2

8th International School and Conference "Saint-Petersburg OPEN 2021" on Optoelectronics, Photonics, Engineering and Nanostructures

M V Maximov[1,2], E I Moiseev[2,1], N V Kryzhanovskaya[2,1], A E Zhukov[2,1]¶

[1] Alferov Federal State Budgetary Institution of Higher Education and Science Saint Petersburg National Research Academic University of the Russian Academy of Sciences
[2] National Research University "Higher School of Economics"

moiseev@spbau.com, aezhukov@hse.ru

8th International School and Conference "Saint-Petersburg OPEN 2021" on Optoelectronics, Photonics, Engineering and Nanostructures was held on May 25 - 28, 2021. The Organizers of the conference are the National Research University "Higher School of Economics" and Alferov Saint Petersburg National Research Academic University of the Russian Academy of Sciences. More than 300 participants from Russia, CIS countries, Europe and America took part in the conference. The conference consisted of poster reports presented by the young participants (students, PhD students, and young researchers) and topical lectures by invited speakers.

Due to restrictions imposed by the city authorities on holding mass events due to the threat of the spread of the COVID-19 infection, this year's conference was held in a hybrid format at the HSE University (Saint-Petersburg, Russia). Most of the participants were present in person, while the rest could listen to reports and ask their questions on-line. Some of the speakers also made their presentations in a distant format. All performances were recorded and are available for viewing at the link https://www.youtube.com/watch?v=pRoRU6T-uM8&t=5s. Posters and video reports of the participants were posted on the conference website https://spb.hse.ru/spbopen/.

The School and Conference included a series of invited talks given by leading professors with the aim to introduce young scientists with actual problems and major advances in physics and technology. The keynote speakers were:

Nikolay Ledentsov (Vertically Integrated Systems GmbH, Germany)
Elizaveta Semenova (Technical University of Denmark, Denmark)
Matthieu Roussey (University of Eastern Finland, Finland)
Vladimir Gavrilenko (Institute for Physics of Microstructures RAS, Russia)
Vladimir Dubrovsky (St Petersburg University, Russia)
Vsevolod Lundin (Ioffe Institute RAS, Russia)
Rustam Khabibullin (Institute of Ultra High Frequency semiconductor electronics RAS, Russia)
Anton Egorov (Connector Optics, Russia)
Wang Qi (Beijing University of Posts and Telecommunications, PRC)
Levon Asryan (Virginia Polytechnic Institute and State University, USA)

The organizing and editorial/program committee of the SPbOPEN-2021 hopes you enjoy the reading of this volume of the JPCS. Besides, we wish to thank all authors and participants for providing their valuable contributions for this proceeding as well as the reviewers for their constructive recommendations and criticism aiding to improve the presented articles. Likewise, we are confident that this issue will serve as a reference for optics, spectroscopy and photonics research worldwide.

The School and Conference is supported by the Russian Science Foundation and Peter the Great St. Petersburg Polytechnic University. Мне казалось, что кто-то еще деньги давал It is a continuation of the annual schools and seminars for youth on topical problems of physics and technology that is organized by the Academic University since 2009.

We invite all the students and young scientists to attend “Saint Petersburg OPEN” in 2022!
More details at https://spb.hse.ru/spbopen/.

Committees

Head of Program Committee
corr. member of RAS Alexey Zhukov (Professor, HSE - St. Petersburg & Alferov University RAS)
Program Committee
Prof. Andrey A. Lipovskii (Peter the Great St. Petersburg Polytechnic University, Russia)
Prof. Valentina Zhurikhina (Peter the Great St. Petersburg Polytechnic University, Russia)
Prof. George Cirlin (Alferov University RAS, Russia)
Prof. Alexander Omelchenko (HSE - St. Petersburg, Russia)
Prof. Pavel Brunkov (Ioffe Institute, Russia)
Prof. Nataliya Kryzhanovskaya (HSE - St. Petersburg, Russia)
Dr. Anna Baldycheva (University of Exeter, UK)
Dr. Kate Berseneva (University of Exeter, UK)

Head of Organizing Committee
corr. member of RAS Alexey Zhukov (Professor, HSE - St. Petersburg & Alferov University RAS)
Organizing Committee
Prof. Mikhail Maximov (Head of Nanophotonics Lab at Alferov University RAS, Russia)
Prof. Andrey Lipovskii (Peter the Great St.Petersburg Polytechnic University, Russia)
Prof. George Cirlin (Alferov University RAS, Russia)
Prof. Valentina Zhurikhina (Peter the Great St.Petersburg Polytechnic University, Russia)
Dr. Vladimir Korenev (Alferov University RAS, Russia)
Mr. Eduard Moiseev (Alferov University RAS, Russia)
Ms. Anna Dragunova (Alferov University RAS, Russia)
Ms. Svetlana Kadinskaya (Alferov University RAS, Russia)
Ms. Balezina Yulia (Alferov University RAS, Russia)
Ms. Kotko Elizaveta (HSE - St. Petersburg, Russia)

Journal of Physics: Conference Series **2086** (2021) 011001 doi:10.1088/1742-6596/2086/1/011001

Peer review declaration

- **Type of peer review: closed, collective.**

  At the first stage, candidates sent their short abstracts when registering on the conference website. Totally, 483 candidates have been registered. The Program Committee pre-selected the submitted abstracts on the basis of the peer review procedure with the involvement of external experts. 286 selected participants were admitted to the conference. Upon the end of the conference, the participants sent in extended versions of their works in the format required for publication in the Journal of Physics: Conference Series. The program committee finally selected 221 articles for publication in the Journal of Physics: Conference Series.

- **Conference submission management system:**

  Articles were downloaded through the conference website.

- **Number of submissions received: 483**

- **Number of submissions sent for review: 483**

- **Number of submissions accepted: 221**

- **Acceptance Rate (Number of Submissions Accepted / Number of Submissions Received X 100): 46**

- **Average number of reviews per paper: 2**

- **Total number of reviewers involved: 10**

- **Contact person for queries:**

Prof. Alexey E. Zhukov,
HSE – St Petersburg,
3(1) Kantemirovskaya str, St Petersburg, 194100, Russia
aezhukov@hse.ru

Nanostructuring at oblique incidence deposition of cobalt

A A Akulov[1], O S Trushin[2], A A Popov[2], A N Pestova[2], L A Mazaletsky[2] and A A Lomov[3]

[1] Department of Nanotechnology in Electronics, P.G. Demidov Yaroslavl State University, Sovetskaya Street 14, Yaroslavl, 150003, Russia
[2] Valiev institute of physics and technology of RAS, Yaroslavl Branch, Universitetskaya Street 21, Yaroslavl, 150007, Russia
[3] Valiev institute of physics and technology of RAS, Nakhimovsky Prospect 36, Moscow, 117218, Russia

alekseyakulov@mail.ru, otrushin@gmail.com

Abstract. Nanocolumnar Co thin films growth by oblique angle deposition on Si substrate is experimentally studied. Formation of regular arrays of tilted Co nanocolumns has been observed at incidence angles more than 70°. It was found that the optimal conditions for nanostructuring are realized at the inclination angle 85°. As obtained films have magnetic anisotropy axis inclined to the substrate surface and oriented along nanocolumns. Such films might be perspective material for applications as a magnetic recording media for next generations of hard disks.

1. Introduction

The progress in the field of microelectronics is largely due to the improvement of the technology for producing thin films. A promising method for the formation of films with special properties is their nanostructuring during growth. One of the known technological methods allowing the growth of nanostructures is oblique angle deposition [1,2]. It is well known that by varying the angle of incidence one can change the direction of magnetic anisotropy [3-5]. Magnetic thin films with easy axis tilted to the substrate surface are of considerable interest as a promising material for ultra high density magnetic recording [6]. Main goal of this work was to study the mechanisms of the growth of nanostructured Co thin films by oblique angle deposition and finding optimal conditions for the formation of perpendicular magnetic anisotropy (magnetization vector tilted to the film surface).

2. Experiment

Experiments on the deposition of Co films on an inclined substrate were carried out on an Oratoria-9 electron beam evaporation unit. The deposition conditions were as follows: base vacuum 4 10^{-6} Torr, electron beam voltage 8 kV, current 0.5 A. A standard single-crystal silicon wafer with a thermal oxide layer 300 nm thick was used as a substrate. As a substrate, we used a piece of a standard single-crystal silicon wafer with orientation (001) of rectangular shape 20*15 mm in size with a 300-nm-thick thermal oxide layer. In order to ensure the same conditions for deposition at different angles of inclination of the substrate, a special tooling was made, which made it possible to simultaneously position 6 samples with variable angles. The distance from the evaporator to the substrate was about 1

Journal of Physics: Conference Series **2086** (2021) 012001 doi:10.1088/1742-6596/2086/1/012001

m. This significant distance from the source of the evaporated material provided a high level of flow uniformity. The schematic of the experiment is shown in Figure 1.

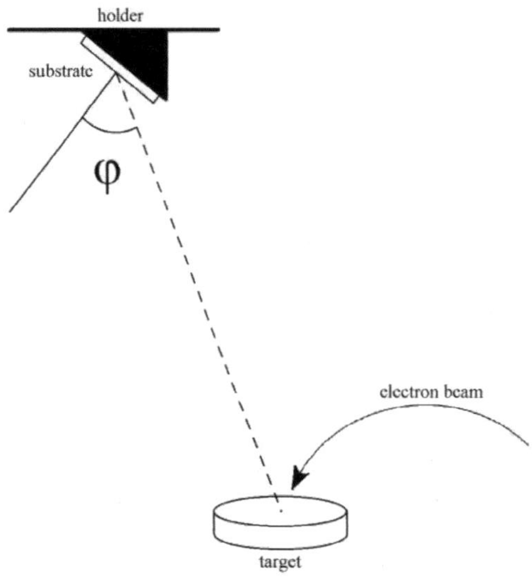

Figure 1. Schematic of the experiment for the oblique angle deposition.

All experiments were done at room temperature. The growth rate of the films varied from 1.5 nm/sec to 0.9 nm/sec with a change in the angle of inclination of the substrate from 50° to 85°.The surface morphology of the obtained films was investigated by scanning electron microscopy (SEM) (Supra 40). Morphology of the film surface was studied using atomic force microscopy (AFM) (SMM-2000). Magnetization reversal curves were measured using vibromagnetometer LakeShore VSM.

3. Results

Cross-section of the film deposited at incidence angle φ=85° is shown at Figure 2(a). One can see that tilted fibrous microstructure is formed at these conditions. More complete information about the morphology of the film is obtained from the analysis of top view on its surface shown at Figure 2(b). One can see that separate Co fibers have shape of narrow and thin strips with the width less than 40 nm and the thickness of several nanometers.

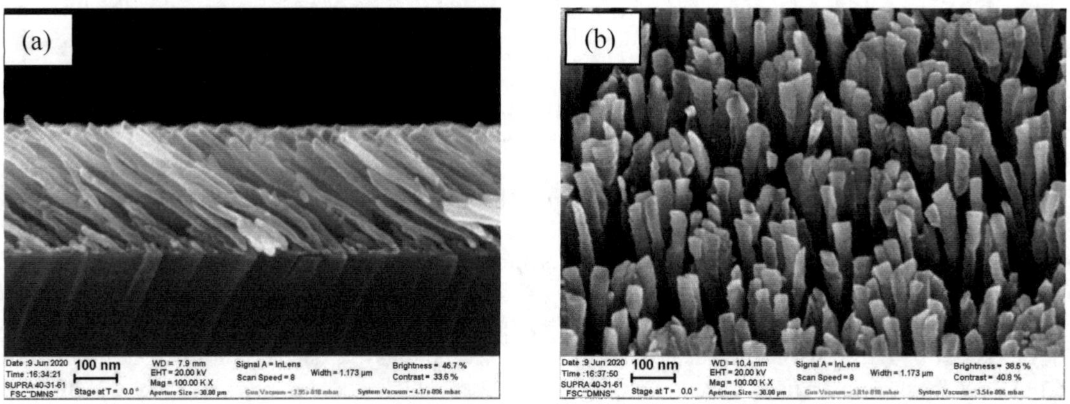

Figure 2. Cross-section (a) of the Co film and top view (b) on its surface. The images were obtained using electron scanning microscopy (SUPRA-40).

These observations are also confirmed by the results of atomic force microscopy, shown at Figure 3. Similar results for nanostructuring at oblique angle deposition of Al were reported earlier [7].

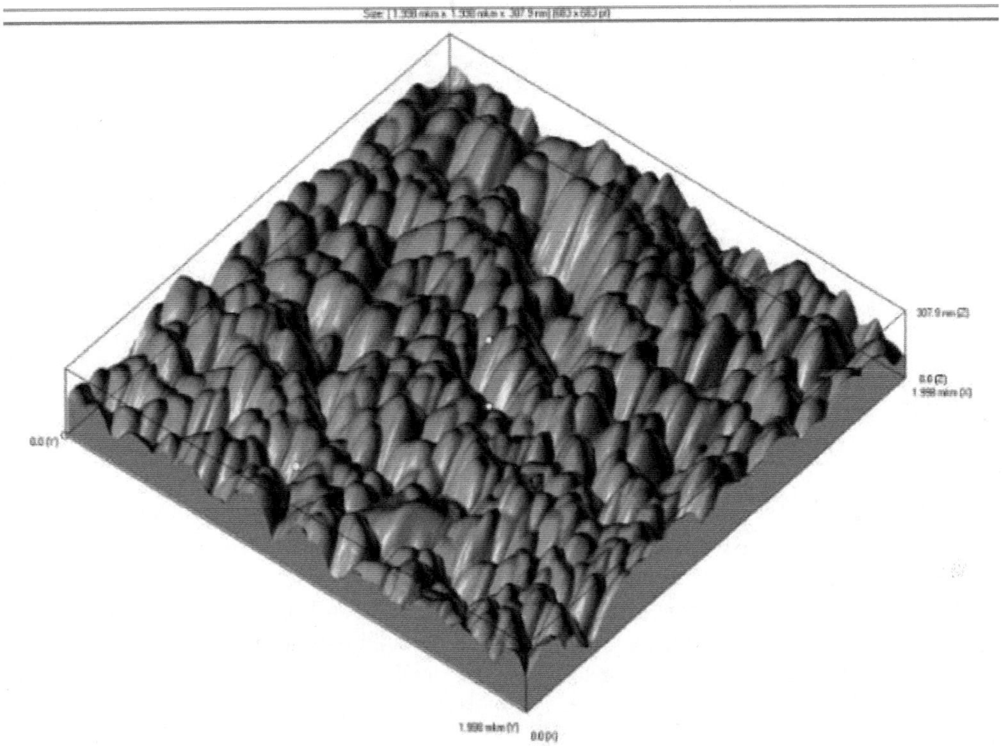

Figure 3. The morphology of Co film obtained using atomic force microscopy (SMM-2000).

Magnetization reversal for as obtained Co films were measured using vibromagnetometer LakeShore VSM at room temperature. Corresponding curves are shown in Figure 4(a,b) and Figure 5(a,b).

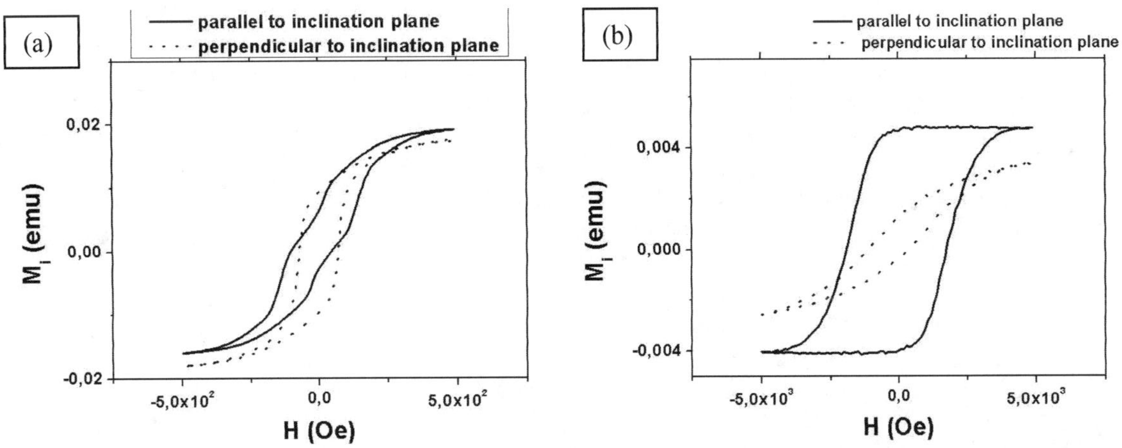

Figure 4. Magnetization reversal curves for the external field parallel to the substrate surface with two possible orientations with respect to the inclination plane. Angle of inclination of the substrate (a) φ=50° and (b) φ=80°.

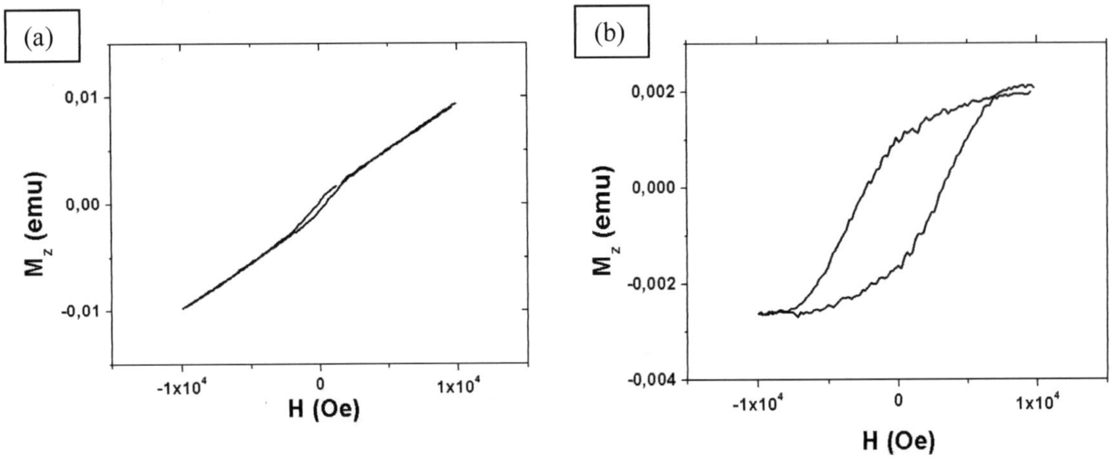

Figure 5. Magnetization reversal curves for the external field normal to the substrate surface. Angle of inclination of the substrate (a) φ=50° and (b) φ=80°.

From the analysis of Figure 4(a,b) one can see that with the increase of inclination angle from 50° to 80° orientation of the easy axis is changed. In case of φ=50° easy axis is oriented perpendicular to the inclination plane, while for φ=80° it is parallel. From other side Figure 5(b) shows that at large inclination angle nonzero component of magnetization normal to the substrate surface is appearing. Thus. it may be concluded that manipulation with the inclination angle may result in obtaining magnetic media having perpendicular anisotropy component.

4. Conclusion

In summary, as the results of experimental studies of the growth of Co thin films by oblique angle deposition it was found that optimal conditions for nanostructuring are achieved at the angle of 85°. As obtained films have magnetic anisotropy axis inclined to the substrate surface and oriented along nanocolumns. Such films might be perspective material for applications as a magnetic recording media for next generations of hard disks.

Acknowledgments

Reported study was carried out under State Programs #0066-2019-0003 and # 0856-2020-0006 of the Ministry of Science and Higher Education of Russia on the equipment of the center for collective use of the scientific equipment «Diagnostics of micro- and nanostructures».

References

[1] Barranco A, Borras A, Gonzalez-Elipe A and Palmero A 2016 *Perspectives on oblique angle deposition of thin films: From fundamentals to devices* (*Progress in Materials Science* vol 76) pp 59–153
[2] Hawkeye M, Taschuk M and Brett M 2014 *Glancing Angle Deposition of Thin Films* (UK: John Wiley & Sons Ltd) pp 1-299.
[3] Kondorsky E and Denisov P 1970 *J. IEEE Trans. on Magn.* MAG-**6** 167
[4] Alameda J, Carmona F, Salas F, Alvarez-Prado L, Morales R, Perez G 1996 *J. Magn. and Magn. Mat.* **154** 249-53
[5] Tang F, Liu D-L, Ye D-X, Zhao Y-P, Lu T-M, Wang G-C, and Vijayaraghavan A 2003 *J. Appl. Phys.* **93** 4194
[6] Piramanayagam S N 2007 *J. Appl.Phys.* **102** 011301
[7] Trushin O, Popov A, Pestova A, Mazaletsky L and Akulov A 2021 *J. Tech. Phys. Lett.* **47** 31

Synthesis of gold nanoparticles by the spark discharge method for visible plasmonics

V I Borisov[1], A A Lizunova[1], D Malo[1], E I Kameneva[1], A A Ramanenka[2] and V V Ivanov[1]

Moscow Institute of Physics and Technology, National Research University,
[1]Institutsky lane 9, Dolgoprudny, Moscow region, 141700, Russia
[2]National Academy of Sciences of Belarus, B.I. Stepanov Institute of Physics, Minsk, 220072, Belarus

Corresponding author's e-mail: borisov.vi@mipt.ru

Abstract. This work demonstrates synthesis of metal Au nanoparticles with a plasmon resonance in the visible optical region by the spark discharge method in atmosphere of argon of purity 6.0. With raising of sintering temperature from 25 to 950 °C, the morphology of synthesized Au nanoparticles changed from agglomerates to individual particles with decreasing the median size from 270 to 90 nm according to aerosol spectrometer. While by transmission electron microscopy primary nanoparticles with a gold crystalline structure with sizes in range from 5 to 120 nm were observed. Synthesized nanoparticles ensembles had broad absorption peaks with maximum in the visible optical region with peak positions approximately at 490 nm. High temperature sintered particles had a spherical shape and an additional absorption peak at approximately 640 nm.

1. Introduction

The plasmon enhancement effect is interest for purposes such as sunlight energy conversion, catalysis, chemical sensors, cell imaging and Raman spectroscopy. In the last two decades Au nanoparticles characterized by plasmon resonance in a visible range became a subject of intense research, due to variety of optical properties related to surface plasmons, high chemical and physical stability and biocompatibility and ease process of surface functionalization with organic and biological molecules [1].

It is known that the size, shape, presence of pores, the distance between the Au particles in the array and their agglomeration play a crucial role to the position of Surface Plasmon Resonance (SPR) peak. The absorption spectra of particles depend on the thickness and chemical composition of the shell around the particle and also the refractive index of the medium. For spherical Au particles with diameter of 20 nm SPR band position changes from 510 to 590 nm with an increase in the refractive index of surrounding media from 1 to 2 [2]. Whereas for nanorods the maximum of absorption peak is located at yellow-red region with wavelength more than 600 nm and can shift to the near-IR region up to 1000 nm with an increase in aspect ratio (length to diameter) from 2 to 6 [3]. Likewise the redshift of the absorption peak during agglomeration are applied in colorimetric sensing based on plasmon coupling [4].

Au nanoparticles are successfully applied in Surface-Enhanced Raman Scattering imaging techniques because it simultaneously ensures fingerprint recognition of chemical species and high detection sensitivity, down to single-molecule level. These facts have tend to application of SERS in many fields as electronics, analytical and bio-chemistry, nanomedicine and cultural heritage [5,6]. The important factors that significantly affect the optical properties of plasmonic structures and the Raman enhancement factor are the aggregation degree and particle shape [6]. In the research [7] the SERS effect of rhodamine was observed at excitation at a wavelength of 785 nm near Au nanoparticles about 150 nm and was found that the enhancement factor systematically increases in order: spherical particles, aggregates of spherical particles, triangular nanoparticles and particles in the form of stars. All of this makes the investigation of new methods of the Au nanoparticles with different size, shape and morphology synthesis urgent.

Different chemical and physical methods are used to produce Au nanoparticles, the main synthetic routes based on reduction of chloroauric acid using sodium citrate or sodium borohydrate with various additional ligand exchange modifications [8]. The physical methods include laser ablation in liquid [9], ionizing and microwave radiation and spark discharge synthesis [10,11]. The spark discharge synthesis allows to synthesize spherical nanoparticles [12] with variable morphology and optical parameters [13], which can be used for the production of microstructures by aerosol jet printing [14]. Aggregates of Au nanoparticles and their reshaped under thermal treatment in tube furnace were obtained by different groups, but optical properties of produced nanoparticles were not discussed [15].

Hence the goal of this research is to produce Au particles in the spark discharge and investigate the influence of temperature treatment on the shape, morphology, size and optical properties of obtained aerosol nanoparticles.

2. Methods and Materials

Custom-built spark discharge generator (SDG), the similar scheme of the setup was presented in [16], was used with the following conditions: capacitor 107 nF, discharge voltage 0.6 kV and pulse repetition rate 0.5 kHz. Nanoparticles synthesis was carried out in the discharge chamber that was made from glass (Duran glass, KF50, Millab) in which two hollow Au electrodes (mass fraction of gold is 0.9999) were opposite mounted. Gas path that was made of stainless steel vacuum fittings KF was pumped to a pressure of 0.4 torr before gas supplying. Further argon of purity 6.0 (oxygen volume fraction lesser than $3\cdot10^{-5}$) at a pressure of 1.5 bar and a flow rate of 1 L/min was fed to the discharge chamber through the hollow electrode, that ensured effective nanoparticles removing from the discharge gap. Synthesized nanoparticles has been passed for 2 seconds through a custom-built tube furnace at temperatures from 25 to 950 °C [17] and were collected on TEM copper grid with a carbon film and a cellulose filter.

The particle size distribution in the flow was measured by using a TSI SMPS 3936 Aerosol spectrometer. These measurements give an equivalent size of the primary nanoparticles agglomerates. The images of primary nanoparticles, their size and crystal structure was received by microscopy studies on transmission electron microscope (TEM) Jeol JEM 2100 (200 kV) with energy dispersive X-ray spectrometer X-MAXN OXFORD Instruments. The UV-vis-NIR spectra of nanoparticles on a cellulose filter were obtained using JASCO V–770 spectrophotometer.

3. Results and Discussion

TEM and electron diffraction images show that the particles have crystalline structure, the average primary particle size in ten samples synthesized at 25, 100, 150, 180, 210, 350, 500, 650, 800 and 950 °C varies from 5 to 120 nm. With an increase in sintering temperature (from 210 °C) the median agglomerates diameter decreases more than twice, that means the sintering of nanoparticles occurs (Figure 1a). However, according to TEM images complete sintering with the formation of spherical particles occurs at temperatures above 950 °C. All obtained nanoparticles nevertheless of its morphology have only face-centered gold cubic crystal structure (Figure 1c). The average diameter of primary nanoparticles is estimated to be 8.9 ± 4.6 nm (Figure 1d), while after sintering at 950 °C the agglomerates reshaped into individual near-spherical nanoparticles with mean size of 56 ± 23 nm and a broad size distribution (Figure 1e).

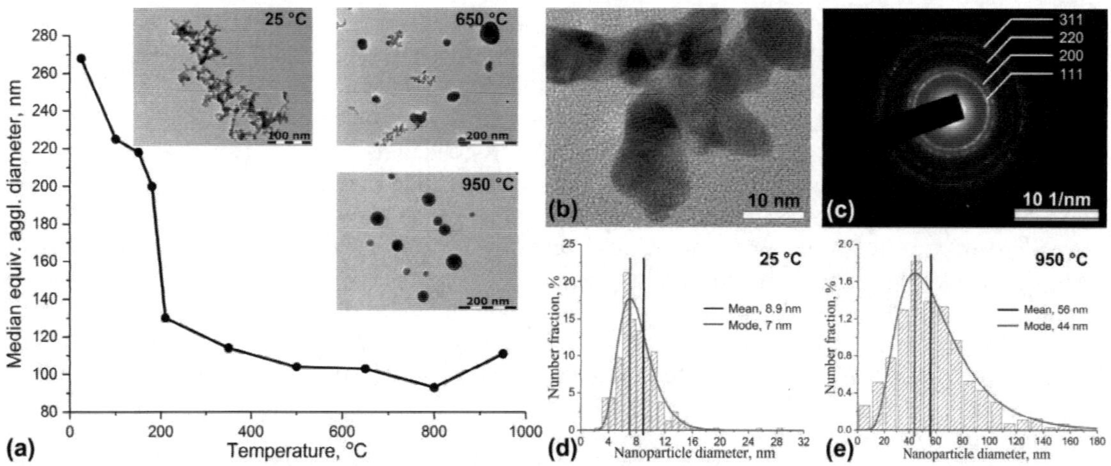

Figure 1. **(a)** Dependence of median agglomerates diameter of the Au nanoparticles on the sintering temperature; Insets: TEM image of typical Au particles at 25, 650 and 950 °C. **(b)** High-resolution TEM image of nanoparticles in aggregate, sintered at 25 °C, **(c)** Selected Area Diffraction Pattern from Au nanoparticles aggregate, produced in spark discharge without thermal treatment and histograms of the particle size distribution of primary Au nanoparticles sintered at **(d)** 25 and **(e)** 950 °C.

The measurements of the Au nanoparticles absorption spectra showed the presence of broad extinction peaks in the visible optical region with peak positions approximately at 490 nm for all sintering temperatures. There are two samples for particles sintereted at 950 °C with collection time 30 and 45 minutes. The second sample has an additional extinction peak at 640 nm (Figure 2a), which can be associated with agglomeration of nanoparticles deposited on the filter due to the long collection time [1].

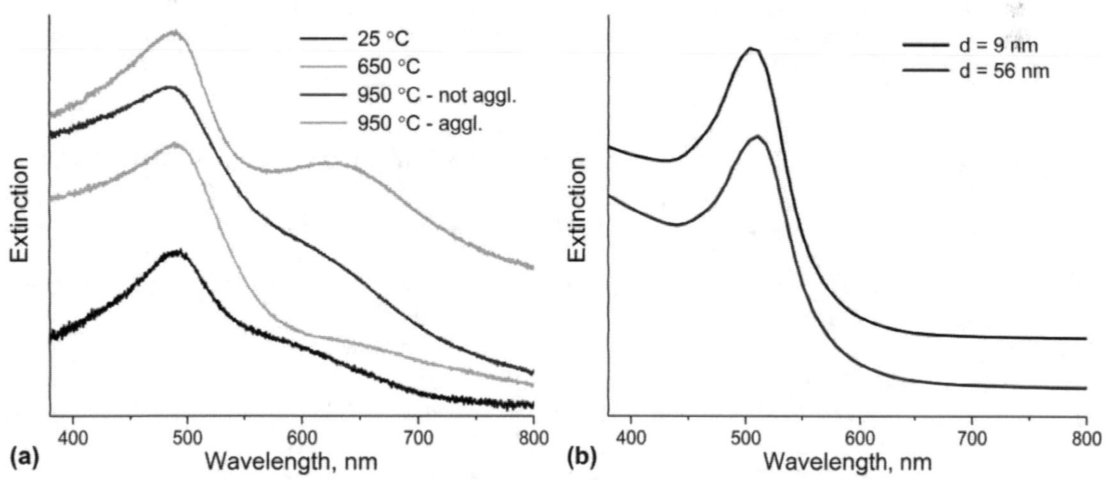

Figure 2. **(a)** Extinction spectra of synthesized Au nanoparticles sintered at 25, 650 and 950 °C, **(b)** simulated extinction curves according to the Mie theory for individual monodispersed Au nanoparticles in argon media with the sizes of 9 and 56 nm.

According to the Mie theory (Figure 2b) with an increase in the primary nanoparticles diameter from 9 to 56 nm there is a slight shift of the extinction peak from 503 to 510 nm, which is absent in the experimental extinction spectra. The mismatch is caused by the broad size distribution of sintered nanoparticles. In future experiments the size distribution of primary nanoparticles can be narrowed by extraction of monodisperse aerosol flow using a differential mobility column, which separates nanoparticles by their electrical mobility.

Thus, a simple synthesis of metal Au nanoparticles with plasmon resonance in the visible optical region by the spark discharge method has been shown. By changing sintering temperature from 25 to 950 °C one can vary the average size of primary nanoparticles and their agglomerates from 9 to 56 nm and from 270 to 90 nm, respectively. However, regardless of the sintering temperature, nanoparticles have a broad extinction peak of 490 nm, which caused by a broad size distribution of sintered primary nanoparticles. To vary the extinction peak, in addition to thermal treatment it is necessary to obtain nanoparticles with a narrow size distribution by separating sintered nanoparticles by size. It is possible to focus nanoparticles on substrate or collect them for production nano-inks [17], that give capacity to use this method as a nanoparticles source in the aerosol jet printing [16].

Acknowledgments

The reported study was funded by RFBR, project number 20-53-00042, and BRFBR, project number T20P-140.

References

[1] Amendola V, Pilot R, Frasconi M, Maragò O M and Iatì M A 2017 *J. Phys.: Condens. Matter.* **29** 203002
[2] Amendola V 2016 *Phys. Chem. Chem. Phys.* **18** 2230–41
[3] Zijlstra P, Chon JWM and Gu M 2009 *Nature* **459** 410–3
[4] Guo L, Jackman J A, Yang H-, Chen P, Cho N- and Kim D- 2015 *Nano Today* **10** 213–39
[5] Pilot R, Signorini R, Durante C, Orian L, Bhamidipati M and Fabris L 2019 *Biosensors* **9** 57
[6] Simonenko N P, Solovey V R, Shumikhin K V, Lizunova A A, Lisovskii S V, Liubavskaya E A, Seregina T V, Basova I G, Dyakonova Y B, Simonenko T L, Simonenko E P, Volkov I A, Khalturin Y A and Ivanov V V 2020 *Heritage Science* **8** 6
[7] Tian F, Bonnier F, Casey A, Shanahan A E and Byrne H J 2014 *Anal. Methods* **6** 9116–23
[8] Trotsiuk L, Antanovich A, Lizunova A and Kulakovich O 2020 *Colloids Interface Sci. Commun.* **37** 100289
[9] Correard F, Maximova K, Estève M-, Villard C, Roy M, Al-Kattan A, Sentis M, Gingras M, Kabashin A V and Braguer D 2014 *Int. J. Nanomed.* **9** 5415–30
[10] Tabrizi N S, Ullmann M, Vons V A, Lafont U and Schmidt-Ott A 2009 *J. Nanopart. Res.* **11** 315–32
[11] Ivanov V V, Efimov A A, Mylnikov D A and Lizunova A A 2019 *Russ. J. Phys. Chem.* **92** 607–12
[12] Lizunova A A, Mylnikov D A, Efimov A A and Ivanov V V 2017 *J. Phys.: Conf. Ser.* **917** 032031
[13] Borisov V I, Lizunova A A, Mazharenko A K, Malo D, Ramanenka A A, Shuklov I A and Ivanov V V 2020 *J. Phys.: Conf. Ser.* **1695** 012021
[14] Efimov A, Arsenov P, Kornyushin D, Lizunova A, Volkov I and Ivanov V 2020 *Materials* **13** 730
[15] Messing M E, Dick K A, Wallenberg L R and Deppert K 2009 *Gold Bull.* **42** 20–6
[16] Efimov A A, Arsenov P V, Borisov V I, Buchnev A I, Lizunova A A, Kornyushin D V, Tikhonov S S, Musaev A G, Urazov M N, Shcherbakov M I, Spirin D V and Ivanov V V 2021 *Nanomaterials* **11** 234
[17] Lizunova A, Mazharenko A, Masnaviev B, Khramov E, Efimov A, Ramanenka A, Shuklov I and Ivanov V 2020 *Materials* **13** 1–10

The influence of temperature and arsenic molecular form at crystallization stage on the InAs nanostructure growth during droplet epitaxy

N E Chernenko, S V Balakirev, M M Eremenko and M S Solodovnik

Institute of Nanotechnologies, Electronics and Equipment Engineering, Southern Federal University, Taganrog 347922, Russia

nchernenko@sfedu.ru

Abstract. In this work, experimental studies of the influence of the arsenic molecular form (di- or tetramers) and substrate temperature on the crystallization of In/GaAs droplet nanostructures during droplet epitaxy have been carried out. We have shown the critical influence of the temperature and arsenic molecular form on the reproducibility of the characteristics of an array of self-organizing InAs nanostructures during crystallization. We also showed that a decrease in the initial In droplet size has a positive effect on the reproducibility of the parameters of the InAs nanostructures arrays. Our work also showed important role of substrate temperature at the crystallization stage.

1. Introduction

Droplet epitaxy is currently one of the most interesting methods of molecular beam epitaxy due to its capabilities in the formation of semiconductor and hybrid nanostructures [1 – 4]. In turn, these structures have great potential for the creation of nanophotonic elements and devices, sources of single and entangled photons, and promising computing platforms on their basis [5 – 7]. Quantum-dimensional structures obtained by the droplet epitaxy technique differ from nanostructures obtained by the Stransky-Krastanov growth method not only in the variety of shapes, but also in functional characteristics, which is due to a number of specific processes underlying the mechanism of droplet nanostructure formation [8 – 12]. At the same time, despite a number of advantages, the use of this technology is still very limited due to its complexity and a huge number of control parameters [4, 13 – 19]. This is especially true of the crystallization stage when metal nanodroplets transform into crystalline semiconductor nanostructures [20 – 23]. It should also be noted separately that the GaAs/AlGaAs system remains the most studied system, while In-containing systems have practically not been studied.

The goal of this work was experimental study the effect of droplet crystallization temperature and arsenic molecular form on the crystallization of indium nanodroplet structures with different initial sizes and their transformation into InAs nanostructures during droplet epitaxy.

2. Experiment

For experimental study, we used GaAs (100) epi-ready wafers. After oxide native removal at substrate temperature of 580°C and growth of buffer layer with thickness of 250 nm we closed all sources. Then substrate temperature was cooled down to droplet deposition temperature. The droplet formation

Journal of Physics: Conference Series **2086** (2021) 012003 doi:10.1088/1742-6596/2086/1/012003

temperature was 150°C and 300°C. Effective thickness of indium deposition was 3.0 for all samples to guaranteed obtain droplet structures in the entire range of conditions under consideration.

Then we exposed obtained samples under identical conditions in equivalent arsenic fluxes at two different cracking zone temperatures of the valve source, providing arsenic supply in the form of dimers (As_2) or tetramers (As_4) and different substrate temperature during crystallization. Additionally, we also changed the temperature of the substrate at the crystallization stage to establish the relationship between the morphological characteristics of the arrays of droplet and crystalline nanostructures.

After the completion of the nanostructure formation all samples was examination by atomic force and electron microscopy techniques to obtain information about morphology and density of InAs nanostructures.

3. Results and discussion

Our experimental studies have shown that the crystallization temperature has a crucial effect on InAs nanostructure formation. As it can be seen from Figure 1, decreasing of temperature during crystallization stage from 500°C to 400°C leads to to suppression of separate InAs quantum dot synthesis and the stage-by-stage formation of the nanodots + ring + diffusion disk and diffusion disk (without any dots) systems at lowest sample temperature. Moreover, it should be noted that a decrease in temperature at the stage of crystallization does not lead to a change in the density of InAs/GaAs structures, only their morphology.

At the same time, in the high-temperature region, there are no diffuse regions around the nanostructures, which may indicate that most of the indium remained in the droplet and transformed into the InAs nanocrystals. In this case, the initial In droplet size is practically irrelevant – the results obtained for various thicknesses of indium deposition correlate well with each other. We believe that this behavior of the system can be due to both thermodynamic and kinetic factors that directly affect the reaction between the metal and arsenic.

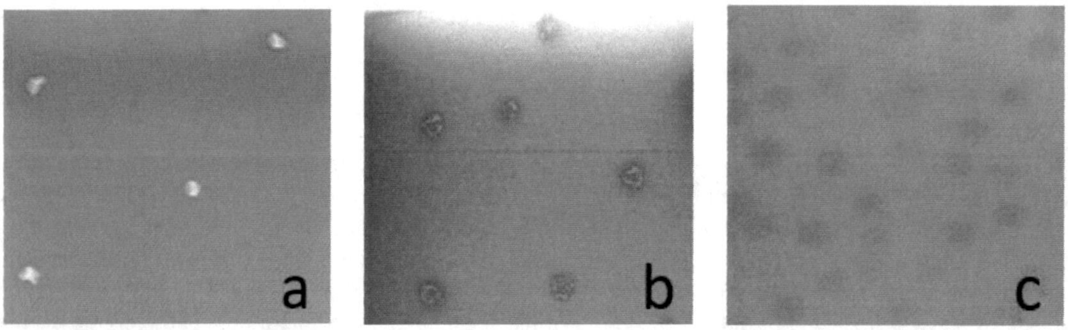

Figure 1. SEM images of InAs nanostructures, obtained during crystallization of In nanodroplets using As_4 flux at different temperatures: *(a)* 500°C, *(b)* 450°C, *(c)* 400°C.

The study of arsenic molecular form influence has shown that the use of tetramers (As_4) makes it possible to obtain crystal InAs nanostructures which correlate in their parameters (size, density) with their predecessors – an ensemble of In nanodroplets. This is true over the entire range of temperatures under consideration.

It is important to note that with a decrease in the initial size of In droplet nanostructures, the degree of their correlation with the final size of InAs crystal nanostructures increases, and their morphology also improves. As it can be seen from comparative analysis of images shown in the Figure 2*a* and Figure 3*b* InAs structures obtained at low temperature and having a small initial size have a pyramidal shape with clear edges, which is characteristic of quantum dots. Whereas large metal nanodroplet obtained at high temperature transform to flower-like nanostructures during crystallization stage. It is important to note that this result is very similar to the results obtained in the previous step. Therefore, apparently, we can say that the droplet crystallization process is largely limited precisely by kinetic factors.

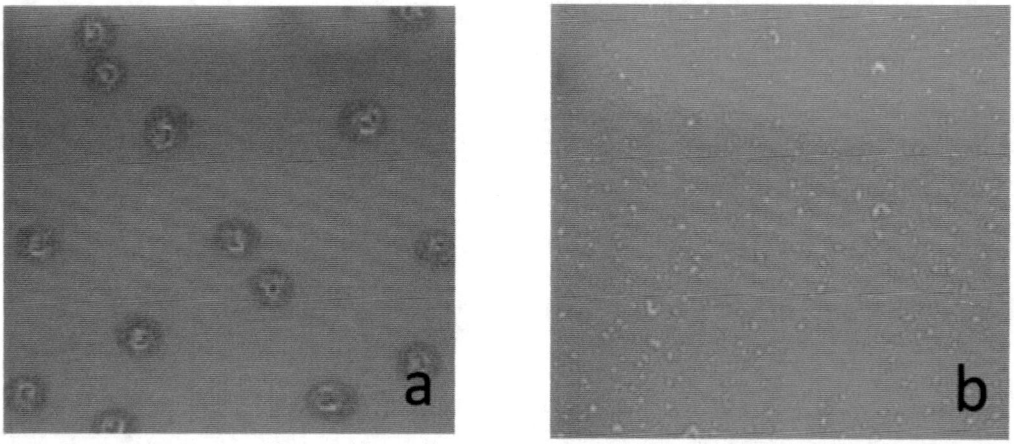

Figure 2. SEM images of InAs nanostructures obtained by crystallization of In droplets in a fluxes of arsenic tetramers *(a)* and dimers *(b)* after deposition of 3.0 ML at growth temperature of 300°C.

In the case of dimers, the situation is completely different. On the one hand, the use of a flux of arsenic dimers leads to the formation of structures with a diameter of 20 – 50 nm in the entire range of considered conditions (Figure 2*b* and Figure 3*b*).

At the same time, the correlation between the parameters of the ensemble of metal nanodroplets and the final crystal structures is practically not observed either in density or in size – regardless of the initial density and the initial size of the structures after crystallization, approximately the same picture is observed. The difference between InAs crystal structures obtained using different arsenic forms increases with increasing growth temperature. As it can be seen from SEM images in Figure 2 and Figure 3, for low-temperature samples (small initial droplet size), the indium structures are not very different in size and density for the crystal structures obtained using tetramers (Figure 3*a*). Although they have a wide variation in size and shapes.

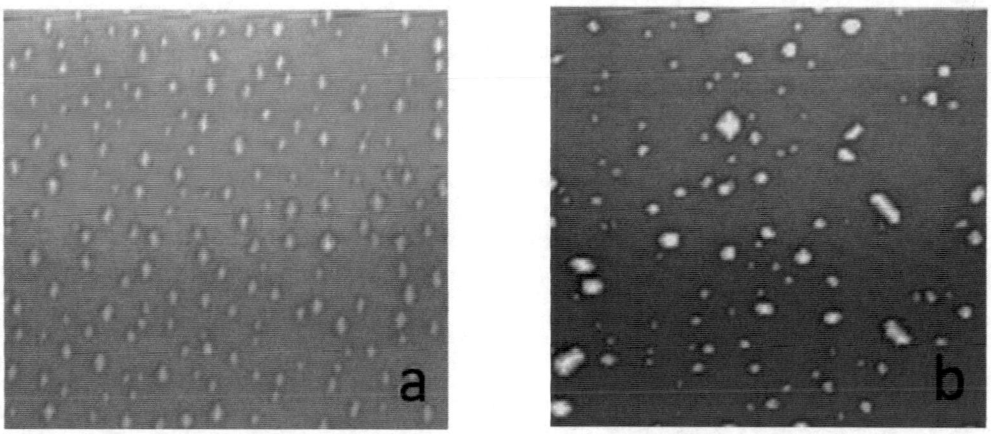

Figure 2. SEM images of InAs nanostructures obtained by crystallization of In droplets in a fluxes of arsenic tetramers *(a)* and dimers *(b)* after deposition of 3.0 ML at growth temperature of 150°C.

At the same time, high-temperature samples are characterized by a more uniform size distribution with a much larger discrepancy between the droplet and crystal structure sizes. This effect of the molecular form of arsenic may be due to the different kinetics of the reactions occurring during the crystallization process using di- and tetramers.

4. Conclusion

Thus, we have shown that the crystallization temperature is extremely important for the controlled and reproducible production of crystalline InAs/GaAs nanostructures in droplet epitaxy. It is shown that an increase in crystallization temperature leads to an improvement in the correlation of parameters between the arrays of droplet and crystalline nanostructures. It was also shown that the use of dimers significantly alters the crystallization processes, in contrast to tetramers. This leads to the fact that during crystallization in the case of using arsenic dimer fluxes, there is no correlation in size and density between droplets and crystal nanostructures. Therefore, the use of arsenic tetramers at the crystallization stage seems to be preferable in InAs droplet epitaxy.

Acknowledgments

This work was supported by the Russian Science Foundation Grant No. №19-79-10099.

References

[1] Mano T, Abbarchi M, Kuroda T, Mastrandrea C, Vinattieri A, Sanguinetti S, Sakoda K and Gurioli M 2009 *Nanotechnology* **20** 395601
[2] Wu J, Wang Z M, Li X, Mazur Y I and Salamo G J 2017 *J. Mater. Res.* **32** 4095–101
[3] Alonso-González P, Martín-Sánchez J, González Y, Alén B, Fuster D and González L 2009 *Cryst. Growth Des.* **9** 2525–8
[4] Balakirev S V, Chernenko N E, Eremenko M M, Ageev O A and Solodovnik M S 2021 *Nanomaterials* **11** 1184
[5] Heindel T et al. 2012 *New J. Phys.* **14** 083001
[6] Bayer M, Stern O, Hawrylak P, Fafard S and Forchel A 2000 *Nature* **405** 923–6
[7] Lochmann A, Stock E, Töfflinger J, Unrau W, Toropov A, Bakarov A, Haisler V and Bimberg D 2009 *Electron. Lett.* **45** 1
[8] Koguchi N, Takahashi S and Chikyow T 1991 *J. Cryst. Growth* **111** 688–92
[9] Balakirev S V, Solodovnik M S, Ageev O A 2018 *Phys. Status Solidi B* **255(4)** 1700360
[10] Mikhaylin I A, Balakirev S V, Eremenko M M, Chernenko N E and Solodovnik M S 2019 *J. Phys. Conf. Ser.* **1410** 012051
[11] Balakirev S V, Eremenko M M, Chernenko N E, Ageev O A and Solodovnik M S 2019 *J. Phys. Conf. Ser.* **1410** 012059
[12] Eremenko M M, Balakirev S V, Chernenko N E, Ageev O A and Solodovnik M S 2019 *J. Phys. Conf. Ser.* **1410** 012045
[13] Sanguinetti S, Watanabe K, Tateno T, Gurioli M, Werner P, Wakaki M and Koguchi N 2003 *J. Cryst. Growth* **253** 71–6
[14] Mano T, Watanabe K, Tsukamoto S, Imanaka Y, Takamasu T, Fujioka H, Kido G, Oshima M and Koguchi N 2000 *Jpn. J. Appl. Phys.* **39** 4580
[15] Mano T, Watanabe K, Tsukamoto S, Fujioka H, Oshima M and Koguchi N 2000 *J. Cryst. Growth.* **209** 504
[16] Watanabe K, Koguchi N and Gotoh Y 2000 *Jpn. J. Appl. Phys.* **39** L79
[17] Mantovani V, Sanguinetti S, Guzzi M, Grilli E, Gurioli M, Watanabe K and Koguchi N 2004 *J. Appl. Phys.* **96** 4416–20
[18] Balakirev S V, Solodovnik M S, Eremenko M M, Konoplev B G, Ageev O A 2019 *Nanotechnology* **30** 505601
[19] Balakirev S V, Solodovnik M S, Eremenko M M, Konoplev B G and Ageev O A 2020 *Nanotechnology* **31** 485604
[20] Schuck C, Boutelle R, Silverman K, Moody G and Simmonds P 2021 *J. Phys. Photonics* **3** 024012
[21] Li X, Wu J, Wang Z, Liang B, Lee J, Kim E and Salamo G 2014 *Nanoscale* **6** 2675
[22] Lee J H, Wang Z M and Salamo G J 2009 *IEEE Transact. Nanotechnol.* **8** 431–36
[23] Bietti S, Basset F B, Tuktamyshev A, Bonera E, Fedorov A and Sanguinetti S 2020 *Sc. Rep.* **10** 6532

Estimation of thermodynamic stability of isoperiodic epitaxial structures whith GaInSbAs and GaInAsP solid solutions

E M Danilina and A S Paschenko

Federal research centre the Southern Scientific Centre RAS, 41, Chekhov st., Rostov-on-Don, 344006, Russia

Abstract. The work studied the thermodynamic stability of GaInSbAs, GaInAsP heterosystems on different substrates. The isotherms of spinodal decomposition caused by chemical changes in the internal energy of the alloy and by elastic stresses at the layer-substrate interface are obtained with the model of quasiregular solutions. It has been found that elastic stresses lead to an expansion of the region of thermodynamic stability of isoperiodic solid solutions for GaSb substrates and a decrease in the critical temperature. The developed model can be using to selection of the technological modes and parameters of epitaxial growth.

1. Introduction

Interest in multicomponent semiconductor solid solutions of III-V compounds [1-5] is due to the need to expand the spectral range and improve the structural perfection of heterointerfaces by the lattice matching of epitaxial layers and thermal expansion coefficient, which opens up possibilities for varying the photoelectric properties of the element base of devices [2].

However, in multicomponent liquid and solid solutions under suitable conditions, the formation of unstable component phases can be observed, leading to decomposition into two coexistent liquid or solid phases of different composition, respectively [3]. The experimental study of the process is labor-intensive and intends the use of precision methods of phase analysis. At the same time, information on the solubility gap of components in the solid phase is extremely important, because the immiscibility and instability of solid solutions significantly limit the range of compositions with targeted properties for use in manufacturing optoelectronic devices [4, 5].

Therefore, theoretical studies of equilibria between the liquid and solid phases in the epitaxy process are important.

2. Materials and Methods

The analysis of thermodynamic stability between the coexisting phases will be carried out using of the Prigogine-Defey criterion [6]. According to which the diffusion stability in isobaric-isothermal processes of multicomponent systems is determined by the system of inequalities

$$\begin{cases} \dfrac{\partial^2 F}{\partial x^2} > 0; \\ \dfrac{\partial^2 F}{\partial x^2}\dfrac{\partial^2 F}{\partial y^2} - \dfrac{\partial^2 F}{\partial x \partial y} \geq 0, \end{cases}$$

where F – Helmholtz free energy;

x, y – molar fraction of elements of III and V groups in solution, respectively.

In accordance with the approximation of a regular solution, a multicomponent solution can be represented as a set of binary compounds. Then the free energy is given by the sum of: the free energy of the mole fraction of binary components, the free energy of mixing, and the excess energy of mixing with the interaction of binary components

$$F = F_0 + \Delta F_M^{Sid} + \Delta F_M^{Sex}. \tag{1}$$

The free energy of the molar fractions of binary components determines the relationship between the nearest pairs of atoms in the first coordinate sphere

$$F_0 = \omega_{AC}(1-x)(1-y) + \omega_{AD}(1-x)y + \omega_{BC}x(1-y) + \omega_{BD}xy,$$

where

A, B – elements of III groups;

C, D – elements of V groups;

ω – pair interaction energy between the nearest tetrahedral bond atoms in different sublattices.

The free mixing energy, model of an ideal solid solution, when the interaction energy between components is negligible

$$\Delta F_M^{Sid} = RT\left(x\ln x + (1-x)\ln(1-x) + (1-y)\ln(1-y) + y\ln y\right),$$

where

R – gas constant.

The excessive mixing energy that takes into account the interaction between the components of the solution within the quasi-chemical approximation

$$\Delta F_M^{Sex} = \alpha^s_{AC-BC}xy(1-x) + \alpha^s_{AD-BD}x(1-x)(1-y) + \alpha^s_{AC-AD}xy(1-y) + \\ + \alpha^s_{BC-BD}y(1-x)(1-y),$$

where

α^s – solid-phase pseudo-ternary system interaction parameter.

According to [7], the sum of the energies of pair interactions can be represented in the following form, which is convenient for experimental determination

$$\omega_{AC} - \omega_{AD} - \omega_{BC} + \omega_{BD} = \Delta S^F_{AD}\left(T^F_{AD} - T\right) + \Delta S^F_{AC}\left(T^F_{AC} - T\right) + \\ + \Delta S^F_{BD}\left(T^F_{BD} - T\right) + \Delta S^F_{BC}\left(T^F_{BC} - T\right) + \frac{1}{2}\left(\alpha^l_{AC} + \alpha^l_{AD} - \alpha^l_{BC} - \alpha^l_{BD}\right),$$

where

ΔS^F, T^F – entropy and melting point of binary compound;

α^l – liquid-phase interatomic interaction parameter.

It suffices to solve the equation to find the region of spinodal decomposition

$$\frac{\partial^2 F}{\partial x^2}\frac{\partial^2 F}{\partial y^2} - \frac{\partial^2 F}{\partial x \partial y} = 0. \tag{2}$$

Equation (2) defines the boundaries of the regions of instability, which reflects the internal interatomic interaction in the lattice of the solid solution [8]. Such a spinodal is called a chemical spinodal.

Another type is a coherent spinodal, which represents the limit of metastability of a supersaturated solid solution. To calculate the coherent spinodal, it is necessary to take into account the energy of heterogeneous deformation F_{st}, the appearance of which is a consequence of the introduction of nonperiodic microinclusions inside the unstable zone

The energy of elastic stresses from the side of the substrate when the layer is oriented in the {100} plane in the absence of misfit dislocations is given by the expression

$$F_{st} = \frac{(c_{11}+c_{12})(c_{11}-2c_{12})}{c_{11}} \frac{N_A a_l^{3/4}}{a_{sub}^2}(a_l - a_{sub})^2, \quad (3)$$

where

$$c_{1j} = \left(c_{1jAC}(1-x)(1-y) + c_{1jAD}(1-x)y + c_{1jBC}x(1-y) + c_{1jBD}xy\right),\ j = 1,2$$

– elastic coefficients of the solid solution layer;

$$a_l = \left(a_{AC}(1-x)(1-y) + a_{AD}(1-x)y + a_{BC}x(1-y) + a_{BD}xy\right)$$

– lattice parameter of the solid solution layer;

a_{sub} – substrate lattice parameter;

N_A – Avogadro's number.

The total free energy of a solid solution is the sum of the thermodynamic energy (1) and the energy of elastic stresses (3)

$$F_{tot} = F + F_{st}.$$

The energy of elastic stresses by changing the solid solution free energy, leads to a change in the limits of solubility of the components and the boundaries of the regions of instability, which are determined by a similar formula (2).

3. Results and Discussion

We calculated the regions of existence of solid solutions of quaternary compounds on GaSb, InSb, GaAs, GaP substrates.

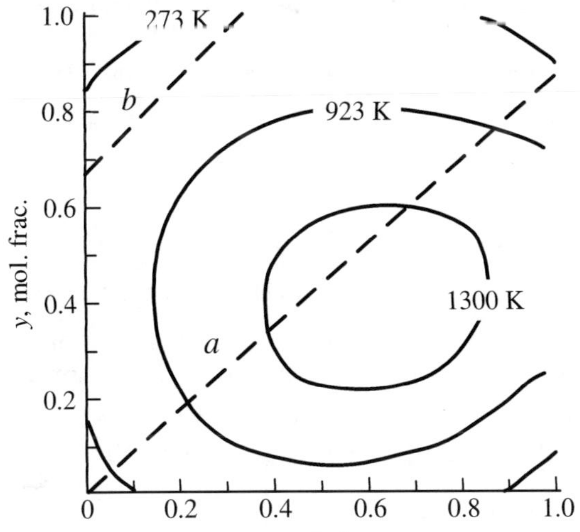

Figure 1. Isotherms of spinodal decomposition of $In_xGa_{1-x}As_ySb_{1-y}$ solution on GaSb substrate without elastic stresses, (*a*) $\Delta a = 0\%$ alloys lattice-matched to GaSb; (*b*) $\Delta a = 5\%$ to GaSb.

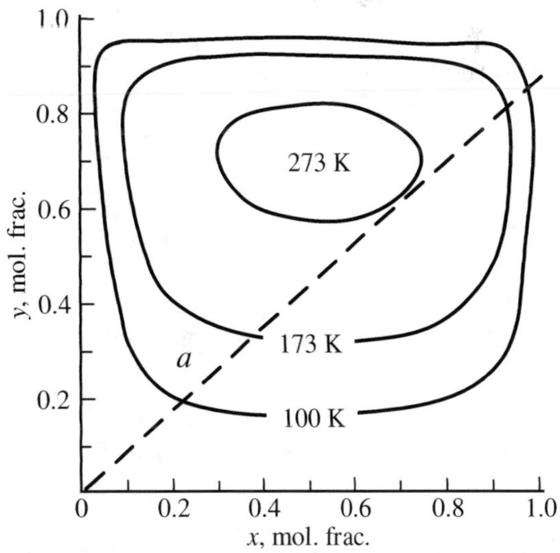

Figure 2. Isotherms of spinodal decomposition of $In_xGa_{1-x}As_ySb_{1-y}$ solution on GaSb substrate with elastic stresses, (*a*) $\Delta a = 0\%$ alloys lattice-matched to GaSb.

An analysis of heterophase equilibria in the Ga–In–Sb–As and Ga–In–As–P systems was carried out within the framework of the model of quasi-regular solid solutions. We also elucidated the regions of thermodynamic stability of solid solutions to spinodal decomposition caused by chemical changes in the internal energy of the alloy and elastic stresses at the layer-substrate interface.

It is shown that for $In_xGa_{1-x}As_ySb_{1-y}$ heterosystem on GaSb substrate, compositions with a lattice mismatch of the solid solution and the substrate can exist (Figure 1 (a), (b)). In addition, it is found that $Ga_xIn_{1-x}As_yP_{1-y}$ / GaAs heterosystem exists in a lattice-matched composition (Figure 3 (a)) and $Ga_xIn_{1-x}As_yP_{1-y}$ / GaP heterosystem exists in a lattice-mismatched composition by 5% (Figure 3 (b)).

A characteristic feature is that elastic stresses lead to an expansion of the thermodynamic stability region of isoperiodic solid solutions for GaSb substrates and a decrease in the critical temperature (Figure 2). No isoperiodic compositions were found among the considered quaternary solid solutions for InSb substrates at room temperature.

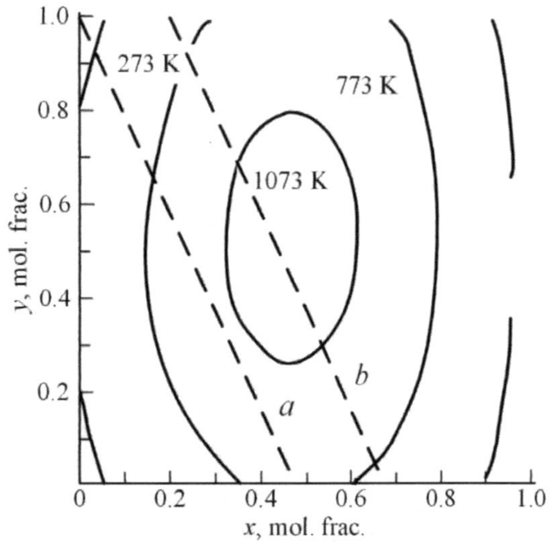

Figure 3. Isotherms of spinodal decomposition of $Ga_xIn_{1-x}As_yP_{1-y}$ solution on GaAs and GaP substrates without elastic stresses, (*a*) $\Delta a = 0\%$ alloys lattice-matched to GaAs; (*b*) $\Delta a = 5\%$ to GaP.

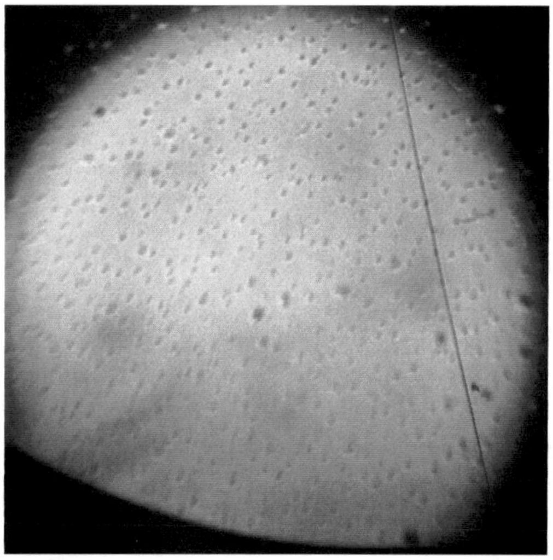

Figure 4. The grown layer of $Ga_xIn_{1-x}As_yP_{1-y}$ solution lattice-matched to GaAs substrate with spinodal decomposition areas.

The isotherms of spinodal decomposition for $Ga_xIn_{1-x}As_yP_{1-y}$ solution were also calculated (Figure 3). The results were compared with the technological data obtained during the growth of $Ga_xIn_{1-x}As_yP_{1-y}$ solid solution film on GaAs substrate (Figure 4). It shows a photograph of a heterosystem with pronounced inhomogeneities caused by spinodal decay. It is established that the growth parameters (temperature, composition) coincide with the data obtained during the simulation.

4. Conclusions

In the course of carrying out numerical experiments, it was found that the presence of elastic stresses in solid solutions leads to an expansion of the region of thermodynamic stability and a decrease in the critical temperature. Comparison of the simulation results with experimental data showed good agreement, which allows us to speak about the correctness of the calculations and the adequacy of the model, as well as the possibility of using the calculation results to substantiate the technological modes of epitaxial growth.

Acknowledgments
This work was funded by as a part of a state order to the Southern Scientific Center of the Russian Academy of Sciences, projects nos. 01201354240. This work was carried out with financial support by the Russian Science Foundation grant 19-79-10024.

References

[1] Weih R, Bauer A, Kamp M and Höfling S 2013 *J. Optical Materials Express* **3** 1624
[2] Bychinskii V A, Tupitsyn A A, Chudnenko K V, Mukhetdinova A V, Fomichev S V and Krenev V A 2013 *Russian Journal of Inorganic Chemistry* **58** 1197
[3] An N, Ma L, Wen G, Liang Zh, Zhang H, Gao T and Fan C 2019 *J. Appl. Sci.* **9** 162
[4] Craig A P, Marshall A R J, Tian Z B and Krishna S 2014 *Infrared Physics & Technology.* **67** 210
[5] Ilahi S, Yacoubi N and Genty F 2017 *Optical Materials* **69**
[6] Prigogine I, Defay R 1954 *Chemical Thermodynamics* (London: Longmans Green and Co)
[7] Onabe K 1982 *J. Phys. Chem. Solids*. **43** 1071
[8] Elyukhin V A 2019 *Journal of Crystal Growth* **515** 53

Study of In/GaAs nanodroplet formation in conditions of non-stationary supersaturation during droplet epitaxy

D D Dukhan, S V Balakirev, N E Chernenko, M M Eremenko and M S Solodovnik

Institute of Nanotechnologies, Electronics and Equipment Engineering, Southern Federal University, Taganrog 347922, Russia

Email: duhan@sfedu.ru

Abstract. In this paper, we present the results of kinetic Monte Carlo study of the In/GaAs growth by droplet epitaxy in conditions of non-stationary vapor supersaturation. These conditions allow achievement of the independent control of size and surface density of nanostructures. The material redistribution is realized on the surface when indium deposition is interrupted and leads to a decrease in the critical thickness of droplet formation. An average droplet size increases with increase in interruption time whereas the surface density decreases. However, additional nucleation within the wetting layer can also be observed during the growth interruptions, which makes it possible to increase the surface density of droplets.

1. Introduction

Independent control of size and surface density of quantum dots is one of the most challenging issues, especially in the field of quantum cryptography and computing where parameters of nanostructures and their mutual arrangement are both important [1-6]. The method of droplet epitaxy has a lot of advantages over the conventional molecular beam epitaxy due to the two-stage nanostructure formation and possibility to control the nanostructure parameters on both stages [7-13]. In order to achieve an independent control of surface density and size of nanostructures, various techniques were used, such as growth on metal-stabilized surfaces [14,15], pre-patterning of the substrates [16,17], use of subcritical deposition thicknesses [18-21], etc. However, these approaches normally do not make it possible to simultaneously reduce the wetting layer thickness. At the same time, the wetting layer has a negative influence on device properties of epitaxial heterostructures [22,23].

It is possible to reduce the thickness of the wetting layer and to improve control over the size and surface density of nanostructures by means of non-stationary supersaturation of vapor. This approach consists in a periodic interruption of the deposition in order to allow deposited material to redistribute over the surface. During this process, the growth system tends to occupy the most energetically favorable positions resulting in the formation of higher-quality nanostructures with required characteristics.

In this paper, we carry out kinetic Monte Carlo simulations to study the effect of non-stationary supersaturation of indium vapor on the parameter of nanostructures formed by droplet epitaxy on the GaAs surface.

2. Description of the model

Previously developed kinetic Monte Carlo model [18,24] was upgraded to take into account the droplet epitaxial growth in conditions of non-stationary vapor supersaturation. We introduced new parameters into the model: the duration of the growth interruption, the frequency and the total number of growth interruptions. These parameters characterize the degree of non-stationary vapor supersaturation. Depending on the values of these parameters, deposition of atoms onto the surface was interrupted for a specified time. During this period, further ripening of the growth system occurred and appeared in the migration of atoms both within the wetting layer and in the region of the formed droplets.

3. Results and discussion

The simulation results demonstrate that the material redistribution occurs on the surface during the growth interruption (Figure 1). Due to the fact that the probability of decay of a stable supercritical island is extremely small, atoms mainly migrate from the wetting layer into the droplet areas. As a result of the material redistribution, the average droplet size increases, while the thickness of the wetting layer decreases. This process has a more pronounced effect when the exposure time is increased from 1 to 3 seconds (Figure 2a). Furthermore, the difference in droplet size increases with increase in substrate temperature. At a temperature of 150 °C, the average diameter of droplets deposited in a continuous mode and with a double interruption for a period of 3 seconds increases from 18 to 22 nm (by 18%), while at a temperature of 300 °C, a change is from 107 to 144 nm (by 26%).

a
b
c

Figure 1. Morphology of nanostructure arrays after deposition of 3 ML of indium at a temperature $T = 150°C$ and growth rate $v = 0.25$ ML/s under various conditions of vapor supersaturation: a) stationary supersaturation, b) with 1-second growth interruption after deposition of 1 ML of In, c) with 3-second growth interruption after deposition of 1 ML of In. The width of the simulation area is 500 nm.

The dependence of critical thickness of droplet formation on the growth temperature (Figure 2b) clearly indicate an increase in the probability of island nucleation in the case of growth with non-stationary vapor supersaturation. This phenomenon is attributed to an increase in the probability of island nucleation as a result of the ripening of the wetting layer, which is taken into account in the simulation through the formation of a critical region of a three-dimensional island.

It was also found that the non-stationary supersaturation has an influence on the intensity of island nucleation, which is caused by an increase in the intensity of surface diffusion of adatoms due to a decrease in the deposition rate of indium atoms. As a result, the final surface density of droplets at a temperature of 300 °C decreases from $1.8 \cdot 10^8$ to $5.2 \cdot 10^7$ cm^{-2} with a two-fold interruption of growth for 3 seconds as compared with continuous deposition (Figure 3). In this case, a two-fold interruption of growth at intervals of 1 second did not lead to a significant change in the surface density of nanostructures.

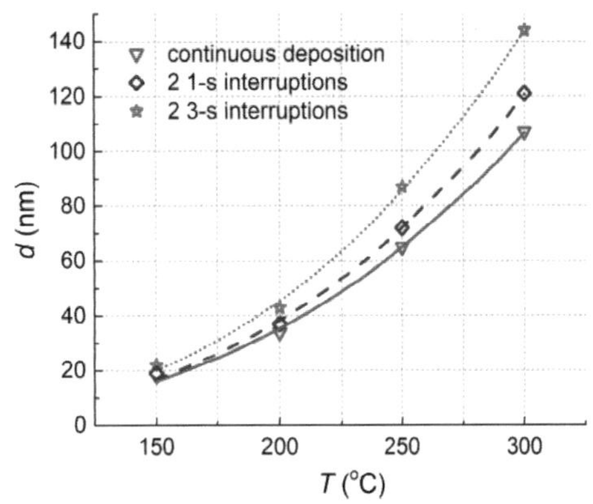

a

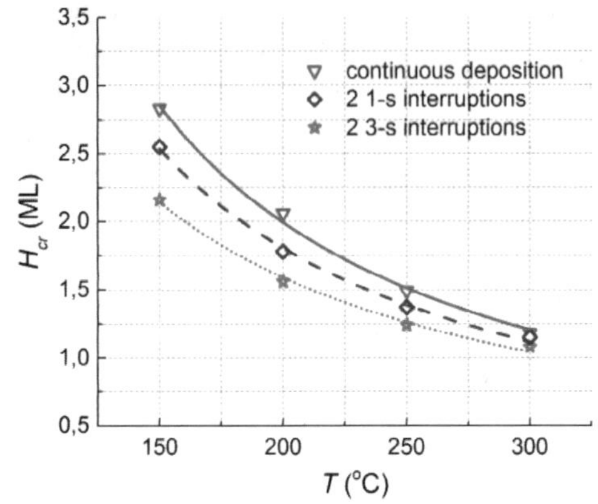

b

Figure 2. Temperature dependences of the a) average droplet diameter and b) critical thickness of droplet formation under various conditions of vapor supersaturation.

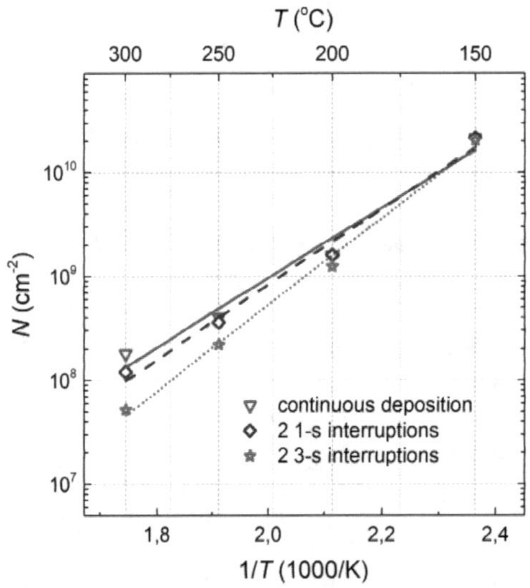

Figure 3. Temperature dependences of surface density of droplets after deposition of 3 ML of indium under various conditions of vapor supersaturation.

As in the case of the average size and critical thickness of droplet formation, the effect of interruption of deposition on the surface density of droplets becomes less significant with decreasing temperature. Thus, at a temperature of 150 °C, the surface density of droplets decreases by only 9% – from $2.2 \cdot 10^{10}$ to $2.0 \cdot 10^{10}$ cm^{-2}.

4. Conclusions

The use of non-stationary indium vapor supersaturation during droplet epitaxy allows an additional control of droplet size and surface density and makes it possible to reduce indium wetting layer thickness as well as the critical amount of deposited material to have droplets formed. During the growth interruptions, indium atoms migrate within the wetting layer and formed droplets resulting in the material redistribution. Because the probability of droplet formation increases, the critical thickness becomes smaller, which is clearer observed in case of low temperatures. At the same time, the growth interruption influence on the droplet size and surface density is more pronounced at high temperatures. The droplet size tends to increase with increase in interruption time while the surface density of droplets decreases.

Acknowledgments
This work was supported by the Russian Science Foundation Grant No. №19-79-10099.

References

[1] Bayer M, Stern O, Hawrylak P, Fafard S and Forchel A 2000 *Nature* **405** 923–6
[2] Bimberg D et al. 2009 *IEEE Photonics J.* **1** 58–68
[3] Lochmann A, Stock E, Töfflinger J, Unrau W, Toropov A, Bakarov A, Haisler V and Bimberg D 2009 *Electron. Lett.* **45** 1
[4] Moreau E, Robert I, Gérard J, Abram I, Manin L and Thierry-Mieg V. 2001 *Appl. Phys. Lett.* **79** 2865–7
[5] Heindel T et al. 2012 *New J. Phys.* **14** 083001
[6] Santori C, Fattal D and Yamamoto Y 2010 *Single-photon devices and applications* (Chichester: John Wiley & Sons)
[7] Koguchi N, Takahashi S and Chikyow T 1991 *J. Cryst. Growth* **111** 688–92
[8] Chikyow T and Koguchi N 1990 *Jpn. J. Appl. Phys.* **29** L2093
[9] Sanguinetti S, Watanabe K, Tateno T, Gurioli M, Werner P, Wakaki M and Koguchi N 2003 *J. Cryst. Growth* **253** 71–6
[10] Mano T, Watanabe K, Tsukamoto S, Imanaka Y, Takamasu T, Fujioka H, Kido G, Oshima M and Koguchi N 2000 *Jpn. J. Appl. Phys.* **39** 4580
[11] Watanabe K, Koguchi N and Gotoh Y 2000 *Jpn. J. Appl. Phys.* **39** L79
[12] Balakirev S V, Chernenko N E, Eremenko M M, Ageev O A and Solodovnik M S 2021 *Nanomaterials* **11** 1184
[13] Balakirev S V, Solodovnik M S, Eremenko M M, Konoplev B G and Ageev O A 2020 *Nanotechnology* **31** 485604
[14] Mano T, Watanabe K, Tsukamoto S, Fujioka H, Oshima M and Koguchi N 2000 *J. Cryst. Growth.* **209** 504
[15] Benyoucef M, Zuerbig V, Reithmaier J P, Kroh T, Schell A W, Aichele T and Benson O 2012 *Nanoscale Res. Lett.* **7** 493
[16] Kiravittaya S, Rastelli A and Schmidt O G 2005 *Appl. Phys. Lett.* **87** 243112
[17] Kim J S, Kawabe M and Koguchi N 2006 *Appl. Phys. Lett.* **88** 072107
[18] Balakirev S V, Solodovnik M S, Eremenko M M, Konoplev B G, Ageev O A 2019 *Nanotechnology* **30** 505601
[19] Lee J H, Wang Z M and Salamo G J 2007 *J. Phys. Condens. Mat.* **19** 176223
[20] Song H Z, Usuki T, Nakata Y, Yokoyama N, Sasakura H and Muto S 2006 *Phys. Rev. B* **73** 115327
[21] Seravalli L, Trevisi G, Frigeri P, Rossi F, Buffagni E and Ferrari C 2013 *J. Phys. D Appl. Phys.* **46** 315101
[22] Duboz J Y, Liu H C, Wasilewski Z R, Byloss M and Dudek R 2003 *J. Appl. Phys.* **93** 1320
[23] Deppe D G and Huffaker D L 2000 *Appl. Phys. Lett.* **77** 3325
[24] Balakirev S V, Solodovnik M S, Ageev O A 2018 *Phys. Status Solidi B* **255(4)** 1700360

Elastic properties of porous silicon layer of hybrid SiC/Si substrates

I A Eremeev[1,2], A A Koryakin[2] and S A Kukushkin[2]

[1] Alferov St. Petersburg Academic University, St. Petersburg 194021, Russia
[2] Institute for Problems of Mechanical Engineering, Russian Academy of Sciences, St. Petersburg 199178, Russia

Abstract. Elastic properties of porous silicon layer of hybrid SiC/Si substrates grown by the atomic substitution method are investigated. The feature of the growth method is the formation of the macroporous silicon layer at the SiC/Si interface during growth. The elastic properties of the layer are studied using the finite element method. The biaxial modulus of the porous silicon is obtained as a function of porosity considering the different shape of the pores and presence of thin SiC boundary layer. The presence of the pores in the silicon layer adjacent to SiC results in the decrease of the elastic moduli by about 35%. However, this leads to a negligibly small change of the substrate curvature.

1. Introduction

Epitaxial silicon carbide thin films synthesized on silicon substrates by the atomic substitution method enable the growth of crack-free thick AlN, GaN and AlGaN layers using HVPE technique [1]. Mentioned wide gap semiconductors are widely used for optoelectronic devices, HEMT transistors, GaN power electronics, etc. Defect formation and the growth mechanism of AlGaN films are influenced by the residual mechanical stress in the hybrid SiC/Si substrate. The latter contains a porous silicon layer that emerged during SiC growth by the method of atomic substitution. There have been relatively few works regarding the stress relaxation mechanisms in the SiC/Si-porous/Si substrates. Telyatnik et al. [2] estimated the residual mechanical strain in the SiC layer (about 1.7×10^{-3} that corresponds to the stress about 1 GPa) using the dependence of the substrate curvature on temperature. Also, different mechanisms of stress relaxation in AlN/SiC/Si were investigated in [2]. However, the real structure of porous silicon was not considered.

Osipov [3] analyzed the SiC/Si interface using UV ellipsometry. It is found that the structure of the hybrid SiC/Si(111) substrate obtained at 1250°C after 15 min of annealing of Si(111) in the CO atmosphere contains a Si-porous layer having ~20% of porosity and ~100 nm of 3C-SiC(111) film. The cubic polytype of SiC is predominantly formed under this condition. The porosity is determined in the frame of Bruggeman's effective medium approximation. The thickness of the porous layer is clearly seen from the SEM micrographs and equals about 2-4um having the structure of dendritic cavities with a thin SiC boundary layer [4]. Consequently, the model geometry of the structure contains Si(111) wafer (400um), porous-Si layer (4um), 3C-SiC film (100nm). Elastic constants of Si wafer and SiC film are known both in cases of monocrystal and polycrystal [5, 6]. The elastic constants of the porous silicon layer were estimated as a function of porosity using the model of spherical pores in [7]. This report continues the investigation of the porous silicon layer considering the octahedral form of the pores and the presence of a thin SiC layer on the surface of pores. Also, the stress relaxation mechanism in the SiC layer is suggested.

2. Dependence of biaxial modulus on porosity

The calculation of elastic tensor of porous Si is performed by the finite element method (FEM) using Comsol Multiphysics software (Cell Periodicity node) [7]. The method is based on the prescribed deformation of the opposite faces of representative volume element (RVE), three normal and three shear traction components. The six strains are set on the opposite boundaries of RVE successively accounting the periodic displacement boundary condition (PDBC).

$$u_1(r_1) = u_2(r_2) + \varepsilon_0(r_1 - r_2) \tag{1}$$

where $u_1(r_1)$ and $u_2(r_2)$ are the displacements of the opposite faces of RVE, $H = r_1 - r_2$ is the characteristic size of RVE geometry, ε_0 is the prescribed deformation tensor. Combination of principal of virtual work with PDBC enables calculation of the displacement field over the RVE [8].

The elastic tensor $\mathbf{C}$ of the heterogeneous material is obtained from the Hook's law using homogenization approach.

$$\boldsymbol{\sigma} = \mathbf{C} \times \boldsymbol{\varepsilon} \tag{2}$$

where $\boldsymbol{\sigma} = \{\sigma_{11}; \sigma_{22}; \sigma_{33}; \sigma_{12}; \sigma_{23}; \sigma_{13}\}$ and $\boldsymbol{\varepsilon}$ are the stress and deformation tensors correspondingly obtained by averaging over the volume of RVE. The method enables the independent calculation of components of the elastic tensor.

RVE of the porous silicon is represented in the form of a cube with pores added to the structure (see figure 1). In this work, the octahedral pores having boundaries in (111) plane and spherical pores are considered. The pores have fixed characteristic size R, which is the height of the half of the regular octahedron and the radius of the pores for the spherical ones. The spherical pores have a thin SiC boundary layer. The positions of the pores are randomly chosen and generated at least 10 times to minimize statistical fluctuations.

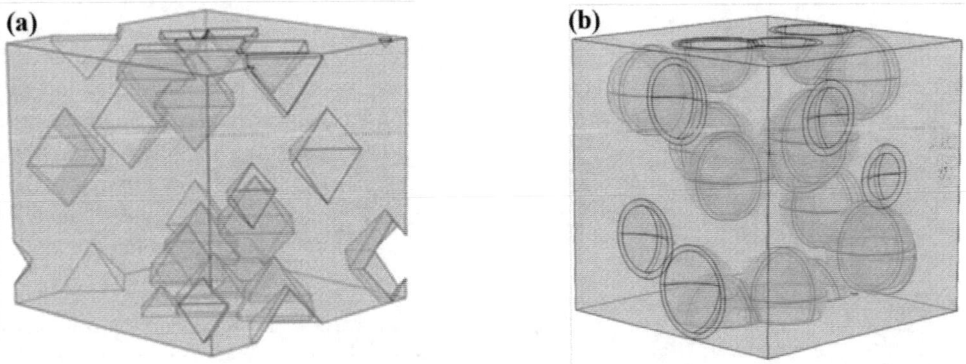

Figure 1. The geometries of the model for the form of the pores of regular octahedron (a) and of the spheres with tSiC/R=0.04 (b).

The three types of errors are minimized during computation: the finite size effect, discretization error, and statistical fluctuations. The characteristic size of the system should be large enough to describe the properties of homogeneous material. The optimization via H/R parameter for the fixed R=0.25 um is carried out using calculation of elastic constants. The procedure shows less than 2% deviation from the converged value for H/R greater than 6 in the case of spherical pores and greater than 4 in the case of octahedral pores. The second source of errors is analyzed via decreasing the size of the maximum partitioning element Δ having constant relation to the minimum element equals to 5.56. It was shown that the variation of Δ/H in the range from 0.05 to 0.45 leads to less than 2% deviation from the average value within the range. The partitioning elements for the discretization are tetrahedra. The remaining parameters for the discretization in Comsol are the resolution of narrow regions (0.5), the curvature factor (0.6) and the maximum element growth rate (1.5). The statistical fluctuations for every calculated elastic constant lead to less than 3% of standard deviation.

The calculation shows three independent constants of the elastic tensor (C_{11}, C_{12}, C_{44}) for the porous silicon. The dependence of biaxial modulus in the [111] direction $M_{(111)}$ on the porosity of porous silicon layer is shown in figure 2. $M_{(111)}$ obtained for the model with octahedral pores is slightly lower than that of the spherical form (filled circles and daggers respectively [7]). Solid lines show the linear approximation of the results, namely $M_{(111)}$=229.7-282.6×φ in the case of octahedra. The value of $M_{(111)}$ increases with the thickness of SiC layer (tSiC). The results calculated with tSiC/R =0.04 and 0.12 are shown by the circles. It is seen from SEM that the tSiC/R is less than 0.04 for the most part of the porous silicon layer, which allows us to omit the influence of SiC boundary layer [4]. The calculated value of Young modulus of porous silicon with octahedral form of the pores for 30% of porosity E_{111}=119 GPa is consistent with the experimental data E_{111}=110 GPa [9].

Figure 2 shows $M_{(111)}$ to be close to silicon value in the case of tSiC/R=0.12. Ellipsometry and Auger electron spectroscopy show the layer to be adjusted to SiC film at the SiC/Si interface for the different polytypes of SiC [3,10]. The dependence of elastic constants on the porosity for each geometry is shown in table 1. Monocrystalline and polycrystalline elastic constants are considered for the thin SiC layer.

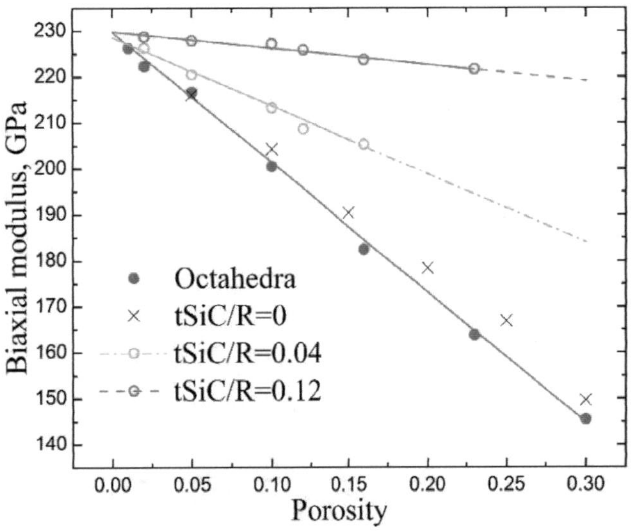

Figure 2. $M_{(111)}$ Dependence on the porosity of porous silicon. Filled circles depict the case of the octahedra. Dashed lines show the linear extrapolation of $M_{(111)}$ to 30% of porosity for tSiC/R=0.04 and 0.12 in the case of monocrystalline SiC. The data obtained in the case of spherical pores is shown by daggers [7].

Table 1. The dependence of the elastic constants of the porous silicon on the porosity φ.

	C11	C12	C44
Octahedra	166-195.4×φ	64.6-85×φ	79.5-96×φ
Monocrystalline 3C-SiC: C11=395 GPa, C12= 132 GPa, C44=236 GPa [6]			
tSiC/R=0.04	165.3-91.1×φ	64.9-55×φ	79.3-51.6×φ
tSiC/R=0.12	166.3-17.8×φ	64.1-15.5×φ	79.9-10.9×φ
tSiC/R=0.28	164.2+87.2×φ	63.4+24×φ	78.7+47.4×φ
Polycrystalline SiC: C11= 500 GPa, C33=535 GPa, C44=165 GPa, C66=210 GPa, C13=50 GPa [6]			
tSiC/R=0.04	166.5-83.4×φ	64-63.3×φ	79.2-55.7×φ
tSiC/R=0.12	166.3-1.7×φ	64.4-31.8×φ	79.8-23.9×φ
tSiC/R=0.28	165.4+104×φ	64-1.3×φ	79.5+20.6×φ

3. Stress relaxation in the SiC film

Using density functional method, Kukushkin et al. [11] have shown that the 3C-SiC/Si(111) interface is the conjugation of 5 elementary cells of 3C-SiC(111) and 4 cells of Si(111). Such configuration corresponds to the minimum energy of the system. This feature of the structure grown by the atomic substitution method greatly decreases the lattice mismatch leading to a compressive strain of SiC film to $|\varepsilon_m|$=0.34%.

The influence of the porous layer has been analyzed using FEM. The lateral component of the stress tensor in the SiC film is investigated. The following relaxation mechanisms are considered: the reduction of the elastic moduli of the porous silicon layer; the folding of the 3C-SiC(111) film; the bending of the substrate. The geometry of the model to be analyzed: 400um of the Si substrate, 4um of the Si-porous layer and 100 nm of the SiC film. The stress in the SiC film for $|\varepsilon_m|$=0.34% equals to 1.97 GPa. The reduction of the elastic moduli of the porous silicon layer leads to less than 1% of relaxation of the stress. Moreover, the contact surface decrease shows about 1% of the relaxation. The latter has been studied via adding the ring-shaped cavities under the surface of 3C-SiC(111) film. The curvature influence is considered during FEM calculation.

4. Conclusion

The dependence of elastic constants of the porous silicon on the porosity is analyzed using FEM. It was found that the porosity reduces the elastic moduli of the porous silicon by about 35%. However, the presence of the porous layer at the SiC/Si interface decreases the stress in the SiC film by less than 1%. The obtained value of the Young modulus of porous layer is consistent with experimental data.

Acknowledgments

This study was carried out within the framework of the Russian Science Foundation, project no. 20-12-00193.

References

[1] Kukushkin S A and Sharofidinov S S 2019 *Phys. Solid State* **61** 2342
[2] Telyatnik R S, Osipov A V and Kukushkin S A 2015 *Phys. Solid State* **57** 162
[3] Kukushkin S A and Osipov A V 2016 *Tech. Phys. Lett.* **42** 175
[4] Redkov A V, Grashenko A S, Kukushkin S A, Osipov A V, Kotlyar K P, Likhachev A I, Nashchekin A V and Soshnikov I P 2019 *Phys. Solid State* **61** 299
[5] Hopcroft M A, Nix W D and Kenny T W 2010 *J. Microelectromechanical* Systems **19** 229
[6] Dirras G F, Djemia P, Roussigne Y and Jackson K M 2004 *Mat. Science and Eng.* **387** 302
[7] Koryakin A A, Eremeev Yu A, Osipov A V and Kukushkin S A 2021 *Tech. Phys. Lett.* **47** 25
[8] Gao L, Wang C, Liu Z and Zhuang Z 2017 *J. Aeronaut.* **30** 1417
[9] Grashchenko A S, Kukushkin S A and Osipov A V 2014 *Tech. Phys. Lett.* **40** 1114
[10] Kukushkin S A, Osipov A V and Feoktistov N A 2014 *Phys. Solid State* **56** 1507
[11] Kukushkin S A and Osipov A V 2020 *Tech. Phys. Lett.* **46** 1103

Study of the initial stage of GaAs growth on FIB-modified silicon substrates

M M Eremenko[1], N A Shandyba[1], N E Chernenko[1], M S Solodovnik[1], S V Balakirev[1] and O A Ageev[1,2]

[1] Institute of Nanotechnologies, Electronics and Equipment Engineering, Southern Federal University, Taganrog 347922, Russia
[2] Research and Education Center "Nanotechnologies", Southern Federal University, Taganrog 347922, Russia

eryomenko@sfedu.ru

Abstract. In this work, we studied the effect of the deposition thickness, growth rate, arsenic flux, and implantation dose on the morphology of the GaAs nanostructures grown on modified Si areas. It is shown that an increase in the growth rate at the initial stages of the growth process leads to the transition of the growth regime from layered-like to one-dimensional with the formation of nanowires. Studies of the effect of As_4 pressure have shown that a change in the equivalent As_4 flux in the range of 3.7 - 5.0 ML/s does not lead to any significant change in the structure of the GaAs layer in the modified areas. An increase in the implantation dose during processing with a focused ion beam led to disordering of the directions of the grown nanowires due to the degradation of the substrate crystal structure.

1. Introduction

In recent years, more and more attention has been attracted to the epitaxial growth of III-V semiconductor compounds on silicon substrates, in particular, the monolithic integration method, which is supposed to allow combining light emission sources with high-speed information processing devices on one chip [1-3]. Therefore, the formation of high-quality III-V light sources on silicon remains one of the important modern problems of optoelectronics and nanophotonics [1-8].

However, there are a number of problems in the monolithic integration of III-V on Si: the growth of a polar III-V semiconductor on a non-polar Si substrate leads to the formation of antiphase domains [7, 8], the difference in the lattice constants of substrate and growth layer contributes to the formation of misfit dislocations and, as a result, threading dislocations [7-11], and the difference in thermal expansion coefficients of materials leads to the formation of thermal cracks [12]. Despite the fact that a wide range of techniques is used to improve the structural and functional characteristics of the III-V optical components on Si, completely new approaches are required to reduce the defectiveness of the resulting structures.

In this work, we investigated the effect of the implantation dose during Si substrate modification by focused ion beams, growth rate, and arsenic flux on the morphology of the grown GaAs structures.

2. Experimental procedure

The GaAs epitaxial materials were fabricated using a SemiTEq STE 35 molecular beam epitaxy system with solid-state sources. The studies were carried out on on-axis Si(100) substrates with local modification areas. The surface modification was carried out by local implantation of Ga ions into the Si substrate by the method of focused ion beams. The implantation dose varied from 1 to 21 pC/μm^2 and the current was 1 pA. The growth rate varied from 0.1 ML/s to 1 ML/s. The removal of native oxide from the Si surface was carried out at 900°C in vacuum for 30 minutes [13]. Then, a high temperature (600°C) GaAs buffer layer was deposited.

3. Results and discussion

Studies have shown that an increase in the GaAs deposition rate from 0.1 to 1.0 ML/s with a constant As_4 flux leads to an intensification of the nucleation and growth processes of GaAs nanowires (Figure 1), regardless of the implantation dose. This behaviour is due to the local accumulation of uncompensated Ga.

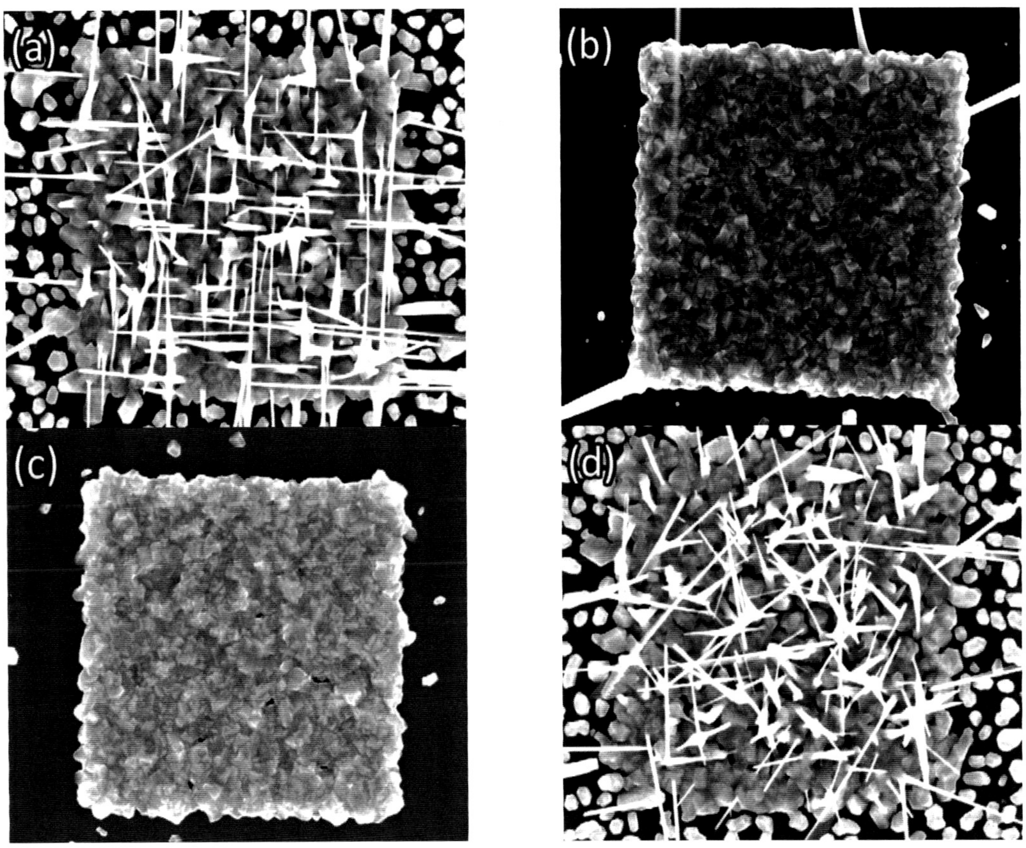

Figure 1. SEM images of modified Si areas after deposition of $H = 200$ nm GaAs at $T = 600$°C with different deposition rates: (a), (d) $v = 1$ ML/s, (b) $v = 0.25$ ML/s, (c) $v = 0.1$ ML/s. The implantation dose and current were 7 pC/μm^2 and 1 pA for (a)-(c), and 21 pC/μm^2 and 1 pA for (d) respectively.

We believe that this may be due to, on the one hand, the presence of an excess of Ga atoms in the modified areas of Si, which arose as a result of the introduction of ions into the substrate, and, on the other hand, to the possible delay of GaAs nucleation processes on the Si surface, which can also lead to a local change in the ratio of the Ga and As_4 fluxes towards the metal component even under conditions of the initial excess of As_4. It should also be noted that such a local shift of the flux ratio towards excess Ga was maintained throughout the entire growth period - the total length of GaAs nanowires reaches 15 – 20 μm.

It is important to note that with an increase in the implantation dose and, as a consequence, the degree of amorphization of the Si crystal lattice, the degree of influence of the crystal lattice of the substrate on the growth of GaAs nanocrystals decreases: from those oriented in accordance with the structure of the substrate in the directions [110], [1-10], [-110] and [-1-10] (the case of actually inheriting the structure of the substrate) to randomly oriented - the case of the absence of the influence of the structure of the substrate.

At the same time outside of the modification areas, an increase in the GaAs deposition rate leads to a sharp increase in the density and size of GaAs crystallites, reducing the selectivity of nucleation and growth processes, which, nevertheless, does not lead to the formation of a continuous film. Separate GaAs crystallites are formed on unmodified Si areas, the density of which is expected to increase from 3.8×10^6 cm^{-2} to 5.38×10^8 cm^{-2} as the GaAs deposition rate increases from 0.1 to 1.0 ML/s, which is due to a decrease in the diffusion length of adatoms.

An increase in the equivalent thickness of GaAs deposition while keeping the remaining parameters unchanged leads to an increase in the degree of filling of the modified areas (up to the formation of a continuous polycrystalline film) and the total height of the GaAs structure (Figure 2). Moreover, the analysis shows that the increase in the filling of the modified areas occurs both due to the growth of GaAs crystallites and due to an increase in their total number. In this case, an increase in the GaAs deposition thickness on unmodified areas of the Si surface did not lead to the formation of continuous films in the entire range of growth conditions under consideration.

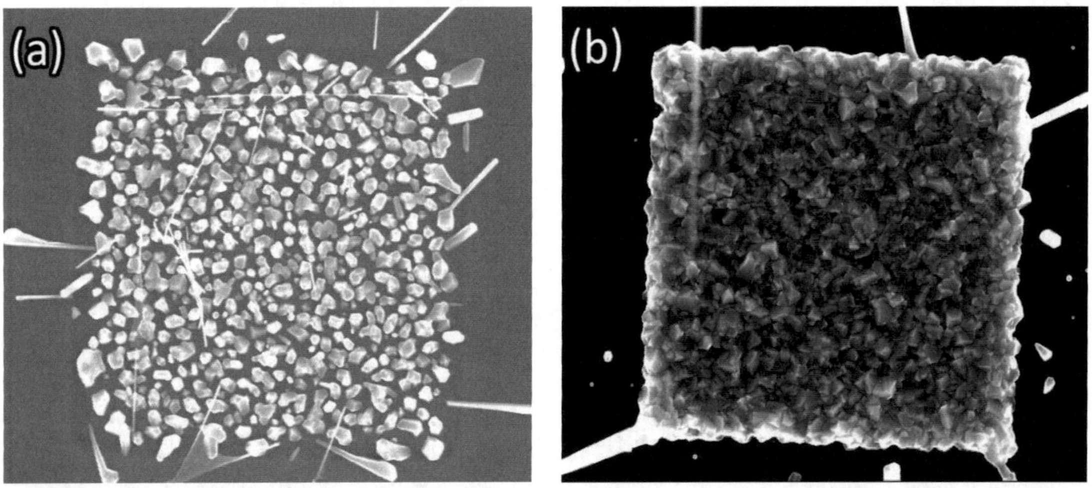

Figure 2. SEM images of modified Si areas after GaAs deposition at $T = 600^{\circ}$C and $v = 0.25$ ML/s with different deposition thicknesses: (a) 50 nm, (b) 200 nm. The implantation dose and current were 7 pC/μm^2 and 1 pA, respectively.

Studies of the effect of As_4 pressure have shown (Figure 3) that a change in the equivalent As_4 flux in the range of 3.7 - 5.0 ML/s does not lead to any significant change in the structure of the GaAs layer in the modified areas, but leads to a sharp increase in the density and size of crystallites GaAs outside of the modified areas. Analysis of the set of experimental data shows that this effect is similar to an increase in the rate of Ga entry to the surface. Thus, the assumption of a local displacement of the flow ratios towards Ga is valid not only for modified, but also unmodified areas. Hence, it can be concluded that, at the initial stage of GaAs growth, the pressure of the Group V component is an important parameter from the point of view of not only the selectivity of epitaxial synthesis processes, but also the suppression, in our case, of the unwanted growth of nanowire nanocrystals.

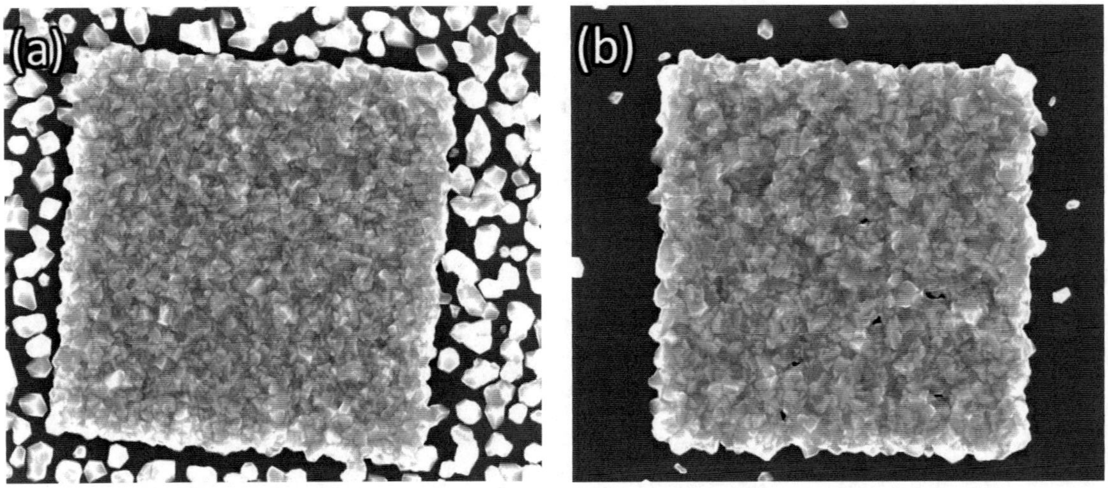

Figure 3. SEM images of modified Si areas after deposition of 200 nm GaAs at $T = 600^{o}C$ and $v = 0.1$ ML/s with various equivalent As_4 fluxes: (a) 3.7 ML/s, (b) 5 ML/s. The implantation dose and current were 7 pC/μm^2 and 1 pA, respectively.

4. Conclusions

Thus, the performed experimental studies show that the formation of GaAs structures on modified Si areas by Ga^+ ions makes it possible to localize the growth of the grown structure. The regularities of the influence are established and it is shown that a change in the main control parameters of the growth of GaAs on Si, such as the arsenic flux, the equivalent deposition rate and thickness, critically changes not only the type of growth, but also the morphology of the resulting nanostructures.

Acknowledgments
This work was supported by the Russian Science Foundation Grant No. 20-69-46076. The results were obtained using the equipment of the Research and Education Center "Nanotechnologies" of Southern Federal University.

References

[1] Wan Y, Li Q, Geng Y, Shi B and Lau K M 2015 *Appl. Phys. Lett.* **107** 081106
[2] Shifren L *et al.* 2014 *IEEE Trans. Electron Devices* **61** 2271–7
[3] Vahala K J 2003 *Nature* **424** 839
[4] Balakirev S V, Solodovnik M S, Eremenko M M, Chernenko N E and Ageev O A 2020 *Nanotechnology* **31** 48
[5] Balakirev S V, Solodovnik M S, Eremenko M M, Konoplev B G and Ageev O A 2019 *Nanotechnology* **30** 50
[6] Chen S *et al.* 2016 *Nat. Photonics* **10** 307–11
[7] Tang M, Park J-S, Wang Z, Chen S, Jurczak P, Seeds A and Liu H 2019 *Prog. Quant. Electron.* **66** 1-18
[8] Barrett C S C, Atassi A, Kennon E L, Weinrich Z, Haynes K, Bao X-Y, Martin P, Jones K S 2019 *J. Mater. Sci.* **54** 7028–34
[9] Balakirev S V, Chernenko N E, Eremenko M M, Ageev O A and Solodovnik M S 2021 *Nanomaterials* **11**(5) 1184
[10] Luxmoore I J *et al.* 2013 *Sci-Rep-UK* **3** 1239
[11] Balakirev S V, Solodovnik M S and Ageev O A 2018 *Phys. Status Solidi B* **255**(4) 1700360
[12] Park J-S, Tang M., Chen S and Liu H 2020 *Crystals* **10** 1063
[13] Eremenko M M, Solodovnik M S, Balakirev S V, Chernenko N E, Kots I N and Ageev O A 2020 *J. Phys.: Conf. Ser.* **1695** 012013

Self-consistent modeling of MBE self-catalyzed GaAs nanowire growth

S V Fedina[1], A A Koryakin[2], V V Fedorov[1,3], G A Sapunov[1] and I S Mukhin[1,3]

[1] Nanotechnology Research and Education Centre of the Russian Academy of Sciences, Alferov University, St. Petersburg 194021, Russia
[2] St. Petersburg State University, St. Petersburg 199034, Russia
[3] Peter the Great St. Petersburg Polytechnic University, 29 Politekhnicheskaya, St. Petersburg 195251, Russia

fedina.serg@yandex.ru

Abstract: Self-catalyzed GaAs nanowires are synthesized by molecular beam epitaxy at various arsenic fluxes and growth temperatures. The growth of GaAs nanowires is simulated considering the kinetics of material transport inside the catalyst droplet. The re-evaporation coefficient of arsenic is estimated for the given growth conditions. Calculated nanowire growth rate is in satisfactory agreement with the experimental data.

1. Introduction

Epitaxial arrays of III-V semiconductor nanowires (NWs) are a promising material for modern optoelectronic and electronic technologies. Particularly, GaAs NWs can be used for fabrication of novel photovoltaic devices [1]. When forming NW by the vapor-liquid-solid (VLS) mechanism, the most extensive possibilities for controlling the morphology of NWs arrays are opened when using catalytic particles of foreign materials (usually Au droplets pre-deposited on the growth substrate). However, the catalyst material can be embedded into the lattice of growing NWs, resulting in deep-level defects that reduce the lifetime of charge carriers and increase the probability of non-radiative recombination. The NW formation by the self-catalytic mechanism [1], when using a group III element of the NWs itself as a catalyst (namely, gallium for GaAs), excludes contamination of III-V compounds, but imposes restrictions on the range of possible growth parameters and, as a result, the morphology of the formed NWs arrays [2].

In this paper, the growth rate of self-catalyzed GaAs NWs is found at different arsenic fluxes and temperatures both experimentally and theoretically. The model proposed in [3] within the "As-only" approach [4] is used for the theoretical assessment of the GaAs NWs growth rate. The value of the re-evaporation coefficient of arsenic is calculated for the given growth conditions by fitting the experimental growth rate.

2. Theory

In the model [3], the nucleation-limited growth rate of NWs which is given by the Zeldovich nucleation rate times the surface area of the top NW facet is considered. It is assumed that the growth rate of 2D islands on the NW top facet is limited by the diffusion flux of arsenic inside the gallium droplet. The simplified geometry of the system is presented in the inset in figure 1b. The growth modeling is performed by two steps. Firstly, for each monolayer, the duration of the droplet refilling stage t_S is calculated. The time t_S is estimated by using the expression for the steady-state nucleation rate:

$$t_S \approx (t + \tau_N(t))_{\min} . \quad (1)$$

Here $\tau_N(t)$ is the time between two consecutive nucleation events calculated at the moment of time t, $\tau_N = 1/(\pi R_0^2 I)$, where I is the Zeldovich nucleation rate, R_0 is the NW radius. Self-consistent modeling is achieved by considering the material balance equation for arsenic in the droplet:

$$\frac{dN_{As}}{dt} = J_{As}^{dir} + J_{As}^{re} - J_{As}^{des} , \quad (2)$$

where N_{As} is the number of arsenic atoms in the droplet; J_{As}^{dir} and J_{As}^{re} are the direct and re-evaporated arsenic flux in the vapor phase, J_{As}^{des} is the desorption arsenic flux. Following the approach [4], the sum of the direct and re-evaporated arsenic fluxes is given by the formula:

$$J_{As}^{dir} + J_{As}^{re} = (1 + \varepsilon_{As}) j_{As}^{dir} S , \quad (3)$$

where ε_{As} is the arsenic re-evaporation coefficient, j_{As}^{dir} is the arsenic flux density, S is the cross section of the droplet. Secondly, the lateral growth of the GaAs nucleus (GaAs island) is modeled. The incorporation flux J_{As}^{inc} should be introduced in the material balance equation:

$$\frac{dN_{As}}{dt} = J_{As}^{dir} + J_{As}^{re} - J_{As}^{des} - J_{As}^{inc} . \quad (4)$$

The island growth rate is determined by the equation:

$$\frac{dR}{dt} = \frac{\Omega}{2\pi h R} J_{As}^{inc} , \quad (5)$$

where Ω is the volume of GaAs pair in solid state, h is the GaAs monolayer height, R is the island radius. The analytical expression derived in [3] is used for the incorporation flux. This expression was obtained by solving a steady-state diffusion problem, and it was assumed that the island has the shape of a disk. It was supposed that the equilibrium concentration of arsenic is maintained at the island edge. Also, the attachment rate of arsenic atoms to the critical island introduced in the nucleation rate I is derived in [3].

3. Experiment

Epitaxial GaAs NWs arrays were synthesized on the vicinal Si (111) substrate with a 4° miscut orientation in $[11\bar{2}]$ direction using the MBE (molecular beam epitaxy) Veeco GEN-III machine with Ga effusion cell and valved arsenide cracker for As_4 molecular flux. Before loading into the MBE chamber, the silicon substrates were cleaned using a modified RCA cleaning and Shiraki etch procedures [5, 6]. Oxidation was carried out by boiling in "base piranha" mixture or RCA-1 cleanser (aqueous solution of ammonia and hydrogen peroxide NH_4OH:H_2O_2:H_2O with the volume ratio of 1:1:3) [7]. The substrate temperature was monitored with a thermocouple calibrated using the silicon oxide decomposition and Si (111) 7x7 to 1x1 phase transition temperatures [8]. Bayard-Alpert vacuum gauge was used to control molecular fluxes of groups III and V and BEP (beam equivalent pressure).

Silicon substrates with oxide were preliminarily degassed in the intro and buffer chambers, and then annealed in the growth chamber, at a temperature lower than the surface oxide decomposition temperature by 20 degrees, to create defects in the oxide layer, which are necessary for the growth of vertical NW. Thereafter, before the growth process, gallium was pre-deposited, which self-organized into droplets. The growth was carried out at the temperatures of 570°C and 600°C (for 10 min and 30 min, respectively) and at the As_4 molecular fluxes of 1.5×10^{-6} Torr and 3×10^{-6} Torr. In this case, the gallium flux was constant (1.25×10^{-7} Torr) which corresponds to the growth rate of planar GaAs 313 nm/h.

The morphology of synthesized GaAs NWs was studied with scanning-electron microscope (SEM) (Zeiss SUPRA 25-30-63).

4. Results and discussion

The re-evaporation coefficient of arsenic ε_{As} that characterizes the arsenic evaporation flux from the substrate and the NW sidewalls is used as the sole fitting parameter. The estimated value of ε_{As} equals 6.5 (two times larger than that obtained in [4]). The value of ε_{As} depends strongly on the surface density of NWs and their morphology. Also, the following value of the conversion factor η connecting the arsenic BEP and the arsenic flux density (j_{As}^{dir} =η×BEP) is found 1.28×10^{24} m^{-2} s^{-1} $Torr^{-1}$. This value is about two times smaller than that used by Glas et al. [4]. because it depends on the configuration of MBE machine. However, the value of the sum $J_{As}^{dir} + J_{As}^{re}$ (equation (3)) is in agreement with the results of the paper [4]. The calculation is performed at the interphase energy of the GaAs island side facet (0.309 J m^{-2}) [3]. The dependences of the NW growth rate on the group V flux calculated at different growth temperatures are shown in figure 1b.

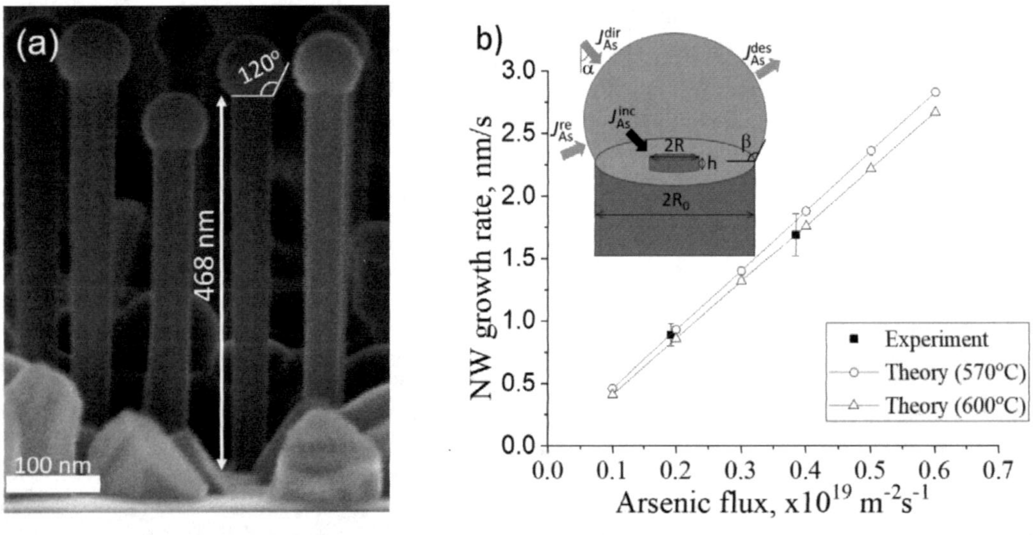

Figure 1. a) Cross-sectional SEM image of GaAs NW array. b) Calculated (lines) and experimental (dots) dependencies of NW growth rate on arsenic flux (red and blue lines are dependencies simulated at 570°C and 600°C, respectively). Schematic representation of self-catalyzed VLS growth of GaAs NW is shown in the inset. The values of NW radius R_0 =51 nm, contact angle β=120° and incident angle of the arsenic flux α=30° are used in computation.

Table 1. NW length calculated within the model [3] and measured in the experiment.

Sample	Calculated length (μm)	Measured length (μm)
S1	3.042	2.998±0.3
S2	0.534	0.502±0.04

It is found that the growth rate obtained in the model is consistent with the real growth rate of synthesized GaAs NWs with an accuracy of about 80-85 percent.

5. Conclusions

In summary, self-catalyzed GaAs NW formation was studied both experimentally and theoretically. The values of the growth rate of self-catalyzed GaAs NWs predicted by the model are in a good agreement with the experimental data. The re-evaporation coefficient of arsenic was used as a sole fitting parameter.

Acknowledgments

The NW samples were grown and characterized under the support of the Ministry of Science and Higher Education of the Russian Federation (project No. 0791-2020-0005). NW growth modeling was performed under financial support of the Russian Science Foundation (grant 19–72–30004).

References

[1] Matteini F, Dubrovskii V G, Rüffer D, Tütüncüoglu G, Fontana Y and Fontcuberta I Morral A 2015 *Nanotechnology* **26** 105603
[2] Dubrovskii V G, Cirlin G E and Ustinov V M 2009 *Semiconductors* **43** 1539
[3] Koryakin A A and Kukushkin S A 2021 *Phys. Status Solidi B* 2000604
[4] Glas F, Ramdani,M R, Patriarche G and Harmand J C 2013 *Phys. Rev. B* **88** 195304
[5] Okumura H, Akane T, Tsubo Y and Matsumoto S 1997 *J. Electrochem. Soc.* **144 (11)** 3765-3768.
[6] Ishizaka A and Shiraki Y 1986 *J. Electrochem. Soc.* **133 (4)** 666–671
[7] Kern W 1990 *J. Electrochem. Soc.* **137 (6)** 1887–1892
[8] Telieps W and Bauer E 1986 *Berichte der Bunsengesellschaft für physikalische Chemie* **90** 197–200

Influence of magnetron sputtering modes on the parameters of ZnO:Ga films

A A Geldash[1,2], E Yu Gusev[1], V N Dzhuplin[2] and O A Ageev[1,2]

[1] Department of Nanotechnology and Microsystems, Southern Federal University, Taganrog 347922, Russia
[2] Research and Education Centre «Nanotechnology» Southern Federal University, Taganrog 347922, Russia

Abstract. The aim of this work is to study the effect of magnetron sputtering modes - its technological parameters (pressure, substrate temperature and DC power source) on the morphology and electrophysical parameters of nanocrystalline ZnO:Ga films, which can be used as contact layers to nanostructures of photovoltaic converters. It was found that with an increase in the substrate temperature, the grain size decreases from 80 to 30 nm and the film surface roughness, as well as the resistivity from $1.68 \cdot 10^{-1}$ to $1.2 \cdot 10^{-2}$ Ω·cm and the mobility of charge carriers with 18.12 to 5.59 cm²/(V·s). In this case, the concentration of charge carriers increases from $5.59 \cdot 10^{18}$ to $3.31 \cdot 10^{20}$ cm^{-3}. With an increase in the power of the DC source, the grain size increases from 35 to 90 nm and the surface roughness of the ZnO:Ga films, as well as the concentration of charge carriers from $5.91 \cdot 10^{18}$ to $3.35 \cdot 10^{20}$ cm^{-3}. In this case, the resistivity decreases from $1.42 \cdot 10^{-2}$ to $1.3 \cdot 10^{-2}$ Ω·cm, and the mobility of charge carriers from 6.74 to 3.22 cm²/(V·s). The results obtained can be used in the development of technological processes for the manufacture of highly efficient photoelectric converters.

1. Introduction
The development of efficient photoconverters of energy is a topical direction in the development of alternative energy. The task is not only to find and study materials suitable for energy conversion, but also materials that play the role of upper contact layers [1, 2]. One of the most common materials for contact layers in photovoltaic converters is conductive indium tin oxide (ITO) [3,4]. However, there is a need to search for alternative materials in order to achieve better parameters (reflection coefficient, surface resistance). Therefore, more and more attention is directed to the study of methods for the formation and post-processing of conductive oxide films. One of these materials is the conductive oxide ZnO:Ga, which is a direct-gap semiconductor with a wurtzite structure and a band gap of 3.36 eV [5-8]. In this case, thin ZnO films have good optical quality, stability, and piezoelectric properties, which makes it possible to use them in various fields of micro- and nanoelectronics [9,10]. The aim of this work is to study the effect of magnetron sputtering modes (pressure, heating temperature, and power of a direct current source) on the morphology and electrophysical parameters of ZnO:Ga films, which can be used as contact layers to nanostructures of photovoltaic converters.

2. Experiment
A series of experiments to study the effect of technological modes during magnetron sputtering of nanocrystalline ZnO:Ga films began with the preparation of substrates. They were glass-ceramic plates with dimensions of 10x10 mm and a thickness of 0.6 mm, as well as p-Si samples with a charge

carrier concentration of the order of 10^{15} cm^{-3}. Si substrates were also obtained by scribing from the original wafer. The obtained substrates were subjected to chemical cleaning from impurities and oxides. The formation of thin ZnO:Ga films was carried out by DC magnetron sputtering using an Auto 500 (Boc Edwards) sputtering device. The source of the vaporized material was a ZnO:Ga target (GIRMET Ltd). Three series of experiments were planned. The first series of experiments was aimed at investigating the effect of the substrate temperature on the morphology and electrophysical parameters of ZnO:Ga films. As a result, samples were obtained that were formed at a substrate temperature of 30-180 °C. In this case, the pressure and power of the direct current source were constant and amounted to 0.135 Pa and 100 W, respectively. The choice of such boundary and interval values of heating temperatures is primarily due to past experimental studies [11, 12], the technological capabilities of the Auto 500 magnetron sputtering facility, as well as temperature constraints during the sputtering of the ZnO:Ga target. The second series of experiments was aimed at investigating the effect of the power of a direct current source on the morphology and electrophysical parameters of the films. Four samples were formed with ZnO:Ga films in the power range of 80-170 W, while the pressure and temperature were constant and were 0.135 Pa and 140 °C, respectively. The third series of experiments was aimed at studying the effect of pressure on the morphology and electrophysical parameters of the films; four samples were formed with ZnO:Ga films in the pressure range of 0.1-0.2 Pa. In this case, the temperature and power of the direct current source were constant and amounted to 140 °C and 100 W, respectively. The thickness control of the deposited films at the stage of their formation was carried out with an integrated quartz thickness gauge.

3. Results and discussion

The study of samples with ZnO:Ga films obtained by magnetron sputtering began with the study of their morphology, which was carried out by scanning electron microscopy (SEM), using a scanning electron microscope Nova Nanolab 600 (FEI.Co, Netherlands), as well as atomic force microscopy (AFM), using the Ntegra probe nanolaboratory (CJSC NT-MDT, Russia). Typical images of the film surface are shown in Figure 1 (a) and Figure 1 (b).

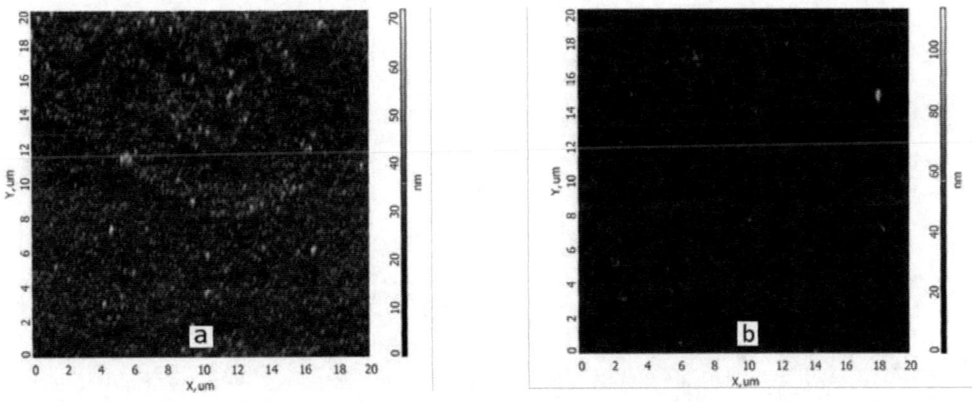

Figure 1. AFM images of the surface of ZnO:Ga films formed at different substrate temperatures: (a) at 130 °C; (b) at 180 °C.

As a result of the analysis of AFM scans, it was found that with an increase in the heating temperature from 30 to 180 °C, the size and height of grains on the surface of the ZnO:Ga film decrease. SEM morphology studies allowed us to obtain not only images of the surface at right angles (Figure 2 (a)), but also at an inclination of 45° (Figure 2 (b)), which provided information on the morphology of individual grains on the film surface - their size varies from 35 to 65 nm. Thus, with an increase in the substrate temperature during the formation of films from 30 to 180 °C, the grain size and hence the roughness decreased from 80 to 30 nm

Figure 2. SEM images of the morphology of ZnO:Ga films obtained at 180 °C: (a) the surface of the film at an angle of 90°; (b) at an angle of 45° (with grain sizes); (c) cleavage of the film with grain sizes.

Also, Figure 2 (c) shows an SEM image of the ZnO:Ga cleavage obtained as a result of cleavage of the sample, which shows that the film consists of individual columnar crystallites with sizes of about 30-60 nm. The thickness of the films was 250-300 nm. The results obtained are similar to the results obtained in [13] and [14], in particular, the dimensions were 50-75 nm, and at 130-180 °C, about 35-65 nm. With increasing pressure in the chamber, the grain size and roughness of the ZnO:Ga film change insignificantly, due to the small value of the pressure difference in the experimental series. However, an increase in the power of the constant current source leads to an increase in the grain size and film roughness. So with an increase in power from 100 to 170 W, the grain size increases from 35-60 nm to 70-90 nm. The study of the electrophysical parameters of the ZnO:Ga films was carried out by measuring the Hall emf using an Ecopia HMS-3000 setup (Ecopia Co., Korea). In Figure 3 shows the dependences of the electrophysical parameters on the substrate temperature.

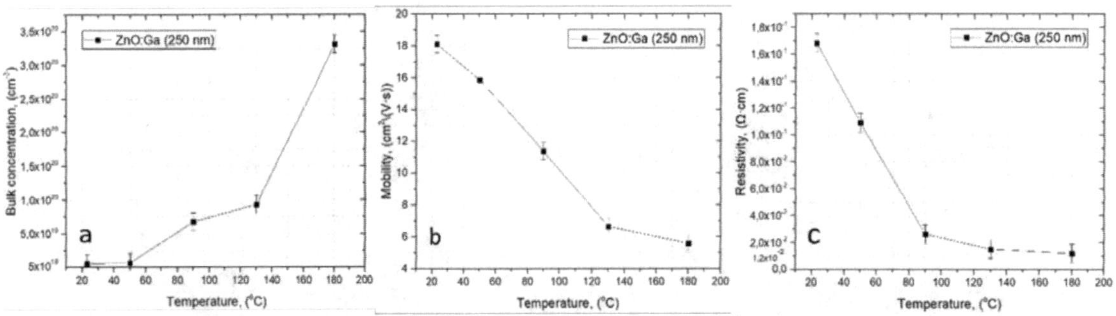

Figure 3. Dependences of the electrophysical parameters of ZnO:Ga films on the substrate temperature: (a) bulk concentration; (b) mobility of charge carriers; (c) resistivity.

From the dependence in Figure 3 (a) shows that the greatest increase in the concentration of charge carriers (by an order of magnitude) in the ZnO:Ga films is observed with an increase in the substrate temperature from 130 to 180 °C. However, from the dependences in Figure 3 (b) and 3 (c) it can also be seen that within the temperature range of 30-100 °C, a pronounced decrease in the mobility of charge carriers and resistivity is observed. From the dependence in Figure 4, it can be seen that with an increase in pressure from 0.135 Pa to 0.235 Pa, the mobility (Figure 4 (b)) and resistivity (Figure 4 (c)) decrease linearly, while the concentration increases linearly by half an order of magnitude.

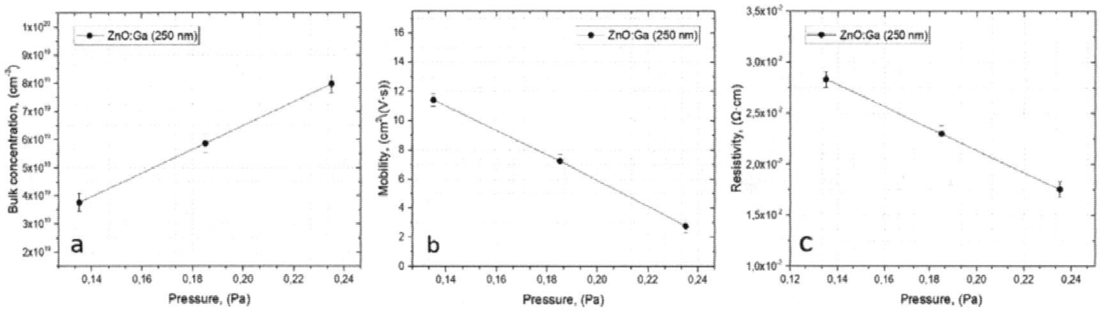

Figure 4. Dependences of the electrophysical parameters of ZnO:Ga films on pressure: (a) bulk concentration; (b) mobility of charge carriers; (c) resistivity.

From the dependencies in Figure 5, it can be seen that with an increase in the power of the dc source from 100 W to 170 W, the mobility (Figure 5 (b)) and resistivity (Figure 5 (c)) decrease almost linearly, while the concentration of charge carriers in the film increases from $6{,}5 \cdot 10^{19}$ cm^{-3} to $1.57 \cdot 10^{20}$ cm^{-3}. The main effect on the decrease in the resistivity value is exerted by an increase in the concentration of charge carriers.

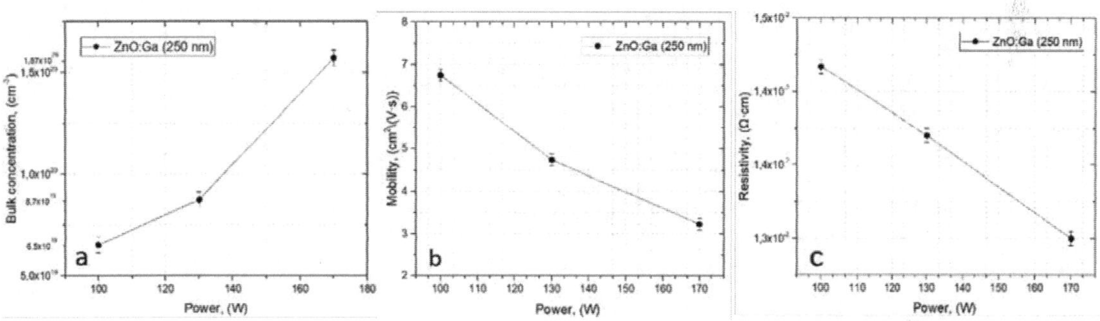

Figure 5. Dependences of the electrophysical parameters of ZnO:Ga films on the power of a direct current source: (a) bulk concentration; (b) mobility of charge carriers; (c) resistivity.

As a result of the studies carried out, it was found that an increase in the substrate temperature, pressure, and power leads to a decrease in the resistivity of ZnO:Ga films, due to a decrease in the number of defects arising during its formation. Its structure plays an important role in changing the electrophysical parameters of ZnO:Ga films. In our case, it is polycrystalline, consisting of individual columnar vertically oriented crystals. This is due to the very method of creating films - magnetron sputtering, characterized by intense sputtering of a large amount of target material per unit time. The experimental results show that a change in one or several technological parameters of magnetron sputtering strongly affects the change in the morphology and size of crystals in the ZnO:Ga film. This change in the morphology and phase composition of the films affects the electrophysical characteristics and optical properties of the resulting films. The results obtained can be used in the development of technological processes for the manufacture of highly efficient photoelectric converters.

Acknowledgments
The reported study was funded by RFBR, projects numbers 19-37-90139. The work was done on the equipment of the Research and Education Centre «Nanotechnology» Southern Federal University.

References

[1] Meyer F, Savoy A, Juan J, Leon J D, Persoz M, Niquille X, Allebe C, Nicolay S, Haug F J, Ingenito A and Ballif C 2021 *Solar Energy Materials and Solar Cells* **110815** 219

[2] Ahmed M, Bakry A, Qasem A and Dalir H 2021 *Optical Materials* **110866** 113

[3] Mondal P and Das D 2017 *Applied Surface Science* **411** 315

[4] Berger L I 1997 *Semiconductor Materials* (Boca Raton: CRC Press) p 496

[5] Ageev O A and Konoplev B G (eds.). 2019 *Nanotechnology in microelectronics* (Moscow: Nauka) p 511 ISBN 978-5-02-040201-0

[6] Saive R, Borsuk A M, Emmer H S, Bukowsky C R, Lloyd J V, Yalamanchili S and Atwater H A 2016 *Advanced Optical Materials* **4(10)** 1470-1474

[7] Minami T 2008 *Thin Solid Films* **516(17)** 5822-5828

[8] Aderhold J, Davydov V Yu, Fedler F, Klausing H, Mistele D, Rotter T, Semchinova O, Stemmer J and Graul J 2001 *J. Cryst. Growth* **222** 701

[9] Ellmer K *Journal of Applied Physics* 2000 **33 17**

[10] Xuhu Yu, Jin M, Feng Ji,Wang Y, Zhang X and Honglei Ma *Thin Solid Films* **483** 296-300

[11] Gusev E Yu, Ageev O A, Gamaleev V A, Mikhno A S, Mironenko O O and Pronin E A 2014 *Advanced Materials Research* **893** 539-542

[12] Geldash A A, Levenets L E, Gusev E Yu, Dzhuplin V N and Ageev O A 2020 *Izvestiya SFedU. Engineering Sciences* **2(2020)** 228-237

[13] Kelly P J and Arnell R D 2020 *Magnetron sputtering: a review of recent developments and applications, Vacuum* **56** 159-172

[14] Khranovskyy V, Grossnerb U, Lazorenko V, Lashkarev G, Svenssonb B G and Yakimova R 2007 *Superlattices and Microstructures* **42** 379-386

Studying of a sensitive material based on Ecoflex and CNTs for flexible strain sensors

N A Demidenko[1], A V Kuksin[1], E S Davydova[1], V A Zaborova[2,3], L P Ichkitidze[1,4], S P Bordovsky[4] and A Yu Gerasimenko[1,4,5]

[1] Institute of biomedical systems, National Research University of Electronic Technology, Zelenograd, Moscow, Shokin Square 1, Russia
[2] Department of Sports Medicine and Medical Rehabilitation, I.M. Sechenov First Moscow State Medical University, Moscow, Bolshaya Pirogovskaya street 2-4, Russia
[3] Sports Adaptology Lab, Moscow Institute of Physics and Technology (National Research University), Moscow, Institutskiy per. 9, Russia
[4] Institute for Bionic Technologies and Engineering, I.M. Sechenov First Moscow State Medical University, Moscow, Bolshaya Pirogovskaya street 2-4, Russia
[5] Institute of Nanotechnology of Microelectronics of the Russian Academy of Sciences, Moscow, Leninsky Prospekt 32A, Russia

Abstract. Nowadays there is a great need for the development of flexible strain sensors that can register human body's movements. In the field of wearable and smart electronics such sensors are actively being developed. Resistive-type flexible sensors are the easiest to manufacture. Their mechanism of sensitivity to deformations is based on a change in electrical resistance during deformations. In this work, we have developed the functional material for strain sensor with high tensile properties, strength and electrical conductivity. This material based on a matrix of silicone elastomer and a multi-walled carbon nanotubes (MCNTs) filler. The material showed a high elongation of 950 % with a tensile strength of 1.437 MPa. The manufacturing process included laser structuring of MCNTs to form an electrically conductive network. The linear gauge factor was 3.4, and the angular gauge factor was 0.26.

1. Introduction

The sensors are widely used both in industry and in the biomedicine. Recently, wearable flexible devices aimed at non-invasive monitoring of human health and physical activities were especially interesting. Flexibility of traditional metal and semiconductor sensors is not flexible enough for this purposes. Traditional sensors usually have high values of the gauge factor only in a limited range of deformation (up to 1 %). For example, in semiconductor materials based on ZnO, the gauge factor of 1250 was obtained at a deformation of less than 1 % [1].

Resistive and capacitive sensors are most commonly used to register human motion [2,3]. Generally they are made by a combination of elastomeric substrates (polydimethylsiloxane (PDMS), polyurethane, polyetherimide, Ecoflex silicones) and conductive fillers (metal and carbon nanoparticles) [4]. Substrates make these nanocomposite materials stretchable. However, there are some limitations. Incompatibility between rigid conductive components and a sufficiently soft elastomeric matrix can lead to high hysteresis, low sensitivity and fragility [5]. The development of methods that make it possible to obtain materials with high elongation and low hysteresis, high values of sensitivity to deformation and strength is an important task. In addition, for biomedical devices that come into contact with the human body, a high degree of biocompatibility, absence of toxicity, and softness of mechanical properties are extremely important [6-10].

In this work sensitive material using a simple technology for laser formation of electrically conductive networks of carbon nanotubes (CNTs) in an Ecoflex silicone matrix was developed. The sensitive material has a small hysteresis (<5%), high tensile strength and remarkable linear sensitivity to deformations.

2. Materials and methods

2.1. Components of a sensitive material

As a filler, we used MCNTs with an outer diameter of 8-30 nm, an inner diameter of 5-15 nm, a length of ≥20 μm, a specific surface of ≥270 m^2/g, and a bulk density of 0.025-0.06 g/cm^3. The matrix was a silicone elastomer Ecoflex 00-10 on platinum. The dynamic viscosity of silicone in a mixed state is140 Pa·s, modulus of elasticity at 100 % elongation is 125 kPa, Shore hardness is 10 A, density is 1.04 g/cm^3, operating temperature range is from -19 °C to +232 °C. Carbon fiber electrodes were included in the finished sensitive material.

2.2. Manufacturing of a sensitive material

The manufacturing process of the sensitive material is shown in figure 1. The first step was to add MCNTs to Ecoflex silicone in the liquid phase at 3 wt% MCNTs. Solution of components was thoroughly mixed for 5 minutes at least. The solution was placed in a vacuum chamber and degassed to remove air microbubbles. Thereafter, the Ecoflex/MCNTs nanocomposite was prepared by stencil printing. The Ecoflex/MCNTs solution was placed in a plastic U-shaped mold. Electrodes were added so that the solution completely covered them. The solution with the electrodes in the mold was left at room temperature (23±5 °C) until complete polymerization (~4 hours). After complete polymerization, the nanocomposite was structured with laser radiation to reduce the resistivity values, form welded joints between nanotubes, and form a structured conductive CNTs network inside the nanocomposite. The effect of the formation electrically conductive network of nanotubes in polymers and the improvement of mechanical and electrical properties was obtained earlier [7]. The laser irradiation parameters were: wavelength 1064 nm, radiation power 12 W, irradiation time 2 minutes. Finally, the laser-structured Ecoflex/MWCNT nanocomposite with electrodes was sealed on both sides with a layer of pure silicone to create an insulating and fully biocompatible coating.

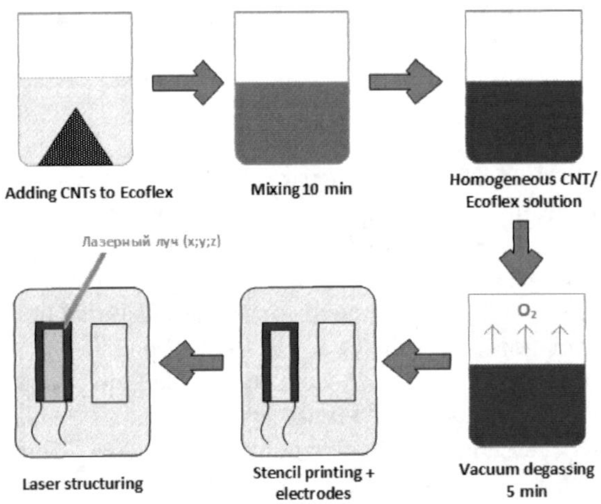

Figure 1. Manufacturing process of the strain-sensitive nanomaterial.

2.3. Gauge Factor Calculation

To determine the gauge factor (*GF*), resistance in comparison with strain (tensile/bending) studies were conducted using a test setup (figure 2).

According to the results of the study, the *GF* were calculated using formulas (1) for tension (linear sensitivity GF_e) and (2) for bending (angular sensitivity GF_a):

$$GF_e = \frac{\Delta R/R}{\Delta l/l} \tag{1}$$

where R is the initial resistance, ΔR is the absolute change in resistance after deformation, l is the initial length of the sensitive element, Δl is the absolute change in its length

$$GF_e = \frac{\Delta R/R}{d/r} \tag{2}$$

where r is the bending radius, d is the coating thickness.

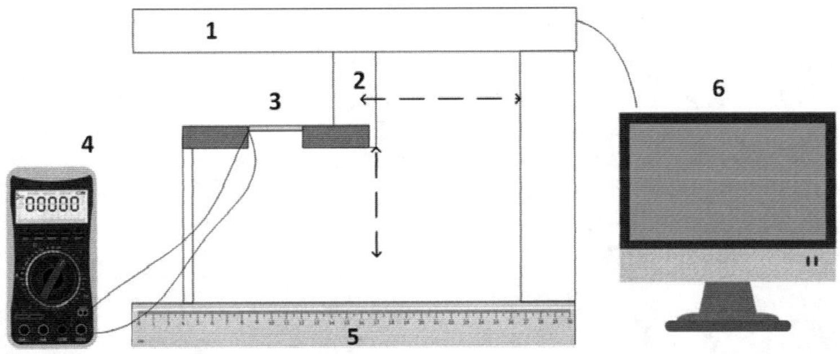

Figure 2. Test setup: 1 - displacement module, 2 - moving head in coordinates (x; y; z) of displacement module, 3 - sample of sensitive material, 4 - digital multimeter connected to the sample of material, 5 - measuring ruler, 6 - personal computer to control the movement module.

2.4. Calculating elastic modulus and Tensile Strength

The research technique was based on the construction of the loading curve and the determination of the elastic deformation limit from its linear region, and the calculation of the elastic modulus (E) was carried out according to the formula (3):

$$E = \frac{F \cdot l}{S \cdot \Delta l} \tag{3}$$

where F is the applied force, l is the initial length of the sample, S is the surface area to which the force is applied, Δl is the elongation of the sample as a result of the application of the force.

Tensile strength σ was calculated by the formula (4):

$$\sigma = \frac{F_m}{A_0} \tag{4}$$

where F_m is the maximum load applied to the sample; A_0 is the initial cross-sectional area of the sample.

3. Results

3.1. Structure of sensitive material

The appearance and microstructure of the developed material are shown in Figure 3. The dimensions of the material were 3.5×1.5×5 mm. Images of the microstructure were obtained using an FEI Helios NanoLab 650 scanning electron microscope. The accelerating voltage of the electron column was 2 kV, the current of the electron probe was 21 pA.

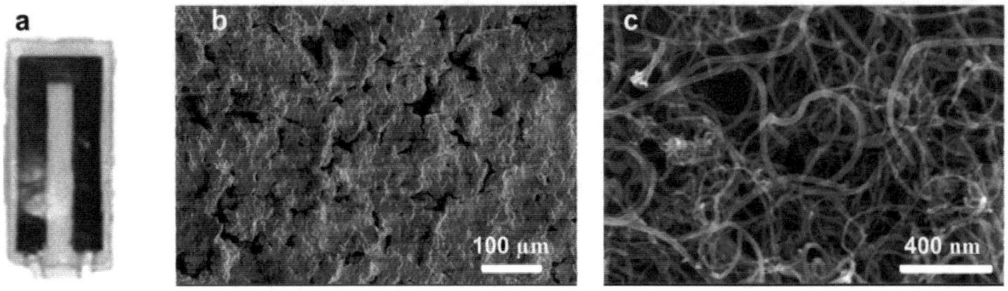

Figure 3. External view of the developed sensitive material (a), SEM images with increase ×60000 (b), ×240000 (c).

During the manufacturing process, the material was irradiated with a pulsed laser, which left the characteristic deepening (figure 3b). As you can see from figure 3c, a network of carbon nanotubes is formed within the material, which provides functional characteristics. Nanotubes form long connections >100 nm throughout the entire volume of the material.

3.2. Mechanism of sensitivity to deformations and Gauge Factor

The loading process (tension and bending) obtained with the test setup is shown in figure 4.

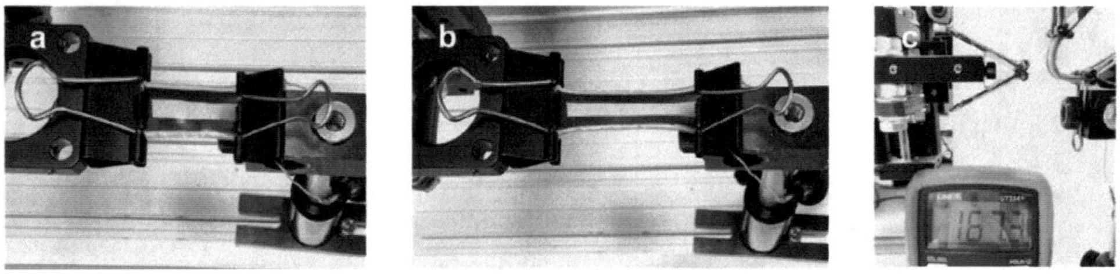

Figure 4. Sensitive material before stretching (a), during stretching (b), in bending (c).

The graphs of the dependence of electrical resistance (figure 5) demonstrate the behavior of the material during deformations. In fact, the material serves as a resistor under an applied voltage (tension). When deformation is applied, the electrical resistance of the conducting network of CNTs in the material changes due to changes in geometric dimensions (length and cross-sectional area), piezoresistive properties of CNTs, tunneling effect or the mechanism of separation (formation and restoration of bonds).

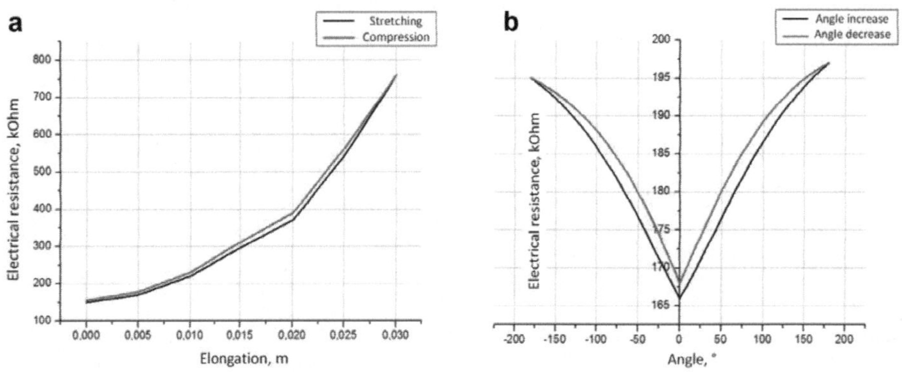

Figure 5. Graph of electrical resistance from elongation (a) according to bending (b).

From the graph in figure 5a, it can be seen that stretching the material by 30 mm (100 % of the original length) results in a linear increase in electrical resistance up to 760 kΩ (black line). Gradually shrinking the material (red line) from 30 mm to 0 mm leads to a decrease in resistance with little hysteresis. The maximum resistance deviation was at the point of tension by 0.025 m and amounted to 20 kΩ. The tensile hysteresis was ~ 4 %. From the graph in figure 5b, it can be seen that bending the sample from 0° to 180° (black line) resulted in a gradual increase in resistance to 195 kΩ. Bending the material from 180° to 0° (red line) resulted in a decrease in resistance. Reverse bending of the material from 0° to -180° (black line) resulted in a gradual increase in resistance to 195 kΩ. Reverse bending of the material from -180° to 0° (black line) resulted in a decrease in resistance. The largest resistance deviation was 5 kΩ. Thus, the bending hysteresis was ~3 %. The following coefficients were obtained: linear GF_e=3.4, angular GF_a=0.26. Low values of the GF_a indicate a low sensitivity of the material to bending.

3.3. Elastic modulus and Tensile Strength

It can be seen from the tension diagram (figure 6) that the linear section of elastic deformation occupies a significant part of the total graph (53 %). The elastic deformation of the sensitive material is in the range of elongation from 0-0.1 m. The elastic deformation limit is in the elongation region of 0.1 m at a load force of 16.9 N. The average value of the elastic modulus was 166.85 kPa. The calculated tensile strength is 1.437 MPa. The maximum elongation of the material from the original length to irreversible damage was 950 %.

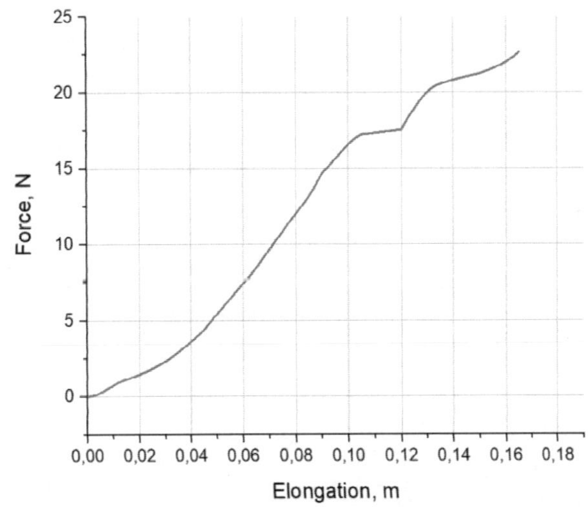

Figure 6. Tensile diagram of the sensor.

4. Conclusion

A simple technology has been developed for manufacturing a strain-sensitive material for flexible sensors. The use of laser radiation made it possible to form a tridimensional network of nanotubes inside a silicone dielectric, which provided the necessary mechanical and electrical characteristics. Due to its high strength, elongation, flexibility and sensitivity, this material opens up broad prospects for the development of wearable, sensitive to deformations (tension, bending, pressure) resistive sensors in the field of flexible electronics.

Acknowledgments

This study was supported by the Ministry of Science and Higher Education of the Russian Federation No. 075-03-2020-216 from 27.12.2019.

References

[1] Kamišalic A, Fister I, Turkanovic M, and Karakatic S 2018 *Sensors* **18** 6 1714
[2] Zhou J, Gu Y, Fei P, Mai W, Gao Y, Yang R, Bao G and Wang Z 2008 *Nano Lett.* **8** 9 3035
[3] Zhang R, Ding J, Liu C and Yang E 2018 *ACS Appl. Energy Mater.* **1** 5 2048
[4] Song J, Do K, Koo J, Son D and Kim D 2019 *Mater. Research Soc.* **44** 643
[5] Sun Z, Yang S, Zhao P, Zhang J, Yang Y, Ye X, Zhao X, Cui N, Tong Y, Liu Y, Chen X and Tang Q 2020 *Appl. Mater. & Interfaces* **12** 11 13287
[6] Markov A, Wördenweber R, Ichkitidze L P, Gerasimenko A Yu, Kurilova U E, Suetina I A, Mezentseva M V, Offenhäusser A and Telyshev D V 2020 *Nanomaterials* **10** 12 2492
[7] Demidenko N A, Kuksin A V, Murashko D T, Cherepanova N G, Semak A E, Bychkov V N and Gerasimenko A Yu 2020 *3D Print. Optics and Additive Photonic Manufacturing II* **11349** 113490U
[8] Gerasimenko A Yu, Kurilova U E, Savelyev M S, Murashko D T and Glukhova O E 2021 *Compos. Struct.* **260** 113517
[9] Zhurbina N N, Kurilova U E, Ichkitidze L P, Podgaetsky V M, Selishchev S V, Suetina I A, Mezentseva M V, Eganova E M, Pavlov A A and Gerasimenko A Yu 2016 *Proc. Of SPIE* **9917** 991718
[10] Gerasimenko A Y, Ten G N, Ryabkin D I, Shcherbakova N E, Morozova E A and Ichkitidze L P 2020 *Spectrochim. Acta. Part A Mol. Biomol. Spectrosc.* **227** 117682

Electrically conductive CNT networks formed by laser

A V Kuksin[1], U E Kurilova[1], O E Glukhova[2,3], E M Eganova[4], A A Dudin[4] and A Yu Gerasimenko[1,3,4]

[1] Institute of biomedical systems, National Research University of Electronic Technology, Zelenograd, Moscow, Shokin Square 1, Russia
[2] Department of Physics, Saratov State University, Saratov, Astrakhanskaya Street 83, Russia
[3] Institute for Bionic Technologies and Engineering, I.M. Sechenov First Moscow State Medical University, Moscow, Bolshaya Pirogovskaya street 2-4, Russia
[4] Institute of Nanotechnology of Microelectronics of the Russian Academy of Sciences, Moscow, Leninsky Prospekt 32A, Russia

Abstract. Materials containing conductive CNT networks were investigated in this work. Binding of CNTs to each other was carried out using laser radiation in scanning mode. As a result of experimental studies, radiation energy density at which the effect of SWCNT binding occured was found – 0.061 J/cm^2. Mechanism by which formation of porous materials with SWCNTs in composition takes place has been established. Materials from CNTs and biopolymers with controllable pore sizes that make up more than 60% of nanocomposite volume have been made.

1. Introduction

Nowadays laser radiation is widely used for modification of nanomaterials made of metals, dielectrics, and semiconductors. One of laser radiation applications is to improve the characteristics of materials for bioelectronics, namely, electrically conductive materials for biointerfaces. Such materials can be used for targeted stimulation of cells and tissues with high spatial resolution. One of the promising materials is carbon nanotubes (CNT) [1]. They are durable [2] and have high biocompatibility [3].

It is known that laser radiation in pulsed and continuous modes can bind carbon nanoparticles (carbon nanotubes, graphene) with each other [4]. This effect is achieved due to two main mechanisms: sublimation of the most defective nanotubes or graphene sheets under the influence of laser radiation and subsequent condensation of sublimation products at nanotubes and graphene junctions; and due to the formation of covalent bonds in defect regions of nanoparticles during laser heating. Defective areas of nanoparticles have the least thermal conductivity and they are the most chemically active. This leads to appearance of chemical bonds in defective areas [5,6]. It is known that carbon nanoparticles have semiconductive and electrically conductive properties. When nanotubes and graphene are bonded to each other, electrically conductive nodes (percolation nodes) appear. It is necessary to create a complex of percolation nodes in given area or in given volume for fabrication of electrically conductive compounds for nanoelectronics or functional 3D materials in bioelectronics [7,8].

Materials such as biopolymers are used to enhance the biocompatibility of interface materials [9]. Albumin under the action of laser forms a nutrient-rich matrix for cell growth. Collagen increases flexibility of biointerface, chitosan has an antibacterial effect [10, 11]. Networks made of nanotubes in matrix, formed under the laser action, imparts electrical conductivity to matrix materials.

In this work, fabrication of SWCNT networks based materials was carried out by depositing disordered system of single-walled carbon nanotubes (SWCNT) on a substrate. Forming of SWCNT network inside its structure helps to achieve high strength and electrical conductivity of material. This is of great relevance in the field of creating bioelectronic devices, for example, when creating electrically conductive implants. In this regard, we have developed technology for creating electrically conductive biopolymer nanocomposites consisting of SWCNT network in biopolymers matrix: albumin, collagen and chitosan.

2. Materials and methods

First stage in preparation of samples was creation of homogeneous dispersion of SWCNT in solvent (dimethylformamide - 90% and methylethylkitone - 10%). Resulting dispersion was applied using a nitrogen sputtering system onto a polished silicon wafer for nano- and microelectronics located on a heating stage for solvent evaporation. Application was carried out layer-by-layer. After that, disordered system of SWCNTs, consisting of 500 layers up to 2 nm thick each was treated with laser action in scanning mode. Laser setup consisted of irradiation system represented by pulsed Yb laser with wavelength 1064 nm, pulse frequency 30 kHz. Positioning of laser beam was carried out using scanner with two mirrors and focusing lens. Sample was uniformly coated with laser pulses with the same energy density. Laser scanning was carried out using computer program along given trajectory.

Nanocomposites with biopolymers were created by the action of same Yb laser radiation on homogeneous dispersion made of albumin, collagen, chitosan and with SWCNTs. Dispersions were prepared from distilled water and nanotubes with concentration 0.01%. Components were mixed on magnetic stirrer for 30 minutes, then agglomerates of nanotubes were separated using ultrasonic homogenizer and ultrasonic bath. Biopolymers were added to the resulting dispersion, treatment was performed with magnetic stirrer for 1 hour and with ultrasonic bath for 1 hour. Biopolymers concentrations were 25% of albumin, 1% of collagen, and 2% of chitosan. Dispersion was deposited layer-by-layer on silicon substrate using the similar method as for SWCNT depositions, but they were irradiated with pulsed laser radiation immediately after each deposition.

All the obtained samples were examined in detail using scanning electron microscope (SEM) FEI Helios NanoLab 650. Accelerating voltage of electron column was 2 kV, electron probe current was 21 pA.

3. Results and discussion

3.1. SWCNT network, formed by laser

As a result of laser exposure, SWCNT network was formed. Experimental studies have made it possible to determine the energy density that led to formation of networks. It was 0.061 J/cm^2. Figure 1a shows optical scheme of laser setup for influencing SWCNT and for creating nanocomposites. An ordered network of connected SWCNTs is shown on pictures, made by scanning electron microscopy (figure 1b-e).

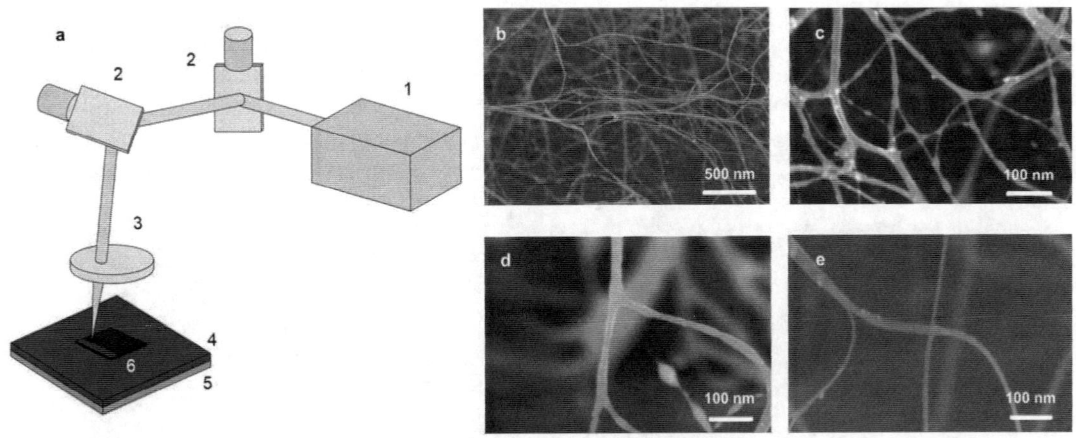

Figure 1. Optical scheme of laser setup for SWCNTs irradiation and for creating nanocomposites (a): 1 – Yb laser, 2 – scanner mirrors, 3 – focusing lens, 4 – SWCNT, 5 – Si substrate, 6 – area of laser influence with laser beam trajectory, SEM images of SWCNT networks (b,c) and places where SWCNTs are welded together (d,e).

It is known that high temperatures promote destruction of nanotubes and C-C bonds, but also leads to subsequent formation of new bonds. High-intensity laser radiation heats nanotubes to high temperatures around 1800 °C. This facilitates the formation of welded SWCNT networks. In addition to nanowelding effect, new chemical bonds are forming on the contact surfaces of heated nanotubes. Formed SWCNT network had high electrical conductivity. In table 1, data on electrical conductivity of obtained SWCNT networks, formed with different modes of laser exposure, are presented.

Table 1. Electrical conductivity of SWCNT network formed by laser action.

	Disordored SWCNTs	SWCNT network, obtained by laser irradiation with energy density E/S			
Laser energy density, J/cm^2		0.002	0.028	0.044	0.061
Number of passes		10	1	1	1
Electrical conductivity, S/m	2.3×10^4	3.8×10^5	6.5×10^5	8.7×10^5	1.1×10^7

3.2. Nanocomposite consisting of SWCNT network and biopolymers

It is known from nonlinear optical studies of materials with carbon nanotubes that nonlinear absorption and nonlinear scattering of radiation occur in liquid dispersed medium exposed to pulsed laser radiation (with high pulse energy). In this case, energy of laser pulse is absorbed by carbon nanotube and converted into heat. Further, gas bubbles appear around the carbon nanotubes because of heating. The resulting bubbles scatter the rest of the pulse and subsequent laser pulses. Nonlinear processes in dispersion with carbon nanotubes are triggered if laser pulse has sufficient threshold energy density. At the same time, threshold energy density was established, at which formation of nanocomposites with a given structure (porosity) took place.

By determining the threshold radiation energy density, it is possible to control pore size in solid composite. The energy density at which porosity of biopolymer composites is more than 60% was determined in this work. Biopolymer composites of albumin, collagen, and chitosan with SWCNT had clearly visible porosity, the sizes of pores were 1-5 μm and 100-200 μm at the threshold energy densities of 0.051, 0.054, and 0.057 J/cm2, respectively. Large pores should promote cell adhesion and proliferation. Small pores are necessary for vascularization and innervation processes. Created nanocomposites can be used as electrically conductive interfaces for restoration of nerve and muscle tissues of the body. Figure 2 shows SEM images of nanocomposites structure, formed by laser radiation with different energy densities: 0.051 J/cm2 (figure 2a) and 0.057 J/cm2 (figure 2b).

Morphology of nanocomposite layer based on SWCNT network can be seen in figure 2c,d. Yellow arrows show the areas of SWCNT connections in network, formed under the influence of laser radiation.

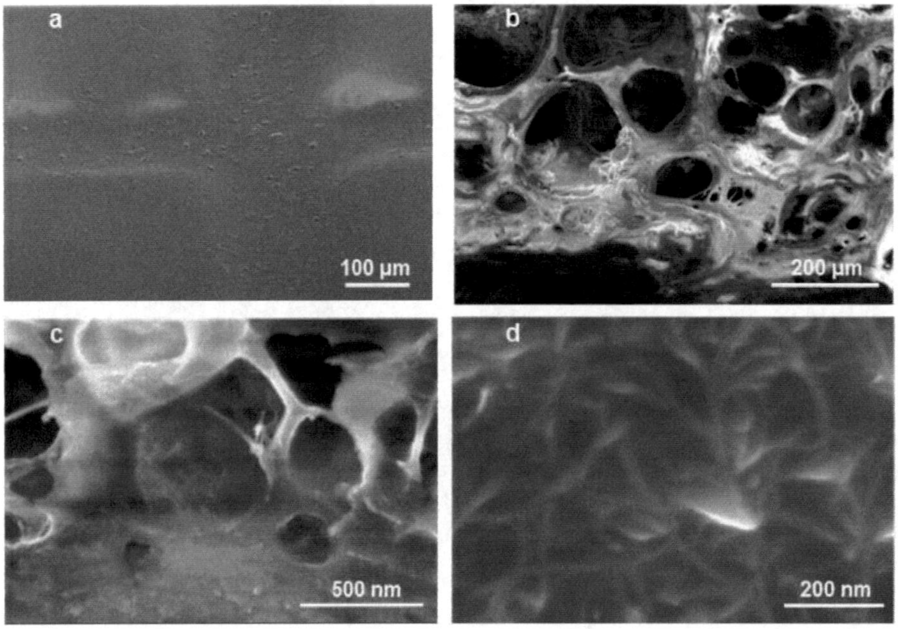

Figure 2. SEM images of nanocomposites structures formed by laser radiation with different energy densities: 0.051 J/cm^2 (a), 0.057 J/cm^2 (b), SWCNT network based on nanocomposite layer with biopolymers (c,d).

It can be seen from the images that the size of network elements exceeds the diameter of single nanotubes. An increase in the diameter of SWCNTs may be due to their coating with biopolymers. Under the influence of laser radiation, elongated biopolymer molecules unfold and attach to SWCNT due to side amino groups [7]. When water molecules are removed, biopolymers completely cover the walls of nanotubes.

4. Conclusion

Materials based on conductive CNT networks were investigated in this work. Creating of CNTs to each other was carried out using laser radiation in scanning mode. As a result of experimental studies, radiation energy density at which the effect of SWCNT binding occured was found – 0.061 J/cm^2. Mechanism by which formation of porous materials with SWCNTs in composition takes place has been established. Materials from CNTs and biopolymers with controllable pore sizes that make up more than 60% of nanocomposite volume have been made. Materials can be created in various forms to create independent implantable structures or coatings for cardiovascular or neural devices.

Acknowledgments

The research was supported by grant of the Russian Science Foundation No. 21-19-00226.

References

[1] Rastogi S K, Kalmykov A, Johnson N and Cohen-Karni T 2018. *J. Mater. Chem. B* **6** 7159-7178

[2] Takakura A, Beppu K, Nishihara T, Fukui A, Kozeki T, Namazu T and Itami K 2019 *Nat. Commun.* **10** 1-7

[3] Vashist A, Kaushik A, Vashist A, Sagar V, Ghosal A, Gupta Y K and Nair M 2018 *Adv. Healthc. Mater.* **7** 1701213

[4] Liu Y-T, Yao T-T, Zhang W-S and Wu G-P 2019 *Mater. Lett.* **236** 244–7

[5] Ozturk Z, Baykasoglu C, Celebi A T, Kirca M, Mugan A and To A C 2015 *Int. J. Hydrog. Energy* **40** 403–11

[6] Piper N M, Fu Y, Tao J, Yang X and To A C 2011 *Chem. Phys. Lett.* **502** 231–4

[7] Gerasimenko A Yu, Kurilova U E, Savelyev M S, Murashko D T and Glukhova O E 2021 *Compos. Struct.* **260** 113517

[8] Markov A, Wördenweber R, Ichkitidze L P, Gerasimenko A Yu, Kurilova U E, Suetina I A, Mezentseva M V, Offenhäusser A and Telyshev D V 2020 *Nanomat.* **10** 2492

[9] Gerasimenko A Y, Ten G N, Ryabkin D I, Shcherbakova N E, Morozova E A and Ichkitidze L P 2020 *Spectrochim. Acta. Part A Mol. Biomol. Spectrosc.* **227** 117682

[10] Zhurbina N N, Kurilova U E, Ichkitidze L P, Podgaetsky V M, Selishchev S V, Suetina I A, Mezentseva M V, Eganova E M, Pavlov A A and Gerasimenko A Yu 2016 *Proc. Of SPIE* **9917** 991718-1

[11] Gerasimenko A Yu, Zhurbina N N, Cherepanova N G, Semak A E, Zar V V, Fedorova Y O, Eganova E M, Pavlov A A, Telyshev D V, Selishchev S V and Glukhova O E 2020 *Int. J. Mol. Sci.* **21** 6163

Method for the modification of graphite subsurface layer to a solid mixture of SiC and graphite

A S Grashchenko[1], S A Kukushkin[1], A V Osipov[1] and A V Redkov[2]

[1] Institute of Problems of Mechanical Engineering, Russian Academy of Sciences, 199178, Saint Petersburg, Russia
[2] Saint Petersburg State University, 199034, Saint Petersburg, Russia

Asgrashchenko@bk.ru

Abstract. We studied the interaction of molten Si and graphite surface during annealing in different atmospheres (CO, vacuum, Ar). The studies have shown that during annealing in CO atmosphere a composite material of SiC and graphite in a thick subsurface layer of the graphite is being formed, whereas at vacuum and Ar atmosphere the modified layer is either thin or absent. The composition and structure of both the composite material itself and the interface between the composite material and the graphite matrix were investigated using the methods of scanning electron microscopy and Raman spectroscopy. Studies have shown that the composite material obtained by this method has a branched fibrous structure consisting of small tubular layers of silicon carbide interspersed with large monocrystalline grains of silicon carbide of the cubic polytype, which leads to significant strengthening of the material. Thus the proposed method can be used to form a thermal protective, chemically resistant coating on graphite surface.

1. Introduction

Composite materials are widely used in dentistry, aerospace, construction, automotive, shipbuilding and many other fields of industry. There are many methods for obtaining different composite materials. In this work, we studied the interaction of molten Si and SiC with the graphite matrix in different atmospheres (CO, vacuum, Ar) and show that during annealing in CO atmosphere a thick subsurface layer of composite material is being formed consisting of small SiC particles having nano sizes interspersed with large single-crystal SiC particles of about 10 μm in size inside the graphite basement. Large single-crystal SiC particles act as reinforcing elements that provide the necessary mechanical characteristics of the material, and the graphite matrix acts as a binder. At the same time annealing in Ar or vacuum atmosphere leads to formation of significantly thinner layers or absence of SiC coating.

The proposed technology of annealing of graphite with molten silicon in CO can be used both to increase the chemical inertness and operating temperature of graphite products, and to obtain a single-crystal SiC powder. Thus, studying the processes occurring during modification and studying the physical characteristics of the resulting mixture of SiC and graphite is a relevant task.

2. Experimental methods

The graphite modification method [1] is partially based on the technology of synthesis of epitaxial SiC films inside a silicon (Si) matrix due to the replacement of Si atoms by carbon atoms [2-4]. It should be noted that the conversion of Si to SiC occurs uniformly over the entire area of the Si substrate. This process is realized as a result of a topochemical reaction:

$$2Si_{solid} + CO_{gas} \rightarrow SiC_{solid} + SiO_{gas}, \quad (1)$$

The synthesis of SiC films occurs at a temperature of 1100–1300°C in an atmosphere of carbon monoxide (CO). It should be noted that the thickness of SiC films obtained by atomic substitution usually does not exceed 250 nanometers. Upon further heating to 1400–1450°C, the silicon base melts and if it is placed on a graphite surface, then graphite is being modified to a mixture of graphite and SiC crystals, and a monocrystalline silicon carbide film is formed on the surface. To study the structure obtained by this method, a graphite plate was taken, on which a piece of a silicon plate with an area of 1 cm^2 was put. The system was heated in a vacuum chamber to 1410°C in an atmosphere of CO 5 minutes. The pressure in the chamber during heating was constant 2 torr. The resulting structure was investigated using a scanning electron microscope (SEM). Studies of the surface and cross-section of the modified region were carried out. Also, the resulting structure was investigated by Raman spectroscopy. The Raman spectra were recorded both in the region of a graphite plate over which Si was initially put and far from this region.

In order to confirm effect of CO on the SiC formation process, within the framework of this work experiments were carried out also on the melting of Si on a graphite plate in argon (Ar) atmosphere and in vacuum. The experiment on melting in Ar was carried out at a pressure inside the chamber of 1.5–2 Torr.

3. Results and discussion

Figure 1 shows the SEM-images of surface of the modified in the presence of CO graphite plate and it's cross-section. One can note on the image of the surface that it is covered with crystals of different sizes. The obtained crystals do not have a common crystallographic direction. Visually, the area of the graphite plate over which the modification occurred in this case has a greenish tint. From the cross-section image, one can determine the depth to which graphite was modified, that is about 1 mm. A further magnification of the cross-section shows that the entire modified region consists of individual faceted crystals within the graphite matrix.

Figure 1. SEM image of a modified region of surface **(a)**, cross-section **(b)**, microcrystal SiC on the cross-section **(c)**.

Figure 2 (a,b) shows the Raman spectra taken from the surface of the graphite plate in the modified region and far from it. Far from the region in which the modifications took place, the Raman spectrum is typical for graphite [5], all its main bands are clearly visible (1351 cm^{-1}, 1571 cm^{-1} и 2692 cm^{-1}). In the case of the modified region, the spectrum radically differs from the typical graphite spectrum, the main bands of the spectrum becomes the SiC band of the cubic polytype (796 cm^{-1}) [6], and graphite bands disappear at all. The obtained result suggests that mainly cubic SiC crystals lie on the surface of the modified region.

One should note that when Si was melted on a graphite plate in argon (Ar) atmosphere and in vacuum, the Si molten mass has not penetrated into the graphite effectively. In these cases, at a temperature of 1410 ° C, a common interface is formed between Si and graphite, and a piece of a silicon wafer turns into a rounded droplet on top of the graphite. Raman spectra taken from the surface of the obtained droplets unambiguously show that it consists of solid Si with very thin layer of SiC (see Figure 2c).

We also note, that the method of Si melting on a graphite plate presented in this work can be used in various modifications. So, for example, in [1], variants of Si preliminarily deposited on graphite in the form of a powder or a thin film were considered. Also in [1], a technique is described which includes dipping a graphite part into liquid silicon, which is in a heat-resistant crucible inside a vacuum reactor, followed by pulling it out of the melt in CO atmosphere. Thus, the method of graphite modification by melting Si in an atmosphere of carbon oxide is a new way to obtain a promising composite material that can be used in various fields of industry.

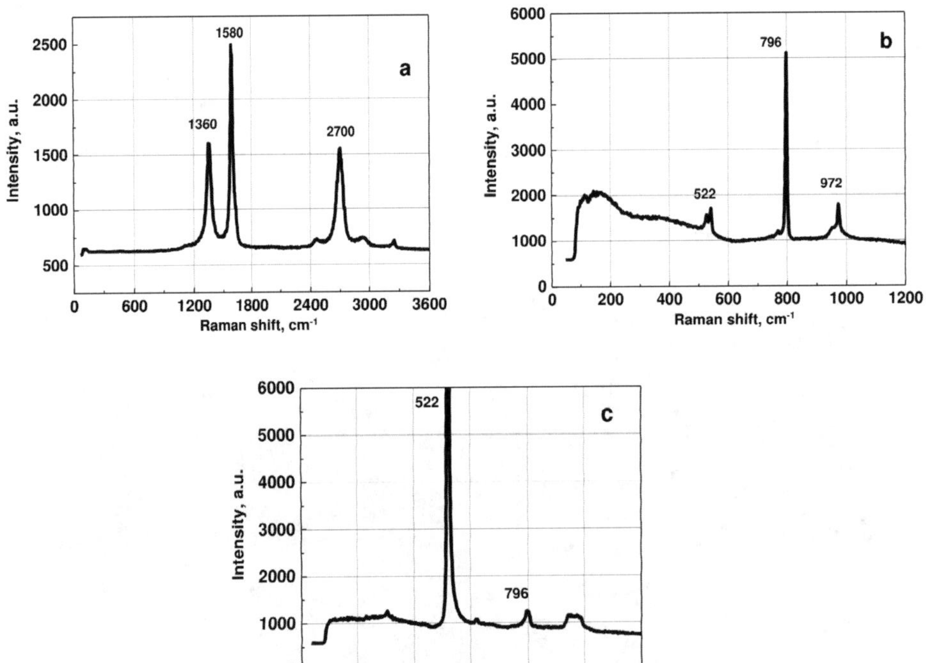

Figure 2. Micro-Raman spectrum of the initial surface of a graphite plate **(a)**, and the surface of the Si on graphite modified in CO atmosphere **(b)**, and Si on graphite modified in argon atmosphere **(c)**.

4. Conclusions

A method of graphite modification into crystalline SiC by melting Si substrates on the surface of the graphite plate in a CO atmosphere is presented. Structural studies of the modified region were carried out. It is shown that the modification occurs not only on the surface, but extends to the depths up to 1 mm. It was also found that the modified region partially consists of separate non-textured misoriented SiC crystals. Using the Raman technique, an analysis of the composition of the modified region was carried out. We also have shown, that the graphite subsurface modification by molten Si takes place actively in presence of CO, whereas in an Ar atmosphere or in the vacuum it does not occur.

Acknowledgments

Grashchenko A S, Kukushkin S A and Osipov A V performed their part of the work (preparation of the samples, formulation of the experiment) within the framework of the state assignment of the Ministry of Science and Higher Education of the Russian Federation for Institute of Problems of Mechanical Engineering RAS (contract No. FFNF-2021-0001), A V Redkov performed his part of the work (Raman investigation, SEM-measurements) within the framework of the state contract of St. Petersburg State University No. 61520973.

The studies were carried out using the equipment of the unique scientific setup "Physics, chemistry and mechanics of crystals and thin films" of the Institute for Problems in Mechanical Engineering of the Russian Academy of Sciences, St. Petersburg.

References

[1] Grashchenko A S, Kukushkin S A, Osipov A V, Redkov A V and Feoktistov N A 2018 Patent of invention № RU 2695423
[2] Kukushkin S A and Osipov A V 2008 *Physics of the Solid State* **50** 1238–1245
[3] Kukushkin S A and Osipov A V 2013 *Journal of Applied Physics* **113** 024909
[4] Grashchenko A S, Kukushkin S A and Osipov A V 2014 *Technical Physics Letters* **40** 1114–1116
[5] Reich S and Thomsen C 2004 *Mathematical, Physical and Engineering Sciences* **362** 2271–2288
[6] Wasyluk J, Perova T S, Kukushkin S A, Osipov A V, Feoktistov N A and Grudinkin S A 2010 *Materials Science Forum* **645** 359–362

Morphology transformation of InGaN nanowires grown on Si substrate by PA-MBE

V O Gridchin[1-3], A S Dragunova[1,4], K P Kotlyar[2], R R Reznik[1-3,5], S D Komarov[1], N V Kryzhanovskaya[1,4], I P Soshnikov[1,3,6], Yu B Samsonenko[3] and G E Cirlin[1-3,5]

[1] Alferov University, Saint-Petersburg 194021, Russia
[2] Saint-Petersburg State University, Saint-Petersburg 198504, Russia
[3] Institute for Analytical Instrumentation RAS, Saint-Petersburg 190103, Russia
[4] HSE University, Saint-Petersburg 190008, Russia
[5] ITMO University, Saint-Petersburg 197101, Russia
[6] Ioffe Institute, St. Petersburg 194021, Russia

gridchinvo@gmail.com

Abstract. The influence of the growth time on the structural properties of InGaN nanowires grown on Si substrate by plasma-assisted molecular beam epitaxy are studied. Under appropriate other growth conditions, the growth for 2h leads to the formation of separated nanowires, whereas the growth for 2h 30min and 3h leads to the formation of nanostructures such as nano-umbrellas. The separated NWs exhibit a photoluminescence spectrum with maxima at about 590 nm, whereas the nano-umbrellas show two pronounced photoluminescence lines at 421 and 619 nm.

1. Introduction

Nowadays, InGaN ternary compounds have proven themselves well for the creation of light-emitting devices and renewable energy sources due to their unique properties [1]. First of all, a band gap energy of InGaN can vary from 0.7 to 3.43 eV depending on the In content inside the compound [2]. However, the lack of the lattice-matched substrates for InGaN makes it impossible to grow high-quality epitaxial layers with a homogeneously chemical composition. Another challenge is the wide miscibility gap of InGaN [3] significantly complicating the formation of InGaN ternary compounds with high In content. A promising approach to overcome these limitations is the growth of nanowire (NWs) structures. It has been demonstrated that the elastic strain relaxation on the nanowire sidewalls can help to overcome the InGaN miscibility gap and achieve a better band gap tunability than in a bulk alloy [2,4,5]. Another advantage of the elastic strain relaxation on the nanowire sidewalls is the possibility of the growth of practically defect-free NWs on substrates with different lattice constants and different thermal expansion coefficients [6,7]. In this regard, the growth of NWs on silicon substrates is one the most promising approach from the point of view of wide applications ones and low cost.

Recently, we have shown that during a plasma-assisted molecular beam epitaxy the InGaN nanowires with spontaneously formed core-shell structures are grown at a narrow growth temperature range, and the core contains high In mole fraction (about 35 %), whereas the shell contains no more than 4 % of In [8,9]. Also, it was obtained that at the lower growth temperature in comparison with ones of InGaN NWs the InGaN nanostructures with three-dimensional morphology can be grown. However, there is still a lack of information about the influence of the growth conditions on the physical properties of these nanostructures. In this work, the influence of the growth time on the formation of InGaN nanowires is studied.

2. Experimental

The experiments of the InGaN NWs growth were carried out using Riber Compact 12 MBE system equipped with indium (In) and gallium (Ga) effusion cells and a nitrogen source. To grow the NWs, p-type Si (111) substrates were used. Initially, the unprepared substrate was transferred into the growth chamber and the substrate temperature was set to 950 °C for thermal treatment. Next, the substrate temperature was reduced to the actual growth temperature set at 600 °C and the nitrogen plasma source was ignited and the N flux was set to 0.4 sccm and 450 W. Finally, Ga and In sources were simultaneously opened. The beam equivalent pressures of Ga and In were equal to each other and set to 10^{-7} Torr. A series of experiments was carried out with growth time in the range from 2h to 3h.

The structural properties of the samples are examined using scanning electron microscopy SUPRA 25 (C. Zeiss SMT Ltd). Photoluminescence (PL) was excited with a helium–cadmium (He-Cd) metal-vapor laser with a wavelength of 325 nm. The PL signal was detected using a DK480 Spectral products monochromator and a single-channel silicon detector using synchronous detection (SRS 510 "Stanford Research Systems"). The measurements were performed at room temperature.

3. Results and discussion

Figure 1 shows SEM images of the InGaN nanowires grown in (a) 2 hours, (b) 2 hours 30 minutes, (c) 3 hours. The insertions below demonstrate plan-view images of corresponding samples.

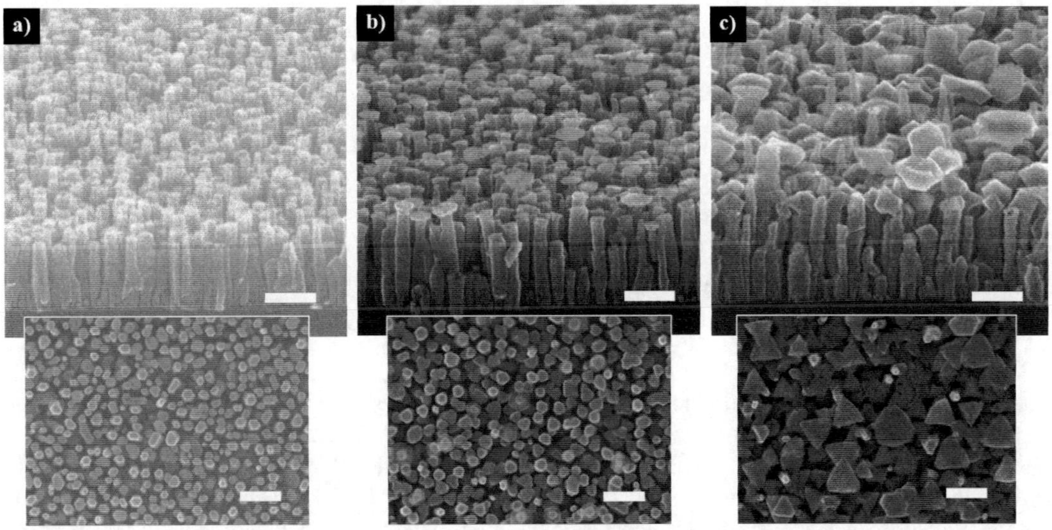

Figure 1. SEM images of InGaN NWs: **(a)** bird's eye view of sample grown at 2h; **(b)** bird's eye view of sample grown at 2h 30min; **(c)** bird's eye view of sample grown at 3h. The insertions below demonstrate plan-view images of corresponding samples. The scale bars correspond to 200 nm.

As can be seen from figure 1(a), the NWs with an average length of 350 nm and an average diameter of 40 nm are grown for 2h. When the growth time was increased to 2h 30min, the upper part of NWs was transformed to nano-umbrellas. The average length of these NWs is 400 nm, and the average diameter is 50 nm. When the growth time was increased to 3h, the morphology of the NWs is changed significantly: a nanotubes layer with an average length 350 nm and average lateral size of 50 nm was formed; above the nanotube layer the InGaN nanostructures with developed morphology, an average length of 100 nm and an average diameter of 150 nm are grown. The nature of these morphology transformations is most likely the local In-rich conditions (In accumulation) in the upper part of NWs [10].

Figure 2(a) shows the results of PL measurements of samples grown for 2h and 2h 30min. The insert in figure 2(a) demonstrates the dependence of the PL maxima on the growth time. Samples grown for 2h and 2h 30min exhibit similar PL spectra with an intense emission of about 590 nm and a weak emission band in the range from 700 to 900 nm, which is apparently explained by the In accumulation in the upper part of NWs. It should be noted that the small deviation of PL maxima of these samples is associated with the In fluctuation within the NWs.

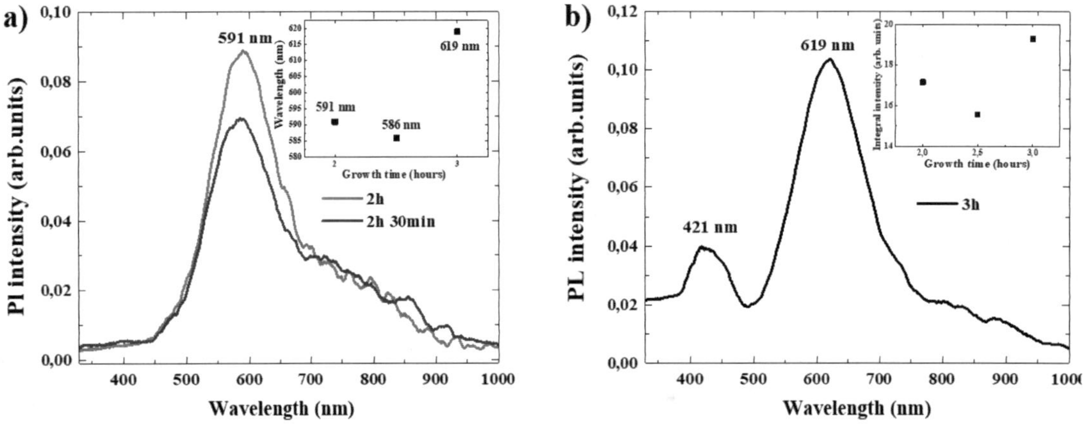

Figure 2. PL spectra of the samples, measured at room temperature: **(a)** PL spectra of samples grown for 2h and 2h 30min. The insert shows the dependence of the PL maxima on the growth time; **(b)** PL spectra of sample grown at 3h. The insert shows the dependence of the integral intensity on the growth time.

Figure 2(b) demonstrates the PL spectrum of sample grown for 3h and the insert with the dependence the integral intensity on the growth time. The main photoluminescence line is shifted by ~30 nm and located at 619 nm, which is apparently associated with the accumulation of In atoms in the upper part of NWs. Also, the addition emission line is detected at 421 nm, which indicates spinodal decomposition in grown NWs and is often observed in InGaN ternary compounds with 0.2-0.8 In mole fraction [3].

4. Conclusion

In the work, we studied the influence of the growth time on the formation of InGaN nanowires during plasma-assisted molecular beam epitaxy. It has been shown that the growth time can significantly affect the structural properties of NWs. In particular, under appropriate growth conditions, the growth for 2h leads to the formation of separated NWs, whereas the growth for 2h 30min and 3h leads to the formation of nanostructures such as nano-umbrellas. Optical studies have shown that samples grown for 2h and 2h 30min exhibit intense PL area of about 590 nm and a weak emission band in the range from 700 to 900 nm, while the sample grown for 3h exhibits two pronounced PL maxima at 421 and 619 nm, which indicated spinodal decomposition in these nano-umbrellas.

Acknowledgements

The samples were grown under the support of the Ministry of Science and Higher Education of the Russian Federation (state task № 0791-2020-0003). The morphology properties were studied under the financial support of the RSF grant (project № 19-72-30010. The optical studies were supported by the Basic Research Program of the National Research University Higher School of Economics.

References

[1] Ambacher O 1998 Growth and applications of group III-nitrides *J. Phys. D* **31** 2653

[2] Roche E, Andre Y, Avit G, Bougerol C, Castelluci D, Réveret F, Gil E, Médard F, Leymarie J and Jean T 2018 Circumventing the miscibility gap in InGaN nanowires emitting from blue to red *Nanotechnology* **29** 465602

[3] Ho I and Stringfellow G 1996 Solid phase immiscibility in GaInN *Appl. Phys. Lett.* **69** 2701–3

[4] Xiang H, Wei S-H, Da Silva J L and Li J 2008 Strain relaxation and band-gap tunability in ternary In x Ga 1− x N nanowires *Phys. Rev. B* **78** 193301

[5] Kuykendall T, Ulrich P, Aloni S and Yang P 2007 Complete composition tunability of InGaN nanowires using a combinatorial approach *Nat. Mater.* **6** 951–6

[6] Hugues M, Shields P, Sacconi F, Mexis M, Auf der Maur M, Cooke M, Dineen M, Di Carlo A, Allsopp D and Zúñiga-Pérez J 2013 Strain evolution in GaN nanowires: From free-surface objects to coalesced templates *J. Appl. Phys.* **114** 084307

[7] Dubrovskii V, Cirlin G and Ustinov V 2009 Semiconductor nanowhiskers: synthesis, properties, and applications *Semiconductors* **43** 1539–84

[8] Gridchin V, Kotlyar K, Reznik R, Dragunova A, Kryzhanovskaya N, Lendyashova V, Kirilenko D, Soshnikov I, Shevchuk D and Cirlin G E 2021 Multi-colour light emission from InGaN nanowires monolithically grown on Si substrate by MBE *Nanotechnology* **32**(33) 335604

[9] Gridchin V, Kotlyar K, Reznik R, Shevchuk D, Kirilenko D, Bert N, Soshnikov I and Cirlin G 2020 Peculiarities of the structural properties of InxGa1-xN polytype nanostructures grown by molecular-beam epitaxy *J. Phys. Conf. Ser.* **1482**(1) 012014

[10] Zhang X, Lourenço-Martins H, Meuret S, Kociak M, Haas B, Rouvière J-L, Jouneau P-H, Bougerol C, Auzelle T and Jalabert D 2016 InGaN nanowires with high InN molar fraction: growth, structural and optical properties *Nanotechnology* **27** 195704

The scalable production of high-quality nanographite by organic radical-assisted electrochemical exfoliation.

E Grushevski[1], D Savelev[1], L Mazaletski[1], N Savinski[2] and D Puhov[2]

[1] The Basic Department of Nanotechnology in Electronics of the P.G. Demidov Yaroslavl State University, Sovetskaya Street 14, 150003, Yaroslavl, Russia.
[2] Valiev Institute of Physics and Technology of Russian Academy of Sciences, Yaroslavl Branch, 150007, Universitetskaya Street 21, Yaroslavl, Russia.

yaregor@mail.ru

Abstract: One of the promising ways to produce graphene is the technology of graphite splitting or exfoliation, both by physical or mechanical and chemical, including electrochemical methods. The product of electro exfoliation is nanographite, which is transformed into multigraphene at the subsequent stage of liquid-phase mechanical and ultrasonic disintegration. This approach demonstrates a successful method of obtaining multigraphene from available graphite raw materials. Since, already at a potential of 1.23V, during the electrolysis of water on a graphite anode, the hydroxyl anion is discharged with the formation of a very active hydroxyl radical oxidizer, it is not surprising that when the graphite electro exfoliation process is overvolted at 10V, graphite oxidation products are formed. In order to control the defectiveness of the graphene lattice by oxidation products, we carried out processes of graphite exfoliation in the presence of both a number of reducing agents ascorbic acid, sodium borohydride, hydrazine hydrate, and in the presence of industrial antioxidants radical traps (2,2,6,6-tetramethylpiperidine-1-il)oxyl (TEMPO), (2,2,6,6-tetramethyl-4 oxo-piperidine-1-yl)oxyl (IPON), a mixture of 5,8,9-bis isomers[(2,2,6,6-tetramethyl - 4 oxo-piperidine-1-yl)]–{5,8,9-[1,1'- bi(cyclopentylidene)]-2,2',4,4'- tetraene}(YARSIM-0215). It should be noted, that the best result of preventing the oxidation of nanographite in electro exfoliation technology in our studies is the ratio of carbon to oxygen (C/O) about 69.

1. Introduction

The uniquely high mobility of carriers, exceptional mechanical and optical properties, and high thermal conductivity have advanced graphene as a precursor material for new opto - and nanoelectronics [1]. One of the limiting factors in the use of graphene in electronics and other areas of use is the lack of industrial scalable, inexpensive methods for the production of graphene derivatives (multigraphenes) with low defects. One of the promising ways to produce graphene is the technology of graphite splitting or exfoliation, both by physical or mechanical and chemical, including electrochemical methods.

At the initial stage of the technology, graphite undergoes electrochemical exfoliation in an aqueous solution of ammonium sulfate at DC and a voltage of 10 V in a potentiostatic mode. The product of electro exfoliation is nanographite, which is transformed into multigraphene at the subsequent stage of liquid-phase mechanical and ultrasonic disintegration. This approach demonstrates a successful method of obtaining multigraphene from available graphite raw materials [2]. Such physical methods of graphite exfoliation as: grinding in ball mills, cleavage with adhesive tape, make it possible to obtain relatively low-defect graphene, but with a very low yield at the level of units of percent. Chemical exfoliation methods are usually based on modified Hammer methods and are used to produce graphene oxides. These methods are easily scaled, the output of products is close to quantitative. However, these chemical methods have a number of significant drawbacks. The result of exfoliation is oxidized graphene derivatives that do not possess graphene properties due to the strong defectiveness of the graphene lattice with oxygen. Even after the reduction stage, the so-called reduced graphene oxide contains a significant amount of oxygen, which is not completely removed.

2. Experiment

2.1. Electrochemical exfoliation

The exfoliation process was carried out on a homemade installation in which GRAFLEX graphite foil was used as the anode, and platinum foil was used as the cathode at a constant current, voltage of 10 volts in a potentiostatic mode. A 3M aqueous solution of ammonium sulfate electrolyte with antioxidant additives was used as the electrolyte. Such as ascorbic acid, hydrazine, sodium borohydride, dimethyl sulfoxide (DMSO), (2,2,6,6-tetramethylpiperidin-1-yl)oxyl (TEMPO), 2,2,6,6-tetramethyl-4-one-piperidin-1-yl)oxyl (IPON), Dimer (2,2,6,6- tetramethylpiperidin-1-yl)oxyl-fulvalene (YARSIM-0215). In acidic electrolytes, anions intercalate into the graphite anode, causing its cleavage. Along with splitting, the processes of decorating the graphite structure with oxygen are underway, including radical processes involving hydroxyl radical [3].

The exfoliated graphene sheets were then transferred onto a Si/SiO_2 substrate using the Langmuir–Blodgett method. Scanning electron microscopy (SEM) and energy-dispersive X-ray spectroscopy (EMF) by means electron microscopes SUPRA-40 and QUANTA 300i were used to study the morphology of graphene flakes on the SiO_2 surface, the size of hundreds of micrometers, and the carbon and oxygen content in the sample. In general, the efficient intercalation of electrolyte into graphite interlayer and the subsequent gas eruption are the key factors governing graphite exfoliation. Since, already at a potential of 1.23V, during the electrolysis of water on a graphite anode, the hydroxyl anion is discharged with the formation of a very active hydroxyl radical oxidizer, it is not surprising that when the graphite electro exfoliation process is overvolted at 10V, graphite oxidation products are formed.

3. Results and disscution

3.1. A mechanism of anodic exfoliation

The mechanism of exfoliation of anodic graphite in an aqueous solution of (NH4)2SO4 was proposed by Parvez et al. in 2014 [3] (figure 1). Hydroxyl anions formed during electrolysis of water at the anode (OH-) are a strong nucleophile that enters into nucleophilic substitution of hydrogen at the sp2 carbon atoms of the aromatic ring at the grain boundaries, forming two vicinal phenolic OH (1) groups. The series-parallel reactions 2 and 3 are dehydration reactions with the formation of oxyrane cycles and carbonyl groups, respectively. Reactions 4 - 6 are channels of gas formation of oxygen and carbon dioxide contribute to the expansion of the interlayer space of graphite layers and the advancement of ions deep into the graphite crystal.

The scheme of the graphite exfoliation process in the presence of a radical scavenger (nitroxyl radical trap TEMPO.) was illustrated in figure 2.

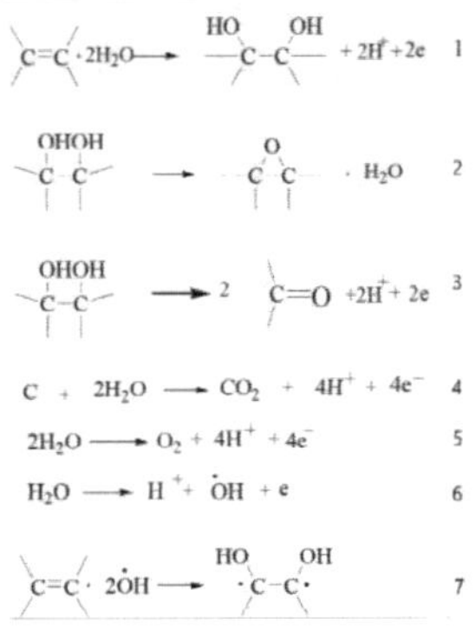

Figure 1. Carbon oxidation reactions.

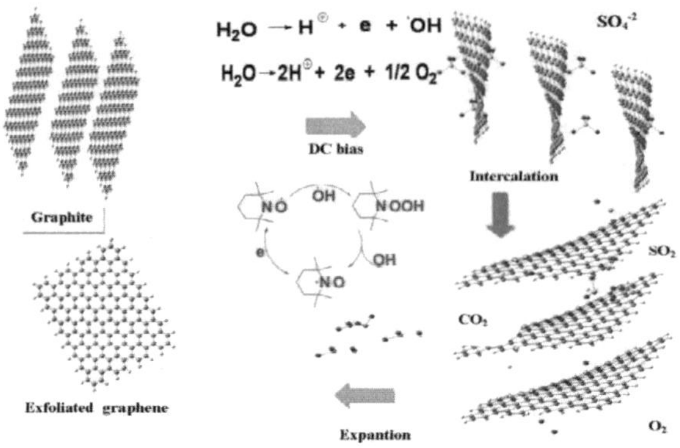

Figure 2. Scheme of the mechanism of anionic electrochemical exfoliation of graphite with an aqueous solution of ammonium sulfate with TEMPO [3].

These gases also assist in the exfoliation of the graphite layers [3]. Along with reaction 1, a hydroxyl radical strong oxidizer (SHE) = -2.8 V was generated at the anode at an overvoltage of 10 V by reaction 6. Reaction 7 of radical addition of hydroxyl radical by the double bond of the aromatic core lead to products similar to reaction 1. In our previous study [4], we identified the products of electrochemical exfoliation of graphite as oxidized nanographite with 20-30 graphene layers. In order to control the defectiveness of the graphene lattice by oxidation products, we carried out processes of graphite exfoliation in the presence of both a number of reducing agents ascorbic acid, sodium borohydride, hydrazine hydrate, and in the presence of radical traps (2,2,6,6-tetramethylpiperidine-1-il)oxyl (TEMPO), (2,2,6,6-tetramethyl-4 oxo-piperidine-1-yl)oxyl (IPON), a mixture of 5,8,9-bis isomers[(2,2,6,6-tetramethyl - 4 oxo-piperidine-1-yl)]–{5,8,9-[1,1'- bi(cyclopentylidene)]-2,2',4,4'-tetraene}(YARSIM-0215).

3.2. Experimental data

The G-band of Raman scattering spectra for sp2 carbon atoms is observed as a single or multiplicative peak near the frequency of 1585 cm^{-1} for all carbon structures of sp2 hybridization. The G^1 - band is also defined by sp2 carbon atoms as a single peak or a multi-peak structure in the range of 2500-2800 cm^{-1} in figure 3.

Figure 3. Raman spectra were obtained using a spectrometer ENSPECTR R532.

The sensitivity to the details of the sp2 structure makes this band a powerful indicator for determining the number of layers of graphene layers in a stack. The Raman spectrum of disordered graphene shows two new sharp peaks around 1350 cm^{-1} peak D and 1620 cm^{-1} peak D^1. These inter-valley D and intra-valley D^1 defect-induced resonant scattering processes are mainly responsible for electron decoherence in the optical and transport phenomena of sp2 carbon atoms. These bands are also blurred and, therefore, their actual positions depend on the laser excitation energy. D-band basic Raman frequency sp2, attributed to disorder and defects. Since absolute intensity is rarely used in Raman spectroscopy, the normalized I_D/I_G intensity is widely used for quantitative measurements of the degree of defect (disorder). Equation 1, obtained for determining the size of the crystal regions of graphene L_a and the defect density, or in other words, the distance between the defects, depending on the energy of the exciting radiation used in the visible range [5,6]

$$L_a(nm)=\frac{560}{E^4_{laser}}(I_D/I_G)^{-1}=(2.4\times10^{-10})\lambda^4_{laser}(I_D/I_G)^{-1} \quad (1)$$

For a laser with a wavelength of 532 nm, the defect density n_D can be represented by equation 2.

$$n_D\left(cm^{-2}\right)=(2.16\times10^{11})\,I(D)/I(G) \quad (2)$$

For a honeycomb ideal lattice, the number of carbon atoms is $N_c = 3.82 \times 10^{15}$ cm^{-2}. If you normalize the number of defects in equation 2 by N_c, you can get the concentration of defects per cm^{-2} in millionths (ppm). The table 1 shows the results of processing the spectra of graphite samples obtained under various exfoliation conditions.

Table 1. Intensity of the general bands of the Raman spectrum of graphite exfoliation products.

№	D/ *λmax 1346 cm⁻¹*	G/ *λmax 1581 cm⁻¹*	G^1/ *λmax 2711 cm⁻¹*	I_D/I_G	$I_G{}^1/I_G$	N_d/N_c (ppm)[a]	Reaction conditions 0.2 M Ammonium sulfate$(NH_4)_2SO_4$, Pt cathode, potential 10V + additives
1	495	523	152	0.95	0.29	53	no additives
2	368	391	120	0.94	0.31	53	Ascorbic acid 0.4% by weight
3	506	472	165	1.07	0.35	60	Hydrazine Hydrate 0.4% w
4	495	523	152	0.95	0.29	53	Sodium borohydride 0.4% w
5	210	518	177	0.41	0.34	23	TEMPO 0.4% w
6	367	806	261	0.46	0.32	26	DMSO 0.4% w

[a]-The density of graphene structure defects according to Equation 3 in ppm

The analysis of the results of Table 1 clearly demonstrates the dependence of the conditions for the electro-exfoliation of graphite on the size and density of defects in the crystal lattice (the basal plane of graphene) of graphite. Thus, carrying out the process without oxidation inhibitors, but with reducing agents: ascorbic acid, hydrazine hydrate, sodium borohydride leads to approximately the same result. That is, it does not significantly affect the density of defects, in contrast to the fact that the use of TEMPO and DMSO significantly reduces the density of defects by 2.5 to 3 times. Probably, a noticeable decrease in the defect density was associated with the blocking of the hydroxyl radical as an oxidizing agent of the basal plane of the graphene crystal lattice by scavengers of radicals by DMSO or TEMPO molecules.

Table 2. Carbon/Oxygen C/O ratio in various graphene products.

Graphene products	Approaches	C/O ratio
Original graphite foil «graflex»		48.11
Exfoliated graphene Electrochemical	Pt electrode 0.05M $(NH_4)_2SO_4$	16.95

Exfoliated graphene Electrochemical	Pt electrode 0.05M $(NH_4)_2SO_4$ + with 1mg /mL TEMPO	17.0
Exfoliated graphene Electrochemical	Pt electrode 0.05M $(NH_4)_2SO_4$ + 1mg /mL IPON	11.5
Exfoliated graphene Electrochemical	Pt electrode 3.M$(NH_4)_2SO_4$ 0.08%w DMSO	69

The main mechanism of hydroxyl radical formation is the neutralization of the hydroxyl anion on the graphite anode, and since it has a high oxidation potential, it is necessary to use radical traps to bind the attacking radical.

It should be noted, if the molecules of industrial antioxidants (TEMPO, IPON) interaction with hydroxyl radical in accordance with the scheme shown in Figure 2 and are practically not consumed in the oxidation reduction cycle, whereas DMSO is consumed during the reaction, forming a neutral non-radical compound methanesulphinic acid through a series of successive stages.

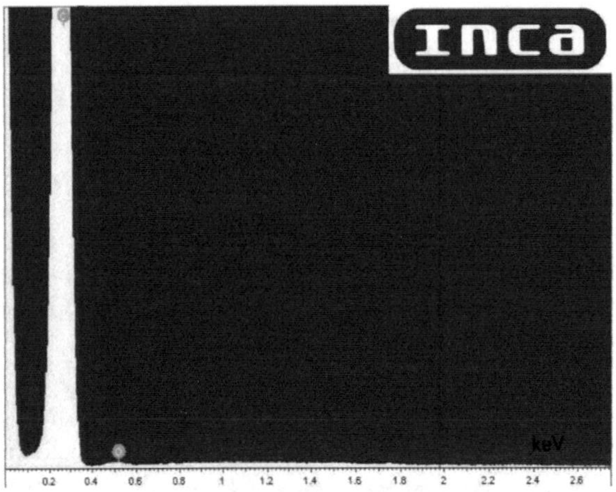

Figure 4. Elemental analysis of exfoliation products using the energy-dispersive X-ray spectroscopy.

4. Conclusion

A method of electrochemical exfoliation of graphite in an aqueous solution of an electrolyte of 3M ammonium sulfate by DC at a voltage of 10 V in a potentiostatic mode has been developed. The exfoliation product is a high-quality nanographite with a good yield close to quantitative. The use of antioxidants in the process allows you to regulate the degree of oxidation and bring the carbon/oxygen ratio to 69. Commercially available and inexpensive reagents, a fast process, and the absence of by-products make such a process promising for industrial scale.

Acknowledgements

This research was financially supported within the framework of the state assignment of the Valiev Institute of Physics and Technology of the Russian Academy of Sciences Yaroslavl Branch of the Ministry of Education and Science of the Russian Federation on the topic No. 0066-2019-0003 and the state assignment of P.G. Demidov Yaroslavl State University No. 0856-2020-0006. Some experimental results were obtained on the equipment of the Center for Collective Use "Diagnostics of Micro-and nanostructures" with the support of the Ministry of Education and Science of the Russian Federation.

We would like to thank Dr. Sergey Salnikov for providing samples of industrial antioxidants of JSC «Yarsintez» Yaroslavl and Professor Evgeny Pliss for practical advice on the use of antioxidants.

References

[1] Novoselov K S, Geim A K, Morozov S V, Jiang D, Zhang Y, Dubonos S V, Grigorieva I V and Firsov A A 2004 *J. Science* **306** 666

[2] Parvez K, Wu Z-S, Li R, Liu X, Graf R, Feng X and Müllen K 2014 *J Am Chem Soc* **136** 6083-91.

[3] Yang S, Brüller S, Shuai Wu Z, Liu Z, Parvez K, Dong R, Richard F, Samori P, Feng X and Müllen K 2015 *J. Am. Chem. Soc.* **137** 13927–32.

[4] Savinsky N G, Melesov N S, Parshin E O, Vasiliev S V, Bachurin V I and Churilov A B 2020 *Bulletin of the Russian Academy of Sciences: Physics* **84** N 6 732–735.

[5] Paton K R *et al* 2014 *Nat. Mater.* **13** 624.

[6] Eckmann A, Felten A, Mishchenko A, Britnell L, Krupke R, Novoselov K S and Casiraghi C 2012 *Nano Lett.* **12** 3925–30.

Effect of Yb^{3+} doping level on the structure and spectroscopic properties of ZnO optical ceramics

E Gorokhova[1], I Venevtsev[2], I Alekseeva[1], A Khubetsov[1], O Dymshits[1], L Basyrova[3], E Oreschenko[1], S Eron'ko[1], F Muktepavela[4], K. Kundzins[4], A Zhilin[1] and P Loiko[3]

[1] S.I. Vavilov State Optical Institute, 36 Babushkina St., 192171 St. Petersburg, Russia
[2] Peter the Great St. Petersburg Polytechnic University, 29 Polytechnic St., 195251 St. Petersburg, Russia
[3] Centre de Recherche sur les Ions, les Matériaux et la Photonique (CIMAP), UMR 6252 CEA-CNRS-ENSICAEN, Université de Caen Normandie, 6 Boulevard du Maréchal Juin, 14050 Caen Cedex 4, France
[4] Institute of Solid State Physics, University of Latvia, 8 Kengaraga St., LV1063, Riga, Latvia

e.gorokhova@rambler.ru; khubezov@gmail.com;

Abstract. Zinc oxide optical ceramics with hexagonal structure doped with 0.6 –5.0 wt% Yb were fabricated by uniaxial hot pressing of commercial oxide powders at 1180 °C in vacuum. The ceramics were characterized by X-ray diffraction, SEM, EDX, X-ray and optical spectroscopy. It is shown that Yb^{3+} ions are distributed between C-type Yb_2O_3 sesquioxide crystals and ZnO grain boundaries. The Yb^{3+} doping of ZnO ceramics enhances the near-band-edge emission of zinc oxide. ZnO:Yb optical ceramics are promising for optoelectronic applications.

1. Introduction

Zinc oxide (ZnO) is a direct wide band gap semiconductor with unique optical and luminescent properties coupled with high radiation tolerance. Ytterbium (Yb^{3+}) doped ZnO nanostructures and films demonstrate efficient energy transfer between the defect states in the ZnO structure and the dopant Yb^{3+} ions [1], making them promising for down-conversion layers of enhanced solar cells [2]. It has been shown that the efficient energy transfer from ZnO to Yb^{3+} ions in the nanocomposites is due to the formation of a tight contact between the Yb_2O_3 and ZnO nanoparticles (i.e., the Yb_2O_3-ZnO heterojunction) and it arises mainly from the quantum cutting mechanism [1]. In the present work, the effect of Yb doping level on the structure, optical and spectral-luminescent properties of hexagonal ZnO ceramics was studied for the first time.

2. Experimental

ZnO optical ceramics with an Yb doping level of 0.6– 5.0 wt% were fabricated from commercial reagent grade ZnO and Yb_2O_3 powders by hot pressing in vacuum at the temperature of 1180 °C. After polishing, ceramic disks had a diameter of 25 mm and a thickness of about 0.5 mm. The undoped ZnO ceramic was pale yellow colored. With Yb^{3+} doping, the color of ceramics changed to slight grey and their transparency decreased, Figure 1.

The X-ray powder diffraction (XRD) patterns of ceramics were measured with a Shimadzu XRD-6000 diffractometer with Ni-filtered Cu Kα radiation. The ZnO unit cell parameters were determined by the Rietveld method. The texture coefficients were calculated from the XRD data using the approach described in Ref. [3]. The microstructure of the polished and etched surfaces of ceramics was characterized by scanning electron microscopy (SEM) using a MERLIN microscope (Carl Zeiss). The same microscope equipped with an Inca Energy X-Max analyzer was employed for the EDX-based element mapping. The fractured surfaces of ceramic disks were studied using a scanning electron microscope (SEM, TESCAN Lyra 3) equipped with an energy dispersive X-ray spectrometer EDS (Oxford, AZtec) in back scattering electron (BSE) mode.

The total transmittance spectra in the visible and near-IR were measured using a SPECORD 200 PLUS spectrophotometer. The absorption spectra at 300 – 3300 nm were measured with a Shimadzu UV-3600 spectrophotometer. The X-ray induced luminescence (XRIL) spectra were recorded in the reflection geometry upon continuous excitation by an X-ray tube with a tungsten anode (40 kV, 10 mA). The decay kinetics of XRIL was recorded using a pulsed X-ray setup with a pulse duration of ~1 ns at a voltage of 30 kV and a current amplitude of 500 mA.

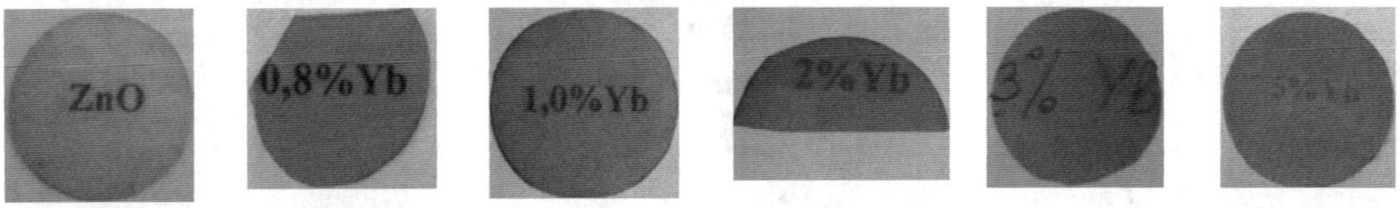

Figure 1. Photographs of undoped and Yb^{3+}-doped ZnO transparent ceramics with increasing Yb doping level (from left to right). The labels indicate the Yb concentration (in wt%).

3. Results and discussion

A typical XRD pattern of Yb^{3+}-doped ZnO ceramic is shown in Figure 2(a). The XRD patterns of the Yb:ZnO ceramics evidence crystallization of hexagonal (sp. gr. $P6_3mc$) phase of ZnO with wurtzite structure and Yb_2O_3 with a cubic (sp. gr. $Ia\bar{3}$) C-type bixbyite structure. The calculated ZnO unit cell parameters for the Yb^{3+}-doped ceramics nearly coincide with those for undoped ZnO (a = 3.249 Å, c = 5.205 Å, International Centre for Diffraction Data (ISDD) card No. 5-0664). They do not change with increasing the Yb content indicating that Yb^{3+} ions do not enter the ZnO structure. The fragment of the XRD pattern in the range of diffraction angles 2θ of 29.0 - 30.3° contains an unambiguously assigned intense diffraction peak of Yb_2O_3 with the Miller's index (*hkl*) = (222), Figure 2(b). The volume fraction of Yb_2O_3 in ceramics gradually increases with Yb content, Figure 2(c). The EDX mapping indicates that Yb^{3+} ions are also localized at the ZnO grain boundaries, Figure 2(d-f).

We performed the SEM study of fractured surfaces of Yb^{3+}-doped ZnO ceramics with the aim to determine the arrangement and location of Yb_2O_3 crystals. The corresponding SEM images for the ZnO: 0.6wt% Yb and ZnO: 3wt% Yb ceramics, cf. Figure 3(a,b), indicate the intragranular fracture mechanism. We failed to find any evidence of brittle fracture on the grain boundaries. This testifies a sufficient strength of ceramics that ensures their reliability in use. The comparison of images presented in Figure 3(a,b) demonstrates that with increasing the ytterbium doping level, the volume fraction of Yb_2O_3 crystals significantly increases.

The Yb_2O_3 crystals are formed in different places within the ZnO ceramic, namely (i) at the grain boundaries (GB) with highest probability, Figure 3(c), (ii) in triple joints of the ZnO grains, which will be shown below, and between the ZnO planes, Figure 3(d). The Yb_2O_3 crystals form regions with sizes ranging from 2-5 to 15-20 μm. These regions are built from smaller grains with sizes of 0.2 -1.0 μm. Their destruction occurs by the quasi-brittle cleavage. The Yb_2O_3 crystals located at the GBs of ZnO grains have sufficient adhesion to ZnO, which indicates strong interaction at the GBs of these oxides. This is evidenced by the SEM-EDX studies of the ZnO:Yb ceramics with different locations of Yb_2O_3

crystals (at the GBs and inside the grains). As an example, Figure 4 shows the SEM-EDX data for the ZnO:0.6wt% Yb ceramic, wherein the grain of Yb_2O_3 is located at the triple joint and partially inside the ZnO grain.

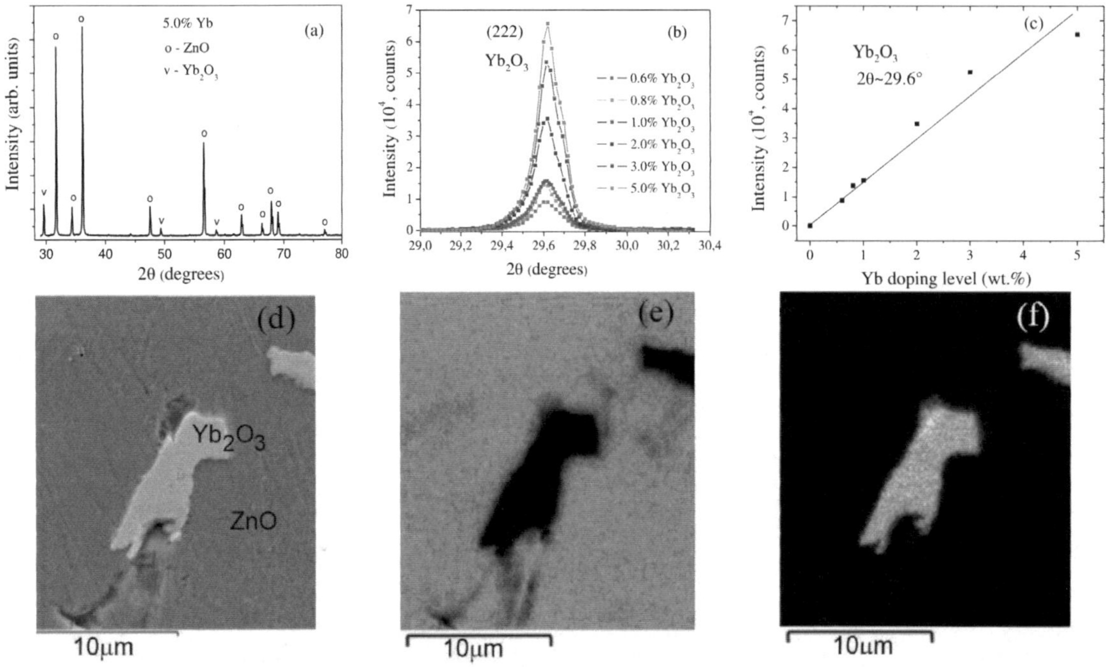

Figure 2. **(a)** XRD pattern of ZnO:5 wt% Yb ceramic; **(b)** XRD patterns of ZnO:Yb ceramics with different Yb doping level in the 2θ range of 29.0 - 30.3° corresponding to the (222) reflection of cubic Yb_2O_3; **(c)** the dependence of the XRD peak intensity at 2θ ~29.6° on the Yb^{3+} concentration; **(d)** SEM image of the polished and etched surface of the ZnO:2 wt% Yb ceramic; **(e,f)** EDX element mapping of **(e)** Zn (Lα) and **(f)** Yb (Mα).

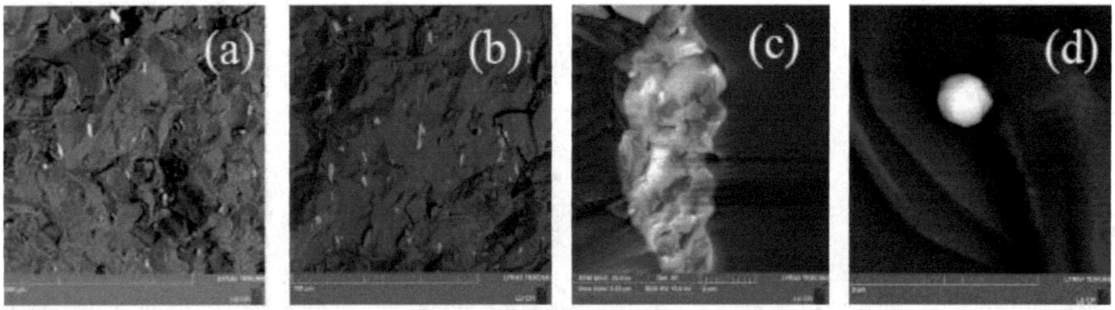

Figure 3. SEM images of the fractured surfaces of ZnO:0.6 wt% Yb **(a)** and ZnO:3 wt% Yb ceramics **(b)**; various localizations of Yb_2O_3 crystals in the ZnO ceramics **(c,d)**.

According to the EDX data for the ZnO:0.6 wt% Yb ceramic, 2.6 wt% Zn is found in the Yb_2O_3 crystal. It seems likely to appear from the surrounding ZnO ceramic body. There are no Yb ions within the ZnO grain. Perhaps this is due to the small ytterbium doping level in this sample, since for the ZnO:3.0 wt% Yb ceramic, 0.2 wt% Yb were detected within the ZnO grain. The EDX data prove the interaction at the GBs of the two oxide phases despite a significant difference in their unit cell parameters (for ZnO, a = 3.249 Å, c = 5.205 Å; for Yb_2O_3, a = 10.441 Å) and ionic radii (ionic radius of $^{IV}Zn^{2+}$ = 0.600 Å, while that of $^{VI}Yb^{3+}$ = 0.868 Å). Considering the high adhesion at the interfaces of

ZnO and Yb_2O_3 grains, we can assume the leading role of the special (nonequilibrium) state of the electronic structure of the GBs in the processes of interaction and transfer of excitation energy from ZnO to Yb^{3+} ions. The previous study by Qi *et al.* [4] of the properties of boundaries in nonequilibrium conditions confirms this possibility.

The analysis of the texture of the ZnO:Yb ceramics indicated that the main components of the texture are the (100) and (110) planes of the hexagonal prism and the (101) planes of the hexagonal pyramid, whose texture coefficient values are significantly higher than those of pure ZnO ceramic.

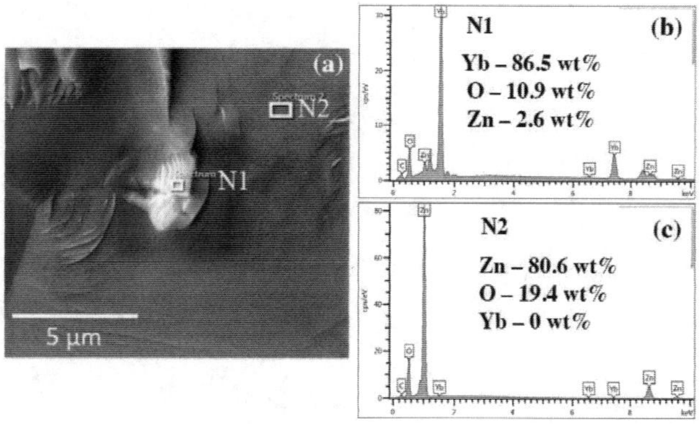

Figure 4. SEM image of a fractured surface of the ZnO:0.6wt% Yb ceramic with the indicated places of spectral analysis **(a)**; spectrum 1 from the Yb_2O_3 grain **(b)**, spectrum 2 from the ZnO grain **(c)**, insets in **(b,c)** - the weight percentages of Zn, O and Yb elements.

The average ZnO grain size of ZnO:Yb ceramics (10-15 µm) practically does not depend on the Yb concentration and is much smaller than that of the undoped ceramic. The inhibitory effect of Yb on the growth of ZnO grains is due to the fact that this impurity is not incorporated into the zinc oxide lattice but is adsorbed at the active surfaces of the crystals, as well as forms crystals of Yb_2O_3 preventing the excessive growth of ZnO crystals.

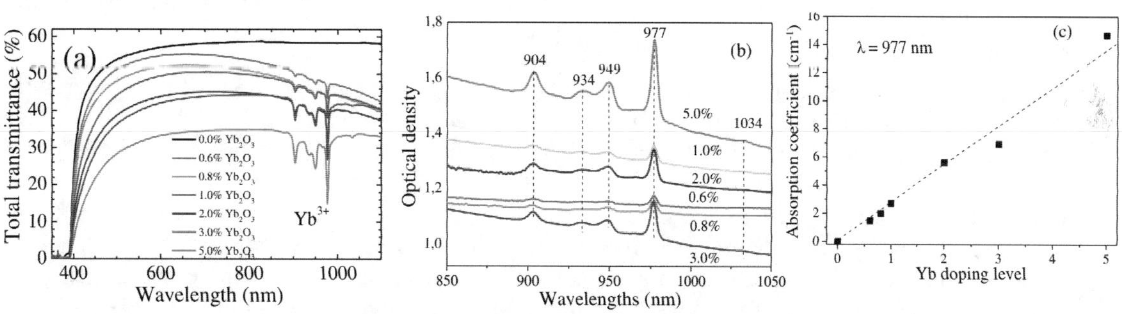

Figure 5 (a-c). Optical spectroscopy of ZnO:Yb optical ceramics: **(a)** total transmittance spectra, thickness: ~0.5 mm; absorption spectra in the range of 850 - 1050 nm **(b)**, dependence of the decimal absorption coefficient at the wavelength of 977 nm on the Yb doping level (**c**).

The total transmittance spectra of ZnO and ZnO:Yb ceramics are shown in Figure 5(a). The spectra of Yb:ZnO ceramics contain the characteristic Yb^{3+} absorption band due to the $^2F_{7/2} \rightarrow {}^2F_{5/2}$ transition. The maximum value of total transmittance in the visible decreases from ~55% to 35% with increasing the Yb content from 0.6 to 5.0 wt%. The addition of Yb also results in shifts of the UV and IR absorption edges of ceramics in opposite directions. There is a shift by 10 nm, from 0.38 to 0.39 µm in the UV and from 4.0 to 2.5-2.7 µm, in the IR spectral range. The latter effect is assigned to increased concentration of free carriers in ZnO, from 3.1×10^{18} cm^{-3} to $7.4–8.0\times10^{18}$ cm^{-3}, as calculated using the approach described in Ref. [5]. Figure 5(b) focuses on the Yb^{3+} absorption. The shape of the absorption band of Yb^{3+} ions does not depend on the doping concentration, which suggests that the site symmetry for these ions does not change with their concentration. The almost linear dependence of the absorption coefficient on the concentration of Yb^{3+} ions at the wavelength of 977 nm indicates the implementation of the Beer's law, Figure 5(c).

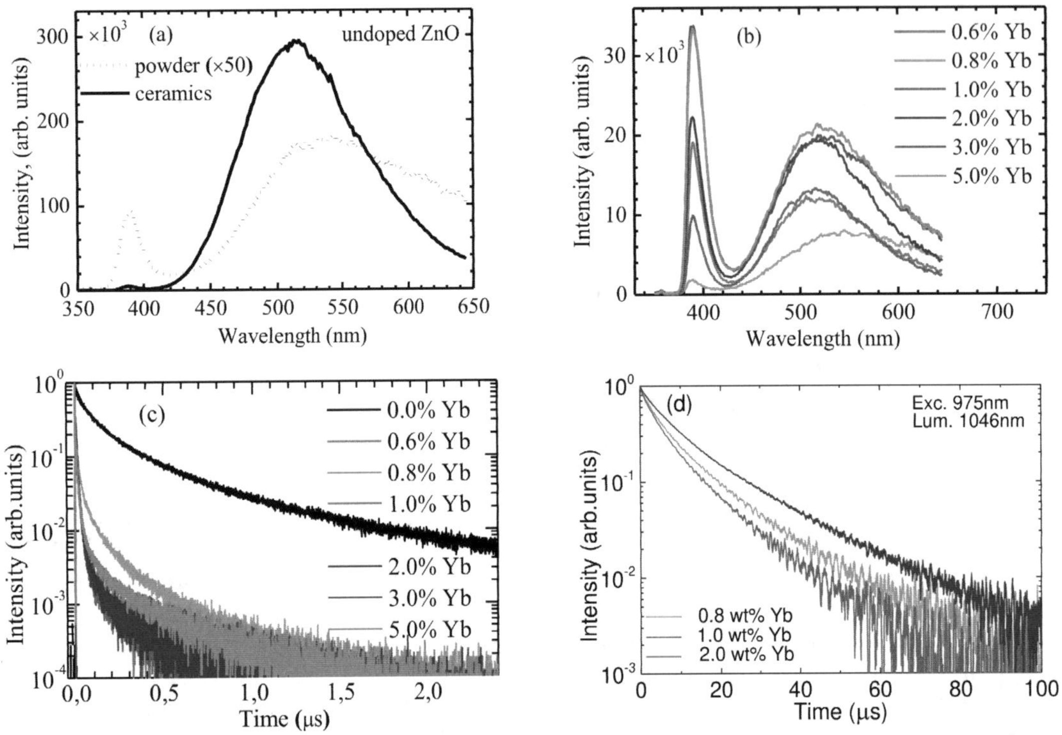

Figure 6(a-d). The spectra of X-ray luminescence of **(a)** the initial ZnO powder and undoped ZnO ceramics and **(b)** Yb^{3+}-doped ZnO ceramics; **(c)** X-ray luminescence decay curves; **(d)** luminescence decay curves for Yb^{3+} ions, $\lambda_{exc} = 975$ nm, $\lambda_{lum} = 1046$ nm.

The X-ray luminescence spectra of the initial ZnO powder, undoped ZnO and ZnO:Yb ceramics contain two emission bands, Figure 6(a,b). One of them is a narrow UV band located near the absorption edge, called the near-band-edge (NBE) emission of an excitonic nature. Another broad band located in the green region of the spectrum is the so-called deep level (DL) emission. The DL emission is usually associated with intrinsic point defects, the exact type of which is still a matter of debate. The DL luminescence is often ascribed to oxygen vacancies. The spectrum of the initial ZnO powder exhibits a weak NBE band with a maximum at 388 nm and a two-times more intense DL band centered at 530 nm. Intensity of the NBE band of the undoped ZnO ceramic is comparable to that of the initial powder. On the contrary, intensity of the DL band of undoped ceramic with a maximum at 513 nm is ~80 times higher than that of the powder. Most likely, this increase in intensity is due to an increase in the concentration of oxide vacancies under the conditions of high-temperature deformation during uniaxial hot pressing in vacuum. It has been widely reported that exposure to high pressure can cause a high concentration of vacancies in ZnO [5]. Intensity of the DL luminescence with a maximum at 520-521 nm for ZnO:Yb ceramics is by an order of magnitude weaker than that for the undoped ceramic. In contrast, intensity of the NBE emission bands of the ZnO:Yb ceramics is higher than that for the undoped ZnO. With increasing the Yb concentration, intensity of the NBE emission increases and saturates for 5 wt% Yb doping. The position of its maximum slightly shifts from 390 to 391 nm probably due to a change in the energy band structure. A similar effect of increasing intensity of the exciton luminescence and decreasing intensity of the DL emission associated with oxide defects upon the addition of Yb^{3+} ions to the ZnO nanopowders was observed by Kabongo *et al.* [6]. The authors assigned these data either to a possible energy transfer from Yb^{3+} ions to ZnO or to a reduction of oxygen related defects.

The Yb doping accelerates the decay of ZnO luminescence, Figure 6(c), although no clear correlation between its content and the shape of the decay curves can be found. With increasing the Yb doping level, the shape of the XRIL kinetics of ZnO:Yb ceramics changes only slightly. The decay curves are not single exponential. For all the studied samples, they can be divided into two time domains: 0-50 ns and 200-2000 ns. The first range mainly contains a "fast" component with a decay time of 7-15 ns for all the studied Yb^{3+}-doped ceramics. For the undoped sample, the characteristic decay time in this region is longer, ~ 50 ns. The second range corresponds to a "slow" decay time of ~ 390-430 ns in all the samples. The undoped ceramic and ceramic doped with 0.8% Yb are distinguished from other ceramics by a much slower decay recorded in the range of 50-500 ns, which correlates with smaller intensities of NBE emissions in comparison with those of the DL, as shown in Figure 6(a,b). It should be noted that the DL emission in the sample with 0.8% Yb is also shifted to longer wavelengths compared to that in other samples. One can assume that the main change in decay kinetics occurs for small (<0.6%) Yb contents.

The luminescence lifetimes of Yb^{3+} ions (the $^2F_{5/2}$ state) are about 10 μs and they are weakly dependent on the Yb doping level, Figure 6(d).

4. Conclusions

ZnO:Yb optical ceramics were fabricated for the first time. Yb^{3+} ions do not enter the ZnO structure while they are located on the ZnO grain surface and form the Yb_2O_3 crystals at the grain boundaries. The optical properties of ZnO:Yb ceramics indicate the ZnO ↔ Yb^{3+} energy-transfer. The study of the fracture mechanisms, phase composition and spectral data allowed us to reveal the role of the nonequilibrium state of the electronic structure of grain boundaries on the excitation energy transfer from ZnO to Yb^{3+} ions, which resulted in an increase of the near-band-edge emission intensity. The developed ceramics are promising for optoelectronic applications.

Acknowledgment

This work was partly supported by the RFBR (Grant 19-03-00855).

References

[1] Shestakov M, Baranov A, Tikhomirov V, Zubavichus Y, Kuznetsov A, Veligzhanin A, Kharin A, Rösslhuber R, Timoshenko V and Moshchalkov V 2012 *RSC Adv.* **2** 8783

[2] Balestrieri M, Ferblantier G, Colis S, Schmerber G, Ulhaq-Bouillet C, Muller D, Slaoui A and Dinia A 2013 *Sol. Energy Mater. Sol. Cells* **117** 363

[3] Márquez J A R , Rodríguez C M B , Herrera C M, Rosas E R, Angel O Z and Pozos O T **2011** *J. Electrochem. Sci.* **6** 4059

[4] Qi Y, Kauffmann Y, Kosinova A, Kilmametov A R, Straumal B B and Rabkin E 2021 *Mater. Res. Lett.* **9** 58

[5] Yang X, Electrical and optical properties of zinc oxide for scintillator applications, 2008 *Graduate Theses, Dissertations, and Problem Reports* 2729 https://researchrepository.wvu.edu/etd/2729 106

[6] Kabongo G L, Mhlongo G H, Mothudi B M, Hillie K T, Swart H C and Dhlamini M S 2014 *Mater. Lett.* **119** 71

Research and calculation of dynamic characteristics of a microelectromechanical device

I E Lysenko[1], N F Kidyaev[1], O A Ezhova[2] D Y Sevostyanov[1],

[1] Southern Federal University, Taganrog 347900, Russia
[2] LLC «MEMS design», Iniciativnaya str., 60, Taganrog, Russia, 347936

Abstract. This paper presents a model of a three-axis micromechanical gyroscope-accelerometer. The stiffness of the suspension of a micromechanical gyroscope-accelerometer is calculated and the dependence of the change in the frequency of vibration of the micromechanical mass on the change in the thickness of the beams of the cruciform suspension element is obtained.

1. Introduction

In the modern electronics market, micromechanical devices (MEMS) have a growing demand. The development of MEMS sensors has significantly influenced the implementa-tion of fundamentally new applications and methods of human interaction with them. sensors of small accelerations and linear speeds are firmly established in the market of smartphones, game consoles, tablets. For a greater pace of development of MEMS devices, it is necessary to develop more advanced sensor models or work on improving existing ones [1, 2]. The work on the model of a three-axis micromechanical gyroscope-accelerometer presented in this article will improve its characteristics, and, consequently, increase the demand for Russian microelectronic products on the world market [3, 4].

2. Design and calculation

The developed model of a micromechanical gyroscope-accelerometer with three axis of sensitivity is shown in Figure 1 [5].

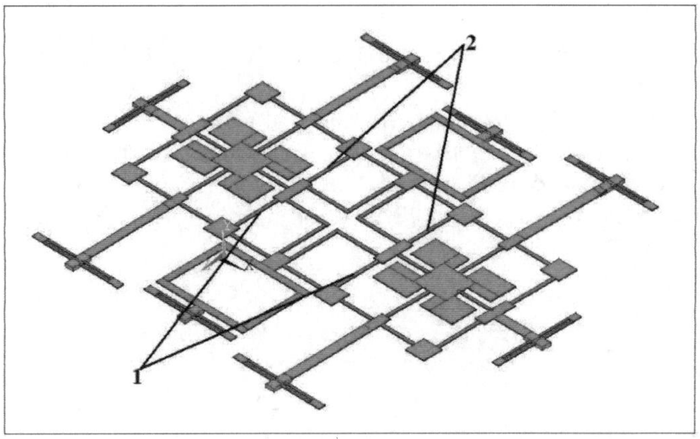

Figure 1. Design of the developed micromechanical gyroscope-accelerometer.

This device allows you to measure linear accelerations along three axis and angular velocity along the X and Z axis located parallel to the substrate plane. A distinctive feature of this device is the presence of a cross-shaped element of rigidity, which is in the center of the structure. It allows the electrostatic actuators to simultaneously interact with the inertial masses of the gyroscope-accelerometer.

To understand the effect of the accelerometer gyroscope suspension design, several types of suspension with different elastic beam widths around the cruciform stiffener were modeled. The design was modeled in the ANSYS software environment. Figures 2-4 show the obtained models of the suspensionof a micromechanical gyroscope-accelerometer.

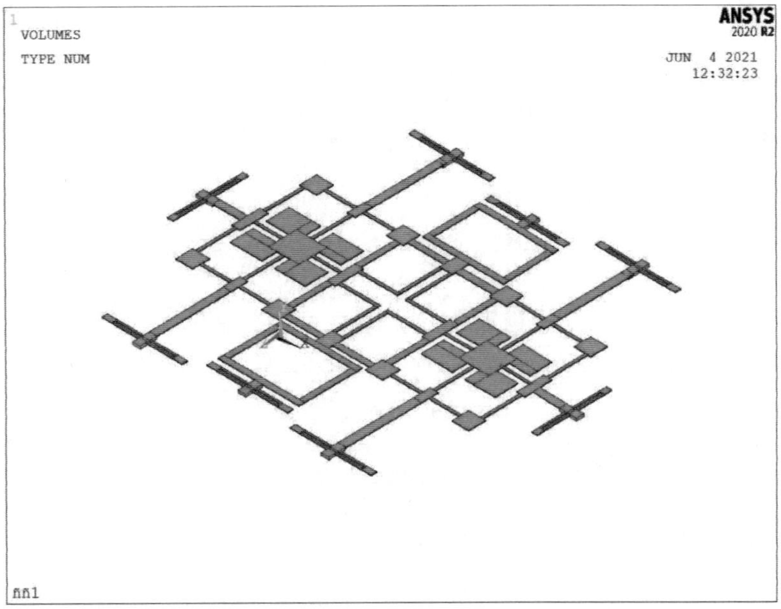

Figure 2. Design of a suspension with a cross-shaped element beam width of 10 microns.

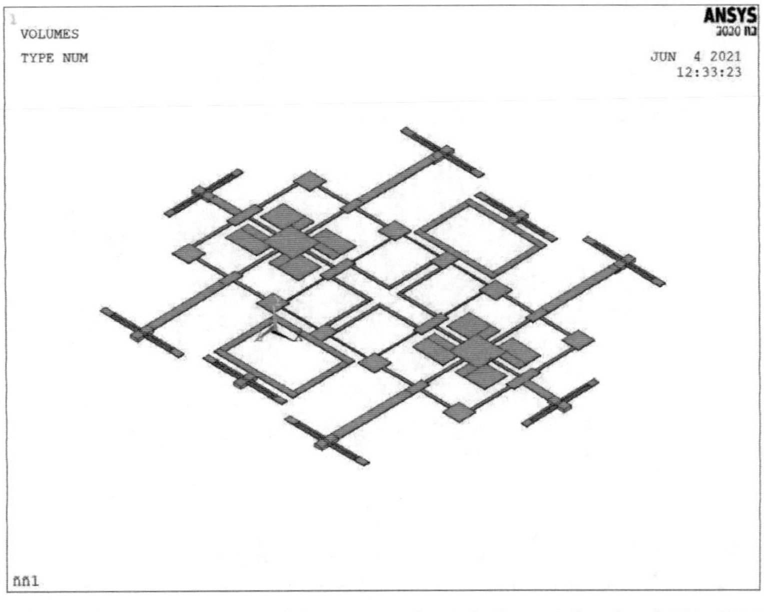

Figure 3. Design of a suspension with a cross-shaped element beam width of 2.5 microns.

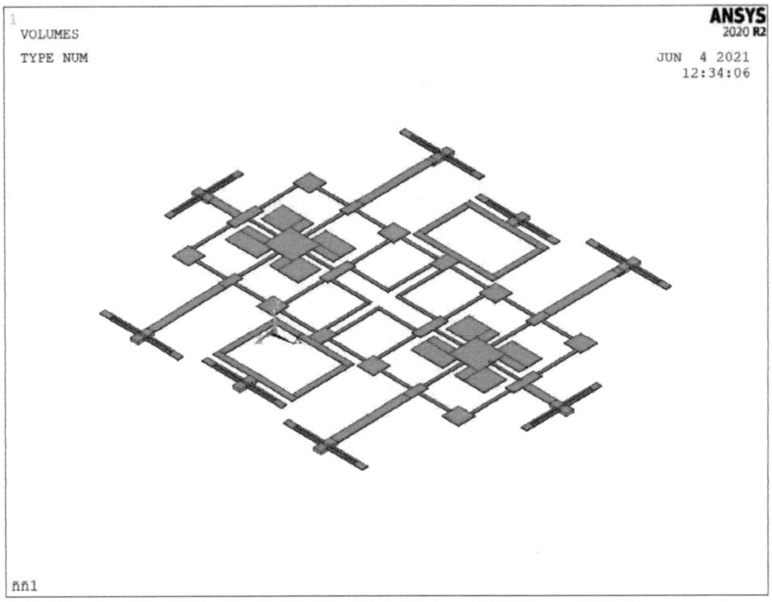

Figure 4. Design of a suspension with a cross-shaped element beam width of 5 microns.

To understand the influence of the cross-shaped element design on the change in the frequency of inertial mass oscillations, the reduced design was calculated in the MATLAB system environment. The calculation is made for the design of a micromechanical gyroscope-accelerometer with different widths of elastic beams 1, 2 in the cruciform suspension element. The dependence of the frequency of natural oscillations of the inertial masses on the X-axis is shown in Figure 5.

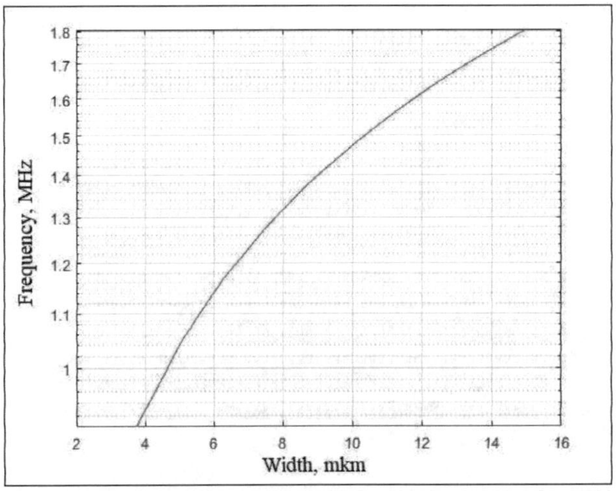

Figure 5. The dependence of the frequency of natural oscillations of the inertial masses along the X-axis.

3. Conclusion

This paper presents a model of a three-axis micromechanical gyroscope-accelerometer. There are also 2 additional models of elastic suspension for the gyroscope-accelerometer with different widths of elastic beams around the cruciform stiffener. The results of calculations of the dynamic characteristics of the presented models, carried out using the MATLAB software, showed a linear dependence of the frequency of natural oscillations of the inertial masses on the change in the width of the elastic beams of the cruciform stiffener of the rake. From the results obtained, it should be concluded that this model of the suspension of a micromechanical gyroscope-accelerometer needs to be improved, or rather, to reduce the rigidity of the suspension to reduce the natural frequencies of oscillations of the inertial masses.

4. Acknowledgments
This work was financially supported by the grant of President of Russian Federation (project number MK-4577.2021.4).

References
[1] Raspopov V Ja 2007 *Micromechanical devices* (Tula: Tula state university)
[2] Lysenko I E, Naumenko D V and Ezhova O A 2020 Design and simulation high aspect ratio torsion suspension of MEMS z-axis accelerometer *Book of Abstracts of 7th International School and Conference «Saint Petersburg OPEN 2020» on Optoelectronics, Photonics, Engineering and Nanostructures* (St. Petersburg: St. Petersburg Academic University) pp. 478–79
[3] Lysenko I E, Ezhova O A, Tkachenko A V, Naumenko D V, Guha K and Rao K S 2019 The results of modelling of a micromechanical accelerometer *Journal of Physics: Conference Series* **1409** *243-86*
[4] Lysenko I E 2019 The results of modelling of a micromechanical accelerometer *Journal of Physics: Conference Series* **1410** 226-*32*
[5] Lysenko I E, Konoplev B G, Kidyaev N F and Shafrostova S I 2019 Integrated micromechanical gyroscope-accelerometer Patent of invention No. 2683810.

Effect of the ultra-low arsenic flux on characteristics of In(As) nanostructures formed during droplet epitaxy

D V Kirichenko, S V Balakirev, N E Chernenko, M M Eremenko and M S Solodovnik

Institute of Nanotechnologies, Electronics and Equipment Engineering, Southern Federal University, Taganrog 347922, Russia

dankir@sfedu.ru

Abstract. In this paper, we present the results of an experimental study of the influence of the ultra-low arsenic flux on the parameters of In nanodroplets obtained by droplet epitaxy on the GaAs substrate. We demonstrate that the arsenic flux can be used to alter the size of droplets without changing their surface density. An increase in the arsenic flux leads to a reduction of the nanostructure size or their complete decay. However, we demonstrate that certain growth conditions allow providing saturation of the size of nanostructures (~30 nm) which ensures good reproducibility of the process. The mechanism of ring and hole formation at various arsenic fluxes is also discussed.

1. Introduction

One of the most significant challenges in modern electronic and photonic devices is a possibility to control characteristics of single nanostructures [1-6]. Applications of quantum communication and computing require operation with single quantum dots with defined parameters [7-9]. Traditional methods of quantum dot formation suffer from an inter-dependence of quantum dot size and surface density which makes it difficult to fabricate small-size nanostructures sufficiently distant from each other [10-13].

Droplet epitaxy is a method of A3B5 nanostructure growth which allows two-stage formation of nanostructures, namely deposition of metal and its further crystallization in the group-V flux [7-9,14-21]. Due to the possibility to control nanostructure parameters at both stages, this method allows formation of low-density droplet arrays with the subsequent altering of their size before the crystallization process. However, a value of the arsenic flux is important to prevent premature fixation of the nanostructure size or their spreading over the surface.

In this paper, we carry out an experimental study of the ultra-low arsenic flux effect on the characteristics of indium droplets formed on the GaAs(001) surface.

2. Experiment

For experimental study, we first prepared GaAs(001) epi-ready substrates by standard oxide removal and further 400-nm buffer layer growth. Then we closed the arsenic valve and cooled down the substrate to a deposition temperature of 300°C. After the background pressure reduced to a value below $1\cdot10^{-7}$ Pa (P_0), we formed droplet arrays on GaAs(001) substrates by deposition of 3.0 and 1.5

equivalent monolayers (ML) of indium. Then we exposed the samples to arsenic fluxes of various ultra-low values.

A clear (2 × 4) reconstruction was observed on the surface by the reflection high-energy electron diffraction (RHEED) prior to the indium deposition. After the indium deposition, a RHEED pattern became hazy and spotty indicating the appearance of metal on the surface [22,23]. No significant changes in the RHEED pattern were observed with further exposure of indium droplets to the ultra-low arsenic flux.

After shutting the arsenic valve, we additionally held the samples in the growth chamber during 5 minutes while cooling down. Then, we transferred the substrates out of the growth chamber and sent them to a scanning electron microscope (SEM) Nova NanoLab 600 and atomic force microscope (AFM) NTEGRA to measure the geometrical parameters of nanostructures.

3. Results and discussion

As we observed previously [24,25], deposition of 3.0 ML and 1.5 ML of indium on the GaAs(001) surface at a temperature of 300°C leads to the formation of arrays of droplets with an average diameter of about 150 nm and 100 nm, respectively, and a surface density at a level of $7 \cdot 10^7$ cm^{-2} in both cases. Exposure of In droplets to the arsenic flux is normally used to crystallize them into InAs nanostructures requiring further capping for the quantum dot formation. However, we used arsenic fluxes of ultra-low values in order to alter the droplet size without their transformation into InAs nanodots, nanorings or nanoholes.

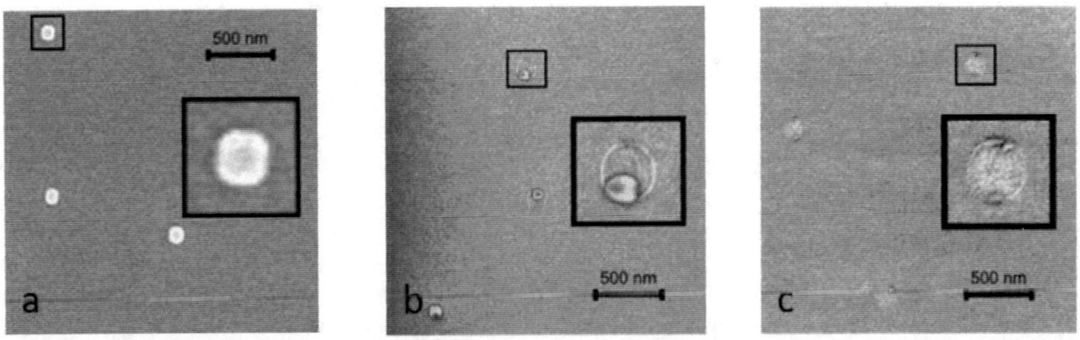

Figure 1. SEM images of nanostructures obtained after deposition of 3 ML of indium and further exposure to the ultra-low arsenic flux: (**a**) $P/P_0 = 4$; (**b**) $P/P_0 = 6$; (**c**) $P/P_0 = 8.1$. The 250 × 250 nm^2 insets demonstrate scaled-up nanostructures.

We demonstrate that an increase in the As pressure ratio leads first to a simple reduction of the average droplet size (Figure 1a). With further increasing arsenic flux, the formation of a ring is observed along the droplet initial perimeter while the droplet is still decreasing in size (Figure 1b,c). It is commonly known that droplet parameters change as a result of the arsenic irradiation and successfully used for the formation of nanorings [26,27] and nanoholes [28,29]. However, we observe that the droplet size can be altered before the final crystallization into InAs nanostructure.

Figure 2 demonstrates As flux dependences of the average diameter of droplets, rings and holes after deposition of 3.0 ML and 1.5 ML of In. While arsenic flux increases, droplets tend to decrease in size. Droplets obtained after deposition of 3.0 ML and 1.5 ML transform into droplet-ring complexes at $P/P_0 = 6$ and 1.4, respectively, so that the ring diameter is approximately equal to the initial droplet size. The ring forms at the interface of three phases due to an increase in the As concentration and, hence, In to InAs crystallization intensity.

It is worth noting that a reduction of droplets without the formation of rings and holes is possible with increasing arsenic flux (Figure 2, 3.0 ML, P/P_0 from 1 to 6). Furthermore, the droplet size is saturated and becomes equal to approximately 30 nm after reaching a certain arsenic flux value

($P/P_0 = 8$). This behavior of the growth system enables a controllable reduction of the droplet size and good reproducibility of this process.

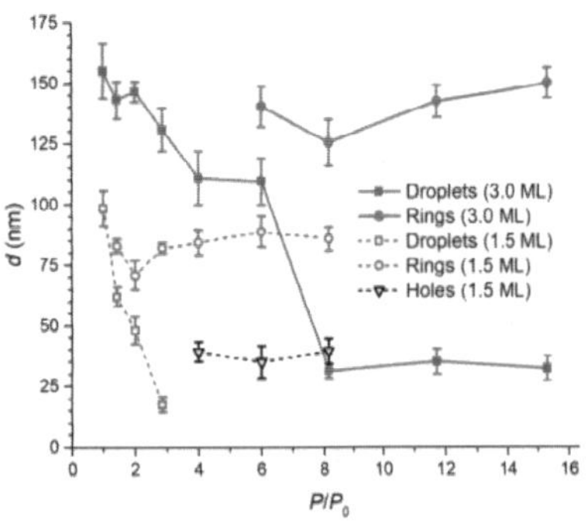

Figure 2. Arsenic pressure ratio dependences of the average diameter of droplets obtained after deposition of 1.5 and 3 ML of indium and their further exposure to ultra-low arsenic flux. The caps represent standard deviations of the diameter of nanostructures.

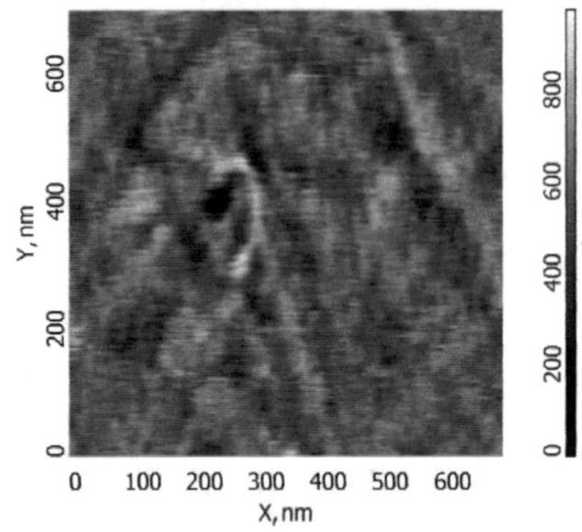

Figure 3. AFM image of a typical nanostructure after the exposure of In droplets obtained after the deposition of 1.5 ML of indium to the ultra-low arsenic flux.

Despite the fact that the study of droplets obtained after deposition of 3 ML of indium demonstrates a clear picture of what occurs on the surface during their exposure to the ultra-low arsenic flux, this amount of deposited material leads to the formation of large droplets which are difficult to transform into optically efficient quantum dots. Therefore, we also exposed droplets formed after the deposition of 1.5 ML (critical thickness for droplet formation at a temperature of 300°C [24]) of indium to the arsenic flux. The surface density of droplets was found to be at the level of 7×10^7 cm^{-2} as in the case of 3.0 ML. An increase in the arsenic flux also led to the shrinkage of droplets (Figure 2). However, droplets obtained at 1.5 ML are initially smaller in size and less stable, as a result of which the droplet reduction was more abrupt and an arsenic pressure at which droplet-ring complexes formed was shifted to the left (to $P/P_0 = 1.4$).

Although the substrate temperature in a series of experiments remained unchanged (300°C), in the case of exposure of 1.5-ML droplets to the arsenic flux, the hole formation was observed (Figure 2). Whereas the ring diameter was still equal to the initial droplet diameter, the diameter of holes tended to saturate as well as droplets in the case of 3.0-ML droplets to a value of about 30 nm (Figure 2). A typical ring-hole complex formed after exposure of droplets obtained at 1.5 ML to the arsenic flux of value above $P/P_0 = 4$ is presented in Figure 3.

The observation of etching phenomena during the arsenic irradiation of droplets is associated with a significant influence of arsenic flux on the chemical composition of a droplet. As a result of the diffusion of arsenic into the droplet, the equilibrium concentration of atoms in the droplet changes. In order to restore the initial volume atom ratio, compensation of deficient atoms of metal is needed. Because In atoms diffused out of the droplet spreading over the surface, their places are occupied by the nearest Ga atoms belonging to the substrate [30]. Thus, Ga atoms migrate from the substrate leading to the formation of the hole at the place of the droplet (Figure 3).

4. Conclusion

Thus, an independent control of the size of nanostructures can be achieved by using an exposure of droplets to the ultra-low As flux. In order to obtain droplets of a small size to further crystallize them into optically efficient InAs quantum dots, a minimum value of the ultra-low As flux and long exposure times should be used. Otherwise, the formation of rings and holes is possible which can in turn be useful for some specific applications.

Acknowledgments
This work was supported by the Russian Science Foundation Grant No. №19-79-10099.

References

[1] Heindel T et al. 2012 *New J. Phys.* **14** 083001
[2] Bimberg D et al. 2009 *IEEE Photonics J.* **1** 58–68
[3] Bayer M, Stern O, Hawrylak P, Fafard S and Forchel A 2000 *Nature* **405** 923–6
[4] Santori C, Fattal D and Yamamoto Y 2010 *Single-photon devices and applications* (Chichester: John Wiley & Sons)
[5] Moreau E, Robert I, Gérard J, Abram I, Manin L and Thierry-Mieg V. 2001 *Appl. Phys. Lett.* **79** 2865–7
[6] Lochmann A, Stock E, Töfflinger J, Unrau W, Toropov A, Bakarov A, Haisler V and Bimberg D 2009 *Electron. Lett.* **45** 1
[7] Mano T, Abbarchi M, Kuroda T, Mastrandrea C, Vinattieri A, Sanguinetti S, Sakoda K and Gurioli M 2009 *Nanotechnology* **20** 395601
[8] Wu J, Wang Z M, Li X, Mazur Y I and Salamo G J 2017 *J. Mater. Res.* **32** 4095–101
[9] Alonso-González P, Martín-Sánchez J, González Y, Alén B, Fuster D and González L 2009 *Cryst. Growth Des.* **9** 2525–8
[10] Gaan S, He G, Feenstra R M, Walker J and Towe E 2010 *J. Appl. Phys.* **108** 114315.
[11] Heitz R, Stier O, Mukhametzhanov I, Madhukar A and Bimberg 2000 *Phys. Rev. B* **62** 11017–28
[12] Mukhametzhanov I, Wei Z, Heitz R and Madhukar A 1999 *Appl. Phys. Lett.* **75** 85–7.
[13] Heitz R, Ramachandran T R, Kalburge A, Xie Q, Mukhametzhanov I, Chen P and Madhukar A 1997 *Phys. Rev. Lett.* **78** 4071–4
[14] Koguchi N, Takahashi S and Chikyow T 1991 *J. Cryst. Growth* **111** 688–92
[15] Balakirev S V, Solodovnik M S, Ageev O A 2018 *Phys. Status Solidi B* **255(4)** 1700360
[16] Sanguinetti S, Watanabe K, Tateno T, Gurioli M, Werner P, Wakaki M and Koguchi N 2003 *J. Cryst. Growth* **253** 71–6
[17] Mano T, Watanabe K, Tsukamoto S, Imanaka Y, Takamasu T, Fujioka H, Kido G, Oshima M and Koguchi N 2000 *Jpn. J. Appl. Phys.* **39** 4580
[18] Mano T, Watanabe K, Tsukamoto S, Fujioka H, Oshima M and Koguchi N 2000 *J. Cryst. Growth.* **209** 504
[19] Watanabe K, Koguchi N and Gotoh Y 2000 *Jpn. J. Appl. Phys.* **39** L79
[20] Mantovani V, Sanguinetti S, Guzzi M, Grilli E, Gurioli M, Watanabe K and Koguchi N 2004 *J. Appl. Phys.* **96** 4416–20
[21] Balakirev S V, Chernenko N E, Eremenko M M, Ageev O A and Solodovnik M S 2021 *Nanomaterials* **11** 1184
[22] Huang S, Kim S J, Levy R, Pan X Q and Goldman R S 2013 *Appl. Phys. Lett.* **103** 132104
[23] Urbańczyk A, Hamhuis G J and Nötzel R 2010 *J. Appl. Phys.* **107** 014312
[24] Balakirev S V, Solodovnik M S, Eremenko M M, Konoplev B G, Ageev O A 2019 *Nanotechnology* **30** 505601
[25] Balakirev S V, Solodovnik M S, Eremenko M M, Konoplev B G and Ageev O A 2020 *Nanotechnology* **31** 485604
[26] Lee J H, Wang Z M, Ware M E, Wijesundara K C, Garrido M, Stinaff E A and Salamo G J 2008 *Cryst. Growth Des.* **8** 1945
[27] Zhao C, Chen Y H, Xu B, Tang C G, Wang Z and Ding F 2008 *Appl. Phys. Lett.* **92** 063122
[28] Li X, Wu J, Wang Z M, Liang B, Lee J, Kim E S and Salamo G J 2014 *Nanoscale* **6** 2675–81
[29] Stemmann A, Heyn C, Köppen T, Kipp T and Hansen W 2008 *Appl. Phys. Lett.* **93** 123108
[30] Hugues M, Teisseire M, Chauveau J M, Vinter B, Damilano B, Duboz J Y and Massies J 2007 *Phys. Rev. B.* **76** 075335

Magnetron sputtered TiO_2 with metal NPs for plasmonic applications

D Konev[2], A Kazakin[1,2], A Vorobyev[1], Y Enns[1,2], A Kuznetsov[1], A Kondrateva[1,2] and M Mishin[1]

[1] St. Petersburg Academic University, St. Petersburg, Russia
[2] Peter the Great Saint-Petersburg Polytechnic University, St. Petersburg, Russia

Abstract. This work is aimed at finding the optimal technological modes for the deposition of TiO2 films by reactive magnetron sputtering for use as a photoactive material in plasmonic applications. The structural and optical parameters of films obtained at different ratios of oxygen in a mixture of working gases were studied. On the basis of the obtained experimental results, a numerical simulation of the spectral characteristics of the TiO2 / NP structure of the metals silver and gold was carried out.

1. Introduction

Currently, much attention is paid to optoelectronic devices based on oxide semiconductors [1-4]. In particular, titanium dioxide (TiO_2) has excellent physical, chemical and optical properties and is widely used in applications such as UV photodetectors, gas sensors and photocatalysts [5-8]. TiO_2 is a well-studied metal oxide wide-gap n-type semiconductor (3.2 eV for anatase and 3.0 eV for rutile phases). Such wide band gap makes TiO_2 a suitable for UV detection applications because of its high sensitivity to background visible and infrared light [6].

The electronic and optical properties of TiO_2 films are provided due to the formation of metal vacancies or interstitial O atoms [9]. Magnetron sputtering technique with an adjustable oxygen ratio and optimised annealing conditions can be used to improve the electrical and optical parameters of TiO_2 films [6]. In addition, compared to other methods used to obtain TiO_2 films [10–13], high-frequency magnetron sputtering has a number of advantages, such as high stoichiometry and significant film uniformity and density, possibility of using different substrates, and low growth temperatures [14, 15].

Plasmonic behaviour of metal nanoparticles (NPs) is a way to alter the optoelectronic devices characteristics and adjust their sensitivity range by changing the particle arrays geometric parameters [1-4]. Silver, aluminum, platinum, and gold are promising materials for distributed arrays of NPs [16]. Plasmonic effects in NPs of these materials lead to trapping of light in localized regions near metal nanoparticles, which adduces to an increase in the electromagnetic field. However, NPs are also significantly influenced by the host matrix material properties.

The aim of this work was the developing of optimal technological modes for the TiO_2 reactive magnetron sputtering in terms of the oxygen content effect on structure and optical properties of formed thin films. The results obtained were used to numerically simulate plasmonic behaviour with silver and gold NPs.

2. Experiment

In this work, TiO_2 films were obtained by reactive magnetron sputtering (DC) with a Ti target in an O/Ar gas mixture on amorphous quartz and Si substrates. Before sputtering, the samples were exposed to plasma cleaning with Ar ions. Spraying was carried out using 0.22-1 gases ratios at an operating pressure of 1.8 mTorr. Process parameters are shown in table 1.

Table 1. TiO_2 magnetron sputtering parameters.

	Current, A	Voltage, V	Work pressure, mTorr	O/Ar	Time, min
Sample 1	1	480	1.8	0.22	120
Sample 2	1	460	1.8	0.5	120
Sample 3	1	470	1.8	1	120
Sample 4	2	480	2.1	1	40

Thickness and refractive index were measured using ellipsometry on a Horiba Jobin Yvon setup with a 638.2 nm HeNe laser. The morphology of the resulting films was analyzed by scanning electron microscopy (SEM) using a JEOL JSM-7001F.

Structure of the films was studied by Raman spectroscopy on a Horib Jobin-Yvon LabRam HR800 instrument with a 1800 g/mm lattice in the backscattering geometry. The laser radiation wavelength is 532 nm, the power is 50 mW and the focal spot aperture is 2 μm. Measurements were carried out in spectral range of the Raman shift (100–1200) cm^{-1} with averaging over 10 measurements to improve signal-to-noise ratio. For data analysis, reference samples of Si and quartz were measured.

Spectral characteristics of transmission and reflection for samples on a quartz substrate and reflection for samples on a silicon substrate were measured in the wavelength range of 250 - 1150 nm using a xenon radiation source. The optical properties of film and substrate were expressed by Fresnel matrices, product of which determines the reflection and transmission of the sample [17]. Based on the data obtained, a system of equations was compiled, the solution of which allows calculating the refractive index and extinction coefficient in the entire spectrum:

$$\begin{cases} R(n,k,\lambda) - R_{exp}(\lambda) = 0 \\ T(n,k,\lambda) - T_{oxp}(\lambda) = 0 \end{cases}, \quad (1)$$

where n is the refractive index, k is the extinction coefficient, λ is the wavelength, and R and T are the reflection and transmission, respectively. The optical band gap was determined from the measured absorption coefficient.

Numerical simulation of the optical characteristics of quartz / TiO_2 / metal NP structures was carried out using the Comsol Multiphysics package in a two-dimensional approximation and a full-wave formulation of the problem. For modeling, the technique described in [18] was used. For the calculation, an assumption was made about the uniform distribution of metal nanoparticles on the surface of the deposited TiO_2. Simulation was carried out on one cell of the array (quartz / TiO_2 / metal NP / air) using a periodic boundary condition on the side surfaces of the cell. The TiO_2 layer thickness is 100 nm. The metal nanoparticle was defined as a hemisphere with a diameter of 20 nm. The refractive index and extinction coefficient of TiO_2 were determined from the spectral characteristics of the prepared films. Simulation was performed for gold and silver nanoparticles described by the Drude-Lorenz model [19].

3. Results

Surface of TiO_2 films deposited on the monocrystalline silicon is shown in figure 1. Magnetron deposition allows produce smooth homogeneous coatings over large working areas. Thus, the images of the formed layers show that all films have a granular structure, the average grain size is about 10 nm, regardless of the formation conditions. The results of ellipsometry in accordance with the deposition modes are shown in table 2. It can be seen that for the samples deposited in different modes, the range of film thicknesses was 24-70 nm. In this case, there is a difference between rate of deposition on Si and quartz substrates, which can be caused by the accumulation of charge on the surface of the quartz plate and a slowdown in the deposition process.

Table 2. Thickness and refractive index of TiO_2 films with Si and quartz substrates.

	Thickness, nm		Velocity, nm/min.		Refractive index (λ = 638.2 nm)	
	Wafer Si	Wafer quartz	Wafer Si	Wafer quartz	Wafer Si	Wafer quartz
Sample 1	70.54 ± 0.04	65.79 ± 0.02	0.59	0.55	2.3929	2.4569
Sample 2	58.6 ± 0.06	54.45 ± 0.03	0.49	0.45	2.3773	2.4907
Sample 3	58.51± 0.03	43.67 ± 0.03	0.49	0.36	2.4085	2.8398
Sample 4	43.37 ± 0.03	24.35 ± 0.06	1.0	0.57	2.3576	2.9252

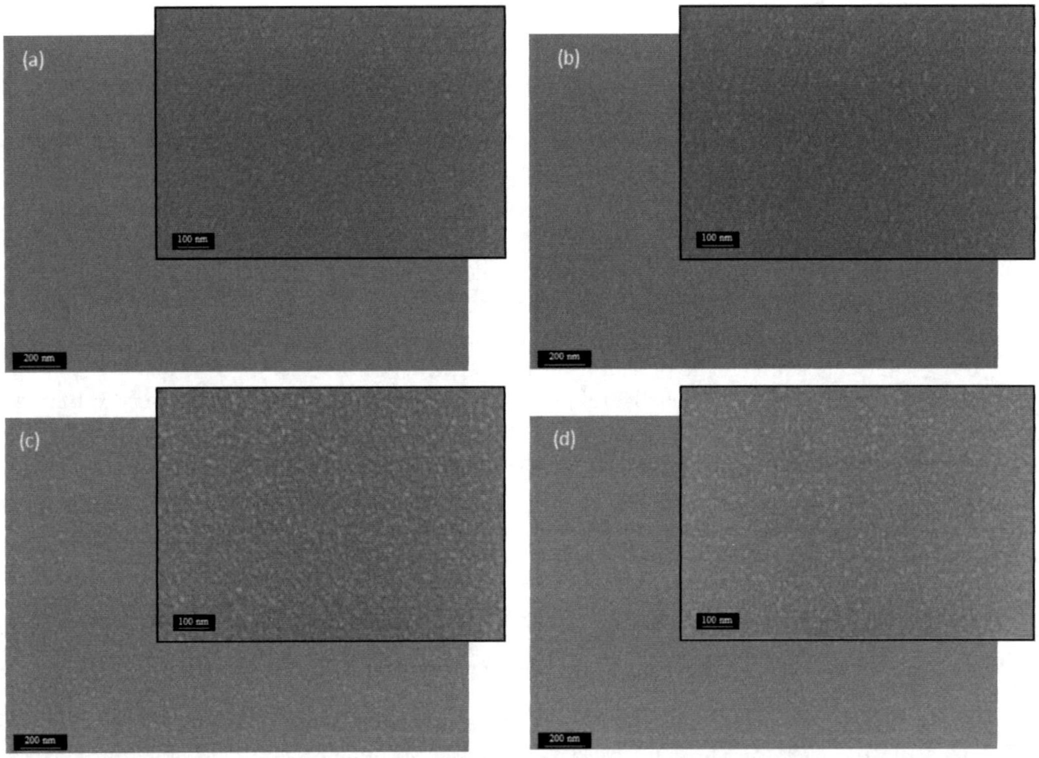

Figure 1. SEM images of TiO_2/Si surface: sample 1(a); sample 2(b); sample 3(c); sample 4(d).

The results of the Raman scattering study for TiO_2 films on Si and quartz substrates are shown in figure 2. In each obtained spectrum, unmarked peaks refer exclusively to silicon or quartz substrates. One peak was recorded at a frequency of 149 cm^{-1}, which is characteristic for both the anatase and rutile TiO_2 structures (figure 2a). However, no peaks (Eg (445 cm^{-1}) and A1g (610 cm^{-1}) related to rutile are observed in the spectral dependences. Also, no peak is observed at a frequency of 235 cm^{-1}, which corresponds to the multi-phonon interaction. At the same time, for the anatase structure, all the peaks, except for 149 cm^{-1} (Eg), are extremely low in intensity because of the small thickness of the films, due to which they could be completely absent or be covered by the peaks of substrate. For TiO_2/quartz structures, a single peak at 149 cm^{-1} was also recorded. Thus, it can be assumed that the type of substrate did not affect structure of the films. The selected technological parameters made it possible to form titanium dioxide in the form of anatase, which is suitable for use in optoelectronic devices.

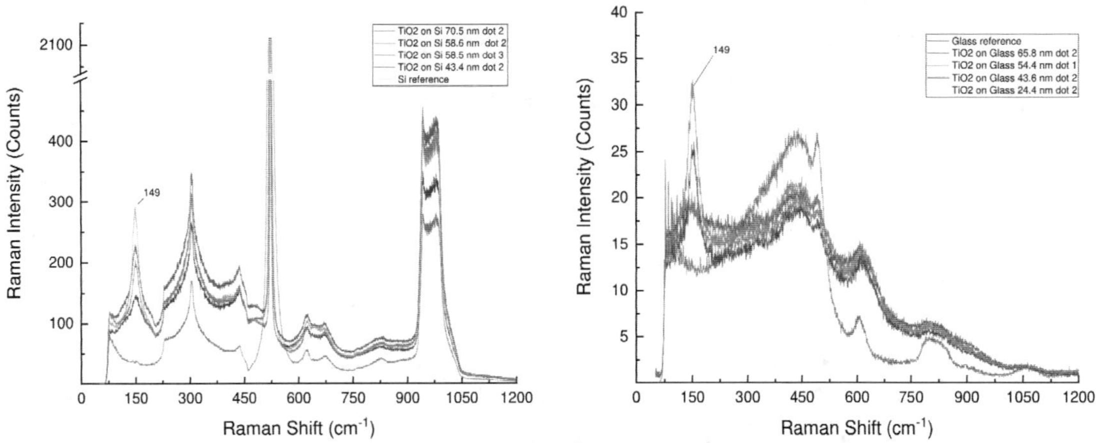

Figure 2. Raman scattering spectra of TiO_2 films with Si (a) and quartz (b) substrate.

Measured spectral characteristics of reflection for films formed on a silicon substrate and spectral characteristics of reflection, transmission and absorption for samples on a quartz substrate are shown in figures 3 and 4, respectively. The obtained characteristics are in good agreement with the literature data for TiO_2 [20].

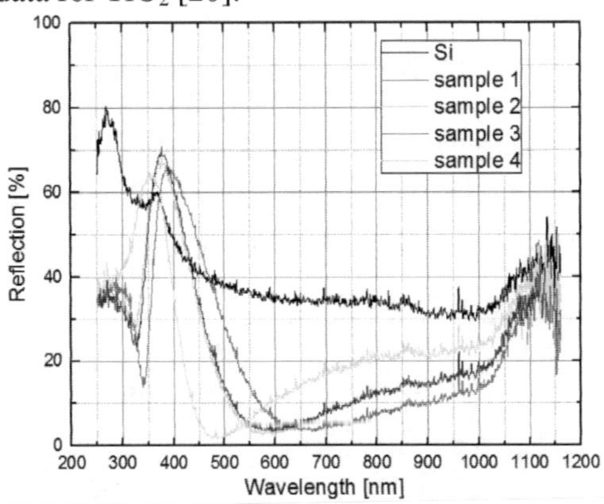

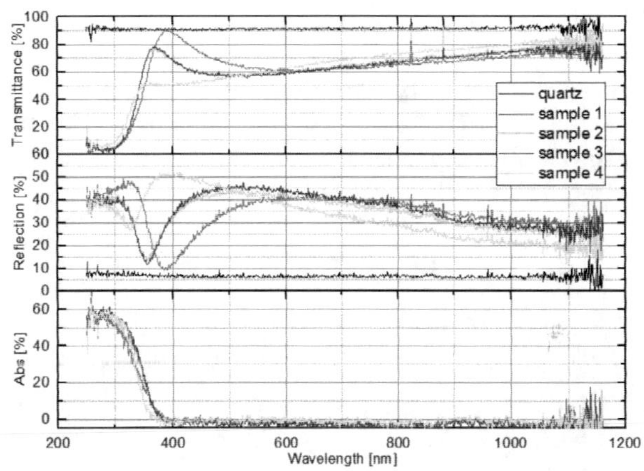

Figure 3. TiO_2/Si and reference substrate reflectance spectrum.

Figure 4. TiO_2 / quartz and reference substrate spectral characteristics.

Based on the data obtained, refractive index and extinction coefficient were determined (Figure 5). Dependences show that the amount of oxygen in the working gas mixture affects the optical parameters of the films. This is primarily pronounced in the UV spectrum of wavelengths. Though, there is no clear relationship between oxygen content in the working gas mixture and calculated values of the optical bandgap of TiO_2. It can be noted that the refractive index of films obtained in this work using a DC magnetron is higher than for films obtained under similar conditions using RF magnetron sputtering [20].

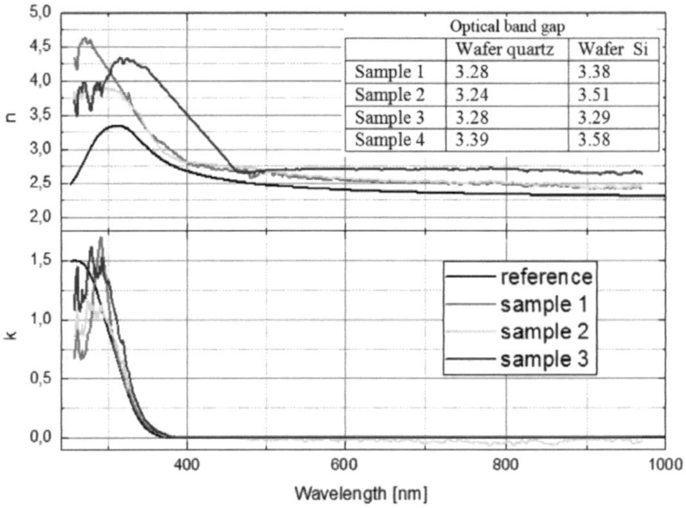

Figure 5. Refractive index and extinction coefficient spectrum for TiO_2 films in comparison with the literature data [20], as well as the calculated value of the optical band gap.

Based on the data obtained, a numerical simulation of the spectral characteristics of the AuNP/TiO_2/Quartz and AgNP/TiO_2/Quartz structures was carried out, the results are shown in figure 6 a and b, respectively. The obtained simulation results agree with the published experimental data for particles with a diameter of 20 nm [21]. It can be seen from the obtained dependences that the maximum difference occurs only for films formed in oxygen plasma without the addition of Ar. This is primarily pronounced in the shift of the plasmon resonance peak of metal nanoparticles to the infrared part of the spectrum.

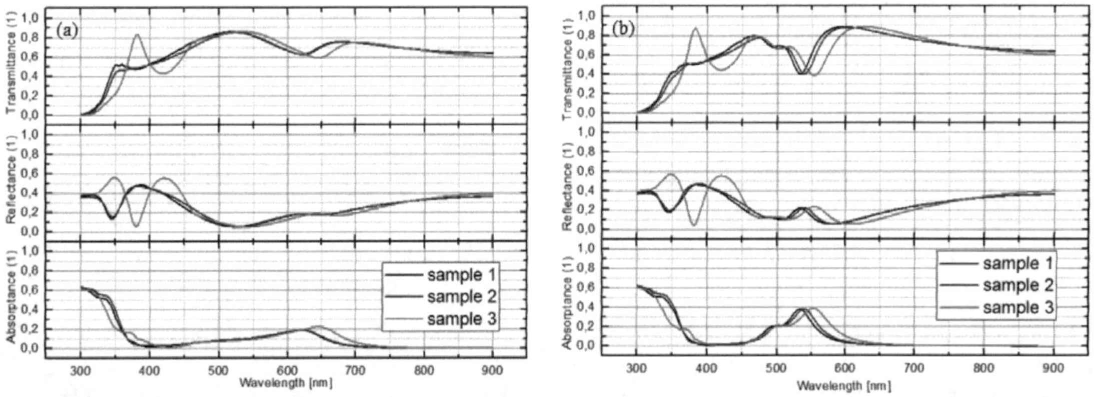

Figure 6. Calculated spectral characteristics for the AuNP / TiO_2 / Quartz (a) and AgNP / TiO_2 / Quartz (b) structures for films obtained under different modes.

4. Conclusion

In the course of this work, TiO_2 films were obtained using DC magnetron sputtering at different oxygen content in the working gas mixture and different sputtering power. The study of the morphology and structure of the obtained films did not show a significant effect of the used deposition parameters. However, the oxygen content, power and type of substrate have a significant impact on the deposition rate. Calculation of the optical characteristics of AuNP/TiO_2/Quartz and AgNP/TiO_2/Quartz structures showed a significant change in characteristics only for films obtained in an oxygen atmosphere.

Acknowledgments
Authors wishing to acknowledge the Ministry of Education foundation for financial support assistance (№ FSRM-2020-0011).

References

[1] Shu J, Qiu Z L, Lv S Z, Zhang K Y and Tang D P 2018 *Anal. Chem.* **90** 2425-29
[2] Wang H L, Zhang L S, Chen Z G, Hu J Q, Li S J, Wang Z H, Liu J S and Wang X C 2014 *Chem. Soc. Rev.* **43** 5234-44
[3] Chen B, Zhang Z, Baek M, Kim S, Kim W and Yong K 2018 *Appl. Catal.* B **237** 763-771
[4] Yang R Y, Yan X X, Li Y M, Zhang X H and Chen J H 2017 *ACS Appl. Mater. Interfaces* **9** 42482-91
[5] Lajvardi M M and Jahangiri M 2016 *IOP Conf. Series: Materials Science and Engineering* **108** 012031
[6] Huanga H, Xiea Y, Zhanga Z, Zhangc F, Xua Q and Wu Z 2014 *Appl. Surface Science* **293** 248-254
[7] Kumar S G and Devi L G 2011 *J. Phys. Chem.* A **115** 13211–41.
[8] Shehzad N, Tahir M, Johari K, Murugesan T and Hussain M 2018 *J. CO2 Util.* **26** 98–122
[9] Liu Z 2015 *Sci. Rep.* **5** 14420
[10] Colgan M J, Djurfors B, Ivey D G and Brett M J 2004 *Thin Solid Films* **466** 92–96
[11] Modes T, Scheffel B, Metzner C, Zywitzki O and Reinhold E 2005 *Surf. Coat. Technol.* **200** 306
[12] Domaradzki J, Kaczmarek D, Prociow E L, Borkowska A, Schmeisser D and Beuckert G 2006 *Thin Solid Films* **513** 269–274
[13] Arconada N, Duran A, Suarez S, Portela R, Coronado J M, Sanchez B and Castro Y 2009 *Appl. Catal.* B **86** 1–7
[14] Heo C H, Lee S B and Boo J H 2005 *Thin Solid Films* **475** 183–188
[15] Kang S H, Kang M S, Kim H S, Kim J Y, Chung Y H, Smyrl W H and Sung Y E 2008 *J. Power Sources* **184** 331–335
[16] Goswami L, Aggarwal N, Shibin K, Singh M, Vashishtha P and Govind G 2020 *ACS Omega* **5** 14535−42
[17] Martin-Palma R J, Martinez-Duart J M and Macleod A 2000 IEEE Transactions on **43** 63-68
[18] Enns Y, Kondrateva A and Mishin M 2020 *J. Phys. Conf. Series* **1695**(1) 012115
[19] Rakic´ D, Djurišic´ A B, Elazar J M and Majewski M L 1998 *J. Appl. Optics* **37**(22) 5271–83
[20] Siefke T, Kroker S and Pfeiffer K *et al.* 2016 *Adv. Optical Mater.* **4** 1780
[21] Kondrateva A, Enns Y, Kazakin A, Morozov I, Karaseov P and Mishin M 2020 *Semiconductors* **54**(14) 1885–88

Laser sintering of oxidized copper nanoparticles deposited by dry aerosol printing

D V Kornyushin, A A Efimov, A I Buchnev, E I Kameneva and V V Ivanov

Moscow Institute of Physics and Technology (National Research University), Dolgoprudny 141701, Russia

kornyushin.d@phystech.edu

Abstract. Sintering of oxidized copper nanoparticles arrays in the form of lines by the local treatment of laser radiation with wavelengths of 527 nm and 980 nm was studied. To form lines with a width equal to 40–150 μm and a thickness equal to 0.5–4 μm focused aerosol jets with average nanoparticle sizes of 110 and 65 nm were used. The production of semiconductor arrays from oxidized copper nanoparticles with a specific electrical resistance of $2 \cdot 10^{-3}$ Ω·m using laser radiation with a wavelength of 980 nm was demonstrated.

1. Introduction

Nowadays, the field of printing technologies for the production of electronic devices (solar cells [1], gas sensors [2]) is actively developing. New methods for the formation of functional microstructures are emerging, which are aimed at simplifying, reducing the cost and minimizing hazardous waste. One of these methods is aerosol printing. It is based on the deposition of aerodynamically focused beams of nanoparticles on substrates. Earlier, the authors of [3-4] demonstrated the possibility of forming arrays of nanoparticles using the dry aerosol printing method.

It is advisable to sinter arrays of nanoparticles formed on a substrate with laser radiation, since this method of energy transfer is local and does not damage the substrate material [5]. Therefore, an urgent task is to find the optimal conditions for laser sintering, which depends on the shape and size of nanoparticles and the laser radiation parameters. In this work, we explore the optimal conditions for sintering by laser radiation of oxidized copper nanoparticles arrays formed using the method of dry aerosol printing.

2. Experimental

To form the nanoparticles arrays in the form of lines used the experimental setup shown in figure 1a. It includes the following key elements: a spark discharge generator of nanoparticles [6-10], a thermal nanoparticles optimizer [11] to control the size and shape of nanoparticles, a coaxial micro-nozzle to deposit nanoparticles on a substrate, a coordinate table to move the substrate at a given speed. Nanoparticles arrays are formed in dry form without the use of solvents and surfactants [12-14]. Nanoparticles obtained as a result of electrical erosion of copper electrodes are generated and deposited on the substrate in an atmosphere of gas Ar+H2(5%).

The morphology and elemental composition of nanoparticles are studied using a transmission electron microscope (TEM) and an aerosol spectrometer. The structural properties of nanoparticles arrays are studied using a scanning electron microscope (SEM) and an optical profilometer.

Nanoparticles arrays are formed in the form of lines with a width equal to 40-170 μm and thickness equal to 0.4–6 μm. The control of size and shape of the nanoparticles is realized by varying the operating modes of the thermal nanoparticles optimizer.

The processes of sintering by laser radiation were carried out using a fixed optical stand, in which the formed arrays were attached to the same coordinate table [15]. The continuous laser radiation with wavelengths of 527 nm and 980 nm used to study the sintering of the arrays. The laser spot moving over the arrays of nanoparticles and the laser radiation power were also varied in the experiments. The diameter of laser spot in the sample plane and the power of incident radiation on the sample in the sintering processes were controlled using a Duma Optronics BeamOn HR laser beam profile meter and an Ophir F150A-BB-26 power meter, respectively.

The estimation of the sintering processes was realized by the measuring of the electrical resistivity of the lines on a heating plate with temperature of 300°C. The absolute resistance and the cross section profile were measured using a Fluke 287 multimeter and an optical profilometer Sensofar S-neox, respectively.

3. Results

Microstructures formed by primary nanoparticles (PNPs) and thermally optimized nanoparticles (TNPs) have different porosity. It was found that microstructures formed by TNPs have a higher density of contacts between the particles and lower porosity than microstructures consisting of PNPs. The TEM images of PNPs and TNP are shown in figure 1(b, c).

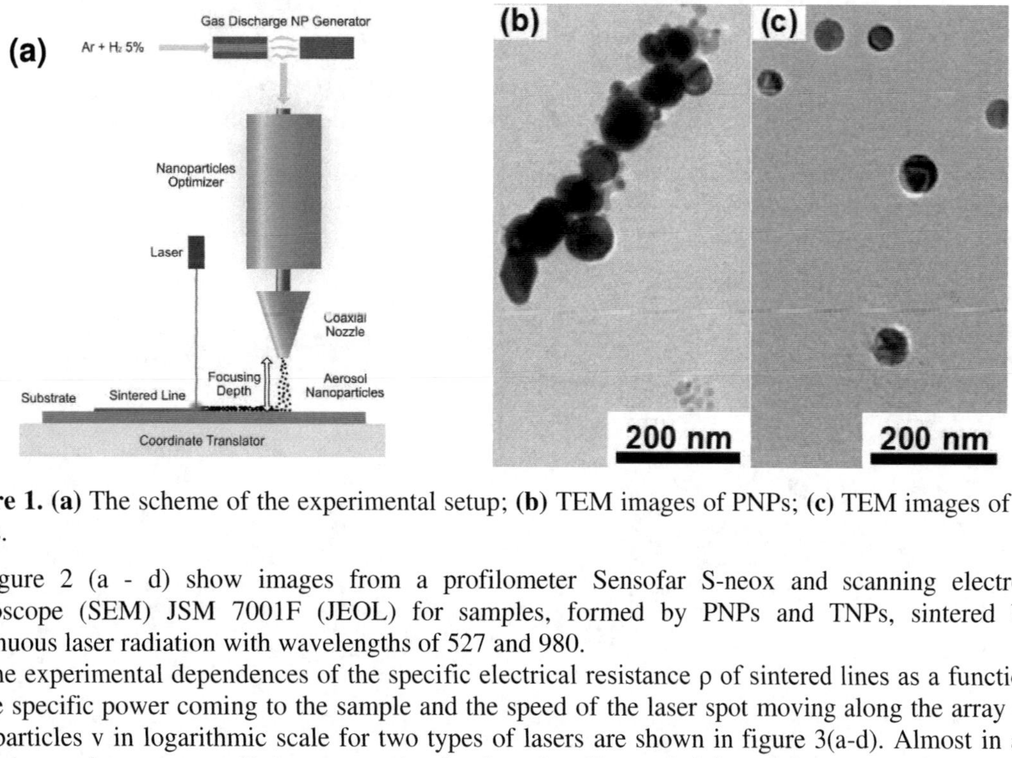

Figure 1. **(a)** The scheme of the experimental setup; **(b)** TEM images of PNPs; **(c)** TEM images of TNPs.

Figure 2 (a - d) show images from a profilometer Sensofar S-neox and scanning electron microscope (SEM) JSM 7001F (JEOL) for samples, formed by PNPs and TNPs, sintered by continuous laser radiation with wavelengths of 527 and 980.

The experimental dependences of the specific electrical resistance ρ of sintered lines as a function of the specific power coming to the sample and the speed of the laser spot moving along the array of nanoparticles v in logarithmic scale for two types of lasers are shown in figure 3(a-d). Almost in all dependences for a given radiation power density there is a line resistivity minimum near the moving speed of the laser spot along an array of nanoparticles of about 10 mm/s. Sintering regimes in this area are close to optimal. The best result of resistivity was obtained at the scanning speed of laser spot about 0.28 mm/s as it shown in figure 3(d). There is a resistivity increasing of the lines to the left of the minimum at low scanning speed in figure 3. This is due to overheating and evaporation of the sintered material as the dose of delivered laser radiation increased. High scanning speeds, where the dose of delivered laser radiation decreases, lead to the growing lines resistivity due to the incomplete sintering of nanoparticles.

Sintering by laser radiation with a wavelength of 527 nm was less energy intensive led to 100-500% greater values of resistivity, than by radiation with wavelengths of 980 nm.

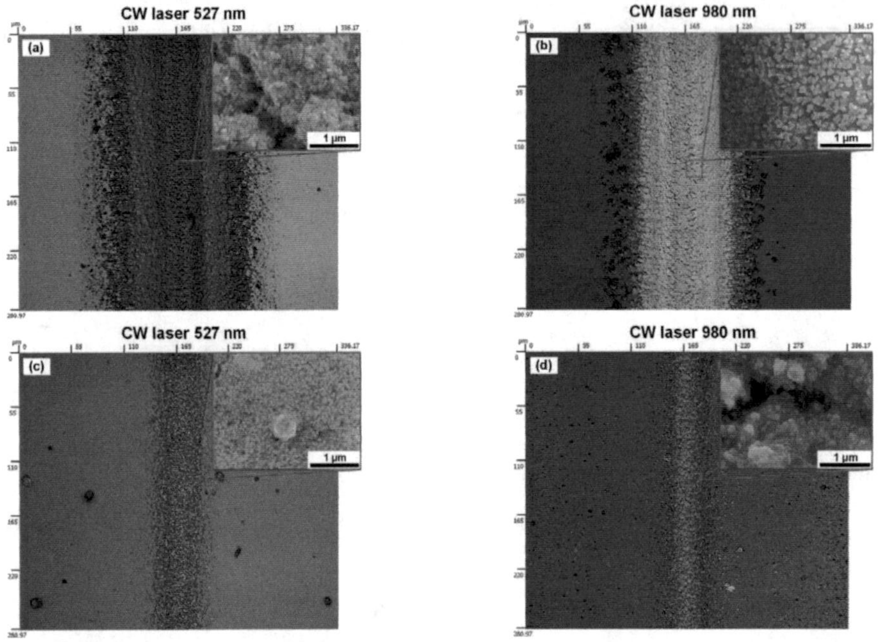

Figure 2. Images from an optical profilometer of microstructures formed by PNPs **(a, b)** and TNPs **(c, d)**, sintered at optimal sintering modes of two lasers. Inset: SEM images of the surface of microstructures.

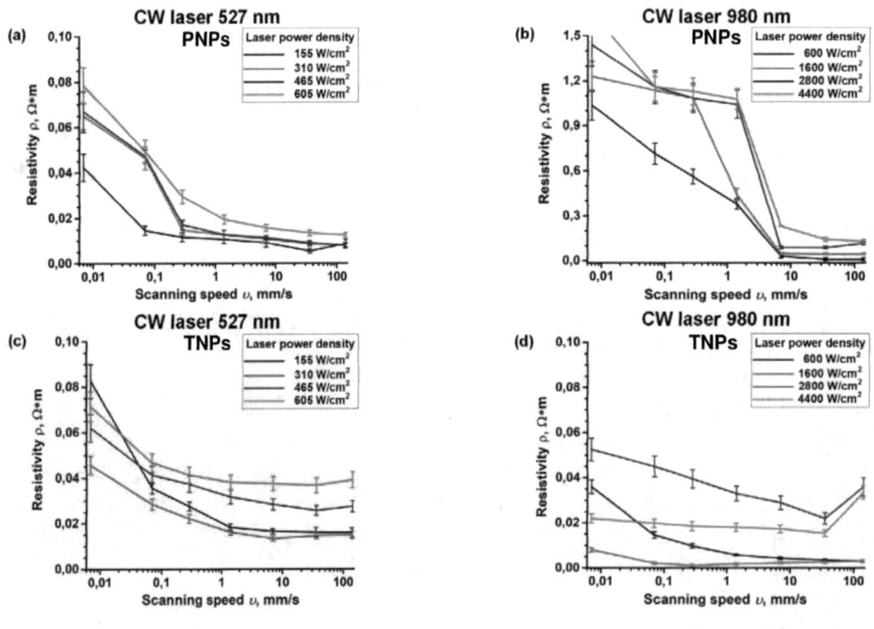

Figure 3. Dependencies of specific electrical resistivity as a function of the specific power coming to the sample and the speed of the laser spot moving along the array of **(a, b)** PNPs and **(c, d)** TNPs for two types of lasers.

4. Conclusion

The advantage of combining the method of dry aerosol printing and laser sintering was demonstrated by the example of sintering arrays of oxidized copper nanoparticles. The minimum resistivity at a heating plate temperature of 300°C was observed for arrays formed from TNPs and sintered by laser radiation with wavelength 980 nm near the speed of 0.28 mm/s and equal to $2\cdot10^{-3}$ Ω·m. In addition, the results obtained show that continuous radiation sinters arrays of nanoparticles more efficiently than pulsed radiation, and radiation with a wavelength of 980 nm is more preferable than with a wavelength of 527 nm for obtaining lines with minimal resistivity. It was found that with an increase in the scanning velocity of the laser spot, a monotonic decrease in the specific resistance of the printed lines occurs.

Acknowledgments

This work was supported by the Russian Science Foundation (project # 19-79-00375).

References

[1] Arsenov P V, Vlasov I S, Efimov A A, Minkov K N and Ivanov V V 2019 *IOP Conf. Ser.: Mater. Sci.Eng.* **473** 012042

[2] Yang C, Zhou E, Miyanishi S, Hashimoto K and Tajima K 2011 *ACS Appl. Mater. Interfaces* **3** 4053–58

[3] Khabarov K, Kornyushin D, Masnaviev B, Tuzhilin D, Saprykin D, Efimov A and Ivanov V 2020 *Appl. Sci.* **10(1)** 246

[4] Kornyushin D V, Efimov A A, Khabarov K M and Ivanov V V 2020 *J. Phys.: Conf. Ser.* **1695** 012032

[5] Kumpulainen T, Pekkanen J, Valkama J, Laakso J, Tuokko R and Mäntysalo M 2011*Optics & Laser Technology* **43** 570-6

[6] Arsenov P V, Efimov A A, Protas N V and Ivanov V V 2018 *IOP Conf. Ser.: Mater. Sci. Eng.* **324** 012016

[7] Mylnikov D, Efimov A and Ivanov V 2019 *Aerosol Sci. Tech.* **53** 1393–403

[8] Ivanov V V, Efimov A A, Myl'nikov D A and Lizunova A A 2018 *Russ. J. Phys. Chem.* **92** 607–12

[9] Efimov A A *et al.* 2021 *Nanomaterials* **11** 234

[10] Khabarov K M, Kornyushin D V, Masnaviev B I, Tuzhilin D N, Efimov A A, Saprykin D L and Ivanov V V 2019 *J. Phys.: Conf. Ser.* **1410** 012217

[11] Lizunova A A, Efimov A A, Arsenov P V and Ivanov V V 2018 *IOP Conf. Ser.: Mater. Sci. Eng.* **307** 012081

[12] Theodorakos I, Zacharatos F, Geremia R, Karnakis D and Zergioti I 2015 *Appl. Surface Sci.* **336** 157-62

[13] Efimov A, Arsenov P, Kornyushin D, Lizunova A, Volkov I and Ivanov V 2020 *Materials* **13** 730

[14] Arsenov P V, Efimov A A, Khabarov K M, Kornyushin D V and Ivanov V V 2020 *Key Eng. Mater.* **834** 37-41

[15] Khabarov K, Kornyushin D, Masnaviev B, Tuzhilin D, Saprykin D, Efimov A and Ivanov V 2020 *Appl. Sci.* **10** 246

Quantitative analysis of metastable wurtzite phase into the self-catalyzed GaP NWs

O Yu Koval[1], V V Fedorov[1,2], I E Eliseev[1], A D Bolshakov[1,3], D A Kirilenko[4], R G Burkovsky[2] and I S Mukhin[1,3]

[1] Nanotechnology Research and Education Centre of the Russian Academy of Sciences, Alferov University, Khlopina 8/3, 194021, St. Petersburg, Russia
[2] Institute of Physics, Nanotechnology and Telecommunications, Peter the Great St.Petersburg Polytechnic University, Polytechnicheskaya 29, 195251, Saint Petersburg, Russia
[3] School of photonics, ITMO University, Kronverksky prospekt 49, 197101, Saint Petersburg, Russia
[4] Ioffe Institute, Politekhnicheskaya 26, 194021, Saint Petersburg, Russia

o.yu.koval@gmail.com

Abstract. In this letter, we report the growth of the self-catalyzed GaP nanowires with a high concentration of wurtzite phase by molecular beam epitaxy. Formation of rotational twins and wurtzite polymorph in vertical nanowires was observed by the developed a complex approach based on the transmission electron microscopy and X-ray diffraction method. Microstructural analysis performed by high resolution transmission electron microscopy and micro-Raman spectroscopy gives us insights on the nanowire formation mechanism and vibrational properties of nanowires with mixed crystal phase. We obtained wurtzite polytype segments with thicknesses lying in the range from several tens up to 500 nm. The results of the work open new perspectives for high phase purity phosphide NWs synthesis and its fast investigation with XRD technique using a laboratory X-Ray source.

1. Introduction

Nowadays, there are no reports on the formation of the wurtzite GaP phase in self-catalyzed nanowires (NW) by means of molecular beam epitaxy (MBE), only grown Au-catalyzed WZ GaP NW by CBE or MOVPE methods are reported [1, 2]. Self-catalyzed nanowires presented in this report are free from foreign catalyst impurities acting as deep trapping levels for charge carriers into synthesized NW. Control and analysis of the crystal phase in semiconductor NW are of high importance owing to the high potential of the phase bandgap engineering [3].

The scientific soundness of the presented paper is based on the presentation of both novel material (wurtzite GaP) and original approach for quantitative characterization of this type of nanostructures. Commonly used crystal phase characterization techniques (i.e. transmission electron microscopy (TEM) or X-ray diffraction (XRD)) are lacking both integral and microscopic data. Thus, the TEM studies reveal only microscopic data and do not allow to evaluate lattice parameters with high accuracy. Thus, the use of Rietveld refinement [4, 5] approach to XRD data handling to perform crystal structure solution, as well as the estimation of volume fraction of the crystalline phases in epitaxial nanoheterostructures are suggested to be inaccurate.

2. GaP nanowires synthesis

GaP NW arrays were synthesized on vicinal silicon (111) substrates with a 4° miscut in <11$\bar{2}$> direction, using Veeco GEN-III MBE machine equipped with Ga effusion cell and valved phosphorus cracker for P_2 molecular flux ($T_{cracker}$ = 900°C). To establish the influence of the growth temperature on NW crystal structure, three samples were grown at the substrate temperature of 610°C. In this work, a two-stage growth with a change of V/III element fluxes ratio, namely 12 and 18, was presented.

3. Result and discussion

3.1. Raman scattering

Raman microspectroscopy (μ-Raman) technique was used as a complementary approach to spatially distinguished WZ-phase NW segments. Polarization–dependent μ-Raman spectroscopy studies were performed at room temperature (300K) using Horiba LabRAM HR800 single monochromator with 1800 gr/mm grating (Kyoto, Japan) and an optical microscope equipped with a x100 0.9 NA lens for excitation and collection of scattered light in backscattering geometry. Single NWs were "dry-transferred" onto transparent quartz substrate mounted on piezo-based scanning stage of Veeco BioScope Catalyst (USA) atomic force microscope, which allowed spatial mapping of μ-Raman signal. The measurements were carried out with a λ = 532 nm (2.331 eV) frequency-doubled diode-pumped solid-state (DPSS) YAG:Nd (neodymium-doped yttrium aluminium garnet) laser providing the above bandgap excitation of ZB and WZ phases of GaP NWs [1, 2, 6].

The μ-Raman measurements of individual NWs were performed at 300K to investigate the spatial distribution of the NW phase composition. Raman spectra were obtained in *-z(x,x)z* backscattering configurations, as depicted on the inset of Figure 1, with *x, y,* and *z* axes are aligned along the [111], [11$\bar{2}$] and [1$\bar{1}$0] GaP ZB phase crystallographic directions. The map of Raman spectra across NW length is presented in Figure 1 (a) inset. The sharp increase of optical scattering signal can be observed at NW top and bottom facets due to the more efficient light scattering and coupling at the NW edges. As can be seen from Figure 1 b) Raman spectra acquired along the NW side facets demonstrate longitudinal optical (LO) mode with low intensity since it is symmetry forbidden. In contrast, Raman spectra obtained at NW top and bottom facets demonstrate bright LO- and surface optical (SO) Raman bands due to selection rules break-down caused by finite size effect [7]. As reported previously, SO mode can be observed in high surface-to-volume ratio nanostructures, this mode is positioned between LO- and TO- modes [1, 8], and can be activated via breakdown of translational symmetry at surfaces under near-resonant laser excitation condition.

It is clearly seen those additional vibrational modes, labelled with blue stars in Figure 1, appear near one of the NW edges, noteworthy LO- and TO- modes are doublet splitted which indicate lower symmetry of the uniaxial WZ crystal structure. In contrast, spectra obtained at the opposite NW edge present only TO-, LO- and SO modes located at 365, 402 and ~ 395 cm^{-1}, respectively, which is typical for ZB-GaP phase [1, 8]. According to the group theory, the following Raman-active bands for WZ phase III-V semiconductors belonging to $P6_3mc$ space group can be excited: A_1 and E_1 modes, where phonons are polarized in *z* and *xy* directions respectively, low and high (E_2^H) frequency E_2 modes [9, 10]. Thus, in a chosen *-z(x,x)z* backscattering geometry in a Raman spectrum, taken from GaP top facet (see Figure 1(c)), we can identify the following 7 Raman-active modes: namely E_2^H positioned at 357 cm^{-1}, A_1(TO) - 363 cm^{-1}, E_1(TO) - 365.5 cm^{-1}, B_1^H - 380 cm^{-1}, SO - 395 cm^{-1}, A_1(LO) - 400 and E_1(LO) - 402 cm^{-1}, as was previously reported by Bruno C. da Silva [6] and Kidong Park [11]. The position of the E_2^H peak of the studied sample is indeed very close to the energy of the TO phonons in GaP at the critical point L on the edge of the Brillouin zone – TO (L) [12, 13]. In addition, the appearance of the Raman-forbidden ("silent") B_1^H mode can be resolved at NW side facets. It should be noted that E_2^H

mode is absent in ZB phase despite the large density of rotational twinning defects, as clearly seen in Figure 1 (b) [14 – 16]. The approximation and visualization of Raman vibrational modes was performed using Lorentz function. Thus, μ-Raman spectroscopy provides clear evidence of the WZ phase in the GaP NW structure.

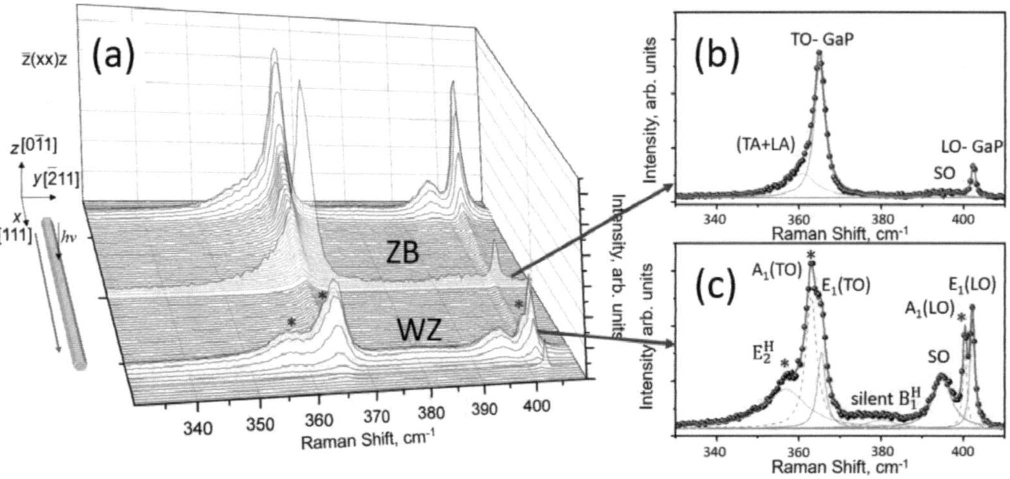

Figure 1. Profile of the Raman spectra across the NW length of the sample with highest WZ crystal phase content, inset: experiment geometry, b) Raman spectrum obtained from GaP NW middle (ZB phase), and c) Raman spectrum obtained from GaP NW base (WZ phase)

3.2. TEM study

Crystal structure is studied with TEM using JEOL JEM–2100F field emission gun TEM (Tokyo, Japan) operating at 200 kV (with point-to-point resolution of 0.19 nm in TEM mode).

According TEM study (Figure 2(a)) there is the existence of 2 crystal phases – cubic ZB and hexagonal WZ. According to X-ray diffraction reciprocal space mapping analysis, there are the existence of ZB phase twinning and absence of WZ phase twinning (not presented here).

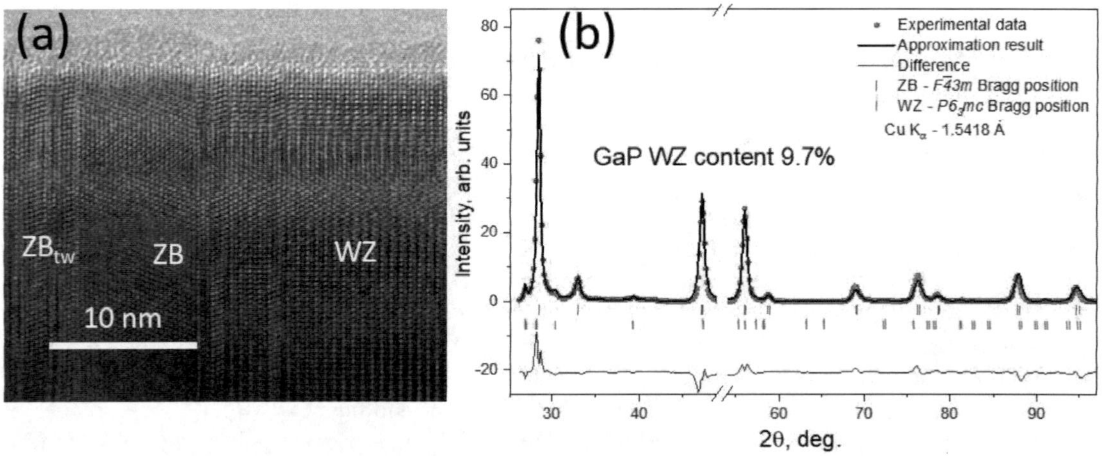

Figure 2. a) HRTEM image of individual studied GaP NW, b) Approximated XRD powder diffraction pattern obtained from disordered GaP NWs into the scotch glue, which was grown at 610° C

3.3. XRD study

The study is performed using SuperNova (Rigaku Oxford Diffraction, Japan) diffractometer, radiation generated by IμS microfocus X-ray tube with Cu Kα-radiation (λ = 1.5418 Å, ~8 keV). NW synthesis by the MBE method is always accompanied by the formation of a two-dimensional parasitic island. That is why, we detached the NW array from the substrate using "scotch-taping" technique for avoiding XRD signal from parasitically grown islands.

As clearly in Figure 2 (b) there are many Bragg reflexes, the highest intensity corresponding to the cubic ZB GaP phase with $F\overline{4}3m$ space group (positions depicted with green sticks) [17] and reflexes with lower intensity can be attributed to WZ phase with $P6_3mc$ space group (positions depicted with magenta sticks) [18]. Quantitative phase analysis and evaluate the WZ GaP phase lattice parameters (a = 3.839 Å, c = 6.344 Å), we present a simple express Rietveld refinement method [5]. The WZ phase content in studied sample was 9.7±1.5 percent. The obtained data provide statistics across a large number of disordered NWs in one measurement which makes it more practically useful compared to TEM analysis.

4. Conclusions

We report on the controllable stabilization of the polymorphic wurtzite crystal phase in the MBE grown self-catalyzed GaP NWs and refine the crystal lattice parameters of metastable WZ GaP phase. We observed the appearance of additional Raman bands, which can be attributed to the hexagonal WZ GaP lattice and, thus, we demonstrate features of Raman scattering signal into phase heterostructured individual GaP NW. We employ TEM that enables direct observations of the WZ phase stabilization up to 500 nm long GaP NW segment, that correlated to powder XRD data. The proposed approach can be used for a quantitative evaluation of the mean volume fraction of polytypic phase segments in heterostructured nanowires highly desirable for optimization of growth technologies.

Acknowledgments

The authors thank the Ministry of Science and Higher Education of the Russian Federation (FSRM-2020-0005). O. Yu. Koval thanked the Russian Foundation for Basic Research (RFBR project № 19-32-90232).

TEM characterizations were performed using equipment of the Federal Joint Research Center "Material science and characterization in advanced technology".

References

[1] Panda J K *et al.* 2013 *Applied Physics Letters* **103** (2) 023108
[2] Assali S *et al. Nano Letters* 2013 **13** (4) 1559–1563
[3] Yang X, Shu H, Liang P, Cao D and Chen X 2015 *Journal of Physics Chemistry C* **119** (21) 12030–12036
[4] Clavier N *et al.* 2013 *Journal of Nuclear Materials* **441** (1–3) 73–83
[5] Wenk H- R, Lutterotti L, Kaercher P, Kanitpanyacharoen W, Miyagi L and Vasin R 2014 *Powder Diffraction* **29** (3) 220–232
[6] Da Silva B C *et al.* 2020 *Scientific Reports* **10** (1) 7904
[7] Spirkoska D, Abstreiter G and Fontcuberta i Morral A 2008 *Nanotechnology* **19** (43) 435704
[8] De Luca M and Zardo I 2017 *Raman Spectroscopy and Applications* InTech
[9] Arguello C A, Rousseau D L and Porto S P S 1969 *Physical Review* **181** (3) 1351–1363
[10] Ketterer B, Heiss M, Uccelli E, Arbiol J and Fontcuberta i Morral A 2011 *ACS Nano* **5** (9) 7585–7592
[11] Park K *et al.* 2019 *ACS Omega* **4** (2) 3098–3104
[12] Dobrovolsky A, Sukrittanon S, Kuang Y J, Tu C W, Chen W M and Buyanova I A 2014 *Applied Physics Letters* **105** (19) 193102
[13] Zardo I *et al.* 2009 *Physical Review B* **80** (24) 245324
[14] Steidl M *et al.*2019 *Nanotechnology* **30** (10) 104002
[15] Aggarwal R, Ingale A A and Dixit V K 2018 *Applied Surface Science* **427** 754–762
[16] Wu J, Zhang D, Lu Q, Gutierrez H R and Eklund P C 2021 *Physical Review B* **81** (16) 165415
[17] L. M. Foster 1974 *Journal of the Electrochemical Society* **121** (12) 1662
[18] Maliakkal C B *et al.* 2019 *Nanotechnology* **30** (25) 254002

Vibrational dynamics in 2D crystal lattices of borophene

I M Kuimov, I O Raikov and D A Parshin

Alferov University, 199034 Saint-Petersburg, Russia

001meridian@mail.ru

Abstract. Vibration dynamics of crystalline borophene is considered in the framework of the Born–von Karman model. The vibrations perpendicular to the plane of 2D borophen lattice (flexural modes) are studied.

1. Introduction

One of the most interesting topics in the modern condensed matter physics is studying of crystalline graphene. As it is well known the crystalline graphene, due to the unusual geometry of the band structure, has exceptional mechanical [1,2], physical [3,4] and optical properties [5], which find applications [6] in various fields. Note here its record high thermal conductivity. For graphene, it was found, that it has unique properties of electrical conductivity and thermal conductivity, therefore, it is of great importance in microelectronics. The discovery of graphene sparked intense research activity. And later a whole class of new materials with no less unique properties appeared. One of these materials is borophene.

2. Borophene

The graphene lattice has a hexagonal honeycomb structure with 2 atoms in a unit cell. Unlike graphene, borophene has a triangular lattice with a single atom in the unit cell, but many elements of graphene symmetry are present in borophene.

Since borophene, unlike graphene, has only one atom in the unit cell, it has 3 acoustic modes like graphene, but no optical modes. Thus we have in borophene LA, TA and ZA modes. We have calculated the frequency spectrum $g(\omega)$ in borofen.

3. Calculations

3.1. Vibrations of a triangular lattice

Consider a triangular lattice, where each atom has 6 neighbors. Length of all links is a, the angle between the links is 60°. Let there be displacement vectors $\mathbf{e_1}$, $\mathbf{e_2}$, $\mathbf{e_3} = \mathbf{e_2} - \mathbf{e_1}$ and three vectors opposite to them. Then $\mathbf{e_1}$, $\mathbf{e_2}$ is elementary lattice vectors, and there is only one atom in the unit cell. Let an atom be bound to its neighbors by springs that have rigidity k. In the Born–von Karman model, the equation of motion is

$$x''(\mathbf{r}) = \frac{\kappa}{m}\big(x(\mathbf{r}{+}\mathbf{e_1}) + x(\mathbf{r}{-}\mathbf{e_1}) - 2x(r)\big) + x(\mathbf{r}{+}\mathbf{e_2}) + x(\mathbf{r}{-}\mathbf{e_2}) - 2x(r) +$$

$$+x(\mathbf{r}{+}\mathbf{e_3}) + x(\mathbf{r}{-}\mathbf{e_3}) - 2x(r)$$

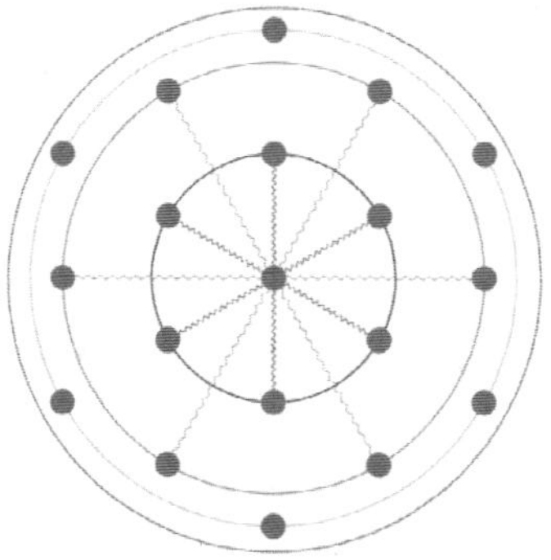

Figure 1. The triangular lattice. The first, second and third coordination sphere are marked in blue, orange, and green, respectively, and the connections to the third sphere are omitted.

Note that in fact it consists of three second-order finite differences. This circumstance will be useful to us. Indeed, we will look for solutions in the form of waves:

$$x(\mathbf{r}) = e^{-i\omega t + i\mathbf{k}\cdot\mathbf{r}}$$

And note that then for any vector **h**:

$$x(\mathbf{r}+\mathbf{h}) + x(\mathbf{r}-\mathbf{h}) - 2x(r) = x(\mathbf{r})\left(e^{i\mathbf{k}\cdot\mathbf{h}} + e^{-i\mathbf{k}\cdot\mathbf{h}} - 2\right) = -4x(\mathbf{r})\sin^2(\frac{\mathbf{k}\cdot\mathbf{h}}{2})$$

And then we get the dispersion equation:

$$\omega^2 = 4\Omega_1^2\left(sin^2\left(\frac{\mathbf{k}\cdot\mathbf{e}_1}{2}\right) + sin^2\left(\frac{\mathbf{k}\cdot\mathbf{e}_2}{2}\right) + sin^2\left(\frac{\mathbf{k}\cdot\mathbf{e}_3}{2}\right)\right)$$

where $\Omega_1^2 = \frac{\kappa}{m}$.

Moreover, note that we did not use the angle between the bonds to derive this equation. We used only the fact that the atom is the center of the lattice inversion, and therefore 6 bonds are split into three pairs, each of which gives sin^2. For a square lattice, for example, there will be a similar formula, but with two sin^2 - one for each pair of vectors. In addition, we will be able to apply this formula for subsequent coordination spheres.

3.2. General formula of dispersion for spheres with even N_i

Let us introduce the following notation: S_i is the set of position vectors drawn from some atom to the atoms of its i-th coordination sphere. For example, for triangular lattice $S_1 = \{\mathbf{e}_1, \mathbf{e}_2, \mathbf{e}_3, -\mathbf{e}_1, -\mathbf{e}_2, -\mathbf{e}_3\}$.

D_i is the set of position vectors drawn to one from each pares of symmetric atoms of the i-th coordination sphere. Such a choice of vectors is ambiguous (one of the vectors can be arbitrarily chosen from each pair), but, as will be seen below, the dispersion equations do not depend on a particular choice. Of course, D_i is determined only when the number N_i of atoms in the i-th coordination sphere is even. For triangular lattice all of N_i are even (and equal to 6). D_1 for it can be chosen, for example, as $\{\mathbf{e}_1, \mathbf{e}_2, \mathbf{e}_3\}$.

For small k:

$$sin^2\left(\frac{\mathbf{k}\cdot\mathbf{h}}{2}\right) = \frac{(\mathbf{k}\cdot\mathbf{h})^2}{4} + O(|k|^4).$$

Let us write using the new notation

$$\omega^2 \approx \Omega_1^2 \sum_{i=1,2,3} (\mathbf{k}\cdot\mathbf{e}_i)^2 = \frac{\Omega_1^2}{2}\sum_{e\in S_1}(\mathbf{k}\cdot\mathbf{e})^2 = \frac{3}{2}\Omega_1^2 a^2|\mathbf{k}|\text{^}2$$

where we went from the sum over three vectors to the sum over all six, the value of which is given to us by the previously obtained formula.

Moreover, it can be noted that all the considerations given here are also true for the following coordination spheres, if they have an even number of atoms. Then the exact formula of the dispersion of N spheres is

$$\omega^2 = 2\sum_i^N \frac{\kappa_i}{m}\sum_{e\in S_1}\sin^2\left(\frac{\mathbf{k}\cdot\mathbf{e}}{2}\right)$$

where κ_i is the stiffness of the bond with the atoms of the ith sphere, and the second summation is carried out over all vectors of the sphere.

Therefore, the condition for the dispersion to be non-linear at zero:

$$\sum_i N_i\Omega_i^2 r_i^2 = 0$$

In particular, for a triangular lattice:

$$\Omega_1^2 + 3\Omega_2^2 + 4\Omega_3^2 = 0$$

3.3 Numerical results

Figure 2 shows the dependence of the ω dispersion law on k in the Brillouin zone, taking into account one coordination sphere.

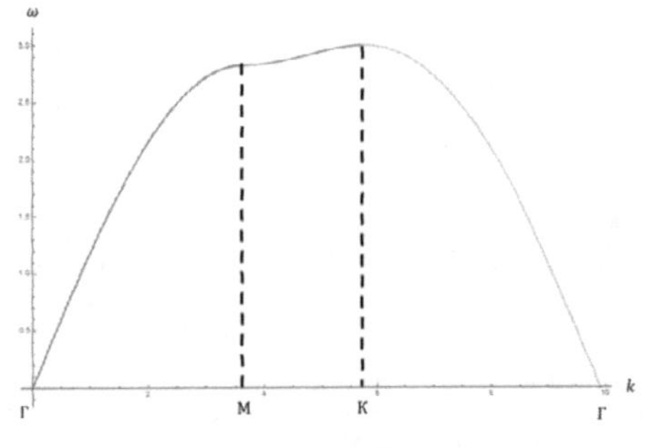

Figure 2. The dependence of the ω dispersion law on k in the Brillouin zone.

The linear dependence at the Γ points indicates that these are phonons. To obtain the quadratic dispersion, it is necessary to take into account the neighbors from the second and third coordination spheres.

Figure 3 shows the dependence of the density of states $g(\omega)$. Linearity means that for small ω, there are phonons with a linear law of dispersion, and the peak is the Van Hove singularity.

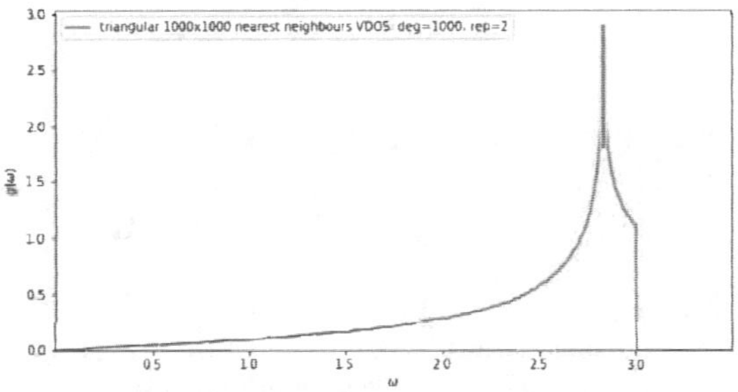

Figure 3. The dependence of the density of states $g(\omega)$.

Acknowledgments
We are grateful very much to Andrey Ipatov and Yury Galperin for discussions of our results.

Appendix
Let us prove the following lemma.

Let us find the next sum that will be needed further

$$\sum (\mathbf{k}\cdot\mathbf{e}_i)^2$$

where the summation is over vectors extended from the center to the vertices of a regular n-gon inscribed in a circle of radius r (that is, in fact, over a coordination sphere with n atoms).

Let us denote by φ the angle between k and the first of the vectors. Then the sum is

$$|\mathbf{k}|^2 r^2 \sum_{i=0}^{n-1} \cos^2\left(\frac{2\pi i}{n}+\varphi\right) = \frac{n}{2}|\mathbf{k}|^2 r^2$$

so the direction of the vector k does not matter.

Moreover:

$$\sum (\mathbf{k}\cdot\mathbf{e})^4 = \frac{3n}{8}|\mathbf{k}|^4 r^4$$

$$\sum \mathbf{k}\cdot\mathbf{e} = 0$$

References

[1] Wirtz L and Rubio A 2004 *Sol. State Comm.* **131** 141
[2] Hobson J P and Nierenberg W A 1953 *Phys. Rev.* **89** 662
[3] Maradudin A A, Montroll E W, Weiss G H and Ipatova I P 1971 *Theory of Lattice Dynamics in the Harmonic Approximation* (New York: Academic Press)
[4] Weige A, Wellein G, Alvermann A and Fehske H 2006 *Rev. Mod. Phys.* **78** 275
[5] Beltukov Y M, Kozub V I and Parshin D A 2013 *Phys. Rev.* B **87** 134203
[6] Beltukov Y M and Parshin D A 2016 *JETP Letters* **104** 552
[7] Falkovsky L A 2012 *JETP* **142** 560
[8] Raikov I O *et al* 2020 *J. Phys.: Conf. Ser.* **1695** 012179

Nanostructured electrodes for supercapacitors

I O Yavtushenko, M Yu Makhmud-Akhunov and A A Adamovich

Ulyanovsk State University, Ulyanovsk 432017, Russia

Abstract. The paper presents the results of studies on the formation of planar capacitive systems based on nanoporous anodic oxide films and a conducting polymer. According to voltammetry data, the capacity of the systems under study was determined. The structure of the porous layers was judged by impedance spectroscopy and electron microscopy.

1. Introduction

The design and creation of elements and systems with increased capacity is a relevant and actively developed area of modern electronics. Improvement of the electrical characteristics of such objects is achieved both through the use of various materials of electrodes and electrolytes, and through the formation of developed surfaces. From this point of view, metals and semiconductors nanostructured due to anodic treatment are promising [1-3]. The advantage of this technique lies in the simplicity of controlling the parameters of the structures being formed.

In this paper, we will consider supercapacitors based on nanoporous titanium and aluminum oxides.

2. Experimental

As capacitor plates, we used metal plates based on titanium and aluminum with surface porous anodic oxide layers. Layers of nanoporous titanium oxide films were produced by electrochemical anodic treatment of titanium foil in aqueous 0.6 wt. % HF solution at 20 V for 50 min [4]. The porous alumina was formed using a standard two-stage anodic treatment at 4 wt. % aqueous solution of $H_2C_2O_4$ at U = 40 V. Each electrode was covered with a layer of the emeraldine form of polyaniline (PANI). A solution of polyvinyl alcohol with 3.4 M orthophosphoric acid served as an interlayer between the electrodes. The parameters of the system under study were estimated by the method of cyclic voltammetry on a pulse potentiostat-galvanostat P-45X in the range from –0.5 to 0.5 V at a potential sweep rate of 20 and 100 mV / s. The impedance spectra were recorded in the frequency range from 0.1 Hz to 50 kHz with an amplitude of 50 mV. The parameters of porous structures were monitored by scanning electron microscopy (SEM) using a PHENOM PRO-X setup.

3. Results and discussion

A typical view of voltammograms for the structures under study is shown in the figure 1. The results of processing showed a capacitance value of $5.6 \cdot 10^{-3}$ F for a structure based on titanium oxide (sweep rate 20 mV / s), which is three orders of magnitude higher than that for a capacitor with electrodes based on porous alumina.

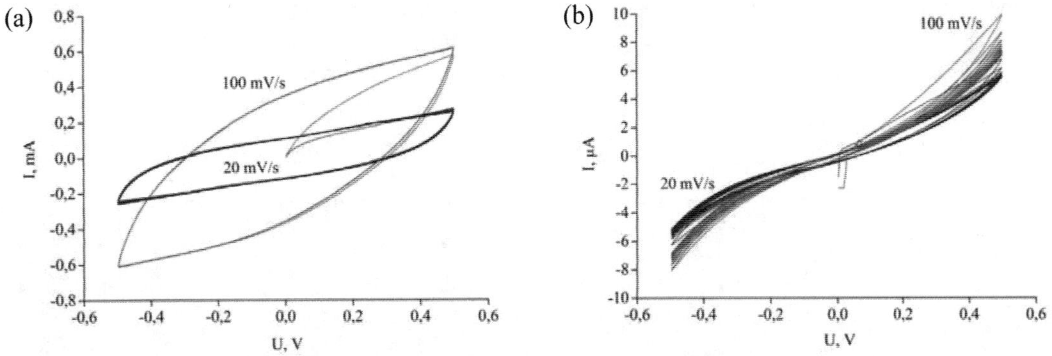

Figure 1. Cyclic voltammograms of supercapacitors based on porous films and PANI: a - titanium oxide; b - aluminum oxide.

A specific feature of the formation of porous titanium oxide films in aqueous solutions is the limitation in thickness, which, as a rule, does not exceed 500–600 nm [5]. Also, anodic titanium oxides, in comparison with oxide films on aluminum, in a number of cases can have dielectric permittivity values by an order of magnitude higher. All this determines the best capacitive properties of the titanium oxide-based system. However, the deviation of the VAG in both cases from an ideally rectangular shape may indicate the influence of the sequential capacitive component of the electrolyte resistance, in this case, polyvinyl alcohol with orthophosphoric acid, as well as the Faraday processes that occur in parallel due to the small thickness of the barrier layer in the pore covered with a thin layer of PANI.

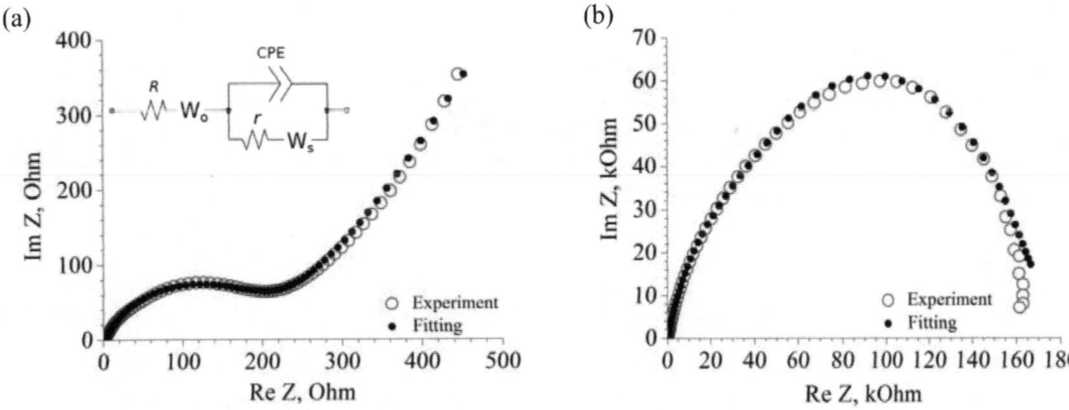

Figure 2. Impedance spectra of supercapacitors based on porous titanium (a) and aluminum (b) oxides. Solid lines correspond to the theoretical dependence of the equivalent circuit, shown in the inset.

The features of the structure of the functional layers of the formed multilayer capacitive systems were judged from the data of impedance spectroscopy. Dependences Im Z (Re Z) (figure 2) are typical for capacitors and have sections with a shape close to a semicircle, which, as noted above, characterizes the value of the Faraday resistance. For the system with aluminum oxide, this parameter is three orders of magnitude higher in comparison with titanium oxide.

The use of electrodes with a developed surface (pore size ~ 80-100 nm, figure 3) determines the shape of the impedance at low frequencies and characterizes the capacitance of the electric double layer (the element with a constant phase is – CPE [6]) and an almost ideal semicircle is observed for systems based on aluminum oxide, characterized, as noted above, by low capacitance values.

$$Z_{CPE} = C_{\alpha}^{-1}(j\omega)^{-\alpha} \tag{1}$$

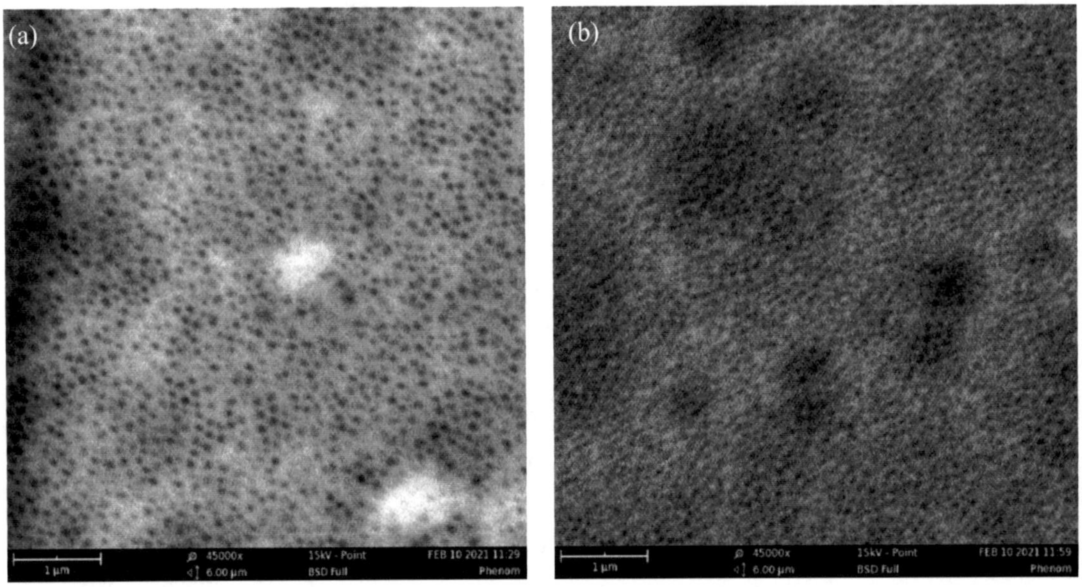

Figure 3. SEM images of Ti (a) and Al (b) surfaces after anodic treatment.

The equivalent circuit used to describe the impedance spectra is shown in the figure 2. The deviation of the obtained curves from the ideal is due not only to the nanostructural nature of the electrodes used, but also to diffusion processes occurring in the interelectrode space. Warburg elements were introduced to describe these components. The W_S element is characterized by the impedance of finite diffusion with a transmissive boundary, the W_O element - with a reflective boundary:

$$Z_{W_S} = \frac{W_S}{\sqrt{\omega}}(1-j)\tanh\left(b_S\sqrt{j\omega}\right) \tag{2}$$

$$Z_{W_O} = \frac{W_O}{\sqrt{\omega}}(1-j)\coth\left(b_O\sqrt{j\omega}\right) \tag{3}$$

where W_S and W_O – are the Warburg coefficients, and $b_{S,O} = d/\sqrt{D}$ (d – is the thickness of the Nernst diffusion layer, D – is the diffusion coefficient of ions).

Table 1. Equivalent circuit parameters.

Parameter	Porous titanium oxide	Porous alumina oxide
R, Ohm	1.51	185.26
r, Ohm	208.61	$81 \cdot 10^3$
C_α^{-1}	$6.03 \cdot 10^{-5}$	$5.88 \cdot 10^{-7}$
α	0.74	0.8
W_S	200	$1.54 \cdot 10^5$
b_S	10	0.39
W_O	$4.5 \cdot 10^{-4}$	$3.5 \cdot 10^3$
b_O	$1.03 \cdot 10^{-5}$	9.7

The parameters of the used structured electrodes established according to the equivalent equivalent circuit (figure 2) are presented in table 1. For this, the EIS Spectrum Analyzer software was used. The presented results were obtained using the Levenberg-Marquard algorithm with amplitude minimization. According to the data obtained, the diffusion coefficient of ions in titanium oxide is several orders of magnitude higher in comparison with aluminum (table 1). Overestimated W_S parameters for aluminum oxide indicate its low diffusion permeability.

4. Conclusions
Thus, the paper considered the features of using nanostructured electrodes based on porous oxide films of titanium and aluminum to create capacitive systems with improved characteristics. The capacity of the system based on titanium oxide turned out to be three orders of magnitude higher in comparison with the structure of electrodes with aluminum oxide and amounted to $5.6 \cdot 10^{-3}$ F. This character is due to the difference in the dielectric constant, the thickness of the formed porous films, as well as the parameters of ion diffusion. Nevertheless, the use of porous aluminum oxide remains promising in the case of its modification with various conductive components, as well as the use of other types of electrolytes in its synthesis.

Acknowledgments
This work is supported by the Russian Science Foundation (project no. 19-71-10063).

References
[1] Rekha S, Subrata M, Shrestha K and Ariga K 2020 *Nanomaterials* **10** 3–27
[2] Sharma V, Singh I and Chandra A 2018 *Scientific Reports* **8** 1–12
[3] Cuihua A, Zhang Y, Guo H and Wang Y 2019 *Nanoscale Advances* **1** 4644–58
[4] Peng Z and Ni J 2019 *Royal Society Open Science* **6** 1–15
[5] Macak J, Tsuchiya H, Ghicov A, Yasuda K, Hahn R, Bauer S and Schmuki P 2007 *Current Opinion in Solid State and Materials Science* **11** 3–8
[6] Sibatov R T and Uchaikin V V 2019 *Appl. Eng. Life Soc. Sci* **87** 87–116

Study of nonlinear optical phenomena in silicon nanowires

V A Mastalieva[1,4], V Neplokh[1,3], A I Morozov[1], A A Nikolaeva[2], A S Gudovskikh[1], I S Mukhin[1,2] and S V Makarov[2]

[1] Department of Physics and Astronomy, Renewable Energy Laboratory, Saint Petersburg National Research Academic University, 8/3 Khlopina, SP 194021, Russia
[2] Faculty of Photonics ITMO University, 49 Kronverk Sky pr., SP 197101, Russia
[3] Graduate School of Physics and Materials Technology, Peter the Great St. Petersburg Polytechnic University, 29 Politekhnicheskaya, SP 195251, Russia

strindberg76@mail.ru

Abstract. This work studies generation of second and third harmonics in arrays of vertically oriented silicon nanowires (SiNWs) encapsulated into a silicone membrane and separated from the growth substrate. The structures were produced by plasma-chemical etching of silicon substrate resulting in a formation of homogeneous arrays of SiNWs. Such SiNW-based membranes demonstrated efficient infrared-to-visible light conversion by generation of second and third harmonic signals visible by a naked eye. This study contributes to the development of technology of optical devices based on silicon and presents a new route for visualization of infrared radiation.

1. Introduction

Converters of infrared (IR) radiation to the visible range are in demand for various optoelectronic devices [1, 2], visualizers of IR radiation, and other nonlinear optical applications. Commercial IR visualizers are based on ceramics which are very sensitive and do not require activation by light [3,4]. However, ceramic visualizers have limited optical transparency and mechanical flexibility and emit light at a fixed wavelength, making it impossible to distinguish between different wavelengths of incident IR radiation. A new alternative approach for visualization of IR radiation based on the generation of the second harmonic (SHG) in arrays of GaP nanowires (NWs) transferred into transparent silicone membrane was presented in our article [2].

The presented GaP NWs-based silicone membranes double the frequency of incident radiation allowing visualization of IR radiation in the range of 800-1400 nm [5]. However, the fabrication of such membranes requires expensive molecular beam epitaxy. In this work, we propose an approach similar to [2] based on SiNWs. Bulk silicon has an efficient third harmonic generation (THG), while only the silicon surface has SHG due to a crystal symmetry violation. Thus, SiNW arrays having large surface area and encapsulated into optically transparent silicone represent a promising object for studying nonlinear optical phenomena. Due to the developed surface of NWs, such membranes allow the study of SHG, while SiNW volume provides a significant THG signal. SiNW membranes can be used for integration with silicon electronics optical devices, including development of optical information transmission technologies [6].

2.The process of obtaining structures

2.1 Photoconverting structures based on vertically oriented structures on silicon obtained by dry plasma-chemical etching at cryogenic temperatures

The substrates with the applied mask were placed in the etching chamber and cooled to -140°C. The substrates were kept for 20 minutes to stabilize the temperature. Then, silicon was etched under the next conditions: ECP power - 30 W, ICP power - 1000 W, pressure - 5 mTorr, oxygen O_2 - 11 ssm^3min^{-1}, SF_6 flow - 50 ssm^3min^{-1}, bias voltage - -100 V

SEM image of the surface of a silicon wafer covered with a mask of latex spheres and the result of cryogenic etching is shown in Figure 1 [1].

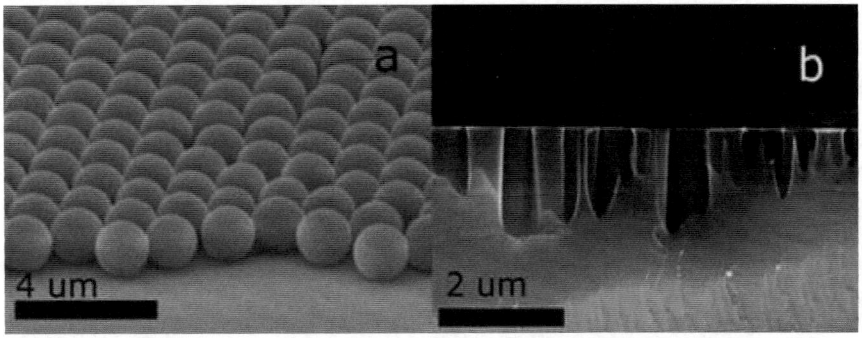

Figure 1. SEM image of the surface of a silicon wafer covered with a mask of latex spheres (a) and after cryogenic etching (b).

To reduce the diameter of the spheres, the spheres were etched in oxygen. The etching result is shown in Figure 2a.

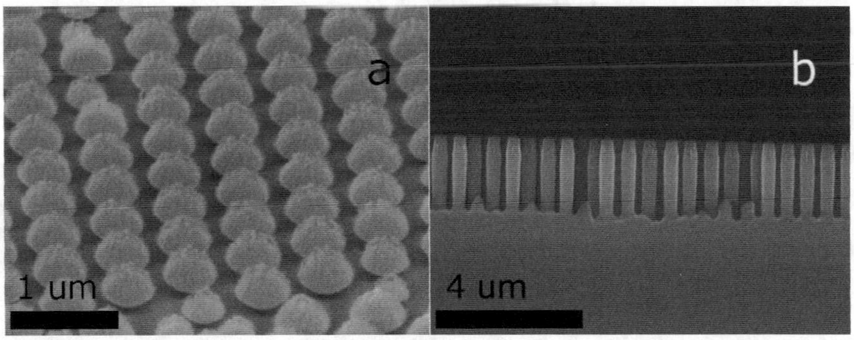

Figure 2. SEM image of the surface of a silicon wafer covered with a mask of latex spheres etched in oxygen plasma (a) and after cryogenic etching under such a mask (b).

The cross-sectional SEM image (Fig.2b) demonstrates that etching with such a mask makes it possible to obtain vertically ordered structures from silicon fibers.[1]

2.2 Nanospheric lithography 0. 9 µm for silicon oxide

For the deposition of spheres, an initial 10% by mass solution of 0.9 µm spheres in water was taken. The solution requires preliminary preparation. First, the solution is placed in an ultrasonic bath to increase the homogeneity of the solution. Next, 500 μl of the solution is withdrawn using a mechanical pipette dispenser and placed in a centrifugal tube. Then the solution is centrifuged at a speed of 13,400 revolutions for 4 minutes. As a result, the latex spheres settle to the bottom of the tube and you can take out excess water using a dispenser. Next, isopropyl alcohol is added. By varying the ratio of alcohol / water, you can control the viscosity of the solution, surface tension and potential. The dose was selected in such a way that 1000 μL of the solution was applied to the substrate. After adding alcohol, the solution is additionally kept in an ultrasonic bath at 50 ° C. Then, the following parameters were used on the centrifuge to apply the resist: the sphere's deposition rate was 300 rpm; drying speed 900 rpm; low acceleration [1].

A four-inch Si substrate completely covered with a single monolayer of close-packed polystyrene spheres is shown in Figure 3. A further increase in the concentration of isopropyl leads to the formation of two or more monolayers. An image of the resulting structure can be seen in Figure 3 [1].

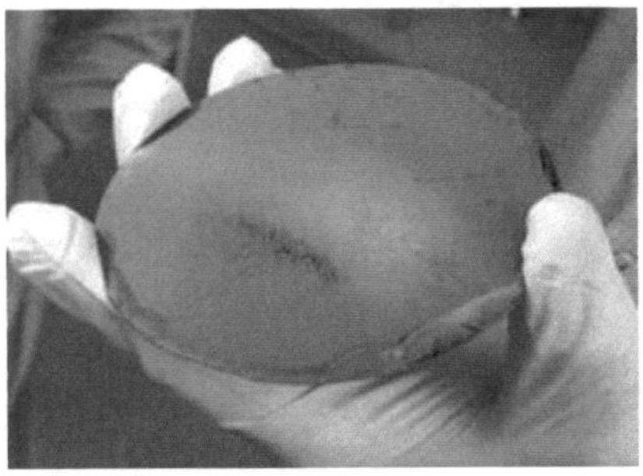

Figure 3. Si substrate 100 mm in diameter completely covered with one monolayer of spheres.

Using this optimal water / isopropyl solution allows us to obtain one monolayer in a tightly packed package for complete coverage of the substrate. In addition, this method makes it possible to regulate the surface concentration of the spheres by changing the concentration of isopropyl in the solution [1].

3. Study of SHG and THG signal

3.1. Samples fabrication

The SHG and THG has been studied in arrays of vertically oriented SiNWs encapsulated into a silicone membrane and peeled from the growth substrate. NWs were produced by plasma-chemical etching of bulk silicon substrates resulting in arrays of vertical-oriented nanostructures. Dense monolayers of SiO_2 spheres served as masks for microsphere lithography. The length of obtained NWs was about 6 μm, while diameter was about 1.5 μm. SEM image of structures is shown in the picture 4.

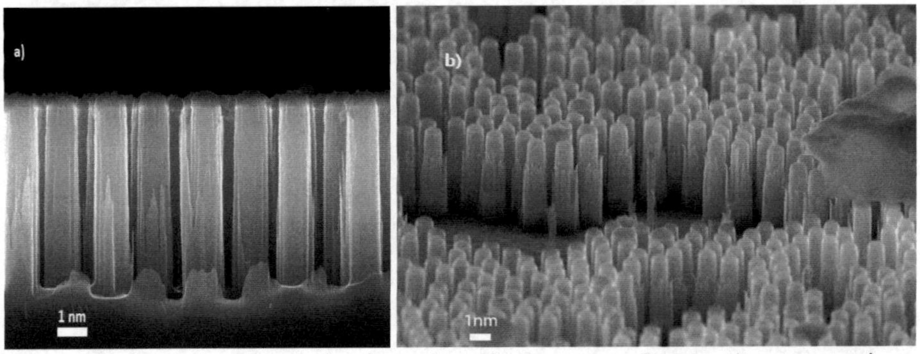

Figure 4. SEM image of the investigated structures. Image a) cross-section whiskers obtained by etching and b) an array of whiskers on a substrate at an angle of 45 degrees.

Further, the NW arrays were encapsulated into a polymer matrix and separated from the growth substrate. This technology enables not only reuse of growth substrates, but also combination of various material systems in order to create ultra-thin (up to several microns thickness) flexible transparent functional structures, and fabrication of large area (up to 50 sq. cm.) samples.The harmonic generation indicators are shown in the Figure 5.

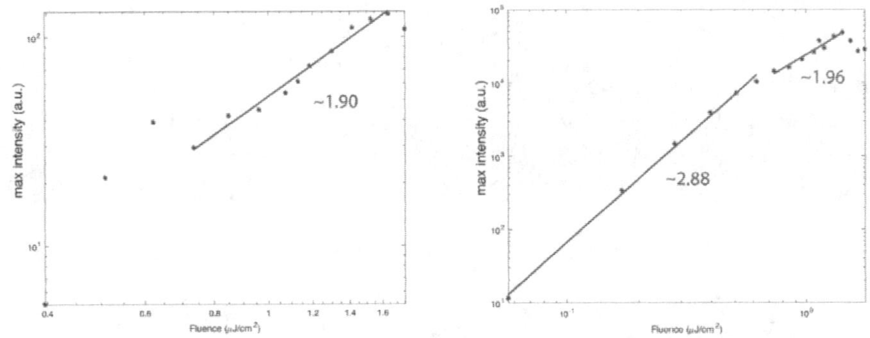

Figure 5. Images of the measurement results – a) generation of the second and b) third harmonics on the sample.

3.2. Measurements of SHG and THG

The SHG and THG signal was measured using a setup of IR femtosecond laser and spectrometer operating in the visible spectral range. The laser power before the objective was 30 mW, and 3 mW after the objective, the fluence was $12*10^{-3}$ mJ / cm^2. Under these excitation conditions the samples generated a THG signal visible by a naked eye in the range of 400-600 nm (1200-1800 nm of incident IR radiation, respectively). In Figure 6, you can see the graphs of the dependence of the wavelength on the intensity with the indicators of the third harmonic generation on the surface of the sample. Figure 7 shows the glow of points on the surface of the sample with different wavelengths.

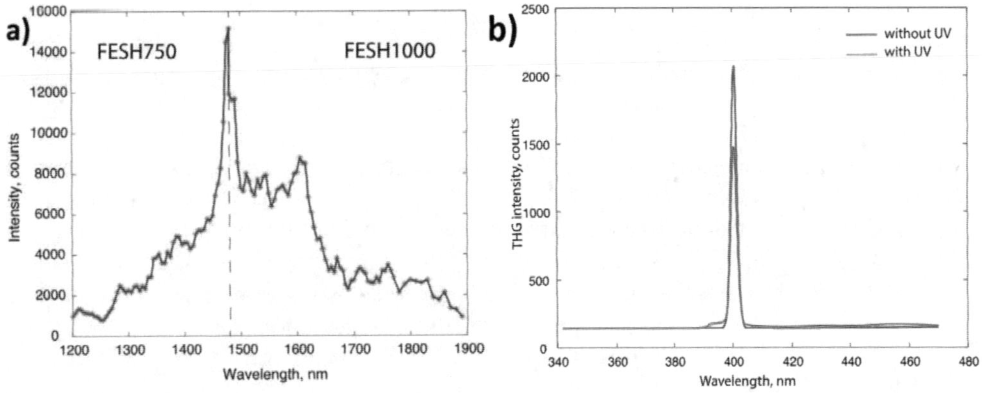

Figure 6. Graph with the third harmonic generation index of 490 nanometers a) without ultraviolet radiation, b) with ultraviolet light glow (third harmonic range from 400-660 nanometers).

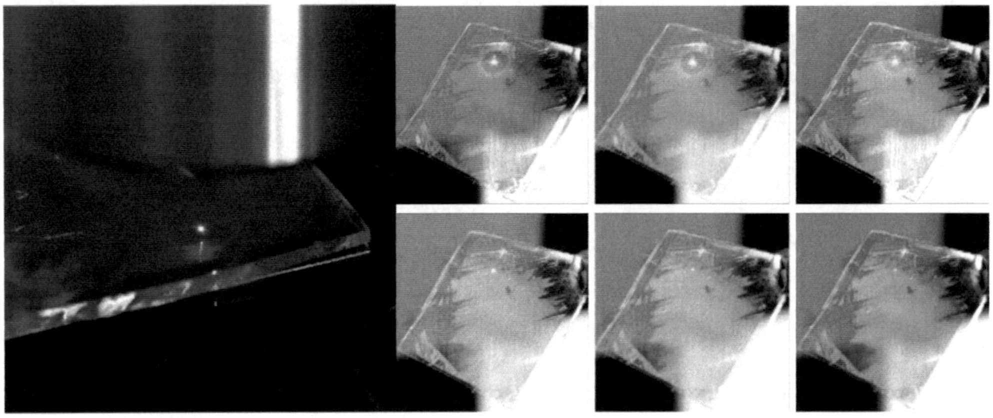

Figure 7. Images with luminescence on the cut membrane surface.

The measured spectra also registered a strong SHG response of two orders of magnitude weaker in comparison to the THG, this may be associated with less efficient processes on the NW surface and a relatively large ratio of the NW volume to its surface area.

4. Summary

In conclusion, we demonstrated efficient IR visualizers based on SiNW/silicone membranes having THG signal visible to a naked eye. The studied structures also demonstrated a strong SHG signal, which can be further enhanced by increasing the contribution of the NW surface (via NW diameter reducing) and surface treatment. This study contributes to the further development of the physics and technology of optical devices based on silicon

Acknowledgments

VN thanks the Russian Foundation for Basic Research (RFBR project № 19-32-60040) for PDMS/NW membrane fabrication.

References

[1] Morozov I 2020 Physico-technological foundations of microstructuring processes for the creation of vertically oriented photoconversion structures based on silicon: *Dis. Cand. Tech. Sciences* 161

[2] Cazzanelli M and Pavesi L 2012 Second-harmonic generation in silicon waveguides strained by silicon nitride *J Nat. Mater* **11** 148

[3] weblink:https://www.thorlabs.com/newgrouppage9.cfm?objectgroup_id=296, last accessed 16/02/2021

[4] weblink: https://www.alphalas.com/products/laser-diagnostic-tools/infrared-to-visible-converters- ir-laser-beam-visualizers-ir-detectors-ir-vis series.html?gclid=Cj0KCQiAgomBBhDXARIsAFNyUqPq76KpwEcZf99j8kRnwNFnr8VHRF3c-RL3YMIy_m36inblJXCKt0EaAsVpEALw_wcB, last accessed 16/02/2021

[5] Fedorov V and Mukhin I 2020 Gallium phosphide nanowires in a free-standing, flexible, and semitransparent membrane for large-scale infrared-to-visible light conversion *J ACS Nano* **14** 10624

[6] Castellan C and Pavesi L 2019 On the origin of second harmonic generation in silicon waveguides with silicon nitride cladding *J Sci. Rep.* **9** 1088

Laser-induced optical nonlinearity in a Li-rich glass

A Yu Moroz[1], E S Babich[2], V P Kaasik[3], A A Lipovskii[3], V G Melekhin[4], A V Redkov[5] and D K Tagantsev[3]

[1] Department of Physics and Technology of Nanostructures, Alferov University, St. Petersburg 194021, Russia
[2] Institute of Physics, Nanotechnology and Telecommunications, Peter the Great St. Petersburg Polytechnic University, St. Petersburg 195251, Russia
[3] Sector of Optics of Heterogeneous Structures and Optical Materials, Alferov University, St. Petersburg 194021, Russia
[4] A. F. Ioffe Physical-Technical Institute, Polytechnicheskaja 26, 194021, Saint-Petersburg, Russia
[5] Laboratory of Phase Transitions,Institute of Problems of Mechanical Engineering, St. Petersburg 199178, Russia

Abstract. The selective crystallization of a photosensibilized lithium silicate glass after a light emitting diode (280 nm) and a nanosecond laser (355 nm) UV irradiation and subsequent heat treatment was studied. Using a focused beam of the laser allows successive drawing regions in which the formation of silver nanoparticles after 500°C heat treatment of the glass co-doped with silver and cerium ions takes place. It is shown that this treatment is a necessary step for the growth of sodium metasilicate crystals under the second thermal treatment of the glass at 600 °C, while these regions demonstrate second optical harmonic generation of 1064 nm nanosecond laser radiation.

1. Introduction

Formation of a given pattern of the second order optical nonlinearity in transparent materials is of interest, first of all because of the possibility to provide quasi-phase-matched generation of optical harmonics in nonlinear media which, by different reason, are not capable of providing phase synchronism at required wavelength. An example of this is, e.g., periodically poled lithium niobate crystal[1]. Another problem relates to the formation of periodic second order nonlinearity in initially centrosymmetricalglasses which does not possess the second-order nonlinearity. To solve this, optical poling [2] and periodic thermo-electric poling [3] of glasseswere proposed. The latter technique requires the formation of periodic anodic electrode to be used for poling at elevated temperatures. That is why there is an interest to search for other (simpler)techniques which could provide prescribed pattern of the second order optical nonlinearity in glassy media.For this, a possibility of optical recording such patterns could look attractive. The approach we consider in this paper supposes using photosensitive glasses, which optically initiated nanocrystallization, contrary to spontaneous formation of crystallites in glasses under thermal treatment for, e.g., nonlinear optical absorbers [4], could provide formation on optically nonlinear glass-ceramics pattern. In the experiments we used a foturan[5]like glass. Foturanlithium-silicate glasses were designed for the formation of relief structures by means of hydrofluoric acid etching of UV-irradiated regions of these glasses after thermal processing [6].These glasses are photosensitive due to their doping with Ag^+ and Ce^{3+} ions. Under UV irradiation the reaction Ce^{3+}+ hν +Ag^+→Ce^{4+}+ Ag^0 takes place, subsequent and thermal processing of the glass first results in silver atoms clustering and then in the glass crystallization,the silver clusters behaving as the crystallization nuclei. The type of crystals formed in a glass depends on the glass composition and structure in accordance with general rules [7]. In the foturan glass, the growth of lithium metasilicate microcrystals (LMS), Li_2SiO_3, takes place. Being non-centrosymmetrical, LMS crystallites should possess quadratic optical nonlinearity.Here we present the studies of the crystallization of a synthesized foturan-like glass and show that the local laser irradiation of the glass with UV nanosecond laser allows "drawing" optically non-linear glass regions.

2. Experiments and results

In the experiments, we used synthesized glass which composition is presented in Table 1. 1 mm thick polished plates of the glass were irradiated either with light emitting diode (LED) at 280 nm wavelength or the 3rd harmonic of aLitron laser (6 ns pulses, maximal pulse energy 14 mJ at 1064 nm and 0.7mJ at 355 nm wavelength). Both LED and laser exposing resulted in similar transformation of optical absorption spectra. The irradiated samples were annealed at the temperature of 500°C for 1 h and then at 600 °C for 10 min - 1 h. The influence of the first thermal treatment on optical absorption spectrum of the glass plates is illustrated with figure 1b.

Table 1. Composition of the glass in weight % of oxides.

Component	SiO_2	Li_2O	Al_2O_3	K_2O	Na_2O	ZnO	Ag_2O	CeO_2	Sb_2O_3
Weight %	75.55	10.44	5.33	5.74	1.17	1.13	0.236	0.085	0.33

We also locally exposed the glass with aunfocused to 120 μm spot laser beam at 355 nm wavelengthand measured optical absorption of the irradiated spots after the first thermal treatment (figure 2a). Prepared samples were characterized with optical microscopy, microRaman spectroscopy(figure 2b) andX-ray diffractometry (figure 3). Finally, we checked the presence of 532 nm radiation (the second harmonic generation (SHG) of the nanosecond laser radiation at the wavelength of 1064 nm) in a light beam transmitted through the sample at normal incidence and measured the dependence of the second harmonic intensity in the transmitted lighton the incidence angle of the fundamental beam- figure 4.

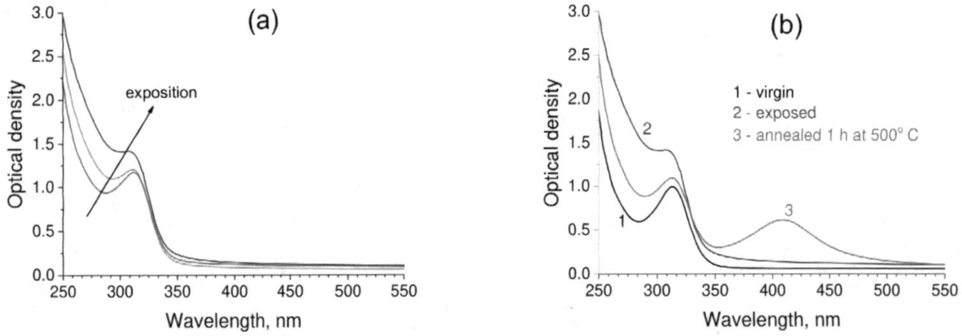

Figure 1. (a) Transformation of optical absorption spectra under successive UV expositionand (b) after UV exposition (curve 2) 1 h anneal of the irradiated glass sampleat 500° C.

3. Discussion

The appearance of the left wing in the cerium optical absorption peak at 280 nm after the irradiation and its growth with the increase of the exposition (figure 1a) indicates the formation of neutral silver atoms in the glass via taking electrons from Ce^{3+} ions by Ag^{+} ions. This process saturates because of limited concentrations of these entities in the glass. First annealing at 500° C results in clustering of silver atoms and in subsequent formation of silver nanoparticles in the subsurface layer of the glass, which thickness corresponds to the penetration depth of exposing UV irradiation. The formation of silver nanoparticles is reflected in appearance of optical absorption peak at 415 nm, which coincides with the wavelength of surface plasmon resonance in silver nanoparticles[8]. This absorption also resulted in yellowish coloration of the specimens. The amplitude of the absorption peak, and, respectively, the number of nanoparticles depends on the UV exposition. We verified this by locally irradiating the specimenswith UV laser beam unfocused to ~ 120μm spot at different laser intensityand annealing them at 500° C for 1 h. Optical microscopy showed that the size of the colored spot formed after this treatment corresponded to the size of the irradiation spot. Measurements of local absorption spectra shown in figure 2a, demonstrate the dependence of the optical density of the irradiated regions after such treatment on the exposition.

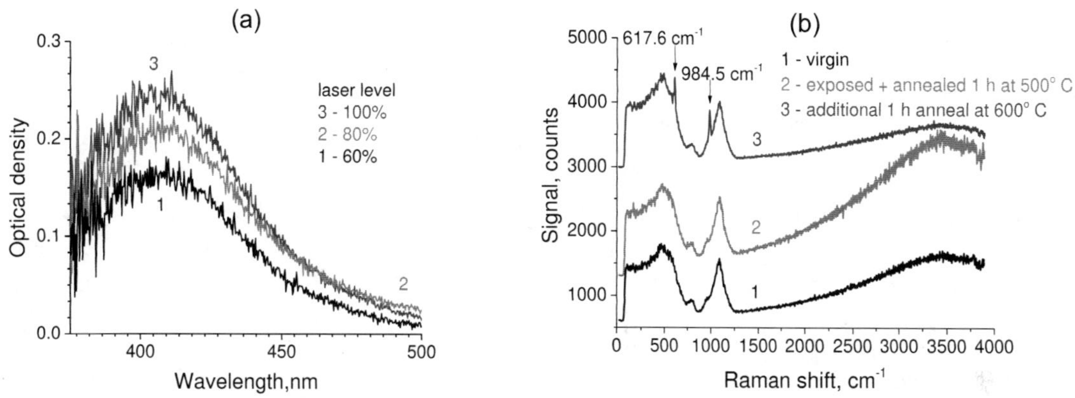

Figure 2. (a) Influence of the UVexposition on optical absorption spectra after the firstthermal treatment and (b) Raman spectra of virgin glass and UV exposed glass after the first and the second thermal treatment.

In the Raman spectra of the UV-irradiated and annealed samples(figure 2b) one can see the increase of the signal in the range 2700-3700 cm^{-1}, which can be attributed to the luminescence of atomic silver clusters, Ag_n, n=2-4,[9,10] and a small peak at 962 cm^{-1}, possibly indicating the glass pre-crystallization. Probably, silver atomic clusters formed in parallel with the formation of silver nanoparticles during the thermal treatment at 500° C, for the luminescence of a virgin glass sample was essentially weaker. The second annealing has resulted in the appearance of narrow lines in the Raman spectrum of the corresponding specimen (indicated with arrows in figure 2b), which indicates the formation of a crystalline phase in the glass. Positions of these lines coincide with ones of orthorhombic lithium metasilicate, Li_2SiO_3[11]. However, due to relatively “soft” second thermal treatment of the glass, which wasused to avoid strong scattering of light in the SHG experiments, only two lines were registered with assurance, and we used X-ray diffractometry to reliably identify the crystalline phase. The diffractometry returned a set of lines perfectly corresponding to orthorhombiclithium metasilicate (see figure 3).

Increasing the duration of the second heat treatment led to the growth of the diffraction peaks as shown in figure 3. The size of the crystalline grainsevaluated in accordance with the diffraction peaks width is about 30-35 nm. Besides, detailed studies of the X-ray diffraction spectra allowed identifying diffraction peaks corresponding to both Li_2SiO_3 and crystalline silver (see inset in figure 3). The latter confirmed the formation of silver nanoparticles after the first thermal processing of the UV-exposed glass and preservation of the nanoparticles under the second thermal treatment. This corresponds to the situation when the nanoparticles behave as nuclei in the glass crystallization.

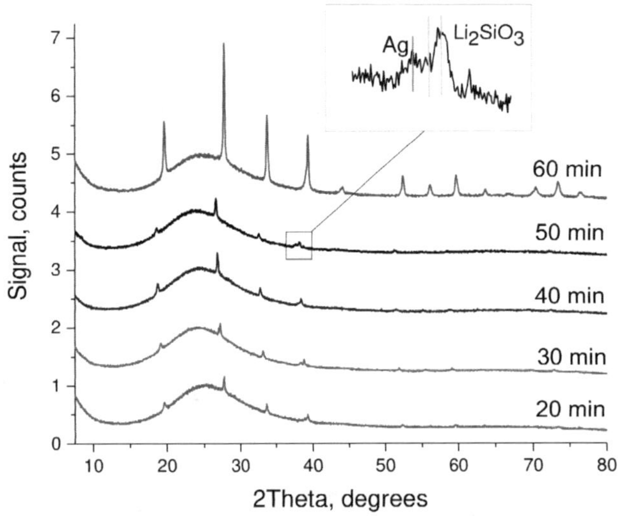

Figure 3. X-ray diffractometry of the light emitting diode exposed specimens after different durations of the second heat treatment (indicated near the curves). Inset: Lithium metasilicate and crystalline silver diffraction peaks at 38.2°.

We were sure that the low symmetry of the grown nanocrystals of orthorhombic lithium metasilicate allows SHG, and chose a specimenthat was 40-min secondary heat-treated to characterize SHG.This choice was based on two criteria: low light scattering andhigh crystalline phase content.The appearance of the second optical harmonic radiation in a light beam passed through the sample confirmed the ability of lithium metasilicate nanocrystals to generate the second harmonic.Figure 4 shows a monotonic decrease in the SHG intensity with an increase in the angle of incidence of the fundamental beam on the sample surface. This allows concluding that there is no oriented crystallization of Li_2SiO_3during the second heat treatment and, as a consequence, powder-like SHG by the glass matrix – nanocrystal system. The observed angular dependence of the SHG intensity is due to an increase in the reflection of waves of both the fundamental and second harmonics with an increase in the angle of incidence of light.

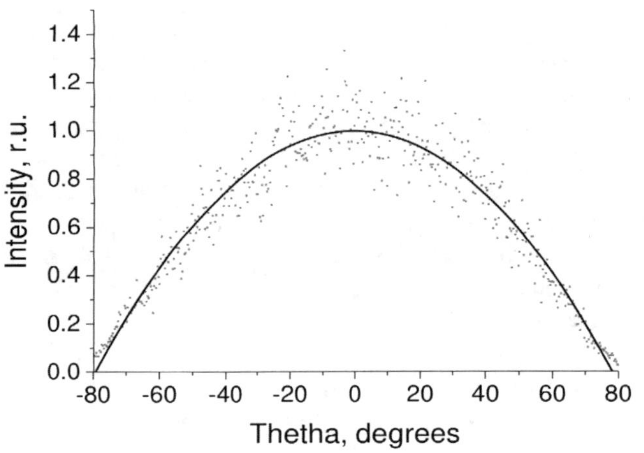

Figure 4. Intensity of 532 nm radiation (second harmonic of 1064 nm laser) in the transmitted beam generated by the UV-exposed and twice thermally processed glass specimen vs the angle of incidence of the fundamental wave.

4. Conclusions

Finally, to form a two-dimensional distribution of optical nonlinearity in the synthesized glass, we used a procedure similar to the traditional one for foturan glasses to form a 2D patternresistantto chemicaletching.Silver nanoparticles formed in the near-surface region of UV-irradiated and heat-treated glass specimens behaved as nuclei of orthorhombic Li_2SiO_3 nanocrystals that grew to 35 nm in size during the second heat treatment.Exposure of the glass to a focused UV laser beam ensured local reduction of silver and,accordingly, 2D "drawing" nanocrystals-containing glass regions. The low symmetry of these nanocrystals provided their second order optical nonlinearity which allowed the second optical harmonic generationby the regions of the glass which contained the nanocrystals. Angular dependence of the second harmonic signal evidenced a random distribution of nanocrystals in the glass subjected to UV irradiation.

Acknowledgments

The study was supported by the Ministry of Science and Higher Education of Russian Federation, project FSRM-2020-001. E.B. thanks the Council for Grants of the President of the Russian Federation (project SP-1491.2021.4).

References

[1] Myers L E and Bosenberg W R 1997 Periodically poled lithium niobate and quasi-phase-matched optical parametric oscillators *IEEE J. Quantum Electron.* **33** 1663–72

[2] Antonyuk B P, Novikova N N, Didenko N V and Aktsipetrov O A 2001 All optical poling and second harmonic generation in glasses: theory and experiment *Phys. Lett. A* **287** 161–8

[3] Li G, Winick K A, Said A A, Dugan M and Bado P 2009 Quasi-phase matched second-harmonic generation through thermal poling in femtosecond laser-written glass waveguides *Opt. Express* **17** 9442

[4] Malyarevich A M, Prokoshin P V, Demchuk M I, Yumashev K V and Lipovskii A A 2001 Passively Q-switched Ho3+:Y3Al5O12 laser using a PbSe-doped glass *Appl. Phys. Lett.* **78** 572

[5] Stookey S D 1949 Photosensitive Glass *Ind. Eng. Chem.* **41** 856–61

[6] Stookey S D 1953 Chemical Machining of Photosensitive Glass *Ind. Eng. Chem.* **45** 115–8

[7] Lipovskii A A, Tagantsev D K, Tatarintsev B V and Vetrov A A 2003 The origin of electrooptical sensitivity of glassy materials: Crystal motifs in glasses *J. Non. Cryst. Solids* **318** 268–83

[8] Zhao J, Lin J, Zhang W, Wei H and Chen Y 2015 SERS-active Ag nanoparticles embedded in glass prepared by a two-step electric field-assisted diffusion *Opt. Mater. (Amst).* **39** 97–102

[9] Eichelbaum M, Rademann K, Hoell A, Tatchev D M, Weigel W, Stößer R and Pacchioni G 2008 Photoluminescence of atomic gold and silver particles in soda-lime silicate glasses *Nanotechnology* **19** 135701

[10] Dubrovin V D, Ignatiev A I, Nikonorov N V, Sidorov A I, Shakhverdov T A and Agafonova D S 2014 Luminescence of silver molecular clusters in photo-thermo-refractive glasses *Opt. Mater. (Amst).* **36** 753–9

[11] Brawer S A and White W B 1975 Raman spectroscopic investigation of the structure of silicate glasses. I. The binary alkali silicates *J. Chem. Phys.* **63** 2421–32

Investigation of the features of the porous morphology of anodic alumina films at the initial stage of anodization

K V Chernyakova[1], E N Muratova[2], I A Vrublevsky[3] and N V Lushpa[3]

[1] State Scientific Institute Center for Physical Sciences and Technologies, Savanoriu Ave. 231, Vilnius, LT-02300, Lithuania
[2] Saint-Petersburg state electrotechnical university «LETI», 5 Prof. Popova Street, Saint Petersburg, 197376, Russia
[3] Belarusian State University of Informatics and Radioelectronics, 6 P. Brovki Street, Minsk, 220013, Belarus

Abstract. The work is devoted to the study of the porous structure formation of anodic alumina films at the initial stage of aluminium anodizing. SEM images of the surface morphology of the oxalic acid anodic films were analyzed. It was shown that at the initial stage, both major and minor pores are formed, the diameter ratio of which is about 1.16 and does not depend on the anodizing voltage. The results obtained indicate that the minor pores in the anodic films are located inside hexagonal cells composed of the major pores.

1. Introduction

One of the effective methods of forming a material with a controlled porous structure is the process of anodic oxidation of aluminum [1, 2]. Figure 1 shows a laboratory complex for the implementation of electrochemical anodizing processes. Films of porous anodic alumina (PAA) obtained as a result of this process are characterized by a regular porous structure with pore sizes proportional to the anodizing voltage.

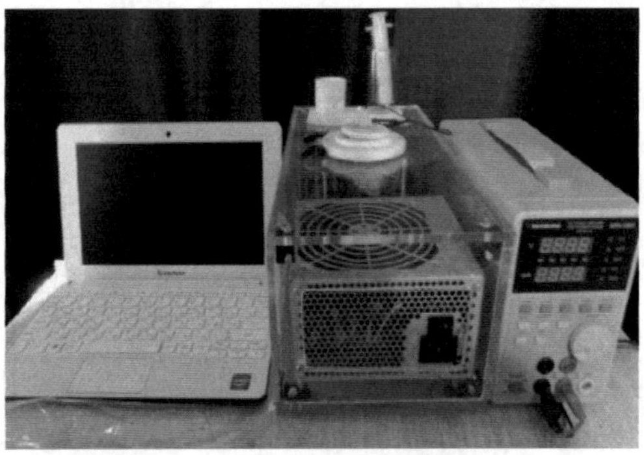

Figure 1. Porous anodic aluminum.

The structure of PAA can be represented as a hexagonal close packing of cylindrical pores perpendicular to the film surface (Figure 2). For the widespread use of PAA films as nanostructures in optics, sensors, and nanotechnologies [3], an important condition is the production of anodic films with high ordering and a given pore diameter. The key role in this is played by the initial stage of the formation of the porous structure of anodic films during anodizing of aluminum, which is poorly studied.

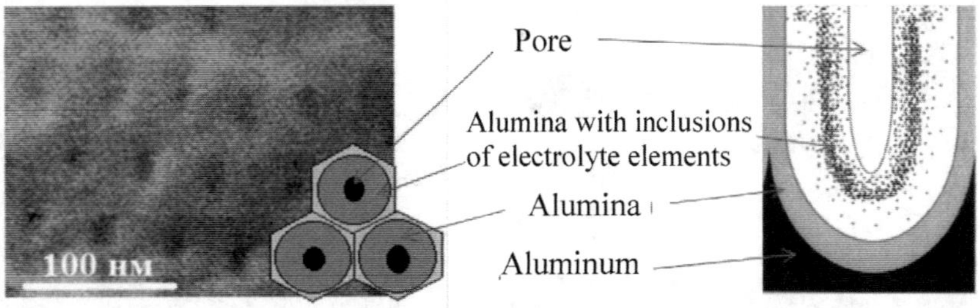

Figure 2. General view of the structure of the porous layer of aluminum oxide: SEM image and schematic representation of the pore profile [4].

At present, the following mechanism has been adopted to explain the pore nucleation process. In the first stage of anodizing (Figure 3), the aluminum surface is covered with a layer of barrier aluminum oxide. In this case, in the depressions and defects of the oxide film, the electric field strength increases sharply. This leads to local dissolution of the oxide layer due to temperature rise and dissolution under the action of an electric field. Due to the competition of neighboring points for charge drain, some of the pores stop growing and, as a result, a uniform growth of the PAA film begins with a constant distance between the pore centers. At the same time, it can be assumed that the development of pores in the initial region of disordered growth has its own characteristics.

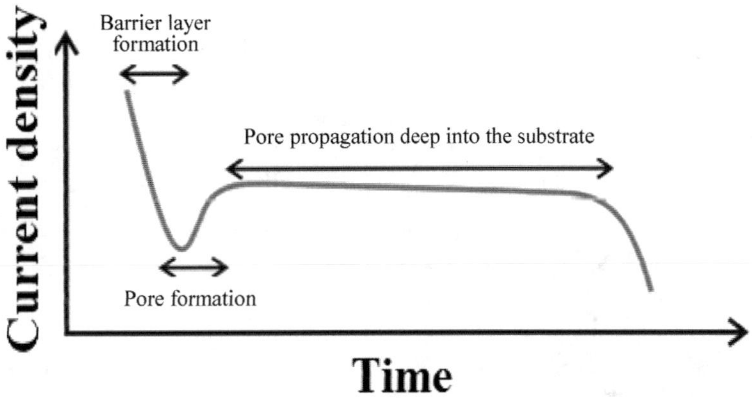

Figure 3. Kinetic dependence of the pore formation process.

The aim of this work was to study the features of the formation of the porous morphology of PAA films at the initial stage of aluminum anodization.

2. Results

Figure 4 shows an appearance of the thin film of PAA and AFM study of anodic alumina membrane synthesized in electrolyte solution based on oxalic acid.

a b

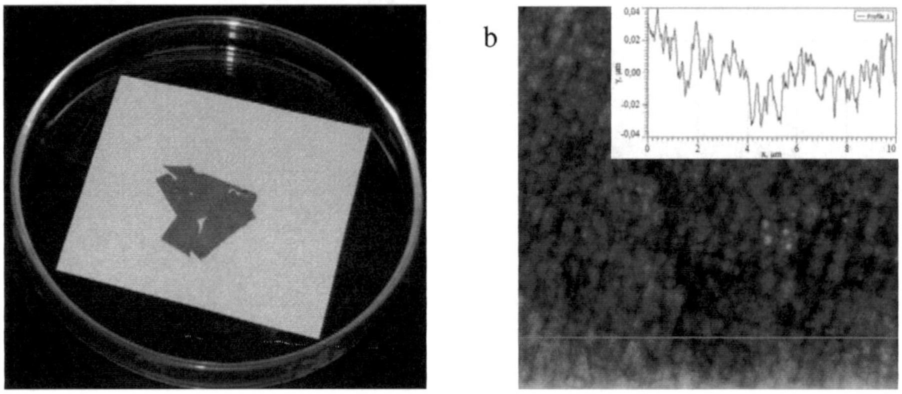

Figure 4. Photo (a) and AFM image (a) of the thin film of PAA formed in a solution of oxalic acid (the scanning area size is 10 x 10 μm, with profile line).

Figure 5 (a) shows SEM image of the surface morphology of a porous alumina film on an aluminum substrate obtained in a potentiostatic mode at 40 V in oxalic acid. To determine the pore diameter, computer processing of the surface morphology of the anodic film was carried out using the ImageJ program. Then, according to the processing results, a graph of the pore diameter distribution was plotted (Fig. 5b). As can be clearly seen from the graph, the pore diameter distribution has two distinct peaks at 22.5 ± 0.2 and 26.4 ± 0.2 nm. The major pores have a larger diameter of 26.4 nm, and the minor pores in the PAA film have a diameter of 22.5 nm. This indicates that at the initial stage of aluminum anodization, the simultaneous development of both major pores and minor pores with similar diameters is observed. The results obtained indicate that the ratio of the diameters of such pores in the anodic films does not depend on the anodizing voltage and is equal to approximately 1.16. It should be noted that the number of minor pores is of the same order as the major pores. Therefore, it can be argued with high probability that the minor pores do not have a chaotic distribution in the anodic film, but are located inside hexagonal cells composed of the major pores. At the same time, at the initial stage of anodizing, the shape of such hexagonal cells is strongly distorted.

a b

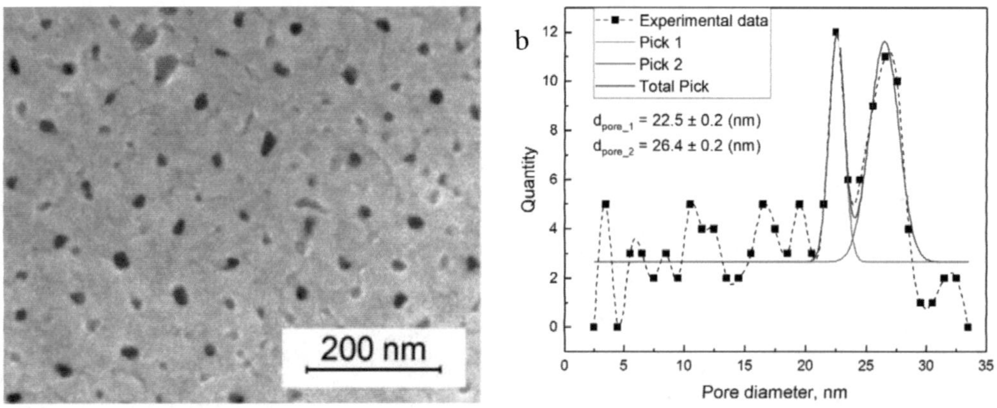

Figure 5. SEM image of the surface morphology (a) and pore diameter distribution (b) for a thin film of porous PAA formed on aluminum in oxalic acid at 40 V for 390 s

The results obtained indicating the presence of major pores and minor pores with similar diameters and the same density in the anodic film are in good agreement with the data presented in [5] for the pore size distribution determined by the method of capillary condensation of isopropanol vapors.

3. Conclusion

It was found that at the initial stage of aluminum anodization, the porous morphology of the anodic films is characterized by the simultaneous development of both major and minor pores of a certain diameter. The ratio of the sizes of the major and minor pores is about 1.16 and does not depend on the anodizing voltage in oxalic acid. The results obtained indicate that the minor pores do not have a chaotic distribution in the anodic film, but are located inside hexagonal cells composed of the major pores.

Acknowledgments

This work was supported by the grant of the President of the Russian Federation: MK-2268.2020.8: agreement No. 075-15-2020-520 dated 04/13/2020 and BRFFR project No. F19VTNG-001.

References

[1] Muratova E N, Spivak Yu M, Moshnikov V A, Petrov D V, Shemukhin A A and Shimanova V V 2013 *Glass Phys. and Chem.* **39** (3) 320–8
[2] Lushpa N V, Dinh Huu Tai, Chernyakova K V, Vrublevsky I A, Muratova E N, Spivak Yu M and Moshnikov V A 2020 Using the method of SEM image processing to study the morphology of porous anodic alumina films *New Materials: Preparation, Properties and Applications in the Aspect of Nanotechnology*, ed. by A G Syrkov and K L Levine (NY: Nova Science Publishers) chapter 13 pp 125–32.
[3] Moshnikov V A, Muratova E N and Spivak Y M 2017 *Altern. Ener. and Ecol.* **42** (34) 21817–21
[4] Muratova E N, et.al. 2017 *Glass Phys. and Chem.* **43** 163–9
[5] Casanova F, et.al. 2008 *Nanotech.* **19** (31) 315709

Nanoobject mass measurement using the node displacement of the second mode of the nanomechanical resonator

N A Solomonov[1], K N Novikova[1], I V Nadoyan[1], A M Mozharov[1], V A Shkoldin[1,2], Y S Berdnikov[3] and I S Mukhin[1,2]

[1] Alferov University, Khlopina 8/3, 194021, St. Petersburg, Russia
[2] ITMO University, Kronverkskii 49, 197101, St. Petersburg, Russia
[3] St. Petersburg State University, Universitetskaya Emb. 13B, 199034, St. Petersburg, Russia

imukhin@spbau.ru, nadoyan@spbau.ru

Abstract. This work suggests a new approach to weighting the nanoscale objects placed at the tip of cantilever vibrating inside the camera of scanning electron microscope. In contrast to traditional approach to mass determination, we suggest tracing the shift of the node of the second vibration mode as an alternative to frequency shift measurement. We demonstrate the applicability of our approach to carbon nanowhisker cantilevers grown on tungsten needles by focused electron beam induced deposition. We compare experimentally the performance of the suggested approach with the traditional frequency shift-based method.

1. Introduction

Nanomechanical resonators are used in design of new class of mass sensors with exceptional precision and fast response [1–3]. Focused electron beam induced deposition (FEBID) is fast versatile technique enabling fabrication of carbon nanowhisker (CNW) cantilevers for nanomass sensing inside a scanning electron microscope (SEM) [4,5]. Typically, the mass measurement mechanism implies determination of the resonant frequency shift induced by a nanoparticle placed on cantilever. This approach inevitably requires recording an amplitude frequency response (AFR) for a specific resonant mode. The amplitude is measured from SEM visualization of the cantilever vibration during frequency scanning. In this process requires long exposure of the CNW cantilever to electron beam and unavoidably leads to parasitic deposition, which limits the precision of the mass sensing [5,6]. This work suggests tracing the shift of the node as an alternative to frequency shift measurement. We demonstrate the applicability of the new approach and compare its performance with the AFR-based method.

2. Fabrication of the mass sensor

The resonant mass sensor consists of vibration generator and CNW cantilever grown on the tip of a tungsten needle. As a generator of vibrations, we used the tubular piezoelectric transducer with generator of signals of special form «АКИП - 3413/3». The sharpened tungsten needle was prepared using the electrochemical etching in a 5% KOH solution as schematically illustrated in Figure 1 (a). The optimal tip sharpness lies in the range of 300-800 nm. Figure 1 (b) show the schematic for FEBID growth of CNW in FEI Quanta Inspect SEM. In our experiments the FEBID process employs the hydrocarbon groups from the residual atmosphere of the SEM camera. Decomposition of hydrocarbon precursors is driven by secondary electrons in the area exposed to primary electron beam [7]. Therefore, the trajectory of the electron beam defines the shape of the deposited carbon nanostructure, thus allowing to control the CNW geometry. Figure 1 (c) sketches the mass sensor consisting of W-needle with the grown CNW placed in piezotube of the vibration generator.

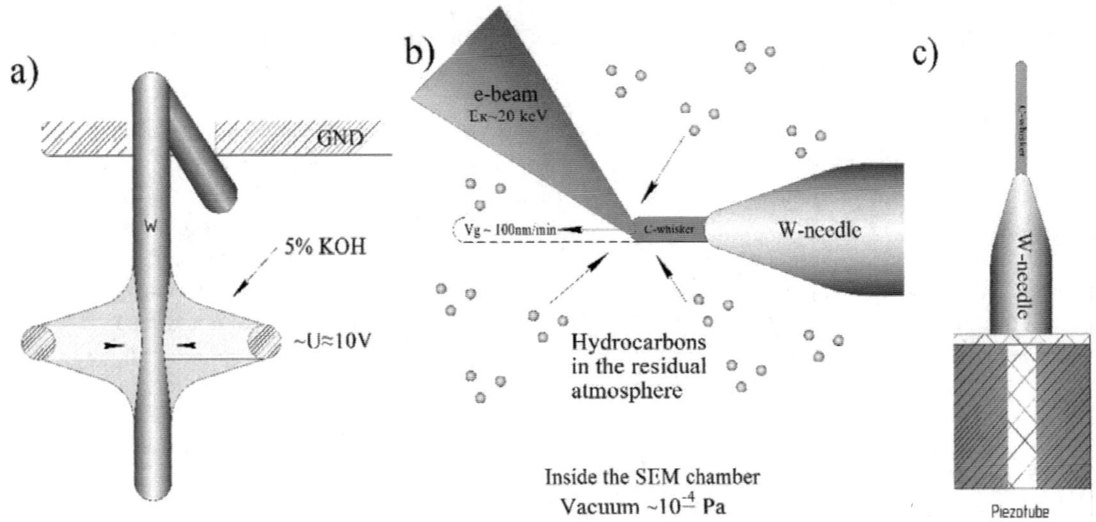

Figure 1. (a) Electrochemical etching of the W-needle; (b) FEBID process in the SEM camera; (c) Mass sensor geometry.

For obtaining the AFR of the CNW cantilever we used the most easy-to-implement method based on visualization of vibrations at the investigated frequency band by means of SEM. When the oscillator frequencies match the resonant frequencies of the nanoresonator, the image of whisker is blurred on the screen and takes the form of a fan or figure-of-eight for the first and second resonances, respectively, as sketched in Figure 2 (a) and (b).

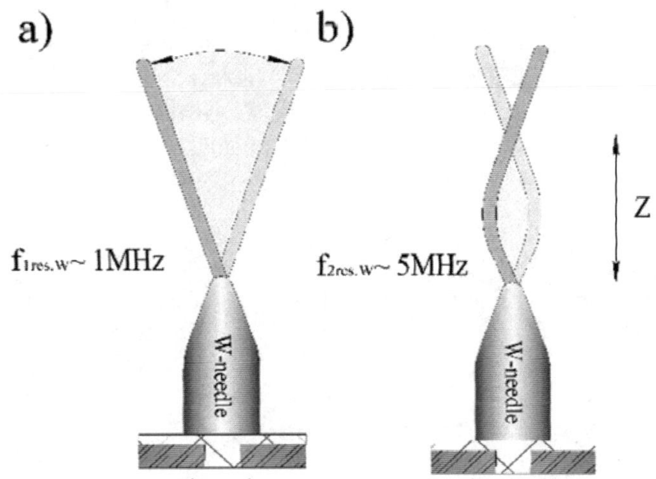

Figure 2. (a) Visualization of the first resonance; (b) Visualization of the second resonance.

Generally, the placement of a nanoobject at the tip of the whisker can be performed using the developed "pick-and-place" method, which consists of picking up a nanoparticle with the tip of a nanomanipulator located in the SEM chamber and transferring it to the target place. The particle can be securely fixed by carbon welding or special nanotraps [5]. However, in this work we consider the test structure grown at the tip of the CNW via the FEBID process to demonstrate the applicability of the suggested mass sensing mechanism.

3. Sensing mechanism

Figure 3 (a) illustrates the standard method of weighing nanoobjects using mechanical nanoresonators consists of determining the shift of the first resonance frequency when the mass under study is localized at the cantilever tip. The AFR of the system is studied by visualization of the vibration shape in a scanning electron microscope camera and requires a long exposure of the sample to the electron beam, which leads to an uncontrolled increase in the whisker mass and an undesirable shift of the resonance frequency. As a solution to this problem, we propose an alternative approach to measure the mass of a nano-object by tracing the displacement of the second resonance node as shown in Figure 3 (b). In this case, it is sufficient to find the resonance with a "package" of frequencies and measure the distance between the needle tip and the resonance node, which significantly reduces the mass measurement time. In addition, the node position depends only on the linear distribution of mass along the length of the whisker [3], and therefore is weakly sensitive to parasitic deposition homogeneous along the CNW length.

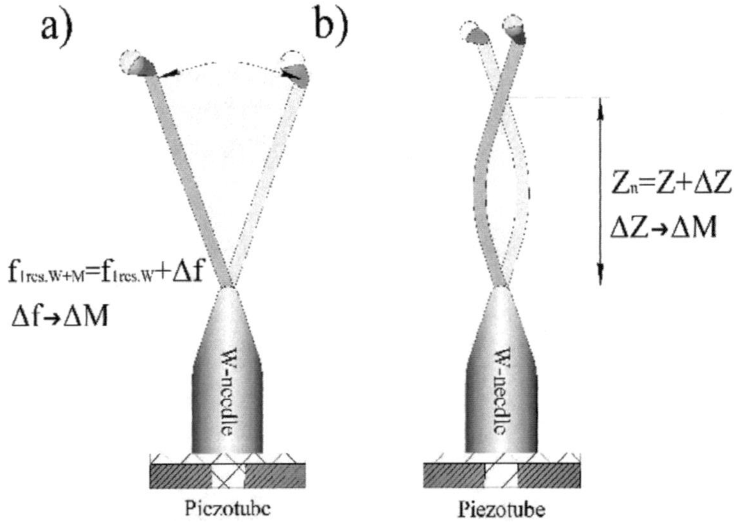

Figure 3. Methods of weighting of nanoobjects; a) by the shift of the first resonance frequency b) by displacement of the second resonance node.

4. Results and discussion

Amorphous CNW with the length of 6.84 μm and diameter of 120 nm was grown at the tip of a sharp tungsten needle with a sharpening radius of about 500 nm. The measured resonant frequencies of the first and second harmonics were 0.95 MHz and 4.5 MHz, respectively.

Figure 4 (a) shows the CNW with the test structure grown at the tip via the FEBID process. The additional mass at the whisker tip M_f can be calculated from the shift of the first resonance frequency to $f_{1res.W+M}$ as [5]:

$$M_f = \frac{3E\left(\frac{d}{2}\right)^4}{16\pi \cdot f_{1res.W+M} \cdot L^3} - 0{,}227 \cdot m_1, \tag{1}$$

where E is the Young's modulus, d is the whisker diameter, L is the whisker length, m_1 is the whisker mass. The frequency shift from 0.95 to 0.8 MHz, according to the eq. (1), corresponds to the additional mass of $5.2 \cdot 10^{-17}$ kg. Note that this value is close to the value of $2.5 \cdot 10^{-17}$ kg, calculated using the density of FEBID grown carbon structures of 2200 kg/m^3 [5] and the measured in SEM sizes of the grown cylindric mass (118 nm in diameter and 650 nm in length). The difference in the results may be due to the fact that the studied object was not a point mass.

The node shift after adding the point mass at the CNW tip can be estimated as [8]

$$M_z \approx \frac{m_1 \cdot \Delta Z}{z_1}, \tag{2}$$

where m_1 is the mass of the whisker, z_1 is the position of the tug without mass, ΔZ is the node shift when the mass is added. The shift of the second resonance node position of ~300 nm was experimentally measured (Figure 4 b, c), which, according to eq. (2), corresponds to a localized mass of $1.04 \cdot 10^{-17}$ kg. Thus, the mass measured from node shift gives the value close to the mass values calculated from the frequency shift and the nanoobject size.

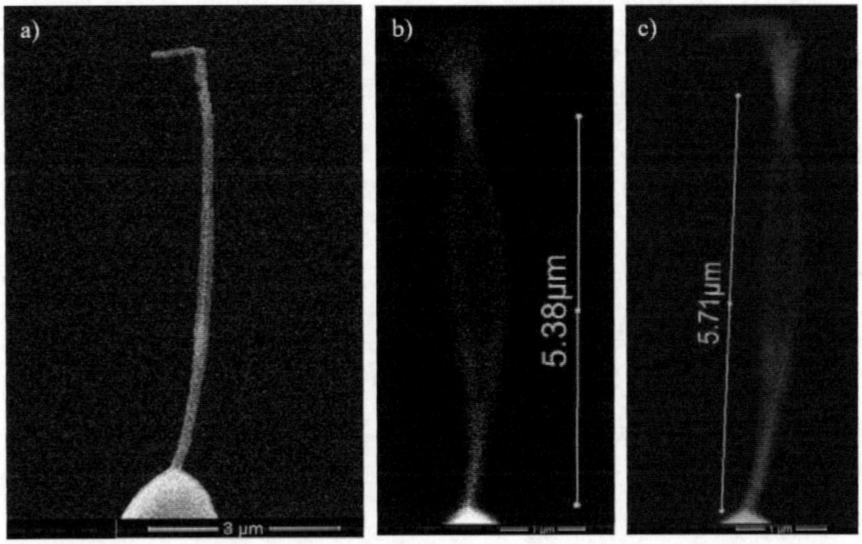

Figure 4. (a) CNW with the added mass at the tip; CNW vibrations at the second resonance before (b) and after (c) additional mass deposition.

5. Conclusions

This study presents a new method of mass measurement based on tracing the node shift at the second resonance of CNW vibrations. This method does not require a long exposure in SEM and excludes errors in determining the whisker geometry size and the influence of uncontrolled frequency drift on the weighing accuracy. We performed weighing of a test carbon mass localized at the tip of the whisker. We observe the first resonance frequency shift of ~150 kHz which corresponds to a localized mass of $5.2 \cdot 10^{-17}$ kg. For the same system experimentally obtained ~300 nm shift of the node position at second resonance, which corresponds to a localized mass of $1.04 \cdot 10^{-17}$ kg. Mass of investigated object determined by geometry is equal to $2.5 \cdot 10^{-17}$ kg. Thus, the suggested method has the accuracy in mass sensing similar to traditional approach but allowing much shorter exposure of the weighting object to the electron beam.

Acknowledgments

The work was financially supported by the Ministry of Science and Higher Education of the Russian Federation (FSRM-2020-0005). YB acknowledges St. Petersburg State University for the research grant 75746688.

References

[1] Yamaguchi H 2017 GaAs-based micro/nanomechanical resonators *Semicond Sci. Technol.* **32** 103003

[2] Chaste J, Eichler A, Moser J, Ceballos G, Rurali R and Bachtold A 2012 A nanomechanical mass sensor with yoctogram resolution *Nat. Nanotechnol.* **7** 301–4

[3] Malvar O, Ruz J J, Kosaka P M, Domínguez C M, Gil-Santos E, Calleja M and Tamayo J 2016 Mass and stiffness spectrometry of nanoparticles and whole intact bacteria by multimode nanomechanical resonators *Nat. Commun.* **7** 1–8

[4] Gruber G, Urgell C, Tavernarakis A, Stavrinadis A, Tepsic S, Magén C, Sangiao S, De Teresa J M, Verlot P and Bachtold A 2019 Mass Sensing for the Advanced Fabrication of Nanomechanical Resonators *Nano Lett.* **19** 6987–92

[5] Lukashenko S Y, Mukhin I S, Komissarenko F E, Gorbenko O M, Sapozhnikov I D, Felshtyn M L, Uskov A V. and Golubok A O 2018 Resonant Mass Detector Based on Carbon Nanowhiskers with Traps for Nanoobjects Weighing *Phys. Status Solidi Appl. Mater. Sci.* **215** 1–5

[6] Lukashenko S Y, Mukhin I S, Veniaminov A V., Sapozhnikov I D, Lysak V V and Golubok A O 2016 Q-factor study of nanomechanical system "metal tip – carbon nanowhisker" at low and ambient pressure *Phys. Status Solidi Appl. Mater. Sci.* **213** 2375–79

[7] Mukhin I S, Fadeev I V, Zhukov M V, Dubrovskii V G and Golubok A O 2015 Framed carbon nanostructures: Synthesis and applications in functional SPM tips *Ultramicroscopy* **148** 151–7

[8] Tymoshenko S P, Yang D Kh and Weaver W 1985 Fluctuations in engineering *Mashinostroenie* **472**

Investigation of the effect of annealing on Si(100) substrate modified by Ga^+ focused ion beam

M M Eremenko[1], N A Shandyba[1], N E Chernenko[1], S V Balakirev[1], L S Nikitina[1], M S Solodovnik[1] and O A Ageev[1,2]

[1] Institute of Nanotechnologies, Electronics and Equipment Engineering, Southern Federal University, Taganrog 347922, Russia
[2] Research and Education Center "Nanotechnologies", Southern Federal University, Taganrog 347922, Russia

larnikitina@sfedu.ru

Abstract. In this work, we studied the effect of annealing the silicon surface on the morphology of focused ion beam modified areas. It was found that an increase in the ion beam accelerating voltage during surface treatment significantly affects the morphology and the appearance of the implanted material on the surface or its absence/evaporation during annealing. It is shown that an increase in number of ion beam passes leads to the formation of holes on the surface of the modified areas, which is a sign that significant damage to the substrate material has occurred.

1. Introduction
Today, silicon technology is the leader in the global micro- and nanoelectronics market. The use of silicon substrates makes it possible not only to obtain structures with a high performance, but also to reduce the cost of manufacturing devices based on them. However, obtaining high-quality emitting devices based on silicon still remains an unsolved problem due to the indirect gap of the semiconductor [1-4]. Therefore, recently, interest has increased in possible solutions to this problem, namely, monolithic integration of III-V on Si, which would allow combining the advantage of both technologies - high mobility of charge carriers (data reception and transmission) and high-performance data processing [3, 5-8].

This hybrid solution is expected to increase energy efficiency, package density, and reduce the final cost of the devices [6, 8]. However, achieving the desired low level of defectiveness and, as a consequence, high-performance structures remains a challenge. To date, a large number of methods are used to reduce the number of defects during monolithic integration of III-V on Si [1-4, 9-12], but studies of the initial stage are still crucial for understanding the formation of subsequent grown layers, and will also make it possible to obtain modes with the lowest density defects, most of which are assumed to be localized in the lower layers.

2. Experimental procedure
In this work, we investigated the effect of initial annealing on modified silicon areas. Silicon substrates were modified with a focused ion beam (FIB) (Ga^+) at various accelerating voltages and beam passages. The accelerating voltages varied from 5 kV to 30 kV, the number of passes was 5 and 200. Annealing

was carried out at temperatures of 600 and 800°C for 30 minutes with and without an additional arsenic flux.

3. Results and discussion

Analysis of the results showed that a decrease in the accelerating voltage from 30kV to 5kV (Figure 1) led to the disappearance of Ga droplets in almost all ranges of the considered modifications and temperatures, except for 200 beam passes at a temperature of 800°C (Figure 1b). We believe that this is due to the intense evaporation of the material from the surface, which does not lie deeply, and this process is more advantageous when modifying the surface with beams with a low accelerating voltage and annealing at high temperatures. On the other hand, with an increase in the accelerating voltage on the surface of the modified areas, the implanted material emerges on the surface already from 5 beam passes. We explain its presence by the fact that at a high accelerating voltage, the depth of the material is high and the material gradually comes out to the surface and does not have time to completely desorb from the surface during annealing. However, it should be noted that during processing with an accelerating voltage of 30kV and 200 beam passes (Figure 1d), holes were left on the surface of the modified areas, without the presence of droplets of implanted Ga. The presence of such holes is apparently caused by a large deformation of the Si surface during processing by focused ion beams.

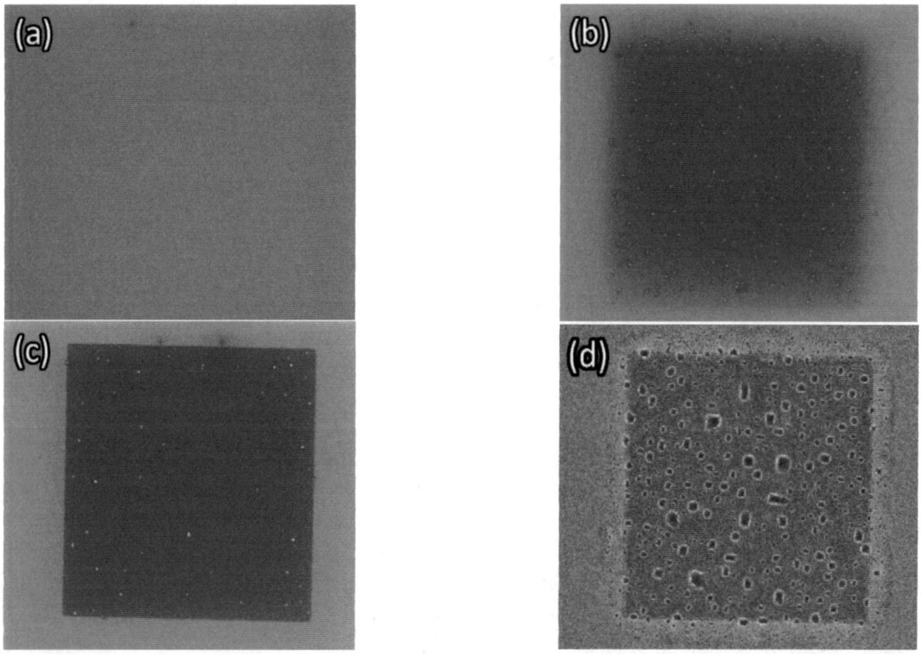

Figure 1. SEM images of modified Si areas after annealing at $T = 800^{o}$C and various accelerating voltages and beam passes, respectively: (a) 5kV and 5, (b) 5kV and 200, (c) 30kV and 5, (d) 30kV and 200.

However, a study of the effect of annealing temperature showed that a decrease in the annealing temperature to 600°C (Figure 2) resulted in no droplets of implanted Ga on the modified area (Figure 2a-c). The observed situation is similar to the mechanism of the evaporation of FIB-implanted Ga during annealing at 800°C, except that Ga droplets are observed only in the areas treated at 30kV and 200 passes (Figure 2d). With a "softer" processing of the substrate than 30kV and 200 passes, the absence of droplets on the surface is explained by the fact that it is not energetically favorable for the material to come to the surface and evaporate at 600°C, when, as a large damage to the surface during processing of 30kV and 200 passes, implanted Ga is collected in large droplets, and part of it evaporates leaving holes in its place

Figure 2. SEM images of modified Si areas after annealing at $T = 600^{o}C$ and various accelerating voltages and beam passes, respectively: (a) 5kV and 5, (b) 5kV and 200, (c) 30kV and 5, (d) 30kV and 200.

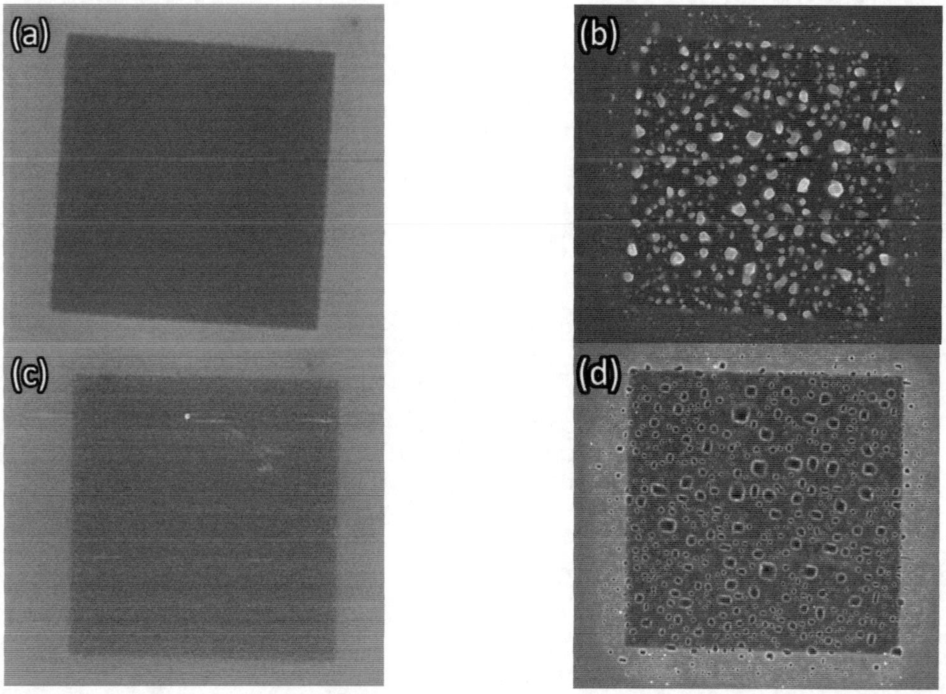

Figure 3. SEM images of modified Si areas with Ga^{+} ions at an accelerating voltage of 30 kV and after annealing with additional arsenic flux at various temperatures and beam passes, respectively: (a) $T = 600^{o}C$ and 5, (b) $T = 600^{o}C$ and 200, (c) $T = 800^{o}C$ and 5, (d) $T = 800^{o}C$ and 200.

Next, we studied the effect of an additional arsenic flux on the annealing of modified silicon areas (Figure 3). It is shown that crystallization of implanted Ga occurs. Moreover, a larger number of small crystallized nanostructures is observed than in the case of annealing without arsenic. This suggests that arsenic reduces the evaporation of Ga from the surface and reduces the probability of droplet coalescence (Figure 1c and Figure 3c). From the cases presented in Figure 2d and Figure 3b, it can be understood that the additional flux of arsenic during annealing at a temperature of $T = 600^{\circ}C$ favorably influenced the formation of crystalline GaAs nanostructures, which can subsequently act as nucleation centers for subsequent grown GaAs layers. However, the use of an additional arsenic flux did not help to reduce evaporation from the surface of the "hard" modified silicon area (Figure 3d). This surface morphology was influenced by two factors: a high annealing temperature, at which GaAs evaporates from the surface with a high probability, and severe damage to the surface during FIB-treatment.

4. Conclusions

Thus, according to the studies carried out, it can be concluded that not only the initial FIB treatment of the silicon surface with Ga ions affects the morphology of the resulting areas, but also such annealing parameters as the temperature and arsenic flux. It is assumed that obtaining optimized annealing parameters together with nucleation layers will make it possible to localize the largest number of defects in the lower layers of structures required for monolithic integration of GaAs on Si. Therefore, in the future, it is planned to study these samples using Raman spectroscopy in order to confirm or disprove our theories.

Acknowledgments

This work was supported by the Russian Science Foundation Grant No. 20-69-46076. The results were obtained using the equipment of the Research and Education Center "Nanotechnologies" of Southern Federal University.

References

[1] Park J-S, Tang M., Chen S and Liu H 2020 *Crystals* **10** 1063
[2] Eremenko M M, Solodovnik M S, Balakirev S V, Chernenko N E, Kots I N and Ageev O A 2020 *J. Phys.: Conf. Ser.* **1695** 012013
[3] Tang M, Park J-S, Wang Z, Chen S, Jurczak P, Seeds A and Liu H 2019 *Prog. Quant. Electron* **66** 1-18
[4] Wang Z *et al.* 2013 *Nano Lett.* **13** 5063–9
[5] Balakirev S V, Chernenko N E, Eremenko M M, Ageev O A and Solodovnik M S 2021 *Nanomaterials* **11**(5) 1184
[6] Baumgartner Y *et al.* 2021 *Opt. Express* **29** 509
[7] Balakirev S V, Solodovnik M S, Eremenko M M, Chernenko N E and Ageev O A 2020 *Nanotechnology* **31** 48
[8] Seifried M *et al.* 2018 *IEEE J. Sel. Top. Quant.* **24** 8200709
[9] Balakirev S V, Solodovnik M S and Ageev O A 2018 *Phys. Status Solidi B* **255**(4) 1700360
[10] Han Y, Li Q, Chang S-P, Hsu W-D and Lau K M 2016 *Appl. Phys. Lett.* **108** 242105
[11] Balakirev S V, Solodovnik M S, Eremenko M M, Konoplev B G and Ageev O A 2019 *Nanotechnology* **30** 50
[12] Chen S *et al.* 2016 *Nat. Photonics* **10** 307–11

The influence of the surface density of oriented nickel networks on the conducting electrode's optical transparency

I R Nizameev [1,2], G R Nizameeva [3] and M K Kadirov [1]

[1] Arbuzov Institute of Organic and Physical Chemistry, FRC Kazan Scientific Center, Russian Academy of Sciences, Kazan 420088, Russia
[2] Kazan National Research Technical University named after A.N. Tupolev - KAI, Kazan 420111, Russia
[3] Kazan National Research Technological University, Kazan 420015, Russia

inizameyev@iopc.ru

Abstract. This study is part of the work on the creation of a transparent conductive coating based on oriented nanonetwork and submicron nickel fibres. It is devoted to finding the optimal values of electrical conductivity and optical transparency of the developed coating. In this work, we study the transmission spectra of oriented nickel networks on a glass substrate in the UV, visible and near-IR regions at different amounts of deposited metal. An exciting feature of the coating was discovered: in the range of 950 nm and above, there is a "bend" of the transmission spectrum downward. This bend (increased absorption of radiation in the near-IR region) is observed only in the presence of nickel nanonetwork and is not typical for a pure submicron network.

1. Introduction

The use of conductive optically transparent coatings has expanded significantly thanks to the development of optoelectronic devices operating on transparent electrodes, such as touch displays and sensors, smart windows and solar panels. The creation of a conductive transparent coating is one of the most urgent problems of modern science, devoted to nanomaterials, composite materials, solar energy, optoelectronics, etc. [1-2]. Indium Tin Oxide (ITO) is now widely used as the primary material in this role because it has low surface resistance and high optical transmittance. ITO is an electrically conductive coating on a glass surface. However, the amount of indium in nature is limited, and the production of ITO is a complex and expensive technological process [2-4]. In this work, we present part of the work devoted to developing and improving an alternative to indium tin oxide (ITO). There are many known methods for producing the transparent conductive coating (TCC). For this, transparent conductive oxides [5-9], conducting polymers [10-13], graphene layers [14-17], carbon nanotubes and nanostructures [18-20], metal nanowires and networks are used. [21-24]. Metals are excellent electrical conductors. Recently, nanowires made of silver, gold and copper are very actively used to create TCC [25-28]. The most relevant direction is the use of structured systems of metal wires: networks and meshes [25-26].

In [29], we developed a technique for creating a transparent electrode based on oriented nickel networks. The basis of the technique is precipitation from the liquid phase by chemical reduction of the nickel salt [30-31]. The forming and adjustment of the transparent conductive layer is carried out using a micellar template of a surfactant (CTAB) and a magnetic field. We used submicron nickel fibres to implement unity to an array of individual-oriented nickel nano-networks. Oriented nickel nanowire networks and a framework of submicron nickel fibres create an electrically conductive and transparent coating.

Within this study's framework, we investigate the effect of the amount of used nickel on the transmission spectra of the developed coating on a glass substrate in the UV, visible and near-IR regions. The surface density of the metal deposition (mass referred to the substrate's surface area) was taken as the value that determines the amount of nickel used.

2. Results and discussion

In this work, samples of the developed coating on glass with different amounts of metallic nickel were obtained: $0.6 \cdot 10^{-6}$ - $311 \cdot 10^{-6}$ g / cm^2. The smallest value in this range corresponds to a coating consisting only of nickel nanowires. Microscopy images of the coating on a glass surface with different deposition densities of nickel are shown in figure 1.

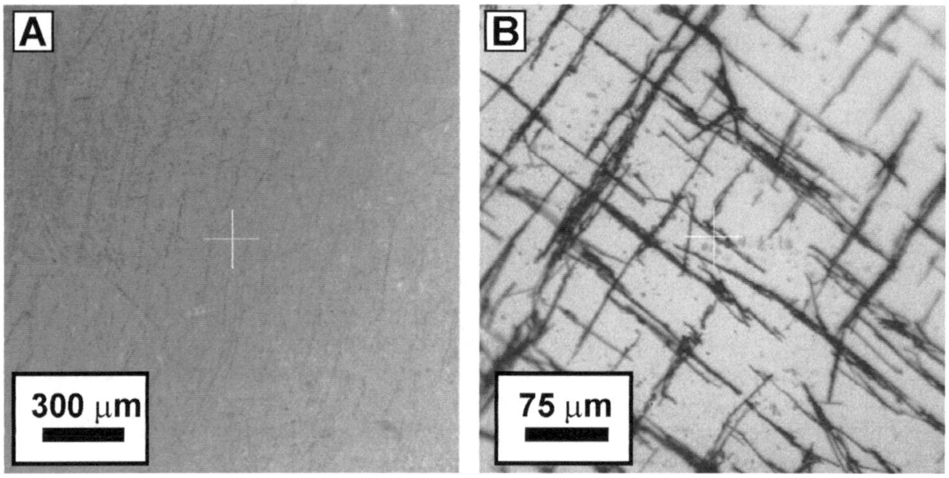

Figure 1. Microscopy images of nickel network coatings on the surface of a glass substrate with different deposition densities: 0.6 μg / cm^2 (A), 62 μg / cm^2 (B).

All obtained samples were examined on a spectrophotometer in the wavelength range of 290-1100 nm. SPECORD 50 (Analytik Jena) was used for all measurements. The averaged transmission spectra for some deposition densities of nickel networks are shown in figure 2. As expected, with an increase in the density of nickel deposition on the glass surface, a monotonic decrease in the transparency coefficient is observed over the entire optical range.

In the research, we discovered an interesting feature of the coating: in the range of 950 nm and above, there is a "bend" of the transmission spectrum downward. This bend (increased absorption of radiation in the near-IR region) is observed only in the presence of nickel nanonetwork and is not typical for a pure submicron network (figure 3).

This phenomenon may be due to the peculiarities of the transmission of radiation through subwavelength apertures. Bethe's criterion shows that when the aperture size is less than half the wavelength of the radiation passing through it, its strong attenuation is observed already in the near field.

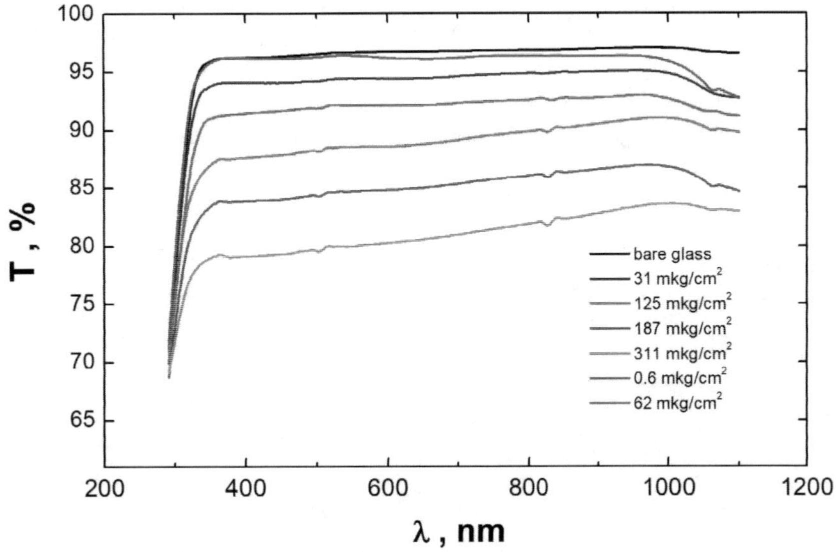

Figure 2. Transmission spectra of oriented nickel networks on a glass substrate in the UV, visible, and near-IR regions at different amounts of deposited metal.

For our case, this means that the material's absorption coefficient increases when the critical value of the wavelength is reached. It should be noted that such an effect is possible only for regular structures such that the scatter of transverse dimensions is small. We believe that the discovered effect may be of interest for applications in the development of nanosensors and near-infrared sensors.

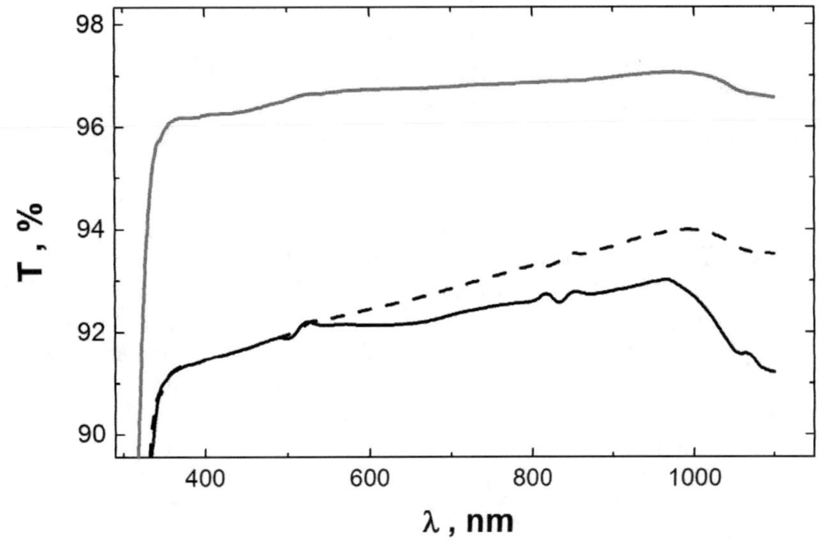

Figure 3. Transmission spectra of the bare glass substrate (red line) and the oriented nickel networks on a glass substrate in the UV, visible, and near-IR regions (the surface density of the metal deposition is 62 mkg/cm^2); the dashed line corresponds to the coating without the use of a nanonetwork.

The work studies transmission spectra of oriented nickel networks on a glass substrate in the UV, visible and near-IR regions at different amounts of deposited metal. The results demonstrate a monotonic decrease in the transparency coefficient at a wavelength of 550 nm with an increase in the metal deposition density. An exciting feature of the coating was discovered: in the range of 950 nm and above, there is a "bend" of the transmission spectrum downward. This bend (increased absorption of radiation in the near-IR region) is observed only in the presence of nickel nanonetwork and is not typical for a pure submicron network.

Acknowledgments
The reported study was funded by RFBR according to the research project № 19-03-00345.

References

[1] Kim H et al 2014 *Small* **10** 3767
[2] Wu H et al 2013 *Nature nanotechnology* **8** 421
[3] Kang M and Guo L 2007 *Advanced Materials* **19** 1391
[4] Lee J, Shin H, Na S and Kim H 2013 *Solar energy materials and solar cells* **109** 192
[5] Lee S, Choo M, Jung S and Hong W 2018 *Applied Sciences* **8** 901
[6] Ali N, Misran N, Mansor M and Jamlos M 2017 *Journal of Physics: Conference Series* **852** 012036
[7] Thampy A and Dhamodharan S 2015 *Physica E: Low-dimensional Systems and Nanostructures* **66** 52
[8] Colombel F, Castel X, Himdi M, Legeay G, Vigneron S and Cruz E 2009 *IET science, measurement & technology* **3** 229
[9] Choi K, Nam H, Jeong J, Cho S, Kim H, Kang J, Kim D and Cho W 2008 *Applied Physics Letters* **92** 194
[10] Heeger A 2001 *The Journal of Physical Chemistry B* **105** 8475
[11] Shi H, Liu C, Jiang Q and Xu J 2015 *Advanced Electronic Materials* **1** 1500017
[12] Zhang X et al 2016 *Solar Energy Materials and Solar Cells* **144** 143
[13] Xia Y, Sun K and Ouyang J 2012 *Advanced materials* **24** 2436
[14] Park H, Brown P, Bulović V and Kong J 2012 *Nano letters* **12** 133
[15] Liu L et al 2020 *ACS Applied Energy Materials* **3** 4068
[16] Seo J, Lee J, Jang A, Choi Y, Kim U, Shin H and Park H 2017 *Chemistry of Materials* **29** 4202
[17] Bae S et al 2010 *Nature nanotechnology* **5** 574
[18] Tung V, Chen L, Allen M, Wassei J, Nelson K, Kaner R and Yang Y 2009 *Nano letters* **9** 1949
[19] Shimoni A, Azoubel S and Magdassi S 2014 *Nanoscale* **6** 11084
[20] Plonska-Brzezinska M, Lewandowski M, Błaszyk M, Molina-Ontoria A, Luciński T and Echegoyen L 2012 *ChemPhysChem* **13** 4134
[21] Kim J, Kim S, Kil H, Kim Y and Park J 2018 *Nano Letters* **18** 4531
[22] Lee H, Jin W, Ovhal M, Kumar N and Kang J 2019 *Journal of Materials Chemistry C* **7** 1087
[23] Francis M, Bhargav P, Ahmed N, Chandra B, Gnanapraksh D, Thyagarajan N and Racchana R 2020 *ChemistrySelect* **5** 1370
[24] Nian Q, Saei M, Xu Y, Sabyasachi G, Deng B, Chen Y and Cheng G 2015 *ACS nano* **9** 10018.
[25] Bellew A, Manning H, Rocha C, Ferreira M and Boland J 2015 *ACS Nano* **9** 11422
[26] Nirmalraj P et al 2012 *Nano letters* **12** 5966
[27] Lee S, Choo M, Jung S and Hong W 2018 *Applied Sciences* **8** 901
[28] Sannicolo T, Lagrange M, Cabos A, Celle C, Simonato J and Bellet D 2016 *Small* **12** 6052
[29] Nizameev I, Nizameeva G, Faizullin R and Kadirov M 2021 *ChemPhysChem* **22** 288
[30] Nizameev I, Nizameeva G and Kadirov M 2019 *ChemistrySelect* **4** 13564
[31] Nizameev I, Nizameeva G and Kadirov M 2019 *Journal of Physics: Conference Series* **1409** 012038

Transmission spectra of transparent electrodes based on oriented platinum nanowires at various concentrations of the metal used

G R Nizameeva [1], I R Nizameev [1,2] and M K Kadirov [2]

[1] Kazan National Research Technological University, Kazan 420015, Russia
[2] Arbuzov Institute of Organic and Physical Chemistry, FRC Kazan Scientific Center, Russian Academy of Sciences, Kazan 420088, Russia

guliya.riv@gmail.com

Abstract. The paper investigates the optical transparency dependence of a coating based on an oriented network of platinum nanowires on the amount of metal used. Oriented platinum nanowires on a glass surface are produced by chemical deposition from an aqueous hexachloroplatinic acid solution. The topography of the deposited metal layer on glass is visualized using atomic force microscopy. Optical transparency was investigated with a spectrophotometer. In almost the entire region of optical radiation, the transparency is approximately 98%. The absorption spectrum shows that the absorption coefficient increases sharply in the near UV region.

1. Introduction

Transparent electrodes have received significant attention due to their importance in electronics. They are an integral part of many optoelectronic devices such as touch displays and sensors, bright windows, solar panels, etc. Today, transparent electrodes are made of solid indium and tin oxides. The material is highly transparent and conductive. However, the material has such disadvantages as fragility, complexity of production technology and high cost. Therefore, scientists from all over the world are looking for an inexpensive alternative to indium tin oxide. Alternative transparent electrodes are available in a wide range of materials. These are films of carbon nanotubes [1], graphene [2] and conductive polymers.

The graphene monolayer is of particular interest. The graphene monolayer has high optical transparency and conductivity. However, due to the problems of scaling the production and quality of the obtained graphene layers, the mass use of graphene in optoelectronics is not yet possible. As you know, for the production of transparent electrodes with large diagonals (for example, in solar cells), large sheets of transparent conductive material without scratches and cracks are required. However, there is currently no suitable technology for producing graphene sheets over large areas. Another disadvantage of graphene production technology is that first graphene is synthesized on catalytic substrates. A copper plate is usually used as such a substrate. This suggests that after synthesis, graphene must be transferred to other transparent substrates. This two-stage synthesis greatly complicates the process. When transferred from catalytic surfaces to transparent substrates, thin graphene layers are destroyed, deformed, and become unsuitable for use as transparent electrodes [3-8].

Transparent conductive polymers are also considered a good alternative in the field of transparent electrodes. These polymers is poly (3,4-ethylenedioxythiophene) polystyrene sulfonate (PEDOT: PSS), polyaniline, polyacetylene, etc. However, these polymers are not independently used as transparent electrodes, since they have low electrical conductivity. Typically, transparent conductive polymers are used in combination with other conductive materials. For example, these is used with metal nanowires to increase mechanical strength and adhesion to the substrate.

Carbon nanotubes are also of interest in the field of optoelectronics. In addition to transparency in the optical range and electrical conductivity, they are flexible. The ability to synthesize directly on flexible substrates is the main advantage of carbon nanotubes. However, such synthesis methods are energy intensive, complex and not scalable. Therefore, it is very important to find a simpler method for obtaining transparent electrodes based on carbon nanotubes or to use other materials.

From a functional point of view, oriented metal nanowires are considered the best candidates for the development of transparent electrodes due to the combination of low surface resistance, high transparency [9], resistance to mechanical deformation and high operating temperatures. In addition, transparent electrodes based on metal nanowires are the most economically efficient.

At the moment, the main idea of creating flexible optically transparent coatings is the use of metal nanowires in conjunction with this or other transparent polymer. One of the most bright representatives of such a polymer is poly (3,4-ethylenedioxythiophene) (common abbreviated name - PEDOT) or its modification with higher electrical conductivity - poly (3,4-ethylenedioxythiophene) polystyrene sulfonate (PEDOT: PSS) [10-13]. PEDOT: PSS is optically clear and has good conductivity.

The use of metals that are weakly susceptible to the effect of oxidation is promising. Of interest is platinum, more precisely, platinum nanobands. Anti-corrosion properties and excellent electrical conductivity make them promising alternatives for creating durable, chemically resistant optically transparent coatings.

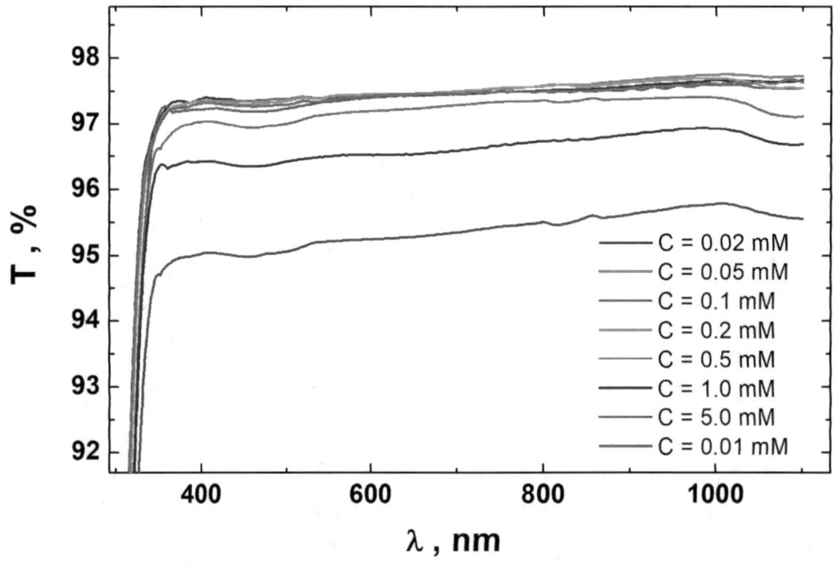

Figure 1. Transmission spectra of oriented platinum networks on a glass substrate in the UV, visible, and near-IR regions at different amounts of deposited metal.

 2086 (2021) 012029 doi:10.1088/1742-6596/2086/1/012029

Previously, we developed a technique for obtaining an optically transparent coating based on oriented grid of platinum nanowires [14, 15]. Within this study's framework, we investigate the effect of the amount of platinum used on the transparency of the resulting coating on a glass substrate in the UV, visible and near-IR regions.

2. Results and discussion

In this work, samples of coating on glass with different amounts of platinum metal were synthesized. The change in the amount of platinum on the glass surface was carried out by adjusting hexachloroplatinic acid concentration ($H_2PtCl_6 \cdot 6H_2O$). The concentration of hydrochloric platinum acid in the aqueous solution varied in the range 0.01 - 5 mM.

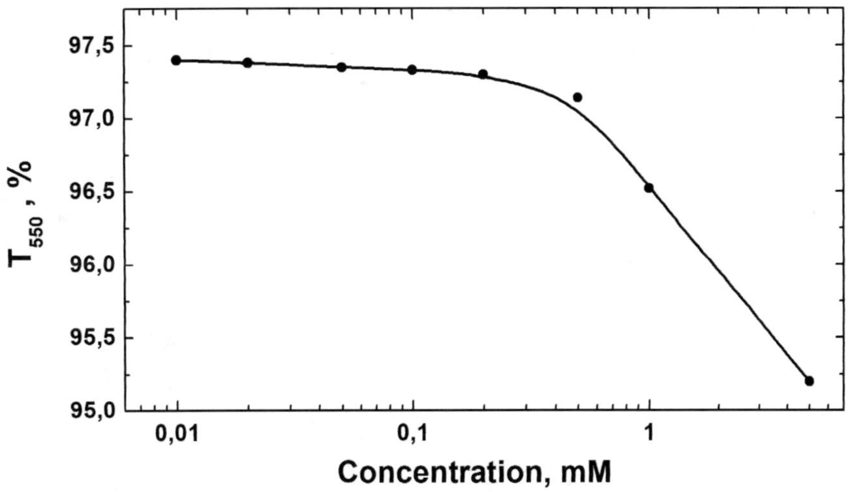

Figure 2. Dependency ratio transparency at a wavelength of 550 nm from initial concentration values hexachloroplatinic acid.

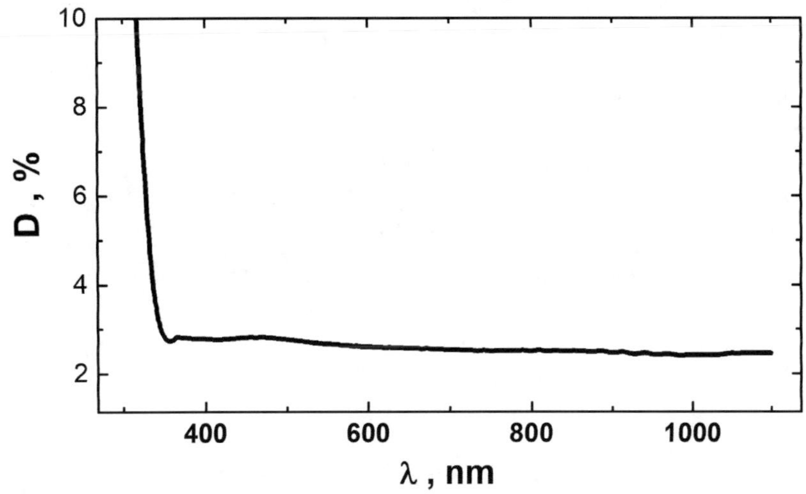

Figure 3. Absorption spectrum oriented nets of platinum on a glass substrate in UV, visible and near-infrared regions, corresponding to the value of the original concentration of hexachloroplatinic acid 0.1 · 10-3 mol/l.

All obtained samples were examined on a spectrophotometer in the wavelength range of 290-1100 nm. The averaged transmission spectra for some deposition densities of platinum nanowires are shown in figure 1. As expected, with an increase in the density of platinum deposition on the glass surface, a monotonic decrease in the transparency coefficient is observed over the entire optical range.

Length dependence of the transparency coefficient waves 550 nm from the value of the initial concentration hexachloroplatinic acid is shown in figure 2. Monotonic decrease of T550 has two linear sections on a logarithmic scale (black line in figure 2). It can be seen in the absorption spectrum that the absorption coefficient increases sharply in the near UV region (figure 3). This is explained properties of the glass substrate: it strongly absorbs UV radiation.

As a result of the studies carried out, the dependences of the coating's transparency on the amount of platinum used were obtained. In almost the entire region of optical radiation, the transparency is approximately 98%. The absorption spectrum shows that the absorption coefficient increases sharply in the near UV region. This phenomenon is due to the properties of the glass substrate: it strongly absorbs UV radiation.

Acknowledgments
This work was supported by the President of Russian Federation grant for the governmental support of young Russian scientists [MK-2264.2019.3].

References
[1] Rathmell A R, Nguyen M, Chi M and Wiley B J 2012 *Nano letters* **12** 6
[2] Kim K S, Zhao Y, Jang H, Lee S Y, Kim J M, Kim Kwang S, Ahn J H, Kim P, Choi J Y and Hong B H 2009 *Nature* **457** 7230
[3] Cai C, Jia F, Li A, Huang F, Xu Z, Qiu L, Chen Y, Fei G and Wang M 2016 *Carbon* **98** 457
[4] Gee C M, Tseng C C, Wu F Y, Chang H P, Li L J, Hsieh Y P, Lin C T and Chen J C 2013 *Displays* **34** 315
[5] Pang S, Chin S F and Yin V 2011 *Advanced Materials* **23** 2779
[6] Allen M J, Tung V C and Kaner R B 2010 *Chemical reviews* **110** 132
[7] Rana K, Singh L and Ahn J H 2014 *Journal of Materials Chemistry C* **2** 2646
[8] Konatham D and Striolo A 2008 *Nano Letters* **8** 4630
[9] Guo C F and Ren Z 2015 *Materials Today* **18** 3
[10] Zhang L, Shen W, He R, Liu X, Tang X, Yang Y and Li M 2016 *Organic Electronics* **32** 134
[11] Kim J, Kim H, Kim S H and Park C E 2016 *The Journal of Physical Chemistry C* **120.2** 956
[12] Chou T R, Chen S H, Chiang Y T, Chang T T, Lin C W and Chao C Y 2017 *Organic Electronics* **48** 223
[13] Xiong Zh and Changqing L 2012 *Organic Electronics* **13.9** 1532
[14] Nizameev I, Nizameeva G and Kadirov M 2019 *ChemistrySelect* **4** 13564
[15] Nizameeva G, Nizameev I, Nefedev E and Kadirov M 2020 *Journal of Physics: Conference Series* **1695** 12007

Modelling electroforming process under constant bias voltage conditions

O O Permyakova[1, 2] and A E Rogozhin[1]

[1] Laboratory of Microstructuring and Submicron Devices, Valiev Institute of Physics and Technology RAS, Moscow 117218, Russia
[2] The School of Electronics, Photonics and Molecular Physics, Moscow Institute of Physics and Technology, Dolgoprudny 141701, Russia

o.permyakova@phystech.edu

Abstract. In order to understand changes in defect concentration during the electroforming process, we modelled the electroforming process in Ta/HfO_2/Pt under constant bias voltage. For this purpose, kinetic Monte-Carlo and finite element methods were utilized. Vacancy profiles were obtained for forming voltages from 3 V to 5 V; modelling of lower stresses is time-consuming. It was found that with decreasing voltage, vacancies begin to accumulate near the inert electrode. When the voltage was dropped from 5 to 3 V, the thickness of such a layer increased by 1 nm, and electroforming time exponentially increase.

1. Introduction

The most promising type of non-volatile memory is resistive random access memory (RRAM) [1]. The main advantage of RRAM is low switching time of fewer than 10 ns [2] and scalability of less than 10 nm [3]. Additionally, the RRAM element can be used for processing information [4]. The main problem of RRAM is the low repeatability of characteristics between elements [1].

A ReRAM element usually has two possible states which differ in resistance value: low resistance state (LRS) and high resistance state (HRS). The transition from one state to another occurs when charge flows through the structure, this process is called resistive switching. Eventually, the most stable resistive switching indicated in RRAM elements based on transition metal oxides (TMOs) such as HfO_X [2, 3] and TaO_X [2].

Resistive switching in RRAM elements based on TMOs mainly occurs due to reduction and disruption of conductive filament (CF). So, RRAM variability strongly depends on the shape and properties of CF [5]. The initial CF formation takes place during electroforming at the voltage values, significantly higher than the operating voltage; this process is called electroforming. Electroforming is a stochastic process; therefore, using the same electroforming process for various elements of the same structure, it is difficult to get similar CF. A structure that does not require an electroforming process is called forming-free structures; such structures usually have low variability [6].

In this work, we present the result of modelling the electroforming process under various constant bias voltage conditions from 3 V to 5 V for structures Ta/HfO_2/Pt. We use kinetic Monte-Carlo and finite element techniques to investigate the formation of CF in hafnium oxide under constant voltage conditions. Results of such a model can be used in the development of forming-free memristor technology.

2. Model description

To examine the growth of CF under the constant-voltage conditions in oxide with thickness 5 nm, we utilize kinetic Monte-Carlo and finite element methods. Monte-Carlo approach was used for simulation of the oxygen vacancy motion and electric current passing through the structure. The finite element method was adopted to calculate the destitution of potential and temperature of modelled device. According to the model, vacancies formation only takes place near the active electrode. After that, vacancies drifted to the inert electrode under an electric field. Vacancy movements were simulated till the entire filament short-circuited the electrodes. We employ our previous model [7] with two exceptions: definition of the cell size and definition of the power dissipated in the storage media during current flow.

The Poisson equation was used to estimate the potential distribution in the device, which was then solved using the finite element method realized via the FEniCS package. Dirichlet boundary conditions were set at the boundary with the electrodes. The potentials at the electrodes were equal to the voltage applied to them. Neumann boundary condition was applied to all other boundaries of the computational grid. It is also considered that vacancies reaching the electrode have a potential equal to that of the electrode, intend that such vacancies represent a potential source inside the storage media.

The nonstationary heat equation was used to estimate the temperature distribution inside the device. Dirichlet boundary condition was set at the boundary with the electrodes, and for other boundaries, Neumann boundary conditions were used. Similar to the phonon-assisted tunnelling model, the power dissipated in the storage media during current flow was based on the difference between the energy levels at which the electron was located and to which it passed.

In order to obtain mesh period d, we define the volume per molecule of hafnium oxide:

$$d = \frac{M}{N \cdot \rho_{ox} \cdot N_a}$$

where M — molar mass, N — number of oxygen atoms per mesh cell, ρ_{ox} — density of storage media, N_a — Avogadro's constant. So, during modelling d was equal to 0.33 nm.

3. Result and Discussion

Typical current-voltage characteristics of the electroforming process under constant 4 V bias voltage are presented in figure 1. The change in conduction was observed. In the time interval from 0 to 3.3 s mean current remain constant under constant bias voltage condition. During that time, part of the vacancies migrated to an inert electrode and moved along it. After 3.3 s, a critical concentration of vacancies at the inert electrode was reached, conductive filament began to form, and current start to rise. An abrupt increase of current after 7.6 s indicates the formation of conductive filament (figure 2b).

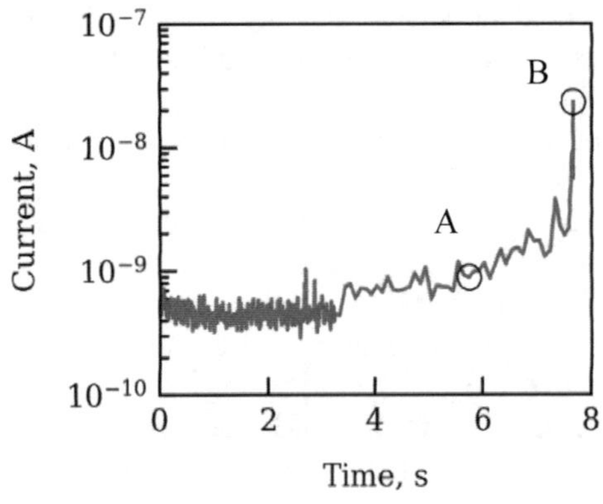

Figure 1. Simulated current-voltage characteristics of electroforming process under 4 V bias voltage. The Electroforming process lasted for 7.66 s.

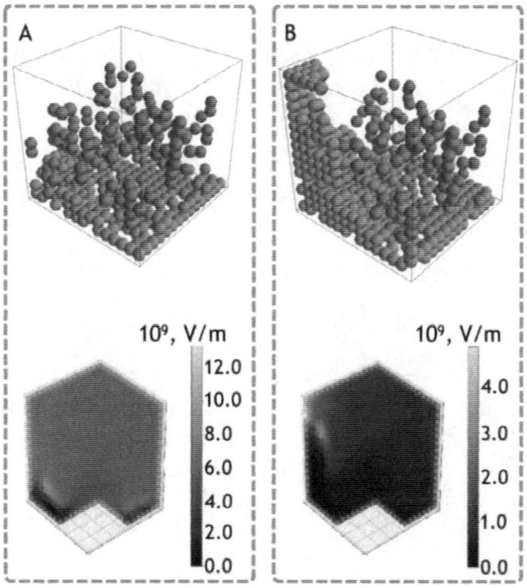

Figure 2. Simulated vacancy distribution and electric field of storage medium. A and B correspond to points A and B in Figure 1. On the distribution of vacancies the vacancies adjacent to the electrode as well as the vacancies associated with them are marked in gray.

It was found that an increase of bias voltage leads to a more significant concentration of vacancies near the inert electrode. The thickness of a vacancy layer near an inert electrode grew by 1 nm when the voltage was reduced from 5 to 3 V (figure 3). Modelled vacancy concentration profiles are presented in Figure 2. We consider that the vacancy concentration profile is determined by the ratio of vacancy motion rates and generation rates. The rate of vacancy generation decreases with stress bias voltage as the rate of vacancy diffusion, but the generation activation energies for the vacancy (1.1 eV) were much lower than diffusion one (0.7 eV). The exponential increase of electroforming time (inserted in figure 3) can be associated with the accumulation of vacancies near the inert electrode. The dependency of forming time on stress, represented on a logarithmic scale, has the shape of a straight line with a slope of 0.77.

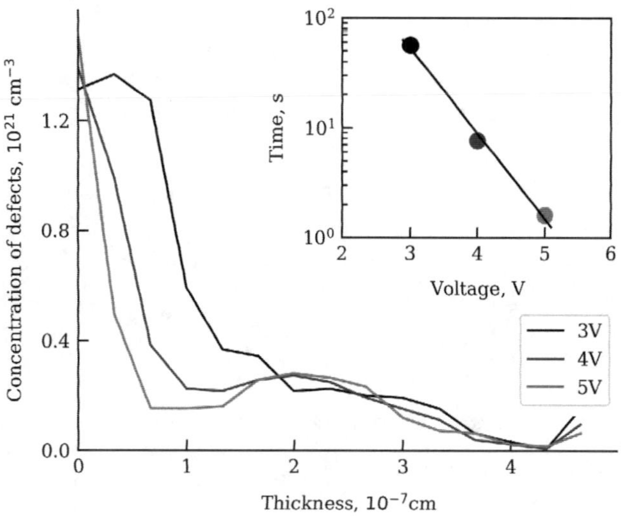

Figure 3. Simulated distribution of vacancy concentration along with the oxide thickness, with an active electrode in the right and an inert electrode in the left. The inserted figure shows the dependence of filament formation time on the constant bias voltage. The slope of the trend line is 0.77.

4. Conclusion

During the electroforming process, vacancies tend to accumulate near inert electrodes with an increase of the applied voltage the thickness of such vacancy layer decreases. Additionally, the time of electroforming exponentially decreases with increasing the forming voltage. Understanding the initial concentration of oxygen vacancies in the storage medium might open the way for forming-free memristor technology, which is based on the setting of a particular initial concentration of oxygen vacancies inside transition metal oxide.

Acknowledgments

The investigation was supported by program №0066-2019-0004 of the Ministry of Science and Higher Education of Russia for Valiev Institute of Physics and Technology of RAS.

References

[1] Lanza M et al. 2019 *Adv. Electron. Mater.* **5** 1800143
[2] Wong H-S P, Lee H Y, Yu S, Chen Y S, Wu Y, Chen P S, Lee B, Chen F T and Tsai M J 2012 *Proc. IEEE* **100** 1951
[3] Govoreanu B et al. 2011 *IEEE Electron Device Lett.* **31** 6
[4] Zahoor F, Zulkifli T Z A and Khanday F A 2020 *Nanoscale Res Lett* **15** 90
[5] Gonzalez M B, Raf J M, Beldarrain O, Zabala M and Campabadal F 2014 *IEEE Transactions on Device and Materials Reliability* **14** 769
[6] Zhang H, Liu L, Gao B, Qiu Y, Liu X, Lu J, Han R, Kang J and Yu B 2011 *Appl. Phys. Lett* **98** 042105
[7] Permiakova O, Miakonkikh A, Rudenko K and Rogozhin A 2020 *Int. Conf. on Information Technology and Nanotechnology,* 26-29 May, Samara, Russia, (New York: IEEE) p 1

Investigation of near-surface layer dislocation density of lithium niobate single crystal wafers using chemical etching

R S Ponomarev[1,2], A V Sosunov[1] , O R Semenova[1] , N P Prokhorov[1] , M Kuneva[3]

[1] Nanotechnology department, Perm State University, Perm 614990, Bukireva St. 15, Russia
[2] Perm Federal Research Center UB RAS, Perm 614990, Lenina St.,13a, Russia
[3] Institute of Solid State Physics, Tzarigradsko Chaussee 72, Blvd., 1784 Sofia, Bulgaria

rsponomarev@gmail.com

Abstract. Using chemical etching it was shown that the density of dislocation in lithium niobate (LN) single crystal wafers is higher near the surface in depth about 20 um than in the depth of crystal. It caused to change of diffusion coefficient during the waveguide formation with proton exchange (PE) method and can increase DC-drift of intensity optical modulators based on PE-waveguides.

1. Introduction

Lithium niobate (LN) crystal is a key material in the manufacture of basic elements of integrated optical circuits, in particular phase and intensity modulators. Distinctive features of LN single crystals are high electro-optical coefficients, high Curie temperature and the possibility of forming waveguide structures with low optical losses [1].

When considering the surface properties of LN single crystals of various cuts, additional opportunities are found for their fundamental research and applications [2]. It is known that the composition and structure of the near-surface layer of the LN crystal differ significantly from the rest of the material volume, which is clearly shown in [3]. These features of the structure of the near-surface layer have a deformation nature and arise during cutting and subsequent polishing of NL plates [3,4]. Many aspects of the formation and state of this layer were discussed repeatedly, but the real structure of this layer was not presented. In this case, the influence of inhomogeneities in the structure of the nearsurface layer of the LN is the main factor associated with the instability of the optical characteristics of integrated-optical schemes [5] and the manufacture of proton-exchange waveguides [6]. Thus, the aim of this work is to reveal the real (dislocation) structure of the surface layer of LN single crystals.

2. Research methodology

In this work, we investigated LN single crystals of congruent composition in X- and Z-cuts. The identification of the real structure of the crystals under study was carried out by the method of selective etching of the working surface of single crystals, as well as the cross section (end) of the LN plates after fracture. The fracture was carried out according to the method described earlier in [3].

Etching of pre-purified samples was carried out in a potassium hydroxide (KOH) melt with the addition of 35 wt% NaOH, as described in [7]. The wet selective etching regime was identical for all samples.

The study of the etching pits on the working surface was carried out by the method of optical microscopy; the fracture structure, i.e. the cross-section of the plates - by scanning electron microscopy. Optical measurements were carried out using a Leica DMi8 microscope in reflected light in bright field mode.

Electron microscopic studies were carried out in the mode of secondary electrons at an accelerating voltage of 20 kV using a Hitachi S3400N microscope. Before the start of electron microscopic studies, a thin (~ 2 nm) metallic Au film was preliminarily deposited on all experimental samples using a compact SC7620 magnetron sputter.

3. Research methodology

It is known that during selective etching of single crystals, etching pits are formed on their surface, which have a certain geometric shape, depending on the crystal structure of the material and the crystallographic orientation of the etched plane [8]. Thus, in [9], it was shown that for Z-cut NL plates when etching the plane (0001), the pits have the form of regular triangles (symmetry axis of the 3-rd order), for planes perpendicular to the $[10\bar{1}0]$ direction, i.e. for the X-cut, the etch pits are rhombic. In the course of etching the working surfaces of the studied NL plates, numerous etching pits were found, corresponding to the dislocation structure of the NL surface layer. Etching patterns are shown in Figure 1 for X- and Z-slices left and right, respectively.

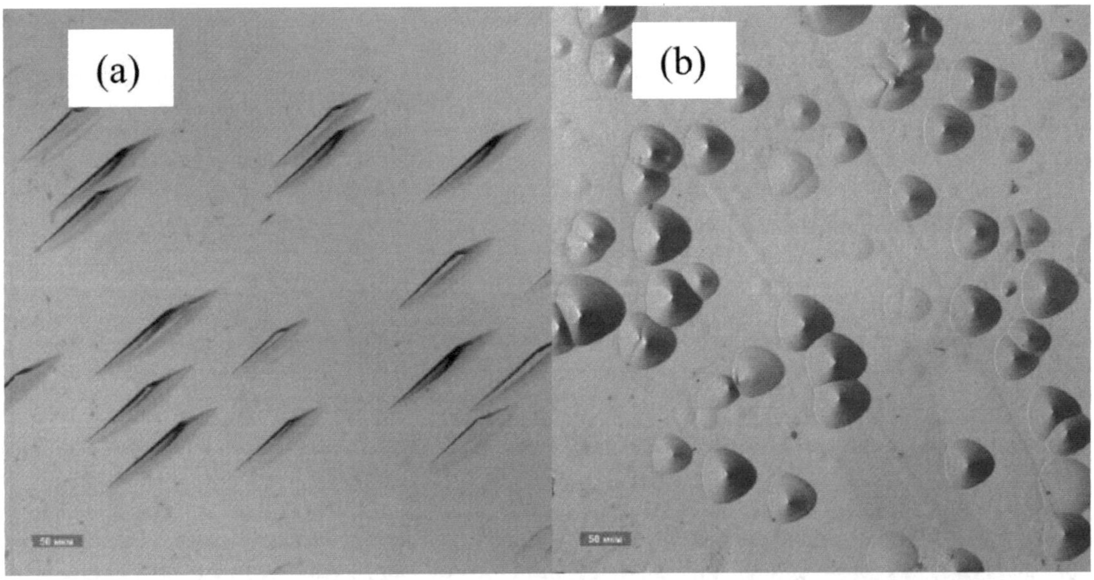

Figure 1. Etching pits for NL X-cut (a) and Z-cut (b). Optical microscopy results.

Figure1 shows that on the surface of the plates of Z-cut shape of most of the etching pits has a symmetry of the third order, which is in good agreement with [9].

The displacement of the top of the pit indicates a deviation of the dislocation axis from the [0001] direction due to the presence of deformations. For an X-cut crystal, etching pits were obtained in the form of highly elongated and distorted rhombuses with long axes parallel to the crystallographic direction $[01\bar{1}\bar{1}]$. The strong distortion of the shape of the etching pits is also associated with the deformation of the crystal lattice in the near-surface layer of the crystal. The density of the etching pits is 8×10^4 cm^{-2} for an X-cut crystal and 2×10^4 cm^{-2} for a Z-cut.

The advantage of this technique for revealing the real structure of the near-surface layer of an LN single crystal is the simplicity and speed of the experiment, however, the fundamental disadvantage of this research method is that it does not allow one to get an idea of the distribution of dislocations in the bulk of the crystal. For this reason, the etching of fresh transverse chips of X- and Z-cut LN plates was carried out. The plates were split according to the method described in [3], thus, the breaks correspond to the spalls close to the planes $(2\bar{1}\bar{1}1)$ and $(01\bar{1}0)$. The indicated planes were etched.

In figure 2 shows the results of revealing the dislocation structure of the LN plates with X- and Z-cuts after fracture (cross section). Investigation of the cross-section of the LN plates after fracture makes it possible to show the real structure of the entire LN crystal as representatively as possible.

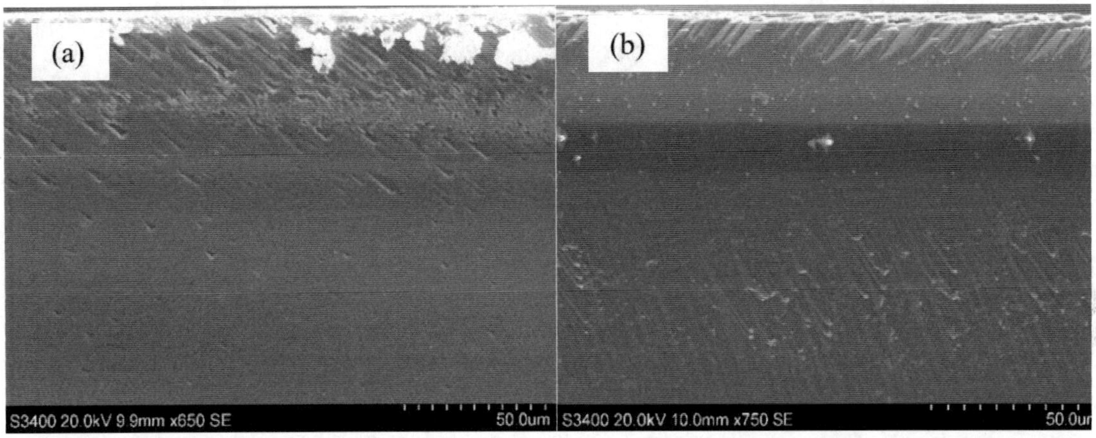

Figure 2. Etching pits of the LN after fracture (cross section): plane $(2\bar{1}\bar{1}0)$ (a) and $(01\bar{1}0)$ (b).

It can be seen that the near-surface layer has an increased density of not just point defects, but dislocations, in contrast to the rest of the material volume.

The depth of the defect layer is up to 50 μm for LN Z-cut (etching plane $(2\bar{1}\bar{1}0)$) and 10 μm for LN X-cut (etching plane $(01\bar{1}0)$). The formation of this defect layer occurs in the process of cutting and subsequent grinding of a single crystal boule. Similar features are typical for other single-crystal wafers, for example, for silicon carbide [10], which is also used to create integrated-optical devices.

4. Summary

Thus, by the method of selective etching, it was possible to reveal the real - dislocation structure of the damaged near-surface layer of LN crystals, which significantly differs in increased defectiveness from the rest of the volume of a single-crystal LN plate.

The results obtained are important, first of all, from a practical point of view. The presence of a defect layer can lead to both a deterioration in the characteristics of devices and a decrease in the stability of the operation of integrated optical devices based on single crystals of LN. In the near-surface layer, proton-exchange, titanium-diffuse, ion and laser waveguides are formed. The optical characteristics of waveguides are determined by the structure of a given subsurface layer. Undoubtedly, in the process of manufacturing and operating various integrated optical components based on LN, it is necessary to take into account the increased density of dislocations in its near-surface layer. In the future, it is planned to study the real structure of NL plates of various manufacturers and search for a method to reduce the defectiveness of the near-surface layer or to remove it.

Acknowledgments

The reported study was funded by RFBR and Perm Territory, project number 20-42-596001 and Ministry of Education and Science of Perm region under Contract No. C-26/848.

References

[1] Chen A and Murphy E 2012 *Broadband Optical Modulators: Science, Technology, and Applications* (CRC Press)
[2] Sanna S and Schmidt W G 2017 *J. Phys.: Condens. Matter* **29** 413001
[3] Sosunov A, Volyntsev A, Tsiberkin K, Yuriev V and Ponomarev R 2017 *Ferroelectrics* **506** (1) 24–31
[4] Galinetto P, Marinone M, Grando D, Samoggia G, Caccavale F, Morbiato A and Musolino M 2007 *Opt. Lasers Eng.* **45** (3) 380–384
[5] Salvestrini J, Guilbert L, Fontana M, Abarkan M and Gille S 2011 *J. Light. Technol* **29** (10) 1522–1534
[6] Kostritskii S, Korkishko Y, Fedorov V, Mitrokhin V, Sevostianov O, Chirkova I, Stepanenko O and De Micheli M 2014 *J. Eur. Opt. Soc. Rap. Public* **9** 14055
[7] Franko N, Ped'ko B and Sorokina I 2004 *Crystallogr. Rep.* **49** 94–99
[8] Sangwal K 1987 *Etching of Crystals: Theory, Experiment and Application* (North Holland)
[9] Nassau K, Levinstein H J and Loiacono G M 1966 *J. Phys. Chem. Solids.* **27** (6-7) 983–988
[10] Friedrichs P 2011 *Silicon carbide: Growth, defects, and novel applications.* (John Wiley & Sons)

Features of the MBE growth of nanowires with quantum dots on the silicon surface

R R Reznik[1], K P Kotlyar[1,2], A I Khrebtov[2] and G E Cirlin[1-4]

[1] Saint-Petersburg State University, Universitetskaya Emb. 13B, 199034 St. Petersburg, Russia
[2] Alferov University, ul. Khlopina 8/3, 194021 St.-Petersburg, Russia
[3] ITMO University, Kronverkskiy pr. 49, 197101, St-Petersburg, Russia
[4] Institute for Analytical Instrumentation RAS, Rizhsky 26, 190103, St-Petersburg, Russia

moment92@mail.ru

Abstract. The development of a new semiconductor element base is necessary to create a new generation of applications. At present time, the synthesis of high-quality hybrid nanostructures based on III-V quantum dots in the body of nanowires of a wide range of material systems is an urgent and important task. In work hybrid III-V nanostructures based on QDs in the body of NWs in GaP/GaAs and AlGaP/InGaP material systems were synthesized in on silicon substrates and their physical properties were investigated.

1. Introduction

Nowadays, the development of a new semiconductor element base is necessary to create a new generation of applications. Semiconductor quantum wells (QWs), nanowires (NWs) and quantum dots (QDs) based on III-V compounds are quantum objects and are promising for creating new devices in the field of optoelectronics, photonics and renewable energy sources and e.g. [1]. Moreover, modern synthesis methods make it possible to form nanostructures based on combinations of such materials, for example, QDs in the body of NWs [2]. Compared to QDs synthesized by other methods, it is an easier task to control the geometric parameters and location of such QDs. In turn, the unique morphology of NWs makes it possible to grow such hybrid structures on a silicon platform [3, 4] due to the unique morphological properties of NWs [5-7], which is an important task for most researchers in the field of micro- and nanoelectronics. In previous works, we have demonstrated that it is possible to synthesize by molecular beam epitaxy (MBE) III-V QDs in the body of NWs based on InP/InAsP [8, 9] and AlGaAs/GaAs [10-15] material systems on silicon wafers. The synthesized nanostructures have high crystallographic and optical qualities. The research results have shown that nanostructures based on InP/InAsP material system exhibit photoluminescence (PL) spectra up to room temperature in a wide range of wavelengths, including strategically important 1.3 and 1.5 μm, and nanostructures based on AlGaAs/GaAs exhibit record-breaking low photoluminescence line width at half maximum (FWHM) and have the properties of single photon sources, which promising for quantum informatics and cryptography. Nevertheless, the synthesis of high-quality hybrid nanostructures based on III-V QDs in the body of NWs in a wide range of material systems remains an urgent and important task.

Within the framework of this work, hybrid III-V nanostructures based on QDs in the body of NWs in GaP/GaAs and AlGaP/InGaP material systems were synthesized in on silicon substrates and their physical properties were investigated.

2. Experimental procedures

Experiments on epitaxial growth of hybrid III-V nanostructures based on QDs in the body of NWs in GaP/GaAs and AlGaP/InGaP material systems were carried out using a unique MBE machine "Compact 21T" from "Riber" company, equipped with effusion sources of III-V semiconductor materials and additional chambers for cleaning and deposition of metals. Monolithic silicon wafers doped with boron atoms and polished on one side were used as substrates for epitaxial synthesis. Growth technology contained a number of fundamental stages. First, the substrate was subjected to chemical cleaning in hydrofluoric acid diluted with water in a ratio of 1 to 4. In addition to cleaning the surface, this procedure allows to partially remove the layer of natural silicon oxide layer on the surface of the substrate. Immediately after chemical cleaning, the substrates were loaded into the lock chamber of the MBE setup, where they were heated to a temperature of 400°C for preliminary thermal cleaning. After 10 minutes of cleaning, the substrate temperature was reduced to 100°C and the substrate was transferred to the metallization chamber. In this chamber, the substrate was heated to 900°C for the purpose of complete thermal cleaning and subsequent metallization of the surface. After 30 minutes of thermal annealing, the substrate temperature decreased to 550°C. After stabilizing the temperature, a few nm thick gold layer was deposited on the substrate surface. Holding at the same temperature for 1 minute after gold deposition allows the formation of homogeneous drops on the substrate surface, which serve as catalysts for the growth of NWs by the vapor-liquid-solid (VLS) mechanism. [1]. At the next stage, the temperature of the substrates was reduced to 100°C and transferred in ultra-high vacuum conditions into the chamber for the growth of nanostructures. Then to grow the GaP NWs the substrate temperature was increased to the 580°C and Ga and P sources were opened for 16 minutes. An the next step Ga and P shutters were close for 1 minute so that phosphorus was pumped out of the chamber and did not participate in the growth of the QDs. After that, Ga and As shutters were pen for 7 second for GaAs QD formation. Finally, after 1 minute pause for pumping of As, Ga and P shutters were again opened for 5 minutes to form final part of GaP NWs. In case of AlGaP NWs with InGaP QDs the growth method was not fundamentally different, but substrate temperature during the growth was set at 530°C due to the specificity of the aluminum and indium atoms participation in the growth.

For determination of the morphological parameters of the synthesized arrays of NWs with QDs, we used the technique of scanning electron microscopy SEM using Supra 25 Zeiss microscope.

3. Results

Figure 1 shows the typical SEM image of GaP NWs array with GaAs QDs inside grown on silicon substrate. The direction of NWs growth <111> indicates epitaxial growth and bond with the substrate. The grown sample demonstrates the surface density of NWs about $3\cdot10^{7}cm^{-2}$, which corresponds to the density of gold droplets. We note that the height and diameter of the grown nanostructures is highly inhomogeneous over the surface which can lead to the difference in the optical properties. The possible reason may be different sizes of the Au particles formed prior the growth of GaP NWs and to investigate the optical properties of this array needs an additional optimization.

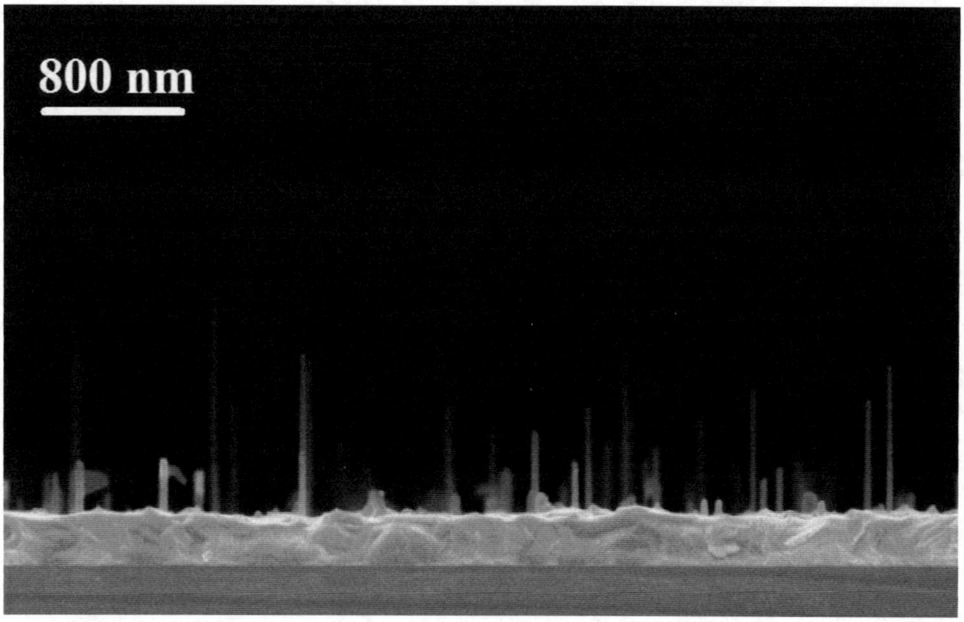

Figure 1. Typical SEM image of GaP NWs with GaAs QDs grown on Si(111) substrate.

Typical SEM images of AlGaP NWs with InGaP QDs are shown in figure 2. Similar to the case of GaP NWs, AlGaP NWs were formed mainly in the <111> direction. It seen that overage high of grown NWs is 900 nm and diameter – 140 nm. It is important to note, that some of the NWs began to bend and grow in a different direction. This is due to the usage of indium as a part of quantum dot material. According to [16] with a sufficient content of indium in the gold catalyst droplet, such a mixed droplet can etch the facets at the top of the NWs, thereby descending to the side of the NWs, and change the direction of NWs growth.

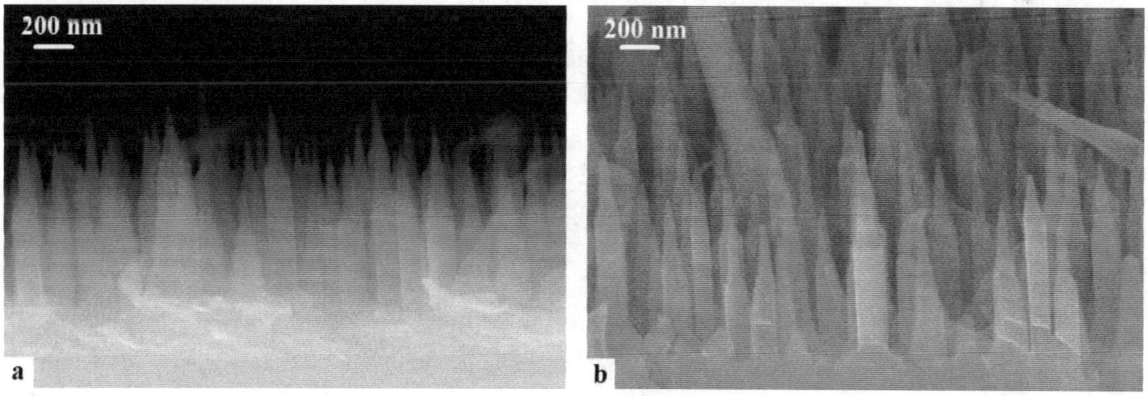

Figure 2. Typical SEM image of AlGaP NWs with InGaP QDs grown on Si(111) substrate.

4. Conclusions

In conclusion, we have demonstrated the principal growth possibility of GaP NWs with GaAs QDs and AlGaP NWs with InGaP QDs on silicon substrates. Morphological parameters of grown arrays of nanostructures were studied using SEM technique, which confirmed their epitaxial formation. Expansion of the range of material systems and the composition of semiconductor III-V QDs in the body of NWs can lead to the creation of new applications based on them. Thus, the research results open up new opportunities for creating quantum applications in the field of computer science and cryptography on silicon.

Acknowledgments
The samples were grown under support of St. Petersburg State University research grant 75746688. Morphological properties measurements were done under support of Russian Science Foundation (project 19-72-30004).

References

[1] Dubrovskii V G, Cirlin G E and Ustinov V M 2009 *Semiconductors* **43** 1539
[2] De La Mata M, Zhou X, Furtmayr F, Teubert J, Gradečak S, Eickhoff M, Fontcuberta i Morral A and Arbiol J 2013 *Journal of Materials Chemistry C* **1** 4300.
[3] Hyun J K, Zhang S and Lauhon L J 2013 *Ann. Rev. Mat. Res.* **43** 451
[4] Ng K W, Ko W S, Tran T-T D, Chen R, Nazarenko M V, Lu F, Dubrovskii V G, Kamp M, Forchel A and Chang Hasnain C J 2013 *ACS Nano* **7** 100
[5] Glas F 2006 *Physical Review B* **74** 121302
[6] Chuang L C, Moewe M, Chase C, Kobayashi N P, Chang-Hasnain C and Crankshaw S 2007 *Appl. Phys. Lett.* **90** 043115
[7] Moewe M, Chuang L C, Dubrovskii V G and Chang-Hasnain C 2008 *J. Appl. Phys.* **104** 044313
[8] Reznik R R, Cirlin G E, Shtrom I V, Khrebtov A I, Soshnikov I P, Kryzhanovskaya N V, Moiseev E I and Zhukov A E 2018 *Technical Physics Letters* **44** 112
[9] Cirlin G E, Reznik R R, Samsonenko Y B, Khrebtov A I, Kotlyar K P, Ilkiv I V, Soshnikov I P, Kirilenko D A and Kryzhanovskaya N V 2018 *Semiconductors* **52** 1416
[10] Leandro L, Gunnarsson C P, Reznik R, Jöns K D, Shtrom I, Khrebtov A, Kasama T, Zwiller V, Cirlin G and Akopian N 2018 *Nano Lett.* **18** 7217
[11] Leandro L, Reznik R, Clement J D, Repän J, Reynolds M, Ubyivovk E V, Shtrom I V, Cirlin G and Akopian N 2020 *Sci. Rep.* **10** 1
[12] Cirlin G E, Reznik R R, Shtrom I V, Khrebtov A I, Soshnikov I P, Kukushkin S A, Leandro L, Kasama T and Akopian N 2017 *J. Phys. D* **50** 484003
[13] Cirlin G E, Reznik R R, Shtrom I V, Khrebtov A I, Samsonenko Yu B, Kukushkin S A, Kasama T, Akopian N and Leonardo L 2018 *Semiconductors* **52** 462
[14] Cirlin G E, Shtrom I V, Reznik R R, Samsonenko Yu B, Khrebtov A I, Bouravleuv A D and Soshnikov I P 2016 *Semiconductors* **50** 1421
[15] Reznik R R, Shtrom I V, Samsonenko Yu B, Khrebtov A I, Soshnikov I P and Cirlin G E 2017 *J. Phys. Conf. Ser.* **929** 012047
[16] Wang J, Plissard S.R, Verheijen M A, Feiner L F, Cavalli A and Bakkers E P 2013 *Nano letters* **13** 3802

Plasma surface treatment of local modify silicon plates

A A Rezvan[1], J V Morozova[1] and V S Klimin[1,2]

[1] Department of Nanotechnology and Microsystems, Southern Federal University, Taganrog 347922, Russia
[2] Research and Education Center "Nanotechnologies", Southern Federal University, Taganrog 347922, Russia

arezvan@sfedu.ru

Abstract. This paper presents a study of the use of silicon Si for element base manufacture of micro- and nanoelectronics by using combined methods of focused ion beams and atomic layer plasma chemical etching. This technology makes it possible to modify surface of Si substrates in the required topology and geometry, followed by removal of atoms to obtain nanoscale elements. The influence of parameters of method of focused ion beams and plasma chemical etching on parameters of the formed structures is analyzed. So, for example, for formation of structures with maximum roughness, it is necessary to increase values of parameters responsible for reactive ion etching, these are such parameters as: the power of capacitive plasma source, the mixing voltage, and the flow rate of an inert gas (argon).

1. Introduction

Silicon nanotechnology is the foundation of the entire world micro- and nanoelectronics. Silicon (Si) in its various forms is used as a base-substrate for further formation of working surfaces of nanoscale devices and an element base based on it. Silicon nanostructures have become so common due to the distinctive physicochemical parameters and the relative ease of obtaining, forming and processing the material, as well as an ability to apply various types of materials and substances on its surface. However, the main principle of nanotechnology is miniaturization of element base in order to increase efficiency, reduce energy consumption and structural dimensions, which at this stage of development on silicon is becoming an increasingly difficult and expensive task, and soon it will become completely impossible, because of it is impossible to create a workable on the basis of silicon transistor with a gate length less than a few nanometers.

After transition to the 28 nm process technology in the industry, for the first time, a situation arose when the next generation transistor was more expensive, not cheaper, than the previous generation transistor. To date, there are only three companies that continue to miniaturize below the 28 nm norm - Intel, TSMC and Samsung. The entire world community is busy solving this problem, actively looking for new materials and modern manufacturing technologies.

This paper presents a study of the use silicon Si for element base manufacture of micro- and nanoelectronics by using combined methods of focused ion beams and atomic layer plasma chemical etching. This technology makes it possible to modify the surface of substrates in the required topology and geometry, followed by removal of atoms to obtain nanoscale elements. The essence of the technique is that embedded ions are used as positive photoresist masks for plasma treatment in the sense that this area, due to the specificity of interatomic interaction, will interfere with chemical processes of material removal.

2. Experiment

On the surface of silicon substrates, local modification of 5x5 μm regions was carried out with a focused Ga^+ ion beam at an accelerating voltage U_{acc} = 30 keV and an ion beam current I_{ib} = 10 pA with a different number of beam passes (10 - 100) and different doses (2.5 - 25 pC/μm^2), respectively. For this, samples were placed in the vacuum chamber of the FIB Nova NanoLab 600 module and oriented in such a way that the flow of accelerated ions hit the substrate in the normal direction. The working vacuum during the exposure to the beams was maintained at a level of $1 \div 2 \times 10^{-4}$ Pa.

Plasma treatment of samples with modified regions was carried out in fluorine-containing plasma with the following parameters: fluorine-containing gas flow N_{SF6} = 15 cm^3/min, argon flow N_{Ar} = 100 cm^3/min, pressure in the reactor P = 2 Pa, power of the inductively coupled plasma source W_{ICP} = 200 W , the power of the capacitive plasma source W_{RIE} = 35 W, the bias voltage U_{bias} = 24 V, the processing time is 30 - 120 seconds.

An analysis of the obtained results shows that, depending on the combination of modes of focused ion beams and plasma chemical etching, two types of action can be realized: masking of substrate surface and activation of substrate etching during etching.

3. Results and Discussion

Based on the results of experimental studies, the dependences of the height of obtained structures on number of passes of ion beam and different etching times in fluorine-containing plasma have been obtained.

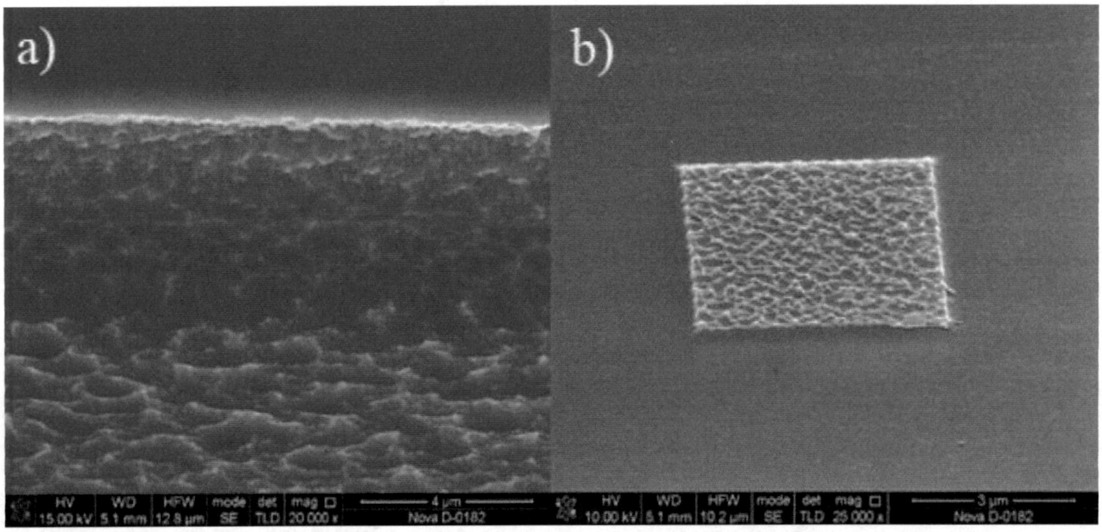

Figure 1 (a, b). SEM - image of the structure of silicon surface after local modification by focused ion beam method at 100 passes of ion beam: a) side face, b) top view

The analysis shows that with the specified combination of modes, the locally modified region is characterized by a lower etching rate (V_{moSi} = 0.019 nm/sec) than the unmodified region of the substrate surface (V_{Si} = 2.21 nm/sec). The masking effect can be explained by the formation of a layer saturated with embedded Ga^+ ions in the near-surface, locally modified region of the silicon substrate, which are characterized by a greater inertness with respect to fluorine-containing plasma ions than silicon atoms.

The obtained dependences show that with an increase in the number of ion beam passes from 10 to 100 at an etching time in plasma of 120 seconds, height of resulting structures varied from 7 to 266 nm. This is explained by the higher exposure dose to the sample surface and, as a consequence, by the different etching rates of the modified regions. Also, with an increase in etching time in plasma from 30 to 120 seconds with the same number of ion beam passes, height of structures after plasma-chemical etching increases from 10 to 266 nm.

Figure 2. AFM - image of the structure of silicon surface after plasma chemical treatment in a fluorine-containing plasma for 120 seconds

Thus, with plasma chemical etching for 120 seconds with an increase in the number of passes from 10 to 100, the height of structures increases from 7 to 266 nm, respectively. With plasma chemical treatment for 30 seconds with 10 passes of ion beam, the etching of modified region is stimulated and a deep structure of 6 nm is formed. This is explained by the appearance of defects due to implantation of gallium ions into the near-surface layers of locally modified region and formation of a damaged layer, which is removed faster during plasma chemical etching than unmodified surface. However, at 100 passes of ion beam with plasma etching for 30 seconds, the modified regions show a masking effect.

AFM images of structures were obtained using the Ntegra probe nanolaboratory (NT-MDT, Russia), and ion processing and acquisition of SEM images of the high-tech complex Nova NanoLab 600. The etching process was carried out on a device for plasma-chemical processing of materials in a combined capacitive and inductive discharge plasma STE ICPe68 (Scientific and Technological Equipment, Russia).

4. Conclusions

The experimental studies carried out made it possible to establish the dependences of parameters influence and modes of the process of focused ion beams and plasma chemical etching on parameters of the formed nanoscale structures, made it possible to establish dependence of rate and time of plasma chemical treatment on the morphology of structures on silicon surface. So, for example, for formation of structures with maximum roughness, it is necessary to increase values of parameters responsible for reactive ion etching, these are such parameters as: power of the capacitive plasma source, mixing voltage, and flow rate of an inert gas (argon). To obtain structures with minimal roughness, it is necessary to increase values of parameters responsible for plasma chemical etching, these are such parameters as: power of the inductively coupled plasma source, flow rate of a fluorine-containing gas (sulfur hexafluoride). Also, an increase in plasma chemical etching speed leads to an increase in the roughness parameter and an increase in the angle of inclination of the structure.

Acknowledgments
This work was supported by a grant from the Russian Science Foundation № 20-69-46076 and the Russian Foundation for Basic Research, project number 19-38-60052. The results were obtained using the equipment of the Scientific and Educational Center and the Center for Collective Use "Nanotechnology" of the Southern Federal University. The obtained research data will be included in the project "Development of an ionization gas sensor with a sensitive element based on a graphene-like film for chemical and environmental safety means" of the winner of the contest "Participant of the youth scientific and innovative competition" ("UMNIK").

References

[1] Tok E S, Neave J H, Zhang J, Joyce B A and Jones T S 1997 *Surf. Sci.* **374** 397
[2] Shiraishi K and Ito T 1998 *Phys. Rev. B* **57** 6301
[3] Kley A, Ruggerone P and Scheffler M 1997 *Phys. Rev. Lett.* **79** 5278
[4] Huffaker D L, Park G, Zou Z, Shchekin O B and Deppe D G 1998 *Appl Phys. Lett.* **73** 2564
[5] Vakulov Z E, Klimin V S, Rezvan A A, Tominov R V, Korzun K, Kots I N, Polyakova V V and Ageev O.A. 2019 *J. Phys.: Conf. Ser.* **1410** 012042
[6] Klimin V S, Rezvan A A and Morozova J V 2019 *J. Phys.: Conf. Ser.* **1410** 012035
[7] Wong P S, Liang B and Huffaker D L 2010 *J. Nanosci. Nanotechnol.* **10** 1537–1550
[8] Grundmann M, Stier O and Bimberg D 1995 *Phys. Rev. B* **52** 11969
[9] Yamaguchi K, Yujobo K and Kaizu T 2000 *Jpn. J. Appl. Phys.* **39** L1245–L1248
[10] Gusev E Y, Jityaeva J Y, Avdeev S P and Ageev O A 2018 *J. Phys.: Conf. Ser.* **1124** 0220345.
[11] Romashko R V, Kulchin Y N, Bezruk M N and Ermolaev S A 2016 *Quantum Electronics* **46(3)** 277
[12] Ergun S, Yaralioglu G and Khuri-Yakub T 2003 *J. Aerosp. Eng.* **16** 76.
[13] Klimin V S, Kots I N, Tominov R V, Rezvan A A, Varzarev Y N and Ageev O A 2019 *J. Phys.: Conf. Ser.* **1410** 012041
[14] Qiu Y, Wang H, Demore C E M, Hughes D A, Glynne-Jones P, Gebhardt S, Bolhovitins A, Poltarjonoks R, Weijer K, Schönecker A, Hill M and Cochran S 2014 *Sensors* **14** 14806.
[15] Judy J W 2001 *Smart Mater. Struct.* **10** 1115.
[16] Michael De Volder F L, Tawfick S H, Baughman R H and Hart A J 2013 *Science* **339** 535—539
[17] Franklin A D et al 2012 *Nano Lett* **12** 758—762
[18] Atul Dhale, Fahim Khan 2013 I*nternational Journal of Engineering Research and Development* **7** 80.
[19] Seymour J P, Wu F, Wise K D and Yoon E 2017 *Microsystems & Nanoengineering* **3** 16066
[20] HajjHassan M, Chodavarapu V and Musallam S 2008 *Sensors* **8** 10 6704-6762
[21] Avdeev S P and Gusev E Yu 2019 *Proc. of SPIE* **11022** 11022-69
[22] Klimin V S, Tominov R V, Avilov V I, Dukhan D D, Rezvan A A, Zamburg E G, Smirnov V A and Ageev O A 2019 *Proc. of SPIE* **11022** 110220E-1
[23] Klimin V S, Rezvan A A, Abramovich T S, Zubova T A, Tominov R V, Vakulov Z E, Kots I N and Ageev O A 2020 *J. Phys.: Conf. Ser* **1695** 012199
[24] Balakirev S V, Eremenko M M, Mikhaylin I A and Solodovnik M S 2018 *J. Phys.: Conf. Ser.* **1124** 022025
[25] Filippov S V, Popov E O, Kolosko A G and Romanov P A 2015 *J. of Phys.: Conf. Ser.* **643** 012101

ZnO nanorods coating modified with $AgInS_2$ quantum dots

A A Ryabko, O A Korepanov, A A Bobkov, O A Aleksandrova, V A Moshnikov

Saint-Petersburg State Electrotechnical University «LETI», 5 Prof. Popova Street, Saint Petersburg, 197376, Russia

a.a.ryabko93@yandex.ru

Abstract. This paper presents the effect on the visible light photoresponse caused by the modification of ZnO nanorod coating with colloidal $AgInS_2$ quantum dots. The modification of ZnO nanorods by immersion in colloidal solution results in the enhancement of visible absorption and photoresponse. Obtained results indicate the possibility of local heterojunctions between ZnO and $AgInS_2$.

1. Introduction

Zinc oxide (ZnO) is an n-type semiconductor with a direct bandgap (3.37 eV at room temperature). ZnO nanorods could be applied in photovoltaic structures [1,2], resistive gas sensors [3], photocatalysis [4]. In the case of photocatalyst, modification of ZnO nanorods with QDs shifts the photoactivation region of the photocatalysis process from the ultraviolet to the visible region [5]. In the case of adsorption resistive gas sensors, ZnO nanorods sensibilization with colloidal QDs could provide gas analysis response at room temperature (no high-temperature heating requirement) [6].

Zinc oxide particles and ternary metal chalcogenides QDs ($AgInS_2$ and $CuInS_2$) are thought to be biocompatible, non-toxic materials. Ternary metal chalcogenides QDs are usually compared with binary ones containing lead, cadmium, or tellurium [7]. The composition of ZnO and ternary metal chalcogenide QDs could be a convenient combination from the toxicology point of view. This work aims at the production and study of ZnO nanorods coating modified with $AgInS_2$ QDs.

2. Experiment

The ZnO nanorod arrays were formed in two stages [8]. At the first stage, a seed layer of ZnO was deposited by ultrasonic spray pyrolysis of a zinc acetate aqueous solution. At the second stage, ZnO nanorods were grown-up in an aqueous solution of zinc nitrate and hexamethylenetetramine by a low-temperature hydrothermal method. After synthesis, the coating of ZnO nanorods was annealed at a temperature of 500°C for 5 minutes to remove organic molecules from the surface. $AgInS_2$ QDs stabilized by 3-mercaptopropionic acid were synthesized in aqueous medium by the heating-up method [9]. 3-mercaptopropionic acid has a sufficiently short molecular chain to ensure the charge carriers' transport between the $AgInS_2$ colloidal QD and the ZnO nanorod [10]. The modification of ZnO nanorods with $AgInS_2$ QDs was performed by substrate immersion in the colloidal solution.

The modified arrays of zinc oxide nanorods were studied by measuring the photocurrent under irradiation in the visible region. For this, nanorods were grown on ceramic substrates with interdigitated electrodes (Sensor Platform, Tesla Blatna, figure 1). The width of the interdigitated electrodes and the distance between them is about 25 μm, the electrode materials are NiCr/Ni/Au. Blue

LED (λ = 460 nm) was used as a visible light source (figure 1). The current through the sample was recorded using Keithley 6485 at the applied voltage of 5 V. Before measuring the photocurrent, the samples were preheated at 150 °C for 10 minutes to reduce the effect of adsorbed moisture.

Figure 1. Ceramic substrate with interdigitated electrodes (Sensor Platform, Tesla Blatna).

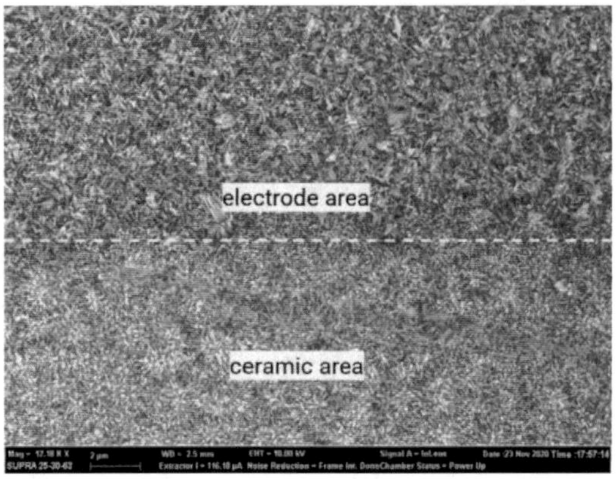

Figure 2. SEM image of ZnO nanorod arrays grown on a ceramic substrate with interdigitated electrodes.

ZnO nanorods cover the entire ceramic substrate with interdigitated electrodes. The area between the electrodes in the SEM image looks brighter due to the surface charging during the scan with an electron beam (figure 2).

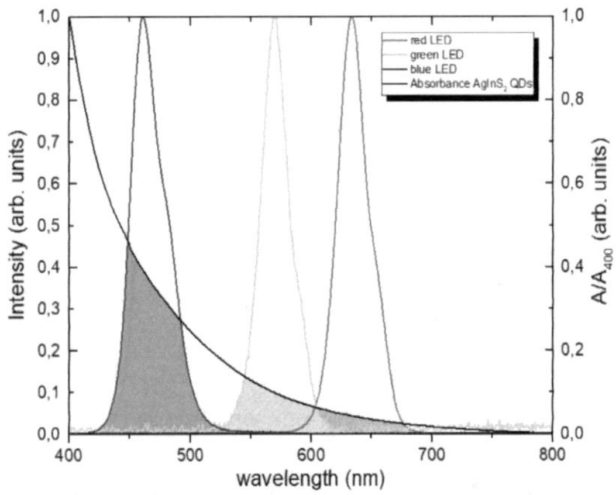

Figure 3. LED emission spectra and $AgInS_2$ QDs absorption spectrum.

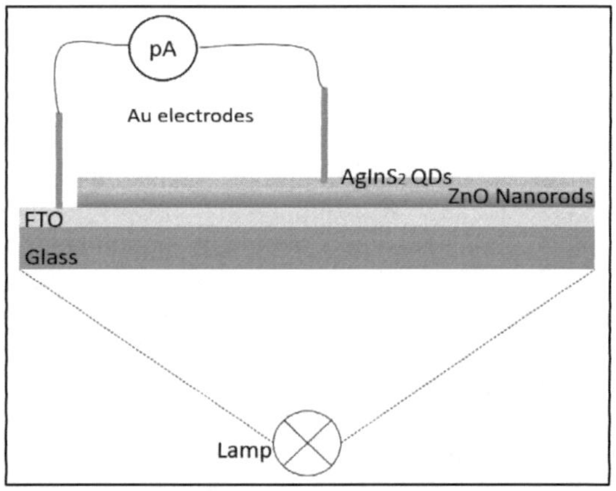

Figure 4. Sketch of the measurement of the short-circuit current of an FTO/ZnO/$AgInS_2$ structure.

Blue LED was chosen as a light source due to the high overlap between its emission spectrum and $AgInS_2$ QDs absorption spectrum.

The surface presence of QDs after immersion was controlled by absorption spectroscopy. For this, nanorods were grown on a quartz substrate.

Finally, the nanorods were grown on an FTO (fluorine-doped tin oxide) coated glass substrate. On which the layer of QDs from aqueous-alcoholic colloidal solution was deposited by the spin-coating method (2500 rpm, 30 sec, 20 cycles). Then, the short-circuit current was measured with gold pressure contact when irradiated with a lamp (figure 4).

3. Results and discussion

As it can be seen from figure 5, after modification the absorption increases due to additives of ZnO and QDs absorption. Photoresponse of a ZnO nanorods array on a ceramic substrate with interdigitated electrodes before and after $AgInS_2$ QDs modification are presented in figure 6. Nanorods without QDs modification show ~3.7 times conductivity change (I_{light}/I_{dark} ratio) under blue light irradiation (for ~5 min), which is due to the influence of intrinsic defects [11]. Modification of nanorods with $AgInS_2$ QDs leads to an increase in the photoresponse of the sample up to ~ 13.4 times (for ~ 5 minutes), which could point to the formation of local heterojunctions between ZnO and $AgInS_2$. ZnO nanorods resistivity after QDs modification could decrease due to the appearance of additional electrically conductive paths, which could arise with the embedding of colloidal QDs between the zinc oxide nanorods at their base.

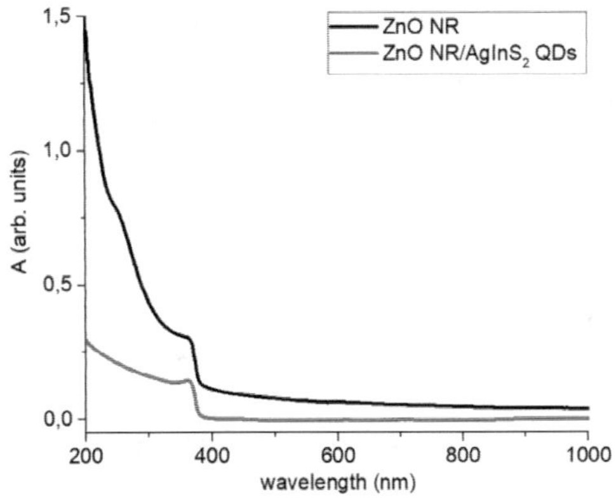

Figure 5. Absorption spectra of ZnO nanorod array on quartz substrate before and after modification with $AgInS_2$ QDs

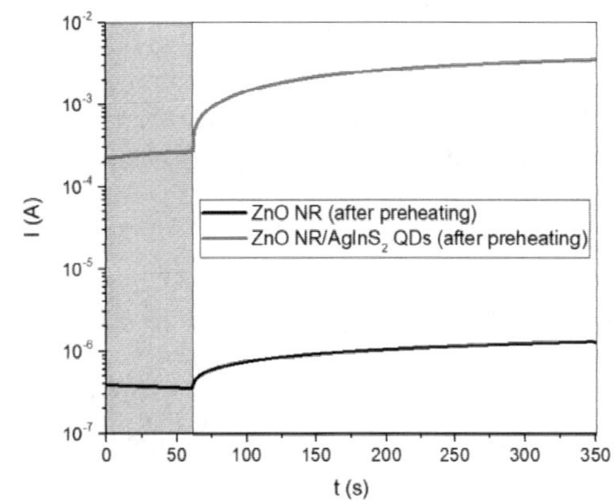

Figure 6. Photoresponse of a ZnO nanorods array on a ceramic substrate with interdigitated electrodes before and after $AgInS_2$ QDs modification. The radiation source is a blue LED (460 nm). The area without irradiation is highlighted with grey.

The photo of ZnO nanorods coating on an FTO coated glass substrate before and after modification with the $AgInS_2$ QDs, and corresponding absorption spectra are shown in Figure 7.

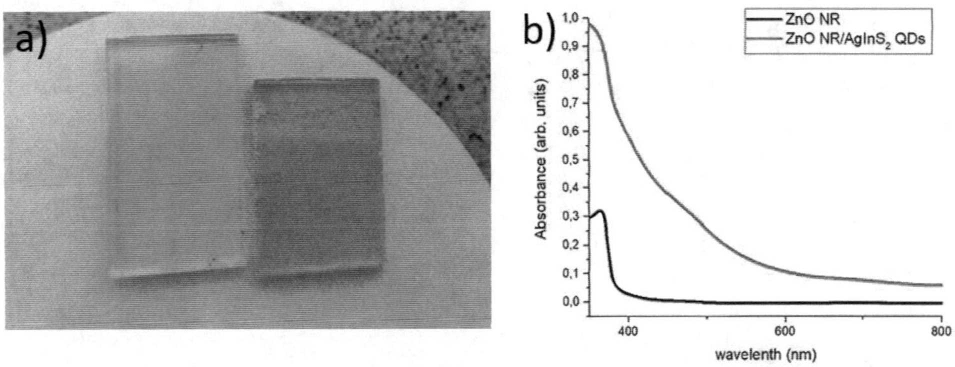

Figure 7. The evolution of an $FTO/ZnO/AgInS_2$ absorption: a) the photo of the sample before and after modification with $AgInS_2$ QDs, b) the corresponding absorption spectra.

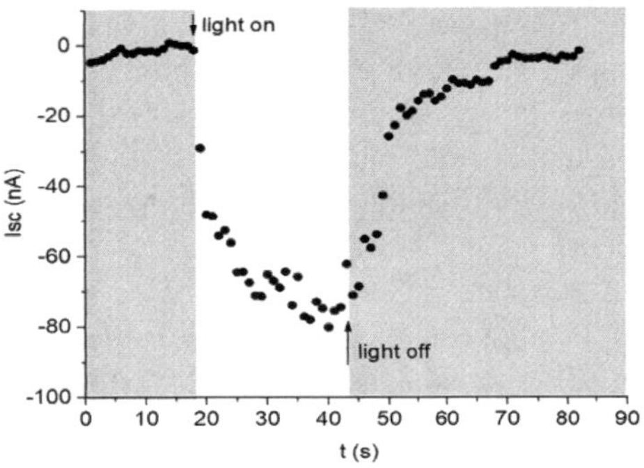

Figure 8. Short-circuit current of FTO/ZnO/$AgInS_2$ by fluorescent lamp irradiation. The area without irradiation is highlighted with grey.

As can be seen from Figure 7, after 20 centrifugation cycles, a layer of $AgInS_2$ QDs is formed, which ensures the absorption of the sample in the visible region (more than 380 nm). In this case, the registration of the short-circuit current during irradiation of the structure indicates the formation of a ZnO/$AgInS_2$ heterojunction (Figure 8).

4. Conclusion

Modification of ZnO nanorod array with $AgInS_2$ QDs leads to an increase in the photoresponse to visible light irradiation. This may indicate the formation of local ZnO/$AgInS_2$ heterojunctions. Short-term annealing at 150 °C does not destroy the colloidal nanocrystals, but allows partial removal of moisture from the surface of the nanostructured coating. Obtained results indicate the possibility to create gas sensors, photocatalysts, and photovoltaic structures based on the ZnO nanorods with $AgInS_2$ colloidal QDs.

Acknowledgments

The reported study was funded by RFBR, project number 19-38-90088.

References

[1] Lashkova N A, Maximov A I, Ryabko A A, Bobkov A A, Moshnikov V A and Terukov E I 2016 *Semiconductors* **9** 1254-1260
[2] Zang Sh, Wang Y, Su W, Zhu H, Li G, Chang X and Liu Y 2016 *Phys. Status Solidi RRL* **10** 745–748
[3] Bobkov A, Varezhnikov A, Plugin I, Fedorov F, Trouillet V, Geckle U, Sommer M, Go V, Moshnikov V and Sysoev V 2019 *Sensors* **19** 4265
[4] Yukhnovets O, Semenova A A, Levkevich E A, Maximov A I and Moshnikov V A 2018 *IOP Conf. Ser. J. Phys. Conf. Ser.* **993** 012009
[5] Yang Y, Que W, Zhang X, Xing Y, Yin X and Du Y 2016 *Journal of Hazardous Materials* **317** 430–439
[6] Nalimova S S, Maximov A I, Moshnikov V A, Bobkov A A, Mazing D S, Ryabko A A, Levkevich E A and Semenova A A 2019 *IEEE International Conference on Electrical Engineering and Photonics (EExPolytech)* **8906789** 223-225
[7] Hong S P, Park h K, Oh J H, Yang H and Do Y R 2012 *Journal of Materials Chemistry.* **22** 18939
[8] Ryabko A A, Maximov A I, Verbitskii, Levitskii V S, Moshnikov V A and Terukov E I 2020 *Semiconductors* **54** (11) 1257-1263
[9] Korepanov O A, Mazing D S, Aleksandrova O A and Moshnikov V A 2019 *Journal of Physics: Conference Series* **1410** (1) 012024
[10] Carey G H, Abdelhady A L, Ning Zh, Thon S M, Bakr O M and Sargent E H 2015 *Chem. Rev.* **115** 12732
[11] Khokhra R, Bharti B, Lee H-N and Kumar R 2017 *Scientific Reports* **7** 15032

Influence of ultrasonic treatment on the change of monocrystalline silicon defective region

A A Solovyev, V V Rybin and A V Kulagin

Ulyanovsk State University, 42 Leo Tolstoy Str., 432017 Ulyanovsk, Russia

vladrib0880@gmail.com

Abstract. The article presents the results of an experimental study of ultrasonic action on monocrystalline silicon samples. The influence of the processing modes on the surface strength of the material under study was found.

1. Introduction

It is known that ultrasonic treatment of silicon crystals leads to a change in the dislocation movement speed at a constant mechanical load [1]. In this case, the nature of the dislocation velocity change is determined by the sign of the external stresses acting on the sample. Compressive forces lead to a decrease, and tensile forces lead to an increase in the dislocation velocity. In addition, the dislocation movement activation energy decreases and the electroplastic effect is enhanced.

It should be emphasized that by now the optical phenomena associated with dislocations movement in the field of intense ultrasound have been well studied [2]. The formation of a surface hardened layer in dislocation materials, including semiconductors, during ultrasonic treatment of the sample surface is also a known effect. The standard explanation for this hardening is an increase in the concentration of dislocations in the surface layer [3].

An interesting aspect of dislocations mobility studying is to reveal the relationship between electronic excitation of dislocations and ultrasonic excitation.

There are practically no data in the literature on the effect of ultrasound on the dislocations mobility in silicon single crystals in the internal stresses field. Therefore, this work is devoted to the study of the ultrasonic treatment effect of silicon single crystals on the defect region dynamics in the field of internal stresses.

2. Experimental technique

In this work, p-Si samples (ρ = 1 Ohm·cm, surface orientation (100)) with formed stress concentrators – scratches (load during scribing 0.5–2.0 N), being in distilled water, were subjected to ultrasonic processing with a signal from the ultrasonic generator at a frequency of 63.4 kHz. The processing time varied from 30 min to 5 hours, after which the samples were annealed at a temperature of T = 650 °C for t = 180 min. At a given annealing time, the introduced internal stresses completely relax and dislocation transport stops. To experimentally study the distribution of dislocations in crystals, we used the method of dislocations selective etching.

3. Results and discussion

3.1 Influence of internal stresses on dislocation movement in silicon

The indentation of the crystal surface leads to the appearance of stresses. Their relaxation upon high-temperature annealing can lead to the generation and displacement of linear defects in the absence of external mechanical disturbances. Indeed, the dislocations movement lowers the internal stresses formed during the indentation of the surface. Therefore, one of the primary tasks was to study the linear defects transport in silicon crystals in the field of internal stresses in the absence of ultrasonic treatment.

For this, the prepared samples with stress concentrators were subjected to high-temperature isothermal annealing (650 °C) for 10 - 240 minutes. The research results showed (figure 1) an increase in the depth of propagation of linear defects from the starting position during annealing. Their movement rate, determined by the profile of elastic stresses, gradually decreased and after ~ 80 minutes of high-temperature isothermal holding practically stopped. In this case, the maximum dislocation path from the stress concentrator was limited to 25 - 30 μm.

Another feature of this studies series is a clearly recorded delay in transport processes at the initial stages of isothermal annealing. Thus, within the first 10 min, it was not possible to detect any dislocations displacement; however, in the next 5 min, the leading dislocations path reached already 10 μm. The dislocation transport start delay on Si and Ge (within 10 - 1200 s) under external loading of the crystal was observed earlier by other authors [4-7] and was associated with the conditions for removing linear defects to the starting position, temperature pretreatment regimes, and experimental conditions.

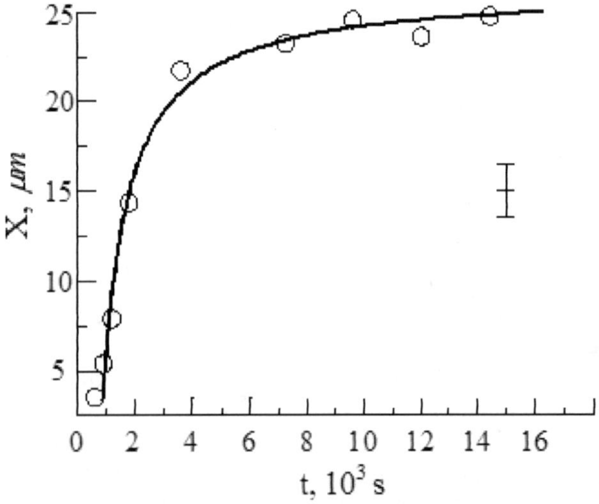

Figure 1. Dependence of the distance traveled by the head dislocations from the scratch (P = 1.75 N) on the high-temperature (T = 650 °C) treatment time.

The results obtained made it possible to describe the temporal changes in the linear defects transport in the field of internal stresses in the form of an empirical dependence:

$$X = \alpha X_\infty \left(1 - \frac{t^0}{t}\right) \quad (1)$$

Here X_∞ – the distance to which dislocations move from the crack at $t \to \infty$; t^0 – time delay of defects movement start; *a* – step function meeting the requirements:

$$\alpha = \frac{f + |f|}{2f} = \begin{cases} 1, \ t > t^0 \\ 0, \ t^0 \geq t > 0 \end{cases}, \text{ where } f = \frac{t - t^0}{t}.$$

Approximation of the experimental data by equation (1) (solid line in Figure 1) made it possible to calculate the values of the dislocations maximum path X_∞ = 24.4 μm at $t \rightarrow \infty$ and the delay time of the onset of their displacement t^0 = 656.9 s. Consequently, at a larger X_∞ distance, the relaxation of residual elastic stresses cannot be realized by squeezing out dislocations from a scratch.

Thus, to determine the ultrasonic treatment effect on the movement of linear defects, the prepared samples were annealed at a fixed temperature T = 650 °C for t = 180 min. At a given annealing time, the introduced internal stresses completely relax and dislocation transport stops. In this case, the maximum distance of dislocations from stress concentrators is determined only by the microhardness of the surface layers.

3.2 Influence of ultrasonic treatment on the transport of linear defects in silicon

The research results showed a decrease in dislocation paths from stress concentrators with processing time (figure 2). In addition, a decrease in the number of etch pits in the selected slip lines is observed (figure 3).

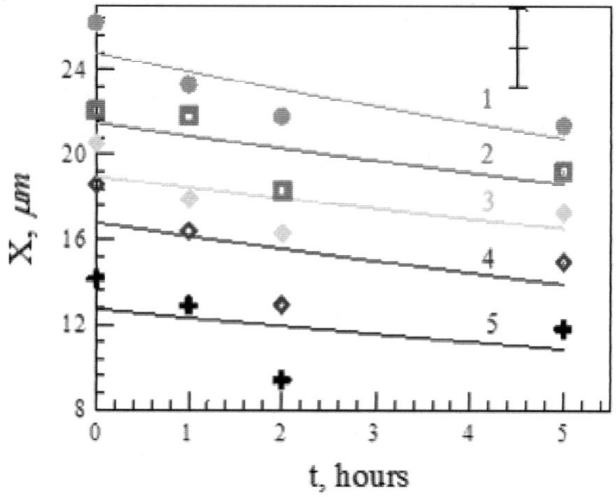

Figure 2. Dependence of the dislocations maximum path in slip lines on the ultrasonic treatment time at various scribing loads: 1 - 2 N; 2 - 1.75 N; 3 - 1.5 N; 4 - 1 H; 5 - 0.5 N. Annealing at T = 650 ° C, t = 3 h.

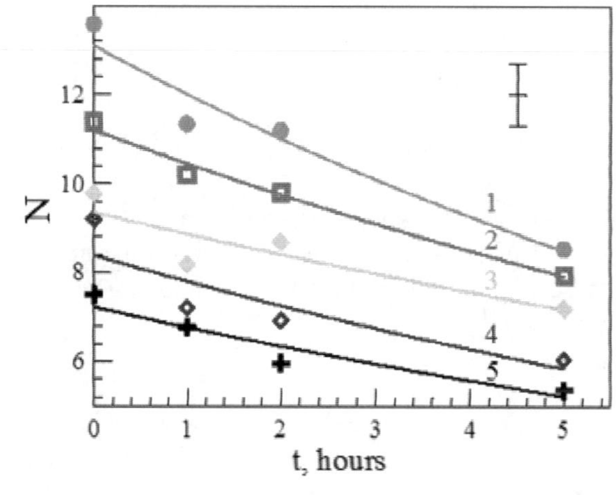

Figure 3. Dependence of the number of locations in slip lines on the ultrasonic treatment time at various scribing loads: 1 - 2 N; 2 - 1.75 N; 3 - 1.5 N; 4 - 1 H; 5 - 0.5 N. Annealing at T = 650 ° C, t = 3 h.

In this case, a regular increase in the controlled parameters with the load on the indenter is clearly recorded when the crystal surface is scribed. The observed decrease in the length of dislocation paths after ultrasonic treatment indicates a decrease in the mobility of dislocations, i.e. the phenomenon of surface hardening is observed.

It should be noted that the action of ultrasonic vibrations by itself did not lead to displacement of dislocations. Clear changes in the dynamics of dislocations are recorded only in the case when isothermal annealing of crystals was carried out after UST. The obtained experimental results can be explained by the influence of a number of factors.

It is known that stimulated diffusion of vacancies and impurities occurs under the action of an ultrasonic wave [8-10]. Taking into account the fact that in our experiments the dislocation half-loop consists of short segments with their ends reaching the surface, as well as the fact that the surface and dislocations play the role of an almost infinite sink of point defects, it can be assumed that the observed effects are associated with an increase in the concentration of point defects under the influence of ultrasonic treatment.

The physical processes of ultrasonic treatment can be associated with the occurrence of acousto-stimulated diffusion of impurities in the samples and an increase in the concentration of impurities in the near-surface layer of silicon. This leads to the predominant nucleation of double kinks on impurities and to an increase in the activation energy of dislocation motion. In addition, the presence of point defects and impurities on the surface and in the bulk, as is known, affects the electronic processes in crystals. That reasons of dislocations motion velocities and the surface microhardness change of the crystal being treated [11].

To check the change in the microhardness of the surface of a single crystal under various modes of ultrasonic processing, studies were carried out to measure the microhardness of crystals. The research results recorded a regular increase in the surface microhardness with the time of ultrasonic treatment (figure 4).

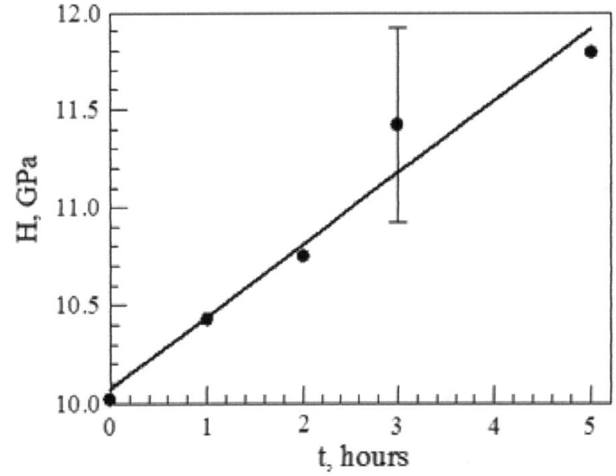

Figure 4. Dependence of the microhardness of crystal surface on the ultrasonic treatment time. Load 100 g at 10 s.

Thus, the ultrasonic treatment of silicon samples containing short near-surface dislocations leads both to a change in the maximum displacement of dislocations with the processing time and to an increase in the microhardness of the crystal surface. It caused by the strengthening of the silicon crystals under study after UST.

4. Conclusion

To check the change in the surface hardness of single crystals under various modes of ultrasonic treatment, a series of experiments was carried out to measure the microhardness of the test samples. The research results recorded a regular increase in the surface microhardness with the time of ultrasonic treatment, which is consistent with the assumptions made on the strengthening of the silicon crystals.

References

[1] Olikh O Y 2018 *Superlattices Microstruct.* **117** 173–188.
[2] Shi C, Shen K, Mao D, Zhou Y and Li F 2018 *Mater. Sci. Technol.* **34**(12) 1511–18.

[3] Mizutani K, Wakatsuki N and Ebihara T 2016 *Jpn J Appl Phys (2008)* **55**(7) 07KA02.
[4] Fan H, Wang Q, El-Awady J A, Raabe D and Zaiser M 2021 *Nat. Commun.* **12**(1) 1845.
[5] Rost H-J, Buchovska I, Dadzis K, Renner M and Menzel R 2020 *J. Cryst. Growth* **552** 125842.
[6] Patel J R 1956 *Phys. Rev. Lett.* **101** 1436–37.
[7] Scandian C, Azzouzi H, Maloufi N Michot G and George A 1999 *Phys. Status Solidi (A) Appl. Mater. Sci.* **171** 67–82.
[8] Ostrovskii I V, Korotchekov O A, Goto T and Grimmeiss H G 1999 *Phys. Reports* **311**(1) 1–46.
[9] Pfahl V, Ma C, Arnold W and Samwer K 2018 *J. Appl. Phys.* **123**(3) 035301.
[10] Olikh O and Voytenko K 2016 *Ultrasonics* **66** 1–3.
[11] Peleshchak R M, Kuzyk O V and Dan'kiv O O 2016 *Ukr. J. Phys.* **61**(8) 741–6.

Effect of wet chemical treatment on the properties of GaAs FIB-modified surface

N A Shandyba, N E Chernenko, J Y Zhityaeva, O I Osotova, M M Eremenko, S V Balakirev and M S Solodovnik

Institute of Nanotechnologies, Electronics and Equipment Engineering, Southern Federal University, Taganrog 347922, Russia

shandyba@sfedu.ru

Abstract. We present the results of studies of the effect of wet chemical treatment on the properties of a GaAs surface modified by a gallium focused ion beam. Our studies based on results of AFM, KpAFM and Raman spectroscopy measurements have shown that, during wet chemical treatment, the damaged areas disappear completely in the case of low accelerating voltages and small doses of ions. At the same time, large accelerating voltages lead to the formation of extended damaged regions, the complete removal of which requires a longer treatment or additional processing.

1. Introduction

Despite active research recently, the problem of positioning self-organizing nanostructures (quantum dots, nanowires, etc.) is still urgent [1 – 4]. Accurate control of the position of nanostructures, as well as their size, is necessary for a wide range of promising devices in nanoelectronics and nanophotonics [5 – 13]. The complexity of solving this problem is largely because in epitaxy the most common approach for ordering structures has become an approach based on surface modification by various methods: from conventional lithography to local techniques based on ion beam treatment and local anodic oxidation [2, 14 – 17].

However, any treatment of growth surface inevitably leads to degradation of its properties, including damage of crystal structure of surface and subsurface layer and surface contamination with unwanted impurities. At the same time, epitaxial synthesis is extremely demanding on the quality of the substrate surface. This requires the development of methods for the pre-growth treatment of modified surfaces, which would improve their quality for subsequent epitaxial synthesis. This is very important for obtaining electrically and optically active nanostructures.

In this paper we present the results of studies of the effect of selective wet chemical etching on the properties of GaAs (001) surfaces modified by the Ga^+ focused ion beams (FIB).

2. Experiment

For experimental studies we formed regular arrays of nanoholes on the GaAs (001) surface (epi-ready substrate) using by the FIB technique at various modification mode. The accelerating voltage of the ion beam was 10 and 30 kV, the ion beam current was 0.3 pA for voltage of 10 kV and 10 pA – 30 kV which is due to the peculiarities of the FIB system [2]. The number of FIB passes was varied in the range from 1 to 300. After that, the samples were treated (etched) in an HCl solution for 3 min. Then all the samples, including the control one (with only FIB modification), were study by atomic force microscopy (AFM), Kelvin probe AFM (KpAFM) and Raman spectroscopy to obtain information about changes in surface morphology, surface potential and crystal structure, respectively.

3. Results and discussion

The results of the AFM study showed that an increase in the accelerating voltage does not lead to significant changes in the depths of the nanoholes. Surface profiles at the Figure 1 show the depth of the nanoholes for the limiting FIB treatment modes is comparable (30 – 35 nm). Perhaps this is due to the significantly larger aspect ratio of the surface structures obtained at higher accelerating voltages and, accordingly, the limited capabilities of the AFM (probe radius). At the same time, the diameter of the nanoholes increases insignificantly, which is due to the large diameter of the ion beam at 30 kV. However, after wet chemical etching of prepared samples, the character of sample morphology is different for different FIB voltage. As can be seen from Figure 1, at low accelerating voltages, the diameter of the nanoholes did not change, but the depth decreased. At the same time, at 30 kV, the diameter decreased, and the depth increased. We attribute this to the fact that at low voltages, implanted Ga^+ ions are accumulated in a thin subsurface area around nanohole.

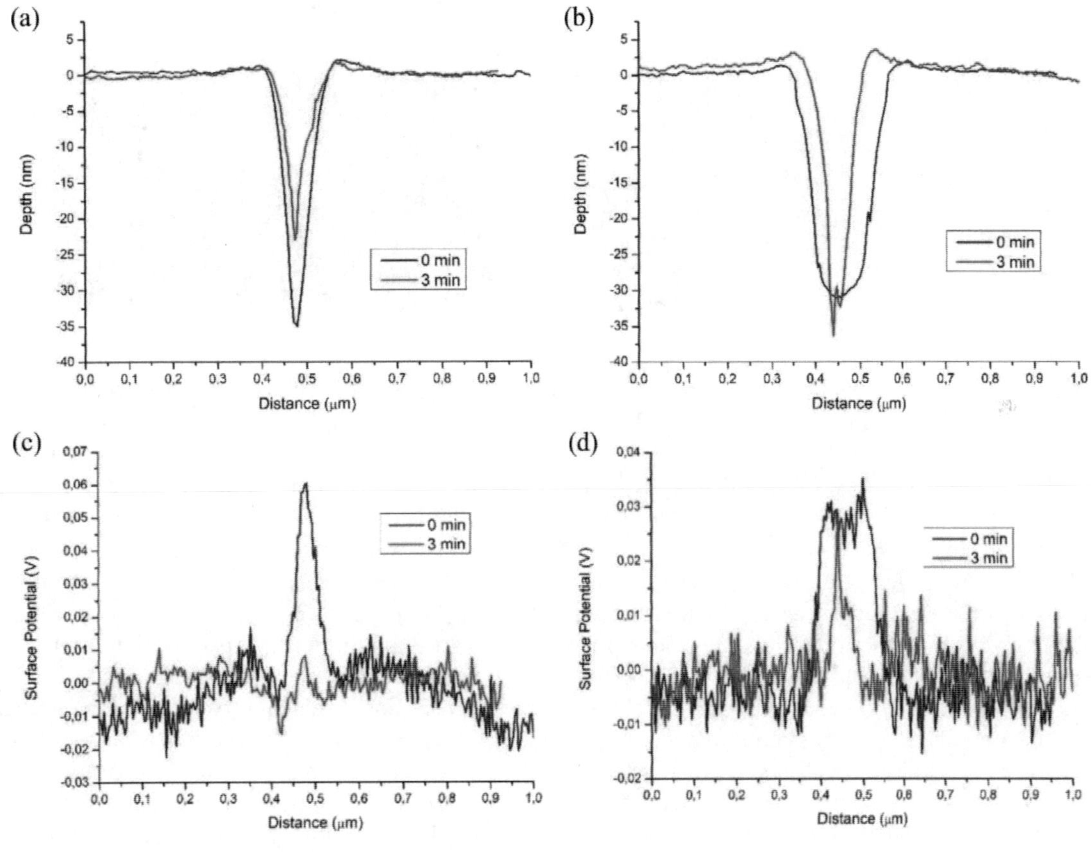

Figure 1. AFM profiles of GaAs surface morphology (a, b) and of potential distribution (c, d) for samples before (0 min) and after wet chemical treatment (3 min) obtained at different FIB-modes: 10 keV, 0.3 pA (a, c) and 30 keV, 10 pA (b, d).

Considering the inevitable process of chemical etching of the unmodified GaAs surface, this leads to a decrease in nanohole depth. At the same time, at high voltages, the Ga^+ ions penetration depth increases. This leads to the formation of extended damaged areas, which significantly increase the surface structure depth during chemical etching. The unequal character of changes in the morphology of nanoholes formed at different voltages may indicate a different etching rate of damaged and original surface areas.

Analysis of the surface potential distribution by the KpAFM (Figure 1 (c, d)) showed that in the case of low accelerating voltages, the implanted Ga^+ ions are localized at the near surface layer, which leads to higher surface potential values (+0,06 V) in comparison with the points formed at higher voltages (+0,03 V), when ions are distributed over a larger volume of the substrate, but with a lower volume density. This leads to the fact that after wet chemical treatment in areas obtained at low accelerating voltages and implantation doses, potential peaks disappear, while at points with a high accelerating voltage, the potential value practically does not change staying at +0.03 V.

It should be noted that, in all cases, a system of concentric areas with different values and signs of the surface potential is formed around the nanohole. The size of the regions reaches 0.5 μm in diameter and weakly depends on the FIB modes and the morphology of the nanohole. We attribute this to the redistribution of mobile charge carriers in the near-surface GaAs layer due to the formation of a charged region of implanted Ga^+ ions. This assumption is supported by the fact that, after wet chemical etching, these regions disappear around the nanoholes formed by the ion beam at low accelerating voltages and remain around ones formed at high voltages (Figure 2). This also additionally confirms the formation of more extended damaged areas around nanoholes during processing with high-energy ion beams.

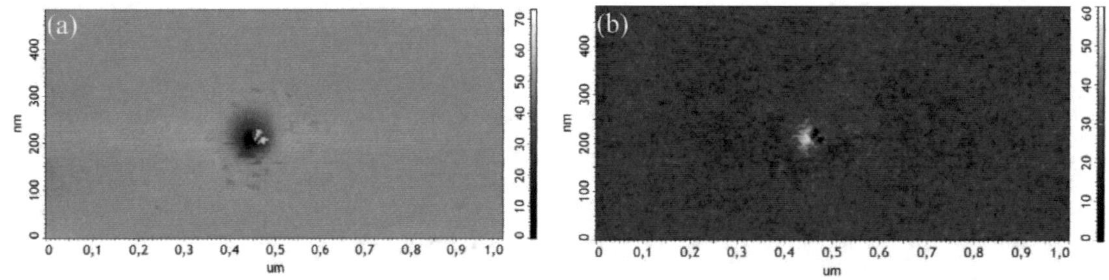

Figure 2. AFM images of GaAs surface morphology (a) and lateral distribution of surface potential (b) for sample after FIB treatment at 30 kV and subsequent wet chemical etching.

The KpAFM results are in good agreement with the Raman spectroscopic data. As can be seen from the Raman spectra in Figure 3, an increase in the Ga^+ ion implantation dose leads to a decrease in the intensity of the GaAs LO-phonon peak and the appearance of an extended region (shoulder) in the range 240 – 280 cm^{-1}. The formation of this region may be due to the presence of components of amorphous arsenic (peaks at 204, 205, and 252 cm^{-1}) and amorphous GaAs (peaks at 245 and 247 cm^{-1}), as well as TO-phonons of crystalline GaAs (265 cm^{-1}) [18 – 21]. It is also important to note that after FIB treatment, the intensity of the GaAs LO-phonon peak is practically equal to the intensity of the multicomponent shoulder at 240 – 280 cm^{-1}. This also indicates the presence of extended areas with a high degree of defectiveness around nanoholes. After wet chemical treatment, the relative intensity of the LO-phonon peak in GaAs sharply increases, and the GaAs TO-phonon peak clearly appears in the structure of the multicomponent shoulder. We attribute this to the predominant removal of the damaged area around nanoholes during the wet chemical etching. At the same time, the extent of the damaged GaAs layer is such that it is not completely removed during etching, leaving a significant contribution to the Raman signal even after chemical treatment.

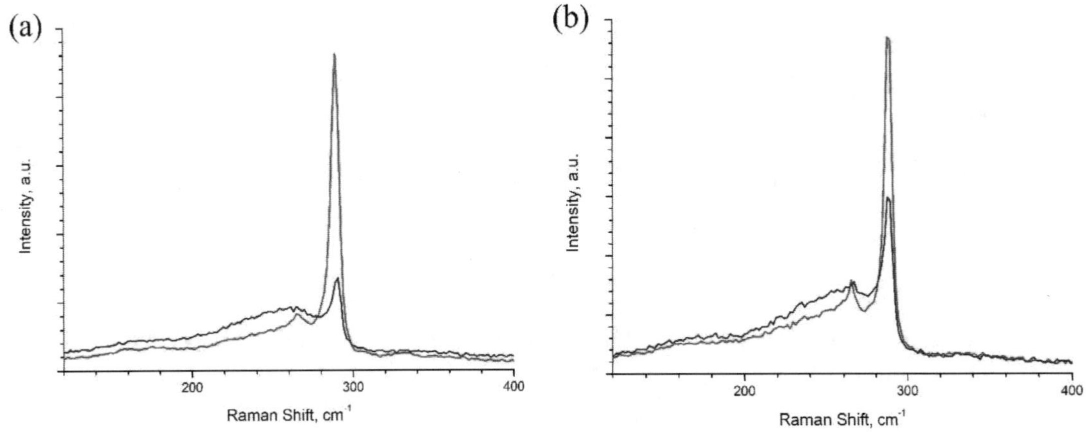

Figure 3. Raman spectra from GaAs area with nanoholes obtained at 30 kV before (a) and after (b) wet chemical etching.

4. Conclusions

Thus, experimental studies of the effect of liquid chemical treatment and FIB modes of nanoholes formation on the morphology, surface potential, and defectiveness of near-surface layer in the modification region of GaAs substrate have been carried out. Based on the results of the analysis of samples by KpAFM and Raman spectroscopy, we have shown the possibility of selectively removing damaged regions around GaAs nanoholes while maintaining its shape and size correlation using low accelerating voltages of the ion beam at the stage of surface structure formation. At the same time, the use of high-energy ion beams leads to the formation of extended regions around nanoholes while maintaining a relatively high defectiveness and built-in charge in GaAs substrate.

Acknowledgments

This work was supported by Ministry of Science and Higher Education of the Russian Federation; the state task in the field of scientific activity No. 0852-2020-0015.

References

[1] Schneider C, Heindel T, Huggenberger A, Weinmann P, Kistner C, Kamp M, Reitzenstein S, Höflinga S and Forchel A 2009 *Appl. Phys. Lett.* **94** 111111
[2] Lisitsyn S A, Balakirev S V, Avilov V I, Kolomiytsev A S, Klimin V S, Solodovnik M S, Konoplev B G and Ageev O A 2018 *Nanotechnol. in Russia* **13** 26 – 33
[3] Ageev O A, Smirnov V A, Solodovnik M S, Rukomoikin A V and Avilov V I 2012 *Semiconductors* **46** 1616 – 1621
[4] Balakirev S V, Solodovnik M S, Eremenko M M, Chernenko N E and Ageev O A 2020 *J. Phys. Conf. Ser.* **1695** 012016
[5] Balakirev S V, Chernenko N E, Eremenko M M, Ageev O A and Solodovnik M S 2021 *Nanomaterials* **11**(5) 1184
[6] Dousse A, Suffczyn'ski J, Beveratos A, Krebs O, Lemaıtre A, Sagnes I, Bloch J, Voisin P and Senellart P 2010 *Nature* **466** 217 – 220
[7] Balakirev S V, Solodovnik M S, Eremenko M M, Chernenko N E and Ageev O A 2020 *Nanotechnology* **31** 48
[8] Khosroshahyab M B, Moaiyeri M H, Angizi S, Bagherzadeh N and Navi K 2017 *Microprocessors and Microsystems* **50** 154 – 163
[9] Cavigli L, Abbarchi M, Bietti S, Somaschini C, Sanguinetti S, Koguchi N, Vinattieri A and Gurioli M 2011 *Appl. Phys. Lett.* **98** 103104

[10] Balakirev S V, Solodovnik M S and Ageev O A 2018 *Phys. Status Solidi B* **255**(4) 1700360
[11] Dias N L, Garg A, Reddy U, Young J D, Verma V B, Mirin R P and Coleman J J 2011 *Appl. Phys. Lett.* **98** 141112
[12] Balakirev S V, Solodovnik M S, Eremenko M M, Konoplev B G and Ageev O A 2019 *Nanotechnology* **30** 50
[13] Kim J S, Kawabe M and Koguchi N 2006 *Appl. Phys. Lett* **88** 072107
[14] Klimin V S, Rezvan A A, Solodovnik M S, Tominov R V and Ageev A O 2019 *J. Phys. Conf. Ser.* **1410** 012036
[15] Chernenko N E, Balakirev S V, Eremenko M M and Solodovnik M S 2019 *J. Phys. Conf. Ser.* **1410** 012007
[16] Balakirev S V, Solodovnik M S, Eremenko M M, Mikhaylin I A and Ageev O A 2018 *J. Phys. Conf. Ser.* **1124** 022001
[17] Avilov V I, Ageev O A, Smirnov V A, Solodovnik M S and Tsukanova O G 2015 *Nanotechnol. in Russia* **10** 214 – 219
[18] Salminen T, Dahl J, Tuominen M, Laukkanen P, Arola E and Niemi T 2012 *Optical materials Express* **2** 799
[19] Burns G, Dacol F H, Wie C R, Burstein E and Cardona M 1987 *Solid State Commun.* **62**(7) 449 – 454
[20] Desnica I D, Ivanda M, Kranjček M, Murri R and Pinto N 1994 *J. Non-Cryst. Solids* **170**(3) 263 – 269
[21] Takagaki Y, Wiebicke E, Ramsteiner M, Däweritz L and Ploog K H 2003 *Appl. Phys. A Mater. Sci. Process.* **76**(5) 837 – 840

Separation of AlN layers from silicon substrates by KOH etching

K Yu Shubina[1], D V Mokhov[1], T N Berezovskaya[1], E V Pirogov[1], A V Nashchekin[2], Sh Sh Sharofidinov[2] and A M Mizerov[1]

[1] Nanoelectronics Lab, Alferov University, St. Petersburg 194021, Russia
[2] Ioffe Physical-Technical Institute RAS, St. Petersburg 194021, Russia

rein.raus.2010@gmail.com

Abstract. In this work, the AlN/Si(111) epitaxial structures grown consistently by plasma assisted molecular beam epitaxy (PA MBE) and hydride vapour phase epitaxy (HVPE) methods were studied. The PA MBE AlN buffer layers were synthesized via coalescence overgrowth of self-catalyzed AlN nanocolumns on Si(111) substrates and were used as templates for further HVPE growth of thick AlN layer. It was shown that described approaches can be used to obtain AlN layers with sufficiently smooth morphology. It was found that HVPE AlN inherited crystallographic polarity of the AlN layer grown by PA MBE. It was demonstrated that the etching of such AlN/Si(111) epitaxial structures results in partial separation of the AlN epilayers from the Si(111) substrate and allows to form suspended structures. Moreover, the avoidance of surface damage and backside overetching was achieved by use thin Cr film as surface protective coating and by increasing the layer thickness accordingly.

1. Introduction

Due to expansion of the semiconductor electronics application area, as well as the growing requirements for compactness, energy efficiency and performance of electronic devices, it becomes necessary to update the component base of modern electronics. Up to date, III-N semiconductors are one of the most promising materials for solving this problem [1]. These materials have unique optical and electrical properties [2]. In addition, they have extremely high thermal and chemical stability [3, 4]. Therefore, the development of III-N materials technology is especially important for the creation of devices operating in harsh environments.

Despite the obvious advantages of using III-N semiconductors for creating new electronic devices, there are still problems that hinder the rapid development and implementation of nitride-based electronics in everyday life. The first and the main obstacle is the lack of low-cost native III-N substrates. Therefore, heteroepitaxy of nitrides on silicon substrates is of increased interest.

At the same time, heteroepitaxy of high quality III-N layers on silicon substrates is quite challenging due to the several reasons. Large lattice mismatch (16.9 % for GaN/Si(111) and 19 % for AlN/Si(111)) leads to the high density of dislocations, and the difference in the thermal expansion coefficients of the materials increases the risk of epitaxial structures cracking. Moreover, there are several growth peculiarities of polar III-N materials on non-polar Si(111) substrates [5]. Furthermore, Al, Ga and Si interdiffusion can cause unintentional doping of III-N layers and silicon substrate [6, 7].

It is well-known, that AlN can be used as a buffer layer for heteroepitaxy of GaN on silicon to solve some of the problems above. However, it is reported that it may have quite rough surface morphology and poor crystalline quality [8]. One of the promising approaches to improve the crystalline quality of III-N-based heterostructures grown on Si(111) substrates can be formation of GaN and AlN epitaxial layers using coalescence overgrowth of GaN and AlN nanocolumns respectively [9].

In our previous work [10], it was shown that etching of such structures in a KOH solution can lead to the gradual separation of AlN layers from the substrate. At the same time, several damage of the sample surface was observed. In addition, a possibility of overetching of the AlN layer backside was found. This work is a continuation of research on the separation of AlN layers from the substrate by KOH etching.

2. Samples and experimental details

The AlN/Si(111) samples were synthesized consistently by PA MBE and then by HVPE method (see figure 1).

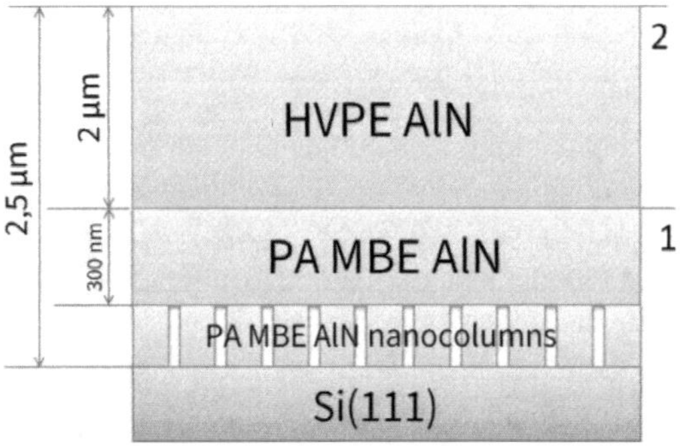

Figure 1. Schematic view of the two types obtained samples.

First, the AlN films with nanocolumnar seed layers were grown on the semi-insulating ($R > 10000$ Ohm×cm) Si(111) substratres using Veeco Gen 200 MBE system. The pre-epitaxial preparation of the substrates was carried out according to the modified Shiraki method [12]. Before the growth, the Si(111) substrates were annealed for 30 min at $T = 850$ °C in order to remove SiO_2 layer. . The growth procedure was similar to the described in papers [10, 11]. The obtained by PA MBE AlN/Si(111) samples were used as templates for HVPE growth of thick AlN layer. At the first stage of the process, the growth zone was heated up to $T = 1080$ °C, then the PA MBE template was moved from the cold zone into the growth zone and annealed in an Ar atmosphere for $t = 10$ min. After annealing, AlN layers were grown on the AlN/Si (111) templates. Aluminium chloride and ammonia were used as reagents. High purity Ar was used as a transport gas. The parameters of the growth flows are arranged as follows: total argon flow $V_{Ar} = 4$ l/min, ammonia flow $V_{NH3} = 1$ l/min, aluminum chloride flow $V_{HCl(Al)} = 200$ ml/min. The growth time was $t = 5$ min.

Both types obtained samples were studied via scanning electron microscopy (SEM) and X-ray diffraction (XRD) analysis. The identification of the crystallographic polarity of the AlN epilayers, as well as the separation process of AlN from Si(111) substrates, were carried out using chemical etching in KOH:H_2O (1:5) solution. To avoid surface damage during the AlN separation by etching in KOH, the Cr mask was formed on the sample surface by e-beam evaporation using BOC Edwards AUTO 500 vacuum coating system.

3. Results and discussion

3.1. Morphology studies

Characteristic SEM images of the two types obtained AlN/Si(111) structures are presented in figure 2. As it can be clearly seen in figure 2(a), AlN nanocolumn coalescence allowed the formation of the continuous AlN epitaxial layer, as it was shown previously [10].

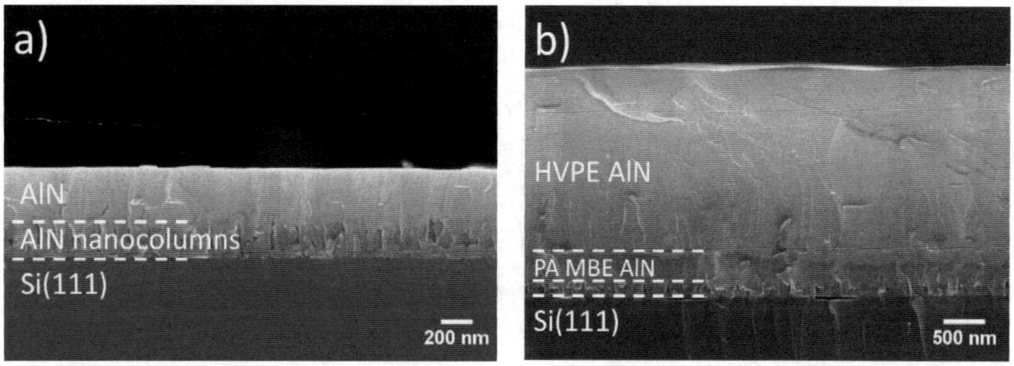

Figure 2. Cross-section SEM images of the AlN/Si(111) samples obtained by PA MBE (a) and both PA MBE and HVPE (b).

The growth of AlN by HVPE on the AlN/Si(111) templates synthesized by PA MBE led to the formation of 2.5 μm thick continuous AlN layers (see figure 2(b)).

3.2. XRD analysis

Next, the samples were studied with XRD measurements using DRON-8 X-ray diffractometer. The characteristic XRD curves for both types of obtained samples are shown in figure 3.

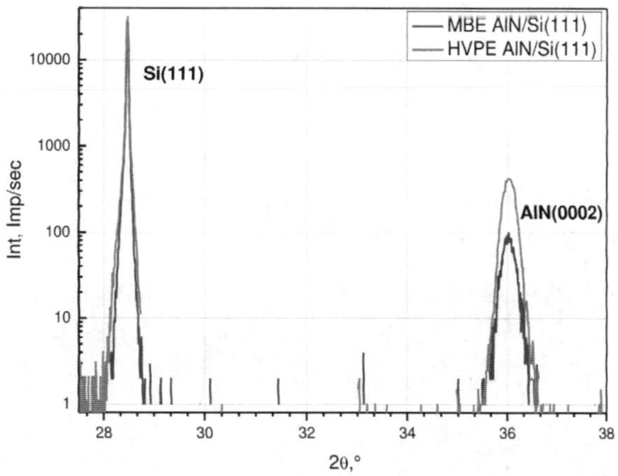

Figure 3. XRD curves of the obtained samples.

The obtained XRD curves clearly show the peaks corresponding to Si(111) substrate and the AlN layers. From the position of the AlN (0002) reflex using the Bragg condition:

$$2d\sin\theta = n\lambda, \tag{1}$$

the lattice constant c of the synthesized AlN layers was calculated, $c_{AlN} \approx 4.980$ Å. The obtained value is consistent with the reference values ($c_{AlN} = 4.980 - 4.982$ Å).

3.3. Crystallographic polarity

It is well known that crystallographic polarity is an essential feature of wurtzite III-N materials, which affects their properties [13]. In particular, III-nitrides with metal-face polarity are the most chemically stable. The crystallographic polarity of the obtained by PA MBE and HVPE methods AlN layers was identified using express technique based on wet chemical etching in 40 °C KOH solution (KOH:H_2O = 1:5) for $t = 5$ min. It was found, that all the samples are Al-polar because after etching the height of the AlN layers and the surface remained unchanged (see figure 4).

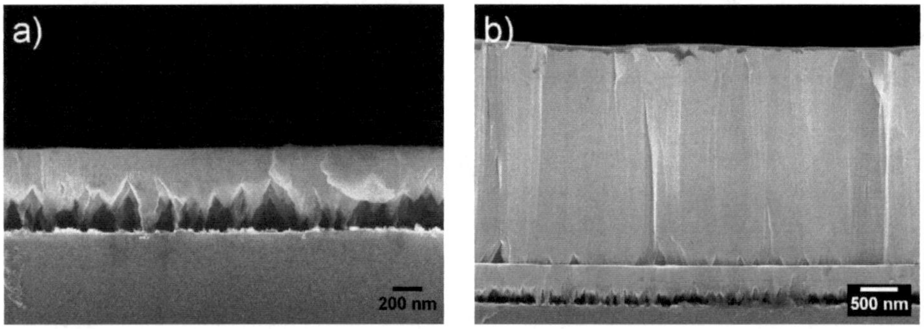

Figure 4. SEM images of the AlN/Si(111) samples grown by PA MBE (a) and both PA MBE and HVPE (b) after etching in KOH KOH:H_2O (1:5), $T = 40$ °C, $t = 5$ min.

Moreover, the experiment showed that HVPE AlN inherited crystallographic polarity of the AlN layer grown by PA MBE in contrast to the N-face AlN layers grown by HVPE on Ga-face GaN/Si(111) PA MBE template, that was shown in our previous work [14]. Besides this, it can be seen in figure 4(b) that not only PA MBE AlN is separating from the substrate, but HVPE AlN tends to separate on the interface HVPE AlN/PA MBE AlN.

3.4. Separation of AlN epilayers by wet chemical etching in KOH

As it was shown in our previous work [10], the approach to separation of AlN layers from Si(111) substrates by KOH etching has two serious disadvantages: surface damage and the risk of backside overetching. In this work, to protect the surface of the AlN epilayers from defect-selective etching, the samples were coated with a 55 nm Cr layer. As it can be seen in figure 5, thin Cr layer indeed allowed to avoid surface damage during the etching in KOH.

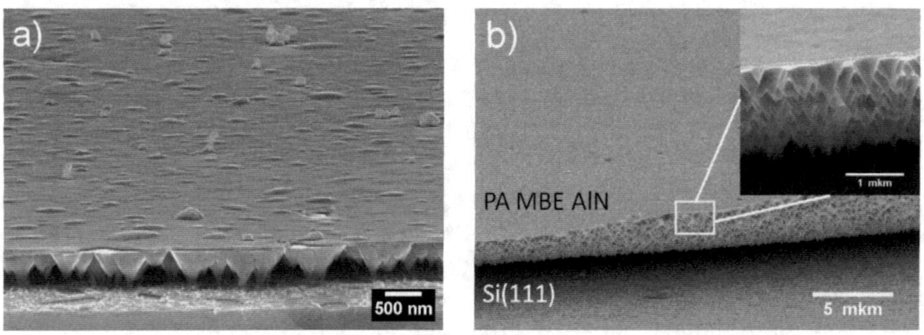

Figure 5. SEM images of the AlN/Si(111) structures after etching in KOH KOH:H_2O (1:5), $T = 40$ °C, $t = 10$ min (a), $T = 70$ °C, $t = 5$ min (b).

Moreover, it was found that the Cr film can hold the separated AlN layer and prevent its crumbling (see figure 5(b)). The slight upward curvature of the Cr/AlN layers after its separation could be due to gradient in the strain of the materials in growth direction [15].

AlN layers grown by HVPE also were partially separated from the substrate by KOH etching (see figure 6.)

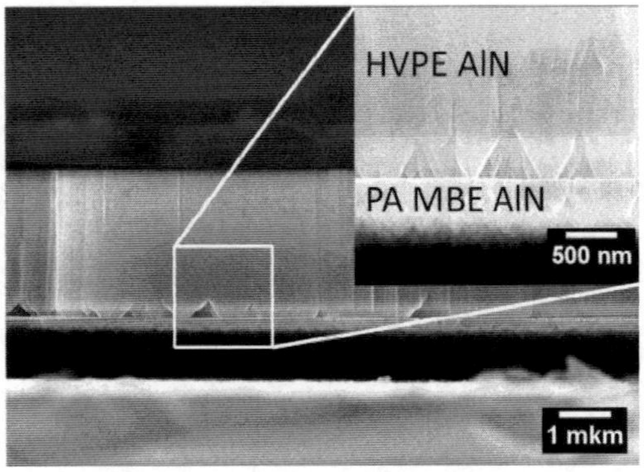

Figure 6. SEM images of the HVPE AlN/Si(111) structures after etching in KOH KOH:H_2O (1:5), $T = 40$ °C, $t = 10$ min.

The increase of the layer thickness up to 2.5 μm indeed allowed to avoid the backside overetching and the film crumbling during the separation of AlN by KOH etching.

4. Conclusion

In this work, the AlN/Si(111) epitaxial structures grown consistently by PA MBE and HVPE methods were studied. The PA MBE AlN layers were synthesized via coalescence overgrowth of self-catalyzed AlN nanocolumns on Si(111) substrates and were used as templates for HVPE AlN growth. Morphology of the obtained epitaxial structures was studied. It was shown that the described approaches allow obtaining sufficiently smooth AlN layers and the design of the PA MBE AlN epilayers with transition from nanocolumnar to 2D morphology can be used as a buffer layer for the heteroepitaxy of high quality III-N layers on silicon substrates. It was identified by etching in KOH that all the obtained samples have Al-face polarity. The experiment showed that HVPE AlN inherited crystallographic polarity of the AlN layer grown by PA MBE. Moreover, it was demonstrated that the KOH etching of such AlN/Si(111) epitaxial structures results in partial separation of the AlN epilayers from the Si(111) substrate. Moreover, the avoidance of surface damage and backside overetching was achieved by use 55 nm Cr film as surface protective coating and by increasing the layer thickness accordingly. Thus, it is shown that etching in KOH can also be promising method to form suspended structures based on AlN/Si(111) or to obtain free-standing III-N structures.

Acknowledgments

This work was supported by the the Ministry of Education and Science of the Russian Federation (grant № FSRM-2020-0008). SEM characterization was partially performed using equipment owned by the Federal Joint Research Center "Material science and characterization in advanced technology" with financial support by Ministry of Education and Science of the Russian Federation.

References

[1] Amano H et al. 2018 *J. Phys. D: Appl. Phys.* **51** 163001
[2] Morkoç H 2008 *Handbook of Nitride Semiconductors and Devices Vol. 1: Materials Properties, Physics and Growth* (Weinheim: Wiley-VCH)

[3] Rais-Zadeh M, Gokhale V J, Ansari A, Faucher M, Théron D, Cordier Y and Buchaillot L 2014 *J. Microelectromech. Syst.* **23** 1252-71
[4] Foster C M, Collazo R, Sitar Z and Ivanisevic A 2013 *Langmuir* **29** 216–220
[5] Lucci I et al. 2018 *Phys. Rev. Mater.* **2** 060401(R)
[6] Zheng Y, Agrawal M, Dharmarasu N, Radhakrishnan K and Patwal S 2019 *Appl. Surf. Sci.* **481** 319-326
[7] Wei L et al. 2020 *Appl. Phys. Lett.* **116** 232105
[8] Lin Y, Yang M, Wang W, Lin Z, Gao J and Li G 2016 *Cryst. Eng. Comm.* **18** 2446
[9] Kusakabe K, Kikuchi A and Kishino K 2002 *J. Cryst. Growth* **237-239** 988-992
[10] Shubina K Yu, Mokhov D V, Berezovskaya T N, Nikitina E V, Mizerov A M and Bouravleuv A D 2020 *J. Phys. Conf. Ser.* **1695** 012042
[11] Mizerov A M, Kladko P N, Nikitina E V and Egorov A Yu 2015 *Semiconductors* **49** 274
[12] Ishizaka A and Shiraki Y 1986 *J. Eltrochem. Soc.* **133** 666
[13] Hellman E S 1998 *MRS Internet J. Nitride Semicond. Res.* **3** e11
[14] Mizerov A M, Kukushkin S A, Sharofidinov Sh Sh, Osipov A V, Timoshnev S N, Shubina K Yu, Berezovskaya T N, Mokhov D V and Buravlev A D 2019 *Phys. Solid State* **61** 2277-2281
[15] Stonas A R, MacDonald N C, Turner K L, DenBaars S P and Hu E L 2001 *J. Vac. Sci. Technol. B* **19** 2838

Journal of Physics: Conference Series **2086** (2021) 012038 doi:10.1088/1742-6596/2086/1/012038

Effects of plasma treatment parameters on the adsorption properties of tin dioxide-based nanomaterials

A P Sigaev, I A Averin, A A Karmanov, I A Pronin and N D Yakushova

Department of Nano- and Microelectronics, Penza State University, 440026, Penza, Russia

Abstract. We have studied the effect of nitrogen plasma exposure on the adsorption properties of SnO_2-based nanomaterials synthesized by the sol-gel method. We have established the correlation between the adsorption properties and plasma treatment parameters. The scanning electron microscopy has confirmed integrity violation of the original structure of nanomaterials and an increase in their specific surface area with high-power plasma source.

1. Introduction

The use of nanomaterials having a developed hierarchical structure with a fractal type of spatial organization is promising in the development of adsorption-type gas sensors. Due to high specific surface area and a pore network system, such materials provide high sensitivity to the analyzed gases. Despite numerous fundamental and applied studies [1, 2], the problem of designing a simple and cheap technique to manufacture nanomaterials with high selectivity, due to similar changes in the material conductivity caused by the adsorption of particles of the analyzed different type gases, is still urgent [3].

One of the possible ways to modify the adsorption properties of sensitive nanomaterials is plasma treatment [4-6]. In this work, we investigate the effect of various nitrogen plasma powers on the adsorption properties of SnO_2-based nanomaterials synthesized by the sol-gel method.

2. Experimental Technique

SnO_2-based hierarchical nanomaterials with a fractal type of spatial organization, being a dielectric matrix with tin dioxide clusters embedded therein were synthesized by the sol-gel method using tetraethoxysilane [7]. A film-forming sol, having been aged for a day before ripening, was deposited onto glass substrates by centrifugation at 2750 rpm. The final stage of the formation of nanomaterials was their thermal annealing at a temperature of 575°C for 50 minutes in air. The samples of nanomaterials to be investigated with a scanning electron microscope (SEM) were formed on an aluminum foil under the same modes.

Inductively coupled plasma processing of nanomaterials was performed with an etching machine under various modes of nitrogen plasma treatment. The modes were selected so that the ions of the process gas break the chemical bonds between atoms only located in the surface layer of the SnO_2-based nanomaterial for redistribution of adsorption centers, rather than etching the film material. The power of inductively coupled plasma source P_1, which determines the energy and density of plasma, and the power of supplementary high-frequency source P_2, which characterizes the ion flux intensity and direction, were chosen as variable parameters.

An analysis of adsorption site distribution on the surface of synthesized samples was carried out by the adsorption method of acid-base indicators in combination with spectrophotometry prior to and following additional processing. The main investigated parameter in the analysis of adsorption sites on the surface of solid-phase systems is acid strength pKa. If adsorption is observed, then the centers are assigned the acid strength corresponding to the indicator-adsorbate molecule, that is, the acid strength of the center corresponds to the acid strength of the indicator. Ten indicators having acid strength values pKa in the range from 0.8 to 11.1 were selected for the experiment.

3. Results and Discussion

To analyze the adsorption site concentration of certain acid strength, three solutions were prepared for each sample and the selected indicator. The spectrophotometer was used to measure the transmittances of the solutions, which were then recalculated into the absorption factors according to the formula:

$$A = \lg\left(\frac{100}{T}\right), \tag{1}$$

where T is the prepared solution transmittance, %. Based on the obtained results, the concentration of adsorption sites was estimated for each pKa value according to the following equation:

$$Q(pKa) = \frac{C_{\text{ind}} \cdot V_{\text{ind}}}{A_0} \cdot \left| \frac{|A_1 - A_0|}{m_1} \pm \frac{|A_2 - A_0|}{m_2} \right|, \tag{2}$$

where C_{ind} is the concentration of the indicator in the solution, V_{ind} is the volume of the indicator solution, m_1 and m_2 are the masses of the studied nanomaterials, A_0, A_1, A_2 are the corresponding absorption factors based on the calculated transmittances with a spectrophotometer.

The surface of a solid is bifunctional, since it is a combination of Lewis and Brönsted acid and base sites. Brönsted acid and base adsorption sites are formed due to the adsorption of water molecules or its fragments at the corresponding Lewis sites.

The histogram shown in Figure 1 confirms that Brönsted acid sites prevail on the surface of the studied plasma-untreated SnO_2-based nanomaterials. These sites are protons bound to surface oxygen atoms in various configurations. The main Brönsted centers, represented by hydroxyl groups, have the Sn–OH structure in various configurations, and are identified on the surface of the investigated materials.

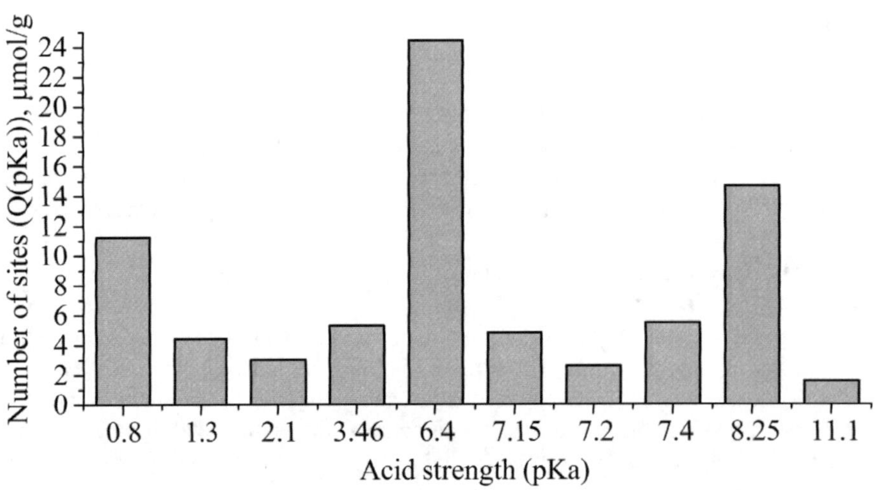

Figure 1. Distribution of adsorption sites on the surface of SnO_2-based nanomaterials.

The total number of adsorption sites on the surface of nanomaterials has decreased after plasma treatment due to the formation of defects, being a result of ionic bombardment, and due to thermal action (Figure 2). Besides, this phenomenon is explained by the transformation of some Brönsted sites into Lewis ones due to plasma action.

An adsorption site with an acid strength of pKa = 6.4 was selected within the research into the dependence of distribution of the adsorption sites of SnO_2-based nanomaterials on the parameters of plasma treatment by the indicator method. The choice is explained by the fact that such adsorption site is characterized by a high concentration on the surface of the films, as follows from Figure 2.

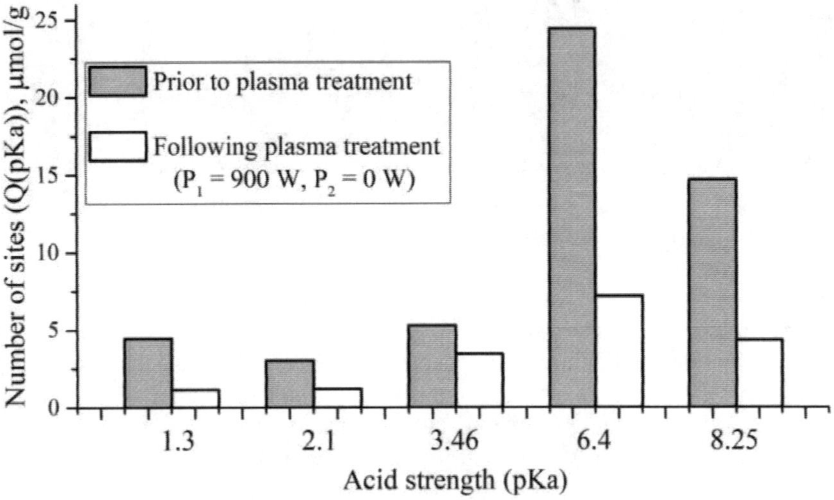

Figure 2. Distribution of adsorption sites on the surface of SnO_2-based nanomaterials prior to and following nitrogen plasma treatment.

An analysis of the results in Figure 3 has showed that the simultaneous use of inductively coupled plasma source, and supplementary high-frequency source leads to a decrease in the number of adsorption sites (pKa = 6.4), located on the film surface.

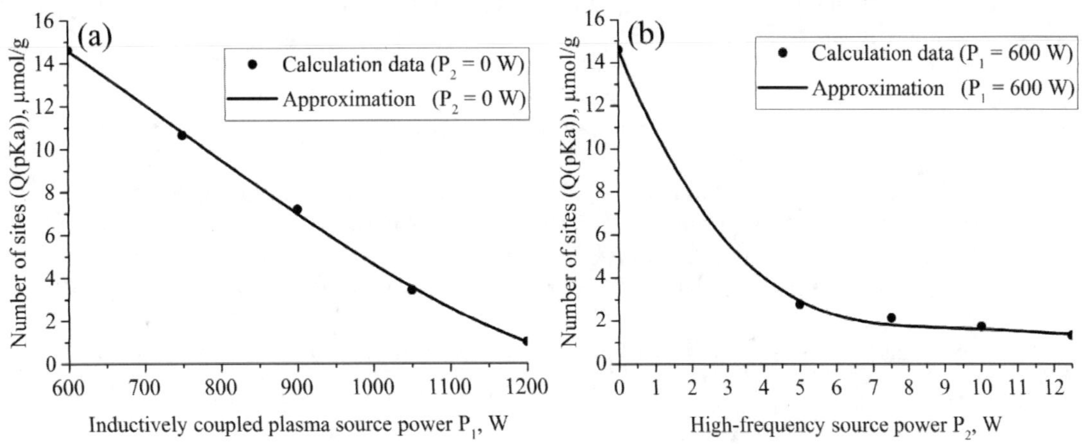

Figure 3. Dependence of the concentration of adsorption site, having acid strength of pKa = 6.4, on the power of an inductively coupled plasma source **(a)**, and on the power of a high-frequency source **(b)**.

Degradation in the structure of nanomaterials at high plasma power P_1 values is confirmed by the data of scanning electron microscopy (SEM) (Figure 4).

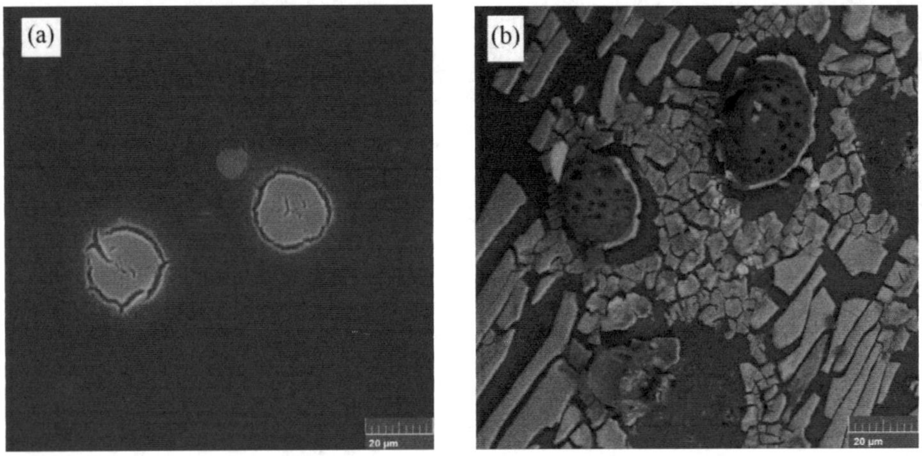

Figure 4. SEM images of the surface of SnO_2-based nanomaterials prior to **(a)** and following **(b)** nitrogen plasma treatment.

4. Conclusions

Nitrogen plasma treatment of SnO_2-based nanomaterials leads to modification of their surface properties due to partial destruction of quasi-spherical aggregates (Figure 4, a), typical for nucleophilic increase in sol-gel systems. This ensures the formation of supplementary macro-, meso-, and micropores in nanomaterials (Figure 4, b), thus increasing their specific surface area. With high-power plasma sources, the overall integrity of the structure of nanomaterials is violated due to the effect of high-energy ions.

Acknowledgments

The study was carried out with the financial support of the Russian Foundation for Basic Research (projects No. 20-38-90155 and No. 19-08-00924 A).

References

[1] Sun G, Qi F, Zhang S, Li Y, Wang Y, Cao J, Bala H, Wang X, Jia T and Zhang Z 2014 *J. Alloys Compd.* **617** 192
[2] Ellert O G, Tsodikov M V and Novotortsev V M 2010 *Rus. Chem. Rev.* **79** 693–712
[3] Shikunov D A, Simakov V V, Sinev I V, Shcherbakova E A and Timoshenko D A 2019 *Physical and chemical aspects of the study of clusters, nanostructures and nanomaterials* **11** 665–71
[4] Dao H T and Makino H 2019 *Materials Science in Semiconductor Processing* **96** 46–52
[5] Sigaev A P, Averin I A, Karmanov A A, Pronin I A and Yakushova N D 2020 *Physical and chemical aspects of the study of clusters, nanostructures and nanomaterials* **12** 162–9
[6] Danilova M N, Pylinina A I, Platonov E A and Yagodovskii V D 2015 *Russian Journal of Physical Chemistry A* **89** 1339–42
[7] Pronin I A and Goryacheva M V 2013 *Surface and Coatings Technology* **235** 835–40

MBE growth of AlGaAs/Ge/AlGaAs core-shell nanowire

A N Terpitskiy[1], I V Ilkiv[2], K P Kotlyar[1,2], D A Kirilenko[3] and G E Cirlin[1]

[1] Alferov University, St. Petersburg, 194021, Russia
[2] St. Petersburg State University, St. Petersburg, 199034, Russia
[3] Ioffe Institute, St. Petersburg, 194021, Russia

terpiczkij@mail.ru

Abstract. Heterostructured AlGaAs/Ge/AlGaAs core-multishell nanowires having hexagonal crystal structure were synthesized by molecular beam epitaxy. Formation of 2-3 nm Ge quantum well structure was demonstrated. Raman characterization revealed a 200 cm^{-1} peak corresponded to hexagonal phases of germanium.

1. Introduction

While germanium (Ge) based heterostructures are one of the most successful for creating sensitive near-infrared detectors [1], their implementation in active photonic devices remains one of the greatest challenges due to the indirect bandgap of Ge. To date, various strategies to facilitate the emissivity ability of Ge have been proposed, including synthesis of SiGe quantum dots [2] and tensile strained layer [3], as well as Ge/GaAs heterostructures [4].

At the same time, Ge-based nanostructures synthesized in the hexagonal phase can demonstrate sufficiently effective emissivity due to the modified direct bandgap [5]. Unfortunately, the methods commonly used typically result in the formation of hexagonal Ge clusters in amorphous matrixes. [6-9]. In turn, alternative approach based on the growth of radial layers on the surfaces of wurtzite A_3B_5 nanowires (NWs) was proposed [10]. Furthermore, light emission in the range 1.8-3.5 μm of was achieved in hexagonal $Si_{1-x}Ge_x$ bulk layers has been demonstrated recently [11]. On the other hand, the growth of Ge hexagonal quantum wells of 1-10 nm thick can be used to tune the emission range up to 1 μm.

Here we report on the growth of hexagonal AlGaAs/Ge/AlGaAs core-multishell nanowires by molecular-beam epitaxy (MBE).

2. Experiment

Heterostructured NWs were grown step-by-step in the same growth session using Riber Compact EB200 MBE set up equipped with A_3B_5 sources, as well as Ge, Si electron-beam evaporators.

The synthesis of $Al_{0.3}Ga_{0.7}As$-core NWs was carried out via "vapor-liquid-solid" mechanism using gold as a catalyst at a substrate temperature of 500°C. Ga and Al deposition rate was equal to 1.7 Å/s and 0.56 Å/s, respectively. It should be noted that AlGaAs NWs grown in such way applying the same growth recipes were used as references in all subsequent experiments. After the NW formation, a technological pause to pump out the residual As species was taken (until the pressure in the growth chamber reached 10^{-9} Torr). Afterwards, the deposition of Ge at a rate of 0.2 Å/s and temperature of 320°C was performed. The growth temperature chosen typically leads to smooth radial Ge layers forming, as previously shown [12]. After the Ge deposition, the growth of AlGaAs cap layer was conducted. At the end of the synthesis, the samples were cooled to room temperature and unloaded from the growth chamber.

The samples obtained were studied by scanning electron microscopy (SEM) using a Zeiss Supra 25 microscope. Structural analysis of NWs was performed by transmission electron microscopy (TEM) using a Jeol JEM-2100F microscope equipped with an EDX spectroscope (Bruker QUANTAX XFlash 6/60). For TEM analysis NWs were mechanically separated from the substrates and transferred onto a carbon-film-coated copper mesh. Raman spectra were acquired using a Witec Alpha 300R microscope. A 532 nm Nd:YAG laser was used for excitation.

3. Results

SEM images of samples obtained are shown in figure 1. As it can be seen, deposition of Ge and AlGaAs layers at the same temperature of 320°C resulted in the formation of NWs with flag-like terminal (figure 1(a)). The average length of such NWs was 250 nm, and the diameter was about 50 nm. At the same time, the flag had a length of about 400 nm and diameter, which gradually decreased from 40 to 10 nm at the tip.

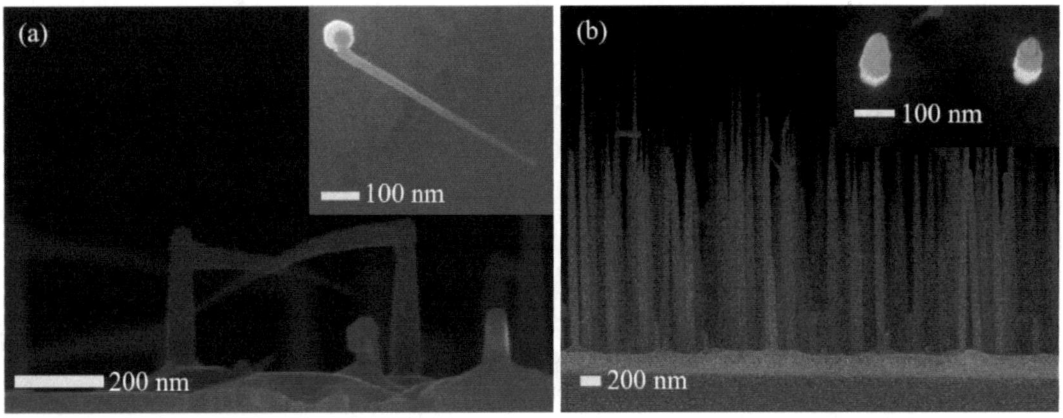

Figure 1. SEM images of the obtained samples: (a) – growth temperature 320°C of AlGaAs cap layer; (b) – growth temperature 500°C of AlGaAs cap layer.

The formation of flags was apparently caused by a migration of a catalyst drop onto the NW side facet during the growth of shells. It should be noted that deposition of Ge at temperatures less than 360°C typically leads to the formation of straight NWs, as it was shown previously [12]. Thus, the droplet migration was occurred in the process of AlGaAs cap layer growth. In this regard, the temperature of cap layer growth was varied in the range of 360-500°C. It was found that the ratio of vertical NWs in the arrays linearly magnifies with increasing temperature. In fact, arrays of straight NWs with a diameter of 70 nm and length of 2.5 μm were obtained at the AlGaAs growth temperature of about 500°C (figure 1(b)).

Combined with EDX map scanning TEM images of straight NWs are presented in figure 2. It is interesting that Au droplet was overgrown. Apparently, Ge droplet formed on the top of the NW could play the role of the catalyst (see figure 2). Unquestionably, it requires more detailed investigations.

TEM images of the NW oriented in the direction of $[1\bar{1}00]$ are shown in figure 3. The formation of about 2 nm thick radial Ge layer was observed. High-resolution TEM images showed that the Ge layer and the covering AlGaAs layers exhibited hexagonal crystal structure of AlGaAs core. This fact was also confirmed by the microdiffraction pattern obtained from the near-surface region of the NW.

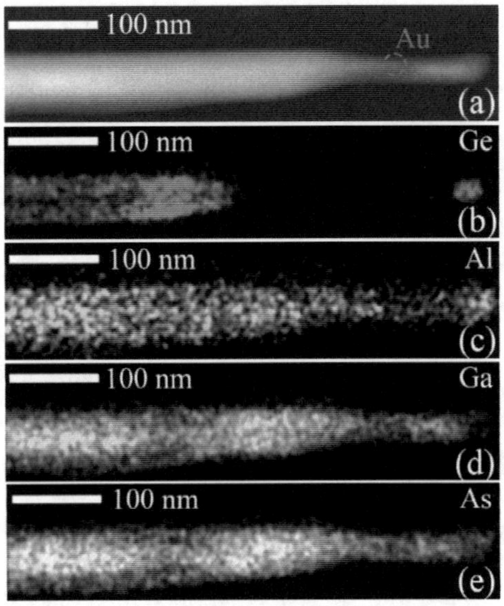

Figure 2. STEM (a) and EDX (b-e) images of a 500°C AlGaAs cap layer growth temperature sample.

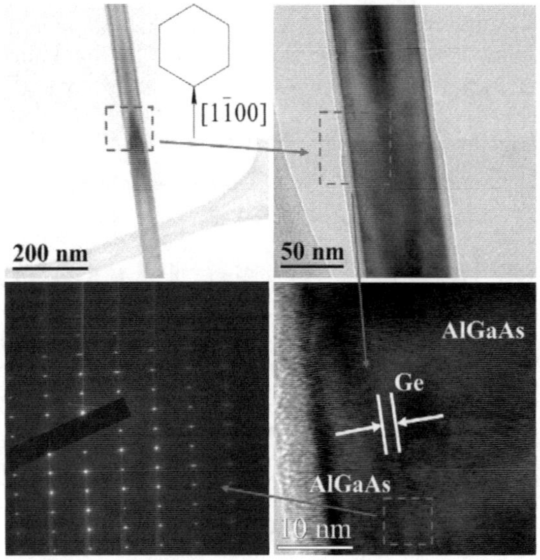

Figure 3. TEM image of NW with a 500°C AlGaAs cap layer growth temperature.

The hexagonal structure of Ge was additionally confirmed by Raman spectroscopy. Several peaks belonging to Ge can be distinguished in the obtained spectra (figure 4). The peak in the range of 300 cm^{-1} is associated with various Ge polytypes [13], while, 200 cm^{-1} peak is attributed to hexagonal phases [13].

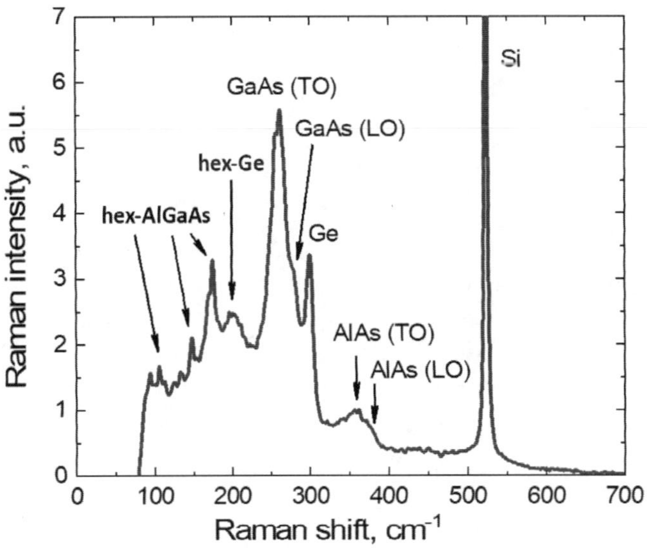

Figure 4. Raman spectra of a sample with a 320°C Ge growth temperature.

4. Conclusion

As a result, heterostructured AlGaAs/Ge/AlGaAs core-multishell NWs were synthesized by MBE. Thin Ge quantum well in AlGaAs barrier layers having radial geometry was demonstrated by TEM. A 200 cm^{-1} peak corresponded to the hexagonal phase of germanium was detected on the Raman spectra. The possibility to growth flag-like NWs by tuning the AlGaAs cap layer growth temperature was shown.

Acknowledgments
This work was financially supported by the Ministry of Science and Higher Education of the Russian Federation (0791-2020-0003). The samples were grown under support of RSF grant (№ 19-72-30010).

References

[1] Aliane A et al 2020 *Semicond. Sci. Technol.* **35** 035013
[2] Wan J, Luo Y H, Jin G L, Jiang Z M, Liu J L, Liao X Z, Zou J and Wang K L 2001 *Mat. Res. Soc. Symp. Proc.* **667** G6.4.1-G6.4.6
[3] Gatti E et al 2014 *J. Appl. Phys.* **116** 043518
[4] Aleshkin V Ya et al 2014 *J. Appl. Phys.* **115** 043512
[5] Rödl C, Furthmüller J, Suckert J R, Armuzza V, Bechstedt F and Botti S 2019 *Phys. Rev. Mater.* **3** 034602
[6] Parsons J R and Hoelke C W 1985 *Philosophical Magazine A* **50** (3) 329-337
[7] Zhang Y, Iqbal Z, Vijayalakshmi S, Qadri S and Grebel H 2000 *Solid State Commun* **115** 657-660
[8] Vincent L, Patriarche G, Hallais G, Renard C, Gardes C, Troadec D and Bouchier D 2014 *Nano Lett.* **14** (8) 4828-4836
[9] Okugawa M, Nakamura R, Ishimaru M, Yasuda H and Numakura H 2016 *J. Appl. Phys.* **120** 134308
[10] Hauge H I T et al 2015 *Nano Lett.* **15** (9) 5855-5860
[11] Fadaly E M T et al 2020 *Nature* **580** 205-209
[12] Ilkiv I V, Kotlyar K P, Kirilenko D A, Osipov A V, Soshnikov I P, Terpitsky A N and Cirlin G E 2021 *Semiconductors* **55** (8) 621-624
[13] Fasolato C et al 2018 *Nano Lett.* **18** (11) 7075-7084

Raman analysis of the crystallinity degree for the local regions in $Ge_2Sb_2Te_5$ films after laser exposure at different parameters

N M Tolkach[1, 2], N V Vishnyakov[2], A O Yakubov[1], A U Sudakova[2], E S Trofimov[2] and A A Sherchenkov[1]

[1] National Research University of Electronic Technology (MIET), Bld. 1, Shokin Square, Zelenograd, Moscow, 124498, Russia
[2] Ryazan State Radio Engineering University named after V.F. Utkin, Bld. 59/1, Gagarina street, Ryazan, 390005, Russia

Abstract. The evaluation of the crystallinity degree for the local regions in $Ge_2Sb_2Te_5$ (GST) thin films after phase state transformation by laser pulses at 405 nm wavelength was analyzed using the Raman spectroscopy. The modes of laser radiation for controlling the reflectivity and transmissivity at 1550 nm telecommunication wavelength of the local regions in the GST film were established. The results obtained make it possible to implement the method of the completely optical control of the multilevel modulation of the optical signals for the integrated optics devices.

1. Introduction

PCM (Phase Change Memory) materials with phase state transformations are successful in this regard, in particular, chalcogenide glassy semiconductors, $Ge_2Sb_2Te_5$ material (GST). GST has a fast and low-energy phase transition. The sizes of the GST active regions are limited (~1 μm) by the wavelengths of the used laser radiation. Such active regions can be integrated into waveguide paths comparable to these sizes. This will allow good levels of integration, miniaturization, and performance in integrated optic and network-on-chip systems [1, 2]. In addition to the amorphous and crystalline phase states, there can be additional intermediate states due to the presence of partially crystalline and amorphous nanosized fractions of the material. Controlling the ratio of the fractional composition of GST opens wide possibilities in fine-tuning of its optical parameters, but at the same time, it requires studying the processes that occur when laser radiation is applied to the GST material. The ratio of amorphous and crystalline fractions in the bulk of the material is characterized by the crystallinity degree. The aim of this work is to study the crystallinity degree of the local regions of the GST film to select the optimal modes of laser exposure for the phase state transformations in the GST material.

2. Methods and materials

The GST thin film with the thickness of 24 nm were deposited by magnetron sputtering. The measurement setup (see figure 1) was used that allowed to carry out investigations using methods of Raman spectroscopy, atomic force microscopy (AFM), optical microscopy, photometry, laser modification. The phase state of the local regions of the GST film (see figure 2a) was changed from crystalline to amorphous phase state by the pulsed laser with 405 nm wavelength, with the focused spot diameter of about 1 μm, pulse durations from 5 to 30 ns and energy exposure from 2 to 4 nJ/μm^2. The radiation of 405 nm wavelength was used due to its high absorption by GST. The crystallinity degree was estimated from the Raman spectra (see figure 3) in the same way as in the work [3].

Raman spectra and AFM scans were obtained using the Ntegra-Spectra probe nanolaboratory (NT-MDT SI, Russia). Photometry of the local regions was used to measure the reflectivity and transmissivity of the film at 1550 nm wavelength.

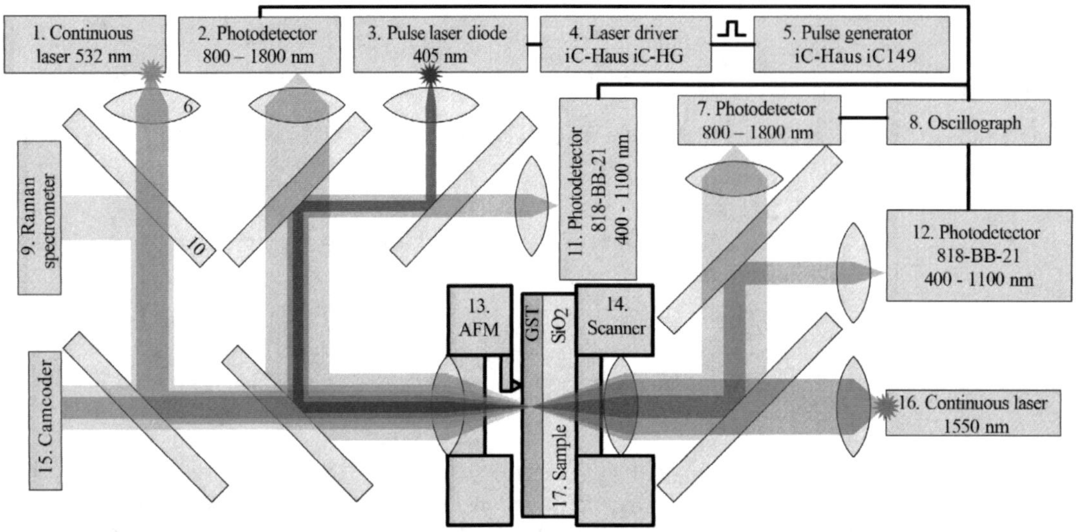

Figure 1. Structure of the measurement setup.

3. Estimation of the film crystallinity degree

Estimation of the fraction of amorphous phase in the local range of the GST film volume is characterized by the crystallinity degree *C*. The crystallinity degree C of the material can be estimated by Raman spectroscopy as follows. Intensity of the Raman scattering from the local region of the GST is equal to [3]:

$$I_{m,n} = 8\times 9^{-1}\pi\omega^4 c^{-4} N I_0 \sum_{\rho\sigma} \left|\left(\alpha_{\rho\sigma}\right)_{m,n}\right|^2, \tag{1}$$

$$K = 8\times 9^{-1}\pi\omega^4 c^{-4}, \tag{2}$$

$$\xi = \sum_{\rho\sigma} \left|\left(\alpha_{\rho\sigma}\right)_{m,n}\right|^2, \tag{3}$$

where I_0 is intensity of the incident laser radiation; ω is a frequency of the scattered light; $\alpha_{\rho\sigma}$ is a polarization tensor component with indices *r* and *s*; *m* and *n* are initial and final electronic states of the molecule; *c* is a velocity of light; *N* is a number of scatterers in the focal volume.

The number of scatterers changes from N_C to N_A in the focal volume due to the amortization of the crystalline film, which influences the intensity of the Raman peaks. In this case following equations can be obtained:

$$I_C = K N_C I_0 \xi_C, \tag{4}$$

$$I_A = K N_A I_0 \xi_A, \tag{5}$$

where I_C, I_A are intensities of the Raman scattering from the local regions of GST film for the amorphous and crystalline phase states, respectively.

Crystallinity degree is expressed as follows:

$$C = N_C\left(N_C + N_A\right)^{-1} = I_C\left(I_C + I_A\xi_C\xi_A^{-1}\right)^{-1}. \tag{6}$$

4. Results and discussion

Figure 2 shows AFM images of local regions in the GST film irradiated by the pulsed laser with 405 nm wavelength and various energy exposure. Figure 3 shows Raman spectra in local regions showed by numbers from 1 to 11 in figure 2a.

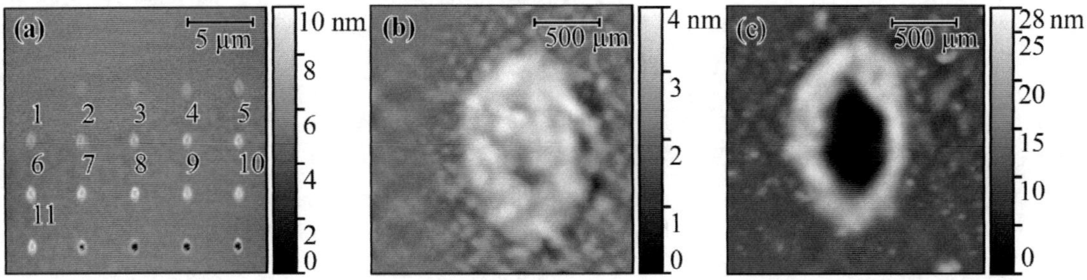

Figure 2. AFM images of the local regions in the GST film irradiated by the laser with various energy exposure from 2.92 nJ/µm² (No. spectra 1, crystalline local region of GST film) to 3.26 nJ/µm² (No. spectra 11, amorphous local region of GST film) (a), energy exposure of 3.26 nJ/µm² (b), energy exposure of 3.57 nJ/µm² (c).

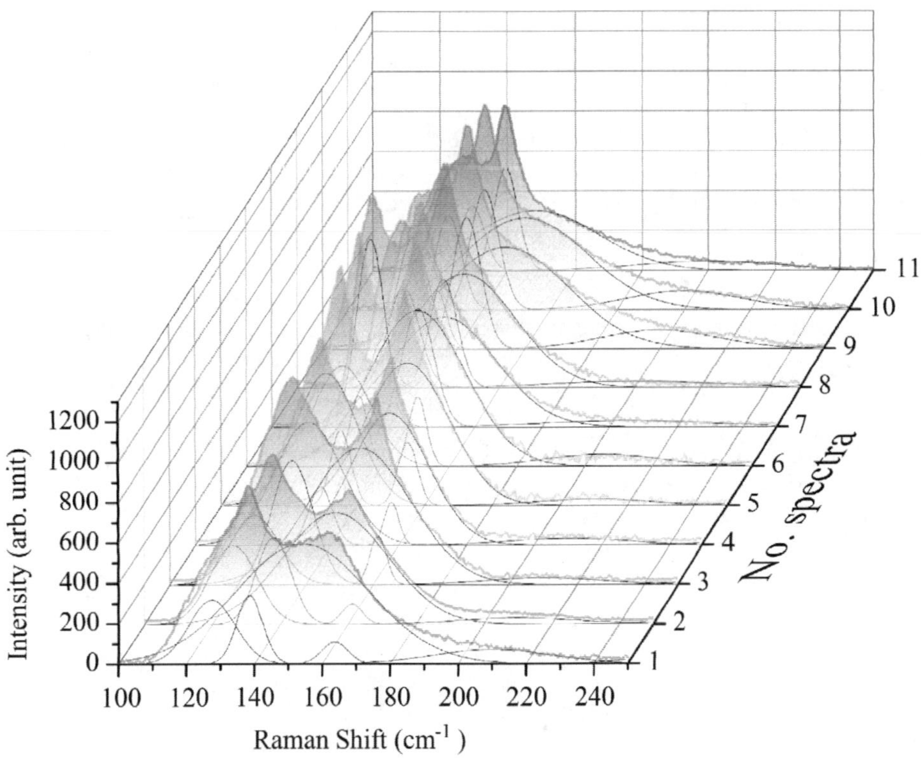

Figure 3. Raman spectra local regions in the GST film irradiated by the laser with various energy exposure from 2.92 nJ/µm² (No. spectra 1) to 3.26 nJ/µm² (No. spectra 11).

The evaluation of crystallinity degree showed that amorphization of the crystalline GST film is observed when exposed to laser with a wavelength of 405 nm at pulse energy exposure of 3.26 nJ/μm^2 (see figure 2b), with pulse duration of 10 ns. Film ablation is observed (see figure 2c) at pulse energy exposures greater than 3.4 nJ/μm^2. The conclusion about film ablation was made from the change in the GST film thickness in the local region using the AFM method. The GST film thickness without ablation was 24 ± 2 nm; during ablation, the thickness changed to the thickness of 0 ± 2 nm.

The change of the spectra from 1 to 11 (see figure 3) obtained after irradiation of local region in the crystalline film at 405 nm wavelength with radiant exposures from 2.92 to 3.26 nJ/μm^2 is due to the evolution of the crystallinity degree. Measurement of the reflectivity and transmissivity for 1550 nm radiation for the local regions of GST films with different crystallinity degree showed the change in these optical properties (see table 1), which allows multilevel amplitude modulation of an optical signal with a modulation depth of 45%.

Obtained values of the energy exposure allow selection of the operation modes of the GST film for the reflection and transmission of 1550 nm radiation in specified ratios. As can be seen table 1, it is possible to perform separation of radiation, partly block it or let it pass.

Table 1. Change of the optical properties of local GST regions.

No. Raman spectra	1	2	3	4	5	6	7	8	9	10	11
Energy exposure for 405 nm, nJ/μm^2	2.92	2.95	2.99	3.02	3.06	3.09	3.13	3.16	3.19	3.23	3.26
Crystallinity degree	0.98	0.83	0.62	0.58	0.55	0.59	0.61	0.50	0.41	0.29	0.04
Reflectivity for 1550 nm	0.57	0.55	0.54	0.51	0.48	0.46	0.41	0.35	0.32	0.27	0.23
Transmissivity for 1550 nm	0.23	0.26	0.29	0.32	0.36	0.38	0.43	0.49	0.53	0.6	0.68

5. Conclusion

Thus, it was found that the change in the energy exposure of a laser pulse irradiation leads to the change of reflectivity, transmissivity, due to the evolution of the crystallinity degree. These results demonstrate the possibility of multilevel modulation of the optical signal at the 1550 nm wavelength and make it possible to establish the optimal modes of laser radiation at the 405 nm wavelength to initiate the phase state transformations of the GST material with necessary crystallinity degree.

Acknowledgments

The reported study was funded by RFBR according to the research project # 19-37-60023 and used the equipment of Region Centre of probe microscopy of Ryazan State Radio Engineering University named after V.F. Utkin.

References

[1] Wen S, Meng Y, Jiang M and Wang J 2018 *Scientific Reports* **8** 4979
[2] Tolkach N M, Vishnyakov N V, Lazarenko P I, Sherchenkov A A, Sudakova A U and Nazimov D R 2020 *J Phys: Conf Ser* **1695** 012075
[3] Samuel A Z 2020 *Spectrochim Acta A Mol Biomol Spectrosc* **224** 117431

Conformality of a-Si:H deposited by low temperature PECVD for solar cells application

A V Uvarov[1,2], A I Baranov[1,2], I A Morozov[1,2], D A Kudryashov[1], A A Maksimova[1,2], E A Vyacheslavova[1,2] and A S Gudovskikh[1,2]

[1] St. Petersburg Alferov University, St. Petersburg, 194021, Russia
[2] St. Petersburg Electrotechnical University "LETI", St. Petersburg, 197376, Russia

Abstract. This work is devoted to the study of the deposition of a-Si:H by the PECVD method in continuous and pulsed regime on test structures in the form of trenches with a high aspect ratio. The thicknesses of the layers obtained in these modes were investigated by the method of scanning electron microscopy. It was shown that the layers obtained in the pulsed mode, as compared with the continuous one, have better conformity.

1. Introduction

Amorphous silicon (a-Si:H) is currently used in such areas of the semiconductor industry as the creation of solar cells, thin-film transistors and TFT displays. PECVD deposition method allows one to obtain best quality amorphous silicon. This deposition method allows one to obtain thin layers of amorphous silicon at temperatures of 150-250 °C, which effectively passivates defects on the surface of silicon substrates. Most solar cells are manufactured using a-Si:H/c-Si heterojunction. However, to reduce the amount of reflected light from solar cells, a textured surface of the substrates is required. In this case, the larger the aspect ratio (height/width) in the texture, the less will be the reflection loss and the higher the resulting efficiency of the solar cell [1]. In addition, textured silicon substrates can provide current matching in silicon-based two-junction systems. This is possible due to enhanced absorption in the p-i-n structure due to a significant increase in the effective area compared to planar layers. It is known that during the deposition of layers by the PECVD method on developed surfaces, conformity significantly depends both on the deposition parameters and on the aspect ratio [2]. Previously, a-Si: H layers were formed by the PECVD method on the surface of silicon nanowires to create photovoltaic converters [3]. However, the nanowires obtained by the VLS method were dispersed and did not have a preferential directionality, which greatly simplifies the deposition of layers by the PECVD method. We have previously shown that an array of vertically oriented silicon wires with a high aspect ratio can be obtained by cryogenic ICP[4]. However, preliminary studies on the formation of thin-film p-i-n structures on such substrates revealed limitations associated with the deposition parameters of amorphous silicon (figure 1). When a-Si:H is deposited onto the surface of an array of vertically oriented silicon wires with a high aspect ratio, a significant inhomogeneity of the deposited layer thickness was observed.

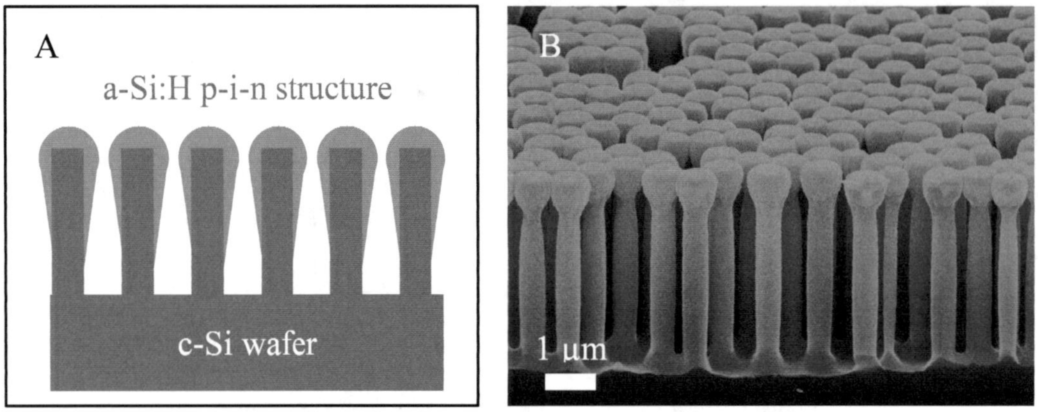

Figure 1. Schematic representation (A) and scanning electron microscopy image (B) of the a-Si:H p-i-n structure deposited on vertically aligned silicon wires.

This may be due to factors such as low ion energy or insufficient circulation of precursors. To increase the conformity of layers in the PECVD method, it is usually recommend increasing the substrate temperature and plasma power, which increases the migration of adsorbed atoms on the surface. However, this can also lead to an increase in the concentration of defects and a decrease in the lifetime of charge carriers in the silicon substrate. Therefore, it is necessary to look for alternative approaches to increase the conformity of the a-Si:H deposition on silicon substrates with vertically aligned silicon wires. One such approach is pulsed deposition, which has proven itself in the deposition of SiO_2 layers by the SA-CVD method for the formation of dielectric coatings in trenches with a high aspect ratio [5]. At this moment, it is of great interest to study the formation of thin layers of amorphous silicon on textured silicon substrates with a high aspect ratio. In this work, we investigate layers of amorphous silicon deposited with continuous and pulsed modes on silicon test structures with a high aspect ratio trenches.

2. Experimental study

To develop the test structure, we use method of cryogenic reactive ion etching of silicon with capacitively and inductively coupled plasma at the temperature of -120 °C using Oxford instruments Plasmalab 100 ICP (figure 2). The shadow mask was formed using the methods of photolithography and electron-beam sputtering of Cr.

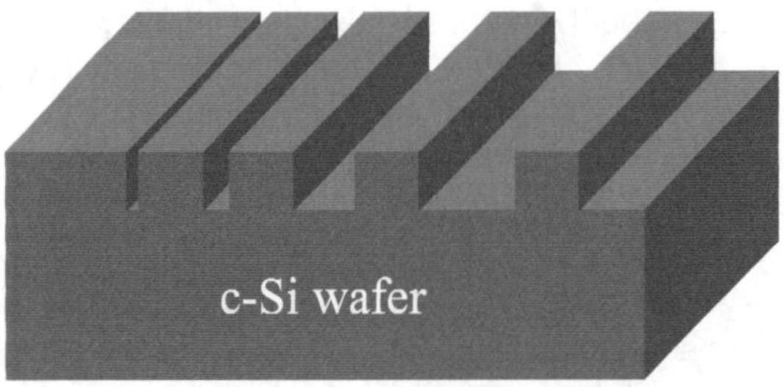

Figure 2. Schematic representation of the test silicon structure with different aspect ratio trenches.

The geometry of the pattern on the silicon substrate was an array of stripes 5 mm long and 1 to 10 µm wide with a distance between them from 1 to 10 µm. The depth of anisotropic etching of silicon by cryo ICP was 10 µm. After the plasma-chemical etching process, the chromium mask was removed from the surface using selective liquid chemical etching. Amorphous silicon layers 140 nm thick were deposited onto test structures by plasma-enhanced chemical vapour deposition at a temperature of 250 °C in continuous and pulsed modes. The layers of a-Si:H were deposited using an Oxford instruments Plasmalab 100 PECVD setup with a 13.56 MHz RF generator from SiH_4 and H_2 precursors. The pressure during deposition was 350 mTorr, the power density of the RF plasma was 8.8 mW / cm2, and the deposition time was 15 minutes. These are typical parameters of layer deposition during the formation of p-i-n structures by the PECVD method. The pulsed deposition mode consisted of periodic purge of the chamber to improve the change in the gas composition in the trenches. The process consisted of 40 pulses of 20 seconds of deposition and 15 seconds of purging the chamber. The resulting structures were cut across the strips and were investigated with a scanning electron microscope.

3. Results and discussion

The SEM images clearly show the formed test structures with deposited layers of amorphous silicon (figure 3). In this case, the side walls of the trenches have a low roughness and located strictly vertically. However, in the areas of the test structure with grooves with a high aspect ratio, the etching depth turned out to be less than 10 μm and ranged from 8 to 9 μm, which is associated with the effect of aspect-dependent etching.

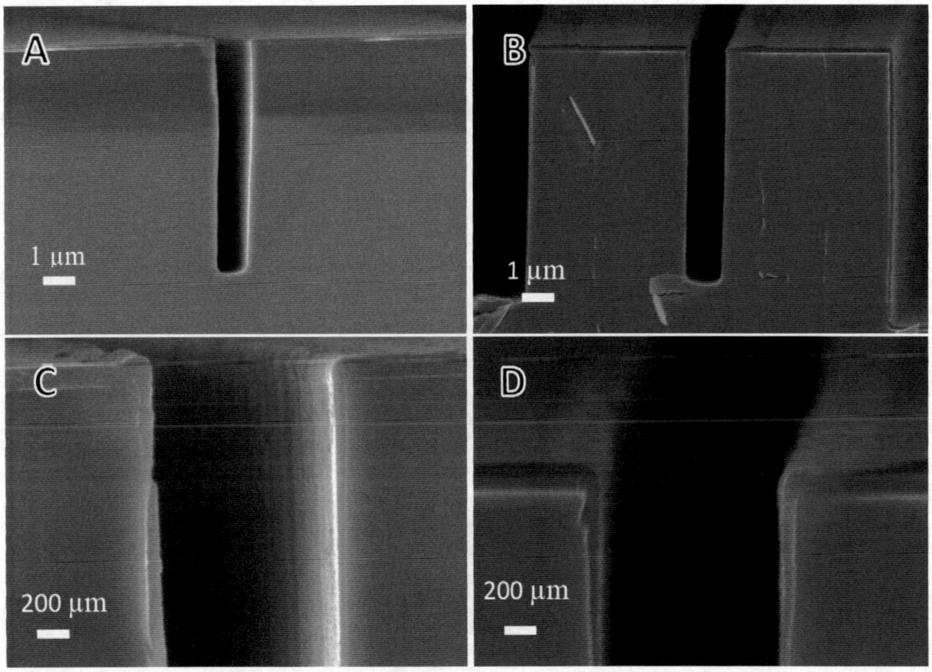

Figure 3. SEM cross-section images of test silicon structures with layers of amorphous silicon deposited in continuous (A) and pulsed (B) modes, as well as their enlarged images (C and D, respectively).

Layers of amorphous silicon have a uniform structure and conformally cover planar areas and trenches. However, in the case of trenches with a high aspect ratio, there is a significant decrease in the thickness of the deposited layer as it deepens. Nevertheless, despite the smaller thickness, the layers have a homogeneous structure with low roughness. Layer thicknesses along the sidewalls were measured for a 1.5 μm wide and 8 μm deep trench.

Table 1. Thickness values of a-Si: H layer deposited on a trench 1.5 μm wide and 8 μm deep in continuous and pulsed mode.

	a-Si:H cont., nm (%)	a-Si:H pulsed, nm (%)
Planar surface	135 (100)	149 (100)
Side wall top	75 (55,5)	110 (73,8)
Side wall at 1 μm depth	53 (39,2)	74 (49,6)
Side wall near the bottom	18 (13,3)	24 (16,1)
Bottom	27 (20)	42 (28,1)

It was noted that in both deposition modes, the thickness values gradually decrease with deepening in trench (table 1). However, in the case of a pulsed deposition mode, the layer thickness along the side wall surface decreases much more slowly. It should be noted that at the bottom of the trench, the layer thickness slightly increases in both cases. For continuous mode, the minimum thickness is 13.3% of the initial, while for pulsed mode this value is 16.1%. Thus, the pulsed deposition method makes it possible to slightly improve the deposition conformity on a surface with a high aspect ratio.

4. Conclusions

In this work, a study was carried out on the deposition of a-Si:H by the PECVD method in continuous and pulsed regime on test structures with the form of trenches with a high aspect ratio. In this case, test structures with depth of 10 μm were etched by cryogenic ICP etching. With a trench width of 1.5 μm, a decrease in the etching depth were observed, which is associated with the aspect-dependent effect in the ICP method. Both in continuous and pulsed regime, the surface of the layers is smooth both on the planar and on the vertical surface. In this case, the thickness of the layers along the side wall of the trenches in these regimes were significantly different. In continuous mode, the minimum depth thickness is 13.3% of the initial, and in pulsed mode, this value reaches 16.1%. This indicates the possibility of using the pulsed mode to improve the conformity of the a-Si:H layers during the deposition of p-i-n structures of photovoltaic converters on silicon substrates with vertically aligned wires with a high aspect ratio.

Acknowledgements

This work was supported by the RFBR project № 21-58-46001.

References

[1] Päivikki R, Haarahiltunen A, Sainiemi L, Yli-Koski M, Talvitie H, Schubert M C and Savin H 2013 *IEEE Journal of Photovoltaics* **3** (1) 90–94
[2] Pan Y, Liu H, Wang Z, Jia J and Zhao J 2021 *Coatings* **11** 510
[3] Yu L, O'Donnell B, Foldyna M and Roca i Cabarrocas P 2012 *Nanotechnology* **23**(19):194011
[4] Morozov I, Gudovskikh A, Uvarov A, Baranov A, Sivakov V and Kudryashov D 2020 *Phys Status Solidi A* **217**(4):2070017
[5] Lisker M, Marschmeyer S, Kaynak M and Tekin I 2019 *ECS Trans* **35**(2):95–104

Structural properties of TiO_2-SnO_2 thin films prepared by new pyrolysis solid-phase method

M G Volkova[1], V Yu Storozhenko[1], V V Petrov[2] and E M Bayan[1]

[1] Department of Chemistry, Southern Federal University, Rostov-on-Don, 344090, Russia,
[2] Research and Education Centre “Microsystem technics and multisensory monitoring systems”, Southern Federal University, Taganrog, 347922, Russia

mvol@sfedu.ru

Abstract. Nanoscale TiO_2-SnO_2 films with the Ti:Sn ratio 1:99, 3:97 and 5:95 mol%, respectively, were obtained by solid-phase low-temperature pyrolysis method. The synthesized materials were studied by X-ray phase analysis and scanning electron microscopy (SEM) analysis. Regardless of the modified agents’ concentration, the structure of cassiterite was observed for all synthesized materials. When studying the effect of synthesis parameters on the materials properties, it was shown that both an increase in the Ti^{4+} concentration and in the calcination temperature leads to an increase in the particle size.

1. Introduction

Nanomaterials based on tin dioxide, doped with different modifying agents, are commonly used to produce optical coatings, gas-sensitive sensors, photocatalysts, etc. [1-2]. Composite materials based on mixed oxides exhibit better optical, electrophysical, and other properties than pure oxide materials, that could be both due to the close ionic radii of tin and titanium and the equal higher valences of these elements. Various studies contain information that the properties of mixed oxides are generally better than those of individual oxides. For example, titanium dioxide nanomaterials with a small addition of tin have higher photocatalytic properties than pure titanium dioxide [3]. The mixture of TiO_2-SnO_2 oxides is very perspective due to the structural analogy between TiO_2 and SnO_2: both oxides crystallize in the rutile structure. For example, it was shown [4] that an increase in the tin concentration up to 40% for TiO_2-SnO_2 materials leads to an increase in the rutile phase of TiO_2 due to the similarity of crystal structures, while pure materials crystallize in the cassiterite and anatase phase, respectively. In [5] it was shown that small additions of Ti^{4+} in SnO_2 reduce the degradation of SnO_2 materials characteristic and increase the density for chemisorption active centers.

To obtain perspective materials with controlled properties different techniques are applied, among them doping and different temperature proceedings are used. The change in the materials films properties is significantly influenced not only by the calcination temperature and the concentration of the additives introduced, but also by the number of layers applied. For example, in paper [6], changes in the optical properties depending on the number of layers (5, 10, 15) of SnO_2-TiO_2 thin films (Sn:Ti=1:1) synthesized by the sol-gel method were shown. According to the authors, the deterioration of the films transparency with an increase in the number of layers can be associated simultaneously with a change in the phase composition of the materials (anatase, anatase-rutile and rutile for 5, 10 and 15 layers, respectively), both an increase in the particle size and in the coating density (231, 340, 413 nm).

The synthesis of nanoscale oxide film materials is carried out by various methods, but some of them have certain disadvantages, such as poor reproducibility of the process (sol-gel method), expensive equipment (hydrothermal synthesis, magnetron sputtering, laser sputtering), etc. The spray pyrolysis [7] is the simplest and most economically feasible, widely used method for materials synthesis in large scale. Another perspective method of thin film synthesis is solid-phase pyrolysis [2, 8], which allows one to obtain thin films with a controlled thickness and specified optimal physicochemical properties. Also, solid-phase pyrolysis technique does not include substrates heat treatment (before precursors' layers applying) in comparison to spray pyrolysis and this technique allows one to obtain sufficiently crystallize films in comparison to sol-gel technology. Comprehensive study of the materials properties synthesized by new technique is an important element for finding methods of their further application.

Thus, the aim of this work was to choose optimal synthesis conditions to obtain high-quality nanoscale films based on tin dioxide containing titanium dioxide in low concentrations by solid-phase low-temperature pyrolysis and to study the structural and morphological properties of the synthesized materials.

2. Experiment

To obtain thin film materials by solid-phase pyrolysis method, stannic chloride pentahydrate $SnCl_4 \cdot 5H_2O$, titanium butoxide $(C_4H_9O)_4Ti$, organic acid, and 1,4-dioxane as a solvent were used as precursors. All chemicals used were of analytical grade or of the highest purity available, were purchased from "ECROS", Russia. The synthesis was carried out in two stages. At the first stage, intermediate products (organic salts of titanium and tin) were obtained from the $SnCl_4 \cdot 5H_2O$ and $(C_4H_9O)_4Ti$ in melt. The preparation of the intermediate product was carried out with the introduction of tin tetrachloride required amount (molar ratio Ti:Sn was 1:99, 3:97, 5:95 for materials 1, 2, 3, respectively) in the melt of organic acid and titanium butoxide. After cooling, the synthesized melt was shredded.

The second stage involved dissolving the resulting product in an organic solvent and then applying it to the pre-prepared substrates (soda-lime glass and silicon). To prepare the substrates, chemical cleaning was applied, which include treatment with a hot mixture of $K_2Cr_2O_7$ and HNO_3 for 10 minutes, three times washing with distilled water and treatment with alcohol. To remove the alcohol residue, the substrates were washed twice in distilled water and dried. Before applying the obtained solution, the substrate was additionally cleaned with acetone for degreasing. The method allows one to obtain films of different thicknesses by applying a precursor's solution several times using the following technology. Each applied layer was dried in air and then at 120 °C in a drying cabinet. The final heat treatment was carried out in a muffle furnace at 550 °C for two hours, the heating rate was 10°/min. After heat treatment all obtained materials were cooled to the room temperature together with a muffle furnace. This cooling method allows one to obtain uniform coatings without cracks. In this paper, three-layer films applied on the pre-prepared silicon and soda-lime glass substrates were synthesized and investigated.

The phase composition of the obtained films was studied by X-ray diffraction (XRD) using an ARLX'TRA diffractometer, Thermo ARL (Switzerland) with CuKα X-rays. Phase composition analysis performed using Crystallography Open Database. The morphology, quality, particles size in the volume and on the surface as well as thickness of the obtained film materials on the silicon substrates were studied by scanning electron microscopy (SEM) using a scanning electron microscope Nova Nanoab 600. The size of the coherent scattering regions (D) was estimated by XRD analysis data using the Scherrer equation: $D = k \cdot \lambda / (\beta \cdot \cos\Theta)$, where k is the geometric shape coefficient, λ is the X-ray wavelength constant (λ = 0,154 nm), and β is the full width at half the maximum of the diffraction line and Θ-diffraction angle.

3. Results and discussion

The results of the XRD analysis showed that all synthesized materials are nanoscale, sufficiently crystallize and contain cassiterite phases (Figure 1). The absence of additional peaks on the XRD pattern suggests a high quality of the films obtained by low-temperature solid-phase pyrolysis technique. With an increasing of the modifying agents concentration the peaks intensity also increased.

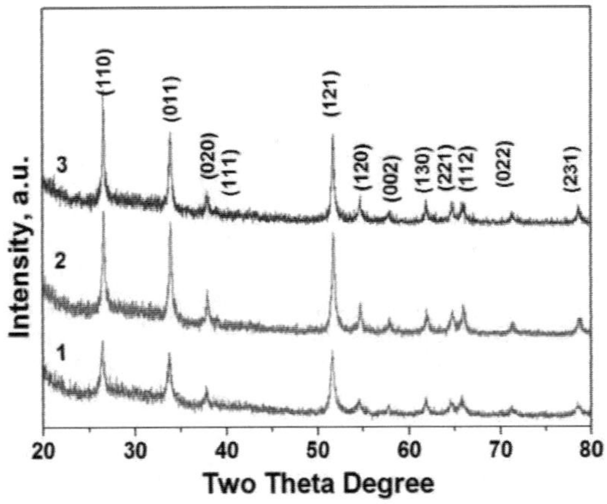

Figure 1. XRD patterns of SnO_2-TiO_2 materials calcined at 550 °C (explanations are presented in text) and SnO_2 [9].

With increasing of the modifying agents' concentration, an increase in the particle size calculated according to the Scherrer equation was also observed (19, 23 and 29 nm for materials 1, 2, 3, respectively). According to the calculations, for materials 1, 2, and 3 the crystallinity was 47, 45, and 43%, respectively. This change may be related to the fact that films containing a high tin ions concentration crystallize even at room temperature in rutile or cassiterite type, while an increase in the titanium dioxide concentration leads to an increase of the amorphous phase [5].

According to SEM analysis (Figure 2a), the films obtained by pyrolysis have relatively smooth surfaces, homogeneous in the volume and solid. The thickness of the films is about 70 nm for three-time application. The presented SEM images do not have clear borders despite the three layers application. This indicates a possibility to synthesize high-quality thin oxide films of the controlled thickness by low-temperature pyrolysis technique.

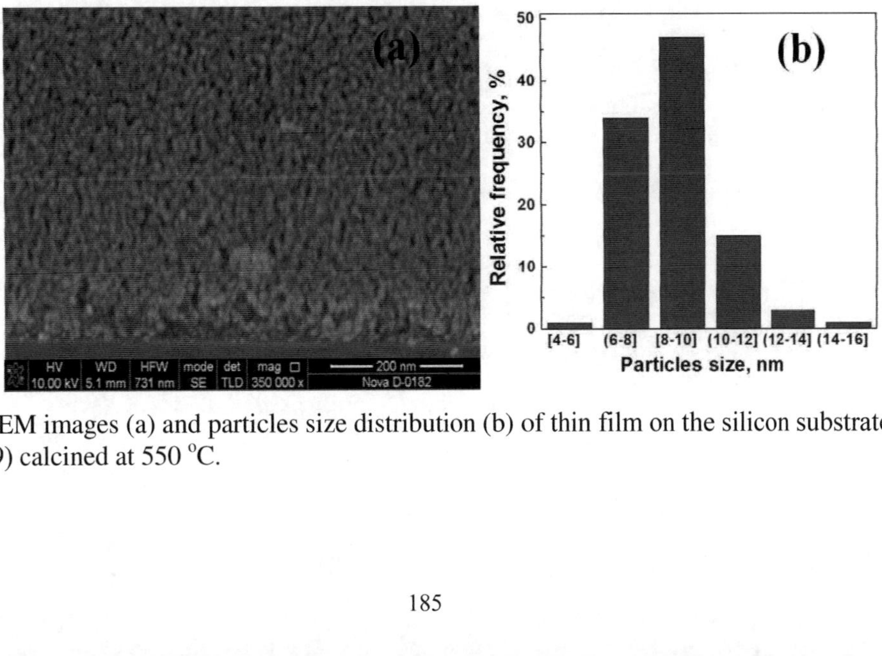

Figure 2. SEM images (a) and particles size distribution (b) of thin film on the silicon substrate (Ti:Sn=1:99) calcined at 550 °C.

All of obtained films materials are consisted from particles almost round in their shape. Statistical analysis of the SEM data showed that the average particle size is 8 and 13 nm, for materials 1 and 3, respectively. For the material Ti:Sn=1:99 (Figure 2b), most of the grains are in the size range of 8-10 nm.

The slightly excessive data obtained by the Scherrer equation compared to the data obtained from the statistical SEM analysis may be due to differences in the determination methods. The XRD method estimates the entire volume of the material, whereas the statistical SEM analysis is a more accurate method, estimating the surface of the synthesized film (table 1).

Table 1. Materials characteristics.

Materials	Particles size, nm (XRD)	Particles size, nm (SEM)	Crystallinity, %
1 (Ti:Sn=1:99)	19	8	47
2 (Ti:Sn=3:97)	23	10	45
3 (Ti:Sn=5:95)	29	13	43

4. Conclusion

Homogeneous thin film nanomaterials TiO_2-SnO_2 were obtained by solid-phase low-temperature pyrolysis technique. The described method allows one to obtain film coatings of different thicknesses by varying the applied layers number of the precursor's solution with good reproducibility. It was shown that the particle size increases with both increasing of the modifying agents' concentration and the calcination temperature. With an increase in the concentration of the introduced additives, the synthesized materials crystallinity decreases. The best crystallinity degree was shown for film material doped with the minimum Ti^{4+} concentration (1 mol.%). The presented method makes it possible to obtain homogeneous films of controlled thickness with uniform and smooth surface, which allows one to recommend it to produce semiconductor film materials as promising electrode or gas-sensing material.

Acknowledgments

This work was financially supported by the RFBR, project 20-07-00653 A.

References

[1] Al-Hamdi A M, Rinner U and Sillanpää M 2017 *Process Safety and Environmental Protection* **107** 190-205

[2] Petrov V V, Bayan E M, Khubezhov S A, Varzarev Y N and Volkova M G 2020 *Chemosensors* **8(2)** 40

[3] Bayan E M, Lupeiko T G, Pustovaya L E and Volkova M G 2020 *Journal of Advanced Dielectrics* **10 (1,2)** 2060018

[4] Medjaldi F, Bouabellou A, Bouachiba Y, Taabouche A, Bouatia K, and Serrar H 2020 *Mater. Res.Express* **7** 016439

[5] Zakrzewska K, Radecka M, Przewoźnik J, Kowalski K and Czuba P 2005 *Thin Solid Films* **490 (1)** 101-7

[6] Sönmezoğlu S, Arslan A, Serin T and Serin N 2011 *Physica Scripta*, **84(6)** 065602

[7] Leng J, Wang Z, Wang J, Wu H-H, Yan G, Li X, H Guo, Liu Y, Zhang Q and Guo Z 2019 *Chem. Soc. Rev.* **48** 3015-72

[8] Volkova M G, Storozhenko V Y, Petrov V V and Bayan E M 2020 *J. of Phys.: Conf. Ser.* **1695(1)** 012023

[9] Downs R T and Hall-Wallace M 2003 *Am. Mineral.* **88** 247-250

Investigation of resistive switching in Ag/Ge/Si(001) stack by conductive atomic force microscopy

V A Vorontsov, D A Antonov, A V Kruglov, I N Antonov, M E Shenina, V E Kotomina, V G Shengurov, S A Denisov, V Yu Chalkov, D A Pavlov, D O Filatov and O N Gorshkov

Lobachevskii University of Nizhnii Novgorod, 603950 Nizhnii Novgorod, Russia

vladislav.vorontsov1@gmail.com

Abstract. We report on an experimental study of resistive switching (RS) of individual dislocations in Ag/Ge/Si(001) memristors by combined grazing incidence ion sputtering of the Ag electrodes and application of Conductive Atomic Force Microscopy to provide an electrical contact to individual Ag-filled dislocations in the Ge layer. Two types of RS were observed corresponding to two different RS mechanisms: (i) drift of Ag^+ ions inside the dislocation cores and (ii) RedOx reactions in residual GeO_x in the etch pits on the Ge layer surface.

1. Introduction

Recently, resistive switching (RS) was studied extensively [1]. The RS is a bi-stable (multistable) reversible change in the resistance of a thin dielectric layer sandwiched between two conducting electrodes under a voltage applied to the ones [2]. The electronic devices, the operation of which is based on RS are called *memristors* [3]. Memristors are considered to be promising for application in the next-generation non-volatile Resistive Random Access Memory (RRAM) [4], neuromorphic computers and neural networks [5], etc.

In most cases, RS mechanism is based on the rupture and restoring of conductive filaments (CFs) shortcutting the memristor stack electrodes under the electric field between the ones. In so-called conducting bridge (CB) memristors, the CFs consist of metal atoms injected into the functional dielectric layer [6]. Recently, a novel type of CB RRAM (so-called Epitaxial RRAM or EpiRRAM) based on relaxed SiGe/Si(001) epitaxial layers (ELs) was introduced [7]. The authors of [7] proposed to use threading dislocations growing through the SiGe ELs as templates for the Ag CFs. In such a device, the Ag atoms are confined within the dislocations cores, which limit the lateral spreading of the Ag atoms inside the SiGe EL. This way, an improved stability of the RS parameters and durability were achieved. However, many details of RS mechanism in EpiRRAM remain unclear to date.

Conductive Atomic Force Microscopy (CAFM) has been proven to be a powerful tool for the investigations of RS at the nanometer scale [8]. CAFM allows visualizing individual CFs on the surface of the functional dielectric films and measuring the electric current flowing through individual CFs. Earlier, CAFM was applied to study RS in individual dislocations in $SrTiO_3$ single crystal films filled with Fe atoms [9].

In the present study, we have applied CAFM combined with grazing incidence ion sputtering (to remove the top Ag electrode) to investigate RS in individual threading dislocations in a Ag/Ge/Si(001)-based EpiRRAM stack.

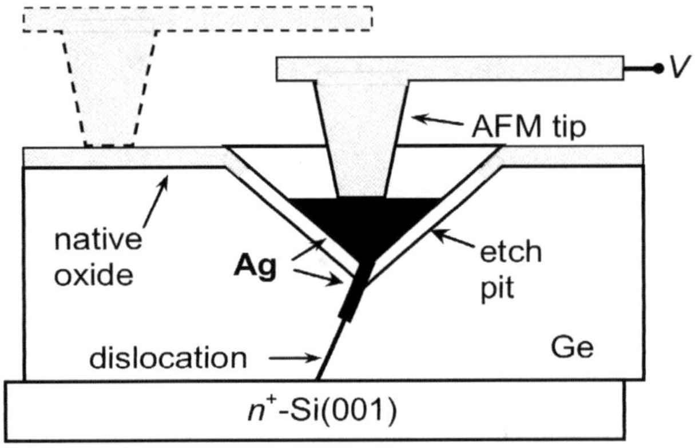

Figure 1. Schematic representation of the experiment of measuring the RS of individual dislocations in the Ge/n^+-Si(001) using CAJM. The dashed contour shows the sample/probe configuration when the CAFM tip is positioned between the etch pits.

2. Experimental

The 190-nm thick Ge EL was grown on n^+-Si(001) substrate by low temperature Hot Wire Chemical Vapor Deposition (HW CVD). In this method, monogermane GeH_4 is introduced into the growth chamber up to the partial pressure from ~10^{-4} Torr up to ~10^{-3} Torr through a leak valve. The flow of the Ge atoms onto the growing Ge EL surface is generated by the thermal decomposition (pyrolysis) of GeH_4 on a Ta strip (playing a role of the hot wire) heated up to ~1400 °C by passing an electric current. The substrate temperature was 325 °C. This method has been proven to allow growing smooth single crystal Ge/Si(001) ELs with controllable density of the threading dislocations, which can be varied in the range from ~10^5 cm^{-2} to ~10^9 cm^{-2} by varying the growth process parameters [10]. Intentionally undoped Ge ELs had a *p*-type conductivity with the background hole concentration ~10^{17} cm^{-3}. The fabricated prototype EpiRRAM devices have demonstrated a stable bipolar RS with the ratio of the currents in the low resistance state (LRS) and the high resistance state (HRS) – I_{ON} and I_{OFF}, respectively – up to 15 [10–13].

The prototype EpiRRAM devices were fabricated by the deposition of the top 40-nm thick Ag electrodes of ~1 mm in sizes on the Ge/Si(001) EL surface by direct current magnetron sputtering through a shadow mask. Prior to the deposition of the top Ag electrodes, the Ge EL was subjected to selective wet chemical etching in HF : H_2O_2 : CH_3COOH 1 : 2 : 3 solution to form the etch pits on the EL surface decorating the threading dislocations growing through the Ge EL. The Ag-filled etch pits stimulated the drift of the Ag^+ ions into the threading dislocation cores in the Ge EL acting as the electric field concentrators [7]. Also, the Ag-filled etch pits served as the reservoirs of the Ag atoms to fill the threading dislocations that is necessary for the EpiRRAM functioning [7]. More details on the growth of the Ge/Si(001) EpiRRAM stacks by HW CVD as well as on the structure and electrical properties of the Ag/Ge/Si(001) prototype memristors fabricated on the basis of these shacks can be found elsewhere [11–13].

For the CAFM investigations, the top Ag electrodes were removed by small-angle (3°) grazing incidence low energy (2 keV) Ar^+ ion sputtering using Gatan® PIPS ion thinning system for sample preparation for Transmission Electron Microscopy. At that, small droplets of Ag remained inside the etch pits (see figure 1). The electric contacts to the Ag droplets were provided by NT-MDT® NSG-11 DCP CAFM probes coated with conductive diamond-like films positioned onto individual Ag droplets inside selected Ag-filled etch pits (figure 1). The second (Ohmic) contact was provided from the back side of the n^+-Si(001) substrate by alloying a foil made from $Sn_{0.9}Sb_{0.1}$ ally by a spark discharge. This way, we were able to measure the cyclic current-voltage (*I–V*) curves of the Ag CFs inside individual threading dislocation cores. The measurements were carried out in ambient air at room temperature using NT-MDT® Solver Pro™ atomic force microscope (AFM) in Contact Mode.

Also, the current images (i. e. the maps of the electric current flowing through the CAFM probe I_t as a function of the probe coordinates x, y in the sample surface plane) were recorded simultaneously with the topographic AFM images of the sample surface by raster scanning with a constant voltage V_g applied to the CAFM probe with respect to the n^+-Si(001) substrate.

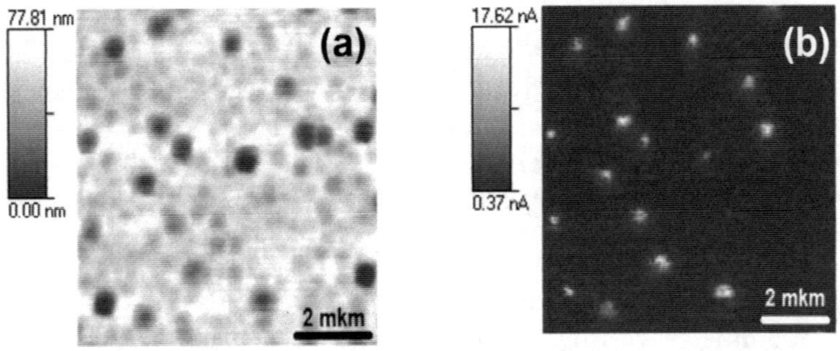

Figure 2. Morphology (a) and current image (V_g = –4 V) (b) of Ge/Si(001) surface after removing the top Ag electrodes.

3. Results and Discussion

Figure 2a presents a topographic AFM image of the Ge/Si(001) stack after removing the top Ag electrode. Numerous etch pits were observed on the Ge EL surface. The surface density of the etch pits (associated with the threading dislocation density in the Ge EL) was ~10^7 cm–2.

Figure 2b presents a current image of the sample surface recorded at V_g = –4 V simultaneously with the topographic image shown in figure 2a. The areas of increased I_t matching the etch pits in figure 2a were associated with the Ag droplets resident inside the etch pits. Every time the CAFM tip encounters an Ag droplet, I_t increases due to increased contact area of the Ag droplet to the inner etch pit surface as compared to the one of the CAFM tip to the Ge EL (< 100 nm^2 [14]). Note that $V_g < 0$ corresponds to a forward biased Schottky barrier of Ag to *p*-Ge EL. Note also that no electroforming has been performed prior to measuring the CAFM images shown in figure 2.

Figure 3 shows typical cyclic *I–V* curves $I_t(V_g)$ recorded with the CAFM tip positioned onto the Ag droplets inside different etch pits after forming. The cyclic *I–V* curves demonstrate pronounced hysteresis loops typical for bipolar RS. Two types of the *I–V* curves were observed that was attributed to different RS mechanisms. The wide hysteresis loops (see, for example, the blue curve in figure 3) was associated with formation and destruction of the Ag CFs in the individual threading dislocations beneath the etch pit due to the drift of the Ag^+ ions inside the dislocation core (so-called CB mechanism [5]). The narrower nonlinear *I–V* curves (see, for example, the red curve in figure 3) were attributed to the reduction/oxidation processes (so called RedOx RS mechanism [1]) in the thin residual GeO_x native oxide layer between the Ag droplet and the inner etch pit surface (see figure 1). Similar *I–V* curves were observed when the CAFM probe was positioned onto the Ge EL surface (also covered with the native oxide) between the etch pits (figure 1, dashed contour). In this case, the hysteresis was attributed to the RS in the GeO_x native oxide layer [15]. At that, the conductive CAFM tip played a role of a nanometer-sized movable top electrode composing a virtual memristive device together with the thin GeO_x film on the surface of the *p*-Ge EL.

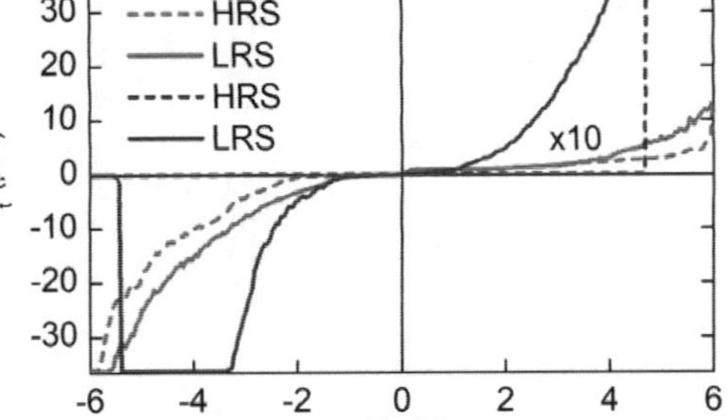

Figure 3. Cyclic *I–V* curves of the contacts of the CAFM probe to the Ag droplets inside the etch pits typical for two types of RS: for the CB RS in the Ag-filled dislocation in the Ge EL (blue curve) and for the RedOx RS tin the native oxide GeO_x between the Ag droplet and the Ge EL surface inside the etch pit (red curve).

4. Conclusion

The results of the present study demonstrate experimentally a possibility to study resistive switching in individual dislocations in Epitaxial Resistive Random Access Memory devices combining the grazing incidence ion sputtering of the top metal electrodes and Conductive Atomic Force Microscopy to provide an electrical contact to individual dislocations. Using this novel technique, we have studied resistive switching in individual threading dislocations in the Ag/Ge/n^+-Si(001) memristor stacks based on the Ge/n^+-Si(001) epitaxial layers grown by low-temperature Hot Wire Chemical Vapor Deposition. We have observed two different types of the current-voltage curves of the conductive atomic force microscope probe to the Ag droplets inside the Ag-filled etch pits on the Ge layer surface corresponding to two different resistive switching mechanisms. One was related to the Conducting Bridge-type switching associated with the drift of the Ag^+ ions along the threading dislocations in the Ge/n^+-Si(001) epitaxial layer. Another one was related to the RedOx-type switching in the GeO_x native oxide layers between the Ag droplets and the surfaces of the etch pits on the Ge epitaxial layers.

Acknowledgments

The present study was supported by Russian Foundation for Basic Research (19-29-03026). The experiments were carried out using the shared research facilities of Research and Educational Centre for Physics of Solid State Nanostructures at Lobachevskii University of Nizhnii Novgorod.

References

[1] Ielmini D and Waser R 2016 *Resistive Switching: From Fundamentals of Nanoionic Redox Processes to Memristive Device Applications* (Hoboken: Wiley–VCH)
[2] Waser R and Aono M 2007 *Nature Materials* **6** 833
[3] Strukov D B, Snider G S, Stewart D R and Williams R S 2008 *Nature* **453** 80
[4] Zhang Y *et al* 2020 *Appl. Phys. Rev.* **7** 011308
[5] Ielmini D 2016 *Semicond. Sci. Technol.* **31** 063002
[6] Lee S H, Zhu X and Lu W D 2020 *Nano Research* **13** 1228
[7] Choi S, Tan S H, Li Z, Kim Y, Choi C, Chen P Y, Yeon H, Yu S and Kim J 2018 *Nature Materials* **17** 335
[8] Lanza M 2017 *Conductive Atomic Force Microscopy: Applications in Nanomaterials* (Hoboken: Wiley–VCH)
[9] Szot K, Bihlmayer G and Speier W 2014 *Solid State Phys.* **65** 353
[10] Denisov S A, Matveev S A, Chalkov V Yu and Shengurov V G 2016 *J. Phys.: Conference Series* **690** 012014
[11] Gorshkov O N, Shengurov V G, Denisov S A, Chalkov V Yu, Antonov I N, Kruglov A V, Shenina M E, Kotomina V E, Filatov D O and Serov D A 2020 *Tech. Phys. Lett.* **46** 91
[12] Filatov D O, Shenina M E, Shengurov V G, Denisov S A, Chalkov V Yu, Kruglov A V, Vorontsov V A, Pavlov D A and Gorshkov O N 2020 *Semicond.* **54** 1833
[13] Gorshkov O *et al* 2020 *J. Phys.: Conference Series* **1695** 01215
[14] Filatov D O, Antonov D A, Gorshkov O N, Kasatkin A P, Pavlov D A, Trushin V N, Antonov I N and Shenina M E 2014 *Atomic Force Microscopy (AFM): Principles, Modes of Operation and Limitations* ed H Yang (New York: Nova Science) pp 335–355
[15] Cheng C H, Chin A and Yeh F S 2011 *Appl. Phys. Lett.* **98** 052905

Crystallization of Bi-substituted iron garnet bi-layers

T V Mikhailova[1], Yu E Vysokikh[2], A N Shaposhnikov[1], V N Berzhansky[1], S Yu Krasnoborodko[2], A S Lutovinov[2], M F Bulatov[2] and D V Churikov[2]

[1] V.I. Vernadsky Crimean Federal University, Simferopol, 295007, Russia
[2] Scientific and Technological Centre of Unique Instrumentation of the RAS, Moscow, 117342, Russia

visokihy@gmail.com

Abstract. Magneto-optical (MO) structures are widely used for different application in the fields of magnetoplasmonics, magneto-optics, photonics e.t.c. Bi-substituted iron garnet (Bi:IG) is high-performance MO material. Integration of Bi:IG films to silicon semiconductor technology gives new opportunities to create nanoscale hight performance MO devices. Vacuum sputtering deposition allows to fabricate Bi:IG structures on different substrate types. Authors investigate crystallization process of Bi:IG bi-layers in a different process parameter (different layers composition and its thickness, temperature and time of annealing) using gadolinium gallium garnet GGG and fused quartz SiO_2 substrates to determine dependences which impact on crystallization.

1. Introduction

MO devices gives opportunities to create high performance isolators, modulators and switches, nanophotonic devices, plasmonic circuit elements [1−5]. The bi-layers films given successful results in magnetophotonic crystals and Tamm structures design [6−12]. Crystallization process is one of most important factors which determine films' quality. Integration of Bi:IG films to silicon semiconductor process is important due to possibility of fabricating new nanoscale and high-performance devices. The synthesis of Bi: IG films on substrates without garnet structure is often performed by using two-layer films structure. [6, 9, 13]. Present work summarized the results on synthesis of Bi:IG bi-layers by reactive ion-beam sputtering with subsequent annealing at air. The deviations of microstructure of films were investigated depending on grown conditions.

2. Experimental

Bi:IG bi-layers were fabricated by reactive ion beam sputtering method (RIBS) at a URM 3-279.014 setup in oxygen–argon mixture using an ion-beam Kholodok-1 source. The films were deposited from ceramic target on gadolinium gallium garnet (GGG) or fused quartz SiO_2 substrates [6−9]. The garnet phase was formed by crystallization annealing at the temperature range from 650°C to 710°C. Annealing time was varied from 20 min to 60 min. Different compositions of Bi:IG were used:

- G1 – $Bi_{1.0}Lu_{0.5}Gd_{1.5}Fe_{4.2}Al_{0.8}O_{12}$
- G8 – $Bi_{2.5}Gd_{0.5}Fe_{3.8}Al_{1.2}O_{12}$
- GN – $Bi_{2.8}Y_{0.2}Fe_5O_{12}$

All investigated samples have two layers with thicknesses as described in table 1.

Table 1. Investigated bi-layers with thicknesses.

Composition	G1/G1	G1/G8	G1/GN
Thicknesses of the layers, nm	80 / 208	80 / 195	80 / 177

In this paper we made the comparison between two process modes (methods) of crystallization annealing.

First method (dipping) description:

- Both layers of the structure were crystallized separately.
- The sample was almost immediately dipped in a thermal field with a temperature from 600 °C to 710 °C.
- Crystallization annealing of the sample occurred within 20 min after reaching thermodynamic equilibrium.

Second method (slow) description:

- Both layers of the structure were crystallized separately.
- The synthesis was carried out as a result of very slow heating (heating rate of about 2-3 °C / min) of the sample to the optimum temperature for a given composition from 600 °C to 710 °C and subsequent crystallization at a constant temperature for 60 min.

Surface structure of bi-layers reveals crystallization features which depends on process parameters (type of substrate, thickness of layers, temperature and time of annealing). Atomic force microscopy (NTEGRA NT-MDT) was used to measure topography parameters of Bi:IG thin films. Active vibroisolating platform was used to protect the system from vibration noises. Average roughness of each film was measured to determine the regime which allows to minimize this parameter. Tapping mode of AFM was used for investigations because it allows one to get the best value of resolution and minimize the negative impacts of the probe and the sample one each other. In this mode, the cantilever console oscillates with self-resonant frequency and high amplitude (approximately 50-100 nm). At these amplitudes, the needle contacts the sample surface at the moment of maximum deflection of the console down from its equilibrium position, which significantly changes the frequency, phase and amplitude of its oscillations. A tapping mode has a higher horizontal resolution compared to the contact mode. Cantilevers NSG10 (TipNano) were used with resonant frequency of 240 kHz and force constant of 11.8 Nm.

3. Results

Figures 1 and 2 show AFM images (15x15 microns and 2x2 microns area) of the surface of two-layer G1/G8 films which were annealed by the first method on GGG and SiO2 substrates respectively. It is showed that the surface texture of the samples is different, although both films exhibit a polycrystalline structure. Thus, the surface of the film on GGG is characterized by a uniform distribution of crystallites. The grains are approximately the same size, even on large-area frames. Average size of the crystallites is about 90-100 nm. The surface of the SiO2 film shows the presence of regions with different crystal structures – regions with large crystallites of about 200-500 nm of a clear rectangular or triangular shape and smooth regions with small "round" crystallites of the same size. A fragment of a 2x2 μm smooth region and its analysis is shown in Figure 2. Such smooth areas show first stage of film formation on a non-granate substrate. Most often, the regions are formed as round shapes of micron sizes. At longer annealing time (or other modes) the regions can be re-formed into large smooth polycrystalline. The size of the crystallites in the homogeneous region of the film is about 60 nm.

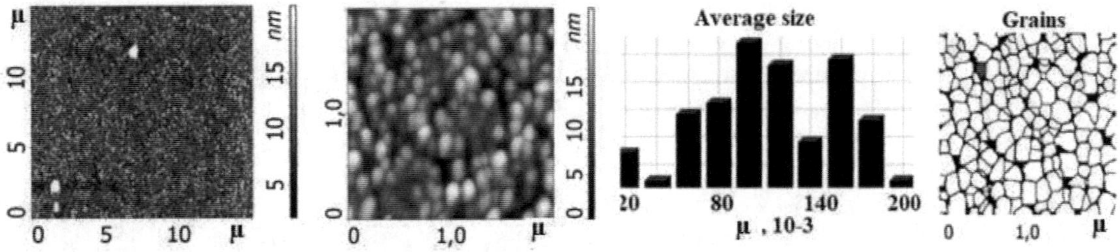

Figure 1. Topography and grain analysis of the GGG/G1/G8 structure annealed by 1 method (dipping).

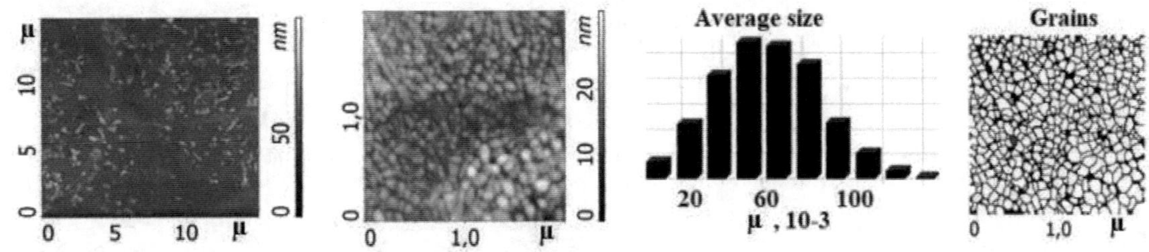

Figure 2. Topography and grain analysis of the SiO2/G1/G8 structure annealed by 1 method (dipping).

We should notice that studies other samples showed different results. The second annealing method led to the formation of a uniformly distributed polycrystalline structure for the SiO2 substrate and the formation of crystallites of various shapes for the GGG substrate.

So in order to synthesize a smoother film, we need to choose a faster synthesis mode (the first method) for growth on GGG substrates and a slower synthesis mode (the second method) for growth on SiO2 substrates. Table 2 shows the morphological parameters of all studied two-layer samples.

Table 2. Comparison of investigated bi-layers annealed by different methods.

Process method	Composition		
	G1/G1	G1/G8	G1/GN
RMS, nm			
1 method (dipping)	4,6	3,4	7,8
2 method (slow)	5,9	9,8	5,2
Peak to peak, nm			
1 method (dipping)	89	51	52
2 method (slow)	65	60	80
Average grain size, μm			
1 method (dipping)	0,087	0,121	0,114
2 method (slow)	-	0,104	0,098

4. Conclusion

Thus, the influence of the crystallization annealing mode on the formation of the Bi:IG bi-layers crystal structure is studied. Two methods, called briefly by the authors "dipping" and " slow ", were compared. First Method is most suitable for crystallization of films on GGG substrates. When using the first method, the surface morphology is more uniform, while the second method leads to the formation of polycrystalline grains of different sizes. However, the second method is better suited for the formation of garnet films on fused quartz SiO_2 substrates which do not have the initial garnet structure.

Acknowledgments

The authors from V.I. Vernadsky Crimean Federal University TVM, ANSh, VNB acknowledge support by the Russian Ministry of Education and Science (Megagrant project N 075-15-2019-1934) for the synthesis of film samples with optimized magnetic parameters.

The authors from Scientific and Technological Centre of Unique Instrumentation of the RAS YuEV, SYuK, MFB, DVCh acknowledge support by the Ministry of Science and Higher Education of the Russian Federation under the State contract No. 0069-2019-0009 for the films measurements. This work was performed using the equipment of the Shared Research Facilities of the STC UI RAS.

References

[1] Inoue M, Baryshev A V, Goto T, Baek S M, Mito S, Takagi H and Lim P B 2013 Magnetophotonic Crystals: Experimental Realization and Applications *Magnetophotonics* eds. Inoue M, Levy M, Baryshev A V (Springer-Verlag Berlin Heidelberg) **7** 163

[2] Armelles G, Cebollada A, García-Martín A and González M U 2013 *Adv. Optical Mater* **1** 10

[3] Bi L, Hu J, Jiang P, Kim H S, Kim D H, Onbasli M C, Dionne G F and Ross C A 2013 *Materials* 6 5094–117

[4] Karpasyuk V K and Bulatov M F 1994 *IEEE Transactions on Magnetics* **30(6)** 4344-4346

[5] Bulatov M F and Parkhomenko Y N 2006 *Semiconductors* **40** 169–171

[6] Berzhansky V N, Karavainikov A V, Mikhailova T V, Prokopov A R, Shaposhnikov A N, Shumilov A G, Lugovskoy N V, Semuk E Y, Kharchenko M F, Lukienko I M, Kharchenko Yu M and Belotelov V I 2017 *JMMM* **440** 175

[7] Berzhansky V, Mikhailova T, Shaposhnikov A, Prokopov A, Karavainikov A, Kotov V, Balabanov D and Burkov V 2013 *Appl. Opt.* **52** 6599-6606

[8] Berzhansky V N, Karavainikov A V, Milyukova E T, Mikhailova T V, Prokopov A R and Shaposhnikov A N 2010 *Functional materials* **17 (1)** 120-126

[9] Berzhansky V N, Shaposhnikov A N, Prokopov A R, Karavainikov A V, Mikhailova T V, Lukienko I N, Kharchenko Yu N, Golub V O, Salyuk O Yu and Belotelov V I 2016 *Journal of Experimental and Theoretical Physics* **123** 744-751

[10] Mikhailova T V, Berzhansky V N, Shaposhnikov A N, Karavainikov A V, Prokopov A R, Kharchenko Yu M, Lukienko I M, Miloslavskaya O V and Kharchenko M F 2018 *Opt. Mater.* **78** 521

[11] Mikhailova T V, Shaposhnikov A N, Prokopov A R, Karavainikov A V, Tomilin S V, Lyashko S D and Berzhansky V N 2018 *EPJ Web of Conferences* **185** 02016

[12] Mikhailova T V, Lyashko S D, Tomilin S V, Karavainikov A V, Prokopov A R, Shaposhnikov A N and Berzhansky V N 2017 *J. Phys. : Conf. Ser.* **917** 062053

[13] Wehlus T, Korner T, Leitenmeier S, Heinrich A and Stritzker B 2011 *Phys. Status Solidi A* **208(2)** 252

Development of epitaxial PbS layers obtaining method for photoelectric transducers

T O Zinchenko, E A Pecherskaya, A V Volik, O A Timohina, V S Alexandrov and V V Antipenko

Department «Information-measuring equipment and metrology», Penza State University, Krasnaya str., 40, Penza, Russia

Abstract. Photoelectric transducers are a semiconductor device that converts photonic energy into electrical energy. This paper describes obtained by the hotwall epitaxy method epitaxial PbS layers technology. Materials, methods, technological parameters of synthesis were selected and substantiated. A theoretical model of the p-n transition has been developed. The calculation of the main parameters has been done. The hotwall epitaxy method was chosen for the synthesis, because it allows to obtain layers with required properties in a single technological cycle with an economical consumption of material. BaF2 was chosen as the substrate, because in this case a smaller difference in the identity periods and the layer and the substrate thermal expansion coefficients is achieved.

1. Introduction

The semiconductor industry is actively developing in the modern world. One of the uses is photonics. Semiconductors have come out on top in terms of breadth of implementation and opportunities for improvement. Such spheres as silicon nanophotonics, glass coatings, two-dimensional crystals for photonics and optoelectronics, and spectral precision lasers are also actively developing [1-3]. This article is devoted to the methods study and the techniques development for obtaining epitaxial layers of photovoltaic cells or photoelectric converters that convert the energy of photons into electricity. At the present stage, converters based on gallium and silicon arsenide are actively used [4]. The development and research of new materials naturally leads to a rethinking of the use of certain structures. Today sulphide-lead, sulphide-cadmium and CIGS (copper, indium, gallium, selenium) structures come to the fore. PbS was chosen because this material has high exponential photosensitivity. Many different photonic devices are based on it. Also, with a decrease in grain size, lead sulfide changes properties due to the size effect, but the structure of the entire material also changes [5, 7].

2. Selection and justification of techniques, materials, technological parameters

Layers of PbS and solid solutions are deposited on substrates from fresh cleavages along the (111) and (100) planes of BaF2 and NaCl single crystals. BaF2 was chosen as the substrate, because in this case a smaller difference in the identity periods and the layer and the substrate thermal expansion coefficients is achieved. The discrepancy between the identity periods between the BaF2 substrate and the epitaxial layers is 4.5-1.3% when the X composition changes from 0 to 1. In turn, the NaCl substrates are hygroscopic, which leads to a violation of the crystal surface perfection. BaF2 (100) substrates are preliminarily subjected to chemical-mechanical polishing in aqueous solutions of hydrochloric and sulfuric acids. After polishing, it is necessary to carry out a pre-epitaxial treatment in a mixture of hydrobromic acid and glycerin.

After processing, the surface relief of the BaF2 plates is characterized by a uniform distribution of pits with an average size of 0.2 μm in diameter and a depth of up to 18 nm, the elevations of which form a labyrinthine morphology. Surface roughness Ra averages 12 ± 1 nm.

The hotwall epitaxy method was chosen because it offers all the advantages of vacuum deposition methods. Moreover, this method makes it possible to obtain epitaxial layers of compounds and based on them solid solutions under conditions as close as possible to thermodynamically equilibrium. It is precisely the observance of the equilibrium conditions for the synthesis of epitaxial layers that is more important than the use of oil-free pumping systems requiring ultrahigh vacuum or ensuring a high degree of purification of the initial components [6].

To optimize the conditions for obtaining layers of high structural perfection, the following rules must be observed:

- The synthesis of layers should be carried out at as close as possible the evaporation temperatures of the initial charge and the substrate.
- Avoid re-evaporation of the deposited layer from the substrate, which leads to a deterioration in the quality of the layer surface.
- The substrate temperature must be high enough to ensure the diffusion of absorbed atoms and molecules over the layer surface and recrystallization and annealing of layer defects.

It is necessary to calculate the deposition time in order to set the n- and p-conductivity layers. The deposition time determines the layer thickness. In turn, the layer thickness is determined based on the fact that the mean free path of charge carriers is comparable to the mean free path. This is required in order for the charge carriers to reach the pn junction. It is also necessary to take into account that a situation of diffusion of defects may arise and the p-n junction is blurred. To select the conditions for obtaining, state diagrams are used, but in addition to them, high-quality homogeneous layers with a structure and properties close to a single crystal are required. For this, kinetic studies are used.

It is taken on the basis that the free passage of charge carriers is comparable to the width of the p-n transition.

The kinetic regularities of obtaining layers on BaF_2 substrates (Figure 1, (a), (b)) and the dependence of the charge carriers concentration on pressure (Figure 2) should be investigated to select the synthesis conditions that make it possible to obtain lead sulfide epitaxial layers of high structural perfection [8].

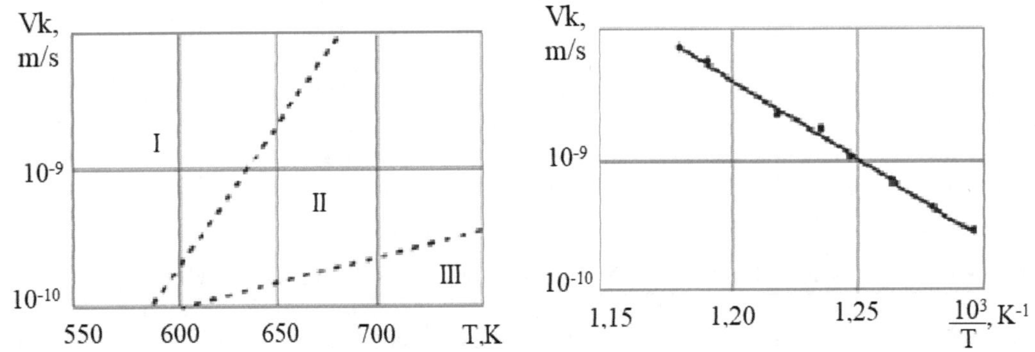

Figure 1. PbS kinetic regularities.

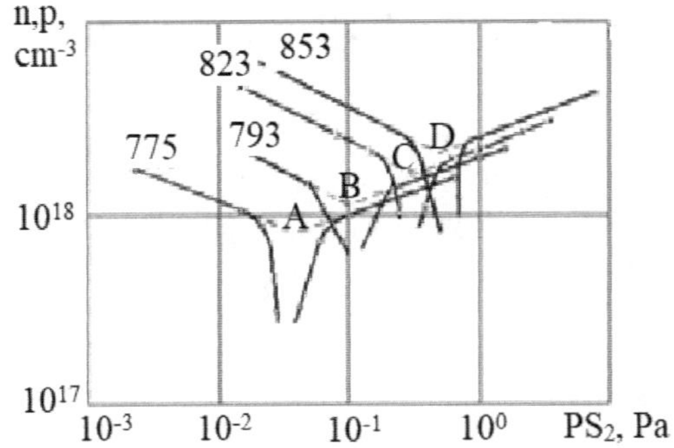

Figure 2. Dependence of the charge carriers concentration on pressure.

As a result of the experimental data analysis, it was concluded that it is advisable to choose a condensation rate of 2.143·10-9 m/s and a substrate temperature of 700 K. The evaporation temperature will be 793 K. For 793 K, the pressure P = 3·10-2 Pa and P = 2·10-1 Pa, with impurity concentration n = 2.6·1018 cm-3 and p = 1018 cm-3.

3. Calculation of the parameters of the p-n transition

The calculation of the intrinsic concentration of electrons and holes was made according to the following formulas:

$$N_c = 2\left(\frac{2\cdot\pi\cdot m_n\cdot k\cdot T}{h^2}\right)^{\frac{3}{2}} = 1.397\cdot 10^{23}\,cm^{-3} \tag{1}$$

$$N_v = 2\left(\frac{2\cdot\pi\cdot m_p\cdot k\cdot T}{h^2}\right)^{\frac{3}{2}} = 2.828\cdot 10^{23}\,cm^{-3} \tag{2}$$

$$n_i = p_i = (N_c\cdot N_v)^{\frac{1}{2}}\cdot\exp\left(-\frac{\Delta E}{2\cdot k\cdot T}\right) = 5.938\cdot 10^{19} cm^{-3} \tag{3}$$

where Nc – effective number of states in the conduction band reduced to the band bottom, Nv – effective number of states in the valence band reduced to the band ceiling, mn – effective electron mass, mp – effective hole mass, k – Boltzmann's constant, T – temperature, h – Planck's constant, ΔE – value of band gap, ni,pi – intrinsic concentration of electrons and holes [9].

The height of the potential barrier is determined as follows:

$$\varphi_0 = k\cdot T\cdot ln\left(\frac{p\cdot n}{(n_i)^2}\right) = 3.603\cdot 10^{-20} J \tag{4}$$

where n – electron concentration, p – hole concentration, φ0 – potential barrier height.

The calculated value of the space charge layer thickness refers to an abrupt p-n transition, in which the impurity concentration changes abruptly. The thickness of the space charge layer of the abrupt p-n transition is:

$$d = \left(\frac{2\cdot\varepsilon\cdot\varepsilon_0\cdot\varphi_0}{n\cdot q^2}\right)^{\frac{1}{2}} = 1.216\cdot 10^{-5} cm \tag{5}$$

where ε0 – permittivity of vacuum, ε – dielectric constant, q – elementary charge.

Contact-potential difference is:

$$U_k = \frac{\varphi_0}{q} = 0.225 \tag{6}$$

Performed calculations showed that to obtain a p-n transition under the given conditions, the time t = 56.783 s is required. It is necessary to take into account the phenomenon of self-diffusion. To calculate the diffusion process, the equation is solved for the variant of diffusion from a constant source in a body of finite dimensions. As a body of finite dimensions, the space charge region of the p-n transition is considered. It was found that the width of self-diffusion processes is in the range from 0.3 to 0.5 μm.

Therefore, in order not to blur the transition, the thickness of the growing layer must be made larger. The choice of thickness should be in the range from 1.5 to 3 microns (one layer). Taking into account the above conclusions, the thickness should be 2.25 μm.

4. Calculation of volt - ampere characteristics of the p-n junction

Volt-ampere characteristics are shown in Figure 3.

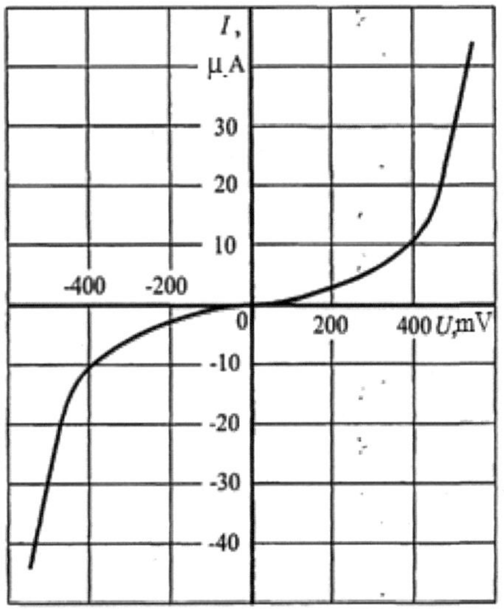

Figure 3. Volt - ampere characteristic.

These volt-ampere characteristics (Figure 3) are similar to the volt-ampere characteristics of contacts without the rectification effect, but with a resistance that depends on the applied voltage. The nonlinearity of the current-voltage characteristic is due to the phenomena at the point contact between the semiconductor microcrystals. With an increase in both forward and reverse voltages, the resistance of the structure decreases due to the tunneling of charge carriers through surface potential barriers, due to the presence of strong electric fields in them, arising in point contacts between microcrystals. The current-voltage characteristics of the structures remain stable when the electric current is repeatedly passed in the forward and reverse directions. However, the current-voltage characteristics of PbS films with contacts change when a current is passed through them. This is explained by the presence of slow surface states, which are localized on the surfaces of microcrystals and in the oxide layer of intercrystalline gaps. Slow surface states are due to absorbed gas molecules, which are formed during the oxidation and decomposition of PbS during the heat treatment of lead sulfide films.

Development of the topology of p-n-junctions manufactured in a single technological cycle, based on epitaxial layers of phases of variable composition

The topology is shown in Figure 4.

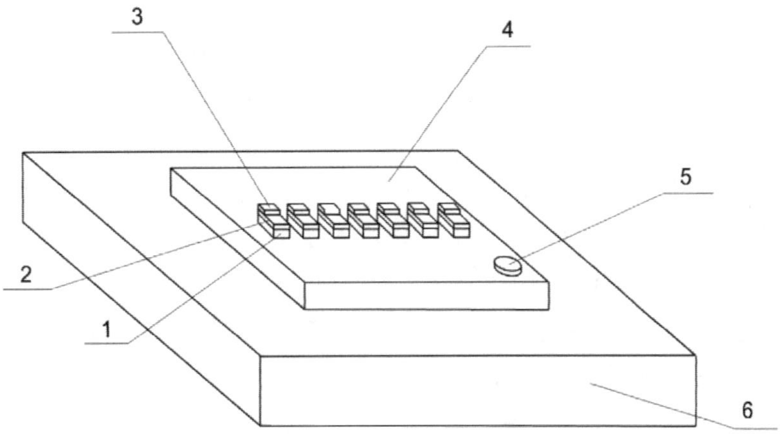

Figure 4. Topology of p-, n- junction based on epitaxial layers of phases of variable composition: 1 and 2 are p- and n- type epitaxial layers of electrical conductivity, respectively; 3 - ohmic contact to n-type layers of electrical conductivity; 4 - epitaxial layer used to form a p-n-junction; 5 - ohmic contact to p-type layers of electrical conductivity; 6 - BaF2 substrate.

5. Conclusion

The obtained indicators prove that the chosen hotwall epitaxy method is one of the most suitable for the use of such structures. Secondly, the selected substrate materials and layers for the p-n transition have the required characteristics for use in photovoltaic converters. Thirdly, the main indicators of the p-n transition are calculated, which make it possible to resolve the topology of the photocell.

6. Acknowledgements

The reported study was funded by RFBR according to the research project № 20-38-90044.

References

[1] Jalali B and Fathpour S 2006 Silicon Photonics *Journal of lightwave technology* **24** (12)

[2] Zinchenko T, Pecherskaya E and Artamonov D 2019 The properties study of transparent conductive oxides (TCO) of tin dioxide (ATO) doped by antimony obtained by spray pyrolysis *AIMS Materials Science* **6** (2)

[3] Wang Q, Lai J and Sun D 2016 *Review of photo response in semiconductor transition metal dichalcogenides based photosensitive devices* Opt. Mater. Express **6**(7)

[4] Gaddy E 1996 Cost trade between multijunction, gallium arsenide and silicon solar cells *Progress in photovoltaics: research and applications* 4

[5] Weiming Q, Mingsheng X, Fei C, Xi Y, Yaxiong N and Hongzheng C 2011 Morphology evolution route of PbS crystals via environment-friendly electrochemical deposition *CrystEngComm* 13

[6] Samoylov A, Kuzminykh O, Synorov Yu, Belonogov E, Belenko S and Agapov B Growth kinetics and microstructure of PbTe films produced on Si and BaF2 substrates by a modified hot-wall method *Inorganic Materials* 2018 **54** (4)

[7] Zinchenko T, Kondrashin V, Pecherskaya E, Kozlyakov A, Nikolaev K and Shepeleva J 2017 Electrical properties of transparent conductive ATO coatings obtained by spray pyrolysis *IOP Conference Series: Materials Science and Engineering* **225**(1) 012255

[8] Averin I 2007 Controlled synthesis of heterogeneous systems: technology and properties *Dissertation for the degree of Doctor of Technical Sciences* (Penza State University, Penza)

[9] Michael Sh and Jasprit S 1990 Physics of Semiconductor Devices *Phys. Today* 43(10)

Optical and structural properties of the GaAs heterostructures grown using AlGaAs superlattice buffer layer on compliant Si(100) substrates with the preformed porous-Si (por-Si) layer.

D Zolotukhin[1], P Seredin[1], A Lenshin[1], D Goloshchapov [1], Y Hudyakov[1], O Radam Ali[1], I Arsentyev[2] and H Leiste[3]

[1] Voronezh State University, Universitetskaya pl. 1, 394018, Voronezh, Russia
[2] Ioffe Institute, Politekhnicheskaya 26, 194021, St. Petersburg, Russia
[3] Karlsruhe Nano Micro Facility H.-von-Helmholtz-Platz 1, 76344 Eggenstein-Leopoldshafen, Germany

seredin@phys.vsu.ru

Abstract. 360 nm and 700 nm thick GaAs layers were grown by MO MOCVD growth technique directly on compliant Si (100) substrate and on supper-lattice (SL) AlGaAs buffer layer. The XRD study revealed better structural quality for the sample grown on SL / por-Si buffer. AFM study revealed a smoother sample surface with blocks of more regular rectangular shape and larger size as well. Photoluminescence spectra of the samples revealed an energy shift of PL maximum intensity for both samples. Sample grown on SL buffer also showed higher PL intensity corresponding to better crystalline perfection.

1. Introduction

The integration of the III-V technology with the most developed Si technology is an actual problem for the state of art science. III-V / Si heterostructures will combine the advantages of both material systems and help to create high-efficient, low cost optoelectronic and power devices. However, large lattice and thermal expansion coefficient (TEC) mismatch between Si and GaAs (~120.4% and ~4.1%, respectively) [1] leading to the generation of high density of threading dislocations (TD) and other defects during the growth procedure, which acts as nonradiative recombination centers and affect on terminal-device efficiency and reliability. The difference in thermal expansion coefficients can led to bending of heterostructures which significantly complicates its post-growth processing, and in some cases can lead to delamination of the films from substrates. As it was shown in our previous work [2] the usage of por-Si layer helps to suppress stress generation and TD propagation due to unique properties of nanoporus material. Due to the presence of nanopores, Si layer can be deformed and obtains less lattice mismatch with overlaying heterolayer. In this work we compare optical and structural properties of GaAs layers grown by MO MOCVD technique on por-Si/Si(100) (sample **A**) and SL/por-Si/Si(100) (sample **B**) templates. The presence of stresses of different signs in SL layers as well as a number of interfaces between GaAs and AlAs layers helps to suppress threading dislocations propagation by their inclination and annihilation in the compressively strained fields [3]. GaAs layer grown on AlGaAs-SL buffer demonstrated more intense PL and XRD spectra corresponding to better structural quality. Besides, AFM revealed smoother surface with blocks having a more pronounced rectangular shape and larger size for the layer grown on SL buffer.

2. Experiment

Porous layer of silicon, por-Si, as a “compliant” virtual substrate was formed on a single-crystalline c-Si (100) plate by its electrochemical etching in the alcohol solution of the fluoric acid according to the standard procedure [4]. Growth procedures were performed on EMCORE GS3100 MO MOCVD setup under the pressure of 77 Torr and substrate rotation speed of 1000 rpm. Trimethylgallium ($(Al(CH_3)_3)$) and Arsine (AsH_3), were used as source gases and hydrogen was applied as a carrier gas. Before growth, both substrates were annealed for 20 min in AsH_3 flow and substrate temperature (T_s) of 750 °C. After annealing the 10-nm-thick AlAs overlaid by 20-nm-think GaAs buffer layer was grown at the T_s=450 °C. GaAs layer with a thickness of 390 nm was grown for sample **A** and the SL buffer layer with a thickness of 100 nm and overlaying 700-nm-thick GaAs layer was grown for sample **B** at the T_s=550 °C.

3. Results and discussion

The AFM study of both samples (Figure 1.**a** and **1.b**) revealed mosaic surface morphology, but for sample **B** the surface blocks have larger size and the shape more close to a rectangular corresponding to better crystalline quality. Well known that larger lateral sizes of the nucleation blocs led to lower generation of threading dislocations during misorientated blocs coalescence.

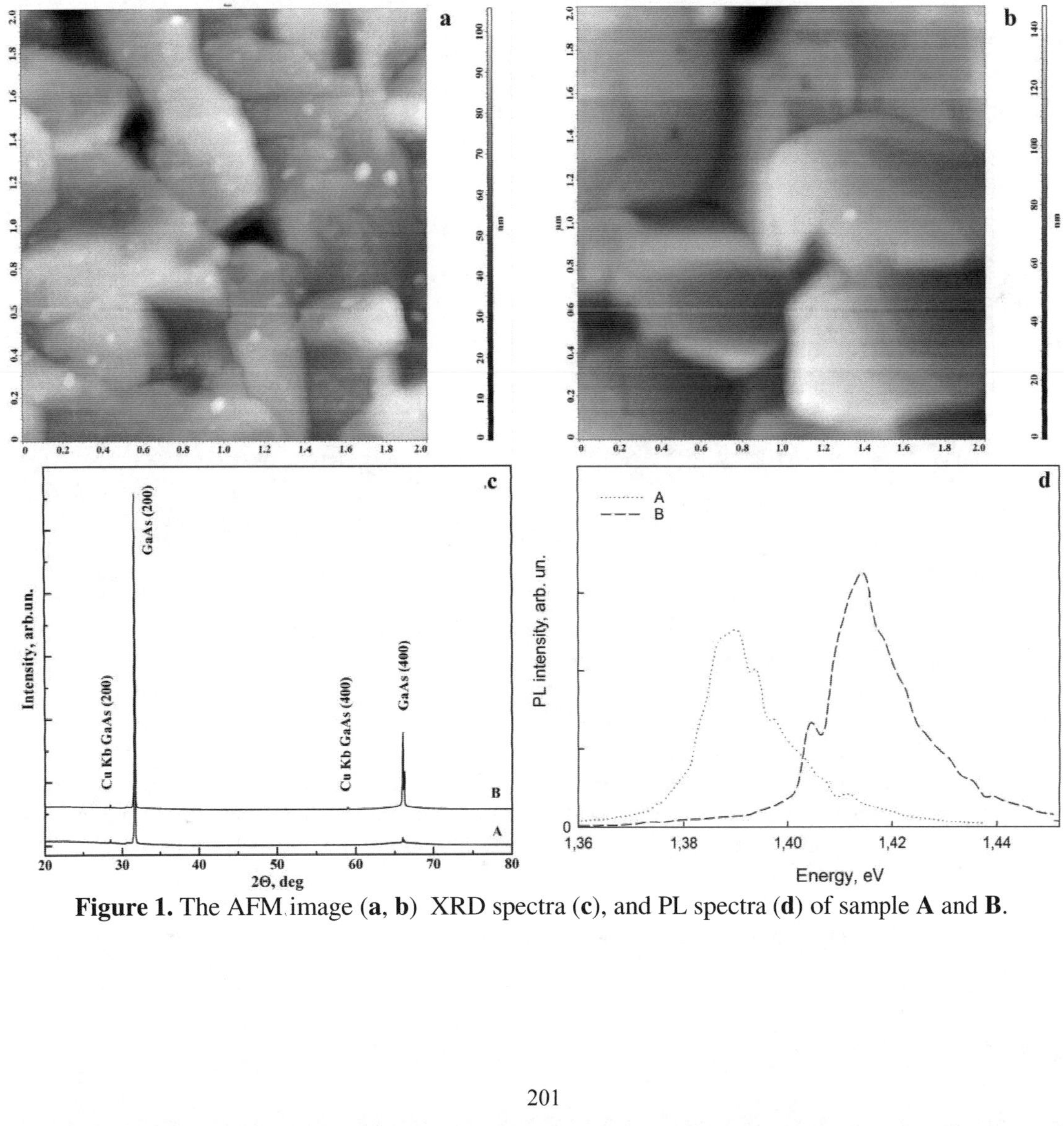

Figure 1. The AFM image (**a**, **b**) XRD spectra (**c**), and PL spectra (**d**) of sample **A** and **B**.

Moreover the 3D morphology of the samples surface corresponding to mosaic structure of all film, thus the TD's can annihilate on blocs free surfaces. This fact is also confirmed by XRD study of the samples (Figure **1.c**) where a more intensive (400) peak can be observed for the sample **B**. As well, no other peaks except (200) and (400) were observed, corresponding to monocrystalline structure of the GaAs epitaxial layer for the both samples.

The PL spectra of the samples (Figure 1.d) demonstrate bright emission at ~1.39 eV for sample A and 1.5 eV for sample B shifted from normal value for GaAs 1.42 eV to the low energy range. The nature of this redshift is under study. The higher PL intensity for sample B confirms the better structural quality of the GaAs layer grown on SL buffer.

Using high-resolution XRD (HRXRD) high-resolution X-ray diffraction on the basis of inverse q-space mapping of the samples (Figure 3) we have calculated the lattice parameters, strain and stress levels in both heterostructures. The images of XRD mapping across symmetric (004) and skew symmetric (511) reflections are listed in Figure 3. It shall be noted that in the Figure 3b (511) reflection from SL can be observed. Using these data we calculate the lattice parameters as a_{GaAs}=5.6528 Å, a_{SL}=5.5791 Å. It should be also noted that the calculated SL lattice parameter is significantly different from that one in AlGaAs binary alloy of the same composition. This difference is associated with atomic self-ordering which is typical for three-component III-V semiconductor alloys with the composition (x ~ 0.50) [5].

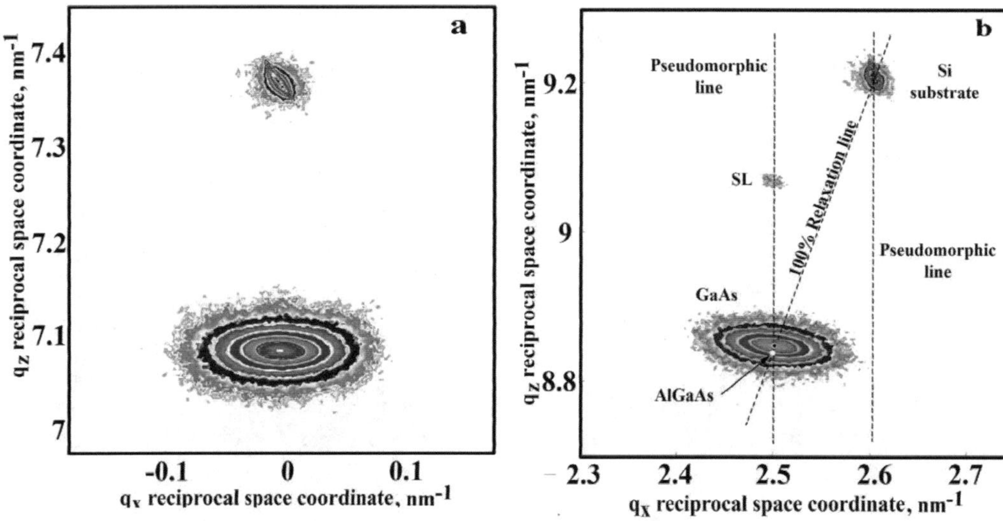

Figure 2. XRD mapping images symmetric (004) (**a**) and skew-symmetric (511) (**b**) reflections of GaAs/SL/proto-Si(100) heterostructures.

However, aAlGaAs lattice parameter calculated in our previous work [6] is different from the value calculated in this work as well as from that one theoretically obtained by S. Laref et. al. in [7]. We believe that this decrease in the lattice parameter of SL in comparison with the values that we observed in our previous works is related to features of the low-temperature growth process, which been used for heterostructure formation.

Our previous researches [8-11] revealed that the growth of GaAs and III-N based epitaxial layers on compliant substrates presented by preformed nanoporous Si layer have advantages in comparison with growth on traditional c-Si substrates: lower TD density, higher PL intensity and less stress level at the RT.

In this work, we confirmed the sufficient improving of the crystalline and optical quality of AlGaAs epitaxial layers grown on substrates with preformed nanoporous Si layer and SL-ALGaAs buffer layer. The AFM study revealed better surface morphology, XRD analysis – better structural quality and PL spectra revealed higher intensity for sample grown on complex substrate SL-AlGaAs/por-Si/Si. The results obtained in this work confirm a positive influence of the usage of compliant Si substrates with porous Si layer and slight misorientations from (111) direction as well as usage SL-AlGaAs buffer layer.

Acknowledgments
The study was financially supported by Russian Science Foundation (grant no. 19-72-10007).
P.V.S. carried out his part of the study with the support from the Ministry of Science and Higher Education of Russian Federation (grant no. FZGU-2020-0036) under the State assignment to higher school institutions.

References
[1] Kukushkin S, Osipov A, Bessolov V, Medvedev B, Nevolin V and Tcarik K. 2008 *Rev Adv Mater Sci n.d.* **17** 1
[2] Seredin P, Goloshchapov D, Lenshin A, Mizerov A and Zolotukhin D 2018 *Phys. E Low-Dimens. Syst. Nanostructures*, **104**, 101
[3] Romanov E, Beltz G, Cantu P, Wu F, Keller S, DenBaars S P and Speck J S 2006 *Appl. Phys. Lett.* **89**, 161922.
[4] Lenshin A, Seredin P, Agapov B, Minakov D and Kashkarov V 2015 *Mater. Sci. Semicond. Proc.* **30**, 25
[5] A. Zunger A, 1997 MRS Bull. **22** 20
[6] Seredin P, Domashevskaya E, Arsentyev I, Vinokurov D, Stankevich A and Prutskij T 2013 *Semiconductors* **47** 1
[7] Laref S, Meçabih S, Abbar B, Bouhafs B, and Laref A 2007 *Phys. B Condens. Matter* **396** 169
[8] Seredin P, Lenshin A, Zolotukhin D, Arsentyev I, Nikolaev A and Zhabotinskiy A 2018 *Phys. B Condens. Matter* **530** 218
[9] Seredin P, Lenshin A, Zolotukhin D, Arsentyev I, Zhabotinskiy A and Nikolaev D 2018 *Phys. E Low-Dimens. Syst. Nanostructures* **97** 218
[10] Seredin P, Lenshin A, Mizerov A, Leiste H and Rinke M 2019 *Appl. Surf. Sci.* **476** 1049
[11] Seredin P, Leiste H, Lenshin A and Mizerov A 2020 *Appl. Surf. Sci.* **508** 145267

Double slot aerosol jet printed antenna for X-band applications

P V Arsenov, A S Sobolev, A A Efimov and V V Ivanov

Moscow Institute of Physics and Technology (National Research University), Dolgoprudny 141701, Russia

arsenov@phystech.edu

Abstract. A double slot antenna for X-band applications was designed and aerosol jet printing technology was used to fabricate the prototype with silver nano-ink on a flexible polyimide substrate. We investigated the microwave losses of printed antennas in the range from 100 kHz to 27 GHz, obtained at sintering temperatures of 200 °C and 250 °C. Double slot X-band antennas have been calculated and measured. It was found that an operating bandwidth of the printed antenna is 10% in the region of the central frequency of 10.5 GHz. Thus, the possibility of forming antennas on flexible polymer substrates with high functional characteristics by aerosol jet printing method has been demonstrated.

1. Introduction

The currently developing printed electronics is attractive not only for the low cost of printing and simplicity of prototyping, but also for the possibility of creating flexible electronic circuits and devices [1–3]. Among the existing printing methods, aerosol jet printing is the most promising technology, since it allows the formation of functional elements with a minimum lateral size of up to 10 microns on different types of substrates, including flexible polymer substrates [4,5]. The aerosol jet printing method is in demand for various fields of application, including the manufacture of transistors, sensors, solar cells, microheaters, microwave structures and interconnections [6–9].

Flexible planar microwave structures are in demand for creating conformal low-profile antennas and quasi-optical filters based on frequency selective surfaces that can be placed on curved aerodynamic surfaces of ground vehicles and aircraft. Fiberglass hulls and keels of unmanned aerial vehicles (UAVs) allow placing antennas on their bearing surfaces without using the ground plane, made in one metal layer. Antennas such as the double slot [10,11], are based on high-impedance transmission lines and, unlike patch antennas, have high antenna efficiency. Single flexible antennas can be combined into conformal antenna arrays that cover large areas on the UAV body and have a high collective gain. We optimized the printing parameters for silver ink on polyimide substrate and printed several samples of the X-band double slot antenna. This range can be used not only for high-speed communication channels, but also for airborne radars and proximity sensors.

2. Experimental

A double slot antenna in the form of a silver film of a given geometry was fabricated by focused deposition of aerosol microdroplets from silver nano-ink onto a polyimide substrate by pneumatic method using an AJ 15XE aerosol jet printer (Neotech AMT GmbH). The nano-ink was a mixture of silver nano-ink PG-007 (Paru Co. Ltd.) based on solvents 1-methoxy-2-propanol (MOP) and ethylene glycol (EG) in a volume ratio of 3:1, respectively [12]. Pure nitrogen N_2 (99.9999%) was used as a carrier gas. A nozzle with an outlet diameter of 300 μm was used to form the antenna. The antenna was formed at the values of aerosol and sheath flow rates, and printing speed equal to 50 sccm, 80 sccm, and 500 mm/min, respectively. After deposition, a silver film was sintered at temperatures of 200 °C and 250 °C for 60 min to form a conductive structure.

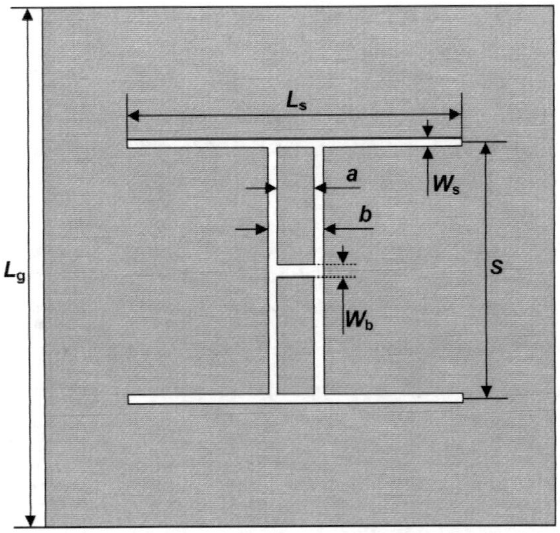

Figure 1. The design of the double slot antenna (L_g = 20 mm, L_s = 11 mm, W_s = 0.1 mm, W_b = 0.4 mm, S = 8.4 mm, a = 1.2 mm, b = 1.8 mm).

The design of the double slot antenna is presented in figure 1. Antenna geometry is designed and optimized for 10.5 GHz using high frequency simulation software (HFSS). The antenna is designed on a flexible polyimide dielectric substrate 120 μm thick and dielectric constant ε = 3.5. The antenna was a silver film 20x20 mm^2 in size and has thickness about 4 μm. Two narrow 0.1 mm holes in the form of rectangular slots were made in the silver film. Functionally, the antenna consisted of two magnetic dipoles formed by half-wave λ/2 resonators based on a slot line. The dipoles are also located at a distance of λ/2 and are interconnected by a coplanar line in the center of which is the power point.

The vector analyzer (Keysight N5242A-200) was used to measure the return loss, S11-parameter, of antennas sintered at 200 °C and 250 °C. Microstrip transmission lines with a characteristic impedance of 50 Ohm were investigated to study microwave losses in deposited silver films in a wide frequency range. In such lines, the current density in the conducting layers has maximum values. Two silver electrodes with a width of 10 mm and 0.38 mm were printed on opposite sides of flexible substrate. SMA connectors were soldered at the ends of the 10 cm long microstrip transmission lines section, see figure 2. The morphology of printed antennas was investigated using a scanning electron microscope (SEM) (JSM-7001F, JEOL Ltd.)

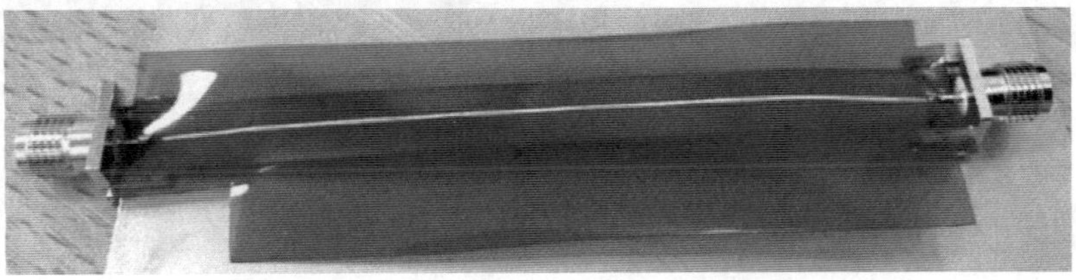

Figure 2. Aerosol jet printed microstrip transmission line on flexible polyimide substrate.

3. Results and discussion

The vector analyzer was used to measure the transmission coefficient S21 through the printed microstrip transmission line in the range from 100 kHz to 26.5 GHz. The measurement results for Ag films with sintering temperatures of 200 °C and 250 °C are presented in figure 3a. The curves show equidistant resonance dips, which are present due to non-optimal contact at the place where the microstrip transmission line is soldered to the coaxial connector. A numerical model of the microstrip transmission line with losses was created, in which the frequency-dependent resistance per square (Ohm sq^{-1}) of the area of the silver films was set. Comparison of the results of numerical simulation by the FDTD method with the experimental curves S21 allows one to obtain the frequency dependence of resistance, see figure 3b. It can be seen that up to 15 GHz the resistance of Ag films with sintering temperature of 200 °C is higher than that for Ag films with temperature of 250 °C. This could be explained by the fact that at higher temperatures printed films have less binding agent and silver nanoparticles has better contact to each other. At frequencies above 15 GHz, the capacitance of the gap between nanoparticles makes a significant contribution to the conductivity. Therefore, the values of resistance are the same for Ag films with different sintering temperatures.

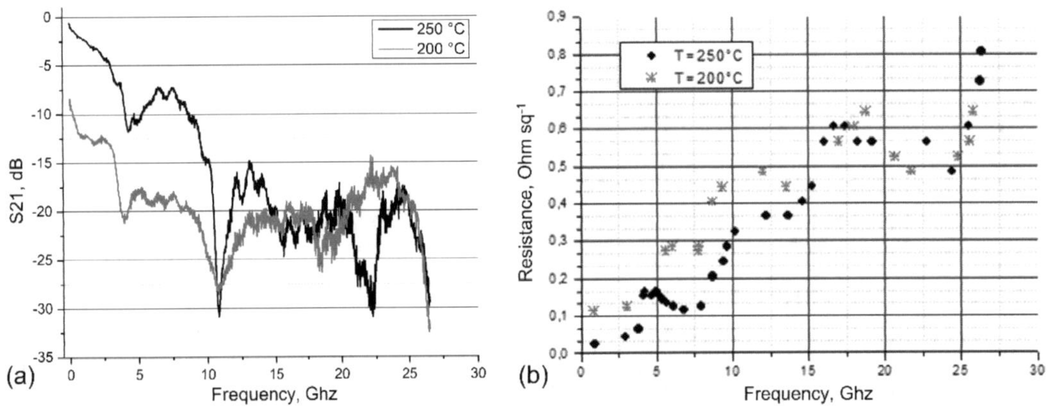

Figure 3. (a) Transmission coefficient S21 through the printed microstrip transmission line in the range from 100 kHz to 26.5 GHz and (b) surface resistance of silver films sintered at 200 °C and 250 °C, calculated from measurements of the S21 parameter.

Figure 4. (a) Photo of the double slot aerosol jet printed antenna on flexible polyimide substrate and (b) typical SEM image of the surface.

Samples of microantennas with sintering temperatures of 200 °C and 250 °C were prepared. A typical photo of an antenna printed on a flexible polyimide substrate is shown in figure 4a. Figure 4b shows SEM image of the antenna surface sintered at 250 °C. The measured return loss, S11-parameter, of antennas sintered at 200 °C and 250 °C is shown in Figure 5. Also figure 5 shows the results of an

FDTD simulation of an antenna made of a Perfect Electric Conductor (PEC) and having resistances of 0.3 Ohm sq^{-1} and 0.5 Ohm sq^{-1}, which corresponds to the frequency of 10.5 GHz.

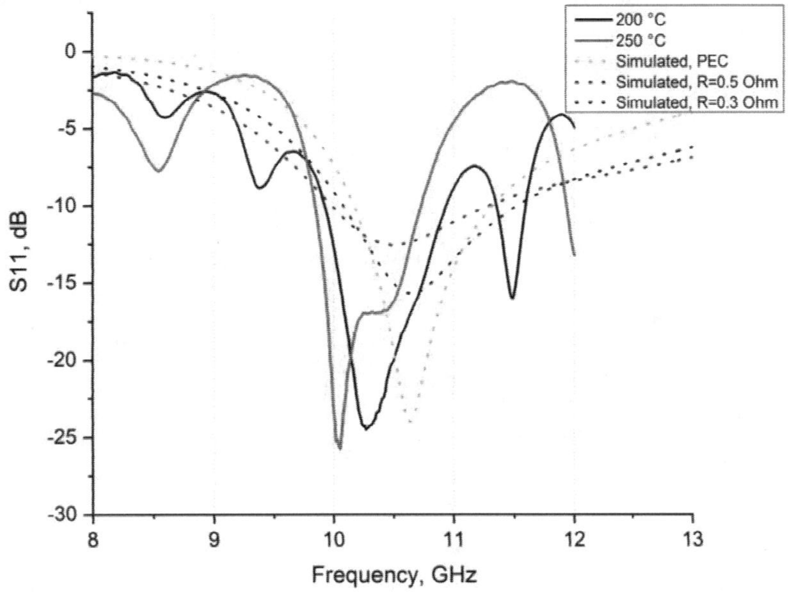

Figure 5. Measured return loss of the aerosol jet printed antenna and FDTD simulation.

4. Conclusion

Aerosol jet printing method can be used to form antennas on flexible polymer substrates. The agreement between microwave losses of printed antennas and the FDTD simulation is observed, taking into account the parasitic interaction of the antenna with the measuring cable. The measured bandwidth of the antennas is more than 1 GHz at a level of −10 dB, which corresponds to 10% of the center frequency. Thus, the possibility of forming antennas on flexible polymer substrates with high functional characteristics by aerosol jet printing method has been demonstrated.

Acknowledgments
The reported study was funded by RFBR, project number 20-33-90058.

References

[1] Ćatić N, Wells L, Al Nahas K, Smith M, Jing Q, Keyser U F, Cama J and Kar-Narayan S 2020 Aerosol-jet printing facilitates the rapid prototyping of microfluidic devices with versatile geometries and precise channel functionalization *Appl. Mater. Today* **19** 100618

[2] Huang Q and Zhu Y 2019 Printing Conductive Nanomaterials for Flexible and Stretchable Electronics: A Review of Materials, Processes, and Applications *Adv. Mater. Technol.* **4** 1800546

[3] Abutarboush H F and Shamim A 2012 Paper-Based Inkjet-Printed Tri-Band U-Slot Monopole Antenna for Wireless Applications *IEEE Antennas Wirel. Propag. Lett.* **11** 1234–7

[4] Arsenov P V, Efimov A A and Ivanov V V 2021 Optimizing Aerosol Jet Printing Process of Platinum Ink for High-Resolution Conductive Microstructures on Ceramic and Polymer Substrates *Polymers* **13** 918

[5] Volkov I A *et al* 2021 Platinum based nanoparticles produced by a pulsed spark discharge as a promising material for gas sensors *Appl. Sci. Switz.* **11** 1–20

[6] Borghetti M, Serpelloni M and Sardini E 2019 Printed Strain Gauge on 3D and Low-Melting Point Plastic Surface by Aerosol Jet Printing and Photonic Curing *Sensors* **19** 4220
[7] Marketing O 2008 Aerosol Jet Printing for High Efficiency Solar Cells to be Showcased at 23rd European Photovoltaic Conference in Valencia, Spain *Optomec*
[8] Tortorich R P, Shamkhalichenar H and Choi J-W 2018 Inkjet-Printed and Paper-Based Electrochemical Sensors *Appl. Sci.* **8** 288
[9] Efimov A A *et al* 2021 Synthesis of Nanoparticles by Spark Discharge as a Facile and Versatile Technique of Preparing Highly Conductive Pt Nano-Ink for Printed Electronics *Nanomaterials* **11** 234
[10] Filipovic D F, Gearhart S S and Rebeiz G M 1993 Double-slot antennas on extended hemispherical and elliptical silicon dielectric lenses *IEEE Trans. Microw. Theory Tech.* **41** 1738–49
[11] Farooqui M F and Shamim A 2013 Dual band inkjet printed bow-tie slot antenna on leather *2013 7th European Conference on Antennas and Propagation (EuCAP)* 2013 7th European Conference on Antennas and Propagation (EuCAP) pp 3287–90
[12] Efimov A, Arsenov P, Kornyushin D, Lizunova A, Volkov I and Ivanov V 2020 Aerosol Jet Printing of Silver Lines with A High Aspect Ratio on A Heated Silicon Substrate *Materials* **13** 730

Terahertz detector utilizing a SiO_2/Graphene/SiO_2 sandwich suspended at the feed of a planar antenna

I Belikov[1], M Rybin[2], A Prikhodko[1,3], D Mikhailov[1,3], I Gayduchenko[1], A Shurakov[1] and G Goltsman[1,3]

[1] Moscow Pedagogical State University, Moscow 119435, Russia
[2] Prokhorov General Physics Institute, RAS, Moscow 119991, Russia
[3] National Research University Higher School of Economics, Moscow 101000, Russia

nknki@mail.ru

Abstract. We report on the fabrication of a terahertz detector utilizing a SiO_2/graphene/SiO_2 sandwich suspended at the planar antenna feed. This design of the detector aims to enhance its sensitivity via weakening of the heat sink between graphene's phonons and those of the substrate. We achieve complete suspension of the sandwich only in case of a low fill-factor of the antenna feed area. Evaluated DC parameters of the samples are consistent with those reported in the literature. The fabrication process developed is suitable for implementing the detector proposed for signal frequencies up to several terahertz.

1. Introduction

Terahertz frequency range is gradually mastered, and terahertz receivers are currently in demand in numerous applications. Despite worse sensitivity compared to that of superconducting competitors, terahertz detectors utilizing graphene produced by the means of chemical vapor deposition (CVD) have certain advantages. Namely, a) they are naturally compliant with integrated circuits making use of various semiconductor platforms, b) recent advances in their development suggest the performance competitive with that of planar Schottky diodes (PSD) [1], c) this performance can be achieved at less technological expenses. For that to happen, however, certain technical issues have to be resolved.

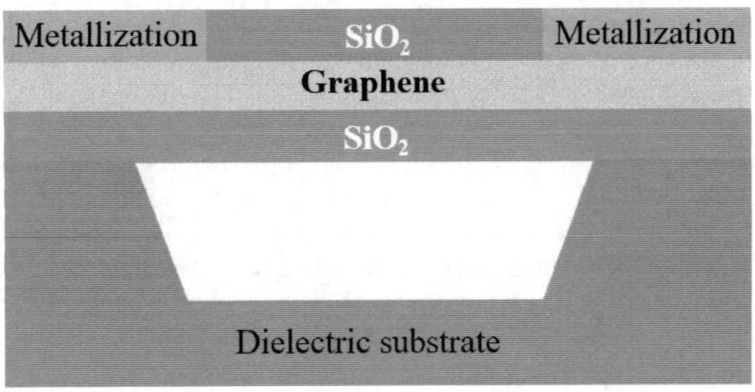

Figure 1. Layer diagram of the graphene based structure proposed.

In this paper we report on the fabrication of a terahertz detector utilizing a SiO_2/graphene/SiO_2 sandwich suspended at the feed of a planar antenna (figure 1). Such a design of the detector is meant to enhance its sensitivity by enlarging the time constant through weakening of the heat sink between graphene's phonons [2] and a phonon reservoir presented by bulk SI-GaAs under the sandwich.

2. Fabrication process

To produce the detector, we relied on a 450 µm thick SI-GaAs (100) wafer to make the fabrication process partially compliant with that of the PSD with a Γ-shaped anode suspended bridge earlier developed by us [3]. Referring to figure 2, the modified technological route contains six key steps:

- fabrication of alignment marks with further deposition of a SiO_2 sublayer (I),
- fabrication of the outer part of terahertz antenna followed by transfer of a CVD graphene (II),
- glancing angle deposition of ohmic and rectifying contacts to graphene (III),
- fabrication of the inner part of terahertz antenna (IV),
- encapsulation of graphene with a SiO_2 cover and plasma etching of unprotected regions (V),
- wet etching of GaAs (VI).

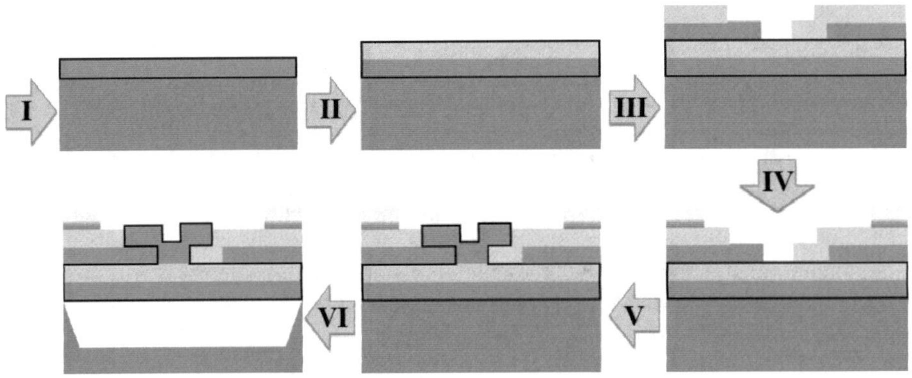

Figure 2. Technological route employed for fabrication of the detector.

To remove organic residuals, GaAs wafer surface was pretreated by O_2 plasma. For steps I-II relying on a lift-off photolithography, we chose AZ 1512 HS positive photoresist, which was spread by centrifugation on a GaAs wafer at 6000 rpm. HMDS was used as an auxiliary adhesive layer. Thermoresistive evaporation was employed to produce the alignment marks and the outer part of terahertz planar spiral antenna presented by Ti/Au (5/200 nm) bilayers. A 50 nm thick SiO_2 sublayer that eventually serves as a foundation for graphene microbridge was sputtered through opening in e-beam resist by e-beam evaporation technique ensuring precise control of the thickness. Lift-off e-beam lithography within steps I, III-V was carried out with the aid of PMMA/MMA simultaneously providing high resolution and thickness required. Footprint of the SiO_2 sublayer was similar to that of the antenna inner part with a spatially-spanned load at its feed.

As shown in the literature [4], electrical properties of graphene significantly change if it comes into contact with photoresist during processing. And removal of photoresist after lift-off process can cause additional damage to graphene monolayer film. Thus, we had to avoid using AZ 1512 HS after transferring CVD graphene to the wafer basic transfer process is described elsewhere [5]. Moreover, fabrication of the outer part of terahertz antenna commonly performed closer to the end of technological route. Inability to use positive photoresist at step IV forced us to organize the order of steps as it is. Raman spectroscopy tests revealed that 30 % of the wafer area was suitable for further processing, i.e. it contained undamaged graphene monolayer film after the transfer.

We chose V and Au as materials for the ohmic and rectifying contacts to graphene. Glancing angle deposition technique [6] was used to implement a micro scale relative shift of the V and Au layers within a V/Au (10/50 nm) metallization system deposited on graphene's top surface. For the deposition, we employed two-element composition of e-beam resists MMA/PMMA providing negative angle of inclination after development. Due to relatively small spacing between the contacts (0.2 - 0.6 μm) it took certain effort to fabricate them. Thus, the parameters of lithography process accompanying the deposition were optimized to ensure implementation of the geometry required. The sufficient dose was found by fixing the development time and varying cathode current along with sleep-in-point time of e-beam. Details of the glancing angle deposition process conducted are shown in figure 3(a, b).

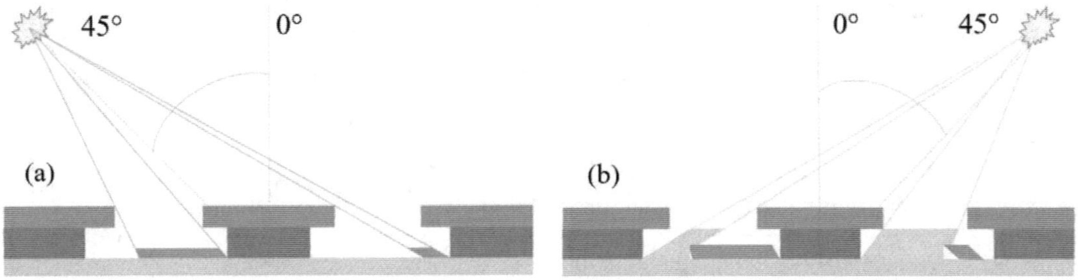

Figure 3. Diagram of glancing angle deposition process: (a) Fabrication of ohmic contact to graphene (V/graphene); (b) Fabrication of rectifying contact to graphene (Au/graphene).

Prior to the fabrication of the antenna inner part, we processed the wafer by O_2 plasma to etch the graphene within its footprint (through opening developed in e-beam resist) and, consequently, to exclude possible AC shunting of the antenna metallization during operation of the detector. Further fabrication of a 75 nm thick SiO_2 cover overlapping the antenna feed area was followed by the removal of unprotected graphene regions in O_2 plasma. Both the SiO_2 sublayer and SiO_2 cover acted as etch-stoppers during the implementation of a SiO_2/graphene/SiO_2 sandwich suspended at the antenna feed.

During step VI, the local removal of the SI-GaAs underneath the sandwich was maintained through a dual sided opening (on both sides of the line connecting the antenna feed points) in PMMA positive e-beam resist intended for the access of wet etchant based on ammonia solution.

The fabrication process described hereinabove was employed to produce a set of terahertz detectors utilizing the suspended SiO_2/graphene/SiO_2 sandwich.

3. Results and discussion

Figures 4(a,c) provide images from a scanning electron microscope (SEM) for the detectors with different areas of graphene microbridge, the labels are denoted in figure 4(b).

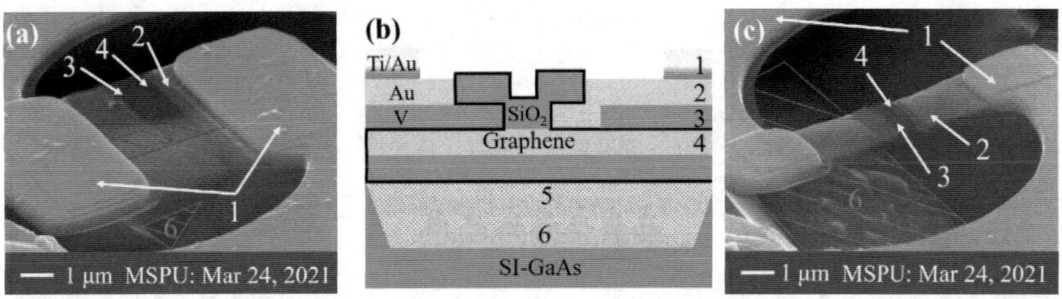

Figure 4. (a) SEM image of sample #1; (b) Layer diagram of the sample; **(c)** SEM image of sample #2.

As one can clearly see, sample #1 (with a 0.6×6 μm^2 area of the microbridge) does not utilize a fully suspended SiO_2/graphene/SiO_2 sandwich, whose central part contacts the SI-GaAs substrate noticeably.

The total contact area is 1.3×4 μm^2, whereas the contact of SI-GaAs to the region underneath graphene microbridge has significantly smaller area of 1.3×0.6 μm^2. We achieved the complete suspension of a SiO_2/graphene/SiO_2 sandwich only in case of a low fill-factor of the antenna feed area (sample #2), i.e. for graphene microbridge of 0.2×2 μm^2. Indeed, this geometry enables bigger opening in the e-beam resist which enhances access of wet etchant to the SI-GaAs being removed. If we had tried to make bigger openings in the resist and/or to increase etching time we would have risked to excessively remove SI-GaAs underneath the antenna's metallization and, consequently, to cause collapse or critical sagging of the microbridge. It is worth mentioning that the microbridge dimensions of 0.2×2μm^2 and 0.6×6μm^2 were also chosen to avoid discontinuities along its length, since irregularities in graphene can be of a submicron scale [7].

To evaluate efficiency of the fabrication process developed, we characterized DC transport properties of the samples produced. Details on the samples' geometry and values of DC resistance measured are summarized in table 1.

Table 1. Parameters of the samples used in the DC characterization.

S_g [μm^2]	$N_{\square}$	S_c [μm^2]	R_{meas} [Ω]
0.6×6	0.6 μm/6 μm	26.7	{370; 350}
0.2×2	0.2 μm/2 μm	9.6	{900; 840}

Here S_g is the area of a graphene microbridge, $N_{\square}$ is the number of squares in the microbridge, S_c is the area of a V/graphene junction and R_{meas} is the value of DC resistance of the sample measured. Graphene's sheet resistance (R_{sheet}) and contact resistivity of a V/graphene junction (ρ_c) can be evaluated by equation (1). Making use of the two sets of data presented in the table, one can calculate R_{sheet} and ρ_c values.

$$R_{meas} = N_{\square} R_{sheet} + \rho_c \,/\, S_c \tag{1}$$

Evaluated DC parameters of the samples produced are consistent with those reported in the literature [8]: graphene's sheet resistance of 728±38 Ω/square and the contact resistivity of a V/graphene junction of 77±9μΩ·cm^2 were obtained. We observed no drastic impact of the finishing removal of the SI-GaAs under SiO_2/graphene/SiO_2 on the samples' DC transport properties.

To observe response to terahertz radiation at zero bias condition, it is vital to create asymmetry in graphene-based detector. In this work, the asymmetry is achieved via the implementation of an asymmetric contact doping, when graphene's opposite sides are in contact with metals possessing different work functions [9]. For the samples produced, we measured the noise equivalent power (*NEP*) values of 9.4 and 50.7 nW/Hz$^{0.5}$ at signal frequencies of 150 and 450 GHz, respectively. No bias voltage or current were applied to the samples during the measurements. Experimental setup was based on a solid state source utilizing microwave frequency synthesizer followed by a series of frequency multipliers and was similar to that previously employed by us in [10]. When mismatch with antenna is corrected for (the detector impedance is estimated based on R_{meas} and RC-model of a graphene/metal junction [11]), the *NEP* values can be reduced to 1 nW/Hz$^{0.5}$ at 150 GHz and 6 nW/Hz$^{0.5}$ at 450 GHz.

4. Conclusion

We report on the fabrication of a terahertz detector utilizing a SiO_2/graphene/SiO_2 sandwich suspended at the planar antenna feed. Complete suspension of the sandwich was achieved only in case of a low fill-factor of the antenna feed area. It is known that the deviation of the spiral antenna inner part from a self-complementary structure leads to the decrease of antenna efficiency at the upper edge of its input frequency range [12]. The minimum dimension, at which this deviation can still be tolerated, is ultimately limited by the resolution requirements imposed on the fabrication of graphene microbridge (equipped with an ohmic contact) with 50-100 Ω impedance. Therefore, we suggest that the proposed design within the fabrication process developed is suitable for the implementation of a graphene-based detector effective at signal frequencies up to several terahertz. Results of our study of DC and AC parameters of samples produced within the fabrication process developed are compliant with this suggestion.

Acknowledgments
We acknowledge support of the Russian Science Foundation grant No. 19-72-10156.

References

[1] Castilla S et al 2019 Fast and sensitive terahertz detection using an antenna-integrated graphene pn junction *Nano letters* **19(5)** 2765–2773
[2] Tong J, Conte M C, Goldstein T, Yngvesson S K, Bardin J C and Yan J 2018 Asymmetric two-terminal graphene detector for broadband radiofrequency heterodyne- and self-mixing *Nano Letters* **18(6)** 3516-3522
[3] Shurakov A et al 2020 Planar Schottky diode with a Γ-shaped anode suspended bridge *J. Phys.: Conf. Ser.* **1695** 012154
[4] Onejae S, Kyumin K, Eunseok C, Joonpyo K, Wanjun P and Seung-Beck L 2016 Reduction of hole doping of chemical vapor deposition grown graphene by photoresist selection and thermal treatment *Nanotechnology* **27(50)** 505205
[5] Kondrashov I, Rybin M and Obraztsova E 2019 Controlled graphene synthesis from solid carbon sources *Phys. Status Solidi B* **256** 1800688
[6] Gayduchenko I et al 2018 Manifestation of plasmonic response in the detection of sub-terahertz radiation by graphene based devices *Nanotechnology* **29(24)** 245204
[7] Rybin M, Islamova V and Obraztsova E 2018 Modification of graphene electronic properties via controllable gas-phase doping with copper chloride *Applied Phys. Letters* **112** 033107
[8] Kim H-Y, Missous M, Monteverde U, Dawood O M, Sexton J, Li Z, Britnell L, Young R and Migliorato M 2015 Multilayer stacking and metal deposition effects on large area graphene on GaAs *Carbon* **96** 83-90
[9] Cai X, Sushkov A, Suess R, Jenkins G, Yan J, Murphy T, Drew H and Fuhrer M 2013 Sensitive room-temperature terahertz detection via photothermoelectric effect in graphene *Nature Nanotechnology* **9(10)** 814-819
[10] Shurakov A et al 2018 Ti/Au / n-GaAs planar Schottky diode with a moderately Si-doped matching sublayer *Microelectronic Engineering* **195** 26-31
[11] Chang J, Huzayyin A, Lian K and Dawson F 2015 Quantum capacitance of graphene in contact with metal *Applied Phys. Letters* **107** 193902
[12] Shurakov A, Seliverstov S, Kaurova N, Finkel M, Voronov B and Goltsman G 2012 Input bandwidth of hot electron bolometer with spiral antenna *IEEE Transactions on Terahertz Science and Technology* **2** 400-405

Monolithic transistor switch for microwave radiometry

V G Tikhomirov[1], Yu V Solov'ev[2,3], A G Gudkov[3], M K Popov[1] and S V Chizhikov[3]

[1] Saint-Petersburg State Electrotechnical University "LETI", St. Petersburg 197376, Russia
[2] Closed Joint Stock Company "Svetlana-Electronpribor", St. Petersburg 194156, Russia
[3] Bauman Moscow State Technical University, Moscow 105005, Russia

Abstract. Modern medical microwave diagnostic equipment requires the application of solutions related to the compactness of the developed devices and high performance. Ensuring these requirements is possible by using a modern semiconductor component base based on A^3B^5 compounds. One of the promising materials for this purpose is gallium nitride. The paper presents the design and manufacturing technology of one of the main control elements of the microwave signal in microwave radiothermometer - monolithic AlGaN/GaN/SiC HEMT SPDT transistor switch.

1. Introduction

Recently, the method of microwave radiometry, which has a number of advantages over traditional diagnostic methods, has been rapidly developing for the use in various fields of human activity. Today, a promising method for measuring soil moisture, based on the use of microwave radiometric sensors of various wavelength ranges, is in the process of implementation into commercial operation. In addition, microwave radiothermometry is actively developing for the functional study of human tissues and organs, which allows non-invasive detecting thermal heterogeneities in them, performing early diagnosis of cancer and correcting the treatment process by changing the parameters of electromagnetic radiation of tissues and organs [1-2].

However, the further development of the radiometry method is hampered by a number of scientific and technological limitations that need to be overcome. Combining the principles of multi-channel, multi-frequency and microminiature in one radiometric complex will lead to a significant reduction in the size of the radiometric receiver and to the need to develop fundamentally new design and technological solutions, namely, its implementation in the form of a single module, which implies the use of a monolithic integrated design. The results of work in this direction are shown in [3-5].

Thus, modern microwave equipment requires technical solutions that ensure the compactness of the developed devices and high performance. The use of the element base, made in the form of monolithic integrated circuits, will increase the variability of circuit solutions in the development of modern microwave radiometers and, thereby, ensure their optimal characteristics and expand the functionality of the devices. In [6-8] the possibilities of using integrated circuits in medical microwave radiothermometers are considered in order to improve the characteristics and functionality of the devices.

The aim of the work is to develop a monolithic switch for designing a miniature microwave radiometer that provides stable operation of the device with an acceptable measurement error.

2. Results

Recently, there has been considerable interest in the implementation of a microwave semiconductor component base based on wide-band materials of the A3B5 group. The use of HEMT AlGaN/GaN heterostructures on semi-insulating SiC substrates in comparison with GaAs-based heterostructures, which are traditional for the microwave application, makes it possible to ensure minimal crystal sizes, efficient heat removal, and work with high input signal power values. In this regard, the use of such a component base in various microwave devices is quite promising.

The design of the monolithic switch was determined by the scope of the developed radiometer. The main requirements for switches are miniaturization and the provision of low insertion loss values and high decoupling values over a wide frequency range. The radiometers most sensitive to the response of the microwave signal from the biological object of research operate in the frequency range of 0.5-3 GHz [9]. Another essential requirement for the developed element base is to provide the radiometer with the ability to receive a signal and calibrate for the" hot "and" cold " temperature load. Based on these requirements, the design of the SPDT type transistor switch (one input-two outputs of the microwave signal) with a minimum topological size of 0.4 microns was selected to ensure reliable operation in the specified frequency range.

A domestic epitaxial HEMT structure grown by gas-phase epitaxy on a semi-insulating silicon carbide substrate with a diameter of 76.2 mm produced by Closed Joint Stock Company "Svetlana-Electronpribor" was used when developing the manufacturing technology of the transistor switch.

In the framework of this work, numerical modeling and experimental optimization of the design of heterostructures of field-effect microwave transistors (HEMTs) based on AlGaN/GaN material systems were carried out. The need to predict the most important output characteristics of transistors required the construction of models for the statistics of carrier distribution in all areas of the heterostructure, a model of differential mobility in the two main materials (GaN, AlGaN) of the heterostructure with possible doping of each, as well as careful tuning of the model of the total energy balance of carrier transport in the HEMT transistor channel to account for the heating of the main carriers and an adequate description of the current transmission processes at the submicron gate length required for the operation of the transistor as part of the switch. Widely known software packages for numerical simulation of semiconductor devices with connected modules, which implement calibrations and original numerical algorithms for calculating heterostructures, were used as basic tools for modeling and optimizing HEMT transistors [10-12]. The simulation results were compared with the experimental results of measuring the electrophysical parameters of heterostructures and HEMT transistors.

The developed heterostructures of HEMTs (figure 1) consisted of an AlGaN germ layer for SiC substrates, an insulating buffer layer GaN with a thickness of 2 microns, an AlN layer with a thickness of 1 nm, and an unalloyed barrier layer AlGaN. The thickness and composition of the AlGaN layer varied in different calculation variants in the range of 15-35 nm and 20-35%, respectively.

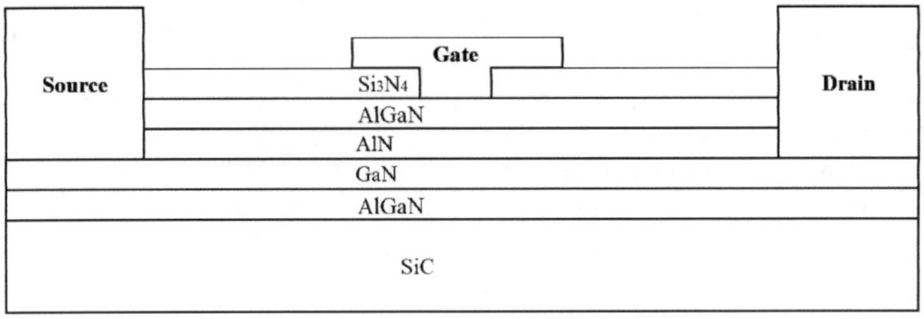

Figure 1. Schematic representation of the developed HEMT heterostructure for use as a switch.

Achieving the required basic parameters of the heterostructure - a high concentration of two-dimensional electron gas and electron mobility in the channel - is the basis for obtaining the necessary characteristics of the transistors in the switch and, accordingly, the switch itself. However, many obstacles must be overcome to achieve the desired characteristics of transistors. Let us briefly consider one that relates directly to the heterostructure. For example, it is well known that an increase in the molar fraction of aluminum in the barrier layer should radically increase the electron concentration in a two-dimensional electron gas [13]. However, such an obvious way to increase the saturation current of the transistor channel has its limitations, which is well shown by numerical modeling. The calculations performed using the Monte Carlo method, which take into account the essentially nonlinear field-velocity characteristic of the barrier layer material and include a modified mobility model [13, 14], suggest the existence of an optimal aluminum content for the selected barrier layer thickness, exceeding which significantly worsens the characteristics of the transistor due to a decrease in the electron mobility in the channel.

Figure 2 shows the values of the electron concentration in a two-dimensional electron gas and the local electron mobility under the gate, obtained by numerical calculation, taking into account the above-mentioned effects, for the barrier layer thickness of 20 nm. The obtained values are in good agreement with the experiment and serve as the basis for creating an effective transistor design for operation as part of the AlGaN/GaN/SiC HEMT SPDT switch.

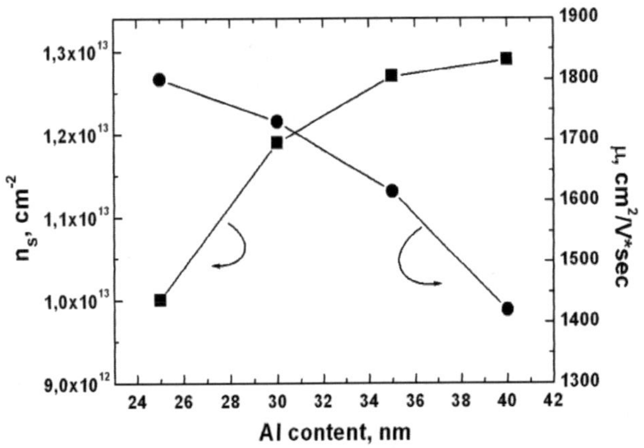

Figure 2. Calculated dependences of the electron concentration in the channel and the local electron mobility under the gate on the aluminum content in the barrier layer for the developed HEMT AlGaN/GaN SPDT switch heterostructure on the plate.

As a result of numerical optimization, the HEMT structure provided the following characteristics: the concentration of two-dimensional electron gas in the channel 9×10^{12} cm^{-2} and the mobility of 1750 $cm^2/V\times sec$.

The formation of metallization of ohmic contacts and wiring was carried out using the method of "explosive" photolithography. The Ti/Al/Ni/Au system was used as the metallization of ohmic contacts, followed by rapid thermal annealing at a temperature of 800 °C in a nitrogen medium. Interstitial isolation was performed by ion implantation. The formation of Ni/Au gate metallization (gate length is 0.4 microns and width is 2 mm) was carried out using electronic lithography. To avoid static breakdown and increase reliability, a triple gate was provided in the switch design. The dielectric protection was provided by applying a layer of silicon nitride. Through-ground holes were formed by plasma-chemical etching of the silicon carbide substrate followed by the process of magnetron sputtering of Ti/Au metallization. Figure 3 shows the HEMT AlGaN/GaN SPDT switch crystal on the plate.

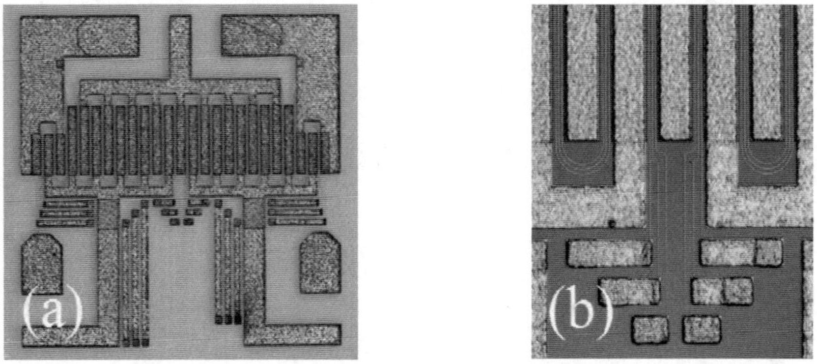

Figure 3. A fragment of the HEMT AlGaN/GaN SPDT crystals of the switch on the plate (a) and its gate part (b).

The CVC of the switch was monitored by using the Cascade MicroTech PA-200 probe station. Figure 4 shows the typical CVC of a HEMT AlGaN/GaN SPDT switch crystal.

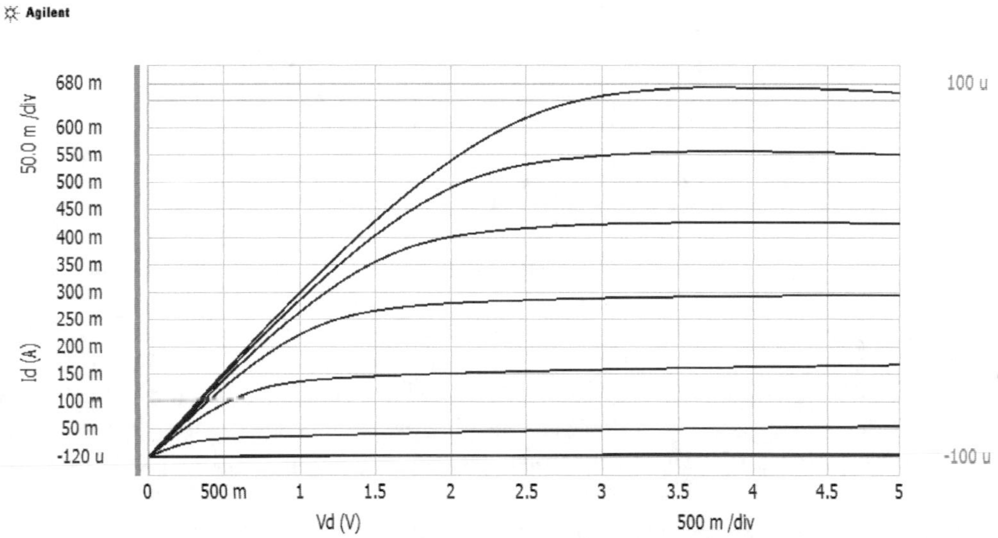

Figure 4. The CVC of the switch HEMT AlGaN/GaN SPDT crystal.

The microwave characteristics were measured by using a test bench consisting of the Cascade PM5 probe station and the Rohde&Schwarz ZVB-20 vector analyzer directly on the chip without using a microwave test board. Figure 5 shows the key characteristics of the developed switch-the loss of passage and the isolation of the closed channel.

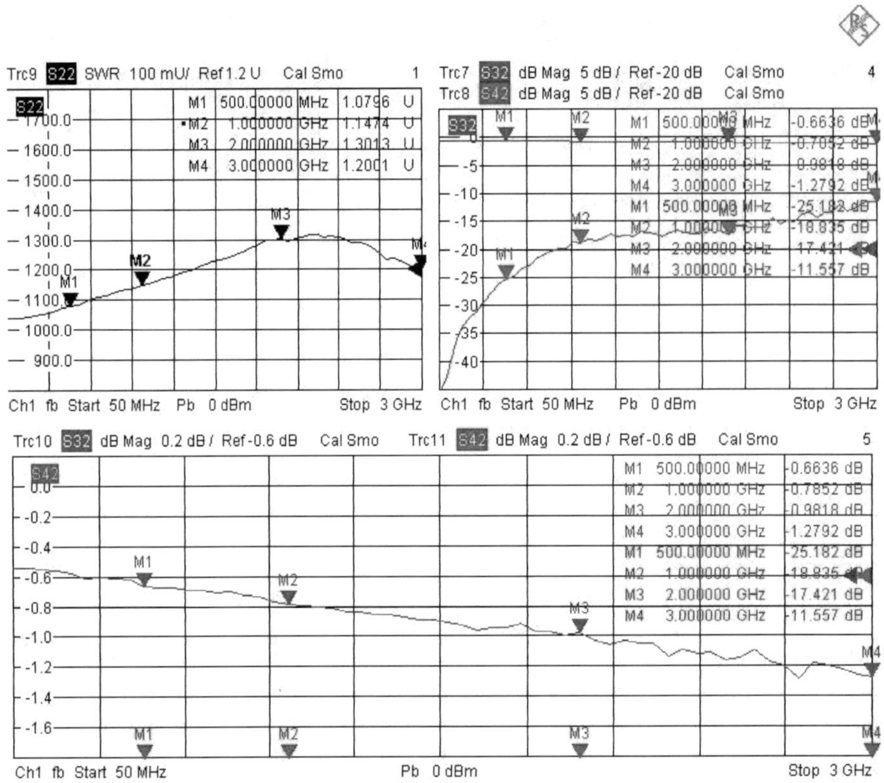

Figure 5. Low-signal microwave characteristics of the HEMT AlGaN / GaN SPDT switch.

According to the presented data, the developed switch provides losses in one channel for the passage of a microwave signal of 0.66-1.27 dB and isolation of no worse than 25-11.5 dB in the frequency range of 0.5-3 GHz. When matching the developed switch on the test board (figure 6), there is an improvement in the key parameter - the loss of the microwave signal per pass of 0.39-0.9 dB, the VSWR value does not exceed 1.33 dB, and the isolation changes in the specified frequency range from 26 dB at 0.5 GHz to 14 dB at 3 GHz, that should be taken into account when matching the switch crystal when mounting in the housing of the microwave radiometer.

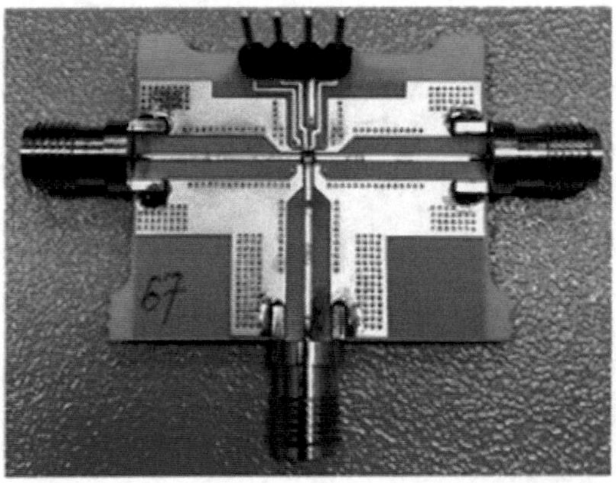

Figure 6. HEMT AlGaN / GaN SPDT switch on test board.

3. Conclusion

Thus, the design and manufacturing technology of the HEMT AlGaN/GaN SPDT switch on a domestic heterostructure on a semi-insulating silicon carbide substrate are presented. The measurements of the microwave parameters of the developed switch show the possibility of using this element component base as part of microwave radiometers, which will allow combining the principles of multi-channel, multi-frequency and miniaturization in one radiometric complex and will lead to the expansion of its functionality and a significant reduction in size.

Acknowledgments
Research was supported by a grant of the Russian Science Foundation (project No. 19-19-00349).

References

[1] Gulyaev Yu V, Leushin V Yu, Gudkov A G, Schukin S I, Vesnin S G, Kublanov V S, Porohov I O, Sedankin M K and Sidorov I A 2017 Pribory dlya diagnostiki patologicheskih izmeneniy v organizme cheloveka metodami mikrovolnovoy radiometrii *Nanotekhnologii: razrabotka, primenenie – Xxi vek* **9** (2) 27-45 (In Russian)

[2] Sidorov I, Gudkov A, Novichikhin E, Taradin A, Haarbrink R and Sergey C 2019 The specific of 3D passive radars sensing alive and non-alive objects *2019 EnT* 1-4

[3] Gudkov A *et al* 2015 Prospects for application of radio-frequency identification technology with passive tags in invasive biosensor systems *Biomed. Eng.* **49** (2) 98-101

[4] Novichikhin E P, Sidorov I A, Leushin V Yu, Agasieva S V, Chizhikov S V 2020 Detection of a local source of heat in the depths of the human body by volumetric radiothermography, *RENSIT* **12**(2) 305-12

[5] Emtsev V V *et al* 2016 The relationship between the reliability of transistors with 2D AlGaN/GaN channel and organization type of nanomaterial *Tech. Phys. Letters* **42** (7) 701-3

[6] Gudkov A G *et al* 2019 Use of multichannel microwave radiometry for functional diagnostics of the brain *Biomed. Eng.* **53** (2) 108-11

[7] Gudkov A G *et al* 2020 Studies of a Microwave Radiometer Based on Integrated Circuits *Biomed. Eng.* **53** (6) 413–16

[8] Chizhikov S V, Solov'ev Yu V 2020 Element base of microwave MIC for microwave radiothermometry *Nanotechnology: development and applications – XXI century* **12** (2) 48-57 (In Russian)

[9] Vesnin S, Sedankin M, Leushin V, Skuratov V, Nelin I, Konovalova A 2019 Research of a microwave radiometer for monitoring of internal temperature of biological tissues *Eastern-European Journal of Enterprise Technologies* **4** (5) 6-15

[10] Tikhomirov V, Zemlyakov V, Volkov V, Parnes Ya, Vyuginov V, Lundin W, Sakharov A, Zavarin E, Tsatsulnikov A, Cherkashin N, Mizerov M and Ustinov V 2016 *Semiconductors* **50** (2) 244–48

[11] Agasieva S *et al* 2017 Dependence analysis of the GaN HEMT parameters for space application on the thickness AlGaN barrier layer by numerical simulation *2nd Int. Conf. on Opto-Elect. Inform. Proc.* 79-82

[12] Tikhomirov V G, Gudkov A G, Agasieva S V, Dynaiev D D, Popov M K, Chizhikov S V 2019 Increasing efficiency of GaN HEMT transistors in equipment for radiometry using numerical simulation *J. of Phys. Conf. Series* **1410**:012191

[13] Ambacher O, Eickhoff M, Steinhoff G, Hermann M, Gorgens L and Werss V 2002 *Proc. ECS* vol 214 (Electrochemical Society, Pennington, NJ) p 27

[14] Shirahata M, Kusano H, Kotani N, Kusanoki S and Akasaka Y 1992 A Mobility Model Including the Screening Effect in MOS Inversion Layer *IEEE Trans. Computer-Aided Design* **11** (9) 1114

Features of modulation of high-power laser radiation for optical space control systems

S A Rodin[1], B K Reznikov[1,2], V V Davydov[3,4] and V Yu Rud[4,5]

[1] The Bonch-Bruevich Saint-Petersburg State University of Telecommunications, Saint-Petersburg 191186, Russia
[2] National Research University of Information Technologies, Mechanics and Optics, Saint-Petersburg 197101, Russia
[3] Peter the Great Saint-Petersburg Polytechnic University, Saint-Petersburg 195251, Russia
[4] All-Russian Research Institute of Phytopathology, Moscow Region 143050, Russia

psnp.174.sut@gmail.com

Abstract. In this work we justify the necessity of using shutter-type modulators to create laser radiation with the required parameters in fiber-optic systems.This is necessary to detect the fact of making changes in the optical system (unauthorized connection with the restoration of the function of communication channels). We have developed a shutter-type optical radiation modulator design. This modulator design has no fundamental limitations on the power of laser radiation, which is used to solve various problems. The investigation results of different work regime to the modulator are presented.

1. Introduction

The development of scientific and technological progress has affected various areas of information transmission and control systems for the parameters of various signals [1-12]. A special place among communication systems is occupied by fiber-optic communication lines and various sensors and systems for converting and monitoring the parameters of optical radiation [12-19]. The main advantage of using optical systems is passivity to electromagnetic radiation, which is growing every year [15-17, 20-26]. In addition, the power of electromagnetic interference increases, which leads to malfunctions in different systems. One of the problems that arise during the maintenance of fiber-optic communication lines is unauthorized access to the system [27-29]. Such connections are used to solve various tasks. They must be installed promptly. Various methods and techniques have been developed to detect such intrusions into the optical system. One of them is associated with the modulation of laser radiation at different frequencies.

In communication systems, electro-optical modulators (EOM) are used to modulate laser radiation. This is an expensive technology. And its use for the previously described tasks is irrational, because the cost of the system with, for example, 20 channels will increase very significantly. In addition, in some cases, for space control systems, it is necessary to use high-power laser radiation, which is not always combined with the basic design of the EOM, which is being developed for fiber-optic communication lines, in which the laser radiation power is quite limited. It is also not always advisable to use a laser transmitting module with internal modulation, since the basic designs of these devices have power limitations and their service life during long-term operation is much less than that of standard laser systems. Therefore, the development of simple and reliable designs of optical radiation modulators, which do not have fundamental limitations when operating in terms of laser radiation power and can operate in a wide frequency range, is an urgent task.

2. Shutter-type modulator design and experimental setup

In this work we propose a shutter-type modulator design. This construction is a universal for realization of the modulator, since it does not require any unique or special components, and it is also compatible with most of the standard optical components used. Figure 1 shows one of the versions for its design.

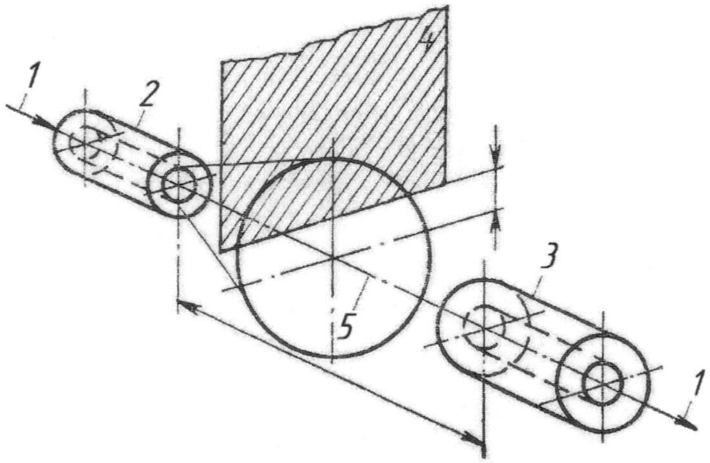

Figure 1. Design of a shutter-type optical radiation modulator: 1 – laser radiation; 2, 3 – optical fibers; 4 – movable shutter; 5 – a gap between optical fibers.

We have constructed an experimental setup for the research of a shutter-type modulator. The block diagram is shown in figure 2.

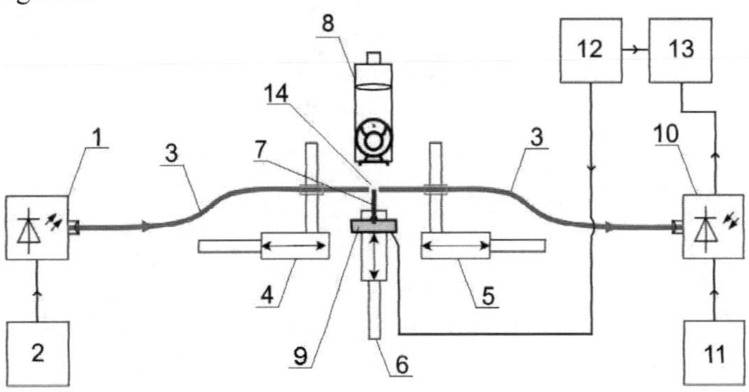

Figure 2. The block diagram of the experimental setup: 1 – LED; 2, 11 – power supply; 3 – multimode optical fibers; 4, 5, 6 – optical translator system; 7 – shutter; 8 – microscope; 9 – piezoceramic; 10 – photodiode; 12 – signal source; 13 – oscilloscope; 14 – a gap between optical fibers.

There is a gap between a pair of optical fibers – "input" and "output" – (3). With the use of an oscillating shutter, placed in this gap, modulation of the laser radiation from the input fiber to the output is performed. The shutter is made of steel and has a size of 37 mm in length and 70 μm in diameter. It is fixed on a piezoceramic plate (9) connected to a signal generator, the frequency of which can change up to tens of MHz. The fibers and plate are placed in optical translator system, which consists of several independent mechanisms (4-6). They are all placed on one base (small optical plate). This system provides accurate positioning along three axes. The distance between the fibers and the position of the shutter is determined by a specially applied scale using a microscope (8). The modulation process was monitored using an oscilloscope, the input of which was supplied with voltage from the photodetector module. For a more complete representation of the operation of the modulator, standard laser radiation with wavelength of 650 nm (power of laser radiation was changed from 0.2 to 20 mW) was supplied to the input fiber. We have performed a series of experiments to detect the optimal mode of the modulator.

3. Experiment results and discussion

An important characteristic in the operation of the modulator is the dependence of the change in the amplitude of the modulated laser radiation on the amplitude of the voltage of the oscillations applied to the piezoceramic plate. In figure 3, as an example, several such dependences are presented for a frequency F=1742 Hz with a change in the distance between optical fibers from 125 μm to 375 μm.

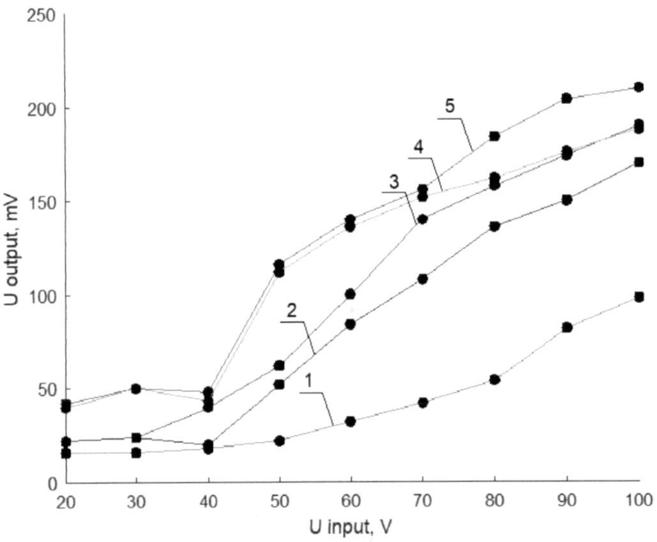

Figure 3. Dependence of the voltage on the photodetector on the shutter oscillation voltage for different distances between the ends of the fibers: 1 – 125 μm; 2 – 187,5 μm; 3 – 250 μm; 4 – 312,5 μm; 5 – 375 μm.

The results show that as the distance between fibers decreases, the efficiency of the modulation process increases. This is due to an increase in the amount of light entering the output fiber, since the light spot has a cone shape, and the ratio of the cone cross-sectional areas in the plane of the end of the output fiber and the core of the output fiber increases.

Another important characteristic of the operation of the modulator is the dependence of the change in the amplitude of the modulated laser radiation on the change in the value of the frequency of oscillations entering the piezoceramic plate. In figure 4, as an example, several such dependences are presented for the value of the oscillation amplitude U=6 V for cases of different distances between the fibers (125 μm-375 μm).

Journal of Physics: Conference Series **2086** (2021) 012050 doi:10.1088/1742-6596/2086/1/012050

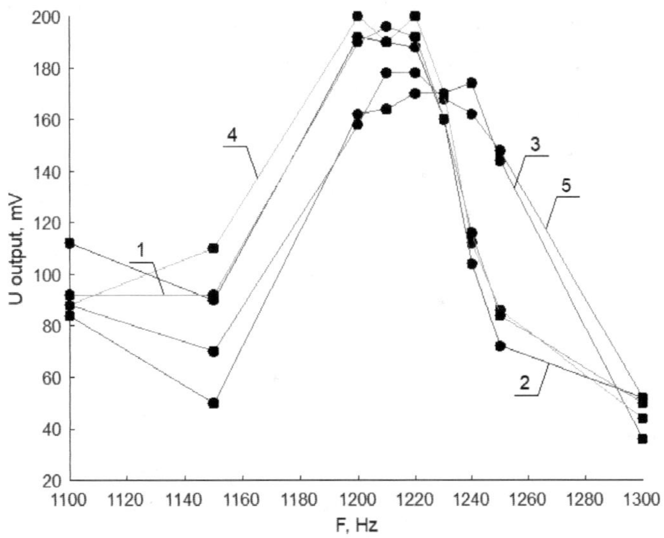

Figure 4. Dependence of the voltage on the photodetector on the oscillation frequency of the shutter for different distances between the ends of the fibers: 1 – 125 μm; 2 – 187,5 μm; 3 – 250 μm; 4 – 312,5 μm; 5 – 375 μm.

The obtained results show that the optimal shutter oscillation frequency is in the range from 1200 to 1230 Hz. This is due to the most linear expansion of the piezoceramic element, on which the shutter is fixed, in the range of specified frequencies.

The last characteristic of the modulator operation is the dependence of the change in the amplitude of the modulated radiation on the position of the shutter between the fibers. Figure 5 shows a few of such characteristics for different values of the depth of introduction of the shutter (from partial to full coverage of fiber cores).

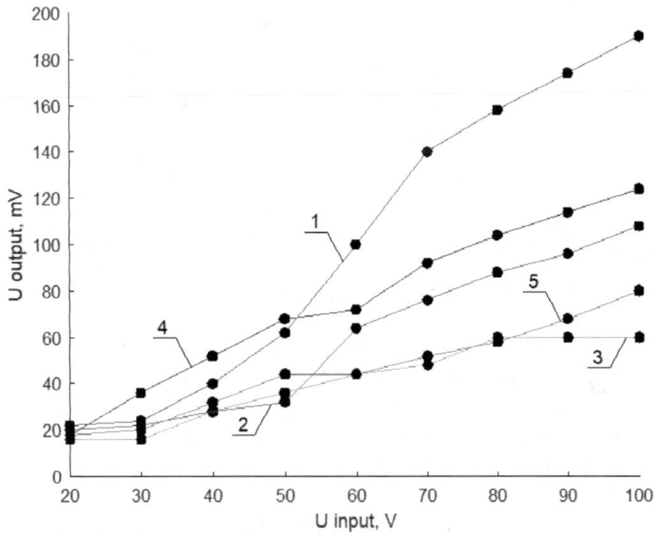

Figure 5. Dependence of the voltage on the photodetector on the shutter oscillation voltage for different values of the depth of introduction of the shutter: 1 – the shutter is in the middle between fibers; 2-5 – shift in steps of 10 μm.

As a result, the most optimal mode of the modulator is the mode when shutter is located exactly in the middle between the optical fibers.

Also as a result of the experiment, it was found that the shutter-type modulator does not distort the signal. It is shown in figure 6. This allows to detect the changes in phase shift in case of unauthorized connection to the system.

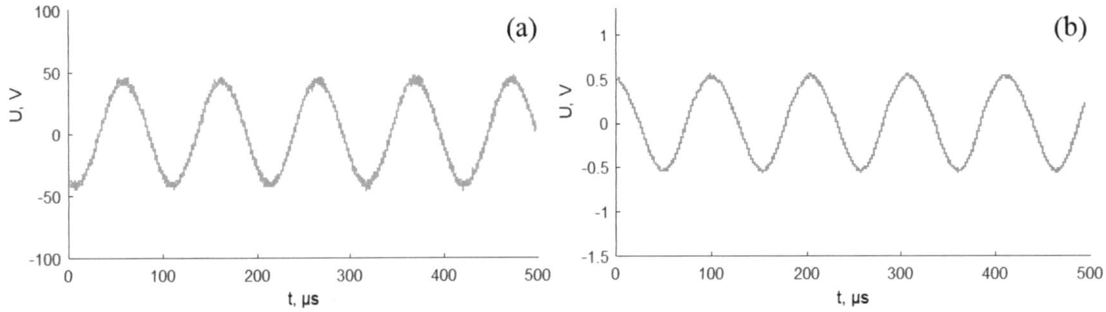

Figure 6. Oscillograms of voltages: (a) – signal from the generator arriving at the piezoceramic plate, (b) – signal from the photodetector.

Additional experiments have shown that up to 500 mW of laser radiation power in a continuous mode, the modulation mode in the developed design remains almost unchanged (the shutter heats up slowly). Disruption of the modulation process (this is shown in figure 7) in our design with an increase in the power of laser radiation will most likely be associated with heating of the shutter surface (thermal expansion). This fact requires additional research with more heat-resistance materials, which will be the result of our further work.

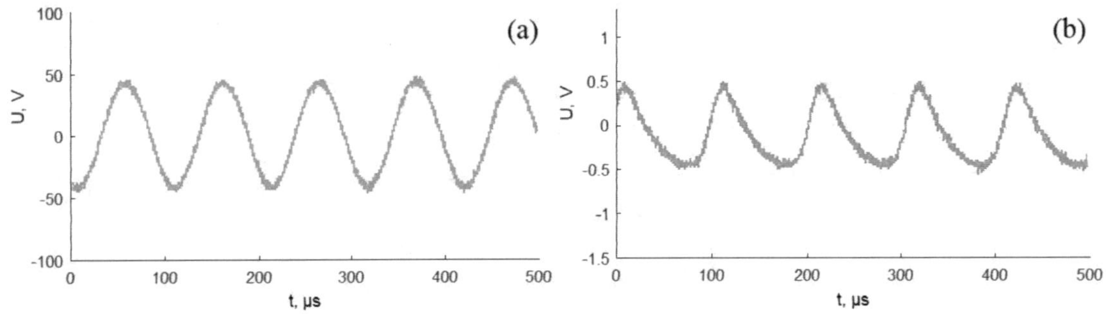

Figure 7. Oscillograms of voltages: (a) – signal from the generator arriving at the piezoceramic plate, (b) – signal from the photodetector.

4. Conclusions

The obtained experimental results showed the presence of modulation in the laser radiation detected by the photodetector. It was found that the optical radiation is modulated with a frequency F. The piezoceramic plate oscillates at this frequency. The optimal range of the laser radiation power is established at which the modulation process is not disturbed.

References

[1] Skidanov R V and Ganchevskaya S V 2015 *Computer Optics* **39(2)** 262-268
[2] Kozlova E S, Kotlyar V V and Nalimov A G 2015 *Computer Optics* **39(5)** 654-660
[3] Kuzmin M S and Rogov S A 2019 *Computer Optics* **43(3)** 391-396
[4] Davydov R, Antonov V, Makeev S and Batov Y 2019 *E3S Web of Conferences* **140** 02001
[5] Davydov R, Antonov V and Moroz A 2019 *IEEE International Conference on Electrical*

Engineering and Photonics (EExPolytech-2019), vol. 8906791 (IEEE), p. 295-297
[6] Andreeva E I and Potapov I A 2020 *Journal of Physics: Conference Series* **1695(1)** 012125
[7] Andreeva E I and Potapov I A 2021 *Springer Proceedings in Physics* **255** 241–245
[8] Andropov A V, Kuzmin S V and Korovin K O 2021 *Springer Proceedings in Physics* **255** 675–681
[9] Raimzhanov T R, Kuzmin S V and Korovin K O 2021 *Springer Proceedings in Physics* **255** 629–635
[10] Makolkina M, Pham V, Kirichek R, Gogol A and Koucheryavy A 2018 *Lecture Notes in Computer Science (including subseries Lecture Notes in Artificial Intelligence and Lecture Notes in Bioinformatics)* **11118 LNCS** 547-559
[11] Hoang T, Kirichek R, Paramonov A and Koucheryavy A 2016 *Lecture Notes in Electrical Engineering* **376** 1249-1259
[12] Davydov R V, Dmitrieva D S, Pilipova V M, Dudkin V I and Andreeva E I *2020 Proceedings of International conference of Laser Optics ICLO-2020 (Saint-Petersburg),* vol. 9285820 (IEEE), p. 171
[13] Smirnov K J, Medzakovskiy V I, Vysoczky M G and Glagolev S F 2017 *Journal of Physics: Conference Series* **917(6)** 062019
[14] Smirnov K J, Glagolev S F and Tushavin G V 2018 *Journal of Physics: Conference Series* **1124(2)** 022014
[15] Davydov R V, Saveliev I V, Lenets V A, Tarasenko M Yu and Yalunina T R 2017 *Lecture Notes in Computer Science (including subseries Lecture Notes in Artificial Intelligence and Lecture Notes in Bioinformatics)* **10531 LNCS** 177-183
[16] Petrov A A and Myazin N S 2017 *Lecture Notes in Computer Science (including subseries Lecture Notes in Artificial Intelligence and Lecture Notes in Bioinformatics)* **10531 LNCS** 561-568
[17] Podstrigaev A S, Smolyakov A V, Myazin N S, Grebenikova N M and Davydov R V 2019 *Lecture Notes in Computer Science (including subseries Lecture Notes in Artificial Intelligence and Lecture Notes in Bioinformatics)* **11660 LNCS** 525-533
[18] Smirnov K J and Batov Y V 2019 *Journal of Physics: Conference Series* **1368(2)** 022073
[19] Lukashev N A, Petrov A A, Grebenikova N M and Valov A P 2018 *Proceedings of 18th International conference of Laser Optics ICLO-2018 (Saint-Petersburg)*, vol. 8435889 (IEEE), p. 271
[20] Moroz A V and Davydov R V 2019 *Lecture Notes in Computer Science (including subseries Lecture Notes in Artificial Intelligence and Lecture Notes in Bioinformatics)* **11660 LNCS** 710-718
[21] Filatov D L, Galichina A A, Vysoczky M G and Yalunina T R 2017 *Journal of Physics: Conference Series* **917**(8) 082005
[22] Fadeenko V B and Pchelkin G A 2019 *Journal of Physics: Conference Series* **1400(4)** 044010
[23] Moroz A V 2019 *Journal of Physics: Conference Series* **1410(1)** 012212
[24] Moroz A V, Davydov V V, Malanin K Y, Krasnov A A and Rud V Yu 2019 *Journal of Physics: Conference Series* **1400(4)** 044009
[25] Fadeenko V B, Pchelkin G A and Beloshapkina O O 2019 *Journal of Physics: Conference Series* **1410(1)** 012238
[26] Fadeenko V B, Kuts V A and Vasiliev D A 2018 *Journal Physics: Conference Series* **1135**(1) 012053
[27] Sinicyna E A, Galichina A A, Lukyanov A S and Podstrigaev A S 2019 *Journal of Physics: Conference Series* **1236(1)** 012075
[28] Grebenikova N, Moroz A, Bylina M and Kuzmin M 2019 *IOP Conference Series: Materials Science and Engineering* **497(1)** 012109
[29] Logunov S E, Rud V Y, Davydov R V, Moroz A V and Smirnov K J 2019 *Journal of Physics: Conference Series* **1326(1)** 012024

LK-5 glass surface modification by glass blowing method based on microsystem technology

A Dzhinikashvili[1], Y Enns[1,2], R Kleimanov[1,2] and A Kazakin[1,2]

[1] Peter the Great Saint-Petersburg Polytechnic University, St. Petersburg, Russia
[2] St. Petersburg Academic University, St. Petersburg, Russia

Abstract. This paper considers the possibility of modifying the surface LK-5 glass by glass blowing methods to solve the problems of creating an atomic clock and ensuring a controlled distribution of metal nanoparticles on its surface. The evolution of the formation of a spherical profile of a glass surface on a hermetically welded cavity in a glass-silicon system was studied experimentally and theoretically. Also, the possibility of modifying the spatial parameters of arrays of nanoparticles distributed on the glass surface has been demonstrated. The obtained experimental and theoretical data demonstrate sufficient convergence.

1. Introduction

The applying of glass substrates as a structural material has become widespread in various sensor applications used in biomedicine, optics, photonics, plasmonics, and microelectromechanical systems (MEMS) [1]. Various technological methods have been developed for the integral fabrication of microfluidic channels, photonic waveguides, and microlenses, such as laser ablation [2], abrasive-jet blasting [3], reactive ion etching (RIE) [4], and femtosecond laser irradiation and chemical etching (FLICE) [5]. Recent studies in the field of creating integral cells of atomic clocks have shown the possibility of using glass blowing methods to produce spherical cells at the wafer level [6]. This technology is based on the formation of glass spheres under the influence of excessive pressure at glass softening temperatures. Excess pressure is formed when the gas is heated in a hermetically sealed cavity. It is a cylindrical hole etched in silicon using deep plasma-chemical etching, sealed with a glass substrate (Pyrex) using an anodic bonding. This paper considers the possibility of modifying the surface LK-5 glass by glass blowing methods to solve the problems of the produced atomic clock and ensuring a controlled distribution of metal nanoparticles on its surface.

The change in the profile of the glass surface is accompanied by a thinning of its thickness and an increase in the surface area. This can be used to modify the distribution of arrays of metal nanoparticles deposited on the glass surface. At the same time, the technology of creating these arrays, which consists in high-temperature annealing of thin metal films deposited on such surfaces as SiO_2, TiO_2, NiO, etc. [7], is compatible with the technology described above. Such arrays of Au, Ag, Al nanoparticles have become widespread due to the surface plasmon resonance effect. Parameters as the particle size, shape, and density of their arrangement largely determine the parameters of the plasmon resonance: wavelength, intensity of photon absorption and reflection of photons. Increasing the area of the upper glass surface should lead to stretching of the array of nanoparticles, which can be used as a method for modifying the array parameters.

This paper considers the possibility of modifying the surface LK-5 glass by glass blowing methods to solve the problems of the produced atomic clock and ensuring a controlled distribution of metal nanoparticles on its surface.

2. Experimental results

2.1. Fabrication

The structures of the sealed cavities were fabricated on standard silicon substrates 350 μm thick and LK-5 glass 485 μm thick. The technological route is shown in Figure 1. Cylindrical cavities in a silicon substrate were formed by deep plasma-chemical etching in the BOSH mode (Figure 1b). The etching depth was 200 μm with a variation in the cavity diameters in the range of 500-1500 μm. In contrast to the works given in [6], where non-standard thick silicon substrates (more than 1 mm) were used, the creation of cavities with a greater depth is impossible due to the limitations of the stiffness of the formed silicon membrane at the bottom of the cavity, so that its thickness would be about 150 μm. The anode bonding of the wafer was carried out at a temperature of 400 °C in the air (Figure 1c). To study the formation of a spherical glass surface, the wafer was divided into 1x1 cm chips by a circular saw. The study was carried out during high-temperature annealing in a muffle oven in the temperature range of 850-950 °C and the time range from 5 to 60 minutes (Figure 1d). The samples cooled at the end of annealing at room temperature. The tendency of deformation of glass with a thickness of 200 μm was also investigated. The glass was thinned by wet etching methods. The profile of the formed structures was studied using the AmbiosXP-1 profilometer.

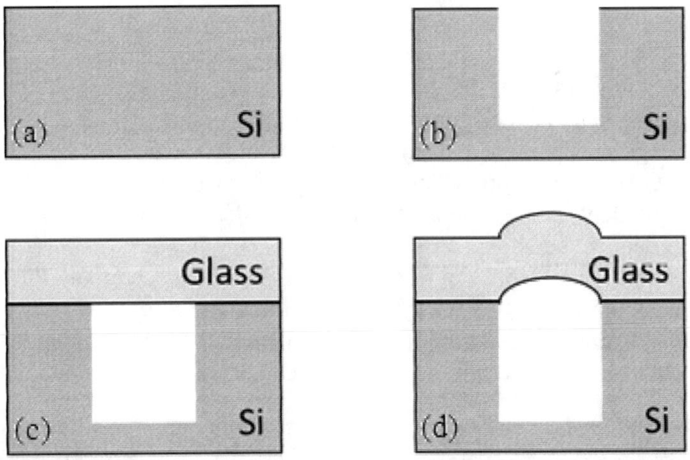

Figure 1. Process of the technological route. (a) silicon substrate; (b) deep plasma chemical etching; (c) anode bonding; (d) high temperature annealing.

To investigate the possibility of ensuring a controlled distribution of metal nanoparticles, an Au film 2 nm thick was deposited on the glass surface before annealing. The gold film was applied using a vacuum thermal evaporation unit. After applying the gold film, the samples were also annealed at a temperature of 950 ° C with a duration of 15 minutes. The distribution of nanoparticles was monitored using SEM.

2.2. FEA of viscoelastic deformation of glass.

The glass blowing process was also analyzed using numerical simulation methods. Heating LK-5 glass to temperatures above 820 °C leads to glass softening and a significant decrease in its viscosity to values below 10^7 Pa·s [8]. This means that the deformation of the glass under the influence of pressure can no longer be determined only by elastic deformation and must be described using viscoelastic models. To describe this behavior, the Oldroyd-B model was used, which includes the viscosity and relaxation time. It should be noted that in this formulation, the internal stresses of the glass after the anode bonding of the wafer, caused by the difference in the temperature coefficients of expansion of the glass and silicon, were not taken into account.

The numerical modeling was carried out in a two-dimensional axisymmetric formulation, which can significantly reduce the cost of computing resources. The geometry of the calculated model includes a cross-section of the glass above the cavity, which can deform under the influence of pressure in the cavity, and a fixed cross-section in the zone of connection with silicon. The fixing corresponding to the welding zone was set along the lower border of the glass. The upper boundary of the glass is defined by a free surface with the possibility of deformation and atmospheric pressure acting on it. The pressure in the cavity was determined based on the equation of state for an ideal gas with allowance for glass deformation. The initial value is determined by the increase in pressure in the sealed cavity due to a temperature change. It is assumed that at the time of the anode bonding, the pressure in the cavity is equal to atmospheric pressure. Under the influence of overpressure occurs glass viscoelastic deformation and volume change of the cavity. The resulting pressure on the lower boundary of the glass above the cavity is determined as the difference between the external pressure and the initial pressure in the cavity during heating, and further due to the deformation of the glass, the volume of the cavity increases and the pressure decreases.

Numerical simulations were carried out in the time domain in the range from 0 to 60 minutes. The load is set as a smooth function of time to reduce non-linearity at the initial stages of the process.

3. Results

After high-temperature annealing, deformation of the upper glass surface in the region above the cavity was observed on the samples (Figure 2a). During prolonged annealing (time more than 30 minutes), the edges of the glass were shattered and a rim was formed along the perimeter of the chip, which made it impossible to carry out further anode bonding (Figure 2b). This can be explained by the surface tension of the glass. The height of the rise of the glass surface was 19.1-70.9 μm, and there was a tendency to increase the deformation of the glass with increasing temperature and annealing time. In the study of the dependence of the deformation on the annealing time, the onset of stationary equilibrium was recorded, in which the difference in glass deformation at 30 minutes and 60 minutes was minimal. This can be caused by a significant decrease in excess pressure during the formation of the spherical surface of the glass. In addition, at a temperature of 950 °C and prolonged annealing (more than 30 minutes), the formation of a crystallizated film was observed on the glass surface, which significantly changes its properties. This limits the choice of temperature and annealing time.

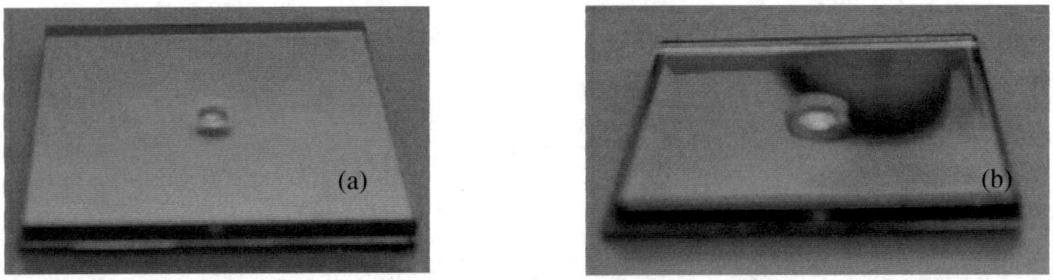

Figure 2. Photo of samples after high-temperature annealing: 5 minutes of annealing time; (b) 2 minutes of annealing time.

Numerical modeling also showed the presence of glass deformation in the region above the cavity and the formation of a spherical profile (Figure 3). The calculation results show good agreement with the experimental results. Figures 4 (a, b) show the numerical and experimental results of the study of the development of glass deformation and changes in excess pressure in the time domain. The obtained theoretical results of changes in the pressure difference between the environment and the chamber confirm the assumption of the onset of stationary equilibrium during a long annealing process. The creation of deeper cavities could increase the maximum deformation, but this requires the use of non-standard silicon substrates and longer plasma-chemical etching. The use of glass with a thinner thickness can also increase the maximum deformation, which confirmed by studies carried out on samples with thinned glass (Figure 4b). Thus, with a smaller diameter of the cavity, it was possible to achieve large deformations (71 μm). However, for this case, there is a greater discrepancy between the theoretical model and the experiment. It can be assumed that this is caused by the greater influence of the shrinkage of the glass during cooling due to the thinner glass wall. In addition, the choice of glass thickness is also limited by the thinning of the glass wall due to the elongation of the spherical profile. This is clearly seen in Figure 4 (a, b), where the change in the initial thickness Δh was 70 μm and 45 μm, respectively.

The difference between the upper and lower surfaces indicates an inhomogeneous thickness, which may be important for the use of this method in optics devices. The most promising way to increase the maximum deformation is to increase the initial pressure in the sealed cavity.

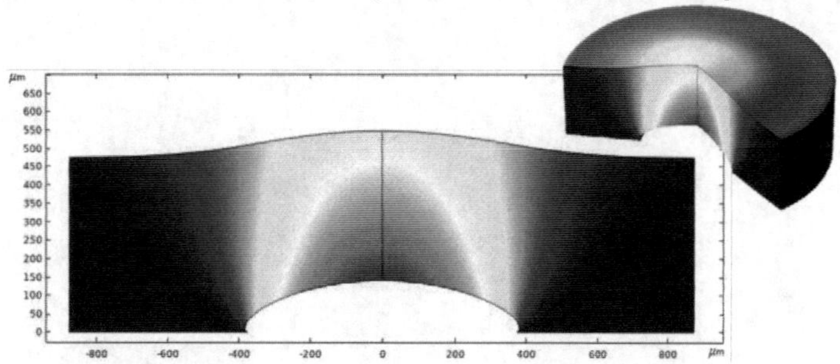

Figure 3. Example of a deformation map with a cavity diameter of 750 microns, a glass thickness of 485 microns, and a heating time of 60 minutes.

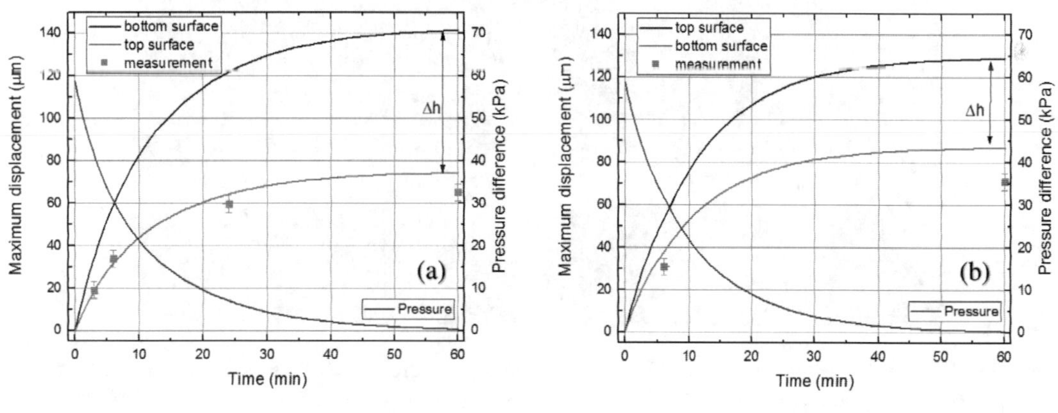

Figure 4. Theoretical and experimental results of studying glass deformation and pressure changes in the time domain: (a) cavity diameter 750 μm, glass thickness 485 μm; (b) cavity diameter 500 μm, glass thickness 200 μm.

On samples with a gold film deposited after annealing, glass deformation above the silicon cavity was also observed. For a cavity diameter of 500 μm, the lifting height of the upper surface was 48 μm, which corresponds to the theoretical results. After annealing, a change in the reflection spectrum was recorded, which indicates the presence of gold in the form of nanoparticles. In addition, a color gradient was observed in the area above the cavity. The EDS results show the presence of gold near the glass surface. The influence of this method of modifying the glass surface on the distribution of nanoparticles was studied in areas close to air bubbles in the glass. These regions had a diameter of the deformed area of about 24 μm, which makes it possible to observe the change in the distribution of nanoparticles using SEM.

Figure 5 (a) shows the SEM image of a glass micro-bubble with a nanoparticle distribution. This image shows a change in the size of the nanoparticles and the distance between them; in the center of the deformed region, nanoparticles of the largest size (about 200 nm in diameter) were observed, and the array had the lowest density (Figure 5b). When moving away from the center, the size of the nanoparticles decreased to a diameter of about 75 μm.

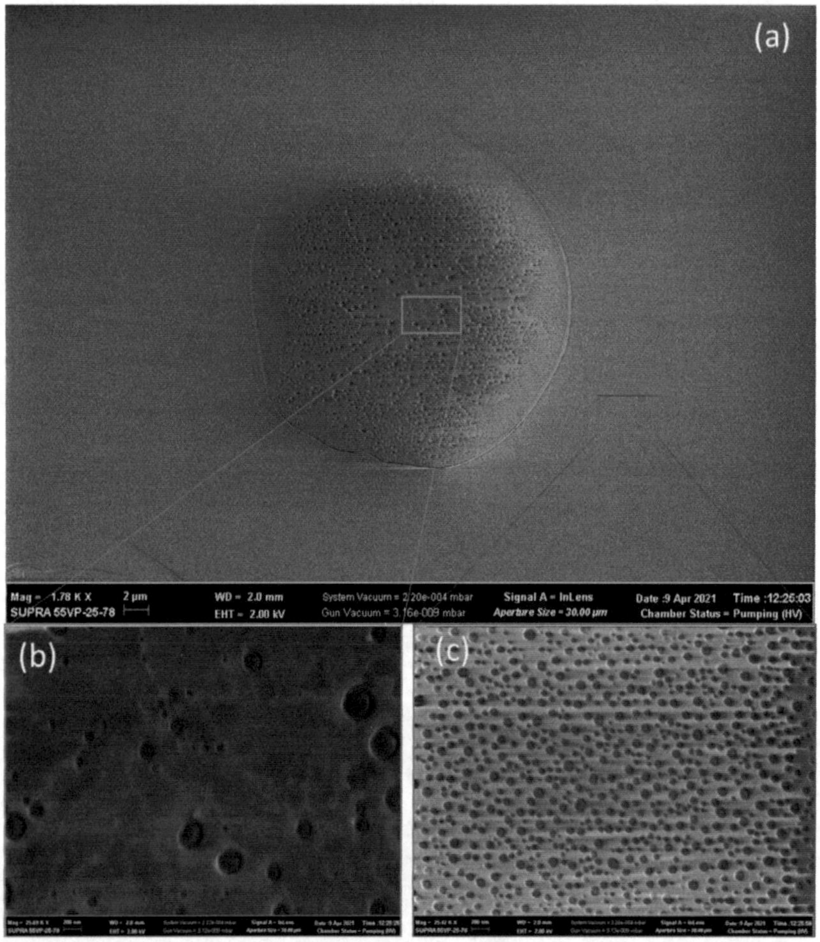

Figure 5. SEM image of the glass surface with nano gold particles (a); area near the center of the formed bubble (b); area on the surface of the undeformed glass (c).

It is worth noting that despite the low adhesive force of gold nanoparticles to glass, mechanical action on the glass surface did not lead to the destruction of the nanoparticle array. It can be assumed that the nanoparticles were immersed in the glass after annealing.

4. Conclusion

The study shows the possibility of modifying the surface of LK-5 glass by glass blowing methods to solve the problems of manufacturing quantum frequency standards and ensuring a controlled distribution of metal nanoparticles on its surface. It is worth noting that the obtained results of glass deformation are significantly lower than the similar work using Pyrex glass [6]. An increase in deformation can be achieved by optimizing the modes of the anode bonding to provide a greater residual pressure in the sealed cavity. Thus, the results of the work provide a broad basis for further research.

Acknowledgments
This work was carried out in Peter the Great St. Petersburg Polytechnic University and was supported by a grant of Russian Science Foundation (project № 20-19-00146) and part of the work on the creation of nanoparticles was done with the support of the Ministry of Education (№FSRM-2020-0011).

References
[1] Asadian M H, Noor R M and Shkel A M 2019 *Proc. 2019 IEEE SENSORS* 1-4
[2] Niino H, Kawaguchi Y, Sato T, Narazaki A, Gumpenberger T and Kurosaki R 2006 *Applied surface science* **252** (13) 4387–91
[3] Belloy E, Sayah A and Gijs M 2000 *Sensors and Actuators A: Physical* **86** (3) 231–237
[4] Li X, Abe T, Liu Y and Esashi M 2002 *J. JMEMS* **11**(6) 625–630
[5] Bellouard Y, Said A, Dugan M and Bado P 2004 *J. Optics express* **12**(10) 2120–29
[6] Eklund E J, Shkel A M, Knappe S, Donley E and Kitching J 2008 *J. Sensors and Actuators A:Physical* **143** (1) pp. 175–180
[7] Kondrateva A, Enns Y, Komarevtsev I, Kudryashov D and Mishin M 2020 J. *Phys.: Conf.* Ser.**1695** 012002
[8] Datasheet for LK5- glass. http://www.ipz.com.ua/datasheet/LK5_rus.pdf

Characterization of focusing grating couplers for telecom wavelengths in the first and second diffraction order

A Elmanova[1], I Elmanov[1], P An[1,2], V Kovalyuk[1,2], A Kuzin[1,4], A Golikov[3] and G Goltsman[1,2,3]

[1] Department of Physics, Moscow Pedagogical State University, 119992, Russia
[2] FRC Institute of Applied Physics RAS, 603950 Nizhny Novgorod, Russia
[3] National Research University Higher School of Economics, Moscow 101000, Russia
[4] Skolkovo Institute of Science and Technology, 121205, Russia

Abstract. In this work we studied how focusing grating couplers, developed for telecommunication C-band wavelength range, can be applied in the near infrared range. In the paper we presented prospects of usage of both first and second diffraction maxima of theoretically computed diffraction grating couplers for photonic aims. The dependence of the central wavelength of the grating on the etching depth of the photonic layer, on the period and filling factor of the grating was studied. We have compared our experimental results with numerical study, performed using finite elements method of solving differential equations. The work is important for different photonic applications and introduces new prospects in application of the already fabricated devices, developed for telecommunication wavelengths.

1. Introduction

Focusing grating couplers (FGCs) have found their wide application for input/output of radiation due to the well-developed CMOS processing together with waveguides, as well as flexible footprint in any place of the nanophotonic chip [1]. Such couplers have efficient filtering of light in the selected spectral range, usually corresponding to the first order of interference [2]. FGCs are considered one of the most compact systems for light coupling from an optical fiber to a photonic integrated circuit (PIC) [3]. However, for a number of quantum applications, for example, quantum key distribution or linear optical quantum computing, in order to reduce parasitic illumination, the coupler efficiency in higher diffraction orders should be known [4].

In this work, the spectra of FGCs, developed for operation at a telecom wavelength of 1550 nm, in the first and second diffraction order was studied. The approximate positions of the first and second diffraction maxima were verified by numerical simulation and experimentally. We used silicon nitride as one of the most popular platforms for developing PICs [1]. However, thanks to the wide bandgap, silicon nitride can be successfully used in both C-band and near IR wavelengths [5]. At the same time, lower optical contrast makes such devices less sensitive to fabrication inaccuracies, and also reduces the scattering loss due to the sidewall roughness [6]. Several works on the development and use of grating couplers have already been presented on this platform, with the coupling efficiency reaching -4.2 dB [7] for a wavelength of 1550 nm. For the FGCs fabrication, one of the most common methods is electron beam (e-beam) lithography [8]. It allows to precisely vary such device parameters as the period p, width w of the grating, hence and the filling factor (ff):

$$ff = \frac{w}{p}. \quad (1)$$

The dependence of the characteristics of focusing devices on them has been actively studied, both theoretically and experimentally [9]. After the e-beam lithography process, the next important step is reactive ion etching structures. Here semi-etched design is used, because it provides a better coupling through evanescent field between different devices on photonic scheme and waveguides than full etching (like optical ring resonators, directional couplers, etc) [10]. In our work, we investigated the performance of the FGCs at different etching depths and wavelength region.

2. Simulation

Before experimental study, it was necessary to determine what diffraction orders we are working with and in what approximately wavelength range one should expect the maximum transmission on diffractive couplers. When positioning an optical fiber with a coupler grating, we consider the fiber inclined at 8 degrees. This angle is shown in Figure 1a. In this case, it is possible to approximately calculate the position of the diffraction maxima of the order of m using the well-known formula [11]:

$$n_{eff} = n_{cover} cos\varphi_c + m \lambda/p, \quad (2)$$

here n_{eff} – effective refractive index, n_{cover} – refractive index for the waveguide layer (considering the structure as fully etched), φ_c – coupling angle (in our case it is equal to 90° – 8° = 82°), m – diffraction order, λ – central wavelength in vacuum and p – period of the diffraction grating (for non-apodized case). Thus, when calculating a structure with a central wavelength of 1550 nm, a grating period of approximately 1.086 μm and a refractive index of the waveguide layer of 1.9963 (for a given wavelength), we obtain an effective refractive index of near 1.839. For wavelengths around 850 nm, the effective refractive index will not be 1.839, but rather (according to the proportional ratio) 1.8663. Then the wavelength of the second maximum falls on even longer wavelengths. Thus, for a device designed for wavelength of 1550 nm, a second maximum is expected in the region of 850-900 nm.

For numerical simulation we used the finite element method (FEM) in the COMSOL Muiltiphysics. A 2D model of a grating coupler was developed. We optimized our parameters for C-band of IR radiation and used parameters possible for fabrication (etching depth h_{etch} =450 nm, filling factor ff =0.86, period of the grating p = 1.086 μm, silicon substrates with SiO2=2.6 μm and Si3N4=450 nm layers atop). First the distribution of normalized electric field and the mode fill diameter were found (Figure 1b). The optimal parameters were determined for a Gaussian beam and a central wavelength of 1550 nm (Figure 1c). To do this, the dependence of the normalized electric field on the coordinate was also determined (Figure 1e). The dependence of the coupling efficiency on the thickness of SiO2 was (Figure 1d). After it was determined, the highest efficiency was found to be for silicon oxide layer thickness 2.6 μm, therefore the fabricated structures had this parameter. Simulation of the propagation of an electric field (TE mode) was carried out for a given wavelength with the parameters of a fully-etched device, and in the wavelength range of about 800 nm. The dependence of the transmission on the Gaussian beam position was also found (Figure 1f). Transmission peaks were found and their effectiveness was evaluated. The maximum efficiencies for the 1520–1570 nm range is 17.8% at a central wavelength of 1558.86 nm, as well as for the 800–810 nm range is 0.87% at a central wavelength of 807 nm were found.

3. Device design and fabrication

For the device fabrication we used commercially available silicon substrates with SiO_2=2.6 μm and Si_3N_4=450 nm layers atop. The fabrication process consisted of one e-beam lithography step. First, ZEP 520A resist was deposited by spin coating procedure. Then, the electron beam lithography was done by Crestec CABL-9050C and resist was developed in O-Xylene, protecting covered silicon nitride under reactive ion etching (RIE) in Ar and CHF_3 gas mixture. Finally, using oxygen plasma the resist was removed. Overall view of the fabricated device is shown in Figure 2a. The device consists of two FGCs, connected via silicon nitride waveguide at a distance of 250 μm. Thus, one device acts as an input port, the other as an output port. More photographs are given on Figure 2d,e. For the primary study, we used a sample with array of structures with period 1.086 μm, filling factor 0.7 and etching depth 225 nm. Then we used five samples with equal arrays of structures, but with a different etch depth (Figure 2f). All devices have a taper angle of 45°, the number of gratings is 20. A 2D array of structures with devices with different filling factors in a range of 0.84 to 0.88 with step 0.01, with different periods, varied from 1.058 to 1.088 with step 0.004 μm was studied. Each was etched to different depths (447, 435, 345, 281, and 170 nm).

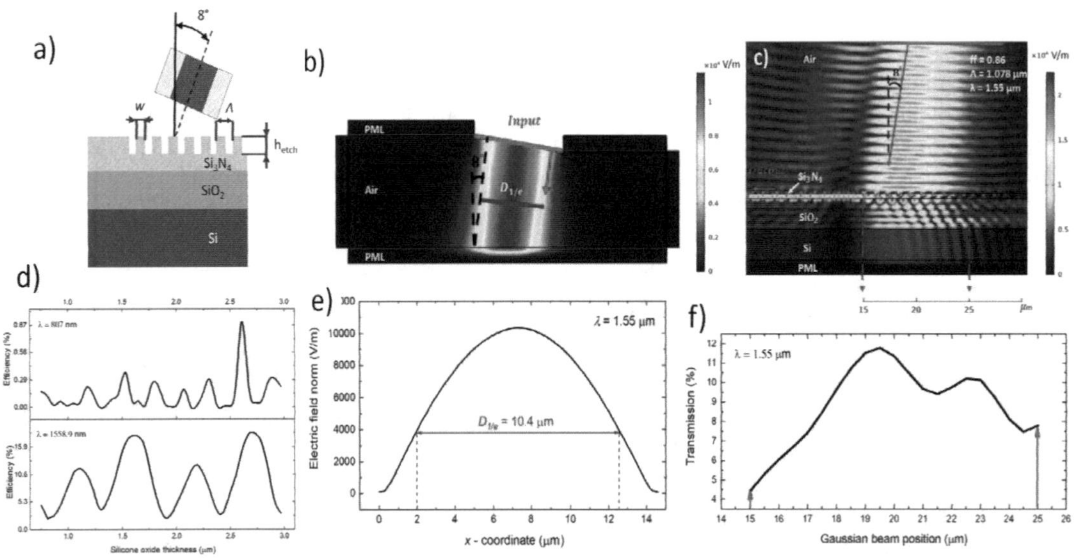

Figure 1. (a) Schematic view of the FGC and optical fiber above at an angle of 8° to the normal. The width of the filled part *w*, period *p* and etching depth h_{etch} are marked. **(b)** Distribution of normalized electric field for Gaussian beam at λ = 1550 nm with an angle of 8° to the grating, $D_{1/e}$ – mode field diameter. The perfect match layer (PML) on boundaries. **(c)** Normalized *E* field for Gaussian beam shown in (b). The period is *p* =1078 nm; the filling factor *ff* = 0.86. **(d)** Dependence of the coupling efficiency on the silicon oxide thickness for the two diffraction orders. **(e)** Dependence of the normalized electric field on the inside Gaussian beam. The mode field diameter is equal to 10.4 μm at a wavelength of 1550 nm. **(f)** Dependence of intensity transmission on the Gaussian beam position.

4. Device design and fabrication

In this work, we used two experimental setups for measuring the transmission spectra of nanophotonic devices. The first setup for measurements at telecom wavelengths consists of a tunable laser (NewFocus TLB-6600 with tune range 1510-1620 nm), a polarization controller, 3D stage with piezo motors, a fast photodetector, as well as a fast analog-to-digital converter (Figure 2b). The fiber array was optimized to work in the wavelength range of the C-band. The second setup (Figure 2c) consisted of the broadband fiber-coupled supercontinuum source (Leucos Rock 400 4) and the spectrum analyzer (Maya 2000 Pro). Light from source pass to the fiber array for near infrared spectra, which is fixed with angle 8°. The results obtained by set up for telecom wavelength (1.55 μm) and is shown in Figure 2g. The results were approximated using Lorentz Fit. Therefore, for infrared range we got FWHM = 35.428 nm and for visible FWHM = 20.193 nm (Figure 2g, h). For the input (P_{in}) and output power (P_{out}), the coupler efficiency (under the condition of low losses in the next waveguide and the same input and output) can be found as:

$$C = \sqrt{\frac{P_{out}}{P_{in}}}. \quad (3)$$

For all of the devices, spectra were obtained, the dependences of the efficiency and position of the central wavelength on the period, filling factor, and etching depth were investigated. Based on the data, color contour maps were compiled (Figure 3) for both ranges (near 1550 nm and 850 nm). The dependences of the central wavelengths of the devices on the etching depth were obtained at a constant filling factor and period.

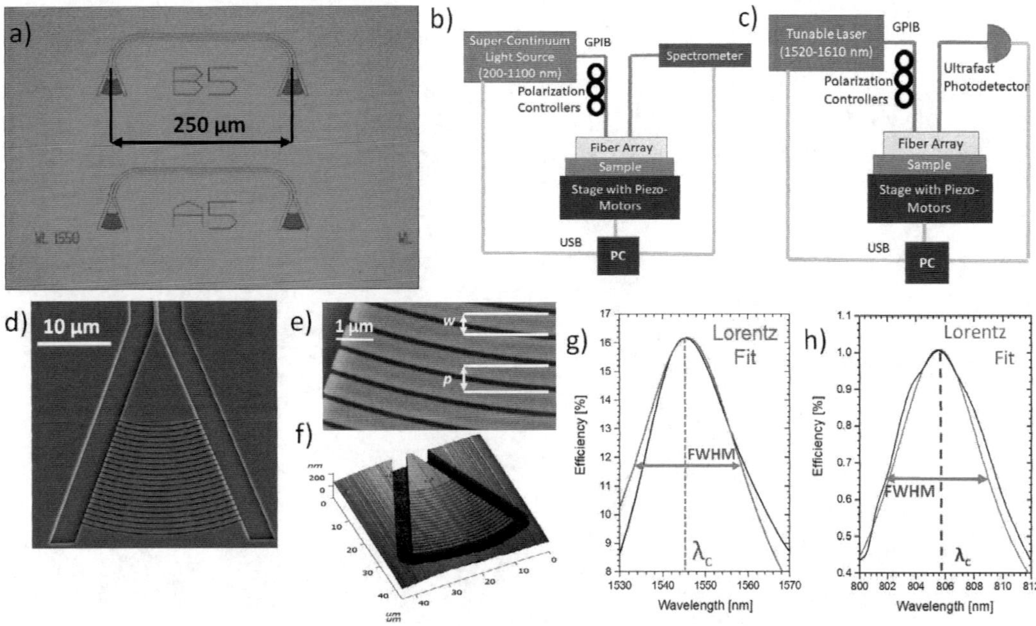

Figure 2. (a) SEM photographs of the fabricated devices. **(b)** Schematically depicted experimental setup for wavelengths range 200-1100 nm. **(c)** Schematically depicted experimental setup for wavelengths range 1520-1610 nm. **(d)** SEM photograph of one of the FGCs. **(e)** SEM photograph with the grating with period and the grating teeth marked. **(f)** Atomic force microscope image of one of the fabricated couplers. **(g)** Experimental transmission spectrum and Lorentz fit for the fabricated device in telecomm wavelength range. Full width at half maximum and notch wavelength are noted. **(h)** Same as **(g)** in near IR wavelength range.

5. Discussion

The obtained contour maps for two wavelength ranges can be compared (Figure 3 a,c). The position of the central wavelength changes similarly, and the efficiency also. Therefore, the notch wavelength of the GC be shifted by changing period or filling factor, as this is a predictable mechanism. The efficiency peaks for both ranges are in the area of low periods (less than 1.07 µm) and filling factors in between 0.85 and 0.87 (Figure 3b,d). It is important to study of the dependence of the wavelength of devices on the etching depth (Figure 4a,c). So, for the least fabricated period of 1.058 µm for various filling factors, the average value of the ratio of increments of the wavelength to the etching depth was -0.1789 and for near IR it was 0.1749. By taking derivatives over the etching depth from both sides of the formula (2), one can obtain the relation for $m = 1$:

$$\frac{dn_{eff}}{dh_{etch}} = \frac{d\lambda}{dh_{etch}} \cdot \frac{1}{p} \tag{4}$$

From where one can get the ratio of the increment of the effective refractive index to the change in the etching depth. On average, this increment for the fabricated structures $1.672 \cdot 10^{-4}$ nm^{-1}. The transmission patterns of fabricated and simulated structures were also compared (Figure 4 b, d).

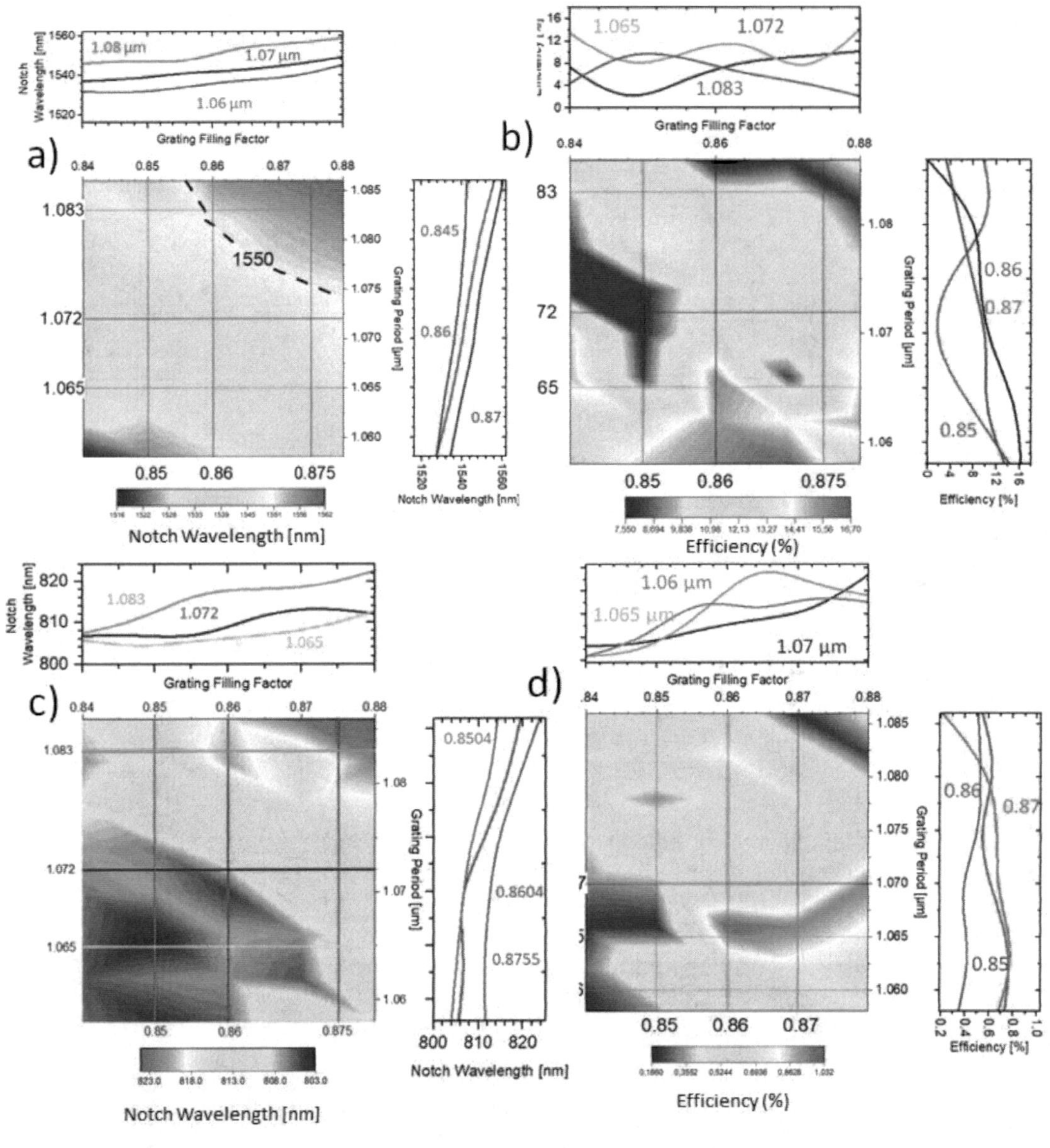

Figure 3. (a) Dependence of the central wavelengths for FGCs on their parameters in telecom range. **(b)** Dependence of the coupling efficiency on λ = 1550 nm for FGCs on parameters in the telecom range. **(c)** Dependence of the central wavelengths for FGCs on their parameters in the near IR range. **(d)** Dependence of the coupling efficiency on λ = 820 nm for FGCs on their parameters in the near IR range.

6. Conclusions

We have studied the performance of FGCs designed for telecommunications wavelengths in the visible range. We have found the same correspondence in the efficiency and notch wavelength depending on period and filling factor of the grating for the two wavelengths ranges. We also determined the dependence of the central wavelength on the etching depth for both diffraction orders, and, therefore, the dependence of the effective refractive index on the etching depth. We have compared the experimental transmission pictures with the simulated curves. We found that the central wavelength is shifted for several nm and the efficiency is of the same order for the simulated and experimental data.

This work is of significant importance for research and work with quantum emitters, for using pumping at a wavelength of shorter than radiation, thus filtering pumping from radiation, and studying the emission spectra of nanophotonic devices in two different wavelength ranges.

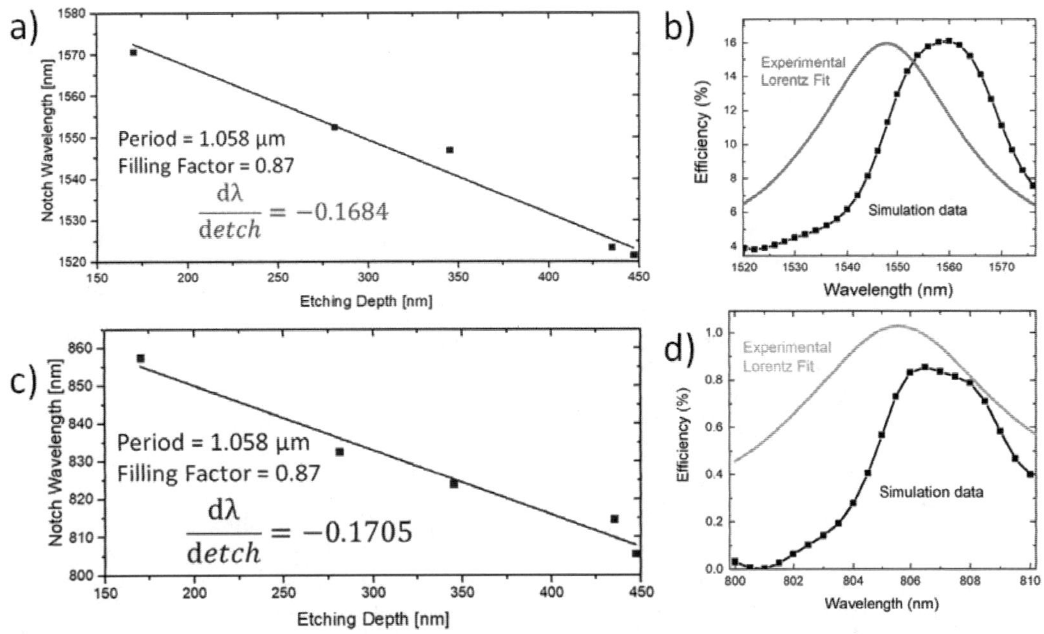

Figure 4. (a) Dependence of the central wavelength on the etching depth in the telecom range for the devices. **(b)** Comparison of the experimental and simulated curves for the device with period 1078 nm, *ff* 0.86 and etching depth 280 nm in the telecom range. **(c)** Dependence of the central wavelength on the etching depth in the near IR range. **(d)** Comparison of the experimental and simulated transmission for the FGC with period 1078 nm, *ff* 0.86 and etching depth 280 nm in the near IR range.

Acknowledgements
The research was performed by support of the Russian Science Foundation grant No. 19-72-10156.

References

[1] Quaranta G, Basset G, Martin O J F and Gallinet B 2018 *Laser Photonics Rev.* **12** 1800017
[2] Song S H and Lee E H 1995 *Appl. Opt.* **34** 5913-5919
[3] Wood M, Sun P and Reano R M 2012 *Opt. Express* **20** 164-172
[4] Ding Y, Bacco D, Cai X, Zhou X, Rottwitt K and Oxenløwe L F 2017 *Quantum Inf.* **3** 25
[5] Zou J, Yu Y, Ye M, Liu L, Deng S and Zhang X 2015 *Opt. Express* **23**(20) 26305-26312
[6] Zhang H, Li C, Tu X, Song J, Zhou H, Luo X, Huang Y, Yu M and Lo G Q 2014 *Opt. Express* **22**(18) 21800-21805
[7] Doerr C R 2010 IEEE *Photonics Technology Letters* **22** 19 461-1463
[8] Song J H, Kongnyuy T D, Troia B, Saseendran S S, Soussan P, Jansen R and Rottenberg X 2019 *OSA Continuum* **2** 1155-1165
[9] Laere F, Bogaerts W, Taillaert D, Dumon P, Van Thourhout D and Baets R 2007 *Conf. on Optical Fiber Communication and the National Fiber Optic Engineers Conf. Proc.* 1-3
[10] Rath P, Kahl O, Ferrari S, Sproll F, Lewes-Malandrakis G, Brink D, Ilin K, Siegel M, Nebel C and Pernice W 2015 *Light: Science& Applications* **4** 338
[11] Waldhäusl R, Schnabel B, Dannberg P, Klez E B, Bräuer A and Karthe W et al 1997 *Appl. Opt.* **36** 9383-9390

Analysis of the possibility of creating an acoustic velocity sensor using GaN epitaxial films

Y Enns[1,2], A Kazakin[1,2], Y Akulshin[2], A Mizerov[1] and R Kleimanov[1,2]

[1] Alferov University, 8 Khlopina, St. Petersburg, 194021, Russia
[2] Peter the Great Saint-Petersburg Polytechnic University, 29 Polytechnicheskaya, St. Petersburg, 195251, Russia

ennsjb@gmail.com

Abstract. This paper results in results of analyzing the possibility of creating an acoustic velocity sensor using epitaxial GaN films. Technology for the fabrication of a microelectromechanical acoustic velocity sensor was developed and a prototype of the sensor was produced. The simulation of the characteristics of the obtained acoustic velocity sensors was carried out on the basis of the measured electrical characteristics, where the sensitivity and the directional pattern were determined.

1. Introduction

Vibration control is of high importance for the diagnosis of industrial equipment, marine, aerospace and nuclear industries. The high number and density of equipment requires the creation of a distributed system of vibration sensors. The overwhelming majority of dangerous vibrations have frequencies in the audio range, therefore, acoustic velocity sensors have received great prospects for vibration diagnostics [1]. The sensors comprise a microelectromechanical system (MEMS) of two short, thin and closely spaced platinum beams that are heated to about 300 ° C [2]. The resistance of the beams depends on the temperature. The action of the acoustic flow instantly changes the temperature distribution, forming a difference in resistance across the sensor beams. This difference in resistance is measured with a bridge circuit that provides a signal proportional to the acoustic flux. Despite the fact that these sensors are proven, their applying in harsh environments is very limited. Therefore, it is necessary to search for new materials that are stable during operation conditions of high temperatures and corrosive gases.

Due to its high chemical, temperature and radiation resistance, GaN is a promising material for electronics and sensors for harsh environments [3]. GaN sensitive elements can be used in miniature gas sensors, temperature and pressure sensors, thermal flow sensors and other MEMS. This work is aimed at creating a design and production technology for acoustic velocity sensors based on GaN epitaxial films.

2. Finite elemental analysis

The results obtained in the course of previous studies of the technology of fabrication microbeam elements, structural parameters and mechanical parameters, as well as the values of the temperature coefficient of resistance (TCR) and gauge factor (GF), show the possibility of creating a MEMS thermal sensor based on epitaxial GaN films. [4,5]. Methods of finite elemental analysis were used to assess the performance of the developed design of thermoanimometers. The simulation was carried out in the COMSOL Multiphysics software. Estimation of the characteristics of the GaN anemometer required an interconnected solution to the problem in the mechanical and electrical domains, as well as in the areas of heat and mass flow. The effect of laminar airflow at low flow rates (up to 1 cm / s) on the sensor output signal was investigated taking into account the effect of the incident angle, i.e. flow velocity vector relative to the plane of thermoelements.

The GaN sensor model consists of two parallel planar beam elements 0.5 μm thick, 250 μm long and 20 μm wide. The beam elements are separated from the silicon substrate by a cavity of 50 μm and fixed at the ends (Figure 1). Such thin-film structures are characterized by the curvature of the profile of the beams after separation from the substrate. The profile of the beam can have a high impact on the distribution of gas flow rates in the vicinity of the threads. The profile was determined based on the previously obtained values of internal stresses for epitaxial GaN films [4]. The mechanical parameters were determined using a modulus of elasticity E 200 GPa and internal stresses 1.39 GPa. It was determined that the magnitude of internal stresses leads to a tensile deformation of 0.006 and the formation of an arc-shaped profile with a deflection height of 12 μm.

The solution of interrelated problems was carried out by finding the self-heating temperature of the beams due to the flow of current through them and changing the convection conditions due to the action of the laminar flow. A potential difference for each of the beam elements equal to 35 V was set on the contact pads with one common ground output. The electrical conductivity parameters for epitaxial GaN layers were set from the previously experimentally determined temperature dependence [5]. In the model, an assumption was made about the ideal isolation of the epitaxial layer by the Si_3N_4 layer from the silicon substrate. At the bottom edge of the silicon chip, a fixed room temperature was set, which corresponds to the ideal heat dissipation from the sensor. The calculation of convective heat removal from the beam elements was carried out on the basis of the atmospheric gas flow in the laminar approximation. The flow rate was set at 10 cm / s. In this case, the incident angle of the flow with respect to the sensor was taken into account, which was set in the range from 5° to 95°.

The resistance of the beam elements for normal conditions and low power was 69 kΩ. Figure 2 shows the results of numerical modeling of the influence of the incident angle of the gas flow on the change in the resistance of the beams. It can be seen from the polar diagram that the maximum difference in resistance occurs at incident angle of about 45°. This is caused by the maximum displacement of the gas flow relative to the beam elements and the redistribution of the heating area (Figure 3). The results of finite elemental analysis show the possibility of creating a sensor that is selective in the direction of action of the air flow. However, it should be noted that the difference in resistance between the beam elements will also be proportional to the flow amplitude. This signal must be isolated using a non-vector flow velocity sensor.

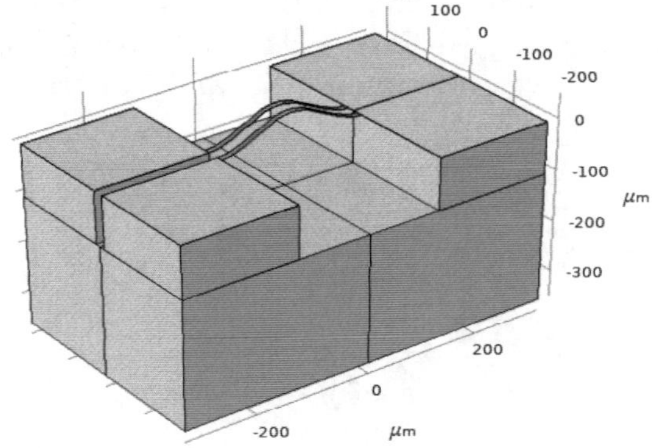

Figure 1. Acoustic velocity vector sensor prototype model.

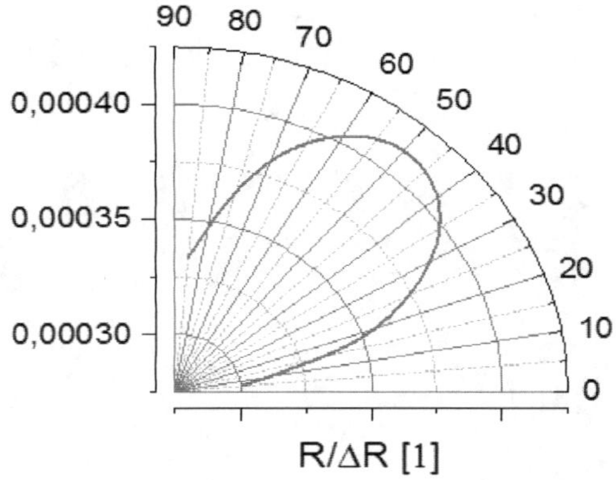

Figure 2. Dependence of the relative change of resistance on the incident angle.

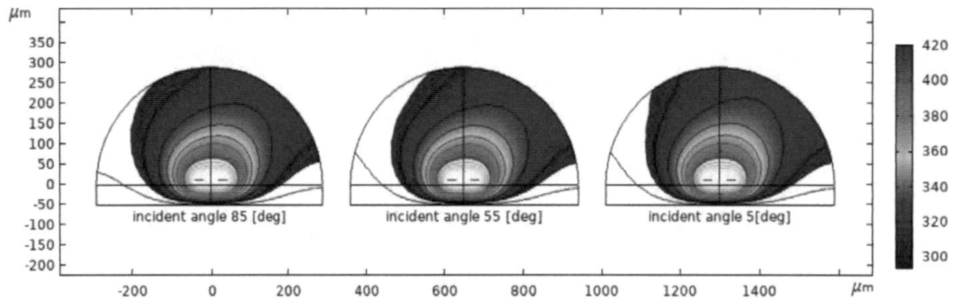

Figure 3. Temperature distribution under the flow rate of 10 cm / s.

3. Experiment

In this work, epitaxial n-type GaN films were grown on high-resistance Si (111) wafers (R> 10000 Ohm cm) by molecular beam epitaxy on a Veeco Gen 200 industrial setup equipped with a high-frequency (13.56 MHz) plasma source Riber RFN- 50/63 [6]. Before growing GaN, silicon substrates were annealed at T = 850 ° C to remove the SiO_2 film. The growth procedure began with the deposition of several Ga monolayers on the silicon surface to prevent Si_xN_y formation. Then an array of GaN nanocolumns was formed as a seed layer. To obtain a continuous GaN layer, the coalescence of GaN nanocolumns was carried out. In order to obtain continuous GaN layer the coalescence overgrowth of GaN nanocolumns was carried out. The samples have a fairly smooth surface morphology; the total GaN height is about 550 nm.

Postgrowth technology has been developed to form the micromechanical structures of the GaN velocity acoustics sensor. Gold platforms chrome adhesive layer were applied to the GaN layer by magnetron sputtering. This stage was followed by the application of a SiO2 mask and the formation of a lithographic patternThe topology of GaN beam heaters was formed using methods of anisotropic dry etching of a GaN layer on an Oxford Plasma Lab ICP 380 setup with a Cl_2 / BCl_3 based reactive plasma-chemical etching mode [7]. The release of micromechanical structures was carried out by isotropic etching of silicon to a depth of 50 μm in SF_6 plasma. The last step was dry removal of the silica mask. The length of the formed beam heaters was 500 μm, 0.5 μm thickness and 10 μm width.

The configuration of the prototypes produced was examined using a SEM Supra 25 Zeiss. The electrical characteristics were measured using a half-bridge circuit using the AKIP 4109/2 data acquisition system. Temperature measurements were carried out in the temperature range of 20 - 140 ˚C by heating the sensor chip with a Heraeus Pt6.8 M1020 microheater and controlling the temperature with a Pt-1000 microthermometer. The limiting values of the gas flow were measured by feeding a narrow flow from a micro-blower through a needle with a diameter of 1.2 mm.

4. Results and discussions

SEM of the produced prototypes is shown in Figure 4. After the release of the beam elements, an arcuate deformation was observed. The lifting height of the beams was 14 μm, which advises the design value.

Figure 4. SEM image of a prototype acoustic velocity sensor.

The measured resistance under normal conditions was 60 kΩ. In this case, the absence of electrical insulation of the beam elements from the silicon substrate was observed. The silicon resistance was measured between contacts without beam elements and was 300 kΩ. Based on the obtained temperature dependences of the resistance of the sensor with and without beam elements, the measurement of the resistance of an individual beam was selected. The results of measuring the temperature dependence of the resistance of the beam are shown in Figure 5.

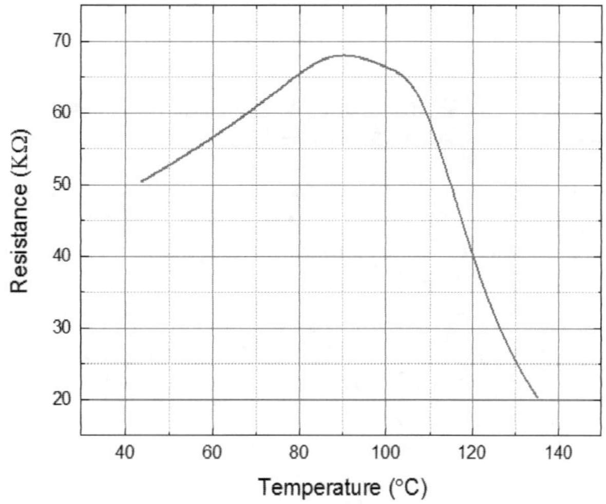

Figure 5. Temperature dependence of the resistance of the beam element.

5. Conclusion

The results of finite elemental analysis showed the fundamental possibility of creating an acoustic velocity vector sensor based on epitaxial GaN layers. The simulation of the characteristics of the obtained acoustic velocity sensors was carried out on the basis of the measured electrical characteristics, where the sensitivity and the directional pattern were determined. The developed technology of epitaxial layer growth and post-process has been successfully demonstrated. It should be noted that the problem of the lack of isolation of the working layer from the substrate can be solved using SOI plates.

Acknowledgments

This work was carried out in Alferov University and was supported by the Ministry of Education and Science of the Russian Federation (grant № FSRM-2020-0008).

References

[1] Guiot M, Comesaña D F and Pousa G C 2015 *Proc. Inter-Noise* 1-8
[2] Bree H-E, Leussink P, Korthorst T, Jansen H, Lammerink T S J and Elwenspoek M 1996 *Sensors and Actuators* **54** 552
[3] Rais-Zadeh M, Gokhale V J, Ansari A, Faucher M, Cordier Y and Buchaillot L 2014 *J. Microelectromech. Syst.* **23** 1252-71
[4] Enns Y, Kazakin A, Mizerov A, Shubina K, Morozov I, Mokhov D and A Bouravleuv 2020 *J. Phys.: Conf. Ser.* **1695**(1) 012019
[5] Kazakin A, Enns Y, Mizerov A, Kleimanov R and Bouravleuv A 2019 *J. Phys.: Conf. Ser.* **1410** 012214
[6] Timoshnev S N, Mizerov A M, Sobolev M S and Nikitina E V 2018 *J. Semiconductors* **52** 660
[7] Shubina K, Berezovskaya T N, Mokhov D V and Morozov I 2018 *J. Semiconductors* **52** 2117

Graphene two terminal detector as THz mixer

A Sh Gazaliev[1,2], M V Moskotin[3], V V Belosevich[1,4], M G Rybin[5], I A Gayduchenko[1,7] and G N Goltsman[1,6]

[1] Moscow Pedagogical State University, Moscow, 119435, Russia
[2] NRC «Kurchatov Institute», Moscow, 123182, Russia
[3] Moscow Institute of Physics and Technology (National Research University), Dolgoprudny, 141700, Russia
[4] National Research University Higher School of Economics, Moscow, 101000, Russia
[5] Prokhorov General Physics Institute, RAS, Moscow, 119991, Russia
[6] NTI Center for Quantum Communications, National University of Science and Technology MISiS, Moscow 119049, Russia
[7] Author to whom any correspondence should be addressed

igorandg@gmail.com

Abstract. The growing requirements for mobile communication networks (data transfer rates over 100 Gbps) makes it necessary to use carrier signal with a frequency of at least 100 GHz. This requires the development of cheap and broadband sub-terahertz (sub-THz) detectors. Here we report on our recent efforts toward the development of a heterodyne sub-THz detector based on a single layer graphene two-terminal device integrated with a bowtie antenna on a sapphire substrate. Our detector operates at frequency of 140 GHz, which corresponds to the maximum transmission of THz radiation in the Earth's atmosphere. The heterodyne detection is achieved by quasi-optical coupling of signals from two sub-THz radiation sources to the same detector. The measured frequency bandwidth is 5.8 GHz.

1. Introduction

The unprecedented increase in wireless data traffic makes it necessary to expand the carrier frequency range in order to increase the bandwidth of communication channels. One of the promising approaches for creating high-performance (over 100 Gb / s) wireless data transmission channels is the use of the terahertz frequency range (0.1-10 THz) [1]. For the practical implementation of such systems, it is necessary to develop an element base based on fast, cheap and energy-efficient THz detectors and photo-mixers.

Due to the gapless band structure and high mobility, graphene is considered a promising material for creating ultrafast photodetectors in a wide spectral range: from visible to terahertz [2]. A weak electron-phonon coupling together with a small specific heat capacity of graphene leads to strong heating of its electronic subsystem under the incident electromagnetic radiation and, as a result, to a thermo-emf signal [3]. The second effect which leads to the appearance of photovoltage is the rectification of THz radiation at the graphene-metal interface (photovoltaic effect) [4]. The photo-thermoelectric effect in graphene, as well as the photovoltaic effect, potentially leads to achieving high volt-watt sensitivity and speeding up to picoseconds even at room temperature [2]. In this work, we demonstrate that an asymmetric two-terminal graphene device can act as an efficient mixer of sub-

THz radiation at a practically important frequency of 140 GHz, which corresponds to the maximum transmission of THz radiation in the Earth's atmosphere.

2. Measurements

Our devices are made in a two-terminal configuration, in which graphene synthesized by chemical vapor deposition is in contact with two electrodes made of metals with different work functions [4] (Figure 1). This geometry leads to non-uniform distribution of charge carrier concentration in the graphene channel due to contact doping. In such systems, photovoltage response occurs due to combination of photo-thermoelectric [5] and photovoltaic effects [4].

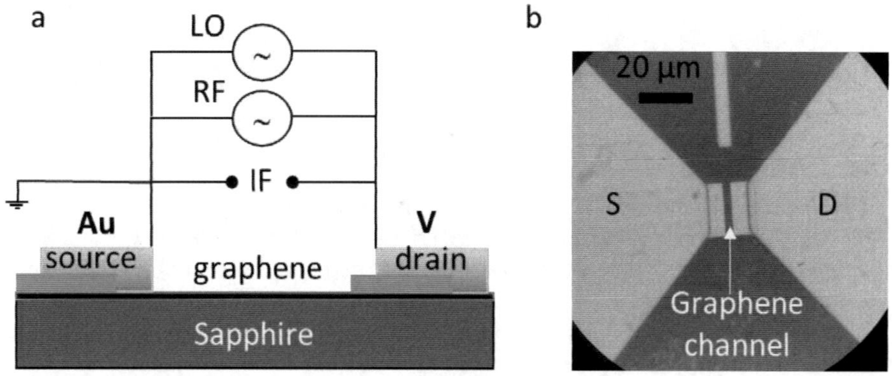

Figure 1. Schematic representation of the experimental device: (a) - electric circuit diagram of the graphene detector, (b) – optical image of the graphene detector integrated with a bow-tie antenna.

The contact electrodes were made in the form of a bow-tie antenna to couple THz radiation to the sensing element using standard e-beam lithography technique (Figure 1). To reduce the parasitic capacitances limiting the performance of the device, we used sapphire substrates. The fabricated device was located in the center of the flat surface of hemispherical Si lens. The source and drain electrodes of the detector were connected to coplanar strip line to form intermediate frequency (IF) transmission line [6].

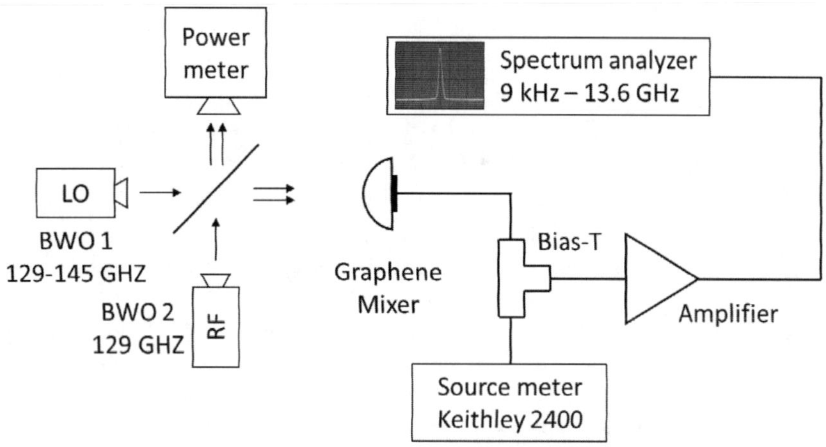

Figure 2. Schematic representation of the experimental setup for measuring the bandwidth of detectors.

The measurement technique used in this study is based on the mixing of both local oscillator (LO) radiation and radio frequency (RF) radiation and described in ref. [6] (see figure 2). The setup includes two backward-wave oscillators (BWO), one as a fixed signal source. The second BWO was used as a tunable LO source which corresponds to an IF up to 13 GHz. The power of both BWOs was controlled by an RF thermistor power meter. Two signals from the specified sources are coupled to the Si lens on which graphene devices are located. The IF signal is amplified with a wideband amplifier that is connected to an FSV 9 kHz – 13.5 GHz Rohde & Schwarz spectrum analyzer. To provide photoresponse measurements under different bias conditions we connect the source meter through a bias tee.

3. Results

The heterodyne measurements for the IF band ranging from 0.1 – 13 GHz are presented in Figure 3. The measured IF gain data was fitted with function [7]:

$$G = G_0 - 20\log(1 + \omega^2\tau^2) \tag{1}$$

where G_0 is IF gain at low frequency, ω is the angular frequency and τ is the response time. The -3 dB bandwidth of the graphene-based mixer is 5.8 GHz that might be limited by bonding wires. The measurements under zero bias and constant bias demonstrate no significant changes in photoresponse value illustrating the non-bolometric origin of the photoresponse. To clarify the detection mechanism we performed the same measurements at temperature of 77 K. Our experimental data show no significant change in the responsivity and operation speed of the detectors. These results are in a good agreement with photo-thermoelectric model for highly doped graphene devices [8].

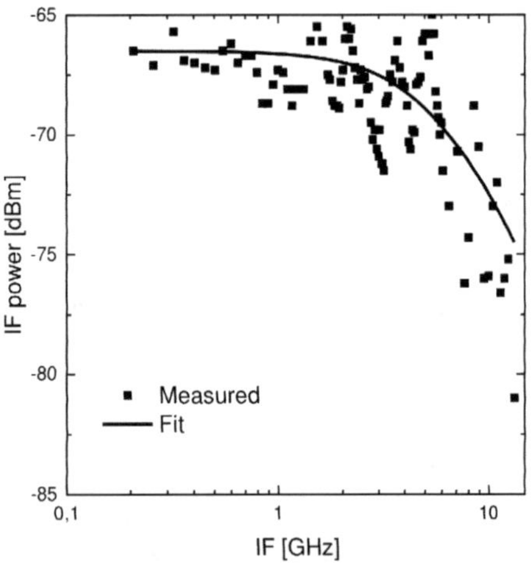

Figure 3. IF bandwidth of the graphene based heterodyne detector. Black dots are experimental data, black line is the result of fitting using formula (1).

4. Conclusion

We have demonstrated that a two-terminal device based on graphene can be used as a THz radiation mixer with a 5.8 GHz bandwidth. Our previous work on the direct detection of THz radiation with graphene detectors showed that the equivalent noise power of such detectors does not exceed 1 $nW/Hz^{0.5}$. These characteristics are close to the best commercial Schottky diodes, but still inferior to them. We also note that the characteristics of our detectors are determined not by internal processes in graphene (which occur at much shorter times of the order of ps), but by the external connections and parasitic capacitances. Further research on optimization and identification of physical mechanisms limiting the characteristics of graphene THz detectors may lead to the creation of fast mixers for next-generation wireless networks.

Acknowledgments
The reported study was funded by RFBR and DFG, project number 21-52-12041(DFG project number 449506295).

References

[1] Elayan H, Amin O, Shubair R M and Alouini M S 2018 *2018 International Conference on Advanced Communication Technologies and Networking (CommNet)* 1-5
[2] Koppens F, Mueller T, Avouris P, Ferrari A, Vitiello M and Polini M 2014 *Nature nanotechnology* **9(10)** 780-793
[3] Sun D, Aivazian G, Jones A, Ross J, Yao W, Cobden D and Xu X 2012 *Nature nanotechnology* **7(2)** 114-118
[4] Gayduchenko I et al 2018 *Nanotechnology* **29(24)** 245204
[5] Cai X et al 2014 *Nature nanotechnology* **9(10)** 814
[6] Krause S, Mityashkin V, Antipov S, Gol'tsman G, Meledin D, Desmaris V, Belitsky V and Rudziński M 2016 *IEEE Transactions on Terahertz Science and Technology* **7(1)** 53-59
[7] Generalov A, Andersson M, Yang X, Vorobiev A and Stake J 2017 *42nd International Conference on Infrared, Millimeter, and Terahertz Waves (IRMMW-THz)* 1-2
[8] Bandurin D A, Gayduchenko I, Cao Y, Moskotin M, Principi A, Grigorieva I V, Goltsman G, Fedorov G and Svintsov D 2018 *Applied Physics Letters* **112(14)** 141101

Features of the formation of the frequency of the microwave excitation signal in the quantum frequency standard on rubidium atoms - 87

A S Grevtseva[1], R A Dmitriev[2], V V Davydov[1, 3] and V Yu Rud[3, 4]

[1] Peter the Great Saint Petersburg Polytechnical University, Russia, St. Petersburg, 195251
[2] Russian Institute of Radionavigation and Time, Russian, Saint Petersburg, 192012
[3] All-Russian Research Institute of Phytopathology, Moscow Region 143050, Russia
[4] A.F. Ioffe Physicotechnical Institute, St. Petersburg, 195152, Russia

annagrevtseva@mail.ru

Abstract. The article discusses the main disadvantages of the current design of a quantum frequency standard based on rubidium-87 atoms. The main disadvantages of the current design of the quantum frequency standard on rubidium-87 atoms are considered. It is noted that the processes associated with light shifts in the optical part contribute to the greatest instability in the long-term operation of the quantum frequency standard. A solution is proposed to improve the design of the rubidium standard. A forecast for improving its metrological characteristics is presented. The results of experimental investigations are presented.

1. Introduction

In modern navigation systems, without determining the exact time and frequency, it is impossible to achieve the required accuracy in determining the coordinates [1-10]. Data on the coordinates of the object is necessary for solving problems of environmental monitoring, carrying out various works, etc. [10-19]. The operating satellite navigation constellations (Russian GLONASS, European GALILLEO, American (USA) GPS and Chinese BSD) actively use quantum frequency standards (QFS) to determine the exact time [1, 2, 8, 20-24]. Among quantum standards in satellite communication systems, rubidium QFSs are most widely used due to their small size and low cost in comparison with other types of standards [25-32].

At present, with the development of electronic equipment, the requirements for the accuracy of satellite navigation systems are constantly increasing, which makes the task of their modernization especially urgent [1, 2, 6, 7, 20-22, 31, 33]. Especially interesting are the areas of modernization using laser radiation, new models of photodetectors and additional feedback systems [34-39]. The modernization of frequency standards is no exception, this process includes changes in the dimensions and weight of the structure, as well as the improvement of metrological characteristics. It should be noted that for the QFS, modernization can be carried out not for its entire structure, but only for individual units [4, 8, 17, 25-29]. This paper discusses one of the possible solutions for modernizing the design of a quantum frequency standard based on rubidium-87 atoms.

2. Features of the formation of the microwave excitation signal in the design of the rubidium frequency standard

To date, domestic and foreign manufacturers successfully represent different models of QFS based on rubidium-87 atoms, however, the basic principles of operation remain unchanged [25-31]. Figure 1 shows a diagram of the current rubidium frequency standard (RFC).

The principle of the QFS operation is based on automatic tuning of the crystal oscillator (CO) frequency to the value of the quantum transition frequency in optically oriented Rb-87 atoms. For the optical orientation of rubidium - 87 atoms in satellite systems, a special pump lamp with a filter cell is used [25-31].

To implement the tuning of the CO frequency, the working cell of the atomic discriminator (DA) is irradiated with a microwave signal, the frequency of which corresponds to the frequency of the quantum transition of excited rubidium-87 atoms. In the case of a deviation of the frequency of the microwave signal from the value of the frequency of the resonant transition, an error signal (ES) is generated, according to which the CO is adjusted. Therefore, one of the important points in the functioning of the RFC is the formation of a microwave signal.

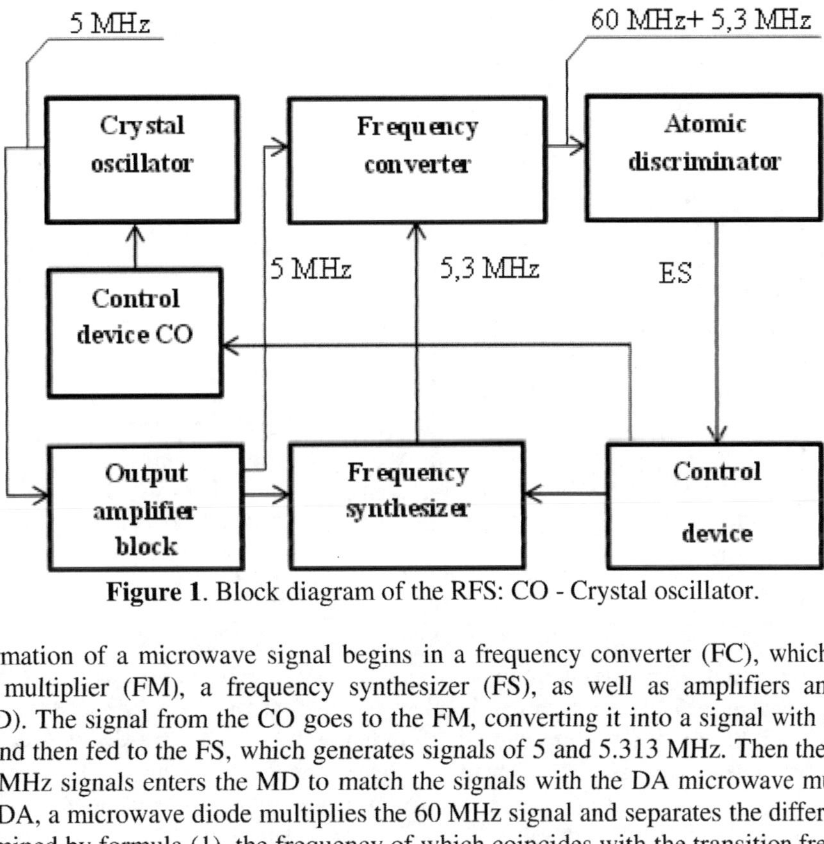

Figure 1. Block diagram of the RFS: CO - Crystal oscillator.

The formation of a microwave signal begins in a frequency converter (FC), which consists of a frequency multiplier (FM), a frequency synthesizer (FS), as well as amplifiers and a matching device (MD). The signal from the CO goes to the FM, converting it into a signal with a frequency of 60 MHz, and then fed to the FS, which generates signals of 5 and 5.313 MHz. Then the mixture of 60 and 5.313 MHz signals enters the MD to match the signals with the DA microwave multiplier diode. Inside the DA, a microwave diode multiplies the 60 MHz signal and separates the difference harmonic f_{dif}, determined by formula (1), the frequency of which coincides with the transition frequency of Rb-87 atoms.

$$f_{dif} = 60 \cdot 114 - 5{,}313 = 6834{,}7 \text{ MHz}, \quad (1)$$

The method of forming a microwave signal, discussed in detail above, has the following disadvantages. The spectrum of the output signal with a frequency of 5.313 MHz contains side amplitude components. The presence of lateral components can lead to an error in establishing the value of the frequency of the output signal of the rubidium standard. Another disadvantage is that the final formation of the microwave signal occurs inside the atomic discriminator. The coincidence of the frequencies of the optical quantum transition and the microwave signal can be judged only by indirect signs, namely, by the magnitude of the error signal, which is formed by the difference in the intensities of the recorded optical signals on the photodetector. Moreover, for different designs of photodetectors [34-36], there may be a small variation in values. For this reason, it was decided to modernize one of the RFS units.

3. Modernization a optical part to design of the rubidium frequency standard

In the new design of the RFC, the microwave signal with a frequency of 6834.7 MHz is proposed to be synthesized using a system of two ring phase-locked loop (PLL). The synthesis scheme is shown in Figure 2. The main elements of the PLL system are a phase detector (PD), one of the inputs of which is fed with a signal from a voltage-controlled oscillator (VCO). Another PD input is connected to a reference signal source with a frequency f_{ref}. A phase detector compares the signals at both inputs and generates an error signal, which, after filtering and amplifying (if necessary), adjusts the VCO frequency. As part of the PLL system, a low-pass filter (LPF) is also used, which is connected between the output of the PD and the input of the VCO and determines its frequency properties.

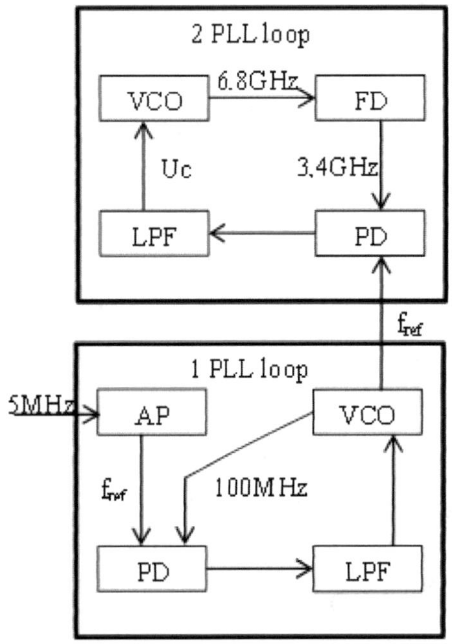

Figure 2. Block diagram of two PLL loop systems: VCO – voltage controlled generator; FD – frequency divider; PD – phase detector; LPF – low-pass filter; AP – analog part; F – filter.

The microwave signal is synthesized in two stages. The first loop of the PLL adjusts the VCO with an output frequency of 100 MHz. The main element of the first ring is a synthesizer microcircuit, which contains the following units: a prescaler, a main divider, a reference frequency divider, an input frequency amplifier, and also the main element - a phase-frequency detector. The principle of PD operation is shown in Figure (3). If the slave generator lags behind the reference, then the Down signal is generated in accordance with Figure (3.a), if the slave generator is ahead of the reference, then the Up signal is generated, in accordance with Figure (3.b), if the signals of the slave and reference oscillators coincide in frequency and phase, the Up Down signals are generated as shown in Figure (3.c).

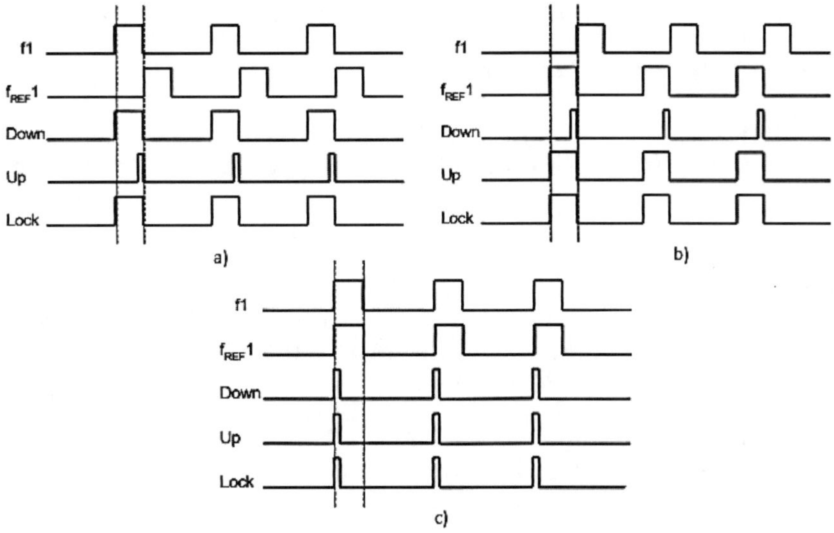

Figure 3. Time diagrams of signals. a) The signals of the slave generator are lagging behind the signals of the reference generator, b) The signals of the slave generator are ahead of the signals of the reference generator, c) The signals of the slave and the reference generator are the same.

To obtain the required output frequency according to formula (2), the division factors of the reference and input frequencies are selected.

$$f_{in} = \frac{f_{ref} * K}{K_{ref}}, \quad (2)$$

where $f_{ref} = 5$ MHz – is the frequency of the reference signal, K is the division factor of the input frequency, K_{ref} – is the division factor of the reference frequency, f_{out} – is the output frequency of thc VCO.

After the formation of the required frequency of 100 MHz at the output of the first ring, the signal enters the second PLL ring, where it is used as a reference. In the second PLL, the VCO with an output frequency of 6.8 GHz is adjusted in a similar manner. At the same time, the implementation of the VCO with a PLL in the form of one compact module was chosen, which will reduce the weight and size characteristics.

To create an accurate value for the frequency of the quantum transition at the final stage, it is necessary to add a fractional component equal to 34.7 MHz to the output signal. Fractional part of 34.7 MHz is created using a special synthesizer. Further, the required signal of 6.8347 GHz is immediately fed to the atomic discriminator input, the microwave diode is excluded from the circuit.

The developed design of the RFS has several significant advantages. First, the use of an indirect synthesis method, namely a phase-locked loop system, allows a cleaner spectrum of the output signal to be obtained. Secondly, the formation of the microwave signal at all stages of the new circuit is controlled; it is possible to precisely adjust the signal frequency to the frequency of the quantum transition of rubidium-87 atoms. In the previous scheme, this was not realized, since the final formation of the microwave signal took place inside the atomic discriminator and the presence of resonance was judged by indirect signs.

4. Conclusion

The developed design of the QFS unit has a number of main advantages: first, a cleaner spectrum of the output optical signal arriving at the photodetector. Second, the possibility of direct control of the microwave signal, and, consequently, its precise tuning to the resonant frequency of rubidium atoms. This makes it possible to reduce the error in establishing the actual value of the frequency of the output signal of the rubidium QFS, which improves the stability of the device.

References

[1] Batori E, Almat N, Affolderbach C and Mileti G 2020 *GNSS-grade space atomic frequency standards: Current status and ongoing developments.* (New York: Advances in Space Research)
[2] Semenov V V , Nikiforov N F and Ermak S V 1991 *Soviet journal of Communications Technology and Electronics* **36(4)** 59 – 63
[3] Petrov A A, Grebenikova N M, Lukashev N A, Ivanova N V, Rodygina N S and Moroz A V 2018 *Journal of Physics: Conference Series* **1038 (1)** 012032
[4] Lukashev N A, Petrov A A, Grebenikova N M and Valov A P 2018 *Proceedings of 18th International conference of Laser Optics ICLO-2018 (Saint-Petersburg),* vol. 8435889 (IEEE), p. 271
[5] Lukashev N A, Glinushkin A P and Lukyantsev V S 2019 *Journal of Physics: Conference Series* **1410(1)** 012211
[6] Lukashev N A, Davydov R V and Glinushkin A P 2019 *Journal of Physics: Conference Series* **1326(1)** 012046
[7] Davydov R, Antonov V and Moroz A 2019 *International Conference on Electrical Engineering and Photonics EExPolytech-2019 (Saint-Petersburg)*, vol. 8906791 (IEEE), p. 295-297
[8] Davydov R, Antonov V, Makeev S, Batov Y, Dudkin V and Myazin N 2020 *E3S Web Conference* **140** 02001
[9] Fadeenko V, Fadeenko I, Dudkin V and Nikolaev D 2019 *IOP Conference Series: Earth and Environmental Science* **390(1)** 012022
[10] Fadeenko V V, Kuts V A and Vasiliev D A 2018 *Journal Physics: Conference Series* **1135(1)** 012053
[11] Gryznova E, Batov Y and Myazin N 2020 *E3S Web of Conferences* **140** 09001
[12] Gryznova E, Grebenikova N, Ivanov D and Bykov V 2019 *IOP Conference Series: Earth and Environmental Science* **390(1)** 012044
[13] Nikitina M, Grebenikova N, Dudkin V and Batov Y 2019 *IOP Conference Series: Earth and Environmental Science* **390(1)** 012024
[14] Yushkova V, Kostin G, Davydov R, Dudkin V and Valiullin L 2019 *IOP Conference Series: Earth and Environmental Science* **390(1)** 012016
[15] Van S, Cheremisin A, Chusov A, Switala F and Davydov R 2019 *IOP Conference Series: Earth and Environmental Science* **390(1)** 012011
[16] Van S, Cheremisin A, Glinushkin A, Davydov R and Yushkova V 2019 *E3S Web of Conferences* **140** 09008
[17] Myazin N S, Yushkova V V and Taranda N I 2019 *Journal of Physics: Conference Series* **1410(1)** 012130
[18] Rukin E V , Myazin N S and Dudkin V I 2019 *Journal of Physics: Conference Series* **1368(4)** 042011
[19] Myazin N S, Yushkova V V and Dudkin V I 2019 *Journal of Physics: Conference Series* **1400(6)** 066008
[20] Petrov A A, Shabanov V E, Zalyotov D V, Bulyanitsa A L and Shapovalov D V 2018 *International Conference on Electrical Engineering and Photonics EExPolytech 2018 (Saint-Petersburg)*, vol. 8564389 (IEEE), p. 52-55
[21] Petrov A A, Vologdin V A and Zalyotov D V 2015 *Journal of Physics: Conference Series* **643(1)** 012087
[22] Petrov A A and Grebennikova N M 2018 *Journal of Communications Technology and*

Electronics **63(11**) 1281–1285
[23] Lukashev N A and Moroz A V 2019 *Journal of Physics: Conference Series* **1236(1)** 012068
[24] Petrov A A and Grebenikova N M 2018 *Lecture Notes in Computer Science (including subseries Lecture Notes in Artificial Intelligence and Lecture Notes in Bioinformatics)* **11118 LNCS** 641-648
[25] Grevtseva A, Davydov V and Rud V 2020 *CEUR Workshop Proceedings* **2667** 15–18
[26] Valov A P, Davydov R V, Rud V Y and Grevtseva A S 2019 *Journal of Physics: Conference Series* **1326(1)** 012040
[27] Grevtseva A S, Davydov R V, Dudkin V I and Rud' V Y 2019 *Journal of Physics: Conference Series* **1326(1)** 012043
[28] Petrov A A and Myazin N S 2017 *Lecture Notes in Computer Science (including subseries Lecture Notes in Artificial Intelligence and Lecture Notes in Bioinformatics)* **10531 LNCS** 561-568
[29] Valov A P 2019 *Journal of Physics: Conference Series* **1410(1)** 012246
[30] Valov A 2020 *CEUR Workshop Proceedings* **2667** 102–105
[31] Valov A P, Nagornaya A I, Kulikova M E and Glinushkin A P 2020 Journal of Physics: Conference Series **1695(1)** 012169
[32] Petrov A A, Zaletov D V, Davydov V V and Shapovalov D V 2021 *Journal of Communications Technology and Electronics* **66(3)** 295–299
[33] Petrov A A, Zalyotov D V, Shabanov V E and Shapovalov D V 2018 *Journal Physics: Conference Series* **1124(1)** 041004
[34] Smirnov K J, Davydov V V and Batov Y V 2019 *Journal of Physics: Conference Series* **1368 (2)** 022073
[35] Smirnov K J, Glagolev S F, Rodygina N S and Ivanova N V 2018 *Journal of Physics: Conference Series* **1038(1)** 012102
[36] Smirnov K J, Glagolev S F and Tushavin G V 2018 *Journal of Physics: Conference Series* **1124(1)** 022014
[37] Davydov R V and Antonov V I 2018 *Journal of Physics: Conference Series* **1124(8)** 081037
[38] Davydov R V and Antonov V I 2018 *Journal of Physics: Conference Series* **1135(1)** 012087
[39] Logunov S E, Rud V Y, Davydov R V, Moroz A V and Smirnov K J 2019 *Journal of Physics: Conference Series* **1326(1)** 012024

Piezoelectric response of aligned carbon nanotubes depending on sublayer material

M V Il'ina[1], O I Il'in[1], O I Osotova[1], N N Rudyk[1] and O A Ageev[2]

[1] Southern Federal University, Institute of Nanotechnologies, Electronics and Electronic Equipment Engineering, Taganrog, 347922, Russia
[2] Southern Federal University, REC "Nanotechnologies", 347922, Taganrog, Russia

mailina@sfedu.ru

Abstract. The results of experimental studies of the effect of the sublayer material on the piezoelectric response and sensitivity to mechanical deformations of aligned carbon nanotubes (CNTs) are presented. It is shown that the highest piezoelectric response (136 nA at a pressing force of 4 μN) and best sensitivity are demonstrated by CNTs grown on a Mo sublayer. This dependence is probably due to the geometric parameters of CNTs and the structure of the CNT array as a whole. The results obtained can be used to develop energy-efficient nanogenerators based on CNT arrays.

1. Introduction

Currently, there is an active development of wearable electronics, which requires the search for suitable energy sources capable of ensuring their autonomous operation. An attractive direction in this area is the ability to use mechanical energy of the environment (ambient noise, vibration, human body activity, etc.) as power sources. One of the devices capable of converting nanoscale deformations and vibrations into electrical energy is a nanogenerator [1, 2]. In this connection, there is an active search for functional nanostructures capable of providing the highest sensitivity to external nanoscale deformations and maximum power [3].

Recent studies show that carbon nanostructures are capable of exhibiting anomalous piezoelectric properties as a result of violation of their centrosymmetric structure through the introduction of defects or the formation of non-uniform deformations [4-7]. Particular attention is attracted by vertically aligned carbon nanotubes (CNTs) due to the ability to transform nanoscale deformations into a surface potential, which leads to the generation of an electric current [7-9]. This makes vertically aligned CNTs one of the most attractive materials for creating energy efficient nanogenerators [10]. However, work in this direction is at an early stage and requires wide research. The aim of this work is to study the effect of the sublayer material of the CNTs on their piezoelectric response and sensitivity to mechanical deformations.

2. Experimental studies

To establish this dependence, a series of 4 arrays of vertically aligned CNTs was grown by the plasma-chemical vapor deposition method [11]. As sublayer materials were used Ti, Mo, TiN and Cr. The growth temperature was 645 °C. The thickness of the catalytic nickel layer was 5 nm. Scanning electron microscopy (SEM) images are shown in Figure 1. Experimental studies of the piezoelectric response of CNT arrays were performed by atomic force microscopy (AFM) using a previously developed technique [8, 12]. As a piezoelectric response, we detected the current arising as a result of the formation of the surface potential of CNTs during their deformation upon pressing the AFM probe with a force of 4 μN. The top electrode was a commercial NSG10/Pt probe with platinum coating and with force constant 8 N/m. It should be noted that the AFM ammeter operates linearly in the range from 10 to -10 nA, and for large values, the logarithmic transformation is turned on. The values of the piezoelectric response of CNTs are recalculated taking into account this transformation.

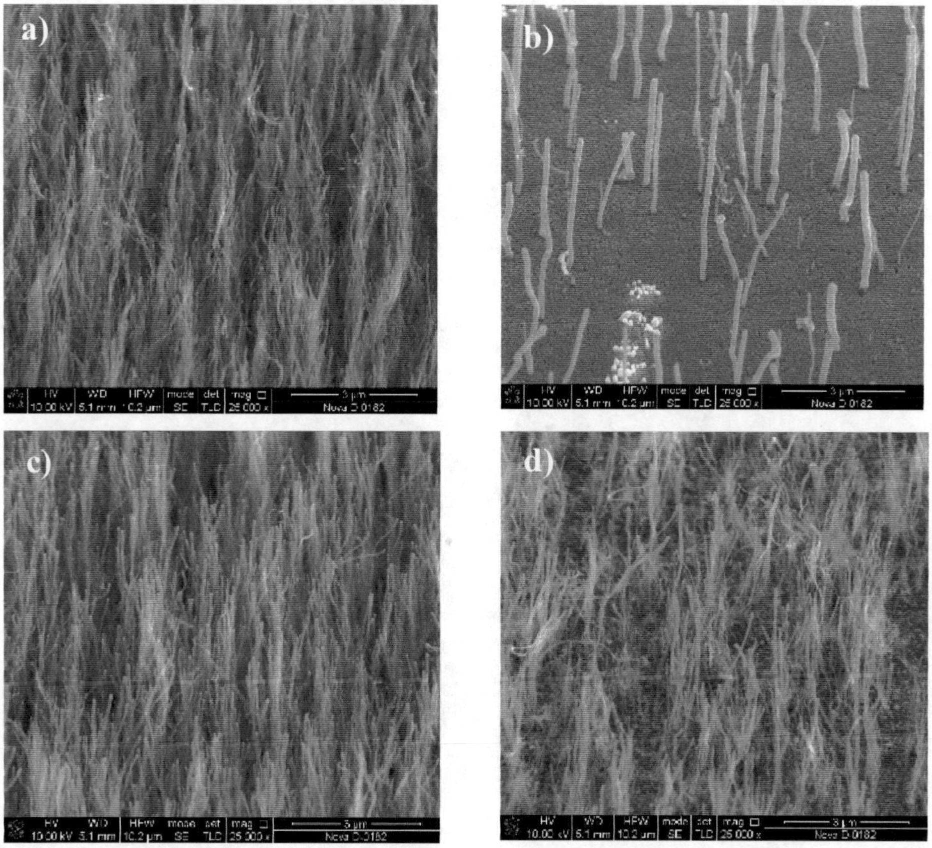

Figure 1. SEM images of the CNT arrays grown on a Ti (a), Mo (b), TiN (c) and Cr (d) sublayer materials.

3. Results and discussion

Analysis of the geometric parameters of CNTs showed that the sublayer material affects the length, diameter and density of CNTs in the array (Figure 1). The average heights of CNTs ranged from 6.6 to 12.9 μm. The average diameters of CNT ranged from 35 to 72 nm. On the Mo sublayer, the growth of separately standing vertically aligned CNTs with a density of 1 μm^{-2} was observed. The growth of vertically aligned CNTs with a high density in the array (26 and 22 μm^{-2}) was observed on the TiN and Ti sublayer respectively. On the other hand, on the Cr sublayer CNTs grew combined into bundles, with a density of 10 μm^{-2}.

Results of the AFM studies have shown that the geometric parameters of CNTs, in turn, have a significant effect on their piezoelectric response of CNTs. So, the value of the piezoelectric response increased from 2 to 39 nA with an increase in the pressing force from 0 to 4 μN for the CNT array grown on a Ti sublayer. For the CNT array grown on a Mo electrode, the piezoelectric response increased from

40 to 136 nA under the same influences. For the CNT array grown on a TiN sublayer, on the contrary, an insignificant decrease in the piezoelectric response from 18 to 13 nA was observed. The CNT array grown on a Cr sublayer showed the smallest piezoelectric response (10 ± 2 nA), which weakly depends on the pressing force. These dependences are probably due to the influence of the geometric parameters of CNTs and the structure of the CNT array as a whole. Thus, a decrease in the piezoelectric response of the CNT array grown on a TiN sublayer is due to a high aspect ratio of the CNT length to diameter (~360), which led to large deviations of the CNT from its axis and the formation of bundles during the measurement. The formation of CNT bundles, in turn, caused a decrease in the total piezoelectric potential due to the compensation of positive and negative charges with each other. A similar situation was observed for the CNT array grown on a Cr sublayer, which was initially combined into CNT bundles (Figure 1d). A gradual increase in the piezoelectric response for CNT arrays grown on Ti and W sublayers is due to an increase in the CNT deformation. In this case, a greater value of the response for CNTs on a Mo sublayer is associated with a lower value of the aspect ratio (~92) compared to CNTs on a Ti sublayer (~223), which led to a greater relative deformation of the CNTs at a given pressing force. It should be noted that the work function of electrons from Ti, Cr, and Mo is similar (4.5 - 4.6 eV) and could not have a decisive influence on the obtained results.

Further, studies of the influence of the sublayer material on the "sensitivity" (the ratio of the piezoelectric response to the amount of deformation) to external mechanical influences were carried out (Figure 2). For this, the AFM probe was gradually shifted into the CNT array to a depth of 500 nm, which led to the formation of 3.5, 7.5, 1.5, and 7% deformation of CNTs grown on Ti, Mo, TiN, and Cr sublayers, respectively.

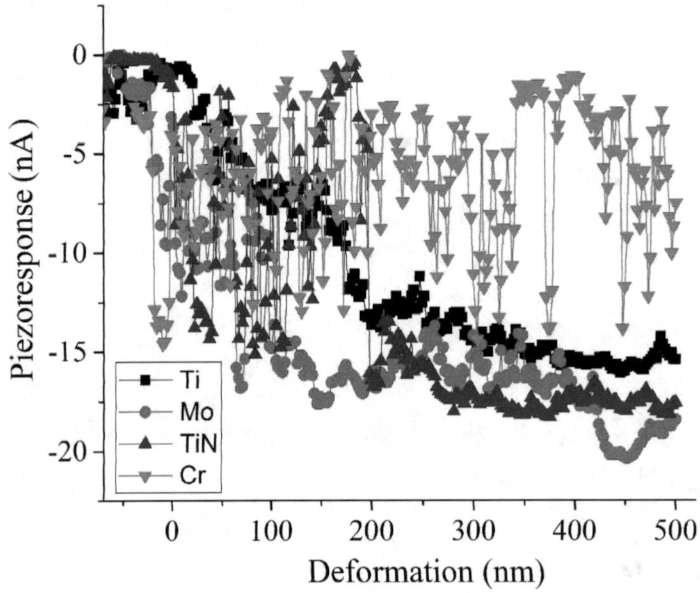

Figure 2. Dependence of the piezoelectric response of CNTs grown on various sublayers on deformation. The graph shows the values obtained directly from an AFM ammeter with a built-in logarithmic converter for currents greater than ± 10 nA.

As a result, it was found that nanotubes grown on a Cr sublayer are not very sensitive to deformations up to 7%. Thus, with an increase in deformation from 0 to 500 nm, the generated current signal was quite noisy and amounted to -7 ± 5 nA in the entire deformation range (Figure 2). This dependence is associated with the fact that nanotubes were combined into bundles during growth, which significantly reduced their "sensitivity" to external mechanical influences. For CNTs grown on a TiN sublayer, at a deformation of less than 200 nm (0.6%), a noisy current signal of -8 ± 5 nA was also observed. With a further increase in deformation, a stable current signal was observed, increasing from -23.3 to -49.7 nA with an increase in deformation from 200 to 500 nm (from 0.6 to 1.5%) (Figure 2). For CNTs grown on a Ti sublayer, an increase in current from 0 to -27.35 nA was observed in the entire range of deformations from 0 to 3.5% (Figure 2). With a deformation of 1.5%, the current was -15.34 nA. The highest current was detected for CNTs grown on the Mo sublayer, both in the region of small and large deformations (Figure 2). The magnitude of the current increased from 0 to -89.2 nA with deformation from 0 to 7.5%, respectively. In this case, at a deformation of 1.5%, the current reached -49.7 nA, similar to the value for CNTs grown on a TiN sublayer. However, at a deformation of less than 1.5%, nanotubes grown on the Mo sublayer, in contrast to those grown on the TiN sublayer, also demonstrated a stable current signal, which increased with increasing deformation.

Analysis of the results obtained allows us to conclude that the "sensitivity" of CNTs to external mechanical influences is determined by their aspect ratio of length to diameter and vertical orientation in the array. Thus, in the region of small deformations (up to 1.5%), the highest "sensitivity" is demonstrated by not bundled nanotubes with an aspect ratio of no more than 200. In the region of large deformations (more than 1.5%), the magnitude of the piezoelectric response increased with an increase in the aspect ratio. At the same time, the tendency of CNTs to combine into bundles significantly reduces their "sensitivity" to external mechanical influences.

4. Conclusion

Thus, it has been shown that vertically aligned carbon nanotubes grown on the Mo sublayer exhibit the maximum piezoelectric response with the best «sensitivity». However, further research is required to investigation of the effect of the sublayer material on the structure of CNTs and their piezoelectric response. The results obtained can be used to develop energy efficient nanogenerators based on vertically aligned CNTs. The results were obtained using the equipment of the Research and Education Center "Nanotechnologies" of Southern Federal University.

Acknowledgments

This study was financially supported by the Russian Science Foundation (project No. 20-79-00284).

References

[1] Hu Y and Wang Z L 2014 *Nano Energy.* **14** 3–14
[2] Briscoe J and Dunn S 2014 *Nano Energy.* **14** 15–29
[3] Dakua I and Afzulpurkar N 2013 *Nanomater. Nanotechnol.* **3** 1–16
[4] El-Kelany K E, Carbonnière P, Erba A, Sotiropoulos J M and Rérat M P 2016 *J. Phys. Chem. C* **120** (14)
[5] Ong M T and Reed E J 2012 *ACS Nano* **6**(2) 1387–1394
[6] Kundalwal S I, Meguid S A and Weng G J 2017 *Carbon* **117** 462–472
[7] Il'ina M V, Il'in O I, Blinov Yu F, Smirnov V A, Kolomiytsev A S, Fedotov A A, Konoplev B G and Ageev O A 2017 *Carbon* **123** 514–524
[8] Il'ina M V, Il'in O I, Blinov Yu F, Konshin A A, Konoplev B G and Ageev O A 2018 *Materials* **11** 638
[9] Il'ina M V, Il'in O I, Guryanov A V, Osotova O I, Blinov Yu F, Fedotov A A and Ageev O A 2021 *J. Mater. Chem. C* **9** 6014 – 6021
[10] Il'ina M V, Konshin A A and Solomin E G 2018 *IOP Conf. Series: Journal of Physics: Conf. Series* **1124** 071010
[11] Il'in O I, Il'ina M V, Rudyk N N, Fedotov A A and Ageev O A 2019 Vertically Aligned Carbon Nanotubes Production by PECVD, *Perspective of Carbon Nanotubes* eds H El-Din Saleh and S M M El-Sheikh (IntechOpen, UK). DOI: 10.5772/intechopen.84732
[12] Il'ina M V, Il'in O I, Smirnov V A, Blinov Yu F, Konoplev B G and Ageev O A 2018 Scanning Probe Techniques for Characterization of Vertically Aligned Carbon Nanotubes, *Atomic-force Microscopy and Its Applications* eds T Tański, M Staszuk and B Ziębowicz (IntechOpen, UK) DOI: 10.5772/intechopen.78061

Study of characteristics of n-p-n type bipolar power transistor in small-sized metalpolymeric package type SOT-89

D A Knyaginin, E A Kulchenkov, S B Rybalka and A A Demidov

Bryansk State Technical University, Bryansk, 50 let Oktyabrya 7, Russia

Abstract. In this study the input, output and current gain characteristics of silicon n-p-n type medium power bipolar junction transistors KT242A91 made by the "GRUPPA KREMNY EL" in modern small-sized metalpolymeric package type (SOT-89) have been obtained. The SPICE model that allows simulating realistic transistor behaviour of n-p-n type transistor KT242A91 has been proposed. It is shown that established experimental characteristics for KT242A91 transistor correspond to similar transistor's type characteristics.

1. Introduction

Widely known that the bipolar power junction transistors are key component of power semiconductor electronics devices that is primarily used as a switch [1]. It should be noted that at present the modern electronics industry produces a wide range of bipolar transistors of various ratings and in various packages type [2].

Moreover, modern production tends to miniaturize the component base without losing its power characteristics and therefore power electronic industry comes down to use of small type of metalpolymeric package such as SOT (Small Outline Transistor), QFN (Quad Flat No-leads) and others [2]. However, in Russia at this time moment there is no serial production of bipolar junction transistors in small type of metalpolymeric package (SOT-89, SOT-23 etc.). Therefore, recently by electronic company the JSC «GRUPPA KREMNY EL» (Bryansk, Russia) the production of power electronics components (SiC Schottky diodes, bipolar junction transistors etc.) in small type of metalpolymeric packages type began within the framework of import substitution program. For instance, in our previous studies it is established that characteristics of the SiC Schottky type diodes made in small type of metalpolymeric packages [3,4] are comparable with the same similar types.

In presented study the main goal is establish characteristics of silicon n-p-n type medium power bipolar junction transistor in modern small-sized (SOT-89) type of metalpolymeric package made by «GRUPPA KREMNY EL» [5].

2. Materials and methods

In experiments was tested the following silicon medium power n-p-n type bipolar junction transistor KT242A91 (JSC «GRUPPA KREMNY EL», Bryansk, Russia) in small-sized SOT-89 package type [5] which is analogous to a similar transistor BCX56 (Nexperia, Netherlands [6]). The main parameters for KT242A91 n-p-n type bipolar junction transistor were calculated with using of TCAD modelling methods by analogy with SiC Schottky type diodes in small type of metalpolymeric packages that was described in our previous works [7-9]. For measuring the characteristics were used a programmable source AKIP 1144-160-40, Tektronix MDO3102 two-channel oscillograph (bandwidth 1 GHz, refresh rate 5 GS/s) and Fluke 8845A digital multimeter.

3. Results and discussion

Figure 1 displays the photo of n-p-n type bipolar junction transistor in small-sized SOT-89 type of metalpolymeric package KT242A91 that is produced by «GROUP KREMNY EL» [5].

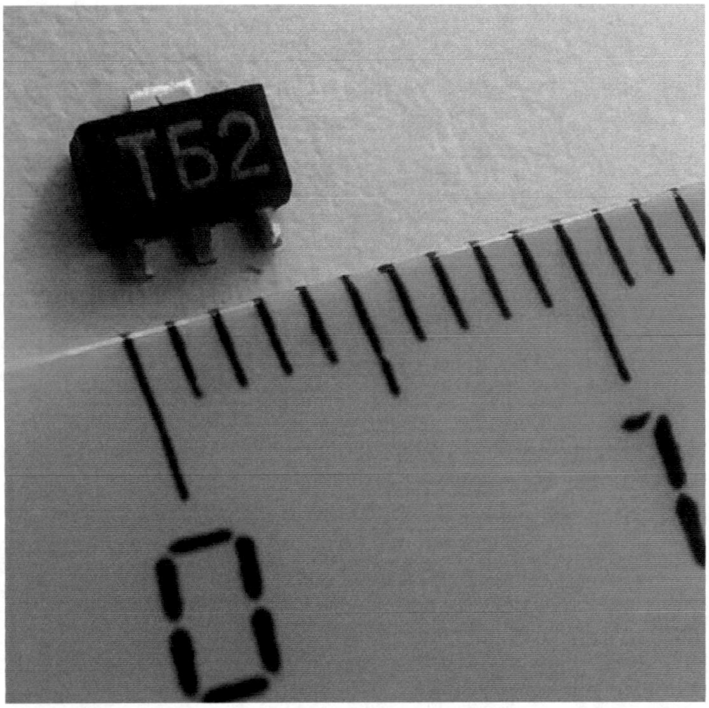

Figure 1. Photo of n-p-n type bipolar junction transistor KT242A91 in small-sized SOT-89 type package [5].

The output characteristic for silicon n-p-n type medium power bipolar junction transistor KT242A91 in small-sized (SOT-89) type package obtained at temperature of 20°C is shown in Figure 2.

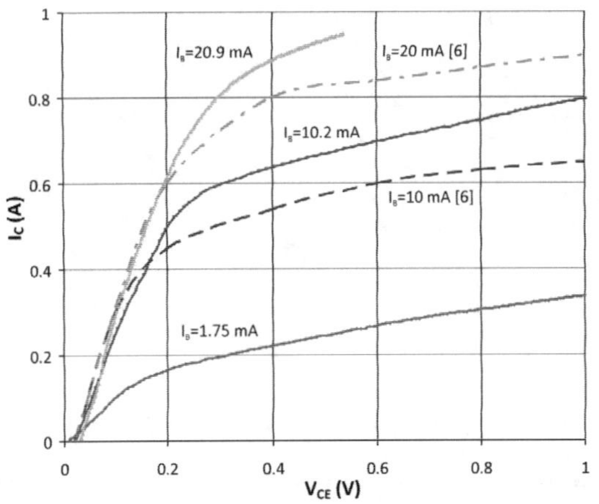

Figure 2. Output characteristics of n-p-n type bipolar junction transistor KT242A91 in small-sized SOT-89 type package.

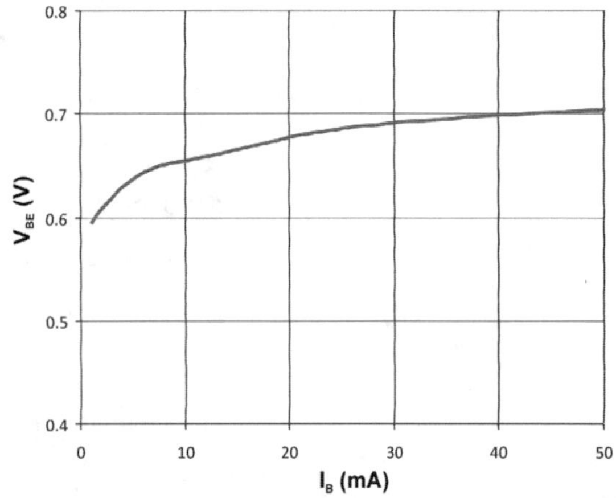

Figure 3. Input characteristics of n-p-n type bipolar junction transistor KT242A91 in small-sized SOT-89 type package (V_{CE}=0 V).

As can be seen from Figure 2 the produced transistor KT242A91 operates with collector current up to 1 A that corresponds to a such type similar transistor BCX56 (1 A [6]). In Figure 2 for comparison also presented data for original medium power n-p-n type bipolar junction transistor BCX56 (Nexperia [6]) for analogous base currents (I_B=10 mA and I_B=20 mA). In Figure 3 presents the experimental input characteristic of n-p-n type bipolar junction transistor KT242A91.

Table 1. The main SPICE model parameters for a modeling of the behavior of the of n-p-n type bipolar junction transistor KT242A91 in small-sized SOT-89 packages type.

Parameters	Description	Value	Units
IS	Transport saturation current	6.2×10^{-14}	A
RC	Collector ohmic resistance	285.417×10^{-3}	Ohm
EG	Bandgap voltage (barrier height)	1.11	eV
VAF	Forward Early voltage	230.45	V
BF	Ideal maximum forward beta	315	–
BR	Ideal maximum reverse beta	20	–
CJC	Base-collector zero-bias p-n capacitance	15.9×10^{-12}	F
FC	Forward-bias depletion capacitor coefficient	495×10^{6}	–
IKF	Corner for forward-beta high-current roll-off	0.9	A
IKR	Corner for reverse-beta high-current roll-off	0.5	A
ISC	Base-collector leakage saturation current	1.2×10^{-13}	A
ISE	Base-emitter leakage saturation current	1×10^{-14}	A
MJC	Base-collector p-n grading factor	0.4	–
MJE	Base-emitter p-n grading factor	0.35	–
NC	Base-collector leakage emission coefficient	1.2	–
NE	Base-emitter leakage emission coefficient	1.2	–
NF	Forward current emission coefficient	0.99	–
NK	High-current roll-off coefficient	0.7	–
NR	Reverse current emission coefficient	0.98	–
RB	Zero-bias (maximum) base resistance	0.2	Ohm
RE	Emitter ohmic resistance	0.08	Ohm
TF	Ideal forward transit time	0.8×10^{-9}	s
TR	Ideal reverse transit time	53×10^{-9}	s
VAR	Reverse early voltage	25	V
VJC	Base-collector built-in potential	0.53	V
VJE	Base-emitter built-in potential	0.67	V
VJS	Collector-substrate built-in potential	729×10^{6}	V
XTB	Forward and reverse beta temperature coefficient	1.3	–

Then, with practical point of view in case when electronic circuits consist from thousands components (diodes, transistors etc.) as a rule for discrete components it makes sense to use of their models created in SPICE (simulation program with integrated circuit emphasis) circuit program that simulates electronic circuits [10].

With this aim the obtained experimental data then were analyzed in Micro-Cap 12 model editor SPICE circuit program [11] that gives us to obtain a SPICE model of the KT242A91 transistor. The main SPICE model parameters for a modeling of the behavior of the of n-p-n type bipolar junction transistor KT242A91 in small-sized SOT-89 package are summarized in the Table 1.

Further, on the base of SPICE model parameters from Table 1 then were calculated in LTspice simulator [12,13] the output and input characteristics for silicon n-p-n type medium power bipolar junction transistor KT242A91 in small-sized (SOT-89) type package (20°C). The output and input characteristics for transistor KT242A91 are shown in Figure 4 and Figure 5. As can be seen, obtained SPICE model for KT242A91 transistor well describes the output and input characteristics with sufficient accuracy.

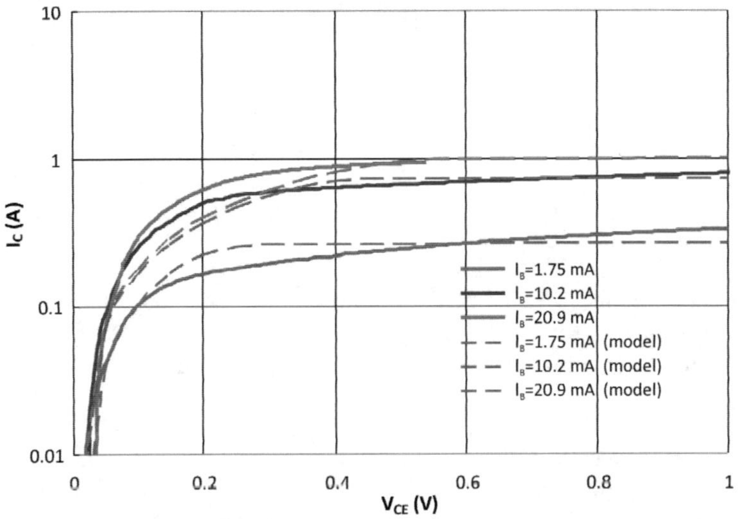

Figure 4. Output characteristics of n-p-n type bipolar junction transistor KT242A91 in small-sized SOT-89 type package calculated in LTspice (dashed line) by model and experimental data (solid line).

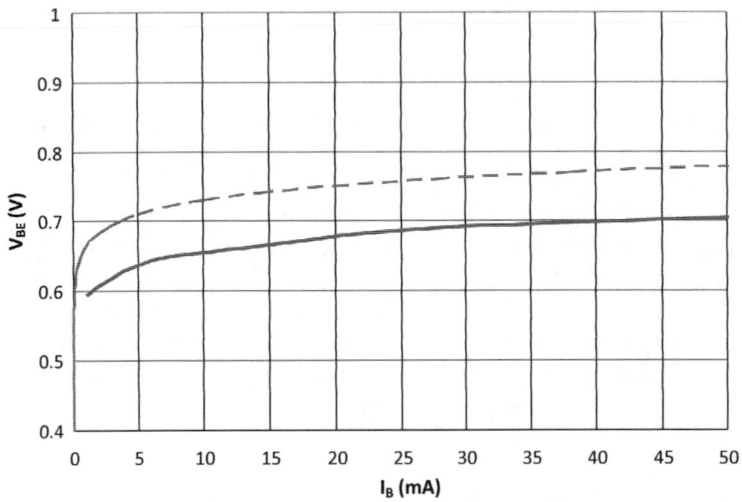

Figure 5. Calculated in LTspice (dashed line) by model and experimental data (solid line) the input characteristics of n-p-n type bipolar junction transistor KT242A91 in small-sized SOT-89 type package (V_{CE}=0 V).

The another important parameter is DC current gain for transistor [14,15] and therefore for KT242A91 transistor this characteristic was obtained experimentally (see Figure 6). Because of this, on the basis of SPICE model for KT242A91 transistor such characteristic was calculated in LTspice simulator that also presented in Figure 6. As follows from Figure 6 the obtained experimental and simulation data of the current gain dependences for KT242A91 transistor (V_{CE}=1 V) are very close and varying in interval of 91-202 that corresponds to a similar BCX56 transistor type, where current gain values equal of 40-250 [6].

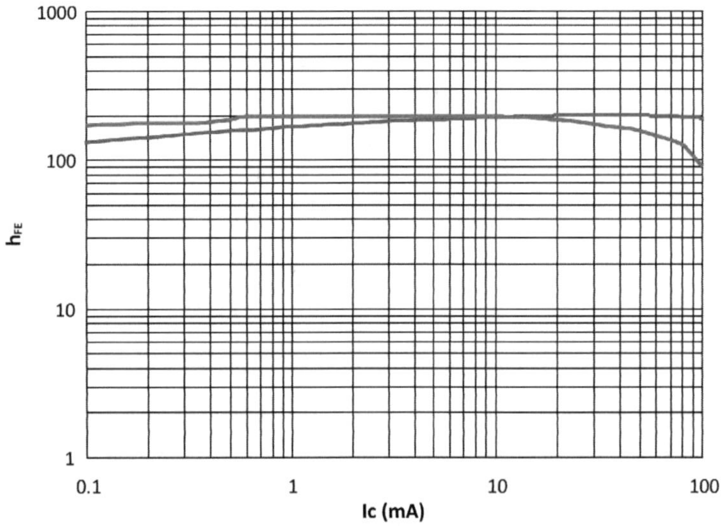

Figure 6. DC current gain dependence on collector current for n-p-n bipolar transistor KT242A91 (small-sized SOT-89 type package) calculated in LTspice (blue line) by model and experimental data (red line) (V_{CE}=1 V).

Thus, analysis of experimental results for silicon medium power n-p-n type bipolar junction transistor KT242A91 made by «GRUPPA KREMNY EL» in small-sized SOT-89 type package shows that obtained characteristics corresponds to data for similar BCX56 [8] transistor type.

4. Conclusions

Finally, the input, output and current gain characteristics of silicon medium power n-p-n type bipolar junction transistor KT242A91 made by «GRUPPA KREMNY EL» in small-sized (SOT-89) type package within the framework of import substitution program have been obtained. In addition, for silicon medium power n-p-n type bipolar junction transistor KT242A91 has been obtained SPICE model that allows simulating realistic transistor behaviour.

It has been established that the studied characteristics for KT242A91 transistor correspond to similar transistor's type and therefore modern small-sized package of this type for transistors can be used in Russian electronic industry within the framework of import substitution program.

Acknowledgements

This work was carried out with financial support of the Russian Ministry of Science and High Education within the framework of complex project by creation of highly technological industry «Creation of highly technological industry of silicon and silicon carbide microelectronic technics products in small-sized metal-polymeric packages of the SOT, SO and QFN types» (agreement of 29 November No. 075-11-2019-035) at the organization of the leading performer of RDDTE (Research and Development Design and Technological Engineering) the Bryansk State Technical University.

References

[1] Baliga B J 2019 *Fundamentals of power semiconductor devices* (Berlin: Springer International Publishing AG)
[2] Lu D and Wong C P 2017 *Materials for advanced packaging* (Cham: Springer International Publishing)

[3] Knyaginin D A, Rybalka S B, Kulchenkov E A, Demidov A A, Zhemoedov N A and Drakin A Yu 2020 *J. Phys.: Conf. Ser.* **1695** 012160
[4] Rybalka S B, Kulchenkov E A, Demidov A A, Zhemoedov N A, Drakin A Yu, Zotin V F and Shishkina O A 2020 *J. Phys.: Conf. Ser.* **1679** 022045
[5] TB2 the N-P-N type bipolar junction transistor, VUZPROMEXPO 2020. https://vuzpromexpo.ru/portfolio-item/tranzistor-bipolyarnyj-n-p-n-tb2
[6] BCP56; BCX56; BC56PA 80 V, 1 A NPN medium power transistors, Rev. 9, 25 October 2011. https://www.nxp.com/docs/en/data-sheet/BCP56_BCX56_BC56PA.pdf
[7] Panchenko P V, Rybalka S B, Malakhanov A A, A.A. Demidov A A, Krayushkina E Yu and Shishkina O A 2017 *J. Phys.: Conf. Ser.* **917** 082010
[8] Panchenko P, Rybalka S, Malakhanov A, Krayushkina E and Radkov A 2016 *Proc. SPIE* **10224** 102240Y-1
[9] Sedykh S, Rybalka S, Drakin A, Demidov A, Ponomaryova N and Shishkina O 2018 *J. Phys.: Conf. Ser.* **1124** 071012
[10] Rashid M H 2012 *SPICE for Power Electronics and Electric* Power (Boca Raton-London-New York: CRC Press)
[11] Spectrum Software - Micro-Cap 12. Analog simulation, mixed mode simulation, and digital simulation software. SPICE and PSpice compatible circuit simulator. http://www.spectrum-soft.com/index.shtm
[12] LTspice Simulator | Analog Devices. https://www.analog.com/en/design-center/design-tools-and-calculators/ltspice-simulator.html
[13] May C 2020 *Passive Circuit Analysis with LTspice®: An Interactive Approach* (Cham: Springer Nature Switzerland AG)
[14] Sedra A S and Smith K C 2015 *Microelectronic Circuits* (New York-Oxford: Oxford University Press)
[15] Sze S M and Kwok K Ng 2007 *Physics of Semiconductor Devices* (Hoboken: John Wiley & Sons Inc)

GaN power IC normally-on and normally-off transistors technology and simulation

V I Egorkin[1], V A Bespalov[1], O B Kukhtyaeva[1], V E Zemlyakov[1], V V Kapaev[1,2] and A A Zaitsev[1]

[1] National Research University of Electronic Technology, Moscow, Russia, 124498
[2] P.N. Lebedev Physical Institute of the Russian Academy of Sciences, Moscow, Russia, 119991

Abstract. GaN technology has been waiting to be widely adopted because of its specific technical requirements. Integration of transistor and driver in a single die will enable to overcome problems with gate driving, high cost of circuit and low device reliability. This paper demonstrates technology of GaN-on-Si normally-on and normally-off transistor with different p-GaN cap-layer thickness as well as simulation of these devices. The simulation data confirm experimental results. P-GaN cap-layer thickness affects the current channel density: the more p-GaN thickness, the less channel density. The fabricated transistors have a maximum drain current in open state of about 800 mA/mm.

1. Introduction

Gallium nitride (GaN) is excellent material for high switching speed and high-power semiconductor device applications due to its properties namely maximum electric field strength, wide band gap and high saturation electron drift velocity. GaN/Si heterostructure advantages are the ability to use large diameter wafers in the manufacture of nitride group devices, reduction of the device cost, relatively high thermal conductivity and the ability to integrate Si and GaN technology on the same substrate. Present GaN-on-Si technology allows to fabricate improved integral E-mode GaN transistors with high breakdown voltages. Due to the modern level of integration many active and passive devices can be manufactured on the same die as in a planar integral circuit (IC) process (Figure 1). These microcircuits have applications as AC/DC and DC/DC converters, switching power supplies etc. Monolithically driver integration solves problem with parasitic inductance in the gate driver loop. The parasitic inductance results in negative and positive gate spikes that can compromise the power FET reliability over time. Thus a monolithically integrated driver is so far the best solution.

There are many methods to form normally-off transistor namely recessed gate, fluorine treatment of the region under the gate, use of NiOx intermediate layer which after thermal treatment acquires p-type semiconductor properties and cascade configuration. In our opinion the perspective method is use of p-GaN or p-AlGaN layer under the gate. There are some methods to fabricate p-GaN layer: traditional selective RF etching, selective growth and hydrogen plasma treatment of p-GaN area not under the gate. In our previous work [1] we have managed to develop RF etching mode to etch p-GaN layer, heterostructure with additional AlN layer and the design process. AlN is a stop-layer at p-GaN etching and achieves a high energy barrier layer to suppress the leakage current induced by thermionic emission. Opportunity to form normally-on and normally-off transistors in a single die will enable to extend GaN electronic component base.

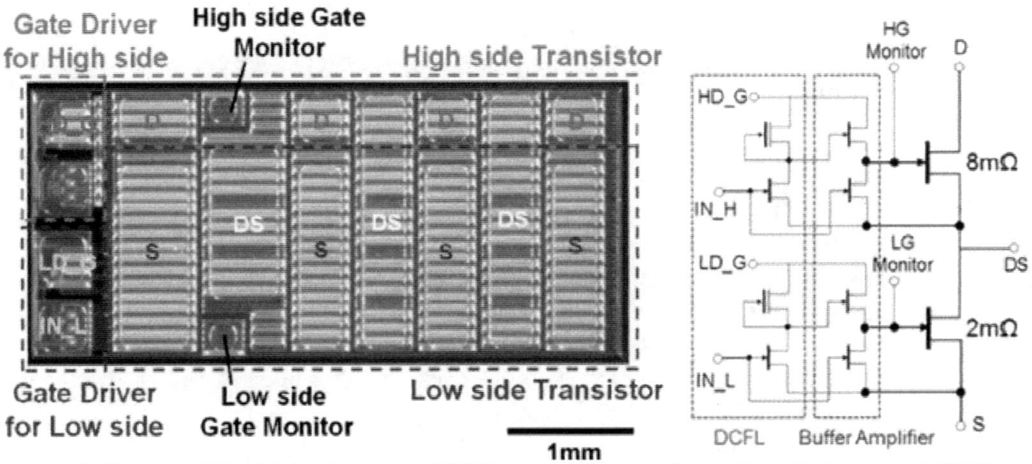

Figure 1. GaN power IC with high voltage HFET, gate driver, under-voltage lock-out and dV/dt control [2].

2. Experiment and simulation

For the experiment 40 nm p-GaN/1nm AlN/12nm AlGaN/1nm AlN/GaN/Si heterostructures were grown by MOCVD. The design process includes formation of alignment marks and isolation, then selective etching of p-GaN to form ohmic contacts and source and drain formation, gate metal deposition and then selective etching of p-GaN, passivation and first level metallisation (Figure 2). There are two ways to form isolation namely by ion implantation or by mesa-etching. Advantage of ion implantation is planarity that facilitates the transistor formation. In the design process gate metal is not heat treated that positively affects the gate region characteristics as p-GaN layer doesn't degrade. Ti-Al-Ni-Au (20/100/40/30 nm) metallization is used for ohmic contacts. The gate metallization is V-Au. To fabricate p-GaN gate ICP etching process has been worked out. Also we have managed to learn to control of etched p-GaN thickness. Gas mixture containing chlorine and fluorine must be used in etching process (BCl_3+SF_6, BCl_3+CF_4 etc.). In this case AlF_3 and AlO_x are formed at which etching process stops. To delete this dielectric layer the waver should be treated in solution of HF/NH_4OH or in hydrochloric acid solution.

P-GaN layer provides normally-off behavior of the device but for certain construction. Based on [3] the constructions have been simulated. Simulation has been done with Sentaurus Technology Computer Aided Design (TCAD). Heterostructure with next parameters has been simulated: thicknesses of AlN layers are 1 nm, thickness of AlGaN barrier layer is 12 nm and Al mole fraction in it is 25%, thickness of p-GaN cap-layer is 40 nm (20nm and without p-GaN layer) and Mg concentration in it is about 10^{18}sm^{-2}. The Al mole fraction is more than in [3] as another task to form transistor with high drain current. Transistors with 1 um gate length, 1.5 um gate-to-source length, 3.5 um gate-to-drain length and different p-GaN thicknesses (40 nm, 20nm and 0nm) were created and simulated. It has been noticed p-GaN cap-layer thickness influences on the current channel density and on type of transistor accordingly (Figure 3, 4).

3. Results

According to the simulation and experiment it has been established the more p-GaN thickness is, the less channel density is (Figures 4, 5). Respectively p-GaN thickness affects the drain current at 0 V gate voltage (Figure 5). Figure 4 shows that minimal channel density is about $2.37 \cdot 10^{13}$ sm^{-3} which corresponds to normally-on behaviour of device. Figure 3 shows dependence of threshold voltage

Journal of Physics: Conference Series **2086** (2021) 012058 doi:10.1088/1742-6596/2086/1/012058

(Vth) on p-GaN layer thickness (t_p). Vth is a key parameter that determines transistor behaviour. Our task is to form normally-off transistor with threshold voltage of about 3 V.

The fabricated transistors are normally-on and have a maximum drain current 800 mA/mm in open state, different threshold voltages (from -3 to -2 V) and different drain currents in open state at 0V gate voltage. The experimental data confirm simulation ones (Figures 6-11). Therefore we have simulated the heterostructure with other parameters of layers to manufacture normally-off transistor.

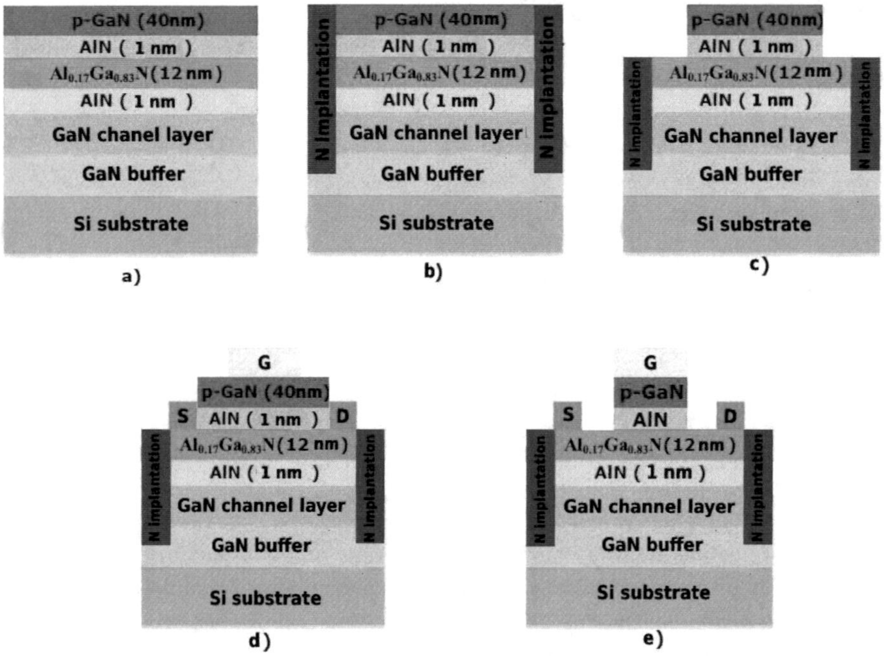

Figure 2. Design process and schematic cross-sectional view of p-GaN gate HEMT: a) GaN-on-Si heterostructure, b) isolation by ion implantation, c) selective etching of p-GaN to form ohmic contacts, d) formation of ohmic contacts and deposition of gate metal, e) etching of p-GaN to form gate.

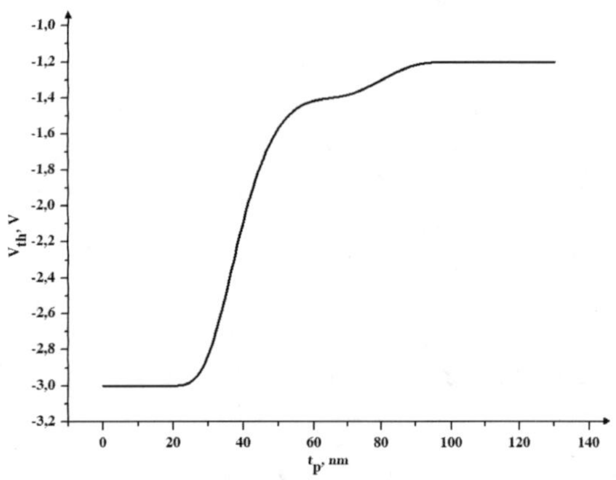

Figure 3. Dependence of threshold voltage (Vth) on p-GaN layer thickness (t_p).

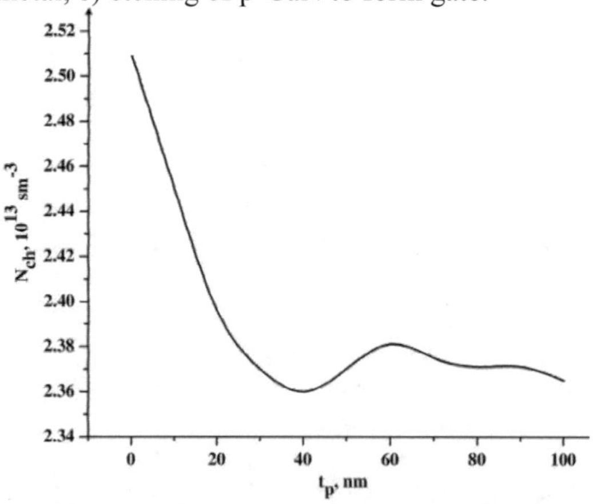

Figure 4. Dependence of channel carrier density (N_{ch}) on p-GaN layer thickness (t_p).

 2086 (2021) 012058 doi:10.1088/1742-6596/2086/1/012058

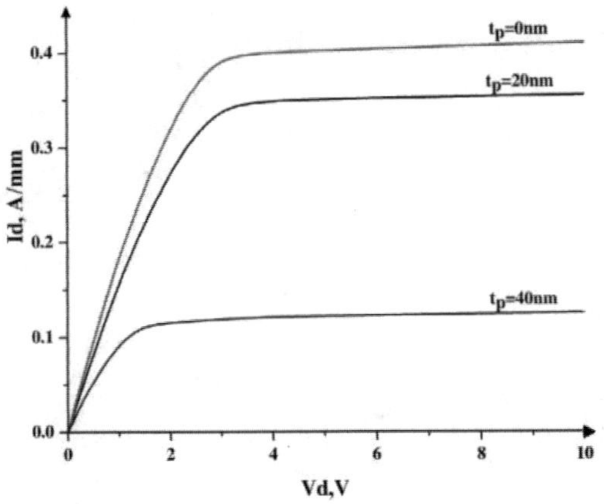

Figure 5. Current-voltage characteristics of HEMTs with different p-GaN layer thicknesses at gate voltage of 0 V.

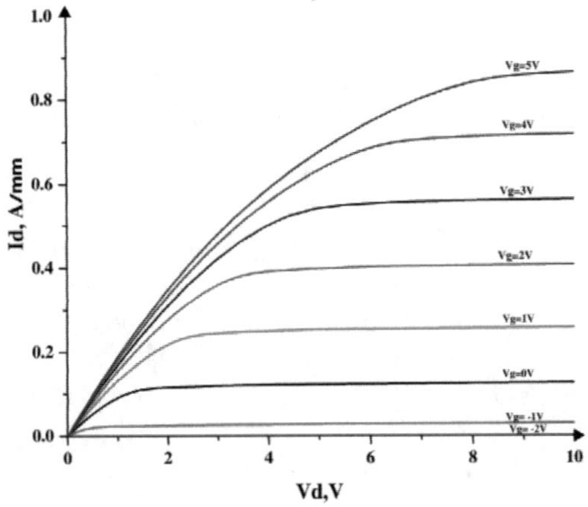

Figure 6. TCAD current-voltage characteristics of the HEMT with 40 nm p-GaN.

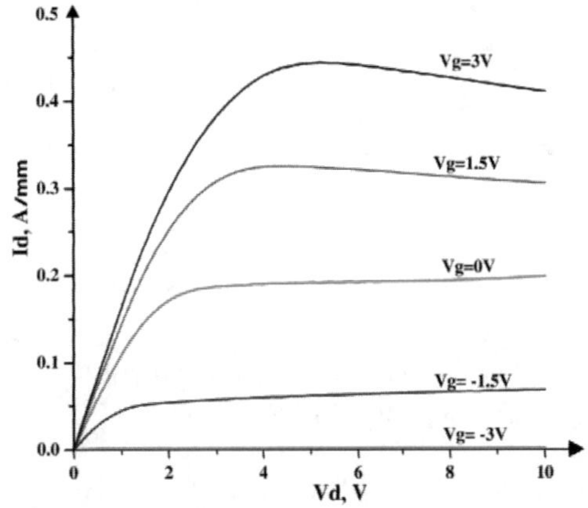

Figure 7. Experimental current-voltage characteristics of the HEMT with 40 nm p-GaN.

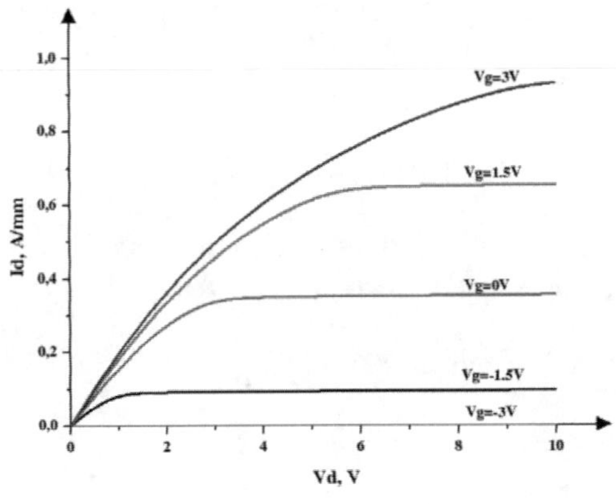

Figure 8. TCAD current-voltage characteristics of the HEMT with 20 nm p-GaN.

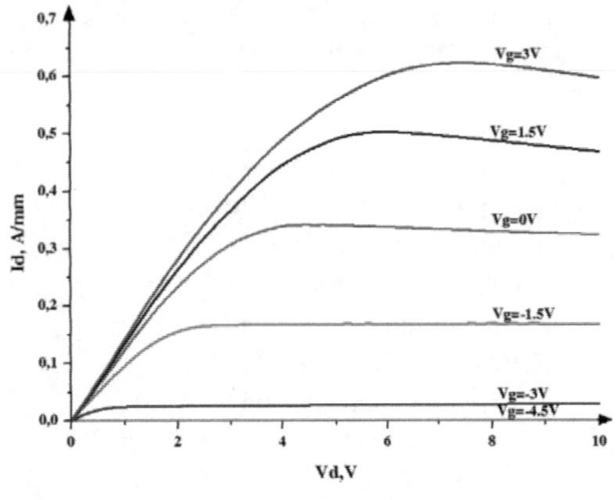

Figure 9. Experimental current-voltage characteristics of the HEMT with 20 nm p-GaN.

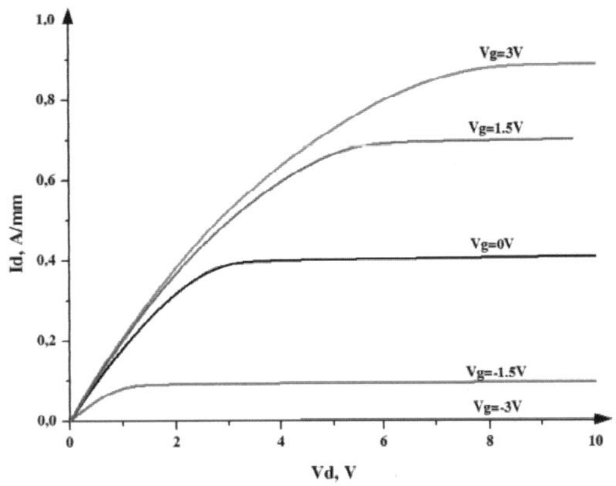

Figure 10. TCAD current-voltage characteristics of the HEMT without nm p-GaN.

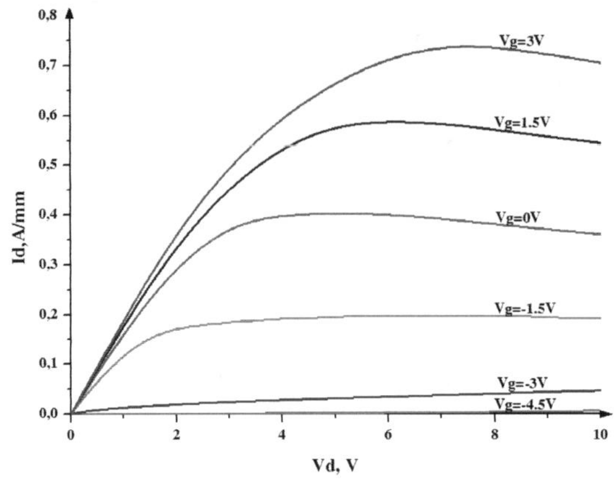

Figure 11. Experimental current-voltage characteristics of the HEMT without nm p-GaN.

4. Conclusion
This work demonstrates the first step to create GaN-on-Si planar IC. We have managed to form normally-off transistor in previous works and today we have managed to simulate and form normally-on transistors with various p-GaN gate thicknesses on a single substrate with our RF etching process. Our simulation results confirm experimental ones. As the simulation is right we have simulated different constructions of heterostructure for normally-off GaN transistor and we are going to form transistors of two types on a single die. Thus we have developed fabrication technology of monolithically integrated driver (normally-on transistor) and normally-off transistor.

Acknowledgments
This work was supported by the Ministry of education and science of the Russian Federation ("Technology development and technological preparation to manufacture transistor crystals based on gallium nitride heterostructures on a silicon substrate with diameter of 150 mm for power converter modules", grant agreement № 075-11-2019-068 from November 26, 2019).

References
[1] Kukhtyaeva O B, Egorkin V I *et al* .2019 *Journal of Physics: Conference Series* vol. 1410 (Saint-Petersburg, Russia) p.012192

[2] Giandalia M and Kinzer D 2019 *PCIM Europe 2019; International Exhibition and Conference for Power Electronics, Intelligent Motion, Renewable Energy and Energy Management* (Nuremberg, Germany)

[3] Chen K J, Haberlen O, Lidow A, Tsai Ch,Ueda T, Uemoto Y and Wu Y 2017 *IEEE Trans. Electron Devices* **64** 779

[4] You S, Li X *et al* 2019 ESSDERC *2019 - 49th European Solid-State Device Research Conf. (ESSDERC)* (Cracow, Poland) pp 158-61

System for monitoring the parameters of overhead power lines

V A Listyuhin, E A Pecherskaya, O A Timokhina and V V Smogunov

Department of Information and measuring equipment and metrology, Penza State University, Penza 440026, Russia

Abstract. The analysis of disturbances (accidents) on overhead power transmission lines and methods and means of the overhead power transmission lines parameters monitoring was carried out. It is shown that the most promising direction for ensuring the required indicators of the reliability of power supply is the development and implementation of information and measurement systems at the facilities of the power supply network complex. The structure of an intelligent information-measuring system for measuring parameters characterizing the state of wires, external factors influencing the state of power lines is proposed. The proposed decision support system will make it possible to promptly identify defective sections of overhead lines, increase the level of technological management of electrical networks.

1. Introduction

Nowadays, the overhead power lines (OHL) parameters monitoring is the most important task to ensure the reliability of power supply, which is implemented through the development and implementation of information and measurement systems (IMS). For example, in [1], it was proposed to introduce an all-optical information and measurement control system for overhead lines based on acousto-optic switching of fiber-optic cables; in [2], it was proposed to use a "smart" electromechanical system (EMS) to solve problems of operational recognition of the emergency mode of a line. Some studies are aimed at using IMS together with high-speed digital protection of transformers against short circuits in the network [3-4]. The authors have developed a version of IMS for measuring and monitoring the overhead lines state, combined with a neuro-fuzzy block of information processing (decision support), which will allow to identify defective sections of overhead lines quickly and increase the level of electrical networks technological control.

2. Requirements for the developed IMS

In order to reduce accidents on overhead lines, as well as to ensure timely maintenance and repair of power grid facilities according to the actual technical level state, it is necessary to develop an overhead lines operational parameters IMS, which makes it possible to identify defects at the stage of formation and predict pre-emergency modes under extreme weather conditions (mainly wind and ice loads).

The following requirements are imposed on the developed IMS:

- non-contact measurement method (the sensor is powered by a "parasitic" method due to the electromagnetic wire field);

- the ability to measure wind speed and ambient temperature as an input device (integrated hot-wire anemometer into the device under development);
- measurement results analysis using a neuro-fuzzy controller (information support for decision support);
- durability to the impact of external natural phenomena.

3. The operating principle and the main parts of the IMS

The operating principle of this device is realized by measuring the distance from the wire to the ground with a laser distance sensor (rangefinder). In parallel, the wind speed and ambient temperature are measured. The measurement results in real time (online) through the module built into the GSM measuring device are transmitted to the control room, where the data is processed in the neuro-fuzzy controller. Based on the measurement results, the decision support system analyzes the data and displays the following messages on the personnel PC, characterizing the state of the controlled object:

- normal mode (horizontal wire position and the wire position, taking into account the deviation from the horizontal position when exposed to non-critical wind loads);
- the initial stage of wire misalignment (non-critical sag of the wire relative to the normal position);
- pre-emergency mode (critical sag of the wire due to possible collision with other wires and the ground);
- emergency mode (wire snagging with other wires or contact of a wire with the ground).

Figure 1 shows a block diagram of the developed IMS.

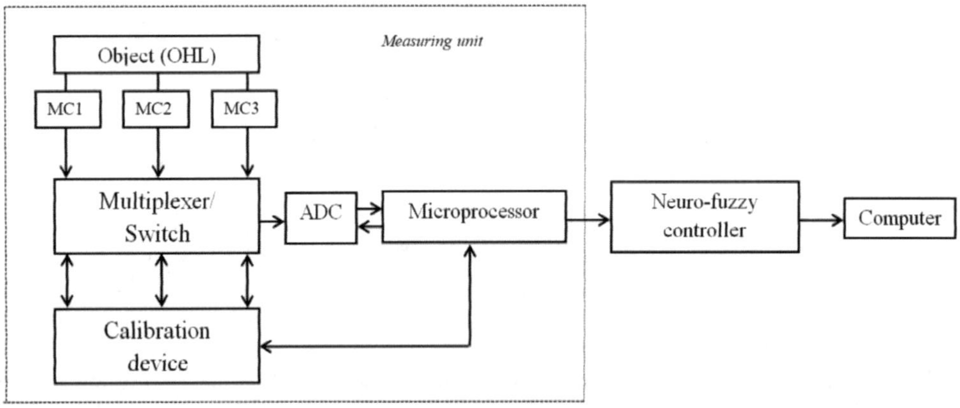

Figure 1. Block diagram of the developed IMS: MC1, MC2, MC3 - respectively, the measuring channels of the distances from the wire to the ground, wind speed, temperature; ADC - voltage-to-digital converter.

A phase laser rangefinder and a hot-wire anemometer are used as primary measuring transducers in the developed IMS. Improving the accuracy and speed of laser devices is a topical scientific and technical issue [5-6]. To date, there is a sufficient number of studies aimed at improving the metrological characteristics of laser devices designed to determine the distance. So in work [7] the principles of operation and characteristics of active-pulse television measuring systems are described, the circuit is considered and the characteristics of a high-precision laser rangefinder with a crystal light modulator are described. The measuring device proposed in [8] has high metrological characteristics, but a small measurement range (2.7 m) is a significant disadvantage of the device. Thus, the use of a phase laser rangefinder (PLR) is due to the fact that, in comparison with other types of rangefinders, PLRs have the required measurement range (up to 50 m) and a higher measurement accuracy. Thus,

PLRs meet the requirements for the developed IMS. The principle of PLR operation is based on determining the distance by comparing the phase of the modulating signal at the output from the radiation receiver (the phase of the radiation that has passed the distance to the object and back) with the phase of the reference signal (the phase of the signal at the radiation source). The distance traveled by the light wave in time *t* is equal to:

$$l = ct,$$

where *c* — speed of light.

During the same time, the phase of the modulated laser radiation that has traveled from the source to the object and back will change by the following value:

$$\varphi = 2\pi f_{\text{M}} t,$$

where f_{M} – radiation modulation frequency.

Thus, to calculate the distance to the object, the following formula is used:

$$l = c\frac{\varphi}{2\pi f_{\text{M}} t}.$$

Figure 2 shows a block diagram of a two-frequency laser phase rangefinder, which implements the classical principle of operation.

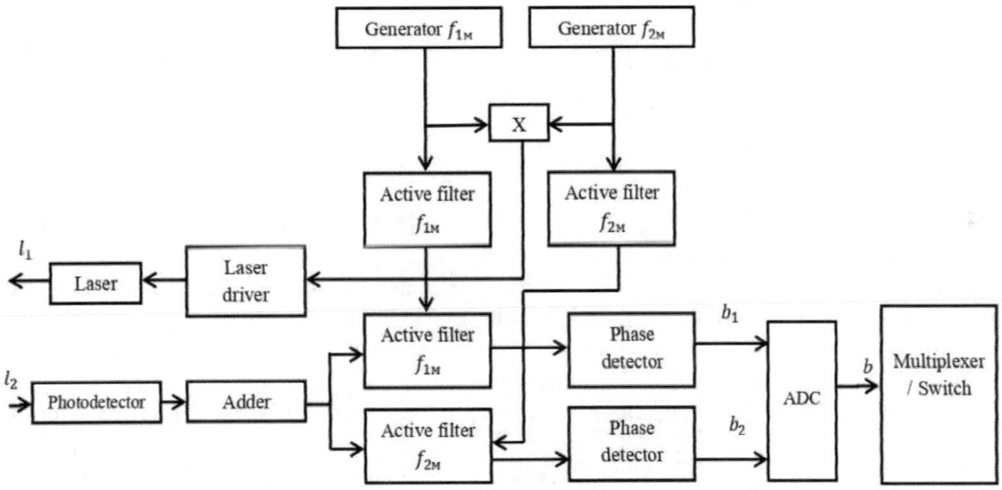

Figure 2. Block diagram of a phase laser rangefinder: l_1 and l_2 – laser beam directed to and from the object; b_1 and b_2 – the number of a roughly defined range and a refined range number to the value of the required error.

At the output of the generators, two signals are generated with a frequency of $f_{1\text{M}}$ and $f_{2\text{M}}$. These signals are logically summed and fed to the laser driver, which modulates the excitation current of the semiconductor laser. Also, these signals are fed through narrow-band filters to two phase detectors as reference voltages. After reflection from the controlled object (in the case of a device under development, from the earth's surface), the phase-changed laser radiation modulated simultaneously by two frequencies is recorded by a photodetector device (PDD). The voltage from the PDD output is amplified by a broadband amplifier and fed to narrowband active filters tuned to frequencies $f_{1\text{M}}$ and $f_{2\text{M}}$. Thus, from the received signal of a complex shape, two harmonics are extracted, which are fed to the working inputs of the corresponding phase detectors. At the outputs of the phase detectors, voltages are generated that are proportional to the phase shift φ_1 and φ_2, which are subsequently converted into a digital signal in the ADC. The digitization of the signals results in two binary numbers with the widths b_1 and b_2. The first number is a roughly determined range, and the second number is a refinement of the range to the value of the required error. To obtain the final result, these numbers are summed up into one number with the digit capacity $b = b_1 + b_2$, which will correspond to the distance to the object (with an error not exceeding the specified value).

To measure the values of wind speed and ambient air temperature in the developed IMS, a hot-wire anemometer is used, which has a remote probe with a built-in hot-wire sensor. The principle of operation of the hot-wire anemometer is based on measuring the temperature resistance of a heated thermistor cooled by an air flow. Miniature platinum thermistors are used as sensitive elements to measure ambient temperature and wind speed.

Implementation of the proposed IMS will allow: timely identification of defective areas by detecting the initial defect formation stage; to raise the level of technological management of power grid complex distribution networks; repair and maintain overhead lines according to their technical condition.

References

[1] Mukhammadieva A Aida and Salakhutdinov I and Yasoveev K. Vasikh 2020 Optical System for Monitoring Power Lines *Proc. Int. Conf. on Electrotechnical Complexes and Systems* pp 1–4

[2] Shilin A N and Shilin A A and Dementiev S S 2019 Smart Electromechanical Systems in Electric Power Engineering: Concept, Technical Realization, Prospects *Smart Electro - mechanical Systems (Studies in Systems Decision and Control vol 174)* eds A Gorodetskiy, I Tarasova (Cham: Springer) pp 251–62

[3] Khrennikov A Y 2016 *Diagnostics of electrical equipment faults and power overhead transmission line condition by monitoring systems (Smart Grid): Short-circuit testing of power transformers* (New York: Nova Science Publishers Inc) p 150

[4] Khrennikov A Y 2012 New "Intellectual Networks" (Smart Grid) for Detecting Electrical Equipment Faults, Defects and Weaknesses *Smart Grid and Renewable Energy* **3** 159–164

[5] Bragynets Y O, Masjurenko Yu O 2020 Phase light rangefinder with adaptive optical calibration system *Technical Electrodynamics* **3** 83–88

[6] Hayrapetyan Yegisabeth, Petrosyan Hovseph, Hunanyan Hovnan, Tsaturyan Artyom 2019 High-precision two-phase laser rangefinder PFSD-1,2 *E3S Web of Conferences* **97** 06042

[7] Chaldina E S and Movchan A K and Kapustin V V 2019 Study of the Time-of-Flight Method for Measuring Distances to Objects Using an Active-Pulse Television Measuring System *Proc. Int. Multi-Conference on Engineering, Computer and Information Sciences* pp 0457–61

[8] Tap H and Moutai L Gatete and Mulliez B 2018 Embedded System for Distance Measurement and Surface Discrimination Applications *Proc. Int. Conf. on Design of Circuits and Integrated Systems* pp 1–6

Modified quantum frequency standard on Hg-199 ions

N A Lukashev[1], V V Davydov[2, 3] and V Y Rud[3]

[1] The Bonch-Bruevich Saint-Petersburg State University of Telecommunications, Saint Petersburg 195251, Russia
[2] Peter the Great Saint-Petersburg Polytechnic University, Saint Petersburg, Russia, 195251
[3] All-Russian Research Institute of Phytopathology, Moscow Region 143050, Russia

n-lukash@list.ru

Abstract. The world around us depends on devices capable of producing or maintaining a signal with an extreme precision. Quantum frequency standards are the answer to this problem. This article presents a modified version of newly developed quantum frequency standard based on trapping Hg-199 ions by magnetic field. The new prototype was developed a while ago and now it was modified due to algorithm improvements and renewed digital hardware, analog and digital circuitry being reordered. Results for Allan deviation show 3 % improvement for long-term frequency stability and more than 5 % for short-term stability

1. Introduction
The Carrying out various scientific studies, transmitting information, determining the coordinates of an object strictly requires the use of time synchronization devices [1-12]. These devices have different principles of physical work, especially when applied at moving objects [4-10, 13-16]. In some cases, instruments based on highly stable quartz are enough to ensure the operation of on-board navigation systems, radio monitoring and radar stations [17-23]. In the case of operation of these devices for a long time in difficult conditions, a number of problems arise [4-9, 24-30], especially in case of large congestion. The accuracy of determining the position of the object is reduced. The use of optical fibers and optical information processing methods [18, 19, 31-36] do not solve this problem.

The modernization process of standards' construction is not a trivial task, since the work of various blocks is related to each other. Therefore, the task of modernizing the frequency standard must be solved in a complex manner [1-5, 10, 24-30, 37]. By reducing the size and weight of the main units, the introduction of new algorithms for the automatic control system of the magnetic field and the photon registration system, the development of a system for automatic tuning of the resonant frequency, this task was realized within the framework of this work before the commissioning of the existing prototype).

2. Supplying scheme
The trap is the power consumer and the main operating unit of the standard. For a new small-sized design of the trap and driver, a switching circuit was developed.

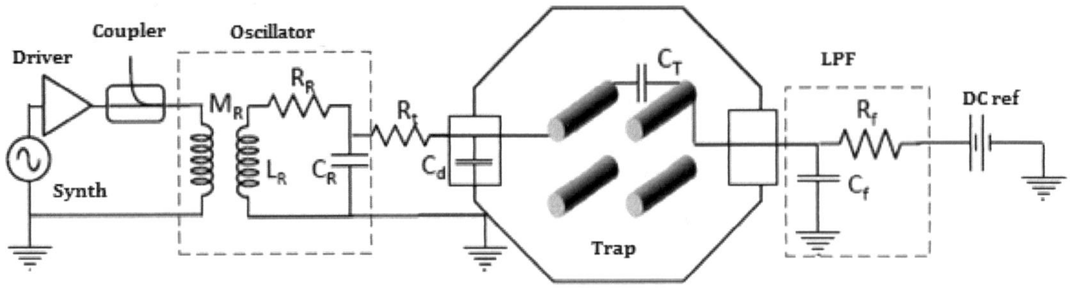

Figure 1. New supplying scheme.

To generate a signal of a certain type and control a power stage with a transformer, we have designed a miniature version of the driver. The signal is pre-amplified and sent through 2 single-shot channels to the key circuit, which accumulates energy in the primary winding of the transformer, feeding the trap electrodes. High response speed and good temperature parameters provide stable control over the RF signal.

The signals of the modulation frequency, clock frequency, as well as the input data received from the microcontroller are received at the counter input. The purpose of the counter is to generate a signal with variable count and pause windows, as well as to count photon pulses over the corresponding time interval. The counting of pulses is carried out with a positive or negative sign, depending on the half-period of the signal of the frequency synthesizer in which the photons were recorded. The resulting counting result characterizes the frequency detuning of the crystal oscillator, on the basis of which the microcontroller generates a control voltage that changes the frequency of the crystal oscillator.

After the assembly, experimental studies of the characteristics of the new prototype, as well as separately designed parts, were carried out. The signal from the driver was tested on a breadboard, which showed that there was no influence of parasitic components on the output signal spectrum. A review of residual oscillations after switching off the device was also carried out, which indicated a good tuning of the output circuit due to the absence of parasitic oscillations and dependences other than ~ exp (-t). The software was pre-tested in the ModelSim environment and then incorporated into the Standard.

3. Computing part

It was obvious that to implement a high-performance counter, the architecture of the entire block had to be changed. Simplification of the circuit consisted in dismembering the module into separate functions and transferring the latter to specific electronic devices. So, the detection function is now performed by a comparator or an optical detector, the microcontroller is engaged in the processing of the output data, and the conversion from the bit sequence to the analog signal of the DAC is performed. These measures significantly reduced the load on the counter processor, but the speed of its operation still did not meet the needs of the consumer. Unfortunately, with all the advantages, microcontrollers are much inferior in speed to programmable logic integrated circuits (FPGAs).

However despite all the advantages of programmable logic integrated circuits, they cannot completely replace the operation of a microcontroller. The fact is that FPGAs have less processing power, and also have no memory, which makes the operation of these devices very limited. Based on the experience of developing such devices, in this project it was decided to implement photon counting on FPGAs, and to process and transmit output data to peripheral devices on the basis of microcontroller technology.

In this work, an Altera FPGA of the MAX 7000S family, model EPM7128SLI84-10N, was chosen as the key element of the photon counting unit. This microcircuit operates at frequencies up to 147.1 MHz, and also has a low latency of 6 ns. Thanks to these parameters, the selected microcircuit will be able to quickly process large input data, and 8 logical blocks will not clutter the program code and output intermediate results to check the FPGA operation algorithm. The choice of a programming tool in our case fell on the VHDL language. The latter surpasses Verilog in readability, but loses significantly in the amount of code.

For FPGAs from Altera, the company of the same name created the Quartus development tool, which was successfully modified by Intel specialists. Quartus software provides the ability to work with FPGAs from design to debugging within one CAD system. Within the framework of this project, the Quartus II version was used, this software supports operation for the selected FPGA, and also has some advantages over later versions: block design technology and hardware debugging capabilities using a logic analyzer. To check the hardware implementation and time characteristics of the project on the FPGA, the HDL-modeling system for digital devices - ModelSim was used.

4. Photon counting

The structure of the photon counting functional block, its structural diagram is shown in Figure 2. Initially, the signals of the modulation frequency, clock frequency, as well as input data received from the microcontroller are sent to the counter input. The task of the counter is to generate a signal with variable counting and pause windows, as well as to count photon pulses over the corresponding time interval.

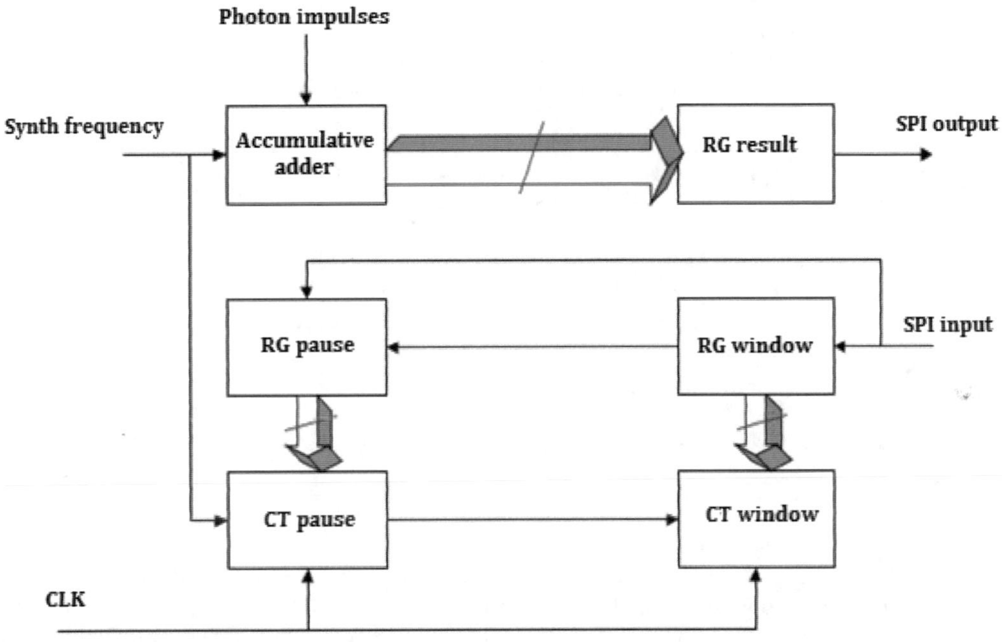

Figure 2. Block diagram of the photon counting functional block

Pulses counting is carried out with a positive or negative sign, depending on the half-period of the signal of the frequency synthesizer, in which the photons were recorded. The resulting counting result characterizes the frequency detuning of the crystal oscillator, on the basis of which the microcontroller forms a control voltage that changes the frequency of the crystal oscillator.

The transmission of the signal parameters, during which the photons are counted, is realized via the serial peripheral interface SPI. This data transmission standard was created for simple and high-speed communication between the microcontroller and peripherals in full duplex mode - the mode of receiving and transmitting information. The SPI bus is designed according to the master-slave principle, and there can be several slaves. Data transfer using this protocol is carried out bit by bit, and the synchronization frequency is generated by the master device, most often the microcontroller processor.

The operation of the block for transmitting data to the FPGA input via SPI is illustrated in the timing diagram (Figure 3). Input signals data00 and data01 set the parameters of the time of the counting and pause windows, the data for transmission are presented in the form of binary numbers. These values are transmitted bit by bit via the SPI interface; for this, special synchronization signals sdvig and strob are used. The synchronized operation of the FPGA and the microcontroller can be checked using the analog signal data1, the level of which at the leading edge of the sdvig pulses corresponds to the input data bits. Zeroing of the data1 signal on the strob edge occurs due to the end of the transmission of the input value data00, with the beginning of the next edge of sdvig, the number of the pause window (data01) will be transmitted.

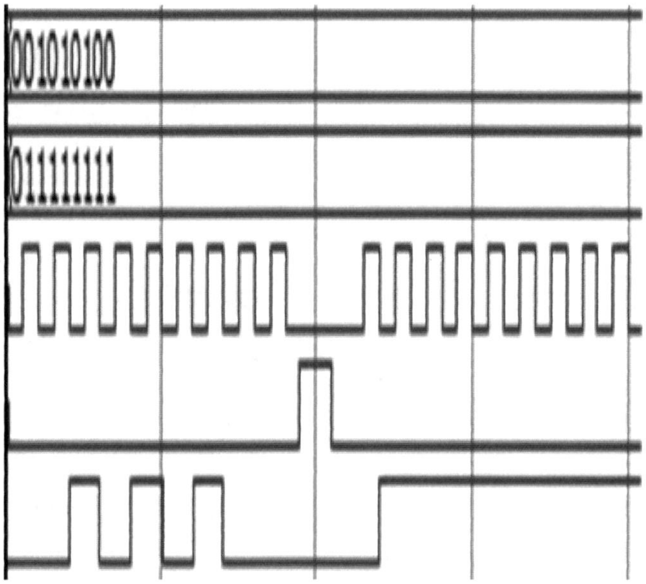

Figure 3. Input data transfer time diagram

Filling the registers of the counting and pause windows with input values is carried out using the overwrite signal S. As the recording ends, namely on the strob front, the next block of the program checks the first two bits of the signal S. After filling the registers with input data, it is necessary to generate a photon counting signal, and its period should be constant and equal to the period of the modulation signal freq_m1. The freq_m1 front serves as the beginning of the digitization of the error signal. Since the work of the counter is performed on digital equipment, it is necessary to take into account the time of switching the FPGA from one logic signal level to another. If you do not take into account the effect of transient processes at the output, you can get the so-called signal race - a situation in which, due to the delay of the input signal relative to the output, false signals appear, and, accordingly, the device malfunctions. One of the options for eliminating the races at the counter output, the formation of the counting window should be shifted relative to the front of the modulation signal. For the same reasons, the photon counting does not stop when freq_m1 decays.

The counting of photon pulses is carried out during the counting windows in the positive and negative half-periods of the modulation signal; this data with the corresponding signs is transferred to the accumulating adder.

The final stage is transfering the count value to the microcontroller for generating the feedback signal of the automatic frequency control system. Output data, like input data, is transmitted via the SPI serial interface. To ensure the transfer of the result, the binary count number is converted into an analog signal using the auxiliary signals sdvig_out_filt and strob_out_filt, the functionality of the latter is similar to the input sdvig and strob, respectively. The difference in the implementation of information transfer is additional protection against FPGA transients.

5. Conclusion

The use of newly developed blocks and the use of new algorithms for automatic tuning made it possible, while reducing the size of the structure of the standard on mercury ions - 199 by three times and the weight by more than 65% compared to the previously operated ground sample and surpassing the accuracy characteristics of the standard applied on rubidium atoms: short-term (from 100 to 10,000 seconds) by 5%, and at long intervals (1 day) by 3% in comparison with previously used designs.

References

[1] Petrov A A, Shabanov V E, Zalyotov D V, Bulyanitsa A L and Shapovalov D V 2018 *International Conference on Electrical Engineering and Photonics EExPolytech 2018 (Saint-Petersburg)*, vol. 8564389 (IEEE), p. 52-55

[2] Petrov A A, Vologdin V A and Zalyotov D V 2015 *Journal of Physics: Conference Series* **643(1)** 012087

[3] Petrov A A and Grebennikova N M 2018 *Journal of Communications Technology and Electronics* **63(11)** 1281–1285

[4] Lukashev N A and Moroz A V 2019 *Journal of Physics: Conference Series* **1236(1)** 012068

[5] Petrov A A and Grebenikova N M 2018 Lecture Notes in Computer Science (including subseries Lecture Notes in Artificial Intelligence and Lecture Notes in Bioinformatics) **11118 LNCS** 641-648

[6] Petrov A A and Davydov V V 2015 *Lecture Notes in Computer Science (including subseries Lecture Notes in Artificial Intelligence and Lecture Notes in Bioinformatics)* **9247** 739-744

[7] Valov A P, Davydov R V and Grevtseva A S 2019 *Journal of Physics: Conference Series* **1326(1)** 012040

[8] Grevtseva A S, Davydov R V, and Dudkin V I 2019 *Journal of Physics: Conference Series* **1326(1)** 012043

[9] Grevtseva A, Davydov V and Rud V 2020 *CEUR Workshop Proceedings* **2667** 15–18

[10] Petrov A A and N S Myazin 2017 *Lecture Notes in Computer Science (including subseries Lecture Notes in Artificial Intelligence and Lecture Notes in Bioinformatics)* **10531 LNCS** 561-568

[11] Semenov V V , Nikiforov N F and Ermak S V 1991 *Soviet journal of Communications Technology and Electronics* **36(4)** 59 – 63

[12] Fadeenko V, Fadeenko I, Dudkin V and Nikolaev D 2019 *IOP Conference Series: Earth and Environmental Science* **390(1)** 012022

[13] Logunov S E, Koshkin A Yu and Petrov A A 2016 *Journal of Physics: Conference Series* **741(1)** 012092

[14] Davydov R V and Antonov V I 2018 *Journal of Physics: Conference Series* **1124(8)** 081037

[15] Davydov R V and Antonov V I 2018 *Journal of Physics: Conference Series* **1135(1)** 012087

[16] Logunov S E, Vysocky M G and Koshkin A Yu 2017 *Journal of Physics: Conference Series* **917(5)** 052058

[17] Moroz A V and Davydov R V 2019 *Lecture Notes in Computer Science (including subseries Lecture Notes in Artificial Intelligence and Lecture Notes in Bioinformatics)* **11660 LNCS** 710 – 718

[18] Filatov D L, Galichina A A, Vysoczky M G and Yalunina T R 2017 *Journal of Physics: Conference Series* **917**(8) 082005

[19] Davydov R V, Saveliev I V, Lenets V A and Tarasenko M Yu 2017 *Lecture Notes in Computer Science (including subseries Lecture Notes in Artificial Intelligence and Lecture Notes in Bioinformatics)* **10531 LNCS** 177-183

[20] Andropov A V, Kuzmin S V and Korovin K O 2021 *Springer Proceedings in Physics* **255** 675–681

[21] Shannikov D V and Kuzmin S V 2003 *Technical Physics Letters* **29(11)** 941–94
[22] Raimzhanov T R, Kuzmin S V and Korovin K O 2021 *Springer Proceedings in Physics* **255** 629–635
[23] Shannikov D V and Kuzmin S V 2002 *1st IEEE International Conference on Circuits and Systems for Communications (ICCSC 2002)*, vol. 1029113 (IEEE), p. 350-353
[24] Petrov A A, Grebenikova N M, Lukashev N A, Ivanova N V, Rodygina N S and Moroz A V 2018 *Journal of Physics: Conference Series* **1038 (1)** 012032
[25] Lukashev N A, Petrov A A, Grebenikova N M and Valov A P *2018 Proceedings of 18th International conference of Laser Optics ICLO-2018 (Saint-Petersburg),* vol. 8435889 (IEEE), p. 271
[26] Valov A P 2019 *Journal of Physics: Conference Series* **1410(1)** 012246
[27] Valov A 2020 *CEUR Workshop Proceedings* **2667** 102–105
[28] Valov A P, Nagornaya A I, Kulikova M E and Glinushkin A P 2020 *Journal of Physics: Conference Series* **1695(1)** 012169
[29] Petrov A A, Zaletov D V, Davydov V V and Shapovalov D V 2021 *Journal of Communications Technology and Electronics* **66(3)** 295–299
[30] Petrov A A, Zalyotov D V, Shabanov V E and Shapovalov D V 2018 *Journal Physics: Conference Series* **1124(1)** 041004
[31] Davydov R, Antonov V and Moroz A 2019 *International Conference on Electrical Engineering and Photonics EExPolytech-2019 (Saint-Petersburg)*, vol. 8906791 (IEEE), p. 295-297
[32] Podstrigaev A S, Smolyakov A V, Grebenikova N M and Davydov R V 2019 *Lecture Notes in Computer Science (including subseries Lecture Notes in Artificial Intelligence and Lecture Notes in Bioinformatics)* **11660 LNCS** 525–533
[33] Davydov R, Antonov V, Makeev S, Batov Y, Dudkin V and Myazin N 2020 *E3S Web Conference* **140** 02001
[34] Saveleiv I K, Sharova N V and Tarasenko M Y 2017 *Journal of Physics: Conference Series* **917(5)** 052020
[35] Tarasenko M Y, Lenets V A, Akulich N V and Yalunina T R 2017 *Lecture Notes in Computer Science (including subseries Lecture Notes in Artificial Intelligence and Lecture Notes in Bioinformatics)* **10531 LNCS** 227–232
[36] Tarasenko M Yu, Lenets V A, Malanin K Yu and Akulich N V 2018 *Journal of Physics: Conference Series* **1038(1)** 012035
[37] Petrov A and Shapovalov D 2019 *Journal of Physics: Conference Series* **1400(4)** 044008

Development of a fiber-optic microwave signal transmissionsystem for an X-band receiving module with dual frequencyconversion

A V Moroz[1, 2], V V Davydov[1, 3], D V Gubareva[2, 4] and V Yu Rud'[3]

[1] Peter the Great Saint Petersburg Polytechnic University, Saint Petersburg, 195251,Russia
[2] NTC JSC "Zaslon", Saint Petersburg 196084, Russia
[3] All Russian Research Institute of Phytopathology, Moscow Region 143050, Russia
[4] The Bonch-Bruevich St. Petersburg State University of Telecommunications, SaintPetersburg 193232, Russia

moroz.com3844@gmail.com

Abstract. The article discusses the use of double frequency conversion in the receiving moduleof the radar. To solve the resulting interference in the feeder paths, it is proposed to use a fiber- optic transmission system. The results of experimental investigations are presented.

1. Introduction

Currently, the world pays great attention to the development of radar systems for various purposes [1- 10]. The number of applications to radars for solving different problems in determining the coordinatesof an object, in environmental monitoring, etc. is constantly increasing [3, 6, 7, 11-19]. Radar stations (RLS) with active phased array antennas have a special place among all the radar [7, 10, 20-26]. Them are placed both on mobile objects and on stationary complexes [6, 7, 14, 15, 27-35].

The design and study of receiving modules is an important cycle in the development of devices for radar operation [21, 24-27, 33, 36-40]. These modules are part of the receiving / receiving unit, which receives a signal from the APAA transducer, generates a probing signal and amplifies it to the level required for the APAA operation.

The development and improvement of the design of receiving modules is an urgent problem. Over the decades of work, various modernized versions of the receiver have been developed, which, in turn,require refinement and optimization [27, 28, 35, 37, 40, 41].

The receiving module is a superheterodyne radio receiver - one of the types of radio receivers basedon the principle of converting the received signal into a signal of a fixed intermediate frequency (IF) with its subsequent amplification. The main advantage of a superheterodyne over a direct-amplified radio receiver is that the parts of the receiving path that are most critical for the reception quality (narrow-band filter, IF amplifier and demodulator) should not be tuned to different frequencies, which allows them to be performed with significantly better characteristics [38, 40, 42].

2. Superheterodyne radio receiver with double frequency conversion and FOTS

The main disadvantage of a direct gain receiver is the complexity of tuning from one frequency to another. It is almost impossible to make a filter with stable parameters when it is tuned in the frequencyrange [21-23, 38].

In order to solve this problem, they began to divide the task into two stages - tuning in the frequency range and ensuring selectivity in the adjacent channel. For tuning in the frequency range, they began touse the spectrum transfer to a certain (usually rather low) intermediate frequency.

Using the mixer, you can easily move the spectrum of the input signal in frequency by changing thefrequency of the local oscillator.

Unfortunately, the superheterodyne receiver can receive both working and mirror channels. Therefore, when implementing a superheterodyne high frequency radio receiver, a contradiction arisesbetween the requirement to reduce the intermediate frequency to facilitate cancellation of the adjacent channel in the intermediate frequency path and the requirement to increase the intermediate frequency to facilitate cancellation of the image channel. In this case, double frequency conversion is used.

One of the most vulnerable spots in APAA for interference is the paths from the reference signal generator and the local oscillator, as well as to the control and signal preprocessing unit, since they have high requirements for frequency stability, signal-to-noise ratio and phase noise. One of our proposed solutions to this problem is the replacement of feeder paths transmitting these signals with fiber-optic transmission system, since optical fiber is not affected by high-frequency interference and has good noise characteristics when transmitting high-frequency signals, has high flexibility and low weight. Figure 1 shows a diagram of a superheterodyne receiver with a fiber-optic transmission system. Figure 2 shows a diagram of a fiber-optic transmission system.

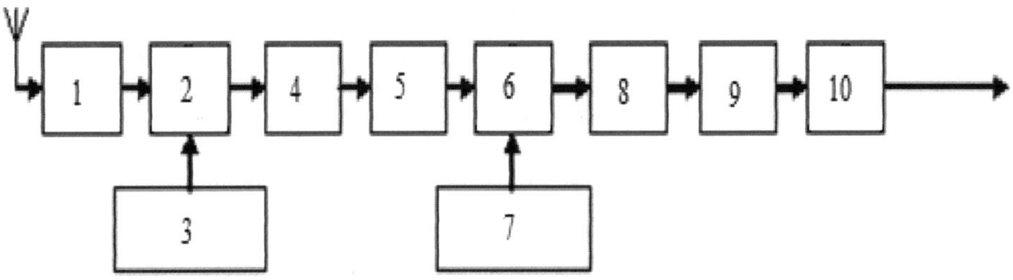

Figure 1. Block diagram of a superheterodyne radio receiver with double frequency conversion and FOTS, where 1 is an amplifier, 2 is a mixer, 3 is a local oscillator, 4 is a fiber-optic transmission system, 5 is an amplifier, 6 is a mixer, 7 is a local oscillator, 8 is a filter, 9 is an amplifier, 10 is a demodulator.

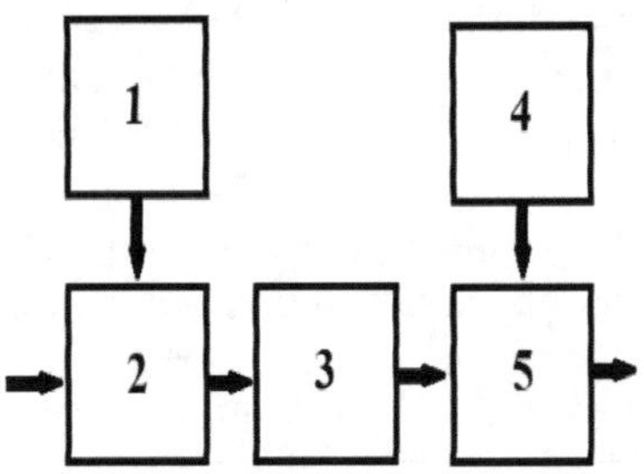

Figure 2. Block diagram of a fiber-optic transmission system: 1 - power driver; 2 - transmitting laser module; 3 - optical isolator; 4 - power driver; 5 - receiving optical module.

The location of the FOTS not only in the receiving module itself, but also at the output can significantly improve the quality of the transmitted signal and facilitate subsequent work with it.

3. Results of using FOTS

The most important characteristic that needs to be investigated when transmitting a heterodyne signal over a FOTS is the dependence of power on frequency. Determine the presence of distortions in the signal spectrum at the FOTS output. Possible distortions in the spectrum during the transmission of the heterodyne signal via the FOTS at an operating frequency of 10 MHz were investigated on the laboratory model of the FOTS developed by us. Figure 3 shows one of the results of these studies as an example.

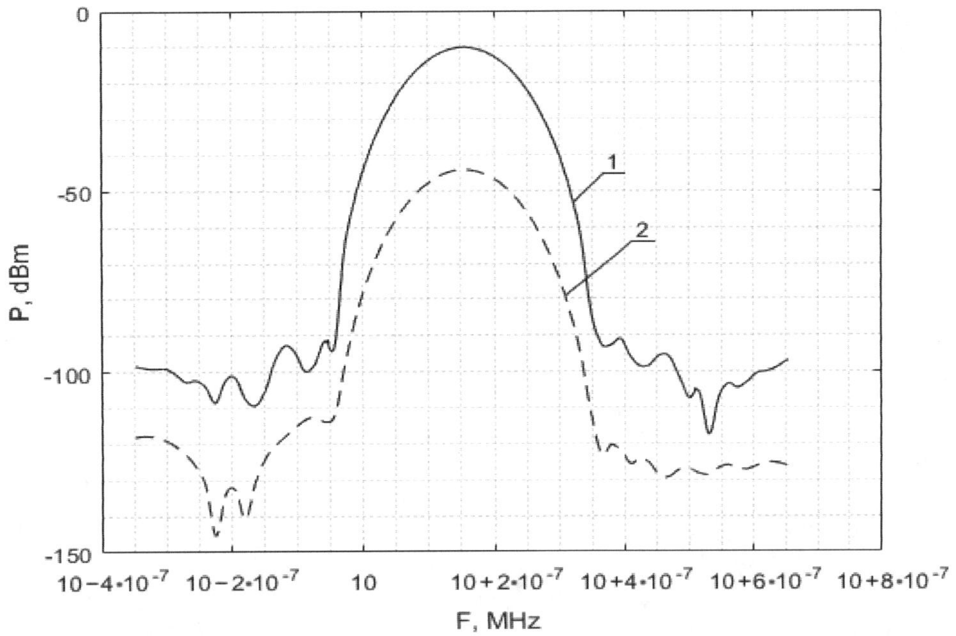

Figure 3. Spectra of the heterodyne signal at the input (graph 1) and output (graph 2) of the FOTS.

Comparison of the obtained spectra shows the high efficiency of the transmission of the heterodyne signal at the carrier frequency of 10 MHz over the FOTS. Distortions in the spectrum are present only on the side components, which does not affect the accuracy of determining the distance to the target in the radar.

Since the developed FOTS is intended for APAA, which are placed on aircraft, it will be operated in different temperature conditions. A change in ambient temperature causes both a change in the refractive index of the fiber and additional elongation of the fiber due to thermal expansion or contraction. This leads to a change in the phase of the light and, accordingly, to a change in the modulation phase of the radiation transmitted through the fiber. Therefore, an experimental estimate was made of the temperature drift of the modulation phase during propagation in the fiber. Figure 4 shows the experimental dependence of the phase shift of light modulation $\Delta\varphi_m$ on the ambient temperature T.

The use of the developed design allows to reduce the number of feeder paths. Provide complete isolation between receiving devices and signal processing systems.

4. Conclusion

As a result of testing the laboratory model of the FOTS developed by us, it was found that the effect of the phase incursion, which is associated with the stretching and compression of the optical fiber under the influence of temperature, on the frequency and amplitude characteristics of the transmitted optical signal is insignificant. Even with temperature fluctuations from 243 to 323 K, since the distance over which the optical signal is transmitted in the APAA is less than 20 m. This, as well as the absence of significant distortions in the signal spectrum, will make it possible to successfully use the receiving module with FOTS in airborne radars. The developed design will meet the high requirements for the elements of stations of this type.

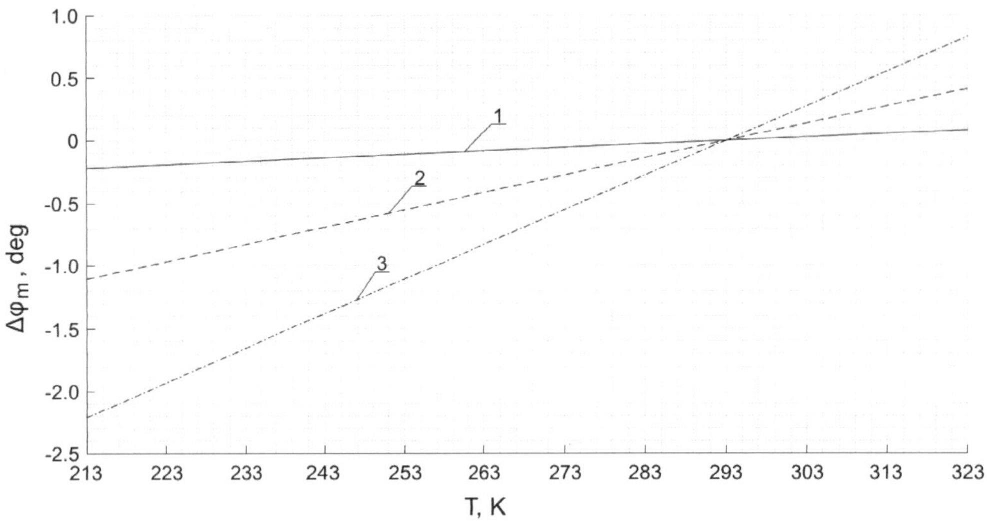

Figure 4. Shift of the modulation phase from the temperature of the medium. Graphs 1, 2 and 3 correspond to the frequency of the heterodyne signal in MHz: 10; 50; 100.

References

[1] Petrov A A, Grebenikova N M, Lukashev N A, Ivanova N V, Rodygina N S and Moroz A V 2018 *Journal of Physics: Conference Series* **1038 (1)** 012032

[2] Davydov R, Antonov V and Moroz A 2019 *International Conference on Electrical Engineeringand Photonics EExPolytech-2019* (Saint-Petersburg) **8906791** pp 295–297

[3] Shannikov D V and Kuzmin S V 2002 *1st IEEE International Conference on Circuits and Systems forCommunications* (ICCSC 2002) **1029113** pp 350–353

[4] Shannikov D V and Kuzmin S V 2003 *Technical Physics Letters* **29(11)** 941–943

[5] Raimzhanov T R, Kuzmin S V and Korovin K O 2021 *Springer Proceedings in Physics* **255** pp 629–635

[6] Fokin G and Lazarev V 2019 *International Congress on Ultra Modern Telecommunications andControl Systems and Workshops* **8970939** pp 123–127

[7] Fokin G 2020 *International Conference on Advanced Communication Technology* (ICACT – 2020) **9061461** pp 1193–1197

[8] Davydov R, Antonov V, Makeev S, Batov Y, Dudkin V and Myazin N 2020 *E3S Web Conference* **140** 02001

[9] Van S, Cheremisin A, Chusov A, Switala F and Davydov R 2019 *IOP Conference Series: Earthand Environmental Science* **390(1)** 012011

[10] Lukashev N A, Davydov R V and Glinushkin A P 2019 *Journal of Physics: Conference Series* **1326(1)** 012046

[11] Gryznova E, Batov Y and Myazin N 2020 *E3S Web of Conferences* **140** 09001

[12] Gryznova E, Grebenikova N, Ivanov D and Bykov V 2019 *IOP Conference Series: Earth andEnvironmental Science* **390(1)** 012044

[13] Nikitina M, Grebenikova N, Dudkin V and Batov Y 2019 *IOP Conference Series: Earth andEnvironmental Science* **390(1)** 012024

[14] Petrov A A, Shabanov V E, Zalyotov D V, Bulyanitsa A L and Shapovalov D V 2018 *International Conference on Electrical Engineering and Photonics EExPolytech 2018* (Saint-Petersburg) **8564389** pp 52–55

[15] Petrov A A, Vologdin V A and Zalyotov D V 2015 *Journal of Physics: Conference Series* **643(1)** 012087

[16] Petrov A A and Grebennikova N M 2018 *Journal of Communications Technology and Electronics* **63(11**) 1281–1285
[17] Grevtseva A S, Davydov R V, Dudkin V I and Rud' V Y 2019 *Journal of Physics: Conference Series* **1326(1)** 012043
[18] Petrov A A and N S Myazin 2017 *Lecture Notes in Computer Science (including subseries Lecture Notes in Artificial Intelligence and Lecture Notes in Bioinformatics)* **10531 LNCS** 561–568
[19] Valov A P 2019 *Journal of Physics: Conference Series* **1410(1)** 012246
[20] Andropov A V, Kuzmin S V and Korovin K O 2021 *Springer Proceedings in Physics* **255** pp 675–81
[21] Podstrigaev A S, Smolyakov A V, Grebenikova N M and Davydov R V 2019 *Lecture Notes in Computer Science (including subseries Lecture Notes in Artificial Intelligence and Lecture Notes in Bioinformatics)* **11660 LNCS** 525–533
[22] Moroz A V, Davydov R V and Davydov V V 2019 *Lecture Notes in Computer Science (including subseries Lecture Notes in Artificial Intelligence and Lecture Notes in Bioinformatics)* **11660 LNCS** 710–718
[23] Davydov R V, Saveleiv I K and Lenets V A 2017 *Lecture Notes in Computer Science (including subseries Lecture Notes in Artificial Intelligence and Lecture Notes in Bioinformatics)* **10531 LNCS** 177–183
[24] Saveleiv I K, Sharova N V and Tarasenko M Y 2017 *Journal of Physics: Conference Series* **917(5)** 052020
[25] Tarasenko M Y, Lenets V A, Akulich N V and Yalunina T R 2017 *Lecture Notes in Computer Science (including subseries Lecture Notes in Artificial Intelligence and Lecture Notes in Bioinformatics)* **10531 LNCS** 227–232
[26] Tarasenko M Yu, Lenets V A, Malanin K Yu and Akulich N V 2018 *Journal of Physics: Conference Series* **1038(1)** 012035
[27] Kuzmin S V 2004 *Technical Physics Letters* **30(11)** 947–949
[28] Shannikov D V, Surikov V V and Kuz'min S V 2005 *Technical Physics* **50(11)** 1531–1534
[29] Fokin G and Lazarev V 2019 *Proceedings of 2019 IEEE Microwave Theory and Techniques inWireless Communications,* (MTTW 2019) **8897228** pp 9–12
[30] Bachevsky S V, Fokin G A, Simonov A N and Sevidov V V 2019 *Journal of Physics: ConferenceSeries* **1368(4)** 042040
[31] Petrov A and Shapovalov D 2019 *Journal of Physics: Conference Series* **1400(4)** 044008
[32] Logunov S E, Rud V Y, Davydov R V, Moroz A V and Smirnov K J 2019 *Journal of Physics: Conference Series* **1326(1)** 012024
[33] Fokin G, Bachevsky S and Sevidov V 2020 *Proceedings of the 2020 IEEE International Conference on Electrical Engineering and Photonics* (EExPolytech 2020) **9243970** pp 94–97
[34] Lazarev V and Fokin G 2021 *Springer Proceedings in Physics* **255** pp 709–717
[35] Stepanets I, Fokin G and Odoevskii S 2021 *Springer Proceedings in Physics* **255** pp 637–645
[36] Fadeenko V, Fadeenko I, Dudkin V and Nikolaev D 2019 *IOP Conference Series: Earth and Environmental Science* **390(1)** 012022
[37] Fadeenko V V, Kuts V A and Vasiliev D A 2018 *Journal Physics: Conference Series* **1135(1)** 012053
[38] Filatov D L, Galichina A A, Vysoczky M G and Yalunina T R 2017 *Journal of Physics: Conference Series* **917(8)** 082005
[39] Moroz A V and Davydov V V 2019 *Journal of Physics: Conference Series* **1410(1)** 012212
[40] Moroz A V, Davydov V V, Malanin K Y, Krasnov A A and Rud V Yu 2019 *Journal of Physics: Conference Series* **1400(4)** 044009
[41] Fadeenko V B, Pchelkin G A and Beloshapkina O O 2019 *Journal of Physics: Conference Series* **1410(1)** 012238
[42] Fadeenko V B and Pchelkin G A 2019 *Journal of Physics: Conference Series* **1400(4)** 044010

Four-point probe stand for magnetoresistance measurement of unpatterned sample

A N Pestova and O S Trushin

Valiev Institute of Physics and Technology of RAS, Yaroslavl Branch, 150007, Yaroslavl, Universitetskaya, 21, Russia

yablokowa-anastasia@yandex.ru, otrushin@gmail.com

Abstract. An experimental stand for express diagnostics of multilayer spin tunnel structures has been developed. The current-in-plane tunnelling method (CIPT) requires no processing, is fast, and provides reliable data which are reflective of the deposition only. The stand is based on the four-probe method for measuring resistance at external alternating magnetic field. This technique can be applied after only a short processing route, thereby saving time and resources, and reducing the potential for damaging the junction.

1. Introduction

Magnetic Tunnel Junction (MTJ) is an important element of modern MRAM cell. It is believed that MRAM might be the next generation of computer memory [1]. Therefore optimization of the technology for MTJ fabrication is very important task. Standard spin-tunnel structure used to make MTJ consists of two magnetic layers separated by the layer of tunnel dielectric. MTJ fabrication involves complex technological route consisting of deposition of multilayer structure and subsequent MTJ pattern formation through lithography. Achievement of high functional characteristics of MTJ requires intensive quality control at different stages of the technological route [2]. This article describes a non-destructive quality control method for a multilayer spin tunnel structure capable of weeding out rejected wafers prior to production. This method is also suitable for controlling spin-valve structures, measuring AMR, etc. Typically, to control the magnetoresistive effect, the spin tunnel structure is prepared by providing contact pads to the lower conductive bus and the upper layer. The process consists of more than ten technological steps and this preparation may be unsuccessful. Possible solution to that problem is the current-in-plane tunneling method (CIPT) being proposed in the paper [3]. This method is fast, requires no processing, and provides reliable data which are reflective of the deposition only and is currently widely used in modern MRAM production. In this work simplified version of the method is proposed. Unlike the original version of CIPT method present realization uses relatively large probe spacing (about 1 mm) and high amplitude probe current (about 1A).

2. Measuring stand

Magnetoresistive effect represents a change in the resistance of a structure in an alternating magnetic field. The resistance of such structure depends on the mutual magnetization of the magnetic layers. For example, if the magnetizations of two layers are codirectional, the resistance of such structure is low, and if the magnetizations are oppositely directed, the resistance is large [1].

The measuring stand consists of two electrical circuits (Figure 1). The first circuit is designed to create an alternating magnetic field and includes a frequency generator, an amplifier, a capacitor, and Helmholtz coil. To ensure the maximum amplitude of the current in the coils, the generator must provide a resonance in the LC circuit, which occurs at a frequency of 20 Hz. At this conditions Helmholtz coil creates a field of about 10 mT. The second circuit is designed for four-probe resistance measurement and consists of a DC source and a measuring head with probes. Standard spring-loaded pogo pins are used as the probes.

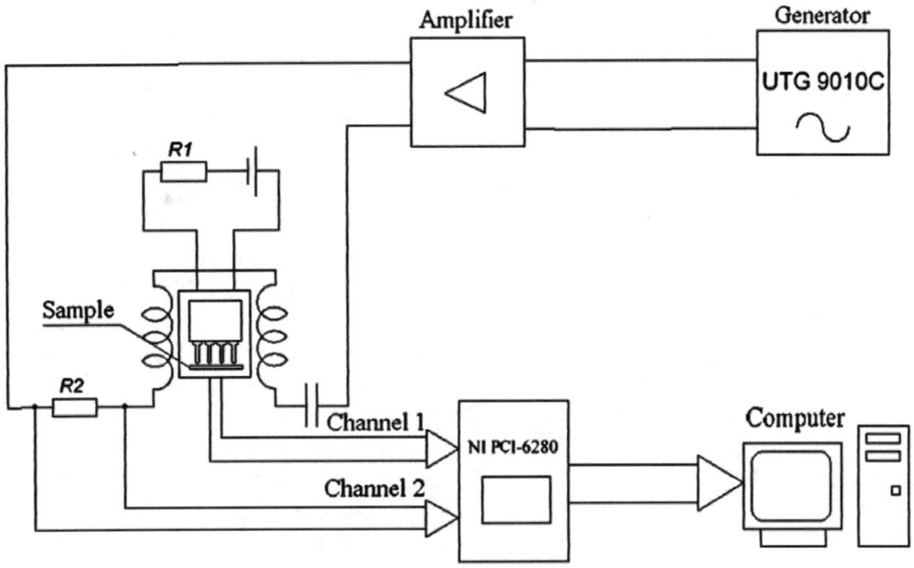

Figure 1. Scheme of the stand.

The measuring stand includes a 4-point probe being in a contact to the sample surface placed in an alternating magnetic field. A probe current is applied to two outer contacts, and the voltage is measured on the two middle contacts. At this stage of development, three probe heads with contact distances of 2.5 mm, 2 mm, 1.6 mm have been prepared. Work is underway to switch to submillimeter sizes. The measured signal is processed on a computer. For this, a program written in the LabVIEW development environment is used.

Figure 2 shows a diagram of the current flow through the sample and a simplified equivalent circuit. Two horizontal resistors represent currents passing in layers and two vertical resistors represent interlayer currents passing through tunnel barrier. The depth of current flow depends on the distance between the contacts. The signal proportional to the interlayer current increases as the probes approach each other. The effect of tunneling magnetoresistance is manifested in a change in the value of the vertical resistors in the equivalent circuit in an external magnetic field.

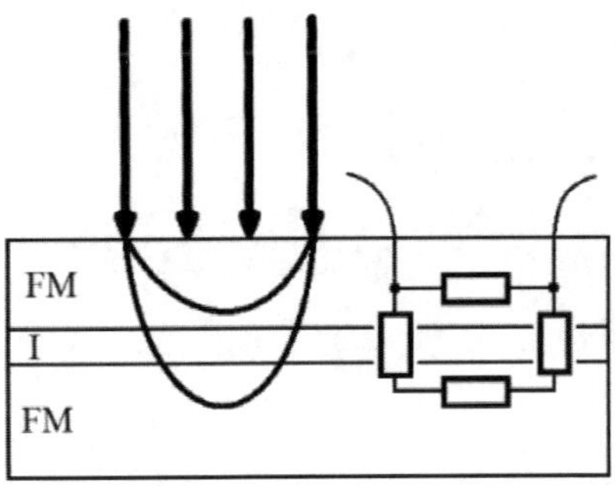

Figure 2. Diagram of the current flow through the sample and the equivalent circuit.

In the papers [3, 4] it has been shown that magnetoresistance effect can be observed if the distance between contacts is around 5-10 microns. But at present version of the described measuring stand the effect can be observed even for the distance between the contacts of about 1 mm. This is achieved by the use of relatively high probe current about 1 A. However, such a large current imposes a limit on the measurement time to prevent overheating of the sample.

There are two assumptions that can explain the observation of the magnetoresistive effect at such a large distance between the contacts. The first assumption is related to the phenomenon of electric current spreading over the surface and over the layers. The second is the heating of the sample when a large current flows through it, which changes the resistance of the layers and makes it possible to observe the effect at large distances.

3. Results

Measurements to control the magnetoresistive effects were carried out using probes with different distances between the contacts. The stand has three measuring heads with contact distances of 2.5 mm, 2 mm, 1.5 mm. Each of them is able to detect the change in resistance in an alternating magnetic field. The samples for measurements were standard spin tunnel structures provided by Crocus Nanoelectronics [2]. Their tunneling magnetoresistance (TMR) value was estimated using CIPT method on CAPRES equipment to be around 140%.

Figures 3, 4 show resistivity versus magnetic field curves for the distances between the contacts of 2.5 mm measured for different orientations of external magnetic field. During the measurement in Figure 3, magnetization occurs along the easy magnetization axis. As a result of measuring the change in resistance in an external magnetic field, a hysteresis loop was obtained for a soft magnetic layer. For complete magnetization reversal of the soft magnetic layer, it is necessary to apply an external magnetic field of about 6 mT. This result is confirmed by earlier studies of the magneto-optical Kerr effect of the same sample. To make sure that the loop obtained in the figure is the effect of tunneling magnetoresistance, it is enough to rotate the sample by 90°. When turning, the external magnetic field becomes directed along the hard axis of magnetization. In this case, magnetization reversal curve change shape. The result of this measurement is shown in Figure 4.

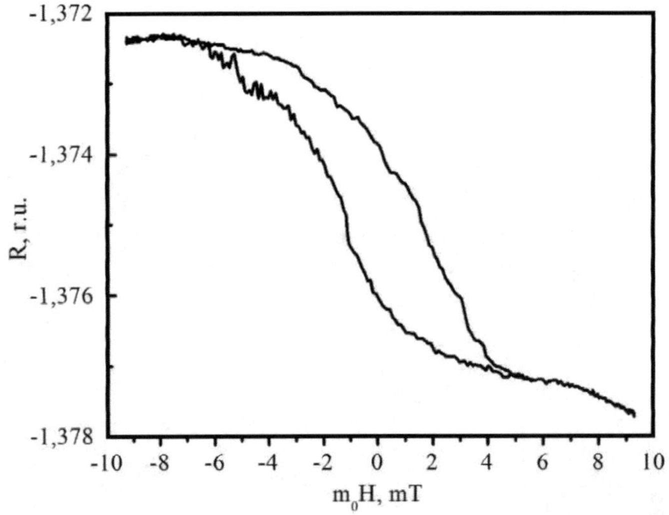

Figure 3. Hysteresis loop for external magnetic field along the easy axis.

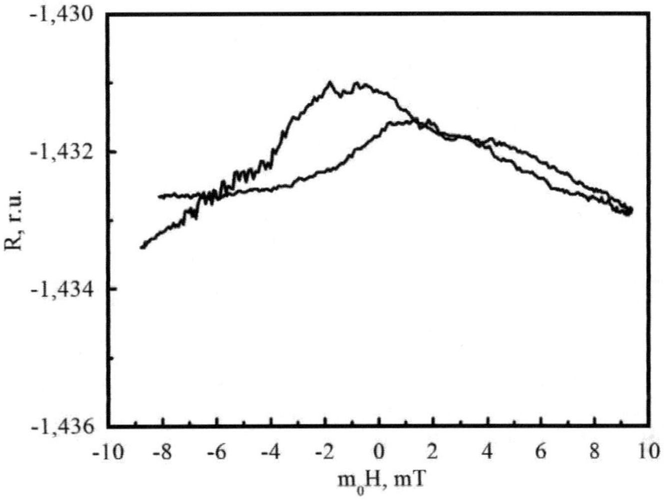

Figure 4. Hysteresis loop for external magnetic field along the hard axis.

4. Conclusion

At the moment, the stand is designed, assembled and tested on a spin-tunnel structure. This stand has shown its efficiency. Thanks to it, it is possible to carry out express diagnostics of structures for the presence of magnetoresistive effects. Unlike the original version of CIPT method present realization uses relatively large probe spacing (about 1 mm) and high amplitude probe current (about 1A). The measurement of the signal for different sample orientations, clearly shows the direction of the easy axis of the structure. It confirms possibility of magnetoresistance measurement at such conditions. The work on the stand improvement and debugging is still underway.

Acknowledgments

This work was carried out on the equipment of the centre for collective use of scientific equipment "Diagnostics of micro- and nanostructures" within the framework of the State assignment of the P.I. K.A. Valiev RAS Ministry of Education and Science of the Russian Federation on topic No. 0066-2019-0003 "Fundamental and applied research in the field of creating promising instrument nanostructures for storing information on new physical principles".

The work was carried out with the support of the Federal State Budgetary Institution "Fund for Assistance to the Development of Small Forms of Enterprises in the Scientific and Technical Sphere" (contract No. 14947ГУ/2019 dated 20.12.2019).

References

[1] Fert A 2008 *UFN* **178** 12 1336
[2] Trushin O S, Simakin S G, Vasiliev S V and Smirnov E A 2018 *Russian Microelectronics* **47** 6 381
[3] Worledge D and Trouilloud D L 2003 *J. App. Phys.* Lett. **83** 84
[4] Christian L Peterseny, Rong Linz, Dirch H Petersenz and Peter F Nielsenz 2006 *2006 14th IEEE International Conference on Advanced Thermal Processing of Semiconductors* 9431599

Towards multipixel THz Schottky diode detector with a single RF output line

A Prikhodko[1,2], **I Belikov**[1], **D Mikhailov**[1,2,3], **A Shurakov**[1] **and G Goltsman**[1,2]

[1] Moscow Pedagogical State University, Moscow 119435, Russia
[2] National Research University Higher School of Economics, Moscow 101000, Russia
[3] All-Russia Research Institute of Physicotechnical and Radio Measurements (VNIIFTRI), Mendeleevo, Russia

Abstract. We propose the design of a dual-pixel array of Schottky diodes. Each diode is fixed between the bow-tie antenna arms on top of a SI-GaAs membrane acting as a waveguide backshort for efficient coupling of the antenna to the feedline of high-directivity horn. The detector utilizes a single RF output line: microwave reflectometer is used for the readout. The pixels are equipped with dual-mode resonator filters to eliminate the cross-talk. We evaluate the design proposed via numerical simulation and performance tests of the array subunits: *NEP* of 300 pW/Hz$^{0.5}$ and dynamic range of 24 dB are demonstrated at 137.5 GHz.

1. Introduction

Terahertz coherent and direct detectors are in demand in numerous applications nowadays. The advantages of GaAs planar Schottky diode (PSD) direct detector are mainly related to a) its implementation-dependent time constant, which is determined only by the diode intrinsic RC-circuitry and output lines RF losses; b) decent sensitivity at room temperature; c) the ease of integration with complex RF circuits ensured by the strict control and reproducibility of the diode parameters. PSD detector also possesses a wide dynamic range beneficial in both the terahertz data transfer and medical imaging applications. However, it is difficult to combine high barrier diodes in array due to the need of DC biasing. This issue is traditionally solved via the use of Mott diodes providing low barriers resulting in high sensitivities if operated with zero bias at room temperature. Alternatively, it is possible to use a microwave frequency comb to readout an array of Schottky diodes [1].

In this paper we report on the evaluated performance of a multipixel terahertz Schottky diode detector with a single RF output line.

2. Detector design and performance tests

2.1. Dual pixel detector design

In accordance with the design proposed, a dual-pixel array of Schottky diodes is arranged as shown in figure 1. Each diode is fixed between the bow-tie antenna arms on top of a back metalized SI-GaAs membrane acting as a waveguide backshort for efficient coupling of the antenna to the feedline of high-directivity horn. There is only one RF output line intended for the injection of a probing microwave dual-frequency comb to readout the pixels [1] electrically connected by the Wilkinson power combiner (WPC). The pixels are equipped with miniaturized dual-mode resonator filters [2] to eliminate the cross-talk. We evaluate the design proposed via numerical simulation and performance tests of the array subunits.

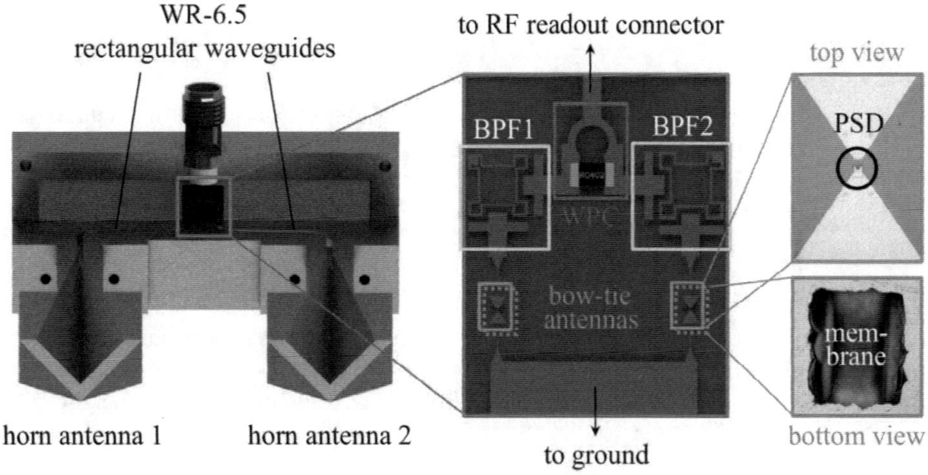

Figure 1. Dual pixel detector design.

The footprint of the miniaturized band-pass filters (BPF) developed for GaAs substrate by the means of EM modeling is within 1.36×1.36 mm^2, which corresponds to a nearly quarter-wave linear dimension given their central frequencies of 18.5 and 20.3 GHz; the fractional bandwidths of ~6 % ensure the pixels isolation of 19.6 dB. The insertion losses introduced by WPC are within 2.2-2.3 dB for each frequency channel, and the return losses are in the range of 15.7-19.8 dB. Please be advised that the inset with a microstrip circuitry of the detector chip is provided to scale, and a 100 Ω 0402 SMD resistor can be used to estimate actual linear dimensions of the microstrips.

Referring to figure 1, the bow-tie antennas are centered with respect to WR-6.5 rectangular waveguides in a way that their arms are oriented along the waveguides' E-planes. For such transitions, EM modeling suggests input return loss of less than -15 dB in the frequency range of 140-160 GHz and insertion loss as low as -3 dB. The diagonal horn antennas [3] possess gain of 23.5 dB and the half power beam width of 12°. Rigidness of the detector chip is due to thick SI-GaAs frame surrounding the membrane.

2.2. Readout method and its evaluation

The detector design adopts planar diodes with a Schottky contact diameter of 3 μm and series resistance of <30 Ω (our fabrication process is described elsewhere [4]) as sensors. It is known that Schottky diode features non-quadratic rectification characteristics at certain operating conditions [5]. Since the probing microwave power can be set for each pixel individually to tune the performance, we theoretically and experimentally investigated the impact of various bias conditions (BC) of the diode on input dynamic range (*IDR*) and the optical noise equivalent power (*NEP*) of the detector.

According to the thermionic emission model for an ideal Schottky diode, the forward bias current-voltage characteristic (IVC) can be described by

$$I_{ideal} = I_s\left(\exp\left(qV_b\left(\eta k_B T\right)^{-1}\right)-1\right). \quad (1)$$

In case of a real Schottky diode, one has to account for the impact of series resistance (R_s) on the shape of IVC at high values of forward bias voltage (V_b) as

$$I_{real} = I_s \left(\exp\left(q\left(V_b - I_{real}R_s\right)\left(\eta k_B T\right)^{-1} \right) - 1 \right). \tag{2}$$

The saturation current (I_s) is defined as

$$I_s = AA_R T^2 \exp\left(-q\Phi_{b0}\left(k_B T\right)^{-1} \right). \tag{3}$$

Here η is the ideality factor, q is the elementary electron charge, k_B is the Boltzmann constant, T is the physical temperature of the diode measured in Kelvin, A is the Schottky contact area, A_R is the Richardson constant, and Φ_{b0} is the zero bias Schottky barrier height. Introducing the Lambert W function (W_0), one can explicitly represent equation (2) as [6,7]

$$I_{real} = \eta k_B T\left(qR_s\right)^{-1} W_0\left(qR_s\left(\eta k_B T\right)^{-1} I_s \exp\left(q\left(V_b + I_s R_s\right)\left(\eta k_B T\right)^{-1} \right) \right) - I_s . \tag{4}$$

The values of η, Φ_{b0}, R_s are extracted upon approximation of the experimental IVC by equation (4) with the aid of the least squares method. In our studies, we rely on symbolic calculations and make use of a cross-platform computer algebra system "Maxima". Flat band barrier height (Φ_{bf}) is further calculated as [8]

$$\Phi_{bf} = \eta\Phi_{b0} - \left(\eta - 1\right)k_B T q^{-1} \ln\left(N_c N_d^{-1}\right),$$

where N_d is the semiconductor impurity concentration, N_c is the effective density of states in the conduction band.

In our experiments, Schottky diode with $\eta = 1.56$, $\Phi_{b0} = 0.577$ eV, $R_s = 14.1\ \Omega$, $\Phi_{bf} = 0.891$ eV is simultaneously biased by a source-meter and microwave frequency synthesizer. The latter also acts as a source of probing signal to readout the diode response to a sub-THz incident power. The output/input power ratio (P_{out}/P_{in}) is measured with the aid of a spectrum analyzer and a sub-THz CW source equipped with precision waveguide attenuator. Referring to figure 2, the slope of a "linear-response" region gradually grows from 1.34 to 1.88 along with a BC set's serial number. This corresponds to the gradual increase of the percentage of AC component (δI_{mw}) induced by the probing signal of 1.4 GHz in the diode bias current (I_b) from 1.25 % for BC#1 to 217.5 % for BC#9. We fix the DC current component (I_{DC}) to 200 μA in all the measurements and observed that $\Delta P_{out}/\Delta P_{in}$ reaches its maximum of 1.91 for BC#10, when the diode is biased by direct current solely and conventional readout scheme is employed. $IDR > 30$ dB can be achieved in both of the above readout options. The current components are related as $I_b = I_{DC}\,(1 + \delta I_{mw})$.

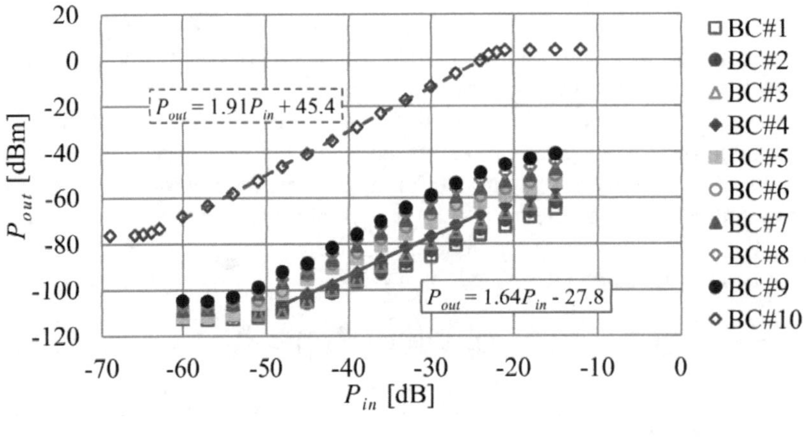

Figure 2. Linearity curves measured for a Schottky diode at different bias conditions.

To analyze the trends in the changes of linearity of a Schottky diode upon increase of the percentage of AC component in its bias current, we calculated a number of time series $[t, I_{real}]$ with the aid of Monte-Carlo method. Analytic approach was inconvenient for the calculation due to the need to deal with modified Bessel functions of the first kind naturally appearing because of $\exp(\xi_0 \sin(\xi(t))$ under integral operator. Indeed, when the diode is under microwaves, its bias voltage is defined as

$$V_b = V_{DC} + V_{AC,0}\left(\sin \omega t\right), \quad (5)$$

where V_{DC} is the DC voltage from a source-meter, $V_{AC,0}$ is the amplitude of AC voltage generated in the diode by frequency synthesizer, and ω is the angular frequency of AC power provided by the synthesizer. Combining equations (4) and (5), one obtains time-dependent current through the diode as

$$I_{real}(t) = \eta k_B T\left(qR_s\right)^{-1} W_0\left(qR_s\left(\eta k_B T\right)^{-1} I_s \exp\left(q\left(\left(V_{DC} + V_{AC,0}\left(\sin \omega t\right)\right) + I_s R_s\right)\left(\eta k_B T\right)^{-1}\right)\right) - I_s. \quad (6)$$

For further analysis, we divide the $I_{real}(t)$ dependence obtained in 32 sections of small-signal response. Sufficiency of such a linear interpolation is justified by the analysis of linearity curves provided in figure 3. Average current value ($\langle I_{real}\rangle$) over the period of microwave probing signal is further calculated. Equating this value to I_b measured for a given BC set's serial number, one acquires the corresponding $V_{AC,0}$ value.

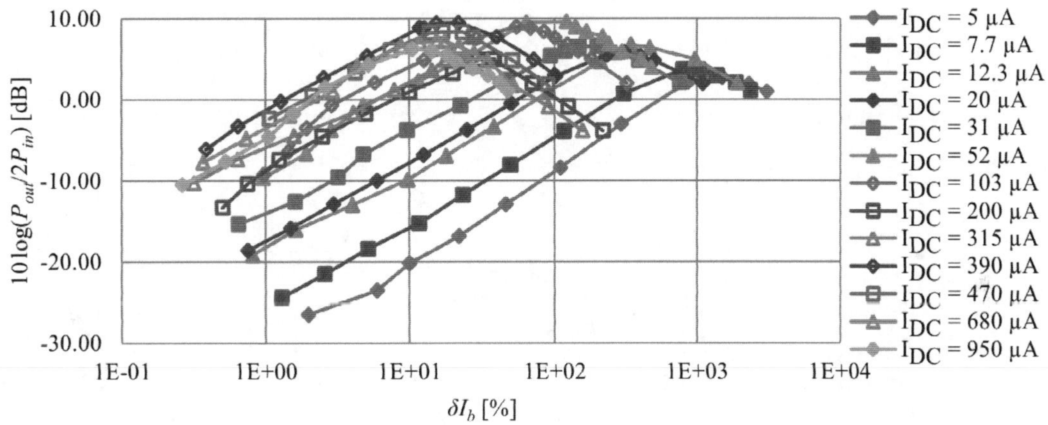

Figure 3. Linearity of the terahertz current generation as a function of DC bias.

Reflected microwave power at a 50 Ω load is further calculated as [9]

$$P_{50}(t) = \left|V_{AC}(t)\right|^2 50\left(\left(50 + \mathrm{Re}\left(Z_{PSD}(t)\right)\right)^2 + \mathrm{Im}\left(Z_{PSD}(t)\right)^2\right)^{-1} \quad (7)$$

with

$$Z_{PSD}(t) = R_s + \left(\left(R_{diff}(t)\right)^{-1} + i\omega C_{tot}\right)^{-1}. \quad (8)$$

Here $R_{diff}(t)$ is the diode time-dependent differential resistance and $Z_{PSD}(t)$ is the diode time-dependent impedance. For its calculation, we assume that total capacitance of the diode element (C_{tot}) is solely determined by the parasitic capacitance value, which is estimated as 4 fF for our diodes [1]. The junction capacitance (C_j) is neglected, since the diode operates as a varistor (figure 4), and instantaneous value of its bias voltage stays within the range from 0.55 to 0.81 V during all the measurements. Such voltages significantly exceed built-in potential (V_{bi}) for the diodes studied.

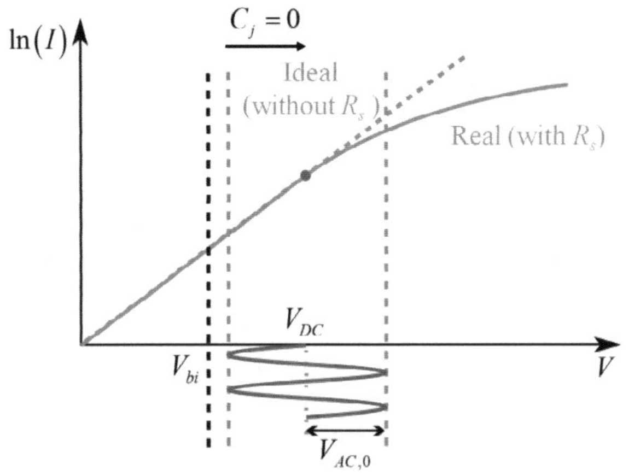

Figure 4. Real and ideal IVCs of a varistor diode.

Referring to figure 5, linearity of the diode response to microwaves gradually grows along with a BC set's serial number. The trend observed is qualitatively consistent with the experimental results.

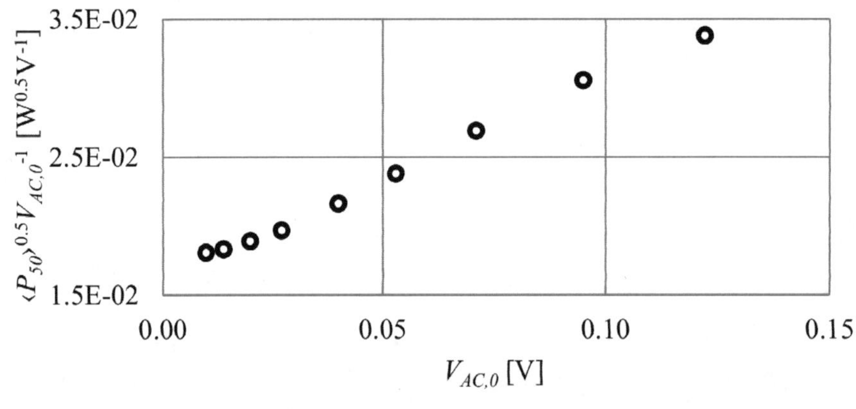

Figure 5. Normalized reflected microwave voltage at a 50 Ω load averaged over the period of microwave probing signal as a function of the amplitude of AC voltage generated by it in a Schottky diode (i.e. as a function of a BC set's serial number).

As a sweet spot of the detector's linearity and sensitivity, we measured $\Delta P_{out}/\Delta P_{in} = 1.64$, $IDR = 24$ dB and $NEP = 300$ pW/Hz$^{0.5}$ at signal frequency of 137.5 GHz for the diode biased by $I_{DC} = 200$ μA with $\delta I_{mw} = 10$ % (figure 2: BC#4).

3. Conclusion

We propose the design of a dual-pixel array of Schottky diodes. The detector utilizes a single RF output line: microwave reflectometer is used for the readout. We evaluate the design proposed via numerical simulation and performance tests of the array subunits. Given the capabilities of the input terahertz optics and the microwave distribution network developed, it seems feasible to expand the design of a dual-pixel array of Schottky diodes to that of the linear array with a noticeably bigger number of elements.

Acknowledgments
We acknowledge support of the Russian Science Foundation grant No. 19-72-10156.

References
[1] Shurakov A, Prikhodko A, Mikhailov D, Belikov I, Kaurova N, Voronov B, and Goltsman G 2020 *Journal of Physics: Conference Series* **1695** 012156
[2] Hong J S and Lancaster M J 1995 *IEEE Microwave and Guided Wave Letters* **5** 371-2
[3] Balanis C A 2005 *Antenna theory: analysis and design* (Hoboken: John wiley & sons)
[4] Shurakov A et al. 2018 *Microelectron. Eng.* **195** 26-31
[5] Harter T et al 2020 *Nature Photonics* **14** 601-6.
[6] Jung W and Guziewicz M 2009 *Materials Science and Engineering: B* **165** 57-59
[7] Latry O, Divay A, Fadil D and Dherbécourt P 2017 *Journal of Semiconductors* **38** 014007
[8] Wagner L F, Young R W and Sugerman A 1983 *IEEE electron device letters* **4** 320-22
[9] Pozar D M 2004 *Microwave Engineering* (Hoboken: John wiley & sons)

The electromagnetic absorbers based nano - structured granular polymer-composites at gigahertz frequencies.

D Savelev[1], E Grushevski[1], N Savinski[2], M Soloviev, V Turov[1] and V Krenev[1]

[1] Department of Nanotechnology in Electronics P.G. Demidov Yaroslavl State University, Sovetskaya Street 14, 150003, Yaroslavl, Russia.
[2] Valiev Institute of Physics and Technology of Russian Academy of Sciences, Yaroslavl Branch, 150007, Universitetskaya Street 21, Yaroslavl, Russia.

dmitriy.savelev9@mail.ru

Abstract: The rapid increase in electromagnetic interference has received a serious attention from researchers who responded by producing a variety of radar absorbing materials especially at high gigahertz frequencies. Ongoing investigation is being carried out in order to find the best absorbing materials which can fulfill the requirements for smart absorbing materials which are lightweight, broad bandwidth absorption, stronger absorption etc. Therefore, this article introduces the electromagnetic wave absorption mechanisms and then reveals and reviews those parameters that enhance the absorption performance.

1. Introduction

With the development of electromagnetic technology, the problem of electromagnetic hazard is becoming more and more serious. So, the roles of electromagnetic absorption are attracting more attention in the military and civilian. Electromagnetic wave absorption coating, as an important means of radar wave absorption material, has a wide range of application. Because few materials in nature can respond to THz frequency, and because there is a lack of effective THz sources and detectors, it still remains a big challenge to effectively manipulate THz waves. Recently, graphene has been demonstrated as a viable alternative in realizing THz absorbers because of its high THz absorption and tenability of surface conductivity [1]. The absorption of electromagnetic energy (EME) is due to dielectric, magnetic and conductivity losses, which are trying to maximize to achieve the maximum efficiency of shielding. Composite materials appear to provide the most attractive and promising solution to demands for lightweight, high-performance structures. Since composite materials have high specific strength and stiffness compared with bulk materials, the range of their applications is rapidly expanding from aerospace and military use to commercial activities. On the other hand, the use of polymers to protect the electronic devices from EME is popular due to the light weight, flexibility and cost-effectiveness. Here we report a method based on generation of materials with a wide range of adjustable electrical and magnetic properties: conductive polymers (polyanilines), flaky nanographite, nano-structured granular composites.

2. Experiment

2.1. Materials

2.1.1. Polyaniline (PANI). The main problem of PANI is that it is poorly soluble in most common solvents, hampering experimental studies and limiting its industrial exploitation. The matrix polymerization of aniline neutralized by polymeric acids is of prime significance as a method of preparing soluble materials combining electrophysical properties of polyaniline (PANI) and the mechanical properties of polymer materials. We used poly (2-acryl-amino-2-methyl-1-propane) sulpha acid (PAMPSA) [2], are distinguished by the regular distribution of sulpha acid groups along the main chain of macromolecules. It was assumed that, upon protonation of aniline by the above polyacids, the resulting PANI macromolecules would be located along polyacid macromolecules and form a two-strand macromolecular structure of the interpolymer complex. In this case, the conformations of polyacid and PANI macromolecules are changed to adjust each other. The synthesis was performed at room temperature while successively mixing solutions of PAMPSA, aniline, and ammonium persulphate in water. The ratio of PAMPSA to aniline in the reaction mixture was 1 : 1 mole-unit/mole, and that of aniline to oxidant 1 : 1.25 mole/mole. The concentration of aniline and PAMPSA in the reaction mixture 0.034 M/L. The conducting emeraldine salt (ES) form can be obtained by oxidative doping of leucoemeraldin base (LB), or by protonation of EB. The electroconductivity of film determined by of polyaniline synthesized by the method of template synthesis with PAMPSA, which was spinning on transparent glass electrode ITO from an aqueous solution at 3500 rpm. An electroconductivity, was measuring by the four-probe method at room temperature, was equal to 0.25 S cm^1.

2.1.2. Ferrites. Ferrites are considered to be the best magnetic material for electromagnetic wave absorbers due to their excellent magnetic and dielectric properties, but they are expensive and heavy. Moreover, nowadays, research interest is focusing on and towards higher gigahertz frequency, nano-particles, ferrite incorporated into polymers, over a single layer, wide bandwidth and to achieve higher and higher absorption. In this study, ferrites produced by JSC «Ferros» Yaroslavl were used. The barium ferrite was used in the form of small magnetized balls, with a diameter of 3 mm and in the form of a powder with a dispersion of 30 microns.

2.1.3. Nanographite Graphite exfoliation. Typically, graphite exfoliation was performed in a two-electrode system, whereby graphite foils were used as working anodes (commercial graphite foil Graflex RF) and platinum foils as counter electrodes. To avoid short circuits (connections between the graphite and Pt foils) during exfoliation, graphite foils were rolled like rods. The Pt foils were placed parallel to the graphite electrodes at a fixed distance of 2 cm. The electrolyte for the graphite exfoliation was prepared by dispersing ammonium sulfate crystals (3.0 M) and DMSO (dimethyl sulfoxide) in DI water. When the electrodes were immersed in the electrolyte, a static potential of 10 V was applied to start the electrochemical exfoliation process. When the exfoliation was complete, the suspended graphene sheets were collected with a 0.2 μm PTFE membrane filter and washed alternatively with DI water and ethanol by vacuum filtration. The washing process was repeated three times to clear any chemical residues. Later, the dark-gray product was dispersed in N, N - dimethylformamide (DMF) via mild sonication (15 min). This homogeneous suspension sat overnight to remove un-exfoliated flakes and large particles. Then, the supernatant was taken out for further measurement and ultimately device fabrication.

2.2. Preparation of microwave-absorbing coatings

As a target, fiberglass was used in the form of an isosceles triangle with an edge 300 mm long, 0.8 mm thick, with an area of ~390 cm^2, and a mass of 33 g. The coating was applied in bulk and with a brush, in the form of several layers with successive drying in a drying cabinet at 80^0C . As a binder, a 30%

(by dry residue) aqueous solution of polyvinyl acetate emulsion DF50/5H was used. Dispersion of solid substances in the emulsion was carried out using a laboratory mechanical dispersant for 20 minutes. If necessary, the emulsion was diluted with deionized water. In the case of sample 5, the dispersant Synthanol 10 at a concentration of 1% by weight was used to improve wetting. The thickness of the coatings was 50-170 microns.

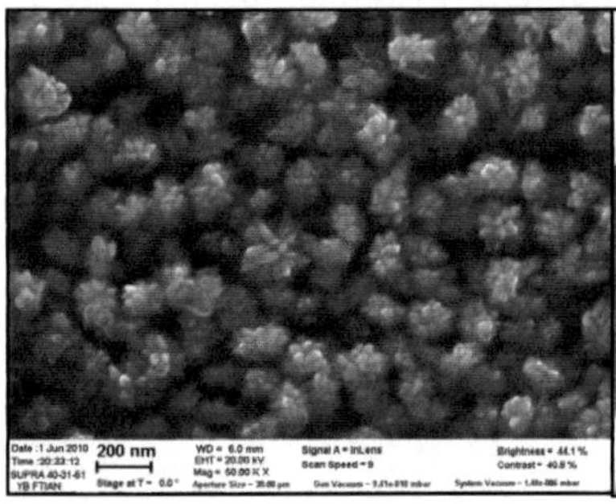

Figure 1. Image of the nanographite after the exfoliation process.

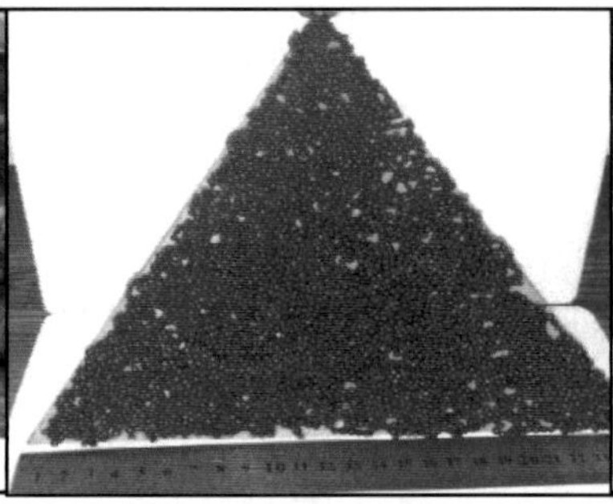

Figure 2. Photo of a target with spherical particles of 3 mm barium ferrite.

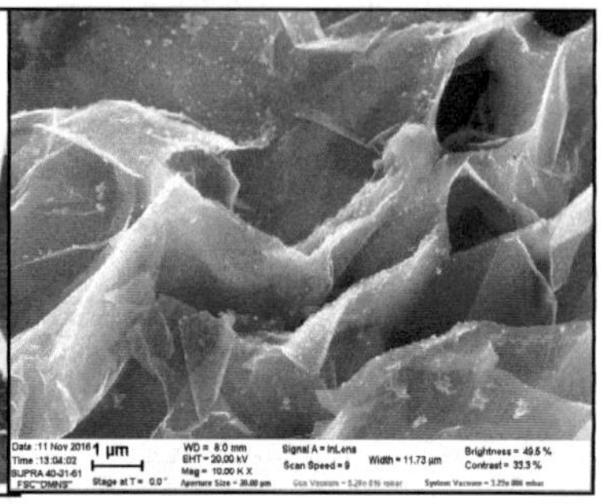

Figure 3. Image of the nanographite after the exfoliation process.

2.3. Measurement

Microcosmic morphology of nano - structured granular polymer-composites and cross section of polymer matrix filled with and GNs were performed using scanning electron microscopy (SEM, SUPRA - 40). The polyaniline particles have a structure similar to «raspberries», small grains have a size of 27-29 nm, large «berry» have a size of 150-160 nm. Phase analysis of the products of exfoliation graphite was done via X-ray diffraction on an ARLX'Tra X-ray diffractometer under the following conditions and parameters: CuKα radiation with a wavelength of nm; λ = 0.15418an accelerating voltage of 35 kV; and an incandescent current of the X-ray tube of 40 mA. The elemental composition of the samples was determined via Rutherford backscattering spectrometry. Three tablets of a compressed product of exfoliation were used as samples and a graphite plate was used as a reference. The measurements were made on the λ = 0.15418 K2MV (HVEE) facility. We used He+ ions with an energy of 1.0 MeV at normal beam incidence. Scattered ions were detected at angle θ = 150° using a surface barrier semiconductor detector with a resolution of 13 keV. The spatial angle of the detector was 3.415 msr. The radiation dose in all experiments was 54 μCl at an ion current of 40 nA on each sample. X-ray diffractometry of our graphite foil samples and products of electrochemical exfoliation revealed the presence of two crystallographic phases: hexagonal 2H and orthorhombic 3R. X-ray diffractometry and electron microscopy showed the exfoliated graphite samples contained crystallites of virtually half the normal size that had 30–50 graphene layers each and could be classified as nanographites. Our Rutherford backscattering spectrometry study confirmed the substantial oxidation of graphite samples during electrochemical exfoliation, due to the generation of a strong oxidizer (a hydroxyl radical) [3,4]

3. Results and disscution

3.1 Factors influence the electromagnetic wave absorption

The general scheme of the processes occurring when electromagnetic radiation falls on the coating is Illustrated in Figure 4.

3.2 Measurements of absorption properties

The portable radar was used to study the absorption properties of electromagnetic radiation. The radar signal was processed using the consistent filtering method. The consistent filtering assumes finding the filter response in the form of convolution of the radar input signal x(t) and the pulse characteristic of the filter $h(t_0\text{-}t)$, consistent with the probe radar signal. With this processing in the output signal of the matched filter (SF) z(t), the maximum signal-to-noise ratio is achieved by the accumulation of the useful signal energy and suppression of out-of-band noise. Additional measures may be used to suppress possible interference [6]. The test sample (target) was located so as to be in the Fraunhofer zone (far field) at a distance from two pairs of broadband radar antennas with a circular aperture of 5 cm and a beam half-width of 20^0. The reflection measurements were carried out at incidence angles of 0^0, 60^0, 120^0 to the normal. For each sample, values were averaged over 100 parallel signals; the obtained data were averaged over 32 more polarizations and over 3 angles of incidence. The received data were normalized by the average reflection value of the sample of a fiberglass substrate without an absorbing coating.

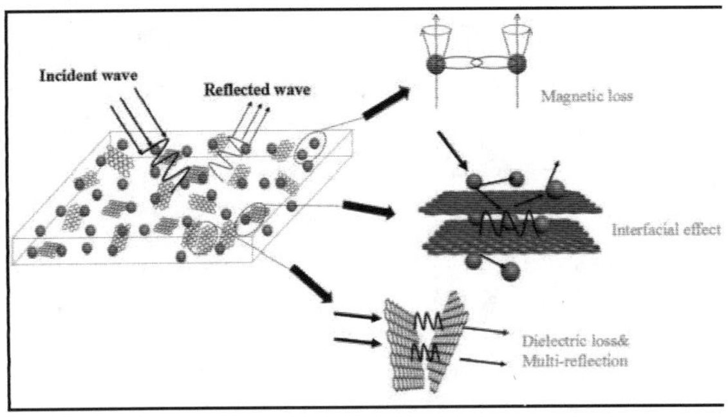

Figure 4. Schematic illustrations of the absorption mechanisms when incident wave enters into coating [5].

Figure 5. Photo of a portable radar for measuring the electromagnetic characteristics of coatings.

3.3. Grain sizes (nanometer vs micrometer and the effects of particle size upon the EM wave properties)

The relationship between thickness, permeability and frequency of minimum reflection loss can be explained using the following equation

$$f_{\mathrm{m}} = \frac{C}{2\pi\mu'' d} \quad (1)$$

where f_m : matching frequency with minimum reflection loss; c : velocity of light; d : sample thickness; μ' : imaginary part of permeability.

Thus, from this equation it is easy to tell that the matching frequency f_m shifts towards lower frequency with increasing sample thickness [7]. More results have been reported on the fact that increasing thickness causes the resonance frequency to move towards lower frequency [8].

Nanomaterials which have a new wave absorption mechanism lead to an excellent electromagnetic wave absorber performance in the GHz range with the effects of small size, surface and shape anisotropy. According to the natural –resonance equations [9],

$$2\pi f_{\mathrm{r}} = H\, r_{\mathrm{a}} \quad (2)$$

where H_a :

$$H_{\mathrm{a}} = \frac{4|K_1|}{3\mu_0 M_S} \quad (3)$$

where r_a is the gyromagnetic ratio, H_a is the anisotropy field and K_1 is the anisotropy coefficient. It demonstrates that the resonance frequency of the magnetic materials shifts to higher frequency when H_a increases due to the small size effect. It is believed that the anisotropy energy of small size materials, especially on nanometer scale, would be remarkably increased due to the increased surface anisotropic field induced by the small size effect.

3.4 Electromagnetic characterist

Table 1. Types of polymer matrix, filler, filler's particle size and its absorbing properties.

№	Structure	Filler's particle size	Coating density (g/cm^2 10^{-2})	Reflection loss (dB) at 10 GHz	Absorbed energy %
1	$BaFe_{12}O_{19}$ Ferrite in the form of small balls, 3 mm in one layer (magnetized balls)	Barium ferrite 3 mm.	94	-14.3	96,3
2	$BaFe_{12}O_{19}$ (1: 1 by mass to mass of 30% emulsion)	Barium ferrite 30μ	10	-11.2	92,4
3	Nanographite:1: 1 by mass to mass of 30% emulsion , 0.1% Syntanol-10	Nanographite 10-100μ, thickness 12 ± 3 nm	4.7	-14.4	96,4
4	Aluminum Powder PAP2 1: 1 by mass to mass of 30% emulsion	Aluminum powder 20-30μ	4.6	-11.8	93,4
5	PANI (Emeraldine salt, sulfonic acids)	Polyaniline 27-29 nm	3.7	-13.3	95,3
6	Ultra-wide flexible radar absorbing material based on nanostructured ferromagnetic microwire [10]	JSC «CDB RPM» МРПК-1Л	10	-17	98

4. Conclusion

As a result, the nanographite coating has a power reflection coefficient of -14.4 dB. This value is close to the material МРПК-1Л (- 17 dB) [10]. But the advantage of nano-structured granular composites, particularly, nanographite is that coating thickness and weight less than competitors with the same degree of reflection and absorption an electromagnetic waves. It is shown that on the basis of granular nanostructures it is possible to create magnetic materials with close complex magnetic and dielectric permittivity at a magnitude of magnetic losses several times higher than traditional ferrite magnetic materials. These materials can be the basis for the creation of effective broadband electromagnetic wave absorbers with a thickness of hundreds of microns.

Acknowledgements

This research was financially supported within the framework of the state assignment of the Valiev Institute of Physics and Technology of the Russian Academy of Sciences Yaroslavl Branch of the Ministry of Education and Science of the Russian Federation on the topic No. 0066-2019-0003 and the state assignment of P.G. Demidov Yaroslavl State University No. 0856-2020-0006. Some experimental results were obtained on the equipment of the Center for Collective Use "Diagnostics of Micro-and nanostructures" with the support of the Ministry of Education and Science of the Russian Federation.

We would like to thank A. Dubov an application ferrites of JSC «Ferros» Yaroslavl.

References

[1] Moradiani F, Farmani A, Yavarian M, Mir A and Behzadfar F 2020 *Physica E: Low-dimensional Syst. and Nanostructures* **122** 114159

[2] Savinsky N G 2012 *Bulletin of the Yaroslavl State University P. G. Demidov Natural and Technical Sciences series* **4** 53-64.

[3] Solov'ev M E, Raukhvarber A B, Savinskii N G and Irzhak V I 2017 *Russian journal of general chemistry* **87** N 4 805-811.

[4] Savinsky N G, Melesov N S, Parshin E O, Vasiliev S V, Bachurin V I and Churilov A B 2020 *Bulletin of the Russian Academy of Sciences: Physics* **84** N 6 732–735.

[5] Duana Y, Liua Y, Cuia Y, Mab G and Tongmina W 2018 *Progress in Organic Coatings* **125** 89–98.

[6] Turov V E 2011 *Radar warfare. Construction and noise protection of basic correlation systems of passive location: Monograph.-M* (*Vuzovskaya kniga*) 205

[7] Mohd F, Idris, Hashim M, Abbas Z, Nazlan R and Ibrahim I 2016 *Journal of Magnetism and Magnetic Materials* **405** 197–208.

[8] Liu L D, Duan Y P, Ma L X, Liu S H and Yu Z 2010 *Appl. Surf. Sci.* **257** 842–846.

[9] Kong J, Liu J, Wang F, Luan L, Itoh M and Machida K 2011 *Appl. Phys.* **105** 351–354.

[10] Patent RU 2322735 C1, H01Q 17/00 (2006.01) *electromagnetic wave absorber* / Ustimenko L G , Vladimirov D N, Smirnov G A, Suslov L M and Khandogina E N; OAO "TsKB RM"; Applic. 2006125655/09 ; Date of public. 20.04.2008 Bull. 11

Silicon carbide of 4H-SiC type Schottky diode current-voltage characteristics in small-sized type metal-polymeric package SOT-89

S V Sedykh, S B Rybalka, A A Demidov and E A Kulchenkov

Bryansk State Technical University, Bryansk, 50 let Oktyabrya 7, Russia

Abstract. The forward and reverse current–voltage characteristics of Ti/Al/4H-SiC Schottky diode type DDSH411A91 in modern small-sized (SOT-89) type metal-polymeric package have been obtained. In forward direction (current up to 2 A) on the basis of analysis it is shown that Schottky diode corresponds to the "ideal" diode with ideality factor n=1.12 and effective Schottky barrier height φ_B=1.2 eV. It is shown that reverse current-voltage characteristics (breakdown voltage 1200 V) can be well approximated by mechanism of field dependence of barrier height lowering by the presence of the intermediate layer in the form of oxide on the 4H-SiC surface.

1. Introduction

It is known that the silicon carbide (SiC) Schottky diodes are key component of power semiconductor electronics devices for high-temperature device applications because of its high breakdown voltage, low series resistance and stability under high temperature conditions [1]. Now modern power electronic industry comes down to use of small type of metalpolymeric package such as SOT (Small Outline Transistor), QFN (Quad Flat No-leads) and others [2], but in Russia manufacture of the main electronic components earlier was produced in standard large-sized package (DIP (Dual In-line Package), TO (Transistor Outline) and others) type. Therefore, recently by electronic company the «GRUPPA KREMNY EL» (Bryansk, Russia) the production of SiC Schottky diodes for power electronics began within the framework of import substitution program. In our previous studies were investigated some characteristics of the SiC Schottky type diodes made in small type of metalpolymeric packages [3]. In this study the main goal is establish current-voltage characteristics of Ti/Al/4H-SiC Schottky diode [4] made in small-sized (SOT-89) type of metalpolymeric package.

2. Materials and methods

In experiments was tested the following Ti/Al/4H-SiC type Schottky diode DDSH411A91 [4] produced by JSC «GRUPPA KREMNY EL» (Bryansk, Russia) in small-sized SOT package type (SOT-89, package dimensions – 4.6×2.6 mm). The main parameters for Ti/Al/4H-SiC type Schottky diode were calculated in our previous works [5-7] with using of TCAD modelling methods. The parameters of Ti/Al/4H-SiC type Schottky diode were the following: the concentration of donors (nitrogen) in the substrate equals 10^{18} cm^{-3} (thickness of substrate is 300 μm), concentration of donors in the n-type epitaxial layer (nitrogen) equals 4.75×10^{15} cm^{-3}, concentration of p+ donors in the guard rings (boron, depth of guard about 2 μm) regions equals 1×10^{18} cm^{-3}, the guard p+ rings consist from one of big guard ring with width of 30 μm and five small guard rings with width of 5 μm (the distance between guard rings was 5 μm), JTE (Junction Terminate Extension) layer formed by boron implantation with p+ concentration 1×10^{17} cm^{-3} (extending 15 μm beyond the edge of the last p+ guard ring), anode material is Ti/Al (titanium/aluminum), the thickness of the epitaxial layer (4H-SiC) equals 14 μm, the radius of the diode equals r=680 μm. For measuring the direct and reverse current-voltage characteristic were used a programmable source AKIP 1144-160-40, Tektronix MDO3102 two-channel oscillograph (bandwidth 1 GHz, refresh rate 5 GS/s) and Fluke 8845A digital multimeter.

3. Results and discussion

Figure 1 presents the photo of Ti/Al/4H-SiC type Schottky diode DDSH411A91 in small-sized SOT-89 type of metalpolymeric package produced by «GRUPPA KREMNY EL» [4].

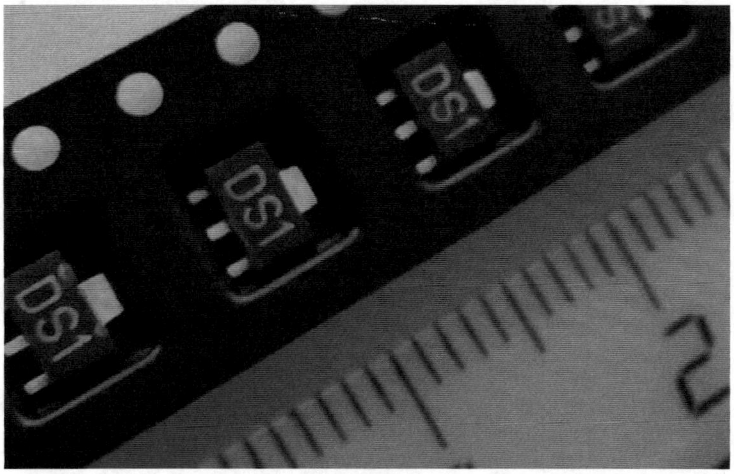

Figure 1. Photo of Ti/Al/4H-SiC type Schottky diode DDSH411A91 in small-sized SOT-89 type package [4].

In Figure 2 is shown the forward current-voltage characteristic for DDSH411A91 Schottky diode in small-sized SOT-89 type package obtained at temperature of 293 K. Further, in Figure 3 presents the reverse current-voltage characteristics for DDSH411A91 Schottky diode (293 K). As follows from current-voltage characteristic the fabricated DDSH411A91 Schottky diode in small-sized package operate with forward current up to 2 A and breakdown voltage 1200 V in reverse direction mode.

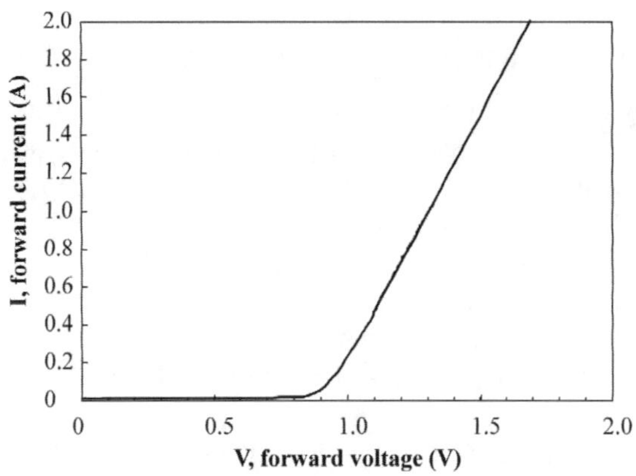

Figure 2. Ti/Al/4H-SiC Schottky diode DDSH411A91 in small-sized SOT-89 type package forward current-voltage characteristic.

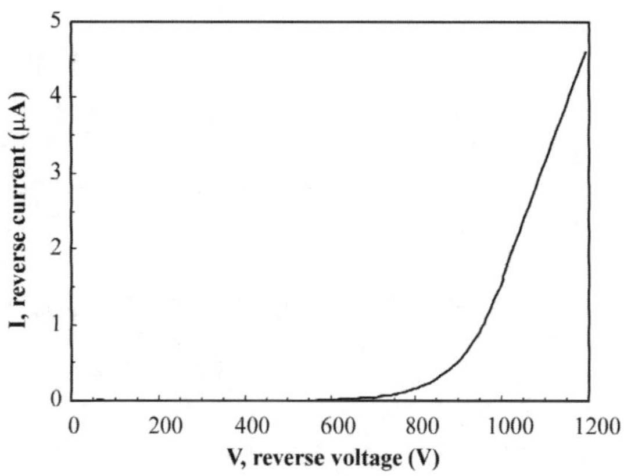

Figure 3. Ti/Al/4H-SiC Schottky diode DDSH411A91 in small-sized SOT-89 type package reverse current-voltage characteristics.

Further, the obtained experimental forward current-voltage characteristics of the DDSH411A91 diode were approximated on the basis of the model of thermionic emission from a semiconductor into metal [1,8] with taking into account the effect of the series resistance R_s (0.35 Ω) of the diode, according to which the diode current in the forward direction can be described by the equations:

$$I = SA^*T^2 \exp\left(-\frac{\varphi_B}{kT}\right)\exp\left(\frac{qV_{SCR}}{nkT}\right), \quad (1)$$

$$V = V_{SCR} + IR_s, \quad (2)$$

where S is the total area of the Schottky regions ($S\approx 0.615$ cm^2), q is the elementary charge, A^* is the effective Richardson constant (A^*=35 A·K^{-2}·cm^{-2}) [9]), φ_B is the Schottky barrier height, V_{SCR} is the voltage drop across the depleted layer of the Schottky contact, k is the Boltzmann constant, T is the absolute temperature and n is the ideality factor of diode.

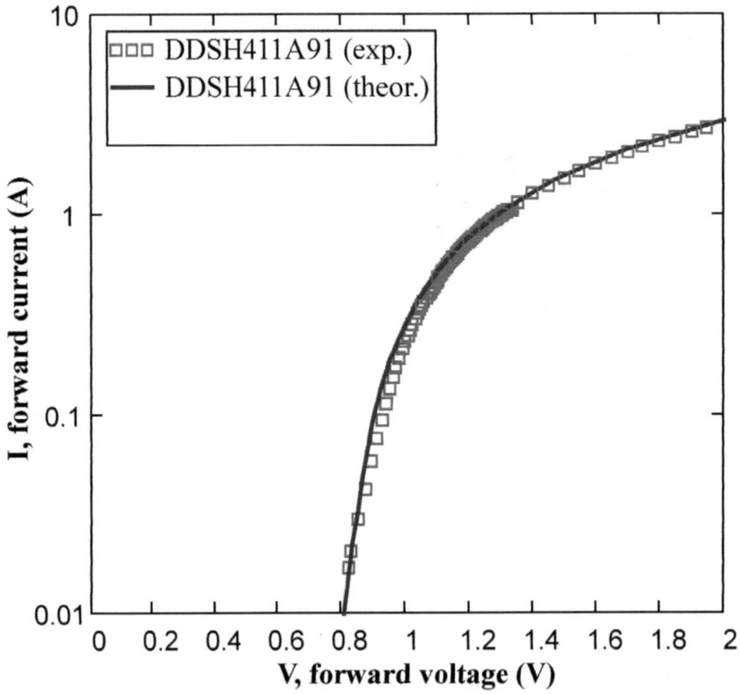

Figure 4. The forward current-voltage characteristic for Ti/Al/4H-SiC Schottky diode DDSH411A91 (T=293 K) in small-sized SOT-89 type package: experimental data (points) and calculated (solid line) by equations (1) and (2) with taken into account the effect of the series resistance R_s (0.35 Ω) and barrier height φ_B (1.2 eV).

It is established that well approximation yielded the following values of the adjustable parameters: barrier height φ_B=1.2 eV and the ideality factor n=1.12 as can be seen from Figure 4. It should also be noted that [10] obtained the barrier height is similar with experimental barrier height for Ti/Al/4H-SiC type Schottky contact (1.2 eV).

In case of the reverse current-voltage characteristic description for Schottky diode type [11] there are the following mechanisms: electron-tunnel transmission from metal to semiconductor; thermal generation of carriers; field dependence of barrier height. For mechanism of field dependence of barrier height the first mechanism is the effect of image forces on the shape and height of the potential barrier (the Schottky effect [12]) and second mechanism is the presence of a thin intermediate insulator layer between metal and semiconductor, across which there is a low voltage drop decreasing the barrier height [11,13]. In particular, as it was established experimentally in RhSi, $ZrSi_2$ and PtSi Schottky diodes [14] formed in n-type silicon the reverse characteristics can be well described by second mechanism where there is the presence of a thin intermediate insulator layer between metal and semiconductor. The above-mentioned mechanism also takes place in case of 4H-SiC type power diodes [15].

Thus, taking into account the field dependence of the barrier height, the approximating formula for the reverse current I_R can be written [11,14,15] in the following form:

$$I_R = I_S \exp\left[\frac{\Delta\varphi_B(V_R)}{kT}\right]\left[1-\exp\left(-\frac{qV_R}{kT}\right)\right], \tag{3}$$

where I_S is the saturation current, $\Delta\varphi_B$ is the barrier lowering caused by the presence of the intermediate layer, V_R is the reverse voltage, q is the elementary charge, k is the Boltzmann constant, T is the absolute temperature.

The saturation current I_S can be written by following equation:

$$I_S = SA^*T^2 \exp\left[-\frac{\varphi_B}{kT}\right], \tag{4}$$

where S is the total area of the Schottky regions ($S\approx 0.615$ cm^2), q is the elementary charge, A^* is the effective Richardson constant ($A^*=35$ A·K^{-2}·cm^{-2}) [9]), φ_B is the Schottky barrier height (1.2 eV).

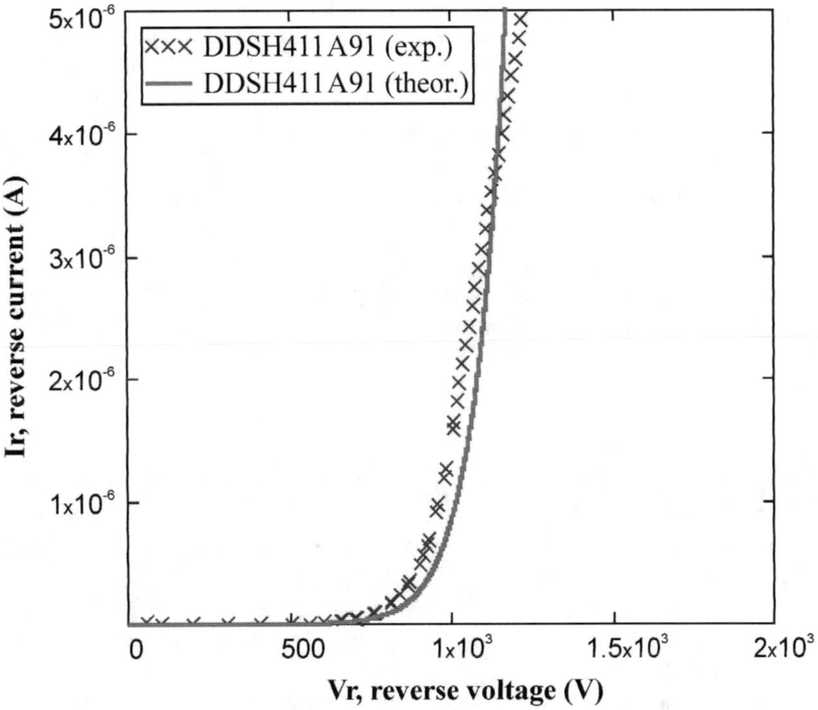

Figure 5. The reverse current-voltage characteristic for Ti/Al/4H-SiC Schottky diode DDSH411A91 (T=293 K) in small-sized SOT-89 type package: experimental data (points) and calculated (solid line) by equations (3)-(7).

Then, the barrier lowering caused by the presence of the intermediate layer $\Delta\varphi_B$ can be written [11,14,15] in the following form:

$$\Delta\varphi_B = \sqrt{\frac{qE_m}{4\pi\varepsilon_o\varepsilon_r}} + \alpha E_m, \tag{5}$$

where ε_r is the relative permittivity (ε_r =9.7 for 4H-SiC [1,16]), ε_o is the permittivity in vacuum (8.85×10^{-12} F/m), E_m is the maximum electric field existing at the metal-semiconductor interface, α is the adjustable empirical parameter. In accordance with [11,14,15] parameter α can be defined:

$$\alpha = \frac{\delta\varepsilon_o\varepsilon_r}{\varepsilon_o\varepsilon_{ir} + q\delta D_s}, \tag{6}$$

where ε_r is the relative permittivity for 4H-SiC (ε_r =9.7), ε_{ir} is the relative permittivity for the SiO_x layer (ε_{ir} =2 [12]), δ is the thicknesses of SiO_x layer, D_s is the energy density of surface states at the Fermi level in metal.

Further, for case when the reverse voltage substantially exceeding the band bending in the semiconductor at the zero bias, the electric field E_m can be expressed in terms of the reverse voltage V_r in the following form:

$$E_m = \sqrt{\frac{2qN_D V_r}{\varepsilon_o\varepsilon_r}}, \tag{7}$$

where N_D is the doping concentration (4.75×10^{21} m^{-3}).

In Figure 5 is presented the reverse current-voltage characteristic for Ti/Al/4H-SiC Schottky diode DDSH411A91 in small-sized SOT-89 type package calculated at 293 K by equations (3)-(7).

As can be seen from Figure 5, the experimental results are very well described by equation (3) for adjustable empirical parameter α=3.159×10^{-9} m. Then, for the SiO_x layer with ε_{ir}=2 and the thicknesses of SiO_x layer δ=3 nm (equal to real our diode structure) we can calculate by equation (6) the energy density of surface states at the Fermi level in metal D_s=1.33×10^{17} $V^{-1}\cdot m^{-2}$. It should be noted that above-mentioned natural oxide thicknesses δ and the values of density of states D_s not only are quite reasonable for the 4H-SiC real surface, but also very similar with values for such type of experimental high voltage 4H-SiC Schottky diode [15]. Because of this, it can be noticed that package's size miniaturization not lead to current-voltage characteristic degradation, moreover in respect of the *dV/dt* and *dI/dt* characteristics for DDSH411A91 Schottky diode [3,17] above-mentioned characteristics are comparable with same values for 4H-SiC Schottky diodes in large-sized (TO-220-F2) package type.

4. Conclusions

Forward and reverse current-voltage characteristic for DDSH411A91 (Ti/Al/4H-SiC) Schottky type diode in small-sized SOT-89 metalpolymeric package (JSC «GRUPPA KREMNY EL», Bryansk, Russia) were obtained. It is shown that fabricated Ti/Al/4H-SiC Schottky diode demonstrate current up to 2 A in forward direction and breakdown voltage 1200 V.

Analysis of experimental results in framework of classical diode theory shows that ideality factor of DDSH411A91 diode equals 1.12 that is close to good Schottky "ideal" diode and effective Schottky barrier height φ_B=1.2 eV.

For the reverse characteristics it is shown that current-voltage characteristics can be well approximated by mechanism of field dependence of barrier height lowering by the presence of the intermediate layer in the form of natural oxide on the 4H-SiC surface.

Thus, on the basis of investigation of forward and reverse current-voltage characteristics for DDSH411A91 (Ti/Al/4H-SiC) Schottky type diode in small-sized SOT-89 metalpolymeric package (JSC «GRUPPA KREMNY EL», Bryansk) it is established that miniaturization of package not lead to current-voltage characteristic degradation and small-sized package of this type can be used in Russian electronic industry.

Acknowledgements
This work was carried out with financial support of the Russian Ministry of Science and High Education within the framework of complex project by creation of highly technological industry «Creation of highly technological industry of silicon and silicon carbide microelectronic technics products in small-sized metal-polymeric packages of the SOT, SO and QFN types» (agreement of 29 November No. 075-11-2019-035) at the organization of the leading performer of RDDTE (Research and Development Design and Technological Engineering) the Bryansk State Technical University.

References
[1] Baliga B J 2019 *Wide Bandgap Semiconductor Power Devices: Materials, Physics, Design, and Applications* (Cambridge: Woodhead Publishing–Elsevier)
[2] Lu D and Wong C P 2017 *Materials for advanced packaging* (Cham: Springer International Publishing)
[3] Rybalka S B, Kulchenkov E A, Demidov A A, Zhemoedov N A, Drakin A Yu, Zotin V F and Shishkina O A 2020 *J. Phys.: Conf. Ser.* **1679** 022045
[4] DSH1 the Schottky diode on the basis of silicon carbide, VUZPROMEXPO 2020. https://vuzpromexpo.ru/portfolio-item/diod-shottki-na-osnove-karbida-kremniya-dsh1/
[5] Panchenko P V, Rybalka S B, Malakhanov A A, A.A. Demidov A A, Krayushkina E Yu and Shishkina O A 2017 *J. Phys.: Conf. Ser.* **917** 082010
[6] Panchenko P, Rybalka S, Malakhanov A, Krayushkina E and Radkov A 2016 *Proc. SPIE* **10224** 102240Y-1
[7] Sedykh S, Rybalka S, Drakin A, Demidov A, Ponomaryova N and Shishkina O 2018 *J. Phys.: Conf. Ser.* **1124** 071012
[8] Ivanov P A, Grekhov I V, Il'inskaya N D, Kon'kov O I, Potapov A S, Samsonova T P and Serebrennikova O U 2011 *Semiconductors* **45**(5) 668
[9] Ivanov P A, Grekhov I V, Potapov A S, Samsonova T P, Il'inskaya N D, Kon'kov O I and Serebrennikova O Yu 2010 *Semiconductors* **44** 653
[10] Perrone D, Ferrero S, Scaltrito L, Naretto M, Celasco E and Pirri C F 2011 *Materials Science Forum* **679-680** 453
[11] Rhoderick E H 1978 *Metal–Semiconductor Contacts* (Oxford: Clarendon Press)
[12] Sze S M and Kwok K Ng 2007 *Physics of Semiconductor Devices* (Hoboken: John Wiley & Sons Inc)
[13] Rhoderick E H 1982 *IEE Proceedings I (Solid-State and Electron Devices)* **129**(1) 1
[14] Andrews J M and Lepselter M P 1970 *Solid-State Electron.* **13** 1011
[15] Ivanov P A, Grekhov I V, Kon'kov O I, Potapov A S, Samsonova T P and Semenov T V 2011 *Semiconductors* **45**(10) 1374
[16] Demidov A A and Rybalka S B 2021 *Applied Mathematics & Physics* **53**(1) 53
[17] Knyaginin D A, Rybalka S B, Kulchenkov E A, Demidov A A, Zhemoedov N A and Drakin A Yu 2020 *J. Phys.: Conf. Ser.* **1695** 012160

Design and simulation of the compact MEMS energy harvester based on aluminium nitride

P S Shlepakov and I V Uvarov

Valiev Institute of Physics and Technology RAS, Yaroslavl Branch, Universitetskaya 21, 150007 Yaroslavl, Russia

p.shlepakov@bk.ru

Abstract. A device for converting the energy of mechanical vibrations to electricity by the piezoelectric effect is presented. A main part of the transducer is a multilayer cantilever with the inertial mass at the tip. A piezoelectric layer is made of 0.5 μm thick aluminum nitride. A feature of the device is the compact lateral size of about 1 mm, which is 10 times smaller in comparison with conventional harvesters. The device is fully compatible with microelectromechanical systems (MEMS) technology. The cantilever has a natural frequency of 45-160 Hz, depending on the size and inertial mass. The transducer generates the output voltage of 0.35 V, which is high enough for rectifying by the diode bridge. The output power of 2.7 nW is relatively low due to the small size. Nevertheless, the figure of merit is higher than that for conventional AlN-based energy harvesters.

1. Introduction

Modern portable devices consume rather low power down to several tens of microwatts [1]. Typically, they are powered by a battery, which requires periodic replacement or charging. In some cases, like GPS trackers or strain gauges in buildings, these operations are inconvenient or impossible. Instead of the battery, portable devices can use energy of environmental sources, e.g. sunlight or wind. A promising and widespread energy source is mechanical vibration. Oscillation of household appliances or human body can be converted to electricity by the harvester, which is a miniature device fabricated by MEMS technology. Among several working principles [2], piezoelectric transduction is the most popular one. The piezoelectric harvester has a simple design compatible with microtechology and demonstrates the highest efficiency [3]. It consists of a resonator with a piezo-material sandwiched between two electrodes. The commonly used material for today is lead zirconate titanate (PZT) due to high piezoelectric coefficient ~100 pm/V [4]. However, lead and its derivatives are toxic and globally considered as hazardous materials. Their usage is expected to be limited in the future, and PZT will be replaced by the lead-free materials like aluminium nitride or zinc oxide [5]. They have lower piezoelectric coefficient (2 pm/V and 5 pm/V for AlN and ZnO, respectively [6]), but they are not toxic and more suitable for MEMS technology. In recent decades, a lot of research are focused at the AlN-based transducers [2,7,8]. These devices have a footprint of about 1 cm2 and generate power up to 64 μW [9]. Their disadvantage is a narrow bandwidth of 1-2 Hz, which limits the application of the harvester by a specific vibrating item. The bandwidth can be expanded by combining several resonators at a single chip, but the price is the significant increase of the harvester size. In this work, we propose a compact piezoelectric transducer of about 1 mm in size. Performance of the device is estimated by the finite element simulation.

2. Methods

A key part of the harvester is a multilayer cantilever illustrated in figure 1. The basic layer is a 0.9 μm thick thermally grown SiO_2. A piezoelectric AlN film has a thickness of 0.5 μm. It is located between two 0.1 μm thick chromium electrodes. The length l and width w of the cantilever are in the range of 500-1000 μm and 200-1000 μm, respectively. The beam is equipped by the inertial mass (IM), which is made of 500 μm thick silicon. The bottom electrode covers the cantilever and IM. It serves as a mask for etching of SiO_2. Piezoelectric layer and upper electrode are located at the beam only, because no mechanical stress is expected above IM. The harvester is attached to the energy source. When the source moves with acceleration, the cantilever deflects from its initial position. Mechanical stress polarizes AlN, which results in a voltage between the electrodes and a current in the external circuit. The current is rectified by diodes and the electric energy is accumulated in a supercapacitor.

Two operating regimes of the harvester are considered. In the static regime, the acceleration is constant and equals to $g = 9.8$ m/s^2. The deflection of the cantilever tip, output voltage and natural frequency are studied. In the vibrating regime, the acceleration oscillates with time as $a = a_0 \cdot \sin(2\pi f t)$, where $a_0 = 0.2$ m/s^2 is the minimal amplitude of acceleration for household devises, f is the frequency of external vibrations, and t is time. The harvester operates in air under normal conditions. An ideal resistor is connected to the electrodes. The resonant frequency, tip deflection and voltage amplitude, as well as the generated power are investigated.

Modeling of the harvester is performed by a finite element method (FEM). The device is divided into small parts called finite elements, as demonstrated in figure 2. Formulation of boundary conditions results in algebraic equations for each element. These equations include laws of elasticity, electrostatics and piezoelectricity. The Navier-Stokes equation is used to find the quality factor. Solution for large system can be found by assembling approximation function for small elements. The base of the cantilever is fixed, while the whole device is located in the acceleration field. One of the electrodes is grounded. The elements have a shape of rectangular tetrahedron. Their amount depends on the harvester size and varies from 7.0×10^4 to 3.2×10^5.

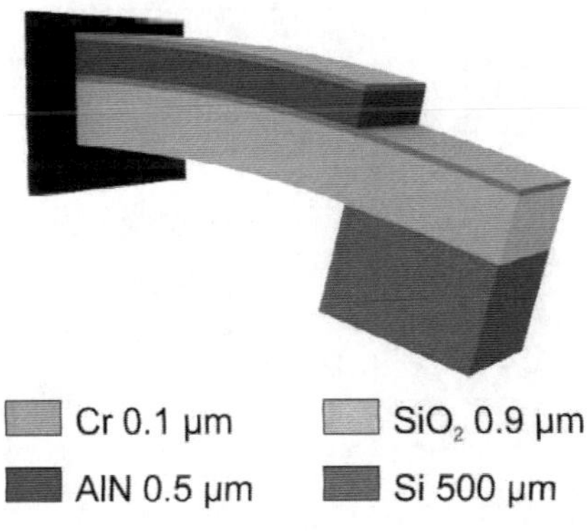

Figure 1. Design of the energy harvester.

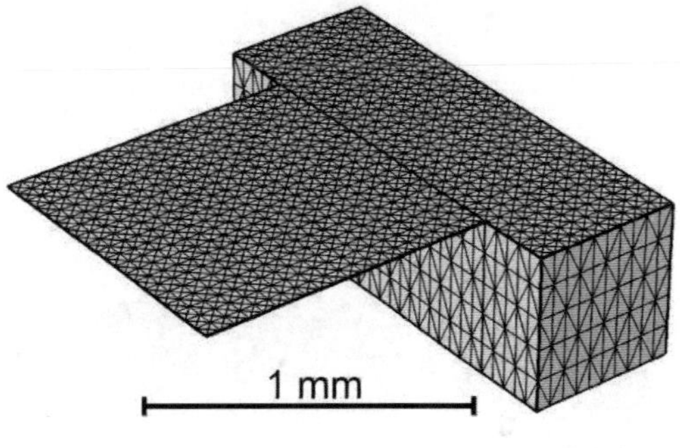

Figure 2. Model of the harvester divided by finite elements.

3. Results and discussion

Cantilever without inertial mass in static regime was investigated. The wider the beam, the more charge is accumulated at the electrodes and the higher output current can be generated. This makes wider cantilevers preferable. The tip deflection increases with the beam length and takes the values from 10 to 130 nm. Static output voltage is of 0.04-0.15 mV, depending on l, which is too low to pass the rectifying circuit that typically requires 0.2-0.3 V [10]. Low voltage is explained by a small stress in AlN. Eigenfrequency f_0 of the cantilever without IM is in the kilohertz range, while the external vibrations typically have a frequency below 200 Hz. An addition of the IM with a length l_m of 100-500 μm and width w_m of 1000 1500 μm lowers the frequency to the desired range. The cantilever of 1000x1000 μm in size demonstrates f_0 of 45-160 Hz, as shown in figure 3. Thus, the operating frequency can be adjusted by changing the IM size. At the same time, the inertial mass increases the tip deflection, mechanical stress and output voltage, see figure 4. These values are three orders of magnitude higher than those for the beam without IM.

The distribution of stress in the cantilever is presented in figure 5. The largest stress is located at the fixed end. Among the materials used in the harvester, AlN has the lowest tensile strength of 300 MPa [11]. A dependence of maximal stress value on w_m (figure 6) demonstrates that the stress is an order of magnitude lower than the tensile strength. Therefore, the damage of the cantilever during operation is not expected. Extrapolation of the dependence to 300 MPa provides maximal output voltage of 0.9 V.

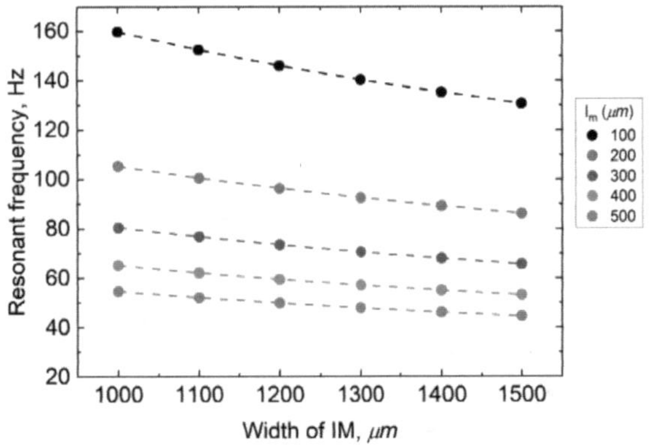

Figure 3. Cantilever eigenfrequency dependance on width of IM for different IM length.

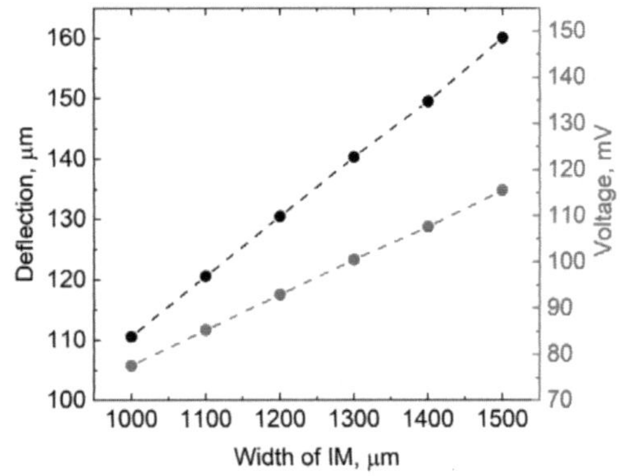

Figure 4. Cantilever deflection and output voltage dependance on width of IM.

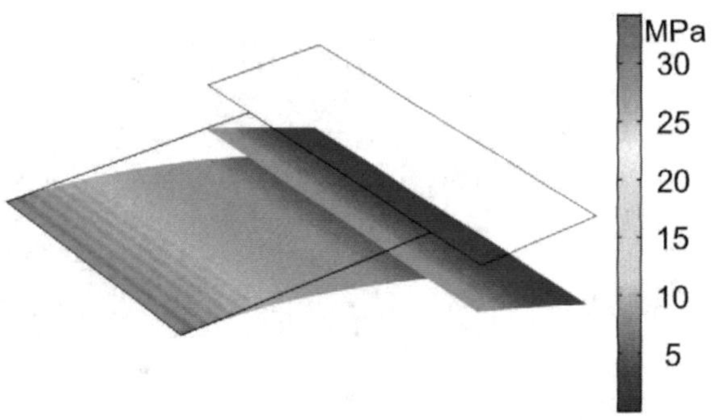

Figure 5. Stress distribution in the cantilever with l_m = 500 μm and w_m = 1500 μm (IM is not shown).

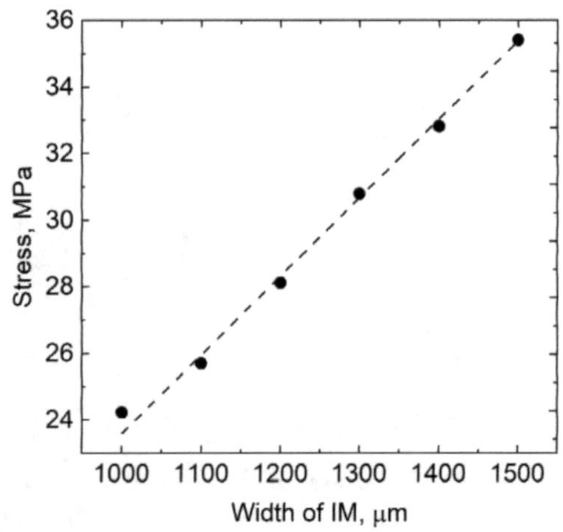

Figure 6. The maximal stress value as a function of w_m at l_m = 500 μm.

The vibrating regime is considered for the cantilever of 1000x1000 μm in size, which is equipped by IM with the size of 500x1000 μm. A load resistance is assumed to be $R = 10^{12}$ Ω. According to FEM simulation, the device has a quality factor $Q = 398$ and a resonant frequency $f_r = 55.05$ Hz. The amplitude of the tip deflection and output voltage as a function of external vibration frequency are shown in figure 7. The bandwidth Δf of the harvester is calculated as a width of the resonant peak at a half of its height and equals to 0.14 Hz. In resonance, the device provides the output voltage of 689 mV. The maximal stress in AlN reaches 100 MPa, which is still lower than the tensile strength.

The output power depends on the load resistance, as demonstrated in figure 8. The power and voltage are calculated for R in range from 10^4 to 10^{10} Ω. The current flowing through the resistor has a sinusoidal shape. Therefore, the power is determined as $P = I_R U_R/2$, where I_R and U_R are the amplitude values of the current and voltage at the resistor. The power has a peak of 2.7 nW at the optimal resistance of 25 MΩ. For this resistance, the voltage is 350 mV, which is high enough for rectifying and energy storage. The output characteristics may be improved by optimizing the harvester design. The base layer can be fabricated from the more flexible material [12] or piezo material [13].

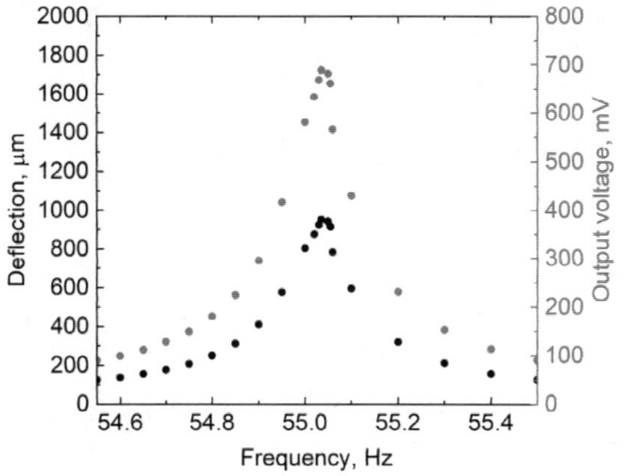

Figure 7. Deflection of the tip and output voltage as a function of excitation frequency.

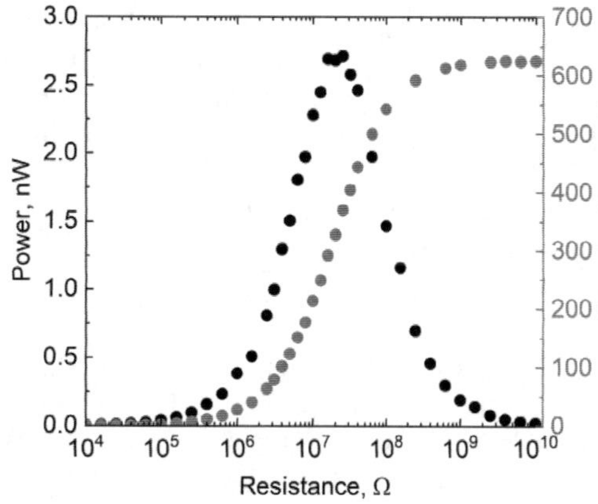

Figure 8. Output power and voltage as a function of the load resistance.

Performance of the harvester is convenient to estimate by a figure of merit F:

$$F = \frac{P}{(V_b + V_m)\cdot a^2}\frac{\Delta f}{f_r}, \quad (1)$$

where $V_b = 1.6\cdot 10^{-6}$ cm^3 and $V_m = 2.5\cdot 10^{-4}$ cm^3 are volumes of cantilever and IM, respectively, a is the acceleration amplitude measured in the units of gravity g. Piezoelectric harvesters based on AlN typically provide F from 1.5 to 6 μW/(g^2·cm^3) [14-16], while the proposed device demonstrates 66 μW/(g^2·cm^3). The enhancement is achieved due to a small size and high quality factor. Thus, the compact device outperforms conventional energy harvesters.

4. Conclusions

A piezoelectric energy harvester based on AlN is presented. A key part of the device is a multilayer cantilever with an inertial mass at the tip. Working characteristics of the harvester are calculated by a finite element method. The cantilevers with large lateral size provide higher performance. The beam itself is unable to generate sufficiently high voltage for rectifying and has a resonant frequency in the kHz range. Addition an inertial mass to the cantilever decreases the frequency to required level and increases mechanical stress in AlN, thereby providing higher output voltage. It is worth noting that the stress does not exceed tensile strength of the material. The resonant frequency can be adjusted by varying the size of inertial mass. The proposed harvester demonstrates several times higher high figure of merit than conventional devices, but the output power is rather low. It can be increased by optimizing the cantilever design.

Acknowledgments
This work was supported by Program no. 0066-2019-0002 of the Ministry of Science and Higher Education of Russia for Valiev Institute of Physics and Technology of RAS.

References
[1] Chandrakasan A, Amirtharajah R, Goodman J and Rabiner W 1998 *IEEE Int. Symp. Circ. S.* **4** 604
[2] Beeby S P, Tudor M J and White N M 2006 *Meas. Sci. Technol.* **17** R175
[3] Roundy S and Wright P K 2004. *Smart Mater. Struct.* **13** 1131
[4] Butt Z, Pasha R A, Qayyum F, Anjum Z, Ahmad N and Elahi H 2016 *J. Mech. Sci. Technol.* **30** 3553
[5] Zhang Y, Kim H, Wang Q, Jo W, Kingon A, Kim S and Jeong C K 2020 *Nanoscale Adv.* **2** 3131
[6] Fraga M A, Furlan H, Pessoa R S and Massi M 2014 *Microsyst. technol.* **20** 9
[7] Elahi H, Eugeni M and Gaudenzi P 2018 *Energies.* **11** 1850
[8] Saadon S and Sidek O 2011 *Energy Convers. Manage.* **52** 500
[9] Andosca R, McDonald T G, Genova V, Rosenberg S, Keating J, Benedixen C and Wu J 2012 *Sens. Actuator A-Phys.* **178** 76
[10] Zhao X, Shang Z, Luo G and Deng L 2015 *Microelectron. Eng.* **142** 47
[11] Gang Zong D, Wo Ong C, Aravind M, Po Tsang M, Loong Choy C, Lu D and Ma D 2004 *Philos. Mag.* **84** 3353
[12] Tsukamoto T, Umino Y, Shiomi S, Yamada K and Suzuki T 2018 *Technol. Adv. Mater.* **19** 660
[13] Ghoddus H and Kordrostami Z 2018 IEEE Sens. J. **18** 8667
[14] Gablech I, Klempa J, Pekáre J, Vyroubal P, Hrabina J, Holá M and Neužil P 2020 *Micromachines.* **11** 143
[15] Elfrink R *et al* 2009 *IEDM: Technical Digest—Int. Electron Devices Mtg.* 1
[16] Cao Z, Zhang J and Kuwano H 2012 *Sens. Actuator A-Phys.* **179** 178

Supercapacitor with electrodes based on high-purity single-walled carbon nanotubes

A Shumilin[1], N Gorshkov[1], A Fomin[1], A Aman[2] and S Palis[2]

[1] Yuri Gagarin State Technical University of Saratov, Saratov 410054, Russia
[2] Otto von Guericke University, Magdeburg 39106, Universitätsplatz, 2

Shumilinai@yandex.ru

Abstract. We studied a supercapacitor with high purity carbon paper electrodes [1] with a specific gravity of 20 g/m^2. The specific capacitance of the electrode during assembly in a coin cell housing with an aqueous 6M KOH electrolyte, at a scan rate of 1 mV/s, is 52 F/g. The specific power is 195.42 W/kg and the specific energy is 0.19 W·h/kg at a scan rate of 100 mV/s, which is included in the region of supercapacitors in the Ragone plot.

1. Introduction

Modern technologies involve significant waste of electrical energy and place high demands on storage elements. One type of device that allows you to accumulate and save electrical energy are capacitors, including ionistors or supercapacitors [2,3,4].

A promising direction for improving supercapacitors is the use of porous carbon materials as electrodes, in which charges accumulate in the form of a double electric layer on a developed porous surface of carbon structures [3]. The issue of increasing the specific capacitance of supercapacitors remains relevant at present. New materials for electrodes and their combinations, alloying, surface modification, selection of the composition of electrolytes - these are the main directions for improving the characteristics of supercapacitors [4,5].

Therefore, it is important to study any, even already well-known material, if it has distinctive features. In this work, a preliminary study of the electrodes for a supercapacitor of single-walled carbon nanotubes with a high degree of purification was carried out. A review of the literature showed that the electrodes of activated carbon and nanotubes can have a specific capacity of up to 100 and 280 F/g, respectively [2], while in ordinary electrolytic capacitors this value is several orders of magnitude lower. Important characteristics in the description of supercapacitors are energy density and power density, the value of the potential window. Typical values for supercapacitors with carbon electrodes are: energy densities up to 80 W/kg and power densities 6–30 kWh/kg, potential window values of the order of $\pm$ 0.8 V when using aqueous electrolytes and symmetric electrode schemes [2].

2. Methodology

In the manufacture of electrodes used carbon paper from nanotubes or SWN – buckypaper. Carbon paper consists of weaving high-strength nanotubes into agglomerates and agglomerates into sheet material, where the individual elements are firmly held together by mechanical weaving and the inter-molecular forces of Van der Waals (Figure 1). For the manufacture of electrodes used paper with different specific weighing 3 and 20 g/m^2.

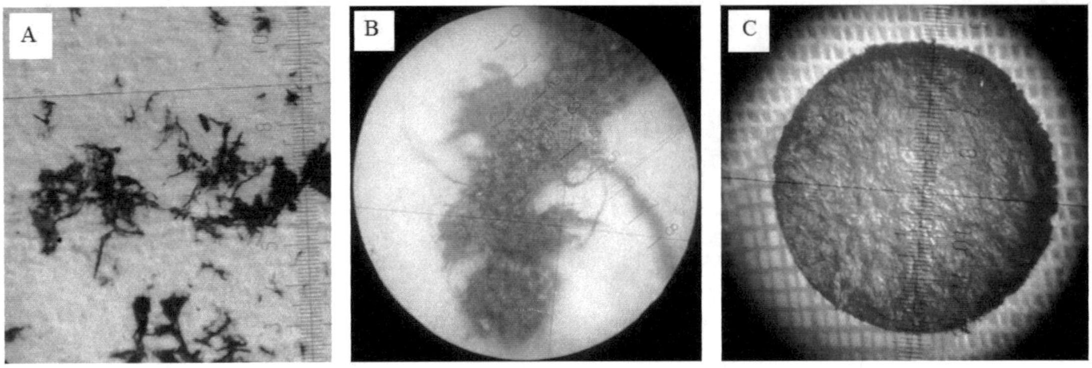

Figure 1. Samples of SWN agglomerates – buckypaper. A – Individual agglomerates of nanotubes with an increase of × 16, B – agglomerates with an increase of × 32, C – a sample of buckypaper (20 g/m^2) at magnification × 8.

Pairs of carbon paper electrodes should be of the same mass, if possible, so it is difficult to produce them by mechanical cutting. To increase the manufacturing accuracy, a laser cutting method was used on an installation of the LRS–50M model, with a coordinate table, a radiation wavelength of 1.064 µm, an average pulse power of 10 W, and a pulse repetition rate of 10 Hz. The carbon paper blank was placed on a metal lining, pressing weights were installed on top to prevent deformation during cutting and holding the blank in the focus of the laser beam (Figure 2A).

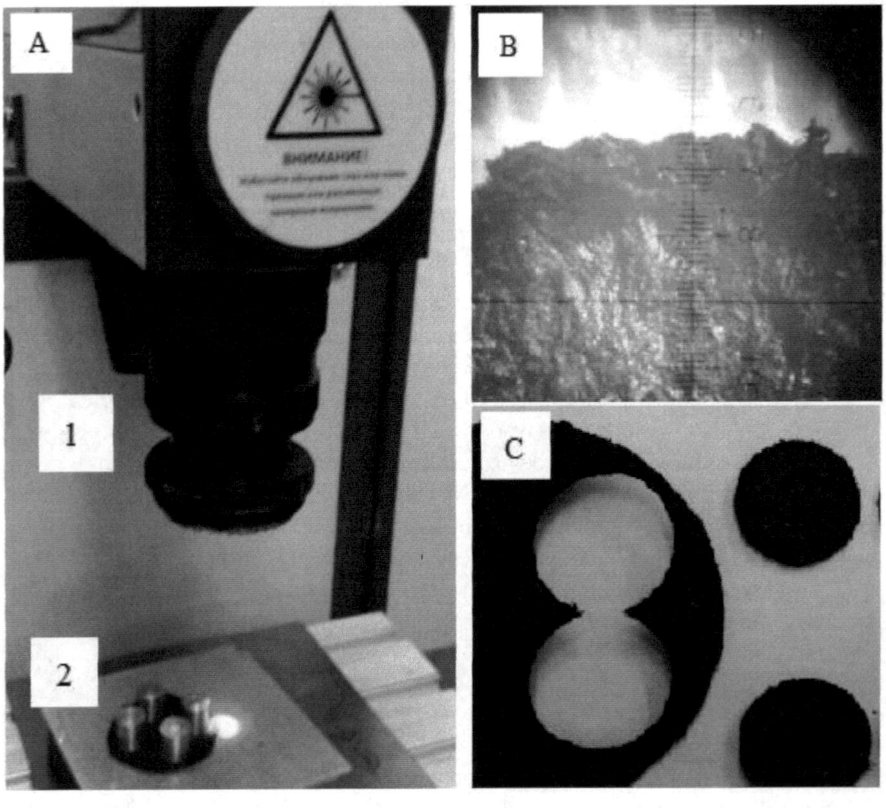

Figure 2. Laser cutting of electrodes from buckypaper (A), paper cut from nanotubes at magnification × 32 (B), appearance of electrodes (C).

In the area of the cut, the buckypaper is stratified, however, in this case, the cut-out electrode retains the shape of the regular circle of Figure 1C and Figure 2C. For the convenience of measurements, samples of electrodes from a high degree of purity SWN-buckypaper up to 99.9 % [1] were placed in a standard coin cell CR2025 battery case. A 6 M aqueous KOH solution with a volume of 50 μl per well was used as an electrolyte. In Figure 3 shows blanks for assembling electrodes in a housing. The electrodes 3 of the SWN-buckypaper accumulate charge in a double electric layer, the metal collectors 2 provide reliable electrical contact between the carbon material and the housing 1. The dielectric separator 4 made of polymer non-woven fiber is an insulator between the electrodes and does not interfere with the free movement of ions in the electrolyte.

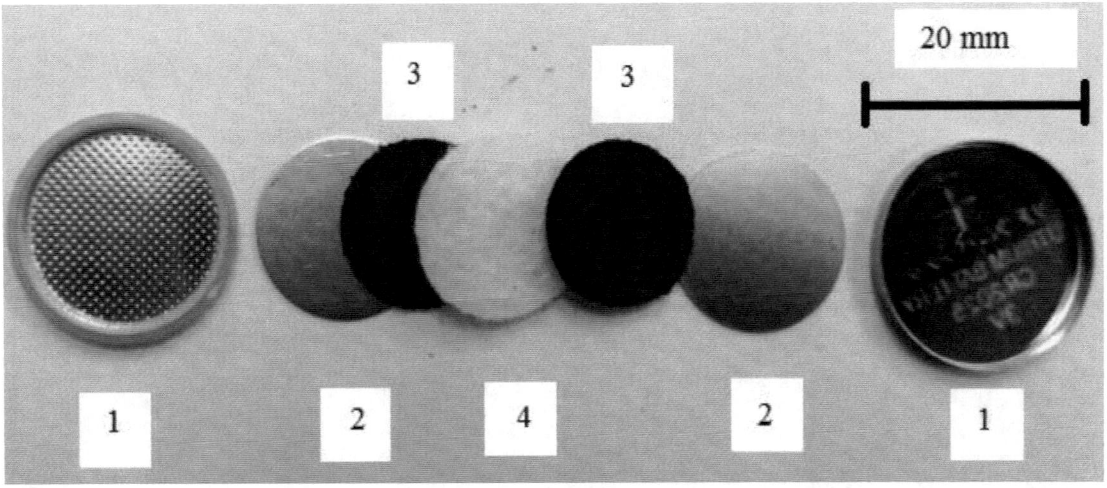

Figure 3. Elements for assembling a coin cell CR2025: 1 – two parts of the outer case made of stainless steel, 2 – metal electrodes-collectors, 3 – electrodes from high-purity SWN-buckypaper, 4 – separator.

Cyclic voltammograms (CV) of samples were obtained according to a two-electrode circuit using a P-50 PRO potentiostat-galvanostat (LLC Elins, Russia) in the scan rate 1–100 mV/s, with a potential window of ± 500 mV, to determine the specific electric capacitance of the electrode material.

3. Results

Preliminary measurements of the assembled coin cell showed the promise of research on the use of electrodes from high-purity buckypaper as plates of capacitors of increased capacity. In Figure 4 shows cyclic voltammograms (CV) for carbon paper of various specific gravity at a sweep speed of 100 mV/s.

The area of the figure in a limited graph corresponds to the capacitor power for one charge – discharge cycle. For a sample with a specific gravity of the electrode of 3 g/m^2 and 20 g/m^2 (Figure 4), the calculated specific capacities were 1.7 F/g and 3.2 F/g, respectively. Graph (Figure 4B) has a shape close to rectangular, which is typical for reversible processes of charge and discharge of a double electric layer [7]. For a sample with a large specific gravity of the electrode (Figure 4B), the calculated specific capacitance is almost twice as large. An increase in the specific capacity with a larger mass of the electrode, ceteris paribus, indicates the influence of factors not related to the material of the electrodes. For example, a low specific capacitance may be due to poor electrical contact between the thinner carbon paper and the metal collector electrode or an improperly selected electrolyte, it is also possible that the separator material does not provide sufficient ion permeability. For the sample, the general laws characteristic of electrodes made of a porous carbon material are observed [6,7]. With a decrease in the potential scan rate from 100 to 1 mV/s (for the sample 20 g/m^2), the specific capacitance of the electrodes increases from 3.2 to 52 F/g, respectively, which is also characteristic of supercapacitors [7]. In Figure 5 shows the CV for the assembled capacitor at voltages U = ± 450 mV, the charge – discharge current reaches 70 μA.

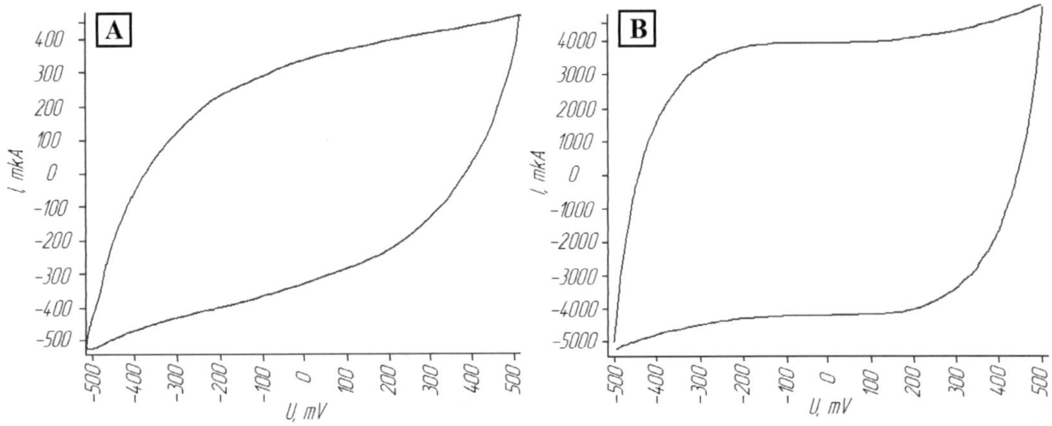

Figure 4. CV plot scan rate of 100 mV/s. Electrodes with density A – 20 g/m², B – 3 g/m².

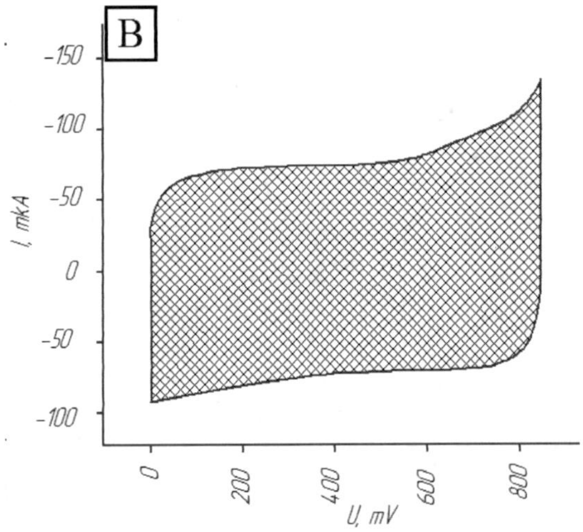

Figure 5. CV plot, scan rate 1 mV/s. Electrodes with a density of 20 g/m².

The stored energy in the capacitor under these conditions is of the order of 220 mJ per cycle. Given the mass of carbon fiber electrodes m_1 = 9.37 mg, m_2 = 9.09 mg, the specific capacity of this material was calculated S_{sp} = 52 F/g. Specific power and energy at a sweep speed of 100 mV/s reach 195.42 W/kg and 0.19 Wh/kg, respectively, these values are in the region of supercapacitors in the Ragonc plot [2]. However, this is not a high enough indicator for carbon materials and additional experiments with different types of separators and coin cell sizes are needed to ensure better electrical contact and greater mobility of electrolyte ions.

4. Conclusion

As a result of the work done, the laser cutting mode of carbon paper buckypaper is set. Mode parameters: average pulse power of 10 W, pulse repetition rate of 10 Hz. The specific capacitance of the electrode during assembly in a coin cell housing with an aqueous 6M KOH electrolyte, at a potentiostatic sweep speed of 1 mV/s, is 52 F/g. A strong dependence of the electrical characteristics of the cell on the specific gravity of the carbon electrode is observed, which is associated with the peculiarities of the assembly of the coin cell. The specific power is 195.42 W/kg and the specific energy is 0.19 Wh/kg at a scan rate of 100 mV/s, which is included in the region of supercapacitors in the Ragone plot.

Acknowledgments
The research was supported by the Michael Lomonosov Program - Line B, 2021. Agreement with the Ministry of Education and Science № 075-03-2021-304. Project executor Shumilin A.

References
[1] https://nanofract.com/
[2] Zhang Y, Feng H, Wu X, Wang L, Zhang Aiqin, Xia T, Dong H, Li X and Zhang L 2009 *Int. J. Hydrogen Energy* **34** 4889
[3] Vivekchand S.R.C, Chandra Sekhar Rout, Subrahmanyam K.S, Govindaraj A and Rao C N R 2008 *J. Chem. Sci.* **120** 9
[4] Béguin F, Raymundo-Piñeiro E and Frackowiak E 2008 *Handbook: Carbons for Electrochemical Energy Storage and Conversion Systems. CRC Press*. 329
[5] Gryzlov L, Rychagov A, Skundin A and Kulova T 2015 *Electrochemical Energetics* **15** 160
[6] Pandolfo A and Hollenkamp A 2006 *J. Pow. Sources.* **157** 11
[7] Frakowiak E and Beguin F 2001*Carbon* **39** 937

High-performance inline RF MEMS switch for application in 5G mobile networks

A V Tkachenko[1], I E Lysenko[1], A V Kovalev[1] and D V Vertyanov[2]

[1] Southern Federal University, Rostov-on-Don region, Taganrog, 347922, Russia
[2] National Research University of Electronic Technology, Zelenograd, Moscow, 124498, Russia

Abstract. This article presents the results of the design and analysis of a radio-frequency switch made using microelectromechanical systems technology. The device is the capacitive switch with a hybrid type of contact, in which the movable electrode of the structure – the metal membrane is part of the microwave signal line of the coplanar waveguide. The switch design is characterized by a high capacitance ratio and low contact resistance. The zig-zag elastic suspension is used to reduce the value of the pull-down voltage – 2 V and the switching time ~ 7 us. The central resonant frequency of the switch is 3.8 GHz. In this case, in the open state, the value of the insertion loss is not more than -0.2 dB and the isolation value in the close state is not less than -55 dB. The effective frequency range is the S-band, as well as the C-, X- and Ku-band, in which the isolation value is at least -30 dB. The presented inline RF MEMS switch is suitable for use in various types of ground and satellite communications, in particular for devices and systems of 5G mobile networks.

1. Introduction

Microelectromechanical systems (MEMS), with their extremely small size and full integration into the radio-frequency interfaces of various telecommunications devices, have become the most promising technology in the last few decades. There is a growing demand for devices with extremely high performance, low power consumption, and compact dimensions in a variety of satellite, defence and communications systems. To date one of the most promising areas of application of MEMS technology is associated with circuits and devices for ultra-high frequency (UHF) (microwave) and radio-frequency (RF) communication. In recent years, the introduction and dissemination of new wireless communication standards, in particular the fifth generation of mobile radio – 5G, has set new challenges in the development of hardware for transceivers. Key RF subsystems in 5G RF transceivers include antennas, configurable filters, power amplifiers and multiple-input multiple-output (MIMO) antennas. At the same time, it can be noted that in all these 5G RF systems or subsystems, from a hardware point of view, the RF switch is one of the most fundamental and important components that is used to route signals along RF transmission paths with a high degree of efficiency; its RF characteristics, switching time, RF signal power and their reliability can directly affect the corresponding properties and performance of 5G applications. RF MEMS switches can effectively replace the currently used active semiconductor analogies based on PIN diodes and FET transistors at frequencies up to 10 GHz and compete with coaxial switches in the range up to 40 GHz or more.

RF MEMS switches are classified according to the actuation mechanism, the movement of the suspended parts, the type of contact, the type of mechanism used, and the electrical configuration. Today electrostatic RF MEMS switches are the most common among micromechanical switches. This is due to the almost zero power consumption in the switched-on state, the small size of the element, the compatibility of the device manufacturing process with the technological processes of manufacturing integrated circuits (IC) using silicon technology and technology based on $A^{III}B^{V}$ elements, the relative ease of manufacture and the short switching time. Electrostatic RF MEMS switches are divided into two types – cantilever and membrane switches with metal-metal and metal-dielectric-metal contacts, respectively. Resistive RF MEMS switches have some advantages over capacitive RF MEMS switches, as they are characterized by a lower pull-out voltage value and a higher switching speed. On the other hand, capacitive RF MEMS switches are characterized by greater reliability and are subject to fewer failure mechanisms. In addition, for the design of RF MEMS switches of medium and low power, switches with a capacitive contact type have an important advantage-the ability to design switch designs with a low value of the pull-down voltage.

Currently, capacitive RF MEMS switches with metal-dielectric-metal contact are the most common in the technology in RF MEMS devices [1, 2]. The main problem of capacitive RF MEMS switches is the imperfect roughness of the contacting layers, which leads to a decrease in the capacitance ratio. Decrease in the capacitance ratio, in turn, leads to a shift in the resonant frequency from the required one and a decrease in the isolation value in the closed state of the RF MEMS switch.

There are a number of studies that have been conducted to achieve a high value of the capacitance ratio of capacitive RF MEMS switches and a low value of the pull-down voltage. In [3], a design of a capacitive RF MEMS switch using a ceramic dielectric layer with a high permittivity is proposed. In [4], a high capacitance ratio was achieved by using the curved design of the metal membrane of the capacitive RF MEMS switch. Another method used to achieve a high value of the capacitance ratio is to increase the air gap between the metal membrane and the dielectric layer [5, 6]. However, there are some obvious disadvantages of these methods, which are that the charge problem of the dielectric layer becomes more significant the smaller the thickness of the applied dielectric layer and the electromechanical parameters of the switch change when the air gap changes.

Meanwhile, a number of studies have proposed some approaches to obtaining a high value of the capacitance ratio, which consist in the use of dielectric materials with high permittivity [3, 7-9]. However, the value of the capacitance ratio is limited by the minimum thickness of the dielectric layer, the maximum value of the permittivity, and the maximum value of the air gap between the metal membrane switch and the RF transmission line. In this regard, the methods used in [4-6] are not the most suitable.

In this paper we present the developed the design of a high-performance capacitive inline RF MEMS switch, which is characterized by a high capacitance ratio, in which the disadvantages of the methods previously used to increase the capacitance ratio are eliminated. Also, the designed RF MEMS switch should be characterized by good radio-frequency characteristics, a low value of the pull-down voltage and a short switching time. In addition, the developed design of the RF MEMS switch should be characterized by a small form factor for use in RF devices and systems of 5G mobile networks.

2. Design and methodology

The proposed capacitive RF MEMS switch consists of a dielectric substrate with a coplanar waveguide (CPW) located on its surface, areas for the contact pads for applying a constant (DC) pull-down voltage to fixed pull-down electrodes, and a metal membrane built into the CPW signal line (t-line) between the RF input and output ports and is part of it, fixed to the anchor areas by means of zig-zag elastic suspensions, as shown in Figure 1 (a)-(d). The choice of materials is made on the basis of the results [10-12].

When a DC pull-down voltage is applied to the fixed pull-down electrodes relative to the metal membrane (positive potential on the fixed pull-down electrodes, negative potential on the metal membrane), an electrostatic force occurs that exceeds the mechanical elastic force, and as a result, the metal membrane sinks to the down position – it is attracted to the fixed pull-down electrodes and the upper metal layer of an additional fixed metal-dielectric-metal (MIM) capacitor, which is formed by this metal layer, a dielectric layer with a high permittivity – high-k (TiO_2) and a metal layer of the signal line CPW. As a result, the incoming RF signal along the signal line of the CPW is shunted to the ground line CPW. In this RF MEMS switch design, an additional fixed MIM capacitor is connected to a shunt capacitor with metal-air-metal (MAM) plates formed by the upper metal layer of the MIM capacitor, a metal membrane and the air space between them. In the case when the metal membrane is in the down position, the MAM capacitor changes to the resistance in the electrical circuit.

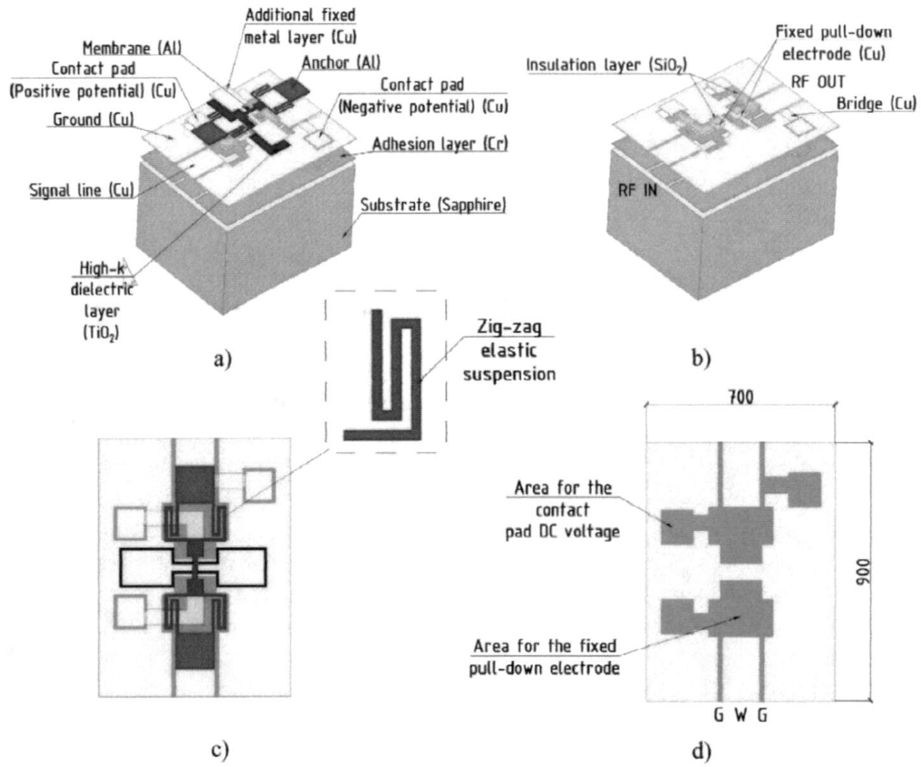

Figure 1. Schematic view of the proposed capacitive inline RF MEMS switch.

Table 1 shows the geometric dimensions of the main structural elements of the proposed capacitive inline RF MEMS switch: CPW, metal membrane, via holes, anchor regions, contact pads for supplying the DC pull-down voltage, fixed pull-down electrodes, air gap, contact pads, additional fixed MIM capacitor and dielectric layers.

In Figure 2 schematically shows the developed zig-zag elastic suspension with an indication of the geometric dimensions. Each elastic suspension consists of five successively connected elastic suspensions with two-fold symmetry, forming the shape of a meander, as well as one connecting elastic suspension having the shape of an elastic beam.

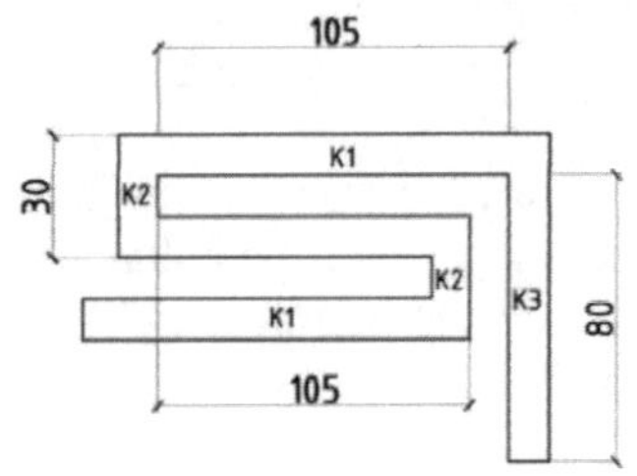

Figure 2. Schematic view and geometric dimensions of the zig-zag elastic suspension.

Table 1. Dimensions of the proposed inline capacitive RF MEMS switch.

Parameter	Dimensions	Material
Substrate	900x700x500 um	Sapphire
CPW		Copper
–G W G	15x140x6 um	
Membrane		
– right/left part	60x50x1 um	Aluminum
– central part	80x20x1 um	
Via holes	4x4x1 um	–
Anchors	140x130x8.2 um	Aluminum
Contact pads	100x100x6 um	Copper
Fixed pull-down electrode	60x60x12 um	Copper
Air gap, g_0	1 um	–
Dielectric film (MIM)		
– right/left part	200x140x0.2 um	TiO_2 ($\varepsilon_r = 90$)
– central part	150x40x0.2 um	
Dielectric film (on fixed pull-down electrode)	60x60x0.1um	SiO_2 ($\varepsilon_r = 4$)

3. Performance numerical analysis

3.1 Electromechanical modeling

In numerical modeling of continuum mechanics problems, the quality of the calculated finite element grid is an important parameter. The question of grid quality is of critical importance in contact problems, since, on the one hand, small elements reflect the behaviour of the structure with higher accuracy when loads are applied to it, and on the other hand, such problems, due to their nonlinearity, require significant computational resources.

Figure 3 (a) shows the results of the distribution of the potential energy in the movable structure of the inline capacitive RF MEMS switch during electrostatic activation. Figure 3 (b) shows the switching time (closure time) inline capacitive RF MEMS switch.

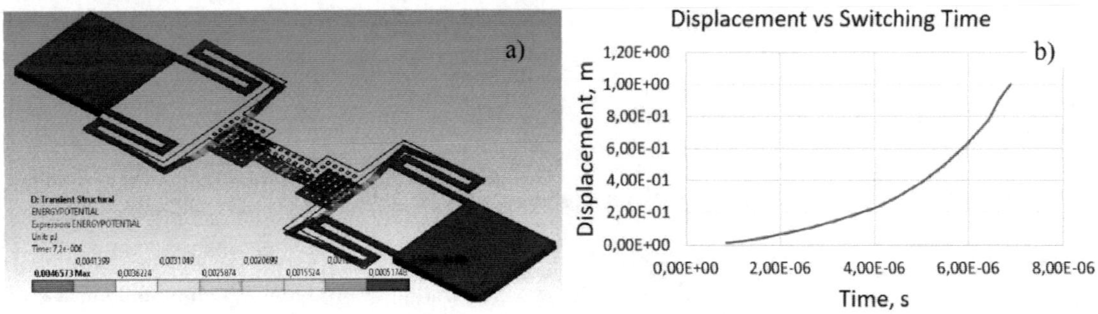

Figure 3. Simulation of switching time of the inline capacitive RF MEMS switch.

According to the results of the numerical simulation of the electromechanical model of the developed inline capacitive RF MEMS switch design, it follows that the value of the pull-down voltage is 2 V and the switching time to the down-state of the metal membrane is 6.9 us.

3.2 Electromagnetic modeling

Figure 4 shows the results of numerical simulation of S-parameters (scattering parameters) of the developed inline capacitive RF MEMS switch. Based on the results of the analysis of the scattering

Journal of Physics: Conference Series **2086** (2021) 012068 doi:10.1088/1742-6596/2086/1/012068

parameters, it follows that the central resonant frequency of the switch is a frequency of 3.8 GHz. In this case, in the open-state, the value of the insertion loss is not more than -0.2 dB and the isolation value in the down-state is not less than -55 dB. The effective frequency range is the S-band, as well as the C-, X- and Ku-band, in which the isolation value is at least -30 dB.

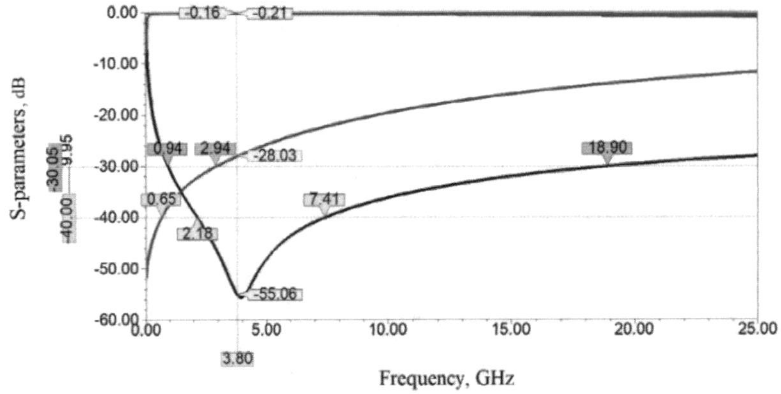

Figure 4. Simulation of the S-parameters the inline capacitive RF MEMS switch.

Figure 5 shows the results of numerical coupled electromagnetic and thermal calculations of the developed capacitive inline RF MEMS switch design when switching an RF signal with a frequency of 3.8 GHz and a power of 1 W in the open-state. The maximum heating in the developed design of the inline capacitive RF MEMS switch occurs in the contact region – the region of the metal membrane. The metal membrane also acts as a heat sink from the contact area. The maximum heating temperature does not exceed the mark of 115°C.

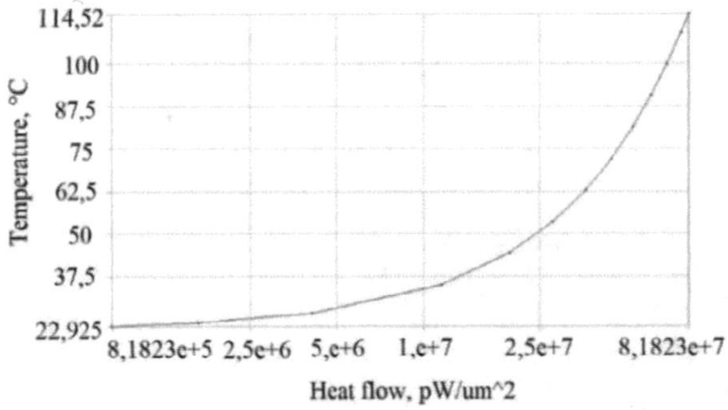

Figure 5. Thermal calculations of the developed capacitive inline RF MEMS switch design.

The value of the capacitance ratio of the developed inline capacitive RF MEMS switch is 5930. When using, for example, Si_3N_4 ($\varepsilon_r = 7.5$), which is the most common material of the dielectric layer of shunt RF MEMS switches, as the material of the dielectric layer, the value of the capacitance ratio will be 545.

The value of the contact resistance in the closed position of the developed inline capacitive RF MEMS switch does not exceed 0.8 Ω.

4. Conclusion

This article presents the results of the design of the capacitive inline RF MEMS switch with a hybrid contact type. The device is the capacitive switch with a hybrid type of contact, in which the movable electrode of the structure – the metal membrane is part of the microwave signal line of the CPW and the shunt connection with the ground line of the CPW is applied directly to the surface of the dielectric substrate. In addition, the switch design is characterized by a high capacitance ratio, obtained by using an additional fixed MIM capacitor. To reduce the value of the pull-down voltage and increase the speed of the switch the zig-zag elastic suspension is used. According to the results of the finite element analysis, the value of the pull-down voltage is 2 V with a short switching time to the down-state (closure position) – no more than 7 us (switching voltage 5 V). Based on the results of the analysis of the scattering parameters, it follows that the central resonant frequency of the switch is a frequency of 3.8 GHz. In this case, in the open-state, the value of the insertion loss is not more than -0.2 dB and the isolation value in the close-state is not less than -55 dB. The effective frequency range is the S-band, as well as the C-, X- and Ku-band, in which the isolation value is at least -30 dB. The value of the capacitance ratio of the developed inline capacitive RF MEMS switch is 5930 and the value of the contact resistance in the close-state does not exceed 0.8 Ω.

The presented inline RF MEMS switch is suitable for use in various types of ground and satellite communications, in particular for devices and systems of 5G mobile networks.

Acknowledgments
The work was carried out at the expense of funds, task No. FENW-2020-0022 for the implementation of scientific research carrying out scientific research at the expense of the Federal budget, in terms of scientific activities on the topic "Development and research of methods and means of monitoring, diagnostics and forecasting state of engineering objects based on artificial intelligence".

References

[1] Karthick R and Babu S P K 2020 Review on Radio Frequency Micro Electro Mechanical Systems (RF-MEMS) Switch *Int. Conf. on Communication, Computing and Electronics Systems* (Singapore/Springer) pp 437–453.
[2] Tongtong C, Tengjiang H and Yulong Z 2020 Research Status and Development Trend of MEMS Switches: A Review. *J. Micromachines* **11(7)** 438–453.
[3] Park J Y, et al 2000 Fully integrated micromachined capacitive switches for RF applications *Proc. of the 2000 IEEE MTT-S Int. Microwave Symp. Dig.* vol 1 (Boston, MA, USA/IEEE) pp 283–286.
[4] Al-Dahleh R and Mansour R R 2010 High-capacitance-ratio warped-beam capacitive MEMS switch designs *J. of Microelectromechanical Syst.* **19** 538–547.
[5] Chang C and Chang P 2010 Innovative micromachined microwave switch with very low insertion loss *J. Sensors and Actuators A: Physical* **79(1)** 71–75.
[6] Bozler C, et al 2000 MEMS microswitch arrays for reconfigurable distributed microwave components *Proc. of the IEEE Antennas and Prop. Soc. Int. Symp.* vol 2 (Salt Lake City, UT, USA/IEEE) pp 587–591.
[7] Angira M, et al 2013 On the investigation of an interdigitated, high capacitance ratio shunt RF-MEMS switch for X-band applications *Proc. of NSTI Nanotech.* vol 2 (Washington D.C, USA/Nano Science and Technology Institute (NSTI)) pp 189–192.
[8] Persano A, et al 2010 Ta2O5 thin films for capacitive RF MEMS switches *J. of Sensors* **10** 1–5.
[9] Liu Y, et al 2002 High-isolation BST-MEMS switches *Proc. of the 2002 IEEE MTT-S Int. Microwave Symp. Dig.* vol 1 (Seattle, WA, USA/IEEE) pp 227–230.
[10] Lysenko I E, Tkachenko A V and Ezhova O A 2019 Research of the microelectromechanical switch with different materials of metal membrane *Proc. of SPIE 11022, Int. Conf. on Micro- and Nano-Electronics 2018* vol 10226 (Zvenigorod, Russian Federation/SPIE) pp 1–12.
[11] Lysenko I E, Tkachenko A V, Ezhova O A, Konoplev B G, Ryndin E A and Sherova E V 2020 The Mechanical Effects Influencing on the Design of RF MEMS Switches *J. Electronics* **9(2)** 1–26.
[12] Lysenko I E, Tkachenko A V, Sherova E V and Nikitin A V 2018 Analytical Approach in the Development of RF MEMS Switches *J. Electronics* **7(12)** 1–23.

A seesaw-type MEMS switch with enhanced contact force: the first results

I V Uvarov[1] **and N V Marukhin**[2]

[1] Valiev Institute of Physics and Technology of Russian Academy of Sciences, Yaroslavl Branch, Universitetskaya 21, 150007 Yaroslavl, Russia
[2] P.G. Demidov Yaroslavl State University, Sovetskaya 14, 150003 Yaroslavl, Russia

Abstract. Outstanding working characteristics make microelectromechanical systems (MEMS) switches attractive for many applications. However, the lack of reliability prevents their commercial success. Due to the small size, MEMS switches develop low contact force compared to their macroscopic counterparts, which leads to instability and fast increase of the contact resistance. This work describes the switch providing significantly larger force than the previously reported device. The enlargement is achieved by the modified shape of the beam and electrodes with the same footprint and lower actuation voltage. Design, simulation, fabrication and first experimental results for the switch are presented.

1. Introduction

In recent time, an intensive growth of radar and wireless communication systems takes place. The growth is achieved due to advanced electronic components, including switches for processing and routing of RF signals. But the enhancement of conventional relays has exhausted, so the new approaches have to be found. The devices provided by MEMS industry are good candidates to substitute conventionally used products. They demonstrate high integration capability in combination with great radio frequency performance, small size and low energy consumption [1]. A significant characteristic of the MEMS switch is the contact resistance that strongly depends on the contact force F_c [2]. The device provides rather low F_c ~10 μN due to the tiny size. This makes the contact spot small and leads to instability and fast increase of the contact resistance. In addition, the switch becomes unable to overcome contamination. Obviously, the force has to be increased in order to ensure reliable operation. Here we present the switch that develops much higher F_c compared to the previous device [3]. High force is achieved at the same footprint and lower pull-in voltage. Performance of the switch is calculated using finite element method (FEM). Samples ready for the test are fabricated and analyzed.

2. Design of the device

The switch is schematically shown in figure 1. A movable electrode (source) is a 100 μm long beam suspended by torsion hinges. The device has two pairs of gate and drain electrodes placed under each arm of the beam. The beam is separated from the electrodes by 1.5 μm gap. It contains 0.5 μm high contact bumps at the bottom side. The material of bumps and electrodes is platinum, while the beam is made of aluminium. Driving voltage fed to the gate attracts the beam to the corresponding drain, turning the device on. The contact breaks under the elastic force of torsion hinges when the voltage is removed.

The previously reported switch, which is further called basic switch, provides quite low contact force $F_c = 12$ μN at the gate voltage $V_g = 60$ V [4]. In order to increase F_c, one has to make the electrostatic force higher. This may be achieved by raising the driving voltage, enlarging the overlap of the electrodes, or reducing the air gap. But these methods increase the switch size and worsen its performance. In this work, we modify the design in such a way that the overall footprint of the working part does not change. The modified shape of the beam and electrodes compared to the basic design is shown in figure 2. The gate is widened to 45 μm and surrounds the contact point. The beam is expanded according to the gate. The length of the electrode, in turn, is reduced. Such configuration increases the area for electrostatic force and transfers large portion of this force to the contact [5]. In addition, the switch withstands higher driving voltage. Thorough information on the modified design is given in our theoretical work [4].

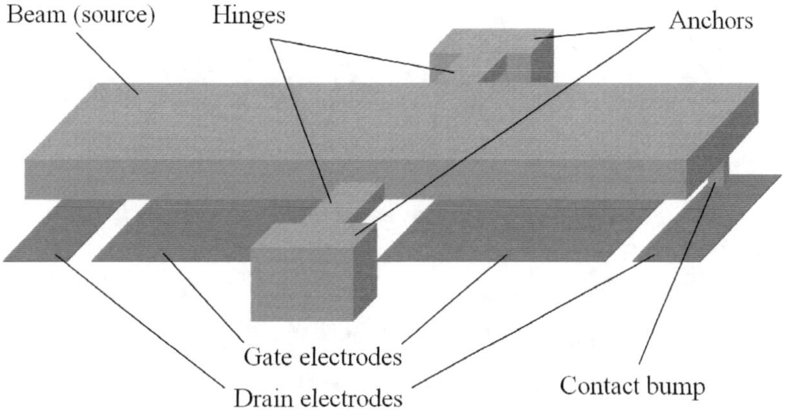

Figure 1. Schematic illustration of the switch.

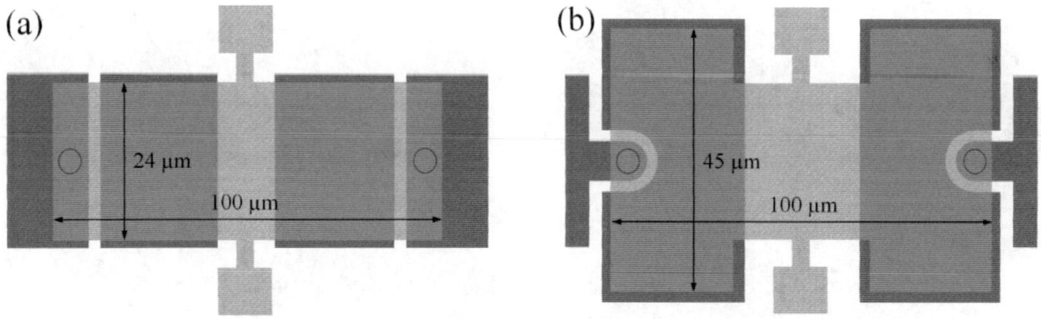

Figure 2. The basic design (a) and modified structure (b), top view.

3. Calculation of working characteristics

Pull-in voltage and resonant frequencies of the modified switch are calculated using validated FEM software according to the previously described procedure [6]. The displacement of the contact bump as a function of the gate voltage is shown in figure 3. The bump sharply falls to the drain at $V_g = 27$ V, while for the basic design the pull-in takes place at $V_g = 45$ V. Thus, the modified switch is actuated by lower voltage that reduces power consumption.

Vibration modes of the beam are shown in figure 4. Notice that we add perforation holes in order to remove the sacrificial layer faster. The first mode has a torsional shape and a frequency $f_1 = 144.4$ kHz. This mode determines the switching time [1]:

$$t_{\rm on} = \frac{3.67}{2\pi f_1} \frac{V_{\rm g}}{V_{\rm pi}}, \tag{1}$$

where $V_{\rm pi}$ is the voltage of electrostatic pull-in. Assuming $V_{\rm g} = V_{\rm pi}$, the switch should go to the on-state in 4 μs. This value is about 1 μs longer than $t_{\rm on}$ of the basic device, which has higher eigenfrequency [3]. The second mode corresponds to bending oscillations and has a frequency of 415.0 kHz. Further we will compare these values with experimental data.

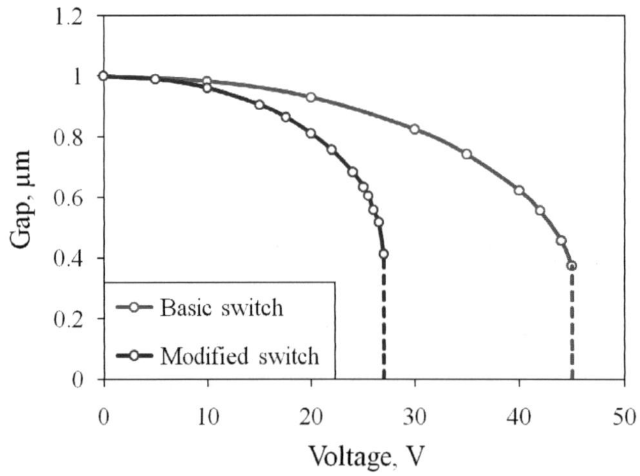

Figure 3. The gap between the contact bump and drain as a function of the gate voltage.

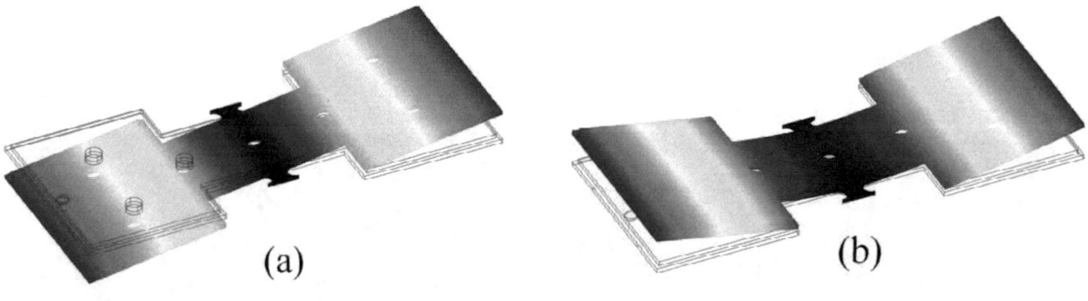

Figure 4. Eigenmodes of the beam: the first (a) and second (b) mode.

A significant characteristic is the dependence of the contact force on the gate voltage. A contact of two bodies is a non-trivial task for FEM simulation, so we use the analytical approach described previously [4]. The result is shown in figure 5. For the same $V_{\rm g}$, the modified switch develops 3.5 times higher force than the classical device. At the voltage of 85-100 V the force exceeds 100 μN. The contact resistance is in proportion to $F_{\rm c}^{-1/3}$ [2]. Thus, increasing the force is expected to lower the resistance by about 30%. Besides, one can expect an improved ability to break contamination films and, therefore, more stable resistance.

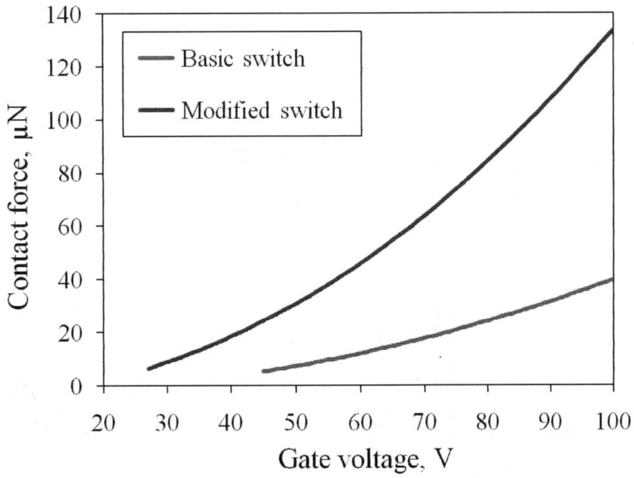

Figure 5. Contact force as a function of the gate voltage.

4. Fabrication

The fabrication sequence is illustrated in figure 6 and thoroughly described in our recent paper [7]. In short, it is carried out on thermally oxidized silicon wafers. The gates and drains are fabricated from Cr/Pt 10/100 nm film by lift-off technique. The beam is reinforced by 1 μm thick frame (figure 6e) and has a total thickness of 2 μm, while the hinges have a thickness of 1 μm. Amorphous silicon (a-Si) is used as a sacrificial material. It is removed from under the beam by plasma etching. Dry release ensures the absence of the fabrication-induced stiction under the capillary forces. In addition, the use of a-Si as a sacrificial material reduces contamination of contacts with organic residues, which is typically observed for the sacrificial photoresist layers. Thus, the proposed route is rather simple and ensures high yield. The fabricated switch is demonstrated in figure 7. One can see main parts of the device, including the contact area.

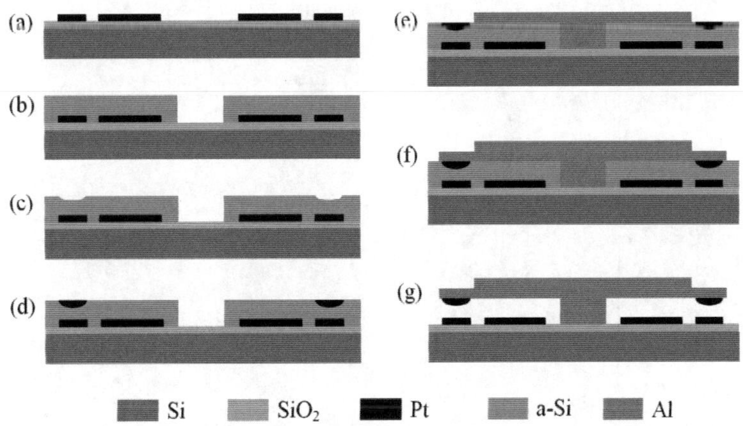

Figure 6. Fabrication sequence: (a) the bottom electrodes are patterned; (b) the sacrificial layer is deposited and etched to form the openings for anchors; (c) the dimples are etched; (d) the dimples are filled with platinum; (e) the first layer of aluminum is patterned to form the reinforcing frame; (f) the second layer of aluminum is patterned to form the anchors, hinges and beam; (g) the beam is released.

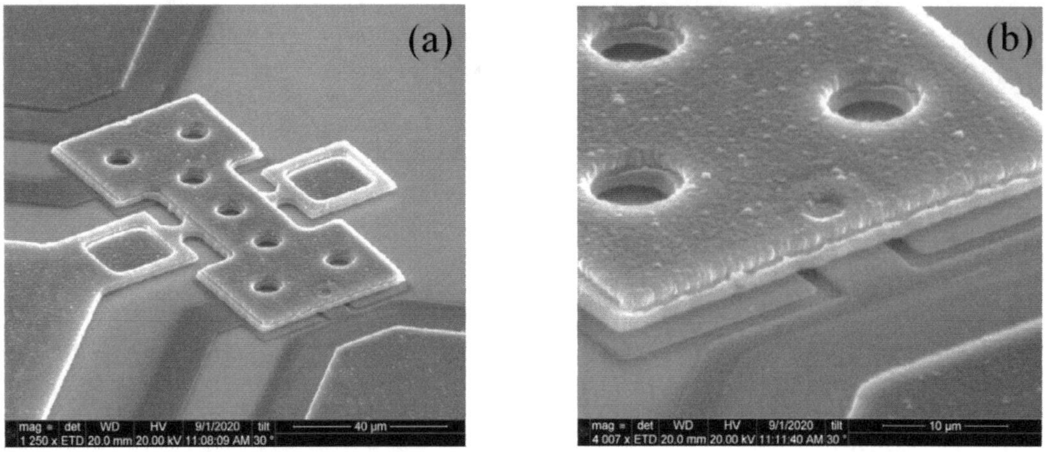

Figure 7. SEM pictures of the switch: (a) general view; (b) close-up view of the contact area.

5. Testing
After fabrication, one has to verify that the resonant frequencies and pull-in voltage of samples correspond to theoretical predictions. This will mean that the switch develops the contact force according to the curve depicted in figure 5. The resonant frequencies are measured with a homemade laser vibrometer [8] under the air pressure of 10^{-1} mbar. In order to excite oscillations of the beam, a sinusoidal voltage with the amplitude of 1 V is fed to the gate. The oscillations are detected by an optical lever unit. Laser light is focused at the beam and reflected to the photo-detector. The resonant curve contains two peaks corresponding to the first and second vibration modes, as shown in figure 8. The resonant frequencies are 15-20% lower than those predicted by FEM. A probable reason is the mismatch between the actual and nominal lateral size of the beam due to excessive undercut of aluminum layers, as well as the difference in elastic properties between thin-film and bulk aluminum. Further, we plan to measure the pull-in voltage and contact resistance in order to estimate the advantages of the modified design.

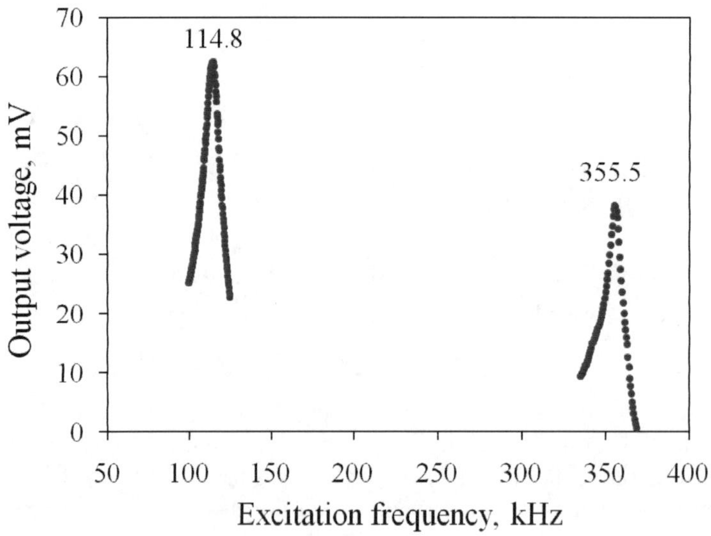

Figure 8. Output voltage of the photo-detector as a function of the excitation frequency.

6. Conclusion

The paper describes electrostatically actuated MEMS switch with protection against stiction. In comparison with the previously demonstrated switch, it provides 3.5 times higher contact force due to the modified design of the beam and electrodes. The modification increases the area for electrostatic force and transfers a large portion of this force to the contact point. The overall footprint of the device is not enlarged, while the pull-in voltage becomes lower. According to FEM simulation, the switch can be actuated by the voltage as low as 27 V. The device is successfully fabricated by surface micromachining using the well established route. Fabricated samples demonstrate 15-20% lower resonant frequencies compared to FEM predictions. The probable reasons are the excessive undercut of aluminum layers during the wet etching and the difference in elastic properties between thin-film and bulk material. Further, it is planned to determine the pull-in voltage and perform long-term measurements of the contact resistance in order to confirm the advantages of the modified design.

Acknowledgments

This work was supported by the grant of President of the Russian Federation № MK-945.2021.4 and the program No. 0066-2019-0002 of the Ministry of Science and Higher Education of Russia for Valiev Institute of Physics and Technology of RAS. The authors thank the Facilities Sharing Centre "Diagnostics of Micro- and Nanostructures" for technical assistance.

References

[1] Rebeiz G M 2003 *RF MEMS: Theory, Design, and Technology* (Hoboken, New Jersey: John Wiley & Sons)
[2] Toler B F, Coutu R A and McBride J W 2013 *J. Micromech. Microeng.* **23** 103001
[3] Uvarov I V and Kupriyanov A N 2019 *Microsyst. Technol.* **25** 3243
[4] Marukhin N V and Uvarov I V 2020 *J. Phys.: Conf. Ser.* **1695** 012157
[5] Stefanini R, Chatras M, Blondy P and Rebeiz G.M. 2011 *J. Microelectromech. Syst.* **20** 1324
[6] Uvarov I V, Naumov V V, Koroleva O M, Vaganova E I and Amirov I I 2016 *Proc. SPIE* **10224** 102241A
[7] Uvarov I V 2021 *Microelectron. Reliab.* 1**25** 114372
[8] Uvarov I V, Naumov V V, Amirov I I 2012 *Proc. SPIE* **8700** 87000S-1

The use of digital data processing to improve the metrological characteristics of the rubidium frequency standard

A P Valov[1], K G Arinushkina[2], V V Davydov[2, 3] and V Yu Rud[3]

[1] The Bonch-Bruevich Saint - Petersburg State University of Telecommunications (Russia, St. Petersburg, 193232, 22 Bolshevikov)
[2] Peter the Great Saint Petersburg Polytechnical University (Russia, St. Petersburg, 195251, 29 Polytechniheskya str)
[3] All Russian Research Institute of Phytopathology (Russia, Moscow Region, 143050, B.Vyazyomy, Institute str., ownership 5)

tony.valov2015@yandex.ru

Abstract. The necessity of improving the metrological characteristics of a quantum frequency standard when determining the coordinates of an object on the Earth's surface or long-term transmission of large amounts of data in satellite communication systems is substantiated. A new algorithm for digital processing of optical signals in the frequency standard is proposed. The results of experimental studies of the metrological characteristics of a quantum frequency standard based on rubidium atoms - 87 are presented

1. Introduction

In the modern world, devices and systems for determining the time are given increased attention [1-10]. These devices are used in communication systems, radar, navigation and scientific experiments [10-21]. Various devices have been developed to solve the problem of determining the exact value of time or frequency [1-3, 14, 15, 22-29].

A special place among devices for determining the frequency and time is occupied by frequency standards [2, 3, 14, 27, 28, 30-36]. The main advantage of the frequency standards over other devices is the use of laser radiation frequency stabilization systems and optical elements in it for stable operation [2, 3, 26-28, 32-39]. A slight deviation of the frequency from the nominal value leads to large errors, especially when transmitting large data streams and determining the coordinates of an object. One of the main problems of the satellite system is the mutual synchronization of the time scales of satellites down to nanoseconds or less. For example, an error in navigation signals emitted by different satellites with a time mismatch of 10 ns causes an additional error in determining the location of an object of 10-15 meters. The solution of new problems in the transmission of information and determination of coordinates requires constant modernization of satellite navigation systems, including quantum frequency standards [32-41]. In this paper, we consider one of the options for upgrading the rubidium frequency standard (RFS) to improve its metrological characteristics.

2. Principle of operation of the rubidium frequency standard

For many years, the general structural scheme of RFS has not undergone fundamental changes for various models of this type of RFS. In the design of rubidium RFS, individual elements or blocks are mainly changed, as well as control systems for various parameters to improve metrological characteristics [41, 42].

The operation of the RFS is based on the principle of tuning a highly stable voltage-controlled crystal oscillator (VCXO) to the rubidium quantum-frequency transition. To implement the noted frequency tuning of the quartz oscillator, the microwave signal from the frequency synthesizer (FS) is fed into a glass vacuum cell filled with rubidium-87 atoms and a buffer gas. When the frequency of the microwave signal coincides with the rubidium frequency of the quantum transition, stimulated emission of photons occurs, which is registered by the photodetector. In this case, the maximum signal-to-noise ratio (S / N) is observed. If the frequency of the microwave signal f_{mw} deviates from the resonant transition frequency, the S / N ratio decreases and the automatic frequency control circuit generates an error signal. This signal is used to adjust the frequency of the voltage-controlled crystal oscillator (VCXO). One of the key aspects of the rubidium RFS operation is the formation of a microwave signal. The process of forming a microwave signal is carried out in a frequency synthesizer.

Therefore, it is extremely important to develop a method that provides, on the one hand, high accuracy of the output frequency when it is tuned in an autonomous mode, regardless of communication with a ground station. In the methods considered in for the RFS, the frequency tuning step is more than 1 Hz using the voltage setting of the voltage-controlled crystal oscillator. This is not enough to meet the new precision requirements. Therefore, the new method uses the design of a frequency synthesizer that provides a fractional conversion factor to a hundredth of a hertz.

The frequency tuning of the crystal oscillator is controlled by the automatic frequency tuning system, which includes a crystal oscillator control unit (CO CU) and a control unit that converts the signal coming from the quantum discriminator and calculates a tuning code for further sending it to the CO CU.

In the previous version of the software for the RFS control device, a simple accumulation of all received data at different frequencies was used, followed by the calculation of the value of the microwave signal error signal. The new version proposes to use the median filtering method as one of the data filtering methods.

The median filtering method uses the ordering of several elements received at the input of the control unit, and the subsequent selection of a value equidistant from the beginning and the end of the ordered row of elements.

The median filtering algorithm has a pronounced selectivity to array elements with a non-monotonic component of the sequence of numbers within the aperture and most effectively excludes single outliers, negative and positive, from signals.

The basic principle of operation of the modified automatic frequency control algorithm of the frequency standard using median filtering of the input data is as follows:

1. Accumulation of data at the first frequency f1. In this case, accumulation begins with a delay, since switching the frequency causes a transient process of approximately $1/5F_m$. Samples are sampled synchronously with a $256F_m$ signal with a frequency of 9984 Hz fed to the microcontroller from a frequency synthesizer.
2. Ordering in ascending order of the received samples and further filtering by the median filtering method.

3. Experimental studies and discussion

During long-term operation of the FS, especially in space, there is an imbalance in the parameters of its operation. To adjust them, it is necessary to calculate the values of the error signal. This signal is greatly influenced by various noises that are present in optical signals. Using median filtering of data from optical signals can reduce the effects of random noise in the error signal, thus reducing phase noise. When using optical light signals for recording resonance conditions on photodetectors, an important characteristic is the power spectral density $S\varphi$. The optical signal arriving at the photodiode is digitized and fed to a control device based on a microcontroller. The data received by the control device are filtered and on their basis the control signal of the crystal oscillator is calculated to adjust to a more highly stable frequency of the rubidium junction. Figure 1 shows as an example the spectral density of noise in the frequency range of tuning the resonant frequency of a quantum transition.

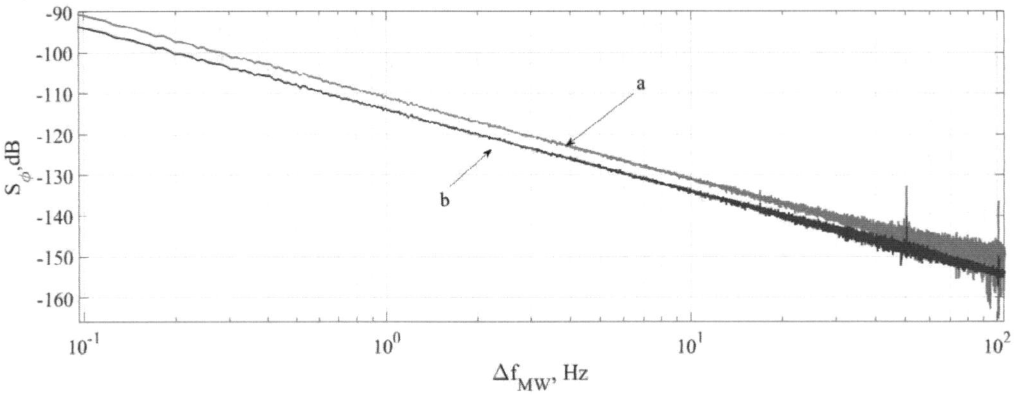

Figure 1. The phase noise spectral power density of error signal: a - the previous version of the software; b - the new version.

In figure 2 as an example, the dependences of the change in the error signal when using median filtering and previously applied methods are presented.

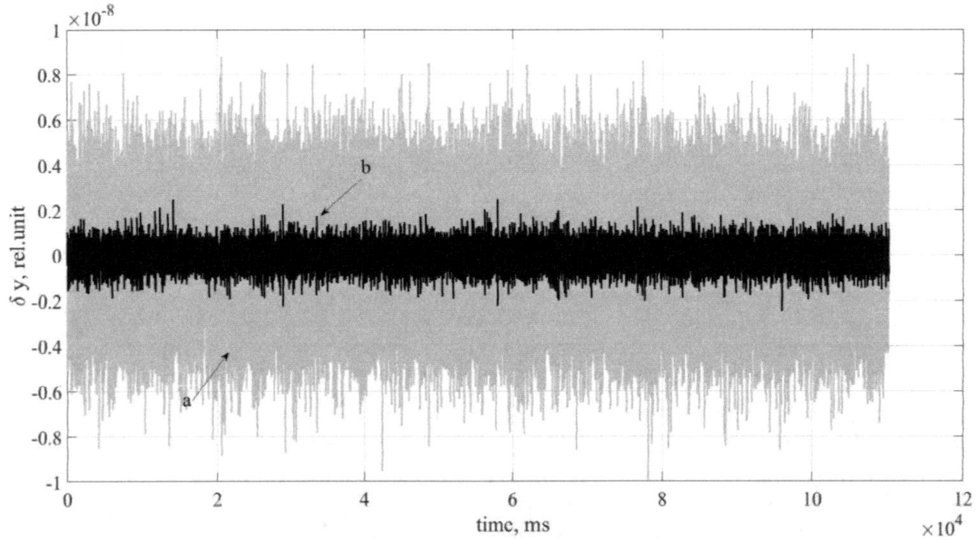

Figure 2: The values of the reference signals δy. Graph (a) corresponds to the earlier value of the error signal with simple accumulation, graph (b) - to the error signal with median filtering.

Analysis of the results obtained shows that the error signal decreases several times. When using optical light signals to register resonance conditions on photodetectors, an important characteristic is the spectral density Sφ [28, 37, 41, 42]. The Sφ value has a significant impact on the S / N ratio. Figure 3 shows the spectral densities of phase noise of the previously used design of the QFS and the new one, which uses the method of improving the parameters of the microwave signal.

The analysis of the results of experimental studies obtained in figure 1 showed that the use of median filtering and a microcontroller for control in the FS allows to reduce the power of phase noise in the output signal spectrum. The use of median data filtering in the automatic frequency control ring of the microwave excitation signal allows one to reduce one of the most important disturbing factors affecting the short-term frequency stability - the spectral noise density.

4. Conclusion

The analysis of the obtained results showed the expediency of using the developed method of automatic frequency control of the RFS microwave excitation signal. The proposed solutions for digital processing of optical signals made it possible to obtain an additional improvement in the metrological characteristics of the RFS.

Studies have shown that an improvement in long-term frequency stability by 5% using the new algorithm reduces the number of bit errors during long-term data transmission by at least 3%. It also increases the accuracy of determining the coordinates of an object by at least 15%. The obtained improvements in the short-term and long-term frequency stability make it possible to increase the degree of reliability of the satellite coordinate determination systems.

References

[1] Lukashev N A, Petrov A A, Grebenikova N M and Valov A P 2018 *Proceedings of 18th International conference of Laser Optics ICLO-2018 (Saint-Petersburg)*, vol. 8435889 (IEEE), p. 271

[2] Polyakov V M, Viktorov E A, Kovalev A V and Orlov O A 2014 *Proceedings of 16th International conference of Laser Optics ICLO-2014 (Saint-Petersburg)*, vol. 6886216 (IEEE), p. 123

[3] Kovalev A V, Polyakov V M and Mak A A 2016 *Proceedings of 17th International conference of Laser Optics ICLO-2014 (Saint-Petersburg)*, vol. 6886216 (IEEE), p. 201

[4] Moroz A V and Davydov R V 2019 *Lecture Notes in Computer Science (including subseries Lecture Notes in Artificial Intelligence and Lecture Notes in Bioinformatics)* **11660 LNCS** 710-718

[5] Davydov R V, Dmitrieva D S, Pilipova V M, Dudkin V I and Andreeva E I *2020 Proceedings of International conference of Laser Optics ICLO-2020 (Saint-Petersburg),* vol. 9285820 (IEEE), p. 171

[6] Smirnov K J, Medzakovskiy V I, Vysoczky M G and Glagolev S F 2017 *Journal of Physics: Conference Series* **917(6)** 062019

[7] Smirnov K J, Glagolev S F and Tushavin G V 2018 *Journal of Physics: Conference Series* **1124(2)** 022014

[8] Hoang T, Kirichek R, Paramonov A and Koucheryavy A 2016 *Lecture Notes in Electrical Engineering* **376** 1249-1259

[9] Davydov R, Antonov V, Makeev S, and Batov Y 2019 *E3S Web of Conferences* **140** 02001

[10] Makolkina M, Pham V, Kirichek R, Gogol A and Koucheryavy A 2018 *Lecture Notes in Computer Science (including subseries Lecture Notes in Artificial Intelligence and Lecture Notes in Bioinformatics)* **11118 LNCS** 547-559

[11] Andropov A V, Kuzmin S V and Korovin K O 2021 *Springer Proceedings in Physics* **255** 675–681

[12] Raimzhanov T R, Kuzmin S V and Korovin K O 2021 *Springer Proceedings in Physics* **255** 629–635

[13] Davydov R V, Saveliev I V, Lenets V A, Tarasenko M Yu and Yalunina T R 2017 *Lecture Notes in Computer Science (including subseries Lecture Notes in Artificial Intelligence and Lecture Notes in Bioinformatics)* **10531 LNCS** 177-183

[14] Petrov A A and Myazin N S 2017 *Lecture Notes in Computer Science (including subseries Lecture Notes in Artificial Intelligence and Lecture Notes in Bioinformatics)* **10531 LNCS** 561-568

[15] Podstrigaev A S, Smolyakov A V, Myazin N S, Grebenikova N M and Davydov R V 2019 *Lecture Notes in Computer Science (including subseries Lecture Notes in Artificial*

Intelligence and Lecture Notes in Bioinformatics) **11660 LNCS** 525-533
[16] Fadeenko V B and Pchelkin G A 2019 *Journal of Physics: Conference Series* **1400(4)** 044010
[17] Moroz A V 2019 *Journal of Physics: Conference Series* **1410(1)** 012212
[18] Moroz A V, Davydov V V, Malanin K Y, Krasnov A A and Rud V Yu 2019 *Journal of Physics: Conference Series* **1400(4)** 044009
[19] Fadeenko V B, Pchelkin G A and Beloshapkina O O 2019 *Journal of Physics: Conference Series* **1410(1)** 012238
[20] Fadeenko V B, Kuts V A and Vasiliev D A 2018 *Journal Physics: Conference Series* **1135**(1) 012053
[21] Sinicyna E A, Galichina A A, Lukyanov A S and Podstrigaev A S 2019 Journal of Physics: Conference Series **1236(1)** 012075
[22] Grebenikova N, Moroz A, Bylina M and Kuzmin M 2019 *IOP Conference Series: Materials Science and Engineering* **497(1)** 012109
[23] Kuzmin M S and Rogov S A 2019 *Computer Optics* **43(3)** 391-396
[24] Fadeenko I, Fadeenko V, Reznik V, Moroz A, Popovskiy N and Nikolaev D 2019 *IOP Conference Series: Earth and Environmental Science* **390(1)** 012022
[25] Ermolaev A N, Krishpents G P and Vysoczkiy M G 2016 *Journal Physics: Conference Series* **741(1)** 012071
[26] Semenov V V, Nikiforov N F and Ermak S V 1991 *Soviet journal of Communications Technology and Electronics* **36(4)** 59 – 63
[27] Petrov A A, Grebenikova N M, Lukashev N A, Ivanova N V, Rodygina N S and Moroz A V 2018 *Journal of Physics: Conference Series* **1038 (1)** 012032
[28] Grevtseva A S, Davydov R V, Dudkin V I and Rud' V Y 2019 *Journal of Physics: Conference Series* **1326(1)** 012043
[29] Logunov S E, Rud V Y, Davydov R V, Moroz A V and Smirnov K J 2019 *Journal of Physics: Conference Series* **1326(1)** 012024
[30] Lukashev N A, Glinushkin A P and Lukyantsev V S 2019 *Journal of Physics: Conference Series* **1410(1)** 012211
[31] Lukashev N A, Davydov R V and Glinushkin A P 2019 *Journal of Physics: Conference Series* **1326(1)** 012046
[32] Petrov A A, Shabanov V E, Zalyotov D V, Bulyanitsa A L and Shapovalov D V 2018 *International Conference on Electrical Engineering and Photonics EExPolytech 2018 (Saint-Petersburg)*, vol. 8564389 (IEEE), p. 52-55
[33] Petrov A A, Vologdin V A and Zalyotov D V 2015 *Journal of Physics: Conference Series* **643(1)** 012087
[34] Petrov A A and Grebennikova N M 2018 *Journal of Communications Technology and Electronics* **63(11)** 1281–1285
[35] Lukashev N A and Moroz A V 2019 *Journal of Physics: Conference Series* **1236(1)** 012068
[36] Petrov A A and Grebenikova N M 2018 *Lecture Notes in Computer Science (including subseries Lecture Notes in Artificial Intelligence and Lecture Notes in Bioinformatics)* **11118 LNCS** 641-648
[37] Valov A P 2019 *Journal of Physics: Conference Series* **1410(1)** 012246
[38] Petrov A A, Zaletov D V, Davydov V V and Shapovalov D V 2021 *Journal of Communications Technology and Electronics* **66(3)** 295–299
[39] Petrov A A, Zalyotov D V, Shabanov V E and Shapovalov D V 2018 *Journal Physics: Conference Series* **1124(1)** 041004
[40] Petrov A and Shapovalov D 2019 *Journal of Physics: Conference Series* **1400(4)** 044008
[41] Valov A 2020 *CEUR Workshop Proceedings* **2667** 102–105
[42] Riehle F 2004 *Frequency standards. Basics and Applications* (New Jersy: Wiley-VCH)

Method for reducing phase fluctuations of a precision frequency response meter for microwave quantum generators

M A Vodopyanov[1], K A Menzorov[1], V V Davydov[1, 2] and V Yu Rud[2]

[1] Peter the Great Saint Petersburg Polytechnic University, St. Petersburg, 195251, Russia
[2] All Russian Research Institute of Phytopathology, Moscow Region, 143050, Russia

vodopyanov.maks@yandex.ru

Abstract: The principle of operation of the device for measuring frequency characteristics is considered. Block diagram of a device for measuring the frequency characteristics of a digital frequency comparator with correlation quadrature processing and description of their components is illustrated. The block diagram of the frequency multiplier is given. The accuracy characteristics of the developed meter design are evaluated. They are compared with the characteristics of previously used devices. The Improve accuracy characteristics and increased stability of operation were found.

1. Introduction

In modern navigation systems, without determining the exact time and frequency, it is impossible to achieve the required accuracy in determining the coordinates of the object [1-10]. The data on the coordinates of the object are necessary when solving problems of environmental monitoring, carrying out various works, etc. [10-19]. In operating satellite navigation constellations, quantum frequency standards (QFS) are actively used to determine the exact time [1, 2, 8, 20-24]. There are a number of problems with these devices [23-27]. One of them is the work of a precision meter. A precision meter helps to solve the problem of the occurrence of phase fluctuations introduced by the frequency multiplier [20-22, 25, 27-31]. In the precision meters of the old generation, there were fluctuations in the signal phase, when it exposed to noise, pulsation of power sources, spurious signals at the I input and noise on the reactive circuits of frequency multipliers. Phase fluctuation is a form of linear distortion caused by the deviation of the transmitted signal from the carrier frequency.

This frequency comparator в прецизионных счетчиках is designed to measure frequency instability and estimate the spectral density of phase fluctuations of signals of quantum frequency standards [30-37]. Precision meters have been developed for a long time, but still need to be upgraded. Device development is important to test the onboard system with higher accuracy.

2. Methods of measuring frequency characteristics

2.1 The method of electronic counting frequency meter

The principle of operation of digital meters is to count the number of pulses by an electronic counter, which is a function of the measured value, during the counting of t.

The device is a device that registers, depending on the operating mode, the number of cycles of the test signal or the sample frequency signal for the time set by one of these signals. The main elements of the electronic counting frequency meters are an electronic pulse counter, a forming device, a time selector, and a selector control device [20-29, 30-34, 38].

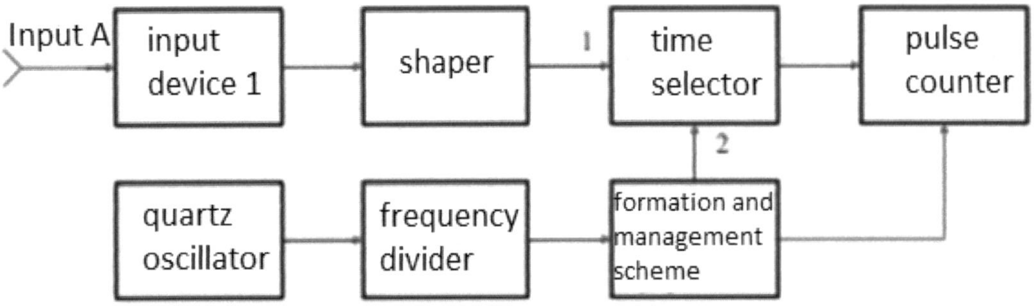

Figure 1. Block diagram of an electronic counting frequency meter.

2.2 The prototype
An analog-to-digital conversion of the measured signal is performed with a time sampling interval determined by the frequency fs of the sampling signal formed from a reference signal with a frequency f0 in accordance with the expression $f_s=4 \cdot f_0$.

A device for measuring the frequency of a harmonic signal, adopted as a prototype, implementing the prototype method, contains a series - connected generator of the measured signal, an analog-to-digital converter, a random-access memory device and a digital signal processing processor connected by a data exchange bus with a personal computer. The synchronization input of the analog-to-digital converter is connected to the output of the frequency synthesizer of the quantization signal, the input of which is connected to the output of the reference signal generator. The synchronization input of the analog-to-digital converter is connected to the output of the frequency synthesizer of the quantization signal, the input of which is connected to the output of the reference signal generator. In the process of implementing the prototype method, the measured signal generator generates a harmonic (sinusoidal) signal at its output, which is fed to the signal input of the analog-to-digital converter. The synchronization input of the analog-to-digital converter receives a sampling signal (quantizing pulses). Quantizing pulses are formed using a quantization signal frequency synthesizer, which performs the operation of multiplying "by four" the frequency f_0 of the reference signal coming from the output of the reference signal generator to the signal input of the quantization signal frequency synthesizer. The analog-to-digital converter performs analog-to-digital conversion of the measured signal with a time sampling interval determined by the sampling signal that is, $t_S = \frac{1}{4*f_0}$. This ensures a phase shift between adjacent samples of the measured signal equal to $\frac{\pi}{2}$.

The disadvantage of the prototype method and the prototype device is a significant limitation of the frequency range of the compared signals. Indeed, let be the maximum operating frequency of analog-to-digital conversion and digital sample storage operations. Then the frequency range of the amplified signals is limited from above by the following relation: $f_x \leq \frac{f_{вых}}{4}$.

In difference to the prototype, an analog-to-digital conversion of the reference signal is additionally performed with a time sampling interval determined by the specified frequency fs of the sampling signal formed in accordance with the expression: $f_S = \frac{4*f_0}{5*(2*j+1)}$, where $j = -\frac{2}{5}, 0,1,2,3 \ldots$ it depends on the frequency range of the compared signals.

For example: let the maximum operating frequency of the hardware implementing the prototype method and the claimed method be 20 MHz. When comparing signals with a nominal frequency of 5

MHz, as is done in the example implementation of the prototype method, the sampling signal frequency f_s=20 MHz. Exactly the same sampling frequency can be selected for the claimed method at $j = -\frac{2}{5}$. Suppose you want to measure the frequency difference with nominal values of 100 MHz. For the prototype method, f_s=400 MHz $\geq\geq$ 20 MHz, which makes the prototype inoperable with the specified implementing hardware. The claimed method remains workable for measuring frequencies with such a nominal value, since, taking j=2, we get f_s=16 MHz $\leq$ 20 MHz.

The frequency response measurement device is designed to measure the frequency characteristics of quartz oscillator signals, frequency and time standards, and frequency synthesizer signals [1, 2]. An integral part of the operation of this device is the formation of a time discredit signal synchronized with one of the input matched signals, and the conversion of the input signal voltages into digital codes of quadrature components, storing these codes in the receiving computing device [2, 3]. As well as the power supply device of the board converts the AC voltage (220±220) V 50 Hz into a DC reference voltage power supply plus 24 V.

The reliable stability of the master generator is 10^{-15}, therefore, the precision meter must be at least 3 times more accurate to detect problems [2, 3].

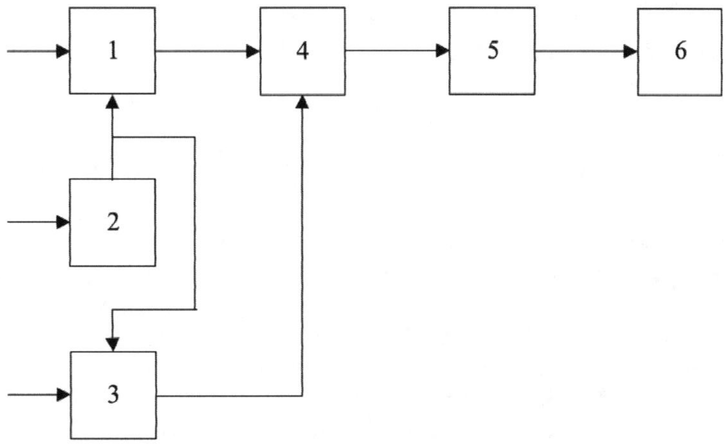

Figure 2. Block diagram of frequency response measurement device of a digital frequency comparator with correlation quadrature processing, where 1 and 3 are an ADC, 2 – frequency multiplier by 4, 4 – digital signal processor (DSP), 5 – field-programmable gate array (FPGA), 6 - Micro Controller Unit (MCU).

The PLL circuit is designed for coherent synthesis of the generator output signal with the sinusoidal signal of the measuring channel. Switching scheme – for controlling the modes of operation of the frequency response measurement device. ADC of the reference channel – for converting the voltage of a reference sinusoidal signal with the clock frequency of a quartz oscillator coherent with the measured signal into a hexadecimal bit code. ADC of the measuring channel - for continuous conversion of the voltage of the measured sinusoidal signal with the clock frequency of a quartz oscillator coherent with the measured signal into a hexadecimal bit code. FPGA – for managing the modes of operation of the frequency response measurement device.

The frequency of one of the matched signals is multiplied by four and the signal from the output of the frequency multiplier is used to take digital samples of the harmonic input signals. The digital signal processor receives digital codes of voltage samples of the measured signal, averages them and filters them to form a code proportional to the phase difference and frequency of the compared signals. A time sampling signal with a nominal frequency of 20 MHz allows you to calculate phase samples of the measured signal every 50 ns. at the same time it forms the quadrature components of the compared signals. Therefore, at the data inputs of the digital signal processing processor (DSP), there are four types of samples: ai; bi; ci; di, following with a nominal period of 0.2 microseconds in each channel. Depending on the processor speed, the DSP processes the codes of these samples, for example, after a time interval of 1 microsecond, it depends on the processor speed.

- S_a, 1 мкс = { a_1, a_2,... a_N)
- S_b, 1 мкс = { b_1, b_2,... b_N)
- S_c, 1 мкс = { c_1, c_2,... c_N)
- S_d, 1 мкс = { d_1, d_2,... d_N)

The DSP processor converts the sample data into a phase difference and a frequency difference.

3. The 1321HD1U chip

The radio receiver chip 1321HD1 allows to receive and process a radio signal at frequencies in the range of 10-300 MHz. To amplify the received signal and transfer it to an intermediate frequency, a built-in low-noise amplifier, a mixer and a PLL system with a fractional division coefficient are used. The received signal is divided into a direct and a quadrature signal and converted into a digital code by a band-pass ΣΔ-ADC and a digital quadrature demodulator. The chip implements an AGC system with a range of 24 dB, as well as the ability to manually adjust the gain of the LNA in the range of up to 15 dB. The ADC and the digital filter block are clocked from an internal PLL with a built-in active part of the VCO. Also, the chip allows to process a signal with a minimum noise factor in the band up to 25 kHz and in the band up to 200 kHz with some degradation of the noise factor.

3.1 Description of the operation of the chip

The input signal at a carrier frequency from 10 MHz to 300 MHz, passing through an external 50-ohm matching chain of the signal source, with a 500-ohm input of the built-in LNA, is amplified with a gain of 17 to 0 dB, depending on the user's settings, and converted to an IF frequency, which is Fclk/8, where Fclk is the clock frequency of the ADC. The frequency of the local oscillator can be supplied either from an external oscillator or generated using an internal PLL with a fractional division coefficient. To eliminate blockers, the chip provides two tunable first-order filters, which also act as anti-reflection filters before the ADC input. The cut-off bands of these filters are also configured via the control registers. To eliminate quantization noise and obtain the required sampling frequency of the direct and quadrature channels, the signal from the output of the digital quadrature demodulator is fed to the input of digital decimator filters with an adjustable decimation coefficient. After filtering, the signal is formatted according to the specified format, and transmitted to the user via the SSI interface. The values are written to the control registers via the SPI interface.

3.2 The composition of the 1321HD1U chip

The integrated circuit consists of:

- low-noise amplifier (LNA), with the possibility of reducing consumption while reducing the gain in steps of 3-4 dB;
- mixer, with the possibility of reducing consumption by reducing IIP3;
- active low-pass filter (LPF), with tunable cut-off band;
- Bandpass Delta-Sigma ADC;
- digital demodulator;
- a block of digital decimator filters, with a programmable decimation coefficient;
- PLL of the local oscillator, with the fractional and integer parts of the output frequency divider;
- reference frequency buffers;

- PLL of the clock signal synthesiser;
- SPI interface, for IC programming;
- SSI interface, for transmitting digital samples for further processing;
- reference voltage and current sources

Figure 3. Receiver module.

4. Conclusion

Our review shows that the claimed group of inventions is feasible and provides the achievement of a technical result, which consists in increasing the accuracy of measuring the frequency of a harmonic signal while expanding the frequency range of the compared signals.

References

[1] Batori E, Almat N, Affolderbach C and Mileti G 2020 *GNSS-grade space atomic frequency standards: Current status and ongoing developments* (New York: Advances in Space Research)

[2] Semenov V V, Nikiforov N F and Ermak S V 1991 *Soviet journal of Communications Technology and Electronics* **36(4)** 59 – 63

[3] Petrov A A, Grebenikova N M, Lukashev N A, Ivanova N V, Rodygina N S and Moroz A V 2018 *Journal of Physics: Conference Series* **1038 (1)** 012032

[4] Lukashev N A, Petrov A A, Grebenikova N M and Valov A P 2018 *Proceedings of 18th International conference of Laser Optics ICLO-2018 (Saint-Petersburg),* vol. 8435889 (IEEE), p. 271

[5] Lukashev N A, Glinushkin A P and V. S. Lukyantsev V S 2019 *Journal of Physics: Conference Series* **1410(1)** 012211

[6] Lukashev N A, Davydov R V and Glinushkin A P 2019 *Journal of Physics: Conference Series* **1326(1)** 012046

[7] Davydov R, Antonov V and Moroz A 2019 *International Conference on Electrical Engineering and Photonics EExPolytech-2019 (Saint-Petersburg)*, vol. 8906791 (IEEE), p. 295-297

[8] Davydov R, Antonov V, Makeev S, Batov Y, Dudkin V and Myazin N 2020 *E3S Web Conference* **140** 02001

[9] Fadeenko V, Fadeenko I, Dudkin V and Nikolaev D 2019 *IOP Conference Series: Earth and Environmental Science* **390(1)** 012022

[10] Fadeenko V V, Kuts V A and Vasiliev D A 2018 *Journal Physics: Conference Series* **1135(1)** 012053

[11] Gryznova E, Batov Y and Myazin N 2020 *E3S Web of Conferences* **140** 09001
[12] Gryznova E, Grebenikova N, D. Ivanov D and Bykov V 2019 *IOP Conference Series: Earth and Environmental Science* **390(1)** 012044
[13] Nikitina M, Grebenikova N, Dudkin V and Batov Y 2019 *IOP Conference Series: Earth and Environmental Science* **390(1)** 012024
[14] Yushkova V, Kostin G, Davydov R, Dudkin V and Valiullin L 2019 *IOP Conference Series: Earth and Environmental Science* **390(1)** 012016
[15] Van S, Cheremisin A, Chusov A, Switala F and Davydov R 2019 *IOP Conference Series: Earth and Environmental Science* **390(1)** 012011
[16] Van S, Cheremisin A, Glinushkin A, Davydov R and Yushkova V 2019 *E3S Web of Conferences* **140** 09008
[17] Myazin N S, Yushkova V V and Taranda N I 2019 *Journal of Physics: Conference Series* **1410(1)** 012130
[18] Rukin E V, Myazin N S and Dudkin V I 2019 *Journal of Physics: Conference Series* **1368(4)** 042011
[19] Myazin N S, Yushkova V V and Dudkin V I 2019 *Journal of Physics: Conference Series* **1400(6)** 066008
[20] Petrov A A, Shabanov V E, Zalyotov D V, Bulyanitsa A L and Shapovalov D V 2018 *International Conference on Electrical Engineering and Photonics EExPolytech 2018 (Saint-Petersburg)*, vol. 8564389 (IEEE), p. 52-55
[21] Petrov A A, Vologdin V A and Zalyotov D V 2015 *Journal of Physics: Conference Series* **643(1)** 012087
[22] Petrov A A and Grebennikova N M 2018 *Journal of Communications Technology and Electronics* **63(11)** 1281–1285
[23] Lukashev N A and Moroz A V 2019 *Journal of Physics: Conference Series* **1236(1)** 012068
[24] Petrov A A and Grebenikova N M 2018 *Lecture Notes in Computer Science (including subseries Lecture Notes in Artificial Intelligence and Lecture Notes in Bioinformatics)* **11118 LNCS** 641-648
[25] Petrov A A and Davydov V V 2015 *Lecture Notes in Computer Science (including subseries Lecture Notes in Artificial Intelligence and Lecture Notes in Bioinformatics)* **9247** 739-744
[26] Valov A P, Davydov R V and Grevtseva A S 2019 *Journal of Physics: Conference Series* **1326(1)** 012040
[27] Grevtseva A S, Davydov R V, and Dudkin V I 2019 *Journal of Physics: Conference Series* **1326(1)** 012043
[28] Grevtseva A, Davydov V and Rud V 2020 *CEUR Workshop Proceedings* **2667** 15–18
[29] Petrov A A and N S Myazin 2017 *Lecture Notes in Computer Science (including subseries Lecture Notes in Artificial Intelligence and Lecture Notes in Bioinformatics)* **10531 LNCS** 561-568
[30] Valov A P 2019 *Journal of Physics: Conference Series* **1410(1)** 012246
[31] Valov A 2020 *CEUR Workshop Proceedings* **2667** 102–105
[32] Valov A P, Nagornaya A I, Kulikova M E and Glinushkin A P 2020 *Journal of Physics: Conference Series* **1695(1)** 012169
[33] Petrov A A, Zaletov D V, Davydov V V and Shapovalov D V 2021 *Journal of Communications Technology and Electronics* **66(3)** 295–299
[34] Petrov A A, Zalyotov D V, Shabanov V E and Shapovalov D V 2018 *Journal Physics: Conference Series* **1124(1)** 041004
[35] Davydov R V and Antonov V I 2018 *Journal of Physics: Conference Series* **1124(8)** 081037
[36] Davydov R V and Antonov V I 2018 *Journal of Physics: Conference Series* **1135(1)** 012087
[37] Logunov S E, Rud V Y, Davydov R V, Moroz A V and Smirnov K J 2019 *Journal of Physics: Conference Series* **1326(1)** 012024
Petrov A and Shapovalov D 2019 *Journal of Physics: Conference Series* **1400(4)** 044008

Metrological aspects of an automated method for measuring electrophysical parameters of soft magnetic materials

A V Volik, E A Pecherskaya, Yu A Varenik, T O Zinchenko, D V Artamonov and O A Timohina

Department of Information and measuring equipment and metrology, Penza State University, Penza 440026, Russia

Abstract. The structure of an automated system for measuring magnetic-hysteresis loops, normal magnetization curve, magnetic permeability with an error of no more than ± 1% is proposed. The measuring principle is based on the inferential measurements of the magnetic induction and the coercive force by integrating the secondary voltage and the excitation current. As a result of metrological analysis, an increase in the measurements accuracy is achieved both by improving the hardware implementation and calibrating the measuring channels, by introducing a correction for the systematic component of the error.

1. Introduction
The relevance of the study of the electrophysical parameters of magnetic materials is due to their wide application in power electronics devices, in motors, generators [1], as well as in micromechanical systems [2]. Experimental studies of the dependences of the parameters of magnetic materials make it possible to discover new directions for their use. For example, in [3], the results of measurements of hysteresis loops of a high-temperature magnetocaloric alloy are presented, which confirm the appearance of an additional phase transition near a temperature of 315 K. The article [5] proposes a non-destructive microwave method for measuring the electrophysical properties of materials, based on the method of transmission and reflection of a plane monochromatic wave through layered materials with a measurement error of ± 2%. The analysis of the listed and other known instruments for measuring the electrophysical parameters of magnetic materials confirms the relevance of improving the principles and methods of measurements in order to increase their accuracy by performing metrological analysis.

2. Relationships between the parameters of soft magnetic materials
Magnetically soft materials are capable of magnetizing in weak magnetic fields (voltage less than 800 A / m). Soft magnets have low coercive force, low hysteresis loss and high magnetic permeability. A measure that determines the softness of a magnet can be the value of the static relative magnetic permeability μ: its initial values are in the range from 102 to 105; the maximum values belong to the interval from 103 to 106, and the losses due to magnetic hysteresis do not exceed 102 J / m^3 per one magnetization reversal cycle.

The behavior of a ferromagnet is described by the dependence of the polarization J (magnetization M) on the magnetic field strength H. The magnetic induction is related by the relationship to the quantities M, J, and H as follows:

$$B = \mu_0 H + \mu_0 M = \mu_0 H + J \quad (1)$$

where $\mu_0 = 4\pi \times 10^{-7}$ H / A^2 (H / m) - magnetic constant (also called vacuum magnetic permeability).

In fact, expression (1) is a macroscopic result of an extremely complex sequence of microscopic processes in which, by a combination of displacements of domain walls, rearrangement of the domain structure, and rotations of the magnetic moments, the system reacts to a change in the applied field H, making a transition through a sequence of metastable states with minimum energy. These processes are associated with irreversibility and losses, as well as with various behavior of the J (H) dependence, for which the phenomenon of hysteresis is inherent. Due to hysteresis, any point on the plane (J, H) can be crossed by an infinite number of trajectories, depending on the history of the process. For magnetic materials, two basic states are determined: saturation, when all domains are oriented in the direction of the external field, and demagnetized state (H = 0, J = 0). The latter is achieved either by starting from saturation ($J = J_s$) and a smooth decrease in the amplitude of the applied ac field to zero, or by cooling the sample to the Curie temperature in the absence of any field. The curve obtained after thermal demagnetization is called the primary curve. Most often, the demagnetization process is performed using a decreasing alternating field, and then the initial magnetization curve is measured. Figure 1 shows the hysteresis loop of a ferromagnet, measured using a prototype of the automated installation under development for measuring the parameters of magnetic materials.

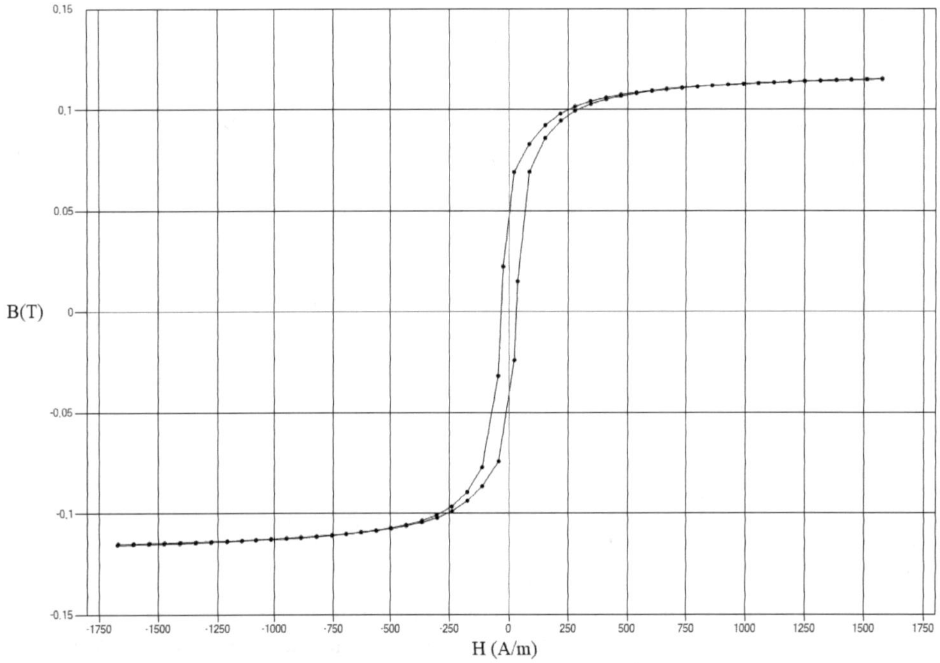

Figure 1. The result of measuring the dependence B (H) in the form of a hysteresis loop using a model of an automated setup for measuring magnetic parameters.

The hysteresis loop contains magnetic properties and materials can be classified according to the value of its parameters. Coercive field H_c, residual polarization J_r, maximum permeability $\mu_p = B_p / H_p$, initial permeability and energy loss W are the main parameters by which the properties of materials can be estimated. The composition of the material actually determines the values of the so-called internal magnetic parameters, such as the Curie temperature, saturation magnetization, magnetic anisotropy constants and magnetostriction constants, which, in turn, affect the magnetization process depending on the structure of the material (for example, crystallographic texture, grain size, foreign phases and lattice defects). Through the correct choice of composition and suitable metallurgical and heat treatment, super soft magnets can be obtained, in which the coercive field can reach values below 1 A / m with a relative permeability of the order of 105. But it should be emphasized that it is necessary to take into account a number of additional properties, such as thermal and structural stability, sensitivity of magnetic parameters to stress, mechanical properties and workability, thermal conductivity and flexibility of reaction to thermomagnetic treatment.

The final conclusion about the applicability of a magnetic material in solving a specific problem will be the assessment of all these properties.

The most common structural material, iron, exhibits soft magnetic properties. The addition of a few percent of Si impurity causes noticeable changes in the physical, mechanical and magnetic properties of iron. Such an alloy is called silicon electric steel, in which iron is the base base. The silicon content in electrical steel provides high electrical resistance and increases the stability of the alloy, helps to increase the magnetic permeability, due to the prevention of the formation of an undesirable phase of cementite Fe_3C. In addition to improving the characteristics, the silicon impurity also has a negative effect on the alloy. Among the negative effects are an increase in the brittleness of the alloy and a decrease in the saturation magnetization. Despite this, electric steel is the main soft magnetic material for mass consumption due to its low cost.

Ni + Fe alloys with a predominant concentration of Ni are called permalloy. The presence in the composition of nickel provides this alloy of a magnetic material with resistance to corrosion and good ductility, which makes it possible to produce thin sheets of this metal that are easy to machine. The magnetic properties of permalloy are largely determined by the chemical composition; therefore, two main types of alloy can be distinguished: high-nickel, containing up to 80% nickel, and low-nickel. The main qualities of this alloy are high values of resistivity and magnetic permeability. Permalloy alloy is quite difficult to manufacture, which is reflected in its high cost.

3. Method for determining static magnetic characteristics

The standard described in GOST 8.377-80 applies to precision soft magnetic materials with a coercive force up to 4 kA / m and establishes a measurement procedure for determining the main magnetization curve and magnetic hysteresis loop for samples of these materials and their parameters; maximum magnetic permeability; the squareness coefficient of the magnetic hysteresis loop; coercive force on magnetization; temperature coefficients of the above parameters, as well as the requirements for these samples and measuring equipment.

The statistical magnetic characteristics and parameters of the samples are determined using an installation (Figure 2), consisting of measuring instruments and devices, the requirements for which are set out below.

The main magnetization curve is determined starting from the lowest required field strength, gradually moving to higher values. Returning from larger values to smaller values is not allowed. The coordinates of each point of the main magnetization curve (magnetic field strength and magnetic induction) are measured according to the following sequence:
The text of your paper should be formatted as follows:

- Calculate the values of the magnetizing current I in amperes, corresponding to the values of H in amperes per meter, at which the magnetic induction will be determined, according to the formula

$$I = \frac{\pi H(D_H + D_B)}{2\omega_1} \quad (2)$$

  instead of ω_1, the number of turns of the magnetizing device (if used) is substituted.
- Connect the magnetizing winding to the power source using the S_2 switch; the measuring winding is disconnected from the galvanometer (webmeter).
- Switch S_1 switch on the current and with the help of adjusting devices R_1 and R_2 (switch K open) set the lowest current value obtained by the formula (4) according to any of the ammeters.
- Change the direction of the current at least 10 times with the S_1 switch. The switching frequency should be no more than two operations per second.

Switch S_3 turn on the measuring winding and, by changing the direction of the current, determine at this moment the deflection of the galvanometer pointer or the weemeter reading. The measurements are repeated many times and the results are averaged to reduce the random component of the measurement result.

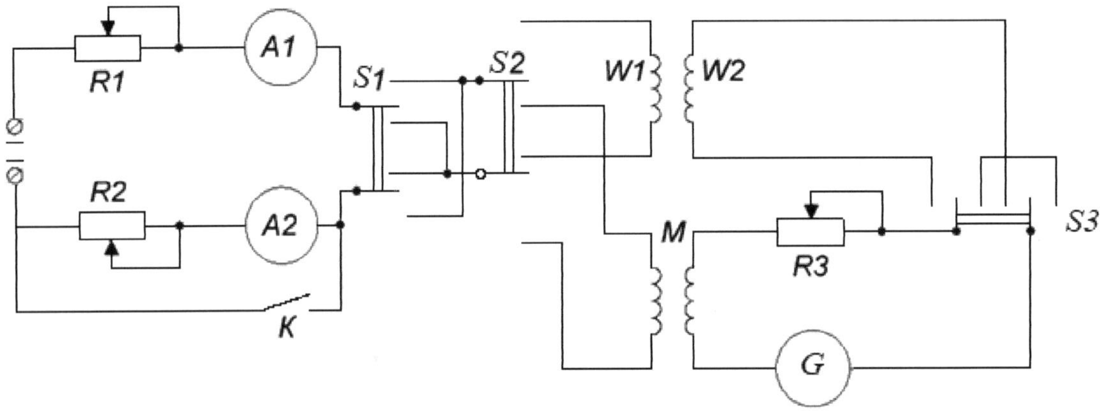

Figure 2. Connection diagram of the installation elements for determining the static magnetic characteristics and parameters of the samples: *R*1, *R*2 - adjusting devices; *A*1, *A*2 - ammeters; *S*1, *S*2, *S*3 - double-pole switches; K - key; W1 - magnetizing winding of the sample or magnetizing device; *M* - exemplary coil of mutual inductance; *R*3 - resistance box; *G* - ballistic galvanometer or webmeter; *W*2 - measuring winding of the sample or measuring coil.

4. The structure of the system for the magnetic materials parameters measuring

The authors have developed an automated information and measurement system (IMS) for studying the parameters of magnetic materials (Figure 3), based on inferential measurements of the magnetic induction and the coercive force by integrating the secondary voltage and the excitation current, respectively.

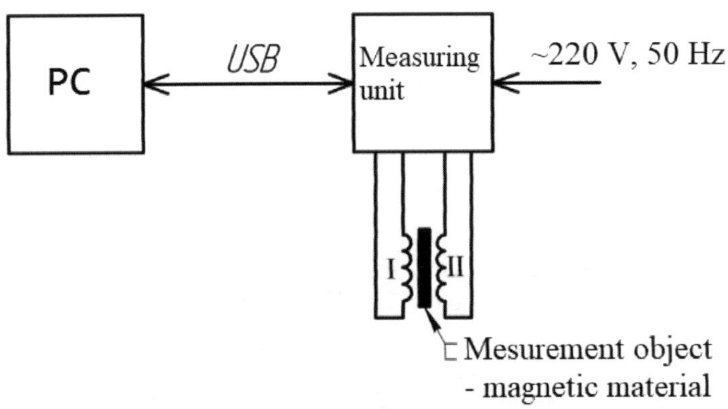

Figure 3. Block diagram of IMS for studying the soft-magnetic materials parameters.

A toroid core is made of the investigated soft-magnetic material. The measuring unit generates the excitation currentin the winding I and converts the EMF of the winding II into a code proportional to the magnetic induction in the investigated soft-magnetic material.

5. Magnetic induction transfer function

Installation control and processing of measurement results is carried out according to unique methods, which are implemented in software using a computer (PC). In particular, both the Preisach model and the modeling methods previously tested by the authors on the hysteresis loops of ferroelectric materials were used to simulate magnetic-hysteresis loops [6, 7]. The value of the desired magnetic induction B is determined according to the following expression:

$$B = \frac{\int \varepsilon(t)dt}{S \cdot n_{II}} \quad (3)$$

where B – magnetic flux; S – the core cross-sectional area; n_{II} – number of magnetizing coil rings; $\boldsymbol{\varepsilon}$ - EMF of the measuring winding, which is converted into a digital code by means of an analog-to-digital converter (ADC) N_B:

$$N_B = \mathrm{Ent}\left[\frac{\frac{\varepsilon t_p}{\tau} + 0{,}5q}{q}\right] \quad (4)$$

where Ent[·] - integer symbol; τ- integrating time constant; t_p- pulse duration at the input of the integrator as part of the measuring unit; q – ADC quantization step.

As a result of the metrological analysis of the developed IMS, it was found that the main instrumental error in the magnetic induction measuring is due to the errors of the integrator based on the operational amplifier, the error in measuring the cross-sectional area S, and the metrological characteristics of the ADC. In order to minimize the integration error, it is proposed to use an integrator, in which the capacitor discharge is realized with a voltage that is formed by a discharge stage on an operational amplifier at the integrator output.

6. Conclusion

The proposed automated IMS performs automated measurements of magnetic-hysteresis loops, normal magnetization curve, magnetic permeability with an error of no more than ± 1%, as well as processing and displaying the results obtained on the monitor screen in graphical and tabular form, saving them in a database for further research. An increase in the measurement accuracy is achieved both by improving the hardware implementation and by calibrating the measuring channels by introducing a correction for the error systematic component.

References

[1] Josefina M Silveyra, Enzo Ferrara and Dale L Huber and Todd C Monson 2018 Soft magnetic materials for a sustainable and electrified world *Science* **362** (6413) eaao0195

[2] Mikhailin N Y, Romanov S G, Kumzerov Y A. et al 2018 Superconducting Properties of Indium Nanostructured in Pores of Thin Films of SiO_2 Microspheres *Phys. Solid State* **60** 1942–1947

[3] Gozdur Roman and Najgebauer Mariusz 2019 Scaling analysis of phase transitions in magnetocaloric alloys *Journal of Magnetism and Magnetic Materials* **499** 166239

[4] Skresanov V N and Shubnyj A I 2006 Autogenerator Spectral Method for Testing of Electrophysical Properties of Materials in Millimeter-Wave Band *16th International Crimean Microwave and Telecommunication Technology* pp 741–742

[5] Gorst A, Zavyalova K, Shipilov S and Yakubov V and Mironchev A 2020 Microwave Method for Measuring Electrical Properties of the Materials *Appl. Sci.* **10** 8936

[6] Artamonov D V, Baranov V A, Pecherskaya E A, Pushkareva A V and Tsypin B V and Fimin A V 2019 Application of a Hyper-Complex Impedance Model for Indirect Measurements of Materials Parameters of Functional Electronics *Proc. International Conference of Young Specialists on Micro/Nanotechnologies and Electron Devices* pp 760–764

[7] Golubkov P E, Pecherskaya E A, Golovyashkin A N, Golovyashkin A A and Pecherskiy A V and Shepeleva Y V 2018 Analysis of the technological parameters influence on the reproducibility of the active dielectrics properties *Proc. International Conference of Young Specialists on Micro/Nanotechnologies and Electron Devices* pp 57–61

Prospective directions for the development of microwave frequency standards for satellite navigation systems

Ding Wang[1, 2], V V Davydov[1, 3] and V Yu Rud[3, 4]

[1] Peter the Great Saint-Petersburg Polytechnic University, Saint Petersburg, Russia, 195251
[2] Jiangsu Normal University, Xuzhou, China, 221116
[3] All-Russian Research Institute of Phytopathology, Moscow Region 143050, Russia
[4] A.F. Ioffe Physicotechnical Institute, St. Petersburg, 195152, Russia

jssdwang06@mail.ru

Abstract. The state of essential various quantum standards of GNSS frequencies for today are collected and presented, the results of analysis in the direction of modernization of time synchronization systems in global navigation satellite systems are presented. The most perspective directions of modernization of global navigation satellite systems are mentioned – the development of new atomic clocks on the mercury ions -199. The data on experimental satellite gives encouraging results.

1. Introduction

Currently, global navigation satellite systems (GNSS) are constantly being developed and modernized [1-9]. The modernization process is determined by the structure of GNSS construction and by the tasks solved by it [8-14]. Four satellite constellations are currently deployed in outer space: the American system - GPS, the Russian system - Glonass, the European system - Galileo and the Chinese system - BDS [13-16].

The most important function of GNSS is to determine the location of an object and provide users with three-dimensional information about its coordinates. For positioning an object using the GNSS system, the time and coordinate system are of critical importance [5-8, 10, 17-22]. Any deviation in the value of time between satellites can have a huge impact on the accuracy of positioning [16, 23].

The coordinate system in GNSS is a reference system. It consists of the origin O, the X-axis, the Y-axis and the Z-axis, in the usual case, the origin O is in the center of mass of the Earth, the X-axis and the Z-axis are directed in different directions, the Y-axis makes a rectangular coordinate system with the X-axis and the Z-axis. For proper operation of the positioning system, while there is no single system, they can replace each other with complex mathematical calculations, this is not very convenient for everyone. This often creates problems. Therefore, it is important to determine the promising directions for the development of GNSS in order to reduce these problems to a minimum in the future.

In addition, the orbits of many satellite navigation systems are filled with space debris (decommissioned satellites). There are more and more of them every year. These satellites create problems both in the operation of guidance systems and in the placement of new satellites in orbits [29, 30, 38-40]. This must also be taken into account when determining the promising directions for the development of satellite navigation systems. The increase in the service life of the satellite, which is now determined by the life of the quantum frequency standard, is a reduction in the rate at which the orbits are filled with "space" debris. This will gain time before an effective system for clearing orbits from "space" debris appears in space. The number of failures in the GNSS work every year due to the "space" garbage increases.

2. Analysis of characteristics of satellite space constellations for navigation systems

All navigation satellites GPS, Glonass, Galileo operate in Medium Earth Orbits (MEO), but each system has a different number of satellites in orbit and a different period of satellite rotation, BDS-3 uses a constellation of 3 orbits: Inclined Geosynchronous Orbits (IGSO), Geostationary Earth Orbits (GEO), and Medium Earth Orbits (MEO), the number of satellites and the rotation period of satellites in each orbit is also different [15, 16, 24-28]. In table 1 shows the data on the different systems.

Table 1. Data on the different system.

System	Orbit altitude (km)	Satellite orbit	Orbit inclination	Satellite rotation period
GPS	20200	MEO	55^0	11 h. 58 m.
Glonass	19100	MEO	64.8^0	11 h. 51 m.
Galileo	23222	MEO	56^0	14 h. 4 m. 45 s.
BSD-3	21528/35768/35768	MEO/GEO/IGSO	$55^0/0^0/55^0$	12 h. 53 m./ 24 h./ 24 h.

Each navigation system uses a different time system: GPS - GPST, Glonass - Glonass-ST, Galileo - GST, BDS - BDT. These time systems are linked together by Coordinated Universal Time (UTC). The UTC time system is based on international atomic time (TAI). There is no extra second in GPST, GST, and BDT after time starts, and there is an extra second in Glonass-ST. Each navigation system has its own coordinate system: GPS coordinate system (WGS-84), Glonass coordinate system (PZ-90), Galileo coordinate system (GTRF), BDS coordinate system (BDCS).

To transmit navigation signals, GPS, Galileo and BDS use code division multiple access (CDMA) technology, while traditional Glonass uses frequency division multiple access (FDMA), currently Glonass supports CDMA. Each system operates in different bands and has different carrier frequencies, in table 2 shows the carrier frequency data of different systems.

From the analysis of the data in table 2, it follows that GPS (L1), Galileo (E1), BDS-3 (B1) have the same carrier frequency 1575.42MHz, GPS (L5) and Galileo (E5a) have the same carrier frequency 1176.45MHz. This shows that the systems can easily be integrated with each other in case of a merger. This is one of the possible directions of modernization of the entire GNSS.

Table 2. Formatting sections, subsections and subsubsections.

System	Range	Carrier frequency (MHz)
GPS	L1, L2, L5	1572.42/1227.60/1176.45
Glonass	G1, G2, G3	1600.995/1248.06/1202.025
Galileo	E1, E5a, E5b, E6	1575.42/1176.45/1207.14/1278.65
BSD-3	B1, B2, B3	1575.42/1191.259/1268.52

The time system in GNSS is linked to the atomic clock of the satellite, which can also be called the heart of the whole satellite system. The GPS satellites use atomic clocks with rubidium (Rb-87) and passive hydrogen maser (PHM), the Glonass satellites use atomic clocks with cesium (Cs-133). The Galileo and BDS satellites (BDS-3) use two types of atomic clocks: passive hydrogen maser (PHM) and rubidium atomic frequency standard (RAFS). There is an experimental satellite with a mercury ion atomic clock (Hg-199) in space. In Figure 1 shows a diagram of the change in the number of atomic clocks in orbit over 40 years. The total number of atomic clocks on active satellites is shown in parentheses.

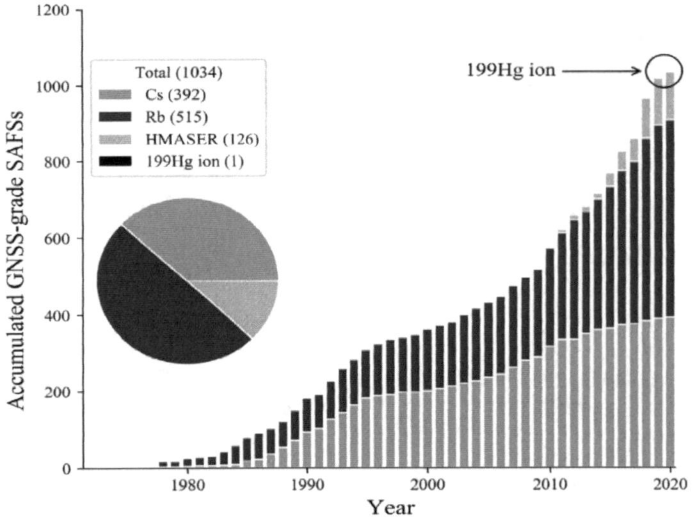

Figure 1. The estimated number of GNSS-grade or better SAFSs sent in space as of 2 March 2021.

Operational data have shown that the atomic clock PHM (Galileo) has the greatest long-term stability of the output characteristics (it does not require constant adjustment of the scale during a communication session with the Earth) [23, 29-33].

3. Analysis of metrological characteristics of quantum standard frequencies

Frequency stability is an important metrological characteristic for quantum standard frequencies, usually described using Allan deviation, Figure 2 and Figure 3 shows 2020 data on Allan deviation for various models of atomic clocks placed on satellites, including new designs of passive hydrogen masers and atomic clocks on mercury ions -199.

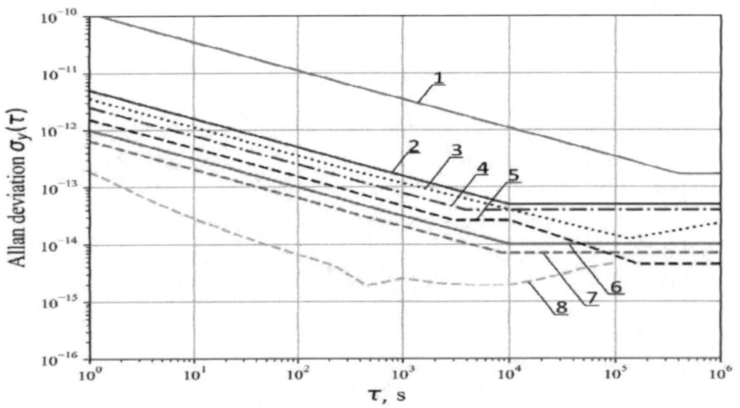

Figure 2. The Allan deviations of different current space atomic clocks**.** Graphs 1, 2, 3, 4, 5, 6, 7 and 8 correspond to the following atomic clock: Cs-133 – GPS; RAFS spec. – Galileo; RAFS – BDS; RAFS – Galileo; PHM–BDS; Cs-133 – Glonass; PHM – Galileo; Radioastron H-maser – GPS and Galileo.

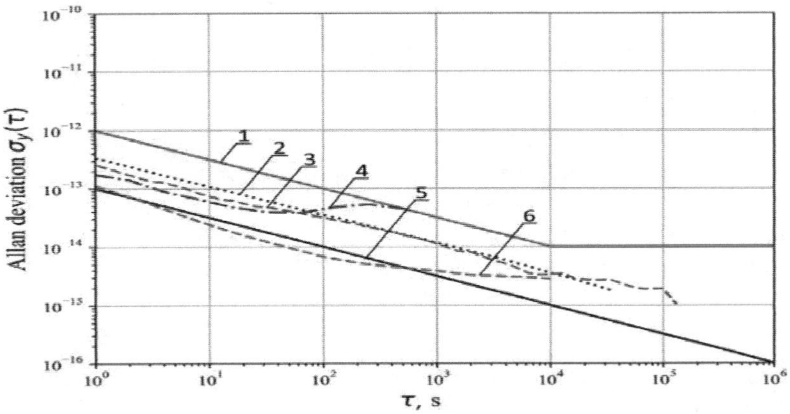

Figure 3. The Allan deviations of different current space atomic clocks and station communication on Earth. Graphs 1, 2, 3, 4, 5, and 6 correspond to the following atomic clock: Cs-133 – Glonass; PHM and RAFS – GPS; JPL 199Hg ion – GPS experimental – USA; Leonardo RAFS – Galileo; PHM – ship communication station; PHM – communication station on Earth.

Analysis of Allan deviation data for different models of atomic clocks (Figure 2), shows that PHM have the best values than other models of atomic clocks. Their disadvantages are limited operating life, high requirements for temperature stabilization and large size compared to atomic clocks on rubidium atoms.

The most prospective direction in the modernization of GNSS is the development of atomic clocks on mercury ions -199 [34-37]. These standards do not have the disadvantages of PHM noted above. The only problem is to work out the design of the Pauli trap while reducing its size and weight for the satellite system. Data on the experimental satellite give encouraging results.

4. Conclusion

The analysis of all the data obtained shows that the two directions of GNSS modernization are the most promising. Creation of a single GNSS system on Glonass orbit using two satellite orbits of the Chinese BDS system. This will solve many of the problems discussed in this article. Especially with the removal of space debris from orbit. The Glonass orbit is the lowest. Forced removal (currently implemented technology on the space robot Ullius) from this orbit of spent satellites will not create problems for other satellite navigation systems.

Or the replacement of all the satellites by new ones with atomic clocks on mercury ions -199. This will reduce the number of satellites in orbit, since the atomic clocks on mercury ions -199 needs almost no adjustment from Earth. The resource of the optical system of this clock is very long (longer than the electronics of the satellite). It is necessary to note that orbits are very cluttered with satellites, taken out of operation and there is less and less place on them. It will also reduce the rate at which space debris fill up orbits.

From this point of view, and taking into account the control of satellite constellation compared to global satellite navigation systems the Chinese system - BDS is the most reliable.

References

[1] Batori E, Almat N, Affolderbach C and Mileti G 2020 *Advances in Space Research (New York: Wiley–Interscience)*

[2] Semenov V V, Nikiforov N F and Ermak S V 1991 *Soviet journal of Communications Technology and Electronics* **36(4)** 59 – 63

[3] Petrov A, Vologdin A and Zalyotov V 2015 *Journal of Physics: Conference Series* **643(1)** 012087

[4] Petrov A A 2017 *Journal of Communications Technology and Electronics* **62(3)** 289–293

[5] Petrov A A 2016 *Journal of Physics: Conference Series* **769(1)** 012065
[6] Hagimoto K 2009 *IEEE Trans. Instrum. Meas.* **48** 496–499
[7] Petrov A 2017 *Lecture Notes in Computer Science (including subseries Lecture Notes in Artificial Intelligence and Lecture Notes in Bioinformatics)* **10531 LNCS** 561-568
[8] Petrov A A, Zalyotov D V, Shabanov V E and Shapovalov D V 2018 *Journal of Physics: Conference Series* **1124(4)** 041004
[9] Lukashev N A, Petrov A A, Grebenikova N M and Valov A P *2018 Proceedings of 18th International conference of Laser Optics ICLO-2018 (Saint-Petersburg),* vol. 8435889 (IEEE), p. 271
[10] Hudson A and Camparo J 2020 *Physical Review Applied* **13(6)** 064007
[11] Chan Y, Johnson W, Karuza S, Young A and Camparo J 2010 *IEEE Transactions on Instrumentation and Measurement* **59(2)** 330–334
[12] Valov A P 2019 *Journal of Physics: Conference Series* **1410(1)** 012246
[13] Petrov A A and Grebenikova N M 2018 *Lecture Notes in Computer Science (including subseries Lecture Notes in Artificial Intelligence and Lecture Notes in Bioinformatics)* **11118 LNCS** 641-648
[14] Petrov A A, Grebenikova N M, Lukashev N A, Ivanova N V, Rodygina N S and Moroz A V 2018 *Journal of Physics: Conference Series* **1038 (1)** 012032
[15] Hegarty C J and Chatre E 2008 In.: Proceedings of the IEEE, vol. 96(12), p. 1902-1917
[16] Hein G W 2020 Satell Navig **1** 22
[17] Petrov A A and Grebennikova N M 2018 *Journal of Communications Technology and Electronics* **63(11)** 1281–1285
[18] Petrov A A, Shabanov V E, Zalyotov D V, Bulyanitsa A L and Shapovalov D V 2018 *International Conference on Electrical Engineering and Photonics EExPolytech 2018 (Saint-Petersburg)*, vol. 8564389 (IEEE), p. 52-55
[19] Formichella V, Camparo J and Tavella P 2017 *Applied Physics Letters* **110(4)** 043506
[20] Martinez F 2014 *Performance of new GNSS satellite clocks* (Scientific Publishing, Karlsruher Institut fur Technologie)
[21] Petrov A and Shapovalov D 2019 *Journal of Physics: Conference Series* **1400(4)** 044008
[22] Lukashev N A, Davydov R V and Glinushkin A P 2019 *Journal of Physics: Conference Series* **1326(1)** 012046
[23] Tavella P and Petit G 2020 *Satell Navig* **1** 10
[24] Jaduszliwer B and Camparo J 2021 *GPS Solut* **25** 27
[25] Hofmann-Wellenhof B 2008 *GNSS - Global navigation satellite systems. GPS, GLONASS, GALILEO and more* (Springer Wien, N. Y.)
[26] Vannicola F 2010 *Proceeding of the 42th Ann. Precise Time and Time Interval Meeting*, p.181–195
[27] Cash P, Krzewick W, MacHado P, et al. 2018 In: 2018 European Frequency and Time Forum, EFTF 2018 (Institute of Electrical and Electronics Engineers Inc.), p. 65–71
[28] Valov A 2020 *CEUR Workshop Proceedings* **2667** 102–105
[29] Cao Y, Huang G, Xie W, et al. 2021 *Acta Geod Geophys* **3** 112
[30] Yang Y, Gao W, Guo S, Mao Yand Yang Y 2019 *NAVIGATION* **66** 7– 18
[31] Grevtseva A S, Davydov R V, Dudkin V I and Rud' V Y 2019 *Journal of Physics: Conference Series* **1326(1)** 012043
[32] Grevtseva A, Davydov V and Rud V 2020 *CEUR Workshop Proceedings* **2667** 15–18
[33] Micalizio S, Calosso C E, Levi F, et al. 2019 *In.: Proc. 2019 IEEE Int. Work. Metrol. AeroSpace, Metroaerosp.*, p. 682–686
[34] Lukashev N A and Moroz A V 2019 *Journal of Physics: Conference Series* **1236(1)** 012068
[35] Lukashev N A 2019 *Journal of Physics: Conference Series* **1400(2)** 022050
[36] Lukashev N A 2020 *Journal of Physics: Conference Series* **1695(1)** 012168
[37] Kuster J 1999 *Joint Meeting of the European Frequency and Time Forum and the IEEE International Frequency control Symposium*, vol. 87(5), p.156 – 168
[38] Petrov A A, Zaletov D V, Davydov V V and Shapovalov D V 2021 *Journal of Communications*

Technology and Electronics **66(3)** 295–299
[39] Batori E, Almat N, Affolderbach C, Gruet F and Mileti G 2020 Proceedings of European Navigation Conference, ENC 2020, (Dresden, Germany, Institute of Electrical and Electronics Engineers Inc), 9317388
[40] Logunov S E, Rud V Y, Davydov R V, Moroz A V and Smirnov K J 2019 *Journal of Physics: Conference Series* **1326(1)** 012024

Scanning ion-conductance microscope with modulation of the sample position along the Z-coordinate and separate Z-axial and lateral (X, Y) scanning

M V Zhukov, S Yu Lukashenko, I D Sapozhnikov, M L Felshtyn, O M Gorbenko and A O Golubok

Laboratory of scanning probe microscopy and spectroscopy, Institute for Analytical Instrumentation RAS, St. Petersburg 198095, Russia

cloudjyk@yandex.ru

Abstract. Scanning ion-conductance microscope with independent piezoscanners in the lateral scanning plane XY and Z axis was designed and tested. For precise, fast and safe approach of the nanopipette to the sample surface, a coarse approach system based on a piezoinertial mover was used. Measurements of test periodic polymer structures were carried out using nanopipettes with an inner pipette diameter of about 100-150 nm. The optimal geometric parameters of the nanopipette were found and the resolution of the method was estimated. To increase the stability and reproducibility of SICM images, the Z-modulation of the position of the substrate with the sample was realized using a bimorph piezomembrane.

1. Introduction

The interest of measuring ion channels in cell membranes and studying of the mechanisms of cell functioning gradually outgrown into the necessity for a qualitative spatial study of the topography of cells and other biological objects at the micro- and nanometer level. This resulted in creation of a scanning ion-conductance microscope (SICM), which allows mapping the surface by monitoring of ion currents using micropipettes with an electrode as a probe [1]. The method based on maintenance of a given value of a constant ionic current by the feedback system, similar to a scanning tunneling microscope, only with an inverted dependence I (z). However, this approach, as it turned out, have instability due to potential drift on the electrodes and external electrical interference, as well as instability when scanning soft objects of complex geometry, which led to probe breakdowns and artifacts [2].

To increase the stability of SICM images, amplitude modulation (AM) [3] of the voltage is used, as well as the method of phase modulation [4], which is less inclined to electronic noise and DC drift. Nevertheless, modulation methods do not solve the problem of studying objects with a complex relief shape and high surface roughness. The hopping mode is most preferable when measuring soft rough surfaces, but its use is limited, because the imaging speed is relatively slow due to the constant movement of the pipette up and down [5]. It was suggested to improve the speed of the formation of an SICM images by means of preliminary fast scanning [6] and prediction of the height at the each next point of the image using hybrid scanning [7]; however, these approaches are suitable only for smooth surfaces.

In [8] the AC (alternating ionic current) SICM method with vibrating probe was developed. In this method the amplitude of the AC ionic current detected by lock-in amplifier is more sensitive to probe-sample distance and is decreased when probe-sample gap is increased unlike to DC (direct ionic current) SICM. This technique is more stable for study of soft objects with complex geometry and has better space resolution.

Another interesting approach to improving the SICM images is to increase the speed of the pipette in the Z-direction by improving the mechanical properties of the piezoelectric mover [9], as well as using high-frequency piezoelectric drives [10].

In this paper, we have explored the design of the scanning ion-conductance microscope by introducing separate scanning scheme in the XY plane, using piezo-membranes, and in the Z coordinate, using a rigid piezo-package, as well as independent Z-modulation of the sample using a piezo-membrane.

2. Experimental setup

The development of technology for the manufacture of piezoceramic elements has made it possible to create mechanisms with a high speed response in comparison with electromagnetic devices. However, they are not widely used due to the fact that with a small contact area of hard surfaces the resulting pressure quickly destroys the flat surface and leads to scratches and disruption of movement. Earlier, the authors created a granular surface for friction pairs, the so-called "orange peel" [11], in which the products of friction are brought to the surface in the intervals between the individual grains. The use of friction pairs with granular surfaces made it possible to create a fairly rigid small-sized SPM unit, in which the same piezoelectric package is used both in a coarse approach system based on a piezo-inertial mover and for separate smooth scanning along the Z axis.

Figure 1 shows a schematic diagram of a scanning ion-conductance microscope. The stacked multilayer piezoelectric element (2) is attached to the fixed base (1) on one side. A titanium guide rod (4) is attached to the opposite side of the piezoelectric element. The titanium carriage (3) is pressed against the surface of the titanium guide rod (4) by a spring. Under the action of an electric pulse, the piezoelectric actuator (2) is sharply stretched or compressed, depending on the sign of the electric voltage. If the inertial force acting on the carriage (3) is greater than the friction force acting between the surface of the guide rod (4) and the surface of the carriage (3), that slides over the surface of the guide rod (4). The direction of movement of the carriage up or down can be changed by changing the sign of the control voltage.

The stacked multilayer piezoelectric element provides both a precision stepwise approach (removal) of a fragile borosilicate nanopipette (NP) to a sample in the range of ~ 5 mm, which protects the NP from destruction, and smooth displacement along the Z coordinate in the range of ~ 7 μm.

The NP (6) fixed on an inertial Z-piezoscanner by a spring clamp (5) and a Petri dish (7) was filled with a 1% NaCl solution and had AgCl electrodes inside. Scanning range of bimorph membrane of piezoelectric elements in the lateral XY plane ~ 100x100 μm^2. To reduce mechanical noise active vibration protection was used. The ionic current I_i at the AgCl electrodes in (6, 7) was converted into an electronic current I_e, which was converted into an electric voltage proportional to the ionic current $U = R \cdot I_e$ through converter (8). In our case, R = (0.1-1) GΩ. This electric voltage U is compared on a differential amplifier (9) with a reference voltage U_0 and fed to the input of the integrator (10) and then to high-voltage amplifier (11) connected to the Z-piezoscanner (5). Such servo system with a negative feedback loop is necessary to maintain a constant average value of the ion current when scanning the sample.

In order to increase the stability of scanning, a mechanical Z-modulation of the substrate with the sample was added through a bimorph piezoelectric membrane. We suppose, modulation leads to a better redistribution of charges and prevents the formation of an electric double layer on the surface of the pipette.

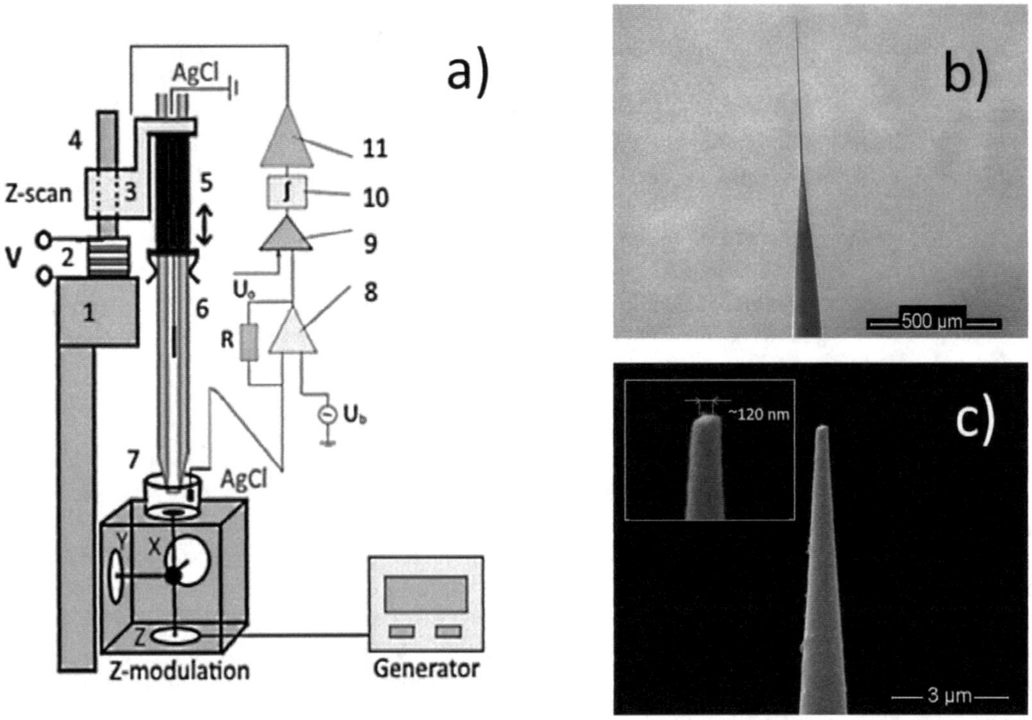

Figure 1. Scheme of the experimental setup (a): 1 - fixed basis, 2 - stacked multilayer piezoelectric element (approach and scanning), 3 - titanium carriage, 4 - titanium guide rod, 5 - NP clamp, 6 - NP with AgCl electrode, 7 - Petri dish with AgCl electrode, 8 - ion current-to-voltage converter, 9 - differential amplifier, 10 - integrator, 11 - high-voltage amplifier, U_b - bios voltage applied to the nanocontact, U_0 - set point voltage signal. SEM image with overview of NP shape (b) and magnification area (c) with approximate diameter of the NP tip is about 100-150 nm.

3. Results and Discussions

In scanning process the set point during scanning was selected based on the curves of the current versus distance (I-Z). In figure 2 the dependences I-Z in the SICM-mode at positive and negative polarity of the voltage applied to electrodes are shown. The set point is selected at the sharpest part of the I-Z curve that is shown by green line. The most stable area for scanning is lying closer to the point where the current starts to fall. In figure 3 a sharp I-Z dependence for a small-diameter pipette about 100-150 nm located near the surface with modulation of Z coordinate of the sample is shown. To control the characteristics of the pipette and estimate the radius of the pipette, the current-voltage characteristics (CVC) were measured (Figure 3, b). The slope of the I-U curve can be used to determine the resistance of the NP and relate this to the radius of the NP outlet. The red vertical lines show the working voltage region for the SICM-mode.

The geometric parameters of the NP were estimated in a scanning electron microscope (SEM) (figure 1, b, c) before and after measurements. NP can be considered as good for scanning with parameters:

- the angle of sharpening of the NP at the apex relative to the axis is about $6\text{-}9^0$;
- the diameter of the NP outlet is about 100-300 nm;
- the unevenness of the NP tip edge from the horizontal plane is no more than $6\text{-}7^0$.

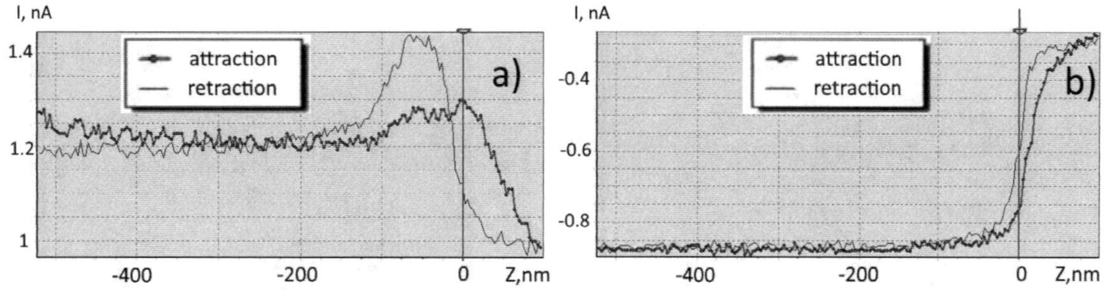

Figure 2. Typical I-Z curves in the SICM-mode when the NP was in contact with the surface at positive (a) and negative (b) polarity.

As you can see in figure 2a, a current leap is observed near the sample surface on the I-z attraction/retraction curves. This phenomenon, as well as a small hysteresis of the attraction/retraction curves, can be associated with an increase in the electric field strength that occurs when the pipette approaches the sample surface. As a result, the flow of ions moving towards the pipette outlet increases. A decrease in the «tip-sample» gap with further approach leads to a decrease in the ion flux. It is assumed that this phenomenon is associated with other phenomena that appear on the CVC and observed earlier - the pipette rectification effect [12, 13].

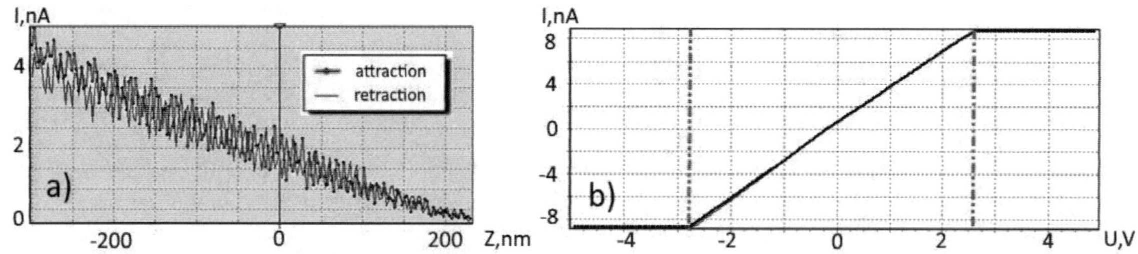

Figure 3. Typical I-Z curves in the SICM-mode of the small tip NP close to surface with modulation of Z coordinate of the sample (a) and the CVC of the NP in the liquid near the surface (b), where red dotted line marks the working voltage region for the SICM-mode.

The SICM-mode was carried out when a sinusoidal signal of about 1 kHz was applied to a Z-piezoscanner of a bimorph membrane, fixed under the bath with the sample. Piezomodules for moving the NP and the sample are set up separately, which excludes the influence of sample and NaCl buffer vibrations at the NP during scanning. The modulated signal allows the redistribution of charges from the contact area and ex decrease the charge migration on the pipette surface. In figure 4a, b an SICM image of a test sample in the form of polymer replica of a rectangular periodic grid with and without the Z-modulation mode is shown. On the standard SICM image one can see the slip lines and Z-scanner jumps at the boundaries of objects. The modulation mode makes it possible to increase the scanning stability. However, a small noise in feedback system was spotted at the imaging due to oscillations. In our case, the sensitivity of the piezomembrane was about 200 nm/V and swing amplitude along the Z axis was about 50 nm. In figure 4c an SICM image of a polymer replica at lower swing amplitude of about 20 nm is shown, at which noise reduction is seen. This result makes possible to distinguish typical small-sized objects on the grid surface (figure 4 c, d, section 1) with typical lateral size ~ 500 nm and height ~200-250 nm and amounted to a minimum lateral size of about 200 nm and a height of about 40 nm (figure 4c, marker 2). Consequently, it is possible to estimate the resolution of the SICM when using borosilicate probe blanks as double diameter of NP aperture in the lateral direction (X, Y) and double amplitude of Z-modulation in the vertical direction.

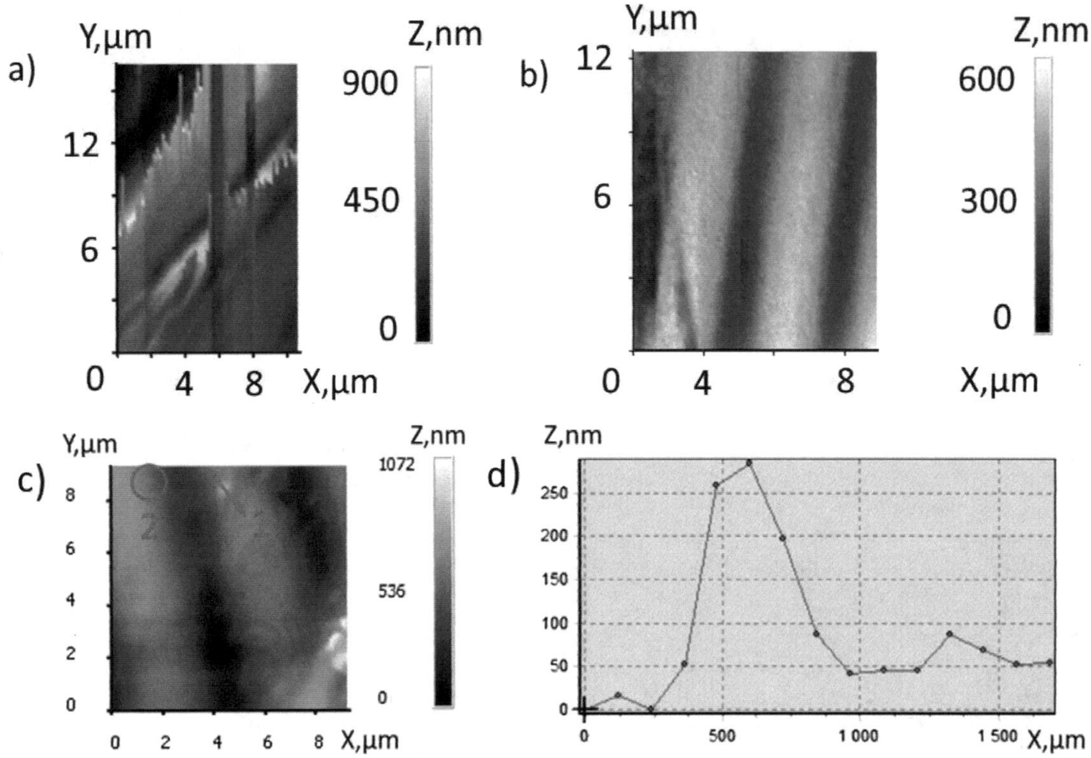

Figure 4. SICM-images of the test sample; a – standard SICM mode, b- AC SICM-mode, c- AC SICM image, small objects on the surface are seen, d-cross section (section 1) of the (c) image.

4. Conclusions

Thus, the scanning ion-conductance microscopy with an independent piezoscanner along the Z axis was created and tested. A new design of the piezoinertial drive was tested with step motion range ~ 5 mm and smooth scanning along the Z coordinate in the range of ~ 7 µm. Such construction makes it possible to locate a Z-piezoelectric membrane under the sample and to provide Z-modulation of the substrate with sample to eliminate scanning artifacts. Measurement of test periodic polymer structure are carried out using borosilicate nanopipettes with an internal pipette diameter on the tip of about 100-150 nm. The optimal geometric parameters of the nanopipette are determined. Small-sized objects with a minimum lateral size of about 200 nm and a height of about 40 nm have been studied. To conclude, the design with independent Z-axis scanning and modulation of the sample position along the Z coordinate improves the quality and eliminating image artifacts in comparison with the standard scanning mode.

Acknowledgments

Work carried out with the support of the state assignment № AAAA-A19-119053190067-8 of the Ministry of Science and Higher Education of the Russian Federation.

References

[1] Hansma P K, Drake B, Marti O, Gould S A C and Prater C B 1989 *Science* **243** 641-643
[2] Edwards M A, Williams C G, Whitworth A L and Unwin P R 2009 *Anal. Chem.* **81** 4482–4492
[3] Peng Li, Lianqing Liu, Yang Yang, Lei Zhou, Dong Wang, Yuechao Wang and Guangyong Li 2015 *Journal of Laboratory Automation* **20** 457–462

[4] Peng Li, Lianqing Liu, Yuechao Wang, Yang Yang, Changlin Zhang and Guangyong Li 2014 *Applied physics letters* **105** 053113
[5] Klenerman D, Shevchuk A, Novak P, Korchev Y E and Davis S J 2013 *Philos. Trans. R. Soc. B* **368** 20120027
[6] Wang Z, Zhuang J, Gao Z and Liao X 2018 *Rev. Sci. Instrum.* **89** 113709
[7] Zhuang J, Jiao Y and Mugabo V 2017 *Micron* **101** 177-185
[8] David Pastre, Hideki Iwamoto, Jie Liu, Gabor Szabo Zhifeng Shao 2001 *Ultramicroscopy* **90** 13–19
[9] Gesper A, Hagemann P and Happel P 2017 *Nanoscale* **9** 14172–14183
[10] Watanabe S and Ando T 2017 *Appl. Phys. Lett.* **111** 113106
[11] Sapozhnikov I D, Gorbenko O M, Felshtyn M L, Zhukov M V and Golubok A O 2019 *IOP Conference Series: Materials Science and Engineering* **699** 012040
[12] Wei C, Bard A J and Feldberg S W 1997 *Anal. Chem.* **69** 4627–4633
[13] Zhukov M V, Lukashenko S Yu, Sapozhnikov I D and Golubok A O 2020 *J. Phys.: Conf. Ser.* **1695** 012167

Determination of hole diffusion length in n-GaN

D S Arteev[1], A V Sakharov[1], A E Nikolaev[1], E E Zavarin[1], W V Lundin[1] and A F Tsatsulnikov[2]

[1] Ioffe Institute, 26 Politekhnicheskaya, 194021, St. Petersburg, Russia
[2] Submicon Heterostructures for Microelectronics, Research & Engineering Center, RAS, 26 Politekhnicheskaya, 194021, St. Petersburg, Russia

ArteevDS@mail.ioffe.ru

Abstract. The paper presents the derivation of a model for minority carriers collection based on the reciprocity theorem and its application for determination of hole diffusion length in n-GaN by means of photoluminescence. The estimated hole diffusion lengths at room temperature are 110 nm and 194 nm in the case of low and high excitation, respectively, which could be explained by saturation of non-radiative recombination centers in bulk GaN and at the surface with photogenerated carriers.

1. Introduction

III-nitride optoelectronic devices, like light-emitting diodes and lasers or photodetectors, are of great interest due to their high efficiency and broad spectral range. Carrier transport in such structures is an important issue. Minority carrier diffusion length (usually holes, as even unintentionally doped GaN is n-type with a relatively high electron concentration of 10^{15}-10^{17} cm^{-3}) is one of the key parameters that directly affects the device performance. Different techniques can be used to determine diffusion length, including electron beam induced current [1], photoluminescence [2, 3], cathodoluminescence [3], etc. However, for correct determination of diffusion length, an appropriate processing of experimental data is required.

In this paper, we derive a simple analytical model for minority carrier collection into quantum well (QW) based on the reciprocity theorem and apply it to determine the minority hole diffusion length in n-GaN/InGaN structures from continuous-wave photoluminescence measurements.

2. Experimental details

A set of GaN/InGaN/GaN single quantum well (SQW) structures with different capping layer thickness (i.e. different distance from QW to the surface) was grown by metalorganic chemical vapor deposition on *c*-face sapphire substrates in our in-house Dragon-125 epitaxial system with a horizontal reactor. Conventional precursors, i.e. trimethylgallium, triethylgallium, trimethylindium and ammonia were used. The growth conditions were exactly the same for all the samples. In order to mitigate the effect of internal spontaneous and piezoelectric polarization fields, GaN layers were doped with [Si]~$5*10^{18}$ cm^{-3} using silane as a precursor. Photoluminescence studies were performed using a continuous wave He-Cd laser (325 nm, ~10 mW) in low and high excitation regimes by changing focused spot diameter from ~5 mm to ~0.5 mm. An AvaSpec-2048 Fiber Optic Spectrometer was used for the spectra measurements.

3. Model derivation

Generally, simulation of charge transport in semiconductors requires self-consistent solving of Poisson and electron- and hole-continuity equations. In simpler cases, when the excess minority carrier density is much smaller than that of the majority one and the electric field is neglected, only diffusion equation for minority carriers can be considered:

$$D\frac{\partial^2 p(z)}{\partial z^2} - \frac{p(z)}{\tau} + G(z) = 0 \tag{1}$$

where D is diffusion coefficient, τ is bulk lifetime, $G(z)$ is carrier generation profile. However, it also requires using of numerical methods.

A reciprocity theorem formulated and proven by Donolato [4] states that charge collection can be described as a solution of the homogeneous version of equation (1) with appropriate boundary conditions (BCs). Introducing $P_{QW}(z)$ as "collection probability" instead of $p(z)$ we get:

$$D\frac{\partial^2 P_{QW}(z)}{\partial z^2} - \frac{P_{QW}(z)}{\tau} = 0 \tag{2}$$

Originally, this model was proposed for charge collection in electron beam induced current (EBIC) analysis. We apply this model to derive an analytical expression for carrier collection into QW.

A general solution of equation (2) is:

$$P_{QW}(z) = A \cdot e^{-\frac{z}{L}} + B \cdot e^{\frac{z}{L}} \tag{3}$$

where $L = \sqrt{D \cdot \tau}$ is diffusion length and coefficients A and B should be found from appropriate BCs. Considering QW of zero thickness located at $z = z_{QW}$ as a perfect collector, the surface at $z = 0$ and infinite buffer thickness, the BCs are:

$$\left.\frac{\partial P_{QW}(z)}{\partial z}\right|_{z=0} = \frac{S}{D} \cdot P_{QW}(0) = \frac{S\tau}{L^2} \cdot P_{QW}(0) \tag{4a}$$

$$P_{QW}(z_{QW}) = 1 \tag{4b}$$

$$P_{QW}(\infty) = 0 \tag{4c}$$

where S is surface recombination velocity. The obtained expression is:

$$P_{QW}(z) = \begin{cases} \dfrac{\cosh\left(\frac{z}{L}\right) + \frac{S\tau}{L}\sinh\left(\frac{z}{L}\right)}{\cosh\left(\frac{z_{QW}}{L}\right) + \frac{S\tau}{L}\sinh\left(\frac{z_{QW}}{L}\right)}, & \text{for } z \le z_{QW} \\ \exp\left(-\dfrac{z - z_{QW}}{L}\right), & \text{for } z \ge z_{QW} \end{cases} \tag{5}$$

Examples of this curve for different conditions are shown in figure 1.
The intensity of photoluminescence from the QW can be expressed as:

$$I_{QW} = \eta_{QW}\int_0^{\infty} G(z)P_{QW}(z)dz \propto \eta_{QW}\int_0^{\infty} e^{-\alpha z}P_{QW}(z)dz \tag{6}$$

where η is a quantum efficiency of QW (portion of radiative recombination acts), $G(z)$ is a photogenerated carriers profile, α=1.2·10^5 cm^{-1} [5] is the absorption coefficient of GaN at 325 nm.

It is also interesting to find PL intensity not only of QW but of GaN as well. It is not just directly proportional to the integral of $G(z)(1-P_{QW})$, because both radiative and non-radiative recombinations take place in bulk, but surface recombination is non-radiative one only. It is therefore necessary to determine the probabilities to recombine on surface and in bulk separately. Let us first

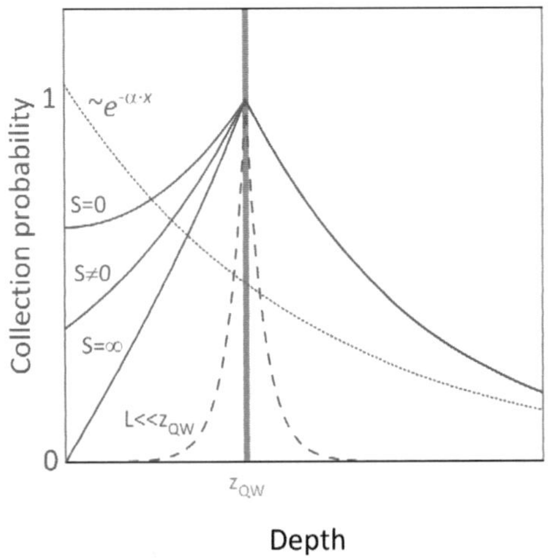

Figure 1. Examples of the collection probability curves for the cases of zero, non-zero and infinite surface recombination velocity S (solid lines) and small diffusion length (dashed line). The dotted line indicates photogenerated carriers profile.

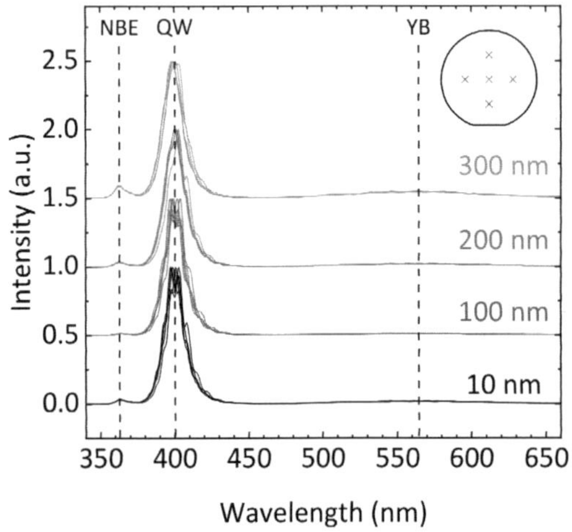

Figure 2. Normalized PL spectra at low excitation level measured at 5 points over the wafer as shown in the inset. Spectra are shifted vertically for clarity.

find the probability P_S to recombine on surface. We start with the same equations as (2) and (3), but with different BCs. The first BC is trivial – the probability to recombine on the surface for the carriers at $z = z_{QW}$ is zero because the probability to get into QW is unity. The second BC is not so obvious. If we used BC similar to (4a), we would get $P_s(z) = 0$ for all z. According to Donolato [6], the correct boundary condition for the surface with finite collection velocity is:

$$D\frac{\partial P_S(z)}{\partial z}\bigg|_{z=0} = S \cdot P_S(0) - S \tag{7}$$

and the corresponding solution for the probability to recombine on the surface is:

$$P_S(z) = \begin{cases} \dfrac{S\tau \cdot \left[exp\left(-\frac{z}{L}\right) - exp\left(-\frac{2z_{QW}-z}{L}\right)\right]}{L + S\tau + L \cdot exp\left(-\frac{2z_{QW}}{L}\right) - S\tau \cdot exp\left(-\frac{2z_{QW}}{L}\right)}, & \text{for } z \le z_{QW} \\ 0, & \text{for } z \ge z_{QW} \end{cases} \tag{8}$$

Hence, the probability P_B for carriers to recombine in bulk can be expressed as:

$$P_B(z) = 1 - P_S(z) - P_{QW}(z) \tag{9}$$

and PL intensity of bulk GaN including near-band-edge (NBE) and 'defect' yellow band (YB) (see figure 2) emission is:

$$I_B = \eta_B \int_0^\infty G(z)P_B(z)dz \propto \eta_B \int_0^\infty e^{-\alpha z}P_B(z)dz \tag{10}$$

where η_B is a portion of radiative recombination acts.

This model can be easily extended to multiple quantum well (MQW) structures (with appropriate BCs) and/or another excitation type. It should be noted that the model ignores carrier overshoot effect and carrier thermal escape effect, which can be significant for shallow QWs.

4. Results and discussion

Internal polarization fields inside InGaN are strongly concentration-dependent. The change in the magnitude of electric field leads to the change of electron-hole wavefunction overlap, which in turn leads to the change in recombination rates and transition wavelength. Therefore, for the correct comparison between the samples one needs to make sure that the indium content is the same. As one can see from PL spectra in figure 2, the samples demonstrate excellent uniformity both over the individual wafer and between the samples.

Then, integrated intensities of QW and GaN emission (including NBE and YB) were obtained from PL spectra and averaged over 5 points. The results normalized to the intensity of QW of the sample with 10 nm cap are shown in figure 3 as symbols. As one can see, the intensity of QW becomes weaker as the cap thickness is increased in both low and high excitation regimes. However, the intensity drop with cap layer thickness increase is smaller in the case of high excitation. The dependence of the intensity of GaN is more interesting with non-monotonic behavior.

The model contains several parameters, i.e. τ, D, L, S, η_{QW} and η_B which are not fully independent, and, moreover, the model does not allow to determine all of them. For convenience, we chose $L = \sqrt{D\tau}$ and $S\tau$ as fitting model parameters. Parameters η_{QW} and η_B were not used, but the dependencies were normalized to the corresponding intensities of the sample with 10 nm cap layer. Then,the dependences of the intensity of QW versus cap layer thickness were fitted with the model. The obtained dependencies for QW and GaN intensity are shown as solid and dashed lines in figure 3, respectively. As one can see, an excellent agreement for QW intensity was achieved. As for GaN intensity, it agrees "semi-quantitatively" with the experiment, probably, because of non-linear

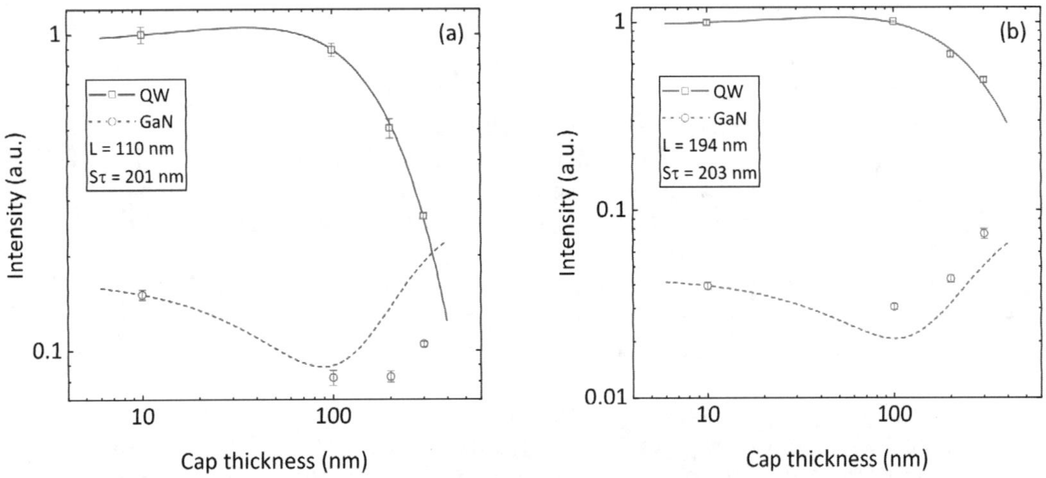

Figure 3. Integrated PL intensities of QW and GaN versus GaN cap layer thickness for the case of low (a) and high (b) excitation. Symbols are experimental data, averaged over 5 points across each wafer and normalized to the intensity of QW of the structure with 10 nm cap layer; error bars indicate standard deviation. Lines are model fit.

dependence of YB emission intensity on excitation or self-absorption effect. Anyway, the model is able to predict such non-monotonous behavior with a minimum at ~100 nm of cap thickness.

The values of the fitting parameters should be discussed. We obtain L=110 nm and $S\tau$=201 nm for the case of low excitation and L=194 nm and $S\tau$=203 nm for the case of high one, i.e. the diffusion length almost doubled when the excitation power density was increased by two orders of magnitude while $S\tau$ didn't change. There could be two situations. In first, S and τ are independent on excitation density, then the mobility of the minority carriers should increase almost fourfold since $D = \mu \frac{kT}{q}$. Possible reasons for increase in mobility are decrease of charged impurity scattering rate or switch from unipolar to ambipolar diffusion. However, none of these can increase the mobility (diffusivity) so much [7]. A second, more viable, hypothesis is fourfold increase in non-radiative lifetime τ in bulk with simultaneous decrease in surface recombination velocity. Both could result from the saturation of non-radiative recombination centers in bulk and at the surface by photogenerated carriers, respectively. The result is in line with [2]. Similar effect of increasing non-radiative lifetime was also reported for iron-, carbon- [8] and magnesium-doped [9] bulk GaN under long-term electron beam irradiation. However, it cannot be excluded that a combination of both effects takes place.

5. Conclusion

In this paper, a model for minority carriers' collection into quantum well was derived based on the reciprocity theorem. The model was applied to the data of photoluminescence of InGaN/GaN single-quantum well structure to determine a hole diffusion length in n-GaN. The estimated hole diffusion lengths was found to be 110 nm and 194 nm in the case of low and high excitation, respectively, which could be explained by saturation of non-radiative recombination centers in bulk GaN and at the surface with photogenerated carriers.

References

[1] Chernyak L, Osinsky A, Temkin H, Yang J W, Chen Q and Asif Khan M 1996 Electron beam induced current measurements of minority carrier diffusion length in gallium nitride *Appl. Phys. Lett.* **69** 2531–3

[2] Netzel C, Hoffmann V, Tomm J W, Mahler F, Einfeldt S and Weyers M 2020 Temperature-Dependent Charge Carrier Diffusion in [000$\bar{1}$] Direction of GaN Determined by Luminescence Evaluation of Buried InGaN Quantum Wells *phys. status solidi B* **257** 2000016

[3] Hafiz S, Zhang F, Monavarian M, Avrutin V, Morkoç H, Özgür Ü, Metzner S, Bertram F, Christen J and Gil B 2015 Determination of carrier diffusion length in GaN *J. Appl. Phys.* **117** 013106

[4] Donolato C 1985 A reciprocity theorem for charge collection *Appl. Phys. Lett.* **46** 270–2

[5] Muth J F, Lee J H, Shmagin I K, Kolbas R M, Casey H C, Keller B P, Mishra U K and DenBaars S P 1997 Absorption coefficient, energy gap, exciton binding energy, and recombination lifetime of GaN obtained from transmission measurements *Appl. Phys. Lett.* **71** 2572–4

[6] Donolato C 1994 Reciprocity theorem for charge collection by a surface with finite collection velocity: Application to grain boundaries *J. Appl. Phys.* **76** 959–66

[7] Ščajev P, Jarašiūnas K, Okur S, Özgür Ü and Morkoç H 2012 Carrier dynamics in bulk GaN *J. Appl. Phys.* **111** 023702

[8] Lopatiuk O, Osinsky A, Dabiran A, Gartsman K, Feldman I and Chernyak L 2005 Electron trapping effects in C- and Fe-doped GaN and AlGaN *Solid-State Electron.* **49** 1662–8

[9] Chernyak L, Burdett W, Klimov M and Osinsky A 2003 Cathodoluminescence studies of the electron injection-induced effects in GaN *Appl. Phys. Lett.* **82** 3680–2

Internal optical loss and light-current characteristic in injection lasers

Zinaida N Sokolova[1], **Nikita A Pikhtin**[1], **Sergey O Slipchenko**[1] **and Levon V Asryan**[2]

[1] Ioffe Institute, St. Petersburg, 194021, Russia
[2] Virginia Polytechnic Institute and State University, Blacksburg, VA 24061, USA

asryan@vt.edu

Abstract. Possibility of tailoring the light-current characteristic (LCC) shape in quantum dot (QD) lasers by varying uniformity of QDs is discussed. Making the QD ensemble less uniform results in roll-over in the LCC. The second branch in the LCC appears with making the QD ensemble even less uniform.

1. Introduction

Internal optical loss that is inherent in semiconductor lasers combines different processes of unwanted photon loss as opposed to mirror loss – on purpose output of photons through the cavity mirrors. While the mirror loss presents the useful optical output from semiconductor lasers, the internal loss is undesired as it consumes the photons emitted during the laser action.

In this paper, the effect of uniformity of quantum dots (QDs) on the light-current characteristic (LCC) in QD lasers is discussed in the presence of internal optical loss. While diode lasers have been extensively studied both experimentally and theoretically before (see, e.g., a review article [1]), a new interesting feature has been however predicted in [2]-[4]: in the presence of internal optical loss that depends on the charge carrier density, roll-over of the LCC can occur. The more so, the second branch can appear in the LCC in addition to the first 'conventional' branch. Here we discuss the possibility of tailoring the LCC shape in QD lasers in the presence of internal optical loss by varying uniformity of QDs.

2. Theoretical model

Our theoretical analysis is based on rate equations that describe the dynamics of charge carries and photons in the laser. In the simplest case, we use a set of three equations – two for carriers [free carriers in the optical confinement layer (OCL) and confined carriers in QDs] and one for photons.

The equations for carriers are

$$b\frac{\partial n_{\mathrm{OCL}}}{\partial t}=\frac{j}{e}+N_{\mathrm{S}}\frac{f_{\mathrm{n}}}{\tau_{\mathrm{esc}}}-\mathrm{v}_{\mathrm{capt,0}}(1-f_{\mathrm{n}})\,n_{\mathrm{OCL}}-bB_{\mathrm{3D}}n_{\mathrm{OCL}}^{2}, \quad (1)$$

$$N_{\mathrm{S}}\frac{\partial f_{\mathrm{n}}}{\partial t}=\mathrm{v}_{\mathrm{capt,0}}(1-f_{\mathrm{n}})\,n_{\mathrm{OCL}}-N_{\mathrm{S}}\frac{f_{\mathrm{n}}}{\tau_{\mathrm{esc}}}-N_{\mathrm{S}}\frac{f_{\mathrm{n}}^{2}}{\tau_{\mathrm{QD}}}-c_{\mathrm{g}}\,g^{\mathrm{max}}(2f_{\mathrm{n}}-1)n_{\mathrm{ph}}. \quad (2)$$

Table 1. Physical quantities entering in eqs. (1)-(3).

Notation	Meaning
$n_{\rm OCL}$	Free carrier density in the OCL
$f_{\rm n}$	Occupancy of the energy level in a QD by a charge carrier
$n_{\rm ph}$	Density of emitted photons (per unit area of the laser heterointerface)
j	Injection current density
e	Electron charge
$N_{\rm S}$	Surface density of QDs
$\tau_{\rm esc}$	Thermal escape time of charge carriers from an individual QD to the OCL
$\rm v_{capt,0}$	Velocity of carrier capture into an unoccupied (at $f_{\rm n} = 0$) QD ensemble (in cm/s)
b	Thickness of the OCL
$B_{\rm 3D}$	Spontaneous radiative recombination constant for the bulk (OCL) region (in $\rm cm^3/s$)
$\tau_{\rm QD}$	Spontaneous radiative lifetime in a QD
$c_{\rm g}$	Group velocity of light
$g^{\rm max}$	Maximum modal gain ($\rm cm^{-1}$)
β	Mirror loss coefficient (in $\rm cm^{-1}$)
$\alpha_{\rm int}$	Internal optical loss coefficient (in $\rm cm^{-1}$)

The equation for photons is

$$\frac{\partial n_{\rm ph}}{\partial t} = c_{\rm g} g^{\rm max}(2f_{\rm n}-1)n_{\rm ph} - c_{\rm g}(\beta+\alpha_{\rm int})n_{\rm ph} . \tag{3}$$

The parameters entering in eqs. (1)-(3) are described in table 1.

The internal optical loss coefficient in (3) is presented as

$$\alpha_{\rm int} = \alpha_0 + \sigma_{\rm int} n_{\rm OCL} , \tag{4}$$

where α_0 is its constant component and $\sigma_{\rm int}$ is its the cross-section.

Solving the set of eqs. (1)-(3) at the steady-state, we then find the emitted optical power as follows:

$$P = \hbar\omega\, c_{\rm g}\beta\, S n_{\rm ph} , \tag{5}$$

where $\hbar\omega$ is the emitted photon energy and S is the cross-section area of the laser heterointerface.

Calculations show that in the presence of internal optical loss that depends on the charge carrier density in the OCL [eq. (4)], roll-over of the LCC can occur at high pump currents and the second branch can appear in the LCC in addition to the conventional branch.

The second branch appears in the LCC if the following condition satisfies:

$$\frac{\sigma_{\rm int} {\rm v}_{\rm capt,0}}{2bB_{\rm 3D} g^{\rm max}} > 1 . \tag{6}$$

Varying each of the parameters entering in (6) can result in single-to-double branch transformation of the LCC. In [4], the cross-section of internal optical loss $\sigma_{\rm int}$ was varied in QD lasers; in [5], the thickness of the OCL b was varied in quantum well (QW) lasers; in [6], the capture velocity $\rm v_{capt,0}$ was varied in QW lasers; and in [7], the temperature was varied in QW lasers (the spontaneous radiative recombination constant B_{3D} entering in (6) is a function of temperature – see [8, 9]). In all these cases, the transformation of the LCC shape from single-to-double branch was shown to occur.

Here, we discuss the transformation of the LCC shape in QD lasers as the QD uniformity is changed.

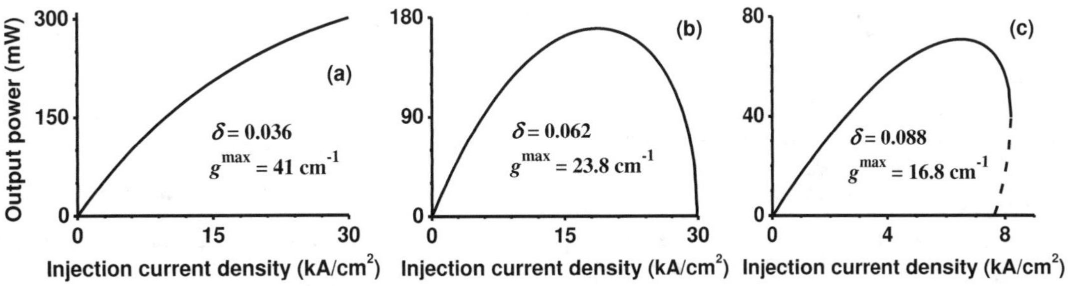

Figure 1. LCC of the QD laser for different values of the RMS of relative QD-size fluctuations δ. The corresponding values of the modal gain g^{max} are also presented. InGaAsP-based structure emitting near 1.55 μm is considered [11, 13]. The internal loss cross-section $\sigma_{int} = 2 \times 10^{-18}$ cm^2, and the constant component of the internal loss $\alpha_0 = 2$ cm^{-1}.

3. Effect of uniformity of QDs on the LCC shape

Self-assembled QDs are never identical in laser structures – they are primarily different in their sizes. The QD-size dispersion causes fluctuations in the quantized energy levels and leads to broadening in the optical transition energy (inhomogeneous line broadening [8]). Due to nonuniformity of QDs the maximum gain of the laser decreases [8], threshold current increases [8] and becomes more sensitive to temperature (the characteristic temperature decreases) [10], multimode generation threshold decreases [11], internal differential efficiency and output power both decrease [12].

Denoting δ the root mean square (RMS) of relative QD-size fluctuations, the maximum modal gain of a QD laser can be presented as follows [8]:

$$g^{max} \propto \frac{1}{\delta}. \quad (7)$$

As seen from (7), g^{max} decreases with increasing δ. Hence it follows from (6) and (7) that the condition for appearing the second branch in the LCC should become satisfied with increasing RMS of QD-size fluctuations, i.e., making the QD ensemble less uniform.

As shown in figure 1(a), at small fluctuations in QD sizes [the case of high modal gain – see (7)], the output power increases with the pump current over the entire range of interest.

At larger fluctuations in QD sizes, the slope efficiency changes from positive to negative with increasing pump current [figure 1(b)]; the output power goes continuously to zero, which marks the stop of lasing. In what follows, such a shape of the LCC is referred to as a 'conventional' roll-over to distinguish from the case of strongly nonuniform QDs.

With further increasing RMS of relative QD-size fluctuations up to a certain value, the LCC retains a conventional roll-over shape. Above that value of the RMS, the LCC transforms qualitatively [figure 1(c)]. The stop of lasing with increasing injection current occurs now at a nonvanishing output power: at the maximum operating current, the conventional branch of the LCC (solid portion of the curve) is detached from the X-axis. This upper branch of the LCC is now continued by the second, lower, branch (dashed portion of the curve); in general, with reducing the pump current the output power can follow either the upper (conventional) or lower branch. Experimental observation of the quenching of lasing at a nonvanishing output power in diode lasers can suggest the presence of carrier-density-dependent internal loss and, in the case of QD lasers, large QD-size dispersion.

4. Conclusions

Possibility of tailoring the LCC shape in QD lasers by varying uniformity of QDs has been discussed. Making the QD ensemble less uniform has been shown to result in roll-over in the LCC. The second branch in the LCC has been shown to appear with making the QD ensemble even less uniform.

Acknowledgments

Z.N.S., N.A.P., and S.O.S. thank the federal program of the Ioffe Institute, and L.V.A. thanks the U.S. Army Research Office (Grant No. W911NF-17-1-0432) for support of this work.

References

[1] Alferov Zh I 2001 Nobel Lecture: The double heterostructure concept and its applications in physics, electronics, and technology *Rev. Mod. Phys.* **73** 767-82
[2] Asryan L V and Luryi S 2003 Two lasing thresholds in semiconductor lasers with a quantum-confined active region *Appl. Phys. Lett.* **83** 5368-70
[3] Asryan L V and Luryi S 2004 Effect of internal optical loss on threshold characteristics of semiconductor lasers with a quantum-confined active region *IEEE J. Quantum Electron.* **40** 833-43
[4] Asryan L V 2006 Maximum power of quantum dot laser versus internal loss *Appl. Phys. Lett.* **88** 073107
[5] Sokolova Z N, Pikhtin N A and Asryan L V 2018 Two-valued characteristics in semiconductor quantum well lasers *J. Lightw. Technol.* **36** 2295-300
[6] Sokolova Z N, Pikhtin N A and Asryan L V 2019 Evolution of light-current characteristic shape in high-power semiconductor quantum well lasers *Electron. Lett.* **55** 550-2
[7] Sokolova Z N, Pikhtin N A, Slipchenko S O and Asryan L V 2020 Temperature-induced single-to-double branch transformation of operating characteristics in semiconductor lasers with a low-dimensional active region *Proc. SPIE* **11301** 113010D
[8] Asryan L V and Suris R A 1996 Inhomogeneous line broadening and the threshold current density of a semiconductor quantum dot laser *Semicond. Sci. Technol.* **11** 554-67
[9] Asryan L V 2005 Spontaneous radiative recombination and nonradiative Auger recombination in quantum-confined heterostructures *Quantum Electron.* **35** 1117-20
[10] Asryan L V and Suris R A 1997 Characteristic temperature of quantum dot laser *Electron. Lett.* **33** 1871-2
[11] Asryan L V and Suris R A 1999 Spatial hole burning and multimode generation threshold in quantum-dot lasers *Appl. Phys. Lett.* **74** 1215-7
[12] Asryan L V, Luryi S and Suris R A 2002 Intrinsic nonlinearity of the light-current characteristic of semiconductor lasers with a quantum-confined active region *Appl. Phys. Lett.* **81** 2154-6
[13] Jiang L and Asryan L V 2006 Excited-state-mediated capture of carriers into the ground state and the saturation of optical power in quantum-dot lasers *IEEE Photon. Technol. Lett.* **18** 2611-3

Surface reconstruction post-processing method for 3D-scanned objects

P D Badillo, V A Parfenov and N L Shchegoleva

Saint Petersburg Electrotechnical University (SPbETU "LETI"), 197376 St. Petersburg, Russia

pbadillo@gmail.com

Abstract. 3D scanning is widely used in multiple applications to obtain high precision / non-destructive documentation of real-life objects, which is especially important in Cultural Heritage (CH) preservation. However, some issues (in particular missing parts which are commonly known as "holes") affect the accuracy of the obtained 3D model after the scanning procedure and requires time-consuming post-processing procedures, which include manual editing by highly-trained personnel. In this article an automatic method to reconstruct the obtained surface of 3D models is proposed, improving previously obtained results for high-density point clouds.

1. Introduction

3D scanning is widely used in multiple applications thanks to its high precision measurements of real-life objects. It has particular importance in Cultural Heritage (CH) preservation as a non-destructive documentation method for artworks, which has multiple uses like art conservation tracing and replication.

Although the 3D scanning procedure is getting faster and easier with the use of laser-based techniques, it still requires to be performed by highly qualified personnel to reduce acquisition-related problems (mainly connected with missing parts in the surface of the object seen as holes in the final created 3D model) and complex and time-consuming post-processing procedures (often requiring manually editing the surface) to improve the accuracy of the final result of the created model. All these factors affect the required quality of such procedures used for documentation of CH objects.

As was above mentioned, one of the most recurrent problems in the application of 3D scanning techniques is the appearance of "holes", due to the absence of measurement data of some parts of the scanned object during the acquisition process. This problem decreases the overall accuracy of the model created and its faithful representation, both problems of particular interest in the CH field.

In this work, a method to improve the accuracy of the computer digital 3D models is described. Its use will allow to increase the speed, reduce the computing costs and the time involved in the post-processing phase. The inverse distance method, developed and used to improve the reconstruction quality of the human face for biometry applications [1], showed good results in the reconstruction of 3D point clouds without the need of a previous mesh creation process [2]. The results are further improved in terms of accuracy and speed by subsampling the surface and applying the corrections in the areas where acquisition data is missing.

2. Method used

Most of the 3D-scanning techniques are based on the response of the object under study to a light source. In some cases, the reflected light is measured to determine the distance to the object (triangulation and time-of-flight techniques), in other cases the deformation of a well-known pattern is evaluated (structured light), in other cases photographs taken from different angles are used to obtain a 3D model using special software reconstruction tools (photogrammetry). The common problem that appears in more or less degree in the use of all these techniques is the presence of areas of the surface in which the sensor cannot measure, generating "holes" in the measurement. This problem could be caused by different factors, but mainly it can be seen when the reflected light is not able to reach the sensor of the device (due to occlusions caused by other parts of the object, highly reflective surfaces like glass or highly light-absorbing like black leather [3-5]). As the sensor is not able to obtain information about the object under study in this area, the device identifies this portion as a "hole" or an empty area in the obtained point cloud. This problem, of course, reduces the overall accuracy of the created digital model, which can compromise the goal of the whole procedure.

To correct these acquisition errors post-processing techniques are used [6]. These techniques include manual procedures, where highly trained personnel reconstruct the missing area trying to keep the shape as close as possible to the original object, and automatic procedures. Automatic procedures [7] use different techniques (from simply "connecting" the mesh using geometrical figures to more advanced techniques that follow the local curvature of the shape) but all have in common a high computing cost and that they are time-consuming processes.

The method of inverse distances was developed for the reconstruction of the 3D surface of a human face in biometry applications. The method uses a small number of points of an irregular grid of measurements to generate the missing points, creating a regular grid along the area. A more detailed description of the method could be found in [2]. In the cited article, it was proven the performance of the method in small areas (300 x 300 pixels), comparable with other more complex methods and requiring less computing power, being able to fix successfully surfaces with missing parts ranging from 3% to 30% of the total area. The method showed a lower performance with a larger number of points (15000 points or more).

To solve this issue, the proposed technique tested along this article detects the holes in the surface of the obtained scan as a first step, then selects an appropriate area around the hole to applies the method. Finally, the processed parts are added to the surface to obtain the final result.

This newly proposed technique shows all the advantages of the method of inverse distances applied to 3D-model correction and improves its performance in high-density point clouds, obtained by modern 3D scanning devices.

3. Results obtained

To test the proposed method, it was developed a script written in Matlab language (R2020b version). In order to properly assess the accuracy of the tested method, the script generates "holes" in the tested scan in a well-known area without any surrounding gaps. Sample areas with different curvatures were used to check the behaviour of the method. The script uses different area sizes and shapes around the hole to calculate the missing information and save the new point cloud in .PLY format. Also, the script calculates the distance between the points of the original point cloud and the new point cloud with the calculated points and obtains its mean value and its standard deviation. Following this procedure, software CloudCompare (2.11.1 version), Meshlab (2020.12 version) and Geomagic (2019 version) were used to compare the original scan and the generated .PLY file, helping to graphically check the location of the maximum difference and the surroundings of the generated testing hole.

The scan used for these experiments was obtained using a 3D-laser triangulation based scanner (Konica Vivid 910). The scanned object was a marble sculpture called "Primavera" from the collection of the "Tsarskoe Selo State Museum and Heritage Site" (Tsarskoe Selo, Russia). The selected scan shows parts of the left side of the face and hair of the original object. To test the effectiveness of the method where the curvature of the zone is different, three different points were selected: In the first place, a point in the forehead/hair of the figure, where the surface is almost plain; in second place, a point in the eye, where the curvature is higher and finally in third place a point in the central part of a flower that forms part of the hair ornament, where the curvature was higher than in any other region of the scan. In the tests were used areas around the hole with square and circle shapes, and the numbers of points used for calculation were between 0.5 and 80 times the number of points contained in the original removed area (sizes were calculated using the distance between the centre of the hole and the border of it adding a percentage between 10% and 700%). Figure 1 shows the generated hole and the area used for calculation for the area named "Flower" (higher curvature) and the obtained results. In the represented case, the calculation area used contained 2 times the number of points of the original number of points removed from the hole.

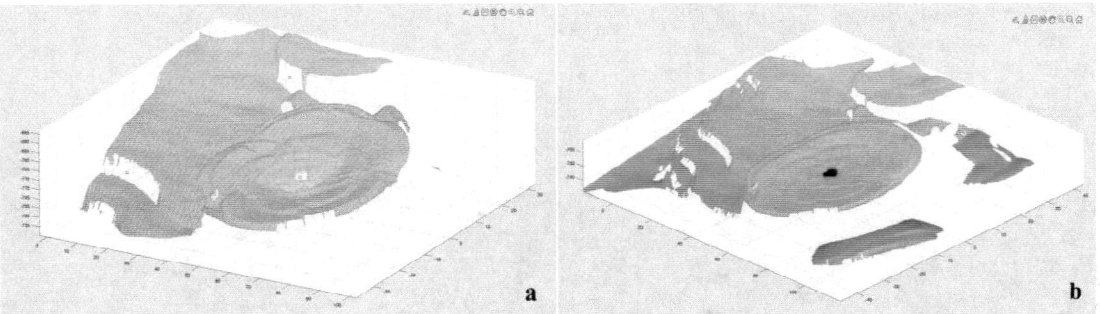

Figure 1. Processing of the data using the Matlab script. In (a) the area used to calculate the missing information is shown (in cyan colour). In (b) the generated points using the method (in black colour).

In all the cases, the method was able to calculate the needed points covering the missing information. As we can observe in Table 1, the results are closer to the original information when the number of points used for calculation is larger than 5 times the number of removed points (two times the distance between the centre and the borders of the hole). Also, we can observe that the results oscillate around this value when the number of points used is larger than 15 times (around 3.5 times the hole centre-border distance). The table also shows better results with fewer points in the zones with lower curvature values (Named "Hair" in the table). The calculation area shape didn't significantly affect the results in our experiments, keeping the same results when comparing areas with the same number of points. The method uses a grid to generate the X and Y coordinates of the new points: different grid sizes were used and the results were almost the same.

Table 1. Results obtained in different zones and with different calculation areas.

	Calculation area size (Original hole point number multiplied *N* times)									
	2		5		10		15		20	
	Mean (x 10^{-3})	Std Dev (x 10^{-3})	Mean (x 10^{-3})	Std Dev (x 10^{-3})	Mean (x 10^{-3})	Std Dev (x 10^{-3})	Mean (x 10^{-3})	Std Dev (x 10^{-3})	Mean (x 10^{-3})	Std Dev (x 10^{-3})
Hair	1.19	59.3	0.52	24.5	0.5	23.6	0.5	24	0.52	24.3
Eye	1.22	67.3	0.55	29.3	0.43	22.7	0.39	21.1	0.38	20.6
Flower	1.33	75.3	0.46	24.1	0.41	21.2	0.41	21.3	0.42	21.5

Using cloud comparing software the results were confirmed. The areas with higher curvature show a higher maximum difference, located in the centre of the calculated area (see the results for the different areas in Table 2). Figure 2 shows the results obtained in the area with higher curvature (named "Flower") using a calculation area size of 5 times the number of points. In this case, the maximum observed difference is nearly 2.25 but as we can see in the standard deviation value (Table 1) and confirm in the image, the number of points that have a high difference with the original information is low. In Table 2 we can observe that the maximum difference tends to stabilize around 2 with larger calculation areas, no matter the curvature of the region used.

Table 2. Maximum distance between calculation and reference with different calculation areas.

	Calculation area size (Original hole point number multiplied *N* times)				
	2	5	10	15	20
	Max distance	Max distance	Max distance	Max distance	Max distance
Hair	5.00	2.12	2.05	2.09	2.10
Eye	5.84	2.61	2.17	1.97	1.96
Flower	8.44	2.28	2.08	2.04	2.02

Figure 2. Comparison of the calculated point cloud information and the original.

4. Conclusions

The method is able to calculate and restore the missing information using the coordinates of the points in the surrounding area of the region of interest in all the tested cases. The procedure showed good results when used with calculation areas containing at least 5 times the number of points of the original area. It is interesting to note that the results (mean distance value between reference and newly calculated region, standard deviation of the same distance and maximum of the same distance) start to oscillate when the number of points used for calculation is larger than 15 times the number of points of the created "hole" in the surface. When the method is applied to a flatter surface, the maximum distance is smaller with less points used, but using around 10 times or more the number of points the results are almost the same for zones with different curvatures. Neither the shape of the used area for calculation nor the grid size of the X-Y coordinates used in the method seem to affect the results obtained. A future improvement will be to find a correlation function between the number of points and the curvature of the area, searching for the ideal number of points in each case.

References

[1] Shchegoleva N 2012 *J Theor Appl Comput Sci* **6(4)** 37-50
[2] Shchegoleva N, Parfenov V and Smolina S 2020 *WSEAS Transactions on Computers* **19(4)** 26-30
[3] Matusik W, Hanspeter P, Ziegler R, Ngan A and McMillan L 2002 *Proc. of the 13th Eurographics Workshop on Rendering* (Pisa) ed P Debevec and S Gibson (New York: Association for Computing Machinery) pp 267–78
[4] Mériaudeau F et al 2012 *Depth Map and 3D Imaging Applications: Algorithms and Technologies* ed A S Malik et al (Hershey: IGI Global) pp 229-43
[5] Zaimović-Uzunović N and Lemeš S 2010 *IMEKO Conf. Proc.: 10th Int. Symp. on Measurement and Quality Control* (Osaka) ed Y Takaya (Tokyo: Japan Society for Precision Engineering) pp 408–11
[6] Berger M, Tagliasacchi A, Seversky L, Alliez P, Levine J, Sharf A and Silva C 2014 *Eurographics 2014 - State of the Art Reports* (Strasbourg) ed S Lefebvre and M Spagnuolo (The Eurographics Association) pp 161-85
[7] Pérez E, Salamanca S, Merchán P and Adán A 2016 *Int. J. Appl. Math. Comput. Sci.* **26(4)** 885-903

The SU-8 spin-coating on silicon nanowires formed by cryogenic dry etching

A I Baranov[1,2], D A Kudyashov[1,2], I A Morozov[1,2], K Yu Shugurov[1], A V Uvarov[1,2], A A Maximova[1,2], E A Vyacheslavova[1,2] and A S Gudovskikh[1,2]

[1] Alferov University, 194021 St Petersburg, Russia.
[2] St Petersburg Electrotechnical University "LETI", 197376 St Petersburg, Russia.

baranov_art@spbau.ru

Abstract. Arrays of vertically aligned silicon nanowires were fabricated by cryogenic dry etching. The post-processing technology was developed to full coating of arrays of NWs by SU-8 and release the top side of SiNWs. The Schottky diodes were fabricated on arrays of SiNWs with and without SU-8 by gold evaporation. The cryogenic dry etching leads to defect formation with E_a=0.28 eV and concentration lower $5\cdot10^{12}$ cm^{-3} in near-surface area in silicon, and no defect are detected in bulk silicon. However, oxygen plasma treatment used to release top side of SiNWs leads to increase of its concentration by two order and formation of defect with E_a=0.39 eV, $\sigma = 1\cdot10^{-16}$ cm^2 and a concentration of $5\cdot10^{14}$ cm^{-3} in a bulk of SiNWs deeper than 1 μm.

1. Introduction

Nowadays, modern types of semiconductor solar cells (SC) like silicon single-junction one and multi-junction ones based on the III-V compounds have almost reached their theoretical limit of efficiency, and their further optimization leads to complication and cost rise. Therefore, fabrication of SC based on low dimensional structures are considered as perspective concepts to increase its efficiency with using of standard processing methods, which can be transferred to industry in the future [1]. Recently, it was shown that usage of vertically aligned silicon nanowires (SiNWs) formed with nanosphere lithography allows one to enhance the absorption of solar radiation in active layers of a-Si:H based top subcell in full silicon a-Si:H/c-Si double-junction SC. In our previous work, cryogenic dry etching allows one to form arrays of vertical aligned nanowires with height of 5 μm and diameter of 700 nm with smooth side surface of NWs using latex spheres with 1 μm diameter [2]. However, such diameter is not sufficient to fabricate SC due to full depletion of NWs in silicon wafer with n=10^{15} cm^{-3}. Furthermore, measurments of deep-level transient spectroscopy (DLTS) shown formation of defects in near-surface area due to plasma influence but it can be partially reduced after wet etching in KOH solution [3]. Here, we use latex spheres with higher diameter of 2 μm which enlarges diameter of SiNWs and can improve performance of SC device. Influence of cryogenic plasma etching and post-growth coating of SiNWs arrays by SU-8 will be studied.

2. Experiments and results

Silicon wafers (n=10^{16} cm^{-3}) were etched in ICP mode in gases flow of SF_6 and O_2 in Oxford Plasma Lab ICP 380 at -140 °C [2] with mask of 2 μm latex spheres to form arrays of vertical aligned nanowires with height of 6 μm and diameter of 1.7 μm. Further, the coating technology of the SiNWs array with a polymer layer with the required thickness has been developed. In future, it will allow to fill SCs based on SiNWs, and then tear off form wafer, so flexible SC will be fabricated. Photoresist SU-8 was chosen as a filler in the project, which is most suitable for the tasks: the height of SiNWs is from 3 to 10 μm, so the photoresist should provide filling in a wide range of thicknesses, and SU-8 was adopted specifically for these purposes and its thickness can be varied by dilution from 100 nm to 50 μm.

The technology has been tested at arrays of SiNWs with height of 5 μm and diameter of 700 nm. Initially, SU-8 with thickness 3.1 and 5.1 μm was spin-coated for 1 minute at of 4000 rpm (figure 1). Further, the sample with the photoresist is subjected to the following procedures: initial annealing (soft bake) at a temperature of 95 °C for 1 min, then exposure under UV illumination for 10 min at room temperature and final annealing (hard bake) at 170 °C for 10 min. In result, the array of SiNWs was fully coated, but the thicknesses of photoresist layer over the SiNWs are 1.6 and 4.5 μm, respectively. Probably, for the studied geometry of the SiNWs array, used photoresist are too viscous, which does not allow them to effectively spread over the sample surface during spin-coating and leads to the formation of an upper layer. The distances between SiNWs are comparable with their submicron diameters, so they are very densely arranged, and, unlike NWs with a diameter of 50-200 nm, have a significant effect on the spin-coating of the SU-8.

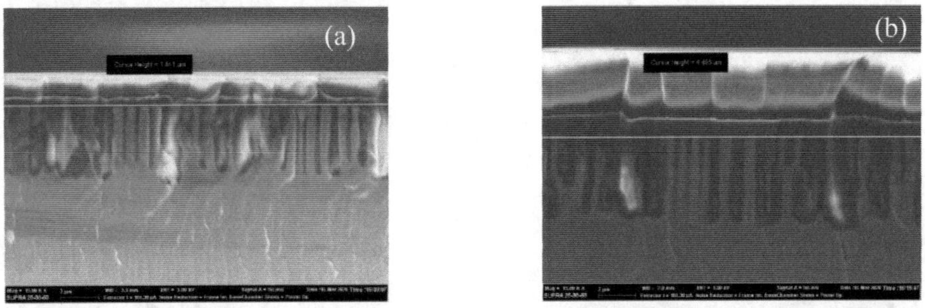

Figure 1. SEM image of structure coated by SU-8 with thickness 3.1 μm (a) and 5.1 μm (b).

Thus, 580 nm SU-8 photoresist was spin-coated at 4000, 2000, 1000 and 800 rpm (figure 2), and a wait of 2 minutes was added before spin-coating, further procedure remained the same. For 4000 rpm (figure 2a), no upper layer of the photoresist is observed, and drops of 30 nm thick settled on the upper faces of the SiNWs. The coating is non-uniform leading to the absence of SU-8 on the sidewall surface of some NWs and between them (figure 2b). For 2000 rpm, the quality of coating improved, but still there are uncoated areas, especially where the SiNWs were are not close-packing, and also SU-8 drops are distinguished on top side of SiNWs. However, for 1000 and 800 rpm, quality of spin-coating significantly changed: total surface of wafer with array of SiNWs is coated by continuous smooth layer of SU-8, and thickness of photoresist on the top faces of the SiNWs is 120 nm. So arrays of new SiNWs with height of 6 μm and diameter 1.7 μm were spin-coated at 800 rpm (figure 3a) with further baked steps. SiNWs were treated in PiNK Plasma system V15-G equipment during 3 minutes in oxygen plasma, and 300 nm of SU-8 was removed (figure 3b). Then, gold was deposited in BOC Edwards Auto500 on the structures to form a Schottky barrier to the top side of the SiNWs (figure 4a), and ohmic contact was fabricated to rear side of wafer. Also, a Schottky barrier and an ohmic contact were formed to the initial sample with an array of SiNWs on a wafer without SU-8 (figure 4b). The obtained samples were used to study defect formation after cryogenic etching in the initial wafer and after the procedure of spin-coating by SU-8 and treatment of top side of SiNWs.

Then, deep-level transient spectroscopy was performed to both samples at different applied bias voltage and pulse amplitude to spatially separate responses from different defects: defects in near-posurface area (V_{init}=0 V, V_{pulse}=+2 V, Forward in figure 5) and far from interface Au/n-Si (V_{init}=-2 V,

V_{pulse}=0 V, Reverse in figure 5). Figure 4 shows DLTS spectra for rate window of 20 s^{-1} measured for samples with an array of SiNWs on a wafer 0.2 Ω·cm with and without SU-8 coating for both modes.

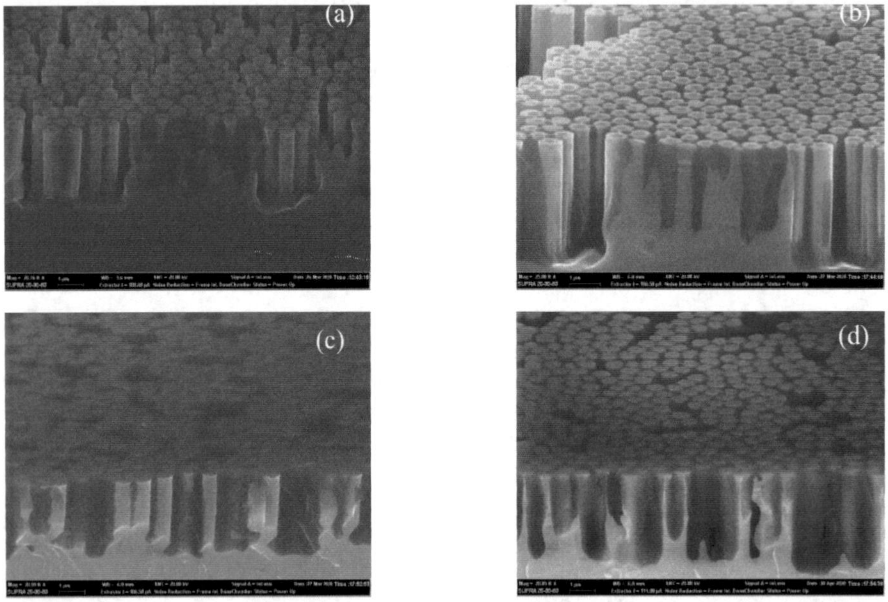

Figure 2. SEM image of array of SiNWs with height of 5 μm and diameter of 700 nm coated by SU-8 with thickness 580 nm at 4000 (a), 2000 (b), 1000 (c), 800 rpm(d).

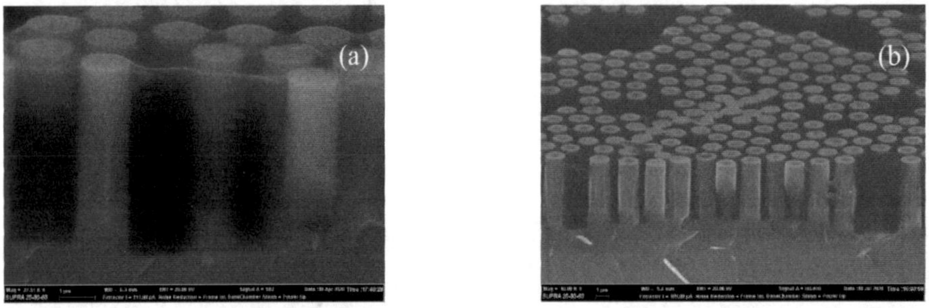

Figure 3. SEM image of array of SiNWs with height of 6 μm and diameter of 1.7 μm coated by SU-8 before (a) and after (b) treatment in oxygen plasma.

Figure 4. SEM image of structure covered by gold metallization: initial SiNWs (a) and SiNWs spin-coated by SU-8 (b).

It is shown that a defect level with E_a=0.28 eV, $\sigma = 1\cdot10^{-17}$ cm^2 and a concentration of less than $5\cdot10^{12}$ cm^{-3} was detected in the near-surface region in the initial array of SiNWs, and there are no responses at a depth of more than 500 nm so used cryogenic dry etching does not lead to high defect formation in bulk silicon. After spin-coating of SU-8 and treatment in oxygen plasma, the concentration of this defect increases by more than two orders of magnitude in the near-surface region. In addition, a defect is formed with E_a=0.39 eV, $\sigma = 1{\cdot}10^{-16}$ cm^2 and a concentration of $5{\cdot}10^{14}$ cm^{-3} at a bulk of SiNWs, and deeper than 1 μm, which is a divacancy arising due to oxygen plasma [4]. Thus, cryogenic etching does not produce defect leading to decreasing lifetimes of charge carriers in silicon wafers, and the standard post-growth treatment in oxygen plasma catastrophically affects on it.

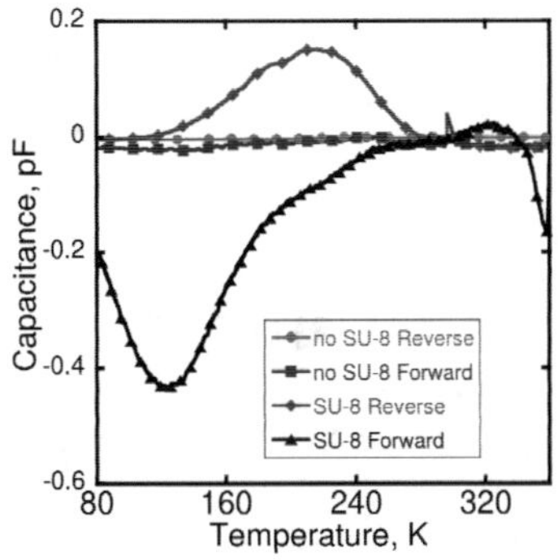

Figure 5. DLTS spectra for window rate of 20 s^{-1} array of SiNWs with and without coating by SU-8 at different mode.

3. Conclusion

Cryogenic dry etching was applied to fabricate array of vertically aligned silicon nanowires with height of 6 μm and diameter of 1.7 μm using mask of latex spheres with diameter of 2 μm. Array of SiNWs was covered by continuous smooth layer of photoresist obtained by spin-coating of SU-8 with thickness of 580 nm thick at 800 rpm. Cryogenic dry etching leads to defect formation with concentration lower $5{\cdot}10^{12}$ cm^{-3} in near-surface area and no defects in bulk silicon. However, defect concentration grows by two order, and new defects as divacancies arising in SiNWs after release of top side of SiNWs by etching of 300 nm SU-8 in oxygen plasma.

Acknowledgments

Post-growth processing (SU-8 spin coating) was carried out within the framework of the project № 075-01024-21-00, which was carried out within the framework of the state assignment of Ministry of Science and Higher Education of the Russian Federation. DLTS measurements were carried out in frame of the grant from the President of Russian Federation No СП-1207.2019.1 awarded to Baranov Artem.

References

[1] Al-Ghzaiwat M, Foldyna M, Fuyuki T, Chen W, Johnson E, Meaot J, Cabaroccas P R 2018 Scientific Reports **8** 1651

[2] Morozov I, Gudovskikh A, Uvarov A, Baranov A, Sivakov V and Kudryashov D 2020 Phys. status solidi **217** 1900535

[3] Baranov A, Kudryashov D, Morozov I, Uvarov A, Shugurov K, Gudovskikh A 2020 J. Phys.: Conf. Ser. **1697**, 012060

[4] Mikelsen M, Monakhov E, Alfieri G, Avset B, Härkönen J, Svensson B. 2005 J. Phys.: Condens. Matter **17** S2247

Features of Degradation of Silicon-based Solar Photovoltavic Cells

Linda Boudjemila[1, 2], **V V Davydov**[1, 3], **V G Malyshkin**[4] **and V Y Rud**[3, 4]

[1] Peter the Great St. Petersburg Polytechnic University, St. Petersburg 195251, Russia
[2] Department of Mechanical Engineering, University of science and technology Houari Boumedien Bab Ezzouar 16111, Algiers, Algeria
[3] All-Russian Research Institute of Phytopathology, Moscow Region 143050, Russia
[4] Ioffe Institute, St. Petersburg, 195256, Russia

lariessai21@gmail.com

Abstract. This work intends to present a method for efficient analysis of the phenomenon of degradation in photovoltaic cells in Saint Petersburg (Russia) and Algiers (Algeria). The data used contain weather parameters (T°, wind, solar radiation) for 10 years. The rates of degradation of Si- photovoltaic cells were obtained from the data. It was noted from the results of comparing various data that the main factor in the formation of the degradation process is the technology of manufacturing photovoltaic cells.

1. Introduction

The development of the technology in the last two centuries required more and more the use of the electrical energy [1-10]. As it's known the production of the electricity from fossils resources has caused many problems of pollution [10-20]. That's why scientists are interested to the alternative energies such as solar energy [21-27]. Recently the solar cell of first generation has known a rapid development and a certain maturity which had led the price to get lower [28-34]. Unlucky these cells are very delicate and many factors such as the shadow, humidity, the hot weather, the wind can cause the degradation in a short time. One of the important elements of studies about the degradation is the modeling of the operation of a solar power plant, taking into account the climatic conditions of operation of photovoltaic converters, the processes of their degradation, etc.

2. Solar cells degradation data analysis

The data obtained from solar power station is a time series. The number of recorded parameters can vary from three (time, current and voltage) to dozens (include temperature, wind speed, humidity, etc.) this depends on power station data acquisition system. Obtained data is typically very noisy, thus their analysis is a non—trivial task. The algorithms of data analysis can be broadly divided on "sensor based" which use sensor-measured solar irradiation and "clear sky", where irradiation and cells temperature is calculated from generated power using mathematical models [35]. One might think that the accuracy of "clear sky" approach will be inferior to the "sensor based", however this is not actually the case, clear sky avoids problems due to drift or recalibration of ground-based sensors. With recent versions of the RDTools software two methods are actually on par with each other. Once the data is collected – a model for data analysis should be build. Current models [35, 36] include the following three stages: 1. Normalization: do nominal power, and temperature normalization. 2. Filtering: remove the data collected during periods of poor or variable solar resource conditions as well as nonrepresentative or biasing data. 3. Analysis: There several approaches, the most commonly used is YOY(Year Over Year) that is typically more robust than SLS (Standard Least Squares) and quantile regression. The result is the distribution of degradation rates. This distribution can be quite noisy due to random and systematic errors. One of the key factors affection the results is the cleanless of solar cells. In [37] the devices which are regularly cleaned (control device) are compared with the devices left to naturally soil (soiled device); the Soiling Loss can contribute up to 10 % to solar station performance reduction. In current work we did not take soiling loss into account. Data noise in for generated power is clearly observed in Figure1 and 2 below along with seasonal power dynamics.

3. Features of the degradation of solar photovoltaic cells

The data base used when simulating the operation of a solar power plant in Algiers and saint-Petersburg is the NASA data on climatic factors illumination of a horizontal surface in energy units per day, temperature and wind speed; these last two parameters are measured at 2 meters altitude.

The RdTools is used in this work as well. It evaluates PV production data logged over several years to obtain rates of performance degradation over time. RdTools can detect the frequency of data automatically and therefore, handle both high frequency (hourly or less) and lower frequency (daily or higher) PV production data. RdTools is an open source tool developed in Python language by NREL (National Renewable Energy Laboratory) in USA based on the improved YOY method [36].

An analysis of the climatic conditions between the two cities shows large differences in illumination and temperature and little difference in wind speed. The first two factors have a significant impact on the power generated by a solar power plant throughout the year. Wind speed has a different effect on the rate of degradation of photovoltaic converters (warm or cold wind and humidity are of great importance). In Saint-Petersburg the low temperatures, strong winds and humidity are a determining stressor in the mechanisms of corrosion and delamination. In Algiers, the degradation process is affected by the high temperatures and humidity which caused the discoloration, hot spots and bubbles. The type of PV cells chosen is mono-crystalline silicon in Saint Petersburg and Algiers. This type is very common within easy reach.

The preferred method of degradation estimation is the annual (YOY) approach [35]. The algorithm includes: Normalizing the measured PV power, by site irradiance and temperature, Filtering the data (the most common filter is a low irradiance cutoff of about 200 W/m²), and Data Analysis step. There are actually two major categories of degradation calculation methods: with an independent measurement of solar irradiation and without it. Advances in algorithm processing [35, 38] allows to obtain sufficiently accurate value of the degradation rate even without independent irradiation measurements. Data availability is one of the major issues. In Figures 1 and (a, b) show the results of calculations of the rate of distribution of the generated power by a solar power plant and the confidence interval for the degree of degradation of photovoltaic stations for Algeria (figure 1) and St. Petersburg (figure 2).

The obtained results of modeling the operation of a solar power plant show that degradation process are more actively occurring at the solar power plant located in the region of St. Petersburg. For the research, data was taken on the operation of SES, in which photovoltaic converters of one company are used.

To obtain a more accurate forecast of the influence of climatic factors on the degradation of solar cells, it is necessary to carry out these measurements at various points in the region. There should be more than 20 such points. This will allow calculating the error in assessing the degree of degradation and the confidence interval.

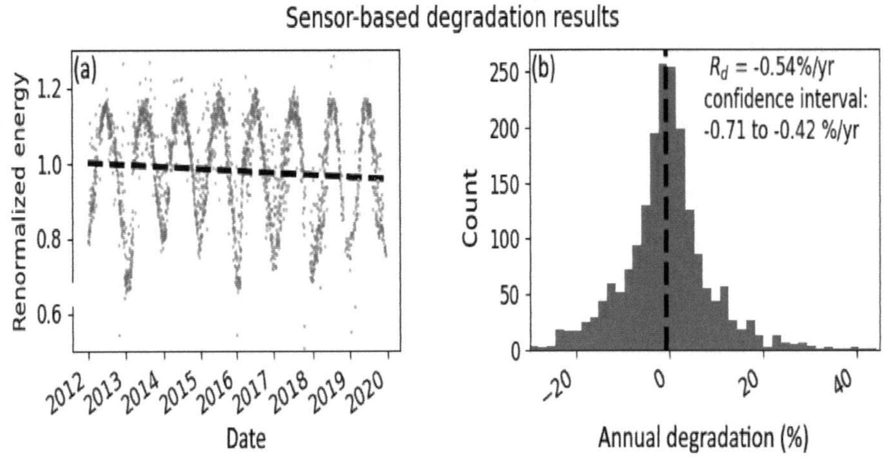

Figure 1. The results of research a fonctionality of solar power plants over the past 8 years in Algeria: (a) is calculations of the rate of distribution of the generated power by a solar power plant, (b) is confidence interval for the degree of degradation of photovoltaic stations.

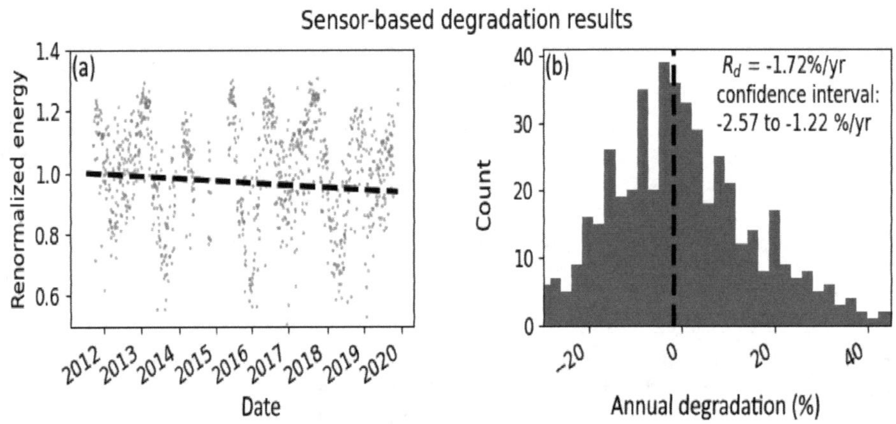

Figure 2. The results of research a fonctionality of solar power plants over the past 8 years in Saint-Petersburg: (a) is calculations of the rate of distribution of the generated power by a solar power plant, (b) is confidence interval for the degree of degradation of photovoltaic stations.

4. Conclusion

The modeling done in this work showed that the rate of degradation in solar cells is not very high. Degradation occurs over years of use. In other words, the efficiency of the solar panel is not affected. Harsh climatic conditions (severe frosts and high temperatures), leads to increase the rate of degradations visible to the naked eye. The results obtained shows that the location of the solar power plan is very important. The rate of degradation in Saint Petersburg (with rainy, snowy and windy weather) is quite high because of the presence of humidity which is much higher than the city Algiers which also has a humid climate but the weather is warmer. Another factor may be responsible is the different defects in solar cells because during the transportation which can cause invisible cracks that accelerate the degradation.

References

[1] Fadeenko I, Fadeenko V, Reznik V, Moroz A, Popovskiy N and Nikolaev D 2019 *IOP Conference Series: Earth and Environmental Science* **390(1)** 012022

[2] Myazin N S 2018 *Journal Physics: Conference Series* **1124(1)** 031004
[3] Davydov R, Antonov V, Makeev S, and Batov Y 2019 *E3S Web of Conferences* **140** 02001
[4] Davydov R, Antonov V and Moroz A 2019 *International Conference on Electrical Engineering and Photonics EExPolytech-2019 (Saint-Petersburg)*, vol. 8906791 (IEEE), p. 295-297
[5] Van S, Cheremisin A, Chusov A, Switala F and Davydov R 2019 *IOP Conference Series: Earth and Environmental Science* **390(1)** 012011
[6] Davydov V V 1999 *Russian Physics Journal* **42(9)** 822–825
[7] D'yachenko S V and Zhernovoi A I 2016 *Technical Physics* **61(12)** 1835–1837
[8] Davydov V V and Kiryukhin A V 2020 *Atomic Energy* **127(5)** 274–279
[9] D'yachenko S V, Kondrashkova I S and Zhernovoi A I 2017 *Technical Physics* **62(1)** 1602-1604
[10] Smirnov K J and Batov Y V 2019 *Journal of Physics: Conference Series* **1368(2)** 022073
[11] Davydov V V, Velichko E N, Dudkin V I and Karseev A Y 2015 *Instruments and Experimental Techniques* **58(2)** 234–238
[12] Myazin N, Neronov Y, Dudkin V and Yushkova V 2018 *MATEC Web of Conferences* **245** 11013
[13] Rukin E V, Myazin N S and Dudkin V I 2019 *Journal of Physics: Conference Series* **1368(4)** 042011
[14] Myazin N S, Yushkova V V and Taranda N I 2019 *Journal of Physics: Conference Series* **1410(1)** 012130
[15] Nikitina M, Grebenikova N and Batov Y 2019 *IOP Conference Series: Earth and Environmental Science* **390(1)** 012024
[16] Gryznova E and Batov Y 2019 *E3S Web of Conferences* **140** 09001
[17] Gryznova E, Grebenikova N, Ivanov D and Bykov V 2019 *IOP Conference Series: Earth and Environmental Science* **390(1)** 012044
[18] Davydov R V, Mazing M S, Yushkova V V and Stirmanov A V 2019 *Journal of Physics: Conference Series* **1410(1)** 012067
[19] Davydov R V, Antonov V I, Yushkova V V and Smirnov K J 2019 *Journal of Physics: Conference Series* **1400(6)** 066037
[20] Davydov V, Fadeenko V, Fadeenko I., Popovskiy N and Rud V 2019 *E3S Web of Conferences* **140** 07006
[21] Kozlova E S and Kotlyar V V 2014 *Computer Optics* **38(1)** 132-138
[22] Kuzmin M S and Rogov S A 2019 *Computer Optics* **43(3)** 391-396
[23] Logunov S E, Davydov R V, Vysotsky M G, Dudkin V I and V Yu Rud 2019 *Journal of Physics: Conference Series* **1368(2)** 022056
[24] Smirnov K J, Glagolev S F, Rodygina N S and Ivanova N V 2018 *Journal of Physics: Conference Series* **1038(1)** 012102
[25] Smirnov K J, Glagolev S F and Tushavin G V 2018 *Journal of Physics: Conference Series* **1124(1)** 022014
[26] Smirnov K J, Medzakovskiy V I, Vysoczky M G and Glagolev S F 2017 *Journal of Physics: Conference Series* **917(6)** 062019
[27] Sachenko A V, Kryuchenko Yu V, Kostylyov V P, Bobyl A V, Terukov E I, Abolmasov S N, Abramov A S, Andronikov D A, Shvarts M Z, Sokolovskyi I O and Evstigneev M 2016 *J. Appl. Phys.* **119(22)** 225702
[28] Boudjemila L, Mahmoudi B, Khenfer K, Davydov V V and Valiullin L R 2020 *Journal of Physics: Conference Series* **1695(1)** 012204
[29] Davydov V V, Bobyl A V, Zabrodskii A G and Malyshkin V G 2020 *International Conference on Electrical Engineering and Photonics EExPolytech-2020 (Saint-Petersburg)*, vol. 9243994 (IEEE), p. 238-241
[30] Bobyl A, Boudjemila L, Malyshkin V, Diuldin M and Cheremisin A 2020 *ACM International Conference Proceeding Series 2020 International Scientific Conference on Digital Trans-*

formation on Manufacturing, Infrastructure and Service, DTMIS 2020 (Saint-Petersburg).
[31] Htein L, Bobyl A, Malyshkin V, Cheremisin A and Gerner A 2019 *IOP Conference Series: Earth and Environmental Science* **390(1)** 012048
[32] Natorkhin M I, Bobyl A V, Cheremisin A V and Sokolov M S 2019 *Journal of Physics: Conference Series* **1236(1)** 012011
[33] Boudjemila L 2021 *Springer Proceedings in Physics* **255** 177–184
[34] Kompan M E and Malyshkin V G 2019 *Technical Physics Letters* **45(1)** 45–47
[35] Jordan D C, Deline Ch, Kurtz S R, Kimball G M and Anderson M 2018 *IEEE Journal of Photovoltaics* 99-105
[36] Jordan D C, Kurtz S R, VanSant K T and Newmiller J 2016 *Prog. Photovolt. Res. Appl.* **24(7)** 978–989
[37] Deceglie M G, Micheli L and Muller M 2018 *IEEE Journal of Photovoltaics* **8(2)** 547-551
[38] John J, Alnuaim A, Elnosh A and Banda S M 2018 *IEEE 7th World Conference on Photovoltaic Energy Conversion (WCPEC)* 712-714

Journal of Physics: Conference Series **2086** (2021) 012080 doi:10.1088/1742-6596/2086/1/012080

Development of visual display and data transmission system for patients with chronic disorders of consciousness

S A Degtiareva[1], D S Shiryaev[1], Y S Andreev[1], I S Polukhin[1], E A Kondratieva[2], I G Smirnova[1] and V E Bougrov[1]

[1] Faculty of Laser Photonics and Optoelectronics, Saint-Petersburg, Russia
[2] Polenov Neurosurgical Institute, the Branch of Almazov National Medical Research Centre, Saint-Petersburg, Russia

svetlanadegtiareva@itmo.ru

Abstract. The monitoring is integral part for patients with chronic disorders, as such cases require serious attention to save their life and predict recovery. Physiological signs such as heart rate, hemodynamic, temperature, saturation are collected from biomedical sensors to bedside monitors that medical staff could detect unexpected life-threatening conditions being around the patients. However, it is a tough job to being stuck at monitor for every person. Furthermore, such continuous supervision may lead to difficulties due to human error. Thus there is a need to capture, display all physiological changes visually and give a medical emergency about health conditions when they are out of the normal range. The main objective is the development of the optoelectronic system for visual monitoring and data transmission of patients in a coma by optical wireless communication.

1. Introduction

A coma is a deep state of prolonged unconsciousness as a result of traumatic brain injury, stroke, brain tumor, diabetes, hypothermia or an infection. The number of accidents happens very high when the patients are admitted in the Intensive Care Unit (ICU) [1]. Coma usually lasts for less than 2 to 4 weeks in itself, then a patient may wake up into a vegetative state/unresponsive wakefulness state (VS/UWS) or minimally conscious state (MCS). It requires operative action to preserve their life and brain function as well as visual controlling of basic physical cues.

Diagnosis of chronic disorders of consciousness is a threat to the patient's life and requires a continuous monitor of the values on bedside monitors, even in the face of the increased workload of medical personnel. In the meantime, patients are admitted to the rehabilitation unit after without any monitoring system after ICU, cause he or she is considered stable. Nevertheless, the need to monitor the physiological parameters remains.

Imaging techniques, including functional magnetic resonance imaging, EEG, and FDG positron emission tomography have great potential to predict the diagnosis of VS/UWS and MSC [2-5]. They can be used to investigate metabolic and functional impairment of the brain. However, diagnosis of the VS/UWS and MSC should be based on a patient’s clinical history and on simple visual observations as well. Therefore, monitoring systems should be applied.

Recently developments provide more convenient and comprehensive medical monitoring. Some of them are focused on system that collect health parameters [6-7]. Other developments are dedicated to wireless patient monitoring systems [8-9]. However visual module is not implemented to monitor physical changes in real-time. Thus, there is a need to develop a system that helps to read vital physiological parameters, display their changes in real-time and indicate a deterioration or improvement of the patient's internal state. The development of a visual display object with established light-emitting diodes allows to represent physiological changes by controlling parameters and send them by optical wireless communication immediately.

2. Methods

A current system has 2-3 sensors on the patients. It is designed to collect the values of physiological parameters from sensors, display received data according to specified scenarios of light-emitting diodes on a visual object, and transmit data to a server or personal computer using optical wireless communication technology for viewing. The system will help to read the state of patients with chronically impaired consciousness, such as the vegetative state (such vital functions as heart activity and breathing are preserved), the state of minimal consciousness (pain stimuli, environmental eye-tracking persist) the syndrome of unresponsive wakefulness (cognitive functions are preserved).

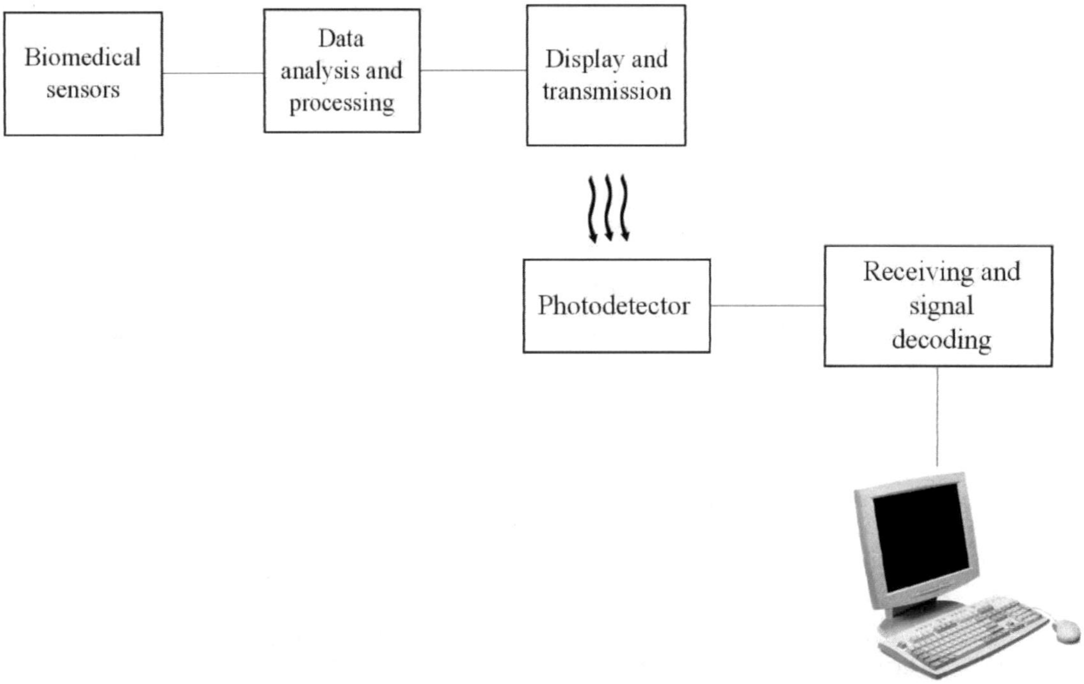

Figure 1. Diagram of the visual display and data transmission system.

The most important indicators of patient physiological state as measured signals were selected: heart rate (sensor working range 60-120 beats per minute), arterial oxygen saturation (sensor working range 95-99%), body temperature (sensor working range 36 - 38 degrees).

Under signs of these types of depression of consciousness, the most important indicators of the physiological state of patients as pulse rate, arterial blood oxygen saturation, and body temperature were selected as measured signals. The block diagram is presented in Figure 1 below.

A working principle is shown in Figure 2.

A bracelet with established light-emitting diodes is currently offer as a visual object. It visualizes the physiological parameters of patients in several scenarios: the values of physiological parameters are normal, the values of physiological parameters are lower than normal values, the values of physiological parameters are higher than normal values. If the values are in the range of normal values, the scenario is implemented in the form of pulsating green color. When the values are outside the normal range on the lower bound, the LEDs have a chromaticity in the range of cold colors. When the values are outside the normal range on the upper bound, the LEDs have a chromaticity in the range of warm colors.

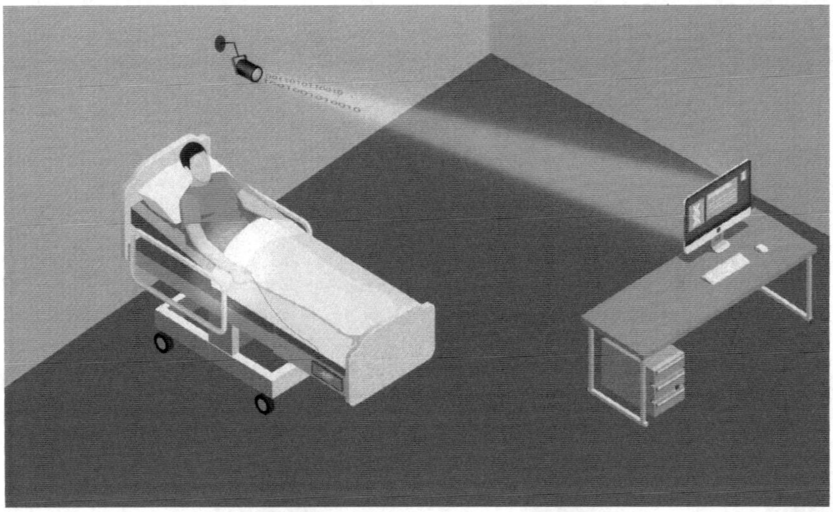

Figure 2. Working principle of the visual display and data transmission system.

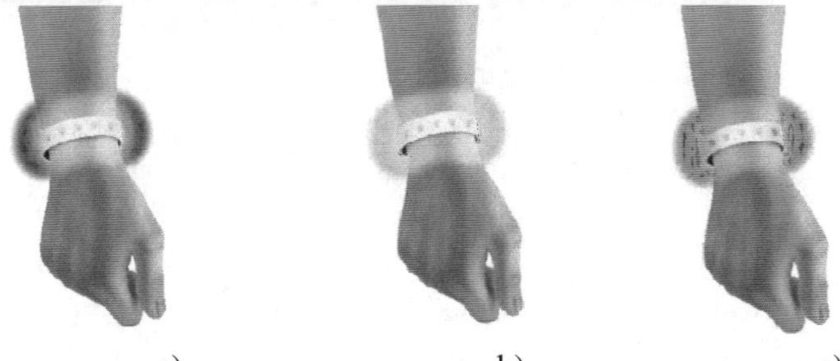

a) b) c)

Figure 3. Scenarios of light-emitting diodes on the visual display object: a) The values are below the normal range, b) The values are in the normal range, c) The values are above normal range

Figure 4 presents a color scale of possible scenarios of light-emitting diodes for given parameters which is bounded by a triangle on the CIE color space diagram.

It has vertices in the chromaticity coordinates, reflecting a wide range of colors for specifying all necessary parameters: $x_r = 0.71, y_r = 0.29$; $x_g = 0.18, y_g = 0.72$; $x_b = 0.16, y_b = 0.03$.

Technical parameters are presented on a table 1.

Arlight light-emitting diodes in the wavelength range of 930-940 nm were selected as a data transmission source. The main criterion for choosing was the spectral sensitivity since the peaks of the LEDs and the photodetector must coincide. A FDS100 silicon photodiode with a spectral sensitivity oяf 950 nm was selected as a receiver. The spectral sensitivity R (λ) is shown in Figure 5.

3. Conclusion

The healthcare industry has perpetually been on the forefront in the adoption and utilization of information and communication technologies for the efficient healthcare treatment. Most of the patients with chronic disorders are young people who were injured, got sunstroke, or hypothermia. The question of their recovery is extremely important, although the chance is quite low, but it is there. The probability is dependent on the extent of structural brain damage, which is difficult to assess by clinical, laboratory, or functional tests. The development of visual display and data transmission system is a critical step towards regaining consciousness, because the goal of this project is to learn how to catch and notice some glimmers and influence them so that the brain wakes up. .The use of visual display and data transmission system will help medical staff monitor the comatose regularly to see if there are any changes in emergency cases.

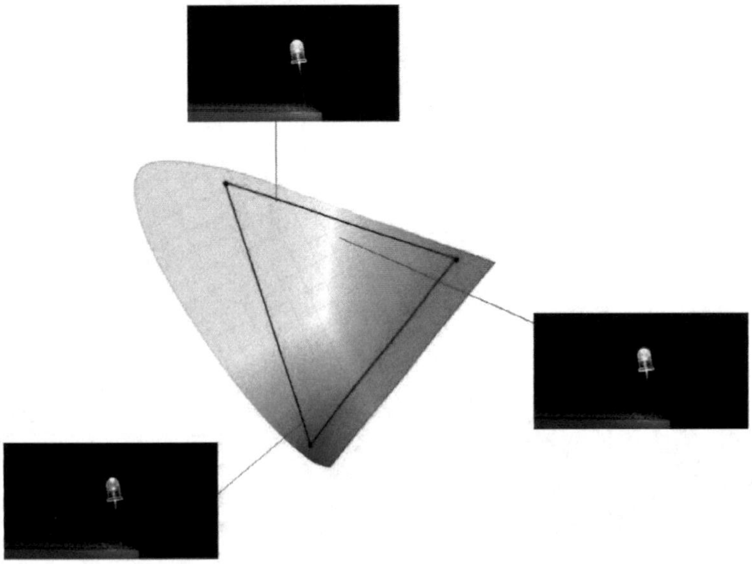

Figure 4. Light-emitting diodes chromaticity scenarios on color space diagram.

Table 1. Technical parameters of light-emitting diodes.

Parameter	Value
The emission angle of RGB LEDs, deg	120
Luminous flux of RGB LED strips, lm/m	150
LED strip power supply voltage, V	up to 12
Peak wavelength of red chip radiation, nm	600-670
Peak wavelength of the green chip, nm	510-570
Peak wavelength of blue chip radiation, nm	420-480
Coordinates of the radiation chromaticity, the normal range	x-0.1-0.4, y-0.5-0.8
Coordinates of the emission chromaticity, below the normal range	x - 0,1-0,3, y - 0,0-0,3
Coordinates of the emission chromaticity, above the normal range	x - 0,5-0,7, y - 0,2-0,4

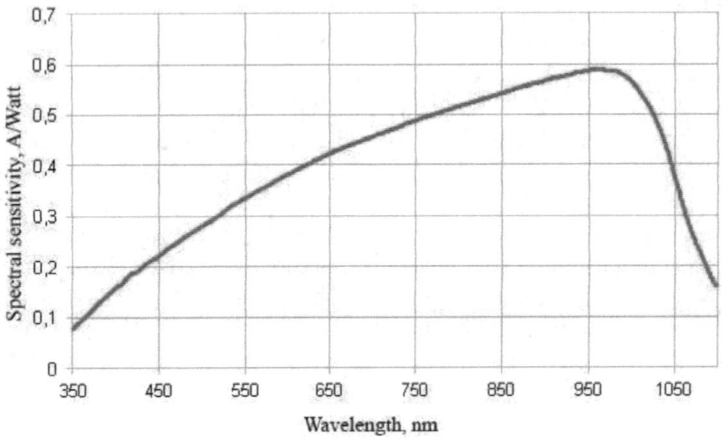

Figure 5. The spectral sensitivity of photodiode.

Acknowledgments
We thank Almazov Research Medical Centre for its great contribution and consulting support. This work was supported by RFBR grant № 19-29-01066/2020

References
[1] Manonmani A, Arivalagan M, Lavanya M and Sellakumar S 2020 Coma patient monitoring using brain computer interface *Int. J. of Psychosocial Rehabilitation* **24** 3701-10
[2] Sutter E 1992 The brain response interface: Communication through visually-induced electricalbrain responses *J. Microcomput. Appl.* **15** 31–45
[3] Edlow B L, Wu O and Giacino J 2013 Functional MRI and Outcome in Traumatic coma *Curr.Neurol. and Neurosci. Rep.* **13** 375
[4] Bates D 2001 The prognosis of medical coma *J. of Neurology, Neurosurger and Psychiatry* **71** 20-23
[5] Heiss W D 2012 Pet in Coma and in vegetative state *Eur. J. of Neurol.* **19** 207-211
[6] Kansal N, Dhillon H 2011 Advanced Coma Patient Monitoring system *Internation Journal ofScientific and Engineering Research* **2**
[7] Kamel M, George L E 2013 Remote Patients Tracking and Monitoring System *InternationalJ.of Computer Sci. and Mobile Computing* **2**
[8] Muneeswaran D, Praveen T 2020 Implementation of body movement detection system for comapatients using WSN *Int. J. of Grid and Distributed Computing* **13** 161-166
[9] Malika, Rana C, 2013 An Indoor Wireless Zigbee based Patient Monitoring system for Hospitals *Int. J. of Engineering Sci. Research-IJESR* **4**

Output power of multilayered InGaAs/GaAs quantum well-dot microdisk lasers

N A Fominykh[1,2], E I Moiseev[2,1], Ju A Guseva[3], M V Maximov[1,2], A I Lihachev[3], N V Kryzhanovskaya[2,1] and A E Zhukov[2,1]

[1] Nanophotonics Laboratory, St. Petersburg Academic University RAS, St. Petersburg 194021, Russia
[2] International Laboratory of Quantum Optoelectronics, Higher School of Economics, St. Petersburg 194100, Russia
[3] Ioffe Institute, St. Petersburg 194021, Russia

fominy-nikita@yandex.ru

Abstract. We studied the output optical power of microdisk lasers with InGaAs/GaAs quantum dots active region. An increase in the number of layers in the active region in the waveguide from 2 to 6 leads to increase in the peak output optical power due probably to increase of the gain. We also observe a corresponding increase of the threshold current due to the increase on the transparence current. The maximal optical power is achieved for structure with 6 layers at approximately 60 mA injection current. Further increase of the number of the QD layers to 10 results in increase of the threshold current and sudden drop of the output power.

1. Introduction

In recent years, much attention has been paid to the development of microlasers based on III-V materials that could be integrated with elements of silicon microelectronics. Injection III-V microdisk (MD) and microring lasers with quantum dots (QDs) show noticeable success: lasing was demonstrated in a continuous wave mode at a temperature of 60 °C in MD lasers synthesized directly on silicon substrates [1], the possibility of direct modulation with a frequency of up to 6.5 GHz [2] was shown, a threshold current density of 0.36 kA/cm^2 was achieved [3], the estimated mean time to failure is more than 80,000 hours [4].

The use of QDs makes it possible to achieve low values of the threshold current density at room and elevated temperatures, significantly reduce the effect of nonradiative recombination on the side walls of the resonator due to the short lengths of lateral diffusion of charge carriers. The disadvantages of the QD material are associated with the low optical gain of QD arrays and the relatively slow relaxation of carriers, as well as their accumulation at excited levels and in the matrix at high pump levels. It was found that the output power of the microlasers can be increased by using dense arrays of QDs in active region deposited by special technological methods (InGaAs/GaAs quantum-well dots (QWD) [5]), or arrays consisting of several layers of such QWDs. When using several active layers in a waveguide, a situation can be realized that only a certain part of the QD layers quantum dots will be in the immediate vicinity of the lightwave maximum. As can be seen from Figure 1, the side layers of

QDs are located outside the maximum of the mode, so that the electric field strength of the optical mode in position of the i-th layer is less than the electric field strength of the optical mode at the center of the waveguide (ε(zi)<ε(z0)) and, therefore modal gain increases sublinearly with the number of QD layers.

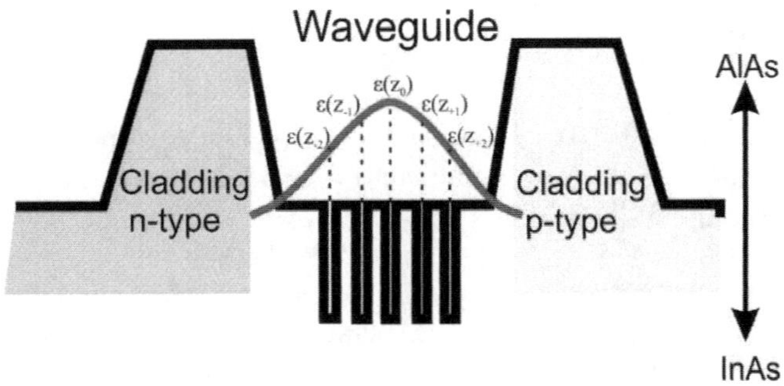

Figure 1. Schematics of the laser structure in the AlGaInAs material system.

In this work we study optical output power of MD lasers containing different number of QD active region layers, while the waveguide thickness is the same and is optimized to maintain only the fundamental mode of the laser vertical waveguide.

2. Experiment

The laser structures were grown by metal-organic vapor phase epitaxy (MOVPE) on an n+-GaAs substrate misoriented by 6° toward [111] direction. The laser active region consists of several QWDs layers. Each QWDs layer was formed by the deposition of 8 monolayers of $In_{0.4}Ga_{0.6}As$. The QWDs represent a dense array of indium-rich regions formed inside In-depleted residual quantum well and can be also considered as shallow QDs of high density. The QWD layers were separated from each other with 40 nm thick GaAs spacer layers and placed in the center of the undoped 0.78 μm thick GaAs waveguide. The waveguide was sandwiched between p- and n-type doped 1.5 μm thick $Al_{0.4}Ga_{0.6}As$ claddings. Five structures (A2, A4, A5, A6 and A10), having 2, 4, 5, 6 or 10 layers of the active region were studied. MD lasers were fabricated by dry etching of 30 μm in diameter circular mesas through both cladding layers and the active region/waveguide layer to the depth of about 5 μm. For structural characterization scanning electron microscope JSM 7001F (JEOL, Japan) was used (Figure 2). Round-shaped AgMn/Ni/Au contacts were formed individually to each MD on top of p+-GaAs cap layers. After that the GaAs substrate was thinned down to approximately 100 μm, and a n-type contact was deposited onto the back surface of the GaAs substrate. Then, the wafer was cleaved into chips containing single microdisk.

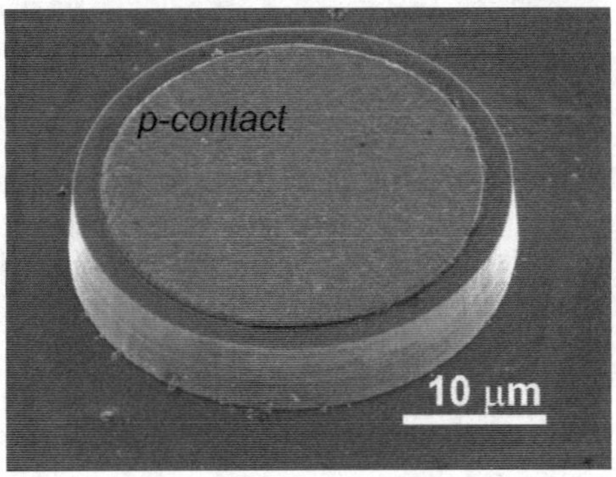

Figure 2. SEM image of 30 μm in diameter QD microdisk laser.

Electroluminescence was excited by a stabilized DC current with a Keithley 2400 Series SourceMeter® power supply. During measurements of MDLs in CW regime the devices were pumped using BeCu probe with Ø20 μm tip. During spectral measurements light emitted by an individual microlaser was collected with a Mitutoyo M Plan Apo NIR objective with 50× magnification. Emission spectra were acquired with a Yokogawa AQ6370C optical spectrum analyzer. The spectral resolution was 20 pm. To measure the optical power emitted into free space, the radiation was collected with Thorlabs S132C photodiode placed closely to the MDL under test.

3. Results and discussion

All the microlasers operate in continuous wave (CW) regime at room temperature. The typical lasing spectra (Figure 3) contains narrow lasing line, corresponding to whispering gallery mode of the resonator, the intensity of which increases sharply at a current above the threshold. The lasing wavelength was about 1100 nm for the lasers under consideration. When the MD laser operates in CW regime, its temperature increases because of the self-heating. The MD laser overheating with increase in the injection current results in a redshift of the WGM lines $\Delta\lambda$. One can estimate the temperature of the active region versus injection current of the microlasers (Figure 3, right axis) taking into account the temperature coefficient $\Delta\lambda/\Delta T$ of about 0.08 nm/K, where ΔT is temperature increase of the MD laser active region. The data presented in the inset to Figure 3 allows determining the thermal resistance R_{T} of the device as:

$$R_{\mathrm{T}} = \frac{\Delta T}{\Delta Q} = \frac{\Delta\lambda/\Delta Q}{\Delta\lambda/\Delta T} \quad (1)$$

where ΔQ is the Joule heat. Since in MD lasers emitting into free space the wall-plug efficiency is low, the Joule heat ΔQ can be estimated as the total consumed electric power, i.e. the product of the bias voltage and the injection current. The thermal resistance RT of the lasers structure was calculated to be ~ 0.55 K/mW.

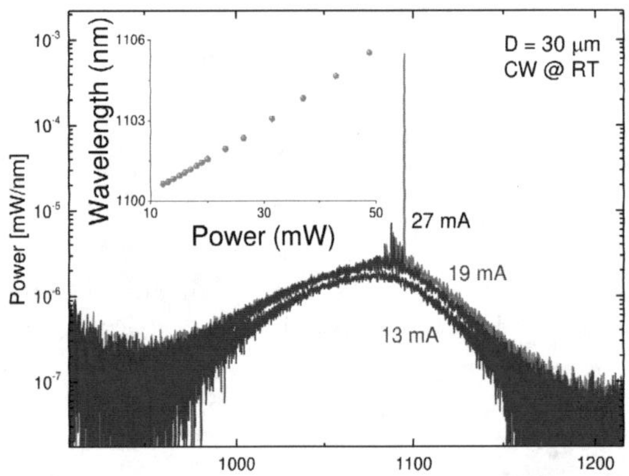

Figure 3. Example of room temperature spectra of 30 μm in diameter QWD microdisk laser (taken using A5 structure). Inset: emission wavelength versus injection current.

The dependencies of the optical power of the structures A1-A10 are presented at Figure 4. Each light-current dependence has three characteristic parts. In the first part at low (sub-threshold) currents the optical power linearly increase with pumping. Above the threshold (second part) we observe rapid increase of the output power. It should be noted, that near the threshold the emission is predominantly of single-mode kind. Increase of the current and subsequent heating of the structure results in the involvement to the lasing resonances with longer wavelengths. We observe this transition to multimode lasing as smoothed steps appeared at the light-power dependence. And next at certain current value the optical power saturates and starts to drop down (third part of the dependence) due to the thermal rollover. Increase of the number of QD layers from 2 to 6 leads to an increase in the peak output optical power due probably to increase of the gain. We also observe a corresponding increase of the threshold current due to the increase on the transparence current. The maximal optical power is achieved for A6 structure at approximately 60 mA injection current. Further increase of the number of the QD layers to 10 (structure A10) results in increase of the threshold current and sudden drop of the output power. Possible reason is that the threshold current in this case becomes too high leading to strong microlaser overheat already below the threshold. What is essential is that for lasers based on self-organizing QDs one can found such a design that allows you to achieve optimal threshold current density and sufficient output power.

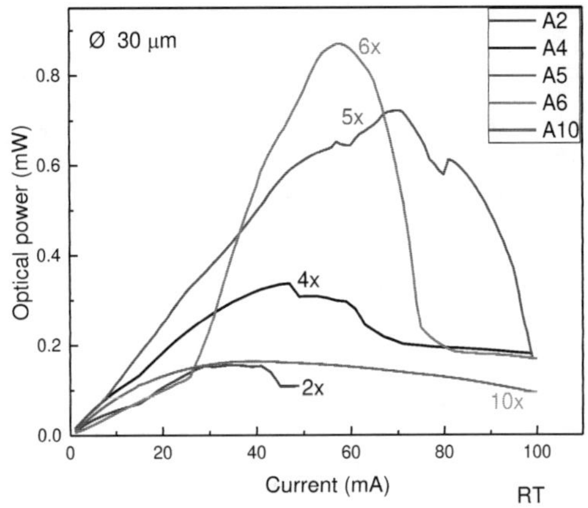

Figure 4. Room temperature light-current dependencies of the 30μm in diameter microdisk lasers with different number of QWDs layers.

4. Conclusions

Microlasers with multilayered InGaAs/GaAs quantum dots active region operating in CW regime are studied at room temperature. Increase in the injection current results in saturation of the output power due to the MD laser overheating (thermal roll-over) and gain saturation. Increase of the number of QD layers from 2 to 6 leads to an increase in the peak output optical power due to increase of the gain. We also observe a corresponding increase of the threshold current due to the increase on the transparence current. The maximal optical power is achieved for structure with 6 layers. To achieve optimal threshold current density and sufficient output power in the microlasers one should manage heat conductance.

Acknowledgement

This study was supported by Russian Science Foundation grant 18-12-00287, https://rscf.ru/project/18-12-00287/. SEM characterization were performed using equipment owned by the Federal Joint Research Center "Material science and characterization in advanced technology".

References

[1] Kryzhanovskaya N V et al 2018 *Laser Physics Letters* **15** 015802

[2] Wan Y, Inoue D, Jung D, Norman J C, Shang C, Gossard A C and Bowers J E 2018 *Photon. Res.* **6** 776-81

[3] Moiseev E I, Kryzhanovskaya N V, Zubov F I, Mikhailovskii M S, Abramov A N, Maximov M V, Kulagina M M, Guseva Yu A, Livshits D A and Zhukov A E 2019 *Semiconductors* **53** 1888-90

[4] Zhukov A E et al 2020 *Materials* **13** 2315

[5] Maximov M V et al 2020 *Appl. Sci.* **10** 1038

Temperature stability of small-signal modulation response of WGM microlasers with InGaAs/GaAs quantum well-dots in the active region

N K Isaev[1,3], E I Moiseev[2,1], N A Fominykh[1,2], N V Kryzhanovskaya[2,1], F I Zubov[1,2], K A Ivanov[2], I S Makhov[2], S A Mintairov[3], N A Kalyuzhnyi[3], Yu A Guseva[3], M V Maximov[1,2] and A E Zhukov[2,1]

[1] Alferov University, St Petersburg, 194021 Russia
[2] HSE University, St Petersburg, 190008 Russia
[3] Ioffe Institute, St Petersburg, 194021 Russia

nabi-isaev@yandex.ru

Abstract. In this paper, results are presented on III-V quantum well dot microring diode lasers tested at elevated temperatures. To the best of our knowledge, the first uncooled microdisk lasers with diameter of 40 μm with 3-dB bandwidth above 2 GHz at 55°C are demonstrated.

1. Introduction

Small-footprint lasers with narrow linewidth and low power consumption are a promising way of developing optical data transmission technology [1, 2]. Such lasers can be based on semiconductor resonators supporting whispering gallery modes (WGM), such as microdisks and microrings. Besides low power consumption and high Q-factor, these microlasers allow direct high-frequency signal modulation. Recently, we have proposed microlasers with active region based on quantum well-dots (QWDs) [3]. High output optical power was demonstrated for a 30 μm in diameter microdisk laser [4] tested at room temperature. High operation temperature and (or) self-heating leads to a decrease in the differential efficiency due to a decrease in the inverted population and an increase in the threshold pumping power due to additional optical losses on the free carriers. An increase in the internal temperature of the laser leads to saturation not only of the output optical power, but also of the resonance frequency. For example, in [5] the WGM microlasers were tested only at 287 K. In [6] microring operation at 20°C was studied. In [7] the 3-dB bandwidth decreases from 12.1 to 10.9 GHz as the temperature rises from 290 to 298 K.

Meanwhile, the reports on the dynamic characteristics of QWD-based microdisk/microring lasers at different temperatures are practically absent. In this work, we studied temperature dependence of small-signal modulation response of the microlasers with InGaAs/GaAs quantum well-dots in the active region.

2. Experiment

A laser heterostructure was grown by low pressure MOVPE on an n^+-GaAs substrate with a surface off-axis oriented by 6° from the (100) plane. An active region represents 5 planes of quantum well-dots formed by the deposition of 8 monolayers of $In_{0.4}Ga_{0.6}As$ and separated with 40 nm-thick GaAs spacers. The active region was placed in the middle of a 0.8 μm-thick GaAs waveguide. GaAs waveguide was inserted between n- and p- doped $Al_{0.39}Ga_{0.61}As$ cladding layers. Microdisk and microring lasers with outer diameters D_{out} varied from 10 to 50 μm were formed by photolithography and dry etching. Mesa height was chosen about 5 μm to ensure sufficient optical confinement of the whispering gallery modes. Sidewall verticality was controlled with the precision of 5 degrees. AgMn/NiAu and AuGe/Ni/Au metallization was used for formation of ohmic contacts to the p^+-GaAs cap layer and the n-doped GaAs substrate, respectively.

Microlasers were then planarized using SU-8 epoxy resist and contact pads were made over the epoxy planarization. A chip comprising a group of MDs of different diameters was soldered on a high-frequency submount suitable for connection using a ground-signal-ground (GSG) radio frequency probe. The top contact of the individual microlaser was connected to a signal line of the submount using a 50 μm-thick gold wire. High-frequency submount were soldered on a copper heat sink with indium and mounted on a heater to be tested at various temperatures. The emission from the investigated microdisk was collected by an Olympus LMPlan IR 100x objective lens, transmitted via an optical fiber, and detected by a Yokogawa AQ6370C optical spectrum analyzer. Dynamic characteristics were measured in the range from 50 MHz to 20 GHz with a New Focus 1434 25 GHz photodetector and an Agilent E8364B network analyzer (schematic layout is shown in figure 1).

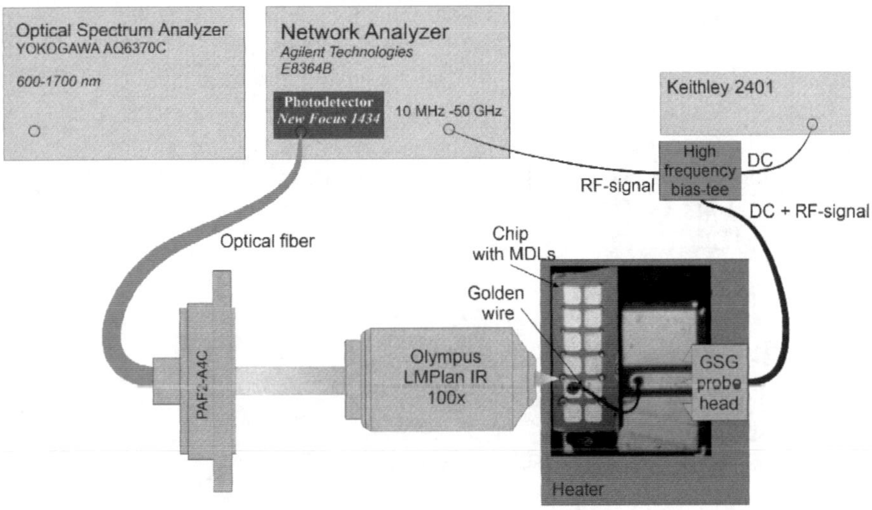

Figure 1. Schematic representation of setup for high-frequency measurement

3. Results and discussion

3.1. Light-current and current-voltage characteristics

Figure 2 presents the light-current and current-voltage (I-V) characteristics of a microring laser with 20μm outer diameter obtained at room temperature. All the lasers demonstrated the diode behaviour with I-V characteristics fitted by equation $U=IR_s+U_0$, where R_s – is series resistance and U_0 – is the turn-on voltage. A turn-on voltage values U_0 approximately correspond to the emission wavelength of the lasers near the threshold (~1.09μm). The series resistance $R_S \approx 50\ \Omega$. At the light-current characteristics the lasing behavior is observed with the threshold current of ~7 mA.

The electroluminescence spectra were measured at injection currents below and above the threshold (spectra obtained at 5 and 21 mA are presented at Figure 3). At the spectra below the threshold one can observe broad line in 950-1150 nm spectral range, corresponding to spontaneous emission of InGaAs/GaAs QWDs. At the spectra above the threshold a dominant WGM lasing line is observed near 1100 nm.

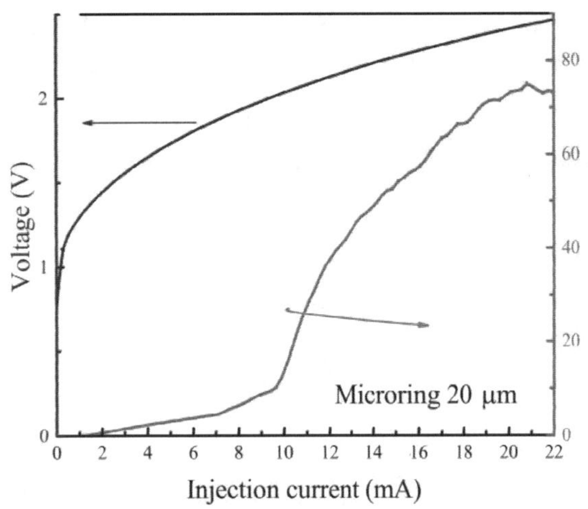

Figure 2. Current-voltage and light-current characteristics of a microring laser with outer diameter 20 μm and inner diameter 6 μm.

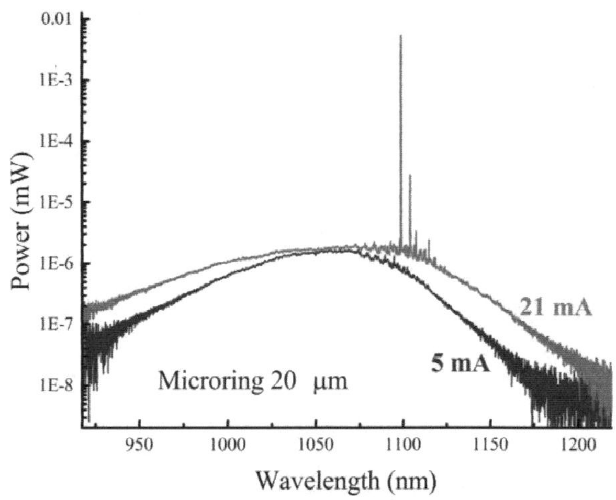

Figure 3. Electroluminescence spectra of a microring laser with outer diameter 20 μm and inner diameter 6 μm.

3.2. Small-signal modulation response at different temperatures

Small-signal modulation response of the microlasers was measured in temperature range from 13°C to 55°C. Modulation response for a microring laser with 40μm outer diameter is presented in Figure 4. At the injection current of 16 mA the 3-dB bandwidth (f_{3dB},) drops from 3.7 GHz to 2.2 GHz when the sample is heated up on 42°C.

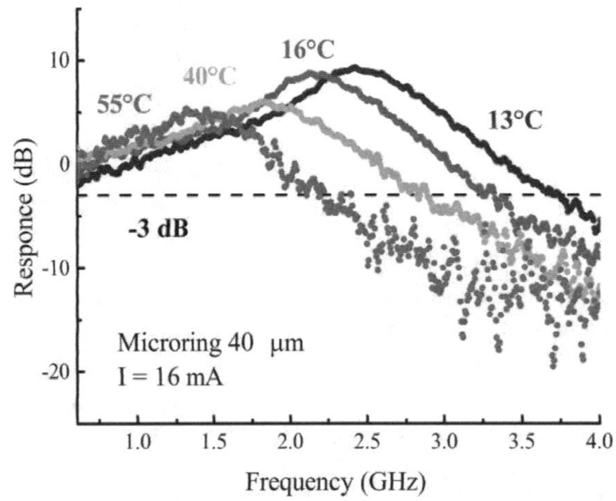

Figure 4. Small-signal modulation response of the microring laser with outer diameter 40 μm and inner diameter 22 μm measured at temperatures 13°C, 16°C, 40°C and 55°C.

The most direct way to increase the modulation frequency is associated with an increase in the photon density in the cavity by increasing the injection current. However, self-heating of the laser with increasing the current leads to a decrease in the active area gain. As a result, the maximum achievable modulation frequency is limited by thermal effects. The 3-dB bandwidth as a function of injection current is presented in Figure 5. Increasing the current the maximal 3-dB bandwidth value of 3.7 GHz for 20-μm in diameter microlaser and 3.1 GHz for 40-μm in diameter microlaser is achieved. Near the threshold, the experimental data can be described well as f_{3dB} = MCEF $(I - I_{th})^{1/2}$ using the modulation current efficiency factor value (MCEF) as fitting parameter (solid lines in Figure 5). The value of MCEF drops from to1.5 GHz/mA$^{1/2}$ at 13°C to only 1 GHz/mA$^{1/2}$ at 55°C for the 40-μm in diameter microring. It can be seen that a change in temperature does not lead to a dramatic deterioration in the characteristics of the laser, although, of course, an improvement in heat dissipation would allow, in general, to increase the frequency characteristics.

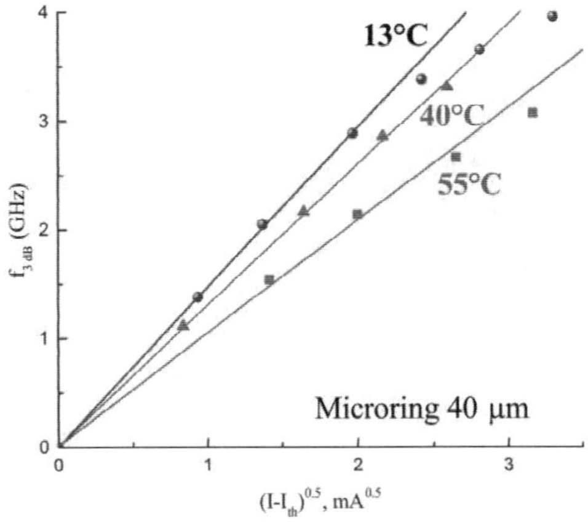

Figure 5. The 3-dB bandwidth for the microring with outer diameter 40 μm at temperatures 13°C, 40°C and 55°C.

MCEFs for microrings with outer diameter 40 μm and 20 μm are presented in Figure 6 as functions of temperature. The dependencies show that a decrease of the laser diameter improves the MCEF parameter due to reduced mode volume. The decrease in MCEF with temperature is caused by a decrease in the differential gain and the obtained results may be used to find dependence of the differential gain on the carrier concentration dg/dn(T) for a given type of active region, as this information is extremely useful for optimization laser design.

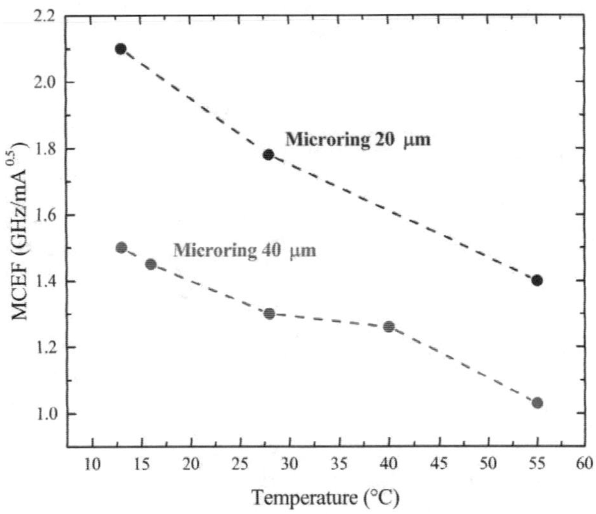

Figure 6. Dependency of MCEF vs temperature for the microring lasers with 40 μm and 20 μm outer diameter.

4. Conclusion

In summary, we examine uncooled QWD-based microdisk lasers at elevated temperatures. We demonstrate, for the first time, small-signal modulation response of the microring lasers at 55°C. Under these operation conditions the 3-dB bandwidth value of 3.1 GHz for 40-μm in diameter microlaser and 3.7 GHz for 20-μm in diameter microlaser is achieved. An increase in temperature leads to a decrease in the modulation current efficiency factor. Obviously, for increasing the modulation frequency of the laser further, it is necessary to reduce the thermal resistance and dissipation level. The results obtained can be useful for the development of components for future photonic circuits for data transmission over short distances.

Acknowledgments

This study was supported by Russian Science Foundation grant 18-12-00287, https://rscf.ru/project/18-12-00287/.

References

[1] Choi S J, Djordjov K, Choi S, Coleman J and Dapkus P 2003 *IEEE Photonics Technol. Lett.* **15** 1330–32
[2] Zou L-X, Lv X-M, Huang Y-Zh, Long H, Yao Q-F, Xiao J-L and Du Y 2013 *Opt. Lett.* **38** 3807
[3] Mintairov S A, Kalyuzhnyy N A, Maximov M V, Nadtochiy A M, Rouvimov S and Zhukov A E 2015 *Electron. Lett.* **51** 1602
[4] Moiseev E, Kryzhanovskaya N, Maximov M, Zubov F, Nadtochiy A, Kulagina M, Zadiranov Y, Kalyuzhnyy N, Mintairov S and Zhukov A 2018 *Opt. Lett.* **43** 4554
[5] Lv X-M, Huanga Y-Z, Yang Y-D, Zou L-X, Long H, Liu B-W, Xiao J-L and Du Y 2014 *Appl. Phys. Lett.* **104** 161101
[6] Yang Y-D, Zhang Y, Huang Y-Zh and Poon A W 2014 *Opt. Express* **22** 824-38
[7] Lv X-M, Huang Y-Z, Zou L-X, Long H and Du Y 2013 *Laser and Photonics Reviews* **7** 818-29

Investigation of the dynamic parameters of electroluminescence in different parts of the spectrum in local regions of the light-emitting heterostructure

I V Frolov[1], O A Radaev[1] and V A Sergeev[1,2]

[1] Ulyanovsk Branch of Kotel'nikov Institute of Radio-Engineering and Electronics of Russian Academy of Sciences, Ulyanovsk 432071, Russia
[2] Ulyanovsk State Technical University, Ulyanovsk 432027, Russia

ilya-frolov88@mail.ru

Abstract. A method for measuring the distribution of the differential charge carriers lifetime over energy levels in the local regions of a light-emitting heterostructure is presented. The method has been tested on commercial green InGaN-based LEDs. It has been determined that with an increase in the energy level (with decreasing wavelength), the charge carriers lifetime decreases. It is shown that the relative inhomogeneity of the distribution of the charge carriers lifetimes in local regions of the die in the long-wavelength part of the electroluminescence spectrum is higher than in the short-wavelength part of the spectrum.

1. Introduction

The inhomogeneity of the distribution of electroluminescence parameters over the area of the InGaN-based light-emitting heterostructures active region is due to the inhomogeneous distribution of nonradiative recombination centers and the inhomogeneous distribution of indium concentration [1-3]. According to the model of inhomogeneous distribution of indium concentration in the heterostructure [4], the short-wave wing of the radiation spectrum is formed by radiation from local regions with a low concentration of indium, and the long-wave one is formed by radiation from local regions with a high concentration of indium. It was shown in [5] that the differential lifetime of charge carriers occupying low energy levels in the quantum well is longer than the differential lifetime of charge carriers occupying high energy levels. This means that the process of radiative recombination in the regions of the heterostructure with a low indium concentration proceeds more intensively, which is the reason for the increase in the current density in this region. Existing measurement methods make it possible to determine the distribution of the differential charge carriers lifetime over energy levels, averaged over the entire area of the structure. The report presents a method for measuring the distribution of the differential charge carriers lifetime over energy levels in local regions of a LED die.

2. Results and discussion

The method consists in sequentially measuring the 3dB frequencies $f_{3\mathrm{dB}}$ distribution profiles of the electroluminescence over the LED die area on the installation presented in [6], when LED emission is passed through i-th optical filters with different central wavelengths of the transmission spectrum λ_{max_i} and with a bandwidth several times smaller than the width of emission spectrum.

A digital microscope based on a CMOS camera with a maximum resolution of 5472×3648 and a DG4162 pulse generator that sets the LED current is used to measure the profile of the radiation power distribution and the 3dB frequencies of electroluminescence modulation over the crystal area. The optical system of the microscope allows you to achieve an image magnification up to 1600x. At currents $I_1 = 20$ μA and $I_2 = 50$ μA, the profile of the distribution of the electroluminescence intensity was measured by storing images of the LED die. To eliminate the influence of non-linear effects, the maximum value of the image brightness should correspond to the linear section of the camera's brightness characteristic. This is achieved by changing the exposure time of the digital camera in the range of 30 ms–2 s. For each pixel of the image, the value of the parameter P_2/P_1 and the 3dB frequency were calculated. The paper [7] presents a technique for measuring the 3dB frequencies of electroluminescence modulation when the LED is supplied with a pulsed current with a duty cycle of 50%. The measuring installation has the following metrological characteristics: the measured frequency range is 1 kHz–40 MHz; the spatial resolution is 0.65 μm; the integration time is 2 ms–3600 s; the amplitude range of current pulses is 20 μA–50 mA; the 3dB frequencies measurement error is 2%.

The differential charge carriers lifetime τ was calculated according to formula [8]: $\tau = \sqrt{3}/(2\pi f_{3dB})$. The energy level is determined by the formula: $E_i[eV] = 1240/\lambda_{\max\_i}[nm]$.

The measurement method was tested on the example of Oasistek green commercial InGaN-based LEDs with a die size of 200 × 130 μm and a center wavelength of the radiation spectrum of 535 nm. The figure 1 shows the electroluminescenc spectrum of the LED and the transmission spectra of the filters.

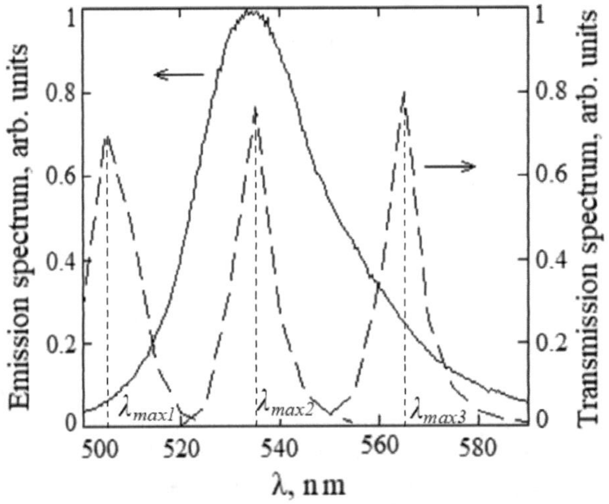

Figure 1. The electroluminescenc spectrum of the LED and the transmission spectra of the optical filters.

The distribution profiles of the 3dB frequencies were measured in turn by passing the LED emission through optical filters with the central wavelengths of the transmission spectrum of 505 nm, 535 nm, and 565 nm and the transmission spectrum width of 8 nm. Based on the results of measurements of the 3dB frequencies distribution profiles, the charge carriers lifetimes distribution profiles over the area of the die were calculated. Then, the die area was conventionally divided into local regions 5 × 5 μm in size. For each optical filter, the average lifetime τ in local regions was calculated. Figure 2 shows the distribution profiles of the differential charge carriers lifetime over the LED active region, measured by transmitting the LED emission through various optical filters.

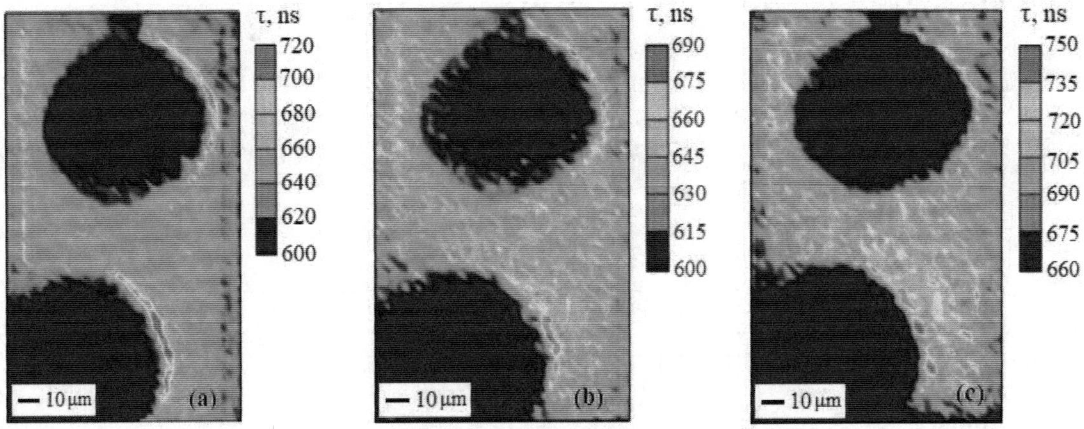

Figure 2. The distribution profiles of the differential charge carriers lifetime over the active region of the LED, measured with optical filters 505 nm (a), 535 nm (b), and 565 nm (c).

The histograms of the scattering of the number of the LED die local regions by the value of the differential charge carriers lifetime are shown in figure 3.

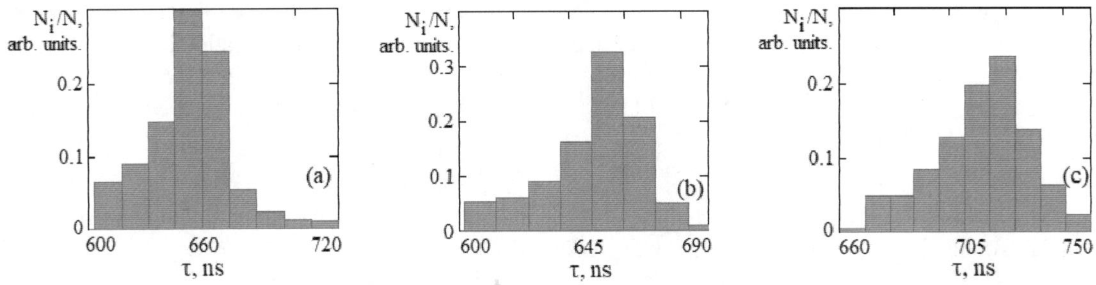

Figure 3. Histograms of the scattering of the number of the LED die local regions with dimensions of 5 × 5 μm by the value of the differential charge carriers lifetime measured with optical filters of 505 nm (a), 535 nm (b), and 565 nm (c).

It has been determined that with an increase in the energy level (with decreasing wavelength), the charge carriers lifetime decreases (figure 4).

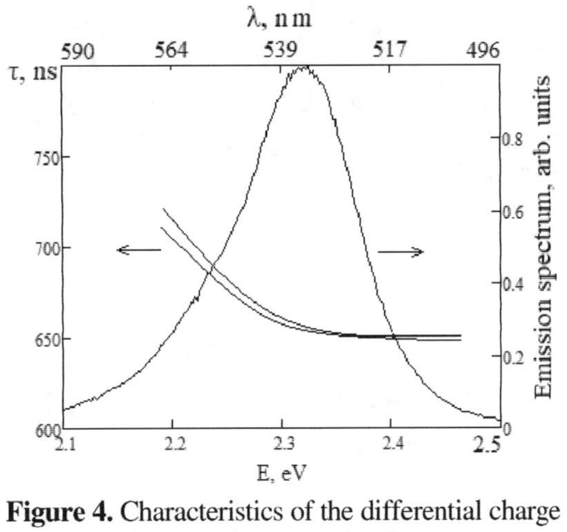

Figure 4. Characteristics of the differential charge carriers lifetime on the energy level.

In the energy range 2.3 – 2.5 eV (short-wavelength part of the emission spectrum), the $\tau(E)$ dependence is weakly expressed. The relative inhomogeneity of the distribution of the τ over the LED die, defined as the ratio of the standard deviation to the mean value, is about 1.7%. In the energy range 2.2 – 2.3 eV (long-wavelength part of the emission spectrum), the dependence $\tau(E)$ is significantly pronounced, and the relative inhomogeneity of the distribution of the τ over the LED die is about 2.7 % (Table 1).

Table 1. Parameters of the distribution of the LED die local areas by the value of the differential charge carriers lifetime.

	Filter center wavelength 505 nm	Filter center wavelength 535 nm	Filter center wavelength 565 nm
$\bar{\tau}(ns)$	645	648	709
$\sigma(ns)$	11	12	19
$\sigma / \bar{\tau}(\%)$	1.7	1.8	2.7

3. Conclusions

The method for measuring the distribution of the differential charge carriers lifetime over energy levels in the local region of a light-emitting heterostructure is developed. The method has been tested on commercial green InGaN-based LEDs. The obtained results indicate that in the investigated structure, the degree of heterogeneity of the distribution of local areas with a high indium concentration is higher than the degree of heterogeneity of the distribution of local areas with a low indium concentration. The presented measurement method can be used as a tool for diagnosing the lateral homogeneity of the composition of light-emitting heterostructures.

Acknowledgments

The work was carried out within the framework of the state task and partially was supported by the Russian Foundation for Basic Research, the project No. 19-07-00562 A.

References

[1] Tian P *et al* 2017 *Journal of Physics D: Applied Physics* **50(7)** 075101
[2] Gelžinytė K, Ivanov R, Marcinkevičius S, Zhao Y, Becerra D L, Nakamura S, DenBaars S P and Speck J S 2015 *Journal of Applied Physics* **117** 023111
[3] Peng Z, Lu Y, Gao Y, Chen G, Zheng J, Guo Z, Lin Y and Chen Z 2018 *IEEE Photonics Journal* **10** 8201908
[4] Manyakhin F I 2018 *Physics of Semiconductor Devices* **52** 359
[5] Frolov I V, Radaev O A and Sergeev V A 2020 *Journal of Physics: Conference Series* **1695** 012076
[6] Frolov I V, Sergeev V A and Radaev O A 2021 *Instruments and Experimental Techniques* **64(2)** 259
[7] Sergeev V A and Frolov I V 2020 *Patent Russia* 2725613
[8] Schubert E F 2006 *Light Emitting Diodes* (Cambridge University Press)

A comprehensive study of current-crowding effect in high power vertical AlInGaN LEDs under high pulsed current

A E Ivanov[1,2], A V Aladov[1], N ATalnishnikh[1], A E Chernyakov[1,2] and A L Zakgeim[1]

[1] Submicron Heterostructures for Microelectronics Research and Engineering Center RAS, St Petersburg 194021, Russia
[2] Saint-Petersburg Electrotechnical University "LETI", St. Petersburg 197376, Russia

a-e-ivano-v@yandex.ru

Abstract. The goal of the study is examination of current-crowding effect in high power AlInGaN LEDs. This effect was presented by mapping of EL (electroluminescence) near filed under high pulse current. LED chip of vertical design was study in high range of current ($10^{-9} \div 70$A). This operating mode of LEDs are interesting for different applications, such as pumping lasers, VLC and LiFi, as well as for investigation accelerated degradation process of LEDs.

1. Introduction

The scope of application of light-emitting diodes (LEDs) keeps expanding. The study of the operation of LEDs in pulse mode to exclude self-heating at high excitation levels is interesting for establishing the injection and recombination mechanisms and identifying the reasons that limit the energy capabilities of devices [1-4]. In this work were studied the vertical design LED chip in the regime of short pulses $\tau = 100$ ns at ultrahigh operating currents – up to 70 A.

In practice, the results of the study can be used to create special designs of AlInGaN visible-range LEDs for working with physical receivers, for example: in VLC (Visible Light Communication) systems, for pumping solid-state lasers [3]. New applications require the operation of LEDs in short pulse modes (from tens of nanoseconds to units of microseconds) when the maximum radiation power (energy per pulse) is reached.

2. Experimental

Commercial non-casted Enhanced Vertical LED chip EV-B40A produced by SemiLEDs [5] was chosen for our study. Chip peak emission wavelength $\lambda_{peak} = 460$ nm, the emission area of 1100x1100 μm^2 and a simple contact geometry Figure 1.

The optical parameters of the LEDs were measured in pulse regime. To ensure the predetermined pulse regime, an Agilent 8114A pulse generator was used with a PicoLAS LDP-V 80-100 V3.3 external amplifier. The pulse width and duty cycle were monitored with a Tektronix TDS3044B oscilloscope. Pulse duration was τ=100 ns and repetition rate was f = 50 Hz to avoid self-heating.

Near-field distribution of EL parameters was monitored by using Mitutoyo optical microscope with Avantes AvaSpec-2048. The minimum and maximum field of view of the optical system was 536×357 μm^2 and 5362×3520 μm^2, respectively, with the best spatial resolution of 25 μm.

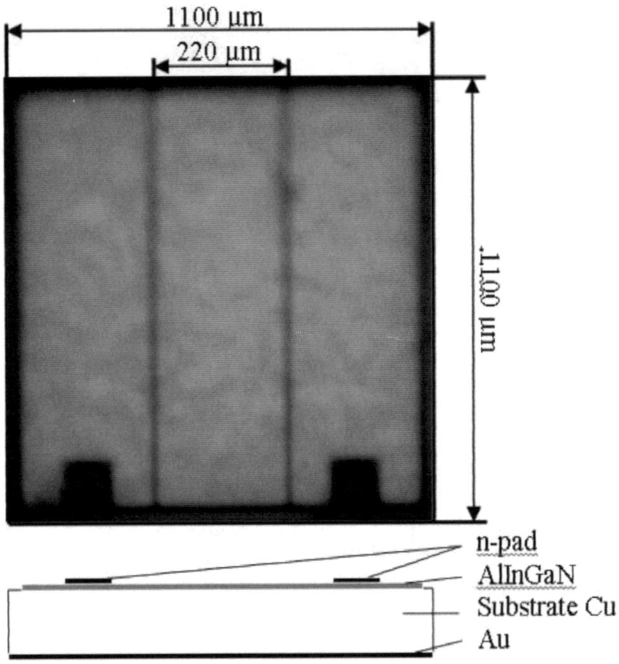

Figure 1. Self emission image and cross-section view of chip.

The optical system collected the light within the 15° cone around the normal direction.

The infrared (IR) thermal imaging was used to determine the surface temperature of the LED chip. It allows measurement of the temperature directly and thereby obtaining more detailed temperature data. The IR thermal radiation in the spectral range 2.5–3.0 µm was mapped by a specially designed IR microscope [6]. The main methodological problems with thermal imaging of AlInGaN structures are the transparency epitaxial layers for IR wavelengths and a large difference in the emissivity of the materials utilized in the LED, i.e. semiconductor layers, metallic electrodes, reflective coatings, mounting elements, etc. [7]. Therefore, extraction of the correct temperature distributions from the IR images requires preliminary calibration of data for every object. Such a calibration was made with the temperature control by means of an external heater in the range 5–100 ^{0} C via recording the IR radiation from the LED chip at zero current. Using the approach mentioned above it was possible to measure the temperature with the accuracy better than 2 K. The calibration procedure is time-consuming, moreover, leads to additional errors in obtaining of temperature maps, especially at points of low emissivity [8].

3. Results and discussion

The current dependences of the power and spectral characteristics of AlInGaN LEDs of vertical design, including their distribution (mapping) over the radiating surface, were studied in a wide range of operating currents up to ~70A. As can be seen from figure 2, the dependences of the λ_{peak} and the external efficiency (*EQE*) on the current have areas of sharp change (approximately up to 10 A) and then the tendency to saturation. We assume that the effect of current crowding under the contacts becomes the main factor in reducing the internal quantum and the radiation-extraction coefficient η_{ext}, limiting the energy capabilities of the LED. To clarify the current distribution, the dependence of the spectra on the coordinate on the chip surface at different operating currents was studied.

It is well known, that at low operating current there is a uniform distribution of LED emission, and the distribution of emission is proportional to the current density over the area of the p-n junction [9]. But at a current of 2A and above, the near field of EL confirms a significant effect of current crowding.

It should be noted that the light intensity distribution only qualitatively reflects the current density distribution. Due to the dependence of the *IQE* on the current, and partly due to the scattering of light inside the LED chip, light distribution is more smoothed compared to the current density distribution. To further refine the current distribution, as well as the temperature over the crystal area, we used spectral distribution mapping, shown at figures 3-4.

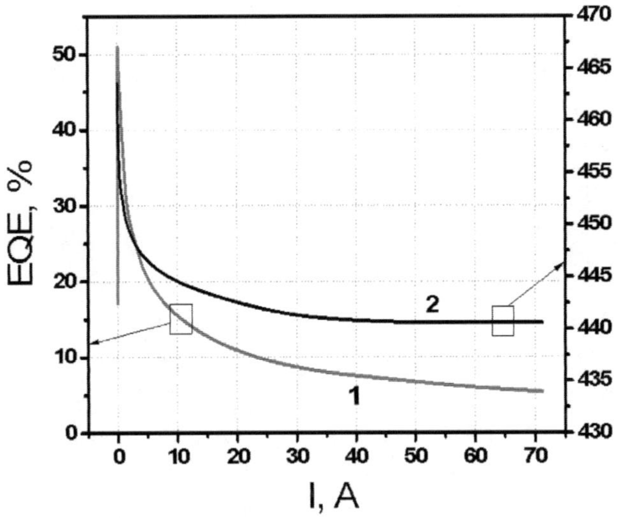

Figure 2. Measured EQE (1) λ_{peak} (2) dependencies on current.

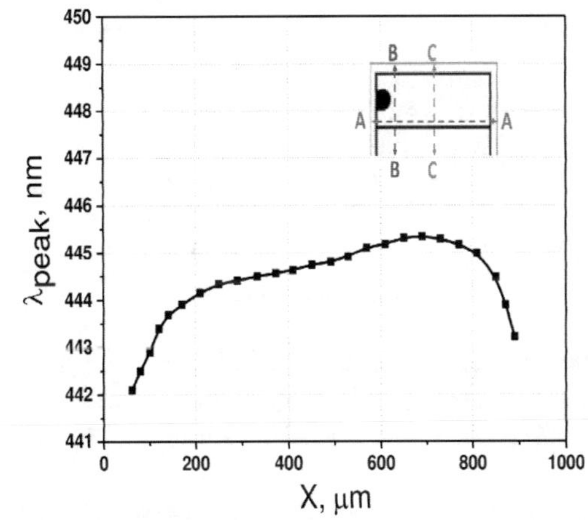

Figure 3. Distribution of λ_{peak} as a function of the coordinate in cross section A.

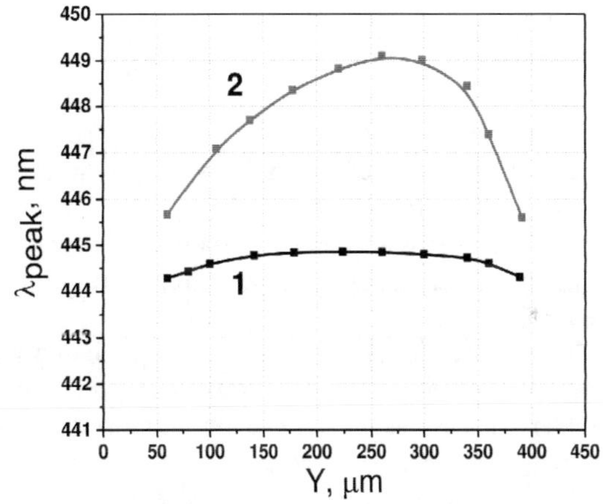

Figure 4. Distribution of λ_{peak} as a function of the coordinate in cross section B (1) and C (2).

Noticeable short-wave shift of the spectrum emitted near the left contact pad (x = 100) compared to the spectrum emitted near the center of chip (x = 600) λ_{peak} = (445 – 442) = 3 nm, is a direct confirmation of the difference in current densities. The gradient was observed both along and transverse the metal strip contacts. The gradient transverse the contacts (figure 4) is associated with poor current flow along the GaN p-layer, and the gradient along the contacts indicates a high resistance of the strip contacts.

Since another parameter – temperature can effect on the peak wavelength the heating of the chip was monitored using IR microscopy, shown at figure 5 The graph shows that there is no increase in the intensity of infrared radiation, even at the highest pulse current used. The slight mismatch of the curves is due to the influence of noise.

In addition, the analysis of the LED radiation spectrum showed no shift in the short-wave arm, which also indicates no increase ~~in~~ of the temperature of the active region.

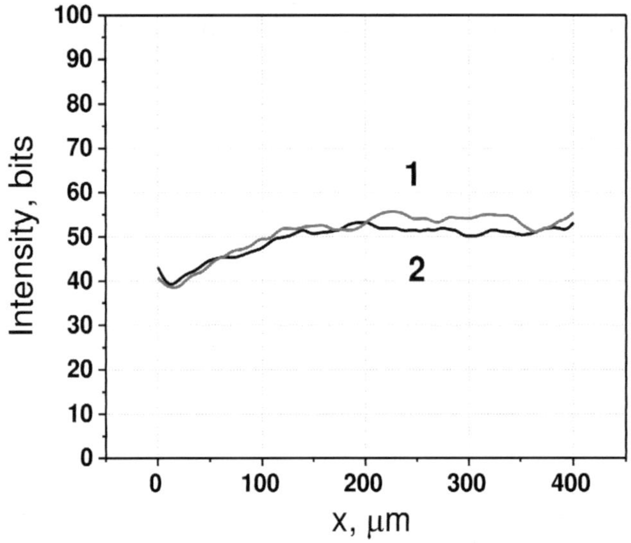

Figure 5. Profile of IR intensity in cross section C (insert in figure 3).

The study of the distribution pattern of the peak wavelength as a function of the coordinate on the chip allowed us to identify the presence of regions with different current densities. It is assumed that the addition to the research methodology of studying the dependence of full width at half maximum of the spectrum.

4. Conclusion

A vertical 1x1 mm^2 LED chip was studied in the current range $I = 10^{-9} \div 70$ A.

There is a sharp change in *EQE* and λ_{peak} in the current range up to 10A, then a saturation of the dependence of these parameters on the current. The explanation for this can be the uneven current flow, as a result of most of the current flows under the strip contacts and increases only non-radiative recombination.

The near-field mapping of the λ_{peak} showed noticeable gradients both across and along the metal strip contacts. Since the λ_{peak} is a current- sensitive parameter. Consequently the gradient of across the contacts is associated with poor current flow along the GaN n-layer, and the gradient along the contacts indicates a high resistance of the strip contacts.

To explain the type of dependence of the spectral parameters on the current, it is necessary to take into account the above-mentioned inhomogeneity of the current distribution.

The mapping of the position of the peak wavelength is the clearest indicator of the current density distribution, since λ_{peak} is not subject to mixing, in contrast to the radiation intensity.

Acknowledgments

This work was carried out at the Center of Multi-User Facilities «Element Base of Microwave Photonics and Nanoelectronics: Technology, Diagnostics, and Metrology».

References

[1] Yadav A, Titkov I E, Sokolovskii, Karpov S Y, Dudelev V V, Soboleva K K, Strassburg M, Pietzonka I, Lugauer H-J and Rafailov E U 2016 *Proc. SPIE* **9768** *LED:Materials, Devices & App. For Solid State Lightning XX p 1*

[2] Titkov I E, Karpov S Y, Yadav A, Zerova V L, Zulonas M, Galler B, Strassburg M, Pietzonka I, Lugauer H-J, Rafailov E U 2014 *IEEE J. of Quantum Electronics* **50** (11) 6910236 pp 911-20

[3] Aladov A V, Chernyakov A E, Ivanov A E, Zakgeim A L and Tsatsulnikov A F 2020 *ICLO p 1*

[4] Chernyakov A E, Zakgeim A L, Bulashevich K A and Karpov S Y 2015 *16th Int.Conf. on Thermal, Mechanical and Multi-Physics Simulation and Experiments in Microelectronics and Microsystems, EuroSimE 2015* **7103132** pp 1-5
[5] LEDs datasheet blue https://www.semileds.com/system/files/EV-B45A.pdf
[6] Zakgeim A L, Kuryshev G L, Mizerov M N, Polovinkin V G, Rozhansky I V and Chernyakov A E 2010 *Semiconductors* **44** 373-9
[7] Hopper R H, Haneef I, Ali S Z, Undrea F and Oxley C H 2010 *Meas. Science and Technol.* **21** (4) 1-6
[8] Chernyakov A E, Aladov A V, Ivanov A E and Zakgeim A L *2018 Conf. 24rd Int. Workshop on Thermal Investigations of ICs and Systems (THERMINIC)* pp 1-6
[9] Aladov A V, Kuptsov V, Chernyakov A E, Zakgeim A L and Valyukhov V P 2016 *Conf.: SPIE/COS Photonics Asia* **10021**

Optimization of the profile and material of wire contacts for an IR photodetector

A V Kamarchuk, D A Bauman and A I Marchenko

Saint-Petersburg National Research University of Information Technologies, Mechanics and Optics, Saint-Petersburg 197101, Russia

annakamarchuck@itmo.ru

Abstract. The study is devoted to the influence of the choice of geometry and materials of wire contacts on the reflection coefficient and thermal characteristics of the photodetector and the quality of the device design. The process of diffusion of materials of wire contacts and contact pads on a photodetector crystal is investigated. The studies were carried out on samples that are rather small in size (250x250x400 um). During the experiment, 4 main types of loop geometry were selected (main loop, reverse loop, double reverse loop, long loop). The loops were formed using a gold wire 25 μm in diameter. The quality of microwelds was investigated in 3 ways: shear and pull-off tests, optical observation using a scanning electron microscope (SEM), and contact resistance measurement. The aim of the work is to create a high-quality design of an IR photodetector, which allows achieving a high sensitivity (at least 0.5 A / W), a large dynamic range (at least 40 dB) and low indicators of dark current values. The developed technology ensures high quality of the photodetector design. Due to the low costs of this technological process (wire material, the number of operations required for installation), relative to other technologies, which allows maintaining high performance in the technical component of the photodetector, the installation method may be of practical interest in production.

1. Introduction

Microwave (MW) photonics is a rapidly developing area of high-speed communication systems with a spectral range of 1.3–1.55 μm. One of the key components of radiophotonics is high-speed photodetectors.[2] Many radio-photonic systems today use PIN-type photodiodes. The desirable characteristics of photodetectors are high sensitivity, wide dynamic range and large bandwidth, but in practice these parameters cannot be achieved simultaneously due to the physical foundations of the photodiode operation and design solutions. It should be noted that in terms of designs and evolving technologies that allow compromise solutions to achieve high bandwidth and high operating power, there are a large number of solutions. Optimization of technologies for creating a photodetector design is one of the main solutions to ensure high technical characteristics of the device.

2. Description of the model

For the study, 4 types of loop geometry were formed on a TPT HB16 semi-automatic microwelding unit. Micro-welds were formed using ultrasonic vibrations. The qualities of the "ball-wedge" microwelding were investigated. Figures 1, 2 show 2 types of formed geometric loops used to create a photodetector.

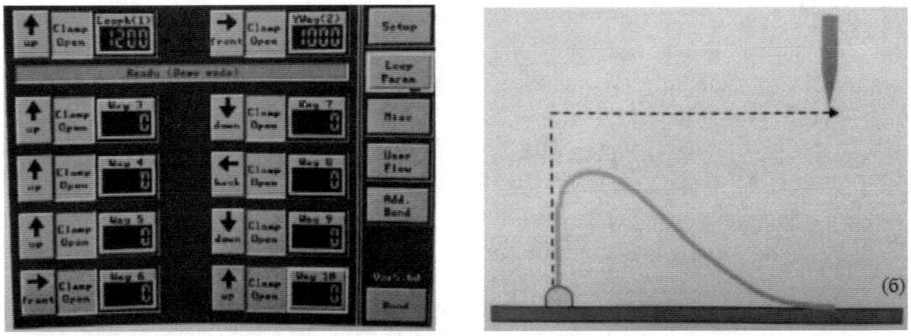

Figure 1. (a) – Loop menu formation parameters; (b) – diagram of the loop geometry and the direction of movement of the tool.

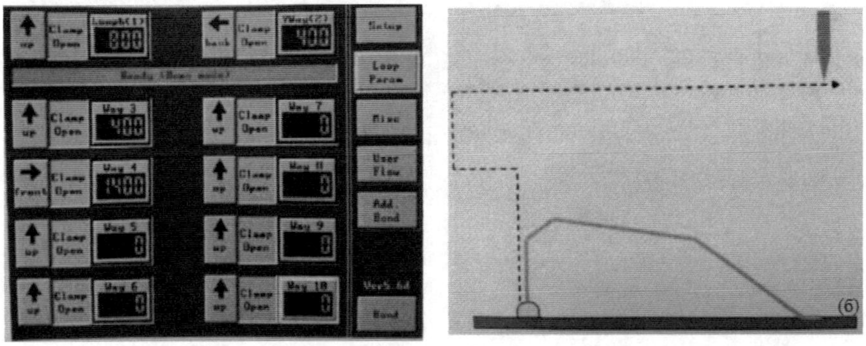

Figure 2. (a) – Reverse loop formation parameters, (b) - a diagram of the loop geometry and the direction of the tool movement.

Figure 3 shows a series of successful microweldings of 4 types of loop geometry.

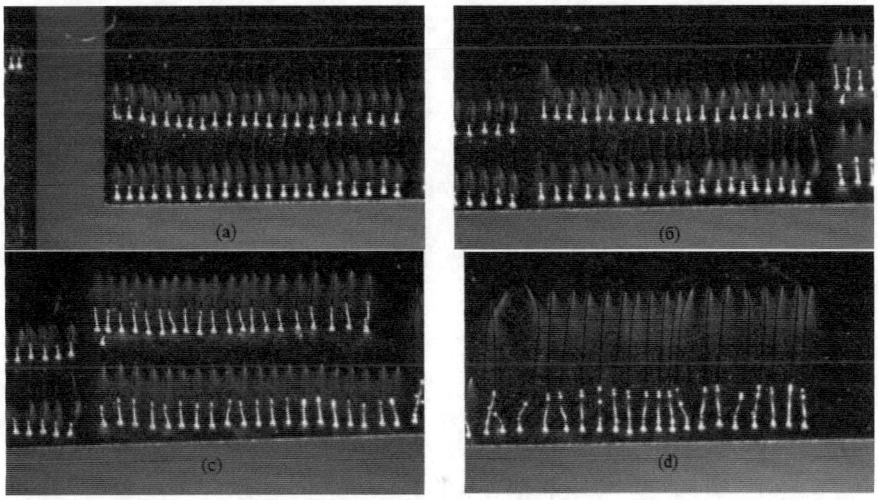

Figure 3. Series of microweldings with different loop geometries: (a) – base loop, (b) – reverse loop, (c) – double reverse loop, (d) – long loop.

3. Results and discussion

Three main research methods are used to analyze the quality of a structure: shear test, optical observation with a scanning electron microscope and contact resistance measurement.

The study according to the first method was carried out using a special tester, on which a force of 0.03 grams was set. The quality hinge profile describes the break at the center of the wire connection. As a result of testing, we obtained 80% of the loops with a wire break in the center.[1]

Figure 4 shows the compounds tested in SEM. An analysis of the diffusion process of the first and second points of microwelding was carried out.

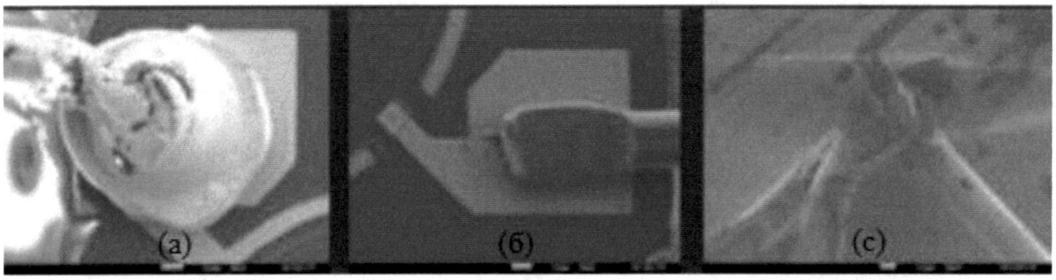

Figure 4. Examples of connections of two points of microwelding a ball and a wedge.

As a result of the research, it was revealed that a high-quality design solution for creating photodetector components in the IR range is the choice of the basic geometry of the loop, which slightly exceeds the dimensions of the photodetector crystal. Table 1 shows the parameters and test results of the practical part of the study.

Table 1. Parameters and test results of the practical part of the study.

Technological process number	First microwelding point			Second point of microwelding			Test result
	Ultrasound	Time	Clamping	Ultrasound	Time	Clamping	
1	180	180	200	180	200	200	separation of the wedge (second point of microwelding)
2	200	180	200	200	200	200	
3	220	180	220	200	200	200	
4	220	180	220	220	200	200	
5	220	180	220	220	200	220	separation of the wedge 50 %
6	220	180	220	220	200	240	break in the center of the wire connection

We considered the contribution of wire bonding, microstrip discontinuities and microstrip-coaxial junction for accurate modelling of reflection coefficient of the matched and biased PD chips. The layout of the PCBs was optimized to minimize reflection coefficient from the PD output. The critical parameters for high-frequency operation were wire-bonding length.

In figure 5 shows the reflection coefficient R from the MB. In the figure, we see a matched board-todie design with bandwidth up to 20 GHz.

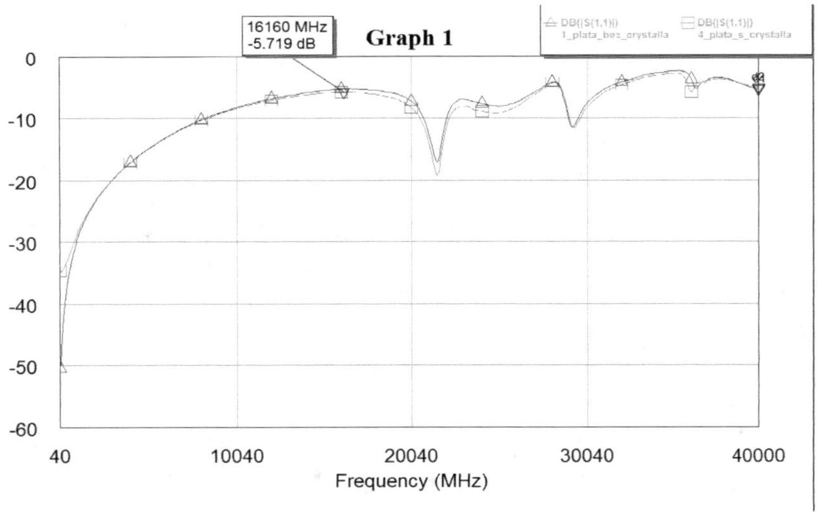

Figure 5. Reflectance value of a board without a crystal and a board with a crystal. After mounting the crystal on the board, the indicator changed by 0.5 dB. Changes up to 3 dB are allowed.

4. Conclusion

Thus, optimization of design solutions (formation of certain wire profiles and material selection), optimization of technological solutions in the process of mounting wire contacts made it possible to create a better design of a photodetector from a structural point of view (analysis of the diffusion process of materials, creation of a consistent design) and technical (achieving a reflection coefficient below -10 dB).

References

[1] Harman G Wire bonding in microelectronics. Third edition. McGraw Hill. 2010
[2] Kurushin A A Solution of multiphysics microwave problems using CAD COMSOL-M., One-Book, 2016. 376 pages.

Optimization of THz quantum cascade lasers with an active module based on two-quantum wells for high-temperature operation

R A Khabibullin[1], D S Ponomarev[1], D V Ushakov[2] and A A Afonenko[2]

[1] V.G. Mokerov Institute of Ultra High Frequency Semiconductor Electronics, Russian Academy of Sciences, Moscow 117105, Russia
[2] Belarusian State University, Minsk 220030, Belarus

khabibullin@isvch.ru

Abstract. Over the past two decades, the operation temperature of terahertz quantum cascade lasers (THz QCLs) has continuously increased from cryogenic level to the current record value of 250 K (about -23°C) [1]. Here we review the state-of-the-art and future prospects of high-temperature THz QCL designs with two-quantum wells in active module based on conventional heterojunction GaAs/AlGaAs and alternative material system HgCdTe. We have analyzed the temperature dependence of the peak gain and predicted the maximum operation temperatures of the given designs.

1. Introduction

Nowadays, there are many approaches to the generation of electromagnetic waves in the terahertz (THz) range, including down-conversion of optical femtosecond pulses [2], photomixing of two laser beams [3] and resonant-tunneling diodes [4,5]. Quantum cascade laser (QCL) is one of the spectrally brightest solid-state source of THz waves with a potentially wide range of practical applications in high-resolution spectroscopy [6,7] and imaging systems [8]. The main barrier to the use of THz QCL "outside the laboratory" is their low operating temperatures [9,10]. The limiting factors for increasing the operation temperatures of THz QCLs are associated with strong optical phonon scattering, the presence of parasitic current channels and the formation of electric field domains as was shown in [11,12]. In addition, the MBE growth of multilayered semiconductor structures for high-temperature THz QCL requires improvement [13,14]. Recently, the mode loss spectra for THz QCLs with double metal waveguide (DMW) were demonstrated in [15]. It was shown the high level of propagation loss of THz radiation in DMW, which exceeds 30 cm^{-1} for room temperature. Thus, to improve the high-temperature performance of THz QCLs it is needed to develop new concepts of active region designs and to reduce losses in DMW.

2. Optimization of the 2-well GaAs/AlGaAs active module

To calculate the THz QCL characteristics we have used the balance equation method with "tight-binding" wave-function basis. This basis has been obtained as superposition of eigenstates of the Schrödinger equation for the entire active region of a THz QCL by minimization of the spatial extension of wave functions of tunnel-coupled states. The localized ("tight-binding") basis is more stable for dephasing impact, as degenerate basis states with $\Delta E \lesssim 3$ meV have a small overlap of wavefunctions and, correspondingly, a low self-scattering rate.

We have analyzed the 2-well designs based on $GaAs/Al_xGa_{1-x}As$ and optimized the height of potential barriers with different Al content (see Table 1). In Figure 1 we demonstrate the operation principle of the given designs and the current flow through the energy levels, indicated by red arrows. It is easy to notice that there are several current channels between electronic levels, including the 3-2 transition between the upper laser and lower laser levels, as well as the parasitic transitions 3-1. In Figure 2, we compare the temperature dependence of the peak gain at 3.9 THz and show the ability to operate above 200 K for designs with Al content x>0.2. For comparison, we demonstrate the calculated "gain vs temperature" dependence for 3-well design proposed in [16] with maximum operation temperature T_{max} = 210,5 K.

Table 1. Parameters of optimized 2-well designs of THz QCL based on $GaAs/Al_xGa_{1-x}As$.

	x	Barrier height U_0, (meV)	Layer sequence in period: **potential barrier**/quantum well, (nm)
Bosco design [16]	0.25	212.5	**1.98**/16.37/**3.39**/7.91
Design A	0.20	178.5	**2.3**/16.4/**4.2**/7.3
Design B	0.25	212.5	**2.0**/17.2/**3.4**/7.9
Design C	0.30	244.5	**1.7**/17.0/**3.1**/7.9

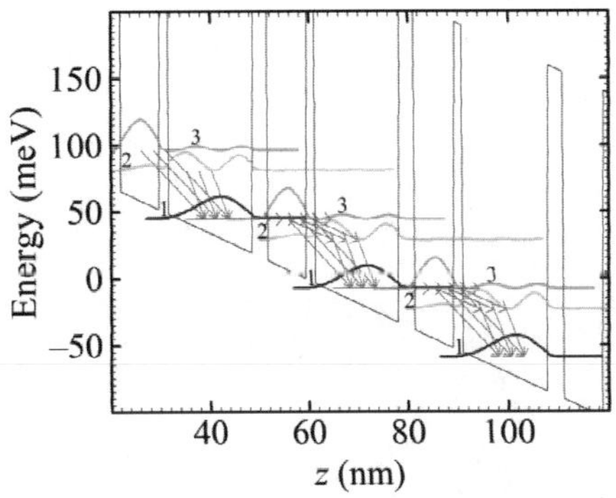

Figure 1. Conduction band diagram and squared modules of wavefunctions for $GaAs/Al_{0.3}Ga_{0.7}As$ design.

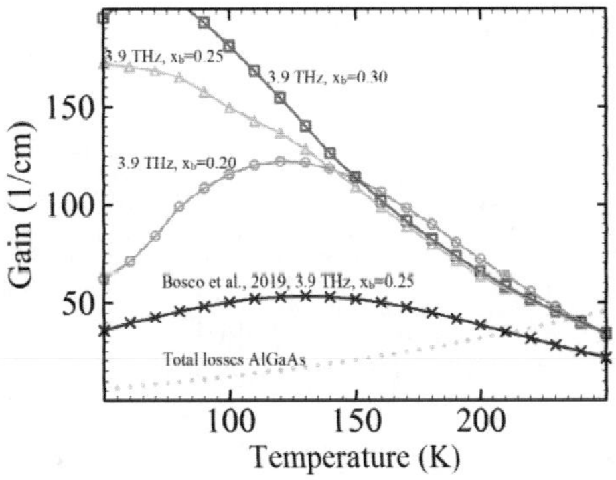

Figure 2. The temperature dependencies of peak gain are shown for GaAs/AlGaAs design with various Al content in barriers.

Optimization of designs with maximum gain at high temperatures (210–240 K) was obtained by scanning the active module thickness in the range from 24 to 36 nm. In this case, the thicknesses of the $Al_xGa_{1-x}As$ barrier layers varied in the range 1.4–4 nm, and GaAs quantum wells in the range 6.3–20 nm with a step equal to half the GaAs constant lattice. The parameters of the optimized active modules for various compositions of the barrier layers (Al contents) are presented in Table 1. In the layer sequence of active module, the GaAs quantum wells are shown in bold. The central part of the wide GaAs quantum well is Si-doped with a layer concentration of $4.5\ 10^{10}$ cm^{-2} as in [16].

The main factors affecting the gain in QCLs are: the electron effective mass m_c, the population inversion N_{32} of laser levels and matrix elements of dipole transitions Z_{32}. Thus, the gain of QCL active module can be represented as $G \sim m_c^{-3/2}\ N_{32}\ Z_{32}^2$. It means that the effective QCL amplification requires a low electron effective mass and a large population inversion and dipole matrix elements of the transitions. The temperature dependences of these values for the optimized active modules in comparison with the Bosco design [16] are shown in Figure 3. It should be noted that the electron effective mass increases in designs with higher potential barriers (with higher Al content in $Al_xGa_{1-x}As$) due to the non-parabolicity effect as shown in Figure 3a. On the other hand, the m_c decreases at elevated temperatures for all designs. In the optimized designs B and C (Figure 3b), the population inverted is higher at temperatures above 200 K. As can be seen from Figure 3c, the matrix elements of dipole transitions of the given designs increase with an increase in temperature from 50 to 250 K: from 4.9 to 5.2 (Bosco 2019), from 4.7 to 5.0 (design A), from 4.4 to 4.7 (design B), and from 4.7 to 5.0 nm (design C). Thus, the combination of all 3 factors for all optimized designs A (x_b = 0.20), B (x_b = 0.25), and C (x_b = 0.30) gives the same predicted maximum operating temperature of 238 K.

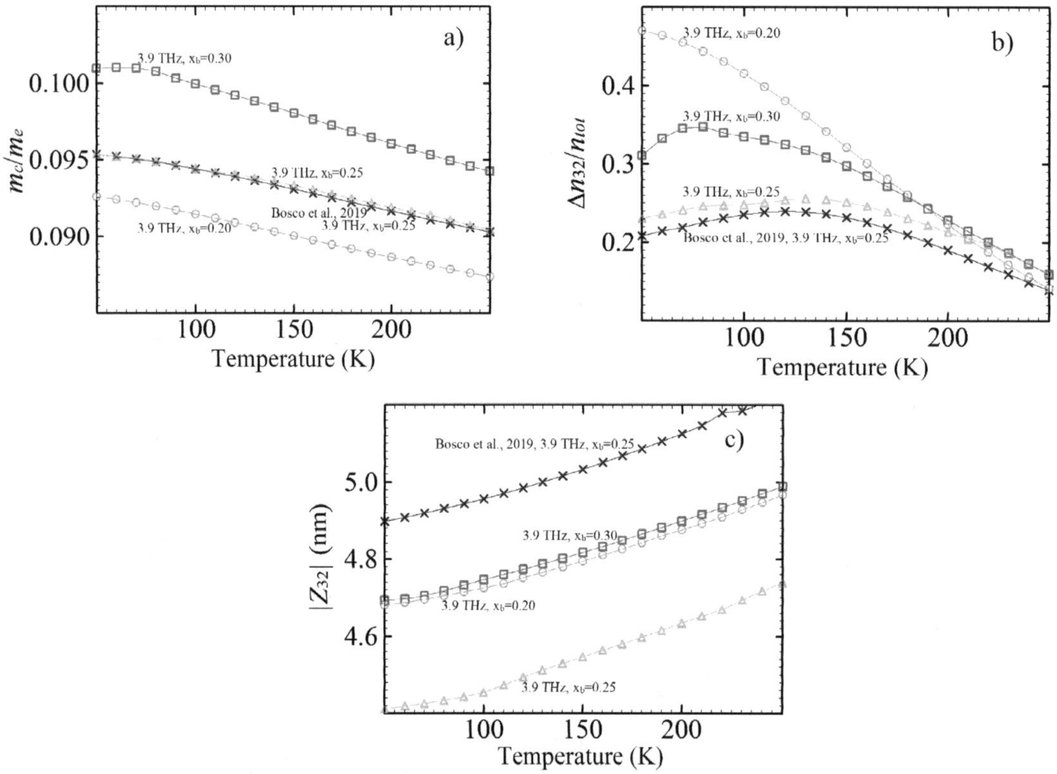

Figure 3. The temperature dependencies of a) electron effective mass m_c; b) population inversion Δn_{32} and c) matrix element of dipole transitions Z_{32} for optimized designs A (x_b = 0.20), B (x_b = 0.25), and C (x_b = 0.30).

3. Optimization of the 2-well HgCdTe active module

We simulate THz QCLs on the basis of narrow-gap HgCdTe quantum wells and calculate its characteristics. The characteristics of a single period, including current through localized states, leakage currents to continuum states, gain spectra and losses, were found by solving a system of balanced equations. We have proposed the 3-well and 2-well HgCdTe designs with maximum operation temperatures of 170 K and 225 K, respectively (see Figure 4) [17].

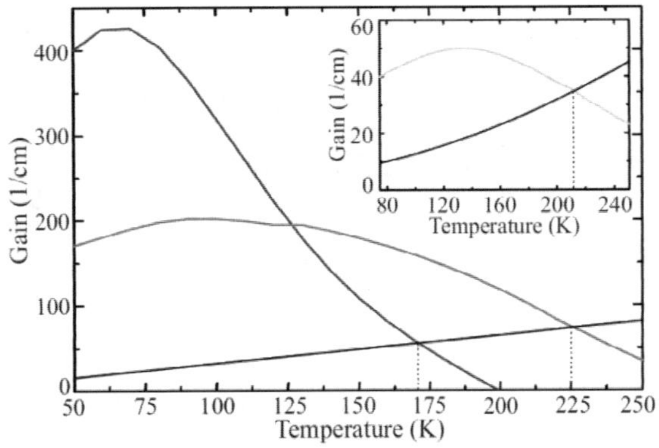

Figure 4. The temperature dependencies of peak gain are shown for the HgCdTe designs based on 3-well (blue line) and 2-well (red line), and for the GaAs/AlGaAs design based on 2-well of Bosco et al. [16] (green line in the inset). Cavity losses of a Cu-Cu waveguide is indicated by the black curves.

Acknowledgments
This work was supported by the Russian Science Foundation Grant No. 21-72-30020.

References
[1] Khalatpour A, Paulsen A K, Deimert C, Wasilewski Z R and Hu Q 2020 *Nat. Photonics* **15** 16
[2] Ilyakov I E, Shishkin B V, Malevich V L, Ponomarev D S, Galiev R R, Pavlov A Yu, Yachmenev A E, Kovalev S P, Chen M, Akhmedzhanov R A and Khabibullin R A 2021 *Opt. Lett.* **46** 3360
[3] Khabibullin R A, Morozov O G, Sakhabutdinov A Z and Nureev I I 2017 *Systems of Signal Synchronization, Generating and Processing in Telecommunications (SINKHROINFO)* 17080288
[4] Rasulova G K, Pentin I V, Vakhtomin Yu B, Smirnov K V, Khabibullin R A, Klimov E A, Klochkov A N and Goltsman G N 2020 *J. Appl. Phys.* **128** 224303
[5] Sobolev A S, Zaitsev-Zotov S V, Maytama M V, Klimov E A, Pavlov A Y, Ponomarev D S and Khabibullin R A 2021 *Opt. Eng.* **60** 082018
[6] Hubers H-W, Richter H and Wienold M 2019 *J. Appl. Phys.* **125** 151401
[7] Volkov O, Pavlovskiy V, Gundareva I, Khabibullin R and Divin Y 2021 *IEEE Trans. Terahertz Sci. Technol.* **11** 330
[8] Sterczewski L A, Westberg J, Yang Y, Burghoff D, Reno J, Hu Q and Wysocki G 2019 *Optica* **6** 766
[9] Khabibullin R A *et al* 2018 *Semiconductors* **52** 1380
[10] Volkov O Yu, Dyuzhikov I N, Logunov M V, Nikitov S A, Pavlovskii V V, Shchavruk N V, Pavlov A Yu and Khabibullin R A 2018 *J. Commun. Technol. Electron.* **63** 1042
[11] Khabibullin R A, Shchavruk N V, Ponomarev D S, Ushakov D V, Afonenko A A, Maremyanin K V, Volkov O Yu, Pavlopskiy V V and Dubinov A A 2019 *Opto-Electronics Review* **27** 329.
[12] Ushakov D V, Afonenko A A, Dubinov A A, Gavrilenko V I, Volkov O Yu, Shchavruk N V, Ponomarev D S and Khabibullin R A 2019 *Quantum Electronics* **49** 913
[13] Cirlin G E, Reznik R R, Zhukov A E, Khabibullin R A, Maremyanin K V, Gavrilenko V I and Morozov S V 2020 *Semiconductors* **54** 1092
[14] Reznik R R, Kryzhanovskaya N V, Zubov F I, Zhukov A E, Khabibullin R A, Morozov S V and Cirlin G E 2017 *J. Phys.: Conf. Ser.* **917** 052012
[15] Ushakov D V, Afonenko A A, Dubinov A A, Gavrilenko V I, Vasil'evskii I S, Shchavruk N V, Ponomarev D S and Khabibullin R A 2018 *Quantum Electronics* **48** 1005
[16] Bosco L, Franckie M, Scalari G, Beck M, Wacker A and Faist J 2019 *Appl. Phys. Lett.* **115** 010601
[17] Ushakov D, Afonenko A, Khabibullin R, Ponomarev D, Aleshkin V, Morozov S and Dubinov A 2020 *Opt. Express* **28** 25371

Computer simulations of solar cells based on silicon/boron phosphide selective contacts

S Y Kiyanitsyn[1] **and A S Gudovskih**[1,2]

[1] Department of Photonics, St. Petersburg Electrotechnical University ("LETI"), 197376 St. Petersburg, Russia
[2] St. Petersburg Academic University of RAS, 194021 St. Petersburg, Russia

serg-kianitsyn@mail.ru

Abstract. Silicon solar cells with selective contacts based on boron phosphide (BP) demonstrate a high potential according to simulation. However, the influence of defects created at the BP/Si interface during BP deposition is a critical issue for solar cell performance. The computer simulations were performed to understand how the defects in the near-surface region and at the interface affect the photovoltaic properties. Calculations of the dependence of the characteristics of solar cells on parameters such as the density of interface states, the concentration of defects in the near-surface region, and its width were made.

1. Introduction

Currently, an intensive search is underway for ways to improve the efficiency of Si solar cells being the main material for solar energy. The theoretical limit of the conversion efficiency of Si solar cells is 29.1% [1], but the real efficiency of such solar cell is below 26%. One of the most promising ways to increase the efficiency is the use of selective contacts to Si, which should provide selective charge carriers transport, a low level of recombination at the interface, and have a high optical transparency for solar radiation. From this point of view, a completely new hole-selective contact based on boron phosphide Si/BP may be of great interest [2]. BP has tremendous potential because it is an indirect gap $E_g \sim 2.1$ eV and has a negative (-0.3 ± 0.1 eV) valence band offset ΔE_V at the BP/Si interface. In addition, this material can be formed at low temperatures using plasma deposition, which is attractive in terms of mass production. However, the use of plasma can potentially lead to the formation of defects both at the BP/Si interface and in the near-surface region of Si. In this work, computer simulation study of the dependence of the characteristics of solar cells on parameters such as the density of surface states, the concentration of defects in the near-surface region and its width was carried out.

2. Experiments and results

The AFORS-HET v2.5 software package was used to calculate the characteristics of solar cells based on the p-BP/n-Si heterojunction. In this work, the element based on p-BP/n-Si heterojunction is investigated. **Figure 1 (a)** shows the structure of solar cell. Initial data for BP were taken from the literature [3-5] and summarized in Table 2.1. Due to the lack of data on the effective density of states typical values for GaP were used [6].

Table 1. Input data for selective contact BP.

dk	χ	E_g	N_c	N_v	μ_n	μ_l	N_a	v_n	v_p	D
-	(eV)		(cm^{-3})		$(cm^{-2}\cdot(V\cdot s)^{-1})$		(cm^{-3})	$(cm\cdot s^{-1})$		(cm)
11,1	2,8	2,1	$2.84\cdot10^{19}$	$2.68\cdot10^{19}$	25	25	10^{19}	10^{7}	10^{7}	$5\cdot10^{-6}$

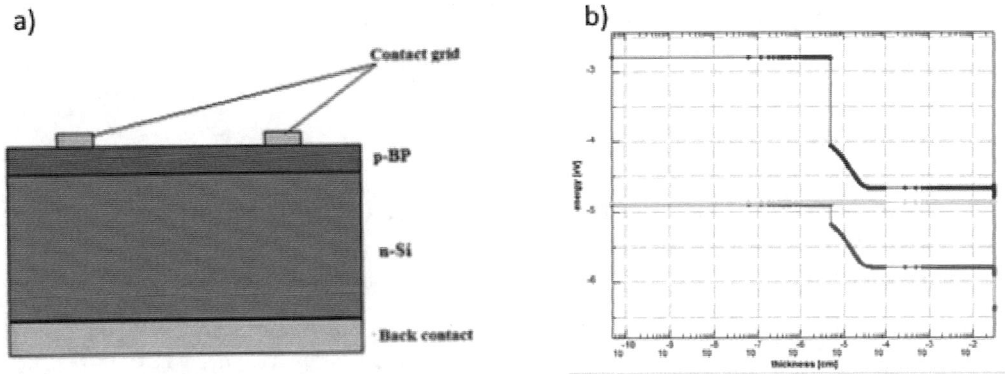

Figure 1. Structure of a solar cell based on a p-BP / n-Si heterojunction **(a)** and band diagram **(b).**

At the first stage, we studied the effect of the density of surface states (D_{it}) at the BP/Si interface on the characteristics of a solar cell. The D_{it} value was varied in the range of 10^9 - 10^{13} cm^{-2} eV. The calculation results are shown in **Figure 2**.

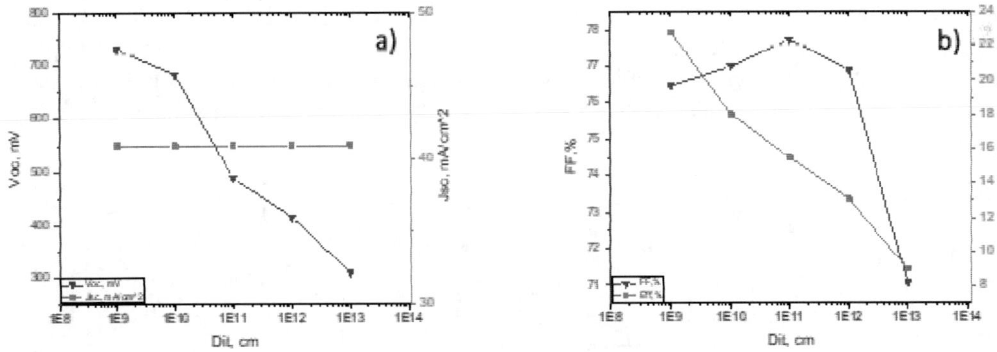

Figure 2. Dependences of Voc, Jsc **(a)**, FF and efficiency **(b)** on the density of interface states (D_{it}) *at the BP/Si interface.*

It can be seen that the open circuit voltage (V_{oc}) strongly reduces with increasing D_{it}, while the short circuit current (Jsc) remains practically constant. As a consequence, there is a drop of efficiency with an increase in D_{it}. Fill factor is weakly dependent on D_{it}.

However, it is known that the sensitivity to interface states can be significantly influenced by the value of the band discontinuity at the interface [7]. Since there is a fairly strong spreading in the band offset values published for the BP/Si interface [5], a series of calculations were performed with a

 2086 (2021) 012087 doi:10.1088/1742-6596/2086/1/012087

variation of this parameter in a wide range from **–0.9** eV to **0.7** eV. D_{it} was taken equal to 10^{11} $cm^{-2}\cdot eV^{-1}$ and similar dependences were plotted. The research results are shown in **Figure 3.**

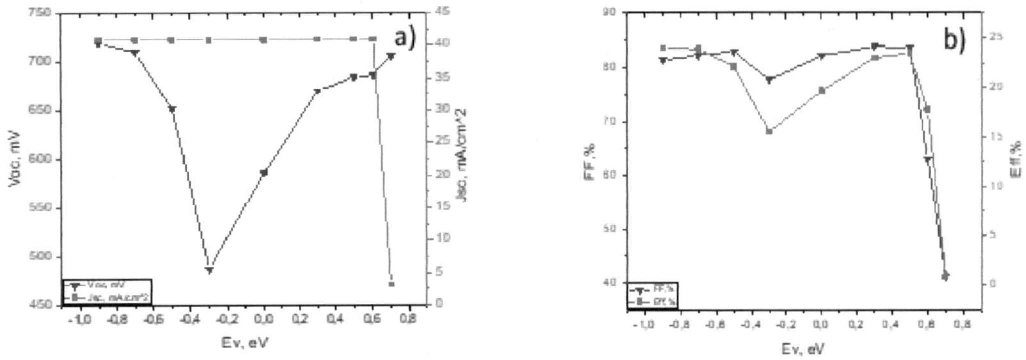

Figure 3. Dependences of Voc, Jsc (a), FF and efficiency (b) on the discontinuity of the valence bands ΔE_v for $D_{it} = 10^{11}$ $см^{-2}$ $эВ^{-1}$.

The valence band offset ΔEv significantly affects the sensitivity of Voc to interface states. As shown in Figure 3, the dependence of Voc on ΔEv has a minimum at ΔEv = -0.3 eV. At this value of ΔEv, the surface states have the most significant effect. With increasing ΔEv the value of Voc is less dependent on Dit, which is related to the increasing band bending in Si (**Figure 4**). The recombination level in this case is determined by the electron concentration, which decreases with increasing ΔEv. It should be noted that for values of ΔEv >0.55 eV, a decrease in FF is observed (**Figure 3**), which is associated with an increase in the potential barrier preventing the transport of holes from n-Si to BP (**Figure 4**). At ΔEv =0.7 eV this barrier leads to a sharp drop of the short-circuit current.

However, the sensitivity to D_{it} also decreases with a significant increase in the negative band gap (ΔEv <-0.3 eV). This effect is due to the fact that at such large negative values of ΔEv the bands, on the contrary, are almost flat and the recombination level is determined by the hole concentration, which decreases with an increase in the negative value of ΔEv. This effect is of practical interest due to reports in the literature of experimentally determined negative ΔEv for BP/Si. Thus, to predict the sensitivity of p-BP/n-Si-based solar cells performance to D_{it}, the value of ΔEv must be determined experimentally.

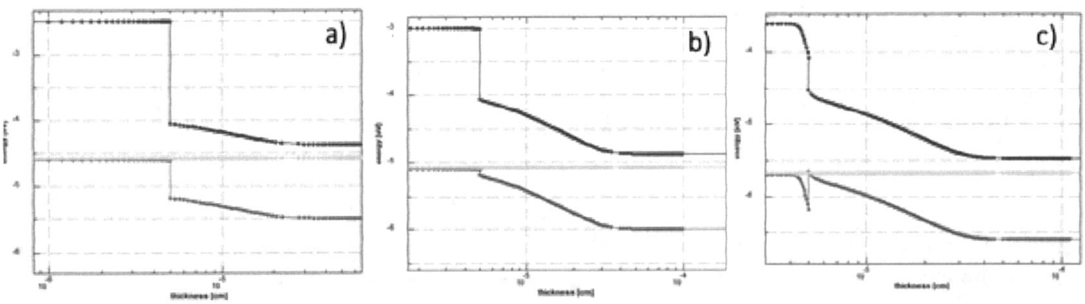

Figure 4. Band diagrams of a solar cell based on p-BP/n-Si heterojunction for ΔEv equal to -0.5 eV **(a)** 0.05 eV **(b)** and 0.5 eV **(c)**.

At the second stage, we studied the effect of defects in the near-surface region of Si. Defects in the near-surface region of silicon can be created during plasma deposition of BP. To study the effect of defect formation in the near-surface region on the characteristics of solar cells, a defective Si layer between BP and Si substrate was introduced into the model. The simulation was performed with two variable parameters: the effective lifetime of the charge carriers in the defect layer (τ) and the thickness of the defect layer (**d**).

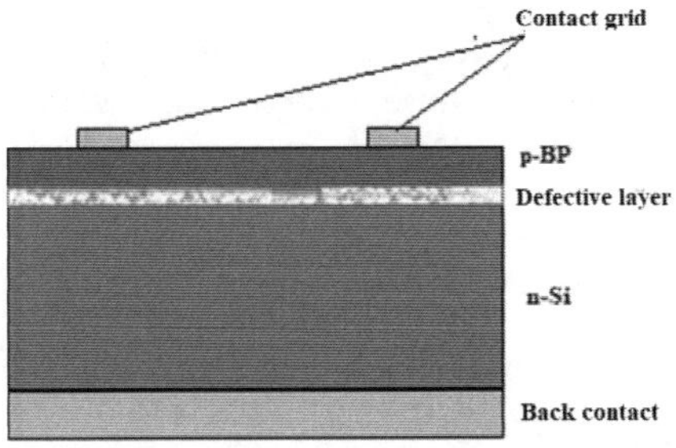

Figure 5. Introduction of a defective layer into the structure of a solar cell based on p-BP/n-Si heterojunction.

A model with a defect layer thickness (d) equal to 100 nm was used to study the influence of lifetime. The obtained dependences of short circuit current, open circuit voltage, fill factor and efficiency are shown in Figure 6. With increasing concentration of defects resulting in decrease of charge carrier lifetime, a decrease of open circuit voltage (Voc) is observed, but the current remains almost unchanged. Due to the drop in voltage, the efficiency of the solar cell decreases also.

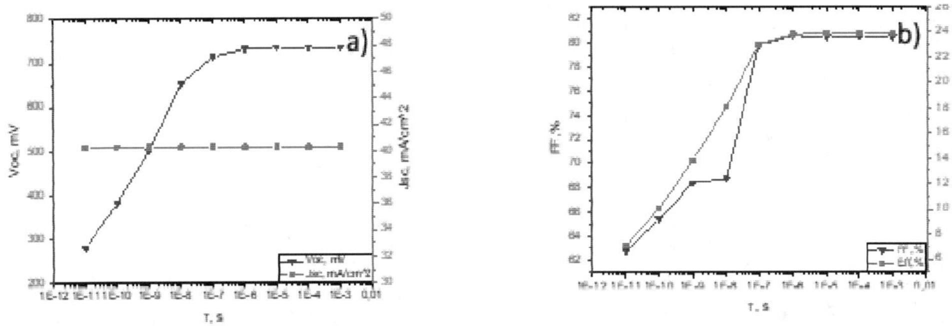

Figure 6. Dependences of Voc, Jsc (a), FF, efficiency (b) on lifetime of charge carriers in the near-surface region at d=100 nm.

The dependence of the defect layer thickness with a lifetime (τ) equal to 10^{-8} s was also investigated. The results are shown in (**Figure 7**). It can be seen that the extension of the defect region near the BP/Si interface also significantly affects the Voc. An increase of the defect region thickness more than 10 nm leads to a drastic drop of Voc.

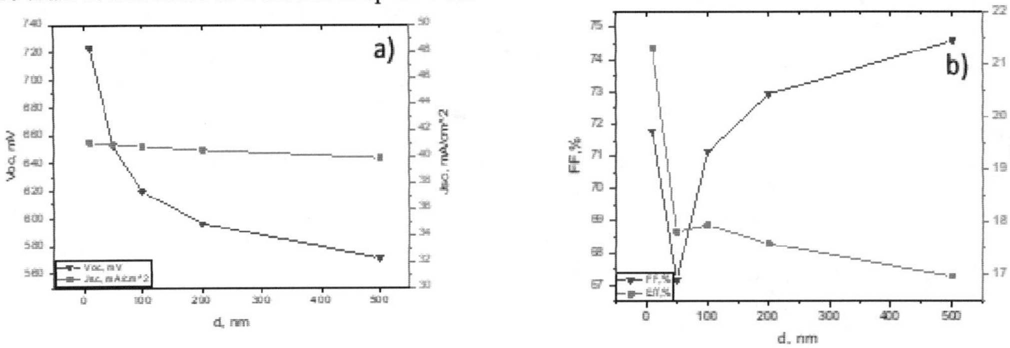

Figure 7. Dependences of Voc, Jsc (a), FF and efficiency (b) on defect layer thickness with $\tau = 10^{-8}$ s.

3. Conclusions

P-type boron phosphide can be very effectively used as hole-selective contacts for n-Si heterojunction solar cells due to the fact that it is a wide-gap material (2.1 eV). Defect states at the BP/Si interface are shown to have a strong influence on the open circuit voltage. The sensitivity to interface states strongly depends on the value of the valence band offset at the BP/Si interface. The study of the influence of defect formation in the near-surface region of Si showed that at a defect depth of 100 nm the decrease in efficiency is observed only when the carrier lifetime in this region is less than 10^{-7} s. If the lifetime of charge carriers in the defect region decreases to 10^{-8} s, the depth of the defect layer should not exceed 10 nm in order the defects do not affect the solar cell efficiency.

Acknowledgments

This work was supported by Ministry of Science and Higher Education of the Russian Federation (research project 0791-2020-0004).

References

[1] Yoshikawa K *et al* 2017 *Nature Energy* **2** 17032
[2] Gudovskikh A S, Kudryashov D A, Baranov A I ,Uvarov A V and Morozov I A 2021 *Tech. Phys. Lett.* **47** 96–98.
[3] Basic Parameters of Gallium Phosphide (GaP) 2021 (Matprop); http://www.matprop.ru/GaP_basic
[4] Dalui S, Hussain S, Varma S, Paramanik D and Pal A K 2008 *Thin Solid Films* **516** 4958–65
[5] Neuberger M 1971 III-V Semiconducting Compounds pp 32-34
[6] Varley J B, Miglio A Ha V-A, Setten M J, Rignanese G-M and Hautier G 2017 *Chemistry of Materials* **29** 2568–73.
[7] Kleider J P, Alvarez J, Ankudinov and A V Gudovskikh A S 2011 *Nanoscale Res Lett* **6** 152.

Optimization of (p)a-Si:H/p-Si ohmic contact for solar cells

D A Kudryashov[1], A I Baranov[1], A V Uvarov[1], I A Morozov[1], A O Monastyrenko and A S Gudovskikh[1,2]

[1] Alferov University, St. Petersburg 194021, Russia
[2] Department of Photonics, Saint Petersburg Electrotechnical University "LETI", St. Petersburg, 197376 Russia

kudryashovda@spbau.ru

Abstract. In this work last results on optimization of (p)a-Si:H/p-Si ohmic contact made by PECVD method are presented. A strong effect of growth temperature and trimethylboron flow on charge carriers transport was demonstrated. An optimized (p)a-Si:H layer also was successfully applied as an emitter layer to a heterojunction solar cell fabrication where a measured open circuit voltage of 0.65 V was obtained.

1. Introduction

Nowadays solar cells have become an important part of our life. We can find them everywhere starting from portative charge devices to large solar plants which collect and transform solar energy to electricity. It is all because of well developed technology and low price of raw materials such as silicon. The highest efficiency for the silicon based solar cells of 26% was reached [1]. Further efficiency increase may be realized with an optimization of individual layers and contacts. For example, there are still lack of data related to an effective (p)a-Si:H/p-Si ohmic contact used for solar cells fabrication by PECVD method [2]. The point is that trimethylboron (TMB) used as a doping component for p-type a-Si:H formation is a very sensitive to process parameters which are individual for different deposition systems [3]. Boron concentration in amorphous silicon depends on plasma power, substrate temperature, pressure and other parameters. This papers presents last results on process parameters optimizations for a (p)a-Si:H/p-Si ohmic contact formation.

2. Experimental detail

CZ p-type silicon wafers (100) with a resistivity of 1-5 Ohm·cm were used as a substrate. Prior the deposition silicon wafer was immersed in HF:H_2O solution to remove native oxide. Amorphous silicon layers were deposited by Oxford plasmalab 100 PECVD system at a power of 5 W and a temperature of 250 °C. Silane, TMB and hydrogen were used as source gases. After the deposition silver contacts were formed by thermal deposition in TLM geometry. Also (p)a-Si:H layers in the same conditions were deposited on quartz substrates for conductivity and activation energy measurements. To check the possibility of usage (p)a-Si:H layer as an emitter layer a heterojunction a-Si:H solar cell was fabricated where n-type CZ silicon wafer with a doping level of 10^{16} cm^{-3} was used. Quantum efficiency spectra were measured using a Solar Laser M266 monochromator. Current-voltage characteristics under AM1.5G simulator (Abet Technology SunLite) were measured using a Keithley 2400 source meter.

3. Results

A series of (p)a-Si:H films with a variation of deposition temperature was grown. TMB flow of 50 sccm was fixed. All the I-V curves measured demonstrate a linear behavior in a voltage range from -10 to 10 V. Figure 1 shows a temperature dependence of current at a voltage value of 10 V for (p)a-Si:H layer deposited on quartz substrate at temperature of 250 °C and TMB flow of 50 sccm. From the slope activation energy of 0.42 eV was calculated. Based on contact geometry a dark conductivity of $5.96 \cdot 10^{-6}$ Sm/cm^2 was also calculated.

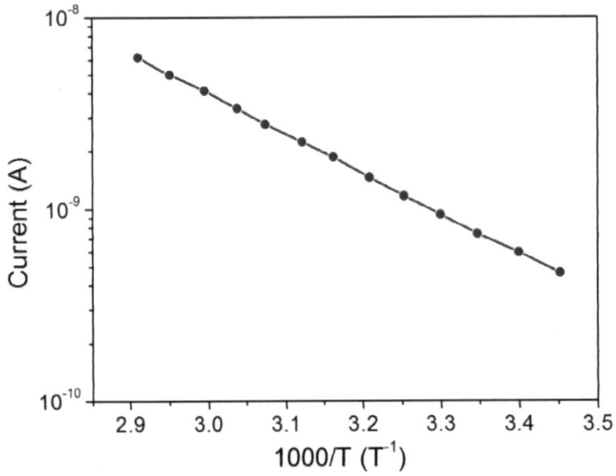

Figure 1. Current values measured at a voltage of 10 V vs temperature for (p)a-Si:H thin film grown at TMB flow of 50 sccm.

The same manner activation energies for (p)a-Si:H thin films deposited at temperatures of 160 and 200 °C were calculated (figure 2). It is clearly seen that a temperature increase leads to a layer conductivity raise. Activation energy shows the opposite behaviour. Both characteristics point on that a deposition temperature of 250 °C is more preferable for (p)a-Si:H growth.

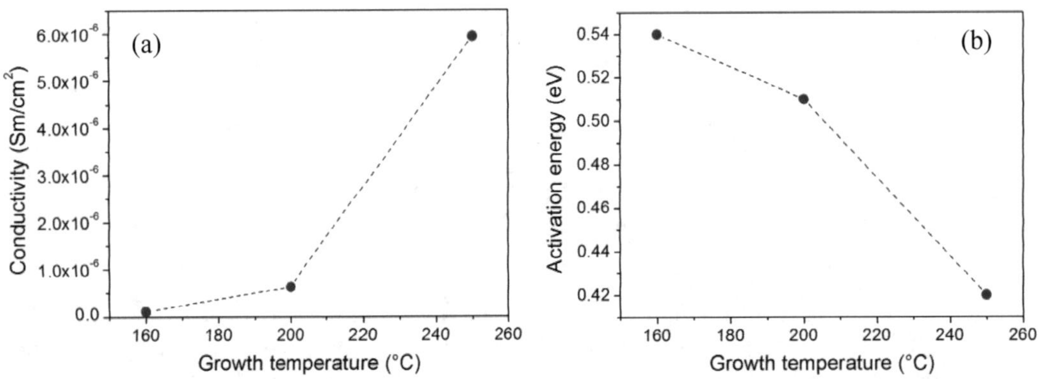

Figure 2. Conductivity (a) and activation energy (b) calculated for (p)a-Si:H grown at TMB flow of 50 sccm.

Next, an activation energy for (p)a-Si:H layers grown at different TMB flows was calculated (figure 3). Optimal activation energy of 0.42 eV at a TMB flow of 50 sccm was found.

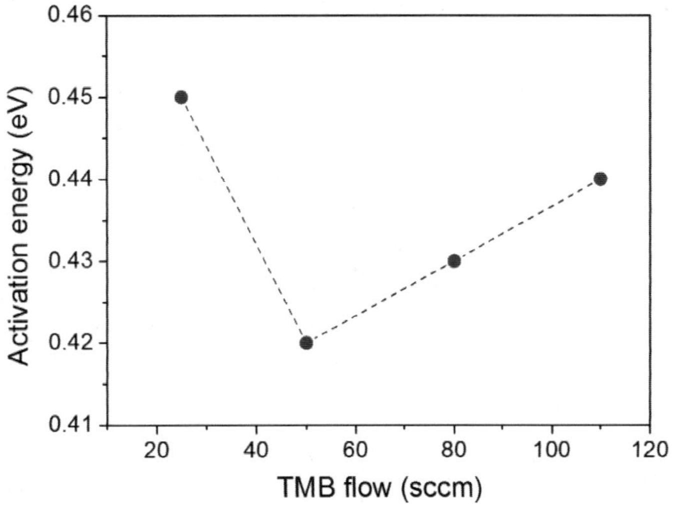

Figure 3. An activation energy for (p)a-Si:H films grown at a temperature of 250 °C and different TMB flows.

Figure 4a shows I-V curves for (p)a-Si:H layers formed on p-type Si wafer with different TMB flows at a fixed SiH_4 flow. It is clearly seen a huge difference in current values. Ultimately such low current values will restrict solar cells characteristics. Figure 4b shows an extremum at TMB concentration of 50 sccm. The values less the TMB concentration lead to a constant current values of approximately 1 mA at 1 V. An increase of TMB more than 50 sccm leads to a sharp current drop. The same dependence was observed for (p)a-Si:H films deposited on quartz substrates.

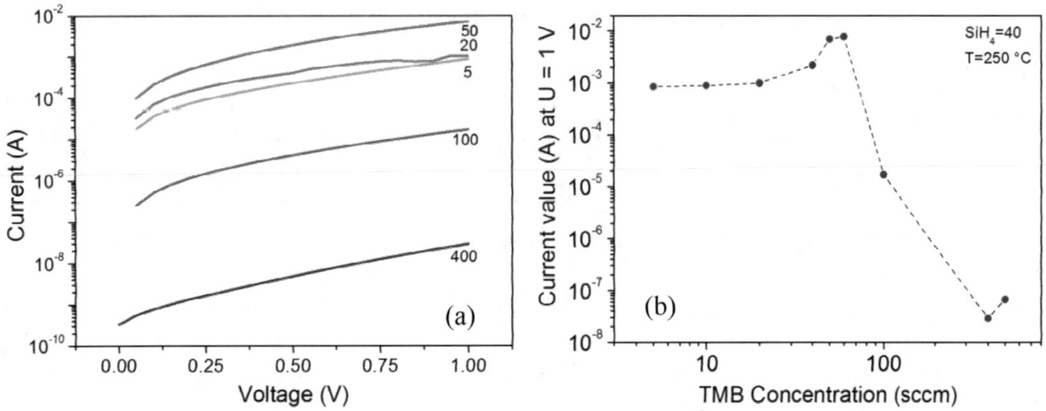

Figure 4. I-V curves for (p)a-Si:H/p-Si contact formed at fixed silane flow and different TMB flow (a) and extracted current values measured for a voltage of 1 V vs TMB flow (b).

The I-V curves behavior may be related to a difference in boron incorporation degree into amorphous silicon and its subsequent activation. Low TMB flow contributes an insufficient impact to the layer conductivity, whereas high boron flow leads to a material degradation.

According to results obtained the optimized (p)a-Si:H layer is characterized by high doping level and a low activation energy, thus it can be also used as an emitter layer in heterojunction a-Si:H solar cell. Figure 5 shows I-V curve and EQE spectra for a heterojunction a-Si:H solar cell fabricated with obtained (p)a-Si:H layer. An open circuit voltage of 0.65 V and a short circuit current of 31 mA/cm^2 were obtained.

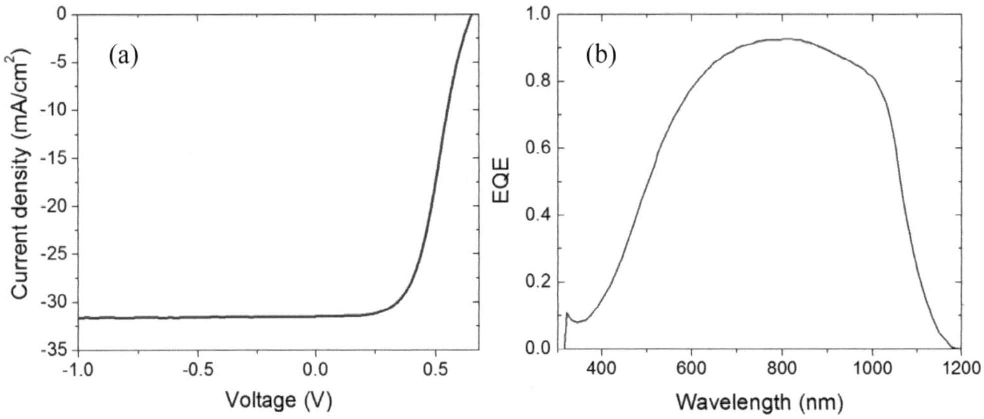

Figure 5. I-V curve of a heterojunction a-Si:H solar cell formed with an obtained (p)a-Si:H layer measured under AM1.5 solar spectrum (a) and its EQE spectra (b).

4. Conclusion

Based on results obtained an ohmic contact for low doped p-Si was fabricated. An optimization of deposition regimes led to a (p)a-Si:H film with a high doping level formation. An application of the contact to a heterojunction a-Si:H solar cell fabrication allowed to reach an open circuit voltage of 0.65 V and a short circuit current of 31 mA/cm^2.

Acknowledgments

This work was supported by Ministry of Science and Higher Education of the Russian Federation (research project FSRM-2020-0004)

References

[1] Yoshikawa K, Kawasaki H, Yoshida W, Irie T, Konishi K, Nakano K, Uto T, Adachi D, Kanematsu M, Uzu H and Yamamoto K 2017 *Nat. Energy* **2** 17032
[2] Froitzheim A, Brendel K, Elstner L, Fuhs W and Schmidt M 2002 *J. Non-Cryst. Solids* **663** 299
[3] Mazzarella L, Kirner S, Mews M, Conrad E, Korte L, Stannowski B, Rech B and Schlatmann R 2014 *Energy Procedia* **60** 123

A new device for autonomous space devices based on a three-component quantum variometer

S E Logunov[1, 2], V V Davydov[2, 3] and V Yu Rud[3, 4]

[1] Northwest Open Technical University, Saint Petersburg 195027, Russia
[2] Peter the Great Saint-Petersburg Polytechnic University, Saint Petersburg 195251, Russia
[3] All Russian Research Institute of Phytopathology, Moscow Region 143050, Russia
[4] A.F. Ioffe Physicotechnical Institute, St. Petersburg, 195256, Russia

sema-logunov@ya.ru

Abstract. The necessity of additional study of magnetic field variations in the magnetic transition layer, in the magnetopause, as well as in the plasma layer and in the tail of the magnetosphere in a wide range of distances from the Earth is substantiated. To obtain additional information in comparison with the studies that are being carried out in outer space at the present time, it is necessary to simultaneously monitor the magnetic field at various points in outer space. It is also necessary to register the dynamics of changes in the magnetic field in time in space by three components. To accomplish this task, a small-sized three-component quantum variometer with autonomous power supply has been developed for space devices, which can be lost in the course of short-term research. The results of the operation of a quantum variometer are presented.

1. Introduction

The Earth's magnetosphere is the most important component of the ecological state of the planet [1-7]. It protects all life on Earth from streams of ionizing radiation (solar wind). An important element of the Earth's magnetosphere is the boundary (magnetopause), at which the pressure from the Earth's magnetic field is equal to the pressure of the solar wind shock wave. There are regions in the magnetosphere that are geomagnetic traps that hold particles in a limited volume. This forms the Earth's radiation belts. The magnetic field, as shown by numerous studies [8-18], carries a huge amount of information that allows you to predict many situations.

Therefore, various areas of magnetic fields, both on Earth and in outer space in near-Earth orbits, must be investigated to predict the evolution of the universe, which has a significant impact on the Earth [1, 4, 19-23]. Information about the magnetic field from several magnetometers located on the autonomous spacecraft Ulysses is insufficient for spatial monitoring of various zones of the magnetosphere and much more. This truncated information does not allow comparing the experimental data with the calculations obtained in various theoretical models to study the dynamics of various physical processes. It is also impossible to check or substantiate forecasts on the influence of magnetic storms in space on processes in the Earth's magnetosphere and on its magnetic field. This requires information about the magnetic field from a large number of points located in outer space. This information should be received continuously (for a long time) to the station, which is in near-earth orbit [24-26]. One of the options for solving this problem can be associated with the use of autonomous small-sized meters of magnetic field variations with autonomous power supply from solar cells, which will fly in outer space. The use of a large number of simple and inexpensive devices (they can be irretrievably lost in space) will make it possible to provide studies of magnetic fields in the required areas of outer space. Therefore, the development of magnetic field meters that can be used for space exploration is an extremely urgent task, especially in the face of changes in the Earth's magnetic field. In our work, we present one of the possible options for an optical meter for magnetic field variations for space applications.

2. Three-component quantum meter of magnetic field variations

At present, a large number of models of various magnetometers have been developed to measure the magnetic field and its components [27-29]. The best precision characteristics are possessed by quantum magnetometers [27-33], which use optical pumping based on laser radiation. These are optical measuring devices based on the principle of operation, which have high stability of accuracy characteristics. But their work requires significant energy resources. In addition, they are large and very expensive. It should also be noted that in space, in some cases, the parameters of laser radiation can change under the influence of gravitational factors and temperature jumps [9-13, 29, 34-37]. Changing the wavelength λ of laser radiation has an extremely negative effect on the operation of the magnetometer. Therefore, these instruments are currently only used on the Ulysses spacecraft. In systems performing an autonomous flight in space for the study of magnetic fields, it is more expedient to use optical systems, measurements in which are less critical to the change in λ.

In our works on the study of magnetic field variations from a moving object at sea depths, we proposed a design of an autonomous sensor based on a ferrofluid cell with a magnetic fluid [34-37]. The variation of the magnetic field in the direction of the force lines of the Earth (one component) was measured, since the moving object creates the greatest distortions in this direction. In the case of solar wind impact (the Sun's magnetic field is 100 times greater than the Earth's magnetic field - about 5 mT), the design of the optical sensor changes. In figure 1 shows the design of an optical sensor developed by us for flight in outer space as part of an autonomous module.

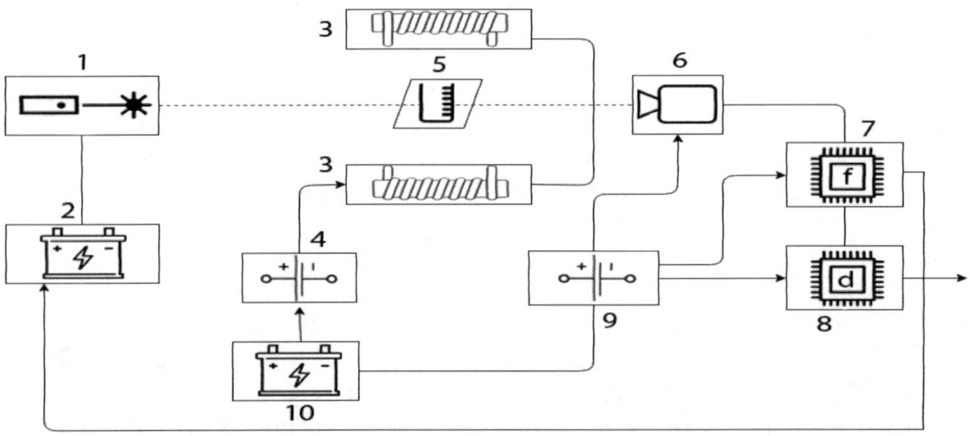

Figure 1. Block diagram of an optical sensor for measuring the magnetic field component and its variation: 1 - semiconductor laser; 2 - laser power supply; 3 - the coil of inductance; 4- stabilized current source; 5 - ferrofluid cell; 6 - a specialized camera; 7 - processing scheme; 8 - data transmission scheme; 9 - power supply; 10 - the battery.

Unlike the designs of optical sensors developed by us earlier [33-36], almost all optical elements (lenses, prims, etc.) are excluded from its composition. This, on the one hand, makes the new design more compact. On the other hand, the number of reflections between optical elements decreases, which increases the contrast of the bands in the recorded diffraction pattern. Radiation from semiconductor laser 1 with λ = 613 nm arrives at the transparent wall of ferrofluid cell 5. The cell contains a magnetic fluid - an aqueous solution of single-domain hematite nanoparticles with a volume concentration of 0.054 with a surfactant (surfactant) tetramethylammonium hydroxide. The ferrofluid cell is located in a weak uniform magnetic field (for example, with an induction B_0 = 0.9 mT and an inhomogeneity of 10-5 cm-1), which is created by special inductors 3 (specialized solenoid). The value of induction B0 can vary from 0 to 4.2 mT.

Previous studies made it possible to establish that in the case of placing a ferrofluid cell in a magnetic field, the nanoparticles of a ferromagnetic liquid are located on the force lines of the magnetic field. For laser radiation passing through a ferromagnetic liquid placed in a cell, this configuration of nanoparticles is similar to a diffraction grating. The period of this grating is determined by the distance between the force lines of the magnetic field [34-37]. The diffraction pattern from the laser radiation passed through the cell is recorded by a specialized camera. The principle of measuring the magnetic field and its variations in the developed design of the optical sensor is implemented as follows. In the case of a high uniformity of the magnetic field created by solenoid 3, in which the ferrofluid cell is located, the photodetector (camera 6) registers a diffraction pattern in the form of stripes with symmetry from the transmitted laser radiation through the ferrofluid cell. Up to a certain value, variations in the magnetic field that will be present in the vicinity of solenoid 3 do not cause visible distortions in the diffraction pattern.

In this case, by decreasing the field of the solenoid, it is possible to obtain distortions in the diffraction pattern due to the presence of variations that worsen the uniformity of the field. Since variations in outer space are non-periodic and non-constant, after a certain period of time (when their influence weakens), it will be possible to register a symmetric diffraction image. Having performed a preliminary calibration of the optical sensor on Earth (before sending it into outer space), it is possible to determine, with an error of no more than 20%, the values of the magnetic field variations. A similar process only with an increase in the induction of the magnetic field of the solenoid can be considered with an increase in the amplitude of the variation of the magnetic field from the solar wind. The restructuring of the magnetic field in these cases does not exceed 0.2-1.0 mT. It doesn't take long. Almost everything happens automatically.

Another case is associated with a change in the induction Bs of the magnetic field of the solar wind (with respect to this value, further variations in the magnetic field arise). This process is longer in time and, as studies have shown, is associated with cosmic magnetic storms or other strong anomalies. In this case, each component of the magnetic field Bs from a certain level will make changes in the magnetic field of the solenoid. The configuration and density of the magnetic field lines in the area of the ferrofluid cell location will change. The distance between adjacent maxima in the diffraction pattern changes. This means that the density of the magnetic field lines has changed per unit area. Having carried out a preliminary calibration on Earth, it is possible to estimate the value of induction B in the area of the ferrofluid cell with an error of less than 20% from the measured values of the density of the magnetic field lines.

$$B = \int j_B(S)dS \tag{1}$$

where j_B - the density of the force lines.

It should be noted that $j_B(S) \sim N$ (the number of force lines per area element). Therefore, it becomes necessary to estimate N, as well as the distance between the force lines

Using a preliminary calibration performed in a stationary laboratory on Earth by the number of lines, the value of jB is determined. Next, we define the induction.

3. Results of experimental studies and their discussion

Figure 2 shows, as an example, diffraction images from the transmitted laser radiation through a ferrofluid cell for various values of the magnetic field inhomogeneity in the area of its location.

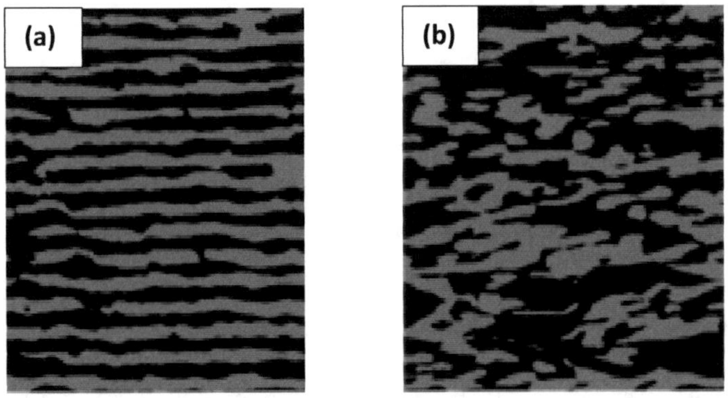

Figure 2. The diffraction pattern of the laser radiation in the case of the magnetic liquid placing: (a) in a uniform magnetic field; (b) in an inhomogeneous magnetic field.

An analysis of the results obtained (Figure 2) shows that the inhomogeneity of the magnetic field caused by the presence of variations significantly distorts the recorded diffraction pattern. It is rather difficult to measure the magnetic induction using the field line density based on the recorded image (Figure 2.b). To determine the number N, which is necessary for calculating the induction value B using (1), the intensity distribution I of the laser radiation is plotted over the width L or height h of the ferrofluid cell (row selection). In figure 3 as an example for the registered images (Figure 2), these lines are presented.

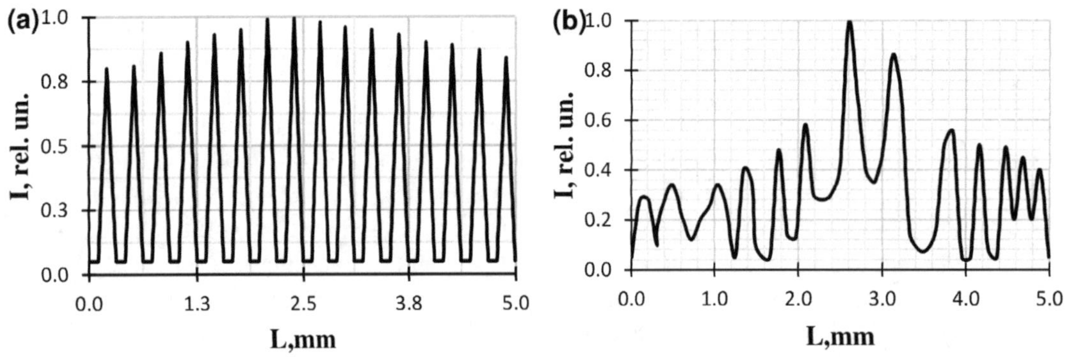

Figure 3. The dependence of the intensity I on the distance between the force lines of the magnetic field: a) in a homogeneous magnetic field; b) in a unhomogeneous magnetic field.

The obtained result shows the validity of using this method. By controlling the magnetic field of the solenoid, it is possible to obtain a change in the diffraction image in the presence of variations in the magnetic field in the area of the ferrofluid cell, which previously led to distortions in the diffraction pattern (Figure 2.b). In figure 4, as an example, we show some possible options for reconstructing the diffraction pattern using the adjustment of the field of solenoid 3 (Figure 1).

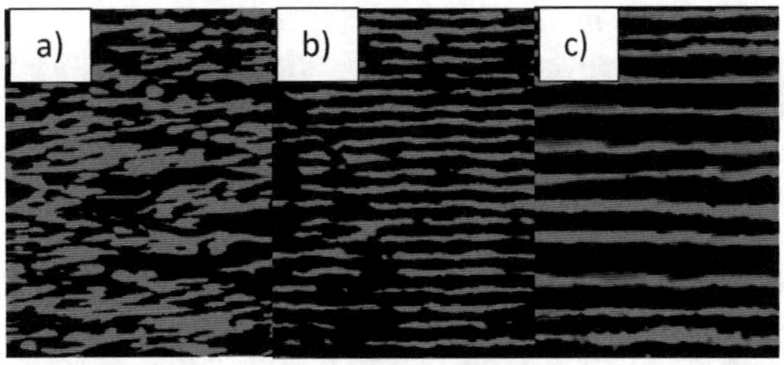

Figure 4. The diffraction pattern of the laser radiation in the case of the magnetic liquid placing: (a) in a uniform magnetic field; (b) in an inhomogeneous magnetic field.

An analysis of the results obtained (Figure 4) shows that by adjusting the induction B0 as a result of tuning, the symmetry in the arrangement of the maxima in the diffraction pattern recorded from the transmitted laser radiation is restored. And it is possible to measure and determine the variation of the magnetic field and the value of Bs. Similarly, the diffraction pattern is adjusted for the case of a change in the constant component in the induction of the magnetic field of the solar wind.

4. Conclusion

The obtained experimental results confirmed the validity of the developed design of the optical sensor for space-based autonomous systems. It should be noted that laser radiation in the visible range is used to obtain a diffraction pattern. In the case of temperature fluctuations, a change in the wavelength λ within the red range of the spectrum by 3-5 nm does not significantly affect the contrast of the recorded diffraction pattern and the error in measuring the number of field lines N. This fundamentally distinguishes the design of the optical sensor developed by us from the quantum magnetometer with optical pumping with using laser radiation.

If three such sensors are placed in an autonomous spacecraft with the field B0 oriented along the x, y, and z axes, it is possible to study the three components of the magnetic field and their variations. The experiments have shown that the service life of such sensors with an autonomous power source is more than 1 year, with the use of solar cells it can significantly increase (until the protective layers of an autonomous object made of non-magnetic material are destroyed).

References

[1] Batori E, Almat N, Affolderbach C and Mileti G 2020 *GNSS-grade space atomic frequency standards: Current status and ongoing developments.* (New York: Advances in Space Research)

[2] Semenov V V , Nikiforov N F and Ermak S V 1991 *Soviet journal of Communications Technology and Electronics* **36(4)** 59 – 63

[3] Petrov A A, Grebenikova N M, Lukashev N A, Ivanova N V, Rodygina N S and Moroz A V 2018 *Journal of Physics: Conference Series* **1038 (1)** 012032

[4] Lukashev N A, Petrov A A, Grebenikova N M and Valov A P *2018 Proceedings of 18th International conference of Laser Optics ICLO-2018 (Saint-Petersburg),* vol. 8435889 (IEEE), p. 271

[5] Lukashev N A, Glinushkin A P and Lukyantsev V S 2019 *Journal of Physics: Conference Series* **1410(1)** 012211

[6] Lukashev N A, Davydov R V and Glinushkin A P 2019 *Journal of Physics: Conference Series* **1326(1)** 012046

[7] Fadeenko V B, Fadeenko I V, and Vasiliev D. A., "Investigation of radiation formation (plasmoid) in the air environment by radar method," Journal of Physics: Conference Series, vol. 1697(1), 012057, December 2020
[8] Davydov V V 1999 *Russian Physics Journal* **42(9)** 822–825
[9] Davydov V V, Dudkin V I and Karseev A Y 2015 *Russian Physics Journal* **58(2)** 146–152
[10] Davydov V V, Dudkin V I and Karseev A Y 2015 *Measurement Techniques* **58(3)** 317–322
[11] Davydov V V, Dudkin V I and Karseev A Y 2015 *Instruments and Experimental Techniques* **58(6)** 787–793
[12] Davydov V V, Dudkin V I and Karseev A Y 2015 *Technical Physics Letters* **41(4)** 355–358
[13] Davydov V V, Dudkin V I and Karseev A Y 2015 *Technical Physics* **60(3)** 456–460
[14] Gizatullin B, Gafurov M, Vakhin A, Mattea C and Stapf S 2019 *Energy and Fuels* **33(11)** 10923–10932
[15] Davydov R, Antonov V, Makeev S, Batov Y and Dudkin V 2019 *E3S Web of Conferences* **140** 02001
[16] Myazin N, Neronov Y, Dudkin V and Yushkova V 2018 *MATEC Web of Conferences* **245** 11013
[17] Neronov Yu I 2017 *Measurement Techniques* **60(1)** 96-102
[18] Myazin N, Neronov Y, Dudkin V and Petrov A 2019 *IOP Conference Series: Materials Science and Engineering* **497(1)** 012111
[19] Petrov A A, Vologdin V A and Zalyotov D V 2015 *Journal of Physics: Conference Series* **643(1)** 012087
[20] Petrov A A and Myazin N S 2017 *Lecture Notes in Computer Science (including subseries Lecture Notes in Artificial Intelligence and Lecture Notes in Bioinformatics)* **10531 LNCS** 561-568
[21] Valov A P 2019 Journal of Physics: Conference Series **1410(1)** 012246
[22] Valov A 2020 *CEUR Workshop Proceedings* **2667** 102–105
[23] Petrov A A, Zalyotov D V, Shabanov V E and Shapovalov D V 2018 *Journal Physics: Conference Series* **1124(1)** 041004
[24] Petrov A A and_Grebenikova N M 2018 Lecture Notes in Computer Science (including subseries Lecture Notes in Artificial Intelligence and Lecture Notes in Bioinformatics) **11118 LNCS** 641-648
[25] Petrov A A and Grebennikova N M 2018 *Journal of Communications Technology and Electronics* **63(11)** 1281–1285
[26] Lukashev N A and Moroz A V 2019 *Journal of Physics: Conference Series* **1236(1)** 012068
[27] Vershovskii A K, Dmitriev S P and Pazgalev A S 2013 Technical Physics Journal **83** 879
[28] Aleksandrov E B and Vershovskii A K 2009 Physics-Uspekhi 52 573-601
[29] Logunov S E, Koshkin A Yu, Davydov V V and Petrov A A 2016 *Journal of Physics: Conference Series* **741(1)** 012092
[30] D'yachenko S V and Zhernovoi A I 2016 *Technical Physics* **61(12)** 1835–1837
[31] D'yachenko S V, Kondrashkova I S and Zhernovoi A I 2017 *Technical Physics* **62(1)** 1602-1604
[32] Davydov V V, Dudkin V I and Petrov A A 2016 *Technical Physics Letters* **42(7)** 692-696
[33] Davydov V V 2017 *Measurement Techniques* **60(2)** 183–189
[34] Logunov S E, Davydov V V, Vysoczky M G and Mazing M S 2018 *Journal of Physics: Conference Series* **1038(1)** 012093
[35] Logunov S E, Davydov V V, Glinuchkin A P, Podkovyrov I Yu and Rud V Yu 2019 *Journal of Physics: Conference Series* **1410(1)** 012113
[36] Logunov S E, Davydov V V and Rud V Yu 2021 *Journal of Physics: Conference Series* **1745(1)** 012007
[37] Logunov S E, Vysocky M G, Davydov V V, Koshkin A Yu and Rud V Yu 2017 *Journal of Physics: Conference Series* **917(5)** 052058

Numerical simulation of effective light transmission through a photonic memory cell

M E Makarov [1,2], A A Sapegin [1,3] and R T Minnullin [1,3]

[1] JSC Molecular Electronics Research Institute, Moscow 124460, Russia
[2] Institute of Microelectronics Technology and High-Purity Materials of Russian Academy of Sciences, Chernogolovka 142432, Russia
[3] Moscow Institute of Physics and Technology, Dolgoprudny 141701, Russia

Abstract. This paper is devoted to numerical simulation of a non-volatile photonic memory cell based on phase-change material $Ge_2Sb_2Te_5$. The parameters of light propagation are presented for both crystalline and amorphous $Ge_2Sb_2Te_5$ phases. The cell structure is optimized for a single TE-mode regime that is suitable for short- and long-distance communication lines.

1. Introduction
Nowadays growth of communication streams leads to increasing requirements for computing, storage, and transmitting devices [1,2]. Despite this fact, most computers are based on von Neumann architecture which has a separate central processing unit (CPU) and memory unit. Therefore, data has to transfer between the CPU and the memory unit in each calculation cycle. The memory unit often uses various types of storage devices for different purposes. Frequently used data temporarily keeps in fast volatile SRAM and DRAM and the remaining data is stored in slow non-volatile SSD and HDD. Different storage locations slow down data processing because of the relatively long transfer time. The proper way out of this situation is to use non-volatile storage devices [3] and the neural architecture [4]. Such devices may be based on a wide range of phase-change materials (PCM). PCM has been studied and used for a long time. For example, these materials have already been used in the field of memory devices on DVDs [5].

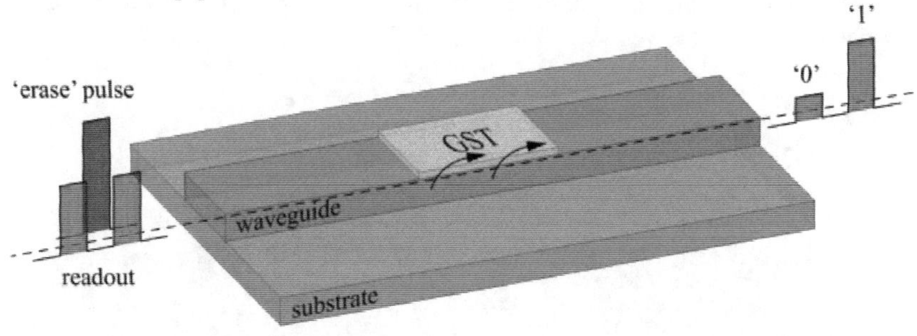

Figure 1. Schematic view of the non-volatile storage element with optical control.

However, the use of PCM in the storage devices alone is not enough to overcome the growth of requirements, because well-known solutions to this problem by the microelectronics methods are reaching physical limits [6]. In the same time, silicon photonics offers several promising solutions [7-10].

This paper is devoted to optimization parameters of a non-volatile photonic memory cell with optical recording and readout based on phase-change material $Ge_2Sb_2Te_5$ (GST).

2. Operation principle and purpose

The photonic memory cell consists of a Si strip waveguide with a 10 nm thick GST layer on its top. The operation principle of the photonic memory cell is similar to PCM memory. It is well known that GST has amorphous and crystalline phase states that are both stable at room temperature and have different optical and electrical properties. Switching between these phase states is performed by heating. Usually, GST crystallization is considered an erasing operation, and amorphization after melting is considered a recording one. The difference between the photonic memory cell and PCM memory is in the heating source and reading operation. Indeed, PCM memory uses Joule heating and difference in electrical resistance of GST phase states while the photonic memory cell uses optical sources and difference in refractive indices.

Consider a photonic storage element. It is worth mentioning that GST has a 20 times larger imaginary part of the refractive index in the crystalline phase state than in the amorphous one. Consider that the GST layer is thin enough to be regarded as a slight nonuniformity for the waveguide mode and the waveguide is supposed to be ideal. In this case, all waveguide mode losses are due to internal absorption in GST. Consequently, there will be two different transmission levels of the signal which correspond to "0" and "1" in binary notation. Switching between them can be performed in different ways by changing the source of heating power. An all-optical configuration was chosen for the calculations. This means that data recording, erasing and reading is performed by laser pulses of different amplitude and duration [11]. Erasing signal consists of a sequence of pulses that have enough power to heat GST to the state near the melting point and have enough delay for the subsequent crystallization. Data recording operation is a short high-energy pulse that melts GST with a short delay for GST amorphization. The optical readout is performed for a low energy pulse which does not affect the GST phase. The goal of the present work is to find the best size ratio of the waveguide to provide a clear difference in the transmission levels of the photonic memory cell.

3. Results and discussion

To reach the goal, two stages of numerical calculations were performed. In the first stage 2D, calculations were carried out. The simulated structure had infinite size in the XY plane. The thickness of buried SIO2 was selected 2 μm to ensure low losses to the substrate [12]. The distance from the waveguide side walls to the boundaries of the simulated area was set to 3 μm; the distance from the GST layer to the top boundary of the simulated area was 2 μm. An approximate Sommerfeld condition (1) was used for the case of plane waves:

$$n(\Delta E_z) + ik_0 E_z = 0 \quad (1)$$

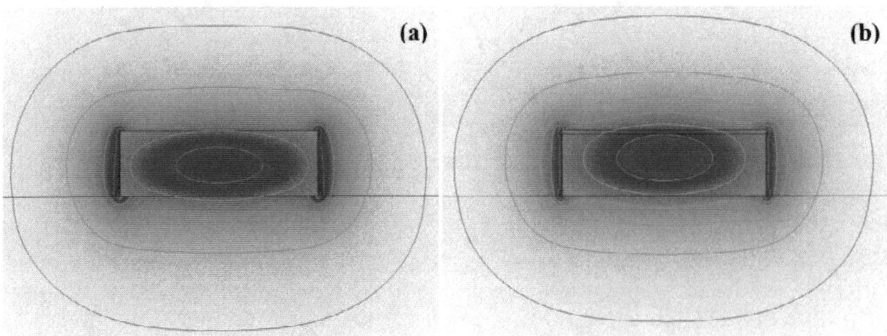

Figure 2. TE fundamental mode profile of the different structures: (a) a simple Si strip waveguide, (b) the photonic storage element with crystalline GST.

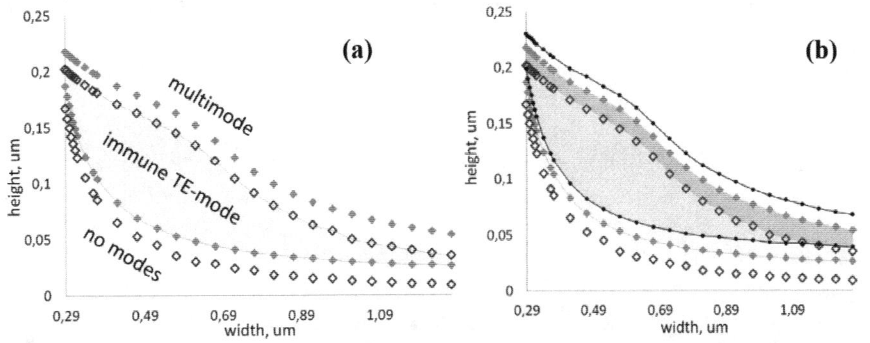

Figure 3. (a) Optical mode regime of the photonic storage element. Green and hollow dots correspond to amorphous and crystalline phase states. The yellow area corresponds to the "immune" regime. (b) The overlap between mode distributions of the photonic storage element and Si strip waveguide (the area inside the black outline). The light green area provides single TE-mode regime and the blue one corresponds to both TE and TM modes.

As a result, mode distributions of the photonic storage element were obtained for both GST phase states and various waveguide geometry. The resulting mode structure is very similar to the mode structure of a simple Si strip waveguide (figure 2a). It is worth mentioning that the difference between these two mode structures may be explained by Snell's law. Indeed, a slight upward shift of the waveguide mode core appears due to a large difference between refractive indexes of crystalline GST and Si (figure 2b). Hence there is a lower critical angle of total internal reflection on the air-GST boundary than on the air-Si boundary. It means that fewer light rays will pass through the top border of the waveguide with the GST layer. Therefore, the waveguide mode exists in the photonic storage element if it satisfies the following condition for the simple Si strip waveguide:

$$n_{buried\_oxide} < n_{eff} < n_{core} \quad (2)$$

where n_{buried_oxide} is SiO_2 refractive index and n_{core} is Si refractive index respectively. This allows finding a certain range of the waveguide width and height where single a TE-mode regime is insensitive (so-called "immune") to switching GST phase state (figure 3a). Furthermore, it was determined that the lower limit of single the TE-mode regime for the crystalline GST phase may be approximated sufficiently accurate by the simple power functions (3):

$$h = \alpha w^{-\beta} \quad (3)$$

where w is the width of the waveguide, h is the total height of the waveguide with the GST layer, α and β are real positive numbers. However, light generally enters the photonic memory cell through a Si strip waveguide and its mode distribution should also be taken into account. The overlap between the "immune" regime and the single TE-mode regime of the Si strip waveguide guarantees the single TE-mode regime in the entire photonic memory cell (figure 3b). Also, this overlap eliminates the losses caused by the appearance of TM-mode and its leakage at the end of the GST layer.

Single TE-mode regime provides propagation of only the fundamental TE mode and the rest of the modes, in this case, are unstable due to the limited size of the waveguide structure. The choice of this regime is explained by the following facts. Firstly, many silicon photonics devices are polarization-sensitive (e.g., MMI, MZI, DC, PhC/Grating, AWG, MRR) [13] and the single TE-mode regime prevents the appearance of parasite high-order modes by definition. Therefore, this regime may be useful for the development of silicon photonics integrated circuits. Secondly, both short- and long-distance transmission lines commonly use signals of one polarization to reduce losses. Usually, it is TE polarization [14]. Bragg gratings are used to connect fiber transmission lines and silicon photonics devices. Bragg condition for a given wavelength and angle of incidence can only be applied for one polarization. As a rule, TE polarization is selected because it corresponds to the fundamental waveguide mode when the width of the waveguide is greater than its height [15]. The possibility of creating Bragg grating for TM polarization was shown, but its efficiency factor was 20% lower than for TE polarization [15,16].

In the second stage of the calculations, transmission characteristics of the photonic memory cell were analyzed. It was determined at what waveguide dimensions the maximum difference between transmission levels is reached.

For this purpose, 2D and 3D numerical calculations were carried out which produced fairly similar results. The structure length was 20 μm to avoid taking into account scattering light. The length of the GST layer was 1 μm. All the other dimensions were the same as in the first stage. A longitudinal section of the photonic storage element was selected for the 2D simulations. Perfectly matched layers (PML) were used in both 2D and 3D numerical calculations as boundary conditions. The key feature of the PML is almost no reflection of the waves that incidents from non-PML material upon the PML. Thus, it was assumed that all light scattered on the GST layer absorbed on the excessively long boundaries of the structure covered by PML.

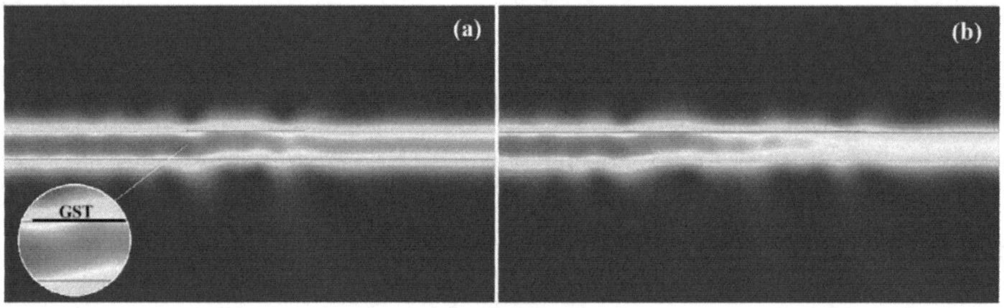

Figure 4. 2D simulations of the photonic storage element with different lengths of crystalline GST. Signs (a) and (b) corresponds to 1 μm and 2 μm respectively.

2D numerical calculations of the guiding light in the photonic memory cell showed the main trend: the difference between transmission levels decreases with the waveguide height. Furthermore, these calculations are consistent with the mode structure obtained in the first stage. Figure 4 shows that the waveguide mode gradually attenuates in the crystalline GST layer and shifts closer to the upper boundary as was demonstrated (figure 2b).

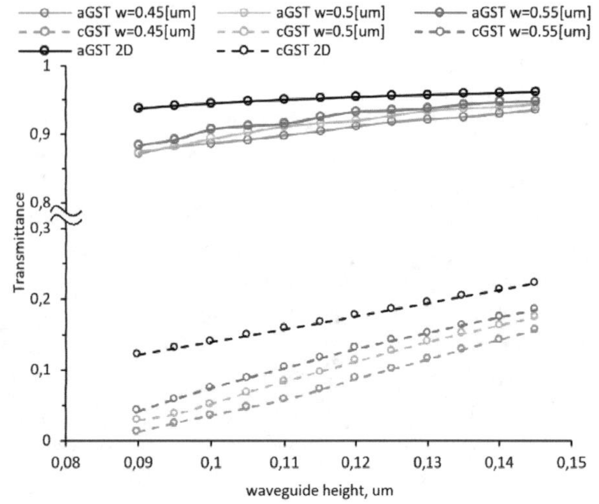

Figure 5. Transmittance versus waveguide height. Dash and solid lines correspond to the crystalline and amorphous GST phases.

3D numerical calculations provided several curves of transmittance versus height for various widths of the waveguide (figure 5). It should be noted that matching between graphs obtained from 2D and 3D calculations of the guiding light is better for wider waveguides. This is completely logical because the strip waveguide is closer to the thin-film waveguide, the wider it is. Also, only one length of the GST layer was considered in this paper. But figure 4 demonstrates that an increase of the GST layer length leads to a stronger attenuation which may cause a greater difference between the transmission levels of the photonic memory cell.

Summing up the above, the largest difference between the transmission levels of the photonic memory cell appears on the bottom border of the overlap between the "immune" zone of the photonic memory cell and the single TE-mode regime of the Si strip waveguide (bottom black line on figure 3b). It reaches about 80 percentage points in the center of the overlap.

4. Conclusion
In conclusion, the optimal geometry of the photonic memory cell suitable for long- and short-distance communication was determined. Furthermore, the approach of designing single TE-mode devices with a switching refractive index, which may be useful for creating silicon photonics devices or photonic integrated circuits, was shown.

Acknowledgments
The reported study was funded by RFBR according to the research project № 19-29-03040

References
[1] Editorial 2018 *Nature* **554** 145
[2] Krasnikov G Y, Zaytsev N A and Krasnikov A G 2015 *Journal of nano- and microsystem technique* **4** 63
[3] Wong HS and Salahuddin S 2015 *Nature Nanotech.* **10** 191-4
[4] Zidan M A, Strachan J P and Lu W D 2018 *Nat. Electron.* **1** 22-29
[5] Noboru Yamad 2000 *Journal of Applied Physics* **88** 7020
[6] Krasnikov G Y and Zaytsev N A 2009 *Journal of nano- and microsystem technique* **1** 2
[7] Eiichi Kuramochi, Kengo Nozaki, Akihiko Shinya, Koji Takeda, Tomonari Sato, Shinji Matsuo, Hideaki Taniyama, Hisashi Sumikura and Masaya Notomi 2014 *Nature Photon.* **8** 474-81
[8] Carlos Ríos, Matthias Stegmaier, Peiman Hosseini, Di Wang, Torsten Scherer, C David Wright, Harish Bhaskaran and Wolfram H P Pernice 2015 *Nature Photon.* **9** 725-32
[9] Jun-Feng Song *et al.* 2016 *Sci. Rep.* **6** 22616
[10] Abel S *et al* 2019 *Nat. Mater.* **18** 42-7
[11] Makarov M, Sapegin A, Barabanenkov M and Italyantsev A 2020 *J. Phys.: Conf. Ser.* **1695** 012114
[12] Reed G T and Knights A P 2004 *Silicon photonics an introduction* (UK: Chichester) p 76
[13] Dai D and Bowers J 2014 *Nanophotonics* **3** 283
[14] Dan-Xia Xu, Schmid J, Reed G, Mashanovich G, Thomson D, Nedeljkovic M, Chen X, Dries Van Thourhout, Keyvaninia S and Selvaraja S 2014 *IEEE Journal of Selected Topics in Quantum Electronics* **20** 189-205
[15] Bogaerts W and Diedrik Vermeulen 2013 *Handbook of silicon photonics* ed Laurent Vivien and Lorenzo Pavesi (US: Boca Raton, Florida) p 121
[16] Vermeulen D *et al* 2010 *Proc. European Conf. on Integrated Optics* (UK: Cambridge University) WeP16

Study of GaP/Si electron-selective contact deposited by plasma

A A Maksimova[1,2], A I Baranov[1], A V Uvarov[1], A S Gudovskikh[1,2], D A Kudryashov[1], I A Morozov[1] and M V Bogdanova[1]

[1] St. Petersburg Academic University, St. Petersburg 194021, Russia
[2] Department of Photonics, Saint Petersburg Electrotechnical University "LETI", St. Petersburg, 197376 Russia

maksimova_alina@spbau.ru

Abstract. The article is based on an important characterization task to accurately evaluate the properties of the layers, their interfaces with c-Si, and to select the best candidates to integrate them into a c-Si-based solar cell. The work has shown that GaP could be doped with n-type doping, thus providing a selective contact for the electrons, and has a significant valence band offset with c-Si, making it an excellent candidate as a selective contact, without requiring an additional ITO layer.

1. Introduction

Silicon-based solar cells are still of great interest due to the novel concept of selective contacts. Solar cells based on a-Si:H/c-Si heterojunction is a fair example of successful selective contacts application. These use a stack consisting of a very thin layers of a-Si:H ensuring a good chemical passivation of the surface of the c-Si, and a layer of doped a-Si:H, n-type or p-type, for contact selectivity to extract electrons or holes, respectively. However, even a thin a-Si:H layer absorbs part of solar light, which would be converted.

In the proposed work, a new path based on phosphide-type materials, which have attractive characteristics for this application, is explored. Thus, gallium (GaP) phosphide has a band gap of 2.26 eV significantly larger than c-Si and higher than a-Si:H, thus limiting the absorption of the solar spectrum in these layers. This fact augurs well for a gain on the short-circuit current compared to the use of a-Si:H. Preliminary work has shown that GaP can be highly doped with n-type doping, and has a significant valence band offset with c-Si. The band discontinuity of the GaP/Si interface values for the conduction and valence band offset are within the range of 0.09–0.35 eV and 0.8–1.05 eV, respectively. Therefore, GaP is an excellent candidate as a selective electron contact [1].

In this article, the capabilities of GaP/Si heterojunctions fabricated by a low temperature plasma-enhanced atomic layer deposition (PE-ALD) for use as selective contacts are explored. This is a reliable, industry-relevant technique that allows one to grow electronic-quality films at low temperature on large areas. The low-temperature (250-380 °C) plasma-chemical synthesis technologies of thin GaP layers on Si substrates [2], and the possibility of their donor doping [3], were recently developed. Thus small (0.09–0.35 eV) conduction band gap (ΔE_C) for GaP/Si [4, 5] provides necessary selectivity based on the heterojunction band diagram (figure 1).

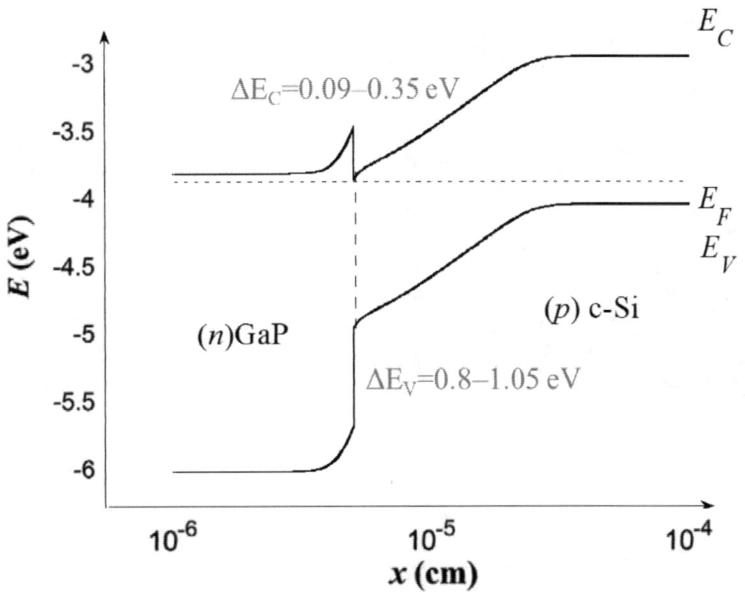

Figure 1. Band diagram of n-GaP/p-Si heterojunction.

2. Experiment details

The GaP layers are grown by plasma technique using the Oxford Plasmalab 100 PECVD setup. First, the potential for passivation of the Si surface due to the deposition of amorphous GaP layers was estimated. For the deposition of amorphous layers, the deposition took place with constant combustion of hydrogen plasma at a reduced plasma power [3]. Also, a process with intermediate passivation of the Si surface by deposition of a thin (3 nm) undoped a-Si:H layer and subsequent deposition of amorphous GaP layer at a temperature of 250°C was carried out. The quality of passivation was tested by Quasi-Steady State Photoconductance (QSSPC) techniques. The measurements were carried out on specially formed structures with double-sided passivation using n-Si substrates with high values of the charge carrier lifetime (1 ms). The contact resistivity of n-GaP/n-Si system was measured by TLM method using Ag evaporated contacts.

Second Si doped GaP layers were grown on B-doped CZ p-Si (10^{16} cm^{-3}) substrate to form n-GaP/p-Si heterojunction. Au/Ge contacts were formed to the n-GaP layer to study photoelectrical and electrical properties of n-GaP/p-Si heterostructures. Indium ohmic contacts were formed to the backside of p-Si. Standard I–V curves under AM1.5G simulator (Abet Technology) were measured using a Keithley 2400 source-meter. Space charge capacitance technique such as capacitance–voltage measurements (C–V) were performed using an Agilent 4980A LCR meter.

3. Result and discussion

The obtained GaP films had amorphous structure as shown by the transmission electron microscopy (Figure 2). The measured by QSSPC dependences of effective lifetime as function of charge carrier concentration for Si substrate passivated by amorphous GaP is presented in figure 3. The maximum value of lifetime does not exceed 60 µs, being extremely low. However, when intermediate ultrathin a-Si:H layer is used this value increased by one order of magnitude (figure 4) to 320 µs, which could be used in photovoltaic devices. The TLM measurements of contact resistivity of n-GaP/n-Si system are presented in figure 5. Quite low value of 0.075 Ohm $^{-1}$ cm^{-2} was obtained.

Current-voltage characteristics of n-GaP/p-Si heterostructures based on amorphous GaP shown in figure 6. Dark I-V curves exhibit rectifying behavior, while I-V curves under illumination demonstrate photovoltaic behavior with open circuit voltage of 0,44 V.

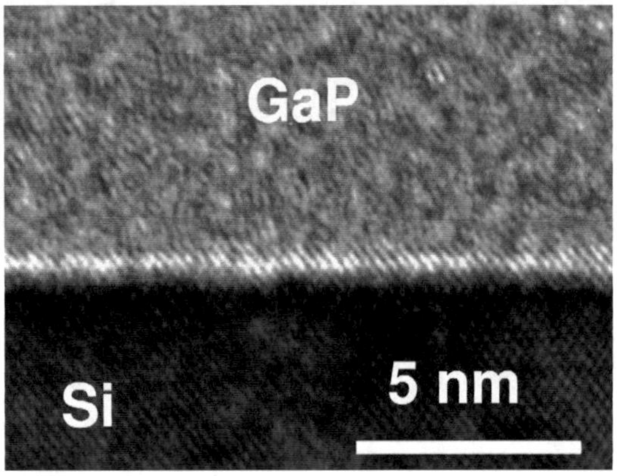

Figure 2. HRTEM cross section bright-field images of GaP/Si interfaces.

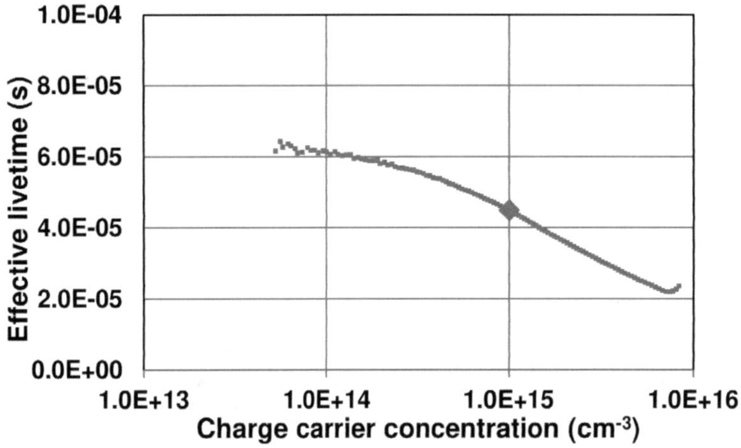

Figure 3. QSSPC measurements of effective lifetime versus charge carrier concentration for n-Si substrate passivated by amorphous GaP.

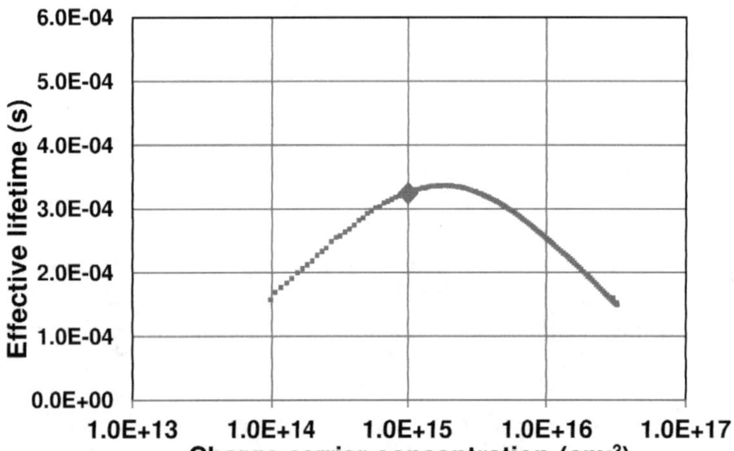

Figure 4. QSSPC measurements of effective lifetime versus charge carrier concentration for n-Si substrate passivated by amorphous GaP with 3 nm a-Si:H interlayer.

The temperature dependences of IV characteristics plotted in semi-logarithmic coordinates are shown on figure 7 a. Lowering the temperature leads to currents shift of the curves toward higher voltage being in agreement with normal p-n junction behavior. An influence of strong series resistance is clearly seen in the figure. However, all curves demonstrate high current down to the temperature of 80 K. The difference in absolute current values in the temperature range of 80-340 K is about 3 orders. The temperature dependence of the current at V=1V is presented in Arrhenius plot (see figure 7 b). The estimated activation energy of about 0.1 eV is a maximum potential barrier to be overpassed by charge carriers.

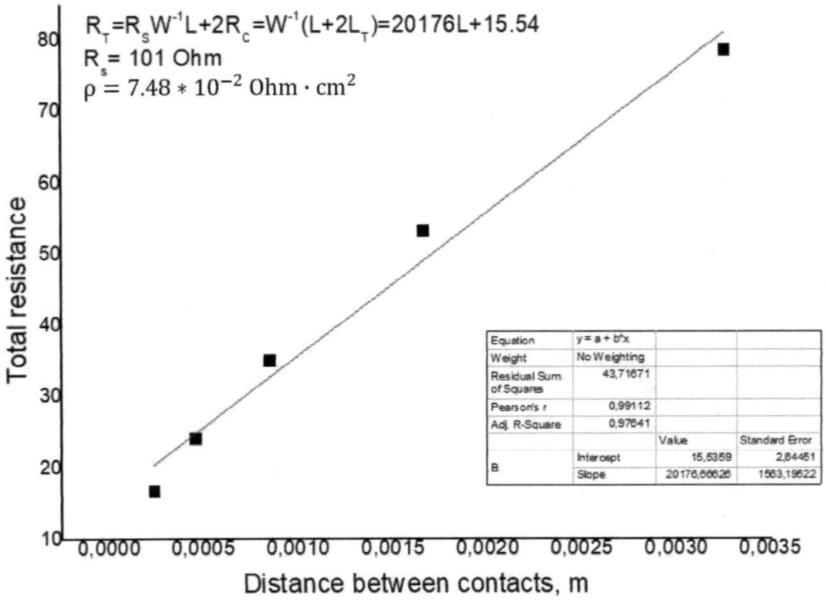

Figure 5. TLM measurements for n-GaP/n-Si contact.

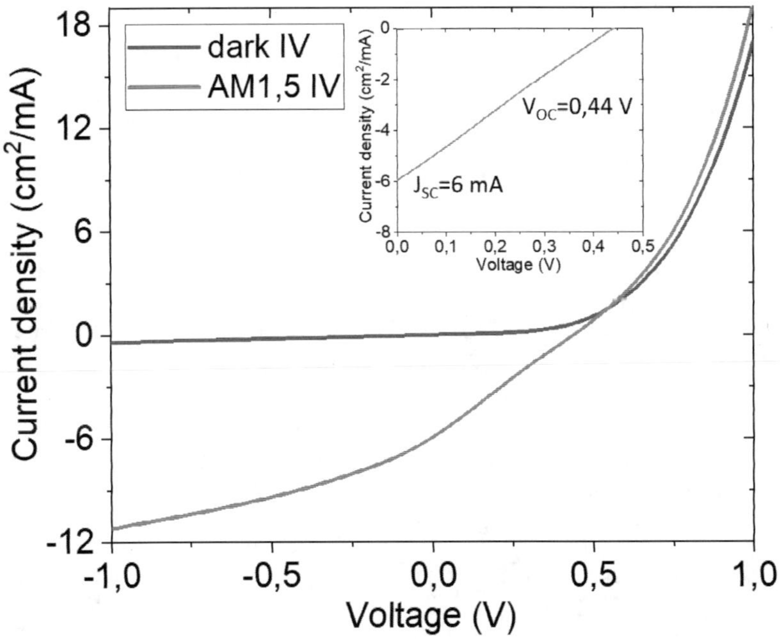

Figure 6. I-V curves of GaP/p-Si in the dark and under AM1.5 illumination

Figure 8 shows the dependence of capacitance on voltage in the Mott-Schottky coordinates. Linear behavior can be seen from the graph. The general trend of the capacitance decreasing with applied voltage could indicate a presence of space charge region in c-Si. Cutoff voltage estimated from the dependences is about 0.85 V is in agreement with band diagram presented in figure 1. Further detailed studies carried out by applying admittance spectroscopy did not show any responses indicating low defect density at the GaP/Si interface.

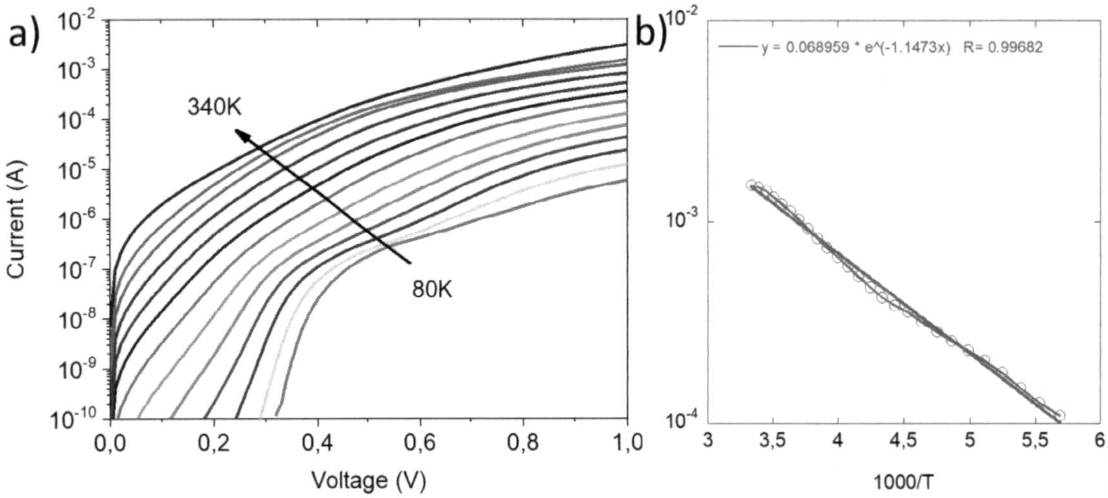

Figure 7. I-V characteristics as a function of temperature (a); Arrhenius plot at T=80K (b)

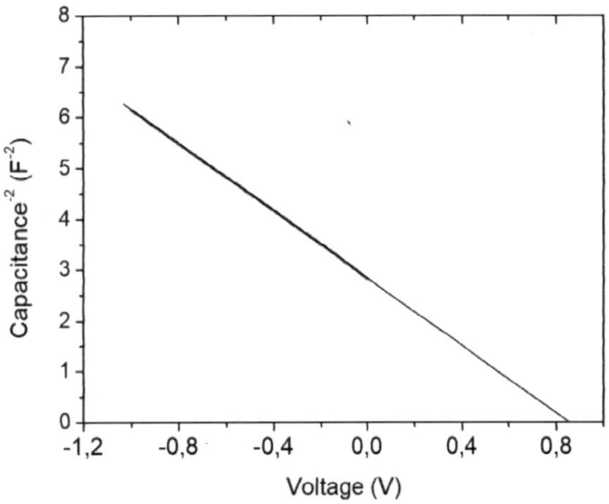

Figure 8. $1/C^2$ versus voltage in Mott-Schottky coordinates.

Thus the work has shown that GaP could be doped with n-type doping, providing a selective contact for the electrons, and has a significant valence band offset with c-Si, making it an excellent candidate as a selective electron contact.

Acknowledgments
This work was supported by the Russian Scientific Foundation under Grant № 17-19-01482

References
[1] Wang X L, Wakahara A and Sasaki A 1996 *J. Crystal Growth* **158(1-2)** 49–52
[2] Gudovskikh A S, Morozov I A, Uvarov A V, Kudryashov D A, Nikitina E V, Bukatin A S and Kleider J P. 2018 *J. Vac. Sci. Technol.* A*: Vacuum, Surfaces, and Films* **36(2)**
[3] Gudovskikh A S et al 2019 *Phys. Status Solidi* A **216(20)**1800617
[4] Wagner H, Ohrdes T, Dastgheib-Shirazi A, Puthen-Veettil B, König D and Altermatt P P 2014 *J. Appl. Phys.* **115(4)** 044508
[5] Sakata I and Kawanami H 2008 *Appl. Phys. Express* **1** 091201

Polarization control algorithm for QKD systems

E E Mekhtiev[1,2,4], **I S Gerasin**[1,2,4], **N V Rudavin**[1,5], **A V Duplinsky**[1,3,4] **and Y V Kurochkin**[1,2,3,4]

[1] QRate, Novaya av. 100, Moscow, Russia
[2] Moscow Institute of Physics and Technology, 9 Institutskiy per., Dolgoprudny, Moscow Region, 141701, Russian Federation
[3] NTI Center for Quantum Communications, National University of Science and Technology MISiS, Leninsky prospekt 4, Moscow 119049, Russia
[4] Russian Quantum Center, Bolshoy Boulevard 30, bld. 1, Skolkovo, Moscow 121205, Russia
[5] Federal State Budget Educational Institution of Higher Education, MIREA—Russian Technological University, 78, Vernadskogo avenye, 119454 Moscow, Russia

e.mehtiev@goqrate.com

Abstract. The crucial task for polarization-encoding fiber QKD is to compensate polarization drift occurring in a quantum channel. To solve this problem, the receiver usually uses a polarization controller. For proper operation, this device must be efficiently managed in real-time. In this work, a gradient-descent-based algorithm is proposed to solve this problem. The algorithm was implemented and tested on a QRate commercial QKD fiber system, that utilizes BB84-protocol. Low and stable QBER has been obtained during a day of continuous operation.

1. Introduction

Today commercial implementations of quantum key distribution (QKD) mostly use phase or polarization encoding. It is well-known that the polarization encoding approach suffers from polarization disturbance occurring in the quantum channel (QC) between sender (Alice) and receiver (Bob) [1]. Moreover, since polarization distortion is not constant in time different approaches were proposed to monitor its changes (see [2] for further references). Regardless of monitoring protocol, however, an active compensating device and an algorithm that controls its operation are needed to recover polarization states of optical pulses sent to QC. Electronic polarization controllers (EPC) based on fiber-squeezers are common devices used in QKD systems to solve this task profiting from their low insertion loss. These devices are usually driven by the number of applied voltages on its channels which are chosen to minimize polarization misalignment (expressed in terms of quantum bit error rate (QBER) in a quantum case). Mathematically this task can be considered as an optimization problem, namely finding the minimum of polarization misalignment or QBER as a function of voltages applied to the polarization controller.

The standard approach to the task stated above is to use gradient-descent-like algorithms. In [3] authors use the Hill-Climbing-based minimization algorithm. In [4] authors calculate first-order approximations of derivatives to make gradient-descent steps on each channel. Unlike previous considerations, we take advantage of the target function structure to improve algorithm convergence.

2. Problem statement

Single-mode optical fiber in the QC does not preserve the polarization state of the input laser pulse due to random birefringence occurring in an optical fiber. There are two main physical sources for this. The first one is a temperature changing of the environment since field-deployed fibers are not thermally stabilized in most cases. The second source is time-varying mechanical stresses occurring in the optical fiber due to different mechanical actions.

EPC is aimed to compensate polarization disturbance occurring in the optical fiber. Such EPC devices have a few piezo-electric squeezers located at the different angles to the optical fiber allowing to perform the most general polarization transformation. Each squeezer is driven by voltage applied to its pins. So, the state of the EPC is characterized by set of voltages applied to each squeezer $\vec{U}$, moreover the vector of voltages at a certain point in time define all polarization changes from Alice's laser to Bob's detectors.

Key indicator of proper operation of the QKD device is QBER. In ideal case after sifting procedure Alice and Bob should have completely identical bit strings which are their secret key. Different kinds of impairments occurring during the key distribution (optical error, eavesdropping) lead to mismatch in Alice's and Bob's bit strings. QBER is essentially the ratio of errors in the bit strings which can be expressed in terms of polarization misalignment angle.

Using EPC allows to minimize polarization misalignment angle and hence QBER. Thus, we consider EPC in a target state when observed QBER is minimal. Mathematically it is an optimization task of QBER as function of $\vec{U}$.

3. Algorithm

Applying the voltage to each channel of EPC causes circular rotation of the polarization vector on the Poincare sphere around a certain axis. The angle of rotation is proportional to the applied voltage on the fiber-squeezer. If misalignment is minimal, the increase or decrease by the same voltage will give the same QBER. Consequently, QBER around the minimum is a symmetric function of voltage on each channel. Moreover, the rotation angle has a period equal to 2π, thus the voltage on each channel has period $T_i \sim 2\pi$. And the last thing to notice is that the maximum QBER is achieved when the polarization vector on the Poincare sphere is opposite to the vector that gives the minimum QBER (Actually one can show that QBER is a harmonic function for the fiber-squeezer EPC). Given all the above, algorithm for QBER minimization works as follows:

1. choose channel $i = 1$ and initialize $\vec{U}$
2. measure $q_c = QBER(U_i)$,
   $q_l = QBER(U_i - h)$,
   $q_r = QBER(U_i + h)$
3. calculate $\partial_i Q = \frac{\partial QBER}{\partial U_i} = \frac{q_r - q_l}{2h}$
4. calculate $\partial_i^2 Q = \frac{\partial^2 QBER}{\partial^2 U} = \frac{q_r - 2q_c + q_l}{h^2}$
5. calculate step $s_i = min\left(\frac{\partial_i Q}{\partial_i^2 Q}, \frac{T_i}{4}\right)$
6. if $\partial_i^2 Q > 0$, then:
   $U_i = U_i - s_i$
7. else:
   $$U_i = U_i - s_i + \frac{T_i}{2}$$
8. $i = i \, mod \, 4 + 1$, then goto step 2

Near the inflection point, the second derivative can be too small for division, thus comparison with quarter-period is introduced at step 5. In case of the second derivative is greater than zero, step 5 yields the position of maximum instead of minimum for QBER function, so the additional half-period voltage bias is required as shown in step 7.

4. Numerical and experimental results

Firstly, we construct a model of the QKD fiber system in terms of John's matrices. The simulation (see Figure 1) shows that gradient-descent for each channel with the second-order first derivative approximation (GD sec order) is more stable in the minimum and is able to achieve lower QBER than the first-order approximation for the first derivative with constant step (GD), but has the same convergence speed. Using Newton's method with a second-order approximation for the first derivative (Adaptive) we achieve faster convergence compared to mentioned above algorithms preserving the stability and the lowest QBER level of the gradient descent with the second-order first derivative approximation.

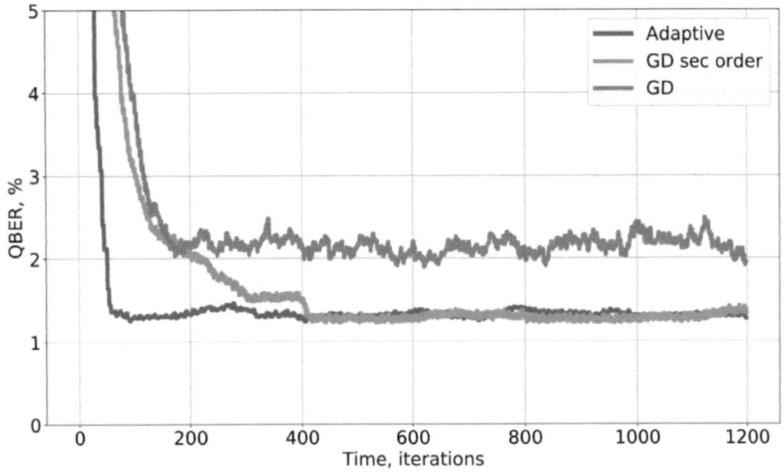

Figure 1. Simulation results.

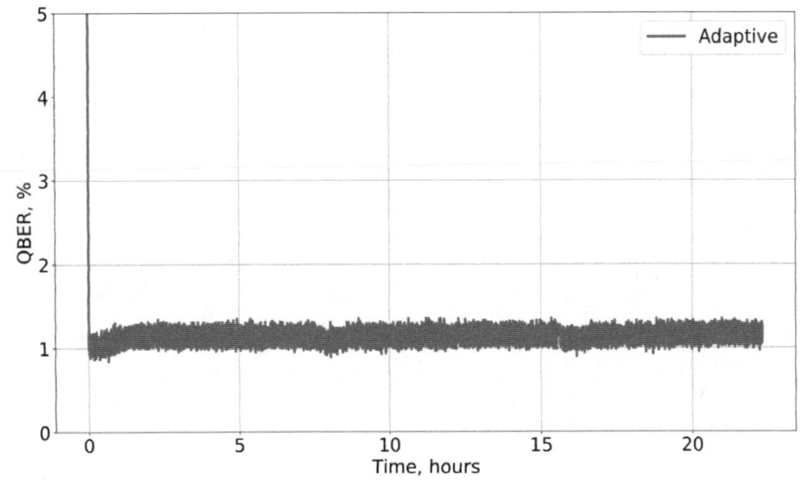

Figure 2. Experimental results.

The proposed algorithm uses only one hyperparameter h - step for derivative estimation. To find an optimal value of this parameter we run the simulation with the numbers of parameters h and find the optimal. Lastly, we implement the proposed algorithm in the QRate commercial QKD scheme [5] and measure the stability of the QBER during a day (Figure 2). We can see stable minimum with small perturbations, which happens due to non-stable optical error in the experimental setup.

For real-time polarization compensation QKD schemes our approach allows us to maintain QBER at a low level. For QKD systems that work in generation and polarization setup regimes, the proposed algorithm allows to achieve fast and stable convergence.

Acknowledgments
We would like to thank many colleagues at the Russian Quantum Center, QRate, for motivating discussions, management, and coordination. This work was supported by The Russian Science Foundation (Grant No. 17-71-20146).

References
[1] Chen J, Wu G, Xu L, Gu X, Wu E and Zeng H 2009 Stable quantum key distribution with active polarization control based on time-division multiplexing *New J.Phys.* **11** 065004
[2] Ding Y-Y, Chen W, Chen H, Wang C, Li Y-P, Wang S, Yin Z-Q, Guo G-C and Han Z-F 2017 Polarization-basis tracking scheme for quantum key distribution using sifted hey bits *Opt.Lett.* **42** 1023-26
[3] Almeida Á, Muga N, Nuno, Silva N, Prata J, André P and Pinto A 2016 Continuous control of random polarization rotations for quantum communications *J. Lightwave Technol.* **34** 3914–22
[4] Chen H, Li M, Liang W-Y, Wang D, He D-Y, Wang S, Yin Z-Q, Chen W, Guo G-C and Han Z 2016 New scheme for finite-retardation limitations of linear retarders with fixed axes in polarization control *Opt. Commun.* **358** 208–14
[5] A. Duplinskiy, V. Ustimchik, A. Kanapin, V. Kurochkin, and Y. Kurochkin 2017 Low loss QKD optical scheme for fast polarization encoding *Optics express* **25** pp 28886-97

Processing and characterization of GaP nanowires encapsulated into a PDMS large-scale membrane for flexible optoelectronics

S M Mukhangali[1], V Neplokh[1,2], F M Kochetkov[1], V V Fedorov[1], A G Nasibulin[3], S V Makarov[4], R M Islamova[5] and I S Mukhin[1,4]

[1] Saint Petersburg Academic University, 8/3 Khlopina, St. Petersburg 194021, Russia
[2] Peter the Great St. Petersburg Polytechnic University, 29 Politekhnicheskaya, St. Petersburg 195251, Russia
[3] Skolkovo Institute of Science and Technology, 30 Bolshoy Boulevard, bld. 1, Moscow 121205, Russia
[4] ITMO University, 49 Kronverksky pr., St. Petersburg 197101, Russia
[5] Saint Petersburg State University, 7/9 Universitetskaya Emb., St Petersburg 199034, Russia

sungat15004@gmail.com

Abstract. This paper presents the methods of fabricating arrays of semiconductor III-V nanowires transferred into a flexible polydimethylsiloxane membrane. Molecular beam epitaxy was used to synthesize GaP nanowires. The synthesized nanowire arrays were encapsulated into a silicone membrane by a heavy load swinging-bucket centrifuge. For optoelectronic applications, the nanowire/polydimethylsiloxane membranes were contacted with single-walled carbon nanotubes, peeled from the substrate, then the second carbon nanotubes contact was formed. For optical experiments, the nanowire/polydimethylsiloxane membranes were bonded to supporting polydimethylsiloxane films by oxygen plasma treatment, and then easily released from the substrate by unsticking. The obtained membranes have a high practical potential in flexible optoelectronics.

1. Introduction

Nowadays, the demand for flexible optoelectronics is growing every year. Flexible optoelectronic devices allow integration with complex form surfaces and biomedical implants. For example, paper-like displays, foldable touch screens, wearable biosensors, antennas, and surgeon implants are extensively studied [1-3]. Today, the most developed flexible technology is based on organic materials [4]. For instance, organic light-emitting diodes (OLEDs) that are commonly used, have a low manufacturing price and a relatively good performance. However, OLEDs are still behind inorganic materials in terms of values of quantum efficiency (EQE) of 2-30% and brightness of 10^2-10^4 cd/m^2, and a limited lifespan [5,6]. Meanwhile, flexible devices based on inorganic thin films are challenged by a highly complicated transfer to flexible carriers, since it requires laborious operations of etching, structuring, and separation from the growth substrate. The use of nanowires (NWs) array embedded in a flexible polymer is a promising way to solve this problem since it combines the efficiency of inorganic materials and simplicity to manufacture large-scale flexible devices.

2. Fabrication

Self-catalyzed GaP NWs arrays were synthesized by molecular beam epitaxy on 3" Si(111) wafers via a vapor-liquid-solid growth mechanism. GaP and Si have a low lattice mismatch (0.37%) that facilitates epitaxial growth of NWs with a high crystallinity on widely spread Si substrates. However, GaP is an indirect bandgap semiconductor with Eg=2.27 eV at 300K [7], and direct bandgap GaPAs or GaPN(As) films grown on silicon are not widely spread due to the high lattice mismatch between such alloys and GaP buffer layer grown on Si. NW geometry, however, allows a combination of very different crystalline materials of high quality because of the relaxation of stress at NW sidewalls.

To obtain n-conductivity, NWs were intentionally doped by introducing a silicon flux during the growth. It was shown, that despite the amphoteric nature of Si dopant in III–V compounds predominantly n-type conductivity is observed in III-phosphide alloy NWs [8]. More detailed information about the synthesis procedure can be found elsewhere [9,10].

In order to embed GaP NWs into flexible silicone, the graft-copolymers of polydimethylsiloxane and polystyrene (PDMS-st) were G-coated in a swinging-bucket centrifuge instead of a standard spinner [11]. The common spin-coating method is not suitable for dense NW array encapsulation due to a high capillary effect preventing pre-polymer gel penetration between NWs. Thus, G-coating is more suitable, it presses the PDMS-st pre-polymer gel between NWs thanks to the centrifugal force direction along the NWs. Moreover, this method preserves NW verticality and allows achieving the record-thin polymer films.

First, the Sylgard 184 PDMS and PDMS-St were mixed in the ratio of 10:1, perfectly stirred, and left in a desiccator for 20 minutes to extract air bubbles. Second, the array of the NWs was covered by the PDMS-st pre-polymer gel and placed into the heavy load centrifuge (Beckman Coulter Avanti J-HC with a Rotor JS-4.2 A) at 4200 rpm for 40 minutes. After PDMS-St deposition, the samples were crosslinked in a muffle oven at 80 ◦C for 4 h. It should be noted that the thickness of the coated PDMS must be lower than the NWs length for application of single-walled carbon nanotubes (SWCNT) contacts to GaP NW top parts. To control the length of the revealed parts of the nanowires the scanning electron microscopy (SEM) was used (Zeiss SUPRA 25). The achieved length of the NWs top parts can be adjusted during the coating procedure in the centrifuge machine and by the plasma treatment. The membranes were etched in a radio frequency (RF) plasma mixture of O_2 / CF_2 gases. The ratio of the O_2/CF_4 fluxes was 20/40 ml/s respectively, the period of etching was about 20 seconds, and the gas discharge power was 500 W.

SWCNT films were used as flexible, conductive, and transparent electrodes because of their high conductivity, transparency, and elastic properties [12]. SWCNT films of 90 nm thickness were synthesized by floating catalytic chemical vapor deposition (FC-CVD) on nitrocellulose filter paper [13, 14]. The SWCNT contact pads of appropriate size were cut and directly applied onto the sample. After the upper contact fabrication, the NW/PDMS membrane was peeled from the substrate by a steel razor blade (figure 1). After the peeling, NWs on the rear surface of the membrane stuck out by approximately 0.5 µm. Hence, SWCNT contact pads may be applied immediately. Figure 1(f) presents an optical image of a released GaP NWs/PDMS-St membrane of a 3 square inch area.

Large area PDMS membranes with embedded III-V NWs can be also used for optical applications such as IR-to-Vis converters based on second harmonic generation [15]. To fabricate such a converter, the NW/PDMS membranes and 50-150 µm thick PDMS cap films were treated in O_2 plasma to modify the surface of PDMS. Then the cap film was put gently on the PDMS/NW membrane forming strong Si-O-Si bonds due to the O_2 plasma surface modification and the whole structure was unstuck from the substrate (figure 2).

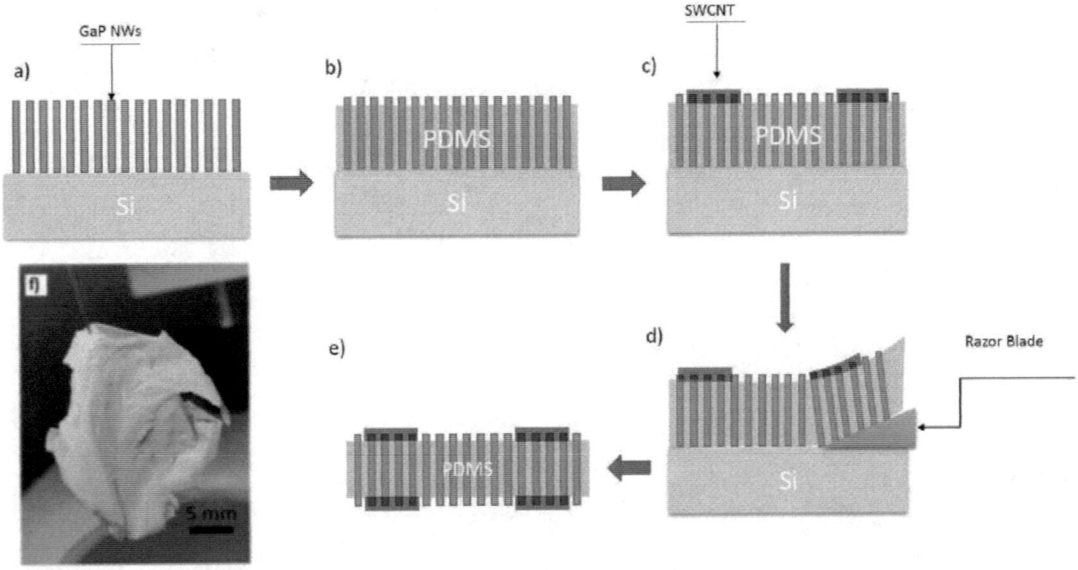

Figure 1. The scheme and of the fabrication procedure of PDMS/GaP NWs membrane (a-e), photo of a representative NW/PDMS-st sample (f).

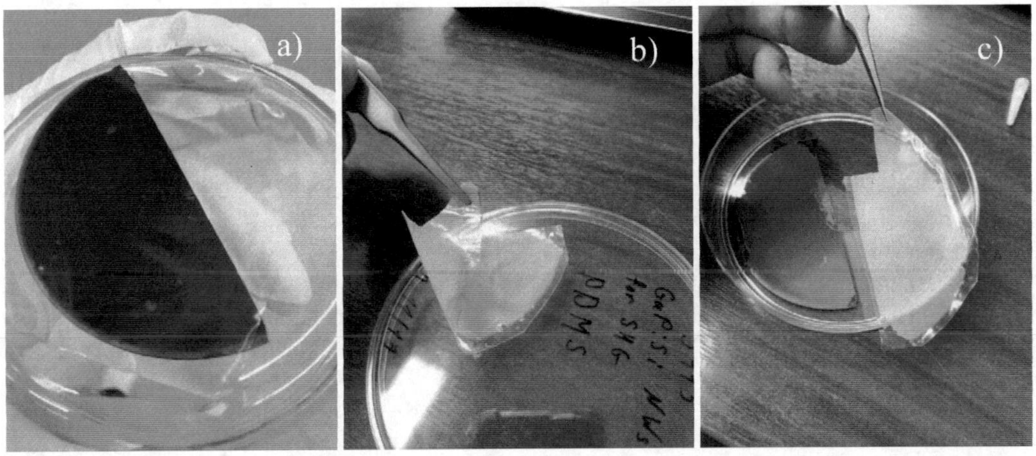

Figure 2. Photos of a PDMS cap film applied directly on a PDMS/GaP NWs substrate (a) and the peeled-off flexible membrane (b,c).

3. Characterization

After the fabrication procedure, the I-V measurements of flexible membranes were performed using a Keithley source meter. The results of measurements for several randomly positioned contact pads are presented in figure 3. The representative I-V results are presented in figure 3. The size of the contact pads was about 0.25 mm^2.

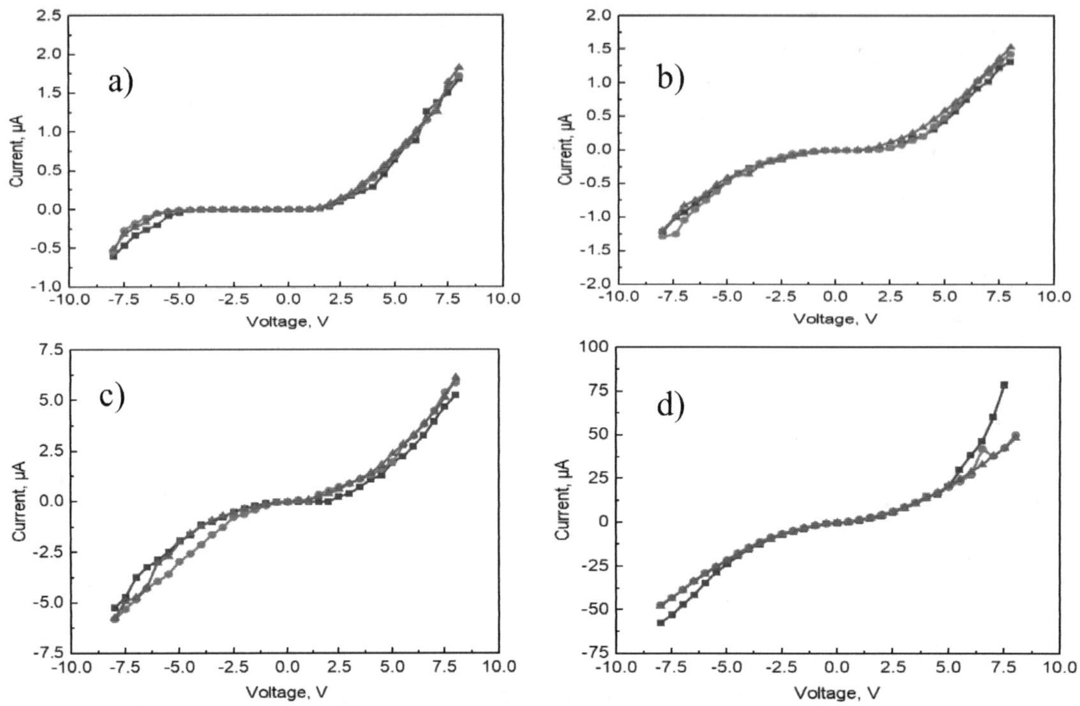

Figure 3. I-V characteristics of the flexible large-scale membrane were measured on the different contact pads (a-d). Colored lines describe the number of measurements.

The form of the current-voltage curve can be explained by the Schottky barrier presence between SWCNT and n-GaP NWs. We consider that SWCNT has a p-type of conductivity [16]. Also, the Schottky shape of the curve indicates that current goes through the NWs and is collected on the SWCNT pads avoiding the shunting effect in the PDMS layer between contiguous NWs. In addition, after 2 months the obtained characteristics were reproduced, and it showed the stability of the electrical properties of the membrane (figure 4) up to 8 V with the current density of 3 A/cm^2.

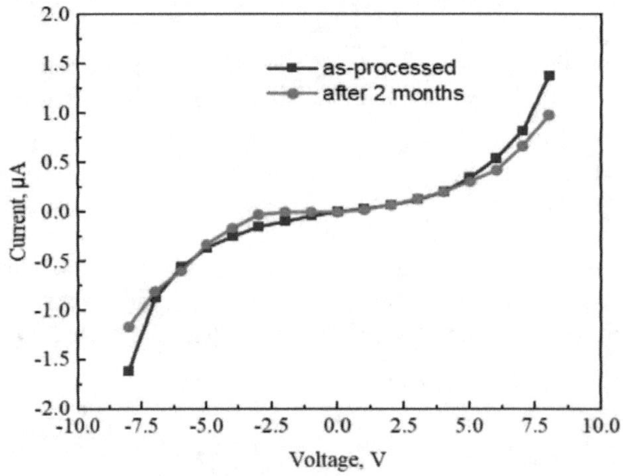

Figure 4. Measured I-V characteristics of the contact pads of the PDMS/GaP flexible membrane recently after the fabrication (blue) and after 2 months (red).

As for the NW/PDMS IR-to-Vis converters, optical characterization of the flexible PDMS/GaP membrane was performed to determine specifically spectral dependence of the SHG in GaP NWs having a high value of the quadratic nonlinear susceptibility ($\chi^2 \approx 70$ pm/V at approximately 1 μm wavelength) and the broad range of transparency. Because of almost similar refractive indexes of GaP (n ≈ 3.4) and Si (n ≈ 3.9), it results in leakage of optical power in NWs that is transferred into the Si substrate. It might be a limiting factor of this design. However, embedded GaP NWs into the flexible PDMS polymer solves this problem and provides new applications of functional optoelectronics [15]. The results are presented in figure 5.

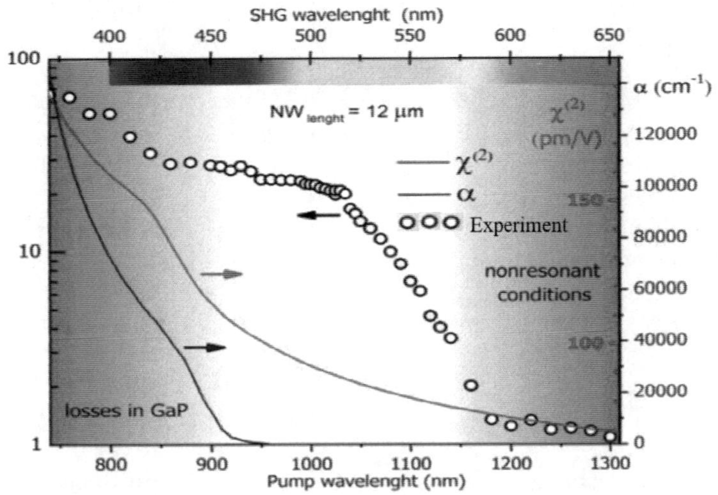

Figure 5. The experimental spectral dependence of the GaP SHG (white dots, left axis). Red and blue curves describe the second-order susceptibility and absorption coefficient of the bulk GaP [17] respectively.

4. Conclusions

In conclusion, this work demonstrated a functional large area flexible membrane based on vertically oriented GaP NWs in the silicon matrix that can be used for LEDs or IR-to-Vis light converters. The developed method has a high potential for applications in flexible optoelectronics.

Acknowledgements

SMM, VN, FMK, VVF, ISM thank the Ministry of Science and Higher Education of the Russian Federation (Grant 0791-2020-0005) for support of the NW synthesis and morphology characterization. VN thanks the Russian Foundation for Basic Research (RFBR project № 19-32-60040) for PDMS/NW membrane fabrication.

References

[1] Li H, Cao Y, Wang Z and Feng X 2019 *Opt. Mater. Express* **9**(10) 4023-4049
[2] Seo J H, Swinnich E, Zhang Y Y and Kim M 2020 *Mater. Res. Lett.* **8**(4) 123-144
[3] Zgierski-Johnston C M, Ayub S, Fernández M C, Rog-Zielinska E A, Barz F, Paul O, Kohl P and Ruther P 2020 *Prog. Biophys. Mol. Biol.* **154** 51-61
[4] Huh B IHS Technology 2015 OLED Display Market Tracker https://cdn.ihs.com/www/pdf/1118/abstract-oled-displaymarket-tracker.pdf (Accessed 15 April 2021)
[5] Salehi A, Fu X, Shin D H and So F 2019 *Adv. Funct. Mater.* **29**(15) 1808803
[6] Dai X, Messanvi A, Zhang H, Durand C, Eymer J, Bougerol C, Julien H and Tchernycheva M 2015 *Nano Lett.* **15**(10) 6958-6964
[7] Fedorov V V, Bolshakov A D, Dvoretckaia L N, Sapunov G A, Kirilenko D A, Mozharov A M,

Shugurov K Yu, Shkoldin V A, Cirlin G E and Mukhin I S 2018 *Semiconductors* **52**(16) 2092-2095

[8] Kochetkov F M, Neplokh V, Fedorov V V, Bolshakov A D, Sharov V A, Eliseev I E, Tchernycheva M Cirlin G E, Nasibulin A G, Islamova R M and Mukhin I S 2020 *Nanotechnology* **31**(46) 46LT01

[9] Bolshakov A D, Fedorov V V, Sibirev N V, Fetisova M V, Moiseev E I, Kryzhanovskaya N V, Koval O Yu, Ubyivovk E V, Mozharov A M, Cirlin G E and Mukhin I S 2019 *Phys. Status Solidi RRL* **13**(11) 1900350

[10] Trofimov P, Pushkarev A P, Sinev I S, Fedorov V V, Bruyere S, Bolshakov A, Mukhin I S and Makarov S V 2020 *ACS Nano* **14**(7) 8126-8134

[11] Neplokh V, Kochetkov F M, Deriabin K V, Fedorov V V, Bolshakov A D, Eliseev I E, Mikhailovskii V Y, Ilatovskii D A, Krsnikov D V, Tchernycheva M, Cirlin G E, Nasibulin A G, Mukhin I S and Islamova R M 2020 *J. Mater. Chem.* C **8**(11) 3764-3772

[12] Khabushev E M, Krasnikov D V, Zaremba O T, Tsapenko A P, Goldt A E and Nasibulin A G 2019 *J. Phys. Chem. Lett.* **10**(21) 6962-6966

[13] Moisala A, Nasibulin A G, Brown D P, Jiang H, Khriachtchev L, and Kauppinen E I 2006 *Chem. Eng. Sci.* **61**(13) 4393-4402

[14] Nasibulin A G, Ollikainen A, Anisimov A S, Brown D P, Pikhitsa P V, Holopainen S, Penttila J S, Helisto P, Ruokolainen J, Choi M and Kauppinen E I 2008 *Chem. Eng. Sci.* **136**(2-3) 409-413

[15] Fedorov V V, Bolshakov A, Sergaeva O, Neplokh V, Markina D, Bruyere S, Saerens G, Petrov M I, Grange R, Timofeeva M, Makarov S V and Mukhin I S 2020 *ACS Nano* **14**(8) 10624-10632

[16] Tune D D, Flavel B S, Krupke R and Shapter J G 2012 *Adv. Energy Mater.* **2**(9) 1043-1055

[17] Hughes J L and Sipe J E 1996 *Phys. Rev. B* **53**(16) 10751

Simulation of double-junction III-phosphides/silicon solar cells

A A Rakitina[1] **and A S Gudovskikh**[1,2]

[1] St. Petersburg Electrotechnical University "LETI", 197376 St Petersburg, Russia.
[2] St. Petersburg Academic University of RAS, 194021 St Petersburg, Russia.

aarakitina@stud.etu.ru

Abstract. The characteristics of two-junction solar cells were calculated in this work, in which phosphides of group III are used as the top junction, and Si as the bottom junction. The following ternary phosphide compounds $Ga_{0.52}In_{0.48}P$ (1.85 eV), $GaPN_{0.02}$ (1.9 eV), and $GaPN_{0.04}$ (1.7 eV) were considered as the active material of the top junction. The use of $GaPN_{0.04}$ allowed one to reach the current matching with the bottom Si junction and an efficiency of 30% was achieved. In addition, the influence of the layer thickness, the lifetime of minority charge carriers and the electronic affinity on the efficiency of solar energy conversion were considered.

1. Introduction

The development of solar energy requires a further increase in conversion efficiency. However, silicon single-junction solar cells (SCs), which are the basis of terrestrial photovoltaics, have practically reached the efficiency limit [1]. A further increase in efficiency is possible with an increase in the number of junctions. One of the most promising ways to create two-junction SC based on Si is an integration with III-V compounds. The formation of III-V layers on the silicon surface is a complex technological task that requires the use of expensive epitaxial methods, high growth temperatures, and also imposes restrictions on the choice of materials in terms of the crystal lattice parameter, which together is a significant obstacle to the development of this path. However, it has recently been shown that epitaxial and microcrystalline layers of III-phosphides can be obtained by plasma deposition at low temperatures [2]. The possibility of mass production of this technology is promising for photovoltaics. In this work, using computer simulation, an assessment was made of the potential possibility of using III-phosphides as the top junction of a double-junction SC with a bottom junction based on a Si substrate. The key feature of this simulation is that potentially defective phosphide layers (for instance due to microcrystalline structure) were taken into account. Therefore, the p-i-n structure was considered, and the effect of defects was also taken into account.

2. Simulation of top junction

Computer simulations were carried out using the AFORS-HET v2.5 software. The modeling was carried out in two stages. At the first stage of modeling, the calculation was carried out without taking into account the effect of defects, at the second stage the defects were taken into account. The following phosphides were considered as active materials of the top junction: $Ga_{0.52}In_{0.48}P$ with a band gap of

1.85 eV, in which $Al_{0.53}In_{0.47}P$ and GaP layers were used as a wide-gap window and a back potential barrier (BSF). For subsequent calculations, the doping level was taken equal to 10^{18} cm^{-3} for the n- and p-layers and 10^{12} cm^{-3} for the i-layer. Schematic representations of these structures are presented in Figure 1 a, b.

a) b)

Figure 1. Schematic representation of the top junction on GaP (a) and $Al_{0.53}In_{0.47}P$ (b).

The radiative recombination rate was taken equal to $1\cdot10^{-13}$ cm^3/s, the values of the Auger recombination coefficients for electrons and holes were $1\cdot10^{-30}$ cm^6/s. The phosphides $GaPN_{0.02}$ and $GaPN_{0.04}$ were also considered for the top junction. The given nitrogen concentrations of 2% and 4% provide the band gap equal to 1.9 eV and 1.7 eV, respectively. Schematic representations of these structures are presented in Figure 2 a, b.

a) b)

Figure 2. Schematic representation of the top junction with an i-layer based on $GaPN_{0.04}$ (a) and $GaPN_{0.02}$ (b).

The main parameters of the materials used in the calculation are shown in Table 1.

Table 1. Material parameters of the upper transition layers.

Layers	ε_0	E_g, eV	χ, eV	μ_n/μ_p, $cm^2V^{-1}s^{-1}$	N_C, cm^{-3}	N_V, cm^{-3}
n-GaP	11.1	2.26 [3]	3.8	250/150	$1.8\cdot10^{19}$	$1.9\cdot10^{19}$
n- $Al_{0.53}In_{0.47}P$	11.9	2.35 [4]	3.78 [6]	100/50	$2\cdot10^{19}$	$2\cdot10^{19}$
i- $Ga_{0.52}In_{0.48}P$	11.8	1.9 [5]	3.8 [5]	500/30 [7, 8]	$1.9\cdot10^{19}$	$1.9\cdot10^{19}$
i - $GaPN_{0.04}$	11.1	1.7[3]	4.08	50/50 [3]	$1.8\cdot10^{19}$	$1.9\cdot10^{19}$
i - $GaPN_{0.02}$	11.1	1.9[3]	4.08	50/50 [3]	$1.8\cdot10^{19}$	$1.9\cdot10^{19}$
p-GaP	11.1	2.26[3]	3.8	250/150	$1.8\cdot10^{19}$	$1.9\cdot10^{19}$
p- $Al_{0.53}In_{0.47}P$	11.9	2.35 [4]	3.78 [6]	100/50	$2\cdot10^{19}$	$2\cdot10^{19}$

Based on the results of the I-V curve calculations and the performed variation in thickness, it was found that the structure in which n and p layers based on GaP, and an i-layer based on $Ga_{0.52}In_{0.48}P$ have a high FF, which indicates that there is no limitation for the transport of charge carriers. In addition, high V_{OC} = 1651 mV, J_{SC} = 13.95 mA/cm2, FF = 91.55% and efficiency = 20% are provided. Also, according to the results of calculations, it was found that the structure based on $GaPN_{0.02}$ is significantly inferior in output parameters to the structure of $GaPN_{0.04}$. In particular, this affects the maximum achievable value of the short-circuit current density. It is significantly lower compared that for the structure based on $GaPN_{0.04}$. With a thickness of 1.4 μm, a current density of 17.5 mA/cm^2 is achieved for $GaPN_{0.02}$. For $GaPN_{0.04}$, at the same thickness, a current density of 22.5 mA/cm^2 was reached. This is due to the band gap difference, for the higher gap less solar radiation is absorbed in the material, therefore, the current density is lower. As a result, the efficiency value is also lower. For $GaPN_{0.04}$ the efficiency was 22.34%, for $GaPN_{0.02}$ - 20.79%.

At the second stage of the simulation, the calculation was carried out for heterostructures in which the $Ga_{0.52}In_{0.48}P$ (1.85 eV) and $GaPN_{0.04}$ (1.7 eV) layers were used as the top junction. The peculiarity of the second stage of modeling was the introduction of defects. An acceptor-type defect with a point distribution was introduced into the $Ga_{0.52}In_{0.48}P$ layer. The energy position of the defects was 1.1 eV, and the concentration was $1\cdot10^{14}$ cm^{-3}. The calculation of the dependence of the SC parameters on the lifetime of charge carriers in $Ga_{0.52}In_{0.48}P$ (which was varied in the range of 10^{-7} - 10^{-12} s) due to the variation of the capture cross-section for electrons and holes (10^{-14} ... 10^{-9} cm^2) was performed. The resulting dependencies are shown in Figure 3 (a). The V_{OC} monotonically decreases with lifetime decrease, while J_{SC} drastically drops for lifetime below 10^{-9} s.

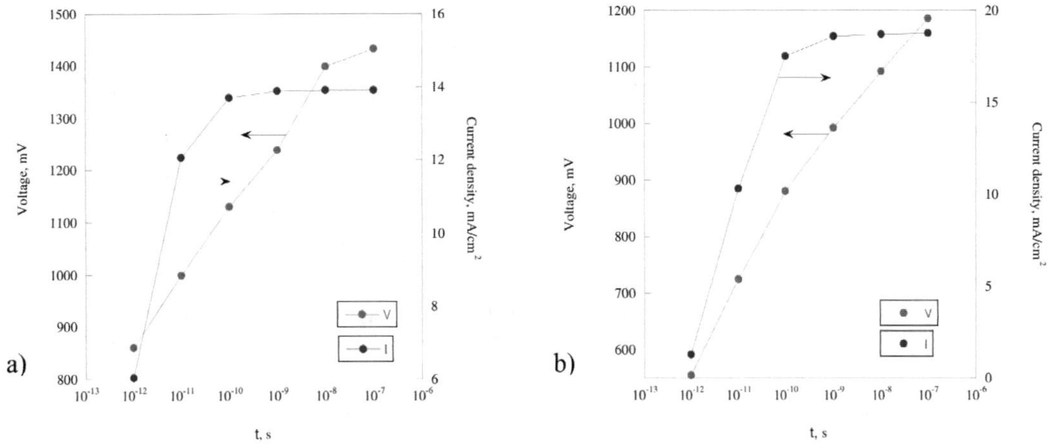

Figure 3. Influence of the lifetime on the current density and the voltage of $Ga_{0.52}In_{0.48}P$ (a) and $GaPN_{0.04}$ (b).

An acceptor-type defect with a point distribution was introduced into the $GaPN_{0.04}$ layer. The energy position of the defects was 0.85 eV. The influence of the lifetime of charge carriers in $GaPN_{0.04}$ as well as the valence band offset (ΔE_V) at the GaPN/GaP interface on the SC performance was explored. The variation of the lifetime is shown in Figure 3 (b). It showed that the values should be at least 10^{-9} s to achieve the current matching with bottom Si junction. The value of ΔE_V at the GaPN/GaP interface, which is not exactly known, has a significant influence on transport properties. An increase of ΔE_V leads to the formation of a potential barrier at the i-GaPN/p-GaP interface that prevents the transport of holes (Figure 4 a) and, as a consequence, leads to a significant decrease in the fill factor of the I–V curve. Indeed, for $\chi = 4.08$ eV ($\Delta E_V = 0.3$ eV) FF is equal to 85.12%, while for $\chi = 3.8$ eV ($\Delta E_V = 0.6$ eV) FF= 63.69%. This transport limitation due to large ΔE_V significantly manifested in the dependence of the short-circuit current on the layer thickness, at which a sharp drop of Jsc is observed when thickness reaches 0.8 μm (Figure 4 b). Thus, the influence of band discontinuities must be taken into account when analyzing experimental data. For the further calculations, ΔE_V at the interface was taken equal to 0.3 eV ($\chi = 4.08$ eV).

3. Simulation of double-junction solar cell

To achieve maximum efficiency of double-junction SC it is necessary to fit current matching condition. The top junction based on $Ga_{0.52}In_{0.48}P$ or $GaPN_{0.04}$ (1.7 eV) should have practically the same short-circuit current as the bottom junction based on n-GaP/p-Si. The performed variation in thickness for $GaPN_{0.04}$ showed that the thickness at which the current matching is ensured is 0.48 μm, while for $Ga_{0.52}In_{0.48}P$ this thickness is 0.3 μm.

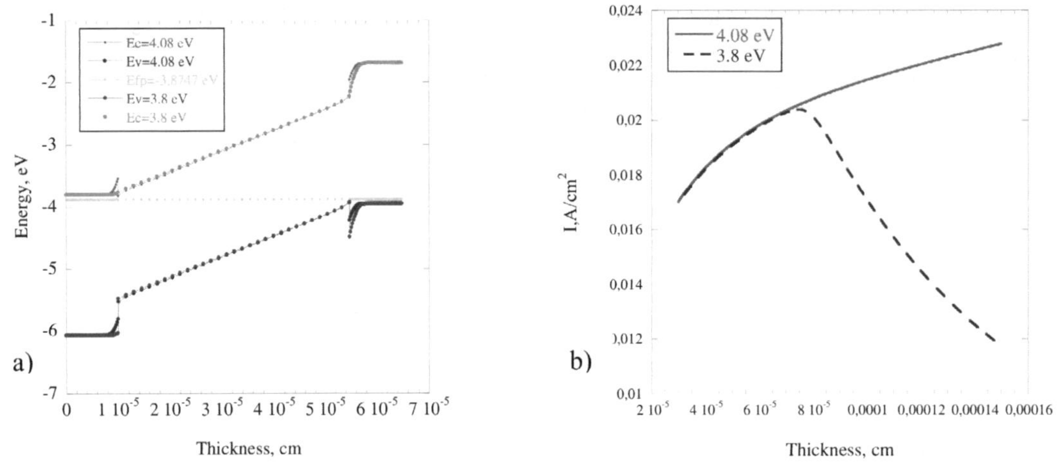

Figure 4. $GaPN_{0.04}$ band diagram (a) and I-V characteristic (b).

The final stage of the calculation is the calculation of the characteristics of double-junction SCs based on group III phosphides - $Ga_{0.52}InP_{0.48}$ and $GaPN_{0.04}$ (1.7 eV) and silicon. A schematic representation of the GaPN/Si and the I-V curves are shown in Figure 5 (a, b). The main parameters of the materials used in the calculation of bottom junction are given in Table 2. The values of doping levels and layer thicknesses for $Ga_{0.52}In_{0.48}P$/Si and $GaPN_{0.04}$/Si double-junction SC are given in Table 3, 4, respectively.

Table 2. Material parameters of heterostructure layers.

Layers	ε_0	E_g, eV	χ, eV	μ_n/μ_p, $cm^2V^{-1}s^{-1}$	N_C, cm^{-3}	N_V, cm^{-3}
p-Si	11.7	1.124	4.05	1107/424.6	$3.2\cdot10^{19}$	$1.8\cdot10^{19}$
(p) a-Si	11.9	1.6	3.85	10/1	$2\cdot10^{20}$	$2\cdot10^{20}$

Table 3. $Ga_{0.52}In_{0.48}P$/Si 2J SC structure.

Layers	Thickness, μm	Doping level, cm^{-3}
n-GaP	0.1	10^{18}
i- $Ga_{0.52}In_{0.48}P$	0.3	10^{12}
p-GaP	0.1	10^{18}
Tunnel junction		
n-GaP	0.05	10^{19}
p - Si	300	10^{16}
(p) a-Si	0.03	10^{19}

Table 4. $GaPN_{0.04}$/Si 2J SC structure.

Layers	Thickness, μm	Doping level, cm^{-3}
n-GaP	0.1	10^{18}
i- $GaPN_{0.04}$	0.448	10^{12}
p-GaP	0.1	10^{18}
Tunnel junction		
n-GaP	0.05	10^{19}
p - Si	300	10^{16}
(p) a-Si	0.03	10^{19}

The obtained I-V curves of the heterojunction structure demonstrates the best current matching and also provides high values of the open circuit voltage. Thus, it is possible to achieve a high value of the short-circuit current density J_{SC} = 18.75 mA/cm^2 and V_{OC} = 1.92 V. As a result, the reaches 30%. A schematic representation of the structure based on GaInP/Si and the I-V characteristic of the heterojunction are shown in Figure 6 a, b, respectively.

According to the calculation results, it was found that the top junction based on $GaPN_{0.04}$ (1.7 eV) provides higher values of the short-circuit current compared to $Ga_{0.52}In_{0.48}P$. For $GaPN_{0.04}$ (1.7 eV), V_{OC} = 1.92 V, and J_{SC} = 18.8 mA, while for $Ga_{0.52}In_{0.48}P$ the same parameters are 2.14 V and 15.47 mA respectively. As a result, the efficiency is also higher: 30.8% for $GaPN_{0.04}$ (1.7 eV) versus 29.5% for $Ga_{0.52}In_{0.48}P$. The calculation results are presented in Table 5.

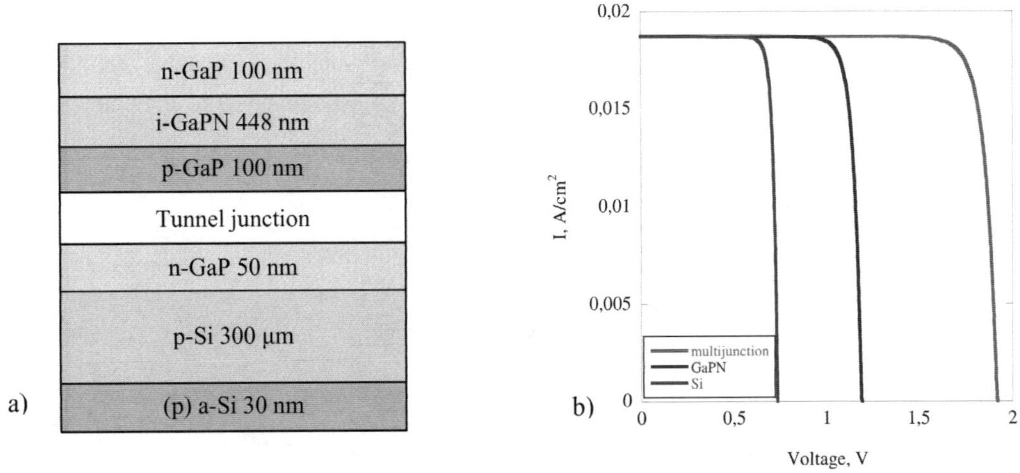

Figure 5. Schematic representation of the construction of $GaPN_{0.04}$/Si (a) and I-V characteristic (b).

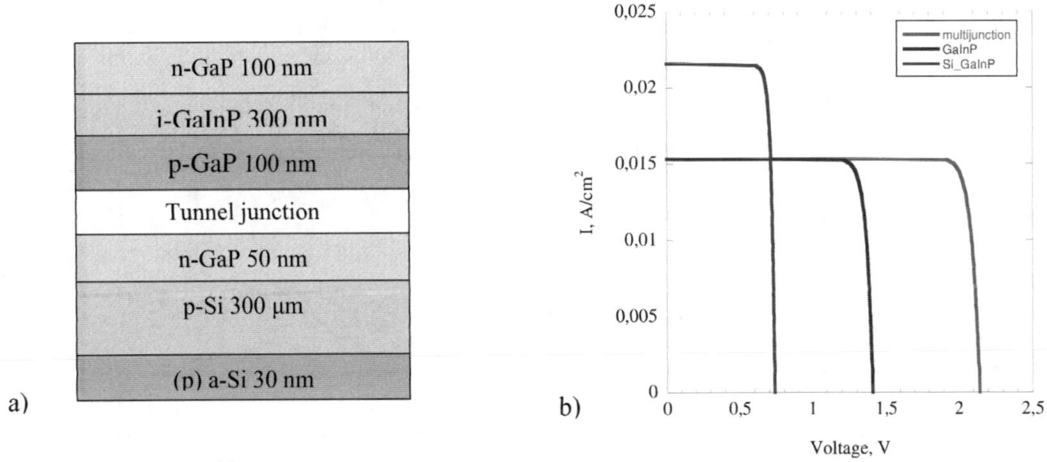

Figure 6. Schematic representation of the construction of $Ga_{0.52}InP_{0.48}$/Si (a) and I-V characteristic (b).

Table 5. Calculation results.

Structure	V_{OC}, mV	J_{SC}, mA/cm^2	FF, %	Eff, %
$GaPN_{0.04}$/Si	1.92	18.75	85.6	30.82
$Ga_{0.52}In_{0.48}P$/Si	2.14	15.47	89.2	29.5

The calculation of the spectral characteristics was also carried out. The spectral dependences of the quantum efficiency of two-junction SC based on GaPN/Si and GaInP/Si are shown in Figure 7.

The spectral characteristic also shows that the structure based on $GaPN_{0.04}$ makes it possible to achieve a higher short-circuit current due to a wider spectrum of absorbed photons. However, the influence of defects in the layers and band offset at the $GaP/GaPN_{0.04}$ interface can minimize this advantage. Thus, further experimental studies of the properties of layers and structures based on group III phosphides grown on silicon are required.

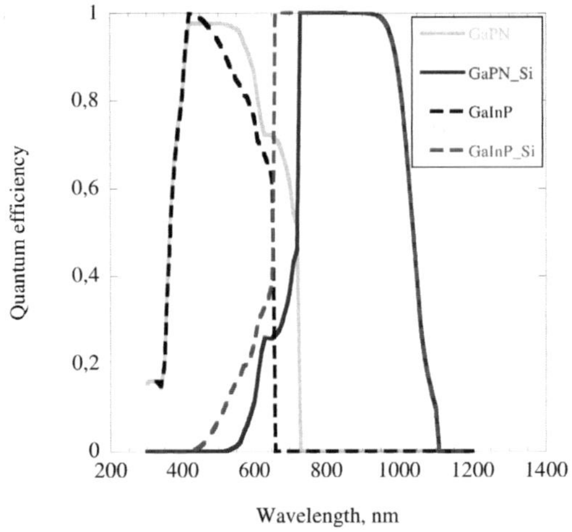

Figure 7. Spectral characteristics of two-junction SCs based on $GaPN_{0.04}$ (1.7 eV) and $Ga_{0.52}In_{0.48}P$.

4. Conclusion

The simulation showed that the use of SCs based on group III phosphides and silicon such as $GaPN_{0.04}$/Si and $Ga_{0.52}InP_{0.48}$/Si is a promising way to reach high efficiency. The influence of the thickness of the i-layer of the upper junction on the filling factor of the I–V characteristic is shown. The influence of the lifetime of charge carriers on the efficiency of solar energy conversion was also considered, and it was found that in order to achieve the best matching in terms of current density, this value should reach at least 10^{-9} s for both structures. In addition, for the $GaPN_{0.04}$ structure, the influence of band discontinuity on the transport of charge carriers was considered. Based on the results of calculations, it was found that the most promising application can find a structure based on $GaPN_{0.04}$, since higher J_{SC} values are provided for it and the efficiency reaches 30%.

Acknowledgments

This work was supported by RFBR research project #20-08-00870.

References

[1] Yoshikawa K et al 2017 *Nat. Energy* **2** 17032
[2] Uvarov A V, Gudovskikh A S, Nevedomskiy V N, Baranov A I, Kudryashov D A, Morozov I A and Kleider J-P 2020 *J. Phys. D: Appl. Phys.* **53** 345105
[3] Kudryashov D A, Gudovskikh A S, Nikitina E V and Egorov A Yu 2014 Development of the design of multi-junction solar cells based on GaPNAs / Si heterostructures by the method of computer simulation *Semiconductor Phys. and Engineering* **48** 396-401
[4] Bour D P, Shealy J R, Wicks G W and Schaff W J 1987 Optical properties of Al x In1− x P grown by organometallic vapor phase epitaxy *Appl. Phys. Lett.* **50** 615-7
[5] Bewley W W, Lindle J R, Kim C S, Kim M, Canedy C L, Vurgaftman I and Meyer J R 2008 Lifetimes and Auger coefficients in type-II W interband cascade lasers *Appl. Phys. Lett.* **93** 041118
[6] Gudovskikh A S, Kaluzhniy N A, Lantratov V M, Mintairov S A, Shvarts M Z and Andreev V M 2008 Numerical modelling of GaInP solar cells with AlInP and AlGaAs windows *Thin Solid Films* **516** 6739-43
[7] Shitara T and Eberl K 1994 Electronic properties of InGaP grown by solid-source molecular-beam epitaxy with a GaP decomposition source *Appl. Phys. Lett.* **65** 356-8
[8] Ikeda M and Kaneko K 1989 *J. Appl. Phys.* **66** 5285-89

Numerical modelling of non-planar GaN LED with CNT top contact

S A Raudik[1, 3], D M Mitin[2, 1], A A Vorobyev[1], A M Mozharov[1], A G Nasibulin[3] and I S Mukhin[1]

[1] Saint Petersburg Academic University, 8/3 Khlopina, St. Petersburg 194021, Russia
[2] Peter the Great St. Petersburg Polytechnic University, 29 Politekhnicheskaya, St. Petersburg 195251, Russia
[3] Skolkovo Institute of Science and Technology, 30 Bolshoy Boulevard, bld. 1, Moscow 121205, Russia

raudik_sa@mail.ru

Abstract. In this paper, the theoretical study of LED based on GaN NWs with carbon nanotubes (CNT) top contact has been presented. The main electrical and optical characteristics of LED have been numerically calculated. In a 0.5 x 0.5 mm NWs array, the Ohmic losses in CNTs were 2.7% with an operating current density of 50 A/cm^2. It proves the possibility of using CNTs as transparent contact.

1. Introduction

In the last decades, III-Nitrides attract attention as materials for optoelectronic devices due to their outstanding properties. AlN, GaN, and InN and their alloys have a direct bandgap varied from 0.7 eV to 6.2 eV [1]. It allows producing light emission diodes (LEDs) with a broad range of emitted wavelengths.

Recent progress in the development of wearable and bio-compatible devices faced a challenge to get stretchability and flexibility of electronics and, in particular, LEDs. These properties can be achieved with III-Nitrides by a transition from conventional planar geometry to 1D structures such as nanorods or nanowires (NWs) [2].

Using NWs solves many inherent problems of thin-film LEDs. Firstly, dislocation density declines in GaN nanowires with a high surface-to-volume ratio. Also, the more intensive strain relaxation was observed in comparison with planar GaN, which led to decreasing internal electrical fields induced by lattice mismatch with traditional substrates such as sapphire and Si [3]. Secondly, such structure changes suppress the waveguide effect on interfaces GaN/air (n=2.4) and promote light extraction [4]. Moreover, such geometry gives additional advantages, the main of which is the ability to control a radiation pattern [5] and wavelength of emitted light [6] by changing the structure sizes and shape.

To take all advantages mentioned before, the contacts to a LED should be transparent and retain their properties upon deformation. The literature reports on the use of conductive composites (silver nanowires/nanoparticles [7] or reduced graphene oxide penetrating into an elastic polymer [8]), woven conductive fibers/filaments [9], ionic hydrogels [10] and liquid metal as stretchable electrodes [11]. However, all of these solutions have their drawbacks. The resistance of silver nanowires increases dramatically when stretched due to the high Young's modulus. Reduced graphene oxide also loses its conductivity as it stretches. Some ionic hydrogels have good stretchability, but insufficient electrical conductivity can result in poor device performance. The gallium-based liquid metal exhibits excellent tensile and recyclability, however, it is extremely susceptible to oxidation and has a high surface tension, which affects its fluidity and makes it difficult to use in nanoscale systems.

Carbon nanotubes (CNTs) can be considered as rolled nanowire graphene sheets, combining the advantages of graphene and silver nanowires, providing low cost, production scalability, ease of use, and improved current collection [11]. The latter properties, together with antireflection characteristics and high stability, make CNTs one of the most promising transparent flexible and extensible contact materials. Despite the fact that the literature contains data on preliminary studies of CNT layers in optoelectronics, their potential has not yet been fully disclosed. This last statement is especially true for high performance stretchable LEDs based on GaN NWs. Therefore, this work is devoted to the numerical modeling of GaN NWs-based LED with CNT top contact to determine its optoelectronic characteristics.

2. Modeling

The considered LED is an array of axial heterojunction GaN NWs with inserted InGaN quantum well. In this system, the current flows mostly vertically due to the geometrical restriction. Thus, each NW separately could be modeled as a 1D structure instead of 3D or 2D with axial symmetry. The modeled NWs contain the n-doped GaN bottom barrier, InGaN quantum well, undoped GaN top barrier, p-doped AlGaN layer, and p-doped GaN.

In the area of InGaN quantum well, the Schroedinger-Poisson equation has been solved to determine the density of states in the quantum well and calculate charge tunneling through it. The modeled energy diagram is presented in Figure 1.

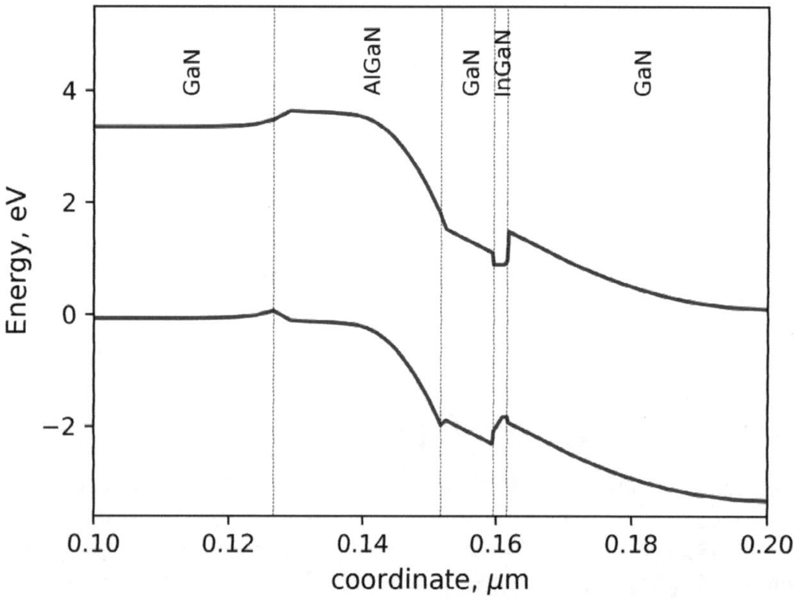

Figure 1. Band diagram of the NWs active area.

Also the current-voltage characteristic and emission spectra of a single NW have been obtained in the calculation process. The calculated at different voltages light emission spectra are presented in Figure 2. It have a maximum power density at the wavelength of 459-460 nm, which matches with experimental data [12].

In the next step, an array of NWs has been modeled by integrating currents over the surface, taking into account the optical losses in the CNT layer. The voltage drop in the CNT was taken into consideration to calculate the dependency of current flows through the NWs on coordinate. The resulting currents spreading in CNT are presented in Figure 3. The Joule heat was calculated by the

Ohm laws using the obtained current distribution. The 0.5×0.5mm CNT layer generates heat of 15.4 mW. The input power was calculated from applied voltage and integrated current. Based on the previous data, we calculated relative Ohmic losses in the CNT layer, which were 2.7% at the current density of 50 A/cm^2.

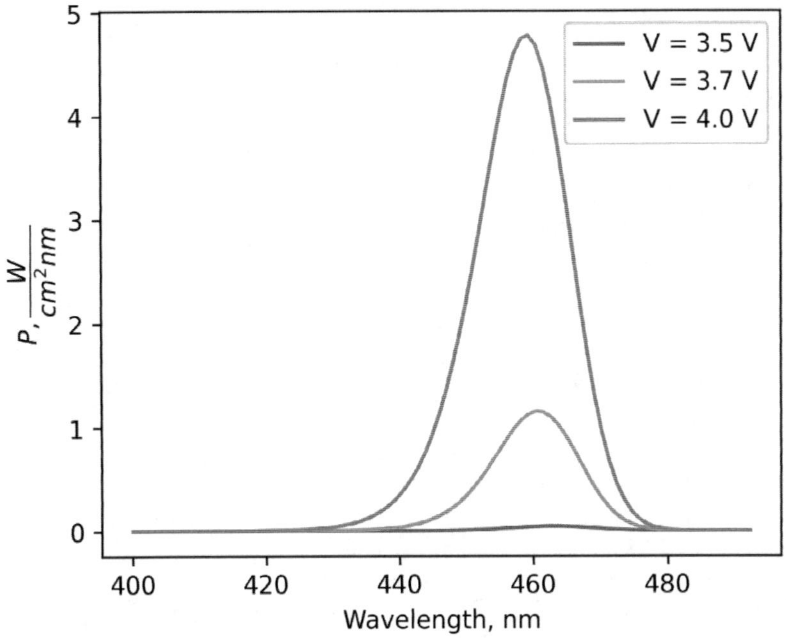

Figure 2. Photoluminescence spectra for the different applied voltages.

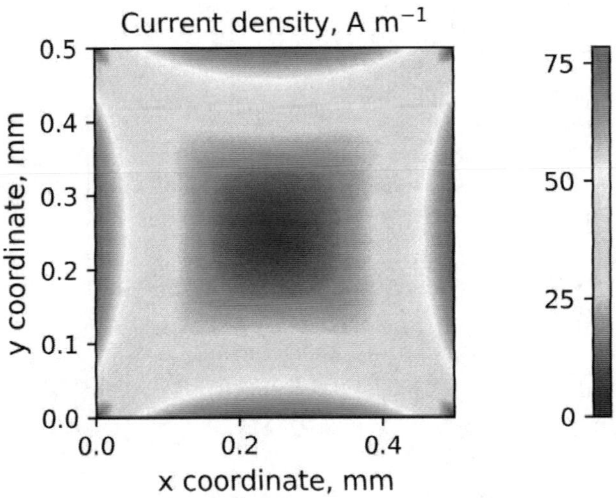

Figure 3. Distribution of currents in the CNT layer.

3. Conclusions

The numerical model of GaN NWs-based LED with CNT contact has been implemented. The main characteristics of LEDs have been obtained. The calculated at different voltages emission spectra show a maximum power density at the wavelength of 459-460 nm. In a 0.5×0.5 mm NWs array, the Ohmic losses in CNTs were 2.7% with an operating current density of 50 A/cm^2. The obtained results demonstrate the possibility of using CNTs as a material for contacts to NWs LEDs.

Acknowledgments
The reported study was funded by RFBR, project number 19-38-60008 (modelling of energy diagram), the Ministry of Science and Higher Education of the Russian Federation (grant 0791-2020-0005).

References

[1] Adachi S 2009 *Properties of semiconductor alloys: group-IV, III-V and II-VI semiconductors* vol 28 (John Wiley & Sons)
[2] Dai X, Messanvi A, Zhang H, Durand C, Eymery J, Bougerol C, Julien F H and Tchernycheva M 2015 *Nano letters* **15** 6958–6964
[3] Bai J, Wang Q and Wang T 2012 Journal of Applied Physics **111** 113103
[4] Tsai Y L, Lai K Y, Lee M J, Liao Y K, Ooi B S, Kuo H C and He J H 2016 *Progress in Quantum Electronics* **49** 1–25
[5] Soshnikov B, Lysak V, Kotlyar K and Soshnikov I 2016 *Journal of Physics: Conference Series* vol 741 (IOP Publishing) p 012154
[6] Sekiguchi H, Kishino K and Kikuchi A 2010 *Applied physics letters* **96** 231104
[7] Kang H, Zhao C, Huang J, Ho D H, Megra Y T, Suk J W, Sun J, Wang Z L, Sun Q and Cho J H 2019 *Advanced Functional Materials* **29** 1903580
[8] Kim S, Gupta M K, Lee K Y, Sohn A, Kim T Y, Shin K S, Kim D, Kim S K, Lee K H, Shin H J *et al.* 2014 *Advanced materials* **26** 3918–3925
[9] Kwak S S, Kim H, Seung W, Kim J, Hinchet R and Kim S W 2017 *ACS nano* **11** 10733–10741
[10] Liu T, Liu M, Dou S, Sun J, Cong Z, Jiang C, Du C, Pu X, Hu W and Wang Z L 2018 *ACS nano* **12** 2818–2826
[11] Yang Y, Han J, Huang J, Sun J, Wang Z L, Seo S and Sun Q 2020 *Advanced Functional Materials* **30** 1909652
[12] Tsatsulnikov A F, Zavarin E E, Kryzhanovskaya N V, Lundin V V, Saharov A V, Usov S O, Brunkov P N, Goncharov V V, Cherkashin N A and Hytch M 2010 *Physics and technology of semiconductors* **44** 1382–1386

Photon counting statistics with imperfect detectors

Aleksei Reutov[4,5] **and Denis Sych**[1,2,3]

[1] P.N.Lebedev Physical Institute, Russian Academy of Sciences, Moscow, 119991, Russia
[2] NTI Center for Quantum Communications, National University of Science and Technology MISiS, Leninsky prospekt 4, Moscow 119049, Russia
[3] Sirius University of Science and Technology, 1 Olympic Ave, 354340, Sochi, Russia
[4] Moscow Institute of Physics and Technology, Dolgoprudny 141700, Russia
[5] QRate, Skolkovo, Moscow 143025, Russia

E-mail: aleksey.reutov@phystech.edu

Abstract. Measurement of photon statistics is an important tool for the verification of quantum properties of light. Due to the various imperfections of real single photon detectors, the observed statistics of photon counts deviates from the underlying statistics of photons. Here we analyze statistical properties of coherent states, and investigate a connection between Poissonian distribution of photons and sub-Poissonian distribution of photon counts due to the detector dead-time corrections. We derive a functional dependence between the mean number of photons and the mean number of photon counts, as well as connection between higher-order statistical moments, for the pulsed or continuous wave coherent light sources, and confirm the results by numerical simulations.

1. Introduction

Quantum properties of light play the key role in various research areas of modern quantum technologies, for example, in quantum communications. There is a class of modern quantum key distribution protocols which security relies on the use of pure coherent states [1, 2]. The modern implementations of various single-photon quantum key distribution protocols, including the original BB84 protocol, also rely on phase-randomized weak coherent state [3–7]. In order to verify applicability of theoretical concepts, such as single photon states or weak coherent states, to the real world applications, such as quantum key distribution, one has to certify the experimental sources of light. Thus the measurement and experimental validation of coherent states is an important practical task. The standard widely employed measurement method for continuous variables is homodyne detection [8, 9]. In particular, one can estimate purity of coherent and single-photon states by measuring two conjugated variables [10]. But the modern developments of novel measurement and control methods employ single photon detection [11–14].

One of the simplest ways to verify a quantum state of light is to measure its photon counting statistics. For the single photon states, it should lead to the well known anti-bunching effect, while for the coherent states we should observe Poissonian distribution of photon counts. However, all the realistic measurements suffer from various imperfections which affect the photon counting statistics and modify the statistical distributions. In particular, detector's dead-time and jitter lead to either sub-Poissonian or super-Poissonian distributions [15, 16].

In this work we show how the detector's dead-time corrections affect the Poissonian distribution for the pulsed or continuous wave coherent light sources. We calculate the mean value of photons, its variance, and high-order moments and determine a connection between the mean number of photons and the mean number of photon counts. Our statistical analysis allows to include the effects of a time-varying repetition rate of laser pulses, source intensity fluctuations and a photo detector jitter. The results can be directly applied to the security enhancement of quantum key distribution protocols, as well as the development of novel protocols that rely on photon counting detectors.

2. Theory and methods

2.1. Coherent source and Poissonian statistics

We can consider laser source emission as coherent states. The mean photon number is determined $\lambda = |\alpha_\lambda|^2$ for every coherent state $|\alpha_\lambda\rangle$. Consider T as the detection time of emission. The distribution of photocounts (photons, detected after the collapse of the wave function) is given by the Poisson distribution:

$$P_{Pois}(n, T) = \frac{\lambda^n}{n!} e^{-\lambda},$$

where n is the number of detected photons.

2.2. Dead-time model

Real devices often has dead-time, time interval during which the detector can not register events. We consider constant dead-time τ and non-paralysable case of dead-time model. Define R and R_d as true rate and detected rate of events, respectively. Then the fraction of all time that the detector is dead is $R_d\tau$. Therefore, the rate decreases by the amount $RR_d\tau$ and it leads to:

$$R_d = \frac{R}{1 + R\tau}.$$

The last equation provides representation of probabilities that any moment of time can be dead ($\mathcal{P}_{dead} = R_d\tau$) or non-dead ($\mathcal{P}_{non-dead} = 1 - R_d\tau$).

2.3. Dead-time correction

We used the cumulative distribution function (CDF), a convenient tool for the statistical analysis of photon count distribution [16]. CDF for a non-dead initial moment of time window of measurement is given:

$$F_k(\lambda,\ T,\ \tau) = \frac{1}{k!}\gamma(k+1,\ \lambda(1 - k\tau/T)) \cdot \theta(T - k\tau), \tag{1}$$

where τ, T, λ are dead-time, time window of measurement and mean photon number during time interval T respectively and

$$\gamma(k+1,\ \beta) = \int_0^\beta z^k\, e^{-z}\, dz$$

is incomplete gamma function, $\theta(T - k\tau)$ is unit step function. Probability function for dead-time corrected distribution can be calculated as:

$$p_k(\lambda,\ T,\ \tau) = F_{k-1}(\lambda,\ T,\ \tau) - F_k(\lambda,\ T,\ \tau). \tag{2}$$

The initial moment of time window of measurement may be dead (Figure 1) due to previous photo count events (dark count rate or registration of counts at previous window). We modify

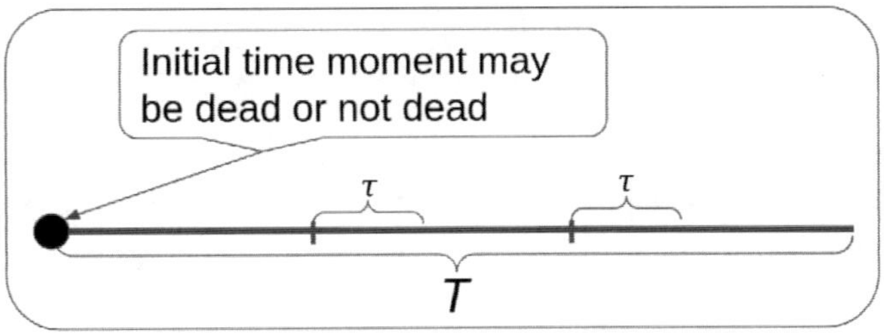

Figure 1. Out-window counts may leads to the dead initial moment of time window. Probabilities of dead or non-dead initial moments are determined by the part of all measurement time that is dead.

the probability function with corrections of possible dead initial time moments. It leads to new formula:

$$\widetilde{p_k}(\kappa,\,\lambda) = \frac{p_k(\lambda,\,1,\,\kappa)}{1+\lambda\kappa} + \frac{\lambda}{1+\lambda\kappa}\int_0^{\kappa} p_k(\lambda,\,1-y,\,\kappa)\,dy, \tag{3}$$

with condition $\kappa = \tau/T$. Series expansion of an incomplete gamma function provides another expression for CDF:

$$F_k(\lambda,\,1-y,\,\kappa) = \theta(1-y-k\kappa)\cdot\left(1-\sum_{n=0}^{k}\frac{e^{-\lambda(1-y-k\kappa)}\lambda^n(1-y-k\kappa)^n}{n!}\right) \tag{4}$$

An integration of expression (4) can be easily completed due to symbolic computation. As a consequence, a well-computed representation of the dead-time corrected distribution is provided by the equations (2), (3) and (4). This result can be naturally generalized for arbitrary sequences of measurement time windows and arbitrary source of coherent lighter.

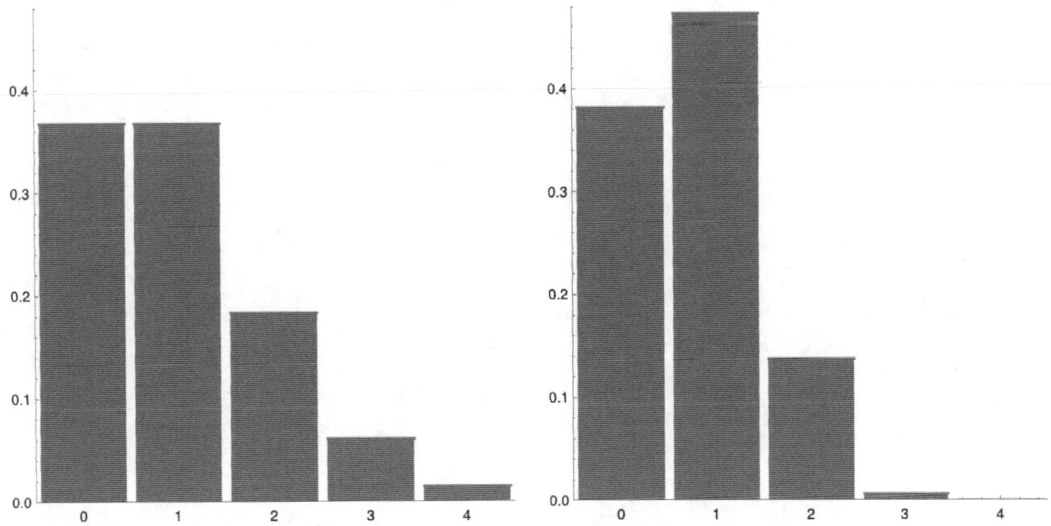

Figure 2. Left: Poisson distribution of photons. Mean number of photons $\lambda = 1$. Right: sub-Poissonian distribution of photon counts. Mean number of photons $\lambda = 1$, dead time $\tau = 0.3\cdot T$, where T is time window of measurement.

3. Results and discussion

The original Poisson distribution and simulated sub-Poissonian distribution are shown in Figure 2 (left and right, respectively). Calculated moments of the corrected distribution for different values of simulated distribution parameters (e.g. dead-time, photon mean number) is used for machine learning analysis and calculation of the photon mean number and the dead-time as functions of mean value and variance.

4. Acknowledgement

The reported study was funded by RFBR, Sirius University of Science and Technology, JSC Russian Railways and Educational Fund "Talent and success", project number 20-32-51004.

References

[1] Sych D and Leuchs G 2010 *New Journal of Physics* **12** 053019
[2] Sych D and Leuchs G 2010 *Optics and Spectroscopy* **108** 326–330
[3] Bennett C H and Brassard G 1984 Quantum cryptography: Public key distribution and coin tossing *Proc. IEEE Int. Conf. on Computers, Systems ad Signal Processing* (New York: IEEE) p 175
[4] Hwang W Y 2003 *Phys. Rev. Lett.* **91**(5) 057901
[5] Sych D V, Grishanin B A and Zadkov V N 2004 *Phys. Rev. A* **70**(5) 052331
[6] Sych D, Grishanin B and Zadkov V 2004 *Laser Physics* **14** 1314–1321
[7] Sych D V, Grishanin B A and Zadkov V N 2005 *Quantum Electronics* **35** 80–84
[8] Leuchs G, Dong R and Sych D 2009 *New J. Phys.* **11** 113040
[9] Sych D, Řeháček J, Hradil Z, Leuchs G and Sánchez-Soto L L 2012 *Phys. Rev. A* **86**(5) 052123
[10] Averchenko V, Reiß D A, Sych D and Leuchs G 2020 *Physica Scripta* **95** 034012
[11] Sych D and Leuchs G 2016 *Phys. Rev. Lett.* **117**(20) 200501
[12] Shcherbatenko M L, Elezov M S, Goltsman G N and Sych D V 2020 *Phys. Rev. A* **101**(3) 032306
[13] Sych D, Averchenko V and Leuchs G 2017 *Phys. Rev. A* **96**(5) 053847
[14] Averchenko V, Sych D, Marquardt C and Leuchs G 2020 *Phys. Rev. A* **101**(1) 013808
[15] Mandel L, Sudarshan E C G and Wolf E 1964 *Proceedings of the Physical Society* **84** 435–444
[16] Bédard G 1967 *Proceedings of the Physical Society* **90** 131–141

Spray deposited thin uniform NiO/Spiro-OMeTAD composite hole transport layer with top carbon nanotube electrode

A V Romashkin, Yu A Polikarpov, G O Silakov and E V Alexandrov

National Research University of Electronic Technology, Moscow 124498, Russia

romaleval@gmail.com

Abstract. Thin Spiro-OMeTAD and NiO nanoparticles layers, as well as their composite layer was formed by layer-by-layer spray deposition as hole-transport layer (HTL), with followed carbon nanotubes (CNT) deposition to form Ti/TiO_2/HTL/CNT structures. Layers' uniformity was estimated by Raman intensity maps, AFM and current-voltage characteristics of the CNT layer and between CNT and Ti contacts. The possibility of formation of thin, less than 100 nm, pinhole-free uniform composite NiO/Spiro-OMeTAD layer by spray-deposition was shown, which manifests itself as continuous HTL even after top CNT layer deposition.

1. Introduction

Due to the insufficient time stability at work under ambient conditions of solar cells based on organic materials, especially in the case of perovskite solar cells [1], new materials and formation principles of the hole transport layers (HTL) are developing, which characterized by dopant-free conception, using composites based on oxide nanoparticles and low molecular weight organic compounds or polymers [2]. Although it is possible to achieve relatively acceptable quantum efficiency even using the dopant-free thin Spiro-OMeTAD layer formed by vacuum deposition, the best values are achieved only with dopant, however, it significantly decreases time stability at ambient conditions [3]. Lifetime stability can be also increased by inorganic compounds adding to the HTL composition [4]. Thus, the development of thin composite HTL based on undoped Spiro-OMeTAD or other molecules with nanoparticles, formed by low temperature spray deposition is a relevant task, which is the subject of present work. Another important aspect of present work is the possibility of using carbon nanotubes (CNT) and their composites with polymers as the top conductive layer (and also, possibly, transparent), which allows improving the perovskite cells stability [5]. Moreover, the replacement of metal and/or ITO with CNTs provides a weak dependence of the electrode conductivity on the concentration of I^- and CN^- anions, which additionally increasing the stability of the structures [6].

2. Materials and methods

Cover glasses 24x24 mm with magnetron deposited thin layers: Ti (about 30 nm) and then TiO_2 (about 3-7 nm with roughness less than 1 nm), were used as substrates. TiO_2 sputtered by Ti target in argon-oxygen gas mixture at rate about 0.5 nm per minute. Substrates with TiO_2/Ti were used for spray-deposition and investigation HTL and top CNT layer (used functionalized single-wall CNT P3-SWNT, Carbon Solutions Inc., USA). Pure Spiro-OMeTAD (99 % (HPLC), Sigma Aldrich, USA), Spiro-OMeTAD using of nickel oxide nanoparticles (nNiO) sublayer (diameter < 80 nm, but with large conglomerates ~ 150 nm presence, 99.9 %, Simplex LLC, Russia), as well as simply nNiO layer were investigated as a HTL. As a result, four types of structures were formed on the substrates with a TiO_2/Ti sublayer to analyse the possibility of uniform HTL and CNT layer forming: A) CNT, B) nNiO/Spiro-OMeTAD/CNT, C) nNiO/CNT and D) Spiro-OMeTAD/CNT. The deposition of both CNTs and Spiro-OMeTAD, as well as nNiO, was carried out by spray-deposition from various solvents: dimethylacetamide (DMA, > 99.5 % (HPLC), Applichem GmbH, Germany) for Spiro-OMeTAD, propyleneglycol with cyclohexanol (PG and CHL, > 99 %, Component-Reactive Ltd., Russia) for CNT, and PG with DMA for nNiO by a method that similar to described earlier [7]. The deposition was carried out on heated substrate (about 90 °C) by own designed spray-coating system with use shadow mask on 2 mm diameter round area for HTL deposition, and then on rectangular area 4x1.5 mm for top CNT layer deposition (centred in HTL area). To estimate the mixing/dipping of CNT into the Spiro-OMeTAD, nNiO and nNiO/Spiro-OMeTAD sublayers atomic force microscope (AFM, Solver Pro, NT-MDT, Russia) was used. Also for this and to measure electrical characteristics of the formed CNT layers (with two Ag-paste contact pads at short sides of CNT rectangular area that not covers HTL area) and structures as a whole (between CNT and Ti layers as contacts) semiconductor parameters analyser IPPP 1/5 (OJSC "MNIPI", Belarus) was used. The nonuniformity degree of Spiro-OMeTAD and CNT materials in appropriate layers were analysed by Raman mapping at 20x20 points (Centaur HR, Nano Scan Technology Ltd., Russia).

3. Results and discussions

Spray deposition of undoped Spiro-OMeTAD on a substrate does not provide a thin (< 150 nm) uniform layer, which was revealed both by AFM and Raman 1615 cm^{-1} region intensity map [3] (figure 1a, where shaded area is the Raman spectrometer background signal < 180 arb.units with Spiro-OMeTAD absence, according to AFM). The substrate was heated to 90 °C during deposition, so Spiro-OMeTAD was crystallized, which was confirmed by Raman spectroscopy. The use of a uniformly deposited nNiO (uniform, but island-like sublayer), which have an energy levels similar to Spiro-OMeTAD [4, 6] and suitable for HTL formation, significantly improved the Spiro-OMeTAD uniformity in sequentially spray-deposited uniform nNiO/Spiro-OMeTAD composite layer when nNiO used as sublayer (figure 1b,c) especially when optimal nNiO substrate coverage was achieved. The value of layer nonuniformity was determined by ratio of standard deviation to median value of Raman intensity [8], that was 0.8 for Spiro-OMeTAD in composite layer, which, although higher than for spray deposited CNTs (ratio is 0.2-0.3), but much lower than for pure Spiro-OMeTAD layer, only partially covering the surface (ratio is 3.3). Spiro-OMeTAD uniformity significantly depends on the nNiO density on the substrate: low nanoparticles covering degree (< 20-30 %) still forms areas without Spiro-OMeTAD (compare figure 1a (corresponds to structure D) and 1b). Therefore significant percent of the surface was still not covered by Spiro-OMeTAD (value of layer nonuniformity is about 1.7, but still higher than 1.0), and that percent was significantly more than in the case of 60-70 % surface coverage of the substrate by nNiO formed sublayer. That percent coverage by nNiO (60-70 %) determined as optimal because acceptable substrate coverage by Spiro-OMeTAD as part of the composite layer was obtained (figure 1c). After obtaining a uniform Spiro-OMeTAD layer, CNTs were deposited on the structure of this type (structure B), as well as on structures of other types to compare uniformity and resistivity of CNT layer. The results of AFM control of the top CNT layer are shown in figure 2. To compare the AFM data in figure 2 with the initial nNiO layer the sublayers of nNiO were also investigated by AFM. Figure 3a,b demonstrated the cases of the substrate covered by nNiO formed island-like underlayer with different percent coverage that lower than 100 % even for figure 3b taking into account deconvolution parameters of AFM tip and true nNiO diameter that equal to height on AFM image. Non-full coverage by nNiO sublayer was used to obtain the composite structure of HTL as nNiO/Spiro-OMeTAD and to form the possibility of charge transport through Spiro-OMeTAD as well as nNiO material or both. Measured CNT layer resistance (R_{CNT}) was: A - 35.7 kΩ, B - 37.6 kΩ, C - 116.9 kΩ, D - 51.8 kΩ; while the CNT surface density differences were < 15 % (control by Raman spectroscopy). According to the AFM data there is a smoother and denser CNT layer on top of composite nNiO/Spiro-OMeTAD with less roughness and a smaller amount of observed deepened areas in comparison to on nNiO sublayer. And finally CNTs weren't observed on top of pure Spiro-OMeTAD (complete CNTs dipping into or mixing with covered by pure Spiro-OMeTAD areas in sublayer is realized). But composite structure provides the absence of deeply dipped into the HTL sublayer CNTs (in contrast to structure D), with nonlinear current-voltage characteristics (CVC) between CNT and Ti (at positive voltage $U_{CNT\text{-}Ti}$, that described below) and therefore the pinhole absence and no shunting of HTL (through formation of shunt CNT-TiO_2/Ti contacts), in contrast to the CVC for structures C and D, which indicates to better sublayer quality and continuous and uniform HTL structure realizing by nNiO/Spiro-OMeTAD composite sublayer.

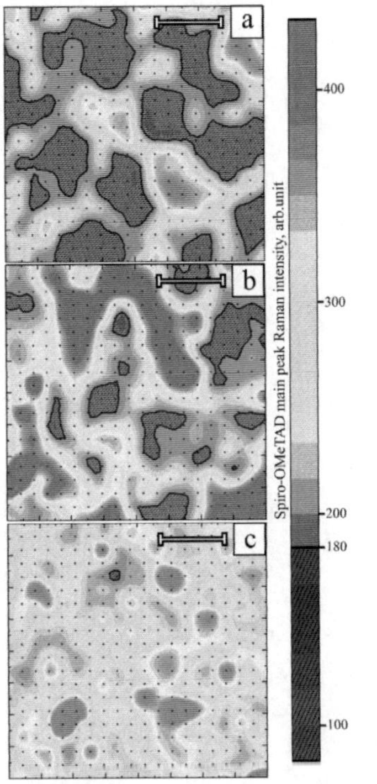

Figure 1. Raman map of Spiro-OMeTAD peak for: a) pure Spiro-OMeTAD; b,c) composite layer nNiO/Spiro-OMeTAD: with ~30 % surface coverage by nNiO sublayer (b), and with ~70 % coverage (c). Scale bar is 10 μm.

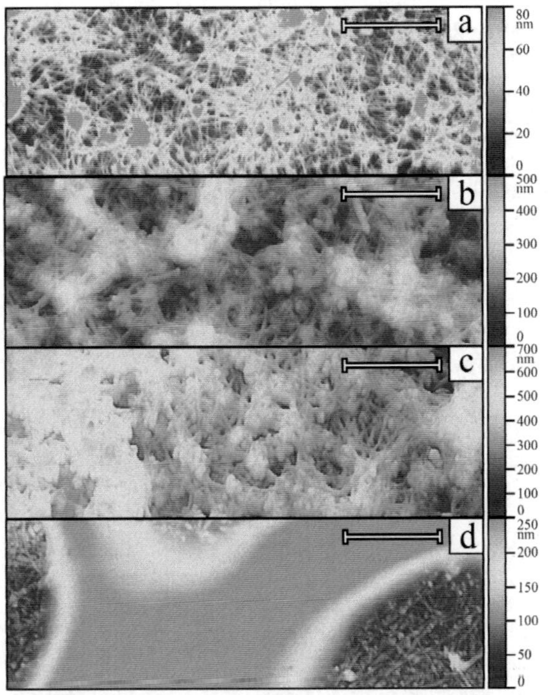

Figure 2. AFM data of structures A-D: a) CNT, b) nNiO/Spiro-OMeTAD/CNT, c) nNiO/CNT, d) Spiro-OMeTAD/CNT. Scale bar is 1 μm.

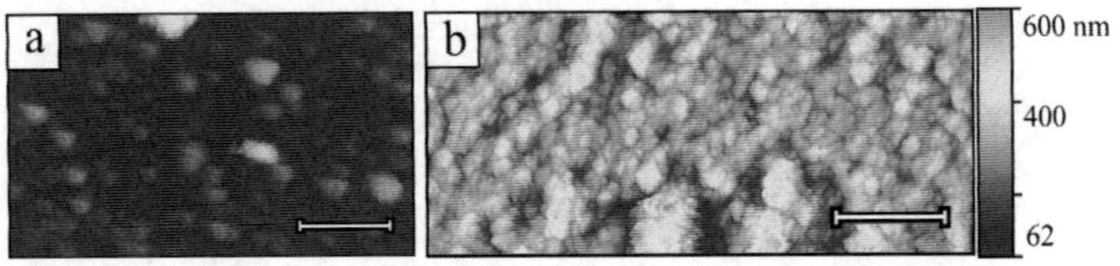

Figure 3. AFM data of nNiO underlayer obtained with: a) 20-30 % covering substrate, b) 60-70 % covering. Scale bar is 2 μm.

That behavior also confirmed by similar R_{CNT} for structure B in comparison with A, having smooth substrate, in contrast to the increased R_{CNT} on nNiO sublayer of structure C, where some CNTs were dipped between nanoparticles (according to AFM there is lower CNTs density on the top of structure C than for A or B). Despite CNTs to Spiro-OMeTAD dipping for structure D, it weakly affects the uniformity of presumably thin and dense CNT layer (according to Raman map) but it couldn't be detected under Spiro-OMeTAD layer by AFM. So R_{CNT} for structure D insignificantly differs from structure A, despite alcohols used for CNTs deposition, partially dissolve Spiro-OMeTAD. Despite to partially dissolving of Spiro-OMeTAD lower influence of that to R_{CNT} for composite sublayer in comparison to structure A was obtained. But differences in R_{CNT} between structures A and B significantly lower than between structures A and D. So Spiro-OMeTAD mixing with CNT in composite HTL layer of structure B was insignificant in comparison to structure D (compare figure 2b and 2d). So Spiro-OMeTAD uniformity insignificantly changed even after the top CNT layer deposition as for composite layer (structure B) as structure D (compare figure 4 and figure 1). The additionally obtained Raman maps of the ratio (I_{Spiro}/I_{CNT}) of the main peak of Spiro-OMeTAD (1615 cm^{-1}) to CNT G-band (1590 cm^{-1}, figure 4a,b) and the CNT G-band (figure 4c,d) demonstrated that even after the CNT deposition on top of nNiO/Spiro-OMeTAD composite sublayer, the distribution of CNTs and the nonuniformity degree of CNT layer was insignificantly changed (value of layer nonuniformity is 0.25) relative to the case of CNTs on the initial TiO_2/Ti substrate (nonuniformity value is 0.24, compare figure 4c and 4d). In addition, in the case of the nNiO/Spiro-OMeTAD composite sublayer, the redistribution of Spiro-OMeTAD with the formation of not covered by Spiro-OMeTAD areas didn't observe, which was determined by comparing the I_{Spiro}/I_{CNT} intensity ratio for structure B in comparison with the values obtained for pure CNT layer (structure A). Estimation of layer nonuniformity value for Spiro-OMeTAD in composite HTL by this ratio was 0.58, which lower than for initial HTL.

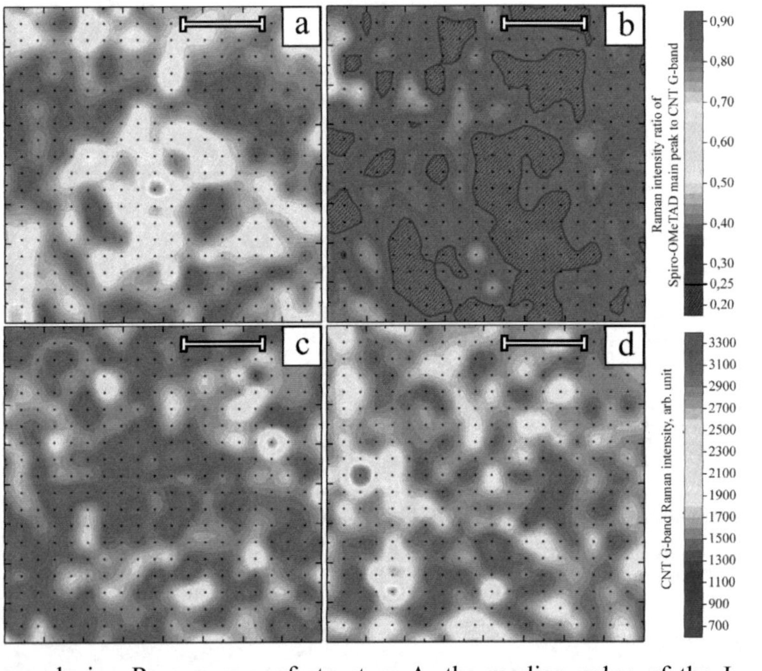

Figure 4. Raman map of: a,b) intensity ratio of Spiro-OMeTAD peak to CNT peak for: structure B with composite nNiO/Spiro-OMeTAD sublayer (a), structure D with pure Spiro-OMeTAD sublayer (b); c, d) CNT peak intensity for: structure B (c), structure A (d). Scale bar is 10 μm.

By analysing Raman map of structure A, the median value of the I_{Spiro}/I_{CNT} ratio was obtained, which when adding standard deviation of this ratio (threshold value 0.23+0.02) characterized situation when Spiro-OMeTAD is absent (shaded areas on figure 4a,b). Therefore values above this limit pointed to some amount Spiro-OMeTAD on the measuring map point. Despite CNTs dipping into or mixing with Spiro-OMeTAD, structure D is characterized by a high uniform CNT distribution (value of layer nonuniformity is 0.23). Comparison of Raman map on figures 4a and 4b can be used for one more evidence of Spiro-OMeTAD presence at almost all areas of the multilayer structure B since for it I_{Spiro}/I_{CNT} intensity ratio isn't decreased lower than the characteristic threshold value for pure CNTs or for areas of structure D with CNTs presence and Spiro-OMeTAD absence. For structure D, this map of I_{Spiro}/I_{CNT} ratio minus the threshold value characteristic for the Spiro-OMeTAD absence gave an estimation of the value of Spiro-OMeTAD sublayer nonuniformity for this structure about 2.7, that also slightly lower than 3.3 for initial pure Spiro-OMeTAD layer, so some smoothing of Spiro-OMeTAD during CNT deposition was observed. Also, to compare the conductivity between CNT and Ti layers for structures B-D, I–V characteristics (CVC) were measured (figure 5). Significant differences between them were observed only for the region of positive values of the U_{CNT-Ti} for structure B. Small hysteresis observed for all structures at CVC cyclic measuring is apparently due to some traps at the HTL-TiO_2 interface. And, taking into account the incomplete coverage of the substrate surface by the HTL layer in the case of pure Spiro-OMeTAD and also pure nNiO, a significant part of CNTs (20-40 %) was in direct contact with the TiO_2/Ti substrate even in the area where the HTL sublayer was deposited, which makes CVCs of these structures (C and D) are similar (especially at "+" $U_{CNT\text{-}Ti}$). And in the case of structure D, the thickness in the regions where CNTs are separated from the substrate by Spiro-OMeTAD can be more than 200-300 nm (see figure 2d and also based on the R_{CNT} in comparison with structure A, CNTs are predominantly concentrated near the top of Spiro-OMeTAD regions), which eventually forms a rather smaller contact area with the substrate (in the round HTL area), as a result of which a slightly lower conductivity is observed both at low "-" and at "+" $U_{CNT\text{-}Ti}$ in comparison with structure C.

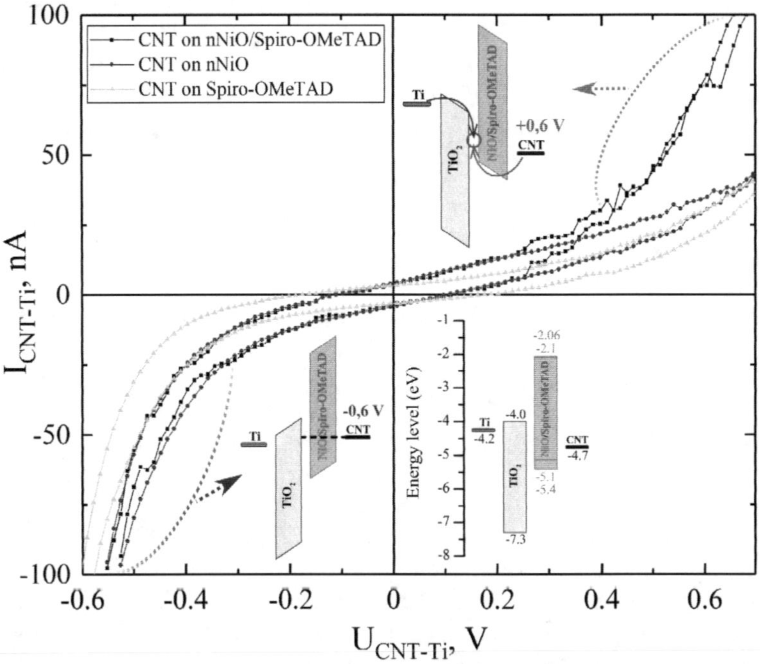

Figure 5. CVC between CNT and Ti contacts for: structure B (black squares), structure C (blue circles) and structure D (green triangles). Ti contact for all structures was grounded. The insets schematically show changes in energy levels in the case of zero and "-" $U_{CNT\text{-}Ti}$ for all structures and "+" $U_{CNT\text{-}Ti}$ for structure B.

And in the case of structure C, which has lower conductivity over the CNT layer and, accordingly, a greater CNTs redistribution over the nNiO layer thickness, CNT contact area directly with the substrate can be some higher than for structure D due to the nNiO island layer. Even despite thin Spiro-OMeTAD layer separates CNTs from TiO_2 and reduces charge transport efficiency through TiO_2 in the case of composite HTL, however, due to fairly high effective area where CNTs locate still near TiO_2, the similarity of CVC in range -0.6 V to +0.2 V was observed, especially for structures B and C. As a result, for structure C (and also for B) the possibility of charge transport through TiO_2 (at high fields formed at "-" $U_{CNT\text{-}Ti}$, taking into account the small thickness of TiO_2) is higher than for structure D. This also can explain the increasing leakage currents through TiO_2 even at "+" $U_{CNT\text{-}Ti}$, when possibility of tunnelling or effective charge injection through TiO_2 cannot be achieved, and only TiO_2 breakdown becomes a possible mechanism of current flow at large positive $U_{CNT\text{-}Ti}$ (except for structure B). The presence of regions in all structures of direct contact of CNTs with TiO_2/Ti outside deposited round region of HTL sublayer (2 mm in diameter) with a region's area of about half of the total CNT layer area (4x1.5 mm), determines the CVCs similarity and presence of current at "-" $U_{CNT\text{-}Ti}$ by charge transport (possibly by tunnelling or field injection) through the triangular barrier of direct CNT-TiO_2/Ti contacts. That marked in the left inset of figure 5 as a dotted line inside the NiO or Spiro-OMeTAD material. Although, directly in the region where HTL forms a continuous layer, it doesn't realize (structure B). Additionally, the shunting of the small HTL islands by CNT, as well as the large HTL thickness in regions where there is no direct CNT-TiO_2/Ti contact, explains the lack of involvement of such HTL layer in charge transport in CNT-Ti contacts in structures C and D. However, in the case of structure B with continuous composite nNiO/Spiro-OMeTAD layer formation, both a relatively small effective thickness of the HTL layer and its sufficient continuity were apparently achieved, which makes possible certain mechanism of current flow at "+" U_{CNT-Ti}, also with a pronounced nonlinear CVC in contrast to the structures C and D. Such a mechanism may be the charges injection (when a sufficient "+" U_{CNT-Ti} voltage is applied) through narrowed barriers of the TiO_2 layer (it plays the role of an electron transport layer, ETL) and HTL: electrons through ETL and holes through HTL to the ETL-HTL interface with their subsequent recombination (figure 5, right inset). This mechanism forms the possibility of effective current flow through the ETL-HTL interface with a nonlinear CVC at "+" U_{CNT-Ti}, and also apparently indicates the formation of a thin continuous HTL in structure B with a significant effective area of the HTL (in contrast to structures C and D). In view of significant difference in LUMO and HOMO levels between ETL and HTL respectively (more than 1 eV), as well as some difference in the CNT work function and the nNiO/Spiro-OMeTAD HOMO level, recombination does not occur quite efficiently at low "+" U_{CNT-Ti}, that was revealed in noticeable CVC nonlinearity in this region and high threshold voltage for type B structures.

4. Conclusion

Spray deposition of only pure Spiro-OMeTAD doesn't provide thin continuous layer formation; however, the use of a sublayer of nNiO with an average size of less than 80 nm and with their filling of the substrate surface already from 70 % allows one to form a thin (less than 100 nm by Spiro-OMeTAD data from AFM, Raman) continuous composite nNiO/Spiro-OMeTAD layer acting as HTL. This manifests itself both on the AFM as a larger number of observed CNTs on top of that HTL and a smoother surface of the structure, as well as on the resistance of the top CNT layer, which only slightly different from the pure CNT layer on a flat TiO_2/Ti substrate, but which significantly higher in the case of island sublayers of nNiO or Spiro-OMeTAD. By obtaining Raman maps, the acceptable continuity and uniformity degree of both the CNT layer deposited on top and the Spiro-OMeTAD itself, and the small Spiro-OMeTAD redistribution with improving uniformity degree in the nNiO/Spiro-OMeTAD composite layer were confirmed, despite on subsequent CNTs deposition from the dispersion of CHL, which partially dissolving Spiro-OMeTAD. Also, this manifests itself when comparing the CVC of the CNT-Ti contacts for a structure with composite HTL, that demonstrates nonlinear CVC at a positive U_{CNT-Ti} voltage, that associated with the possibility of recombination at the TiO_2-HTL interface, in contrast with the CVC of structures where the HTL is formed by an island layer of only nNiO or pure Spiro-OMeTAD, which both don't manifest itself as continuous HTL.

Acknowledgments

This work was supported by grant of the President of the Russian Federation for young Russian scientists support (grant No. MK-1024.2020.8, agreement No. 075-15-2020-439 of 16.03.2020).

References

[1] Meng L, You J and Yang Y 2018 *Nat. Commun.* **9** 1
[2] Guo Y, Lei H, Xiong L, Li B and Fang G 2018 *J. Mater. Chem. A* **6** 2157
[3] Barranco A et al 2020 *Adv. Energy Mater.* **10** 1901524
[4] Jin I S, Park S H, Kim K S and Jung J W 2020 *J. Alloys Compd.* **847** 156512
[5] Ahn N, Jeon I, Yoon J, Kauppinen E I, Matsuo Y, Maruyama S and Choi M 2018 *J. Mater. Chem. A* **6** 1382
[6] Zhou Y et al 2018 *ACS Appl. Mater. Interfaces* **10** 31384
[7] Romashkin A V, Polikarpov Y A, Alexandrov E V and Nevolin V K 2020 *J. Phys. Conf. Ser.* **1695** 012086
[8] Polikarpov Y A, Romashkin A V, Struchkov N S, and Levin D D 2019 IEEE Conf. of Russian Young Researchers in Electrical and Electronic Engineering (Moscow) (IEEE) p 1980

QKD key generation control protocol

N V Rudavin[1,5], I S Gerasin[1,2,4], E E Mekhtiev[1,2,4], A V Duplinsky[1,3,4] and Y V Kurochkin[1,2,3,4]

[1] QRate, Novaya av. 100, Moscow, Russia
[2] Moscow Institute of Physics and Technology, 9 Institutskiy per., Dolgoprudny, Moscow Region, 141701, Russian Federation
[3] NTI Center for Quantum Communications, National University of Science and Technology MISiS, Leninsky prospekt 4, Moscow 119049, Russia
[4] Russian Quantum Center, Bolshoy Boulevard 30, bld. 1, Skolkovo, Moscow 121205, Russia
[5] Federal State Budget Educational Institution of Higher Education, MIREA—Russian Technological University, 78, Vernadskogo avenye, 119454 Moscow, Russia

n.rudavin@goqrate.com

Abstract. Polarization-encoding fiber QKD requires compensation of polarization distortion caused by birefringence in optical fiber. Solving this task inevitably requires losing some effectiveness in terms of the final key rate. In this work, a time-division multiplexing protocol for polarisation calibration is suggested. This protocol was implemented in a QRate commercial QKD fiber system, utilizing BB84-protocol. Parameters of the protocol were optimized to maximize the secret key rate.

1. Approaches to QKD key generation control

Quantum technologies evolve rapidly and now the quantum key distribution (QKD) fiber systems become more and more available. The fiber QKD systems have the advantage that they can be directly deployed on existing telecom optical fiber infrastructure. However, for polarization-encoding protocols polarization control is required since in single-mode optical fiber any given input polarization of light passing through optical fiber suffers time-dependent disturbance. To provide a high secure key-rate efficient polarization calibration protocol should be used.

There are two main approaches to organize polarisation calibration protocols: (i) interrupting scheme (time-division multiplexing, TDM) and (ii) real-time scheme. The first approach requires stopping the key generation process in order to send a sequence of calibration pulses (either quantum or classical) [1]. The second method implies utilizing wave-division multiplexing (WDM) for calibration signals [2] or using some part of the secret key to estimate QBER [3]. WDM systems require some additional equipment to multiplex signals which leads to the cost increase. Interrupting schemes allow to send high-intensity calibration pulses that can gather statistics for error estimation faster than from the weak pulses used for secret key generation. Another advantage of the TDM approach, that the calibration pulses pass the same optical path, which is used by signal pulses. In this work, we propose interrupting calibration protocol with flexible switch criteria.

2. Protocol description

2.1. Problem statement
During the calibration mode, key generation is impossible so one needs to decide when to switch into key generation regime and back. Consider the work of some iteration algorithm that minimizes QBER (e.g. gradient descent). Usually, iteration algorithms do not have well-defined stop criteria. On the one hand, if protocol switches to the key-generation mode too early during the convergence, the final QBER value is not optimal and the maximum secure key rate is not achieved. On the other hand, if protocol switches to the generation too late, the key rate is maximized but key generation time is decreased. Thus, for interrupting protocol, a crucial thing is to choose switching criteria efficiently in terms of secret key length. We aim to develop adaptive switching criteria, which can be tuned by some parameters. Varying the parameters manually or automatically, the optimal ones can be found. Moreover, we trying to avoid the adding of some additional equipment.

2.2. Flexible criteria
On each iteration of the convergence algorithm, we measure M values of QBER (each point is measured during the same time interval τ, see Figure 1). Parameter M is introduced to change the accuracy of the gathered QBER values. The longer the measured time (larger M), the more accurate measurement. Then mean $\overline{QBER}$ and the standard deviation σ_{CM} is calculated for each iteration that lasts $M\tau$. $\overline{QBER}$ is passed to convergence algorithm. Switching criteria is based on QBER linear interpolation as a function of time via MSE (N last mean $\overline{QBER}$ points are used) and a comparison of the slope with some threshold, which is dynamically calculated on each iteration of the convergence algorithm. The threshold is calculated as $k_0 = -\sigma_{CM}/NM\tau$. The k_0 is an estimation of the minimal slope of the MSE line. Increasing M with fixed N allows manipulating threshold value. Thus, by changing M and N we can operate the time needed for calibration and the minimal QBER value achieved during calibration.

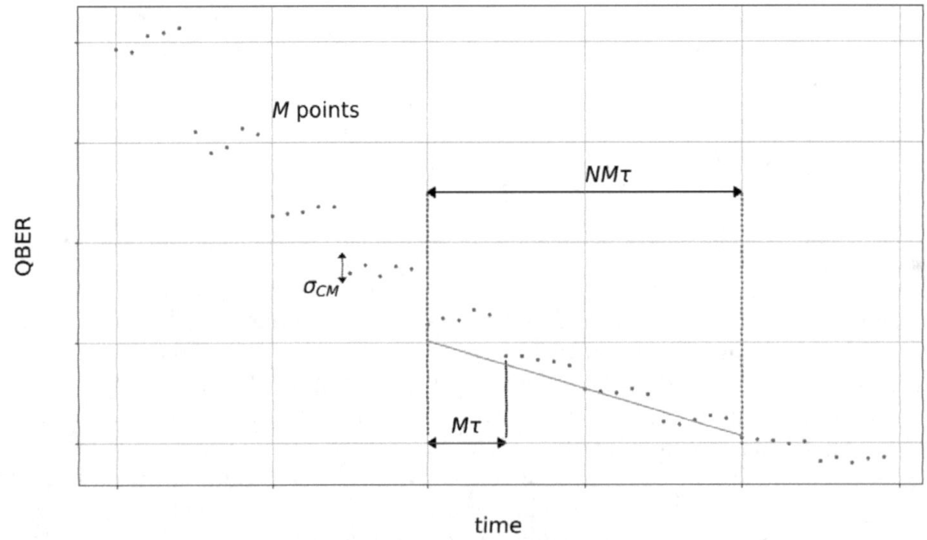

Figure 1. Criteria principle.

2.3. Proposed protocol
The whole process of QKD is divided into two parts: key-generation mode (GM) and calibration mode (CM). During GM $\overline{QBER}$ is calculated from decoy-state clicks and σ_{GM} is estimated as the standard deviation of the ratio of two random values $N_{wrong\ clicks}$ / $N_{total\ clicks}$ with the Poissonian distribution. Generally, σ_{GM} can be chosen in other ways depending on QBER estimation technique

during key generation. Moving back to the CM is regulated with a certain threshold based on QBER statistics analysis. The proposed protocol works as follows:

1) enable polarization calibration mode
2) wait until the convergence algorithm makes N steps
3) using the last N points $(\overline{QBER_i}, t_i)$ approximate $QBER(t)$ as a linear function through MSE method: $QBER = kt + b$
4) calculate the threshold slope $k_0 = -\frac{\sigma_{CM}}{NM\tau}$
5) if $k < k_0$ then
   make another iteration of the convergence algorithm
   go to step 3
6) else if $k \geq k_0$ then
   set $QBER_{thrs} = \overline{QBER_N} + 3\sigma_{GM}$
7) enable key-generation mode
8) monitor $\overline{QBER}$ until $\overline{QBER} > QBER_{thrs}$
9) go to step 1

3. Experimental results

The proposed protocol was implemented and tested on QRate commercial QKD fiber system [4]. This system, utilizing BB84 protocol, uses one-way polarization encoding optical scheme [5]. In the experimental setup, transmitter (Alice) and receiver (Bob) was connected by 50 kilometers of standard single-mode telecommunication optical fiber SMF-28e. As polarization controller was used PolaRITE III, developed by General Photonics. M and N are hyperparameters to be optimized. We run the key distribution process with the fixed parameters $(M, N) \in \{4, 6, 8\} \otimes \{5, 7, 9\}$ and measure the average secret key rate during 3 hours for each pair. The obtained results are represented in Table 1. We can notice that choice of hyperparameters does not influence much. Thus, we choose $(M, N) = (6,7)$ as optimal pair of parameters for our system.

Table 1. Average key rate for fixed parameters during.

N \M	4	6	8
5	14.76	14.84	14.86
7	14.87	14.97	14.85
9	15.02	14.70	14.40

Acknowledgments

We thank many colleagues at the Russian Quantum Center, QRate, for motivating discussions, management, and coordination. This work was supported by The Russian Science Foundation (Grant No. 17-71-20146)

References

[1] Chen J, Wu G, Li Y, Wu E, and Zeng H 2007 Active polarization stabilization in optical fibers suitable for quantum key distribution *Optics express* **15** pp 17928-17936
[2] Xavier G B, Vilela de Faria G, Temporão G P, and von der Weid J P 2008 Full polarization control for fiber optical quantum communication systems using polarization encoding *Optics express* **16** 1867-1873
[3] Ding Y-Y, Chen W, Chen H, Wang C, Li Y-P, Yin Z-Q, Wang S, Guo G-C, Han Z-F 2017 Polarization basis tracking scheme for quantum key distribution with revealed sifted key bits *Optics Letters* **42** 1023-1026

[4] Duplinskiy A, Ustimchik V, Kanapin A, Kurochkin V, and Kurochkin Y 2017 Low loss QKD optical scheme for fast polarization encoding *Optics express* **25** 28886-28897
[5] Bennett C H, Brassard G, and Mermin N D 1992 Quantum cryptography without Bell's theorem *Physical Review Letters* **68** 577

Development of a light control system using an optical aerial information transmission system

D D Savin[1], V V Davydov[1, 2] and V Yu Rud[2, 3]

[1] Peter the Great Saint Petersburg Polytechnic University, 29 Politechnicheskaya str., Saint Petersburg, Russia, 195251
[2] All-Russian Research Institute of Phytopathology, 5 Institutsky Ave., P. B. Vyazemy, Moscow region, Russia, 143050
[3] FTI A. F. Ioffe, 26 Politechnicheskaya str., Saint Petersburg, Russia, 194021

savin.dd@edu.spbstu.ru

Abstract. The inefficiency of the use of electric energy for indoor lighting is shown. The necessity of modernization of lighting control systems in the workplace area is justified. The method of automatic control and adjustment of illumination of workplaces and zones where it is necessary is offered. A control system has been developed, which is integrated into the air optical information transmission system (Internet). The simulation of the system operation is carried out, its parameters are calculated. The main characteristics of the fiber-optic communication line with the connected device were measured. The efficiency of the developed device was evaluated.

1. Introduction

The development of scientific and technological progress, the increase in the population and the demand for consumption requires more and more electric energy [1-10]. The increase in the volume of electricity production has an extremely adverse effect on the ecological state of the planet [11-19]. In addition, natural sources of fuel and raw materials for electricity production are being depleted more and more rapidly [4-8, 10-12, 20-23]. Alternative energy sources cannot fully compensate for the growing energy consumption [24-27].

It should also be noted that during the processing of raw materials, a large amount of impurities is released into the atmosphere [9, 11, 28-32]. These pollutants lead to a deterioration in the quality of natural light sources (smog is present in the atmosphere) [32-34]. It is necessary to compensate for the loss of light by increasing the intensity of artificial lighting, that is, to spend more electricity. Therefore, solving the problems of efficient use of the generated electricity is a very urgent task for humanity, which affects several areas.

One of the ways to reduce electricity consumption is to reduce its lighting costs, especially in office and educational premises with windows. This problem is still very relevant, despite the large number of its technical solutions. The main disadvantage of these systems is that they are aimed at controlling the illumination from artificial lighting in some points of the room with automatic adjustment of the power of lighting devices. It is due to the lack of natural light metering in these systems that leads to an unjustified consumption of electrical energy.

2. Automatic light control system with optical communication channel

Nowadays, information is transmitted in various systems (including lighting control systems) using Wi-Fi. This leads to some difficulties: with a large number of users, there is a large congestion of the airspace and an increase in the amount of electromagnetic interference [35-40]. An effective solution to this problem was an optical channel using fiber-optic communication lines (FOCL). Optical systems have greater information transfer capabilities than Wi-Fi, at least by an order of magnitude [36-40]. In addition, various electromagnetic interference and congestion of the air space do not affect the transmitted information [35-40].

Figure 1 shows a block diagram of the automatic lighting control system of workplaces integrated into the optical air information transmission system (Internet).

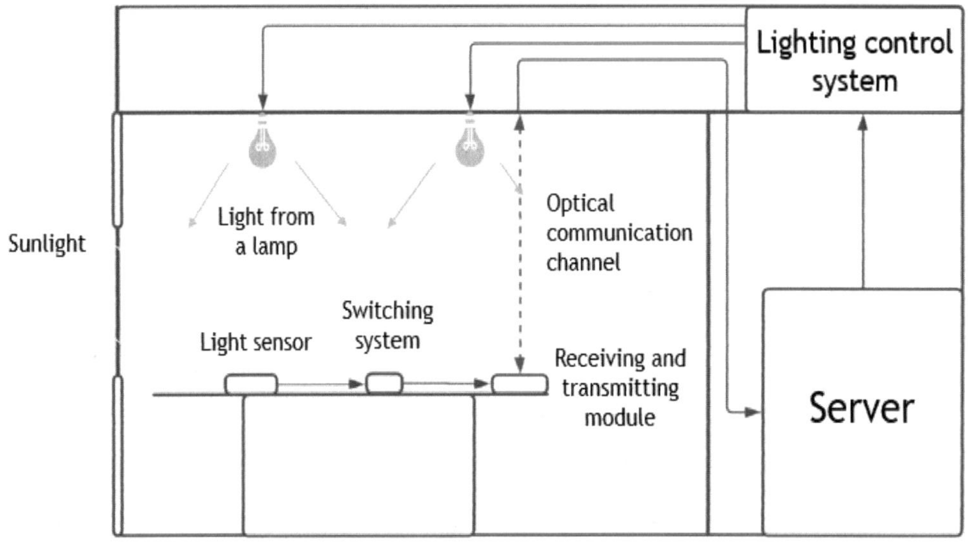

Figure 1. Diagram of an automatic illumination monitoring system using an optical aerial information transmission system.

The information from the light sensors is sent to the information processing unit developed by us, which is located in the same building as the computer connection system. In this block, data packets are formed based on the received data. In this case, the channel switching device selects the least loaded channel and receives a time interval during which information about the light level at control points in the workplace will be transmitted over the optical Internet. This creates several features when operating an aerial optical information transmission system. One of them is associated with an additional displacement of the operating point of the electro-optical modulator when switching from one channel to another, since additional temperature stabilization is not provided in such systems.

After the receiving and transmitting module, the light data packets are transmitted over the optical channel to the server, where they are processed. Based on the information received, the server sends commands to the lighting control system to increase or decrease the voltage on a particular lamp.

In addition, to increase the efficiency of the system (in terms of time), the issue of additional charging of the optical communication system batteries from the energy of solar cells located in a number of zones of work desks and lighting devices powered by light was considered.

3. System results and discussion

The conducted simulation of the operation of the lighting control system showed that the Internet speed does not decrease significantly. The additional energy required for the stable operation of the system is not more than 5 % of the electrical energy consumed for the operation of the optical communication channel.

Also, the main parameters of the optical communication channel were analyzed, since when connecting a lighting control system in a room, the main problem is the deterioration of these parameters.

One of the important parameters that characterize the operation of fiber-optic communication systems (FOCL), which include an air optical communication channel, is the amplitude-frequency response (AFC). Figure 2 shows the frequency response of a fiber-optic network in the frequency range from 3 to 7 GHz, which operates over the aerial optical channel of the Internet.

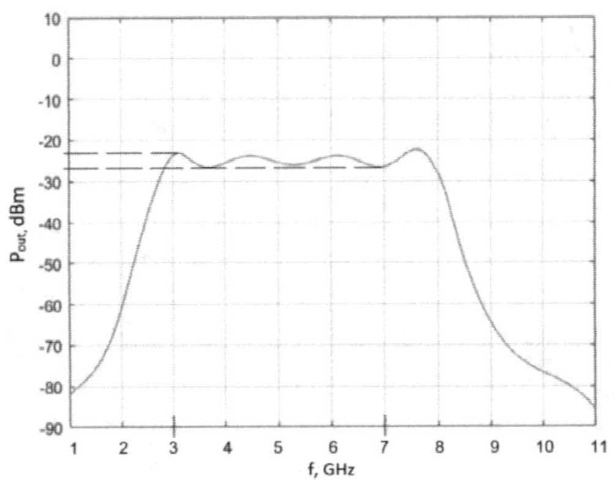

Figure 2. The amplitude-frequency response of an air optical channel with a connected light control device at T=294.2 K.

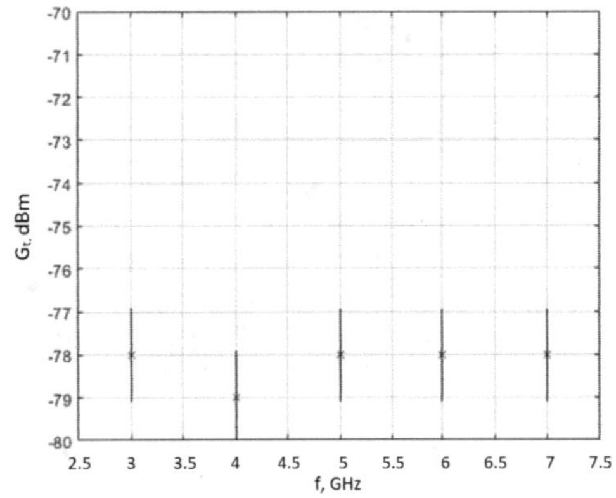

Figure 3. Dependence of the tangential sensitivity of the receiving path on the frequency of the transmitted signal at T=293.7 K.

The obtained result shows that the unevenness of the frequency response of a fiber-optic communication system in the frequency range (3-7 GHz) is about 4-5 dB. This allows you to realize a stable signal transmission with a large amount of information after connecting the lighting control device developed by us.

Another important characteristic of the operation of a fiber-optic cable with an air optical channel is the tangential sensitivity G of the receiving path, which is placed on the users ' table. Figure 3 shows the dependence of the tangential sensitivity G of the receiving path on the frequency of the transmitted signal.

When the lighting control system is connected to the path, the G value has not changed significantly and is about 78 dBm. Therefore, it is not necessary to include additional low-noise amplifiers in the receiving path circuit after connecting the lighting control system developed by us. That is, the main equipment does not need to be reworked.

Let's define the dynamic range of our system. Figure 4 shows the experimental dependence of the power of the P_{out} output signal on the power of the P_{in} input signal.

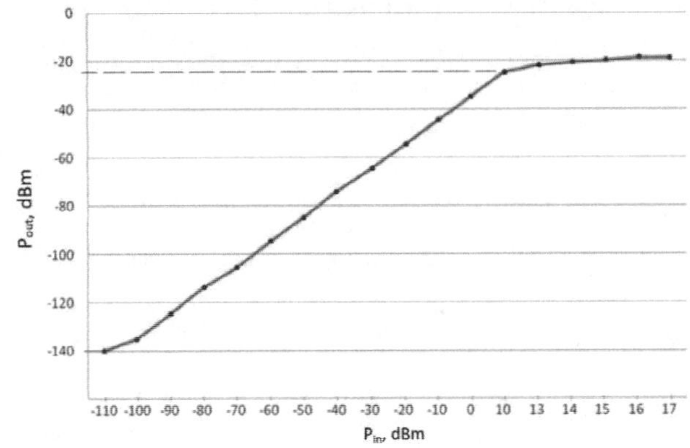

Figure 4. Dynamic characteristic of a fiber optic cable with an air optical channel with a connected light control system at T=294.2 K.

The analysis of the obtained results shows that after connecting the lighting control system, the dynamic range of stable operation of the fiber-optic cable with an air optical channel was about 112 dBm. This value of the linearity of the characteristic allows reliable transmission of large amounts of information and corresponds to the values in the classical backbone fiber-optic lines.

4. Conclusion

The simulation of the developed light control system and the calculation of its parameters showed that this system does not introduce significant distortions in the transmitted information over the optical channel for Internet users. For different seasons of the year, preliminary estimates have shown that the use of this system allows you to reduce the electricity consumption for indoor lighting from 10 to 20 %.

References

[1] Davydov V V, Dudkin V I and Karseev A Yu 2014 *Optical Memory & Neural Networks (Information Optics)* **23(4)** 259 – 264
[2] Davydov V V, Dudkin V I and Karseev A Yu 2014 *Optical Memory & Neural Networks (Information Optics)* **23(3)** 170 – 176
[3] Davydov V V, Dudkin V I and Karseev A Yu 2013 *Optical Memory & Neural Networks (Information Optics)* **22(2)** 215 – 218
[4] Nikitina M, Grebenikova N and Batov Y 2019 *IOP Conference Series: Earth and Environmental Science* **390(1)** 012024
[5] Gryznova E and Batov Y 2019 *E3S Web of Conferences* **140** 09001
[6] Gryznova E, Grebenikova N, Ivanov D and Bykov V 2019 *IOP Conference Series: Earth and Environmental Science* **390(1)** 012044
[7] Davydov R V, Mazing M S, Yushkova V V and Stirmanov A V 2019 *Journal of Physics: Conference Series* **1410(1)** 012067
[8] Davydov R V, Antonov V I, Yushkova V V and Smirnov K J 2019 *Journal of Physics: Conference Series* **1400(6)** 066037
[9] Van S, Cheremisin A, Chusov A, Switala F and Davydov R 2019 *IOP Conference Series: Earth and Environmental Science* **390(1)** 012011
[10] Davydov V V 1999 *Russian Physics Journal* **42(9)** 822–825
[11] Fadeenko I, Fadeenko V, Reznik V, Moroz A, Popovskiy N and Nikolaev D 2019 *IOP Conference Series: Earth and Environmental Science* **390(1)** 012022
[12] Myazin N S 2018 *Journal Physics: Conference Series* **1124(1)** 031004
[13] Davydov R, Antonov V, Makeev S, and Batov Y 2019 *E3S Web of Conferences* **140** 02001
[14] Davydov R, Antonov V and Moroz A 2019 *International Conference on Electrical Engineering and Photonics EExPolytech-2019 (Saint-Petersburg)*, vol. 8906791 (IEEE), p. 295-297
[15] D'yachenko S V and Zhernovoi A I 2016 *Technical Physics* **61(12)** 1835–1837
[16] Davydov V V and Kiryukhin A V 2020 *Atomic Energy* **127(5)** 274–279
[17] Myazin N S 2018 *Journal Physics: Conference Series* **1135(1)** 012061
[18] Davydov R V and Antonov V I 2018 *Journal of Physics: Conference Series* **1124(8)** 081037
[19] Davydov R V and Antonov V I 2018 *Journal of Physics: Conference Series* **1135(1)** 012087
[20] Myazin N, Neronov Y, Dudkin V and Yushkova V 2018 *MATEC Web of Conferences* **245** 11013
[21] Rukin E V, Myazin N S and Dudkin V I 2019 *Journal of Physics: Conference Series* **1368(4)** 042011
[22] Myazin N S, Yushkova V V and Taranda N I 2019 *Journal of Physics: Conference Series* **1410(1)** 012130
[23] Gizatullin B, Gafurov M, Vakhin A, Mattea C and Stapf S 2019 *Energy and Fuels* **33(11)** 10923–10932
[24] Neronov Y I 2021 *Biomedical Engineering* **54(5)** 333–336

[25] Myazin N S and Yushkova V V 2019 *Environmental Research, Engineering and Management* **75(2)** 28–35
[26] Myazin N, Neronov Y, Dudkin V and Petrov A 2019 *IOP Conference Series: Materials Science and Engineering* **497(1)** 012111
[27] Smirnov K J and Batov Y V 2019 *Journal of Physics: Conference Series* **1368(2)** 022073
[28] Grevtseva A S, Smirnov K J, Greshnevikov K V and Glinushkin A P 2019 *Journal of Physics: Conference Series* **1368(2)** 022072
[29] Grevtseva A S, Smirnov K J, Davydov V V and Rud V Yu 2018 *Journal of Physics: Conference Series* **1135(1)** 012056
[30] Kozlova E S and Kotlyar V V 2014 *Computer Optics* **38(1)** 132-138
[31] Kuzmin M S and Rogov S A 2019 *Computer Optics* **43(3)** 391-396
[32] Smirnov K J, Glagolev S F, Rodygina N S and Ivanova N V 2018 *Journal of Physics: Conference Series* **1038(1)** 012102
[33] Smirnov K J, Glagolev S F and Tushavin G V 2018 *Journal of Physics: Conference Series* **1124(1)** 022014
[34] Smirnov K J, Medzakovskiy V I, Vysoczky M G and Glagolev S F 2017 *Journal of Physics: Conference Series* **917(6)** 062019
[35] Myazin N S, Dudkin V I, Grebenikova N M, Rud' V Y and Podstrigaev A S 2019 *Lecture Notes in Computer Science (including subseries Lecture Notes in Artificial Intelligence and Lecture Notes in Bioinformatics)* **11660 LNCS** 744–756
[36] Fadeenko V B, Kuts V A and Vasiliev D A 2018 *Journal Physics: Conference Series* **1135(1)** 012053
[37] Davydov R V, Saveliev I V, Lenets V A and Tarasenko M Yu 2017 *Lecture Notes in Computer Science (including subseries Lecture Notes in Artificial Intelligence and Lecture Notes in Bioinformatics)* **10531 LNCS** 177-183
[38] Moroz A V and Davydov R V 2019 *Lecture Notes in Computer Science (including subseries Lecture Notes in Artificial Intelligence and Lecture Notes in Bioinformatics)* **11660 LNCS** 710 – 718
[39] Moroz A V 2019 *Journal of Physics: Conference Series* **1410(1)** 012212
[40] Moroz A V, Davydov V V, Malanin K Y, Krasnov A A and Rud V Yu 2019 *Journal of Physics: Conference Series* **1400(4)** 044009

Influence of dielectric overlayers on self-heating of a microdisk laser

S A Scherbak[1,2,3], E I Moiseev[2,1], I A Melnichenko[2,1], Ju A Guseva[4], M V Maximov[1,2], A I Lihachev[4], N V Kryzhanovskaya[2,1], A A Lipovskii[1,3] and A E Zhukov[2]

[1] Alferov University, St. Petersburg 194021, Russia
[2] International Laboratory of Quantum Optoelectronics, Higher School of Economics, St. Petersburg 194100, Russia
[3] Institute of Physics, Nanotechnology and Telecommunications, Peter the Great St. Petersburg Polytechnic University, St. Petersburg 195251, Russia
[4] Ioffe Institute, St. Petersburg 194021, Russia

Abstract. We studied experimentally and numerically self-heating of a microdisk laser developed in the AlGaInAs material system and covered with dielectric layers. By experiments, we found that planarization of the microlaser with SU-8 photoresist significantly (almost, 2-fold) decreases the microlaser thermal resistance. Calculations demonstrate that a downward heat flux through the substrate to the heat sink is a dominant way of heat dissipation, and upward convection is much less relevant. Also, the calculations showed that covering microlaser with a TiO_2 layer barely affects microdisk temperature but decreases heat localization in the structure.

1. Introduction

Increase of internal temperature (self-heating) of a microlaser negatively affects its performance [1]. In optical microdisk (MD) lasers, the self-heating effect with increasing pump power is pronounced due to their small size (a few microns) and low thermal conductivity of materials, including Al_2O_3, which is often used for optical confinement on the substrate side. Self-heating leads to a decrease in the differential efficiency of the lasers due to a decrease in the inverted population and an increase in the threshold pumping power due to additional optical losses through scattering by the free carriers. Also, internal heat of the laser causes saturation not only of the output optical power, but also of the modulation bandwidth [2]. These impede using such microlasers as light sources in all-optical interchip communications. Thus, ways to improve microlasers thermal properties and studies of these properties are of interest. One of the most effective solutions of the problem of overheating of a microlaser is using an additional heat sink with high thermal conductivity.

Recently, a novel strip-loaded horizontal slot waveguide (SLSW) based on dielectric $TiO_2/SiO_2/TiO_2$ layers have been proposed for routing microdisk lasers emission [3]. The substrate with a microdisk semiconductor laser is planarized with the use of SU-8 photoresist, and on the top of this structure a planar (horizontal) slot optical waveguide composed of $TiO_2/SiO_2/TiO_2$ layers is deposited. Both SU-8 and waveguide layers influence thermal behavior of the structure. In this paper we analyze the influence of covering dielectric layers on self-heating of a microdisk laser.

2. Experiment and methods

A semiconductor heterostructure was grown using molecular beam epitaxy on a semi-insulating GaAs (100) substrate. An active region comprises five layers of InAs/$In_{0.15}Ga_{0.85}As$/GaAs quantum dots (QDs) with a 30-nm-thick GaAs spacer between the layers. The spectral position of the QD ground-state transition is located around λ = 1.3 μm at room temperature. The active region is placed in the middle of the GaAs layer confined from both sides with 10 nm thick $Al_{0.3}Ga_{0.7}As$ barriers. The total thickness of the vertical waveguide is 360 nm. This waveguide layer was grown on the top of a 400-nm-thick $Al_{0.98}Ga_{0.02}As$ layer. The $Al_{0.98}Ga_{0.02}As$ layer was then transformed into an $(AlGa)_xO_y$ layer by selective oxidation to ensure the optical confinement on the substrate side. MDs of different diameters from 4 to 11.5 μm were fabricated using photolithography and argon-ion beam etching. The scanning electron microscopy (SEM) image of a typical MD laser fabricated is shown in figure 1. The microdisk lasers were studied at room temperature under optical pumping with a cw-operating YLF: Nd laser (λ = 527 nm). When the microdisk laser operates in continuous wave regime, its temperature increases due to self-heating, and the released Joule heat is dissipated through the ambient of limited thermal conductivity. The microlaser heating with increase in the optical pump power results in a redshift of the lasing resonance lines. Lasing spectra were measured with a following optical setup: the signal was collected by a microobjective x100 and detected by ANDOR iDus multi-channel detector and a confocal optical spectroscopy setup (Integra Spectra, NT MDT). The redshift of the lines is mainly determined by the temperature-induced change of the refractive index and is about 0.08 nm/K for GaAs-based microlasers in the temperature range of 20–80 °C [4]. Using the dependence of the resonance wavelength on the optical pump power one can estimate the active region temperature.

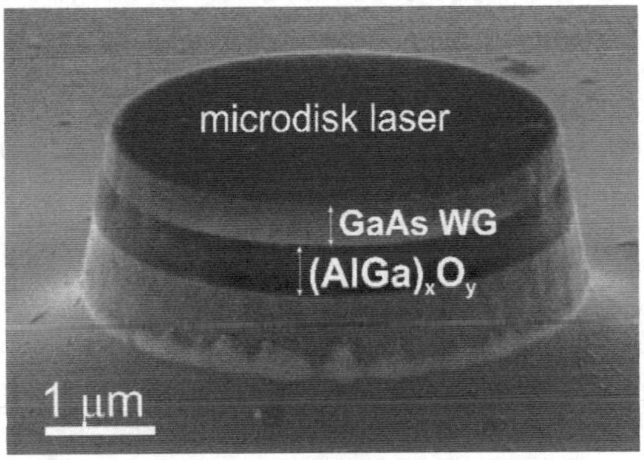

Figure 1. SEM image of a MD laser.

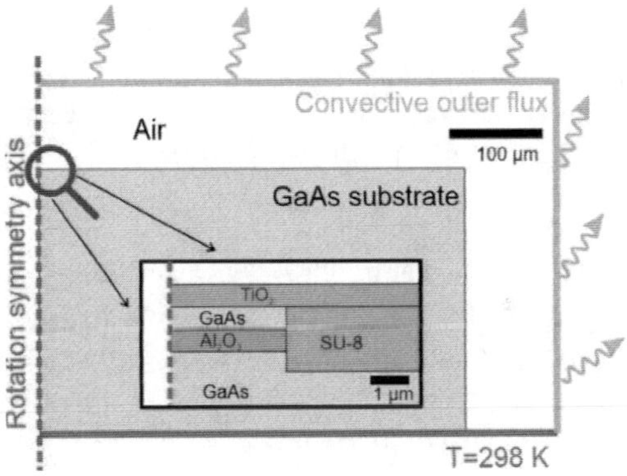

Figure 2. Scheme of numerical simulations (half cross-section of the structure).

To analyze microdisk self-heating we used numerical simulation in COMSOL Multiphysics environment via the finite element method. The scheme of the problem is shown in figure 2. We considered axially symmetric problem. We used a boundary condition of convective heat flux at upper and side air boundaries and a constant (room) temperature condition at a lower boundary. The latter corresponds to an ideal heat sink beneath the structure. Also, we assumed that the resonators heating power is equal to the power of optical pump, since free space emission of the MD laser is negligible in this case. The material properties used in the simulation are presented in table 1.

Table 1. Material properties used in simulations

	Air	GaAs [5]	Al_2O_3 [6]	SU-8 [7]	TiO_2 [8]
Thermal conductivity (W $m^{-1}K^{-1}$)	0.0026	51	34	0.2	8
Heat capacity (J kg^{-1} K^{-1})	1005	327	780	1200	500
Density (kg m^{-3})	1.2	5300	4000	1200	4200

3. Results and discussion

3.1 Self heating vs. pump power

We evaluated temperature increase of the 6-μm MD surrounded by air in accordance with a measured shift of a lasing wavelength for different pumping power. The threshold pump power for this laser is about 0.4 mW. Also, we carried out corresponding numerical simulation of temperature distribution. The results of experimental and calculated heating of the microlaser vs pump power are presented in figure 3.

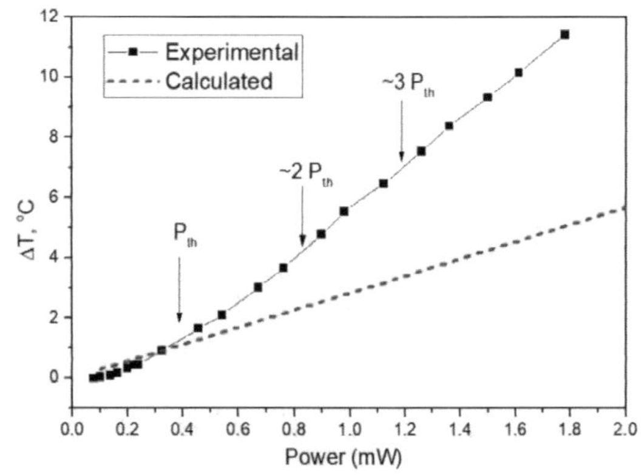

Figure 3. Measured (dots) and calculated (dashed line) heating of the MD laser of 6 μm in diameter and surrounded by air; P_{th} – threshold power.

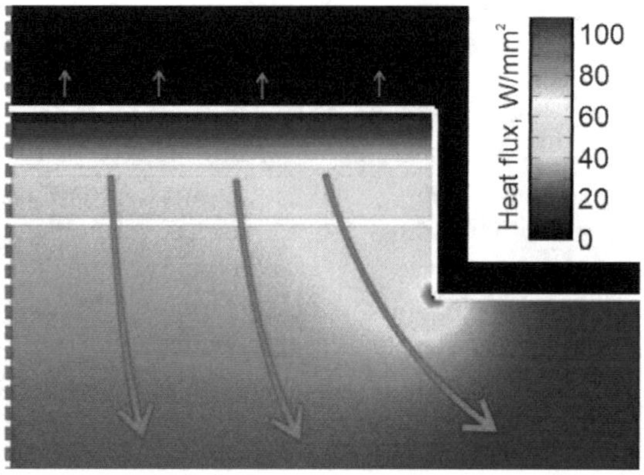

Figure 4. Distribution of heat flux magnitude in the MD vicinity; MD diameter is 6 μm, pump power – 1.2 μW. Arrows qualitatively indicate the heat flux direction.

The dependences in figure 3 are almost linear as expected and their slope is called thermal resistance. One can find that self-heating of the MD laser under consideration is moderate – about 7 °C at pump power 3-times exceeding the threshold. Moreover, experimentally evaluated heating noticeably exceeds the calculated one. Probably, this is because of imperfection of the real heat sink and the imperfect thermal contact between it and the structure. In figure 4 we demonstrate calculated heat flux distribution in the structure, arrows there qualitatively indicate the heat flux direction. A bright spot in the corner in figure 4 is an artifact of numerical modeling, since the flux suffers discontinuity at cutting edges. Important to note that heat flux through the substrate and further to the heat sink significantly dominates the heat flux to the upper surrounding. Particularly, the overall downward heat flux exceeds the overall upward heat flux by about 4 orders. This supports abovementioned reasonings: since the downward heat flux to the heat sink is dominant, imperfection of the thermal contact can significantly increase the experimental heating compared to the ideal simulated one.

3.2 Thermal resistance vs. disk diameter and SU-8 influence

Thermal resistance shows a change of the structure temperature relatively to a change of the input power. In accordance with the described above method, we determined thermal resistance of MD lasers with different diameters under air surrounding and planarized with SU-8 photoresist. Experimental results are demonstrated in figure 5a. Also, we carried out corresponding calculations, results of which are presented in figure 5b.

In figure 5a we observe that thermal resistance is expectedly lower for larger MDs, i.e. thermal resistance is inversely proportional to a MD diameter. Basic difference between measured and calculated results in figure 5a and b is the same as in figure 3: heating and, therefore, thermal resistance of the real structure are higher than ones of the ideal modeled structure. Also, the measurements showed that planarization of a microlaser with the use of SU-8 leads to a noticeable decrease in thermal resistance, whereas, according to the calculations, the influence of SU-8 is almost negligible (two curves in figure 5b coincide). The result of the calculations is explained by dominance of the downward heat flux, for it makes influence of cover layers insignificant. The experimental result, in turn, offers that the downward heat flux is somehow restricted, probably, by imperfect thermal contact between the structure and the heat sink, and the upward heat flux becomes relevant. The SU-8 cover increase the upward heat flux, since its thermal conductivity, albeit low, is still higher than the one of air (see table 1). Thus, planarization with SU-8 noticeably improves thermal resistivity.

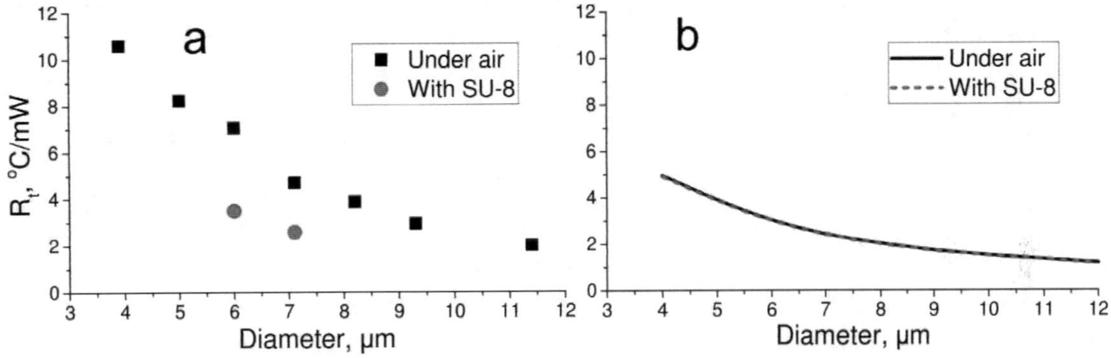

Figure 5. Experimental (a) and calculated (b) dependences of thermal resistance on MD diameter for microlasers under air surrounding and with SU-8 planarization.

3.3 Influence of TiO_2 cover

We calculated and compared temperature distribution in the MD laser under air surrounding and with upper TiO_2 cover. An actual SLSW consists of layers $TiO_2/SiO_2/TiO_2$, however, the inner SiO_2 layer is much thinner than the clad TiO_2 layers. Therefore, in the calculations we considered the covering layer as uniform TiO_2. We analyzed the MD of 6 μm in diameter under pump power of 1.2 mW (~3 P_{th}). We calculated temperature distribution in the structure with and without the TiO_2 cover, particularly, temperature dependence along a vertical axis. The result of the calculations is presented in figure 6.

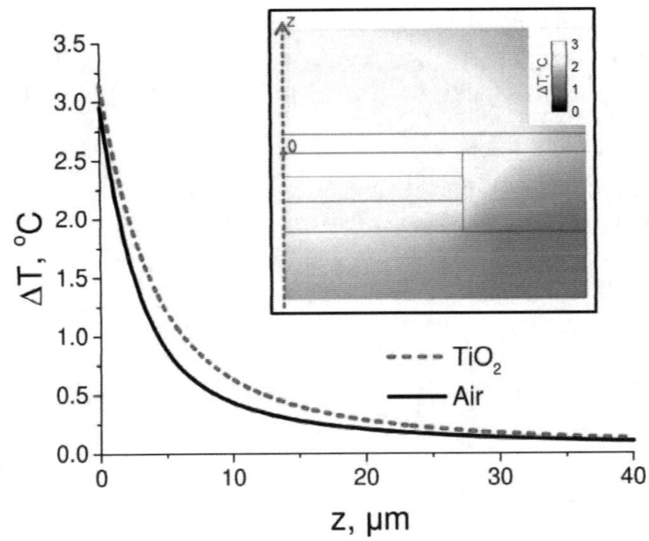

Figure 6. Temperature distribution along a vertical axis in the microlaser structure with and without (under air) the TiO_2 cover. The MD diameter is 6 μm, pump power – 1.2 mW. Inset – spatial temperature distribution in case of the TiO_2 cover, vertical axis *z* is designated.

The average MD self-heating in both cases is similar: ~2.9 °C with the TiO_2 cover and ~3.1 °C without cover. The reason is the one discussed above – dominance of the downward heat flux makes upper covers almost irrelevant. However, temperature distribution in the structure with TiO_2 cover becomes less localized compared to the air surrounding. This, possibly, can make the structure with the SLSW cover more sensitive to any kind of additional active air cooling – less localized heat is easier to dissipate.

Conclusion

We studied experimentally and numerically thermal resistance of microdisk lasing structures. We determined self-heating of MDs of different diameters against pump power via measurements of the lasing wavelength shifting. Thermal resistance, which is a slope of the dependences obtained, was lesser for larger diameter of the MD, i.e. inversely proportional to diameter. Qualitatively, this was supported by numerical calculations of temperature distribution in the structures. Though, modeled ideal structures had lesser thermal resistance than real ones. This is probably because of the imperfect thermal contact between the substrate and the heat sink. Modeling showed that the downward heat flux through the substrate to the heat think was much greater than the upward convective heat flux, therefore, good quality of the heat sink and the thermal contact there is crucial. We also experimentally demonstrated that planarization of the structure with SU-8 noticeably decreases thermal resistance of the structure. By numerical modeling, we showed that the TiO_2 cover of the MD slightly decreases its temperature and reduces heat localization. The latter makes the structure with a dielectric cover more sensitive to convective heat dissipation, e.g. by active air cooling.

Acknowledgement

This work was supported by the Russian Foundation for Basic Research (project 20-02-00334). The optical studies were supported by the Ministry of Science and Higher Education of the Russian Federation (state task № 0791-2020-0002). SEM characterization were performed using equipment owned by the Federal Joint Research Center "Material science and characterization in advanced technology".

References

[1] Zhukov A E *et al* 2020 *IEEE J. Quantum Electron.* **56** 1–8
[2] Zhukov A E, Moiseev E I, Nadtochii A M, Kryzhanovskaya N V, Kulagina M M, Mintairov S A, Kalyuzhnyi N A, Zubov F I and Maximov M V 2020 *Tech. Phys. Lett.* **46** 515–9
[3] Fetisova M *et al* 2020 *J. Opt. Soc. Am. B* **37** 1878
[4] Zhukov A E, Kryzhanovskaya N V, Maximov M V, Lipovskii A A, Savelyev A V, Shostak I I, Moiseev E I, Kudashova Y V, Kulagina M M and Troshkov S I 2015 *Semiconductors* **49** 674–8
[5] Blakemore J S 1982 *J. Appl. Phys.* **53** 123–81
[6] https://www.makeitfrom.com/material-properties/Alumina-Aluminum-Oxide-Al2O3
[7] Oh S H, Lee K-C, Chun J, Kim M and Lee S S 2001 *J. Micromechanics Microengineering* **11** 221–5
[8] Thurber W R and Mante A J H 1965 *Phys. Rev.* **139** 1655–65

Simulation optical parameters nanowire structures of various shapes.

I A Shishkin, D A Shishkina and N V Latukhina

Samara National Research University, Samara 443086, Russia

Abstract. In this study, three-dimensional fragments of nanowires structures were simulated in which the thread cross sections had a square, triangular and round shape. The theoretical data were compared with the experimental results, described in detail in [1]. The influence of the shape of the filaments on the theoretical course of the reflection curve is also shown.

1. Introduction

Silicon nanowires are a modern and promising nanomaterial of great interest for various fields of science [2-4]. Due to the complex surface, silicon nanowires have unique properties that allow them to be used as an anti-reflective coating for solar cells. Silicon nanowires also have advantageous optical and structural properties that differ from porous silicon. Silicon nanowires created on the surface have micro-nano textures that contribute to the hydrophobicity of the surface. These hydrophobic characteristics can be used in self-cleaning applications. These surfaces can be used for solar energy, in dusty environments [5]. Dust settles on surfaces and partially prevents solar radiation from reaching energy-saving devices such as photovoltaic panels and volumetric receivers, which reduces the performance of the device over time [6]. The spectral characteristics of nanostructures are modelled. At the moment, there are several models that can reliably describe the real course of the reflection curve for nanostructured materials, including silicon nanowires. In this paper, it is shown that the theoretical reflection coefficient depends on the shape of the wires (square, triangular, pyramidal), and one of the simulated structures coincides with the experimental results described in the article [1].

Figure 1. SEM-image of the real structure of nanowires and one of the structures, which is used in the Comsol Multiphysics software package.

2. The method of sample production

Silicon nanowires (SiNW) were manufactured by a two-stage metal-assisted chemical etching method (MACE). A metal-assisted chemical etching method was used, and AgNO3 (99.8%), H2O2 (30% in water), H2SO4 98%, and HF (48%) were used as the chemical etchants. The first step was to deposit silver on the surface of the silicon wafers, and then etching was carried out in an etching solution of HF / H2O2 5 M / 30% of the etchant. The remaining metal particles were removed from the surface with a solution of H2O: HCl: HNO3 for one hour [7-9].

For silicon nanowires of crystalline quality with transverse dimensions from 10 nm and higher, the so-called quantum-size effect for charge carriers is realized, which leads to an increase in the band gap and a shift of the optical absorption edge to the high-energy region. The optical properties of silicon nanostructures with large transverse dimensions will significantly depend on the effects associated with the spatial distribution of local electric fields and caused by light scattering by both individual nanoobjects and their ensembles. The study of such effects in nanostructures in the form of ensembles of silicon nanowires is particularly relevant, since they are easily integrated with microelectronics and sensor devices and materials for various applications, including biophotonics and medicine.

There are a number of practical tasks for creating light-sensitive structures, for which it is necessary to reduce the reflection coefficient at the boundary of two media. Modelling the distribution of the electric field on a filamentous surface allows us to describe physical processes with non-standard surface geometry.

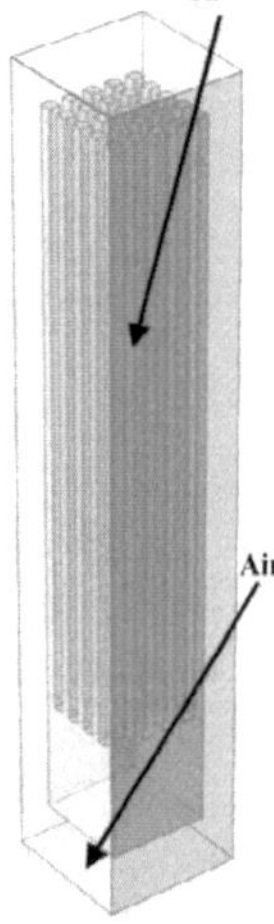

Figure 2. Scheme of representation of silicon nanowires in the air environment.

3. Method of simulation

The Comsol Multiphysics software package was used to describe the passage of electromagnetic waves in a nanowire structure consisting of 10 wires with a thickness of 0.05 microns [10]. To solve this problem, a system was chosen in which a part of the silicon substrate is in the air (Fig. 2), so a wave equation was adopted based on an alternating electric field for two media:

$$\nabla \times (\nabla \times \boldsymbol{E}) - k_0^2 e_r \boldsymbol{E} = 0$$

By default, the software package already contains the basic optical parameters of the two media for solving the equation. In our case, such media are air and silicon. Simulation of the electric field propagation, carried out in the software package in the optical range (300-1000 nm), allows you to calculate the distribution of the near field.

The boundary conditions under which the condition $n \times E = 0$ is satisfied in this system are the side faces of the bounded air medium. The upper face of this scheme is the plane of the energy flux density of the incident radiation (Fig. 3). It is also necessary to determine the scattering parameters (S-parameter) in terms of the electric field. To convert the picture of the electric field on the "in" port into a scalar complex number corresponding to the voltage in transmission line theory, it is necessary to decompose the electromagnetic fields on the ports according to their own modes. We assume that the fields are normalized with respect to the integral of the power flow through each port cross-section, respectively. This normalization depends on the frequency. The port excitation is applied using the main eigenmode. The calculated electric field ***E*** at the input "in" consists of the excitation and the reflected field. S-parameters are set:

$$S_{in} = \frac{\int_{\partial\Omega}(E - E_1)E_1}{\int_{\partial\Omega} E_1 \cdot E_1}$$

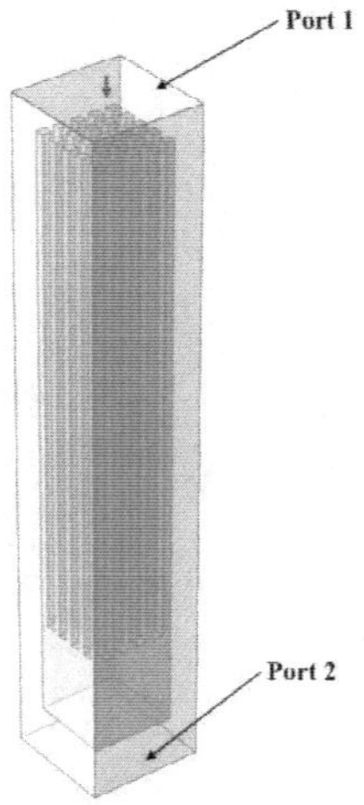

Figure 3. The plane of incidence of the electric field through the input port in the air environment of the Comsol Multiphysics software package.

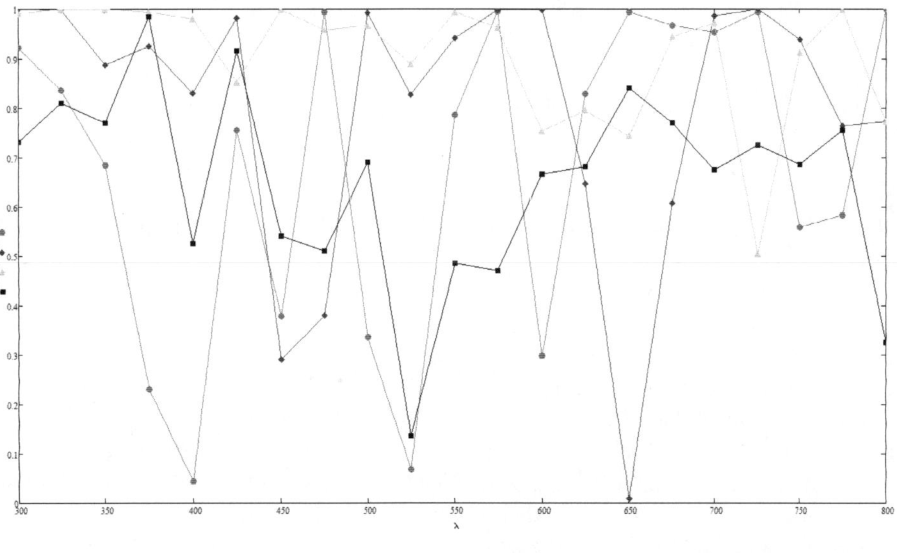

Figure 4. Graphs of the dependence of the reflection coefficient on the wavelength obtained in various theoretical models (square, triangular, pyramidal) and practical data: red line – triangular nanowires, blue line - square nanowires, green line - pyramidal nanowires, black line - practical data.

The numerical method for solving this problem was to construct a finite element grid for the entire system. The dimensions of one grid element ranged from 0.1 to 5 microns. The wave optics module has a special periodic state. The periodic condition can identify simple mappings at the boundaries of the source and destination plane of equal form. The destination can also be rotated relative to the source. In this paper, a continuous type of periodic conditions was chosen. The tangential components of the solution variables are equal to the source and destination.

4. Results and discussion

The description of the passage of electromagnetic waves in the structure of nanowires consisting of 25 nanowires with a thickness of 0.2 microns with different shapes (pyramidal, triangular and square) was carried out in the Comsol Multiphysics software package.

Figure 4 shows the results of modelling the dependence of the light reflection coefficient on the length of the incident wave. Electromagnetic radiation passed through a structure with a nanowire, in which the distance between each element of the nanowire system was 0.2 microns. The analysis shows that in the wavelength range from 300 to 800 nm, the minimum reflection coefficient for theoretical models is in different values, but it is the triangular nanowires (Fig.5) are as close as possible to the practical values obtained in [1].

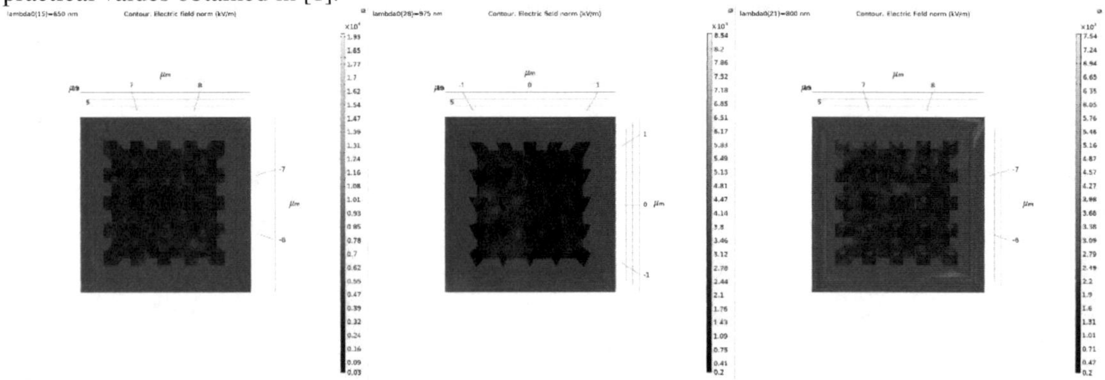

Figure 5. 3D map of light distribution in structures of various shapes nanowires (pyramidal, triangular, square). Matrix 5x5 with a minimum reflection coefficient.

Acknowledgments

This work was supported by the Foundation for Assistance to Small Innovative Enterprises, grant № 12980GU/2018.

References

[1] Yilbas B S, Salhi B, Muhammad Y, Al-Sulaiman F, Haider A and Nasser A 2016 *Scientific Reports.* **6.1** 1-13
[2] Li X 2012 *Curr. Opin. Solid State Mater. Sci.* **16** 71–81
[3] Noor M O and Krull U J 2014 *Anal. Chim. Acta* **825** 1–25
[4] Chen L J 2007 *J. Mater. Chem.* **17** 4639–4643
[5] Liu K and Jiang L 2012 *Annu. Rev. Mater. Res.* **42** 231–263
[6] Mani M and Pillai R 2010 *Renew. Sust. Energ. Rev.* **14** 3124–3131
[7] Ma D D D, Lee C S, Au F C, Tong S Y and Lee S T 2003 *Science* **299** 1874–1877
[8] Wu S, Shao Y M, Nie T, Xu L, Jiang Z and Yang X 2015 *Nanoscale Res. Lett.* **10** 325–333
[9] Han H, Huang Z and Lee W 2014 *Nanotoday* **9** 271–304
[10] Hochbaum A I 2005 *Nano Letters.* **5** 457

Photovoltaic characteristics of structures with porous silicon obtained by various technological plans

D A Shishkina[1,2], N A Poluektova[1] and I A Shishkin[2]

[1] Faculty of Electronics and Instrument Engineering, Samara National Research University, Samara, 443086, Russia
[2] Department of Physics, Samara National Research University, Samara, 443086, Russia

Abstract. In this work, the influence of technological parameters on the current-voltage characteristics of solar cells with porous silicon is investigated. It is shown that for photosensitive structures with a porous layer, the optimal mode is pore formation by electrochemical etching followed by diffusion. The effect of etching modes affects the character of the photosensitivity curve.

1. Introduction
Currently, alternative energy sources are becoming more and more relevant. This is due to many factors. In particular, the main sources of energy, such as oil, gas, coal, etc., are non-renewable. Their reserves are gradually being depleted, and sooner or later the time will come when they will have to be abandoned. In addition, their extraction is dangerous for the environment. For example, the high consumption of oil and petroleum products has led to the need to extract raw materials such as bituminous sands. But their development causes great harm to the environment: oil obtained from bituminous sands produces three times more carbon dioxide than with traditional drilling; the open-pit method involves deforestation and the removal of up to several tons of soil and peat; the waste water of the quarry can seep into the environment and pollute the groundwater.

Most coal-fired power plants are located in the immediate vicinity of the mines. This is due not only to the destruction of the natural environment by the hectares of workings created by the mines, but also to the air pollution in this region with carbon dioxide, sulfur dioxide, nitrogen oxide and dust. The electrostatic filters and chimneys of powerful power plants do not stop the release of toxic chemical compounds that cause damage to agricultural land and forests in nearby regions. Creating such large open spaces also requires appropriate adaptation of the terrain and the environment.

The process of hydraulic fracturing (FRACKING) raises the greatest doubts about the entire technology of shale gas production and is criticized not only by environmentalists. Fracking involves injecting a mixture of water, sand, and chemicals under high pressure into a well and into a closed rock containing natural gas. Energy companies are protected from disclosing what chemicals are used in the fracking process.

As a result, alternative energy sources are used, such as the movement of water in rivers (hydroelectric power plants), kinetic wind energy (wind power plants) and electromagnetic radiation from the Sun (solar power plants). In the field of solar energy, solar photovoltaic installations are considered more profitable, due to the direct conversion of solar radiation into electricity using solar panels, which are a combination of semiconductor photovoltaic converters. It is shown in [1] that the use of porous silicon as a working layer can increase the efficiency of solar cells. However, the different order of technological processes can negatively affect the characteristics of structures and, as a result, their efficiency.

2. Materials and methods

Samples of photosensitive structures with a porous layer were divided into two groups.

Before creating photosensitive structures, p-type silicon wafers with a textured surface were cleaned in a peroxide-ammonia solution and divided into two groups. In one group of samples, the first stage was diffusion at 1000C°, the exposure time was 1 hour.

The porous layer was created in two ways: by electrochemical etching in a horizontal cell in an alcoholic solution of hydrofluoric acid [2] and by metal-stimulated chemical etching (MACE) [3]. Electrochemical etching modes: time 15 minutes, current density j=15 mA/cm^2.

A typical MAC etching procedure involves a noble metal that partly covers a Si substrate and an etchant composed of HF and an oxidative agent, such as $Fe(NO_3)_3$, $KAuCl_4$, K_2PtCl_6, and H_2O_2[4]. The former serves as a local cathode catalyzing oxidant reduction and injecting holes into the Si substrate. The holes transferred from the "cathode" oxidize the Si substrate. The silicon oxide is then dissolved by the acidic etchant. As the process goes on, the Si beneath the noble metal is etched away while the Si surface without any metal is nearly impacted. As a result, the noble metal "drill" into the Si substrate, forming various Si nanostructures.

The chemistry of Si dissolution reactions in the presence of HF and H_2O_2 include cathodic reactions at the metal interface, which result in hole (h) injection into Si via the following chemical reaction of H_2O_2 decomposition in HF[5]:

$$H_2O_2 + 2H^+ = 2H_2O + 2h^+ \tag{1}$$

and anodic reactions at the Si interface resulting in Si dissolution in HF

$$Si + 6HF + nh^+ = H_2SiF_6 + nH^+ + [(4-n)/2]H_2. \tag{2}$$

If n = 2, anisotropic etching (Si Back-bond theory— [100] preferential etching)

$$\begin{gathered} Si + 2H^+ + 2h^+ = Si^{4+} + H_2 \\ Si^{4+} + 6HF = SiF_6^{2-} + 6H^+ \end{gathered} . \tag{3}$$

If n = 4, isotropic etching (or electropolishing)

$$\begin{gathered} Si + 2H_2O + 4h^+ = SiO_2 + 4H^+ \\ SiO_2 + 6HF = H_2SiF_6 + 2H_2O \end{gathered} , \tag{4}$$

where n is number of holes per dissolved Si atom.

From the overall dissolution reactions

$$Si + 6HF + H_2O_2 = H_2SiF_6 + 2H_2O + H_2 \tag{5}$$

for anisotropic etching,

$$Si + 6HF + 2H_2O_2 = H_2SiF_6 + 4H_2O \tag{6}$$

for isotropic etching.

As shown in figure 1, the metal particles serve as a redox center and function as a short-circuited galvanic cell with a flux of electrons inside the metal particles, while protons would migrate from the anode to the cathode site outside the metal particles.

In horizontal electrochemical etching, passivation of the silicon surface occurs during anodizing in an aqueous solution of HF and pore formation is observed[6].

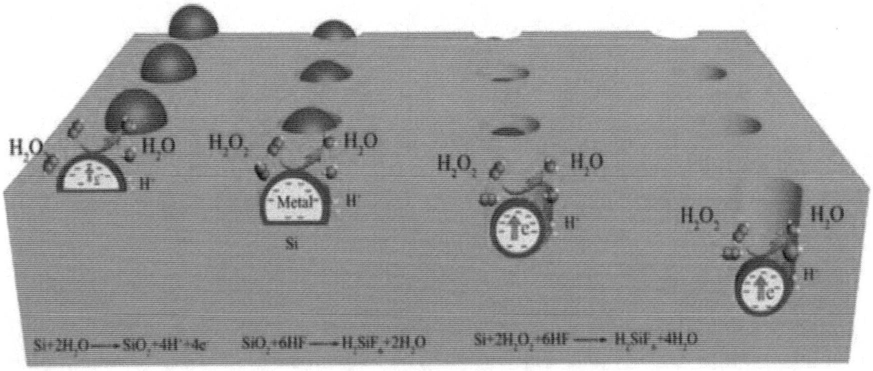

Figure 1. Metal-assisted catalytic reactions on bulk Si surface [4].

Figure 2 shows some possible pathways for the reaction involved in porous silicon formation.

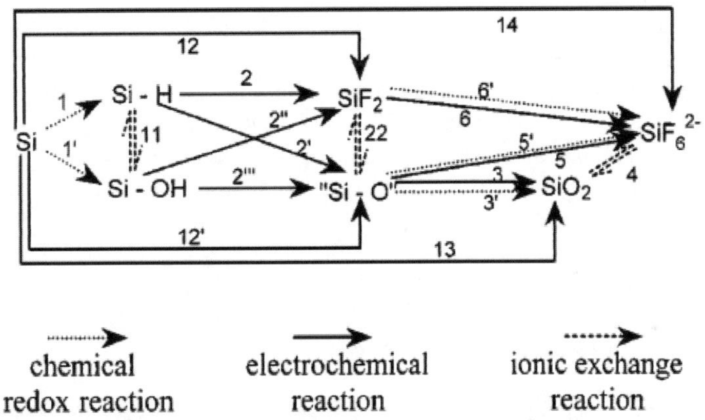

Figure 2. Possible reaction pathways during porous silicon formation [6].

During anodic treatment in electrolytes containing hydrofluoric acid, the following reactions occur on the surface of the electrode:

electrochemical reaction of the formation of silicon fluoride

$$Si + 2HF + 2e^{+} = SiF_2 + H_2F, \quad (7)$$

chemical reduction of silicon from silicon bifluoride

$$\begin{gathered} 2SiF = Si + SiF_4 \\ SiF_4 + 2HF = H_2SiF_6' \end{gathered} \quad (8)$$

chemical oxidation of silicon bifluoride to silicon dioxide and its dissolution in hydrofluoric acid

$$\begin{gathered} SiF_2 + 2H_2O = SiO_2 + 2HF + H_2 \\ SiO_2 + 4HF = SiF_4 + 2H_2O \quad . \\ SiF_4 + 2HF = H_2SiF_6 \end{gathered} \quad (9)$$

Depending on the treatment conditions, one of the reactions (8 or 9) prevails, resulting in either the formation of por-Si (reactions 7 and 8) or electropolishing (reactions 7 and 9).

Figure 3 schematically shows the process of horizontal electrochemical etching of silicon.

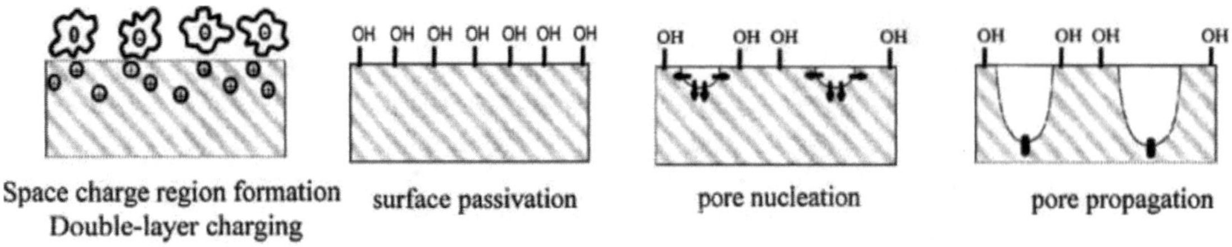

Figure 3. Scheme of horizontal electrochemical etching [6]

The first stage of metal-stimulated chemical etching was $AgNO_3$ deposition. For this purpose, silver nitrate weighing 0.2 g was dissolved in 45.5 ml of hydrofluoric acid HF. The samples were fixed on the bottom of the bath in a horizontal position and then filled with the solution for 3 minutes. Then the samples with the deposited silver on the surface of the plates were placed in a mixture of hydrogen peroxide H_2O_2 and hydrofluoric acid for subsequent etching for 5 minutes. After etching, the samples are placed in concentrated nitric acid to remove the silver nanoparticles.

The other group had the reverse order of technological processes.

Photosensitivity measurements were made using the MDR-3 diffraction monochromator. The radiation source is a xenon arc lamp of the DKSEL 100 type. The light from the lamp hits the collimator, which creates a focused beam of light that enters the input slot of the monochromator. At the output of the monochromator, we have light of the required wavelength, which, falling on the second collimator, is focused and directed to the sample located in the holder. The signal that occurs when the structure is illuminated is recorded. To register the signal, a voltammeter B7-21A was used, which operated in the ammeter mode (nA).

3. Results and discussions

Figure 4 shows SEM images of structures made by the MACE method (figure 4a) and the electrochemical method (figure 4b).

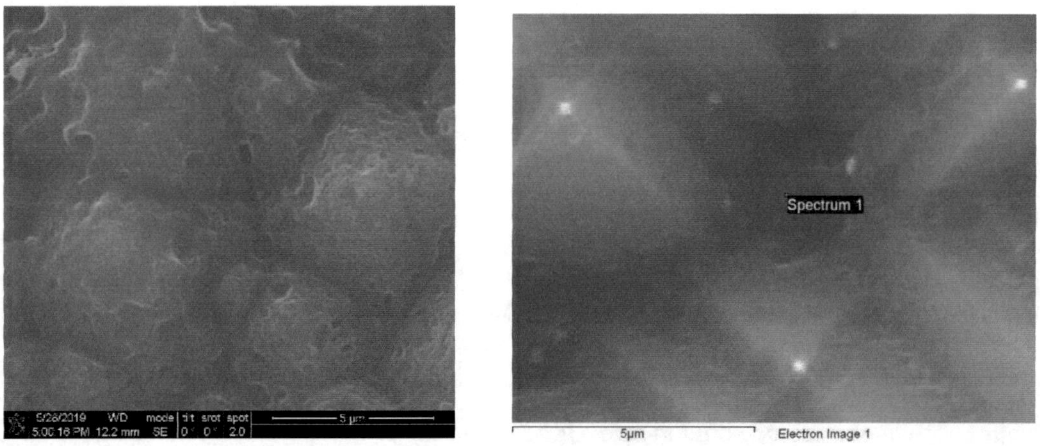

Figure 4. SEM images of a porous layer obtained by (a) MACE, (b) horizontal electrochemical etching.

In the electrochemical method, pores are formed mainly at the joints of the pyramids. With MACE, etching occurs on the entire surface of the pyramids, which allows for better radiation capture.

Figure 5 shows the results of measurements of the current-voltage characteristics of porous structures performed by various methods before diffusion (purple curve) and after diffusion (green curve).

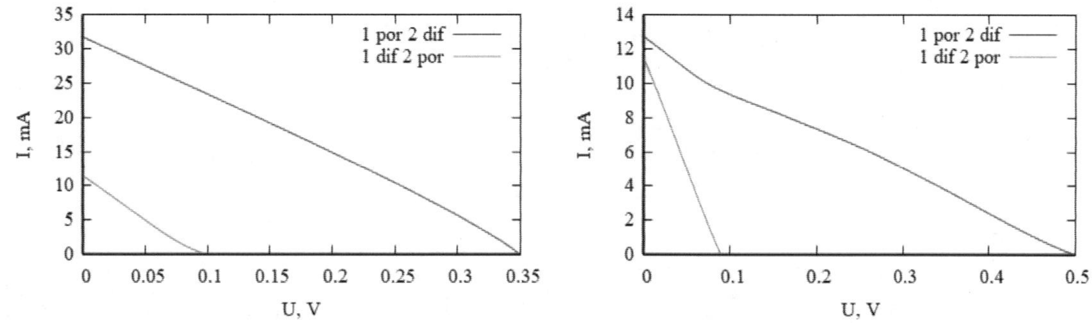

Figure 5. The current-voltage characteristic of photosensitive structures with a porous layer was performed (a) by electrochemical etching in a horizontal cell, (b) by the MACE method, before diffusion (purple curve), after diffusion (green curve).

Figure 6 shows the results of measuring the photosensitivity of porous structures in which pore formation was carried out before diffusion (purple curve) and after diffusion (green curve).

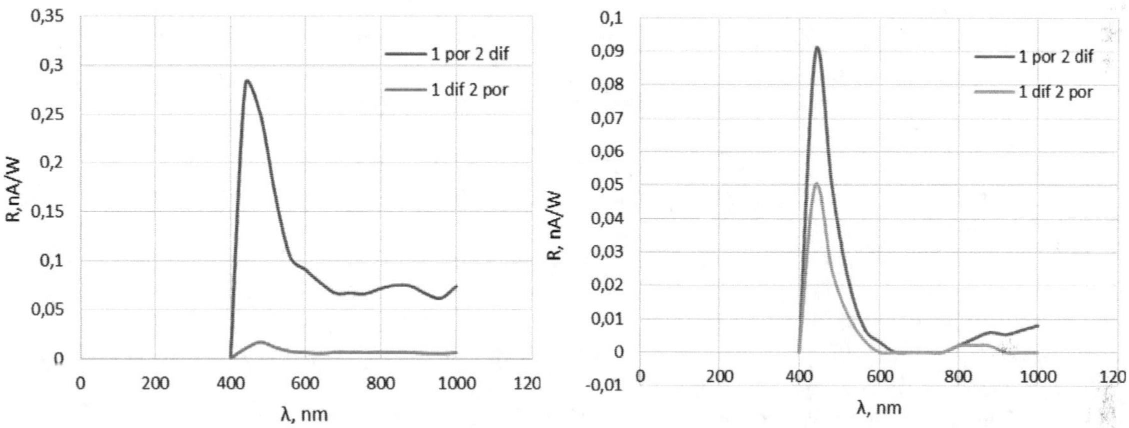

Figure 6. The spectral characteristics of photosensitive structures with a porous layer were performed (a) by electrochemical etching in a horizontal cell, (b) by the MACE method, before diffusion (purple curve) and after diffusion (green curve).

The analysis of the graphs shows that for both methods of manufacturing porous layers, the route by which the porous layer is first made, and then diffusion is carried out, is more effective. If the diffusion is carried out before pore formation, then as a result of etching, active etching of the n-type occurs from the substrate, which negatively affects the main parameters of the solar cell: short-circuit current and no-load voltage. Such a technological route also reduces the spectral sensitivity of the samples.

It was also observed that the method of creating a porous layer affects the characteristics of the solar cell (figure 7).

This is primarily due to the characteristics of the resulting porous layer on the textured surface (figure 7). With the help of MACE, a porous layer is formed over the entire surface of the pyramid, providing better light capture, which affects a higher no-load voltage.

However, in horizontal etching, where the porous layer passes only along the bases of the pyramids, the short-circuit current is higher than in the MACE.

 2086 (2021) 012102 doi:10.1088/1742-6596/2086/1/012102

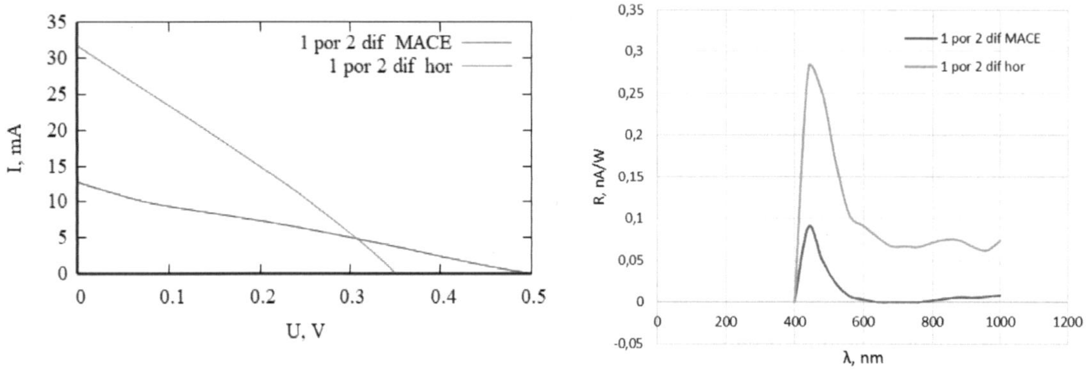

Figure 7. Comparison of (a) current-voltage and (b) spectral characteristics of structures with porous silicon obtained by horizontal etching (green curve) and MACE (purple curve).

The lower short-circuit current in photosensitive structures manufactured by the MACE method is associated with current losses caused by poor contact.

4. Conclusion

Thus, the conducted studies have shown that the optimal technological route is pore formation and subsequent diffusion. However, this technological route requires further refinement, namely, the selection of the optimal mode of diffusion and the reduction of recombination losses.

References

[1] Kirsanov N Y, Latukhina N V, Lizunkova D A, Rogozhina G A and Stepikhova M V 2017 *Semiconductors* **51** p 3
[2] Sze S M and Kwok K Ng 2006 *Physics of Semiconductor Devices*
[3] Bandarenka H V, Girel K V, Niauzorau S A, Gonchar K A and Timoshenko V U 2016 *Doklady BGUIR* **2** p 74
[4] Peng Yu, Jiang Wu, Shenting Liu, Jie Xiong, Chennupati Jagadish and Zhiming M Wang 2016 *Nano Today* **11** pp 704–737
[5] Fatima Toor 2016 *Nanotechnology* **27** p 412003
[6] Ronkel F and Schultze J W 2000 *Journal of Porous Materials* **7** pp 11-16

STM Light Emission and I(V) study of single gold nanoantenna

V A Shkoldin[1,2], D V Levedev[1,3,4], A M Mozharov[1], D V Permyakov[2], L N Dvoretckaia[1], A A Bogdanov[2], A K Samusev[2], A O Golubok[4] and I S Mukhin[1,2]

[1] Alferov University, 8/3 Khlopina str., St. Petersburg, 194021 Russia
[2] ITMO University, 9 Kronverksky pr., St. Petersburg, 197101 Russia
[3] Saint Petersburg State University, 7/9 Universitetskaya nab., St. Petersburg, 199034 Russia
[4] Institute for analytical instrumentation RAS, 26 Rizhskii pr., St. Petersburg, 190103 Russia

shkoldin@spbau.ru

Abstract. Hight-speed optical nanoemitters are of importance for on-chip optical data processing. A tunnel junctions can be a base for such light emitters, however such structures suffer from low quantum efficiency. One of the ways to improve efficiency of tunneling electron energy to photon generation conversion is the increase of the local density of optical states by using of optical nanoantennas. In this work, we study optoelectronic properties of single gold nanodisc with high spatial resolution. We show nonuniform distribution of electromagnetic near-fields of nanodisk, which is consistent with nanoantenna optical modes. And we demonstrate direct correlation between nanoantenna optical states and features on current-voltage characteristics of tunnel junction between metal tip and nanodisk.

1. Introduction

The optical communication transfer enormous amount of information and this volume grows every year at the rate of 20-30% [1]. Optical links are one of the best ways to transfer information on large distance. At the same time metal wires are commonly used for data transfer within the parts of integrated circuits. Optical signal processing has many advantages, including the absence of ohmic losses and big speed of signal propagation. Transition to photonic can be next step of computing systems' development. Currently, the size mismatch between nanophotonic components, CMOS-components and optical sources is a technological bottle neck. Available today light emitting sources, such as VECSELs, Fabry-Perot lasers or microring and microdiscs lasers, have resonators with minimal sizes significantly exceeding their emission wavelengths. Also, new type of optical sources is required to high frequency modulation. One of ways to overcome this size mismatch can be based on the use of inelastic electron tunneling process causing light emission. This phenomenon enables to create subwavelength light sources.

A process of light emission in the metal-insulator-metal (MIM) structures was discovered by the first time by Lambe and McCarthy in 1976 [2]. However, quantum efficiency (QE) of this process is very low (10e-6 – 10e-4 photon/electron). In recent years, there has been renewed interest in light emission from tunnel junctions. Many articles were focused on the improvement of light emission from a tunnel junction [3–5]. It was shown that QE can be enhanced by increasing of local density of optical states (LDOS) in the tunnel junction. Theoretically, the value of QE may reach 10% [6,7] due to a large LDOS enabled by plasmonic confinement. Using of nanoantennas is a promising way to increase LDOS in the tunnel junction.

2. Experiment

Scanning tunnel microscopy (STM) is powerful tool for studying tunnel-junction light emission[8,9]. In this article, we used an ultra-high vacuum STM combined with a home-made optical setup. The experiment was produced in the STM Omicron VT650 with a vacuum not worse than 5e-9 mBar. The optical system consists of a small lens (d = 4 mm, N.A. = 0.32) positioned near the sample inside an STM vacuum chamber and the avalanche photodiode (IDQuantique ID120) with mirror system outside the vacuum chamber. The output signal from photodiode was directed to STM controller. It makes possible to acquire STM topography and obtain light emission map simultaneously.

We used STM to study the LDOS distribution of a single gold nanodisc by a Pt/Ir STM probe (DPT10, Bruker). The array of nanodiscs was made on thin gold film on mica substrate with low roughness[4]. The period of nanodiscs' array was 1 μm. The nanodiscs produced by e-beam lithography have 50 nm height and 180 nm diameter.

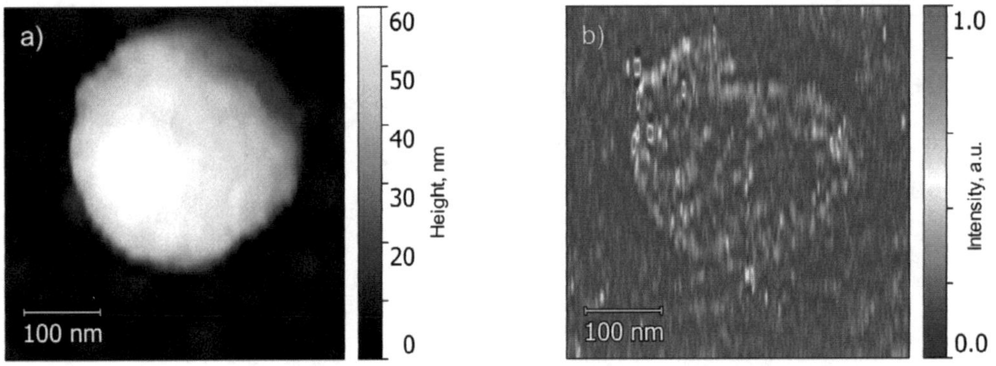

Figure 1. (a) Typical STM topography image of a nanodisc and (b) acquired light emission map.

We experimentally observed the increase of light emission under STM tip on edges of nanodiscs (Figure 1). As well known, the inelastic tunnel current is proportional to the integral of the density of optical states of the sample[9,10]. Therefore, we investigated current-voltage (I-V) characteristics at points on the line crossing the disk (Figure 2a). Figure 2b demonstrates I-V curves acquired at the edge of nanodisk. One can see the features on I-V dependence near 1.6 V and 2.2 V. The fingerprint map presented in Figure 2 c) shows d^2I/d^2V dependence for all points denoted at Figure 2 a). The features, which appear in the form of puddles on the color map, can be associated with inelastic tunneling and consequent light emission process [4]. This map allows to estimate the area of nanodisc with enhanced LDOS.

Also, we performed numerical simulation of the electromagnetic field distribution excited in a plasmonic nanodisc by a point vertical dipole, located between nanodisk and STM tip. The numerical simulation shows that two optical modes at 668 nm and 566 nm can be excited in the nanoantenna, when the STM tip located at the edge of nanodisk. At the same time with the probe positioned in the center of the nanoantenna, only one mode at 668 nm can be excited. Both modes have a «hot spot» at the edge of the nanoantenna, which confirms the increased LDOS at the periphery of the structure. This explains the increasing of the emission process efficiency from the tunnel contact, which was observed in the experiment (Figure 1 b). IV-curves analysis also shows a change of LDOS at the edge of the nanoantenna (Figure 2).

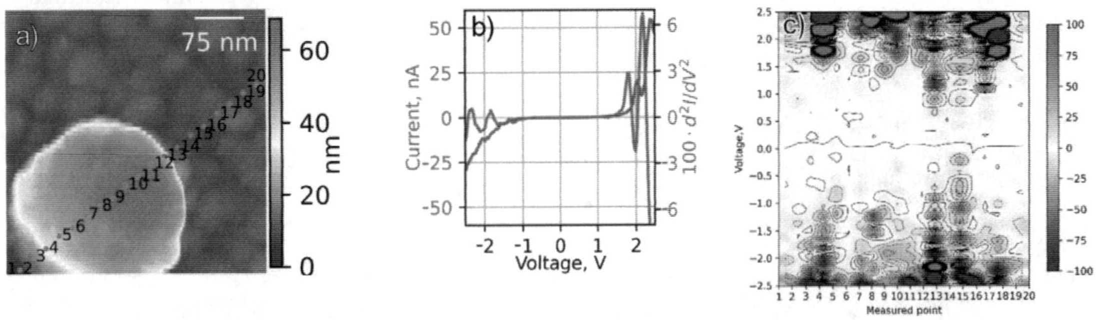

Figure 2. (a) STM topography image of a nanodisc with the markers enumerating the points of I(V) study. (b) Typical single I-V, d^2I/d^2V dependences measured at the edge of the disc (point #4). (c) d^2I/dV^2 "fingerprint" map of the gold disc.

Also, optical spectra were measured on the same disks at ambient conditions. We use SmartSPM (AIST-NT) coupled with spectrometer Princeton Instruments SpectraPro-2500i (USA) equipped with PyLoN:400BR_eXcelon detector with liquid nitrogen cooling.

The optical spectra of the STM-induced nanoantenna emission contain two clear bands with maxima at 670 nm and 765 nm (Figure 3). The long-wavelength band dominates in the spectra obtained on the substrate between nanodisks (see green lines in Figure 3). While the short-wavelength band dominates in the spectra acquired on the nanoantenna (see blue lines in Figure 3). When the STM tip was located at the edge of nanodisk, optical spectra had the enhanced short-wavelength component. These experimental results qualitatively correspond to the results of numerical simulations, at the periphery of nanodisk the mode with shorter wavelength can be excited. We suggest that the difference of the peak spectral positions predicted by modelling can be caused by the changing of inelastic tunneling process compared to vacuum conditions, as well as the imperfections of disk fabrication. The simulation was performed for an ideal cylinder in vacuum, while in air the sample surface is covered with a thin layer of water, which affects the parameters of the tunnel contact, including the energy position of Fermi level. The low selectivity of mode excitation is most likely related to the relatively large diameter of the STM tip used, which is comparable with the size of the nanoantenna[11].

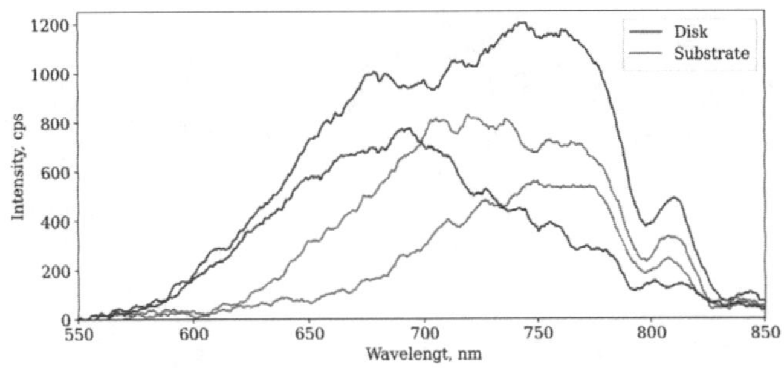

Figure 3. Radiation spectra of the tunnel contact obtained at different positions. The blue lines are spectra obtained from the tunnel contact with a nanoantenna. The green lines are typical spectra obtained from the gold substrate.

3. Results

The I-V and its second-order derivative curves of the tunnel contact allow to investigate optical modes nanoantennas with subwavelength resolution. It may be used for simplifying of the development of nanostructures for local optical sources based on inelastic electron tunneling phenomena. In that case the use of optical detection system is not required. The proposed technique is essential for the development of electrically-driven plasmonic optical sources for on-chip optical data processing.

Acknowledgments
This work was done with the financial support of the Russian Federation President Council (MK-2428.2020.2) and the Russian Foundation for Basic Research (project 19-32-90028).

References
[1] "Cisco Annual Internet Report (2018–2023) White Paper" https://www.cisco.com/c/en/us/solutions/collateral/executive-perspectives/annual-internet-report/white-paper-c11-741490.html (access data: 01.03.2021)
[2] Lambe J and McCarthy S L 1976 Light Emission from Inelastic Electron Tunneling *Phys. Rev. Lett.* **37** 923–5
[3] Kern J, Kullock R, Prangsma J, Emmerling M, Kamp M and Hecht B 2015 Electrically driven optical antennas *Nat. Photonics* **9** 582–6
[4] Lebedev D V, Mozharov A M, Bolshakov A D, Shkoldin V A, Permyakov D V, Golubok A O, Samusev A K and Mukhin I S 2020 Indirect Detection of the Light Emission in the Local Tunnel Junction *Phys. Status Solidi - Rapid Res. Lett.* **14**
[5] Shalem G, Erez-Cohen O, Mahalu D and Bar-Joseph I 2021 Light Emission in Metal – Semiconductor Tunnel Junctions: Direct Evidence for Electron Heating by Plasmon Decay *Nano Lett.* **21** 1–6
[6] Doderer M and et al. 2019 Light Emission from a Waveguide Integrated MOS Tunnel Junction 2019 *Conf. Lasers Electro-Optics, CLEO 2019 - Proc.* 3–4
[7] Parzefall M and Novotny L 2019 Optical antennas driven by quantum tunneling: a key issues review *Reports Prog. Phys.* **82** 112401
[8] Shkoldin V A and et al. S 2019 Crucial Role of Metal Surface Morphology in Photon Emission from a Tunnel Junction at Ambient Conditions *J. Phys. Chem. C* **123** 8813–7
[9] Lebedev D V, Shkoldin V A, Mozharov A M, Permyakov D V, Dvoretckaia L N, Bogdanov A A, Samusev A K, Golubok A O and Mukhin I S 2021 Scanning Tunneling Microscopy-Induced Light Emission and I(V) Study of Optical Near-Field Properties of Single Plasmonic Nanoantennas *J. Phys. Chem. Lett.* **12** 501–7
[10] Sinev I S, Voroshilov P M, Mukhin I S, Denisyuk A I, Guzhva M E, Samusev A K, Belov P A and Simovski C R 2015 Demonstration of unusual nanoantenna array modes through direct reconstruction of the near-field signal *Nanoscale* **7** 765–70
[11] Mukhin I S, Fadeev I V, Zhukov M V, Dubrovskii V G and Golubok A O 2015 Framed carbon nanostructures: Synthesis and applications in functional SPM tips *Ultramicroscopy* **148** 151–7

Several processes participating in a decrease and the droop of external quantum efficiency in green InGaN/GaN MQW structures

N A Talnishnikh[1], E I Shabunina[2], N M Shmidt[2] and A E Ivanov[1,3]

[1] Submicron Heterostructures for Microelectronics Research and Engineering Center RAS, St Petersburg 194021, Russia
[2] Ioffe Physical Technical Institute, St Petersburg 194021, Russia
[3] Saint-Petersburg Electrotechnical University ETU "LETI", St. Petersburg 197376, Russia

nadya.fel@mail.ru

Abstract. The obtained experimental results allow us to clarify the nature of mechanisms related to the presence of cations in disordered InGaN alloy and hetero-interfaces. The capture of charge carriers by cations reduces the external quantum efficiency (EQE) in green MQWs at $j \leq 10$ A/cm^2. The EQE droop phenomenon caused by smoothed out lateral potential fluctuations occurs at $j > 10$ A/cm^2. At $j > 40$ A/cm^2 the droop associated with interactions between charge carriers and dislocations and grain boundaries takes place.

1. Introduction

InGaN/GaN MQW structures are the foundation for light-emitting diodes (LEDs) and lasers. Phenomena such as low values of the external quantum efficiency (EQE, η) at maximum and the EQE droop at $j > 10$ A/cm^2 in green LEDs have been under intense discussions for the last decade and have yet to find a commonly accepted explanation [1-3]. Current-voltage (I-U) characteristics and EQE dependences on current in green LEDs are known to show more complex behavior which could not be described by the commonly used ABC-model [2]. To clarify the processes resulting in low EQE values at maximum and the EQE droop in green MQWs we considered our experimental results in the framework of atomic [4], localization landscape [5], and trap-assisted tunneling [6] models and took into account the tunneling radiative recombination at $j \leq 10$ A/cm^2 [7,8]. The atomic and localization landscape models are based on the assumption that cations in disordered InGaN alloy and hetero-interfaces impacts the properties of nitride-based MQWs

2. Experimental

We studied commercially available green LEDs based on MQW InGaN/GaN nanostructures having EQE in the range of 15-30 % at 520-530 nm. We should note that η values ~ 30 % at 530 nm in commercially available green LEDs can be considered reasonably high. We also used reference blue LEDs having EQE in the range of 50-55 % at 455-460 nm. All LEDs have 1 mm^2 active area. We studied I-U characteristics, electroluminescence spectra (EL), η dependences on direct current (DC) and in pulse mode (PM) (100 ns pulse at 50 Hz) at 50-350 K and the distribution of both $\eta_{j,U}$ and full width at half maximum of EL (FWHM) over wavelengths.

3. Results and discussion

Experimental η(j) dependences at DC and in PM at 300 K in the green and blue (reference) LEDs shown in a semi-logarithmic plot demonstrate the η droop and a typical difference between η values in green and blue LEDs (Figure 1a). The red circles show η value in PM. Different shapes of η(j) dependences in regions I, II, III, and IV (Figure 1a) indicate the presence of several mechanisms. Red lines mark wavelengths corresponding to values of threshold voltage (V_{th} =2.62 V at 2 A/cm ^{2} in blue LEDs, V_{th} = 2.65 V at 3.5 A/cm ^{2} in green LEDs with high η and V_{th} = 3.4 V at 20 A/cm ^{2} in green LEDs with low η). Thus, η(j) dependences in regions I, II indicate mechanisms occurring in QWs situated at a depletion layer (DL) around p-n junction. At the same time, η(j) in regions III and IV shows processes in QWs outside of DL, when the p-n junction opens. It was shown earlier [7, 8] that $\eta(j) \sim j^{1/2}$ at j < 10 A/cm ^{2} is due to the tunneling radiative recombination. The same η(j) shape is observed in region I in blue (Figure 1a, curve 1) and green (Figure 1a, curve 2 and 3) LEDs. This allows us to assume that the process of radiative recombination of localized tunneling carriers in QWs situated in DL around p-n junction takes place. However, $\eta(j) \sim j^{1/2}$ in green LEDs is observed only within a small range of j values, practically by an order of magnitude shorter than that in blue LEDs (Figure 1a, curve 2). $\eta(j) \sim j^{1/2}$ is absent in green LEDs with low η (Figure 1a curve 3). To clarify the causes behind the difference, we studied η distribution over wavelengths in green LEDs (Figure 1b). It seems, that the η distribution over wavelengths in green LEDs in the same range of the j as in Figure 1 (a) allows us to visually show several processes developed in regions I-IV. We assume that radiative recombination of localized tunneling carriers in QWs in DL is responsible for increasing η values at the fixed wavelength 532,5 nm at region I (Figure 1b, curve 1).

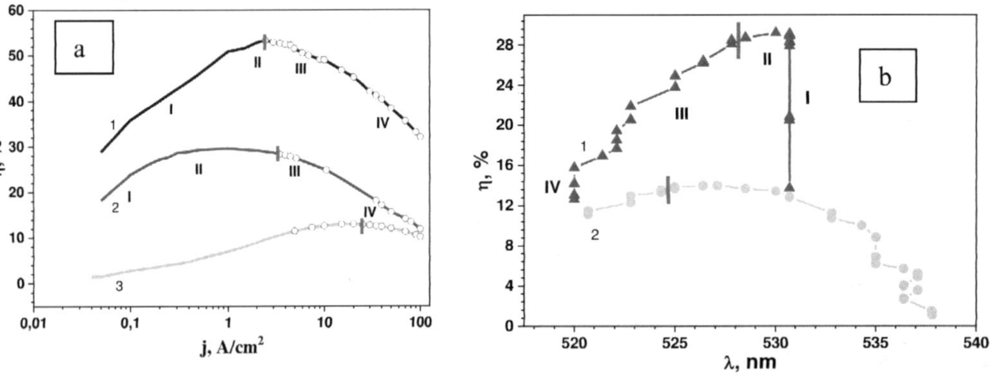

Figure 1. (a) EQE dependences on j in blue (curve 1) and green (curve 2 and 3) LEDs. Red curves show EQE dependences in green LEDs in PM. (b) η distribution over wavelengths in green LEDs. Red lines mark the wavelengths corresponding V_{th} values.

Prevalence of tunneling transport of charge carriers in region I and II is confirmed by the absence of Lorentzians in noise spectra [9], the ideality factor n > 2, and shapes of reverse and forward I-U characteristics [6], also their temperature dependences. I-U characteristics in green LEDs that we observed experimentally are presented in Figure 2a (the insert shows reverse I-U characteristics). (Figure 2b) shows the dependences of current on temperature at U = 2, 2.5 and 3 V. Forward I-U characteristics in green LEDs have n > 2. Their shape is in good agreement with simulations obtained in the framework of the multi-phonon trap-assisted tunneling (TAT) model [6]. The shape of current temperature dependences, however, is considerably weaker than that in case of predominant Shockley-Read-Hall recombination in DL. Peculiarities of reverse I-U characteristics indicates the presence of cations in DL leading to the carrier transport via electron hopping among localized states within the band gap. High reverse currents (insert to Figure 2a) as well as the absence of effective radiative recombination at fixed wavelengths in green LEDs with low η (Figure 1b, curve 2) is observed.

Moreover, a wide change in wavelengths in the range of 10 nm and instable low η values with a longer wavelength tile at λ > 530 nm on η (j) dependence in DL is visually shown.

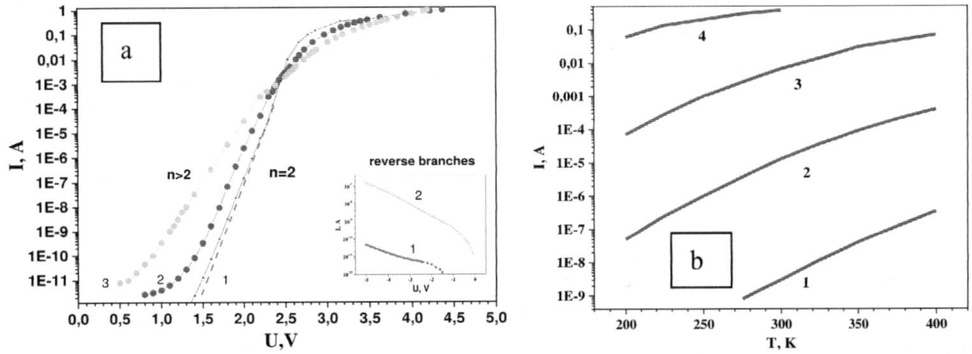

Figure 2. (a) Forward I-U characteristics in: 1 – blue (reference) LEDs, 2 – green LEDs with η =28%, 3 – green LEDs with η =14 %. On the insert: reverse I-U characteristics in green LEDs with different values of η: 1-28 %, 2-14 %. (b) Dependences of current on temperature at: 1-1.5 V, 3-2.5 V, 4-3 V.

Similar tile was observed earlier in green LED structures with disordered nanomaterial enriched by In [10]. We assume that the shape of η distribution over wavelengths (Figure 1b, curve 2) depicts random cations' distribution in disordered hetero-interface and InGaN alloy. Our assumption correlates well with the atomic and localization landscape models [4, 5]. It was found out that a decrease in η values in region II in both types of green LEDs is due to the capture of charge carriers by cations in DL. This process is reflected in the narrowing of FWHM values of EL followed by the increasing bias and injection current (Figure 3a (the falling branches)).The relative mistake of FWHM value determination is ±0.6nm. This effect is observed not only in LEDs with η = 30 % (Figure 3a, curve 1), but also in LEDs with different η values (Figure 3a). The more prominent the effect is, the lower η values in green LEDs are. The falling branches are not observed in blue LEDs with EQE > 40 %. The filling of quantum states at U < U_{th} in QWs in DL is equilibrium.

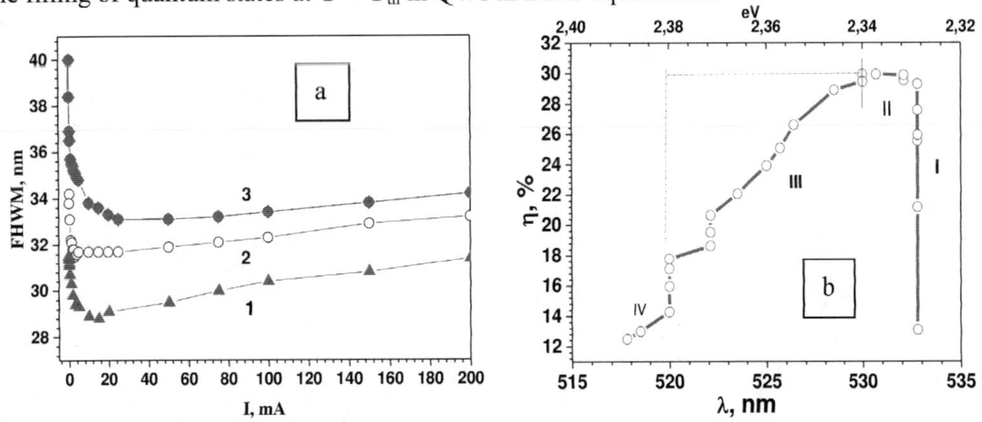

Figure 3. (a) FWHM dependences on j in green LEDs with different η values: 28 % (curve 1), 20 % (curve 2), and 14 % (curve 3). (b) η distribution over wavelengths in green LED with η = 28 %.

On the contrary, at U > Uth (part III) the filling of quantum states in QWs located outside of DL is non-equilibrium since FWHM values grow steadily over wavelengths with a current increase, which is accompanied by the EQE droop in both types of LEDs (Figure 3b). The opening of p-n junction is accompanied by an increase in the concentration of delocalized carriers activated by injection. They fill all active region of LED outside of DL, including the lateral alloy fluctuations each InGaN layer. It is known that lateral dimensions of random alloy fluctuations may vary from several nanometres to several microns [5]. Lateral alloy fluctuations and rough hetero-interfaces form lateral potential fluctuations (LPF) along InGaN layers, according to the atomic and lateral landscape models. LPF is revealed at the η distribution over wavelengths (Figure 3b). Experimental value of LPF is about 40 meV, which is in good agreement with the values predicted by landscape model [4]. We considered each QW as bonded lateral nano- and micro-wells. The tunneling (migration) of injected carriers along InGaN layers and hetero-interface get smoothed out LPF. This process is followed by carriers filling in nano- and micro-wells and partially getting captured in rough hetero-interfaces and cations in local regions of disordered alloy. As a result, the η droop and the shift of EL maximum is observed at $j < 30$ A/cm^2 (region IV). The droop starts at $j > 30$ A/cm^2 in PM as well. The concentration of injected carriers (N) is higher than 10^{18} cm^3 and their interactions with all defects including extended ones start. The amplification of non-radiative recombination in dislocations and grain boundaries in LED structures with an increase in carrier concentration was found out by cathodoluminescence and AFM [12]. Aleksiejūnas R. et al. [13] observed charge carriers' diffusion through grain boundaries in LED structures and GaN substrates over growing carrier concentration $N > 10^{18}$ cm^3. It is important that the spatial mechanism of charge carrier transport changes from tunnelling and migration to the prevalent diffusion. However, the diffusion length is very short in nitrides and can be estimated by the size of grains [14]. As a result, the crowding effect occurs.

4. Conclusion

The results obtained confirm the importance of contribution of randomly distributed cations and rough hetero-interface to LED parameters predicted by the localization landscape model. It was clarified that the mechanisms occurring in QWs depend significantly on their placement inside or outside of DL around a p-n junction. Vertical tunneling of charge carriers accompanied by the filling of QWs situated in DL in green LEDs takes place at $j < 10$ A/cm^2. EQE reduction at $j < 10$ A/cm^2 in green LEDs is due to the capture of tunneling carriers by cations in rough hetero-interface and local regions enriched by In (cations). The droop, starting at $j > 20$ A/cm^2 when the p-n junction opens, is related to the loss of carriers tunneling in low LPF caused by lateral disorder of hetero–interface and InGaN layers. At $j > 40$ A/cm^2 the diffusion transport of carriers becomes predominant. The carrier loss is amplified by dislocations and grain boundaries when the injection current increases and the crowding effect occurs.

Acknowledgments

Measurements of near-field EL intensity distribution were made at the Center of Multi-User Facilities "Element Base of Microwave Photonics and Nanoelectronics: Technology, Diagnostics, Metrology".

References

[1] Aurelien D, Young N, Lund C and Craven M 2020 *ECS Journal of Solid State Science and Technology* 9 016021
[2] Titkov I E, Karpov S Yu, Yadav A, Mamedov D, Zerova V L and Rafailov E 2017 *Materials* 10 1323 doi:10.3390/ma10111323
[3] Piprek J, Römer F and B Witzigmann 2015 *Applied Physics Letters* 106 101101
[4] Schulz S, Caro M A, Coughlan C and O'Reilly E P 2015 *PHYSICAL REVIEW* B **91** 035439
[5] Li C-K, Piccardo M, Lu L-S, Mayboroda S, Martinelli L, Peretti J, Speck J S, Weisbuch C, Filoche M and Wu Y-R 2017 *PHYSICAL REVIEW* B **95** 144206 .
[6] Mandurrino M, Goano M, Vallone M, Bertazzi F, Ghione G, Verzellesi G, Meneghini M, Meneghesso G and Zanoni E 2015 *J. Comput.* 14 2 444-455
[7] Voronenkov V, Bochkareva N, Gorbunov R, Latyshev P, Lelikov Y, Rebane Y, Tsyuk A, Zubrilov A and Shreter Y 2013. *J. Appl. Phys.* **52** 08JE14
[8] Shmidt N M et al. 2020 *Technical Physics Letters* 46 1253
[9] Chernyakov A E, Levinshtein M A, Talnishnikh N A, Shabunina E I, Shmidt N M 2014 *J. Cryst. Growth* 401 302
[10] Shmidt N M, Chernyakov A E, Talnishnikh N A, Nikolaev A E, Sakharov A V, Petrov V N

2019 *J. Crys. Growth* 520 82
[11] Zhou S, Lu Z, Wu Y and Zhang Y 2018 Japanese *Journal of Applied Physics* 57 051003
[12] Pozina G, Ciechonski R, Bi Z, Samuelson L, Monemaral B 2015 *APL* 251106
[13] Aleksiejūnas R, Nomeika K, Miasojedovas S, Nargelas S, Malinauskas T, Jarašiūnas K, Tuna Ö and Heuken M 2015 *Phys. Stat. Sol. (B)* 252 977
[14] Scˇajev P, Jarasiunas K, Okur S, Özgür Ü and Morkoç H 2012 *J. Appl. Phys*. 111 023702.

Laser treatment of ITO thin films with Carbon Nanotubes for Liquid Crystal Devices

A S Toikka[1,2,3]

[1] Department of Photonics, St. Petersburg Electrotechnical University ("LETI"), 197376 St. Petersburg, Russia
[2] Lab for Photophysics of media with nanoobjects, Vavilov State Optical Institute, 199053 St. Petersburg, Russia
[3] Advanced Development Division, Petersburg Nuclear Physics Institute, 188300 Gatchina, Russia

Abstract. In this paper, the influence of laser ablation on the refractive properties of indium tin oxides (ITO) thin films with deposited single-wall Carbon Nanotubes (CNTs) was considered. Sputtering of CNTs was preliminary based on the laser-oriented method with application of the external electric field. The laser ablation of ITO-CNTs coatings allows changing the electric, optical and mechanical properties dramatically. Moreover, this technical operation permit to switch the topology of the surface, thus it leads to the conversion of the refractive index. The possibility of index-matching due to the laser treatment contributes to the expansion of the technical capabilities of LC devices.

1. Introduction

Liquid Crystal – is a mesophase that combines the optical anisotropy and properties of liquids. These molecules have no regularity of the centre mass; however, their orientation can be controlled with external exposure (optical beam, electric and magnetic field, acoustic waves). These features allow constructing the devices for the optical modulation issues such as electro-optical and spatial light modulators, optical switching elements, lenses with the variable focus point, etc. In the common case, electro-optical LC devices require high optical transmittance in the operating spectra range and low electric consumption. Spectral properties of mesophase depend on the LC modifications and the materials of sensitizers. One of the most perspective ways for optical optimizing of LC is structuring with the nanoparticles, such as fullerenes, nanotubes, quantum dots, and others [1-3]. Moreover, they are able to increase the local polarizability of media which correlated with the temporal characteristics of devices [4-5]. Another approach for tuning of parameters is connected with the topology and composition of the adjacent layers to the LC phase in the construction of the device. In order to decrease the response time of LC molecules, variable approaches can be used. The deposition of single-wall carbon nanotubes (CNTs) permits not to use high-resistive orienting layers and increases the laser strength [6]. It should be noticed, that these nanoparticles can realize anti-reflection functions that depend on the deposition parameters, such as density, penetration depth, and tilt angle.

In the current research, we processed via the CO_2 laser the surface of ITO with the deposited CNTs. Its influences the topology of films, thus refractive properties can be modified under irradiation. In this paper, we compare the refractive indices of two groups of samples — with the laser ablation and without the laser ablation.

2. Materials and methods

For deposition of the CNTs on the ITO surface, we used a laser-oriented deposition (LOD) method based on a CO_2 laser (λ = 10.6 µm, *P*= 30 W, *d*= 5 mm and speed of treatment *v* was 1-3 cm×s^{-1}). The supply grid with the electric field 100 V×cm^{-1} was in LOD scheme as well. The laser ablation was performed with the laser system «C-Marker C30» (λ = 10.6 µm, *P*= 5 W, the beam diameter is 100 µm, *f*= 1 kHz, *v*=3 cm×s^{-1}, the distance between laser grooves was 500 µm). The refractive index of structures was measured with the ellipsometry method (J.A. Woollam M-2000RCE) in 26 points for each sample per aperture 20 mm. The thickness of ITO was approximately 100 nm. The diagnostic of relief was performed through the atomic-force microscope (AFM) Solver Next NT-MDT in semi-tapping mode. The aperture was 50 µm and the scan rate was 1 Hz.

3. Results

During the laser ablation process, besides the absorption of the beam in media, the part of optical flux can propagate along the surface. In this case, induced surface electromagnetic wave (SEW) has the same wavelength. However, the intensity of SEW is dramatically decreased in comparison with the incident wave. According to these features, it is possible to realize the secondary laser ablation technique with SEW (Figure 1). This phenomenon was considered in detail by M.N. Libenson in papers [7-8].

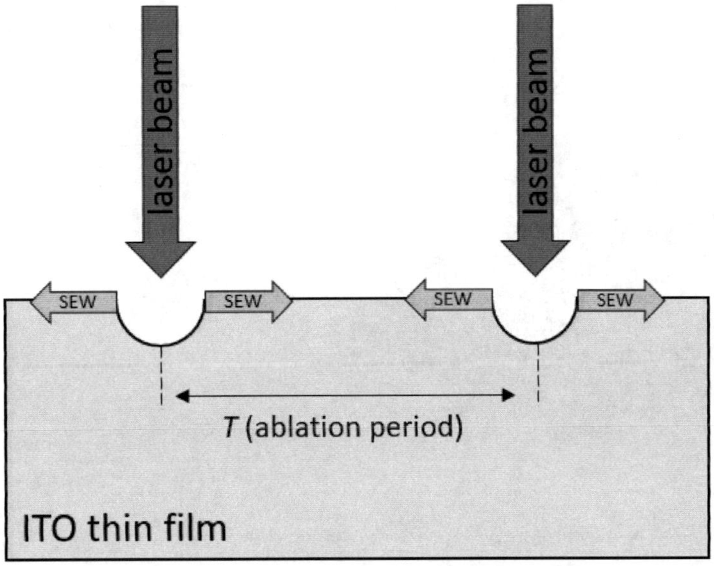

Figure 1. The scheme of laser ablation of ITO surface through the SEW propagation

In the case of the smooth surface of pure ITO, the described waves produce periodic lattice in relief with the following period coinciding with the wavelength. The diffuse reflection becomes higher and the measured effective refractive index for ITO with SEW processing is higher in range the from 1280 nm to 1700 nm (Figure 2a).

When the ITO structure is modified with the CNTs, the relief roughness is increased. Two against mechanisms exist. The first one is connected with the material aspect and based on the antireflective properties of CNTs due to the less value of the refractive index. However, the new geometry of the surface can provide higher reflection losses in comparison with the smooth structure.

The properties of the new interface can be changed and improved via the SEW processing with the further decrease of refractive index and rise of transmittance (Figure 2b).

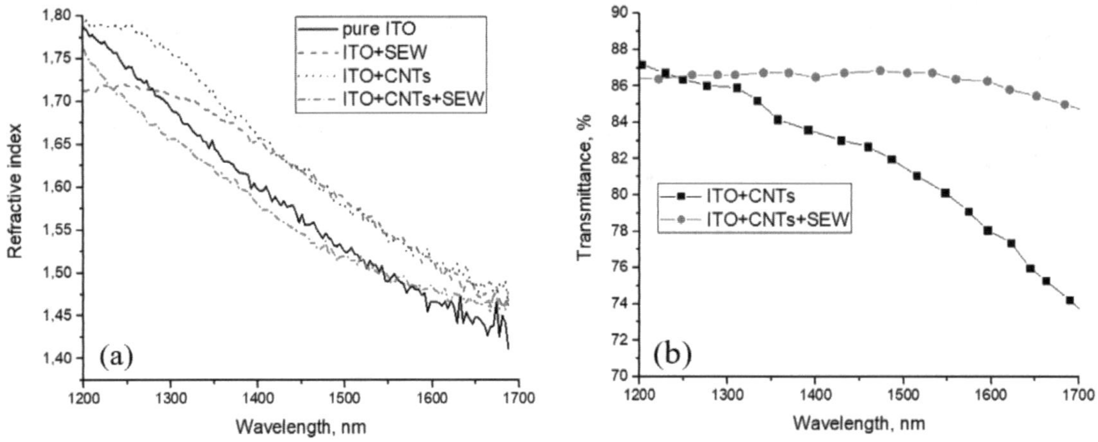

Figure 2. (a) Spectral dependences of refractive index for ITO modification on glass substrate; **(b)** Optical transmittance for ITO with CNTs before and after laser treatment.

Moreover, SEW treatment of ITO with CNTs is an additional mechanism for liquid crystal orientation. The order of cavities height is tens of nm and the size of the LC molecule is approximately 1.5 nm (Figure 3). These features allow to control and to set the initial orientation of mesophase molecules in the construction of LC cell.

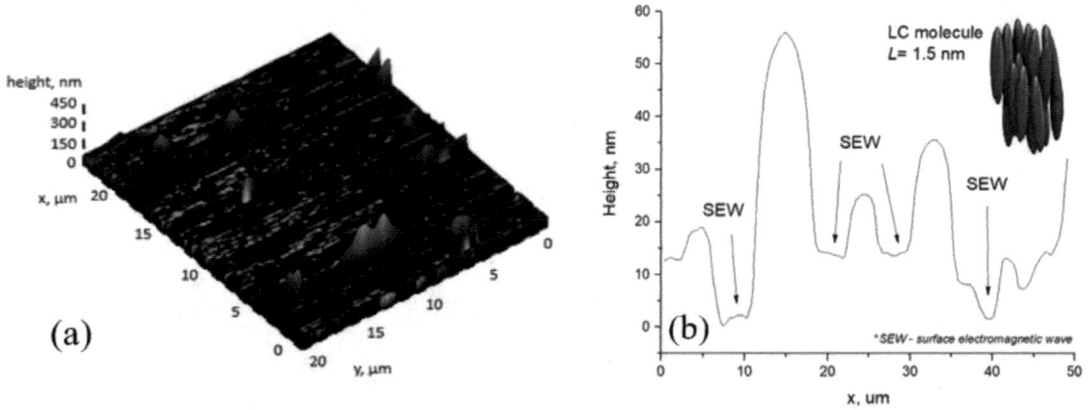

Figure 3. (a) 3D AFM image of ITO with deposited CNTs after the laser treatment; **(b)** 1D image of ITO with deposited CNTs after the laser treatment

4. Conclusion

In this article, the refractive properties of the various modifications of ITO thin films were discussed. During the interaction of the incident laser beam and the ITO surface, the SEW is produced. Induced waves have the same wavelength, however, the intensity is dramatically decreased. This technological process has not contradicted with the one shown previously in paper [9] and can be applied for several reasons in the construction of LC cell:

1). Probability to high precise control the effective refractive index of ITO films via the geometry of relief. It allows creating an index-matched ITO-LC interface with small reflection losses.

2). SEW processing is an additional mechanism for the orientation of LC molecules. In this case, we influence the initial conditions for the director of LC, thus temporal parameters of LC cells can be improved.

3). In addition, the cavities after SEW can perform the orientation functions for CNTs deposition inside the laser-oriented deposition technique. Its influences the tilt angles of CNTs and their penetration depth. The discussions of this phenomenon will demonstrate in the further articles.

The described features of SEW treatment perform to increase the transmittance of LC devices via the index-matching of conductive layers and LC mesophase.

Acknowledgments
The author would like to thank his supervisor N.V. Kamanina (Dr.Sci.,PhD, head of Lab for Photophysics of media with nanoobjects at Vavilov State Optical Institute) for productive discussions and seminars.

References
[1] Kamanina N V 2005 *Physics-Uspekhi* **48(4)** 419
[2] Chausov D N, Kurilov A D, Kazak A V, Smirnova A I, Velichko V K, Gevorkyan E V, Rozhkova N N and Usol'tseva N V 2019 *Liquid Crystals* **46(9)** 1345
[3] Petrescu E and Cirtoaje C 2018 *Beilstein J. Nanotechnol.* **9** 233
[4] Kamanina N V, Zubtsova Yu A, Toikka A S, Likhomanova S V, Zak A and Tenne R. 2020 *Liq. Cryst. and their Appl.* **20(1)** 34
[5] Toikka A S and Kamanina N V 2020 *J. Phys.: Conf. Ser.* **1695** 012071
[6] Kamanina N, Toikka A and Gladysheva I 2021 *Nano Express* **2** 1
[7] Bonch-Bruevich A M, Libenson M N, Makin V S and Trubaev V V 1992 *Optical Engineering* **31(4)** 718
[8] Libenson M N, Makin V S, Shiryaev V A and Soileau M J 1994 *Proc. SPIE* **2428** 676
[9] Vasilyev P Ya and Kamanina N V 2007 *Technical Physics Letters* **33(1)** 8

Radiofrequency ablation of biological tissue and determination of the moment of achievement of the transmural effect: the influence of applied power on tissue resistance

V V Antipenko[1], E A Pecherskaya[1], S A Antipenko[2], O A Timokhina[1], D V Artamonov[1] and A I Levin[1]

[1] Department of Information and measuring equipment and metrology, Penza State University, Penza 440026, Russia
[2] JSC PO Elektropribor, Penza 440011, Russia

Abstract. Radiofrequency ablation (RFA) is used to treat both congenital and acquired heart diseases, including arrhythmias of various types. In this work, experimental studies are carried out on biological tissues under the influence of the applied power on the tissue. Based on the experiments carried out, it was concluded that the effectiveness of the effect on biological tissue and the quality of the result obtained clearly depend on the correct choice of power applied to the treated tissue.

1. Introduction

The RFA method is currently used in the treatment of congenital and acquired heart defects, ischemic heart disease complicated by arrhythmias, as well as in isolated forms of arrhythmia.

The purpose of the method is to destroy the areas of the myocardium through which pathological impulse activity is conducted. Destruction is carried out by applying radiofrequency energy to specific areas of the myocardium from the apparatus using a consumable instrument containing electrical contacts. After the destruction process, the abnormal impulses propagation that excite the myocardium stops, which allows maintaining the correct heart rhythm.

The currently used equipment and consumable tools have significant drawbacks. The main disadvantage is inaccurate determination of the achieving transmural lesion effect in the exposure process. This indicates the relevance of improving the hardware, methods of exposure and determining the achievement of the transmurality effect, to which this study is devoted.

2. The biological tissue impedance measurement while applied power exposed to tissue

This method is based on the biological tissue impedance measuring when the power is applied to the tissue. The graph of the impedance change on time dependence (Figure 1) while applied power exposed to tissue consists of three main sections [1-3]:

- 1. Impedance decrease.
- 2. The impedance level has reached a minimum value, possibly constant for some time, or changes insignificantly.
- 3. Impedance increase.

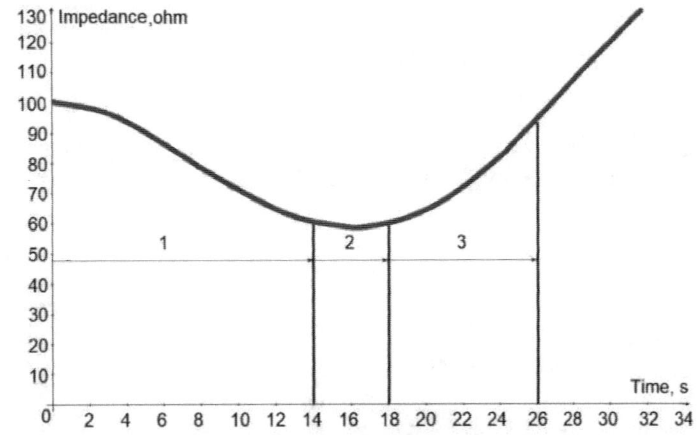

Figure 1. A graph of the change in impedance on time dependence

Based on the data on the tissue impedance and the "operating point" position on the graph of the impedance change on time dependence, it is necessary to investigate and draw conclusions about the course of the ablation process and the achievement of transmural damage to the myocardial tissue [4].

The task of obtaining a transmural zone of myocardial damage is the need for the shortest possible period of time to heat the required area of the myocardium [5].

When analyzing a number of scientific studies [6 - 9], it was found that after passing the flat section of the graph and reaching the section of impedance growth, there is a certain impedance threshold, the achievement of which may indicate the achievement of transmurality. The impedance threshold is individual in nature and for each specific situation it is different. Experimental studies carried out by the authors on biological tissues have shown that after passing the point on the graph of the impedance change on time dependence, corresponding to the minimum tissue impedance, there is a rapid, often avalanche-like increase in the tissue impedance.

The impedance change rate depends on a number of parameters: thickness of myocardial tissue; the state of myocardial tissue; power applied to the tissue, the area (or line length) of the instrument's contact with the tissue, etc. When high power is applied to the biological tissue, excessive heating of the tissue surface layers and insufficient heating of the inner layers occur. The impedance of the tissue region will be the sum of the impedances of the surface layers and the inner layers.

3. Influence of applied power on tissue impedance and determination of the moment of transmural effect achievement

The task of determining the moment of achieving the effect of transmurality is the need for the shortest possible period of time to heat the desired part of the myocardium by applying power. At the same time, insufficient power supply to the myocardial site will lead to the absence of transmural damage. Excessive power supply to the myocardial site can lead to destruction of myocardial tissue and loss of mechanical integrity of the organ.

Experimental studies were carried out by the authors on areas of the myocardium up to 10 mm thick. For a successful experiment, a different constant power was applied to different parts of the tissue. The objective of the study is to formulate conclusions about the course of the ablation process and the achievement of transmural damage to myocardial tissue, based on data on tissue conductivity and the position of the so-called working point on the characteristic curve of tissue impedance.

When a power of 20 W is applied in the first experiment (Figure 2), there is a rapid heating of the surface layers of the tissue and insufficient heating of the inner layers. The conductivity of a tissue site is influenced by both the conductivity of the surface and inner layers. In this case, the surface layers of the fabric are characterized by high conductivity, the inner layers have low conductivity. In this case, the total conductivity is the sum of the conductivity of the inner and surface layers of the tissue, which will indicate the achievement of transmurality in its absence.

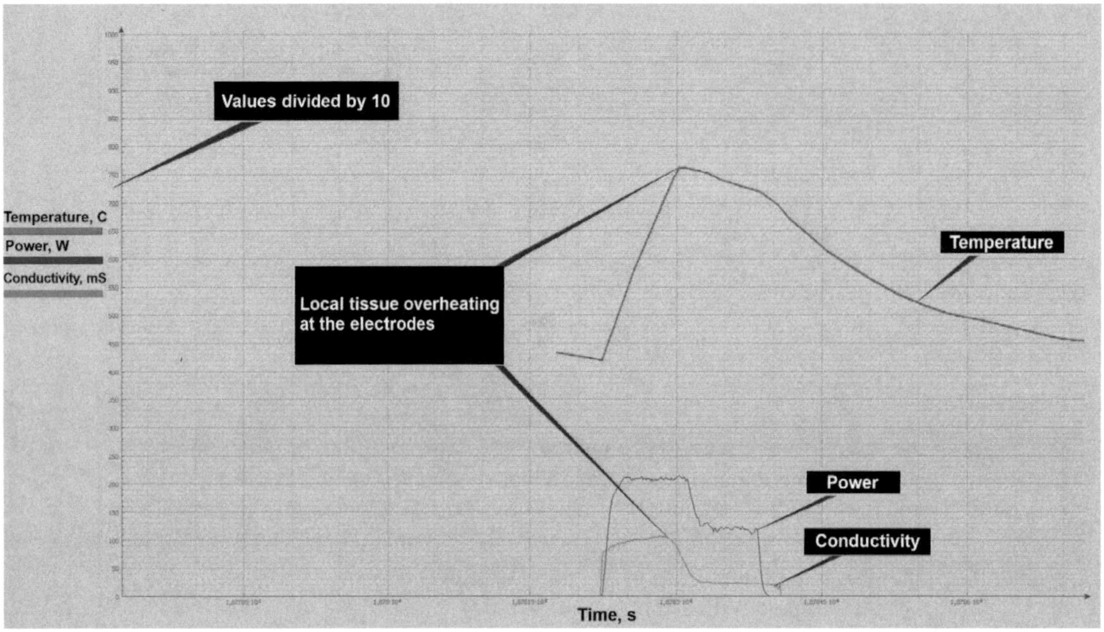

Figure 2. Graph of the dependence of the conductivity of biological tissue under RF exposure with a power of 20 W.

In the experiment under consideration, the excess power, local tissue overheating at the electrodes and a drop in conductivity and power began at a temperature of 76°C in the center of the tissue. Heating to such a temperature is not necessary, since during rapid heating, the temperature in the center of the tissue differs from the temperature in the area applied to the tissue of the electrodes (Figure 3). In this case, transmurality is not achieved, the inner layers are not completely included, and overheating takes place on the surface layers.

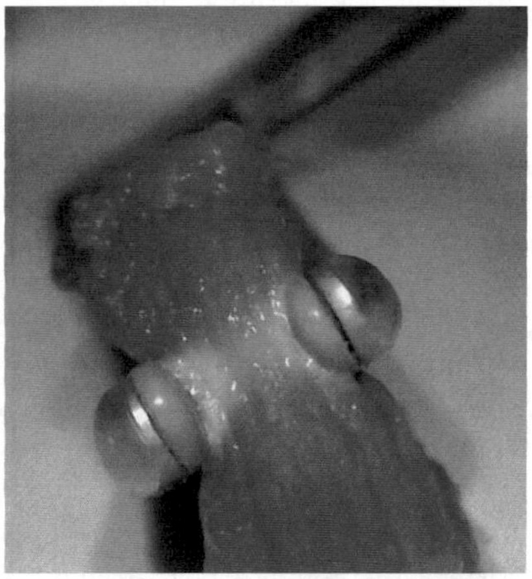

Figure 3. Experiment with a section of the tissue myocardium under HF exposure with a power of 20 W.

In the second experiment (Figure 4), when a power of 15 W is applied, the inner layer is uniformly heated. The part of the myocardium had time to warm up in the center before the conductivity and power of the electrodes decreased due to local overheating of the tissue. This power level in relation to the tissue sample under study provided the rates of impedance change close to the required values in the corresponding sections of the characteristic tissue impedance curve.

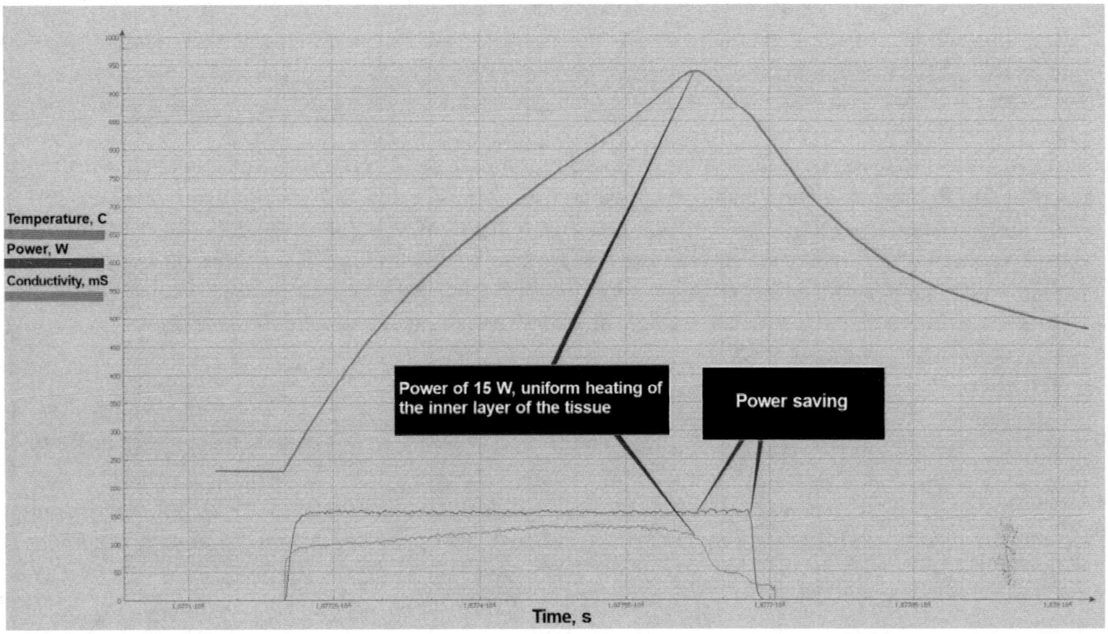

Figure 4. Graph of the dependence of the conductivity of biological tissue under RF exposure with a power of 15 W.

In Figure 5, it is visually visible that the area of the myocardium sandwiched between the clamp contacts has become a white hue, which indicates transmural damage. In this case, the zone of the ablated tissue does not go beyond the projection of the clamp onto the tissue, it is approximately the same throughout the entire thickness of the tissue.

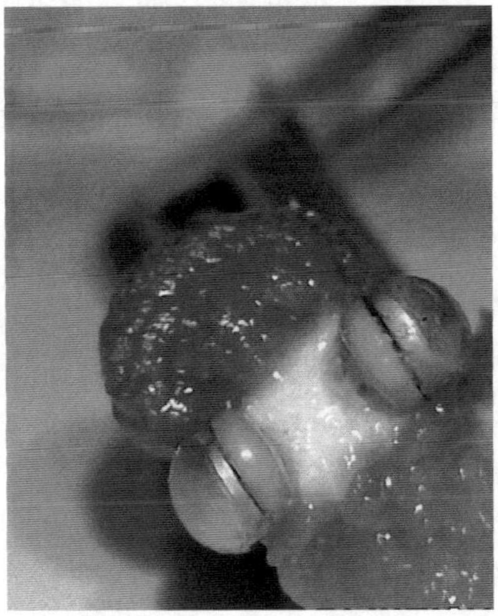

Figure 5. Experiment with a section of the tissue myocardium under HF exposure with a power of 15 W.

Figure 4 shows that the power applied to the tissue is maintained throughout the duration of exposure. This is because the voltage rises to maintain power. For practical purposes, the power level can and should be varied throughout the exposure time. Of key importance is the average rate of change of impedance in the corresponding section of the characteristic curve of tissue impedance. Therefore, to achieve the effect of transmurality, it is necessary to control and regulate the rate of tissue heating, assess the volume of the processed tissue, calculate the required amount of energy to heat the calculated volume of tissue, and monitor the correspondence of the actually consumed amount of energy to the calculated one.

4. Conclusion

In the course of experimental studies, it was proved that the effectiveness of the effect on biological tissue and the quality of the result obtained clearly depend on the correct choice of power applied to the treated tissue, which was used in the development of a software and hardware complex that implements the RFA method and achieves the effect of transmurality. Taken together, the achieved results contribute to a reduction in the time of cardiac surgery and an increase in the percentage of cured arrhythmias both immediately after the operation and in subsequent periods.

References

[1] Artyukhina E A, Taymasova I A and Revishvili A S 2020 Catheter ablation of atrial arrhythmias in patients after thoracoscopic ablation of persistent atrial fibrillation *Russian Journal of Cardiology* **25**(7) 3655

[2] Mulder B A, Luermans J G L M and Hindricks G and Blaauw Y 2021 Innovations and paradigm shifts in atrial fibrillation ablation *Europace: European pacing, arrhythmias, and cardiac electrophysiology journal of the working groups on cardiac pacing, arrhythmias, and cardiac cellular electrophysiology of the European Society of Cardiology* **23**(2) ii23-ii27

[3] Van de Sande W W P J, Ali A and Radaelli G 2021 Design and Evaluation of a Passive Constant Force Mechanism for a Cardiac Ablation Catheter *Journal of Medical Devices, Transactions of the ASME* **15**(2) 021003

[4] Hooks D A and et al 2021 Automated rhythm-based control of radiofrequency ablation close to the atrioventricular node: Preclinical, animal, and first-in-human testing *Heart Rhythm* **18**(5) 734–742

[5] Ghannam M and et al. 2021 Catheter ablation of the left and right atrial appendages without isolation in persistent atrial fibrillation *Heart Rhythm* **18**(5) 694–701

[6] Antipenko V V, Pecherskaya E A, Zinchenko T O, Artamonov D V and Spitsina K Yu and Pecherskiy A V 2020 Development of an automated bioimpendance analyzer for monitoring the clinical condition and diagnosis of human body diseases *J. Phys. Conf. Ser.* **1515** (5) 052075

[7] Antipenko V V, Pecherskaya E A, Zinchenko T O, Melnikov O A and Fimin A V and Zaryvahina S A 2020 Analysis of methodological errors in measuring a digital automated bio-impedance meter *J. Phys.: Conf. Ser.* **1695** 012051

[8] Golubkov P E, Pecherskaya E A, Karpanin O V, Kraynova K Y and Artamonov D V and Shepeleva Y V 2018 Automated system for bioimpedance measuring *Proc. 19th International Conference of Young Specialists on Micro/Nanotechnologies and Electron Devices* pp 641 – 644

[9] Starchenko E I 2009 Low-Voltage Precision Analogue Multiplier with wide Frequency Range, *Proc. 2nd IEEE International Conference on Circuits and Systems for Communications* p 156

Chemical characterization of extracellular vesicles of mesenchymal stromal cells: TOF-SIMS and BCARS approach

A V Aybush[1], A A Gulin[1], A A Kuzoiatova[2], M V Gubina[1], F E Gostev[1], M S Syrchina[1], A S Ermakov[2,3], E A Suprunenko[2] and V A Nadtochenko[1,4]

[1] N.N. Semenov Federal Research Center for Chemical Physics RAS, Moscow 119991, Russia
[2] Lomonosov Moscow State University, Faculty of Biology, Department of Embryology, 1, b. 12 Lenin Hills, Moscow, 119991, Russia
[3] Koltsov Institute of Developmental Biology, Russian Academy of Sciences, Vavilova str., 26, Moscow, 119334 Russia
[4] Lomonosov Moscow State University, Faculty of Chemistry, 1-3 Lenin Hills, Moscow, 119991, Russia

Abstract. Paracrine functions of mesenchymal stem (stromal) cells (MSCs) rely, at least partly, on membrane-bound extracellular vesicles (EVs) with rich composition of lipids, nucleic acids and signaling proteins. Elucidation the underlying chemistry could potentially lead to MSCs-free therapy. However, the secretome of MSCs (EVs' composition) is non-static and depends on many other factors including surrounding cells and medium. Thus, the research techniques must be able to provide not only bulk but microscopy-scale data within a reasonable time frame. Two of these label-free techniques are subject of this work toward the question of chemical composition of the EVs.

1. Introduction

Mesenchymal stem/stromal cells (MSC) are known as a subset of pluripotent cells that can differentiate into various cell types. Earlier experimental studies have shown the ability of MSCs to osteogenic, chondrogenic and adipogenic differentiation, which is currently one of the obligate criteria of MSCs [1]. Recent studies also show MSCs differentiating capability into cardiomyocytes, cells of skeletal, muscle and nervous tissues [2-4]. MSCs of multiple tissues, i.e. bone marrow, adipose tissue, umbilical cord can be isolated, amenable to ex vivo cultivation and subsequent cells transplantation [5] with initial idea to augment the structure and function of damaged or diseased tissues via direct cell replacement. Recent studies suggest that, in addition to direct differentiation, stem cells can regulate the reparative process through paracrine secretion of various growth factors, glycosaminoglycans, cytokines and extracellular vesicles (EVs) modulating immune response, apoptosis, proliferation, differentiation, angiogenesis [6,7].

In recent years, interest in the field of EVs has grown tremendously leading to some paradigm shift that EVs could be one of the main components of the cell's secretome responsible for paracrine functions. Secreted by the majority of cell types, EVs can be divided into exosomes, microvesicles, apoptotic bodies, and other subsets of extracellular vesicles [8]. Smaller sized EVs, i.e. exosomes (30–100 nm) and microvesicles (up to 1000 nm), derived from MSCs have recently become known as promising agents for tissue regeneration and immunomodulation while their remarkable handling advantages can possibly lead to so-called 'cell therapy without cells' [9,10]. Exosomes and microvesicles differ based on their origin within the cell. The formation of exosomes is associated with endosome biogenesis and directly depends on the multivesicular body while endosome maturation [11]. Unlike exosomes, microvesicles formed by direct cell membrane budding and contain the same characteristic lipids as exosomes [12]. Both exosomes and microvesicles carry functional cargo composed of nucleic acids (including microRNAs, messenger RNAs and DNAs) and proteins [13].

Quality and reproducibility of current EVs isolation methods are crucial for understanding the actual source of regenerative potential in the secretome of MSCs. Several isolation methods have been described to isolate EVs [14], including differential ultracentrifugation, density gradient ultracentrifugation with centrifugal forces more than 100,000×g, isolation reagents, exclusion chromatography. Some recommendations for the characterization of EV samples were given by International Society for Extracellular Vesicles (ISEV) [15]. In addition, the subtypes of EVs obtained from different tissues using several methods were summarized [9,16]. In view of importance of potential applications, various techniques were already employed for EVs characterization, including electron microscopy (SEM, TEM) [10]; PCR [17]; dynamic light scattering (DLS), spontaneous Raman scattering [18,19]; narrow band CARS, SHG [20]; Raman scattering with multispectral optical tweezers [21]; mass spectrometry [22]. In this work we study EVs using multimodal approach based on laser multicolor confocal fluorescence scanning microscopy (LSM) and modern label-free chemical imaging techniques which are: broadband CARS microspectroscopy (BCARS) of the fingerprint range and time -of-flight secondary ion mass spectrometry (TOF-SIMS). Particularly, both BCARS and TOF-SIMS are especially good at lipid imaging/characterization. The latter are essential molecular components of EVs, but the role of lipid content for EVs functioning is still not fully elucidated [23].

2. Materials and methods

2.1. Samples
Adipose tissue derived MSCs (d122, kindly provided by Koltsov Institute of Developmental Biology, Moscow, Russia) were prepared using a standard methodology [24]. Culturing continued until ~90-95% confluence density was reached, followed by serum-free cultivation for 2 days. EVs isolation procedure included multi-stage centrifugation to obtain debris-free media followed by ultracentrifugation. The pellet containing exosomes/microvesicles was re-suspended in DPBS (Gibco).

2.2. TOF-SIMS measurements
Mass spectrometry studies of EVs were performed on the TOF.SIMS.5 instrument (ION-TOF, Germany). Bismuth cluster ions (Bi_3^+) with the energy 30 keV were used as an ion source. An analysis area was 300×300 μm (64×64 pixels) and the primary ion dose density was kept below 6×10^{11} ions/cm^2 for each measurement. Electron flood gun was activated to avoid charging. Since TOF-SIMS requires high vacuum conditions, EVs have to be analyzed in the dehydrated state. Dehydration procedure was as follows. 10 μL of EVs solution in mQ water was applied on the surface of cleaned silicon wafer and dried in the argon atmosphere.

2.3. LSM measurements
LSM studies of live MSCs cells were performed on the LSM-980 multicolor scanning microscope with stage top incubator (Zeiss, Germany). Live cell staining fluorescent dyes used in experiments: 1) CellMask™ Deep Red Plasma membrane Stain (5 mg/ml, excitation at 639 nm, emission range 650-680 nm); 2) Hoechst 33342 (1 mg/ml, excitation at 405 nm, emission range 415-480 nm); 3) BioTracker 488 Green Microtubule Cytoskeleton Dye (1 mg/ml, excitation at 488 nm, emission range 510-550 nm). Incubation time with dyes was 30 min at 37 ºC followed by multiple washing with DMEM. Finally, nutrient mixture DMEM/F-12 (Gibco) was added.

2.4. Broadband CARS measurements
Broadband coherent anti-Stokes Raman scattering (BCARS) microspectrometer implemented as a part of femtosecond laser complex [25] representing two pulses collinear scheme for CARS generation.

Narrowband (FWHM < 10 1/cm) sub-picosecond pump pulses centred at 730 nm to tune for vibrational bands of the fingerprint range: 700-1800 1/cm. Bell-shaped Stokes pulse (FWHM ~500 1/cm) at 800 nm was transformed to flat-top pulse (FWHM ~1000 1/cm) using real-time 1D-SLM/spectrometer optimization algorithm; additionally, negatively chirped was supplied to 1D-SLM to provide dispersion compensated pulse (~50 fs) at microscope area of the setup. Measured spectra of biological samples were related to ones of sapphire glass to exclude non-resonant component of CARS. Finally, the related spectra transformed to Raman bands using CARS-variation of MEM described earlier [26].

3. Results and discussion

The result of NTA (nanoparticles tracking analysis) shows (figure 1a) the isolated EVs solution contains both exosomes (< 100 nm) and microvesicles (< 1000 nm) with average particle size ~120 nm. Although bright field microscopy also shows large particles (> 1000 nm), they generally tend to uneven distribution over the thickness of the solution and accumulate mostly in the upper part of the solution (as shown in figure 1b). In general, the fraction of such particles is small. Overall concentration of all EVs, according to NTA is ~10^{10} particles/ml.

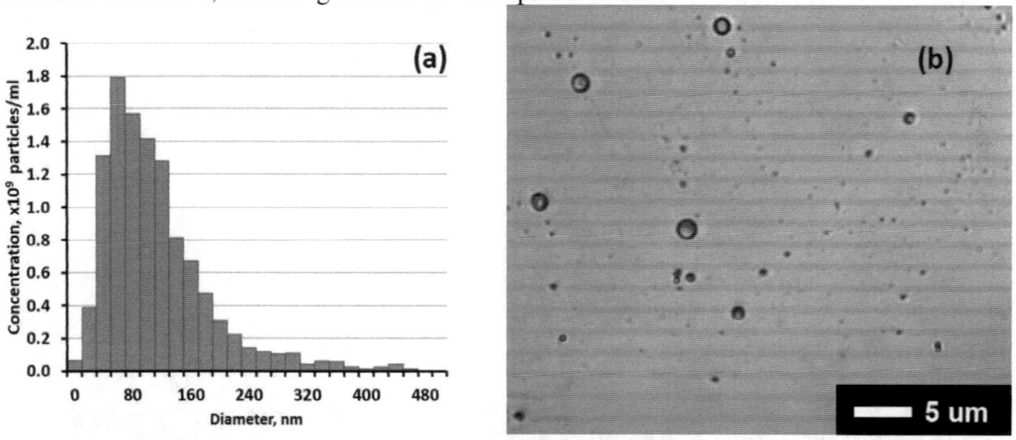

Figure 1. Size distribution of isolated EVs (a) bright field microscopy of EVs (b).

Figure 2 represents typical lipid components presented in the mass spectrum of EVs. While lipid ions are clearly seen, they are not dominant in mass spectrum. The majority of intense peaks are likely attributed to the inorganic salts presented in EVs solutions. However, despite significant complication of mass spectra caused by presence of the salts the same lipid peaks were prominent for all samples. Phosphatidylcholine (PC), sphingomyelin (SM) and cholesterol fingerprint fragment ions were detected in the positive ion mode. Negative ion spectra show peaks of myristic (C14:0), palmitoleic (C16:1), palmitic (C16:0), linolenic (C18:3), linoleic (C18:2), oleic (C18:1), stearic (C18:0) acids. It should be noted that fatty acids ions likely formed under bismuth bombardment of phospholipids (at least PC and SM) or acylglycerols. In general, formation of these lipid ions corresponds to the lipid content of the cell plasma membrane [27,28]. However, peaks intensity was significantly lower due to low concentration of EVs. Therefore, only the most intense bands of the corresponding lipids are clearly visible (Figure 2).

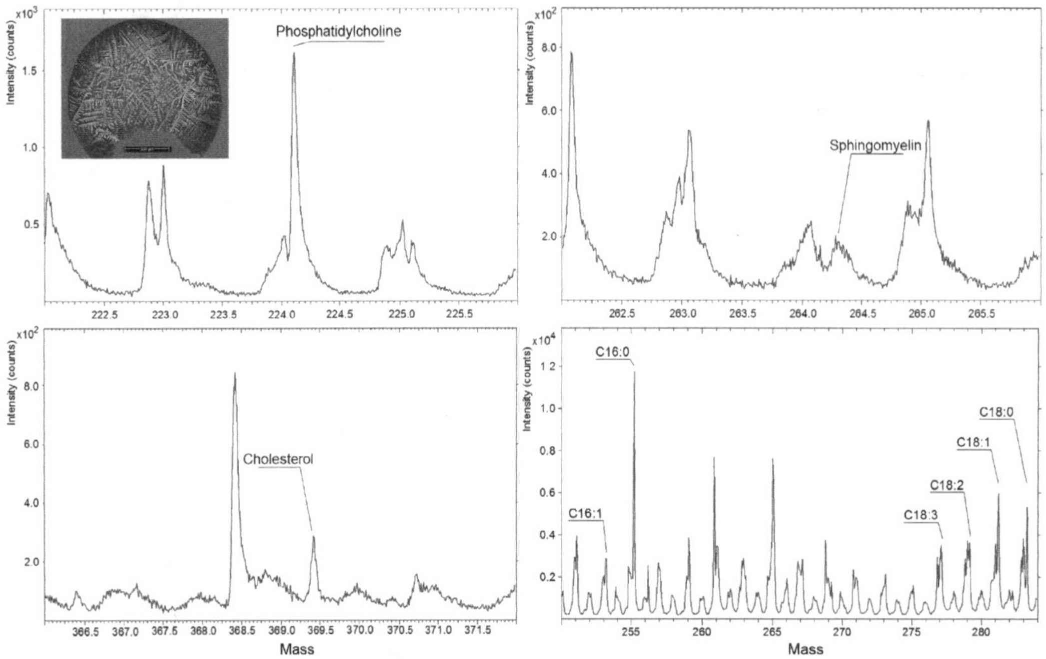

Figure 2. TOF-SIMS spectrum of isolated EVs.

An attempt to increase the concentration by applying several drops to the same site showed no improvement in lipid sensitivity due to the presence of salts in the solution. Besides lipids, the detection of following phosphate ion peaks ($H_4PO_4^+$, PO_3^-, PO_2^-), as well as many fragments of carbohydrate ions (CH_3O^+, $C_2H_5O^+$, $C_3H_5O^+$, $C_5H_5O^+$, $C_5H_7O^+$, $C_5H_9O_3^+$) can also be an evidence of nucleic acid (NA) content of EVs [29].

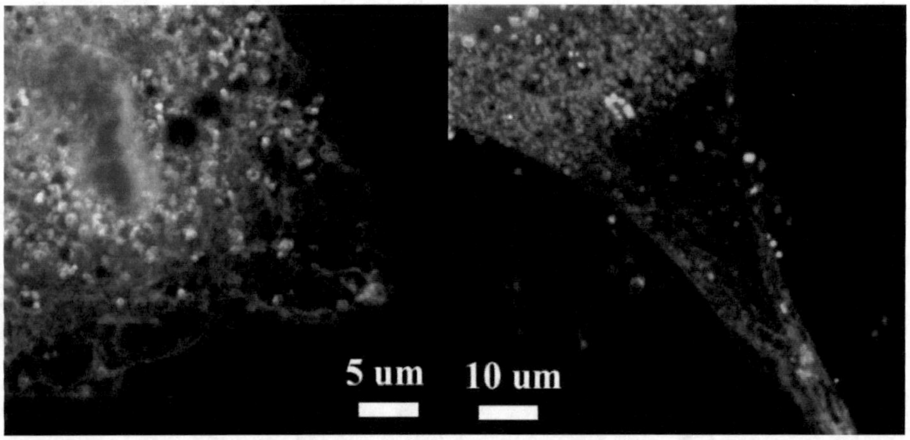

Figure 3. LSM images of live MSCs (Blue: Hoechst 33342, Green: cytoskeleton dye; Red: Deep Red Plasma membrane stain)

Spatial resolution of LSM images shown in Figure 3 is of order of 200 nm (Airy scan mode of LSM-980). Since concentration of lipid content in the vesicles is locally high, the red (membrane dye) spots of the scan can be treated as microvesicles (exosomes are beyond the device resolution). Additionally, we believe that the shape of the microvesicles should be close to spherical due to the membrane structure of the vesicles [10]. In this sense, one can say the surface density of EVs over the cell membrane can have essentially changes, forming large EVs clusters in some areas over the cell (Figure 3b). One of that clusters is probably also detected during label-free BCARS measurements (with band coverage reduced for image mode). Each point of the BCARS scan show integral value over the band of ~1440 1/cm. Locally increased value of the integral (by the factor of 5) can indicate that molecules of the same chemical class are concentrated at the corresponding scan points. The band of ~1440 1/cm is one of the main bands of lipids in the fingerprint range [30].

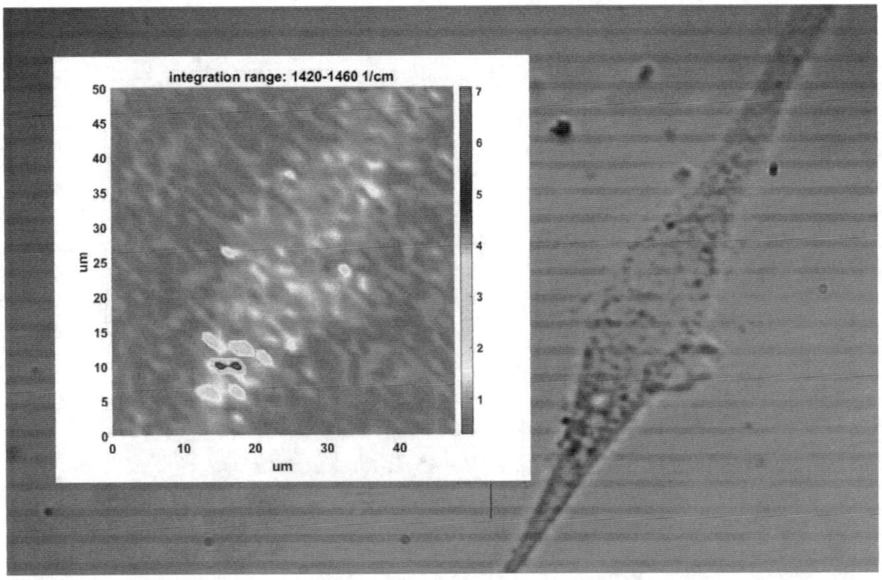

Figure 4. BCARS image of live MSC: integration over the band of ~1440 1/cm.

The lipid content of isolated EVs samples can be recognized by a typical spectral profile which is reflected in following main bands of lipids: ~1442 1/cm (CH2 scissoring mode), ~1660 1/cm (C=C stretching mode), ~1300 1/cm (in phase CH3 twisting mode), 1060–1090 and 1110–1180 1/cm (C-C stretching vibrations) including gauche C–C stretching (~1086 1/cm) and trans C–C (~1120 1/cm), ~1260 1/cm (=C-H deformations), ~1740 (C=O stretching of ester groups) [30].

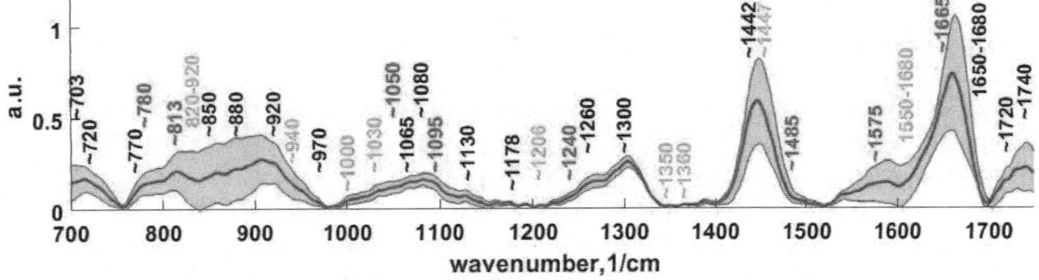

Figure 5. BCARS (transformed to Raman bands) spectrum of isolated EVs. Blue, magenta and green markers represent typical bands of lipids, nucleic acids and proteins, respectively.

The band of ~1740 1/cm is characteristic of triacylglycerols, cholesteryl esters and complex phospholipids like phosphatidylcholine (PC) and phosphatidylethanolamine (PE) while stretching vibrations of the choline $N+(CH_3)_3$ group (PC and sphingomyelin, SM) is at ~720 1/cm; cholesterol (CHL) and cholesteryl esters band at ~703 1/cm is also detected (figure 5). The bands of nucleic acids are mainly seen at ~1575 1/cm (ring breathing modes), ~780 1/cm (various breathing modes of U,T,C) [31].

4. Conclusion

EVs derived from MSCs have recently become known as promising agents for tissue regeneration and immunomodulation while their remarkable handling advantages can possibly lead to new type of therapy based on paracrine functions of the EVs. Elucidating relationship between cell secretome chemistry and response of the target objects can significantly expand our knowledge of the MSCs paracrine status. Advanced femtosecond microspectroscopy and mass spectrometry techniques like BCARS and TOF-SIMS can be effectively used to determine the main features of complex chemical composition of the EVs as well as for MSCs imaging.

Acknowledgement

This work was supported by RFBR Project No. 19-29-04136 mk in part of the sample preparation and TOF-SIMS experiments. The work was supported by RFBR Project No 19-53-52007 in part of laser spectroscopy and microscopy measurements. Experiments were performed using shared research facilities of FRCCP RAS (no. 1440743). The authors express their gratitude to Evgeniy G. Evtushenko, Lomonosov Moscow State University, Department of Chemical Enzymology for the kindly provided NTA analysis of MSC-derived EVs.

References

[1] Dominici M, Blanc K L et al 2006 *Cytotherapy* **8** 315
[2] Yang W, Zheng H, Wang Y, Lian F, Hu Z, Xue S 2015 *Mol. Med. Rep.* **11(1)** 133
[3] Rastegar F, Shenaq D, Huang J et al 2010 *World J. Stem Cells* Aug 26 **2(4)** 67
[4] Ulrich H, Nascimento I C, Bocsi J, Tarnok A 2015 *Stem Cell Reviews and Reports* **11(3)** 474
[5] Koga H, Engebretsen L et al 2009 *Knee Surg. Sports Traumatol. Arthrosc.* **17(11)** 1289
[6] Pankajakshan D, Agrawal D K 2014 *J. Biomed. Technol. Res.* **1** 107
[7] Suprunenko E A, Sazonova E A, Vasiliev A V 2021 *Russ. J. Dev. Biol.* **52**, 129
[8] Raposo G, Stahl P D 2019 *Nat. Rev. Mol. Cell. Biol.* **20(9)** 509
[9] Phinney D G, Pittenger M F, 2017 *Stem Cells* **35** 851
[10] Teixeira J H et al 2016 *Science* **55** 53
[11] Zhang Y, Liu Y, Liu H, Tang W H 2019 *Cell Biosci.* **15(9)** 19
[12] Gyorgy B, Szabo T G et al 2011 *Cell Mol. Life Sci.* **68** 2667
[13] Muralidharan-Chari V, Clancy J W, Sedgwick A, D'Souza-Schorey C 2010 *J. Cell Sci* **123** 1603
[14] Gardiner C, Di Vizio D, Sahoo S, Thery C, Witwer K W, Wauben M, Hill A F 2016 *J. Extracell. Vesicles* Oct 31 **5** 32945
[15] Lener T, Gimona M, Aigner L et al 2015 *J. Extracell. Vesicles* **4** 1
[16] Rani S, Ryan A E, Griffin M D, Ritter T 2015 *Mol. Ther.* May **23(5)** 812
[17] Hunter M P, Ismail N et al 2008 *PLoS ONE* **3(11)** e3694
[18] Zhang H, Silva A C, Zhang W, Rutigliano H, Zhou A 2020 *PLoS ONE* **15(7)** e0235214
[19] Autefage H, Gentleman E et al 2015 *Proc. Natl. Acad. Sci. USA* Apr 7 **112(14)** 4280
[20] Parodi V, Jacchetti E et al 2020 *Int. J. Mol. Sci.* **21(22)** 8498
[21] Carney R P, Hazary S et al 2017 *Anal. Chem.* May 16 **89(10)** 5357
[22] Pocsfalvi G, Stanly C et al 2016 *Mass Spec. Rev.* **35** 3
[23] Skotland T, Sagini K, Sandvig K, Llorente A 2020 *Adv. Drug Delivery Rev.* **159** 308
[24] Minteer D, Marra K G, Rubin J P 2013 *Adv. Biochem. Eng. Biotechnol.* **129** 59
[25] Nadtochenko V A, Aybush A V et al 2017 *Nanomaterials* **7** 371
[26] Vartiainen E, Rinia H, Muller M, Bonn M 2006 *Optics Express* **14** 8 3622
[27] Draude F, Korsgen M, Pelster A, Schwerdtle T, Muthing J, Arlinghaus H F 2014 *Anal. Bioanal. Chem.* **407** 2203
[28] Gulin A A, Pavlyukov M S et al 2015 *Biochem. Moscow Suppl. Ser. A* **9** 202
[29] May C, Canavan H, Castner D 2004 *Anal. Chem.* **76** 1114
[30] Czamara K, Majzner K, et al 2015 *J. Raman Spec.* **46** 4
[31] Movasaghi Z, Rehman S, Rehman I 2007 *App. Spec. Rev.* **42** 5 493

Investigation of the cytotoxicity of silver nitrate and silver-cysteine nanocomplexes

Bogdanov An A[1,2,3,4], Shmakov S V[5], Verlov N A[1,2,4], Klimenko V V[1,2], Knyazev N A[1,2], Terterov I N[1,2] and Bogdanov A A[1,2]

[1] Saint-Petersburg clinical scientific and practical center for specialised types of medical care (oncological), St. Petersburg, Russian Federation
[2] State Research Institute of Highly Pure Biopreparations, St Petersburg, Russian Federation
[3] Institute for Analytical Instrumentation of the Russian Academy of Sciences, St Petersburg, Russian Federation
[4] Petersburg Nuclear Physics Institute NRC Kurchatov Institute, Gatchina, Russian Federation
[5] St Petersburg Academic University, St Petersburg, Russian Federation

vip.nasa@bk.ru

Abstract: Currently, a large number of studies are devoted to the investigation of the antitumor activity of silver nanoparticles and compounds, one of which is silver nitrate. However, silver nitrate has systemic and local toxic effects. In this work, a method was proposed for the synthesis of non-metallic complexes that do not contain toxic nitrate ions, and the cytotoxicity of silver nitrate and silver-amino acid nanocomplexes was investigated.

1. Introduction

The study of the antitumor activity of silver nanoparticles and compounds in vitro and in vivo, as well as the study of the mechanisms underlying these effects, is relevant today. There is a theory that the main active agent is not the nanoparticles themselves, but silver ions Ag +, dissociating from the surface [1-4]. However, the molecular mechanisms underlying the antitumor activity of silver ions and the molecular triggers responsible for the activation of these mechanisms have not yet been identified [5]. In addition, there are no studies in which the in vivo effect of silver is studied in, presumably, the most active, ionic form, where the most soluble salt, silver nitrate, acts as an ion donor. This can be explained by the high general and local toxicity of silver nitrate when administered into the bloodstream, in particular, due to the formation of nitric acid during the interaction of silver ions with sulfide groups of proteins and amino acids, toxicity of nitrate ion, as well as the formation of insoluble particles of silver chloride during the interaction of Ag + with Cl- ions present in the blood. One of the problems of silver nanoparticles that prevent their systemic administration is the presence of a metal core that prevents their biodegradation and excretion from the body.

The aim of this work is to study the toxicity of silver nitrate and silver-cysteine nanocomplexes, as well as to develop a method for the synthesis of these nanocomplexes.

2. Materials and methods

In this work, we used cell cultures: K-562, human chronic myelogenous leukemia, human cervical carcinoma cell line - HeLa, Capan-2 - human pancreatic adenocarcinoma, CT-26 - mouse colon carcinoma, Mouse embryonic fibroblasts 3T3b and transformed embryonic 3T3 fibroblasts -SV40 / All cell cultures were obtained from the Russian collection of vertebrate cell cultures, Institute of Cytology, Russian Academy of Sciences. HeLa, 3T3b, and 3T3-SV40 cells were cultured in DMEM medium (HyClone, USA), and K-562, CT-26, Capan-2 cells were cultured in RPMI-1640 medium (HyClone) supplemented with 10% fetal bovine serum (HyClone, USA), in the presence of 40 μg ml-1 gentamicin (Sigma Aldrich, USA) at 37 ° C in an atmosphere of 5% CO2 Human peripheral blood mononuclear cells (PBMC) of peripheral blood of healthy donors were obtained by centrifugation in a density gradient Ficoll-Paque PLUS (GE Healthcare , USA) according to the standard method. with the addition of 10% fetal bovine serum in the presence of 100 units / ml penicillin, 100 μg / ml streptomycin (BioLot, Russia) and 2 mM glutamine.

To determine the cytotoxic activity of the studied substances in vitro, flow cytometry and colorimetric MTS test were used. To determine the cell viability by the colorimetric method, the MTS Cell Proliferation Colorimetric Assay Kit (BioVision, USA) was used in accordance with the following protocol: cells were pre-scattered in a 96-well plate in the amount of 10 thousand per well, incubated for 24 hours, exposed to silver nitrate and incubated for another 24 hours. Then, 20 μl of MTS reagent was added to each well and the plate was incubated for another 2 hours. The optical absorption was measured at a wavelength of 490 nm using a Multiskan ™ GO Microplate Spectrophotometer (Thermo Scientific, USA). Cell survival was determined using the following formula:

$$Cell\ viability = \frac{A_{sample} - A_{medium}}{A_{control} - A_{medium}} \times 100\%$$

where A_{sample} is the optical absorption value in the well, A_{medium} is the optical absorption value of the medium, $A_{control}$ is the control optical absorption value.

The study of the size and morphology of the obtained nanoparticles was carried out using dynamic light scattering methods. The hydrodynamic diameter of these complexes for various ratios (R) was measured by dynamic light scattering on a ZetaSizer Nano ZS device, and the electrokinetic potential of the obtained complexes was also determined. The electrokinetic potential of nanoparticles was measured by electrophoretic light scattering.

3. Results

Earlier in the study of the cytotoxic and antitumor activity of silver nitrate, we showed that silver nitrate has a cytotoxic effect, as a result of the study, an IC50 value for the HeLa and K-562 tumor lines was obtained, which is 20 times higher than the IC50 for of human peripheral blood mononuclear cells. It has also been shown that silver nitrate has an antitumor effect against Ehrlich's solid tumor, but the compound also has systemic and local toxic effects [6]. Presumably, one of the mechanisms of local toxicity is the release of nitric acid during the interaction of silver ions with sulfide-containing compounds, such as cysteine, the concentration in the blood of which is normally 166.6-249.9 μM, which, with the introduction of silver nitrate (excluding buffer capacity of blood), can cause a local decrease in blood pH to 4 at a normal value of 7.5. The nitrate ion, in turn, is also a toxic compound. A possible way to reduce the systemic and local toxicity observed with the introduction of pure silver nitrate is the synthesis of non-metallic complexes that do not contain toxic nitrate ions. The simplest, most accessible and physiological complexing substance can be the amino acid cysteine, which can effectively bind Ag + ions. In the course of the work, a general synthesis scheme was developed, and the optimal conditions for the synthesis and purification of such complexes were selected.

When an aqueous solution of cysteine interacts with silver nitrate at different ratios R = Cys: Ag, nanosized complexes are formed. The hydrodynamic diameter of these complexes for different R was measured by the method of dynamic light scattering, and the values of the electrokinetic potential of the obtained complexes were also determined. Figure 1 shows the position of the distribution peak in terms of scattering intensity, volume, z-average value, as well as the value of the electrokinetic potential.

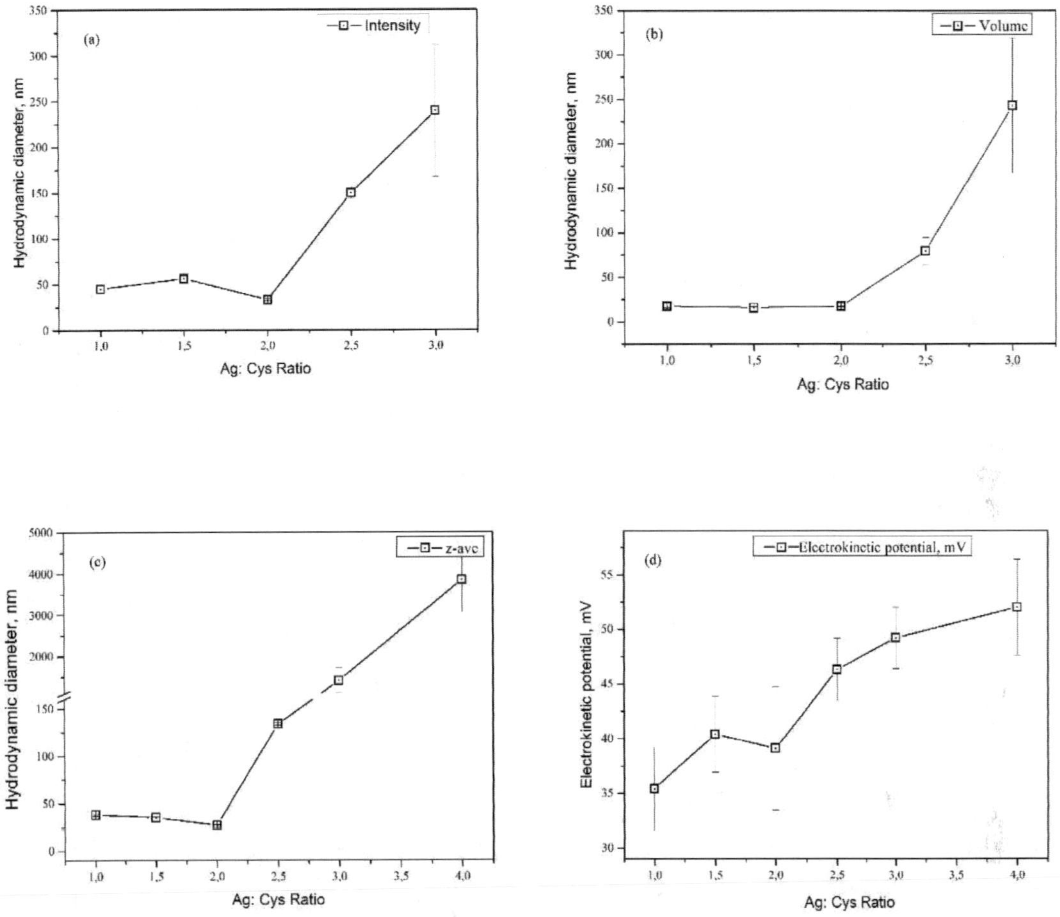

Figure 1.(a) - the position of the maximum on the distribution of the hydrodynamic diameter by the intensity of the scattered light; (b) - the position of the maximum on the distribution of the hydrodynamic diameter over the volume occupied by the scattering dispersed phase; (c) - z-average value, (d) - electrokinetic potential values. The values are given for the crude complexes.

The complexes under study have a characteristic size of several tens of nanometers up to the ratio R = 1: 2; with an increase in the mole fraction of silver nitrate in the initial mixture to 2.5, the hydrodynamic diameter increases to ~ 100 nm, and upon reaching R = 1: 4 it reaches a micron size. The electrokinetic potential of the investigated particles increases uniformly with an increase in R, when the ratio Ag: Cys = 1: 1, its value is 35 ± 3.78 mV, with an increase in R to 1: 4, the value of the electrokinetic potential increases to 52 ± 4.4 mV.

When measuring the spectra of ultraviolet and optical absorption of the obtained complexes, it was shown that the absorption values increase with an increase in the ratio R over the entire investigated wavelength interval. The absence of a broad plasmon resonance peak at λ = 410 nm, characteristic of metallic silver nanoparticles, confirms the non-metallic nature of the synthesized complexes (Figure 2).

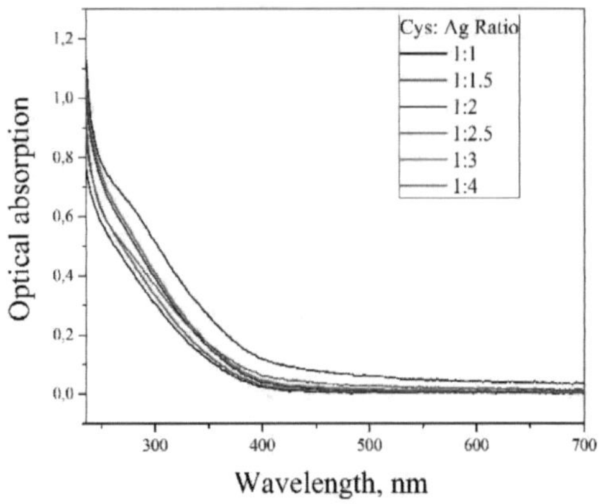

Figure 2. UV and optical absorption spectra of AgCys complexes before purification

The measured pH value of the solution after the formation of AgCys complexes was 1.5, which makes the crude complexes unsuitable for both in vivo and in vitro studies.

After the obtained complexes were purified from nitric acid as well as the residues of unreacted starting compounds, the pH value of the resulting solution was 7, which indicates the absence of nitric acid and the potential applicability of the complexes for in vitro and in vivo studies. The purified complexes were analyzed on a ZetaSizer Nano ZS device, the position of the peak of the distribution by scattering intensity, volume, z-average value, as well as the values of the electrokinetic potential are shown in Figure 3.

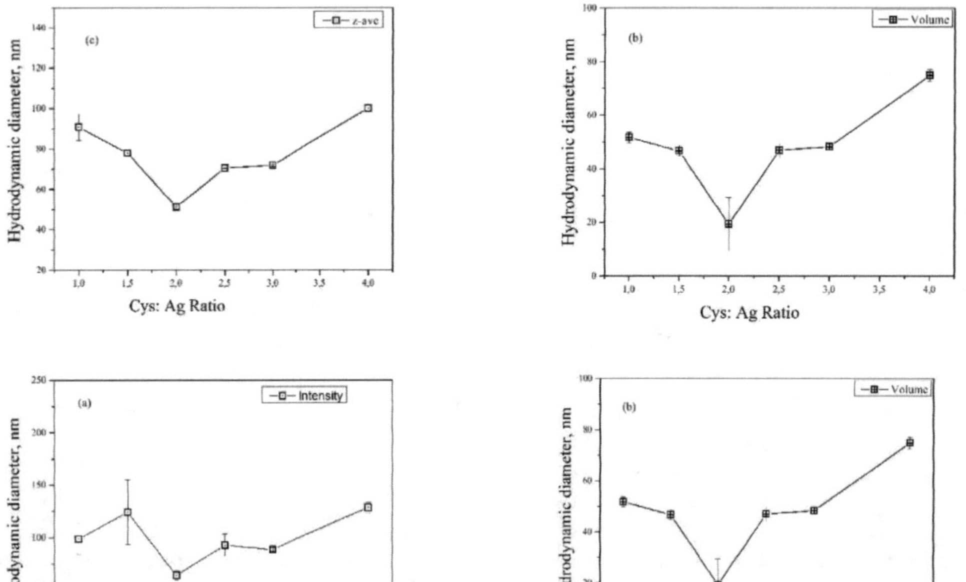

Figure 3. (a) - the position of the maximum on the distribution of the hydrodynamic diameter by the intensity of the scattered light; (b) - the position of the maximum on the distribution of the hydrodynamic diameter over the volume occupied by the scattering dispersed phase; (c) - z-average value, (d) - electrokinetic potential values. The values are given for the purified complexes.

In contrast to the crude complexes, the size dependence on the Ag: Cys ratio is not directly dependent. At the ratio R = 2, the minimum value of the hydrodynamic diameter is observed (z-average = 50 nm), there is no increase in size up to hundreds of nanometers and microns at the ratios R> 2. Apparently, the initial complexes aggregate and lose their individuality during centrifugation; further reconstitution by ultrasonication leads to the formation of new particles, the size of which depends on both the parameters of the acting ultrasound and the internal physicochemical properties. Figure 4 shows the data of the spectrophotometric study of the obtained complexes.

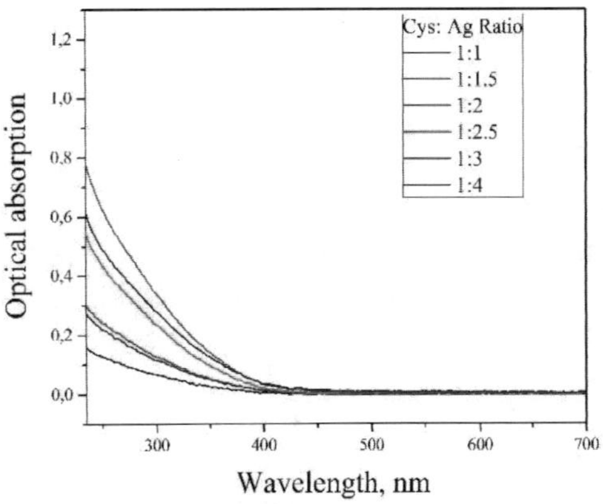

Figure 4. Ultraviolet and optical absorption spectra of the purified complexes.

As in the case with the initial complexes, there is no broad plasmon resonance peak at a wavelength of λ = 410 nm. Thus, the larger size of the initial complexes formed at ratios R> 2 is an advantage, allowing the isolation of a larger amount of the purified product. This is due to the almost complete sedimentation of particles larger than 100 nm in comparison with small particles formed at low ratios of silver and cysteine ions.

Based on the results of the synthesis and purification of silver-cysteine complexes, we carried out work on the study of the antitumor activity of these complexes in vitro on various tumor and normal, human and mouse cell cultures. As in the case of using silver nitrate, it was shown that cell lines with the highest proliferation rate (Figure 5)

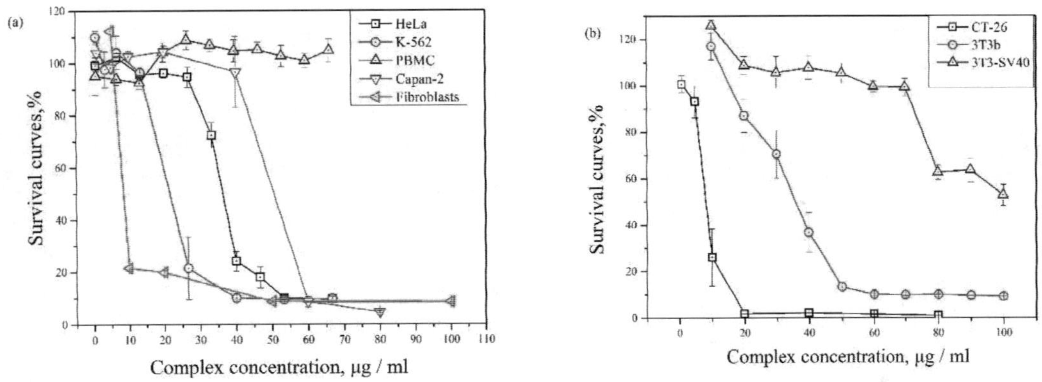

Figure 5. Survival curves of various cell cultures exposed to AgCys complexes with R = 2.5. (a) survival curves for human cell lines, (b) survival curves for mice.

As a result of the study, the concentration values of the half-inhibition of the AgCys complexes were obtained for various cell cultures (Table 1).

Table 1. Concentration values of the half-inhibition of AgCys complexes for various cell cultures.

Cell culture	IC_{50}, µg / ml
K-562	22.2±1.61
HeLa	36±2.37
Human peripheral blood mononuclear cells	-
CT-26	9±2.17
Capan-2	50±1.94
3T3b	40±2.46
3T3-SV40	-
Human fibroblasts	10±0.94

In the range of the investigated concentrations, no toxicity for of human peripheral blood mononuclear cells was observed, which indicates the selectivity of the action in relation to tumor cells

4. Conclusion

As a result of the work performed, a method for the synthesis of silver-cysteine nonmetallic nanocomplexes was proposed, and the optimal conditions and concentrations were investigated and selected. The antitumor activity of silver-cysteine complexes was shown in various normal and tumor cell cultures of humans and mice.

References

[1] Kim S, Choi, J E, Choi J 2009 *Toxicology in vitro*. **23(6)** 1076-1084
[2] Jannathul F M, Lalitha P 2015 *Prog Biomater*. **4(2-4)** 113-121
[3] Jeyaraj M, Rajesh M, Arun R 2013 *Colloids Surf B Biointerfaces*. **102(1)** 708–717
[4] Rutberg F G, Dubina M V, Kolikov V A 2008 *Dokl Biochem Biophys*.421 191
[5] Gurunathan S, Lee K J, Kalishwaralal K, 2009 *Biomaterials*. **30(31)** 6341–6350
[6] Bogdanov A A, Verlov N A , Shmakov S V 2020 *Collection of scientific papers on materials XII International Scientific Conference*. 4-10

Analysis of the limiting behavior of a biological neurons system with delay

E G Fedorov and I Yu Popov

Faculty of Control Systems and Robotics, ITMO University, St. Petersburg, 197101, Russia

fedevg95@gmail.com

Abstract. In this work an analytical and numerical analysis of the limiting behaviors of a system consisting of a pair of biological neurons was carried out. In this case connection between neurons will occur with a delay. As a neuron model, the FitzHugh-Nagumo model was chosen as a model that can reproduce many dynamic behaviors of a real neuron and, at the same time, is not very complex computationally.

1. Introduction

For different neurons and in different external conditions, the time for impulse transmission can differ significantly. In this case the behavior of the entire network as a whole can significantly depend on the strength of the connection and the delay time during the transmission of the impulse. These parameters correspond to the properties of the synapse and the speed of excitation propagation along the axon (see, e.g., [1]). By identifying dynamic behavior and controlling the dynamic behavior of the network, it is possible to identify and predict unwanted behaviors, as well as to restore the normal rhythms of the network's functioning.

To describe a biological neuron, one of the most complete models at the moment is the Hodgkin-Huxley model [2]. Later many other models were proposed to simplify the Hodgkin-Huxley model. One of these models was the FitzHugh-Nagumo model [3, 4], which retains the basic types of neuron behavior, but is much simpler from a computational point of view. In this work this model is used to describe a neuron and an analysis of possible types of behavior of a network of two connected neurons is carried out, depending on the main parameters of the model.

2. Model description

We will consider two neurons connected to each other through a sigmoidal connection (independent of the potential of the postsynaptic neuron) with a delay. Each neuron is described by the FitzHugh-Nagumo model. Then we can write down the final model of the considered system of neurons [5]:

$$\begin{cases} \dot{u}_1 = -a\,u_1 + (a+1)\,u_1^2 - u_1^3 - v_1 + c\tanh\left(u_2^\tau\right) \\ \dot{v}_1 = b\,u_1 - \gamma\,v_1 \\ \dot{u}_2 = -a\,u_2 + (a+1)\,u_2^2 - u_2^3 - v_2 + c\tanh\left(u_1^\tau\right) \\ \dot{v}_2 = b\,u_2 - \gamma\,v_2 \end{cases},$$

where $i = 1, 2$, functions $u_i = u_i(t)$ and $v_i = v_i(t)$ describe the state of the corresponding neuron at time t, a, b, γ are positive parameters, $u_i^\tau = u_i(t - \tau)$, c -- a positive parameter that corresponds to the strength of the connection between neurons, τ -- time delay in the transmission of an impulse between two neurons.

3. Analytical results

Equilibrium points are analyzed for the described system. As a result, the number (from 1 to 9) of equilibrium points was determined depending on the parameter values. Also stability areas are determined for the found equilibrium points.

According to the results obtained, if $a \leq 1$ and $3b \geq \gamma$, then there are no more than 3 equilibrium states in the system (in this case the neurons have the same potential values in the equilibrium state). Otherwise the system can have up to 9 states of equilibrium. In what follows, we will consider only the case $a \leq 1$ and $3b \geq \gamma$.

Now if $c > a + {}^{b}/_{\gamma}$, then the system will have exactly 3 equilibrium states (with a potential greater than 0, equal to 0 and less than 0). If $c < a + {}^{b}/_{\gamma}$, then there is one equilibrium state in the system before the critical value of c, and then three (with a potential equal to 0 and two states with a positive potential).

Further the analysis of local bifurcations in the system is carried out. As a result many transitions were found from stable equilibrium points to stable periodic synchronous and asynchronous oscillations.

Series were obtained for the value of the delay, candidates for a change in the type of behavior. These values distinguish stable equilibrium from oscillation.

$$\begin{cases} \omega^2 = \dfrac{1}{2}\left(-(\alpha^2 + \gamma^2 - c'^2 - 2b) \pm \left(\left(\alpha^2 - \gamma^2 - c'^2\right)^2 - 4b\left((\alpha+\gamma)^2 - c'^2\right)\right)^{0.5}\right) \\ \tau = \dfrac{1}{\omega}\arctan\left(\left(-\omega\,(\omega^2 + \gamma^2) + b\omega\right)\left(\alpha\,(\omega^2+\gamma^2) + b\gamma\right)^{-1}\right) + \dfrac{\pi k}{\omega}, \quad k \in Z \end{cases},$$

where $\alpha = 3u^2 - 2u(a+1) + a, c' = c\cosh(u)^{-2}$, u -- potential of the corresponding state of equilibrium.

For candidates for local bifurcations a criterion was obtained to determine the effect on the number of eigenvalues with positive real parts for the linearized system:

$$(\omega^2 + \gamma^2) > b^2 + 2b\gamma(a+1).$$

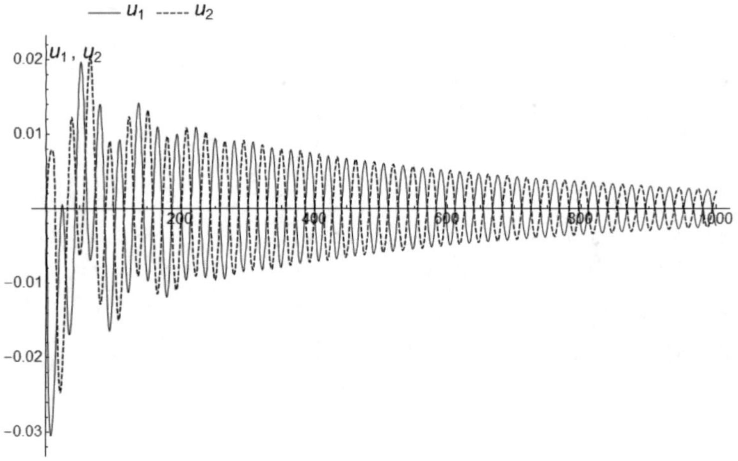

Figure 1. $\{u_1(t)\}, \{u_2(t)\}$ for $a = 0.15, b = \gamma = 0.02, c = 0.2, \tau = 11, u_1(t, t < 0) = 0.01, u_2(t, t < 0) = 0.$

4. Numerical result

All analytical results are confirmed numerically. The global bifurcations in the system are also numerically researched. As a result, for this system of a pair of neurons without external influences, such areas of parameter values have been found for which global stability, periodic oscillations, multistability, simultaneous synchronous and asynchronous stable oscillations, and chaotic oscillations are observed in the system.

Below are examples for these types. Global stability (figure 1); periodic oscillations (figure 2); multistability, simultaneous synchronous and asynchronous stable oscillations (figure 3); chaotic oscillations (figure 4).

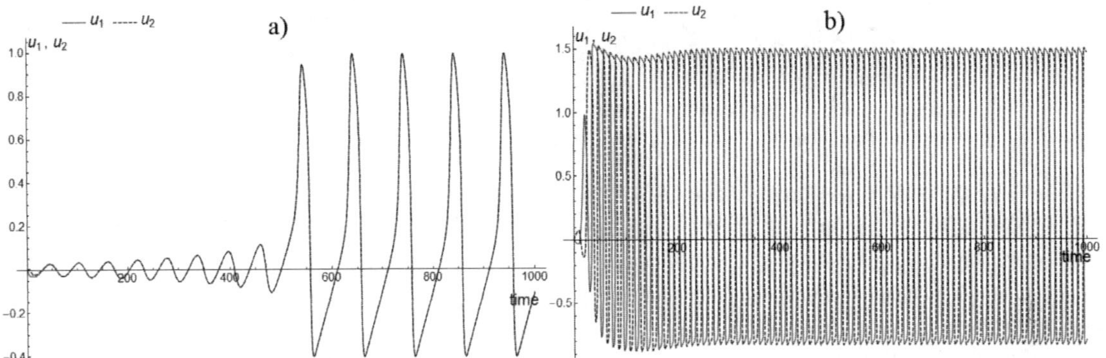

Figure 2. **a)** $\{u_1(t)\}$, $\{u_2(t)\}$ for $a = 0.15$, $b = \gamma = 0.02$, $c = 0.2$, $\tau = 3$, $u_1(t, t < 0) = 0.01$, $u_2(t, t < 0) = 0$; **b)** $\{u_1(t)\}, \{u_2(t)\}$ for $a = 0.15, b = \gamma = 0.02, c = 1.5, \tau = 10, u_1(t, t < 0) = 0.01, u_2(t, t < 0) = 0$.

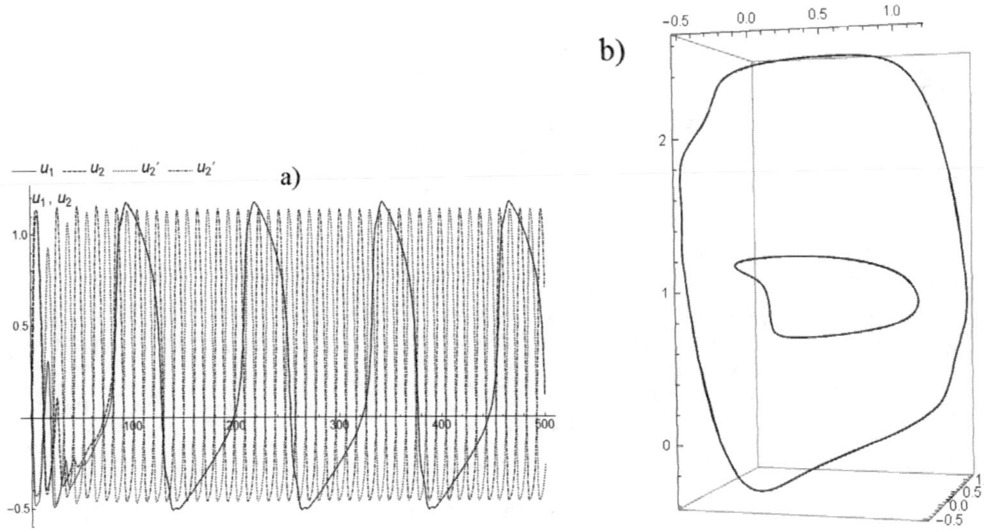

Figure 3. **a)** $\{u_1(t)\}$, $\{u_2(t)\}$, $\{u_1'(t)\}$, $\{u_2'(t)\}$ for $a = 0.15$, $b = \gamma = 0.02$, $c = 0.5$, $\tau = 7$, $u_1(t, t < 0) = 0.5$, $u_2(t, t < 0) = 0$, $u_1'(t, t < 0) = 0.4$, $u_2'(t, t < 0) = 0$; **b)** $\{u_1(t), u_2(t), 5v_1(t)\}$, $\{u_1'(t), u_2'(t), 5v_2'(t)\}$ for $a = 0.15, b = \gamma = 0.02, c = 0.5$, $\tau = 7$, $u_1(t, t < 0) = 0.5$, $u_2(t, t < 0) = 0$, $u_1'(t, t < 0) = 0.4$, $u_2'(t, t < 0) = 0$.

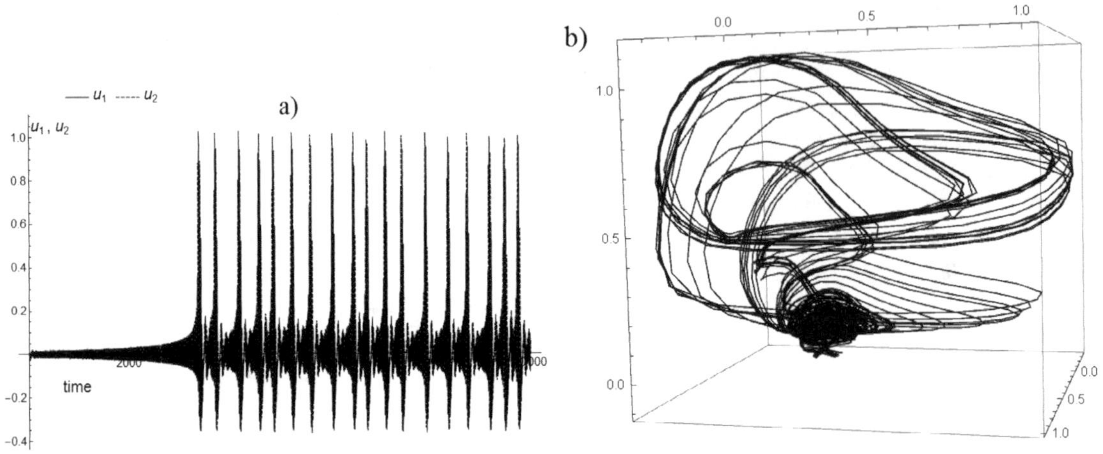

Figure 4. a) $\{u_1(t)\}, \{u_2(t)\}$ for $a = 0.15, b = \gamma = 0.02, c = 0.2, \tau = 11.7, u_1(t, t < 0) = 0.01, u_2(t, t < 0) = 0$; **b)** $\{u_1(t), u_2(t), 5v_1(t)\}$ for $a = 0.15$, $b = \gamma = 0.02$, $c = 0.2$, $\tau = 11.7$, $u_1(t, t < 0) = 0.01, u_2(t, t < 0) = 0$.

5. Conclusion

A model of a pair of neurons was built based on the FitzHugh-Nagumo model. Analytical and numerical studies have been carried out for this model. This made it possible to identify the main types of dynamic behavior, as well as criteria for determining the types.

The main controllable parameters in such systems are the strength of the connection and the transmission rate of the impulse or the delay in transmission between neurons. Thus, based on the results obtained, undesirable behaviors in the system can be diagnosed, and then, by influencing the strength of the connection and the delay in transmission, the system can be brought to the desired state.

Acknowledgments
The reported study was funded by RFBR, project number 20-31-90036

References

[1] Fedorov E G, Popov A I and Popov I Y 2019 Metric graph version of the FitzHugh-Nagumo model *Nanosystems: Physics, Chemistry, Mathematics* **10** (6) 623–626
[2] Hodgkin A L and Huxley A F 1952 A quantitative description of membrane current and its application to conduction and excitation in nerve *The Journal of physiology* **117** (4) 500-544
[3] FitzHugh R 1961 Impulses and physiological states in theoretical models of nerve membrane *Biophysical journal* **1** (6) 445-466
[4] Nagumo J, Arimoto S and Yoshizawa S 1962 An active pulse transmission line simulating nerve axon *Proceedings of the IRE* **50** (10) 2061-70
[5] Burić N, Grozdanović I and Vasović N 2005 Type I vs. type II excitable systems with delayed coupling *Chaos, Solitons & Fractals* **23** (4) 1221-33

AUTHOR INDEX

8th International School and Conference "Saint Petersburg OPEN 2021": Optoelectronics, Photonics, Engineering and Nanostructures (SPbOPEN 2021)

Journal of Physics: Conference Series Volume 2086

St. Petersburg, Russia
25 – 28 May 2021

Part 2 of 2

ISBN: 978-1-7138-5057-1
ISSN: 1742-6588

Printed from e-media with permission by:

Curran Associates, Inc.
57 Morehouse Lane
Red Hook, NY 12571

Some format issues inherent in the e-media version may also appear in this print version.

Institute of Physics
Dirac House, Temple Back
Bristol BS1 6BE UK

Phone: 44 1 17 929 7481
Fax: 44 1 17 920 0979

techtracking@iop.org

Additional copies of this publication are available from:

Curran Associates, Inc.
57 Morehouse Lane
Red Hook, NY 12571 USA
Phone: 845-758-0400
Fax: 845-758-2633
Email: curran@proceedings.com
Web: www.proceedings.com

TABLE OF CONTENTS

PART 1

PART 2

The effect of optogenetic activation of astrocytes on the hippocampal neurons activity

E I Gerasimov[1], A I Erofeev[1], S A Pushkareva[1], A V Bol'shakova[1], A A Borodinova[3], P M Balaban[3], I B Bezprozvanny[1,2] and O L Vlasova[1]

[1] Laboratory of molecular neurodegeneration, Peter the Great St. Petersburg polytechnic university, Saint-Petersburg, 195251, Russia
[2] Department of Physiology, University of Texas Southwestern Medical Center, Dallas, TX, USA
[3] Cellular Neurobiology of Learning Lab, Institute of Higher Nervous Activity and Neurophysiology of the Russian Academy of Science, Moscow 117485, Russia

evgeniigerasimov1997@gmail.com, olvlasova@yandex.ru

Abstract. The method of optogenetics has spread widely in neurobiology over the past 10 years and has found extensive application in various fields of this sciences. It allows to control and regulate cellular activity with high spatial and temporal resolution. In this study, optogenetic activation was applied to astrocytes expressing ChR2. Optogenetic stimulation parameters were determined, in which the frequency of spontaneous currents of hippocampal pyramidal neurons significantly changed. In the future, it is planned to use the obtained data on the modes of optogenetic stimulation of astrocytes to normalize the functions of the hippocampus in mice-models of Alzheimer's disease.

1. Introduction

Astrocytes play an integral role in the maintenance and regulation of neural networks in the brain. They are able to influence neuronal activity by regulating the extracellular concentration of potassium ions, as well as neurotransmitters, due to the expression on their membrane of a large number of transporters of electrogenic mediators (glutamate [1], gamma-aminobutyric acid [2], glycine [3]). By releasing gliotransmitters, astrocytes act on neuronal receptors, modulating neuronal excitability, synaptic transmission, and plasticity. Astrocytes do not generate an action potential in response to a stimulus, but respond with intracellular calcium waves [Ca^{2+}] [4]. When this wave propagates, serine, cytokines, and lactate are released, which modulate the activity of neurons [5]. The ability of astrocytes to release glutamate allows to regulate the function of NMDA receptors, thereby controlling the excitation in the network of neurons [6]. Astrocytes are closely related to the pathogenesis and pathological processes of neurodegenerative diseases [7], so the ability to control their activity becomes an urgent and necessary task in the treatment of neurodegenerative diseases. As the main method for the activation of astrocytes, the method of optogenetics is used in this work, which allows selectively, with physiological space-time resolution, to regulate the activity of hippocampal astrocytes.

2. Materials and methods

2.1. Plasmids used in the study
For expression of channelrhodopsin selectively in astrocytes we used the available plasmid pZac2.1 GfaABC1D_ChR2(H134R)-mCherry (Addgene #112496) driven by short version of astrocyte-specific GFAP promoter. The payload was packaged using rAAV5 serotype (Addgene # 104964) to generate recombinant AAV2/5 viruses.

2.2. AAV production and purification
The production of viral particles was carried out according to the standard protocols used for AAV preparation. Briefly, HEK293T cell were subjected to the triple-plasmid transfection using PEI reagent. At day 3 post-transfection, media and cells were collected and processed separately. Cells were harvested and subjected to the 3 freeze/thaw cycles in liquid nitrogen. Supernatants were treated with polyethylene glycol (PEG) 8000, the PEG-precipitated AAVs were collected by centrifugation. AAVs from cells and supernatant were mixed and treated with benzonase nuclease to destroy any unpacked DNA. Then rAAV particles were purified by iodixanol gradient ultracentrifugation. The required fraction, containing AAVs, was collected, filtered and transferred to the Amicon Ultra-15 centrifugal filter unit for buffer exchange and concentration of virus suspension to the final volume of 110-130 mkl.

Virus titer was then determined by quantitative PCR using primer pair targeting AAV2 ITR sequence (Forward: GGAACCCCTAGTGATGGAGTT; Reverse: CGGCCTCAGTGAGCGA). To remove any extra-viral DNA before qPCR measurements, virus aliquotes were treated with DNase I. The protocol was run twice to obtain two different batches of rAAVs with the titers equal to 9,10E+14 vg/ml and 2,63E+14 vg/ml.

2.3. Stereotaxic surgery and immunohistochemical staining
Testing of the viral construct (AAV5 pZac2.1 GfaABC1D_ChR2(H134R)-mCherry), selectivity and light stimulation modes was performed in the hippocampus of wild-type mice. After the animal was anesthetized, the viral constructs were administered according to the standard protocol [9] at the following coordinates: AP -2.1, DV -1.8, ML ±2.4, with a volume of 1.5 μl at a rate of 0.1 μl/min. The specificity of the injected virus was controlled on fixed sections of mouse brain tissue by immunohistochemical staining (Figure 1).

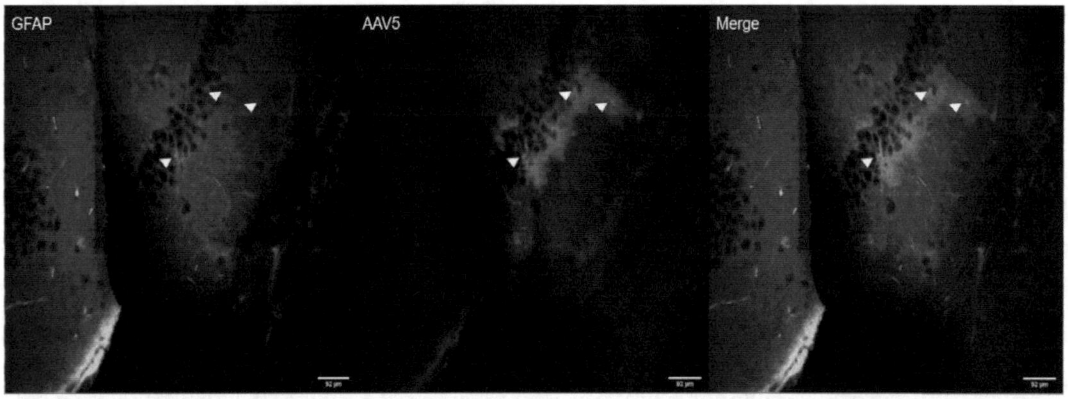

Figure 1. Confocal images of fixed slices of mouse brain tissue with a thickness of 20 microns three weeks after unilateral administration of AAV5 pZac2. 1 GfaABC1D_ChR2(H134R) - mCherry, magnification x40. On the left - immunohistochemical staining of brain tissue with serum to the glial fibrillary acid protein of astrocytes (GFAP, λ = 488 nm), in the center - localization of ChR2-mCherry in astrocytes (AAV5, mCherry, λ = 561 nm), on the right - superimposition of the first two images.

2.4. Hippocampal acute brain slices
Preparations of acute hippocampus brain slices of mice were carried out after anesthesia of the animals by intraperitoneal administration of a mixture that consists of zoletil and xylazine (40 μl of zoletil [Zoletil Virbac] + 150 μl [NaCl+xylazine] + 160 μl of 0.9% NaCl). After a control check of the depth of anesthesia in the animal, the preparation of slices was carried out according to a modified protocol [10].

2.5. Patch-clamp method on neurons of mice hippocampal acute brain slices
Electrophysiological experiments were performed three weeks after the introduction of the AAV5 pZac2.1 GfaABC1D_ChR2(H134R)-mCherry virus construct into the mouse hippocampus. Registration was carried out in the CA1 region of the hippocampus using glass microelectrodes (2-3 MOm) filled with a solution of: (120 mM K-gluconate, 20 mM KCl, 10 mM HEPES, 0.2 mM EGTA, 2 mM $MgCl_2$, 0.3 mM Na_2GTP, 2 mM MgATP). The EPSC was recorded in the Voltage-clamp mode in the whole-cell configuration at a holding potential of -70 mV using a Multiclamp 700B amplifier (Molecular devices) and a Digidata 1440A series digitizer (Molecular devices) using MultiClamp Commander and pClamp 10.7 software (Axon Instruments). Before each recording, a test registration of the currents was performed for 5 minutes, which was not taken into account in the calculation of the data, but was used to check the stability of the patch clamp and the viability of the cell.

2.6. Method of optogenetics
EPSC registration was carried out within 5 minutes before the light stimulation, as well as during and after it. The ThorLabs DC4401 LED (λ = 470 nm, I = 35 MW / mm2) was used as the light excitation source. In the case of a successful test recording, the baseline was recorded for 10 minutes (EPSC before light stimulation), then EPSCs were recorded during light stimulation in the pulse mode (5 minutes) using one of the following modes: t = 20 ms, T = 20 ms; t = 20 ms, T = 1 s; t = 100 ms, T = 1 s) and continuous mode (t = 5c), after which EPSC was recorded for another 10 minutes after light stimulation
(Figure 2).

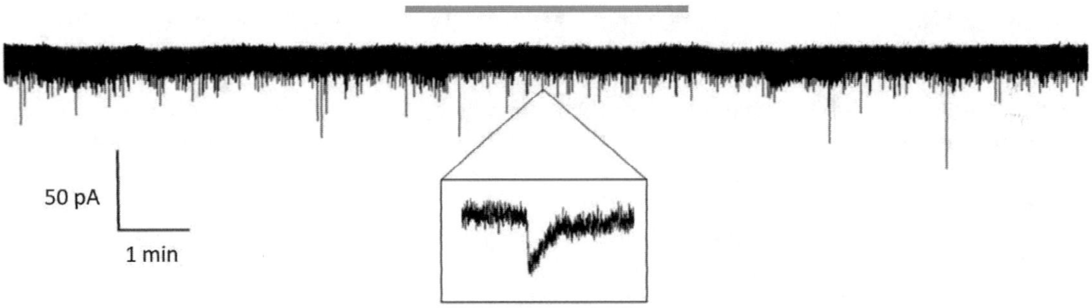

Figure 2. Example of registered EPSCs of the hippocampal CA1 neuron in the Voltage-clamp mode in the whole-cell configuration before light stimulation (λ = 470 nm) of astrocytes (10 minutes), as well as during (5 minutes pulse mode, 5 seconds continuous) and after it (10 minutes). Registration was performed three weeks after the unilateral administration of AAV5 pZac2. 1 GfaABC1D_ChR2 (H134R) - mCherry into the mouse hippocampus. The blue line indicates the moment of light stimulation for the pulse mode.

2.7. Methods of statistical processing
The obtained data were processed in the Mini Analysis V6. 0. 3 program. The processing included EPSCs recorded from the 5th to the 10th minute of the baseline and the last 5 minutes after optical stimulation. The EPSC frequency of each record was normalized to the EPSC frequency of the baseline. Obtained results are shown in Figure 3.

3. Results

The results of the studies showed an increase in the activity of hippocampal pyramidal neurons in response to optical stimulation of astroglial cells containing the AAV5 pZac2.1 GfaABC1D_ChR2(H134R)-mCherry virus and allowed us to determine the mode of light stimulation of astroglia. At the pulse mode t = 100 ms, T = 1 s and continuous mode t = 5 s, a significant change in the EPSC frequency relative to the base level value was observed.

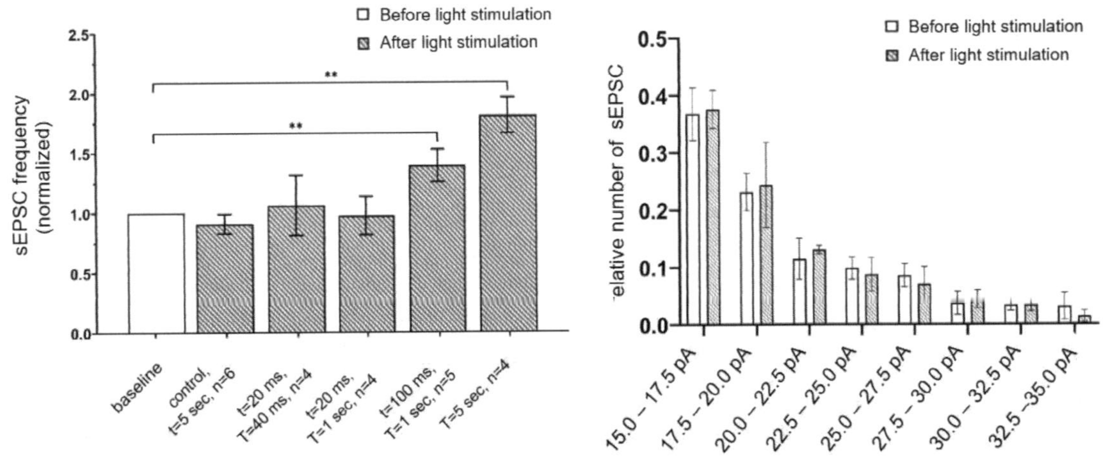

Figure 3. Values of the normalized EPSC frequency of hippocampal neurons in the CA1 region after light activation (t = 20 ms, T = 20 ms; t = 20 ms, T = 1c; t = 100 ms, T = 1 s; continuously t = 5 s) of astrocytes. The data are presented as an average of ± SEM, **: p <0.01, the confidence is calculated by the Mann-Whitney test.

4. Conclusions

In this work, duration and frequency of light excitation of astroglia were determined. At these parameters of optogenetic stimulation (pulse mode t=100 ms T=1 sec and continuous excitation t=5 sec) significant change in activity of pyramidal neurons in acute hippocampal slices was observed. In the future, it is planned to use the obtained data on the modes of optogenetic stimulation of astrocytes for their excitation and registration of fEPSP in mice-models of Alzheimer's disease, long-term potentiation and conduct behavioral tests.

Acknowledgments
The work was supported by the grant of the Russian Science Foundation 20-65-46004 (OLV).

References

[1] Huang Y H and Bergles D 2004 *Curr. Opin. Neurobiol.* **14** 346–52
[2] Schousboe A 2000 *Neurochem. Res.* **25** 1241–4
[3] Jursky F and Nelson N 1996 *J. Neurochem.* **67** 336–44
[4] Kuga N, Sasaki T, Takahara Y, Matsuki N and Ikegaya Y 2011 *Neurosci.* **31** 2607–14
[5] Salmina A B, Gorina Y V, Erofeev A I, Balaban P M, Bezprozvanny I B and Vlasova O L 2021 *Rev. Neurosci.* **32** 459-79
[6] Halassa M M and Haydon P G 2009 *Annu. Rev. Physiol.* **72** 335–55
[7] Toniolo S, Sen A and Husain M 2020 *Int. J. Mol. Sci.* **21** 1–37
[8] Osten P, Cetin A, Komai S, Eliava M and Seeburg P H 2007 *Nat. Protoc.* **1** 3166–73
[9] Ting J T, Lee B R, Chong P, Soler-Llavina G, Cobbs C, Koch C, Zeng H and Lein E 2018 *J. Vis. Exp.* 1–13

Model of 'implant-host' neural circuits in a microfluidic chip in vitro

V N Kolpakov[1], Y I Pigareva[1], A A Gladkov[1,2], A S Bukatin[3,4], V B Kazantsev[1,5,6], I V Mukhina[1,2] and A S Pimashkin[1]

[1] Neurotechnology Department, Lobachevsky State University of Nizhny Novgorod, 603950 Nizhny Novgorod, Russia
[2] Cell Technology Department, Central Research Laboratory, Privolzhsky Research Medical University, 603005 Nizhny Novgorod, Russia
[3]The Laboratory of Renewable Energy Sources, Alferov Saint-Petersburg National Research Academic University of the Russian Academy of Sciences, 194021 Saint-Petersburg, Russia
[4] The Laboratory of Bio and Chemosensor Microsystems, Institute for Analytical Instrumentation of the RAS, 198095 Saint-Petersburg, Russia
[5] Neuroscience and Cognitive Technology Laboratory, Center for Technologies in Robotics and Mechatronics Components, Innopolis University, 1 Universitetskaya Str., 420500 Innopolis, Russia
[6] Center for Neurotechnology and Machine Learning, Immanuel Kant Baltic Federal University, 14 Nevsky Str., 236016 Kaliningrad, Russia

pimashkin@neuro.nnov.ru

Abstract. In this study, we developed a new model of neuronal cells plating into a developed neural network to study functional integration using microfluidic methods. The integration was modeled in a three-chamber microfluidic chip by growing two weakly coupled neuronal networks and enhancing its connectivity by plating new dissociated cells. The direction of connections was formed by the asymmetric design of the chip. Such technology can be used to develop a new type of scaffold to recover the modular structure of the network.

1. Introduction

Neural tissue regeneration in the brain injury area can be performed with the implantation of scaffold structures composed of various biocompatible and biodegradable materials (polymers, hydrogels, and hyaluronic acid) that carry neuronal cells to restore the cellular structure. Such recovery must also restore a heterogeneous or modular network architecture and its functional structure. A neurite outgrowth direction can be defined by scaffold structure to form a synaptic connection with the host brain tissue and restore signaling pathways that form functions.

However, a method to integrate neuronal cells in the network and localize in precise spatial structure in order to restore its functionality remains one of the main technical challenges. Recent advances in microfluidic methods combined with microelectrode arrays allow the development of an in vitro model of such network monitoring and spatial reconfiguration to study the formation of new functional connections which enhances network functionality. Guidance of axon growth by microchannels between separate networks enabled modeling of the complex biologically inspired architecture of hippocampal regions using primary rodent cell cultures. In this study, we developed a microfluidic chip with three chambers connected by microchannels with the asymmetric design which allows integration of new cells between unidirectionally and synaptically weakly coupled networks in order to enhance these functional connections and provide spike pattern propagation in a predefined direction.

2. Materials and methods

2.1. Microfluidic Device Fabrication

Microfluidic chips were fabricated using a 10:1 mix of PDMS (polydimethylsiloxane) silicon elastomer and its curing agent (Sylgard 184, Dow Corning, UK). Liquid PDMS was poured onto a master mold and cured in an oven at 70°C at least for two hours. The wells for cell plating were cut with a puncher (2 mm in diameter) at two opposite sides of each chamber.

The chip consisted of three chambers: a Source, an Implant, and a Target (figure 1). The Source chamber and the Implant chamber were connected by 16 microchannels, each of which was 400 μm long and consisted of two triangular sections. The Implant chamber and the Target chamber were connected by 16 asymmetric microchannels 600 μm in length, each consisted of three sections. The second and the third sections contained two 'traps' that prevented axons growing in the Target - Implant direction.

2.2. Cell Culturing

Mouse hippocampal cells (E18) with a density of 5500 cells/mm2 were cultured in the Source and the Target chambers of the chip. On 7th day of culture development in vitro (DIV) in the Source and the Target chambers, the new dissociated hippocampal cells were plated into the Implant chamber, modeling the cell integration of new cells into the host neuronal network (figure 1). Details of the culturing protocol can be found in a previous study [1].

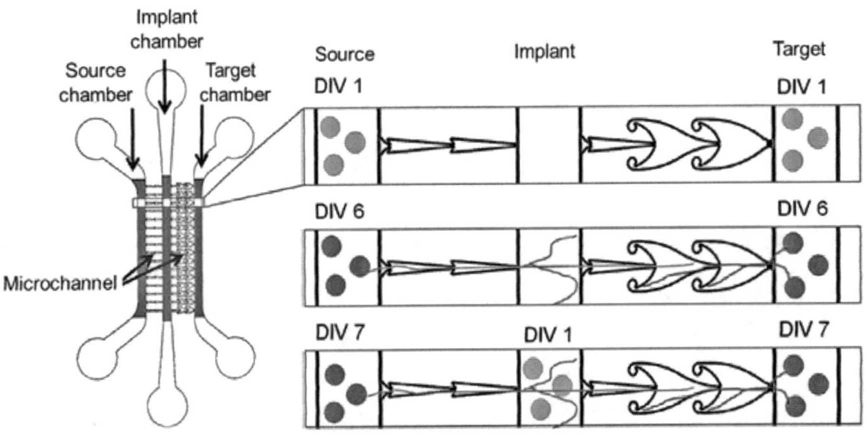

Figure 1. Microfluidic chip design and scheme of the experiment. Dissociated hippocampal cells were plated in the Source and the Target chamber, the Implant chamber remained empty (DIV1). The axons of the Source network grew into the Implant chamber within 6 days after plating (DIV6). Axons from the Target mostly grew into the "trap" section. Dissociated hippocampal cells were plated into the Implant chamber 7 days after plating in the host chambers (DIV7).

2.3. Neurite Analysis

The axon growth cone regions in microchannels were observed using bright-field microscopy with

Axio Observer.A1, Carl Zeiss, Germany. We analyzed neurite growing each day from 1 to 6 DIV (n = 4 chips, 64 microchannels). We estimated the percent of the microchannels, in which neurites grew from the Source to the Target chambers on 5 DIV. The axons from the Source and Target chambers were visually difficult to separate within the same microchannel on later days of culture development.

2.4. Extracellular Spike Recording and Stimulation protocol

Microfluidic chips were manually mounted onto the surface of a planar microelectrode array (MEA, Multichannel Systems, Germany). The electrophysiological activity was recorded from 59 (1 reference) TiN electrodes of the MEA system USB-MEA120-Inv-2-BC-System (Multichannel Systems, Germany) at a sample rate of 20 kHz. Electrode diameter was 30 µm with a 200 µm interelectrode distance.

Stimulation of the electrodes on the MEA was performed using the STG-4004 stimulator (Multichannel Systems, Reutlingen, Germany). Analysis of the recorded bioelectrical activity was performed with custom-made software in Matlab [1]. We used a low-frequency stimulation which consisted of biphasic voltage pulses ±800 mV, 260 µs per phase, positive first, with 3s interstimulus intervals. Series of 60 stimuli were applied separately on two electrodes in the Source and the Target chambers.

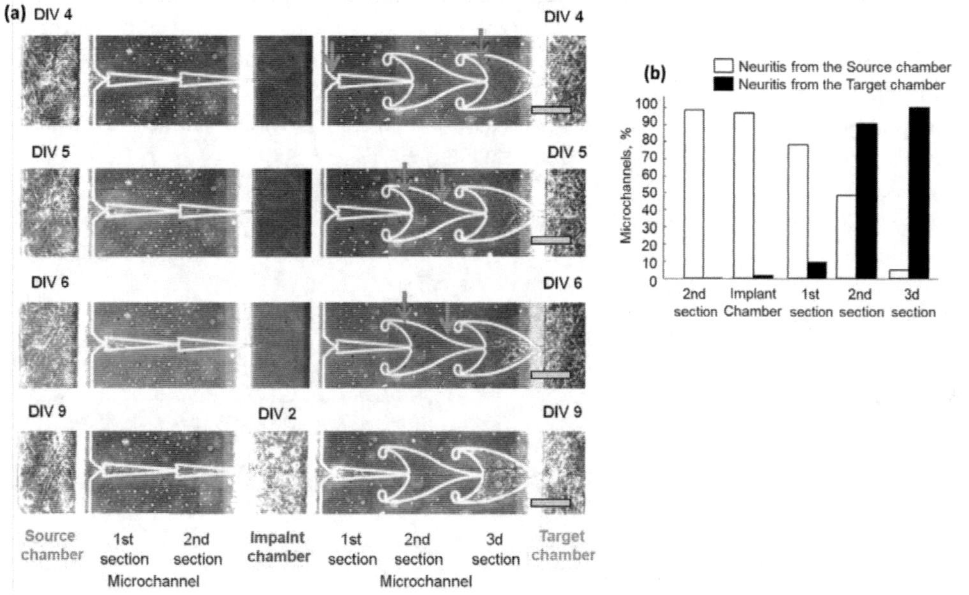

Figure 2. Axon growth in three-chamber microfluidic chip. (a) Microphotographs of the axons on hippocampal cells in the microchannels on 4, 5, 6 and 9 DIV. The maximum distances to which the neurites grew from the Source chamber are marked by green arrows, the neurites grew from the Target chamber are marked by blue arrows. The cells were plated into the Implant chamber on 7 DIV. Scale bar = 100 µm. (b) The number of the microchannels (%), where neurites grew to the corresponding section from the Source (white) and the Target (black) chambers at 5 DIV (64 microchannels, 4 chips).

2.5. Analysis of network activity

Spike and burst detection methods were described in detail in a previous study [2]. Spontanous spiking activity was recorded on 10, 14, 18, 21, 27 DIV.

The probability of the burst propagation in the forward direction was estimated as a number of bursts propagated [4] from the Source to the Implant normalized to the burst number in the Source. The *probability of the burst propagation* in the opposite direction was estimated as the number of bursts propagated from the Implant to the Source normalized to the burst number in the Implant. The characteristics were also estimated between the Implant and the Target. *Burst propagation delay* was estimated as a mean time delay between the maximums of array-wide spiking rate with a 1 ms time bin found for each pair of propagated bursts.

Electrical stimulation applied to the electrodes in the chambers evoked a response in the form of a network burst recorded by a group of electrodes. We used a *post-stimulus time histogram* (PSTH) tocharacterize the evoked network bursts. PSTH values on the electrodes of the Source, Implant, and Target chambers were analyzed separately and were normalized on the number of analyzed electrodes. The PSTH was estimated at 10–50 ms intervals after stimulation with a 2 ms time-bin and calculated by averaging the responses to 60 stimuli.

3. Results and discussion

Axons of cells from the Source chamber grew into the Implant chamber at a distance of 400-600 μmon 5 DIV in more than 95% of microchannels (n=64 microchannels from 4 chips, figure 2b). The microchannel length 400 μm between the Source and the Implant chambers did not allow cell dendrites from the Source chamber to grow into the Implant chamber and reduced the effect of the implanted cells on cells in the Source chamber. These axons of the Source neurons formed the basis for directed synaptic connections with the new dissociated implanted cells in the Implant chamber. In addition, the axons continued to grow further than the Implant chamber, entering the next channel connecting the Implant chamber and the Target chamber, where they met the neurites (usually dendrites) from the Target chamber culture at 5 DIV in 48% of microchannels. A weak unidirectional connection could be formed between the Source and the Target cultures before the new dissociated cells were plated into the Implant chamber. Axons of cells from the Target chamber grew at a distance of 600-800 μm on 5 DIV only in 1.6% of all microchannels..

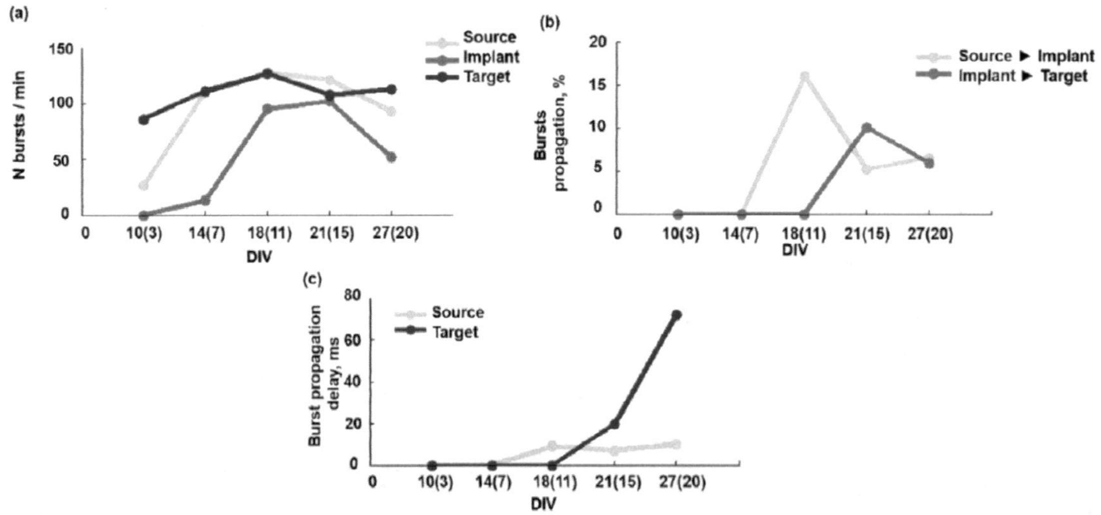

Figure 3. Spontaneous activity of neuronal culture in a three-chamber microfluidic chip. (a) The number of bursts per minute in the Source (green), Implant (red) and Target (blue) chambers. (b) Percentage of bursts propagation from chamber Source to chamber Implant (green), from Implant chamber to Target chamber (blue). (c) Burst propagation delay from the Source chamber to the Implant chamber (green), from the Implant chamber to the Target chamber.

We evaluated the characteristics of spontaneous network activity during culture development and after plating new cells (figure 3a). The *number of bursts per minute* was 264 in the Source chamber and 859 in the Target chamber on 10 DIV while the culture in the Implant chamber remained inactive (see Methods). The bursting activity appeared on 14 DIV (7 days after new cell plating). The *burst propagation* was not observed between the Source and the Implant chambers until 18 DIV and between the Implant and the Target chambers until 21 DIV (figure 3b). The p*robability of the burst propagation* in the forward direction was in the range of 5-16.6% between the Source and the Implant chambers as well as between the Implant and the Target chambers on 18-27 DIV. There was no propagation in the opposite direction.

The *burst propagation delay* (figure 3a) from the Source chamber to the Implant chamber on 18 DIV was equal to 9 ms, and 10 ms on 27 DIV. The *burst propagation delay* from the Implant chamberto the Target chamber was 19.7 ms on 21 DIV, and 72 ms on 27 DIV..

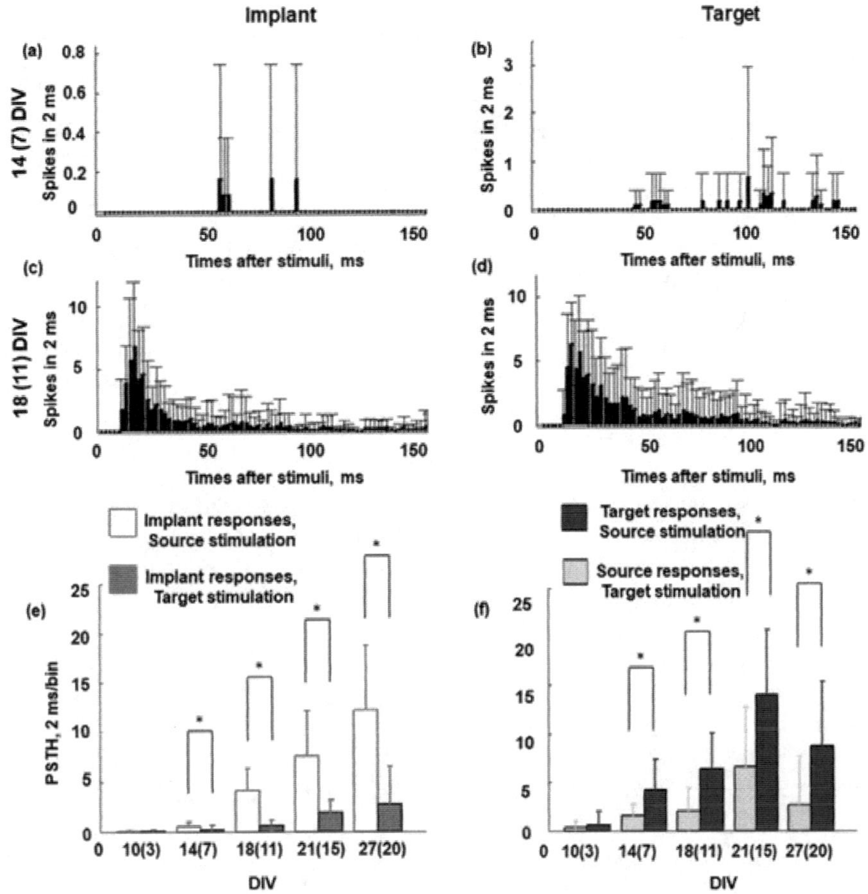

Figure 4. The responses induced by low-frequency stimulation applied to the electrode in the Source chamber. The number of spikes after stimuli in the Implant (a) and the Target (b) chambers on 3 DIV after cell plating in the Implant chamber (10 DIV from first culture plating). The number of spikes after stimuli in the Implant (c) and the Target (d) chambers on 11 DIV after cell plating in the Implant chamber (18 DIV from first plating). (e) PSTH of the responses in the Implant chamber induced by the Source and the Target chamber stimulation. (f) PSTH of the responses in the Sourcechamber induced by the Target chamber stimulation and the responses in the Target chamber induced by the Source chamber stimulation. Mann-Whitney Rank Sum Test, *$p < 0.001$.

The postsynaptic responses were estimated in the Implant and the Target chambers in response to electrical stimulation of the Source chamber electrodes. The stimulation evoked spikes in the Implant and the Target chambers after 3 days from cell plating in the Implant chamber (figure 4 a - d, 10 DIV from the plating of first cultures). The average number of stimulus-evoked spikes (PSTH) on the Implant chamber electrodes increased from 0.01 ± 0.08 on 3 DIV to 12.2 ± 6.5 on 20 DIV after cell plating in the Implant chamber (Kruskal-Wallis test, $p < 0.001$). The electrical stimulation of the

Target chamber electrodes evoked spikes in the Source and the Implant chambers. The PSTH in the Implant chamber was significantly less to stimulation of the Target than the Source chamber during days: 7, 11, 15, and 20 DIV (Mann-Whitney Rank Sum Test, $p < 0.001$). The PSTH in the Source chamber (stimulation in the Target) was significantly less than the PSTH in the Target chamber (stimulation in the Source) during days: 7, 11, 15, and 20 DIV (Mann-Whitney Rank Sum Test, $p < 0.001$).

4. Discussion

Many studies have demonstrated the convenience of platforms based on MEA and microfluidic chips for modeling cellular and neurites disorders. It was shown that neurons derived from stem (pluripotent) cells can functionally interact with primary cultures of neurons [3] and enhance information processing performance [4]. In this work, we investigated the integration of cells between unidirectionally and weakly coupled two subnetworks to enhance its connectivity. We applied electrical stimulation to evoke spiking activity and analyzed functional connectivity between subnetworks. Network bursts in the Implant and Target chambers, recorded with delays of 10-50 ms in response to the stimulus in the Source chamber, indicated the formation of synaptic connections with the cells in the corresponding chambers. Stimulation in the opposite direction to axon growth also demonstrated a spiking response which might be explained by back-propagation of action potentials through the axons and then could evoke a network response in the chamber. Nonetheless, the responses in integrated cell culture and further connected with it were significantly greater than to opposite suggesting activitypropagation dominantly in a predefined direction.

5. Conclusion

In this study, we investigated a three-chamber chip platform to study spiking activity changes after the integration of the cells into weakly coupled cultured neural networks. The method was based on a previously proposed experimental scheme. We showed that integrated cells formed synaptic connectivity in the form of a "bridge" to transfer spontaneous and stimulus-evoked spikes. Such an experimental platform allows to model the recovery of the functional structure of neural networks and can be further developed using IPS cells for recovery of the brain injury.

Acknowledgments

The microfluidic device design and fabrication were funded by RFBR (project № 18-34-00639); the neurite growth analysis and the spontaneous activity analysis were funded by RFBR (project № 18-29-10068); the study of functional connections between networks was funded by Russian Science Foundation (project № 21-75-10154).

References

[1] Gladkov A, Pigareva Y, Kutyina D, Kolpakov V, Bukatin A, Mukhina I, Kazantsev V andPimashkin A 2017 *Sci. Rep.* **7(1)** 15625

[2] Pimashkin A, Gladkov A, Agrba E, Mukhina I and Kazantsev V 2016 *Cogn. Neurodyn.* **10** 287–299

[3] Shimba K, Sakai K, Takayama Y, Kotani K and Jimbo Y 2015 *Biomed. Microdevices* **17** 1–10

[4] Fumika M, Kenta S, Kiyoshi K and Yasuhiko J 2021 *J Neural Eng.* 1–19

Study of the supramolecularly ordered layered structure of chitosan gel films

A A Konduktorova, V A Kurochkina, T S Babicheva, S L Shmakov and A B Shipovskaya

Educational and Research Institute of Nanostructures and Biosystems, Saratov State University 83 Astrakhanskaya St., Saratov 410012, Russian Federation

konduktorova.anastasiya@gmail.com

Abstract. Structural and morphological features of chitosan gel films with a radially periodic structure, obtained by neutralizing the salt form of the polymer with sodium hydroxide or triethanolamine, were visualized by scanning electron microscopy. The formation of such supramolecularly ordered layered structures was found to obey diffusion kinetics and the regularities of Liesegang periodic precipitation. The revealed dependence of the morphostructure of our chitosan gel films on the neutralizing reagent used is due to differences in the diffusion rate of inorganic and organic substance, as well as some spatio-temporal features of the mass transfer process.

1. Introduction

Studies of the physicochemical regularities of mass transfer during the formation of supramolecularly ordered structures will make it possible to control the process of layered ordering and obtain multilayer materials (capsules, plates, and films) with desired properties. An urgent solution to the problem of the short duration of drug substance in the biological environment, the uncontrolled release of the biologically active compound in the human body is the use of systems with targeted drug delivery to the zone of planned localization [1]. The targeted prolonged release of a drug from the polymer system makes it possible to effectively carry out therapy, simultaneously eliminating any negative effect of the drug upon the gastrointestinal tract and other side effects. Multilayer chitosan-containing gel films promising for targeted drug delivery were obtained in our work [2]. The method for their preparation is based on the interfacial reaction of polymer-analogous chitosan salt → base transformation [2]. Strong alkali (NaOH) was used as a neutralizing reagent. To increase the variety of layered structures and, at the same time, increase the environmental friendliness of the process of their preparation, it is advisable to utilize an organic alkaline reagent, e.g., triethanolamine. In this work, we performed a comparative analysis of the morphostructure of chitosan gel films obtained with an organic and inorganic base.

2. Materials

We used chitosan with a viscosity-average molecular weight of 700 kDa, a deacetylation degree of 80 mol% produced by Bioprogress Ltd. (RF) and a 1.5% aqueous solution of glycolic acid (Sigma-Aldrich, USA). Based on these reagents, a polymer solution with a concentration within 1.5–4 wt% was prepared, wherein chitosan was in the salt form (chitosan glycolate). Chitosan was converted into its basic form with a 5% aqueous solution of NaOH (Khimreakiv Corp., RF) or 50% aqueous solution of triethanolamine ("Baza No. 1 Khimreakiv" Corp., RF).

3. Experimental methods

The objects of our study were gel films obtained from a chitosan solution in glycolic acid by converting the polymer into its basic form. The neutralizing agent effect on the surface and internal morphostructure of the samples was evaluated by scanning electron microscopy on a MIRA \\ LMU microscope at a voltage of 15 kV and a conducting current of 400 pkA. To explore the surface structure, the samples were prepared as follows: a glass mold with removable sides was filled with a chitosan solution, a solution of NaOH or triethanolamine was applied to the center of its surface until the salt groups were completely neutralized. The gel film sample was removed from the mold, washed with distilled water until neutral pH, and dried in air. To obtain a sample for exploring its internal structure, a chitosan solution in the mold was poured with an excess of a neutralizing agent, after the formation of the basic chitosan form, the sample was immersed in ethyl alcohol to remove water, then removed from the mold, cut into two parts with a medical scalpel, and dried in air. Before microscopic studies, a 5 nm thick layer of gold was deposited on the samples using a K450XCarbonCoater.

4. Results

During our experiment, it was found that the obtained chitosan-containing gel films were formed in the form of supramolecularly ordered radial periodic structures with bulky circular zones (Figure1 и 2). An increase in the initial polymer concentration increased the number of circular zones, but insignificantly affected the width of rings with a common serial number.

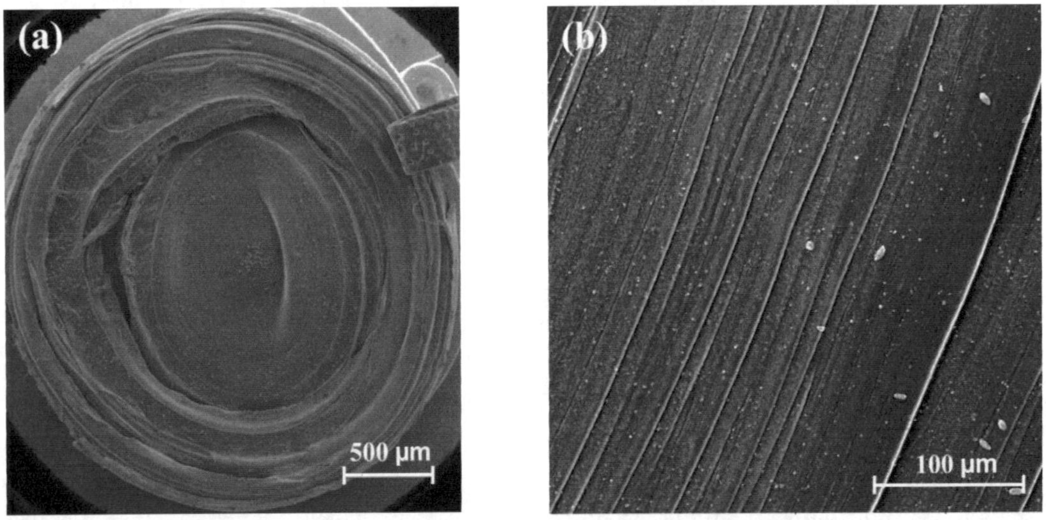

Figure 1. SEM photos of the surface of a gel film obtained from a 4 wt% chitosan solution in glycolic acid after NaOH neutralization: (a) – general view; (b) – central concentric zones.

Earlier, the surface structure of a chitosan gel film was examined [2], in the preparation of which sodium hydroxide was used as a neutralizing agent. Three regions differing in morphostructure were identified, namely: the primary interface of an almost circular shape with spherulite formations, central concentric zones and peripheral edge rings, whose structural features and orientational ordering depend on the polymer concentration (Figure 1 a). Presumably, the formation of spherulite structures depends on the "ring" growth of crystallites, induced by the nuclei of a new phase formed upon contact of chitosan glycolate with NaOH, which is characteristic of the process of aggregation of spherulites during crystallization in high-viscosity polymer systems. The topographic morphology of the central circular zones of the gel film is represented by an array of circularly symmetric micro-rings (Figure1 b). The width of these micro-rings varies in the range of ~2–6 μm. The circular shape of the mutual orientation of the supramolecular formations is also preserved at the periphery, however, some compaction of the material structure is observed.

When using an organic neutralizing agent (triethanolamine), three regions differing in morphostructure were also observed (Figure 2 a). However, there were a number of significant differences. First, the primary gel-film interface is represented by supramolecular formations as scale-like particles arranged in ribbon-like structures (Figure 2 b). Possibly, this orientation is formed as a result of the presence in the sample of a complex salt of tris (2-hydroxyethyl) ammonium glycolate, which has a pseudotricyclic structure [3]. The width of micro-rings of circular symmetry in the central zone differs insignificantly and varies in the range of 2–7 µm, and it is 400–700 nm at the periphery (Figure 2 c, d). Some differences in the supramolecular ordering of the central and edge ring regions of the periodic structure may be due to some kinetic features of mass transfer [4], which is currently being investigated.

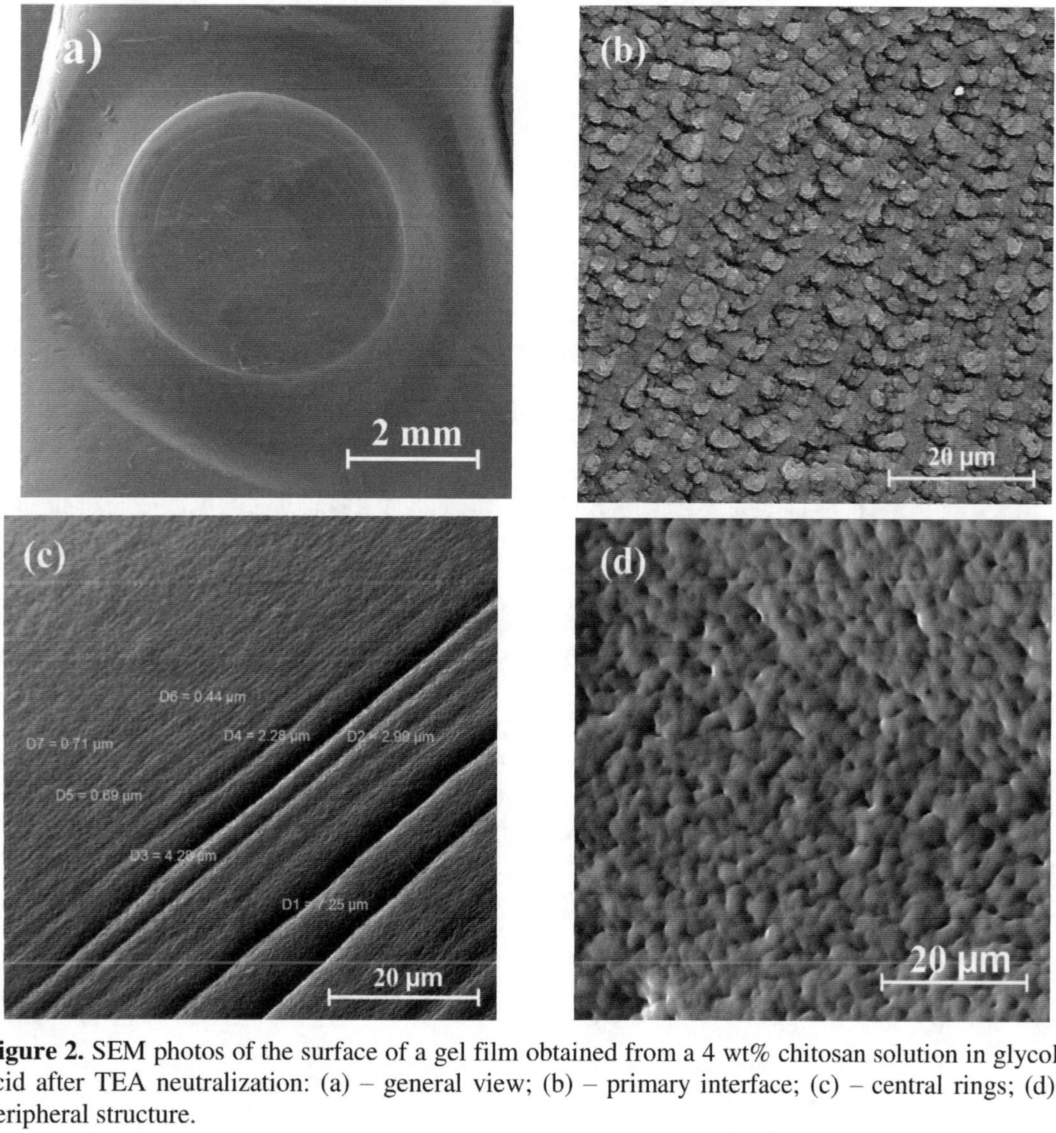

Figure 2. SEM photos of the surface of a gel film obtained from a 4 wt% chitosan solution in glycolic acid after TEA neutralization: (a) – general view; (b) – primary interface; (c) – central rings; (d) – peripheral structure.

In addition, the morphology of the internal structure on a perpendicular section of the sample was studied. It was found that the formation of circular zones occurred not only on the surface, but throughout the entire volume of the sample (Figure 3).

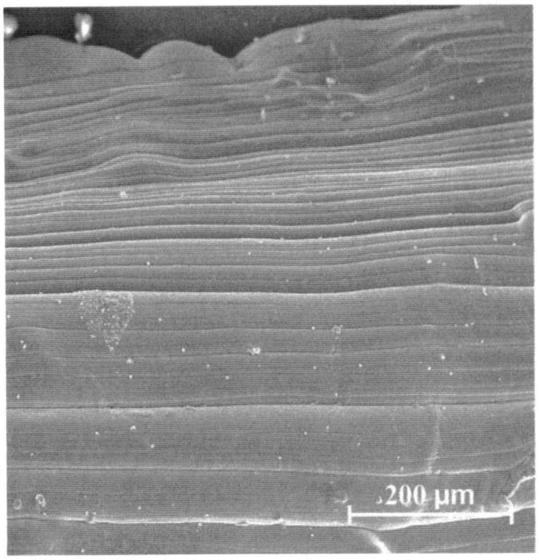

Figure 3. SEM photo of a perpendicular section of a gel film obtained from a 4 wt% chitosan solution in glycolic acid after NaOH neutralization.

The results obtained for the formation of spatially separated concentric rings in the structure of our chitosan gel-film show common features with the main regularities of the Liesegang phenomenon [5–8]. To confirm this assumption, the spatiotemporal evolution of the formation of the chitosan solid phase in the gel film in the course of the studied reaction of polymer-analogous transformation was analyzed by the method from Ref. [2]. It was proved that for both neutralizing reagents the formation of a chitosan base layer obeyed diffusion kinetics and the regularities of Liesegang periodic sedimentation.

Thus, as a result of using the method of scanning electron microscopy, features of the structure of gel films of the chitosan base obtained using an organic and inorganic alkaline reagent were visualized. It was found that the supramolecular ordering of the samples depended on the neutralizing reagent nature. This makes it possible to obtain gel films with a predetermined structure. The use of triethanolamine makes it possible to prepare multilayer structures with a large number of periodic formations.

References

[1] Sarode A, Annapragada A, Guo J and Mitragotri S 2020 *Biomaterials* **242** 119929
[2] Babicheva T S, Konduktorova A A, Shmakov S L and Shipovskaya A B 2020 Journal of Physical Chemistry 124 419255–66
[3] Dain I A, Loginov S V, Rybakov V B, Ofitserov E N, Kulikov E A, Storozhenko P A 2017 *Butlerov Communications* **50 (6)** 1–12
[4] Nie J, Lu W, Ma J, Yang L, Wang Z, Qin A and Hu Q 2015 *Scientific reports* **5(1)** 1–7
[5] Rajurkar N and Ambekar B 2015 *Journal of Molecular Liquids* **204** 205–9
[6] Molnar Jr F, Izsák F and Lagzi I 2008 *Physical Chemistry Chemical Physics* **10(17)** 2368–73
[7] Droz M 2000 *Journal of Statistical Physics* **101** 509–19
[8] George J and Varghese G 2002 *Colloid and Polymer Science* **280(12)** 1131–36

Application of an automated complex resistance and phase difference measuring method for rheographic studies of human cardiovascular system

A I Levin , E A Pecherskaya, Yu A Varenik, V V Antipenko, D V Artamonov and O A Timokhina

Department of Information and measuring equipment and metrology, Penza State University, Penza 440026, Russia

Abstract. The structure of an automated system for measuring the complex resistance of the human body and the phase difference is proposed. It is used in the rheographic method for diagnosing human cardiovascular system diseases using a multilevel microcontroller structure. The metrological analysis of the automated system made it possible to achieve an increase in the measurement accuracy both by reducing the methodological and instrumental components of measurement errors. The relative basic measurement error does not exceed ± 0.5%.

1. Introduction

The relevance of the human cardiovascular system study is conditioned by a sharp increase in diseases associated with it and high mortality from them. There are several methods and well-known instruments for non-invasive measurement of the human body parameters, in particular the cardiovascular system. One of these methods is rheography [1]. Using this method, the following parameters of the cardiovascular system can be measured: blood volume (OS) for the recorded areas, cardiac contractile ability (SPS), determined by the amplitude of the systolic wave on the thoracic differential rheogram, stroke volume of the heart (VOS), reduction in heart rate (HR) is determined by processing rheograms (contractions per minute), blood minute volume (MCV), blood volume index or blood volume (TC) of a region, and several dozen other parameters of a person's cardiovascular system.

Such an informative method of medical research confirms the relevance of improving its principles and measurement methods in order to increase their accuracy by conducting metrological analysis. This method allows you to examine large vessels of the human circulatory system for clogging with blood clots. To do this, it is required to place at least four electrodes on one limb (two input and two output) on the upper and lower parts of the limb.

2. The structure of the software and hardware rheographic complex system

The authors have developed an automated information system and a system for complex resistance and phase difference measuring for studying the human cardiovascular system parameters (Figure 1) [2]. The central node in it is the master microcontroller, which controls the slave microcontrollers via the I^2C interface [3, 4]. Slave microcontrollers (of which there can be up to 127 units in the system) generate a digital signal of various frequencies, which is fed to a digital-to-analog converter (DAC).

There it is converted into an analog signal with a given frequency and amplitude. The signal frequency can be adjusted by the user from 1 to 100 kHz. It is known that the rheography method works only at frequencies up to 100 kHz, since the complex impedance reactive component weakens at frequencies above 10 kHz, and at frequencies above 100 kHz it has no effect [5]. The received analog signal enters the stage of operational amplifiers (OA) and through the electrodes passes through the human body, where at the receiving electrode the signal goes to the receiving stage OA, where it is amplified. Then, the amplified signal passes through a low-pass filter (LPF) and enters the input of a digital-to-analog converter (ADC), from where the value of the difference between the input and output voltage of a signal of different frequencies is formed in the registers in the slave microcontroller. Data is transmitted over the I^2C interface to the master microcontroller from each slave microcontroller. The host microcontroller processes the received packets and transmits them via the USB interface to a personal computer (PC) for displaying the obtained complex resistance and phase difference values in the dialog menu.

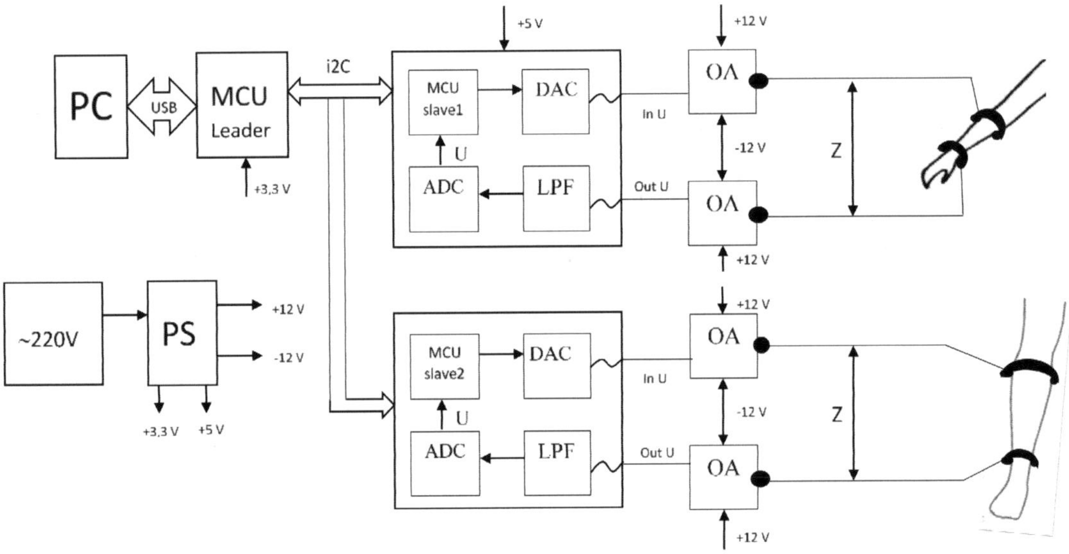

Figure 1. Block diagram of the proposed rheographic complex.

The results obtained from medical practice and scientific developments show that at the moment a multichannel measuring system manifests itself best in the method of the human body composition impedance analysis. In this regard, the use of several devices of the same type, the control of which will be based on the I^2C data bus, will provide a multichannel mode of simultaneous measurements with a minimum arrangement of devices in a miniature device [6, 7].

The use of the multichannel measurement mode provides a more accurate heart rate and stroke volume calculation by averaging the data obtained in different parts of the human body. The use of the device multichannel operation mode will be useful when carrying out surgical operations on the limbs, on different body parts to let the surgeon have information about the amount of blood flowing through the vessels in parallel with the control of the patient's pulse in front of him. In emergency situations, this method is able to calculate the volume of blood in case of closed extremities injuries, which does not enter the injured area if blood flow is impaired.

The use of external operational amplifiers can reduce matching errors. The main purpose of the external interface is to expand the ADC dynamic range by bringing the voltage offset of the output measuring signal to the input OA, and to reduce the output impedance of the measuring circuit, since it is added to the measured one and can introduce a significant error [8].

3. Functioning of the measuring complex

The control of the rheographic hardware-software complex and the measurement results processing is carried out according to the method, which is implemented in software using a PC, master and slave microcontrollers [5, 9]. The master microcontroller contains a control program that controls the slave microcontrollers, sends control signals to each of them, and receives the obtained values. Then it transfers them to the PC. In this case, the leading microcontroller transmits the obtained results of the complex resistance calculation to the program dialog box on the PC. The slave microcontrollers generate voltage signals at different frequencies to measure the complex impedance of the human body part and process the measured values of the voltage passing through the human body to calculate the complex resistance and phase difference. An integrated transducer of the impedance spectral composition in a wide frequency range is built on the basis of a slave microcontroller, in which the active (*R*) and reactive (*X*) components of the impedance *Z* are calculated for each frequency, from which the impedance modulus is then calculated:

$$|Z| = \sqrt{R^2 + X^2}$$

and its phase:

$$\varphi = arctg\left(\frac{X}{R}\right),$$

and packets of the received data on the complex resistance and the difference of the phase value are formed for transmission to the master microcontroller. In turn, the main component of the complex resistance is X_c – the capacitive resistance of the human body. It is determined by the formula:

$$X_C = \frac{1}{2f\pi C},$$

where f – probing signal frequency, C – capacity of the human body. The inductive component contribution to the final measured impedance is extremely small.

Measurement of the phase difference between the probe current source signal and the recorded signal allows to formalize the effect of the skin resistance capacitive component on the impedance through the measured signal mathematical processing, which helps to reduce the error of measurement results.

By the calculated difference in the blood volumes readings through the measured areas at the same test signal frequency and the same pulse, one can judge about the weakening of the blood flow. This value will depend on the blood vessel length, the degree of clogging with blood clots, the vessel elasticity [9, 10]. The implementation of this technique will make it possible to formulate conclusions about the state of large blood vessels, which, in fact, is an alternative to such generally accepted diagnostic methods as computed tomography and magnetic resonance imaging [12].

4. Conclusion

The structure of an automated system for the human body complex resistance and the phase difference measuring is proposed. It is used in the rheographic method for diagnosing human cardiovascular system diseases with an error of no more than ± 0.5%. The measurement results allow to indirectly calculate such parameters as the stroke volume of the heart, the amount of blood rolled through the vessels, etc. Processing and display of the results obtained using the automated system can be displayed in graphical and tabular form, as well as stored in a database for further research.

An increase in the measurement accuracy is also achieved by improving the hardware implementation and the method of the human body conductivity multifrequency measurement, followed by the capacitive parasitic component of the measured signal calculation.

References

[1] Malakhov A I, Tikhomirov A N, Shchukin S I, Kobelev I A, Kobelev AV, Belenkov Yu N, Shakaryants G A and Kozhevnikova M V and Kaplunova V Yu 2016 Electroimpedance

Methods of Investigation of Cardiac Activity *Journal Cardiology* **56** (12) 33–39

[2] Golubkov P E, Pecherskaya E A and Karpanin O V and Kraynova K Y 2018 Automated system for bioimpedance measuring *Proc. International Conference of Young Specialists on Micro/Nanotechnologies and Electron Devices* pp 641–644

[3] Tsai B, Birgersson E, Birgersson U 2018 Mechanistic multilayer model for non-invasive bioimpedance of intact skin *Journal of Electrical Bioimpedance* **9** 31–38

[4] Artamonov D V, Baranov V A, Pecherskaya E A, Pushkareva A V and Tsypin B V and Fimin A V 2019 Application of a Hyper-Complex Impedance Model for Indirect Measurements of Materials Parameters of Functional Electronics *Proc. International Conference of Young Specialists on Micro/Nanotechnologies and Electron Devices* pp 760–764

[5] Kawala A, Khoma V and Zmarły Y D and Sovyn Y 2008 Use of Wavelet Transform for Qualification of Rheograms Characteristic Points *Journal Plezegland Elektrotechniczy* **84** (3) 132–133

[6] Leonov S D, Obrazcov S A and Troickiy Y V 2011 A Bioimpedance Meter Measuring both Active and Reactive Components *Journal Biomedical Engineering* **45**(4) 128–131

[7] Podyacheva E , Zemlyanukhina T and Shadrin L and Baranova T 2020 Features of Hemodynamics of Pulmonary Circulation During the Diving Reflex *Journal Biological Communications* **65** (3) 244–251

[8] Didenko V I and Ivanov A V 2009 A Metrological Approach to the Investigation of the Quantization Noize of Delta-Sigma ADCs *Measurement Techniques* **52** (5) 521–527

[9] Righetti X and Thalmann D 2010 Proposition of modular I2C-Based wearable Architecture *Proc. 15th IEEE Mediterranean Electrotechnical Conference "MELECON"* pp 802–805

[10] Dobrynina L A, Gnedovskaya E V, Shabalina A A, Sergeeva A N and Kravchenko M A and Nikolaeva N S 2018 Biomarkers and mechanisms of early vascular damage *Journal of Neurology and Psychiatry* **12** (2) 23–32

[11] Balkanov A S ,Gaganov L E and Stashuk G A and Sherman L A 2008 To the mechanism of edema around malignant brain glioma *Vestnik rentgenologii i radiologii* **4–6** 55–58

[12] Zakharova N E, Danilov G V, Potapov A A, Pronin I N, Alexandrova E V, Kravchuk A D, Oshorov A V, Sychev A A and Polupan A A and Savin I A 2019 The prognostic value of MRI-classification of traumatic brain lesions level and localization depending on neuroimaging timing *Zhurnal Voprosy Nejrokhirurgii Imeni N N Burdenko* **83**(4) 46–55

Medical applications of porous silicon

E M Loginova[1], **D A Shishkina**[1] **and M A Zhuravleva**[2]

[1] Department of Electronics and Instrument Engineering, Samara University, Samara 443086, Russia
[2] Department of Biology, Samara University, Samara 443086, Russia

Abstract. Materials based on porous silicon are extremely attractive for biomedical applications due to their simple and flexible production, biocompatibility, biodegradability, controlled morphology, and multiple ways of introducing the drug into the body. This paper presents the results of studies of porous structures with a broad spectrum antibiotic ceftriaxone. It has been shown that the characteristics of the porous structure change upon saturation of the pores with the drug. It was shown that solutions of porous silicon + ceftriaxone have a characteristic peak at a wavelength of 1070 nm with increasing sonication time

1. Introduction

Silicon is a component of all cells of the human body, its average daily consumption is about 20-50 mg. Depending on the degree of porosity, silicon particles can be biologically active, bioinert or biodegradable [1]. Currently, the possibility of using porous silicon powders as materials for nanocontainers for transporting drugs is being actively studied. It is possible to regulate the rate of release of the drug and the rate of resorption of the container in the body by controlling the development of the surface and porosity [2]. Porous silicon is an extremely attractive candidate for nanomedicine due to its interesting properties: 1) the process of making porous silicon is simple; 2) hydrophobic drugs with low solubility in a biological medium can be loaded into the pores of porous silicon particles after dissolution in an organic solvent; 3) the large surface area and the widely adaptable surface composition of the porous silicon carriers contribute to a high drug loading [3]; 4) retention of drug molecules in the pores to prevent crystallization; 5) high drug loading (up to 60% by weight) and concentration (up to 400 times); 6) surface chemistry for attaching target groups to porous silicon particles deliver playloads to diseased organs [4]. High interest is the absorption capacity of porous silicon, which allows it to be used for the treatment of cancer [5]. Porous silicon particles are capable of protecting sensitive proteins from degradation and enzymatic denaturation in stomach [6].

The main goal of drug delivery systems based on nanocontainers is the selective effect of the loaded drug on the focus of the disease with minimal side effects and toxicity associated with the distribution of the drug throughout the body during normal administration. Now the concept of nanomedicine has gone further, the direction of combining a therapeutic and diagnostic agent into a single delivery system is developing. Thus, many opportunities open up not only in the treatment of diseases, but also in early diagnosis and understanding of the biological response to certain treatments [7]. Porous silicon nanoparticles can also act as time-synchronized imaging tools by exploiting the optoelectronic properties of porous silicon [8-10].

It was determined that orthosilicic acid, which is a breakdown product of porous silicon, is readily absorbed in the gastrointestinal tract and subsequently excreted through the urinary system. Orthosilicic acid is also associated with homeostasis and regulation of key body processes such as collagen synthesis, prevention of atherosclerosis, Alzheimer's disease, maintenance of bone health, and improvement of the immune system [11].

The ability of porous silicon to support cell growth [12] enables ex-vivo tissue and organ regeneration [13] with further potential for the growth of the patient's own skin for subsequent autotransplantation or the use of donor cells for allotransplantation on burns or other serious wounds

2. Experiment

Pretreatment of samples of silicon substrates was carried out in order to clean them from organic contaminants to ensure the uniformity of the etching process. For this, a solution of ammonium-ammonia peroxide was prepared in following proportions H_2O_2:NH_4OH:H_2O 1:1:4. Samples were immersed in a boiling solution for 7 minutes, then dipped in three waters.

19 samples of porous silicon were prepared by the electrochemical method in a horizontal cell [14] in an alcoholic solution of hydrofluoric acid in the ratio HF: C_2H_5OH 1: 1. The current density was j = 30 mA/cm^2, the etching time was 20 minutes. After etching, the samples were washed in distilled water and dried on filter paper.

To determine the porosity, a semi-destructive gravimetric method was used [15], which included sequential weighing of the samples before etching, after etching. To evaluate the obtained nanostructures, some of the samples were examined with a NEOPHOT 21 photomicroscope. Pieces were cut off with a scalpel and their cleavage was examined to determine the thickness to tell if it corresponded to the average thickness of the nanostructured substrates. The spectral characteristics of the absorption coefficient were studied using a Shimadzu UV-2450 spectrophotometer in the wavelength range from 190 to 1100 nm. The scanning speed is slow, the slit width is 5 nm.

3. Results

Figure 1 shows a cleavage of a sample with porous silicon. Under these conditions, the thickness of the porous layer was 50 μm.

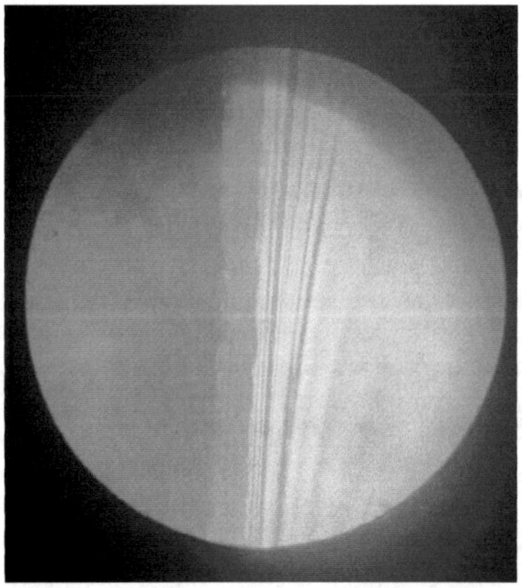

Figure 1. Image of the cleavage obtained with an optical microscope

Porosity was determined by the formula:

$$P = \frac{m1-m2}{\rho \cdot S \cdot d} \cdot 100\%, \quad (1)$$

there m_1 – mass of the silicon sample before anodizing, m_2 – mass of the sample after anodizing, ρ – density of silicon, S – area of the porous layer, d – thickness of the porous layer

Table 1 shows the results of determining the porosity of the samples.

Table 1. Results of determining the porosity of samples

№	m_1 (g)	m_2 (g)	Δm (g)	d (μm)	S, cm^2	P (%)
1	0,5666	0,5480	0,0186	49,65	3,8	42,31
2	0,5045	0,4868	0,0177	48,94		40,26
3	0,6058	0,5870	0,0188	50,32		42,76
4	0,7817	0,7622	0,0195	52,06		44,35
5	0,5997	0,5799	0,0198	51,44		45,04

Under these etching modes, the porosity of the structures varied from 40 to 45%. Further, some of the porous silicon samples were saturated with a broad-spectrum antibiotic ceftriaxone (Table 2).

Table 2. Sample weights before and after saturation

№	Weight to saturation (g)	Weight after saturation (g)
1	0,5453	0,5720
2	04441	0,4585
3	02708	0,3019
4	0,5936	0,6245
5	0,5524	0,5807
6	0,5843	0,6004

Table 2 shows that after saturation with ceftriaxone, the weights of the porous silicon samples increased. To confirm the presence of ceftriaxone in the pores, the volt-ampere characteristics of the samples were examined before and after the application of the drug (Figure 2).

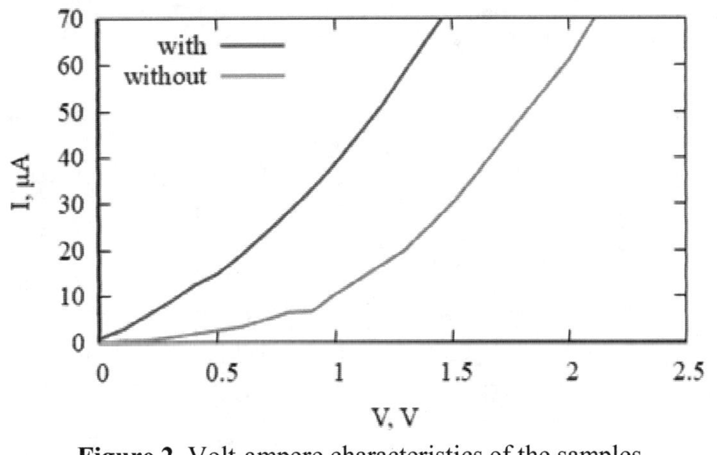

Figure 2. Volt-ampere characteristics of the samples.

As can be seen from Figure 2, after saturation of porous silicon with ceftriaxone, the current-voltage characteristic changes, which confirms the filling of pores.

Then, porous silicon with Ceftriaxone was removed from the surface to create colloidal solutions and dissolved in 2 ml of water. Then 5 samples of the dissolved substance were subjected to grinding in an ultrasonic bath for 5, 10, 20, 30 and 35 minutes. A drug-free porous silicon solution was used as a control.

Figure 3. Samples grinded in the ultrasonic bath.

Figure 4 shows the spectral dependences of the absorption coefficient of aqueous solutions of porous silicon with Ceftriaxone.

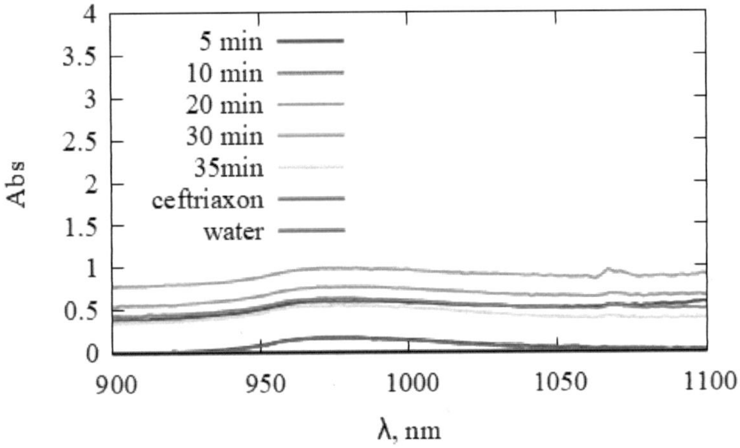

Figure 4. Absorption spectra of aqueous solutions of porous silicon with ceftriaxone .

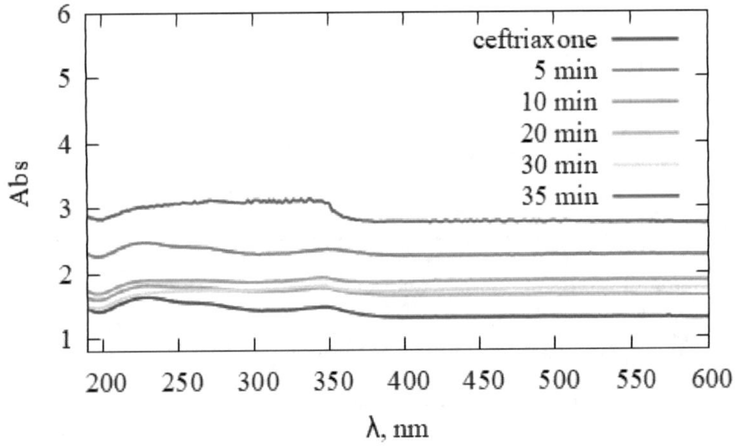

Figure 5. Absorption spectra of dried porous silicon samples with ceftriaxone .

Since Ceftriaxone was diluted with water, the effect of the latter on the spectra was very significant. Therefore, drops of solutions were applied to a glass slide and dried for 2 days to obtain a dry silicon powder with an antibiotic in its pores (Figure 5).

A noticeable absorption peak can be seen at 340 nm, which increases with increasing sonication time on the sample. This is due to a decrease in the particle size of porous silicon with the drug and a more thorough mixing of the solution.

4. Conclusion
Thus, as a result of the work by the method of electrochemical etching, porous structures with a porosity of the order of 40-45% were obtained. When ceftriaxone enters the pores, the characteristics of porous silicon change. The size of the nanocontainer affects the absorption spectrum of an aqueous solution of porous silicon with ceftriaxone. The possibility of using porous silicon nanoparticles as a means of targeted delivery of drugs of various pharmacological actions has been shown.

Acknowledgments
This work was supported by the Foundation for Assistance to Small Innovative Enterprises, grant № 12980GU/2018.

References
[1] Canham L 2014 *Handbook of Porous Silicon*
[2] Anglin E J, Cheng L Y, Freeman W R and Sailor M J 2008 *Adv. Drug Delivery Rev.* **60** pp 1266-77
[3] Zhang D, Esser L, Vasani R, Thissen H and Voelcker N 2019 *Future Medicine* **14** pp 3213-30
[4] Kumeria T, Mclnnes S, Maher S and Santos A 2017 *Expert Opinion on Drug Del.* **14** pp 1407-22
[5] Chen S *et al.* 2020 *Materials Today Adv.* **6** p 100066
[6] Shrestha N *et al.* 2016 *Adv. Funct. Mat.* **26** pp 3405-16
[7] Kelkar S S and Reineke T M 2011 *Bioconjugate Chem.* **22** pp 1879-1903
[8] Gu L, Hall D, Zhengtao Q, Anglin A, Joo J, Mooney D, Howell S and Sailor M 2013 *Nat. Comm.* **4** p 2326
[9] Secret E *et al.* 2014 *Adv. Materials* **26** pp 7643-48
[10] Joo J, Liu X, Kotamraju V R, Ruoslahti E, Nam Y and Sailor M J 2015 *ACS Nano* **9** pp 6233-41
[11] Jurkić L M, Capanec I, Pavelić S K and Pavelić K 2013 *Nutrition & Metabolism* **10** p 2
[12] Sweetman M J, Ronci M, Ghaemi S R, Craig J E and Voelcker N 2012 *Adv. Funct. Mat.* **22** pp 1158-66
[13] Schwartz M P, Derfus A M, Alvarez S D, Bhatia S N and Sailor M J 2006 *Langmuir* **22** pp 7084-90
[14] Spivak Y M, Belorus A O, Somov P A, Tulenin S S, Bespalova K and Moshnikov V A 2015 *Journal of Phys.: Conference Series* **643** p 012022
[15] Zavatski S A and Bondarenko A V 2017 *Doklady BGUIR* **110** pp 21-25

Influence of the nature of a polysaccharide on the surface-morphological and physicomechanical properties of sol–gel plates

O N Malinkina and A B Shipovskaya

Department of High Molecular Compounds, Saratov State University, Saratov 410012, Russia

Abstract. Thin-film sol–gel plates based on chitosan and glucomannan polysaccharides in a wide range of compositions were obtained. The effect of the polysaccharide component on the SEM surface morphology and physicomechanical properties was evaluated. It was found that the introduction of the polysaccharide glucomannan (and/or *Aloe Vera* juice) into the composition of chitosan sol–gel plates increased the elastoplastic characteristics of the samples and had a significant effect on the surface relief of their solid phase. The obtained structures can be used for the design of bioadhesive sensor coatings with a high therapeutic effect, which is an urgent area for theranostics and regenerative medicine.

1. Introduction

The use of biopolymers, e.g. natural polysaccharides, makes it possible to form photonic materials in the form of thin-film coatings and plates for use as selectively sensitive biosensors in functional nanomaterials [1]. In addition, the polysaccharide matrix contributes to the provision of an active biointerface of bionanocomposites with body tissues.

Sol–gel synthesis is a technique for obtaining such materials of varying degrees of structuring (particles, layers, and 3D structures). It has a number of advantages over traditional methods and makes it possible to obtain optically transparent organic–inorganic composites with the possibility of incorporating functional additives of various natures into the matrix [3]. The use of silicon tetraglycerolate $Si(OGly)_4$, a pharmacologically active gel precursor, allows one-pot sol–gel synthesis of hybrid hydrogels under conditions close to physiological ones [4–5]. The use of biologically active polysaccharides such as cationic chitosan (CS) and neutral glucomannan (GM), as well as their mixtures as a matrix, contributes to the expansion of the application fields of the sol–gel materials obtained. These polymers are thermodynamically compatible, have the ability to gelation due to the formation of intermolecular contacts [6–8] and, importantly, are biocompatible and capable of biodegradation. In our recent study [9], the effect of the plant-derived therapeutic agent of *Aloe Vera* juice (AV, dry extract contains ~55% GM), a functional additive, on the surface morphology of sol–gel plates based on CS ascorbate, promising for the design of specific optical sensors and optodes of biomedical applications, was found. In this study, AV was used as the source of GM.

The microrelief and surface roughness parameters, as well as some physicochemical properties (surface energy, sorption capacity, adhesion, etc.) of hydrogel systems are known to largely determine the useful properties of the material [10]. In this connection, the control of the morphostructure of organic–inorganic hydrogels is topical when designing materials with desired properties for new technological applications. SEM imaging is most commonly used for surface inspection due to its accessibility, relative ease of sample preparation and result interpretation.

This study examines the effect of the nature of the polysaccharide (cationic chitosan or neutral glucomannan) on the surface-morphological, sorption and physicomechanical properties of the sol-gel plates.

2. Methodology

The following reagents were used: chitosan (CS) with a viscosity average molecular weight 32 kDa, a degree of deacetylation 80 mol.% (Golden Shell, CN); glucomannan (GM, Uspekh Corp, RU); poly(vinyl alcohol) with a weight average molecular weight of 80 kDa a degree of hydrolysis 85 mol.% (PVA, Titan Corp, RU); glycerol solution of $Si(OGly)_4$ (IOS UB RAS, RU); ascorbic and aminocaproic acids (AscA and AmA, Khimreaktiv Corp, RU); 95% ethyl alcohol (RFK Corp, RU); *Aloe Vera* leaves and distilled water.

To prepare 15 wt% CS stock solutions in 3% AscA and 3% AmA, a sample of polymer powder was suspended in water on a magnetic stirrer, followed by addition of an air-dry AscA and AmA powders. The system was left at 20±2°C for 1 day in the absence of natural light until complete dissolution.

Aqueous GM solutions of a concentration of 0.5 wt% were prepared by suspending a sample of polymer powder in water on a magnetic stirrer at 80±5ºC during 1 hour.

Aloe Vera juice (AV) was obtained in the laboratory by mechanical extraction of previously washed leaves of a 3-year-old plant, followed by filtration. The obtained AV was used directly immediately or stored until 7 days at 4±0.5°C in the absence of natural light.

Aqueous PVA solutions (bioinert structure-forming component) of a concentration of 8 wt% were prepared by suspending a sample of polymer powder in the calculated amount of water on a magnetic stirrer for 5 min, followed by 850 W microwave treatment in a laboratory microwave system Mars-5 (CEM Corporation, US) during 60 s.

Sol–gel plates were prepared by mixing the polysaccaride component, the PVA solution and $Si(OGly)_4$ in a 6 : 1 : 1 weight ratio (Figure 1). The composition of the polysaccharide component was varied by the ratio of the initial solutions CS and/or GM and/or AV in the mixture. The mixtures were thoroughly stirred for 4–5 min until homogeneity, placed on a horizontal silicone substrate at a rate of 0.5±0.1 mL/cm^2 and allowed for sol–gel process to proceed at 20±2°C. Gelation was carried out in a thin (4±1 mm) layer within 1–3 days. The solid phase (xerogels) was isolated by exhaustive extraction with ethyl alcohol within 7 days followed by drying at 20±5ºC.

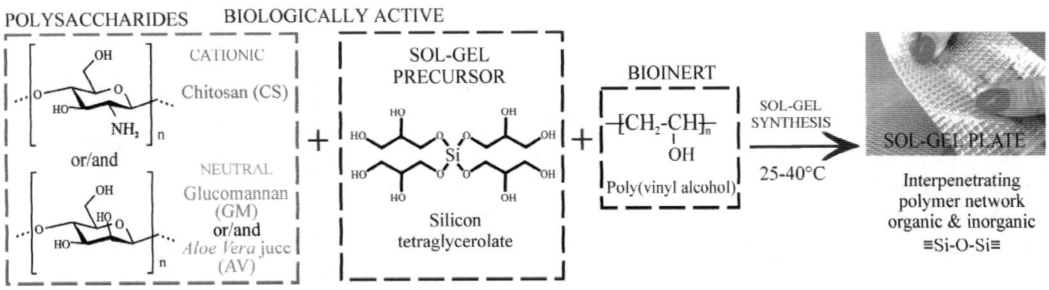

Figure 1. Scheme of the sol–gel plates preparation.

Sorption properties were assessed gravimetrically (Ohaus Adventurer AR 1530 scales, US) in H_2O vapor at 22±1°C. The sorption degree of water vapor (S_w, wt%) and the diffusion coefficients of the initial and final stages of sorption (D_1 and D_2, cm^2/s) were determined according to the standard method. The resorption rate was estimated by the immersion method based on the loss of structure integrity upon exposure to model media with pH = 1.1, 6.5, and 7.4 at 37±0.5°C.

Elastoplastic properties were evaluated on an Instron 3342 tensile testing machine (DE) used a 0.5 kN load cell in a uniaxial tensile mode at a speed of 10 mm/min. The breaking stress (σ_b, MPa),

Young's modulus (E, MPa) and the elongation at break (ε_b, %) were calculated from the averaged data «stress–strain» curves ($\sigma - \varepsilon$) of 10 replicate measurements by standard procedures. Statistical data processing was performed using BLUEHILL-3 INSTRON software. Physicomechanical characteristics were calculated taking into account the initial length and thickness of the sample. The thickness of plates was measured with an Electronic Digital Outside Micrometer CT 200-521 (CN) with a 10 μm division and was expressed as the arithmetic mean of 10 measurements.

SEM images were taken on a MIRA\\LMU scanning electron microscope (Tescan, CZ) at a voltage of 7 kV and a conducting current of 400 pA. A 5 nm thick gold layer was deposited on each sample using a K450X Carbon Coater (DE).

3. Results and discussion

It has been shown that as a result of gelation of the precursor $Si(OGly)_4$, flexible monolithic hydrogel coatings are formed on the polysaccharide matrix (Figure 1). They possess a pH-dependent behavior characteristic of physically cross-linked hydrogels: the sorption capacity and the resorption rate of hydrogel plates at neutral pH are higher than in media with pH = 1.1 and 7.4. Complete resorption of samples in model biological media was observed after 3 days on average.

The assessment of the sorption properties of our sol–gel plates has shown that the water vapor diffusion in the polymer matrix does not obey Fick's law and proceeds in an anomalous mode. The diffusion coefficients calculated from the initial (D_1) and final regions (D_2) of the kinetic curves of sorption differ by an order of magnitude (Table 1). For the plates based on a neutral polysaccharide (AV and/or GM), higher values of both diffusion coefficients and the sorption degree are observed in comparison with the compositions containing the amino polysaccharide CS. It is possible that the diffusion characteristics are affected by the change in the supramolecular structure of the samples due to variation in the polysaccharide component composition [11].

Table 1. Properties of sol–gel plates based on CS and/or GM and/or AV.

Polysaccharide component CS : AV : GM	Water vapor sorption parameter			Physicomechanical parameter		
	Diffusion coefficients (cm²/s)		S_w (wt%)	σ_b (MPa)	ε_b (%)	E (MPa)
	$D_1 \cdot 10^9$	$D_2 \cdot 10^{10}$				
0 : 0 : 6	5.71	13.5	77	0.12	6.54	0.011
3 : 0 : 3	3.47	9.3	72	0.36	23.41	0.025
0 : 3 : 3	5.11	11.4	77	0.15	17.05	0.013
3 : 1.5 : 1.5	4.23	10.8	71	0.39	30.20	0.029
0 : 6 : 0	4.52	12.4	76	0.09	5.29	0.006
3 : 3 : 0	3.38	11.3	74	0.33	26.10	0.023

In the study of the physicomechanical properties of our sol–gel plates under uniaxial tension, it was found that deformation curves characteristic of plastic polymer materials which do not reach the yield point, e.g., xenopericardium or cartilaginous tissue, were recorded for all samples [12]. Plates based on AV and/or GM (no CS) show low elastoplastic properties in comparison with similar CS-containing compositions (Table 1). This indicates that CS makes a positive contribution to the formation of the structure of the plates with a high fraction of the elastic component of deformation. This behavior can be explained by the formation of a large number of intermolecular contacts between the components of the gel-forming mixture. This, in turn, leads to an increase in the number of physical nodes of the resulting network and contributes to the formation of a stronger and more elastic sol–gel plate.

The results of SEM microscopy of the solid phase of our sol–gel plates also showed a positive effect of the combined use of polysaccharides of different nature on the surface-morphological properties of the hybrid matrix. E.g., our analysis of SEM images revealed that the surface of all samples was represented by a non-porous structure with a small number of relatively uniform aggregates and structural inhomogeneities (Figure 2).

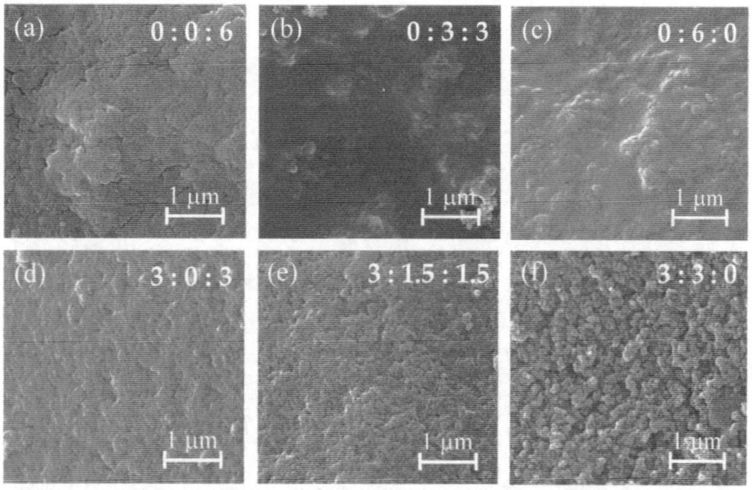

Figure 2. SEM images of the xerogel's morphostructure isolated from sol–gel plates based on polysaccharide component containing GM **(a, b, d, e)**, AV **(b, c, e, f)** and CS **(d–f)**.

The plates based on the neutral polysaccharide GM have a smooth morphostructure, those based on AV have a more granular one. It is worth noting that the joint presence of GM and AV enhances the structuring of the surface. For all the studied samples, the presence of the aminopolysaccharide CS in the composition contributes to the formation of a rougher surface relief.

Our studies have shown a significant effect of the polysaccharide's nature on the morphological, sorption and elastoplastic properties of the sol–gel plates. According to SEM, the resulting material belongs to composites, AV and GM acts as a structure-forming component. The obtained structures can be used for the design of bioadhesive sensor coatings with a high therapeutic effect, which is an urgent area for theranostics and regenerative medicine.

References

[1] Momtaz M and Chen J 2020 *ACS Appl. Mater. Interfaces* **12** 54104
[2] Singh V, Malviya T, Gupta S, Dwivedi L M, Baranwal K and Prabha M 2020 *In AdvancedBiopolymeric Systems for Drug Delivery* (Cham: Springer). pp. 151-181
[3] Xie Y T, Chen J R, Chen Y T, Jiang B C, Sie Z H, Hsu H Y and Hsueh H Y 2021 *Chem. Eng. J.* **405** 126572
[4] Shipovskaya A B, Zhuravleva Yu Yu, Khonina T G, Malinkina O N and Gegel N O 2019 *J. Sol-Gel Sci. Technol.* **92** 349
[5] Gegel N O, Zhuravleva Yu Yu, Shipovskaya A B Malinkina O N and Zudina I V 2018 *Polymers* **10** 259
[6] Yang D, Yuan Y, Wang L, Wang X, Mu R, Pang J, Xiao J and Zheng Y 2017 *Int. J. Mol. Sci.* **18** 2250
[7] Jegina S, Kukle S and Gravitis J 2018 *In IOP Conf. Ser.: Mater. Sci. Eng.* **459** 012016
[8] Hikmawati D, Rohmadanik A R and Putra A P 2018 *In IOP: Conf. Ser.* **1120** 012096
[9] Zhuravleva Yu Yu, Malinkina O N, Zudina I V and Shipovskaya A B 2020 *IOP Conf. Ser.: J. Phys.* **1695** 012099
[10] Xu Y, Zeng S, Xian W, Lin L, Ding H, Liu J and Sun L 2021 *Macromol. Rapid Commun.* **42** 2000446
[11] Shurshina A, Bazunova M, Chernova V, Galina A, Titlova A and Kulish E 2021 *J. Drug Deliv. Sci. Technol.* **63** 102560
[12] Halfwerk F R, Rouwkema J, Gossen J A and Grandjean J G 2018 *J. Mech. Behav. Biomed. Mater.* **77** 400

Application of the Kohonen neural network for monitoring tissue oxygen supply under hypoxic conditions

M S Mazing[1,2], A Y Zaitceva[1] and R V Davydov[2]

[1] Institute for Analytical Instrumentation, Russian Academy of Sciences, 190103, Saint Petersburg, Russia
[2] Peter the Great Saint Petersburg Polytechnic University, 195251, Saint Petersburg, Russia

Abstract. The article presents the results of application of the Kohonen artificial neural network (KANN) in assessing the oxygen status of human tissues, as well as for studying the adaptive-compensatory response of the body to functional load. In the experiment, the registered digital oxygen images of the tissue of 31 subjects were distributed into three classes using the KANN. Each group is characterised by different resistance of the organism to hypoxia. The research results have shown the effectiveness of using an artificial neural network structure and the possibility of its implementation for recognition of the functional state of a person under conditions of metabolic hypoxia; it seems relevant and has theoretical and practical significance in the framework of ecological physiology.

1. Introduction

In the modern world, there is a constant deterioration of the ecological situation [1-10]. Deterioration of the environment has a negative impact on human health [11-16]. For this reason, it is required to pay increased attention to the control of health at different time intervals, especially in moments of poor health [17-23]. In this situation, the development of devices for express diagnostics of the state of human health, as well as support systems for making medical decisions is extremely relevant [21-28]. During the operation of such systems, a number of problems arise [21, 22, 25-27, 29, 30], one of which is associated with obtaining a large amount of unstructured data, which often require automated processing. One of the possible options for solving this problem is the intellectualization of information processing using data mining methods: cluster analysis algorithms and artificial neural networks [31-35]. The task of data mining is to identify latent rules and patterns in data sets, and data mining methods are a direction that complements classical statistical research methods. To date, algorithms built on the basis of neural networks show decent results in various fields of human activity and fields of science [15-18, 20, 33, 34, 36]. In addition, ANNs have many advantages over traditional computational methods. The advantages of neural networks, which include the ability to solve problems with unknown patterns, high resistance to noisy data, low error rate, potentially high speed, continuous improvement and optimization of various network learning algorithms, make neural networks an increasingly promising direction in data mining. Today there are many different artificial neural networks, differing in many parameters for learning, generalization, abstraction. One of the popular classes of neural networks is the Kohonen neural network in the form of Self-organizing Kohonen maps (SOM). Self-organizing process is a learning process without a teacher. With such training, the training set consists of the values of the input variables, and in the training process there is no comparison of the outputs of neurons with the desired values. Self-organizing Kohonen maps are used for visualization, forecasting, searching for patterns in large data sets, identifying independent features and reducing the dimension of the feature space. In particular, the capabilities of SOM can be used to solve a wide range of non-formalized problems in the field of medical diagnostics, as well as for clustering and classification of various medical indicators [37, 38].

Most of the known diseases are directly or indirectly associated with oxygen deficiency in the organs and tissues of the body. Hypoxia is one of the main biologically significant factors that a person encounters throughout his life. Oxygen deficiency can occur in tissues due to dysfunction of the cardiovascular and respiratory systems, and negatively affect the development of various pathological processes in the human body. In this regard, the assessment of the oxygen supply of tissues with oxygen is of great importance both in express diagnostics of the general functional state of the human body and in the therapy of critical conditions.

The proposed work describes the application of an artificial neural network of the SOM for the analysis of oxygen digital tissue images containing information about the oxygen status of human tissues and recorded using the developed optical non-invasive complex, in order to project data into a space of lower dimension, identify the cluster structure and classify them.

2. Method

Registration of oxygen digital tissue images, which contain information about the oxygen status of tissues, was carried out using the developed non-invasive optical medical diagnostic complex [37]. The complex is based on a six-channel analyzer of optical spectra of the visible range, which detects backscattered light in tissues. The principle of operation of the instrumental complex is based on the ability of hemoglobin to absorb light of a certain wavelength, while the degree of light absorption depends on the percentage of oxyhemoglobin in the blood. Each oxygen digital tissue image represents the readings of the optical sensors of the diagnostic system, recorded at different points in time during the functional load. Operations related to system control, recording of results, processing and analysis of the obtained data are performed by the information module of the developed optical complex.

The obtained measurement results are analyzed using Kohonen Self-Organizing Maps. SOM are one of the various neural network algorithms. This algorithm solves clustering problems, and during training, the unsupervised learning method is used, that is, the training result depends only on the structure of the input data. An important feature of the algorithm for the functioning of self-learning maps is that all neurons are ordered into a certain structure (two-dimensional grid). At the same time, in the course of training, not only the winning neuron is modified, but also its neighbors in a certain radius, albeit to a lesser extent. Before starting training the map, it is necessary to initialize the weights of the neurons. Initialization occurs with random values, when all weights are given small random values. Self-organizing Kohonen maps are trained by the method of successive approximations using iterations in which neurons-vectors are corrected. At each iteration, one vector is randomly selected from the initial sample, and then a search is made for the vector of neuron coefficients most similar to it. The neuron most similar to the input vector is declared the winning neuron. As a criterion for the proximity and determination of the winning neuron, the criterion of the minimality of the Euclidean distance between the points specified by the coordinates of the vectors was chosen:

$$\|x - w_{ij}\| = \sqrt{\sum_{t=0}^{n}\left(x(t) - w_{ij}(t)\right)^2} \tag{1}$$

$$d = \min_{i,j}\|x - w_{ij}\| \tag{2}$$

where, x is the input vector, t is the current iteration, $x(t)$ is the input vector's instance at iteration t, $w_{ij}(t)$ is the weight of the connection between the nodes i,j in the grid, and the input vector's instance at the iteration t, d is the distance between vectors x and w.

After that, all the weights of the neural network are corrected, so that the winning neuron and its neighboring vectors in the grid are shifted towards the input vector, and both the winning neuron and its neighbors are modified. The weighting coefficients of the winning neurons and neurons lying in its vicinity are corrected by the following formula:

$$w_{ij}(t+1) = w_{ij}(t) + \alpha_i(t) \cdot \beta_{ij}(t) \cdot \left(x(t) - w_{ij}(t)\right) \quad (3)$$

where, $\alpha_i(t)$ is the learning rate, decreasing with time in the interval [0,1], to ensure the network converges, $\beta_{ij}(t)$) is the neighbourhood function, monotonically decreasing and representing a node *i*, *j*'s distance from the winning neuron.

Thus, the change in the value of the weighting coefficient of the *i*-th neuron is the less, the further this neuron is located from the winner's neuron. If the neuron does not belong to the neighborhood $S_w(t)$, then its weighting coefficient will not change. The execution of the algorithm will continue until the output values appear on the feature map. After the final organization of neurons into a specific structure, the resulting map is displayed, which is an n-dimensional grid. Further, for each cell, statistical characteristics are determined, which make it possible to assign certain color to a cluster of cells.

To assess the stability of a cluster solution, one can compare the clustering result that was obtained using SOM with the results of solutions obtained using other clustering algorithms (non-hierarchical, hierarchical).

In an experimental study, 31 healthy young people took part on a voluntary basis. All subjects performed the proposed functional load, which was holding the breath for the maximum possible time. During all periods of functional load (before breath holding, during breath holding and after breath holding), the readings were recorded using a non-invasive optical diagnostic system. During the measurements, each subject was in a sitting position; measurements were taken on the cyst of the left hand. The obtained oxygen digital images of each subject were multivariate data, which were further processed using SOM.

3. Results

To process the oxygen digital images of subjects obtained as a result of the study using SOM, the software environment R was used. The analysis used the readings that were recorded 4 minutes after the functional load (the recovery period after holding the breath). Prior to the application of the Kohonen neural network, the readings of each sensor were standardized with respect to the readings of each sensor, which were recorded before the functional load (resting state). As a result of using the neural network, the subjects were divided into 3 clusters. The result of grouping nodes of the Kohonen map is shown in figure 1.

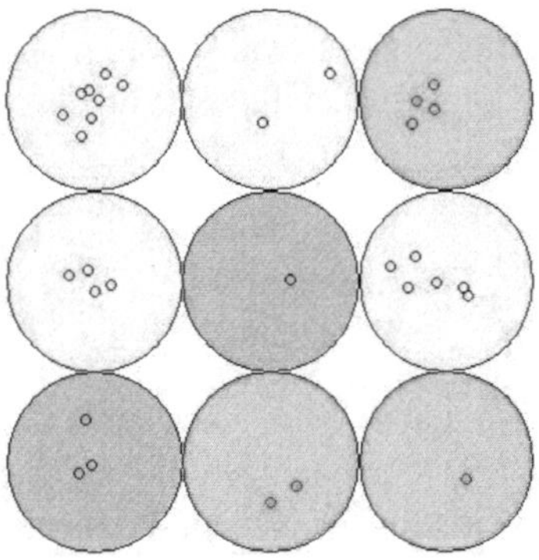

Figure 1. Cluster nodes of the Kohonen map with the distribution of subjects by nodes.

Node clusters are highlighted in different colors: yellow, green, and blue. Each node contains a different number of objects (test subjects).

In addition, to assess the stability of the cluster solution, the same data was processed by the hierarchical clustering method. The created cluster solution coincided with the result of clustering by the hierarchical complete-linkage clustering method (figure 2), which indicates the statistical significance and stability of the cluster solution.

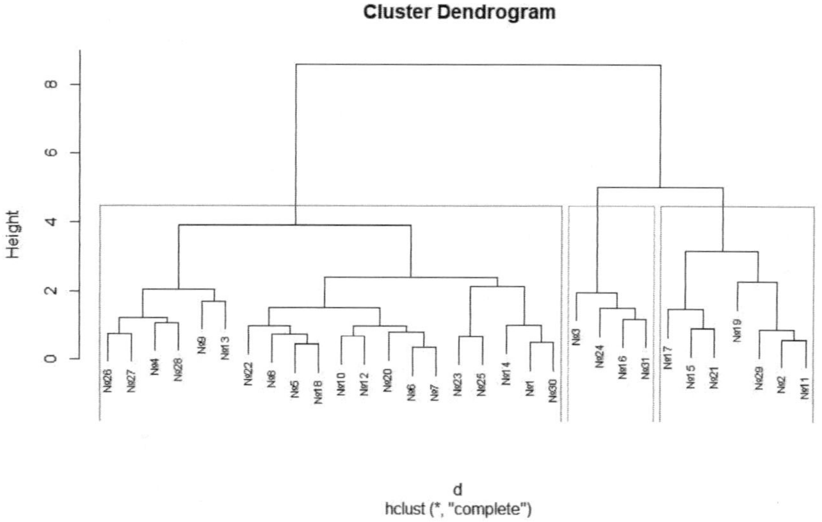

Figure 2. The result of cluster analysis using the complete-linkage clustering with splitting the tree into three groups.

Each cluster is characterized by both a different level of the body's resistance to hypoxia and a different general functional state of the subjects. Each cluster contains 20, 4 and 7 people, respectively. The results obtained are confirmed by independent biomedical research.

4. Conclusion

Thus, evaluating the result obtained using the network, we can conclude that self-organizing Kohonen maps can be used to solve the problem posed by us. The proposed approach to assessing the oxygen status of human tissues and recognizing the general functional state of a person has shown prognostic ability and can be used to create an intelligent information system for express diagnostics.

Acknowledgments

The study was carried out within the framework of State Assignment No. 075-00280-21-00 SU NIR 0074-2019-0013.

References

[1] Davydov V V, Cheremiskina A V, Velichko E N and Karseev A Yu 2014 *Journal of Physics: Conference Series* **541(1)** 012006
[2] Kozlova E S and Kotlyar V V 2014 *Computer Optics* **38(1)** 132-138
[3] Kuzmin M S and Rogov S A 2019 *Computer Optics* **43(3)** 391-396
[4] Smirnov K J and Batov Y V 2019 *Journal of Physics: Conference Series* **1368(2)** 022073
[5] Smirnov K J, Glagolev S F, Rodygina N S and Ivanova N V 2018 *Journal of Physics: Conference Series* **1038(1)** 012102
[6] Gryznova E, Batov Y and Myazin N 2020 *E3S Web of Conferences* **140** 09001
[7] Gryznova E, Grebenikova N, D. Ivanov D and Bykov V 2019 *IOP Conference Series: Earth and Environmental Science* **390(1)** 012044
[8] Nikitina M, Grebenikova N, Dudkin V and Batov Y 2019 *IOP Conference Series: Earth and Environmental Science* **390(1)** 012024
[9] Davydov V V 1999 *Russian Physics Journal* **42(9)** 822–825

[10] Myazin N S and Yushkova V V 2019 *Environmental Research, Engineering and Management* **75(2)** 28–35
[11] Davydov V V, Velichko E N, Dudkin V I and Karseev A Y 2014 *Measurement Techniques* **57(6)** 684–689
[12] Davydov V V, Dudkin V I and Karseev A Y 2015 *Technical Physics Letters* **41(4)** 355–358
[13] Davydov V V, Dudkin V I and Karseev A Y 2015 *Technical Physics* **60(3)** 456–460
[14] D'yachenko S V and Zhernovoi A I 2016 *Technical Physics* **61(12)** 1835–1837
[15] Myazin N S, Yushkova V V and Dudkin V I 2019 *Journal of Physics: Conference Series* **1400(6)** 066008
[16] Rukin E V, Myazin N S and Dudkin V I 2019 *Journal of Physics: Conference Series* **1368(4)** 042011
[17] Neronov Y I 2021 *Biomedical Engineering* **54(5)** 333–336
[18] D'yachenko S V and Zhernovoi A I 2016 *Technical Physics* **61(12)** 1835–1837
[19] Davydov V V, Dudkin V I and Karseev A Y 2015 *Instruments and Experimental Techniques* **58(6)** 787–793
[20] Makeev S S, Grevtzeva A S, Glinushkin A P and Matorin D N 2020 *Journal of Physics: Conference Series* **1695(1)** 012112
[21] Myazin N S, Yushkova V V and Taranda N I 2019 *Journal of Physics: Conference Series* **1410(1)** 012130
[22] Marusina M Y and Karaseva E A 2018 *Asian Pacific Journal of Cancer Prevention* **19(10)** 2771–2775
[23] Marusina M Y, Fedorov A V, Prokhorovich V E, Tkacheva N V and Mayorov A L 2018 *Measurement Techniques* **61(3)** 297–302
[24] Myazin N, Neronov Y, Dudkin V and Petrov A 2019 *IOP Conference Series: Materials Science and Engineering* **497(1)** 012111
[25] Davydov R V, Mazing M S, Yushkova V V and Stirmanov A V 2019 *Journal of Physics: Conference Series* **1410(1)** 012067
[26] Grevtseva A S, Smirnov K J, Greshnevikov K V and Glinushkin A P 2019 *Journal of Physics: Conference Series* **1368(2)** 022072
[27] Grevtseva A S, Smirnov K J, Davydov V V and Rud V Yu 2018 *Journal of Physics: Conference Series* **1135(1)** 012056
[28] Myazin N, Neronov Y, Dudkin V and Yushkova V 2018 *MATEC Web of Conferences* **245** 11013
[29] Myazin N S 2018 *Journal Physics: Conference Series* **1124(1)** 031004
[30] Myazin N S 2018 *Journal Physics: Conference Series* **1135(1)** 012061
[31] Logunov S E, Davydov V V, Glinuchkin A P, Podkovyrov I Yu and Rud V Yu 2019 *Journal of Physics: Conference Series* **1410(1)** 012113
[32] Tuckova J 2013 *11th International Workshop of Electronics, Control, Measurement, Signals and their application to Mechatronics* (IEEE), p. 1-6
[33] Logunov S E, Rud V Y, Davydov R V, Moroz A V and Smirnov K J 2019 *Journal of Physics: Conference Series* **1326(1)** 012024
[34] Grevtseva A S, Davydov V V and Rud V Yu 2020 *Journal of Physics: Conference Series* **1697(1)** 012079
[35] Valov A P., Davydov R V, Rud V Y and Grevtseva A S 2019 *Journal of Physics: Conference Series* **1326(1)** 012040
[36] Davydov V V, Logunov S E and Fadeenko V B 2018 *Russian Journal of Nondestructive Testing* **54(3)** 213–221
[37] Mazing M S, Zaitceva A Y and Kislyakov Y J 2021 *Springer Proceedings in Physics* **255** 233–239
[38] Mazing M S, Zaitceva A Y, Kislyakov Y Y, Avdyushenko S A and Davydov V V 2020 *International Journal of Pharmaceutical Research* **12** 1974–1978

Thermostability of lysozyme amyloid fibrils

N M Melnikova[1], M I Sulatsky[2], Yu D Diordienko[1] and A I Sulatskaya[1]

[1] Laboratory of Structural Dynamics, Stability and Folding of Proteins, Institute of Cytology, Russian Academy of Sciences, 4 Tikhoretsky avenue, 194064 Saint Petersburg, Russia
[2] Laboratory of Cell Morphology, Russian Academy of Sciences, 4 Tikhoretsky avenue, 194064 Saint Petersburg, Russia

melnikova07nm@gmail.com

Abstract. Ordered protein aggregates, amyloid fibrils, are a marker of many serious diseases, such as Alzheimer's, Parkinson's, prion diseases, etc. At present, special attention is paid to the study of external influences that can affect the structure and stability of mature amyloid fibrils, which may be in demand in the development of approaches to the therapy of amyloidosis, as well as in the creation of new high-strength materials on the basis of these protein aggregates. An external factor, the influence of which on fibrils was studied in this work, was temperature denaturation. It was shown that heating lysozyme amyloid fibrils to 60 °C does not lead to their degradation, but leads only to a reversible increase in the intramolecular mobility of amyloid-forming proteins, but does not change their morphology. At the same time, boiling of lysozyme amyloids leads to their irreversible degradation, which occurs at least 5 days after exposure: fibrils that form larger clusters change their secondary structure, and fibrils with a lesser degree of clustering are divided into separate fibers. Obtained data about the factors that change the stability and structure of amyloids can be applied in biotechnology for creating new high-strength nanomaterials on their basis.

1. Introduction

Highly ordered protein aggregates, amyloid fibrils, are a marker of many serious diseases, such as Alzheimer's, Parkinson's, prion diseases, and others [1, 2]. At the same time, physiological amyloid fibrils in bacteria, unicellular eukaryotes, fungi, plants, insects and mammals were found, which are necessary for the normal functioning of cells and tissues, which emphasizes the biological significance of amyloid formation [3, 4]. Due to the widespread occurrence of amyloid fibrils, as well as the variety of their functions (mechanical protection and modification of the cell surface; biotic or abiotic surface adhesion; cellular internalization; ensuring resistance to various external influences; biosynthesis of pigments, control of homeostasis, storage and the release of hormones, signal transduction, etc. [5, 6]) the study of amyloids is a very urgent task. The relevance of amyloid fibrils study is also due to their unique mechanical properties, such as high strength and extensibility, which makes them an attractive object for the creation of new nanostructures and nanomaterials [7]. The aim of this work was to analyze the resistance of amyloid fibrils to high temperatures. The object of the study was model amyloid fibrils formed from the globular protein lysozyme with different morphology.

2. Experimental techniques

Fibrillogenesis of lysozyme was carried out using two different protocols, making it possible to obtain aggregates with different morphologies [8]. According to one of these protocols, lysozyme was incubated in 20% acetic acid in the presence of 100 mM NaCl (pH 2) at 37 °C. According to the second protocol, the protein was incubated in 0.05 M KH2PO4-NaOH buffer in the presence of 3 M GdnHCl (pH 7) at 57 °C. Both samples were incubated with constant stirring for 2 days. Then the fibrils were transferred to distilled water. The prepared amyloid fibrils were studied by the absorption and fluorescence spectroscopy, method of circular dichroism and electron microscopy. In addition, using a specially developed approach based on the samples preparation by equilibrium microdialysis [9, 10, and 11], the interaction with amyloids of a specific fluorescent probe thioflavin T (ThT) was studied. The recorded fluorescence intensity of the samples was corrected on the primary inner filter effect according to a specially elaborated protocol [12].

3. Results

According to electron microscopy data, amyloid fibrils prepared at pH 2 predominantly are thin fibers, having a tendency to form clots. At the same time, in the buffer solution with pH 7 shorter fibrils with a significantly higher degree of clustering are formed (Figure 1).

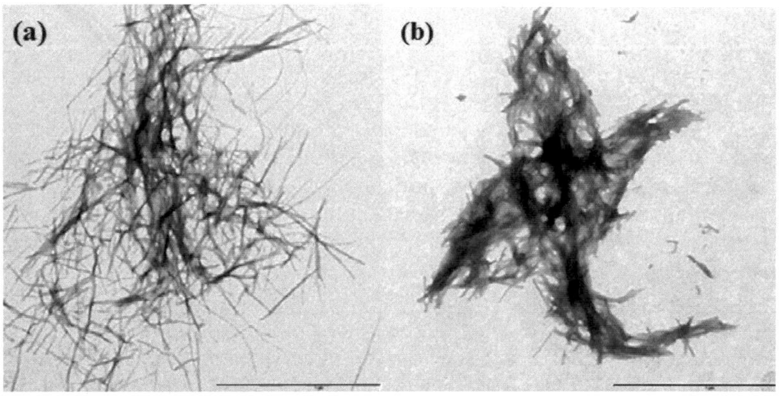

Figure 1. Visualization of lysozyme amyloid fibrils prepared at pH 2 (a) and pH 7 (b) and transferred to water. The scale bar is 1 μm.

First of all, we investigated the stability of the prepared amyloid fibrils when changing temperature in the range from 0 to 60 ºC. The fluorescence spectra of the sample (free and bound to fibrils ThT) and reference (free ThT) solutions obtained by the equilibrium microdialysis were recorded when they were heated from 0 to 60 °C. It turned out that heating the prepared samples leads to a decrease in the fluorescence intensity of ThT bound to fibrils. In order to check whether this decrease is reversible, we cooled the samples to 0 °C. The curve characterizing the heating of the samples almost completely coincided with the curve characterizing their cooling for both samples (Figure 2), which made it possible to conclude that the processes occurring in the samples when they are heated to 60 °C are reversible.

In order to understand what exactly is responsible for the decrease in the ThT fluorescence intensity with heating the sample, we analyzed images of amyloid fibrils obtained by electron microscopy before and after external exposure. We found no changes in the structure or degradation of amyloids after their heating to 60 °C (Figures 3, 4).

It can be assumed that an increase in the sample temperature increases the probability of the ThT molecule transition to the conformation with an angle between the benzothiazole and aminobenzene rings equal to 90 °C, which, in turn, leads to nonradiative deactivation of the excited state of the dye and a decrease in the quantum yield of its fluorescence. This can occur due to a reversible change in the intramolecular mobility of side radicals of amyloid-forming proteins and fragments of the ThT molecule relative to each other, without changing the structure or degradation of amyloid fibrils. Thus, obtained results indicate that lysozyme amyloid fibrils prepared under different conditions are stable when heated to 60 °C, which confirms the idea of a high resistance of amyloids to some extreme external influences.

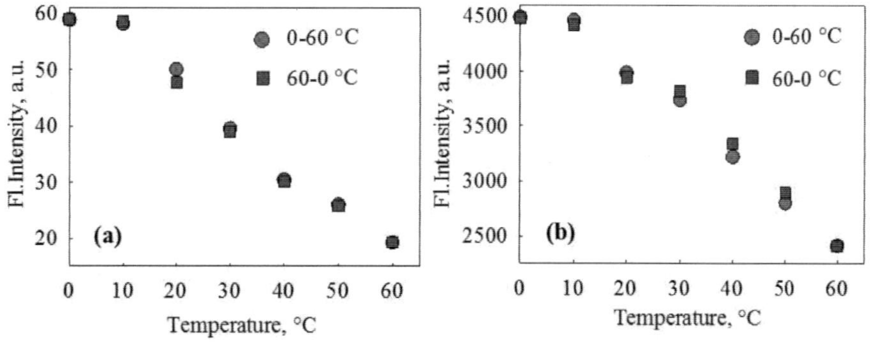

Figure 2. Corrected on the primary inner filter effect the fluorescence intensity of ThT (at 480 nm, maximum of fluorescence spectrum) bound to lysozyme amyloid fibrils, prepared at pH 2 (a) and pH 7 (b). Changes in fluorescence intensity are shown with increasing (0-60 °C) and decreasing (60-0 °C) temperature.

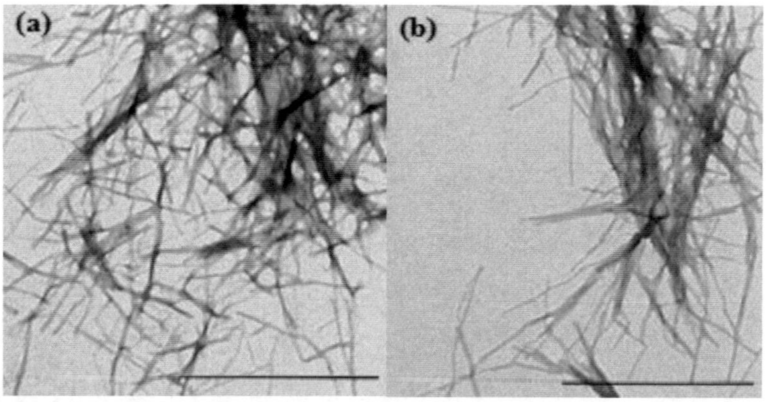

Figure 3. Visualization of lysozyme amyloid fibrils prepared at pH 2, before (a) and after (b) heating to 60 °C. The scale bar is 0.25 μm.

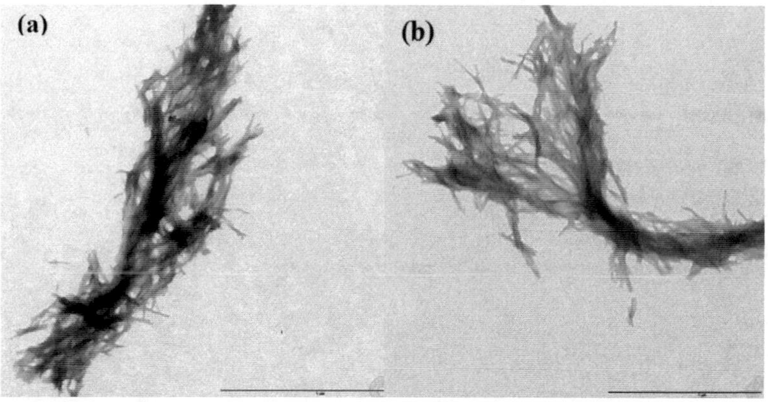

Figure 4. Visualization of lysozyme amyloid fibrils prepared at pH 7, before (a) and after (b) heating to 60 °C. The scale bar is 1 μm.

Then, we used even more extreme environmental conditions and analyzed whether boiling affects the structure and stability of lysozyme amyloid fibrils. Registration of the intrinsic tryptophan fluorescence spectra of amyloid fibril samples prepared under different conditions made it possible to conclude that changes in the sample with lysozyme amyloid fibrils continue for at least 5 days after their boiling and are irreversible (Figure 5).

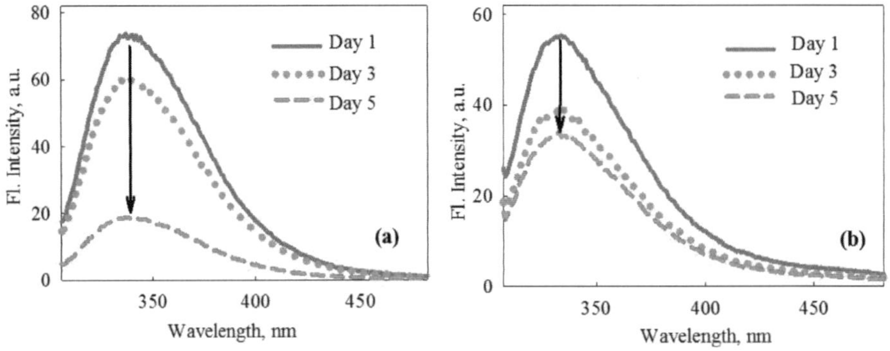

Figure 5. Spectra of intrinsic tryptophan fluorescence of lysozyme amyloid fibrils prepared at pH 2 (a) and pH 7 (b) on the first, third and fifth days after boiling.

Electron microscopy data of the samples a week after boiling indicate that boiling lysozyme amyloid fibrils, prepared at pH 2, leads to their partial degradation: declusterization of not numerous fibrils clots into the separate thin filaments (Figure 6) and to a slight change in their secondary structure, which was evaluated by the recorded circular dichroism (CD) spectra of the samples in the far UV region (Figure 7 a).

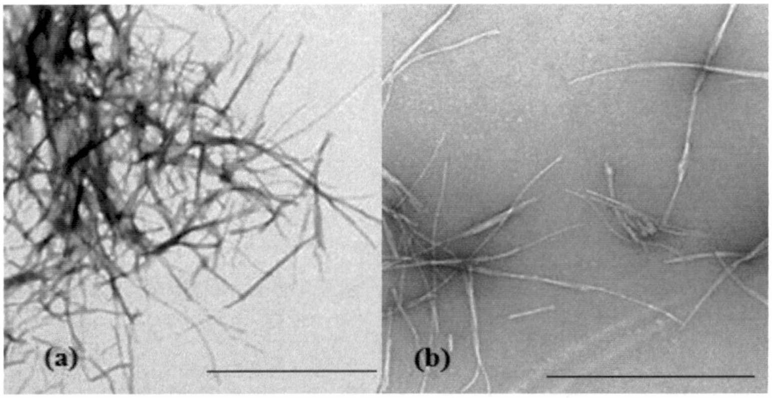

Figure 6. Visualization of lysozyme amyloid fibrils prepared at pH 2, before (a) and after (b) boiling. The scale bar is 0.25 μm.

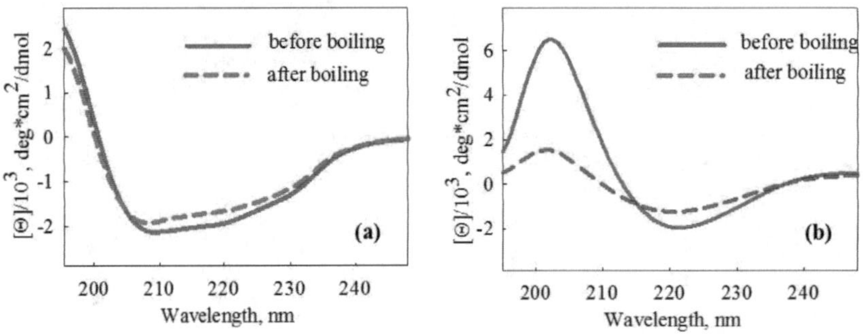

Figure 7. CD spectra of lysozyme amyloid fibrils prepared at pH 2 (a) and pH 7 (b), before and after boiling.

The boiling of lysozyme amyloid fibrils prepared at pH 7 also led to their degradation, which, however, was more manifested in a change in the secondary structure of amyloid-forming proteins (Figure 7 b), but practically did not lead to declusterization of fibrillar clots (Figure 8).

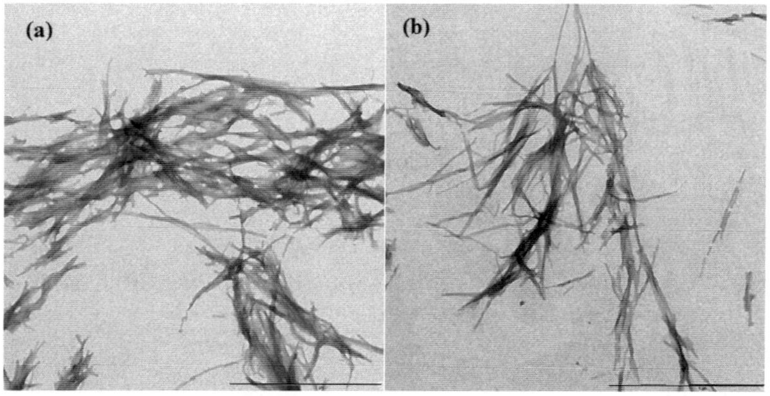

Figure 8. Visualization of lysozyme amyloid fibrils prepared at pH 7, before (a) and after (b) boiling. The scale bar is 1 μm.

4. Conclusion

Based on the obtained results, it was concluded that heating lysozyme amyloid fibrils with different morphology up to 60 °C causes only a reversible increase in the intramolecular mobility of amyloid-forming proteins. At the same time, boiling of lysozyme amyloids leads to their irreversible degradation: amyloids that form larger clusters change their secondary structure, while fibrils with a lesser degree of clustering are divided into separate fibers. New data about the factors that influence the stability of amyloids can be applied in biotechnology for creating new high-strength nanomaterials on their basis.

Acknowledgments

This work was supported by the Russian Science Foundation (project № 18-74-10100).

References

[1] Iadanza M G, Jackson M P, Hewitt E W, Ranson N A and Radford S E 2018 *Nat. Rev. Mol. Cell. Biol.* **19** 755-73
[2] Eisenberg D and Jucker M 2012 *Cell* **148** 1188–1203
[3] Chiti F and Dobson C M 2006 *Annual Rev. Biochem.* **75** 333-66
[4] Fersht A 1998 *Structure and Mechanism in Protein Science: A Guide to Enzyme Catalysis and Protein Folding* vol 42, ed W H Freeman and Company (New York: W H Freeman) p 631
[5] Fowler D M, Koulov A V, Balch W E and Kelly J W 2007 *Trends Biochem. Sci.* **32** 217-24
[6] Pham C L L, Kwan A H and Sunde M 2014 *Essays Biochem.* **56** 207–19
[7] Cherny I and Gazit E 2008 *Angewandte Chemie International Edition* **47** 4062-69
[8] Sulatskaya A I, Rodina N P, Povarova O I, Kuznetsova I M and Turoverov K K 2017 *Journ. Of Mol. Struct.***1140** 52-8
[9] Kuznetsova I M, Sulatskaya A I, Uversky V N and Turoverov K K 2012 *Mol. Neurobiol.* **45** 488-49
[10] Sulatskaya A I, Kuznetsova I M and Turoverov K K 2011 *The Journal of phys. chemistry B.* **115** 11519-24
[11] Kuznetsova I M, Sulatskaya A I and Uversky V N, Turoverov K K 2012 *Plos One* **7**(2) 30724
[12] Fonin A V, Sulatskaya A I, Kuznetsova I M and Turoverov K K 2014 *Plos One* **9**(7) 103878

IOP Publishing
2086 (2021) 012118 doi:10.1088/1742-6596/2086/1/012118

The study of molecular composition in biomimetic interface of biocomposite/dentin

D Goloshchapov[1], V Kashkarov[1], I Ippolitov[2], Yu Ippolitov[2], K Nikitkov[1], J Vongsvivut[3] and P Seredin[1]

[1] Department of Solid State Physics and Nanostructures, Voronezh State University, Voronezh, Russia
[2] Department of Pediatric Dentistry with Orthodontia, Voronezh State Medical University, Voronezh, Russia
[3] Australian Synchrotron (Synchrotron Light Source Australia Pty LTD), 800 Blackburn Rd Clayton, VIC 3168, Australia

goloshchapov@phys.vsu.ru

Abstract. The aim of the study is the problem of formation of the biomimetic interface between the dental product and dentin of the human tooth as well as the investigations of molecular-chemical features in biointerface with the use of molecular multivariate IR-visualization. The data on synchrotron IR-mapping made enabled to differentiate the regions of sound dentin tissue and biomimetic transition layer and also to determine molecular groups responsible for the process of integration.

1. Introduction

Biomimetic remodeling of enamel and dentin of teeth demonstrates that in order to improve integration of dental products with the hard dental tissue it is required to form nano-filled biomimetic buffer layers. Furthermore, the composition of buffer layers should involve nanocrystalline hydroxyapatite (n-HAP) and amino acid matrix[1]. Incorporation of nano-HAP with various defect structure into the composition of a hybrid layer, as well as a number of polar amino acid help to resemble characteristic for the enamel matrix [1–4]. Such strategy makes it possible to improve integration of synthetic materials and natural tissue [5–7], while the transition layer ceases to be a weak link of the restoration dentistry [8]. Therefore, investigations of formation of the biomimetic biocomposite/dentin interface and determination of its molecular composition with the use of FTIR-microspectroscopic visualization technique was the aim of our study.

2. Materials and methods

The study was performed with tooth samples removed from patients in the age of 20-40 according to the orthodontic criteria. With the use of Er:YAG laser the hard dental tissue was prepared up to dentin. Next, dentin conditioner, bio-primer, universal adhesive and compomer were deposited layer by layer on the prepared surface of dentin[6]. Treatment processes were performed in the conditions of alkaline environment with pH > 11 for the formation of the buffer hybrid layer. To promote formation of the bond with the apatite of dentin polar amino acids – L-lysine hydrochloride and L-arginine hydrochloride were additionally introduced in the composition of bio-primer. Nanocrystalline carbonate-substituted hydroxyapatite (n-CHAP) was introduced as a filler in the universal adhesive [9].

In accordance with the requirements of the microspectroscopic research techniques for the geometry of the samples, we prepared plane-parallel segments of the teeth. Molecular composition of the interface samples was studied with the use of synchrotron ATR FTIR-microspectroscopy[10].The synchrotron FTIR experiment was performed at Infrared Microspectroscopy (IRM) beamline (Australian synchrotron, Victoria, Australia), using a Bruker Vertex 80v spectrometer coupled with a Hyperion 3000 FTIR microscope and a liquid nitrogen-cooled narrow-band mercury cadmium telluride (MCT) detector (Bruker Optik GmbH, Ettlingen, Germany). The synchrotron FTIR spectra were recorded within a spectral range of 3800–700 cm^{-1} using 4-cm^{-1} spectral resolution. Blackman-Harris 3-Term apodization, Mertz phase correction, and zero-filling factor of 2 were set as default acquisition parameters using OPUS 8.0 software suite (Bruker Optik GmbH, Ettlingen, Germany). A subsequent synchrotron macro ATR-FTIR mapping measurements were performed on specific areas of interest found in the prior overview map where a good contact with the Ge ATR crystal was achieved, using a smaller step interval of 1 μm and 32 co-added scans.

3. Results

The proposed technique of molecular analysis mapping of biointerface was applied for the region between biomimetic apatite layer and dentin tissue of the human tooth (60×30 μm) (Figure 1,2) with a spatial resolution of 1 μm. Basing on the color coding one-dimensional IR-maps were designed that demonstrate the distribution of molecular group of ether –$COOCH_3$ (1762-1701 cm^{-1}) associated with compomer (Figure 1), vibrations of molecular groups CH_2, CH_3, the bands of Amid III and collagen which are present in dentin (Figure 2), the distribution of molecular group Amid I (1685-1611 cm^{-1}) (Figure 3) and mineral (phosphate) PO_4 component (1113-961 cm^{-1}) (Figure 4).

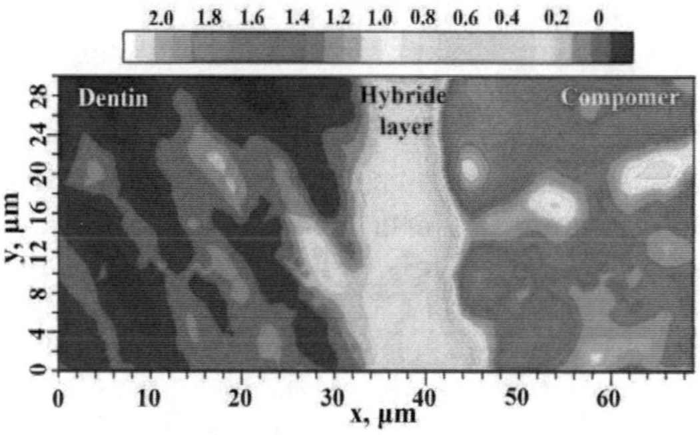

Figure 1. The FTIR map of dentin\composite interface calculated on the basis of color coding of the absorption band 1762-1701 cm^{-1} intensity.

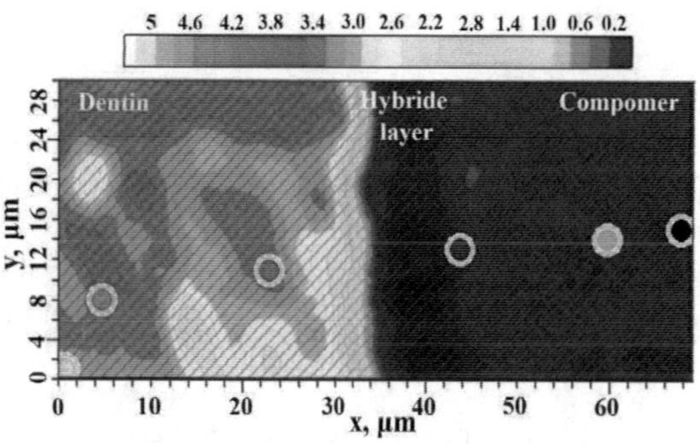

Figure 2. The FTIR map of dentin\composite interface calculated on the basis of color coding of the absorption band 1485-1376 cm^{-1} intensity.

The obtained FTIR maps (see Figure 1,2) provide information on the spatial distribution of the etheric (bioprimer-adhesive-compomer) and protein (collagen) components in the region of dentin/compomer. In the area of dentin one can easily observe the bands with the average color intensity (Figure 1). Their appearance in the map is caused by the presence of dentin tubules. Analysis of FTIR map of the distribution for Amid I band in the range of 1685-1611 cm^{-1} (Figure 3) does also support this statement. It means penetration of bioprimer components and universal adhesive into dentin. This coincides with the results of works concerned with the microscopic investigations of dentin/adhesive interface [11]. The detailed analysis of the data on FTIR-micromapping of the interface areas revealed the presence of a hybrid layer between the area of a sound dentin and the area of adhesive/dental material. It should be noted that the hybrid layer is composed both of the components of bioprimer and adhesive as of the collagen fibers and enamel apatite, that follows from the analysis of spatial distribution of the vibration bands intensities: 1762-1701 cm^{-1}, 1485-1376 cm^{-1} and 1113-961 cm^{-1}, respectively (Figure 1,2,4).

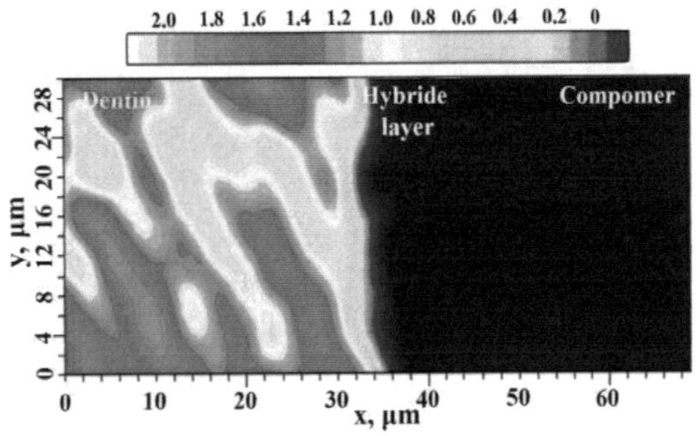

Figure 3. The FTIR map of dentin\composite interface calculated on the basis of color coding of the absorption band 1685-1611 cm^{-1} intensity.

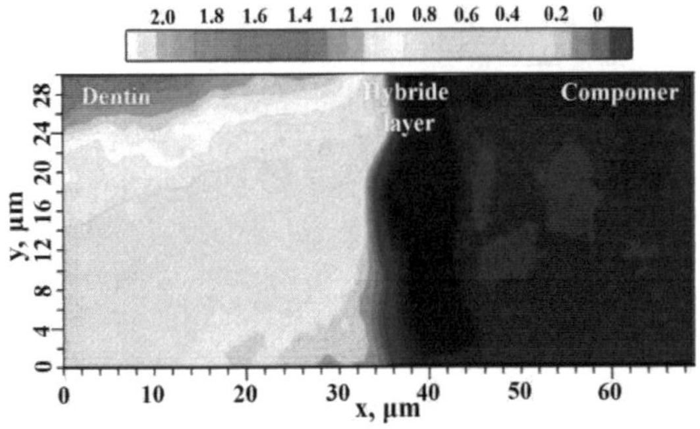

Figure 4. The FTIR map of dentin\composite interface calculated on the basis of color coding of the absorption band 1113-961 cm^{-1} intensity.

Joint analysis of the spectral data array (position and intensity of Amid I and phosphate bands) in the area of hybrid layer (Figure 2,4) indicates at the appearance of the interphase interaction [12]. Considering the characteristics of preparation of the dentin tissue with the erbium laser this interaction is realized as a result of formation of the demineralized and disorganized area in dentin and the following penetration of bioprimer and adhesive components into dentin. The following hierarchical cluster analysis of the spectral data [13] demonstrated that the interaction between dental product and the natural hard tissue of dentin is realized by the biomimetic buffer layer that is bound with partially demineralized dentin matrix (Figure 2,3), and organic-mineral interaction occurs within the 10 µm region. Analysis of FTIR map of the mineral (phosphate) component (Figure 4) in the enamel apatite demonstrated correctness of this assumption. Decrease of intensity in the area of dentin means demineralization of the hard dental tissue in the direction of dentin tubules. At the same time non-zero intensity of the phosphate component within the area of hybrid layer (Figure 4) is explained by the presence of nanocrystalline hydroxyapatite in the adhesive. It is known [14], that inclusion of various remineralization agents in the adhesive systems is just a factor which already allows formation of the hybrid interface based on polymer, collagen and resin substituting native tissue of dentin at the boundary between dentin and adhesive [1]. Such approach provides positive results and it enables to form stable bounds between adhesive and dentin.

3. Conclusion

Applying synchrotron IR-mapping it was shown that our elaborated biomimetic buffer layer involving n-CHAP and the complex of the main polar amino acids which are characteristic for the natural dental tissue can produce the functional bond as with the dental compomer as with natural dentin. The use of biomimetic buffer layers makes it possible to achieve an excellent biocompatibility during repair of dental structure and it can form the basis for the new therapeutic approach in dentistry.

Acknowledgments

This work was supported by the grant of Russian Science Foundation, grant number 21-15-00026. The part of this research was undertaken with The Infrared Microspectroscopy (IRM) beamline at the Australian Synchrotron.

References

[1] Comeau P and Willett T 2018 *Sci. Rep.* **8** 12700-1-11

[2] Goloshchapov D L, Ippolitov Yu A and Seredin P V 2020 *Results Phys.* **18** 103277-1-3

[3] Yanyan S, Guangxin W, Guoqing S, Yaming W, Wuhui L and Osaka A 2020 *RSC Adv.* **10** 37005–13

[4] Li Z, Ren Q, Cui J, Hu D, Tian T, He T, Wang K, Jiang W and Zhang L 2020 *CrystEngComm* **22** 3814–23

[5] Barandehfard F, Kianpour Rad M, Hosseinnia A, Khoshroo K, Tahriri M, Jazayeri H E, Moharamzadeh K and Tayebi L 2016 *Ceram. Int.* **42** 17866–75

[6] Seredin P V, Goloshchapov D L, Prutskij T and Ippolitov Yu A 2017 *Results Phys.* **7** 1086–94

[7] Provenzi C, Leitune V C, Collares F M, Trommer R, Bergmann C P and Samuel S M 2014 *Appl. Adhes. Sci.* **2** 1-5

[8] Delgado A H and Young A M 2021 *Mater. Basel Switz.* **14** 760 1-18

[9] Goloshchapov D L, Lenshin A S, Savchenko D V and Seredin P V 2019 I *Results Phys.* **13** 102158-1-8

[10] Seredin P, Goloshchapov D, Ippolitov Y and Vongsvivut J 2020 *Sci. Rep.* **10** 20891-1-12

[11] Takamizawa T, Imai A, Hirokane E, Tsujimoto A, Barkmeier W W, Erickson R L, Latta M A and Miyazaki M 2019 *Dent. Mater.* **35** 1791–804

[12] Spencer P, Wang Y, Katz J L and Misra A 2005 *J. Biomed. Opt.* **10** 031104-1-11

[13] Savić D, Joković N and Topisirović L 2008 *Dairy Sci. Technol.* **88** 273–90

[14] Kim K-I, Kim D-A, Patel K D, Shin U S, Kim H-W, Lee J-H and Lee H-H 2019 *Sci. Rep.* **9** 4921-1-11

Comparison of the effectiveness of blood transfusion and reinfusion

E V Pimakhina[1], A A Pimakhin[1], N V Vishnykov[2], N M Tolkach[2] and S B Arsentiev[3]

[1] Ryazan State Medical University, Ryazan 390026, Russian Federation
[2] Ryazan State Radio 2Engineering University, Ryazan 390005, Russian Federation
[3] Regional Clinical Hospital, Ryazan 390039, Russian Federation

elenapim@yandex.ru, +7 (920) 960 03 39

Abstract. Studies by atomic force and scanning electron microscopy have shown that erythrocytes of foreign blood have morphological defects. The sequestration of foreign erythrocytes makes it difficult for the erythron to selfrepair. The actual solution to this problem is the application of the blood reinfusion technique using the Cell Saver apparatus. Transfusion of autoerythra suspension, harvested using the Cell Saver apparatus during the operation, stabilizes red blood counts in the early post-transfusion period in patients and reduces the manifestation of massive hemotransfusion syndrome. Hardware reinfusion of erythrocytes is effective and safe for massive blood loss in obstetrics. Reinfusion dictates the need for parallel correction of all blood parameters. Application of the principles of patient blood management can reduce the transfusion load, improve the quality of medical care.

1. Introduction
Donor blood transfusion in obstetric practice is characterized by large volumes. Transfused erythrocytes, especially those of long shelf life, are rapidly destroyed in the recipient's bloodstream, which contributes to the development of postoperative anemia.[1]

2. Experiment and methods
To explain the complications from the transfusion of large doses of donor erythrocytes, we studied the structure of normal and donor erythrocytes membranes using atomic force microscopy (afm, surface topography and morphometry) with large scanning fields on the equipment of the Regional Center of Probe Microscopy of Collective Use of the Ryazan State Radio Engineering University named after V. F. Utkin at the probe laboratory "Ntegra -Aura" (NT MDT SI, Russia). AFM processing was performed using builtin software. Various forms of erythrocytes have been analyzed.

Atomic force microscopy allows analisys of the object with a probe, with the of nanometer curvature radius. The atomic force microscope can obtain a truly three-dimensional relief of the surface.[2] This method opens up opportunities for studying adhesion, molecular binding, electrostaticity, elasticity and mobility of cell membranes. Blood cells were one of the first biological objects studied by the AFM method. Usage of large scanning fields helps to obtain a comprehensive stereometric characteristic of whole erythrocytes. In the taping mode with nanometer resolution protruding protein globules on the membrane surface were differentiated.[3] The differences in the elastic properties of individual sections of the membrane of these shaped elements were revealed with phase contrast method. A change in the geometric properties of erythrocytes was found when they were exposed to various substances.

Innovative microscopy methods make it possible to obtain new data in the pathomorphology of massive blood transfusion syndrome. It was found that under the action of EDTA (Ethylenediaminetetraacetic acid), which is used as an anticoagulant [4], the erythrocyte membrane changes. there are outgrowths in the form of thorns on the erythrocyte, which loses the elastic function of the membrane. It is very important when the erythrocyte passes through the lumen of the capillary. Similar violations occur during the procurement and storage of donated blood. It was found that transfused erythrocytes, especially long-term storage, have uneven areas, broken edges, which affects the transfer of oxygen and disrupts the gas transport function of the red blood cell. The increased moisture content of the erythrocyte leads to the symptom of "wet erythrocyte" [5], which leads to sticking together and sludge syndrome, closure of the capillar by microconglomerates with the subsequent ischemia of cells and tissues. The described changes in erythrocytes were studied in women whose pregnancy was complicated by preeclampsia.

To explain the complications arising from the transfusion of large doses of donor erythrocytes, we studied the structure of the membrane of normal and donor erythrocytes using AFM (surface topography and morphometry) methods using large scanning fields. The experiments were carried out at the Ntegra-Aura probe laboratory (NT-MDT SI, Russia). Processing and construction of AFM images was carried out using built-in software. Various forms of erythrocytes have been studied in the experiments.

According to the results of hemoscanning, the majority of erythrocytes had a rounded shape. The diameter of the erythrocyte, calculated by studying the profile of the cell, corresponded to the average known values. The depth of the depression of normocytes was 0.25 ± 0.06 nm. When studying the ratio of the diameter of the erythrocyte to the diameter of the cavity, the value was 21 ± 2 units.

When studying donor erythrocytes, differences in the shape of erythrocytes were found. Elongated cells were observed, which is associated with a violation of the elasticity of the membrane. The passage of such an erythrocyte through the microvasculature is difficult. An increased content of cytoplasmic bridges between cells was also found, which promotes stasis, sludge and then - thrombosis. Discocytes with a crest, discocytes with outgrowth, discocytes with multiple outgrowths, and discocytes in the form of an irregular oval were found. The curvature and depth of the central platform is different. The depth of the depression of normocytes was 0.15 ± 0.06 nm. When studying the ratio of the diameter of the erythrocyte to the diameter of the cavity, the value was 11 +- 2 units. These types of altered erythrocytes cannot be restored and are irreversibly deformed. They cannot take part in the process of oxygen transportation and are destroyed in the microvasculature.

The sequestration of foreign erythrocytes makes it difficult for the erythron to self-repair.[6] The actual solution to this problem is the application of the blood reinfusion technique using the Cell Saver apparatus. Comparative analysis of the effectiveness of reinfusion in comparison with blood transfusion was carried out using the example of typical cases.

3. Results and discussion

Blood transfusion for blood loss of 2700 ml. Blood counts before transfusion: Hb = 54g/l (liters), erythrocytes = $1.65 * 10^{12}$/l, Ht = 14.9%, leukocytes – $16.0 * 10^9$/l, thrombocytes = $70.6 * 10^9$/l. Transfusion: ffp = 1780ml, erythrocyte suspension = 870ml.

Blood parameters after transfusion: Hb = 84g/l, erythrocytes = $2.7 * 10^{12}$/l, Ht = 23.6%, leukocytes = $14.7 * 10^9$/l. Blood parameters values for different groups are presented in table 1.

Blood transfusion with blood loss 4000 ml. Blood parameters before transfusion: Hb = 39g/l, erythrocytes = $1.3 * 10^{12}$/l, Ht = 11.5%. Transfusion: ffp = 2980 ml, erythrocyte suspension = 1825 ml, reinfusion of washed erythrocytes using the Haemonetics Cell Saver-5 apparatus was 982ml.

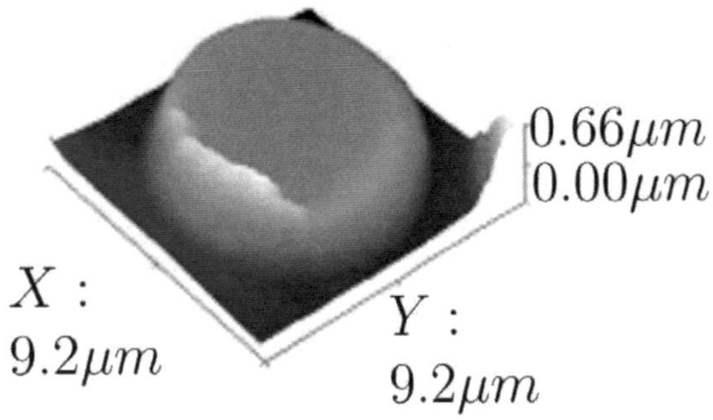

Figure 1. Normal erythrocyte size

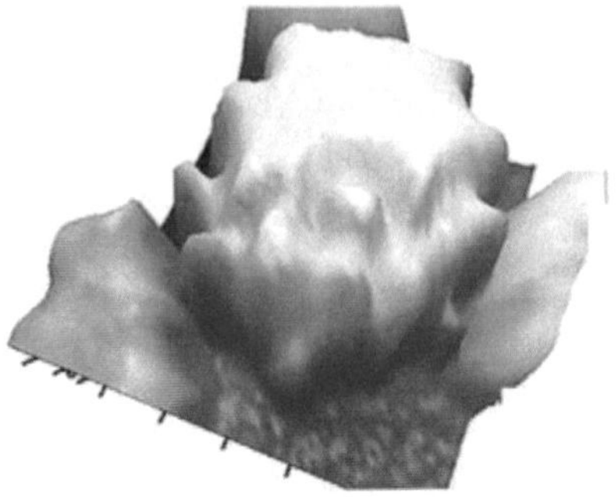

Figure 2. Canned erythrocyte (25-30 days of storage)

Table 1. Dynamics of blood parameters during reinfusion.

Group	Before transfusion	Immediately after transfusion	18 hours after transfusion	24 hours after transfusion	34 hours after transfusion
Hemoglobin g/l	54	84	80	58	50
Erythrocytes 10^{12} /l	1.65	2.72	2.5	1.8	1.62
Leukocytes10^{9} /l	14.9	14.7	18.1	11.1	11
Thrombocytes10^{9} /l	70.6	90	86	62	63

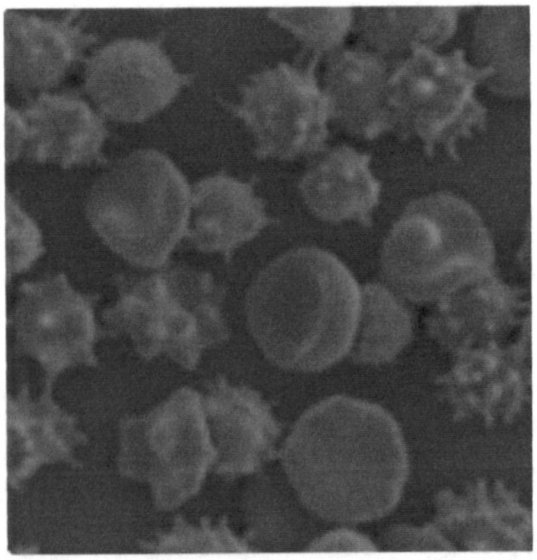

Figure 3. Donated erythrocytes before storage

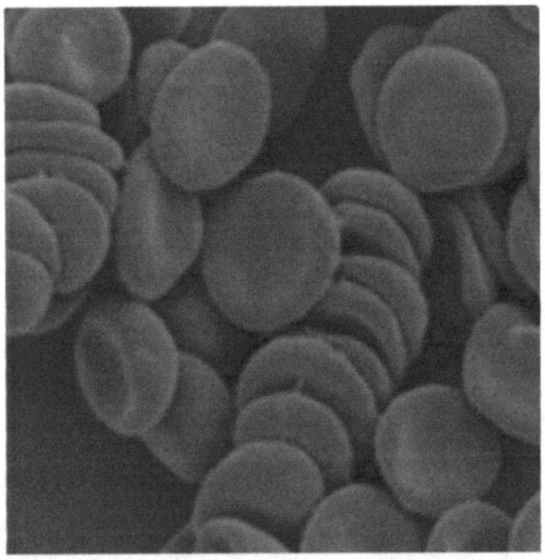

Figure 4. Donor erythrocytes after 42 days of storage

Table 2. Dynamics of blood parameters during reinfusion.

Group	Before reinfusion	Reinfusion	7 hours after reinfusion	10 hours after reinfusion	24 hours after reinfusion	34 hours after reinfusion
Hemoglobin g/l	39	186	95	84	74	71
Erythrocytes 10^{12} /l	1.3	5.2	3.2	2.8	2.53	2.18
Leukocytes 10^{9} /l	20.1	11.5	11.5	9.6	4.9	-
Thrombocytes 10^{9} /l	258	154	159	103	100	-

Blood counts after transfusion: Hb = 186g/l, erythrocytes = $5.2 * 10^{12}$/l, Ht = 55.5%, leukocytes = $20.1 * 10^{9}$/l, thrombocytes = $258 * 10^{9}$/l. Blood parameters values for different groups are presented in table 2.Reinfusion is characterized by a decrease in hemoglobin (by 9g/l, 10g/l and 3g/l), smooth and slow, in contrast to the methods of introducing donor erythrocyte mass (by 22g/l, 8g/l). 72 hours after the transfusion of blood products, a significant decrease in hemoglobin, erythrocytes, leukocytes and thrombocytes was noted. However, with reinfusion, the decrease was smoother and did not reach the initial low values.

Reinfusion of donor erythrocytes is cost-effective as well. So 1 dose of fresh frozen plasma costs 18,752 rubles, 1 dose of erythromass — 15,337 rubles, 1 dose of autoerythrocytes corresponds to 2-3 doses of donor hermass.

4. Conclusion

Transfusion of autoerythra suspension, harvested using the Cell Saver apparatus, stabilizes red blood counts in the early post-transfusion period and reduces the manifestation of massive blood transfusion syndrome. Hardware reinfusion of erythrocytes is effective and safe for massive blood loss in obstetrics.Reinfusion dictates the need for parallel correction of all blood parameters. Application of the principles of patient blood management can reduce the transfusion load, improve the quality of medical care.

Acknowledgments

The work was performed on the equipment Regional center of probe microscopy collective use Ryazan State Radio Engineering University named after V. F. Utkin.

References

[1] Yang Yu et al 2013 *Exp. and therapeutic medicine* **5** 1466-70
[2] De Oliveira RRL, Albuquerque DAC, Cruz TGS, Yamaji FM and Leite FL 2012 *Atomic force microscopy-imaging measuring and manipulating surfaces at the atomic scale* **3**
[3] Li J, Liu Y, Yuan Y and Huang B 2021 *Frontiers of Medicine* **15** 43-52
[4] Ross EH and Dickinson A 2021 *J. of the American Animal Hospital Association* **57** 199-204
[5] Pape EH 2018 *Structure and Properties of Cell Membranes* 167-93 (CRC Press)
[6] Weiler S and Nairz M 2021 *Frontiers in Oncology* **11** 831

Using of «bubble sensors» to control the quality of sequencing by the Illumina / Solexa method

V S Reznik[1, 2], V A Kruglov[1, 2] and V V Davydov[2, 3]

[1] Institute of Analytical Instrumentation of the Russian Academy of Sciences, St. Petersburg,190103, Russia
[2] Peter the Great Saint Petersburg Polytechnic University, Saint Petersburg, 195251, Russia
[3] All Russian Research Institute of Phytopathology, Moscow Region 143050, Russia

vlreznik97@gmail.com

Abstract. In the modern world, sequencing is an integral part of medicine, biology and other scientific fields. The Illimina / Solexa method is a new generation method and relates to methods of mass parallel sequencing. One of the features of using this method is the sequential pumping of various chemicals through the flow cell in which the reaction occurs. For uniformity and high quality of DNA sequencing, it is necessary that the amount of gas in liquids be minimized. Because many it can adversely affect both during chemical reactions and at the stage of recording reaction results. This article will examine the sequencing system using the Illumina\Solexa method using bubble sensors. An algorithm was developed that periodically receives information from bubble sensors in a microfluidic tube. The information received is processed and allows at certain stages to report deviations from the normal conditions for sequencing. The experimental results are presented.

1. Introduction

Deterioration of the ecological situation [1-5], man-made disasters that lead to mutations [6-9], required the development of various methods for research. Currently, sequencing of the sequence of nucleotides that enter DNA and RNA has become a familiar and necessary phenomenon in various fields [10-14].

This method has several advantages in terms of informativeness of research in comparison with other methods, for example, optical spectroscopy [15 - 20] or nuclear magnetic spectroscopy [21-27]. The nuclear magnetic spectroscopy has a higher measurement accuracy compared to other methods [26-33]. But the information contained in the NMR spectrum is not enough for a genetic analysis with a unambiguous result. The using of the various methods for optical signal processing techniques [15-20, 34-37] in the optical spectroscopy does not allow to obtain a complete DNA chain. Therefore, DNA sequencing methods remain the most reliable.

One of the widely used methods for carrying out this process is the use of the Illumina / Solexa method. The result of sequencing is data on the sequence of nucleotides that make up the sample. One of the important features of the implementation of this sequencing method is the use of an optical system during detection.

Illumina / Solexa sequencing uses a large number of different chemicals. For ease of implementation, these reagents are placed in test tubes, and then through the flow channels enter the reaction cell.

In this connection, the problem of getting into the system of bubbles of gas dissolved in reagents arises. In addition, air can enter the system at the junction of the flow channel with the reaction cell.

The ingress of gas bubbles into the reaction cell at the stage of carrying out amplification can lead to a decrease in the number of clusters, which in turn will negatively affect the quality of further analysis.

On the other hand, the ingress of gas bubbles at the detection stage leads to a decrease in the total number of clusters in the field of view of the detection system. The used of other industrial devices for determining bubbles is excluded [33-35].

To solve this problem, it was proposed to use the so-called “bubble sensor”, which allows counting the number of gas bubbles passing through the reaction cell at all stages of sequencing.

2. Working process of bubble counting system

To count the number of bubbles, it is advisable to use an optical sensor as a sensitive element. This fact is justified primarily by the fact that flow tubes, due to their transparency, transmit light. The radiation intensity will be highly dependent on the medium. Accordingly, if a gas bubble passes through the flow channels, the value of the radiation intensity will change its value.

Therefore, it is necessary to use a combination of a radiation source and a photodetector. The obvious solution in this case will be the use of optocouplers.

However, this measuring circuit will be sufficiently effective only if the gas bubble enters the field of view of the optocoupler. In the event that the bubble is smaller than the diameter of the tube or moves along the walls of the tube, the value of the radiation intensity recorded by the photodetector may not be high enough to determine the passage of the bubble.

To solve this problem, it was decided to use two optocouplers located perpendicular to each other. The block diagram of the total operation of the bubble detection system is shown in Figure 1.

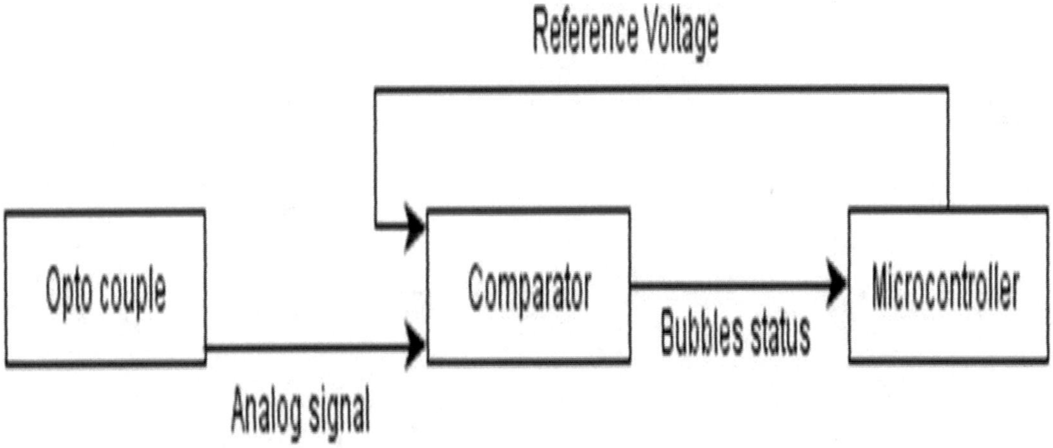

Figure 1. Block-diagram of signals in using of «bubble sensors».

A feature of the work is the use of a comparator. After the bubble passes through one of the optocouplers, the level of radiation received by the photodetector changes. The radiation level of the photodetector is converted into an electrical signal.

The converted electrical signal is fed to the input of the comparator, the second input of which receives the reference signal. The level of this signal corresponds to the minimum level of radiation arising from the passage of a bubble. This signal is generated at the output of the digital-to-analog microcontroller converter. As a result of this, the level of this signal can be adjusted to different types of reagents and bubble sizes and is usually determined experimentally.

If the signal level from the optocoupler exceeds the value of the reference signal, a voltage is generated at the output of the comparator, which corresponds to the level of a logical unit, which in turn is interpreted by the microcontroller as the presence of a bubble.

Similarly, the processing of signals from the second optocoupler occurs. It is also important to note that optocouplers are located behind the reaction cell, since this allows one to take into account the possible ingress of gas bubbles at the junction of the reaction cell with the flow channels.

The converted electrical signal is fed to the input of the comparator, the second input of which receives the reference signal. The level of this signal corresponds to the minimum level of radiation arising from the passage of a bubble. This signal is generated at the output of the digital-to-analog microcontroller converter. As a result of this, the level of this signal can be adjusted to different types of reagents and bubble sizes and is usually determined experimentally.

If the signal level from the optocoupler exceeds the value of the reference signal, a voltage is generated at the output of the comparator, which corresponds to the level of a logical unit, which in turn is interpreted by the microcontroller as the presence of a bubble. Similarly, the processing of signals from the second optocoupler occurs.

3. Analysis data from «bubble sensors»

During sequencing, the control program sends requests to the microcontroller, to which the sensors are connected, requesting the number of registered gas bubbles every 5 seconds (see Figure 2).
Further analysis is carried out in two ways:

- The number of bubbles for a period of 5 seconds should not exceed 500
- The number of bubbles for the entire duration of the sequencing should not exceed 25,000

If the number of bubbles is exceeded in one of the two above parameters, the general sequencing algorithm proceeds to interrupt the system.

These threshold values of the number of bubbles were obtained experimentally and are justified by a significant deterioration in the sequencing results if these values are exceeded.

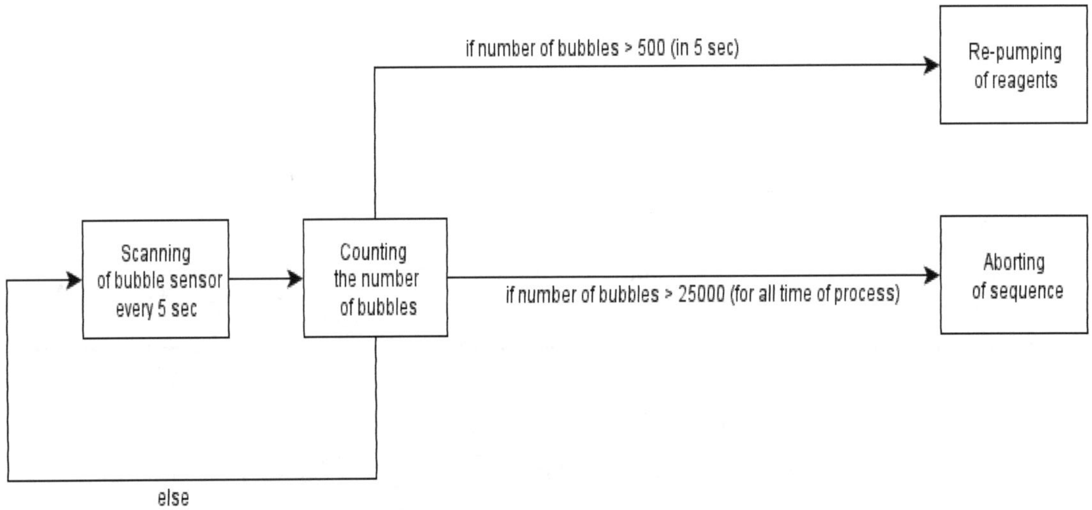

Figure 2. Block diagram of the main algorithm with using «bubble sensors».

4. Conclusion

Thus, the developed system for analyzing the number of gas bubbles passing through the reaction cell during mass parallel sequencing using the Illumina / Solexa method can significantly improve the quality of the output data of the analysis.

References

[1] Davydov R, Antonov V and Moroz A 2019 *International Conference on Electrical Engineering and Photonics EExPolytech-2019 (Saint-Petersburg)*, vol. 8906791 (IEEE), p. 295-297

[2] Davydov R, Antonov V, Makeev S, Batov Y, Dudkin V and Myazin N 2020 *E3S Web Conference* **140** 02001

[3] Nikitina M, Grebenikova N, Dudkin V and Batov Y 2019 *IOP Conference Series: Earth and Environmental Science* **390(1)** 012024

[4] Yushkova V, Kostin G, Davydov R, Dudkin V and Valiullin L 2019 *IOP Conference Series: Earth and Environmental Science* **390(1)** 012016

[5] Van S, Cheremisin A, Chusov A, Switala F and Davydov R 2019 *IOP Conference Series: Earth and Environmental Science* **390(1)** 012011

[6] Van S, Cheremisin A, Glinushkin A, Davydov R and Yushkova V 2019 *E3S Web of Conferences* **140** 09008

[7] Gryznova E, Batov Y and Myazin N 2020 *E3S Web of Conferences* **140** 09001

[8] Gryznova E, Grebenikova N, D. Ivanov D and Bykov V 2019 *IOP Conference Series: Earth and Environmental Science* **390(1)** 012044

[9] Davydov R V, Mazing M S, Yushkova V V, Stirmanov A V and Rud V Yu 2019 *Journal of Physics: Conference Series* **1410(1)** 012067

[10] Belov Yu V, Petrov A I, Ivanov V V and Kurochkin V E 2011 *Scientific instrumentation* **21(1)** 44-49

[11] Alekseev Ya I, Belov Yu V, Malyuchenko O P, Monakhova Yu A, Natyrov A N, Orekhov V A, Konovalov S V, Kurochkin V E and Petrov A I 2012 *Scientific instrumentation* **22(4)** 86-92

[12] Antonova O S, Rudnidskya G E, Tupik A N, Bylianitsa A L and Evstrapov V E 2011 *Scientific instrumentation* **21(4)** 5-21

[13] Reznik V S, Kruglov V A, Petrov A I, Glinuchkin A P and Rud V Y 2019 *Journal of Physics: Conference Series* **1410(1)** 012078

[14] Davydov V V, Reznik V S, Kruglov V A and Rud V Yu 2019 *International Conference on Electrical Engineering and Photonics EExPolytech-2019 (Saint-Petersburg)*, vol. 8906872 (IEEE), p. 69-71

[15] Logunov S E, Vysoczky M G and Mazing M S 2018 *Journal of Physics: Conference Series* **1038(1)** 012093

[16] Logunov S E, Davydov V V, Glinuchkin A P, Podkovyrov I Yu and Rud V Yu 2019 *Journal of Physics: Conference Series* **1410(1)** 012113

[17] Logunov S E and Rud V Yu 2021 *Journal of Physics: Conference Series* **1745(1)** 012007

[18] Logunov S E, Vysocky M G, Koshkin A Yu and Rud V Yu 2017 *Journal of Physics: Conference Series* **917(5)** 052058

[19] Grevtseva A S, Smirnov K J, Greshnevikov K V, Rud V Yu and Glinushkin A P 2019 *Journal of Physics: Conference Series* **1368(2)** 022072

[20] Grevtseva A S and Smirnov K J 2018 *Journal of Physics: Conference Series* **1135(1)** 012056

[21] Neronov Yu I 2017 *Measurement Techniques* **60(1)** 96-102

[22] Myazin N S, Yushkova V V and Taranda N I 2019 *Journal of Physics: Conference Series* **1410(1)** 012130

[23] Rukin E V , Myazin N S and Dudkin V I 2019 *Journal of Physics: Conference Series* **1368(4)** 042011

[24] Myazin N S, Yushkova V V and Dudkin V I 2019 *Journal of Physics: Conference Series* **1400(6)** 066008

[25] Neronov Y I 2021 *Biomedical Engineering* **54(5)** 333–336

[26] Rutberg F G, Kolikov V A, Snetov V N, Cherepkova I A and Dyachenko S V 2016 *High Temperature* **54(2)** 170–174

[27] Myazin N S, and Yushkova V V 2019 *Environmental Research, Engineering and Management* **75(2)** 28–35

[28] Gizatullin B, Gafurov M, Vakhin A, Mattea C and Stapf S 2019 *Energy and Fuels* **33(11)** 10923–10932
[29] Myazin N, Neronov Y, Dudkin V and Petrov A 2019 *IOP Conference Series: Materials Science and Engineering* **497(1)** 012111
[30] D'yachenko S V, Kondrashkova I S and Zhernovoi A I 2017 *Technical Physics* **62(1)** 1602-1604
[31] Davydov V V, Dudkin V I and Petrov A A 2016 *Technical Physics Letters* **42(7)** 692-696
[32] Davydov R V and Antonov V I 2018 *Journal of Physics: Conference Series* **1124(8)** 081037
[33] Davydov R V and Antonov V I 2018 *Journal of Physics: Conference Series* **1135(1)** 012087
[34] Kozlova E S and Kotlyar V V 2014 *Computer Optics* **38(1)** 132-138
[35] Kuzmin M S and Rogov S A 2019 *Computer Optics* **43(3)** 391-396
[36] Smirnov K J and Batov Y V 2019 *Journal of Physics: Conference Series* **1368(2)** 022073
[37] Grebenikova N M and Davydov R V 2019 *Journal of Physics: Conference Series* **1326(1)** 012012

Femtosecond laser synthesis and comparative analysis of fluorescent carbon dots from L-lysine aqueous solution

A A Astafiev, A M Shakhov, A A Gulin, A A Vasin, M V Gubina, M S Syrchina and V A Nadtochenko

N.N. Semenov Federal Research Center of Chemical Physics, Russian Academy of Sciences, Kosygina 4, Moscow, Russia

Abstract. Laser synthesis of fluorescent species from biomolecules in living cells and tissues offers unique capabilities for fluorescent bioimaging, yet little is known about its mechanisms and characteristics of products. We examine synthesis of fluorescent products from water solution of L-lysine upon irradiation by trains of femtosecond laser pulses with varying parameters. We demonstrate that irradiation products contain nanoscale carbon-based fluorescent particles (carbon dots) with multi-colour and excitation-dependent emission. Morphology, chemical composition and fluorescent characteristics of irradiation products strongly depend on laser pulses parameters.

1. Introduction

Production of fluorescent species in living cells and tissues by femtosecond laser irradiation is a novel approach to fluorescent bioimaging which offers a unique advantage of non-invasive, rapid, highly-localized and targeted fluorescent labelling [1-2]. Exact mechanisms of laser synthesis and chemical nature of fluorescent products remain debatable [2-3] and their analysis is complicated by heterogeneity of cells and tissues and difficulty of isolating irradiation products. A possible role of synthesis of fluorescent carbon dots (CDs) in laser production of fluorescent species was highlighted recently [2,4].

In order to clarify synthesis mechanisms and characteristics of products we examined femtosecond laser production of fluorescent species from a model biological molecule. A common proteinogenic amino acid L-lysine, which is also highly effective as precursor for hydrothermal or microwave synthesis of CDs, was chosen as a model compound. In order to understand the role of various physicochemical effects of nonlinear absorption and the influence of laser parameters the water solution of l-lysine was irradiated with femtosecond laser pulses in three different regimes: one corresponding to laser breakdown in aqueous media and two sub-breakdown regimes with different laser wavelength, and comparative analysis of irradiation products was performed suing a combination of optical spectroscopy, energy-dispersive X-ray spectroscopy, TOF-SIMS mass spectroscopy, and FTIR spectroscopy.

2. Materials and methods

2.1. Laser synthesis and isolation of products.

Lysine solution of 500 mg/mL in water (Sigma, HPLC) was prepared from L-lysine powder (Sigma). 100 μL of the solution was placed in 100 μL quartz cuvette for irradiation by femtosecond pulses at 80 MHz repetition rate and 2 mL was placed into 10 mL serum vial for irradiation at 1 kHz repetition rate. Lysine solutions were irradiated in 3 regimes: 1) femtosecond laser pulses at 1 kHz repetition rate, wavelength of 800 nm, 1.4 mJ pulse energy, 50 fs pulse duration and 20 h irradiation time; 2) femtosecond laser pulses at 80 MHz repetition rate, wavelength of 740 nm, 1.8 W average optical power, 50 fs pulse duration and 70 h irradiation time; 3) femtosecond laser pulses at 80 MHz repetition rate, wavelength of 370 nm (SHG), 0.3 W average optical power, 100 fs pulse duration and 70 h irradiation time, resulting in samples referred as **1-3**. After irradiation samples were dialyzed for 72 hours in 2,000 MWCO dialysis units in order to remove unreacted L-lysine.

2.2. Samples characterization
UV-VIS absorption and photoluminescence spectra of aqueous solutions of **1-3** in a 3.8-ml quartz cuvette were recorded with Shimadzu UV-3600 spectrophotometer and Shimadzu RF-5031PC spectrofluorometer. Fluorescence quantum yields of **1-3** solutions in ethanol were estimated by the slope method using excitation at 356 nm wavelength and anthracene solution in ethanol (Φ-27%) as a fluorescence standard. For measurement of fluorescence anisotropy decay of aqueous solutions of **1-3** we recorded two fluorescence decay curves $I_{\|}(t)$ and $I_{\perp}(t)$ for emission with polarization parallel and perpendicular to the excitation laser. Fluorescence was excited with frequency-doubled femtosecond laser pulses (355 nm central wavelength, 100 fs pulse duration). Decay curves were recorded at 450 nm emission wavelength with a Beckr-Hickl SPC-150N time-correlated single photon counting system. The fluorescence anisotropy was calculated by the formula $r(t)=(I_{\|}-\alpha(\lambda)I_{\perp})/(I_{\|}+2\alpha(\lambda)I_{\perp})$, where α is the empirical correction factor. The average hydrodynamic volume of luminescent particles was estimated by the formula: $V = k_B T \tau_{rot} / \eta$, where τ_{rot} is the characteristic anisotropy decay time, k_B is the Boltzmann constant, T is the temperature, and η is the dynamic viscosity of the water.

Samples diluted in ethanol and dried on a cover slip were analyzed with an atomic-force microscopy unit (SMENA-B, NT MDT) in an intermittent contact mode. FTIR infrared spectra were recorded using the Lumos I FTIR microscope-spectrometer in attenuated total reflection (ATR) mode from a drop of water solution dried on a surface of a CaF_2 window. Elemental analysis with energy-dispersive X-ray spectroscopy was performed using a Prisma-E electron microscope (Thermo-Fisher), using L-lysine with a known C:N:O ratio (3:1:1) as a calibration sample. Mass spectroscopy analysis of dried samples on borosilicate cover slip in positive and negative ions was performed with TOF.SIMS 5 mass spectrometer (ION-TOF). Mass spectra of l-lysine film on a cover slip were used as a reference.

3. Results and discussion
Exposure to femtosecond laser pulses resulted in yellowing of originally colorless lysine solution and appearance of appreciable visible luminescence in all irradiation regimes. Analysis of UV-Vis absorption spectra demonstrated a large difference between samples. **1** absorbed mostly in the UV region with a conspicuous shoulder at 330 nm and a weak absorption tail above 400 nm, **2** had appreciable absorption only in the deep UV region, **3** had a strong absorption peak at 315 nm, but very small visible absorption (Figure **1**). Absorption features in the near-UV and visible range demonstrates effective formation of chromophores with conjugated systems in **1** and **3**, and the kilohertz irradiation regime was especially effective for producing chromophores absorbing in the visible range.

All three samples exhibited multicomponent near-UV and visible photoluminescence with at least two separate emission peaks: a near UV peak with an emission maximum at ca. 380 nm and excitation maximum at 310 nm, and a blue peak with emission maximum at 420-430 nm and excitation maximum at ca. 350 nm. The second peak was excitation-dependent: its maximum shifted to a longer wavelength with increase of the excitation wavelength parallel to the attenuation of emission intensity.

Such excitation dependence is typical for photoluminescence of carbon nanodots [5] and is interpreted as a result of multicomponent character of their emission [6, 7]. Importantly, the proportion between emission components was different, and **1** had the strongest visible luminescence component, whereas **3** – the strongest near-UV component.

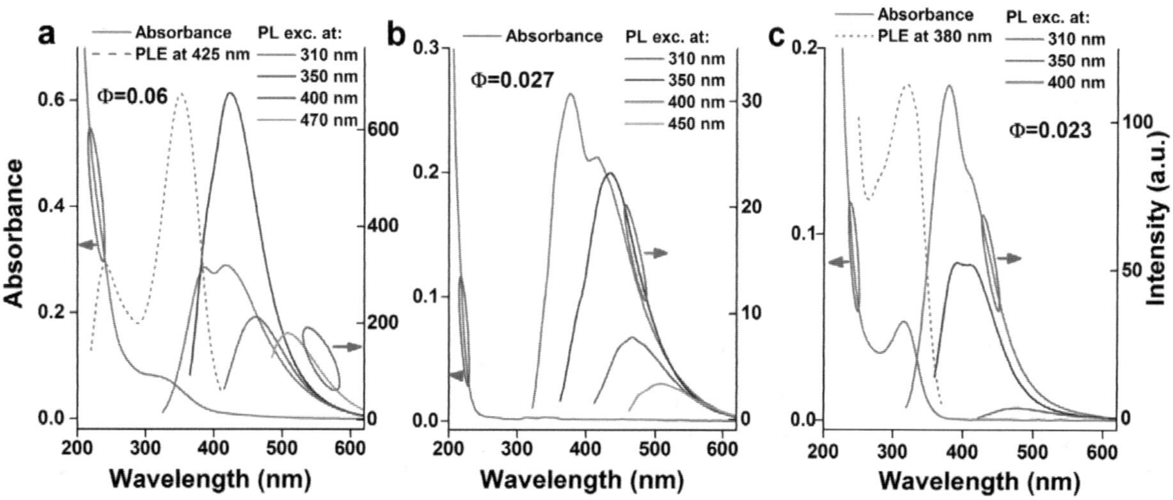

Figure 1. Absorbance, photoluminescence emission (PL) and photoluminescence excitation (PLE) spectra of aqueous solutions of **1** (a), **2** (b) and **3** (c).

Additionally, under 470 nm **1** excitation exhibited a distinct green emission peaks with emission maximum at ca. 510 nm. Estimated quantum yield of the blue luminescence component was small for all the three samples, the highest yield of 6% belonged to **1**. In summary, femtosecond laser irradiation produced several fluorescent moieties from l-lysine, their composition and relative content was strongly dependent on the irradiation regime, and kilohertz irradiation was the most effective for producing moieties with visible luminescence.

In order to remove unreacted l-lysine for further chemical analysis the samples **1**-**3** were subjected to dialysis, which also removed irradiation products with a molecular mass smaller than appr. 2 kDa. This removal resulted in attenuation of absorption and luminescence intensity which was the strongest for **3** and the weakest for **1** (Figure 2a). This indicates that absorption and photoluminescence was only partly attributable to products with large molecular mass (nanoparticles, oligomers, aggregates etc), while a large part of luminescence belonged to smaller molecular weight fragments which were removed by dialysis. From the attenuation ratio it can be concluded that the kilohertz regime (**1**) produced products with the largest mass, whereas the second harmonic irradiation (**3**) – with the smallest. Absorption and photoluminescence spectra of **1** were virtually unchanged by dialysis, whereas for the **2** and **3** these changes were strong and their spectra became nearly identical (Figure 2b). UV-Vis spectra of the dialyzed **2** and **3** were similar to **1** with a characteristic absorption shoulder at ca. 320 nm and a tail above 400 nm. The relative strength of the near-UV luminescence peak at 380 nm also strongly decreased. Thus, UV-emitting fluorophores in **2**-**3** tended to be associated with relatively small molecules or aggregates, whereas fluorophores with blue and green emission - with heavy weight products. We further elucidated characteristics of dialyzed samples **1**-**3** using measurements of emission anisotropy kinetics in ethanol solution under pulsed laser excitation at 356 nm (Figure 2c). Anisotropy decay from an initial value of approximately 0.3 on a nanosecond and sub-nanosecond time scale is attributed to rotational diffusion of fluorescent moieties, which changed orientation of the emission dipole with time and depolarized emission. Characteristic rotation time calculated from the multiexponential peak yields effective hydrodynamic volumes of **1**-**3** as 10.6, 0.7 and 1.4 nm^3 respectively. The last two values are unexpectedly small, since particles with volumes smaller than appr. 2.5 nm^3 can pass through pores of the dialysis membrane and are mostly filtered out during the dialysis.

We explain anomalously fast rotational diffusion with a flexible structure of heavy-weight products in **2**-**3**, e.g. flexible polymer chains, where segments bearing fluorophore can change orientation independently of other segments, leading to fast depolarization (Figure 2d). By a contrast, **3** has fluorophores are bound to rigid nanoparticles and can change orientation only with rotation of the nanoparticle as a whole which occurs on the scale of nanoseconds.

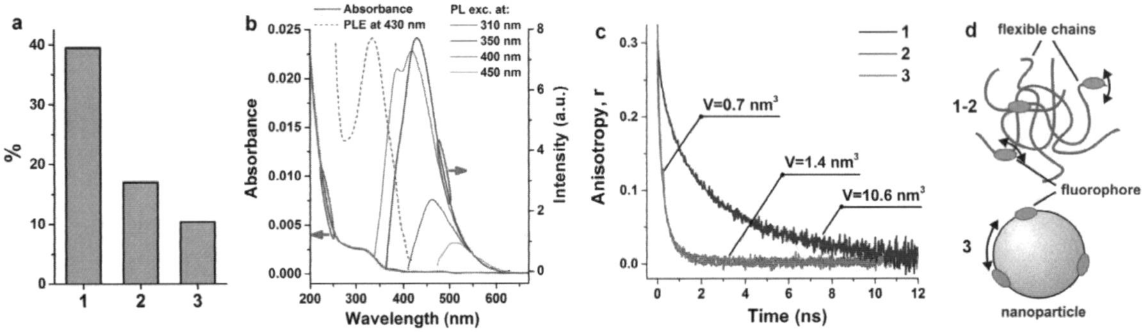

Figure 2. (**a**) Percentage of the original PL intensity remaining after dialysis for 1-3, excitation at 350 nm. (**b**) Absorbance, PLE and PL spectra of the dialyzed samples 2-3. (**c**) Emission anisotropy decay kinetics of **1-3** in aqueous solution. Excitation with femtosecond laser pulses at 356 nm. (**d**) Proposed structure of heavy-weight luminescent products in **1-3**.

According to the EDS measurements (Table 1) the content of carbon in irradiation products was larger than in l-lysine, especially in **1**, which indicates an onset of laser-driven carbonization. The O:N ratio was larger than 1:1 in l-lysine. Additional oxygen apparently resulted from oxidation of products during the irradiation in presence of air dissolved in aqueous solution.

Table 1. Elemental composition of **1-3** compared with L-lysine.

Sample	C (at. %)	N (at. %)	O (at. %)
L-lysine	60	20	20
1	67.6	14.5	17.9
2	62.1	16.4	21.5
3	61.1	17.2	22.4

FTIR spectra demonstrate that chemical composition of **2-3** was similar and at the same time much different from **1** (Figure 3a). Spectrum of **1** included typical amide peaks (amide I peak at 1630 cm^{-1}, amide II peak of secondary amides at 1560 cm^{-1}, amide III peak of primary amides at 1390 cm^{-1}), which indicates that formation of heavy weight products under kilohertz irradiation proceeded via polymerization of l-lysine. Curiously, presence of primary amides indicates that the resulting poly-lysine nanoparticles underwent reorganization, typical for carbon dots formation in polar solvent, which transferred amide groups from the interior to the nanoparticle surface [8]. CH_2 stretching and bending peaks at 2930, 2860 and 1450 cm^{-1} show presence of alkyl chains derived from l-lysine, meaning that carbonization of nanoparticles was incomplete. Broad OH and CO stretching peaks at 3400 and 1070 cm^{-1} indicate that oxidation of nanoparticles resulted in formation of surface hydroxyl groups. **2** and **3** exhibited strong asymmetric and symmetric bending vibrations of NH_3^+ at 1605 and 1510 cm^{-1}, asymmetric and symmetric stretching vibrations of COO^- at 1580 and 1405 cm^{-1}, a broad NH_3^+ stretching peak at 3000 cm^{-1} and COO^- bending at 620 cm^{-1}, which are characteristic of the zwitterionic form of l-lysine. They indicate that the size of polymerized fragments was much smaller than in 1, at best short olygomer chains were formed leaving many unreacted COO^- groups. At the same time strong C-O-C asymmetric stretching peak at 1080 cm^{-1} and a weaker symmetric stretching at 860 cm^{-1} reveal presence of ether groups. Thus, lysine monomers or olygomers in **2-3** are connected by ether bonds.

Mass-spectra measurements confirm polymerization of L-lysine under femtosecond laser exposure: the relative intensity of lysine monomer ion at m/z=145 decreased for **2** and practically disappeared for **1** parallel to decrease of NH^- and COO^- signals, while stronger signal of CN^- and CNO^- attributable to peptide bonds appeared (Figure 3b). **1** yielded especially strong peptide signals, confirming large scale of polymerization.

Thus, chemical analysis demonstrates difference in structure and formation mechanism of macromolecular products for kilohertz and megahertz laser irradiation regimes (Figure 3c). Kilohertz irradiation with high-energy pulses promotes polymerization and formation polymer clusters. Cross-linking of polymer chains gives these clusters a rigid structure.

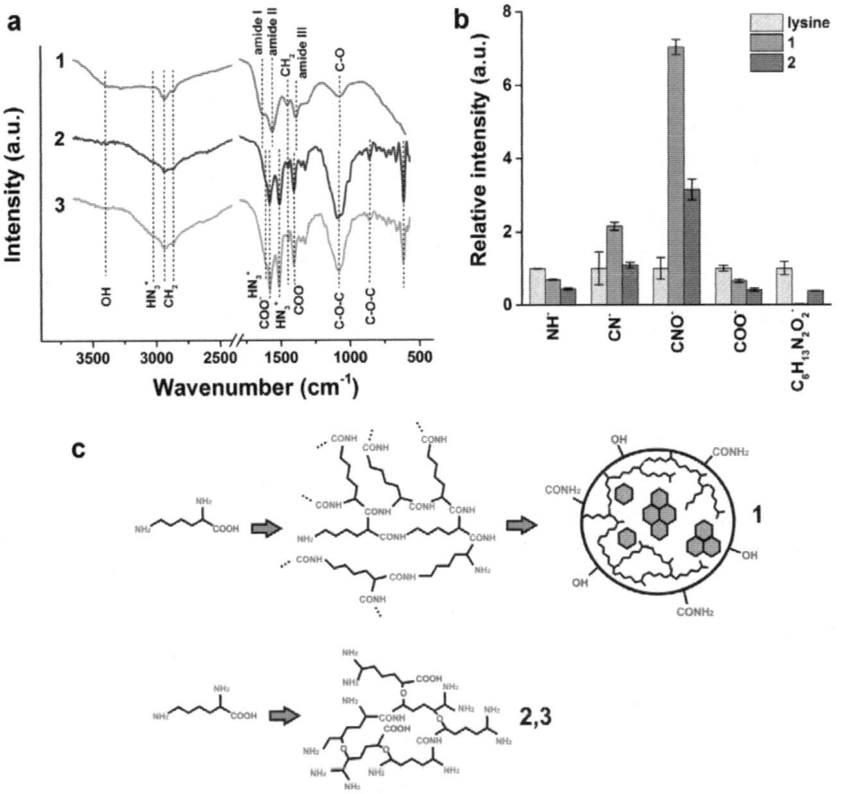

Figure 3. (**a**) ATR FTIR spectra of **1-3**, (**b**) Relative intensity of TOF-SIMS negative ions signals with m/z=15 (NH^-), 26 (CN^-), 42 (CNO^-), 44 (COO^-) 145 ($C_6H_{13}N_2O_2^-$) for l-lysine, **1** and **2**. (**c**) Proposed structures of L-lysine heavy-weight irradiation products in **1-3**.

Partial carbonization produces partly carbonized or even aromatic domains in nanoparticles, while oxidation and reorganization of amide groups forms a shell of polar amide and hydroxyl surface groups. This process is similar to synthesis of carbon nanodots by solvothermal or microwave methods from amino acids [9, 10], and resulting nanoparticles can be described as carbonized polymer dots [8]. Exposure to high repetition nanojoule femtosecond pulses produces flexible marcomolecules composed of lysine monomers or short oligomers connected by ether bonds. Curiously, despite this difference in structure both types of products exhibit excitation-dependent visible photoluminescence typical for carbon nanodots. It can be hypothesized that fluorophores are produced by partial carbonization and formation of conjugated structures from sections of the polymer/olygomer chains.

4. Conclusion

We demonstrated that under exposure to femtosecond laser pulses chemical products with visible absorption and photoluminescence are formed from l-lysine aqueous solution independently of the Maillard reaction. Molecular weight, structure and chemical composition and absorption and photoluminescence characteristics of products strongly depend on the irradiation regime. Exposure to high-energy pulses in the laser breakdown regime is especially effective for obtaining carbon nanodots-like luminescent products. Our results shed new light on possible mechanism of formation of luminescent products in material of living cells and tissues under pulsed laser irradiation.

Acknowledgments
The work has financial support from the Russian Scientific Foundation grant No. 21-72-20169. The measurements were performed in the FRCCP shared research facilities No. 506694 and large-scale research facilities No. 1440743.

References

[1] Sun Q, Qin Z, Wu W, Lin Y, Chen C, He S, Li X, Wu Z, Luo Y and Qu J 2018 *Biomed. Opt. Express* **9** 581-90
[2] Astafiev A A, Shakhov A M, Osychenko A A, Syrchina M S, Karmenyan A V, Tochilo U A and Nadtochenko V A 2020 *ACS OMEGA* **5** 21 12527-38
[3] Qin Z, Sun Q, Lin Y, He S, Li X, Chen C, Wu W, Luo Y and Qu J 2018 *Biomed. Opt. Express* **9** 3373-3390
[4] Astafiev A A, Shakhov A M, Kritchenkov A S, Khrustalev V N, Shepel D V, Nadtochenko V A and Tskhovrebov A. G 2021 *Dyes and Pigments* **188** 109176
[5] Zhu S, Song Y, Zhao X, Shao J, Zhang J, Yang B 2015 Nano Research **8** 355-381
[6] Van Dam B, Nie H, Ju B, Marino E., Paulusse J M J, Schall P, Li M, Dohnalova K 2017 Small **13** 1702098
[7] Sharma A, Gadly T, Gupta A, Ballal A, Ghosh S K, Kumbhakar 2016 J.Phys.Chem.Lett **7** 3695-3702
[8] Xia C, Zhu S, Feng T, Yang M, Yang B 2019 Advanced Science **6** 1901316
[9] Choi Y, Thongsai N, Chae A, Jo S, Kang E B, Paoprasert P, Park S Y, In I 2017 J. Ind. & Eng. Chem. **47** 329-335
[10] Sahiner N, Suner S S, Sahiner M, Silan C, 2019 J. Fluorescence **29** 1191-1200

Computational modeling of schedule-specific chemotherapy outcomes in mouse tumor model

I N Terterov[1], V A Chubenko[1], N A Knyazev[1,2], V V Klimenko[1,2], A A Bogdanov[1,2], V M Moiseyenko[1] and A A Bogdanov[1,2]

[1] Saint-Petersburg Clinical Scientific and Practical Center of Specialized Types of Medical Care (Oncological), Leningradskaya str., 68A., Pesochny, St. Petersburg 197758, Russia
[2] Research Institute of Ultra Pure Biologicals, Pudozhsakya str., 7, St. Petersburg 197110, Russia

Abstract. Despite the rapid development of new innovative strategies in cancer treatment like immunotherapy, chemotherapy still remains a common choice in many cases. Standard protocols of chemotherapeutic administration rely on a maximal tolerated dose paradigm, however there is growing evidence that this approach is not always optimal. Alternative scheduling, like metronomic - low dose continuous drug administration - were recently proved their efficacy. The space of available variants of drug administration protocols is prohibitively large to be explored empirically, and there is an urgent need of predictive mathematical models for rational chemotherapeutic scheduling design. In this work we tested the ability of the minimal pharmacokinetic-pharmacodynamics model to describe schedule-specific tumor volume time evolution in different mouse tumor models.

1. Introduction

In the standard maximal tolerated dose (MTD) protocol, chemotherapeutics are administered in high doses with the aim to affect the maximal number of cancer cells. However, in some cases this does not lead to tumor eradication and the drug-resistance emerges. Alternative approaches were recently proposed, like metronomic chemotherapy (MC), which consists in low dose continuous drug administration [1]. MC is characterized with lower side effects while demonstrating promising efficacy. Tumor growth deceleration during MC therapy is attributed to anti-angiogenic, immunostimulation and cytostatic effects, which all take a minor role compared to direct toxicity in high dose MTD [1,2]. This diversity of biological mechanisms involved complicates the rational design of the optimal chemotherapeutic administration protocols in MC.

To date several predictive computational models have been developed to assist schedule optimization of MC protocol, because the huge number of possible variants makes empirical search impractical. Many of these rely on the pharmacokinetic-pharmacodynamics (PK/PD) modeling approach, which computationally links drug concentration in the body with its therapeutic action [2,3]. Such models were successfully applied in some cases [2,3,4], however they often contain a large number of adjustable parameters, and uncertainties in their estimation limits the model performance. Previously we developed the PK/PD Minimal model that is defined by a minimal set of parameters, which is sufficient to capture schedule-specific outcomes [5]. In this work we test performance and transferability of the Minimal model on the data for mice bearing different tumor types.

2. Materials and Methods

In this work two different tumor models were used, namely Ehrlich ascites carcinoma model in SHR mice (2.5 106 cells per mouse) and the model of colon cancer - CT26 in Balb/C mice (0.5 106 cells per mouse). Gemcitabine in different scheduling was used, as it is a chemotherapeutic known to be effective in MC [5]. For each tumor model, mice were divided into four groups: Control, MTD (25 mg/kg i.p.), MC (0.5 mg/kg i.p daily, 50 days) and Comb (combined protocol: 25 mg/kg+0.5 mg/kg daily). All animal experiments were approved by the local animal ethics committee in accordance with institutional guidelines for the welfare of animals ("N.N. Petrov National Research Center of Oncology``, St.Petersburg, Russia).

General scheme and differential equations of the Minimal model are depicted in the bottom panel of figure 1. Unperturbed growth is defined by the linear model with three parameters (λ0, λ1 and Ψ), among which one is fixed (Ψ=20) and the remaining two together with the value of initial volume V0 were obtained by the model fit to mean tumor volume data in untreated group. Anti-tumor drug action in the Minimal model is coupled to the on-site effective drug concentration (Ceff(t)) and defined by direct cytotoxicity (k) and antiangiogenic effect (IC50). Parameter α in the second equation (figure 1) describes drug partitioning rate between plasma concentration (C(t)) and effective compartment concentration (Ceff(t)). These three parameters (k, IC50, α) together define drug action and are typically obtained by model fitting to tumor volume data in specific treatment groups.

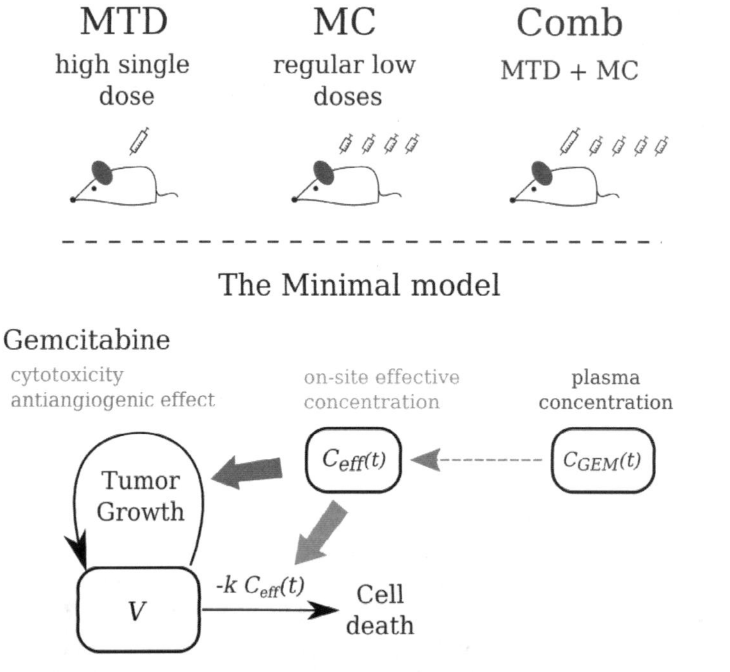

$$\frac{dV}{dt} = \frac{\lambda_0 \cdot V}{[1 + (\lambda_0/\lambda_1 \cdot V)^{\Psi}]^{(1/\Psi)}} \cdot \left(\frac{IC_{50}}{C_{eff}(t) + IC_{50}}\right) - k \cdot C_{eff}(t) \cdot V$$

$$\frac{dC_{eff}(t)}{dt} = -\alpha \cdot (C_{eff}(t) - C_{Gem}(t))$$

Figure 1. Treatment protocols applied in the experiments (top) and basic description of the Minimal model (bottom).

3. Results and discussion

Previously we have shown that the Minimal model is able to efficiently and simultaneously describe tumor response of MTD and MC gemcitabine based on Ehrich carcinoma mouse model data [5]. Here we systematically test and compare the model results on the data of two different mouse tumor models (Ehrich carcinoma and CT26) subjected to the same treatment protocols (see figure 1 top panel). It is known that plasma gemcitabine concentration follows a one compartment pharmacokinetic model, whose parameters were obtained by fit to experimental data from ref. [7]. For each tumor model firstly parameters of unperturbed growth were obtained by the weighted (1/y) non-linear least square fit of the model to the time evolution curve of mean volume (see table 1, λ_0 , λ_1 ,V_0). With these parameters for free tumor growth the model was either simulations fitted to the data from MTD and MC groups or fitted to data from Comb group to obtain parameter estimates (see table 1. k, IC_{50}, α). With parameters from each model fit tumor volume time evolution was simulated for all treatment protocols. Resulted time curves together with experimental data points are presented in figure 2 and 3.

For Ehrich carcinoma with parameters from simultaneous fit to MTD and MC data the model well reproduces experimental data trend of combined protocol (figure 2 B). Conversely, when fitted to Comb data the model successfully predicts trends for MTD and MC protocols (figure 2 C, D). These results demonstrate that the antitumor effect of gemcitabine is adequately described by the mathematical form of the Minimal model as we have shown in detail previously [5]. It should be noted that fit results presented here slightly differ from those published in [5] due to slight fitting protocol difference.

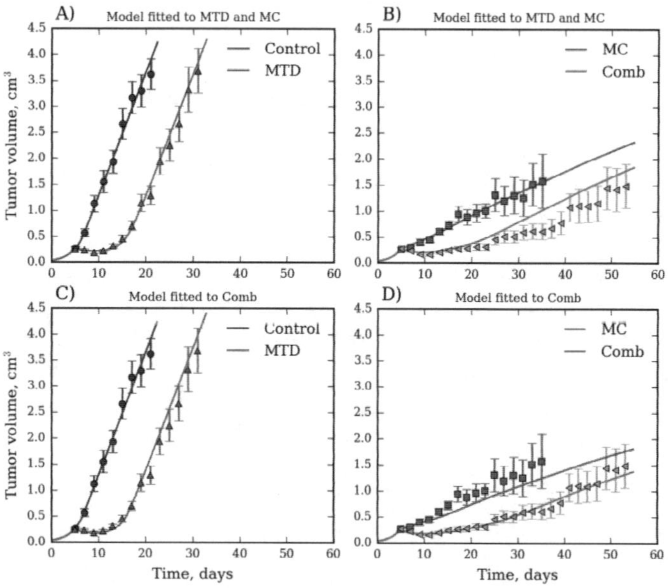

Figure 2. Ehrlich carcinoma data modeling. Lines - simulated curves. Circles - Control , triangles -MTD, squares - MC, diamonds - Comb.

Results of the analogous fitting procedure in case of CT26 tumor model are presented in figure 3. When parameters are obtained by the simultaneous model fit to MTD and MC data, the model adequately predicts trend for Comb protocol (figure 3 B), as in the previous case. Value of root mean square deviation (RMSE) for simulated Comb curve is even less than that for MTD and MC groups which were fitted in this case (see table 2). When the model is fitted to the Comb group data (figure 3 C, D), the predicted tumor volume curves go close to the experimental points of MTD and MC schedules. While it seems that the predicted curve for MTD scheduling becomes parallel to Control data too early, the curve lies within the uncertainty region and corresponding RMSE does not exceed the value for intrinsic experimental data variation (see table 2). The above observations for CT26 tumor (figure 3) are qualitatively the same as for Ehrich carcinoma (figure 2). Taken together these results indicate that the Minimal model is valid for description of outcomes of different gemcitabine scheduling in different tumor models.

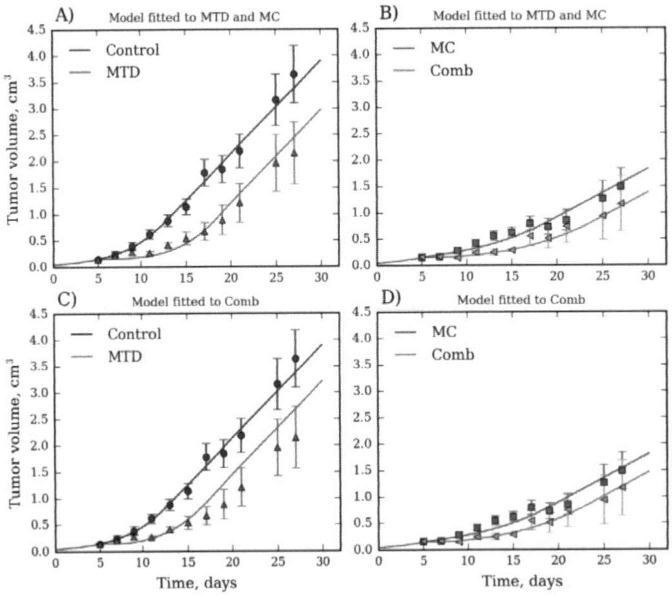

Figure 3. CT26 tumor data modeling. Lines - simulated curves. Circles - Control, triangles -MTD, squares - MC, diamonds - Comb.

Parameter values for description of unperturbed growth are accurately estimated (with low coefficient of variance CV, table 1) and differ for Ehrlich and CT26 tumor models, as expected. Model fit to Comb data or simultaneous fit to MTD and MC data resulted in slightly different but close parameter estimates in case of Ehrich carcinoma. In case of CT26 tumor estimate of parameter k tends to zero in simultaneous fit, while converges to non-zero value with large CV when model is fitted to Comb data (table 1). It seems that the cytotoxicity term ($-k\ C_{eff}(t)$) may be unimportant in this case.

Table 1. Parameter estimates.

	fit to Control			fit to MTD and MC		fit to Comb	
Param.	Values	CV(%)	Param.	Values	CV(%)	Values	CV(%)
			Ehrlich carcinoma				
$V_0\,[cm^3]$	0.036	27	$k\,[(day\,\mu M)^{-1}]$	0.28	18	0.43	11
$\lambda_0\,[1/day]$	0.40	11	$\alpha\,[1/day]$	0.48	4	0.58	30
$\lambda_1\,[cm^3/day]$	0.24	4	$IC_{50}\,[\mu M]$	0.0059	5	0.0048	7
			CT26 tumor model				
$V_0\,[cm^3]$	0.048	12	$k\,[(day\,\mu M)^{-1}]$	0.0	N/A	0.041	326
$\lambda_0\,[1/day]$	0.23	6	$\alpha\,[1/day]$	0.66	28	1.07	89
$\lambda_1\,[cm^3/day]$	0.18	6	$IC_{50}\,[\mu M]$	0.0231	16	0.0234	26

The Minimal model is able to describe tumor response to a different scheduling for both tumor models. Because the model prediction looks qualitatively similar for such a different experimental setting like Ehrlich carcinoma and CT26 tumor, we may conclude that the mathematical components of the model capture universal gemcitabine biological effect. Surprisingly, for both Ehrlich and CT26 tumors when the model is fitted to Comb data, predictions for other schedules are characterized by relatively low RMSE comparable with intrinsic data variance (table 2). Parameter estimates with which the model best describes the data differs in case of Ehrlich or CT26 tumor. This is not surprising because the tumor growth in Control groups differs in these two cases (cf. figures 2 and 3 and table 1). We expect that there may be a relationship between parameters for unperturbed tumor growth and parameters for anticancer drug action. The ability to estimate the treatment response based on the native tumor growth data with such a relationship will substantially increase transferability of the Minimal model results in preclinical settings and may be a promising tool for future personalized medicine. Further work with various tumor mouse models is needed to establish such interrelations.

Table 2. Goodness of fit estimation.

	RMSE [cm^3]			
	Control	MTD	MC	Comb
fit to MTD and MC (Ehrlich carcinoma)	143.4	117.7	74.2	240.0
fit to Comb (Ehrlich carcinoma)	143.4	186.6	209.5	86.5
Exp.[a] (Ehrlich carcinoma)	233.3	226.5	246.7	211.1
fit to MTD and MC (CT26 tumor model)	122.6	132.6	100.4	76.0
fit to Comb (CT26 tumor model)	122.6	276.3	101.0	58.0
Exp.[a] (CT26 tumor model)	272.6	285.0	174.6	246.7

[a] Experimental RMSE are estimated as a root of mean square standard deviations of mean tumour volumes

References

[1] Pasquier E, Kavallaris M and Andre N 2010 *Nat. Rev. Clin. Oncol.* **7** 455

[2] Benzekry S, Pasquier E, Barbolosi D, Lacarelle B, Barlesi F, Andre N and Ciccolini J 2015 *Semin. Cancer Biol.* **35** 53-61

[3] Barbolosi D, Ciccolini J, Lacarelle B, Barlési F and Andre N 2016 *Nat. Rev. Clin. Oncol.* **13** 242

[4] Ciccolini J, Barbolosi D, Meille C, Lombard A, Serdjebi C, Giacometti S, Padovani L, Pasquier E, and Andre N 2017 *Cancer Res.* **77** 4723–33

[5] Terterov I, Chubenko V, Knyazev N, Klimenko V, Bogdanov A, Moiseyenko V and Bogdanov A 2021 Minimal pk/pd model for simultaneous description of the maximal tolerated dose and metronomic treatment outcomes in mouse tumor models *Cancer Chemother. Pharmacol.* https://doi.org/10.1007/s00280-021-04326-x

[6] Cham K *et al* 2010 *Br. J. Cancer* **103** 52–60

[7] Veerman G, van Haperen V, Vermorken J, Noordhuis P, Braakhuis B, Pinedo H, Peters G 1996 *Cancer Chemother. Pharmacol.* **38** 335–42

Identification of new silk-like protein from *B. magister* and development of functional materials based on it

A A Vronskaia, A D Mikushina and I E Eliseev

Alferov University, St. Petersburg 194021, Russia

vrons.ann@gmail.com

Abstract. Tandem repeat proteins have composite structure and unique properties, which allow them to be used in multiple fields, such as soft photonics, drug delivery and textile industry. The recent discovery of squid ring teeth (SRT) proteins have expanded the existing repertoire of repetitive polypeptides. We chose previously unexplored squid *B. magister* for our research, isolated and analyzed a new protein forming its ring teeth and hooks, and amplified the corresponding gene. Finally, we used this new isolated SRT protein to fabricate transparent thin films and microspheres.

1. Introduction

Conventional hard materials, including synthetic polymers, are typical for manufacturing photonic devices [1], however, they often do not meet the biodegradability and biocompatibility requirements necessary to work with live systems. On the other hand, natural proteins can be optimized to create new biocompatible materials with diverse functionalities and complex structure [2].

Squid ring teeth are located inside suction cups on tentacles of a squid and serve to seize prey [3]. These teeth are built of a native protein complex, which appeared a novel silk-like fibrillar protein with tandem repeat structure [1] that allows its use in the polymer industry [4] and other fields [1]. Although it has been shown that SRT proteins meet the requirements of biodegradability and possess remarkable physical properties [4], the knowledge of clear structure-property relationships remains elusive [1]. This fact together with unique properties of SRT proteins demands further research to design materials with previously unattainable characteristics [2].

2. Results and discussion

For our research we have chosen another cephalopod — a squid *Berryteuthis magister.* It is spread in high latitude waters of the North Pacific and anatomically similar to the European common squid *Loligo vulgaris* previously studied in the works [2, 4-6]. The limbs of this squid also have suction cups with characteristic ring teeth made of a hard semitransparent material. However, compared to *Loligo vulgaris*, arms of a female *B. magister* are additionally equipped with small tubercles bearing massive hooks. Microscopic images of collected squid ring teeth and squid hooks (SH) are shown in figures 1a and 1b. To enable further application of biochemical methods, we first developed protein extraction protocols with formic acid or DMSO. The initial analysis of solubilized SRT and SH protein complexes with SDS-PAGE identified one major component which we called "SRT-37" (see figure 1c).

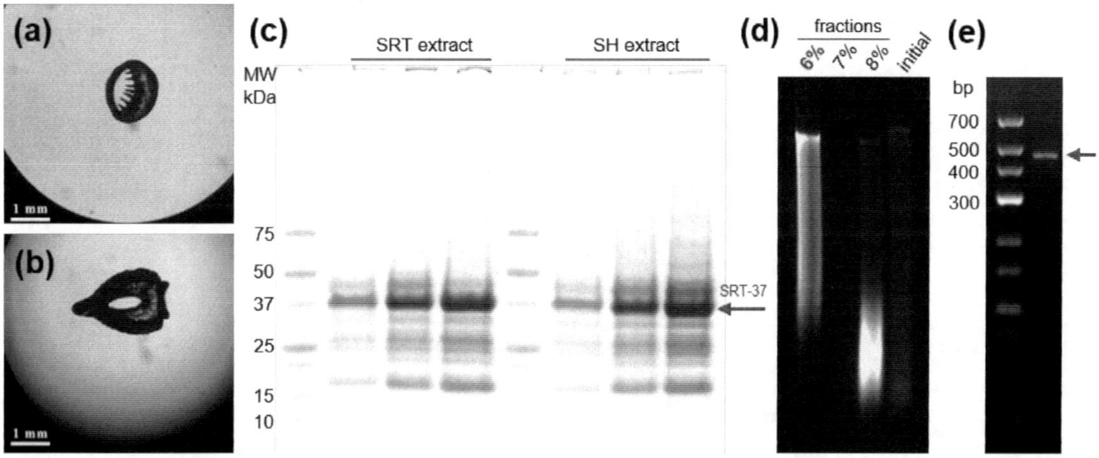

Figure 1. Collected SRT (a) and SH (b) from *B. magister* were examined using bright-field microscopy. They consist of a solid translucent protein material. (c) SDS-PAGE of the isolated protein complexes from *B. magister,* the arrow marks the major component, SRT-37. (d) Electrophoretic analysis of extracted gDNA after fractionation by PEG precipitation. (e) Electrophoresis of the purified PCR product.

Based on the protein electrophoresis results, we hypothesized that SRT-37 is homologous to the 32 kDa SRT protein from *Loligo vulgaris* [4]. The presumably large number of histidine residues and associated positive charge of the SRT-37 based on homology encouraged us to use ion exchange chromatography (IEX) and metal-affinity chromatography (IMAC) for its purification. Indeed, we were able to purify SRT-37 by IEX and IMAC and then studied it by mass-spectrometry. The protein was excised from the gel and examined by ESI-MS after trypsin digestion. Surprisingly, a single 8 residue peptide from the Suckerin-6 belonging to the organism with sequenced genome — *Sepioteuthis lessoniana* — was identified (see figure 2a). Peptides from the main components of muscle (actin and tropomyosin) were also found in SRT-37 sample.

Identification of SRT-37 implies studying of the gene encoding the protein and consequently requires its amplification. The genomic DNA (gDNA) was extracted from tissue of *B. magister*, purified and fractionated. Electrophoresis in agarose gel revealed high fragmentation of gDNA (figure 1d), therefore different fractionation methods were examined to enrich high molecular weight gDNA in the sample. Precipitation series by ethanol and polyethylene glycol (PEG) turned out to be the most successful (figure 1d).

The identification of the peptide by mass-spectrometry played a key role in designing primers. The sequence analysis allowed to find homologous proteins Suckerin-4 and Suckerin-6 from a closely related species *S. esculenta.* BLAST search with the amino acid sequence of the peptide did not yield any results among eukaryotes, as did the search with the nucleotide sequence, which is remarkable for such a short peptide consisting of only 8 amino acids. This observation can be explained by the hydrophobicity of the peptide due to the high content of the nonpolar (hydrophobic) amino acid valine (V) and rare amino acid proline (P). Using multiple sequence alignment (MSA), a similar peptide differing by one amino acid was identified in *S. esculenta* Suckerin-6. To design a forward primer, we analyzed the signal sequences at the beginning of the proteins since they tend to be very conservative. The MSA results demonstrated high similarity of signal peptides among Suckerin-6 proteins from different organisms. Especially conservative was the peptide "LAVIALS", which is identical in all three proteins. It is reasonable to assume that its conservatism is also caused by the hydrophobicity of the signal peptide core (all residues except serine are hydrophobic).

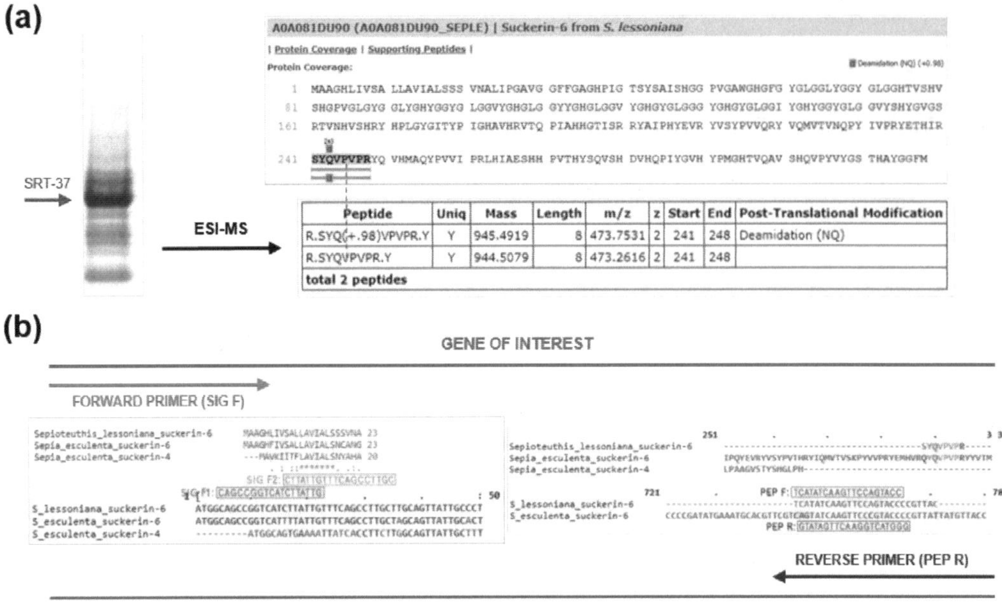

Figure 2. (a) Mass-spectrometry results after trypsin digestion: the identified in SRT-37 component peptide is similar to *S. lessoniana* Suckerin-6 protein at amino acid position No. 241-248. (b) Scheme of designing primers for PCR: MSA of signal and peptide amino acid and nucleotide sequences of Suckerin-6 protein from *S. lessoniana* and Suckerin-4 and Suckerin-6 proteins from *S. esculenta*.

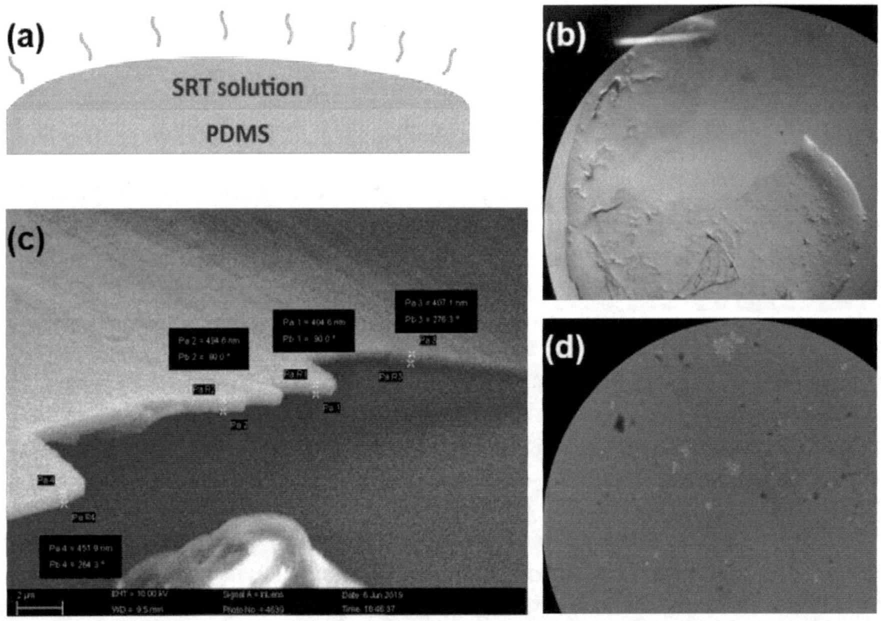

Figure 3. (a) Thin films casting scheme: a solution of high concentration in a volatile compound is placed on a substrate. (b) SRT film, 5x magnification. The film is transparent and elastic. (c) SEM of the film at an accelerating voltage of 10 kV. Film thickness is about 0.4-0.5 μm. (d) SRT particles, 40x magnification. The obtained particles were observed using phase-contrast microscopy.

Based on the analysis of the signal and peptide sequences, primers for amplification were designed (see figure 2b). After gradient polymerase chain reaction (PCR) and purification of PCR products, electrophoretic analysis with ethidium bromide showed a single product (figure 1e). Having successfully sequenced the amplicon, it is possible to sequence the whole gene by chromosome walking techniques and then produce the recombinant protein.

After the extraction and characterization of SRT-37 we used it to create thin plastic films by solvent evaporation. One of the key steps in making films was selecting a smooth planar substrate with low adhesion. Two substrate candidates have been identified: polytetrafluoroethylene (PTFE, Teflon) and polydimethylsiloxane (PDMS). With PDMS, a smooth hydrophobic surface is readily formed after polymerization. SRT protein solutions were evaporated on PDMS at different temperatures to get films, the scheme of the experiment is shown in the figure 3a. Microscopic photography and SEM of the film are shown in figures 3b and 3c. We also obtained microsphere particles that were formed by adding the protein solution to the NaCl solution (see figure 3d).

3. Conclusion

In summary, here we describe a research on tandem repeat proteins from previously unexplored squid *B. magister*. SRT and SH protein complexes were analyzed in detail and turned out to contain only one major component SRT-37. It was experimentally shown that SRT-37 is a basic protein and is saturated with histidine similarly to other homologous SRT proteins. A short peptide identified in SRT-37 by mass-spectrometry was used to design primers and amplify the gene encoding the protein of interest. Thin films and colloids were also synthesized, demonstrating reversibly agglomeration after removing the solvent. Further sequencing will allow its further structural and functional research and possible application in the development of new biomimetic materials.

4. Materials and methods

4.1. SRT and SH collection

Individual squid hooks (SH) and squid ring teeth (SRT) were removed from the defrosted squid tentacles and separated from the connective tissue using a toothpick, a scalpel and tweezers. Collected SRT and SH were soaked in 70% (v/v) ethanol in deionized water and freeze-dried.

4.2. SDS-PAGE

For casting two gels in Mini-PROTEAN Tetra Cell (Bio-Rad) 5 ml of 5% stacking gel and 10 ml of 12.5% resolving gel were used. Before electrophoretic analysis samples were prepared by dissolving solid SRT in 20 μl Laemmli Buffer (20% glycerol, 2x Tris-HCl pH 6.8, 5 M urea, 4% sodium dodecyl sulfate (SDS), 0.01% bromophenol blue) and heating at 95°C for 3 min. Sodium dodecyl sulfate polyacrylamide gel electrophoresis (SDS-PAGE) was performed under 100-150 V for about 60 min. For the protein gels staining Coomassie blue dye was used.

4.3. Ion exchange chromatography (IEX) purification

200 ml of buffer A (25 mM disodium ethylenediaminetetraacetate dihydrate (EDTA-Na_2) in dimethyl sulfoxide (DMSO), pH 4) and 200 ml of buffer B (25 mM EDTA-Na_2, 1 M ammonium nitrate in DMSO, pH 4) were prepared and filtered using polytetrafluoroethylene membrane filter. Before IEX purification solid SRT were dissolved in 1 ml DMSO up to 23 mg/ml, centrifuged, the supernatant was transferred to fresh tube and 1 mM EDTA-Na_2 was added. After equilibration with buffer A for 20 minutes, SRT proteins were separated on a HiTrap SP HP 1 ml column (GE Healthcare) with a 20-min linear gradient from buffer A to buffer B. 20 fractions of 1 ml were sampled after the 22nd minute.

4.4. Immobilized metal affinity chromatography (IMAC) purification

50 ml of buffer A (50 mM HEPES, 1 M ammonium nitrate, 3 μl imidazole in DMSO, pH 8) and 50 ml of buffer B (50 mM HEPES, 1 M ammonium nitrate, 0.5 M imidazole in DMSO, pH 8) were

prepared. For IMAC separation fractions 5-8 after IEX purification were chosen, 20 μl of buffer B and 20 μl of nickel chloride were added to them. IMAC was performed using a peristaltic pump on a HisTrap HP 1 ml column (GE Healthcare). After washing with deionized water and equilibration with buffer A the flow-through was collected. For elution 15 ml of buffer A and 15 ml of buffer B were used, four fractions of 1 ml were sampled.

4.5. Genomic DNA (gDNA) extraction and fractionation with ethanol and polyethylene glycol (PEG)

The squid tentacles were thawed and sliced into small pieces. The pieces were equally distributed among the tubes (about 30-40 mg per tube). PureLink Genomic DNA Mini Kit (Invitrogen) reagents were used for lysis: 20 μl of Proteinase K and 180 μl of preheated to 55°C Genomic Digestion Buffer were added to each tube. The tubes were incubated at 55°C for 16 hours in water bath. For DNA purification was used salt/chloroform extraction method [7].

We used a PEG precipitation protocol from paper [8]. Based on the results of electrophoresis, the most optimal method of fractionation was selected: precipitation with 5% PEG, then transfer of the supernatant to a new tube and precipitation with 6% PEG. DNA obtained after supernatant collection and sediment permeabilization was used in polymerase chain reaction (PCR).

4.6. DNA electrophoresis and amplification

0.5% agarose gel was prepared using 0.125 g of agarose, 25 ml of 1x TAE buffer and casted in Mini-Sub cell GT (Bio-Rad). 6x Sybr Green and 6x Loading Dye (Bio-Rad) were added to samples. Electrophoresis was carried out under 75 V until the front of the dye reached the edge of the gel.

For gradient polymerase chain reaction (PCR) Phusion High-Fidelity DNA Polymerase (NEB) was used, total volume of reaction is 25 μl. The thermal cycling conditions were composed of an initial denaturation step at 98°C for 3 min, 35 cycles at 98°C for 30s, gradient in range from 55.5°C to 59.5°C for 30s and 72°C for 30s.

4.7. SRT films and particles

To cast thin films solid SRT were dissolved in 10% formic acid in acetonitrile, a concentration of 11.4 mg/ml was obtained. 20 μl of SRT solution was applied to the PDMS with a size of 0.5x0.5 cm^2.

For particles fabrication solid SRT were dissolved in 5% formic acid aqueous solution. 720 μl of 220 mM sodium chloride was added to 120 μl of the SRT solution. After vortexing for 2 min 30 μl of the mixture was applied to a glass slide and examined under a microscope.

References

[1] Pena-Francesch A and Demirel M C 2019 *Front. Chem.* **7** 69
[2] Yılmaz H, Pena-Francesch A, Shreiner R, Jung H, Belay Z, Demirel M C, Özdemir S K and Yang L 2017 *ACS Photonics* **4** 2179-86
[3] Jung H, Pena-Francesch A, Saadat A, Sebastian A, Kim D H, Hamilton R F, Albert I, Allen B D and Demirel M C 2016 *PNAS* **113** 6478-83
[4] Pena-Francesch A, Florez S, Jung H, Sebastian A, Albert I, Curtis W and Demirel M C 2014 *Adv. Funct. Mater.* **24** 7401-09
[5] Pena-Francesch A, Akgun B, Miserez A, Zhu W, Gao H and Demirel M C 2014 *Adv. Funct. Mater.* **24** 6227-33
[6] Jung H, Szwejkowski C J, Pena-Francesch A, Tomko J A, Allen B, Özdemir S K, Hopkins P and Demirel M C 2017 *Analyst* **142** 1434-41
[7] Kanai N, Fujii T, Saito K and Tokoyama T 1994 *J Clin Pathol.* **47** 1043-44
[8] Paithankar K R and Prasad K S 1991 *Nucleic Acids Res.* **19** 1346

Microfluidic chips for real-time PCR

A N Zubik, G E Rudnitskaya, A L Bulyanitsa, T A Lukashenko and A A Evstrapov

Laboratory of Bio-&Chemosensors microsystems, Institute for Analytical instrumentation RAS, Saint Petersburg 190103, Russia

tunix@yandex.ru

Abstract. The results of real-time PCR in single-chamber microfluidic chips made of silicon-glass materials and optically transparent polymethyl methacrylate are presented. Conditions for efficient thermal cycling in microchip devices with several reaction chambers are discussed. A simplified theoretical estimation of the duration of heating a liquid in a polymer microchip is proposed, the results of which correlate with experimental data.

1. Introduction

The development of micro-fabrication technologies and the improvement of bioanalytical methods have led to the emergence of various instruments and devices for studying biological samples at the "point-of-care" [1, 2], in particular, for the detection of nucleic acids by amplification methods (polymerase chain reaction (PCR), loop mediated isothermal amplification (LAMP), etc.). An efficient PCR requires a high accuracy of the set temperatures. Prolonged heating at the stage of denaturation leads to a decrease in the activity of the DNA polymerase enzyme and a decrease in the efficiency of the reaction [3]. Due to multiple (more than 35) heating-cooling cycles, the duration of the temperature change makes a significant contribution to the total duration of the analysis. The heating-cooling rate is determined by both the characteristics of the heater and the inertia of the reaction volume. Therefore, the use of planar systems (microfluidic chips) with a high surface area to volume ratio can increase the heat transfer rate, and therefore reduce the analysis time compared to traditional equipment and polymer tubes.

At present, the relatively high cost of microfluidic chips in comparison with traditional test tubes or microplates is a significant obstacle to their widespread use. The cost of microchips depends on the cost of the material and components, the manufacturing and sealing technologies used, in some cases – on the cost and laboriousness of additional technological operations (surface modification, integration of elements, lyophilization of reagents, etc.), serial production.

The high thermal conductivity of silicon (160 $W\cdot(m^2\cdot°C)^{-1}$) is an advantage of microfluidic chips for PCR. However, the silicon surface reduces the efficiency of PCR, and the opacity of the material in the UV and visible spectral (UV-VIS) ranges does not allow the implementation of optical methods for the analysis of PCR products on an all-silicon microchip. The combination of silicon with glass materials has become more widespread, allowing you to maintain high thermal conductivity and use the advantages of optical detection [4]. For silicon surfaces, various options for protective coatings have been proposed, for example, by applying inorganic layers (silicon oxide) or the formation of organic layers (silanizing agents, etc.) [5].

Manufacturing microfluidic chips from polymeric materials can reduce their cost. For polymers, many available technologies for the formation of microstructures have been developed. However, polymers have a low thermal conductivity ($0.1 \div 0.3$ W·(m^2·°C)$^{-1}$), therefore, they require more careful control of heating to carry out the PCR.

At the IAI RAS, microfluidic chips made of various materials were developed: glass, polymethyl methacrylate (PMMA), epoxy resins, etc. In this paper, a theoretical and experimental evaluation of the performance characteristics of microfluidic PCR chips with the same topology, but made of different materials (silicon-glass and PMMA) is carried out.

2. Materials and methods

Silicon-glass microfluidic chips were manufactured by "Svetlana-Electronpribor" LLC (St. Petersburg, Russia). Reaction chambers with a depth of 250 μm and a volume of 25 μL were formed in a silicon wafer 0.55 mm thick by photolithography and liquid etching. For bonding glass and silicon plates, an anodic bonding method was used.

Polymer microchips consisted of two plates glued together. The top plate (with inlet / outlet holes) was made of PMMA grade ACRYMA®72 with a thickness of 1 mm; the bottom plate (with the reaction chambers) was 1 mm or 0.8 mm thick from PMMA grades TOSN or SO-120-K, respectively. Reaction chambers with a depth of 300 μm and a volume of about 30 μL were engraved by laser ablation at "Laser center" LLC (St. Petersburg, Russia).

In a polymer microchip, the heat transfer process under heating (from 30 to 95 °C) was described by a parabolic differential equation with the following simplifications:

- It was a spatially one-dimensional model with heat fluxes in the vertical plane.
- The temperature change was carried out by setting a temperature jump by 65 °C at the lower boundary of the microchip, at the other boundary, the condition of the Newtonian heat flux was set.
- Instead of jointly solving problems of thermal diffusivity for several media with different characteristics and subsequent smooth gluing of solutions, we considered an equivalent medium with averaged parameters: the coefficients of thermal diffusivity, thermal conductivity, heat transfer, and specific heat. The geometric mean is taken as the average value (by analogy with the models of averaging the dielectric constant for composite media).
- Based on the characteristics and dimensions of the polymer, the Bio criterion (which determines the ratio of heat transfer and thermal conductivity) for various microchip designs ranged from 0.4 to 0.8.
- The calculated value of the heating time was considered when the average temperature in the reaction chamber reached a value that differed from the steady-state (asymptotic) temperature value by no more than 1 °C.

Experimentally, the time required for the liquid to reach a predetermined temperature can be estimated using a test solution with fluorescent heat-sensitive probes [6]. When this solution is heated, the fluorescence signal increases, and the temperature dependence of fluorescence is nonlinear. When the fluorescence signal reaches a plateau, it can indicate that a predetermined temperature has been reached, which means uniform heating in the reaction chamber. Fluorescence measurement was performed in real time every 7 seconds. The heating time was determined in 3-5 repetitions by the output of the fluorescence signal of the test solution to the maximum value. The obtained values were averaged over all chambers of the microchip.

For real-time PCR, a kit of reagents was used to detect the DNA of the pathogen of potato ring rot (Cat. No. RN-001, "Syntol" LLC, Russia). Calibrators provided by "Syntol" LLC with known concentrations were used as the target DNA. Bovine serum albumin was added to the reaction mixture for these microchips to prevent the sorption of the DNA polymerase enzyme on the silicon surface. The reaction chambers were filled with a laboratory dispenser. After filling, mineral oil was introduced into the microchip channels, and the inlets were sealed with adhesive films for PCR microplates. Real-time PCR was performed on devices developed at the IAI RAS: in test tubes – on ANK-32 and in microchips – on the prototype experimental device for microchips.

3. Results and discussion

The results of real-time PCR on microfluidic chips with a single chamber are shown in figure 1 (a). An increasing or decreasing background signal on the graphs can be corrected using mathematical processing. The thermal cycling mode for these microchips was the same: denaturation at 94 °C – 20 s, annealing and elongation at 65 °C – 70 s. The curve for a polymer microchip (with a 1 mm bottom plate) is comparable to that obtained in conventional test tubes. The curve for a silicon glass microchip has a steeper slope and an earlier plateau. This is probably due to better heat transfer through the silicon bottom plate and, as a consequence, an increase in the reaction efficiency.

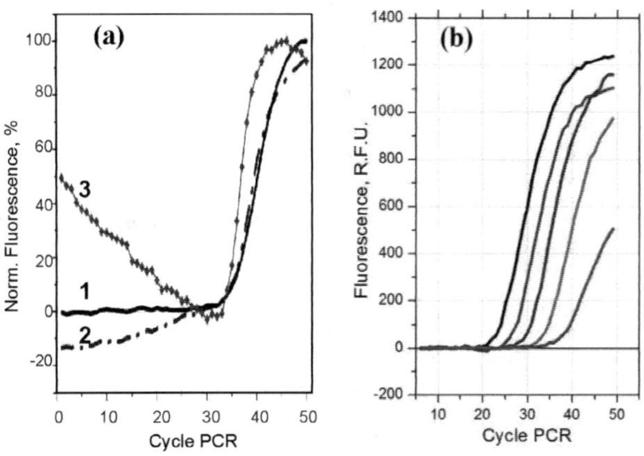

Figure 1. (a) Amplification curves of real-time PCR: 1 – in test tubes on ANK-32 (solid line); 2 – in a PMMA microchip with a single chamber on the prototype device for microchips (dash line with dot); 3 – in a microchip made of silicon-glass materials with a single chamber on the prototype device for microchips (line with dot). The concentration of the target DNA was $5 \cdot 10^4$ copies/μl. **(b)** Amplification curves of real-time PCR in a multichamber chip from PMMA with sequential ten-fold dilution of the target DNA from 10^6 to 10^2 copies/μl.

For the transition from single-chamber microchips to multichamber chips, the heating conditions were studied in more detail. The overall dimensions of the five-chamber microfluidic chips were (60 × 25) mm. The rate of temperature change in the reaction chamber of the microchip, in addition to the characteristics of the heater, is influenced by the following factors: the thermal conductivity and the thickness of the bottom of the microchip, the mass of heated objects, and the thermal contact between the surface of heater and microchip. The latter depends on the design of the clamping device and the materials of the contacting surfaces. After PCR in microchips made of PMMA, an imprint of the heater was observed on the lower surface of the microchip, which indicates a tight fit of the microchip to the heater. A fragile (with such dimensions) multichamber silicon-glass chip with significant clamping force can break if random solid irregularities (or microparticles) are found on the contacting surfaces. The absence of a clamp can lead to insufficiently tight contact with the heater; therefore, earlier [7] we used a thin elastic substrate (0.2 mm thick, the thermal conductivity of the substrate is much lower than that of silicon) that provides contact with surfaces and is suitable for repeated use.

The duration of heating the fluorescent test solution in a silicon-glass multichamber chip on an elastic substrate was (6 ± 2) s, which is comparable with the measurement discreteness. Comparison of the calculated and experimental heating duration in polymer multichamber chips with different bottom thicknesses are presented in Table 1. The difference between the calculated values and the experimental ones can be caused by losses during heat transfer in a real heater-microchip system, as well as by uneven heating over the heater area. It is also possible that averaging the thermal diffusivity leads to some overestimation of the intensity of the heat transfer process, which leads to underestimated times. According to the calculated data, a further decrease in the thickness of the bottom of the microchip down to 0.1 mm results in a decrease in the heating duration to 8 seconds.

Table 1. Heating duration (30 – 95 °C) of fluorescent test solution in multichamber chips made of PMMA.

Bottom thickness (mm)	Heat duration 30 – 95 °C (s)	
	calculated	experimental
0.7	29	35±3
0.5	18	26±4

The obtained values can be used to determine the duration of the denaturation stage for PCR in polymer chips. It was taken into account that the temperature difference during thermal cycling is almost two times less than the temperature difference when using a test fluorescent solution (35 °C versus 65 °C). For multichamber polymer microchip with a 0.8 mm thick bottom plate, the following durations were set: the denaturation stage was 22 s, the annealing and elongation stage was 65 s. These times were consistent with the PCR program for tubes in ANK-32. The results of real-time PCR are shown in figure 1 (b). The systematic delay of the threshold cycles of PCR in the multichamber polymer microchip on the prototype of the experimental device was (0.76 ± 0.27) cycles in comparison with the test tubes on the ANK-32, while the definition of the threshold cycles was carried out according to various algorithms using the software of these devices. The value of the PCR efficiency determined by the calibration dependence for the microchip is comparable to the results in polymer tubes and amounted to more than 95 %.

4. Conclusions

The advantage of microfluidic devices is the ability to create analytical systems that combine several stages of analysis on a single microfluidic chip. To implement a polymer chain reaction on a chip, it is necessary to carefully select the material, as this affects the reaction conditions. The use of silicon-glass microchips (with an oxide layer) provides a rapid temperature change in the reaction chamber. However, polymer materials are simpler and cheaper to manufacture, are suitable for one-time use, and the technologies used make it possible to quickly establish the serial production of chips.

For polymer microchips with a bottom thickness of 0.7 and 0.5 mm, the duration of heating up to 65 degrees (from 30 to 95 °C) is 18 to 35 s. The duration of the thermal cycling program for these microarrays is comparable to the duration of the analysis in polymer tubes, which is confirmed by the obtained PCR results. According to the calculated data, reducing the thickness of the bottom of the polymer microchip to 0.1 mm will reduce the heating duration to 8 s, which will provide the possibility of a faster temperature change and reduce the total duration of PCR analysis.

Acknowledgments

The research was carried out within the state assignment of Ministry of Science and Higher Education of the Russian Federation (theme No. 075-00280-21-00).

References

[1] Zhu H, Fohlerova Z, Pekarek J, Basova E and Neuzil P 2020 *Biosens. Bioelectron.* **153** 112041
[2] Zhang L, Ding B, Chen Q, Feng Q, Lin L and Sun J 2017 *TrAC* **94** 106–16
[3] Grunenwald H 2003 Optimization of Polymerase Chain Reactions *PCR Protocols* ed J M S Bartlett and D Stirling (Totowa: Humana Press) chapter 20 pp 89–99
[4] Shoffner M A, Cheng J, Hvichia G E, Kricka L J and Wilding P 1996 *Nucleic Acids Res.* **24** 375–9
[5] Zhang C, Xu J, Ma W and Zheng W 2006 *Biotechnol. Adv.* **24** 243–84
[6] Sochivko D, Varlamov D, Fedorov A and Kurochkin V 2016 *Tech. Phys. Lett.* **42** 362–4
[7] Afonicheva P, Zubik A, Bulyanitsa A, Rudnitskaya G and Evstrapov A 2020 *J. Phys. Conf. Ser.* **1695** 012060

Combination of cytostatic chemotherapy with cisplatin and photodynamic therapy with Radachlorin reduced resistance of K562 and Hela cell lines

V V Klimenko[1,2], S V Shmakov[3], N A Knyazev[1,2], N A Verlov[1,2,4], An A Bogdanov[1,2,4], I N Terterov[1] and A A Bogdanov[1,2]

[1] Saint-Petersburg clinical scientific and practical center for specialised types of medical care (oncological)», 197758, St Petersburg, Russia
[2] State Research Institute of Highly Pure Biopreparations, 197110, St Petersburg, Russia
[3] St Petersburg Academic University, 194021, St Petersburg, Russia
[4] Petersburg Nuclear Physics Institute NRC Kurchatov Institute, 188300, Gatchina, Russia

klimenko.vova@gmail.com

Abstract. In this work were study combination effect of photodynamic therapy and cisplatin on the proliferation activity of K562 human leukemia cells and Hela cervical carcinoma cells. A decrease in cell viability and an increase the fraction of apoptotic cells for combination treatment compared with single therapy were observed. It has been shown that the G2/M-phase of cell cycle decreases compared with cisplatin treatment alone, which demonstrates an increase anti-proliferative effect. The combination index of the photodynamic therapy with Radachlorin and cisplatin was calculated and indicates a synergistic effect.

1. Introduction
Photodynamic therapy (PDT) is an actively developing method for the treatment of oncological diseases of various localization. However, in clinical practice PDT is not often used as a monotherapy and is usually used in combination with chemotherapy as an additional antitumor effect. Currently, there are no unambiguous protocols for the combined use of cytostatic drugs and PDT. However, in a number of works [1, 2, 3, 4], the effectiveness of the combined use of various chemotherapeutic drugs and PDT with the photosensitizer Photofrin in vitro was shown and a synergistic effect of the combination treatment was revealed. Chlorin e6 based photosensitizers (Radachlorin and Photoditazin) have less dark toxicity than Photofrin and widely used in clinical practice in Russian Federation. The aim of this study was to investigation the combination effect of photodynamic therapy with Radachlorin and the cytostatic drug cisplatin on human erythroleukemia K562 cell line and human cervical carcinoma Hela cell line. K562 cells are cisplatin-resistant ($IC_{50} > 40$ μM), whereas Hela cells are sensitive to cisplatin ($IC_{50} < 15$ μM). This work were evaluated the cytotoxic and cytostatic effects in combination cisplatin treatment and PDT with the identification of the cell death mechanisms and the cell cycle changes.

2. Materials and methods

2.1. *Cell cultures*
The human erythroleukemia K562 cells and human cervical cancer HeLa cells were obtained from the bank of cell cultures, Institute of Cytology of the Russian Academy of Sciences, Russia. The K562 cells were cultured in RPMI-1640 and the Hela cells were cultured in DMEM medium (HyClone, USA) supplemented with 10% fetal bovine serum (FBS; HyClone, USA) and 40 μg/mL of gentamicin (Sigma, USA) in a humidified incubator with 5% CO_2 at 37°C.

2.2. *Photosensitizer*
Radachlorin (Rada-Pharma, Russia) was used as photosensitizer drag for photodynamic treatment. Radachlorin is a water soluble chlorin e6 complex with the molar extinction coefficient 28300 $M^{-1}cm^{-1}$ at 654 nm in dPBS and 34000 $M^{-1}cm^{-1}$ at 662 nm in PBS + 10% FBS. The stock solution of Radachlorin has concentration 3.5 mg/mL. The Radachlorin was diluted in dPBS and work concentration was evaluated by spectrophotometer Nanodrop 2000c (Thermo Fisher Scientific, USA).

2.3. *Photodynamic therapy*
The medical semiconductor laser Lakhta-MILON (Milon Lasers, Russia) were applied for photodynamic treatment at wavelength of 662 nm. The effect of single PDT on cell viability was studied for K562 cells and Hela cells. Cells were seeded in a 12-well plate at density $1x10^5$ cells per well in 1 ml cultured medium. At 24-hours photosensitizer Radachlorin was diluted in 100 □l dPBS and added to experimental wells at final concentration 2 μM and incubated for 24 hours in dark. Intracellular photosensitizer uptake was controlled by confocal microscopy and Radachlorin concentrations were equal to 8.7 ± 0.7 μM for K562 cells and 7.0 ± 0.5 μM for Hela cells [5]. Cells were treated by laser irradiation with light doses 0.125, 0.25, 0.5, 1.0, 1.5, 2.0, 2.5, 3.0, 4.0 J/cm^2 and 5 J/cm^2 at fluency rate 20 mW/cm^2. Cell viability and cell cycle analysis were obtained by flow cytometry for 24 hours after irradiation (72 hours after cell seeding).

2.4. *Cisplatin treatment*
Cisplatin (Kemoplat, Germany) was used as a chemotherapy drug. The effect of single cisplatin treated on cell viability K562 cells and Hela cells were studied. Cells were seeded in 12-well plate at density $1x10^5$ cells/well in cultured medium. At 24 hours the cisplatin was added to experimental wells at concentrations of 1–166 μM for K562 cells and 1–25 μM for Hela cells. Cell viability and cell cycle analysis were obtained by flow cytometry for 24 hours and 48 hours after cisplatin treatment.

2.5. *Combination treatment of cisplatin and photodynamic therapy*
The study the combination effect of cisplatin and PDT on cell viability was used a cisplatin at concentrations of 15 μM for K562 cells and 5 μM for Hela cells. These cisplatin concentrations were not caused the cytotoxic effect, but significantly affected on the cell cycle and inhibit cell division. The experiments of combination treatment were carried out according to the following scheme: cells were seeded in a 12-well plate at $1x10^5$ cells per well; at 0 hours (24 h after seeding) cisplatin 15 μM (K562) or cisplatin 5 μM (Hela) was added to cells; at 24 h after cisplatin addition (48 h after seeding) PDT was performed with light doses 0.25, 0.5, 1.0, 1.5, 2.0, 2.5, 3.0 and 4.0 J/cm^2 at fluency rate 20 mW/cm^2; at 48 hours after cisplatin addition (72h after seeding) the cell viability and the cell cycle analysis were taken.

2.6. *Cell viability by flow cytometry*
Cell viability and dead cells (early apoptosis and necrosis) were detected by FITC Annexin V Apoptosis Detection Kit I (BD Bioscience, USA) and Propidium Iodid (PI, Sigma-Aldrich, USA) for 24 h after PDT and for 24 h, 48 h after cisplatin treatment. The cell viability was evaluated by a flow cytometry EPICS XL (Beckman Coulter, USA). Cell viability was presented as presents of live cell

count in experimental groups relative to live cell count in untreated control and was named cell viability (% of control). Data of cell viability were presented as mean ± SD for three independent experiments. Apoptosis of cells was presented as presents of cells with FITC-Annexin V positive FL1 signal and PI negative FL3 signal in individual samples. Data of apoptosis were presented as mean ± SD for three independent experiments.

2.7. Cell cycle analysis

The distribution of cells between phases of the cell cycle (G0/G1-, S- and G2/M-phases) was assessed by determining the relative DNA content in cells. Cells were washed three times with phosphate-buffered saline (PBS) and then 0.05 mg/ml propidium iodid (Invitrogen, USA), 0.2 mg/ml saponin (Sigma-Aldrich, USA), 0.25 mg/mL RNase (Sigma-Aldrich, USA) were added to each sample. After incubation for 30 min at room temperature, the samples were analyzed by a flow cytometry EPICS XL (Beckman Coulter, USA). More than 20,000 cells were recorded for each sample. SubG1-phase presented in percent from total cell count and was not included in cell cycle distribution. Data were analyzed with ModFit LT 4.1 software (Verity Software, Topsham, ME, USA). Cell cycle distribution were obtained for control samples, single cisplatin treatment for 24 h and 48 h, single PDT for 24 h after PDT and combination treatment of cisplatin and PDT for 48 h.

3. Results

3.1. Cisplatin treatment of K562 cells and Hela cells

The cell viability of K562 cells and Hela cells at cisplatin treatment were investigated. Figure 1 shown the relative cell viability of K562 cells from cisplatin 10-166 μM for 24 h and 48 h treatment intervals. The cisplatin concentrations which inhibited cell viability to 50% were determined and equal $LD50_{K562(24h)}$ = 75 μM for 24 h and $LD50_{K562(48h)}$ = 40 μM for 48 h. Figure 2 shown the relative cell viability of Hela cells from cisplatin concentration for 24 h and 48 h treatment intervals. The cisplatin concentrations inhibited cell viability to 50% were equal to $LD50_{Hela(24h)}$ = 14 μM for 24 h and $LD50_{Hela(48h)}$ = 10 μM for 48 h.

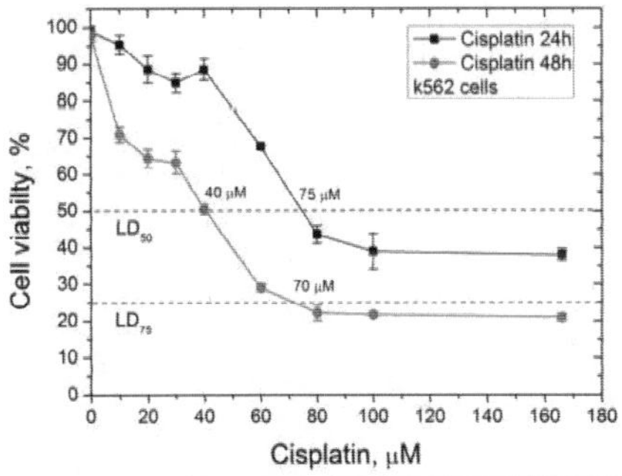

Figure 1. Cisplatin-indused cytotoxity in human erythroleukemia K562 cells. Cell viability (% to control) in K562 cells treated with cisplatin at 10-166 mM for 24 h and 48 h.

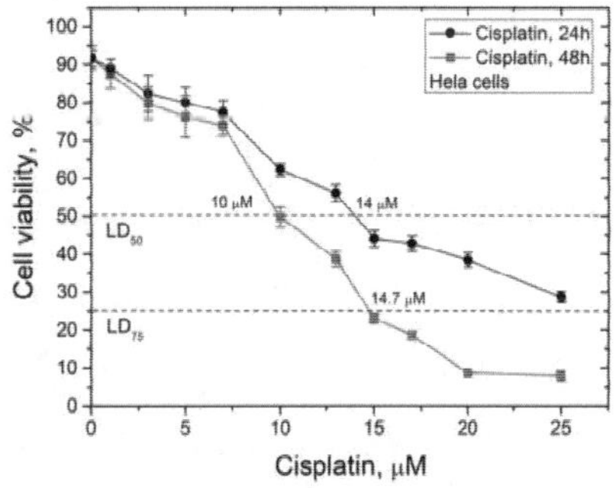

Figure 2. Cisplatin-indused cytotoxity in human cervical cancer HeLa cells. Cell viability (% to control) in Hela cells treated with cisplatin at 1-25 mM for 24 h and 48 h.

The investigation of cytostatic effect of cisplatin on the distribution cell cycle phases was carried out. On the Figure 3(A, B) shown the dependences of the cell cycle phases distribution from cisplatin concentration for K562 cells for 24 h and 48 h after cisplatin addition. The significant changes of the cell cycle were observed at cisplatin concentrations from 5 to 30 μM, in which the S-phase increased up to 90% without strongly cytotoxic effect. The maximal cytostatic effect of K562 cells was observed at cisplatin concentration about 20-30 μM due to the blocking in G0/G1- and the S-phase of the cell cycle for 48 h after treatment. The blocking of cell cycle in the G0/G1-phase is observed up to 50%, the S-phase decreases to 40% at increasing cisplatin concentration more than 30 μM. This effect of cisplatin practically inhibited the cell division and causes an increases in the sub-G1 phase up to 40 %, (DNA fragmentation in apoptotic cells), which well correlates with the cell viability data. The presence of the G2/M phase at high concentration of cisplatin (60 and 80 μM) may be associated with the rapid DNA intercalation by cisplatin and cell cycle arrest at the G2/M phase that indicates the damage of intracellular DNA is difficult to repair. It is also worth noting that at low concentrations, up to 7 μM, there is an increase in the cell population up to 40 % in the G2/M phase, followed by a decrease to 0% with increasing concentrations, that also indicates the damage DNA but is possible to repair.

On the Figure 3 (C, D) shown the dependences of the cell cycle phases distribution from cisplatin concentration for Hela cells for 24 h and 48 h after cisplatin addition. The maximal cytostatic effect of Hela cells was observed at cisplatin concentration about 7-10 μM due to the blocking in G2/M- and the S-phase of the cell cycle for 48 h after treatment.

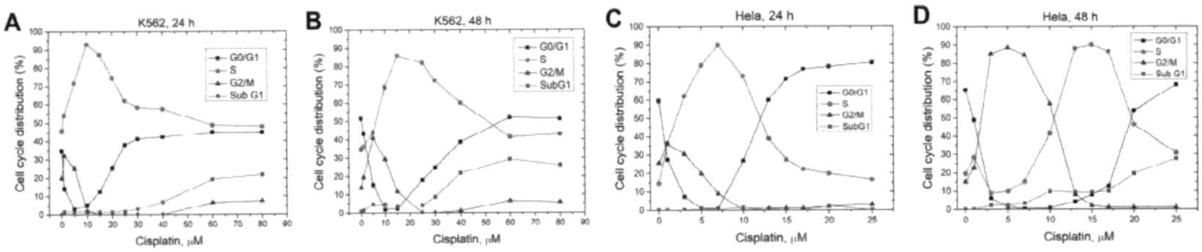

Figure 3. The dependensis of cell cycle phases (G0/G1, S, G2/M) from cisplatin concentration of K562 cells for 24 h (A) and 48 h (B) and Hela cells for 24 h (C) and 48 h (D).

3.2 Cell cycle analysis Combination treatment of cisplatin and PDT

Combination effect of cisplatin treatment and PDT was investigated by analysis of cell viability in experimental groups: single "PDT" (light doses 0.125-5 J/cm^2, Radachlorin 2 μM) and "cisplatin + PDT" (treated with cisplatin 15 μM - K562 cells and 5 μM-Hela cells; 24 h later cells were threated with PDT at light doses 0.25-5 J/cm^2, Radachlorin 2 μM). Cell viability was analyzed for 48 h after cisplatin addition and 24 h after PDT as described in material and methods.

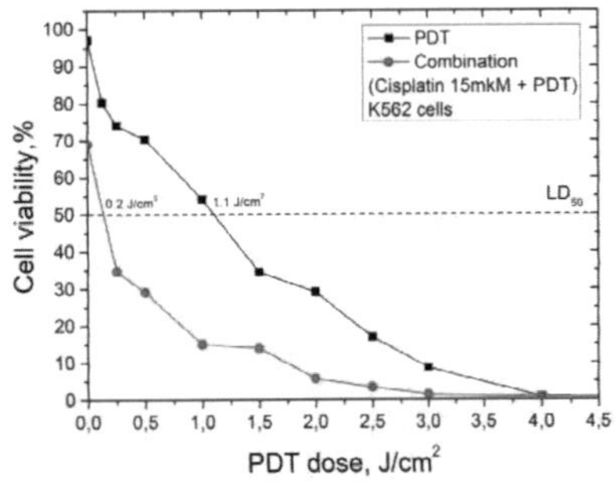

Figure 4. Cell viability (% to control) of K562 cells from light dose at single PDT and combination treatment of cisplatin 15 μM with PDT.

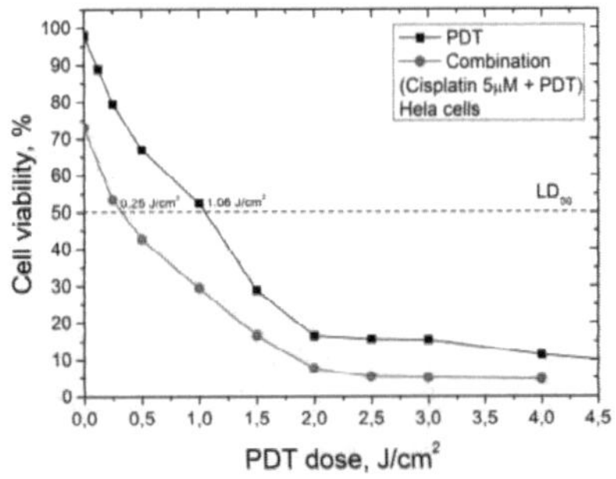

Figure 5. Cell viability (% to control) of Hela cells from light dose at single PDT and combination treatment of cisplatin 5 μM with PDT.

Figure 4 shows the cell viability of K562 cells from PDT light dose under single PDT and combination treatment of cisplatin 15 μM and PDT for 48 h after cisplatin addition. The light dose leading to inhibit of cell viability to 50% were determined by exponential approximation of cell viability decay. For PDT the light dose was $LD50_{PDT}$ = 1.10 ± 0.02 J/cm^2, for the combination treatment $LD50_{comb}$ = 0.20 ± 0.04 J/cm^2. Figure 5 shows the cell viability of Hela cells from light dose under single PDT and combination treatment of cisplatin 5 μM and PDT for 48 h after cisplatin addition. The irradiation dose leading to cell viability 50% were determined. For PDT, the light dose is $LD50_{PDT} = 1.06 \pm 0.04$ J/cm^2, for the combination treatment $LD50_{comb} = 0.25 \pm 0.05$ J/cm^2.

Based on the obtained data was calculated combination index (CI) using the formula: CI = D1/D50(1)+D2/D50(2)+D1 x D2/(D50(1) x D50(2)), where D1 and D2 are the values of the cisplatin concentration and PDT light dose giving a 50% of cell viability in combination "cisplatin + PDT" treatment, and D50(1) and D50(2) are the values of the actions required to achieve 50% of cell viability for "single cisplatin" or "single PDT" when these treatment are applied separately (CI < 1, CI = 1 and CI> 1 show, respectively, synergism, additive effect or antagonism of the considered influences) [4,6]. The combination index was calculated and in the case of K562 cells - CI = 0.62, in the case of HeLa cells - CI = 0.87. Thus, we can talk about the synergistic effect of cisplatin and PDT under the conditions of the experiments. This data was demonstrated the presence of a significant synergistic effect with the combination treatment of cisplatin and photodynamic therapy on K562 and Hela cells.

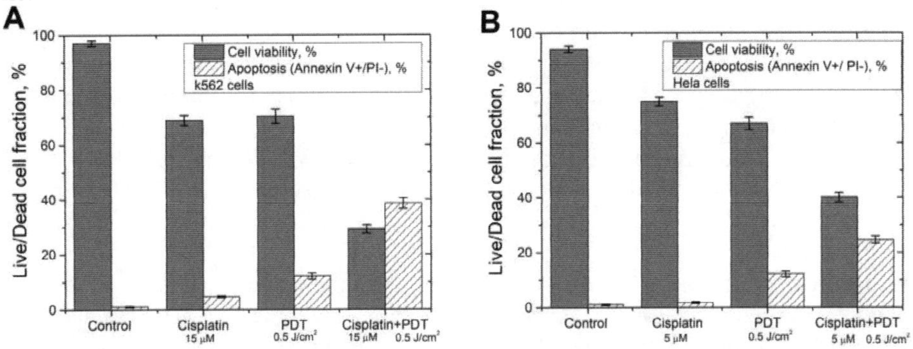

Figure 6. Effect of combination treatment cisplatin and PDT in human erythroleukemia K562 cells (A) and human cervical cancer Hela cells (B). Cell viability and % value of apoptotic cells value were obtained for 48h after treatment.

Figure 6 demonstrated cell viability in % relative to untreted control and dead cell fraction in early apoptosis (Annexin V+/PI-, %) for experimental groups: control, single cisplatin treatment, single PDT, combination treatment of cisplatin and PDT. It has been shown (Figure 6A) that the combination treatment of cisplatin 15 µM and PDT with a light dose 0.5 J/cm^2 on K562 cell line leads to a decreased cell viability to 29.2% and increased apoptosis percentage greater than 38.6 %, which significantly exceeded the single treatment and indicated the presence of a synergistic effect. The same study for Hela cells (Figure 6B) demonstrated that the combination effect of cisplatin 5 µM and PDT with a light dose 0.5 J/cm^2 which lead to decreased cell viability to 40.1% and increased apoptosis percentage up to 24.6%, which also demonstrated a synergistic effect.

3.3 Cell cycle distribution at combination treatment

On the Table 1 demonstrated the distribution between phases of the cell cycle at control samples, single cisplatin treatment, single PDT and combination treatment of cisplatin and PDT for K562 cells. The effect of cisplatin at 15 µM leads to blocking of the cell cycle in the S-phase. Single PDT does not lead to strong changes in the cell cycle, however, the proliferative index (G2/M+S)/(G0/G1) and G2/M phase were increased for survived cells. The percentages sub-G1 phase were increased from 1.0% to 9.6%. The combination treatment of cisplatin and PDT was lead to the accumulation of cells in the S-phase cell cycle from 93.3% to 98.9 % and a strong decrease in the G2/M-phase from 6.7% to 1.1% compere to single cisplatin treatment. The cell fraction in the sub-G1 phase for combination treatment was 21.2% (1.7% - single cisplatin; 9.6% - single PDT). The sub-G1 fraction demonstrated the percentage of cells with DNA fragmentation which occurs at later phase of apoptosis. This values have a good correlation with present apoptosis for combination treatment 38.6% (4.7% - single cisplatin; 12.1% - single PDT) obtained by flow cytometry using the Annexin V-FITC/PI test.

Table 1. Cell cycle distribution of K562 cells and Hela cells at combination treatment for 48 h.

Cell cycle phase	K562				Hela			
	Control	Cisplatin 15 μM	PDT 0.5 J/cm^2	Cisplatin 15 μM + PDT 0.5 J/cm^2	Control	Cisplatin 5 μM	PDT 0.5 J/cm^2	Cisplatin 5 μM + PDT 0.5 J/cm^2
G0/G1, %	47.1	0.0	28.8	0.0	65.4	1.6	61.3	11.4
S, %	46.0	93.3	45.2	98.9	19.6	9.8	25.9	17.4
G2/M, %	6.9	6.7	26.0	1.1	15.0	88.6	12.8	71.2
Sub G1, %	1.0	1.7	9.6	21.2	0.0	2.0	6.9	12.7

On the table 1 demonstrated the cell cycle distribution for Hela cells. The effect of cisplatin at 5 μM leads to blocking of the cell cycle in the G2/M-phase. Single PDT does not lead to strong changes in the cell cycle compere to control cells, only increasing the sub-G1 phase was observed from 0.0% (control) to 6.9% (PDT). The combination treatment of cisplatin and PDT lead to the accumulation of cells in the S-phase cell cycle from 9.8% to 25.9 % and a decrease in the G2/M-phase from 88.6% to 71.2% compere to single cisplatin treatment. The cell fraction in the sub-G1 phase (DNA fragmentation) for combination treatment was 12.7% (2.0% - single cisplatin; 6.9% - single PDT) and have a good correlation with percentage of apoptosis at combination treatment 38.6% (1.7% - single cisplatin; 12.1% - single PDT) obtained by flow cytometry using the Annexin V-FITC/PI test.

4. Conclusion

In this work, new data were obtained on the cell viability and the cell cycle distribution of K562 and Hela cells under the combination treatment of cisplatin and PDT. The obtained data were demonstrated a significant synergistic effect in the combination treatment of cisplatin and photodynamic therapy with CI=0.62 for K562 cells (cisplatin-resistant cell line) and CI=0.87 for Hela cells (cisplatin-sensitive cell line). It can be assumed that the demonstrated effect may be associated with a photodynamic effect on the mitochondria of cells [7], leading to membrane disruption, release of cytochrome C into the cytoplasm, followed by activation of caspase-3, which triggers the apoptosis cycle on the one hand, and on the other, with an increase in expression of the Fas/Fas ligand, which also activates the apoptosis mechanism, when exposed to cisplatin [8]. It was found that PDT with Radachlorin in combination with cisplatin treatment significantly increased apoptotic cell death up to 40% for K562 cells and up to 25% for Hela cells at low cisplatin concentration. PDT did not significantly affect the cell cycle distribution during treatment with low-dose cisplatin, but increases cisplatin-cytotoxic effect on proliferating cancer cells. Further investigation of the intracellular mechanisms under the combination treatment of cisplatin and PDT will possibly reveal the reasons for the synergistic effect and find more effective treatment regimens for cisplatin.

References

[1] Nonaka M, Ikeda H and Inokuchi T 2002 *Cancer Lett.* **184** 171-178
[2] Compagnin C, Mognato M, Celotti L, Canti G, Palumbo G and Reddi E 2010 *Cell Prolif.* **43** 262-274
[3] Biswas R, Mondal A and Ahn JC 2015 *J Photochem Photobiol B* **148** 118-127
[4] Crescenzi E, Chiaviello A, Canti G, Reddi E, Veneziani B M and Palumbo G 2006 *Mol Cancer Ther.* **5(3)** 776-85
[5] Klimenko V V, Shmakov S V, Kaydanov N E, Knyazev N A, Kazakov N V, Rusanov A A, Bogdanov A A and Dubina M V 2017 *Proc. SPIE* **10417**, 1041703
[6] Chou T and Talalay P 1984 *Adv Enzyme Regul* **22** 27-55
[7] Kessel D, Luo Y, Deng Y and Chang C K 1997 *Photochem. Photobiol.* **65** (3) 422–426
[8] Micheau O, Solary E, Hammann A, Martin F and Dimanche-Boitrel M T 1997 *J. Natl. Cancer Inst.* **89** 783–789

Method development to secure the stability of the parameters of pulsed and continuous radiation in laser systems with semiconductor pumping

S R Abdurakhmanova[1, 2], G D Bukharov[1], Z S Gheisser[2], V V Davydov[1, 3] and V Yu Rud[3]

[1] Peter the Great Saint-Petersburg Polytechnic University, Saint Petersburg, Russia, 195251
[2] LLC FEDAL, Saint Petersburg, Russia, 197342
[3] All-Russian Research Institute of Phytopathology, Moscow Region 143050, Russia

sofya.abdurakhmanova@mail.ru

Abstract. The necessity of increasing the stability of the parameters of pulsed and cw laser radiation when operating at peak powers is substantiated. Particular attention is paid to the parameters of pulsed laser radiation in systems with semiconductor pumping. The main requirement in this case is the repeatability of the parameters of the output pulses with high accuracy. This condition is ensured by the stability of the diode supply drivers, especially at peak powers, which are required for scientific research. It was found that even a short-term excess (about 1 ms) of the upper range of the permissible current leads to damage to the laser diode, which instantly degrades the parameters of laser radiation. The developed technique can significantly reduce the likelihood of these phenomena and increase the reliability of the laser.

1. Introduction

At present, lasers are used in all areas of nanophotonics and in many areas of optics [1–10]. The most widely used are diode-pumped semiconductor lasers operating in both cw and pulsed modes [10-18]. When carrying out many scientific research and practical work, especially in industry, high power of laser radiation with a high degree of stability of its parameters is required [7-10, 13-15, 19-24]. The most critical is the operation in a pulsed mode [25-29]. Special requirements for the stability of pulsed laser radiation arise when it is used in radars for various purposes or in lidar installations when working with moving objects [30–37].

Any failure in the operation of laser systems at high powers is highly undesirable. Taking into account the fact that constantly to solve new problems it is necessary to increase the peak power of laser radiation [38-41], the need to develop devices that will control, adjust and maintain expensive laser equipment is extremely urgent. The paper presents a developed technique for monitoring the parameters of power supplies for diode drivers that perform semiconductor pumping of a laser and its practical implementation.

2. Experimental setup and research method

In laser systems currently in operation, there are allowable offsets in the wavelength λ for laser diodes, which differ in operating ranges. For example, in IR for λ = 808 nm and higher, the shift to longer wavelengths with increasing temperature is 0.2-0.3 nm per degree. The presence of such tolerances allows the stabilization system to restore the temperature regime and control the process of changing λ. With large sharp temperature effects or an increase in current (operation at peak powers), which leads to heating, mode effects appear in laser radiation. The pulse duration changes, the edges are distorted, etc. It becomes impractical to use such radiation.

The current stabilization circuits of the pump diode operating mode when operating at peak powers with short pulses (on the order of microseconds or less) and high duty cycle do not provide the necessary stability of the edges, etc. The classic circuit for comparing currents is malfunctioning. Therefore, in the method developed by us to stabilize the parameters of laser radiation, it is proposed to use a modernized current comparison circuit. Calculations and experiments have shown that the use of the bridge method with a voltage difference of 10 V between the arms of the circuit allows you to work out a sharp increase in voltage and start the stabilization process at the initial distortion section, in contrast to other circuits that work out the result of the completed process or at the stage of its completion.

To implement our proposed technique in the current comparison circuit, we developed a voltage converter supplied to the input of the laser diode power supply to a constant voltage of 10 V. The circuit of the new converter is shown in figure 1.

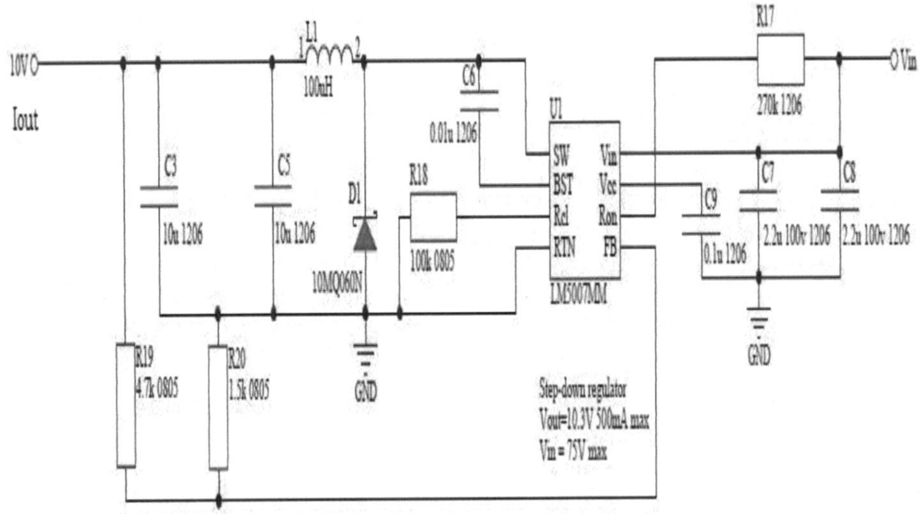

Figure 1. The voltage converter.

The main control element in the LM5007 converter is the feedback comparator. The output voltage of the regulator is measured at the feedback pin (FB) and compared to an internal 2.5 V reference. If the FB signal is below the reference voltage, the comparator switch is turned on for a fixed time pulse determined by the input voltage and resistor R17. The minimum recommended opening time is 300 ns, at the maximum input voltage. Accordingly, the resistor is selected so that at 55 V at the input, the opening time is at least 300 ns. If the voltage at the FB pin remains below the reference value, the comparator will turn on again. This switching will continue until the voltage at the FB pin reaches the reference voltage level.

Also, the LM5007 has a built-in comparator designed to protect the output of the circuit from overvoltage due to a surge in the input voltage or a change in the load at the output. The overvoltage comparator monitors the FB pin against an internal 2.875 V reference. If the voltage on the FB rises above 2.875 V, the comparator shuts down the regulator immediately.

Thus, a voltage equal to the driver supply voltage is supplied to the input of the converter, and a stable voltage of 10 V is obtained at the output of the circuit, which is fed to the voltage adjustment pin of the LM5116 controller. In figure 2 shows a new circuit for comparing currents developed by us.

In the current comparison circuit, the target current supplied to the UPR input is compared with the actual load current supplied to the left arm of the circuit at the Iout input. The principle of operation of the circuit is as follows. If the sum of the voltages in the left and right arms of the circuit is 10 V, then the system is stabilized, a signal of 1.215 V is applied to the FB input of the LM5116 microcircuit, equal to the reference voltage of the error amplifier in the microcircuit, in this case, the driver works in a steady state.

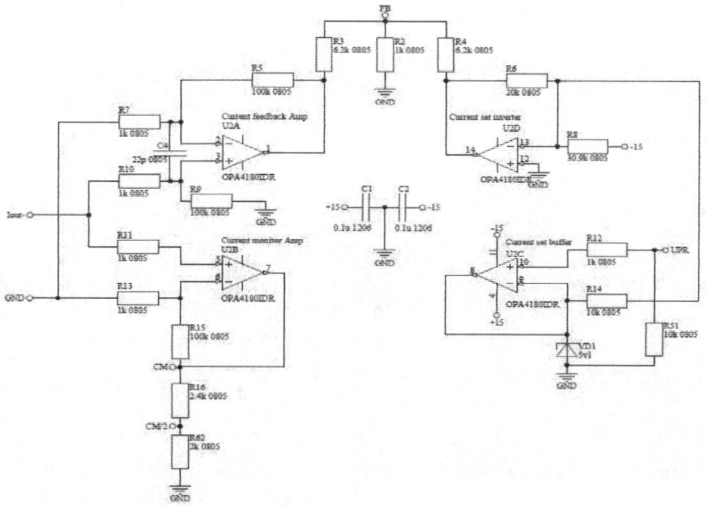

Figure 2. The current comparison circuit.

If the sum of the voltages is more than 10 V, then a signal is supplied to the input of the LM5116 FB microcircuit that is greater than the reference voltage of the error amplifier. In this case, the error amplifier will reduce the duty cycle, and, accordingly, reduce the input voltage, thereby the current supplying the load will also be less.

If the FB signal is less than the reference, then the amplifier will increase the output voltage, thereby increasing the output current, respectively, the signal supplied to the I_{out} input of the comparison circuit will increase, while in the left arm of the circuit, the voltage will increase until the sum of the voltages becomes 10 V, and the FB signal equals the reference signal of the error amplifier in the microcircuit.

3. Result of experimental investigations

In figure 3 the results of the work of a new power driver circuit developed by us with the current comparison method implemented in it are shown.

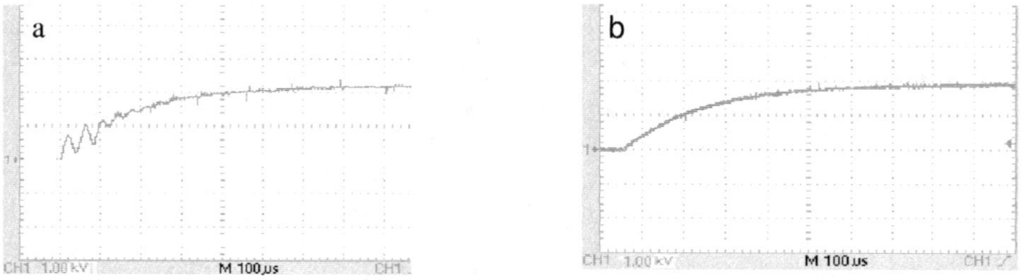

Figure 3. Power driver output signal: (a) - unstable mode is present; (b) - compensation of signal distortions.

The results obtained show that the use of the new circuit makes the distortion values insignificant in relation to the signal amplitude at the output of the power driver.

As an optical result, figure 4 shows the directional diagram of laser radiation, where curve 1 was taken after stabilization of the power supply, curve 2 - before stabilization of the noise of the power supply at the initial stage.

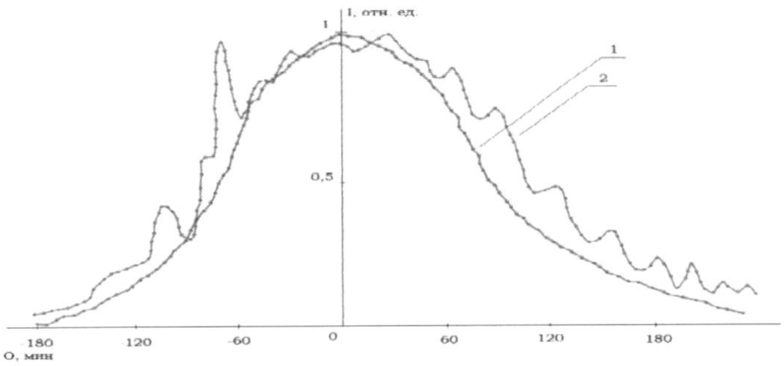

Figure 4. The laser radiation pattern.

Analysis presented in figure 4 of the measurement results shows that even with small noise in the power driver, distortions occur in the laser radiation pattern, which do not allow the use of this radiation for solving many practical and scientific problems.

4. Conclusion

The obtained experimental results of the operation of the stabilization scheme with the developed technique showed that most of the distortions in the fronts of laser pulses associated with the initial section of overheating of the laser diode becomes insignificant, even at its high steepness. There is also no mode distortion in laser radiation compared to the use for stabilization of other techniques.

References

[1] Kozlova E S, Kotlyar V V and Nalimov A G 2015 *Computer Optics* **39(5)** 654-660
[2] Kozlova E S and Kotlyar V V 2014 *Computer Optics* **38(1)** 132-138
[3] Kuzmin M S and Rogov S A 2019 *Computer Optics* **43(3)** 391-396
[4] Skidanov R V, Blank V A and Morozov A A 2015 *Computer Optics* **39(3)** 384-392
[5] Davydov R V and Antonov V I 2018 *Journal of Physics: Conference Series* **1124(8)** 081037
[6] Davydov R V and Antonov V I 2018 *Journal of Physics: Conference Series* **1135(1)** 012087
[7] Logunov S E, Rud V Y, Davydov R V, Moroz A V and Smirnov K J 2019 *Journal of Physics: Conference Series* **1326(1)** 012024
[8] Davydov R, Antonov V and Moroz A 2019 *International Conference on Electrical Engineering and Photonics EExPolytech-2019 (Saint-Petersburg)*, vol. 8906791 (IEEE), p. 295-297
[9] Davydov R, Antonov V, Makeev S, Batov Y, Dudkin V and Myazin N 2020 *E3S Web Conference* **140** 02001
[10] Logunov S E, Vysoczky M G and Titova O A 2018 *International Conference on Electrical Engineering and Photonics EExPolytech-2018 (Saint-Petersburg)*, vol. 8564402 (IEEE), p. 282-284
[11] Andreeva E I and Potapov I A 2020 *Journal of Physics: Conference Series* **1695(1)** 012125
[12] Andreeva E I and Potapov I A 2021 *Springer Proceedings in Physics* **255** 241–245
[13] Davydov R V, Dmitrieva D S, Pilipova V M, Dudkin V I and Andreeva E I 2020 *Proceedings of International conference of Laser Optics ICLO-2020 (Saint-Petersburg),* vol. 9285820 (IEEE), p. 171
[14] Dmitrieva D S, Pilipova V M, Dudkin V I, Davydov V V and Rud V Yu 2020 *Journal of Physics: Conference Series* **1697(1)** 012145
[15] Dmitrieva D, Pilipova V, Andreeva E, Dudkin V and Davydov V 2020 Proceedings of ITNT 2020 – 6th IEEE International Conference on Information Technology and Nanotechnology, vol. 925334 (IEEE), p. 123-127.
[16] Dmitrieva D S, Pilipova V M, Davydov V V and Valiullin L R 2020 *Journal of Physics: Conference Series* **1695(1)** 012130
[17] Grevtseva A S, Davydov R V, Dudkin V I and Rud' V Y 2019 *Journal of Physics: Conference Series* **1326(1)** 012043
[18] Valov A P, Davydov R V, Rud V Y and Grevtseva A S 2019 *Journal of Physics: Conference*

Series **1326(1)** 012040

[19] Lukashev N A, Petrov A A, Grebenikova N M and Valov A P 2018 *Proceedings of 18th International conference of Laser Optics ICLO-2018 (Saint-Petersburg),* vol. 8435889 (IEEE), p. 271

[20] Lukashev N A, Glinushkin A P and V. S. Lukyantsev V S 2019 *Journal of Physics: Conference Series* **1410(1)** 012211

[21] Lukashev N A, Davydov R V and Glinushkin A P 2019 *Journal of Physics: Conference Series* **1326(1)** 012046

[22] Petrov A A, Grebenikova N M, Lukashev N A, Ivanova N V, Rodygina N S and Moroz A V 2018 *Journal of Physics: Conference Series* **1038 (1)** 012032

[23] Podstrigaev A S, Smolyakov A V, Myazin N S and Slobodyan M G 2018 *Lecture Notes in Computer Science (including subseries Lecture Notes in Artificial Intelligence and Lecture Notes in Bioinformatics)* **11118 LNCS** 509-515

[24] Valov A P 2019 *Journal of Physics: Conference Series* **1410(1)** 012246

[25] Petrov A A and N S Myazin 2017 *Lecture Notes in Computer Science (including subseries Lecture Notes in Artificial Intelligence and Lecture Notes in Bioinformatics)* **10531 LNCS** 561-568

[26] Petrov A A, Vologdin V A and Zalyotov D V 2015 *Journal of Physics: Conference Series* **643(1)** 012087

[27] Petrov A A, Zalyotov D V, Shabanov V E and Shapovalov D V 2018 *Journal Physics: Conference Series* **1124(1)** 041004

[28] Davydov R, Antonov V and Moroz A 2019 IEEE International Conference on Electrical Engineering and Photonics (EExPolytech-2019), vol. 8906791, (IEEE), p. 295-297

[29] Davydov V V, Dudkin V I and Karseev A Yu 2014 *Optical Memory & Neural Networks (Information Optics)* **23(4)** 259 – 264

[30] Petrov A and Shapovalov D 2019 *Journal of Physics: Conference Series* **1400(4)** 044008

[31] Fadeenko V B, Kuts V A and Vasiliev D A 2018 *Journal Physics: Conference Series* **1135(1)** 012053

[32] Davydov R V, Saveliev I V, Lenets V A and Tarasenko M Yu 2017 *Lecture Notes in Computer Science (including subseries Lecture Notes in Artificial Intelligence and Lecture Notes in Bioinformatics)* **10531 LNCS** 177-183

[33] Davydov V, Fadeenko V, Fadeenko I., Popovskiy N and Rud V 2019 *E3S Web of Conferences* **140** 07006

[34] Fadeenko I, Fadeenko V, Reznik V, Moroz A, Popovskiy N and Nikolaev D 2019 IOP Conference Series: Earth and Environmental Science **390(1)** 012022

[35] Moroz A V and Davydov R V 2019 *Lecture Notes in Computer Science (including subseries Lecture Notes in Artificial Intelligence and Lecture Notes in Bioinformatics)* **11660 LNCS** 710 – 718

[36] Filatov D L, Galichina A A, Vysoczky M G and Yalunina T R 2017 *Journal of Physics: Conference Series* **917**(8) 082005

[37] Podstrigaev A S, Smolyakov A V, Myazin N S, Grebenikova N M and Davydov R V 2019 *Lecture Notes in Computer Science (including subseries Lecture Notes in Artificial Intelligence and Lecture Notes in Bioinformatics)* **11660 LNCS** 525-533

[38] Pavlova Z S and Fedorov A V 2019 *Photonics* **13(1)** 42-47

[39] Smirnov K J, Glagolev S F and Tushavin G V 2018 *Journal of Physics: Conference Series* **1124(1)** 022014

[40] Myazin N S, Dudkin V I, Grebenikova N M, Rud' V Y and Podstrigaev A S 2019 *Lecture Notes in Computer Science (including subseries Lecture Notes in Artificial Intelligence and Lecture Notes in Bioinformatics)* **11660 LNCS** 744–756

[41] Davydov V V, Nikolaev D I, Bukharov G D and Pavlova Z S 2020 IEEE International Conference on Electrical Engineering and Photonics (EExPolytech-2020), vol. 9243948, (IEEE), p. 227-229

Optical properties of plasmonic metal nanoparticles on GaN surface

P Alay[1], Y Enns[1,2], A Kazakin[1,2], A Mizerov[1], E Nikitina[1], A Kondrateva[1,2], E Vyacheslavova[1], P Karaseov[2], M Mishin[1]

[1] Alferov University, St. Petersburg, Russia
[2] Peter the Great Saint-Petersburg Polytechnic University, St. Petersburg, Russia

Abstract. Effect of the plasmonic resonant absorption in metal nanoparticles formed on the GaN surface on optical properties of samples is studied. Silver and gold nanoparticles are formed by solid-state dewetting on epitaxial GaN grown by molecular beam epitaxy (MBE) on c-sapphire substrate. Theoretical and experimental optical characteristics show the appearance characteristic absorption of the surface plasmon resonance. The results of the work show the possibility of increasing the efficiency of GaN-based optoelectronic devices.

1. Introduction

Gallium nitride (GaN) is widely used in modern innovative electronics as the base material for light-emitting diodes (LED), lasers, photo-detectors and other high-power and high-frequency devices [1, 2]. Attention to this material is due to the combination of its physical properties such as wide and direct bandgap, high breakdown voltage, and high effective mobility of charge carriers, good thermal conductivity, and chemical inertness [3]. GaN based photo detectors are considered as promising devices in UV spectral range (wavelength $\lambda<365$ nm). The wide bandgap provides high selective photosensitivity in UV which makes it possible to avoid the installation of any additional optical filters absorbing visible and infrared spectral parts.

Recently, metal nanoparticles (NPs) have been proposed as a viable tool to improve the performance of optoelectronic devices due to their plasmonic behavior [4-6]. Such structures are capable of exhibiting the effect of localized surface plasmon resonance (LSPR), which is widely used for tuning of optoelectronic processes in photovoltaics [7], photocatalysis [8], and photodetectors [9] due to improved optical absorption and / or charge transfer [10]. The most common materials used to make nanoparticles are silver [11], aluminum [4], platinum [5] and gold [6] because of their high optical activity.

In this study, the plasmon effect leading to the capture of photons in the process of localized scattering is studied in detail, which increases the time of interaction of light and material. The efficiency of the electro-optical conversion can be increased by enhanced interaction stimulating by scattering.

2. Experimental

Highly conductive GaN layers were synthesized by molecular beam epitaxy with plasma nitrogen activation (MBE PA) on semi-insulating 2 um thick [0001] GaN layer pre-grown by MOCVD on c-Al_2O_3 substrate. The surface GaN/Al_2O_3 plate was cleaned according to the protocol developed by us (see details in [12]). Samples were placed to the Veeco Gen 200 industrial-type machine immediately after the cleaning. 300 nm thick silicon-doped GaN layers were grown by the PA MBE at constant values of the substrate temperature T_s = 700 °C. The gallium flux F_{Ga} was ~ 0.25 μm/h, and an activated nitrogen flux F_N was ~ 0.05 μm/h. The silicon flux used during the MBE PA synthesis made it possible to obtain the electron concentration in GaN layer of ~ 3.7×10^{18} cm^{-2}, as it was previously established in [12]. The formation of metal nanoparticles was carried out by the solid state dewetting using the technique described in [13]. Metallic films of Au and Ag were deposited on the surface of the epitaxial GaN layer by vacuum thermal evaporation. The thickness of the films was 1.5 and 3 nm for both metals. The formation of nanoparticles was carried out by high-temperature annealing at 550 °C in an air ambient for 60 minutes.

The simulation of the optical characteristics of an array of metal nanoparticles on the GaN surface was carried out using Comsol Multiphysics software by the full-wave equation solution as described in [14]. Uniform distribution of metal nanoparticles on the GaN surface was assumed in order to somehow simplify calculation. Thus, the simulation was carried out on one hexagonal-shaped simulation cell containing the stack Sapphire/GaN/metal NP/air with periodic boundary conditions applied in all lateral directions of the cell. The metal nanoparticle was modelled as a 30 nm diameter hemisphere, which is consistent with the structures made under similar technological conditions [15]. GaN optical parameters (refractive index, extinction coefficient) were taken using data for similar epitaxial layers [16, 17]. Metal nanoparticles were described using the Drude-Lorenz model [18].

The morphology of the epitaxial GaN film, metal films and the distribution of metal nanoparticles were studied by SEM using Supra 55VP setup. The elemental composition of the samples was determined by EDS technique. The photoluminescence spectra of the structures were measured using an Accent RPM Sigma setup with 266 nm laser pumping at low power. Optical transmittance and reflectance spectra were measured using AvaSpec-ULS2048XL-EVO-RS spectrometer with a xenon light source. The rough back surface of the c-Al_2O_3 substrate was coated with immersion liquid GEM-A 1.79 R.I to eliminate undesirable light scattering. The equality of the refractive indices of the liquid and the sapphire substrate makes it possible to exclude the rough interface of the sapphire substrate.

3. Results

Figure 1 presents plan-view and cleavage edge images of GaN samples obtained by SEM. It can be seen in these images the absence of a pronounced interface between the surface of the GaN/c-Al_2O_3 substrate and the upper GaN layer grown by the PA MBE. At the same time, on the surface of the GaN layer synthesized by the MBE PA method, a relatively high density of V-shaped defects is observed. These defects have a characteristic shape of inverted hexagonal pyramids, which can be formed around dislocations seeded at the sapphire substrate and threading through all GaN layer. Also, the data obtained using the SEM indicated that the total thickness of the GaN epitaxial layer is about 2.1 - 2.4 μm. The upper GaN layer grown by the PA MBE method has the thickness of ~ 300 nm over the entire substrate area.

The spectral characteristics of the transmission, reflection and absorption of the sample with epitaxial layers before metal deposition are shown in the Figure 2. The resulting films have high transparency in the visible and IR spectral ranges and high absorption at wavelengths below 370 nm. Analysis of the interference peaks of the spectral characteristics confirmed the thickness of the epitaxial layers. The numerical simulation results showed good agreement with the measured spectra. However, the obtained samples show a sharper absorption edge.

EDX analysis of the elemental composition after the samples after thin Au and Ag film deposition confirmed their presence on the GaN surface. At the same time, depending on the thickness, the quantitative ratio differed twofold, which correlated with the parameters of the technological process. Figure 3 shows SEM images of samples after metal deposition. For samples with a metallization thickness of 1.5 nm, both for Au (Figure 3a) and Ag (Figure 3c), a nonuniform metal distribution over the GaN surface is seen. Formation of individual nanoparticles is clearly observed. At the same time, for samples with a thickness of 3 nm, coverage is better and a sort of cluster structure instead of nanoparticles is seen (Au Figure 3b; Ag Figure 3c).

Figure 1. Plan-view and cleavaged edge SEM images of the GaN/c-Al_2O_3

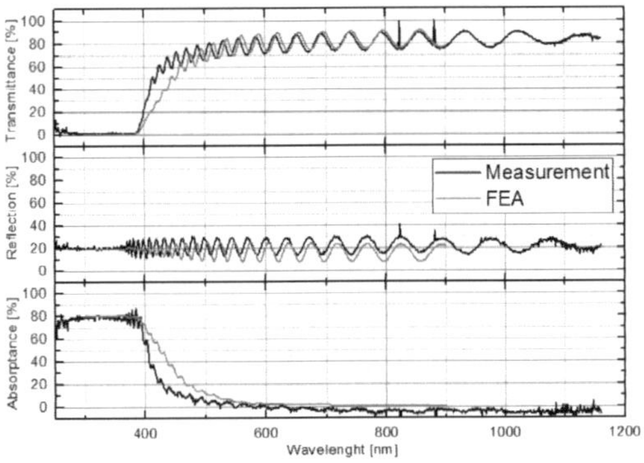

Figure 2. Experimental and theoretical reflection and absorption transmission spectra of GaN / c-Al_2O_3 substrate

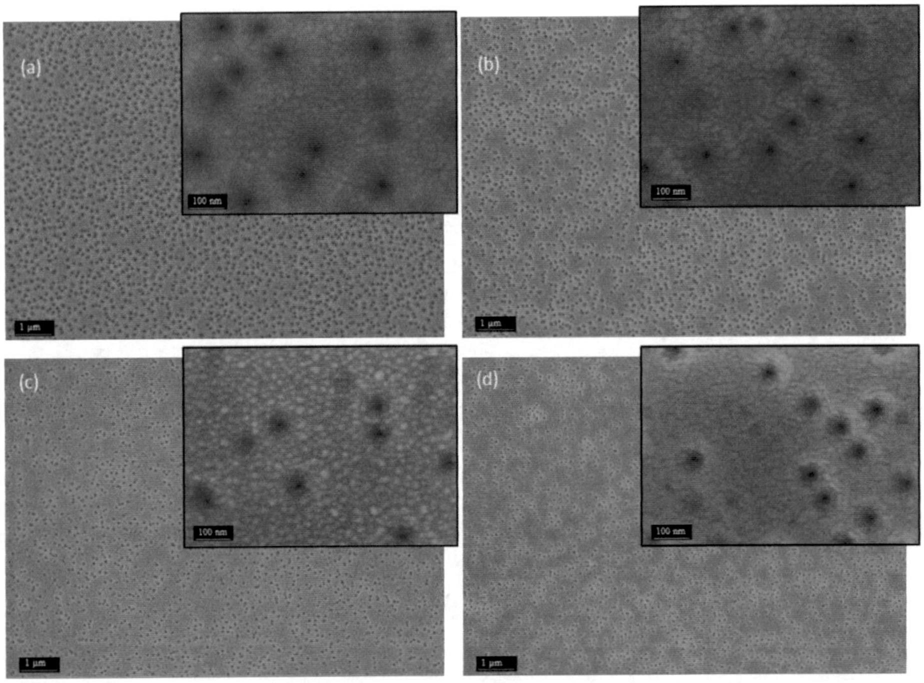

Figure 3. SEM image of GaN surface after metal film deposition: **(a)** Au 1.5 nm; **(b)** Au 3 nm; **(c)** Ag 1.5 nm; **(d)** Ag 3 nm.

Figure 4 shows the photoluminescence spectra of virgin GaN sample and samples covered with thin metallic films. It can be seen from the Figure 4 (a) that an 1.5 nm Au film has almost no effect on the band-to-band PL intensity, in contrast to a thicker gold film. Silver affects it much effectively, which can be explained by the higher activity of Ag in the UV spectral range. In this case, the dumping of PL intensity is approximately proportional to the film thickness. After high-temperature annealing, a significant decrease in band-to-band PL intensity is observed on all samples with a metal coating (Figure4b). No clear dependence of PL on the film thickness is seen after annealing. At the same time, an increase in the photoluminescence intensity is observed in the yellow spectral range for samples coated with Ag, which is primarily seen for a thinner coating.

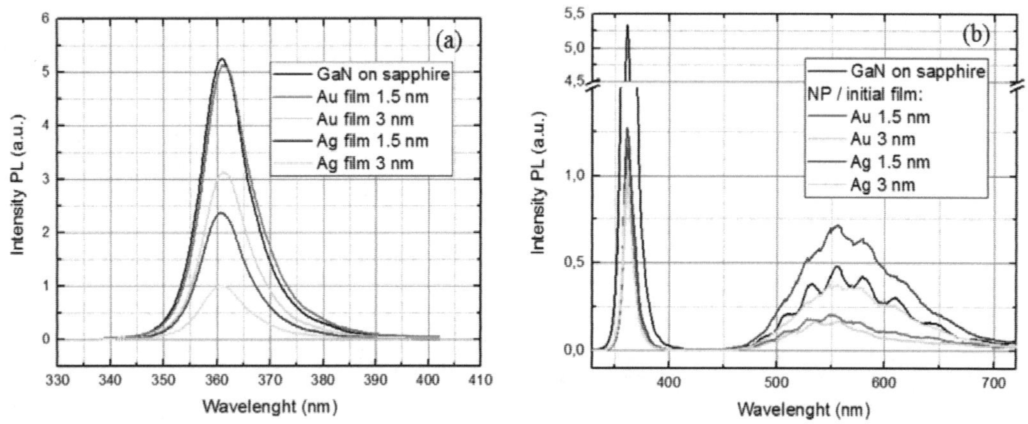

Figure 4. Photoluminescence spectra of virgin epitaxial GaN and samples coated with a thin metal film: **(a)** before; and **(b)** after high temperature annealing.

A preliminary analysis of the spectral characteristics of metal nanoparticles on a GaN/c-Al_2O_3 substrate by numerical methods showed the appearance of their plasmon behavior (Figure 5).

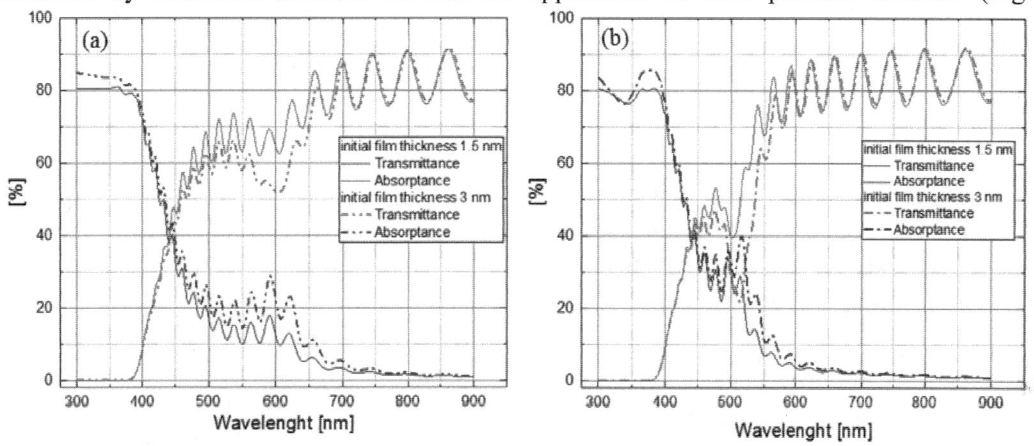

Figure 5. Calculated spectral characteristics of metal nanoparticles on a GaN / c-Al2O3 substrate: **(a)** AuNP; **(b)**AgNP

Figure 6 shows the experimental transmission and reflection spectra of the Au / GaN / c-Al_2O_3 (a) and Ag / GaN / c-Al_2O_3 (b) samples after high-temperature annealing in comparison with the initial GaN / c-Al_2O_3 substrate. For samples with AuNP, a good agreement with the numerical model, where plasmon resonance is observed at wavelengths of 550 and 600 nm (Figure6a) is seen. In addition, an increase in absorption in the UV part of the spectrum is observed. However, no plasmon peaks were observed in the spectral characteristics of Ag-coated samples after annealing (Figure 6b), with the same increase in absorption in the UV part of the spectrum. This can be caused by a change in composition of the silver coating during high-temperature annealing in the air. Silver nanoparticles could be fully oxidized during annealing.

4. Conclusion

The effect of plasmonic resonant scattering in metal nanoparticles formed on the surface of silicon-doped GaN is studied. Epitaxial silicon doped GaN layers are grown on the GaN surface of MOCVD-GaN / c-Al2O3 substrate by the PA MBE. Silver and gold nanoparticles are formed by solid-state dewetting on epitaxial GaN film grown by molecular beam epitaxy (MBE). The morphology of epitaxial GaN layers, thin Au and Au films, and a distributed array of nanoparticles has been studied.

The theoretical and experimental optical characteristics show the appearance of the surface plasmon resonance effect. The results of the work show the possibility of increasing the efficiency of GaN-based optoelectronic devices.

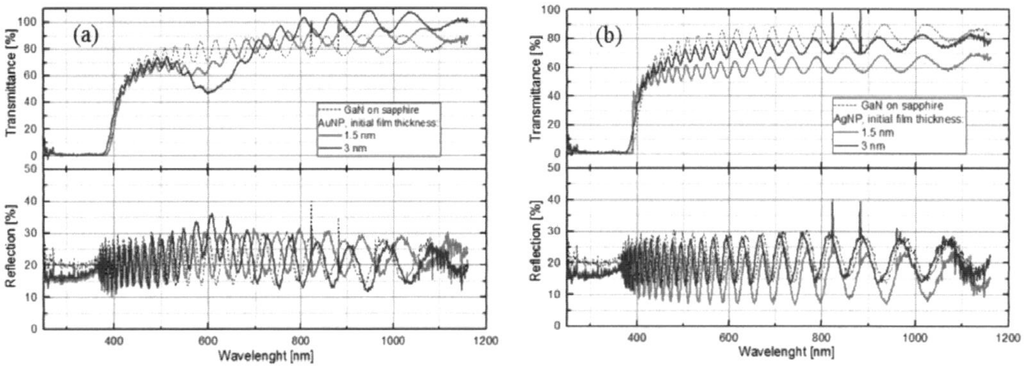

Figure 6. Transmission and reflection spectra of Au/GaN/ c-Al_2O_3(**a**)and Au/GaN/ c-Al_2O_3(**b**) samples after high-temperature annealing.

Acknowledgments
The work was done as a part of the state assignment (№FSRM-2020-0008) of the Ministry of Science and Higher Education of Russian Federation.

References
[1] Wasisto HS, Prades JD, Gülink J, Waag A 2019*Appl. Phys. Rev.***6**.
[2] UedaT2019 *J. Appl. Phys.***58**.
[3] Fu N, Li E, Cui Z, Ma D, Wang W, Zhang Y, Song S, LinJ,2014 *J. Alloys Compd.***59**692-97.
[4] Chang S, Chang M, Yang Y2017*J. Phot.***91−7**.
[5] Zhang X, Liu Q, Liu B, Yang W, Li J, Niu P, JiangX2017 *J. Mater. Chem.* C **5** 4319.
[6] Shetty A, Sundar KJ, Roul B, Mukundan S, Chandan G, Mohan L, Ghosh A, Vinoy K, Krupanidhi JSB2014 *IEEE 2nd International Conference*.
[7] Hwang J D, ChanY D & Chou T C2015 *Nanotechnol***26** 465202-1–465202-6.
[8] Ewa, K, Orlando O P M, Ryu A. & Bunsho O2010 *Chem. Phys.***12** 2344–2355
[9] Narendar G, Arun KS, Sumita S, Santanu M & Samit K R2014 *Sci. Rep.***4** 6483-1–6483-9.
[10] MarkL B, Naomi JH & Peter2015 *Nat. Nanotechnol.***10** 25–34.
[11] Li D, Sun X, Song H, Li Z, Chen Y, Jiang H *Adv. Mater.* 2012 **24** 845−849.
[12] MizerovA M et al 2019 *Semiconductors*5381120–1130.
[13] Kondrateva A, Enns Y, Kazakin A, Morozov I, Karaseov P, Mishin M2020 *Semiconductors* **54**(14) стр. 1885–1888
[14] Enns Y, Kondrateva A, Mishin M 2020 *Journal of Physics: Conference Series***1695**(1) 012115
[15] Sui M., Li Ming-Yu., Pandey P., Zhang Q., Kunwar S., Lee J 2018 *Metals and Materials International*(24) 337–350
[16] Ambacher O et al 1996 *Solid State Communications***97**(5) 365–370
[17] Kawashima T. et al. 1997 *J App. Phys.***82** 3528
[18] Raki ´ A D, Djuris˘i ´ A B, Elazar and Jajewski. L. 998 Appl. Opt. 37(22) 5271-5283

Fiber Optic Attenuator

D P Andreev[1] and E I Andreeva[2]

[1] Institute of Physics, Nanotechnology and Telecommunications, Peter the Great St. Petersburg Polytechnic University, St. Petersburg 194064, Russia
[2] The Bonch-Bruevich Saint-Petersburg State University of Telecommunications, St.Petersburg 193232, Russia

Abstract. A study of a bending-type fiber-optic attenuator has been carried out. It is shown that the use of G.655 - optical fiber provides the largest dynamic range of the attenuator. The spectral dependence of the bending losses was measured in a wide range of wavelengths: from 1270 to 1610 nm.

1. Introduction

At times in a fiber optic network the need arises to reduce or balance the power of the light transmitted from one device to another device. A fiber optic attenuator is used in such cases. Fiber optic attenuator are used for reducing the power of the light in the fiber optic network [1-3]. Attenuators allow tuning the transmitting and receiving equipment without the danger of overloading the signal registration system, measuring optical power and loss level etc. The most used attenuators are adaptor type. They are simple in design and use, but this also has its disadvantages: such attenuators introduce large return losses, which can adversely affect the operation of the transmitting equipment. The lowest level of back reflections is provided by attenuators, made in the form of a fiber-optic patch cord with a macrobend. They can be fixed or variable. For practical use, such characteristics of the device as the achieved dynamic range, linearity of regulation, and a low level of return loss are important.

To study the effect of bending of an optical fiber on its characteristics, a stand was assembled, the block diagram of which is shown in Figure 1.

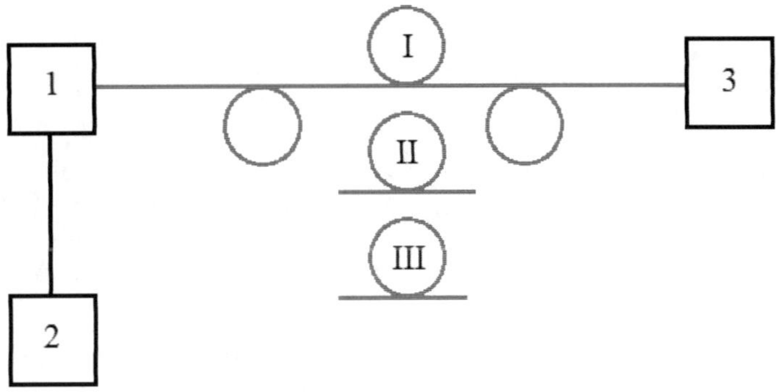

Figure 1. Block diagram of the measuring setup: 1 – laser, 2 – Yokogawa AQ6370C optical spectrum analyzer, 3 – Rubin-300 optical power meter, I – NZDSF, II – SSMF, III – BIF.

An experimental of the bending losses was carried out by the differential method [3]. For optical fibers of the G.652, G.655, and G.657, the spectral dependence of the bending losses was measured in a wide range of wavelengths: from 1270 to 1610 nm. CWDM SFP modules (Small Form-Factor Pluggable) were used as optical sources. The optical signal registration was carried out by the differential method with the spectral sensitivity photodetector correction [3]. Optical power measurements were carried out using a Rubin-300 stationary optical power meter. The source parameters were monitored using a Yokogawa AQ6370C optical spectrum analyzer. The measurement results are shown in Figure 2.

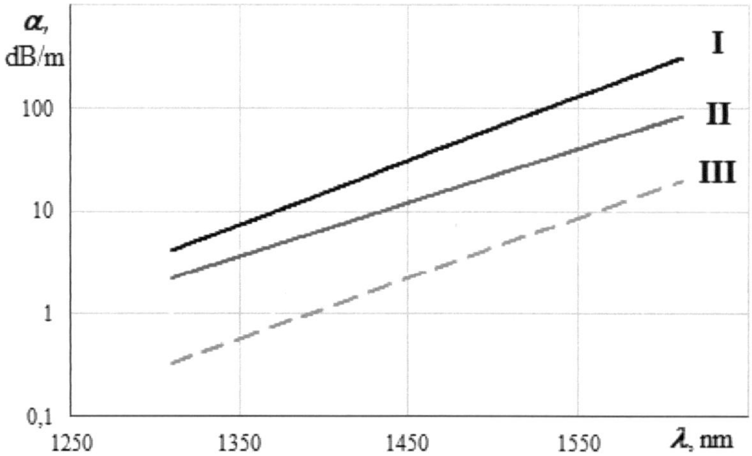

Figure 2. Bending loss $\alpha(\lambda)$ for optical fibers: I – NZDSF (G.655), II – SSMF (G.652), III - BIF (G.657) with the bend radius $R = 6.25$ mm.

The experimental results showed that, regardless of the refractive index profile, the spectral dependence of bending losses $\alpha(\lambda)$ is similar. The NZDS - fiber selected for the optical attenuator layout is characterized by the highest bending sensitivity. The optimal coil diameter was selected: 12 ... 20 mm (Figure 3).

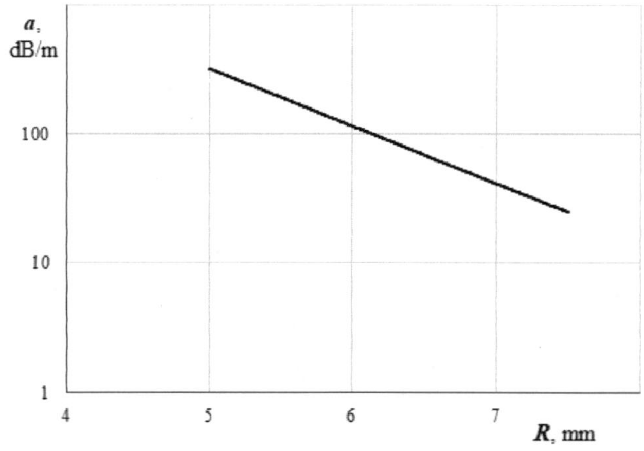

Figure 3. Bending loss dependence $\alpha(R)$ for λ= 1550 nm (NZDSF G.655).

It is known that the axial tension of optical fiber does not introduce additional losses. However, the axial tension of the curved fiber leads to additional bending losses, which can be used to adjust the attenuator. Attenuator block diagram is presented in Figure 4. Additional losses α_F per turn is shown in Figure 5.

Significant tension can lead to damage to the fiber. Thus, when the axial tension of the optical fiber is up to 200 N, it can lead to its rupture. Therefore, it is advisable to apply efforts that are significantly less than this value.

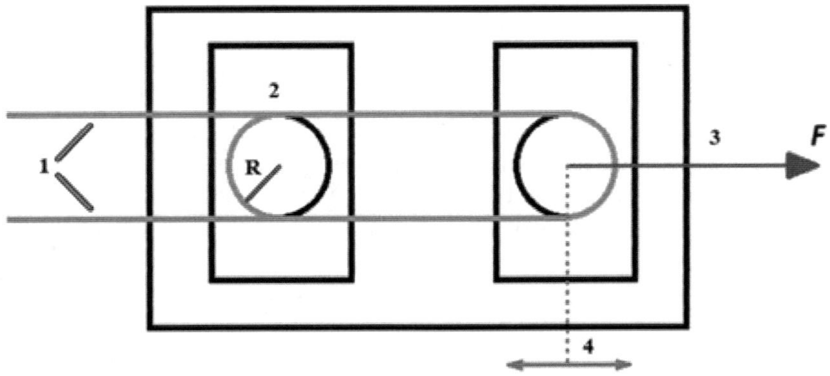

Figure 4. Attenuator block diagram: 1 – optical fiber, 2 – coil with radius R, 3 - tension screw, 4 – moving coil.

In the design of the studied attenuator, one coil with optical fiber was fixed, while the second one could be shifted using a screw. This created a tension of the optical fiber wound on these coils. The number of turns varied (5 ... 10). The output ends of the optical fiber equipped with SC/APC connectors were connected via measuring patch cords to the source and photodetector.

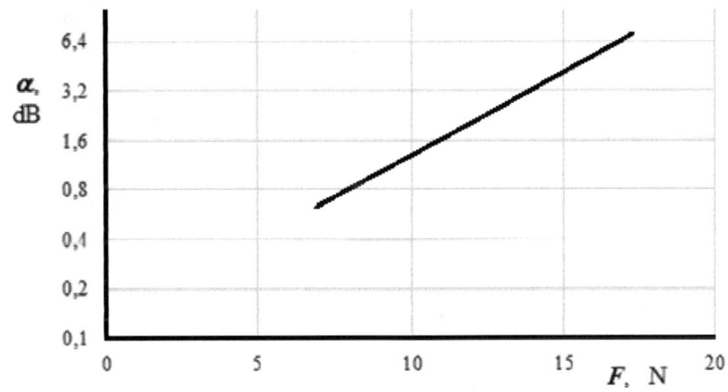

Figure 5. Bending loss $\alpha(F)$ per turn at tension force F (coil 2R = 12.5 mm, G.655) for λ= 1550 nm.

The dynamic range of the studied attenuator was 50 dB at the wavelength λ = 1550 nm. The threshold is about 1 dB. If it is necessary to increase the dynamic range, the number of fiber turns can be increased.

2. Conclusion

A study of a bending-type fiber-optic attenuator has been carried out. The effect of macrobending on the attenuation of the optical signal in the fiber has been studied experimentally. The values of bending losses for optical fibers of various standards are presented: G.652, G.655 and G.657. It is shown that the use of G.655 - optical fiber provides the largest dynamic range for a fixed optical fiber length in the device. Measurement of the spectral dependence of losses in the range of 1270… 1610 nm. It is shown that in the 1550 nm wavelength region the attenuator provides an attenuation of the transmitted signal by more than 100 dB/m, is easy to use.

A bending-type fiber-optic attenuator with a minimum level of back reflections is studied. The effect of macro-bending on the attenuation of an optical signal passing through a light guide has been experimentally studied. The values of bending losses for optical fibers of various standards are presented: G.652, G. 652, G .655, and G.657. It is shown that the use of the NZDS - fiber of the G. 655 standard provides the greatest dynamic range of the device.

The spectral dependence of bending losses in the range of 1270 ... 1610 nm is measured. It is shown that, despite the difference in the refractive index profiles, the spectral dependence of the bending loss $\alpha(\lambda)$ of the NZDS - fiber is like SSMF. The dependence of loss from the bending radius $\alpha(R)$ is also preserved. The dependence of the bending losses with the axial tension force of the fiber is studied. It is shown that in the wavelength range of 1550 nm, the attenuator provides attenuation of the transmitted signal by more than 100 dB/m, increase in use, with low (-55 dB) return losses.

References

[1] Dmitriev S and Slepov N 2010 *Fiber optic technology: current state and prospects* (Moscow: Technosphera) p 608

[2] Zenevich A 2019 Fiber-optic single-mode attenuator based on macro-bends *Vestnik svyazi* **3** pp 55-59

[3] Andreev D, Andreeva E, Sergeev A and Sumkin V 2020 Influence of Optical Cable Bending in Fiber Optic Systems Video Surveillance and Subscriber Access with Spectral Multiplexing *APINO Proc.* **1** pp 57-61

Mn^{4+} concentration effect on spectral properties of lithium-germanate glass-ceramics

A N Babkina, M S Kovova, E V Kulpina, A S Pavliuk, K S Zyryanova, A A Monogarova, A I Bukhvostov, A I Ignatiev and R K Nuryev

Research Center for Optical Material Engineering, ITMO University, Saint Petersburg 199034, Russia

babkina.anastasya@bk.ru

Abstract. A series of lithium-germanate glasses with different manganese concentration is synthesized. $Li_2Ge_7O_{15}$ nanocrystals nucleate in the glass matrix via standard volume crystallization technique thus obtaining lithium-germanate glass-ceramics. The glass-ceramics possess intense emission near 660-670 nm under two-band excitation at 330 and 450 nm. The luminescence lifetime is 550 µs for 0.05 mol.% MnO_2-doped glass-ceramics. The glass-ceramics obtained can be used as a source of deep-red radiation.

1. Introduction

A narrow-band red emission has been major requirement in solid-state lighting technologies in terms of a white-light- emitting diode (w-LED) with an InGaN UV/blue LED chip [1]. Currently, there are two primary strategies to fabricate the w-LED. One is employing the blue LED with yellow- emitting phosphor $Y_3Al_5O_{12}$: Ce (YAG:Ce) and nitride red phosphor [2], the other is combining UV LED chips with tri-color (blue, green, and red) phosphors. Between these two methods, the red-emitting phosphor is the indispensable component to obtain the vivid and warm white light illumination.

There have been mainly reported two different ions, namely, Mn^{4+} and Eu^{2+}, doped as the red-emitting activator ions in various hosts [3]. The general performances of Mn^{4+} (Eu^{2+}) as an activator ion are peak emission at 600–650 nm, excitation peak edge of 450-500 nm, photoluminescence decay lifetime of a few milliseconds (a few microseconds or less), and quantum yield of 80–90%. With this regard and from an economic point of view, it seems that non-rare-earth Mn^{4+} is superior than Eu^{2+}.

Herein, we successfully synthesized novel transparent lithium-germanate glass-ceramics doped with Mn^{4+} with red emission under UV/blue excitation and studied its properties under the influence of manganese concentration.

2. Materials and methods

The initial glasses were synthesized with the following composition: $xMnO_2$-$10Li_2O$-(90-x) GeO_2 (where x=0.005; 0.05; 0.1; 0.25; 0.5; 1; 2) mol.%. The synthesis was carried out in air atmosphere at a temperature of 1250°C using corundum crucibles. For the synthesis of glass-ceramics based on germanate glass the samples were subjected to the heat treatment at 560°C in programmable muffle furnaces (Nabertherm).

The glass absorption spectra were measured by a Lambda 650 spectrophotometer (Perkin Elmer) in the spectral range of 200–900 nm with a step of 1 nm at room temperature. For the luminescence and excitation spectra recording, we used the LS-55 spectrofluorimeter (Perkin Elmer). Based on the luminescence spectra recorded with different delays after the exciting pulse, the luminescence decay curves were obtained, and the luminescence lifetime was determined. The luminescence spectra were corrected considering the spectral sensitivity of the spectrofluorimeter photodetector. For the excitation spectra, no correction was conducted. Glass transition temperature and the crystallization temperatures were determined using an STA 449F1 Jupiter (Netzsch) differential scanning calorimeter (DSC) with an accuracy of ±10 °C in a temperature range of 30 – 700 °C. The occurrence of nanocrystalline phase in the glass bulk was revealed with X-ray diffraction (XRD) method. Diffraction patterns were recorded with Rigaku Ultima IV X-ray diffractometer using CuKα radiation. The diffractogram was taken in the angle range from 10° to 50° in the Bragg-Brentano geometry

3. Experimental results

The absorption spectra shown in Figure 1b are dominated by an intense absorption band with a maximum at 500 nm, which belongs to a series of $^5B_1$ (5E) → $^5A_2$ (^{5}T2), $^5B_1$ (^5E) → $^5B_2$ ($^5T_2$) and $^5B_1$ (^5E) → $^5A_1$ ($^5T_2$) transitions in trivalent manganese ions [4,5]. With an increase in the manganese oxide content, the intensity of the band increases, and the color of the glass becomes dark purple. The initial glasses do not exhibit luminescence in the visible region.

The DSC results show glass transition temperature to be 490°C and two exothermal regions with maxima at 550-560 and 630°C. On the DSC curve of glass with no manganese, the first crystallization peak locates at 550°C, with an increase in the manganese content, it shifts towards 560°C. The position of the second crystallization peak does not change. Moreover, its intensity decreases with increasing MnO_2 content, and the intensity of the former increases. This indicates the possibility of the formation of two types of lithium-germanate crystals in the glass matrix; however, manganese promotes the crystallization of only one of them. Therefore, further synthesis of glass-ceramics was carried out at 560°C for 5 hours.

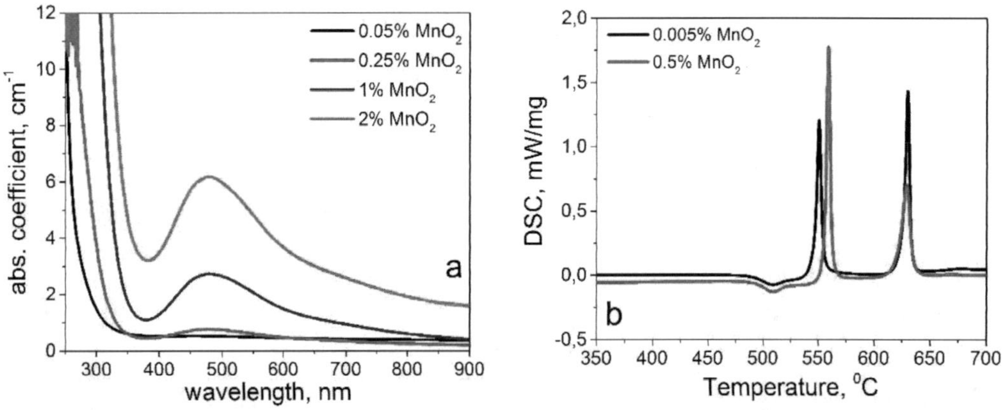

Figure 1. (a) The absorption spectra and (b) the DSC curves of initial lithium-germanate glasses with different MnO_2 content.

The glass ceramics structure was investigated by XRD, which showed the presence of $Li_2Ge_7O_{15}$ crystals in the matrix (Figure 2b). In the sample with 0.5% MnO_2 and after heat treatment at 560°C for 5 hours, the mean size of the nucleated crystals was 8 nm. It is known [4,6] that Mn^{2+} ions have a characteristic absorption at 410-415 nm and 360 nm. Thus, the bands of divalent manganese cannot be identified unambiguously in the absorption spectra before and after heat treatment (Figure 2a). Nevertheless, in the spectrum of glass-ceramics, the maximum of the band in the visible region shifted by 450 nm, and in the region of 330-350 nm the absorption intensity increased. It is known [7,8] that the absorption corresponding to the $^4A_2$→$^4T_1$ and $^4A_2$→$^4T_2$ transitions of tetravalent manganese falls on 330 and 460 nm. It is also possible to determine the oxidation state of manganese ions by studying the luminescent properties.

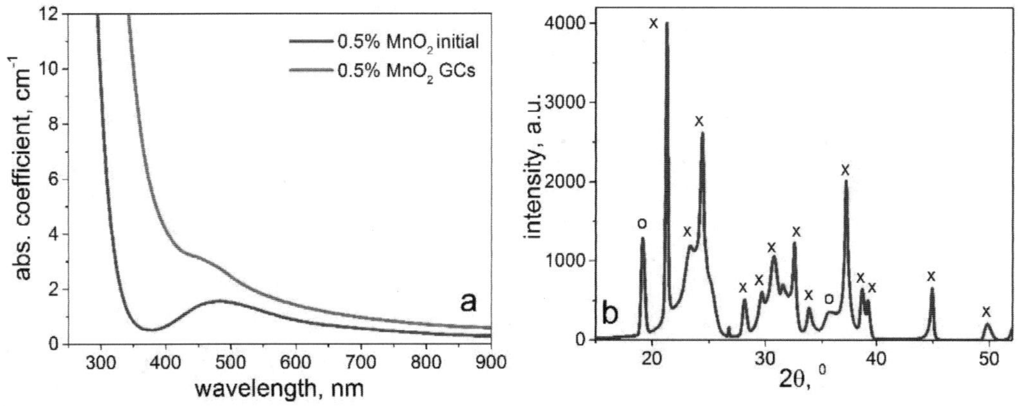

Figure 2. (a) The absorption spectra of initial glass and glass-ceramics with 0.5% MnO_2; (b) the XRD results of lithium-germanate glass-ceramics with 0.5% MnO_2: **x** is denoted for $Li_2Ge_7O_{15}$ phase, **o** is denoted for $Li_{0.115}MnO_2$ phase.

The luminescence of divalent manganese locates at the red-orange region with a maximum at 580-620 nm upon excitation at 410-415 nm [9]. Under the influence of the environment, the luminescence band of Mn^{2+} can shift up to 700 nm, but the large FWHM will remain its characteristic (~100 nm) [10]. The energy level diagram of Mn^{4+} is described by the Tanabe-Sugano diagram for d^3 transition elements in an octahedral field [11]. In this regard, in the crystalline environment, Mn^{4+} ions produce intense luminescence in the red region. There are many crystalline phosphors doped with tetravalent manganese, which have luminescence in the 660-670 nm region [1,3].

In Figure 3b, the luminescence spectra of the obtained glass-ceramics show a maximum at 667 nm, regardless of the activator content (excitation at 332 nm). Photoexcitation spectra are normalized to the level at 330 nm (luminescence at 667 nm). The spectra (Figure 3a) show the presence of three excitation bands with maxima at 332, 400, and 460 nm. With an increase in the manganese content, the contribution of the long-wavelength band increases. So, the specific location and shape of the luminescence and photoexcitation bands indicates that the manganese ions are in 4+ oxidation state.

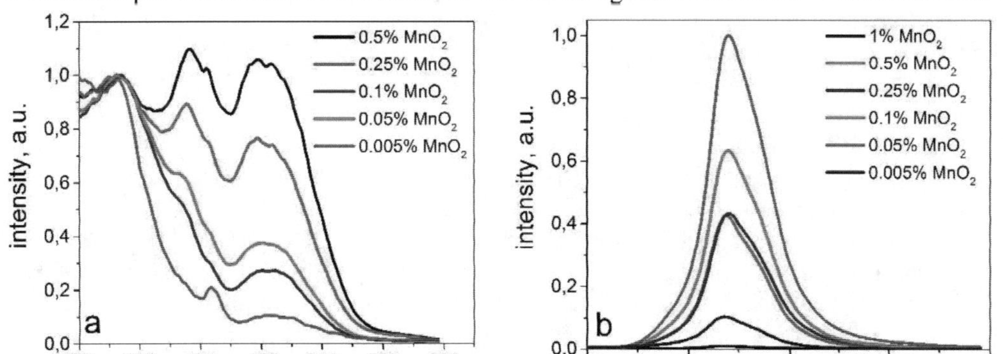

Figure 3. The excitation (a) and luminescence (b) spectra of the lithium-germanate glass-ceramics with different MnO_2 content.

The luminescence lifetime of Mn^{4+} ions depends on the environment; in crystalline phosphors it can vary from 1 to 6 ms at room temperature [12,13]. The decay curves were measured based on the luminescence spectra of glass-ceramics recorded with a delay after the exciting pulse (Figure 4). The luminescence decay curve has a two-exponential character. The luminescence lifetime for a glass-ceramic sample with 0.05% MnO_2 was 550 μsec, which was unexpectedly small. With a further increase in the manganese oxide content, the luminescence lifetime decreases, which indicates concentration quenching.

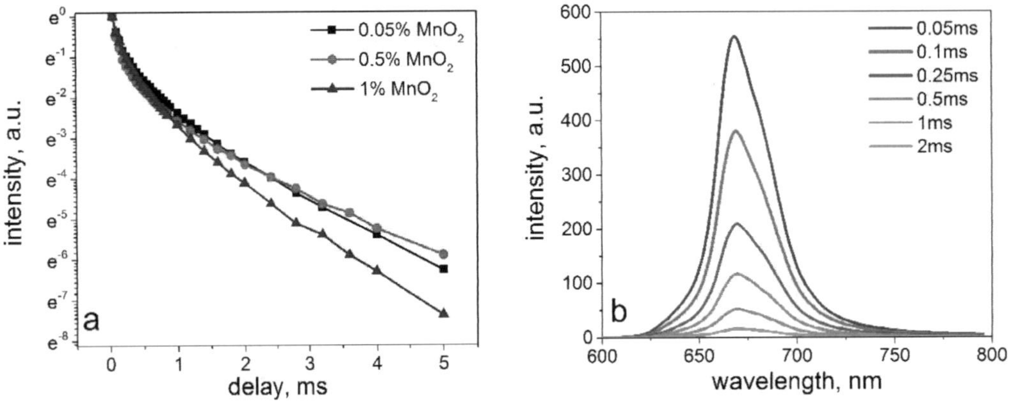

Figure 4. (a) The luminescence decay curve for glass-ceramics with different MnO_2 content (λ_{lum}=667 nm); (b) the glass-ceramics luminescence spectra with 0.5% MnO_2 after different delays from the exciting pulse (λ_{ex}=332 nm).

4. Conclusion

A series of Li_2O-GeO_2 glasses with MnO_2 concentration from 0.005 to 2 mol.% was synthesized. Lithium-germanate glass-ceramics were synthesized using standard volume crystallization technique thus nucleating $Li_2Ge_7O_{15}$ nanocrystals in the glass matrix. The glass-ceramics possessed intense emission near 660-670 nm under two-band excitation at 330 and 450 nm. The luminescence lifetime was 550 μs for 0.05 mol.% MnO2-doped glass-ceramic sample. The specific location and shape of the luminescence and photoexcitation bands indicated that the manganese ions in this matrix were in 4+ oxidation state. Due to this, lithium-germanate glass-ceramics doped with manganese ions were a promising material for the development of deep red radiation sources.

Acknowledgments

Research was funded by Russian Science Foundation (Agreement #19-72-10036).

References

[1] Adachi S 2018 *J. Lumin.* **202** 263–81
[2] Li K, Zhu D and Van Deun R 2017 *Dye. Pigment.* **142** 69–76
[3] Chen D, Zhou Y and Zhong J 2016 *RSC Adv.* **6** 86285–96
[4] Sevastianova I, Aseev V, Tuzova I, Fedorov Y and Nikonorov N 2019 *J. Lumin.* **205** 495–9
[5] Czaja M, Lisiecki R, Chrobak A, Sitko R and Mazurak Z 2018 *Phys. Chem. Miner.* **45** 475–88
[6] Saleh M A, Nugroho A A, Dewi K K, Supandi A R, Onggo D, Kuhn H and Van Loosdrecht P H M 2019 *Key Eng. Mater.* **811 KEM** 179–83
[7] Suzuki R, Kunitomo J, Takahashi Y, Nakamura K, Osada M, Terakado N and Fujiwara T 2015 *J. Ceram. Soc. Japan* **123** 888–91
[8] Kunitomo J, Suzuki R, Takahashi Y, Miyazaki T, Terakado N and Fujiwara T 2014 *J. Ceram. Soc. Japan* **122** 725–7
[9] Reisfeld R, Kisilev A and Jørgensen C K 1984 *Chem. Phys. Lett.* **111** 19–24
[10] Machado I E C, Prado L, Gomes L, Prison J M and Martinelli J R 2004 *J. Non. Cryst. Solids* **348** 113–7
[11] Tanabe Y and Sugano S 1954 *J. Phys. Soc. Japan* **9** 766–79
[12] Senden T, Van DIjk-Moes R J A and Meijerink A 2018 *Light Sci. Appl.* **7** 1–13
[13] Liang S, Shang M, Lian H, Li K, Zhang Y and Lin J 2016 *J. Mater. Chem. C* **4** 6409–16

Unambiguous state discrimination and joint measurement attacks on passive side channel of the light source in BB84 decoy-state protocol

D Babukhin[1,3,4] **and D Sych**[2,3,4,5]

[1] QRate LLC, Moscow 121353, Russia
[2] NTI Center for Quantum Communications, National University of Science and Technology MISiS, Leninsky prospekt 4, Moscow 119049, Russia
[3] National University of Science and Technology MISiS, 119049, Moscow, Russia
[4] Sirius University of Science and Technology, 1 Olympic Ave, 354340, Sochi, Russia
[5] P.N.Lebedev Physical Institute, Russian Academy of Sciences, Moscow, 119991, Russia

Abstract. Quantum key distribution (QKD) has a promise of unconditionally secure communication between the remote sides. The real-world QKD implementations, however, have numerous loopholes, both of engeneering and physical origin, and compromise the security promise. In this work, we investigate two attack strategies on the passive side channel of the light source along with the optimal cloning attack on the BB84 protocol with decoy-states. We calculate an upper bound of a secret key rate for these situations and show that the joint measurement attack on the signal and side channel degree of freedom is more effective to the adversary.

1. Introduction

Quantum communication is the most developed part of quantum technologies up to date. Quantum communication promises to create a secure sequence of bits between two parties (Alice and Bob) oblivious to the third party (Eve). It is possible if legitimate sides use single photons as carriers of secret bits. The quantum no-cloning theorem guarantees that one cannot copy a photon state without introducing disturbance. This change will result in errors in communication between legitimate parties, and thus eavesdropping will be revealed. When the QKD protocol is implemented in hardware, different loopholes and side channels arise in the protocol. These side channels allow the adversary to obtain more information about secret bits without alerting legitimate parties. This oblivious leakage makes it impossible to correct these errors during security amplification, and the promised security is compromised. For example, a sender-side can use several lasers as a source of photons for the protocol. Physical differences in lasers lead to differences in carrier photons. These discrepancies lead to additional distinguishability between photons, and thus, adversary can obtain more information about communication. Investigation of side channels arising from the difference between theory and real-world devices is a subject of practical security of QKD [1]. From a point when the security of theoretical BB84 protocol was established [2], many efforts were to incorporate different discrepancies between theoretical models and real-world devices [3-5].

Here we investigate two attack strategies, which the adversary can use to get additional information about the secret bit from physical differences of carrier photons. We estimate this information using two-photon interference [1] for a phase-covariant optimal cloning attack on the decoy-state BB84 protocol. We provide an upper bound for the secret key rate for these cases and demonstrate, that it is more effective to make a joint measurement of both cloned signal and side channel states rather than make an unambiguous state measurement of the side channel followed by measurement of the cloned signal.

2. The individual unitary attack on the BB84 protocol

During the individual unitary attack, Eve brings a carrier photon into the unitary interaction with an ancillary quantum system to obtain two quantum systems with states which are optimal clones of the initial state. That means states of outcoming two systems have a close to unity fidelity with the initial system. This optimal cloning procedure is equivalent to the individual attack on the BB84 protocol [6]. If we consider photon states from X and Y bases, the following transform does the optimal cloning

$$U|\Psi(\varphi)\rangle_B|0\rangle_E = \frac{1}{\sqrt{2}}(|0\rangle_B|0\rangle_E + cos\eta|1\rangle_B|0\rangle_E + sin\eta|0\rangle_B|1\rangle_E) \quad (1)$$

$$|\Psi(\varphi)\rangle_B = \frac{1}{\sqrt{2}}\left(|0\rangle_B + e^{-i\varphi}|1\rangle_B\right) \quad (2)$$

Here, B denotes the Hilbert space of Bob, E denotes the Hilbert space of Eve, φ is a relative phase of quantum states from XY plane, and η is a control parameter of cloning. This operation creates two quantum states which have a high-fidelity overlap with the quantum state sent by Alice. It is instructive to see the extreme cases of this transform: for η = 0, this transform does nothing, and for η = π/2, it flips the state of Bob and the initial state of Eve.

3. Attack strategies on the passive side channel of the light source

As a model for photon distinguishability, we consider the augmentation of a side channel degree of freedom. Formally, we attached a side channel degree of freedom to the bases of the BB84 protocol:

$$\rho_x = \frac{1}{2}\left(|0_x\rangle\langle 0_x| \otimes \rho^{\Delta}_{x,0} + |1_x\rangle\langle 1_x| \otimes \rho^{\Delta}_{x,1}\right) \quad (3)$$

$$\rho_y = \frac{1}{2}\left(|0_y\rangle\langle 0_y| \otimes \rho^{\Delta}_{y,0} + |1_y\rangle\langle 1_y| \otimes \rho^{\Delta}_{y,0}\right) \quad (4)$$

where states $Tr\left[\rho^{\Delta}_{a,b}\rho^{\Delta}_{a',b'}\right] = \delta_{aa'}\delta_{bb'} + (1-\delta_{aa'})(1-\delta_{bb'})\Delta$ are nonorthogonal to each other in general case. We use this model to demonstrate two strategies of eavesdropping with the use of photon distinguishability.

3.1. Unambiguous discrimination of side channel states (USD)

A USD measurement is a type of optimal quantum measurement, which allows obtaining the full information about the state sent from an ensemble of possible states [7] The trade-off is that this measurement can fail, and no information will be obtained about the measured state. For an ensemble of non-orthogonal and linearly independent states $\{|\varphi_i\rangle$, i = 1,2,3,4$\}$, this measurement is formulated as follows:

$$\sum M_k = I_n \quad (5)$$

$$\langle\varphi_i|M_k|\varphi_i\rangle = p_i\delta_{ik}, 1 \leqslant i, k \leqslant n \quad (6)$$

where M_k are POVM operators, representing quantum measurement device and p_i are probabilities of states in the ensemble. The probability to fail measurement is defined as a probability to obtain an inconsistent result on a particular sent state, it is provided with a formula

$$\langle\varphi_i|M_0|\varphi_i\rangle = 1 - p_i \quad (7)$$

In this scenario of attack, Eve applies the USD measurement to the state of the side channel. If the measurement succeeds, Eve knows exactly what was the state of side channels. She waits until the basis exchange of legitimate sides and afterward knows exactly what was the state of the carrier photon on a corresponding position. However, if the measurement has failed, Eve attacks the signal photon state with the optimal cloning attack as in a usual scenario of the BB84 protocol eavesdropping. As a result, her information about the secret bit is

$$I_{AE} = P_{usd} + (1 - P_{usd})I_{AE}^{ind} \tag{8}$$

where I_{AE}^{ind} is the amount of information about the secret bit, which Eve obtains from a cloning attack on the signal photon state.

3.2. Joint measurement attack on side channel states

In this scenario of attack, Eve attacks a signal photon with an optical phase covariant cloner, thus obtaining a non-ideal copy of the signal photon state. Afterward, she jointly measures a copy of the signal photon with a side channel state with a minimum error measurement. After the cloning attack, Eve (after waiting for the basis exchange step) effectively has an ensemble, consisting of two states of the form:

$$|0_x\rangle\langle 0_x| \otimes \rho_{x,0}^{\Delta} \text{ or } |1_x\rangle\langle 1_x| \otimes \rho_{x,1}^{\Delta} \tag{9}$$

To discriminate two states, Eve uses a Helstrom measurement. A probability to guess which state was sent from the ensemble of states $\rho_1$and ρ_2 with minimal error is

$$P_{guess} = \frac{1}{2}\left(1 + Tr\left[\sqrt{(\rho_1 - \rho_2)(\rho_1 - \rho_2)^{\dagger}}\right]\right) \tag{10}$$

4. Decoy state BB84

Although in theory the BB84 protocol is formulated for single photons, in a real experiment it is hard to incorporate a truly single-photon sources in a QKD setup. For QKD purposes, attenuated weak coherent pulses with randomized phases, which are easily generated with lasers, are usually used. These pulses have a form

$$\rho_x^{\mu} = \sum e^{-\mu}\frac{\mu^2}{k!}|k;x\rangle\langle k;x|, \tag{11}$$

where μ is the intensity of the coherent signal and x is a value of a secret bit. Signal states of this form have an incorporated vulnerability, which consists of two and more photon parts. This kind of state opens the door for another way of oblivious eavesdropping with a photon-number-splitting attack. Because the quantum phase of this state is random, Eve can make a non-demolishing photon number resolving the attack, and if there is more than one photon in the signal, she splits a single photon to herself. Because this photon has the same state, as other photons in the many-photon state, Eve has the same quantum state as Bob has at the end of the communication. After the basis exchange step, both Bob and Eve measure their photons in the proper basis and thus have the same value of a secret bit. Because there is no error in communication between Alice and Bob in this scenario, the QKD protocol is no longer secure.

The mentioned loophole is closed with the addition of decoy states to the conventional BB84 protocol [8,9,10]. Alice sometimes sends special phase-randomized weak coherent pulses with an intensity, different to informational signal pulses. Because Eve cannot discriminate between signals with different intensities, she applies the same attack sequence to every pulse in the channel. After the whole communication process, Alice and Bob reveal, which pulses were informational, and which pulses were decoy-states. They use decoy states statistics to estimate the single-photon component in the whole communication process, a component which Eve cannon attack without introducing disturbance. Thus, a security loophole of many-photon states from coherent pulses is closed.

Let us elaborate on the use of decoy states statistics. If Bob receives a signal with intensity μ, a probability to have a click on the detector is

$$Q_{\mu} = \sum e^{-\mu}\frac{\mu^2}{k!}Y_k, \tag{12}$$

where Y_k is a k-photon yield:

$$Y_k = \sum\sum P_A(a) P_{A|B}(b|A = a), \tag{13}$$

Using this statistics of counts, Bob can estimate the bit error rate

$$E_\mu = \frac{1}{Q_\mu} \sum e^{-\mu} \frac{\mu^2}{k!} Y_k e_k, \tag{14}$$

These detection outcomes are further used to estimate the vacuum yield Y_0, the single-photon yield Y_1, and single-photon error rate e_1. These quantities are used to calculate the single-photon part of all communication process, which is a quantum-secure part. The secret key rate for this component is given by (see [11]):

$$R(e_1) = \frac{1}{2}\left(Q_1 I_{AB}(e_1) - f Q_\mu max\left(I_{AE}(E_\mu), I_{BE}(E_\mu)\right)\right), \tag{15}$$

where f is the efficiency of error correction and e_1 is a single-photon fraction error.

5. The Hong-Ou-Mandel visibility and side channel states

The Hong-Ou-Mandel (HOM) effect is a two photons interference effect when these photons are incident on a balanced beam splitter. When modes of these photons are perfectly matched (i.e., they have the same polarization, spectral and temporal distribution), they exit the beam splitter pairwise. But when the indistinguishability of these photons is not complete, they will sometimes exit different sides of the beam splitter. The degree of photons indistinguishability is the HOM interference visibility, which for phase randomized weak coherent pulses reads (see [1] for details)

$$V_{\rho_1\rho_2} = \frac{1}{2} Tr[\rho_1\rho_2] \tag{16}$$

6. Results and discussion

We have calculated secret key rates for the aforementioned attack strategies with different powers of informational leakage through a source side channel (which correspond to different values of the HOM visibility of the sources). We provide calculation results in figure 1. We used the following parameters for simulations: a postprocessing efficiency f = 1.22, a quantum channels attenuation values α = 0.2 dB/km, efficiency of the Bob detector η_{Bob} = 0.01, background yield Y_0 = 10^5, background error rate e_0 is 0.5 and mean photon number in a pulse μ = 0.5. The critical length of the transmission line is provided with the formula

$$L(e_1) = \frac{-10}{\alpha} log_{10}\left[\frac{Y_0(e_0 - e_1)}{\eta_{Bob}(e_1 - e_{det})}\right], \tag{17}$$

where e_1 is the error in the single-photon fraction and e_{det} is the detection error.

From figure 1 we conclude, that the joint measurement of the signal state after phase-covariant cloner attack and the side channel state allows Eve to gain more information about the secret bits, than using the USD attack on the side channel alone and then proceeding with the attack on the signal degree of freedom if the USD attack fails. We also prove that using the physical distinguishability of photons does not give Eve much additional information about the secret key. For HOM visibility values in the range from 0.4 to 0.5 (which correspond to the state-of-the-art values of contemporary photon sources), we see only a slight decrease in the security of the protocol. We can conclude that physical differences between signal photons, which lead to additional distinguishability at the level of the state-of-the-art photon sources, do not lead to a severe drop in the BB84 protocol secrecy under the considered attacks.

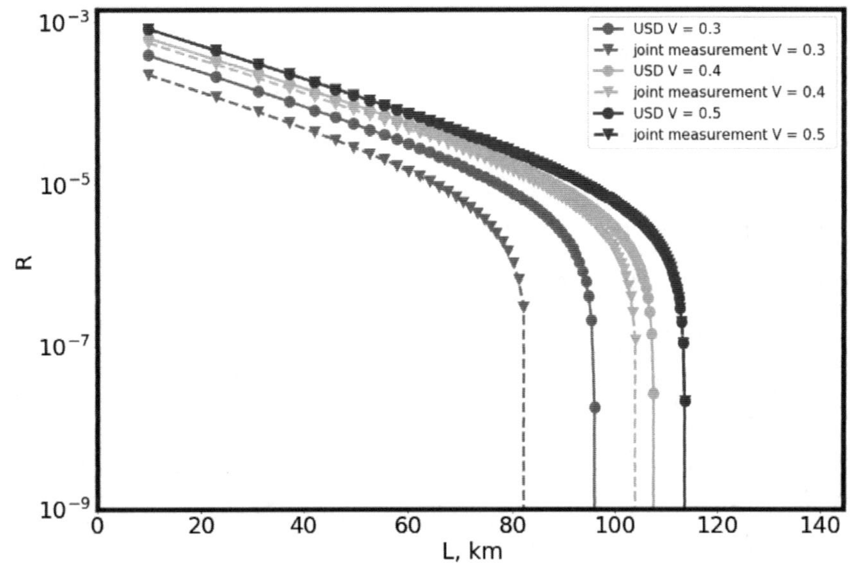

Figure 1. Dependence of the secret key rate R on the length of the transmission channel L for several values of the two-photon interference visibility V. Different curves correspond to the USD measurement of side channel state (circles) and the joint measurement of cloned signal and side channel states (triangles).

Acknowledgements
Denis Sych acknowledges funding by RFBR, Sirius University of Science and Technology, JSC Russian Railways and Educational Fund "Talent and success", project number 20-32-51004. Danila Babukhin acknowledges funding by FASIE.

References
[1] Duplinskiy A and Sych D V 2021 *Phys. Rev. A* **104** 012601
[2] Shor P W and Preskill J 2000 *Phys. Rev. Lett.* **85** 441-444
[3] Gottesman D, Lo H-K, Lutkenhaus N and Preskill J 2004 *Quant. Inf. Comput.* **5** 325-360
[4] Koashi M and Preskill J 2003 *Phys. Rev. Lett.* **90** 057902
[5] Koashi M 2009 *New J. Phys.* **11** 045018
[6] Bruß D, Cinchetti M, D'Ariano G M and Mac-chiavello C 2000 *Phys. Rev. A* **62** 012302
[7] Eldar Y C 2003 *IEEE Trans. Inf Theor.* **49** 446-456
[8] Hwang W-Y 2003 *Phys. Rev. Lett.* **91** 057901
[9] Lo H-K, Ma X and Chen K 2005 *Phys. Rev. Lett.* **94** 230504
[10] Lo H-K 2004 *Proc. IEEE ISIT* (IEEE Press) p 137
[11] Ma X, Qi B, Zhao Y and Lo H-K 2005 *Phys. Rev. A* **72** 012326

Encapsulation of Recrystallized Inorganic Perovskite Quantum Dots in Nonwoven Fluoropolymer Fibers

M Baeva[1], V Neplokh[1], D I Markina[2], A M Pavlov[3], D A Kirilenko[4], I S Mukhin[1], A P Pushkarev[2], S V Makarov[2] and A A Serdobintsev[3]

[1] Department of Physics, Alferov University, Khlopina 8/3, 194021 St. Petersburg, Russia;
[2] Department of Physics and Engineering, ITMO University, Lomonosova 9, 197101 St. Petersburg, Russia;
[3] Education and Research Institute of Nanostructures and Biosystems, Saratov State University, Astrakhanskaya 83, 410012 Saratov, Russia;
[4] Ioffe Institute, Politekhnicheskaya 29, 194021 St. Petersburg, Russia;

baeva@spbau.ru

Abstract. Since emerging Metal Halide Perovskites were attracting elevated attention from scientific and industrial communities. Although, we still yet to see commercial products based on perovskite materials a significant body of work has been already done. Here we present our own method of Metal Halide Perovskites integration into possible industrial applications. Our way of Perovskite Quantum Dots encapsulation in polymer fibers provides exceptional water and ambient stability as well as optimal photoluminescence.

1. Introduction

The promising photoactive materials with a general stoichiometry of ABX3, where A = Cs, MA, FA; B = Pb, Sn; X = Cl, Br, I, - Metal Halide Perovskites (MHPs) - captivate the attention of international industrial and scientific community due to their excellent optical and electrophysical properties. For instance, Inorganic Metal Halide Perovskite (IMHP) Quantum Dots (QDs), CsPbBr3 QDs in particular, were reported to possess not only the tunable with the QDs size photoluminescence (PL) spectra but also the 95% efficient PL quantum yield[1]. Therefore, a significant body of work has already been done to apply CsPbBr3 QDs in photovoltaic (solar cells) or/and optoelectronic devices (LEDs, LASERs, LASER beam detectors, etc.)[2-6] Nevertheless, to introduce any flexible device into the civil use market it is of crucial importance to encapsulate CsPbBr3 QDs to improve their stability and user safety (Pb toxicity concerns).

In this work, we present the synthesis of fluoropolymer nonwoven mats with CsPbBr3 QDs encapsulated in their fibers, where during the polymer fiber electrospinning, we observe the self-organized IMPH QDs recrystallization. Moreover, the perovskite QDs, encapsulated inside of the polymer fibers demonstrate two-photon absorption nonlinear process in addition to excellent water stability.

2. Experimental section

2.1 Synthesis of perovskite nanoparticles

To obtain monodisperse $CsPbBr_3$ QDs we utilize hot injection method[7], where at 140-200 °C under N_2 atmosphere perovskite nanoparticles precipitate from dissolved in 1-octadecene perovskite precursors, Cs-oleate and lead (II) halide ($PbBr_2$). Quantum dots were stabilized by introducing 1:1 mixture of oleylamine and oleic acid into 1-octadecane. The $CsPbBr_3$ QDs nucleation and growth takes mere seconds with resulting size distribution from 4 nm to 15 nm, achievable via reaction temperature varying (140-200 °C). The synthesized QDs were transferred from 1-octadecene to toluene by centrifugation due to the higher stability of nanoparticles in toluene. To be ready for nonwoven mats fabrication QDs/toluene solution was dried resulting in stoichiometrically pure $CsPbBr_3$ material.

2.2 Synthesis of nonwoven mats

Polymer material for nonwoven mats is prepared from 12% (wt) solution of vinylidene fluoride-tetrafluoroethylene copolymer (100-150 KDa, Halopolymer, Russia) in a 1:1 mixture of dimethylformamide (DMF) and butyl acetate (BA). For mats containing perovskite material first $CsPbBr_3$ QDs were dispersed in BA, see Figure 1, a) and b), then combined with DMF, next fluoropolymer was dissolved in perovskite containing solution, see Figure 1, c) and d). A typical procedure involves 12 mL of solvent (BA/DMF 1:1), 3.2 mg of NPs and 1.5 g of fluoropolymer.

A conventional laboratory syringe setup was used to electrospin polymer fibers with needle connected to ground terminal of the power supply (HCP 140-65000, F.u.G. Elektronik GmbH, Germany) and negative (60 kV) terminal being a rectangular (15x20 cm) aluminum plate located at 20 cm distance from the needle tip. The polymer solution was pumped through the needle using the syringe pump at a flow of 10 mL/h for 30 minutes and collected onto the coated paper fixed to the metal electrode surface, see the final sample in Figure 1, e) and f).

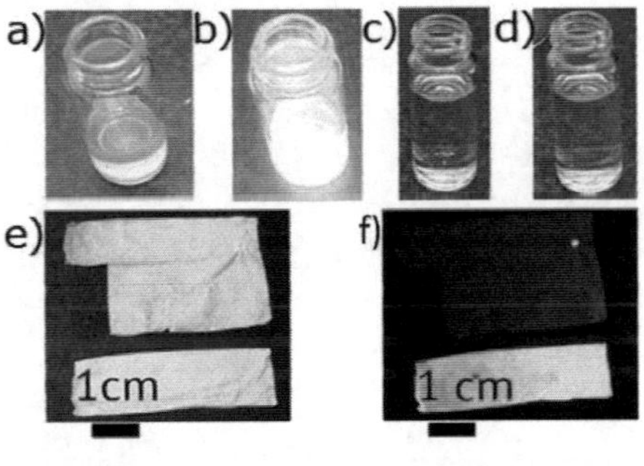

Figure 1. Photographs of $CsPbBr_3$ QDs in BA under (a) fluorescent and UV (b) lighting. $CsPbBr_3$ QDs in 1:1 BA/DMF solution under fluorescent (c) and UV (d) lighting. Nonwoven mats without (up) and with (down) recrystallized $CsPbBr_3$ QDs in the daylight (e) and under UV lighting (f).

2.3 Microscopic and optical characterization

The imaging of the insides of the nonwoven mats' fibers were investigated with Transmission Electron Microscopy (TEM), using JEOL JEM-2100F microscope, with an operating voltage of 200 kV. The $CsPbBr_3$ QDs phase content was studied through powder x-ray diffraction method via measurements of 2D XRD patterns in Kappa Apex II diffractometer (Bruker AXS) using CuK_α (λ = 1.54184 Å) radiation generated by a $I_\mu S$ microfocus X-ray tube. The X-ray powder diffraction patterns were collected from the CsPbBr3 NPs dispersed in the polymer material, and pure polymer material was used as a reference sample.

The continuous PL characterization of nonwoven mats was performed using Axio Imager A2m (Carl Zeiss) microscope with 50× objectives (Carl Zeiss EC Epiplan-NEOFLUAR), 365 nm UV mercury lap and optical fiber spectrometer (Ocean Optics QE Pro) with detection area spot size 2 μm.

To evaluate the IR to visible light up-conversion we used Pharos PH2-SP-20W-2mJ single-unit integrated femtosecond laser system and high power optical parametric amplifier Orpheus-F with laser excitation wavelength of 900 nm, laser pulses repetition rate 100 kHz, pulse duration 220 fs.

The water stability study was carried out in a following manner: the PL signal was first recorded from several points from the dry nonwoven mat sample, next water was added and PL was measured at several time points within a 4-hour time span. Two series of measurements in a one-week interval were performed with samples being stored in a fridge at 2-4 °C.

3. Results and Discussion

The Figure 2 demonstrates the TEM images of reference sample fibers next to fibers with incapsulated $CsPbBr_3$ QDs. According to our data the average size of perovskite inclusions in the polymer fibers is around 10 nm. The inset in Figure 2, b) shows the powder XRD data, with clearly identifiable (100), (200), and (211) $CsPbBr_3$ peaks proving the crystallinity of our recrystallized nanoparticles.

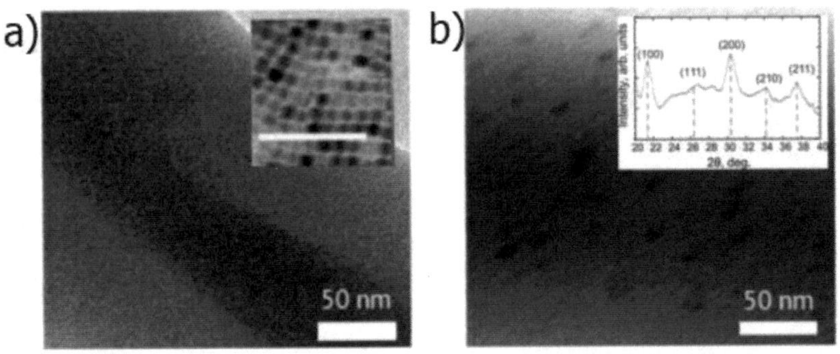

Figure 2. TEM images of the electrospun mats without $CsPbBr_3$ QDs and with them. Inset in a) – TEM image of as-synthesized QDs, scale bar is 100 nm. Inset on b) – X-ray powder diffraction pattern of $CsPbBr_3$ QDs in nonwoven fibers

It can be seen from the Figure 3, a), that recrystallized QDs (mean QD size 10,5 nm) in fibers demonstrate a slightly wider QDs size distribution as well as a PL signal red-shift when compared to the as-synthesized CSPbBr3 QDs (mean QD size 9,5 nm) in colloid solution. We showed experimentally that the CsPbBr3 QDs nonwoven mats demonstrate a visible luminesce coming from the two-photon nonlinear absorption phenomenon, which is essential for the IR LASER light detection, see Figure 3, b).

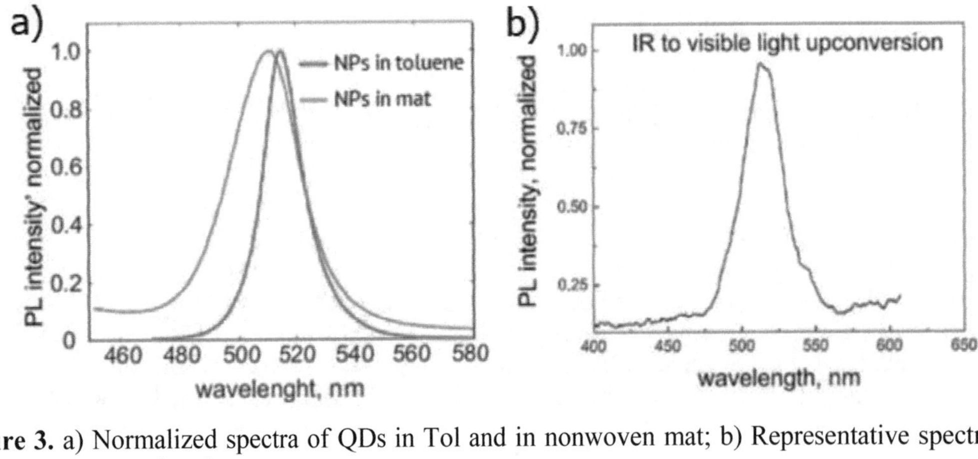

Figure 3. a) Normalized spectra of QDs in Tol and in nonwoven mat; b) Representative spectrum of up-conversion PL signal.

Additionally, we conducted water stability tests of our recrystallized QDs: IMHP QDs in nonwoven mats show only a 40% PL signal intensity drop after a 9-hour immersion test, whereas as-synthesized QDs in colloid solution dissolve immediately when placed into water, see **Figure *4***.

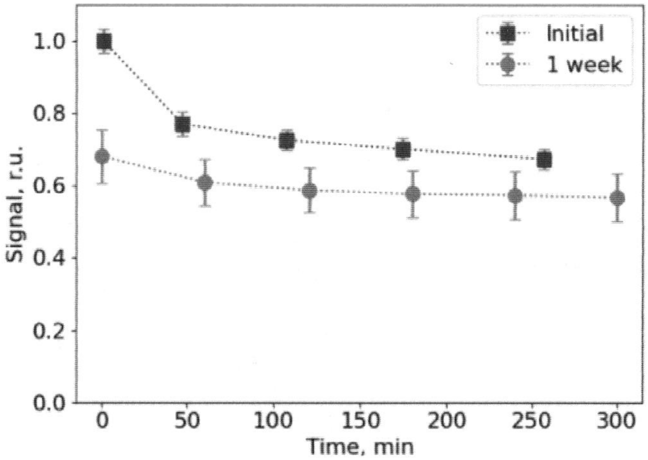

Figure 4. Fluorescence signal from electrospun mats with NPs, intensity related to the initial maximum value. Two series of measurements on a single sample are shown.

4. Conclusion

To conclude, we report spontaneous $CsPbBr_3$ QDs recrystallization during the electrospinning process of polymer fibers. From PL spectra (λ_{PL} = 506nm) deconvolution we calculated that the recrystallized perovskite nanocrystals exhibit mean QD size of 10,5 nm with size distribution range of 5–14 nm, which is proven by TEM imaging. In addition to bright UV light down-conversion, our nonwoven mats display up-conversion at IR fluence 1 mJ/cm^2 to 513 nm green light. Water stability test reveals 60% PL intensity retention over 9 h.

Thus, the composite system studied can be considered an encouraging solution for perovskite QDs integration into feasible industrial applications, such as IR LASER light detection, etc.

Acknowledgments

This work was supported by State assignment of the Russian Ministry of Science and Higher Education № 0791-2020-0005.

References

[1] Sutherland B R and Sargent E H 2016 *Nat. Photonics* **10** 295
[2] Green M A, Ho-Baillie A and Snaith H J 2014 *Nat. Photonics* **8** 506–514
[3] Lu M, Zhang Y, Wang S, Guo J and Yu W W 2019 *Adv. Funct. Mater.* **29** 1902008
[4] Polushkin A S, Tiguntseva E Y and Pushkarev A P 2020 *Nanophotonics* **9** 599–610
[5] Luo W, Yan L, Liu R and Zou T 2019 *Semicond. Sci. Technol.* **34** 074004
[6] Soranyel G C, Galian R E and Pérez-Prieto J 2015 *J. Mater. Chem. A* **3** 39187–39193
[7] Lu M *et al* 2019 *Adv. Funct. Mater.* **29** 1902008

Synthesis and studying properties of the GNPs@Fe_xO_y structure

P G Bespalova [1], A A Vorobyev [2], A B Speshilova [1], V M Studzinsky [1], E A Patirimova [3], P A Karasev [1] and M V Mishin [2]

[1] Peter the Great St. Petersburg Polytechnic University, St. Petersburg 195251, Russia
[2] Saint Petersburg Academic University, St. Petersburg 194021, Russia
[3] Lobachevsky State University of Nizhny Novgorod, Nizhny Novgorod 603950, Russia

polinlin0508@mail.ru

Abstract. An article presents a study of the regularities of the formation of gold up to 8 nm thick, deposited by vacuum thermal deposition on silicon wafers with a natural oxide, its annealing and subsequent deposition of iron oxide by chemical vapor deposition. The stages were accompanied by SEM analysis of the sample surface, as well as fixation of the optical and FTIR spectra, I–V characteristics of the obtained structures.

1. Introduction

Iron oxide is a widespread compound, interest in which is justified by the presence of various phases with significantly different properties. It is obtained in the form of particles and films by chemical vapor deposition, plasma-chemical methods, magnetron sputtering, etc [1]. The selected method allows to obtain a phase with desired properties. At the same time, the subsequent modification of such a material can lead to a significant change in certain properties. Therefore, the study of laws of formation and basic properties (optical and electrical) is quite urgent.

In turn, the patterns of gold formation are of great interest in connection with the manifestation of LSPR, and the combination of surface plasmon resonance and oxide matrices is relevant for a variety of sensors - gas sensors, IR reflectors, SERS etc [2,3].

In work, gold of various thicknesses was deposited on n-Si silicon substrates by vacuum thermal sputtering. Then, by chemical deposition from the gas phase, iron oxide (Fe_2O_3) was applied (the process is described in more detail here [4] at a temperature of 250 °C. At the same time, gold was annealed with different thicknesses. The thickness of the gold was determined by the RBS method. Gold-oxide coatings were studied by scanning electron spectroscopy, spectrophotometry, IR spectroscopy. An I–V characteristics were also measured using two probe method.

2. Methodology

In this work, gold of various thicknesses was deposited on silicon substrates n-Si by vacuum thermal spraying. Then, iron oxide was applied by chemical vapor deposition at a temperature of 250 °C. At the same time, gold of various thicknesses was annealed. The thickness of gold was determined by the RBS method (Rutherford backscattering spectroscopy).

Gold-oxide coatings were studied by scanning electron spectroscopy Zeiss Supra, spectrophotometry (Shimadzu UV-3600 Plus spectrophotometer), IR-spectroscopy (FSM 1201), Reflection high-energy electron diffraction (RHEED). The study was made of electrical properties using I – V measurements.

3. Results

3.1 Surface morphology

The thickness of the coatings was measured by Rutherford backscattering spectroscopy. The gold thicknesses were 1, 1.5, 2.5, 3, and 6 nm. Results are presented in Figure 1. Then the resulting structures were annealed in an argon atmosphere at a reduced pressure P = 1000 Pa and temperature of 250 °C. In one case, they were only annealed, in the other case, they were annealed and iron oxide was applied (the technology is described in more detail in [5].

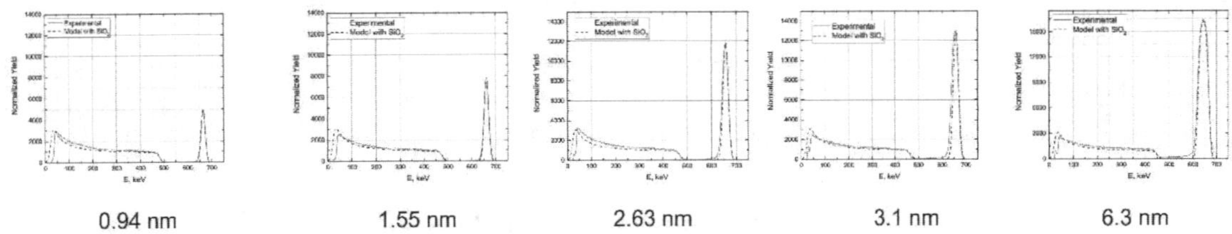

Figure 1. Measurements of gold film thickness by Rutherford backscattering spectroscopy.

Figure 2a-e shows SEM images of the surface of samples coated with gold of various thicknesses. The study of the formation patterns showed that on the natural oxide, gold of a smaller thickness (less 2 nm) forms particles, and with an increase in the thickness of the structure known as "golden islands".

After gold deposition SEM images of the deposited films were recorded (Figure 2 f-j). This series of images is characterized by a sequential increase in the specific surface area occupied by gold, while spherical particles are characteristic for thicknesses less than 2 nm and with an increase in thickness structures called golden islands were formed.

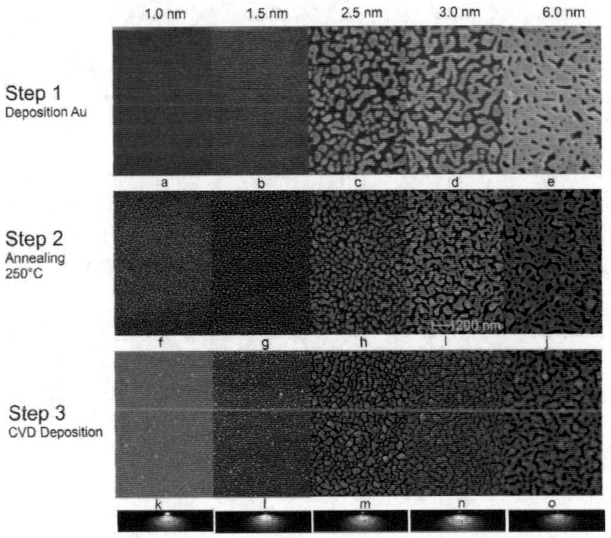

Figure 2. SEM-images during different steps of formation.

At the second step, during the annealing of such structures, the surface morphology of the sample changes; at the same time, at thicknesses less than 2 nm, the particles agglomerate and at large thicknesses, after annealing, the specific area occupied by gold decreases.

At the last step, gold is coated with iron oxide at a temperature 250 °C. A uniform coating of gold particles with iron oxide occurs while maintaining the shape of the original surface (Figure. 2 k-o, while the crystallinity of iron oxide increases with an increase in the initial thickness of gold).

With the simultaneous folding of gold and the application of iron oxide on thin films, particles of two types were formed, the structures of the "golden islands" type were simply covered, while maintaining the topology as in simple annealing. The type of crystals is most characteristic of alpha-iron oxide (as it was proven by RHEED) and have a trend for formation crystals with gold thicknesses increasing).

3.2 Optical measurements

The optical reflection spectra are presented for gold and gold with iron oxide in the range of 200-1600 nm. It can be seen that for both groups of spectra, the maximum reflection is in the region of 800 nm, while for samples without iron oxide, a maximum is observed in the region of 300 nm, which is typical for silicon. 6 nm gold with iron oxide also has a minimum reflection (7%) in the region of 500 nm.

IR-spectra show a decrease in the transmission intensity of gold, while the 6 nm film almost completely absorbs IR-radiation. Films of smaller thickness have peaks characteristic of silicon and its oxide and a lower transmission intensity. Figure 3 shows the optical reflectance spectra of the obtained structures (gold and gold with iron oxide).

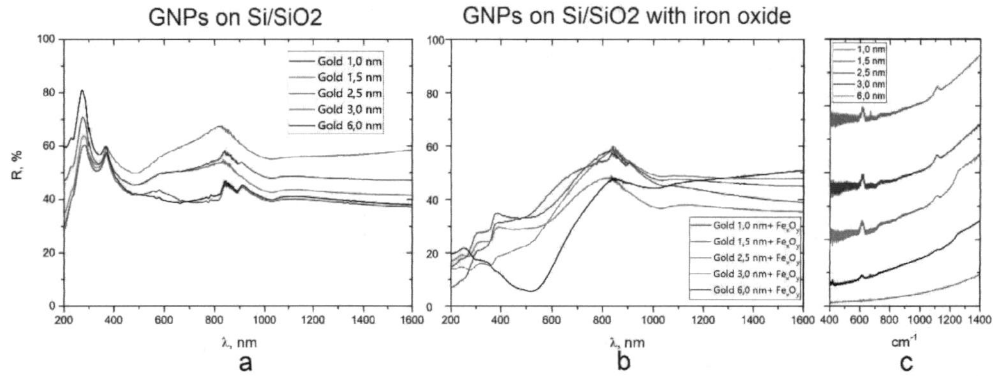

Figure.3 Optic spectra of: a) Si/SiO_2 with gold nanoparticles annealed at 250 °C; b) $Si/SiO_2/GNPs@Fe_xO_y$., c) view of FTIR spectra of $Si/SiO_2/GNPs@Fe_xO_y$ structures.

The nature of the spectra for 1 and 1.5 nm is very similar. In this case, the spectra 1, 1.5, 2.5 nm of gold-iron oxide reflection are characterized by the form of the spectrum, characterized by a maximum reflection at 860, 300 and 400 nm. While the structures GNPs @ Fe_xO_y 3 and 6 nm gold thickness have a gentler bend in the 200-400 nm region, and in the 400-600 region the sample has a pronounced minimum of reflection.

Figure 3c shows a view of the IR spectra of GNPs-Fe_xO_y samples. The IR-spectra were recorded in transmission mode. IR-spectroscopy of gold films is not very common because gold films are good reflectors of IR-radiation. Nevertheless, it can be seen that with decreasing film thickness, the reflection intensity decreases, and at 2.5 and 3.0 nm gold thicknesses have a shoulder typical for silica and thick film when 6 nm nearly absolutely reflect gold.

3.3 I-V measurements

The I–V characteristics of the samples were obtained using a two-probe measurement method. Measurements were made using a Keithley 2400 current source and voltage meter. The I-V measurements of the samples were obtained at a longitudinal displacement. Two gold contacts were deposited on the surface of the structure. A halogen lamp was used as a light source. All measurements were carried out under the same illumination conditions. Figure 4 shows the I-V characteristics of the samples at longitudinal displacement.

From the graphs presented in Figure 4 it can be seen that the nature of the dependences in the absence and presence of irradiation for the samples is different. As the thickness of the initial gold film increases, the value of the photoinduced current decreases and the behavior of the I–V characteristic changes. The ohmic character of the dependence begins to dominate from the barrier character as the film thickness increases, which is associated with an increase in the degree of filling the substrate surface with the deposited materials. To understand the behavior of the I – V characteristics of the samples consider Figure 5.

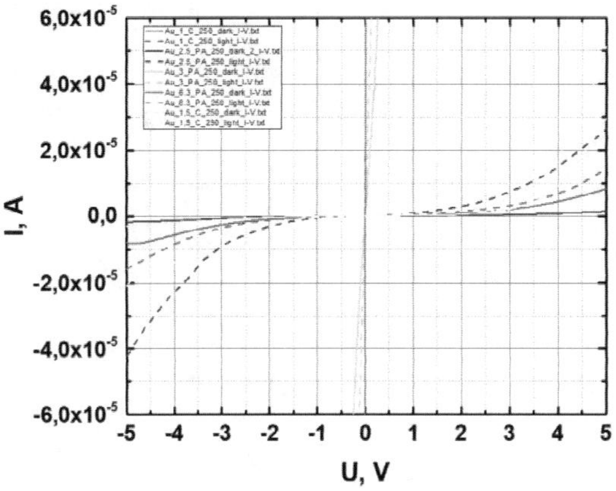

Figure 4. I-V measurements of structures with and without illumination. The inset picture shows zoomed image of I-V measurements.

Based on Figure 5, it can be assumed that the photo-response of the obtained samples changes depending on the initial thickness of the gold film. As the thickness of the initial gold film increased, an extreme value of the photoinduced current was observed. With increases of the film thickness the ohmic nature of the dependence begins to prevail on the barrier character. Presumably, this is due to the thickness of the final coating after the deposition of iron oxide the efficiency of light capture and separation of charge carriers.

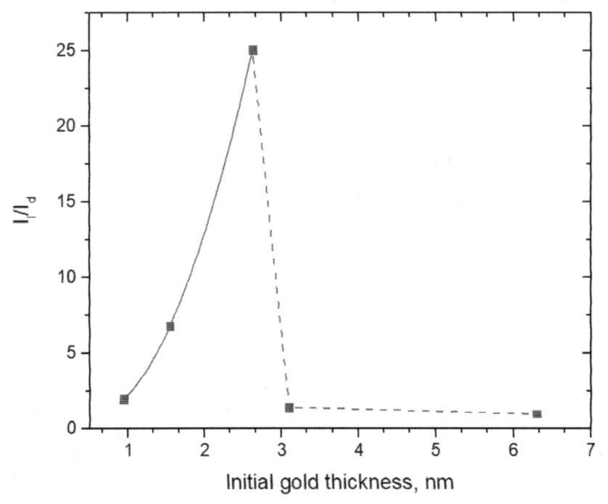

Figure 5. Dependence of ratio of light and dark currents from initial gold thickness.

4. Conclusions

We presented a study of the regularities of the formation of gold up to 8 nm thick, deposited by vacuum thermal deposition on silicon wafers with a natural oxide, its annealing and subsequent deposition of iron oxide by chemical vapor deposition. The stages were accompanied by SEM analysis of the sample surface, as well as fixation of the optical and FTIR spectra, I–V characteristics of the obtained structures.

Acknowledgments
The reported study was partially supported by Ministry of Science and Higher Education of the Russian Federation (research project FSRM-2020-0011). Metal contacts' fabrication and I–V-investigations was funded by RFBR, project number 20-38-90243. Research group thanks Ioffe Institute Ferroics Physics Laboratory for the provided equipment (Shimadzu UV-3600 Plus spectrophotometer).

References
[1] Vorobyev A, Sedov Y, Bespalova P, Shakhmin A, Kondrateva A, Gabdullin P and Mishin M 2020 Controlled formation of iron oxide nanoparticles by pulse-modulated RF discharge at atmospheric pressure *Materials Today: Proceedings* **30** 417-421
[2] Kondrateva A, Enns Y, Kazakin A, Kleimanov R, Morozov I, Karaseov P and Mishin M 2020 Electrooptical Properties of TiO_2 Doped with Gold Nanoparticles *Semiconductors* **54(14)** 1885-1888
[3] Kondrateva A, Enns Y, Komarevtsev I, Kudryashov D and Mishin M 2020 Au Nanoparticle Sub-Monolayers Buried between Magnetron Oxide Thin Layers *Journal of Physics: Conference Series* **1695(1)** 012002
[4] Bespalova P, Enns Y, Kunkel T, Balanov V, Speshilova A, Vorobyev A and Karaseov P 2020 Gold Nanoparticle Array Formation by Low-Temperature Annealing *International Youth Conference on Electronics, Telecommunications and Information Technologies* 255-262
[5] Bespalova P, Vorobyev A, Kunkel, T, Shakhmin A and Mishin M 2020 Characterization of iron oxide coatings prepared by MOCVD method from $Fe(CO)_5$ *Materials Today: Proceedings* **30** 434-438

Coherent tunable diffractional pulse shaping and generation of the 0π-pulse in Rb vapor

S N Bagayev[1], V A Averchenko[2], I A Chekhonin[2], M A Chekhonin[2], I M Balmaev[2] and I B Mekhov[2]

[1] Institute of Laser Physics, Novosibirsk 630090, Russia
[2] St. Petersburg State University, St. Petersburg 199034, Russia

chekhonin@mail.ru

Abstract. We have experimentally studied for the first time a new operation principle of the coherent diffractional pulse shaper (Rabi shaper). In the experiment, we observed an effect of tunable pulse shaping of nanosecond semiconductor laser pulse during the resonant pumping of the D_2 line (780.24 nm) of ^{87}Rb vapor in the range of self-diffraction angles $\varphi = \pm 4^o$. We observed the synthesis of nanosecond 0π-pulses at the small length of the nonlinear interaction 0.1...1 mm. We propose to use the Rabi shaper as an energy efficient tunable shaper of classical and single-photon wave packets. We analyze a possibility of the Rabi shaper operation in quantum systems with feedback.

1. Introduction

A problem of the angular deflection of laser radiation is the most difficult problem of the laser radiation control [1]. In [2,3], we proposed a new principle of the angular deflection of radiation wave vector during the laser pulse diffraction from the atomic resonant diffraction grating with the spatial pitch, which is time dependent – the Rabi deflector or Rabi pulse shaper.

As known, laser pulses with the shaped amplitude, frequency, phase, and wave vector $\boldsymbol{K}$ excite atoms and molecules more efficiently [4], which can be applied for the problems of all-optical signal processing and quantum technology. A review of modern linear methods of pulse shaping is given in [5].

In this report, we present the study of nonlinear transformation of the shape and pulse area during the increase of the pulse self-diffraction angle φ in the resonant medium: positive pulse → 0π-pulse → negative pulse.

The 2π-pulses and 0π-pulses are classical objects of resonant optics of atoms. According to the theory of self-induced transparency, their area is preserved during the propagation in a resonant medium. After the pulse has passed, the atomic system returns to its original state. This property is very important for problems of optical signal processing, because the limiting pulse rate of processed signals is limited by their duration only.

Nevertheless, the 0π-pulses have several important properties, which the 2π-pulses do not have. In the case, where the wavefront of a 2π-pulse is different from the plane wave, it loses its stability during its propagation at long distances. The 0π-pulse can have an arbitrary wavefront and preserve its stability. In addition, 0π-pulses are stable with respect to the degeneration of a resonant transition and to the inhomogeneous broadening.

The theory of transformation of a pulse of a small area to the 0π-pulse was given in [6] and was confirmed by multiple experiments with classical fields, cf. e.g. [7]. Later, the formation of resonant 0π-pulses was observed for single-photon wave packets [8,9]. The transformation of a small area pulse to the 0π-pulse during its propagation in a resonant medium is a quasilinear phase effect and it requires the long interaction length and high density of resonant atoms N_o.

In this report we show a new nonlinear method of pulse shaping using the effects of nonlinear self-diffraction in resonant media [2,3].

In the experiment, the generation of a 0π-pulse took place at a small distance of a nonlinear interaction with the medium of 0.1 ... 1 mm.

2. Experimental

The description of experimental setup and the signal processing method are presented in [3]. A cell with ^{87}Rb vapor was pumped by a pulsed tunable laser diode with the duration of 5.15 ns at the wavelength of transition D_2 (780.24 nm). The density of ^{87}Rb atoms was $N_o = 2.78 \cdot 10^{12}$ cm^{-3}. The cell was filled by buffer gas Ar at the pressure of 1 Torr. Laser pulse power did not exceed 10 mW. The transverse spatial distribution of the pump filed $E_{in}(t,x) = E_{in}(t) \cdot G(x)$ had a Gaussian shape: $G(x) = \exp[-(x/\sigma)^2]$, $\sigma = 0.017$ mm. The full pulse area was $\theta_{in} = 3\pi$.

In the experiment, we analyzed the form and pulse area of the diffraction pulse scattered from the resonant cell at different diffraction angles φ. The pulse form was registered by the single-photon counting method using Single Photon Avalanche Diode with the time resolution of 27 ps. After treating the temporal histogram of detected photons, one can obtain the radiation pulse $E_p(t)$ of the resonant polarization of a medium in the form of a signal $S1(t) - S2(t) \approx 2 \cdot E_{in}(t) \cdot E_p(t)$ [3].

Note that the method of processing the experimental signals we used enables us separating the field of the reemission of the polarization of the medium taking into account its sign.

We performed the calculation, which models the operation of a Rabi shaper prototype. We solved numerically semiclassical Bloch equations in the model of a two-level medium for a transition with the homogeneous line broadening and given spatial profile of the pump field $E_{in}(t,x) = E_{in}(t) \cdot G(x)$.

In Figures 1-6, we show the results of numerical simulations for a realistic pulse shape and the results of the observation of nonlinear transformation of the pulse shape for different angles of self-diffraction in a resonant medium.

The generation of a positive pulse.

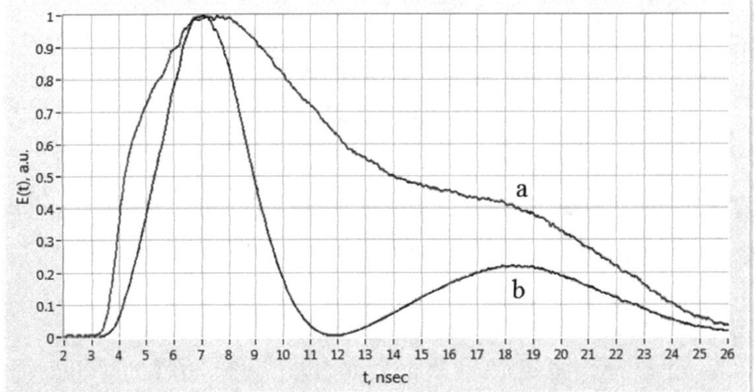

Figure. 1. (a) a realistic pump pulse $E_{in}(t)$; $E_{in}(t) \sim (N_{ph/channel})^{1/2}$.
(b) calculated normalized quantity $2 \cdot E_{in}(t) \cdot E_p(t)$ at the diffraction angle of $\varphi = 0.78^o$.

Figure. 2. The measured value of $2 \cdot E_{in}(t) \cdot E_p(t)$ at the diffraction angle of $\varphi = 0.78^o$.

The generation of a 0π-pulse.

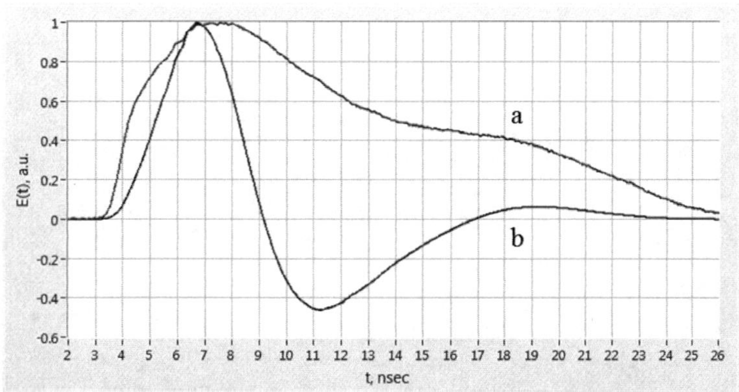

Figure. 3. (a) a realistic pump pulse $E_{in}(t)$; $E_{in}(t) \sim (N_{ph/channel})^{1/2}$.
(b) calculated normalized quantity $2 \cdot E_{in}(t) \cdot E_p(t)$ at the diffraction angle of $\varphi = 1.82^o$.

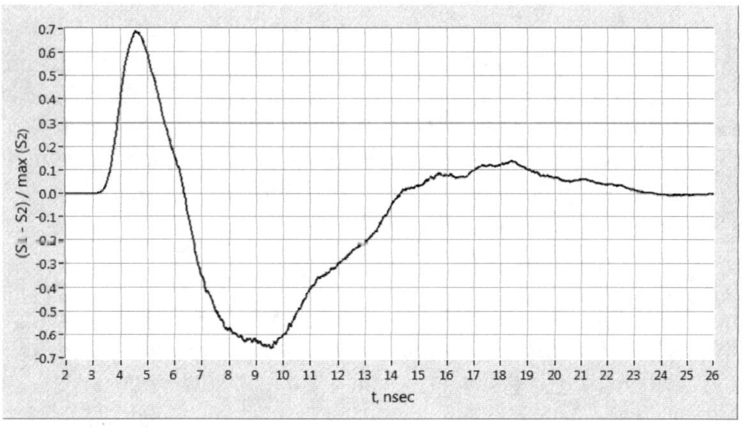

Figure. 4. The measured value of $2 \cdot E_{in}(t) \cdot E_p(t)$ at the diffraction angle of $\varphi = 1.82^o$.

The generation of a negative pulse.

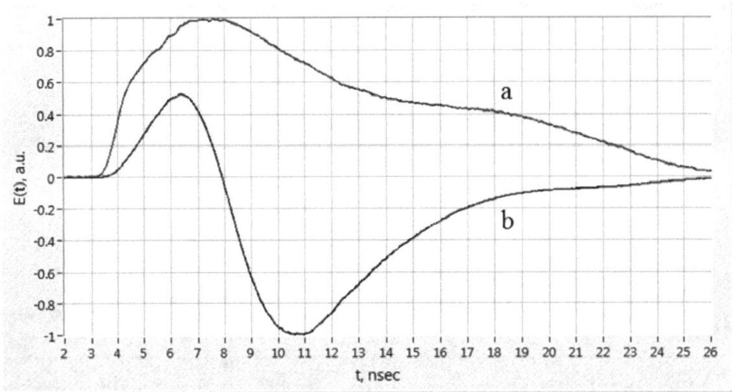

Figure. 5. (a) a realistic pump pulse $E_{in}(t)$; $E_{in}(t) \sim (N_{ph/channel})^{1/2}$. (b) calculated normalized quantity $2{\cdot}E_{in}(t){\cdot}E_p(t)$ at the diffraction angle of $\varphi = 2.86^o$.

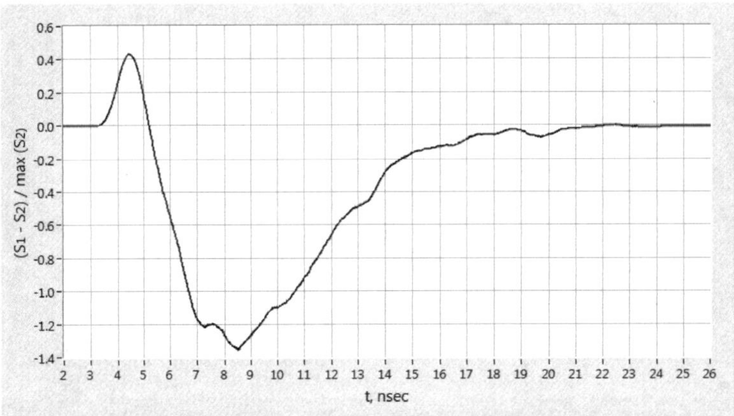

Figure. 6. The measured value of $2{\cdot}E_{in}(t){\cdot}E_p(t)$ at the diffraction angle of $\varphi = 2.86^o$.

3. Conclusion

In this report, we analyze a possibility to use the tunable Rabi shaper of classical and single-photon wave packets. Note that the studied Rabi shaper can perform shaping of nanosecond pulses at the small distances of nonlinear interactions of 0.1 ... 1 mm.

We showed that the nonlinear coherent tunable Rabi shaper has the following degrees of freedom for shaping the classical and single-photon wave packets in a broad range:
- the amplitude of a pump pulse, its spatial shape $E_{in}(x)$;
- the temporal (amplitude and phase) profile of the pump pulse $E_{in}(t)$;
- the diffraction angle φ.

The coherent tunable Rabi shaper can be used as a prototype of perspective quantum systems with a feedback loop [10,11]. Being based on the coherent and collective [12,13] light scattering, it can serve as a prototype for testing theories in atomic [14,15] and molecular [16] media, which are yet difficult to realize.

Acknowledgements
This work was supported by Russian Science Foundation (project № 17-19-01097-P).

References

[1] Sarantos C H and Heebner J E 2010 *Opt. Lett.* **35** (9) 1389
[2] Arkhipov R M, Arkhipov M V, Egorov V S, Chekhonin I A, Chekhonin M A and Bagayev S N 2015 *J. Phys.: Conf. Ser.* **643** 012029
[3] Bagayev S N, Averchenko V A, Chekhonin I A, Chekhonin M A, Balmaev I M and Mekhov I B 2020 *J. Phys.: Conf. Ser.* **1695** 012129
[4] Dudovich N, Oron D and Silberberg Y 2002 *Phys. Rev. Lett.* **88** (12) 123004-1
[5] Monmayrant A, Weber S and Chatel B 2010 *J. Phys. B: At. Mol. Opt. Phys.* **43** 103001
[6] Crisp M D 1970 *Phys. Rev. A* **1**, 1604
[7] Rothenberg J E, Grischkowsky D and Balant A C 1984 *Phys. Rev. Lett.* **53** (6) 552
[8] Costanzo L S *et al.* 2016 *Phys. Rev. Lett.* **116** 023602
[9] Specht H P, Bochmann J, Mücke M, Weber B, Figueroa E, Moehring D L and Rempe G 2009 *Nature Photon* **3** 469
[10] Ivanov D A, Ivanova T Yu, Caballero-Benitez S F and Mekhov I B 2020 *Phys. Rev. Lett.* **124** 010603
[11] Mazzucchi G, Caballero–Benitez S F, Ivanov D A and Mekhov I B 2016 *Optica* **3** 1213
[12] Mekhov I B and Ritsch H 2011 *Laser Phys.* **21** 1486
[13] Kozlowski W, Caballero–Benitez S F and Mekhov I B 2015 *Phys. Rev. A* **92** 013613
[14] Kozlowski W, Caballero–Benitez S F and Mekhov I B 2016 *Phys. Rev. A* **94** 012123
[15] Caballero–Benitez S F and Mekhov I B 2015 *New J. Phys.* **17** 123023
[16] Mekhov I B 2013 *Laser Phys.* **23** 015501

On necessity for analytical solution of the Bloch equations for nuclear magnetic resonance signals at condition express control of liquid medium

M N Davydov[1], **V V Davydov**[1, 2] **and V Yu Rud**[2, 3]

[1] Peter the Great Saint Petersburg Polytechnical University, St. Petersburg, 195251, Russia
[2] All Russian Research Institute of Phytopathology, Moscow Region, 143050, Russia
[3] A.F. Ioffe Physicotechnical Institute, St. Petersburg, 195152, Russia

mdavydov2010@mail.ru

Abstract. The necessity of using express analysis methods to control medium condition is substantiated. It has been shown that the method of express control based on the phenomenon of nuclear magnetic resonance is one of the most preferable. It was found that to increase the information about the medium condition state obtained from the recorded NMR signal, it is necessary to use a mathematical model (based on analytical solutions of the Bloch equations). Two approaches are considered that are used to describe the NMR signal in a liquid medium. It is determined that in the classical approach in the system of Bloch equations it is possible to take into account the peculiarities of using radiotechnical methods of signal registration. The direction of the analytical solution of the Bloch equation is proposed. The experimental data are compared with the numerical solution.

1. Introduction

With the development of scientific and technological progress, the deterioration of the ecological state of the environment and other reasons, express control has become an important element in human life [1-9]. Nuclear magnetic resonance (NMR) is one of the promising methods of express control of liquid media at the sampling site [9-17]. Devices, the principle of operation of which is based on the phenomenon of NMR, do not make irreversible changes in the composition and structure of the studied medium [7, 8, 10-14, 18-26]. Liquid media after express control using NMR instruments can be examined using other instruments (to obtain confirmation of the identified deviations in the state of the medium). New requirements for express control methods create a number of problems. One of them is the low functionality of the NMR method under express control. From the measured values of the times of longitudinal T_1 and transverse T_2 relaxation of a liquid medium, it is possible to establish only the presence of deviations from the standard state in it [20, 22, 26-30]. The rest of the information about the medium, which is contained in the recorded NMR signal using the modulation technique, is extremely difficult to obtain.

One of the main reasons that does not allow this is the lack of an adequate model for describing the recorded NMR signal using a modulation technique. Its development requires an accurate analytical solution of the Bloch equations with respect to absorption and dispersion signals, with coefficients that take into account the registration of an NMR signal in a weak magnetic field using a modulation technique.

Only in this case it is possible to take into account the phase change with time in the recorded NMR signal by an autodyne detector.

2. System of Bloch equations for the case of recording an NMR signal using a modulation technique

The system of Bloch equations in a rotating coordinate system is represented in the following form [31-34]:

$$\begin{cases} \dot{u} + \frac{u}{T^*_2} + v\Delta\omega = 0 \\ \dot{v} + \frac{v}{T^*_2} - u\Delta\omega = -|\gamma|H_1 M_z \\ \dot{M}_z + \frac{M_z}{T_1} - |\gamma| v H_1 = \frac{M}{T_1} \end{cases} \quad (1) \qquad \begin{cases} ב = v + ju \\ \frac{dב}{dt} + \left(\frac{1}{T^*_2} + j\Delta\omega\right) ב = -|\gamma|H_1 M_z \\ \frac{dM_z}{d\tau} + \alpha M_z - v = \alpha M \end{cases} \quad (2)$$

By introducing new functions, system (1) is simplified (2). The general solution is found by the method of variation of arbitrary constants. After mathematical transformations, we obtain the Voltaire integral equation of the second kind for M_z

$$M_z = \frac{1}{T_1} \int_{-\infty}^{t} M(t') e^{-\frac{t-t'}{T_1}} dt' - \gamma^2 {H_1}^2 \int_{-\infty}^{t} M_z(t') K(t,t') dt' \quad (3)$$

This integral equation (3) is solved by the method of successive approximations [31]. The constructed solution can be interpolated. Interpolation of the response waveform by various orthogonal systems of functions and analysis of interpolation errors will be performed. Other approximal solutions of Bloch equations do exist [31, 35, 36].

These approximations were carried out for different conditions of recording the NMR signal. An analysis of these conditions showed their inconsistency with those that arise when the NMR signal is recorded in a weak (magnetic) field. Since the use of a modulation technique when registering an NMR signal in a weak magnetic field is indispensable [22-26, 37].

This circumstance creates problems with the application of previously developed theories, which in their conclusions or solutions of equations used adiabatic (fast) passage through resonance. Therefore, we propose the following solution to the problem of describing the configuration of the NMR signal line recorded in a weak (magnetic) field using a modulation technique. Coefficients are introduced into the Bloch equations, containing information about the use of the modulation technique when registering:

$$\Delta\omega = \gamma H_0 + \gamma H_m \sin(\omega_m t) - \omega_{nmr} \quad (4)$$

where H_0 is the constant magnetic field, H_m is the modulation coil field, ω_m is the modulation frequency.

$$M = \chi_0 (H_0 + H_m \sin(\omega_m t)) \quad (5)$$

Before substituting new coefficients (4) and (5) into equation (1), it is necessary to take into account one more feature of recording an NMR signal in a weak field. The NMR signal should be recorded only at the resonance frequency ($\omega_{nmr} = \omega_0 = \gamma H_0$). This is due to the small volume of the investigated medium, from which the NMR signal is recorded. If the frequency of recording the NMR signal ω_{nmr} is detuned from the resonance frequency ω_0, the signal-to-noise ratio (S/N) can become less than 1.3. Accumulation of the NMR signal at this S / N ratio is impossible. Without accumulation, it is extremely difficult to learn the S / N> 3 ratio and carry out measurements with an error of no more than 1%. When this feature is fulfilled, formula (4) is transformed into the following function:

$$\Delta\omega = \gamma H_m \sin(\omega_m t) \quad (6)$$

After substituting (5) and (6) into (1), the Bloch equations are transformed into:

$$\begin{cases} \frac{du(t)}{dt} + \frac{u(t)}{T_2} + \gamma H_m \sin(\omega_m t)\, \upsilon(t) = 0 \\ \frac{dv(t)}{dt} + \frac{v(t)}{T_2} - \gamma H_m \sin(\omega_m t) u(t) + \gamma H_1 M_z(t) = 0 \\ \frac{dM_z(t)}{dt} + \frac{M_z(t)}{T_1} - \chi_0 \frac{H_0 + H_m \sin(\omega_m t)}{T_1} - \gamma H_1 \upsilon(t) = 0 \end{cases} \quad (7)$$

In Bloch equations (7), there appear coefficients with a time term $\gamma H_m sin(\omega_m t)$. In this case, it becomes difficult to implement the classical solutions of equations (7). Namely, transform (7) by the method of Gvozdover and Magazanik [31, 35] into the integral form of Voltaire. Expansion in a series with the identification of the convergence condition ($\gamma^2 H_1^2 T_1 T_2^* < 1$) is also not acceptable. It should be noted that the condition for the convergence of the series is the implementation in an NMR meter (small-sized) of fast adiabatic passage through resonance. It is extremely difficult to provide it in it when registering an NMR signal in a weak (magnetic) field.

Therefore, a number of the author began to implement new approximations and assumptions, for example $H_0 >> \gamma H_m sin(\omega_m t)$. This assumption assumes that, when considering (2.6), the effect of modulation on the (constant) field H_0 is very small - in principle, it can be neglected. This approach made it possible to solve equations (7). This solution is very similar to the solutions obtained in [31–35]. Further, a term is added to the solution itself. With this term, a number of authors decided to take into account the modulation. If we analyze all such solutions, then we can choose the most adequate of them to describe the configuration of the NMR signal. In [36], this is presented:

$$U = U_0 \exp(-\frac{t}{T_2^*}) \cos(\omega_0 t + \frac{1}{2}\gamma \frac{dH}{dt} t^2) \quad (8)$$

In a weak field, for adequacy, it is necessary to transform (8), taking into account (4), into a new relation:

$$U = U_0 \exp(-\frac{t}{T_2^*}) \cos(\omega_0 t + \frac{1}{2}\gamma H_m \omega_m \cos(\omega_m t)\, t^2) \quad (9)$$

The resulting relationship was also used to describe the recorded NMR signal in a weak field.

3. Description of NMR signal lines and their comparison with experiment

For example, figure1 shows the experimental NMR signal from water (temperature T = 14.4 ° C) and the result of its simulation using (9)

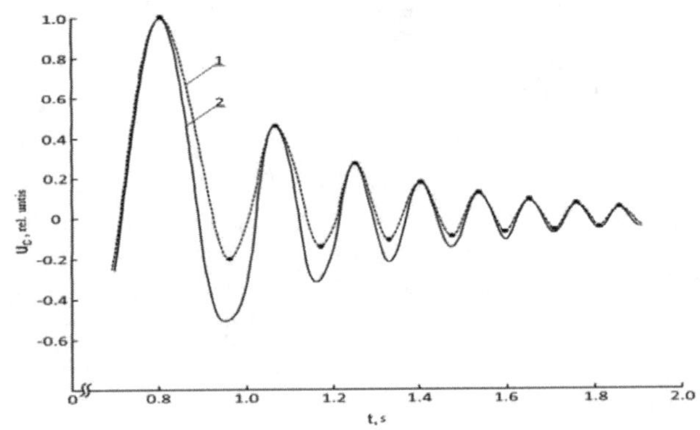

Figure 1. NMR waveform from tap water. Line 1 corresponds to experiment. Line 2 – theoretical analysis

The results obtained show a large difference between the experimental signal and the theoretical calculation. At the bottom of the dependencies, there is a complete mismatch. This is mainly due to the absence of the magnetic field parameter H_1 in ratio (9). This field is generated in the autodyne registration coil. It plays a primary role in determining the conditions for passage through resonance. Without taking it into account, it is impossible to obtain an adequate value of the parameters for the maximum of the S /N ratio at the output of the registration circuit in a weak field.

To determine the adequacy of our proposed Bloch equations (7) for describing the NMR signal in a weak field, a mathematical model was developed for the numerical solution of equations (7). Equations (7) were solved for the components $\upsilon(t)$, $u(t)$, and M_z under the following initial conditions: $u/_{t=0} = \upsilon/_{t=o} = 0$, $M_z/_{t=0} = M_t$. The values of the relaxation constants T_1 and T_2 were determined from the recorded NMR signal. These T_1 and T_2 values were used to calculate the absorption and dispersion signals. Figure2 shows the recorded NMR signal from tap water at a temperature T = 293.3 K.

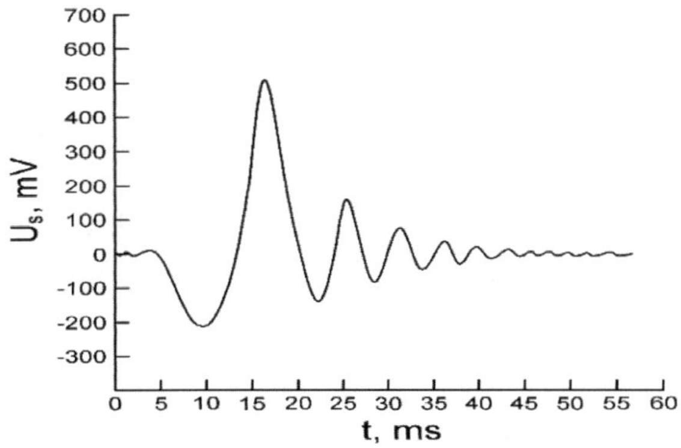

Figure 2. NMR waveform from tap water.

Figure3 and figure4 show the calculations of absorption and dispersion signals for tap water.

Analysis of the obtained dependences figure3 and figure4 show that they reflect the nature of the physical processes corresponding to the experiment. However, additional calculations have shown that with an increase in t, every $T_m/2$ changes in the phase of the components $\upsilon(t)$ and $u(t)$ by 180^0, as well as a sharp decrease in the values of the peak amplitudes. In real experiments, the same NMR signal is recorded by the autodyne every half-period $T_m/2$. Comparison with experiment can be carried out only within the first peak, since the phase is not taken into account.

Difficulties also arise in the interpretation of the resulting NMR signal (experimental), which consists of absorption and dispersion signals. It is necessary to enter a factor that takes into account the phase change. This coefficient has to be selected for each case, which is very difficult. The greatest agreement is achieved at the same contributions of the absorption and dispersion signals to the generated NMR signal. Figure5 shows this signal.

In other cases of modeling, it is necessary to constantly select a coefficient that takes into account the phase of the signal and the coefficients of the contributions of the absorption and dispersion signals. It takes a long time. In the conditions of conducting research in express mode, such time costs are excluded.

This can be excluded only by obtaining an analytical solution to the Bloch equations. This problem is supposed to be solved using the Matlab software package used for technical computing tasks.

4. Conclusion

The preliminary results obtained showed that the direction we have chosen to expand the functionality of using express control using an NMR signal is promising. The analytical solution of the Bloch equations will take into account the phase of the NMR signal and delineate long-term studies of media using the modulation technique.

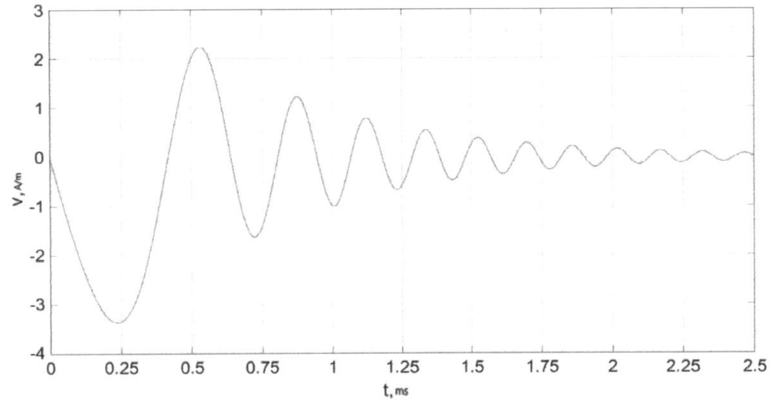

Figure 3. Calculated line shape of the absorption signal from water

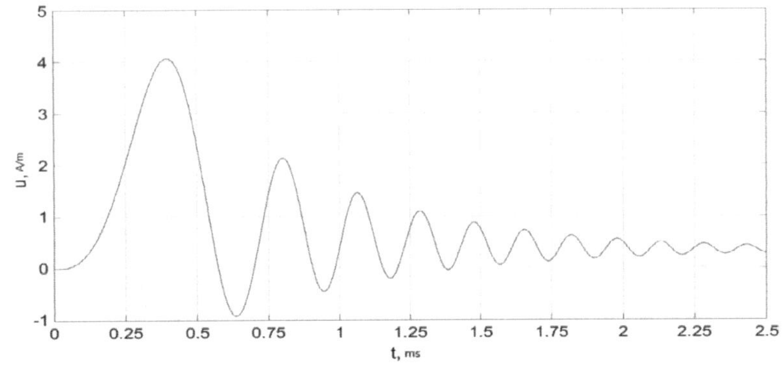

Figure 4. Calculated line shape of the dispersion signal from water.

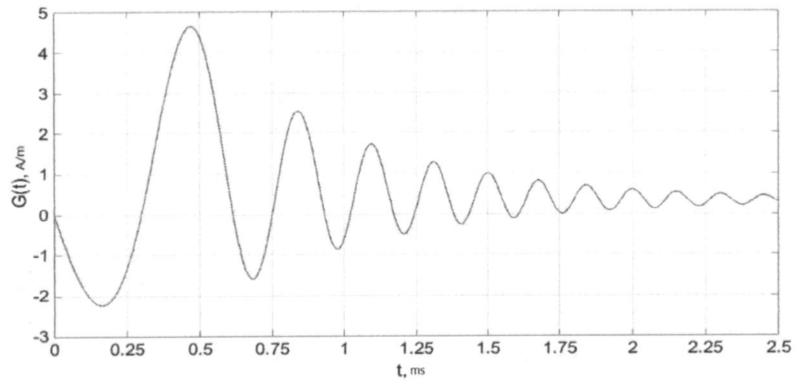

Figure 5. The result of modeling the NMR signal line for water.

References

[1] Logunov S E, Rud V Y, Davydov R V, Moroz A V and Smirnov K J 2019 *Journal of Physics: Conference Series* 1326(1) 012024

[2] Davydov R, Antonov V and Moroz A 2019 *International Conference on Electrical Engineering and Photonics EExPolytech-2019 (Saint-Petersburg)*, vol. 8906791 (IEEE), p. 295-297

[3] Davydov R, Antonov V, Makeev S, Batov Y, Dudkin V and Myazin N 2020 *E3S Web Conference* **140** 02001

[4] Myazin N S, Sviatkina V I and Yushkova V V 2020 *Journal of Physics: Conference Series* 1695(1) 012141

[5] Davydov V V, Moroz A V, Myazin N S, Makeev S S and Dudkin V I 2020 *Optics and Spectroscopy* 128(10) 1678–1685

[6] Davydov V V, Myazin N S, Makeev S S and Dudkin V I 2020 *Measurement Techniques* 63(5) 368–374

[7] Davydov V V, Dudkin V I, Myazin N S and Davydov R V 2020 *Journal of Communications Technology and Electronics* 65(5) 558–564
[8] Davydov V V, Myazin N S and Kiryukhin A V 2020 *Atomic Energy* 127(5) 274–279
[9] Davydov R V, Mazing M S, Yushkova V V and Stirmanov A V 2019 *Journal of Physics: Conference Series* 1410(1) 012067
[10] Myazin N S, Yushkova V V, Davydova T I and V Yu Rud' 2017 *Journal of Physics: Conference Series* **917** 042017
[11] Karseev A Yu, Cheremiskina A V, Davydov V V, Velichko E N 2014 *Journal of Physics: Conference Series* **541(1)** 012006
[12] Rutberg F G, Kolikov V A, Snetov V N, Cherepkova I A and Dyachenko S V 2016 *High Temperature* 54(2) 170–174
[13] D'yachenko S V and Zhernovoi A I 2016 *Technical Physics* 61(12) 1835–1837
[14] D'yachenko S V, Kondrashkova I S and Zhernovoi A I 2017 *Technical Physics* **62(1)** 1602-1604
[15] Neronov Yu I 2017 Measurement Techniques **60(1)** 96-102
[16] Neronov Y I 2021 *Biomedical Engineering* 54(5) 333–336
[17] Myazin N S and Yushkova V V 2019 *Environmental Research, Engineering and Management* 75(2) 28–35
[18] Davydov V V, Dudkin V I and Karseev A Y 2015 *Instruments and Experimental Techniques* 58(6) 787–793
[19] Davydov V V, Dudkin V I and Karseev A U 2013 *Optical Memory and Neural Networks (Information Optics)* 22(2) 112–117
[20] Myazin N S, Yushkova V V and Dudkin V I 2019 *Journal of Physics: Conference Series* 1400(6) 066008
[21] Rukin E V, Myazin N S and Dudkin V I 2019 *Journal of Physics: Conference Series* 1368(4) 042011
[22] Myazin N S, Yushkova V V and Taranda N I 2019 *Journal of Physics: Conference Series* 1410(1) 012130
[23] Myazin N S 2018 *Journal Physics: Conference Series* **1124(1)** 031004
[24] Myazin N S 2018 *Journal Physics: Conference Series* **1135(1)** 012061
[25] Davydov V V 2017 *Measurement Techniques* 60(2) 183 189
[26] Davydov V V, Velichko E N, Dudkin V I and Karseev A Y 2014 *Measurement Techniques* 57(6) 684–689
[27] Davydov V V 1999 *Russian Physics Journal* 42(9) 822–825
[28] Gizatullin B, Gafurov M, Vakhin A, Mattea C and Stapf S 2019 *Energy and Fuels* 33(11) 10923–10932
[29] Davydov V V, Velichko E N, Dudkin V I and Karseev A Y 2015 *Instruments and Experimental Techniques* 58(2) 234–238
[30] Davydov V V, Dudkin V I and Karseev A Y 2015 *Technical Physics* 58(6) 787–793
[31] Leshe A *Nuclear induction* (Berlin: Verlag Der Wissenschaften)
[32] Bloch F 1946 *Phys. Rev.* **70(7)** 460 – 476
[33] Bloch F, Hansen W W and Packard F 1946 *Phys. Rev.* **70(7)** 474 – 488
[34] Packard M E and Varian R H 1954 *Phys. Rev.* **93(8)** 941 952
[35] Abragam A *The principles of nuclear magnetism* (Oxford: Qxford at the Clarendon)
[36] Zhernovoi A I and Latyshev G D 1965 Nuclear magnetic resonance in a flowing liquid (New York: Consultants Bureau)
[37] Myazin N, Neronov Y, Dudkin V and Yushkova V 2018 *MATEC Web of Conferences* 245 11013

Condition monitoring of current media using a differential refractometer

V I Svyatkina[1], V V Davydov[1, 2] and V Yu Rud[2, 3]

[1] Peter the Great Saint-Petersburg Polytechnic University, Saint Petersburg 195251, Russia
[2] All-Russian Research Institute of Phytopathology, Moscow Region 143050, Russia
[3] A.F. Ioffe Physicotechnical Institute, St. Petersburg, 195152, Russia

svyatkina.vi@edu.spbstu.ru

Abstract. A new design of a differential flow refractometer has been developed to monitor the condition of flowing media in a pipeline. A new method of refractive index measurement has been implemented, taking into account the specifics of flowing and closed cuvette arrangement, as well as the angles of incidence of laser radiation on their walls. The effect of changes in the optical density in the flowing liquid on the refractive index measurement result is determined. The results of experimental investigations of different media are presented.

1. Introduction

Currently, the number of tasks associated with monitoring the state of condensed matter is constantly increasing [1-10]. Depending on the tasks to be solved, different methods and instruments are used [10-18]. The greatest difficulties arise while monitoring the liquid flowing through the pipeline, for example, during scientific research or monitoring the operation of cooling systems [13-15, 19-25]. The use of hostile requirements or severization for the stability of the measurement process has led to the preference for non-contact measurement methods when monitoring the state of the current environment [26-32]. In some cases, like food& beverages, chemical or pharma industry, apart from strong requirements for accuracy and reliability of measurements, there is a number of arrangements [33-35]. The measurements shall not change the physical and chemical properties of the environment. [34-38]. All these requirements are met only by nuclear magnetic flowmeters - relaxometers or flow refractometers [8, 13, 15, 24, 34, 35, 39].

Refractometers, unlike nuclear-magnetic flow meters - relaxometers, allow one measuring device to monitor both the current and stationary state of the matter [34, 35, 39-44]. In addition, refractometers are easier to operate, etc. [34, 35, 39-45]. The most frequently used types of fractometers are differential refractometers and refractometers, based on the use of the phenomenon of complete internal reflection of laser radiation from the boundary of two matters [39-41]. A number of problems arise in refractometers with complete internal reflection. As the temperature changes, the refractive indexes of the current liquid and prism do change, so it requires a quick automatic adjustment of the optical part of the device. With fast fluid flows, the refractive index measurement error is impacted by the Hus-Hengchen optical beam shift phenomenon, which complicates the operation of the device and limits the measurement range n.

Therefore, the development of new designs of flow refractometers of the differential type and methods for implementing in them the measurement of the refractive index n with an error of less than 10^{-4} is an urgent task.

2. Differential type refractometer designs and refractive index measurement errors

The study of the effect of the change in optical density on the measurement error of the current fluid index [45] made it possible to develop a new design of a flow refractometer of the differential type (Fig. 1). In this refractometer design, a differential method is used to detect laser radiation. Therefore, the refractometer can be considered differential.

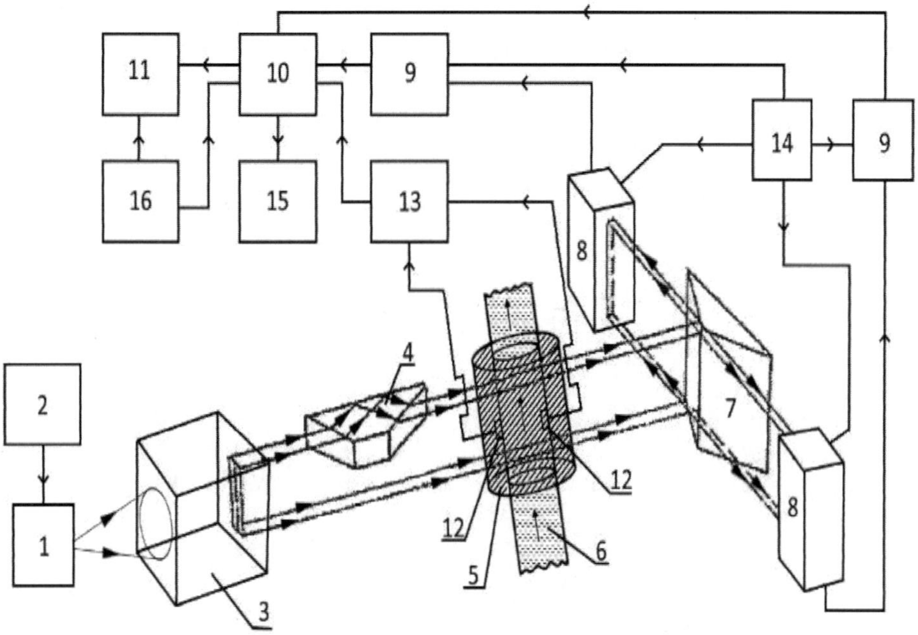

Figure 1. The Block diagram of a laboratory model of a differential refractometer: 1 - semiconductor laser; 2 - power source; 3 - optical system; 4 - prism Dove; 5 - closed cuvette; 6-flowing cuvette; 7 - dividing prism; 8 - photodiode array; 9 - ADC; 10 - processing device; 11 - indicating device; 12 - temperature sensor; 13 - device for processing information with a temperature sensor; 14 - specialized power supply ADC and photodiode arrays; 15 - laptop; 16 - power supply unit for processing and indication devices.

The radiation from the semiconductor laser 1 with a λ = 632.8 nm is transmitted to the optical system 3 (prism collimator). The collimator is an anamorphic system consisting of two prisms. In this system, unlike lens systems, there is no axial symmetry when converting a laser beam. The laser kit includes a long-focus and short-focus lens. Their use allows narrowing the laser beam to 0.4 mm. Overall makes it possible to obtain at the exit of the collimator a parallel beam in the form of a narrow strip measuring 24 mm by 1 mm. The laboratory model of the refractometer we have developed uses cylindrical cuvettes made of ARCUNA glass with n = 1.4953. Flow cell 6 is placed inside closed 5 (Fig. 1). In our design of the refractometer, we have proposed a new method for placing ditches in an upright position (Figure 1). Its use makes it possible to eliminate the formation of voids completely in the flow cell at low liquid flow rates and vortex flows at high flow rates q, which were previously used in the differential cells. The use of this method made it possible to use cells of large diameter (20 and 28 cm) in the design of the refractometer - voids in the flow cells are not formed at low flow rates q.

This arrangement of cells 6 and 5 also provides a more efficient heat exchange between fluids (current and reference) than previously used differential cells designs. This significantly reduces the inmpact of the temperature error contribution. The refractive index value n_f of the current liquid is determined from the ratio:

$$\beta = \Delta n \, \mathrm{tg}(\gamma) \tag{1}$$

where γ is the angle between the incoming beam and the perpendicular to the boundary of the flow and closed cell, $\Delta n = n_f - n_r$ (n_r is the refractive index of the reference liquid located in the closed cell).

The conducted researches showed that the main contribution to the uncertainty value of the measurement δn_f of refraction index will only change the optical density of the current liquid, the impact of other deviations on measurement nf result will be insignificant. In this case, the value of δnf can be determined from the following formula:

$$\delta n_f = 0{,}993 \frac{\alpha^2 L \,\Delta d}{\Delta n} \tag{2}$$

where $\Delta d = d_f - d_0$ is the change of optical density df relative to initial value d_0, a - half-width of light beam, L- distance from wall of cylindrical cell 5 to linear photodiode array 8.

Taking into account the features of the placement of cells and the effect on the current and reference liquid of laser radiation, we have developed a method for measuring the refractive index of n_f using a differential optical signal recording circuit. Magnitude of amplitudes difference of normalized output signals from photodiode lines 8 is proportional to value β and L (distance from cell wall 6 to linear photodiode array 8. In this case, formula (1) for determining the informative parameter β is converted into the following form:

$$\beta = A^1_{ar} - A^2_{ar} = LK_{ar}\,\Delta n = LK_{ar}\,(n_f - n_r), \tag{3}$$

where K_{ar} - proportionality coefficient, A^1_{ar}, A^2_{ar} - normalized output signals from photodiode lines 1 and 2.

Normalization of output signal A from photodiode ruler 8 is performed according to the following principle:

$$A = \frac{\sum_{i=1}^{512} A_i}{A_{max}} \tag{4}$$

where A_{max} is the maximum value of the illumination signal from the element of the photodiode line, A_i is the amplitude of the signal from the element of the photodiode cell, i is the element number.

From the photodiode lines 8, signals are transmitted through analog-to-digital converters to the information processing device 10, developed on the basis of the microcontroller STM32 (ARM Cortex M3 core - STM32F100RBT6B). Information on n_f value is transmitted from this device to indication device 11. In addition, a laptop 15 was used to monitor the operation of the information processing device in the laboratory layout, which received information from the microcontroller. It was designed to determine the value of the n_f and visualize the change in the parameters of the matter during long monitoring of its condition.

3. Results of experimental data and their discussion

The validity of n measurements using the developed refractometer design was tested on several liquid matters for different temperatures T. Figure 2 shows the n (T) dependencies for several current matters.

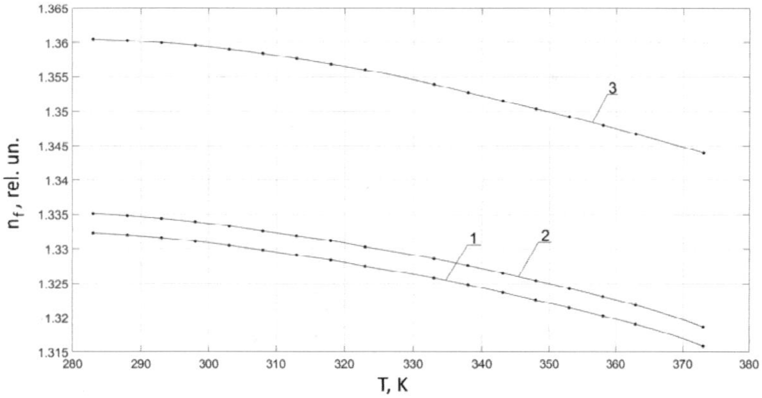

Figure 2. The dependence of the refractive index n_f from the temperature T. Charts 1, 2, 3 correspond to the following environments: distilled water, tap water, ethyl alcohol.

The refractive indices of these matters were also measured using the industrial flow refractometer PRM-100 alpha (company ATAGO, Japan) - measurement deviation 10^{-4}. The obtained refractive index values using two instruments coincided within the measurement deviation range. In addition, the refractive index values of these matters, measured on the laboratory model of the differential-type refractometer developed by us, coincided with the refractive index values obtained earlier by scientists using other refractometer designs [34, 35, 39-41, 44, 45]. In some cases, the liquids were stationary during measurements, so the measurement error n was less than for the flowing liquid.

An important parameter when considering the error associated with the change in optical density of the liquid medium is transparency. Especially many liquid matters with low transparency are used in the food and pharmaceutical industries. Refractometers are often used to monitor the condition of these matters. Figure 3 shows as an example the relationship of the refractive index change n_f different oil grades to the temperature change T.

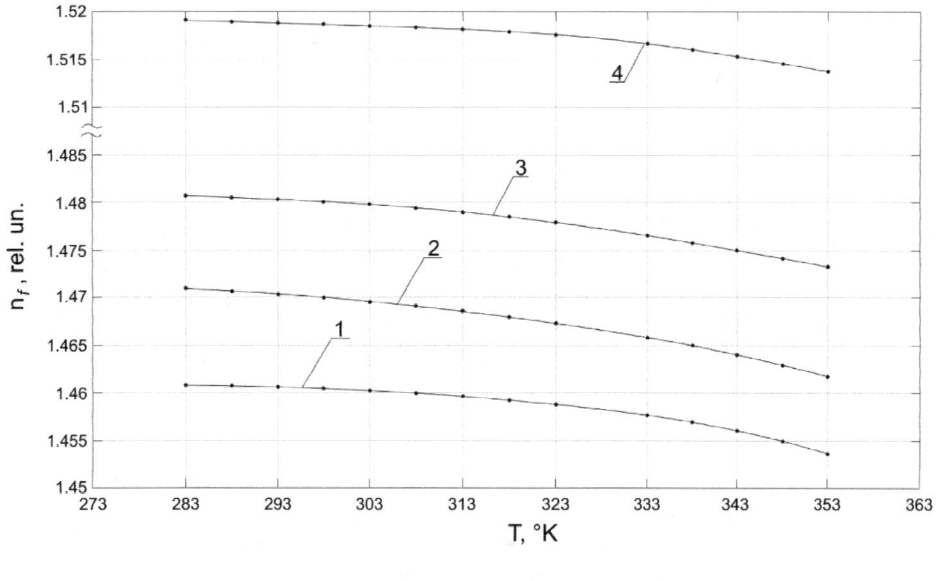

Figure 3. The dependence of the refractive index n_f of edible oils from the temperature T. Charts 1, 2, 3, 4 correspond to the following oils: olive, sunflower, castor, cedar.

Refractive indices of oils were measured using an industrial flow refractometer PRM-100 alpha (company ATAGO, Japan) - measurement error 10^{-4}. The refractive index results obtained coincided within the measurement error range. Besides, comparison of measured values of refractive indices of n_f oils using the design of refractometer developed by us with data obtained on stationary refractometers [46] showed their coincidence within the limits of measurement deviation.

4. Conclusion

The results of the studies showed that the use of the design of a flow refractometer developed by us with a differential method for measuring the refractive index of n_f implemented in it allows us to monitor the state of the current matter in the entire range of changes in its flow q and temperature T, unlike refractometers, the principle of operation of which is based on the use of the total internal reflection phenomenon from the boundary of two matters.

In the newly developed refractometer design, there are no limits on the measuring range n_f in comparison with refractometers operating on the basis of total internal reflection phenomenon, in which there are restrictions on changing the angle of complete internal reflection depending on the temperature of the current matter T, etc. Its use allows you to control, for example, the state of the current flow of phoenylhydrazine $n_f = 1.6105$, used in the medical industry in the production of drugs (for example, antipyrin or amidopyrin). Currently, control of the state of this substance in the flowing stream by industrial refractometers is not implemented. All limitations in the work of the refractometer we have developed are related to the transparency of the liquid and the presence of insoluble particles in it. They were considered in detail in the discussion of the experimental results.

References

[1] Logunov S E, Rud V Y, Davydov R V, Moroz A V and Smirnov K J 2019 *Journal of Physics: Conference Series* **1326(1)** 012024
[2] Davydov R, Antonov V and Moroz A 2019 *International Conference on Electrical Engineering and Photonics EExPolytech-2019 (Saint-Petersburg)*, vol. 8906791 (IEEE), p. 295-297
[3] Davydov R, Antonov V, Makeev S, Batov Y, Dudkin V and Myazin N 2020 *E3S Web Conference* **140** 02001
[4] Myazin N S, Sviatkina V I and Yushkova V V 2020 *Journal of Physics: Conference Series* **1695(1)** 012141
[5] Davydov V V, Moroz A V, Myazin N S, Makeev S S and Dudkin V V 2020 *Optics and Spectroscopy* **128(10)** 1678–1685
[6] Davydov V V, Myazin N S, Makeev S S and Dudkin V I 2020 *Measurement Techniques* **63(5)** 368–374
[7] Davydov V V, Dudkin V I, Myazin N S and Davydov R V 2020 *Journal of Communications Technology and Electronics* **65(5)** 558–564
[8] Davydov V V, Myazin N S and Kiryukhin A V 2020 *Atomic Energy* **127(5)** 274–279
[9] Davydov R V, Mazing M S, Yushkova V V and Stirmanov A V 2019 *Journal of Physics: Conference Series* **1410(1)** 012067
[10] Davydov V V 1999 *Russian Physics Journal* **42(9)** 822–825
[11] Kozlova E S and Kotlyar V V 2014 *Computer Optics* **38(1)** 132-138
[12] Kuzmin M S and Rogov S A 2019 *Computer Optics* **43(3)** 391-396
[13] Davydov V V, Dudkin V I and Karseev A U 2013 *Optical Memory and Neural Networks (Information Optics)* **22(2)** 112–117
[14] Davydov V V, Dudkin V I and Karseev A Y 2014 *Optical Memory and Neural Networks (Information Optics)* **23(3)** 170–176
[15] Davydov V V, Dudkin V I and Karseev A Y 2014 *Optical Memory and Neural Networks (Information Optics)* **23(4)** 259–264
[16] Davydov V V, Velichko E N, Dudkin V I and Karseev A Y 2014 *Measurement Techniques* **57(6)** 684–689
[17] Makeev S S, Grevtzeva A S, Glinushkin A P and Matorin D N 2020 *Journal of Physics:*

Conference Series **1695(1)** 012112
[18] Myazin N S, Yushkova V V and Dudkin V I 2019 *Journal of Physics: Conference Series* **1400(6)** 066008
[19] Rukin E V, Myazin N S and Dudkin V I 2019 *Journal of Physics: Conference Series* **1368(4)** 042011
[20] Myazin N S, and Yushkova V V 2019 *Environmental Research, Engineering and Management* **75(2)** 28–35
[21] Davydov V V, Dudkin V I and Karseev A Y 2015 *Russian Physics Journal* **58(2)** 146–152
[22] Davydov V V, Dudkin V I and Karseev A Y 2015 *Measurement Techniques* **58(3)** 317–322
[23] Davydov V V, Dudkin V I and Karseev A Y 2015 *Instruments and Experimental Techniques* **58(6)** 787–793
[24] Myazin N S, Yushkova V V and Taranda N I 2019 *Journal of Physics: Conference Series* **1410(1)** 012130
[25] Davydov V V, Dudkin V I and Karseev A Y 2015 *Technical Physics* **60(3)** 456–460
[26] Davydov V V, Dudkin V I and Karseev A Y 2015 *Technical Physics Letters* **41(4)** 355–358
[27] Davydov V V, Logunov S E and Fadeenko V B 2018 *Russian Journal of Nondestructive Testing* **54(3)** 213–221
[28] Davydov V V, Fadeenko V B and Logunov S E 2018 *Technical Physics Letters* **44(2)** 153–156
[29] Smirnov K J and Batov Y V 2019 *Journal of Physics: Conference Series* **1368(2)** 022073
[30] Neronov Yu I 2017 *Measurement Techniques* **60(1)** 96-102
[31] Neronov Y I 2021 *Biomedical Engineering* **54(5)** 333–336
[32] Rutberg F G, Kolikov V A, Snetov V N, Cherepkova I A and Dyachenko S V 2016 *High Temperature* **54(2)** 170–174
[33] Velt I D and Mikhailova Yu V 2013 *Measurement Techniques* **56(3)** 283
[34] Davydov V V, Grebenikova N M and Smirnov K J 2019 *Measurement Techniques* **62(6)** 519
[35] Karabegov M A 2012 *Measurement Techniques* **55(11)** 1104
[36] Gizatullin B, Gafurov M, Vakhin A, Mattea C and Stapf S 2019 *Energy and Fuels* **33(11)** 10923–10932
[37] D'yachenko S V and Zhernovoi A I 2016 *Technical Physics* **61(12)** 1835–1837
[38] D'yachenko S V, Kondrashkova I S and Zhernovoi A I 2017 *Technical Physics* **62(1)** 1602-1604
[39] Grebenikova N M and Davydov R V 2019 *Journal of Physics: Conference Series* **1326(1)** 012012
[40] Grebenikova N M, Davydov V V and Rud V Y 2019 *Journal of Physics: Conference Series* **1410(1)** 012186
[41] Karabegov M A 2013 *Measurement Techniques* **55(11)** 1301–1310
[42] Myazin N S 2018 *Journal Physics: Conference Series* **1124(1)** 031004
[43] Myazin N S 2018 *Journal Physics: Conference Series* **1135(1)** 012061
[44] Karabegov M A 2009 *Measurement Techniques* **52(1)** 97–104
[45] Davydov V V and Moroz A V 2020 *Optics and Spectroscopy* **128(9)** 1415–1420
[46] Calhoun W R, Maeta H, Combs A, Bali L M and Bali S 2010 *Optics Letters* **35(8)** 1224–1226

Fiber-optic sensor for remote monitoring the γ-radiation of various powers

D S Dmitrieva[1], V V Davydov[2, 3] and V Y Rud[3, 4]

[1] The Bonch-Bruevich Saint Petersburg State University of Telecommunications, Saint-Petersburg, Russia, 193232
[2] Peter the Great Saint Petersburg Polytechnic University, Saint Petersburg, Russia, 195251
[3] All-Russian Research Institute of Phytopathology, Moscow region, Russia, 14305
[4] Ioffe Institute, Saint Petersburg, Russia, 194021

emilylitov@gmail.com

Abstract. The necessity of improving the metrological characteristics and functional capabilities of the fiber-optic sensor for measurements at the large distances (more than 10 km) is substantiated. The new method for constructing a communication line with the fiber-optic sensor for controlling exposure dose in a large range of its variation (several orders of magnitude) in a remote mode is proposed. The functional capabilities of the sensor are determined; its connection setup and measurement limits are developed. The experimental results are presented.

1. Introduction

The development of scientific and technical progress led to the appearance of large number of facilities, both scientific and industrial, which use radioactive materials [1-4]. Every year the number of these increases, because of which has increased the number of cases of radioactive materials leakage (getting them into water, soil, into the atmosphere etc.) [1, 3, 5-8]. Moreover, the number of radioactive releases (mainly γ-radiation) outside the protective zone (through the protective screens and ceilings) has increased. It has led to a deterioration of the ecological situation in several regions [9-14]. Therefore, great attention is paid to constant control the radiation situation, both in the atmosphere and on territory of various facilities [5, 7, 8, 14-19].

The main interest is the probability of remote control the radiation situation in automatic mode. Such monitoring is necessary for facilities, where the possibility of radioactive materials leakage is high: in various land-based and sea-based systems of nuclear power plants and on the territories of test sites [4, 6, 8, 19-21]. In some cases, remote control should be realized at the distance more than 10 km from the location of the monitoring center. Moreover, the exposure dose D_R in control zone can change by several orders in a short period of time. After the exposure dose is terminated, device takes measurements in normal mode after a while. In such operation conditions, it is extremely difficult to use most of dosimetry devices, because they often go off and can accumulate the radiation on the instrument case and functional units, which prevents normal operation due to a great danger to humans.

One of the perspective directions for such measurements is the use of fiber-optic sensors [22-24]. Fiber-optic sensors developed nowadays, which operation principle is based on the measurements of changes in laser radiation polarization under the γ-radiation influence, allow to register small D_R changes, which cause a decrease in the laser radiation power by 0.05 dB. At a high D_R values laser radiation in them is completely damped and sensor stopped its work for a long time. The relaxation process without interventions can take 10^6 s and more [22-24], because for measurements are used small powers (no more than 5 mW). The use of more powerful radiation can damage photosensitive layer in photodetector module. Therefore, developing of new models of fiber-optic sensors is extremely actual.

2. Principle of operation a fiber-optic sensor for remote monitoring γ-radiation on based the governance to color centers relaxation rate

The color centers appear in an optical fiber after ending the γ-radiation influence [22-24]. The amount of these defects depends on two radiation effects: displacement and ionization. The displacement is due to the atoms shift from a stationary state to an unstable. Due to this shift in the crystal lattice the vacancies are formed, the inter-site introductions of additional atoms occur, etc. In some cases, these defects can be unstable (if the radiation level is low enough for an optical fiber can recover due to relaxation processes to primary state). In the case, when radiation influence was too long or radiation level was too high, the displacement effect leads to the appearance of more difficult defects with the union of various defects (for example, defects which are connected to the presence of alloying additives, etc). As the result the crystal lattice destroys, and an optical fiber does not recover to the primary state.

The ionization effect is based on the formation of electron-hole pairs. An electron with high energy knock oxygen out of OH compounds and takes its place in the crystal lattice. In a case of alloying fiber core by the germanium oxide GeO_2, oxygen is knocked out of this chemical compound. The increase in the amount of ionization atoms changes the electrical properties of the optical medium, that can lead to the change of the refractive index n. The effects caused by ionization disappear after ending the γ-radiation influence, because the free charge carriers have a high mobility. But in some cases, when exposure dose is high or long, the free charge carriers can miss in the deep energetic levels of the optical material and make stable connections. It also leads to the impossibility of return the optical fiber to the initial state.

Relaxation process begins right after the formation of electron-hole pairs, which take part in the formation of color centers. Relaxation velocity depends on the value of exposure dosses of irradiation and the duration of the γ-radiation influence. It is possible to determine the presence of γ-radiation in its placement zone (using the data of preliminary calibration), if measuring the decrease of laser radiation power at the output of an optical fiber. But this process has some difficulties. Previous researches [22-24] showed, that the relaxation time after ending the radiation influence can take 10^6 and more. In this case next powerful γ-radiation influence cannot be recorder, and it will be extremely difficult to estimate the value of exposure dose D_R. As the result the possibility of remote control of radiation situation will be stop for a long period. So, maintain the operability of fiber optical sensors becomes possible only in the case of control the relaxation rate of radiation color centers

3. Experimental setup and research results

It is necessary to provide additional research of γ-radiation influence on the optical fiber for develop a new design of the fiber-optic sensor for remote monitoring the radiation situation. We have assembled earlier an experimental setup for these purposes [23,24]. Its block diagram is presented at the figure 1.

In the experimental setup was used modified transmitting optical module DMPO131-23M (company Dilaz) with a wavelength λ=1550 nm and variation of radiation power from 0.1 to 5.4 mW. It was used as a continuous source of laser radiation. To measure losses, the classical formula was used:

$$\alpha_s = \frac{10}{L} \lg \left(\frac{P_{in}}{P_{out}}\right) \quad (1)$$

where P_{in} – the power input in to the optical fiber, P_{out} – the power output from the optical fiber, L – the length of the optical fiber.

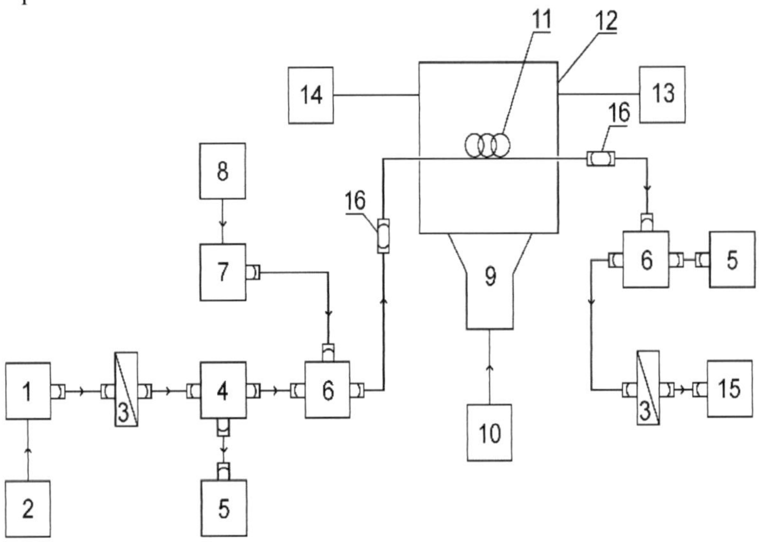

Figure 1. The block diagram of the experimental setup: 1 - laser transmitting module; 2 - power driver; 3 - polarizer; 4 - optical divider; 5 - optical power meter; 6 - мультиплексор; 7 - pulsed semiconductor laser; 8 - pulse power driver; 9 - source of γ - radiation; 10 - control unit of the source of γ - radiation; 11 - optical fiber; 12 - closed volume of a polymeric material; 13 - a device for changing the temperature; 14 - a device for controlling the dose of radiation; 15 - receiving optical module; 16 – optical adapter.

For connecting source, receiver and isolator together with the optical fiber are applied the detachable connections, which have typical losses α_{pc} = 0.2 dB. There are 11 such connections. The losses in the optical adder are 0.46 dB. There are 2 such adders. The losses on the optical polarizer are 0.32 dB. Total losses on the connections and adders are 3.44 dB. For registration of low power of the laser radiation at the high losses in the optical fiber was used the receiving optical module DFDMSH40-16M (company Dilaz), which has highly sensitive in the wavelength range of 980-1650 nm. It is better to use a set of two devices from one company, because it allows to obtain a higher signal-to-noise ratio at a low power of laser radiation.

This experimental setup allows to conduct measurements both at the moment of γ-radiation influence and immediately after its termination, in contrast to the works, which were previously conducted by other scientists. Also, there is no large time interval between the time of the γ-radiation influence and the beginning of measurements. It allows to establish a number of new features for optical fibers, which relate to the color centers relaxation.

At the figure 2 as an example are presented the dependences of the α_s losses in single-mode fibers with the change of exposure dose D_R for various alloying percentage of optical fiber.

Analysis of the results obtained at the figure2 showed, that an increase in alloying percentage increases the sensitivity of this optical fiber to the exposure dose of γ-radiation. It allows to register the changes in small D_R values, which decrease of the laser radiation power at the output of FOCL by 0.2 dB, when exposed to an optical fiber.

At the figure 3 as an example are presented the research results of velocity of the optical properties recovery after γ-radiation influence with a dose of 100 G.

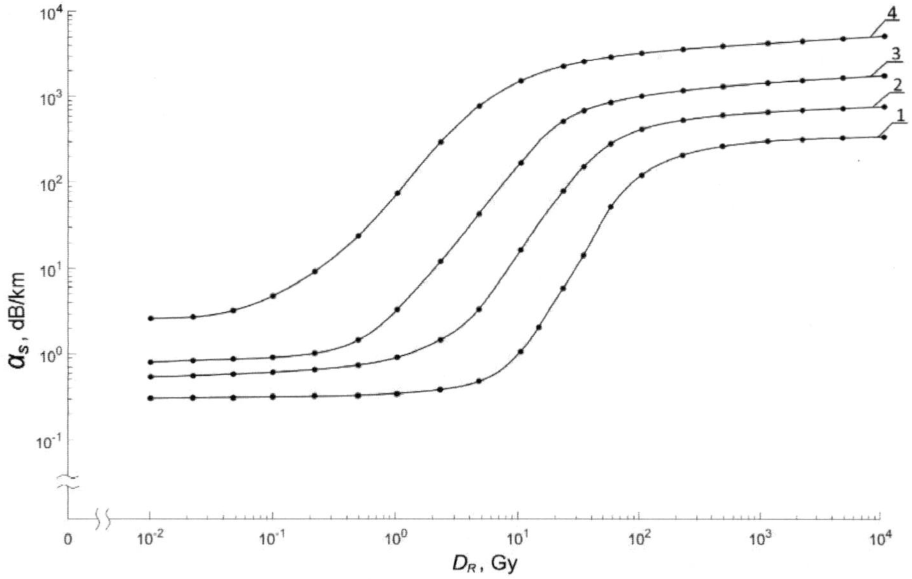

Figure 2. Dependence of the α_s changes at a wavelength λ=1550 nm with irradiation dose D_R for a single-mode fiber with a $SiO_2 - GeO_2$ core at T = 294.2 K. Charts 1, 2, 3 and 4 correspond to different alloying in %: 1.5; 4.0; 10.0 and 20.0

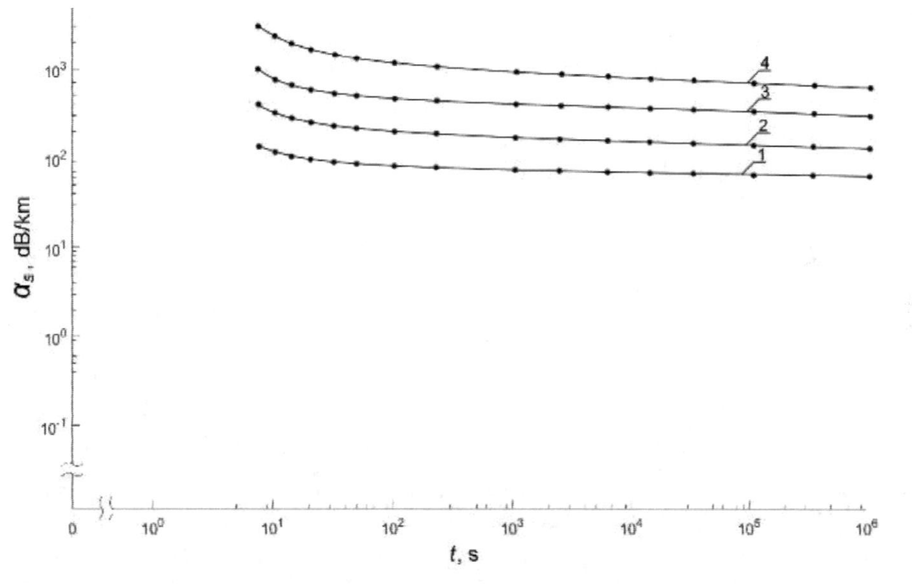

Figure 3. Dependence of the α_s changes at a wavelength λ=1550 nm with irradiation dose D_R for a single-mode fiber with a $SiO_2 - GeO_2$ core at T = 294.2 K. Charts 1, 2, 3 and 4 correspond to different alloying in %: 1.5; 4.0; 10.0 and 20.0.

Obtained results show that relaxation velocity of color centers higher with an increase of the alloying percentage. As a new design of the fiber-optic sensor, FOCL with a pure quartz core is proposed to connect to an optical fiber 200 m long with a $SiO_2 - GeO_2$ core and various alloying percentage (the alloying degree can be changed depending on the solving tasks). To recover an optical fiber, was decided to use the method developed earlier by us [16, 18], which is based on using an additional laser radiation with a wavelength λ=1310 nm and a dose of 100 G for pulsed and continuous laser radiation. For conducting this research was used pulsed laser radiation with a duaration of 0.1 s with various powers during 10 s.

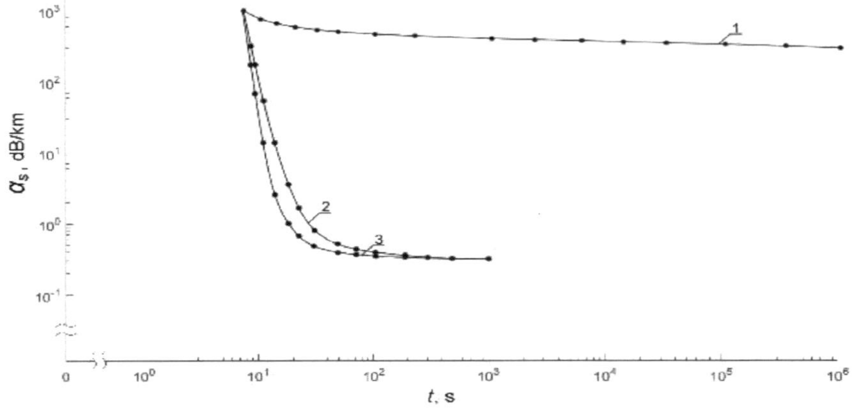

Figure 4. Dependence of the change in losses α_s with time t at a wavelength λ=1550 nm for a single-mode fiber with a $SiO_2 - GeO_2$ core (alloying 10.0 %) and polymer cladding at T= 294.3 K. Charts 1, 2 and 3 are correspond to different laser radiation powers in mW: 0; 40; 80.

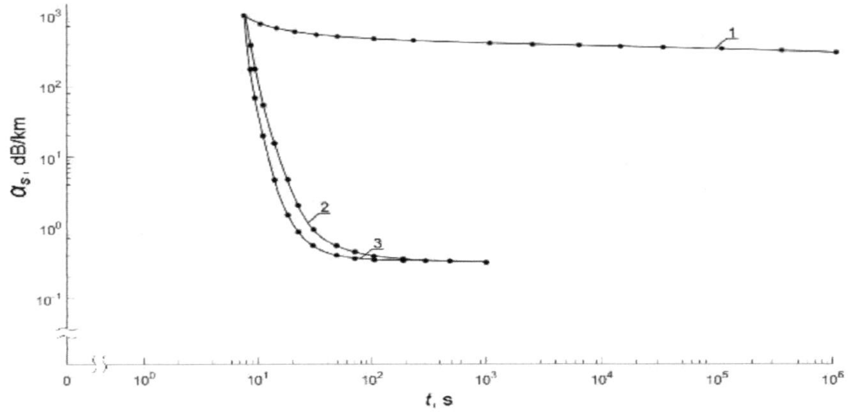

Figure 5. Dependence of the change in losses α_s with time t at a wavelength λ=1550 nm for a single-mode fiber with a $SiO_2 - GeO_2$ core (alloying 10.0 %) and polymer cladding at T=294.3 K. Charts 1, 2 and 3 are correspond to different laser radiation powers in mW: 0; 20; 40.

Analysis of the obtained results shows that the use of additional laser radiation in communication line allows to control the relaxation velocity of color centers (the increase of additional radiation power significantly increase the relaxation velocity). It allows to recover the optical properties of an optical fiber in less 10 s. Moreover, it was found in research, that the use of pulsed laser radiation is more effective than continuous.

4. Conclusion

Obtained experimental results showed a reliable work of the developed sensor for remote control an exposure dose from 0.1 to 1000 G with an opportunity of operation mode recovery with using an additional laser radiation with λ=1310 nm, as in trunk FOCLs developed earlier [21-24]. It is difficult to use the method developed by us in fiber-optic sensors with the measurements, based on the polarization, because there is high probability to failure the photodetectors with high sensitivity used in them.

References

[1] Fadeenko I, Fadeenko V, Reznik V, Moroz A, Popovskiy N and Nikolaev D 2019 *IOP Conference Series: Earth and Environmental Science* **390(1)** 012022
[2] Myazin N S 2018 *Journal Physics: Conference Series* **1124(1)** 031004
[3] Davydov R, Antonov V, Makeev S, and Batov Y 2019 *E3S Web of Conferences* **140** 02001
[4] Davydov V V and Kiryukhin A V 2020 *Atomic Energy* **127(5)** 274–279
[5] Nikitina M, Grebenikova N and Batov Y 2019 *IOP Conference Series: Earth and Environmental Science* **390(1)** 012024
[6] Gryznova E and Batov Y 2019 *E3S Web of Conferences* **140** 09001
[7] Gryznova E, Grebenikova N, Ivanov D and Bykov V 2019 *IOP Conference Series: Earth and Environmental Science* **390(1)** 012044
[8] Davydov R, Antonov V and Moroz A 2019 *International Conference on Electrical Engineering and Photonics EExPolytech-2019 (Saint-Petersburg)*, vol. 8906791 (IEEE) 295-297
[9] Van S, Cheremisin A, Chusov A, Switala F and Davydov R 2019 *IOP Conference Series: Earth and Environmental Science* **390(1)** 012011
[10] Davydov R V, Mazing M S, Yushkova V V and Stirmanov A V 2019 *Journal of Physics: Conference Series* **1410(1)** 012067
[11] Davydov R V, Antonov V I, Yushkova V V and Smirnov K J 2019 *Journal of Physics: Conference Series* **1400(6)** 066037
[12] Davydov V V 1999 *Russian Physics Journal* **42(9)** 822–825
[13] Davydov V V, Dudkin V I and Karseev A Y 2015 *Measurement Techniques* **58(3)** 317–322
[14] Davydov V V, Dudkin V I and Karseev A Y 2015 *Instruments and Experimental Techniques* **58(6)** 787–793
[15] Elokhin A P 2000 *Atomic Energy* **89(6)** 1011-1026
[16] Fadeenko V B, Kuts V A and Vasiliev D A 2018 *Journal Physics: Conference Series* **1135(1)** 012053
[17] Davydov V V, Dudkin V I and Karseev A Yu 2014 *Optical Memory & Neural Networks (Information Optics)* **23(4)** 259 – 264
[18] Davydov R V, Saveliev I V, Lenets V A and Tarasenko M Yu 2017 *Lecture Notes in Computer Science (including subseries Lecture Notes in Artificial Intelligence and Lecture Notes in Bioinformatics)* **10531 LNCS** 177-183
[19] Davydov V, Fadeenko V, Fadeenko I., Popovskiy N and Rud V 2019 *E3S Web of Conferences* **140** 07006
[20] Davydov V V, Velichko E N, Dudkin V I and Karseev A Y 2015 *Instruments and Experimental Techniques* **58(2)** 234–238
[21] Ermolaev A N, Krishpents G P and Vysoczkiy M G 2016 *Journal Physics: Conference Series* **741(1)** 012071
[22] Kashaykin P F, Tomashuk A L, Salgansky M Y, Guryanov A N and Dianov E M Proceedings of SPIE International Society for Optical Engineering **10681** p 1068110
[23] Dmitrieva D S, Pilipova V M, Dudkin V I, Davydov V V and Rud V Yu 2020 *Journal of Physics: Conference Series* **1697(1)** 012145
[24] Dmitrieva D, Pilipova V, Andreeva E, Dudkin V and Davydov V 2020 In.: Proceedings of ITNT 2020 – 6th IEEE International Conference on Information Technology and Nanotechnology **925334**

Stellar calibration of the single-photon receiver for satellite-to-ground quantum key distribution

A V Khmelev[1,2], A V Duplinsky[1,5], V L Kurochkin[1,3,4] and Y V Kurochkin[1,2,3,4]

[1] QRate, Novaya av. 100, Moscow, Russia
[2] Moscow Institute of Physics and Technology, 9 Institutskiy per., Dolgoprudny, Moscow Region, 141701, Russian Federation
[3] NTI Center for Quantum Communications, National University of Science and Technology MISiS, Leninsky prospekt 4, Moscow 119049, Russia
[4] Russian Quantum Center, Bolshoy Boulevard 30, bld. 1, Skolkovo, Moscow 121205, Russia
[5] Bauman Moscow State Technical University, Moscow, Baumanskaya 2-ya, 5/1, 105005, Russian Federation

al.duplinsky@goqrate.com

Abstract. Satellite quantum communication is the technology that allows to deploy large-scale quantum networks with a communication range of thousands kilometres We report the ground receiver for downlink quantum key distribution (QKD) with satellite. An optical part of this system including an active tracking loop is mounted on a 600-mm Ritchey-Chretien telescope and permits to distinguish polarization states to perform QKD between ground and satellite. Moreover, a procedure of calibration the receiver using stars with known brightness is presented. Measurements of the photon count rate of stars in the spectral range of 845 nm - 855 nm are performed and compared with an estimate.

1. Introduction

Quantum key distribution (QKD), the technology that allows secure communication immune against future advances in computational power, is now being deployed on a large geographic scale.

The need for cryptography, that is, the encryption of important messages arose in the ancient times and has only grown over the millennia. The today's most popular encryption algorithm is RSA [1], which is not unbreakable, but hacking it is computationally complex and would take hundreds of years on the best modern supercomputers. Such robustness had been sufficient until in 1994 Shor proposed an algorithm that would have allowed breaking RSA in minutes using a quantum computer [2].

Although the actual large-scale implementation of Shor's algorithm is some time away, quantum computers already show significant success. Not so long ago Google declared quantum supremacy achievement (quantum devices without error correction can perform a well-defined computational task beyond the capabilities of currently available classical computers) on 53 qubits [3].

Moreover, Chinese physicists published an article in December 2020, where they demonstrated a quantum photon processor with 25 qubits. The calculation speed of the computer is 100 trillion times higher than on any existing classical supercomputer [4]. Having achieved quantum superiority, they thus showed the prospects for the development of a quantum computer based on photons.

The idea to create cryptography unbreakable in principle and counteract the "quantum threat" unsurprisingly also emerged from the quantum world. In fact, the concept of quantum cryptography was proposed even earlier, in 1984, by Bennett and Brassard [5]. It is based on the fact that if one measures the quantum of light, it changes its state.

However, the impossibility of cloning the quantum states, being the basis of quantum cryptography, also limits the transmission distance of the signal over the optical fiber to 100–200 km (limited by fiber loss). This imposes a restriction on the length of the communication line.

Hence, satellite quantum key distribution as a new method to share secret information for extra long distances can be the realization of an alternative way to the fiber optic distribution.

In this paper we present a ground receiver for downlink quantum key distribution with satellite that opens up a possibility for useful quantum cryptography. It is crucial to have the highest possible level of useful signal during the entire communication session with the satellite, therefore we performed a stellar calibration for our receiver.

2. Methods

The Zvenigorod observatory is located about 80 km from Moscow. The observatory station consists of a Ritchey–Chretien Alt-Az telescope with an aperture of 0.6 m and a focal length of 4.8 m, a coarse camera with the field of view (FOV) of 0.7° × 0.7° (512 × 512 pixels, frame rate of 10 Hz) and an optical receiver box mounted on the back of the telescope main mirror.

The coarse-tracking system consists of an alt-azimuth mount and a 70mm-coarse camera in a control loop. The coarse camera identifies optical signals from distant objects such as satellites or stars and then the telescope accurately points to it. Successful testing of a similar system is described in reference [6].

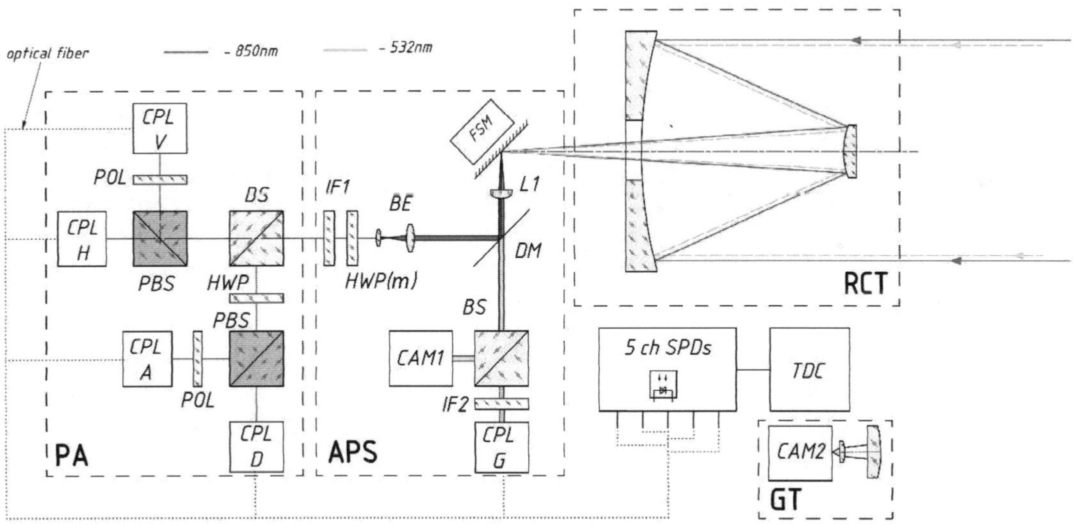

Figure 1. Experimental set-up. CPL – coupler; POL – polarizer; PBS - polarizing beam splitter cube; HWP – half-wave plate; BS – 50:50 beam splitter; IF1 – interference filter (CWL=850nm; FWHM=10nm); HWP-m – motorized half-wave plate; BE – beam expander; DM – dichroic mirror; L1 – lens (f = 75 mm); FSM – fast steering mirror; RCT – 600-mm-aperture Ritchey-Chretien telescope; CAM1 – fine camera (field-of-view of 4 mrad×4 mrad); IF2 – interference filter (CWL=532 nm; FWHM=10 nm); 5 ch SPDs– 5 channel single photon detectors module; TDC – time to digital converter; PA – polarization analyzer; APS - analysis and processing module, GT – guide telescope, CAM2 – guide telescope camera.

The optical receiver box holds the fine-tracking system as well as the 850-nm photon receiver. It primarily consists of a dichroic mirror (DM), a fast-steering mirror (FSM) and a complementary metal-oxide semiconductor (CMOS) imaging sensor or a fine camera with the FOV of 3 mrad × 3 mrad (640 × 640 pixels, frame rate of 100 Hz). The FSM is used to fine-tune the optical direction based on image data gathered by the fine feedback camera (CMOS). The dichroic mirror is used to separate the reference optical beam from the 850-nm photons, with one going to the fine camera for tracking and the other to an 850-nm photon receiver connected to the four single-photon detectors (SPDs).

The 850-nm photons are obtained by a BB84 passive-basis-choice polarization analysis module, which contains a 50:50 beam splitter, half-wave plate, and polarizing beam splitters after going through a motorized half-wave plate (HWP) and an interference filter (central wavelength of 850 nm, bandwidth of 10 nm). The receiver module is coupled to four single-photon detectors by four multimode optical fibers with core diameters of 105 μm. At 850nm, the SPDs have a quantum efficiency of 50%, an active area of 180 μm, and a dead time of 22 ns. The electric output pulses from all SPDs are fed into a time-to-digital converter (TDC), which records the detecting time and detector channel numbers. The acquired data is stored in the computer for further analysis.

The maximal possible deviation of the telescope from the exact position at the object is limited by two optical fiber parameters: numerical aperture (NA) and core diameter (CD). In our tests, they are equal to 0.22 and 105 μm, respectively.

3. Results

To calibrate the ground receiver in the spectral range 850±5 nm, we use stars with known apparent magnitude in infrared diapason (m) and record the count rate (C) or photons per second, which is proportional to the brightness of the stars. By definition

$$m = -2.5lg\frac{C}{C_0}, \qquad (1)$$

where C_0 is the count intensity of a star with a zero apparent magnitude.

Furthermore, the turbulence in the atmosphere causes a deviation in the calculation outcome of each star. Figure 2 depicts an example of the count rate fluctuation for the star Etamin, which has an apparent magnitude of 0.23.

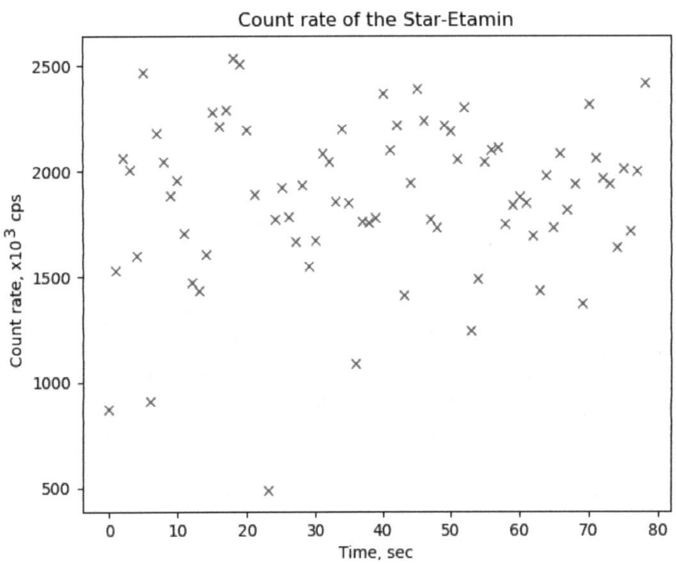

Figure 2. The photon flux of the star Eltanin (Gamma Draconis) obtained by the ground station and recorded by avalanche photodiodes (SPCM-AQRH-14-FC) over an 80-second period. The altitude of the star is 70 degrees. The mean value is 1875x10^{3}cps, and the standard deviation is 374x10^3 cps, indicating a 20% error in photon flux calculation due to atmospheric turbulence.

Thus, we perform 21 measurements of the count rate from the stars, normalized them for 100% transmission of atmosphere and calculated the average value of coefficient C_0 for that case, using equation (1). It equals $3850 \pm 150\text{x}10^3$ cps.

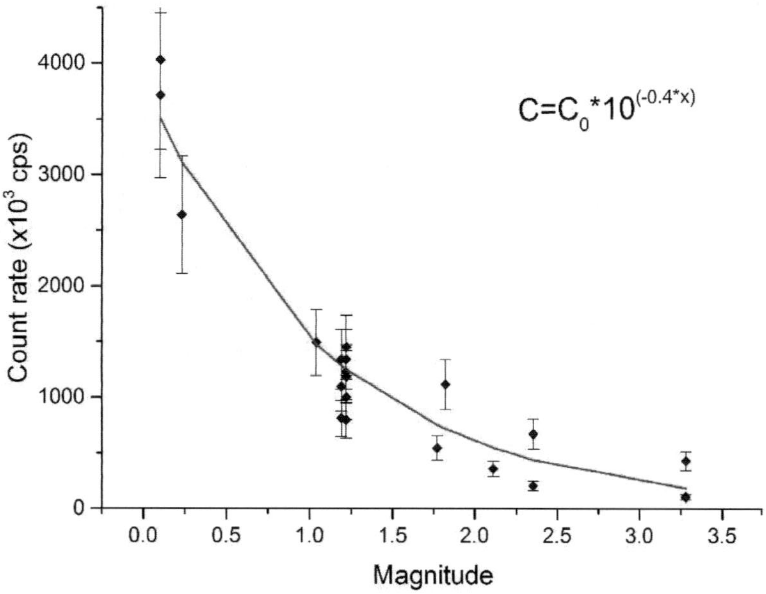

Figure 3. The dependence of the count rate on the magnitude. Measurements were taken on various clear nights of September 2020.

According to the open database, the brightness (L) of a star with a magnitude of zero in the I-spectral range at the Earth's surface is $1.058\text{x}10^{-11}\frac{W}{nm\cdot m^2}$. The incident number of photons (C_0^{th}) predicted is as follows:

$$C_0^{th} = \frac{L \cdot S \cdot \eta \cdot B}{E_{ph}}, \qquad (2)$$

Hence, taking into account $\eta = 0.55$, $S = 0.283\ \text{m}^2$, $B = 10$ nm, C_0^{th} equals $51400\text{x}10^3$ cps.

Finally, we can calculate the transmittance (T) of our entire optical device, including the telescope, using estimated and experimental measurements as follows:

$$T = \frac{C_0}{C_0^{th}} = 7.5 \pm 0.2\% \qquad (3)$$

4. Conclusion

We have shown the transmittance of our receiver using the method of stars measuring in narrow spectral range. This coefficient allows us to assess the quality of the system setup and to improve the system for further experiments. The developed device for satellite QKD opens up a possibility for useful quantum cryptography at a global scale and convenient integration with other quantum networks such as fiber-based networks. It will be interesting to investigate the experimental implementation of space QKD in the backbone network for long-distance transmission.

Acknowledgments
We thank many colleagues at the Russian Quantum Center, QRate, for motivating discussions, management, and coordination. We also thank our colleagues at the Institute of Astronomy of the Russian Academy of Sciences, especially S. Barabanov and M. Nalivkin for their assistance and support. This work is supported by the Russian Science Foundation under project 17-71-20146.

References

[1] Rivest R L, Shamir A and Adleman L 1978 A method for obtaining digital signatures and public - key cryptosystems *Commun. ACM*, **21(2)** pp 120-26.
[2] Shor P W 1994 Proceedings of the 35th Annual Symposium on the Foundations of Computer Science *Computer Society Press* pp 124–34.
[3] Arute F et al Quantum supremacy using a programmable superconducting processor 2019 *Nature* **574** 505-10.
[4] Zhong H S et al 2020 Quantum computational advantage using photons *Science* **370(6523)** 1460-63.
[5] Bennett C H and Brassard G 1984 Proceedings of the international conference on computers, systems and signal processing pp 175–79.
[6] Kurochkin V L, Khmelev A V, Petrov I V, Miller A V, Feimov A A, Mayboroda V F, Balanov M Y, Krushinsky V V, Popov A A and Kurochkin Y V 2020 *Journal of Physics: Conference Series* **1680(1)**.

Structure and spectral properties of Fe:$ZnAl_2O_4$ transparent glass-ceramics and ceramic

K Eremeev[1,3], L Basyrova[2], O Dymshits[3], S Balabanov[4], A Belyaev[4], I Alekseeva[3], A Khubetsov[3], M Tsenter[3], A Volokitina[5], A Zhilin[3], V Popkov[6] and P Loiko[2]

[1] Saint-Petersburg State Institute of Technology, 26 Moscovsky Pr., 190013 St. Petersburg, Russia
[2] Centre de Recherche sur les Ions, les Matériaux et la Photonique (CIMAP), UMR 6252 CEA-CNRS-ENSICAEN, Université de Caen Normandie, 6 Boulevard Maréchal Juin, 14050 Caen Cedex 4, France
[3] S.I. Vavilov State Optical Institute, 36 Babushkina St., 192171 St. Petersburg Russia
[4] G.G. Devyatykh Institute of Chemistry of High-Purity Substances of the Russian Academy of Sciences, 49 Tropinin St., 603950 Nizhny Novgorod Russia
[5] ITMO University, 49 Kronverkskiy Pr., 197101 St. Petersburg, Russia
[6] Ioffe Institute, 26 Politekhnicheskaya St., 194021 St Petersburg, Russia

kirilleremeev42@gmail.com, vodym1959@gmail.com

Abstract. Transparent glass-ceramics based on iron-doped $ZnAl_2O_4$ (gahnite) nanocrystals were prepared by melt-quenching (at 1580 °C) with subsequent heat-treatments (at 720 – 1050 °C). Polycrystalline transparent Fe:$ZnAl_2O_4$ ceramic was synthesized by hot pressing of powders (at 1600 °C, 50 MPa) using the sintering additive, ZnF_2. A comparative study of the structure, Raman spectra and absorption properties of these materials was performed.

1. Introduction
Gahnite ($ZnAl_2O_4$) is a promising optical material belonging to the family of spinels, AB_2O_4. It is very suitable for doping with transitions metal ions [1,2]. Divalent iron ions (Fe^{2+}) are of interest because of their broadband absorption extending into the mid-infrared [3]. The growth of gahnite single crystals is technologically complicated. Thus, it is relevant to develop other $ZnAl_2O_4$-based transparent materials, such as transparent nanophase glass-ceramics (GCs) and transparent ceramics. Recently, first GCs based on Fe^{2+}:$MgAl_2O_4$ [4] and on Fe^{2+}:$ZnAl_2O_4$ [4,6] nanocrystals were prepared and studied. In the present work, we report on the fabrication of such materials doped with iron ions, as well as their structure and optical properties.

2. Experimental

2.1. Synthesis of glass-ceramics
The model glass of the ZnO-Al_2O_3-SiO_2 system nucleated by TiO_2 [1] was doped with 1.0 mol% FeO and melted in a quartz ceramic crucible at 1580 °C for 4 h with stirring, cast onto a metal plate and annealed at 620 °C. The batch weight was 400 g. Transparent gahnite-based GCs were prepared from

the glass by two-stage heat-treatments with the first (nucleation) stage at 720 °C and the second stage in the range from 750 to 1200 °C (the duration of each stage was 6 h).

2.2. Synthesis of ceramic

Metallic aluminum A99 (≥99.99% purity), anhydrous isopropyl alcohol (reagent grade), zinc formate (99.99% purity), zinc fluoride (99.99% purity) and iron (II) formate (99.99% purity) were used as starting materials. Firstly, aluminum isopropoxide $Al[OCH(CH_3)_2]_3$ was synthesized by direct interaction of aluminum with isopropyl alcohol. The obtained aluminum isopropoxide was distilled in vacuum at a temperature of ~160 °C and an absolute pressure ~0.5 kPa. Then, an alcoholic solution of aluminum isopropoxide was hydrolyzed with an aqueous solution of zinc formate, zinc fluoride and iron (II) formate by dropping with intense stirring. The $Zn_{(form)}$ / (Al + Fe) molar ratio was 1:2. The contents of zinc fluoride and iron ions were 3 wt% ZnF_2 and 1 at.% Fe^{2+}, respectively. After vacuum drying at the final temperature of 300°C, the resulted precursor xerogel was ground in a planetary ball mill, and then calcined in air at 700 °C for 20 min. The obtained 3 wt% ZnF_2, 1 at.% Fe^{2+}:$ZnAl_2O_4$ powder was compacted in a steel mold with a diameter of 20 mm at 50 MPa, then wrapped into a graphite foil and hot pressed at a temperature of 1520 °C under uniaxial pressure of 40 MPa for 4 h in a ~10 Pa vacuum. After polishing, transparent Fe^{2+}-doped $ZnAl_2O_4$ ceramic with a thickness of ~2 mm and a diameter of 20 mm was obtained, Figure 1.

Figure 1. Photographs **(a)** of the laser-grade polished 1 at.% Fe:$ZnAl_2O_4$ ceramic (thickness: ~2 mm); **(b)** the 1.0 mol% FeO-doped initial glass and $ZnAl_2O_4$-based GCs Labels 750 – 1050 °C indicate the heat-treatment temperature at the second stage. The heat-treatment at the nucleation stage is at 720 °C for 6 h.

2.3. Characterization methods

The differential scanning calorimetry (DSC) study for glasses was carried out using a simultaneous thermal analyzer NETZSCH STA 449 F3 Jupiter at a heating rate of 10 K/min in a dynamic flow atmosphere of Ar in the temperature range between room temperature and 1250 °C. A bulk sample of about 15 mg in weight was used. The X-ray powder diffraction (XRD) patterns of the ceramic and GCs were measured with a Shimadzu XRD-6000 diffractometer with Ni-filtered Cu Kα radiation. The mean sizes of crystals with spinel structure in the GCs were estimated from broadening of the X-ray diffraction peak at $2\theta \approx 65.0°$ according to the Scherrer equation [1]. Their unit cell parameter a was determined using the same diffraction peak with the Millers' indices (440). The error for the crystal size estimation was ~5%. The Raman spectra were measured using a Renishaw inVia confocal Raman microscope equipped with a Leica ×50 objective and an edge filter. The excitation wavelength was 514 nm (an Ar^+ ion laser line). The absorption spectra were recorded with a Shimadzu 3600 spectrophotometer in the spectral range 250 - 3300 nm with a resolution of 1 nm. Plane-parallel polished plates with a thickness of 1 mm were used.

3. Results and discussion

Let us first describe the properties of transparent GCs. Figure 2 presents the results of the DSC study of the initial glass and glass heat-treated at 720 °C for 6 h (the nucleation stage). An intense peak on the DSC curve of the initial glass with the crystallization onset temperature T_{on1} of 804 °C and the maximum crystallization temperature T_{max} of 840 °C evidences the gahnite crystallization. The peaks of low intensity in the high-temperature part of the DSC curve are assigned to rutile (TiO_2) and probably cristobalite (SiO_2) crystallization. The nucleation affects the kinetics of crystallization: the glass transition temperature T_g decreases by ~10 °C, the temperature of gahnite crystallization decreases by ~35 °C and the exothermal peak broadens, the temperatures of crystallization of high-temperature phases also decrease.

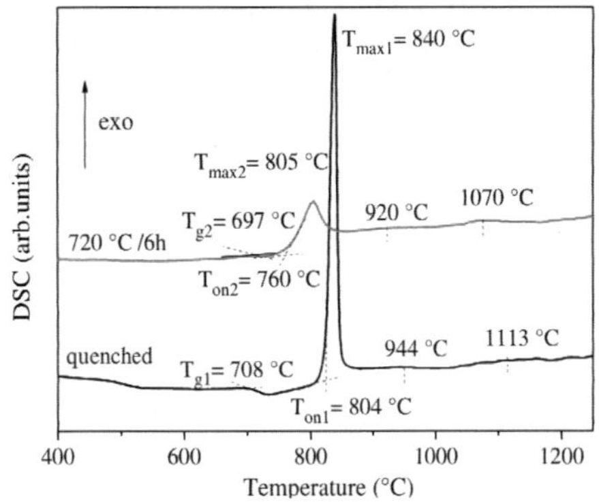

Figure 2. DSC curves of the initial glass and glass heat-treated at 720 °C for 6 h. T_g is the glass transition temperature, T_{on} and T_{max} are the onset and maximum crystallization temperatures.

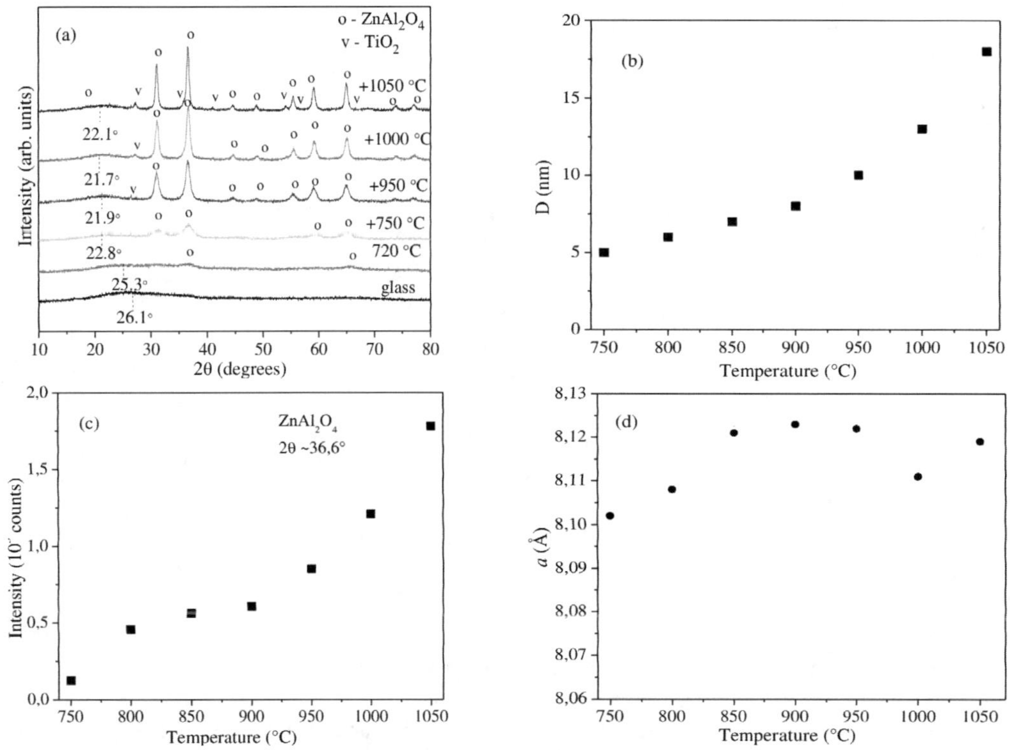

Figure 3. **(a)** XRD patterns of the initial and heat-treated glasses; **(b)** the average size of $ZnAl_2O_4$ crystals; **(c)** the intensity of the $ZnAl_2O_4$ diffraction peak at 2θ=36.6°; **(d)** the unit cell parameter *a* of the $ZnAl_2O_4$ crystals as a function of the heat-treatment temperature of the initial glass. The heat-treatment duration is 6 h. The heat-treatment at the nucleation stage is at 720 °C for 6 h.

The XRD pattern of the initial glass (Figure 3a) shows an amorphous halo with a maximum at 2θ = 26.1° and no signs of any crystalline phases. The traces of gahnite crystals are seen after the heat-treatment at 720 °C. As the temperature of the heat-treatment increases, the Scherrer diameter of gahnite crystals gradually increases from 5 to 19 nm, Figure 3b. The crystallinity of gahnite also increases (Figure 3c). The parameter of the gahnite unit cell *a* changes with increasing the heat-treatment temperature in a complex manner (Figure 3d) and ranges from 8.102 Å (at 750 °C) to 8.119 Å (at 1050 °C). The maximum value of a = 8.123 Å is observed after the heat-treatment at 900 °C. These changes reflect a complex variation in the composition and structure of the crystals with heat-treatment temperature. Rutile crystallizes during the heat-treatment at 1000 °C and the corresponding nanocrystals have a size of 24 nm.

The Raman spectrum of the initial glass, Figure 4, contains two broad bands, one with a maximum at 460 cm^{-1} and another one of a complex shape with two local maxima at 800 and 920 cm^{-1}. The bands at 440 and 800 cm^{-1} are attributed to vibrations of tetrahedra of the aluminosilicate network. The more intense band at 920 cm^{-1} is connected with vibrations of TiO_4 tetrahedra entering the aluminosilicate network [4]. The spectrum changes after the heat-treatment at 720 °C for 6 h. There is a change in the positions of Raman bands and a redistribution of their intensities. The band at ~920 cm^{-1} weakens, and the band at ~800 cm^{-1} shifts to ~790 cm^{-1} and noticeably increases. These changes are associated with the development of liquid phase separation and formation of the amorphous zinc aluminotitanate phase wherein titanium ions form TiO_5 and TiO_6 polyhedrons giving rise to a characteristic band at ~790 cm^{-1} superimposed on the band due to vibrations of tetrahedra of the aluminosilicate network. After further heat-treatments, the phase separation develops with a formation of gahnite, and then, at 950 – 1050 °C, rutile (Figure 3). The multiple vibrations associated with gahnite manifest its partly inversed structure probably caused by Ti^{3+}, Ti^{4+}, Fe^{2+} and Fe^{3+} ions entering its structure.

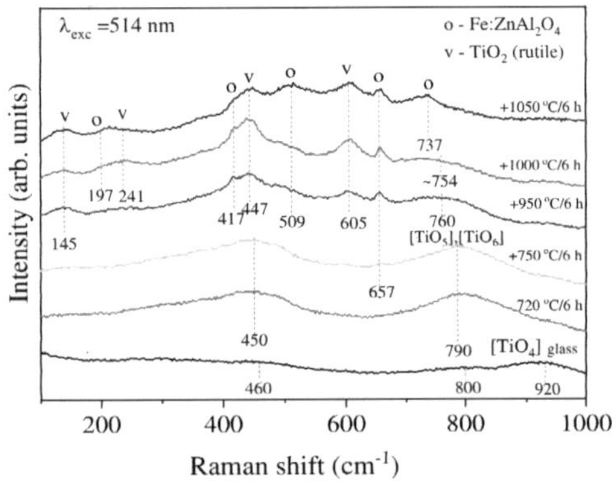

Figure 4. Unpolarized Raman spectra of the initial glass and transparent GCs, the labels 750 – 1050 °C indicate the heat-treatment temperature at the second stage, λ_{exc} = 514 nm. The heat-treatment at the nucleation stage is at 720 °C for 6 h.

The absorption spectrum of the initial glass reveals a short-wave absorption edge (E_g = 2.95 eV), two broad and smooth bands with maxima at about 1070 nm (the $^5T_2 \rightarrow {}^5E$ (^5D) transition) and about 1850 nm (the $^5E \rightarrow {}^5T_2$ (^5D) transition), assigned to the absorption of Fe^{2+} ions in octahedral (predominantly) and tetrahedral coordination, respectively, and a broad band in the region of 2700-3300 nm due to absorption of OH^--groups in the glass (Figure 5). As the temperature of the heat-treatment increases, the short-wavelength absorption edge shifts to longer wavelengths, hindering the absorption band centered at 1000 nm, the intensity of the absorption band at ~2 μm markedly increases, the position of its maximum is shifted to ~2030 nm, the absorption band of OH^--groups is structured and its intensity increases. A sharp increase in absorption in the visible and near-IR regions is connected with the Fe^{2+}-Fe^{3+} intervalence transitions, Fe^{2+} and Fe^{3+} d→d transitions in the zinc aluminotitanate amorphous phase and in gahnite (Fe^{3+} and Fe^{2+} in octahedral site symmetry positions). The absorption at ~2030 nm is associated with incorporation of Fe^{2+} ions into the gahnite nanocrystals in tetrahedral positions (the $^5E \rightarrow {}^5T_2(^5D)$ transition). The structuring of the IR absorption band caused by the OH^-- groups is also connected with gahnite crystallization as the OH^--groups can enter the gahnite structure.

We have performed a comparative study of the structural and optical properties of the transparent GC prepared by the heat-treatment at 720 °C for 6 h and at 1050 °C for 6 h and the ceramic. Scanning electron microscopy (SEM) revealed a close-packed microstructure of ceramic, Figure 6a, while the TEM photograph of GC revealed its multiphase nature. Intense diffraction peaks in the XRD pattern of the ceramic, Figure 6c, belong to gahnite, $ZnAl_2O_4$, with a cubic structure (sp. gr. *Fd*3*m*). The unit cell parameter of gahnite ceramic *a* is 8.12 Å. The increased *a* value as compared to that of undoped

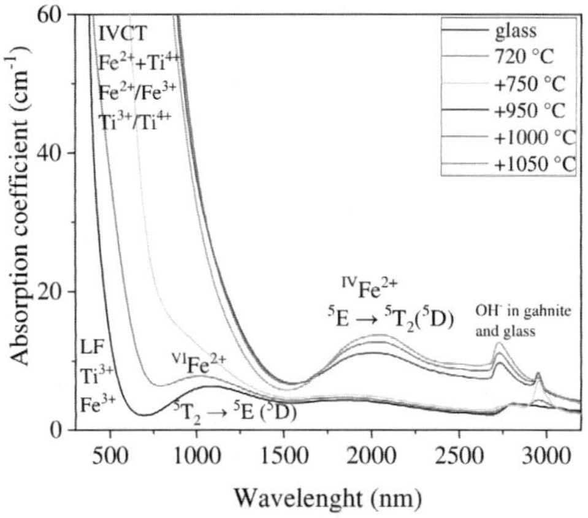

Figure 5. Absorption spectra of the initial glass and transparent GCs. The labels 750 – 1050 °C indicate the heat-treatment temperature at the second stage.

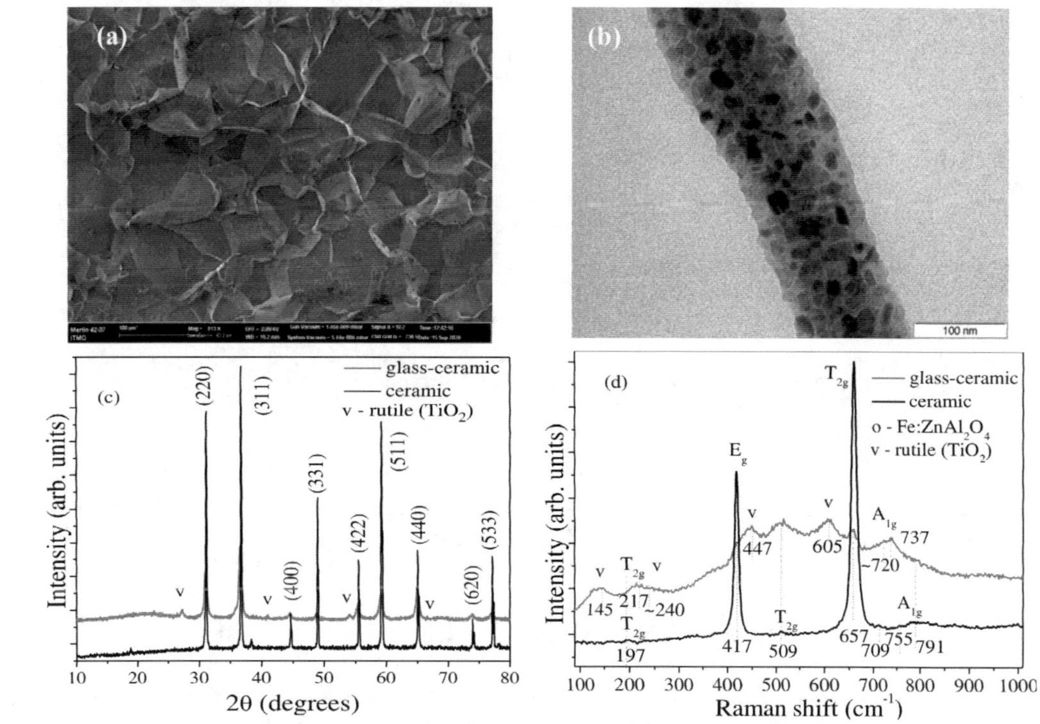

Figure 6. Structural characteristics of the $Fe{:}ZnAl_2O_4$-based transparent GC prepared by the heat-treatment at 720 °C for 6 h and at 1050 °C for 6 h and the $Fe{:}ZnAl_2O_4$ ceramic: **(a)** SEM image of the fractured surface of ceramic; **(b)** TEM image of GC; **(c)** XRD patterns, *numbers* denote the Miller's indices (*hkl*); **(d)** Raman spectra, $\lambda_{exc} = 514$ nm.

$ZnAl_2O_4$ (a = 8.088 Å) indicates a certain degree of inversion. A comparison of the XRD patterns of the GC and ceramic, Figure 6c, revealed that they differ in phase composition, size and fraction of gahnite crystals. The XRD pattern of the GC shows broad peaks of gahnite, broad peaks of low intensity due to rutile, TiO_2, and a weak halo due to the residual highly siliceous glass phase. The mean size of gahnite nanocrystals in the GC is 18 nm and their unit cell parameter a is 8.098 Å, also manifesting a certain degree of inversion. The average size of rutile crystals is 39 nm. The complex phase composition of the GC is also revealed in its Raman spectrum, see Figure 6d.

The absorption spectra of both the GC and ceramic demonstrate absorption of Fe^{2+} ions in tetrahedral (T_d) sites in gahnite crystals due to the $^5E \rightarrow {}^5T_2(^5D)$ transition, Figure 7. The inversion of Fe:$ZnAl_2O_4$ is confirmed by absorption of Fe^{2+} ions in octahedral (O_h) sites (the $^5T_2(^5D) \rightarrow {}^5E$ transition). The profound red-shift of the UV absorption edge for the GC as compared with that of the ceramic is due to intervalent and charge-transfer transitions in multiphase GC.

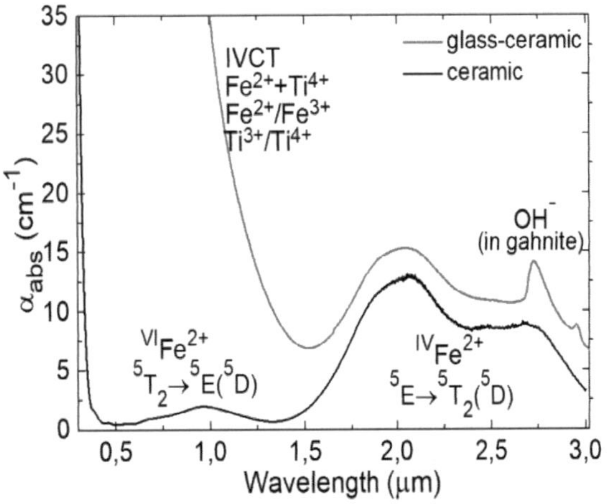

Figure 7. Absorption spectra of Fe:$ZnAl_2O_4$-based transparent GC prepared by the heat-treatment at 720 °C for 6 h and at 1050 °C for 6 h and Fe:$ZnAl_2O_4$ ceramic.

4. Conclusions

Transparent materials based on iron-doped gahnite (glass-ceramics and ceramic) were prepared. Their structure and spectral properties were evaluated and compared. A broad absorption band at 1.5-3 μm due to the $^5E \rightarrow {}^5T_2(^5D)$ transition of Fe^{2+} ions in T_d sites in gahnite crystals makes these materials promising for saturable absorbers of lasers emitting in the short-wave infrared spectral range.

Acknowledgment
This work was partly supported by the RFBR (Grant 19-03-00855).

References

[1] Loiko P, Belyaev A, Dymshits O, Evdokimov I, Vitkin V, Volkova K, Tsenter M, Volokitina A, Baranov M, Vilejshikova E, Baranov A and Zhilin A 2017 *J. Alloys Compd.* **725** 998

[2] Belyaev A, Basyrova L, Sysoev V, Lelet M, Balabanov S, Kalganov V, Mikhailovski V, Baranov M, Stepanidenko E, Vitkin V, Dymshits O and Loiko P 2020 *J. Alloys Compd.* **829** 154514

[3] Mirov S, Fedorov V, Moskalev I, Martyshkin D and Kim C 2010 *Laser Photon Rev.* **4** 21

[4] Basyrova L, Bukina V, Balabanov S, Belyaev A, Drobotenko V, Dymshits O, Alekseeva I, Tsenter M, Zapalova S, Khubetsov A, Zhilin A, Volokitina A, Vitkin V, Mateos X, Serres J M, Camy P and Loiko P 2021 *J. Lumin.* **236** 118090

[5] Eremeev K, Dymshits O, Alekseeva I, Khubetsov A, Zapalova S, Tsenter M, Basyrova L, Loiko P, Zhilin A and Popkov V 2020 J. *Phys. Conf. Ser.* **1697** 012125

[6] Eremeev K, Dymshits O, Alekseeva I, Khubetsov A, Tsenter M, Zhilin A, Basyrova L, Loiko P and Popkov V 2020 *J. Phys. Conf. Ser.* **1965** 012011

Features of the fiber-optics data system using optical solitons

A A Ermolaev, M A Shevchenko and E I Andreeva

The Bonch-Bruevich Saint-Petersburg State University of Telecommunications, St.Petersburg 193232, Russia

Abstract. The analysis of the parameters of a high-speed information data system using dispersion-managed solitons is carried out. It is shown that the quality of information transmission can be increased by choosing the input-output point of the symbol sequence on the dispersion map.

1. Introduction
Soliton information transmission systems are distinguished by high reliability of information transmission [1-8]. However, in real fibers, it is possible to realize the soliton regime at certain approximations. Various variants of the soliton-like regime are used depending on the specific problem. The classical regime of the fundamental soliton provides a high speed in the transmission channel but requires a relatively high peak power of symbol pulses. The use of wavelength division multiplexing is hindered by the effect of the interaction of solitons. Pass-average solitons suitable for long hop systems but has symbol width limitations. For high-speed WDM systems, it is advisable to use the dispersion-managed soliton mode. The distinguishing feature of this mode is the low threshold level forming soliton in an optical fiber, which ensures a low level of inter-channel interference. The soliton is formed as averaged over a long nonlinearity manifestation. To implement this mode, the topology of a dispersion map (Figure 1) with alternating segments with a dispersion β_2 of opposite signs is used. Along the segment length, the parameters of the soliton pulse, such as the pulse duration *T(z)*, the frequency chirp *C(z)*, experience periodic changes, keeping their values T_o and C_o at the segment boundary (Figure 2).

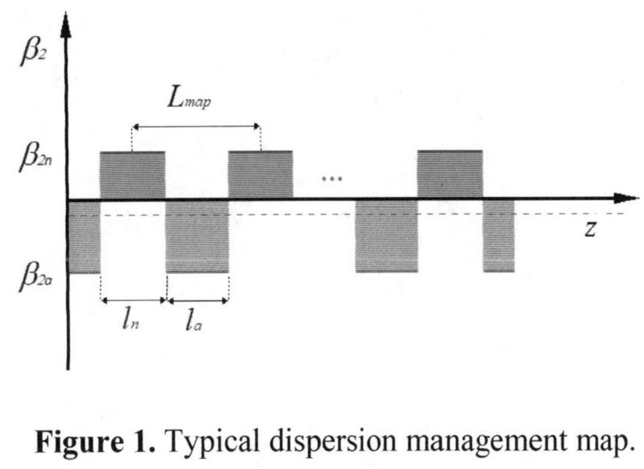

Figure 1. Typical dispersion management map.

Dispersion scheme (Figure 1) consist of the two fibers with the anomalous (negative) dispersion β_{2a} and length l_a and positive dispersion β_{2n} and length l_n [4]:

$$L_{map} = l_a + l_n, \quad (1)$$

where L_{map} - is the period of the dispersion map.

The input pulse width T_o is determined depending on the initial chirping C_o and the parameter T_{map} – dispersion map parameter [4]:

$$T_o = T_{map}\sqrt{\frac{1+C_o^2}{|C_o|}}, \quad (2)$$

where T_o – pulse width and C_o - pulse chirp between the segments with different dispersion, and

$$T_{map} = \sqrt{\frac{|\beta_{2n}\beta_{2a}l_n l_a|}{\beta_{2n}l_n - \beta_{2a}l_a}}, \quad (3)$$

The dispersion map parameters largely determine the bit rate. However, with the given parameters of the dispersion map, it is possible to select the parameters of the symbol pulse (Figure 2), providing a higher bit rate and higher SN-ratio in the system. For this, a differential dispersion map is used, which makes it possible to input / output data at the midpoint of the segment, the so-called point of zero chirp, where the pulse width takes on a minimum value.

The fiber-optic system with a period $l_a = l_n = 48$ km and 12 km was modeled in OptiSystem. In first configuration input soliton pulse width, corresponding to $T_o = 12$ ps and $C_o = 1$, so that the $T_{map} = 8.5$ ps realized bit rate of 10 Gbit/s. Second configuration referred to as the dense dispersion management: $T_o = 6$ ps and $C_o = 1$, so that the $T_{map} = 4.2$ ps (Figure 2 – 4). In this configuration bit rate of 20 Gbit/s can be realized.

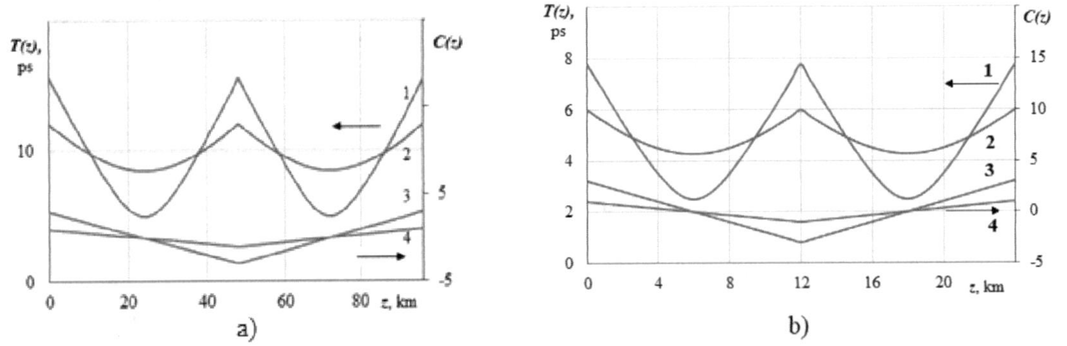

Figure 2. Spatial dynamics of changes in the pulse width *T(z)* (1 - 2) of the soliton pulse and a chirp *C(z)* (3 - 4) at $C_o = 3$ (1, 3) and $C_o = 1$ (2, 4) (a) on the period of the dispersion map $l_a = l_n = 48$ *km*, $|\beta_2| \cong 3$ ps^2/km (a), $l_a = l_n = 12$ *km*, $|\beta_2| \cong 3$ ps^2/km (b).

On the long-distance propagation solitons can be formed if the nonlinear effects are balanced by the path-average dispersion:

$$\overline{\beta_2} = \frac{\beta_{2n}l_n + \beta_{2a}l_a}{l_n + l_a}, \quad (4)$$

As a result, solitons can be formed in an average sense, but the width and power of such solitons oscillate periodically on the dispersion management map.

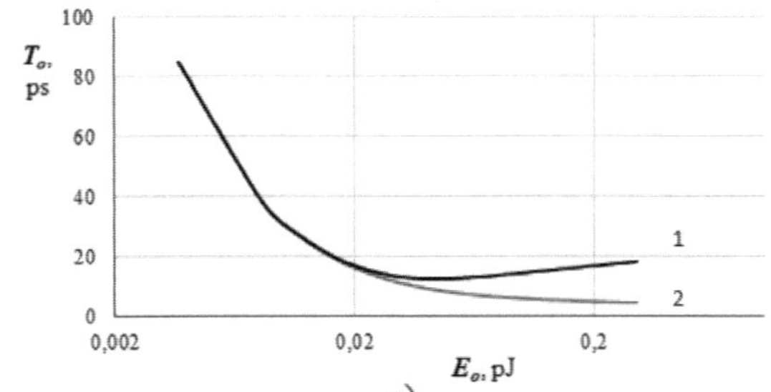

a)

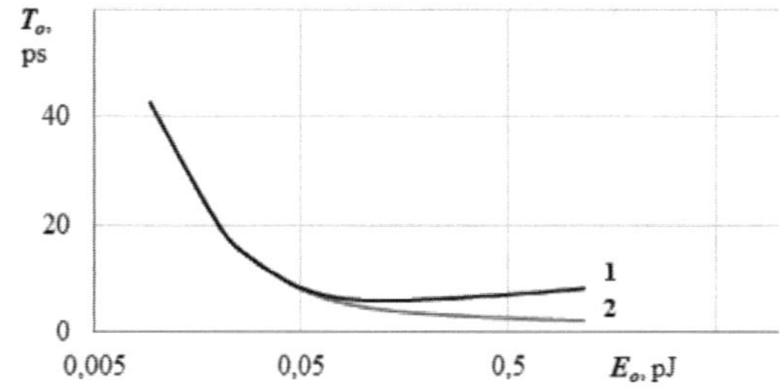

b)

Figure 3. Dependence of the pulse width T_o at the boundary of segments with different dispersion (1) and width T_{min} (2) in the middle of the segment on the initial energy E_o of the soliton pulse (dispersion map parameters are presented in Figure 2).

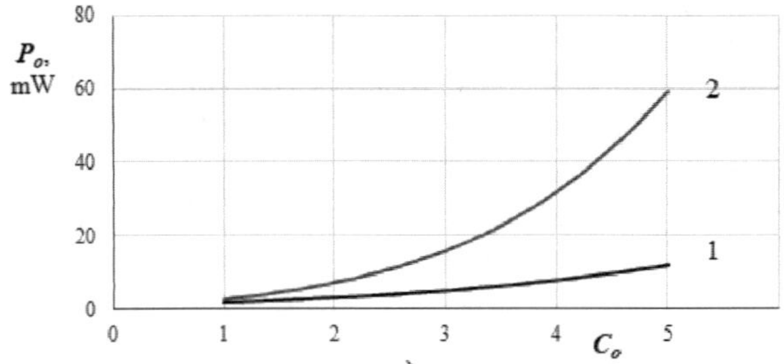

a)

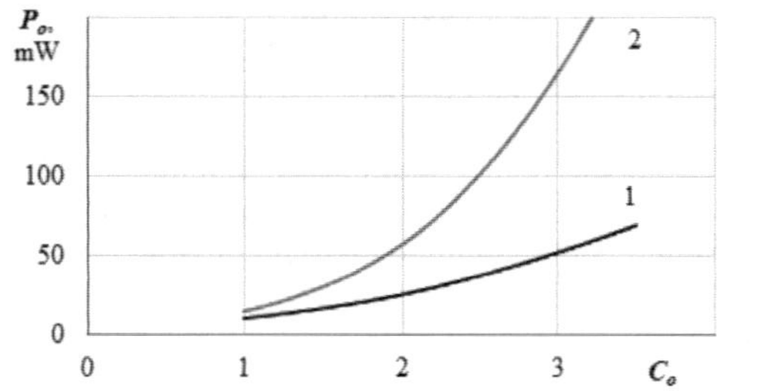

b)

Figure 4. Dependence of the pulse power P_o at the boundary of segments with different dispersion (1) and width P_m (2) in the middle of the segment on the chirp C_o of the soliton pulse (dispersion map parameters are presented in Figure 2).

As shown by computer simulation with the OptiSystem-program, with an initial value of $C_o = 1$ and entering the middle of a segment with anomalous dispersion, the value of the Q- parameter increases by 3 times (Figure 5).

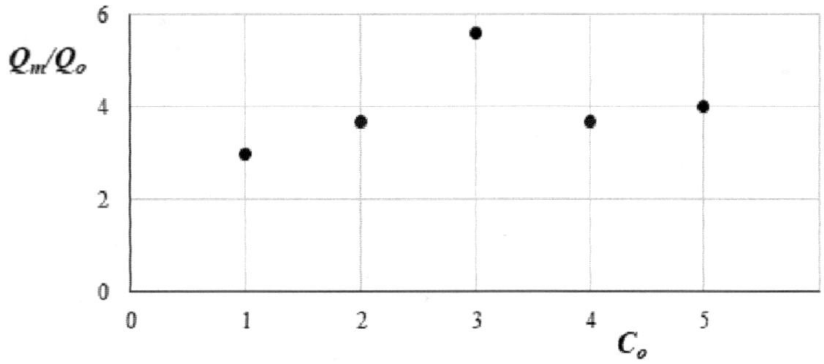

Figure 5. Q-parameter at the output of the system Q_m when entering the middle of a segment with anomalous dispersion (at the point of zero chirp) depending on the value of the C_o chirp at the boundary of segments with variance of different signs, normalized to the value of Q_o - the value of the Q-parameter at $C_o = 1$ and data input at segment boundaries.

An increase in the initial energy E_o of the pulse and, accordingly, an increase in $C_o = 3$ additionally increases the Q- parameter up to 5.6 times (Figure 5). However, a further increase in these values C_o already leads to a decrease in the value of the Q - parameter since the pulse width at the segment boundary and the interaction of solitons significantly increase.

2. Conclusion

The computer simulation of the fiber-optic data transmission system using optical solitons has been carried out. It is shown that when constructing the system with a dispersion map, which assumes input-output of spectrally limited pulses, the quality of data transmission increases. The optimal parameters of such a system can be chosen.

References

[1] Andreeva E, Bylina M, Glagolev S and Chaimardanov P 2018 Properties of Temporary Optical Solitons in Optical Fibers and the Possibility of their Use in Telecommunications. Part 1 *Proceedings of Telecommunication Universities* **4** (1) pp 5–12

[2] Shcherbakov A, Andreeva E 1996 Performance data of lengthy-span soliton transmission system *Optical Fiber Technology* **2** pp 127-133

[3] Agrawal G 2013 *Nonlinear Fiber Optics* (San Diego: 5th Academic Press)

[4] Kivshar Y and Agrawal G 2003 *Optical solitons. From Fibers to Photonic Crystals* (New York: The Institute of Optics University of Rochester)

[5] Andreeva E, Bylina M, Glagolev S and Chaimardanov P 2018 Properties of Temporary Optical Solitons in Optical Fibers and the Possibility of their Use in Telecommunications. Part 2 *Proceedings of Telecommunication Universities* **4** (2) p 26–35

[6] Ferreira M 2011 *Nonlinear Effects in Optical Fibers* (New Jersey: John Wiley & Sons inc.)

[7] Andreeva E, Bylina M, Glagolev S and Chaimardanov P 2019 Properties of Temporary Optical Solitons in Optical Fibers and the Possibility of their Use in Telecommunications. Part 4 Proceedings of Telecommunication Universities **5** (1) pp 15–24

[8] Andreeva E and Potapov I 2021 Possibilities of Using Optical Solitons in High-Speed Systems *Springer Proceedings in Physics* **255** pp 241-245

Characterization of In(Ga,Al)As/GaAs metamorphic heterostructures for mid-IR emitters by FTIR photoreflectance spectroscopy

D D Firsov[1], M Yu Chernov[2], V A Solov'ev[2] and O S Komkov[1]

[1] Micro- and nanoelectronics department, St. Petersburg Electrotechnical University "LETI", Professora Popova 5, Saint Petersburg 197376, Russia
[2] Ioffe Institute, Politekhnicheskaya 26, Saint Petersburg 194021, Russia

Abstract. Infrared photoreflectance (PR) spectra of In(Ga,Al)As/GaAs metamorphic heterostructures have been obtained using a novel photomodulation FTIR spectroscopy technique. An analysis of the PR spectra features allowed us to estimate the critical point energies corresponding to the direct interband transitions in various regions of the In(Ga,Al)As heterostructures, and distinguish the PR signals originating from Fabry-Perot interference. Observation of Franz-Keldysh oscillations originating from the InAlAs virtual substrate and an InGaAs waveguide layer has enabled determination of the built-in electric field intensities within the heterostructures. The obtained results open up possibilities for contactless control of free carrier concentration in In(Ga,Al)As/GaAs metamorphic heterostructures developed for growth of emitters of mid-IR spectral range.

1. Introduction

In(Ga,Al)As/GaAs metamorphic heterostructures have been recently shown to be very promising for high-efficiency semiconductor lasers and light-emitting diodes operating in the 3–5 μm mid-infrared (mid-IR) spectral range [1]. Such emitters are in demand for a variety of applications, including optical gas sensing, chemical process control, noninvasive medical tests, and free-space optical communications [2,3]. A promising approach to their creation is based on the formation of an active region comprising a type-II submonolayer InSb insertion embedded in the middle of a type-I InAs/InGaAs quantum well, which provides good confinement of carriers along with a rather strong overlap of electron and hole wave functions [1]. The key element of such structures is a convex-graded $In_xAl_{1-x}As$ metamorphic buffer layer (MBL) [4], which enables fabrication of low-defect-density active region on strongly lattice mismatched but well-developed GaAs substrates.

This work is devoted to applying a novel photomodulation spectroscopy technique based on a Fourier-transform infrared (FTIR) spectrometer [5] for studying the properties of In(Ga,Al)As/GaAs metamorphic structures, which are playing the role of a "virtual substrate" for the active region of mid-IR emitters. FTIR photoreflectance spectra, obtained both at room and at low temperatures, include a number of features corresponding to direct interband transitions in the InGaAs and InAlAs layers, as well as Franz-Keldysh oscillations, the analysis of which enables contactless determination of internal electric fields within the metamorphic In(Ga,Al)As/GaAs mid-IR emitter structures.

2. Experimental setup

The studied In(Ga,Al)As metamorphic structures were grown by molecular-beam epitaxy (MBE) in a RIBER 32P setup on semi-insulating (SI) GaAs (001) substrates via a convex-graded $In_xAl_{1-x}As$ metamorphic buffer layer [4]. The composition profile of the MBL is described by a dependence $x(z) = x_i + (x_{max} - x_i)(z/L)^{0.5}$, where x_i and x_{max} are the In contents at the bottom and the top of MBL, and L – its total thickness [6]. Atop the MBL an $In_{0.70}Al_{0.30}As$ layer of constant composition was grown (as illustrated in the inset of Fig. 1*b*), which is intended as a "virtual substrate" (VS) for an active region of the mid-IR sources. Some of the studied structures also included a thick $In_{0.75}Ga_{0.25}As$ layer within an $In_{0.75}Al_{0.25}As$ VS (see the inset of Fig. 2), which is a prototype of a waveguide where an emitting quantum well is to be located.

The optical setup for photomodulation measurements was based upon a VERTEX 80 FTIR spectrometer equipped with InSb and CdHgTe detectors, enabling spectra measurements in the 0.9 - 16 μm range. During the photoreflectance (PR) measurements, the samples were excited by a 405 nm or an 809 nm laser diode (P_{las} up to 100 mW), which was modulated at 3.5 kHz. The PR component of the signal was amplified by a SR-830 lock-in amplifier tuned to the frequency of modulation, then digitized by the spectrometer's internal ADC, and processed using an algorithm described elsewhere [7]. Additionally, photoluminescence (PL) spectra of the studied samples were measured with the same spectrometer, excitation laser, and a closed-cycle helium cryostat (Janis CCS-150) as described earlier [8]

3. Results and discussion

The first part of the work was devoted to studying the properties of the basic (In,Al)As/GaAs structures, which contain a single virtual substrate layer atop a metamorphic MBL. The photoreflectance spectra ΔR/R of a typical heterostructure with an $In_{0.70}Al_{0.30}As$ VS layer are presented in Figure 1 *a*,*b*. It can be seen that the PR spectra exhibit a number of features, which differ significantly at cryogenic and room temperature.

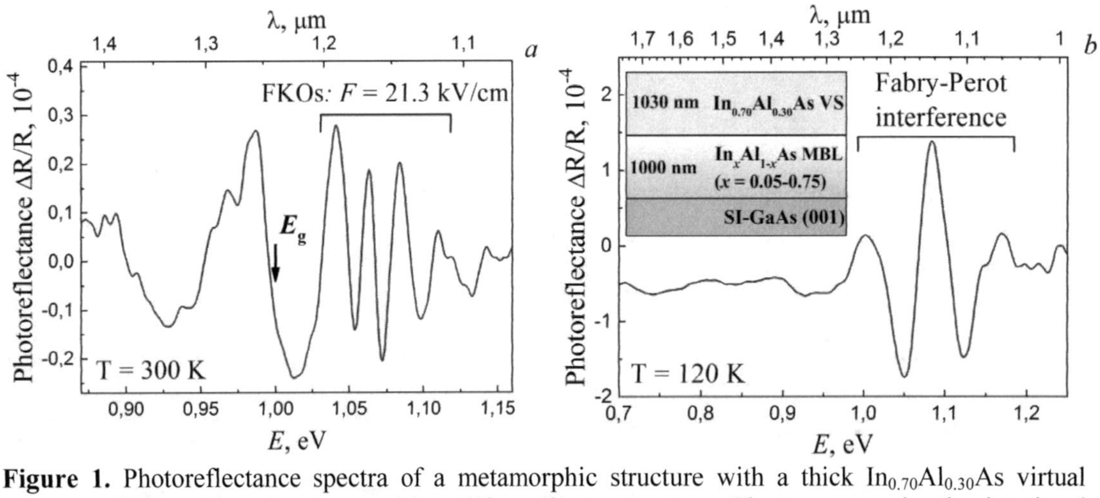

Figure 1. Photoreflectance spectra of a metamorphic structure with a thick $In_{0.70}Al_{0.30}As$ virtual substrate (VS) measured at room (*a*) and low (*b*) temperature. The arrow marks the interband transition energy obtained using the technique from Ref. [9]. The brackets highlight the spectral regions containing Franz-Keldysh oscillations (*a*), and Fabry-Perot interference (*b*), respectively. The inset in (b) shows the sketch of the layers within the heterostructure.

The broad differential-like feature at 0.97-1.03 eV observed PR in the spectrum at 300 K (Fig. 1, *a*) can be attributed to the direct interband transition E_g in the virtual substrate. Indeed, the critical point energy value of 1.00 eV, determined using a modified three-point method from Ref. [9], matches precisely the expected band gap of $In_{0.70}Al_{0.30}As$ (estimated according to data from [10]). The periodic oscillating structure in the same PR spectrum at higher energy (1.03-1.12 eV range) corresponds to

Franz-Keldysh oscillations (FKOs) within the $In_{0.70}Al_{0.30}As$ layer. Analyzing the period of FKOs enables determination of the intensity *F* of the internal electric field, using a technique described in Ref. [11]. The obtained field value for the sample from Fig. 1 is $F = 21.3\pm0.5$ kV/cm. Considering the large thickness (1.03 μm) of the VS layer, which clearly exceeds the penetration depth of the 809 nm laser into $In_{0.70}Al_{0.30}As$, the observed electric field is evidently related to the near-surface region. The intensity of a surface electric field in a semiconductor layer depends on the density of the charged states at the surface (characterized by Fermi level pinning), and the concentration of free carriers within the material (see eg. [5]). Therefore, given that the Fermi level pinning value could be determined independently, the obtained results open up a possibility of determining free carrier concentration in (In,Al)As virtual substrates based on analysis of the period of FKOs in their PR spectra.

The low temperature PR spectrum (Fig. 1, *b*) also exhibits an oscillating structure, but its period is significantly greater than that of FKOs observed at 300K. Considering that the period of FKOs is proportional to the field intensity (and hence to the carrier concentration, which is strongly decreased at 120K), this signal is obviously of another origin. The period of the oscillations matches that of the Fabry-Perot interference signal observed in the reflectance spectra *R* of the (In,Al)As/GaAs structure. Analyzing the distance between the PR peaks to determine the layer thickness *d* using a technique described in [12] with an effective refractive index of n=3.46 (calculated by a model from [13] at $\lambda = 1.15$ μm for an $In_{0.70}Al_{0.30}As$ alloy) results in a value of $d = 2160$ nm, which is close to the combined thickness of the (In,A)As VS and MBL epitaxial layers. Therefore, it can be concluded that the oscillating signal observed in the PR spectra at 120 K corresponds to the photoinduced change in the refractive index (see eg. [14]), which causes the modulation of Fabry-Perot interference in the epitaxial structure.

The photoreflectance spectrum ΔR/R of a complex In(Ga,Al)As/GaAs metamorphic heterostructure with an (In,Ga)As waveguide layer is presented in Figure 2, along with the photoluminescence of the same structure.

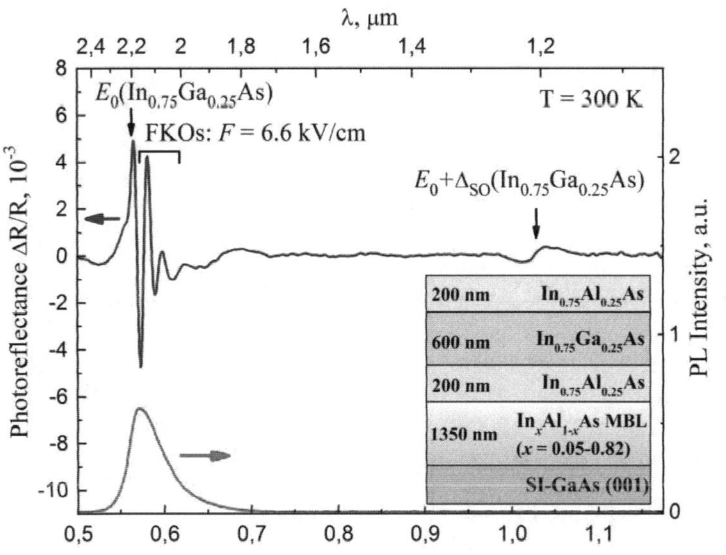

Figure 2. Photoreflectance and photoluminescence spectra of a metamorphic structure with an $In_{0.75}Ga_{0.25}As$ waveguide layer. The arrows mark the transition energies obtained using the techniques from Ref. [16] (E_0) and Ref. [9] ($E_0+\Delta_{SO}$). The bracket shows the spectral region containing Franz-Keldysh oscillations in InGaAs. The inset shows the sketch of the layers within the heterostructure.

It can be seen that in the 0.55-0.65 eV range, where a PL peak from the direct interband transition in $In_{0.75}Ga_{0.25}As$ is present, the PR spectrum exhibits a signal with a number of Franz-Keldysh oscillations, evidently related to the same $In_{0.75}Ga_{0.25}As$ layer. To the best of our knowledge, FKOs have not been reported for $In_xGa_{1-x}As$ with x above 0.53[15], which makes this result the first experimental observation of Franz-Keldysh oscillations for an $In_xGa_{1-x}As$ alloy of such a high indium composition.

The lineshape of the near-band-edge PR signal from the $In_{0.75}Ga_{0.25}As$ waveguide exhibiting Franz-Keldysh oscillations is characteristic of the so-called intermediate-field regime (see eg. [17]). In this case the exact band gap can be determined from the FKO signal according to a model from [16], which has resulted in a value of $E_0 = 0.563$ eV. Meanwhile, an analysis of the differential signal observed in the higher-energy part of the PR spectrum using an aforementioned three-point technique [9] results in a corresponding critical point energy of 1.028 eV. This value is significantly above the band gap of bulk $In_{0.75}Al_{0.25}As$ at 300 K (0.890 eV according to [10]). However, a 0.465 eV difference between this signal and the E_0 peak is rather close to the expected spin-orbit splitting value for an In-rich alloy ($\Delta_{SO} = 0.39$ eV for InAs [10]), and thus the 1.028 eV peak can likely be interpreted as a $E_0+\Delta_{SO}$ transition in $In_{0.75}Ga_{0.25}As$.

Determination of the internal electric field intensity from the period of FKOs for the sample from Fig. 2 results in a value of $F = 6.6\pm0.3$ kV/cm. Since the $In_{0.75}Ga_{0.25}As$ layer is capped with 200 nm of $In_{0.75}Al_{0.25}As$, the observed electric field is not related to the surface states, but rather to a heterointerface within the metamorphic structure. Therefore, an active region grown inside the $In_{0.75}Ga_{0.25}As$ waveguide will be placed into a strong built-in electric field, which will undoubtedly affect the efficiency of future mid-IR emitters.

4. Acknowledgments

D D Firsov acknowledges the support of the Grants Council of the President of Russian Federation, project # SP-4626.2021.3.
The authors would like to deeply thank Prof. S V Ivanov for the original concept of the studied metamorphic heterostructures, and his prolonged guidance of their research.

5. References

[1] Ivanov S V, Chernov M Yu, Solov'ev V A, Brunkov P N, Firsov D D and Komkov O S 2019 *Progress in Crystal Growth and Characterization of Materials* **65** 20
[2] Hoffman A and Gmachl C 2012 *Nat. Photonics* **6** 407
[3] Lu Q, Zhuang Q and Krier A 2015 *Photonics* **2** 414
[4] Chernov M Yu, Komkov O S, Firsov D D, Meltser B Y, Semenov A N, Terent'ev Y V and Solov'ev V A 2017 *Journal of Crystal Growth* **477** 97
[5] Komkov O S, Firsov D D, Lvova T V, Sedova I V, Semenov A N, Solov'ev V A and Ivanov S V 2016 *Physics of the Solid State* **58** 2394
[6] Pobat D B, Solov'ev V A, Chernov M Y and Ivanov S V 2021 *Physics of the Solid State* **63** 84
[7] Firsov D D, Komkov O S 2013 *Technical Physics Letters* **39** 1071
[8] Firsov D D, Komkov O S, Solov'ev V A, Kop'ev P S and Ivanov S V 2016 *Journal of Physics D: Applied Physics* **49** 285108
[9] Hosea T J C 1995 *Physica status solidi b* **189** 531
[10] Vurgaftman I, Meyer J R and Ram-Mohan L R 2001 *Journal of Applied Physics* **89** 5815
[11] Pikhtin A N, Komkov O S and Bazarov K V 2006 *Semiconductors* **40** 592
[12] Komkov O S, Firsov D D, Semenov A N, Meltser B Y, Troshkov S I, Pikhtin A N and Ivanov S V 2013 *Semiconductors* **47** 292
[13] Pikhtin A N and Yas'kov A D 1988 *Sov. Phys. Semicond.* **22** 613
[14] Kallergi N, Roughani B, Aubel J and Sundaram S 1990 *Journal of applied physics* **68** 4656
[15] Chen Y H, Hsu K T, Chen K L, Lin H H and Jan G J 1994 *Japanese journal of applied physics* **33** 2448
[16] Hughes P J, Weiss B L and Hosea T J C 1995 *Journal of applied physics* **77** 6472
[17] Misiewicz J, Sitarek P, Sęk G and Kudrawiec R 2003 *Materials Science* **21** 263

Passive decoy-state quantum key distribution with imperfect source

A Gavrilovich[1,2,3], D Sych[4,5,6] and Y Kurochkin[1,2,6]

[1] QRate, Skolkovo, 143026, Russia
[2] Russian Quantum Center, Skolkovo, 143026, Russia
[3] Moscow Institute of Physics and Technology, Dolgoprudny, 141700, Russia
[4] P.N.Lebedev Physical Institute, Russian Academy of Sciences, Moscow, 119991, Russia
[5] Department of Physics, Moscow State Pedagogical University, Moscow, 119992, Russia
[6] National University of Science and Technology MISiS, Moscow, 119049, Russia
[7] Sirius University of Science and Technology, Sochi, 354340, Russia

E-mail: gavrilovich.aa@phystech.edu

Abstract. Passive generation is a sophisticated way of state preparation in quantum key distribution (QKD) systems which is designed to exploit some internal physical process as a source of randomness. It can be profitable in a wide range of scenarios. However, the original analysis of the passive scheme implies an ideal interference which is almost impossible to assure in practice, therefore utilizing such method potentially compromises the security of the system. Here we develop a general technique to estimate decoy-state parameters for a passive protocol with an arbitrary experimental distribution of intensity. We compare this analysis with the original method and show that the proposed technique can provide higher key generation rates.

1. Introduction

QKD allows establishing a secret key between two distant users in the presence of an eavesdropper with unlimited computational power. Currently, there exist a lot of different QKD protocols [1–8]. Many of them are similar in the sense that they imply performing a random choice of some quantity (intensity, phase, etc.) at the transmitter's side. For the unconditional security of QKD it is crucial to make this choice using quantum randomness. Therefore, conventionally QKD schemes involve a fully-fledged quantum random number generator which alongside a modulator carries out the active state preparation. An alternative approach is a so-called passive generation. It allows exploiting internal physical process of the system to perform a random choice of the necessary quantity. Passive method is expected to be more robust against side-channel information leakage and may significantly simplify the experimental setup.

In general, there are many different ways to realize passive generation. One can, for example, make use of entanglement [9, 10] by operating a spontaneous parametric down-conversion source. However, existing devices are not capable of producing states on demand, therefore this implementation is not widely used. Another evident source of quantum randomness is the phase difference of the independently emitted laser pulses. By interfering them one can get pulses of random intensity and use the result as decoy states in BB84. Since the proposal of such scheme in [11], several works were dedicated to refining the original analyses, for example [12] which takes into account statistical fluctuations. However, recent research [13] shows that besides finite-size effects there are also numerous imperfections of a laser source that significantly change the shape of intensity probability density function (PDF). Any emerging mismatches may provide additional information to an adversary, thus violating the security of the system. To bridge the gap between the original model and a real-world setup one should take into account the combined effect of chirp, jitter, and relaxation oscillations.

In this work, we consider phase-randomized coherent light and develop a method to estimate BB84 decoy-state parameters for any experimental distribution of intensity. We explicitly derive the lower and the upper bound of single-photon yield and error rate. Then we calculate the secret key and show that it can provide higher secret key rates in comparison with the original analysis.

2. Passive generation

The phase difference of the independently emitted laser pulses, being a fundamentally random value, can be used to achieve different intensities via interference as shown in Fig.1. The scheme allows not only to generate pulses of random intensity but also to read out the signal by detecting half of it in the second arm of the interferometer. Assuming that interfering pulses have the same polarization and each pulse shape exhibits a Gaussian temporal profile, the intensity of a signal produced by a passive source can be written the following way:

$$I = I_1 + I_2 + 2\eta\sqrt{I_1 I_2}\cos\Delta\Phi$$

where I_1 and I_2 are the intensities of interfering pulses, $\Delta\Phi$ is the phase difference and η is the visibility. I depends on various real-life imperfections of apparatus, such as intensity fluctuations, laser chirp, jitter, etc. For a detailed review, see [13]. As an illustration, three different cases from [13] are presented in Fig.2

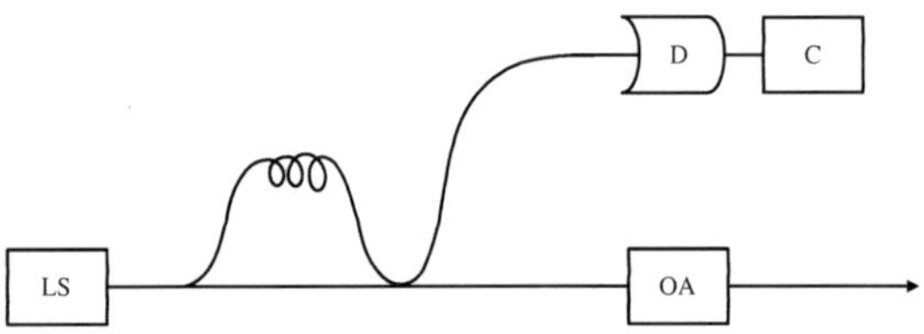

Figure 1. Conceptual scheme of a passive source: LS — laser source, OA — optical attenuator, D — classical detector, C — comparator(s).

The resulting spectre of I can be used to define decoy states by assigning nonoverlapping intervals of intensity to different types of signal as illustrated in Fig 3. After setting the boundary values for each type of signal, one can distinguish them in practice by tuning a system of comparators (see Fig.1).

3. Decoy state BB84

We use H.F. Chau's idea of utilizing an arbitrary number of decoy intensities [14] and expand this method to the passive setup.

Let yield Y_1 be a conditional probability of detection event on Bob's side given that Alice sends i photons, then under the assumption that the light is coherent and phase-randomized after summing the joint probability distribution over all values of $i \in \{0, \infty\}$ one can get the

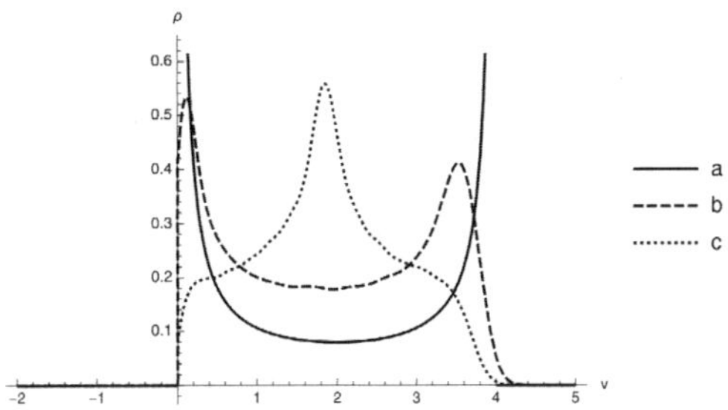

Figure 2. Possible shapes of intensity probability density functions of passively prepared states: a) without fluctuations of interfering pulses b) with fluctuations of interfering pulses c) with fluctuations of interfering pulses and a linear chirp.

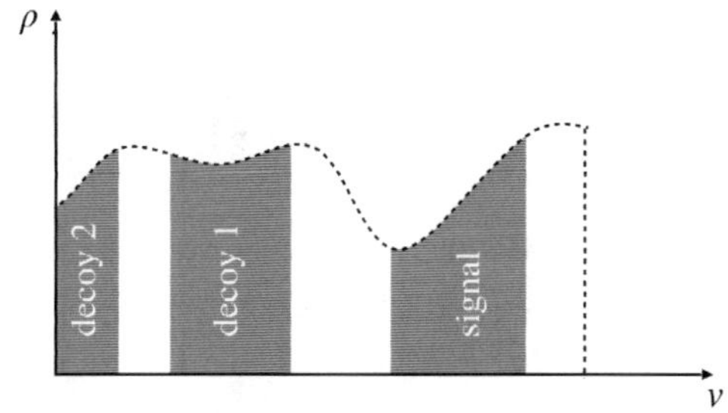

Figure 3. Example of a passive BB84 decoy-state protocol with three types of signals.

unconditional probability Q_v of a click for a signal of intensity v. By averaging it with the posterior distribution $\rho_m(v)$ of intensity within m-th type of signal one can derive the unconditional probability for a pulse of m-th type to be detected. Applying this procedure to all k types of signal gives a system of linear equations for Y_i:

$$\begin{pmatrix} Q_{v_1} \\ Q_{v_2} \\ \vdots \\ Q_{v_k} \end{pmatrix} = \underbrace{\begin{pmatrix} \int \rho_1 e^{-v_1} dv_1 & \dots & \int \rho_1 v_1^{k-1} e^{-v_1} dv_1 \\ \int \rho_2 e^{-v_2} dv_2 & \dots & \int \rho_2 v_2^{k-1} e^{-v_2} dv_2 \\ \vdots & \ddots & \vdots \\ \int \rho_k e^{-v_k} dv_k & \dots & \int \rho_k v_k^{k-1} e^{-v_k} dv_k \end{pmatrix}}_{\tilde{V}} \begin{pmatrix} Y_0 \\ Y_1 \\ \vdots \\ \frac{Y_{k-1}}{(k-1)!} \end{pmatrix} + \begin{pmatrix} \sum_{j=k}^{\infty} \int \rho_1 \frac{v_1^j Y_j}{j!} e^{-v_1} dv_1 \\ \sum_{j=k}^{\infty} \int \rho_2 \frac{v_2^j Y_j}{j!} e^{-v_2} dv_2 \\ \vdots \\ \sum_{j=k}^{\infty} \int \rho_k \frac{v_k^j Y_j}{j!} e^{-v_k} dv_k \end{pmatrix} \quad (1)$$

Analogically, if e_i is a conditional probability of an error for a detected i-photon pulse, then one can write down a system of k equations for unconditional probabilities of having an erroneous bit. Here E_{v_m} is an overall quantum bit error rate for m-th type of signal.

$$E_{v_m} Q_{v_m} = \sum_{j=0}^{k-1} \tilde{V}_{m,j} \frac{e_j Y_j}{j!} + \sum_{j=k}^{\infty} \int \rho_m \frac{v_m^j e_j Y_j}{j!} e^{-v_m} \, dv_m$$

It is secure to use only single photon pulses, therefore combining the GLLP approach with decoy state method one can derive formula for key generation rate:

$$R = q \left\{ Y_1^L [1 - h_2(e_1^U)] \sum_{i=1}^{k} P_{v_i} [v_i e^{-v_i}] - f_{ec}(E_{v_i}) \sum_{i=1}^{k} P_{v_i} Q_{v_i} h_2(E_{v_i}) \right\}$$

where q depends on the implementation (1/2 for the original BB84, ≈ 1 for the efficient BB84 protocol), $h_2(x)$ is binary Shannon information function, f_{ec} is the error-correction efficiency function, P_{v_i} is the probability of choosing i-th type of signal. Q_{v_i} and E_{v_i} are experimentally measured quantities, whereas Y_1^L and e_1^U (superscripts L and U denote a lower and an upper bound respectively) should be theoretically estimated. By introducing a modified Vandermode's matrix $\tilde{V}$ in (1) and deriving the explicit formula for it's inverse we obtain that

$$Y_1^L = \sum_{i=1}^{k} (\tilde{V}^{-1})_{2,i}\, Q_{v_i}\ , \qquad e_1^U = \sum_{i=1}^{k} (\tilde{V}^{-1})_{2,i}\, (E_{v_i} Q_{v_i})/Y_1^L$$

$$(\tilde{V}^{-1})_{ij} = (-1)^{i+j} \frac{\int \prod_{t=1,t\neq j}^{k} \rho_t e^{-\sum_{t=1,t\neq j}^{k} v_t} S_{k-i,j} \prod_{l<m,(l,m\neq j)}^{k} (v_m - v_l) d^{k-1}v}{\int \prod_{t=1}^{k} \rho_t e^{-\sum_{t=1}^{k} v_t} \prod_{1\leq l<m\leq k} (v_m - v_l) d^k v}$$

where $S_{m,i}^k$ is defined as the sum of all possible power products of m (where $1 \leq m \leq k-1$) different intensity values v_{t_j} out of k used in the protocol such that $1 \leq t_1 \leq \ldots \leq t_m \leq m$, $t_j \neq i$ (where $1 \leq i \leq k$).

4. Simulations

We compare the proposed technique with the original method by simulating the performance of a simple passive source in which the intensity spectre is split into two parts with a threshold value $I_{th} = I_{max}/2$.

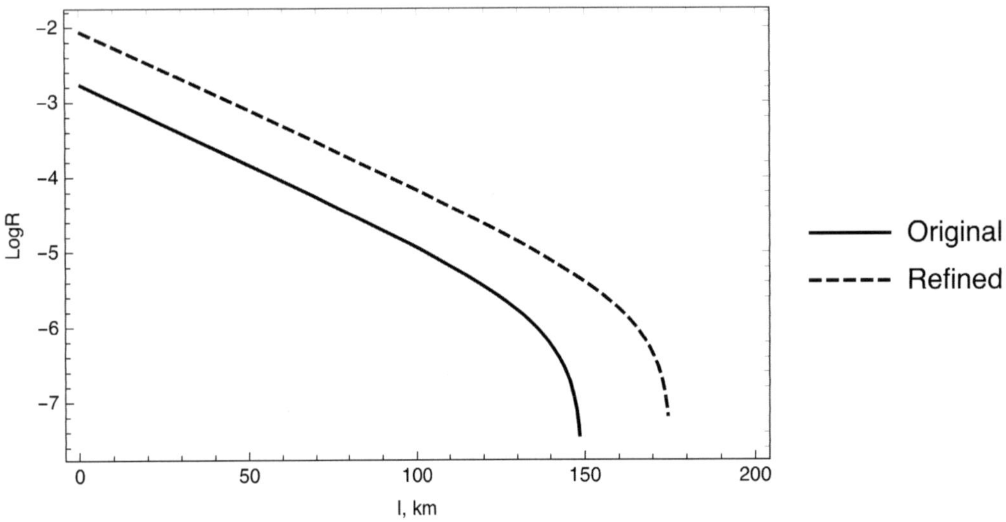

Figure 4. Comparison of the key rate as a function of communication distance for original and refined analysis of a passive BB84 decoy-state protocol with two types of signals.

From the results of the simulation, one can see that the developed method allows to significantly refine a secret key rate.

5. Conclusion

In this paper we developed a straightforward approach for decoy state estimation of a passive setup with an arbitrary distribution of intensity. Simulation results show that the refined analysis provides higher key generation rate, thus demonstrating the practical interest in the proposed analysis.

Acknowledgments

Denis Sych acknowledges funding by RFBR, Sirius University of Science and Technology, JSC Russian Railways and Educational Fund "Talent and success" project number 20-32-51004, Yury Kurochkin acknowledges support from Russian Science Foundation grant 17-71-20146 and Arina Gavrilovich acknowledges support from FASIE.

References

[1] Bennett C H and Brassard G 1984 Quantum cryptography: Public key distribution and coin tossing *Proc. IEEE Int. Conf. on Computers, Systems ad Signal Processing* (New York: IEEE) p 175
[2] Bruß D 1998 *Phys. Rev. Lett.* **81** 3018 – 3021
[3] Bechmann-Pasquinucci H and Tittel W 2000 *Phys. Rev. A* **61**(6) 062308
[4] Sych D V, Grishanin B A and Zadkov V N 2004 *Phys. Rev. A* **70**(5) 052331
[5] Sych D, Grishanin B and Zadkov V 2004 *Laser Physics* **14** 1314–1321
[6] Sych D V, Grishanin B A and Zadkov V N 2005 *Quantum Electronics* **35** 80–84
[7] Sych D and Leuchs G 2010 *New Journal of Physics* **12** 053019
[8] Sych D and Leuchs G 2010 *Optics and Spectroscopy* **108** 326–330
[9] Sych D, Averchenko V and Leuchs G 2017 *Phys. Rev. A* **96**(5) 053847
[10] Averchenko V, Sych D, Marquardt C and Leuchs G 2020 *Phys. Rev. A* **101**(1) 013808
[11] Curty M, Ma X, Qi B and Moroder T 2010 *Phys. Rev. A* **81**(2) 022310
[12] Shao-Jie T and Rong-Zhen J 2010 *Communications in Theoretical Physics* **54** 443–446
[13] Shakhovoy R, Sych D, Sharoglazova V, Udaltsov A, Fedorov A and Kurochkin Y 2020 *Opt. Express* **28** 6209–6224
[14] Chau H F 2018 *Phys. Rev. A* **97**(4) 040301

Investigation of a single-photon hybrid emitting system based on NV-centers in nanodiamonds integrated with GaP NWs

A S Goltaev[1], A M Mozharov[1], V V Yaroshenko[2], D A Zuev[2] and I S Mukhin[1,3]

[1] Nanotechnology Research and Education Centre of the Russian Academy of Sciences, Alferov University, St. Petersburg 194021, Russia
[2] ITMO University, St. Petersburg, 197101 Russia
[3] Peter the Great St. Petersburg Polytechnic University, 29 Politekhnicheskaya, St. Petersburg 195251, Russia

AKBAPNYM@yandex.ru

Abstract: NV-centers can be used for quantum informatics, quantum communication and quantum sensing. The calculation of optical modes formed in a GaP cylindrical nanocavity covered by nanodiamonds has been performed. GaP nanowires have been synthesized with molecular beam epitaxy and played the role of optical resonators for light-emitting centers on the base of nanodiamonds with NV-centers. The optical characteristics of the GaP-based nanocavity were analyzed. The increase in the rate of spontaneous emission of NV-centers optically coupled to the nanocavity was estimated by the time correlated single photon counting method.

1. Introduction

Nanoscale quantum optical emitters are main elements of quantum optics and there are widely used for sensing applications. Nanodiamonds (NDs) with optically active point defects in the crystal lattice can be used as such sources. One of the most studied crystal lattice defects is a nitrogen-substituted vacancy or NV-center. The spins of electrons in NV-centers can be adjusted at room temperature using a magnetic field, electric field or optical radiation [1]. The main problem that limits application of NV-centers in quantum nanophotonic systems is associated with the low brightness of zero-phonon line (ZPL) at room temperature due to broad phonon side-band [2]. A conventional approach to solve the problem of increasing the NV-center ZPL emission rate as well as efficient collection of emitted photons is based on the integration of the radiative nanodiamonds with optical cavities or waveguides [3]. Gallium phosphide (GaP) nanowires (NWs) which have large length-to-diameter ratio are the most promising candidates to be used as optical resonance systems since GaP NWs have an indirect band gap with relatively large energy that makes its transparent in the visible range including ZPL line of NV-centers. Another advantage is the optical refractive index, which is about 3.2 within the visible range which is higher than most optical materials. For example, the refractive index of diamond is 2.4.

2. Numerical calculation

The primary calculation aim is to determine the conditions providing efficient coupling between a nanocavity based on a single GaP NW and nanodiamond with an emitting NV-center. Another main task is related to assess the system resistance to coupling errors. To solve this problem we used numerical simulation of a coupled optical system in the wave optics approximation with Comsol Multiphysics software package. The analysis of the system was led into two steps. At the first step we looked for NW diameters having optical resonance near the ZPL frequency (470 THz). Also we studied the near-field distribution in the cavity and the corresponding far-field radiation pattern. At the second step the nanodiamond was placed to NW sidewall and the Purcell factor and the radiation pattern as a function of the NW diameter and the polarization of the NV-center were determined for ZPL frequency.

In the numerical simulation GaP NW was represented as a 3D elongated objectwith a hexagon cross-section and 5 μm length. The nanodiamond was represented by a cube with a size of 60 nm. The refractive indexes and absorption coefficients for GaP and diamond were taken from [4, 5]. Analyze of the obtained data was carried out for NW diameters which provides eigen frequency of optical resonance equaled to ZPL.

According to the results of numerical simulation we plotted the Q-factor of optical nanocavities based on GaP NWs at ZPL frequency (470 THz), see Figure 1. Figure 2 shows the maps of the electric field distribution inside the resonator and the corresponding radiation patterns for three typical NW diameters. One can see that the increase in the NW diameter leads to the increase in the Q-factor of the resonator. It can be associated with the size-dependence of the energy localization inside the NW, Figure 1.

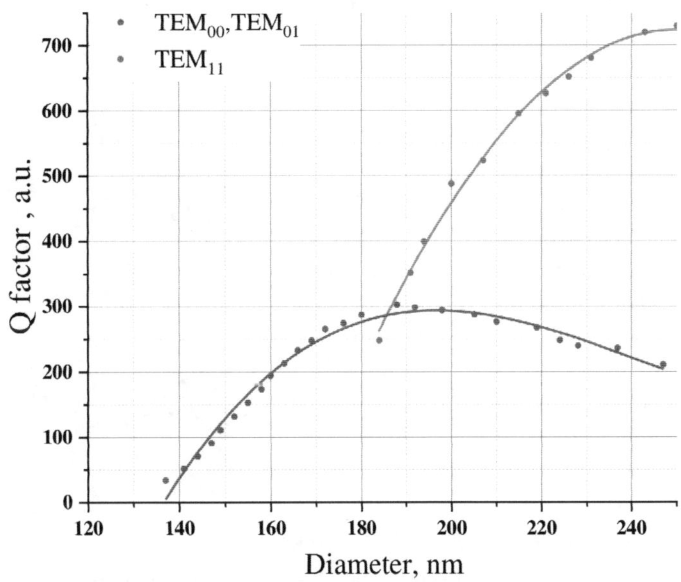

Figure 1. Plot of Q factor of optical modes of GaP NW.

Analysis of the distribution of electromagnetic fields shows that with a diameter lower than 160 nm the resonator modes are mainly represented by the fundamental TEM_{00} mode, Figure 2 (137 nm), which is optimal for obtaining directional (along the NW axis) radiation. A further increase in diameter leads to the appearance of higher order transverse modes such as TEM_{01} and TEM_{11}, which lead to the increase in the contribution of side share to the radiation pattern, Figure 2 (160 nm), (188 nm).

The calculation of composite GaP-NW / NV-center system was carried out for the obtained at previous stage NW diameters. Nanodiamond was placed onto NW sidewall and play a role of the optical dipole source. . The calculation was performed for three polarizations of the dipole source: the one is along the NW axis (Z) as well as two transverse polarizations – along (X) and across (Y) the sidewall. The distribution of the electromagnetic field of a nanocavity with a diameter of 137 nm and the radiation pattern for three orientations of the dipole are presented in Figure 3. As can be seen from the maps of the near field the maximum amplitude of the electromagnetic wave induced in the NW is achieved for the Y polarization which is associated with the penetration of the near field of the NV-center dipole into the core of the NW. The minimum coupling, in turn, is observed for the X orientation of the polarization. Far-field analysis shows that for a NW diameter of 137 nm, Y and Z polarizations of the NV-center provide the light propagation along the NW axis, which can be successfully used to supress the directivity pattern of the NV-center.

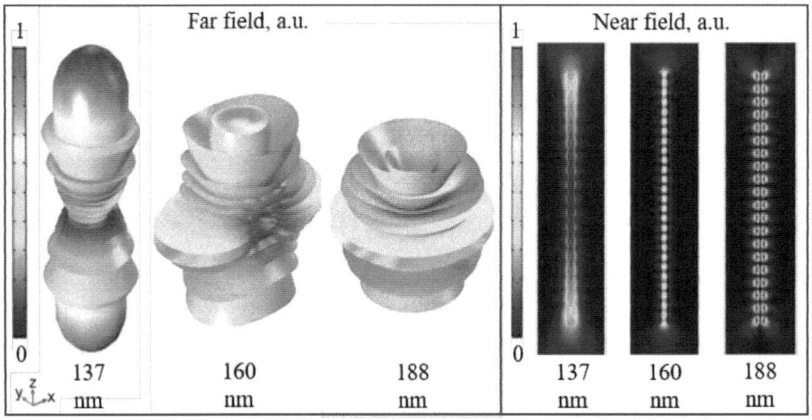

Figure 2. Far and near fields' for three characteristic diameters of GaP NW: b) 137 nm, c) 160 nm, d) 188 nm.

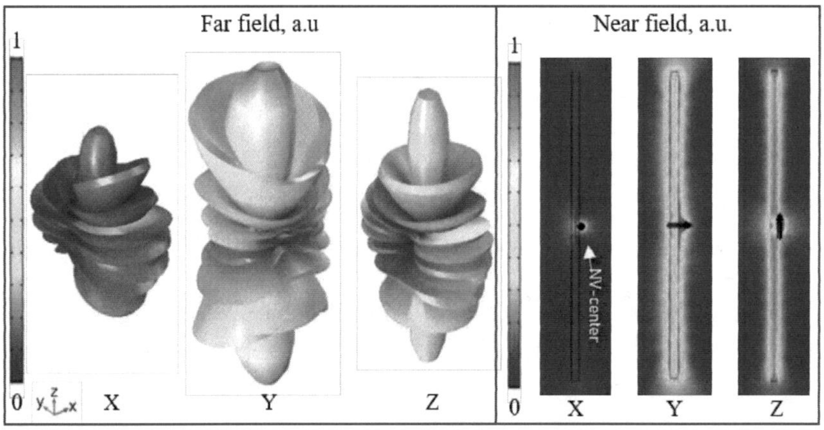

Figure 3. Far and near fields' distributions for GaP NW (137 nm in diameter) for three different polarization of NV-centers.

According to the results of numerical calculation, the dependence of the Purcell factor on the nanocavity diameter was calculated for three different polarizations of the NV-center. Figure 4 shows a graph of the obtained dependencies.

As noted above, for the X polarization (when the dipole weakly interacts with the cavity and the near field strongly goes beyond its limits), we observed the smallest Purcell factor (gray curve). It confirms the weak optical coupling of the NV-center with GaP NWs. For NW diameter about 160 nm, we observed the highest value of the Purcell factor. Presumably, it is associated with the complication of the mode composition of NWs and the enhancement of the optical coupling between NWs and nanodiamonds. Y (red curve) and Z (blue curve) polarizations are characterized by enhanced values of the Purcell factor, since they have a larger optical coupling with the resonator. The dependence of the slope of the curve for Y polarization is related to the fact that with an increase in the diameter the fraction of the energy contained in the evanescent field decreases, while the fraction of the energy inside the resonator increases. For Z orientation, there is also nonlinear dependence and the optimal value is determined by the coincidence of the field lines generated by the dipole with the localization maxima of the modes in the nanocavity. This is due to the fact that with an increase in the diameter of GaP NWs, the effective refractive index of the system increases and the field maxima begin to approach each other. Thus, we obtain the optimal configuration of the field maxima near NW diameter of 150 nm, which leads to the best optical coupling between the NV-center and GaP NWs. Oscillations on the graph are associated with the dipole "hitting" an even or odd localized mode in the near field, Figure 4.

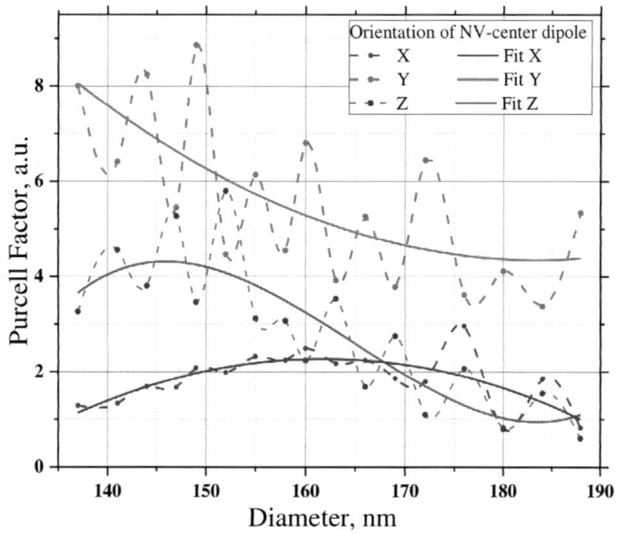

Figure 4. Plot of the Purcell factor on NW diameter for different NV-center orientations. Gray curve – polarization along the lateral face of the NW, Red curve – polarization across the lateral face of the NWs, Blue curve – polarization is directed along the NWs. Solid lines – fitting of the acquired data.

Based on the obtained model, we determined the optimal geometric parameters of a nanocavity based on nanodiamond combined with GaP NWs for three different polarization. The NW diameter near ~ 150-160 nm provides the best characteristics for increasing the luminosity of NV-centers due to the high Purcell factor.

3. Experiment

Arrays of GaP NWs were synthesized on a Si (111) substrate using a molecular beam epitaxy Veeco GEN III machine. The morphology of the grown GaP NWs was studied using a scanning electron microscope (SEM) (Zeiss SUPRA 25-30-63). The obtained NWs had the length of 12 ± 2 um and diameter 155 ± 15 nm. Nanodiamonds were deposited to NWs array from a liquid suspension with microdispenser, Figure 5 (a).

Microspectroscopic studies of the photoluminescence (PL) response and Raman scattering of light (RS) of individual NWs covered with the ensemble of nanodiamonds were performed at a room temperature (300 K) (see Figure 5). Figure 5 (b) demonstrates two typical peaks corresponding to two charge states of the defect, namely: NV^0 (575 nm) and NV^- (637 nm). As shown in Figure 5 (d) the PL spectrum for a single NW with deposited nanodiamonds demonstrates intensity modulation associated with the eigenmodes of the cavity which also confirms the optical coupling between emitting NV-centers and NWs.

The photoluminescence decay time of NV-centers was investigated using the time-correlated single photon counting (TCSPC) method on the sample with single NWs as well as reference glass covered by NDs We used a 532 nm femtosecond laser with a Mitutoyo M Plan 100X NA 0.7 objective. Figure 6 shows the intensity and lifetime maps of NW-NV-center system.

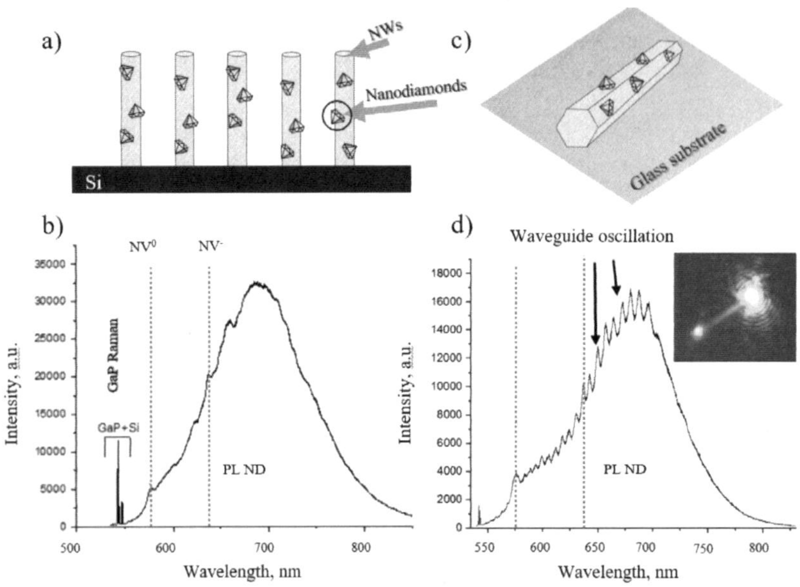

Figure 5. a) Scheme of the synthesized GaP nanowire array covered by nanodiamonds b) PL spectrum of NW array with NDs, c) Scheme of a single NW with NDs transferred on an arbitary glass substrate, d) PL spectrum of a single NW with ND; the inset shows an optical image of NWs and demonstrates the light trapping effect by the NW.

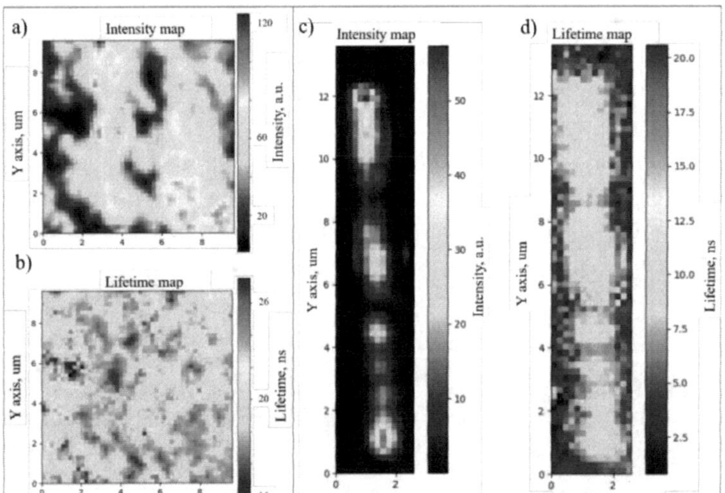

Figure 6. Intensity map and map of lifetime for: a), b) - a reference sample, c), d) - a single NW with a deposited ensemble of nanodiamonds, respectively.

The average lifetime of NV-centers on the reference sample was ~ 18.2 ns, while on the sample with a single GaP NWs, ~ 9.2 ns. The lifetime ratio for these samples gives the Purcell factor corresponding to ~ 2 [6]. This value lower compared to the results of numerical calculation due to many reasons. Firstly, the lifetime value of the reference sample is underestimated due to presence of greater than unity inherent Purcell factor of reference sample. Thus, this leads to decreasing the life-time ratio. Secondly, in our samples there are many NV-centers located at different points inside the diamond particle thus we observe the PL decay rate averaged over the ensemble of emitting NV-centers. It is known that such averaging usually leads to lower values of the Purcell factor [7]. Thirdly, in our optical measurement geometry collection and pumping occur perpendicular to the NW axis and therefore the optical signal is mainly contributed by two dipole components: X and Z which have lower Purcell factor compared to Y part (on average ~ 2-5, see Figure 4). The last conclusion is supported by numerical calculation.

4. Conclusions

In this work, we performed a numerical calculation of the Q-factor of the optical modes of a nanocavity based on GaP NW at the ZPL frequency (470 THz), and obtained diagrams of the electric field distribution inside a cavity based on GaP NW combined with an NV-center. The Purcell factor was analyzed as a function of the NW diameter for three different dipole polarization of the NV-center in nanodiamonds. A method of nanodiamond integration with NV-centers into an array of GaP NWs has been developed. The analysis of the optical properties of the fabricated structure was carried out and the PL and Raman spectra were obtained, which confirmed the waveguide properties of NWs. Increasing the spontaneous emission rate of an NV-center in a nanocavity system in comparison with the emission of reference samples has been experimentally demonstrated.

Acknowledgments

Modeling and development of a nanocavity based on GaP NWs combined with a diamond nanoparticle containing an NV-center was supported by the Russian Science Foundation (grant 19-19-00693).

References

[1] Maurer P C et al 2010 *Nature Phys.* **6** 912-918
[2] Aharonovich I, Castelletto S, Simpson D A, Su C H, Greentree A D and Prawer S 2011 *Prog. Phys.* **74** 076501
[3] Verardo D, Lindberg F W, Anttu N, Niman C S, Lard M, Dabkowska A P and Linke H 2018 *Nano Lett.* **18** 4796-4802
[4] Aspnes D E and Studna A A 1983 *Physical review* **27** 985
[5] Phillip H R and Taft E A 1964 *Physical Review* **136** 1445
[6] E M Purcell 1995 *Confined Electrons and Photons* 839-839
[7] Canet-Ferrer J et al 2012 *Optics Express* **20** 7901–7914

On the possibility of using the optical method for express quality control of fruits

E M Gryaznova[1], V V Davydov[1, 2] and V Y Rud[2, 3]

[1] Peter the Great Saint-Petersburg Polytechnic University, Polytechnicheskya street 29, Saint Petersburg, Russia, 195251
[2] All-Russian Research Institute of Phytopathology, Moscow Region, Russia, 143050
[3] A.F. Ioffe Physicotechnical Institute, St. Petersburg, 195152, Russia

katya.gryaznova@mail.ru

Abstract. The necessity of express control of fruit quality in various situations has been substantiated. Various methods for the implementation of express control of fruit quality in express mode are considered. A method for express control of fruit quality using laser radiation has been developed. The change in the degree of spatial coherence in the recorded reflected laser radiation from defects on the surfaces of fruits was investigated. Experimental results of the study of various defects on fruits, as well as their hardness and degree of ripening, are presented using the method developed by us.

1. Introduction

At present, much attention is paid to monitoring the state of various media (water, air, etc.), as well as materials [1-9]. The greatest difficulties arise while controlling various products, especially in express mode [10-17]. Among the products, one of the most difficult areas is fruit quality control [18-24]. Since the products must have a marketable condition preference for inspection is given to non-destructive various methods, for example, using nuclear magnetic resonance [24-31] or illuminated microscopes to visualize anatomical features using microscopes [32, 33]. In some cases, acoustic or electromagnetic methods are used. All these methods are rather laborious and in most cases require a stationary laboratory. While using them for express control, a large number of problems arise or it is required to take a sample from the object under investigation (for example, a piece of skin from a fruit), which will not allow this fruit to be put up for sale in the future or stored for a long time before processing. Optical non-invasive methods based on color, spectrozonal or luminescence characteristics are ineffective [34–39] if structural changes are not accompanied by significant and unambiguous biochemical processes.

In this paper, we consider the possibility of using the correlation parameters of laser light scattering to register tissue structural rearrangements without significantly affecting the vital activity of organic compounds found in fruits or vegetables. It is extremely important that vitamin complexes in fruits are not destroyed during control.

2. Methodology for determining the quality of fruit and its practical implementation

Our studies allowed us to develop a technique for express control of fruits, based on obtaining the contrast of the interference pattern, which, when the intensities of the interfering beams are equal, is equal to the degree of spatial coherence (DSC). Figure 1 shows a diagram of the setup for obtaining DSC in laser radiation reflected from the surface of the fruit.

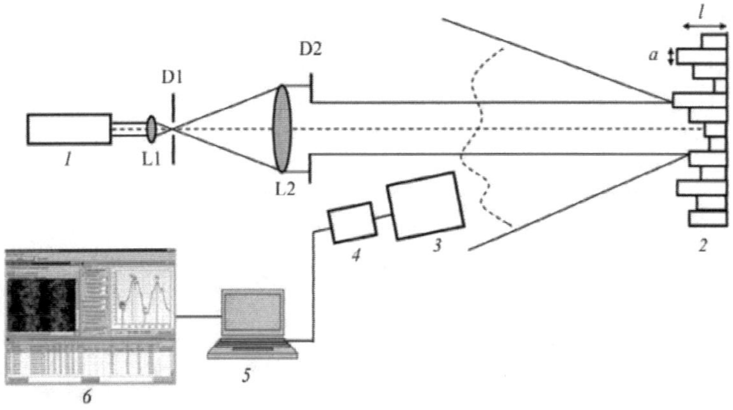

Figure 1. Installation diagram: 1 - helium-neon laser; 2 - biological object; 3 - polarization interferometer; 4 - CCD camera; 5 - PC; 6 - program for processing the results; L1, L2 - collimator lens; D1 - Fourier filter, D2 - aperture diaphragm.

The radiation of a single-mode laser is directed to a collimator objective with a Fourier filter and an aperture diaphragm, after which a weakly diverging coherent Gaussian beam is incident on a biological object. The radiation scattered by it falls into a polarization interferometer, whose interference pattern is recorded by a CCD camera connected to a personal computer. The program installed on it determines the extrema of the interference fringes and calculates the DSC.

In addition, this design of the experimental setup allows, on the basis of this method, to develop a portable device that is connected to a personal computer to determine the quality of the fruits in the express control mode (at the place of storage).

3. Results of experimental studies and its discussion

To confirm the reliability of the methodology developed by us, various fruits with the most typical defects for them were studied. It was found that the DSC for all types of fruits (for example, apples, pears, grapes, etc.) is significantly reduced in the case of defects in them. It should be noted that the possibility of using the developed method does not depend on the biochemical state of the defect and the color of the fetus. The measurement results are presented in table 1.

Table 1. Changes in DSC when reflected from fruits of different quality.

Object of study	The condition of the integumentary tissue	DSC (%)	Defect recognition probability
Apples	Intact	0.480±0.114	0.99
	Peel off the skin	0.090±0.056	
	Puncture	0.030±0.017	
Plums	Intact	0.510±0.041	0.99
	Peel off the skin, puncture	0.093±0.038	
Cherry	Intact	0.560±0.063	0.99
	Crack	0.102±0.037	
Pears	Intact	0.413±0.190	0.99
	Peel off the skin, puncture	0.074±0.064	
	Peel off the skin, puncture	0.074±0.064	
Orange	Intact	0.290±0.018	0.95
	Dry rot	0.103±0.009	

The ripeness of the fruits was also tested with tis method, since the structure of the skin and pulp changes during the ripening process, which leads to a change in the DSC. Many fruits are removed unripe so it is easier to transport and store them. This process takes place during storage. For each type of fruit the data on the hardness of the pulp obtained using a penetrometer coincide with the results of the considered optical method with a correlation coefficient of at least 0.95 (see table 2).

Table 2. Results of measurements of the degree of fruit ripeness.

	Unripe fruits		Ripe fruits	
Object of study	DSC (%)	Hardness (kg/cm^2)	DSC (%) (fruits with skin)	Hardness (kg/cm^2)
Pear:				
Krasavitsa Chernenko	16.2±1.7	6,8±0,4	3,83±1,4	1,2±0,34
Avgustovskaya Rosa	13.9±0.8	4,6±0,4	3,65±0,7	0,65±0,05
Osennee Yakovleva	11.8±2.2	4,4±0,4	6,30±0,7	1,5±0,22
Yanvarskaya	14.5±2.8	5,8±0,4	7,20±0,8	1,2±0,35
Apricot	12,53±0,37	2,9±0,17	7,62±0,57	1,48±0,17

Using the proposed method, it is also possible to determine the dynamics of fruit ripening, which is useful for predicting the keeping ability of fruits or controlling the duration of storage. For this, the degree of spatial coherence of light scattering in the process of accelerated ripening was recorded using the example of pears. A sample of 20 pears with winter ripening in December was removed from storage in a refrigerator with a controlled atmosphere, kept in a room at room temperature for 24 days, and the DSC was measured every 3 days from the equatorial surface (Figure 2).

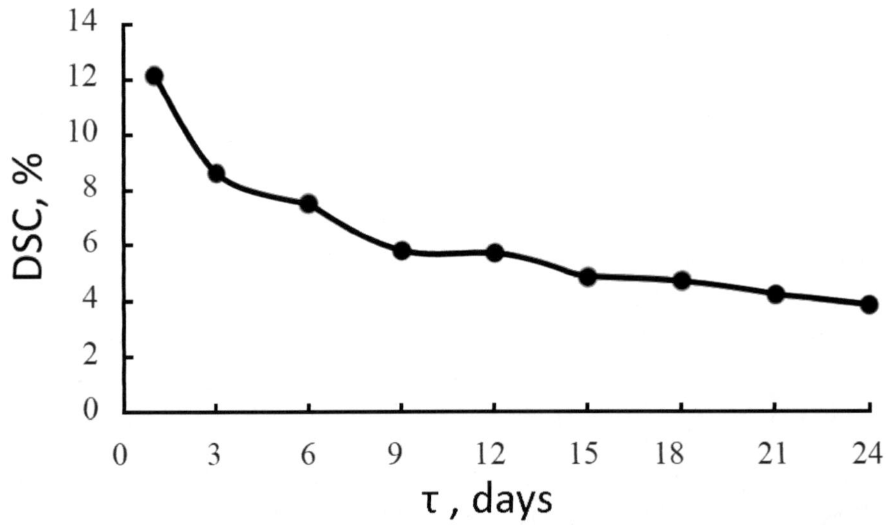

Figure 2. Changes in the DSC of light scattering of pear fruits during ripening.

Fruit hardness was measured at the beginning, in the middle and in the end of ripening (see table 3). Fruit hardness was assessed by two criteria, since DSC is more dependent on the state of the superficial, mainly epidermal layer of cells.

Table 3. Measurement of hardness and DSC of pears in the process of accelerated ripening.

	Destructive method (penetrometer)		Non-destructive method (laser)
Ripening time at room temperature	Hardness of fruit with skin, kg/cm^2	Pulp hardness (kg/cm^2)	DSC (%) (fruits with skin)
1 day	9.45±0.26	6.35±0.18	12.2±1.4
12 days	4.42±0.2	3.17±0.17	5.7±0.7
24 days	4.08±0.16	2.85±0.14	3.77±0.6

The correlation coefficient between the DSC and the hardness of the fruit with the skin is practically equal to the correlation coefficient between the DSC and the hardness of the pulp, which indicates a fairly close relationship between the phase parameters of light scattering and general physiological and physicomechanical changes occurring during ripening in all tissues of the fruit.

4. Conclusion

The data obtained confirm the possibility of using the method under consideration to control the quality of fruit in the express mode. The use of the DSC as an assessment of the characteristics of the state of the fruits has a sufficiently high accuracy, and the non-invasiveness and simplicity of the method make it attractive for the usage. The method allows you to assess not only the defects of fruits and the degree of their ripeness, but also to predict the dynamics of ripening.

References

[1] Davydov R, Antonov V and Moroz A 2019 *International Conference on Electrical Engineering and Photonics EExPolytech-2019 (Saint-Petersburg)* vol 8906791 (IEEE) pp 295-7

[2] Davydov R, Antonov V, Makeev S, Batov Y, Dudkin V and Myazin N 2020 *E3S Web Conference* **140** 02001

[3] Nikitina M, Grebenikova N, Dudkin V and Batov Y 2019 *IOP Conference Series: Earth and Environmental Science* **390(1)** 012024

[4] Yushkova V, Kostin G, Davydov R, Dudkin V and Valiullin L 2019 *IOP Conference Series: Earth and Environmental Science* **390(1)** 012016

[5] Van S, Cheremisin A, Chusov A, Switala F and Davydov R 2019 *IOP Conference Series: Earth and Environmental Science* **390(1)** 012011

[6] Van S, Cheremisin A, Glinushkin A, Davydov R and Yushkova V 2019 *E3S Web of Conferences* **140** 09008

[7] Gryznova E, Batov Y and Myazin N 2020 *E3S Web of Conferences* **140** 09001

[8] Gryznova E, Grebenikova N, D. Ivanov D and Bykov V 2019 *IOP Conference Series: Earth and Environmental Science* **390(1)** 012044

[9] Petrov A A, Vologdin V A and Zalyotov D V 2015 *Journal of Physics: Conference Series* **643(1)** 012087

[10] Davydov V V, Moroz A V, Myazin N S, Makeev S S and Dudkin V I 2020 *Optics and Spectroscopy* **128(10)** pp 1678–85

[11] Davydov V V, Myazin N S, Makeev S S and Dudkin V I 2020 *Measurement Techniques* **63(5)** pp 368–74

[12] Davydov V V, Dudkin V I, Myazin N S and Davydov R V 2020 *Journal of Communications Technology and Electronics* **65(5)** pp 558–64

[13] Kozlova E S and Kotlyar V V 2014 *Computer Optics* **38(1)** pp 132-8

[14] Kuzmin M S and Rogov S A 2019 *Computer Optics* **43(3)** pp 391-6
[15] Davydov V V, Dudkin V I and Karseev A Y 2015 *Russian Physics Journal* **58(2)** pp 146–52
[16] Davydov V V, Dudkin V I and Karseev A Y 2015 *Measurement Techniques* **58(3)** pp 317–22
[17] Davydov V V, Dudkin V I and Karseev A Y 2015 *Instruments and Experimental Techniques* **58(6)** pp 787–93
[18] Neronov Yu I 2017 *Measurement Techniques* **60(1)** pp 96-102
[19] Myazin N S, Yushkova V V and Taranda N I 2019 *Journal of Physics: Conference Series* **1410(1)** 012130
[20] Rukin E V , Myazin N S and Dudkin V I 2019 *Journal of Physics: Conference Series* **1368(4)** 042011
[21] Myazin N S, Yushkova V V and Dudkin V I 2019 *Journal of Physics: Conference Series* **1400(6)** 066008
[22] Neronov Y I 2021 *Biomedical Engineering* **54(5)** pp 333–6
[23] Rutberg F G, Kolikov V A, Snetov V N, Cherepkova I A and Dyachenko S V 2016 *High Temperature* **54(2)** pp 170–4
[24] Myazin N S, and Yushkova V V 2019 *Environmental Research, Engineering and Management* **75(2)** pp 28–35
[25] Gizatullin B, Gafurov M, Vakhin A, Mattea C and Stapf S 2019 *Energy and Fuels* **33(11)** pp 10923–32
[26] Myazin N, Neronov Y, Dudkin V and Petrov A 2019 *IOP Conference Series: Materials Science and Engineering* **497(1)** 012111
[27] Myazin N S 2018 *Journal Physics: Conference Series* **1124(1)** 031004
[28] Myazin N S 2018 *Journal Physics: Conference Series* **1135(1)** 012061
[29] Davydov V V, Dudkin V I, Myazin N S and Davydov R V 2020 *Journal of Communications Technology and Electronics* **65(5)** pp 558–64
[30] Davydov V V, Dudkin V I and Karseev A Y 2015 *Technical Physics* **60(3)** pp 456–60
[31] D'yachenko S V and Zhernovoi A I 2016 *Technical Physics* **61(12)** pp 1835–7
[32] Zagoryan E M 1985 *Biological and chemical sciences* **6** pp 68-70
[33] Golev I M 2013 *Storage and processing agricultural raw materials* **12** pp 28-31
[34] Smirnov K J and Batov Y V 2019 *Journal of Physics: Conference Series* **1368(2)** 022073
[35] Grebenikova N M and Davydov R V 2019 *Journal of Physics: Conference Series* **1326(1)** 012012
[36] Grebenikova N M, Davydov V V and Rud V Y 2019 *Journal of Physics: Conference Series* **1410(1)** 012186
[37] Smirnov K J, Glagolev S F, Rodygina N S and Ivanova N V 2018 *Journal of Physics: Conference Series* **1038(1)** 012102
[38] Smirnov K J, Glagolev S F and Tushavin G V 2018 *Journal of Physics: Conference Series* **1124(1)** 022014
[39] Logunov S E, Rud V Y, Davydov R V, Moroz A V and Smirnov K J 2019 *Journal of Physics: Conference Series* **1326(1)** 012024

The development of an optical system for lighting various rooms with sunlight

I M Gureeva[1], V V Davydov[1, 2] and V Yu Rud[2, 3]

[1] Peter the Great Saint-Petersburg Polytechnic University, Saint Petersburg, Russia, 195251
[2] All-Russian Research Institute of Phytopathology, Moscow Region 143050, Russia
[3] A.F. Ioffe Physicotechnical Institute, St. Petersburg, 195152, Russia

irena-gureeva@mail.ru

Abstract. The necessity of using light in the room, the spectral composition of which corresponds to the spectrum of the sun for a comfortable state of a person, is justified. A passive system that does not consume energy is proposed. An energy-efficient natural light system has been developed, creating a comfortable environment for the eyes, well-being and mood of a person. An experimental study of the system operation was carried out. The dependence of the light intensity on the angle of incidence in the light guide is obtained.

1. Introduction

The deterioration of the environmental situation, the appearance of a large number of negative factors, as well as an increase in stress load negatively affects the state of human health [1-12]. This dramatically reduces its efficiency and contributes to the appearance of various diseases [10-19]. It is extremely difficult to improve the environmental situation, as this is a long process [11, 12, 20-27]. It is possible to exclude other factors or reduce their influence [20, 21, 26-31]. One of them is the level of illumination and the quality of the light wave. Numerous studies of scientists [15-18, 32-38] have shown that the most favorable effect on vision and the psyche is provided by sunlight. Therefore, the developers of various light sources strive to reproduce the spectrum of sunlight in the light stream.

With the increase in the cost of electric energy, the production of most light sources goes in the direction of reducing energy consumption, increasing the resource of its operation, as well as compliance with strict safety measures for operation [32, 33, 38-40]. Less attention is paid to the spectral component in the emitted light from the source, since the implementation of the full spectrum of the sun is very expensive technologies [33, 38, 41].

Our research has shown that almost all currently produced LED lamps have a truncated spectrum of sunlight, especially with a long service life [32, 35, 38, 41]. Currently, only two of the light sources produced fully reflect the spectrum of the sun. Traditional incandescent light bulb. It has a number of disadvantages associated with a low service life and the possibility of an explosion of the glass bulb when the spiral burns out. A lamp consisting of a combination of LEDs of different wavelengths that emit with different power (corresponding to the solar spectrum). The disadvantage of this lamp is the high price. If one or more LEDs fail during power surges, the light spectrum is distorted [41, 42].

Therefore, the search for various solutions related to the possibility of using sunlight, especially in residential areas, is extremely relevant at the present time.

2. Optical system and research methodology

In this work, one of the possible solutions is considered – the design of a system containing a set of optical elements that concentrate daylight, transmit it through a system of quartz light guides and disperse it in the indoor areas of the building. The system shown in Figure 1 includes collecting and scattering lenses, a rotating prism, quartz light guides and a transparent protective panel. The transparent panel is placed on the roof of the house.

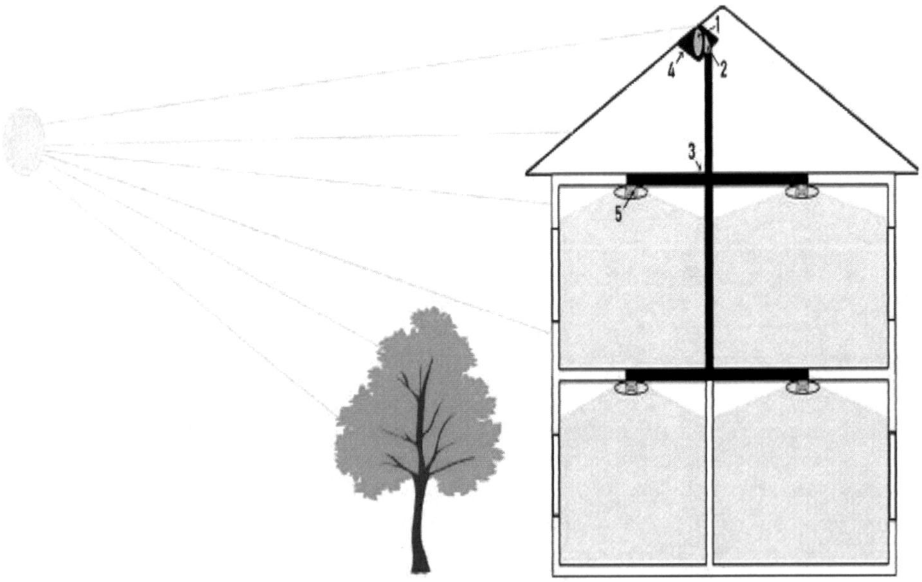

Figure 1. Operation of the optical system for transmitting sunlight: 1-collecting lens; 2-rotating prism; 3-quartz light guide system with couplers; 4-transparent protective panel; 5-scattering lens.

This system will function throughout the daylight hours. The introduction of such a system has a positive impact on the health of people with continuous exposure to the visible spectrum of natural light. Other benefits include a reduction in energy losses and energy inputs from buildings, as well as a positive impact on the planet's ecology by reducing conditional CO2 emissions into the atmosphere. The use of quartz light guides ensures the operation of the system in any weather due to their resistance to low and high temperatures. The presence of a rotating prism eliminates additional light losses due to reflections on the bends and walls of the light guide when transmitting it from the collecting lens to the vertical light guide.

3. Result of experimental investigations and discussion

The results of experimental studies have shown that in the developed optical system, when light is transmitted through optical fibers, its spectral composition does not change. Figure 2 shows as an example the results of the study of the radiation spectrum using a diffraction grating.

It should be noted that there is also no significant redistribution of the light intensity over the wavelengths in the studied radiation after its passage through the optical fibers. An important parameter in the developed design of the lighting system is the dependence of the illumination E in the room on the angle at which the light stream enters the light guide after the prism (Figure 3).

The dependence of the change in illumination E on the bending angle of the light guide φ after the introduction of sunlight into it was also studied (Figure 4).

With an increase in the angle at which the light stream enters the light guide, there is a decrease in illumination. Similarly, with a larger bend of the light guide, the illumination of the rooms decreases.

The conducted studies allowed us to establish that the most effective for transmitting sunlight in the optical system developed by us is the following. For any roof geometry, the optical axis of the focusing lens is perpendicular to the input surface of the light guide. The focus of the lens is placed on this surface of the light guide. The bending angle of the light guide can be minimized by increasing its length. The choice of the bending angle of the optical fiber in accordance with its length, as well as other parameters of the optical system are determined using mathematical modeling. The model is developed taking into account the design of the roof, since different placement of optical elements is possible for different cases. The optimum is determined by the maximum illumination.

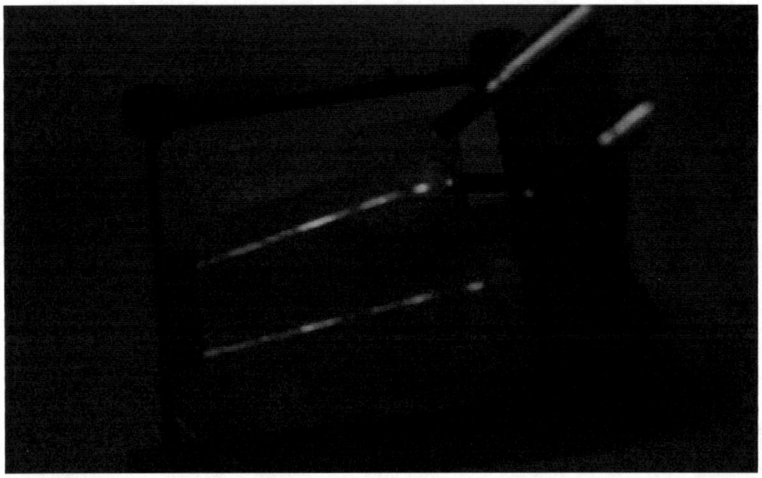

Figure 2. The spectral composition of sunlight at the output of two optical fibers after the divider.

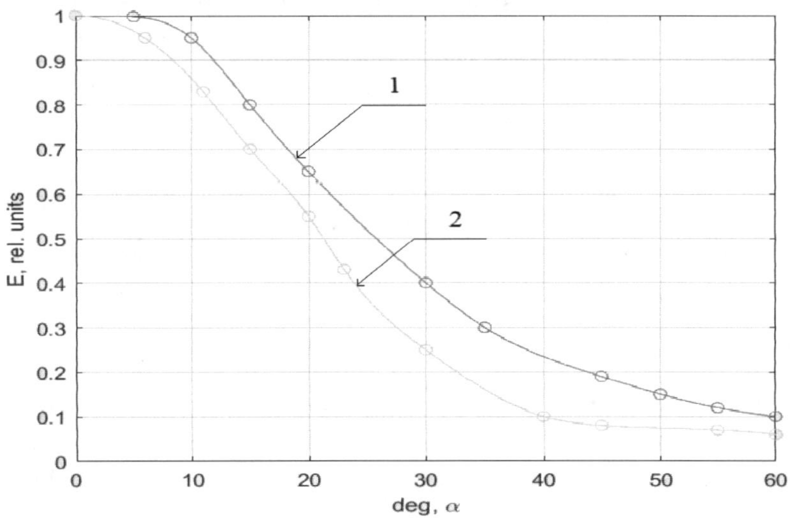

Figure 3. The dependence of the change in light intensity on the angle α at which light enters the light guide for two types of light guides, graph 1 – a single light guide, graph 2-a light guide with two branches.

4. Conclusion

Conducted studies have shown that when implementing the system, it is important to take into account the geometry of the roof and the angle at which the optical axis of the focusing lens will be located. The focus of the lens should be located on the outer surface of the light guide, the optical axis of the focusing lens should be perpendicular to the surface of the light guide on which the radiation falls. For each indoor lighting project, it is necessary to calculate the system and simulate its operation in order to determine the optimal configuration for the maximum intensity of light entering the premises.

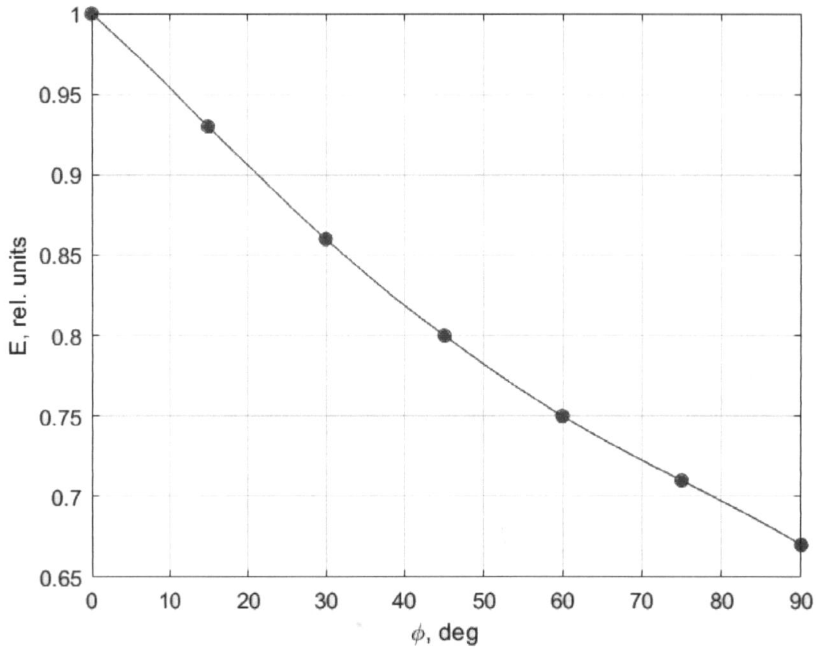

Figure 4. The dependence of the change in illumination E on the bending angle of the light guide φ.

Thanks to its properties, the developed optical lighting system creates an atmosphere of comfort in the premises, and significantly reduces the energy costs for lighting, heating and air conditioning of the buildings in which it is installed.

References

[1] Logunov S E, Rud V Y, Davydov R V, Moroz A V and Smirnov K J 2019 *Journal of Physics: Conference Series* **1326(1)** 012024
[2] Davydov R, Antonov V and Moroz A 2019 *International Conference on Electrical Engineering and Photonics EExPolytech-2019 (Saint-Petersburg)*, vol. 8906791 (IEEE), p. 295-297
[3] Davydov R, Antonov V, Makeev S, Batov Y, Dudkin V and Myazin N 2020 *E3S Web Conference* **140** 02001
[4] Davydov V V 1999 *Russian Physics Journal* **42(9)** 822–825
[5] Davydov V V, Dudkin V I and Karseev A Y 2014 *Optical Memory and Neural Networks (Information Optics)* **23(3)** 170–176
[6] Davydov V V, Dudkin V I and Karseev A Y 2014 *Optical Memory and Neural Networks (Information Optics)* **23(4)** 259–264
[7] Nikitina M, Grebenikova N, Dudkin V and Batov Y 2019 *IOP Conference Series: Earth and Environmental Science* **390(1)** 012024
[8] Yushkova V, Kostin G, Davydov R, Dudkin V and Valiullin L 2019 *IOP Conference Series: Earth and Environmental Science* **390(1)** 012016
[9] Van S, Cheremisin A, Chusov A, Switala F and Davydov R 2019 *IOP Conference Series: Earth and Environmental Science* **390(1)** 012011
[10] Van S, Cheremisin A, Glinushkin A, Davydov R and Yushkova V 2019 *E3S Web of Conferences* **140** 09008
[11] Gryznova E, Batov Y and Myazin N 2020 *E3S Web of Conferences* **140** 09001
[12] Gryznova E, Grebenikova N, D. Ivanov D and Bykov V 2019 *IOP Conference Series: Earth and Environmental Science* **390(1)** 012044
[13] Davydov R V, Mazing M S, Yushkova V V and Stirmanov A V 2019 *Journal of Physics:*

Conference Series **1410(1)** 012067
[14] Myazin N S, and Yushkova V V 2019 *Environmental Research, Engineering and Management* **75(2)** 28–35
[15] Grevtseva A S, Smirnov K J, Greshnevikov K V and Glinushkin A P 2019 *Journal of Physics: Conference Series* **1368(2)** 022072
[16] Grevtseva A S, Smirnov K J, Davydov V V and Rud V Yu 2018 *Journal of Physics: Conference Series* **1135(1)** 012056
[17] Davydov R V, Rud V Y and Yushkova V V 2020 *Journal of Physics: Conference Series* **1695(1)** 012064
[18] Davydov R V, Antonov V I, Yushkova V V, Rud V Yu and Smirnov K J 2019 *Journal of Physics: Conference Series* **1400(6)** 066037
[19] Davydov R V, Antonov V I, Yushkova V V, Grebenikova N M and Dudkin V I 2019 *Journal of Physics: Conference Series* **1236(1)** 012079
[20] Rutberg F G, Kolikov V A, Snetov V N, Cherepkova I A and Dyachenko S V 2016 *High Temperature* **54(2)** 170–174
[21] Myazin N, Neronov Y, Dudkin V and Yushkova V 2018 *MATEC Web of Conferences* **245** 11013
[22] Davydov R V and Antonov V I 2018 *Journal of Physics: Conference Series* **1124(8)** 081037
[23] Davydov R V and Antonov V I 2018 *Journal of Physics: Conference Series* **1135(1)** 012087
[24] Gizatullin B, Gafurov M, Vakhin A, Mattea C and Stapf S 2019 *Energy and Fuels* **33(11)** 10923–10932
[25] Neronov Yu I and Seregin N N 2013 *Measurement Techniques* **55(11)** 1287-1293
[26] Davydov V V, Cheremiskina A V, Velichko E N and Karseev A Yu 2014 *Journal of Physics: Conference Series* **541(1)** 012006
[27] Grebenikova N M and Davydov R V 2019 *Journal of Physics: Conference Series* **1326(1)** 012012
[28] Davydov V V, Velichko E N, Dudkin V I and Karseev A Y 2014 *Measurement Techniques* **57(6)** 684–689
[29] Davydov V V, Dudkin V I and Karseev A Y 2015 *Technical Physics Letters* **41(4)** 355–358
[30] Davydov V V, Dudkin V I and Karseev A Y 2015 *Technical Physics* **60(3)** 456–460
[31] D'yachenko S V and Zhernovoi A I 2016 *Technical Physics* **61(12)** 1835–1837
[32] Münch M, Wirz-Justice A, Brown S, Kantermann T, Martiny K, Stefan O, Vetter C, Wright K, K. Wulff K and Skene D 2020 *Clocks & Sleep* **2(1)** 61-85
[33] McCloughan C L B, Aspinall P A and Webb R S 1999 *International Journal of Lighting Research and Technology* **31(3)** 81-88
[34] Rukin E V, Myazin N S and Dudkin V I 2019 *Journal of Physics: Conference Series* **1368(4)** 042011
[35] Myazin N S, Yushkova V V and Taranda N I 2019 *Journal of Physics: Conference Series* **1410(1)** 012130
[36] Kozlova E S and Kotlyar V V 2014 *Computer Optics* **38(1)** 132-138
[37] Kuzmin M S and Rogov S A 2019 *Computer Optics* **43(3)** 391-396
[38] Smirnov K J and Batov Y V 2019 *Journal of Physics: Conference Series* **1368(2)** 022073
[39] Smirnov K J, Glagolev S F, Rodygina N S and Ivanova N V 2018 *Journal of Physics: Conference Series* **1038(1)** 012102
[40] Smirnov K J, Glagolev S F and Tushavin G V 2018 *Journal of Physics: Conference Series* **1124(1)** 022014
[41] Smirnov K J, Medzakovskiy V I, Vysoczky M G and Glagolev S F 2017 *Journal of Physics: Conference Series* **917(6)** 062019
[42] Davydov V V, Nikolaev D I, Bukharov G D and Pavlova Z S 2020 *International Conference on Electrical Engineering and Photonics EExPolytech-2020 (Saint-Petersburg)*, vol. 9243948 (IEEE), p. 227-229

On the need to control the state of the upper edge of the prism in a flow-through refractometer for measurements with an error less than 10^{-4}

F A Isakov[1], V V Davydov[1, 2] and V Yu Rud[2]

[1] Peter the Great Saint Petersburg Polytechnic University, Saint Petersburg 195251, Russia
[2] All-Russian Research Institute of Phytopathology, Moscow Region 143050, Russia

ifa-spb@yandex.ru

Abstract. The article substantiates the need to use a flow-through refractometer to control the state of a flowing liquid during scientific research, when automating a technological process, etc. The main negative factors that affect an increase in the measurement error of the refractive index n of a flowing liquid are determined. It was found that one of these factors is the formation of a thin film (plaque) on the upper face of the prism, which is in contact with the medium under study. The study of the influence of this film on the measurement error has been carried out. A method has been developed to eliminate plaque from the upper face of the prism during the operation of the refractometer. The results of experimental studies are presented.

1. Introduction

The development of scientific and technological progress has posed a number of difficult tasks for scientists [1-8]. One of which is monitoring the state of the current environment in real time [9-14]. Especially great difficulties arise if the medium flows through the pipeline at high pressure and temperature [12, 13, 15-18]. The diameter of the pipeline and its configuration are very critical in this case [19-25]. In addition, the measurements carried out should not change the flow structure and hydraulic resistance of the system. This is important when using liquid media in various cooling systems, both in scientific research and in industrial installations [26-29].

In most cases, when monitoring the state of liquid media, the measurements carried out should not also change their physical structure and chemical composition [30–34]. This is especially important when conducting scientific experiments, in control systems in the food and pharmaceutical industries, in mechanical engineering when processing materials and others [13, 14, 26-29, 35]. At present, only two types of devices (nuclear magnetic flowmeters-relaxometers and flow-through refractometers) satisfy all these requirements [15-19, 29, 34-37]. The latter, in contrast to nuclear magnetic flowmeters-relaxometers, can monitor the state of the liquid in the pipeline both in the current state and in the stationary state. And they have a number of other advantages: measurements of the refractive index n do not depend on the speed of fluid flow through the pipeline and its size; measurements do not require a large straight section of the pipeline and others [35-40].

In contrast to the operation of stationary or portable refractometers, a number of problems arise during the operation of a flow-through refractometer. One of them is associated with the formation of a deposit from a thin film on the surface of the prism during the operation of the device. This optical part of the refractometer is in contact with the investigated flowing medium. Our studies have shown that the formation of this plaque does not depend on the studied environment. Even water that stagnates during the maintenance of the entire pipeline is capable of forming a deposit on the edge of the prism after a while. In cases where refractometers are used to control viscous, poorly transparent liquids, for example, special lubricants for milling or turning equipment, engine oils, etc., the rate of plaque formation increases. In stationary refractometers, after each measurement of any medium, the prism is cleaned and this problem is not.

2. Experimental setup and scurf destruction method

The reasons for the formation of plaque on the surface of the prism are different. It should only be noted that there are a number of factors that accelerate or reduce the time of plaque formation. These factors include fluid viscosity, temperature, flow rate, etc. Our studies have shown that with a coating layer thickness of about 0.3 mm, changes in the refractive index n in the current medium due to the ingress of impurities and other factors will not affect the measurements of n_m, which are carried out with a refractometer. A refractometer will measure the refractive index of the plaque formed. Control over the state of the current environment will be completely lost. It is also worth noting that the plaque has a higher refractive index than the medium under study. Therefore, in the process of its formation (the device works to control the state of the flowing liquid for a long time), the measurement error n will worsen. All these circumstances were taken into account when developing a method for removing plaque from the upper edge of the prism. In figure 1 shows the design of a refractometer with an ultrasonic transducer, developed by us, for studying the process of formation and elimination of plaque.

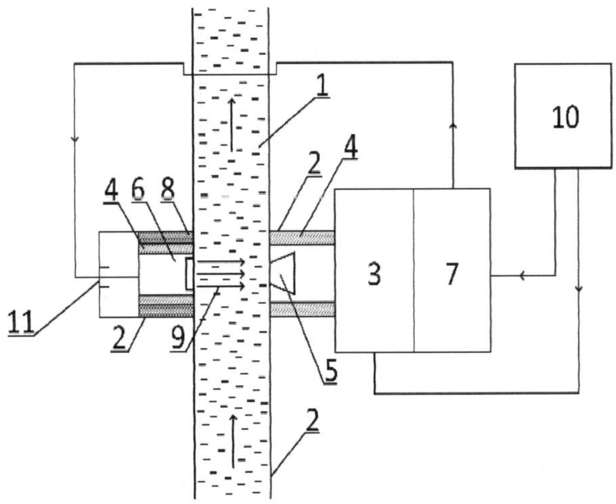

Figure 1. Refractometer design with a pipeline section: 1 - flowing medium, 2 - pipeline wall, 3 - electronic and optical part of the refractometer, 4 - sealing gaskets, 5 - prism, 6 - ultrasonic transducer, 7 - device for monitoring the state of the prism surface, 8 - inserts into pipeline, 9 - ultrasonic wave, 10 - multifunctional power supply, 11 - plug with hydraulic channel for the cable.

Radiation from a semiconductor laser located in the optical part of the refractometer enters in the form of a narrow strip on the upper face of the prism 5 at different angles. At a certain value of the angle of incidence of laser radiation on the face of the prism, total internal reflection occurs [35–37]. On the photodiode array, which is used to register laser radiation, a pronounced light-shadow boundary is formed [35-37]. The position of this boundary is used to determine the angle of total internal reflection α_c and then n_m is calculated. When the value of n_m changes due to the appearance of impurities in it or a change in temperature T_m the angle α_c changes (the light - shadow boundary shifts). The measurement error n_m is mainly determined by the degree of contrast of the light-shadow boundary and the resolution of the photodiode ruler (by recording the position of the light-shadow boundary on one photosensitive sensor). With the formation of plaque, the contrast of the light-shadow border deteriorates (the measurement error n_m increases). There is also a change in the structure of the light-shadow boundary, which is recorded on the photodiode array. This is due to the fact that the plaque on the upper face of the prism is formed unevenly (the refractive index at the upper face of the prism begins to change abruptly - several angles of total internal reflection appear). This is the only sign that allows you to determine the beginning of plaque formation on the upper edge of the prism. The control device 7 establishes this fact of change in the structure of the light-shadow border. The ultrasonic transducer 6 is turned on, which is located opposite the prism 5. The ultrasonic wave 9 destroys the plaque, which is washed off by the current flow. Experiments have shown that a change in the temperature T_m or the flow rate of the medium does not destroy the plaque.

In the case of prolonged immobility of the liquid in the pipeline, a deposit forms on the edge of the prism. Therefore, it is recommended to turn on the ultrasonic transducer during n_m measurements (do not turn on the fluid movement through the pipeline) and monitor the n_m values for a certain period of time. The refractometer is placed in a vertical section of the pipeline, so the plaque will fall down under the influence of gravity and dissolve.

3. Experimental results and discussion

In figure 2 shows the results of the study of the influence of plaque on the control of the state of drinking water and the result of the work of the developed method for cleaning the upper edge of the prism.

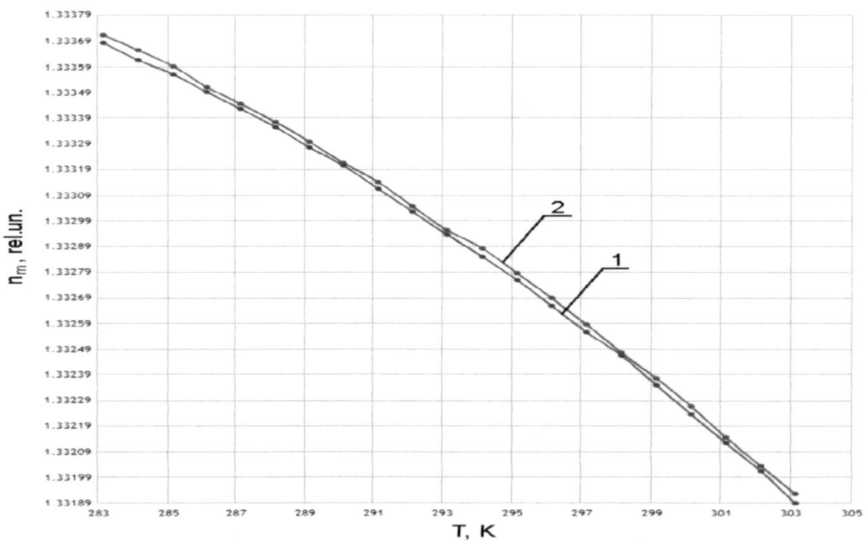

Figure 2. The dependence of the change in the refractive index n_m of drinking water on the temperature T. The graphs correspond to the state of the prism edge: 1 - no plaque, 2 - plaque is formed.

Comparison of the obtained results of studying the state of water with the data in [35-37, 40] showed that the measurement error and n_m values when using an ultrasonic prism cleaner in the refractometer design correspond to the standard. In the presence of plaque, the measurement error n_m is high, monitoring the state of the environment is difficult.

In figure 3 as an example, the results of a study of the influence of plaque on the control of the condition of Rheinol 5W-30 engine oil are presented.

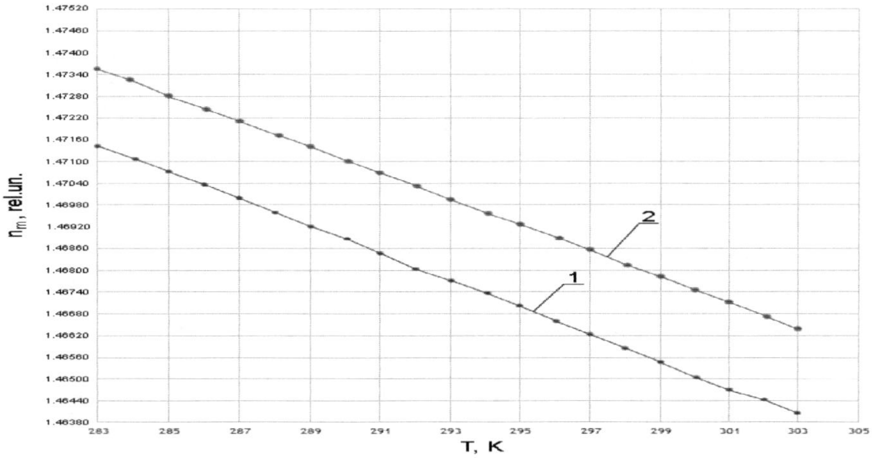

Figure 3. The dependence of the change in the refractive index n_m of engine oil on the temperature T. The graphs correspond to the state of the prism edge: 1 - no plaque, 2 - plaque formed.

Comparison of the obtained results of studying the state of engine oil and water (Figures 2 and 3) showed that the higher the viscosity of the liquid, the greater the difference between the refractive index of the plaque n_s from the value n_m. This leads to a higher measurement error and to an unreliable control of the state of the environment.

In figure 4, as an example, the results of a study of the change in the results of measuring the refractive index n_m from time to time for water and Rheinol 5W-30 engine oil during the formation of a plaque to the value of n_s and after turning on the cleaning system are presented.

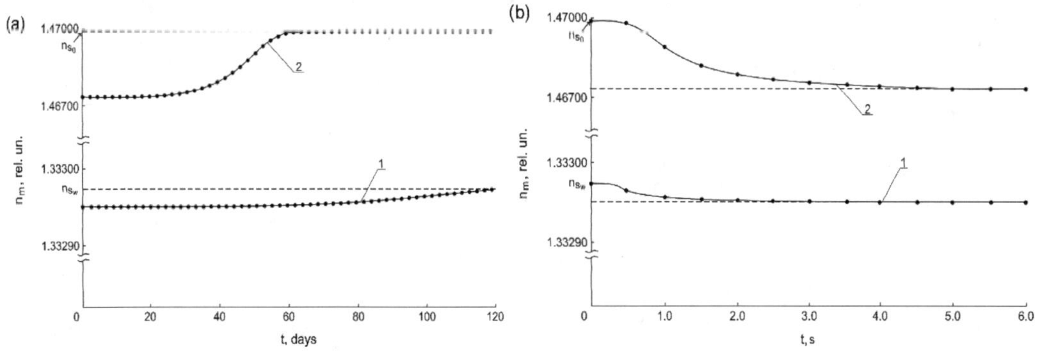

Figure 4. Dependence of the change in the refractive index n_m of water and engine oil on time t at a temperature of T = 293.1 K: (a) – the cleaning system is off, (b) - the cleaning system is on. Graphs 1 and 2 correspond to the presence of water or engine oil in the pipeline.

The results obtained showed that the time of formation of a plaque on the upper face of the prism depends on the parameters of the liquid and can be short enough to begin to introduce changes in the measurement error n_m. The time for cleaning the upper edge of the prism from a dense layer of plaque, which formed during the operation of the refractometer for various flowing media at an ultrasonic vibration frequency f_u = 50 kHz (above the audibility threshold of 20 kHz) is less than 10 seconds.

4. Conclusion

The conducted studies have shown that the use of the developed design for placing a refractometer with an ultrasonic cleaner and placing it on a vertical section of the pipeline completely eliminates the influence of plaque formation on the measurement error n_m.

The results obtained have shown that it is most expedient to use the ultrasonic cleaner autonomously in the case of placing the refractometer structure on the vertical section of the pipeline. Switch it on every 20 minutes for 3 seconds for any type of liquid. The reliability of the mode of operation of the purifier is carried out by a sound signal.

References

[1] Kozlova E S and Kotlyar V V 2014 *Computer Optics* **38(1)** 132-138
[2] Kuzmin M S and Rogov S A 2019 *Computer Optics* **43(3)** 391-396
[3] Skidanov R V, Blank V A and Morozov A A 2015 *Computer Optics* **39(3)** 384-392
[4] Davydov R V and Antonov V I 2018 *Journal of Physics: Conference Series* **1124(8)** 081037
[5] Davydov R V and Antonov V I 2018 *Journal of Physics: Conference Series* **1135(1)** 012087
[6] Logunov S E, Rud V Y, Davydov R V, Moroz A V and Smirnov K J 2019 *Journal of Physics: Conference Series* **1326(1)** 012024
[7] Davydov R, Antonov V and Moroz A 2019 *International Conference on Electrical Engineering and Photonics EExPolytech-2019 (Saint-Petersburg)*, vol. 8906791 (IEEE), p. 295-297
[8] Davydov R, Antonov V, Makeev S, Batov Y, Dudkin V and Myazin N 2020 *E3S Web Conference* **140** 02001
[9] Myazin N S, Sviatkina V I and Yushkova V V 2020 *Journal of Physics: Conference Series* **1695(1)** 012141
[10] Davydov V V, Moroz A V, Myazin N S, Makeev S S and Dudkin V V 2020 *Optics and Spectroscopy* **128(10)** 1678–1685
[11] Davydov V V, Myazin N S, Makeev S S and Dudkin V I 2020 *Measurement Techniques* **63(5)** 368–374
[12] Davydov V V, Dudkin V I, Myazin N S and Davydov R V 2020 *Journal of Communications Technology and Electronics* **65(5)** 558–564
[13] Davydov V V, Myazin N S and Kiryukhin A V 2020 *Atomic Energy* **127(5)** 274–279
[14] Davydov R V, Mazing M S, Yushkova V V and Stirmanov A V 2019 *Journal of Physics: Conference Series* **1410(1)** 012067
[15] Davydov V V 1999 *Russian Physics Journal* **42(9)** 822–825
[16] Davydov V V, Dudkin V I and Karseev A U 2013 *Optical Memory and Neural Networks (Information Optics)* **22(2)** 112–117
[17] Davydov V V, Dudkin V I and Karseev A Y 2014 *Optical Memory and Neural Networks (Information Optics)* **23(3)** 170–176
[18] Davydov V V, Dudkin V I and Karseev A Y 2014 *Optical Memory and Neural Networks (Information Optics)* **23(4)** 259–264
[19] Davydov V V, Velichko E N, Dudkin V I and Karseev A Y 2014 *Measurement Techniques* **57(6)** 684–689
[20] Davydov V V, Dudkin V I and Karseev A Y 2015 *Technical Physics Letters* **41(4)** 355–358
[21] Davydov V V, Dudkin V I and Karseev A Y 2015 *Technical Physics* **60(3)** 456–460
[22] Myazin N S, Yushkova V V and Taranda N I 2019 *Journal of Physics: Conference Series* **1410(1)** 012130
[23] Davydov V V, Dudkin V I and Karseev A Y 2015 *Russian Physics Journal* **58(2)** 146–152
[24] Davydov V V, Dudkin V I and Karseev A Y 2015 *Measurement Techniques* **58(3)** 317–322
[25] Davydov V V, Dudkin V I and Karseev A Y 2015 *Instruments and Experimental Techniques* **58(6)** 787–793
[26] Makeev S S, Grevtzeva A S, Glinushkin A P and Matorin D N 2020 *Journal of Physics: Conference Series* **1695(1)** 012112
[27] Myazin N S, Yushkova V V and Dudkin V I 2019 *Journal of Physics: Conference Series*

1400(6) 066008
[28] Rukin E V, Myazin N S and Dudkin V I 2019 *Journal of Physics: Conference Series* **1368(4)** 042011
[29] Myazin N S, and Yushkova V V 2019 *Environmental Research, Engineering and Management* **75(2)** 28–35
[30] Myazin N, Neronov Y, Dudkin V and Petrov A 2019 *IOP Conference Series: Materials Science and Engineering* **497(1)** 012111
[31] Myazin N S 2018 *Journal of Physics: Conference Series* **1135(1)** 012061
[32] Myazin N, Neronov Y, Dudkin V and Yushkova V 2018 *MATEC Web of Conferences* **245** 11013
[33] Davydov V V, Logunov S E and Fadeenko V B 2018 *Russian Journal of Nondestructive Testing* **54(3)** 213–221
[34] Davydov V V, Fadeenko V B and Logunov S E 2018 *Technical Physics Letters* **44(2)** 153–156
[35] Grebenikova N M, Davydov V V and Rud V Y 2019 *Journal of Physics: Conference Series* **1410(1)** 012186
[36] Karabegov M A 2013 *Measurement Techniques* **55(11)** 1301–1310
[37] Davydov V V, Grebenikova N M and Smirnov K Y 2019 *Measurement Techniques* **62(6)** 519–526
[38] Davydov V V and Moroz A V 2020 *Optics and Spectroscopy* **128(9)** 1415–1420
[39] Davydov V V, Grebenikova N M and Dudkin V I 2020 *Measurement Techniques* **62(12)** 1090–1098
[40] Karabegov M A 2009 *Measurement Techniques* **52(1)** 97–104

Simulation of the photonic nanojet effect for Raman scattering enhancement in the diagnostics of oxide films

A I Ivanina, D S Agafonova, N V Roshina and I A Lamkin

Saint Petersburg Electrotechnical University «LETI», Department of Photonics, Saint Petersburg 197376, Russia

ellieams6@mail.ru, dsagafonova@gmail.com, roshina1997@yandex.ru, ialamkin@mail.ru

Abstract. In this paper, by means of numerical simulations in the COMSOL Multiphysics software it's demonstrated that Raman scattering enhancement can be achieved for the diagnosis of metal oxide films using spherical particles made of barium titanate with a 10-micron diameter sphere. The formation of photonic nanojet in the sphere/film/substrate system at different radiation wavelengths and microsphere refractive index, film, and substrate was studied. The optimal interval of the particle refractive index is n≈1.8-2 was determined, at which the gain occurs directly at the particle/film interface. It is shown that for the UV wavelength range of wavelengths and film thicknesses from 50 to 200 nm, the gain is maximum. For ZnO and PZT films in the perovskite phase, sitall and quartz are preferred as the substrate material, while for PZT in the pyrochlore phase, sapphire is preferred.

1. Introduction

Raman spectroscopy is a method of molecular spectroscopy based on inelastic light scattering in a substance. It can be used to obtain information about the chemical composition and phase transition of the material under study. Due to the small cross-sections of the interaction, the Raman signal is extremely weak and difficult to separate from the much stronger Rayleigh scattering signal [1]. The analysis of nano-sized objects, such as nanoparticles, biomolecules, and proteins, is challenging due to the low intensity of Raman scattering and the small sample size. Various approaches are used to increase the Raman signal, one of them being the photonic nanojet (PNJ) effect. If you illuminate a spherical particle that is larger than the incident radiation wavelength with a well-chosen refractive index, the particle behaves like a microlens, focusing light into" jets " of sub-wavelength length that come out of the shadow surface of microsphere. For a spherical particle, the PNJ is formed within a narrow line (transverse size of the order λ / 3), while its propagation to the depth (lateral size) it can be stretched to a distance greater than λ, depending on the refractive index of the medium and the substrate on which the sphere is placed [2,3]. The inhomogeneity of the refractive index at the interface between the particle and the sample leads to a change in the electromagnetic field distribution, including an increase in its localization. Therefore, Raman scattering can be amplified, since it strongly depends on the magnitude of the electromagnetic field [4,5]. Compared to traditional approaches such as surface-enhanced Raman light scattering (SERS), Raman microsphere-enhanced spectroscopy may be advantageous due to the ease of sample preparation, particle position control, and microsphere size control [6]. In [6], an increase in Raman scattering was demonstrated due to the use of nanostructures on silicon dioxide particles. It is shown that the most effective amplification occurs when the particle diameter is equal to the size of the incident radiation spot. In [7], the Raman scattering amplification using individual polystyrene microspheres was investigated. It was also found that the maximum gain occurs when the sphere is approximately equal to the size of the incident beam.

The Raman spectroscopy method can be used to study metal oxide films. At present, metal oxide films are both scientific and practical interest [8,9,10,11], since their application is possible in a wide field of optoelectronics both in the form of layers in film heterostructures [12] and in the form of composite systems [9,13]. Zinc oxide films are used as conductive transparent and barrier layers in optoelectronics, in particular, in solar energy [11,14], the conductive and optical properties of the films are modified by doping and special surface treatment [15]. Lead zirconate-titanate films are characterized by a wide variability of properties, including refractive index, which are determined by the composition [16] and can crystallize in different phases [14,17], which differ in refractive indices. The limitations of using the Raman spectroscopy method in the study of metal oxide films are associated with small sample thicknesses, and, as a result, small Raman scattering signals. Signal amplification is possible with a local increase in the incident radiation intensity, which can be realized when the PNJ is formed on the surface of the film.

In this paper, the possibility of amplifying Raman scattering using dielectric microspheres for the diagnostics of thin metal oxide films of zinc oxide and lead zirconate-titanate (PZT) in the perovskite phase and the pyrochlore phase is theoretically investigated.

2. Methods and materials

The possibility of amplifying Raman scattering using microspheres was investigated by numerical simulation in the COMSOL Multiphysics software. The physical model consisted of five components: an incident plane wave, the environment (air), a spherical particle, a film, and a substrate. The direction of propagation of the plane wave is perpendicular to the film surface. Schematically, the physical model is shown in figure 1.

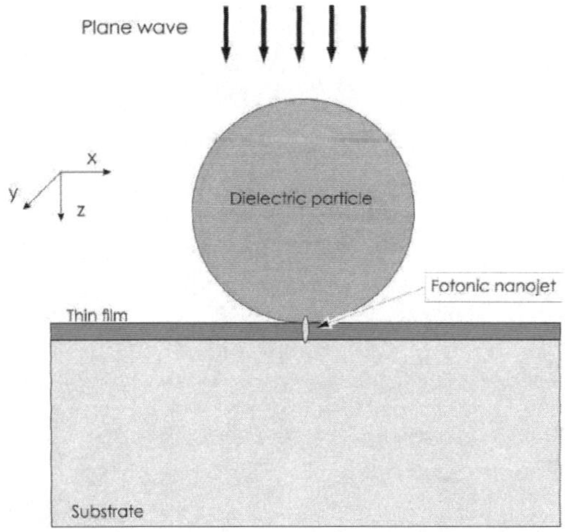

Figure 1. Schematic physical model in COMSOL Multiphysics software.

Barium titanate (BaTiO3) was selected for the microsphere material. Medium – air (n0 = 1). The choice is determined by the dependence of the PNJ size on the particle refractive index. Previous calculations showed that if the refractive index is less than 1.8, the PNJ begins to move away from the shadow surface of the particle and extends in length, and if it is more than 2.1, the PNJ moves inside the particle. Therefore, the optimal range is 1.8> n> 2 BaTiO3 has a refractive index of 1.9, so it is excellent for simulation the effects that occur behind the particle surface at the boundary with the studied film.

The substrate on which the sample is placed can strongly affect the light holding and propagation of the nanostructure, as demonstrated in [18]. Therefore, three different materials, such as sapphire (n=1.77), quartz (n=1.46), sitall (n=1.53), were used as substrates to choose the most effective combination.

In this work, a plane wave was modelled, the wavefront width is equal to the sphere diameter. The sphere diameter is 10 microns. The PNJ effect was achieved by focusing a plane wave through a microsphere. The radiation intensity profiles were calculated along the x-axis at the boundary of the sphere and the film (figure 1), the maximum and average radiation intensity values were determined over the entire depth of the film, as well as the geometric dimensions of the PNJ in the lateral ($_n$) и and transverse (R_τ) directions. The parameters R_n and R_τ are the full width at the half maximum (FWHM) of the PNJ intensity peak in the sections along z and x, respectively, passing through the point of maximum intensity. The simulation was carried out in a wide wavelengths range to select the best value at which the maximum gain in the sample will be achieved. It is assumed that the value of the PNJ radiation intensity is a measure of the Raman signal amplification.

3. Results and Discussion

Figure 2 shows graphs of the dependence of the average and maximum field intensity in the ZnO film on the wavelength. At this thickness, we can assume that the value is the same across the entire depth of the film. It can be seen that for different variants of the substrate and the film thickness, the intensity is maximum in the UV region of the spectrum, and it decreases with increasing wavelength.

According to figure 2, 248 nm wavelength can be chosen for the study of the film in practice, for example, the radiation of an excimer laser (KrF). It is also possible to use radiation with 350-380 nm wavelengths and 470 nm for the visible range (there are peaks at the graph in these areas). For other PZT films, the peaks are observed in the same ranges.

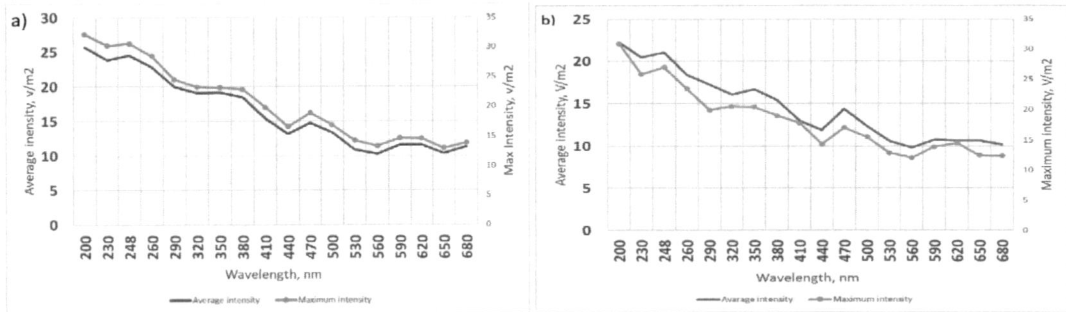

Figure 2. The value of the light intensity in the ZnO film depending on the light source wavelength. Film thickness: **(a)** 100 nm; **(b)** 500 nm. Substrate – sitall.

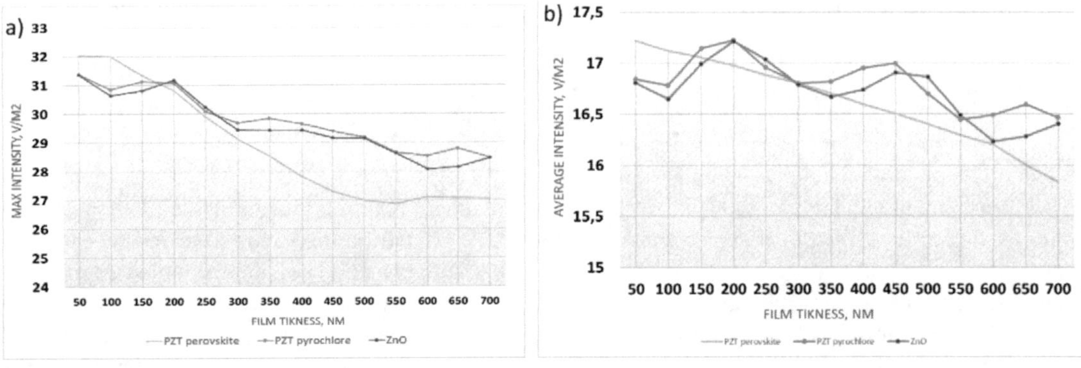

Figure 3 - Dependence of the **(a)** average and **(b)** maximum radiation intensity on the film thickness. Substrate – sitall.

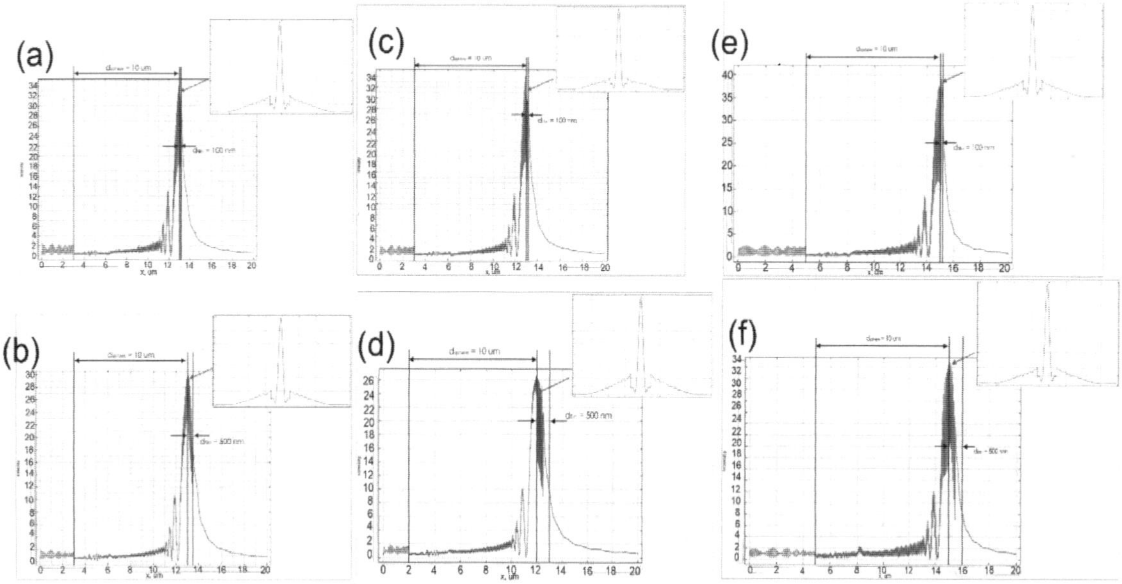

Figure 4. Field intensity distribution along the x-axis: **(a)** BaTiO3/ZnO/sitall $d_{film} = 100$ nm **(b)** $d_{film} = 500$ nm. **(c)** BaTiO3/PZT (perovskite)/sitall $d_{film} = 100$ nm **(d)** $d_{film} = 500$ nm, **(e)** BaTiO3/PZT (pyrochlore)/Al2O3 $d_{film} = 100$ nm, **(f)** $d_{film} = 500$ nm. Radiation wavelength – 248 nm.

Graphs of the dependence of the light intensity on the sample thickness are shown in Figure 3. The samples under study: ZnO and PZT films in the perovskite phase and pyrochlore phase were placed on sital substrates, and the film thickness varied from 50 nm to 700 nm.

It's found that for the PZT in the perovskite phase, the dependence is almost linear, while the PZT in the pyrochlore phase and zinc oxide have well defined peaks. For zinc oxide, it is 200, 450 and 500 nm, for PZT in the pyrochlore phase – 200 nm, 450 and 650 nm. In general, we see a tendency to decrease in intensity with increasing film thickness due to absorption within the film.

Table 1. Full width at half-maximum (FWHM) for various film/substrate combinations, radiation wavelength-248 nm.

Film material	Substrate material	R_τ (FWHM), um	R_n (FWHM), um	R_n (FWHM), um	R_τ (FWHM), um
		Film thickness, nm			
		100		500	
ZnO	Sapphire	0.140	1.046	0.139	1.157
	Quartz	0.142	0.979	0.142	1.153
	Sitall	0.120	0.880	0.132	1.021
PZT (perovskite)	Sapphire	0.135	1.101	0.143	1.237
	Quartz	0.129	0.928	0.123	1.261
	Sitall	0.127	0.884	0.134	1.022
PZT (pyrochlore)	Sapphire	0.141	0.883	0.151	1.166
	Quartz	0.168	0.936	0.152	1.069
	Sitall	0.144	1.025	0.146	1.139

Table 1 shows the results of calculating the main geometric characteristics of the PNJ – "length" R_n (lateral size) and "width" R_τ (transverse size). The PNJ formation for films with 100 and 500 nm thickness was considered. According to Figure 2, maximum increase in intensity was expected on the film with d = 100 nm, and therefore the values of R_τ and n should be less than at d = 500 nm. The maximum increase was observed on the sitall substrate for zinc oxide films and PZT in the perovskite phase, but it is small relative to another substrates 2–3%. For the PZT film, the pyrochlore phase is the highest intensity achieved on a sitall substrate. So, by selection, it was possible to identify the most effective combination of film and substrate.

The increase in intensity at the particle-film interface was considered in more detail. Figure 4 shows the radiation intensity distribution along the x-axis for each case of maximum gain. The position of the particle and the thin film is schematically marked. The intensity peak is located at the boundary between the particle and the film. Also, the peak intensity is shown on the field strength distribution at the point of contact of the particle with the sample along the y axis. This shows that the amplification occurs directly in the sample, which means that the Raman scattering can be enhanced. After the peak, the gain decreases.

4. Conclusions

In this paper, it was demonstrated by means of numerical simulations that it is possible to achieve Raman scattering amplification using microspheres for the diagnostics of metal oxide thin films. The formation of photonic nanojet in air at different radiation wavelengths, the microsphere refractive index and diameter, and the film refractive index are studied. It was found that to enhance Raman scattering by microspheres, it is necessary to choose the refractive index of the particles in the range n ≈ 1.8–2. Then the gain occurs directly at the particle / film boundary. Also, it is necessary to choose the relevant wavelength. It was determined for all test samples that it is best to use the UV range wavelength and the optimal film thickness is 50-200 nm. For ZnO and PZT (pyrochlore phase) films, intensity peaks are observed also in the 450–500 nm thickness range. As a substrate for sample deposition, sitall and quartz can be used for ZnO and PZT films in the perovskite phase. For PZT in the pyrochlore phase, the gain is maximum if one uses the sapphire substrate.

Thus, with the correct choice of the radiation wavelength, the microsphere refractive index, the film refractive index and thickness, it is possible to effectively increase the intensity of the field inside the film. This method is easy to prepare and implement, cost-effective, and can provide field amplification for Raman scattering detection.

Acknowledgments

The research was carried out within the framework of the project № FSEE-2020-0008, which was carried out within the framework № 075-01024-21-00 of the state assignment of Ministry of Science and Higher Education of the Russian Federation

References

[1] Darafsheh A 2021 Photonic nanojets and their applications, *J. Phys. Photonics* **3** 6678-6690

[2] Yang H, Trouillon R, Huszka G, Gijs M A 2016 Super-resolution imaging of a dielectric microsphere is governed by the waist of its photonic nanojet *Nano Lett.* **16** 4862–4870

[3] Chen Z, Taflove A, Backman V 2004 Photonic nanojet enhancement of backscattering of light by nanoparticles: a potential novel visible-light ultramicroscopy technique *Opt. Express.***12** 12141220

[4] Du C L, Kasim J, You Y M 2011 Enhancement of Raman scattering by individual dielectric microspheres *J. Raman Spectrosc.* **47** 42145–42148

[5] Das G M, Laha R, Dantham V R 2016 Photonic nanojet-mediated SERS technique for enhancing the Raman scattering of a few molecules *J. Raman Spectrosc* 47895–900.

[6] Yi K J, Wang H, Lu H, Yang Y F 2007 Enhanced Raman scattering by self-assembled silica spherical microparticles. *Appl. Phys.* **47** 42145–42148

[7] Du C L, Kasim J, You Y M, Shi D N and Shen Z X 2011Enhancement of Raman scattering by individual dielectric microspheres *J. Raman Spectrosc.* **42** 145-146

[8] Afanasjev V P, Mukhin N V, Redka D N 2017 Surface Modification of ZnO by Plasma and Laser Treatment *Ferroelectrics* **508** **1** 124–129

[9] Afanas'ev V P, Vorotilov K A, Mukhin N V 2016 Effect of the Synthesis conditions on the Properties of Polycrystalline Films of Lead Zinconate Titanite of Non-Stoichiometric Composition *Glass Physics and Chemistry* **42** **3** 295–301

[10] Mukhin N V 2014 Diffusion model of intrinsic defects in lead zirconate titanate films on heat treatment in air *Glass Physics and Chemistry* **40** **2** 238–242

[11] Evseenkov A S, Tarasov S A, Lamkin I A, Solomonov A V and Kurin S Y 2015 The efficiency of UV LEDs based on GAN/ALGAN heterostructures *Proceedings of the 2015 IEEE North West Russia Section Young Researchers* 27-29

[12] Mukhin N V 2016 Investigation of the formation kinetics of grain boundary inclusions of lead oxide in lead zirconate-titanate films *Glass Physics and Chemistry* **42** **1** 64–69

[13] Afanasjev V P, Chigirev D A, Mukhin N V and Petrov A A 2019 Formation and properties of PZT-PbO thin heterophase films *Ferroelectrics* **496** **1** 170–176

[14] Redka D N, Mukhin N V, Zakharov I G 2016 Variations in Optical and Structural Properties of Zinc Oxide Films after Laser Processing *Technical Physics* **61** **11** 1744–1746

[15] Agafonova D S and other 2018 Phases in Polycrystalline Films of Ferroelectric Oxides of the Perovskite- Type on the Basis of Bi2SrTa2O9 and Pb (Zr,Ti)O3 *Glass Physics and Chemistry* **44** **1** 15-20

[16] Iakovlev A and other 2019 Laser surface modification of ZnO for solar converters *Proceedings of SPIE* **10062** 100621F

[17] Mukhin, N, Chigirev, D, Bakhchova, L and Tumarkin, A 2019 Microstructure and properties of PZT films with different PbO content-ionic mechanism of built-in fields formation *Materials* **12** **18** 2926

[18] Dantham V R, Bisht P B, Namboodiri C K R 2018 Enhancement of Raman scattering by two orders of magnitude using photonic nanojet of a microsphere *J. Appl. Phys.* **109** 103103

Formation of planar plasmon microstructures by dry aerosol printing

K M Khabarov, **M Nouraldeen**, **A A Lizunova**, **M N Urazov** and **V V Ivanov**

Moscow Institute of Physics and Technology (National Research University), Dolgoprudny 141701, Russia

kirill.khabarov@phystech.edu

Abstract. Optical properties and microstructure of samples formed by dry aerosol printing are studied. Silver nanoparticles flat layers of two types were formed on substrates surfaces and were investigated by a spectrophotometer, a scanning electron microscope, and a transmission electron microscope. It is shown that all microstructures support plasmon resonance on individual nanoparticles with the Q factor depending both on the width of the nanoparticles size distribution in the aerosol and on their tendency to agglomeration and aggregation.

1. Introduction

The optical properties of planar arrays of nanoparticles (NPs) significantly depend on their material, morphology, and degree of ordering in the sample. For example, with the rise of the isolated NP size the wavelength of the radiation scattered by that NP increases. Moreover, the specific geometric shapes of NPs exhibit a certain spectral response, which was studied on colloidal NPs in [1]. The spherical shape of NPs is the most common and allows us to analytically predict the position of spectral features for both individual NPs [2] and periodic lattices with individual spherical NPs located in lattice points [3]. Thus, it was shown in [3] that square lattices of gold spherical NPs may simultaneously support several resonant features in the visible wavelength range associated with the excitation of plasmon oscillations. Other authors [4] studied same features of the extinction spectra depending on the geometry of the periodic silver NPs lattice. The study of such structures makes them indispensable in many areas of biology and medicine for targeted delivery of drugs [5,6], biomarkers and contrast agents based on individual NPs [7–9] and for the manufacturing of biosensors [10,11] and of neurocomputer interfaces [12].

At present, the task of manufacturing such structures is massively solved by the lithography and vacuum deposition methods [13]. At the same time, dry aerosol printing allows to synthesize by the spark discharge method [14] and to deposit chemically pure NPs in real time for similar tasks of manufacturing planar microstructures [15] with resonant features in optical spectra. However, in the transport process of aerosol to the substrate, individual NPs form dendrite-like agglomerates and aggregates that destroy the order and distort the optical spectra of the formed structures. Nevertheless, disordered NPs arrays are still able to support plasmon resonance, and therefore the formation of such structures by dry aerosol printing also looks promising for the applications described above.

In this work, disordered arrays consisting of spherical silver NPs, their agglomerates and aggregates were produced by dry aerosol printing on the surface of glass and silicon substrates. Optical transmission spectra in the range from 320 to 800 nm were studied for the obtained samples. It is shown that all microstructures support plasmon oscillations on individual NPs with the Q factor depending both on the width of the NP size distribution in the aerosol and on their tendency to agglomeration and aggregation.

2. Experiment

The studied silver NP structures were made by dry aerosol printing in the form of 3x3 mm^2 flat square layers on substrates surfaces. Thus, glass and polished silicon substrates were used in the study of optical spectra and in the microscopic study of samples, respectively. Aerosol NPs were depositing through a coaxial nozzle with an outlet hole of 300 μm located at 1 mm from the substrate surface, according to the printing pattern shown in Figure 1 (b) by the red line. The line width was controlled by adjusting the flow rate of the additional shielding gas flow entering the nozzle in accordance with the works [16,17]. In the work the width of the printed line W was set to 150 μm. The printing step between the lines S was chosen experimentally such as, firstly, to minimize the surface waviness (Figure 1 (a)), and secondly, to create a sufficient thickness of the layer H, as the aerosol flow falling on the substrate surface could destroy the already formed lines.

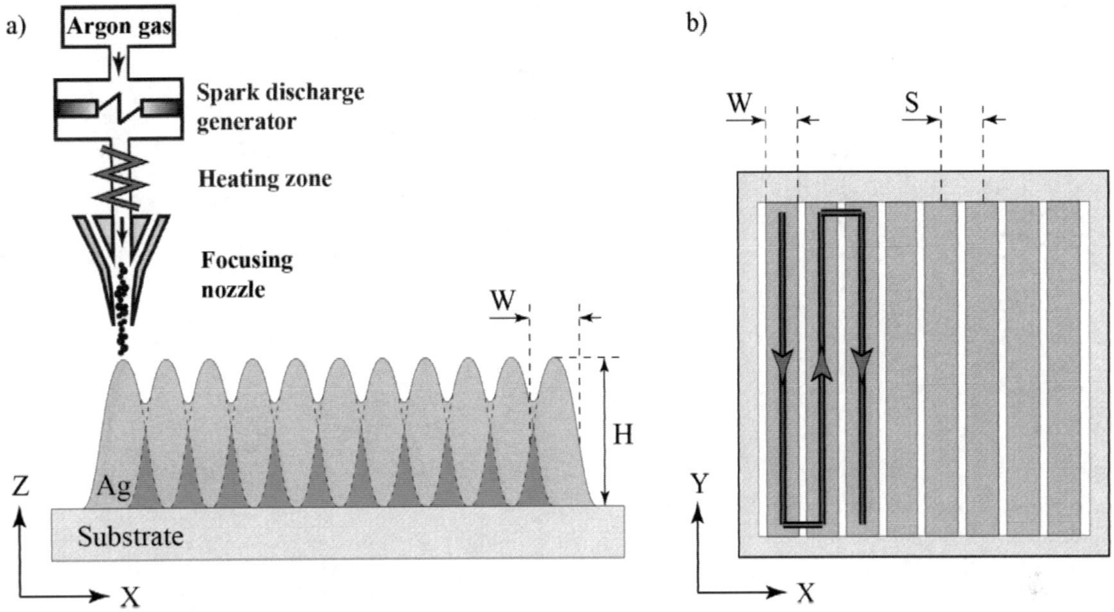

Figure 1. (a) Scheme of the setup for the formation of silver NPs arrays; (a,b) Model image of the printing structure – the side view and the top view with the given template for deposition of NPs on a substrate (red line).

The morphology of NPs was studied by the transmission electron microscope (TEM) JEM-2100 (JEOL), for which they were depositing on copper TEM grids right after the structure was printed. Additionally, the sizes of agglomerates and aggregates of NPs were studied directly in the aerosol stream by the electrodiffusion method based on their differential electrical mobility via the aerosol NPs analyzer SMPS 3936 (TSI Inc.). The microstructure of the sample surface, including the features of NPs stacking, was studied by the scanning electron microscope (SEM) JSM 7001F (JEOL) on samples formed on the surface of polished silicon substrates. The sample thickness measurements were done in the cut of the central part of arrays by the optical profilometer S neox (SENSOFAR) with a 50x confocal lens. For that purpose we used a blade to perform a longitudinal cut of the structure, followed by a transverse shift for the removing of NPs. The optical spectra of samples made on glass substrates were studied by the spectrophotometer V-770 UV-Visible/NIR (JASCO) in the wavelength range from 320 to 800 nm with the radiation beam completely located inside the NPs array.

3. Results and discussion

In this paper we analyzed two types of samples differing in the average diameter of spherical NPs. The first type of microstructures consisted of primary synthesized NPs with an average size of 5 nm, tending to agglomerate in the gas flow into dendride-like structures and forming strong aggregates with an average size of 160 nm (Figure 2 (a,c)). Morphologically such resulting particles are characterized by a low degree of ordering. The second type structures were formed by NPs obtained by modifying aggregates of primary NPs in a section of the gas path heated to 750 °C. The composition of these NPs was characterized by the bimodality of their size distribution with maxima of 40 and 140 nm. The level of their agglomeration in the gas flow was insignificant and aggregation was not observed (Figure 2 (b,d)).

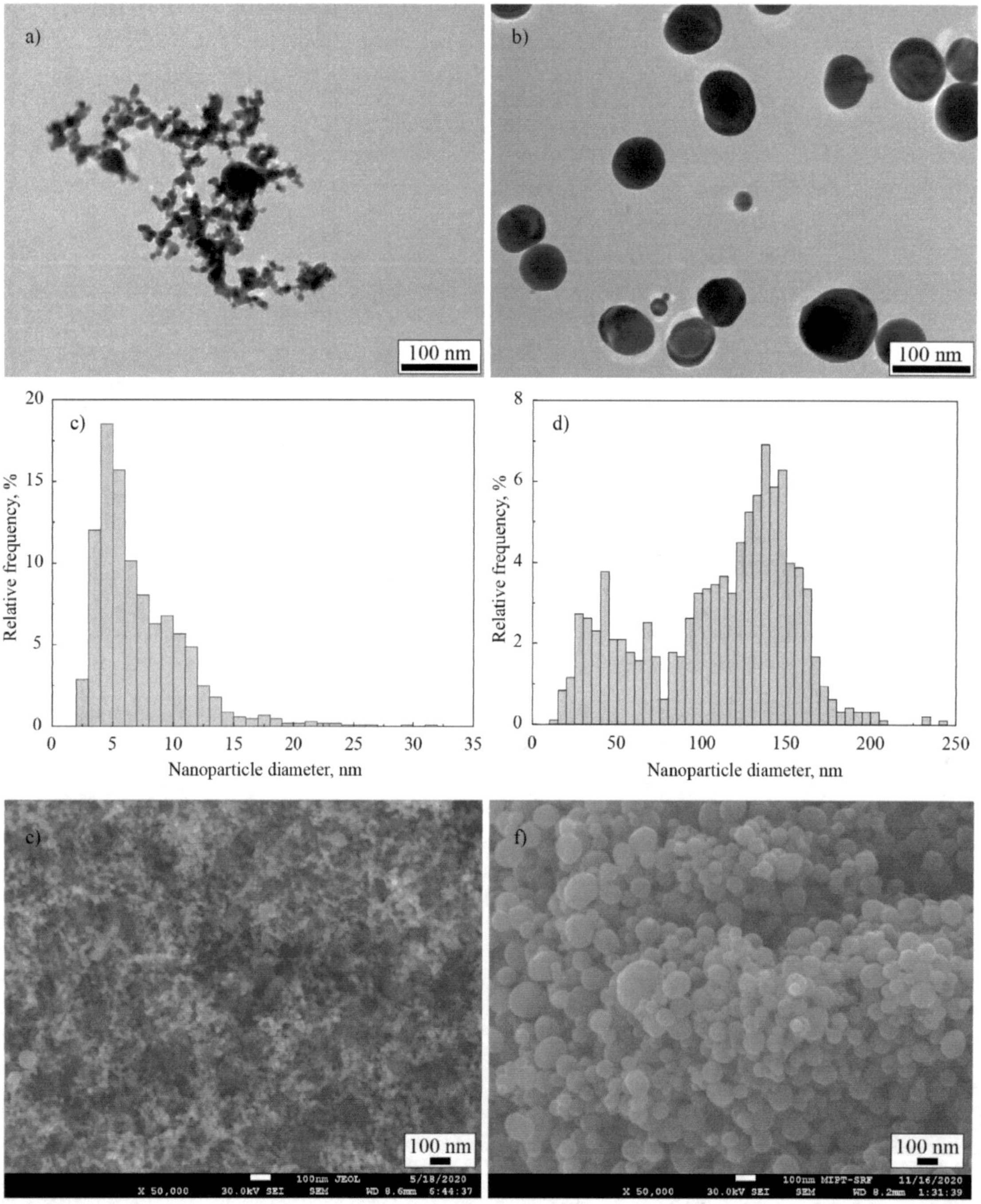

Figure 2. (a,b) TEM images and (c,d) size distributions of NPs forming structures and (e,f) SEM images of arrays surfaces for the first and second type structures.

Scanning electron microscopy of the sample surface allowed us to determine the features of NPs stacking (Figure 2 (e,f)). The surface of the first type structure, consisting of aggregates of primary NPs, turned out to be loose, while NPs in the second type structure were packed more compactly.

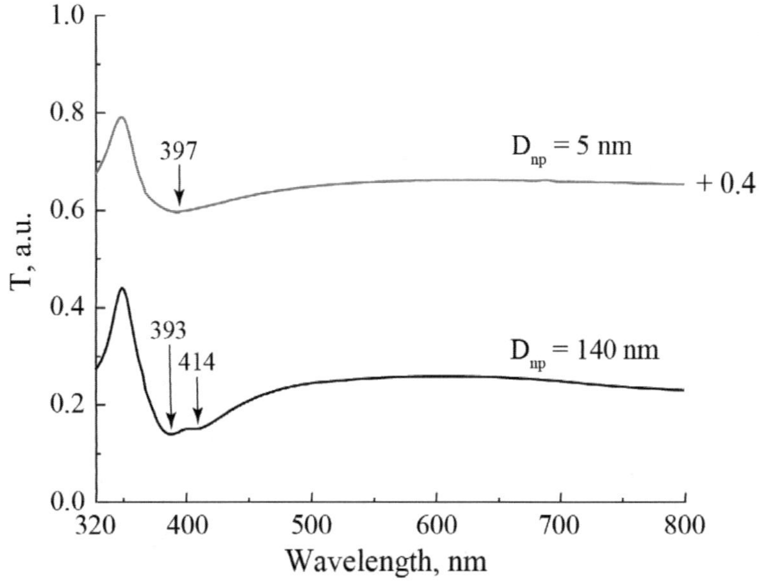

Figure 3. Optical spectra of the first (a) and second type (b) microstructures without substrate.

The optical spectra of the microstructures under the study, with the pre-subtracted spectrum of the glass substrate, demonstrate resonant features in the range of 390-420 nm and are shown in Figure 3. In the transmission spectra for the first type structure only one resonant contour was observed, while for the second type structure this resonant contour split in two. We also noted that the transmission spectrum of the first type structure was higher compared to the transmission spectrum of the second type structure, i.e. the first type structure was more transparent, which corresponds to the thicknesses of the studied samples of 160 and 320 nm, respectively (Figure 4).

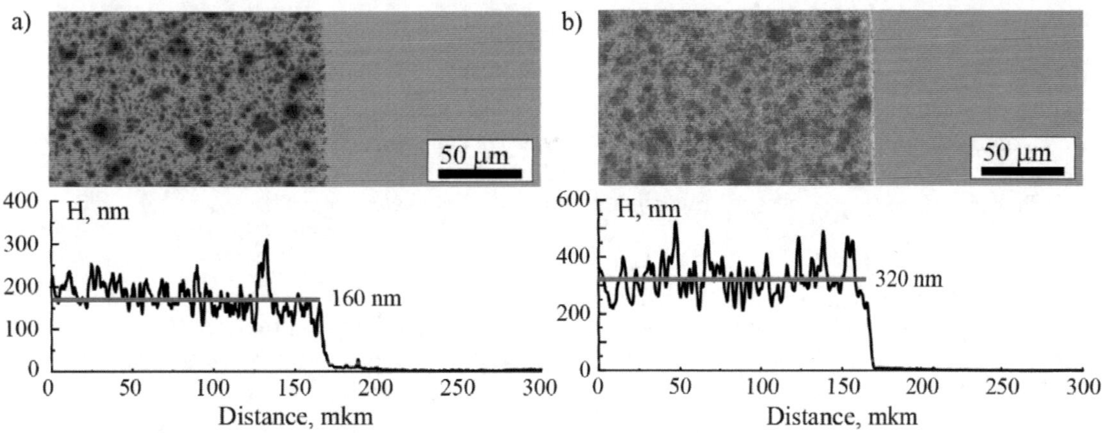

Figure 4. Surface images of the first (a) and second type (b) microstructures in an optical microscope with averaged cross-section profiles corresponding to this region.

The observed resonant features in the optical spectra of the samples are obviously related to plasmon oscillations on NPs and their agglomerates and aggregates in the samples. According to the SEM images, the structures under the study are disordered arrays of point-contacting NPs (for the second type structures) and NPs aggregates (for the first type structures). In the structures of the first type, the degree of ordering is obviously much lower due to the fact that the structure is formed from disordered NPs aggregates of random shape. This explains the difference in the spectra and the lower Q factor of the resonance for the structure of the first type.

The discussed wide absorption resonance in the range of 390-420 nm logically corresponds to the Mie resonance on individual silver particles [2]. Thus, the splitting of the resonance for the second type structure is in good agreement with the positions of the Mie resonance on individual NPs of two particular sizes with diameters of 40 and 140 nm (Figure 2 (d)) at wavelengths of 393 and 414 nm, respectively [18]. However, the position of the resonance for the first type structure at a wavelength of 397 nm is extremely unusual for the position of the resonance on individual silver NPs with a diameter of 5 nm. However, considering this microstructure in the model of aggregated particles with a size of about 160 nm, it can be argued that the NPs in the aggregates work as a whole, showing the properties of extended dendride-like plasmon nanoantennas [19]. In this case, with an increase in the size of the aggregates the resonance redshifts. This is observed in the optical spectra for the first type structure. At the same time, a higher Q factor of the observed resonances can be achieved by reducing the width of the NP size distribution and preventing their agglomeration and aggregation [14].

4. Conclusion

Spherical silver NPs arrays formed by dry aerosol printing support plasmon oscillations, appearing in the optical spectra in the form of wide wavelength areas of radiation absorption. The Q factor of these resonances is small both for arrays of NPs with an average diameter of 5 nm, tending to agglomerate in the gas flow into dendride-like structures and forming strong aggregates with an average size of 160 nm, and for arrays of NPs with an average diameter of 140 nm, mostly not forming agglomerates. This peculiarity of the two types of arrays under the study may be a consequence of a wide range of NP size distribution and may also be associated with their random agglomeration and aggregation. In all cases, the observed resonant features are logically associated with the excitation of Mie oscillations on individual NPs. However, in the case of a structure consisting of primary NPs with a diameter of 5 nm, the resonance redshifts due to their aggregation, when considering the aggregates as extended dendride-like plasmon nanoantennas. The observed resonance splitting for the structure of the second type is explained by the bimodal size distribution of NPs with maxima of 40 and 140 nm. Thus, various sensors can be created on the basis of the manufactured structures, provided that the Q factor of the resonances under consideration increases by reducing the width of the NPs size distribution and their tendency to agglomeration and aggregation.

Acknowledgments

This research was funded by the Ministry of Science and Higher Education of the Russian Federation (state contract no. 075-03-2021-095, project identifier 0714-2021-0007, project title: “Development of functional materials with controlled electrical, chemoresistive and catalytic properties for manufacturing sensor microsystems by using methods of printed electronics”).

References

[1] Mock J J, Barbic M, Smith D R, Schultz D A and Schultz S 2002 Shape effects in plasmon resonance of individual colloidal silver nanoparticles *J. Chem. Phys.* **116** 6755–9
[2] Mie G 1908 Beiträge zur Optik trüber Medien, speziell kolloidaler Metallösungen *Ann. Phys.* **330** 377–445
[3] Shaimanov A N, Khabarov K M, Merzlikin A M, Bykov I V and Baryshev A V 2017 Plasmon resonances in a two-dimensional lattice of metal particles in a dielectric layer: Structural and polarization properties *J. Exp. Theor. Phys.* **124** 584–91
[4] Humphrey A D and Barnes W L 2014 Plasmonic surface lattice resonances on arrays of different lattice symmetry *Phys. Rev. B* **90** 075404
[5] Neuberger T, Schöpf B, Hofmann H, Hofmann M and von Rechenberg B 2005 Superparamagnetic nanoparticles for biomedical applications: Possibilities and limitations of a new drug delivery system *J. Magn. Magn. Mater.* **293** 483–96

[6] V. Zelepukin I, V. Yaremenko A, O. Shipunova V, V. Babenyshev A, V. Balalaeva I, I. Nikitin P, M. Deyev S and P. Nikitin M 2019 Nanoparticle-based drug delivery via RBC-hitchhiking for the inhibition of lung metastases growth *Nanoscale* **11** 1636–46

[7] Parak W J, Gerion D, Pellegrino T, Zanchet D, Micheel C, Williams S C, Boudreau R, Gros M A L, Larabell C A and Alivisatos A P 2003 Biological applications of colloidal nanocrystals *Nanotechnology* **14** R15–27

[8] Reuveni T, Motiei M, Romman Z, Popovtzer A and Popovtzer R 2011 Targeted gold nanoparticles enable molecular CT imaging of cancer: an in vivo study *Int. J. Nanomedicine* **6** 2859–64

[9] Pallares R M, Thanh N T K and Su X 2019 Sensing of circulating cancer biomarkers with metal nanoparticles *Nanoscale* **11** 22152–71

[10] Maier S A 2007 *Plasmonics: Fundamentals and Applications* (Springer Science & Business Media)

[11] Lednický T and Bonyár A 2020 Large Scale Fabrication of Ordered Gold Nanoparticle–Epoxy Surface Nanocomposites and Their Application as Label-Free Plasmonic DNA Biosensors *ACS Appl. Mater. Interfaces* **12** 4804–14

[12] Dipalo M, C. Messina G, Amin H, Rocca R L, Shalabaeva V, Simi A, Maccione A, Zilio P, Berdondini L and Angelis F D 2015 3D plasmonic nanoantennas integrated with MEA biosensors *Nanoscale* **7** 3703–11

[13] Seshan K 2001 *Handbook of Thin Film Deposition* (William Andrew)

[14] Khabarov K, Urazov M, Lizunova A, Kameneva E, Efimov A and Ivanov V 2021 Influence of Ag Electrodes Asymmetry Arrangement on Their Erosion Wear and Nanoparticle Synthesis in Spark Discharge *Appl. Sci.* **11** 4147

[15] Efimov A A, Potapov G N, Nisan A V and Ivanov V V 2017 Controlled focusing of silver nanoparticles beam to form the microstructures on substrates *Results Phys.* **7** 440–3

[16] Protas N V, Efimov A A, Zemlyanoy V K and Ivanov V V 2018 Computer simulation of aerosol nanoparticles focusing and deposition process for functional microstructure fabrication *J. Phys. Conf. Ser.* **1124** 081033

[17] Arsenov P V, Efimov A A and Ivanov V V 2018 Effect of Methods of Changing in Focusing Ratio on Line Geometry in Aerosol Jet Printing *Key Eng. Mater.* **779** 159–64

[18] Bilankohi S M 2015 Optical Scattering and Absorption Characteristics of Silver and Silica/Silver Core/shell Nanoparticles *Orient. J. Chem.* **31** 2259–63

[19] Sattler K D 2016 *Handbook of Nanophysics: Nanoparticles and Quantum Dots* (CRC Press)

Neural networks application to determine the types and magnitude of aberrations from the pattern of the point spread function out of the focal plane

P A Khorin[1], A P Dzyuba[3], P G Serafimovich[1,2] and S N Khonina[1,2]

[1] Samara National Research University, 34, Moskovskoye Shosse, 443086, Samara, Russia
[2] IPSI RAS - branch of the FSRC «Crystallography and Photonics» RAS, 151 Molodogvardejskaya street, 443001, Samara, Russia
[3] Saint-Petersburg National Research University of Information Technologies, Mechanics and Optics, 49, Kronverksky Prospekt, 197101, Saint Petersburg, Russia

paul.95.de@gmail.com

Abstract. Recognition of the types of aberrations corresponding to individual Zernike functions were carried out from the pattern of the intensity of the point spread function (PSF) outside the focal plane using convolutional neural networks. The PSF intensity patterns outside the focal plane are more informative in comparison with the focal plane even for small values/magnitudes of aberrations. The mean prediction errors of the neural network for each type of aberration were obtained for a set of 8 Zernike functions from a dataset of 2 thousand pictures of out-of-focal PSFs. As a result of training, for the considered types of aberrations, the obtained averaged absolute errors do not exceed 0.0053, which corresponds to an almost threefold decrease in the error in comparison with the same result for focal PSFs.

1. Introduction

The task of detecting, recognizing and compensating for wavefront aberrations [1, 2] is closely related to the problem of improving the resolution of optical systems [3-8]. The solution to this problem is relevant in many applications, including microscopy [9-15], astronomy [16-20], and ophthalmology [21-23], therefore it has a long history and many different approaches and methods.

One of the tools for describing wavefronts, in addition to the deviation patterns, is the decomposition of aberrations over various bases. The most famous decomposition bases include Zernike polynomials [24-28], as well as Seidel aberrations. However, direct measurement of aberration coefficients is possible only for low levels of aberrations [28].

Recently, to solve the problem of recognition and compensation of optical system aberrations from the PSF intensity pattern in a certain plane, the means of intellectual analysis and machine learning are increasingly being used [29-35]. Some works [33, 34] showed the problems of using focal PSF at low levels of aberrations, when the intensity pattern is almost indistinguishable from the focal spot.

The purpose of this work is to investigate the possibility to improve the results of using neural networks to solve the problem of recognizing the type and magnitude of wavefront aberration from the PSF pattern outside the focal plane. The architecture of the neural network corresponded to the Xception [36] adapted for the regression problem.

Thus, in this work, we use a neural network to solve the regression problem [36]. In a regression problem, we aim to predict the outcome of a continuous value such as probability or price. This problem differs from the classification problem, where we seek to select a discrete quantity from a list of classes. A numerical experiment showed that, trained on 8 separate types of aberrations, matched with the Zernike functions, the neural network confidently copes with the classification task, i.e. detection in a superposition of two different types of aberrations of their presence. The obtained averaged absolute errors do not exceed 0.0053, which corresponds to an almost threefold decrease in the error as compared to the analogous result for focal PSFs.

2. Theoretical foundations

Wavefront (WF) aberrations are usually described in terms of Zernike functions $Z_{nm}(r,\varphi)$:

$$W(r,\varphi)=\exp\left[i\psi(r,\varphi)\right],\quad \psi(r,\varphi)=2\pi\alpha\sum_{n=0}^{n_{\max}}\sum_{m=0}^{n}C_{nm}Z_{nm}(r,\varphi), \tag{1}$$

where,

$$Z_n^m(r,\phi)=A_nR_n^m(r)\begin{Bmatrix}\cos(m\varphi)\\ \sin(m\varphi)\end{Bmatrix}, \tag{2}$$

$A_n=\sqrt{(n+1)/\pi}$ is the normalization factor, $R_n^m(r)$ are the radial Zernike polynomials [1], and α is the wave aberration level.

PSF outside the focal plane z can be calculated using the Fourier transform with the addition of the defocusing function to the wavefront (1):

$$G(u,v,z)=\int_{-\infty}^{\infty}\int_{-\infty}^{\infty}W(x,y)\exp\left[\pi i/_{\lambda z}(x^2+y^2)\right]\exp\left[-2\pi i(ux+vy)\right]dxdy\cdot \tag{3}$$

Table 1. Fragment of a dataset depicting extrafocal PSFs of various types of individual wave aberrations for λ = 532 nm, f =100 mm is the focal plane.

(n.m)	α	WF(phase) z=0 / PSF(intensity) z=f, z=105 mm	*(n.m)*	α	WF(phase) z=0 / PSF(intensity) z=f, z=105 mm
	0.10	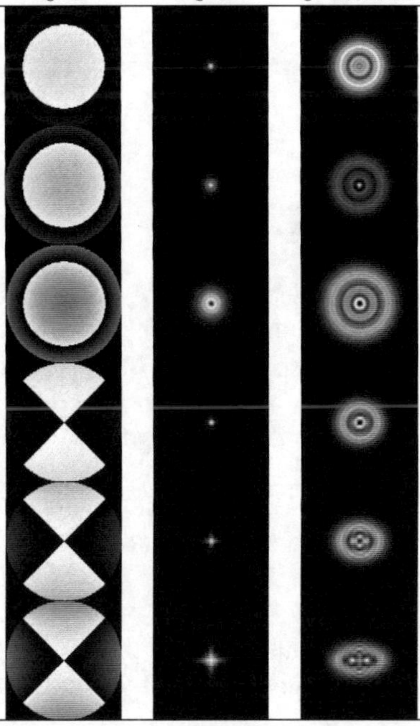		0.10	
(2.0)	0.25		(3.1)	0.25	
	0.50			0.50	
	0.10			0.10	
(2.2)	0.25		(3.3)	0.25	
	0.50			0.50	

Table 1 shows a fragment of the dataset and the stages of its modeling. Similar intensity distributions were calculated for aberrations up to 4 orders in [37, 38]. A phase consisting of one Zernike polynomial (2) with a weight coefficient $C_{nm} = 1$ and a variable level of wave aberration α was applied to the input to the mathematical model of the aberrated wavefront (1). Then, using the Fourier transform with the addition of the defocusing function (3) to the wavefront (1), the PSF outside the focal plane was calculated. In a number of numerical experiments, the most optimal value of the parameter z turned out to be equal to 105 mm at a wavelength of $\lambda = 532$ nm.

It can be seen from the intensity distribution that outside the focal plane the "character" of the distribution is preserved, but it becomes more pronounced, including at small aberrations.

3. Results and discussion

A dataset with 2 thousand images of out-of-focal PSFs (Table 1) was used to train the neural network. It was divided in a 4:1 ratio into training and validation sets.

To train the neural network, it was decided to use the Google Colab service. The Keras library was used as a machine-learning library, which is an add-on over the Deep learning4j, TensorFlow and Theano libraries. Single-channel images with a size of 128 × 128 pixels were fed to the input of the neural network, examples of which are presented in Table 1. The complete architecture of Xception is shown in Figure 1.

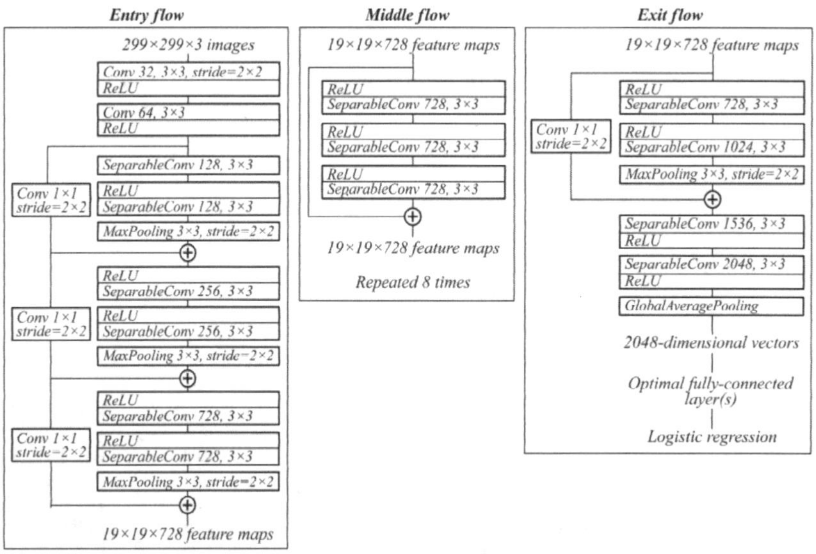

Figure 1. Complete Xception architecture.

To solve the problem of recognizing aberrations, it was decided to use the ready-made Xception architecture, presented in 2015 [36]. A distinctive feature of this model is the use of the Depthwise separable convolution. The essence of this technique is to split the transformations over the input tensor on each layer into two stages:

- at the first stage, the tensor is convoluted with a 1 × 1 kernel. This operation is called "pointwise convolution";
- at the second stage, standard convolution is performed. After that, the results of the convolution of the tensor obtained as a result of applying the "pointwise convolution" operation to the input tensor are summed up with all the filters of the layer and form the output tensor of the layer. This operation is called "depthwise convolution".

This technique allows you to significantly reduce the number of weights in the neural network, which significantly affects the learning rate required for training resources and the weight of the pretrained model. Thus, the Xception architecture is the best choice, providing sufficient training accuracy, while requiring significantly less resources.

Adam was chosen as the optimization algorithm, the initial learning factor of which was 10^{-4}. The loss function was the mean absolute error. The learning process was carried out in stages with a stepwise decrease in the learning coefficient of the Adam algorithm. The number of epochs was 250. Data augmentation using noise and random image rotation in the range from 0 to 45 degrees did not bring any tangible improvement in the results. The achieved indicators of prediction errors for all aberrations of a certain type - mean absolute error (MAE) and mean square error (MSE), are given in Table. 2.

Table 2. Correspondence of Zernike functions to prediction errors.

(n,m)	MAE	MSE
(1,1)	0.0053	0.0020
(2,0)	0.0047	0.0018
(2,2)	0.0042	0.0016
(3,1)	0.0045	0.0017
(3,3)	0.0042	0.0016
(4,0)	0.0046	0.0017
(4,2)	0.0043	0.0016
(4,4)	0.0041	0.0015

The computational experiment showed that the trained neural network confidently copes with the classification task, i.e. identifying two non-zero types of aberrations for each image. The main contribution to the error is made by the error in determining the value of the coefficient of two nonzero aberrations. In addition, the dataset calculated in the out-of-focal plane made it possible to determine the level of aberration about 3 times better than the same result for focal PSF [33].

4. Conclusion

In this work, PSF patterns were simulated outside the focal plane for a coherent case, and a neural network training process was carried out to recognize the type and magnitude of different types of aberrations. The use of out-of-focal PSFs made it possible to almost 3 times reduce the recognition error in comparison with the same result for focal PSFs

The most obvious way to reduce the error in recognizing a single wave aberration is to increase the dataset. In addition, it is planned to consider a dataset calculated in the out-of-focus plane for a superposition of two different types of aberrations. In the future, a full-fledged regression problem will be solved to determine the superposition of aberrations from the PSF pattern.

Acknowledgments

This work was supported by the Russian Foundation for Basic Research (grants 20-37-90129, 19-29-09054).

References

[1] Born M and Wolf E 1968 *Principlies of Optics* (Oxford: Pergamon Press)

[2] Welford W T 1986 *Aberrations of optical systems* (Bristol, Philadelphia: Adam Hilger Press)

[3] Yang W and Kotinski A B 2004 One-sided achromatic phase apodization for imaging of extra solar planets *The Astrophysical Journal* **605** 892-901

[4] Reddy A N K, Sagar D K, Khonina S N 2017 Complex pupil masks for aberrated imaging of closely spaced objects *Optics and Spectroscopy* **123** 940–949

[5] Belov A M, Denisova A Y 2018 Spectral and spatial superresolution method for Earth remote sensing image fusion *Computer Optics* **42** 855-863

[6] Reddy A N K, Khonina S N 2018 Apodization for improving the two-point resolution of

coherent optical systems with defect of focus *Applied Physics B* **124** 229-(9p)

[7] Reddy A N K, Martinez-Corral M, Khonina S N, Karpeev S V 2018 Focusing of light beams with the phase apodization of the optical system *Computer Optics* **42** 620-626

[8] Dzyuba A 2020 Calculation and research of point spread function of apodized optical system in order to compensate for defocusing *Proceedings of ITNT 2020 - 6th IEEE International Conference on Information Technology and Nanotechnology* **1** 9253262

[9] Albert O, Sherman L, Mourou G, Norris T B, Vdovin G 2000 Smart microscope: an adaptive optics learning system for aberration correction in multiphoton confocal microscopy *Opt. Lett.* **25** 52–54

[10] Marsh P N, Burns D and Girkin J M 2003 Practical implementation of adaptive optics in multiphoton microscopy *Opt. Express* **11** 1123

[11] Booth M J 2007 Adaptive optics in microscopy *Phil. Trans. R. Soc.* **365** 2829

[12] Thomas B, Wolstenholme A, Chaudhari S N, Kipreos E T, Kner P 2015 Enhanced resolution through thick tissue with structured illumination and adaptive optics *J. Biomed. Opt.* **20** 026006-026006

[13] Booth M J 2014 Adaptive optical microscopy: the ongoing quest for a perfect image *Light Sci. Appl.* **3** e165

[14] Ji N 2017Adaptive optical fluorescence microscopy *Nat. Methods* **14** 374–380.

[15] Beckers J M 1993 Adaptive optics for astronomy: principles, performance, and applications *Ann. Rev. Astron. Astrophys.* **31** 13-62

[16] Ellerbroek B L and Vogel C R 2009 Inverse problems in astronomical adaptive optics *Inverse Problems* **25** 063001

[17] Davies R and Kasper M 2012 Adaptive optics for astronomy *Ann. Rev. Astron. Astrophys.* **50** 305

[18] Klebanov I M, Karsakov A V, Khonina S N, Davydov A N, Polyakov K A 2017 Wavefront aberration compensation of space telescopes with telescope temperature field adjustment *Computer Optics* **41** 30-36

[19] Rastorguev A A, Kharitonov S I, Kazanskiy N L 2018 Modeling of arrangement tolerances for the optical elements in a spaceborne Offner imaging hyperspectrometer *Computer Optics* **42** 424-431

[20] Atchison D A 2009 Wavefront aberrations and their clinical application *Clin. Exp. Optom.* **92** 171-172

[21] Khorin P A, Khonina S N, Karsakov A V, Branchevskiy S L 2016 Analysis of corneal aberration of the human eye *Computer Optics* **40** 810-817

[22] François Hénault, Alain Spang, Yan Feng and Laura Schreiber 2020 Crossed-sine wavefront sensor for adaptive optics, metrology and ophthalmology applications *Eng. Res. Express* **2** 015042

[23] Love G D 1997 Wavefront correction and production of Zernike modes with a Liquid crystal spatial light modulator *Appl. Opt.* **36** 1517-1525

[24] Khonina S N, Kotlyar V V, Soifer V A, Wang Y, Zhao D 1998 Decomposition of a coherent light field using a phase Zernike filter *Proc. SPIE* **3573** 550-553

[25] Booth M J 2003 Direct measurement of Zernike aberration modes with a modal wavefront sensor *Proc. SPIE* **5162** 79-90

[26] Sheppard C J R 2015 Zernike expansion of pupil filters: optimization of the signal concentration factor *J Opt Soc Am A* **32** 928-933

[27] Khonina S N, Karpeev S V, Porfirev A P 2020 Wavefront aberration sensor based on a multichannel diffractive optical element *Sensors* **20** 3850

[28] Guo H, Korablinova N, Ren Q, Bille J 2006 Wavefront reconstruction with artificial neural networks *Opt Express* **14** 6456-6462

[29] Nikonorov A V, Petrov M V, Bibikov S A, Kutikova V V, Morozov A A, Kazanskiy N L 2017 Image restoration in diffractive optical systems using deep learning and deconvolution

Computer Optics **41** 875-887
[30] Paine S W, Fienup J R 2018 Machine learning for improved image-based wavefront sensing *Opt Lett* **43** 1235-1238
[31] Dzyuba A P 2019 Optical phase retrieval with the image of intensity in the focal plane based on the convolutional neural networks *J Phys Conf Ser* **1368** 022055
[32] Nishizaki Y, Valdivia M, Horisaki R, Kitaguchi K, Saito M, Tanida J, Vera E 2019 Deep learning wavefront sensing *Opt Express* **27** 240-251
[33] Rodin I A, Khonina S N, Serafimovich P G, Popov S B 2020 Recognition of wavefront aberrations types corresponding to single Zernike functions from the pattern of the point spread function in the focal plane using neural networks *Computer Optics* **44** 923-930
[34] Evdokimova V V, Petrov M V, Klyueva M A, Zybin E Y, Kosianchuk V V, Mishchenko I B, Novikov V M, Selvesiuk N I, Ershov E I, Ivliev N A, Skidanov R V, Kazanskiy N L, Nikonorov A V 2021 Deep learning-based video stream reconstruction in mass production diffractive optical systems *Computer Optics* **45** 130-141
[35] Chollet F 2017 Xception: Deep learning with depthwise separable convolutions *Proc IEEE Conf on Comp Vis Pattern Recogn* 1251-1258
[36] Fischer P, Dosovitskiy A, Brox T 2015 Image orientation estimation with convolutional networks *In Book: German Conference on Pattern Recognition* 368-378 (Springer)
[37] Khorin P A 2019 Analysis wavefront propagating in free space based on the Zernike polynomials and Gauss-Laguerre modes expansion *J Phys Conf Ser* **1096** 012104
[38] Khorin P A, Podlopnov V V, Khonina S N 2020 Generation of scalable wavefront for testing optical systems *Proc of SPIE* **11516** 115161K

Thermo-optical properties of nanophotonic devices with carbon nanotube films

S Komrakova[1], P An[2], V Kovalyuk[2], A Golikov[2,3] Y Gladush[4], A Mkrtchan[4], A Nasibulin[4] and G Goltsman[1,2]

[1] National Research University Higher School of Economics, 101000, Moscow, Russia
[2] Department of Physics, Moscow Pedagogical State University, 119992, Russia
[3] Moscow Institute of Physics and Technology (State University), 141700, Russia
[4] Skolkovo Institute of Science and Technology, Nobel Street 3, 43026 Moscow, Russia

Abstract. Here, thermo-optical properties of hybrid nanophotonic circuits SWCNTs/SiN were investigated by studying the temperature dependence of the resonance wavelengths. After experimental and theoretical study, we found the thermo-optical coefficient of SWCNTs film is equal to 2.02 10-6 RIU/0C.

1. Introduction

Integrated optical devices operating in the near-infrared wavelength range have become a popular platform for sensing applications, communication, and optical signal processing, and for studying the interaction of light and matter [1]. One of the main tasks of these schemes is to match the spectral responses of all-optical devices. Despite the advanced fabrication technologies, the inhomogeneity of fabrication leads to a discrepancy between the spectral responses [2]. One of the ways for solving the problem is the use of metallic microheaters [3]. However, metals absorb near-infrared wavelengths, so they must be placed a few microns away from the devices being tuned to avoid degrading the performance of optical devices. Thus, heating is not localized and results in thermal crosstalk between multiple optical devices. Silicon nitride has proven as a platform for thermal tuning. On the one hand, due to its wide bandgap (5 eV) and low optical loss (<0.1 dB/ cm), the silicon nitride platform, is mainly used to create high-end on-chip passive photonic devices [4]. On the other hand, due to its weak nonlinearity, active optical elements should be performed based on additional nonlinear materials. A promising candidate for waveguide integration is single-walled carbon nanotubes (SWCNTs) films because of their unique electrical [5], mechanical [5], and optical [6] properties. Integration of silicon nitride waveguides with SWCNT films for thermal tuning can be eliminated from the disadvantages of metal heaters since the film can be placed directly on the waveguide. The development of photonic devices based on SWCNTs requires knowledge of the thermo-optical properties of films. Since the expected value of the thermo-optical coefficient is small, it cannot be measured by direct methods. Therefore, we made the SiN O-ring resonators (ORRs) with SWCNTfilm cells atop. Since the SiN thermo-optical coefficient is also small enough (dn/dT)Si3N4 =2.51 × 10-5 [7], due to the shifts of resonances in the transmission spectrum of OR, such a device provides a possibility to find refractive index and small thermo-optic coefficient of a hybrid waveguide structure (SiN/SWCNTs).

2. Device design and fabrication

For device fabrication, we used commercially available 540 μm thickness silicon wafers with 2.6 μm thermally oxidized and low-pressure chemical vapor deposited (LPCVD) 450 nm layer of Si_3N_4 atop. The ORRs were formed by direct e-beam lithography. E-beam resist ZEP 520A was used as a sensitive layer. Silicon nitride was etched at 225 nm in a CHF_3 plasma atmosphere. The resonator waveguide width of 1.6 μm, the ring length of 400.417 μm, and the 1.4 μm gap between the ring and bus waveguide were chosen, supporting a free spectral range (FSR) of about 3 nm. Focusing grating couplers are used for the input/output of light and have been optimized for wavelength 1565 nm and provide a coupling efficiency of 18 %.

To measure the properties of SWCNTs films were deposited on the nanophotonic chips. The films were made by the aerosol (floating catalyst) chemical vapor deposition (ACVD) method [8]. In this method, nanotubes grow on aerosol particles in the hot zone of the reactor and collect on a filter at the outlet, forming thin uniform films that do not require further purification. Further, the nanotube film was transferred to photonic device by dry transfer. Fabrication steps are shown in Figure 1(a). The nanotubes were grown so that the diameter corresponded to the S_{11} optical transition in the region of 1550 nm. The thickness of the film was measured as 22±4 nm. The AFM image of the SWCNTs film on the waveguide is shown in Figure 1(a). As can be seen from the image, the nanotubes in the film, intertwining with each other, completely cover the waveguide from all sides. In the end, the rectangular cells of SWCNT films using laser lithography and O_2 plasma etching were finalized (Figure 1 (c)). The SWCNTs cell width is 350 nm.

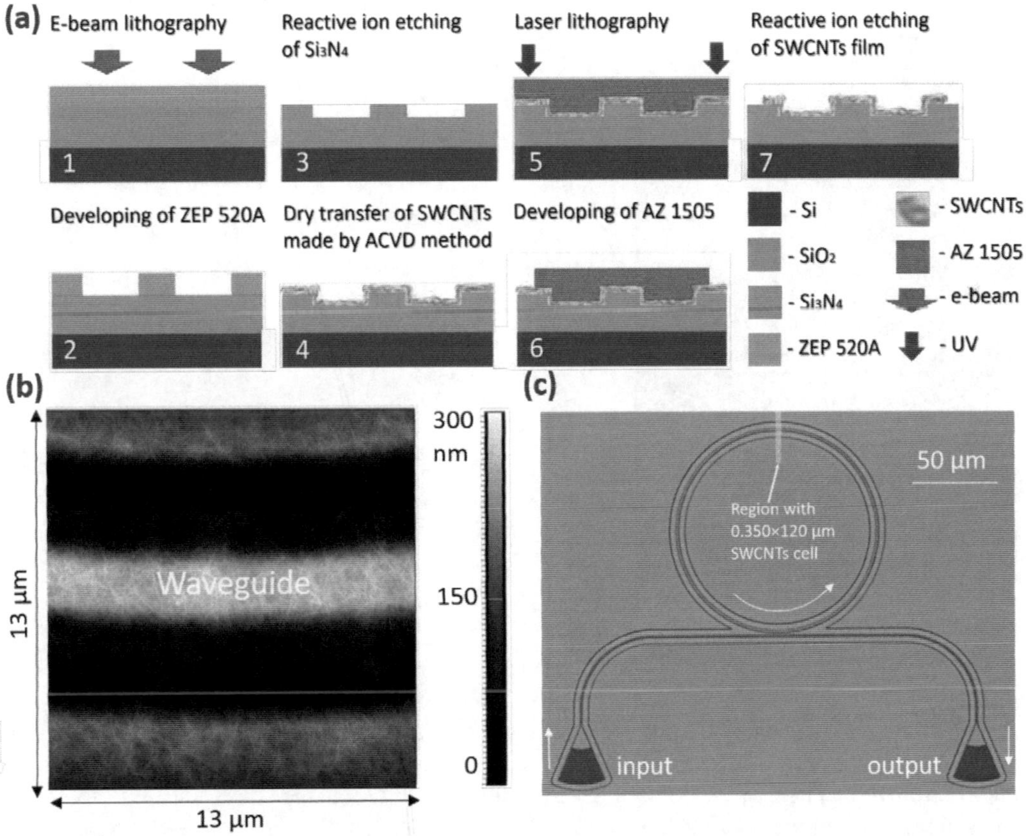

Figure 1. **(a)** Schematic view of the device fabrication process; **(b)** AFM image of SWCNTs film atop of SiN waveguide; **(c)** Optical micrograph of the fabricated device. White arrows show the propagation of light. The transparent light blue area shows the SWCNTs film cell.

3. Experimental results

To study the thermo-optical properties of the SWCNTs film, the transmission spectra of the fabricated devices were measured at different temperatures. The obtained spectra were analyzed andusing finite element method, the thermo-optical coefficient of the SWCNTs film was obtained.

3.1. Experimental setup

The measurements were performed using a setup, that included a tunable laser source (New Focus TLB-6600), an optical Erbium-Doped Fiber Amplifier (EDFA), polarization controller, an array of optical fibers aligned with on-chip focusing grating couplers by motorized x, y, z, θ – stage (Figure 2 (a)) and temperature stubilized using proportional integral derivative (PID) controller. This setup makes it possible to measure the transmission spectra of devices at different temperatures. The optical signal transmitted through the waveguide was recorded with a fast photodetector. One of the transmission spectra of our devices is shown in Figure 2 (b).

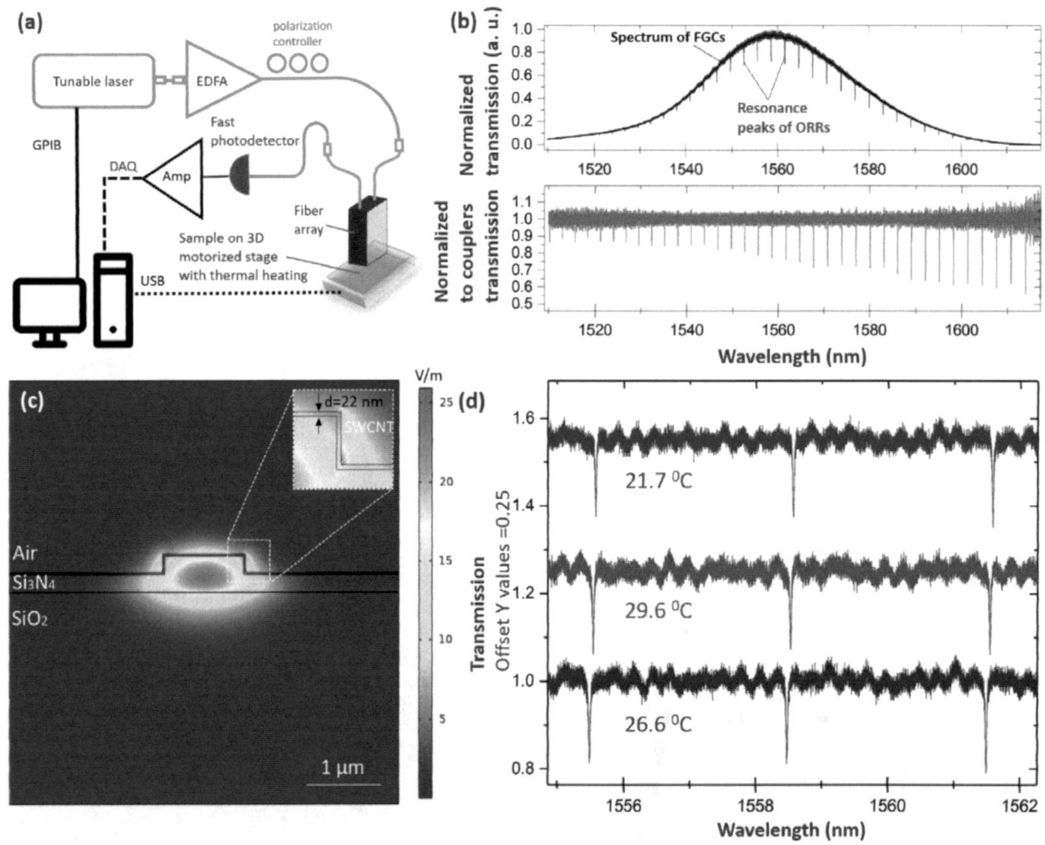

Figure 2. (a) Experimental setup for measurements of the thermo-optical characteristics of fabricated hybrid waveguide devices with SWCNTs. Electric cables are shown by black lines, orange and blue are optical fibers; **(b)** Measured transmission spectrum of one of the fabricated O-ring resonator with SWCNTs film (top), the same spectrum normalized to the transmission of FGC (bottom); **(c)** The cross-section of a waveguide structure with SWCNTs atop. Color change shows field distribution for TE-like mode of waveguide. On the inset is an enlarged image showing the coverage of the film of the waveguide and its thickness; **(d)** Transmission spectra of one of O-ring resonator with SWCNTs for different temperatures. Along the Y-axisan offset of 0.25 is used.

3.2. Calculation method and results

Heating the sample led to a shift of resonance peaks in the transmission spectrum of the ORRs (Figure 2 (d)). Compared to ORR without nanotubes, the shift of the resonance wavelength occurs faster, which indicates that such a scheme is sufficiently sensitive to measuring the thermo-optical coefficient of the SWCNTs film. From the analysis of the transmission spectra, the shifts of the resonance peak wavelength with a temperature change were obtained. The wavelength shift without SWCNTs is equal to 1.922×10^{-2} nm·K^{-1}, as well as the wavelength shift with SWCNTs is equal to 1.941×10^{-2} nm·K^{-1} (Figure 3 (a)). Then using the formula: $\Delta n_{\rm eff} = \Delta\lambda_c(T)/2\pi Rm$, connected the order of interference m, waveguide parameters, and resonance peak wavelength the effective refractive index change from a temperature of the ORRs was extracted. The average value of $dn_{\rm eff}/dT$ extracted from the linear fit was found as value refractive index units per Celsius degree 1.6616×10^{-5} RIU/^{0}C.

The calculation of the thermo-optical coefficient was carried out by the finite element method implemented in the COMSOL Multiphysics® software. For this, the cross-section of a waveguide with nanotubes atop was constructed. In the model, the film covers the waveguide from all sides as a solid object. Then, using the literature data for $(dn/dT)_{\rm Si3N4} = 2.51\times10^{-5}$, $(dn/dT)_{\rm SiO2} = 0.96\times10^{-6}$ [7] and the cross-section simulation of the effective refractive index value, the value $(dn/dT)_{\rm SWCNTs}$ was selected in such a way that the theoretical value of the change in $n_{\rm eff}$ from temperature was consistent with the experimental one. The thermo-optical coefficient of the SWCNTs film is linear within the temperature range from 25 ^{0}C to 70 ^{0}C and equal to $2.02\times10^{-6}\pm0.25\times10^{-6}$ RIU/^{0}C (Figure 3 (b)).

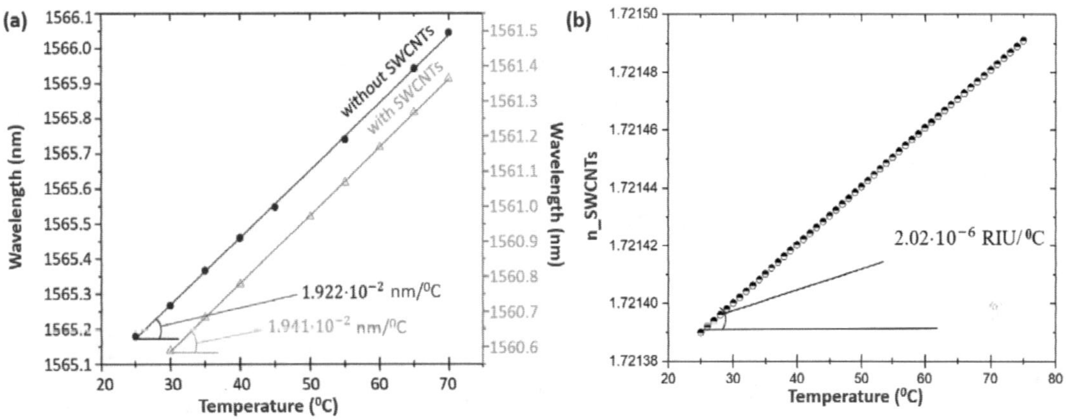

Figure 3. (a) Change in the resonance wavelength of a ring resonator with a change in temperature for a device with nanotubes (circles) and without (triangles). The lines are linear fit for each case, respectively; **(b)** The refractive index of SWCNTs film from temperature.

4. Conclusion

Here we demonstrated easy way for hybrid SiN/SWCNT on-chip integration Using experimental and theoretical studies, we found the thermo-optical coefficient of SWCNTs film as a function of temperature between 25 ^{0}C to 70 ^{0}C. The thermo-optical coefficient of the SWCNTs film is linear within this temperature range and equal to 2.02×10^{-6} RIU/^{0}C. The presented results provide new practical guidelines in designing photonic circuits for studying temperature optical phenomena. The proposed method can also be used to study other materials with a weak dependence of the refractive index on temperature.

Acknowledgments

We acknowledge support of the Russian Science Foundation, grant No. 19-72-10156 (experimental study), as well as the Russian Foundation for Basic Research, grants No. 18-29-27031 (waveguide fabrication).

References

[1] Prasad P, Selvaraja S and Varma M 2016 *Optics Letters* **41** 3153
[2] Selvaraja S, Jaenen P, Bogaerts W, Thourhout D, Dumon P and Baets R 2009 *Journal of Lightwave Technology* **27** 4076
[3] Ovvyan A, Gruhler N, Ferrari S and Pernice W 2016 *Journal of Optics* **18** 064011
[4] Blumenthal D, Heideman R, Geuzebroek D, Leinse A and Roeloffzen C 2018 *Proceedings of the IEEE* 1–23
[5] Sankar J and Udhaya K 2011 *European Journal of Scientific Research* **3** 342-358
[6] Pei S, Ma L and Cheng H 2014 *Advance Materials* **26** 1958−1991
[7] Elshaari A, Zadeh I, Jöns K and Zwiller V 2016 *IEEE Photonics Journal* **3** 1943-0655
[8] Ermolaev G, Tsapenko A, Volkov V, Anisimov A, Gladush Y and Nasibulin A 2020 *APL* **116** 231103

Traffic interception in fiber optical video-systems

Yu E Krivenko and E I Andreeva

The Bonch-Bruevich Saint-Petersburg State University of Telecommunications, St.Petersburg, 193232, Russian Federation

Abstract. In fiber-optic video systems, as well as in optical communication systems, standard single mode optical fibers (SSMF, standard G.652) are usually used. One of the advantages of these fibers is the ability to use CWDM in a wide spectrum. At the same time, more optimal near the wave-length of 1550 nm are provided by non-zero dispersion fiber (NZDSF, standard G.655) fibers. However, as studies have shown, these optical fibers have an increased sensitivity to bending. This fact can be used to traffic interception. It is shown that fiber-optics systems with SSMF have more protection from traffic interception than systems with NZDSF. To transmit a high-confidentiality video signal, special techniques, such as frequency modulation, can be used, or additional noise signals can be added.

1. Introduction

The use of the optical cable in the communication systems provides such advantages as information capacity, reliability, durability, and protection. To achieve these advantages, different types, and standards of fibers are used [1-3]. However, this does not exclude the interception of the transmitted information.

Often the optimization some system parameter is associated with difficulties in maintaining the proper level of other system parameters. One of the key issues in the design of a video surveillance system is the choice of the type of optical fiber as a direct signal transmission medium. The characteristics of the optical fiber must ensure the maximum of the transmission distance on the one hand, and the protection from the interception of the transmitted information on the other. To provide the transmission a high-quality video signal, it is necessary to transmit a wide- spectrum optical signal. Therefore, to transmit such optical signal over a long distance, fibers with a low losses and low dispersion can be used, such as NZDSF – nonzero dispersion shifted fibers (standard G.655). In fiber-optic video surveillance systems, as well as in optical communication lines, standard optical fibers (SSMF, Standard Single Mode Fiber, standard G.652) are most often used. One of the advantages of these fibers is the ability to use the CWDM wavelength division multiplexing in a wide spectrum range (from 1270 to 1610 nm). At the same time, the optimal parameters for losses and dispersion in the third transparency window (near the wavelength of 1550 nm) are provided by NZDSF fibers. However, as studies have shown, these fibers have an increased sensitivity to bending. This can lead to additional losses when laying optical cables and an increased risk of unauthorized access. To mitigate these risks, the Bending Loss Insensitive Fiber (BLIF – fibers, standard G.657) can be used.

The main methods of interception relate to the formation of optical informative signals by outputting a part of the optical information signal, e. g., by removing a part of the radiation when the fiber is bended.

To study the effect of bending of an optical fiber on its characteristics, a stand was assembled, the block diagram of which is shown in Figure 1.

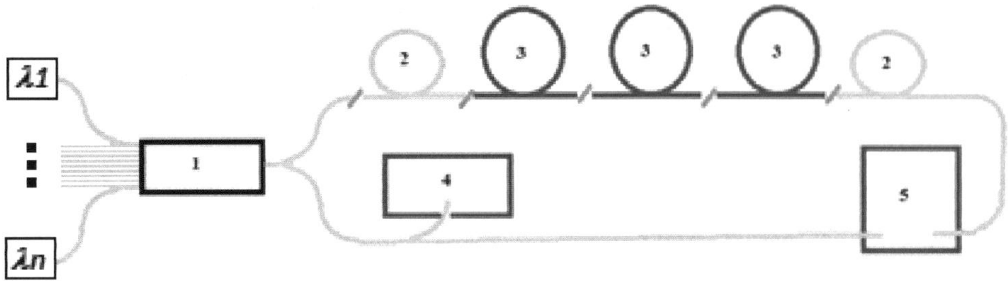

Figure 1. Block diagram of the measuring setup: 1 – WD-multiplexor, 2 - measuring optical fiber, 3 - tested optical fibers, 4 - optical spectrum analyzer, 5 - optical power meter.

ExperimentalAn experimental study of the level of bending losses was carried out by the substitution method. Testing was carried out in 2 stages. At the first stage, the spectral dependence of the level of introduced losses was recorded in a wide range of wavelengths: from 1270 to 1610 nm. CWDM SFP modules (Small Form-Factor Pluggable) were used as radiation sources. Registration of the optical signal was carried out by the differential method considering the spectral sensitivity of the photodetector [2]. Optical power measurements were carried out using a Rubin-300 stationary optical power meter. The source parameters were monitored using a Yokogawa AQ6370C optical spectrum analyzer. Results of the investigation of the bend sensitivity of the different fibers is presented in figure 2.

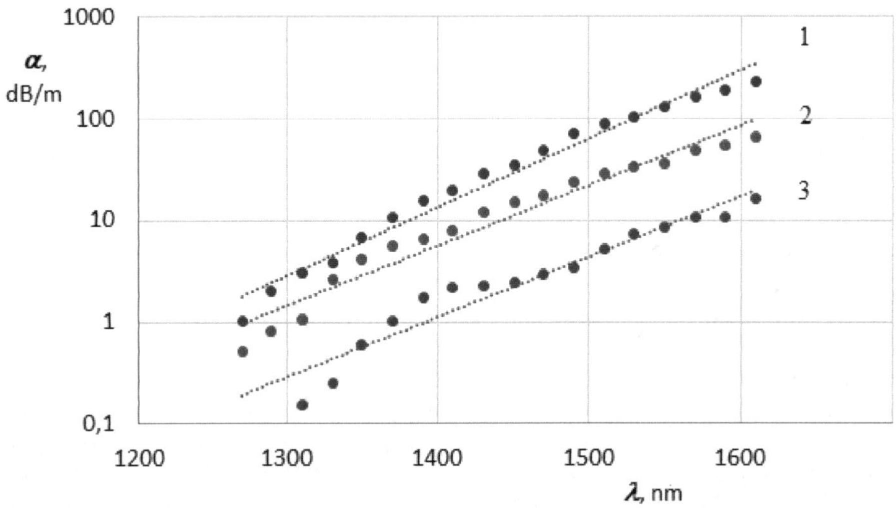

Figure 2. The bend loss α in dependence from wavelength λ. for NZDSF (standard G.655) (1), for SSMF (standard G.652) (2) and BLIF (standard G.657A2) (3) is presented. (The bend diameter is 12.5 mm.)

NZDSF has maximal bend sensitivity within studied fibers. So, this fiber has big risk of the information interception.

In systems with CWDM, traffic interception can be carried out on channel splitting devices. Adjacent channel isolation is nearly 30 dB. Non-adjacent channel isolation is nearly 40 dB. So transmitted signal can be registered in different output channel. To test the possibility of intercepting traffic, a stand was assembled, figure 3. The peart of the experimental setup is shown in figure 4. The possibility of intercepting traffic at the demultiplexor output has been experimentally confirmed.

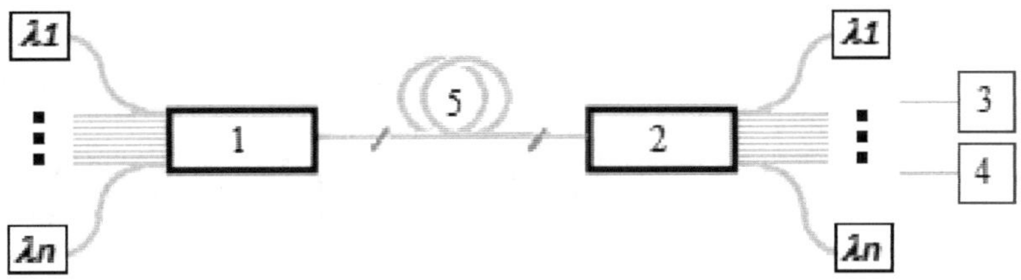

Figure 3. Block diagram of the measuring setup: 1 – WD-multiplexor, 2 - WD-demultiplexor, 3 - optical power meter, 4 - optical spectrum analyzer, 5 - optical fiber.

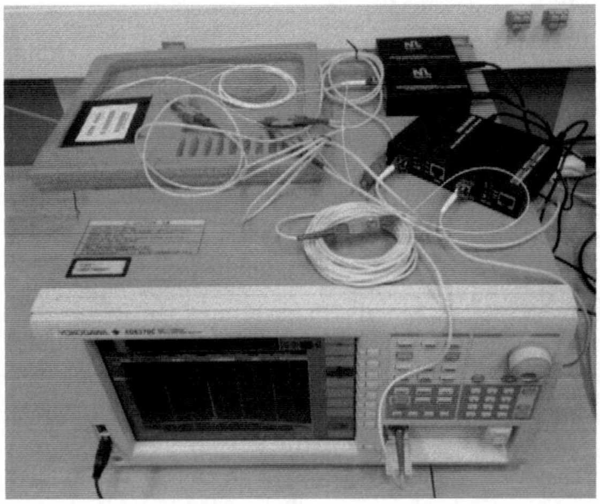

Figure 4. Experimental setup: CWD-multiplexor, CWD-demultiplexor, Yokogawa AQ6370C optical spectrum analyzer, optical fiber.

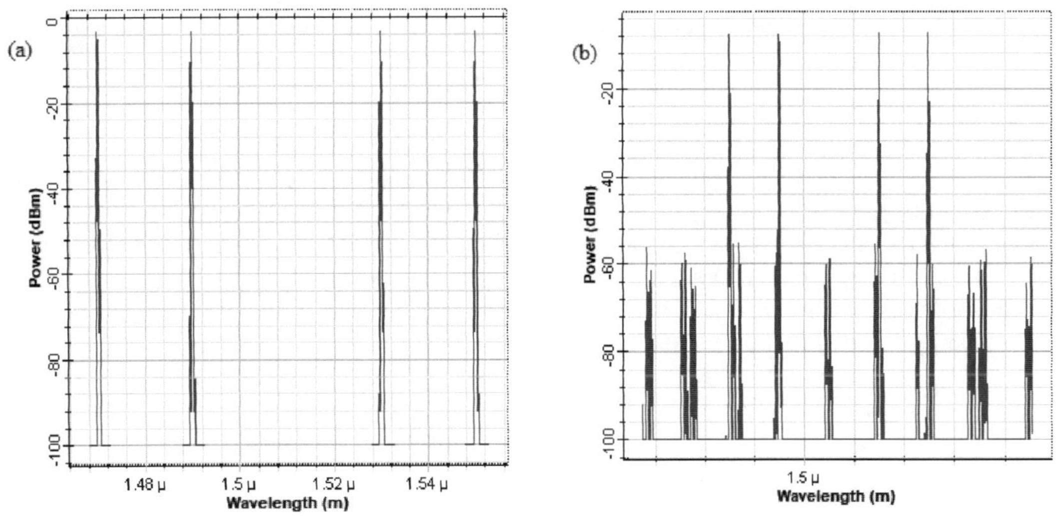

Figure 5. Optical signal at input (a) and output (b) optical fiber with dispersion D = 2 ps/nm/km. Signal wavelength: λ = 1470, 1490, 1530 and 1550 nm. Distance 20 km.

In WDM systems, part of the transmitted signal can fall into the adjacent spectral channel due to FWM (Four Wave Mixing). The simulation was carried out in the OptiSystem program. The optical signal was fed into the spectral channels of the CWDM grid: 1470 and 1490, 1530 and 1550 nm. After passing the effective loss manifestation length L_{eff} = 20 km, signals due to the original ones were observed in the adjacent spectral channels. The level of these signals is 53 ... 55 dB lower than that transmitted in NZDSF and 55 ... 60 dB lower than that transmitted in SSMF (figure 5). This fact reflects the difference in the nonlinearity coefficient of these optical fibers. Standard optical fibers are characterized by greater protection from unauthorized access then NZDS fibers.

Table 1. Nonlinear fiber coefficient.

Fiber	A_{eff}, μm^2	n_2, 10^{-20} m^2/W	n_2/ A_{eff}, 10^{-10} W^{-1}	$\gamma=(2\pi n_2)/(\lambda A_{eff})$, W^{-1}km^{-1}
SSMF	80	2,35	2,94	1,2
NZDSF	55	2,6	8,7	3,5

A_{eff} – the effective core area
n_2 – nonlinear refractive index
γ - nonlinearity coefficient

2. Conclusion

It is experimentally confirmed that NZDS-fibers (with low dispersion) of the G.655-standard are characterized by increased sensitivity to bending, as result of which they require greater security to avoid interception of the transmitted signal traffic. NZDSFs provide more long-distance transmission of video signal. At that time SSMFs provide greater protection against traffic interception.

It has been shown experimentally that interception of the transmitted signal traffic can occur in the CWDM system.

The question of the penetration of the fraction of the transmitted signal into adjacent channels in WDM-system due to nonlinear effects is investigated.

Comparative analysis shows that a standard optical fiber (SSMF) has a higher protection, while the signal transmission range can be increased using a dispersion-shifted fiber (NZDSF).

To transmit a high-privacy video signal, special modulation methods, such as frequency modulation, can be used, or additional noise signals can be added.

References

[1] Andreev D, Andreeva E, Sergeev A and Sumkin V 2020 Influence of Optical Cable Bending in Fiber Optic Systems Video Surveillance and Subscriber Access with Spectral Multiplexing *APINO Proc.* **1** pp 57–61

[2] Andreeva E, Valyukhov V and Kuptsov V 2019 Hight-Quality Video Transmission Over Fiber-Optic Network with CWDM-Technique *APINO Proc.* **1** pp 56–60

[3] Andreeva E, Valyukhov V and Kuptsov V 2018 Fibre Optics CCTV Systems and Security Alarms *APINO Proc.* **1** pp 57–61

Polarization-sensitive terahertz spectroscopy of graphene nanostructures

A Kvitsinskiy[1], M Rybin[2], A Zaitsev[1], K Bogdanov[1], D Zykov[1], and E Obraztsova[2,3]

[1] ITMO University, 49 Kronverkskiy Prospekt, Building A, Saint Petersburg, 197101, Russian Federation

[2] Prokhorov General Physics Institute of the Russian Academy of Sciences, 38 Vavilova Street, Moscow, 119991, Russian Federation

[3] Moscow Institute of Physics and Technology, 4 Nauchny Pereulok, Dolgoprudny, Moscow Region, 141701, Russian Federation

E-mail: anatolykvitsinskiy@gmail.com

Abstract. Efficient devices for control temporal and spatial properties of electromagnetic waves are essential for the development of terahertz (THz) technologies. But despite the great progress achieved in a study of graphene, the influence of the number of graphene layers on its optical and electrical properties in the THz frequency range has not yet been sufficiently studied. In this work, we experimentally studied optical and electrical properties of multilayer graphene (MLG) thin films in the frequency range 0.2–0.8 THz (corresponding to a wavelength range ~1.5–0.37 mm), at a controlled room temperature of 291 K, and a relative humidity of 40 %. Using our custom-made THz time-domain spectroscopic polarimetry system, we obtained spectra of the complex relative permittivity and the electrical conductance of the chemical vapor deposition graphene with ~14, ~40, and ~76 layers of graphene on borosilicate glass substrates. It is shown that the conductance increases nonlinearly with an increase in the graphene layer number and reaches, for ~76 layers, 0.06 S for the real, and 0.03 S for the imaginary part, respectively. Our results show that by using MLG it is possible to create tunable devices that can be used in the advanced areas of THz photonics.
Keywords: graphene, multilayer thin films, chemical vapor deposition, raman spectroscopy, terahertz time-domain spectroscopy, permittivity, electrical conductivity, polarization of light.

1. Introduction

In recent years, there has been a real breakthrough in the study of all possible carbon allotropes—fullerenes, carbon nanotubes, graphene and their heterostructures [1,2]. The latest nanotechnology allows us to engineer materials at the atomic level and create dynamically-controlled nanostructures for their real integration into all areas of our life, from genome editing to hyper-speed wireless Internet networks. And therefore, advanced fundamental and applied science is less and less within the rigid boundaries of the classical fields, and more and more becomes truly interdisciplinary [3].

One of these topical complex areas is the terahertz (THz) photonics of low-dimensional nanostructures. In connection with the development of sources and receivers of THz radiation, it became possible to develop compact systems for data transmission for the next generation of communication networks. But the problem of efficient manipulation of the temporal and spatial parameters of electromagnetic (EM) waves in the THz range in open space remains unresolved.

A promising method for solving this problem is the use of graphene, a material with proven unique optical and electronic properties, as an active medium for THz wave control devices [4, 5]. But despite the great progress achieved in the theoretical and experimental study of graphene-based nanostructures, the question of the influence of the graphene layer number on its fundamental properties has not yet been resolved, and there are still many dark spots in this scientific field.

The aim of our work was to experimentally study a room-temperature layer-number effect of multilayer graphene (MLG)-on-glass structures on its electrical conductance via polarization-resolved THz time-domain spectroscopy.

In this work, we experimentally studied optical and electrical properties of MLG thin films in the frequency range 0.2–0.8 THz (corresponding to a wavelength range ~1.5–0.37 mm), at a controlled room temperature of 291 K, and a relative humidity of 40 %. We found that the conductance of the MLG increases nonlinearly with an increase in the graphene layer number and reaches, for ~76 layers, 0.06 S for the real, and 0.03 S for the imaginary part, respectively. Our results show that by using MLG it is possible to create tunable devices that can be used in the advanced areas of THz photonics.

2. Materials and methods

2.1. Sample preparation

For our study, we selected three samples of MLG thin films with different number of graphene layers. Our graphene thin films were synthesized by the modified chemical vapor deposition (CVD) method [6] on the nickel foil. The samples were synthesized in a specially designed home-made system [7] that allows continuous monitoring of each synthesis parameter, including temperature and pressure. Such control makes it possible to obtain graphene samples with a high repeatability and a certain predetermined number of layers. Then the films were transferred onto the commercially available borosilicate cover glass substrates [8] with a thickness of ~150 µm.

2.2. Raman spectroscopy

To study the structural properties and qualitative composition of graphene films, we performed the Raman spectroscopy [9]. Experimental data were obtained using a standard commercially available confocal Raman microscope 'inVia™' (Renishaw, UK) using an excitation wavelength of 488 nm [10] in the backscattering geometry. The spectra were recorded at a room temperature, in the range from ~103 cm^{-1} to ~3503 cm^{-1}, with a step of ~2 cm^{-1}.

2.3. Polarization-resolved transmission terahertz spectroscopy

To study the optical and electrical properties of graphene thin films in the THz range, we used an upgraded spectroscopy system [11, 12]. This system is based on the THz radiation generation using a InAs crystal in a static magnetic field [13], and the detection using a CdTe crystal with the electro-optic sampling. Using a system of high-precision passive THz polarizers, we have added the ability to perform polarimetric measurements [14–16]. The spectra were recorded at a room temperature, a relative humidity of 40 %, in the range from 0.2 THz to 0.8 THz, with a step of ~0.03 ps.

3. Results and Discussion

3.1. Structural properties

Raman spectroscopy results are shown in Figure 1 (a). The obtained results of the Raman spectroscopy clearly show that with an increase in the number of graphene layers, an enhancement of the G-peak at ~1580 cm^{-1}, and a suppression of the 2D-peak at ~2710 cm^{-1} occurs, which indicates a transformation from two-dimensional carbon-based nanostructures to quasi-bulk graphite-like ones. The absence of a strong D-peak at ~1360 cm^{-1} in the spectrum indicates a high structural integrity of graphene films, despite the relatively large number of layers.

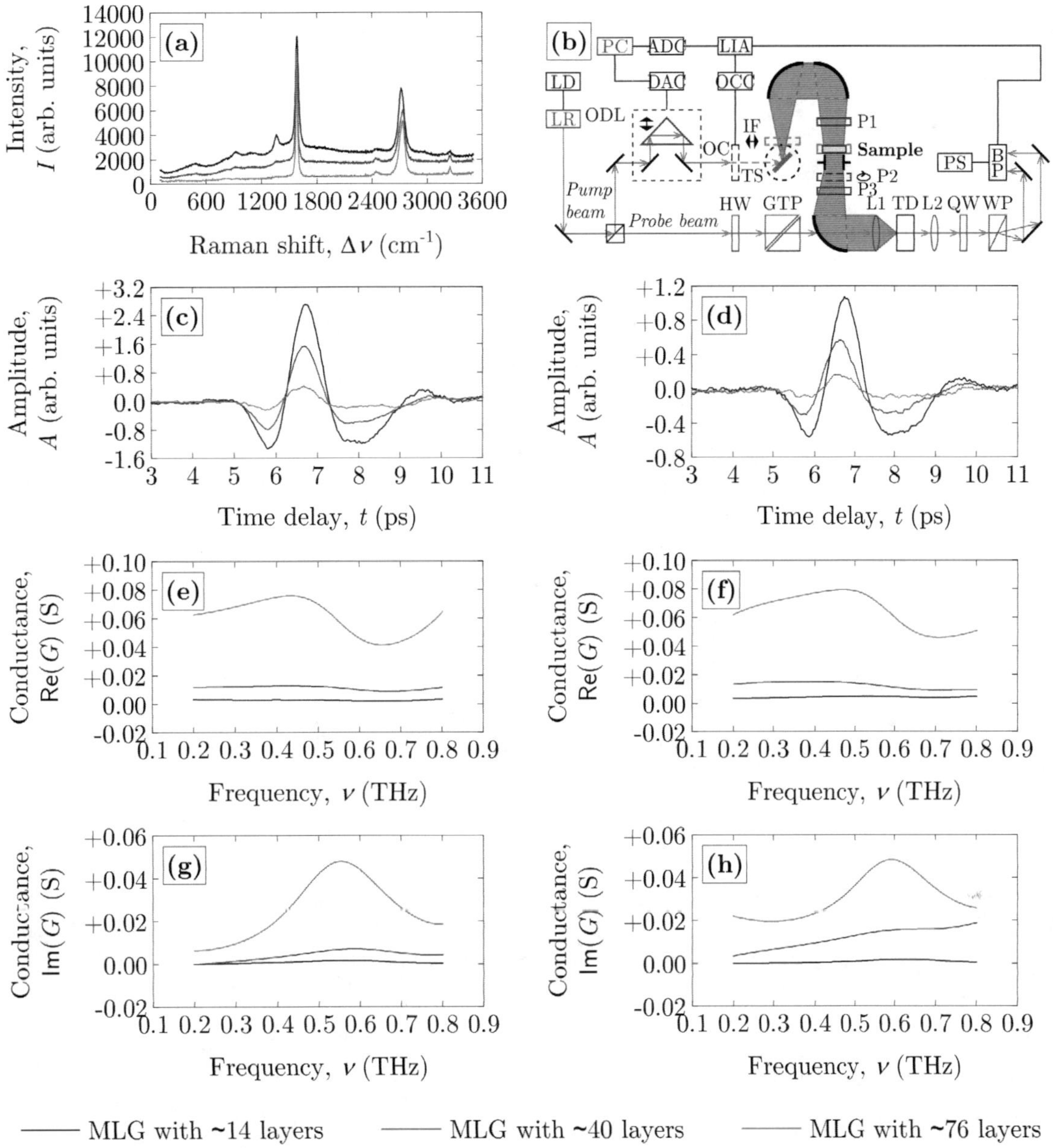

Figure 1. **(a)** Raman spectra of the MLG thin films; **(b)** scheme of the THz-TDSP setup (**ADC**—analog-to-digital converter; **BP**—balanced photodetector; **DAC**—digital-to-analog converter; **GTP**—Glan–Taylor prism; **HW**—half-wave plate; **IF**—poly(tetrafluoroethylene) ((C_2F_4)n) infrared cut-off filter; **L1**, **L2**—lenses; **LD**—laser diode driver; **LIA**—lock-in amplifier; **LR**—femtosecond Yb:KYW 1040 nm laser; **OC**—optical chopper; **OCC**—optical chopper controller; **ODL**—optical delay line; **P1**, **P3**—static polarizers; **P2**—rotatable polarizer; **PC**—personal computer; **PS**—balanced detector power supply; **QW**—quarter-wave plate; **TD**—THz radiation detector based on the CdTe crystal; **TS**—THz radiation source based on the InAs crystal; **WP**—Wollaston prism); typical temporal waveforms of the THz signals at P2 polarizer rotation angle of **(c)** 0°, and **(d)** 45°; the real part of the electrical conductance of the MLG thin films at P2 polarizer rotation angle of **(e)** 0°, and **(f)** 45°; and the imaginary part of the electrical conductance of the MLG thin films at P2 polarizer rotation angle of **(g)** 0°, and **(h)** 45°.

3.2. Terahertz properties
Terahertz spectroscopy results are shown in Figure 1 (c)–(h). For the analysis of the obtained waveforms of the THz waves transmitted through the samples we used a basic Tinkham thin-film calculations [17] and a modernized data processing techniques [18]. The use of THz polarizers (see Figure 1 (b)) with a higher extinction ratio relative to our previous work described in Ref. [19] made it possible to abandon strong signal noise reduction and, thus, to obtain more accurate results. We managed to obtain the electrical and optical properties, including the complex electrical conductance of the MLG thin films [20], based on THz polarimetric measurements. The results show that the conductance of the MLG films increases nonlinearly with an increase in the graphene layer number and reaches, for ~76 layers, 0.06 S for the real, and 0.03 S for the imaginary part, respectively. And the use of calculations at two angles of the P2 polarizer rotation makes it possible to determine the difference of these changes.

4. Conclusions
In this work, we experimentally studied optical and electrical properties of MLG thin films in the frequency range 0.2–0.8 THz (corresponding to a wavelength range ~1.5–0.37 mm), at a controlled room temperature of 291 K, and a relative humidity of 40 %. Using our custom-made THz-TDSP system, we obtained spectra of the complex relative permittivity and the electrical conductance of the CVD graphene with ~14, ~40, and ~76 layers of graphene on borosilicate glass substrates. Our results show that by using MLG it is possible to create tunable devices that can be used in the advanced areas of THz photonics.

Acknowledgments
The reported study was funded by the Russian Foundation for Basic Research (RFBR) according to the research project No. 20-32-90203 ('Terahertz electromagnetic waves polarization modulation in one- and two-dimensional carbon-based nanostructures').

Author contributions
A. K.: Conceptualization (lead); Data curation (lead); Formal analysis (equal); Funding acquisition (lead); Investigation (equal); Methodology (lead); Project administration (lead); Validation (lead); Visualization (lead); Writing—original draft (lead); Writing—review and editing (lead). **M. R.**: Formal analysis (equal); Investigation (equal); Resources (lead). **A. Z.**: Formal analysis (equal); Investigation (equal). Software (lead). **K. B.**: Formal analysis (equal); Investigation (equal). **D. Z.**: Formal analysis (equal); Investigation (equal). **E. O.**: Supervision (lead).

References
[1] Han P, Wang X and Zhang Y 2020 *Adv. Opt. Mater.* **8**(3) 1900533 URL `https://doi.org/10.1002/adom.201900533`
[2] Kužel P and Němec H 2020 *Adv. Opt. Mater.* **8**(3) 1900623 URL `https://doi.org/10.1002/adom.201900623`
[3] Lawler N B, Ho D, Evans C W, Wallace V P and Iyer K S 2020 *J. Mater. Chem. C* **8**(32) 10942–55 URL `https://doi.org/10.1039/D0TC01716G`
[4] Degl'Innocenti R, Kindness S J, Beere H E and Ritchie D A 2018 *Nanophotonics* **7**(1) 127–44 URL `https://doi.org/10.1515/nanoph-2017-0040`
[5] Ma Z T, Geng Z X, Fan Z Y, Liu J and Chen H D 2019 *Research* **2019** 6482975 URL `https://doi.org/10.34133/2019/6482975`
[6] Lee H C, Liu W W, Chai S P, Mohamed A R, Aziz A, Khe C S, Hidayah N M S and Hashim U 2017 *RSC Adv.* **7**(26) 15644–93 URL `https://doi.org/10.1039/C7RA00392G`
[7] Rybin M G, Kondrashov I I, Pozharov A S, Nguyen V C, Phan N M and Obraztsova E D 2018 *Phys. Status Solidi B* **255**(1) 1700414 URL `https://doi.org/10.1002/pssb.201700414`
[8] Uprety P, Junda M M and Podraza N J 2017 *Surf. Sci. Spectra* **24**(2) 026003 URL `https://doi.org/10.1116/1.4997898`

[9] Wu J B, Lin M L, Cong X, Liu H N and Tan P H 2018 *Chem. Soc. Rev.* **47**(5) 1822–73 URL https://doi.org/10.1039/C6CS00915H

[10] Bogdanov K, Fedorov A, Osipov V, Enoki T, Takai K, Hayashi T, Ermakov V, Moshkalev S and Baranov A 2014 *Carbon* **73** 78–86 URL https://doi.org/10.1016/j.carbon.2014.02.041

[11] Kvitsinskiy A, Demchenko P, Litvinov E, Masyukov M, Anoshkin I, Vozianova A and Khodzitsky M 2020 *Opt. Spectrosc.* **128**(8) 1262–71 URL https://doi.org/10.1134/S0030400X20080214

[12] Kvitsinskiy A, Demchenko P, Novoselov M, Anoshkin I, Bogdanov K, Baranov A and Khodzitsky M 2020 *J. Phys. Conf. Ser.* **1695** 012097 URL https://doi.org/10.1088/1742-6596/1695/1/012097

[13] Bespalov V G, Gorodetskiĭ A A, Denisyuk I Y, Kozlov S A, Krylov V N, Lukomskiĭ G V, Petrov N V and Putilin S E 2008 *J. Opt. Technol.* **75**(10) 636–42 URL https://doi.org/10.1364/JOT.75.000636

[14] Watanabe S 2018 *Photonics* **5**(4) 58 URL https://doi.org/10.3390/photonics5040058

[15] Kanda N, Konishi K and Kuwata-Gonokami M 2007 *Opt. Express* **15**(18) 11117–25 URL https://doi.org/10.1364/OE.15.011117

[16] Aschaffenburg D J, Williams M R C, Talbayev D, Santavicca D F, Prober D E and Schmuttenmaer C A 2012 *Appl. Phys. Lett.* **100**(24) 241114 URL https://doi.org/10.1063/1.4729148

[17] Tinkham M 1956 *Phys. Rev.* **104**(3) 845–6 URL https://doi.org/10.1103/PhysRev.104.845

[18] Zaitsev A D, Demchenko P S, Zykov D V, Korotina E A, Makarova E S, Tkhorzhevskiy I L, Tukmakova A S, Kablukova N S, Asach A V, Novotelnova A V and Khodzitsky M K 2020 *Appl. Sci.* **10**(8) 2724 URL https://doi.org/10.3390/app10082724

[19] Kvitsinskiy A, Demchenko P, Grebenchukov A, Litvinov E, Masyukov M, Zaitsev A, Baldycheva A, Kovalska E, Vozianova A and Khodzitsky M 2019 *SN Appl. Sci.* **1**(12) 1714 URL https://doi.org/10.1007/s42452-019-1748-x

[20] Lloyd-Hughes J and Jeon T I 2012 *J. Infrared Millim. Terahertz Waves* **33**(9) 871–925 URL https://doi.org/10.1007/s10762-012-9905-y

Peculiarities of ion-exchange in poled glasses

E A Lubyankina[1,2], D V Raskhodchikov[1,2], E S Babich[1,2], V P Kaasik[1,2] and A A Lipovskii[1,2]

[1] Department of Physics and Technology of Nanostructures, Alferov University, St. Petersburg 194021, Russia
[2] Institute of Physics, Nanotechnology and Telecommunications, Peter the Great St. Petersburg Polytechnic University, St. Petersburg 195251, Russia

katylubyankina@gmail.com

Abstract. We demonstrate for the first time that the results of ion exchange processing of thermally poled soda-lime glass essentially depend on the poling conditions. In particular, the processing of vacuum-poled soda-lime glass in silver-sodium nitrate melt results in the diffusion and reduction of silver ions followed by clustering silver nanoparticles in the subsurface layer of the glass after either ion-exchange or additional heat treatment of the ion-exchanged samples. Poling in air atmosphere with deposited gold film anode prevents silver ions penetration in the glass, but electric field stimulated diffusion of gold in this configuration leads to the formation of gold nanoparticles in the glass after heat treatment. It is also shown that corona poling of the glass in air atmosphere does not completely block silver penetration.

1. Introduction

The study of the processes taking place in thermal poling of glasses is of interest in connection with the possibility to use thermal poling for producing micro- and nanostructures for photonics, microfluidics and integrated optics [1]. It is known that the poling results in the redistribution of positive ions, the main charge carriers in glasses, which leads to the formation of spatial charge under the anodic surface of the poled glass [2]. Thermal poling of multicomponent glasses can differently modify the composition and structure of the subanodic region of glass, the modification being dependent on the poling mode. In the case of closed anode poling, which prevents access of atmospheric species to the anodic surface of glass, the subsurface region of the glass becomes depleted with alkali and alkali earth ions, and the latter form a peak of concentration beneath the depleted layer [3]. If open anode poling is used, the subanodic region of the glass is enriched with hydronium ions generated from atmospheric water vapor in the vicinity of the anodic electrode and penetrating into the glass during the poling [4]. Experimental data [5] indicate that the hydronium ions are more mobile than alkaline earth metal ions, and they occupy positions of alkali ions which shift in the glass bulk towards the cathode, while alkaline earth ions retain their position. Previous studies have shown that poling changes the resistance of glasses to chemical etching [6], allows the formation of phase patterns [7] or dissolves metal nanoparticles in glasses [8]. In this paper, we present the studies of the influence of thermal poling regime on silver ion exchange in soda-lime glass.

2. Experiments and results

We used 1 mm thick soda-lime glass slides "Menzel" purchased from Agar Scientific. The glass composition is presented in Table 1.

Table 1. Composition of Menzel glass in wt % of oxides.

SiO_2	Al_2O_3	Na_2O	K_2O	MgO	CaO	Others
72.2	1.2	14.3	1.2	4.3	6.4	0.33

We thermally poled the glass samples with electrodes presenting 1) pressed glassy-carbon plates, 2) deposited metal (55 nm gold and 200 nm chromium) films on the both sides of the slides, and poled using 3) corona discharge in air atmosphere. This is illustrated with figure 1. The list of the samples and the conditions of their poling, following ion exchange and thermal processing are presented in table 2. Further, we put main attention to the anodic electrodes, for they essentially influence results of glass poling.

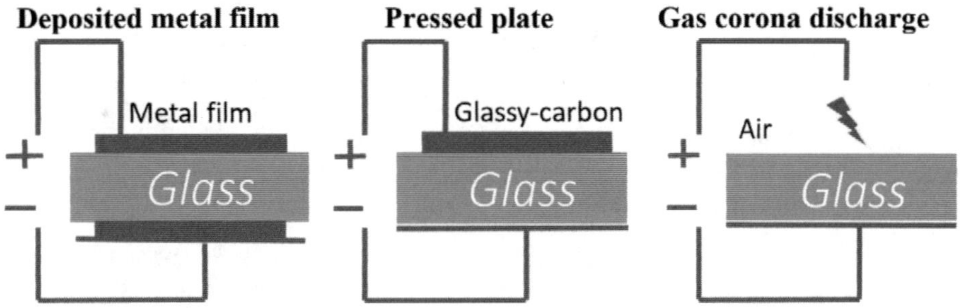

Figure 1. Three different types of anodic electrode.

Table 2. Prepared samples.

	Pressed anode, vacuum poling		Deposited anode, vacuum poling		Deposited anode, air poling		Corona poling
Anode type	Glassy carbon		Chromium	Gold	Chromium	Gold	
Voltage (V)	500	800	800	500	800	500	4000
Ion Exchange $Ag_{0.05}Na_{0.95}NO_3$ (in wt.%)	20 min 325 °C						10 h 325 °C
Annealing, 1 h 600 °C	-	-	-	+	+	+	+

All samples were poled at the temperature of 300 °C. Duration of the poling with anodic electrodes corresponded to 90% drop of the poling current. The corona poling was performed for 3 h. It has been found that vacuum poling with pressed electrodes results in the coloration of the samples after the ion exchange. Measurements of the optical absorption spectra of the sample poled with pressed electrodes (figure 2a) showed wide absorption peaks centered at ~375 nm in 500 V-poled and a narrower peak at ~420 nm in 800 V-poled glass (yellowish-colorated sample). We attributed the first peak at ~375 nm to the absorption of Ag_6-Ag_9 neutral clusters [9] and the second peak - to the surface plasmon resonance absorption of a polydispersed ensemble of silver nanoparticles [10]. Thus, vacuum poling with pressed electrodes resulted in the reduction of silver ions penetrated into the glass during the ion exchange. Vacuum poling with deposited chromium electrodes under 800 V similarly led to the formation of polydispersed silver nanoparticles after the ion exchange, however, their number, which correlates to the height of optical absorption peak (see figure 2b), essentially increased. We associate the reduction of silver with the presence of negatively charged non-bridging oxygen atoms in the subanodic layer of the poled glasses [2]. Supposedly, the non-bridging oxygen atoms donate electrons to silver ions, and thermal diffusion of reduced silver atoms leads to their clustering and the growth of silver nanoparticles. The higher poling voltage and poling in "closed anode" configuration results in the generation of higher spatial charge and, correspondingly, in an increased number of silver atoms, which facilitates the formation of the nanoparticles.

Vacuum poling of the glass using deposited gold electrodes induced lower (compared to pressed and deposited film chromium electrodes) light absorption after ion exchange. This corresponds to fewer silver clusters formed in the subanodic region of this sample glass during ion exchange (figure 2c). However, annealing (600° C, 1 hour) of the poled and ion-exchanged slide in air led to the formation of numerous nanoparticles on the anode surface of the glass. This is evidenced by the appearance of the absorption peak at the wavelength of surface plasmon resonance of silver nanoparticles in the residual spectrum (figure 2c). Here and below residual spectra were obtained by the subtraction of absorption spectra of slides after anode surface cleaning from the spectra of the slides before cleaning and marked with "R" in figures. We attribute lower silver-related optical absorption of this sample to a lower concentration of silver ions penetrating into glass. Probably, this is because of the limitation of their diffusion due to partial generation and entering of gold ions into the glass and occupying of non-bridging oxygen bonds by these ions. In this poling mode, the gold film anode electrode is the source of some amount of gold ions.

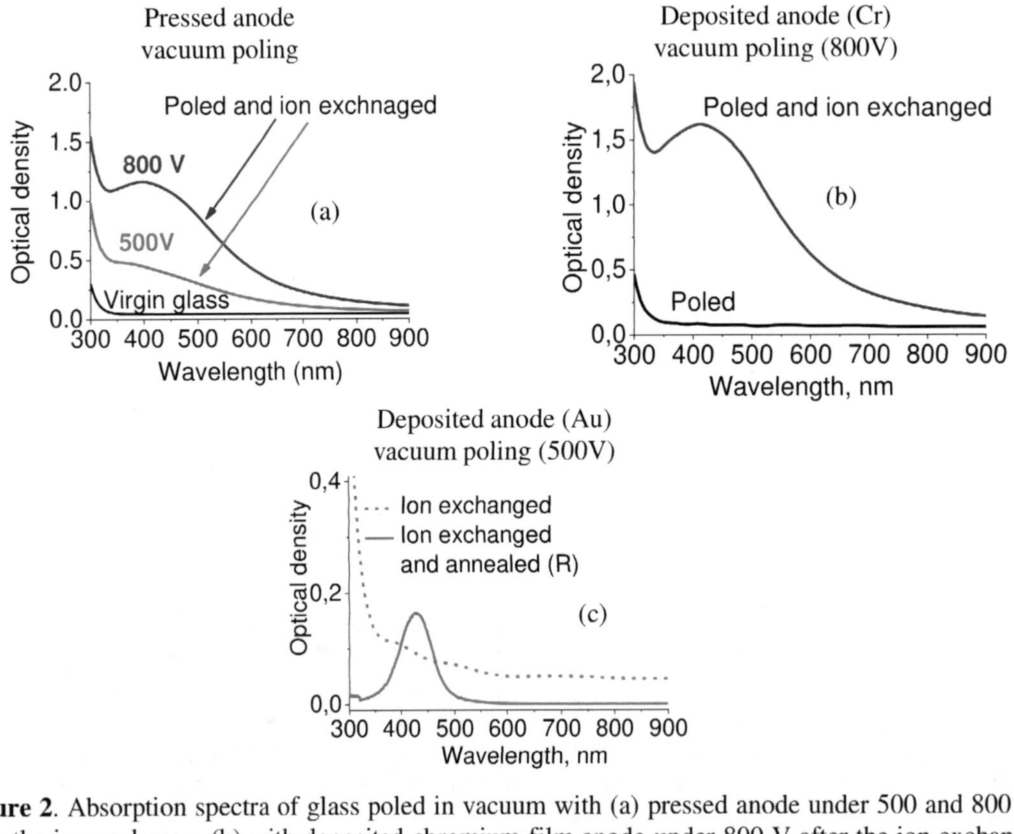

Figure 2. Absorption spectra of glass poled in vacuum with (a) pressed anode under 500 and 800 V after the ion exchange, (b) with deposited chromium film anode under 800 V after the ion exchange and (c) glass poled under 500 V with deposited gold film anode after the ion exchange and annealing (residual spectrum, marked with "R").

Contrary to vacuum poling, poling in air atmosphere using deposited gold film anode (300° C, 500 V) completely prevents silver ions penetration into glass. There is no optical absorption corresponding to silver clusters/nanoparticles in the measured spectrum of the poled glass after both ion exchange and subsequent annealing (also, no surface silver nanoparticles absorption is seen in the residual spectrum in figure 3). We suppose that this is due to blocking of silver penetration by gold ions formed at the anode via some oxidation of the gold film, their drifting into the glass and occupying non-bridging oxygen sites. Occupation of non-bridging oxygen bonds by gold atoms should leave no space for silver ions. Indeed, the heat treatment of the poled sample in air atmosphere at 600° C for 1h has demonstrated the appearance of a peak at ~530 nm in its optical absorption spectrum (see figure 3). This corresponds to the spectral position of surface plasmon resonance in gold nanoparticles [11], and the peak reflects the formation of such nanoparticles inside the glass, not on the surface. The latter was confirmed by the absence of changes in the spectrum after cleaning the slide surface. The appearance of the absorption at the wavelength of surface plasmon resonance of gold nanoparticles indicated the drift of low-mobile gold ions into the glass during poling in this mode. Presumably, thermal poling in the air atmosphere provided some oxidation of the gold film anode, which allowed arising gold ions to penetrate into the glass under the influence of an electric field. In addition, the binding of trivalent gold ions to non-bridging oxygen atoms should block the ion exchange of silver. Gold nanoparticles form at a relatively high temperature, 600° C, at which metal nanoparticles are usually formed in glasses containing noble metal ions [10]. The air poling with deposited chromium film anode (300° C, 800 V) completely prevented silver diffusion into the glass during the ion exchange (see figure 3b). There is no silver clusters/nanoparticles absorption in the spectrum after the ion exchange, and the residual spectrum after additional thermal treatment of the ion exchanged poled glass does not show a presence of nanoparticles on the glass surface. The absence of silver ions penetration into the glass is the result of a strong polymerization of subanodic glass region in this mode of thermal poling.

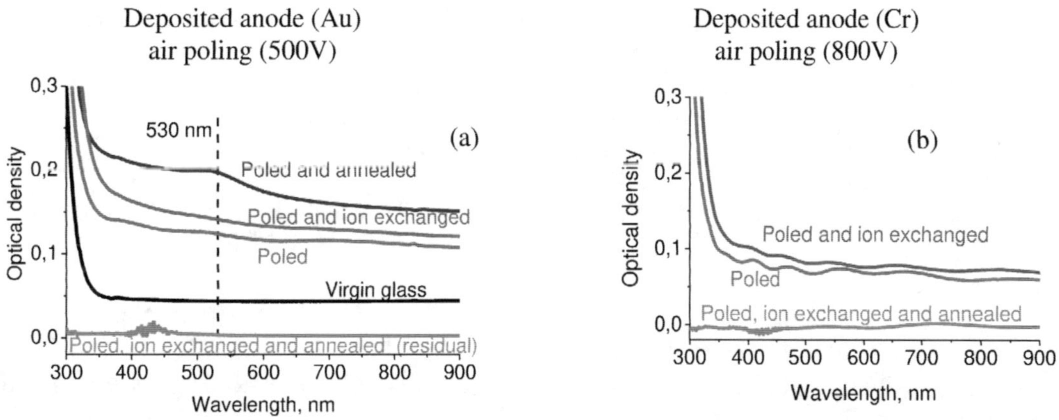

Figure 3. (a) Absorption spectra of the glass poled in air atmosphere under 500 V with deposited gold film anode before and after ion exchange and annealing. (b) Absorption spectra of the glass poled in air atmosphere under 800 V with deposited chromium film before and after ion exchange and annealing. Residual spectra are marked with "R".

3. Conclusions

Finally, we have shown that the mode of thermal poling of the soda-lime glass significantly affects results of subsequent ion exchange processing of glass. The most significant results are the formation of silver nanoparticles in the subanodic region of glasses poled in vacuum using pressed or deposited chromium film anode without an additional heat treatment of the glass after the ion exchange. Poling in vacuum with gold film anode allows some penetration of silver ions in the poled glass, but silver nanoparticles do not form in the glass and additional anneal results in their formation on the glass surface only. Poling in air corona discharge allows silver ion exchange and formation silver nanoparticles after heat treatment of the ion-exchanged glass. Using deposited chromium and gold film anodic electrodes for poling in air prevents silver ions penetration in the poled glasses. However, in the case of gold electrode diffusion of gold in the glass occurred and gold nanoparticles formation took place after additional heat treatment of the glass. Obtained results can be employed in the formation of the metal nanoparticles in glasses, and prescribed 2D distribution of the nanoparticles can be formed if profiled anodic electrodes are used.

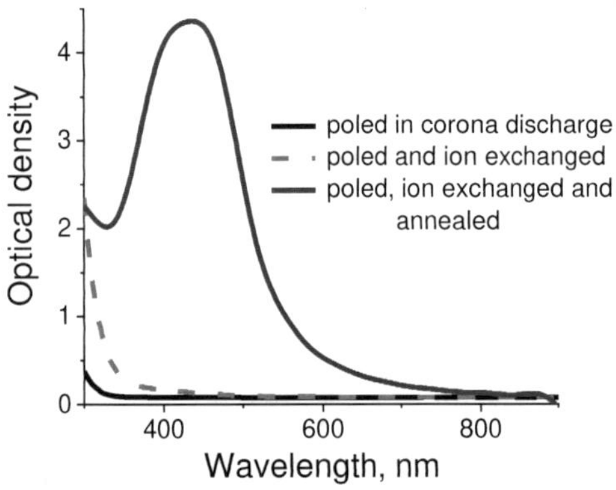

Figure 4. Absorption spectra of the glass poled in air corona discharge under 4000 V before and after ion exchange and annealing.

Acknowledgments
The study was supported by the Ministry of Science and Higher Education of Russian Federation, project FSRM-2020-001.

References

[1] Ikutame N, Kawaguchi K, Ikeda H, Sakai D, Harada K, Funatsu S and Nishii J 2013 *J. Appl. Phys.* **114** 083514
[2] Xu W, Arentoft J, Wong D and Fleming S 1999 *IEEE Photonics Technol. Lett.* **11** 1265–7
[3] Redkov A, Melehin V, Raskhodchikov D, Reshetov I, Tagantsev D, Zhurikhina V and Lipovskii A 2019 *J. Non. Cryst. Solids* **503–504** 279–83
[4] Doremus R 2005 *Appl. Phys. Lett.* **87** 232904
[5] Lepienski C, Giacometti J, Leal Ferreira G, Freire F and Achete C 1993 *J. Non. Cryst. Solids* **159** 204– 12
[6] Margulis W and Laurell F 1996 *Opt. Lett.* **21** 1786
[7] Lipovskii A, Rusan V and Tagantsev D 2010 *Solid State Ionics* **181** 849–55
[8] Lipovskii A, Melehin V and Petrikov V 2006 *Tech. Phys. Lett.* **32** 275–7
[9] Harb M, Rabilloud F, Simon D, Rydlo A, Lecoultre S, Conus F, Rodrigues V and Félix C 2008 *J. Chem. Phys.* **129** 194108
[10] Sancho-Parramon J, Janicki V, Dubček P, Karlušić M, Gracin D, Jakšić M, Bernstorff S, Meljanac D and Juraić K 2010 *Opt. Mater. (Amst).* **32** 510–4
[11] Cattaruzza E, Battaglin G, Gonella F, Ali S, Sada C and Quaranta A 2007 *Mater. Sci. Eng. B* **149** 195– 9

Journal of Physics: Conference Series **2086** (2021) 012153 doi:10.1088/1742-6596/2086/1/012153

Computation of optical waveguide interaction for quantum gates implementation

A A Lytaev and I Yu Popov

Faculty of Control Systems and Robotics, ITMO University, St. Petersburg, 197101, Russia

Abstract. The system of two coupled optical dual-mode waveguides is considered. The coupling of the system is studied to find a circuit for building a control switch for two qubit gates. The classical coupled mode theory is applied and the exact expressions for coupling coefficients are derived. The parameters of the system for performing the desired operations are numerically computed and analysed. The system describing the influence of intermodal interactions is solved numerically. The distortions are analysed.

1. Introduction

The physical implementation of quantum computer is a problem of vital importance since the idea was suggested by Richard Feynman [1]. One of the main obstacles on the way to build a multiple qubit quantum computer is a quickness of the system decoherence. The optical quantum computing is an approach, that allows to reduce the effect of decoherence due to the low effectiveness of photon-photon and photon-environment interaction. The conventional linear optical model is based on the usage of the bulk optics devices such as beamsplitters, polarizing beamsplitters, mirrors, phase shifters etc. However, the practical implementation would require the usage of integrated optics devices, where the bulk optics devices are replaced with waveguides. The universal set of quantum gates based on waveguides interaction is described in [2].

Nevertheless, the existing models of linear optical two-qubit gates are probabilistic with 1/9 probability of success, that can be increased only with a large number of ancilla qubits, thus, making them unscalable for building a complex computational scheme [3]. The alternative is to use nonlinear effects for quantum gate implementation. However, the low effectiveness of Kerr-like nonlinearities would require long optical paths, that could potentially accumulate the computational error due to the waveguide border imperfections [4].

One of the possible solutions was suggested in [5], where it is suggested to encode quantum bits with two mode optical waveguides. This work was devoted to the development of this approach by studying the system of two similar two-mode waveguide that can be used as a control switch for nonlinear CNOT gate.

2. Two mode waveguide optical model

The idea to perform quantum computing using two mode optical waveguides was suggested in [5]. $|0\rangle$ qubit state can be encoded with a TE_0 waveguide mode, while $|1\rangle$ with TE_1. The quantum theory of rays allows one to note the analogy between the transverse field distribution in the waveguide and the quantum mechanical wavefunction, so the system behaves as a quantum bit [6]. One of the main benefits of this approach is the high optical density of the light within the core region of the waveguide that increases the intensity of nonlinear interaction, so it can be used to perform qubit operations, thus making the model closer to possible practical implementations. Note that coding of qubit by two modes is used not only in waveguides but also in free space channels where the Gaussian beam modes play the role of waveguide modes (see, e.g., [7]).

The model of CNOT gate for two mode waveguide qubits is described in [5] (see figure 1). The main part of the circuit is the waveguide Mach-Zehnder interferometer with nonlinearities in its arms with intensity dependent refractive index. The waveguide coupling regions perform the waveguide state dependent mode separation and combine operations. Thus, the closer is the state of the control qubit to $|1\rangle$, the higher is intensity of light, that is transferred to the upper arm of Mach-Zehnder interferometer and the larger is the phase shift the target qubit is subjected to. The necessary condition for the correct operation of the system is, that the coupling region transfers TE_1 mode of the control waveguide to the upper MZI arm, while keeping TE_0 within the control waveguide. The adjustment of the coupling region parameters (the coupling length L and the distance R between the waveguides) is needed to make the circuit perform operations in the desired way. It is our goal.

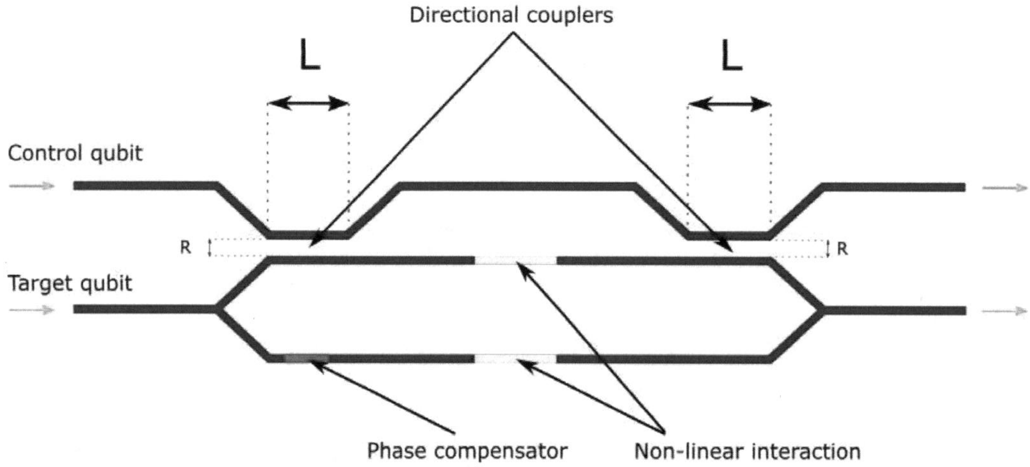

Figure 1. CNOT gate scheme.

3. The coupled mode theory

The interaction between two mode waveguides can be described using the coupled mode theory [8]. This approach is approximate because the field in the system is presented as the fields of the individual waveguides with amplitude coefficients.

$$\boldsymbol{E} = \sum_{v,m=0,1} A_{vm}(z)\,\boldsymbol{E}_{vm}$$
$$\boldsymbol{H} = \sum_{v,m=0,1} A_{vm}(z)\,\boldsymbol{H}_{vm} \tag{1}$$

The fields are substituted into the Maxwell equations.

$$\nabla \times \boldsymbol{H} = \mathrm{i}\omega\varepsilon_0 n^2(x,z)\,\boldsymbol{E}$$
$$\nabla \times \boldsymbol{E} = -\mathrm{i}\omega\mu_0 \boldsymbol{H} \tag{2}$$

The resulting equations for amplitude coefficients are multiplied with conjugate fields and integrated over the cross sections.

$$\sum_{v,m=0,1}\int_{-\infty}^{+\infty} dx(\frac{\partial A_{vm}}{\partial z}\boldsymbol{e}_z(\mathbf{H}_{vm}\times\mathbf{E}_{v'm'}^{-})+\frac{\partial A_{vm}}{\partial z}\mathbf{e}_z(\boldsymbol{E}_{vm}\times\boldsymbol{H}_{v'm'}^{-}) - \mathrm{i}\omega\varepsilon_0(n_{\mathrm{core}}^2-n_{\mathrm{clad}}^2)_{(v+1)mod2}A_{vm}\,\boldsymbol{E}_{vm}\boldsymbol{E}_{v'm'}^{-})=0, \qquad v',m'=0,1 \tag{3}$$

One cannot find the general solution in an analytic form. It can be made numerically. However, it is possible to simplify the analysis by neglecting terms, that describe weak interactions of high orders. The introduction of coupling coefficients c_m and $D_m^{m'}$ yields to the following system of four differential equations.

$$\begin{aligned}
&\frac{\partial A_{00}}{\partial z}c_0 + A_{10}\mathrm{i}D_0^0 + A_{11}\mathrm{i}D_1^0\mathrm{e}^{\mathrm{i}\Delta\beta z} = 0\\
&\frac{\partial A_{01}}{\partial z}c_1 + A_{10}\mathrm{i}D_0^1\mathrm{e}^{-\mathrm{i}\Delta\beta z} + A_{11}\mathrm{i}D_1^1 = 0\\
&\frac{\partial A_{10}}{\partial z}c_0 + A_{00}\mathrm{i}D_0^0 + A_{01}\mathrm{i}D_1^0\mathrm{e}^{\mathrm{i}\Delta\beta z} = 0\\
&\frac{\partial A_{11}}{\partial z}c_1 + A_{00}\mathrm{i}D_0^1\mathrm{e}^{-\mathrm{i}\Delta\beta z} + A_{01}\mathrm{i}D_1^1 = 0
\end{aligned} \tag{4}$$

It is possible to simplify the system further by neglecting the terms of intermodal interaction with $\mathrm{e}^{\mathrm{i}\Delta\beta z}$ as well. The analytical solution can be found for the simplified system.

$$\begin{aligned}
A_{0m}(z) &= A_{0m}(0)\cos\left(\frac{D_m^m}{c_m}z\right) - \mathrm{i}A_{1m}(0)\sin\left(\frac{D_m^m}{c_m}z\right)\\
A_{1m}(z) &= A_{1m}(0)\cos\left(\frac{D_m^m}{c_m}z\right) - \mathrm{i}A_{0m}(0)\sin\left(\frac{D_m^m}{c_m}z\right)
\end{aligned} \tag{5}$$

In this approximation the system transfers TE_1 mode to another waveguide, while keeping TE0 to remain in the same, when the following condition in integers is satisfied.

$$\frac{4v_1+1}{4v_0} = \frac{\mathrm{D}_1^1/c_1}{\mathrm{D}_0^0/c_0} \qquad v_0,\ v_1 \in \mathrm{N} \tag{6}$$

The analytical expressions for D_m^m/c_m were obtained by directly evaluating the integral expressions with TE mode fields given in [8].

$$D_0^0/c_0 = 4\pi^2\left(n_{\mathrm{core}}{}^2 - n_{\mathrm{clad}}{}^2\right)\cos(\chi_0 d)\exp[-\gamma_0(R-d)]\beta_0^{-1}(\lambda n_{\mathrm{clad}})^{-2}\left(\gamma_0^2+\chi_0^2\right)^{-1} * \left[\gamma_0\cos(\chi_0 d)\mathrm{sh}(\gamma_0 d)+\chi_0\sin(\chi_0 d)\,\mathrm{ch}(\gamma_0 d)\right]\left[d+\frac{\sin(2\chi_0 d)}{2\chi_0}+\frac{\cos^2(\chi_0 d)}{\gamma_0}\right]^{-1} \tag{7}$$

$$D_1^1/c_1 = 4\pi^2\left(n_{\mathrm{core}}{}^2 - n_{\mathrm{clad}}{}^2\right)\sin(\chi_1 d)\exp[-\gamma_1(R-d)]\mathrm{ch}(\gamma_1 d)\beta_1^{-1}(\lambda \mathrm{n}_{\mathrm{clad}})^{-2}\left(\gamma_1^2+\chi_1^2\right)^{-1} * \left[\chi_1\cos(\chi_1 d)-\gamma_1\sin(\chi_1 d)\right]\left[d-\frac{\sin(2\chi_1 d)}{2\chi_1}+\frac{\sin^2(\chi_1 d)}{\gamma_1}\right]^{-1}$$

Hence, the ratio (6) can be satisfied by adjusting the coupling distance R to match almost any given pair of v_0 and v_1. The coupling length L, that makes the system perform the desired control operation can be expressed as

$$L = (2D_1^1)^{-1}c_1(4v_1+1) \tag{8}$$

4. Numerical results

The numerical evaluation of the coupling coefficients and various propagation constants as a solution of transcendental equation was implemented for a symmetrical system of coupled waveguides. The system parameters are the light wavelength λ, the waveguide width $2d$ and the refractive indices L, n_{core} and n_{clad} of the core and cladding layers respectively. The algorithm for determination of the distance between waveguides R, so that the coupling coefficient are adjusted to satisfy the condition (6), was implemented. The secondary goal, that is achieved by the algorithm is the minimization of the coupling length L, so the compactness of the device is increased and the potential error due the waveguide boundaries imperfections is reduced. In order to perform that, the algorithm tests a number of parameters R, the one that yields the minimum value of L is selected.

The analysis above was performed for the system of equations (4), that was simplified by neglecting of weak intermodal interactions. In addition, the system (4) without simplifications was solved numerically with the finite difference method in order to estimate the influence of intermodal interaction and the potential error due to them.

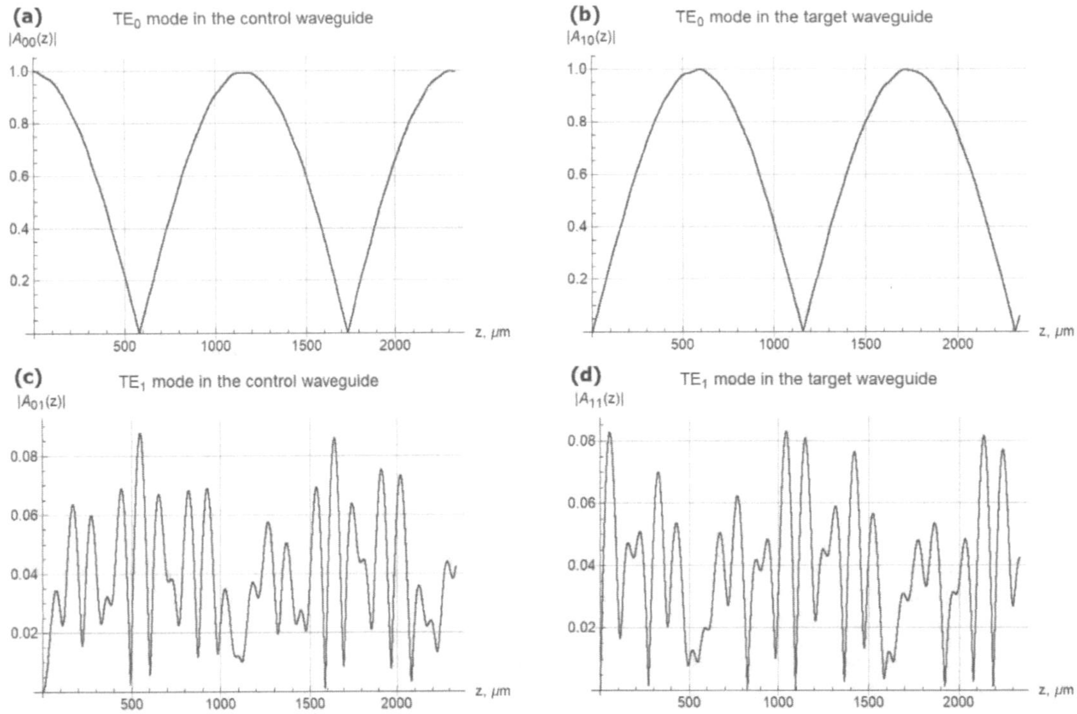

Figure 2. Propagation of waveguide modes for $|0\rangle$ as input parameter in the control qubit. **(a)** TE_0 in the control waveguide. **(b)** TE_0 in the target waveguide. **(c)** TE_1 in the control waveguide. **(d)** TE_1 in the target waveguide.

It was obtained, that including of intermodal interactions distorts the performance of the device notably. For the system with parameters $\lambda = 1.018$ μm, $n_{\text{core}} = 1.57$, $n_{\text{clad}} = 1.55$, $d = 1.83$ μm, $R = 3.96$ μm (figure 2, 3), that would perform ideally in the theoretical model governed by equations (5), it was obtained, that the loss in the signal due to the intermodal interaction is about 0.5 %. The loss is due to the undesired guidance of modes of the different order and the related phase shift in the signal mode. The error of that order should be taken into account, as it would multiply in the complex scheme of several consecutive CNOT gates, however it could be coped with by the application of various quantum error correction algorithms.

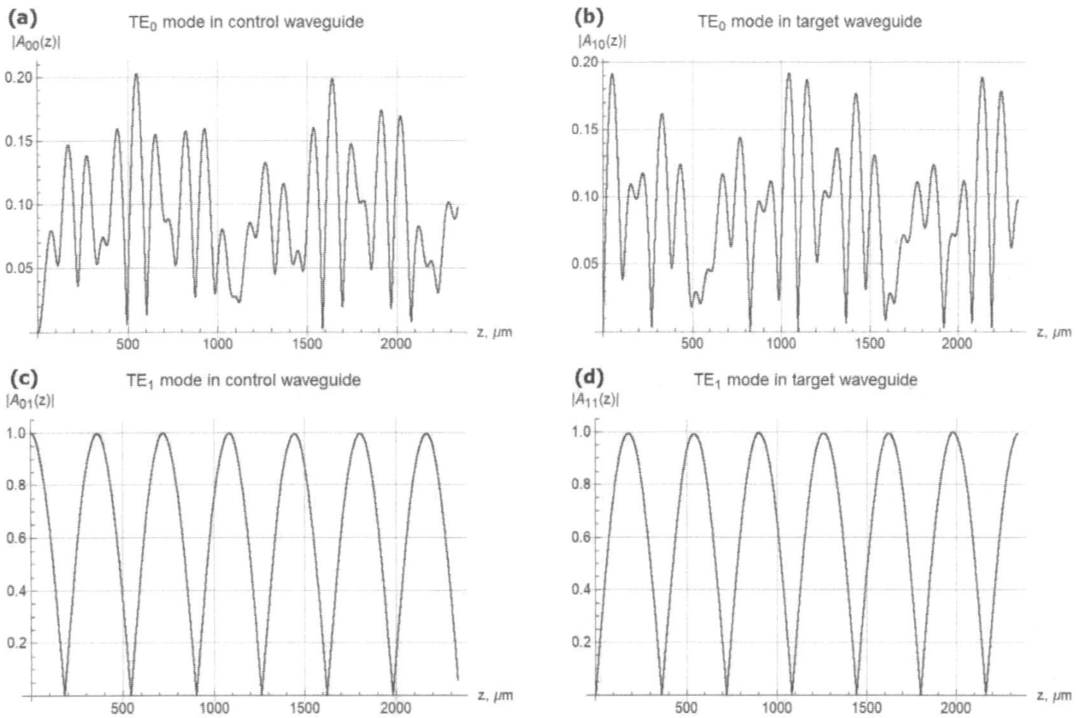

Figure 3. Propagation of waveguide modes for $|1\rangle$ as input parameter in the control qubit. **(a)** TE_0 in the control waveguide. **(b)** TE_0 in the target waveguide. **(c)** TE_1 in the control waveguide. **(d)** TE_1 in the target waveguide.

5. Conclusion

The system of two slab symmetrical two mode coupled optical waveguides was considered with the aim to be used as a switcher for nonlinear CNOT gate. The coupled mode theory approximation was applied to simulate the behaviour of the system. The approximate analytical model was studied in order to obtain the parameters, that would make the coupling perform in the desired way. The extended differential equations system was solved numerically with the finite difference method to consider the influence of the intermodal interactions. It was obtained, that the intermodal interaction could cause the computational error of 0.5 %. However, the possible distortions due to the waveguide bends and boundary imperfections were not considered.

References

[1] Feynman R 1986 *Found Phys.* **16** 507
[2] Gavrilov M, Gortinskaya L, Pestov A, Popov I and Tesovskaya E 2007 *Phys. Part. Nuclei Lett.* **4** 137
[3] Knill E, Laflamme R and Milburn J 1997 *Nature* **409** 46
[4] Milburn G 1989 *Phys. Rev. Lett.* **62** 2124
[5] Fu J and Tang S 2003 *Chinese Phys. Lett.* **20** 1426
[6] Gloge D and Marcuse D 1969 *J. Opt. Soc. Am.* **59** 1629
[7] Faleeva M and Popov I 2020 *Nanosyst. Phys. Chem. Math.* **11** 651
[8] Marcuse D 1982 *Light Transmission Optics* ed S Mitra (New York: van Nostrand Reinhold) chapter 10 pp 519–531

Features of spectral analysis of nuclear magnetic resonance signal for express-control of hydrocarbon media

S S Makeev[1], V V Davydov[1, 2], N S Myazin[1] and V Yu Rud[2]

[1] Peter the Great Saint Petersburg Polytechnic University, Saint Petersburg 195251, Russia
[2] All-Russian Research Institute of Phytopathology, Moscow Region 143050, Russia

st_makeev@mail.ru

Abstract. The article substantiates the need to study the structure of the nuclear magnetic resonance signal recorded using the modulation technique. The features of registration of the nuclear magnetic resonance (NMR) signal using the modulation technique, which are both in the current and in the stationary state, have been established. The influence of the properties of the medium on the features of recording the NMR signal has been determined. A new method is proposed for determining the contributions of absorption and dispersion signals to the recorded NMR signal. The features of the use of spectral analysis in the study of the NMR signal from hydrocarbon media have been determined. The results of experimental studies are presented.

1. Introduction

The development of science and technology has led to the emergence of a large number of problems in applied physics related to the control of the state of condensed media [1-10]. There are a number of difficulties in the development of fast and reliable methods of express control [2, 7, 11-15]. Measurements that are used for express control should not change the physical structure and chemical composition of the investigated medium [2, 7, 11, 16-24]. The fulfillment of this condition is possible when devices are used for express control, the principle of operation of which is based on the phenomenon of nuclear magnetic resonance (NMR) [24-32]. Other types of devices for express control of condensed media (optical, ultrasonic, X-ray, etc.) fulfill this condition when they work with a certain class of media [18, 32-35].

On the other hand, there are a number of limitations when using devices based on the phenomenon of nuclear magnetic resonance. One of them is associated with the registration of an NMR signal in a weak magnetic field. This does not allow obtaining the NMR spectrum from the studied medium [2, 7, 31-33, 36, 37]. All information about the composition and structure of the studied medium is contained in the registered NMR signal in the form of "wiggles". Therefore, the purpose of this work is to develop a method using spectral analysis to obtain the necessary information about the state of the medium from the recorded NMR signal from it using a modulation technique. And the purpose is to determine the features of its application for various conditions of physical experiments and compositions of the medium. Particular attention is paid to hydrocarbon media and their mixtures.

2. Features of the use of spectral analysis to decipher information about the state of the medium in the NMR signal recorded in a weak magnetic field

In [31, 32, 36-41], the Bloch equations obtained after substitution of coefficients that take into account the registration of the NMR signal using the modulation technique are considered in detail. They also substantiate the mathematical relation for constructing the NMR signal waveform recorded using an autodyne detector:

$$G(t) = F(t)\left(\frac{A}{A+B}\upsilon^2(t) + \frac{B}{A+B}u^2(t)\right)^{1/2} \quad (1)$$

where $\upsilon(t)$, $u(t)$ are the absorption and dispersion signals, A and B are the coefficients that determine the contribution to the recorded NMR signal from the absorption and dispersion signals, and F(t) is the coefficient taking into account phase changes.

As a result of the studies carried out, NMR signals from water were recorded, and their shapes were constructed using (1) and (2). This made it possible to establish that the value of the coefficient F(t) depends on many factors (temperature, composition of the medium, etc.). Therefore, for each case of studying a medium or a mixture of media in the express mode, it is necessary to select F(t). This can lead to large time delays or errors in research and interpretation of results. In addition, in the above description of the NMR waveforms with increasing t, every $T_m/2$ changes in the signal phase by 180^0 and a successive decrease in the amplitude of the peaks occurs. In the experiment, the same NMR signal is recorded every half-period $T_m/2$ of the modulation field H_0. This discrepancy between theory and experiment, the previously noted difficulties with the choice of the coefficient F(t) does not allow the results of the experiment to be fully displayed.

Therefore, to solve this problem, we proposed to use spectral analysis, especially for hydrocarbon media, in which there are a number of features. The most important for express control is associated with the fact that when mixed, these media do not enter into a chemical reaction with each other. They form a conglomerate (for example, a mixture of gasoline, kerosene and gasoline, etc.). Each of the components of the mixture exists independently and can be isolated from the general mixture.

Another feature of spectral analysis is that the recorded NMR signal is a non-periodic oscillation in the form of damped peaks. Therefore, we use the discrete Fourier transform to describe it, as well as the calculated absorption and dispersion signals:

$$y_k = \sum_{n=0}^{N-1} x_n e^{-j2\pi kn/N} \quad (2)$$

where n = 0, 1, 2, ..., $N-1$, x_n is the input data sequence, N is the number of elements of the input data sequence x_n.

The harmonics of the spectrum are located on the frequency axis with a discrete $\Delta f = f_s/N$, where f_s is the sampling frequency of the original sequence x. The sampling rate is determined as follows. Let τ be the duration of the NMR signal. Then f_s can be calculated using the following relation:

$$f_s = N/\tau \quad (3)$$

In the case where N is a power of two, DFT is calculated by the FFT algorithm (Fast Fourier Transform), which is significantly faster discrete Fourier transform and requires fewer computational resources.

The discrete Fourier transform is symmetric with respect to the Nyquist frequency equal to $f_s/2$, which allows to combine the harmonic numbers (N/2-k) and (N/2+k). As a result of combining the harmonics, a one-sided complex spectrum is obtained with frequencies from 0 to $fs/2$, which corresponds to the indices k = 0 ... (N/2-1). The scaled-sided complex spectrum of the discrete input sequence x_n is given by:

$$y_k = \begin{cases} y_0 & k = 0 \\ \sqrt{2}\dfrac{y_k}{N} & k = 1,2,\ldots\left\lfloor \dfrac{N}{2} - 1 \right\rfloor \end{cases} \tag{4}$$

In relation (4), the operation in brackets [N/2-1] means rounding to the nearest smallest integer. Accordingly, the amplitude spectrum S is the modulus of the one-sided complex spectrum, the phase spectrum $P(f) = \arg y_k$ is its argument, where $f = k\,\Delta f$.

3. Study of the spectra of the NMR signal taking into account the determination of the contributions of the absorption and dispersion signals

In case is recording an NMR signal in a weak magnetic field, the parameters of the autodyne detector and the modulating magnetic field are adjusted to the maximum S/N. In this case, the signals υ(t) and u(t) take part in the formation of the NMR signal. In figure 1 shows NMR signals from oil and kerosene as an example.

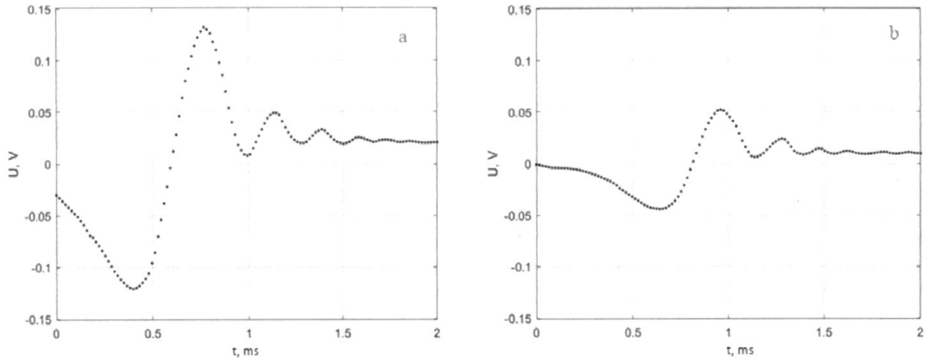

Figure 1. The NMR waveforms from: a) oil, b) kerosene.

Analysis of the received signals shows the absence of a period by the location of the decaying peaks. In figure 2 shows the amplitude spectra of NMR signals recorded from a sample of oil and kerosene, obtained using relations (2) - (4). In figure 2 and 3 shows the phase spectra of the NMR signal from a sample of oil and kerosene.

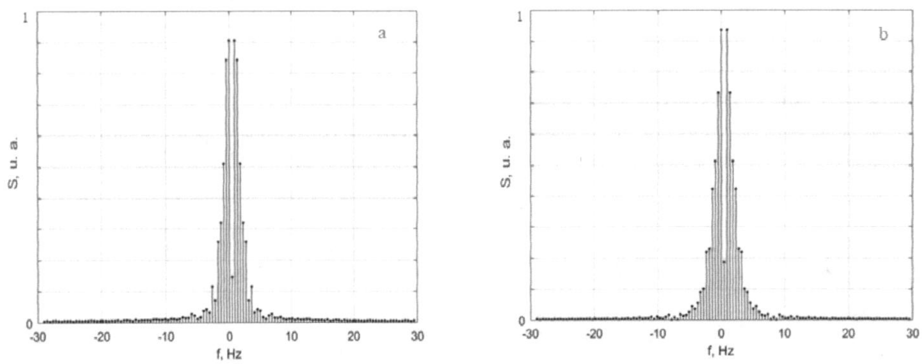

Figure 2. The amplitude spectra from the experimental NMR signal recorded from: a) oil, b) kerosene.

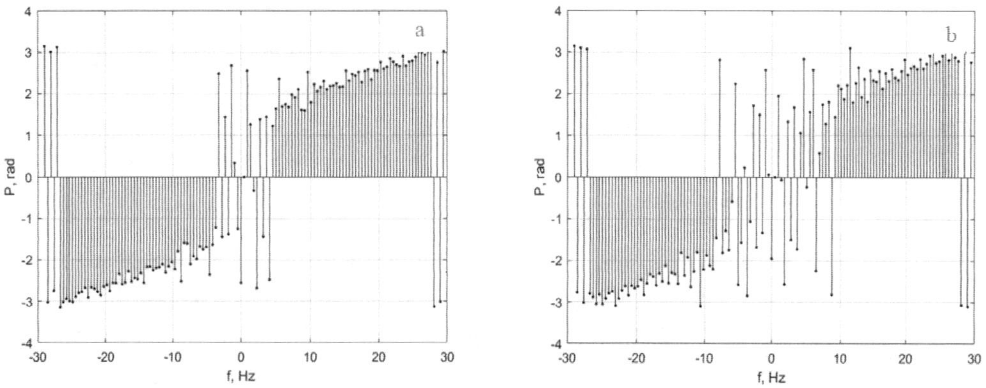

Figure 3. The phase spectra of the NMR signal from: a) oil, b) kerosene.

To obtain information on the contributions of absorption and dispersion signals to the recorded NMR signals, it is necessary to calculate the values of υ(t) and u(t) using the Bloch equations. In figure 4 shows the calculated dependences υ(t) and u(t) for oil as an example.

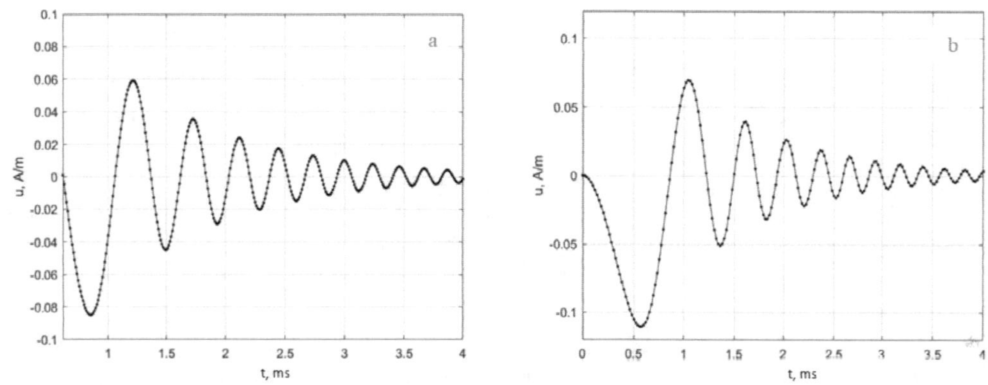

Figure 4. The calculation forms of NMR signals from oil: a) absorption, b) dispersion.

For absorption and dispersion signals, the amplitude and phase spectrum is calculated (figure 5 and 6)

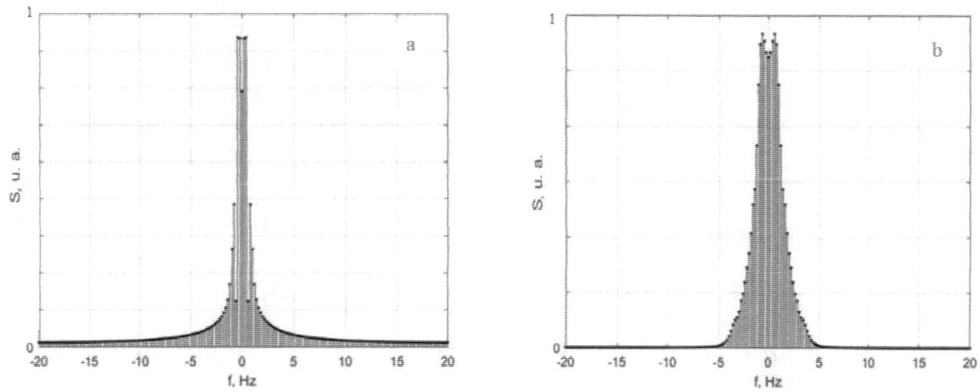

Figure 5. The amplitude spectra of the NMR signal from oil: a) absorption, b) dispersion.

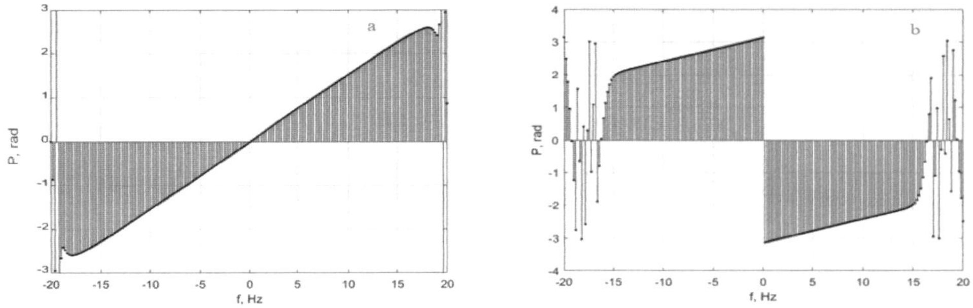

Figure 6. The phase spectra of the NMR signal from oil: a) absorption, b) dispersion.

Comparison of the obtained results showed that the spectral components of the amplitude and phase spectrum can be expressed in terms of the spectral components of the amplitude and phase spectra from the signals υ(t) and u(t). The coefficients that will determine this relationship will be the contribution of these signals to the recorded NMR signal.

4. Conclusion

As a result of experiments, it was found that the spectral method developed by the research has no restrictions on use. For its application must register the NMR signal from the medium containing nuclei with magnetic moments, for example, at the resonant frequency of protons (in the more than 99% of liquid media include protons) and to measure T_1 and T_2 relaxation constants to calculate the dispersion and absorption signals.

It should be noted that the registration of the NMR signal in a weak field is carried out at a maximum signal/noise ratio signal. In this case, the absorption signal is always greater than the dispersion of the signal, the coefficients are determined uniquely. In the study of such mixtures number of coefficients increases, moreover appear coefficients corresponding media concentrations in the mixture. Their determination allows one to identify the composition and concentration of the components in the mixture.

Since in mixtures from hydrocarbon media the contributions to the recorded NMR signal from protons are determined in proportion to the concentrations of the media in the mixture, it is possible to unambiguously determine the coefficients in the spectra and identify the composition and concentration of the components in the mixture.

Acknowledgments

The reported study was funded by RFBR according to the research project № 20-32-90012.

References

[1] Neronov Yu I and Seregin N N 2013 *Measurement Techniques* **55(11)** 1287-1293

[2] Davydov V V, Cheremiskina A V, Velichko E N and Karseev A Yu 2014 *Journal of Physics: Conference Series* **541(1)** 012006

[3] Davydov R V, Mazing M S, Yushkova V V and Stirmanov A V 2019 *Journal of Physics: Conference Series* **1410(1)** 012067

[4] Gizatullin B, Gafurov M, Vakhin A, Mattea C and Stapf S 2019 *Energy and Fuels* **33(11)** 10923–10932

[5] Davydov R, Antonov V, Makeev S, Batov Y and Dudkin V 2019 *E3S Web of Conferences* **140** 02001

[6] Rutberg F G, Kolikov V A, Snetov V N, Cherepkova I A and Dyachenko S V 2016 *High Temperature* **54(2)** 170–174

[7] Myazin N, Neronov Y, Dudkin V and Yushkova V 2018 *MATEC Web of Conferences* **245** 11013

[8] Davydov R V and Antonov V I 2018 *Journal of Physics: Conference Series* **1124(8)** 081037
[9] Davydov R V and Antonov V I 2018 *Journal of Physics: Conference Series* **1135(1)** 012087
[10] Davydov R, Antonov V and Moroz A 2019 *International Conference on Electrical Engineering and Photonics EExPolytech-2019 (Saint-Petersburg)*, vol. 8906791 (IEEE), p. 295-297
[11] Davydov R V, Antonov V I, Yushkova V V and Smirnov K J 2019 *Journal of Physics: Conference Series* **1400(6)** 066037
[12] Davydov V V 1999 *Russian Physics Journal* **42(9)** 822–825
[13] Davydov V V, Dudkin V I and Karseev A Y 2015 *Russian Physics Journal* **58(2)** 146–152
[14] Davydov V V, Dudkin V I and Karseev A Y 2015 *Measurement Techniques* **58(3)** 317–322
[15] Davydov V V, Dudkin V I and Karseev A Y 2015 *Instruments and Experimental Techniques* **58(6)** 787–793
[16] Grebenikova N M and Davydov R V 2019 *Journal of Physics: Conference Series* **1326(1)** 012012
[17] Myazin N S, Dudkin V I, Grebenikova N M and Podstrigaev A S 2019 *Lecture Notes in Computer Science (including subseries Lecture Notes in Artificial Intelligence and Lecture Notes in Bioinformatics)* **11660 LNCS** 744–756
[18] Smirnov K J and Batov Y V 2019 *Journal of Physics: Conference Series* **1368(2)** 022073
[19] Davydov V V 2017 *Measurement Techniques* **60(2)** 183–189
[20] Davydov V V, Velichko E N, Dudkin V I and Karseev A Y 2014 *Measurement Techniques* **57(6)** 684–689
[21] Davydov V V, Dudkin V I and Karseev A Y 2015 *Technical Physics Letters* **41(4)** 355–358
[22] Davydov V V, Dudkin V I and Karseev A Y 2015 *Technical Physics* **60(3)** 456–460
[23] D'yachenko S V and Zhernovoi A I 2016 *Technical Physics* **61(12)** 1835–1837
[24] D'yachenko S V, Kondrashkova I S and Zhernovoi A I 2017 *Technical Physics* **62(1)** 1602-1604
[25] Neronov Yu I 2017 *Measurement Techniques* **60(1)** 96-102
[26] Neronov Y I 2021 *Biomedical Engineering* **54(5)** 333–336
[27] Myazin N S and Yushkova V V 2019 *Environmental Research, Engineering and Management* **75(2)** 28–35
[28] Myazin N, Neronov Y, Dudkin V and Petrov A 2019 *IOP Conference Series: Materials Science and Engineering* **497(1)** 012111
[29] Davydov V V, Dudkin V I and Karseev A Y 2015 *Instruments and Experimental Techniques* **58(6)** 787–793
[30] Davydov V V, Dudkin V I and Karseev A U 2013 *Optical Memory and Neural Networks (Information Optics)* **22(2)** 112–117
[31] Myazin N S, Yushkova V V and Dudkin V I 2019 *Journal of Physics: Conference Series* **1400(6)** 066008
[32] Rukin E V, Myazin N S and Dudkin V I 2019 *Journal of Physics: Conference Series* **1368(4)** 042011
[33] Myazin N S, Yushkova V V and Taranda N I 2019 *Journal of Physics: Conference Series* **1410(1)** 012130
[34] Grebenikova N M 2019 *Journal of Physics: Conference Series* **1410(1)** 012186
[35] Davydov V V, Dudkin V I and Petrov A A 2016 *Technical Physics Letters* **42(7)** 692-696
[36] Myazin N S 2018 *Journal Physics: Conference Series* **1124(1)** 031004
[37] Myazin N S 2018 *Journal Physics: Conference Series* **1135(1)** 012061
[38] Leshe A 1963 *Nuclear induction* (Veb Deustscher Verlag Der Wissenschaften, Berlin)
[39] Myazin N S 2019 *Journal Physics: Conference Series* **1236(1)** 012026
[40] Davydov V V, Logunov S E and Fadeenko V B 2018 *Russian Journal of Nondestructive Testing* **54(3)** 213–221
[41] Makeev S S, Grevtzeva A S, Glinushkin A P and Matorin D N 2020 *Journal of Physics: Conference Series* **1695(1)** 012112

Organic phosphor based fiber-optic sensor for detection of UV radiation

A S Matrosova[1,2,3], **V A Ananyev**[2], **G A Pchelkin**[1,3,4], **D N Shurupov**[1,4], **S K Evstropiev**[1,2,3,5], **V M Volynkin**[1], **V V Demidov**[1,3] **and N V Nikonorov**[2]

[1] R&P Association Vavilov State Optical Institute, St. Petersburg 192171, Russia
[2] ITMO University, St. Petersburg 197101, Russia
[3] Bauman Moscow State Technical University, Moscow 105005, Russia
[4] Peter the Great St. Petersburg Polytechnic University, St. Petersburg 195251, Russia
[5] Saint-Petersburg State Institute of Technology, St. Petersburg 190013, Russia

Abstract. The paper presents a simple and low-cost fiber-optic sensor for detection of UV radiation. A sensor construction consists of a silica capillary with a photoactive composition based on an organic phosphor, organic solvent and epoxyacrylate inside and a multimode optical fiber in contact with each other. By adjusting the proportion of components in a photoactive composition, it is possible to obtain a pronounced optical signal at wavelength near 440 nm which is the luminescence emission wavelength of the chosen organic phosphor. The potential of using the construction as a UV sensor is confirmed by the linear dependence of the optical signal amplitude at the fiber output on the optical power supplied to the fiber input.

1. Introduction

Detection of ultraviolet (UV) radiation is a vital challenge in many spheres of anthropogenic activities because being uncontrolled it can be detrimental to human health and environment [1-5].

World Health Organization specifies the following parts of the UV radiation spectrum according to the impact on human health: UV-C (λ = 100-280 nm), UV-B (λ = 280-315 nm), and UV-A (λ = 315-400 nm) [6]. UV-A radiation is harmful to eyes and skin even at low doses due to its deep penetration. UV-B light damages DNA and causes cancer. Natural UV-C radiation is totally absorbed by ozone in the upper layers of the atmosphere, but under extreme exposure from a laser source it destroys some groups of cells despite a small depth of penetration.

On the other hand, UV light also has positive effects. Particularly, UV-A radiation with moderate doses contributes to the formation of vitamin D in humans, facilitates metabolism, and has a therapeutic effect on skin [3, 4]. Using UV-C radiation, it is possible to disinfect air and water, sterilize objects, surfaces and medicines [2, 4, 7, 8].

Moreover, UV light is widely applied to solve technological and industrial problems. Thus, UV-C radiation assists in detection of electric discharge in the presence of strong electromagnetic interference in high voltage equipment and systems of signal intelligence [8]. UV-A radiation is useful in electrical equipment to measure gas parameters of corona discharge [3, 9]. UV radiation also operates in fire protection devices, control systems of flame processes, production of biodiesel, photolithography, technology of polymer coatings, and technology of laser sources [2, 4].

Detection of UV radiation with the use of fiber optic sensors (FOSs) has a number of advantages compared to other types of devices with analogous functionality, mainly thermoelectric [10]. FOSs are compact, lightweight, electrically safe, immune to electromagnetic interference, and provide an accurate measurement of environmental parameters in a single point or along the whole fiber length. In recent years, considerable attention is paid to the development of luminescent FOSs which convert UV light into visible one due to photoluminescence. Their benefits are fast response and high sensitivity [11]. There are at least 4-5 groups of such devices which incorporate optical fibers made of polymers with various optically active materials in a cladding or outside it [1, 2, 4, 5, 7, 8, 12, 13], phosphate glasses with rare earth metal ions [14, 15], multicomponent silicate, phosphate and borate glasses with molecular clusters [3], silica glasses and a cladding made of polymers or their modifications with organic dyes in a composition [16].

This work aimed to develop and study a prototype of a simple and low-cost luminescent capillary-type FOS for the UV-A radiation detection. The main idea was to use a combination of a photoactive composition which converts UV light into visible one and a silica optical fiber responsible for capturing and transferring of a spectrally converted optical signal inside a silica capillary.

2. Synthesis and characterization of the photoactive composition

Similar to the methodology described in [17], the photoactive composition was synthesized in two stages. As a first step, an organic phosphor 1,3,5-triphenyl-4,5-dihydro-1H-pyrazole (Sigma Aldrich, mass concentration 98%) and an organic solvent dimethylformamide (EKOS-1, mass concentration 99.9%) were stirred vigorously and exposed to the action of ultrasound waves at normal conditions until a homogeneous and low-viscous (10^{-3} Pa·s) solution was formed. At a second step the solution was mixed with a polymer material based on epoxyacrylate DeSolite 3471-3-14 (Royal DSM) under identical conditions in order to provide the viscosity of the solution about 10 Pa·s. The latter requirement regarding a relatively high viscosity of the final three-component solution (Figure 1a) arises from the following reasons. First, a segment of a silica capillary, which constitutes the basic element of the FOS construction, should be filled in a uniform manner. Second, a silica optical fiber should be rigidly fixed in the solution for subsequent UV curing.

Figure 1. Three-component solution based on 1,3,5-triphenyl-4,5-dihydro-1H-pyrazole, dimethylformamide and epoxyacrylate used for the synthesis of the photoactive composition (a) and the photoactive composition UV cured on a silica glass substrate (b).

The main criterion in the choice of a components ratio to synthesize the photoactive composition of the FOS was the condition to register bright visible emission of photoluminescence excited in the coating formed from the composition on a silica glass substrate (Figure 1b). Table 1 shows the characteristics of four synthesized solutions and coatings on their basis. In each case, curing of a solution was performed in the UV furnace under 0.4 J/cm^2 radiation dose during 30 seconds. Photoluminescence excitation and emission spectra were registered using a spectrophotometer LAMBDA 650 UV/Vis (PerkinElmer) and fluorescent/luminescent spectrometer LS-50B (PerkinElmer) respectively.

Table 1. Weight of the components in the solutions used for the synthesis of the photoactive composition and luminescence properties of the coatings formed from these solutions.

N	Weight, g			Photoluminescence emission of the coating in the visible part of the spectra
	1,3,5-triphenyl-4,5-dihydro-1H-pyrazole	dimethylformamide	epoxyacrylate	
1	0.03	10	10	None
2	0.03	5	5	None
3	0.07	5	5	Occurs at λ = 440 nm, the signal cannot be separated from the basic level
4	1	5	5	Occurs at λ = 440 nm, the signal 5 times higher than the basic level

Figure 2 demonstrates photoluminescence excitation and emission spectra of the coating based on the photoactive composition synthesized from the solution 4 with a maximum weight of the organic phosphor. As can be seen from the curves, the measured photoluminescence excitation wavelength locates in the spectral range λ = 365-390 nm, while the photoluminescence emission wavelength falls into the spectral range λ = 425-440 nm. These data correlate with the corresponding bands of pyrazoline derivatives [18]. This fact confirms that noticeable structural conversions of the organic phosphor are absent while stirring the initial components and curing the synthesized solution.

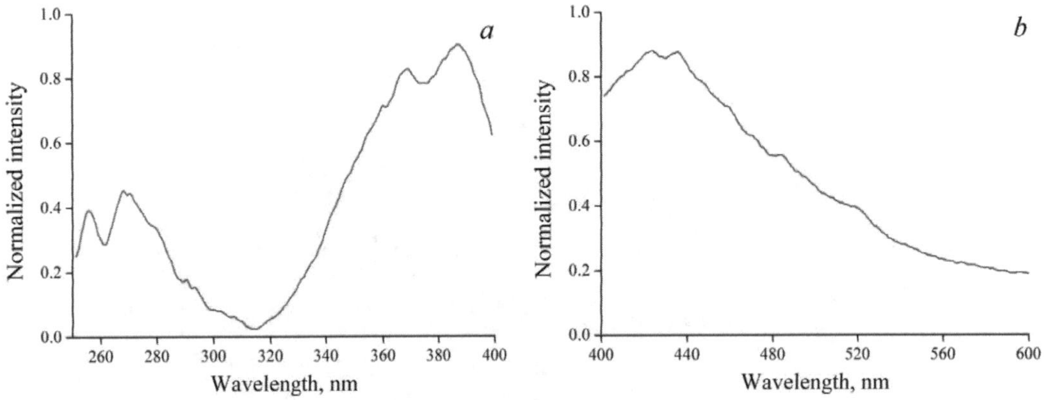

Figure 2. Photoluminescence excitation (a) and emission (b) spectra of the coating based on the photoactive composition synthesized from the solution 4.

Further, the components ratio corresponding to the solution 4 was used to synthesize the photoactive composition for the FOS construction.

3. Construction of the sensor

The FOS prototype consisted of a horizontally oriented silica capillary with 0.8 mm inner diameter and 26 mm length, filled with the photoactive composition from the one end (Figure 3a). The capillary was filled by pumping the composition under pressure through a plastic pipe, hermetically connected to the capillary. The length of the filled capillary segment was about 10 mm. After filling, the pipe was disconnected and the face end of the capillary was sealed by a cap made of the UV resistant plastic to prevent leakage of the composition during its curing. From the other capillary end a silica optical fiber 100 mm in length was inserted close to the photoactive composition. To provide correct operation, the protective polymer coating from the fiber end was removed and the end face was polished. In order to transfer the converted portion of radiation at wavelength λ = 440 nm the multimode fiber with a pure silica core and thin fluorosilicate cladding was used. The transmission of the fiber in the visible range exceeds 99 %/m. The core diameter and numerical aperture of the fiber were equal to 200 μm and 0.22 respectively. Such parameters enabled to simplify focusing of radiation from a xenon lamp on the fiber input end.

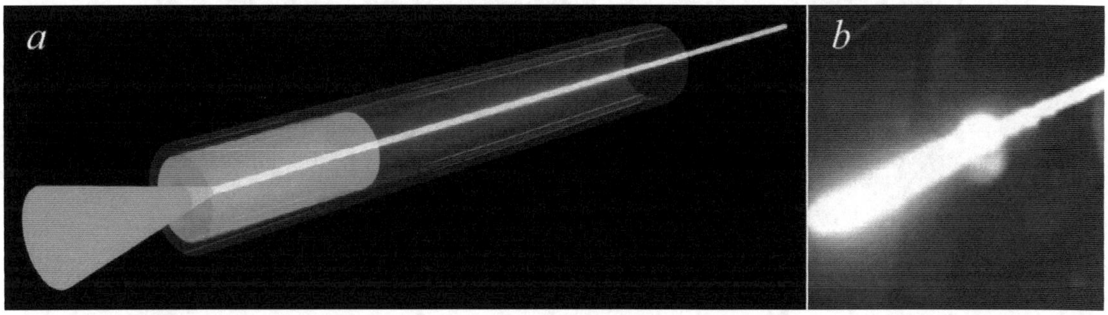

Figure 3. Scheme illustrating the FOS construction with a purple cone of radiation from a xenon lamp at the input (a) and the FOS view right after UV curing of the photoactive composition (b).

Following the contact of the photoactive composition and the optical fiber, the last one was moved towards the end face of the capillary (Figure 3a) with the use of a specially fabricated tool. The tool represented a system of six silica rods 0.26 mm in diameter fixed close to each other on the inner surface of a capillary identical to the basic one. The fiber with outer diameter 0.27 mm was placed in the center of the mentioned rods system, which automatically provided fiber centering relative to the axis of the FOS capillary. The requirement of positioning the fiber close to the end face of the capillary was caused by the necessity to reduce the optical path length – the distance radiation passes through the photoactive composition towards the fiber input end. Due to a relatively high absorbance of the photoactive composition, photoluminescence was excited in a small volume of the material.

Then the construction with the optical fiber rigidly fixed inside the capillary was subjected to the short-term (less than 10 seconds) exposure in the UV furnace under 0.5 J/cm^2 radiation dose. Subsequently, the preliminary installed cap was removed from the end face of the capillary and the centering tool was disconnected from the other one. The capillary once more was placed in the UV furnace for 30 seconds and extracted as the integrated element with the completely cured photoactive composition (Figure 3b).

4. Results and discussion

Figure 4 demonstrates the output end of the optical fiber in the process of the FOS operation. For clarity, two drops of the phosphor-free polymer material PC-3745AP (Fospia Efiron) were applied close to the fiber end that has been cleaned from the protective polymer coating and polished. The FOS was fixed in a mechanical 5-axis positioner in a way that radiation from a xenon lamp XBO 150 W/4 (OSRAM) was focused on the input end of the fiber after passing a monochromator MDR-23 (LOMO) and the capillary filled with the photoactive composition. The use of the 5-axis positioner allowed focusing UV radiation more precisely in order to provide visually intensive blue emission at the fiber output ($\lambda \sim$ 440 nm). As can be seen from Figure 4a, the drop of the fluoroacrylate PC-3745AP transparent in the visible part of the spectra applied close to the fiber output became blue. This indicates that the optical signal at the FOS output has a relatively high intensity. The efforts to focus UV radiation in the plane of the capillary end face or space between the capillary end face and the fiber input led to the absence of the optical signal at the FOS output.

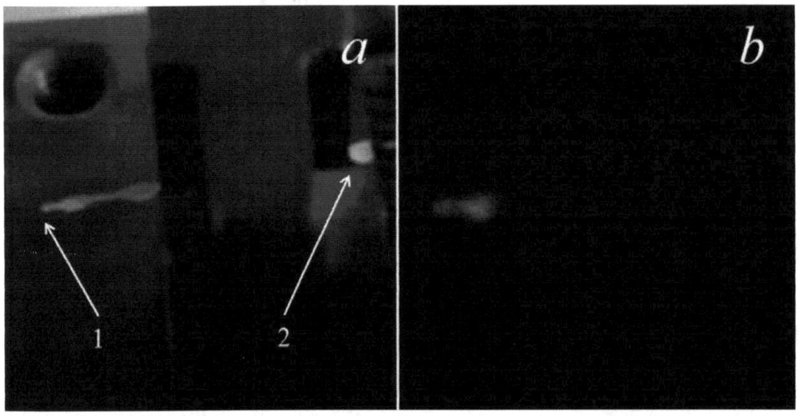

Figure 4. Visualized spectral conversion of the UV-A radiation by the FOS fixed in a mechanical positioner under daylight (a) and in a semi-dark room (b). In Figure 4a: 1 – optical fiber end face, 2 – silica capillary.

To quantitatively confirm the visually registered effect of photoluminescence, spectral characteristic of optical signal at the FOS output was measured. The measurement was performed on the experimental setup described in [17]. The optical scheme was supplemented by an optical amplifier eLockIn 204 (Anfatec Instruments) and an optical power meter AQ-1135E (Ando Electric Corporation) connected with a silicon photodetector FPD510-FV (Menlo Systems). Figure 5a demonstrates the measured characteristic. Note that validity of photodetector indications was influenced by the extrinsic radiation. For that reason, the measurement was performed in a semi-dark room with a minimum noise level.

The character of the curve in Figure 5a indicates an increase of the optical signal at least on 3 dBm above the basic transmission level of the FOS near the photoluminescence emission wavelength excited in the phosphor-based photoactive composition. In the authors' view, this fact confirms the conversion of $\lambda \sim 365$ nm radiation focused on the optical fiber input into $\lambda \sim 440$ nm radiation after passing the capillary segment filled with the photoactive composition. The absence of the optical signal in the wavelength range $\lambda < 400$ nm in the spectral characteristic is related to the sensitivity of the silicon photodetector functioning in the spectral range $\lambda = 400$-900 nm.

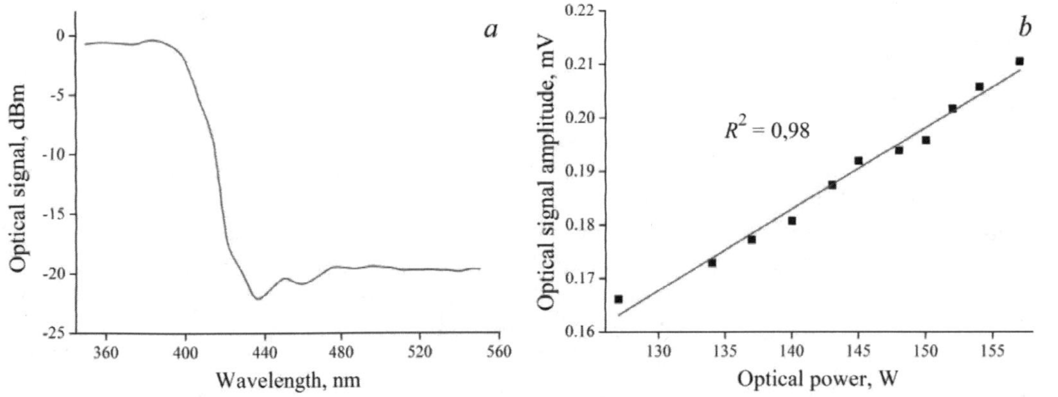

Figure 5. Measured spectral characteristic of the FOS transmission (a) and dependence of the FOS optical signal amplitude on the UV radiation power (b).

Experimentally, the FOS potential to be implemented as a UV-A radiation sensor was evaluated by measurement of the optical signal amplitude at wavelength of 440 nm with an increase of the UV radiation power at wavelength of 365 nm. For this purpose, the optical power meter in the optical scheme was replaced with a digital oscilloscope Infinium HP54830 (Agilent Technologies). As a result of experimental testing, it was revealed that dependence of the FOS optical signal amplitude on the UV radiation power has a linear character with a 98% approximation confidence interval (Figure 5b). Thus, based on the determined linear dependence it is possible not only to detect UV radiation in the environment but also to qualify changes of its relative power.

5. Conclusions

An experimental study of the luminescent sensor for the UV-A radiation detection was reported. The device comprised of a silica capillary partially filled with the organic phosphor 1,3,5-triphenyl-4,5-dihydro-1H-pyrazole containing photoactive composition which contacted the silica multimode optical fiber. Functioning of the sensor was driven by the spectral conversion of the UV-A light ($\lambda = 365$ nm) delivered by the fiber to the phosphor-based photoactive composition. The results demonstrated the possibility to register an optical signal of the sensor at the photoluminescence emission wavelength of the phosphor $\lambda = 440$ nm by choosing a components ratio in the initial solution used for the synthesis of the photoactive composition (1 g organic phosphor, 5 g organic solvent, 5 g epoxyacrylate) and conditions of the fiber input end excitation. The potential of the developed sensor to be used for the UV-A radiation detection was confirmed by the linear dependence of the optical signal amplitude at the sensor output on the UV radiation power. Among the benefits of the sensor there are an availability of raw materials and simplicity of fabrication, no need to change a construction of the fiber by partial removal of the cladding and/or deposition of coatings based on optically active materials, as well as an opportunity to function at considerably long distances far from sources of UV radiation.

Acknowledgements
This research is supported by the Russian Science Foundation (Project No. 19-19-00596).

References

[1] Joza A V, Bajic J S, Stupar D Z, Slankamenac M P, Jelic M and Zivanov M B 2012 *Telfor Journal* **4** 133
[2] Miluski P, Dorosz D, Kochanowicz M, Zmojda J and Dorosz J 2014 *Proc. SPIE* **9290** 92900G
[3] Agafonova D S, Kolobkova E V, Ignatiev A I, Nikonorov N V, Shakhverdov T A, Shirshnev P S, Sidorov A I and Vasiliev V N 2015 *Opt. Eng.* **54** 117107
[4] Miluski P, Kochanowicz M, Zmojda J and Dorosz D 2016 *Metrol. Meas. Syst.* **23** 615
[5] Zolkapli M, Sani A and Saad G 2020 *IEEE 8th International Conference on Photonics* **1** 113
[6] www.who.int
[7] McSherry M, Fitzpatrick C and Lewis E 2004 *IEEE Sens. J.* **4** 619
[8] Yan Y, Zhang X, Li H, Ma Y, Xie T, Qin Z, Liu S, Sun W and Lewis E 2018 *Sensors* **18** 3754
[9] Cho H-T, Seo G-S, Lim O-R, Shin W, Jang H-J, and Ahn T-J 2018 *Curr. Opt. Photon.* **2** 303
[10] Grattan K T V and Sun T 2000 *Sens. Act. A: Phys.* **82** 40
[11] Jackson R G 2004 *Novel sensors and sensing* (London: CRC Press)
[12] Fitzpatrick C, Lewis E, Al-Shamma'a A, Pandithas I, Cullen J and Lucas J 2001 *Opt. Rev.* **8** 459
[13] Lyons W B, Fitzpatrick C, Flanagan C and Lewis E 2004 *Sens. Act. A: Phys.* **115** 267
[14] Zmojda J, Dorosz D, Kochanowicz M, Miluski P, Czajkowski K and Ragin T 2013 *Proc. SPIE* **8903** 890306
[15] Zmojda J, Kochanowicz M, Miluski P and Dorosz D 2014 *Fibers* **2** 150
[16] Agafonova D S and Sidorov A I 2011 *J. Opt. Technol.* **78** 735
[17] Matrosova A S, Evstropiev S K, Mironov L Yu, Nikonorov N V, Komarov A V and Demidov V V 2019 *Opt. Spectr.* **127** 746
[18] Zhang L, Liu J, Gao J, Zhang F and Ding L 2017 *Molecules* **22** 1304

Magneto-Optical Spectra of Magnetic Photonic Crystal with Composite (SiO_2-Au) Layer

T V Mikhailova, S V Osmanov and V O Boyko

V.I. Vernadsky Crimean Federal University, Simferopol 295007, Russia

taciamikh@gmail.com

Abstract. The resonant enhancement of magneto-optical effects due to structure modes arising at the boundary of magnetic photonic crystal $[TiO_2 / SiO_2]^m$ / iron garnet / SiO_2 / (SiO_2-Au), in which the upper layer (SiO_2-Au) is a composite layer of SiO_2 with metallic Au nanoscale inclusions, and iron garnet is a bi-layer of composition $Bi_{1.0}Lu_{0.5}Gd_{1.5}Fe_{4.2}Al_{0.8}O_{12}$ / $Bi_{2.3}Dy_{0.7}Fe_{4.2}Ga_{0.8}O_{12}$, has been considered by modelling of 4×4 transfer matrix method.

1. Introduction

The synthesis and investigation of magnetic photonic crystals (MPC) with one-dimensional (1D), two-dimensional (2D) and three-dimensional (3D) structuring attracts the attention of researchers due to the implementation of new effects associated with the propagation of light waves inside such structures [1]. Structuring opens up the possibility of localizing and amplifying the electromagnetic field of a light wave inside magnetic layers of 1D structures or magnetic elements of various shapes of 2D and 3D structures, creating resonance features in optical and magneto-optical (MO) spectra. The use of metal and composite metal-dielectric components allows transformation of properties due to the excitation of propagating surface plasmon-polaritons, localized plasmons and optical Tamm states (OTSs). It should be noted, that the properties of MPCs based on composites have not been considered previously. The authors present the modelled optical and MO spectra of a similar structure – MPC with a composite layer (SiO_2-Au).

2. Structure model and methods

The structure under consideration can be represented by the general formula

$$[TiO_2 / SiO_2]^4 / G1 / G2 / (SiO_2\text{-Au}). \quad (1)$$

There, $[TiO_2 / SiO_2]^4$ is Bragg mirror made of silicon and titanium dioxide layers with repetition number $m = 4$; G1 is bismuth-substituted garnet layer $Bi_{1.0}Lu_{0.5}Gd_{1.5}Fe_{4.2}Al_{0.8}O_{12}$; G2 is bismuth-substituted garnet layer $Bi_{2.3}Dy_{0.7}Fe_{4.2}Ga_{0.8}O_{12}$; SiO_2 is a buffer layer of silicon dioxide and (SiO_2-Au) is a composite layer with nanoparticles of Au.

As a comparison, we also present the MO spectra of MPC with an upper continuous layer of Au:

$$[TiO_2 / SiO_2]^4 / G1 / G2 / SiO_2 / Au. \quad (2)$$

In order to simulate the optical and MO properties of MPC and numerically solve Maxwell's equations, a software algorithm was implemented using the transfer matrix method 4×4. The software algorithm takes into account the angle of incidence of light on the structure θ, incident light wave polarization state Ψ, anisotropy and gyration of structure layers [2]. The dielectric constant of composite (SiO_2-Au) layer was calculated based on the Maxwell-Garnett model [3]. Nanocomposite is an uniaxial crystal. Its effective permittivity is represented by a diagonal tensor with components $\varepsilon_x = \varepsilon_{\|}$ and $\varepsilon_{y,z} = \varepsilon_{\perp}$ (the subscripts $\|$ and $\perp$ refer to two orientations of electric field vector of electromagnetic wave: parallel and perpendicular to optical axis of metal nanoparticles, respectively). The components can be calculated using the following expression:

$$\varepsilon_{\perp,\|} = \varepsilon_m \left(1 + \frac{f\left(\varepsilon_p - \varepsilon_m\right)}{\varepsilon_m + (1-f)(\varepsilon_p - \varepsilon_m) g_{\perp,\|}} \right), \quad (3)$$

where ε_m and ε_p are dielectric constants of metal inclusions and dielectric matrix, respectively; f is volume fraction of inclusions; $g_{\perp,\|}$ are geometric factors that take into account the effect of nanoparticles shape on the induced dipole moment of nanoparticles. We assumed that $g_{\perp} = g_{\|} = 1/3$. That is, the nanoparticles are spherical. The sizes of nanoparticles in the matrix should be much less than the wavelength, that is, should not exceed 30 nm.

The structures under consideration are shown schematically in figure 1. The substrate of fused quarts is also take into account. All optical and MO parameters of the layers were determined as a result of measurements and fitting of optical and MO spectra of single layers forming the structures. The procedures are described in more detail in the works [4, 5].

The following general invariant parameters of the layers were chosen. The thicknesses of layers in Bragg mirror are $h(TiO_2) = 76$ nm and $h(SiO_2) = 108$ nm. The iron garnet layers have the thicknesses $h(G1) = 80$ nm and $h(G2) = 250$ nm, respectively. Normal incidence $\theta = 0$ deg of light was taken into account.

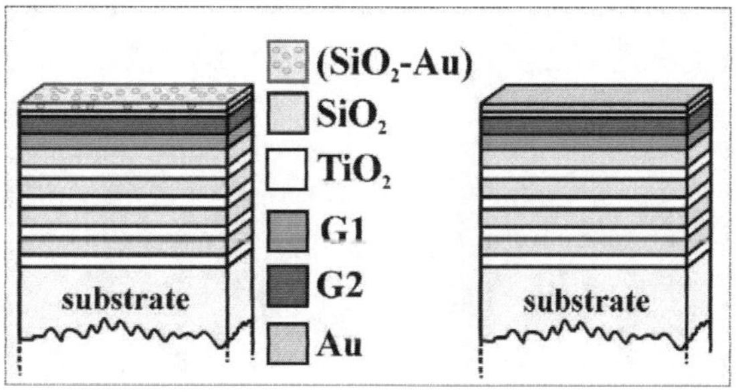

Figure 1. Schematic representation of structure (1) (a) based on composite (SiO_2-Au) and (2) (b) based on a layer of Au.

3. Spectral dependences of Faraday Effect and transmittance for structural modes

Figure 2 shows the formation of optical and MO spectra and resonances for structure (1) in the case of $f = 0.2$. The composite possess only positive ε component at these fraction in investigated wavelength region. As a result, all resonances I, II and III should be attributed to resonance defect mode (RM) corresponding of G1 / G2 / SiO_2 / (SiO_2-Au) layer. A change in the thickness of any layer of upper part of structure leads to a shift in resonant wavelength to longer wavelengths. The greatest enhancement of Faraday Effect for any of configurations will be observed in 600 nm wavelengths range. This region is characterized by the highest values of the components ε of the composite.

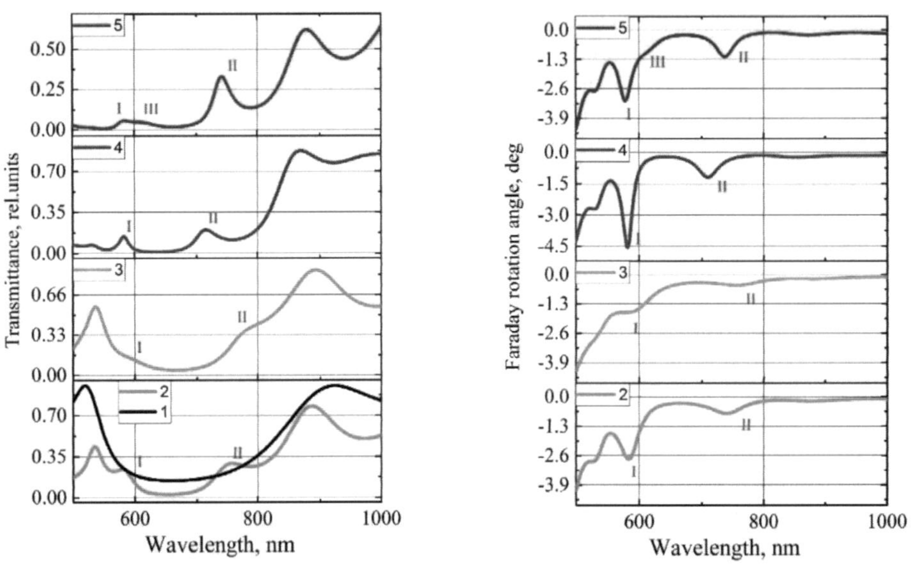

Figure 2. Optical (a) and magneto-optical (b) spectra for different configuration of structure (1): spectra 1 for a Bragg mirror; 2 for Bragg mirror / G1 / G2; 3 for Bragg mirror / G1 / G2 / SiO_2 with $h(SiO_2) = 80$ nm; 4 for the whole structure with $h(SiO_2) = 80$ nm, $h(SiO_2\text{-Au}) = 55$ nm and $f = 0.2$; 5 for the whole structure with $h(SiO_2) = 80$ nm, $h(SiO_2\text{-Au}) = 120$ nm and $f = 0.2$.

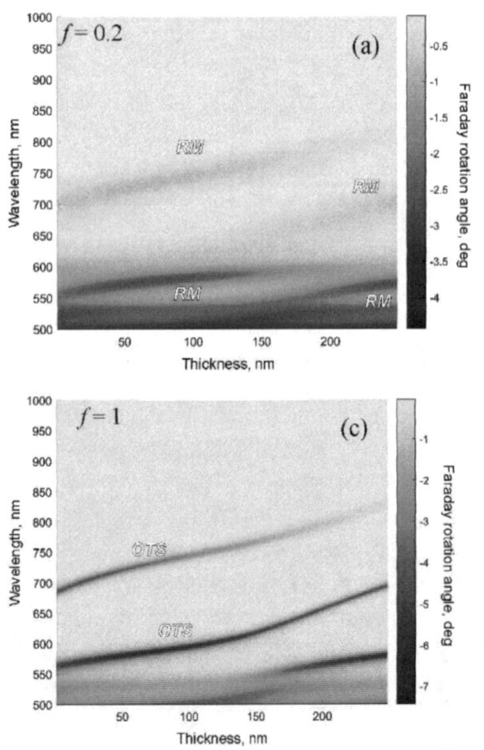

Figure 3. Dependences of magneto-optical spectra on the thickness of top layer SiO_2 and fraction f: structure (1) $f = 0.2$ (a), structure (1) $f = 0.4$ (b), structure (2) or structure (1) $f = 1$ (c). The thicknesses of composite and Au layers were fixed $h(SiO_2\text{-Au}) = 120$ nm and $h(\text{Au}) = 55$ nm, respectively.

A break in the vicinity of 630-650 nm for the spectral dependences of structure (1) on the thickness of top SiO_2 layer (or with a change in the thickness of any layer of upper part of the structure, G1 or

G2) becomes characteristic with an increase in fraction f above 0.35. This discontinuity is due to the fact that in this region ε of the composite passes through zero point. On the left (in the region of short wavelengths) ε has negative values. On the right (in the region of long waves) ε has positive values.

In this case, the excitation of OTS and RM is possible (figure 3). OTS exists only in a certain area – in the area, where the composite is similar to metal [3]. The discontinuity is most pronounced in the spectra of Faraday rotation angle; therefore, we present here only these spectra. At the same time, we still observe the greatest increase in MO rotation for the short wavelength range, where in this case there already exists a OTS.

It should be noted that the existence of OTS depends only on ε-component of composite and fraction. However, we can create several modes simultaneously by changing the thickness of composite layer $h(SiO_2\text{-}Au)$. Figure 4 shows the formation of RM within (SiO_2-Au) layer, which can exist simultaneously with OTS for higher composite concentrations.

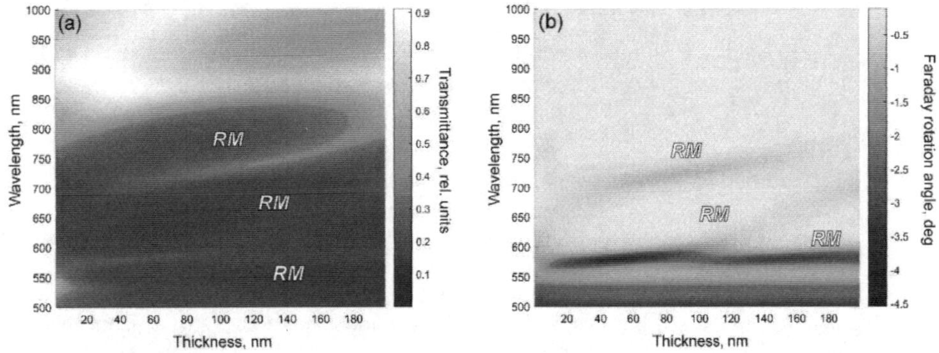

Figure 4. Dependences of magneto-optical spectra on the thickness of composite layer (SiO_2-Au). The thickness of top SiO_2 $h(SiO_2)$ = 80 nm and fraction f = 0.2.

4. Conclusions

As a result of modelling of MO properties of magnetophotonic crystal with a composite layer (SiO_2-Au), it was shown that the highest values of Faraday rotation can be achieved in the short spectral region. The results are interesting for shortwave applications: design of multicolor optical filter / green and blue light modulators for displays.

Acknowledgments

This work was supported by Russian Science Foundation (project No.19-72-20154). Structure development and layer analysis were carried out using the equipment of Common Research Center «Physics and technology of micro- and nanostructures» of the IPM RAS.

References

[1] 2013 *Magnetophotonics*, ed Inoue M, Levy M and Baryshev A V (Heidelberg: Springer-Verlag Berlin Heidelberg)
[2] Passler N C and Paarmann A 2017 *Journal of the Optical Society of America* B **34** 2128
[3] Vetrov S Ya, Bikbaev R G and Timofeev I V 2013 *JETP* **117** 988
[4] Dzibrou D O and Grishin A M 2009 *J. Appl. Phys.* **106** 043901
[5] Mikhailova T V, Berzhansky V N, Shaposhnikov A N, Karavainikov A V, Prokopov A R, Kharchenko Yu M, Lukienko I M, Miloslavskaya O V and Kharchenko M F 2018 *Opt. Mater.* **78** 521

Mass spectrometry analysis of C-dots produced by femtosecond laser irradiation of L-lysine film.

A A Astafiev, A A Gulin, A A Vasin, A M Shakhov, A D Zalessky, A A Osychenko and V A Nadtochenko

N.N. Semenov Federal Research Center of Chemical Physics, Russian Academy of Sciences, Kosygina 4, Moscow, Russia

alina.chemphys@gmail.com

Abstract. The production of carbon dots (C-dots) by femtosecond lasers within living cells and tissues is a novel approach, which has a great potency for intracellular bioimaging. An exact mechanism of fluorescent particles production as well as their composition still remains unknown. In this work we use L-lysine film as a model system to study the mechanism and the composition of C-dots produced by femtosecond laser irradiation investigated by time-of-flight secondary ion mass spectrometry (ToF-SIMS).

1. Introduction

Fluorescent bioimaging is the key method of living cell structure and functioning investigation. Usually bioimaging requires exogenous fluorophore delivery or insertion of genes, coding special fluorescent proteins. A novel approach for bioimaging has been proposed recently [1-3], which is based on the generation of fluorescent carbon dots (C-dots) within a cell or a tissue from its material by femtosecond laser. But chemical-physical processes, leading to C-dots formation, are poorly described and debatable. One research concluded that fluorescent particles are produced as a result of thermally-induced Maillard reaction from proteins and amino acids [4]. Our results indicate that this fluorescent synthesis implies local carbonization and graphitization of the biomaterial [2, 5], induced by laser-generated plasma [6]. However, both processes are possible or a consequence of biomolecule conversion can take place, and their contribution depends on the irradiation parameters.

Complexity and heterogeneity of the chemical composition of cells and tissues make it difficult to analyse the laser synthesis processes of fluorescent products. In this work we used a film made of L-lysine amino acid as a model system to investigate laser synthesis of C-dots. Being subjected to intensive femtosecond laser radiation, L-lysine is known to be an effective precursor for fluorescent species C-dots synthesis. L-lysine film, exposed to 80 MHz and 50 kHz 800 nm femtosecond laser, was then analysed by ToF-SIMS.

2. Methods

2.1. L-lysine film preparation

10 μl of L-lysine solution (0,5 g/l) was placed at the cover glass (24*24 mm, Heinz Herenz), previously subjected to UV lamp (185 and 254 nm), and then covered with another cover glass, which

was not subjected to UV. This "sandwich" was dried for 24 hours at 40°. Then the upper cover glass was removed, and the smooth L-lysine film remained at the lower glass.

2.2. Laser parameters

Femtosecond laser pulses at 790 nm central wavelength, 40 fs duration and 80 MHz repletion rate were generated by a Ti:sapphire oscillator (Tsunami, Spectra-Physics). Pulses could be optionally frequency-doubled in a nonlinear crystal or their repetition rate could be decreased to 50 KHz by a pulse-picker (Avesta Project). Laser pulses trains were coupled to an optical microscope (Olympus IX71) and focused by an objective lens (40X, 0.7NA). L-lysine film was placed on the microscope sample stage and scanned across the focused laser beam using a piezoelectric scanner (NT-MDT).

2.3. Parameters of irradiation

In these experiments we used following parameters for the irradiation of different areas of L-lysine film: (1) $\lambda = 790$ nm, $\upsilon = 50$ kHz, pulse energy 6 nJ, the area has been irradiated 10 times, area sixe is 100*100 μm, 80*80 points; (2) $\lambda = 790$ nm, $\upsilon = 50$ kHz, pulse energy 6 nJ, the area has been irradiated only once (single irradiation), area sixe is 100*100 μm, 80*80 points; (3) $\lambda = 790$ nm, $\upsilon = 80$ MHz, pulse energy 1,9 nJ, single irradiation, area sixe is 80*80 μm, 30*30 points; (4) $\lambda = 390$ nm, $\upsilon = 80$ MHz, pulse energy 1,3 nJ, single irradiation, area sixe is 100*60 μm, 40*40 points; (5) $\lambda = 390$ nm, $\upsilon = 80$ MHz, pulse energy 0,17 nJ, single irradiation, area sixe is 100*60 μm, 40*40 points. We used 40x objective with NA = 0.55.

2.4. Mass spectrometry imaging by ToF-SIMS

L-lysine film was analyzed by mass spectrometer ToF-SIMS 5 (ION-TOF, Germany). Bi_3^+ cluster ions were used as primary ions. Primary ions dose density was 4 x 10^{11} ions/cm^2. Scanning area was 400 x 400 μm with pixels size ~ 0.8 μm. Chemical maps were obtained for both positive and negative secondary ions.

2.5. Fluorescent spectra measurement

Fluorescence of L-lysine film was locally excited by two-photon absorption of femtosecond laser pulses (790 nm, 40 fs, 80 MHz, 0.5 nJ) focused by the microscope objective lens, then collected by the same lens, coupled to the monocrhomator (Acton SP300i), and its spectra were registered by an intensified CCD camera (PI MAX 2, Roper Scientific).

2.6. Confocal imaging

Confocal images were obtained by LSM 980 confocal system (Carl Zeiss Microscopy, Jena, Germany), 20× Plan-Apochromat objective (NA = 0.8). Excitation of the fluorescence was performed by one-photon mode by 405 nm laser. Emission recorded at 450/600 nm range.

3. Results

Fluorescent products were formed within each area after femtosecond laser irradiation (figure 1). Fluorescent spectra of irradiated and non-irradiated L-lysine film are shown at figure 2 and it has maximum at nearly 510 nm.

The chemical composition of irradiated L-lysine film was studied by ToF-SIMS mass spectrometer. Figure 3 represents chemical maps of irradiated areas for different secondary ions. Lysine ($C_6H_{15}N_2O_2^+$) signal (d) decreases for each irradiated area. The lowest signal was observed in the area 1, which corresponds to the highest irradiation dose. NH_3^+ yield (b) dropped in the area 1 as well, which is likely caused by nitrogen splitting off.

Lysine and NH_3^+ redistribution was clearly observed in the areas 3 and 4 and could evidence lysine diffusion under laser irradiation. Interestingly, sodium signal increased in the area 1. One may suggest that sodium diffused into the film from the surface of substrate (cover glass). However silicon signal was not detected hence the film in the area 1 was not evaporated completely. Ion with m/z = 177 also showed increase yield in the area 1. This ion may be attributed to $C_{12}H_{17}O^+$ - the product of lysine dimerization. On the contrary ion with m/z = 171 (possibly $C_8H_{17}N_3O^+$, $C_9H_{17}NO_2^+$ or $C_9H_{19}N_2O^+$) showed increase intensity in areas 2, 4 and 5.

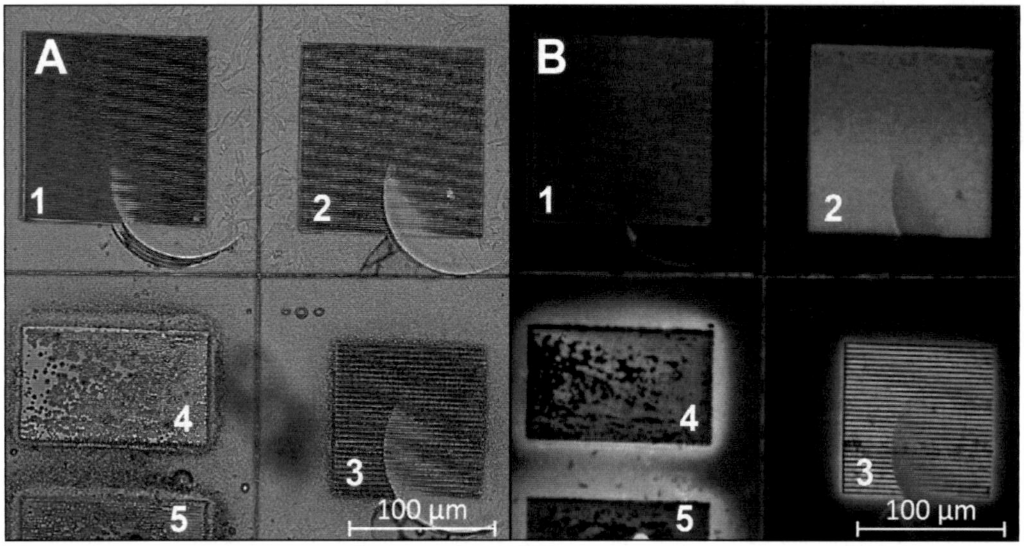

Figure 1. Confocal imaging of irradiated L-lysine film. A – transmitted light image, B – fluorescent image. Numbers indicate areas with different parameters of laser irradiation. Non-irradiated L-lysine is around areas 1-5.

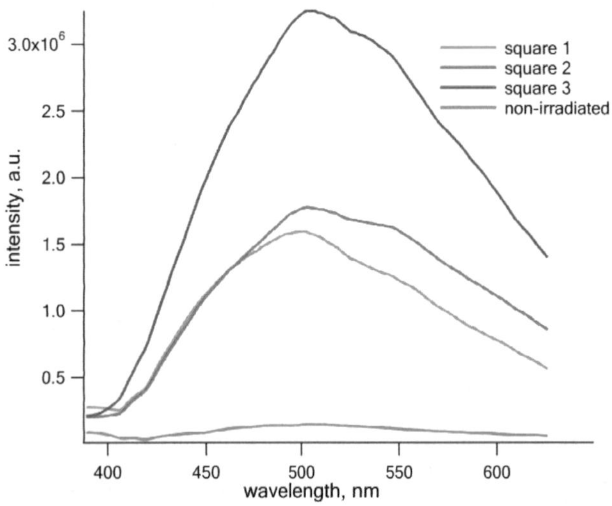

Figure 2. Fluorescence spectra of irradiated and non-irradiated L-lysine.

Alterations in other ion distributions were also observed. The signal of the ion m/z 44 (CO_2^-) dropped in irradiated fields, and the most intensive decrease was observed in the square 1. The intensity of the ion with m/z = 168 is in a good correlation with the irradiation doze, so it is probably the final product of the laser irradiation. Possible formula of this ion are: $C_7H_8N_2O_3^+$, $C_{10}H_6N_3^+$, $C_{12}H_8O^+$, $C_8H_8O_4^+$. As well, ions C_4 (m/z = 48), $C_6H_5^+$ (m/z = 77) and $C_7H_7^+$ (m/z = 91) had an increase in irradiated L-lysine.

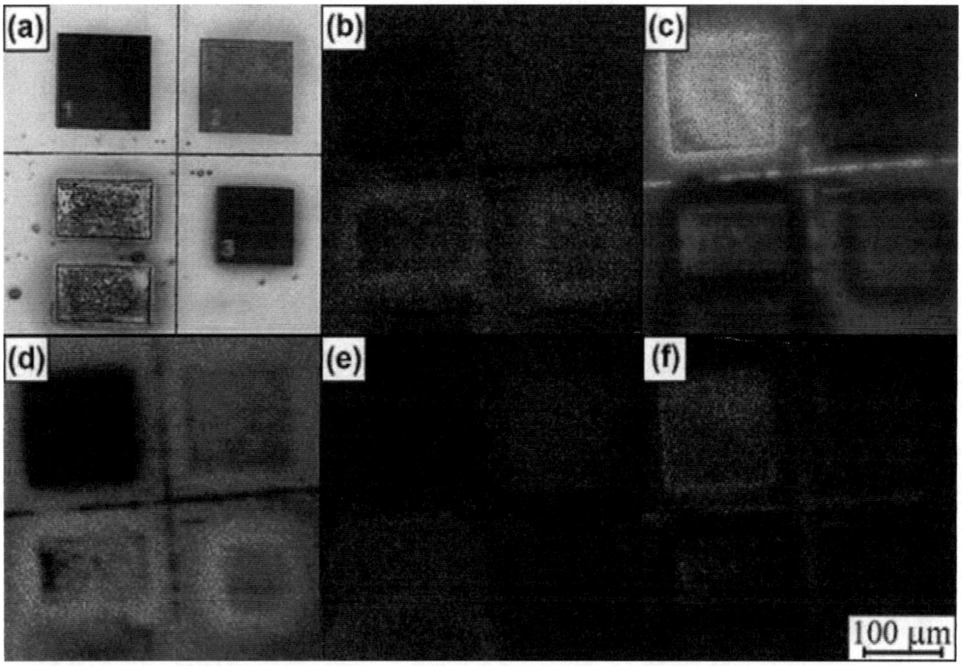

Figure 3. Mass spectrometry imaging of L-lysine film irradiated by femtosecond laser. Numbers indicate areas with different parameters of laser irradiation. A - optical image; B - NH_3^+ distribution, C - Na^+ distribution, D – lysine molecular ion ($C_6H_{15}N_2O_2^+$) distribution, E - distribution of ion with m/z = 171, F - distribution of ion with m/z = 177.

4. Discussion

An average fluorescence intensity in square 1 seems to be lower than in square 2 is spite of 10 times higher irradiation doze (Figure 1, b). It can be explained by photobleaching of fluorophores, produced at the first irradiation, during subsequent irradiations. Thereby, the significant lysine signal decay (figure 3, d) is not a consequence of C-dots formation. The fact that with an increase in the duration of the C-dots synthesis reaction or an increase in the synthesis temperature, the products became less fluorescent, is in a compliance with our results.

C-dots formation under the action of second harmonic laser radiation ($\lambda = 390$ nm) is also possible, but 390 nm radiation has lower order of nonlinear absorption. It is clearly visible by melting of 4 and 5 squares (figure 1).

L-lysine dimers and oligomers are not fluorescent. Fluorescence of irradiated L-lysine could occur as a result of carbonization and aromatization leading to carbon dots-like products. It is confirmed by C_4 (m/z = 48), $C_6H_5^+$ (m/z = 77) and $C_7H_7^+$ (m/z = 91) ions signal increase. Although it is difficult to accurately identify the final product at the moment, it should be noted that the position of mass-spectrometry peaks indicates either a very small amount of hydrogen relative to carbon, or a noticeable content of oxygen and / or nitrogen. This most likely indicates the occurrence of carbonization and oxidation processes during irradiation, and oligo- and poly- L-lysine could be the intermediate products, from which C-dots are produced.

5. Conclusion

C-dots formation from L-lysine film is possible for both 790 and 390 nm femtosecond laser radiation, but 790 nm allows to produce C-dots more precisely. The mass-spectrometry analysis has performed that greater yield of L-lysine polymerization products was achieved by the laser radiation with higher pulse energy. Carbonization and aromatization can be involved in the process of C-dots formation.

Acknowledgments

The work has financial support from the Russian Foundation for Basic Research grant 19-53-52007 in the part of ToF-SIMS analysis. Laser synthesis of C-dots was supported by the Russian Scientific Foundation grant No.21-72-20169. The work was performed on facilities of ACBS Center of the Collective Equipment (no. 506694, FRCCP RAS) and large-scale research facilities No. 1440743.

References

[1] Sun Q, Qin Z, Wu W, Lin Y, Chen C, He S, Li X, Wu Z, Luo Y and Qu J 2018 *Biomed. Opt. Express* **9** 581-90

[2] Astafiev A A, Shakhov A M, Osychenko A A, Syrchina M S, Karmenyan A V, Tochilo U A and Nadtochenko V A 2020 *ACS OMEGA* **5** 21 12527-38

[3] Osychenko A, Zalessky A, Astafiev A, Shakhov A, Kostrov A, Krivokharchenko A and Nadtochenko V 2020 *Exp. Cell Res.* 111887

[4] Qin Z, Sun Q, Lin Y, He S, Li X, Chen C, Wu W, Luo Y and Qu JY 2018 *Biomed. Opt. Express* **9** 3373-3390

[5] Astaf'ev A A, Zalesskii A D, Shakhov A M, Osychenko A A and Nadtochenko V A 2016 *High Energy Chemistry* **50** 421-423

[6] Vogel A, Noack J, Hüttman G and Paltauf G 2005 *Appl. Phys.* **81** 1015-1047

Control of the mode composition of optical radiation in a microstructured fiber

G A Pchelkin[1, 2], V B Fadeenko[1, 3], V V Davydov[1, 4] and V Yu Rud[4]

[1] Peter the Great St. Petersburg Polytechnic University, Polytechnicheskaya street, 29, Saint-Petersburg, Russia, 195251
[2] Joint Stock Company «Scientific and Production Association S. I. Vavilov State Optical Institute», Babushkina street, 36, bldg.1, Saint-Petersburg, Russia, 192171
[3] Joint Stock Company «Concern «Granit-Electron», Gospitalnaya street, 3, Saint-Petersburg, Russia, 191014
[4] All-Russian Research Institute of Phytopathology, Institutsky pr., 5, Moscow region, Russia, 143050

beegrig@mail.ru

Abstract. The construction structure of microstructured fibers is considered. A research scheme of the mode composition and defects control in optical fibers is developed. A microstructured fiber for studying optical vortex fields has been developed and manufactured. The results of studies of the same fiber structure and the distribution of optical radiation depending on the parameters of the technological cycle of its production are presented.

1. Introduction

The development of science and technology is inextricably linked with the increase in the transmitted information volume [1-12]. Fiber-optic lines (FOCL) have the greatest advantages among terrestrial communication networks [1, 2, 7-10, 13-19]. It should be noted that FOCL is also included in the systems of cellular and satellite communication lines, performing an important function in their composition [20-26]. The main problem in the fiber optic communication system modernization is the limit on the speed and volume of data transmitted over a single-mode fiber, both on trunk and local communication lines [1, 2, 7, 9, 27-32].

Spectral compaction technologies in communication channels have reached their limit, and further compaction leads to loss information or a sharp increase in signal attenuation during transmission [33-37]. The use of multimode fiber in trunk communication lines is associated with great difficulties in providing a thermal stabilization mode, the use of which at long distances is not always justified [1, 2, 29-34]. Therefore, various developments and studies are being conducted to solve this problem. One of the promising directions is the use of microstructured fiber with a given mode composition in communication systems [38-41]. The latter is quite a difficult task, especially with changes in temperature, pressure and other negative factors within large limits. Therefore, the technology of manufacturing microstructured fiber and the study of the influence of various factors on the mode composition in the fiber and its stability is given increased attention. It is worth noting that the parameters of the technological production of such fibers are selected at the first stage by modeling, but the main stage is experimental research, since it is impossible to take everything into account in the calculation. The paper presents one of the possible solutions based on the conducted research.

2. Manufacturing technology of microstructured fiber and an experimental monitoring its modal composition stand.

The technology of manufacturing a microstructured fiber with a given modal composition consists in choosing the internal structure, which is assembled with the help of capillaries. Then it is inserted into the quartz tube for further extraction. An example of a billet is shown in Figure 1.

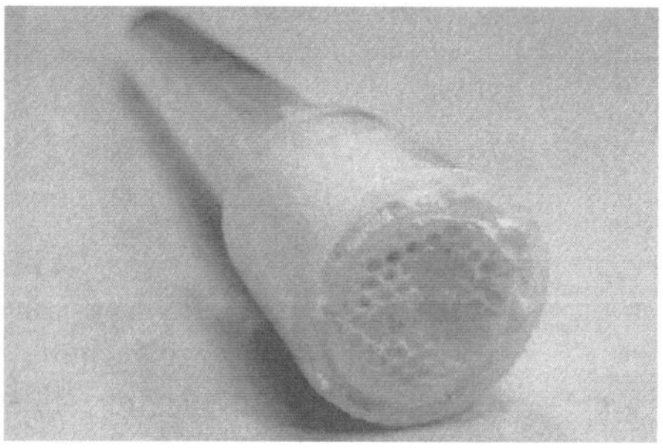

Figure 1. Fiber extraction billet.

Fibers with concentric birefringent or non-concentric cores are twisted at high speed when passing through a miniature furnace in a pulling unit. The furnace temperature and pressure can be adjusted. As a result, an optical fiber consisting of real gratings is formed at the output (Fig. 2).

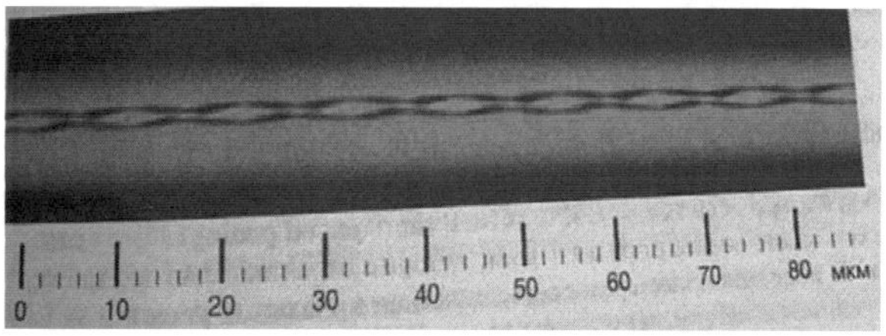

Figure 2. Side view of the double-spiral chiral grid.

A stand was developed for detecting vortexes in chiral microstructured fibers with an annular core geometry and a broken shell symmetry. The previously proposed schemes have large dimensions and increased requirements for temperature stabilization.

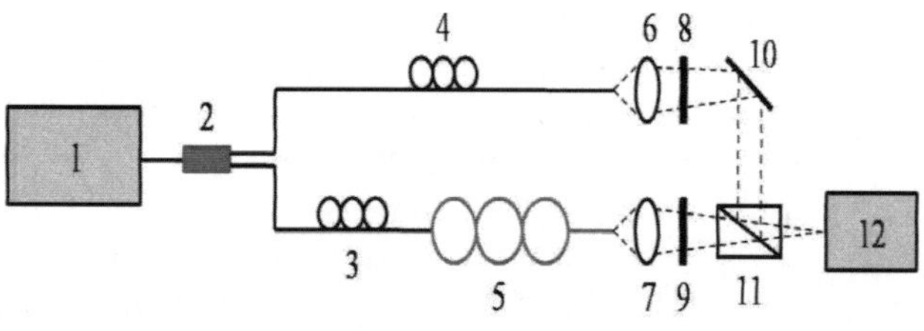

Figure 3. Optical scheme for vortex detection: 1 – semiconductor laser with a wavelength of 1550 nm, 2 – 50/50 coupler, 3, 4 – polarization controller, 5 – chiral photonic-crystal light guide, 6, 7 – objective, 8, 9 – polarizer, 10 – mirror, 11 – beam mixer, 12 – CCD camera.

The semiconductor laser radiation with a wavelength of 1550 nm enters the input of the optical divider 1:2. One of its outputs is connected to the fiber under study 5. Radiation through the system (lens and polarizer) enters the mixer 11 from the output of the fiber 5. The second part of the laser beam is directed to the second input of the mixer. A radial pattern of the field distribution is recorded in chamber 12.

It should be noted that the spatial modes LP_{01}, LP_{11}, LP_{21} have a similar transverse distribution of the radiation intensity. Therefore, when they are summed up on a photodetector, the most stable result is obtained from the transmitted optical signals (there should be a clear ring in the field distribution figure).

3. Results of simulation and experimental research

Before microstructured fiber manufacturing its running characteristics are calculated, since it is necessary to take into account the peculiarities of the technological cycle during its production. The calculation was performed in the COMSOL Multiphysics program for the following parameters for a ring fiber: the number of circular holes orders – 4; the structure pitch (the distance between adjacent holes) – 10.35 microns; the hole diameter – 6.21 microns (the ratio of the hole diameter to the structure pitch – 0.60); the outer diameter of the fiber – 125 microns; the laser wavelength – $\lambda = 1550$ nm.

In Figure 4 shows, for example, the dependence of the value of the effective (waveguide) refractive index for the first twelve spatial modes, the propagation of which is uniquely supported in a microstructured (ring) fiber.

The analysis of the obtained results shows that the spatial modes LP_{01}, LP_{11}, LP_{21}, which we need to obtain in the fiber for efficient information transmission, have similar values n_{eff}.

In Figure 5 shows the results of calculating the distribution of the radiation intensity in the cross section for different modes in a microstructured (ring) fiber, taking into account the obtained n_{eff} values. For the convenience of comparison, all information is presented in the form of a table.

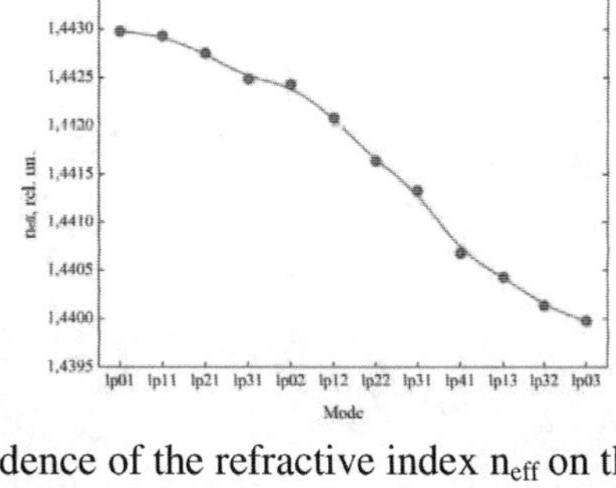

Figure 4. Dependence of the refractive index n_{eff} on the mode number.

Mode	LP_{01}	LP_{11}	LP_{21}
Transverse distribution	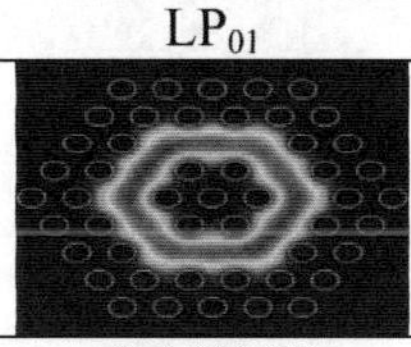	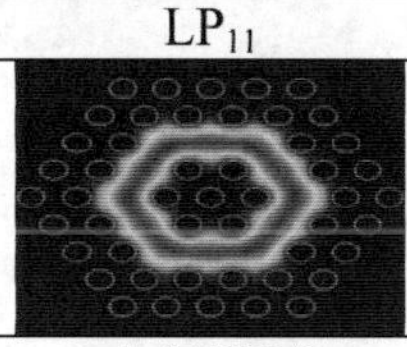	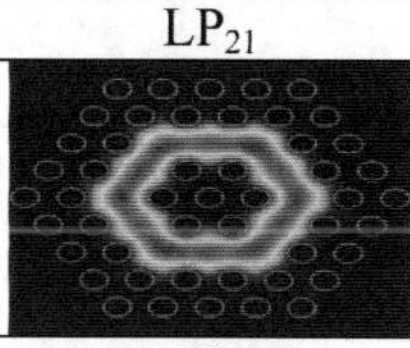
n_{eff}	1,442975	1,442925	1,442745
Mode	LP_{31}	LP_{02}	LP_{12}
Transverse distribution	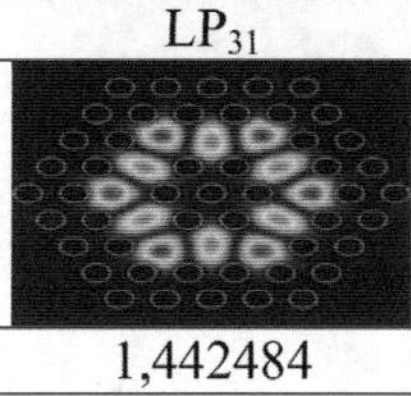	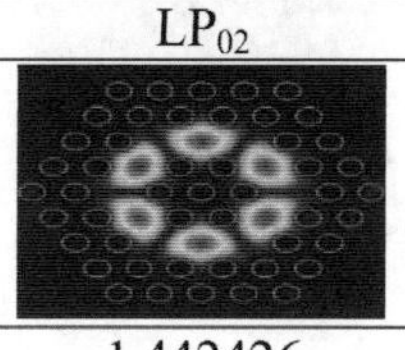	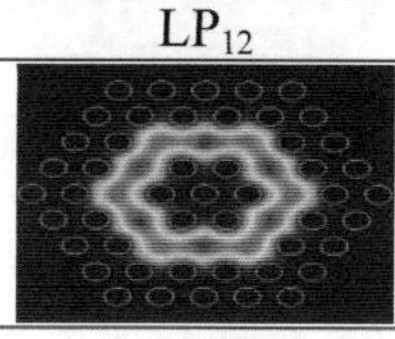
n_{eff}	1,442484	1,442426	1,442077

Mode	LP_{22}	LP_{31}	LP_{41}
Transverse distribution	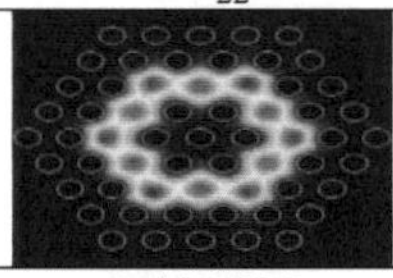	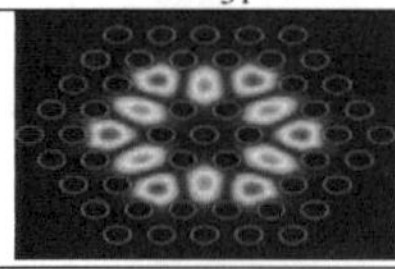	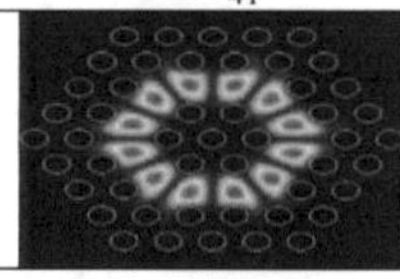
n_{eff}	1,441635	1,441328	1,440678
Mode	LP_{13}	LP_{32}	LP_{03}
Transverse distribution	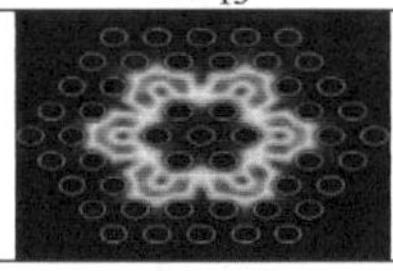	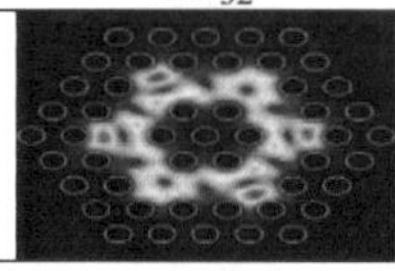	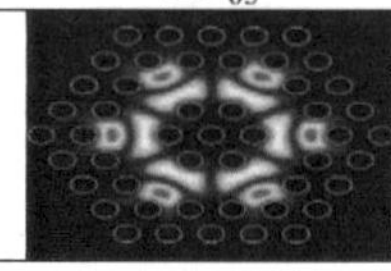
n_{eff}	1,440427	1,440134	1,439976

Figure 5. Calculated spatial distribution of laser radiation in cross-section for different models.

The great possibilities of this optical fiber structure are shown by the obtained theoretical calculations. In Figure 6 shows as an example two topologies of chiral microstructured fibers, for which the topology of the distribution of optical radiation at the output of them was studied for the cases of their manufacture at different twisting frequency ω_c and other parameters of the technological cycle.

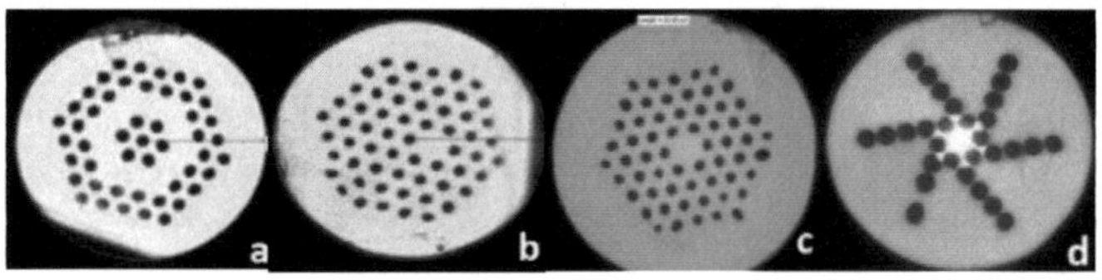

Figure 6. Cross-sectional images of photonic-crystal fibers: a) ring-core fiber , b) displaced core fiber, c) hexagonal structure fiber and d) star-core fiber.

In Figure 7 shows, as an example, the distribution of optical radiation at the microstructured fiber output at different speeds of rotation of the workpiece per meter of fiber length. The fiber topology is shown in Figure 6 (d). The furnace temperature is T = 1700 K, pressure P = 2.34 atm.

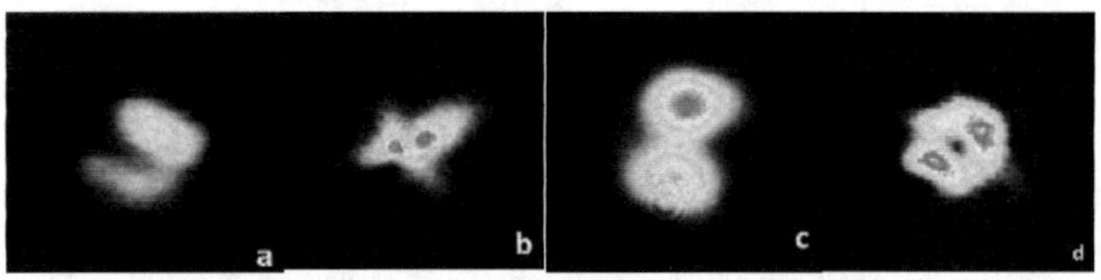

Figure 7. Images of cross-sections of microstructured fibers at different rotational speeds. Images a), b), c) and d) correspond to the following rotation speeds in revolutions per meter: 10; 20; 65 and 200.

The research results showed a significant dependence of the mode composition of the fiber on the speed of rotation of the workpiece. They were also calculated using the developed mathematical model of the mode distribution inside some types of microstructured fibers, which helps in obtaining the desired fiber. It should be noted that the use of spatial modes LP_{01}, LP_{11}, LP_{21} strongly affects the stability of the final signal due to the similar geometry of the intensity distribution.

4. Conclusion

The obtained experimental results (cross-section images) allow us to establish the dependence of the number of modes that are present in the microstructured fiber on the speed of rotation of the workpiece per meter of fiber length, temperature and pressure in the furnace. These data allow to adjust the process of forming the structure of the workpiece, to set the optimal range of changes in the values of ω_c, temperature and pressure during the technological process to obtain clear three modes.

It should be noted that the stand design developed by us allows to determine the mode composition and vortex fields in various microstructured fibers. Increasing the contrast of the recorded images will improve the accuracy of determining the temperature T and pressure P values. The step of changing the operating mode is 5 revolutions per meter on the engine. It should be reduced to 1 turn per meter for more thorough research in the future.

References

[1] Grebenikova N, Moroz A, Bylina M and Kuzmin M 2019 *IOP Conference Series: Materials Science and Engineering* **497(1)** 012109

[2] Burdin V A, Dashkov M V, Demidov V V, Zhukov A E and Bourdine A V 2020 *Fibers* **8(3)** 18

[3] Podstrigaev A S, Smolyakov A V, Myazin N S, Grebenikova N M and Davydov R V 2019 *Lecture Notes in Computer Science (including subseries Lecture Notes in Artificial Intelligence and Lecture Notes in Bioinformatics)* **11660 LNCS** 525-533

[4] Podstrigaev A S, Smolyakov A V, Myazin N S and Slobodyan M G 2018 *Lecture Notes in Computer Science (including subseries Lecture Notes in Artificial Intelligence and Lecture Notes in Bioinformatics)* **11118 LNCS** 509-515

[5] Makolkina M, Pham V, Kirichek R, Gogol A and Koucheryavy A 2018 *Lecture Notes in Computer Science (including subseries Lecture Notes in Artificial Intelligence and Lecture Notes in Bioinformatics)* **11118 LNCS** 547-559

[6] Hoang T, Kirichek R, Paramonov A and Koucheryavy A 2016 *Lecture Notes in Electrical Engineering* **376** 1249-1259

[7] Sivers M and Fokin G 2015 Lecture Notes in Computer Science (including subseries Lecture Notes in Artificial Intelligence and Lecture Notes in Bioinformatics) **9247** 393–396

[8] Andreeva E I and Potapov I A 2020 *Journal of Physics: Conference Series* **1695(1)** 012125

[9] Andreeva E I and Potapov I A 2021 *Springer Proceedings in Physics* **255** 241–245

[10] Davydov R V, Dmitrieva D S, Pilipova V M, Dudkin V I and Andreeva E I *2020 Proceedings of International conference of Laser Optics ICLO-2020 (Saint-Petersburg),* vol. 9285820 (IEEE), p. 171

[11] Davydov R, Antonov V, Makeev S and Batov Y 2019 *E3S Web of Conferences* **140** 02001

[12] Davydov R, Antonov V and Moroz A 2019 IEEE International Conference on Electrical Engineering and Photonics (EExPolytech-2019), vol. 8906791 (IEEE), p. 295-297

[13] Ermolaev A N, Krishpents G P and Vysoczkiy M G 2016 *Journal Physics: Conference Series* **741(1)** 012071

[14] Fadeenko I, Fadeenko V, Reznik V, Moroz A, Popovskiy N and Nikolaev D 2019 *IOP Conference Series: Earth and Environmental Science* **390(1)** 012022

[15] Davydov R V and Antonov V I 2018 *Journal of Physics: Conference Series* **1124(8)** 081037

[16] Davydov R V and Antonov V I 2018 *Journal of Physics: Conference Series* **1135(1)** 012087

[17] Logunov S E, Rud V Y, Davydov R V, Moroz A V and Smirnov K J 2019 *Journal of Physics: Conference Series* **1326(1)** 012024

[18] Moroz A V and Davydov R V 2019 *Lecture Notes in Computer Science (including subseries Lecture Notes in Artificial Intelligence and Lecture Notes in Bioinformatics)* **11660 LNCS** 710 – 718

[19] Filatov D L, Galichina A A, Vysoczky M G and Yalunina T R 2017 *Journal of Physics: Conference Series* **917**(8) 082005

[20] Davydov R V, Saveliev I V, Lenets V A and Tarasenko M Yu 2017 *Lecture Notes in Computer Science (including subseries Lecture Notes in Artificial Intelligence and Lecture Notes in Bioinformatics)* **10531 LNCS** 177-183

[21] Petrov A A 2015 *Lecture Notes in Computer Science (including subseries Lecture Notes in Artificial Intelligence and Lecture Notes in Bioinformatics)* **9247** 739-744

[22] Petrov A A and Myazin N S 2017 *Lecture Notes in Computer Science (including subseries Lecture Notes in Artificial Intelligence and Lecture Notes in Bioinformatics)* **10531 LNCS** 561-568

[23] Petrov A A, Grebenikova N M, Lukashev N A, Ivanova N V, Rodygina N S and Moroz A V 2018 *Journal of Physics: Conference Series* **1038 (1)** 012032

[24] Grevtseva A S, Davydov R V, Dudkin V I and Rud' V Y 2019 *Journal of Physics: Conference Series* **1326(1)** 012043

[25] Valov A P, Davydov R V, Rud V Y and Grevtseva A S 2019 *Journal of Physics: Conference Series* **1326(1)** 012040

[26] Lukashev N A, Petrov A A, Grebenikova N M and Valov A P *2018 Proceedings of 18th International conference of Laser Optics ICLO-2018 (Saint-Petersburg),* vol. 8435889 (IEEE), p. 271

[27] Fadeenko V B and Pchelkin G A 2019 *Journal of Physics: Conference Series* **1400(4)** 044010

[28] Smirnov K J, Glagolev S F and Tushavin G V 2018 *Journal of Physics: Conference Series* **1124(1)** 022014

[29] Dmitrieva D S, Pilipova V M and Valiullin L R 2020 *Journal of Physics: Conference Series* **1695(1)** 012130

[30] Dmitrieva D S, Pilipova V M, Dudkin V I, Davydov V V and Rud V Yu 2020 *Journal of Physics: Conference Series* **1697(1)** 012145

[31] Moroz A V and Davydov V V 2019 *Journal of Physics: Conference Series* **1410(1)** 012212

[32] Moroz A V, Davydov V V, Malanin K Y, Krasnov A A and Rud V Yu 2019 *Journal of Physics: Conference Series* **1400(4)** 044009

[33] Fadeenko V B, Pchelkin G A and Beloshapkina O O 2019 *Journal of Physics: Conference Series* **1410(1)** 012238

[34] Fadeenko V B, Kuts V A and Vasiliev D A 2018 *Journal Physics: Conference Series* **1135**(1) 012053

[35] Andropov A V, Kuzmin S V and Korovin K O 2021 *Springer Proceedings in Physics* 255 675–681

[36] Raimzhanov T R, Kuzmin S V and Korovin K O 2021 *Springer Proceedings in Physics* 255 629–635

[37] Sinicyna E A, Galichina A A, Lukyanov A S and Podstrigaev A S 2019 *Journal of Physics: Conference Series* **1236(1)** 012075

[38] Pidishety S, Pachava S, Gregg P, Ramachandran S, Brambilla G and Srinivasan B 2017 *Opt. Lett.* **42(21)** 4347-4350

[39] Xie Z, Lei T, Li F, Qiu H, Zhang Z, Wang H, Min C, Du L, Li Z and Yuan X 2018 *Light: Sci. Appl.* **7(4)** 18001

[40] Carlon Zambon N, St-Jean P, Milicevic M, Lemaître A, Harouri A, Le Gratiet L, Bleu O, Solnyshkov D, Malpuech G, Sagnes I, Ravets S, Amo A and Bloch J 2019 *Nat. Photonics* **13(4)** 283–288

[41] Shcherbakov A S and Andreeva E I 1996 *Optical Fiber Technology* **2(2)** 127–133

Absorption properties of bromide photo-thermo-refractive glasses doped with Ytterbium

V Pesnyakov, R Kharisova and A Ignatiev

Department of Photonics, ITMO University, St. Petersburg 197101, Russia

Abstract. In this work bromide photo-thermo-refractive glasses with different Ytterbium concentration were explored. It is shown that UV irradiation and subsequent heat treatment of samples with Yb 2 mol.% causes significant red-shift absorption line from 410 nm with temperature lower than T_g to 517 nm with temperature T_g +60 °C. For small concentration of Yb (0,1 mol. %) maximum red-shift in plasmon resonance absorption band is shown as 485 nm and it is the same for heat treatment temperatures from T_g +20 °C to T_g +60 °C.

1. Introduction

Photo-thermo-refractive (PTR) glass is a very promising material that is widely used for photonic applications such as recording the volume Bragg gratings [1], waveguide structures [2], lasers structuring [3] and many others. The main processes in PTR glasses which lead to the formation of Ag nanoparticles and subsequent growth of the AgBr-NaBr shell are the following [4,5,6]. On the first stage samples are exposed under UV irradiation into the Ce^{3+} absorption band (310 nm) which causes electron donation. Most of these electrons are trapped by antimony and only partly trapped by silver ions with formation of neutral silver atom and molecular clusters. The subsequent heat treatment near the glass transition temperature (T_g) leads to the Sb^{5+} donate electrons which are trapped by silver nanoparticles with continuing formation of neutral silver atom and molecular clusters. With subsequent thermal treatment higher than T_g causes a formation of silver nanoparticles and a growth of AgBr-NaBr shell on them.

2. Experiment and results

In this research bromide PTR glasses with Na_2O-ZnO-Al_2O_3-SiO_2 matrix doped with photosensitizers such as CeO_2 (0,007 mol.%), Sb_2O_3 (0,04 mol.%), Ag_2O (0,064 mol.%) and with variable Yb_2O_3 concentration (0,1; 1; 2 mol.%) were studied. On the first stage there was found the transition temperature (T_g) of synthesized glasses. To find the value of glass transition temperature the scanning calorimeter STA 449 F1 Jupiter (Netzsch) was used. The obtained T_g for 0,1 mol.% is 515 °C, for 1 mol.% is 550 °C and for Yb 2 mol.% is 590 °C.

Studied temperatures were chosen according to the following rule: -40, 0, +20, +40, +60 °C from T_g. Hence samples with Yb 2 mol.% were thermally treated with temperatures 550, 590, 610, 630, 650 °C, samples with Yb 1 mol.% temperatures 510, 550, 570, 590, 610 °C were chosen and for Yb 0,1 mol.% temperatures 475, 515, 535, 555, 575 °C were selected. Second, samples were UV irradiated with mercury lamp for 10 minutes and then were thermally treated with described temperatures during 1, 3 and 10 hours in sum. Absorption spectrum were recorded on each step and for each temperature. Obtained results for 3 and 10 hours thermally treated Yb 0,1 mol.% glasses are shown in figure 1. For comparison here and after absorption spectra for sample without UV irradiation (VIR) and after UV irradiation but without thermal treatment (UV) are also shown.

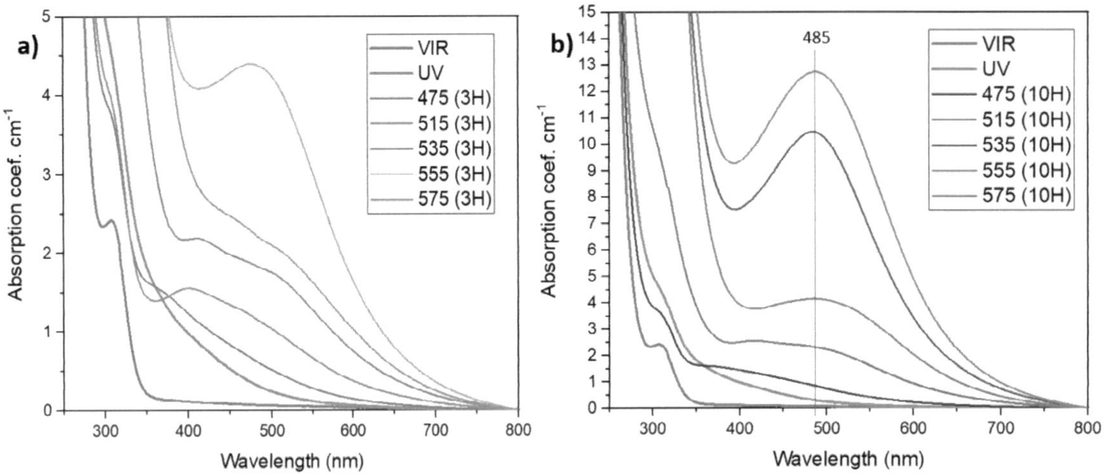

Figure 1. Absorption coefficient spectra for samples with Yb 0,1 mol.% thermally treated for 3 hours **(a)** and 10 hours **(b)**.

With increase of thermal treatment time plasmon resonance absorption band also increases. It should be noticed that for 3 hour thermally treated samples absorption spectra do not have a single plasmon resonance absorption band for temperatures under 555 °C instead it can be seen complex spectrum which commonly says that plasmon nanoparticles are not in a spherical shape, but it was not confirmed by other methods such as XRD analysis. In figure 1 (b) plasmon resonance absorption band does not exceed 485 nm and absorption coefficient decreases significantly after thermal-treatment temperature T_g +40 °C.

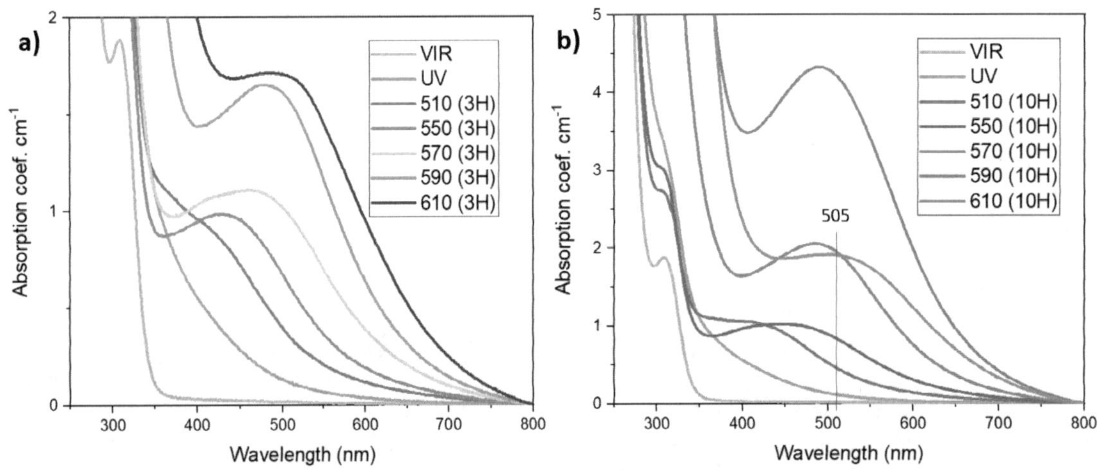

Figure 2. Absorption coefficient spectra for samples with Yb 1,0 mol.% thermally treated for 3 hours **(a)** and 10 hours **(b)**.

In figure 2 is shown absorption coefficient spectra for samples with Yb 1 mol.%. In that case plasmon resonance absorption band red-shifts with the increase of Yb concentration and equal to 505 nm for sample with T_g +60 °C heat treatment temperature. As for Yb 0,1 mol.% heat treatment under the temperature T_g +60 °C causes decreasing in plasmon resonance absorption band coefficient.

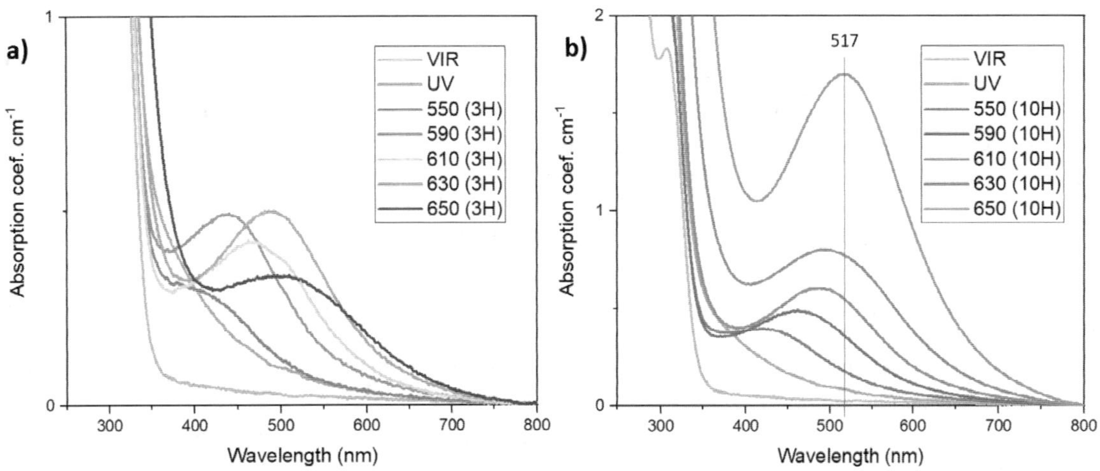

Figure 3. Absorption coefficient spectra for samples with Yb 2,0 mol.% thermally treated with described above temperatures for 3 hours **(a)** and 10 hours **(b)**.

Absorption coefficient spectra for samples with Yb 2,0 mol.% are shown in figure 3. In contrast to absorption coefficient spectra for Yb 1 mol.%. and 0,1 mol.% (for 10h) in which the plasmon resonance absorption band decreases after exceeding the heat treatment temperature T_g +40 °C for the samples with Yb 2 mol.% concentration it increases even under the temperature T_g +60 °C (650 °C) comparing the one for T_g +40 °C (630 °C). Also, the maximum the plasmon resonance absorption band at 517 nm is higher than previously was shown for glass without Yb [5].

Figure 4 shows a photo of Yb 2,0 mol.% samples after 10 hours heat-treatment.

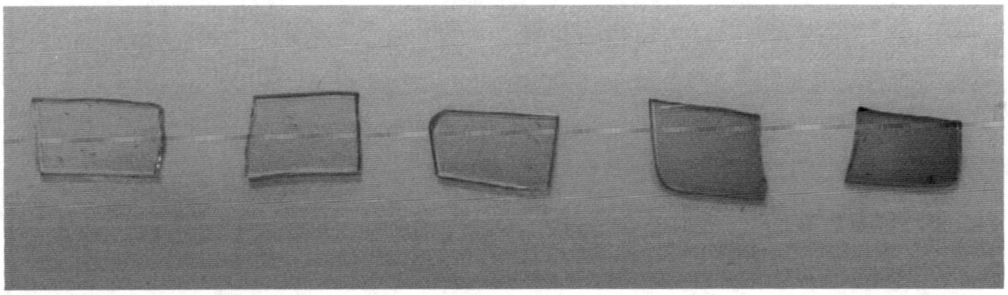

Figure 4. Photo of Yb 2,0 mol.% samples after 10 hours heat-treatment temperatures from left to right: 550, 590, 610, 630, 650 °C.

3. Conclusion

In conclusion, bromide PTR glasses doped with Ytterbium were studied. Absorption spectra for glasses with 0,1; 1 and 2 mol.% Yb were obtained after UV irradiation and subsequent heat-treatment with different temperatures. It was shown that by increasing the Yb concentration it can be achieved a maximum red-shift of plasmon resonance absorption band from 485 nm for 0,1 mol.% Yb to 517 for 2 mol.% Yb but the magnitude of the absorption decreases with increasing the Yb concentration. The magnitude of plasmon resonance absorption band for Yb 2 mol.% does not drops after exceeding the T_g +40 °C heat treatment temperature what is seen for 0,1 and 1 mol.%.

References

[1] Glebov A, Mokhun O, Rapaport A, Vergnole S, Smirnov, V and Glebov, L 2012 Volume bragg gratings as ultra-narrow and multiband optical filters *Proc. SPIE Photonics Europe, Micro Optics* **8428**

[2] Sgibnev Y, Nikonorov N, Vasilev V and Ignatiev A 2015 Optical gradient waveguides in photo thermo-refractive glassformed by ion exchange method. *Journal of lightwave Technology* **33.17** 3730–35

[3] Ivanov S and Aseev V 2014 Resonator free Er-Yb laser based on photo-thermo-refractive (PTR) glass *Solid State Lasers XXIII: Technology and Devices*. **8959** 89591E

[4] Nikonorov N, Sidorov A, Tsekhomskiĭ V and Lazareva K 2009 Effect of a dielectric shell of a silver nanoparticle on the spectral position of the plasmon resonance of the nanoparticle in photochromic glass *Optics and spectroscopy* **107.5** 705–7

[5] Dubrovin V, Nikonorov N and Ignatiev A 2017 Bromide photo-thermo-refractive glass for volume Bragg gratings and waveguide structure recording. *Optical Materials Express* **7.7** 2280–92

[6] Nikonorov N and Sidorov A 2018 Silver nanoparticles in silicate glasses: synthesis, modification and destruction. *Met. nanopart.* 61–123

Development of a fiber-optic system for testing instruments for monitoring nuclear power plants

V M Pilipova[1], V V Davydov[2, 3] and V Y Rud[3, 4]

[1] The Bonch-Bruevich Saint Petersburg State University of Telecommunications, Saint-Petersburg, Russia, 193232
[2] Peter the Great Saint Petersburg Polytechnic University, Saint Petersburg, Russia, 195251
[3] All-Russian Research Institute of Phytopathology, Moscow region, Russia, 14305
[4] Ioffe Institute, Saint Petersburg, Russia, 194021

lera_pilipova@mail.ru

Abstract. The necessity of testing the equipment for monitoring the operation of nuclear power plants is justified. It is proposed to use optical fiber and pulsed laser radiation for these purposes. It is established that optical fiber is more resistant to radiation than other communication systems. The design of a fiber-optic emergency simulator is developed. The calculation of its characteristics is carried out. Their experimental research was carried out. It was found that with an optical signal delay of 98.6 microseconds, the loss is – 26 dB with an uneven frequency response of ± 2 dB. This makes it possible to test the entire set of equipment that uses optical signals used to control a nuclear power plant.

1. Introduction

Reduction of fuel reserves led to the development of nuclear energy [1-7]. Nuclear energy is a dynamically developing high-tech field with broad prospects for progress [8-13]. The environmental problems associated with this atomic energy [14-22], because of its great prospects, are relegated to the background.

The use of nuclear power plants (NPP) also remains relevant on sea vessels, both civil and military, which, in turn, leads to both the emergence of a number of new types of nuclear power plants and the modernization of existing ones [4, 11-13, 20, 23-26].

One of the main problems arising during the operation of nuclear power plants is their safe operation. One of the main conditions for the safe operation of nuclear power plants is the flawless operation of the reactor cooling systems, for the control of the operation of which reliable equipment must be used. For cooling in nuclear power plants, the so-called "heavy" water is most often used, the flow rate of which and its parameters are controlled using flow meters, the principle of which is based on the phenomenon of nuclear magnetic resonance (NMR) [11-13, 25, 27].

In an emergency in a nuclear reactor there is a sharp increase in the level of ionizing radiation, which affects the "heavy" water of the reactor cooling system. It has been experimentally established that in this case there is a shift in the timing of the appearance of the NMR signal in the recording system and a change in signals appearance. This shift happens once. By measuring this one-time shift (time interval) between the recorded NMR signals and comparing it with the one previously determined (data from various accidents) it is possible to make a correct decision much faster in an emergency situation.

To keep the operating parameters of the time domain measurement system in working order, it must be checked and tested regularly [11-13, 19-21, 27, 28]. Since mobile objects, as a rule, have a large amount of various kinds of interference for the transmission of NMR signals, fiber-optic communication lines (FOCL) are most often used in these systems [29-32]. The FOCLs developed at present are insensitive to powerful electromagnetic radiation, contact interference arising during transient processes in ship power plants, etc [29-37]. Therefore, to test the system for measuring time intervals, it is more expedient to use devices based on fiber-optic communication lines and laser radiation. At present, a large number of models of various lasers have been developed, which makes it possible to eliminate many problems in the creation of such devices.

2. Development fiber-optical emergency simulator

A during developing fiber-optic systems for various purposes for objects with nuclear power plants, it is necessary to take into account the effect of γ-radiation, which can have a negative effect on FOCL [33-35]. In this regard, in the developed fiber-optic simulator, it is advisable to use a single-mode fiber with a pure quartz core. At present, relaxation constants are determined in NMR-based flowmeters, and the flow of "heavy" water is also recorded in a wide frequency band from 3 to 40 MHz at a modulation frequency f_m of a constant magnetic field in the range from 10 to 50 Hz. The time interval between the recorded NMR signals is from 0.01 to 0.05 s, and the range of time of the possible delay of the NMR signal when the state of "heavy" water changes is from 40 to 130 μs.

The inclusion of the simulator in the information transmission line can introduce the minimum possible loss for the attenuation of the transmitted NMR signal to ensure the required measurement accuracy when testing the control system of the nuclear power plant, especially when it operates at the "edges" of the dynamic range of measurements. Distortion of the form of the transmitted signal is impossible, since this will instantly degrade the accuracy of measurement of the relaxation constants, as well as the very time interval between signals, which characterizes changes in "heavy" water. In Figure 1 shows as an example the line shape of the NMR signal from "heavy" water, recorded at a frequency of f_{nmr} = 4573 kHz with a modulation frequency of f_m = 16.3 Hz.

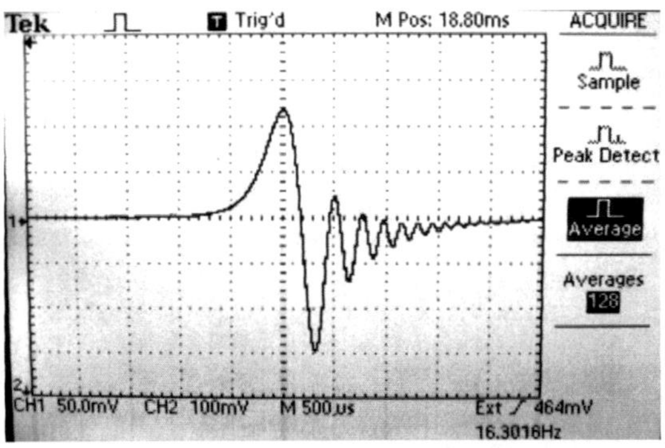

Figure 1. The NMR signal from «heavy» water in wide registration band.

For accurate measurement of relaxation times T1 and T2, an NMR signal with a large number of damped "wiggles" of oscillations is required. Such signals should be transmitted without distortion of the form to the processing device (the calculator of the reactor compartment and to central computers of the ship), in which T1 and T2 will be determined and the corresponding conclusions and recommendations will be issued.

It should also be noted that in order to objectively check the operation of the NPP control system using NMR flowmeters-spectrometers, it is necessary to cover the entire range of variation of the delay times τ_d. As part of solving this problem, the task was to develop six simulators. for different values of τd, which are about 40, 50, 75, 100, 110 and 130 μs. Also, simulators should work stably in the temperature range from 278 to 318 K. The developed simulators can be placed on one rack and in turn connected to the information transmission system. In the presented work, a FOCL-based simulator is developed on the basis of a passive fiber-optic delay line with an optical isolator switched on at the input to exclude reflected signals.

The signal attenuation was calculated for the developed design of the simulator. All intrinsic losses A_{Σ} in the fiber are defined as follows:

$$A_{\Sigma} = a_a + a_d + a_i + a_{ex}, \qquad (1)$$

where: a_a –takeover loss; a_d-scattering loss; a_i- impurity losses; a_{ex}-additional losses.

Absorption losses are defined by the following expression:

$$a_a = 8.69 \frac{\pi\, tg(\delta) n_1}{\lambda}, \qquad (2)$$

where: n_1- refractive index of the core, for the fiber we have chosen is 1.4676; tg(δ) - dielectric loss angle tangent (2.5·10-12), then absorption losses for operating wavelength (λ= 1310 nm) constitute a_a= 0.076 dB/km.

The scattering loss is determined by the following formula:

$$a_d = 4.34 * 10^3 \left(8\pi^3 \frac{n_1^2 - 1}{3\lambda^4} \kappa\beta T\right), \qquad (3)$$

where: T – fiber core manufacturing temperature; β = 8.1·10-11 m^2/H – compressibility factor. The calculated value for operating wavelength is a_d =0.234 dB/km.

To transmit NMR signals in NPP control systems use single-mode optical fibers of the firm Corning SMF-28e+ with a operating wavelength of 1310 nm are used, therefore, for them, the loss of extraneous material and additional losses are very small compared to the estimated a_a and a_d.Then all your own losses will be A_{Σ}= 0.31 dB/km. In addition to these losses, there are back losses on four additional compounds (FC - APC connectors) a_c –4·0.2 = 0.8 dB. Loss in optical insulator 0.41 dB. Then all the losses associated with switching on the fiber-optic simulator will be:

$$a_{im} = (\alpha_a + a_d) * L_{op} + 1.21, \qquad (4)$$

where: Lop – fiber length.

For a system energy margin of 2 dB, let us determine the maximum possible length l_{max} of the regeneration section of the fiber-optic delay line:

$$l_{max} = (A_{max} - M - n_{pc} a_{pc})/(\alpha + \frac{\alpha_{nc}}{l_{ct}}), \qquad (5)$$

where: n_{pc}– number of connectors; a_{pc}– loss in detachable compounds; l_{ct}- face-to-face cable length (l_{ct}= 63000 m); α_{nc}– loss in integral compounds (when welding face-to-face cable lengths α_{nc}= 0,05 dB), α= α_{im} for calculation of L_{op}=l_{ct}; M – system attenuation margin at the regeneration section (M = 2 dB); A_{max} – maximum closured attenuation value of instrumentation.

Thus, the value l_{max}=124900 m. The results obtained indicate that there are great opportunities for using this type of fiber for the production of simulators based on fiber-optic communication lines for signal delay.

3. Experimental results and discussion

To check the correct functioning (measurement of delay times τ_d) of the developed fiber-optic simulators and study their dynamic characteristics, an experimental setup was created, the block diagram of which is shown in Figure 2.

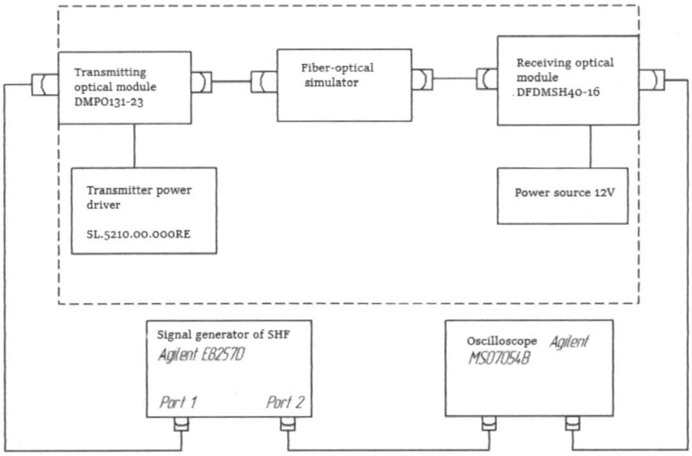

Figure 2. The design Scheme of the Simulator Test Facility.

To measure the dynamic characteristics of simulators in an experimental setup, the Agilent MS07954B oscilloscope in the block diagram is changed to an Agilent PXA N9030A spectrum analyzer. In the case of K_n noise figure measurements, the Agilent MS07954B oscilloscope in the block diagram is replaced by an Agilent PXA N9030A Spectrum Analyzer, and an Agilent E82570 Signal Generator is replaced by an Agilent N4002A SNS Series Noise Source.

In Figure 3 shows, as an example, the results of measurements of the delay times τd for three simulators (values 49 μs, 75 μs and 98.6 μs). The upper signal on the oscillogram was obtained without including the simulator in the measurement circuit shown in Figure 2, the lower one with the simulator turned on.

Analysis of the signals obtained during the experiment shows a decrease in power. Thus, when a FOCL-based simulator is included in a fiber-optic NMR signal transmission system for testing nuclear power plant control systems, the transmitted signals will need to be amplified to ensure the required accuracy during measurements. The measured dynamic characteristics of fiber-optic simulators showed that all the developed designs have a dynamic range of at least 56 dB, which is quite easy to implement in practice.

An important characteristic when transmitting optical signals over fiber-optic lines is the intrinsic noise of an additional device connected to the communication line. The noise figure Kn of the developed simulators was investigated by the Y-factor method in the frequency range from 3 to 44 MHz.

In Figure 4 shows the experimental dependence of the noise figure Kn on the input signal frequency F for three values of τd, obtained during the experiment.

Based on the data obtained during the experiment, it was found that the noise figure varies in the range from 23 to 28 dB. For the frequency range F from 18 to 42 MHz, the nonlinearity of K_n is no more than 1.5 dB. In the specified frequency range, it is most convenient to test the control system for nuclear magnetic meters of NPP parameters. In this regard, when designing nuclear magnetic counters for these control systems, developers are advised to choose the frequency of recording the NMR signal in the range from 18 to 40 MHz in order to be able to qualitatively check its performance.

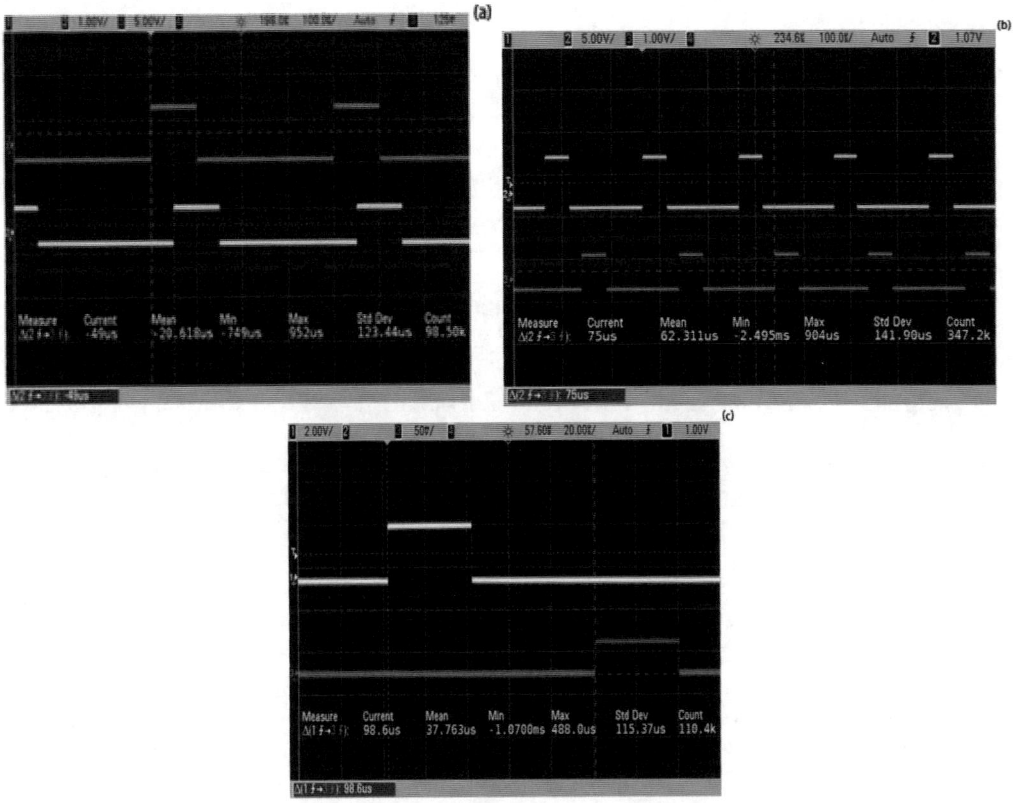

Figure 3. Temporary passage of the optical signal through a simulator with different lengths of fiber – optical L: a) – corresponds to L = 10 000 m (τ_d = 49.6 μs, b) - L = 15 000 m (τ_d = 75.0 μs), c) - L = 20 000 m (τ_d = 98.6 μs).

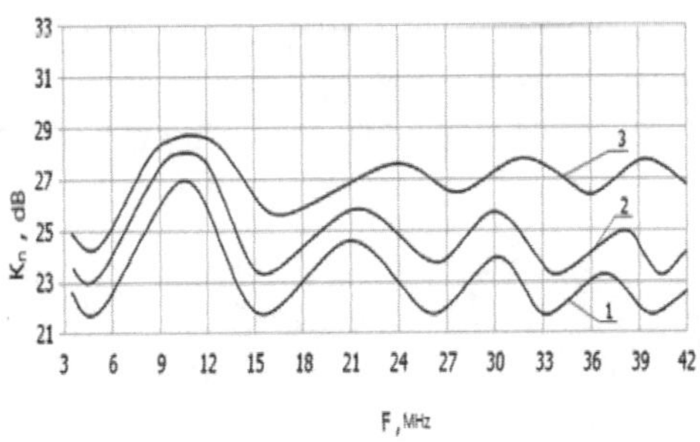

Figure 4. The dependence of noise coefficient K_n on the frequency of the input signal F. Graphs 1, 2 and 3 correspond τ_d in a μs: 49.6; 75; 98.6.

4. Conclusion

The results obtained in the course of the experiment showed sufficient reliability of the developed simulator based on FOCL, intended for testing the operation of time intervals meters in NPP cooling systems.

The conducted studies of the parameters of simulators based on FOCL showed that they can also be successfully used for testing equipment of radar stations, since the noise figure of the developed simulator, as shown by additional studies, in the frequency range from 2 to 18 GHz, varies by no more than 17 dB (the maximum allowable value for these devices is 27 dB). This shows the wide functionality of the developed equipment.

References

[1] Fadeenko I, Fadeenko V, Reznik V, Moroz A, Popovskiy N and Nikolaev D 2019 *IOP Conference Series: Earth and Environmental Science* **390(1)** 012022
[2] Myazin N S 2018 *Journal Physics: Conference Series* **1124(1)** 031004
[3] Davydov R, Antonov V, Makeev S, and Batov Y 2019 *E3S Web of Conferences* **140** 02001
[4] Davydov R, Antonov V and Moroz A 2019 *International Conference on Electrical Engineering and Photonics EExPolytech-2019 (Saint-Petersburg)*, vol. 8906791 (IEEE), p. 295-297
[5] Van S, Cheremisin A, Chusov A, Switala F and Davydov R 2019 *IOP Conference Series: Earth and Environmental Science* **390(1)** 012011
[6] Davydov V V 1999 *Russian Physics Journal* **42(9)** 822–825
[7] D'yachenko S V and Zhernovoi A I 2016 *Technical Physics* **61(12)** 1835–1837
[8] Davydov V V and Kiryukhin A V 2020 *Atomic Energy* **127(5)** 274–279
[9] D'yachenko S V, Kondrashkova I S and Zhernovoi A I 2017 *Technical Physics* **62(1)** 1602-1604
[10] Smirnov K J and Batov Y V 2019 *Journal of Physics: Conference Series* **1368(2)** 022073
[11] Davydov V V, Dudkin V I and Karseev A Yu 2014 *Optical Memory & Neural Networks (Information Optics)* **23(4)** 259 – 264
[12] Davydov V V, Dudkin V I and Karseev A Yu 2014 *Optical Memory & Neural Networks (Information Optics)* **23(3)** 170 – 176
[13] Davydov V V, Dudkin V I and Karseev A Yu 2013 *Optical Memory & Neural Networks (Information Optics)* **22(2)** 215 – 218
[14] Nikitina M, Grebenikova N and Batov Y 2019 *IOP Conference Series: Earth and Environmental Science* **390(1)** 012024
[15] Gryznova E and Batov Y 2019 *E3S Web of Conferences* **140** 09001
[16] Gryznova E, Grebenikova N, Ivanov D and Bykov V 2019 *IOP Conference Series: Earth and Environmental Science* **390(1)** 012044
[17] Davydov R V, Mazing M S, Yushkova V V and Stirmanov A V 2019 *Journal of Physics: Conference Series* **1410(1)** 012067
[18] Davydov R V, Antonov V I, Yushkova V V and Smirnov K J 2019 *Journal of Physics: Conference Series* **1400(6)** 066037
[19] Davydov V V, Dudkin V I and Karseev A Y 2015 *Russian Physics Journal* **58(2)** 146–152
[20] Davydov V V, Dudkin V I and Karseev A Y 2015 *Measurement Techniques* **58(3)** 317–322
[21] Davydov V V, Dudkin V I and Karseev A Y 2015 *Instruments and Experimental Techniques* **58(6)** 787–793
[22] Davydov V V, Velichko E N, Dudkin V I and Karseev A Y 2015 *Instruments and Experimental Techniques* **58(2)** 234–238
[23] Myazin N, Neronov Y, Dudkin V and Yushkova V 2018 *MATEC Web of Conferences* **245** 11013
[24] Rukin E V, Myazin N S and Dudkin V I 2019 *Journal of Physics: Conference Series* **1368(4)** 042011
[25] Myazin N S, Yushkova V V and Taranda N I 2019 *Journal of Physics: Conference Series* **1410(1)** 012130

Raman spectroscopy and optical microscopy of medical infusion solutions for parenteral nutrition

E V Pimakhina[1], A A Pimakhin[1], N V Vishnykov[2], N M Tolkach[2] and S B Arsentiev[3]

[1] Ryazan State Medical University, Ryazan 390026, Russian Federation
[2] Ryazan State Radio Engineering University, Ryazan 390005, Russian Federation
[3] Regional Clinical Hospital, Ryazan 390039, Russian Federation

Elenapim@yandex.ru, +7 (920) 960 03 39

Abstract. Usage of Raman spectroscopy, optical microscopy, and retaining track membranes to find 20% SMOFlipid microparticles with a diameter of 5 to 10 μm and to prevent complications that arise during parenteral nutrition.

1. Introduction

The study was carried out on the equipment of the regional center of probe microscopy for collective use at the Ryazan State Radio Engineering University. We used an optical microscope and a spectrometric probe complex Integra-Spectra (manufactured by NT-MDT, Russia, Moscow).

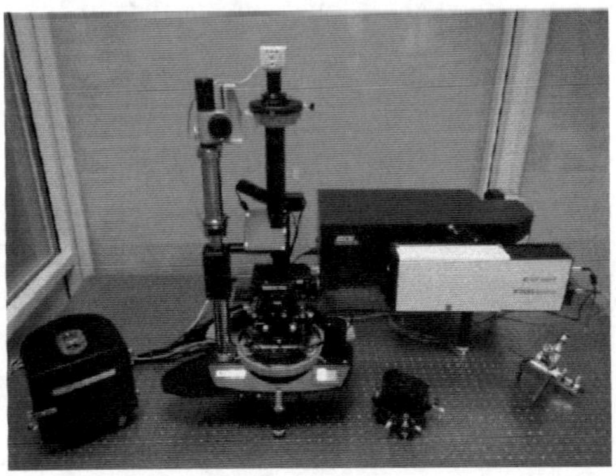

Figure 1. Particle detection equipment Spectrometric probe complex Ntegra Spectra (manufacturer NT-MDT, Zelenograd), located in the Regional Center for Probe Microscopy for Collective Use V.F. Utkin in Ryazan

A focused laser changes the vibrational state of molecules of foreign particles of the solution on the sample, and the molecules begin to emit Raman radiation. But any substance constantly emits Rayleigh radiation, with constant wavelength. Raman radiation (with dynamic wavelength) can be filtered with a dichroic filter.

Monochromatic light created by a laser is used for the initial radiation (for example, Nd: YAG – a laser based on Yttrium – Aluminum Garnets with the addition of Neodymium, with a wavelength of 532 nm – a green visible color).

The dependencies between the intensity of Raman radiation (based on the results obtained on the photosensitive matrix) and wavelength of Raman radiation is plot-ted on a graph. Each molecule has its own shift of Raman radiation, so the amount of molecules depends on the intensity of Raman radiation. Raman spectroscopy allows to analyze a substance in liquid, gaseous and solid states larger than 1 micron. [1]

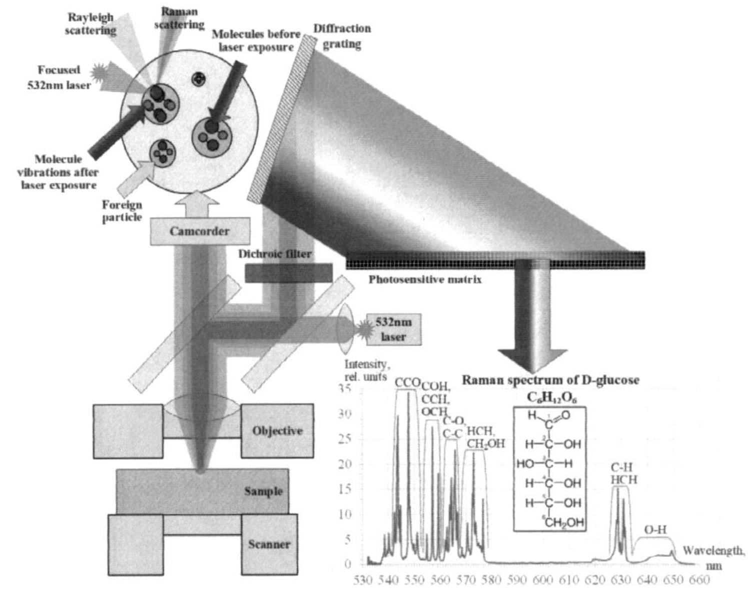

Figure 2. The experimental scheme for detection of foreign particles

Analysis stages:

- Determination of the chemical composition of foreign particles by Raman spectrums;
- Determination of the area of each foreign particle $S_1, S_2 \ldots S_i$;
- Calculation of particle concentration: n = N/(S•d) [*unit./ml*];

Calculation of the volume fraction of particles: $\omega = (100\ (S_1+S_2+\ldots S_i))/S$ [%]; N is the number of found foreign particles; s is the area in [cm^2] occupied by the test solution; d –layer thickness in [cm] occupied by the test solution.

Optical microscopy and Raman spectroscopy in a 20% SMOFlipid solution revealed homogeneous microparticles with sizes from 1 μm to 50 μm (average size ~ 10μm, an average concentration of 10,000 particles / ml, and a volume fraction of $6.4 \cdot 10^{-3}$. Filters of standard medical droppers with a pore diameter of 150 microns do not retain the above described microparticles.

The circulation of microparticles through the bloodstream for 60 minutes was experimentally established when the centrifuged plasma of women aged 19 to 42 was added (10 to 1 by volume) to each of the AB0 / rhesus combinations. The average size of microparticles at 36.6°C did not change within 60 minutes, which is due to the specifics of the enzymatic activity of plasma albumin, liver and kidney enzymes and the properties of cell receptors.

The Doppler method was used to calculate the transit time of the injected solution through the subclavian vein of fullterm newborns – 0.07 s, which is incomparable with the period of lipolysis activation 60 minutes.

When microparticles enter the pulmonary and cerebral bloodstream, they cause microembolism, which leads to a pathological decrease in ventilation-perfusion relations.

2. Experiment and methods

It is proposed to filter microparticles through track membranes with a pore diameter of 3 μm, which retain microparticles with a diameter of 5 to 10 μm, which is confirmed by the homogeneous composition of the filtered microparticles and their absence in the solution during repeated filtration. The presence of microparticles and their absence on the surfaces of the track membranes used for filtration were established by optical microscopy. The homogeneity of the microparticles was established by Raman spectroscopy.

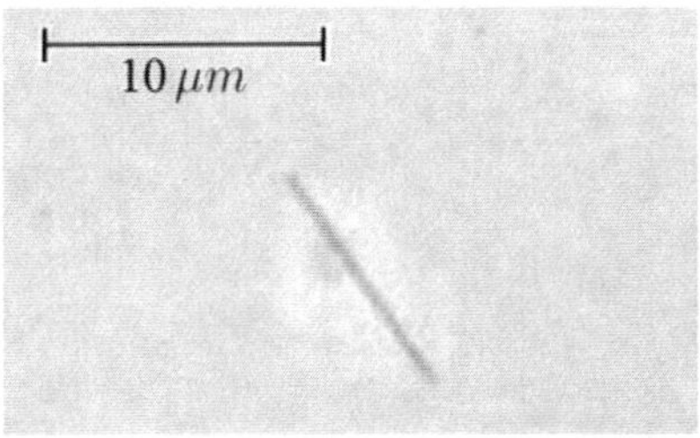

Figure 3. Particles in test solution SMOFlipid (sizes are indicated in micrometers (μm)

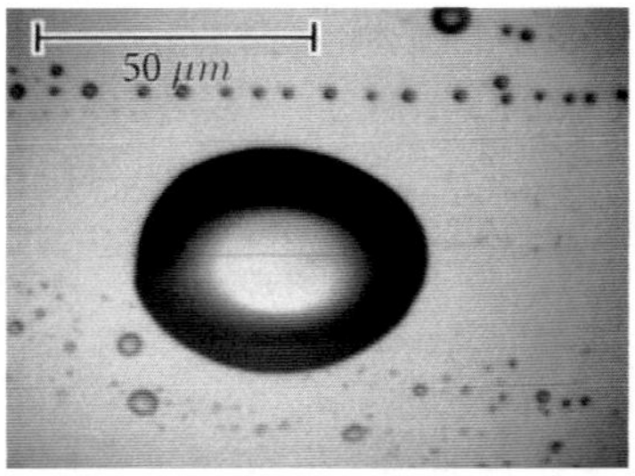

Figure 4. The microparticle of SMOFlipid in the blood plasma at room temperature 24_C. The diameter of the image is 500 μm. Optical microscopy

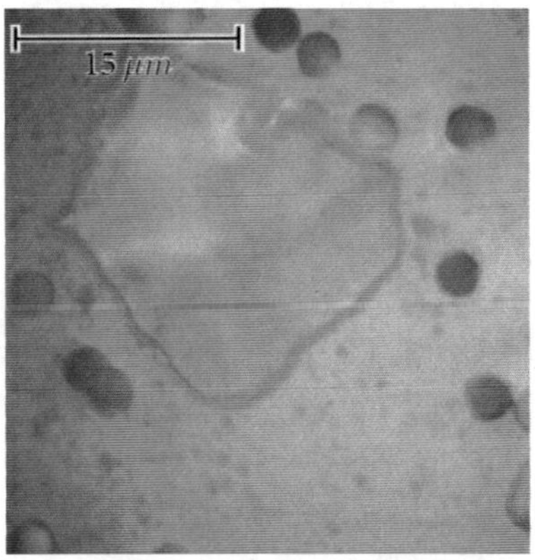

Figure 5. It is proposed to filter microparticles through track membranes with a pore diameter of 3 μm

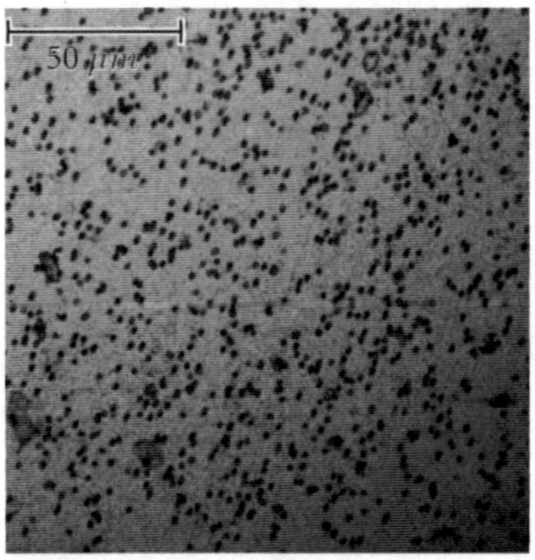

Figure 6. The surfaces of the track membranes

The filtration efficiency was confirmed by testing on laboratory rats – microscopy of lung sections revealed a smaller number of neutrophils in the airspace and a smaller proportion of the airspace containing hyaline membranes, in comparison with sections of the lungs of intact rats. It is caused by a reduction in capillary stasis and a decrease in the intensity of perifocal inflammation.

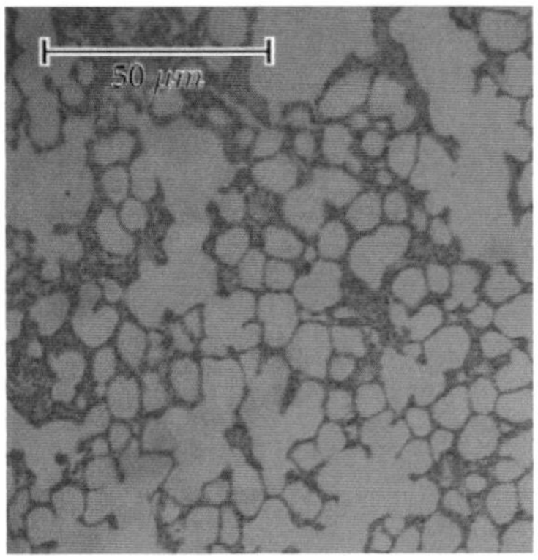

Figure 7. The microscopy of the lungs of intact rats

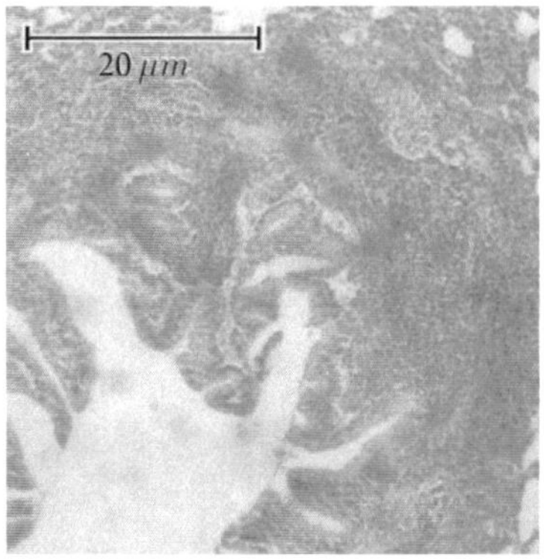

Figure 8. The microscopy of the lungs of intact rats. The experimental model of perifocal inflammation after use of foreign particles with out of the track membranes

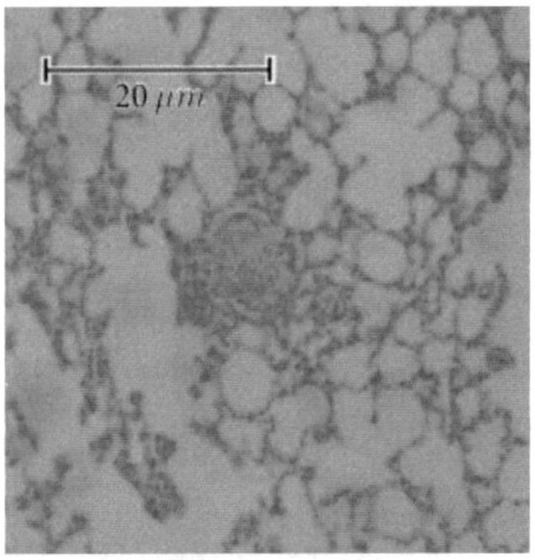

Figure 9. The microscopy of the lungs of intact rats. The experimental model of the thromb after use of foreign particles with out of track membranes

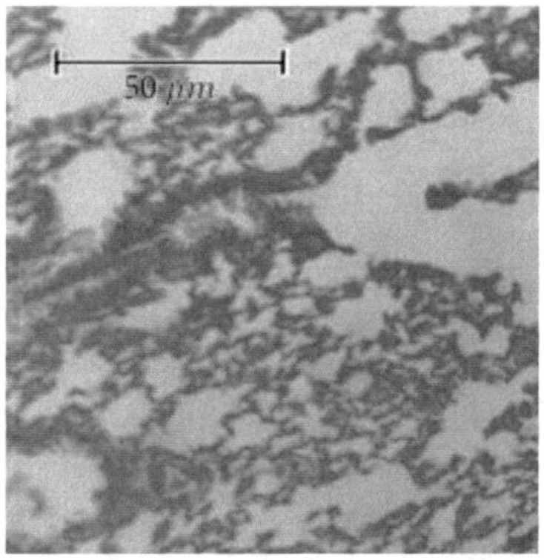

Figure 10. The microscopy of the lungs of rats after use of 20% SMOFlipid with of track membranes

With parenteral nutrition of fullterm newborns with 20% SMOFlipid, the use of additional medical filters of the Octopus brand with a pore diameter of 3 μm did not reveal complications associated with impaired hemostasis, while when using filters of standard medical droppers, increased bleeding was revealed, which is manifested by point intradermal hemorrhages in the area of catheter placement, shortterm bleeding from the umbilical wound against the background of normal values of fibrinogen, antithrombin, 8 coagulation factor.

3. Results and discussion

The first paragraph after a heading is not indented (Bodytext style). Complications that arise during parenteral nutrition using standard medical dropper filters can be explained by the manifestation of hypercoagulation against the background of latent fibrinolysis. The penetration of a large number of microparticles of the infusion emulsion into the peripheral, and then into the pulmonary and cerebral blood flow causes a deviation of hemostasis parameters and activation of the fibrinolytic link. To determine the state of hemostasis accurately, it is necessary to conduct thromboelastometry followed by analysis of the thromboelastogram.

Thus, microparticles in a 20% SMOFlipid solution cause symptoms of hemorrhagic disease of newborns. The introduction of the solution must be carried out with the obligatory use of a filter with a micropore size of up to 5 microns. The infusion rate should correspond to the rate of activation of plasma enzymes and should not exceed the rate recommended in the drug's instructions.

Acknowledgments

The work was performed on the equipment Regional center of probe microscopy collective use Ryazan State Radio Engineering University named after V. F. Utkin.

References

[1] Pardeshi N N *et al* 2017 *Journal of pharmaceutical sciences* **106(2)** 511

Numerical simulation of optical coupling between a microring resonator and a directly connected straight waveguide

V V Pirogov[1], S A Scherbak[1,2,3], A A Lipovskii[1,3], N V Kryzhanovskaya[1,2] and A E Zhukov[1,2]

[1] Department of Physics and Technology of Nanoheterostructures, St. Petersburg Academic University RAS, St. Petersburg 194021, Russia
[2] International Laboratory of Quantum Optoelectronics, Higher School of Economics, St. Petersburg 194100, Russia
[3] Institute of Physics, Nanotechnology and Telecommunications, Peter the Great St. Petersburg Polytechnic University, St. Petersburg 195251, Russia

Abstract. We studied numerically optical coupling of whispering gallery modes of a microring resonator with modes of a radially directed straight waveguide connected to the microring. Optical outcoupling from the resonator through the waveguide was calculated for different widths of the latter. The output was maximized under conditions when the width fits an odd integer number of half-wavelengths of light inside the microring.

1. Introduction

Semiconductor microdisk (MD) and microring (MR) lasers are of great interest for applied photonics. They support optical whispering gallery modes (WGMs) which have extremely high Q-factor [1]. This determines interest in them, particularly, for optical communications and integrated optics [2,3]. Also, they are used in environmental (particularly, biochemical), thermal and stress sensors [4–7] due to high sensitivity of WGMs to changes of environment, and in nonlinear optics because of high magnitude of an electric field of the modes [8]. Direct emission of such microlasers can empower their applications. However, due to extreme Q-factor of WGMs, the modes are highly localized in the resonator volume. Thus, the problem of controllable and directional output of their radiation is actual. A promising way to achieve this is to couple a laser with an optical waveguide (OW), either nearly located [3,9,10] or directly connected one [11,12]. The former requires fine tuning and provides relatively low output. The latter can, principally, ensure more efficient coupling, however such configuration has never been modeled. In this study, we numerically model outcoupling from a MR resonator through a straight optical waveguide directly connected to it in radial direction and reveal some general tendencies of the phenomenon.

2. Approach

To analyze WGMs coupling with the OW we used numerical simulation in COMSOL Multiphysics environment via the finite element method. The scheme of the problem is shown in figure 1. The model includes a MR with inner/outer radii 2.6/3.0 μm with refractive index n_{in}=3.4 (close to GaAs) and an OW with refractive index n_{ow}=1.5 (close to some glasses or polymers), the index of the outer medium is n_{out}=1.0. The OW width d was varied to find the condition of a resonant coupling. We considered 2D geometry (cylinder of infinite width) instead of 3D because this allows reducing required computational power while the coupling properties are qualitatively the same. Also we limited our consideration with TE-polarized (H-field in-plain, E-field out-of-plane) modes only.

We considered WGMs in near infrared spectral region (wavelengths 1300-1350 nm) that is typical for microlasers with an InAs/InGaAs/GaAs quantum dots active region [13]. We used an eigenfrequency solver to find wavelengths of WGMs, their electromagnetic field spatial distributions and Q-factor. Also, we set an imaginary part of the MR index $k_{in}=5\cdot10^{-5}$ to simulate internal losses. This allowed matching modeled Q-factor with Q-factor of typical MD lasing structures, which is of 10^4 order [14].

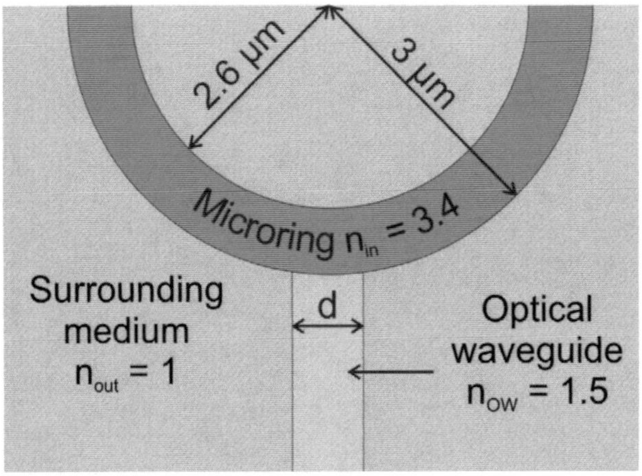

Figure 1. Scheme of a microring resonator coupled with an optical waveguide.

3. Results and discussion

3.1. Optical modes in the structure

We simulated a WGM of the 1st radial and the 42nd azimuthal order, TE-polarized (E-field of the mode is out-of-plane), denoted as TE(1,42). Properties of other modes including TM-polarized would be qualitatively the same. The wavelength of this mode is λ=1344.2 nm. The isolated MR possesses axial symmetry, therefore, WGMs orientation regarding azimuthal angle is arbitrary. However, the OW connected to the MR breaks axial symmetry, and TE(1,42) mode (as well as others) splits into two modes: symmetric (S) and asymmetric (A) one relatively to the mirror symmetry axis (see figure 2). Wavelengths of S- and A-modes slightly differ from each other and from the initial wavelength 1344.2 nm, but the difference does not exceed a few angstroms that indicates a weak influence of the OW on the MR.

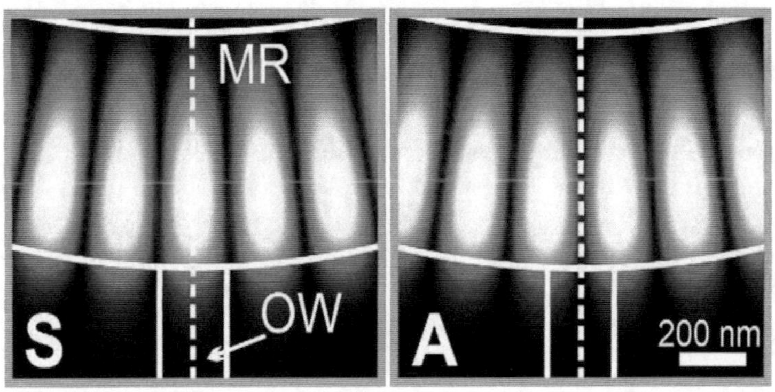

Figure 2. Images of |E|-field of symmetrical (S) and asymmetrical (A) modes. Regions of the microring (MR) and the optical waveguide (OW) are designated. Dashed line is a mirror symmetry axis.

3.2. Optical power output

Evidently, direct connection of the OW to the MR causes additional scattering of light from the conjunction area, because the index of the OW is higher than the index of the surrounding and this disturbs condition of the total internal reflection in that area. Some of this scattered light is directly outcoupled through the waveguide. However, some of this light scatters in the open space. In figure 3a we schematically show direct (to the waveguide), P_{out}^{dir}, and indirect (out of the waveguide), P_{out}^{sc}, scattering of the light. Note that this indirect scattering occurs in addition to natural indirect radiation of the WGM, though it is significantly greater and natural WMG radiation can be neglected at all.

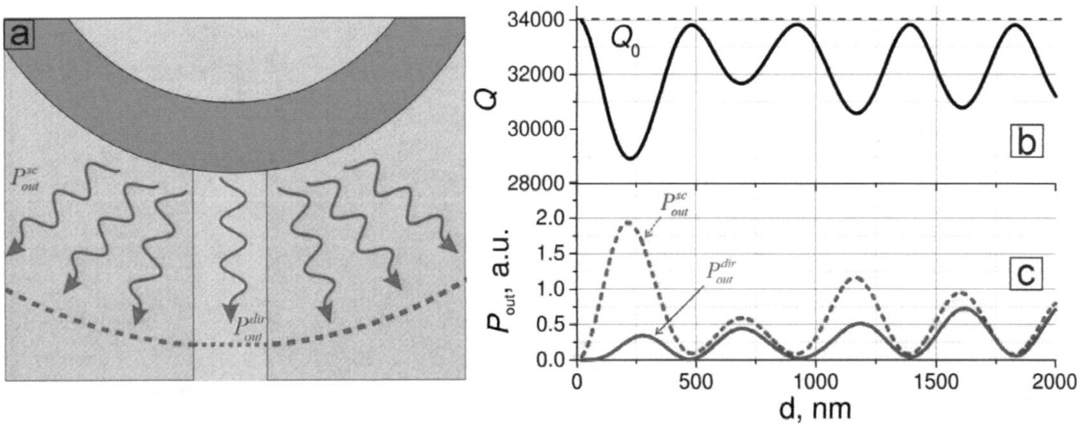

Figure 3. a) Schematic illustration of direct (P_{out}^{dir}) and indirect (P_{out}^{sc}) scattering of the light; dot green line – area of direct scattering integration; dashed red line – area of indirect scattering integration. b) Calculated Q-factor of the S-type of TE(1,42) mode vs OW width; Q_0 – Q-factor of the undisturbed mode in the isolated MR. c) Calculated direct (P_{out}^{dir}) and indirect (P_{out}^{sc}) optical power outputs.

In order to study MR-OW coupling we, as an example, chose the S-type of TE(1,42) mode and varied the OW width *d*. In figure 3b we demonstrate the dependence of calculated quality-factor of the mode on *d*. Q-factor of the undisturbed mode in the isolated MR, Q_0, is also designated. The dependence represents a series of drops in Q-factor, nature of which will be discussed further. Knowing Q-factor we can determine relative radiation losses in the structure [15]:

$$\chi = \frac{1}{Q} - \frac{1}{Q_0}. \tag{1}$$

However, Eq. (1) represents total radiation losses in the structure, both P_{out}^{sc} and P_{out}^{dir}, whereas relative direct radiation losses, which are the MR-OW power outcoupling coefficient, are of interest. We can approximately separate P_{out}^{sc} and P_{out}^{dir} via integrating the Poynting vector along corresponding surfaces, which are schematically designated in figure 3a. Results of these calculations are presented in figure 3c. One can observe that drops of Q-factor correspond to peaks of $P_{\rm out}$ and that peaks of P_{out}^{sc} and P_{out}^{dir} qualitatively coincide. The latter means that maximal directional outcoupling always accompanied by maximal indirect scattering losses. Knowing ratio between P_{out}^{sc} and P_{out}^{dir} one can determine power outcoupling coefficient:

$$\chi_{dir} = \left(\frac{1}{Q} - \frac{1}{Q_0}\right) \frac{P_{out}^{dir}}{P_{out}^{dir} + P_{out}^{sc}}. \tag{2}$$

3.3. Power outcoupling coefficient

We calculated coupling coefficient in accordance with Eq. (2) for S- and A-types of TE(1,42) mode for different OW widths. Also, we calculated spatial distributions of electric field that correspond to maximal outcoupling to determine through which mode of the OW the outcoupling occurs. The results are presented in figure 4. As was mentioned above, the dependences represent series of peaks at certain OW widths: for S-mode the coupling coefficient is maximized at $d \approx 270$, 690, 1170, 1620 nm, etc; for A-mode – at $d \approx 480$, 930, 1380, 1830 nm, etc. Worth to note that minima of A-mode outcoupling correspond to maxima of S-mode outcoupling in terms of waveguides width, and vise-versa. Apparently, periodicity of these maxima (approximately 450 nm) correlates with wavelength of light inside the MR, which is $\lambda/n_{in} \approx 450$ nm. Also, |E|-field distributions in the insets of figure 4 show that the coupling is maximized when the OW width fits an odd integer number of half-wavelengths inside the MR for S-mode (insets in figure 4a) and an even integer number of half-wavelengths for A-mode (insets in figure 4b). This condition can be formulated more generally: OW width should fit integer number (even for A-modes and odd for S-modes) of the WGM intensity peaks. Size of WGM intensity peak is $2\pi R/m$, where R is the resonator radius and m is the WGM azimuthal order. This condition can be applied for WGMs of any radial order. Particularly, for WMGs of the first radial order this condition coincides with the one given above, since $2\pi R/m$ approximately corresponds to the wavelength of light inside the MR (λ/n_{in}) for first radial modes. Such formulation allows an important conclusion: resonant outcoupling of modes with different azimuthal order requires differently wide waveguides. This offers possibility of selective outcoupling: waveguide of given width provides resonant optical output for a WGM of given m.

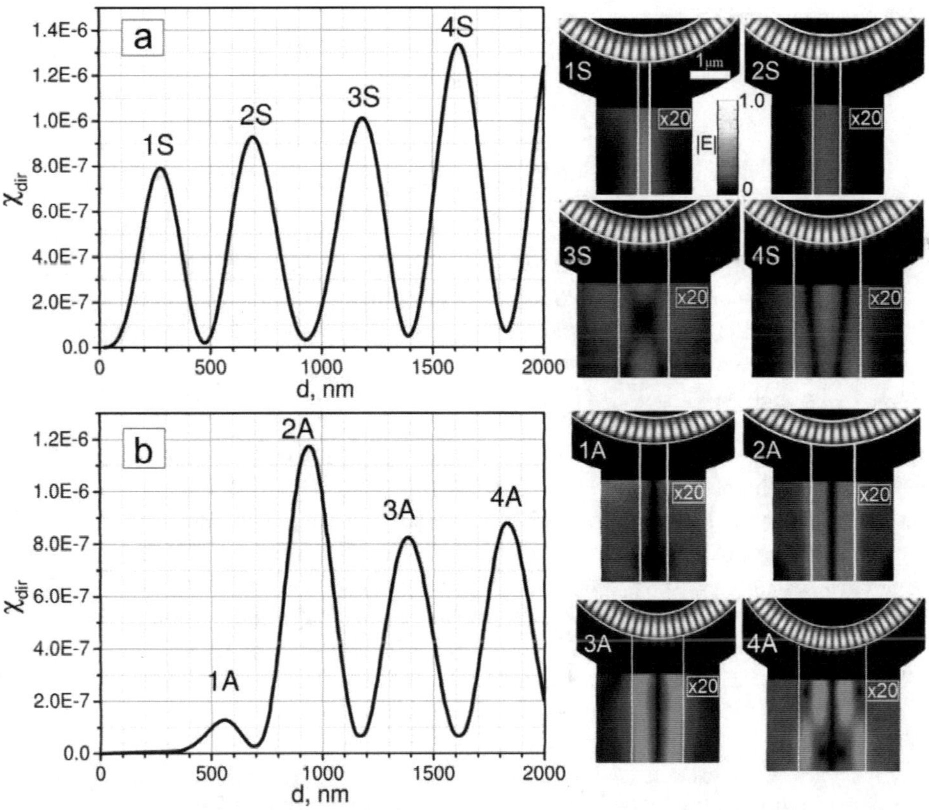

Figure 4. Dependence of the relative coupling coefficient on the OW width for S-mode (a) and A-mode (b). Insets: |E|-field distributions for maxima of the output (d = 270, 690, 1170, 1620 nm for S-mode and d = 480, 930, 1380, 1830 nm for A-mode); |E|-field in the OW and its vicinity is 20-fold artificially enhanced for better visualization.

Distributions in figure 4 also demonstrate through which kind of the OW mode outcoupling occurs. Peaks 1S and 2S in figure 4a correspond to excitation of the fundamental (1st) mode of the OW, since the OWs of the given widths do not support higher odd modes (3rd, 5th, etc.). However, OWs of widths corresponding to peaks 3S and 4S do support the 3rd mode, and in insets 3S and 4S one can observe superposition of the 1st and the 3rd OW modes. Similar situations occurs for the A-mode, though in this case even OW modes (2nd, 4th, etc.) would be under discussion. Peak 1A in figure 4b has a noticeably low magnitude. This is because the OW of that width barely supports the 2nd mode, therefore, this mode turns out to be weakly localized (see inset 1A) and outcoupling efficiency is low. OWs of widths that correspond to peaks 2A and 3A are wide enough to support the 2nd mode, therefore this mode is strongly localized and well-guided (see insets 2A and 3A), and the coupling efficiency is considerably higher. The OW of width d=1830 nm (peak 4A) is wide enough to also support the 4th mode, therefore in inset 4A we observe the superposition of the 2nd and 4th waveguide modes.

4. Conclusion

We numerically simulated optical coupling of whispering gallery modes of a microring resonator with modes of a radial straight waveguide connected to the microring. The calculations showed that conjunction of the OW to the MR causes light scattering partially to the waveguide and partially to the outer space. The optical power output was maximized under conditions when the OW width fits an odd integer number of half-wavelengths of light inside the microring, or, more generally, integer number of the WGM intensity peaks. Evidently, OW widths that fit this condition are different for modes with different azimuthal numbers that offers possibility of selective output. Also, we demonstrated that outcoupling occurs through superposition of all the optical modes supported by the waveguide of a given width. Narrow waveguides support only the first order fundamental mode and the outcoupled electromagnetic field distributed accordingly. Wider waveguides support more modes and distribution of electromagnetic field becomes more complex.

Acknoledgement

This study was supported by Russian Foundation for Basic Research, project #20-02-00334.

References

[1] Savchenkov A A, Matsko A B, Ilchenko V S and Maleki L 2007 Optical resonators with ten million finesse *Opt. Express* **15** 6768

[2] Matsko A B and Ilchenko V S 2006 Optical resonators with whispering-gallery modes-part I: basics *IEEE J. Sel. Top. Quantum Electron.* **12** 3–14

[3] Cai D-P, Lu J-H, Chen C-C, Lee C-C, Lin C-E and Yen T-J 2015 High Q-factor microring resonator wrapped by the curved waveguide *Sci. Rep.* **5** 10078

[4] Khozeymeh F and Razaghi M 2018 Cylindrical optical resonators: fundamental properties and bio-sensing characteristics *J. Opt.* **20** 045301

[5] Ma Q, Rossmann T and Guo Z 2010 Whispering-gallery mode silica microsensors for cryogenic to room temperature measurement *Meas. Sci. Technol.* **21** 025310

[6] Ioppolo T, Kozhevnikov M, Stepaniuk V, Otugen M V and Sheverev V 2008 Micro-optical force sensor concept based on whispering gallery mode resonators. *Appl. Opt.* **47** 3009–14

[7] Fetisova M V., Kryzhanovskaya N V., Reduto I V., Moiseev E I, Blokhin S A, Kotlyar K P, Scherbak S A, Lipovskii A A, Kornev A A, Bukatin A S, Maximov M V. and Zhukov A E 2018 Room temperature lasing from microdisk laser in aqueous medium *J. Phys. Conf. Ser.* **1124** 051007

[8] Braginsky V B, Gorodetsky M L and Ilchenko V S 1989 Quality-factor and nonlinear properties of optical whispering-gallery modes *Phys. Lett. A* **137** 393–7

[9] Rowland D R and Love J D 1993 Evanescent wave coupling of whispering gallery modes of a dielectric cylinder *IEE Proc. J Optoelectron.* **140** 177

[10] Raskhodchikov A V., Scherbak S A, Kryzhanovskaya N V., Lipovskii A A and Zhukov A E

2020 Dielectric surrounding bleaches the optical bond between a microdisk resonator and a straight optical waveguide *J. Phys. Conf. Ser.* **1695**

[11] Ma X-W, Lv X-M, Huang Y-Z, Yang Y-D, Xiao J-L and Du Y 2014 Mode characteristics for unidirectional-emission microring resonator lasers *J. Opt. Soc. Am. B* **31** 2773

[12] Liu S, Sun W, Wang Y, Yu X, Xu K, Huang Y, Xiao S and Song Q 2018 End-fire injection of light into high-Q silicon microdisks *Optica* **5** 612

[13] Kryzhanovskaya N V., Polubavkina Y S, Scherbak S A, Moiseev E I, Zhurikhina V V., Zubov F I, Lipovskii A A, Kulagina M M, Troshkov S I, Zadiranov Y M, Maximov M V. and Zhukov A E 2017 3.5- μ m radius race-track microlasers operating at room temperature with 1.3- μ m quantum dot active region *J. Appl. Phys.* **121** 043104

[14] Reduto I, Fetisova M, Kryzhanovskaya N, Kotlyar K, Raskhodchikov A, Scherbak S, Guseva Y, Lipovskii A, Maximov M and Zhukov A 2019 Lasing in III–V microdisk core–TiO 2 shell lasers *J. Opt. Soc. Am. B* **36** 2285

[15] Raskhodchikov A V., Scherbak S A, Kryzhanovskaya N V., Lipovskii A A and Zhukov A E 2020 Dielectric surrounding bleaches the optical bond between a microdisk resonator and a straight optical waveguide *J. Phys. Conf. Ser.* **1695** 012128

Features of the construction of photonic integrated circuits for communication systems

N I Popovskiy[1], V V Davydov[2, 3] and V Yu Rud[3, 4]

[1] The Bonch-Bruevich Saint-Petersburg State University of Telecommunication, Saint Petersburg 195251, Russia
[2] Peter the Great Saint Petersburg Polytechnic University, Saint Petersburg 195251, Russia
[3] All-Russian Research Institute of Phytopathology, Moscow Region 143050, Russia
[4] A.F. Ioffe Physicotechnical Institute, St. Petersburg, 195152, Russia

nikitanikita24@mail.ru

Abstract. The article discusses the main methods of signal processing in a coherent optical transport network based on a photonic integrated circuit to increase the speed and the onset of the terabit era in optical transport networks, cloud and high-performance computing systems. The properties and operational characteristics of the main material platforms of photonic integrated circuits and their future technological units are considered.

1. Introduction

In modern human life, communication systems play an important role [1-10]. The volume of information and the requirements for its transfer are increasing every year [5-7, 11-18]. For different areas of human activity, different areas of communication are developing [3, 4, 6-8, 17-24]. A special place among communication systems is occupied by fiber-optic communication lines, which have found application in almost all areas of information transmission [16, 18, 22, 24-35]. Optical transport network at development of technology is stimulated by new services, such as cloud connection services of data centers, high-bandwidth video services and mobile 5G network services will drive future optical communications industry development and architecture transformation [5, 6, 13, 34, 36-40]. Service providers have driven higher speed Ethernet solutions for decades. Router connections, client side optics for optical transport networks (OTN) equipment, and wireless backhaul have continually pushed Ethernet to higher rates and distances to meet the demands for wireless connectivity [36-41]. And with global demand by consumers for video, this shows no signs of changing.

The main task of the network is to deliver traffic from one subscriber to another [34-41]. And do it as quickly as possible, both in terms of bandwidth and latency. Accordingly, the main task of the node is to transfer the incoming packet to the correct output interface as quickly as possible, having time to change its headers and apply policies. Therefore, there are pre-populated packet transmission tables - switching tables, routing tables, label tables, and neighborhood tables [36, 38, 40, 41].

The electrical or optical signal, getting to the device through the input physical port, is restored to the bit stream, separate Ethernet frames are isolated from it, then based on the headers (Ethernet, IP, MPLS), a decision is made about which output port this packet should be sent to and with which set of headers. On its way from the input port to the output port, the packet still passes through the Traffic Manager module, where the following things can happen to it: buffering, policing, shaping, priority processing.

2. Optical Communication Systems

In connection with the increase in video traffic on the Internet and with at increase on the transmission speed over fiber-optic data transmission systems up to 100 Gbps, and in the future up to 400 Gbps per channel, there is a problem of packet processing on switches and routers of the aggregation and core levels telecom operators, since the speed there is limited by electronic components, where the maximum frequency is 50 GHz. Photonic integrated circuits can meet this growing demand. Super channels allow you to increase the DWDM bandwidth in terabits in one cycle without any loss in terms of spectral efficiency and with the same optical range as the current generation of coherent transponders.

All major optical functions on all 10 100G line cards are bundled into one PIC pair - one for transmit and one for receive. All 10 carriers can now be brought into service in a single duty cycle, consuming much less power than 10 discrete transponders, and the result is significantly improved service reliability. PICs bring the same engineering usability to super channels that electronic integration brings to multi-core processors or graphics engines.

Within reasonable limits, the more carriers there are in the superchannel, the simpler the electronics and the better the optical performance. PICs remove the complexity limitation of optical components and allow you to choose the right engineering balance. As PIC technology advances, new circuits will emerge that will either allow new applications to reliably combine different technologies or continue to scale the component density for photonic VLSI systems.

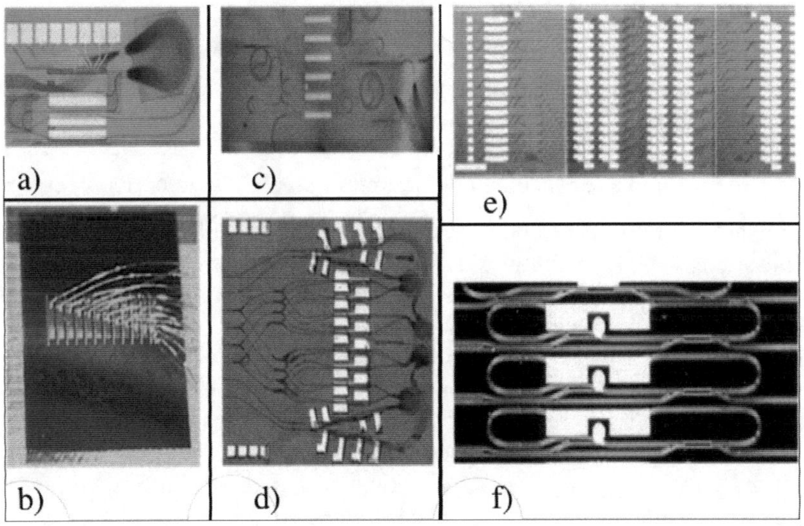

Figure 1. Examples of the platform on PIC are implemented. (a) Configurable discrete laser with nanosecond switching speed based on AWG; (b) Multi-stage SOA switching circuit with a speed of 320 GB/s; (c) multifunctional delay interferometer; (d) 4×4 cross-selective coupling of space and wavelength; (e) 16×16 photon switch for broadband photon packet routing; (f) wide frequency comb laser.

3. Photonic Integrated Circuit

The most productive commercial chip today produces 25.6 Tb/s. This is a major engineering challenge for developers. And there is no reason to believe that hyperscalers and examscalers will moderate their appetites and decide to stop there. The speed will increase. The PHY chip is on the transceivers, and the SerDes chip is on the network chip chip. The signal between them goes through an electrically conductive medium - along a metal track. As the speeds increase, so does the design complexity and the electricity consumed. Sooner or later (rather, sooner) we'll hit something. In the case of silicon photonics, the PHY chip is transferred inside the switching chip itself. Photonic ports are "embedded" in the crystal, allowing communication between chips at the speed of light through the optical medium.

The idea is not new, and only waiting for its time, namely, when the technology will reach the desired level of maturity. The problem was that the materials and processes used to make the photonic chips were fundamentally incompatible with the silicon chip manufacturing process - CMOS. Possible alternative solutions: installing a separate chip on the board that converts the electrical signal into an optical one, or installing it inside the network chip, but not on the crystal itself (still requires converting the medium). But this technological dam is washed away by persistent developments in this direction, and in the near future, microelectronics is waiting for big changes.

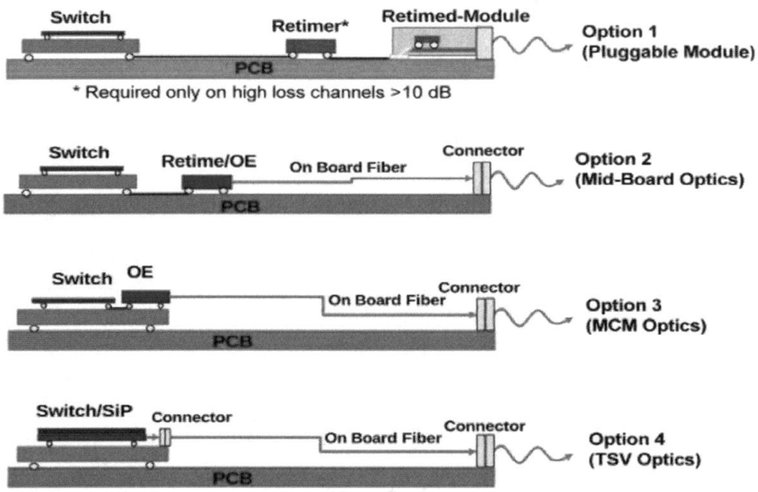

Figure 2. Silicon photonics chip-on-chip layout options.

Figure 3 (a) shows a number of components that we can manufacture using passive waveguides. MMI connectors and AWG demultiplexers are the most important of them. Deeply embedded waveguides allow you to create MMIRE reflectors and compact ring filters. The polarization converter is building block that can be created using a passive waveguide. It is possible to make polarization splitters and unifiers by placing a polarization converter in a Mach-Zehnder interferometer (MZI), placing it half in a polarization-dependent component, its response is independent of polarization. A wide range of functionality is offered by SOA on combination of passive devices. As shown in Figure 3 (b): Fabry-Perot lasers (FP), multiwave lasers, ring lasers and tunable lasers. It is possible to create picosecond pulsed lasers using a short SOA section with a reverse offset as a protected absorber. Figure 3 (c) shows the functionality will be implemented by a block of phase modules with passive devices: amplitude modulators, spatial switches, switches with wavelength selection, such as WDM cross-connections and multiplexers with addition and reset. It is also possible to create ultrafast switches using the nonlinear properties of SOA integrated into MZI.

4. Conclusion

The paper discusses the main ways to help the onset of the terabit era in optical transport networks, in the clouds, as well as in high-performance computing systems. New signal processing technologies using photonic integrated circuits in coherent optical networks are considered. Three main platforms of photonic integrated circuits, their technological components, properties and performance characteristics are considered.

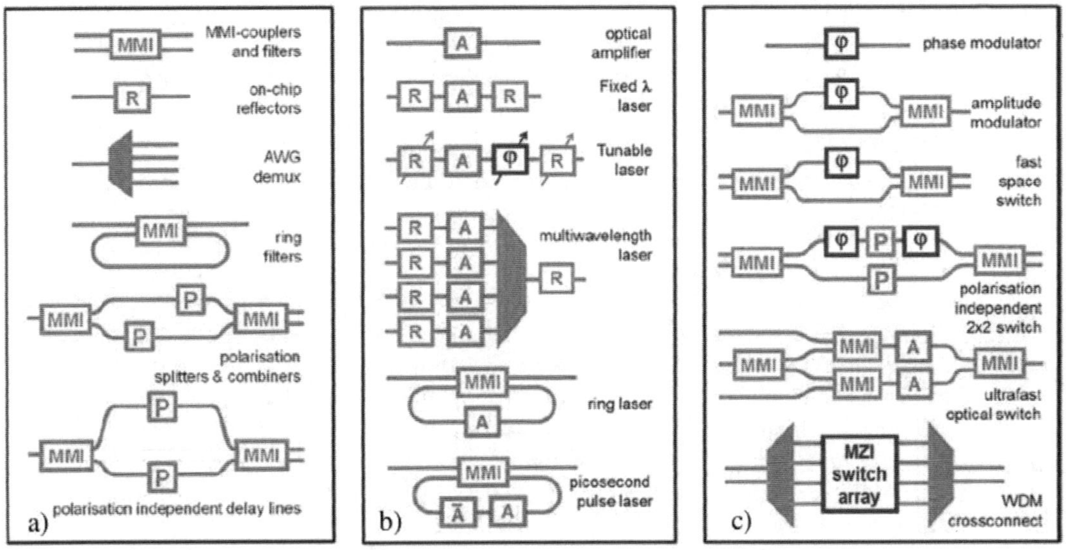

Figure 3. Examples of implementing devices using (a) passive waveguide devices (b) passive waveguide devices in combination with optical amplifiers (c) passive waveguide devices in combination with phase modulators.

References

[1] Grevtseva A S, Davydov R V, Dudkin V I and Rud' V Y 2019 *Journal of Physics: Conference Series* **1326(1)** 012043

[2] Valov A P, Davydov R V, Rud V Y and Grevtseva A S 2019 *Journal of Physics: Conference Series* **1326(1)** 012040

[3] Lukashev N A, Glinushkin A P and V. S. Lukyantsev V S 2019 *Journal of Physics: Conference Series* **1410(1)** 012211

[4] Lukashev N A, Davydov R V and Glinushkin A P 2019 *Journal of Physics: Conference Series* **1326(1)** 012046

[5] Andreeva E I and Potapov I A 2020 *Journal of Physics: Conference Series* **1695(1)** 012125

[6] Andreeva E I and Potapov I A 2021 *Springer Proceedings in Physics* **255** 241–245

[7] Davydov R V, Dmitrieva D S, Pilipova V M, Dudkin V I and Andreeva E I *2020 Proceedings of International conference of Laser Optics ICLO-2020 (Saint-Petersburg),* vol. 9285820 (IEEE), p. 171

[8] Logunov S E, Vysoczky M G and Titova O A 2018 *International Conference on Electrical Engineering and Photonics EExPolytech-2018 (Saint-Petersburg)*, vol. 8564402 (IEEE), p. 282-284

[9] Davydov R V and Antonov V I 2018 *Journal of Physics: Conference Series* **1124(8)** 081037

[10] Davydov R V and Antonov V I 2018 *Journal of Physics: Conference Series* **1135(1)** 012087

[11] Kozlova E S, Kotlyar V V and Nalimov A G 2015 *Computer Optics* **39(5)** 654-660

[12] Kozlova E S and Kotlyar V V 2014 *Computer Optics* **38(1)** 132-138

[13] Kuzmin M S and Rogov S A 2019 *Computer Optics* **43(3)** 391-396

[14] Skidanov R V, Blank V A and Morozov A A 2015 *Computer Optics* **39(3)** 384-392

[15] Petrov A A, Grebenikova N M, Lukashev N A, Ivanova N V, Rodygina N S and Moroz A V 2018 *Journal of Physics: Conference Series* **1038 (1)** 012032

[16] Podstrigaev A S, Smolyakov A V, Myazin N S and Slobodyan M G 2018 *Lecture Notes in Computer Science (including subseries Lecture Notes in Artificial Intelligence and Lecture Notes in Bioinformatics)* **11118 LNCS** 509-515

[17] Valov A P 2019 *Journal of Physics: Conference Series* **1410(1)** 012246

[18] Filatov D L, Galichina A A, Vysoczky M G and Yalunina T R 2017 *Journal of Physics: Conference Series* **917**(8) 082005
[19] Petrov A A and N S Myazin 2017 *Lecture Notes in Computer Science (including subseries Lecture Notes in Artificial Intelligence and Lecture Notes in Bioinformatics)* **10531 LNCS** 561-568
[20] Petrov A A, Vologdin V A and Zalyotov D V 2015 *Journal of Physics: Conference Series* **643(1)** 012087
[21] Petrov A A, Zalyotov D V, Shabanov V E and Shapovalov D V 2018 *Journal Physics: Conference Series* **1124(1)** 041004
[22] Davydov V V, Dudkin V I and Karseev A Yu 2014 *Optical Memory & Neural Networks (Information Optics)* **23(4)** 259 – 264
[23] Petrov A and Shapovalov D 2019 *Journal of Physics: Conference Series* **1400(4)** 044008
[24] Myazin N S, Dudkin V I, Grebenikova N M, Rud' V Y and Podstrigaev A S 2019 *Lecture Notes in Computer Science (including subseries Lecture Notes in Artificial Intelligence and Lecture Notes in Bioinformatics)* **11660 LNCS** 744–756
[25] Davydov R, Antonov V and Moroz A 2019 IEEE International Conference on Electrical Engineering and Photonics (EExPolytech-2019), vol. 8906791, (IEEE), p. 295-297
[26] Davydov R V, Saveliev I V, Lenets V A and Tarasenko M Yu 2017 *Lecture Notes in Computer Science (including subseries Lecture Notes in Artificial Intelligence and Lecture Notes in Bioinformatics)* **10531 LNCS** 177-183
[27] Moroz A V and Davydov R V 2019 *Lecture Notes in Computer Science (including subseries Lecture Notes in Artificial Intelligence and Lecture Notes in Bioinformatics)* **11660 LNCS** 710 – 718
[28] Fadeenko I, Fadeenko V, Reznik V, Moroz A, Popovskiy N and Nikolaev D 2019 IOP Conference Series: Earth and Environmental Science **390(1)** 012022
[29] Davydov V, Fadeenko V, Fadeenko I., Popovskiy N and Rud V 2019 *E3S Web of Conferences* **140** 07006
[30] Smirnov K J, Glagolev S F and Tushavin G V 2018 *Journal of Physics: Conference Series* **1124(1)** 022014
[31] Fadeenko V B, Kuts V A and Vasiliev D A 2018 *Journal Physics: Conference Series* **1135(1)** 012053
[32] Moroz A V 2019 *Journal of Physics: Conference Series* **1410(1)** 012212
[33] Moroz A V, Davydov V V, Malanin K Y, Krasnov A A and Rud V Yu 2019 *Journal of Physics: Conference Series* **1400(4)** 044009
[34] Fadeenko V B, Pchelkin G A and Beloshapkina O O 2019 *Journal of Physics: Conference Series* **1410(1)** 012238
[35] Fadeenko V B and Pchelkin G A 2019 *Journal of Physics: Conference Series* **1400(4)** 044010
[36] Makolkina M, Pham V, Kirichek R, Gogol A and Koucheryavy A 2018 *Lecture Notes in Computer Science (including subseries Lecture Notes in Artificial Intelligence and Lecture Notes in Bioinformatics)* **11118 LNCS** 547-559
[37] Ermolaev A N, Krishpents G P and Vysoczkiy M G 2016 *Journal Physics: Conference Series* **741(1)** 012071
[38] Hoang T, Kirichek R, Paramonov A and Koucheryavy A 2016 *Lecture Notes in Electrical Engineering* **376** 1249-1259
[39] Grebenikova N, Moroz A, Bylina M and Kuzmin M 2019 *IOP Conference Series: Materials Science and Engineering* **497(1)** 012109
[40] Karanov B and Chagnon M 2018 *J. Lightwave Technol* **36 (20)** 4843-4855
[41] Popovskiy N I and Davydov V V and Valiullin L R 2020 *Journal of Physics: Conference Series* **1695(1))** 012120

Thermo-optical properties of silicon nitride Mach-Zehnder interferometer for the on-chip quantum random number generator

A Prokhodtsov[1,2,3], V Kovalyuk[2,3], P An[2,3], A Golikov[2,4], Y Konyshev[1,2], R Shakhovoy[5,6], V Sharoglazova[5], A Udaltsov[5], Y Kurochkin[5,6,7] and G Goltsman[1,6]

[1] National Research University Higher School of Economics, Moscow, 101000, Russia
[2] Department of Physics, Moscow State Pedagogical University, 119992, Russia
[3] Institute of Applied Physics of the Russian Academy of Sciences, 603950, Russia
[4] Moscow Institute of Physics and Technology (State University), 141700, Russia
[5] QRate, 100 Novaya str., Skolkovo, Russia
[6] NTI Center for Quantum Communications, National University of Science and Technology MISiS, 4 Leninsky prospekt, Moscow, Russia
[7] Russian Quantum Center, 45 Skolkovskoye shosse, Moscow, Russia

Abstract. Here we study the thermo-optical properties of an on-chip silicon nitride Mach-Zehnder interferometer (MZI). The spectral shift of the MZI is associated with a change in the chip temperature. This can be explained by a change in the splitting ratio of the directional couplers, as well as a significant change in phase difference between waveguide arms. We experimentally found a phase shift of 2π when heated by 1.67 °C and changes in resonant wavelength at different temperatures ($d\lambda/dT$) equal 12.0 pm/°C, theoretically obtained a formula for an arbitrary splitting ratio of the directional couplers in an MZI, and determined the temperature stability required to the device operation inside a quantum cryptography system.

1. Introduction

High thermo-optical (TO) stability of photonic integrated circuits (PICs) is required in most modern applications, such as quantum key distribution and quantum random number generators, where passive elements, such as Mach-Zehnder interferometers (MZIs), are used to transfer phase-encoded signals into amplitude modulation [1]. In fact, MZIs generally require accurate resonant wavelengths; therefore, the temperature dependence of the effective refractive index n should be taken into account. Inasmuch as the temperature coefficient of the Si_3N_4 refractive index (2.45×10^{-5} K^{-1} [2]) is much smaller than in silicon on an insulator (1.8×10^{-4} K^{-1} [3]), the former is less sensitive to temperature fluctuations. However, due to the large difference between the arms in our Si_3N_4 interferometers [4], even a small change in temperature can significantly affect the operation of the device due to the shift in the optical spectrum. For this reason, the use of such elements in PICs requires thermal control, which is usually solved by packaging. But, for example, maintaining the laser wavelength in high-throughput multiplexing systems requires maintaining the temperature within ± 0.1 °C [5]. Here, we study the temperature dependence of the optical spectrum, visibility at wavelength, change resonance wavelength and the phase shift in the MZI.

2. Theory of a Mach-Zehnder Interferometer

In general case, an interferometer is an optical device that splits a wave into two parts and creates a phase difference between them by passing different lengths of optical paths, and then aligns them and measures their superposition. Figure 1 shows a free space Mach-Zehnder interferometer, which consists of two beam splitters that serve to split and combine waves, and two mirrors that set the optical path lengths of the interferometer arms.

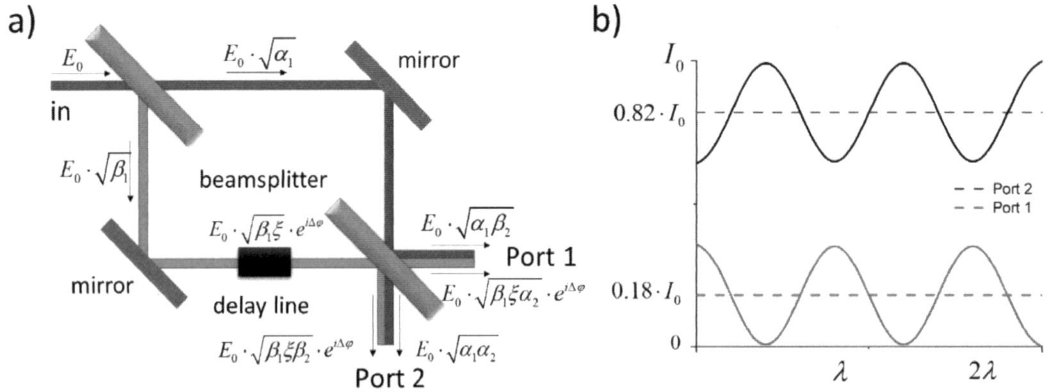

Figure 1. (a) Free space Mach-Zehnder Interferometer; **(b)** Spectral intensities at the two outputs of the interferometer at transmission coefficients of beam splitters equal to 0.9.

Let the transmission and reflection coefficients of the beam splitters be equal to α and β, respectively. In case of low absorption and scattering losses we get $\alpha + \beta \approx 1$. The wave with amplitude E_0 splits at the first beam splitter into $E_0\sqrt{\alpha_1}$ and $E_0\sqrt{\beta_1}$, then these two waves are redirected to the second beam splitter. Due to different optical paths, two waves come to the second beam splitter with a phase difference $\Delta\varphi$; moreover, additional losses may occur in one of the arms, which may be introduced via a transmission coefficient ξ . In fact, when the length of one of the arms is much larger than the other, as in our case, the losses in one arm are greater than in the other. Thus, two waves fall on the second beam splitter $E_0\sqrt{\alpha_1}$, and $E_0\sqrt{\beta_1\xi}\cdot e^{i\Delta\varphi}$. As a result, we get four waves after the second beam splitter: $E_0\sqrt{\alpha_1\cdot\beta_2}$ and $E_0\sqrt{\beta_1\xi\alpha_2}\cdot e^{i\Delta\varphi}$, which interfere in one port, and $E_0\sqrt{\beta_1\xi\beta_2}\cdot e^{i\Delta\varphi}$ and $E_0\sqrt{\alpha_1\alpha_2}$ – in the other. Using the equation (1):

$$I = E\cdot E^* = I_1 + I_2 + 2\sqrt{I_1\cdot I_2}\cdot\cos(\Delta\varphi), \tag{1}$$

where I_1 and I_2 the intensities of the interfering waves, we find the intensities of the optical radiation at the output ports:

$$I_{P_1}(\Delta\varphi) = I_0\big(\alpha_1\beta_2 + \beta_1\xi\alpha_2 + 2\sqrt{\alpha_1\alpha_2\beta_1\beta_2\xi}\cdot\cos(\Delta\varphi)\big) \tag{2}$$

$$I_{P_2}(\Delta\varphi) = I_0\big(\alpha_1\alpha_2 + \beta_1\xi\beta_2 + 2\sqrt{\alpha_1\alpha_2\beta_1\beta_2\xi}\cdot\cos(\Delta\varphi+\pi)\big). \tag{3}$$

Further, using these expressions, the interference visibility colour maps were built, which is defined as:

$$V = \frac{I_{max} - I_{min}}{I_{max} + I_{min}} = \frac{I_p(2\pi) - I_p(\pi)}{I_p(2\pi) + I_p(\pi)}, \tag{4}$$

where I_{max}, I_{min}, are the intensities at the output of the interferometer in the case when the interfering waves arrived in-phase and antiphase, respectively. The maximum visibility of interference is achieved when the waves that interfere are equal in intensity to each other.

2.1. Lossless example

Consider the case when the losses in the interferometer arms are the same ($\xi = 1$). Figure 2 shows the visibility maps for the interferometer outputs. We are interested in two special cases: 1) the beam splitters in the Mach-Zehnder interferometer are the same, which leads to the equality of the coefficients α and β, and 2) the coupling ratio of one of the splitters is 0.5.

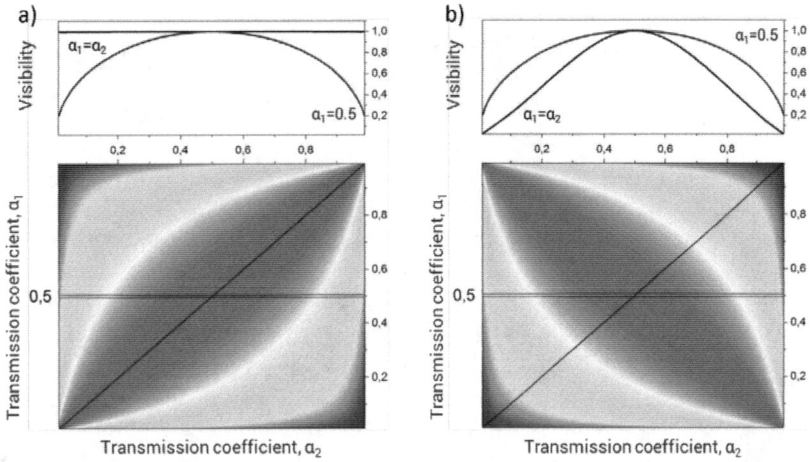

Figure 2. (a) Dependence of the visibility on the splitting ratios of the first port without taking into account losses. **(b)** Dependence of the visibility on the splitting ratios for the second port without taking into account losses.

In case of identical beam splitters, the visibility takes the highest values at the output of the first port. At the output of the second port, the visibility has a maximum only at transmission coefficients close to 0.5. At the same time, with significant deviations of the transmittance $\boldsymbol{\alpha}$ from 0.5, the visibility drops sharply.

To understand why this happens, consider, for example, the case in which the transmittance $\boldsymbol{\alpha}$ is close to unity. The initial radiation I_0 is split into four waves. The most intense of them, which passed the two beam splitters, enters port 2. The wave with the lowest amplitude, which is reflected twice, also enters port 2. The two remaining waves once reflected and passed once through the beam splitter, are identical in intensity and enter port 1. All this leads to the fact that in the case of identical beam splitters and an arbitrary coupling ratio, the radiation intensity in port 1 significantly exceeds the intensity in the second port, while the radiation visibility, on the contrary, is maximal at the output of the second port.

If one beam splitter divides the radiation in half, there are no differences in the intensity and visibility of the interference at the output from the two ports. The outputs differ only in that the interference patterns are shifted relative $\boldsymbol{\pi}$ to each other (Figure. 1 (b)).

2.2. Example with losses

Let us consider the case when the losses in the arm, into which the light enters after reflection at the first beam splitter, are much higher than in the other. In this case, ξ is different from unity. Figure 3 shows the dependence of the visibility on the transmittance of the beam splitters α_1 and α_1 at $\xi = 0.01$.

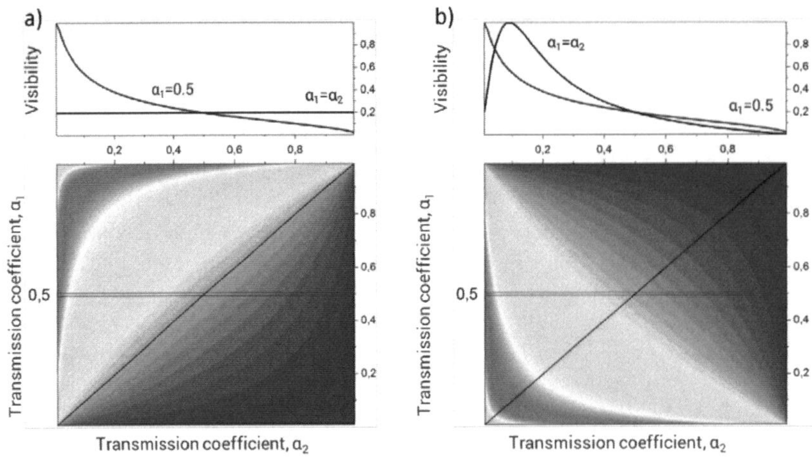

Figure 3. (a) Dependence of the visibility on the splitting ratios of the first port, taking into account the loss. **(b)** Dependence of the visibility on the splitting ratios for the second port, taking into account the loss.

In the case of identical beam splitters, the visibility takes the highest values at the output of the second port at low values of the transmittance. In the case when it is less than 0.5, most of the radiation enters the arm with large losses. If you choose such that it will satisfy the condition $\alpha^2 = \beta^2\xi$ then the maximum visibility will be observed at the output of the second port, since the interfering waves will be equal in intensity. The visibility of the interference in port 1 is not affected by the values of the division factors, as in the case when $\xi = 1$. When ξ decreases, the visibility of the interference also decreases, since one of the waves loses its intensity, and the intensity of the second wave is not affected by ξ, and the difference between the intensities of the interfering waves increases.

Since in our work along delay line is used in one of the arms, ξ differs significantly from one. The loss in the delay line was estimated to be about 20 dB ($\xi = 0.01$). Therefore, to get the highest visibility, the second port was chosen for the operation. In addition, the values of the transmittance of the beam splitters are optimal in the range from 0.05 to 0.15.

3. Device fabrication and experimental setup

To fabricate our device, we use a commercially available 525 μm silicon (Si) wafer with 2.6 μm thick buried silicon oxide (SiO_2) and over a 450 nm silicon nitride (Si_3N_4) layer. To form the geometry of the waveguide, we used single-stage electron-beam lithography with a high-contrast positive resist ZEP 520A and dry reactive-ion-etching (RIE) in a CHF_3-Ar mixture.

We fabricated a 2D array of the MZIs with a delay line, with various parameters of directional couplers interaction lengths in a range of 60 μm to 100 μm and gaps from 0.6 μm to 1.15 μm, the single device is shown in Figure 4 (a). The measurement transmission spectra of a 2D array with gap = 0.85 μm and different interaction length shown in Figure 5 (b).

In the second step, we characterized transmission spectra of on-chip MZI with different temperature substrate. For that, we used the experimental setup shown in Figure 5 (a).

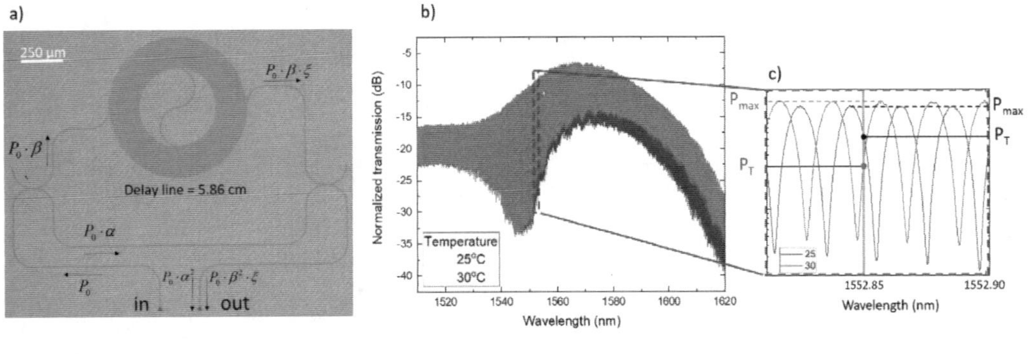

Figure 4. **(a)** Optical micrograph of the on-chip MZI; **(b)** The normalized transmission spectra of MZI without coupling losses with temperature 25 °C and 30 °C; **(c)** Enlarged normalized transmission spectrum of MZI with temperature 25 °C and 30 °C.

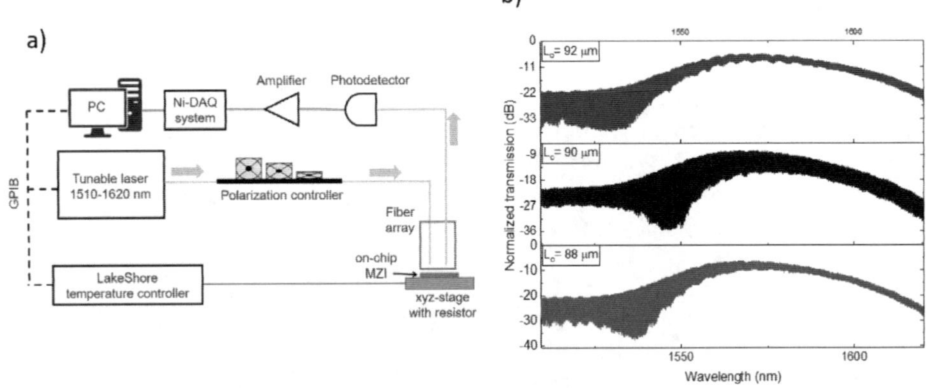

Figure 5. **(a)** Schematic view of the experimental setup for MZI transmission measurements with different temperature of the substrate; **(b)** The measured transmission spectra of MZIs with constant gap = 0.85 µm and different interaction length.

Light from a tunable laser source through polarization controller coupled into the chip using a fibre array separated by with distance of 250 µm SMF-28 optical fibres as well. At the output fibre by a low-noise photodetector and amplifier and recorded by Ni-DAQ system. Chip was situated on rotation stage with *x*,*y*,*z*-motors. The change in the temperature of the stage in the range from 25 °C to 30 °C occurred due to the Joule heating of the resistor, which in turn heated the stage on which the sample was located. The stage temperature was controlled using a PID controller.

We characterized the transmission of MZI in the wavelength range 1510-1620 nm at a wavelength of 1552.85 nm. After the temperature normalization, the transmission spectrum of the MZI was measured. Figure 4 (a) shows the measured normalized transmission spectra of the MZI without coupling losses for two different temperatures (T), equal to 25 °C and 30 °C. In Figure 4 (c) we can see a zoomed image of optical transmission, and the maximum (P_{max}) and the minimum (P_T) optical power at a wavelength of 1552.82 nm for two different temperatures. In the final, we calculated visibility (ν_λ) at wavelength use equation $\nu_\lambda = P_{max} - P_T/P_{max} + P_T$. The results are shown in Figure 6 (a).

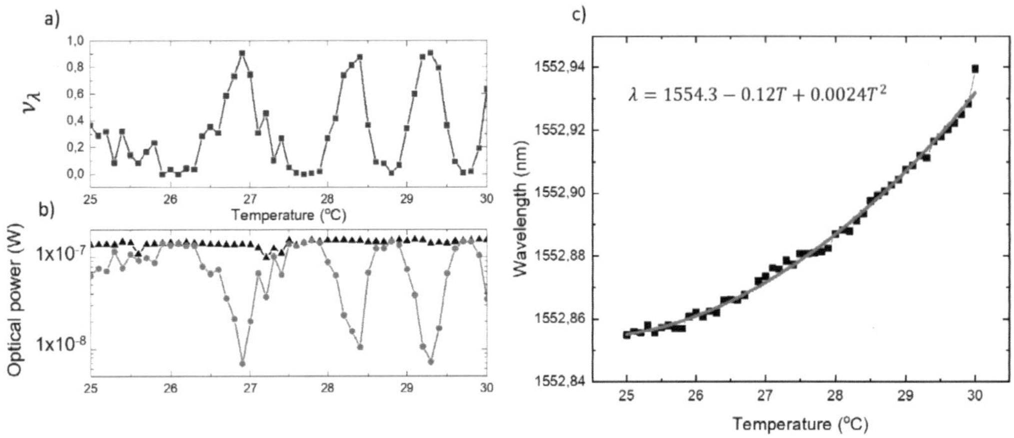

Figure 6. **(a)** Visibility at a wavelength = 1552.82 nm vs temperature; **(b)** Measured maximum optical power (black) and optical power at a wavelength = 1552.82 nm (red); **(c)** Wavelength in the minimum vs temperature.

Based on the measurement data (Figure 6 (c)) we found changes in resonant wavelength at different temperatures ($d\lambda/dT$) equal 12.0 pm/°C as well as a phase shift by 2π with a change in the substrate temperature by 1.67 °C. We also theoretically obtained a formula for an arbitrary division ratio of directional couplers in an interferometer and determined the temperature stability required to operate inside a quantum cryptography system.

4. Conclusion

In this work, we theoretically study the visibility of interference at splitting ratio in Mach-Zehnder interferometer and demonstrate visibility without and with losses in one of the interferometer arm. To study the TO effect, we change the temperature of the substrate in the range from 25 °C to 30 °C. Of the measured transmission spectra in this range we calculated the visibility of interference at wavelength 1552.85 nm. We have demonstrated a change in visibility and saw the change in resonance wavelengths by 12.0 pm/°C, as well as the phase shift by 2π with a change in the substrate temperature by 1.67 °C. In particular, this will allow selecting the optimal temperature for work such device.

Acknowledgments

We acknowledge support of the Russian Science Foundation grant No. 19-72-10156 (waveguide fabrication), grant No. 17-71-20146 (theoretical and numerical calculations).

References

[1] Raffaelli F, Sibson P, Kennard J E, Mahler D H, Thompson M G and Matthews J C F 2018 *Opt. Express* **26** 19730-19741.

[2] Arbabi A and Goddard L L 2013 *Optics letters* **38** (19) 3878-3881.

[3] Komma J, Schwarz C, Hofmann G, Heinert D and Nawrodt R 2012 *Applied Physics Letters* **101** (4) 041905.

[4] Prokhodtsov A, Kovalyuk V, An P, Golikov A, Shakhovoy R, Sharoglazova V, Udaltsov A, Kurochkin Y and Goltsman G 2020 *Journal of Physics: Conference Series* **1695** 012118

[5] Enright R, Lei S, Nolan K, Mathews I, Shen A, Levaufre G and Hernod D 2014 *Bell Labs Technical Journal* **19** 31-45

Helical phase in two-dimensional magnets due to four-spin interactions

G R Rakhmanova[1], D I Ilin[1], A N Osipov[1], I V Shushakova[1], I V Iorsh[1], I A Ado[2,3] and M Titov[2]

[1]School of Physics and Engineering, ITMO University, Saint Petersburg 197101, Russia
[2]Radboud University, Institute for Molecules and Materials, 6525 AJ Nijmegen, The Netherlands
[3]Institute for Theoretical Physics, Utrecht University, 3584 CC Utrecht, The Netherlands

gulnaz.rahmanova@metalab.ifmo.ru

Abstract. We demonstrate that in ferromagnets with the D_{3h} point group of symmetry a possible origin of phase transition from a collinear ferromagnetic state to a non-collinear state can be the fourth order contributions to the free energy density that are allowed by this point group of symmetry. At the same time, Dzyaloshinskii-Moria interaction vanishes in such materials. Via symmetry analysis we derive seven possible fourth order contributions to the free energy density with respect to the unit vector of the local magnetization direction but only two of them can be considered as independent. Moreover, for two-dimensional systems only one survives. Considered symmetry class is essential because a large group of two-dimensional intrinsic ferromagnets belongs to it, for example a monolayer Fe_3GeTe_2. The four-spin chiral exchange does also manifest itself in peculiar magnon spectra and favors spin waves.

1. Introduction

The discovery of single-layer graphene in 2004 attracted intense interest in the field of two-dimensional (2D) materials. The field has been growing ever since and large number of other stable atomically thin crystals have been studied, such as hexagonal hBN, MoS_2 and other dichalcogenides and layered oxides.

Potential applications of monolayers include spin-based computer logic and new ways to store information that require possibility to stabilize and effectively manipulate magnetic order [1]. In particular, noncollinear magnetic structures, as skyrmions, might become the basis for future memory devices [2].

However, 2D magnets have not been realized in experiments for a long time. Early attempts mainly focused on inducing magnetism in 2D nonmagnetic materials by magnetic defects, including vacancies, doping impurities, localized electronic states arising at the edges [3-4]. But the implementation of long-range magnetic order in 2D materials is limited by these methods. Another method for observing magnetism in 2D materials is the effect of magnetic proximity in heterostructures. For example, it was shown in [5] that in the graphene/ferromagnet heterostructure, it is possible to effectively modulate spin current in the graphene layer by changing magnetization of the ferromagnetic thin film. This do 2D magnetism possible, but the complex structure and fabrication at nanoscale severely limit its application [1]. Only in 2017 a long-range magnetic order was discovered in 2D van der Waals materials $Cr_2Ge_2Te_6$ [6] and CrI_3 [7]. Also, intrinsic ferromagnetism was described in atomic thin Fe_3GeTe_2 [8, 9]. Moreover, stable chiral magnetic textures have been observed in two different heterostructures based on Fe_3GeTe_2 [10, 11] and in thin layers of this material [12]. It is known that noncollinear magnetic order requires the existence of an antisymmetric exchange interaction called the Dzyaloshinsky-Moria interaction (DMI) [13, 14]. However, the origin of noncollinear spin states in Fe_3GeTe_2 cannot be correctly explained by DMI [15,16].

In this paper we discuss another possible origin of non-collinear states in ferromagnets with the D_{3h} group symmetry, for which all LI terms in micromagnetic functional are forbidden, while multi-spin exchange interactions are allowed by symmetry [17]. We propose that four-spin chiral exchange interaction in such 2D magnets causes a phase transition from a collinear state to a non-collinear ground state. The paper is presented based on the results of the article [19].

2. Helical phase in magnet with D3h symmetry

Using standard symmetry analysis [18], we identify seven fourth order contributions to the free energy density that are allowed by D_{3h} point group of symmetry. Only two invariants out of seven can be chose as independent. Moreover, for 2D structures we are left with a single fourth order term. So, the free energy functional of ferromagnet with D_{3h} point group of symmetry reads:

$$F[m] = \int d^2r\left[A \textstyle\sum_\alpha (\nabla m_\alpha)^2 + 8Bm_y\left(m_y^2 - 3m_x^2\right)[\nabla \times m]_z + Km_z^2 - \boldsymbol{Hm}\right], \quad (1)$$

where $\boldsymbol{H}$ stands for external magnetic field measured in energy unit, $A > 0$, K – anisotropy constants. The first term represents usual symmetric exchange, the second term describes independent fourth order invariant which corresponds to four-spin chiral interaction.

Next, we consider a non-collinear ansatz for local magnetization field $\boldsymbol{m}(r)$ that can be written in a form:

$$\boldsymbol{m}(\boldsymbol{r}) = \boldsymbol{n}\cos\alpha + [\boldsymbol{m}_1\cos(\boldsymbol{kr}) + \boldsymbol{m}_2\sin(\boldsymbol{kr})]\sin\alpha, \quad (2)$$

where $\boldsymbol{n}$, $\boldsymbol{m_1}$ and $\boldsymbol{m_2}$ are mutually perpendicular unit vectors: $\boldsymbol{m}_1 = (\sin\phi, -\cos\phi, 0)^T$, $\boldsymbol{m}_2 = (\cos\phi\cos\theta, \sin\phi\cos\theta, -\sin\theta)^T$, $\boldsymbol{n} = \boldsymbol{m}_1 \times \boldsymbol{m}_2$.The angle α corresponds to the transition from collinear state, $\alpha = 0$, to a pure helix state, $\alpha = \pi/2$.

We substitute this ansatz (2) into (1) and minimize with respect to wave vector and find the free energy as:

$$\frac{F[m]A}{VB^2} = -\frac{9}{64}(\sin\alpha + 5\sin 3\alpha)^2 \sin^4\theta\cos^2\theta + \frac{KA}{B^2}\cos^2\theta\cos^2\alpha + \frac{KA}{2B^2}\sin^2\theta\sin^2\alpha - \boldsymbol{Hm}. \quad (3)$$

In the absence of the external field the free energy is parametrized by a single dimensionless parameter KA/B^2. The figure 1 provides the dependence of an angle describing a transfer from collinear to helix state (α) on both constant of magnetic anisotropy and external magnetic field parallel to the plane (H_z=H_y=0). It is seen the sharp phase transitions from the collinear to the helical phase and back with K increasing. The corresponding jumps are also seen in the left panel. The maximum achieved value of α is $\alpha = \pi/6$. Naturally for large value of anisotropy or an external magnetic field, their effect suppresses the effect of four-spin interaction and the stable state is collinear. The effect of the four-spin interactions is thus in some way similar to the DMI.

3. Magnon spectra

Let us now investigate how the four-spin chiral interaction may affect the magnon spectra. To develop the magnon dispersion it is needed to solve the Landau-Lifshits equation:

$$\frac{d\mathbf{m}}{dt} = \mathbf{m} \times \mathbf{H_{eff}}, \quad (4)$$

where $\boldsymbol{H_{eff}} = -\delta F/\delta\boldsymbol{m}$ with respect to a small variation $\delta\boldsymbol{m}$.

We consider collinear ground state which is characterized by a constant magnetization vector $\boldsymbol{n}$. For the case of the external magnetic field is $\boldsymbol{H} = H_0(\sin\theta\cos\varphi, \sin\theta\sin\varphi, \cos\theta)$ $\boldsymbol{n}$ reads:

$$\boldsymbol{n} = \left(\lambda\sin\theta\cos\varphi \,/\left(1-\frac{K\lambda}{H_0}\right), \lambda\sin\theta\sin\varphi \,/\left(1-\frac{K\lambda}{H_0}\right), \lambda\cos\theta\right), \tag{5}$$

where $\lambda = f(K/H_0, \theta)$ is the solution of the equation $|\boldsymbol{n}| = 1$ that corresponds to the minimal free energy. It is equal to 1 in the absence of anisotropy, $\lambda = f(0, \theta) = 1$.

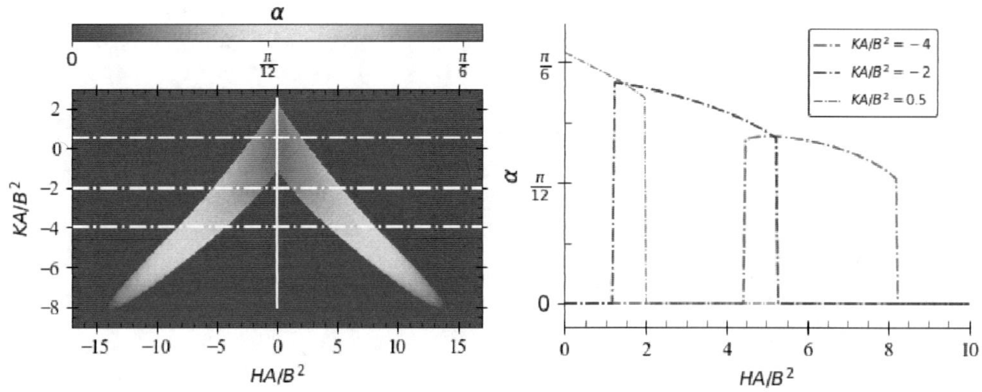

Figure 1. Phase diagram of the collinear and non-collinear phases as a function of anisotropy and external magnetic field in lateral direction. The left panel shows three vertical cross sections for different constants of anisotropy.

Using the ansatz

$$\boldsymbol{m}(\boldsymbol{r}, t) = \boldsymbol{n} + (\boldsymbol{n} \times \delta)e^{i\boldsymbol{q}\boldsymbol{r} - i\omega t}, \tag{6}$$

we linearize the Landau-Lifshits equation with respect to δ and find the magnon dispersion:

$$\omega_q^{\pm} = \pm S(q^2, \lambda) + \frac{3\lambda^3 Bq}{\left(1-\frac{K\lambda}{H_0}\right)^2}\sin^2\theta\cos\theta\sin(3\varphi + \chi), \tag{7}$$

where χ is the angle between the direction of propagation of magnon and the direction of in-plane component of magnetic field. In the case of zero anisotropy, S is just $S|_{K=0} = Aq^2 + H_0$. We notice the symmetry $\omega_q^{\pm} = -\omega_{-q}^{\mp}$. The maximal asymmetry $\omega_q^{+} - \omega_{-q}^{+}$ is reached when $\chi = \pm\pi/2 - 2\phi$. Moreover, the asymmetry is linear with respect to Bq and has non-trivial dependence on θ, the polar angle of the external magnetic field (see figure 2).

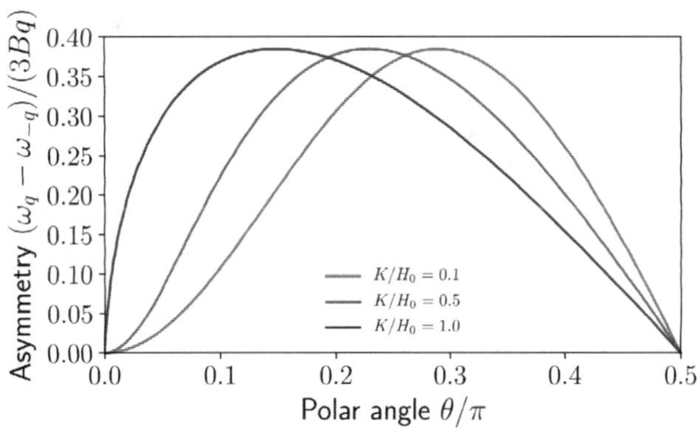

Figure 2. Dependence of the asymmetry on the magnetic field polar angle for different vales of anisotropy strength K/H_0.

4. Conclusion

In conclusion, we demonstrate that fourth order contribution $\boldsymbol{\omega(r)} = \boldsymbol{8Bm_y(m_y^2 - 3m_x^2)[\nabla \times m]_z}$ to the free energy density is responsible for the appearance of a long-range non-collinear magnetic order in ferromagnets with D_{3h} group of symmetry. It looks compatible with spin spirals observed in a recent experiment on Fe_3GeTe_2. Also, we showed how such four-spin chiral interaction in 2D magnets leads to non-reciprocal magnon dispersion.

Acknowledgments

This work was supported by the Ministry of Science and Higher Education of Russian Federation, goszadanie no. 2019-1246.

References

[1] Gong C and Zhang X 2019 *Science* **363** 4450
[2] Parkin S, Hayashi M and Thomas L 2008 *Science* **320** 190
[3] Son Y, Cohen M and Louie S 2006 *Nature* **444** 347
[4] Shidpour R and Manteghian M 2010 *Nanoscale* **2** 1429 -35
[5] Xu R, Liu B, Zou X and Cheng H-M 2017 *ACS Appl. Mater. Interfaces* **9** 38796 -801
[6] Gong C, *et al* 2017 *Nature* **546** 265
[7] Huang B, *et al* 2017 *Nature* **546**, 270
[8] Deng Y, *et al* 2018 *Nature* **563** 94
[9] Fei Z, *et al* 2018 Nat. Mater. **17** 778
[10] Wu Y, *et al* 2020 Nat. Commun. **11** 3860
[11] Park T-E, *et al* 2021 *Phys. Rev.* B **103** 104410
[12] Meijer M, Lucassen J, Duine R, Swagten H, Koopmans B, Lavrijsen R and Guimar M 2020 *Nano Lett.* **20** 8563
[13] Dzyaloshinsky I 1958 *J. Phys. Chem. Solids* **4** 241
[14] Moriya T 1960 *Phys. Rev.* **120** 91
[15] Hals K and Everschor-Sitte K 2019 *Phys. Rev.* B **99** 104422
[16] Laref S, Kim K-W and Manchon A2020 Phys. Rev. B 102 060402(R)
[17] Ado I, Qaiumzadeh A, Brataas A and Titov M 2020 *Phys.Rev.* B **101** 161403
[18] Hals K and Everschor-Sitte K 2017 *Phys. Rev. Lett.* **119** 127203
[19] Ado I A, Rakhmanova G R, Zezyulin D, Iorsh I V, Titov M 2021 arXiv:2105.14495

The investigation of the features optical vortices focusing by ring gratings with the variable height using high-performance computer systems

D A Savelyev[1,2]

[1] Samara National Research University, Moskovskoye Shosse 34, Samara, Russia
[2] Image Processing Systems Institute - Branch of the Federal Scientific Research Centre "Crystallography and Photonics" of Russian Academy of Sciences, Molodogvardeyskaya str. 151, Samara, Russia

dmitrey.savelyev@yandex.ru

Abstract. The diffraction of vortex laser beams with circular polarization by ring gratings with the variable height was investigated in this paper. Modelling of near zone diffraction is numerically investigated by the finite difference time domain (FDTD) method. The changes in the length size of the light needle and focal spot size are shown depending on the type of the ring grating.

1. Introduction

The interesting properties of singular beams make it possible to use them for optical manipulations with micro- and nanoparticles [1], material processing [2] and in microscopy [3]. In addition, the optical vortices are used in wireless communication systems [4], for transmitting information over fiber [5] and in quantum informatics [6], for polarization state recognition [7]. Such beams can be obtained using spiral and twisted axicons [8], spiral phase plates [9], multi-order diffractive optical elements [10].

The introduction of a vortex phase singularity into the input beam makes it possible to enhance the longitudinal component of uniformly polarized laser beams on the optical axis in the focal region [7, 11]. So, it is makes it possible to change the diffraction pattern due to the redistribution of energy between the components of the electromagnetic field [12].

The diffractive axicon (ring grating) as the conical axicon forms an extended light needle along the optical axis [13, 14]. It is also possible to obtain dynamic focus by complementing the optical system with an adjustable lens or axicon [15]. The change in the substrate thickness has a significant effect on the diffraction pattern of a limited plane wave by diffraction axicons with different numerical apertures (NA). This action was shown earlier to change the substrate thickness from 0.2λ to 0.3λ [11].

The features of focusing optical vortices on ring gratings with the variable relief height h are studied in this paper. Numerical calculations (3D) of laser radiation propagation were performed using the finite difference time domain (FDTD) method using high-performance computer systems [16, 17].

2. The investigation of focusing by ring gratings with variable height

Modeling parameters: wavelength λ = 0.532 μm, computational cell size x, y, z ∈ [–5.6λ; 5.6λ]. The thickness of the absorbing layer PML ~ 1.1λ, the sampling step in space is λ/16, the time step is λ/(32c), where c is the light speed. The refractive index of the considered optical elements and the substrate n was equal to 1.5.

The height h was varied as follows: at the center h = 3λ, then the height decreased with each subsequent grating ring with a step of 0.25λ up to h = 1.75λ to the edge of the element (the direct ring grating). The opposite case is also considered: in the center there is a minimum, h = 1.75λ and the height increases to 3λ from the center to the grating edge (the inverse ring grating). The diffractive axicon (h = 3λ) with NA = 0.95 was also considered for comparison. The period of all considered ring gratings was 1.05λ.

The Laguerre-SuperGauss modes (1,0) of degree 6 with circular polarization (polarization is opposite in sign of the introduced vortex phase singularity) and the super-Gaussian beams with the same degree are considered as the input laser radiation (Figure 1).

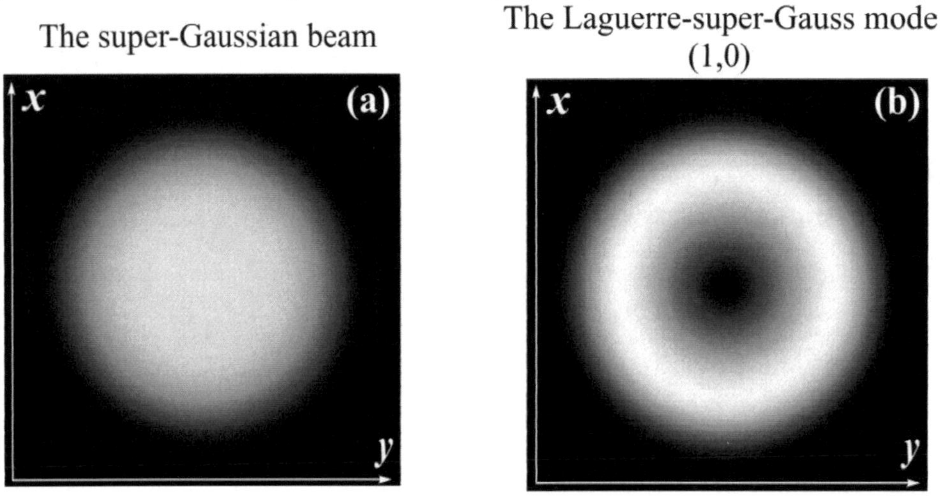

Figure 1. The input beams: (a) the super-Gaussian beam, (b) and the Laguerre-super-Gauss mode (1,0).

The research results are shown in Figure 2.

The size of the focal spot on the optical axis was estimated from the full width at half maximum (FWHM) of the intensity, and the length of the light needle was also estimated from the depth of focus (DOF), also at half maximum.

It should be noted that for the case Laguerre-super-Gauss mode (0,1), the most compact focal spot was obtained for the case of the direct ring grating (FWHM = 0.54λ), and the longest light needle was obtained for the case of the inverse ring grating (DOF = 4.81λ).

When considering the case of a super-Gaussian beam, it should be noted that focusing inside the element is observed for a direct ring grating. For the case of inverse ring grating, one can note a decrease in the size of the focal spot (FWHM = 0.48λ) and an increase in the length needle (DOF = 2.12λ) in comparison with a standard diffractive axicon with the same period (FWHM = 0.92λ and DOF = 1.06λ, respectively).

The graphs with cross-sections for compare direct and inverse ring grating are shown in Figure 3. It should be noted the drop in intensity for the case of direct ring grating (the Laguerre-super-Gauss mode (1,0)).

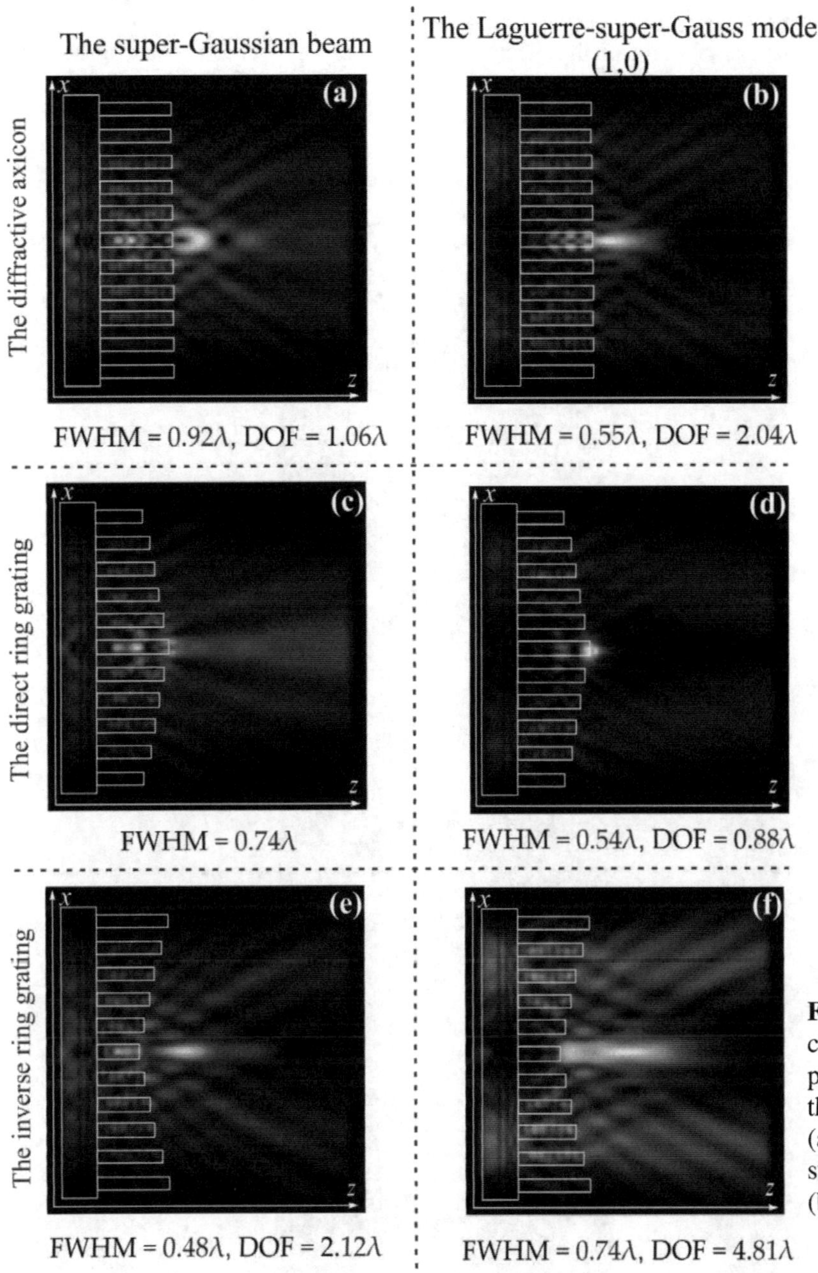

Figure 2. The longitudinal cross section (xz) of propagation laser radiation: the super-Gaussian beam – (a), (c), (e), and the Laguerre-super-Gauss mode (1,0) – (b), (d), (f).

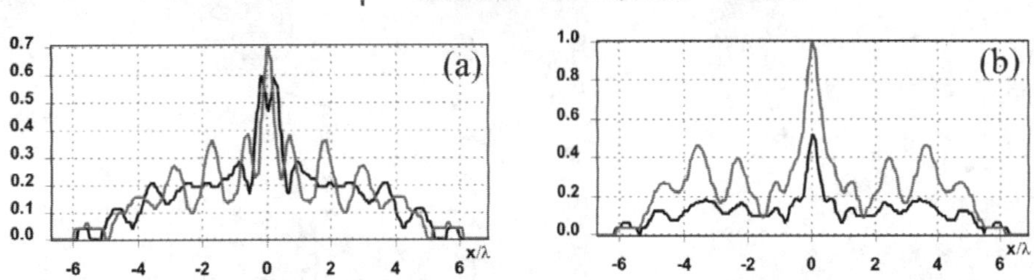

Figure 3. The cross sections for the cases of Figure 2 (c), (d), (e), (f), the direct ring grating (black line) and inverse ring grating (green line): (a) the super-Gaussian beam, (b) the Laguerre-super-Gauss mode (1,0).

3. The propagation of the pulsed Laguerre-super-Gauss mode (0,1)

One of the factors affecting the distribution of the electric field in the focal region is the duration of the laser pulse [18-24]. Let's expand on the research from the previous section for Laguerre-super-Gauss mode (0,1). The results of modeling the propagation of these pulsed beams at different times t are shown in Figure 4.

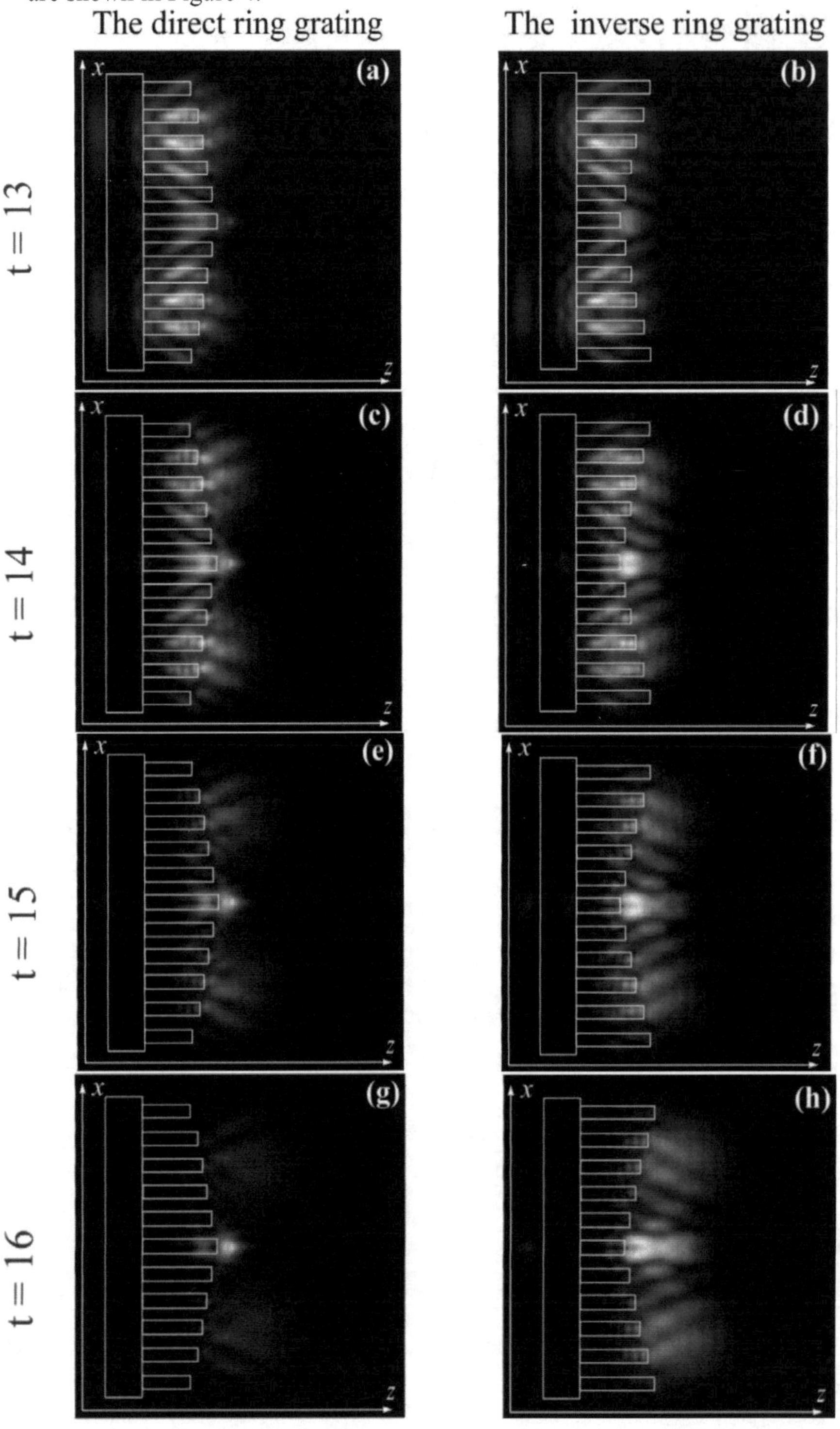

Figure 4. The propagation Laguerre-super-Gauss mode (1,0) pulses through the: (a), (c), (e), (g) the direct ring grating, (b), (d), (f), (h) inverse ring grating.

The graphs with cross-sections for compare direct and inverse ring grating at the output of the pulse from the element are shown in Figure 5. It should be noted the narrower focal spot for the case of direct ring grating than for inverse ring grating (FWHM = 0.5λ and FWHM = 1.06λ, respectively).

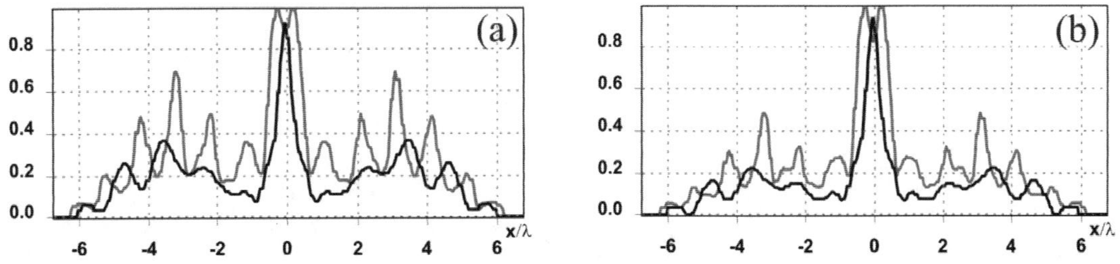

Figure 5. The cross sections for the cases of Figure 4, the direct ring grating (black line) and inverse ring grating (green line): (a) t = 15, (b) t = 16.

4. Conclusion

The formation of the Laguerre-super-Gauss modes (1,0) with circular polarization and a super-Gaussian beam by ring gratings with the variable height were investigated. The FDTD method is used for numerical simulation of diffraction.

It should be noted that a significant increase in the length of the light needle is observed with an increase in the height of the grating rings from the center to the edges in comparison with the standard diffractive axicon.

It is shown that or the case Laguerre-super-Gauss mode (0,1), the most compact focal spot was obtained for the case of the direct ring grating (FWHM = 0.54λ), and the longest light needle was obtained for the case of the inverse ring grating (DOF = 4.81λ). And for the case of a super-Gaussian beam should be noted inverse ring grating: one can note a decrease in the size of the focal spot (FWHM = 0.48λ) and an increase in the length needle (DOF = 2.12λ) in comparison with a standard diffractive axicon with the same period (FWHM = 0.92λ and DOF = 1.06λ, respectively).

Acknowledgments

This research was funded by the Russian Science Foundation (project No. 20-72-00051) in the parts «The investigation of focusing by ring gratings with variable height», «Conclusion»; scholarship of the President of the Russian Federation SP-916.2019.5 in the part «The propagation of the pulsed Laguerre-super-Gauss mode (0,1)» and was performed under the «ERA.Net RUS plus» program and funded by RFBR, project number 20-52-76021 in the part «Introduction».

References

[1] Dennis M R, O'Holleran K and Padgett M J 2009 *Progress in Optics* **53** 293
[2] Nivas J J, He S, Song Z, Rubano A, Vecchione A, Paparo D, Marrucci L, Bruzzese R and Amoruso S 2017 *Appl. Surf. Sci.* **418** 565
[3] Khonina S N and Golub I 2012 *JOSA A* **29** 2242
[4] Li S et al. 2020 *Advanced Optical Materials* **8** 1901666
[5] Pryamikov A, Alagashev G, Falkovich G and Turitsyn S 2020 *Scientific Reports* **10** 1
[6] Li S, Pan X, Ren Y, Liu H, Yu S and Jing J 2020 *Physical Review Letters* 124 083605
[7] Savelyev D and Kazanskiy N 2021 *Sensors* **21** 1
[8] Supp S and Jahns J 2018 *J. Eur. Opt. Soc.-Rapid Publ.* **14** 18
[9] Kotlyar V V, Kovalev A A, Skidanov R V, Khonina S N, Moiseev O Y and Soifer V A 2007 *J. Opt. Technol.* **74** 686
[10] Khonina S N and Ustinov A V 2019 *Applied Optics* **58** 8227
[11] Khonina S N and Savelyev D A 2013 *Journal of Experimental and Theoretical Physics* **117** 623
[12] Savelyev D A, Khonina S N 2015 *Computer Optics* **39** 654

[13] Khonina S N, Savelyev D A, Pustovoy I A, and Serafimovich P G 2012 *Journal of Optical Technology* **79** 626
[14] Savelyev D A 2021 *Computer Optics* **45** 214
[15] Khonina S N, Ustinov A V and Porfirev A P 2019 *JOSA A* **36** 1039
[16] Khonina S and Savelyev D 2015 *Radiophysics & Quantum Electronics* **57** 650
[17] Savelyev D 2020 *International Conference on Information Technology and Nanotechnology (ITNT), IEEE* 1
[18] Li X, Salamin Y I, Hatsagortsyan K Z and Keitel C H 2016 *JOSA B* **33** 405
[19] Savelyev D A and Krasnov S V 2019 *Optical Technologies for Telecommunications 2018, International Society for Optics and Photonics* **11146** 1114609
[20] Savelyev D A and Fomchenkov S A 2019 *Journal of Physics: Conference Series* **1410** 012138
[21] Khonina S N and Golub I 2018 *JOSA A* **35** 985
[22] Krasnov S V, Volotovskiy S G and Savelyev D A 2019 *PhotonIcs & Electromagnetics Research Symposium-Spring (PIERS-Spring) IEEE* 2025
[23] Wong L J, Kärtner F X and Johnson S G 2014 *Opt. Lett.* **39** 1258
[24] Savelyev D and Volotovskiy S 2020 *22nd International Conference on Transparent Optical Networks (ICTON) IEEE* 1

Discrete optical Zeno effect for polarization of light

K O Sedykh[1,2] **and D V Sych**[1,2,3,4]

[1] Moscow State Pedagogical University, Moscow 119992, Russia
[2] National University of Science and Technology MISiS, Moscow 119049, Russia
[3] Sirius University of Science and Technology, 1 Olympic Ave, 354340, Sochi, Russia
[4] P. N. Lebedev Physical Institute, Moscow 119991, Russia

kseniaolegovna98@gmail.com

Abstract. Quantum Zeno effect concerns deterministic dynamics of a quantum system induced by a series of projective quantum measurements. Applying this effect in optics, one can achieve an arbitrary lossless transformation of linear polarization of light with help of linear polarizers. However, to demonstrate this effect in practice, we have to take into account unavoidable losses in each polarizer that limits probability of successful transformations. In this work, we theoretically study a realistic quantum Zeno effect with an optimal discrete set of polarizers and find the maximum success probability

1. Introduction

There is no motion according to Zeno's arrow paradox [1, 2]. Since a flying arrow is at rest at every moment of time in a certain point in space, this means that it is at rest all the time. In other words, it is motionless. The quantum Zeno effect is the inhibition of the evolution of the state of a quantum system by method of repeated measurements [3, 4]. The measurement result in quantum mechanics is intrinsically probabilistic. However, repeated measurement of a physical quantity after measurement should give the same result. Thus, the evolution of a quantum system is inhibited. If there is a particle which tends to abandon its initial quantum state and we try to observe how it changes its quantum state, the probability of the change tends to zero. With quite frequent observation rate, we can keep the state of the quantum system, and the probability for this tends to unity.

The relevance of the research topic is due to the development of the direction of quantum control and quantum interaction-free measurements research [5–7], and finds applications in quantum communication [8-16]. Quantum Zeno effect can be applied in the counterfactual communication in which information is transferred by the phase part of wave function [17].

In our work, we investigate a discrete version of the quantum Zeno effect and want to make an arbitrary deterministic transformation of the quantum polarization state of light with help of only projective measurements. Consider a vertically polarized photon directed through a set of linear polarizers, slightly tilted with respect to each other. Each of polarizers tilts the polarization plane to a certain angle depending on the number of polarizers. In this case, a vertically polarized photon can be transformed into horizontally polarized one at the end of the set of polarizers. This polarization rotation is an example of the evolution of the quantum system that we want to guide by quantum measurements. In another interpretation of quantum Zeno effect, the dynamics of a quantum system can be inhibited by frequent observations [18-23]. In this work, we provide theoretical investigation of the discrete optical Zeno effect on the example of polarization of light and a set of lossy polarizers.

2. Ideal and real cases of Zeno effect

2.1. Ideal Zeno effect

Consider an optical system consisting of a set of linearly polarized light source, a set of polarizers and power meter. If a horizontal polarizer is added to the system, then only horizontal component of light remains. Consider a case, when we put horizontally and vertically oriented polarizers one after another. Apparently, there is no light passes through such system. However, if we add a diagonally oriented polarizer between them, we can observed a part of light that passes through the system. The more polarizers we add to the system, the greater part of light passes through it, if we pay attention how to put these devices to the optical scheme. There is a special relation between the number of added polarizers and the correct angle of their relative polarization. There is a rotation by $\frac{\pi}{2N}$, so the initially polarized light is orthogonally polarized at the end.

The probability of light passing through the set of polarizers depends on the number of additional polarizers:

$$P_N(\theta) = \left(\cos^2 \frac{\pi}{2N}\right)^N \tag{1}$$

The more polarizers are added to the optical system, the more light we can obtain (A1). In the limit of infinitely many polarizers, we obtain 100% light transmission. This is the ideal case of the quantum optical Zeno effect.

$$\lim_{N\to\infty} \left(cos^2 \frac{\pi}{2N}\right)^N = 1 \tag{2}$$

2.2. Real Zeno effect

Consider a realistic optical system linearly polarized light source, a set of lossy polarizers and a power meter. For the real case of quantum Zeno effect, we need to take into account the non-unit transmittance of the polarizers due to various losses, which affects the result. For different transmittance of polarizers, we need to add different number of polarizers to achieve maximum light transmittance. The probability of light passing through a set of polarizers depending on the transmittance coefficient *k* is

$$P_N(\theta) = \left(k \cdot cos^2 \frac{\pi}{2N}\right)^N \tag{3}$$

For example, if the transmittance k=0.9, then the maximum light passing through the system can be obtained using four additional polarizers (B1). In the limit of infinitely many polarizers, we obtain no light, in contrast to the ideal quantum Zeno effect:

$$\lim_{N\to\infty} \left(k \cdot cos^2 \frac{\pi}{2N}\right)^N = 0 \tag{4}$$

In the real case of the Zeno effect, the probability of light passing depends on the transmittance coefficient of the polarizers. Due to losses on each device, instead of 100% light transmission, only a small part (~35%) remains for the case of 10% loss on each polarizer.

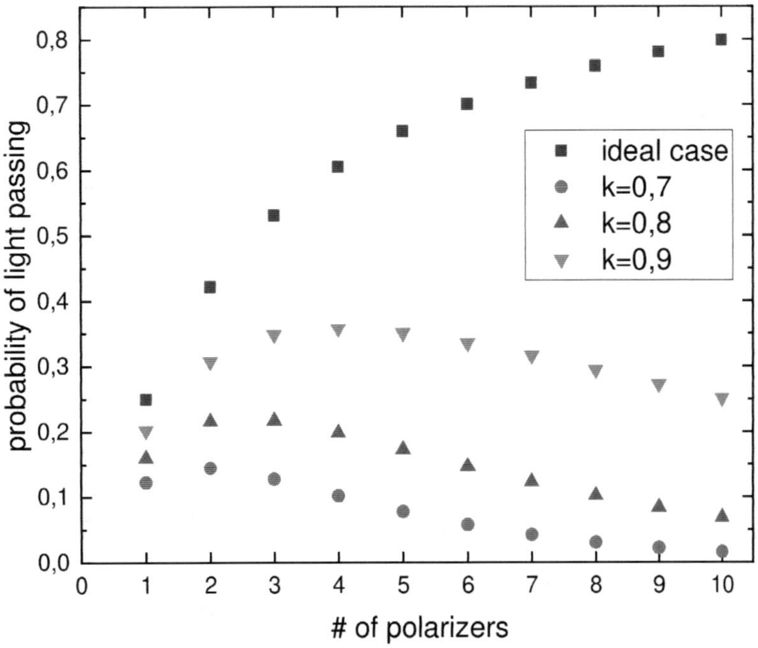

Figure 1. Quantum Zeno effect for the ideal case (black squares) and for the real case with different transmittance coefficients (k).

3. Conclusion

In this work, we consider a realistic version of the quantum optical Zeno effect on the example of changes of polarization of light with a set of polarizers. In the ideal case, the more polarizers we put to the system, the more light we can obtain at the end. In the real case, we have to take into account unavoidable losses in each polarizer, and the maximum amount of light can be obtained with a finite number of polarizers. This effect can be helpful for quantum state preparation via projective measurements, e.g. for quantum communication.

Acknowledgments

The reported study was funded by RFBR, Sirius University of Science and Technology, JSC Russian Railways and Educational Fund "Talent and success", project number 20-32-51004.

Appendices

A1. Probability of light passing through the set of polarizers in ideal case.

# of added polarizers	Angle (°)	$P(\theta)$
1	45	0.25
2	30	0.422
3	22.5	0.531
4	18	0.605
5	15	0.659
6	12.9	0.699
7	11.25	0.733
8	10	0.759
9	9	0.781

B1. Probability of light passing through the set of polarizers in real case with transmittance coefficient k=0.9.

# of added polarizers	Angle (°)	$P(\theta)$
1	45	0.206
2	30	0.308
3	22.5	0.348
4	**18**	**0.356**
5	15	0.351
6	12.9	0.334
7	11.25	0.316
8	10	0.294
9	9	0.272

References

[1] Aristotle *Physics* **VI** 9 239b10
[2] Bathfield M 2018 Foundations of Science **23** 649–679
[3] Misra B and Sudarshan E C G 1979 *J. Math. Phys.* **18** 756
[4] Kwiat P G *Illinois Physics* Interaction Free Measurements https://physics.illinois.edu/people/kwiat/interaction-free-measurements.asp#Zeno
[5] Kwiat P, Weinfurter H, Herzog T and Zeilinger A 1996 *Sci. Am.* **275** 72
[6] Kwiat P, Weinfurter H, Herzog T, Zeilinger A and Kasevich M 1995 *Ann. New York Acad. Sci.* **755** 3
[7] Elitzur A C and Vaidman L 1993 *Found. Phys.* **23** 987
[8] Sych D and Leuchs G 2016 Phys. Rev. Lett. **117** 200501
[9] Shcherbatenko M L, Elezov M S, Goltsman G N and Sych D V 2020 Phys. Rev. A **101** 032306
[10] Sych D V, Grishanin B A and Zadkov V N 2005 Quantum Electronics **35** 80–84
[11] Sych D, Averchenko V and Leuchs G 2017 Phys. Rev. A **96** 053847
[12] Sych D, Grishanin B and Zadkov V 2004 Laser Physics **14** 1314–1321
[13] Averchenko V, Sych D, Marquardt C and Leuchs G 2020 Phys. Rev. A **101** 013808
[14] Sych D and Leuchs G 2010 New J. of Phys. **12** 053019
[15] Sych D and Leuchs G 2010 Optics and Spectroscopy **108** 326–330
[16] Sych D and Leuchs G 2015 Foundations of Physics **45** 1613-1619
[17] Yuan Cao *et al* 2017 *Proc. Natl. Acad. Sci.* **114** (19) 4920-24
[18] Longhi S 2006 *Phys. Rev. Lett.* **97** 110402
[19] Kitano M 1997 *Opt. Commun.* **141** 39
[20] Yamane K, Ito M and Kitano M 2001 *Opt. Commun.* **192** 299
[21] Biagioni P, della Valle G, Ornigotti M, Finazzi M, Duò L, Laporta P and Longhi S 2008 *Opt. Express* **16** 3762
[22] Porras M A, Luis A, Gonzalo I and Sanz A S 2011 *Phys. Rev. A* **84** 052109
[23] Porras M A, Luis A and Gonzalo I 2013 *Phys. Rev. A* **88** 052101

Simulating the spectral characteristics of reflection in planar porous structures with antireflection coatings ZnS/DyF_3

D Shishkina[1,2], I Shishkin[1] and P Tishin[2]

[1] Department of Physics, Samara National Research University, Samara, 443086, Russia
[2] Faculty of Electronics and Instrument Engineering, Samara National Research University, Samara, 443086, Russia

Abstract. This paper presents the results of modeling a planar multilayer structure with layers of porous silicon, ZnS and DyF_3 coatings by the optical matrix method. It was shown that the optical matrix method, taking into account the model of porous silicon with a variable band gap, which takes into account the porosity gradient, allows us to approximate the course of the curve of the real experiment

1. Introduction

Porous silicon is a material of particular interest for solar energy, due to its large absorbing surface area. Due to the porous nature (Figure 1), the effective refractive index of porous silicon is lower than that of state material. Simulating the reflective properties of a multilayer structure allows us to identify the optimal parameters of the layers, which lead to a decrease in the reflection coefficient in the entire spectral range under study. In this paper, the reflective properties of a multilayer structure with a porous layer were modeled using the optical matrix method. This approach was used by the authors of [2,3] for structures with a single-layer antireflection coating of porous silicon and showed good agreement with the experiment.

Figure 1. Cross cleavage of a porous structure [1].

2. Experiment

The porous layer was obtained by electrochemical etching on a silicon substrate in alcoholic solutions of hydrofluoric acid. The thickness of the porous layer of the resulting structure was 10 microns. The

ZnS and DyF_3 coatings were applied by thermal evaporation in vacuum. The spectral characteristics of the reflection coefficient were studied using a Shimadzu UV-2450 spectrophotometer.

3. The optical matrix method

Along with the traditional calculation, there is a matrix method [4-5], allowing to establish a number of relationships between the input and output parameters of the system using a single matrix of the optical system. The choice of one method or another is dictated by the requirements of the task at hand.

Plane wave interacting with a stack of layers

$$E = E_0 e^{ikr-i\omega t};\ H = H_0 e^{ikr-i\omega t} \tag{1}$$

$$\begin{bmatrix} E_{j-1} \\ H_{j-1} \end{bmatrix} = M_{j-1} \begin{bmatrix} E_j \\ H_j \end{bmatrix} = \begin{bmatrix} \cos\varphi_j & \frac{i}{n_j}\sin\varphi_j \\ in_j \sin\varphi_j & \cos\varphi_j \end{bmatrix} \times \begin{bmatrix} E_j \\ H_j \end{bmatrix} \tag{2}$$

$\varphi_j = \frac{2\pi}{\lambda} n_j (z_j - z_{j-1})$ - phase thickness of the layer

Border conditions:

$$\begin{cases} E_0 = E_{0^-}^{(t)} + E_{0^-}^{(r)} \\ H_0 = n_0 E_{0^-}^{(t)} - n_0 E_{0^-}^{(r)}\ (3) \\ E_{(m)^+}^{(t)} = 1 \\ H_m = n_m E_{(m)^+}^{(t)} = n_m \end{cases} \tag{3}$$

Transformation matrix:

$$\begin{bmatrix} E_0 \\ H_0 \end{bmatrix} = M_1 M_2 M_3 \ldots M_{m-1} \begin{bmatrix} E_m \\ H_m \end{bmatrix} \tag{4}$$

Taking into account the boundary conditions:

$$\begin{bmatrix} E_{0^-}^{(t)} + E_{0^-}^{(r)} \\ n_0 E_{0^-}^{(t)} - n_0 E_{0^-}^{(r)} \end{bmatrix} = \begin{bmatrix} m_{11} & im_{12} \\ im_{21} & m_{22} \end{bmatrix} = \begin{bmatrix} 1 \\ n_m \end{bmatrix} \tag{5}$$

Amplitude and energy coefficients of reflection and transmission of a multilayer interference system:

$$r_{0^-} = \frac{E_{0^-}^{(r)}}{E_{0^-}^{(t)}} = \frac{n_0(m_{11}+in_m m_{12}) - (m_{21}+in_m m_{22})}{n_0(m_{11}+in_m m_{12}) + (n_m m_{22}+im_{21})}; \tag{6}$$

$$R = \left|\frac{(E_-)_i}{(E_+)_i}\right|^2 = |r|^2 \tag{7}$$

The main parameters of modeling are shown in Table 1. In all cases, the total thickness of the porous silicon layers was 10 µm, and the refractive index was taken in values between the refractive index of silicon and the refractive index of zinc sulfide [6]. The model [7-8] assumes the formation of nanocrystallites and pores of various sizes in PSi layers. It is supposed that smaller nanocrystallites are located near the surface and their concentration reduces with depth. This causes a porosity gradient and, correspond ingly, macroscopic dielectric function and refractive index gradients. Therefore, PSi can be considered as a semiconductor with a varying forbidden gap (gradedbandgap semiconductors) or, in conformity with our case, as a stack of layers with different refractive indexes increasing towards the inner interface with the silicon substrate/

Table 1. Simulation parameters

	Si	por –Si	ZnS	DyF_3
n	3.42	3	2.2	1.6
D, μm	270	10	0.056	0.13

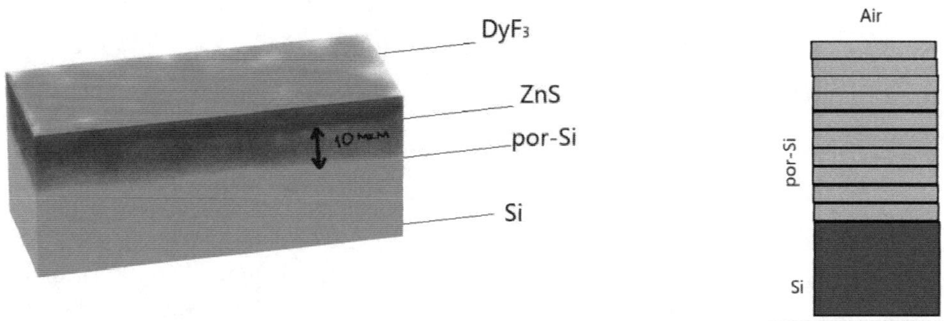

Figure 2. (a) Figure with short caption; **(b)** Figure with short caption.

4. Results and discussion

Taking into account the porosity gradient of the structure (Figure 3), several modifications of the model were considered: 1 porous layer, 6 porous layers, 8 porous layers and 10 porous layers. The control graph was the graph of the real experiment. The simulation results are shown in Figure 4.

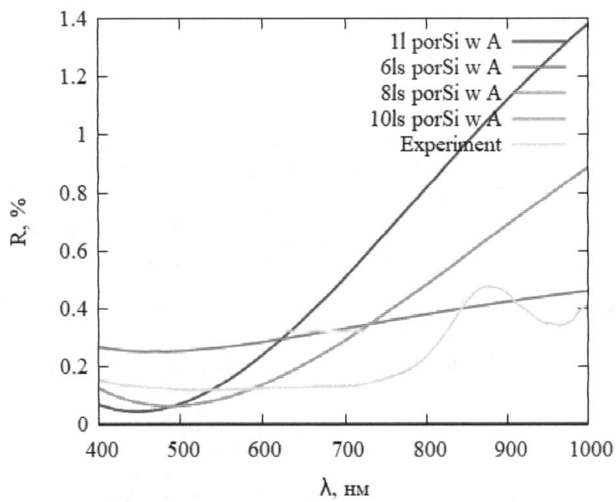

Figure 3. Comparison of the simulation results a multilayer porous structure with a different number of layers and experiment.

By reducing the simulation step, it was possible to achieve some agreement with experiment

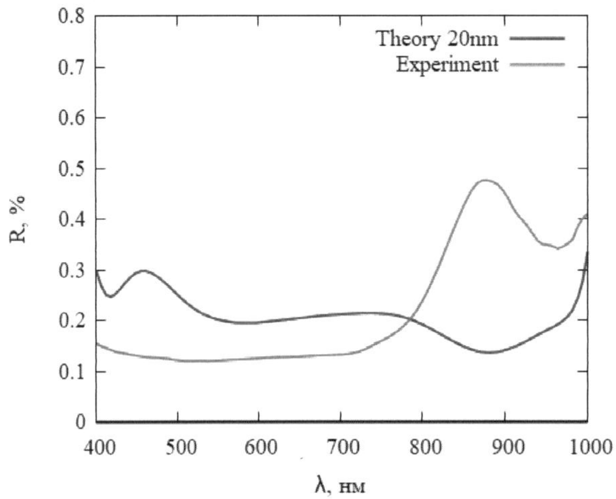

Figure 4. Comparing the simulation results with the multilayer porous structure experiment.

From the analysis of the obtained graphs, it can be seen that the representation of porous silicon in the form of 10 layers of the same thickness with a refractive index varying from silicon to air allows us to approximate the course of the curve of the real experiment. However, the discrepancy between the theory and the experiment can be explained by the features of absorption in real porous structures, which were not taken into account sufficiently fully in this model. Similar results were obtained in the works [3].

5. Conclusion

Thus, the resulting model allows us to determine the optimal parameters of the photosensitive structure without the use of complex mathematical calculations, which will further simplify the technological process of creating multilayer solar cells with a low reflection index.

Acknowledgments

This work was supported by the Foundation for Assistance to Small Innovative Enterprises, grant № 12980GU/2018.

References

[1] Latukhina N, Chepurnov V and Pisarenko G 2013. *ELECTRONICS science technology business* **4** 00126
[2] Aroutiounian V M, Maroutyan K R, Zatikyan A L and Touryan K J 2002 *Thin Solid Films*
[3] Adamian Z N, Hakhoyan A P, Aroutiounian V M, Barseghian R S and Touryan K 2000 *Solar Energy Materials & Solar Cells* **64**
[4] Acquaroli L N 2018 *arXiv*:1809.07708v1
[5] Pérez E X 2007 *Fabrication and characterization of porous silicon multilayer optical devices* p27-57
[6] Pickering C, Beale M I J, Robbins D J, Pearson P J and Greef R, 1984 *J. Phys. C, Solid State. Phys.* **17** 6335
[7] Aroutiounian V M 1995 *Reports of NAS of Armenia (in Russian)* **95** 151, 229
[8] Aroutiounian V M 1998 *Phys. Stat. Sol. (a)* **165** 105

Investigation of laser and thermal sintering processes of silver nanoparticles agglomerates synthesized by spark discharge

S S Tikhonov, M Nouraldeen, K M Khabarov, A A Efimov and V V Ivanov

Moscow Institute of Physics and Technology (National Research University), Dolgoprudny 141701, Russia

sergei.s.tikhonov@phystech.edu

Abstract. Changes in the shape and size of silver nanoparticles (NPs) during their laser and thermal sintering have been studied experimentally and theoretically. Aerosol silver NPs forming dendrid-like agglomerates 180 nm in size were synthesized by spark discharge and exposed to laser radiation and high temperature of 750 °C. The shape and size of the NPs were investigated depending on the power of the laser radiation and the temperature of the gas. It is estimated that, at a power density of laser radiation of the order of 10^3-10^4 W/cm^2, the formation of spherical NPs with an average size of 140 nm is expected. Such particles turn out to be similar to NPs thermally heated in a gas flow at 750 °C for 6 seconds.

1. Introduction

Nanoparticles (NPs) are an interesting and important object for study, due to their high demand for research problems and their promising potential in many technological industries. With their help, point delivery of drugs, detection of hazardous substances, change in the optical and electronic properties of materials is carried out [1-3]. Currently, there are many methods for producing NPs: lithography, synthesis from colloidal solutions, laser ablation, and others [4-6]. However, one of the most effective ways to obtain NPs in terms of mass productivity and synthesis capabilities is their synthesis in a gas discharge. [7,8]. This method makes it possible to control the delivery of NPs and their agglomerates, composition and morphology directly in the aerosol flow using external influences [9].

In this work, we study the effect of continuous laser radiation with wavelengths of 527 and 980 nm and high temperatures on agglomerates of aerosol silver NPs synthesized in a spark discharge in order to obtain non-agglomerated spherical NPs.

2. Experiment setup and research method

The synthesis of NPs was carried out by the method of a spark gas discharge in an argon flow (99.99998%) between coaxially arranged silver cylindrical electrodes. Primary silver NPs with an average size of 10 nm in an aerosol flow agglomerated and formed dendrid-like nanostructures, which were studied in an aerosol NPs analyzer SMPS 3936 (TSI Inc.) and in a transmission electron microscope (TEM) JEM-2100 (JEOL). The gas path connected in series two zones - with a controlled temperature and with laser radiation brought inside, in which agglomerates of NPs were subjected to two types of impact: thermal, in a heated section of the gas path, and laser, in a quartz cell. The quartz cell consisted of two combined tubes, into one of which an aerosol stream was injected, and into the other continuous laser radiation with wavelengths of 527 and 980 nm with the ability to control power in the range of up to 1.5 W for a green laser and up to 6 W for an infrared one (Figure 1). Thermal modification of agglomerates of NPs was carried out in a heated section of the gas path 40 cm long at a temperature of 750 °C.

Determination of the power of laser radiation was carried out by a power detector VEGA (Ophir).

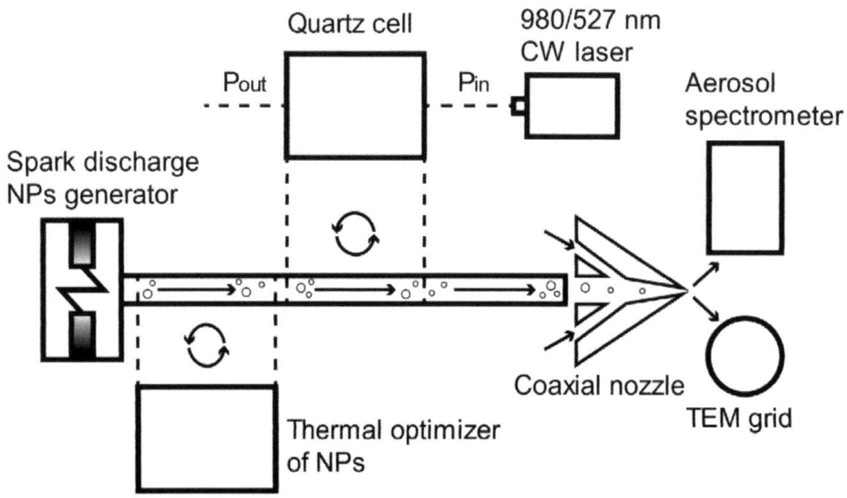

Figure 1. Schematic of an experimental setup for studying the sintering processes of nanoparticle agglomerates obtained in a spark discharge generator of NPs.

Theoretical assessments were carried out in accordance with the model given in the article [10]. The calculations of the heating and melting times of NPs were carried out using the heat transfer equation in spherical coordinates:

$$\rho_m C_m V \frac{\partial T_m}{\partial t} = P - (q_{rad} + q_{cond}), \quad (1)$$

where ρ_m – is the density of a metal nanoparticle, C_m – specific heat, T_m – temperature, V – volume of a nanoparticle, P – power loss of laser radiation on one agglomerate of NPs, q_{rad} and q_{cond} – heat received or given off by a nanoparticle due to radiation and heat transfer respectively. All theoretical calculations were performed for a spherical silver nanoparticle with a size of $d = 180\,nm$. This is much greater than the mean free path in argon gas, which in this case is the carrier gas. This indicates that the change in the nanoparticle temperature depends mainly on heat transfer and is almost independent of radiation; therefore q_{rad} is neglected in this case. Formula for calculating q_{cond} or the case when the temperature of the particles is higher than the temperature of the medium:

$$q_{cond} = SG(T_m - T_0), \quad (2)$$

where S – surface area of the nanoparticle, G – thermal conductivity of the nanoparticle-medium interface, T_0 – temperature of the gas medium. The power of laser radiation that is lost on one agglomerate of NPs can be estimated by knowing the length and diameter of the quartz cell - 160 mm and 3 mm, respectively, the diameter of the laser beam - 2 mm and obtaining data on the concentration of nanoparticle agglomerates in the carrier gas from an aerosol analyzer. Also, in this estimate, the melting point of silver will be taken as a temperature of 750 °C, and the transformation time of NPs agglomerate to an enlarged spherical nanoparticle is taken to be equal to the sum of the times of heating and complete melting of a nanoparticle with a size of 180 nm.

3. Results and discussion

It was experimentally established that at an output power of laser radiation of about 1.2 W for two wavelengths, an infrared laser has a significantly less effect on NPs than green laser radiation. A small change in the size of agglomerates of silver NPs at threshold powers is achieved due to the fact that agglomerates of NPs are modified by absorbing radiation. However, the absorption peak for silver NPs is in the blue region of the spectrum, which is much closer to 527 nm than to 980 nm.

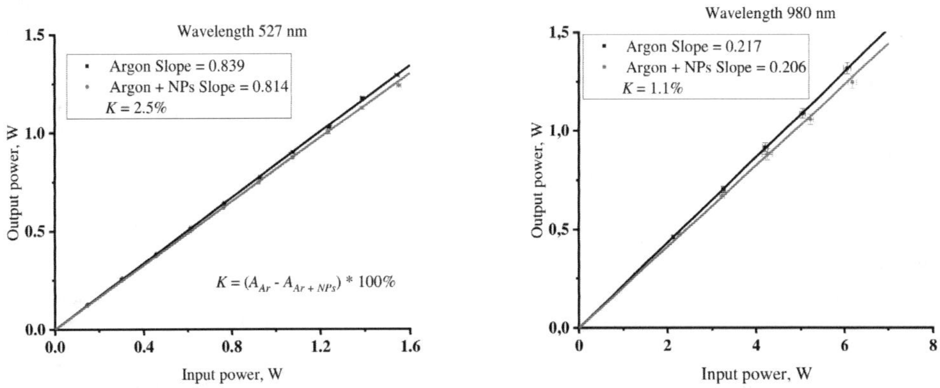

Figure 2. Dependence of the power of laser radiation at the exit from a quartz cell on the power at the entrance with a carrier flux Q=100 ml/min at room temperature for a laser with an operating wavelength of 527 nm (left) and 980 nm (right).

Analysis of the power of laser radiation at the input and output of a quartz cell (Figure 2) showed that NPs absorb laser radiation with a wavelength of 527 nm more efficiently than 980 nm by about 2.5 times. It is worth saying that the coefficient K here reflects all the energy losses of laser radiation on particles, that is, extinction: scattering of radiation on particles and absorption of radiation by them. But in this work, the scattering of energy by particles in theoretical calculations is assumed to be small in comparison with absorption.

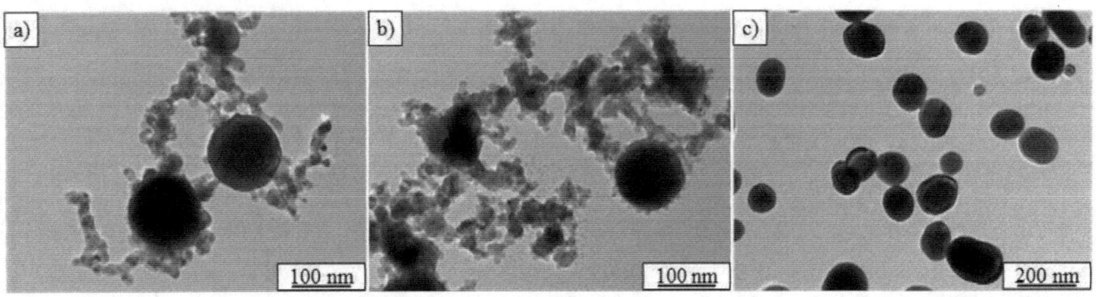

Figure 3. TEM images of silver NPs: primary (a), passed through a quartz cell with laser radiation of 527 nm at a power of 1.2 W (b) and optimized in a heated section of the channel at a temperature of 750 °C (c).

Analysis of the data obtained using an analyzer of aerosol NPs and the TEM showed that at a gas flow of 200 ml/min, agglomerates of NPs significantly change their shape, turning into spheres with an average diameter of 140 nm, determined from TEM images (Figure 3 (c)).

TEM images of NPs (Figure 3) confirm that thermal sintering of agglomerates of NPs at a temperature of 750 °C is more effective than sintering with laser radiation, in which the agglomerates of NPs do not undergo significant changes. However, it can be noted that agglomerates exposed to laser radiation are a structure of individual NPs with more resolvable boundaries, in comparison with agglomerates of primary NPs.

Table 1. Results of theoretical evaluation of comparison of two types of sintering of nanoparticle agglomerates.

	Heating time (s)	Final temperature of NP (°C)	Current power density (W/cm^2)	Estimated power density of complete NPs sintering (kW/cm^2)
Thermal optimizer	6.0	750	-	-
527 nm CW laser	0.7	31	50	7.9
980 nm CW laser	0.7	36	200	15.0

The results of theoretical calculations (Table 1) show that in the case of heating NPs using laser radiation under current conditions, a silver nanoparticle with a size of 180 nm will, on average, heat up by only 4-9 degrees during its flight through a quartz cell. This happens because the cooling factor of NPs upon collision with molecules of an unheated gas medium has a significant effect. Comparing the heating and melting time of a nanoparticle in a thermal element with that for a nanoparticle exposed to laser radiation, it is possible to estimate the power density of laser radiation at which it is possible to achieve the results of sintering agglomerates of NPs similar to those shown in the image (Figure 3 (c)). At the same time, having made similar estimates for a nanoparticle with a size of 10 nm, one can come to the conclusion that the current power density of laser radiation is sufficient for its heating and melting, which is consistent with the experimental results in the image, where one can see sintered small NPs on the surface of large particles in the composition agglomerate (Figure 3 (b)). The estimates made indicate that to improve the efficiency of modifying NPs in a gas cell, the radiation power density or radiation absorption efficiency should be two orders of magnitude higher for green and infrared lasers. This can be achieved, for example, by building a more complex optical system with a narrower laser beam. This development seems to be possible, since agglomerates of NPs moving along the tube in the gas flow are mostly distributed closer to the center of the tube.

4. Conclusion

As a result of a comparison of two types of exposure (thermal and laser), it was found that for sintering agglomerates of NPs to comparable sizes and shapes, it is necessary to increase the radiation power density by two orders of magnitude for wavelengths of 527 nm and 980 nm. In this case, either the absorption efficiency of radiation turns out to be higher for a wavelength of 527 nm, compared with a wavelength of 980 nm, which corresponds to the peculiarities of the absorption spectrum of silver NPs. This allows us to conclude that when changing the radiation wavelength to a closer to the peak of plasmon absorption of the NPs under study, one can also achieve comparable results with those obtained by modifying particles in the heated section of the gas path.

Acknowledgments

The reported study was funded by RFBR and BRFBR, project number 20-53-00042 in part of the investigation of thermal sintering of nanoparticles and by the Russian Science Foundation, project number 19-79-00375 in part of the investigation of laser sintering of nanoparticles.

References

[1] Kuznetsov A I, Miroshnichenko A E, Brongersma M L, Kiyshar Y S and Luk'yanchuk B 2016 Optically resonant dielectric nanostructures *Science* **354**

[2] Devabharathi N, M. Umarji A and Dasgupta S 2020 Fully Inkjet-Printed Mesoporous SnO2-Based Ultrasensitive Gas Sensors for Trace Amount NO2 Detection *ACS Appl. Mater.*

Interfaces **12** 57207-57217

[3] Palani G, Kannan K, Radhika D, Vijayakumar P and Pakiyaraj K 2020 Bioengineered metal and metal oxide nanoparticles for photocatalytic and biological applications (review) *Physics and Chemistry of Solid State* **21** 571-583

[4] Ealia S A M and Saravanakumar M P 2017 A review on the classification, characterisation, synthesis of nanoparticles and their application *IOP Conf. Ser.: Mater. Sci. Eng.* **263** 032019

[5] McDarby S P, Wang S J, King M E and Personick M L 2020 An Integrated Electrochemistry Approach to the Design and Synthesis of Polyhedral Noble Metal Nanoparticles *Journal of the American Chemical Society* **142** 21322-21335

[6] Yu W-J, Geng P, Wen M and Chen Z-G 2020 Gelatin-mediated coprecipitation synthesis of copper sulfide nanoparticles for photothermal ablation of cancer cells *Cailiao Gongcheng/Journal of Materials Engineering* **48** 68-74

[7] Mylnikov D, Efimov A and Ivanov V 2019 Measuring and optimization of energy transfer to the interelectrode gaps during the synthesis of nanoparticles in a spark discharge *Aerosol Science and Technology* **53** 1393-1403

[8] Efimov A A, Arsenov P V, Borisov V I, Buchnev A I, Lizunova A A, Kornyushin D V, Tikhonov S S, Musaev A G, Urazov M N, Shcherbakov M I, Spirin D V and Ivanov V V 2021 Synthesis of Nanoparticles by Spark Discharge as a Facile and Versatile Technique of Preparing Highly Conductive Pt Nano-Ink for Printed Electronics *Nanomaterials* **11** 234

[9] Lizunova A, Mazharenko A, Masnaviev B, Khramov E, Efimov A, Ramanenka A, Shuklov I and Ivanov V 2020 Effects of Temperature on the Morphology and Optical Properties of Spark Discharge Germanium Nanoparticles *Materials* **13** 4431

[10] Xi Q, Li Y, Zhou J, Li B and Liu J 2019 Role of Radiation in Heat Transfer from Nanoparticles to Gas Media in Photothermal Measurements *Int. J. Mod. Phys.* **30** 1950024

Tunable laser induced periodic surface structures in Ge2Sb2Te5 thin films

Irina G Bessonova[1], Pavel I Trofimov[1], Petr I Lazarenko[2], Demid A Kirilenko[3], Nikolay A Bert[3], Sergey A Kozyukhin[4] and Ivan S Sinev[1]

[1] Department of Physics and Engineering, ITMO University, St. Petersburg, Russia
[2] National Research University of Electronic Technology, Zelenograd, Moscow, Russia,
[3] Ioffe Institute, St. Petersburg, Russia,
[4] Kurnakov Institute of General and Inorganic Chemistry of the Russian Academy of Sciences, Moscow, Russia

irina.bessonova@metalab.ifmo.ru

Abstract. Planar photonic structures, such as gratings and metasurfaces, are routinely used for beam steering, waveguide coupling, and light localization. However, conventional fabrication techniques that involve lithography are demanding in terms of time and cost. Much cheaper and simpler methods for surface patterning and formation of periodic surface structures are enabled by direct laser processing. Here, we demonstrate low-cost rapid fabrication of high-quality phase gratings based on the formation of laser induced periodic surface structures (LIPSS, or ripples) in $Ge_2Sb_2Te_5$ (GST) thin films. Due to unique phase change properties of GST, the structures demonstrate strong modulation of refractive index with period controlled by the wavelength of laser irradiation. We study the formation of phase change LIPSS in a broad range of excitation wavelengths and observe transition between regimes with different orientations of generated ripples with respect to laser polarization.

1. Introduction

Chalcogenide phase change materials (PCM) is a unique family of optical materials that can maintain two different phase states at normal conditions – amorphous and crystalline – with large contrast of optical and electrical properties. PCM can be rapidly switched between these two states when exposed to optical, or electrical pulses of nanosecond duration [1]. On the other hand, an important effect that frequently accompanies the action of pulsed laser radiation on the surface of solid material is the formation of laser-induced periodic surface structures (LIPSS or ripples). This direct non-contact surface patterning method is already a viable alternative to lithography and can be used for various applications [2]. The mechanism of the formation of LIPSS is still under discussion and depends on the material of the exposed substrate and the regimes of laser irradiation [3]. Generally, two types of laser-induced periodic surface structures can be observed: high (HSFL) and low (LSFL) spatial frequency LIPSS. In both cases these surface structures reproduce the interference pattern of the incident and scattered light which is imprinted in the sample material through absorption usually followed by ablation of the overheated material.

However, this is not the case for the LIPSS formed in the PCM thin films there the absorption leads to a phase transition accompanied with a change in density and a significant change in the optical properties, in particular, the refractive index. Thus, LIPSS formation in PCM films leads to appearance of both periodic surface relief (due to the material volume change during phase transition) and spatial modulation of the optical properties [4]. Moreover, in line with the majority of phase transitions in PCM, this process is potentially reversible, allowing formation of rewritable gratings with variable period and direction depending on the laser irradiation regimes.

In this work we study the fabrication of LIPSS-based GST phase change gratings using fs-laser pulses. We scale the gratings using line-by-line scanning of the laser beam over the PCM film surface and study different regimes of grating formation with varying period and direction with respect to laser polarization. Finally, we take an important step towards the realization of rewritable phase gratings by demonstrating that LIPSS-based structures can be fabricated not only in as-deposited amorphous GST film but also in a re-amorphized film.

2. Results and discussion

To create 2D phase gratings in 100 nm thick GST films on sapphire substrates we focus a laser beam with tunable wavelength (Light Conversion Pharos + Orpheus optical parametric amplifier, 1MHz repetition rate, 300 fs pulse duration) at the film surface from the substrate side. We then scan the beam waist across the film surface using a piezoscanner to irradiate 50x50 um^2 areas (figure 1a). Making use of the wide tuning range of the laser system, we studied the spectral dependence of LIPSS period and direction of periodic pattern in a wide spectral range from 600 to 2000 nm. To determine the period of the formed LIPSS-based gratings we perform atomic force microscope (AFM) scanning of the laser-imprinted area. Examples of the obtained AFM images are shown in figure 1c,d. The crystallized regions of the film are suppressed below the surface of the initial amorphous film by several nanometers due to a decrease in material volume during the crystallization.

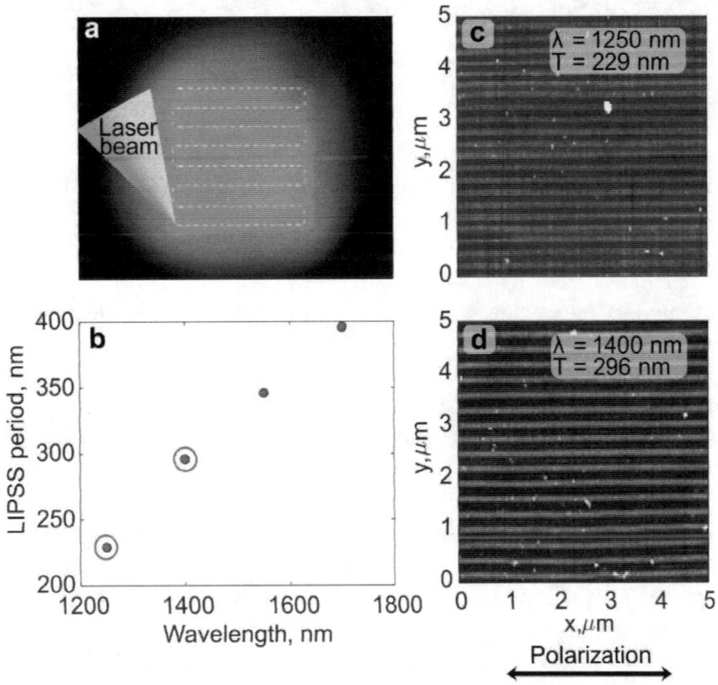

Figure 1. (a) Scheme of phase grating fabrication in GST thin film using femtosecond laser writing. (b) Dependence of the LIPSS period on the wavelength for 100 nm GST film on sapphire. (c, d) AFM images of the obtained gratings in 100 nm GST film at different wavelengths. Data denoted with blue circles in (b) are extracted from AFM images (c, d).

The dependence of LIPSS period on the laser wavelength for 100 nm GST film extracted from the AFM data is shown in figure 1b. For the wavelengths from 1200 to 2000 nm, periodic surface structures are formed parallel to laser polarization, which can be attributed to the "dielectric" type of LIPSS formation [5]. In contrast, for the wavelengths lower than 800 nm the regime of the LIPSS formation is changed. In this wavelength region for GST film of 50 nm the LIPSS perpendicular to the polarization of the laser beam are formed, which in turn can be attributed to the "metallic" regime [6].

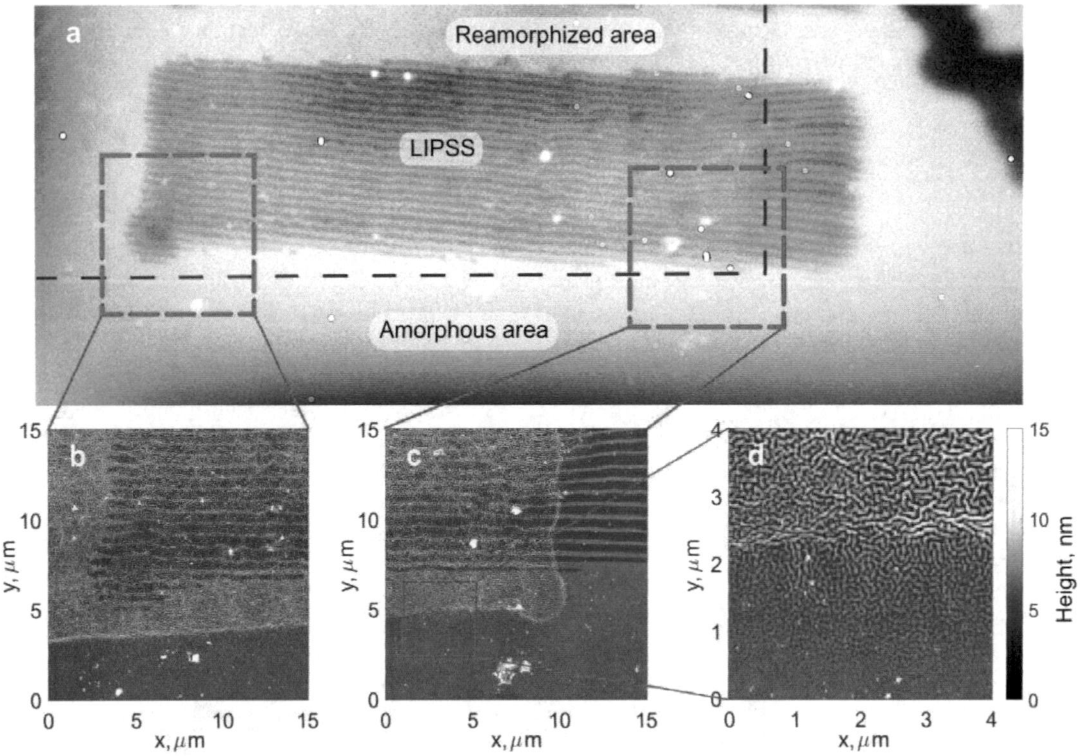

Figure 2. LIPSS-based optical grating imprinted into re-amorphized and as-deposited GST film. (a) Optical image of the GST grating. The edges of the re-amorphized area are marked with blue dashed line. (b, c) AFM images of the imprinted LIPSS grating in re-amorphized region (b) and on the boundary between re-amorphized and as-deposited amorphous GST film (c). (d) Higher resolution AFM image of the boundary between as-deposited (bottom) and phase-cycled (top) GST film showing random surface relief generated during film preprocessing.

As it was mentioned earlier, the phase change nature of the GST film allows to fabricate LIPSS-based gratings that are not only tunable, but also potentially re-writable. However, for realization of rewritable gratings it is crucial that successful imprint of LIPSS phase gratings can be performed both in as-deposited GST films and in phase-cycled (re-amorphized) film. To verify this concept we re-amorphize an area of the as-deposited amorphous 50 nm thick GST film. During the process we first crystallize the initial amorphous substrate with CW laser (532 nm) and subsequently amorphize it with femtosecond laser pulses (1 Hz, 1550 nm, 290 fs). After that we imprint LIPSS-based gratings both in as-deposited amorphous and re-amorphized areas of GST film. As a result, in the optical image we observe the successful formation of periodic structures in both amorphous and phase-cycled areas of the GST film (figure 2a). Further details of GST gratings formation process for the cases of as-deposited and re-amorphized films are revealed by the AFM images of the imprinted gratings. In accordance with the optical images, for both initial states of GST film (as-deposited amorphous and phase cycled) we observe the formation of similar periodic surface reliefs attributed to formation of the LIPSS-based gratings (figure 2b,c). Importantly, for the case of the re-amorphized GST film we also observe the corrugation of film surface. Despite the fact that the optical image shows exactly the same reflectivity modulation between the areas the random surface relief is formed during phase cycling of the GST film (figure 2d). It is worth mentioning that such surface relief could potentially facilitate the formation of LIPSS gratings [7].

In conclusion, we showed the formation of laser-induced phase change gratings in thin $Ge_2Sb_2Te_5$ films on sapphire. We demonstrate the possibility of producing high-quality phase-change gratings flexible in terms of period and direction with respect to laser polarization, are fast and easy to fabricate. In addition, we demonstrate that LIPSS patterning can be performed both on as-deposited amorphous and re-amorphized GST films which shows great promise for the realization of the re-writable GST gratings. All these qualities make LIPSS-based GST gratings strong candidates for realization of rewritable on-chip couplers and beam steering devices.

Acknowledgement
The reported study was funded by RFBR according to the research project № 20-32-90220.

References
[1] Wuttig M and Taubner T 2017 Phase-change materials for non-volatile photonic applications *Nat. Photonics* **11** 465-476
[2] Fuentes-Edfuf Y, Garcia-Lechuga M, Puerto D, Florian C, Garcia-Leis A, Sanchez-Cortes S, Solis J and Siegel J 2017 Coherent scatter-controlled phase-change grating structures in silicon using femtosecond laser pulses *Sci. Rep.* **7** 4594
[3] Bonse J and Graf S 2020 Maxwell meets Marangoni — a review of theories on laser-induced periodic surface structures *Laser Photonics Rev.* **14** 2000215
[4] Kozyukhin S, Smayev M, Sigaev V, Vorobyov Y, Zaytseva Y, Sherchenkov A and Lazarenko P 2020 Specific features of formation of laser-induced periodic surface structures on Ge2Sb2Te5 amorphous thin films under illumination by femtosecond laser pulses *Phys. Status Solidi B* **257** 1900617
[5] Bonse J, Hohm S, Kirner S V, Rosenfeld A and Kruger J 2016 Laser-induced periodic surface structures — a scientific evergreen *IEEE J. sel. top. quan.* **23**
[6] Fuentes-Edfuf Y, S´anchez-Gil J A, Florian C, Giannini V, Solis J and Siegel J 2019 Surface plasmon polaritons on rough metal surfaces: role in the formation of laser-induced periodic surface structures *ACS Omega* **4** 6939–6946
[7] Sipe J E, Young J F, Preston J S and van Driel H M 1983 Laser-induced periodic surface structure I. Theory *Phys. Rev. B* **27** 1141

Simple representation of the quantum entanglement of coupled harmonic oscillators in terms of the reflection coefficient

Yu V Tsykareva and D N Makarov

Department of Fundamental and Applied Physics, Northern Arctic Federal University named after M V Lomonosov, Severnaya Dvina Embankment 17, Arkhangelsk 163002, Russia

aisonoka@gmail.com

Abstract. Quantum entanglement of coupled harmonic oscillators is frequently applied in quantum and non-linear physics, molecular chemistry and biophysics, which is why its study is of a great interest for modern physics. In this work a quantum entanglement of a coupled harmonic oscillator in a simple form was found. This simple form is presented as a single parameter – reflection coefficient R. All parameters of the studied system are included in the R coefficient. It is shown that the derivation of the expression can have applications in quantum optics, in particular in quantum metrology.

1. Introduction

Studying the properties of coupled harmonic oscillators is a relevant area of modern physics. The interest towards it stems primarily from the fact that models of such systems can be encountered in many applications of quantum and non-linear physics, molecular chemistry and biophysics [1-2]. In quantum physics this interest generally originates from quantum entanglement for such a system. In particular, quantum communication protocols such as quantum cryptography, quantum dense coding, quantum calculation algorithms and teleportation of quantum states can be explained using entangled states. On the other hand, physical models of coupled harmonic oscillators are used in biophysics to explain the issue of photosynthesis [3] which is still considered an unsolved puzzle.

The simplest model capable of serving as a foundation for presenting multiple physical processes is the coupled harmonic oscillator model. Because of a large number of system parameters evaluation of quantum entanglement in a simple analytical form is problematic for this model. The goal of this research is to find simple expressions for evaluating the quantum entanglement of the harmonic oscillator in the case of a dynamic system that would only depend on reflection coefficient R. The Schmidt parameter [4] and the von Neumann entropy [5], both of which depend on a common parameter (the Schmidt mode) were considered as measures of quantum entanglement. In the expressions obtained as a result of the research the R coefficient incorporated all other system parameters, which allows the expressions of quantum entanglement evaluation to take simple analytical forms. It is worth noting that this approach can not only be used for evaluating the quantum entanglement, but also for other systems that can be presented in the form of linked harmonic oscillators.

2. Calculating methods

For this purpose we take a look at a system of two coupled harmonic oscillators with Hamiltonian (1)

$$\hat{H} = \frac{1}{2}\left(\frac{1}{m_1}\hat{p}_1^2 + \frac{1}{m_2}\hat{p}_2^2 + Ax_1^2 + Bx_2^2 + Cx_1x_2\right). \tag{1}$$

Work [2] provides a solution for a nonstationary Schrödinger equation for this system under the condition that the system was not linked at zero time (C=0). However, for our case it is more convenient to present the creation and annihilation of quantum states via operators. In this case the Hamiltonian of the system will look as follows [6]

$$\hat{H} = \sum_k^2 \hbar\omega_k\left(\hat{a}_k^\dagger\hat{a}_k + \frac{1}{2}\right) + \hbar\Omega x_1x_2, \tag{2}$$

where $\omega_1 = \sqrt{A/m_1}$; $\omega_2 = \sqrt{B/m_2}$; $\Omega = C/2(m_1m_2AB)^{1/4}$; $\hat{a}_k^\dagger = \frac{1}{\sqrt{2}}\left(x_k - \frac{\partial}{\partial x_k}\right)$, $\hat{a}_k = \frac{1}{\sqrt{2}}\left(x_k + \frac{\partial}{\partial x_k}\right)$ – are the operators of annihilation and creation of quantum states s_k respectively. For the Hamiltonian in this form the following equation solution was found [6]

$$|\Psi(t)\rangle = e^{-i\hat{H}t}|s_1, s_2\rangle = \frac{1}{\sqrt{s_1!s_2!}}e^{-i\hat{H}t}\hat{a}_1^{\dagger\, s_1}\hat{a}_2^{\dagger\, s_2}|0,0\rangle. \tag{3}$$

For further study we decompose the obtained wave function of the system according to the Schmidt theorem

$$|\Psi(t)\rangle = \sum_k \sqrt{\lambda(t)}u_k(x_1,t)v_k(x_2,t), \tag{4}$$

where $u_k(x_1,t)$ и $v_k(x_2,t)$ – are wave functions of a pure state of system's oscillators, $\lambda(t)$ – is the Schmidt mode. Some of the most well-known methods for measuring quantum entanglement are the Schmidt parameter (5) and the von Neuman entropy (6)

$$K = \left(\sum_k {\lambda_k}^2\right)^{-1} \tag{5}$$

$$S_N = -\sum_k \lambda_k\, ln(\lambda_k). \tag{6}$$

For this task these measures of quantum entanglement are basically equal. All main patterns in these measures are identical, though different in absolute meaning. Some measures are sometimes more convenient to use in calculations; in our case it is the Schmidt parameter (5). Expressions become simpler and easier to interpret. As you can see, both methods depend on the Schmidt mode, therefore, the main issue during the measurement of quantum entanglement is the calculation of the Schmidt modes.

Furthermore, to move on to the reflection and transmission coefficients we used well-known transformations (7) for the linear beam splitter [7], where R is the reflection coefficient, T is the transmission coefficient, φ is the phase shift

$$R = \frac{sin^2(\Omega t/2\sqrt{1+\varepsilon^2})}{1+\varepsilon^2};\ T = 1 - R; \tag{7}$$

$$cos\,\varphi = -\varepsilon\sqrt{\frac{R}{T}};\ \varepsilon = \frac{\omega_2^2 - \omega_1^2}{2\sqrt{\omega_1\omega_2}\Omega}.$$

As was demonstrated in work [6], the entire dependence on the multiparameter system for quantum entanglement can be presented as a single parameter – reflection coefficient R

$$\lambda_k(R) = \left|c_{k,s_1+s_2-k}\right|^2, \tag{8}$$

where $c_{k,p} = \sum_{n=0}^{s_1+s_2} A_{n,s_1+s_2-n}^{s_1,s_2} A_{n,s_1+s_2-n}^{*k,p} e^{-2in\,\arccos(\sqrt{1-R}\sin\varphi)}$,

$$A_{n,m}^{k,p} = \frac{\mu^{k+n}\sqrt{m!n!}}{(1+\mu^2)^{(n+m)/2}\sqrt{k!p!}} P_n^{(-(1+m+n),m-k)}\left(-\frac{2+\mu^2}{\mu^2}\right),$$

$$\mu = \sqrt{1+\frac{1-R}{R}\cos^2\varphi} - \cos\varphi\sqrt{\frac{1-R}{R}},$$

where P are Jacobi polynomials, k and p are quantum numbers of the oscillator in the state $|k, p\rangle$, with $k+p = s_1+s_2$. It is worth noting that $\varphi \in \left(0,\frac{\pi}{2}\right)$, yet the dependence on φ is contained in trigonometrical functions. Therefore, the expression form of the dependence on R alone is determined. Despite the fact that quantum entanglement can be calculated, its form is still difficult to interpret and can be compared to numerical computations. Indeed, in order to calculate quantum entanglement it is necessary to conduct a summation of rather complicated special functions. It is essential to use simple expressions that allow to analyze quantum entanglement rather easily. We found such expressions for certain states.

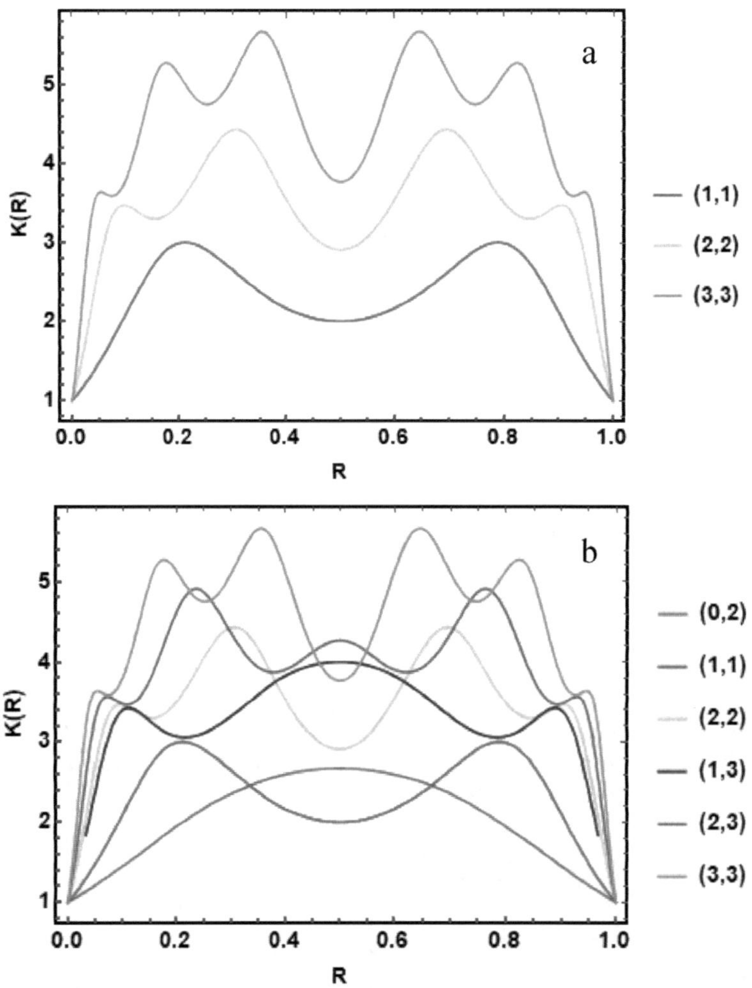

Figure 1. The dependence of the Schmidt parameter as a function of R similar values of quantum number in set (a) and for various sets of quantum numbers (b). In the figures, the dependencies are presented for different initial values of quantum numbers (s_1,s_2).

3. Results and Discussion

The graphic interpretation of the Schmidt parameter depending on R for various pairs of quantum numbers s_1 and s_2 for harmonic oscillators is presented in Figure 1, where R $\in(0,1)$.

The forms of expressions for the Schmidt parameter for various sets of quantum numbers s_1 and s_2:
For $s_1 = 0$ and $s_2 = 2$:

$$K = \frac{1}{1-2R(1-R)\big(2-3R(1-R)\big)} \quad (9)$$

For $s_1 = 2$ and $s_2 = 2$:

$$K = \frac{1}{1-24R(1-R)\Big(1-3R(1-R)\big(4-5R(1-R)(4-7R(1-R))\big)\Big)} \quad (10)$$

For $s_1 = 1$ and $s_2 = 3$:

$$K=\frac{1}{1\text{-}2R(1\text{-}R)\Big(10\text{-}R(1\text{-}R)\big(99\text{-}80R(1\text{-}R)(5\text{-}7R(1\text{-}R))\big)\Big)} \quad (11)$$

Using the expressions (9-11), it is possible to find such values of R for which the quantum entanglement of this pair of quantum numbers reaches its maximum. Table 1 presents the results of such a calculation. The expression for the Schmidt parameter for $s_1 = 1$ and $s_2 = 1$ is presented in work [6].

Table 1. R values at maximum quantum entanglement for various quantum numbers.

(s_1,s_2)	K	R
(1,1)	3	$0.5\left(1\pm 1/\sqrt{3}\right)$
(0,2)	2.67	0.5
(2,2)	4.4	1/3 and 2/3
(1,3)	4	0.5

This is quite an interesting result, since at first glance it seems that the maximum entanglement should be at R = 1/2. Figure 1 shows that quantum entanglement strongly depends on the reflection coefficient R but is always zero at R = 0; 1. The larger the quantum numbers s_1; s_2, the greater the quantum entanglement.

It should be added that the Eq. (8) can be used in quantum optics to calculate the quantum entanglement of photons in a beam splitter (with two input and output ports) with a reflection coefficient R. In this case, ω_1 and ω_2 are the frequencies of 1 and 2 photons, respectively, and is some parameter characterizing the beam splitter. Combining the parameters ω_1 and ω_2; Ω; t, you can always choose the value of R with maximum quantum entanglement using the expressions (7).

4. Conclusions

It was shown that the quantum entanglement of two coupled harmonic oscillators is expressed as a single parameter, this is the reflection coefficient R; It is shown that for certain values of R and a given pair of quantum numbers s_1 и s_2 quantum entanglement has a maximum and can be large; for R = 0 and R = 1 there is no quantum entanglement; the equations for Schmidt mode presented in this work have the most simple analytical form; quantum entanglement, expressed as a single parameter (the reflection coefficient R, was obtained without taking into account the environment that can significantly affect the quantum entanglement. It is worth adding that the search for simple expressions of quantum entanglement is a relevant task, as modern sources generally involve numerical computations. This makes it difficult to see the pattern and even more so the fact that quantum entanglement can depend on a single parameter that includes all parameters of the studied system.

Acknowledgments

The work was supported by Russian Science Foundation grant № 20-72-10151.

References

[1] Fetter A L and Walecka J D 1971 *Quantum Theory of Many Particle Systems* (New York: McGraw-Hill)
[2] Makarov D N 2018 *Scientific Reports* **8** 8204
[3] Romero E, Augulis R, Novoderezhkin V I, Ferretti M, Thieme J, Zigmantas D and van Grondelle R 2014 *Nature Physics* **10** 676–682
[4] Ekert A and Knight P L 1995 *Amer.J.Phys.* **63** 415–423
[5] Casini H and Huerta M 1996 *J. Phys. A: Math.Theor.* **42** (50) 504007
[6] Makarov D N 2020 *Phys. Rev.* E **102** 052213
[7] Zhe-Yu Je Ou 2007 *Multi-Photon Quantum Interference* (New York: Springer)

Identification of pigments of Russian icons by means of X-ray fluorescence spectroscopy

A V Vasilieva and **V A Parfenov**

Photonics Department, Saint Petersburg Electrotechnical University "LETI", Professora Popova str., 5, 197376 St. Petersburg, Russian Federation

anastasiastru@mail.ru

Abstract. In this work the analysis of paint layers of Russian icons of the 16th-19th centuries was carried out by means of X-ray fluorescence spectroscopy. The data obtained on the elemental composition of pigments made it possible to identify them. Main problems of identification were described in this paper. The results of the research will be used to create a database of XRF-spectra of pigments, which were used by Russian icon painters of certain schools and time periods.

1. Introduction

This work is devoted to use of the method of X-ray fluorescence analysis (XRF) in the field of conservation and restoration. Before starting the restoration process of an art object, it is necessary to conduct studies, which mainly rely on experience of restorers and art historians, according to the current practice in Russia, where instrumental methods of artwork analysis started to be used in recent years only. Meanwhile, non-destructive optoelectronic diagnostic methods are actively developed and widely used in museum work all over the world, due to the fact that they are non-invasive and do not require sampling [1]. Currently, one of the most frequently used methods for studying Cultural Heritage objects is the XRF method. It is based on the detection of X-ray fluorescence emission from inorganic substance atoms, which occurs when X-ray radiation interacts with a substance. As a result, obtained XRF-spectra give information about the elemental composition of analyzed substances. XRF analysis of artworks provides information about the time of its origin, modifications that could occur to it, and about the authenticity of objects [2–4]. Furthermore, this technique may be used for quantitative analysis of effectiveness of restoration of artworks [5].

In Russia, icon painting was the main type of fine art from the end of the 10th century to the beginning of the 18th century. Such a long period of development singled out Russian icon painting as a separate branch in fine art, and as a result, many ancient icons have been saved nowadays in the depositories of museums and cathedrals. Most of them are in critical condition and require essential restoration. That is why problems of preservation of icons are very actual for our country. The use of XRF method in studying pigment layers of ancient icons makes it possible to develop a reliable restoration methodology based on evidential proof prerequisites. Exactly for that reason this approach is important for further development of scientific restoration in our country. However, one of most actual tasks of practical use of XRF is creating the database of XRF-spectra of pigments, which were used by Russian icon painters of certain schools and time periods. This is the main goal of researches conducting by our research group, and in this paper we report the preliminary results of our work.

2. Investigation of the paint layer from Russian icons

NitonXL3tGold+ spectrometer (Thermo Fisher Scientific, USA) was used in the experiments. Operating voltage on X-ray tube - 6 ... 50 kV, operating current - up to 200 μA. Icons for investigation were provided by the Department of Easel Painting Restoration of the St.Petersburg State Academic Institute of Painting, Sculpture and Architecture named after I.E. Repin.

2.1. Method

Figure 1 shows a methodology of XRF-spectrum registration and analysis at the example of icon "Five-figure half-length Deisus" (Vologda State Historical-Architectural and Art Museum-Reserve, 18th century).

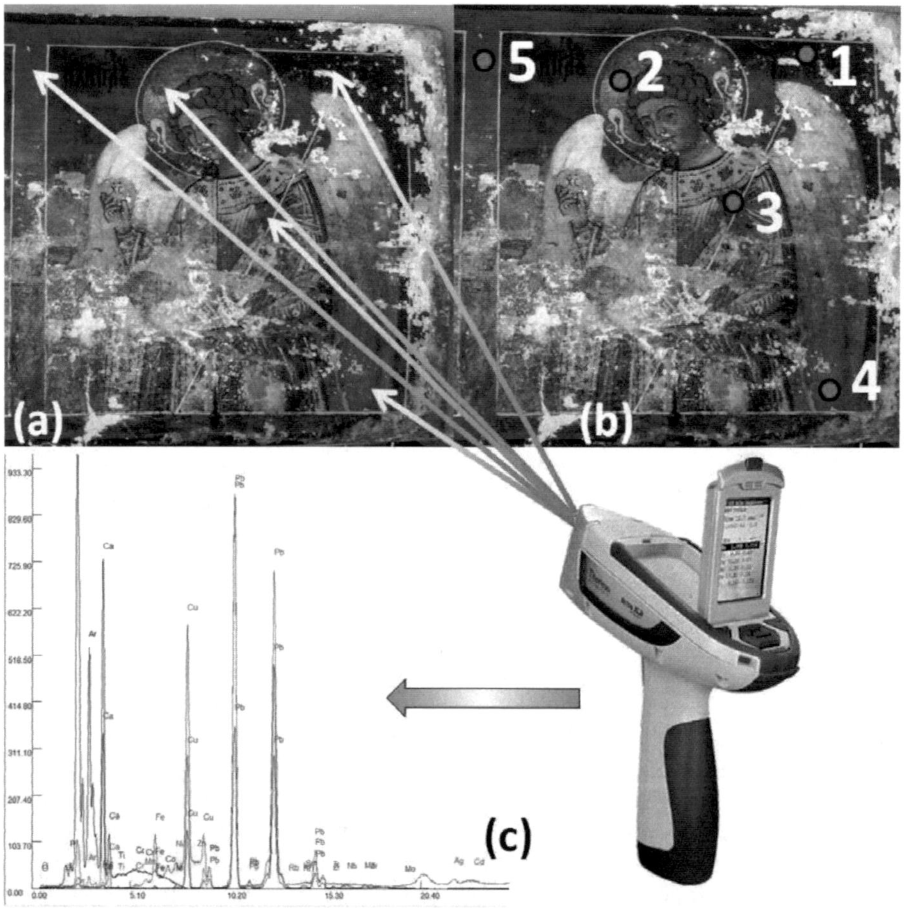

Figure 1. Measurement of XRF-spectra: general view of the icon (a), the same photo with set of reference points (b), example of XRF-spectrum (c).

Measurements were carried out according to a set of reference points (RP), which correspond to the main colors of paints and were chosen by restorers.

2.2 Results

XRF - spectra were obtained at each reference point. The main elements identified from each XRF-spectrum are presented in table 1.

Table 1. Main elements from XRF-spectra in reference points (RP).

№ of RP	**1**	**2**	**3**	**4**	**5**
Main elements	*Pb, Ca, Cu*	*Ca, Fe, Cu, Pb, Ag*	*Cu, Pb, Ca*	*Hg, Ca, Pb*	*Fe, Ca, Pb*
Color on icon	Dark - green	Gold	Dark - green	Red	Ocher

There is a calcium peak in the spectrum of each RP. This is due to the fact that the ground layer of icons – gesso ground, was usually made from a mixture of chalk, which contains calcium, and animal or fish glue. Therefore, calcium peaks will not be taken into account during the pigment identification.

2.2.1 Reference point 1 and 3

The spectra at these points contain the same basic elements and their general view is similar (Figure 2).

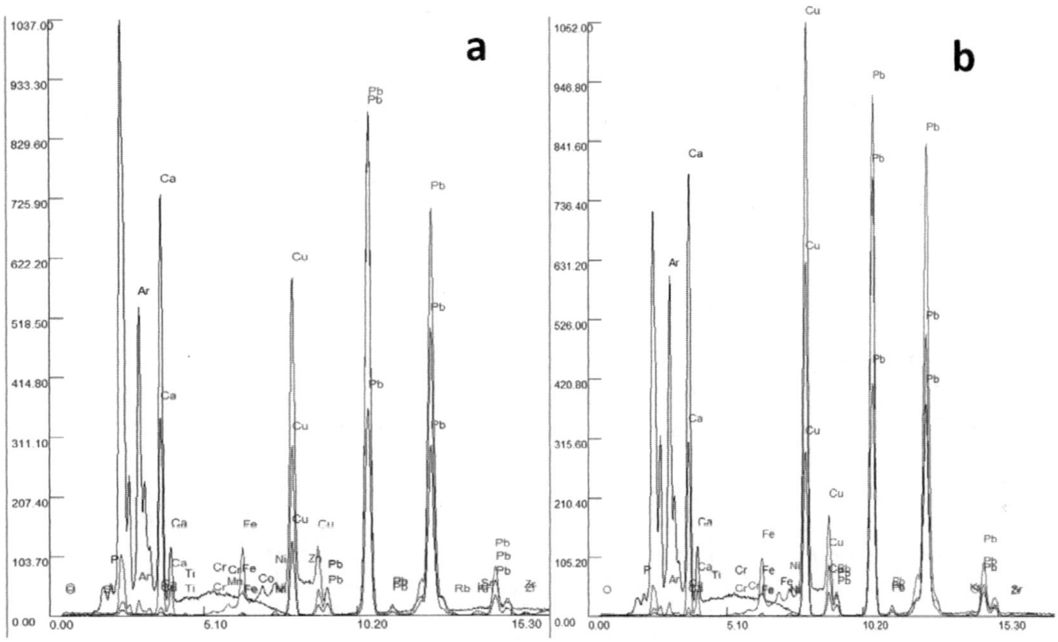

Figure 2. General view of XRF – spectra: RP 1 (a), RP 3 (b).

It is assumed that at both points there is the pigment verdigris ($Cu (CH3COO) \cdot 2Cu(OH)2$), mixed with lead white ($2PbCO3 \cdot Pb(OH)2$) at different concentrations. Verdigris mixed with white lead was widely used in Russian icon painting of the 18th century to paint backgrounds and clothing. As it is seen from Figure 1, RP 1 is laid in the area of background, while RP 3 is in the area of sleeve. Difference in the intensities of Cu peaks may be explained by the change in the concentration of verdigris in these areas. It is possible to obtain different shades of color by changing concentration one of the mixture components.

2.2.2 Reference point 4

The most intensive peak in this point is Hg, the color in the area of RP 4 is red. There is an explicit peak of Pb also (Figure 3).

In the 18th century, a pigment such as cinnabar was often used to highlight something on an icon, for example, an element of the background or clothing. Its color is bright – red, sometimes blood – red. The cinnabar mineral is mercury (II) sulfide (HgS). Because of the fact, that the spectrum contains a slightly less intensive lead peak, this may mean that the cinnabar was mixed with white lead and applied to the icon.

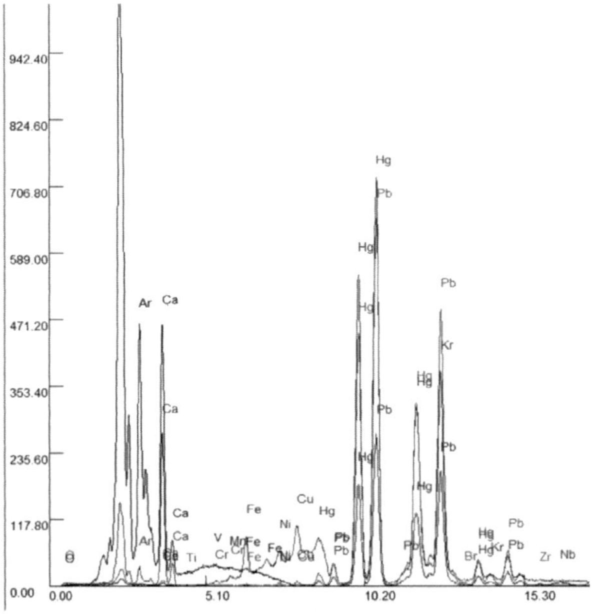

Figure 3. XRF – spectrum in reference point 4.

Another assumption is that the red tint was obtained by mixing the pigments of cinnabar and red lead (Pb_3O_4). Such mixtures have been used since the Roman Empire period to reduce the price of the cinnabar pigment, because its extraction was under the strict control.

2.2.3 Problems of pigment identification
The spectrum at RP 2 showed the absence of explicit peaks of metallic components, and at RP 5 (Figure 4) intensive peak of iron and color in that area indicate that ocher was used in that point - a natural pigment consisting of iron oxide hydrates in combination with clay.

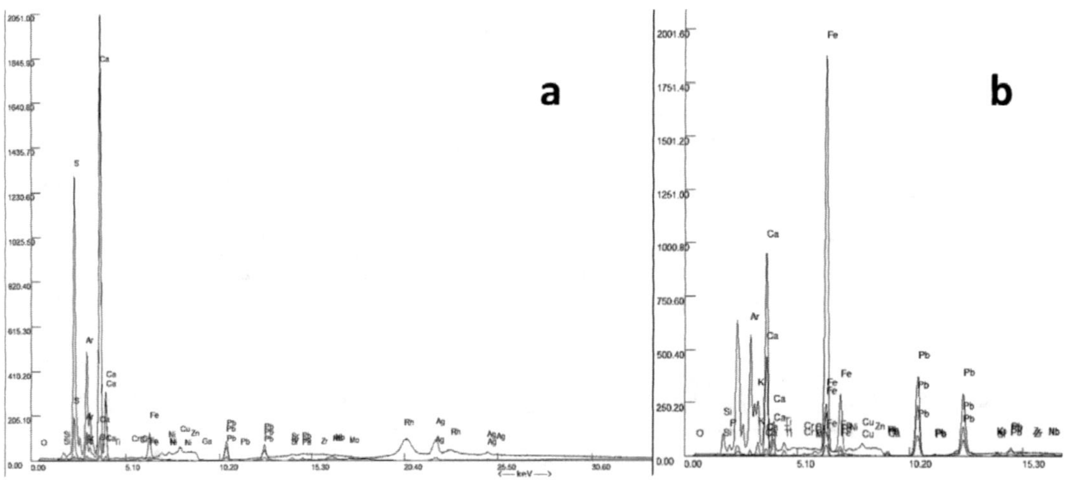

Figure 4. XRF – spectra: RP 2 (a), RP 5 (b).

The RP 2 is on the golden halo. Since no gold was found in the spectrum, the use of gold leaf when writing this icon is excluded. Also, no explicit peaks of other metals were found in the spectrum. This could mean that the golden pigment is made from either organic components or non-metals that are difficult to detect with XRF.

The most common group of pigments containing ferrous components (RP 5) is ocher. There are a lot of ocher shades, which makes it difficult to analyze this pigment. However, such pigments as mustard ocher and limonite were often used to mark the frame of the icon. RP 5 is also situated in the frame area. Though the XRF - data is not enough to determine unequivocally which pigment was used at point 5.

In the experiment, 7 icons were investigated with XRF – method. All icons had gilding or ocher areas. Based on XRF method identification of such pigments was not reliable for all investigated icons.

3. Conclusion

Our studies allowed to identify most typical red and green pigments consisting Cu, Hg and Pb, which were used in Russian icon art created by artists of Russian cities Pskov and Novgorod. Obtained XRF-spectra will be used for systematization of information about specific pigments of different time periods and icon painting schools. It allows creating a database of XRF-spectra of Russian icons of that period, which will become an important information resource for restorers and art historians.

It should be noted that in the experiments it was revealed that some groups of pigments could not be identified with XRF analysis, for example, pigments imitating gilding, or ocher. These pigments require additional study by means of alternative analytical methods to identify impurities in them, including organic ones, which may provide more complete information about their origin and properties.

Acknowledgments

Authors are grateful to the Department of Easel Painting Restoration of the St.Petersburg State Academic Institute of Painting, Sculpture and Architecture named after I. E. Repin for providing access to spectrometer NitonXL3tGold+, as well as Professor F. Bobrov for valuable advice.

References

[1] Janssens K et al 2016 *Top Curr. Chem. Z* 374:81

[2] Gao N and Janssens K 2004 *XRay Spectrometry: Recent Technological Advances* (Chichester: Wiley & Sons)

[3] Moran Th, Kaye A, Rao A and Bueno F 2016 *Journal of Forensic Radiology and Imaging* **5** 38–46

[4] Revenko A and Revenko V 2007 *Methods and Objects of chemical analysis* **2** 1 4–29

[5] Kudryashov V, Serebryakov A and Parfenov V 2010 *Journal of Optical Technology* **77** 8 469 – 472

Performance of microheaters for tunable on-chip interferometer

I O Venediktov[1], M S Elezov[1,2,3], A I Prokhodtsov[1,4,5], V V Kovalyuk[1,5] , P P An[1,5], A D Golikov[1], M L Shcherbatenko[1,2,3], D V Sych[1,2,3,6], G N Goltsman[1,4]

[1] Department of Physics, Moscow State Pedagogical University, 119992, Russia
[2]National University of Science and Technology MISiS, Moscow, 119049, Russia
[3]Sirius University of Science and Technology, Sochi, 354340, Russia
[4]National Research University Higher School of Economics, Moscow, 101000, Russia
[5]Institute of Applied Physics of the Russian Academy of Sciences, Nizhny Novgorod, 603950, Russia
[6]P.N.Lebedev Physical Institute, Russian Academy of Sciences, Moscow, 119991, Russia

ilia1999ven@gmail.com

Abstract. Mach-Zehnder interferometer (MZI) is a valuable practical tool in many optical science areas. In particular, high-contrast MZI are required for experimental realization of displacement-based quantum receivers that can discriminate weak coherent states of light with the minimum error rate. In this work we study phase modulators of tunable on-chip interferometer on silicon nitride (Si_3N_4) platform for telecom wavelength (1550 nm) consisting of several MZI. Phase modulators on one of the arms of MZI consists of microheaters and waveguide. Microheaters heat waveguides changing its refractive index due to thermo-optical effect providing a phase shift. We measure the bandwidth of phase modulators and study their operation in pulse mode.

1. Introduction

High-contrast interferometry is an important prerequisite for experimental demonstration of novel methods for quantum measurements, communications, and control [1-5]. Although the recent progress in realization of ultra-sensitive quantum receivers was achieved with the optical fibers [6-9], the use of integrated optical circuits offers potentially much higher contrast. Photonic integrated circuits based on silicon nitride (Si_3N_4) have many advantages over fiber and free space realizations because of a small footprint of the device itself, less susceptibility to heat and phase fluctuations and higher stability.

One of the technical problems is how to obtain the desired phase shift with the integrated Mach-Zehnder interferometer (MZI). In this work, we use a microheater as a phase modulator. Microheater's work is based on the thermo-optical effect, i.e. the refractive index of waveguides material is changing due to the temperature change [10]. We investigate performance of phase modulators and find their main parameters such as effective bandwidth.

2. Device description

The tunable on-chip interferometer on silicon nitride shown in figure 1. Tunable interferometer consists of 3 MZI: main interferometer MZI 1 and 2 interferometers MZI 2, MZI 3 on its arms, which are used for attenuation of signal. To control MZI 1 - phase modulators PM1 and PM3 are used, to control MZI 2 and MZI 3 - PM4, PM5 are used respectively. As waveguide material, we used Si_3N_4 due to its low absorption in infrared (IR) and good mechanical properties [11] and possibility of integration of SNSPD [12] on chip which allows to fabricate more accurate devices. Phase modulator consists of waveguide and microheater. Microheaters are Ti/Au film with thickness of Au 300 nm. They can be divided in two parts: one part lying on waveguide to heat it and another part is contact plates to apply voltage with contact plate width 225 μm. Part on waveguide has significantly smaller width in comparison with leading contact plates, therefore we may neglect resistance and heating of the contact plates. Four of microheaters are used as phase modulators and one of them (PM2) is used to change splitting ratio of beam splitter dividing optical power between output ports 1 and 2. Interferometer (except contact plates) is covered by layer of SiO_2 of thickness 300 nm to exclude contact of microheaters with atmosphere and prevent burn down at high temperatures. We used FGC (Focusing Grating Couplers) for telecom wavelength with grating period 1.088μm and fill factor 70% (in design). FGC 3 is used to input light, and FGC 1 and FGC 2 are used to output light from the interferometer.

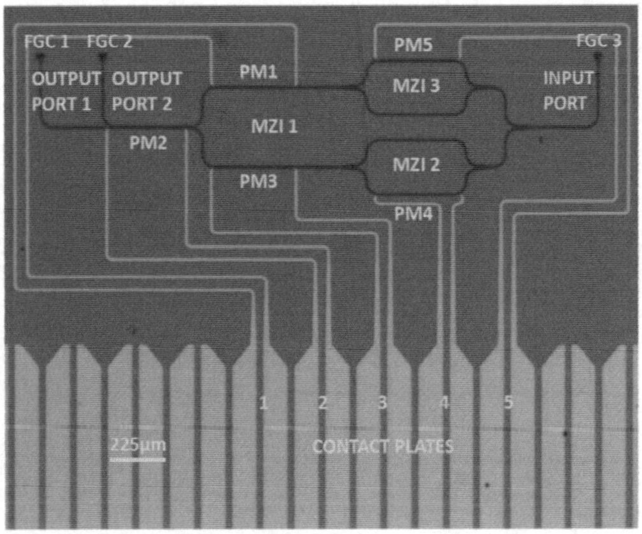

Figure 1. Optical micrograph of interferometer. PM 1, PM 3, PM 4, PM 5 - phase modulators; PM2 - microheater to control splitting ratio of beam splitter for output ports 1 and 2; MZI 1 - main Mach-Zehnder interferometer; MZI 2, MZI 3- interferometers for attenuation of the signal in the arms of MZI 1; FGC 1, FGC 2 - focusing grating couplers for output optical power; FGC 3 - focusing grating coupler for input optical power.

3. Bandwidth of microheaters

To measure bandwidth of phase modulators we used experimental setup shown on Figure 2. We used lock-in amplifier with internal sinusoidal signal generator. Sinusoidal signal was applied to one of the microheater, radiation of 1596 nm from the laser was introduced to interferometer through fiber array and FGC. Simultaneously optical power from output port 1 was measured by InGaAs detector with bandwidth 2 GHz. Sinusoidal voltage of constant amplitude heats microheater which leads to the phase shift in arms of interferometer. Output optical power also has sinusoidal form and measured by detector.

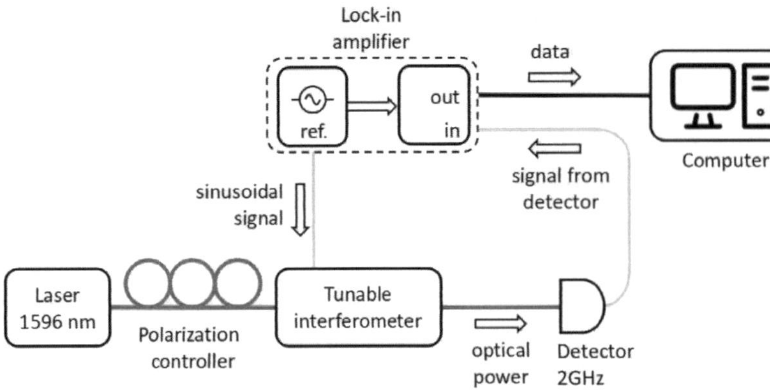

Figure 2. Experimental setup for bandwidth investigation. Laser 1596 nm - New Focus TLB-6600; Lock-in amplifier - Stanford Research System RS830; Detector 2GHz - Hamamatsu G9801; Blue line corresponds to optical fiber; Green line corresponds to electric wires.

The results can be seen on the graph on Figure 3. Obtained curve has flat region, which means phase modulators able to work in this frequency range. Curve goes down after some frequency, that means that phase modulators are not able to work on frequencies higher than some characteristic frequency.

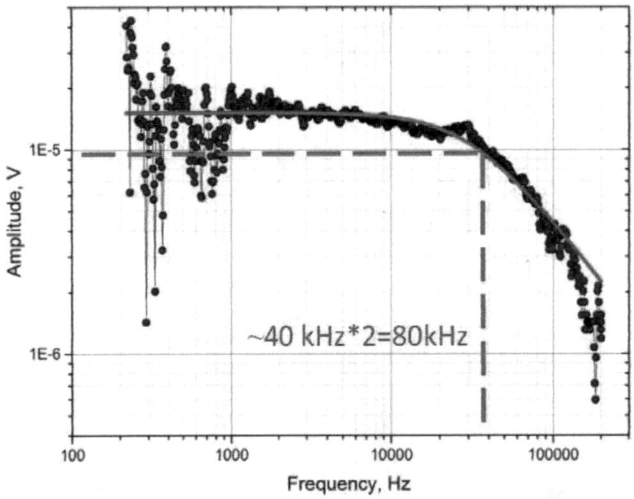

Figure 3. Frequency response curve. Red line corresponds to fitting curve. Characteristic frequency is shown on the graph, but the performance will be twice bigger because heat process is independent on sign of applied voltage.

As characteristic frequency can be taken such a frequency that output amplitude is twice smaller than amplitude in flat region, which corresponds to -3 dB. This frequency can be obtained by fitting frequency response curve by equation 1.

$$U(f)=U\left(1+\left(f/f_{3dB}\right)^{2}\right)^{-0.5}. \qquad (1)$$

A value of 40 kHz for characteristic frequency was obtained after fitting. But as the sinusoidal signal was used, the performance will be twice bigger, because we take into account periods of heat which are independent on voltage sign. Therefore final performance is 80 kHz.

4. Optimal pulse mode for microheater operation.

Performance value of 80 kHz was obtained, but in practice it is often necessary to get a phase difference of 0 or π during some period of time for some switching speed . Therefore we define optimal impulse regime as maximum switching speed between 0 and π. There are 2 restrictions on maximum switching speed. Firstly, thermo-optical effect based phase modulators are relatively slow because of inertia of thermal processes connected with specific heat and heat conductivity of waveguides and microheaters. Switching speed defined not only by speed of heating of waveguides, but also by their speed of cooling. Those parameters are defined by design of structure. Second restriction connected with maximum voltage, which can be applied to microheater. After constant voltage applied, microheater is heating itself and heats waveguide leading to phase shift due to thermo-optical effect. Work region for bias voltage 4-8 V [13]. If constant voltage is high or applied for a long time it leads to burn out of microheaters. Therefore it is better to use impulse regime, periodically apply constant voltage for the short period of time. In other time waveguides and microheaters are cooling down for initial temperature because of heat dissipation into the substrate. To find optimal regime of phase shift experimental setup shown in Figure 4 was used. Radiation of 1596 nm from laser introduced into interferometer through fiber array and focusing grating couplers. Rectangular electrical signal from delay generator was applied to microheater PM4, heating up the waveguide and modulating phase shift. The electrical signal had an amplitude of 7 V and a duration of 500 ms. At the output of the chip, the interfered light was detected by an InGaAs detector with a 5 GHz bandwidth.

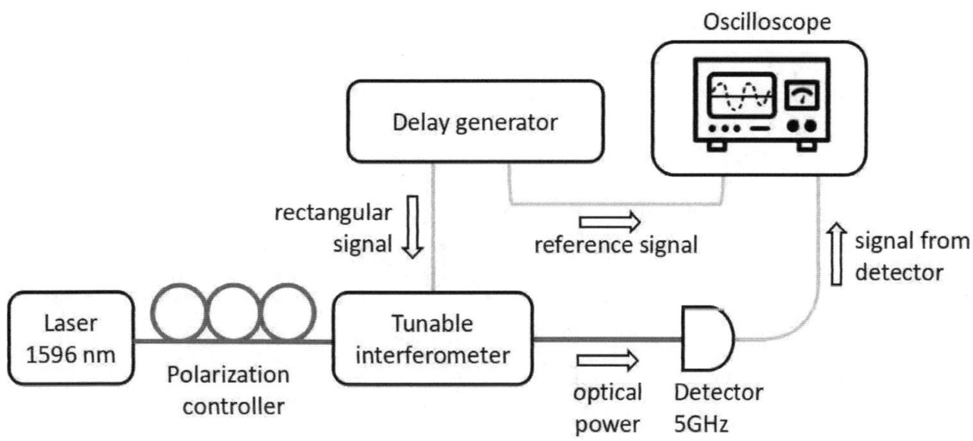

Figure 4. Scheme of the experimental setup for investigation of microheaters in pulse mode. Laser 1596 nm - Venturi Tunable Laser TLB-6600; Delay generator - Highland Technology P400 Digital Delay Generator; Oscilloscope - Rohde & Schwarz RTO 1012; Detector 5GHz - ThorLabs DET08CFC/M; Blue line corresponds to optical fiber; Green line corresponds to electric wires.

Figure 5 represents the response from the detector. The absence of voltage on the microheater corresponds to an arbitrary interference result. At a voltage of 7 V, we get destructive interference, which corresponds to a phase difference of 0 in the interferometer. However, the time of establishing destructive interference is long and is approximately 400 μs. For 100 μs after that, the interference is stable. Further, after the voltage is turned off, the microheater is cooled due to the heat leaving the substrate and returns to its original state. The maximum repetition rate of rectangular pulses is 4 Hz. As the repetition rate increases, the chip overheats, because the microheater and the waveguide do not have time to cool down and the interference result becomes unstable. The same thing happens if you increase the time of applying voltage to the microheater. Accordingly, to obtain constructive interference, the amplitude of the displacement voltage changes by 10.2 V, which corresponds to the phase difference π. Thus, we found the optimal pulse mode for microheaters: a rectangular pulse signal with a repetition frequency of 4Hz and a pulse duration of 500 μs with amplitudes of 7 V and

10.2 V for the phase shift of 0 and π, respectively. The relaxation time for heating is 400 μs. The time with a stable phase is 100 μs.

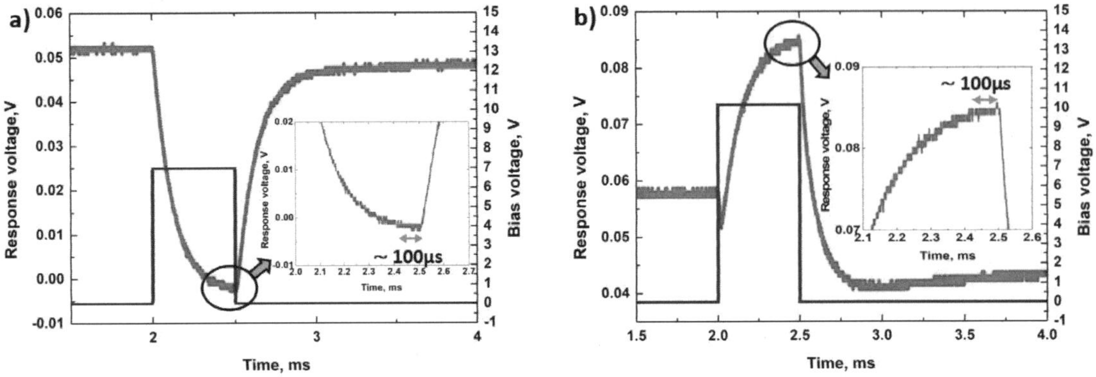

Figure 5. (a) Detector response for destructive interference for pulse with amplitude 7 V. **(b)** Detector response curve for constructive interference for pulse with amplitude 10.2 V. Blue curve corresponds to rectangular bias voltage applied to microheater. Red curve corresponds to detector response to optical power from output port 1. Enlarged graphs for stable interference regions provided at the right.

5. Conclusion

The output frequency response curve was obtained. After fitting bandwidth value 80 kHz was obtained.Optimal pulse mode for the microheaters was found. Detector response curves for rectangular voltage were obtained and analyzed. These results have potential for using this tunable on-chip interferometer as a part of quantum receiver of weak coherent signals.

Acknowledgments

Denis Sych, Mikhail Elezov and Mikhail Shcherbatenko acknowledge Russian Foundation for Basic Research grant 20-32-51004. Iliia Venediktov acknowledges FASIE, agreement № 645ГУЦЭС8-D3/63065.

References

[1] Sych D V and Leuchs G 2016 Phys. Rev. Lett. **117** (20) 200501
[2] Shcherbatenko M L, Elezov M S, Goltsman G N and Sych D V 2020 Phys. Rev. A **101** (3)032306
[3] Sych D V, Grishanin B A and Zadkov V N 2005 Quantum Electronics **35** 80–84
[4] Averchenko V, Sych D V, Marquardt C and Leuchs G 2020 Phys. Rev. A **101** (1) 013808
[5] Sych D V, Averchenko V and Leuchs G 2017 Phys. Rev. A 96 (5) 053847
[6] Shcherbatenko M L, Elezov M S, Sych D V and Goltsman G N 2020 Journal of Physics: Conference Series **1695** 012140
[7] Elezov M S, Scherbatenko M L, Sych D V and Goltsman G N 2018 Journal of Physics: Conference Series **1124** 051014
[8] Elezov M S, Scherbatenko M L, Sych D and Goltsman G N 2019 EPJ Web of Conferences **220** 03011
[9] Elezov M S, Shcherbatenko M L, Sych D V and Goltsman G N 2020 Journal of Physics: Conference Series **1695** 012126
[10] Ovvyan A P, Gruhler N, Ferrari S and Pernice W H P 2016 J. Opt. **18** (6) 064011
[11] Silverstone J V, Bonneau D, O'Brien J L, Thompson M G 2016 IEEE **22** 390
[12] Kovalyuk V V et al 2017 Scientific Reports 7 4812
[13] Venediktov I O, Elezov M S, Prokhodtsov A I, Kovalyuk V V, An P P, Golikov A D, Shcherbatenko M L, Sych D V, Goltsman G N 2020 J. Phys.: Conf. Ser. **1695** 012117

Photoionization of polarized doublet states of Xenon atom

E A Viktorov[1], A A Pastor[1], P Yu Serdobintsev[1], K Miculis[2,3] and N N Bezuglov[1,3]

[1] Department of Optics, Faculty of Physics, St. Petersburg State University, 198504, St. Petersburg, Russia
[2] Moscow State Engineering Physics Institute, Kashirskoe shosse 31, 115409, Moscow, Russia
[3] University of Latvia, Institute of Atomic Physics and Spectroscopy, LV-1586 Riga, Latvia

eaviktorov@yandex.com

Abstract. We studied experimentally and theoretically the ionization of a two-photon excited superposition of 4f states of Xe atom by probe femtosecond pulse laser in a supersonic beam. The revealed oscillation structure in the observed photoionization signals is associated with the interference of the coherently populated components of Xe doublet states. Our results demonstrate the high efficiency of proposed experimental scheme for registering quantum beats in the photoelectron current.

1. Introduction

Analyzing of fluorescence signals after exciting atomic states by polarized light is an important brunch of modern spectroscopy [1]. Observation of magnetic resonances in polarized gas media is the basis of construction and realization of sensitive magnetometers [2]. Diverse types of oscillating structures in detected signals in the presence of magnetic fields allow making precision measures of atomic state parameters beyond limitations due to Doppler broadening [3]. Traditional methods of spectroscopy are associated mainly with bound-bound optical transitions. In the present report, we demonstrate a possibility to examine photocurrent oscillations which occurs due to the interference between coherently excited components of doublet atomic states upon their photoionization by probe polarized photons.

2. Experimental setup

The supersonic beam of Xe atoms crosses a focused laser beam in the presence of a magnetic field, it is necessary to draw the electrons from the ionization region (figure 1). We use a "pump-probe" experiment with two-photon excitation of Xe atoms to nearby states $5p^5(^2P_{3/2})4f[3/2]_2$ and $5p^5(^2P_{3/2})4f[5/2]_2$ using a "pump" pulse. Excitation is performed by a horizontally polarized femtosecond laser pulse at a wavelength of about 220nm. Photoionization of the excited atoms in the J=2 states is performed by a "probe" pulse of a horizontally or circularly polarized Ti: sapphire femtosecond laser at a wavelength of 795nm, the spectral width of the ionization pulse is greater than the distance between these 4f levels.

An electron spectrometer of the "magnetic bottle" type was used to record the electron energy spectra. Analysis of the energy spectra of electrons, obtained in different delay times of the probe pulse relative to the pump pulse, made it possible to obtain data on quantum beats. The energy spectra provide information on the levels from which ionization took place and on the intensity of ionization from these levels.

The Ti-Sapphire femtosecond laser PULSAR 10 (Amplitude Technologies, France) has the following characteristics: main wavelength - 790nm, pulse energy - 5mJ, pulse duration - 50fs, repetition rate - 10Hz. The outcoming pulse was divided into two parts: (i) 70% of the pulse power pumped parametric amplifier TOPAS and was converted to UV radiation for two-photon excitation of atomic states. (ii) The rest 30% of the power was used as probe radiation in the form of the first harmonic for ionization of pump-excited atoms.

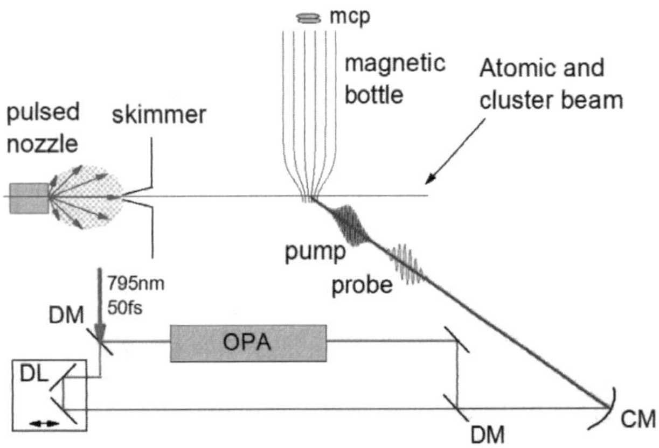

Figure 1. Experimental setup: OPA – optical parametric amplifier; DL – delay line; DM – dichroic mirrors; CM – concave mirror.

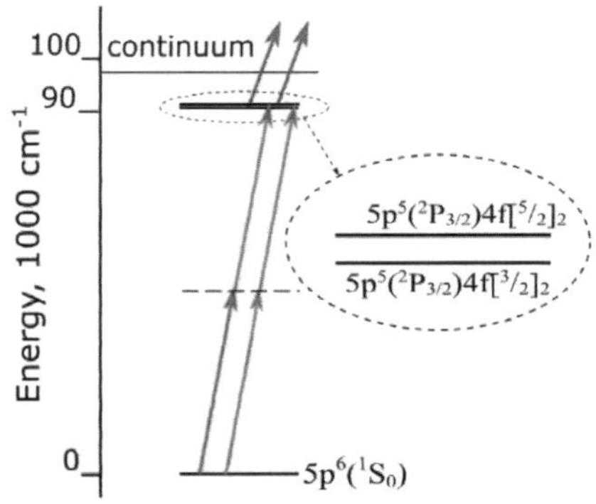

Figure 2. Researched Xe levels.

An optical system was used to control the delay time between pulses. The delay time between laser pulses was varied from 0 to 24ps. With a change in the delay time t, periodic oscillations of the intensity of the photoelectron current appear at the difference frequency of the levels $\Delta\omega$ (figure 3). Which equal to the frequency distance between levels $\Delta\omega=\omega_{5/2}$ - $\omega_{3/2}$. From the experimental data shown in figure 3, we have a beat period of 0.549ps. This is in good agreement with the energy gap between these xenon levels (Xcm^{-1}), which corresponds to a beat period of 0.55ps.

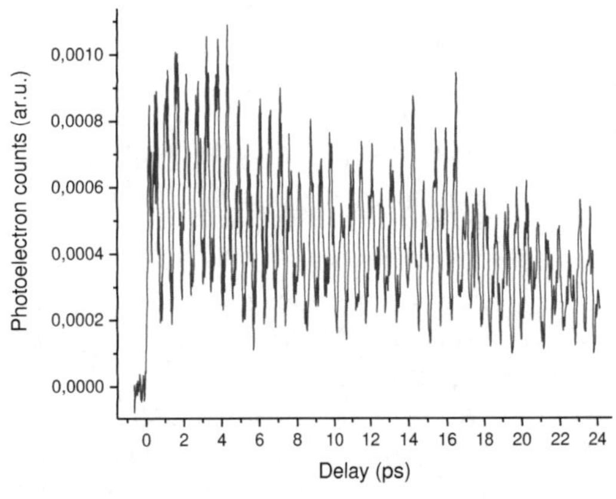

Figure 3. Temporal dependence of the photocurrent I for one-photon ionization. The one-photon ionization occurs together with two-photon ionization due to the strong probe pulse.

3. Theoretical analysis

The effects of the coherent interaction between polarized quantum particles and electromagnetic fields are convenient to analyze in the frame of the irreducible tensor operators technique which deals with atomic and photon polarization moments ρ_q^κ , Φ_q^κ [2,4]. In this section, we focus on a brief overview of our theoretical study of how the interference of the doublet components of the Xe states results in a beating structure (see figure 3) in photoionization signals upon photoionization stimulated by a single photon (single-photon photoionization).

3.1. Partial photoionization signals
The authors of the book [2] obtained an explicit expression for the photoionization cross section ([2], Section 20.2) for a polarized degenerate quantum *J*-state with Zeeman sublevels $M=-J,...,J$. In our recent work [5], we applied the formalism from [2] for the case when external magnetic fields split the corresponding Zeeman *M*-components. A fairly direct reformulation of the previous results makes it possible to describe the photosignal $I(t)$ upon photoionization of the studied doublet state $\Lambda=\{\Lambda_1,\Lambda_2\}$ of Xe, consisting of two sub-states (see figure 2)

$$\Lambda_i=5s^2 5p^5(^2P_{3/2})nl[K_i]_J\ ;\ i=1,2\ ;\ K_1=3/2;K_2=5/2 \tag{1}$$

with the same total atomic orbital moment *J* (*jl*-coupling scheme [4]). The corresponding photo signal looks like this:

$$I_{\Lambda,\Lambda'}(t)/J_{ion}\sim|C_1|^2\Pi_{11}+|C_2|^2\Pi_{22}+2\mathrm{Re}\left(C_1C_2^*\Pi_{12}\exp(i\Delta\omega)\right), \tag{2}$$

$$\Pi_{i\bar{i}}=\sum\nolimits_{\kappa=0}^{2}D_{i\bar{i}}^{\kappa}M_{i\bar{i}}^{\kappa}\Phi_0^{\kappa}\rho_0^{\kappa}\ . \tag{3}$$

Here J_{ion} is the intensity of the luminous flux of the ionizing laser. The indexes in $I(t)$ indicate the quantum numbers $\Lambda=\{\Lambda_1,\Lambda_2\}$ of the polarized bound doublet atomic state in the initial channel and $\Lambda'=3s^2 3p^5(^2P_j)\varepsilon'l'[K']_{J'}$ that's in the final (ionization) channel with the notation ε' for the free photoelectron energy. It is assumed that the quantum numbers (P,j) of atomic core remain unchanged (the absence of intercombination transitions).

The first two terms on the right-hand side of equation (2) determine the contributions to the ionization of the doublet i-sub-states, while the third term is responsible for the effect of their interference. Coefficients C_i correspond to sub-states Λ_i probability amplitudes normalized to the total sub-states population over all their Zeeman sublevels: $n_i=|C_i|^2=\sum_m|C_m^{(i)}|^2$. It is important to note that although the amplitudes C_i may differ, the atomic polarization moments ρ_q^{κ} with $\kappa=0$ (population), $\kappa=1$ (orientation) and $\kappa=2$ (alignment) must be the same, since both components Λ_i of the doublet have the same quantum number *J* and they are excited with the same pump photons. The presence of the atomic moments $\rho_{q=0}^{\kappa}$ in equation (3) is due to the pure polarization (linear or circular) of the probe laser that implies that the corresponding photon moments $\Phi_q^{\kappa}\equiv 0$ if $q\neq 0$ [2].

The factors $M_{i\bar{i}}^{\kappa},D_{i\bar{i}}^{\kappa}$ take into account the influence of the mutual orientation of the orbital moments of atoms and ionizing photons on the photoionization processes efficiency. The value of $M_{i\bar{i}}^{\kappa}$ turns out to be equal to one ($M_{i\bar{i}}^{\kappa}\equiv 1$) while factors $D_{i\bar{i}}^{\kappa}$ are expressed through 6*j*-symbols

$$D_{i\bar{i}}^{\kappa}=(-1)^{J+J'+1}\begin{Bmatrix}1 & 1 & \kappa\\ J & J & J'\end{Bmatrix}(\Lambda_i\|d\|\Lambda')(\Lambda_{\bar{i}}\|d\|\Lambda') \tag{4}$$

and the reduced dipole matrix elements for bound-free optical transitions [2,4].

Expressions (2)-(4) provide information on the partial photoionization cross-section, when the quantum numbers l',K',J' in the final channel turn out to be fixed (i.e. measured in an experiment).

3.2. Aggregate photoionization signal

In our experiment, we do not separate photoelectrons over its quantum numbers l', K', J'. In other words, all partial signals contribute to the final aggregate photocurrent signal $I(t) = \sum_{\Lambda'} I_{\Lambda,\Lambda'}(t)$. Importantly, the sum rule (see in [4] equation (4.85)) for the product of three 6*j*-symbols entering the factor $D^{\kappa}_{i\bar{i}}$ (4) makes it possible to perform explicit summation: $I(t) = \sum_{l'=l\pm1} |(nl \| d \| \varepsilon' l')|^2 \, I_{ll'}$, where the remaining partial signals $I_{ll'}$ relate to the output of a free photoelectron with two possible values $l' = l \pm 1$ of its orbital quantum number. The ratio $I_{ll'} / J_{ion}$ is expressed through a formula similar to equation (1) with a somewhat modified coefficient

$$D^{\kappa}_{i\bar{i}} = \begin{Bmatrix} 1 & 1 & \kappa \\ l & l & l' \end{Bmatrix} \tag{5}$$

and a nontrivial factor

$$M^{\kappa}_{i\bar{i}} = (2J+1)\sqrt{(2K_i+1)}\sqrt{(2K_{\bar{i}}+1)} \begin{Bmatrix} K_i & K_{\bar{i}} & \kappa \\ J & J & 1/2 \end{Bmatrix} \begin{Bmatrix} l & l & \kappa \\ K_i & K_{\bar{i}} & j \end{Bmatrix} (-1)^{\Phi} \tag{6}$$

with a phase $\Phi = K_i + K_{\bar{i}} - 0.5 + J - l - j$.

3.3. Results

We have derived that the observed photocurrent signal $I(t)$ during single-photon photoionization of an atomic doublet state with polarized components is calculated as a sum

$$I(t) = \sum_{l'=l\pm1} \chi_{l'} I_{ll'}(t); \; \chi_{l'} = (nl \| d \| \varepsilon' l')^2 \tag{7}$$

is calculated as a sum of two partial signals $I_{ll'}(t)$

$$I_{ll'}(t) / \Upsilon = |C_1|^2 \, \Pi_{11} + |C_2|^2 \, \Pi_{22} + 2\mathrm{Re}\left(C_1 C_2^* \Pi_{12} \exp(i\Delta\omega)\right), \tag{8}$$

whose statistical weights $\chi_{l'}$ are given by the reduced dipole matrix elements (7) of bound-free optical transitions of the valence electron. The multiplier Υ depends on intens1ities of applied lasers and experimental set-up, while the probability amplitudes C_i of the initial doublet sub-states (1) depend on the parameters of the pump laser, in particular, on its polarization. The factors $\Pi_{i\bar{i}}$ (3) are expressed via 6j-symbols as it is seen from equations (5), (6) and via the polarization moments connected with Wigner 3*j* symbols [2]

$$\Phi^{\kappa}_0 = \sqrt{2\kappa+1}(-1)^{1-m} \begin{pmatrix} 1 & 1 & \kappa \\ m & -m & 0 \end{pmatrix}; \; \rho^{\kappa}_0 = \sqrt{2\kappa+1}(-1)^{J-M} \begin{pmatrix} J & J & \kappa \\ M & -M & 0 \end{pmatrix} \tag{9}$$

which values are presented in table 1.

It is noteworthy that our theoretical results can be extended to the case of two-photon ionization of polarized atomic states. The corresponding formulas are similar to expressions (7), (8), however the factors $\Pi_{i\bar{i}}$ (3) contains five atomic polarization moments ρ^{κ}_0 with $\kappa = 0, \ldots, 4$ and 9j Wigner symbols. Taking into account the presence of a control external magnetic field is reduced to introducing the time dependence of the moments ρ^{κ}_0, as described in detail in [5].

4. Conclusion

The polarization atomic moments (PAM) strongly affect the magnitude of the long-range dipole-dipole interaction between the excited states of the atoms [6]. This interaction plays an important role in modern applied physical problems associated with quantum computing based on an ensemble of cold atoms in magneto-optical traps [7]. For this reason, real-time control and diagnostics of PAM are an urgent problem in the implementation of entangled atomic states (Q-bits) and in the performance of basic quantum operations [8]. Traditional optical diagnostics of PAM is associated with photon-stimulated bound- bound transitions and registration of accompanying absorption [9] or fluorescence [2,3] spectra. The use of bound-free transitions for diagnostic purposes has a number of advantages due to almost 100% registration of photoelectrons, which significantly increases the possibilities of recording subtle physical effects [10, 11]. The presented here and in our previous work [5] results on the study and description of the oscillatory structure of the photocurrent with their subsequent spectral analysis make it possible to find PAM necessary for calculating the dipole-dipole interaction between excited atoms in cold media.

Table 1. Photon Φ_0^κ and ρ_0^κ atomic polarization moments. Φ_0^κ-moments are shown for circular (*m*=1) and linear polarizations (*m*=0) of the probe laser, while ρ_0^κ-moments corresponds to the excitation states with *J*=2 and Zeeman component *M*=2 [2].

	$\kappa=0$	$\kappa=1$	$\kappa=2$
Φ_0^κ; linear	23.56	0	$-\sqrt{2/3}$
Φ_0^κ; circular	34.64	$1/\sqrt{2}$	$1/\sqrt{6}$
ρ_0^κ	23.76	$\sqrt{2/5}$	$\sqrt{2/7}$

Acknowledgments
This work was supported by the Latvian Science Council Grant No lzp-2019/1-0280. We thank Professor M. Auzinsh for useful discussions on physics of polarized states of atoms and photons. The equipment of the Resource Center "Physical Methods of Surface Investigation" of the St. Petersburg State University was used in experiments.

References
[1] Quack M, Merkt F 2011 *Handbook of High-Resolution Spectroscopy* (Wiley-Blackwell, Hoboken)
[2] Auzinsh M, Budker D and Rochester S 2010 *Optically Polarized Atoms. Understanding light atom interactions* (New York: Oxford University Press)
[3] Aleksandrov E B, Chaika M P and Khvostenko G I 1993 *Interference of Atomic States. Springer Series on Atoms and Plasmas* vol 7 (Berlin: Springer-Verlag)
[4] Sobelman I 1992 *Atomic Spectra and Radiative Transitions. Springer Series on Atoms and Plasmas* vol 12 (Berlin Heidelberg: Springer)
[5] Viktorov E A, Dimitrijević M S, Srećković V A, Bezuglov N N, Miculis K, Pastor A A and Serdobintsev P Yu 2021 *Eur. Phys. J.* D **75** 13
[6] Pillet P and Gallagher T F 2016 *J. Phys. B: At., Mol. Opt. Phys.* **49** 174003
[7] Saffman M, Walker T G and Mølmer K 2010 *Rev. Mod. Phys.* **82** 2313
[8] Saffman M 2016 *J. Phys. B: At., Mol. Opt. Phys.* **49** 202001
[9] Sydoryk I, Bezuglov N N, Beterov I I, Miculis K, Saks E, Janovs A, Spels P and Ekers A 2008 *Phys. Rev.* A **77** 042511
[10] Porfido N, Bezuglov N N, Bruvelis M, Shayeganrad G, Birindelli S, Tantussi F, Guerri I, Viteau M, Fioretti A, Ciampini D, Allegrini M, Arimondo E, Ekers A and Fuso F 2015 *Phys. Rev.* A **92** 043408
[11] Efimov D K, Bruvelis M, Bezuglov N N, Dimitriyević M S, Klyucharev A N, Sreković V A, Gnedin Yu N and Fuso F 2017 *Atoms* v **5** p 50

Polarized spectroscopy of electric and magnetic dipole transitions of Europium (III) ions in C_2 sites

A Volokitina[1,2], P Loiko[3], E Dunina[4], A Kornienko[4], J M Serres[2], M Aguiló[2], F Díaz[2], A Pavlyuk[5] and X Mateos[2,6]

[1] ITMO University, 49 Kronverkskiy Pr., 197101 St. Petersburg, Russia
[2] Universitat Rovira i Virgili, Física i Cristal·lografia de Materials i Nanomaterials (FiCMA-FiCNA), Marcel.li Domingo 1, 43007 Tarragona, Spain
[3] Centre de Recherche sur les Ions, les Matériaux et la Photonique (CIMAP), UMR 6252 CEA-CNRS-ENSICAEN, Université de Caen Normandie, 6 Boulevard Maréchal Juin, 14050 Caen Cedex 4, France
[4] Vitebsk State Technological University, 72 Moskovskiy Pr., 210035 Vitebsk, Belarus
[5] A.V. Nikolaev Institute of Inorganic Chemistry, Siberian Branch of Russian Academy of Sciences, 3 Lavrentyev Ave., 630090 Novosibirsk, Russia
[6] Serra Húnter Fellow

anna.itmo@gmail.com

Abstract. Polarization anisotropy of luminescent properties of europium (III) ions in low-symmetry C_2 sites is studied using monoclinic (sp. gr. *C2/c*) tungstate crystal $KY(WO_4)_2$. The $^5D_0 \rightarrow {}^7F_J$ (where J = 0...6) transitions are characterized for the principal light polarizations. Polarization selection rules for the magnetic dipole $^5D_0 \rightarrow {}^7F_1$ transition are presented. The stimulated-emission cross-sections for Eu^{3+} ions relevant for laser operation are determined.

1. Introduction

Trivalent europium ions, Eu^{3+} or Eu (III), possessing an electronic configuration of [Xe]4f^6 are well-known for their emissions in the visible owing to transitions from a metastable (long-living) $^5D_0$ state to a group of lower-lying ^7F_J (J = 0...6) multiplets ($^7F_0$ is the ground-state). Among them, the electric dipole (ED) transition to the $^7F_2$ state falling in the red spectral range (~610 nm) typically dominates in the emission spectrum. This determines the applications of Eu^{3+}-doped materials as red phosphors with high colour purity [1,2], e.g., commercialized Eu^{3+}:Y_2O_3S. Note that europium also exhibits another oxidation state, Eu^{2+}. Eu^{2+}-doped materials are used as blue phosphors [3].

Eu^{3+} ions are also attractive as structural probes [4]. First, the transition $^5D_0 \rightarrow {}^7F_0$ is both electric dipole and magnetic dipole (MD) forbidden. Its observation is due to J–J-mixing in the local field [5]. Both the emitting and the terminal levels are non-degenerate (J = 0), so that the number of emission lines corresponds to the number of non-equivalent Eu^{3+} sites in the host lattice. Second, the transition $^5D_0 \rightarrow {}^7F_1$ is of purely MD nature (so that its probability is weakly dependent on the host), while the spectrally adjacent transition $^5D_0 \rightarrow {}^7F_2$ is of purely ED nature (it is a hypersensitive one, i.e., its probability depends on the local site symmetry and its distortion). The ratio $R = I_{ED}(0\rightarrow2)/I_{MD}(0\rightarrow1)$ is known as an asymmetry parameter [4]: if Eu^{3+} site does not exhibit a centre of inversion, $R >> 1$.

Finally, Eu^{3+}-doped crystals are also suitable for direct generation of red laser emission [6-8]. Recently, Loiko *et al.* reported on a highly-efficient Eu laser in a stoichiometric compound, namely a monoclinic $KEu(WO_4)_2$ crystal, operating on the $^5D_0 \rightarrow {}^7F_4$ transition and delivering watt-level output at ~703 nm (in the deep-red spectral range) [9], which stimulated the study of polarized emission properties of Eu^{3+}-doped crystals.

Among the host matrices intended for Eu^{3+} doping, monoclinic double tungstates (MDTs) with a chemical formula $KRE(WO_4)_2$ (where RE = Gd, Y, Lu) appear very suitable. Their advantages are: (i) high available doping concentrations up to stoichiometric compositions, (ii) long RE^{3+} - RE^{3+} interatomic distances leading to weak concentration quenching of luminescence, and (iii) strong polarization anisotropy of emission properties [9,10]. Considering the great potential of Eu^{3+}-doped MDT crystals for the development of red and deep-red lasers operating on different $^5D_0 \rightarrow {}^7F_J$ transitions, it is relevant to study in detail the anisotropy of transitions in emission.

In the present work, we have studied in detail the polarization anisotropy of the ED and MD transitions of Eu^{3+} ions in the monoclinic $KY(WO_4)_2$ crystal.

2. Experimental

The studied crystal, europium-doped potassium yttrium double tungstate, 2 at.% $Eu:KY(WO_4)_2$, was grown from the flux by Dr. Anatoly Pavlyuk. The so-called Low Temperature Gradient (LTG) method was used. Potassium ditungstate ($K_2W_2O_7$) was used as a solvent. An undoped seed was oriented along the [010] axis. The thermal gradients in the flux were below 0.1 °C/cm. More details can be found elsewhere [10]. A photograph of the as-grown crystal is shown in Figure 1.

Figure 1. Photograph of the 2 at.% $Eu:KY(WO_4)_2$ crystal, the growth direction is along the [010] axis.

$Eu:KY(WO_4)_2$ is monoclinic (sp. gr. C^6_{2h} - $C2/c$) and optically biaxial. The crystal was oriented by single-crystal X-ray diffraction. The orientation of the optical indicatrix frame is the following [11]: the N_p axis is parallel to the ***b***-axis and the N_m and N_g ones are lying in the orthogonal ***a-c*** plane, the angle $N_g \wedge \boldsymbol{c}$ is 18.5° (out of the monoclinic angle). The principal optical directions N_p, N_m and N_g are selected according to the relation for the corresponding refractive indices, $n_p < n_m < n_g$.

In the $KY(WO_4)_2$ lattice, the Eu^{3+} ions are replacing for the Y^{3+} cations in a single type of sites with C_2 symmetry and VIII-fold oxygen coordination (the corresponding ionic radii are R_Y^{VIII} = 1.019 Å and R_{Eu}^{VIII} = 1.066 Å [12]). This coordination depicts a square antiprism polyhedron.

Polarized luminescence spectra of Eu^{3+} ions were measured at room temperature (RT, 293 K) using a confocal Raman microscope (InVia, Renishaw) equipped with an Ar^+ ion laser as an excitation source (λ_{exc} = 488.0 nm), a ×50 Leica microscope objective, a 1800 l/mm grating, a CCD detector, an edge filter and a set of a polarizer and an analyser. The spectral resolution was ~1 cm^{-1}.

3. Results and discussion

An unpolarised overview luminescence spectrum of Eu^{3+} ions was measured as shown in Figure 2. The observed emission bands are due to the $^5D_0 \rightarrow {}^7F_J$ transitions, at 581 nm ($J = 0$), 590-596 nm ($J = 1$), 613-627 nm ($J = 2$), 649-658 nm ($J = 3$), 694-707 nm ($J = 4$), 742-764 nm ($J = 5$) and 803-840 nm ($J = 6$). No spectroscopic evidences of Eu^{2+} species were found. The measured luminescence lifetime of the $^5D_0$ state was 430 μs [10]. The $^5D_0 \rightarrow {}^7F_2$ transition dominated in the spectrum and it determined the red colour of the luminescence (CIE 1931 colour coordinates, $x = 0.670$, $y = 0.329$). The dominant emission wavelength λ_d was calculated to be 613 nm and the colour purity p exceeded 98%.

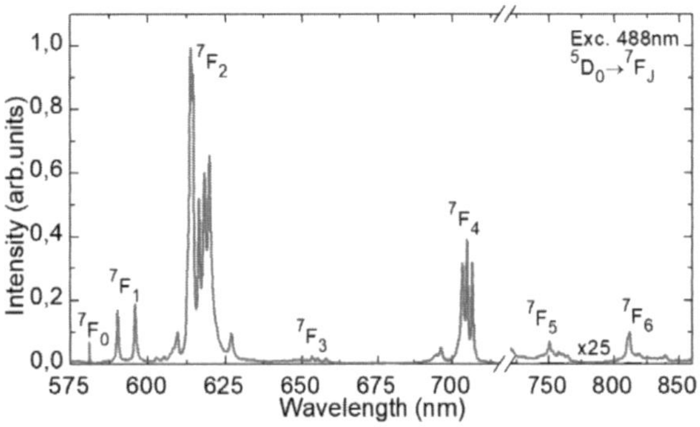

Figure 2. Unpolarized overview luminescence spectrum of Eu^{3+} ions in $KY(WO_4)_2$, $\lambda_{exc} = 488$ nm.

The polarization anisotropy of the emission properties is of particular interest for transitions suitable for laser operation, namely $^5D_0 \rightarrow {}^7F_2$ (red emission) and $^5D_0 \rightarrow {}^7F_4$ (deep-red emission). Both are of pure ED nature. First, polarized luminescence spectra were measured for $\boldsymbol{E} \parallel N_p$, N_m and N_g. Then, stimulated-emission (SE) cross-sections, σ_{SE}, were calculated by the Füchtbauer-Ladenburg (F-L) formula [13]:

$$\sigma_{SE}^{i}(\lambda) = \frac{\lambda^5}{8\pi\langle n\rangle^2 c\tau_{rad}} \frac{W_i(\lambda)B(JJ')}{(1/3)\sum_{j=p,m,g}\int \lambda W_j(\lambda)d\lambda}, \quad (1)$$

where, the i, $j = p$, m, g indices mark the polarization state, λ is the light wavelength, $\langle n\rangle$ is the mean refractive index, c is the speed of light, $\tau_{rad} = 464$ μs [10] is the radiative lifetime of the emitting state ($^5D_0$), $B(JJ') = 83.7\%$ ($0 \rightarrow 2$) and 12.8% ($0 \rightarrow 4$) [10] are the luminescence branching ratios for the J → J' transitions and $W_i(\lambda)$ is the luminescence spectrum for the i-th polarization. The $B(JJ')$ and τ_{rad} values were calculated previously using the Judd-Ofelt theory [10].

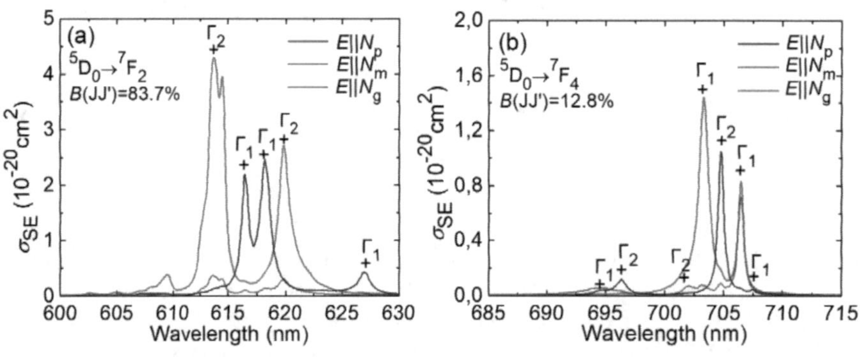

Figure 3. Stimulated-emission (SE) cross-section, σ_{SE}, spectra for electric-dipole (ED) transitions (**a**) $^5D_0 \rightarrow {}^7F_2$ and (**b**) $^5D_0 \rightarrow {}^7F_4$ of Eu^{3+} ions in the monoclinic $KY(WO_4)_2$ crystal for light polarizations $\boldsymbol{E} \parallel N_p$, N_m and N_g. "+" marks electronic transitions, $\Gamma_{1(2)}$ – irreducible representations corresponding to sub-levels of the ^7F_J multiplets.

The σ_{SE} spectra are shown in Figure 3(a,b). They are strongly polarized. For the $^5D_0 \rightarrow {}^7F_2$ transition falling in the red spectral range, the highest SE cross-sections correspond to light polarization $\boldsymbol{E} \parallel N_m$. The two prominent emission peaks are at 613.7 nm (4.31×10^{-20} cm^2) and 619.8 nm (2.72×10^{-20} cm^2). For the former broader and more intense peak, the emission bandwidth (full width at half maximum, FWHM) is 1.6 nm. Another favorable light polarization is $\boldsymbol{E} \parallel N_p$. Again, two intense emission peaks are observed at 616.5 nm (2.20×10^{-20} cm^2) and 618.2 nm (2.46×10^{-20} cm^2). The principal polarization $\boldsymbol{E} \parallel N_g$ corresponds to much lower SE cross-sections and it is less interesting for laser applications.

For the $^5D_0 \rightarrow {}^7F_4$ transition corresponding to deep-red emissions, the highest SE cross-sections also correspond to light polarization $\boldsymbol{E} \parallel N_m$. Two intense emission peaks at 703.3 nm (1.44×10^{-20} cm^2) and 706.5 nm (0.83×10^{-20} cm^2) are observed; the emission bandwidth (FWHM) of the former peak is 1.0 nm. For $\boldsymbol{E} \parallel N_p$, there is only one intense peak at 704.8 nm (1.05×10^{-20} cm^2).

For Eu^{3+} ions in C_2 sites, every $^{2S+1}L_J$ multiplet (with integer J) is split into a total of $2J + 1$ Stark sub-levels with irreducible representations Γ_1 and Γ_2, cf. Table 1.

Table 1. Expected number of Stark sub-levels for Eu^{3+} ions in C_2 sites and the corresponding irreducible representations.

$^{2S+1}L_J$ (Eu^{3+})	Γ $(2J + 1)$	$^{2S+1}L_J$ (Eu^{3+})	Γ $(2J + 1)$
$^7F_0$	Γ_1 (1)	$^7F_4$	$5\Gamma_1+4\Gamma_2$ (9)
$^7F_1$	$2\Gamma_1+\Gamma_2$ (3)	$^7F_5$	$5\Gamma_1+6\Gamma_2$ (11)
$^7F_2$	$3\Gamma_1+2\Gamma_2$ (5)	$^7F_6$	$7\Gamma_1+6\Gamma_2$ (13)
$^7F_3$	$3\Gamma_1+4\Gamma_2$ (7)	$^5D_0$	Γ_1 (1)

Table 2. Polarization dependent selection rules for ED J $\rightarrow$ J' transitions of Eu^{3+} ions in the monoclinic $KY(WO_4)_2$ crystal, C_2 – 2-fold symmetry axis, p, m, g – polarization states.

J $\rightarrow$ J'	Γ_1	Γ_2
Γ_1	$\boldsymbol{E} \parallel C_2$ (p)	$\boldsymbol{E} \perp C_2$ (m, g)
Γ_2	$\boldsymbol{E} \perp C_2$ (m, g)	$\boldsymbol{E} \parallel C_2$ (p)

The polarization selection rules for ED J $\rightarrow$ J' transitions of Eu^{3+} ions are listed in Table 2. For the monoclinic $KY(WO_4)_2$ crystal, the point group is $2/m$. The 2-fold symmetry axis (C_2) is parallel to the crystallographic ***b***-axis ($C_2 \parallel N_p$). It is orthogonal to the mirror plane coinciding with the ***a-c*** plane and, consequently, containing the N_m and N_g axes. The polarization selection rules are different for light polarized along the 2-fold axis ($\boldsymbol{E} \parallel N_p$) and orthogonal to it ($\boldsymbol{E} \parallel N_m$, N_g). This agrees well with the observed polarization anisotropy of ED transitions of Eu^{3+} ions, cf. Figure 3.

Let us now discuss the $^5D_0 \rightarrow {}^7F_{0,1}$ transitions of Eu^{3+} ions in $KY(WO_4)_2$ which correspond to the "structural probe" role of these species. For the ED and MD forbidden $^5D_0 \rightarrow {}^7F_0$ transition, a single narrow emission line was observed at 581.3 nm (17203 cm^{-1}) for light polarization $\boldsymbol{E} \parallel N_p$, Figure 4(a). This agrees with a single type of sites for Eu^{3+} ions. As both the emitting and the terminal levels for this transition are non-degenerated and the irreducible representation corresponding to the single Stark sub-level is Γ_1 (Table 1), according to the polarization selection rules (cf. Table 2), it can be observed only for light polarized along the 2-fold symmetry axis (i.e., along the N_p-axis).

The polarized luminescence spectra corresponding to the $^5D_0 \rightarrow {}^7F_1$ Eu^{3+} transition are shown in Figure 4(b). In our experiments, it was observed that the relative intensity and position of the emission lines depend not only on the polarization of luminescence $\boldsymbol{E}$, but also on its propagation direction $\boldsymbol{k}$ (like for Raman spectra). To clarify this, Porto's notations were used: $\boldsymbol{k}_{exc}(\boldsymbol{E}_{exc}\boldsymbol{E}_{lum})\boldsymbol{k}_{lum}$, where $\boldsymbol{k}_{exc}$ and $\boldsymbol{k}_{lum}$ are the directions of propagation of the excitation light and luminescence, respectively (in our case they correspond to the same direction in the crystal because of the confocal geometry), and $\boldsymbol{E}_{exc}$ and

$\boldsymbol{E}_{lum}$ are the polarization states of the excitation light and luminescence, respectively. In Figure 4(b), the spectra are grouped according to the polarization state of luminescence ($\boldsymbol{E}_{lum}$, *p*, *m* or *g*).

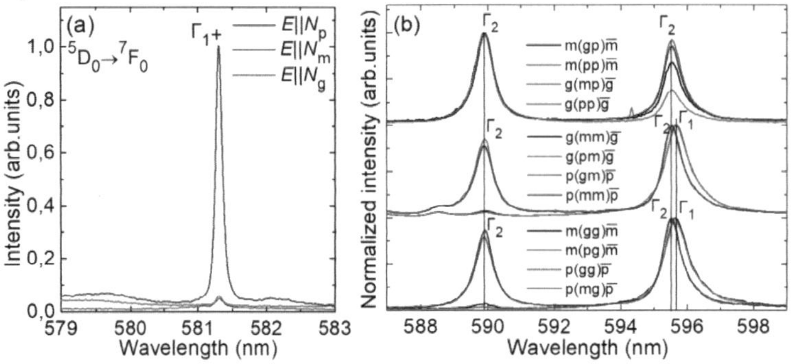

Figure 4. Polarization anisotropy of emission corresponding to the $^5D_0 \rightarrow {}^7F_{0,1}$ transitions of Eu^{3+} ions in $KY(WO_4)_2$: **(a)** ED and MD forbidden $^5D_0 \rightarrow {}^7F_0$ transition; **(b)** MD allowed transition $^5D_0 \rightarrow {}^7F_1$, Porto's notations are used. "+" marks electronic transitions, $\Gamma_{1(2)}$ – irreducible representations corresponding to sub-levels of the ^7F_J multiplets.

Similar behavior was described and interpreted previously for MD transitions in emission of rare-earth ions in optically uniaxial (e.g., tetragonal) and optically biaxial (e.g., orthorhombic) crystals [14]. It has never been observed for monoclinic crystals. A MD transition is caused by an interaction of an active ion (Eu^{3+}, in our case) with the magnetic field component of the light through the magnetic dipole. The orientation of the latter is determined by the vector $\boldsymbol{M}$. Thus, the orientation of the magnetic field vector $\boldsymbol{H}$ of radiation with respect to $\boldsymbol{M}$ becomes relevant.

According to Table 1, the $^5D_0$ multiplet consists of a single sub-level (Γ_1). The $^7F_1$ multiplet contains three sub-levels, denoted as Γ_1 and $\Gamma_2^{(1)}$, $\Gamma_2^{(2)}$. The probability of a MD transition in emission is proportional to $|\boldsymbol{e}_{\rho k}\cdot[\boldsymbol{k}\times\langle f|\boldsymbol{M}|i\rangle]|^2$, where $\boldsymbol{e}_{\rho k}$ is the polarization vector, $\boldsymbol{k}$ is the Poynting vector and *f* and *i* indices mark the initial (*i*) and final (*f*) Stark sub-levels [14]. The vector product in square brackets is maximized when $\boldsymbol{k} \perp \boldsymbol{M}$ and the whole mixed product is maximized when all three vectors are mutually orthogonal. From these considerations, polarization selection rules for the MD $^5D_0 \rightarrow {}^7F_1$ transition are obtained, see Table 3.

Table 3. Polarization selection rules for the pure MD $^5D_0 \rightarrow {}^7F$ transition of Eu^{3+} ions in C_2 sites in the monoclinic $KY(WO_4)_2$ crystal.

$\boldsymbol{M}$ vector	$^5D_0 \rightarrow {}^7F_1$	Preferred $\boldsymbol{k}$	Possible $\boldsymbol{e}_{\rho k}$
$\boldsymbol{M} \parallel C_2$ ($\boldsymbol{M} \parallel p$)	$\Gamma_1 \rightarrow \Gamma_1$	$\parallel g$	$\parallel m$
		$\parallel m$	$\parallel g$
$\boldsymbol{M} \perp C_2$ ($\boldsymbol{M} \parallel g, m$)	$\Gamma_1 \rightarrow \Gamma_2^{(1)},\Gamma_2^{(2)}$	$\parallel p$	$\parallel g, m$
		$\parallel g$	$\parallel p, m$
		$\parallel m$	$\parallel p, g$

Let us analyze Figure 4(b) using Table 3. All the four spectra from the upper group (*m*(*gp*)*m*, etc.) correspond to the $\Gamma_1 \rightarrow \Gamma_2^{(1)},\Gamma_2^{(2)}$ selection rule (two emission lines are observed). For the intermediate group, two spectra (*g*(*mm*)*g* and *g*(*pm*)*g*) correspond to the selection rule $\Gamma_1 \rightarrow \Gamma_1$ (one line) the other two (*p*(*gm*)*p* and *p*(*mm*)*p*) – to the $\Gamma_1 \rightarrow \Gamma_2^{(1)},\Gamma_2^{(2)}$ one (two lines). Finally, for the lower group of spectra, two of them (*m*(*gg*)*m* and *m*(*pg*)*m*) correspond to the selection rule $\Gamma_1 \rightarrow \Gamma_1$ (one line) the other two (*p*(*gg*)*p* and *p*(*mg*)*p*) – to the $\Gamma_1 \rightarrow \Gamma_2^{(1)},\Gamma_2^{(2)}$ one (two lines). Thus, the measured spectra well agree with the constructed table of polarization selection rules for MD transitions.

The analysis of the $^5D_0 \rightarrow {}^7F_J$ transitions in emission of Eu^{3+} ions in $KY(WO_4)_2$ allowed us to construct the energy-level scheme. It is shown partially in Figure 5 focusing on the $^7F_{0\text{-}2}$ and $^5D_0$ states, as well as the corresponding emission wavelengths.

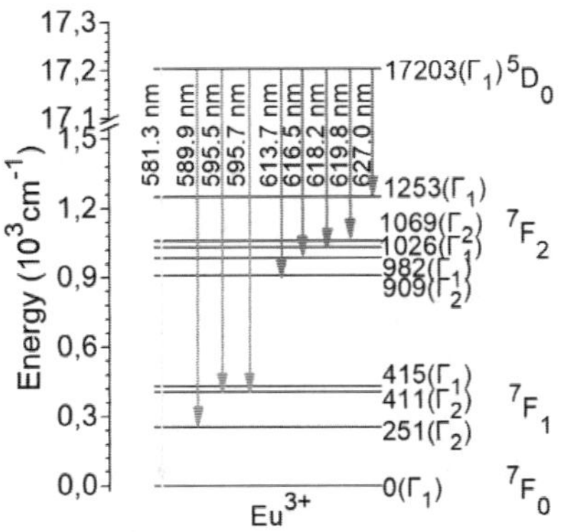

Figure 5. Crystal-field splitting of $^5D_0$ and $^7F_{0\text{-}2}$ Eu^{3+} multiplets for the monoclinic Eu:KY$(WO_4)_2$ crystal, *arrows* indicate the transitions in emission.

4. Conclusion

Eu^{3+} ions embedded in low-symmetry sites (C_2) in monoclinic crystals such as KY$(WO_4)_2$ exhibit strong polarization anisotropy of their emission properties. The nature of the $^5D_0 \rightarrow {}^7F_J$ electronic transitions (ED or MD) has a direct consequence on the polarization selection rules. The anisotropy of the emission due to the pure ED transitions terminating at the $^7F_2$ and $^7F_4$ states is determined by the orientation of the ***E*** vector with respect to the 2-fold symmetry axis of the crystal. These transitions are of interest for laser emission in the red and deep-red ranges. The anisotropic emission properties corresponding to the pure MD transition terminating at the $^7F_1$ state are determined by the orientation of the ***H*** vector. The description of this transition is important for the "structural probe" analysis.

Acknowledgments

The reported study was funded by RFBR, project number 19-32-90199. It was also supported by Spanish Government, Ministry of Science and Innovation (project No. PID2019-108543RB-I00) and by Generalitat de Catalunya (project No. 2017SGR755).

References

[1] Huang X, Guo H, Li B 2017 *J. Alloys Compd.* **720** 29
[2] Zhu H, Fang M, Huang Z, Liu Y G, Chen K, Min X, Mao Y, Wang M 2016 *J. Lumin.* **172** 180
[3] Kim K B, Kim Y I, Chun H G, Cho T Y, Jung J S, Kang J G 2002 *Chem. Mater.* **14** 5045
[4] Binnemans K, and Görller-Walrand C 1996 *J. Rare Earths* **14** 173
[5] Loiko P A, Rachkovskaya G E, Zakharevich G B, Kornienko A A, Dunina E B, Yasukevich A S, Yumashev K V 2014 *J. Non-Cryst. Solids* **392** 39
[6] Dashkevich V, Bagayev S, Orlovich V, Bui A, Loiko P, Yumashev K, Yasukevich A, Kuleshov N, Vatnik S, Pavlyuk A 2015 *Laser Phys. Lett.* **12** 085001
[7] Demesh M, Yasukevich A, Kisel V, Dunina E, Kornienko A, Dashkevich V, Orlovich V, Castellano-Hernández E, Kränkel C, Kuleshov N 2018 *Opt. Lett.* **43** 2364
[8] Bagayev S, Dashkevich V, Orlovich V, Vatnik S, Pavlyuk A, Yurkin A 2011 *Quantum Electron.* **41** 189
[9] Loiko P, Rytz D, Schwung S, Pues P, Jüstel T, Doualan J-L, Camy P 2021 *Opt. Lett.* **46** 2702
[10] Loiko P, Dashkevich V, Bagaev S, Orlovich V, Yasukevich A, Yumashev K, Kuleshov N, Dunina E, Kornienko A, Vatnik S, Pavlyuk A 2014 *J. Lumin* **153** 221
[11] Loiko P A, Yumashev K V, Kuleshov N V, Pavlyuk A A 2012 *Appl. Phys. B* **106** 663
[12] Shannon R D 1976 *Acta Cryst. A* **32** 751
[13] Aull B F, Jenssen H P 1982 *IEEE J. Quantum Electron.* **18** 925
[14] Volokitina A, Loiko P, Vilejshikova E, Mateos X, Dunina E, Kornienko A, Kuleshov N, Pavlyuk A 2018 *J. Alloys Compd.* **762** 786

Design and Simulation of the Two-Axis Micromachined Angular Rate Sensor

I E Lysenko, M A Denisenko and A S Isaeva

Southern Federal University, Taganrog, 347922, Russia

dema@sfedu.ru

Abstract. Micromechanical inertia sensors - accelerometers, gyroscopes, multisensor modules and systems based on them - are widely used in navigation, for compensation of other instruments (accelerometers, inclinometers) or stabilization (gyroscopes). The paper presents the designed construction of a MEMS angular rate sensor with two sensitivity axes, topology of gyroscope is presented; modal and static analysis is performed using ANSYS CAD; simulation results of micromechanical gyroscope operation under the action of angular velocities using VHDL-AMS are presented.

1. Introduction

In many segments of commercial MEMS s market inertial and pressure sensors dominate. In part, this is due to the existing problems. Inertia sensors – accelerometers, gyroscopes, multisensor modules and systems based on them - are widely used in navigation, for compensation of other instruments (accelerometers, inclinometers) or stabilization (gyroscopes). Inertial Measurement Units (IMU) are one of the main applications of inertial sensors in a MEMS design [1-10].

Recently known micromechanical sensors provide possibility to measure linear accelerations and angular velocities by single or double axes. These devices are manufactured in the form of hybrid or integrated microsystems [9-12].

2. Design and simulation

In this work its authors investigated LL-type inertial micromechanical sensor, which structure is presented at figure 1.

Micromechanical gyroscope contains substrate 1, stationary electrodes of capacitive displacement transducers (CDT) 2-11, stationary comb electrodes of electrostatic drives (ESD) 12-23, mobile comb electrodes of electrostatic drives 24-27, spring beams 28-35, 44-47, 52-71, supports 36-41, 72-90, inertia masses 42 and 43, mobile electrodes of capacitive displacement transducers 48-51. Displacements of sensors along Y axis is performed in movement mode, and along X and Z axes – in sensibility mode.

In such a way, developed micromechanical device represents integrated micromechanical gyroscope, providing opportunity to measure values of angular velocity along X axis, located in substrate plane, and Z axis, directed normally to substrate plane of the device.

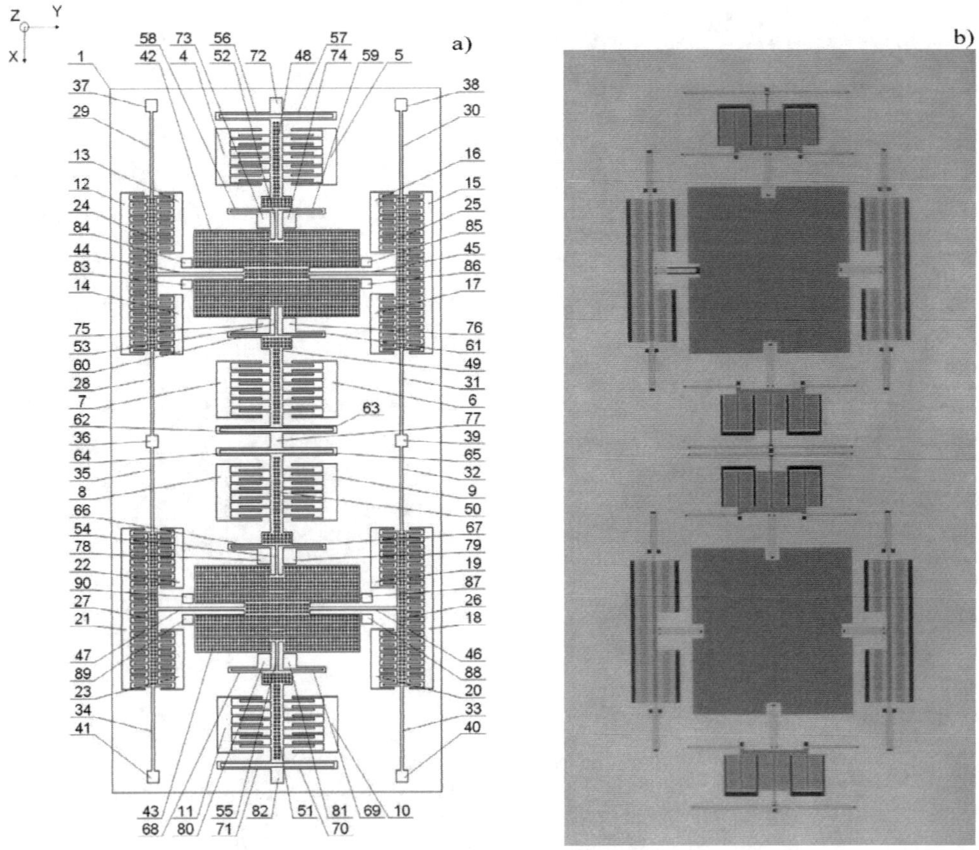

Figure 1. Integrated micromechanical gyroscope with two sensitive axes: a) scheme, b) topology.

The main feature of the developed angular velocities and linear accelerations sensor is that elastic suspensions, inertia mass, mobile and stationary comb electrodes of electrostatic drives, stationary electrodes of capacitive displacement transducers are all made in a single structural layer. Using two suspensions allows to exclude mutual influence of primary and secondary oscillations of MMG sensors.

Within this investigation geometrical and finite-elements models of a single sensor of proposed gyroscope with two sensitive axes were developed. In accordance with proposed simulation method, calculations of natural modes and oscillation frequencies of gyroscope sensor were performed.

Figure 2 presents the geometrical and numerical models and results of micromechanical gyroscope sensor modal analysis along sensitive axes performed in ANSYS CAD.

Modal analysis of micromechanical gyroscope sensor along axis of motion and statical analysis of micromechanical device under the action angular velocities along sensitivity axes also were obtained by using of ANSYS CAD.

Figures 3-5 present the simulation results of micromechanical gyroscope operation under the action of angular velocities performed in hAMSter software on the basis of the developed parametrized high level description language VHDL-AMS.

As can be seen at the figure 3, duration of linear acceleration action constitutes 100 ms, rising and falling edge duration constitutes 30 ms.

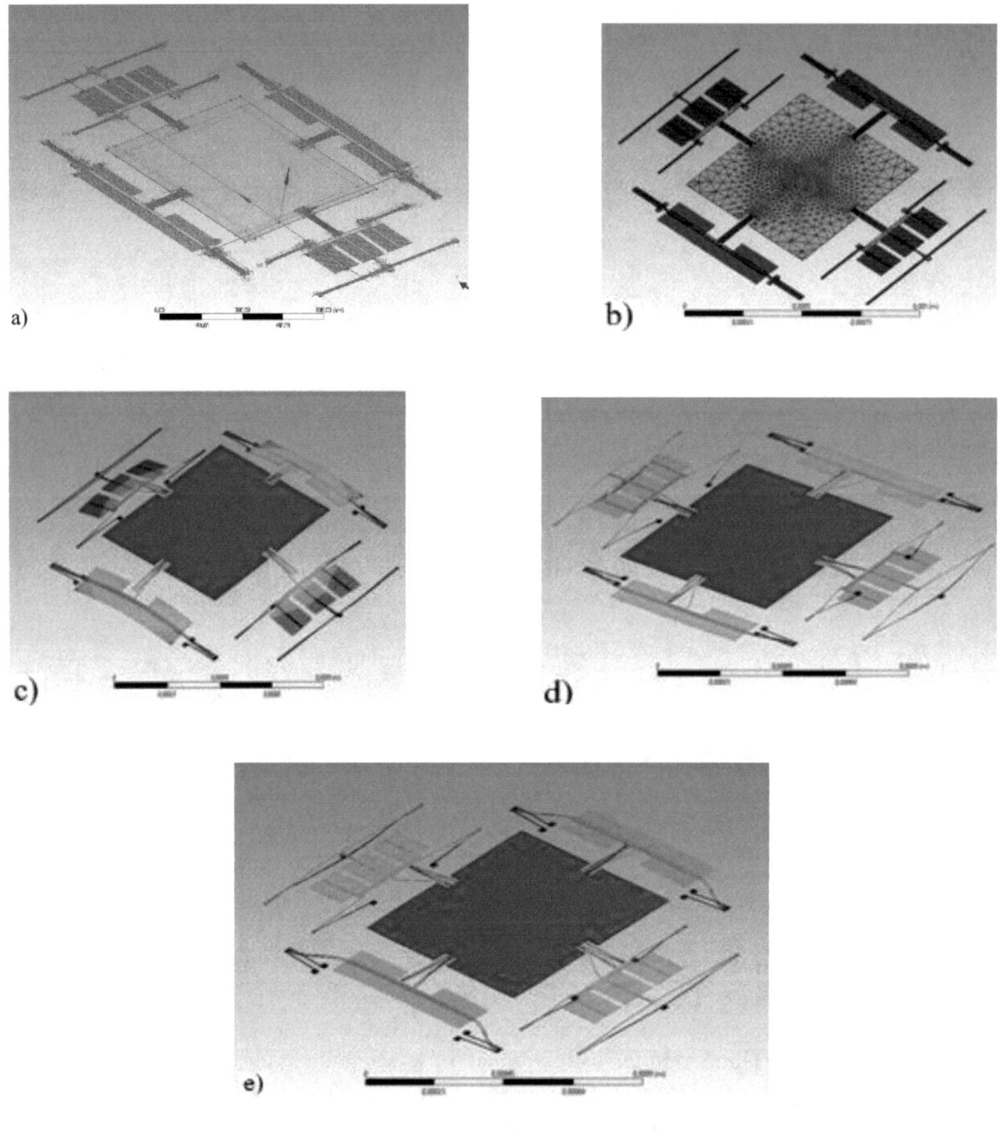

Figure 2. Second natural mode of gyroscope sensitive element: a) geometrical model, b) numerical model, c) corresponding to the mode of sensitivity by axis Z (2.82 kHz), d) corresponding to the mode of sensitivity by axis X (9.2 kHz), e) corresponding to the mode of drive by axis Y (9.2 kHz).

As it was shown by simulation results, initial capacity of displacement transducers along X axis constitutes 90 pF, and along Z axis – 2.2 pF. Under the action of angular velocities differential capacity alteration along sensitivity axis X constitutes 25 pF at 500 degrees/s, along sensitivity axis Z – 0.6 pF at 500 degrees/s.

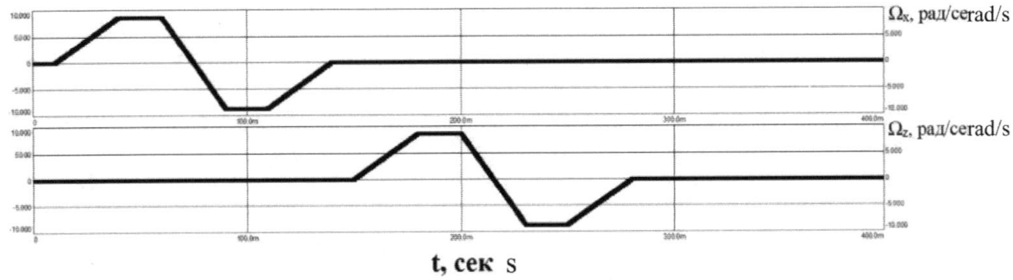

Figure 3. Measurement of angular velocities in the whole dynamical range.

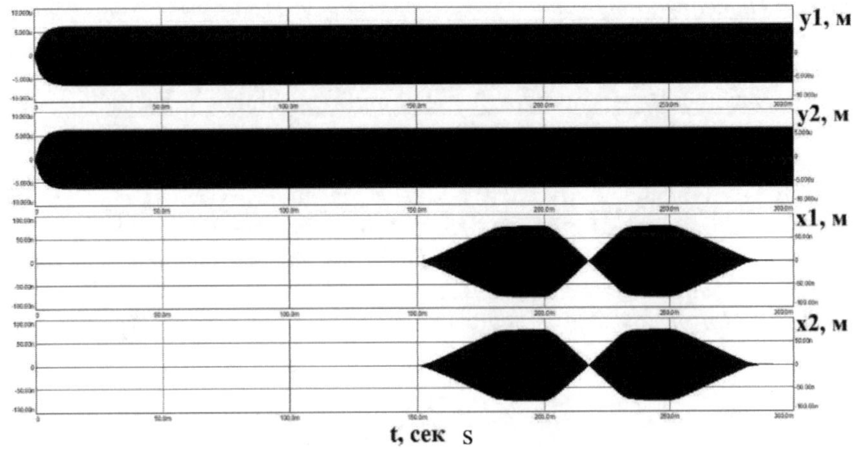

Figure 4. Displacement of gyroscope sensitive element along X, Y, Z axes under the action electrostatic forces and angular velocities.

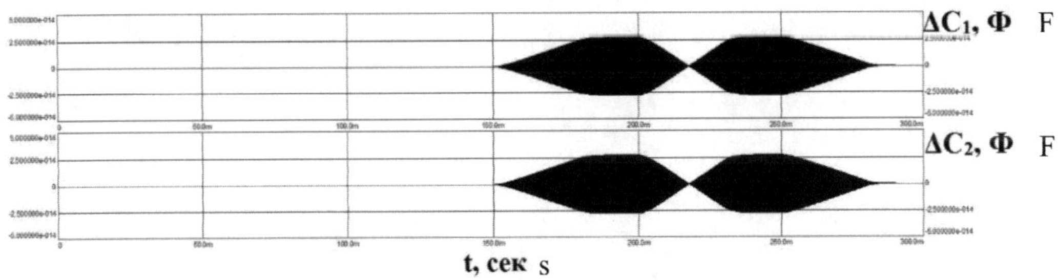

Figure 5. Capacities measurement of gyroscope displacement capacitive transducers.

3. Conclusion

From the results of performed investigations it is clear, that the proposed construction of elastic suspensions of sensitive element of angular velocities micromechanical sensor in the mode of sensitivity along X and Z axes and in the mode of motion along Y axis with sensitivity and capacitance indicators, does not accede to analogues.

Acknowledgments
This work was financially supported by the grant of President of Russian Federation (project number MK-2130.2020.8).

References

[1] Raspopov V Ya 2007 *Micromechanicheskie pribory: uchebnoye posobie* (Tula: Tulskiy gos. Universitet)
[2] Ramadoss R 2013 *MEMS devices for biomedical applications* Solid State Tech. **10** 10–3
[3] Konoplev B G, Lysenko I E and Ezhova O A 2016 *Criteria of equality of modal frequency of micromechanical gyroscopes-accelerometers sensitive elements* Modern Appl. Sci. **10** 52–5
[4] Konoplev B G, Lysenko I E and Ezhova O A 2015 *Evolution Criteria Fingers Hardness Electrode MEMS Comb Converters* Bioscience Biotech. Res. Asia **12** 2345–50
[5] Prokofiev I V and Tikhonov R D 2011 *Nano- i microswistemy dlia monitoringa parametrov dvijeniya transportnykh sredstrv* Nano- i Microsys. Tech **12** 48–50
[6] Pristupchik N K, Konoplev B G, Kulikova I V, Klunnikova Yu V and Lysenko I E 2015 *Quantum Tunneling in Mesoscopic Electromechanical Transducers* Mod. Appl. Sci. **9** 20–9
[7] Sysoeva S 2010 *Kliuchevye segmenty rynka MEMS-komponentov. Inertsyonnye systemy — ot Low-End do High-End* Komp. i tech. **5** 15–9
[8] Anchurin S, Maximov V N, Morozov S, Golovan S and Shilov V F 2011 *Blok inertsyonnykh datchikov* Nano- i microsys. Tech. **1** 50–3
[9] Elwenspoek M and Wiegerink R 2005 *Mechanical microsensors* (New York: Springer).
[10] Konoplev B G, Ryndin and Lysenko I 2007 *Elementnaya baza nanoelectroniki, micro- i nanosistremnoy techniki* (Tatganrog: Izdatelstvo Technologicheskogo instituta YuFO).
[11] Lysenko I E 2013 *Modeling of the micromachined angular rate and linear acceleration sensors LL-type with redirect of drive and sense axis* World Appl. Sci. Journal **27** 759–62
[12] Lysenko I 2010 *Integralnyi sensor uglovykh skorostei I nineinykh uskoreniy* Injenernyi vestnik Dona(2010)3

Determination of measurement fidelity for a superconducting photon-number resolving detector with micron-wide strips

M Dryazgov[1, 2], N Simonov[2], Yu. Korneeva[2] and A Korneev[1, 2]

[1] Higher School of Economics - National Research University, 101000, Russia
[2] Department of Physics, Moscow State Pedagogical University, 119992, Russia

mdryazgov@hse.ru

Abstract. We report a study of multiphoton detection fidelity (or accuracy) for sequential photon-number resolving detectors based on micron-wide superconducting strips. It was found that an increase in the width of the superconducting strips by a factor of 5 leads to an improvement in the measurement accuracy by more than a factor of 10.

1. Introduction

Photon-number resolving detectors (PNR detectors) [1, 2] based on superconducting strips have found applications in the areas where it is necessary not only to validate the fact of photon detecting, but to find out the number of photons: quantum computers [3] and quantum cryptography [4].

The standard architecture of a PNR detector consists of several superconducting single-photon detectors (SSPD) [5] connected either in series with parallel resistors or in parallel with series resistors. Distinguishing the number of photons occurs using spatial multiplexing, when several sections are triggered simultaneously by several photons, and the voltage pulses from the sections are summed up resulting in the amplitude proportional to the number of photons.

For a PNR detector with a series connection of sections (figure 1), the maximum pulse height is limited by the value of the bias current and does not depend on the total number of sections. At a low bias current, as in samples with a superconducting strip width of 100 nm, and a large number of sections, the pulses are difficult to distinguish due to thermal noise. As a consequence, when i photons are absorbed actual voltage may be larger or smaller due to thermal noise, and resulting amplitude will correspond to i+1 or i-1 photon pulse.

Here we address the issue of fidelity (or accuracy) of the photon number identification: due to electric noise the resulting amplitude may be smaller or larger, thus giving the wrong measurement result. We consider the detection fidelity as the probability to detect exactly i photons, when i sections are triggered, when the internal quantum efficiency of the detector is 100%.

The following is a description of the method for determining detection fidelity and its comparison for a PNR detector with a series number of sections with a strip width of 0.5 μm and 100 nm.

2. Measurement procedure

Voltage pulses are obtained for a serial PNR detector with a strip width of 0.5 μm, 7 sections, the length of one section is 240 μm, and the total area is 35x30 μm^2. The pulses pass through 2 Mini-

Circuits ZFL-1000LN amplifiers and are measured using a Rohde&Schwarz RTO1012 digital oscilloscope (figure 2).

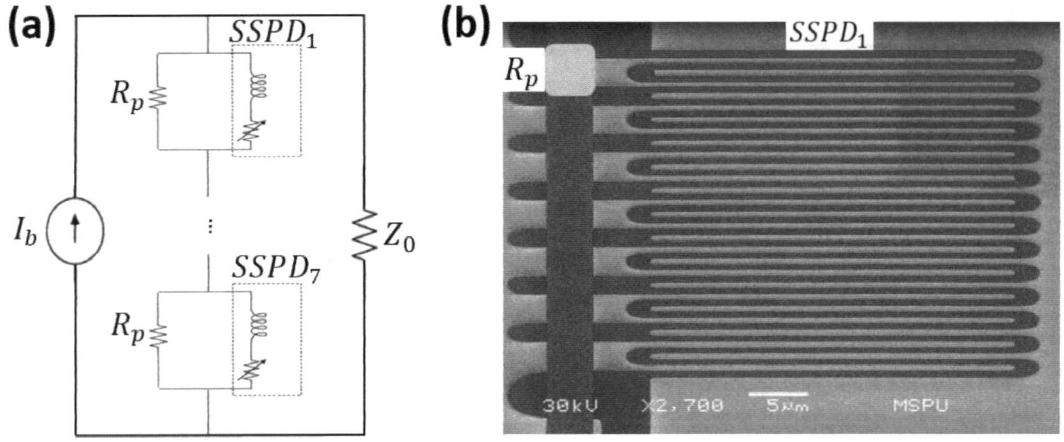

Figure 1. (a) A simplified electrical scheme of the PNR with 7 series connection of sections. Here I_b is bias current, R_p is parallel resistors, Z_0 is resistance of the coaxial line. **(b)** SEM image of PNR detector, the width of the superconducting strips is 0.5 μm. One section and the corresponding resistor are highlighted in color.

We obtain a histogram of the frequencies for the occurrence of a sample of a certain voltage in the noise waveform. This histogram represents the normal distribution (figure 2 (in the inset)). Using the normal distribution formula for relative frequencies, the probability function of detecting noise with amplitude U_{noise} is obtained:

$$P(U_{noise}) = \frac{1}{\sigma\sqrt{2\pi}} \exp\left(\frac{-U_{noise}^2}{2\sigma^2}\right)$$

The mean value for the normal distribution in this case μ = 0, the deviation value σ is selected by the least squares method. The $P(U_{noise})$ values were also plotted with a solid line (figure 2 (in the inset)). The $P(U_{noise})$ values were also plotted with a solid line (figure 2 (in the inset)).

To detect the $i-th$ pulse with the amplitude U_{imp}^i, the trigger value is $U_{tr}^i = (U_{imp}^i - U_{imp}^{i-1})/2$. This means that if the noise value is $U_{noise} \leq U_{tr}^i - U_{imp}^i$, the equipment will fix $i-1$ photon. Similarly, if $U_{noise} \geq U_{tr}^{i+1} - U_{imp}^i$, the equipment will capture $i+1$ photons.

Then we define fidelity as the probability of detecting exactly i pulses, taking into account thermal noise:

$$F = P\left(U_{tr}^i - U_{imp}^i < U_{noise} < U_{tr}^{i+1} - U_{imp}^i\right) = \int_{U_{tr}^i - U_{imp}^i}^{U_{tr}^{i+1} - U_{imp}^i} \frac{1}{\sigma\sqrt{2\pi}} \exp\left(\frac{-U_{noise}^2}{2\sigma^2}\right) dU_{noise}$$

As shown in figure 2, with a superconducting strip width of 0.5 μm and 7 sections, the desired fidelity will be 100% for each pulse. But with an increase in the number of sections, the difference in heights of neighboring pulses decreases, as well as with a decrease in the width of the superconducting strips.

To assess this effect, the classical electro-thermal model [7], rewritten for an SSPD with micron-wide superconducting strips [8], was modified for a PNR-detector with a series connection of sections. The simulation result is shown in figure 2 (dashed red lines). Good agreement with the experimental result is seen.

Using this model, it is possible to obtain voltage pulses for a PNR detector with any number of sections and any width of the superconducting strip. This makes it possible to obtain the fidelity value for each voltage pulse and compare them with each other for a PNR detector based on superconducting strips 0.5 μm and 100 nm wide.

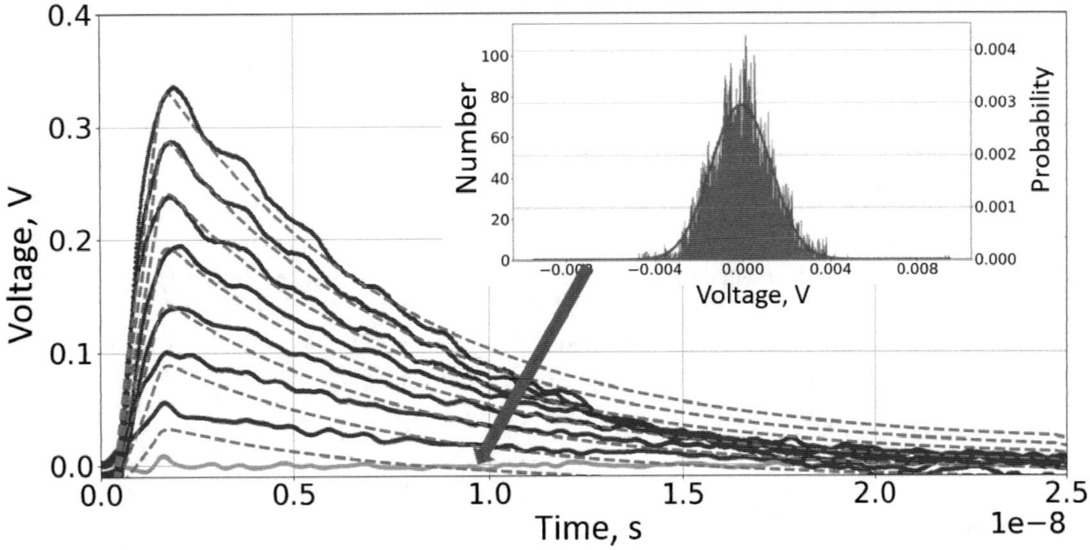

Figure 2. Experimentally obtained voltage pulses for a PNR detector with a 7 series connection of sections, the width of the superconducting strips is 0.5 μm. The noise track is highlighted in purple. Red lines are simulation voltage pulses. The inset shows a histogram of the voltage probability distribution of the noise track, the purple curve is the normal distribution of relative frequencies.

3. Results

In the course of the simulation, it was found that when using 0.5-μm-wide superconducting strips, a significant effect of noise occurs only if the total number of sections of the PNR detector exceeds 40.

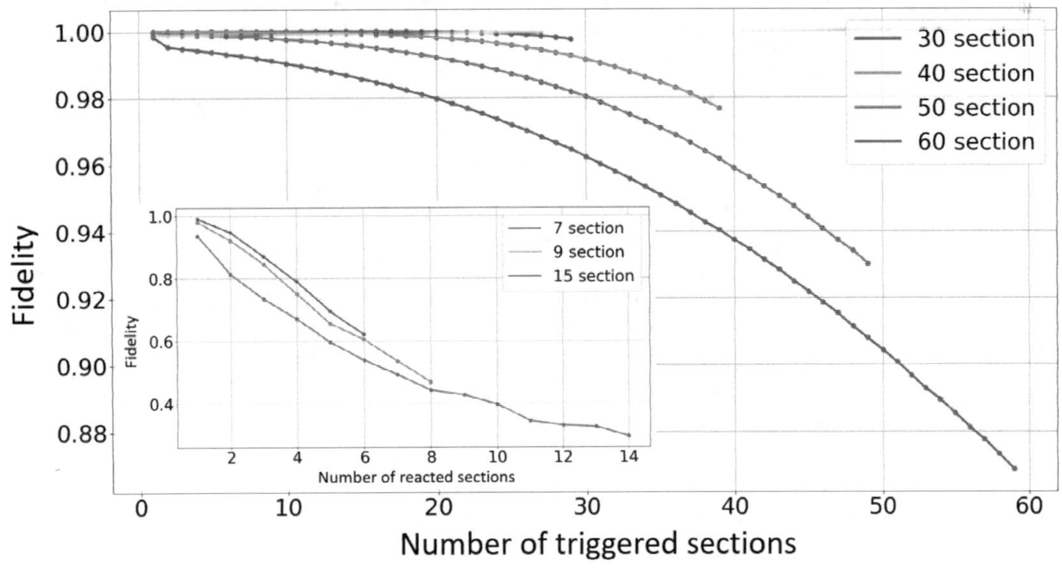

Figure 3. Dependence of fidelity on the number of triggered sections for a PNR detector with a series connection of sections with a width of superconducting strips of 0.5 μm. Inset for 100 nm stripes.

Figure 3 shows the dependence of the calculated fidelity on the number of triggered sections for a PNR detector with a different number of sections. As can be seen from the graphs, with the total number of sections less than or equal to 30, the noise does not affect the probability of detecting N voltage pulses when N sections are triggered. With an increase in the total number of sections, the difference in amplitude of neighboring impulses decreases, which causes a decrease in fidelity.

The inset in figure 3 shows similar plots for a PNR detector with a series connection of sections with a superconducting strip width of 100 nm. It is clearly seen that even with 7 sections, the fidelity is significantly reduced, and with the number of sections more than 9, it becomes less than 50%.

4. Conclusion

We demonstrate a method for determining the fidelity of detection of i-photon pulses of the PNR-detector for different widths of superconducting strips and different total number of sections. The use of micron-wide superconducting strips makes it possible to improve the fidelity of detecting voltage pulses for a PNR-detector with a series connection of sections due to higher critical current. This fact and low kinetic inductance give an advantage to micron-wide superconducting strips.

5. Acknowledgments

This work was supported by the Russian Science Foundation (RSF) Grant No. 20-12-00287.

References

[1] Mattioli1 F, Zhou Z, Gaggero A, Gaudio R, Jahanmirinejad S, Sahin D, Marsili F, Leoni R and Fiore A 2015 *Superconductor Science and Technology* **28** 104001

[2] Divochiy A, Marsili F, Bitauld D, Gaggero A, Leoni R, Mattioli F, Korneev A, Seleznev V, Kaurova N, Minaeva O, Gol'tsman G, Lagoudakis K, Benkhaoul M, Lévy F and Fiore A 2008 *Nature Photonics* **2** 302–306

[3] You L 2020 *Nanophotonics* **9** 2673–2692

[4] Cattaneo M, Paris M and Olivares S 2018 *Phys. Rev. A* **98** 012333

[5] Gol'tsman G, Okunev O, Chulkova G, Lipatov A, Semenov A, Smirnov K, Voronov B, Dzardanov A, Williams C and Sobolewski R 2001 *Applied Physics Letters* **78** 705–707

[6] Zou K, Meng Y, Xu L, Hu N, Wang Z and Hu X 2020 *Physical Review Applied* **14** 044029

[7] Yang J, Kerman A, Dauler E, Anant V, Rosfjord K and Berggren K 2007 IEEE Transactions on Applied Superconductivity **17** 581

[8] Dryazgov M, Semenov A, Manova N, Korneeva Yu and Korneev A 2020 *Journal of Physics: Conference Series* **1695** 012195

Study of the welded joint of VT1-0 titanium with 1.3343 steel

I Egorov and A Fomin

Yuri Gagarin State Technical University of Saratov, Saratov 410054, Russia

Abstract. The work describes a method for obtaining a composite structure of small plates. The resulting plates are a layered structure consisting of a substrate (1.3343 steel) and a titanium coating (VT1-0). A method of resistance welding in the open air was applied to form a layered structure. The resulting titanium-steel compound was tensile tested. The maximum force at break of the welded joint varied in the range from 1.05 to 2.17 kN.

1. Introduction

The development of modern industry is mainly aimed at creating high-tech technologies to increase productivity and reduce the cost of production in oil and gas processing, production of high-tech equipment [1]. One of the innovative directions in these areas is the introduction of technologies involving the use of new materials in the form of composites. In the manufacture of composite materials, depending on the operating parameters, it is required to optimize the technology to achieve the necessary service life.

Production of composite materials largely involves the use of combined methods [2-3]. In many cases, these methods differ a lot from each other. The materials used can be dissimilar in their purpose and mechanical properties as well. In the case of various welding methods in the obtained joints between dissimilar materials, chemical non-homogeneity and intermetallic phases may arise, thus leading to a decrease in physical and mechanical properties [5]. In this regard, an urgent issue in obtaining layered structures is the formation of a high-quality and durable connection that will ensure the reliability and high resource of the finished unit.

The use of titanium-coated steel products has great practical interest. Steel products are used in almost all industries [1,2]. Steel as a structural material has high strength, hardness and elasticity, depending on the alloying elements. However, steel products have low corrosion resistance, which reduces the field of their application and service life [6]. They are covered with various coatings and films in order to increase the latter characteristic.

The known methods of coating deposition frequently require the use of resource-intensive equipment and expensive materials for surface protection. The chemical and mechanical resistance of the resulting coatings in some cases is not high. Resistance welding conducted under certain conditions provides a strong and reliable connection between dissimilar materials [4,5]. In this work, a titanium coating (cp-Ti VT1-0) was applied on a substrate of 1.3343 tool steel (analogous to R6M5). A method of spot resistance welding was used for coating deposition.

2. Methodology

To study the tensile strength of the welded joint, samples of titanium and steel were made. The materials used were 1.3343 steel with a square cross-section with a side of 13 mm, a height of 3.5 mm and a central hole of 4.2 mm, and VT1–0 titanium in the form of a 70×20 mm plate with a thickness of 0.5 mm.

Before welding, the contact surfaces of the samples were processed with abrasive paper to remove the oxide layer and impart the required roughness parameters. Further the surfaces were degreased in 96% ethanol.

The welded joint of steel with titanium was formed using "Telwin Digital Modular 230" resistance spot welding setup. The constant parameter of resistance welding was the compression force of the electrodes of about 600 N (±10 %) and the operating voltage of 2.5 V. The power and the consumed current were the output characteristics and varied in the following ranges: P = 3.6–8.2 kW (power consumption), I = 17–39 A (consumed current), and pulse duration t = 0.1–1.2 s.

Table 1. Resistance welding modes

Sample №	Resistance welding modes			
	Compression force F, N	Pulse time t, s	Consumed current I, A	Consumed power P, kW
1	600	0.4	13	2.9
2	600	0.4	21	4.7
3	600	0.4	29	6.6
4	600	0.4	39	6.6
5	600	0.4	27	5.7
6	600	0.6	28	6.2
7	600	0.6	28	6.2
8	600	0.9	17	3.6
9	600	1.2	23	4.9

Static tensile tests were performed in "Impuls IR5282-100" software-controlled tensile testing machine. The stretching rate was chosen equal to 10 mm/s. When stretching, the maximum force was recorded. After the test, the fracture site was examined using "MBS-10" optical microscope.

3. Results

According to the results of a static tensile test, it was revealed that a stable connection was formed in almost all modes (Table 2). Depending on the applied resistance welding modes, the strength properties had significant differences. When analyzing the surface of sample No.1, the area of thermal action in the electrode contact zone was clearly visible. The surface of the sample remained unchanged and slight traces of contact were observed. When tested for rupture, the sample showed one of the highest results, destruction occurred along the metal interface. At a current I = 39 A and a pulse duration of t = 0.4 s, the surface of the titanium coated sample had macrodefects in the form of dents and metal splashes (Figure 1). This sample was destroyed before the tensile test when clamped.

Table 2. Maximum breaking force

Sample №	1	2	3	4	5	6	7	8	9
Breaking force Fm, kN	1.87	1.37	1.05	-	1.77	1.08	1.83	1.37	2.17

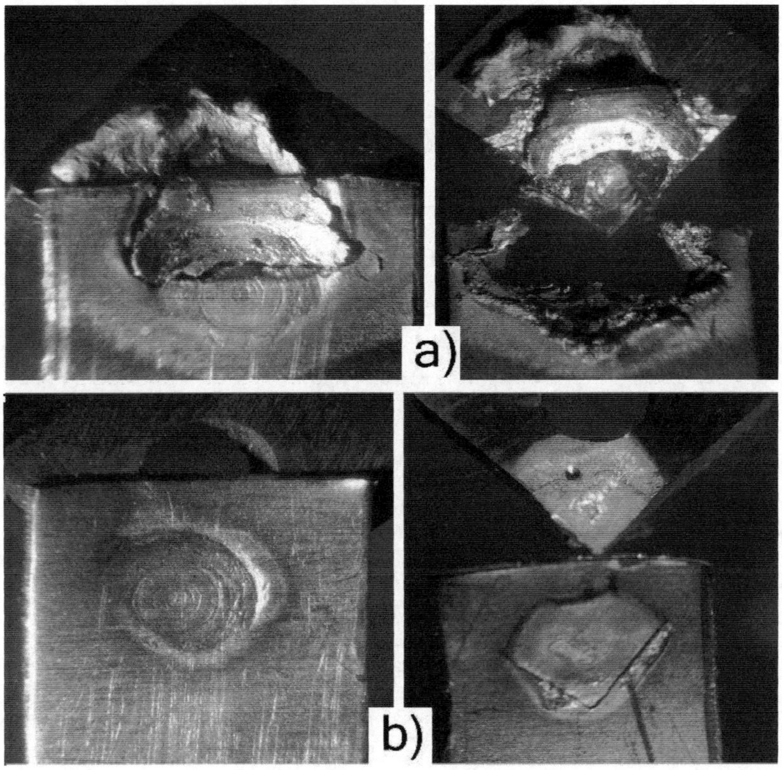

Figure 1. Samples after resistance welding and tensile tests obtained at various modes: I = 39 A, t = 0.4 s (**a**); I = 27 A, t = 0.4 s (**b**)

Sample No.5 obtained at modes I = 27 A, t = 0.4 s at rupture showed a close result to the that of sample No.1 (Fig. 1). When analyzing the surface morphology of the parts to be welded, an intense nature of the thermal effect was revealed, which led to the formation of metal splashes. The titanium surface in the contact area was deformed and a clear trace from the electrode appeared. It was destroyed along the metal interface like the rest of samples. When fractured, chipping of the steel base was also observed. This type of destruction showed that hardening had occurred in the steel base. The increase in hardness ultimately led to stresses in the steel and embrittlement. Sample No.7 with modes I = 28 A, t = 0.6 s in the tensile test showed the same result as sample No.2. However, after welding, the reference planes of the sample surfaces remained undeformed, and traces of a moderate thermal effect of the electrodes were observed. The destruction of the joint occurred along the interface. After breaking, the samples remained intact, which indicated a moderate thermal effect, which did not lead to deterioration in the strength of the metals. Sample No.9 obtained at I = 23 A, t = 1.2 s had optimal tensile strength. The sample surfaces retained their flatness after resistance welding. There was a well-defined uniform contact plane of the electrodes on the surfaces. The heat trail on the steel base was more pronounced compared to other samples. However, this did not lead to metal splashes. The nature of the fracture of the welded joint was the same as for sample No.5 (tearing off the edge part of the steel base). In this case, the steel was in a stressed state (quenching occurred), but the breaking force Fm was 2.17 kN. In samples No.2, 3, 6, 8 the tensile strength was minimal during tests. For a more detailed analysis, the tensile diagrams of the samples with best strength indexes were considered (Fig. 2).

Sample No.1 had a moderate tensile strength result in a test, while the joint was the least ductile in contrast to the other samples. Sample No.7 turned out to be the most ductile of the selected samples.

Tensile strain was maximal. There were no particularly pronounced areas in the graph, which indicated a uniform elongation without any abrupt transitions. Sample No.9 with the highest strength index had significant elongation before fracture.

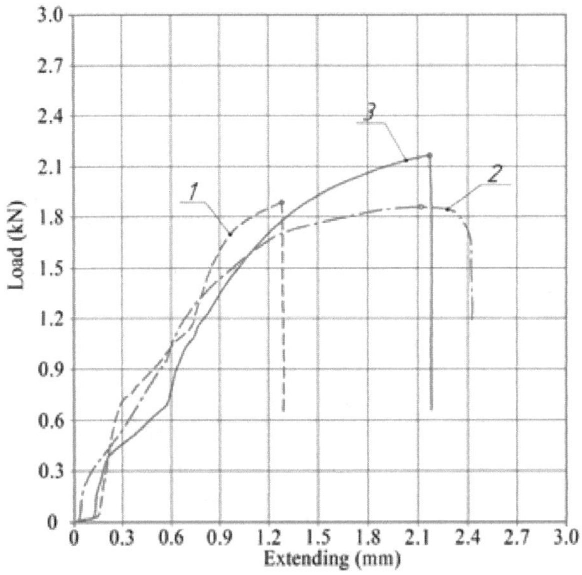

Figure 2. Tension diagrams for samples of welded joints: 1 – sample No.1; 2 – sample No.7; 3 – sample No.9.

4. Conclusions

Thus, the application of the resistance spot welding method for the formation of a welded joint between 1.3343 steel and commercially pure titanium was considered. According to the results of the study, it was determined that a pulse with a current strength of about 39 A led to the destruction of the samples. For the duration of 0.4–0.9 s and the current of about 27 A, overheating of the steel substrate and its embrittlement occurred. Static tensile tests of welded joints showed that under the modes $I = 23$ A, $t = 1.2$ s and $I = 28$ A, $t = 0.6$ s, the joint was strong enough and at the same time it had high ductility and required no subsequent heat treatment (tempering, annealing). At the mode $I = 13$ A, $t = 0.4$ s, a sufficiently strong joint with low plasticity was formed compared to other modes, which had a positive effect on the geometry of the final product.

Acknowledgments

The research was supported by the scholarship from the President of the Russian Federation (SP-571.2019.1). Part of the work (optical analysis of the surface morphology of the welded joint) was performed within the framework of the "U.M.N.I.K." program (project 15686GU/2020).

References

[1] Kuz'min, E V, Korolev M P, Lysak V I, Kuz'min S V, Zarubin M S and L'vov V A 2020 *VolgSTU.* **246** 19-23
[2] Ren D, Jiang Y, Hu X, Zhang X, Xiang X and Ling H 2021 *Intermetallics.* **132** 107115
[3] Li J, Vivek A, Daehn G 2021 *Journal of Materials Science & Technology.* **79** 191-204
[4] Egorov I, Shchelkunov A, Fomin A, Rodionov I 2018 *J. Phys. : Conf. Ser.* **1124** 081015
[5] Fomin A, Egorov I, Shchelkunov A, Fomina M, Koshuro V and Rodionov I 2018 *Compos. Struct.* **206** 467
[6] Chen, T, Ling, Z, Wang, M and Kong L 2020 Journal of Materials Processing Tech **285** 116797

Investigation of the dielectric fatigue on the example of lead titanate films $PbTiO_3$

A V Fimin, E A Pecherskaya, O A Timokhina, V S Aleksandrov, A V Volik and A E Shepeleva

Department of Information and measuring equipment and metrology, Penza State University, Penza 440026, Russia

Abstract. The phenomenon of dielectric fatigue of active dielectrics, which consists in a decrease in the residual polarization depending on the number of switching cycles, is researched. A model of the dependence of the residual polarization of ferroelectric materials on the number of switching cycles is proposed. The model is based on piecewise - linear approximation of the results of measurements of the hysteresis loops of thin films $PbTiO_3$ at a temperature T = 470 (°C), the electric field strength E = 100 (kV/cm). The developed model was used in the development of a technique for studying dielectric fatigue, depending on different modes of material switching.

1. Introduction

Piezoelectrics are characterized by the appearance of electric polarization at certain elastic deformations even in the absence of an electric field (direct piezoelectric effect). A consequence of the direct piezoelectric effect is the inverse piezoelectric effect - the appearance of mechanical deformations under the action of an electric field. In general, we are talking about a linear relationship between mechanical deformation (mechanical stress) and electrical parameters (polarization P, electric field E, electric induction D). This feature allows the use of piezoelectric materials in the sensitive elements of products of sensor equipment, detectors, transducers [1, 2].

Piezoelectric instrumentation is a narrowly focused specific field in which the most important aspect is the study of the parameters of materials, on which the characteristics and modes of operation of elements based on them depend [3].

When operating elements and devices based on piezoelectric materials, it is necessary to take into account the phenomenon of material degradation, which affects the reliability and service life of the device as a whole [4]. The operation of a product that does not take into account such an important factor contributes to obtaining unreliable experimental data.

Thus, the development of the direction for the formation of thin-film sensitive structures of sensors is conditioned not only by the expansion of the technical characteristics of devices in general, but also by the increase in the reliability of such elements. It is important to note the simplest way to improve the quality of measurements in a long time interval by introducing corrections in the measurement results.

2. Basic principles of dielectric fatigue of active dielectrics

The class of active dielectrics is massively used in various products of functional electronics, such as small-sized capacitors, non-volatile memory chips and others, since in the ferroelectric phase at certain values of the external electric field strength, these materials have high values of anisotropic relative permittivity.

Exposure to an external electric field of polarization promotes switching of subminiature domains - memory cells in which information is recorded at this moment. Therefore, an important qualitative characteristic of ferroelectric films is the number of polarization switching cycles (more than 10^{10} times) [5].
From a metrological point of view, the task is to develop the following methods:

- measurement of hysteresis loops of polarization dependences on the electric field strength $P(E)$, determination of characteristic parameters (saturation polarization P_s, residual polarization P_r, coercive field E_s);
- study of the dynamics of changes in the characteristic parameters of the hysteresis loop (their temporal instability) depending on time; depending on the number of switching cycles.

In works [6–8], the basic principles of degradation of thin ferroelectric films are considered.

3. Modeling of the mechanism of dielectric fatigue of films $PbTiO_3$

According to the developed measurement technique, using an automated setup [9], during interval switching of the polarization cycle N, a set of hysteresis loops P(E) is measured, on the basis of which the dependence of the residual polarization on the number of polarization switching cycles is formed.

The graph of the residual polarization dependence on the number of polarization switching cycles $P_r(\lg N)$ for thin films of lead titanate $PbTiO_3$ at a temperature $T = 470°C$, an electric field strength $E =$ 100 (kV / cm) is shown in Figure 1.

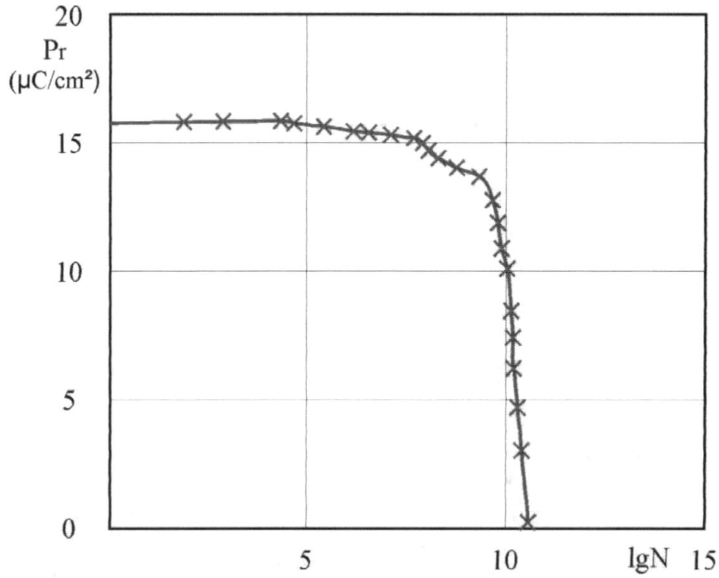

Figure 1. Dependence of the residual polarization of thin films of lead titanate on the number of switches.

The $P_r(\lg N)$ dependence shown in Figure 1 can be represented as linear sections, which is shown in Figure 2.

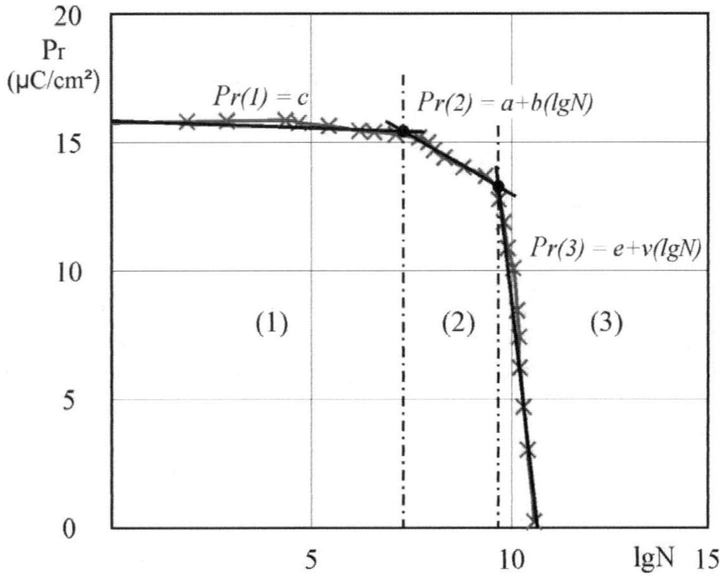

Figure 2. Linear approximation of the dependence of the residual polarization of thin films of lead titanate on the number of switchings.

The approximated three straight lines, obtained in accordance with Figure 2, can be represented in the form of the following expressions:

$$P_r = \begin{cases} c, \text{if } 0 < N \le N_1, \\ a + b \cdot \lg N, \text{if } N_1 < N \le N_2, \\ e + v \cdot \lg N, \text{if } N_2 < N \le N_3, \end{cases} \quad (1)$$

where the values of the coefficients (c, a, b, e, v) in the approximating equations of the straight lines, as well as the intervals themselves, bounded by the values N_1, N_2, are selected in accordance with the least squares method.

Relations (1) make it possible to determine the relative deviation of the residual polarization for the continuous switching mode in the time intervals $\left[\frac{N_1}{f}; \frac{N_2}{f}\right]$ at the switching frequency f according to the formula (2):

$$\delta_P(t_x) = 1 - \frac{c}{a + b \cdot \lg(f t_x)} \quad (2)$$

The maximum permissible value of the number of switchings is related to the distribution of measurement results that do not go beyond the interval of the standard deviation σ. This maximum amount is determined by the formula (3):

$$N_{max} = 10^{\frac{Pr(t_0)(1-\delta_p)-a-\Delta p-3\sigma_{max}}{b}}, \quad (3)$$

where Δ_P is the maximum absolute deviation.

With an idea of the maximum number of operations, the maximum permissible deviation can be compared with the calculated absolute value. The excess of the calculated values relative to the limiting value indicates the need for relaxation of the dielectric material parameters.

The method described above for modeling the $P_r(lg\text{N})$ dependences of active dielectrics was used to develop a technique for investigating dielectric fatigue in three modes of material switching (continuous polarization switching mode, polarization switching mode with alternating long waiting times, and a mode with an uneven polarization switching frequency).

Thus, on the basis of this model, one can calculate the dielectric fatigue of the lead titanate film $PbTiO_3$ and correct the operation of this component of the piezoelectric device. It is important to note that this technique can serve as a basis for determining the dielectric fatigue parameter for other active dielectrics.

4. Conclusion

The presented model allows for a more in-depth presentation of the design of sensitive elements formed by lead titanate films or similar active dielectrics from the composition of various products of functional electronics. The presented technique makes it possible to take into account the degradation of the piezoelectric material, which affects the service life of the sensor as a whole.

Acknowledgements

The reported study was funded by RFBR according to the research project № 20-38-90151.

References

[1] Waseem A, Johar M A, Hassan M A, Bagal I V, Abdullah A, Ha J and Lee J K and Ryu S. 2021 Flexible self-powered piezoelectric pressure sensor based on GaN/p-GaN coaxial nanowires *Journal of Alloys and Compounds* **872** 159661

[2] Esmaeel A and Ahmed K I E and Fathel-Bab A M R 2021 Determination of damping coefficient of soft tissues using piezoelectric transducer *Biomedical Microdevices* **23**(2) 23

[3] Ryabov D V, Pecherskaya E A and Shepeleva J V and Pecherskaya R M 2014 Automated method of measuring the temperature dependences of the dielectric parameters of ferroelectrics with second kind phase transition *Journal of Physics: Conference Series* **541** (1) 012012

[4] Pecherskaya E A, Solov'ev V A, Metal'nikov A M, Varenik Y A and Gladkov I M and Ryabov D V 2013 Controlling the temporal instability of the dielectric parameters of ferroelectrics. *Semiconductors* **47**(13) 1720–1722

[5] Delimova L A and Gushchina E V and Yuferev V S 2016 Peculiarities of Electrical Characteristics of Ferroelectric Memory Elements Based on PZT-Films *Russ Phys J* **58** 1301–1305

[6] Sidorkin A S and Nesterenko L P and Pakhomov A Y 2012 Influence of fatigue processes on the switching currents in lead titanate and lead zirconate-titanate films *Phys. Solid State* **54** 1008–1010

[7] Nguyen H T and Sidorkin A S and Milovidova S D 2020 Dielectric properties of ferroelectric nanocomposites of nanocrystalline cellulose and sodium nitrite *Appl Nanoscience* **10** 499–506

[8] Darinskii B, Sidorkin A and Sigov A and Popravko N 2018 Influence of Depolarizing Fields and Screening Effects on Phase Transitions in Ferroelectric *Composites Materials* **11** 85

[9] Pecherskaya E A, Artamonov D V, Kondrashin V I, Golubkov P E and Karpanin O V and Zinchenko T O 2017 Software - Hardware Complex for Measurement and Control of Ferroelectrics Parameters *IOP Conference Series: Materials Science and Engineering* **225** 012254

Structure and characteristics of a thin-layer "aluminum - carbon nanotubes" sandwich structure

A Fomin[1], V Koshuro[1], M Fomina[1], A Aman[1,2] and S Palis[3]

[1] Yuri Gagarin State Technical University of Saratov, Saratov 410008, Russia
[2] Otto Vollmann GmbH & Co. KG, Gevelsberg, Germany, 58285
[3] Otto-von-Guericke-Universität Magdeburg, Universitätsplatz 2, Magdeburg, Germany 39106

afominalex@gmail.com

Abstract. The study describes a way to produce thin-layer sandwich structures containing layers of aluminum foil and an interlayer filler of carbon nanotubes (CNTs). To obtain a homogeneous structure of the composite, induction heat treatment was used combined with a device providing primary compression of the assembly. To determine the functional characteristics of sandwich structures, the homogeneity of their microstructure and surface conductivity were studied. It was found that the use of CNTs allowed increasing the surface conductivity by a factor of 8.3–9.5 compared to the sandwich structure without filler.

1. Introduction

It is known that carbon nanotubes (CNTs) are a promising and available material that has a unique combination of strength and the ability to control physical properties such as electrical conductivity. This type of filler is widely used in the manufacture of composites. For example, CNTs are used to improve the strength of aluminum; however the limited ductility of high-strength aluminum matrix composites still remains a critical issue [1]. The elongation of Al/CNT composite (with small volume fraction of 1.3 %) sintered at 900 K is enhanced by 82 % compared to that sintered at 800 K, while an increment of 18 % in yield strength (from 127 MPa to 150 MPa) is achieved, as well as 14 % in ultimate tensile strength (from 186 MPa to 212 MPa).

To obtain layers containing CNTs, various types of solvents are used, *e.g.* the *m*-cresol (3-methylphenol, $CH_3C_6H_4(OH)$) weak organic acid [2]. Composites with CNTs have higher mechanical properties and the ability to pre-form compared to the initial material [3]. An important fact is that the amount of filler providing a noticeable strengthening effect is at a low level – not more than 0.5–5 wt.%.

Powder metallurgy, e.g. cold pressing of powder with nanosized particles, is used for the production of Al/CNT nanocomposite (0.5 wt.%) [4]. This material has the stable coefficient of thermal expansion (CTE) in the temperature range of 25–250 °C, which decreases by about 75–85 % compared to that of the aluminum. Aluminum-based composites containing MW-CNTs are produced by a powder metallurgy route involving a ball-milling technique [5]. The composite with individually dispersed nanotubes shows an increased strength with unchanged electrical and thermal conductivities. However, the composite with a network structure of MWCNTs reveals much enhanced electrical conductivity of about 33×10^7 S/m and thermal conductivity of 172 W/m×K (for aluminum – electrical conductivity of 3.77×10^7 S/m and thermal conductivity of 237 W/m×K).

The combined use of CNTs and graphene nanoplatelets (GNPs) synergistically reduced the carbon fiber reinforced plastics surface resistivity by 4 times and increased the thermal conductivity by more than 7 times, thus creating possibilities for the replacement of metallic mesh structures for electromagnetic interference shielding and lighting strike protection [6].

CNTs are many times stronger and lighter than metals (magnesium, aluminum, titanium, iron, copper, etc.), and they appear to be a good candidate for reinforcing them [7]. The integration of CNTs into a metal matrix depends on dispersion uniformity. In this case the composite material properties are a function of the uniformity of CNTs dispersion, interfacial bonding, weight percent of fillers, and the geometric parameters of the matrix itself.

2. Methodology

The samples were pressed and thermally treated layers of aluminum foil (thickness of 12.9–13.0 μm, Fe impurity of 0.5–1 wt.%) (figure 1,*a*). The sandwich structure had from 16 to 64 layers, which was associated with the peculiarities of the formation of test samples. The interlayer filler was applied by staining with a colloidal solution containing carbon nanotubes (5 mg CNTs in 100 ml *m*-cresol). Homogenization of the solution was performed using ultrasonic treatment. Further the solution was applied on the aluminum foil by staining followed by drying at 50–60 °C (figure 1,*b*).

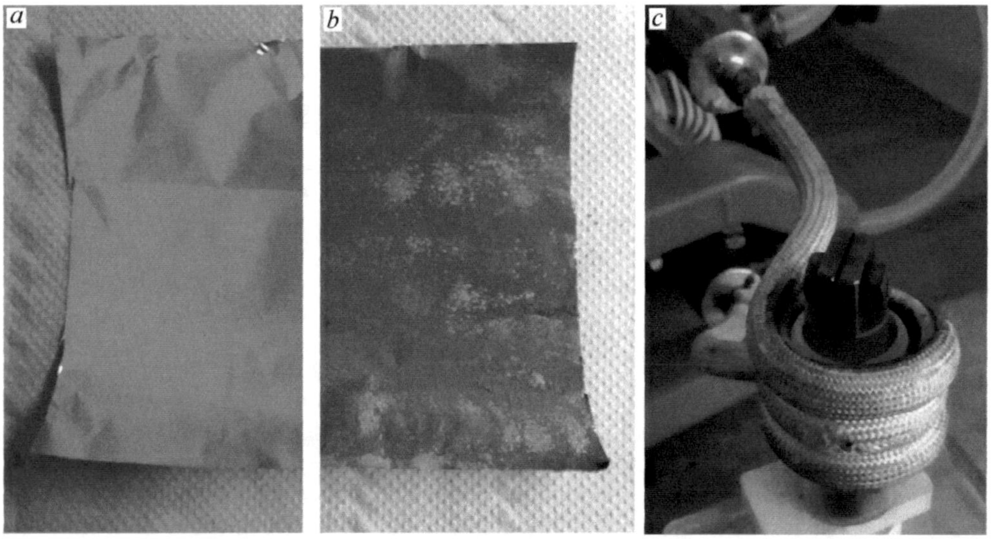

Figure 1. Aluminum foil (**a**); an applied layer of CNTs (**b**); the process of IHT at the sintering temperature (**c**).

The induction treatment of sandwich structures was conducted using a special device that provided primary compression of the assembly. The treatment temperature of the layered assemblies lied in the range of (0.9–0.95)×T_m with a process duration within 0.5–1 min (figure 1,*c*).

When studying the structure of the resulting aluminum foil sandwich, methods of optical and scanning electron microscopy (SEM) were used. Mechanical tests were performed to establish tensile strength and the clamps with soft inserts were used to fix the samples.

Surface electrical resistance was measured by the four-probe method using a picoammeter ("Keithley 6485", Textronix, US) and a DC voltage source ("Agilet 6614C", US). Surface resistance tests were performed in accordance with ASTM D4496–87 standard.

3. Results

During induction treatment of the formed assembly, there was a thermal effect on the system containing a special clamping device (figure 2,*a*). As the temperature grew, the plasticity of the heated material increased as well, and in a number of cases, partial melting was observed. The pressure in the system decreased and a homogeneous structure was formed; however, during mechanical tests (bending), defects appeared, in particular delamination occurred (figure 2,*b*).

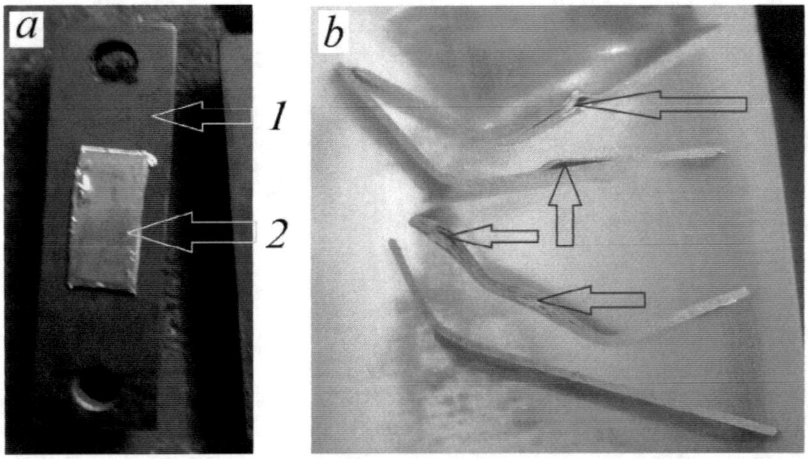

Figure 2. Sandwich structure *1* on the surface of mold half *2* (**b**); cross-section of "Al – CNTs" sandwich structures obtained with different parameters of induction treatment (**b**).

The internal structure of the sandwiches was layered and there were joint areas between the layers (figure 3,*a*). In the enlarged parts, changes in the directional structure of the deformed metal were observed, which was associated with the temperature effect as a result of IHT (figure 3,*b*). A sandwich structure sample having 32 aluminum layers without any melt regions on the lateral surface was characterized by increased strength (about 25–28 %) and high plasticity. The thickness of the resulting composite was at the level of 0.89–0.93 of the theoretical value.

In the presence of CNTs filler, the joint areas were retained; however, a fibrous structure was observed between the layers (figure 3,*c*). The strength of the resulting structure was increased on average by 10–12 % whereas the ductility was unchanged. Thus, CNTs formed fibrous structures between the layers of aluminum foil (figure 3,*d*).

When studying the functional characteristics of the resulting samples, it was also found that the use of CNTs allowed increasing the surface conductivity by a factor of 8.3–9.5 compared to the sandwich structure without filler. This may be due to the influence of CNTs on the integral value of the conductivity of a thin-layer system containing a strengthened aluminum alloy (Fe content of 0.5–1 wt.%).

4. Conclusions

The possibility of creating a homogeneous "Al – CNTs" sandwich structure with increased strength was shown in the study. The use of CNTs filler allowed increasing the surface conductivity by a factor of not less than 8, which can find application in the creation of electromechanical systems.

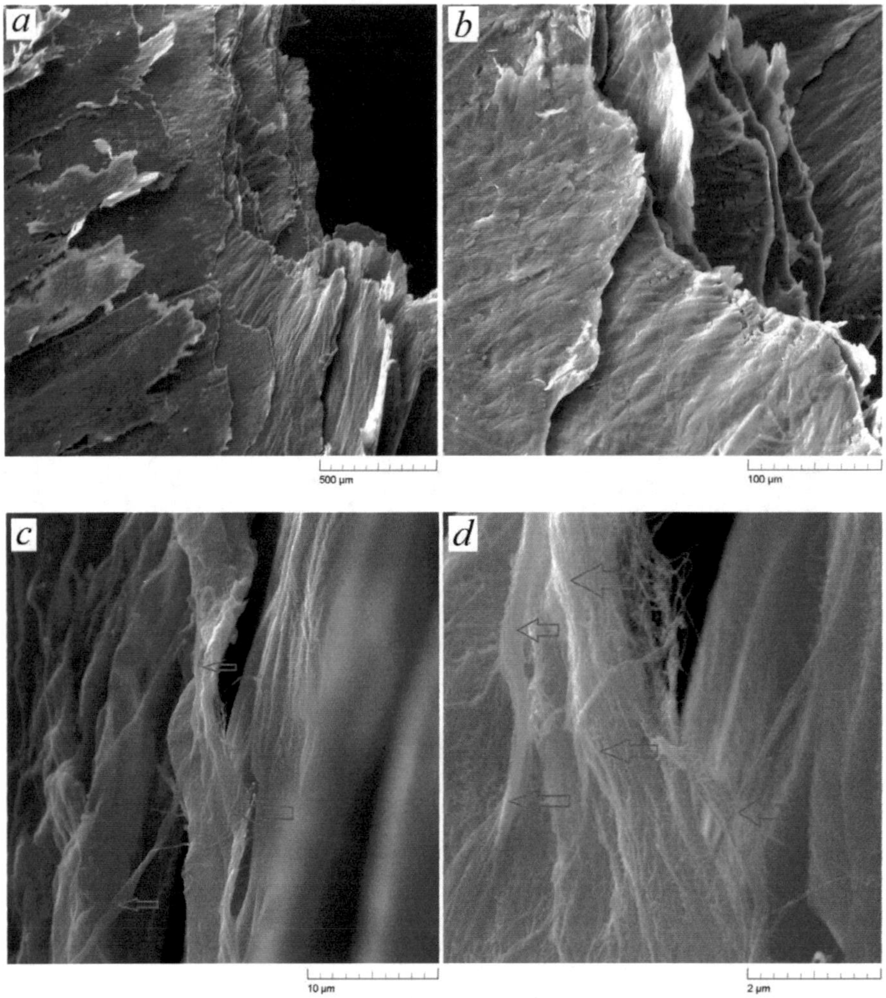

Figure 3. Fracture surface of the aluminum sandwich (**a**); an enlarged fragment of the layered structure (**b**); an area of the interaction between CNTs and the aluminum foil (**c**); an enlarged fragment showing accumulation of fibers of CNTs (**d**).

Acknowledgments
The research was supported by the Ministry of Education and Science of the Russian Federation and German Academic Exchange Service (DAAD) in the framework of the program "Mikhail Lomonosov" (Agreement No. 075-03-2021-304/3 from 21.05.2021; Project Executor Fomin A.A.).

References

[1] Chen B, Kondoh K, Imai H, Umeda J and Takahashi M 2016 *Scr Mater* **113** 158
[2] Chiou K, Byun S, Kim J and Huang J 2018 *PNAS* **115** (22) 5703
[3] Du J-H, Bai J and Cheng H-M 2007 *EXPRESS Polym Lett* **1** 5 253
[4] Sharma M, Pal H and Sharma V 2014 *Int J Chemtech Res* **6** (3) 2057
[5] Shin S E, Choi H J and Bae D H 2012 *J Compos Mater* **47** (18) 2249
[6] Li Y, Zhang H, Liu Y, Wang H, Huang Z, Peijs T and Bilotti E 2018 *Compos Part A Appl Sci Manuf* **105** 9
[7] Radhamani A V, Lau H C and Ramakrishna S 2018 *Compo Part A Appl Sci Manuf* **114** 170

Induction-thermal action effect on the surface area of titanium products

M Fomina, A Shchelkunov and A Shumilin

Yuri Gagarin State Technical University of Saratov, Saratov 410054, Russia

fominamaspb@mail.ru

Abstract. The changes in the surface area of titanium samples occurring during induction heat treatment (IHT) were studied. The dependence of the surface area of titanium samples on the exposure temperature was revealed. When a titanium sample was heated to a temperature of 1000 °C in the air at an exposure time of 60 s, there was a 45-fold area increase. The assessment of the porosity of the formed coating is carried out. The number of pores increased with increasing temperature from 374 to 1029 pieces. Accordingly, the average pore size decreased by 40%.

1. Introduction

Increasing the surface of titanium parts is used in the manufacture of dental implants. Sandblasting and plasma spraying of porous coatings are conventional methods for preparing the surface of implants [1,2]. The developed surface morphology facilitates integration with the surrounding biological tissues [1]. Oxidation of titanium parts is a simple way to obtain a morphologically heterogeneous surface oxide layer [3]. The works on the synthesis of anatase powders with a high specific surface area are also known [4,5]. However, the production of an adhesive strong coating with a controlled surface area is of practical interest.

2. Methodology

The samples for the study were made of VT1-00 commercially pure titanium and had the shape of a cylinder with a diameter of 5 mm and length of 5 mm. Thermal oxidation was carried out using "VCh-15A" induction equipment (Figure 1).

The IHT comprised several stages: accelerated heating, high-temperature exposure and cooling in a quiet mode (on a ceramic plate). The IHT temperature varied in the range of 800–1000 °C with a fixed duration of 60 s. The "Surface Area & Pore Size Analyzer System Quantachrome Nova 2000e" complex was used to measure the surface area. Sample preparation was characterized by the following parameters: outgas time 2 hrs; outgas temp +200 °C; analysis gas – nitrogen. The surface area was measured using the BET (Single Point Surface Area) method at a pressure ratio of $P/P_0 = 0.3$.

The mass of the samples was taken as a conventional unit, since their sizes were the same. With an increase in the P/P_0 ratio to 0.3, the adsorbent pore surface was filled with a monolayer of adsorbate molecules (Figure 1).

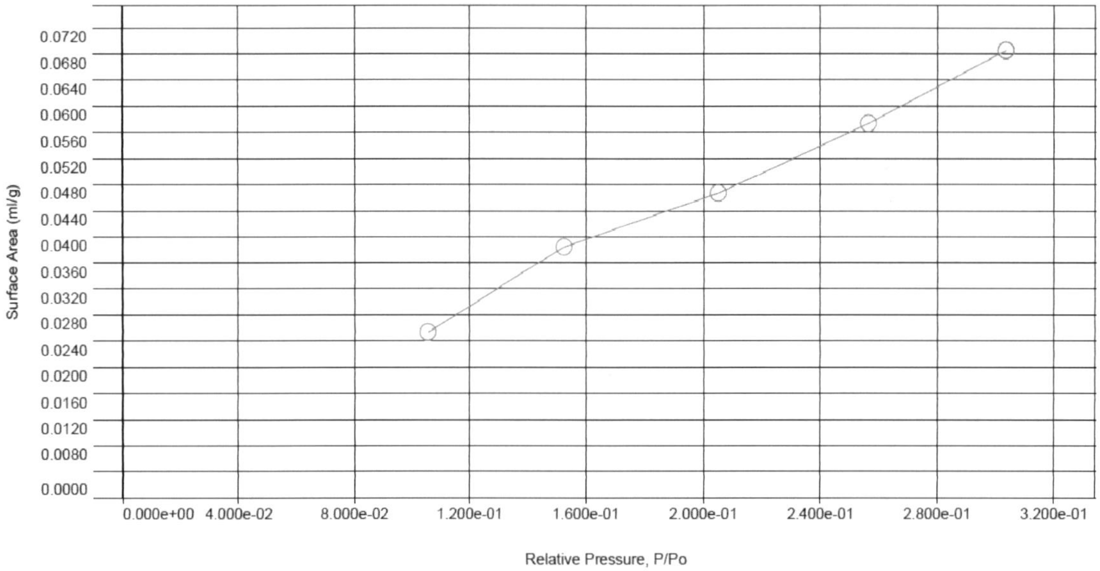

Figure 1. Measurement of surface area at various ratios P/P_0.

The BET model assumes three assumptions: 1) The surface of the adsorbent is homogeneous and all adsorption centers are energetically equivalent. 2) The adsorption centers fill the adsorbate molecules layer by layer and the molecules in all layers are considered to be completely identical. 3) The model takes into account only the forces of interaction between adsorbate and adsorbent molecules and neglects the forces of interaction between adsorbate molecules on the surface in a given adsorption layer. These assumptions make it possible to accurately determine the specific surface area from type II isotherms.

After measuring the surface area, a comparison was made between structural elements and porosity indicators. The study was carried out using the software for the analysis of geometric parameters of micro-objects "Metallograf". The image frame width was 220 μm.

3. Results

The surface areas of geometrically identical samples grew with increasing IHT temperature (Figure 2). When heated to a temperature of 850 °C and above, the surface layers of titanium were actively saturated with oxygen. This led to the appearance of an oxide layer and changes in the morphology of the samples. The areas of two titanium cylindrical samples were measured simultaneously for each heating mode. The measured value of the surface area of the samples (without heat treatment) was at the sensitivity limit of the device.

From the form of the adsorption isotherm, a conclusion was made about the surface area, porosity of the adsorbing body and the nature of the interaction between the adsorbent and adsorbate. Sample No. 4 in was investigated in the range from 0 to 1 for the ratio p / p0. A type II isotherm was obtained (according to Langmuir's classification). Type II isotherms are inherent in non-porous solids or macroporous adsorbents. In our case, we can talk about a macroporous thin layer formed as a result of oxidation.

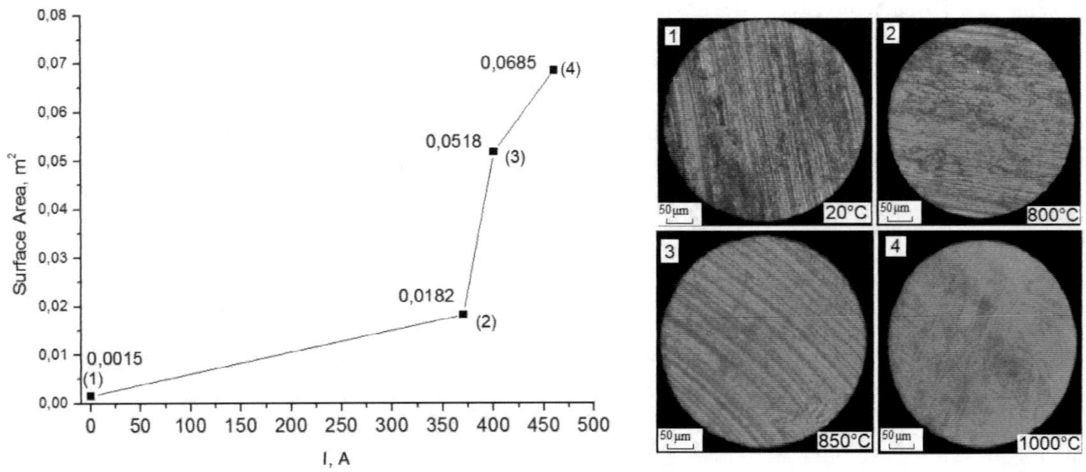

Figure 2. Dependence of the surface area on the inductor current and photomicrographs showing the sample morphology after IHT.

The increase in surface area was explained by the formation of a morphologically non-homogeneous oxide layer [3]. The layer thickness depended on the heating temperature and the exposure time at a given temperature. At the temperatures of 770–800, 850–880 and 1000–1030 °C, there was a 12-, 35- and 45-fold area increase, respectively. Thus, the real area of the oxidized sample was 45 times larger than that of the cylinder, the surface of which was formed only by machining (turning).

The porosity for sample 1 (Figure 3-1) was 47%, this value will be considered a reference after turning. For sample 2 (Figure 3-2), heat treatment of which was about 800 °C, the porosity value was 50%.

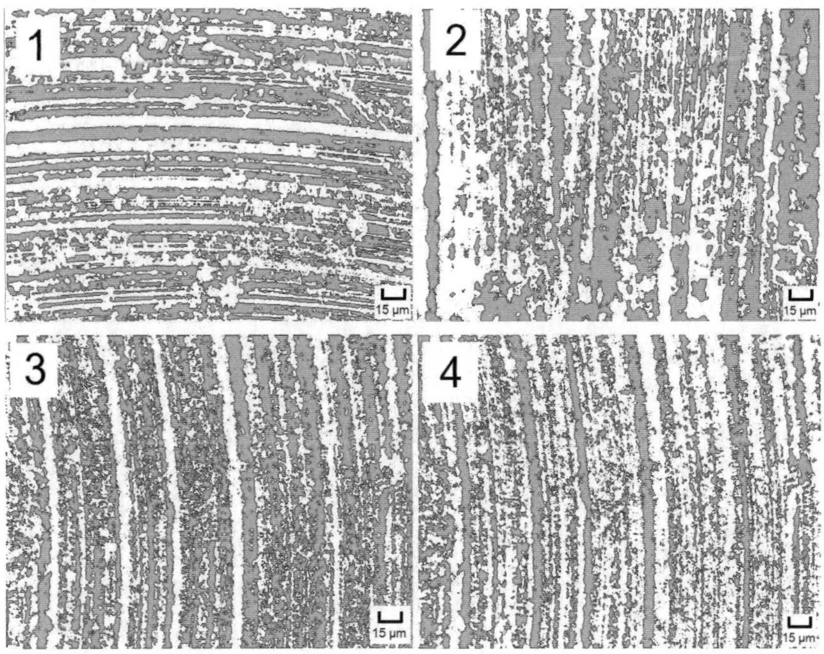

Figure 3. Photomicrographs showing the morphology (porosity) of the sample after IHT: 1 – 47%; 2 – 50%; 3 – 43%; 4 – 37%.

Sample 3 (Figure 3-3) shows a decrease in surface porosity of up to 43%. On sample surface 4 (Figure 3-4), the porosity was 37%.

4. Conclusion

The possibility of changing the functional surface area of titanium products by controlled heating and growth of the oxide coating was shown. IHT in the air at a temperature of about 1000–1030 °C allowed a 45-fold increase in the surface area of the titanium sample compared to the initial value, which, combined with high mechanical strength (hardness, wear resistance), will provide a higher quality of titanium products.

Acknowledgments

The research was supported by the Russian Science Foundation (project No. 18-79-10040).

References

[1] Velasco-Ortega E, Alfonso-Rodríguez C A, Monsalve-Guil L, España-López A, Jiménez-Guerra A, Garzón I, Alaminos M and Gil F J 2016 *Mater Sci Eng C* **64** 1
[2] Zhang W, Gu J, Zhang, Cheng, Xie Y and Zheng X 2019 *Surf Coat Technol* **358** 511
[3] Fomin A, Egorov I and Shchelkunov A 2018 *Proc SPIE* **10716** 107161O
[4] Zhang Z, Brown S, Goodall J B M, Weng X, Thompson K, Gong K, Kellici S, Clark R J H, Evans J R G and Darr J A 2009 *J Alloys Compd* **476** 451
[5] Li J, Wan W, Zhou H, Lia J and Xu D 2011 *Chem Commun* **47** 3439

Solid-state phase transition in n-alkanes of different parity

S A Gureva, A K Borisov, V A Marikhin and V M Egorov

Strength Physics Lab, Ioffe Institute RAS, St. Petersburg 194021, Russia

Abstract. The kinetics of the first-order solid-state structural transition in monodisperse n-alkanes samples of tricosane $C_{23}H_{48}$ and tetracosane $C_{24}H_{50}$ was studied by DSC and FTIR spectroscopy. The initial nuclei location of the new phase was revealed. The process of crystal structure rearrangement is initiated in the interlayers between neighboring lamellar for odd tricosane, while the nanonuclei in even tetracosane arise in the crystalline lamella cores. Thus, the influence of the number evenness of carbon atoms in the n-alkanes chains on the first-order structural phase transition has been proved.

1. Introduction

The purpose of this work was a fundamental study of temperature phase transitions (PTs) in long-chain molecular crystals (LCMCs), using the example of normal alkanes with the general formula $CH_3(CH_2)_{n-2}CH_3$, by differential scanning calorimetry (DSC) and IR Fourier spectroscopy.

N-alkanes are model systems for studying the regularities of PTs in LCMCs and polymers, and also have important practical applications as PCM (phase change materials). The objects of research are monodispersed samples of n-alkanes tricosane $C_{23}H_{48}$ and tetracosane $C_{24}H_{50}$ with a very high purity ~ 99.9% produced by Sigma – Aldrich Company. During crystallization of n-alkanes, thin lamellar crystals are formed, in which depending on the number parity of methylene CH_2 groups molecules stacked in triclinic or orthorhombic subcell for even and odd n-alkanes, respectively.

Under studying of various LCMCs (i.e. n-alkanes, n-alcohols, mono- and dicarboxylic acids) by the DSC technique [1 – 4], it was found that the transition from the solid to the liquid state (and vice versa) is carried out in two stages. The initial stage is a first order solid-phase (structural) transition (PT-1) occurring at temperatures $T_{PT-1} < T_m$. The final stage is a second order phase transition of the order-disorder type or melting at T_m.

The main attention was paid to the study of PT-1, which is associated with a change in the type of the subcell symmetry in the crystalline cores of elementary lamella. It was shown [2] that PT-1 develops according to a heterogeneous mechanism (i.e. with the formation of a new phase nuclei) in a narrow temperature range ΔT ~ 2K in accordance with a theory of the diffuse first order phase transitions [5]. It was found that the transition is due to the change in the initial subcell symmetry to the hexagonal one. According to the theory, nanonuclei of new phases appear in the initial phase volume, and then the interfaces gradually propagate to the entire sample volume by adding similar nuclei to the resulting phase boundary.

In our previous works [6, 7], the development kinetics of structural PT-1 in monodisperse sample of odd tricosane $C_{23}H_{48}$ and even tetracosane $C_{24}H_{50}$ was studied by FTIR spectroscopy. Considering the lamellar structure of LCMCs, it is possible to consider the development of PT-1 in two quasi-independent structural components: in the lamella cores containing methylene sequences and in interlamellar layers formed by the van der Waals interactions between the terminal methyl groups of adjacent lamellae. The study of the transformations in the crystalline lamella cores was carried out on the basis of the frequency temperature dependences analysis of various vibration CH_2 modes in methylene trans-sequences. It was shown that in the cores PT-1 develops in a temperature range $\Delta T = 40.3 - 41.7$°C for tricosane and $\Delta T = 45.5 - 47.6$°C for tetracosane. It was found that in the PT-1 temperature range a hexagonal phase nucleus appear in the volume of the initial triclinic or orthorhombic phase, and the amount of the new phase gradual increase occurs due to the disappearance of the former one.

From a fundamental point of view, it was necessary to find out the location of a first hexagonal nucleus in the volume of the initial phase. The nucleation can occur both in the lamella cores and in the interlamellar space. Thus, considerable interest was a study of changes in the interlamellar layers formed by the van der Waals contacts between the terminal methyl CH_3 groups of the neighboring lamella.

2. Experiments and results

FTIR spectroscopic analysis reveals the kinetics of a first-order solid-state phase transition (PT-1) by the smallest displacements in the molecular vibration spectra, which are due to changes of the subcell symmetry in the crystalline cores of elementary lamella.

The samples with a 10 – 20 μm thickness were prepared by depositing the synthesis products – lamellar flakes of tricosane $C_{23}H_{48}$ and tetracosane $C_{24}H_{50}$ – on NaCl plates, their melting, and subsequent slow cooling until an equilibrium crystalline structure was obtained. The absorption spectra of monodisperse samples were recorded on IFS-88 FTIR spectrometer (Bruker, Germany). Temperature variations of IR spectra were studied in the interval from 19 to 51°C with steps $\Delta T = 2$°C and $\Delta T = 0.1$°C in the PT-1 region and the samples were kept during 10 min until reaching the equilibrium state at the given temperature.

To detect changes in the interlamellar layers, stretching vibrations of C – H bonds in the spectral region $\nu = 2800 - 3000$ cm^{-1} was studied that shown in figure 1. According to [8], eight absorption bands are observed in the spectral range under study, the frequencies of which differ slightly for n-alkanes with various chain lengths. Adopted the following notations bands:

r^+ are the symmetric stretching vibrations in terminal CH_3 groups ($\nu(C_{23}H_{48}) = 2871.0$ cm^{-1} and $\nu(C_{24}H_{50}) = 2870.6$ cm^{-1});

r^-_a, r^-_b are the antisymmetric stretching vibrations in CH_3 groups, in-plane and out-of-plane components, respectively ($\nu(C_{23}H_{48}) = 2960.6$ and 2955.0 cm^{-1}, $\nu(C_{24}H_{50}) = 2961.5$ and 2953.5 cm^{-1});

d^+, d^- are the symmetric and antisymmetric stretching vibrations in CH_2 groups in chain ($\nu(C_{23}H_{48}) = 2849.0$ and 2918.4 cm^{-1}, $\nu(C_{24}H_{50}) = 2848.7$ and 2916.1 cm^{-1});

d^+_w are the symmetric stretching vibrations in w-CH_2 groups adjacent to CH_3 ($\nu(C_{23}H_{48}) = 2858.3$ cm^{-1}, $\nu(C_{24}H_{50}) = 2859.3$ cm^{-1}).

In addition, there are 2 bands (r^+_{FR} and d^+_{FR}) due to the Fermi resonance of symmetric stretching vibrations r^+ and d^+ with an overtone of the CH_2 groups bending vibrations inside the crystalline lamella cores.

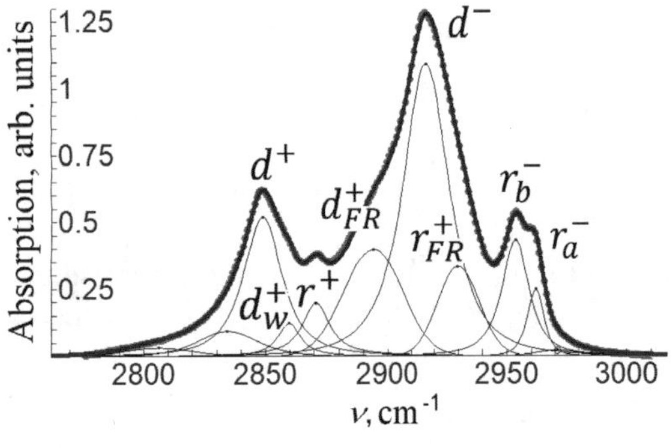

Figure 1. The IR spectrum of tetracosane $C_{24}H_{50}$ in the stretching vibrations region at 19°C. Spectrum decomposition was performed in the Fityk 0.9.8 program.

Samples of tricosane and tetracosane weighing 0.4 – 1.2 mg were studied on a DSC 500 calorimeter with varying scanning speeds in the range 0.2 – 5 K/min. Low sample weights and scan speeds ensures the minimum thermal resistance of the calorimeter cell and minimizes the effect of thermal delay. The dependences of heat capacity on temperature were received for tricosane and tetracosane samples with variable scanning speeds. Figure 2 shows PT-1 on DSC curves obtained at a scan rate of 0.2 K/min. Methodological errors were eliminated by a special method of extrapolating the peak temperatures to zero scanning speed [3].

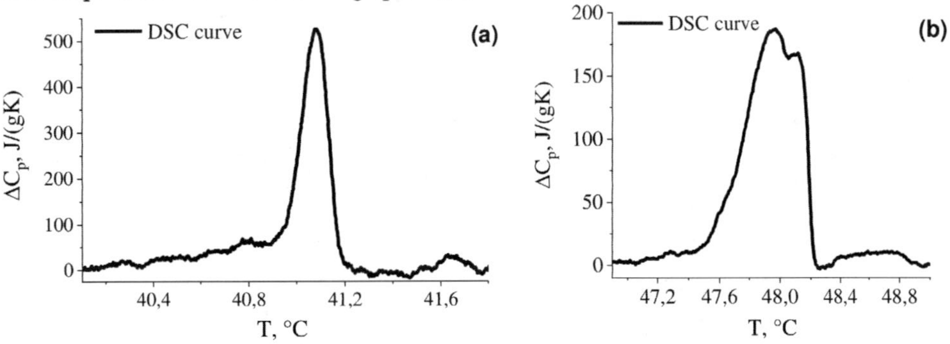

Figure 2. DSC PT-1 curve for tricosane (a) and tetracosane (b).

3. Results and discussions

In the work, we investigated the features of the rearrangement in the interlamellar space on the basis of the temperature dependences of the various C – H bond stretching vibrations modes in the terminal methyl CH_3 groups, where radical changes in the PT-1 range were found (figure 3 and 4).

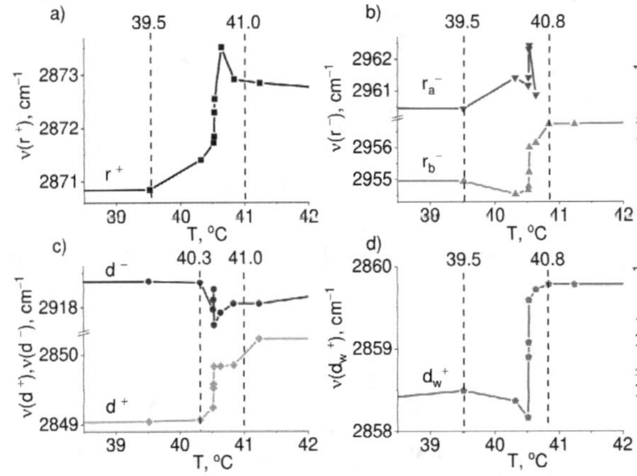

Figure 3. Temperature dependences of the absorption band positions of r^+ (a), r^-_a, r^-_b (b), d^+, d^- (c) and d^+_w (d) stretching vibrations of C – H bonds in the tricosane $C_{23}H_{48}$.

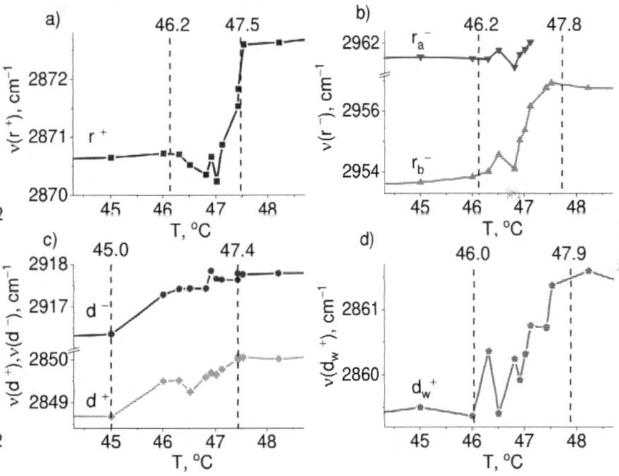

Figure 4. Temperature dependences of the absorption band positions of r^+ (a), r^-_a, r^-_b (b), d^+, d^- (c) and d^+_w (d) stretching vibrations of C – H bonds in the tetracosane $C_{24}H_{50}$.

The temperature study of the CH_2 stretching vibrations $d^{\pm}$ in the lamella cores confirms the results obtained in our previous works [6, 7]. The change of $d^{\pm}$ frequencies in the lamella cores occurs in the temperature range ΔT = 40.3 – 41.0°C for tricosane (figure 3(c)) and ΔT = 45.0 – 47.4°C for tetracosane (figure 4(c)).

It was established that for tricosane PT-1 begins with a change of the stretching vibrations $r^{\pm}$ in the terminal CH_3 groups and d^+_w in the CH_2 groups closest to them at ΔT = 39.5 – 41.0°C (figure 3(a, b, d)). The rearrangement of the tetracosane crystal structure manifests in changes of vibrations in the terminal groups at more higher temperatures ΔT = 46.2 – 47.8°C (figure 4(a, b, d)) as compared to the vibrations inside the lamella. It can be concluded that for even n-alkane $C_{24}H_{50}$ the initial nuclei of a new phase arise precisely in the lamella cores and only then gradually extend over the entire volume. It is assumed that in this case before a complete transition to the hexagonal symmetry it is energetically more favorable for molecules inclined relative to the base plane of the lamellae to occupy a vertical position, and only after that to change the molecules packing and the contact of the terminal groups.

DSC researches were carried out to confirm and refine the results obtained by FTIR spectroscopy. The temperature dependences of the heat capacity in the PT-1 range at a scanning rate of 0.2 K/min for the samples under study are shown on figure 5. It can be seen that the transition covers the interval ΔT = 40.6 – 41.3°C for $C_{23}H_{48}$ and ΔT = 47.2 – 48.4°C for $C_{24}H_{50}$, that is consistent with IR data on the complete transition to the hexagonal phase.

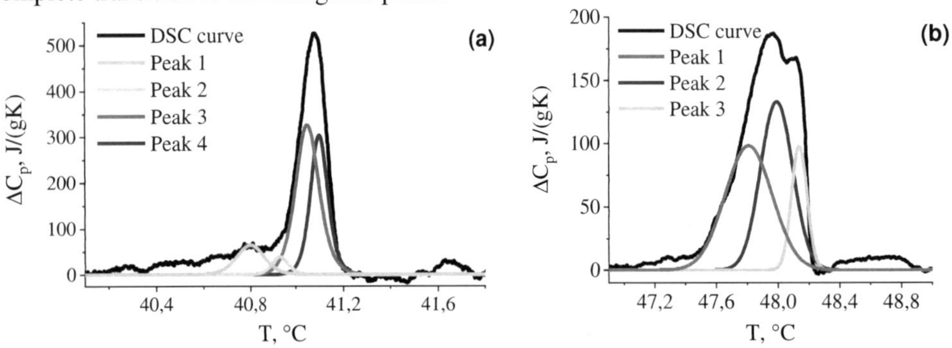

Figure 5. DSC PT-1 curve for tricosane (a) and tetracosane (b).

The peak shapes were analyzed according to the thermodynamic theory of a self-consistent field [5, 9, 10] as applied to symmetric Λ – shaped diffuse PT-1. The theory analyzes the change in the system of interacting particles of different phases, in particular, the formation and subsequent growth of stable nuclei of a new phase in the matrix of the old one. The main idea of the theory is that in the old phase volume, localization of numerous fluctuations occurs in the form of stable nuclei of a new phase, the so-called elementary transition volumes ω. Further new phase propagation is carried out by successive additions of nuclei with a volume of ω to the arising interphase boundary.

The relation for the temperature dependence of the heat capacity at a diffuse phase transition has the form [11]:

$$\Delta C_p(T) = 4\Delta C_{\max} \cdot \exp\left[\frac{B(T-T_0)}{T_0}\right]\cdot\left[1+\exp\left[\frac{B(T-T_0)}{T_0}\right]\right]^{-2}, \quad (1)$$

where T_o is a temperature of the PT-1 peak, K; $\Delta C_{\max}$ is a maximum value of heat capacity at $T = T_o$, J/(g·K); B is an athermal parameter.

As can be seen from figure 5, the shape of the DSC curve is asymmetric in both cases, however, for the self-consistent field theory, the symmetric Λ – shaped peaks are required. Therefore, according to the ratio (1), the separation into symmetric peaks was performed using the previously developed method [12]. As shown in figure 5, the main peak is divided into two Λ – shape peaks (peaks 3 and 4 for tricosane and peaks 1 and 2 for tetracosane). Also, on the tricosane curve it was possible to distinguish additional peaks preceding the main one (peaks 1 and 2), and for tetracosane – peak after the main one (peak 3). The enthalpies for the main and secondary peaks are compared with the calculated values of the transition energies in the lamella cores $\Delta H_{(n-4)CH2} = (n-4) \cdot 3.84$ J/g (where $(n-4)$ – number of CH_2 groups in the lamella cores) and in the interlamellar space $\Delta H_{2CH3+2CH2}$ = 16 J/g (two terminal and nearest to them CH_2 groups) [13, 14]. The experimental values of enthalpies for $C_{23}H_{48}$: $\Delta H_{peak1} + \Delta H_{peak2}$ = 14 J/g, $\Delta H_{peak3} + \Delta H_{peak4}$ = 72.4 J/g; and for $C_{24}H_{50}$: $\Delta H_{peak1} + \Delta H_{peak2}$ = 71.9 J/g, ΔH_{peak3} = 12.6 J/g (figure 5). Comparison of experimental and calculation data is presented in table 1.

Table 1. Enthalpies of interlamellar space and lamella cores of the samples under study.

Samples	Calculation data			Experimental data		
	$\Delta H_{interlam.space}$ (J/g)	$\Delta H_{lam.cores}$ (J/g)	$\frac{\Delta H_{lam.cores}}{\Delta H_{interlam.space}}$	$\Delta H_{interlam.space}$ (J/g)	$\Delta H_{lam.cores}$ (J/g)	$\frac{\Delta H_{lam.cores}}{\Delta H_{interlam.space}}$
$C_{23}H_{48}$	16	73.0	4.56	14	72.4	5.17
$C_{24}H_{50}$	16	76.8	4.8	12.6	71.9	5.71

As can be seen from the table 1, the experimental data correspond to the calculated ones with good accuracy. However, it should be noted that an underestimation of the total enthalpy is observed for the tetracosane sample. This may be due to some measurement error, since the ratio between the enthalpies of the lamella cores and the interlamellar space corresponds to the calculated one, and there is also a tendency towards an increase in the chain length.

4. Conclusion

The kinetics of the first-order solid-state structural transition in monodisperse n-alkane samples of odd tricosane $C_{23}H_{48}$ and even tetracosane $C_{24}H_{50}$ was studied by FTIR spectroscopy and DSC. It is shown that the hexagonal nucleus in the initial phase matrix appears by the fluctuation mechanism in a narrow temperature range $\Delta T \approx 2 - 3$°C for the samples under study. The present investigation made it possible to study in more detail the sequential structure rearrangement of different parity n-alkanes during the phase transition. In addition, the initial nuclei location of the new phase was revealed. The process of crystal structure rearrangement of even n-alkanes is initiated in the crystalline lamella cores, when the molecules tend to line up vertically. Further, the transition includes the terminal groups, and the molecular packing in the layers changes. But in odd n-alkanes, it was shown that PT-1 begins in the interlamellar space with changes in the terminal group contacts, and only then the transition spreads to the lamella cores. FTIR spectroscopy and DSC have demonstrated the influence of the parity effect of n-alkanes depending on the chain length on the development of the structural phase transition.

References

[1] Egorov V M, Marikhin V A and Myasnikova L P 2013 *Phys. Solid State* **55** 1057
[2] Egorov V M and Marikhin V A 2016 *Phys. Solid State* **58** 2574
[3] Egorov V M, Marikhin V A, Myasnikova L P and Yakushev P N 2017 *Phys. Solid State* **59** 2070
[4] Borisov A K, Egorov V M and Marikhin V A 2019 *J. Phys.: Conf. Ser.* **1400** 055020
[5] Malygin G A 2001 *Physics-Uspekhi* **171** 187
[6] Marikhin V A, Gur'eva S A, Myasnikova L P, Volchek B Z and Medvedeva D A 2019 *Phys. Solid State* **61** 1785
[7] Gureva S A, Marikhin V A, Myasnikova L P, Volchek B Z and Medvedeva D A 2019 *J. Phys.: Conf. Ser.* **1400** 055040
[8] MacPhail R A, Straws H L and Snyder R G 1984 *J. Phys. Chem.* **88** 334
[9] Fisher M 1965 *The Nature of Critical Points* (Colorado: Univ. of Colorado)
[10] Rolov B N and Yurkevich V E 1983 *Fizika razmytykh fazovykh perekhodov* (Rostov: Publishing Rostov University)
[11] Malygin G A 2001 *Phys. Solid State* **43** 1909
[12] Egorov V M, Borisov A K and Marikhin V A 2021 *Phys. Solid State* **63** 498
[13] Gureva S A, Borisov A K, Marikhin V A and Egorov V M 2020 *J. Phys.: Conf. Ser.* **1695** 012180
[14] Van Krevelen D W 1972 *Properties of Polymers — Correlation with Chemical Structures* (Amsterdam: Elsevier-North Holland) p 427

Electrophoretic deposition and capacitive characteristics of composite electrode materials based on CNTs and ruthenium oxide for planar supercapacitors

Yu I Kakovkina[1, 2], E A Lebedev[2], E P Kitsyuk[1], D A Kuzmin[2], A V Alekseyev[2] and Yu P Shaman[1]

[1] SMC "Technological Centre", Moscow, 124498, Russia
[2] MIET, Moscow, 124498, Russia

Kitsyuk.e@gmail.com

Abstract. In this work, a method for the formation of planar supercapacitors, which combines electrophoretic deposition of composite electrode materials based on CNTs and ruthenium oxide, and laser engraving, is proposed. The features of electrophoretic deposition are considered and the influence of the main technological modes on the morphology and composition of the formed layers of electrode materials is determined. The conducted studies of the electrophysical characteristics of the formed samples confirmed the possibility of producing planar capacitors with a high capacity and their potential applicability for a number of applications in microelectronics.

1. Introduction

The advance of miniaturized electronics and Internet of Things has a great impact on the development of miniaturized energy storage devices [1]. Planar supercapacitors are considered a promising device in this area due to their significant energy and power density, and long life cycle [2, 3]. It is known that electrochemical capacitors accumulate charge not only due to the electric double layer formation, but also due to the Faraday pseudocapacitance of reversible redox reactions. The combination of these two mechanisms of charge storage at once can significantly increase the capacitive characteristics of the device.

For practical implementation, it is necessary to develop a method for the formation of composite electrode materials in which heterogeneous components will be responsible for different mechanisms of energy storage [4, 5]. Great efforts have been made to maximize their performance with new carbon nanomaterials and to increase their specific capacity and energy density by utilization of transition metal oxides such as RuO_2, MnO_2, NiO, Co_3O_4, SnO_2, ZnO, TiO_2, CuO, $Fe2O_3$, WO_3 [6]. One of the approaches to creating multicomponent electrode materials for supercapacitors is electrophoretic deposition (EPD) - a simple method that allows the formation of layers on conductive substrates with a given composition, morphology, and porosity [7-9].

Therefore, within the framework of this work, the features of the electrophoretic deposition of composite electrode materials based on CNTs and ruthenium oxide will be investigated, their capacitive characteristics are determined, and a method for the formation of planar supercapacitors is proposed.

2. Experiment

In this work, the objects of study were "sandwich-like" and planar supercapacitors, a schematic representation of the route of their production is shown in figure 1. Nickel foil was used as substrates for bulk samples, and sitall plates with a 300 nm thick layer of nickel deposited by magnetron sputtering were used for planar samples. The formation of composite electrode materials on their surface was carried out by the method of electrophoretic deposition on a setup consisting of two electrodes (a substrate and a counter electrode) connected to a power source and a container with a suspension. Preparation of suspensions was carried out by ultrasonic dispersion of components in a solvent. The main components of the suspension in this work were few walls carbon nanotubes (FWCNTs) synthesized by CVD method [10] and ruthenium oxide hydrate produced by Alfa Aesar. Acetone, isopropyl alcohol, ethanol and their combinations were used as solvents. The concentrations of carbon nanotubes and ruthenium oxide were 0.2 mg/ml.

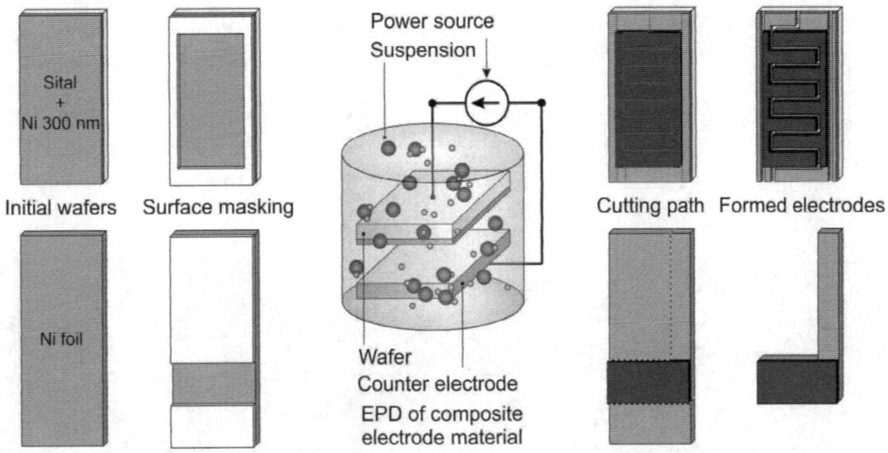

Figure 1. Schematic representation of supercapacitor formation processes.

EPD of composite materials was produced in potentiostatic mode - the electric field strength varied from 50 to 150 V/cm. To determine the mass of the sediment, the samples were weighed before and after the deposition process, and the deposition area was limited using overhead masks-templates. The morphology and composition of the materials were controlled using SEM and energy dispersive X-ray analysis. After the deposition of composite materials, electrodes were formed - for "sandwich-like" samples, they were cut with scissors, and planar supercapacitors were manufactured using laser engraving on a CNC 3018 Pro machine. "Sandwich-like" supercapacitors consisted of two electrodes separated by a separator and placed in a sealed bag filled with electrolyte. For measuring the capacitance characteristics planar supercapacitors were placed in a container with an electrolyte. In both cases, a 1 M aqueous solution of KOH was used as the electrolyte. The capacitance characteristics were measured by the method of cyclic voltammetry in the voltage range from 0 to 1 V with a sweep rate of 10 and 100 mV/s.

3. Results

At the first stage of the study, the dependence of the stability of the suspension on the selected solvent was experimentally determined. Acetone, isopropyl and ethyl alcohols, and their mixtures were used as solvents. The stability assessment was carried out on the basis of photographs of the suspensions taken at 0.5, 1, 2, 24 and 48 hours after preparation. The change in transparency and the formation of sediment at the bottom of the tube were visually assessed. Within the framework of this work, a suspension was considered stable if the time of its visual stability was many times greater than the duration of the electrophoretic deposition process. Figure 2 shows an example of a set of photographs for an acetone solvent. The carried out experimental work showed that the most stable suspensions were obtained with ethyl alcohol - acetone (E-Ac) and isopropyl alcohol - acetone (IPA - Ac) solvents mixed in equal volume fractions.

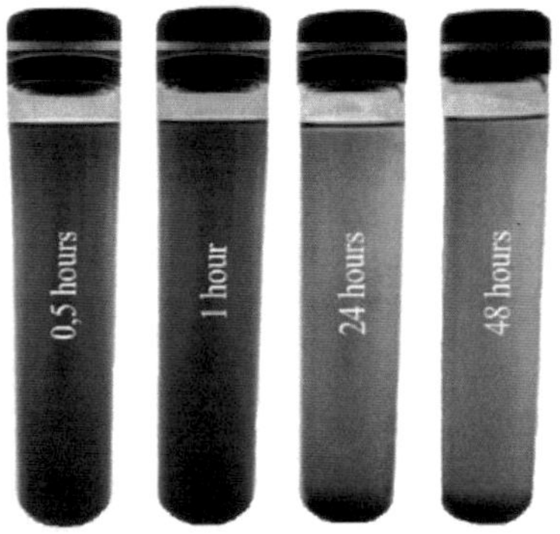

Figure 2. Photographs of the evolution of acetone-based suspensions over time.

At the next stage, the effect of the solvent on the morphology and capacitive characteristics of the formed electrode materials was studied. For this, suspensions were prepared based on E-Ac and IPA-Ac, and deposition was carried out on substrates made of nickel foil at electric field strength of 100 V/cm. Figure 3 shows SEM images of the obtained composites.

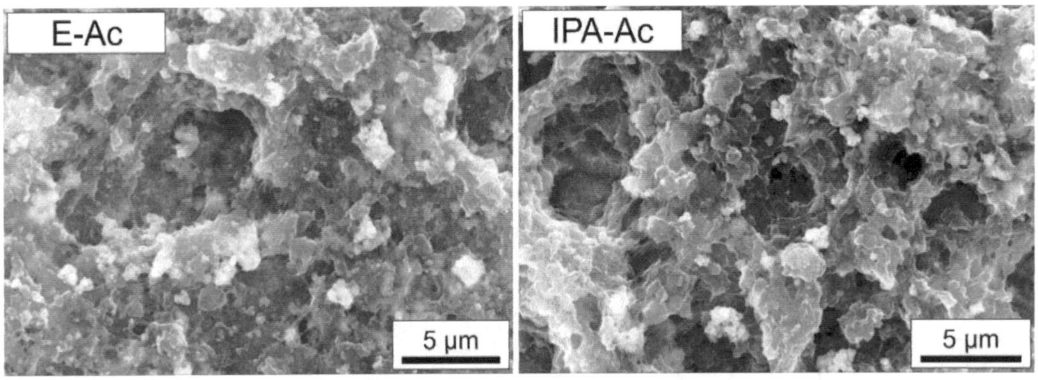

Figure 3. SEM images of FWCNT-RuO_2 composite layer deposited from a suspension based on ethanol-acetone (EA-Ac) (left) and isopropyl alcohol – acetone (IPA-Ac) (right).

The layers obtained from the IPA-Ac based slurry have a more porous structure. This feature can make a positive impact on the capacitive characteristics of a supercapacitor, since the effective electrode-electrolyte contact area will be larger, and, hence, the contribution to the total capacity of the electric double layer. This hypothesis was confirmed by studying the capacitive characteristics of "sandwich-like" supercapacitors with electrodes based on the materials mentioned above. Figure 4 shows cyclic voltammograms obtained with sweep rates of 10 and 100 mV/s.

At a sweep rate of 10 mV/s, the specific capacitance of supercapacitors with electrodes formed from suspensions based on E-Ac and IPA-Ac was 8.1 mF and 10.3 mF and at 100 mV/s – 4.1 and 5.3 mF respectively. Further, the influence of the electric field strength on the deposition rate and the morphology of the formed sediments was investigated. An IPA-Ac suspension was used for the experiments. Figure 5 shows SEM images of the morphology of the formed sediments at electric field strengths of 50 and 150 V/cm. Based on the obtained experimental results, the field strength insignificantly affects the morphology.

The uniformity of the distribution of components in the sediment was monitored using elemental analysis. Figure 6 shows the distribution maps of the components over the surface of the layer of the composite electrode material CNT - ruthenium oxide formed from the IPA-Ac suspension. Obtained results indicate a uniform distribution of elements in the sediment.

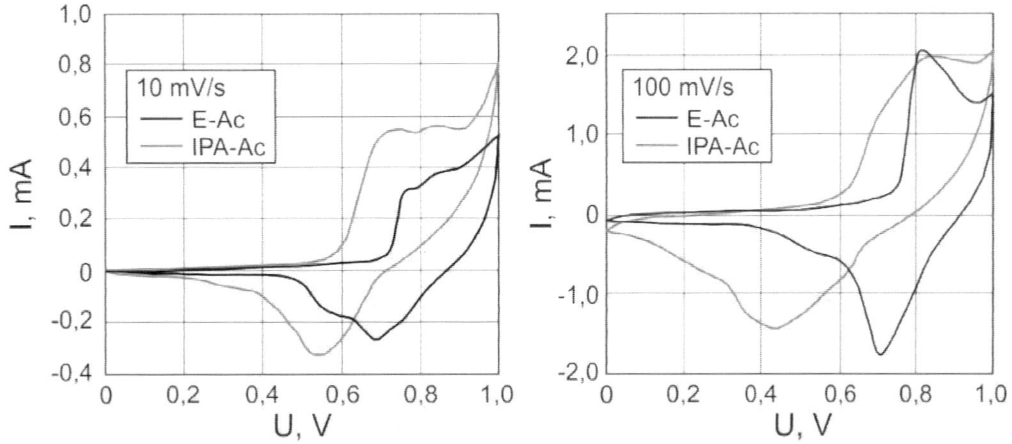

Figure 4. Cyclic sweeps of the samples at 10 mV/s (left) and 100 mV/s (right).

Figure 5. Layer of FWCNT-RuO_2 composite applied from the IPA-Ac suspension at field strengths of 50 V/cm (left) and 150 V/cm (right) respectively.

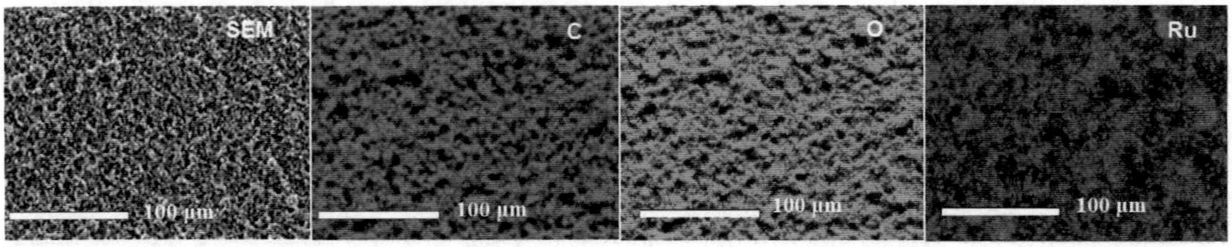

Figure 6. Maps of the distribution of elements over the surface of a composite electrode material.

The presence of dark areas on the maps of the distribution of elements may be due to the fact that after the solvent dried out, the layer cracked and had a porous and rough structure. The presence of micro and macro pores can have a positive effect on the capacitive and power characteristics, since the contact area between the electrode material and the electrolyte will increase.

The deposition rate increases linearly with an increase of electric field strength in the range from 50 to 150 V/cm - the obtained experimental dependence is shown in figure 7.

After examining the “sandwich-like” samples, experiments were carried out with the samples based on sitall substrate having a planar structure. The formation of electrode materials by the EPD method was carried out from IPA-Ac suspensions in similar modes.

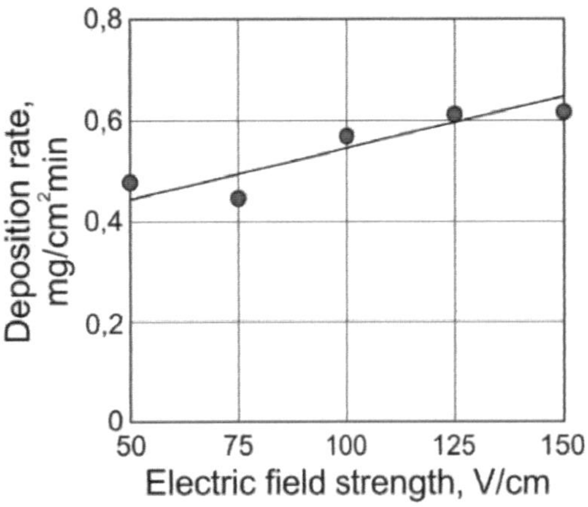

Figure 7. Graph of the dependence of the deposition rate on the electric field strength.

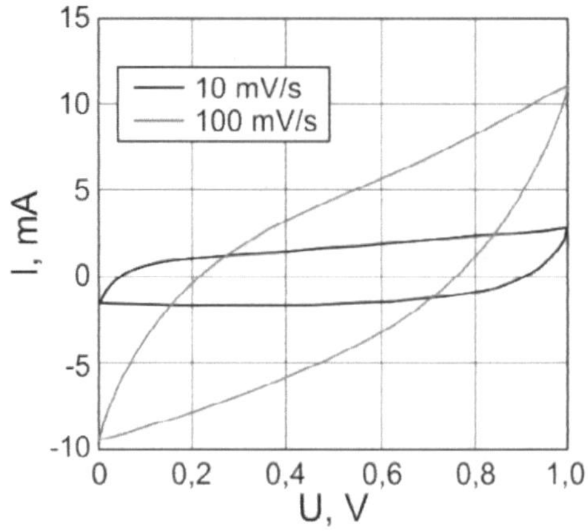

Figure 8. Cyclic sweep of the planar supercapacitor at 10 mV/s and 100 mV/s.

Figure 8 shows cyclic sweeps for a sample of a planar supercapacitor with composite electrode materials based on CNTs and ruthenium oxide, taken at a sweep rate of 10 and 100 mV/s. The average calculated specific gravimetric capacitance values for 10 measurement cycles for sweep rates of 10 and 100 mV/s were 35.3 and 9.2 F/g, respectively. For planar supercapacitors, the more correct specific unit of measurement is "F/cm^2", since their location is implied directly next to the consumer on the chip. The specific capacitance per unit surface area for this sample was 44.5 and 11.6 mF/cm^2.

4. Conclusions

In this work, an approach to create planar supercapacitors with composite electrodes based on carbon nanotubes and ruthenium oxide was proposed. The electrode materials were formed using electrophoretic deposition, and the structure of the planar power source was formed using laser engraving. To optimize the EPD, the influence of the solvent composition on the suspension stability, porosity and morphology of the resulting precipitates was investigated, and the deposition rate of the material was determined. The dependence of the capacitive characteristics of sandwich-like samples and planar supercapacitors on the measurement modes was also investigated. The measured specific values for planar supercapacitors were 44.5 and 11.6 F/cm^2 for sweep rates of 10 and 100 mV/s.

Acknowledgments

This work was supported by the Russian Federation President's grant (MK-1823.2021.4).

References

[1] Zhao J and Burke A F 2021 *J. Energy Chem.* **59** 276–291
[2] Zhou R, Meng C, Zhu F, Li Q, Liu C, Fan S and Jiang K 2010 *Nanotechnology* **21** 345701
[3] Bu F, Zhou W, Xu Y, Du Y, Guan C and Huang W 2020 *Npj Flex. Electron.* **4** 31
[4] Wang Y, Chen J, Cao J, Liu Y, Zhou Y, Ouyang J-H and Jia D 2014 *J. Power Sources* **271** 271
[5] Wang Y, Zhang L, Hou H, Xu W, Duan G, He S, Liu K and Jiang S 2021 *J. Mater. Sci.* **56** 174
[6] Ellis B L, Knauth P and Djenizian T 2014 *Adv. Mater.* **26** 3368-3397
[7] Xu S, Wei G, Li J, Ji Y, Klyui N, Izotov V and Han W 2017 *Chem. Eng. J.* **317** 1026–1036
[8] Argüello J A, Rojo J M and Moreno R 2019 *Electrochim. Acta* **294** 102–109
[9] Tjandra R, Liu W, Zhang M and Yu A 2019 *J. Power Sources* **438** 227009
[10] Pavlov A A, Sysa A V, Shaman Y P, Bazarova M I, Gavrilin I M and Polokhin A A 2017 *Russ. Microelectron.* **46** 82-87

Lithium aluminosilicate glass-ceramics for low-temperature anodic sealing of MEMS sensors

R Kleimanov[1,2], I Komarevtsev[1,2], Y Enns[1,2], Y Akulshin[1], A Korshunov[1], A Shashkin[3], D Arefiev[4] and A Kazakin[1,2]

[1] Peter the Great St Petersburg Polytechnic University, St Petersburg 195251, Russia
[2] Alferov University, St Petersburg 194021, Russia
[3] S.I. Vavilov State Optical Institute, St Petersburg 192171, Russia
[4] Innovative Crystal Systems, 2-1-1 3rd Samotechnyi per., Moscow 127473, Russia

Abstract. The results of a study of the anodic bonding parameters of transparent glass-ceramics based on lithium aluminosilicates which are promising as structural materials of MEMS and MOEMS sensors are presented. A comparison of the optical transmittance of these materials and classical for MEMS industry glasses has been carried out. The glass-ceramics electrical conductivity in a wide temperature range has been measured. The procedure of hermetic sealing of glass-ceramics by the anodic bonding at temperatures of 150 – 250 °C has been worked out. A prototype of glass-ceramic atomic cell has been fabricated.

1. Introduction

For the correct operation of various microelectromechanical (MEMS) sensors, an inert atmosphere or vacuum on the chip level is often required, as well as the optical transparency of the substrate. Such devices include lab-on-chip, quantum magnetometers and atomic clocks, gyroscopes, accelerometers and pressure sensors with optical measurement scheme, bolometers and other MEMS. For hermetic sealing of such devices on a wafer level, various bonding technologies are used, the most widespread of which is the high-temperature anodic bonding of silicon and glass wafers [1]. Materials for the anodic bonding are special borosilicate glasses (in the case of silicon technologies, it is Borofloat 33, Pyrex 7740, Hoya SD-2, etc.). These glasses have high ionic conductivity at anodic bonding temperatures (about 300 – 500 °C), optical transparency, thermomechanical compatibility with other sensor materials [2]. At the same time, high voltages and temperatures required for high-quality anodic bonding of commercial borosilicate glasses, fixed coefficient of thermal expansion (CTE) and a number of other factors limit the application of this technology for some types of MEMS.

Transparent glass-ceramics can become alternative materials for anodic sealing of MEMS because of many advantages over glasses – high working temperature, low gas permeability, low fragility, high strength, thermal shock resistance, durability during mechanical processing [3]. Glass-ceramics based on lithium aluminosilicates (LAS) are promising structural materials for MEMS sealing since the CTE of LAS glass-ceramic can be adjusted within a wide range by changing composition and heat-treatment schedule. Depending on the ratio of the crystallization phases in Li_2O-Al_2O_3-SiO_2 systems, some of which have a negative CTE (eucryptite), others have a positive CTE (spodumene, quartz-like solid solutions), a glass-ceramic material with a negative, zero or positive CTE is formed. Therefore CTE of LAS glass-ceramics may be approximately matched to that of silicon [4]. Successful attempts have been made to synthesize the LAS glass-ceramic composition for anodic bonding with silicon [5] and with fused quartz (using a sputtered conductive layer) [6]. But such materials were not widespread in the MEMS industry because of the presence of lithium is both a positive and a negative property. Lithium ions in the glass and glass-ceramics have an increased ionic mobility compared to sodium and potassium ions. For this reason, LAS materials can be bonded to silicon at lower temperatures than conventional borosilicate glasses. But lithium also has a high rate of penetration into the gate dielectrics of field-effect transistors and into silicon itself, which prohibits the use of such materials in silicon technologies. However there are many MEMS applications where this is not an obstacle – for example, microfluidic chips for bio-diagnostics and gas cells for miniature quantum devices such as magnetometers and atomic clocks [7–9]. For chip-scale atomic clock application, the low gas permeability of the glass-ceramics will allow for an increase in the long-term frequency stability by several orders of magnitude due to a decrease in the helium and neon leak rate [10]. Anodic sealing at temperatures below 170 °C will open the way to the mass production of MEMS quantum frequency standards with organic anti-relaxation wall coatings to reduce the spin relaxation in the alkali vapor cells and increase short-term frequency stability of the atomic clock [11].

This article presents the results of the work carried out to assess the prospects for use of optically transparent glass-ceramics based on lithium aluminosilicates (LAS) for anodic sealing of MEMS.

2. Experimental details

The objects of study were polished wafers with a thickness of 0.5 to 1 mm, made of glass-ceramics synthesized on the basis of lithium aluminosilicates with compositions of 55–65% SiO_2, 15–25% Al_2O_3, 5–15% Li_2O, 5–10% TiO_2, 2–4% CaO. Concentrations of the crystal phase (spodumene and β-eucryptite) varied by the duration and temperature of annealing. TiO_2 was used as compound nucleation agent. The basic glasses were prepared by melting and casting. The reagents of high pure oxides were mixed, and then they were melted in quartz ceramic crucibles with mixing by quartz stirrers at 1550 – 1620 °C for 8 hours in a laboratory furnace with molybdenum disilicide heaters. The melts were cast into the pre-heated metal mold. Subsequently, the glasses were annealed at 650 – 680 °C for 30 minutes and then passively cooled to room temperature. Secondary heat treatment was carried out in a two-stage mode in the temperature range of 680 – 1000 °C.

Temperature dependences of the LAS samples CTE were measured by dilatometer Linseis L75VS1000 (monocrystalline silicon was used as reference sample). Optical transmittance was measured by spectrophotometer Shimadzu uv-3600.

The measurement of the glass and glass-ceramic electrical conductivity was carried out on laboratory equipment designed for anodic bonding of silicon and glass (figure 1a). The wafers were preliminarily divided into square samples with an area of 1 cm^2, and blocking electrodes made of a 500 nm thick aluminum film were deposited on both sample sides by DC-magnetron sputtering. The magnitude of the initial current peak at the moment of switching on the voltage was used to estimate the specific electrical conductivity of glass-ceramics at a given temperature. Bonding of LAS samples to each other was carried out according to the scheme shown in figure 2b. Previously, a thin polysilicon layer was deposited on the sample, which serves as an anode.

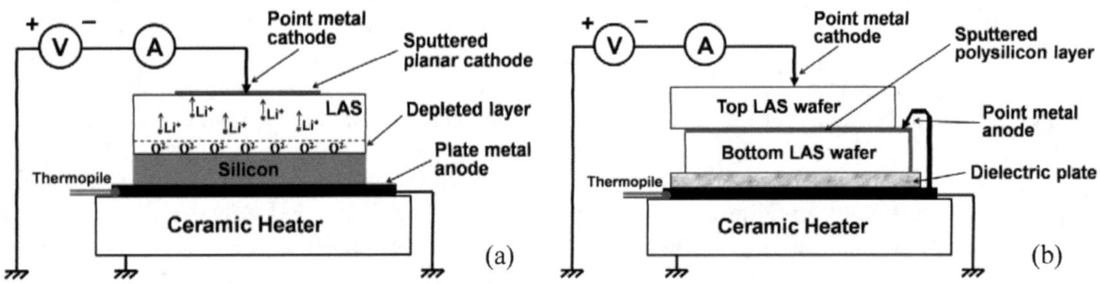

Figure 1. Schemes for Si–to–LAS bonding (a) and LAS–to–LAS bonding using thin Si interlayer (b).

3. Results and discussion

For the anodic bonding development, two types of LAS glass-ceramic material were tested. First type is SO-33M glass-ceramic (russian name is "CO-33M"), which has an average CTE close to zero and is a domestic analog of the Schott Zerodur glass-ceramic. Second type is experimental LAS samples with a CTE close to CTE of silicon. The work on the exact alignment of glass-ceramics and silicon CTE is not yet finished. To date, the best of the already synthesized samples have a slightly larger CTE (~$4\cdot10^{-6}$ 1/K) compared to silicon ($3.3\cdot10^{-6}$ 1/K). Figure 2 shows the difference between the CTE of silicon and the CTE of one of these samples ("LAS24") in the temperature range from 20 to 400 °C.

The measured values of LAS glass-ceramic conductivity were 2 to 3 orders of magnitude higher than that of Borofloat 33 glass in the temperature range of 150 – 250 °C, as shown in figure 3. This case made it possible to successfully carry out a hermetic anodic bonding of some glass-ceramics samples with monocrystalline silicon already at a temperature of 150 °C.

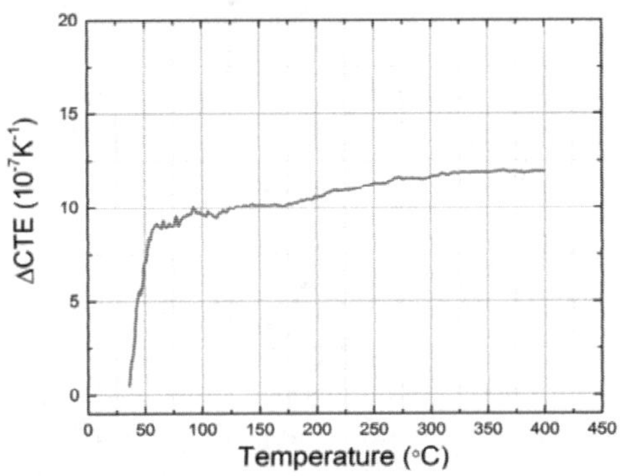

Figure 2. Difference of CTE of sample LAS24 and CTE of silicon vs. temperature.

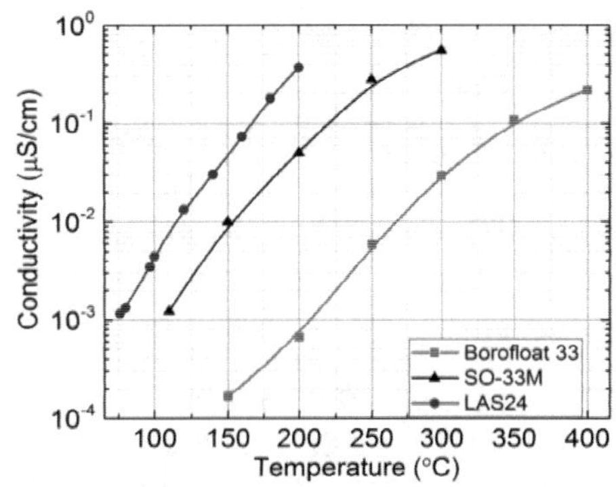

Figure 3. Conductivity of Borofloat 33, SO-33M and sample LAS24 vs. temperature.

The obtained time dependences of the current density during anodic bonding of SO-33M and LAS24 to silicon at different voltage and temperatures are shown in figure 4. The increase in the current in the first minutes of bonding was due to the spreading of the joint spot over the entire area of the silicon sample under the influence of electrostatic force. After that, all current responses showed a purely ionic character of conductivity with exponential decrease in current over time, which indicates the formation of a depleted layer at the glass-ceramic–silicon interface (as shown in figure 1a). The area under the presented curves is equal to the passed charge spent on the depleted layer formation and is a comparative criterion for the bonding strength (minimal charge density sufficient to create a bond between the Si and the glass is 3 mC/cm^2) [6].

All glass-ceramic samples provided uniform high-quality bonding to the entire silicon area at temperatures above 180 °C even when using a point cathode and low voltage (< 500 volts). The passed charge density during 30 minutes under these conditions was approximately 50 – 100 mC/cm^2. Bonding at a lower temperature was accompanied by a negligible charge density, even at elevated voltages (> 1000 V). This fact required the use of a distributed planar cathode. The use of an aluminum thin film electrode sputtered on the cathode side of the glass-ceramic samples made it possible to provide charge densities at the level of 10 – 20 mC/cm^2 when bonding was carried out at 100 – 150 °C. Samples of LAS24 glass-ceramics bonded with silicon at low temperatures and a voltage of 500 volts are shown in figure 5. The minimal temperature of a successful bonding was 110 °C. In this case, a stable sealing is achieved only under the surface of the planar electrode which is clearly visible on the left side of figure 5.

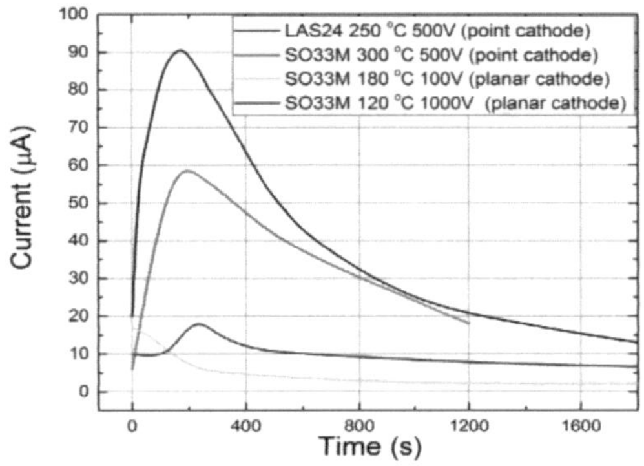

Figure 4. Current during the anodic bonding of silicon to SO-33M and LAS24 (S = 1 cm^2).

Figure 5. Results of silicon-LAS24 bonding at 200 °C, 150 °C, 110 °C (from left to right).

Samples of SO-33M glass-ceramic were bonded to each other. To perform this process a thin polysilicon films with a thickness of 50 – 200 nm was deposited on the top surface of one of samples (bottom wafer). The method of DC-magnetron sputtering from a single-crystal silicon target with a resistivity of 0.1-0.12 Ohm·cm was used. The resistivity of the deposited polycrystalline films was 5000 times higher and was approximately 600 Ohm·cm. This corresponded to the surface resistance of 30 MOhm/□ and 120 MOhm/□ for films with a thickness of 200 nm and 50 nm, respectively.

The bonding results depended on the polysilicon film thickness and the bonding scheme used, and are explained using figure 6, which shows samples bonded under different conditions at 250 °C and 800 V (figure 6a) and the corresponding time dependences of the bonding current (figure 6b). When the standard bonding scheme shown in figure 1a was used, the bottom LAS wafer acted as a silicon wafer, but its conductivity was provided by the lithium ion flow from the side connected to metal anode to the sputtered polysilicon film. Due to the presence of pinholes in the film, some of the lithium ions from the bottom wafer could pass through the barrier polysilicon layer and cause a local electrical breakdown of the depleted layer formed in the top wafer. Applying additional wire anode contact directly to the polysilicon film at the wafer edge did not help to completely neutralize this negative circumstance, especially for very thin films due to the their high resistance comparable to the resistance of bulk LAS glass-ceramics at bonding temperatures. As a result, the depleted layer could not be formed and the bonding interface at the edge of the wafer opposite to the point anode contained unbonded voids (central sample in figure 6a). The successful results on the SO-33M bonding were obtained when the scheme with additional dielectric plate shown in the figure 1b was used. The sample with 200 nm thick interlayer bonded in this way is shown at the left sides of figure 6a.

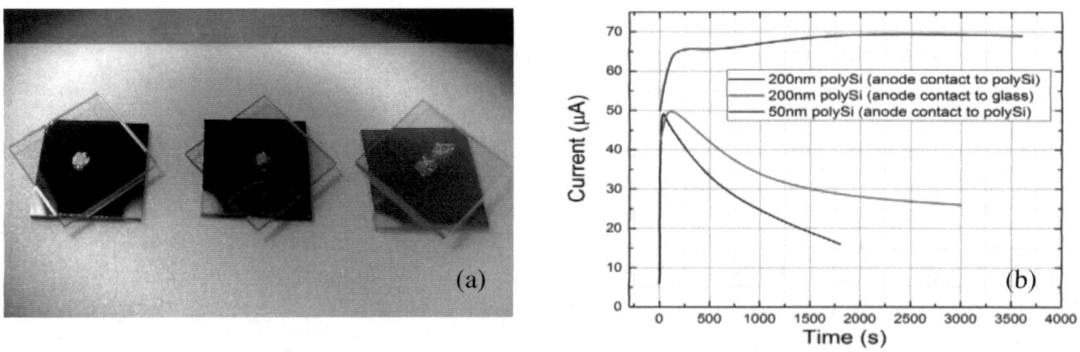

Figure 6. Results of SO-33M–to–SO-33M bonding using polysilicon interlayer with thickness of 200 nm (left and centre) and 50 nm (right) (a) and the time dependences of the bonding current (b).

This configuration of bonding equipment was used to fabricate a prototype of two-chamber gas cell for a chip-scale atomic clock, shown in figure 7. The cell was formed by sandwiching a glass-ceramic wafer with two through-hole cavities and connection channels between two another wafers by anodic bonding using polysilicon interlayers. Cell fabrication process is shown in figure 8.

First wafer micromachining was carried out by drilling with tubular drill bits with diamond grains binded by electroplating. The drill bits diameter was 2 and 3 mm. The wafer was fixed to the movable substrate holder by molten paraffin. The cavities formation was carried out by through-wafer sinking the drill bit at spin rate of 1000 rpm. Formation of two arc-shaped channels with a width of 200 μm was carried out by sinking the tubular drill bit to the depth of 150 μm at the center point between the cavities. After removing the paraffin in toluene, the wafer was cleaned in isopropanol, water and piranha solution. Then 200 nm thick polysilicon films were sputtered on both sides of the wafer.

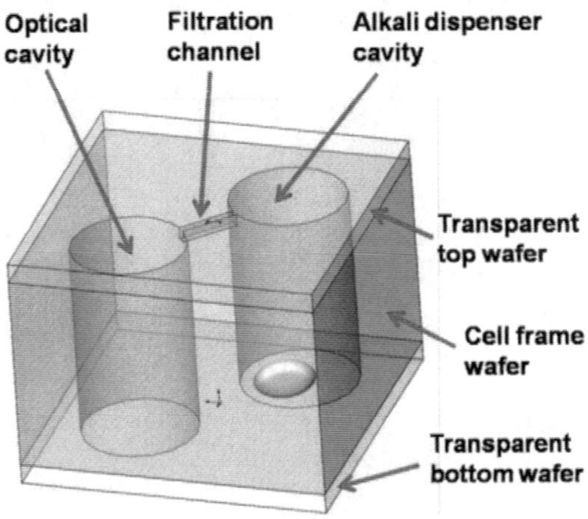

Figure 7. Scheme of two-chamber atomic cell.

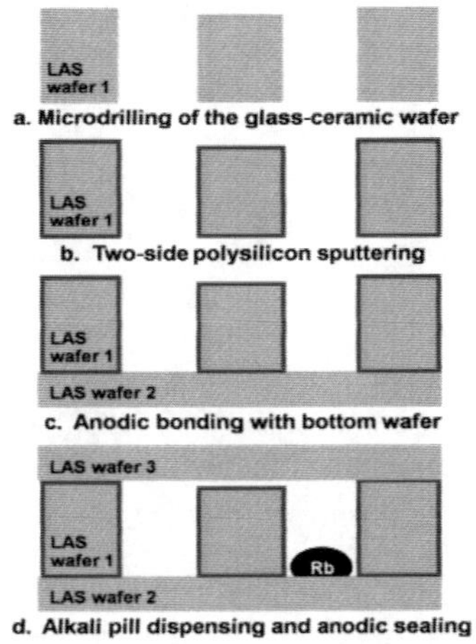

Figure 8. Cell fabrication process.

First anodic bonding at 250 °C and 500 V was carried out to join the micromachined wafer to the bottom glass-ceramic wafer. The operation of the filling the cell with an alkali dispenser was not performed at the prototype manufacturing stage. Second anodic bonding at the same conditions was carried out in vacuum chamber to seal the cell by top glass-ceramic wafer. Exponential drop of the current passed through the interface during the first and second anodic bonding are shown in figure 9. The value of the second bonding current was slightly less due to the lower real temperature near the connected surfaces, as the SO-33M glass-ceramic has a low thermal conductivity. The sandwich of three SO-33M glass-ceramic samples bonded in the described process is shown in figure 10.

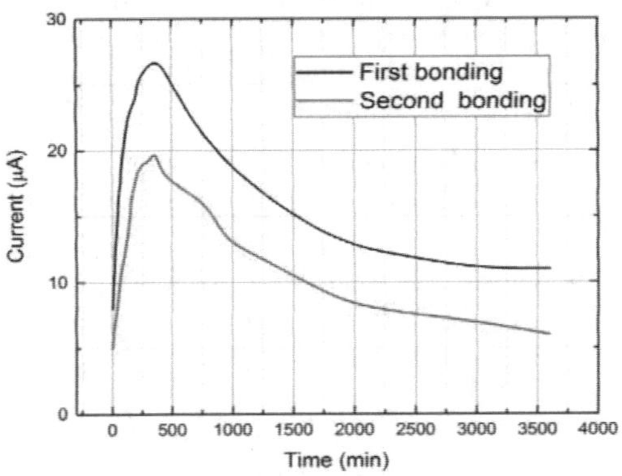

Figure 9. Current response of first and second anodic bonding for atomic cell sealing.

Figure 10. Fabricated atomic cell prototype (bonded samples size: 10 x 10 x 0.5 mm).

The transmittance spectra of Borofloat 33 glass used in the production of chip-scale atomic clock and SO-33M glass-ceramic in the range of 300 – 2500 nm were compared. As shown in figure 11, the SO-33M glass-ceramic transmits light only 2% worse than Borofloat 33 in the entire measured range. However, a near-zero coefficient of thermal expansion of SO-33M glass-ceramic is an advantage for MOEMS application.

The same bonding conditions were applied to seal SO-33M with fused quartz due to the proximity of their CTE. The bonded sandwich of SO-33M glass-ceramic and fused quartz after etching of the polysilicon layer in the optical cavity is shown in the figure 12. This technology may be promising for manufacturing of microfluidic sensors for UV diagnostics of various biomaterials and atomic cells with Hg vapours.

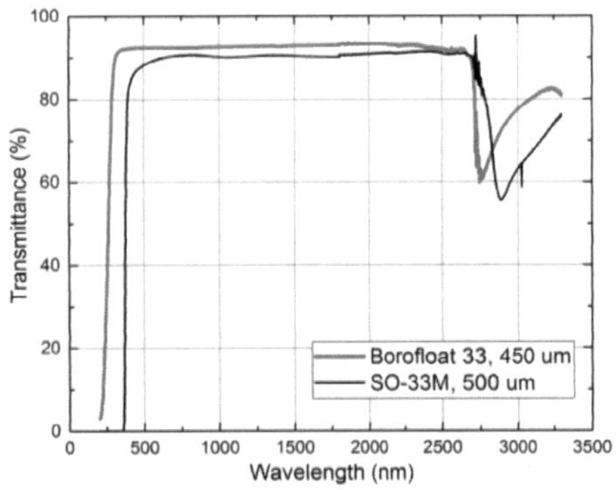

Figure 11. Transmittance of Borofloat 33 glass and SO-33M glass-ceramic.

Figure 12. Result of quartz–to–SO-33M bonding using polysilicon layer with thickness of 200 nm.

4. Conclusion

The hermetic anodic sealing of lithium aluminosilicate glass-ceramic wafers to each other as well as to silicon and fused quartz wafers at temperatures of 150 – 250 °C has been worked out. A prototype of glass-ceramic gas cell for chip-scale atomic clock has been fabricated. The obtained results can be applied in the technology of MEMS atomic cells and microfluidic devices.

Acknowledgments

This work was carried out in Peter the Great St.Petersburg Polytechnic University and was supported by a grant of Russian Science Foundation (project № 20-19-00146).

References

[1] Dziuban J A 2006 *Bonding in Microsystem Technology (Springer Series in Advanced Microelectronics* vol 24) ed K Itoh et al (Netherlands: Springer) pp 204-9
[2] Sinev L S and Ryabov V T 2017 *J. Micro/Nanolith. MEMS MOEMS* **16**(1) 015003
[3] O Dymshits, Shepilov M and Zhilin A 2017 *MRS Bulletin* **42** 200–5
[4] Li H, Du Y, Cheng J and Wang T 2008 *Key Engineering Materials* **368-372** 1436–8
[5] Shoji S, Kikuchi H and Torigoe H 1997 *Proc. 1997 10th Annu. Int. Workshop Micro Electro Mech. Syst.* 482–7
[6] Van Elp J, Giesen P T M and Van der Velde J J 2005 *J. Vac. Sci. Technol.* B **23** 96–8
[7] Sagitov E A, Ermak S V, Petrenko M V and Semenov V V 2016 *J. Phys.: Conf. Ser.* **769** 012044
[8] Kitching J 2018 *Appl. Phys. Rev.* **5** 031302
[9] Gorecki C 2012 *Procedia Engineering* **47** 898–903
[10] Rushton J A, Aldous M and Himsworth M D 2014 *Rev. Sci. Instrum.* **85** 121501
[11] Straessle R, Pellaton M, Affolderbach C, Petremand Y, Briand D, Mileti G and de Rooij N F 2013 *J. Appl. Phys.* **113** 064501

Single-step alkaline etching of deep silicon cavities for chip-scale atomic clock technology

I Komarevtsev[1,2], Y Akulshin[1] and A Kazakin[1,2]

[1] Peter the Great St Petersburg Polytechnic University, St Petersburg 195251, Russia
[2] Alferov University, St Petersburg 194021, Russia

Abstract. This paper presents the results of experiments on the development of the technology of MEMS alkali vapor cells for a miniature quantum frequency standard. The classical design of a two-chamber silicon cell containing an optical chamber, shallow filtration channels and a technical container for a solid-state alkali source was implemented in a single-step process of wet anisotropic silicon etching. To prevent the destruction of the filtration channels during etching of the through silicon cavities, the shapes of the compensating structures at the convex corners of the silicon nitride mask were calculated and the composition of the silicon etchant was experimentally found. The experiments results were used in the manufacture of chip-scale atomic clock cells containing vapors of ^{87}Rb or ^{133}Cs isotopes in the neon atmosphere.

1. Introduction

During last few decades, deep alkaline silicon etching has been extensively used in the microelectromechanical systems (MEMS) industry. The anisotropic nature of silicon micromachining in organic and inorganic alkaline solutions is due to the significant difference in the dissolution rates of different facets of monocrystalline silicon [1,2]. Compared to deep vertical plasma etching of silicon, alkaline etching does not require expensive equipment and allows simultaneous processing of a large number of wafers. Therefore, it is preferred for the mass production of various MEMS containing bulk structures of a simple rectangular topology with a not high aspect ratio. These are pressure sensors, thermal sensors, accelerometers, microfluidics chips, etc. [3]. The same type of MEMS includes gas cells for chip-scale atomic clocks (CSAC) [4].

MEMS atomic cells containing alkali metal vapors are key elements of miniature optical-pumped quantum devices [5], such as magnetometers or frequency standards [6] used in satellite navigation [7]. The small size and energy consumption of CSAC operating on the effect of coherent population trapping (CPT) [8,9] are an undeniable advantage in this field. The atomic vapor cells are typically made of a silicon wafer sandwiched between two transparent borosilicate glass wafers. The main processes of the MEMS cells technology are through-wafer etching of silicon, filling with an alkali metal source and vacuum-tight sealing of the cell in the appropriate buffer atmosphere. Methods for sealing cells and filling them with alkali vapors are described in review [10]. Through-wafer dry or wet etching is used to form absorption cavities in silicon. Moreover, according to [9], it is desirable to use thick silicon substrates (500 – 1000 μm), since the depth of the cavities for optical probing plays an important role to determine the short-term relative frequency stability of CSAC.

The simplest cell designs contain a single absorption chamber, like the first cells produced by the NIST Company [4]. However, single-chamber cells require extremely precise dosing of a liquid or solid alkali metal source, since its excess leads to a violation of transparency. For this reason, for the mass production of CSAC two-chamber cell designs are used. Such cells consist of two sealed cavities with a volume of several cubic millimeters, connected by narrow filtration channels. One of the cavities is used to fill the cell with a solid alkali dispenser. The other contains only pure alkali metal vapors and an inert gas to provide optical absorption at a wavelength corresponding to the atomic line D1 of ^{133}Cs, ^{85}Rb or ^{87}Rb. Filtration channels provides, due to a small cross-section, a transfer of alkali atoms into the optical cavity without byproducts formed during laser activation of the dispenser [10].

The cells of the described design are mainly made by plasma etching of silicon [11]. Alkaline etching of silicon is excellent for single-chamber cells fabrication [9]. However, it is difficult to form compact two-chamber cells with an optical path length of about 1 mm by alkaline etching due to excessive undercutting at the convex corners of the mask, i.e. in places where the filtration channels connect to the cavities [12]. The formation of the internal volume of the cell by alkaline etching can be carried out in two stages to separately form through-wafer cavities and shallow channels. However, this way requires performing of several operations of deposition and removing of mask coatings, precise patterns alignment, and photolithography on the surface with deep relief.

To reduce the cost of atomic cells technology, a single-step alkaline etching for the simultaneous formation of through-wafer cavities and filtration channels has been developed. To reduce channel underetching, the appropriate etching solution has been experimentally determined and the designs of the mask topology with convex corners undercutting compensating structures have been developed.

2. Experimental details

The objects of the study were polished 3″ silicon wafers with (100) orientation and thickness of 600 μm. 200 nm thick layer of silicon nitride was deposited on both sides of the wafer, and a mask pattern was formed on one side by direct photolithography and plasma etching. The mask pattern corresponded to the basic topology of two-chamber atomic cells with a total size of 6 x 6 x 1.6 mm (figure 1). The parameters of cells with ^{87}Rb vapors in the neon atmosphere at pressures of 200 – 400 Torr were calculated on the basis of preliminary experiments carried out in [13]. Their basic design contained two volumetric cavities with sizes of 3 x 1.5 x 0.6 mm, connected by rectangular channels with a length of 1 mm and a width of 100 or 200 μm.

Then, through-wafer etching of the cavities was carried out in tetramethyl ammonium hydroxide (TMAH) or potassium hydroxide (KOH) solutions at 80 °C for a time of about 6 – 10 hours (figure 2(a)). After mask removing, the wafers were used to make the cells as shown in figure 2.

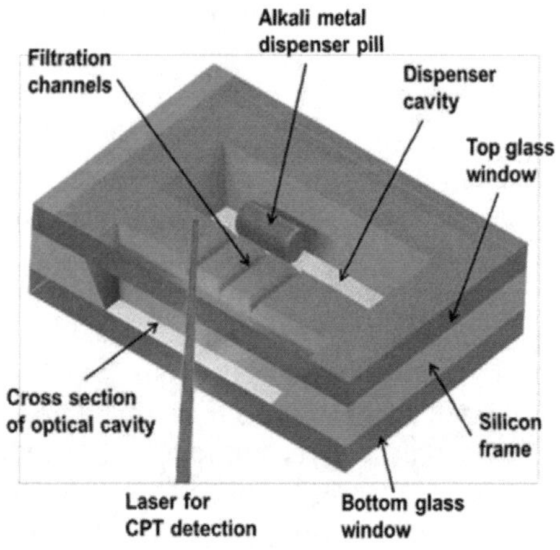

Figure 1. Two-chambers cell scheme.

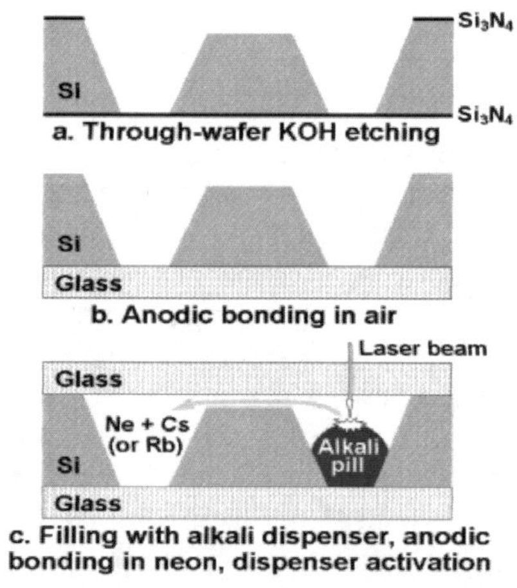

Figure 2. Cell fabrication process.

Depending on the etching conditions, the real appearance of the chips could differ from the basic design (figure 3). The etching rate and the distortion of the rectangular shape of the channels near the convex corners of the mask were measured using optical microscope and SEM (figure 4). For each alkaline composition, at a fixed etching depth, the undercutting length l and the beveled angle α, i.e. an angle between the direction [ml0] of maximum lateral undercutting and mask edges direction [110], were measured. The beveled angle α is a criterion for determining the Miller indexes of the {mlk} facets with the highest rate of dissolution in the etchant used [14]. The criterion of the relative rate of the convex corner undercutting is a dimensionless undercutting ratio $U = l / h$, where h is the etching depth. These values were used to select the type and calculate the size of additional mask elements designed to compensate for the corner undercutting.

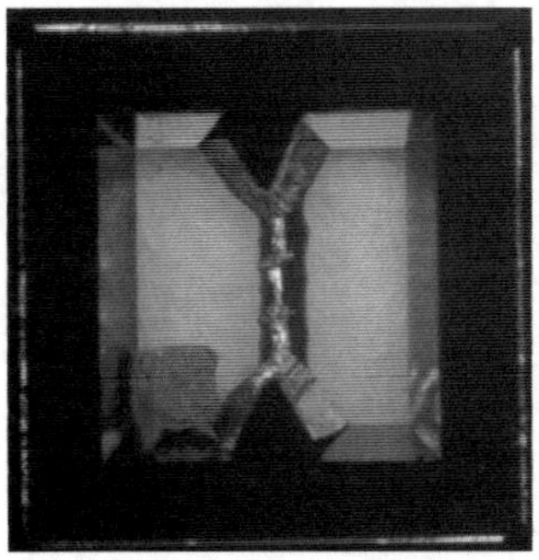

Figure 3. The ^{87}Rb vapor cell, made by etching in 30% KOH (Si chip size: 6 x 6 x 0.4 mm).

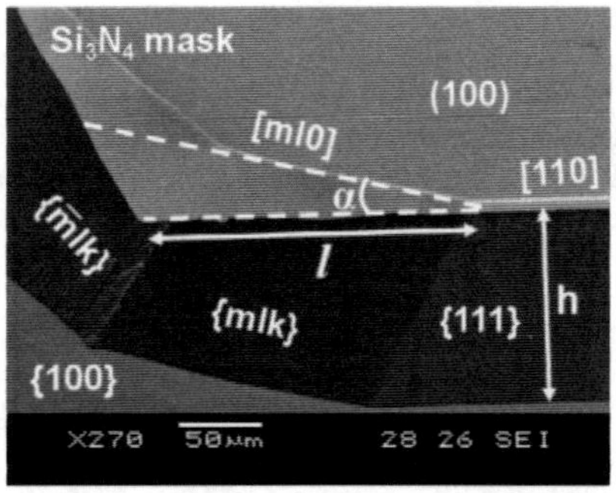

Figure 4. Convex corner undercutting.

3. Results and discussion

During anisotropic alkaline etching of (100) silicon surface through masks oriented in the <110> direction and containing rectangular holes with concave corners, through-wafer cavities or V-groove trenches with smooth walls inclined to the surface at an angle of 54.7 ° are formed, since they are limited by the (111) planes having an almost zero dissolution rate. When etching rectangular convex corners in silicon using anisotropic etchants such as TMAH or KOH, deformation of the edges always occurs (figure 4). This is due to the increased rate of dissolution of silicon facets with high Miller indexes ({211}, {311}, {331}, {411}, {772}, and etc.) with respect to the (100) surface. The specific facets that develop during convex corner undercutting depend on the composition of the etchant, its concentration, temperature, duration of etching, and the accuracy of matching the mask pattern with the [110] direction. For the two-chamber atomic cell design shown in figure 1, the convex corners of the mask are in places where the V-grooves channels connect to the through-wafer cavities. As a result, during the deep Si etching (> 400 μm) at normal conditions (30% aqueous solution of KOH at 80 °C), the cell filtration channels with a length of 1 mm disappear and optical and dispenser cavities completely transform into one large cavity, as can be seen in figure 3.

But it is possible to compensate for the undercutting of convex corners by a suitable mask design. In the corner compensation method, extra features called compensating structures are added at all convex corners in the mask layout design to eliminate the deformation at convex corners during alkaline etching of silicon. Compensating structures of various geometries are used – triangle, square, <110>-oriented beam, <100>-oriented beam, superimposed squares and asymmetric beams and etc. (there are more than a dozen variants in total) [12]. Most types of known compensating structures are not suitable for the design of our cell geometry. The limited space for their placement in the cell cavities near narrow filtration channels does not allow the use of compensating structures in the form of triangle, <100>-oriented beam, superimposed squares and asymmetric beams. Only two variants of corners undercutting compensators can be chosen – <110>-oriented beam and square. The design of the nitride mask with these elements is shown in figure 5(a) and figure 6(a), respectively. Figures 5(b) and 6(b) show a compensation schemes for the fabrication of V-groove trenches connected to cell cavities. The dotted lines designate the shape of the structures as etching proceeds. The consumption of <110>-beam or square takes place by the initiation of undercutting at its convex corners. The evolving etched profile of compensation pattern clearly indicates that the significant beveling will take place at the convex corners of channels. <110> beam type compensation design exhibits smaller beveling, depending on the beam's width. For our design the width b was chosen to be 50 μm.

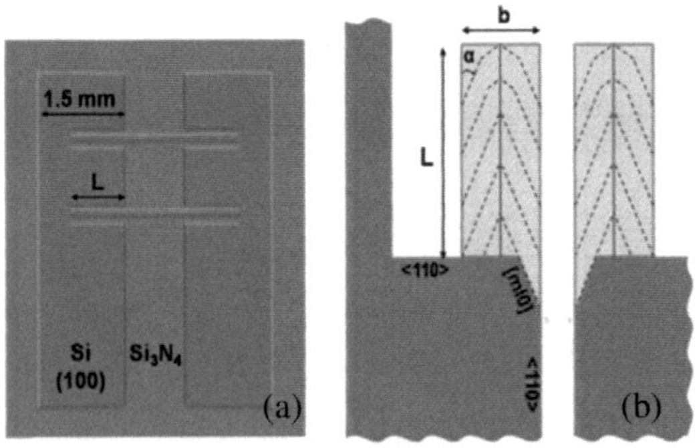

Figure 5. Mask design (a) and compensation scheme (b) for <110>-beam compensators.

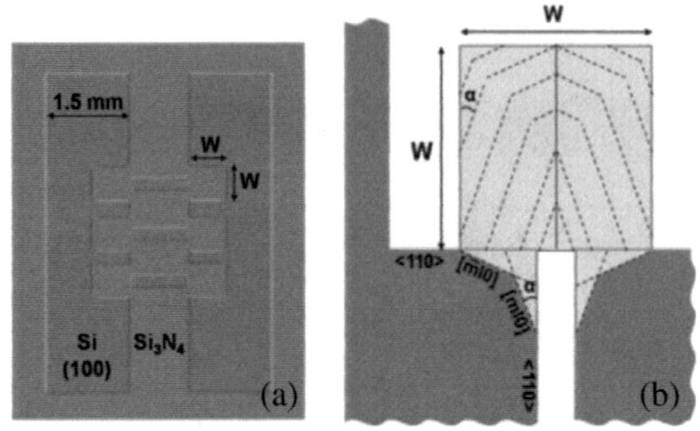

Figure 6. Mask design (a) and compensation scheme (b) for square compensators.

In order to determine the side length of such rectangular compensation geometries for etch depth h, simple formulas can be derived using geometrical relations:

$$L = Uh - 0.5b\text{ctg}\alpha \quad (1)$$

$$w = \frac{2Uh\sin\alpha}{2\sin\alpha + \cos\alpha} \quad (2)$$

Various parameters must be known for the analysis and the accurate design of the compensation geometry: etch rate of wafer surface plane (v_{100}); etch depth (h); undercutting length (l); beveled angle (α); undercutting ratio: ($U = l/h$). These parameters strongly depend on the etching characteristics of the etchant and can only be found experimentally.

For the atomic cells fabrication, etchants based on aqueous solutions of KOH and TMAH of different concentrations were studied. Taking into account the large etching depth, the concentration of solutions in each group was selected to ensure the maximum rate and uniformity of (100) silicon surface dissolution. From this point of view, the best results were obtained with 30% KOH and 15% TMAH solutions. At 80 °C, the etching rates were 1.3 μm/min and 0.7 μm/min, respectively.

For both etchants, the value of convex corner undercutting was investigated and the directions of maximum lateral undercutting were determined. Figures 7(a) and 7(b) show the top views of the <110>-oriented beam structures etched in these etchants at 80 °C to the same depth of 18 μm. The values of beveled angle α, undercutting length l and ratio U and are given in table 1 in the corresponding columns. For both etchants, angle α was close to 30.96 °. According to [14], it means that the maximum lateral undercutting occurs in the [410] direction, and the {mlk} facets developed during etching are {441} for 15% TMAH and {411} for 30% KOH [15,16]. By substituting the values U and α in formulas 1 and 2, the length L of the <110>-oriented beam and side W of the square compensators, which are necessary to protect the channels during through-wafer etching of 600 μm thick substrate, was calculated. These values are shown in the last two columns of table 1. It is clear that TMAH etchant is not suitable for the manufacture of cells because the dimensions of both compensating structures exceed the width of the optical cavity or the distance between adjacent channels. In the case of KOH etchant, the compensating structures would occupy almost the entire space of the optical cavity. The huge size of the rectangular compensators for both etchants is due to the fact that the dissolution rates of the {110}, {441} and {411} facets exceed the etching rate of the (100) plane by one and a half or two times [2,16].

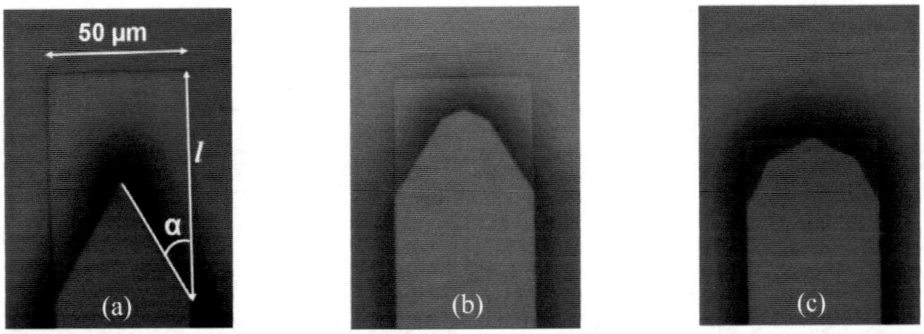

Figure 7. Convex corner undercutting in 15% TMAH (a), 30% KOH (b) and 30% KOH:IPA (c) solutions at 80 °C (beams width is 50 μm, etch depth is 18 μm).

Table 1. Etching parameters for calculating the size of rectangular compensating structures

Etchant	v_{100} (μm/min)	l (μm)	$U = l / h$	α (°)	{mlk}	L_{600} (μm)	W_{600} (μm)
15% TMAH	0.7	75.4	4.19	29.9 ± 2	{441}	2470 ± 4	1345 ± 50
30% KOH	1.3	45.4	2.52	30.7 ± 2	{411}	1470 ± 3	820 ± 30
30% KOH : IPA	1.2	23.1	1.28	25.3 ± 2	{331}	715 ± 5	370 ± 30

To reduce the rate of filtration channels undercutting, aqueous KOH solutions with additives of isopropyl alcohol (IPA) have been used. It is known that alcohol additives in KOH etchants reduce the rates of dissolution of the silicon facets with high Miller indexes [14,15]. The best results for the cells fabrication have been obtained with an etchant consisting of equal volumes of 30% aqueous KOH solution and IPA. Silicon etching has been carried out at a temperature of 80 °C in a container with a cooled lid to prevent the alcohol evaporation. The view of convex corner undercutting after etching under these conditions to a depth of 18 μm is shown in figure 7(c). The beveled angle α is 25.3 ° ± 2 °, which is close to the angle between [110] and [310] directions – 26.57 ° [14]. According to [15], it means that the facets developed during etching are {331}. The (100) plane etching rate in KOH:IPA etchant is 1.2 μm/min and it is comparable to the etching rate of aqueous KOH solution. But undercutting ratio U is two times less. The calculated average sizes of compensating structures were L = 715 μm and W = 370 μm for the <110>-beam and square types, respectively.

Using the ACES program, a simulation of the cell formation process in this etchant with a mask topology containing both of compensating structures was performed. For the simulation, the etching rate V_{100} from table 1 and the ratios V_{331}/V_{100} and V_{110}/V_{100} from [15] were used. For example, the etched silicon relief evolution when using a <110>- beam corner undercutting compensator is shown in figure 8. The achievement of the etching depth of 600 μm was determined by the complete closure of the (111) planes inside a square hole with a side a equal to 850 μm.

After making the appropriate photomask, KOH:IPA etchant was successfully used to fabricate one hundred atomic cells on a single silicon wafer. The optical cavities were etched through, the compensating structures were completely etched at the upper side of silicon relief, leaving the V-groove filtration channels open. However, there were distorted silicon ridges near the convex corners on the bottom side of the relief. It may be due to the development of additional facets during deep etching, which is difficult to simulate in the frame of simple model. To ensure that this phenomenon is not an obstacle to the normal operability of the atomic cells, the formation process was carried out with over-etching. The etching duration was increased to 11 hours, while the compensating structures size was increased to 1 mm for <110>-oriented beam and to 500 μm for square (while the distance between the channels was 700 μm). Figure 9 shows photos of the chips after the final stages of alkali vapor cell manufacturing – filling with a source of rubidium (or caesium) dispenser and anodic sealing with glass in a neon atmosphere. The suitability of cells manufactured using this technology for chip-scale atomic clocks was confirmed by appropriate tests [17].

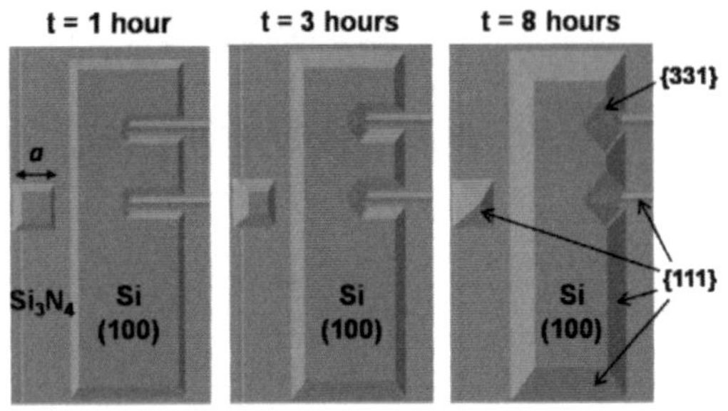

Figure 8. Evolution of etched relief of cell cavity with <110>-beam compensators in 30% KOH:IPA etchant at 80 °C (a = 850 μm).

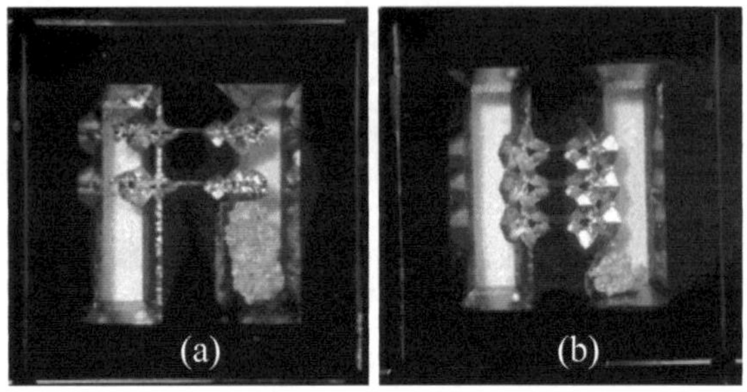

Figure 9. Cells made by etching in 30% KOH:IPA solution with the mask with compensators of the <110>-oriented beam (a) and square (b) types.

4. Conclusion

A method of single-step alkaline etching of silicon for fabrication of two-chamber MEMS atomic cells containing vapor of rubidium-87 or caesium-133 isotopes in a neon atmosphere was developed. To prevent the destruction of the filtration channels during etching of the through-wafer cavities, the <110>-oriented rectangular compensating structures at the convex corners of the silicon nitride mask were calculated and 30% KOH:IPA solution was used as silicon etchant. Developed technology is perspective for mass-production of chip-scale atomic clock.

Acknowledgments

This work was carried out in Peter the Great St.Petersburg Polytechnic University and was supported by a grant of Russian Science Foundation (project № 20-19-00146).

References

[1] Seidel H, Csepregi L, Heuberger A and Baumgartel H 1990 *J. Electrochem. Soc.* **137** 3612–26
[2] Shikida M, Sato K, Tokoro K and Uchikawa D 2000 *Sens. Act.* A **80** 179–88
[3] Petersen K E 1982 *Proceedings of the IEEE* **70**(5) 420–57
[4] Knappe S, Shah V, Schwindt P D D, Hollberg L, Kitching J, Liew L and Moreland J 2004 *Appl. Phys. Lett.* **85** 1460–2
[5] Kitching J 2018 *Appl. Phys. Rev.* **5** 031302
[6] Sagitov E A, Ermak S V, Petrenko M V and Semenov V V 2016 *J. Phys.: Conf. Ser.* **769** 012044
[7] Lozov R K, Baranov A A, Ermak S V and Semenov V V 2019 *J. Phys.: Conf. Ser.* **1236** 012077
[8] Ermak S V, Semenov V V, Piatyshev E N, Kazakin A N, Komarevtsev I M, Velichko E N, Davydov V V and Petrenko M V 2015 *St. Petersb. Polytech. Univ. J. Phys. Math.* **1** 37–41
[9] Knappe S 2008 *Compr. Microsyst.* **3** 571–612
[10] Knapkiewicz P 2019 *Micromachines* **10** 25
[11] Chutani R K, Hasegawa M, Maurice V, Passilly N and Gorecki C 2014 *Sens. Act.* A **208** 66–72
[12] Pal P and Sato K 2015 *Micro and Nano Syst. Lett.* **3** 6
[13] Fedorov M I, Ermak S V, Petrenko M V, Piatyshev E N and Semenov V V, 2016 *J. Phys.: Conf. Ser.* **769** 012046
[14] Zubel I 2000 *Sens. Act.* A **84** 116–25
[15] Zubel I, Barycka I, Kotowska K and Kramkowska M 2001 *Sens. Act.* A **87** 163–71
[16] Pal P, Sato K and Chandra S 2007 *J. Micromech. Microeng.* **17** R111–33
[17] Bobrov M A et al 2019 *J. Phys.: Conf. Ser.* **1400** 077014

III–V nanowires for ammonia detection

V M Kondratev[1], A S Kuznetsov[1], V O Gridchin[1], S V Fedina[1], R R Reznik[1, 5], S S Nalimova[2], V A Moshnikov[2], G E Cirlin[1-4] and A D Bolshakov[1, 3]

[1] Alferov University, St. Petersburg, 194021, Russia
[2] Saint Petersburg Electrotechnical University "LETI", St. Petersburg, 197376, Russia
[3] ITMO University, St. Petersburg, 197101, Russia
[4] Institute for Analytical Instrumentation RAS, St. Petersburg, 190103, Russia
[5] Saint Petersburg State University, St. Petersburg, 199034, Russia
[6]Moscow Institute of Physics and Technology, Moscow, 117303, Russia

Abstract. Ammonia is an inorganic agent found both in nature and in the human body, which is of great interest for modern sensory applications. Here we use GaP, GaN and GaAs epitaxial nanowires as sensitive elements of the ammonia sensors fabricated via a simple protocol on the platform with golden interdigital contacts. Impedancemetry is used to study change of the device properties with addition of ammonia to the reference medium (water). GaP and GaN-based devices exhibit sufficient response to the ammonia presence with the detection limit lower than 10ppm. This work is aimed at fabrication and study of precise, technological and relatively cheap ammonia sensors compatible with a liquid medium, and motivated by the possibility of using this type of adsorption sensors in medical, environmental equipment and biological purposes.

1. Introduction

Ammonia in concentrations of the order of 1 ppm is produced by human and is a biological marker of pathological changes in human body, but concentrations of the order of 100 ppm, ammonia is toxic and hazardous. Among modern methods for detection of small concentrations of inorganic agents in liquids, indicator analysis and voltammetry methods are mostly often used. Classical adsorption sensors based on metal oxides, which have proven themselves in gas sensing, are widely presented [1-5]. To improve the sensitivity features of the adsorption-type device, structures with developed area can be used. Despite the high ratio of the surface area compared to volume of nanowires (NWs) and the pronounced conductivity features of quasi-1D structures compared to bulk materials, the adsorption properties of III–V NWs and their correlation with electrical characteristics are still very poorly studied [6–7]. Here we fabricate and study several ammonia sensors based on different III-V NWs to investigate potential of these nanostructures for sensory applications.

2. Synthesis and study of nanowires

Vertically oriented GaAs, GaN and GaP NWs were synthesized on (111) silicon substrates by molecular beam epitaxy, typical parameters and growth conditions are presented in previous works [8–10]. The synthesized structures were studied by scanning electron microscopy, and transferred to an auxiliary substrate for further characterization with Raman scattering spectroscopy. The results of the measurements are shown in Figure 1.

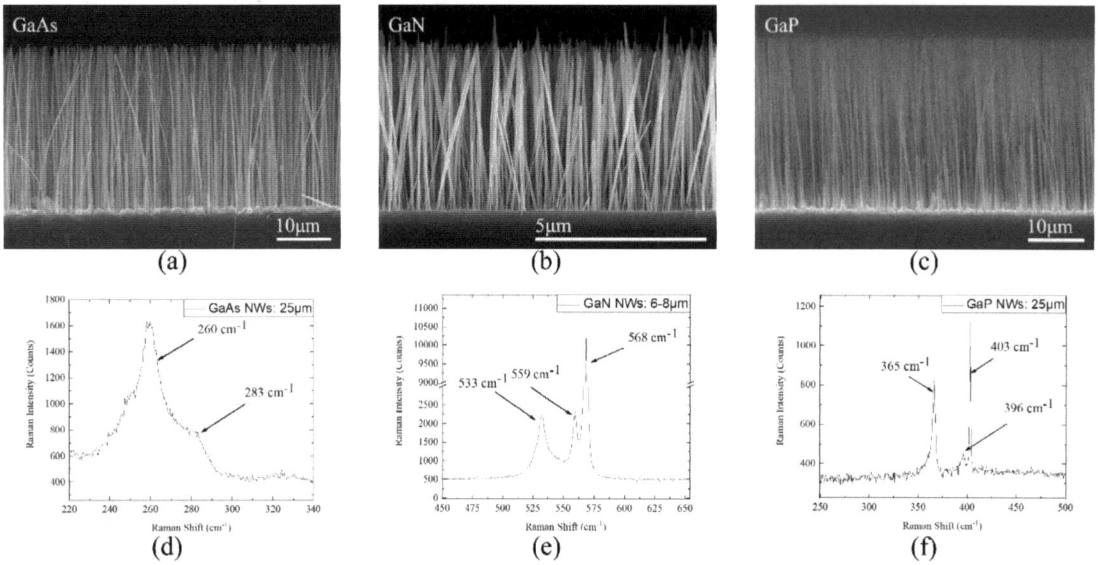

Figure 1. SEM image of the vertically oriented NWs and Raman spectrum: (a, d) – GaAs, (b, i) – GaN, (c, f) – GaP.

GaAs NWs have a length of about 25 μm and a cross section of 100–200 nm, and their Raman scattering spectra show modes that correspond to the zinc blende (ZB) GaAs crystal structure [11–12]. GaN NWs are about 6–8 μm long and 20–80 nm cross–section have sharp features in Raman spectra corresponding to the wurtzite structure [13].

Raman spectrum of self-catalyzed GaP NWs [14] (about 25 μm long and 150–250 nm cross–section) is well consistent with the results of the work which examined high crystal quality ZB GaP [15].

3. Sensor fabrication and ammonia detection

The NWs were separated from the growth substrate by ultrasonication in isopropanol, and transferred to the surface of a sensor platform with gold interdigital contacts (contact pitch of 5 and 10 μm) with subsequent annealing for sensor fabrication. The step–by–step protocol of sensor production is shown in the Figure 2.

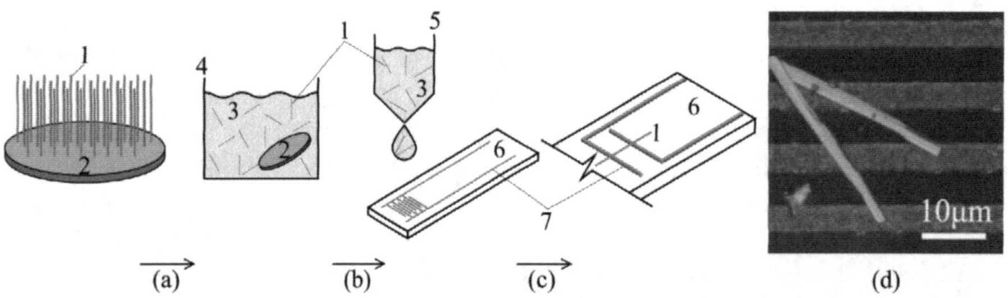

Figure 2. The step–by–step sensor fabrication protocol: 1 – as-synthesized vertical Si NWs, 2 – Si substrate, 3 – isopropyl alcohol, 4 – ultrasonic separation tube, 5 – diffusion system, 6 – sensor platform, 7 – gold interdigital gold contacts; (a) – separation of NWs from the growth substrate by ultrasonication (b) – dispersion of NWs in isopropanol and transfer to the sensor platform, (c) – fabrication of sensor and formation of gold–NWs contacts by annealing, (d) – typical image of a III-V NW on Au interdigital contacts obtained by optical microscopy.

Gold–NWs contacts are found to be of the Schottky type, which is proven by the voltage–current characterization obtained using Keithley 2400. Also, it is necessary to take into account the contribution to the overall conductivity of the sensor from the electrochemical cell formed between the interdigital contacts when the sensor is placed in liquid medium. The impedance of the NWs and electrochemical cell in this sensor configuration are connected in parallel. A schematic electrical diagram of a sensor based on III – V NWs is shown in Figure 3.

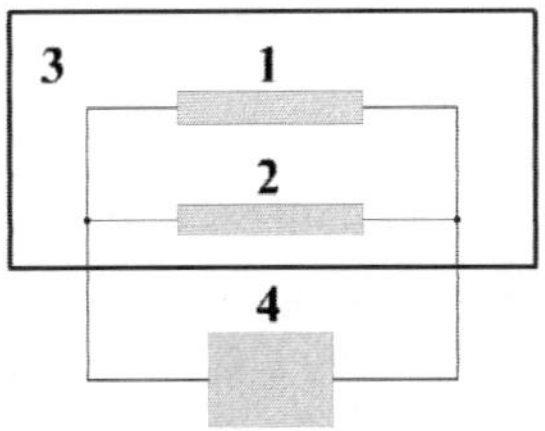

Figure 3. Schematic electrical diagram of a sensor based on III – V NWs, drawn in accordance with a parallel equivalent circuit: 1 – the III-V NWs impedance; 2 – the electrochemical cell impedance, filled with a liquid sample; 3 – the experimental area with sensor and liquid medium; 4 – the impedancemeter Z-500P.

Based on the latter assumption, the signal corresponding to the current passing through the NWs can be obtained via subtraction of the impedance of the electrochemical cell from the total sensor impedance. Total impedance $(Z_\Sigma)_i$ of the sensor for each i-th frequency of the voltage measuring signal can be calculated using the expression 1:

$$\frac{1}{(Z_\Sigma)_i} = \frac{1}{(Z_{H_2O})_i} + \frac{1}{(Z_{NWs})_i}; \tag{1}$$

here $(Z_{H2O})_i$ is the impedance of the electrochemical cell, $(Z_{NWs})_i$ is the impedance of the NWs. To describe the sensitivity of the sensors, the real and imaginary parts of the impedance were considered:

$$Z_i = \mathrm{Re}(Z)_i + j \cdot \mathrm{Im}(Z)_i; \tag{2}$$

Z_i is the impedance, $\mathrm{Re}(Z)_i$ is the real component of the impedance, $\mathrm{Im}(Z)_i$ is the imaginary, $j = (-1)^{1/2}$.Then the expression for signal normalization, which consists in subtracting the contribution from the electrochemical cell from the sensor impedance formed between the interdigitated contacts can be represented in the form:

$$\mathrm{Re}(Z_{NWs})_i = \frac{R}{R^2 + I^2}, \mathrm{Im}(Z_{NWs})_i = \frac{I}{R^2 + I^2}; \tag{3}$$

where

$$R = \left(\frac{\mathrm{Re}(Z_\Sigma)_i}{|Z_\Sigma|_i^2} - \frac{\mathrm{Re}(Z_{H_2O})_i}{|Z_{H_2O}|_i^2} \right), I = \left(\frac{\mathrm{Im}(Z_\Sigma)_i}{|Z_\Sigma|_i^2} - \frac{\mathrm{Im}(Z_{H_2O})_i}{|Z_{H_2O}|_i^2} \right); \tag{4}$$

In accordance with the above, Figure 4 shows the impedance spectra of the sensors based on III-V NWs in a comparative medium – distilled H_2O and in H_2O with the target adsorbate – NH_3. In this paper, we designate the exposure time as an interval during which the sensor is exposed to liquid adsorbates before the measurement of the impedance spectrum begins. At the same time, the exposure of the comparative medium-water was carried out for about 1000 s., in order to achieve an equilibrium of the adsorption processes on the surface of the NWs.

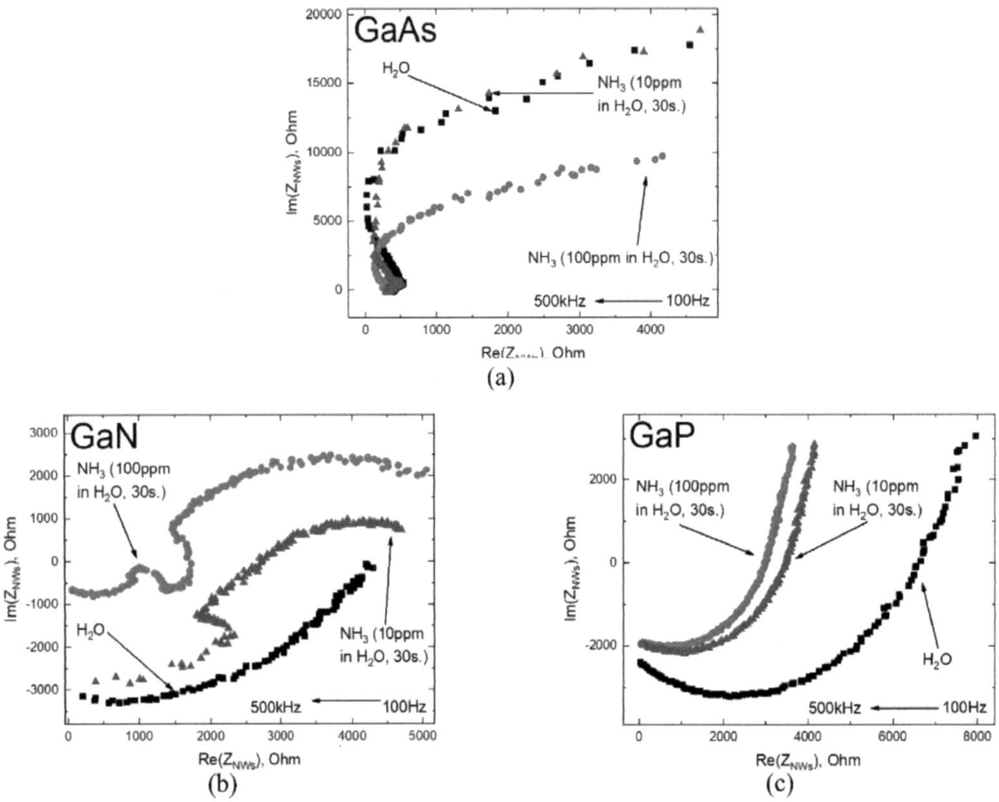

Figure 4. Impedance spectra of the sensors based on: (a) – GaAs, (b) – GaN, (c) – GaP; Plots: black – in pure deionized H_2O with exposure time over 1000s.; red - spectrum in 100ppm NH_3 dissolved in distilled H_2O with adsorbate exposure time 30s.; blue – sensor impedance spectrum in 10ppm NH_3 dissolved in distilled H_2O with adsorbate exposure time 30s.

The GaAs NWs based sensor allowed us to qualitatively determine the presence of ammonia in distilled water followed by reduction of the imaginary part of the impedance over the entire range of voltage frequencies relative to the sensor impedance spectrum in the presence of distilled water. However, no restoration of the original electrical performance was observed. We assume that this may be due to the defective nature of the adsorption centers on the NWs and the degradation of the GaAs crystal surface, which together determines the capacitive nature of the GaAs NWs sensor and makes a significant contribution to the reactive component of the impedance. The optimal frequency range for working with GaAs sensors is 200 Hz–8 kHz, where the strongest separation of the impedance spectra corresponding to water and ammonia are observed.

GaN and GaP NWs demonstrated high chemical stability in liquid medium. The first demonstrated the possibility of reproducible ammonia detection with the restoration of the electrical characteristics of the sensor. In the presence of ammonia, an increase in the imaginary part of the sensor impedance was observed with the real part almost unchanged over the entire frequency range under consideration. The second sensor is suitable for quantitative detection of ammonia with reduction of the imaginary and real part of the impedance during adsorption of ammonia, with the restoration of the original electrical characteristics in water. In this case, the entire considered range of measuring frequencies is found suitable.

4. Conclusion

Here we fabricate and study ammonia sensory properties of the devices based on GaP, GaAs and GaN NWs in liquid medium. The signal normalization is proposed, which makes it possible to study changes in the NWs conductivity due to the ammonia molecules adsorption on the NW sidewalls. Normalization consists in subtracting the electrochemical contribution of the liquid between the interdigitated contacts to the sensor impedance spectrum and allows us to evaluate the electrical properties of NW and their correlation with the adsorption properties.

The fabricated GaN- and GaP-based devices demonstrate ammonia detection limit lower than 10ppm in water. Conclusions are drawn about the biocompatibility of quasi-one-dimensional nanostructures of compounds III-V with respect to slightly alkaline media, which are close in concentration to those of humans.

We demonstrated an effective protocol for manufacturing sensors based on NWs III–V and showed the optimal frequencies for working with this type of sensor when using impedance engineering - as a way to control the effect of adsorption on the electrical properties of NWs in terms of sensitivity to ammonia.

Acknowledgments
V.M.K and A.S.K. thanks for support of the sensor fabrication Ministry of Science and Higher Education of the Russian Federation (grant № FSRM-2020-00011). A.D.B. thanks for support of the spectrometry measurements the Russian Science Foundation (grant 20-72-10192).

V.O.G. and G.E.C. thanks for support of the GaAs and GaN synthesis the Ministry of Science and Higher Education of the Russian Federation as part of state order no. 0791-2020-0003.

S.V.F. thanks for support of the substrates preparation and GaP NWs growth the Ministry of Science and Higher Education of the Russian Federation as part of state order no. 0791-2020-0005.

References

[1] Nalimova S S, Bobkov A A, Kondrat'ev V M, Ryabko A A, Moshnikov V A and Shomakhov Z V 2020 *IEEE Conference of Russian Young Researchers in Electrical and Electronic Engineering (EIConRus)* 991.

[2] Nalimova S S, Kondratev V M, Ryabko A A, Maksimov A I and Moshnikov V A 2020 *Journal of Physics: Conference Serial* **012033** 1658.

[3] Nalimova S S, Kondrat'ev V M 2020 *IEEE Conference of Russian Young Researchers in Electrical and Electronic Engineering (EIConRus)* 987.

[4] Levkevich E A, Maksimov A I, Kirillova S A, Nalimova S S, Kondrat'ev V M and Semenova A A 2020 *IEEE Conference of Russian Young Researchers in Electrical and Electronic Engineering (EIConRus)* 984.

[5] Kondratev V M, Bolshakov A D and Nalimova S S 2021 *IEEE Conference of Russian Young Researchers in Electrical and Electronic Engineering (ElConRus)* 1163.

[6] Kang D, Ko J H, Bae E, Hyun J, Park W, Kim B K, Kim J J and Lee C 2004 *J. Appl. Phys.* **96** 7574.

[7] Arora P, Sindhu A, Dilbaghi N, Chaudhury A 2013 *Appl Nanosci* **3** 363.

[8] Bolshakov A D, Dvoretckaia L N, Fedorov V V, Sapunov G A, Mozharov A M, Shugurov K Yu, Shkoldin V A, Mukhin M S, and Cirlin G E 2018 *Semiconductors* **52(16)** 2088.

[9] Gridchin V O, Kotlyar K P, Reznik R R, Dvoretskaya L N, Parfen'eva A V, Mukhin I S and Cirlin G E 2020 *Tech. Phys. Lett.* **46** 1080.

[10] Fedorov V V, Bolshakov A, Sergaeva O, Neplokh V, Markina D, Bruyere S, Saerens G, Petrov M I, Grange R, Timofeeva M, Makarov S V and Mukhin I S 2020 *ACS Nano* **14 8** 10624.

[11] Volodina V A and Sinyukova M P 2014 *JETP Letters* **99** 396.

[12] D. Spirkoska et al 2009 *PhysRevB* **80** 245325.

[13] Li E, Song S, Ma D, Fu N and Zhang Y 2014 *Journal of Elec Materi* **43** 1379.

[14] Bolshakov A D, Mozharov A M, Sapunov G A, Fedorov V V, Dvoretckaia L N and Mukhin I S 2017 *Mater. Res. Express* **4** 125027.

[15] Dobrovolsky A, Sukrittanon S, Kuang Y J, Tu C W, Chen W M and Buyanova I A 2014 *Appl. Phys. Lett* **105** 193102.

Influence of laser processing conditions on the depth and microhardness of layers formed on titanium

V Koshuro, L Osipova, O Markelova, M Fomina and A Fomin

Yuri Gagarin State Technical University of Saratov, Saratov 410054, Russia

dimirion@mail.ru

Abstract. In this work, we studied the depth and microhardness of the layers formed on titanium during laser processing in a graphite environment. The energy and duration of the laser pulse varied from 0.76 to 10.17 J and from 0.5 to 3 ms, respectively. As a result of processing, the formation of layers characterized by a depth of 8.3-800 μm and a microhardness of 9.68-28.01 GPa took place. Regression models are constructed that describe the effect of laser processing conditions on the indicated characteristics of the layers. It was found that the pulse energy had the greatest effect on the depth and microhardness of hardened titanium.

1. Introduction

At present, carburizing technologies using gas and solid carbon-containing media are widely used to harden the titanium surface. Cementation in a gas atmosphere at a temperature of 1150 °C allows to reduce the coefficient of friction of the titanium surface. Surface properties depend on the composition of the reaction medium [1].

Cementation is also carried out by pressing titanium bases with graphite powder and subsequent heating. Heating up to 800 ° C is carried out by supplying current for 20 minutes. As a result, titanium carbide layers are formed, characterized by a carbon content of up to 36 at.% And a hardness of 6.53 GPa (Knoop, 100 gf). [2]. Heating of titanium products can also be induction. The titanium product is placed in a refractory container filled with graphite. During processing, the samples are heated to temperatures comparable to the temperature of the container. As a result of cementation, TiC layers with a thickness of up to 14 μm and a hardness of up to 20 GPa are formed [3].

It is more rational to carry out local hardening by laser processing of titanium in a graphite environment. Laser treatment of a powder mixture deposited on a titanium sublayer and consisting of titanium with a dispersion of 20-30 μm and CNT, at a radiation power of up to 700 W, allows the formation of carbide layers with a hardness of up to 1125 $HV_{0.5}$ (11 GPa) [4]. After laser processing in a pulsed mode with a pulse duration of 4 to 12 ms and a power of 250 W, layers with a hardness of up to 934 HV (9.2 GPa) are formed. The formed layers were characterized by a dendritic structure [5].

Despite the existence of works on laser carburizing of titanium, the effect of laser processing conditions on the thickness of the hardened layers and their microhardness has not been sufficiently studied.

2. Methodology

In the experimental work, specimens in the form of VT1 titanium plates with dimensions of 10 × 10 × 1 mm were used. On the surface of the plates, a bulk layer with a thickness of 100-300 μm was formed from graphite powder with a dispersion of 10-35 μm. The surface was processed with a diameter of 8 mm with 50% overlap using the "LRS-50A" installation. Laser radiation was focused on the graphite surface into a spot 0.75 mm in diameter (figure 1). The energy and duration of the laser pulse varied from 0.76 to 10.17 J and from 0.5 to 3 ms, respectively.

The structure of the layers was investigated by optical and scanning electron microscopy (SEM) using the optical system of the "PMT-3" microhardness tester and the "MIRA II LMU" electron microscope. The depth of structural changes was measured from the images of microsections according to the well-known method [6]. Hardness of the coatings was evaluated by microindentation using "PMT-3M" (at the load of 0.98 N). The construction of regression models was carried out in the "Data Fit 9" program.

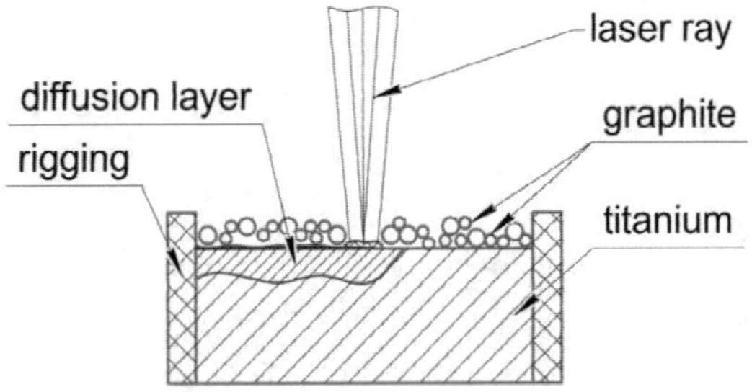

Figure 1. Laser treatment of titanium in a graphite environment.

3. Results

Laser treatment in a graphite medium made it possible to form a layered system on titanium, consisting of a thin, 0.5-6 μm thick coating and diffusion layers with a depth of 8.3-800 μm. These structural areas were characterized by the presence of a fine (from 0.1 μm) dendritic structure (figure 2). The size and shape of the dendritic formations depended on the pulse duration. With an increase in the duration of the laser pulse, the structural elements merged; therefore, the size of individual grains increased. Dendritic structure is visually similar to the structures formed by laser processing in similar works [5].

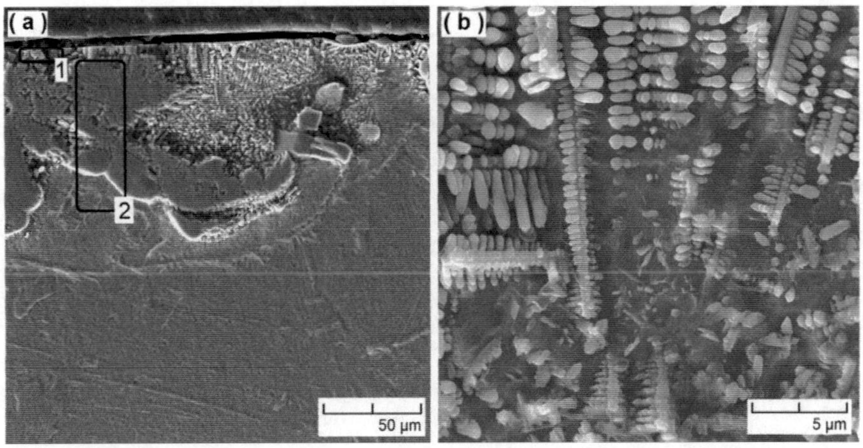

Figure 2. Microstructure of titanium **(a)** after laser treatment at an energy and pulse duration of 1.5 J and 2 ms, respectively, where coating *1* and diffusion layer *2* with a dendritic structure **(b)** are distinguished.

In addition to the coating and diffusion layer, the microstructure was distinguished by the presence of a transition layer and, accordingly, boundaries between them, as well as the boundary of thermal effect on the metal (figure 3).

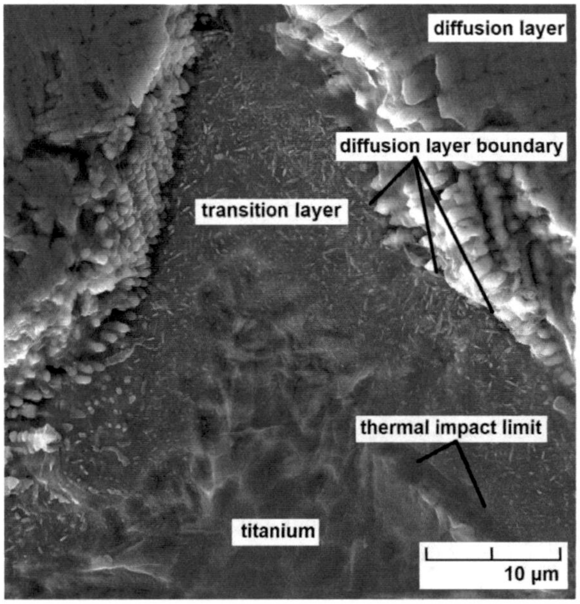

Figure 3. Diffusion layer boundary.

The pulse energy (E, J) had a pronounced effect on the thickness (h, μm) of the diffusion layers. Diffusion layers formed at an energy and pulse duration of 10.17 J and 3 ms, respectively, differed in maximum depth (up to 800 μm) (figure 4a). The pulse duration influenced not only the shape of dendritic formations but also the microhardness of the diffusion layer (H, GPa). The layers obtained with a pulse duration (τ, ms) of 3 ms and average energy values differed in the maximum microhardness, approximately 28 GPa (figure 4b).

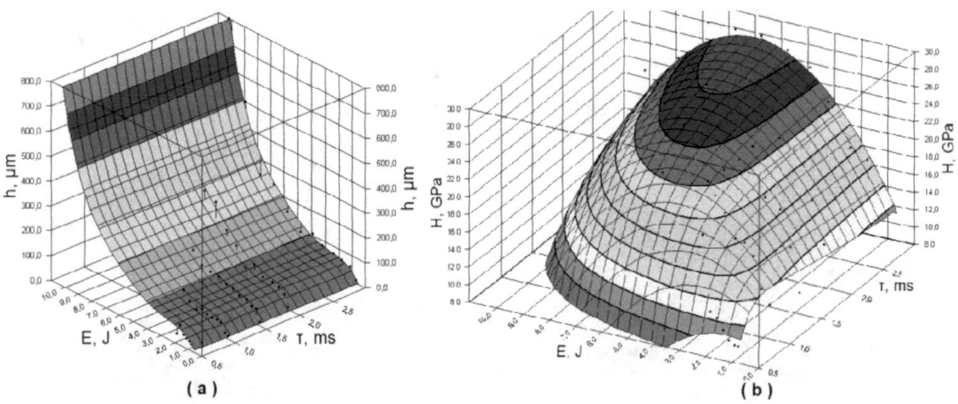

Figure 4. Influence of laser processing conditions (energy and duration of a laser pulse) on the depth **(a)** and microhardness **(b)** of layers formed on titanium.

4. Conclusions

Laser treatment of titanium in a graphite medium made it possible to obtain a layered structure consisting of coatings, a diffusion layer, and a transition layer. The coatings were characterized by a thickness of up to 6 μm. The diffusion layer had a dendritic structure and high microhardness, up to 28 GPa. The size of structural elements and hardness were most influenced by the pulse duration. The depth of the hardened layer reached 800 μm and depended on the pulse energy. The significant thickness of the structures being formed allows surface finishing.

The established dependences make it possible to choose the modes of laser treatment for the formation of structures with specified properties on titanium.

Acknowledgments

The research was supported by the scholarship of the President of the Russian Federation SP-2631.2019.1.

References

[1] Ajikumar P K, Vijayakumar M, Kamruddin M, Kalavathi S, Kumar N, Ravindran T R and Tyagi A K 2012 *Int J Refract Met* H **31** 62
[2] Boonruang C and Thongtem S 2010 *Chiang mai J Sci* **37** 206
[3] Voyko A V, Fomina M A, Koshuro V A, Fomin A A, Rodionov I V, Atkin V S, Galushka V V, Zakharevich A M and Skaptsov A A 2018 *Proc. biomed. opt. imag.* **10716** 107161L
[4] Savalani M M, Ng C C, Li Q H and Man H C 2012 *Appl Surf Sci* **258(7)** 3173
[5] Hamedi M J, Torkamany M J and Sabbaghzadeh 2011 *Opt Laser Eng* **49(4)** 557
[6] Fomina M, Koshuro V, Papshev V, Rodionov I and Fomin A 2018 *Data in Brief* **20** 1409

Study of hardness and morphology of carbide coatings obtained on complex shaped steel items by electro-spark alloying

O Markelova [1], V Koshuro[1], A Fomin[1], A Aman[2,3] and S Palis[3]

[1] Yuri Gagarin State Technical University of Saratov, Saratov 410054, Russia
[2] Otto Vollmann GmbH & Co. KG, Gevelsberg, Germany, 58285
[3] Otto-von-Guericke-Universität Magdeburg, Universitätsplatz 2, Magdeburg, Germany 39106

dimirion@mail.ru

Abstract. The work investigated the possibility of forming carbide coatings by electrospark alloying on steel products of complex shape. It has been established that electrospark alloying at an AC current of 1.0 to 4.5 A makes it possible to form coatings of hard carbide alloy VK6 and T15K6 characterized by microhardness up to 11.5 GPa and by hardness HRA 86.6 and 81.5 with the nitial hardness of the steel product HRA 80.3.

1. Introduction

In most cases, protective coatings on steel products are formed from carbides, nitrides, borides, carbonitrides of a number of metals (Cr, W, Mo, V, Ti, Ta) and transition metal oxides (Zr, Ti, Al). At present, protective layers are formed on steels by CVD, PVD, ion-plasma spraying, etc. [1-2]. Nitride (TiN) coatings formed by CVD and PVD increase the corrosion resistance of the steel surface. In this case, the hardness of PVD coatings of the TiN system formed on steel substrates can reach 20 GPa, and the grain size can be 15-50 nm [3]. Layers of CrC /C, TiC and TiC/CrC/C systems increase the wear resistance of steel substrates [4]. At the same time, these layers are characterized by significant brittleness, small thickness and low values of adhesive strength, which sharply narrows the area of their rational use.

Despite the widespread use of CVD and PVD technologies, there remains a need for methods that allow the formation of wear-resistant layers characterized by high values of hardness, adhesive strength, low porosity, and a given surface morphology. The known method of electrospark alloying (ESA) allows surface alloying of metals and the formation of functional layers from conductive materials, including hard alloys [5]. Despite the knowledge of the ESA method, it remains necessary to study the features of the formation of high-strength coatings on products of complex shape.

2. Methodology

Steel products of complex shape with an initial hardness of HRA 80.3 were used as samples. Local spherical areas of the surface were hardened (figure 1a). For the formation of coatings from hard materials of grades VK6 (WC-94%, Co-6%) and T15K6 (WC-79%, TiC-15%, Co-6%) (ISO 513-75) at values of alternating current from 1.0 to 4.4 A, an "ARKOGRAF" electric spark marker was used.

The contact area of the electrodes and the surface of the samples was 1 ± 0.25 mm2. The contact area of the electrodes and the surface of the samples was 1 ± 0.25 mm^2. Accordingly, the current density during ESA was: 2.8 and 4.0 A / mm^2 when using VK6 electrodes; 2.9 and 3.7 A / mm^2 for T15K6.

Images of the surface morphology of the coatings were obtained using an "MBS-10" microscope and an optical system of a "PMT-3" microhardness tester. The magnitude of open porosity, the distribution of the linear sizes of grains and coating defects were determined from the surface images using the "Metallograph" software, according to a well-known technique. The morphology of the formed coatings was investigated according to the well-known technique [6].

The microhardness was measured using a "PMT-3" microhardness tester with a load on the indenter of 100 gf in accordance with the requirements of known regulatory documents (GOST 9450 - 76, ISO 6507-1: 2005).The measurement of the hardness of the formed carbide layers was carried out using a portable Rockwell hardness tester "TX PHR-2". The qualitative assessment of the adhesion of the applied layers was carried out on the HF1-HF6 scale according to Verein Deutscher Ingenieure Normen VDI 3198.

3. Results

During ESA, a heterogeneous coating was formed on selected curved areas (figure 1b).

Figure 1. General view (a) of a steel sample with areas of ESA coating from T15K6 and the morphology of the formed coating (b).

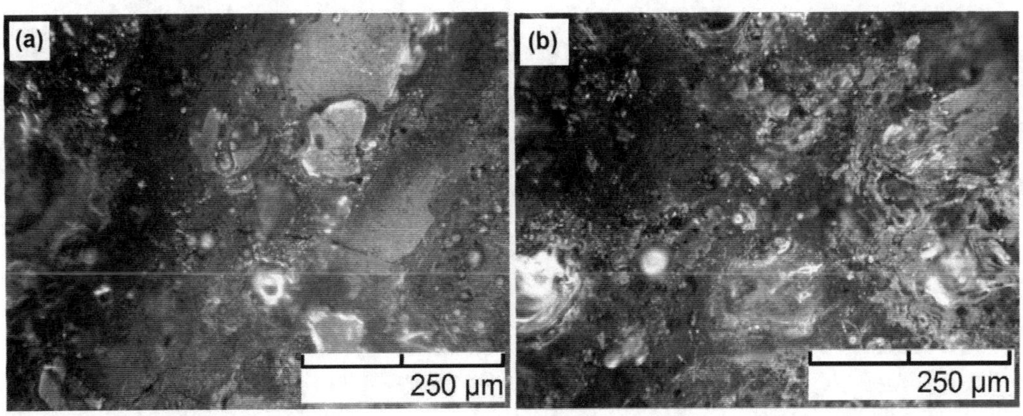

Figure 2. Morphology of coatings formed by ESA electrodes from VK6 (a) and T15K6 (b) at currents of 2.9 and 2.8 A, respectively.

No significant differences in the morphology of coatings formed by ESA from VK6 and T15K6 were observed. Defects in the form of cracks, delamination and significant pores were not visualized (figure 2ab). The porosity of the coating was influenced by the current density at ECA. With an increase in the current density, the open porosity of the formed layers increased from 50-52% to 58%.

The study of the microstructure showed that coatings from VK6 and T15K6 are formed without significant defects, namely, pores and cracks. Coatings obtained on ESA VK6 steel are formed from particles with a size of 500 nm and more (figure 3a). With T15K6, coatings are formed from larger particles, as small as 1 μm (figure 3b). The difference in the micro-structure is probably due to the different structure of the electrodes used. The thickness of the coatings was 15–45 μm.

The maximum microhardness of coatings obtained by ESA with the T15K6 electrode reached 11.5 GPa. When alloying VK6 steel, coatings with a maximum microhardness of 10.4 GPa were formed. Surface hardness increased from HRA 80.3 to 81.5 and 86.6 for T15K6 and VK6, respectively. The difference in hardness is justified by the different thicknesses of the layers being formed. After indentation, no cracks or delamination of the coating was found around the indentations, which corresponds to HF1 (VDI 3198) and is a high indicator of adhesive-cohesive strength.

4. Conclusions
As a result of the research, the possibility of forming carbide coatings on complex-profile steel surfaces has been established. The formed coatings are distinguished by the absence of significant defects in the form of cracks and pores. Carbide layers are formed from sub-micron-sized particles. Electrospark alloying makes it possible to obtain highly hard layers with high values of adhesive-cohesive strength.

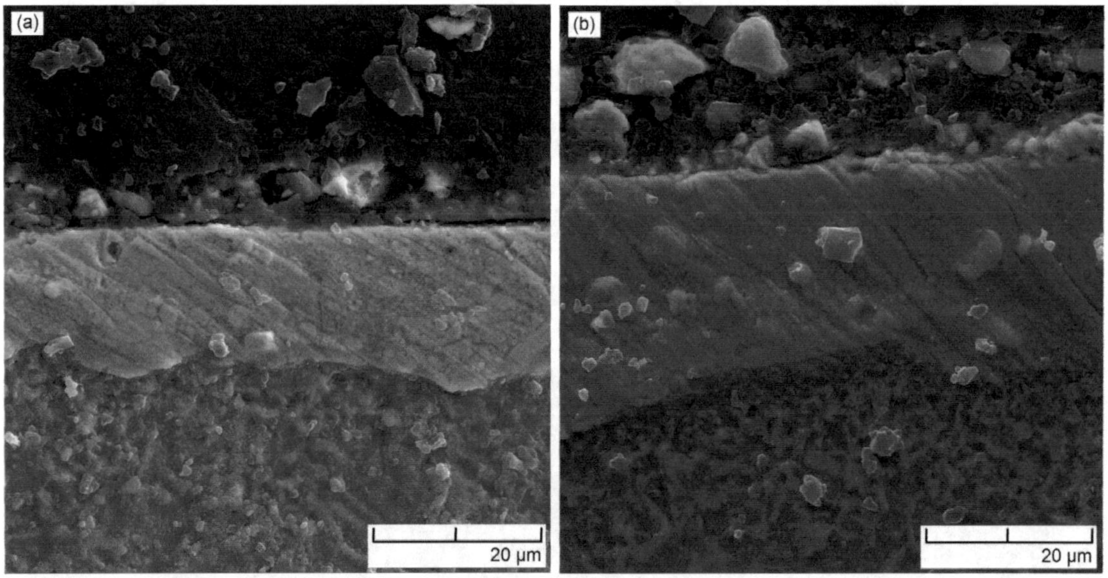

Figure 3. Microstructure of coatings formed by ESA electrodes from VK6 (a) and T15K6 (b) at currents of 2.9 and 2.8 A, respectively.

Acknowledgments
The research was supported by the Ministry of Education and Science of the Russian Federation and German Academic Exchange Service (DAAD) in the framework of the program "Mikhail Lomonosov" (Project Executor Koshuro V.A.).

References
[1] Michalski J, Lunarska E, Wierzchon T and AlGhanem S 1998 *Surf Coat Tech* 72 189

[2] Helmersson U, Johansson B O, Sundgren J E, Hentzell H T G and Billgren P 1985 *J Vac Sci Technol A* **3** 308
[3] Lukaszkowicz K and Dobrzański L A 2008 *J Mater Sci* **43** (10) 3400
[4] Nilsson M and Olsson M 2011 *Wear* **273** (1) 55
[5] Li Q H, Yue T M, Guo Z N and Lin X 2013 *Metall Mater Trans A* **44** (4) 1767
[6] Fomina M, Koshuro V, Papshev V, Rodionov I and Fomin A 2018 *Data in Brief* **20** 1409

Dependence of field-effect biosensor sensitivity on photo-induced processes in Si and its conductivity type

A V Kozlowski and S V Stetsyura

Institute of Physics, Saratov State University, Saratov 410012, Russia

kozlowsky@bk.ru

Abstract. The effect of photoelectron processes in *n*-Si and *p*-Si on the glucose sensitivity of a capacitive field-effect biosensor based on electrolyte/oxide/semiconductor structure was investigated. We obtained that illumination of *n*-Si/SiO_2/PEI structure during the GOx adsorption increases the glucose sensitivity by three times compare to GOx adsorption in the dark. In contrast, *p*-Si illumination during the GOx adsorption led to a decrease in sensor sensitivity from 2.9 mV/mM to 2.2 mV/mM. The result is explained by a change in the density of immobilized GOx molecules due to a change in the electrostatic forces of attraction under illumination and stabilization of the photo-generated charge on the surface electronic states of the Si/SiO_2 and SiO_2/PEI interfaces after illumination.

1. Introduction

Electrolyte/oxide/semiconductor (EOS) structures are widely used to detect nano- and bio-objects ionized in solution. In addition, EOS-structures are used to recognize enzymatic reactions [1], viruses [2], RNA/DNA sequencing. The parameters of enzyme biosensor based on EOS-structures can be improved by increasing the concentration of enzyme molecules adsorbed onto a semiconductor transducer surface. Ionic strength or pH of solution are commonly used to modulate the electrostatic interactions between enzyme molecules and semiconductor surface. However, variation of pH or ionic strength leads to a nonmonotonic and unpredictable change in the surface density of enzyme molecules.

Previously [3], we report on the influence of the white-light illumination of Si substrate during glucose oxidase (GOx) adsorption on the surface concentration of immobilized GOx molecules. This method is called photo-stimulated adsorption (PSA). The PSA efficiency of charged nano- and bio-objects significantly depends both on a Si conductivity type and on a buffer poly(ethylene-imine) (PEI) layer on Si surface [4]. However, studies on the effect of PSA of enzyme molecules onto a semiconductor transducer surface on the sensitivity of a capacitive field-effect biosensor have not existed before. In this work, we studied the effect of photo-stimulated layer-by-layer adsorption of GOx on the glucose sensitivity of field-effect-based capacitive EOS-sensors.

2. Experimental Section

2.1. Materials and Methods

The capacitive field-effect EOS structures were fabricated using single-crystal (100) Si wafers (400±10 μm thickness) of *n*-type (ρ = 8-10 Ω cm) and *p*-type (ρ = 9-15 Ω cm). Initially, the wafers were boiled in a peroxide–ammonia solution and rinsed in deionized water. This treatment leads to "reconstruction" of a native oxide layer to SiO_2 (~2 nm thickness). It can be note that isoelectric point (IEP) of the SiO_2 surface is ca. pH=2-3 [5] and above this pH value the surface is negatively charged. Afterwards, wafers were cut into substrates of 10×10 mm^2.

The enzyme glucose oxidase (GOx) form *Aspergillus niger* were used as enzyme molecules. In a wide pH range of the solution, the GOx molecule have an effective negative charge. The size of the GOx molecule is 6.0×5.2×7.7 nm^3 [6]. Branched poly(ethylene-imine) (PEI) with a molecular weight of 25 kDa was used as cationic polyelectrolyte to increase the adsorption of negatively charged GOx onto the silicon substrates. The PEI molecules were adsorbed onto the silicon substrates from the 1 mg/ml aqueous solution during 10 min followed by rinsing in deionized water (ρ ~18.2 MΩ cm) during 10 min and drying in nitrogen flow. The glucose solutions were prepared by dissolving D-glucose in the working buffer. As working buffer, a 0.2 mM potassium phosphate buffer solution (pH = 7.3) containing 150 mM NaCl as an ionic strength adjuster was used. The photo-stimulated layer-by-layer adsorption technique suggested in [7] was used to adsorb GOx from the 0.5 mg/ml aqueous solution onto silicon substrates covered with PEI. A halogen lamp (Philips 13186 EPX/EPV) was used to activate photoelectric processes in a silicon wafer during adsorption of polyelectrolyte molecules. The Si substrate was either in the dark or under illumination during the GOx adsorption, other things being equal. Thus, the Si/SiO_2/PEI/GOx biosensor structures were fabricated.

2.2. Measurement set-up

Figure 1b schematically shows the experimental set-up, which has been utilised for the characterisation of the EOS sensors with the adsorbed PEI and GOx layers. For the measurements, the EOS sensor was mounted into an electrochemical cell, sealed by an O-ring and contacted on its front side by the electrolyte and Ag/AgCl reference electrode, and on its rear side by a gold-plated pin. The measurements have been performed at room temperature.

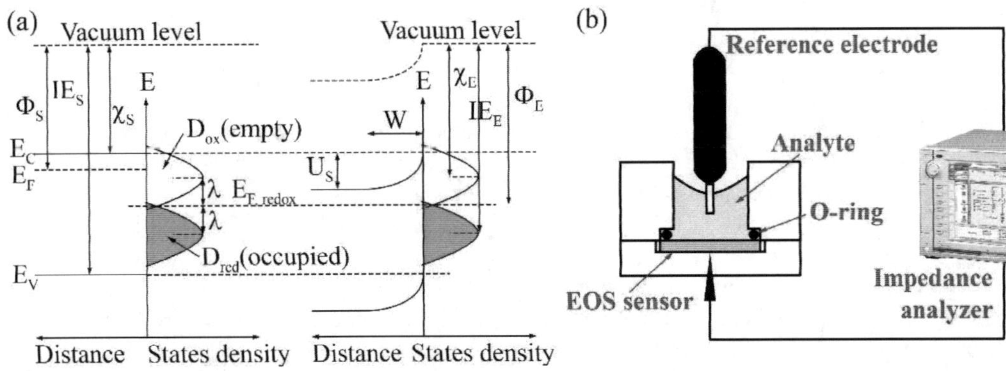

Figure 1. (a) - Schematic energy band diagram of the Si/SiO_2/electrolyte contact before (left) and after (right) equilibrium: D_{red} and D_{ox} - density distribution of states of the oxidation-reduction system in the electrolyte, λ - energy of reorganization, Φ, χ and IE are the work function, electron affinity, and ionization energy, respectively, of the semiconductor (index S) and electrolyte (index E); **(b)** - Schematic of a capacitive EOS-sensor.

EOS glucose biosensors have been characterised in glucose solutions with different content of D-glucose from 1 mM to 10 mM by means of Capacitance-Voltage (C-V) method using a semiconductor device analyzer (Agilent B1500A). The C–V measurements have been performed at a frequency of 1 kHz. Before the adsorption of PEI and GOx, the pH-sensitive behaviour of the as-prepared Si/SiO_2 structure has been investigated in buffer solutions of pH 3.8–7.5 by C–V method. For operation, a DC voltage is applied via the Ag/AgCl reference electrode and a small AC voltage (20 mV) is applied to the system in order to measure the capacitance of the sensor. For the measuring procedure, about 0.5

mL of the working buffer or particular glucose solution was applied to the EOS gate surface, and C-V curve has been read out for after 10 min. After each measurement, the EOS gate region was rinsed with buffer solution.

3. Results and Discussion

3.1. Effects of pH on EOS sensor

As the working principle of the enzyme-modified EOS sensor is based on the detection of pH changes near the Si/SiO_2 surface induced by the enzymatic reaction, the pH sensitivity of the SiO_2 layer has been proven before GOx immobilization. Figure 2 depicts a typical C-V curves (a) and calibration curve (b) of EOS sensor recorded in 0.1 M sodium-acetate (pH 3.8-4.5) and 0.1 M potassium phosphate (5.9-7.5) buffer solution. The EOS sensor shows a nearly linear pH response between pH 3.8 and pH 7.5 and an average sensitivity of about 42-43 mV/pH at 1 kHz. It didn't depend on the silicon conductivity type.

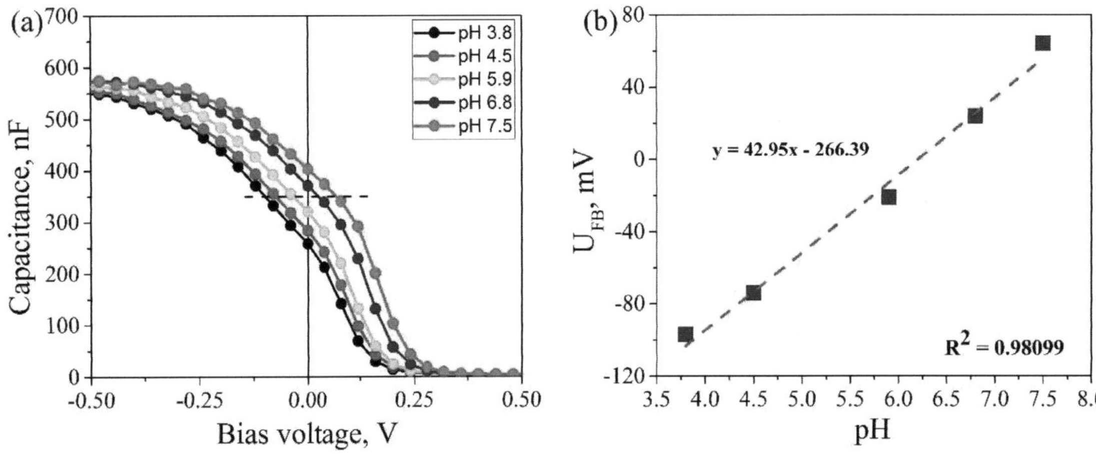

Figure 2. (**a**) - C-V curves of EOS sensor before immobilization of PEI and GOx measured in 0.1 M sodium-acetate (pH 3.8-4.5) and 0.1 M potassium phosphate (5.9-7.5) buffer solutions; (**b**) - the calibration curve derived from the values of the flat band potential shifts.

This change occurs because with an increase in pH of the raster on the semiconductor surface, a negative charge increases. These leads to electrons repelled from the depletion region of Si, which in turn reduces the width of space charge region (SCR) in *p*-Si and increases it for *n*-Si. As a result, a total capacity either increases (*p*-Si) or decreases (*n*-Si) [8].

3.2. Effects of photoelectron processes in Si on the sensitivity of EOS biosensor

The glucose sensitivity of EOS biosensors was studied by means of C–V measurements. Figure 3 shows exemplarily a typical set of C–V curves of the EOS biosensor, where GOx deposited in the dark (Figure 3a) and under illumination of *n*-Si during GOx adsorption (Figure 3b) measured in work buffer and in D-glucose solutions of different concentrations from 1 to 10 mM. The C–V curves exhibit a high-frequency behaviour with typical accumulation, depletion and inversion region. With increasing glucose concentration in the solution, the C–V curves are shifted along the voltage axis in the direction to more negative voltage values. This is due to an increase in the H^+ concentration on the EOS sensor surface after the enzymatic reaction.

It was found that for EOS sensor, where the GOx molecules were deposited in the dark, the sensitivity to glucose is 1.90 and 2.87 mV/mM for *n*-Si and *p*-Si, respectively. In the case of PSA - 5.8 and 2.23 mV/mM for *n*-Si and *p*-Si, respectively. These data are in good agreement with the previously obtained results on PSA and are explained using the model of recharging of surface electron states under illumination of Si and adsorption of charged molecules from a solution, which is partially presented in [3, 4]. The change in the sensitivity of EOS sensor in the case of photo-stimulation of the Si/SiO_2 structure during GOx adsorption occurs due to a change in the density of the adsorbed enzyme.

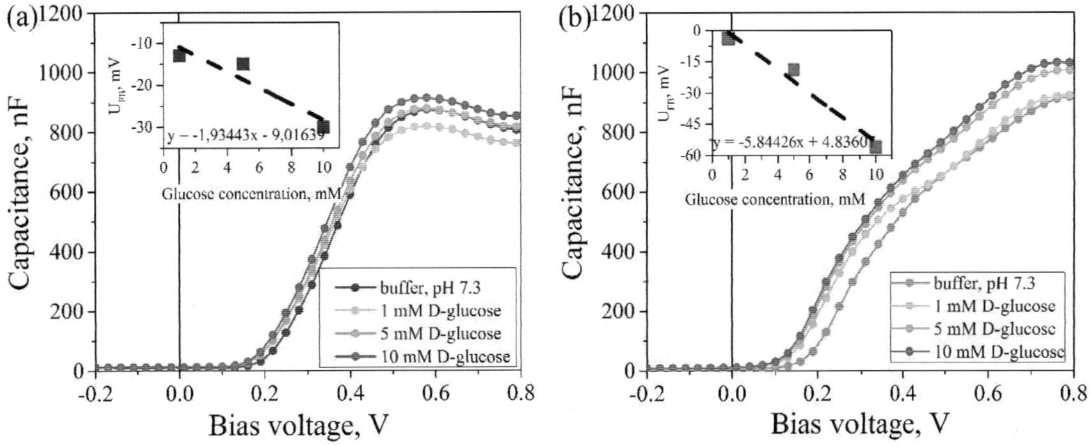

Figure 3. Typical C–V curves of the EOS sensors based on *n*-Si measured in potassium phosphate buffer and in glucose solutions of different concentrations from 1 to 10 mM; **(a)** – GOx adsorbed in the dark, **(b)** – GOx adsorbed under illumination. Inset: calibration curves.

The influence of illumination on the GOx adsorption can be explained as follows. Under equilibrium conditions (i.e., in the absence of illumination and external electric fields), the charge of the space-charge region (SCR) is equal in magnitude and opposite in sign to the charge of surface electronic states (SES). The sign of SES charge usually corresponds to that of main carriers in semiconductor. The negative sign of SES in *n*-Si leads to electrons depleted region in the near-surface layer and the corresponding upward bending of the energy bands. The PEI deposition leads to a decrease in band bending. The total charge of SES, traps in SiO_2, and PEI in structures based on *n*-Si will be close to zero. When Si/SiO_2/PEI structure is immersed into an aqueous solution of anionic PE (e.g., GOx), the negatively charged molecules are adsorbed and held on the PEI surface mostly due to electrostatic interactions. The illumination of Si/SiO_2/PEI structures leads to rectification of their energy bands due to the generation of electron–hole pairs and their separation in the SCR field. As a result, the charge of SES at the Si/SiO_2 interface decreases. In *n*-Si/SiO_2/PEI structure illumination, the SCR field facilitates the drift of holes toward the *n*-Si/SiO_2 interface, which leads to an increase in the effective positive charge in this structure at the PEI/GOx interface and results in increasing amount of adsorbed GOx molecules as compared to that adsorbed in the dark. In [3], a similar model for the PSA of GOx molecules onto *n*-Si/SiO_2/PEI structure was presented and the experimental results showed a significant increase in adsorbed GOx molecule number on *n*-Si substrate compare to GOx deposition in the dark. In the case of *p*-Si/SiO_2/PEI structure illumination, the drift of electrons toward the *p*-Si/SiO_2 interface due to SCR field leads to a decrease in the effective positive charge at the PEI/GOx interface and, as a result, the adsorbed GOx molecule number is decreases compared to adsorption in the dark. The change is not so significant for *p*-Si, since the positive charge of PEI does not change under illumination.

4. Conclusion

Thus, an increase in the sensitivity to glucose of an enzymatic biosensor based on n-Si fabricated with PSA of enzyme molecules onto a semiconductor substrate has been experimentally and theoretically proved. This sensitivity exceeds the sensitivity of biosensors both without PSA and based on p-Si, regardless of the fabrication conditions. Our research is useful in fabrication of capacitive field-effect biosensor since the growth of enzyme molecules density on the surface of the semiconductor transducer increases both sensitivity and selectivity of the sensor.

5. Acknowledgments

The work was supported by the German-Russian Interdisciplinary Science Center (G-RISC), Project no. P-2018a-10.

References

[1] Poghossian A and Schöning M 2020 *Sensors* **20** 5639
[2] Poghossian A, Jablonski M, Molinnus D, Wege C and Schöning M 2020 *Front. Plant Sci.* **11** 598103
[3] Maslennikova A A, Kozlowski A V, Santer S and Stetsyura S V 2019 *Journal of Physics: Conference Series* **1400** 077052
[4] Malyar I V, Gorin D A, Santer S and Stetsyura S V 2017 *Appl. Phys. Lett.* **110** 133104
[5] Cloarec J-P, Chevalier C, Genest J, Beauvais J, Chamas H, Chevolot Y, Baron T and Souif A 2016 *Nanotechnology* **27** 295602
[6] Hecht H J, Kalisz H M, Hendle J, Schmid R D and Schomburg D 1993 *J. Mol. Biol.* **229** 153–72
[7] Malyar I V, Gorin D A, Santer S and Stetsyura S V 2013 *Langmuir* **29** 16058
[8] Garyfallou G, de Smet L and Sudhölter E 2012 *Sensors and Actuators B* **168** 207

Optimization of deep silicon etching process for microstructures fabrication

V Kuzmenko[1,2], A Miakonkikh[1,2] and K Rudenko[1,2]

[1] Valiev Institute of Physics and Technology of Russian Academy of Sciences, Moscow, Russia
[2] Moscow Institute of Physics and Technology, Dolgoprudny, Russia

kuzmenko.vo@phystech.edu

Abstract. The paper presents the study of cyclic process of deep anisotropic silicon etching, called Oxi-Etch, in which the steps of etching and oxidation alternate, allowing deep etching of silicon with an anisotropic profile. This process forms typical for cyclic etching process sidewall profile called scalloping. Opportunities for modification and optimization of the process for specific application were investigated. The effects of optimization of the bias voltage and the duration of the etching step on the parameters of the resulting structures, such as the etching depth, wall roughness, and the accuracy of transferring the lithographic size, are considered. Balance between etch rate and scalloping was established.

1. Introduction

Plenty of microstructures relevant for various applications are fabricated by deep silicon etching. Some devices of droplet microfluidics [1] can be based on thin trenches made in silicon. Array of holes in silicon can form two-dimensional photonic crystal, which can be applied for terahertz communication devices [2]. Another application is microstructures for 3D integrated capacitors using in dynamic random-access memory (DRAM) [3]. In order to increase capacitance and consequently density of the stored information, the area of the capacitor plates is increased, so usually integral capacitors are formed by conformal deposition of films on sidewalls of high aspect ratio structures in silicon. Manufacturing of all these high aspect ratio structures requires technology of deep anisotropic plasma etching of silicon. Different approaches of anisotropic silicon etching have their own features. One of these technologies is well-known Bosch process, in which anisotropy is achieved by protecting sidewalls of structure with fluorocarbon film deposited form plasma. The idea of this process consists in the cyclic repetition of the steps of passivation of the surface with a polymer film in a fluorocarbon plasma and anisotropic etching with a DC bias in a fluorine-containing plasma. Structures etched in this process have typical sidewall profile with periodical notches called "scallops", which could be smoothed by further optimization of the process, i.e. shortening of both steps duration.

Here an alternative room temperature approach of deep anisotropic silicon etching known as Oxidation-Etching (Oxi-Etch approach) was studied. Earlier, it was proposed [4] and patented [5]. The idea is based on the replacement of polymer passivation by plasma oxidation of silicon. Silicon oxide 1–2 nm thick on the sidewall has sufficient plasma etch resistance for one etching cycle. The process studied in this work is a cyclical repetition of the following basic steps: 1) passivation in the O_2 plasma - the formation of a SiO_2 layer several nanometres thick on all open areas of silicon (oxidation step), 2) etching in SF_6 plasma - removal of passivation from the bottom with accelerated ions, and spontaneous isotropic etching (etching step). The method we are investigating has several advantages over the Bosch process. In the etching process, no hard-to-remove polymer film remains on the side walls, which facilitates subsequent technological operations. Furthermore, fluorocarbon gases used in Bosch process for passivation have significant global warming potential, which means that increased content of these gases in atmosphere leads to an increase in the greenhouse effect. Passivation by O_2 plasma is more eco-friendly.

In addition to the above, there are ample opportunities for process optimization. For example, in addition to the basic steps of etching and passivation, intermediate steps can be introduced in which the gas composition in the reactor is changed; in these steps, the value of the input power and bias voltage can also be changed. This is especially true in reactors in which continuous wave plasma is used for reasons of reproducibility. In this case, the consideration of processes in a mixture of gases between steps is also important.

At present work approaches of modification and optimization of the oxidation-etching process are examined. In studied process DC bias is applied for a short time only for breakthrough passivation and then etching of exposed silicon regions is isotropic without bias. As bias-stimulated etching break the passivation layer predominantly on horizontal areas, sidewalls are preserved from isotropic etching. Minimization of duration of DC bias should increase selectivity in process.

2. Experimental

The experiment was carried out in the PlasmaLab 100 plasma etching tool (Oxford Instruments Plasma Technology, UK) on silicon samples 2x2 cm with a SiO_2 mask 2 μm thick. A feature of this reactor equipped with a Faraday shield is that the ignition of an ICP discharge is possible only when a bias voltage is applied to the sample. For this reason, the charge is ignited once at the very beginning of the process and more ICP power is not interrupted.

Mask has round holes 2, 4, 8, and 12 μm in diameter and 16 μm width trenches formed by photolithography and plasma etching. In this work basic Oxi-Etch process, described in details earlier [4], was modified in following way. Typical step duration time, gas flows and applied powers diagram is shown on figure 1.

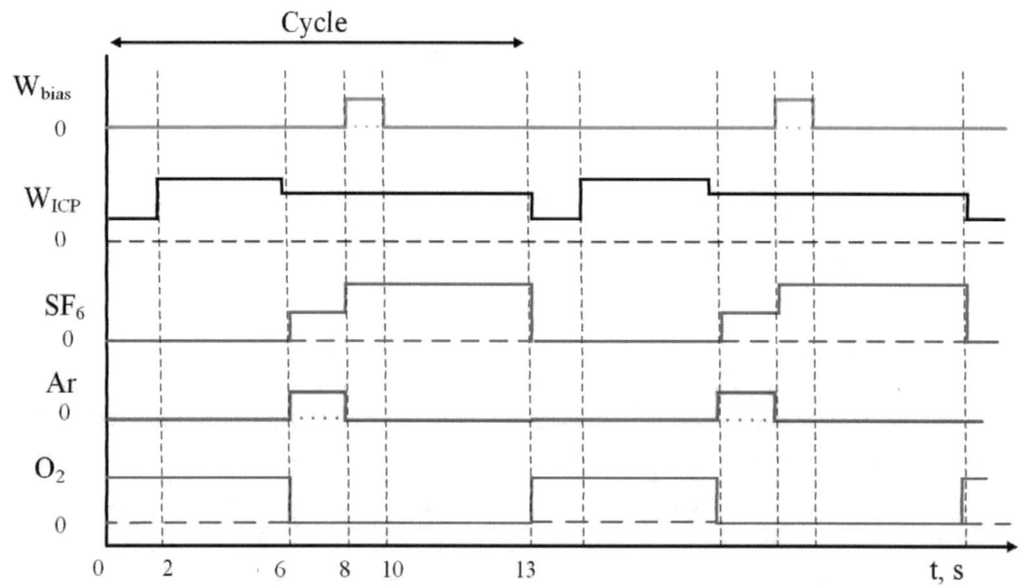

Figure 1. Diagram of variation of parameters in process.

Oxidation step was conducted in O_2 plasma at high ICP power (2000 W) without DC bias and step time was varied 4-8 s. Then was short (2 s) transition step at which oxygen in chamber was replaced by SF_6/Ar mixture. And right after that during 2 s breakthrough step DC bias was applied in plasma mixture and passivation on the bottom of the structure was broken. Then it was 3-6 s etching step, when in SF_6 plasma 1750 W ICP power structures were etched isotropically without DC bias. Finally, it was 2 s transition step from SF_6 to O_2 plasma with low ICP power 1200 W. During the process the sample was glued to the 100 mm silicon wafer fixed in chuck cooled to room temperature. As a result, we obtained etched structures and used SEM (Carl Zeiss Supra 55) to characterize features such as undercut and mentioned above scalloping.

It can be seen that the switching on of the bias power in the process is performed very scarcely. This is due to the fact that ion-stimulated etching is required only to remove passivation from the bottom of the trench. During the rest of the time, the excessive flow of accelerated ions will not stimulate the silicon etching process, but will only lead to additional erosion of the mask. Moreover, if oxygen ions undergo acceleration due to DC bias, this will lead to their deep penetration into the bottom of the silicon etching structure and the creation of a deeply oxidized layer, which can even cause an etching-stop effect.

Duration of both oxidation and etching steps were varied in order to research ability to tune etch rate and scalloping of the process. Effect of DC bias during breakthrough step was studied. SEM pictures of microstructures were obtained after 50 cycles process. Using modified process high aspect ratio structures with vertical sidewall profile were achieved in 530 cycles of etching process.

3. Results and discussion

At first oxidation and etching without bias steps durations were varied. Values of etch rate, undercut and scalloping in 50 cycles process on 16 μm trench are shown in table 1. Etch rate is increasing with decrease of oxidation step duration and with the increase of etching step duration due to increasing the proportion of etching time in the cycle. Undercut grows with decreasing of oxidation time and increasing of etching time due to lack of passivation of the top of structure. Scalloping is decreasing with the decrease of oxidation duration and increase of etching duration probably because of improving the break-up of oxide at the edges of the bottom of the structure. As a result, optimized process with acceptable etch rate and fine undercut and scalloping is process with decreased duration both oxidation and etching steps.

Table 1. Etch rate, and values of undercut, and scalloping, and etching per cycle (EPC) on 16 μm trench for different duration of oxidation and etching step.

Columns: t_{oxid}, s Lines: t_{etch}, s	4	6	8
3	Rate=1.14 μm/min EPC=0.25 μm/cycle Undercut=158 nm Scalloping=64 nm	Rate=1.04 μm/min EPC=0.26 μm/cycle Undercut=100 nm Scalloping=88 nm	
6	Rate=1.31 μm/min EPC=0.35 μm/cycle Undercut=400 nm Scalloping=59 nm	Rate=1.27 μm/min EPC=0.38 μm/cycle Undercut=374 nm Scalloping=67 nm	Rate=1.06 μm/min EPC=0.35 μm/cycle Undercut=247 nm Scalloping=78 nm

DC bias during passivation breakthrough step was varied. On figure 2 results of etching of 2 μm holes in 50 cycles process for different bias voltage in the range of 38-83 V are shown. As one can see at low DC bias with increase of aspect ratio leads to incomplete passivation breakthrough and consequently nanograss formation at the bottom of the structure. Increasing of DC bias leads to more complete breakthrough in the depth of the structure, but structures still have positive sidewalls angle. At high DC bias in first 50 cycles complete breakthrough is achieved which gives vertical sidewalls profile. On the other hand, with growth of DC bias mask etch rate is also increasing. In processes with 38 V, 65 V and 83 V DC bias mask etch rate is growing from ~ 0 nm/min, 4.6 nm/min and 7.4 nm/min respectively. This is consistent with the known concepts that the silicon oxide of which the mask is made is etched in a fluorine-containing plasma in a threshold process after 40 eV, with a bias exceeding 60-80 volts, the etching rate does not change so significantly. It is noteworthy that the effect of the formation of roughness at the bottom does not appear immediately, but after reaching a certain etching depth and aspect ratio of the hole. This means that it is possible to investigate the possibility of changing the bias voltage depending on the cycle number with increasing etching depth.

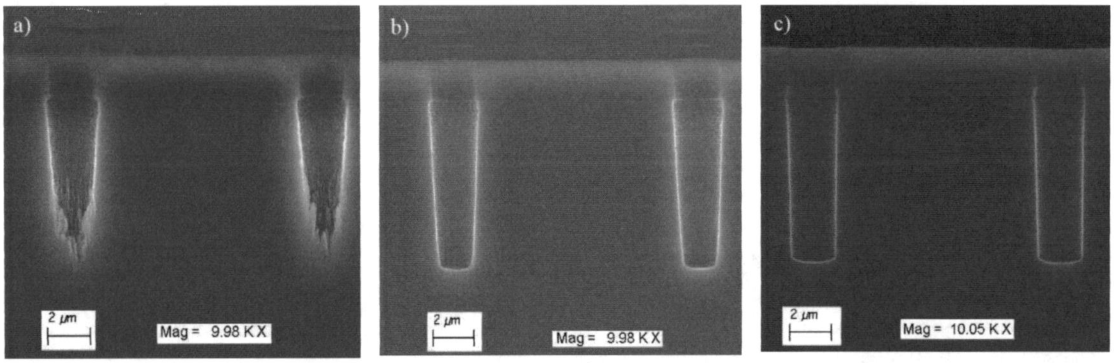

Figure 2. SEM pictures of 2 µm holes after 50 cycles processes with a) 38 V, b) 65 V and c) 83 V DC bias during breakthrough step.

Using modified Oxi-Etch approach deep silicon etching process was carried out to achieve high aspect ratio structures. In order to obtain vertical profile and achieve deep structures with 2 µm mask 530 cycles process with 4 s oxidation time and 3 s etching time was chosen. DC bias during breakthrough step was increased every few cycles from 65 V at start to 130 V in the end. SEM pictures of 16 µm trench and 8 µm hole are shown at figure 3.

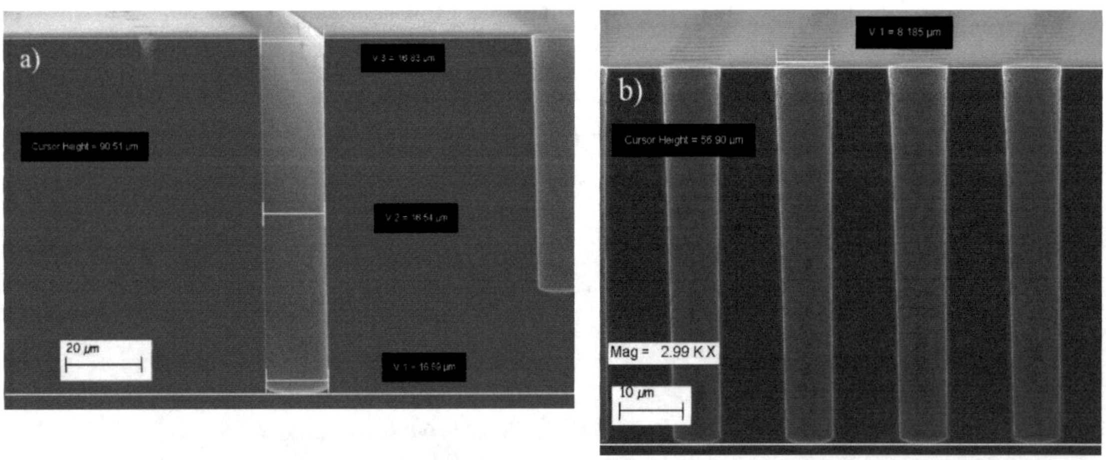

Figure 3. SEM pictures of a) 16 µm trench and b) 8 µm holes after 530 cycles of optimized Oxi-Etch process.

Both structures have vertical sidewalls. For example, trench structure on the figure 3a width variation less than 1% over the entire etching depth. Aspect ratio for trench and hole are 5.6 and 7 respectively. Effect of aspect ratio dependence etching was observed, for example holes with diameters of 4 μm, and 8 μm have depths 45.4 μm, and 56.9 μm respectively. This typical for deep silicon etching effect is that with increasing of aspect ratio transport of active particles to the etching region becomes more complicated. As a result, narrower structures have lower etch rate.

4. Conclusions

In present work cyclic Oxidation-Etching deep silicon etching process was modified, studied and optimized. This approach has some advantages over well-known deep plasma etching processes. Influence of steps duration and DC bias during breakthrough step were studied. It is shown that there is a trade-off between the etching rate and undercutting depending on the ratio of the etching time and the oxidation time in the cycle. It is shown that with a decrease in the bias voltage, a significant increase in the selectivity of the Oxi-Etch process is possible. At the same time, when etching high-aspect-ratio structures, a small bias voltage may be insufficient to break through the passivation layer at the bottom; this can lead to the formation of an inclined wall and even, with incomplete breakthrough of the passivation layer, to the formation of nanograss at the bottom. It is necessary to develop processes with variable characteristics depending on the cycle number to obtain maximum selectivity and vertical profile.

Using the process high aspect ratio structures with acceptable sidewalls profile were achieved. For example, for a process with a duration of 530 cycles, trench coats with strictly vertical walls were obtained. Etch selectivity, i.e. the ratio of the etching depth to the mask consumption was 67 for trench coats with a width of 16 μm, and 42 for holes with a diameter of 8 μm. Researched method can be used for manufacturing structures for different applications.

Acknowledgments

The investigation was partially supported by Program no. 0066-2019-0004 of the Ministry of Science and Higher Education of Russia for Valiev Institute of Physics and Technology of RAS and partially funded by RFBR according to the research projects #20-07-00832.

References

[1] Yadavali S, Lee D and Issadore D 2019 Robust microfabrication of highly parallelized three-dimensional microfluidics on silicon *Sci. Rep.* **9** 12213

[2] Withayachumnankul W, Fujita M and Nagatsuma T 2018 Integrated silicon photonic crystals toward terahertz communications *Adv. Optical Mater.* **6** 1800401

[3] Strambini L et al. 2020 Three-dimensional silicon-integrated capacitor with unprecedented areal capacitance for on-chip energy storage *Nano Energy* **68** 104281

[4] Miakonkikh A V, Averkin S N and Rudenko K V 2019 Anisotropic plasma etching of silicon in gas chopping process by alternating steps of oxidation and etching *J. Phys.: Conf. Ser.* **1243** 012009

[5] Averkin S N, Antipov A P, Lukichev V F, Myakonkikh A V, Rudenko K V, Rylov A A and Semin Yu F 2019 Method of anisotropic plasma etching of silicon microstructures in a cyclic two-step oxidation-etching process Patent RU 2691758 C1

Separation of III-N partially-coalesced nanowire arrays from Si substrate

V V Lendyashova[1,2], K P Kotlyar[3], V O Gridchin[1,3,5], R R Reznik[4], A I Lihachev[2], K Yu Shubina[1], T N Berezovskaya[1], E V Nikitina[1], I P Soshnikov[1,2,5] and G E Cirlin[1,4,5]

[1] Alferov University, St. Petersburg 194021, Russia
[2] Ioffe Institute, St. Petersburg 194021, Russia
[3] Saint-Petersburg State University, St. Petersburg 198504, Russia
[4] ITMO University, St. Petersburg 197101, Russia
[5] IAIRAS, St. Petersburg 198095, Russia

erilerican@gmail.com

Abstract. In modern optoelectronics, arrays or single nanowires (NWs) of III-N materials, in particular InGaN, separated from the original substrates are used to fabricate light-emitting diodes or single photon sources. This work describes a technology of separation super-dense arrays or arrays of partially-coalesced InGaN nanowires and single nanowires from a Si substrate by chemical etching in HF:HNO_3 solution, which allows preserving the optical properties of the structure for further use.

1. Introduction

Nowadays semiconductor nanowires (NWs) are promising for applications in nanophotonics and optoelectronics. One of the priorities is high-efficient light-emitting diodes (LEDs) based on III-N NWs [1]. Flexible LEDs based on the NWs arrays separated from the substrate, especially indium gallium nitride (InGaN) are of great interest. This material is selected because of a direct band gap in the range from 0.7 to 3.43 eV available by tuning the chemical composition. The special class of objects is partially-coalesced arrays of nanowires – aggregate nanostructures – which are also used to fabrication light-emitting devices [2] or act as a buffer layer for the growth of III-N layers on substrates with a large mismatch of lattice parameters [3,4]. For several applications, it is necessary to separate these structures from the original substrate. The methods used for separately located NWs (mechanical separation, the formation of a composite material polymer/NWs and its subsequent separation, etc.), in this case, do not lead to the desired result. In this regard, it is necessary to develop a technology of separating super-dense arrays from the substrate (in particular, Si(111)) to further study the properties of synthesized NWs and to fabricate optoelectronic devices.

2. Experiments and results

Arrays of partially-coalesced NWs of indium gallium nitride (InGaN) on Si(111) substrate were synthesized by molecular beam epitaxy (MBE) at Riber Compact 12 MBE system as an example of III-N NWs [5]. The surface morphology of the samples was studied with scanning electron microscopy (SEM, Zeiss Supra 25 machine). The optical properties of the structure were exanimated by photoluminescence spectroscopy (Pl, Accent RPM Sigma machine) at each stage of separation. Laser has a power of 1 mW with a wavelength of 266 nm.

The method of separation consists of wet chemical etching in HF:HNO_3:H_2O (1:1:3) solution at room temperature [6] (figure 1).

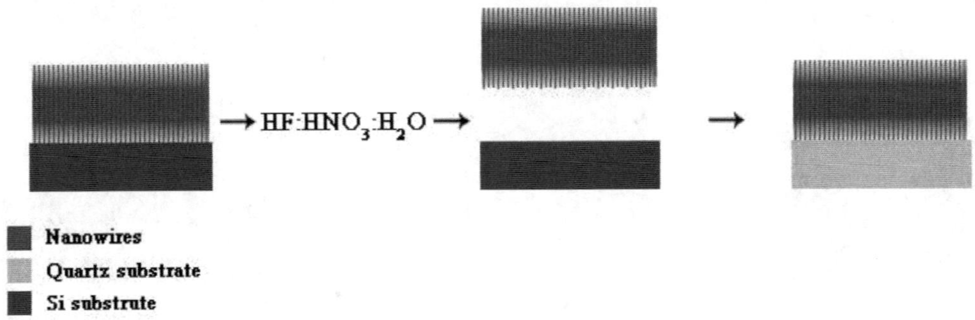

Figure 1. Principle schema of the method of separation.

Initially, the InGaN NWs array was a structure with a total height of about 2 μm, consisting of partially-coalesced in the mid-height zone spindle-shaped nanowires with a diameter from 80-100 at the tops and bases up to 200 nm (figure 2a). After placing the samples in a container with an acid solution etching was carried out in stages with results checked. Initial etching took 2 minutes and further etching stages lasted for 10 seconds. It was observed that the separation of the structure primarily occurs around defects and along the edges of the sample, as shown in figure 2b [7]. This feature is explained by the fact that in these sites it is easier for the etching solution to reach the NWs/Si interface.

Directly, the process of separating the NWs array during etching can be decomposed into two components. First, the large difference between the permanent crystal lattices of silicon and nitride NWs causes large mechanical stresses in the structure. When the bases of nanowires are etched, stress relaxation occurs, which leads to separation of the structure from the substrate. Second, the etching mechanism is based on the formation of gaseous products of the reaction of silicon oxide and silicon substrate with an acid solution penetrating through the space (ex., pores, channels) between NWs. Gas bubbles rising from the base of the structure contribute to its "lift-off" separation.

Consider the chemical mechanism of an etching reaction. To begin with, the oxide film on the silicon surface is etched with hydrofluoric acid:

$$SiO_2 + 6HF \rightarrow H_2SiF_6 + 2H_2O \quad (1)$$

At the same time there is a process of dissociation of nitric acid in water:

$$4HNO_3 \rightarrow 4NO_2\uparrow + O_2\uparrow + H_2O \quad (2)$$

After removing the silicon oxide, the substrate interface becomes available for reaction and is oxidized by the products of nitric acids dissociation with the release of gaseous hydrogen:

$$Si + 2NO_2 + 2H_2O \rightarrow SiO_2 + H_2\uparrow + 2HNO_2 \quad (3)$$

The silicon oxide formed during this reaction reacts with hydrofluoric acid. Thus, the complete equation of the etching reaction is:

$$SiO_2 + HNO_3 + 6HF \rightarrow H_2SiF_6 + H_2O + H_2\uparrow + HNO_2 \quad (4)$$

After separation, the NWs array, cleared of the residual etching solution, can be either immediately transferred to any type of substrate or placed in a cuvette for subsequent ultrasound treatment. Such an impact on the nanostructure allows the layer to be divided into small flaxes, single NWs, or parts of them (individual tops) depending on the morphology of the initial sample. Thereafter, it is possible to deposit parts of the NWs array or single NWs on the substrate (figure 3).

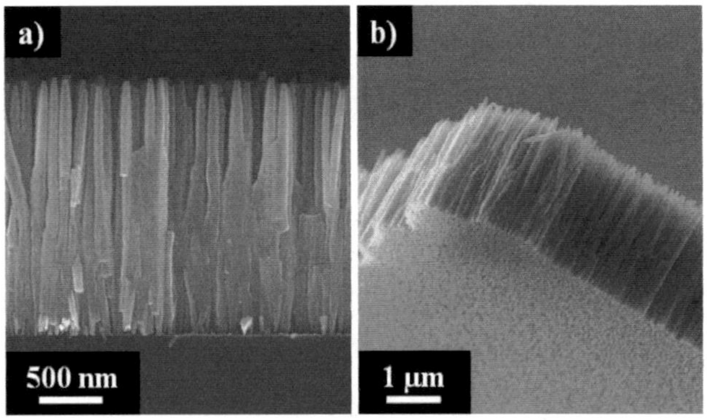

Figure 2. SEM images of (a) InGaN NWs array after MBE growth; (b) NWs array separated from the edge of the sample.

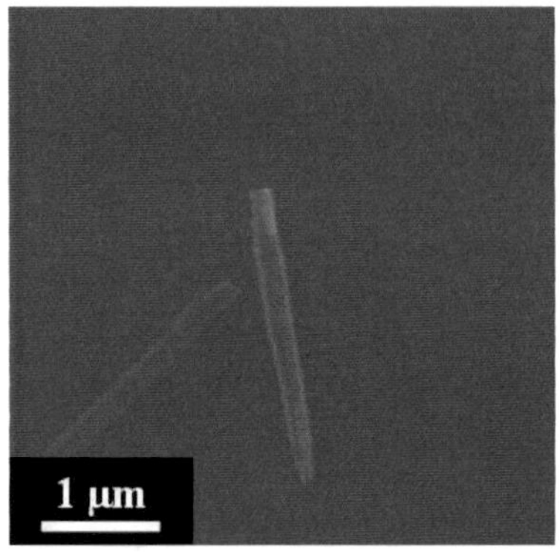

Figure 3. SEM image of single NWs after transferred on glass.

The most interesting was the influence of etching and transfer of the array to another substrate on the optical properties of the structure. Figure 4 illustrates photoluminescence spectra of the initial sample (a), sample after primary etching (b) and transferred to glass InGaN NWs array (c) which did not show any significant changes in optical properties. The only exception is for the shift of the photoluminescence peak caused by relaxation of internal stresses in the structure. The slight decrease in intensity is due to the lack of reflection effect from the substrate in the transferred structure and to the disorientation of the NWs relative to the beam of exciting radiation.

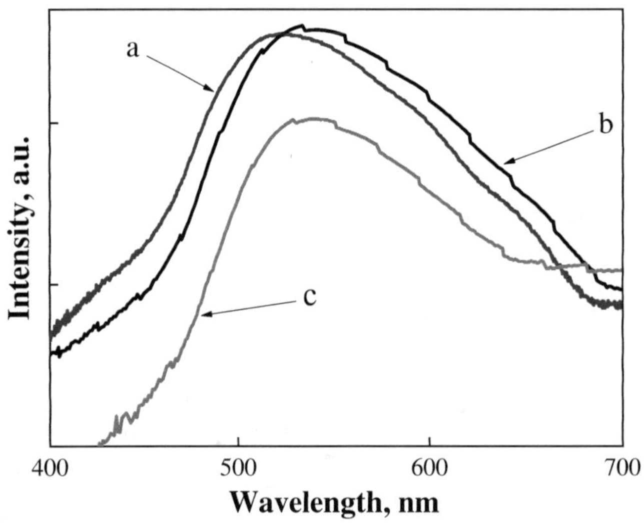

Figure 4. Photoluminescence spectra of (**a**) InGaN NWs on Si(111); (**b**) InGaN NWs on Si(111) after etching; (**c**) InGaN NWs on glass.

This work describes a new method of separation arrays of partially-coalesced III-N NWs from Si(111) substrate by wet chemical etching in an $HF:HNO_3$ solution. It is shown that the proposed method allows preserving the optical properties of the structure during separation and transfer to other types of substrates, as well as separating single III-N NWs from the original substrate.

Acknowledgments
The samples were grown with the support of the Ministry of Science and Higher Education of the Russian Federation (state task №0791-2020-0003). SEM and Pl studies of the samples were supported by RFBR (grant 19-32-90156).

References
[1] Dubrovskii V G, Cirlin G E and Ustinov V M 2009 *Semiconductors* **43**(12) 1539
[2] Kikuchi A, Kawai M, Tada M and Kushino K 2004 *Jpn. J. Appl. Phys.* **43** L1524
[3] Yang T H, Ku J T, Chang J R, Shen S G, Chen Y C, Wong Y Y, Chou W C, Chen C Y and Chang C Y 2009 *J. Cryst. Growth* **311** 1997-2001
[4] Li S and Waag A 2012 *J. Appl. Phys.* **111** 071101
[5] Gridchin V O, Kotlyar K P, Reznik R R, Dragunova A S, Kryzhanovskaya N V, Lendyashova V V, Kirilenko D A, Soshnikov I P, Shevchuk D S and Cirlin G E 2021 *Nanotechnology* **32** 335604
[6] Karar N, Opila R and Beebe T Jr 2011 *J. Electrochem. Soc.* **158**(6) D342-D350
[7] Kukushkin S A, Osipov A V and Red'kov A V 2017 *Semiconductors* **51**(3) 414

Features of Electrophoretic Formation of Local Heat Sources Based on Nanosized Powder Al

E S Leonenko[1], L I Sorokina[1], R M Ryazanov[2] and E A Lebedev[1]

[1] National Research University of Electronic Technology (MIET), 124498, Moscow, Zelenograd, Russia
[2] Scientific Manufacturing Complex "Technological Centre", 124498, Moscow, Zelenograd, Russia

katejustkate86@gmail.com

Abstract. In this study, the features of electrophoretic deposition (EPD) method to form nano-Al based energetic layers were investigated. The influence of solvents and additive components on the EPD process was analyzed, and the optimal composition of the suspension for the best deposition of layers based on nanoscale Al particles was acquired. The obtained layers can be used as an initiator of secondary reactions for on-chip energetic systems and a local heat source for joining surfaces by reactive bonding.

1. Introduction

Nowadays nanoscale energetic materials attract considerable research interest as a local heat source for low-temperature joining of thermosensitive elements [1]. Another one of the fields where these materials can be used is as an initiator of secondary reactions for on-chip energetic systems. Usually thermite composite materials consist of fuel particles and oxidizer particles [2, 3]. The most represented fuel material is aluminum. Among oxidizers, metal oxides such as CuO, Fe_2O_3 and WO_3 are popular. If using more active nanosized aluminum and initiate combustion in air, then an oxidizer is not needed - oxygen from the atmosphere will be used for the reaction.

In order to obtain such coats electrophoretic deposition (EPD) is considered to be the most affordable method that has the ability to deposit thin layers locally [4]. Thus, goal of this study was to investigate how suspension composition affects its stability and how process modes affects the morphology of the layer surface.

2. Methodical

To carry out the experiments it is necessary to prepare suspensions. They consisted of solvent, main and additive components [5]. For solvents, isopropyl alcohol (IPA) and acetone were used solely and in different ratios together. The main component of the suspension was Al nanopowder synthesized by the method of wire electric explosion in the argon atmosphere at Advanced Powder Technologies LLC. The aluminum powder has the particles with the mean size of 90–110 nm and contains at least 85–87 at.% of aluminum and 13–15 at.% of oxygen, the presence of which is associated with the thin layer of native oxide on the surface of the particles. Additive components as nickel (II) nitrate hexahydrate $N_2NiO_6 \cdot 6H_2O$ was used for suspension stabilization. The suspensions were dispersed in the ultrasonic bath. All suspension compositions are presented in the Table 1.

Table 1. Suspension composition.

№	Suspension composition	Al mass, mg	$N_2NiO_6 \cdot 6H_2O$ mass, mg	Solvent ratio, ml
1	IPA/Al		-	50
2	IPA/Acetone/Al		-	25/25
3	IPA/Acetone/Al/$N_2NiO_6 \cdot 6H_2O$	100	10	25/25
4	IPA/Al/$N_2NiO_6 \cdot 6H_2O$		10	50

The EPD process requires an electrophoretic cell. It includes two electrodes connected to a power source and a suspension immersed into a beaker. Titanium foils were used as substrates on which the material was deposited, and stainless steel plates were used as a counter electrode. The area was masked with round template, deposited surface area was 1,5 cm^2. Layers were deposited at potentiostatic mode at an electric field strength from 30 to 150 V/cm at different time limits.

Surface morphology, thickness and stoichiometry of the obtained energetic materials were investigated using scanning electron microscopy (SEM) with an attachment for analysis of energy dispersive X-ray spectroscopy (EDS).

3. Results and discussion

First, the stability of suspension has been investigated. To stabilize it different solvents and additives were used. The change in transparency and the formation of sediment at the bottom of the tube as well as uniformity of deposited layer were visually assessed. Within the framework of this work, a suspension was considered stable if the time of its visual stability was many times greater than the duration of the electrophoretic deposition process.

First, layers obtained from the suspension composition № 1 have uneven surface, because stability of suspension was low (Figure 1, a). Visually suspension sedimented on the bottom of the beaker in approximately 10 minutes. Then it was decided to add acetone to the suspension (composition № 2, Table 1). Visually layers were uniform, which tells about higher stability of the suspension in general (Figure 1, b). Afterwards, composition № 3 was tested. Layers obtained from the suspension have visually bigger deposited surface area although it was masked (Figure 1, c). Therefore, after removing acetone from composition № 3 layers were more even (Figure 1, d).

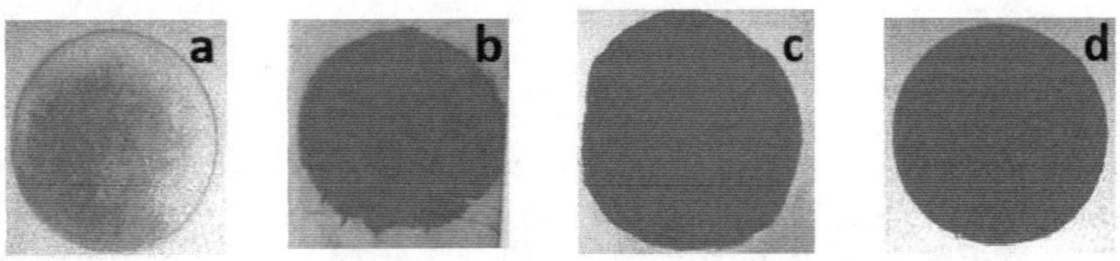

Figure 1. Photos of specimens obtained from suspension composition presented in table 1 respectively: (a) – composition № 1, b – composition № 2, c – composition № 3, d – composition № 4.

According to the experimental results, two compositions of suspensions showed better stability – suspension composition № 2 and № 4 (Table 1). The addition of acetone into the suspension made of Al and isopropyl alcohol allows to increase the stability of the suspension. In another case, the addition of nickel (II) nitrate hexahydrate to the same suspension composition makes it possible to stabilize it. In both cases layers were visually uniform.

Afterwards, the influence of different concentrations of $N_2NiO_6 \cdot 6H_2O$ to the velocity of the EPD process was studied (Figure 2). There were 4 suspension compositions with different concentration of $N_2NiO_6 \cdot 6H_2O$ (Table 2).

Table 2. $N_2NiO_6 \cdot 6H_2O$ concentration in the suspension.

Suspension composition	Al mass, mg	$N_2NiO_6 \cdot 6H_2O$ concentration, mg/ml	Solvent volume, ml
		0	
		0,1	
IPA/Al/$N_2NiO_6 \cdot 6H_2O$	100	0,2	50
		0,3	

Layers were deposited at potentiostatic mode at an electric field strength from 30 to 150 V/cm with 20 V/cm increments. Consequently, seven specimens from each suspension were obtained. Analyzing the dependence of the deposition velocity on $N_2NiO_6 \cdot 6H_2O$ concentration suspension composition with 0,1 mg/ml concentration of the charger has higher velocity. Though it was decided to use the composition of the suspension with 0,2 mg/ml concentration of $N_2NiO_6 \cdot 6H_2O$, because the stability of the suspension was higher.

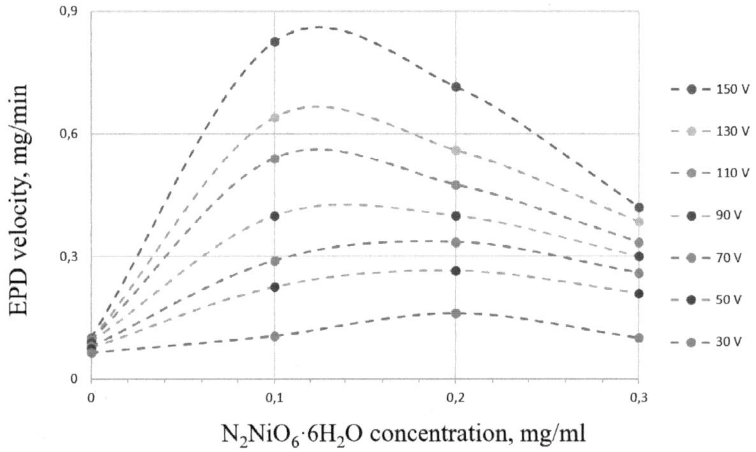

Figure 2. EPD velocity dependence on $N_2NiO_6 \cdot 6H_2O$ mass.

Though stability of the suspension composition № 4 from Table 1 was good, the velocity of the deposition was lower than in the composition № 2 from the same table. Thus it was decided to use isopropyl alcohol/acetone/Al composition. Graph of the deposited weight dependency on electric field strength has linear tendency (Figure 3). Varying such characteristics as electric field strength and time limits it is possible to control deposition velocity.

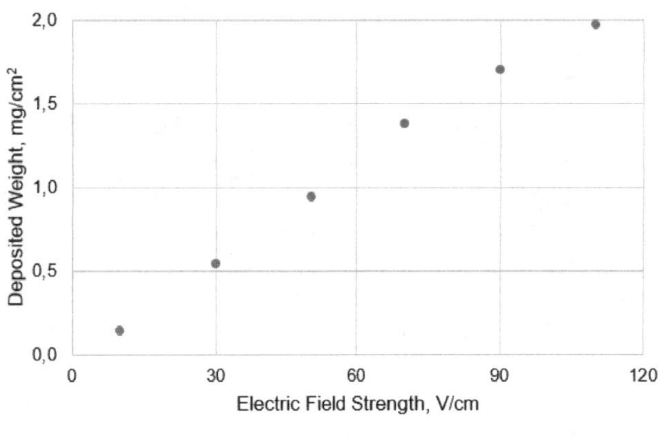

Figure 3. Deposited weight of the specimen dependence on electric field strength, deposited from IPA/Acetone/Al suspension.

Then the morphology of the obtained coats was examined with the scanning electron microscopy. Layers deposited in the electric field of 10 V/cm have big agglomerates and was uneven. The deposition velocity in the electric field of 110 V/cm was higher, thus deposited layers turned out to be more porous, which negatively affected the adhesion of the formed layers, and some areas of the layer desorbed from the substrate during the process. In the course of the research it was revealed that coat deposited at the electric field strength of 70 V/cm is more even than coat deposited at 10 V/cm and less porous than coat deposited at 110 V/cm (Figure 4). In this regard, the electric field strength of 70 V/cm was chosen.

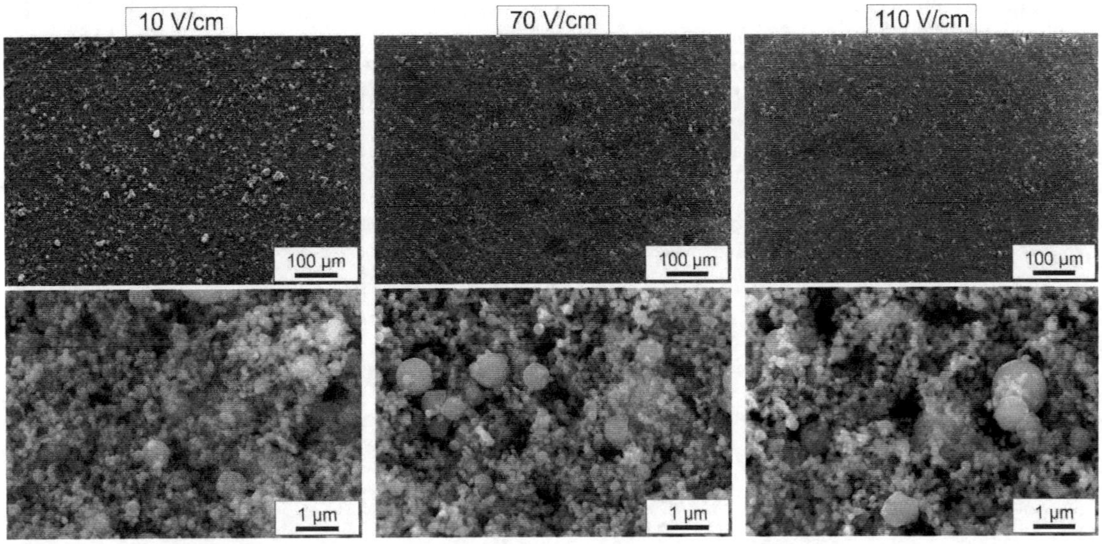

Figure 4. SEM photos of specimens obtained from suspension composition IPA/Acetone/Al.

Therefore, suspension composition of choice is isopropyl alcohol and acetone 1:1 ratio (50 ml), Al (100 mg); EPD mode – 70 V/cm, 5 minutes.

After finding out the electrophoretic deposition mode, examination of self-propagating high-temperature synthesis. The self-propagating reaction was initiated by a lighter. Combustion was propagating evenly through the entire surface area. Complete combustion of a double-sided specimen with the length of 3 cm continued for 4,3 seconds. Combustion rate frames were acquired using VirtualDub software (Figure 5).

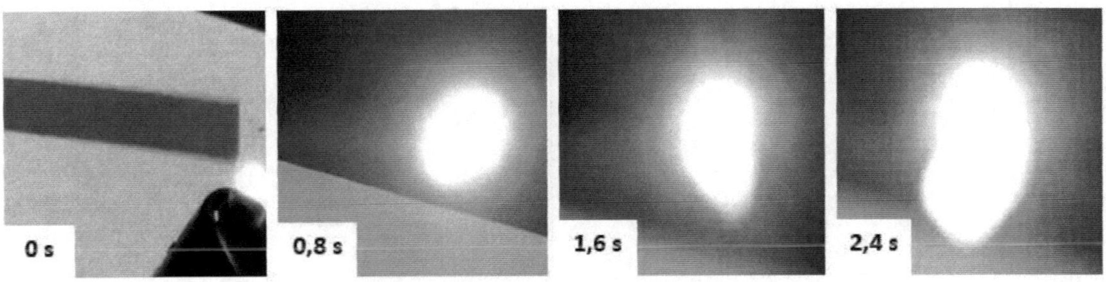

Figure 5. Combustion rate frames.

Knowing length of the specimen and time of self-propagating high-temperature reaction the combustion rate was calculated. It is appeared to be 0,007 m/s.

4. Conclusions

Based on the research results, the influence of solvents and additive components of suspension on the EPD process was analyzed, and the optimal composition of the suspension for deposition of layers based on nanosized Al particles was acquired. The obtained coats can be used as an initiator of secondary reactions for microelectromechanical systems (MEMS), and as a local heat source for reactive bonding of the surfaces.

Acknowledgments
This work was supported by the State assignment 2020-2022 (research project FSMR-2020-0018).

References

[1] Hinterreiter A P, Rebhan B, Flotgen C, Dragoi V and Hingerl K 2018 *Microsystem Technologies* **24** 773-777
[2] Yin Y and Li X 2019 *Vacuum* **163** 216-223
[3] Sullivan K T, Zhu C, Tanaka D J, Kunz J D, Duoss E B and Gash A E 2013 *The Journal of Physical Chemistry* **117(6)** 1686-1693
[4] Sullivan K T, Marcus A W, Kuntz J D and Gash A E 2012 *Combustion and Flame* **159** 2210-2218
[5] Besra L and Liu M 2006 *Progress in Materials Science* **52(1)** 1-61

Influence of active structure parameters on resonant frequency of acoustic transducer membranes

S V Malokhatko[1,2], E Yu Gusev[1] and O A Ageev[1,2]

[1] Department of Nanotechnologies and Microsystems, Southern Federal University, Taganrog 347922, Russia
[2] Research and Education Centre"Nanotechnologies", Southern Federal University, Taganrog 347922, Russia

malohatko@sfedu.ru

Abstract. The paper presents the results of calculations of the resonant frequency of a multilayer square membrane for ultrasonic microelectromechanical sensors. Various combinations of active layer materials and metal electrodes were taken into account. The dependences of the resonant frequency on the side length of membrane, as well as on the thickness of active piezoelectric layer and metal electrode for SiO_2/Ti/ZnO, SiO_2/Al/ZnO, SiO_2/Ti/PZT and SiO_2/Al/PZT structures were ob-tained. According to the calculations, the values of the resonant frequencies are in the ranges of 46.1–498.3 kHz for SiO_2/Ti/ZnO; 45.4–477.3 kHz for SiO_2/Al/ZnO; 39.4–411.4 kHz for SiO_2/Ti/PZT; 38.1-381 kHz for SiO2/Al/PZT. It is shown that the resonant frequency can be increased due to changes in the geometric parameters of the membrane; and the material and dimensions of the piezoelectric layer have the greatest influence. The results of analytical and numerical simulations for particular case of SiO_2(1μm)/Ti(1μm)/ZnO(2μm)/Ti(1μm) membrane with a side length of 600 μm are also compared. The obtained results could be used to optimize the design and process technology of microelectrome-chanical ultrasonic sensors.

1. Introduction

Ultrasonic transducers have found application in various fields of industry and medicine [1, 2]. The key parameter when choosing an ultrasonic transducer for specific applications is the resonant frequency, the choice of which is a compromise between the difference achieved at high frequencies and attenuation, which is less common at lower frequencies [3]. The main element of these sensors, on which such characteristics as sensitivity, electromechanical coupling coefficient and frequency depend, is the membrane. A multilayer membrane is an active layer of piezoelectric material located between metal electrodes [4]. Studies of constructive solutions with different active layers (ZnO, PZT, AlN) [5] and electrodes (Al, Mo, Ti, Pt, Au) are reported [6]. The values of ultrasound frequencies in such transducers start from 20 kHz. An urgent task is to choose the most optimal combination of materials in a multilayer structure and their thicknesses for working at such frequencies. The purpose of the work is to theoretically study the influence of materials and geometric dimensions of the active layer and electrodes on the values of resonant frequencies.

2. Experimental data and results

The paper analyzes the piezoelectric materials and metal electrodes used in ultrasonic transducers. Based on the reference data [5-7] on the properties of materials, the most optimal combinations of the multilayer structure were selected. A square-shaped membrane was chosen as the design for further calculations. The values for the side length of the square membrane was chosen based on previous study [8]. The structure for which the analytical calculations were performed consisted of three layers: SiO_2, piezoelectric (ZnO, PZT) and bottom electrode (Al, Ti) (Figure 1).

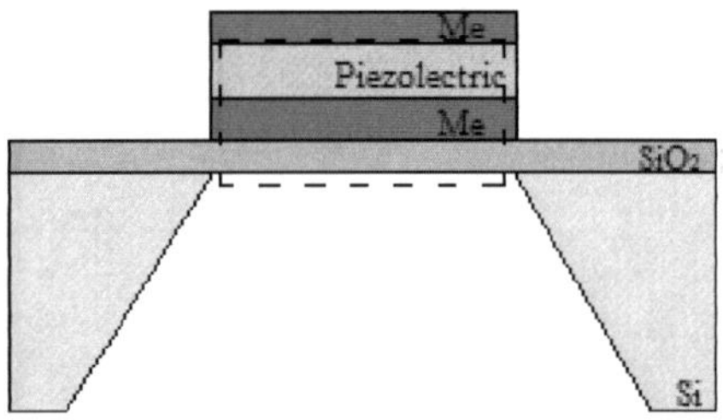

Figure 1. Shematic of the multilayer membrane of an acoustic transducer (dotted line indicates the cell used in analytical calculations).

Since the structure is multilayer, the resonant frequency (f_r) was calculated taking into account the parameters of each layer according to the formulas [7]:

$$f_r = \frac{3{,}19^2}{2\pi a^2}\left(\frac{D}{\mu}\right)^{1/2} \quad (1)$$

where a is half of the side length of the membrane; μ is the mass per area of the membrane; D is the bending stiffness of the membrane:

$$D = \frac{1}{3}\sum \frac{(h_n - z_s)^3 - (h_{n-1} - z_s)^3}{s_{11}^{(n)}(1 - (v_{12}^{(n)})^2)} \quad (2)$$

where v is the Poisson's ratio; z_s is the height above the bottom plane of the membrane; h_n is the total thickness of the layers; s_{11} is the plasticity coefficient.

Calculations were performed for the following combinations of multilayer membrane materials: 1) SiO_2/Ti/ZnO, 2) SiO_2/Al/ZnO, 3) SiO_2/Ti/PZT, 4) SiO_2/Al/PZT. The initial data for the calculations are given in Table 1 [5-7].

Table 1. Parameters of multilayer membrane materials.

	SiO_2	Al	Ti	ZnO	PZT
υ	0.17	0.35	0.32	0.33	0.36
ρ, kg/m^3	2200	2700	4506	5672	7500
s_{11} * 10^{-12}, m^{-2}/N	14.3	14.3	8.62	4.76	16.2
E, GPa	70	70	115	210	61,7
t, μm	1	1-2	1-2	2-4	2-4
a, μm			200-500		

The dependences of the resonant frequency on the thickness of the piezoelectric material and metal, as well as on the length of the side of the square membrane were obtained (Figure 2). The corresponding values of the resonant frequencies are given in Table 2.

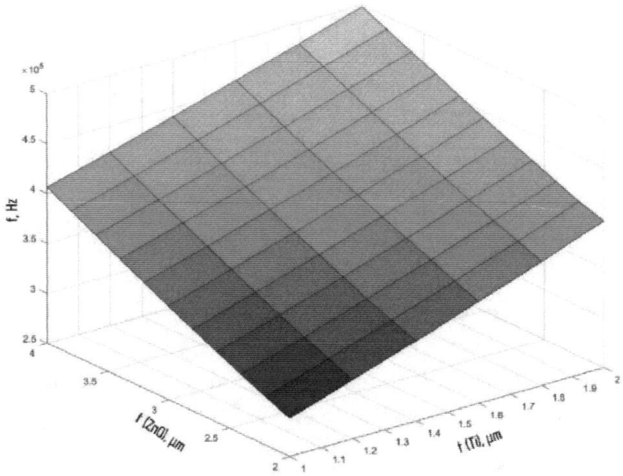

Figure 2. The dependence of the resonant frequency on the thickness of the piezoelectric (ZnO) and the bottom electrode (Ti) for 400 μm side length.

Table 2. Resonant frequency values (kHz) calculated for four type of membrane.

Stucture	Layer thickness, μm	400	600	800	1000
SiO_2/Ti/ZnO	1 / 1 / 2-4	288.2-406.6	128.1-18.7	72-101.6	46.2-65
	1 / 2 / 2-4	396.7-498.3	176.3-221.4	99.2-124.5	63.5-79.7
SiO_2/Al/ZnO	1 / 1 / 2-4	284-396,8	126.2-176.4	71-99.2	45.4-63.5
	1 / 2 / 2-4	387.6-477.3	172.3-212.1	96.9-119.3	62-76.4
SiO_2/Ti/PZT	1 / 1 / 2-4	246.6-305.1	109.6-135.6	61.6-76.3	39.4-48.8
	1 / 2 / 2-4	358.5-411.4	159.3-182.8	89.6-102.8	57.4-65.8
SiO_2/Al/PZT	1 / 1 / 2-4	238.3-288.4	105.9-128.2	59.6-72.1	38.1-46.1
	1 / 2 / 2-4	343.4-381	152.6-169.3	85.9-95.3	48-54.9

To verify the results of analytical calculations, numerical simulation of the frequency characteristics of a multilayer membrane was performed for one of the particular cases: a square membrane of 600×600 μm SiO_2(1 μm)/Ti(1 μm)/ZnO(2 μm)/Ti (1 μm). The resonant frequency calculated by the numerical method for such structure was 144 kHz (Figure 3), and by the analytical method - 128 kHz.

The difference in the values of the resonant frequency is presumably due to the fact that the presence of the top electrode was taken into account in the numerical model.

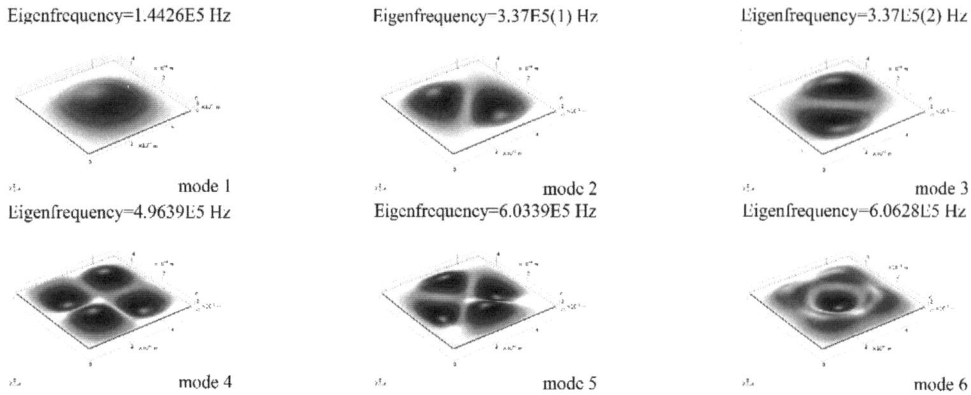

Figure 3. Deformation of the membrane for the first 6 modes of the resonant frequency.

The obtained results show that as the side length of the square membrane increases and the thickness of the layers decreases, the resonant frequency decreases, and that the choice of piezoelectric material makes the greatest contribution to the change in the resonant frequency in comparison with the electrode material. The highest values of the resonant frequency (396.7-498.3 kHz) were obtained for the SiO2/Ti/ZnO case, which makes it the most suitable for applications in devices operating at high frequencies. The obtained results could be used to optimize the design and process technology of microelectromechanical ultrasonic sensors.

Acknowledgment
The reported study was funded by RFBR, project number 20-37-90087. The work was done on the equipment of the Research and Education Centre «Nanotechnology» of Southern Federal University.

References
[1] Kanno I 2018 *Japanese Journal of Applied Physics* **57** 040101
[2] Ageev O A and Konoplev B G (eds.) 2019 *Nanotechnology in microelectronics* (Moscow: Nauka) 511
[3] Hoskins P R, Martin K and Thrush A 2010 *Diagnostic Ultrasound: Physics and Equipment* (Cambridge medicine. Cambridge University Press) 263
[4] Qiu Y, Gigliotti J V, Wallace M, Griggio F, Demore C E M, Cochran S and Trolier-McKinstry S 2015 *Sensors* **15** 8020
[5] Trolier-McKinstry S and Muralt P 2004 *Journal of Electroceramics* **12** 7
[6] Jung J, Lee W, Kang W, Shin E, Ryu J and Choi H 2017 *J. Micromech. Microeng.* **27** 113001
[7] Muralt P, Baborowski J, Barzegar A, Gentil S, Belgacem B, Petitgrand S, Bosseboeuf A and Setter N 2005 *IEEE Transactions on ultrasonics, ferroelectrics, and frequency control* **52** 2276
[8] Malohatko S V, Gusev E Yu, Jityaeva J Y and Ageev O A 2020 *Journal of Physics: Conference Series* **1697** 012100

Modeling of the interaction of porphyrin molecules in a nonpolar solvent

V N Mironyuk[1,5], A J R Al-Hassani[2], A J K Al-Alwani[1], N N Begletsova[1], M V Gavrikov[1], A S Kolesnikova[1,3], M V Pozharov[4] and E G Glukhovskoy[1]

[1] Education and Research Institute of Nanostructures and Biosystems, Saratov State University, Saratov, 410012, Russia
[2] The Iraqi Ministry of Education, 8CH9+9F Baghdad, Iraq
[3] Bogoliubov Laboratory of Theoretical Physics, Joint Institute for Nuclear Research, Dubna, Moscow Oblast, 141980, Russia
[4] Institute of Chemistry, Saratov State University, Saratov, 410012, Russia
[5] E-mail: bestblogger@yandex.ru

Abstract. The article presents the data of a theoretical molecular dynamics study of the interaction between a pair of porphyrin molecules, i.e. symmetrically substituted 5,10,15,20-tetra (4-n-methyloxyphenyl) porphyrin (P) and asymmetrically substituted 5-(4 hydroxy-phenyl)-10,15,20-tris (4-n-methyloxyphenyl) porphyrin (P-OH). We studied three systems, each of which consisted of a pair of porphyrin molecules (P || P, P-OH ↑↑ P-OH and P-OH ↑↓ P-OH) and chloroform molecules as a non-polar solvent. The effect of substitution, different orientations of asymmetrically substituted molecules and temperature on the geometry and energy of the system was investigated. It was shown that all three systems show signs of a true solution with chloroform as a solvent; the distance between asymmetrically substituted P-OH molecules was less than in the case of two P molecules. This may serve as an indirect evidence that the molecules are not prone to aggregation in the presence of chloroform.

1. Introduction and motivation

An important point in the Langmuir-Blodgett technology is the formation of highly organized floating monomolecular layers on the water surface. This process can be seriously disrupted by uncontrolled aggregation of molecules that make up a monolayer. Moreover, such aggregation can occur at the initial stage of preparing a solution for LB technology. In the case of simple surfactant molecules, this problem is not very common. They dissolve quite well in highly volatile solvents such as chloroform, i.e. form true solutions. Such solutions easily and evenly spread out over the water surface between barriers over the entire area of the Langmuir bath, and, after the solvent evaporates, the surfactant molecules are separated by distances commensurate with the sizes of the molecules themselves.

The task of obtaining a true monolayer becomes more complicated when the surface is comprised of molecules with a complex structure and uneven charge distribution or induction of dipoles. Such molecules include derivatives of porphyrin, phthalocyanine, proteins, etc. Some authors often conclude that it is impossible to obtain a true monolayer for such substances [1]. Nevertheless, this problem can be solved by varying the required conditions, i.e. introducing additional spacer substances into the composition of the monolayer substrate (MS), adding nanoparticles, changing the ratio of surfactants and quantum dots, and using various physical fields and radiation [2-4].

One of the most powerful technological factors is temperature. It has a significant impact on the state of multicomponent systems such as "water-air-monolayer". So, when the water temperature changes from 0 to 100 °C, its surface tension of water will change from 75 to 59 mN/m. In the absolute majority of cases, water is used in the LB bath to form MS. Therefore, the temperature during the formation of MS in the LB bath is the most important technological factor [5-7].

The aim of this work is theoretical study of the effect of temperature on various aspects of interaction between molecules of porphyrin derivatives with different substitutions and orientations of asymmetrically substituted molecules surrounded by solvent (chloroform) molecules.

2. Object of research

The objects of the study are molecules of porphyrin derivatives with symmetric substitution – 5,10,15,20-tetra (4-n-methyloxyphenyl) porphyrin (P) – and with asymmetric substitution -5- (4-hydroxyphenyl) -10,15,20- tris (4-n-methyloxyphenyl) porphyrin (P-OH). Their structural formulas are shown in figure 1.

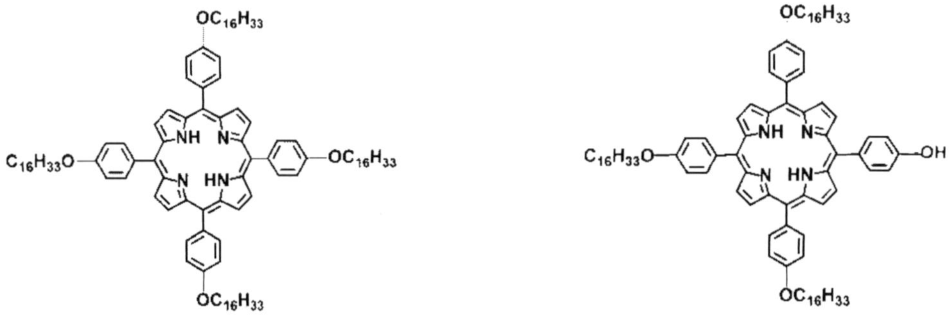

(a) (b)

Figure 1. Structural formula of: (a) symmetric 5,10,15,20-tetra (4-n-methyloxyphenyl)porphyrin (P), (b) asymmetric 5- (4-hydroxyphenyl) -10,15,20-tris (4- n-methyloxyphenyl) porphyrin(P-OH).

Three model systems were investigated. Each (of them) consisted of a pair of identical porphyrin molecules and chloroform molecules. System I contained two molecules of symmetrically substituted porphyrins (P || P); system II contained two molecules with asymmetric substitution with central macrocycles aligned in parallel, so that their dipole moments were oriented in one direction (P-OH ↑↑ P-OH); system III consisted of two molecules with asymmetric substitution with a parallel arrangement of central macrocycles, however, their dipole moments were counter-oriented (P-OH ↑↓ P-OH). System I included 292 chloroform molecules, systems II and III - 244 chloroform molecules each.

Due to the fact that the porphyrin molecule has a dipole moment, the interaction of two electric dipoles with dipole moments μ_1 and μ_2, located at a distance r from each other and oriented relative to each other, produces energy w. According to [8] this energy of the «charge-dipole» interaction can be written as

$$w(r, \theta_1, \theta_2, \varphi) = -\frac{\mu_1 \mu_2}{4\pi\varepsilon_0 \varepsilon r^3}[2\cos\theta_1 \cos\theta_2 - \sin\theta_1 \sin\theta_2 \cos\varphi] \quad (1)$$

where ε_0 is the dielectric permeability of space, ε is the dielectric permeability of molecules, φ is the angle made by two oppositely charged dipoles, while θ_1 and θ_2 are the angles formed by two with respect to the line connecting their centers. From the same equation, it follows that the maximum attraction will be observed when the dipoles are lined up along a single line connecting the centers of mass of the dipoles, and the energy will be determined as:

$$w(r, 0, 0, \varphi) = \frac{-2\mu_1 \mu_2}{4\pi\varepsilon_0 \varepsilon r^3} \quad (2)$$

If the dipoles are located parallel to each other and perpendicular to the line connecting their centers of mass, then the energy of their interaction $w(r, 90°, 90°, 180°)$ at the same distance r will be two times less than in the first case.

In the first variant of the arrangement of such molecules (when $\theta_1 = 0, \theta_2 = 0$), the distances between them are too great for the interaction to be effective The molecules selected for the calculation have a large extension in the plane of the macrocycle. In the first variant of the arrangement of such molecules (when $\theta_1 = 90, \theta_2 = 90$).

3. Research method

The search for the equilibrium state of the systems under study was carried out by the method of molecular mechanics MM+ with the calculation algorithm of Polak - Reiber. In all cases, the simulation began with the same position of the porphyrin molecules: the planes of the macrocycles of the molecules were parallel to each other and located at a distance of 1.55 nm from each other. For this, all three systems were constructed from a single initial model with a pair of symmetrically substituted porphyrin molecules (system I). To create systems II and III, -OH groups were introduced into the molecules P (initial system I) as the replacement of the aliphatic chains in the required fragments so as to obtain asymmetrically substituted molecules with the desired mutual orientation of the dipole moments - parallel (system II) and antiparallel (system III), as shown in Figure 2. This made it possible to keep the same initial positions of the central macrocycles in all molecular systems. The distance between the centers of mass of the central porphyrin macrocycle of two molecules was associated with the distances between the molecules and was subsequently used for analysis.

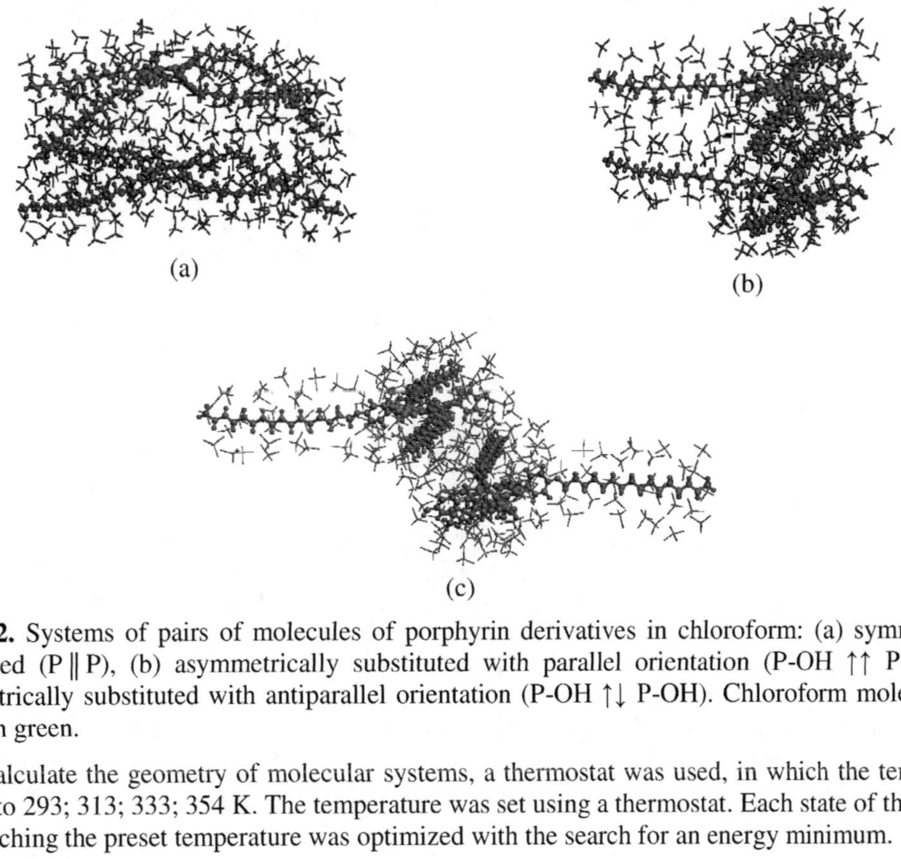

(a) (b) (c)

Figure 2. Systems of pairs of molecules of porphyrin derivatives in chloroform: (a) symmetrically substituted (P || P), (b) asymmetrically substituted with parallel orientation (P-OH ↑↑ P-OH), (c) asymmetrically substituted with antiparallel orientation (P-OH ↑↓ P-OH). Chloroform molecules are shown in green.

To calculate the geometry of molecular systems, a thermostat was used, in which the temperature was set to 293; 313; 333; 354 K. The temperature was set using a thermostat. Each state of the systems after reaching the preset temperature was optimized with the search for an energy minimum.

4. Results and discussion

The molecule of symmetrically substituted 5,10,15,20-tetra (4-n-methyloxyphenyl) porphyrin in vacuum at a temperature of 0 K is practically non-polar and has a dipole moment close to 0 in the case of a symmetric arrangement of aliphatic chains, when they and the macrocycle lie in the same plane.

Figure 3 shows asymmetric 5-(4-hydroxyphenyl)-10,15,20-tris (4-n-methyloxyphenyl) porphyrin (P-OH) optimized by MM+ at 293 K. It can be seen that at this temperature the P-OH has a dipole moment of $\mu_{P-OH} = 1.17$ D. The direction of the vector of the dipole moment is shown in Figure 3 as a segment with the signs «+» and «–» at its ends.

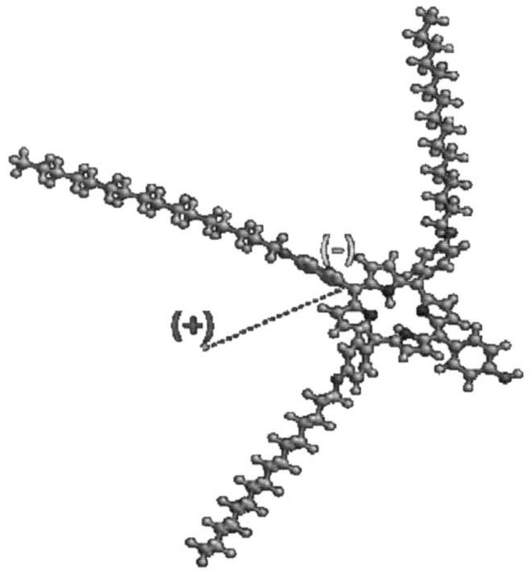

Figure 3. Asymmetric 5-(4-hydroxyphenyl)-10,15,20-tris (4-n-methyloxyphenyl) porphyrin (P-OH) optimized by MM+ at 293K.

Figure 4 demonstrates the temperature dependence of the distance between the centers of mass of molecules and the total energy of the system. As can be seen, at relatively low temperatures (293 and 313 K), the difference in distances is not significant – about 1 Å. Moreover, the intermolecular distance in system I is less than that for system III. There are two possible reasons for this behavior. First, the allotted time might not have been enough for the molecules to find a deeper local energy minimum. On the other hand, the value of the temperature, and, consequently, the intensity of the internal displacements of individual fragments of molecules, might not be sufficient for the system to arrive at a deeper energy minimum. This is also evidenced by the higher energy values at low temperatures for all systems. This is especially evident for system I: its transition from a temperature of 293 to 313 K is accompanied by a noticeable decrease in the total energy by 60 kcal/mol. For systems II and III $\Delta w_{\text{I}} = w_{\text{I}}(293) - w_{\text{I}}(313) = 60$ this decrease was less pronounced, but also took place: $\Delta w_{\text{II}} = 5\ \Delta w_{\text{III}} = 15$.

At higher temperatures, the behavior of systems becomes more logical and expected. It is likely that at such temperatures, the system moving more intensively manages to find a deeper energy minimum in a shorter time.

The difference in the strength of molecular interaction systems II and III are less significant as compared with the system I. Nevertheless, it can be noted here that the distances between molecules with oppositely oriented dipole moments are slightly smaller (and, accordingly, the interaction is greater) than for molecules with dipole vectors directed in one direction.

5. Conclusion

The effect of temperature on the features of the interaction of molecules of porphyrin derivatives surrounded by chloroform molecules as a solvent has been studied. Numerical experiment at temperatures 293; 313; 333; 354 K revealed a non-monotonic effect of temperature on the intermolecular distance.

The total energy of the system with a parallel arrangement of porphyrin molecules (P-OH ↑↑ P-OH) was higher than in the case of an antiparallel arrangement (P-OH ↑↓ P-OH).

The results obtained show that the molecules of symmetrically and asymmetrically substituted porphyrin form a true solution with chloroform. The presence of a weak intermolecular interaction is not critical and does not lead to the formation of aggregates.

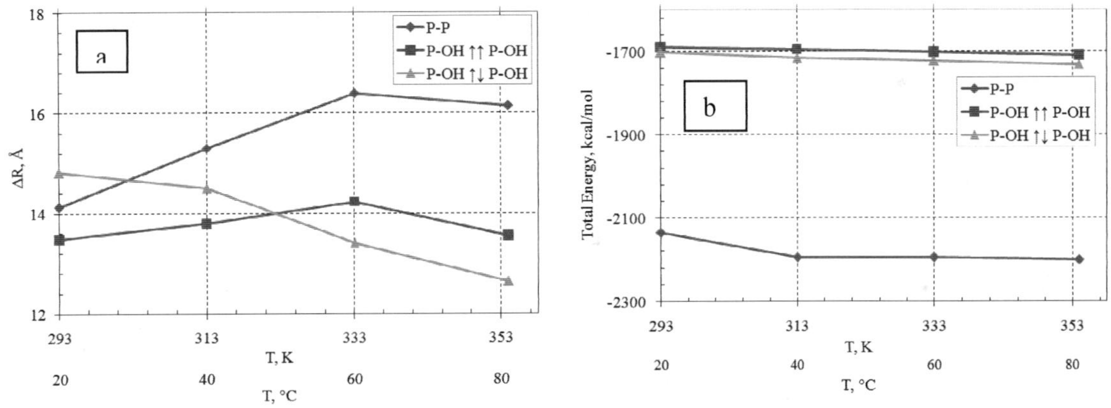

Figure 4. Dependence of the distance between the centers of mass of porphyrin molecules (a), and the total energy of the system (b) on temperature.

Acknowledgments
The study supported by a grant of the Russian Science Foundation (project No. 21-73-20057) and Saratov State University

References

[1] Mildner J and Dynarowicz-Łątka P 2012 β-Carotene does not form a true Langmuir monolayer t the air/water interface *Colloids Surf. B Biointerfaces* **90** 244–7
[2] Kazak A V, Marchenkova M A, Khorkov K S, Kochuev D A, Rogachev A V, Kholodkov I V, Usol'tseva N V, Savelyev M S and Yu. Tolbin A 2021 Ultrathin Langmuir–Schaefer films of slipped-cofacial J-type phthalocyanine dimer: supramolecular organization, UV/Vis/NIR study and nonlinear absorbance of femtosecond laser radiation *Appl. Surf. Sci.* 148993
[3] Al-Alwani A J, Chumakov A S, Gorbachev I A, Kuznetsov N, Kletsov A A and Glukhovskoy E G 2017 Influence of Thermal Separation of Oleic Acid on the Properties of Quantum Dots Solutions and Optoelectronic of Their Langmuir Monolayers *BioNanoScience* **7** 496–500
[4] Chumakov A S, Gorbachev I A, Ermakov A V, Kim V P and Glukhovskoy E G 2013 Langmuir Monolayers in the Electric Field *Izv. Saratov Univ. New Ser. Ser. Phys.* **13** 80-3
[5] Begletsova N N, Mironyuk V N, Santer S, Smirnova A I, Usol'tseva N V and Glukhovskoy E G 2020 Effect of the composition and temperature of the subphase on the surface potential of the Langmuir monolayer of 8CB liquid crystal *J. Phys. Conf. Ser.* **1697** 012112
[6] Chumakov A S, Al-Alwani A J, Gorbachev I A, Ermakov A V, Kletsov A A, Glukhovskoy E G, Kazak A V, Usol'tseva N V and Shtykov S N 2017 Temperature and Mixing Ratio Effects in the Formation of CdSe/CdS/ZnS Quantum Dots with 4′-n-octyl-4-p-Cyanobiphenyl Thin Films *BioNanoScience* **7** 666–71
[7] Begletsova N N, Baimagambetova L T, Mironyuk V N, Smirnova A I, Venig S B, Usol'tseva N V and Glukhovskoy E G 2019 Formation of Gibbs and Langmuir floating layers based on copper nanoparticles: temperature and concentration effects *J. Phys. Conf. Ser.* **1410** 012216
[8] Israelachvili J N 2011 *Intermolecular and Surface Forces* (University of California, Santa Barbara, California, USA: Publisher: Elsevier Academic Press Inc.)

Study of Langmuir monolayers and Langmuir-Schaefer films based on symmetrical *meso*-aryl-substituted porphyrin derivative

V N Mironyuk[1,4], A J K Al-Alwani[1], N N Begletsova[1], M V Gavrikov[1], A S Kolesnikova[1,2], M V Pozharov[3] and E G Glukhovskoy[1]

[1] Education and Research Institute of Nanostructures and Biosystems, Saratov State University, Saratov, 410012, Russia
[2] Bogoliubov Laboratory of Theoretical Physics, Joint Institute for Nuclear Research, Dubna, Moscow Oblast, 141980, Russia
[3] Institute of Chemistry, Saratov State University, Saratov, 410012, Russia
[4] E-mail: bestblogger@yandex.ru

Abstract. This paper presents the results of a study of *meso*-aryl-substituted porphyrin Langmuir monolayers by the method of compression isotherms. Experimental data were used to plot the dependences of the compression modulus (C^{-1}) on the specific area. Monolayers at specific surface pressure were transferred to solid substrates and investigated. The monolayers were transferred to the surface of monocrystalline silicon at surface pressures of 8, 25, and 60 mN/m and examined them by atomic force microscopy (AFM) in a semi-contact mode. It was found that with an increase in the transfer pressure, the coarsening of molecular aggregates occurs. The smallest roughness is observed for a porphyrin film formed and transferred at a pressure of 8 mN/m.

1. Introduction

Thin films of porphyrin derivatives are a promising material for the creation of thin-film photosensitive and photovoltaic devices [1–3]. The properties of porphyrin strongly depend on the presence of peripheral substituents, the inclusion of metals inside the porphyrin ring, and various conformations [4]. All this gives designers great freedom to develop new devices with unique properties.

One of the important technological problems in obtaining multilayer structures is the formation of highly ordered monomolecular layers of porphyrin and their derivatives (PP) on solid substrates. Currently, there are many technologies for producing films and coatings, but it is the Langmuir-Blodgett (LB) technology that combines the possibility of forming monomolecular layers and the ease of implementation. Using the basic principle of LB technology (self-organization of substances at the water-air interface), it is possible to form PP monolayers with different packing densities and molecular orientations in the layer. Using the Langmuir-Schaeffer (LS) transfer technique, such layers are easily transferred to solid substrates, while the crystal structure of the monolayer that is formed on the surface of the water can be preserved on the solid substrate.

To obtain highly oriented defect-free Langmuir monolayers and LB films, it is important to understand the regularities of their self-organization at the air-water interface, which is a template that initiates both the 2D format and the orientation of molecules in the layer. The strict control of technological conditions is important, in particular the amount of substance that is applied to the surface of the water.

The molecule of the symmetrically substituted porphyrin under study has 4 hexadecyloxyphenyl substituents in the mesopositions of the macrocycle [5]. This substitution gives reason to expect that the amphiphilic properties of these molecules will be maximized. This will facilitate better ordering of molecules in the ML on the water surface and more successful technological transfer of it to solid substrates.

One of the non-contact methods for controlling the structure of LB films, with the help of which it is possible to determine the surface morphology and dimensions of the formed nanoobjects, is the method of atomic force microscopy. In the course of previous experiments, in which the effect of the amount of surfactant in a solution of quantum dots on the structure of the resulting films was studied by atomic force microscopy, it was shown that the removal of excess oleic acid leads to a more ordered and dense packing of quantum dots [6]. The AFM method allowed the authors of [7] to show that protonated porphyrins form J-aggregates at the gas-liquid and liquid-liquid interface, the optical properties of which depended on the degree of compression. In study [8], the LB method was used to form chiral mesoscopic structures of two tetraphenylporphyrin derivatives functionalized with a steroid group in each of the four meso-positions. AFM made it possible to reveal the formation of rod-like aggregates intertwined in such films. The AFM method showed the inclusion of copper nanoparticals stabilized by the surfactant sodium dodecyl sulfate leads to an increase in stability, an increase in the value of surface potential, and the relief of the film transferred to a solid substrate has a pronounced macrostructure in the form of relatively large objects with dimensions of 300-500 nm in plan and height up to 80 nm [9].

The aim of the work is: to obtain floating (Langmuir) monolayers based on symmetrically substituted porphyrin and to study the influence of the conditions of their formation and transfer (the amount of solution introduced to the surface of the water subphase, the degree of compression of monolayers) on the morphology of thin films transferred to a solid substrate by the Langmuir-Scheffer method.

2. Materials and methods

Monolayers of porphyrin derivatives were formed and examined with a KSV NIMA 2002 KN LB bath (medium size bath with a compression ratio of about 1:9). Deionized water with a resistivity of 18.2 MΩ×cm, which was obtained on a Smart2Pure unit from ThermoScientific, was used as the aqueous subphase. The studies were carried out at a room temperature of 24 °C. A solution of *meso*-aryl-substituted porphyrin (PP) in chloroform with a concentration of 5×10^{-4} M was used to form a monolayer. This porphyrin derivative (the structural formula of which is shown in Figure 1) was synthesized according to the method described in [5].

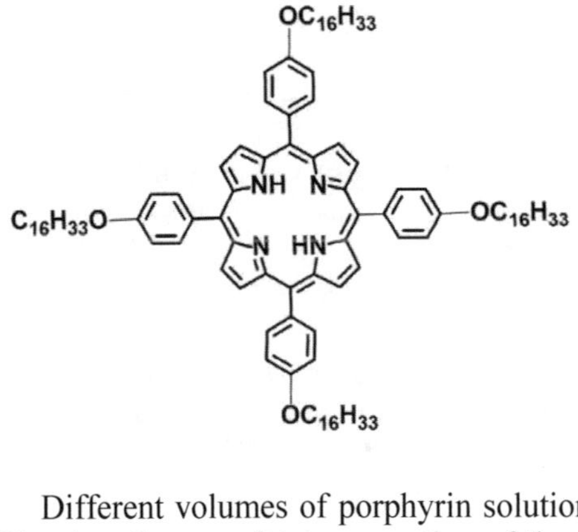

Figure 1. Structural formula of the *meso*-aryl-substituted porphyrin compound.

Different volumes of porphyrin solution were placed on the water surface – 50, 100, and 150 μL. The time for complete evaporation of the solvent was 10 min for each case, after which the ML was compressed by two barriers at a constant rate of 13 mm/min (symmetric compression mode). The surface tension during compression was measured with a Wilhelmy balance.

The monolayer was transferred onto a solid n-type monocrystalline silicon substrate coated with a natural oxide. Immediately before the transfer of ML, the silicon surface was treated with chloroform. After reaching a given pressure (at surface pressures of 8, 25, and 60 mN/m) and relaxation of the monolayer, it was transferred by the Langmuir-Schaeffer method - a horizontal lift. Before scanning, the samples were dried at room temperature.

Then, the surface relief of the samples was examined using a SOLVER NANO scanning probe microscope (AFM NT-MDT). Scanning was carried out in a semicontact mode, with probes of the HA-HRA/W2C series with a resonance frequency of 230 kHz and a radius of curvature of the tip not exceeding 35 nm, at a scan recording speed of 0.8 Hz and an operating point current of 2.5 nm.

The resulting images were processed using the software for visualization and data analysis Gwyddion version 2.58 [10].

3. Results and Discussions

Figure 2 shows the π-A compression isotherms of the *meso*-aryl-substituted porphyrin compound depending on the volume sprayed on the surface of the aqueous subphase.

Compression isotherms showed an implicit dependence on the volume of the solution deposited on the surface of the subphase. The maximum compression pressure was 66 mN/m.

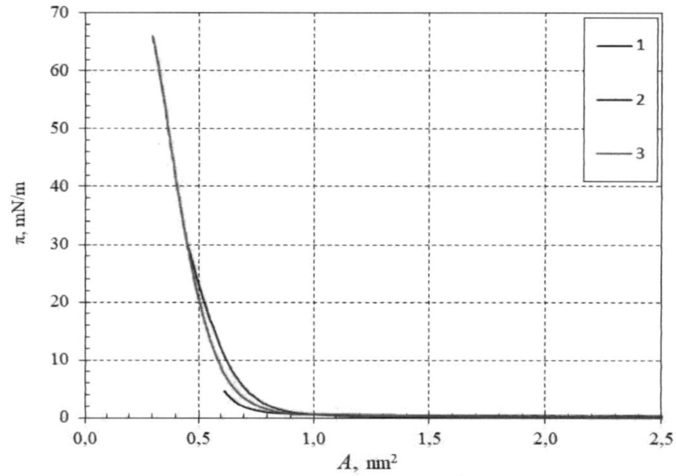

Figure 2. π-A isotherms of the *meso*-aryl-substituted porphyrin compound depending on the volume of the solution (numbers 1, 2, 3 correspond to the volume of the solution introduced onto the water surface - 50; 100; 150 μl, respectively).

The calculated specific area A per molecule immediately after spreading (before compression) of the porphyrin solution was about 5.3; 3.9; 2.6 nm^2 for 50, 100 and 150 μl, respectively. The tight packing state for 50 μL was not achieved. No collapse was observed for volumes of 100 and 150 μL.

The choice of pressure for the transfer of a monolayer to solid substrates was chosen from the following considerations. Were selected 3 areas in the compression isotherm plot for 150 μl. These areas correspond to different phase states of ML. For them, the area per molecule in a monolayer (A_0) was calculated, which was obtained by extrapolating the linear section to the A-axis (Figure 3, one such tangent to the middle section of the isotherm is shown as an example). For each of the selected regions, the modulus of compression, compressibility and area per molecule in a monolayer were calculated. Knowledge of these parameters makes it possible to determine the phase states of a monolayer, to analyze and correct the conditions for deposition of the formed films on solid substrates.

Together with the compression isotherm, a C^{-1}-A dependence was plotted (the dependence of the compression modulus on the specific area). It is noteworthy that this dependence exhibits local maxima in the range of values from 0.47 to 0.37 nm^2. This indicates that achieving closest packing of amphiphilic molecules or fragments of amphiphilic molecules, ML components. Upon further compression, a multilayer structure begins to form.

Table 1 shows the values of the parameters of PP monolayers, calculated from the experimental curves: pressure of close packing of molecules π_{tr}, area A_0, modulus of compression C^{-1} of a monolayer, compressibility K_0.

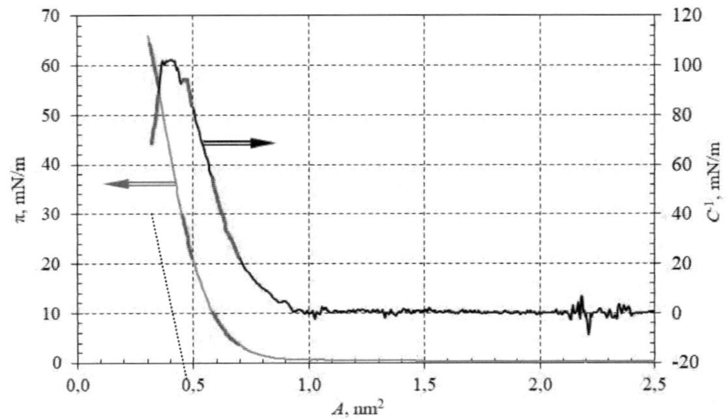

Figure 3. π-A isotherms of *meso*-aryl-substituted porphyrin compound (blue curve) and C^{-1}-A Compression modulus dependence (black curve). Red areas correspond to areas tight molecular packing.

Table 1. Parameters of PP monolayer on the surface of pure water (transfer pressure π_{tr}, area A_0, compression modulus C^{-1} of the monolayer, compressibility K_0) for the selected sections of the isotherm (curve 3 in Figure 2) and films on a solid substrate (lateral dimensions of nanoobjects L, height of nano-objects h, root-mean-square roughness R_a).

π_{tr}, mN/m	A_0, nm^2	C^{-1}, mN/m	K_0, m/mN	L, nm	h, nm	R_a, nm
8	0,76	44	0,022	100-150	5-10	2,3
25	0,62	117	0,0085	200-300	10-15	3,8
60	0,60	134	0,0074	300-500	10-20	6,9

At pressures of about 8 mN/m, the ML state can be classified as liquid-expanded, at which the modulus of compression is not constant due to the changing conditions of intermolecular interactions. At higher pressures, the packing of molecules in the ML changes little and the mechanical properties remain practically constant. This is an indirect indication that in the pressure region of 25 mN/m the ML is in a stressed state, but at these stresses its structure is preserved. At a pressure of about 60 mN/m stresses in the ML begin to form 3D objects, such as multi-layer folds of molecular aggregates.

AFM studies also indicate such rearrangements in the ML. For AFM studies, the formed monolayers were transferred onto solid substrates of monocrystalline silicon at transfer pressures of 8, 25, and 60 mN/m. Figure 4 shows AFM images and elevation profiles of a *meso*-aryl-substituted porphyrin compound. The obtained images of the relief show that on the samples with deposited films, at different transfer pressures, noticeable differences are observed: the shape of the nanoobjects and the features of the relief change. Typical dimensions of nanoobjects were 100-150 nm in plan and up to 10 nm in height for films obtained at a pressure of 8 mN/m, and at flat dimensions of 450-500 nm and a height of up to 20 nm for films obtained at a pressure of 60 mN/m, that is, the height and lateral dimensions of the objects increased with increasing transfer pressure.

Statistical processing of AFM images also showed the expected coarsening of molecular aggregates and an increase in film roughness with an increase in the surface transfer pressure to 60 mN/m. The root-mean-square roughness R_a of the surface of the samples obtained was 2.3; 3.8; 6.9 nm at transfer pressures of 8; 25; 60 mN/m, respectively, i.e. a film of porphyrin molecules formed at a pressure of 8 mN/m has the smallest roughness, which may be associated with a lower aggregation of molecules.

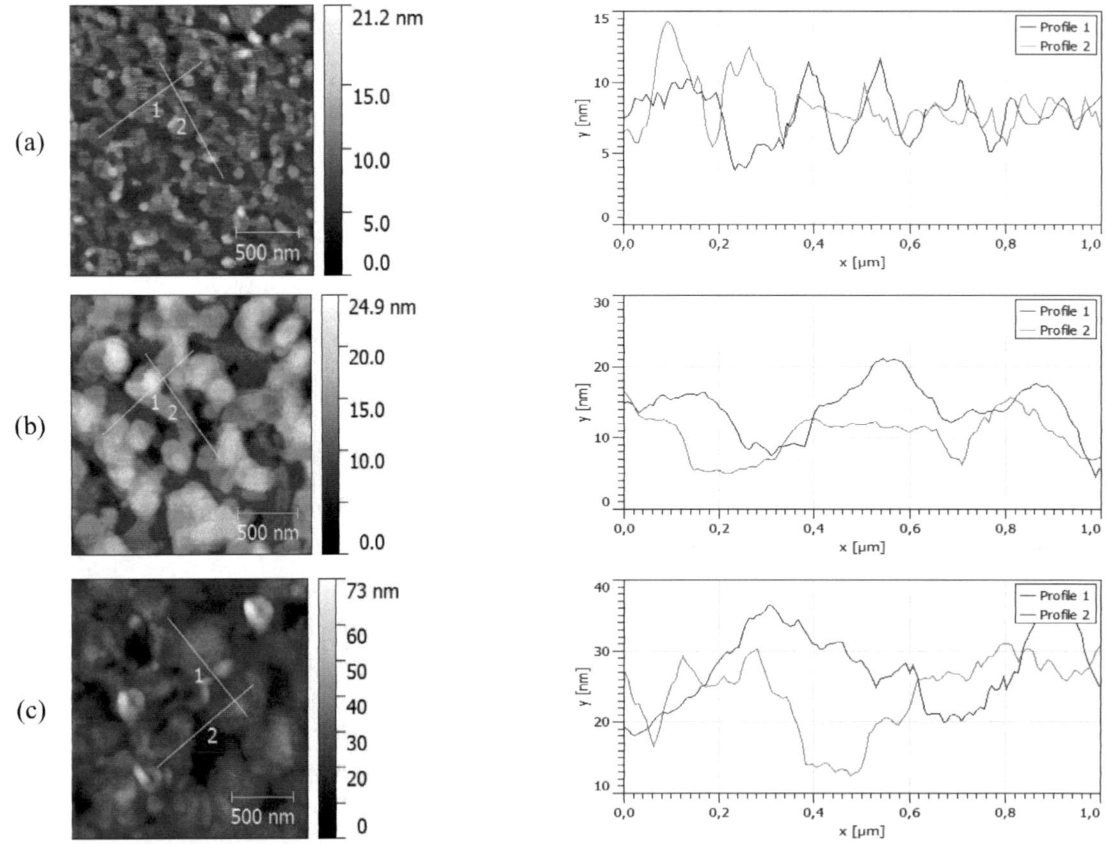

Figure 4. AFM image of *meso*-aryl-substituted porphyrin (left) and their profiles (right) at transfer pressures: (a) 8 mN/m, (b) 25 mN/m, (c) 60 mN/m.

4. Conclusion

Thus, Langmuir monolayers and Langmuir-Schaeffer films of a meso-aryl-substituted porphyrin compound were studied in this work. Compression isotherms showed an implicit dependence on the volume of the solution deposited on the surface of the subphase. The mechanical parameters of the monolayer were determined for different aliquots. By extrapolating the linear region to the axis, the most favorable regions of the close packing of molecules for further transfer to solid substrates were revealed. Analysis of AFM images showed the expected coarsening of molecular aggregates and an increase in film roughness with increasing surface transfer pressure. The smallest roughness is observed in the PP film formed and transferred at a surface pressure of 8 mN/m, which may be associated with a lower aggregation of molecules.

Acknowledgments

The study supported by a grant of the Russian Science Foundation (project No. 21-73-20057) and Saratov State University

References

[1] Zeng K, Tong Z, Ma L, Zhu W-H, Wu W and Xie Y 2020 Molecular engineering strategies for fabricating efficient porphyrin-based dye-sensitized solar cells *Energy Environ. Sci.* **13** 1617–57

[2] Marczak R, Sgobba V, Kutner W, Gadde S, D'Souza F and Guldi D M 2007 Langmuir−Blodgett Films of a Cationic Zinc Porphyrin−Imidazole-Functionalized Fullerene Dyad: Formation and Photoelectrochemical Studies *Langmuir* **23** 1917–23

[3] Usol'tseva N V, Smirnova A I, Kazak A V, Giricheva N I, Galanin N E, Shaposhnikov G P, Bodnarchuk V V and Yablonskii S V 2017 Mix-substituted phthalocyanines of a "push–pull"-type and their metal complexes as prospective nanostructured materials for optoelectronics *Opto-Electron. Rev.* **25** 127–36

[4] Kazak A V *et al* 2020 Floating layers and thin films of mesogenic mix-substituted phthalocyanine holmium complex *Thin Solid Films* **704** 137952

[5] Smirnova A I, Soldatova K M, Ezhov A V, Bragina N A, Giricheva N I and Usol'tseva N V 2019 Experimental and Theoretical Absorption Spectra of A3B-Type Porphyrine Derivatives *Liq. Cryst. Their Appl.* **19** 25–37

[6] Al-Alwani A J, Kosolapova K I, Chumakov A S, Lukyanova V O, Gorbachev I A, Glukhovskoy E G, Kazak A V, Smirnova A I, Usol'tseva N V and Shtykov S N 2018 Studying of Surfactant Excess Separation from Non-aqueous Quantum Dots Solution on its Monolayer Formation Process *BioNanoSci.* **8** 1081–6

[7] Okada S and Segawa H 2003 Substituent-Control Exciton in J-Aggregates of Protonated Water-Insoluble Porphyrins *J. Am. Chem. Soc.* **125** 2792–6

[8] Lettieri R, Cardová L, Gatto E, Mazzuca C, Monti D, Palleschi A, Placidi E, Drašar P and Venanzi M 2017 Hierarchical transfer of chiral information from the molecular to the mesoscopic scale by Langmuir–Blodgett deposition of tetrasteroid-porphyrins *New J. Chem.* **41** 639–49

[9] Begletsova N N, Mironyuk V N, Ezhov A V, Smirnova A I, Usol'tseva N V and Glukhovskoy E G 2020 Features of formation of Langmuir monolayers of porphyrin derivatives on the surface of aqueous solutions of copper nanoparticles *J. Phys. Conf. Ser.* **1697** 012118

[10] Nečas D and Klapetek P 2012 Gwyddion: an open-source software for SPM data analysis *Open Phys.* **10** 181–8

Effect of encapsulation of fullerene in semiconducting zigzagcarbon nanotubes on optical and thermoelectric properties

E V Morozova and D A Timkaeva

Ulyanovsk State University, Ulyanovsk, 432017, Russia

kat-valezhanina@yandex.ru

Abstract. We study the optical and thermoelectric properties of carbon nanotubes (CNTs) with encapsulated C_{60} fullerene molecules. Using ab-initio methods, we calculate optical and thermoelectric paramters for CNT with fullerenes periodically located inside the nanotube at different distances from each other. Dependencies of these parameters on fullerene concentration and diameter of CNT are analyzed.

1. Introduction

Nanopeapods discovered by the author of Ref. [1] are hybrid systems of single-wall carbon nanotubes (CNTs) and fullerenes inside and linked with tubes by van der Waals interactions [2-5]. Nanopeapod systems are promising for use in nanoelectronics: solar cells, nano-emitters, etc. [2-4]. Fullerenes arranged in an orderly manner represent a set of quantum dots in a one-dimensional superlattice. By varying the concentration of fullerenes inside, and the types of fullerenes, one can modulate the band structure of the CNT-fullerene system and control the electronic and phonon characteristics of nanopeapods.

In this work, using ab-initio methods, we calculate the optical and thermoelectric properties of CNTs (n,0) with fullerenes periodically located inside the nanotube at a fixed distance from each other.

2. Optical and thermoelectric properties

Semiconductor CNTs (14,0), (17,0) are considered. The distances between the C_{60} fullerenes in nanopeapods were chosen equal to 10,32 and 22,74 Å. The diameter of the C_{60} fullerene is 7.1 Å.

Table 1. Parameters of the investigated carbon nanotubes

Chirality indice sof CNTs	CNT diameter,D, Å	The smallest distance from the fullerene to the CNT wall, R, Å
(14,0)	1.096	2.03
(17,0)	1.331	3.27

Optimization of the considered systems CNT (14,0) -C_{60}, CNT (17,0) -C_{60} (17,0). was carried out using the density functional method (DFT) implemented in QuantumATK (Synopsys). The cutoff energy of the electron wave functions is 500 eV. To describe the effects of electron exchange and correlation, we used the Purdue-Burke-Ernzerhof approximation generalized to the case of solids (GGA-PBEsol). To generate k points in the Brillouin zone for all nanotubes, the Monkhorst-Pack method with a 1 × 1 × 24 grid was used.

Transfer coefficients were calculated using the nonequilibrium Green's function (NEGF) method, the DFT method, and nonequilibrium molecular dynamics. We used the standard model, in which the central part of the tube is connected to semi-infinite left and right electrodes (see, for example, [5,6]). QuantumATK calculates the indicated thermoelectric coefficients according to linear response theory. The following relationships are used.

$$G_e = \left.\frac{dI}{dV_{\text{bias}}}\right|_{dT=0}, S = -\left.\frac{dV_{\text{bias}}}{dT}\right|_{I=0}, \lambda_e = \left.\frac{dI_Q}{dT}\right|_{I=0}, \Pi = \left.\frac{I_Q}{I}\right|_{dT=0} = SV_{\text{bias}}.$$

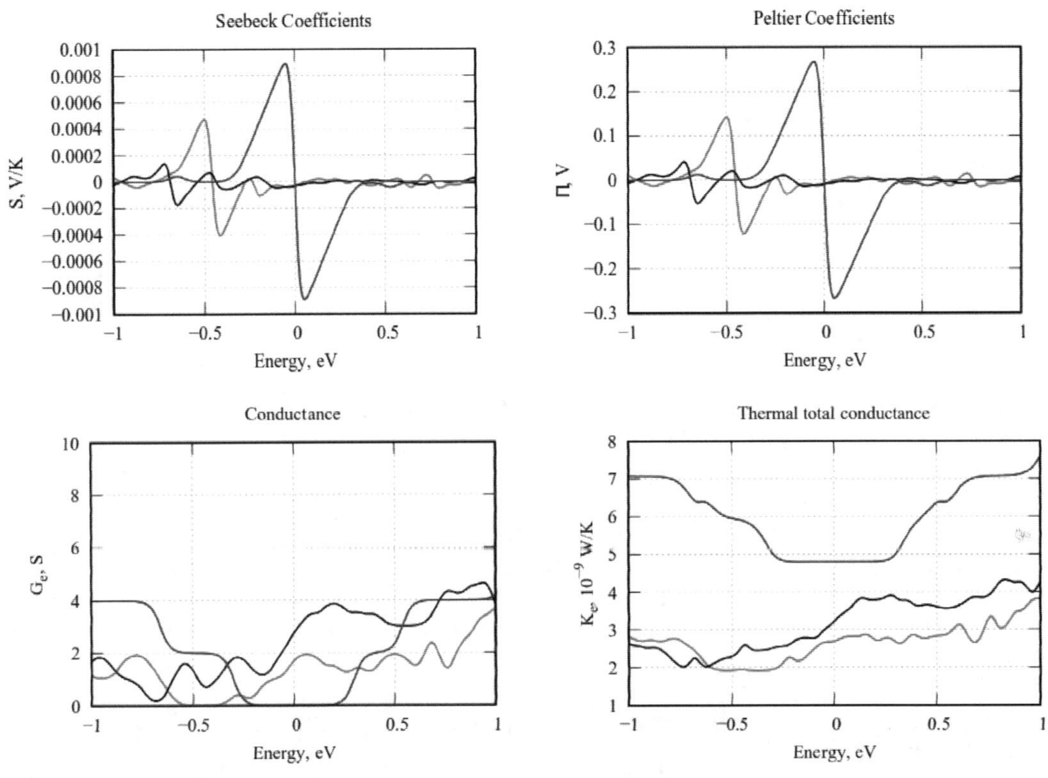

Figure 1. Thermoelectric coefficients of the CNT (14,0) - C_{60} system with distances between fullerene centers R = 22,64 Å (red color), R = 10,32 Å (black color) and pure CNT (14,0) (blue color).

Figure 1 shows the effect of the encapsulation of fullerenes in CNTs, at which a shift in the spectra readings to the range of negative energies is observed, which leads to a decrease in the Seebeck and Peltier coefficients and causes additional fluctuations in the spectra of electrical and thermal conductivity. The readings of the thermal conductivity coefficients decrease, which can be caused by a change in the Fermi level due to the encapsulation of the tube with fullerenes. Also, the created temperature gradient causes the phenomenon of entrainment of electrons by phonons.

The same calculations were performed for a semiconductor nanotube (17,0) with a distance of 10,32 Å between the centers of two neighboring fullerenes.

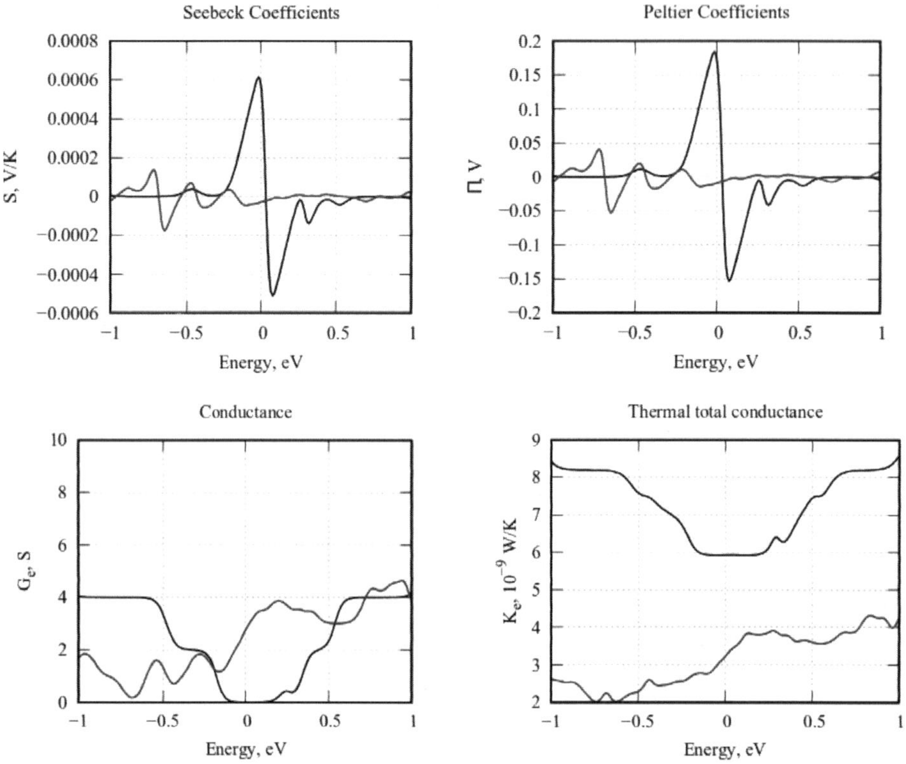

Figure 2. Thermoelectric coefficients of CNT (14,0) - C_{60} systems with distances between fullerene centers R = 10,32 Å (blue color) and CNT (17,0) - C_{60} systems with distances between fullerene centers R = 10,32 Å (black color).

Comparing the readings of the thermoelectric coefficients of CNT (14,0) and CNT (17,0) with the distance between the centers of two neighboring fullerenes 10,32 Å (Figure 2), it can be seen that an increase in the tube diameter leads to a resonance of the Seebeck and Peltier coefficients near zero energy. Also, an increase in the CNT diameter leads to an increase in the thermal conductivity coefficient, which may be associated with a large distance between the fullerene molecules and the tube wall. (see table 1)

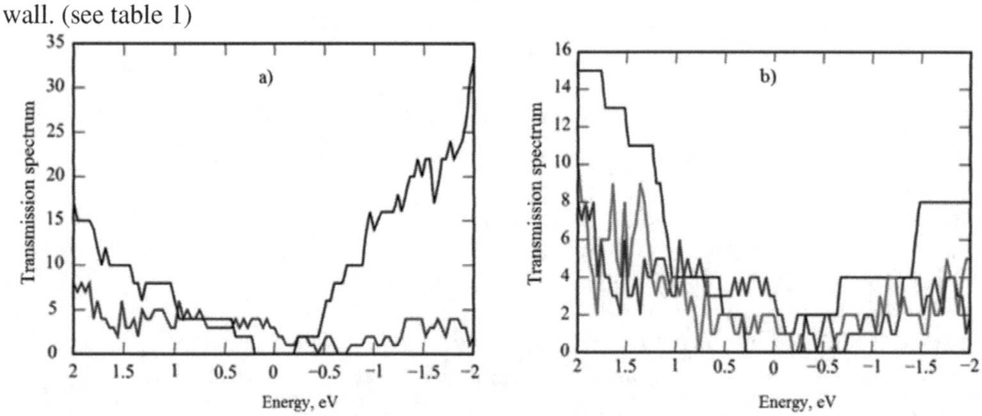

Figure 3. a) Transmission spectrum of CNT (14,0) - C_{60} (blue color) and CNT (17,0) - C_{60} (black color); b) Transmission spectra of CNT (14,0) - C_{60} systems with distances between fullerene centers R=10,32 Å (black color), R = 22,64 Å (red color) and CNT (14,0) (blue color).

The transmission coefficient of a CNT system (17,0) with encapsulated fullerenes at a distance of 10, 32 Å from each other is higher than that of a similar CNT system (14,0) (Figure 3 a). Transmission spectra of CNT systems (14,0) - C_{60} is approximately the same (within fluctuations) for the considered distances between fullerenes (Figure 3 b).

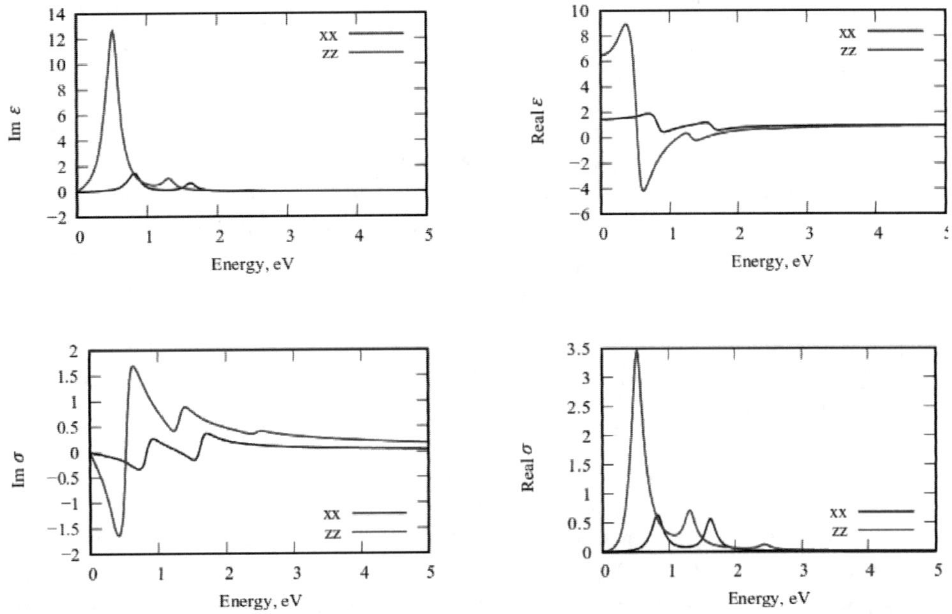

Figure 4. The coefficients of dielectric constant and optical conductivity of CNTs (17,0) with encapsulated fullerenes, the distance between the centers of which is 10,32 Å.

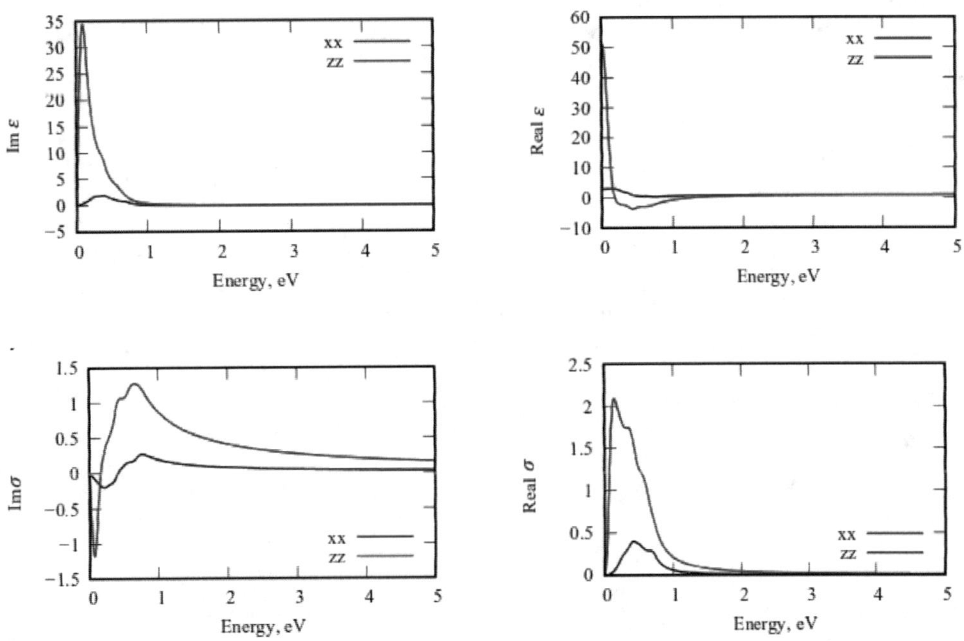

Figure 5. The coefficients of dielectric constant and optical conductivity of CNTs (14,0) with encapsulated fullerenes, the distance between the centers of which is 10,32Å.

The dielectric constant in the short-wavelength spectrum of CNTs with encapsulated fullerenes is higher for the tube (14,0), while an increase in the diameter leads to an increase in the conductivity in the short-wavelength region of the spectrum (Figure 4, Figure 5). Accordingly, we can talk about high readings of the capacity of the considered nanotubes in the short-wavelength region of the spectrum. Also, an increase in the tube diameter leads to the appearance of small fluctuations in the conductivity index in the energy range from 1 to 2,5 eV.

3. Conclusions

Using ab-initio methods, we calculated optical and thermoelectric paramters for CNT (14,0) and (17,0) with fullerenes periodically located inside the nanotube at different distances from each other. Dependencies of these parameters on fullerene concentration and diameter of CNT are analyzed.

With a decrease in the nanotube diameter, the optical conductivity of the CNT-C_{60} system in the short-wavelength spectrum increases. The thermal conductivity of CNT-C_{60} depends on the distance between the encapsulated fullerene molecules and the diameter of the carbon nanotube.

Acknowledgments

This work is supported by the Ministry of Science and Higher Education of the Russian Federation (state program 0830-2020-0009).

References

[1] Smith B W, Monthioux M and Luzzi D E1998 *Nature* **396**
[2] Okazaki T, Okubo S, Nakanishi T, Joung S K, Saito T, Otani M, Okada S, Bandow S and Iijima S 2008 *J. Am. Chem. Soc.* **130**
[3] GlukhovaOE, KolesnikovaAS, NefedovIS and Slepchenkov M M 2014 *JETP Letters* **99**
[4] Kodama T, Ohnishi M, Park W, Shiga T, Park J, Shimada T, Shinohara H, Shiomi J and Goodson K E 2017 *Nature mater.***16**
[5] Kochaev A I, Meftakhutdinov R M, Sibatov R T and Timkaeva D A 2020 *Comput. Mater. Sci.***186**
[6] Meftakhutdinov R M, Sibatov R T and Kochaev A I 2020 *J. Phys. Condens. Matter.***32**

Analysis of frequency response sensor of MEMS gyroscope in vacuum chamber

I E Lysenko, D V Naumenko and O A Ezhova

Southern Federal University, Taganrog 347922, Russia

dvnaumenko@sfedu.ru

Abstract. This article presents a study of the frequency response of a MEMS gyroscope in a vacuum chamber. On the basis of experimental studies by the method of laser Doppler vibrometry, the dependences of the amplitude of oscillations of the inertial mass in the vertical plane at various pressures are obtained. The bandwidth of the MEMS sensor was also measured.As a result of the experiments, the damping factors were determined to compose a more complete mathematical model and for more accurate finite element modeling in ANSYS, and refined parameters of the electrostatic drive and the amplitude of oscillations along the axis of the drive were obtained. These studies will be useful for determining the residual degree of vacuum in the case for further frequency tuning of the device.

1. Introduction

Micromechanical gyroscopes are designed to measure angular velocity. Due to their advantages such as small size, low cost and sufficient accuracy for many tasks, they have found application in many areas, for example, in mobile devices, in vehicles. Experimental studies of MEMS structures are necessary to test analytical and numerical methods of calculation [1], to determine the calculated damping coefficients [2], as well as to determine the optimal operating modes of the device. This article examines the behavior of a micromechanical sensor at various degrees of vacuum.

2. Description of the test sample

A sample of a micromechanical gyroscope (Figure 1) was made using the BDRIE technology [3] with a device layer thickness of 50 μm and an aspect ratio of 1:25. The minimum technological gap is 2 microns. The MEMS sensor of the gyroscope has two perpendicular axes X and Y. The movable inertial mass (1) is fixed on an elastic suspension (2) through intermediate decoupling frames (3, 4), which provide movement along two axes and which, in turn, are fixed on the anchor (5). An electrostatic vibration drive with a variable area (6) ensures the movement of the inertial mass in the driving mode, and a capacitive sensitive element with a variable gap (7) is responsible for the information signal. The MEMS sensor parameters are described in Table 1.

After the samples were made, the dimensions and technological tolerances were controlled using an SEM microscope. SEM image show on Figure 2.

Young's modulus and other parameters of anisotropic silicon were obtained from the factory, manufacturer, laboratory of Center for Microtechnologies Fraunhofer ENAS.

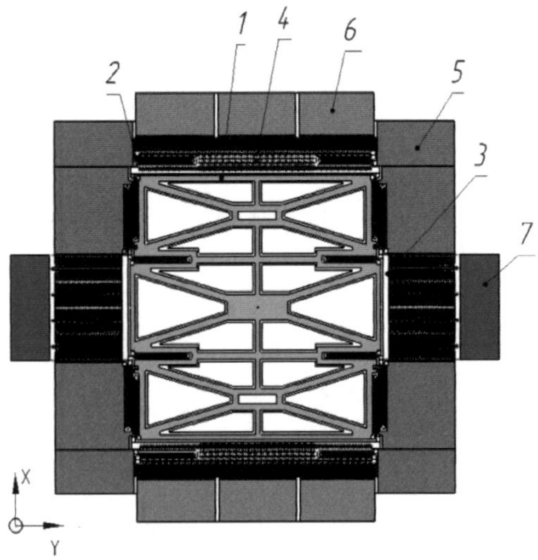

Figure 1. Topology of a micromechanical gyroscope.

Figure 2. SEM image of the test sample.

Table 1. Mechanical parameters of the micromechanical gyroscope.

Parameter	Value
Inertial mass	0,63x10^{-7} kg
Natural frequency of X-axis	10300 Hz
Stiffness coefficient k_x	1.260×10^3 N / m
Young's modulus	142 GPa
Width of elastic beam	6 µm
Device layer thickness	50 µm

3. Mathematical description of the experiment

Understanding the dynamics and response characteristics of a MEMS gyroscope is of great importance in the design of both electrostatic actuators and vibration gyroscope sensing elements. The sample under study is a typical mechanical resonator with one degree of freedom, shown in Figure 3. The equation of motion for a resonator with inertial mass m, stiffness coefficient k, and damping coefficient c has the form (1):

$$m\ddot{x} + c\dot{x} + kx = F_x(t)\#(1)$$

Differential equations were modeled in MATLAB / Simulink , the simulation scheme is shown in the Figure 4. The diagram contains the displacement integrators, a sine load generator, an oscilloscope, and a Bode diagram.

F_x (t) is the force of the electrostatic variable-area actuator, shown in figure 5. The equivalent force of which is calculated by the formula (2) [4]:

$$F_x = -\frac{\varepsilon_0 z_0}{y_0} N V^2 \#(2)$$

The results of calculating the electrostatic drive at different voltage values are shown in the Figure 6. Typical values for electrostatic drive forces are in the range of 0.1 to 1 µN.

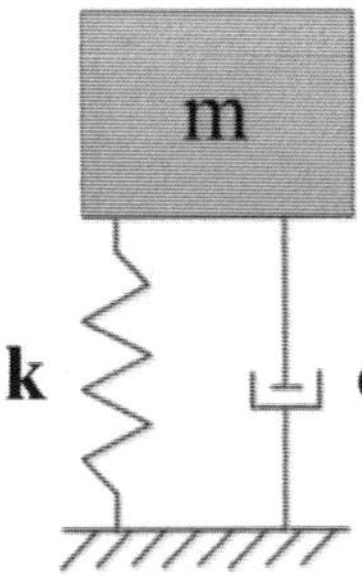

Figure 3. Lumped-element model of mechanical resonator.

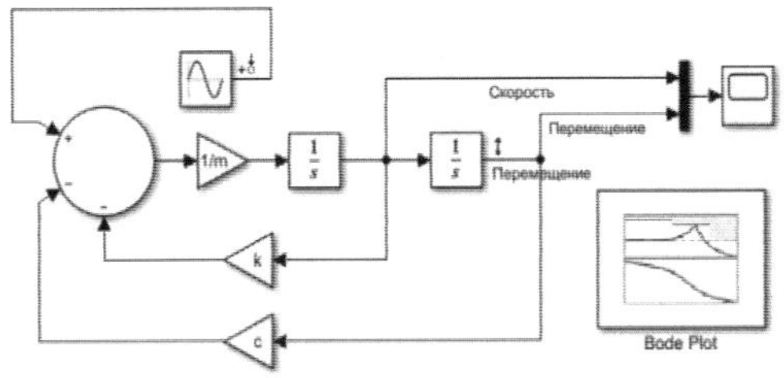

Figure 4. MATLAB / Simulink model.

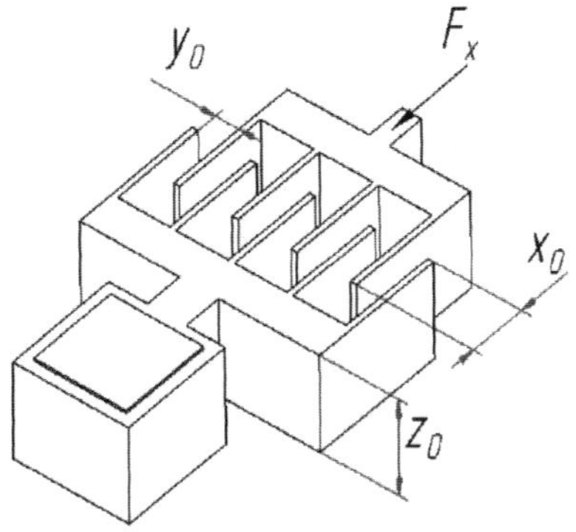

Figure 5. Dimensions of the electrostatic actuator.

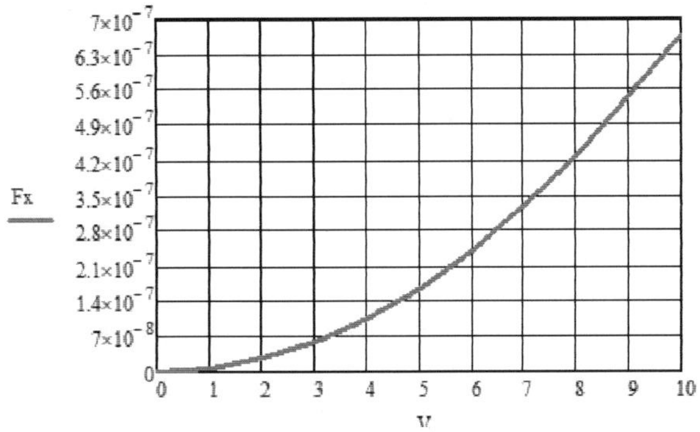

Figure 6. Electrostatic drive force.

For the study, the average values of the electrostatic drive were chosen equal to 5V. Further harmonic analysis was carried out using an equivalent electrostatic drive force of 0.17μN.

4. FEA of frequency response MEMS gyro

As a result of modal analysis, the natural frequencies of the oscillatory mode were obtained. The first mode of vibration with a frequency close to 10 kHz is shown in the figure 7. For research, the first vibration mode is sufficient, since other mods are not involved in this research.

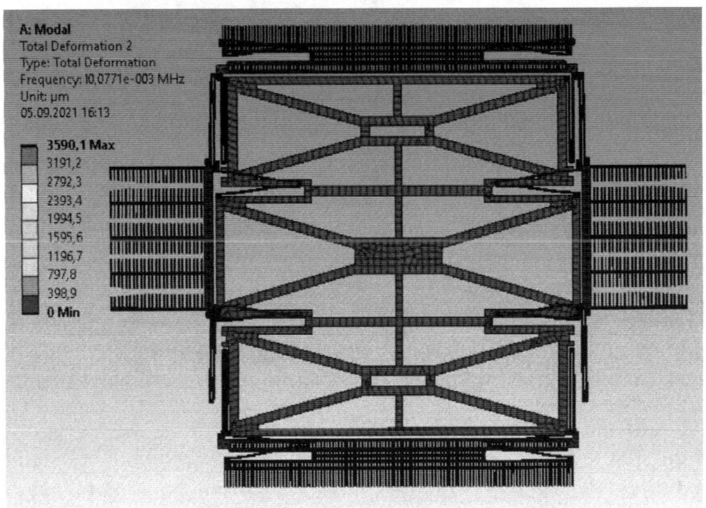

Figure 7. First mode of vibration.

The study of the amplitude frequency response was carried out using harmonic analysis with the support of expanded modes. FEM frequency response is shown in the figure 8.

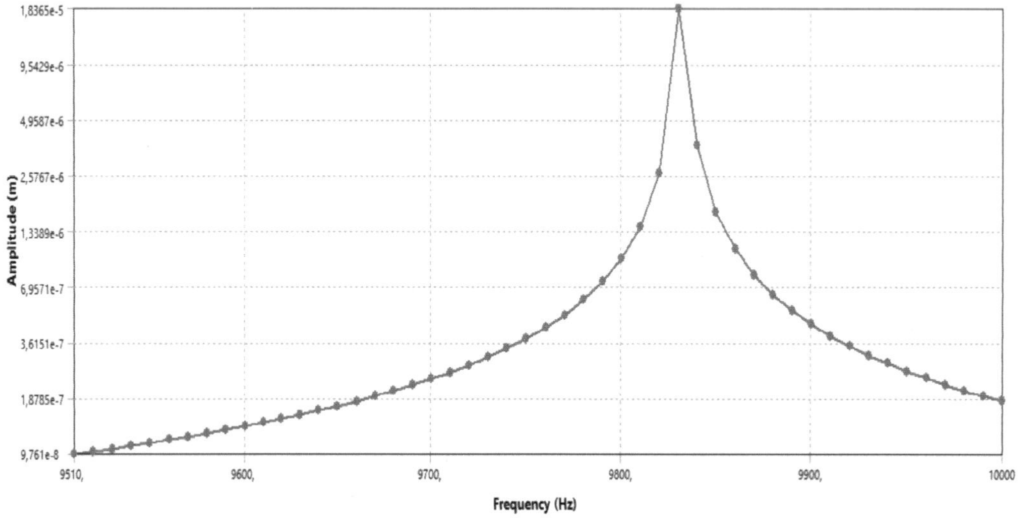

Figure 8. FEM frequency response.

Since we have not determined the coefficients of viscous damping, the amplitude is 18 microns. This is the maximum theoretical amplitude excluding frictional losses in material and viscous media.

5. Description of the experimental research

The examination of samples using scanning laser vibrometry and stroboscopy technologies are standard methods for characterizing micromechanical gyroscopes [5]. These research technologies are combined in the Polytec MSA-500 microsystem analyzer shown in Figure 9. The MEMS sensor installation for measurements is shown in Figure 10. The sample (1) is placed in a vacuum chamber (2), from which a vacuum pump regulates the pressure. The microsystem analyzer (4) makes measurements. The generator (5) provides the movement of the inert mass. the measurement results are displayed on the monitor (6).

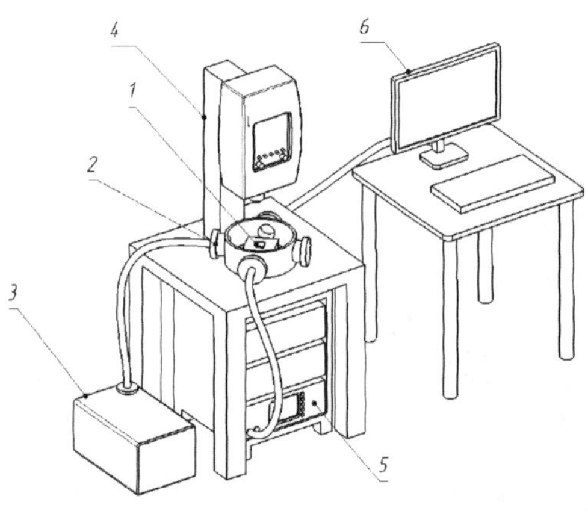

Figure 9. Scheme of the experimental stand.

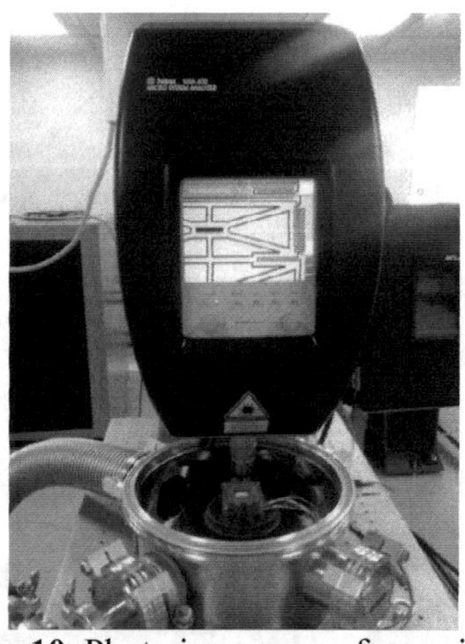

Figure 10. Photo in process of experiment.

The course of the experiment was as follows. A pressure of 2000 Pa was chosen as a starting point, since this pressure provides a stable frequency response during operation of the device. The following two examination modes correspond to different variants of the micromechanical transducer housing. A low vacuum on the order of a few Pa is typical for wafer level packing using a porous titanium oxide getter. An average vacuum of 200 Pa is typical when sealing in sintered bodies under vacuum conditions.

6. Experimental result

In the first experiment, an experimental bode diagram, shown in the Figure 11, was obtained using stroboscopic video microscopy. Using this method, you can get the amplitude frequency characteristics in the horizontal plane.

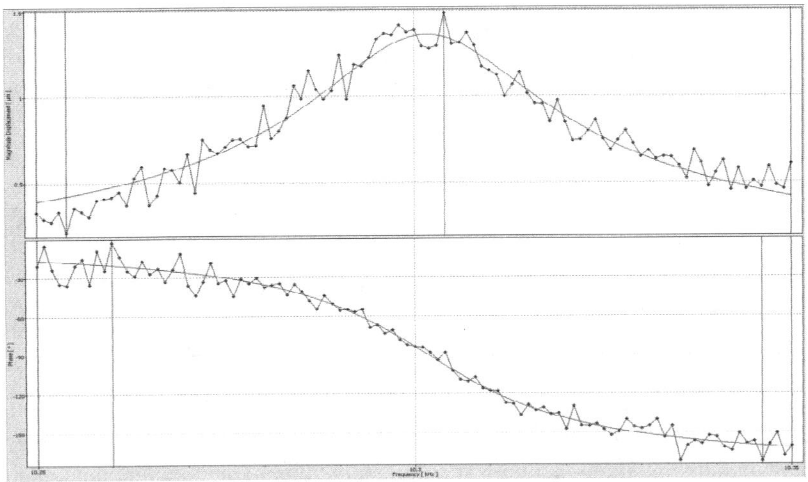

Figure 11. Bode diagram of test structure.

As a result of measurements, the maximum amplitude is 1.485 μm. Using the software of the microsystem analyzer, the numerical characteristics presented in Figure 12 were obtained.

The second part of the experiment was to determine the frequency response along the vertical axis using doppler vibrometry technology. The results of studies at a predetermined pressure are presented in Table 2. To determine the correlation between vertical and horizontal amplitudes, the first measurement was performed at the same pressure equal to 2000 Pa. The results of this measurement are shown in Figure 13.

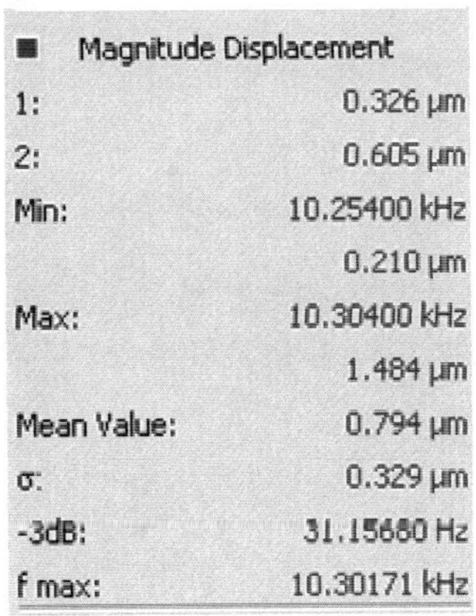

Figure 12. Frequency response in the horizontal plane.

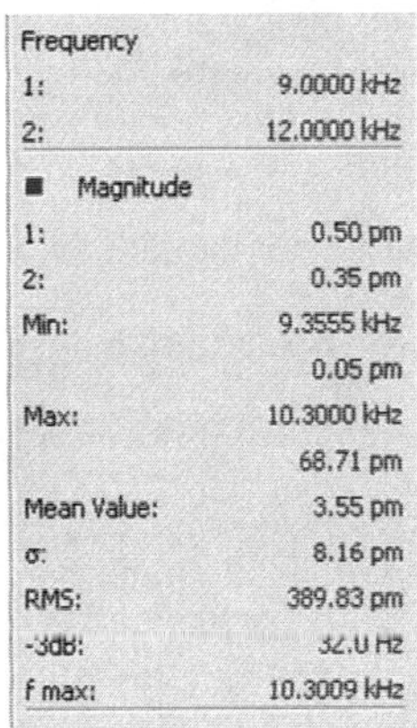

Figure 13. Frequency response in the vertical plane.

With a decrease in pressure from 2000 Pa to 2 Pa, the amplitude increased 5.15 times from 68.71, which corresponds to 1.485 microns in the horizontal plane to 353.7 pm.

Following calculations, the amplitude at a pressure of 2 Pa is 7.65 μm. And it is less than half of the theoretically possible amplitude.

Table 2. Result of experimental research of microgyroscop.

Pressure, Pa	Bandwidth, Hz	Amplitude, pm
2000	31,5	68,71
100	28	243,1
2	20	353,7

7. Conclusion

As a result of the study, the minimum working pressure in the gyroscope body was determined. The experimental dependence of the amplitude of the inert mass on the pressure is obtained. Determination of the bandwidth of the device at various pressures. The results obtained will be used to determine the damping coefficients, and to check the mathematical model. Also, these results are necessary for setting up and debugging the device, determining the residual pressure in the case, and also to improve the accuracy of calculations by the finite element method.

Acknowledgments

This work was supported by RFBR research project No. 19-37-90136 “Development of the fundamental foundations for the construction of high-aspect torsion suspensions for inertial MEMS sensors”.

References

[1] Lysenko I E and Naumenko D V 2019 Calculation of the amplitude-frequency characteristics of the sensitive element of a micromechanical gyroscope using finite element modeling *Izvestiya SFU. Technical science.* **3** 205

[2] Raspopov V Ja 2007 *Micromechanical devices* (Moscow: Mashinostroenie Publishers) p 400

[3] Billep D, Hiller K, Küchler M, Schröter B, Dienel M, Scheibner D and Gessner T 2005 Bonding and deep RIE: a powerful combination for highaspect-ratio sensors and actuators *Proc. of SPIE.* **5715** 80–91

[4] Acar C and Shkel A 2009 *MEMS vibratory gyroscopes – Structural approaches to improve robustness in MEMS* (New York: Springer) p 260.

[5] *IEEE Standard Specification Format Guide and Test Procedure for Coriolis Vibratory Gyros* Dec 20 2004 (IEEE Std 1431-2004) , pp.1–78

Spin-dependent transport through helical Aharonov-Bohm interferometer

R A Niyazov[1,2], **D N Aristov**[2,1] **and V Yu Kachorovskii**[3]

[1] Department of Physics, St. Petersburg State University, St. Petersburg 198504, Russia
[2] NRC "Kurchatov Institute", Petersburg Nuclear Physics Institute, Gatchina 188300, Russia
[3] Ioffe Institute, St. Petersburg 194021, Russia

E-mail: `r.niyazov@spbu.ru`

Abstract. We discuss spin-dependent transport via tunneling Aharonov-Bohm interferometer formed by helical edge states tunnel-coupled to helical leads. We focus on the experimentally relevant high-temperature case as compared to the level spacing and obtain the full 4×4 matrix of transmission coefficients in the presence of magnetic impurities. We show that spin conserving and spin-flip transmission coefficients of the setup can be effectively tuned by the magnetic flux. These features are attractive due to possible applications for spintronics, magnetic field detection, and quantum computing.

1. Introduction

The physics of topological insulators is one of the recently emerged, but already very actively studied areas of condensed matter physics [1, 2]. In these materials, as in conventional insulators, the Fermi level lies in the band gap, so that there is no bulk conductivity. At the same time, conducting states exist on the surface of a topological insulator, and the conduction through these states is not suppressed by disorder scattering due to topological reasons. Surface excitations exist due to strong spin-orbit interaction and have chiral symmetry: the direction of the momentum is rigidly related to the spin projection. These are the so-called helical edge states (HES), in which electrons with different spins propagate in opposite directions.

In two-dimensional topological insulators, interference of electron waves propagating at the edge state along the perimeter of the system is possible. Thus, the transport through the edge states of the topological insulator is, in fact, the transport through the interferometer (see figure 1). The ability to control interference by a magnetic field due to Aharonov-Bohm effect [3] provides an additional potential for experimental and theoretical analysis [4].

Therefore, HES are candidates for efficient creation, control, and transfer of electron polarization. It is very promising from the point of view of the application in quantum spin-sensitive interferometry, especially in the modern and rapidly developing field of spintronics, as well as for the creation of quantum informatics devices using spin degrees of freedom, such as spin filters and quantum qubits.

In this paper, we focus on an experimentally relevant case, when temperature exceeds the level spacing, $T \gg \Delta$. Even for such high T, the Aharonov-Bohm effect is not entirely suppressed and reveals itself in sharp antiresonances in the dependence of conductance, G, on the magnetic flux, Φ [5]. Current progress in the research of helical Aharonov-Bohm interferometer is given in recent review [6]. Here, we generalize previous studies and consider all possible ways of electrons transmission through interferometer including backscattering. Such generalization allows to study case of helical leads instead of metallic leads (see figure 3).

2. Model

We consider a two-dimensional topological insulator and assume that there are HES on its edge (see figure 1). The length of the upper (lower) shoulder is L_1 (L_2) and the full length of the edge is $L = L_1 + L_2$. HES are coupled to metallic leads by tunneling point contacts. Since time-reversal symmetry protects HES from backscattering on non-magnetic impurities than scattering matrix, S_{pc}, which describes the contacts, has a block-diagonal form:

$$S_{pc} = \begin{pmatrix} S_{pc}^{(2)} & 0 \\ 0 & S_{pc}^{(2)} \end{pmatrix}, \quad S_{pc}^{(2)} = \begin{pmatrix} -t & r \\ r & t \end{pmatrix}, \quad r^2 + t^2 = 1, \quad t = e^{-\lambda}. \tag{1}$$

S_{pc} connects incoming states $(\psi_\uparrow, h_\uparrow, \psi_\downarrow, h_\downarrow)$ with outgoing states $(\psi'_\uparrow, h'_\uparrow, \psi'_\downarrow, h'_\downarrow)$ (see figure 2). ψ_α denotes electron at the lead, h_α denotes electron at HES. λ is the transparency of the contact: $\lambda = 0$ ($\lambda = \infty$) corresponds to the closed (open) interferometer.

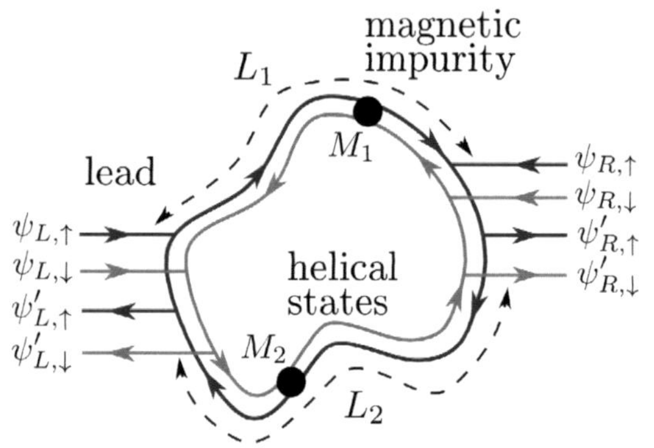

Figure 1. Aharonov-Bohm interferometer formed by helical edge states containing magnetic impurities.

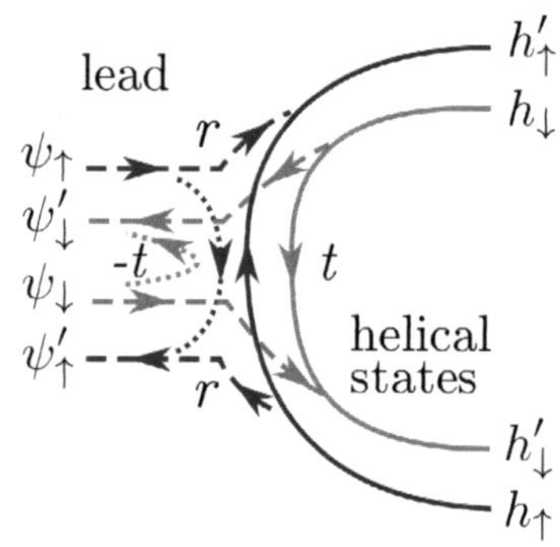

Figure 2. Point contact of helical edge states and metallic leads.

We assume that each shoulders of the interferometer may contain one magnetic impurity (MI) of arbitrary strength. The internal dynamics of MI is neglected. Scattering on such MI is described by the matrix

$$S_M = \begin{pmatrix} e^{i\zeta}\cos\theta & i\sin\theta\, e^{i\varphi} \\ i\sin\theta\, e^{-i\varphi} & e^{-i\zeta}\cos\theta \end{pmatrix}. \tag{2}$$

Here, θ is related to the MI strength: θ is small for a weak MI and $\theta = \pi/2$ for a strong MI – all electrons are fully reflected by it. The forward scattering phase, ζ, can be removed by appropriate shift of ϕ, and we set $\zeta = 0$ below. The corresponding transfer matrix reads

$$M = \frac{1}{\cos\theta} \begin{pmatrix} 1 & i\sin\theta e^{i\xi} \\ -i\sin\theta e^{-i\xi} & 1 \end{pmatrix}, \tag{3}$$

where $\xi = \varphi - 2kx_0$ and k is the electron momentum.

3. Transmission coefficients

Now we can compute the scattering matrix of the whole setup, S, which relates incoming states $(\psi_{L,\uparrow}, \psi_{L,\downarrow}, \psi_{R,\uparrow}, \psi_{R,\downarrow})$ and outgoing states $(\psi'_{L,\uparrow}, \psi'_{L,\downarrow}, \psi'_{R,\uparrow}, \psi'_{R,\downarrow})$, superscript "L" ("R") stands for the electron in the left (right) contact (see figure 1):

$$\begin{aligned}
S &= -\begin{pmatrix} \kappa^2 & 0 \\ 0 & \kappa^{-2} \end{pmatrix} + \frac{r^2}{t} \begin{pmatrix} \kappa^{-1} & 0 \\ 0 & \kappa\Lambda_1 \end{pmatrix} g \begin{pmatrix} \kappa & 0 \\ 0 & (\Lambda_1\kappa)^{-1} \end{pmatrix}, \\
g &= \begin{pmatrix} \kappa^2\Lambda M_2 & 0 \\ 0 & \kappa^{-2} \end{pmatrix} \begin{pmatrix} D & D \\ D & D \end{pmatrix} \begin{pmatrix} \kappa^2 M_1 & 0 \\ 0 & \kappa^2 \end{pmatrix} \begin{pmatrix} \sigma_3 & 0 \\ 0 & \sigma_3 \end{pmatrix}, \\
D &= (1-\kappa^2 M_1 \kappa^2 \Lambda M_2)^{-1}, \; \kappa = \begin{pmatrix} \sqrt{t} & 0 \\ 0 & 1/\sqrt{t} \end{pmatrix}, \; \Lambda = \begin{pmatrix} e^{ikL} & 0 \\ 0 & e^{-ikL} \end{pmatrix} e^{2\pi i\phi}, \\
\Lambda_1 &= \begin{pmatrix} e^{ikL_1} & 0 \\ 0 & e^{-ikL_1} \end{pmatrix} e^{2\pi i\phi\frac{L_1}{L}}, \; \sigma_3 = \begin{pmatrix} 1 & 0 \\ 0 & -1 \end{pmatrix}.
\end{aligned} \tag{4}$$

Here $\phi = \Phi/\Phi_0 = 1/2 + n$ is the dimensionless flux, with $\Phi_0 = hc/e$ the flux quantum. M_1 (M_2) is the transfer matrix for MI, equation (3), in the upper (lower) shoulder.

The matrix of transmission coefficients is defined by $Y_{ij} = \langle |S_{ij}|^2 \rangle_\epsilon$, where the thermal averaging, $\langle \cdots \rangle_\epsilon = -\int d\epsilon(\cdots)\partial_\epsilon f_F(\epsilon)$, is performed with the Fermi function $f_F(\epsilon)$. Computation of Y_{ij} in an analytical way is possible in two cases: either for weak impurities in both shoulders or for one strong impurity in one of the shoulders. We focus on strong MI because the spin polarization of outgoing electrons can reach 100% in this case.

To be specific we assume MI in the upper shoulder, the matrix g can be represented as follows

$$\begin{aligned}
g &= H_0 + \sum_{\alpha=\pm} \frac{1+\alpha\hat{H}}{1-t^2 e^{i(kL+\alpha 2\pi\phi_0)}}, \\
H_0 &= \begin{pmatrix} -1 & i\sin\theta e^{i\xi} & -\cos\theta & 0 \\ 0 & 0 & 0 & 0 \\ 0 & 0 & 0 & 0 \\ 0 & -\cos\theta & \sin\theta i e^{-i\xi} & -1 \end{pmatrix}, \quad H = \begin{pmatrix} a_2 & b_2 e^{i\xi} & a_1^* & b_1^* e^{i\xi} \\ b_2 e^{-i\xi} & -a_2 & b_1^* e^{-i\xi} & -a_1^* \\ a_1 & b_1 e^{i\xi} & a_2 & b_2 e^{i\xi} \\ b_1 e^{-i\xi} & -a_1 & b_2 e^{-i\xi} & -a_2 \end{pmatrix}, \\
a_1 &= i(e^{-2\pi i\phi} - \cos(2\pi\phi_0)\cos\theta)/\sin(2\pi\phi_0), \quad b_1 = e^{-2\pi i\phi}\sin\theta/\sin(2\pi\phi_0), \\
a_2 &= i(e^{-2\pi i\phi}\cos\theta - \cos(2\pi\phi_0))/\sin(2\pi\phi_0), \quad b_2 = \sin\theta\cot(2\pi\phi_0),
\end{aligned} \tag{5}$$

where ϕ_0 obeys $\cos(2\pi\phi_0) = \cos\theta\cos(2\pi\phi)$. The elements of matrix H obey $a_1^2 + b_1^2 = \cos^2\theta$, $a_2^2 + b_2^2 = \cos^2\theta$ and depend on the strength of the impurity and the magnetic flux only, while the dependence on the energy is encoded in the exponents $e^{\pm i\xi}$. In Y_{ij} these exponents are absent. Thus, thermal averaging acts only on the following combinations:

$$\begin{aligned}
&\left\langle \frac{1}{1-t^2 e^{i(kL+\alpha 2\pi\phi_0)}} \frac{1}{1-t^2 e^{-i(kL+\beta 2\pi\phi_0)}} \right\rangle_\epsilon = \frac{1}{1-t^4 e^{(\alpha-\beta)2\pi\phi_0}}, \\
&\left\langle \frac{1}{1-t^2 e^{\pm i(kL+\alpha 2\pi\phi_0)}} \right\rangle_\epsilon = 1.
\end{aligned} \tag{6}$$

Now one is in a position to compute the matrix of transmission coefficients.

The metallic lead allows to study only unpolarized beams of incoming electrons. For example, the transmission coefficient for electrons with spin up and down through the interferometer from the left lead to the right are defined as $\mathcal{T}_\uparrow = Y_{31} + Y_{41}$ and $\mathcal{T}_\downarrow = Y_{32} + Y_{42}$. Then the transmission coefficient for the whole setup is $\mathcal{T} = (\mathcal{T}_\uparrow + \mathcal{T}_\downarrow)/2$. The spin transmission coefficient is defined as $\mathcal{T}_s = (\mathcal{T}_\uparrow - \mathcal{T}_\downarrow)/2$.

However, the matrix Y_{ij} contains information about the contributions of different spins separately. Thus, we can consider the cases of ferromagnetic leads or the situation when HES form leads themselves (see figure 3). In the latter case one has to demand that electrons with opposite spin propagate in different directions. For example, lead "1" consist of incoming state $\psi_{L,\uparrow}$ and outgoing state $\psi'_{L,\downarrow}$. In such a basis the matrix of transmission coefficients has to be rotated $\bar{Y} = R_1 Y R_2$, where

$$R_1 = \begin{pmatrix} 0 & 1 & 0 & 0 \\ 0 & 0 & 1 & 0 \\ 0 & 0 & 0 & 1 \\ 1 & 0 & 0 & 0 \end{pmatrix}, \quad R_2 = \begin{pmatrix} 1 & 0 & 0 & 0 \\ 0 & 0 & 0 & 1 \\ 0 & 0 & 1 & 0 \\ 0 & 1 & 0 & 0 \end{pmatrix}. \tag{7}$$

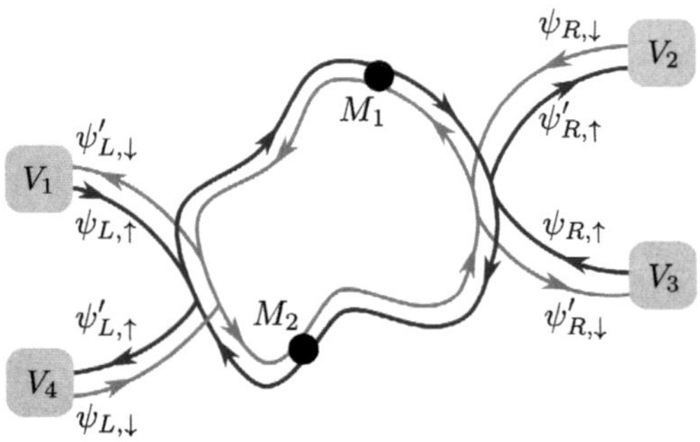

Figure 3. Helical Aharonov-Bohm interferometer with leads formed by helical states.

Let us discuss electron transport. For weak MIs the transmission of electrons from the left leads, 1 and 4, to the right leads, 2 and 3, depends on the transparency of the contacts, λ. For small λ the transmission is small, for large λ the transmission tends to unity. However, for strong MI, $\theta = \pi/2$, the transmission in one of the interferometer shoulders is blocked. In this case for MI in the upper (lower) shoulder the transmission from the lead 1 (4) to the right lead is absent. In the intermediate region of parameters λ and θ, the spin-flip processes have significant contribution into the transmission as shown in figure 4. Incoming electrons from lead 1 have spin "up", outgoing electrons in lead 2 have the same spin, whereas in lead 3 electrons have spin "down". Accordingly, $\bar{Y}_{21}$ describes spin conserving processes, $\bar{Y}_{31}$ describes spin-flip process. It is seen, that these transmission coefficients may be of the same order. Besides, fine-tuning by magnetic flux is possible to make one of the coefficients larger or lower than the other. Transmission coefficients for electron emitted from lead 4 are shown in figure 5. This electron has spin "down" and $\bar{Y}_{23}$ describes spin conserving processes, whereas $\bar{Y}_{24}$ describes the spin-flip process.

Our system can be characterized by the matrix of conductances, $\bar{G}_{ij}$ as $I_i = \bar{G}_{ij} V_j$. Here, I_i is the current flowing in the channel i and V_j is the voltage applied to the channel j. The Kubo formula relates $\bar{G}_{ij}$ with $\bar{Y}_{ij}$, namely $\bar{G}_{ij} = (\delta_{ij} - \bar{Y}_{ij})/2$. Kirchhoff's rules result in constraints on $\bar{G}_{ij}$ such that $\sum_i \bar{G}_{ij} = \sum_j \bar{G}_{ji} = 0$. Consequently, it is possible to define more convenient combinations of I_i and V_j:

$$\begin{aligned} I_R &= (I_1 - I_2 - I_3 + I_4)/2, \quad V_R = (V_1 - V_2 - V_3 + V_4)/2, \\ I_D &= (I_1 + I_2 - I_3 - I_4)/2, \quad V_D = (V_1 + V_2 - V_3 - V_4)/2, \\ I_S &= (I_1 - I_2 + I_3 - I_4)/2, \quad V_S = (V_1 - V_2 + V_3 - V_4)/2, \\ I_\Sigma &= \sum\nolimits_j I_j/2, \quad V_\Sigma = \sum\nolimits_j V_j/2. \end{aligned} \tag{8}$$

Physically, such combinations have the following meanings: I_R is the charge current flowing to the right, I_D is the charge current flowing to the down and I_S is the spin current. In this new

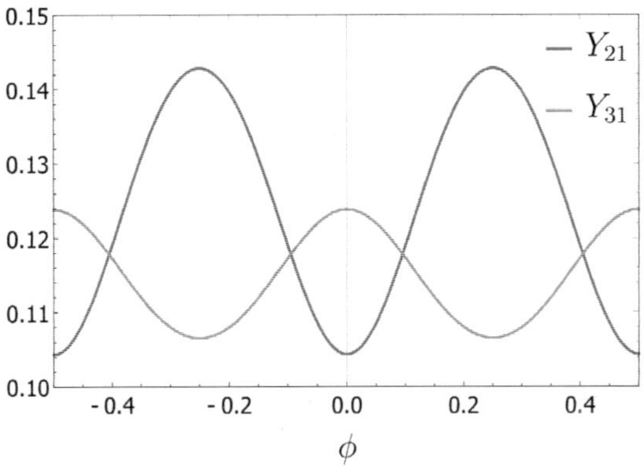

Figure 4. Coefficients of transmission electrons from the left lead 1 (spin "up") to the right leads.

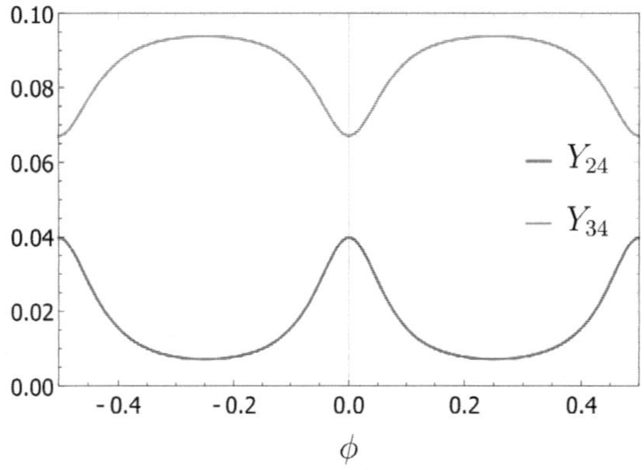

Figure 5. Coefficients of transmission electrons from the left lead 4 (spin "down") to the right leads.

basis, the matrix of conductances has a simple form

$$G = R_0^T \bar{G} R_0 = \begin{pmatrix} \mathcal{T} & 0 & \mathcal{T}_s & 0 \\ 0 & 1-\tanh\lambda & 0 & 0 \\ \mathcal{T}_s & 0 & 1+\mathcal{T}_s/\tanh\lambda & 0 \\ 0 & 0 & 0 & 0 \end{pmatrix}, \quad R_0 = \frac{1}{2}\begin{pmatrix} 1 & 1 & 1 & 1 \\ -1 & 1 & -1 & 1 \\ -1 & -1 & 1 & 1 \\ 1 & -1 & -1 & 1 \end{pmatrix}, \tag{9}$$

$$\mathcal{T} = \tanh\lambda\left(1 - \frac{\sin^2\theta\sinh^2\lambda\cosh(2\lambda)}{\cosh^2(2\lambda)-\cos^2\theta\cos^2(2\pi\phi)}\right), \quad \mathcal{T}_s = -\frac{\sin^2\theta\sinh^2\lambda\cosh(2\lambda)}{\cosh^2(2\lambda)-\cos^2\theta\cos^2(2\pi\phi)}.$$

4. Conclusion

We have studied Aharonov-Bohm interferometer formed by helical edge states of 2D topological insulators and tunnel coupled to helical leads (see figure 3). We calculated the full S-matrix of the interferometer as a whole, including both transmission and reflection processes. In high temperature regime, $T \gg \Delta$, the transmission coefficients, Y_{ij} are oscillatory functions of dimensionless magnetic flux, ϕ, with the period 1/2. Y_{ij} additionally depend only on the transparency of the contact, λ, and the strength of magnetic impurity, θ. Other details, such as the geometry of the device, the relation between L_1 and L_2, the position of the magnetic impurities or the Berry phase, do not affect Y_{ij}. Manipulating by the magnetic flux allows one to tune spin conserving and spin-flip transmission coefficients. For example, it is possible to make one of the coefficients dominant. Our study thus indicates possible directions for expansion of quantum spin-sensitive interferometry research.

Acknowledgments

The work of R.N. was funded by RFBR, project number 19-32-60077. The work of D.A. and V.K. was funded by the Russian Science Foundation, Grant No. 20-12-00147.

References

[1] Hasan M Z and Kane C L 2010 *Rev. Mod. Phys.* **82** 3045
[2] Qi X L and Zhang S C 2011 *Rev. Mod. Phys.* **83** 1057
[3] Aharonov Y and Bohm D 1959 *Phys. Rev.* **115** 485
[4] Aronov A G and Sharvin Y V 1987 *Rev. Mod. Phys.* **59** 755
[5] Dmitriev A P, Gornyi I V, Kachorovskii V Y and Polyakov D G 2010 *Phys. Rev. Lett.* **105** 036402
[6] Niyazov R A, Aristov D N and Kachorovskii V Y 2021 *JETP Letters*

Research of the structure and mechanical properties of gas-thermal coatings after induction heat treatment

E O Osipova, O A Markelova, V A Koshuro and A A Fomin

Yuri Gagarin State Technical University of Saratov, Saratov 410054, Russia

lenusindra@mail.ru

Abstract. In this article the process of induction-thermal modification of titanium coatings formed by electroplasma spraying was considered. The influence of the inductor current on the temperature of processed samples was experimentally established. The research results showed that thermal treatment of the samples with titanium coatings at a temperature of 750–1200 °C and a duration of 300 s led to an increase in porosity from 56±2 to 61±1 % and in microhardness from 1035–1532 to 1825–1883 $HV_{0,98}$, the sprayed layer thickness decreased from 320±30 to 114±15 µm as well. A change of nanoscale structural elements shape was also observed.

1. Introduction

Designs of restorative medicine are often made from titanium based alloys [1]. The element-phase composition and structural-mechanical properties of the surface influence the osseointegration of implantable structure. Hence, a biocompatible coating similar in structure to bone tissue is formed on the implant.

In biomedicine, aircraft construction, mechanical engineering, coatings formed by gas-thermal spraying (GTS), electrochemical deposition and vacuum technologies (CVD and PVD) are widely used [2–4]. Gas-thermal spraying is distinguished by high productivity, technological simplicity, so it has become widespread in production. One of the GTS methods is electroplasma spraying (EPS) that allows the formation of titanium coatings by applying powder material. According to the review results the porosity of these layers is 10–50 %, whereas the adhesion strength between the coating and the base may reach 20 MPa [5–7].

The main disadvantages of the EPS method are the unevenness of the elemental-phase composition, the presence of structural defects in the spayed material, which leads to the coating destruction. To improve the functional properties of plasma coatings clad powders are used and the base is subjected to preliminary induction heating. Besides, modification methods such as microarc oxidation, laser and high-temperature processing are also used. During induction heat treatment (IHT) simultaneous heating of the coating and base occurs. This treatment contributes to a change of porosity, hardness and adhesive-cohesive strength. It is known that IHT enables the formation of nanostructured metal oxide layers on titanium, which increase osteoconductive properties of the implant surface [8].

At the moment the works devoted to IHT of sprayed coatings are not enough for the further development of surface modification technology. The purpose of this work was to study the effect of IHT parameters on the structure and hardness of titanium coatings formed by electroplasma spraying.

2. Methodology

The research samples were titanium disks with diameter of 15 mm and a thickness of 0.5 mm. Prior to the coating deposition the sample surfaces were subjected to abrasive blasting with an electrocorundum powder having a dispersion of 250–300 μm and at a pressure of 4±0.5 MPa on the "Chaika-20" setup. Technological contaminants were removed from the samples in a surfactant solution using an "UZUMI-2" ultrasonic bath. The coating was formed by EPS of PTS grade titanium powder with a dispersion of 100–150 μm (TC 14-22-57-92 Titanium powder) using the "UPN-28" installation. The following spraying modes were set: plasmatron arc current – 350 A (±1 %); spraying distance – 150±10 mm; transporting gas consumption. IHT of samples was carried out on a "VCh-15" installation at a constant exposure of 300 s, an inductor current of 3.2–8.0 kA which provided heating of samples to the temperatures from 650 to 1150 °C (figure 1).

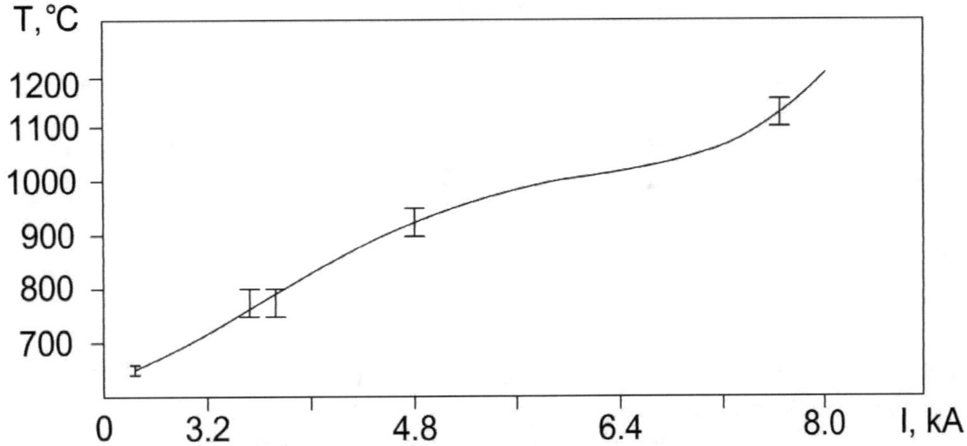

Figure 1. Accordance of the inductor current and the treatment temperature.

The analysis of the coating structure was carried out by scanning electron microscopy (SEM) using a "MIRA 2 LMU" microscope. Size of the structural elements, porosity and thickness of the resulting layers were determined from the images of the coating areas (size – 660 by 495 μm) using the "Metallograph" program for analyzing geometric parameters of microobjects [9]. The coating and titanium base microhardness was assessed according to the manufactured cross-sections with a step of 50 μm from the "coating – base" boundary with the use of "PMT-3" hardness tester at a load on the Vickers indenter equal to 0.98 N (ISO 6507-1:2018).

3. Results

In the course of plasma spraying the coating was formed from individual molten particles arranged in layers and forming agglomerates with a size of 100–200 μm (figure 2a). Deep pores and cracks characterized by a summarized surface porosity of 56±2 % were observed.

Thermal modification in the low-temperature range (T = 750–800 °C) allowed the reduction in the number of cracks. Porosity increased to 61±1 % (figure 2b). The number of spherical particles less than 5 μm has decreased, while the number of larger particles with the size of 10–50 μm was practically unchanged. A growth of modification temperature to 900–950 °C led to a noticeable decrease in the number of pores and cracks, while the porosity was 55±2% (figure 2c). Particles of 30–40 μm in size prevailed in the structure of surface modified coating. A further increase in the inductor current and, accordingly, the treatment temperature did not result in a significant change of porosity as the latter remained equal to 56±1% (figure 2d).

It was revealed that IHT affected the sizes and shapes of the coating structural elements. Nanoparticles with a size of 100–200 nm were observed during EPS (figure 3a). After the coating was modified at a minimal temperature of 750–800 °C prismatic crystals appeared in its structure, whereas the particle size increased to 500–600 nm (figure 3b).

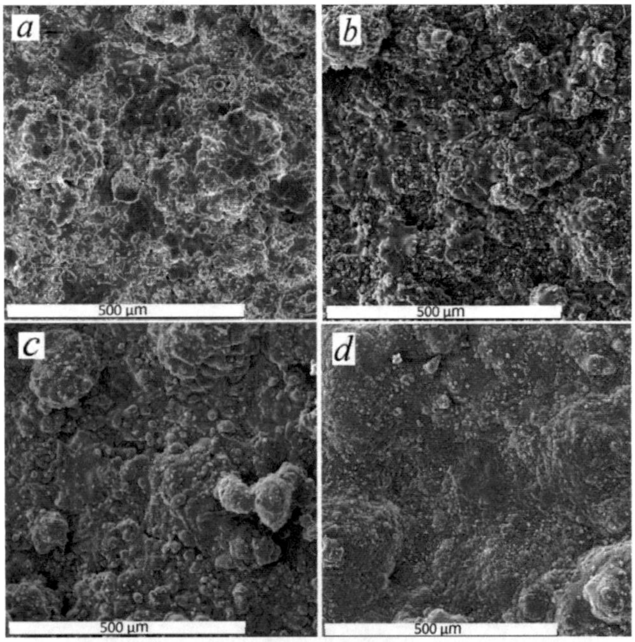

Figure 2. SEM of the sample images: a – plasma sprayed coating (no IHT); b – T = 750–800 °C; c – T = 900–950 °C; d – T = 1150–1200 °C.

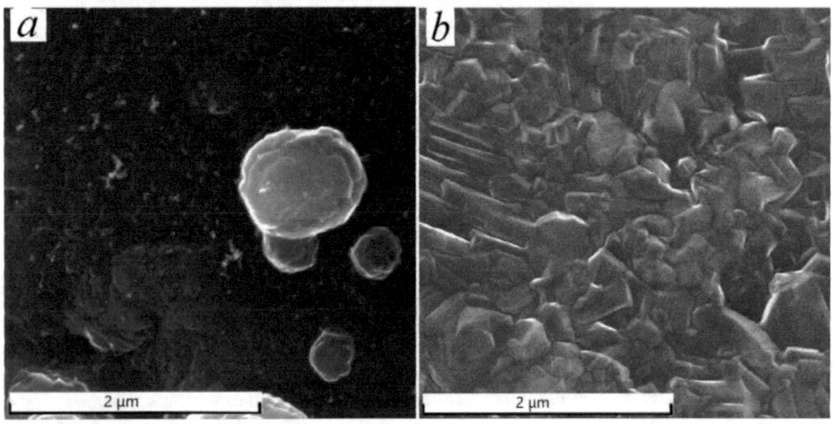

Figure 3. SEM showing the images of samples with nanoparticles: a – coating without IHT; b – T = 750–800 °C.

The thickness of the initial layer obtained by the EPS method was 320±30 μm. Microindentation showed that the resulting porous structure was characterized by a hardness of 1035–1530 $HV_{0.98}$ at a depth of 200 μm from the "coating – base" interface (figure 4). Microhardness of the titanium base of samples did not exceed 145 $HV_{0.98}$. Oxidation processes were likely to occur during low-temperature induction treatment, which led to an increase of the coating thickness to 360±35 μm. The modified layer microhardness at a depth of 100–150 μm reached 1825 $HV_{0.98}$, while at the value of 25 μm (close to Ti base) it was 600 $HV_{0.98}$. Thermal modification in the medium and high temperature ranges (900–950 and 1150–1200 °C) resulted in a decrease of the coating thickness to 115±15 μm and 190±25 μm and stabilization of microhardness at the level of 1600 $HV_{0.98}$ and 1885 $HV_{0.98}$, respectively. After heat treatment at 900–950 °C, the hardness of the titanium base grew to 270 $HV_{0.98}$. At the highest IHT temperature, the hardness of the sample bases reached 440 $HV_{0.98}$. Thus, the change in hardness was associated with oxidation and the formation of strengthened diffusion layers.

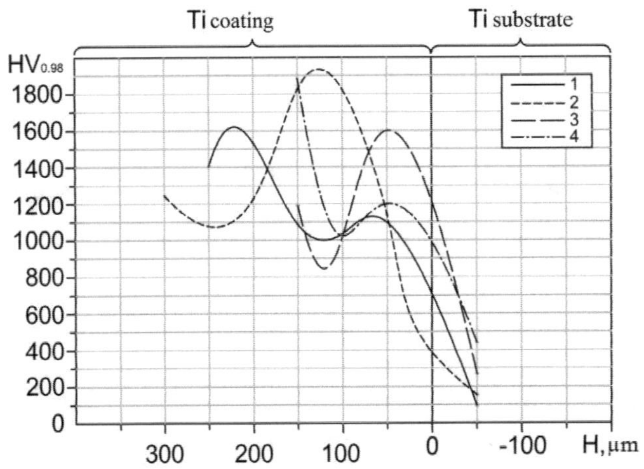

Figure 4. Microhardness distribution over the cross-section of the samples: 1 – EPS coating; 2 – T = 750–800 °C; 3 –T = 900–950 °C; 4 – T = 1150–1200 °C.

4.Conclusions

Thus, the plasma sprayed coatings had a high summarized porosity of about 56±2 % at a thickness of 320±30 μm. In this case, nanoparticles with a size of 100–200 nm were observed. The resulting structure had a high microhardness of 1035–1530 $HV_{0.98}$. The titanium base hardness during this treatment did not exceed 145 $HV_{0.98}$. IHT at an inductor current of 3.5 kA, which corresponded to a temperature of 750–800 °C, contributed to an increase of open porosity up to 61±1 % and a coating thickness to 360±35 μm. Particle size grew to 500–600 nm. The microhardness of this coating was 1100–1825 $HV_{0.98}$ and it gradually fell to 600 $HV_{0.98}$ at the interface of the titanium base. An increase of the inductor current to 4.8 kA and the process temperature to 900–950 °C led to the porosity reduction of 55±2 %. At the maximum value of the inductor current equal to 8.0 kA, and, accordingly, the exposure temperature of about 1150–1200 °C, no significant changes in porosity were revealed (56±1%). Individual structural elements of the coating have acquired a prismatic shape. Treatment at an inductor current of 4.8–8.0 kA led to a decrease in the coating thickness to 114±15 μm and 186±24 μm, the microhardness over the cross-section of the sprayed material did not exceed 1600–1880 $HV_{0.98}$. Thus, it can be concluded that the resulting coatings can be used in the manufacture of titanium designs in the field of restorative medicine.

Acknowledgments

The paper was prepared as a part of the scientific research within the framework of the grant of the President of the Russian Federation № MD-965.2021.4.

References

[1] Pohler O E M 2000 *Injury* **31** 7
[2] Vilardell A M, Cinca N, Garcia-Giralt N, Dosta S, Cano I G, Nogués X and Guilemany J M 2018 *Journal of Materials Science: Materials in Medicine* **29(2)** 1
[3] Jeyachandran Y L, Karunagaran B, Narayandass S K, Mangalaraj D, Jenkins T E and Martin P J 2006 *Materials Science and Engineering: A* **431(1-2)** 277
[4] Cinca N, Barbosa M, Dosta S and Guilemany J M 2010 *Surface and Coatings Technology* **205(4)** 1096
[5] Qian M, Froes F H 2015 *Butterworth-Heinemann* 627
[6] Vogel D, Dempwolf H, Baumann A and Bader R 2018 *Journal of the Mechanical Behavior of Biomedical Materials* **77** 600
[7] Kalita V I, Komlev D I and Radyuk A A 2016 *Inorganic Materials: Applied Research* **7(4)** 536
[8] Fomin A A, Dorozhkin S, Fomina M, Koshuro V, Rodionov I, Zakharevich A, Petrova N and Skaptsov A. 2016 *Ceramics International* **42(9)** 10838
[9] Fomina M, Koshuro V, Papshev V, Rodionov I and Fomin A 2018 *Data in Brief* **20** 1409

Improving the mechanical properties of tool steel by induction chemical-thermal treatment

P Palkanov, V Koshuro and A Fomin

Yuri Gagarin State Technical University of Saratov, Saratov 410054, Russia

Abstract. The study results of the structure and microhardness of the surface layer of high-speed tool steel after induction chemical-thermal treatment in a gaseous nitrogen-containing medium at a temperature of 900–1100 °C were presented. Due to the strengthening treatment of products a gradient diffusion nitride layer with a thickness of about 200 μm and a surface microhardness of 1950±70 $HV_{1.98}$ was formed.

1. Introduction

In modern industry, various types of working media are used for nitriding. The most common of them is a gas mixture of ammonia and propane (1:1). The nitriding process in this environment is performed in shaft or vacuum retort furnaces in the temperature range from 520 to 570 °C for 60–90 hrs. As a result of processing, a strengthened diffusion layer with a thickness of 150±50 μm and a microhardness of 900–1100 HV is obtained [1]. A method of ion-plasma nitriding is often currently applied, which is performed in special installations with a nitrogen-containing rarefied medium consisting of nitrogen or ammonia at temperatures from 300 to 590 °C with cycle duration of about 24 hrs. Ionic nitriding allows obtaining a nitrided layer with the thickness of 300±50 μm and a microhardness of 1000–1300 HV [2].

An alternative heating method for chemical-thermal treatment (CTT) can be induction heating, which reduces the process time [3]. Based on the mentioned above, the most promising way of heating products during the nitriding process can be induction heating [4,5]. In this paper, the mechanical properties of diffusion layers obtained on tool steel using induction chemical thermal treatment (ICTT) were studied.

2. Methodology

Cylindrical samples used in the studies had a diameter of 10 mm and a length of 5 mm and were made of high-speed tool R6M5 steel not subjected to heat treatment.

Nitriding was carried out by placing the samples in a chamber of "VCh-15A" heating device designed for ICTT (Figure 1). Sample *5* was in the active zone of inductor *3*. Before the nitriding took place, air was removed from the chamber by blowing with nitrogen (at least for 1 min), which was supplied through a hole in the lower base of chamber *4*. Then the outlet valve of chamber *1* was closed, as a result of which overpressure was observed.

The ICTT process was carried out at a temperature of 900–1000 °C for 10 min and a pressure of 0.2±0.05 MPa. The heating rate and temperature were controlled by varying the operating current *I* from 300 to 360 A, which corresponded to 1:16 of the inductor current for this device. The temperature of the induction heating was controlled according to the colors of heating of the processed solids. When the exposure was finished, the samples were cooled to a temperature of 550–570 °C by blowing the chamber with gaseous nitrogen.

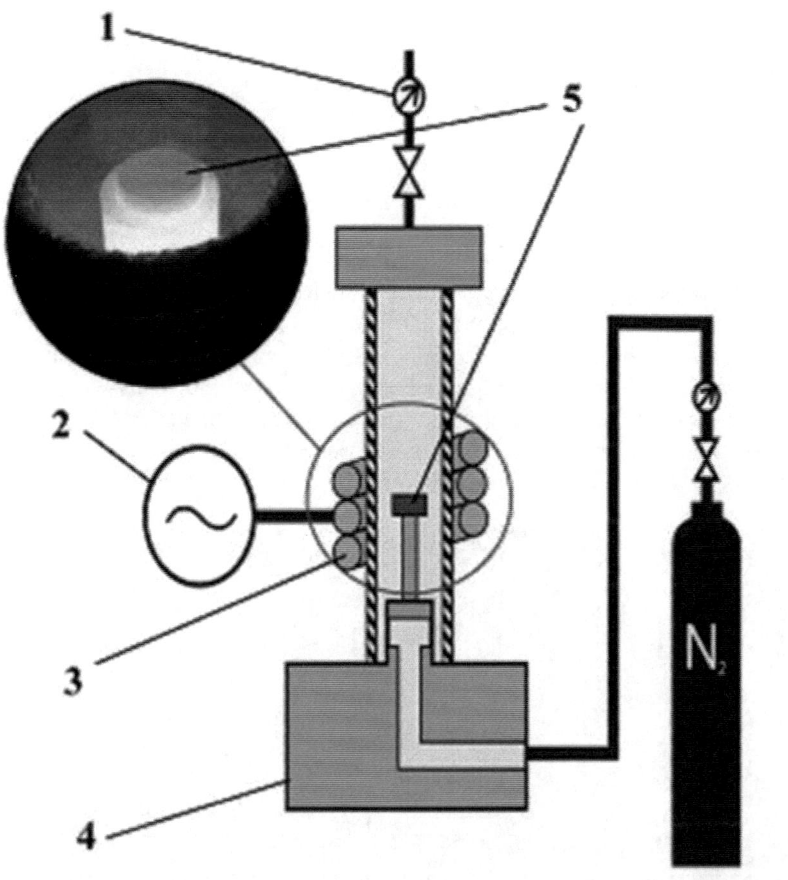

Figure 1. Scheme showing the chamber for ICTT in nitrogen: *1* – outlet pressure reducing valve; *2* – "VCh-15A" setup; *3* – water-cooled inductor; *4* – lower base of the chamber for puffing of a working gaseous medium; *5* – sample.

The structure and depth of the nitrided layer were studied using an "Olympus BX-51" optical microscope. Cross-sections of the samples were made according to a well-known technique using a "CitoPress-1" pressing machine and a "TegraPol-15" grinding and polishing machine. The microhardness was evaluated using a "Durascan-20" microhardness tester with a load on the Vickers indenter equal to 100 gf (0.98 N) for microsections and 200 gf (1.98 N) for the sample surface.

3. Results

As a result of ICTT, the samples acquired a gray-matte shade typical for nitrided products. At a maximum operating current of 360 A, the sample melted. Thus, this current value was taken as the maximum and the exposure time was within 10 min. All samples were distinguished by their uniformity of texture in the central parts and the peripheral areas had a darker shade. This was explained by their uneven heating in the course of treatment with high frequency currents. However, the microhardness over the entire surface was characterized by sufficient uniformity.

The maximum microhardness of the surface of the nitrided layers was stabilized at the level of 1928–1950 $HV_{1.98}$ (Table 1).

Table 1. Microhardness of the sample surfaces.

Designation of a series of samples	Hardness, $HV_{1.98}$
Untreated sample	256±9
Hardened sample (after quenching)	799±34
No. 1, I_P – 300 A	1862±98
No. 2, I_P – 320 A	1950±75
No. 3, I_P – 340 A	1928±70
No. 4, I_P – 360 A	1837±55

With an increase in the operating current of the inductor, an increase in the indices of the surface microhardness was observed. Changing the values of the operating current of the inductor led to a growth of the heating rate of the processed products and the emergence of a pronounced temperature gradient [6]. A significant increase in microhardness indirectly indicated the strengthening of the surface layer as a result of the formation of nitride phases in the course of ICTT.

When studying the microstructure of the nitrided samples, the uniformity of the modified layer with the thickness of 160–200 μm was noted (Figure 2).

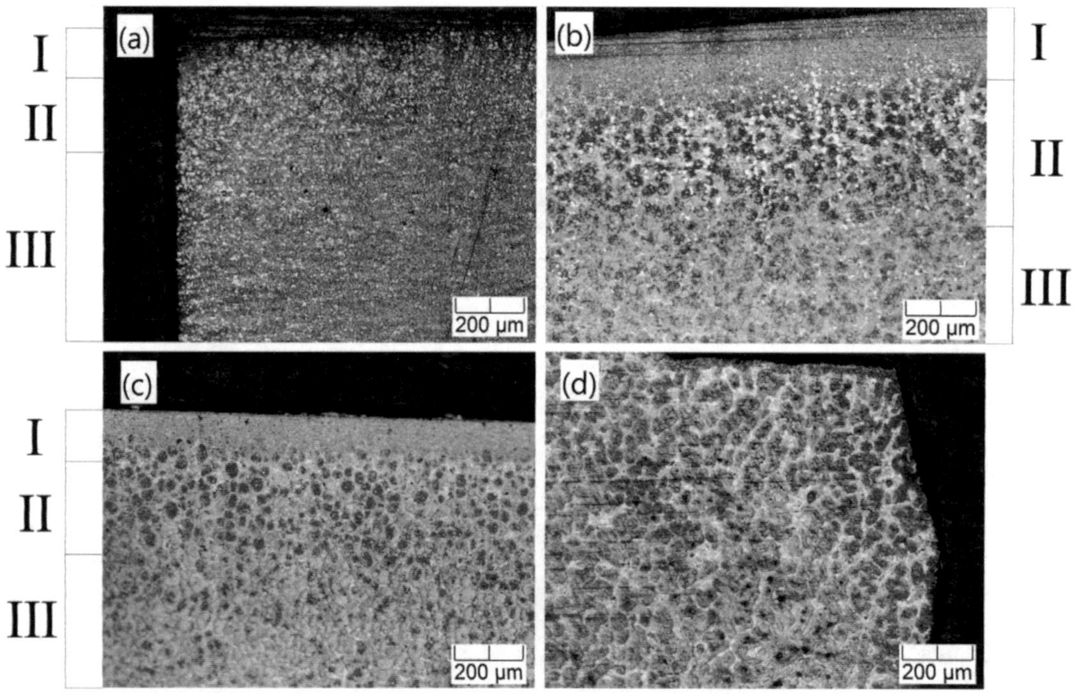

Figure 2. Layered structure of the samples after ICTT containing a strengthened layer (I), a diffusion layer (II) and a base metal (III): *a* – 300 A; *b* – 320 A; *c* – 340 A; *d* – 360 A.

The thickness of the strengthened layer for sample No. 4 was not revealed, which was due to its melting and structure homogenization. The strengthening boundary was determined by microindentation along the section to a depth of about 1 mm (Figure 3). At an operating current of 300–320 A, the layers with a smooth decrease in the microhardness over the cross section to 150 μm were formed. A microhardness gradient within 100 μm was observed in the samples treated at 340–360 A.

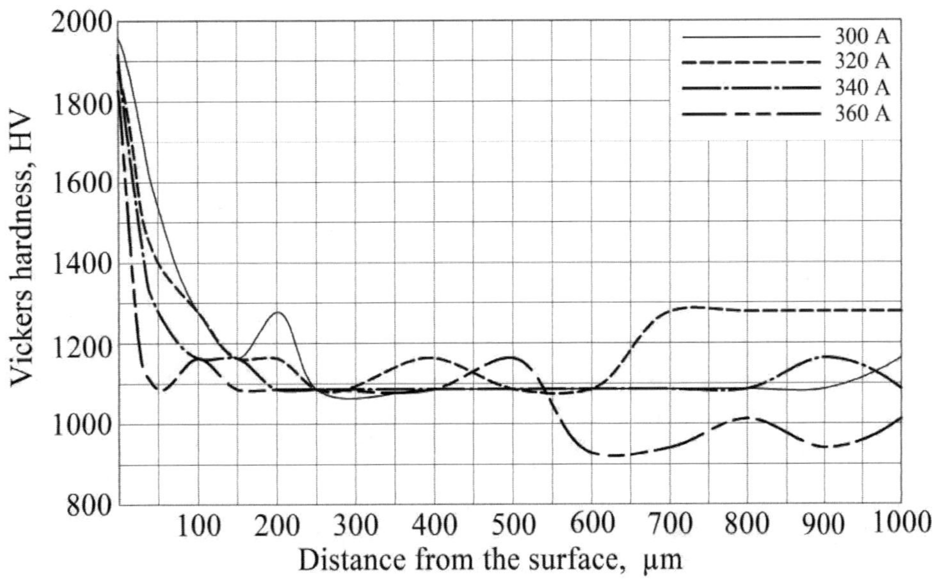

Figure 3. Dependences of the microhardness of nitrided samples on the indentation depth.

The constancy of microhardness was observed from a depth of 250 μm and more. The depth from 30 to 250 μm was characterized by high hardness (1694–1281 $HV_{0.98}$), which was associated with the formation of finely dispersed strengthening phases in this region. In the case of sample No. 4 minimum values of microhardness in deep layers (more than 500 μm) were noted, which was associated with partial fusion of the sample.

4. Conclusion

Thus, the layers obtained on steel products were characterized by the required combination of high microhardness and a large depth of the strengthened layer, which can be effectively used under conditions of friction and dynamic loads. Using the ICTT method, a nitrided layer with a high microhardness of about 1928-1950 $HV_{1.98}$ was formed on the surface of high-speed R6M5 steel.

The depth of the diffusion layer was 160–200 μm in the operating current range of 300–340 A with process duration of 10 min. Treatment in the range of 300–320 A was characterized by higher microhardness values of the strengthened layers with its uniform distribution. Intense nitriding during induction heating was achieved by increasing the localized activity of the gaseous medium, accelerating the adsorption and diffusion processes.

Acknowledgments

The research was supported by the "U.M.N.I.K." program (Project No. 0067326 Executor Palkanov P.A., 2021).

References

[1] Wang B, Sun S, Guo M, Jin G, Zhou Z and Fu W 2015 *Surf Coat Technol* **279** 60
[2] Rocha A, Strohaecker T and Hirsch T 2003 *Surf Coat Technol* **165** 176
[3] Fomin A, Voyko A, Fomina M, Mokrousov S and Koshuro V 2020 *Compos Struct* **245** 112393
[4] Takesue S, Kikuchi S, Misaka Y, Morita T and Komotori J 2020 *Surf Coat Technol* **399** 126160
[5] Fomin A, Fedoseev M, Palkanov P, Voyko A, Fomina M, Koshuro V, Zakharevich A, Kalganova S and Rodionov I 2018 *Proc SPIE* **10716** 107161N
[6] El-Hossary F 2002 *Surf Coat Technol* **150** 277

Research of FIB local milling processes for creation of nanosized field emission structures

I V Panchenko, N A Shandyba and A S Kolomiytsev

Southern Federal University, Taganrog 347924, Russia

ivpanchenko@sfedu.ru

Abstract. The paper presents the results of experimental studies of the influence of the main parameters of a focused ion beam (FIB) during surface profiling on the accuracy of transfer of a pattern to a silicon substrate to create nanoscale field emission structures. In this work, the optimal FIB currents are determined, introducing a minimum amount of distortions during the formation of structures of various sizes. The possibilities of the method of local ion-beam etching of structures in a wide range from 0.1 to 2 μm are shown. The prospects of using this technology for the creation of field emission structures have been demonstrated. It is determived the current-voltage characteristic of the fabricated field-emission cells with a threshold voltage of the onset of emission of ~ 2.5 V and a maximum current of 300 nA at 30 V.

1. Introduction

One of the urgent and important tasks in the development of modern electronics is the creation of miniature vacuum devices operating on the effect of field emission. The development and formation of elements of vacuum field emission nanoelectronics allows achieving high performance, noise immunity and low power consumption in finished devices [1]. To obtain the most advanced devices with improved output parameters and characteristics, it is necessary to use ever smaller mass and size parameters of the structural elements of the device down to nano sizes when creating devices. Also, this approach allows you to increase the density of the placement of elements on a single chip (substrate). For example, a decrease in the interelectrode distance makes it possible to sharply decrease the threshold voltage of the emission start. However, when providing the necessary parameters of the elementary component base, a number of technological difficulties arise. Modern production processes based on the operation of optical lithography, liquid and plasma etching do not allow achieving sufficient accuracy and resolution in the formation of structures of field emission nanoelectronics. The application of the focused ion beam (FIB) method helps to overcome the main limitations of traditional methods and expand the ranges of parameters of the resulting structures.

The FIB method allows performing technological operations of local milling and deposition of materials from the gas phase under high vacuum conditions without the need for the use of resists, masks, and chemical etchants [2]. The main advantage of this method is the possibility of profiling the substrate surface with nanometer accuracy and in a strictly specified area. By creating various templates in the FIB control software, it is possible to achieve almost any shape of the structure.

The aim of this work is experimentally studying the main parameters of FIB during surface profiling, to determine their optimal values during the formation of structures of given sizes, as well as to create nanoscale field emission cells and study their parameters.

2. Surface profiling by FIB method

The work was carried out on a Nova Nanolab 600 electron microscope equipped with a FIB system. One of the main parameters of the FIB, which has the greatest effect on the profile of the final structure during ion-beam milling of the surface, is the ion beam current. It also determines the milling rate and has a great influence on the degree of deviation of the obtained object sizes from those specified in the pattern. At this stage of the work, experimental studies of the accuracy of transfer of the pattern to the silicon substrate during local ion-beam milling for various FIB currents were carried out. For this, a pattern was generated in the software for controlling the ion beam, consisting of 20 circles with different diameters varying in the range from 0.1 μm (left upper structure) to 2 μm (right lower structure) with a step of 100 nm. The main parameters of the FIB: accelerating voltage 30 kV, beam current 0.1-0.5 nA, beam dwell time (Dt) 1 μs and beam overlap (OL) 50%. SEM images of the structures are shown in Figure 1.

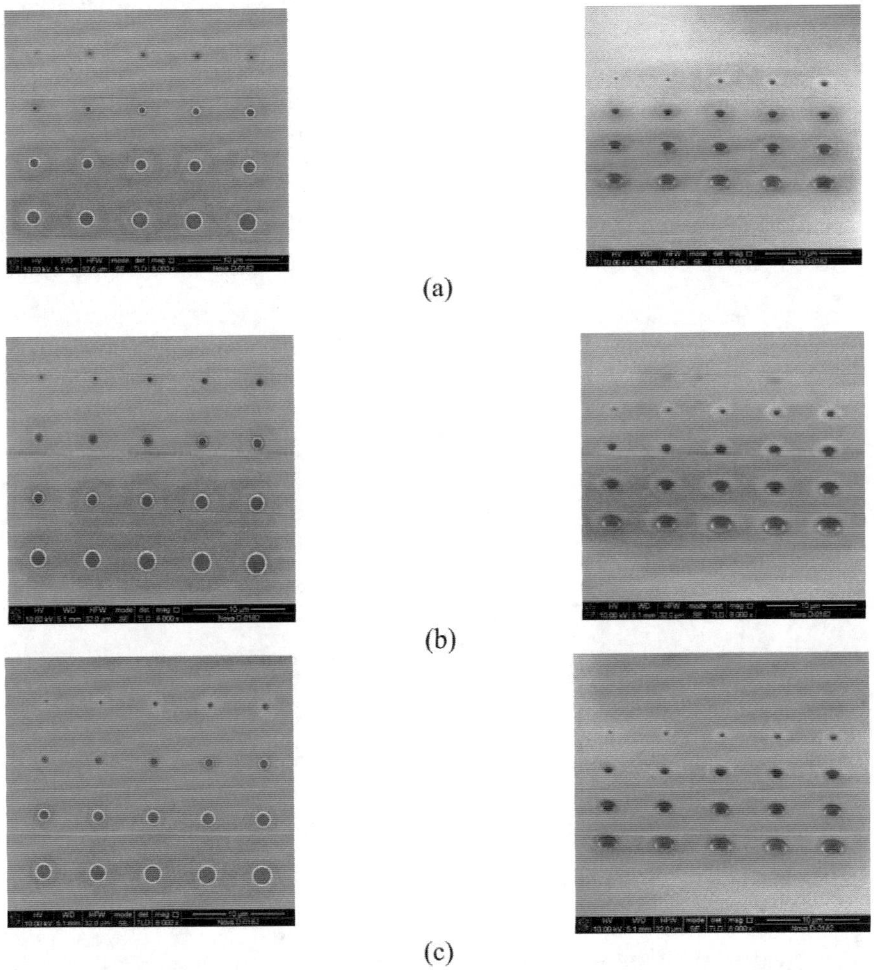

Figure 1. SEM images of an array of structures after milling with a FIB current of 0.1 nA (a), 0.3 nA (b), and 0.5 nA (c) at an angle of 0 ° and 52 °, respectively.

From the analysis of SEM images, it can be seen that the experimentally obtained structures have a surface raster near the circle, which is most noticeable in the image at an angle, and the angle of the lateral surface of the recesses is different from 90 °. This is due to the fact that the FIB has a rather complex beam shape, which can be described by the law of normal distribution or the Gaussian function. Figure 2 shows the simulated beam shape. In this regard, the milling profile of the substrate is very different from that specified in the template and has the indicated inaccuracies when transferring the pattern of the template to the substrate.

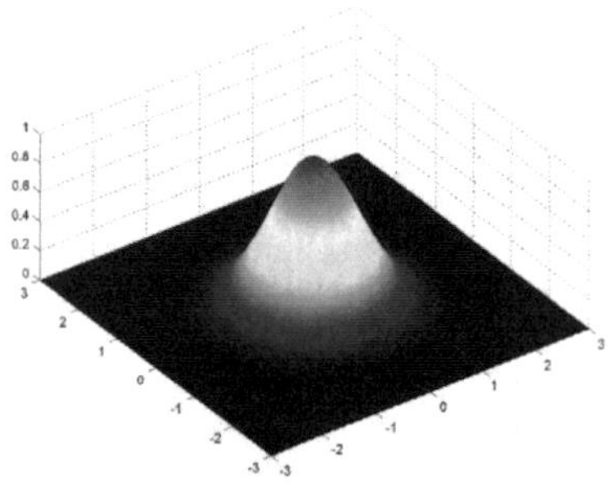

Figure 2. Ion beam shape described in terms of the Gaussian function.

In this regard, to assess the degree of influence of the FIB current on the resulting nanoscale structures, the dependences of the deviation of the specified values in the pattern from the experimentally obtained values were plotted. The measurement of the inner circumference obtained at the bottom of the milled object will be used as the measured value for the diameter of the structure. The use of this plane is justified by the fact that it is this plane that is the "useful" area on which the field-emission cathode will be grown in the future. Thus, the dependences of the deviation of the given value from the practically obtained for the inner diameter of the circle were plotted, presented in Figure 3.

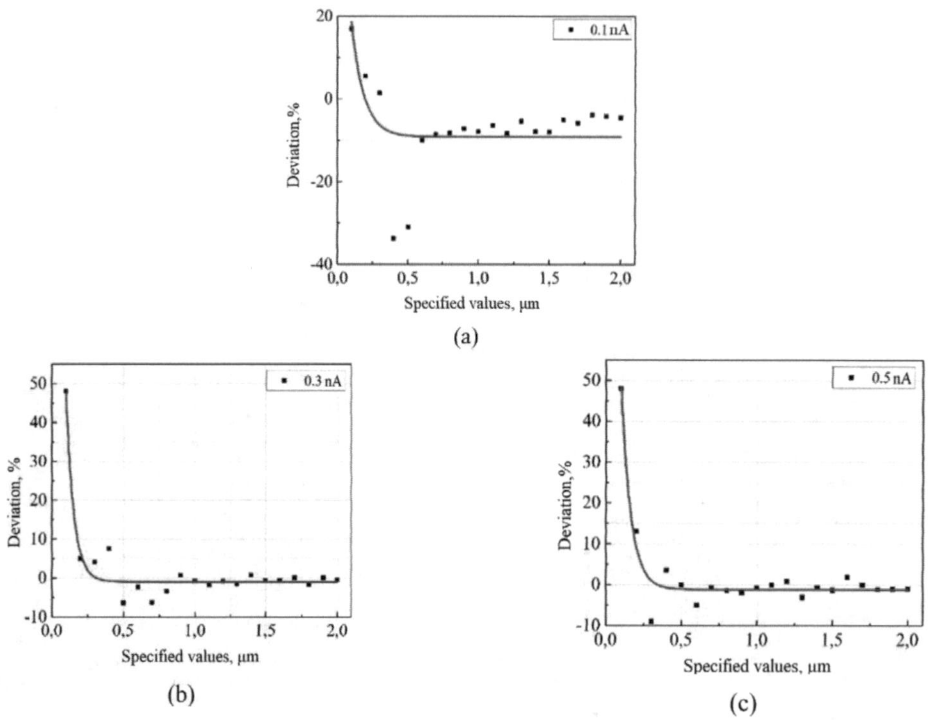

Figure 3. Dependences of the deviation of a given diameter of structures from the experimentally obtained at a FIB current of 0.1 nA (a), 0.3 nA (b), 0.5 nA (c).

From the obtained dependences it follows that, in general, the percentage of deviations has not too large values and their spread. Exponential dependence is caused by the presence of extreme points, such as 100 nm, which slightly distort the real picture and cause some inaccuracies when transferring the pattern to the substrate. The increase in the percentage of deviations for structures with a given diameter of less than 500 nm is associated with the difficulty of obtaining the minimum diameter of the ion beam, due to the Coulomb interaction of like-charged particles. Thus, for the formation of structures with sizes less than 500 nm, it is preferable to use a FIB current of 0.1 nA, since in this region it provides the most accurate transfer of the pattern to the substrate surface. For the formation of structures with dimensions greater than 500 nm, the FIB current of 0.5 nA is optimal, since, with a similar spread in the percentage of deviation with the data obtained in the graph for a current of 0.3 nA, it provides a shorter milling time.

3. Fabrication of a field emission cathode

The obtained experimental results can be used in the formation of nanoscale field emission cells with a combination of methods of local FIB milling and deposition of material from the gas phase. Next, one of the possible options for the technology of cell formation will be presented.

The process of creating a field-emission cell with an emitter grown at the bottom of a depression was carried out in two stages. First, the depression itself was obtained by the method of local ion-beam milling of the FIB, and then an cathode was grown using ion- and electron-stimulated (as a comparison) deposition at the bottom of the structure [3]. The cell was formed on the basis of the data obtained at the previous stage for the parameters of the pattern and FIB. SEM images of fabricated field emission cells are shown in Figure 4.

So, when forming a depression for the deposition W of the FIB emitter (Figure 4, a), the minimum possible cell diameter is 1.5 μm, which, in accordance with the dependence shown in Figure 3, makes it possible to find the diameter at the bottom of the cell equal to ~ 1 microns. The formation of structures with a smaller value of the cell diameter is impossible, since this is due to the complete overgrowing of the depression with the deposition material, caused by a different physics of distribution and interaction of chemically active gases with an ion beam in submicron objects. For an cathode made of C (Figure 4, b) and W (Figure 4, c) formed by FIB and an electron beam, respectively, it was possible to achieve a cell diameter of 930 nm, which is ~ 500 nm at the bottom of the structure. The height of the resulting emitter, as well as the height of the cell for all cases, was about 1.3 μm. The settling time of the points from the W and C FIB is, respectively, 84 s. and 28 s., and an electron beam 45 s. The tip radius of the field emission cathode for the structures in Figure 4 is 65 nm, 75 nm, and 5 nm, in accordance with the captions to the SEM images. It is also seen that during the deposition of an cathode from carbon, a negative factor is found associated with the redeposition of the material on the cell walls, which leads to the shunting of the electrodes and the absence of the field effect.

It is also worth noting that a further decrease in the radius of the tip of the emitter tip by “sharpening” it by ion milling according to a special pattern is impossible, since the material sputtered by the ion beam in such a closed volume will be re-deposited on the cell walls, which will inevitably lead to a short circuit of the entire structure.

The current-voltage characteristics of the vacuum field emission cell were measured on an Analytical Submicron Prober EM-6040A setup using a KEITHLEY 4200-SCS semiconductor parameter measurement system.

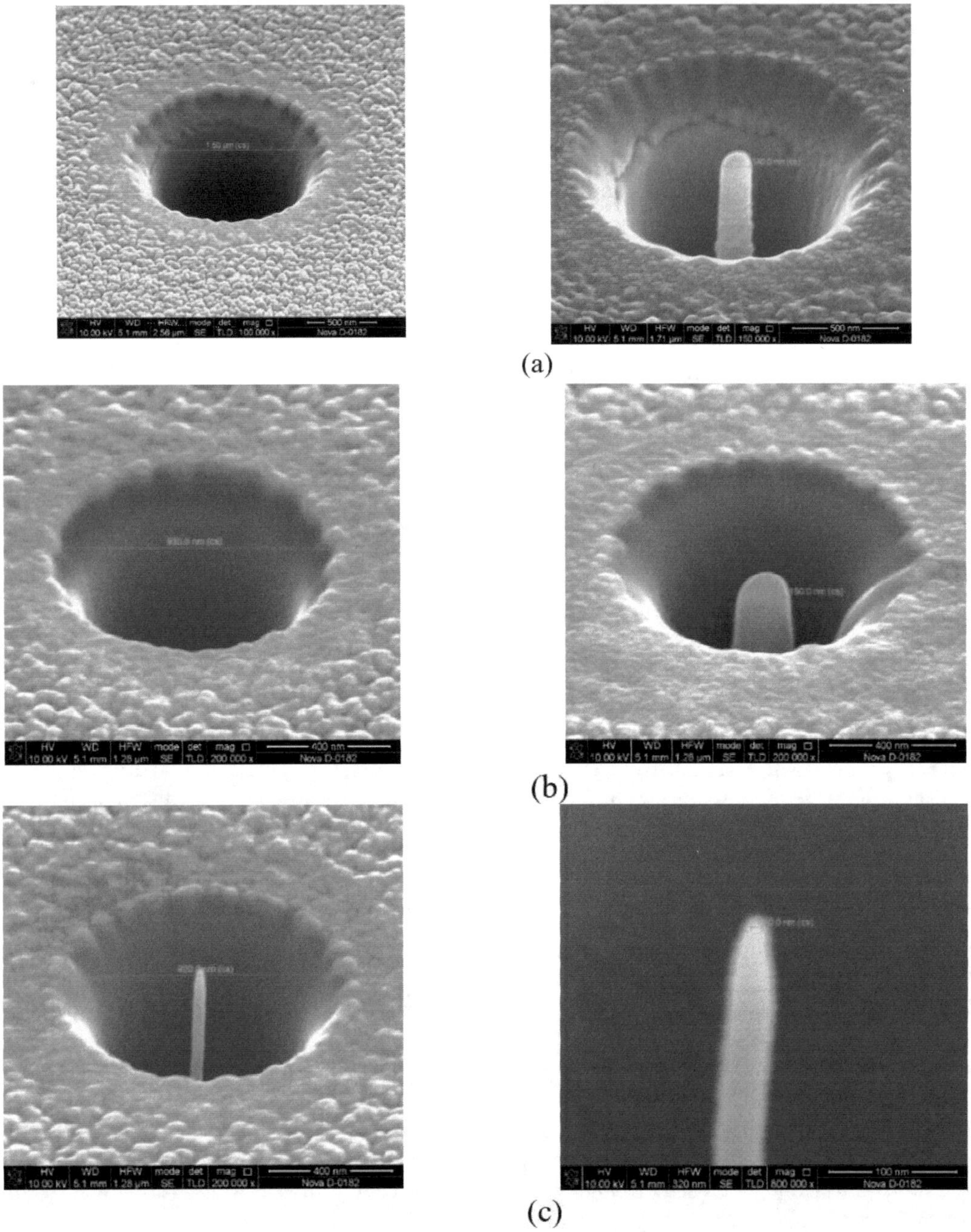

Figure 4. SEM images of the formed field emission cells by the FIB method with a vertically oriented emitter from W (a) and C (b), as well as an electron beam with an cathode from W (c).

One probe, which was at a positive potential, was placed on the upper Ni layer, the other on the contact area obtained by ion-beam milling of the FIB on the lower nickel layer, to which a negative bias was applied. The I - V characteristic was measured in the range from 0 to 30 V. The results of measuring one of the formed structures are shown in Figure 5.

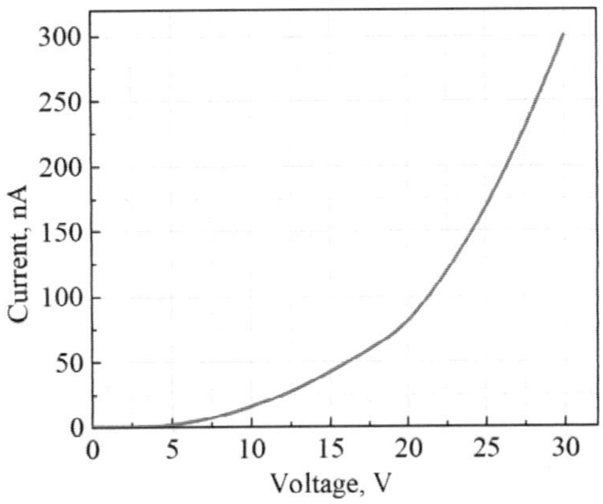

Figure 5. I - V characteristic of a field emission cell with a vertically oriented emitter.

It follows from the analysis of the curve that this field emission cell demonstrates an exponential increase in the anode current depending on the applied voltage, which indicates the presence of a pronounced field effect. The threshold voltage of the onset of emission for this type of construction and cell formation technology is about 2.5 V. At an interelectrode voltage of 30 V, the field emission current is 300 nA.

4. Conclusions

Thus, from the results of experimental studies, the optimal FIB current was determined for the formation of structures with sizes less than 500 nm, equal to 0.1 nA, and for structures larger than 500 nm, equal to 0.5 nA, at which the minimum amount of distortions is introduced into the formed structures and the most accurate transfer is provided. drawing the template onto the surface. To reduce the deviation in the range of values less than 500 nm, it is necessary to additionally investigate such FIB parameters as beam dwell time and the overlap. Based on the data obtained, field emission cells with a vertically oriented emitter were formed. Measurements of the current-voltage characteristic of the obtained cell have been carried out, from the graph of which it follows that the threshold voltage of the onset of emission is about 2.5 V, and the field current at a voltage of 30 V is 300 nA. The prospects of using this technology for the creation of field emission structures have been demonstrated.

Acknowledgments

This work was supported by the Russian Science Foundation Grant No. 20-69-46076. The results were obtained using the equipment of the Scientific-Educational Center and the Center for Shared Use "Nanotechnologies" of the Southern Federal University (Taganrog).

References

[1] Panchenko I V, Shandyba N A, Kolomiytsev A S and Lisitsyn S A 2019 *Journal of Physics: Conference Series* **1400** 044005

[2] Panchenko I V, Shandyba N A, Ballouk A, Rodrigues D, Kolomiytsev A S and Ageev O A 2020 *Journal of Physics: Conference Series* **1695** 012182

[3] Shandyba N A, Kolomiytsev A S, Panchenko I V and Jityaev I L 2020 *Journal of Physics: Conference Series* **1697** 012056

Optimizing self-rotating bristle-bots for active matter implementation with robotic swarms

V A Porvatov[1,2,3], A D Rozenblit[4], A A Dmitriev[4], O I Burmistrov[4], D A Petrova[2,4], G Yu Gritsenko[5], E M Puhtina[4], E I Kretov[4], D S Filonov[1], A Souslov[6] and N A Olekhno[4*]

[1]Moscow Institute of Physics and Technology, 141701 Moscow, Russia
[2]National University of Science and Technology "MISIS", 119991 Moscow, Russia
[3]Public Joint Stock Company "Sberbank of Russia", 117997 Moscow, Russia
[4]ITMO University, 197101 Saint Petersburg, Russia
[5]Lomonosov Moscow State University, 119991 Moscow, Russia
[6]University of Bath, Bath, UK

E-mail: *olekhnon@gmail.com

Abstract. Robotic swarms have been recently explored as a versatile and scalable alternative to traditional microscale platforms for experimental studies of active matter. These robotic setups consist of either self-propelled or self-rotating particles. In the present paper, we develop and experimentally realize a swarm of self-rotating bristle-bots suitable for a wide range of active-matter experiments. We focus on optimizing the bristle-bot design and controlling the sliding friction between individual robots.

1. Introduction

The physics of active matter explores mechanical phenomena in non-equilibrium systems consisting of particles that individually convert their internal energy into mechanical work and motion. Typical examples of such systems include various micro- and nanoscale setups, ranging from bacterial colonies [1] and liquid crystals with molecular motors [2] to self-propelled colloidal particles [3, 4]. However, the complexity of conducting experiments at the microscale has recently resulted in the development of swarm-robotics platforms [5, 6, 7, 8], which offer an unprecedented degree of control over system parameters.

In the present paper, we consider a swarm of self-rotating bristle-bots. Optimizing the bristle-bot design, we define an optimal set of bristle parameters and construct a system with adjustable side friction between individual robots. Such a system can be used to study the mobility of individual robots upon the variation in their packing density and side friction, thus paving the way towards experimental studies of friction-assisted phase transitions in active matter.

2. Experimental setup

We aim to explore bulk active phases and transitions between them. To separate such bulk phenomena from boundary effects, we focus on systems with self-rotating robots in which both the number of robots and the confinement area shape can be varied. We realize experimentally the ensemble of 80 self-rotating bristle-bots vibrating at frequencies of approximately 60 Hz,

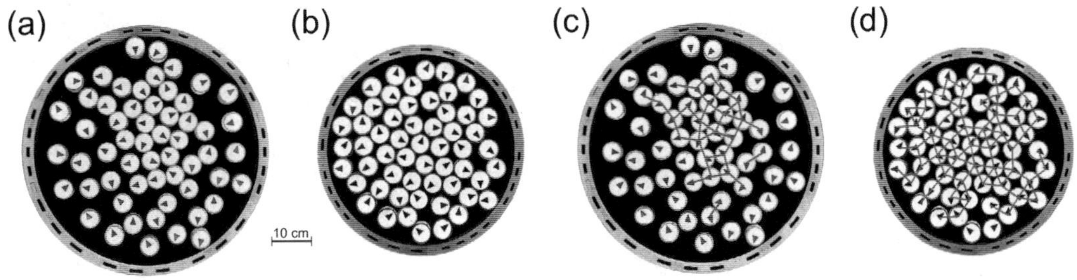

Figure 1. Examples of experimental setup with bristle-bots inside a circular boundary of diameters d = 60 cm (a) and d = 45 cm (b). (c,d) Post-processed images with connected clusters of robots shown by solid lines.

Figure 1(a,b). The vibration of an electric motor inside each robot is converted to a rotation around the robots' center by a circularly shaped layer of elastic bristles at the bottom of each robot. The robots are 50 mm in diameter and consist of a 3D-printed body made of PLA plastic, a vibration motor, a rechargeable 3.7V battery, and a control circuit. The circuit allows turning the robots on and off with an infrared remote control.

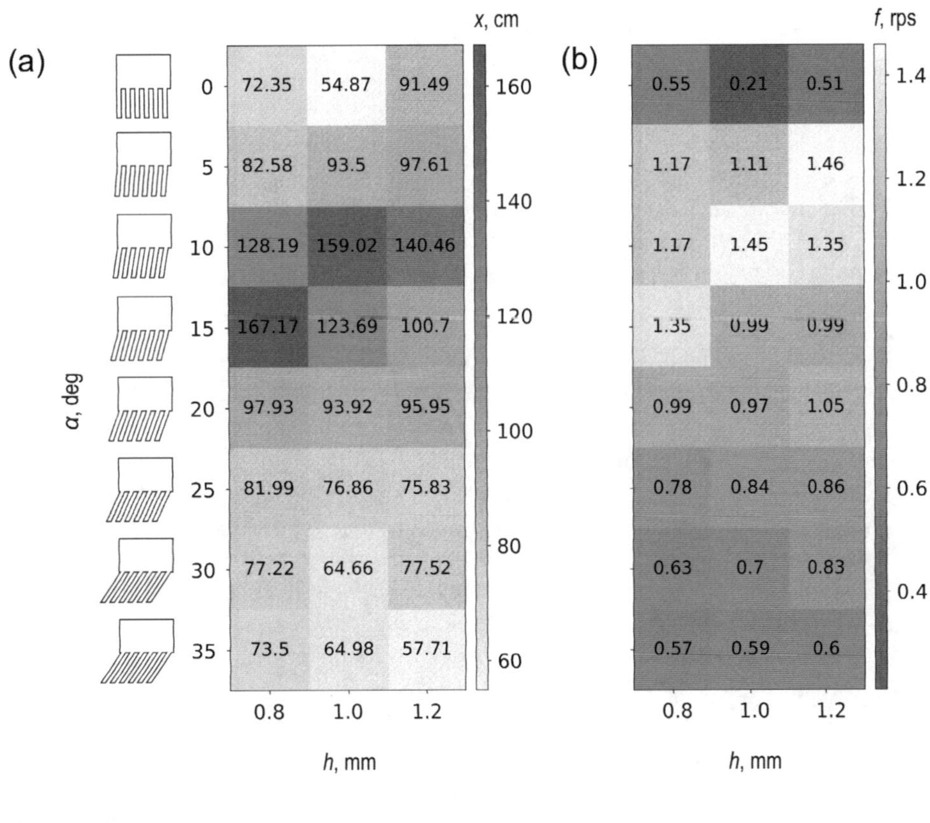

Figure 2. Optimization maps for the Bflex bristles, showing (a) the distance x travelled by the robot and (b) the self-rotation frequency f for different bristle angles α (vertical axis) and thicknesses h (horizontal axis) of elastic bristles.

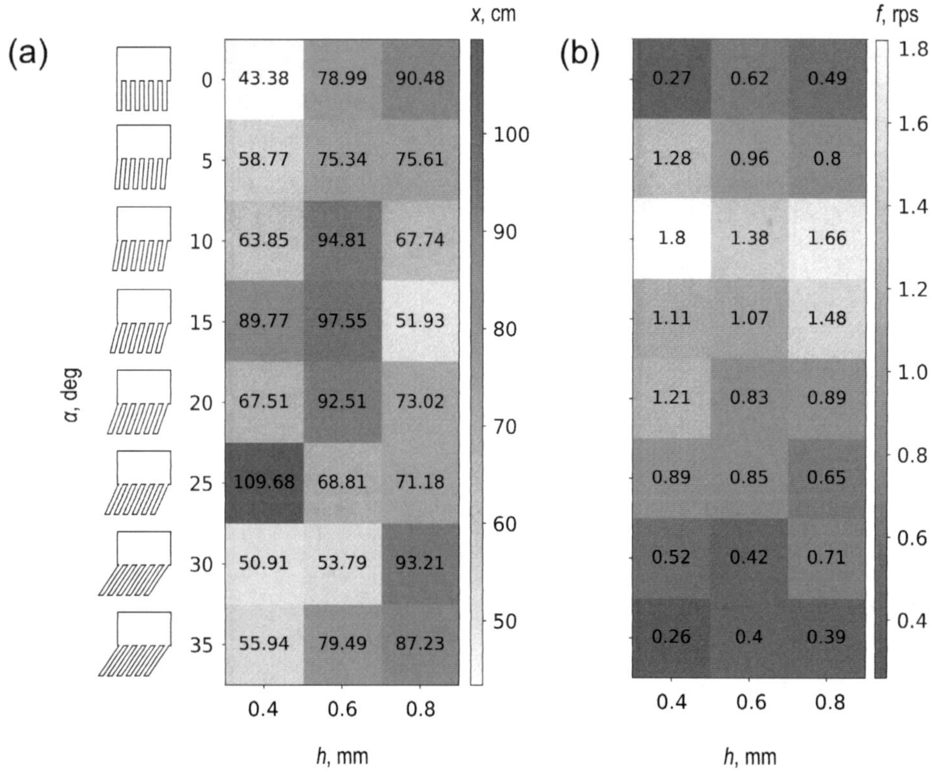

Figure 3. Optimization maps for the PLA bristles, showing (a) the distance x travelled by the robot and (b) the self-rotation frequency f for different bristle angles α (vertical axis) and thicknesses h (horizontal axis) of elastic bristles.

The motion of all the robots is captured by a camera and then analyzed with the help of deterministic algorithms from the OpenCV library in order to simultaneously track the triangle-shaped markers placed on top of every bristle-bot. Such an imaging allows us to capture the detailed kinematics of the system, Figure 1(c,d).

The applied approach to extracting the system's kinematics is motivated by the properties of constructed setup. In particular, the proposed design of self-rotating bristle-bots leads to minor changes in the ensemble's geometry during each time step between the consequent recorded frames. This feature of spatio-temporal locality was used in the nearest neighbor search algorithm in order to recover the robots' trajectories. Moreover, the formation of connected clusters of robots in the ensemble can be extracted from the tracking information using collision graphs and further explored by breadth-first search, as shown in Figure 1(c,d).

3. Bristle-bot design optimization

The effects of bristles shape on the movement pattern of individual bristle-bots have been explored theoretically and experimentally in the case of self-propelled bristle-bots [9] and driven self-rotators [10]. To optimize the bristles for our particular design of self-rotating bristle-bots, we perform a set of experiments varying the shape and thickness of bristles and considering different bristle materials.

We study 48 realizations including straight-line bristles composed of resin-like Bflex material,

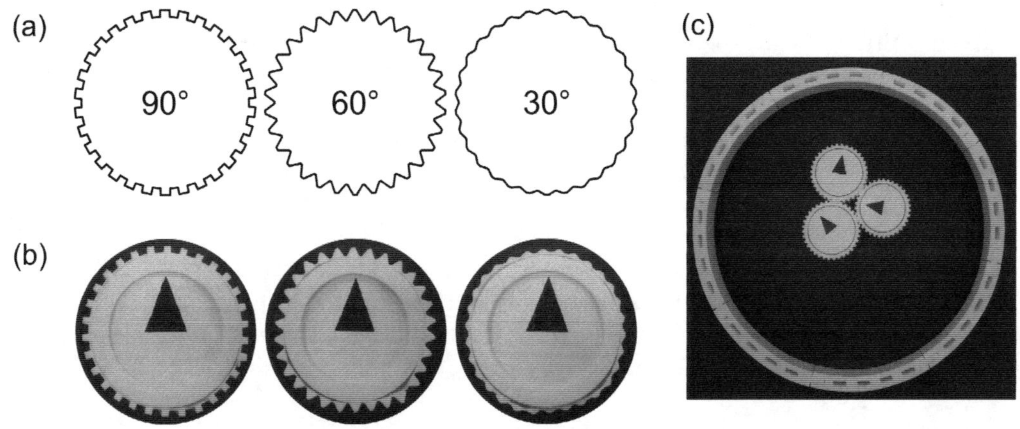

Figure 4. (a) Top view schematics of the bristle-bots covered with different ribbons with angles $\beta = 30^o$, $\beta = 60^o$ and $\beta = 90^o$, respectively. (b) Images of experimentally fabricated bristle-bots corresponding to panel (a). (c) Photo of three interacting bristle-bots with the maximal value of the effective side friction ($\beta = 90^o$).

Figure 2, and PLA plastic, Figure 3, with angles of $\alpha = 0^o$ up to $\alpha = 35^o$ with the step 5°. According to the properties of each material, we use thicknesses $h = 0.4$, 0.6, and 0.8 mm for PLA and $h = 0.8$, 1.0, and 1.2 mm for Bflex. We optimize two parameters: the traveled distance of each robot (which should be minimized for self-rotating robots) and the self-rotation frequency (which should be maximized). In order to obtain more representative results, we perform averaging over 3 independent realizations of each experiment considering different bristle-bots. After this stage, we apply OpenCV library to extract required magnitudes from the captured videos.

As seen from Figure 2, in the case of Bflex bristles the optimal realization with $\alpha = 5^o$ and $h = 1.2$ mm corresponds to 1.46 self-rotations per second at travelled path of 97.61 cm during the 30-second observation time. However, these results are outperformed by the PLA bristles, as seen from Figure 3. Indeed, the PLA bristles with $\alpha = 10^o$ and $h = 0.4$ mm allow achieving 1.8 rotations per second at the displacement of 63.85 cm during the 30-second observation time. Thus, we select this set of bristle parameters for the following realization of the bristle-bot swarm.

4. Adjustment of side friction between robots

Next, we focus on controlling the side friction between two (or more) individual bristle-bots. To proceed, we consider interacting rotating particles in the form of trapezoidal gears, schematically depicted in Figure 4(a). The effective friction coefficient is controlled by the angle at the base of the trapezoid β, with the limit of $\beta = 90^o$ corresponding to infinite effective friction. For the first tests, we fabricated a set of scalloped PLA ribbons with a trapezoidal meander profile wrapped around the side surface of each robot. Creating ribbons with angles β in the range $20^o \leq \beta \leq 90^o$ with step 10°, we optimize the trapezoids' geometry in different ways so as to preserve the period $T \approx 5.7$ mm and equal dimensions of the small base and the meander gap. Thus, for $\beta = 20^o...70^o$, we fix bases $a \approx 1.0$ mm and $b \approx 4.7$ mm along with changing the trapezoid height $H \approx 0.6...5$ mm, whereas for $\beta = 80^o$, we set different dimensions of the bases $a \approx 2.0$ mm, $b \approx 3.7$ mm and the height $H = 5$ mm. Finally, for the rectangle meander ($\beta = 90^o$), we set $a = b \approx 2.5$ mm and $h = 2$ mm. Thus, we make a gap larger than the base of the rectangle since the obtained "gears" otherwise may not fit each other due to imperfections of the 3D-printing process. Figure 4(b) shows the fabricated bristle-bots covered by ribbons of three different types. If the bristle-bots rotate in the same direction, then the ribbons with larger angles β may cause pronounced cluster formation, while the ones with small angles provide weaker interaction between particles. An example of interaction between three bristle-bots is shown in Figure 4(c).

5. Conclusion
We have developed, optimized, and experimentally realized the active matter setup consisting of self-rotating bristle-bots. The implemented setup allows us to study the effects of changing the side friction between particles on phase formation in active matter, an exciting topic which has only recently begun to be explored [11, 12].

Acknowledgments
The experimental part of the work was supported by the Russian Foundation for Basic Research (Grant No. 20-19-00491). A.S. acknowledges the support of the Engineering and Physical Sciences Research Council (EPSRC) through New Investigator Award No. EP/T000961/1.

References
[1] Mathijssen A J T M, Guzmán-Lastra F, Kaiser A and Löwen H 2018 *Phys. Rev. Lett.* **121**, 248101
[2] Duclos G, et al 2020 *Science* **367** 1120-1124
[3] Bricard A, Caussin J-B, Desreumaux N, Dauchot O and Bartolo D 2013 *Nature* **503** 95-98
[4] Ginot F, Theurkauff I, Detcheverry F, Ybert C and Cottin-Bizonne C 2018 *Nat. Commun.* **9** 696
[5] Deblais A, Barois T, Guerin T, Delville P H and Vaudaine R 2018 *Phys. Rev. Lett.* **120**, 188002
[6] Yang X, Ren C, Cheng K and Zhang H P 2020 *Phys. Rev. E* **101**, 022603
[7] Fruchart M, Hanai R, Littlewood P B and Vitelli V 2021 *Nature* **592** 363–369
[8] Wang G et al 2021 *Phys. Rev. Lett.* **126**, 108002
[9] Koumakis N et al 2016 *New J. Phys.* **18** 113046
[10] Scholz C et al 2016 *New J. Phys.* **18** 123001
[11] Liu P et al 2020 *PNAS* **117**, 11901
[12] Li S et al 2021 *Sci. Adv.* **7** eabe8494

Structural, morphological and catalytic characteristics of nanostructured palladium catalysts

P D Pushankina, I S Lutsenko, I S Glazkova, T I Malkov and M A Mukhanov

Department of Physics, Kuban State University, Krasnodar, 350040, Russia

polina_pushankina@mail.ru

Abstract. A new method for the highly active palladium catalyst synthesis on the surface of a Pd-23%Ag film has been developed to increase the material activity with respect to reactions involving hydrogen. Comparison of the electrochemical experiments data of classical palladium black and a new developed nanocatalyst demonstrated a significant increase in catalytic activity in the methanol oxidation reaction (up to 17.09 μA cm^{-2}) for electrodes modified with the latter catalyst. The reason for that is an increase in the number of localized potentially more active surface regions due to the creation of a larger number of active sites in comparison with spherical particles. Estimation of resistance to CO poisoning showed high efficiency of nanocatalysts. Chronoamperometric experiment established the long-term stability and activity of the developed catalyst and confirmed the possibility of its practical use.

1. Introduction

Nanoparticles based on noble metals are of great interest for many branches of modern science and industry [1-6]. Particular attention is drawn to the potential of using palladium-based functional nanomaterials as catalysts for hydrogen separation and purification [6-12]. Such a catalyst must be active, stable, with low overvoltage, which requires effective catalytic properties, such as high electronic conductivity and a large number of active sites [13]. Since the reactions catalyzed by Pd exhibit structural sensitivity, the activity and selectivity depend on the control of the shape and size of metal nanoparticles and the arrangement of atoms on the surface. Therefore, the key to the development of highly efficient nanoscale systems is to understand the effect of nanoparticle morphology (facets with high or low Miller index, defects, bends, adatoms, etc.) on their physicochemical and catalytic properties.

At the moment, various methods for the synthesis of nanostructures have been developed, which made it possible to obtain particles of various shapes, structural organizations, monometallic or bearing synergistic effects from secondary metals [14, 15]. The most advanced of these are particles with one or more fifth-order axes [16-18]. However, these particles have a regular geometric shape (icosahedron, decahedron) close to spherical and therefore have low activity.

Therefore, the aim of the article was to synthesize new types of highly efficient nanostructured palladium-based catalysts that increase the adsorption / desorption activity of the material with respect to reactions involving hydrogen and to investigate their catalytic characteristics.

2. Methods and materials

2.1. Synthesis of nanostructured catalysts
The substrate of a Pd-23% Ag film was prepared by washing in ethyl alcohol (96%) and boiling for 20 minutes to degrease before deposition. Then the film was washed in running distilled water for 5-10 minutes. After that, the prepared film was fixed in the working cell as an anode, the cathode from which the deposition was carried out was fixed parallel to anode. Then the substrate was polarized in sulfuric acid (0.05 M) at a current density of 10 mA/cm^2, after which it was washed with bidistillate. In order to start deposition the electrolytic cell was filled with a palladium chloride solution (2%). Depending on the type of the obtained coating, a surfactant (tetrabutylammonium bromide) in various concentrations was added to the growth solution. The coating deposition on the substrate was carried out at a current density varying for different coatings from 5 to 6 mA/cm^2 for 20 minutes. The highly dispersed palladium coating electrolytic deposition was carried out on a potentiastat-galvanostat (Elins) R-40X. At the end of the process, the film was washed with bidistillate.

2.2. Electrochemical measurements
Electrochemical experiments were carried out using a potentiastat-galvanostat "Elins" P-40X. The experiment was carried out in a standard three-electrode cell filled with a solution of NaOH (1 M) and methanol (0.5 M) and consisting of a working electrode, a counter electrode, and a reference electrode. The working electrode was a sample of palladium-silver films modified with various types of palladium coatings. The counter electrode was a platinum foil comparable in size to the working electrode. A silver chloride (Ag/AgCl) electrode was used as a reference electrode. Before measurements, all electrodes were washed and the platinum counter electrode was degreased.

Measurements of the samples electrocatalytic characteristics were carried out by the method of cyclic voltammetry (CV). The multiscanning was carried out in the operating potential range from -0.9 V to +0.5 V at a scan rate of 50 mV s^{-1}. The currents were normalized to the geometric area of the working electrodes. All potentials were reported with respect to the reference electrode.

Chronoamperometric (CA) measurements of the samples stability were carried out at a constant potential of -0.3 V for 2400 s.

3. Results and discussion
Modified films samples of two series were synthesized by the electrolytic deposition method with variation of some parameters. The synthesis of the first sample series was carried out without surfactant at the current density during deposition of 6 mA cm^{-2}. The synthesis of the second sample series was carried out with tetrabutylammonium bromide as a surfactant at a concentration of 0.005 M at the current density during deposition of 5 mA cm^{-2}.

Modified palladium-silver films surface microphotographs were taken using SEM. The resulting images shown in figure 1 made it possible to study the morphology of the synthesized Pd nanocatalysts. SEM images of the first sample series (classical palladium black) (figure 1a) demonstrate a spherical structure of particles with a characteristic size in the range of 90-100 nm. SEM images of the second sample series (figure 1b) demonstrate a filamentary structure of particles with pointed tips. Probably, the obtained particles morphology was influenced by a change in such synthesis parameters as a rate of the particles deposition on a substrate and a change in the composition of the growth solution by increasing the concentration of surfactant, which also affected the halide ions ratio (Cl^- and Br^-) in the growth solution. The characteristic size of the particles synthesized on the surface of the Pd-Ag film was 100-150 nm.

CV of two series samples were carried out in the potential range from -0.9 V to +0.5 V at a scan rate of 50 mV s^{-1} at room temperature (25 °C). The choice of the potential range is due to the presence of deep hydrogen evolution below the negative threshold and oxygen evolution above the positive threshold. The current peaks observed in the methanol oxidation reaction (MOR) on palladium (figure 2) are caused by the occurrence of competing processes of electrochemical adsorption of methanol and water on the catalytic palladium surface. Electrochemical methanol adsorption occurs with the formation of Pd – C bonds. MOR is a multi-stage process, the result of which is the formation of carbon monoxide tightly bound to the Pd surface. The MOR process limiting stages on Pd are the oxygen chemisorption stage and / or the chemisorbed CO removal stage.

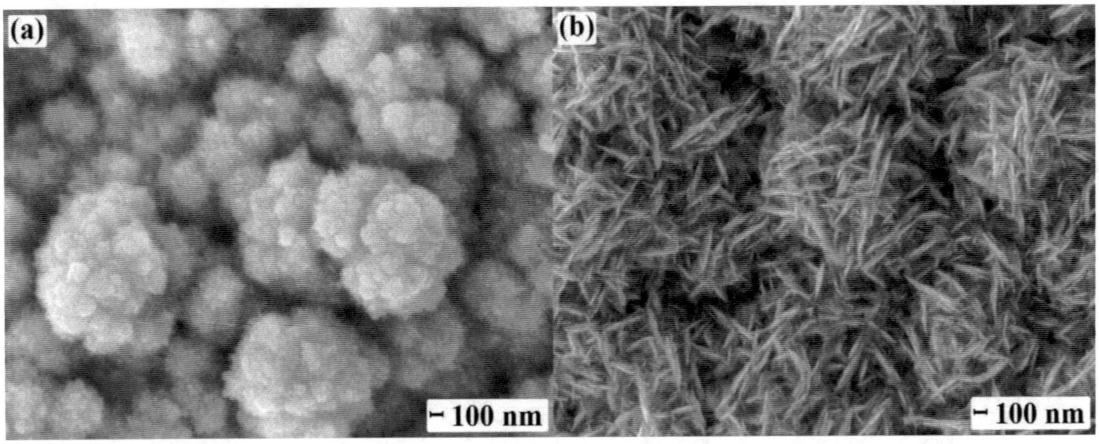

Figure 1. SEM images of the first (a) and second (b) series samples surfaces of Pd-Ag films.

A distinct peak recorded during direct scanning on the anode branch in the potential range from -0.3 V to -0.05 V for all samples is due to processes of electrochemical methanol adsorption on Pd, water adsorption and removal of carbon monoxide bound to the palladium surface. A negative potential peak bias indicates good catalytic performance with respect to MOR. Second series sample had the highest value of the current density peak, and, consequently, the highest activity - 17.09 mA cm^{-2}. Next, the current density peak decrease can be explained by the electrochemical methanol adsorption rate decrease, caused by the modified surface sample blocking by the products of the electrochemical adsorption of water. At higher anodic potentials (over 0.3 V), methanol oxidation proceeds in a parallel reaction with weakly bounded to the surface active particles participation.

A lower intensity peak with the reverse potential sweep in the cathodic direction was observed in the potential range from -0.33 V to -0.4 V, it is due to the methanol oxidation resumption. The oxidation reaction becomes possible when the catalyst surface is reduced. In this case, adsorbate-free palladium atoms are formed, which can participate in the electrochemical methanol adsorption process.

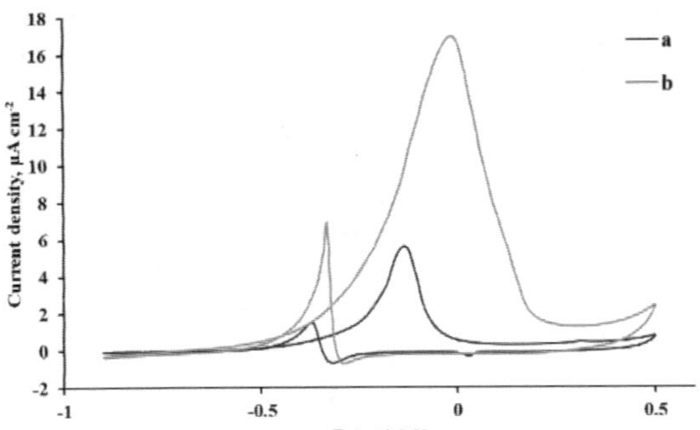

Figure 2. CV of steady-state cycles of the first (a) and second (b) series samples in the alkaline methanol oxidation reaction.

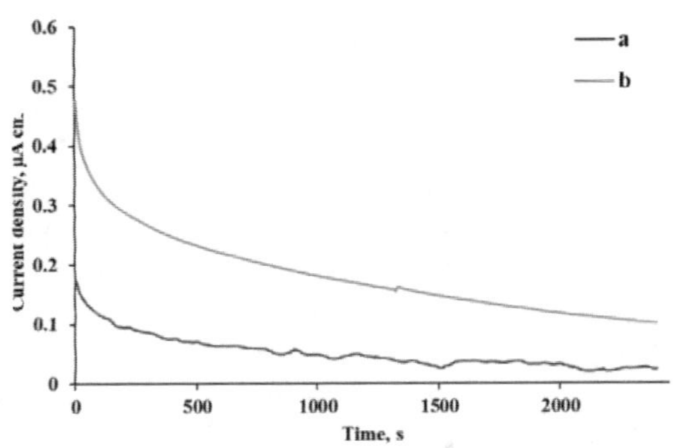

Figure 3. Chronoamperometry of the first (a) and second (b) series of samples in alkaline methanol oxidation.

The catalysts resistance to CO poisoning was assessed through the current peaks i_F / i_B.. ratio [19, 20]. Low values of the ratio indicate an excessive residual forms of carbonaceous products accumulation on the catalyst surface, while higher values indicate effective CO desorption on the studied catalyst. So the entire studied sample series demonstrated a fairly high carbon monoxide poisoning resistance, however, the highest i_F / i_B.. values were achieved for the first sample series, which amounted to 3.83 in numerical equivalent.

The stability of two series of samples was investigated by CV-multiscanning of 100 cycles. The highest current density values were achieved by the samples on average by the 30th cycle, after which a moderate current decrease was observed. Based on the obtained experimental data, the second sample series had the most long-term stability, for which the current density decrease by the 100th cycle was 11.99%.

Further activity and stability assessment of the obtained sample series was carried out by the CA method in MOR (figure 3). According to the experiment results, the second sample series showed the highest stabilized current density value - 0.099 mA cm^{-2} compared to the rest of the samples.

Along with the presented above indicators, the electrochemically active surface area (ECSA) and the roughness factor (RF) are also quite important in the catalyst quality assessing. The values of these indicators were calculated by equation (1):

$$ECSA = \frac{C_{DL}}{C_S}, \tag{1}$$

where C_{DL} is the double layer capacity, C_S is the specific sample capacity or the capacity of an atomically smooth flat surface of the material per unit area under identical electrolyte conditions.

The roughness factor (RF) was calculated by dividing the calculated ECSA value by the geometric area of the electrode as in equation (2):

$$\boldsymbol{R_F} = \frac{\boldsymbol{ECSA}}{\boldsymbol{S}}. \tag{2}$$

The obtained data are presented in table 1. It is shown that the second sample series had the highest ECSA and the highest RF values, which indicates an increased active sites number in comparison with the of the first sample series.

Table 1. Characteristics of the synthesized catalysts.

Sample characteristics	Types of coatings	
	I series	II series
Electrolytic deposition conditions		
Deposition time, min	20.0	20.0
Current density, mA cm^{-2}	6.0	5.0
Surfactant concentration, M	–	0.005
Size characteristics		
Average particle size, nm	90-100	100-150
Layer thickness, µm	2.102	2.224
Electrocatalytic characteristics		
ECSA, cm^2	0.031	0.19
R_F	0.62	3.8
Peak potentials		
E_F, V	-0.13	-0.012
E_B, V	-0.37	-0.33
Current densitiy peaks		
i_F, mA cm^{-2}	5.6	17.09
i_B, mA cm^{-2}	1.45	6.93
i_F / i_B	3.86	2.47

I_{st}, mA cm^{-2}	0.023	0.099
Stability, %	12.8	11.99

The study was demonstrated a new approach to solving the problem of creating a highly efficient nanostructured palladium-based catalysts. These catalysts increase the material adsorption/desorption activity with respect to reactions involving hydrogen. High catalytic activity indicators (up to 17.09 mA cm^{-2}), long-term stability and CO oxidation resistance for the second sample series with a completely new catalyst is due to an increase not only in the surface roughness, but also in the reactive active sites number, compared with classical palladium black.

Acknowledgement
The research was funded by Ministry of Science and Higher Education state assignment of Kuban state university № FZEN-2020-0022 and Kuban Science Foundation in the framework of the Commerciazable scientific and innovation projects competition № NIP-20.1/13.

References

[1] Giannone G, Santi M, Ermini M L, Cassano D and Voliani V 2020 Cost-Effective Approach for Non-PersistentGold Nano-Architectures Production *Nanomaterials* **10** 1600.
[2] Slesiona N, Thamm S, Stolle H L K S, Weißenborn V, Müller P, Csáki A and Fritzsche W 2020 DNA-Biofunctionalization of CTAC-Capped Gold Nanocubes *Nanomaterials* **10** 1119.
[3] Petriev I S, Baryshev M G, Voronin K A, Lutsenko I S, Pushankina P D and Kopytov G F 2020 Gas transmission properties of Pd–Ag membranes coated with modifying layer *Russian Physics Journal* **63(3)** 457-461.
[4] Barbillon G 2020 Latest Novelties on Plasmonic and Non-Plasmonic Nanomaterials for SERS Sensing *Nanomaterials* **10** 1200.
[5] Naik S S, Lee S J, Begildayeva T, Yu Y, Lee H and Choi M Y 2020 Pulsed laser synthesis of reduced graphene oxidesupported ZnO/Au nanostructures in liquid with enhanced solar light photocatalytic activity *Environ* Pollut **266** 115247.
[6] Kopytov G F, Malyshko V V, Elkina A A, Moiseev A V, Dzhimak S S, Baso, A A and Baryshev M G 2020 Sorption capacity of silver nanoparticles on plain polished catgut *Russian Physics Journal* **63(6)** 989-996.
[7] Theerthagiri J, Madhavan J, Lee S J, Murthy A P, Choi M Y 2020 Ashokkumar M and Pollet B G Sonoelectrochemistry for energy and environmental applications *Ultrasonics Sonochem* **63** 104960.
[8] Lytkina A A, Orekhova N V, Ermilova M M, Petriev I S, Baryshev M G and Yaroslavtsev A B 2019 Ru–Rh based catalysts for hydrogen production via methanol steam reforming in conventional and membrane reactors *Int. J. Hydrogen Energy* **44** 13310–13322.
[9] Petriev I S, Bolotin S N, Frolov V Y and Baryshev M G 2019 Synthesis and Gas-Transport Parameters of Membranes Modified by Star-Shaped Palladium *Nanocrystallites Doklady Phys.* **64** 210–213.
[10] Pati S, Jat R A, Anand N S, Derose D J, Karn K N, Mukerjee S K and Parida S C 2017 Pd-Ag-Cu dense metallic membrane for hydrogen isotope purification and recovery at low pressures *J. Membr. Sci.* **522** 151–158.
[11] Lytkina A A, Orekhova N V, Ermilova M M and Yaroslavtsev A B 2018 The influence of the support composition and structure ($M_XZr_{1-X}O_{2-\delta}$) of bimetallic catalysts on the activity in methanol steam reforming *Int. J. of Hydrogen Energy* **43(1)** 198-207.
[12] Petriev I S, Bolotin S N, Frolov V Y, Baryshev M G, Kopytov G F and Isaev V A 2019 Monte carlo simulation of hydrogen absorption in palladium and palladium-silver alloy *Russian Physics J.* **61(10)** 1894-1898.
[13] Peng C, Hu Y, Liu M and Zheng Y 2015 Hollow raspberry-like PdAg alloy nanospheres: High electrocatalytic active for ethanol oxidation in alkaline media *J. Power Sources* **278** 69–75.

[14] Chowdhury S R, Ghosh S, Bhattachrya S K 2017 Improved Catalysis of Green-Synthesized Pd-Ag Alloy-Nanoparticles for Anodic Oxidation of Methanol in Alkali *Electrochim Acta* **225** 310–321.
[15] Petriev I , Pushankina P , Lutsenko I , Shostak N and Baryshev M 2020 Synthesis, electrocatalytic and gas transport characteristics of pentagonally structured star-shaped nanocrystallites of Pd-Ag *Nanomaterials* **10(10)** 1-19.
[16] Mackay A L 1962 A dense non-crystallographic packing of equal spheres *Acta Crystallogr.* **15** 916–918.
[17] Kuo K H 2002 Mackay, anti-mackay, double-mackay, pseudo-mackay, and related icosahedral shell clusters *Struct Chem* **13** 221–230.
[18] Petriev I, Pushankina P, Bolotin S, Lutsenko I, Kukueva E and Baryshev M 2021 The influence of modifying nanoflower and nanostar type Pd coatings on low temperature hydrogen permeability through Pd-containing membranes *J. Membr. Sci.* **620** 503.
[19] Woo S, Lee J, Park S-K, Kim H, Chung T D and Piao Y 2015 Electrochemical codeposition of Pt/graphenecatalyst for improved methanol oxidation *Curr. Appl. Phys.* **15** 219–225.
[20] Petriev I S 2021 Hydrogen permeability and catalytic characteristics of a palladium-silver membrane with a modified surface *J. of Phys.* **1758(1)** 1-7.

Fabrication of probe tips via the FIB method for nanodiagnostics of the surface of solids by atomic force microscopy

D J Rodriguez, A V Kotosonova, H A Ballouk, N A Shandyba, O I Osotova and A S Kolomiytsev

Southern Federal University, Institute of Nanotechnologies, Electronics and Equipment Engineering, Taganrog, 347922, Russia

priviet122@gmail.com

Abstract. In this work, we carried out an investigation of commercial atomic force microscope (AFM) probes for contact and semi-contact modes, which were modified by focused ion beam (FIB). This method was used to modify the original tip shape of silicon AFM probes, by ion-etching and ion-enhance gas deposition. we show a better performance of the FIB-modified probes in contrast with the non-modified commercial probes. These results were obtained after using both probes in semi-contact mode in a calibration grating sample.

1. Introduction

Atomic force microscopy (AFM) is a powerful tool for studying topography and various physical and chemical properties of the surface of solids. The most important factor affecting the quality and accuracy of AFM images is the quality of fabrication and the geometric parameters of the tip of the probe sensor. Standard AFM probes have a limited range of geometric parameters, which often leads to undesirable image distortions [1]. Often, to obtain adequate results of measurements of the relief of nanoscale structures, it is necessary to manufacture probes with a special tip shape, for example, with an increased aspect ratio or a reduced radius of rounding. One of the methods that allows the formation of probe points in a wide range of geometric parameters is the use of a focused ion beam (FIB) for local ion-stimulated etching and deposition of materials from the gas phase [2]. The purpose of this work is to manufacture AFM probes by the FIB method and to study the influence of the geometric parameters of the probes on the accuracy of the study of the relief of nanoscale periodic structures by the AFM method.

2. Modelling

To identify the regularities of the influence of the ion beam parameters on the evolution of the surface relief during ion treatment, numerical modelling was performed using the SRIM software package, which is based on the methods of stochastic Monte Carlo modelling. By simulating the distribution of ion runs in a solid, it is possible to estimate the influence of the ion energy and the direction of their movement on the dynamics of the profile formation on the surface of the solid, and by calculating the sputtering coefficient, it is possible to judge the influence of these parameters on the etching rate or the rate of dissociation of process gases.

For the simulation, a multilayer structure was constructed: the first layer is C10H8 gas with a thickness of 10 nm, the next layer is absorbed C10H8 gas molecules with a thickness of 5 nm, and the last layer is a silicon substrate with a thickness of 200 nm. The choice of the material is determined by the method of the subsequent experiment, which is based on the decomposition of the C10H8 gas, which enters locally through the injector needle, under the action of the Ga+ FIB with an energy of up to 30 keV. Numerical simulations were performed to determine the dependences of the ion penetration depth and the sputtering coefficient (on the angle of incidence of the ion beam (Figure 2, a) and on the value of the Ga+ ion energy (Figure 1, b).

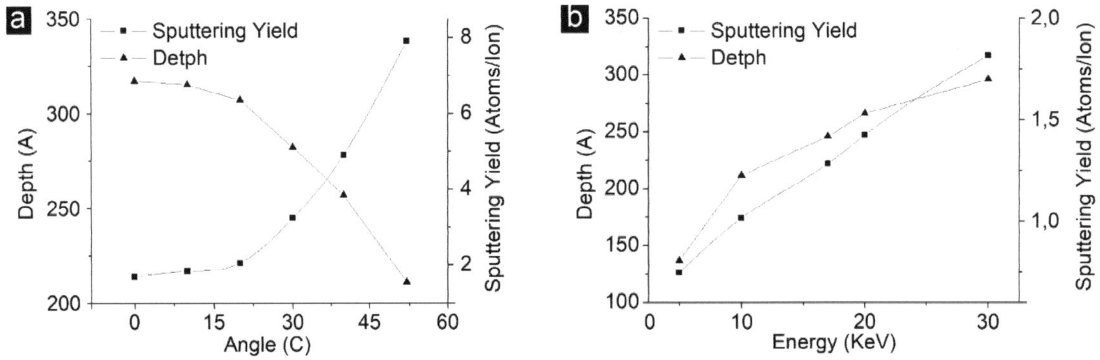

Figure 1. Simulation results of the ion-gas-substrate interaction: the dependence of the ion penetration depth and the sputtering coefficient on (**a**) the energy and (**b**) the angle of incidence of the ions.

The analysis of the obtained dependences showed that with an increase in the angle of incidence of the ion beam, the depth of ion penetration decreases, and at the same time the sputtering coefficient increases. Such a change in the parameters is interrelated: if the ion energy is equal in the case of a normal fall, it will mainly be spent on the gradual penetration of the ion deep into the substrate, and if the ion hits the surface at an angle to the normal, the energy of the primary ions will be transferred mainly to the near-surface atoms and the probability of sputtering will increase. The change in the accelerating voltage similarly affects both the penetration depth and the sputtering coefficient for the same reasons: as the energy of the incident ion increases, both the number of collisions (and, accordingly, the penetration depth) increases until the ion completely stops, and the probability of sputtering of the substrate atoms increases.

These results were used to determine the parameters of ion-stimulated etching and deposition. The depth of ion penetration into the structure will determine the thickness of the amorphized layer and the dose of embedded gallium ions. When planning and conducting experimental studies, it is necessary to take into account that etching and deposition by an ion beam incident at an angle to the surface can lead to distortion of the geometry of the formed structure; and a decrease in the accelerating voltage can significantly reduce the resolution and efficiency of decomposition of precursor gases. The results of the influence of the beam parameters on the sputtering coefficient allow us to estimate the dynamics of the ion-stimulated deposition or ion-beam etching process.

3. Experimental

Experimental studies were carried out using a scanning electron-ion microscope Nova NanoLab 600 (FEI, the Netherlands), equipped with a FIB system. The tip of the probes was formed on the basis of the beams of standard AFM HA_HR cantilevers (NT-MDT), whose own points were previously removed by ion-beam etching FIB. The new probe points were formed by ion-stimulated carbon deposition, when a volatile compound C10H8 was fed into the zone of impact of the ion beam. To study the influence of the probe shape on the accuracy of measuring the surface topology, cantilevers with two types of points were made: cylindrical with a high aspect ratio and conical stepped. The technique was also applied, where ion-stimulated etching was used to modify the formed high-aspect probes.

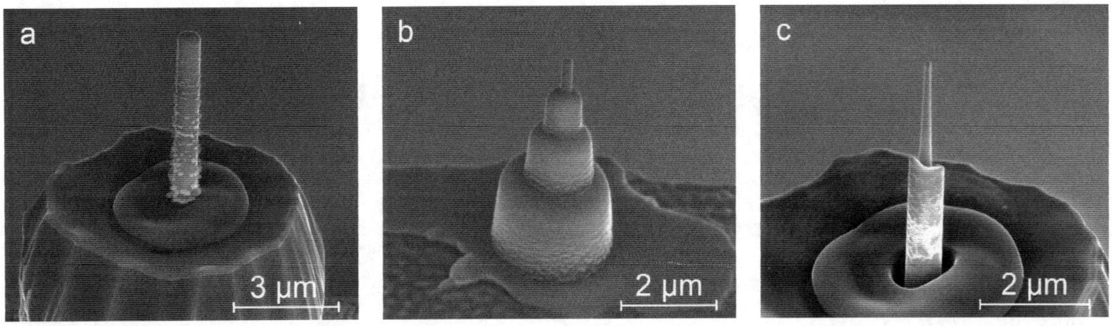

Figure 2. SEM images of (**a**) high-aspect, (**b**) stepped, and (**c**) sharpened probes manufactured using FIB technology.

3.1. High aspect ratio probes
The formation of high-aspect ratio points was carried out by ion-stimulated carbon deposition at FIB current values from 30 to 1 pA, an accelerating voltage of 30 keV, and an ion beam exposure time at a point of 1 µs. The geometric parameters specified in the template were different for each probe. The deposition parameters for each specific tip are shown in Table 1.

Table 1. Deposition parameters of high-aspect probes

	Designation of the cantilever on the beam of which the probe was formed				
	T1	T2	T3	T4	T5
Ion beam current (nA)	30	30	10	1	1
Height (µm)	5.62	5.91	4.92	4.58	5.15
Radius of curvature (nm)	521	330	230	171	125

3.2. Stair-step base probes
The first step probe (S1) was deposited as a structure of five disks with a height of 1.3; 1.2; 0.9; 0.7 и 0.6 microns and diameters of 3; 2.5; 2.0; 1.5 and 0.4 microns, respectively. The ion current during deposition was 30 nA for the first three disks and 1 pA for the last two. The second probe (S2) was deposited as a structure of four disks, the height of which was 1.4 ; 0.7; 0.7 и 0.7 microns, and the diameter of 2; 1; 0.5 and 0.2 microns, respectively. The ion current during deposition was 300 pA for the first two disks, 100 pA for the third, and 1 pA for the last.

3.3. FIB-sharpened probes
The next probe (M1) was made according to the scheme shown in Figure 3. This method uses a combination of ion-stimulated deposition and etching techniques. At the first stage, a tungsten tip probe with a diameter of 800 nm, was deposited at a beam current of 0.10 nA. Next, ion-stimulated deposition of the formed tip was carried out using the raster templates shown in Figure 3. The final radius of rounding of the probe was 81 nm.

In order to assess the influence of the shape of the obtained probes on the accuracy of measurements using the T1-T5, S2 and M1 cantilevers, the surface of the TGZ3 (NT-MDT) calibration grating was studied in the semi-contact AFM mode. For a comparative evaluation of the results of the study, such a scan was also performed using a standard NSG10 cantilever.

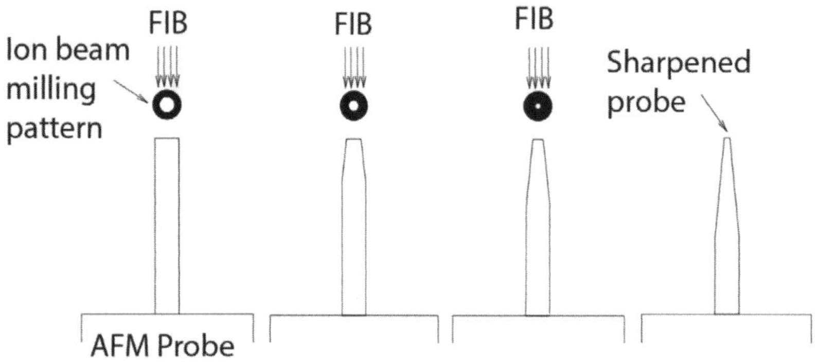

Figure 3. Technology for the formation and sharpening of the probe, including ion-stimulated deposition from gas phase and subsequent ion-stimulated etching by bitmap templates.

4. Results and discussion

The T2-T5 cantilever probes were broken at the initial stage of scanning due to the base being too thin and consequently brittle. The profilograms of the TGZ3 lattice surface obtained for the NSG10, T1, T2, S2, and M1 cantilevers are shown in Figure 4.

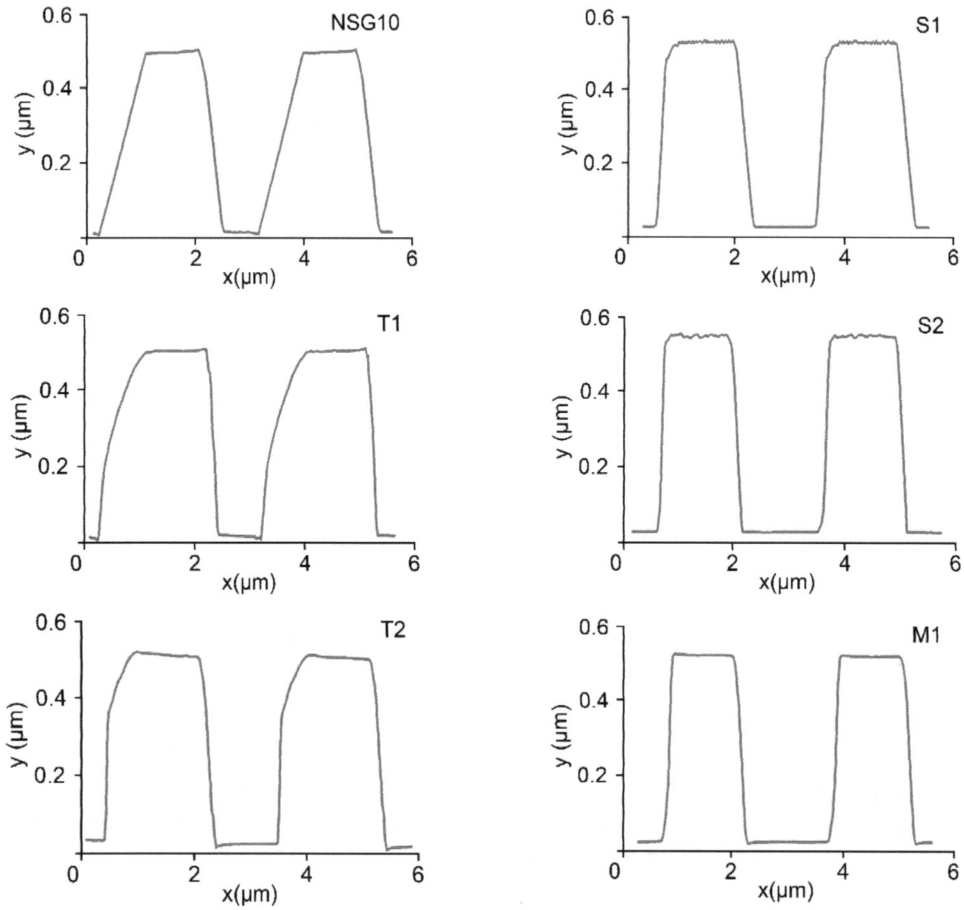

Figure 4. Profilograms obtained by scanning the TGZ3 calibration grating. The cantilevers used for scanning are marked in the figures.

Comparing the obtained profilograms, we can note an increase in the scanning resolution when using modified probes. To numerically evaluate the accuracy of the profilograms, the values of the calibration grid parameters specified in the passport, namely, the width and height of the steps, were calculated. The data obtained with the corresponding errors are presented in Table 2. The increase in the accuracy of AFM studies using FIB-modified probes was estimated by the value of the deviation from the set values, which was determined by the ratio of the measured structure parameter, taking into account its average deviation and measurement error, and its passport value.

Table 2. Parameters of the TGZ3 lattice determined by different cantilevers.

	Width		Height	
	Measurments (microns)	Deviation	Measurments (microns)	Deviation
Technical data	1.10±0.01	-	500±2.5	-
NSG10	1.66±0.68	113.4 %	492.3±3.05	2.66 %
T1	1.79±0.69	113.8 %	489.1±5.3	3.74 %
T2	1.62±0.44	88.6 %	495.6±4.0	2.18 %
S1	1.49±0.40	72.1 %	505.49±2.2	2.10 %
S2	1.36±0.27	49.1 %	521.9±3.1	5.50 %
M1	1.31±0.26	43.6 %	496.3±2.7	1.77 %

For a more visual representation of the obtained data, graphs were constructed (Figure 5). These graphs show the parameter values determined by various cantilevers and show the error during scanning. Also, a blue bar was plotted on the graphs, the width of which corresponds to the passport data of the TGZ3 grid, taking into account the confidence interval presented in the passport.

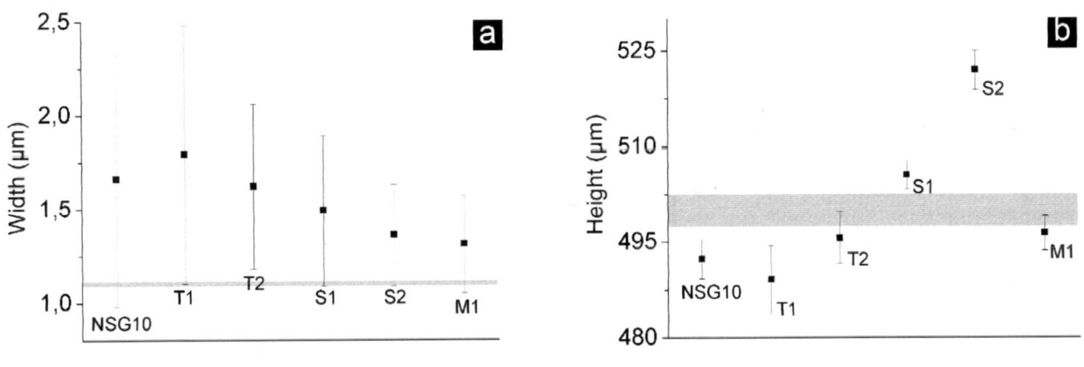

Figure 5. Comparison of the obtained values (**a**) of the width and (**b**) of the height of the steps of the TGZ3 grid with the passport data, taking into account the deviation during scanning.

The only probe that did not allow for improved scanning quality, compared to the standard NSG10 cantilever, was the T1 cantilever probe. This result is due to the too large radius of the rounding. The remaining manufactured cantilevers showed good results in the study. The obtained parameter values are more accurate than for the NSG10 cantilever, and the scanning errors are also significantly reduced, especially in the lateral direction. For the M1 cantilever, the resolution was the highest, which allowed for a 2.4 – fold increase in the accuracy of the width measurement and a 3.55-fold increase in the height measurement.

5. Conclusion

The manufactured probes showed less distortion in the image than the standard AFM probe, which allowed to increase the measurement accuracy several times. Tip probes with a baseless than 330 nm showed low strength during the scanning process. The stepped probes avoid this problem thanks to a stronger base. Sharpening the probe with a focused ion beam allows to get a smaller radius of rounding (compared to unmodified probes), and this operation is performed simultaneously in one stage. The probes made by this method showed the highest scanning accuracy.

The performed studies have shown that the use of the FIB method makes it possible to form the tips of the probes, the parameters of which differ significantly from the parameters of standard probes, thereby increasing the accuracy of the study of the surface relief. The results of this work can be useful in the study of micro-and nanostructures, and can also be used for further development of the technology for manufacturing and modifying probes by the FIB method.

Acknowledgments
This work was supported by the Russian Science Foundation Grant No. 20-69-46076. The results were obtained using the equipment of the Scientific-Educational Center and the Center for Shared Use “Nanotechnologies” of the Southern Federal University (Taganrog).

References.
[1] De Felicis D, Mughal M and Bemporad E 2017 A method to improve the quality of 2.5 dimensional micro-and nanostructures produced by focused ion beam machining. *Micron* **101** 8–12
[2] Vysokik E *et al* 2021 Carbon tip aperture cantilevers: Fabrication & features in near-field magneto-optical imaging *Journal of Magnetism and Magnetic Materials* **529** 167837.

In situ Investigation of Individual Filament Growth in Conducting Bridge Memristor by Contact Scanning Capacitance Microscopy

M A Ryabova, D A Antonov, A V Kruglov, I N Antonov, D O Filatov and O N Gorshkov

Lobachevskii University of Nizhnii Novgorod, 603950 Nizhnii Novgorod, Russia

rmargo01@mail.ru

Abstract. We report on the application of Contact Scanning Capacitance Microscopy (CSCM) to trace the growth of an individual Ni filament in a ZrO_2(Y) film on a Ni sublayer (together with a conductive Atomic Force Microscope probe composing a nanometer-sized virtual memristor). An increasing of the filament length in the course of electro-forming results in an increasing of the capacitance between the probe and the sample, which can be detected by CSCM technique. This way, the filament growth can be monitored in real time *in situ*.

1. Introduction

In recent years, resistive switching (RS) effect in thin dielectric films has attracted much attention due to its potential application in the next-generation non-volatile computer memory (Resistive Random Access Memory, RRAM) devices [1]. The electronic devices, the operation of which is based on the RS effect are called *memristors* [2]. The RS is a reversible switching of resistance of a thin insulator of a capacitor-like stack under a voltage applied to the plates. In most cases, the RS mechanism is understood in terms of forming and rupture of conductive filaments (CFs) formed inside the insulator under the electric field between the plates. In the memristors based on metal oxides (HfO_x, NiO_x, TaO_x, etc.), SiO_x, GeO_x, etc., the CFs are composed of oxygen vacancies (V_Os) [3]. In so-called conducting bridge (CB) memristors, the CFs consist of metal atoms (e.g. Au, Ag, Cu, etc.) injected into the insulator (SiO_x, etc.) [4].

To date, such methods as Cross-Sectional Transmission Electron Microscopy (X-TEM) and Scanning Tunneling Microscopy (STM) [5] as well as Conductive Atomic Force Microscopy (CAFM) [6] have been applied to study the growth/rupture of the CFs *in situ*. In the present work, we applied Contact Scanning Capacitance Microscopy (CSCM) [7] to trace the growth of an individual Ni CF in a ZrO_2(Y)/Ni film.

2. Experimental part

The ZrO_2(Y) film (~12% mol. Y_2O_3) with a thickness $d \approx 10$ nm was deposited by radio-frequency (RF) magnetron sputtering using Torr International® **** vacuum setup onto an n^+-Si(001) substrate with a pre-deposited 10-nm thick Ni sublayer. The CSCM experiment was carried out in ambient air at room temperature with NT-MDT® Solver Pro™ AFM in the contact mode using NT-MDT® HA-HR DCP AFM probes with conducting diamond-like coated tips.

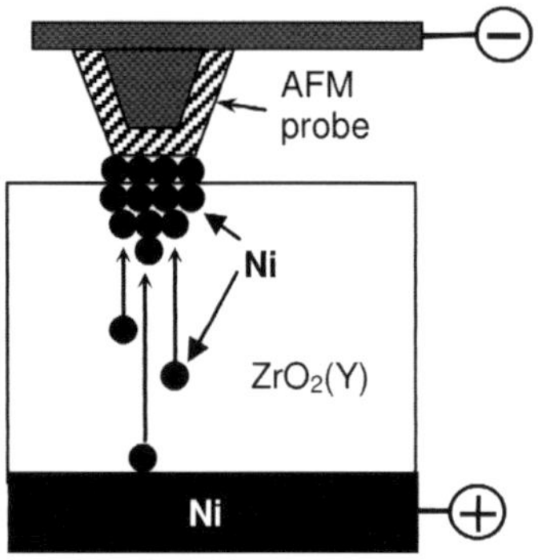

Figure 1. Schematic representation of the filament forming process in the $ZrO_2(Y)/Ni/Si(001)$ film under the biased CAFM probe.

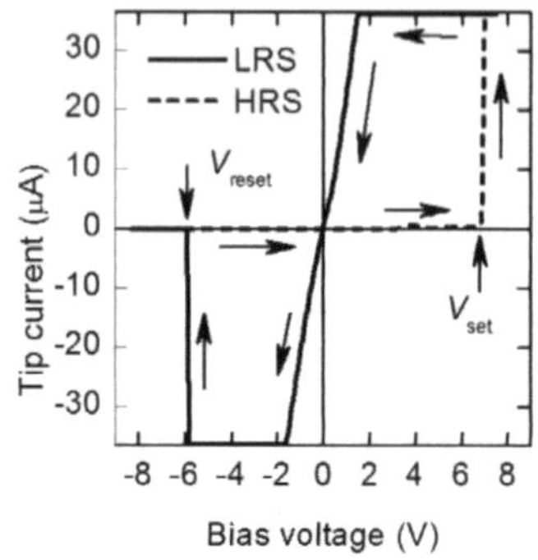

Figure 2. Cyclic *I*–*V* curve of the contact of CAFM probe to the $ZrO_2(Y)/Ni/Si(001)$ film.

The probe was brought into a contact with the $ZrO_2(Y)$ film surface, and a sawtooth voltage V_g (amplitude 6 V, period 6 s) was applied between the CAFM probe and the Ni sublayer. The forming of the Ni CF inside the $ZrO_2(Y)$ film started at a negative V_g applied to the CAFM probe with respect to the sample due to an electrochemical (anodic) oxidation of the Ni sublayer surface at the $Ni/ZrO_2(Y)$ interface, drift of the Ni ions in the electric field between the probe and the Ni sublayer, and reduction of the Ni ions at the growing filament tip (figure 1) [6].

3. Results and Discussion

Figure 2 shows typical cyclic current-voltage (*I*–*V*) curve of the contact of the CAFM probe to the $ZrO_2(Y)/Ni/Si(001)$ film surface. The *I*–*V* curve demonstrated a pronounced hysteresis typical for the CB memristors attributed to the forming and rupture of the Ni filaments inside the $ZrO_2(Y)$ film under the voltage V_g applied between the CAFM probe and the Ni conducting sublayer.

Figure 3 shows a surface topography image (figure 3a) and a current image (a map of the current flowing through the CAFM probe I_t as a function of the in-plane probe coordinates *x*, *y*, figure 3b) acquired simultaneously with the surface topography with the bias voltage V_g = 4 V applied to the CAFM tip with respect to the Ni sublayer. Prior to this measurement, a single Ni filament was formed in the center of the frame by applying a triangular pulse of $V_g < 0$ (amplitude 6 V, duration 3 s) to the CAFM probe positioned in an arbitrary point on the sample surface.

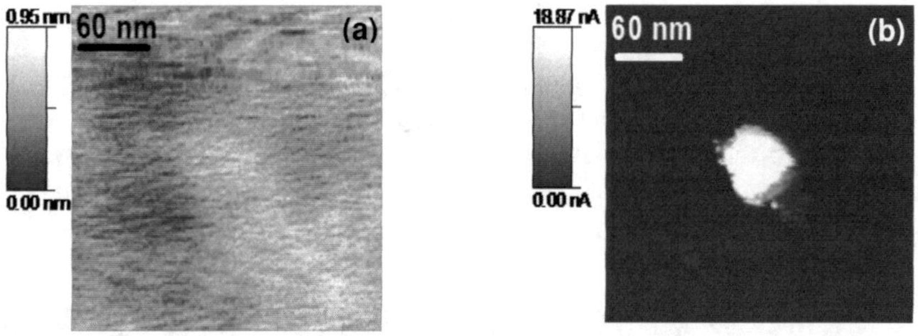

Figure 3. Topography (a) and current image (b) of the $ZrO_2(Y)/Ni/Si(001)$ film surface after forming an individual Ni filament. $V_g = 4$ V.

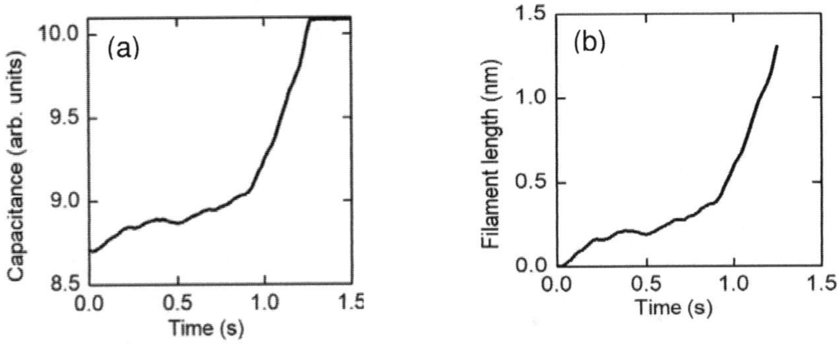

Figure 4. Kinetics of the capacitance $C_{CF}(t)$ (a) and of the CF length l_{CF} (b) extracted from the $C_{CF}(t)$ dependence.

A local increasing of I_t in the center of the frame in figure 3b was attributed to the Ni CF formed between the ZrO_2(Y) film surface and the Ni sublayer: once the CAFM probe encounters the upper end of the Ni CF, an increased current flows between the CAFM probe tip and the Ni sublayer directly through the highly conductive Ni CF. Note that a small hillock (of ~ 1 nm in height) was observed in figure 3a at the place of the local increasing of I_t attributed to the accumulation of the Ni atoms under the CAFM probe tip in the course of the Ni filament forming (as shown schematically in figure 1).

Figure 4a shows the kinetics of the capacitance between the CAFM probe and the sample $C_{CF}(t)$ (t is the time) measured in the course of the forming process using an add-on for CSCM to NT-MDT® Solver Pro™ AFM.

The increasing of C_{CF} was attributed to the increasing of the CF length l_{CF} in the course of forming (as shown in figure 1). A dependence of C_{CF} on l_{CF} can be expressed in the flat capacitor approximation as

$$C_{CF} = \frac{\varepsilon\varepsilon_0 S}{d - l_{CF}}, \qquad (1)$$

where S is the effective filament cross-section area, ε is the dielectric permittivity of the insulator material, and ε_0 is the electric constant. As $l_{CF} \to d$, $C_{CF} \to \infty$. In the experiment, the overload of the CSCM add-on lock-in amplifier took place at $t \approx 1.25$ s (figure 4a).

Figure 4b shows a dependence $l_{CF}(t)$ calculated from the one $C_{CF}(t)$ according to (1).

Once the dependence $l_{CF}(t)$ was obtained, one can calculate many other characteristics of the filament growth process. For example, a time dependence of the filament growth rate $v_g(t) = dl_{CF}(t)/dt$ is shown in figure 5a. The dependence $v_g(t)$ shown in figure 5c was calculated from the one $l_{CF}(t)$ in figure 4b by numerical differentiation with non-linear smoothing (7-point adjacent averaging). One can see v_g to be almost constant at $t < 0.8$ s. Then v_g increased rapidly and saturated at $t > 1$ s.

Figure 5c shows a dependence of v_g on the electric field strength in the gap between the CF tip and the Ni sublayer F. The latter was calculated as

$$F(t) = \frac{V_g}{d - l_{CF}(t)}. \qquad (2)$$

One can see in figure 5b that v_g was almost constant at $F < 3\cdot10^6$ V/cm, increased abruptly at $F \approx 3\cdot10^6$ V/cm, and saturated again at $F > 5\cdot10^6$ V/cm. Such a behavior of the dependence $v_g(F)$ can be interpreted as follows. At low F (that corresponds to low l_{CF} and low V_g) v_g is limited by the rate of arrival of the Ni ions (dispersed inside the ZrO_2(Y) film) to the growing Ni filament tip. However, at higher F (corresponding to higher V_g), the electrochemical reaction of anodic oxidation of the Ni sublayer at the Ni/ZrO_2(Y) interface starts [8]. In this case, v_g is limited by the anodic oxidation rate because all Ni ions generated at the Ni/ZrO_2(Y) readily arrive at the Ni filament tip.

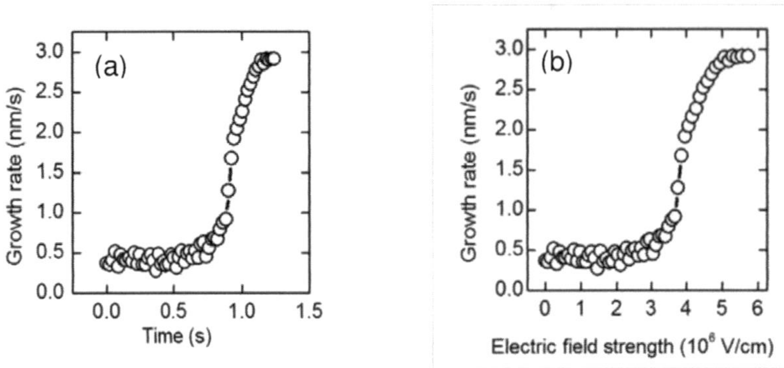

Figure 5. Kinetics of the CF growth rate $v_g(t)$ (a) and dependence of v_g on the electric field strength in the gap between the Ni CF tip and the Ni sublayer F (b) extracted from the $C_{CF}(t)$ dependence.

To date, a lot of mathematical models for the CF growth have been published in the literature starting from the simplest current-driven memristor model [3]. However, very few studies on in situ monitoring CF (ususlly by X-TEM) have been published because such studies are extremely difficult and expensive. The results of the present study show the prospects of the CSCM technique in the *in situ* studies of the CF growth. This technique provides the information on the dynamics of the CF sizes, which can be compared to the modeling results directly. It should be stressed that such experiments can be carried out in ambient air. Also, more realistic models of the CF shape (e. g. a conical CF) can be used in future studies.

4. Conclusion

In the present study, we have demonstrated experimentally a capability of CSCM technique to trace the growth of the CF in the virtual memristors composed of the CAFM probe contact to the functional dielectric film on the conductive substrate and, moreover, to measure the time dependence of the CF dimensions, at least, semi-quantitatively. In this scope, the CSCM technique worked out in the present study seems to be very promising for application in future studies of the CF dynamics during the forming and RS processes in various material systems.

Acknowledgments

The present work was support Russian Foundation for Basic Research (20-02-00830). The CAFM investigations were carried out using the shared research facilities of Research and Educational Centre for Physics of Solid State Nanostructures at Lobachevskii University of Nizhnii Novgorod.

References

[1] Amiri P K (Ed.) 2020 *Emerging Memory and Computing Devices in the Era of Intelligent Machines* (MDPI AG).
[2] Strukov D B, Snider G S, Stewart D R and Stanley-Williams R 2008 *Nature* **453** 80
[3] Riess I 2017 *J. Electroceram.* **39** 61
[4] Lee S H, Zhu X and Lu W D 2020 *Nano Research* **13** 1228
[5] Yang Y, Takahashi Y, Tsurumaki-Fukuchi A et al. 2017 *J. Electroceram.* **39** 73
[6] Lanza M 2017 *Conductive Atomic Force Microscopy: Applications in Nanomaterials* (Wiley-VCH)
[7] Williams C C, Hough W P and Rishton S A 1989 *Appl. Phys. Lett.* **55** 203
[8] Lubben M and Valov I 2019 *Adv. Electron. Mater.* 1800933

Research Of Dynamic Characteristics Of A Three-axis Micromechanical Gyroscope-Accelerometer

I E Lysenko, D Y Sevostyanov, N F Kidyaev and A V Kovalev

Southern Federal University, Taganrog 347900, Russia

Abstract. In this paper, an original design of a micromechanical gyroscope-accelerometer is proposed. A parametrizable geometric and finite element model of a micromechanical device is presented. The dynamic characteristics of the structure are studied and the dependences of the natural oscillation frequencies on the geometric parameters of the gyroscope-accelerometer are obtained.

1. Introduction

The market for modern electronics puts developers in front of the need to improve weight and size characteristics, increase energy efficiency and reduce the cost of devices being developed. Micromechanical (MEMS) devices are actively used in many systems, such as virtual reality devices, medical technology and aerospace navigation systems [1, 2]. The high demand for new devices requires labor-intensive research and development work, the use of new materials, the improvement of geometry, the use of new physical principles. All of this is done to meet the ever-growing customer requirements for increased sensitivity and sensitivity range [3]. Therefore, the analysis of dynamic characteristics and the development of a micromechanical gyroscope-accelerometer with three axes of sensitivity, which makes it possible to measure the linear velocity and angular acceleration, and which is a single device made on one substrate in one technological cycle, is an urgent task [4, 5].

2. Design and simulation

The developed original design of a micromechanical gyroscope-accelerometer with three axis of sensitivity is shown in Figure 1 [5, 6].

This device has the ability to measure the values of angular velocities and linear acceleration along the X- and Y- axis located mutually perpendicular in the plane of the substrate, and the Z-axis directed perpendicular to the plane of the substrate. A special feature of the design is two inertial masses located one inside the other. It is assumed that this will improve the accuracy of recording linear acceleration along the Z-axis and reduce the values of the natural frequencies of oscillations along the X- and Y-axis.

The ANSYS program system was used for the research. In particular, the Mechanical APDL module was used to analyze the dynamic characteristics of the developed gyroscope-accelerometer. In it, a geometric model of a gyroscope-accelerometer was built and a finite element mesh was imposed, as shown in Figure 2.

The dependence of the dynamic characteristics on the geometric parameters of the micromechanical gyroscope-accelerometer is shown in Figure 5. Figure 5a shows the dependence of the natural frequencies of oscillations of the micromechanical gyroscope-accelerometer on the size of the inertial mass. Figure 5b shows the dependence of the natural frequencies of oscillations of the micromechanicalgyroscope-accelerometer on the height of the structural layer.

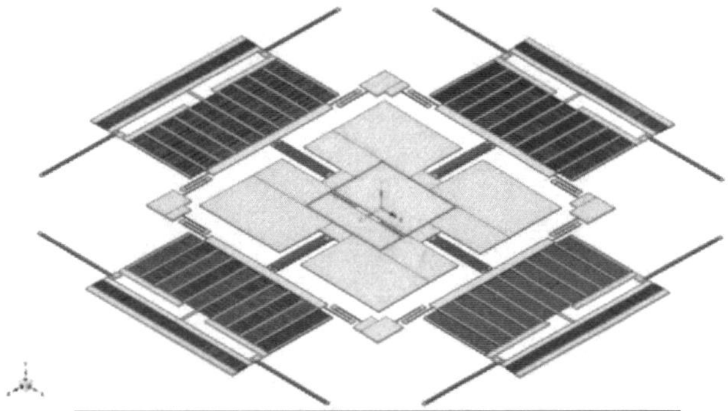

Figure 1. Design of the developed micromechanical gyroscope-accelerometer.

Figure 2. Finite element model of a micromechanical gyroscope-accelerometer.

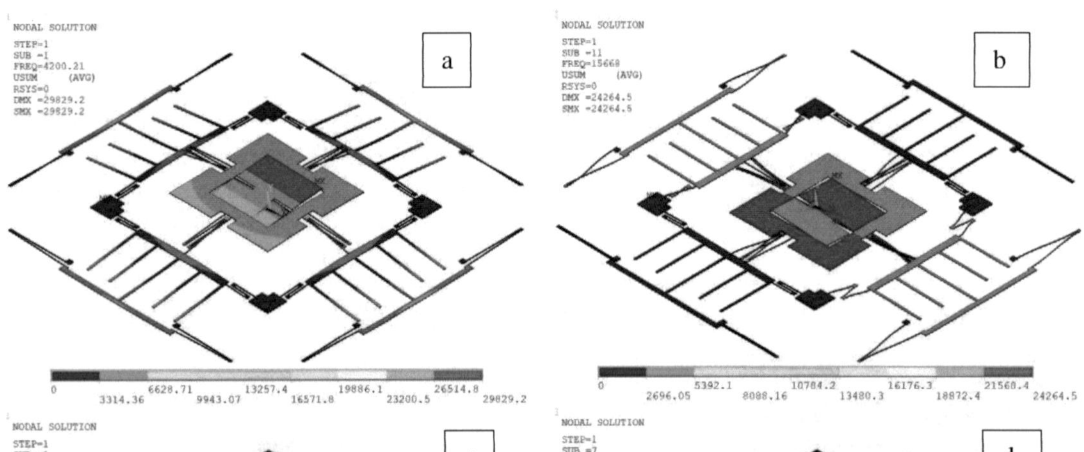

Figure 3. The results of modal analysis of changes in the geometric parameters of the external inertial mass, **a** along the z-axis and **b** along the x-axis with the size of the inertial mass 500x500 mkm, **c** along the z-axis and **d** along the x-axis with the size of the inertial mass 1700x1700mkm

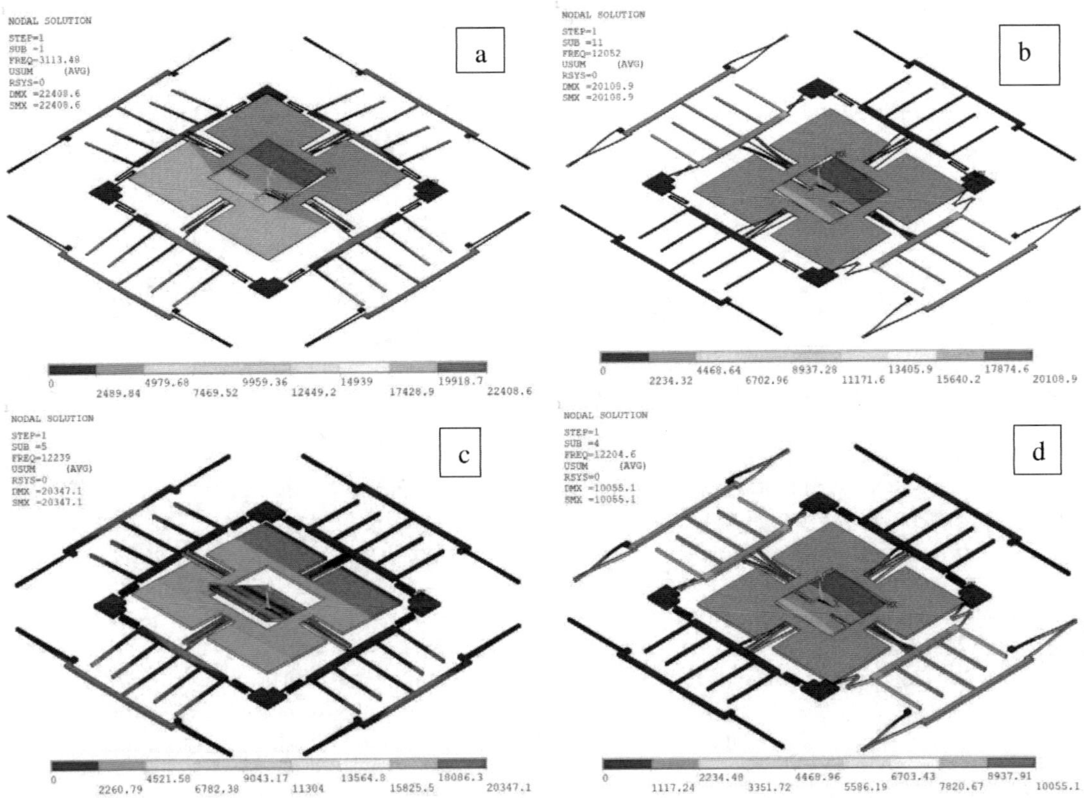

Figure 4. Results of modal analysis of changes in the height of the structural layer, **a** along the Z- axis at a structural layer height of 3 mkm, **b** along the X axis at a structural layer height of 3 mkm, **c** along the Z axis and **d** along the X axis at a structural layer height of 12 mkm

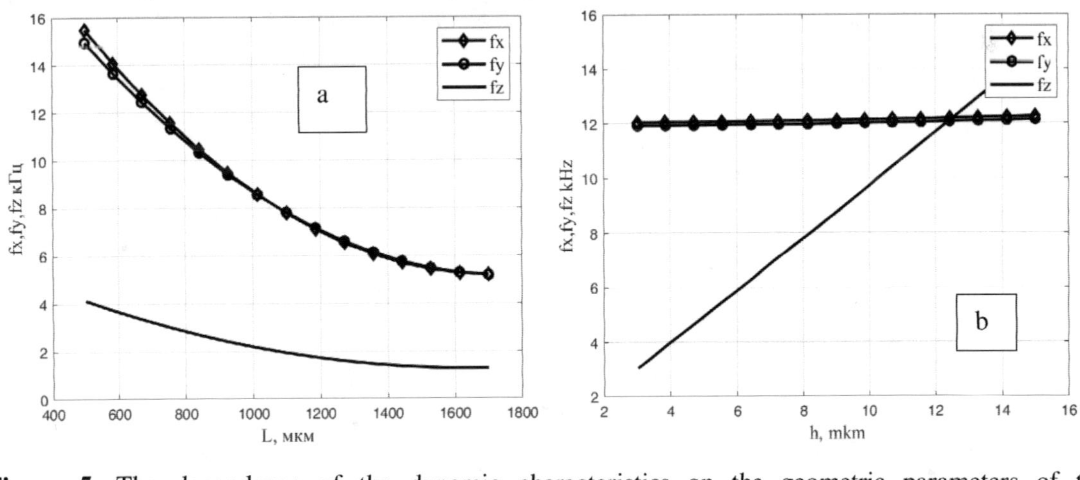

Figure 5. The dependence of the dynamic characteristics on the geometric parameters of the micromechanical gyroscope-accelerometer **a** form the size of the inertial mass, **b** from the height of the structural layer.

3. Simulation results

The results of modal analysis of a micromechanical gyroscope-accelerometer with various geometric parameters of inertial mass are shown in Figure 3.

The results of modal analysis of a micromechanical gyroscope-accelerometer with different values ofthe height of the structural layer are shown in Figure 4.

4. Conclusion

In this paper, the design of a micromechanical sensor for linear accelerations and angular velocities was proposed, and the dynamic characteristics of a gyroscope-accelerometer were modeled. The simulation results revealed the dependence of the natural frequencies of vibrations on the geometric parameters ofthe micromechanical device.

Acknowledgments

The work was carried out at the expense of funds, task No. FENW-2020-0022 for the implementation of scientific research carrying out scientific research at the expense of the federal budget, in terms of scientific activities on the topic "Development and research of methods and means of monitoring, diagnostics and forecasting state of engineering objects based on artificial intelligence".

References

[1] Raspopov V Ja 2007 *Micromechanical devices* (Tula: Tula state university)

[2] Lysenko I E, Ezhova O A, Tkachenko A V and Naumenko D V 2020 Analysis of the micromechanical three-axis accelerometer *Book of Abstracts of 7th International School and Conference «Saint Petersburg OPEN 2020» on Optoelectronics, Photonics, Engineering and Nanostructures* (St. Petersburg: St. Petersburg Academic University) pp 449–50

[3] Lysenko I E, Naumenko D V and Ezhova O A 2020 Design and simulation high aspect ratio torsion suspension of MEMS z-axis accelerometer *Book of Abstracts of 7th International School and Conference «Saint Petersburg OPEN 2020» on Optoelectronics, Photonics, Engineering and Nanostructures* (St. Petersburg: St. Petersburg Academic University) pp 478–79

[4] Lysenko I E 2019 The results of modelling of a micromechanical accelerometer *Journal of Physics: Conference Series* **1410** pp 226-*32*

[5] Lysenko I E, Ezhova O A, Sevostyanov D Y and Konoplev B G 2020 Integral micromechanical gyroscope–accelerometer Patent of invention No. 2716869.

[6] Ageev O A and Konoplev B G 2019 *Nanotechnology in microelectronics* (Moscow: Nauka) p.511

Probing electric properties of GaP nanowires with Kelvin probe force microscopy

V A Sharov[1,2], P A Alekseev[2], V V Fedorov[1] and I S Mukhin[1,3]

[1] Alferov University, Khlopin St. 8/3, Saint-Petersburg 194021, Russia
[2] Peter the Great Saint-Petersburg Polytechnic University, Polytechnicheskaya st. 29, Saint-Petersburg 195251, Russia
[3] ITMO university, Kronverksky avenue 49, Saint-Petersburg 197101, Russia

vl_sharov@mail.ru

Abstract. Surface electronic properties of GaP nanowires were investigated using scanning probe force microscopy. I-V curves of individual free-standing NWs with different doping types were obtained. Surface Fermi level positions in the nanowires of different crystal phases and doping types were extracted using phase-modulated Kelvin probe force microscopy. The results indicate on weak Fermi level pinning in GaP nanowires. The difference between wurtzite and zinc blende GaP work function is observed.

1. Introduction

Semiconductor nanowires (NWs) stay in focus of attention. NW Mechanical strength and the ability to be grown on highly mismatched substrates as well as significant progress in growth and characterization techniques opened paths to fabrication of a wide range semiconductor NW-based devices in nanoelectronics, photovoltaics, sensing and energy harvesting. In this case, comprehensive understanding of NW electronic transport is of interest. Surface phenomena play critical role in the transport properties of low-dimensional semiconductor structures. In the case of nanowires, the surface effects are especially critical due to the high surface to volume ratio. Termination of the lattice periodicity leads to the formation of dangling bonds on the surface. These bonds, coupled with surface defects and native oxide states, accounts for surface electronic states inside energy gap. At a high density of surface states, their influence on the surface conductivity becomes dominant. In particular, presence of surface states leads to Fermi level pinning. In this case, the conductivity of the III-V NWs with unoxidized surface is governed mainly by surface states. Conventionally, III-V NW surface is terminated by (110) facets. Surface electronic properties of III-As semiconductors are widely investigated via Kelvin probe force microscopy (KPFM), scanning tunneling spectroscopy (STS) and other characterization techniques. Fermi level in GaAs and in InAs (110) surface with native oxide layer is pinned in the midgap and near the conduction band respectively, resulting in different transport properties. The surface Fermi level does not coincide with the bulk level, which leads to the transfer of carriers and energy band bending. Midgap-pinned Fermi level leads to the formation of near-surface depleted area whose width is governed by doping level ND, NW diameter a, surface state density, and work function. The difference between E_F in n + and p + (110) GaAs doesn't exceed 200 meV [1] due to strong pinning effect so the conductivity of thin NWs is terminated.

Early studies of electronic surface states on UHV cleaved GaP (110) revealed Fermi level pinning 1.5-2.1 eV above the valence band depending on cleavage-induced surface states. Unlike other III-Vs, the tail of intrinsic surface band lies in the band gap in case of GaP [2]. To quantitatively analyze transport properties of GaP NW, one should know a position of the surface Fermi level pinning (work function) as it is necessary for calculating the Schottky barrier height and optimizing ohmic contacts. A strong Fermi level pinning effect induces high contact resistance and obstructs realization of ohmic contact. However, surface Fermi level position in III-P NWs with native oxide layer is still poorly studied. Gallium phosphide (GaP) NWs are prospective for betavoltaic devices and photodetectors, while wurtzite GaP is promising for green-range optoelectronics due to direct band gap. GaP window is often used as a window layer in AlInGaP LEDs [3]. For designing GaP NW-based devices with controllable electronic parameters, the rate of pinning effect and the work function value have to be obtained. GaP NWs are still poorly studied from this point of view.

2. Results and discussion

In phosphides and arsenides, it has been shown that the appearance of the wurtzite phase is restricted to the nanowire form and is never observed in the bulk. Wurtzite III-V NWs are attracting a great interest in opto- and nanoelectronics. The influence of the crystal phase on transport properties is questionable.

It was shown, that the work function of (110) ZB and (10¯10) WZ InAs NWs differs at least by 50 meV [4]. In this work, we discuss the effect of the crystalline phase on the work function of GaP NWs. We obtain WF values of ZB and WZ i-GaP as well as n- and p-GaP. Fermi level pinning is studied with respect to NW diameter. GaP nanowires were synthesised in self-catalysed mode on Veeco Gen-III MBE setup on (111) Si substrates. Gallium drops were used as catalytic caps instead of Au in order to improve crystal quality. Three samples with different doping type and crystal structure were studied. Sample 1 is 5 μm long i-GaP (see figure 1, a). Sample 2 is silicon-doped (see figure 1, b). Silicon is known to be n-dopant in gap. Sample 3 is p-GaP doped with Be. NW crystal structure of sample 1 was studied with transmission electron microscopy (TEM) and turned out to be polytypic with mainly twinned zinc blende crystal structure and wurtzite base and top (see inset in figure 1, a). Self-catalysed VLS mode, which has been realized mainly in molecular beam epitaxy (MBE), allows one to modify the droplet volume during growth by changing the ratio of the fluxes of group III and group V species. During the end of growth process Ga droplet was expended resulting in the formation of 500 nm long wurtzite (WZ). WZ phase in III-V nanowires is being formed in case of triple line nucleation – at specific volume and contact angle of Ga droplet [5]. Thus, WZ segments in NW bases may be formed due to droplet formation while that in tops emerged due to it consumption. Sample 2 and 3 possess zinc blende crystal structure.

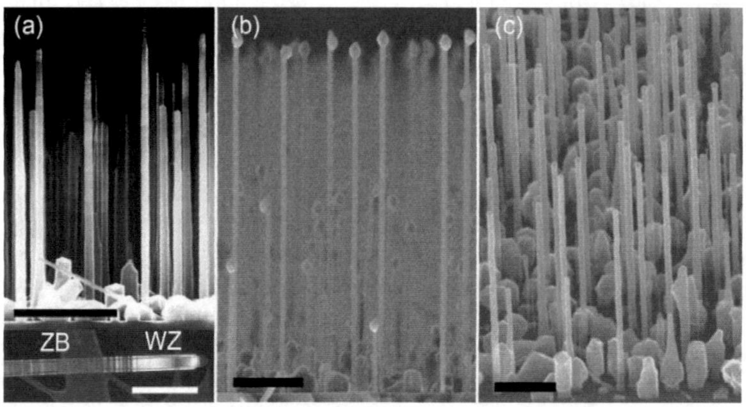

Figure 1. (a) SEM image of sample 1 (polytypic WZ/ZB i-GaP NWs), TEM image of the NW top is shown in the inset; (b) and (c) SEM images of sample 2 (n-GaP ZB NWs) and sample 3 (p-GaP ZB NWs).

WZ inset on a big number of NWs on sample 1 was confirmed by micro-Raman mapping. For mapping we used Horiba Lab RAM setup with 532 nm excitation laser and x100 lens. 1D Raman scan along a NW is shown in figure 2. The 1D profiling revealed strong signal from base and top of NW caused by waveguiding effect. Black and red curves represent Raman spectra from base and top respectively. Left shoulder on transverse optical mode at 357 cm^{-1} corresponds at the presence of WZ phase and is observed only near the NW top. WZ phase is absent relating to samples 2 and 3 because the droplet wasn't consumed because of rapid temperature decreasing in the end of the growth.

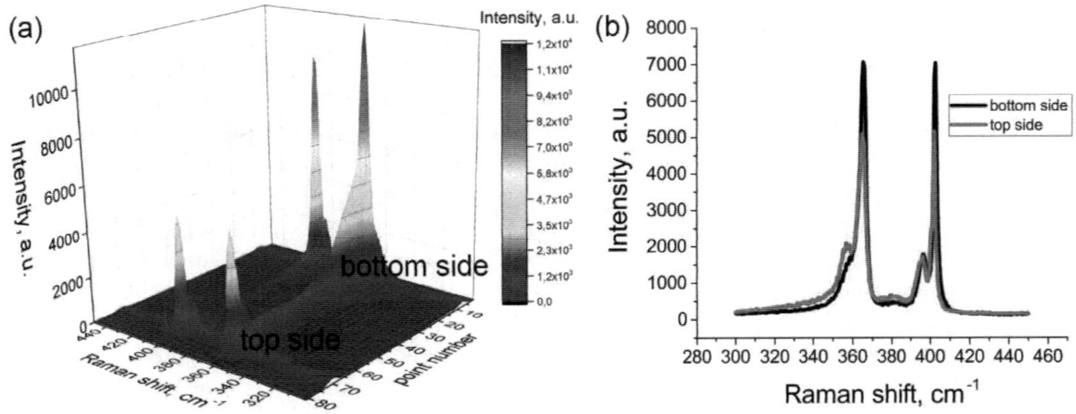

Figure 2. Micro-Raman mapping of ZB i-GaP nanowire with WZ inset. Left: 1D micro-Raman profile along a NW. Right: Raman spectra from bottom (black) and top (red) NW ends.

We check individual NW conductivity with conductive atomic force microscopy (c-AFM) study by obtaining its I-V curve. The first conductive electrode was created using conductive AFM probe being accurately positioned on the NW top. The second one was formed between the highly doped Si growth substrate and an AFM sample holder. scheme of this experiment and the results for sample 2 are shown in figure 3, a. The curve slope as well as knee and breakdown voltage values correspond at n-type doping. The I-V curve from individual NW from sample 3 is shown in figure 3, b.

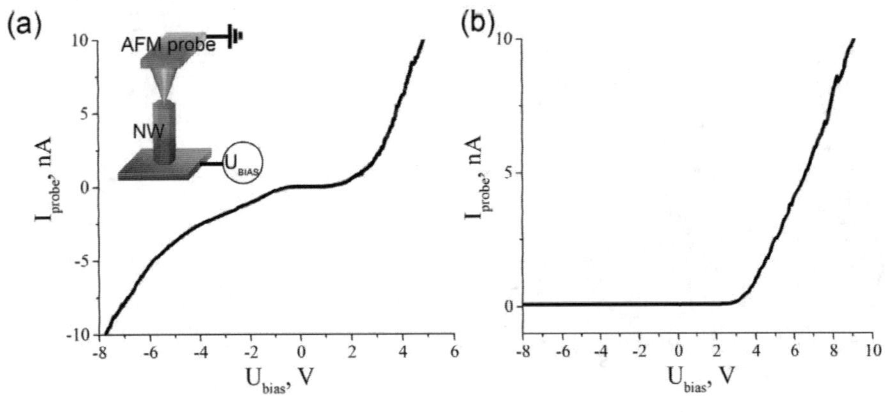

Figure 3. I-V curves from individual n- (a) and p- (b) doped GaP nanowires

Kelvin probe force microscopy (KPFM) is specializing in structural and electronic surface characterization of semiconductors. KPFM allows mapping contact potential difference (CPD) between a sample and a metallic probe. Knowing the probe work function, one can calculate the sample work function from CPD. KPFM is typically utilized for investigating planar structures such as p-n junctions and segnetoelectrics. Recently, a publication activity over nanowire-related KPFM studies increased. GaN and InP NWs with axial p-n junction were investigated [6-8]. It was shown that the Fermi level at oxidized cleavage surfaces of AlGaAs and GaInAs is pinned at the same position, which is attributed to excess surface arsenic [9] CPD value and band bending of individual GaN and Si nanowires were studied with respect to diameter [10]. These studies prove that KPFM spatial resolution allows distinguishing and qualitative analysis of work function of n- and p- regions in a single nanowire. However, the effect of crystal phase on work function is more delicate, and convenient amplitude-modulated (AM) KPFM that was used in these works seems to lack spatial sensitivity for its registration.

KPFM measurements were performed on NT-MDT NTegra Aura setup under low-vacuum ($5 \cdot 10^{-5}$ Bar) conditions. The NW arrays were transferred from the growth substrates on the surface of freshly cleaved highly oriented pyrolytic graphite (HOPG). HOPG work function (4.48 eV) was well defined [11] which is helpful for quantitative data analysis. The main idea of KPFM may be summarized to utilizing relations between electrodynamic force between probe and sample and the probe amplitude and phase. We utilize double-pass phase-modulated (PM) KPFM, which is described in detail in [12]. Firstly, the morphology was acquired using the semi contact mode. Secondly, based on the morphology profile obtained, the tip was retracted from the surface and vibrated at a low amplitude. Then, the dc tip to sample polarization was applied only during the second pass and frequency or phase shifts were detected. Thus, constant distance explorations of long-range force gradients can be achieved and linear conditions can reasonably be obtained. Gradient KPFM techniques have much higher spatial resolution which is crucial for nanostructures analysis. We used rigid TipsNano HA NC/W2C probes (force constant k=12 N/m, resonant frequency f=235 kHz). After sampling several probes, we have found that rigid ones have better spatial resolution and signal-to-noise ratio in KPFM due to higher quality factor.

AFM topography and CPD scan of sample 1 are shown in figure 4. CPD value on HOPG is stabilized at -0.05 V so one can estimate the probe work function as φ_{probe}=4.48-0.05=4.43 eV. The CPD value significantly varies along the NW. Thus, in the bottom side of NW (top side of scan) it amounts to -0.1 V, against 0 V and 0.15 V at the middle and at the very end of NW. We associate the increased value of CPD at the NW top with WZ crystal structure. In this case, KPFM measurements are in agreement with TEM, CL and micro-Raman mapping. Surface potential gradient along the NW may be connected with gradient of flat defects density arising from changed growth conditions. Thus, work function of ZB and WZ GaP are 4.43 and 4.28 eV respectively. The electron affinity in GaP is 3.8 eV, so for both cases surface Fermi level is pinned near the middle of band gap.

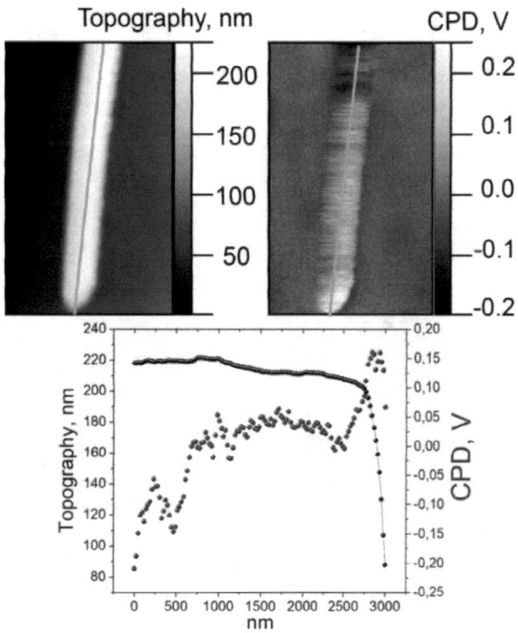

Figure 4. AFM topography and CPD scans of ZB i-GaP nanowire with WZ inset.

KPFM results for sample 2 are shown in figure 5. CPD value on n-GaP is 0.6 V and doesn't significantly change along the NW. Thus, we estimate n-GaP work function as 5.0 eV. The significant difference between i- and n- GaP WF indicates on weak Fermi level pinning in GaP in comparison with GaAs.

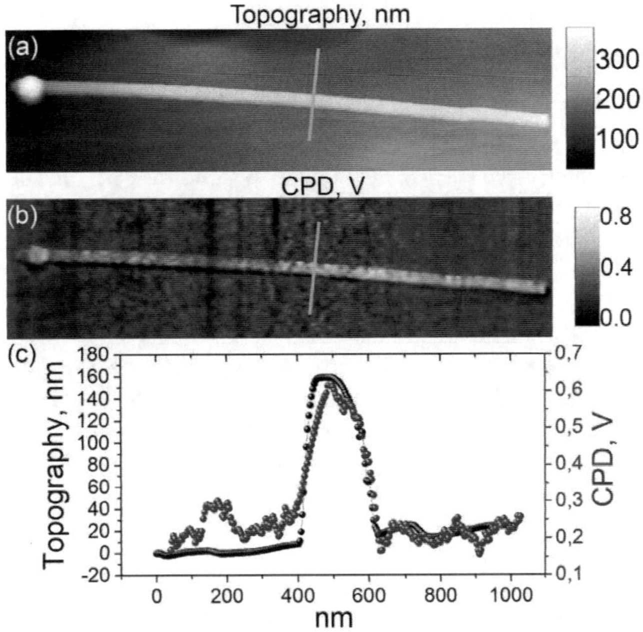

Figure 5. AFM topography and CPD scans of n-GaP nanowire

KPFM results for sample 3 are shown in figure 6. CPD value on p-GaP is 0.05 V and doesn't significantly change along the NW. The increased value of CPD near the NW base in the bottom of the scan may be attributed to growth instability. Thus, we estimate p-GaP work function as 4.33 eV that is close to the data obtained on i-GaP indicating on low Be-doping level.

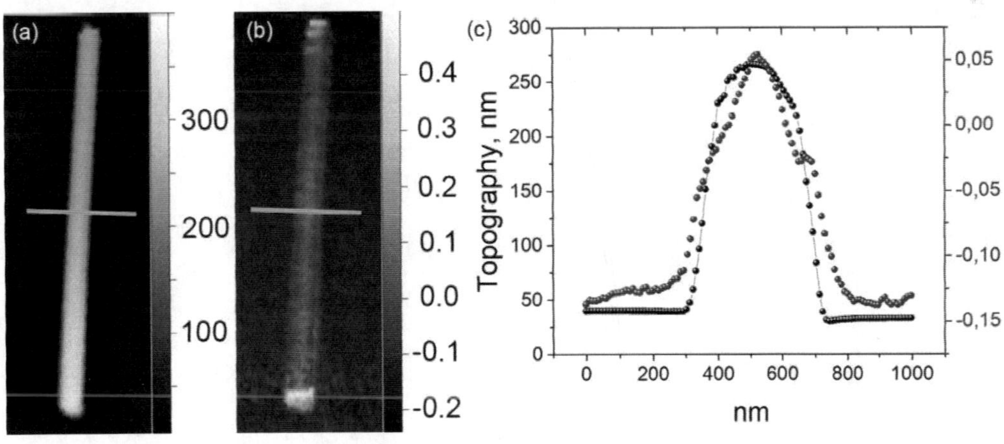

Figure 6. AFM topography and CPD scans of p-GaP nanowire

3. Conclusion

In this work we present a study of surface potential of GaP nanowires with different crystal structure and doping type. We observe different work function between wurtzite and zinc blende GaP (4.43 and 4.28 eV respectively). The response from nanoscale regions possessing different crystal phases within the same nanowire was quantified due to high spatial resolution of force gradient Kelvin probe force microscopy. Work function values for n- and p- doped ZB GaP were 5.0 eV and 4.33 eV respectively. Significant difference between the work function with respect to doping type indicates on relatively weak pinning in gallium phosphide.

Acknowledgments
The work was supported by the ministry of science and higher education of the Russian Federation (0791-2020-0005).

References

[1] Alekseev P, Dunaevskiy M, Cirlin G, Reznik R, Smirnov A., Kirilenko D, Davydov V and Berkovits V 2018 *Nanotechnology* **29** 314003
[2] Guichar G, Sebenne C and Thuault C 1979 *Journal of Vacuum Science and Technology* **16** 1212-15
[3] Tseng M, Chen C, Lai N, Chen S, Hsu T, Peng Y and Horng R 2014 *Optics express* **22** A1862-67.
[4] Hjort M, Lehmann S, Knutsson J, Zakharov A, Du Y, Sakong S, Rainet T, Nylund G, Lundgren E, Kratzer P, Dick K and Mikkelsen A 2014 *ACS nano* **8** 12346-55
[5] Dubrovskii V 2015 *The Journal of chemical physics* **142** 204702.
[6] Minj A, Cros A, Auzelle T, Pernot J and Daudin B 2016 *Nanotechnology* **27** 385202
[7] Sun X, Wang X, Wang P, Sheng B, Li M, Su J, Zhang J, Liu F, Rong X, Xu F, Yang X, Qin Z, Ge W and Shen B 2017 *Optical Materials Express* **7** 904-12
[8] McKibbin S, Colvin J, Troian A, Knutsson J, Webb J, Otnes G, Dirscherl K, Sezen H, Amati M, Gregoratti L, Borgstrom M, Mikkelsen A and Timm R 2019 *Nano letters* **20** 887-95
[9] Parida S, Sahoo A, Madapu K, Jaya S and Dhara S 2020 *Applied Surface Science* **510** 145502
[10] Koren E, Hyun J, Givan U, Hemesath E, Lauhon L and Rosenwaks Y 2011 *Nano letters* **11** 183-7
[11] Hansen W and Hansen G 2001 *Surface science* **481** 172-84
[12] Portes L, Girard P, Arinero R and Ramonda M 2006 *Review of Scientific Instruments* **77** 096101

Study of the hardness distribution after induction heat treatment of titanium over the surface and the cross-section

A Shchelkunov, I Egorov and A Fomin

Yuri Gagarin State Technical University of Saratov, Saratov 410054, Russia

shelkynov.a_94@mail.ru

Abstract. In this work, the mechanical properties (microhardness) of a titanium disk after induction heat treatment (IHT) were studied. The influence of the processing parameters (inductor current and temperature) on the distribution of microhardness over the cross-section of the experimental samples was established.

1. Introduction

Currently, there is a growing demand for the use of titanium and its alloys in various industries, as well as in the field of biomedicine [1]. For this reason there is a need in studying the ways to modify its structure. The problem of using commercially pure titanium in medicine is associated with low values of tribological characteristics [2,3]. One of the options for increasing physical and mechanical parameters, in particular hardness, and reducing the friction coefficient is the formation of oxide-ceramic structures by chemical-thermal treatment [4,5]. Such structures can be obtained by methods of physical or chemical vapor deposition (PVD and CVD), ion implantation, nitriding, carburizing or laser surface treatment in various atmospheres. Examples include carbide, nitride, oxide structures and their combinations.

To obtain nanostructured coatings with enhanced physical and mechanical properties, the Aerosol Assisted Chemical Vapor Deposition (AACVD) method is used [6]. This method can be used to obtain nanostructured morphology. Achieving durable, wear-resistant and biocompatible coatings is also possible using the plasma electrolytic oxidation method.

2. Methodology

The samples to be studied were prepared from commercially pure titanium (VT1-00) and had cylindrical shapes (diameter 14 mm, thickness 2 mm). Prior to the induction heat treatment (IHT), the surface of the samples was processed by stepwise grinding by abrasive paper (P400 – P2500) until the surface roughness Ra = 0.16–0.32 μm was obtained. Another stage preceding IHT included the washing of the samples in ethanol.

IHT of titanium disks was performed using "HF-15A" setup. The treatment comprised three stages: intense heating, exposure and free cooling in the open air. The main processing parameters were the inductor current, temperature and processing duration. The temperature range corresponded to T = 900–1500 °C (±50 °C) with a duration t = 1–3, 60, 120, 300 s. The inductor current value ranged from 2.5 to 3.5 kA.

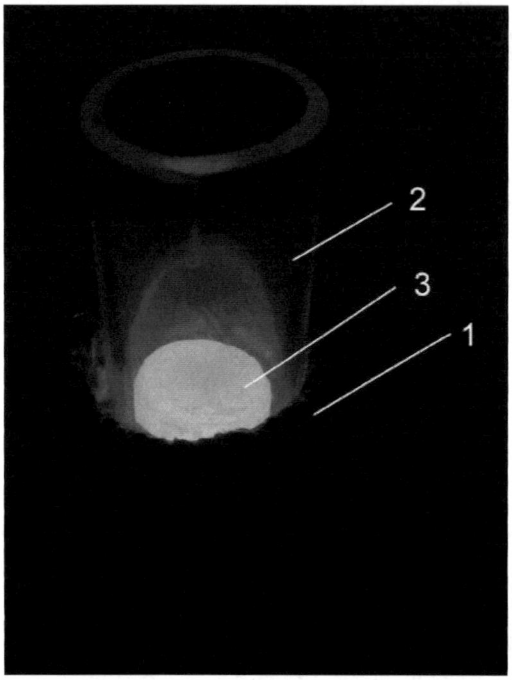

Figure 1. The process of induction heat treatment: 1 – an inductor, 2 – a quartz tube, 3 – a sample.

The microhardness over the surface and the cross-section of the sample was measured using "PMT-3" hardness tester (GOST 9450-76). The load on the indenter was 0.19 N. A microsection was made to measure the microhardness over the cross-section of the sample. The beginning of the measurement (from the surface of active saturation with air oxygen) was performed stepwise (50, 150 200, 300, 400, 500, 750, 1000 μm). The temperature during IHT was controlled by thermal imaging analysis and colorimetric method (using a heat color scale). The operating current was measured using "UNI-T UT205" current clamp.

3. Results

The samples subjected to IHT were divided into three groups: low temperature (900–1100 °C), medium temperature (1150–1250 °C), and high temperature ones (1300–1500 °C). The surface microhardness in the low-temperature samples varied from 5–8 GPa to 8–11 GPa. The cross-sectional data were lower than those obtained on the surface and equaled from 7.6 to 2.6 GPa.

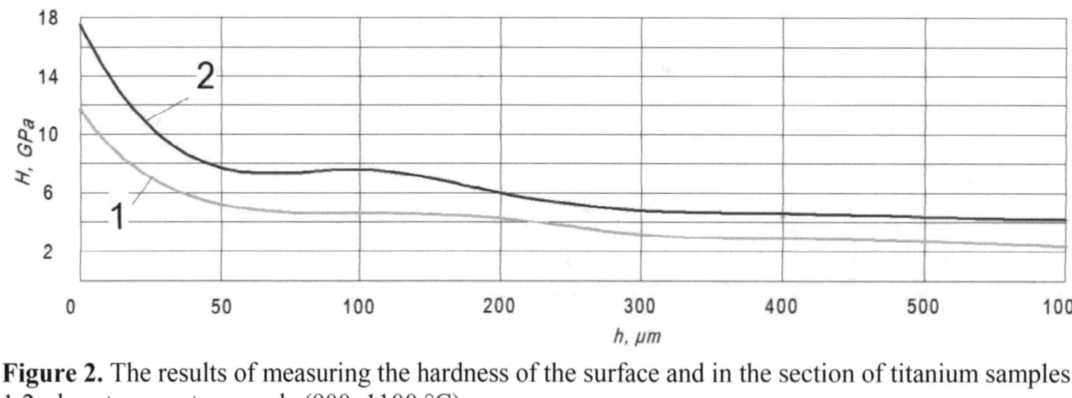

Figure 2. The results of measuring the hardness of the surface and in the section of titanium samples: 1,2 - low-temperature mode (900–1100 °C)

For the medium temperature mode, the surface microhardness varied from 8–9 GPa to 16 GPa. At high temperature IHT (T = 1200±50 °C and t = 300 s), areas with a hardness of up to 20 GPa were found. In the study of the microhardness over the cross section, the values varied from 10.6 to 2.6 GPa.

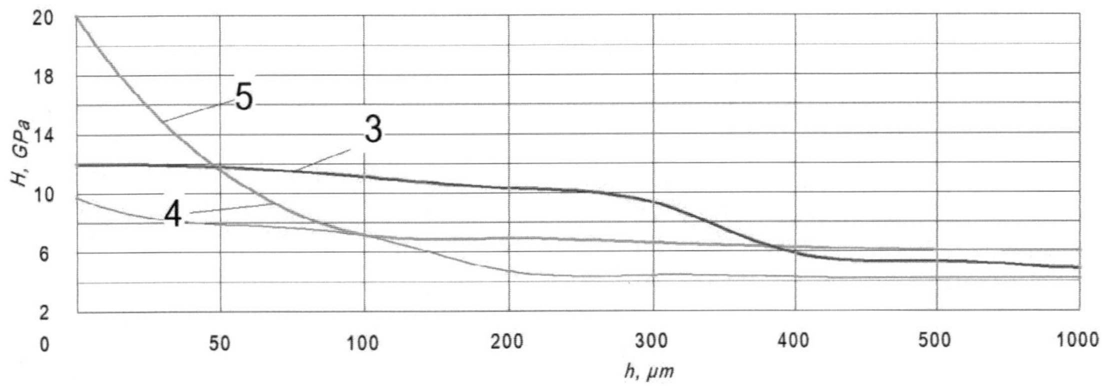

Figure 3. The results of measuring the hardness of the surface and in the section of titanium samples: 3,4,5 - medium-temperature mode (1150–1250 °C)

High temperature data were obtained for the surface measurements in the range of 8–14 GPa with the solid inclusions of 15–18 GPa. For the third group of samples, the regularity of reduced microhardness was retained and corresponded to the range of 4.5–13.6 GPa.

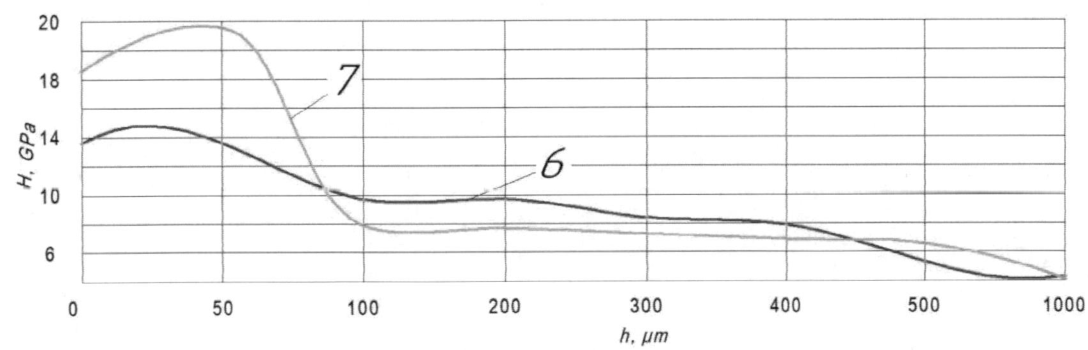

Figure 4. The results of measuring the hardness of the surface and in the section of titanium samples: 6,7 - high-temperature mode (1300–1500 °C)

4. Conclusion

Thus, the average value of microhardness for the low temperature IHT group of samples (depth within 200–250 μm) was 14–18 % higher relative to deep layers (from 300 to 1000 μm), for the medium temperature group – 13–15 %, and for the high temperature group – 15– 22 %. The presence of a highly hard coating (up to 10–20 GPa) and a hard near-surface layer can increase the tribological characteristics of titanium products, which will expand the area of their effective use in the manufacture of medical products.

Acknowledgments
The reported study was funded by RFBR, project number No. 20-33-90053.

References

[1] Sarma J, Kumar R, Sahoo A K and Panda A 2020 *Mater Today Proc* **23** 561
[2] Senthilselvan J, Monisha K, Gunaseelan M, Yamini S, Kumar S Arun, Kanimozhi K, Manonmani J, Shariffc S M and Padmanabham G 2020 *Mater Charact* **160** 110118
[3] Dolgun E, Zemlyakov E, Shalnova S, Gushchina M and Promahov V 2020 *Mater Today* **30** 688
[4] Maytorena-Sánchez A, Hernández-Torres J, López-Huerta F, Hernández-Campos M A, Zamora-Peredo L, Pacio-Castillo M, Serrano-De la Rosa L E and García-González L 2021 *Mater Lett* **282** 128679
[5] Fomin A, Egorov I, Shchelkunov A, Fomina M, Koshuro V and Rodionov I 2018 *Compos Struct* **206** 467
[6] Taylor M, Pullar R C, Parkin I P and Piccirillo C 2020 *J. Photochem. Photobiol.*, A. **400** 112727

Properties of plasma enhanced atomic layer deposited ruthenium thin films from $Ru(EtCp)_2$

E A Smirnova , A V Miakonkikh , A E Rogozhin and K V Rudenko

Valiev Institute of Physics and Technology of Russian Academy of Sciences, Moscow 117218, Russia

smirnova@ftian.ru

Abstract. Ruthenium thin films were deposited by plasma enhanced atomic layer deposition using bis(ethylcyclopentadienyl)ruthenium(II) or $Ru(EtCp)_2$ and oxygen plasma. The growth characteristics have been studied on a silicon substrate with different interfaces in a wide temperature range. On Si and SiO_2, a nucleation delay period has been observed, which can be substantially reduced by the use of a tantalum nitride underlayer of ~ 0.3 nm. The surface analysis shows that the substrate's temperature strongly affects the composition of the film from ruthenium oxide at low temperatures to pure ruthenium film at higher temperatures.

1. Introduction

The development of deposition techniques for thin good-quality ruthenium films has recently become an actual and popular problem due to the prospects for its application in metallization systems of integral systems [1-5]. The interest in ruthenium is partly due to its high technological properties of the noble metal at a relatively low cost, which makes it possible to use it in the mass production of integrated circuits. Ultrathin continuous films without any voids are needed when Ru is used as a diffusion barrier for Cu interconnects. The application of Ru as a conducting interconnection material requires Ru with low resistivity and low impurity contents.

Particular interest in the development of the use of ruthenium as a metal for an IC interconnect system is due to the fact that processes for plasma etching of ruthenium are well known, with the possibility of obtaining a good vertical profile at room temperature and without redeposition [6]. Such processes open up the prospect of abandoning the damascene process, which is used to create copper metallization. In the process of damascene, due to the impossibility of proposing an adequate etching process for copper, it is necessary to use plasma etching of the dielectric, which inevitably causes its degradation, which could not be completely eliminated despite great efforts [7].

There are a few approaches for deposition of metallic films: physical deposition (i.e. magnetron sputtering), chemical vapour deposition (CVD), thermal atomic layer deposition and plasma enhanced atomic layer deposition (PEALD). Although physical deposition methods are the easiest to implement, they cannot be applied to some applications where conformal coating and filling of thin trench and holes is required, which is especially important in IC metallization systems where ruthenium can be applied at levels with a minimum critical size.

PEALD is one of the most attractive methods for producing films with these characteristics, given its inherent ability to easily transition to three-dimensional structures.

A typical ALD process consists of the alternating dosing of precursor and coreactant gas that interact with a substrate through self-limiting surface reactions. In this work, we examine the growth and nucleation of Ru thin films deposited using bis(ethylcyclopentadienyl)ruthenium(II) precursor in PEALD process that includes an oxygen plasma pulse. In particular, we look at how the substrate temperature influences the growth and nucleation of the film and how it can be used to manipulate the material properties.

It was shown earlier [8] that one of the main problems of film growth is also delamination, which is apparently associated with mechanical stresses in the film. This becomes especially critical for films with a thickness of more than 50 nm.

The research of the resistance of Ru-based films obtained in the optimal mode is presented in the previous work [9].

2. Experiment

ALD processes were performed in a commercial FlexAl system (Oxford Instruments Plasma Technology) with a base pressure of 10^{-6} Torr, using a 13.56 MHz ICP remote plasma source. $Ru(EtCp)_2$ was contained in a bubbler, preheated to 70°C and delivered to the reactor by argon flow gas at a flow rate of 150 sccm. The manifold lines were heated to 100 °C and the reactor walls to 120 °C to prevent the condensation of the precursor. The recipe for Ru ALD (Table 1) consisted of a 2 s $Ru(EtCp)_2$ dosing, 1 s pumping, 4 s Ar purge, 2 s step of oxygen pressure stabilization in the chamber, 1 s O_2 plasma ignition and exposure (O_2 pressure of 0.065 Torr), and 4 s Ar purge. In the work, we used gases of a semiconductor grade.

Table 1. Conditions at each step of a single PEALD cycle for deposition of Ru films

	Dose Ru	Ru Purge	Plasma Gas Stabilization	Plasma	Post Plasma Purge
Flow, sccm	150 Ar	150 Ar	60 O_2	60 O_2	150 Ar
Pressure, mTorr	80	80	65	65	80
ICP RF power, W	-	-	-	75	-
Time, s	2	4	2	1	4

Depending on the thickness of the films, adequate *ex situ* methods were used to measure the growth rate: for films with a thickness of less than 10 nm, spectroscopic ellipsometry (Woollam M-2000X) was used, and for thicker films, a cross-section scanning electron microscopy (Carl Zeiss Ultra-55). To apply the ellipsometry method, the dependence of k and n on the wavelength was determined experimentally from a thick ruthenium film obtained by our method (> 30 nm). In this case, it can be assumed that the light reflected from the lower boundary does not leave the sample and the calculation is carried out as for bulk ruthenium. The dependencies were approximated by a B-spline.

Stress analysis of samples was carried out using a Zygo New View 5000 interference microscope.

3. Results and Discussion

We examine the range of growth temperature of 200–400 °C and it turned out that the substrate temperature plays a critical role in determining whether Ru or RuO_2 is deposited. It was noticed that at 375 °C an abrupt change in surface reaction mechanisms takes place, leading to the changing in film composition and structure from polycrystalline RuO_2 at low temperatures to pure Ru film at higher temperatures.

Figure 1a shows how ruthenium is deposited onto an unprepared silicon surface. It can be seen that the growth has an insular character. Even if, during long-term deposition, the islands close together giving a continuous layer, its roughness will be higher than the permissible values. Then we used prepared wafers with a tantalum nitride underlayer of ~ 0.3 nm. Figure 1b shows the character of growth on prepared silicon. Blistering can be seen, which means that the film has significant compressive stresses, and the adhesion to the surface is insufficient. Figure 1c shows the growth of a film on a SiO_2 layer deposited on silicon. In this case, it can be seen that the film is continuous without visible roughness within the resolution of the scanning electron microscope. AFM measurements confirmed this with a roughness of about 1.5 nm.

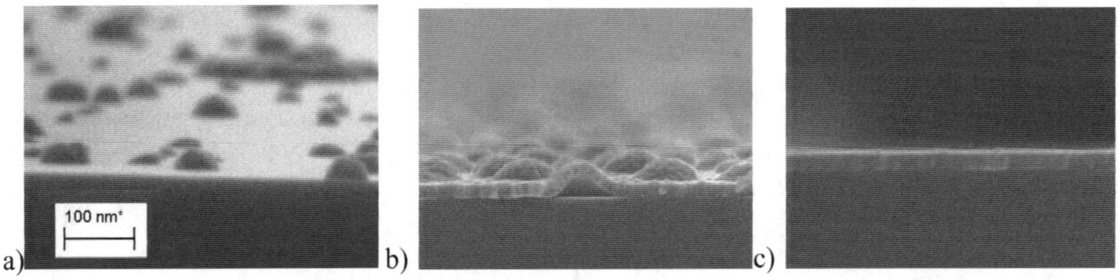

Figure 1. Cross-sectional image of the samples obtained with a scanning electron microscope. a) Island growth of a film on unprepared silicon, b) on a prepared silicon surface, c) on a prepared SiO_2 surface.

For estimation of Ru film stress on the silicon surface, the deflections of single-crystal silicon wafers with a thickness of 380 μm were measured with films deposited on them. The measurement was carried out in a field of 5 mm x 5 mm; the vertical measurement accuracy was about 1 nm. Figure 2 shows the distribution of heights on the investigated central part of the silicon wafer (the arrow shows the section along which the approximation was carried out). Figure 3 shows a cross-sectional profile that was then approximated by a circle giving the radius of curvature of the surface.

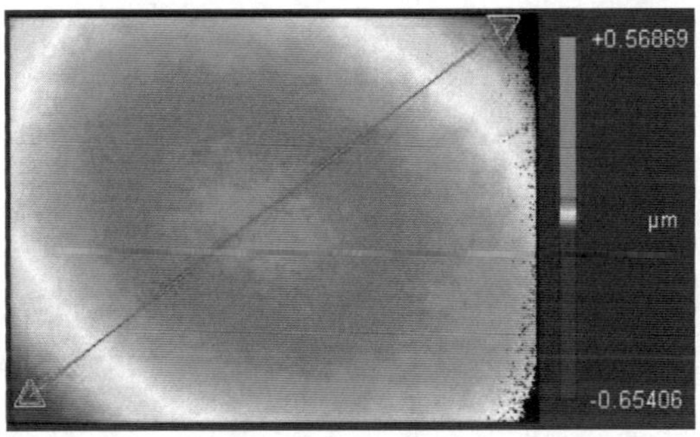

Figure 2. Dependence of the surface stress of the films on the deposition temperature.

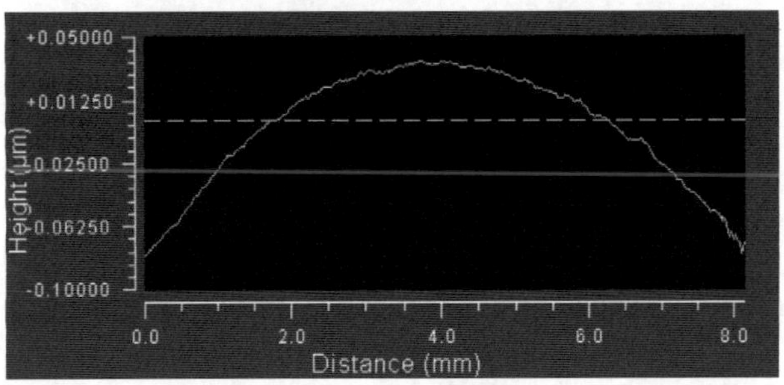

Figure 3. Dependence of the surface stress of films grown at a temperature of 375°C on the number of PEALD cycles.

The obtained distributions were approximated by a spherical profile, and the mechanical stress in the film was calculated using the Stoney formula:

$$\sigma = \frac{E_S}{6(1-v_s)} \frac{h_S^2}{h_f} \left(\frac{1}{R} - \frac{1}{R_0}\right) \quad , \tag{1}$$

where E_S is Young's modulus, v_s is the Poisson's ratio, h_s is the thickness of the substrate, h_f is the thickness of the film, and R and R_0 is the curvature radii of the substrate after and before deposition, respectively.

The application of this approach will allow optimizing the stresses in the films by varying the deposition modes and plasma treatment, which will make it possible to obtain films of metallic ruthenium with better quality.

Mechanical stress data are consistent with the roughness of the films. The mechanism of relaxation of the ruthenium film on a single-crystal silicon surface is associated with the formation of blisters. Stress values of 2.8–4.6 GPa (on Si) were measured.

4. Results and Discussion

In a remote plasma ALD system, we have fabricated Ruthenium thin films by PEALD technology using $Ru(EtCp)_2$ and oxygen plasma on silicon and SiO_2 surfaces with and without modification by TaN 0.3 nm sub-layer (5 cycles). High uniformity and integrity of films as well as reasonable GPC rate were achieved over the area of 100 mm silicon wafer on SiO_2 layer modified by TaN deposition. The fact that such a result was achieved on the silicon oxide layer is explained by the fact that, in contrast to crystalline silicon, it makes it possible to reduce the stresses of the film. For interconnection technology, this is permissible since in it the metal is always deposited on an intermediate dielectric layer. In the future, it is of interest to study the deposition of ruthenium on stop SiOC layers on the surface of a low-k dielectric.

Acknowledgments

The investigation was partially supported by Program no. 0066-2019-0004 of the Ministry of Science and Higher Education of Russia for Valiev Institute of Physics and Technology of RAS and partially funded by RFBR, according to the research project № 18-29-27029.

References

[1] Methaapanon R, Geyer S M, Lee Han-Bo-Ram and Bent S F, *J. Mater. Chem.*, 2012, **22**, 25154
[2] Chopra S N, Vos M F J, Verheijen M A, Ekerdt J G, Kessels W M M and Mackus A J M, *J. Vac. Sci. Technol.*, 2020, A **38**, 062402
[3] Popovici M, Groven B, Marcoen K, Phung Q M, Dutta S, Swerts J, Meersschaut J, Van Den Berg J A, Franquet A, Moussa A, et al. *Chem. Mater.* 2017, **29**, 4654–66
[4] Müller R, Ghazaryan L, Schenk P, Wolleb S, Beladiya V, Otto F, Kaiser N, Tünnermann A, Fritz T and Szeghalmi A, *Coatings*. 2018, 8(**11**):413
[5] Dutta S, Kundu S, Gupta A, Jamieson G, Granados J F G, Bömmels J, Wilson C J, Tokei Z and Adelmann C, *IEEE Electron. Device Lett.* 2017, **38**, 949–951
[6] Hwang S M, Garay A A, Choi J H and Chung C W, *Thin Solid Films* 2016, **615**, 311-317
[7] Miyajima H, Ishikawa K, Sekine M and Hori M, *Plasma Process. Polym.* 2019, **16**, 1–20
[8] Smirnova E A, Miakonkikh A V, Rogozhin A E and Rudenko K V, 2020 *J. Phys.: Conf. Ser.* **1695** 012045
[9] Rogozhin A, Miakonkikh A, Smirnova E, Lomov A, Simakin S and Rudenko K, *Coatings* 2021, **11**, 117

Study of ripple formation on Si surface under Ga ion beam bombardment

M A Smirnova, A S Ivanov, V I Bachurin and A B Churilov

P.G. Demidov Yaroslavl State University, Yaroslavl, Russia

masha_19957@mail.ru

Abstract. The process of microrelief formation on Si (100) surface under 30 keV Ga^+ ion beam bombardment with doses $2 \cdot 10^{17}$ - $4 \cdot 10^{18}$ ion/cm^2 at incident angles $\theta = 0$ - $50°$ was studied. It was found that wave-like structures form on the surface at $\theta = 25°$ - $35°$ and doses $6 \cdot 10^{17}$ - $2 \cdot 10^{18}$ ion/cm^2. The nice ripple formed at $\theta = 30 \pm 2°$ incident angles and irradiation dose 10^{18} ion/cm^2.

1. Introduction

Irradiation of the surface of a solid body by ion bombardment leads to the formation of various types of micro – and nanostructures on the surface. Depending on the type of substrate, ion energy and type of ions, it is possible to form a certain morphology on the surface. The formation of micro- and nanostructures on the surface of semiconductor materials by ion bombardment has recently attracted interest. In particular, a large number of works is devoted to the formation of periodic wavelike structures on a nanometer scale called ripples. The reviews [1, 2] present the main results, devoted to the experimental and theoretical study of the formation of such structures on the surface of various materials irradiated with ions of inert and chemically active gases. The main regularities of the parameters of the forming structures were experimentally established. Thus, ion bombardment may have different effects on the surface, depending on many factors such as incident ion energy, type of ions, angle of incidence, sputtered substrate temperature and material composition and etc. There are differences between using inert and chemically active ion beams in the process of ripple formation. The irradiation dose, which the ripple formation begins with is one of the distinguishing features. In case of using chemically active ions this dose is orders of magnitude less than in case of using inert ions.

This article presents the results of an experimental study of the processes of microrelief formation on the Si surface irradiated by ion beam Ga^+ with a 30 keV. This type of ions is neutral in relation to Si, since the implanted ions don't form chemical compounds with silicon atoms in contradistinction to N_2^+ and O_2^+ ions. The implanted ions of Ga are present in the surface layer in the form of precipitates. The deposition depth of precipitates is comparable to a projective range [3, 4]. This behavior of embedded ions differs from classical rare-gas atoms. At the moment, there is a small number of works devoted to the formation of the microrelief by Ga ion beams on the Si surface [5, 6], on the glass surface [7], on the crystal diamond surface [8]. It should be noted that focused Ga^+ ion beams are a widespread experimental tool in the formation of nanostructures on various surfaces, under different irradiation conditions (angle of incidence, irradiation dose, etc.). Formation of the relief can lead to a change of the sputtering rate [9] and the reproducibility of the experimental results, especially in the case of formation the structures with high aspect ratio [10].

2. Experimental

The irradiation experiments of Si (100) with a 30 keV Ga^+ ion beam were carried out on a Quanta 3D 200i. Beams with a diameter of 4 μm and 300 nm were used at various irradiation parameters (percentage of overlap, delay time, scanning mode). Three series of experiments were carried out.

In the first one the angles of incidence of the ion beam were Θ = 25°, 30°, 35°, 40°, irradiation doses were D = $2{\cdot}10^{17}$, $4{\cdot}10^{17}$, $6{\cdot}10^{17}$, $8{\cdot}10^{17}$, 10^{18}, $2{\cdot}10^{18}$, $4{\cdot}10^{18}$ ion/cm^2, the ion beam current was 5 nA. In the second - the angles of incidence of the ion beam were Θ = 0°, 20°, 30°, 40°, 50°, the irradiation doses were D = 10^{18} ion/cm^2, the ion beam current was 5 nA. The surface topography was explored by using a Quanta 3D 200i. In the third – preliminarily some samples were created at the fixed angle of incidence of ion beam Θ = 30°, irradiation doses were D = $2{\cdot}10^{17}$, $4{\cdot}10^{17}$, $6{\cdot}10^{17}$, $8{\cdot}10^{17}$, 10^{18}, $2{\cdot}10^{18}$, $4{\cdot}10^{18}$ ion/cm^2, ion beam current 5 nA. Under certain experimental conditions of irradiation of the Si surface with Ga^+ ions, a wavelike structure (ripples) was formed on the surface. This structure, which is formed on the Si surface at these parameters, is characterized by a high reproducibility of the result. A platinum band was applied perpendicular to the direction of propagation of the ripples on the surface of each sample for more accurate determination of the parameters of the relief. It was made to protect the wavelike structure from the further exposure of the incident ions. Further, the cross-section along the platinum band was made by ion etching. The formation of samples and the cross-section line was carried out on a Quanta 3D 200i (in-situ). The surface topography of all samples was explored ex situ by Supra 40 electron microscop in detail.

3. Results and discussions

It was found that at irradiation doses and angles of incidence more than 40° the formation of a relief on the surface does not occur. At angles of incidence of the ion beam less than 25° and irradiation doses exceeding $4{\cdot}10^{17}$ ion/cm^2, a granular structure is formed on the surface and it becomes reticulated (characteristic size about 100 nm) with the irradiation dose increasing. At angles of incidence in the range from Θ = 25° to 35° and irradiation doses higher than $4{\cdot}10^{17}$ ion/cm^2, a wavy relief appears. A well reproducible periodic relief (ripples) is formed near the angle of incidence of the ion beam Θ = 30°. This is consistent with the results of the work [6]. However, in contrast to the results presented in [6], the generation of ripples was observed in this work at more then twice irradiation dose (> $2{\cdot}10^{17}$ ion/cm^2). Perhaps it is due to the fact that in [6] the relief parameters were determined by using the AFM method, which is sensitive to the exploration of relief inhomogeneities.

Figure 1 demonstrates micrographs of a Si surface irradiated with a Ga^+ ion beam at an incidence angle 30° and irradiation doses D = $4{\cdot}10^{17}$, $6{\cdot}10^{17}$, 10^{18}, $2{\cdot}10^{18}$, ion/cm^2. One can see that the generation of ripples (D = $4{\cdot}10^{17}$, $6{\cdot}10^{17}$ ion/cm^2) with bright hills (hilloks) are observed on the surface. Those hilloks are Ga drops. It was determinated by the analysis of the irradiated surface by the SIMS method on the Cameca IMS 4f. Previously, in [11] it was experimentally established that at irradiation doses of the order of 10^{17} ion/cm^2 and the angle of incidence of the ion beam Θ = 30°, the Ga distribution in the near-surface layer is inhomogeneous and a separate regions enriched in Ga are observed. For cases Θ = 30°, D = $4{\cdot}10^{17}$, $6{\cdot}10^{17}$, 10^{18} and $2{\cdot}10^{18}$ ion/cm^2 the value of wavelength λ was measured by the SEM method, $\lambda \approx$ 150 nm and 450 nm respectively for cases D = $4{\cdot}10^{17}$ and $2{\cdot}10^{18}$ ion/cm^2. A similar result was obtained in [5] when the Si surface was irradiated by Ga^+ ions at an angle of incidence Θ = 30°. At irradiation doses higher than $2{\cdot}10^{18}$ ion / cm^2, there is an intersection of waves, the destruction of the wave-like structure, and the increase of the wavelength magnitude. The wavelength λ reaches the value 800 nm at the irradiation dose D = $4{\cdot}10^{18}$ ion/cm^2, which is also consistent with the results of work [5].

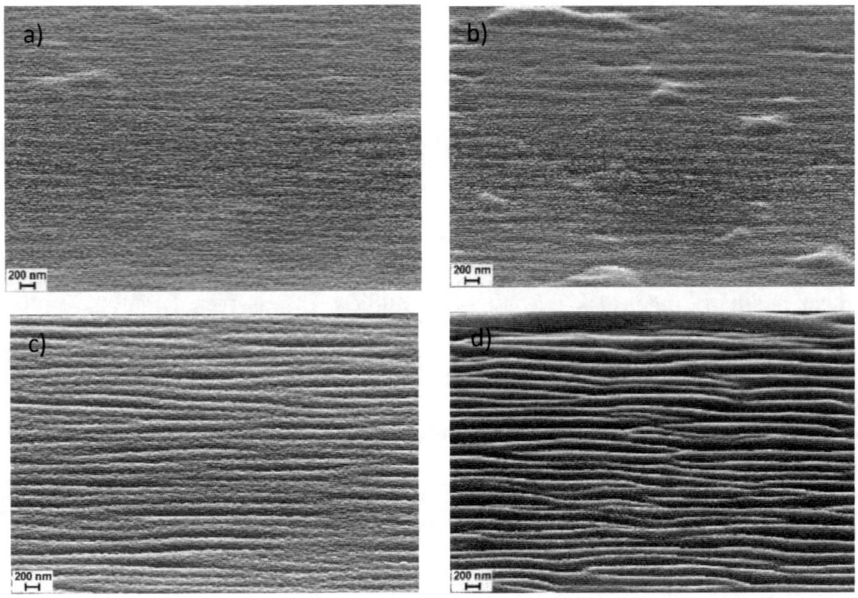

Figure 1. Micrographs of the Si surface. Irradiation doses: a) $4{\cdot}10^{17}$, b) $6{\cdot}10^{17}$, c) 10^{18} and d) $2{\cdot}10^{18}$ ion/cm^2.

Figure 2 shows electron microscopic images of wave-like structures in a cross-section. These structures were formed on the Si surface by with a Ga^+ ion beam at an angle of incidence $\Theta = 30°$, irradiation doses were $4{\cdot}10^{17}$, $8{\cdot}10^{17}$, $2{\cdot}10^{18}$ ion/cm^2.

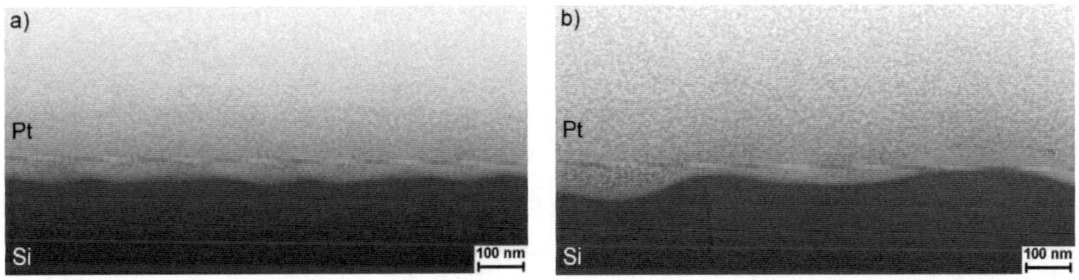

Figure 2. Micrographs of ripples cross-section on the Si surface. Irradiation doses: a) $4{\cdot}10^{17}$ and b) $2{\cdot}10^{18}$ ion/cm^2.

A platinum band was applied over the wave-like structure as a mask before the formation of the cross-section. Thereby the wave-like was saved from further ionic influence. The values of the ripple amplitude were obtained by the SEM method. For the case shown in Figure 2 (a) the average value of the peak-to-valley amplitude h = 32.5 nm, in Figure 2 (b) – h = 70 nm. Thus, the value of the ripple amplitude increases with the rise of the irradiation dose.

As noted above, behavior of Ga ions in the near-surface layer significantly differs from embedded ions of inert gases, for example Ar. The implanted Ga atoms form precipitates near the surface, which are located at the depth of the ions projective range, as shown in [3, 4]. Apparently, the emergence of such irregularities on the surface in the form of Ga drops is the reason of ripple formation. This ripples are formed at Si surface at irradiation doses by two orders of magnitude less than at inert gas ion bombardment (Ar, Xe) with close energies. The sputtering yields of Ga precipitates and substrate material (Si) is different. This can lead to the appearance of topographic irregularities. The creation of these irregularities with the simultaneous sputtering of Si and the deposition of metals, which form silicides, leads to the ripples generation on the Si surface [12]. The creation of a relief of an arbitrary shape on the initial Si surface leads to the impetuous formation of ripples with a wavelength, which is specific to the parameters of ion bombardment, as shown in [13, 14]. The absence of relief at angles of incidence of the ion beam more than 35° is due to a sharp drop in the Ga content in the near-surface layer [11]. The sputtering yield almost does not change [11] at angles of incidence less than 25°. The necessary condition for the appearance of a wavy relief on the surface under ion irradiation is not met [1].

4. Conclusions
The process of topography formation on Si surface under irradiation with 30 keV Ga^+ ion beam bombardment at incident angles $\theta = 0$ - 50° was studied. It was found that the process of ripple formation starting with irradiation dose $2 \cdot 10^{18}$ ion/cm^2 at angles of incidence of the ion beam nearly 30°. The wavelength and amplitude of the microrelief are increase with grows of the irradiation dose from $4 \cdot 10^{17}$ to $2 \cdot 10^{18}$ ion/cm^2. The value of the ripple wavelength is increasing from 150 nm to 500 nm. The value of the ripple amplitude is increasing from 30 nm to 70 nm. The existence of an implanted gallium in the near-surface layer in the form of separate precipitates is a possible reason of the early appearance of the wavelike structures. The difference between sputtering yields leads to the appearance of topographic irregularities. These irregularities are initiate the formation of the wavelike relief.

Acknowledgments
This study was performed with financial support from the Ministry of Education and Science of the Russian Federation within the framework of the state assignment of the P.G. Demidov Yaroslavl State University of topic no. 0856-2020-0006 and using equipment of the Center for Collective Use Diagnostics of Microstructures and Nanostructures

References
[1] Munoz-Garcia J, Vazquez L, Castro M, Gago R, Redono-Cubero A, Moreno-Barrado A and Cuerno R 2014 *Mater. Sci. & Eng.* R **86** 1
[2] Makeev M A, Cuerno R and Barabasi A 2002 *Nucl. Instr. Methods Phys. Res. B* **197** 185
[3] Frey L, Lehrer C and Ryssel H 2003 *Appl. Phys.* A **76** 1017
[4] Rommel M, Spoidi G, Yanev V, Beuer S, Amon B, Jambreck J, Petersen S and Bauer A J 2010 *J. Vac. Sci. Technol. B* **28** 595
[5] Habenicht S, Lieb K P, Koch J and Wieck A D 2002 *Phys. Rev* B **65** 115327
[6] Qian H X and Znou W 2012 *Mater. Lett.* **77** 1
[7] Gnaser H, Reuscher B and Zeuner A 2012 *Nucl. Instr. Methods Phys. Res. B* **285** 142
[8] Adams D P, Vasile M J, Mayer T M and Hodges V C 2003 *J. Vac. Sci. Technol. B* **21** 2334
[9] Carter G 1999 *J. Appl. Phys.* **83** 455
[10] MoberlyChan W 2006 *MRS Proceedings* **960** 0960-N10-02-LL06-02
[11] Bachurin V I, Zhuravlev I V, Pukhov D E, Rudy A S, Simakin S G, Smirnova M A and Churilov A B 2020 *J. Surf. Invest.* **14** 784
[12] Engler M, Frost F, Muller S, Masko S, Will M, Feder R., Spemann D, Hubner R, Fasko S and Michely T 2014 *Nanotechnology* **25** 115303
[13] Smirnov V K, Kibalov D S, Lepshin D A and Bachurin V I 2000 *Izv. Akad. Nauk, Ser. Fiz.* **64** 626
[14] Kamakar P, Mollick S A, Ghose D and Chakrabarti A 2008 *Appl. Phys. Lett.* **93** 103102

The electron transmission properties in a non-planar system of two chained rings

M O Smolkina, I Yu Popov, A S Bagmutov and I V Blinova

Faculty of Control Systems and Robotics, ITMO University, St. Petersburg, 197101, Russia

Abstract. The electron transmission properties in a model of two chained orthogonal quantum rings with input and output wires were investigated by using the quantum graph theory and quantum waveguide theory. The model was obtained for three-dimensional space. It also was shown that changing of the orientation of second ring in the respect to the field is a way to control the electron transmission and reflection.

1. Introduction

There is a number of works about the spectral and electron transmission properties of a system consisting of rings (see, for example, [1, 2]). Manipulating the magnitude of the magnetic field allows one to control the ballistic electron transport in nanostructures [3, 4]. The quantum graphs in these articles are plane. Such models do not allow controlling the electron transport by changing the direction of the external magnetic field. This work proposes a non-planar construction consisting of two chained rings (i.e., the model is not plane) that allows one to control the electron transport due to the direction of the magnetic field or, in other words, due to the orientation of the structure. The example of such a structure can be found among the macromolecules – biphenyl molecules. In particular, discrete model of biphenyl molecules is described in [5] and the electron transmission properties were also studied. This molecule is non-planar and contains connected rings, as the system discussed in this work, although the geometry is different. We construct solvable continuous model in contrast with the discrete model in [5]. This model makes it simpler to choose the system parameters for better control of ballistic electron transport.

2. Theoretical model

Figure 1 shows the proposed non-planar model of two chained rings, described by using the quantum graph theory. One ring is in the ZOY plane and the other is in the XOY plane. These two rings are chained to each other and have a junction point. The system also has one input wire and one output wire. It is also supposed that the right ring can rotate around the Y axis, while the left one is fixed. The magnetic field is co-directed with Z axis.

The functions in the incoming and outgoing leads, denoted by: $\psi_I(x_I), \psi_{II}(x_{II})$, are

$$\psi_I(x_I) = e^{ikx_I} + C_1 e^{-ikx_I}, \psi_{II}(x_{II}) = C_{10} e^{ikx_I}, \quad (1)$$

the functions in the edges on rings:

$$\begin{aligned}
\psi_{L_{up}}(\gamma) &= C_2 e^{ikR_L\gamma} + C_3 e^{-ikR_L\gamma},\\
\psi_{L_{low}}(\gamma) &= C_4 e^{ikR_L\gamma} + C_5 e^{-ikR_L\gamma},\\
\psi_{R_{up}}(\gamma) &= C_6 e^{ikR_R\gamma} + C_7 e^{-ikR_R\gamma},\\
\psi_{R_{low}}(\gamma) &= C_8 e^{ikR_R\gamma} + C_9 e^{-ikR_R\gamma},
\end{aligned} \tag{2}$$

where $R_L(R_R)$ are the radii, γ is an angle between the input(output) wire and the Y axis, k is a wave number.

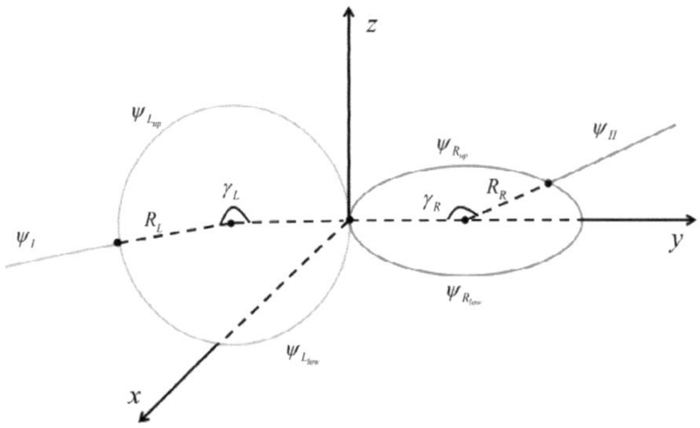

Figure 1. Quantum graph with two semi-infinite leads (input and output quantum wires) and four leads which form two rings. $R_L(R_R)$ are the radii, $\gamma_L(\gamma_R)$ are the angles between the input(output) wires and the Y axis.

The magnetic Kirchhoff conditions in the vertices of the graph are as follows:

$$\begin{cases} e^{(-1)^{\mu} i\Phi_e(\gamma_e)}\psi_e(\gamma_e) = \psi_{e'}(\gamma_e)e^{(-1)^{\mu} i\Phi_{e'}(\gamma_e)} \\ \sum_e (-1)^{[e]}\partial\psi_e(\gamma_e)e^{(-1)^{\mu} i\Phi_e(\gamma_e)} = 0, \end{cases} \tag{3}$$

where $[e]=0$ for the output edge and $[e]=1$ for the input edge, γ_e is an angle between junctions of the rings and wires. There are some points where the edges orientation does not coincide with the parameters on the ring, so we will take into account the negative sign in front of Φ_e, and μ equals 0 or 1, depending on the orientation. Besides

$$\Phi_e = \begin{cases} 0, \\ \int_{\gamma_e} a(\tau)d\tau, \end{cases} \tag{4}$$

where 0 is in the beginning of the edge, $\int_{\gamma_e} a(\tau)d\tau$ in the end of the edge.

According the model, the magnetic field is co-directed with Z axis. This means that $\Phi_1(\gamma) = \frac{1}{2}BR_L{}^2\gamma\sin(\alpha), \Phi_2(\gamma) = \frac{1}{2}BR_R{}^2\gamma\cos(\alpha)$, and $\Phi_1(\gamma)$ refers to the ring in the ZOY plane, $\Phi_2(\gamma)$ refers to the ring in the XOY plane, α is a rotation angle around the Y axis.

Taking into account the magnetic Kirchhoff conditions (3) and the expressions for functions in the incoming and outgoing leads (1), for similar rings (2), we obtain such conditions for the vertices of the graph:

$$\begin{gathered}\psi_I(0)=\psi_{L_{up}}(0)=\psi_{L_{low}}(2\pi)e^{i\Phi_1(2\pi)},\\-\psi_I'(0)+\psi_{L_{up}}'(0)-\psi_{L_{low}}'(2\pi)e^{i\Phi_1(2\pi)}=0,\end{gathered}\tag{5}$$

for the first vertex (the junction point between the incoming wire and left ring).

For the second vertex (the junction point of two rings) the equations are:

$$\begin{gathered}\psi_{L_{up}}(\pi)e^{i\Phi_1(\pi)}=\psi_{L_{low}}(\pi)e^{i\Phi_1(\pi)}=\psi_{R_{up}}(0)=\psi_{R_{low}}(2\pi)e^{i\Phi_2(2\pi)},\\-\psi_{L_{up}}'(\pi)e^{i\Phi_1(\pi)}+\psi_{L_{low}}'(\pi)e^{i\Phi_1(\pi)}+\psi_{R_{up}}'(0)-\psi_{R_{low}}'(2\pi)e^{i\Phi_2(2\pi)}=0.\end{gathered}\tag{6}$$

For the third vertex (the junction point between the outcoming wire and the right ring) the equations are:

$$\begin{gathered}\psi_{R_{up}}(\pi)e^{i\Phi_2(\pi)}=\psi_{R_{low}}(0)=\psi_{II}(0),\\-\psi_{R_{up}}'(\pi)e^{i\Phi_2(\pi)}+\psi_{R_{low}}'(0)+\psi_{II}'(0)=0.\end{gathered}\tag{7}$$

As the left ring is proposed to be fixed relatively to the magnetic field, α for the left ring is 0. This means that $\Phi_1(\gamma)=0$ and the next system of equations takes place:

$$\begin{cases}1+C_1-C_2-C_3=0\\C_2+C_3-C_4e^{ikR_L2\pi}-C_5e^{-ikR_L2\pi}=0,\\-ik+ikC_1+ikR_LC_2-ikR_LC_3-ikR_LC_4e^{ikR_L2\pi}+ikR_LC_5e^{-ikR_L2\pi}=0\\C_2e^{ikR_L\pi}+C_3e^{-ikR_L\pi}-C_4e^{ikR_L\pi}-C_5e^{-ikR_L\pi}=0\\C_4e^{ikR_L\pi}+C_5e^{-ikR_L\pi}-C_6-C_7=0\\C_6+C_7-(C_8e^{ikR_R2\pi}+C_9e^{-ikR_R2\pi})e^{i\Phi_2(2\pi)}=0\\-ikR_LC_2e^{ikR_L\pi}+ikR_LC_3e^{-ikR_L\pi}+ikR_LC_4e^{ikR_L\pi}-ikR_LC_5e^{-ikR_L\pi}+\\+(ikR_RC_6-ikR_RC_7)-(ikR_RC_8e^{ikR_R2\pi}-ikR_RC_9e^{-ikR_R2\pi})e^{i\Phi_2(2\pi)}=0\\(C_6e^{ikR_R\pi}+C_7e^{-ikR_R\pi})e^{i\Phi_2(\pi)}-C_8-C_9=0\\C_8+C_9-C_{10}=0\\-(ikR_RC_6e^{ikR_R\pi}-ikR_RC_7e^{-ikR_R\pi})e^{i\Phi_2(\pi)}+(ikR_RC_8-ikR_RC_9)+ikC_{10}=0\end{cases}\tag{8}$$

It is a time-consuming and difficult task – to find the explicit form of solution of (8) , therefore the most effective way was chosen in the work: to carry out numerical modelling for the fixed parameters of the system. To obtain all the coefficients $C_1,C_2,\ldots C_{10}$ in (8) the Gaussian elimination method was used. Finally, the expression for reflection coefficient is: $\mathrm{R}_{ref}=C_1\cdot C_1^*$, for transmission coefficient: $T=C_{10}\cdot C_{10}^*$.

3. Numerical results

By fixing physical and geometric parameters of the system and employing the numerical methods, the values of coefficients $C_1,C_2,\ldots C_{10}$ were obtained. The effects of rotation angle α, radii R_L,R_R, wave number k, the angles $\gamma_L=\gamma_R$, magnetic induction B, on the transmission coefficient were also obtained.

The dependence of electron transmission coefficient T and reflection coefficient R_{ref} on rotation angle α for values of system parameters: $R_L=R_R=1.14, k=0.18, \gamma_L=\gamma_R=\pi$ is shown in figure

2(a) and $R_L = R_R = 1.2, k = 0.77, \gamma_L = \gamma_R = \pi$ in figure 2(b). One can see that transmission coefficient T and reflection coefficient R_{ref} are not zero. There are values of rotation angle when the coefficient $T = 1$ and $R_{ref} = 0$, so the perfect transmission takes place (the wave completely passes in the output wire). Besides there exist intervals of values of α when R_{ref} is close to 1 while T is close to 0, so the perfect reflection takes place.

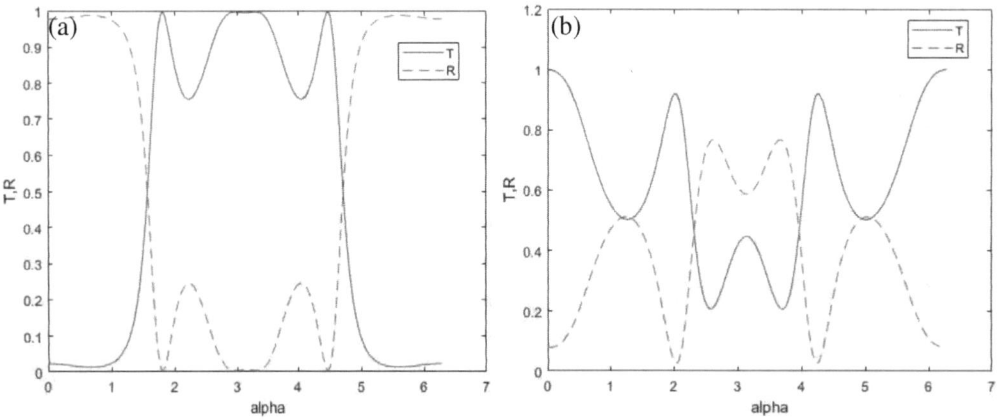

Figure 2. Electron transmission coefficient T and reflection coefficient R_{ref} as functions of rotation angle α, while $R_L = R_R = 1.14, k = 0.18, \gamma_L = \gamma_R = \pi$ (a), $R_L = R_R = 1.2, k = 0.77, \gamma_L = \gamma_R = \pi$ (b), α changes from 0 to 2π.

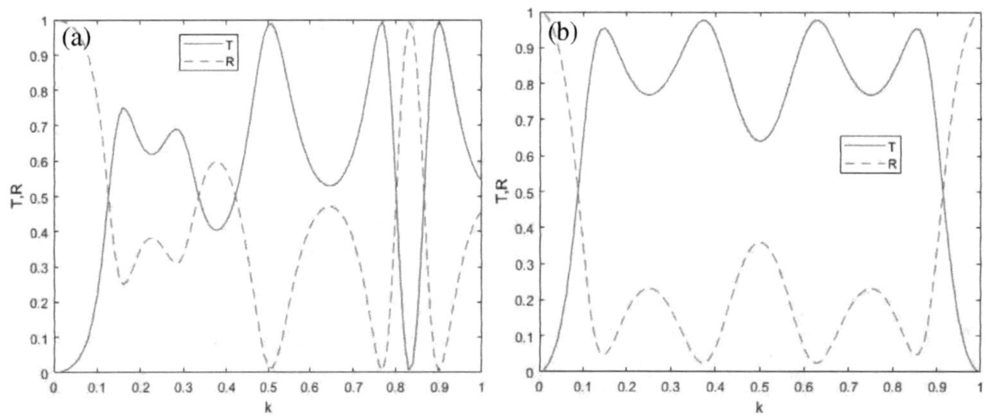

Figure 3. Electron transmission coefficient T and reflection coefficient R_{ref} as functions of wave number k for $\alpha = \pi, \gamma_L = \gamma_R = \pi$, $R_L = R_R = 1$, (a), $R_L = R_R = 1.2$ (b).

Figure 3 shows the dependence of electron transmission coefficient T and reflection coefficient R_{ref} on the wave number k for chosen values of parameters: $R_L = R_R = 1, \alpha = \pi, \gamma_L = \gamma_R = \pi$ in figure 3(a), $R_L = R_R = 1.2, \alpha = \pi, \gamma_L = \gamma_R = \pi$ (b) in figure 3(b). The behaviour of coefficients is similar to figure 2. There exists such a value of wave number, when perfect transmission (reflection) takes place.

When the rotation angle $\alpha = \pi$, the rings are coplanar, there is a very high probability of perfect electron transmission. These results are confirmed by few works [4, 6, 7] about the spin-filtering properties in two coupled Rashba quantum rings.

But changing the geometry of the model is not the best way to control the electron transmission. The better way is a magnetic flux.

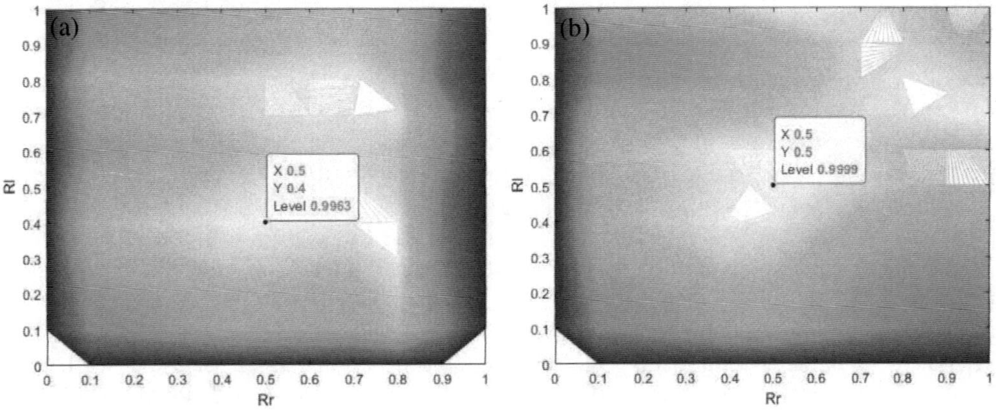

Figure 4. Contour map of electron transmission coefficient T as function of radii of the rings for $k = 0.77, \gamma_L = \gamma_R = \pi$, $\alpha = 1.8$ rad (a), $\alpha = \pi$ (b).

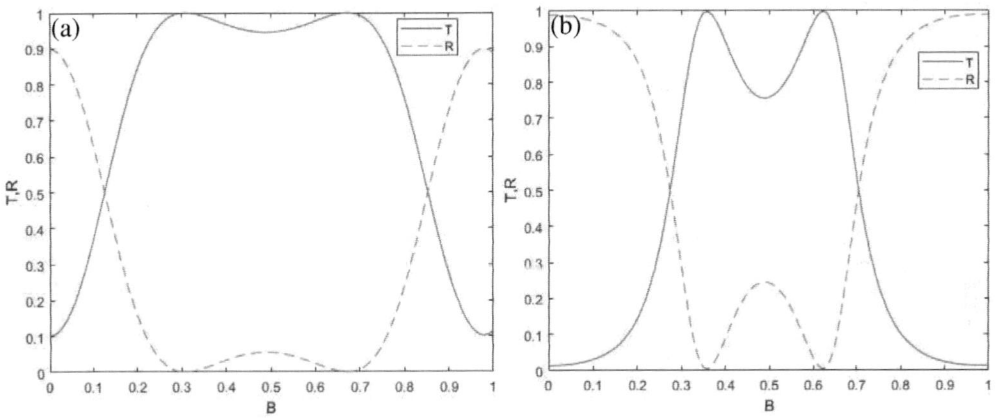

Figure 5. Electron transmission coefficient T and reflection coefficient R_{ref} as functions of B for $R_L = R_R = 1.2, \alpha = 1.8$ $rad, \gamma_L = \gamma_R = \pi$, $k = 0.44$ (a), $k = 0.51$ (b).

Figure 5 shows the dependence of ballistic electron transport on the value of magnetic field. For chosen from figures 2-4 values of rotation angle α, radii of the rings, wave number and junction angles, the electron transmission coefficient T and reflection coefficient R_{ref} are represented as functions of B. Figure 5 shows that perfect transmission can be obtained when B goes to 0.31, 0.66.

4. Conclusion

The value of transmission coefficient T changes from 0 to 1 and depends on physical and geometric parameters of the system: the wave number, the rotation angle, the radii of the rings and the angles between junction points. If the value of transmission coefficient approaches 1, there is a ballistic electron transport without reflection through the system, otherwise, when approaching 0, there is no ballistic electron transport, the wave is being fully reflected. During the work we have found appropriate values of wave number, rotation angle, radii of the rings and the angles between the input(output) wires and the Y axis, when the transmission coefficient T is close to 1, so the perfect electron transmission takes place.

Acknowledgements
The reported study was funded by RFBR, projects number 19-31-90164 and number 20-31-90050.

References

[1] Pavlov B S, Popov I Yu, Geyler V A and O S Pershenko 2000 Possible construction of quantum multiplexer *Europhys. Lett.* **52** (2) 196-202
[2] Geyler V A and Popov I Yu 2001 Quantum interference rectifier *Physica E: Low-dimensional Systems and Nanostructures* **9** (4) 631-634
[3] Dehghana E, Khoshnouda D S and Naeimi A S 2018 Logical spin-filtering in a triangular network of quantum nanorings with a Rashba spin-orbit interaction *Physica B* **529** 21-26
[4] Smolkina M O, Popov I Y and Blinova I V 2020 The spin-filtering properties in two coupled Rashba quantum rings *J. Phys.: Conf. Ser.* **1697** 012198
[5] Kondo H, Nara J, Kino H and Ohno N 2009 Transport properties of a biphenyl-based molecular junction system the electrode metal dependence *J. Phys.: Condens. Matter* **21** 064220
[6] Smolkina M O, Popov I Y and Blinova I V 2020 The spin-filtering properties in two coupled Rashba quantum rings *J. Phys.: Conf. Ser.* **1695** 012174
[7] Popov I Y, Blinova I V, Shamionova E A and Smolkina M O 2021 On the electron transmission control by a direction of magnetic field *International Journal of Numerical Modelling: Electronic Networks, Devices and Fields* **34** (6) e2918

Normal-state transport in superconducting NbN films on r-cut sapphire

M D Soldatenkova[1], A D Triznova[1], E M Baeva[1,2], P I Zolotov[1,2], A I Lomakin[1,2], A I Kardakova[1,2] and G N Goltsman[1,3]

[1] Moscow Pedagogical State University, Moscow 119435, Russia
[2] National Research University Higher School of Economics, Moscow 101000, Russia
[3] NTI Center for Quantum Communications, National University of Science and Technology MISiS, Leninsky prospekt 4, Moscow 119049, Russia

soldatenkova.mariya.99@mail.ru

Abstract. High-quality thin NbN films are very crucial for realizing quantum devices. Here, we investigated electrical transport and noise properties of a series of thin NbN films of various thicknesses grown on r-cut sapphire substrate using a DC magnetron sputtering technique. The films exhibit non-uniform thickness dependences for superconducting transition temperature (T_c) and normal-state resistivity. Morphological characterization of NbN samples of various thicknesses reveals uniform structure in thin films and granular structure in thick films. By measuring transport and noise properties in a normal state, we observe that the granular structure of NbN films does not have a strong effect on resistivity and does not cause an additional source of current noise.

1. Introduction

Niobium nitride (NbN) has become a material that is increasingly important for development of superconducting electronic devices for high-frequency, high-speed, and high-temperature operations because it has a high transition temperature (T_c), and good thermal cycling stability. Thin NbN films are widely used in low-temperature superconducting devices, such as single-photon detectors [1] and hot-electron bolometers [2], and it is considered as a potential material for application in superconducting quantum circuits [3-5]. Great interest on the fundamental level is related to effects of disorder on superconducting and normal state properties of NbN thin films [6-9] as well as a demonstration of thermal and quantum phase slips in NbN nanowires [10], an important phenomenon for understanding one-dimensional superconductivity. Therefore, there has been a growing demand for high-quality thin NbN films for a variety of applications.

Extensive studies of NbN during the past decades [11-14] show that properties of NbN are strongly dependent on deposition-related conditions such as crystal structure and lattice constant. The latter is usually a direct result of the choice of a substrate material. NbN films grown on sapphire (Al_2O_3) substrates, preferably with c-cut and a-cut orientations, demonstrate a prospective performance with a high T_c and low resistivity in a normal state [12, 15, 16]. Here, we focus on study of transport and electronic properties of NbN thin films grown on r-cut sapphire, which is the least studied system. As will be shown further, morphological characterization of NbN samples of various thicknesses on r-cut sapphire reveals uniform structure in thin films and granular patterns in thick films.

2. Fabrication and experiment

Fabrication of samples includes the following steps: (1) NbN film deposition, (2) fabrication of gold metallic pads, and (3) patterning of NbN film. NbN film deposition is performed in an ultra-high vacuum system (AJA Orion series). The r-cut sapphire substrates are preheated up to 400^0C with a resistive heater supplied with a PID controller. After 10-minute temperature stabilization niobium target (2'' diameter, 99.95% purity) is first pre-conditioned in argon atmosphere for 5 minutes followed by 5-minute discharge stabilization in a gas mixture containing 20% of nitrogen. Set current of 600 mA resulted in approx. 300 W power at operating pressure of 3.6 mTorr. The deposition of variable thicknesses is produced by opening the magnetron shutter and controlling the deposition time. The resulting thicknesses are calculated from pre-calibrated deposition rate obtained with a quartz crystal nanobalance detector. The calculated deposition rate is 0.065 nm/s, which is the optimal sputtering rate to control thickness below 10 nm. In the following step, the gold pads are fabricated by means of photolithography, e-beam evaporation, and lift-off processes. To pattern NbN film a negative resist mask is formed with electron-beam lithography, then unwanted NbN areas are removed using plasma-chemical etching in SF_6 atmosphere. The parameters of NbN films are presented in Table 1.

Table 1. Parameters of NbN films. d is the film thickness, R_s is the resistance per square at 300 K, r_R is the resistance ratio taken as the ratio of R_s between 300 K and 20 K, T_c is defined at a temperature where resistance is half of normal-state resistance, D is diffusion coefficient.

d (nm)	R_s(300 K) (Ω/sq)	r_R	T_c (K)	D cm^2/s
5	340	0.83	11.7	-
10	151	0.85	12.6	0.474
30	55.9	0.83	13.2	-
50	37.6	0.8	12.9	0.33
100	20.76	0.74	13.3	-
200	10.56	0.74	14	-
300	7.78	0.72	13.9	0.29

The morphological characterization of the samples is performed using atomic force microscopy (AFM) setup from NT-MDT (INTEGRA series) in classical contact mode. Figure 1 reveals morphologies of NbN films for thicknesses 10, 100 and 300 nm. AFM images, presented in Figure 1(a-c), show a pronounced granular structure in the thickest film. Figure 1(d-e) presents analysis of granular structure of NbN films, measured by AFM. The effective diameter of a grain D_g is defined as diameter of cross-sectional area of a grain. It is measured to be 2.66±0.02, 10.30±0.23, and 33.50±2.1 nm for 10-, 100- and 300-nm-thick NbN films, respectively. The change of film morphology with thickness increase may be related to a change of a growth mode, which is suggested to change from substrate-confined growth at the beginning (NbN phase) to the thermal dynamic growth Nb_4N_3 phase as the deposition time increased [14].

Further, we consider the effect of granular morphology on resistivity in a normal state. Figure 2 (a) shows the temperature dependence of normalized resistance for NbN films of different thicknesses. At decreasing T, the resistance increases and drops to zero at critical temperature T_c. The resistance slope above T_c can be described by resistance ratio $r_R=R_{300}/R_{20}$. The parameters of T_c and r_R are presented in Table 1. Interestingly, we observe a noticeable increase of resistance above the resistive transition with increasing d (see Figure 2 (a)). This effect cannot be explained by weak-localization phenomena, predicting such behaviour for films thinner than L_φ, where $L_\varphi \approx 4 - 10$ nm is the phase-breaking length in NbN at 15-20 K [17]. Normal-state resistivity ρ, obtained as $\rho=R_s d$, as a function of film thickness is presented in Figure 2 (b).

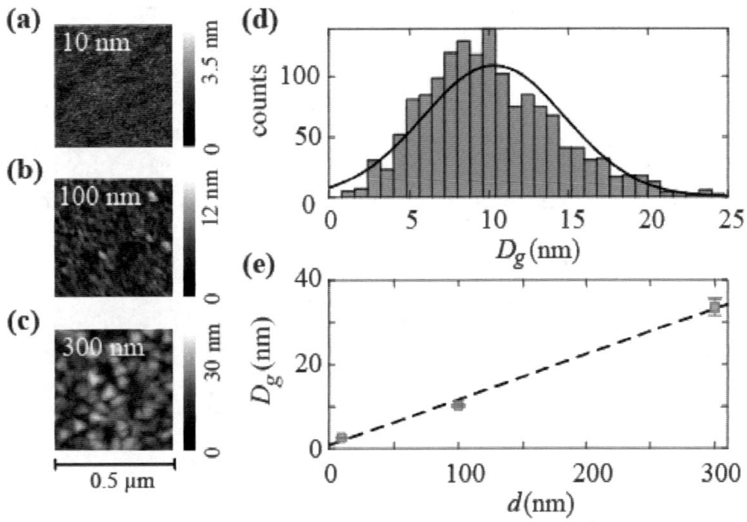

Figure 1. Morphological characterization of NbN films. **(a-c)** AFM images of the NbN films of 10, 100, 300 nm thickness, respectively. **(d)** Grain diameter distribution for one of the studied 100-nm NbN film. Solid line indicates the best-fitting normal grain diameter distribution. **(e)** Average grain diameter D_g as a function of the film thickness d. The experimental dependence is fitted with a linear function (dashed line).

One can see that resistivity of the 5-nm NbN film is higher than for the 10-nm and 30-nm films, meanwhile for films with d > 10 nm it also gradually increases at increasing d. The change in ρ is usually associated with an increase of disorder or emergence of additional scattering mechanisms. For example, the presence of grains can affect resistivity through scattering at grain boundaries. To define the effect of grains on ρ we estimate the mean free path l for electrons. Figure 2 (c) shows the temperature dependence of resistance at different values of an applied perpendicular magnetic field. From the temperature slope of upper critical magnetic field B_{c2} (see inset of Figure 2 (c)) we estimate the diffusion coefficient $D = 4k_B/\pi e\ (dB_{c2}/dT)^{-1}$, where k_B is the Boltzmann constant, e is the electron charge.

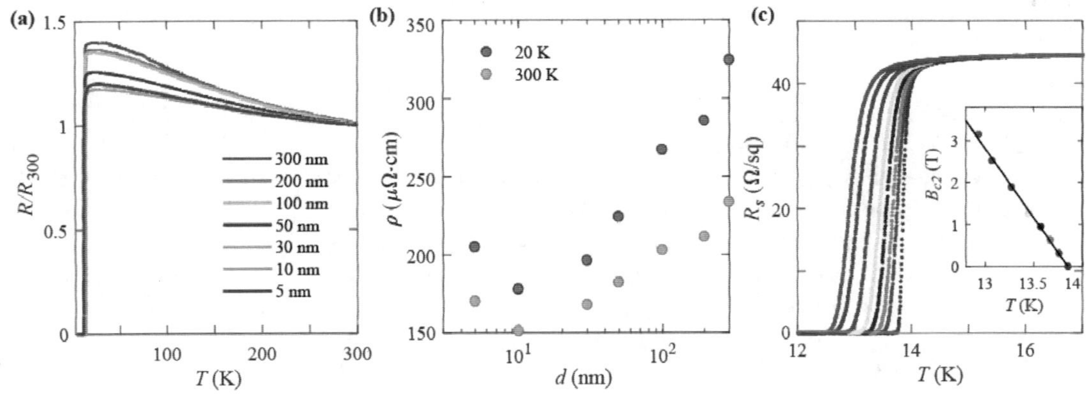

Figure 2. Electrical and transport properties of NbN films in a normal state. **(a)** The temperature dependencies of normalized resistance of NbN films in a wide temperature range. **(b)** The thickness dependence of resistivity. The blue circles correspond to resistivity at 20 K, the orange circles correspond to resistivity at 300 K. **(c)** Main: the temperature dependence of resistance per square for the 50-nm thick film in an external magnetic field B applied perpendicular to the film. Inset: the temperature dependence of the upper critical magnetic field B_{c2} determined at resistance $R=R_s/2$.

The values of D for 10, 50, and 300 nm thick films are presented in Table 1. Here, it is instructive to estimate the mean free path as $l=3D/v_F$, where v_F is Fermi velocity. This yields estimation of mean free path l=0.043-0.067 nm at $v_F=2\times10^6$ m/s in NbN [15], where the shortest value of l is determined for 300 nm film. Since obtained l is much smaller than thickness and average grain size D_g, the Fuchs–Sondheimer's surface scattering model [18-19] and the Mayadas-Shatzkes' grain boundary scattering model [20] are not applicable for describing resistivity data in the studied NbN samples. We assume that an increase in resistivity and a decrease in the mean free path can be related to the growth of film disorder due to a long duration time of film deposition. In fact, a long deposition process is accompanied by absorption of nitrogen by the magnetron target and degassing of the magnetron chamber. Thus, the effect of grains on resistivity can be negligible.

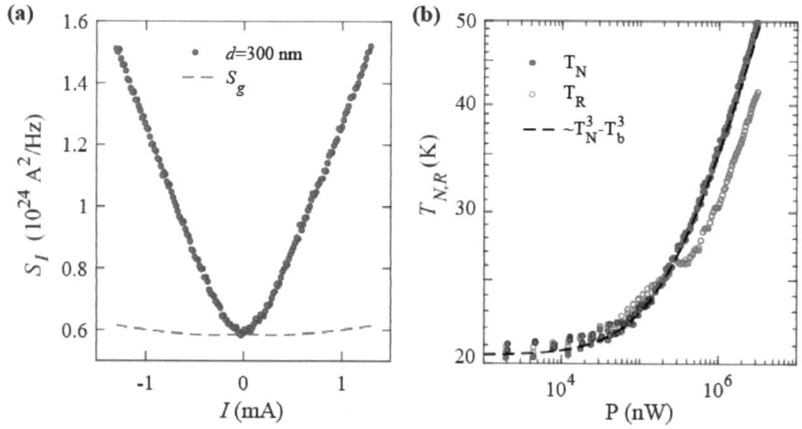

Figure 3. **(a)** The current dependence of current shot noise S_I for the 300-nm NbN sample (blue circles) is in comparison with the noise caused by granular structure S_g (red dashed line). **(b)** Increase of temperatures T_R and T_N versus joule power (P) dissipated in the NbN sample. The T_R is retrieved from the $R(T)$ dependence of the sample. The black dashed line corresponds to the fit by equation $P\sim T_e^3-T_b^3$.

Next, we consider the effect of granular morphology on spectral density of current fluctuations (S_I) in a sample patterned from the 300-nm thick NbN film. The experimental setup for measurement of current fluctuations is described in Ref. [21]. We investigate the current noise in the normal state above T_c. Experiments are performed at a resonance frequency of 45 MHz, high enough to ignore possible $1/f$-like noise contributions. In this measurement, the sample is maintained in the sample holder in a vacuum. The samples are biased with a DC current at bath temperature (T_b) of 20-21 K. The Joule power (P) dissipated in NbN film is defined as $P=IV$, where I is the bias current, V is the measured voltage. In scenario of intrinsically inhomogeneous transport in a granular material, the fluctuations can be dominated by a partition noise of a network of nanoscopic weak links. Weak links could be modeled as tunnel barriers or constrictions, separated by large conductive regions with strong energy relaxation [21-22]. Here we suppose that the spectral density of current fluctuations can be determined as $S_g=2e|I|/N \coth(2e|V|/2k_BT_bN)$, where N is a number of weak links defined as $N=L/D_g$. Figure 3 (a) shows the experimental dependence of S_I(I) for the NbN sample (0.33×64 μm^2) in comparison with an estimate of the noise of weak links S_g at $N=L/D_g=64$ μm/33 nm ~ 2000. One can see that the estimated S_g predicts less noise than we observe here. In contrast, the experimental data can be described by electron heating in NbN film. In noise thermometry, the noise temperature T_N, which is equivalent to electron temperature T_e, can be found with the Johnson-Nyquist formula $T_N=S_I (dV/dI)/4k_B$, where dV/dI is a differential resistance of a sample. Figure 3 (b) shows the P dependence of T_N on a log-log scale. With intense heating, when the regime $T_e \gg T_b$ is achieved, we observe the $P(T_e)$ dependence with the heat power-law $P\sim T_e^3-T_b^3$ (the dashed black line). In addition, we plot the electron temperature of the sample (T_R), which is reestablished from the $R(T)$ dependence and sample resistance $R=V/I$ data. With small heating, the electron temperatures obtained by means of noise thermometry and secondary (resistive) thermometry give similar results. Meanwhile, at intense heating, we observe a small deviation T_R from T_N, which can be related to the fact that the *R(T)*-dependence measured at a low current bias. Thus, we assume that the electron heating effect describes S_I better than the current noise caused by granular structure of the sample.

In conclusion, our preliminary results demonstrate that granular structure of thick NbN films does not have a strong effect on resistivity and does not cause an additional source of current noise.

Acknowledgments

Authors acknowledge the support of RSF project N19-72-10101 for the study of the NbN samples with the noise measurements, and the support of RFBR project N 19-32-60076 for fabrication and preliminary DC measurements of the unpatterned NbN films. The authors also thank V. Khrapai for useful discussion of this work.

References

[1] Goltsman G N, Okunev O, Chulkova G, Lipatov A, Semenov A, Smirnov K, Voronov B, Dzardanov A, Williams C, and Sobolewski R 2001 *Appl. Phys. Lett.* **79** 705
[2] Hajenius M, Baselmans L J A, Gao J R, Klapwijk T M, de Korte P A J, Voronov B, and Goltsman G N 2004 *Supercond. Sci. Technol.* **17** S224
[3] Yamashita T, Kawakami A, and Terai H. 2017 *Phys. Rev. Appl.* **8** 054028 Annunziata A L, Santavicca D F, Frunzio L, Catelani G, Rooks M J, Frydman A, and Prober D E 2010 *Nanotechnology* **21** 445202
[4] Niepce D, Burnett J, and Bylander J 2019 *Phys. Rev. Appl.* **11** 044014
[5] Chand M et al 2012 *Phys. Rev. B* **85** 014508
[6] Noat Y et al 2013 *Phys. Rev. B* **88** 014503
[7] Kamlapure A et al 2013 *Sci. Rep.* **3** 2979
[8] Zolotov P et al 2021 *IEEE Tran. Appl. Supercond.* **31** 2400305
[9] Delacour C, Pannetier B, Villegier J, and Bouchiat V 2012 *Nano Lett.* **12** 3501
[10] Zou Y et al 2016 *Sci. Rep.* **6** 22330
[11] Hazra D, Tsavdaris N, Jebari S, Grimm A, Blanchet F, Mercier F, Blanquet E, Chaoelier C, and Hofheinz M 2016 *Supercond. Sci. Technolog.* **29** 105011
[12] Wang Z, Kawakami A, Uzawa Y, and Komiyama B 1996 *J. Appl. Phys.* **79** 7837
[13] Chang H W, Wang C L, Huang Y R, Chen T J, and Wang M J 2017 *Supercond. Sci. Technol.* **30** 115010
[14] Chockalingam S P, Chand M, Jesudasan J, Tripathi V, and Raychaudhuri P 2008 *Phys. Rev. B* **77** 214503
[15] Semenov A et al 2009 *Phys. Rev. B* **80** 054510
[16] Sidorova M, Semenov A, Hübers H-W, Ilin K, Siegel M, Charaev I, Moshkova M, Kaurova N, Goltsman G N, Zhang X, and Schilling A 2020 *Phys. Rev. B* **102** 054501
[17] Fuchs K *Math. Proc. Cambridge Philos. Soc.* 1938 **34** 100
[18] Sondheimer E H 1952 *Adv. Phys.* 1952 **1** 1
[19] Mayadas A F and Shatzkes M *Phys. Rev. B* 1970 **1** 1382
[20] Baeva E M, Titova N A, Kardakova A I, Piatrusha S U, and Khrapai V S 2020 *JETP Lett.* **111** 104
[21] Gousev Yu, Goltsman G N, Semenov A D, Gershenzon E M, Nebosis R S, Heusinger M A, and Renk K F 1994 *J. Appl. Phys.* **75** 3695
[22] Denisov A O, Tikhonov E S, Piatrusha S U, Khrapach I N, Rossella F, Rocci M, Sorba L, Roddaro S, and Khrapai V S 2020 *Nanotechnology* **31** 324004

Al-CuO_x multilayer nanostructures: formation features and thermal properties of new type of local heat source

A I Novoseltsev[1], L I Sorokina[1], A V Sysa [2], R M Ryazanov[2] and E A Lebedev[1,2]

[1] National Research University of Electronic Technology – MIET, 124498, Moscow, Zelenograd, Russia
[2] Scientific-Manufacturing Complex "Technological Centre", 124498, Moscow, Zelenograd, Russia

larasork@gmail.com

Abstract. In this work, multilayer nanostructured thermite materials are considered - a new type of local heat sources. Aluminium and copper oxide were chosen as components of the thermite mixture. The formation of multilayer structures was carried out on the surface of the substrate by the method of magnetron sputtering. The features of the deposition process as well as the energy properties of the formed materials have been investigated. The results obtained confirm the prospects of using this class of materials as local heat sources.

1. Introduction

Technological progress in micro- and nanoelectronics requires the solution of new problems associated with the technology of surface joining. Reaction bonding or soldering using thin layers of energetic materials, acting as local heat sources, already now make it possible to effectively solve a number of technological problems. Energy materials manufactured in the form of multilayer structures consist of two or more components and are able to maintain in their volume a self-propagating exothermic reaction in the wave combustion mode after initial initiation. The reaction front temperature can significantly exceed 1000 °C and reaches this value in milliseconds. For soldering, a layer of energetic material is placed between the solder layers, and initiation occurs with an electric spark or low-power laser. The heat released during the reaction melts the solder, which wets the surfaces to be joined and, after solidification, forms a reliable joint. Depending on the physical state of the surfaces to be joined, all joining processes can be divided into three groups - solid bonding [1], solder bonding [2] and fusion bonding [3]. In the latter case, melting of the surfaces to be joined is necessary, which, as a rule, is associated with significant heating of the object, which can lead to the loss of its unique properties. And this is precisely the main problem of micro- and nano-assemblies - how to limit and accurately dose the heating, and how to limit the heated volume or how to localize it? Energetic nanoscale multilayer foils for bonding surfaces have been developed and partially commercialized in the last decade. Due to their high reactivity and intense heat release, these functional foils can act as heat sources for joining thermosensitive materials and micro- and nano-sized components [4].

2. Experimental details

Multilayer structures with different thicknesses of individual layers were formed by alternate sputtering of Al and CuO_x targets by magnetron sputtering. Studies of the formed materials using energy-dispersive X-ray spectroscopy and stylus profilometry made it possible to reveal the dependences of the deposition rate and composition of copper oxide layers on the partial pressure of oxygen in the process of target sputtering. Thermal effects were measured by differential scanning calorimetry, and the propagation velocity of the wave combustion front was carried out using high-speed video camera.

3. Results

Multilayer Al-CuO_x structures were formed by magnetron sputtering. The rates of deposition of aluminum and copper oxide were preliminarily determined. The sputtering rate of aluminum was 0.125 nm / s at a sputtering power of 500 W. The copper oxide target was sputtered in a mixture of argon and oxygen at a power of 350 W. Deposition rate of copper oxide as well as the ratio of copper to oxygen in the formed layer varied depending on the partial pressure of oxygen during the sputtering process. The obtained experimental dependences are shown in Figure 1. The decrease in velocity is associated with a decrease in the sputtering efficiency due to the lower mass of oxygen ions in comparison with argon.

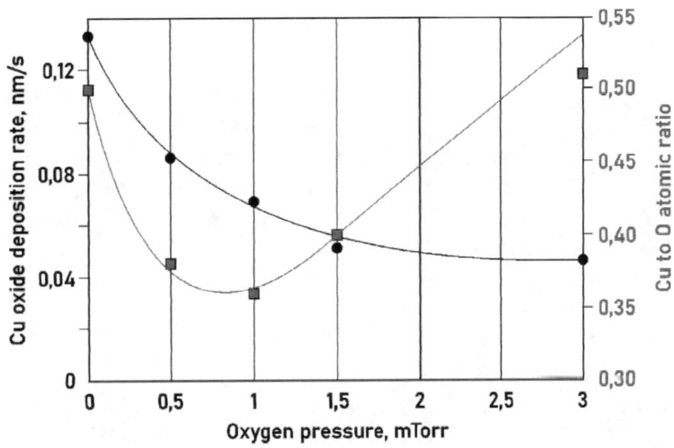

Figure 1. Dependences of the copper oxide deposition rate and the composition of the formed layers on the partial pressure of oxygen during sputtering.

The presence of a minimum for a pressure of 1 mTorr was also revealed on the dependence of the atomic ratio of copper to oxygen. On the one hand, the increased oxygen content in the copper oxide layers will increase the energy properties of the multilayer structure, and on the other hand, additional technological difficulties appear - the deposition rate decreases and it becomes necessary to change the composition of the atmosphere, which in combination significantly increases the total duration of the process. Therefore, for further studies, the copper oxide target was sputtered in an argon atmosphere.

In this work, 3 multilayer structures with different bilayer thicknesses (the total thickness of one aluminum layer and one copper oxide layer) - 75, 150, and 300 nm were formed. The total thickness of the structure was the same in all cases and was equal to 3 μm. The deposition was carried out on various substrates - sitall and Kapton tape. The control of the geometric characteristics of the formed structures was carried out using an SEM and the resulting images are shown in Figure 2. The ratio of the thicknesses of the components was determined based on the chemical reaction $2Al + 3CuO \rightarrow Al_2O_3 + 3Cu + Q$.

The obtained SEM images indicate the presence of a clear periodic structure with the required thicknesses of individual layers and the total thickness of the structure.

At the next stage of the study the wave combustion process for Al-CuO_x structures deposited on sital wafers with different bilayer thickness was investigated using high-speed video at a speed of 15000 fps. The combustion rate of samples with a bilayer thickness of 75, 150, and 300 nm was 6.3, 4.0, and 1.9 m/s, respectively, which is close to a linear dependence.

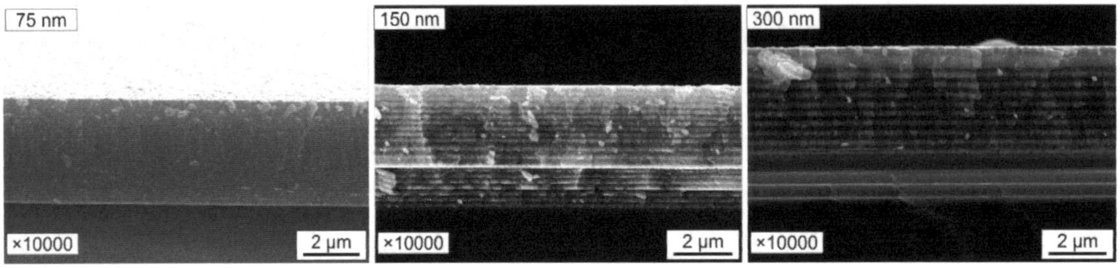

Figure 2. SEM images of cross-sections of Al-CuO_x multilayer structures with different bilayer thicknesses - 75, 150, and 300 nm.

Combustion storyboards are shown in Figure 3. A decrease in the front velocity with an increase in the bilayer thickness is associated with an increase in diffusion lengths in the volume of the multilayer structure and a decrease in the contact area of the reagent layers. Combustion was initiated in all cases using an electric spark from a piezoelectric generator. Combustion took place in a wave mode.

Figure 3. Storyboard of the combustion process of Al-CuO_x.

The combustion products of multilayer Al-CuO_x structures were studied using scanning electron microscopy and elemental analysis. Despite active gas evolution during combustion, most of the reaction products remain on the substrate surface in the form of solidified melt droplets (see Figure 4).

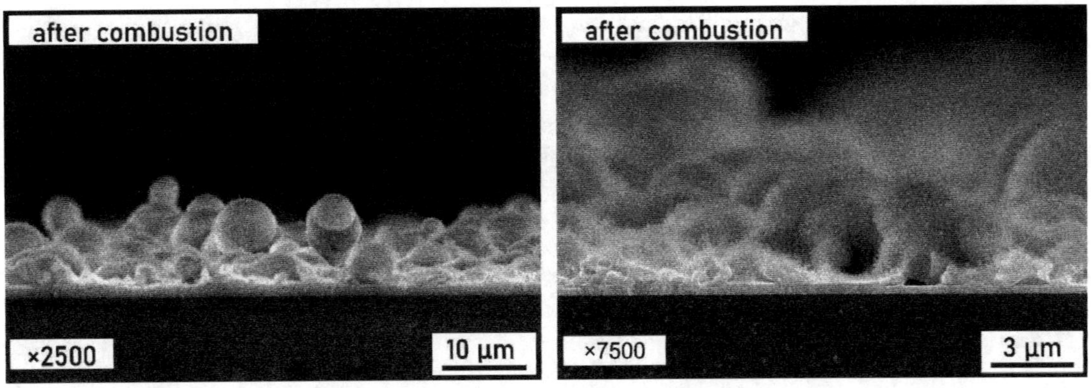

Figure 4. SEM images of combustion reaction products on the substrate surface.

Elemental analysis made it possible to determine the composition of the combustion products (see Figure 5) - these are solidified copper drops, the surface of which is partially covered with aluminum oxide. In addition, nanoscale fibers were found on the particle surface at higher magnification, the nature and properties of which will be studied in further studies.

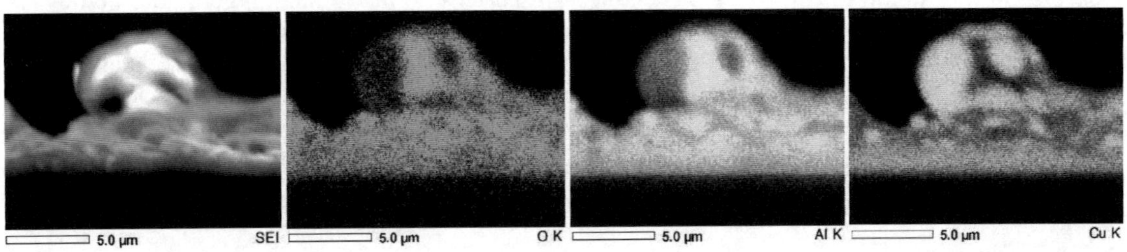

Figure 5. Results of elemental analysis of combustion products of Al-CuO_x multilayer thermite materials.

The energy properties of multilayer materials were studied using DSC and TGA (obtained results presented at Figure 6). For these purposes, multilayer structures, formed on the surface of the Kapton tape, were separated from it with a scraper. The measurements were carried out on a TA Instruments SDT Q600 series instrument in air and in an argon atmosphere in ceramic crucibles with a heating rate of 10 degrees per minute in the range from 150 to 750 °C. Some general patterns were found - the change in the weight of the samples when measured in an argon atmosphere was less than 1%, which indicates the absence of the influence of the atmosphere. When measured in a stream of air, the weight gain in all cases was within 9%. One main exothermic effect was observed on the DSC curves in the temperature range from 475 to 625 °C.

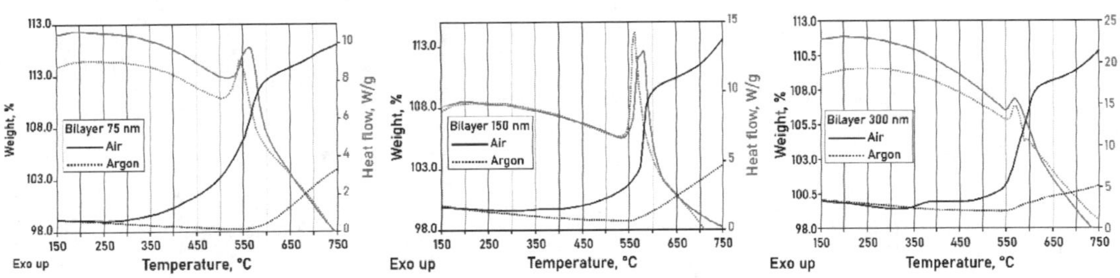

Figure 6. DSC and TGA curves obtained for multilayer Al-CuO_x structures with different values of the bilayer thickness.

The position of the maximum of the peak in measurements in argon was found in the region of lower temperatures. However, a change in the value of the overall thermal effect was observed, which amounted to 730, 1050, and 550 J/g for the bilayer 75, 150, and 300 nm, respectively, for measurements in argon and 1120, 1690, and 870 J/g in argon. The variation in the value with a change in the bilayer thickness can be associated with both the completeness of the reaction in the multilayer material and the contribution of the mixed layer that forms at the interface between aluminum and copper oxide during magnetron sputtering. An increase in the general thermal effect in air indicates the occurrence of the oxidation reaction of aluminum with oxygen from the atmosphere, which is also evidenced by the weight gain.

4. Conclusion

In this work, we investigated the features of the process of the formation of multilayer structures by the method of magnetron sputtering. The effect of oxygen in the process of sputtering a copper oxide target on the deposition rate of layers and their composition has been studied. An experimental study of the size effect on the process of wave combustion of multilayer structures with different bilayer thicknesses has been carried out. A significant increase in the combustion front velocity from 1.9 to 6.3 m / s with a decrease in the bilayer thickness from 300 to 75 nm was found. The results of DSC and TGA measurements of samples of multilayer structures made it possible to evaluate the energy characteristics of the formed materials when they are heated in air and in an argon atmosphere. The overall thermal effect during slow heating is influenced not only by the fraction of the material reacted during the deposition process at the boundary of the layers, but also, to a greater extent, by the completeness of the reaction. The contribution of the reaction of aluminum with oxygen from the atmosphere in all cases was approximately the same.

The studies carried out have confirmed the high potential of using Al-CuO_x multilayer thermite materials in various fields as local heat sources.

Acknowledgments

This work was supported by the Russian Federation President's grant (МК-4750.2021.4).

References

[1] Lu Y, Huang JY, Wang C, Sun S and Lou J 2010 *Nature Nanotechnology* **5(3)** 218
[2] Do J-W, Chang NN, Estrada D, Lian F, Cha H and Duan XJ 2015 *ACS Nano* 15 **9(5)** 4806
[3] Hu A, Peng P, Alarifi H, Zhang XY, Guo JY and Zhou Y 2012 *J. of Laser App.* **24(4)** 042001
[4] Simões S, Viana F, Ramos AS, Vieira MT and Vieira MF 2014 *Micr. and Microanal.* **21(1)** 132

Nanocarbon mesh in the flame of aromatic compounds

O V Vasilyeva[1], S I Ksenofontov[2] and A N Lepaev[3]

[1] Chuvash State University named after I.N. Ulyanov, Cheboksary 428015, Russia
[2] Chuvash State Pedagogical University named after I.Y. Yakovlev, Cheboksary 428000, Russia
[3] Cheboksary Institute (branch) Moscow Polytechnic University, Cheboksary 428000, Russia

dprostokvashino@mail.ru

Abstract. The soot in the flame of the pyrotechnic composition is presented in a variety of shapes and structures. Part of the dispersed particles consists entirely of carbon. The sizes of such particles are in the submicron range. Another part of the particles contains carbon in the form of nanostructures - needles and tubes. A detailed study of the morphology and chemical composition of dispersed particles on optical and scanning electron microscopes showed the existence of yet another unidentified species of particles - tuba. The resulting formations are micron-sized and retain the properties of nanostructures.

1. Introduction

Various ordered carbon nanostructures - linear-chain carbon, fullerenes, single-walled nanotubes - are formed depending on the physicochemical conditions of the experiment. A film of two-dimensionally ordered linear-chain carbon is obtained by sputtering by the method of pulse-plasma evaporation of a graphite cathode. The resulting films have a thickness of up to several microns, high anisotropy of electrophysical and physicochemical characteristics [1]. Fullerenes or sphere-like carbon molecules are closed spherical networks consisting of a combination of 6- and 5-membered carbon rings. They can be obtained by evaporation in an inert atmosphere of a graphite rod in a volt arc [2, 3]. Nanotubes are extended structures consisting of graphite-like layers rolled into single or multilayer tubes with a cavity diameter of up to 6 nm. The distance between the layers is 340 pm, which is close to the interplanar distance in graphite. The length of nanotubes reaches tens of micrometers. The end of such a tube can be open or closed by a fullerene-like hemisphere [3]. Carbon nanotubes can also be produced by arc discharge, laser ablation, or chemical vapor deposition.

Carbon nanostructures obtained during the combustion of magnesium in carbon dioxide at a pressure of 2000 atm are shown in Figure 1. The phases of the formation of nanostructures are clearly visible. First, this is the formation of round or spherical clusters of amorphous carbon (Figure 1a). Then, structuring of these nanoclusters into a "zigzag" system with rather extended linear regions of a loose nanocluster structure (Figure 1b). Further, the compaction of this structure with the almost complete disappearance of round or spherical clusters and the formation of linear or zigzag bamboo-like carbon nanofibers of considerable length (1 - 5 μm), with an outer diameter of up to 40 nm and an inner diameter of up to 20 nm (Figure 1c) [4].

The introduction of solids into the pyrolysis gas medium of hydrocarbons leads to the formation of fibrous carbon on the surface of solids with catalytic activity (SiO_2, Al_2O_3, MgO) [5].

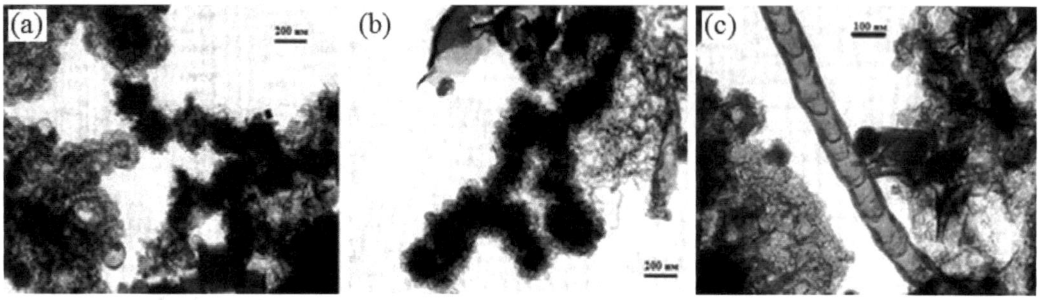

Figure 1. Combustion products of magnesium in CO_2. Formation of carbon clusters of various shapes: (a) circular; (b) spherical; (c) bamboo.

2. Research object

Combustion products in a flame of pyrotechnic compositions based on magnesium powder, inorganic oxidant and aromatic compounds. Powders of the components were thoroughly mixed and pressed into cardboard casings. The combustion of the samples was carried out under room conditions.

3. Research methods

a) sampling of condensed dispersed particles from the flame of pyrotechnic compositions by the method of carrying glass plates; b) examination of samples by methods of optical and scanning electron microscopy to study the morphology of dispersed particles; c) IR – spectrometry for the analysis of organic carbon compounds.

4. Experimental results and their discussion

A high temperature (3100 K) in the flame is achieved by burning a pyrotechnic composition based on magnesium powder, inorganic oxidant. The presence of diphenyl in the composition, as a combustion phlegmatizer, helps to reduce the maximum flame temperature. An excess of the mass concentration of diphenyl of 30% can cause an unstable combustion regime, and subsequently, the cessation of the combustion itself. The oxygen of the oxidizing agent, first of all, is spent on the reaction with the metal, and only then on the reaction with hydrogen and carbon. At the same time, an insignificant part of oxygen goes to the combustion of carbon, as a result of which, compounds with a negative oxygen balance form flames with a large amount of smoke.

The selection of condensed combustion products was carried out by passing a quartz plate through the flame. The particles deposited on the plates were "frozen" and subjected to microscopic studies, IR –spectrometry.

Soot in the flame of a pyrotechnic composition can be represented in different ways [6, 7]. Part of the dispersed particles consists entirely of carbon. The sizes of such particles are in the submicron range. Another part of the particles contains carbon in the form of nanostructures - needles and tubes. A detailed study of the morphology and chemical composition of dispersed particles on optical and scanning electron microscopes showed the existence of yet another unidentified species of particles - tuba. The resulting formations have micron sizes, retaining the properties of nanostructures [8, 9].

The tube is a woven mesh with a diameter of 30 - 40 microns and a length of more than a millimeter (Figure 2). It begins on the surface of a particle that reacts by a heterophase mechanism. The enveloping mass of diphenyl is a supplier of material for the formation of nanotubes. The resulting mesh of a certain length is a pipeline of combustion products. The two-phase medium of combustion products moves along the tube. The gas component freely passes through the tube grid, dispersed particles settle on the inner surface of the tube. The settled particles of micron size deform the surface of the tube and form bulges. Submicron sized particles settle on the inner surface of the tube and plug the tube mesh. In this case, the hydrodynamic resistance increases greatly, and this contributes to the formation of new cells of the tube.

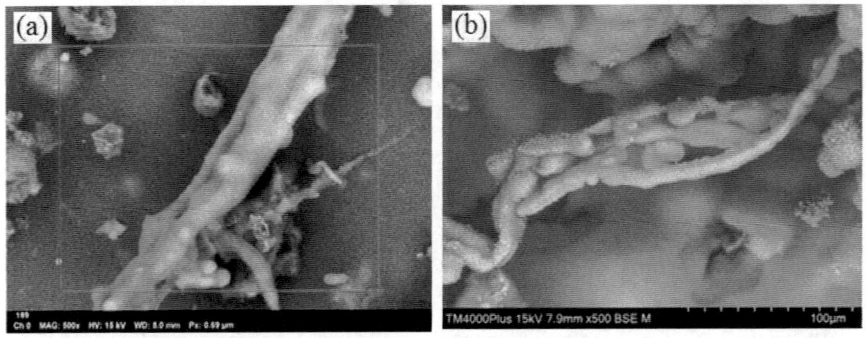

Figure 2. Fragments of carbon tubes. Magnification (a) 200x; (b) 500x.

A model of a tube with dispersed particles moving inside it is shown in figure 3. The growth process of the tube ends with the cessation of combustion. The formation of parallel branches or a change in the directions of growth of the tube itself is possible when the tube is clogged.

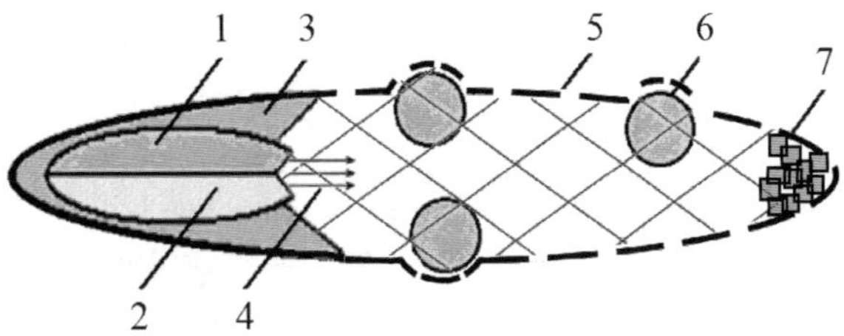

Figure 3. Model of the formation of a carbon tube: 1 - metal, 2 - oxidizer, 3 - diphenyl shell, 4 - two-phase flow of combustion products, 5 - tube, 6, 7 - dispersed particles.

The carbon mesh of the tube burns out during combustion, and a dense layer of metal oxide particles remains, preserving the appearance of the tube. The micrographs show tubes with a partially burnt out carbon mesh (Figure 4).

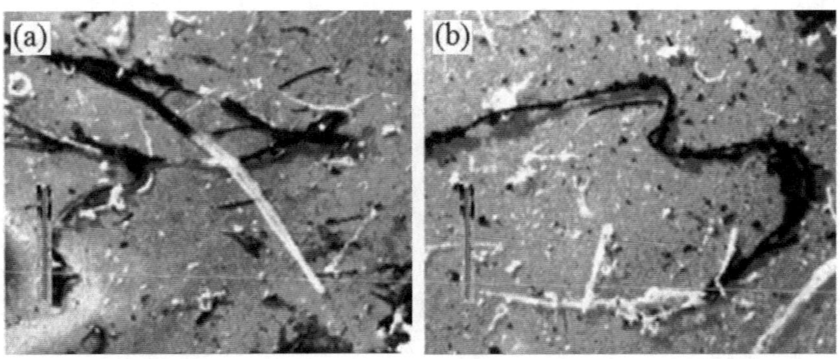

Figure 4. Micrograph of the sample area.

The section of the tube was studied by X-ray spectral analysis. Below are maps of the distribution of individual chemical elements from the analyzed area of the sample (Figure 5). The distribution of carbon, magnesium and oxygen is discrete and coincides with the contours of the tube.

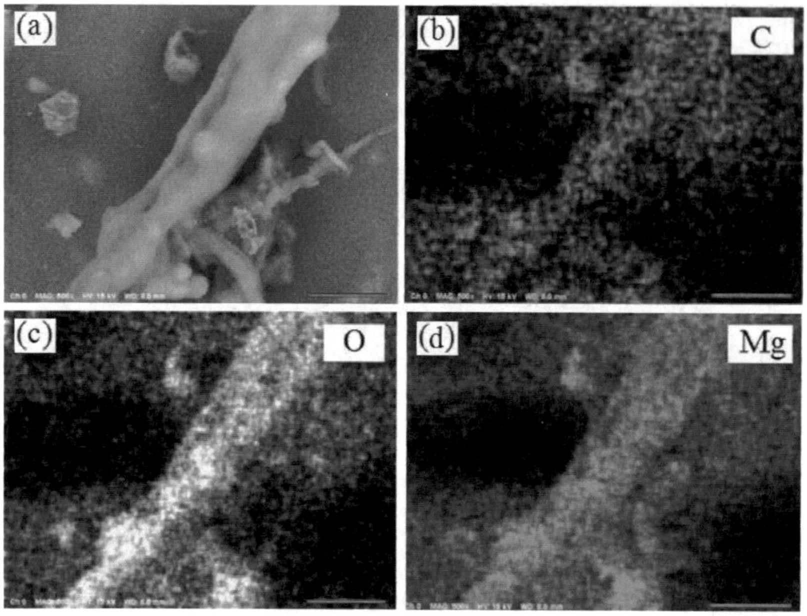

Figure 5. Results of high-quality energy microanalysis: (a) electron microscopic image; (b – d) distribution map of chemical elements.

The IR absorption spectrum indicates the presence of various organic compounds in condensed combustion products (Figure 6). The absorption bands due to stretching vibrations of the OH– group are located in the frequency range 3500 – 3300 cm^{-1} in the form of a broad intense absorption band. Absorption bands related to stretching vibrations of $–CH_2$ and $–CH_3$ – groups are observed in the range of 2980 – 2820 cm^{-1} and 2985 – 2850 cm^{-1}, respectively, bending vibrations of $–CH = CH_2$ groups – 1410 – 1420 cm^{-1} appear as peaks against the background of intense absorption. Stretching vibrations of groups with multiple bonds –C = C– appear in the frequency range of 2167 cm^{-1}. The absorption bands at 1460 cm^{-1} are skeletal vibrations of the benzene ring.

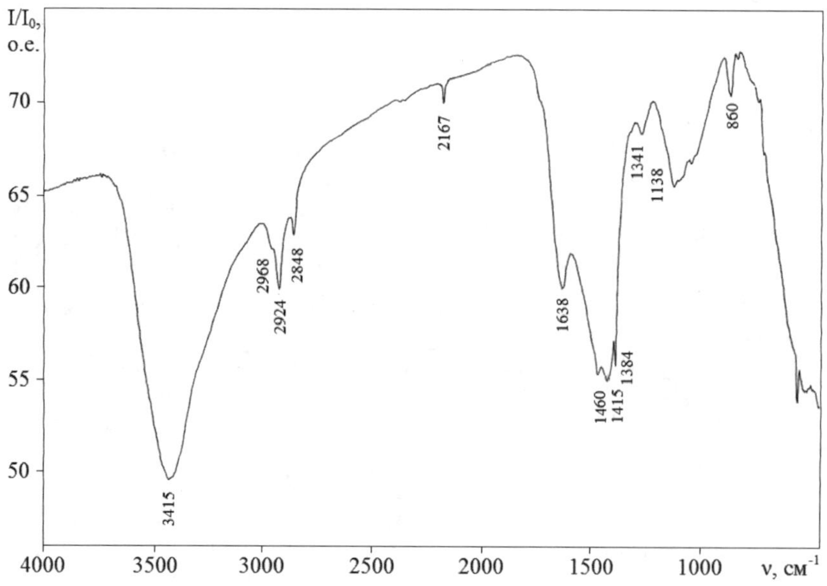

Figure 6. IR - absorption spectrum of combustion products.

5. Conclusion

Thus, in the flame of pyrotechnic compositions, many structures containing carbon are formed. Some structures are identified as nanosized carbon tubes and needles, while other structures are micron-sized tubes in diameter and several millimeters in length. However, there is no description of the mechanism, the rate of formation of submicron carbon tubes, and methods for determining their size factor in the literature.

Acknowledgments

The authors would like to express sincere gratitude and appreciation to Chistyakov I., Kremer E., InterLab company for assistance in conducting research on Hitachi TM–4000Plus scanning electron microscope.

References

[1] Korobova Yu G 2012 *JETP Letters* **95** 524
[2] Mansurov Z A 2012 *J. Physics of Combustion and Explosion* **48** 77
[3] Romanenko A V and Simonov P A 2007 *Industrial catalysis in lectures* (Moscow: Kalvis) p 128
[4] Kartashov Yu I 2008 *Innovation* **116** 92
[5] Likholobov V A 1997 *Soros Educational Journal* **5** 35
[6] Vasilyeva O V, Ksenofontov S I, Baturova G S and Panyshev A N 2017 *Proc. of the V–th All–Russian Scientific Conf. Nanostructured materials and converting devices for solar energy* (Cheboksary: Publisher of Chuvash State University Press) pp 7–9
[7] Ksenofontov S I and Vasilyeva O V 2018 *Proc. of the VI–th All–Russian Scientific Conf. Nanostructured materials and converting devices for solar energy* (Cheboksary: Publisher of Chuvash State University Press) pp 86–90
[8] Vasilyeva O V, Ksenofontov S I, Baturova G S and Kiprova L A 2019 *Proc. Int. Conf. Science – Society – Technology* (Moscow: Moscow Poly) pp 200–205
[9] Vasilyeva O V, Ksenofontov S I and Kokshina A V 2019 *J. of Phys. Conf. Ser.* **1410** 012232

Application of the technology of induction chemical-thermal treatment to improve the physical and mechanical properties of tantalum

A Voyko, A Gerasimov, M Fomina and V Koshuro

Yuri Gagarin State Technical University of Saratov, Saratov 410054, Russia

voyko.leha@mail.ru

Abstract. The results of the chemical thermal treatment (CTT) of tantalum in a solid carbon-containing medium in the temperature range from 1000 to 1300 °C were presented. CTT consisted in heating a workpiece with a working medium in a container made of a refractory material. The induction heating method was used for heating. After strengthening treatment, tantalum samples were characterized by increased hardness, which grew from 140 to 1100 $HV_{0.02}$.

1. Introduction

Tantalum interacts with non-metallic elements that form interstitial solid solutions (H, C, N, O) [1]. For example, the absorption of hydrogen is slow under normal conditions, but at a temperature of about 500 °C, the absorption rate is maximal. As a result of this interaction, not only absorption occurs, but also the formation of chemical compounds – TaH hydrides. Tantalum absorbs nitrogen at a temperature of 600 °C, and at a higher temperature, TaN nitride is formed. Oxidation with atmospheric oxygen begins at a temperature of 200–300 °C, whereas above 500 °C intense oxidation occurs with the formation of the higher oxide Ta2O5. Simultaneous saturation with oxygen and nitrogen leads to the formation of a film with high mechanical strength compared to other coatings of tantalum oxide [2]. Carbon-containing gases (CH_4, CO) at the temperature of 1200–1400 °C react with metal, which results in the formation of hard and refractory carbides, e.g. TaC [3].

A method of chemical-thermal treatment (CTT) allows the production of the necessary compounds on the surface of metals and their alloys. One of the effective ways of CTT of tantalum is carburization in a solid carburizer such as graphite. The use of eddy currents arising during induction heating is effective as a heating source [4].

2. Methodology

For CTT, the prepared sample was placed in a heat-resistant metal container and then carbon-containing powder was poured from all sides. The container with the samples was subjected to induction heating in the air atmosphere at the given electrophysical parameters (power, inductor current) at temperatures of 1000, 1100, 1200, and 1300 °C, with an exposure time of 4 min. After processing, the chemical elemental composition of the coatings was investigated using energy dispersive X-ray fluorescence analysis (EDX), the surface morphology was studied by scanning electron microscopy (SEM), and mechanical properties (microhardness) – by the Vickers method

[5,6]. The hardness was assessed with two loads on the indenter, a low load (20 gf) made it possible to determine the hardness of the coating and a large load (100 gf) was used to determine the hardness of the underlying layer (the effect of the tantalum base).

3. Results

The chemical composition of the surfaces of tantalum samples after CTT was represented by tantalum, carbon and oxygen. The amount of tantalum in the coatings decreased as the treatment temperature grew from 29 to 23 at.% (Figure 1). The amount of carbon increased from 39 to 61 at.% at a temperature of 1200 °C; with a further growth of temperature, the percentage of carbon reached 58 at.%. The amount of oxygen was minimal (11 at.%) at a temperature of 1200 °C, an increase and decrease in the CTT temperature led to a higher percentage of oxygen.

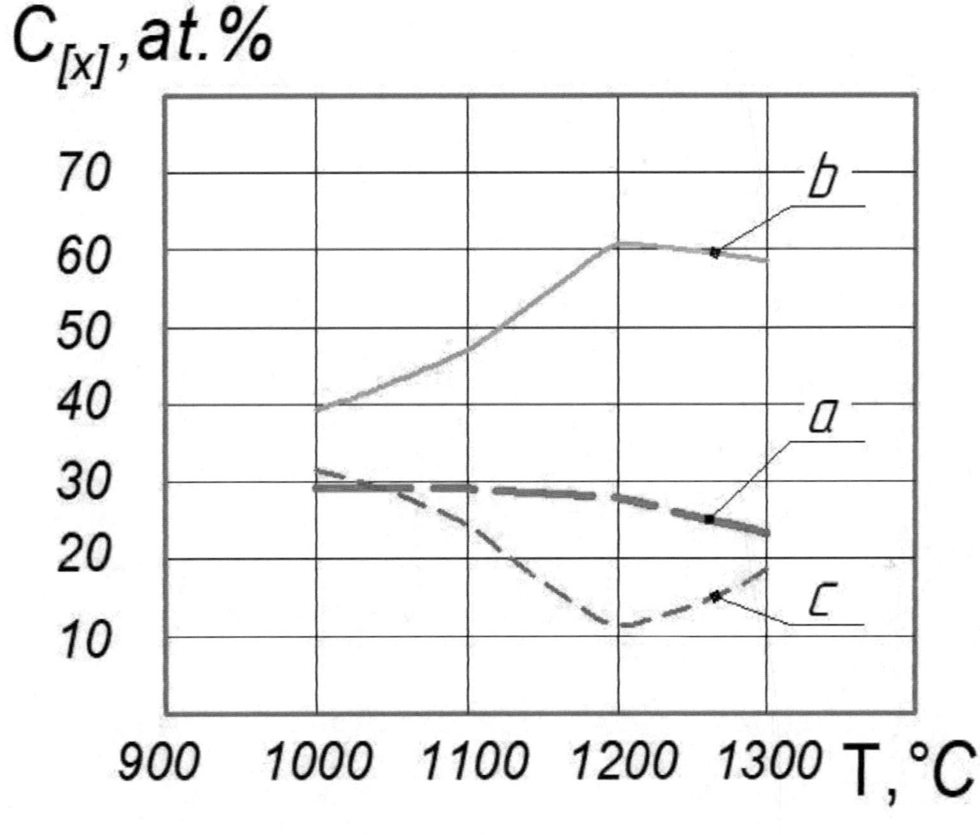

Figure 1. Effect of CTT temperature on the change in the amount of tantalum (a), carbon (b) and oxygen (c) in sample coatings

The surface morphology after CTT had significant changes – coatings with nanosized particles were formed on the surface. The minimum processing temperature ensured the formation of a uniform coating, while the scratches less than 1 μm wide were visible resulting from grinding the tantalum base. The obtained coating had single pores not more than 2 μm in size and nanosized particles in the form of single protrusions 10–20 nm in size (Figure 2a). Raising the temperature to 1100 °C in the course of carburization led to the formation of a coating with an increased number of surface pores with the size up to 3 μm and a decrease in the number of small particles (Figure 2b). In some areas, structural elements with a size of not more than 3 μm were observed. Figure 2c shows the surface morphology of a tantalum sample after CTT at 1200 °C. In this case the nanostructure was similar to the first sample but the particle size increased; there were also areas with pores up to 4 μm in size. The most defective structure is shown in Figure 2d. At a given CTT temperature, pores occupied a significant part of the surface. The particles had a wide spread in size, and their minimum size equaled about 100 nm.

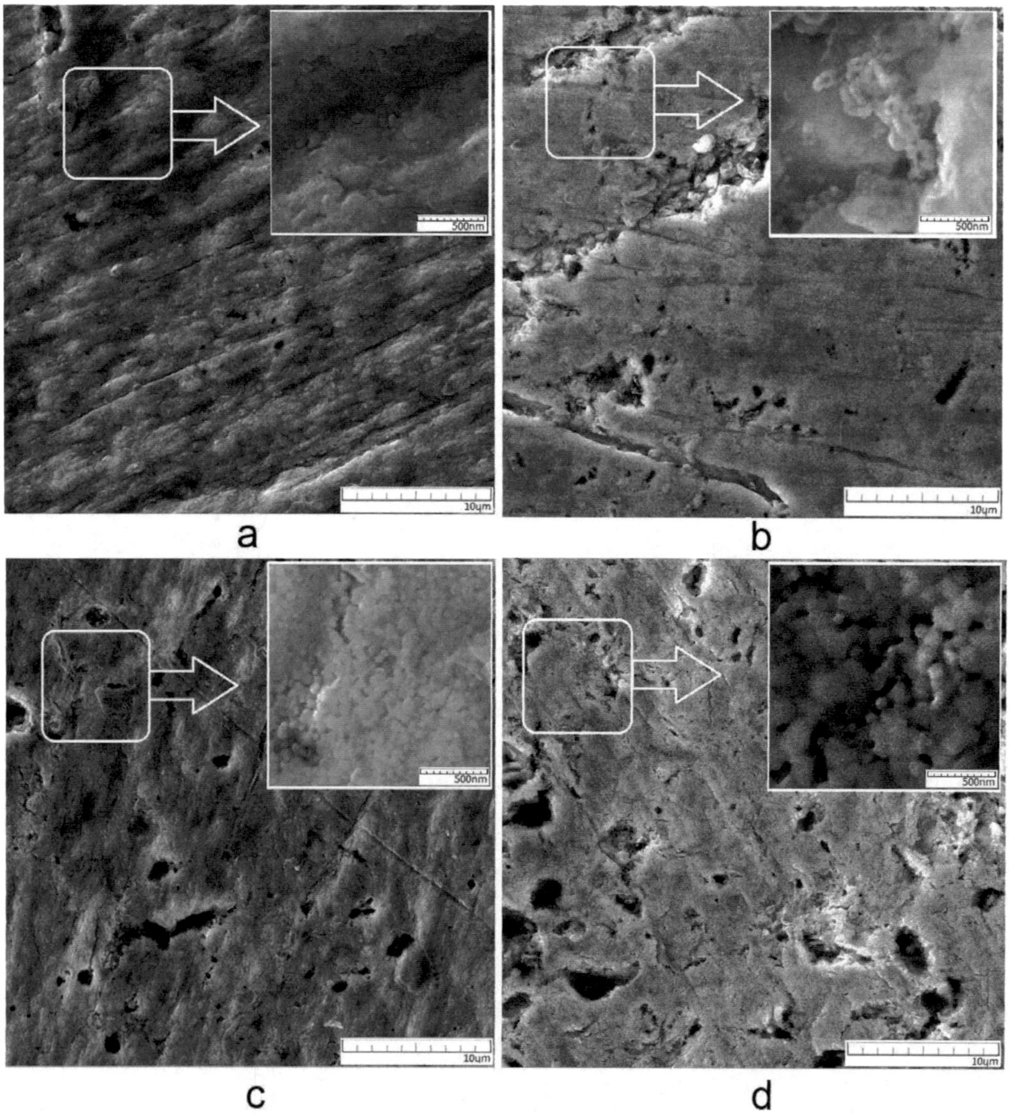

Figure 2. Surface morphology of tantalum samples after CTT at processing temperatures of 1000 °C (**a**), 1100 °C (**b**), 1200 °C (**c**), and 1300 °C (**d**)

The initial hardness of the tantalum samples was 140±7 HV. The summarized measurement results were presented graphically (Figure 3). The obtained hardness curves showed that at the highest CTT temperature (1300 °C) coatings with a high hardness of 1100±64 $HV_{0.02}$ were formed. At this temperature, the hardness of the base increased significantly to 740±57 $HV_{0.01}$. A decrease in the process temperature to 1200 °C led to lower hardness of the coating (900±48 $HV_{0.02}$) and the base (522±70 $HV_{0.02}$).

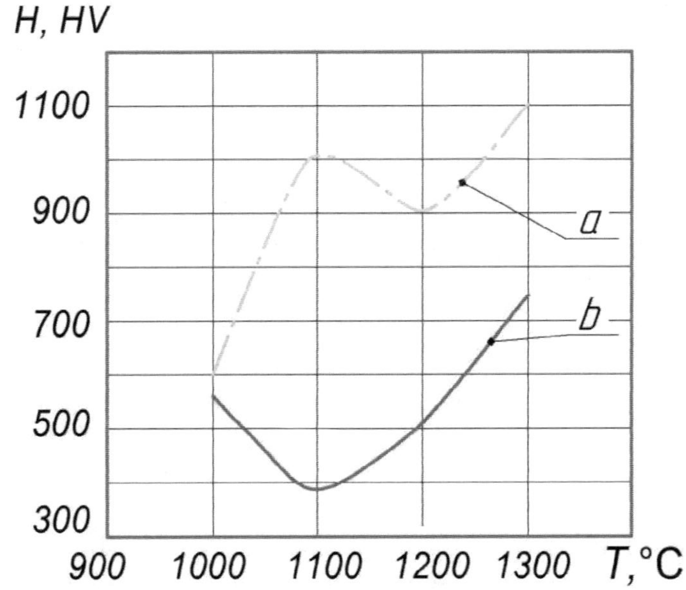

Figure 3. Effect of the CTT treatment temperature on the change in surface hardness of tantalum samples at $HV_{0.02}$ (a) and $HV_{0.1}$ (b)

Hard coating and soft base were observed at temperatures close to the minimum, where the difference between the values of the coating and the base equaled 650 HV. The minimum CTT temperature (1000 °C) contributed to the formation of a coating with a low hardness of 611±64 $HV_{0.02}$. In this case, the hardness of the base was close in value to that of the coating and amounted to 569±59 $HV_{0.1}$.

4. Conclusions

Thus, in the course of the work, it was found that as a result of CTT, three-component coatings were formed consisting of tantalum, carbon and oxygen. The structure of the resulting coatings contained nanograins, the size of which grew with an increase in the temperature of the CTT process. The hardness of the coatings had the highest value at the maximum temperature T = 1300 °C.

Acknowledgments
The research was supported by the Russian Science Foundation (project No. 18-79-10040).

References

[1] Puhr-Westerheide J and Elssner G 1970 *J Less Common Met* **20(4)** 371–374
[2] Hirpara Jignesh G, Chawla V and Chandra R 2020 *Mater Today Commun* **23** 101113
[3] Zhou H, Hu Y, Zhu K, Li F, Long Z and Luo W 2018 *J Vac Sci Technol* **38(1)** 48–52
[4] Fomin A, Voyko A, Fomina M, Mokrousov S and Koshuro V 2020 *Compos Struct* **245** 112393
[5] Fomin A A, Fomina M A, Rodionov I V, Koshuro V A, Poshivalova E Yu, Shchelkunov A Yu, Skaptsov A A, Zakharevich A M and Atkin V S 2015 *Tech Phys Lett* **41** 909–911
[6] Koshuro V, Fomina M, Fomin A and Rodionov I 2018 *Compos Struct* **196** 1–7

Entanglement of two dipole-coupled qubits induced by a thermal field of one-mode lossless cavity with Kerr medium

R K Zakharov and E K Bashkirov

Samara University, 34, Moskovskoye shosse, Samara, 443086, Russia

rodionzhakharov@yandex.ru

Abstract. We studied the dynamics of two qubits interacting with one-mode thermal quantum electromagnetic field of microwave cavity with Kerr medium. Using the exact solution for considered model we derived the qubit-qubit negativity for separa coherent initial qubits states. We showed that initial qubits coherencee interaction can greatly enhance the degree of qubits entanglement in the presence of the Kerr nonlinearity and dipole-dipole interactionyeven for high thermal field intensities.

1. Introduction

Cavity quantum electrodynamics (CQED) deals with the interaction between the natural or artificial atoms (qubits) and quantum electromagnetic field of micro-cavity. In recent years, many schemes have been proposed for generating, controlling, and protecting the entanglement of qubits in CQED systems, such as ions in Paul traps, neutral Rydberg atoms, superconducting Josephson circuits, electron and nuclear spins in solids, quantum dots, nitrogen substituted vacancies in diamond, quantum optomechanical systems, hybrid quantum systems [1]. A system consisting of two two-level atoms (qubits) interacting with one-mode of cavity field is one of the simplest quantum system to exhibit atom-atom entanglement. This system has been actively studied both theoretically and experimentally. Theoretical studies of CQED systems are based on the Jaynes-Cummings model or many-atoms Tavis-Cummings model (TCM) and its generalizations [2] Recently, special attention has been paid to the study of TCM with nonlinearities, in particular with Kerr nonlinearities (see refs in [3]).The Kerr effect is a widely used phenomenon in nonlinear quantum optics, and has been used to generate squeezed states, parametrically convert frequencies, and create ultra-fast pulses. But, in the optical range, the Kerr nonlinearities are small compared to the photon loss rate, which makes it difficult to use this effect to control the entanglement of qubits. However, the situation may be fundamentally different for superconducting circuits in microwave cavity. In the field of quantum optics with microwave circuits, the direct analogue of the Kerr effect is naturally created by the nonlinear inductance of a Josephson junction. In [4] the author experimentally realized the interaction of a qubit with a resonator field in a Kerr medium, by connecting a superconducting qubit (transmon) in a sapphire medium with two three-dimensional high-Q superconducting microwave resonators and achieved the single-photon Kerr regime.

P. Knight and coauthors showed that thermal field of cavity can induce the atom-atom entanglement. They studied this effect for case of two-atom TCM with one-photon transitions [5]. The same for two-atom TCM with degenerate and non-degenerate transitions was investigated in [6, 7].

The result is that atom-atom entanglement induced by a nonlinear interaction is larger than that induced by a linear interaction. It is well known that dipole-dipole interaction is a natural mechanism of atom-atom entanglement. The influence of dipole-dipole interaction between qubits on its entanglement induced by a thermal cavity noise was investigated by Aguiar and coauthors [8]. Hu and co-authors considered the entanglement in a system of two two-level atoms resonantly interacting with a single-mode thermal field for separable coherent initial atomic state [9]. The results show that the initial atomic coherence can enhance the amount of entanglement.

Recently, we investigated the thermal entanglement in two-atom Tavis-Cummings model with Kerr nonlinearity [10, 11]. It is of interest to generalize these results to the case of large thermal field intensity and dipole coupled qubits. In this paper we study the entanglement properties in a system of two dipole–dipole coupled qubits resonantly interacting with a single-mode thermal field of a cavity with additional Kerr medium for coherent initial qubits states.

2. Model and negativity calculation

We consider two identical dipole-coupled qubits A_1 and A_2 resonantly interacting with cavity mode. We suppose that qubit-field coupling constants are equal. The suppose also that there is an additional Kerr medium in the cavity. The interaction Hamiltonian for the system under rotating wave approximation can be written as

$$H = \sum_{i=1}^{2} \hbar g(\sigma_i^+ a + a^+ \sigma_i^-) + X a^{+2} a^2 + \hbar\Omega(\sigma_1^+ \sigma_2^- + \sigma_2^+ \sigma_1^-), \quad (1)$$

where $\sigma_i^+ = |+\rangle_{ii}\langle -|$, and $\sigma_i^- = |-\rangle_{ii}\langle +|$ are the transition operators between the excited $|+\rangle_i$ and the ground states in the i th qubit $(i = 1, 2)$, a^+ and a are the creation and the annihilation operators of photons of the cavity mode, g is the coupling constant between qubits and the cavity mode, X is the dispersive part of the third- order nonlinearity of Kerr medium in the cavity and Ω is dipole-dipole coupling constant. Note for artificial atoms such as superconducting circuits the effective parameter of dipole-dipole interaction Ω may be larger than the qubit-field coupling constant g.

The initial qubits state is assumed to be coherent separable such as

$$|\Psi(0)\rangle_{A_1A_2} = |\Psi(0)\rangle_{A_1} \otimes |\Psi(0)\rangle_{A_2},$$

where

$$|\Psi(0)\rangle_{A_1} = \cos\theta_1 |+\rangle_1 + \sin\theta_1 |-\rangle_1, \quad |\Psi(0)\rangle_{A_2} = \cos\theta_2 |+\rangle_2 + \sin\theta_2 |-\rangle_2. \quad (2)$$

Here θ_1 and θ_2 are the parameters which define the degree of initial coherence of qubits A_1 and A_2

The initial cavity mode state is assumed to be thermal one-mode state

$$\rho_F(0) = \sum_{n=0}^{\infty} p_n |n\rangle\langle n|,$$

where $p_n = \bar{n}^n / (1+\bar{n})^{n+1}$. The average photon number is $\bar{n} = (\exp[\hbar\omega / k_B T] - 1)^{-1}$, where T is the cavity temperature.

Before studying the dynamics of two qubits interacting with thermal field, we investigate evolution of two identical two–level atoms interacting with cavity field prepared in a Fock state. Assume that the excitation number of the system under consideration n $(n \geq 0)$. Then only four basic states $|-,-,n+2\rangle, |+,-,n+1\rangle, |-,+,n+1\rangle, |+,+,n\rangle$ are needed to describe the system evolution. By expanding the state vector of the atom-field system as

$$|\psi_n(t)\rangle = C_{1n}(t)|-,-,n+2\rangle + C_{2n}(t)|+,-,n+1\rangle + C_{3n}(t)|-,+,n+1\rangle + C_{4n}(t)|+,+,n\rangle \quad (3)$$

and using the Schrodinger equation, one gets

$$C_{in}(t)=\sum_{j=1}^{4}A_{ij}^{n}(t)C_{jn}(0),$$

$$A_{ij}^{n}(t)=\sum_{k=1}^{4}\alpha_{in}^{k}\alpha_{jn}^{k}\exp[-i\lambda_k gt],$$

where λ_k are the scaled eigenvalues of the Hamiltonian (1) , and α_{in}^{k} is the ith element of the kth eigenvector . These eigenvalues are

$$\lambda_{1n}=\frac{1}{3}\mathrm{Re}\left(\alpha+2\chi+3n\chi+3n^2\chi+x_n/y_n+y_n\right),$$

$$\lambda_{2n}=\frac{1}{12}\mathrm{Re}\left(4(\alpha+(2+3n(1+n))\chi)-2i\left(-i+\sqrt{3}\right)x_n/y_n+z_n\right),$$

$$\lambda_{3n}=\frac{1}{12}\mathrm{Re}\left(4(\alpha+(2+3n(1+n))\chi)+2i\left(i+\sqrt{3}\right)x_n/y_n-z_n\right),$$

$$\lambda_{4n}=-\alpha+n\chi+n^2\chi,$$

where

$$x_n=18+12n+\alpha^2-2\alpha\chi+4(1+3n(1+n))\chi^2$$

$$z_n=2i\left(i+\sqrt{3}\right)y_n,\ \ y_n=(q_n+s_n)^{1/3},$$

$$q_n=9(3+2n)\alpha+\alpha^3+36n\chi-3\alpha^2\chi-6(1+6n(1+n))\alpha\chi^2+4(2+9n(1+n))\chi^3,$$

$$s_n=\frac{1}{2}(-4x_n^3+4q_n^2)^{1/2},\ \ \alpha=J/g,\ \ \chi=X/g\ .$$

Using the wave-function (1) we can obtain the time-dependent density matrix for considered system

$$\rho(t)=\sum_{n=0}^{\infty}p_n\,|\psi_n(t)\rangle\langle\psi_n(t)|.$$

Taking a trace over the field variables we derived the reduced atomic density operator

$$\rho_A(t)=Tr_F\rho(t)=\begin{pmatrix}(\rho_A)_{11} & (\rho_A)_{12} & (\rho_A)_{13} & (\rho_A)_{14}\\ (\rho_A)_{12}^{*} & (\rho_A)_{22} & (\rho_A)_{23} & (\rho_A)_{24}\\ (\rho_A)_{13}^{*} & (\rho_A)_{23}^{*} & (\rho_A)_{33} & (\rho_A)_{34}\\ (\rho_A)_{14}^{*} & (\rho_A)_{24}^{*} & (\rho_A)_{34}^{*} & (\rho_A)_{44}\end{pmatrix}. \tag{4}$$

We do not give an explicit form of the elements of matrix (4) here due to their excessive cumbersomeness. The partial transpose of a reduced atomic density matrix (4) is

$$\rho_A^{T_1}(t)==\begin{pmatrix}(\rho_A)_{11} & (\rho_A)_{12} & (\rho_A)_{13}^{*} & (\rho_A)_{23}^{*}\\ (\rho_A)_{12}^{*} & (\rho_A)_{22} & (\rho_A)_{14}^{*} & (\rho_A)_{24}^{*}\\ (\rho_A)_{13} & (\rho_A)_{14} & (\rho_A)_{33} & (\rho_A)_{34}\\ (\rho_A)_{23} & (\rho_A)_{24} & (\rho_A)_{34}^{*} & (\rho_A)_{44}\end{pmatrix}. \tag{5}$$

Using the partial transpose of a reduced atomic density matrix (4) we can calculate the two-qubit negativity

$$\varepsilon=-2\sum\mu_i^{-}, \tag{6}$$

where μ_i^- is the negative eigenvalues of the partial transpose of a reduced atomic density matrix $(\rho_A^{T_1})$. All four eigenvalues of matrix (5) can be negative and therefore contribute to negativity (6). In this paper, we do not give an explicit expression for the negative, since the eigenvalues of matrix (4) have a cumbersome form. We show the results of negativity computer simulation in Figure 1.

3. Results and discussion

The negativity for the initial separable coherent qubits state (2) is shown in Figure 1 as a function of the scaled time gt for a model with Kerr nonlinearity and various values of the mean thermal cavity photons. The solid curves correspond to coherent state (2) with $\theta_1 = \pi/4,\ \theta_2 = -\pi/4$ and the dashed curves correspond to incoherent state $|+,-\rangle$.

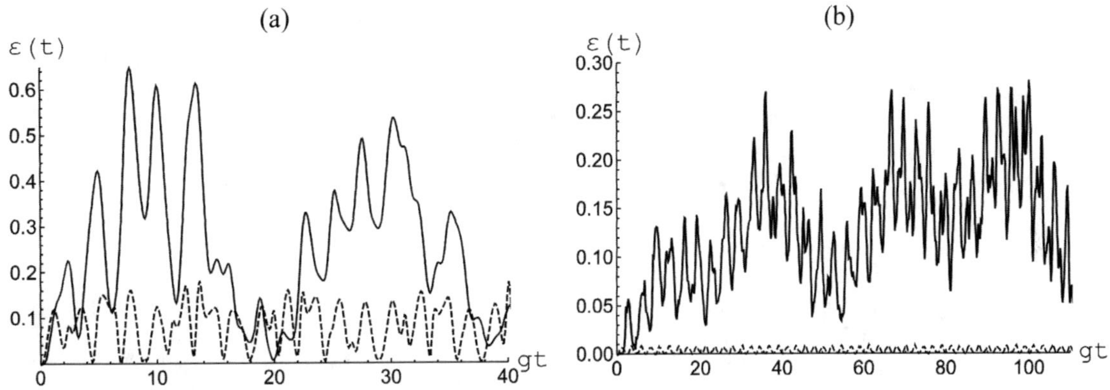

Figure 1. The negativity as a function of a scaled time gt for initial qubits state (2) with $\theta_1 = \pi/4,\ \theta_2 = -\pi/4$ (solid) and $|+,-\rangle$ (dashed). The scaled parameter of Kerr nonlinearity $\chi = 1$. The mean photon number $\bar{n} = 0.5$ (a) and $\bar{n} = 10$ (b). The parameter $\alpha = 0.01$

Negativity revealed by the solid line (coherent state) is much stronger than that revealed by the dashed line (incoherent state). Obviously, the dipole–dipole interaction and Kerr nonlinearity can produce appreciable amount of entanglement for atoms prepared in a coherent states. It is clearly seen from the figure that even for intense thermal fields the initial atomic coherence greatly enhance the maximum degree of qubits entanglement. The maximum amount of entanglement is achieved for the values $\theta_1 = \pi/4,\ \theta_2 = -\pi/4$ of the parameters that determine the initial atomic coherence.

References

[1] Georgescu I M, Ashhab S and Nori F 2014 *Rev. Mod. Phys.* **88** 153
[2] Shore BW and Knight P L 1992 *J. Mod. Opt.* **4** 1195
[3] Alqannas H S and ·Khalil E M 2021 *Optical and Quantum Electronic* **53** 1
[4] Kirchmair G, Vlastakis B, Leghtas Z, Nigg S E, Paik H, Ginossar, E, Mirrahimi M, Frunzio L, Girvin S M and Schoelkopf R J 2013 Nature **495** 205
[5] Kim M S, Lee J, Ahn D and Knight P L 2002 *Phys. Rev.* A**65** 040101(R)
[6] Zhou L and Song H S 2002 *J. Opt.* B**4** 425
[7] Bashkirov E K 2006 *Laser Phys. Lett.* **3** 145
[8] Aguiar L S, Munhoz P P, Vidiella-Barranco A and Roversi J A 2005 *J. Opt.* B**7** S769
[9] Hu Y H, Fang M F, Jiang L and . Zeng K 2008 *Chin. Phys.* **17** 1784
[10] Evseev M M and Bashkirov E K 2020 *Proc. ITNT 2020 - 6th IEEE Int. Conf. Inf. Tech.* Nantech. **9253347** 1
[11] Bashkirov E K 2021 *Proc. SPIE* . **11846** (118460W)

Electrophysical and photocatalytic properties of SnO_2-ZnO thin films prepared by sol-gel method

V Yu Storozhenko[1], M G Volkova[1], Yu N Varzarev[2], A P Starnikova[2], V V Petrov[2] and E M Bayan[1]

[1] Department of Chemistry, Southern Federal University, Rostov-on-Don, 344090, Russia,
[2] Research and Education Centre "Microsystem technics and multisensory monitoring systems", Southern Federal University, Taganrog, 347922, Russia

vstorozhenko@sfedu.ru

Abstract. Electrophysical properties of SnO_2-ZnO thin films prepared by sol-gel method have been studied. The resistance of thin films have a temperature hysteresis, the films resistance decreases up to two times when the temperature reaches 210-300 °C and returns to its initial value when cooling down to 90-30 °C. That phenomenon can be explained by the processes of thermal generation - recombination of electrons, and adsorption - desorption of oxygen on the surface of the films.

1. Introduction

Thin films of semiconductor oxide materials are widely used in microelectromechanical systems, electronic devices, etc [1]. One of the most used is a composite material based on mixed oxides of Zn(II) and Sn(IV) due to the combination of their high chemical stability, non-toxicity, and unique electrophysical properties [2].

In [3,4] thin films of SnO_2-ZnO were applied by ultrasonic spray pyrolysis technique. With an increase in the tin content in the sample, the thickness of the films increased too, and the resistivity decreased. These works showed that the obtained ZnO-SnO_2 films can be used in photovoltaic devices. In [5], ultrasonic spray pyrolysis technique was used to obtain films based on zinc tin oxide. It was shown that an increase in zinc concentration reflects a deterioration in crystallinity in terms of a decrease in grain size. This is explained by an increase in the number of lattice dislocations due to the larger number of dangling bonds formation with an increase in the zinc content. In [6] SnO_2-ZnO thin films prepared by radio-frequency magnetron sputtering. The effect of the substrate temperature on the morphology and properties of the films was shown. It was found that with an increase in temperature, the particle size increases too, and the resistivity decreases.

In addition to unique electrophysical properties, materials based on zinc and tin oxides have photocatalytic activity [7]. The photocatalytic activity of SnO_2-ZnO thin films obtained by the sol-gel method was quantified by the degradation of methylene blue via ultraviolet irradiation. It was found that films containing a mixture of ZnO and SnO_2 exhibit a higher photocatalytic activity in comparison to pure phases SnO_2 and ZnO [8].

In addition to studying the properties of zinc and tin oxides thin films, an important stage is the synthesis of thin-film materials. Several methods of preparation are described in the literature, the main of them are spray pyrolysis, co-precipitation method and sol-gel method [9, 10]. In physical methods of obtaining, complex devices and conditions that differ from standard ones are most often used. Therefore, in recent years, much attention has been paid to chemical methods for obtaining film materials from solutions, since they are simpler, more economical, and does not include expensive equipment using. The sol-gel method is the simplest method for producing films of complex composition [11].

The aim of this work was to synthesize SnO_2-ZnO thin films with a Sn:Zn molar ratio from 0.5:99.5 to 5:95 by the sol-gel method and study their electrophysical and photocatalytic properties.

2. Experiment

To obtain thin film materials by the sol-gel method, $SnCl_4 \cdot 5H_2O$, $Zn(NO_3)_2 \cdot 6H_2O$ and isopropanol as a solvent were used. The Sn(IV):Zn(II) molar ratio in solutions was 0.5:99.5, 1:99, 5:95 by mole. Calcination was carried out for 2 hours at 550 °C. The phase composition of the obtained films was studied by the X-ray diffraction (XRD) using a diffractometer ARLX'TRA, Thermo ARL (Switzerland) with CuKα X-rays. The morphology and thickness of the films studied by scanning electron microscopy (SEM) on a Nova Nanolab microscope. The photocatalytic activity of obtained thin film was studied in the reaction of a model pollutant (methyl orange) photodegradation under the illumination of UV light [12]. The study of the electrophysical properties of the film samples was carried out according to the previously described method [13]. The measurement results were compared with pure ZnO obtained earlier [14].

3. Results and discussion

According to XRD data, all of obtained materials are nanosized, insufficiently crystallized. For all materials, the main phase is the wurtzite. The analysis of diffraction peaks showed the presence of hexagonal wurtzite structure of ZnO for all the thin films. The cassiterite phase appears in samples with a tin content of 0.5 mol% or more. The low intensity of the peaks can be a consequence of both the crystallites small size influence and the insufficiently crystallization of the material. As the content of tin in the sample increases, the degree of crystallinity and the intensity of the peaks increase too. The lowest peak intensity is observed for the sample with the lowest tin addition (0.5 mol. %).

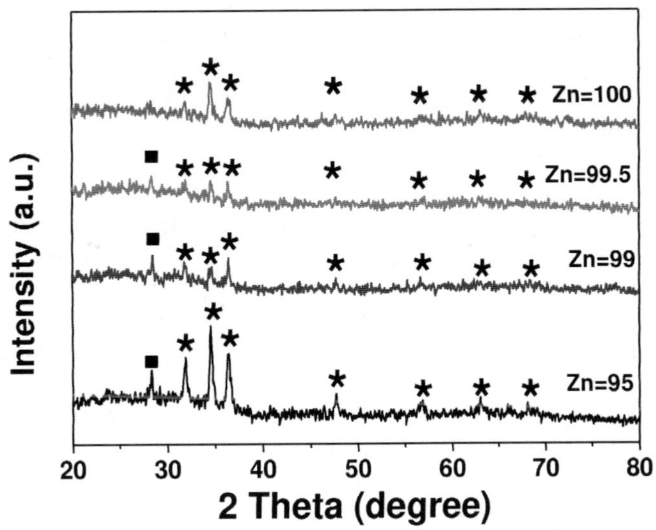

Figure 1. XRD patterns of the synthesized SnO_2-ZnO films, calcined at 550 °C (* – wurtzite structure, ■ – cassiterite structure).

SEM data showed that the films surface and thickness are inhomogeneous. The cross section shows that nanoscale films have a porous structure and an average thickness of about 120-130 nm. The thickness reaches 400-500 nm in some places, which is typical for films formed by the sol-gel technology.

Temperature dependences of the resistance (R) measurements showed that the resistance of pure ZnO is 20% higher than that for films $0.5SnO_2$:99.5 ZnO and $1SnO_2$:99ZnO, and 30% higher than that for material $5SnO_2$:95ZnO (Figure 1), which correspond with the another research works [14, 15].

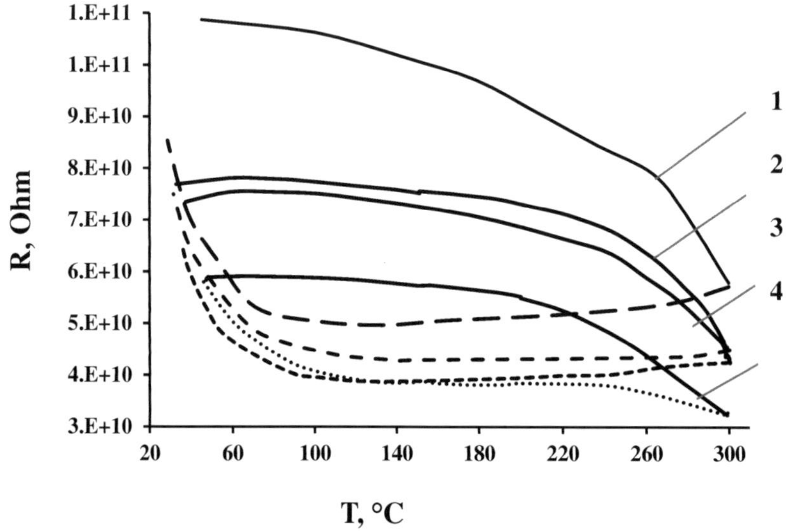

Figure 2. Temperature dependence of the resistance for films materials ZnO (1); $1SnO_2$:99ZnO (2); $0.5SnO_2$:99.5ZnO (3); $5SnO_2$:95ZnO (4) obtained by sol-gel method

The temperature hysteresis is observed in all the presented dependences of resistance on temperature – the heating and cooling curves do not coincide. When the temperature increases from 30 to 210-240 °C, the resistance of the films changes insignificantly. With further heating up to 300 °C, resistance decreases almost twice. When the temperature decreases from 300 °C to 100-90 °C, the resistance reaches the minimum value (and even slightly decreases), and when cooling down to 30°C, increases to the initial value.

This difference in the heating and cooling curves may be due to the competition of two processes – thermal generation and recombination of electrons, on the one hand, and adsorption-desorption processes of oxygen on the films surface, on the other, which also lead to a change in the charge carriers concentration.

The values of the activation energy of conduction estimated by the Arrhenius equation for the temperature ranges in which the resistance changes significantly were 0.15-0.19 eV for the range of 240-300 °C upon heating and 0.07-0.09 eV for the range of 90-30 °C upon cooling, respectively. These energy values correspond to two levels that are electron donors. The fact that they appear separately at the heating and cooling stages can be explained by the intermediate capture of electrons by oxygen adsorbed on the surface at high temperatures. In areas with a slight change in resistance, the activation energy is sufficiently small. However, the presence of a section with negative activation energy on the cooling curve can also indicate the effect of adsorbed oxygen.

The photocatalytic activity study for all materials showed that all materials, regardless of their composition, exhibit photocatalytic properties. It was noted that SnO_2-ZnO films had higher catalytic properties than pure zinc oxide films. At the same time, the material containing the minimum amount of SnO_2 ($0.5SnO_2$:99.5 ZnO) has the best photocatalytic activity. In 60 minutes, when exposed to UV light irradiation, the degradation of the model pollutant methyl orange was 40%.

4. Conclusion

Thin films of SnO_2-ZnO, including the wurtzite and cassiterite phases, were synthesized by the sol-gel method. It was shown that the material containing the minimum amount of tin dioxide (0.5SnO_2:99.5ZnO) exhibits the lowest degree of crystallization. The resistance of thin SnO_2-ZnO films have a temperature hysteresis, the films resistance decreases up to two times when the temperature reaches 210-300 °C and returns to its initial value when cooling down to 90-30 °C. SnO_2-ZnO thin films efficiently utilized for photocatalytic degradation of methyl orange dye under UV light.

Acknowledgments

The present study was performed with financial support of RFBR, project number 20-07-00653 A.

References

[1] Chang Y-J, Lee D-H, Herman G S, Chang C-H 2007 High-Performance, Spin-Coated Zinc Tin Oxide Thin-Film Transistors *Electrochem. Solid-State Lett.* **10(5)** 135-138

[2] Schmidt-Mende L, MacManus-Driscoll J L 2007 ZnO–nanostructures, defects, and devices *Materials today* **10(5)** 40-48

[3] Jazmati A K, Kakhia M, Abdallah B 2018 Physical, optical and sensing properties of sprayed zinc doped tin oxide films *Optik* **158** 1113–1122

[4] Atay F, Akyuz I, Durmaz D, Kose S 2019 Characterization of ZnO-SnO_2 oxide systems produced by ultrasonic spray pyrolysis *Solar Energy* **193** 666–675

[5] Hegazy A R, Salameha B, Mathaic M, Alsmadi A M 2020 Morphology and optical analysis of defect levels in ultrasonically-sprayed zinc tin oxide thin films *Ceramics International* **46** 13151–13158

[6] Zhao J, Ni J, Zhao X, Xiong Y 2011 Preparation and characterization of transparent conductive zinc doped tin oxide thin films prepared by radio-frequency magnetron sputtering *Journal of Wuhan University of Technology-Mater. Sci. Ed.* **26(3)** 388–392

[7] Hamrouni A, Moussa N, Parrino F, Di Paola A, Houas A, Palmisano L 2014 Sol-gel synthesis and photocatalytic activity of ZnO-SnO_2 nanocomposites *Journal of Molecular Catalysis A: Chemical* **390** 133-141

[8] Torres Martínez D Y, Castanedo Pérez R, Torres Delgado G, Zelaya Ángel O 2012 Structural, morphological, optical and photocatalytic characterization of ZnO–SnO_2 thin films prepared by the sol–gel technique *Journal of Photochemistry and Photobiology A: Chemistry* **235** 49–55

[9] Krunks M, Mellikov E 1995 Zinc oxide thin films by the spray pyrolysis method *Thin Solid Films* **270** 33-36

[10] Yang Z, Lv L, Dai Y, Xv Z, Qian D 2010 Synthesis of ZnO–SnO_2 composite oxides by CTAB-assisted co-precipitation and photocatalytic properties *Appl. Surf. Sci.* **256** 2898–2902

[11] Chenb W-J, Liua W-L, Hsieh Sh-H, Hsu Y-G 2012 Synthesis of ZnO:Al Transparent Conductive Thin Films Using Sol-gel Method *Procedia Engineering* **36** 54 – 61

[12] Bayan E M, Lupeiko T G, Pustovaya L E, Volkova M G 2020 Synthesis and photocatalytic properties of Sn–TiO_2 nanomaterials, *Journal of Advanced Dielectrics* **10(01n02)** 2060018

[13] Petrov V V, Plugotarenco N K, Nazarova T N, Korolev A N 2007 Preparation of Sols from Water–Alcohol Solutions of Tetraethyl Orthosilicate and $SnCl_4$ and the Effect of Sol Composition on the Surface Morphology of Sol–Gel Films *Inorganic Materials* **43(9)** 1010–1014

[14] Petrov V V, Bayan E M, Khubezhov S A, Varzarev Yu N, Volkova M G 2020 Investigation of Rapid Gas-Sensitive Properties Degradation of ZnO–SnO_2 Thin Films Grown on the Glass Substrate *Chemosensors* **8(2)** 40

[15] Petrov V V, Varzarev Yu N, Abdullin Kh A 2019 Temperature Dependence of Electrical Properties of ZnO Nanorods Array *IOP Conf. Series: Materials Science and Engineering* **703** 012040

AUTHOR INDEX